CHILTON®

EUROPEAN DIAGNOSTIC SERVICE 2006 EDITION

Audi
BMW
MINI
Saab
VW

Australia • Canada • Mexico • Singapore • Spain • United Kingdom • United States

THOMSON

DELMAR LEARNING

CHILTON®
EUROPEAN
DIAGNOSTIC SERVICE
2006 Edition
Audi, BMW, MINI, Saab, VW

Vice President,
Technology Professional Business Unit:
Gregory L. Clayton

Publisher,
Technology Professional Business Unit:
David Koontz

Director of Marketing:
Beth A. Lutz

Production Director:
Patty Stephan

Editorial Assistant:
Rebecca Rokitowski

Production Manager:
Andrew Crouth

Marketing Manager:
Brian McGrath

Marketing Coordinator:
Jennifer Stall

Publishing Coordinator:
Paula Baillie

Sr. Content Project Manager:
Elizabeth C. Hough

Managing Editor:
Terry Blomquist

Editors:
David Koontz
Ryan Price

Graphical Designer:
Melinda Possinger

Printed in the United States of America
1 2 3 4 5 6 XX 10 09 08 07 06

For more information contact:
Thomson Delmar Learning
Executive Woods
5 Maxwell Drive, PO Box 8007,
Clifton Park, NY 12065-8007
Or find us on the World Wide Web at
www.delmarlearning.com

ISBN: **1-4180-2924-6**

NOTICE TO THE READER

TABLE OF CONTENTS

SECTIONS

USING THIS INFORMATION

Organization

To find where a particular model ction or procedure is located, look in e Table of Contents. Main topics are ted with the page number on which ey may be found. Following the main pics is a listing of all of the subjects thin the section and their page umbers.

Manufacturer and Model Coverage

This product covers 1996-2006 European models that are produced in sufficient quantities to warrant coverage, and which have technical content available from the vehicle manufacturers before our publication date. Although this information is as complete as possible at the time of publication, some manufacturers may make changes which cannot be included here. While striving for total accuracy, the publisher cannot assume responsibility for any errors, changes, or omissions that may occur in the compilation of this data.

Part Numbers & Special Tools

Part numbers and special tools are recommended by the publisher and vehicle manufacturer to perform specific jobs. Before substituting any part or tool for the one recommended, you must be completely satisfied that neither your personal safety, nor the performance of the vehicle will be endangered.

ACKNOWLEDGEMENT

ne publisher would like to express preciation to the following vehicle anufacturers for their assistance in oducing this publication. No further reproduction or distribution of the material in this manual is allowed without the expressed written permission of the vehicle manufacturers and the publisher. Audi of America, Inc., BMW of North America, LLC, MiniUSA, Saab Cars USA, Inc., Volkswagen of America, Inc.

PRECAUTIONS

Before servicing any vehicle, lease be sure to read all of the llowing precautions, which deal with personal safety, prevention of component damage, and important points to take into consideration when servicing a notor vehicle:

- Always wear safety glasses or goggles when drilling, cutting, grinding or prying.
- Steel-toed work shoes should be worn when working with heavy parts. Pockets should not be used for carrying tools. A slip or fall can drive a screwdriver into your body.
- Work surfaces, including tools and the floor should be kept clean of grease, oil or other slippery material.
- When working around moving parts, don't wear loose clothing. Long hair should be tied back under a hat or cap, or in a hair net.
- Always use tools only for the purpose for which they were designed. Never pry with a screwdriver.
- Keep a fire extinguisher and first aid kit handy.
- Always properly support the vehicle with approved stands or lift.
- Always have adequate ventilation when working with chemicals or hazardous material.
- Carbon monoxide is colorless, odorless and dangerous. If it is necessary to operate the engine with vehicle in a closed area such as a garage, always use an exhaust collector to vent the exhaust gases outside the closed area.
- When draining coolant, keep in mind that small children and some pets are attracted by ethylene glycol antifreeze, and are quite likely to drink any left in an open container, or in puddles on the ground. This will prove fatal in sufficient quantity. Always drain the coolant into a sealable container.
- To avoid personal injury, do not remove the coolant pressure relief cap while the engine is operating or hot. The cooling system is under pressure; steam and hot liquid can come out forcefully when the cap is loosened slightly. Failure to follow these instructions may result in personal injury. The coolant must be recovered in a suitable, clean container for reuse. If the coolant is contaminated it must be recycled or disposed of correctly.
- When carrying out maintenance on the starting system be aware that heavy gauge leads are connected directly to the battery. Make sure the protective caps are in place when maintenance is completed. Failure to follow these instructions may result in personal injury.
- Do not remove any part of the engine emission control system. Operating the engine without the engine emission control system will reduce fuel economy and engine ventilation. This will weaken engine performance and shorten engine life. It is also a violation of Federal law.
- Due to environmental concerns, when the air conditioning system is drained, the refrigerant must be collected using refrigerant recovery/recycling equipment. Federal law requires that refrigerant be recovered into appropriate recovery equipment and the process be conducted by qualified technicians who have been certified by an approved organization, such as MACS, ASI, etc. Use of a recovery machine dedicated to the appropriate refrigerant is necessary to reduce the possibility of oil and refrigerant incompatibility concerns. Refer to the instructions provided by the equipment manufacturer when removing refrigerant from or charging the air conditioning system.

• Always disconnect the battery ground when working on or around the electrical system.

• Batteries contain sulfuric acid. Avoid contact with skin, eyes, or clothing. Also, shield your eyes when working near batteries to protect against possible splashing of the acid solution. In case of acid contact with skin or eyes, flush immediately with water for a minimum of 15 minutes and get prompt medical attention. If acid is swallowed, call a physician immediately. Failure to follow these instructions may result in personal injury.

• Batteries normally produce explosive gases. Therefore, do not allow flames, sparks or lighted substances to come near the battery. When charging or working near a battery, always shield your face and protect your eyes. Always provide ventilation. Failure to follow these instructions may result in personal injury.

• When lifting a battery, excessive pressure on the end walls could cause acid to spew through the vent caps, resulting in personal injury, damage to the vehicle or battery. Lift with a battery carrier or with your hands on opposite corners. Failure to follow these instructions may result in personal injury.

• Observe all applicable safety precautions when working around fuel. Whenever servicing the fuel system, always work in a well-ventilated area. Do not allow fuel spray or vapors to come in contact with a spark, open flame, or excessive heat (a hot drop light, for example). Keep a dry chemical fire extinguisher near the work area.
Always keep fuel in a container specifically designed for fuel storage; also, always properly seal fuel containers to avoid the possibility of fire or explosion. Do not smoke or carry lighted tobacco or open flame of any type when working on or near any fuel related components.

• Fuel injection systems often remain pressurized, even after the engine has been turned OFF. The fuel system pressure must be relieved before disconnecting any fuel lines. Failure to do so may result in fire and/or personal injury.

• The evaporative emissions system contains fuel vapor and condensed fuel vapor. Although not present in large quantities, it still presents the danger of explosion or fire. Disconnect the battery ground cable from the battery to minimize the possibility of an electrical spark occurring, possibly causing a fire or explosion if fuel vapor or liquid fuel is present in the area. Failure to follow these instructions can result in personal injury.

• The EPA warns that prolonged contact with used engine oil may cause a number of skin disorders, including cancer! You should make every effort to minimize your exposure to used engine oil. Protective gloves should be worn when changing oil. Wash your hands and any other exposed skin
areas as soon as possible after exposure to used engine oil. Soap and water, or waterless hand cleaner should be used.

• Some vehicles are equipped with an air bag system, often referred to as a Supplemental Restraint System (SRS) or Supplemental Inflatable Restraint (SIR) system. The system must be
disabled before performing service on or around system components, steering column, instrument panel components, wiring and sensors. Failure to follow safety and disabling procedures could result in accidental air bag deployment, possible personal injury and unnecessary system repairs.

• Always wear safety goggles when working with, or around, the air bag system. When carrying a non-deployed air bag, be sure the bag and trim cover are pointed away from your body. When placing a non-deployed air bag on a work surface, always face the bag and trim cover upward, away from the
surface. This will reduce the motion of the module if it is accidentally deployed.

• Electronic modules are sensitive to electrical charges. The ABS module can be damaged if exposed to these charges.

• Brake pads and shoes may contain asbestos, which has been determined to be a cancer-causing agent. Never clean brake surfaces with compressed air. Avoid inhaling brake dust. Clean all brake surfaces with a commercially available brake cleaning fluid.

• When replacing brake pads, shoes, discs or drums, replace them as complete axle sets.

• When servicing drum brakes, disassemble and assemble one side at a time, leaving the remaining side intact for reference.

• Brake fluid often contains polyglycol ethers and polyglycols. Avoid contact with the eyes and wash your hands thoroughly after handling brake fluid. If you do get brake fluid in your eyes, flush your eyes with clean, running water for 15 minutes. If eye irritation persists, or if you have taken brake fluid internally, immediately seek medical assistance.

• Clean, high quality brake fluid from a sealed container is essential to the safe and proper operation of the brake system. You should always buy the correct type of brake fluid for your vehicle. If the brake fluid becomes contaminated, completely flush the system with new fluid. Never reuse any brake fluid. Any brake fluid that is removed from the system should be discarded. Also, do not allow any brake fluid to come in contact with a painted or plastic surface; it will damage the paint.

• Never operate the engine without the proper amount and type of engine oil; doing so will result in severe engine damage.

• Timing belt maintenance is extremely important! Many models utilize an interference-type, non-freewheeling engine. If the timing belt breaks, the valves in the cylinder head may strike the pistons, causing potentially serious (also time-consuming and expensive) engine damage.

• Disconnecting the negative battery cable on some vehicles may interfere with the functions of the on-board computer system(s) and may require the computer to undergo a relearning process once the negative battery cable is reconnected.

• Steering and suspension fasteners are critical parts because they affect performance of vital components and systems and their failure can result in major service expense. They must be replaced with the same grade or part number or an equivalent part if replacement is necessary. Do not use a replacement part of lesser quality or substitute design. Torque values must be used as specified during reassembly to ensure proper retention of these parts.

INTRODUCTION TO OBD SYSTEMS

1

Table of Contents

INTRODUCTION TO OBD

Contents

Notes & Cautions

Before servicing any vehicle, please be sure to read all of the following precautions, which deal with personal safety, prevention of component damage, and important points to take into consideration when servicing a motor vehicle:

- Observe all applicable safety precautions when working around fuel. Whenever servicing the fuel system, always work in a well-ventilated area. Do NOT allow fuel spray or vapors to come in contact with a spark, open flame, or excessive heat (a hot drop light, for example). Keep a dry chemical fire extinguisher near the work area. Always keep fuel in a container specifically designed for fuel storage; also, always properly seal fuel containers to avoid the possibility of fire or explosion. Refer to the additional fuel system precautions that follow.
- Fuel injection systems often remain pressurized, even after the engine has been turned OFF. The fuel system pressure must be relieved before disconnecting any fuel lines. Failure to do so may result in fire and/or personal injury.
- Brake fluid often contains Polyglycol Ethers and Polyglycols. Avoid contact with the eyes and wash your hands thoroughly after handling brake fluid. If you do get brake fluid in your eyes, flush your eyes with clean, running water for 15 minutes. If eye irritation persists, or if you have taken brake fluid internally, IMMEDIATELY seek medical assistance.
- The EPA warns that prolonged contact with used engine oil may cause a number of skin disorders, including cancer. You should make every effort to minimize your exposure to used engine oil. Protective gloves should be worn when changing oil. Wash your hands and any other exposed skin areas as soon as possible after exposure to used engine oil. Soap and water, or waterless hand cleaner should be used.
- The air bag system must be disabled (negative battery cable disconnected and/or air bag system main fuse removed) for at least 30 seconds before performing service on or around system components, steering column, instrument panel components, wiring and sensors. Failure to follow safety and disabling procedures could result in accidental air bag deployment, possible personal injury and unnecessary system repairs.
- Always wear safety goggles when working with, or around, the air bag system. When carrying a non-deployed air bag, be sure the bag and trim cover are pointed away from your body. When placing a non-deployed air bag on a work surface, always face the bag and trim cover upward, away from the surface. This will reduce the motion of the module if it is accidentally deployed. Refer to the additional air bag system precautions later in this section.
- Disconnecting the negative battery cable on some vehicles may interfere with the functions of the on-board computer system(s) and may require the computer to undergo a relearning process once the negative battery cable is reconnected.
- It is critically important to observe all instructions regarding ground disconnects, ignition switch positions, etc., in each diagnostic routine provided. Ignoring these instructions can result in false readings, damage to electronic components or circuits, or personal injury.

Preliminary Diagnostics

HISTORY OF OBD SYSTEMS

Starting in 1978, several vehicle manufacturers introduced a new type of control for several vehicle systems and computer control of engine management systems. These computer-controlled systems included programs to test for problems in the engine mechanical area, electrical fault identification and tests to help diagnose the computer

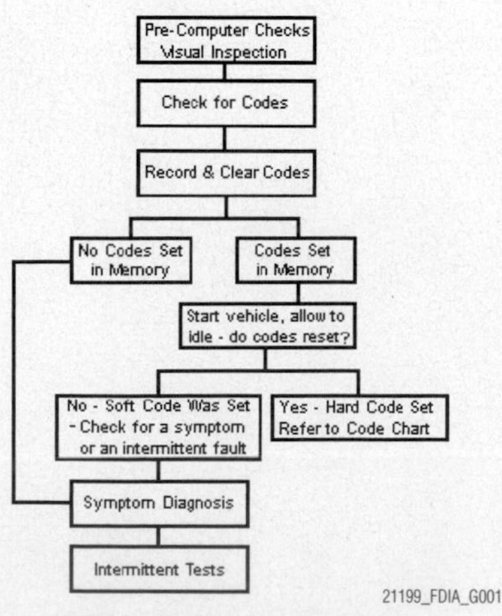

Fig. 1 OBD I diagnostic flow chart

OBD I SYSTEM DIAGNOSTICS

One of the most important things to understand about the automotive repair industry is the fact that you have to continually learn new systems and new diagnostic routines (the test procedures designed to isolate a problem on a vehicle system). For OBD I and II systems, a diagnostic routine can be defined as a procedure (a series of steps) that you follow to find the cause of a problem, make a repair and then verify the problem is fixed.

CHANGES IN DIAGNOSTIC ROUTINES

In some cases, a new Engine Control system may be similar to an earlier system, but it can have more indepth control of vehicle emissions, input and output devices and it may include a diagnostic "monitor" embedded in the engine controller designed to run a thorough set of emission control system tests.

OBD I Diagnostic Flowchart

See Figure 1.

The OBD I Diagnostic Flowchart on this page can be used to find the cause of problems related to Engine Control system trouble codes or driveability symptoms detected on OBD I systems. It includes a step-by-step procedure to use to repair these systems. Compare this flowchart with the one used on OBD II systems.

The steps in this flow chart should be followed as described (from top to bottom).

- Do the Pre-Computer Checks.
- Check for any trouble codes stored in memory.
- Read the trouble codes - If trouble codes are set, record them and then clear the codes.

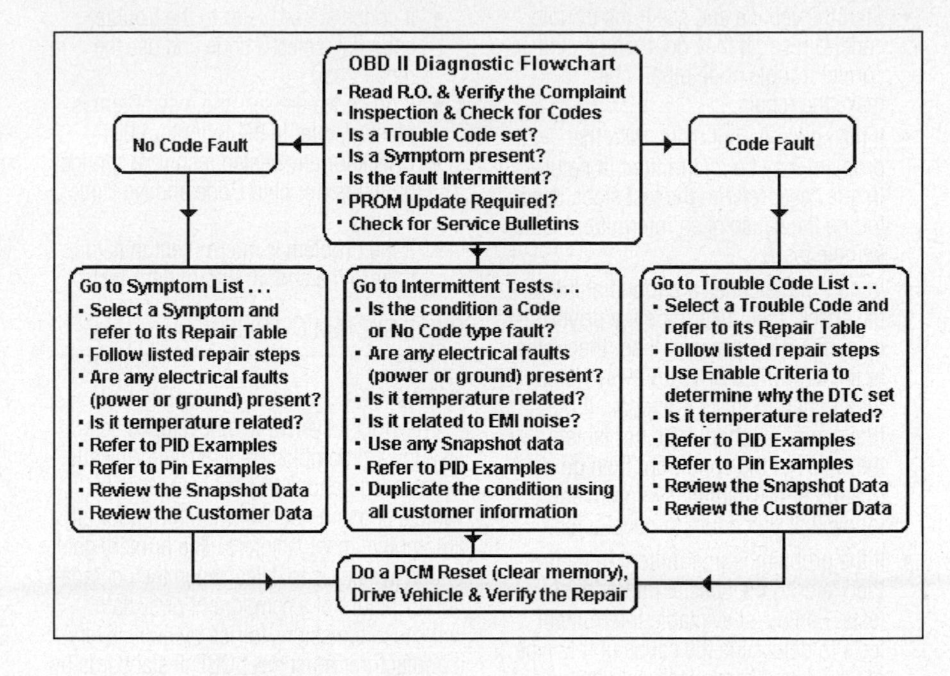

Fig. 2 OBD II diagnostic flow chart

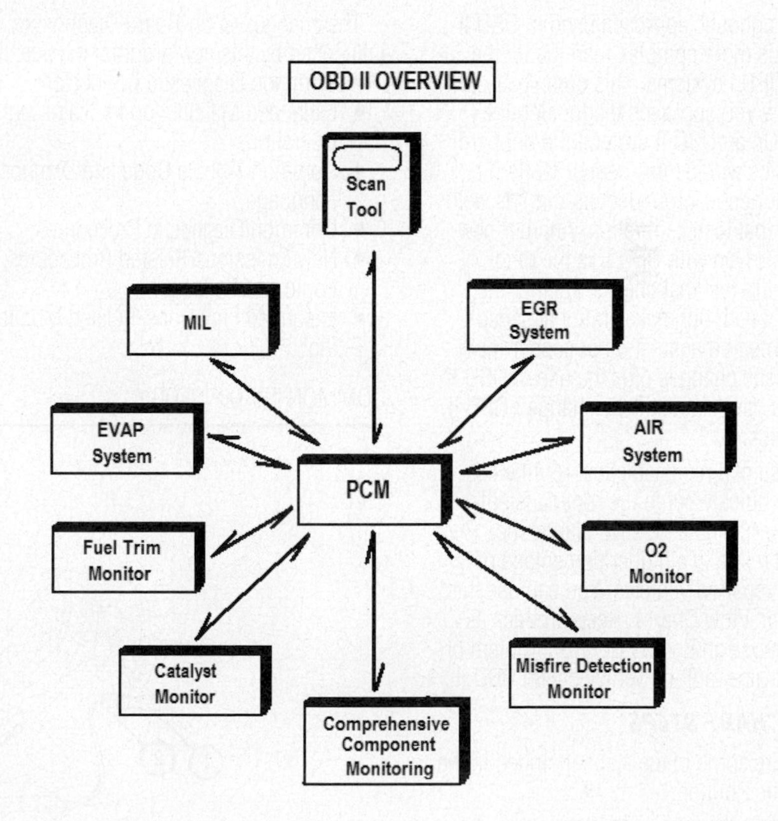

Fig. 3 PCM inputs and outputs

- Start the vehicle and see if the trouble code(s) reset. If they do, then use the correct trouble code repair chart to make the repair.
- If the codes do not reset, than the problem may be intermittent in nature. In this case, refer to the test steps used to find the cause of an intermittent fault (wiggle test).
- In no trouble codes are found at the initial check, then determine if a driveability symptom is present. If so, then refer to the approriate driveability symptom repair chart to make the repair. If the first symptom chart does not isolate the cause of the condition, then go on to another driveability symptom and follow that procedure to conclusion.
- If the problem is intermittent in nature, then refer to the special intermittent tests. Follow all available intermittent tests to determine the cause of this type of fault (usually an electrical connection problem).

OBD II System Diagnostics

See Figure 2.

The diagnostic approach used in OBD II systems is more complex than that of the one for OBD I systems. This complexity will effect how you approach diagnosing the vehicle. On an OBD II system, the onboard diagnostics will identify sensor faults (i.e., open, shorted or grounded circuits) as well as those that lose calibration. Another new test that arrived with OBD II is the rationality test (a test that checks whether the value for one input makes rational sense when compared against other sensor input values). The changes plus the use of OBD II Monitors have dramatically changed OBD II diagnostics.

The use of a repeatable test routine can help you quickly get to the root cause of a customer complaint, save diagnostic time and result in a higher percentage of properly repaired vehicles. You can use this Diagnostic Flow Chart to keep on track as you diagnose an Engine Control problem or a base engine fault on vehicles with OBD II.

FLOW CHART STEPS

Here are some of the steps included in the Diagnostic Routine:

- Review the repair order and verify the customer complaint as described
- Perform a Visual Inspection of underhood or engine related items
- If the engine will not start, refer to No Start Tests
- If codes are set, refer to the trouble code list, select a code and use the repair chart
- If no codes are set, and a symptom is present, refer to the Symptom List
- Check for any related technical service bulletins (for both Code and No Code Faults)
- If the problem is intermittent in nature, refer to the special Intermittent Tests

OBD II SYSTEM OVERVIEW

See Figure 3.

The OBD II system was developed as a step toward compliance with California and Federal regulations that set standards for vehicle emission control monitoring for all automotive manufacturers. The primary goal of this system is to detect when the degradation or failure of a component or system will cause emissions to rise by 50%. Every manufacturer must meet OBD II standards by the 1996 model year. Some manufacturers began programs that were OBD II mandated as early as 1992, but most manufacturers began an OBD II phase-in period starting in 1994.

The changes to On-Board Diagnostics influenced by this new program include:

- Common Diagnostic Connector
- Expanded Malfunction Indicator Light Operation
- Common Trouble Code and Diagnostic Language
- Common Diagnostic Procedures
- New Emissions-Related Procedures, Logic and Sensors
- Expanded Emissions-Related Monitoring

COMMON TERMINOLOGY

OBD II introduces common terms, connectors, diagnostic language and new emissions-related monitoring procedures. The most important benefit of OBD II is that all vehicles will have a common data output system with a common connector. This allows equipment Scan Tool manufacturers to read data from every vehicle and pull codes with common names and similar descriptions of fault conditions. In the future, emissions testing will require the use of an OBD II certifiable Scan Tool.

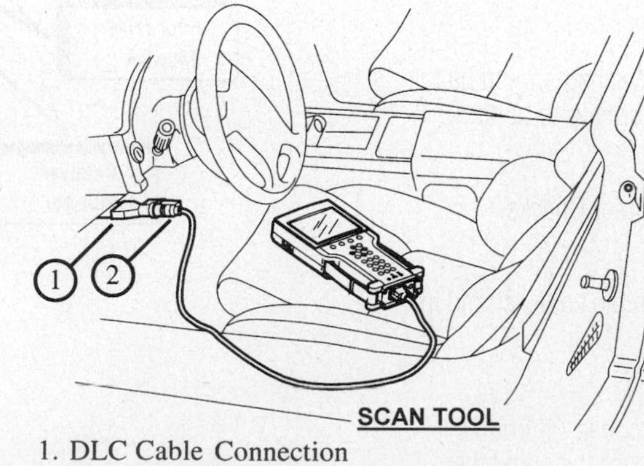

Fig. 4 Typical scan tool hook up

Diagnostic Tools & Circuit Testing

HAND TOOLS & METER OPERATION

To effectively use this or any diagnostic information, you should have a solid understanding of how to operate required tools and test equipment.

SCAN TOOLS

See Figure 4.

Vehicle manufacturers designed their computers to have an accessible data line where a diagnostic tester could retrieve data on sensors and the status of operation for components.

These testers became known in the automotive repair industry as "Scan Tools" because they scanned the data on the computers and provided information for the technician.

The Scan Tool is your basic tool link into the on-board electronic control system of the vehicle. Scan Tools are equipped with, or have separate software cards, for each OEM needed to be diagnosed. In this case, always secure a scan tool that has the latest OEM-specific diagnostic software included. Spend some time in the scan tool user's manual to ensure you know how to properly operate the tool and how to select the necessary programs required for full and proper diagnostics.

MALFUNCTION INDICATOR LAMP

Emission regulations require that a Malfunction Indicator Lamp (MIL) be illuminated when an emissions related fault is detected and that a Diagnostic Trouble Code be stored in the vehicle controller (PCM) memory.

When the MIL is illuminated, it is an indication of a problem within one of the electronic components or circuits. When the scan tool is attached to the Data Link Connector (DLC) in the vehicle, it can access the DTCs. In some situations, without the use of a scan tool, the MIL can be activated to flash a series of long and short flashes, which correspond to the numbering of the DTC.

OBD II guidelines define when an emissions-related fault will cause the MIL to activate and set a Diagnostic Trouble Code (DTC). There are some DTCs that will not cause the MIL to illuminate. OBD II guidelines determine how quickly the onboard diagnostics must be able to identify a fault, set the trouble code in memory and activate the MIL (lamp).

ELECTRONIC CONTROLS

You should have a basic knowledge of electronic controls when performing test procedures to keep from making an incorrect

IDENTIFYING THE PROBLEM

2

Table of Contents

Problem Identification

INTRODUCTION

System Control Modules

See Figures 1 and 2.

Before attempting diagnosis of the Electronic Engine Control system, familiarize yourself with the basics of how the system is designed to operate. It consists of a central processing unit: Powertrain Control Module (PCM), Engine Control Module (ECM), Transmission Control Module (TCM) and/or the Body Control Module (BCM). These units are the "heart" of the electronic control systems on the vehicle. In some cases, these units are integral with one another, and on some applications, they are separate. As you get deeper into actual diagnostic testing, you will find out which units are used on the vehicle you are testing.

The PCM is a digital computer that contains a microprocessor. The PCM receives input signals from various sensors and switches that are referred to as PCM inputs. Based on these inputs, the PCM adjusts various engine and vehicle operations through devices that are referred to as PCM outputs. Examples of the input and output devices are shown in the graphic.

Powertrain Subsystems

A key to the diagnosis of the PCM and its subsystems is to determine which subsystems are on a vehicle. Examples of typical subsystems are:

- Cranking & Charging System
- Emission Control Systems
- Engine Cooling System
- Engine Air/Fuel Controls
- Exhaust System
- Ignition System
- Speed Control System
- Transaxle Controls

WHERE TO BEGIN

See Figure 3.

Diagnosis of engine performance or drivability problems on a vehicle with an onboard computer requires that you have a logical plan on how to approach the problem. The "Six Step Test Procedure" is designed to provide a uniform approach to repair any problems that occur in one or more of the vehicle subsystems.

The diagnostic flow built into this test procedure has been field-tested for several years at dealerships - it is the starting point when a repair is required!

It should be noted that a commonly overlooked part of the "Problem Resolution" step is to check for any related Technical Service Bulletins.

Six-Step Test Procedure

The steps outlined as follows were defined to help you determine how to perform a proper diagnosis. Refer to the flow chart that outlines the Six Step Test Procedure as needed. The recommended steps include:

Verify The Complaint & Check For TSBs

To verify the customer complaint, the technician should understand the normal operation of the system. Conduct a thorough visual and operational inspection, review the service history, detect unusual sounds or odors, and gather diagnostic trouble code (DTC) information resources to achieve an effective repair.

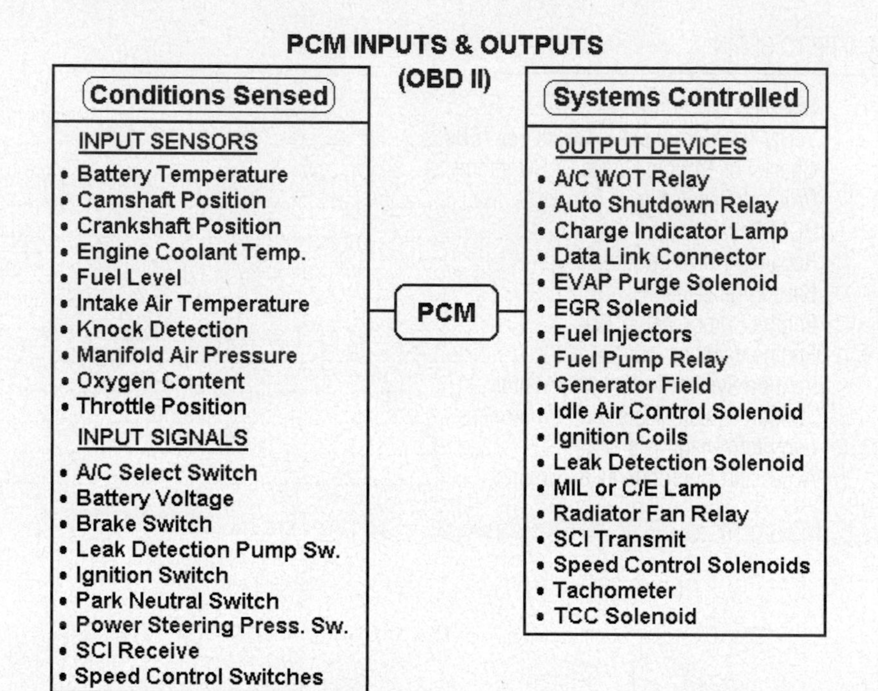

Fig. 1 An example of OBD II input and output devices 21199_FDIA_G005

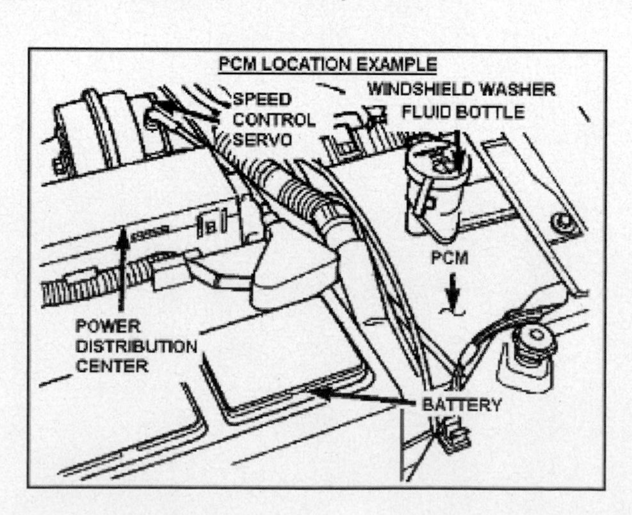

Fig. 2 Typical PCM location 21199_FDIA_G006

This check should include videos, newsletters, and any other information in the form of TSBs or Dealer Service Bulletins. Analyze the complaint and then use the recommended Six Step Test Procedure. Utilize the wiring diagrams and theory of operation articles. Combine your own knowledge with efficient use of the available service information.

Verify the cause of any related symptoms that may or may not be supported by one or more trouble codes. There are various checks that can be performed to Engine Controls that will help verify the cause of a related symptom. This step helps to lead you in an organized diagnostic approach.

Check For Trouble Codes Or Symptoms

Determine if the problem is a Code or a No Code Fault. Then refer to the appropriate published service diagnostic information to make the repair.

Problem Resolution & Repair

Once the problem component or circuit has been properly identified and verified using published diagnostic procedures, make any needed repairs or replacement to restore the vehicle to proper working order. If the condition has set a DTC, follow the designated repair chart to make an effective repair. If there is not a DTC set, but you can determine specific symptoms that are evident during the failure, select the symptom from the symptom tables and follow the diagnostic paths or suggestions to complete the repair or refer to the applicable component or system in service information.

If the vehicle does not set a DTC and has only intermittent operating failures or concerns, to resolve an intermittent fault, perform the following steps:

- Observe trouble codes, DTC modes and freeze frame data.
- Evaluate the symptoms and conditions described by the customer.
- Use a check sheet to identify the circuit or electrical system component.
- Many Aftermarket Scan Tools and Lab Scopes have data capturing features.

PCM Reset

It is a good idea, prior to tracing any faults, to clear the DTCs, attempt to replicate the condition and see if the same DTC resets. Also, once any repairs are made, it will be necessary to clear the DTC(s) - PCM Reset - to ensure the repair has totally resolved the problem. For procedures on PCM Reset, see DIAGNOSTIC TROUBLE CODES.

Repair Verification

Once a repair is completed, the next step is to verify the vehicle operates properly and that the original symptom was corrected. Verification Tests, related to specific DTC diagnostic steps, can be used to verify a repair.

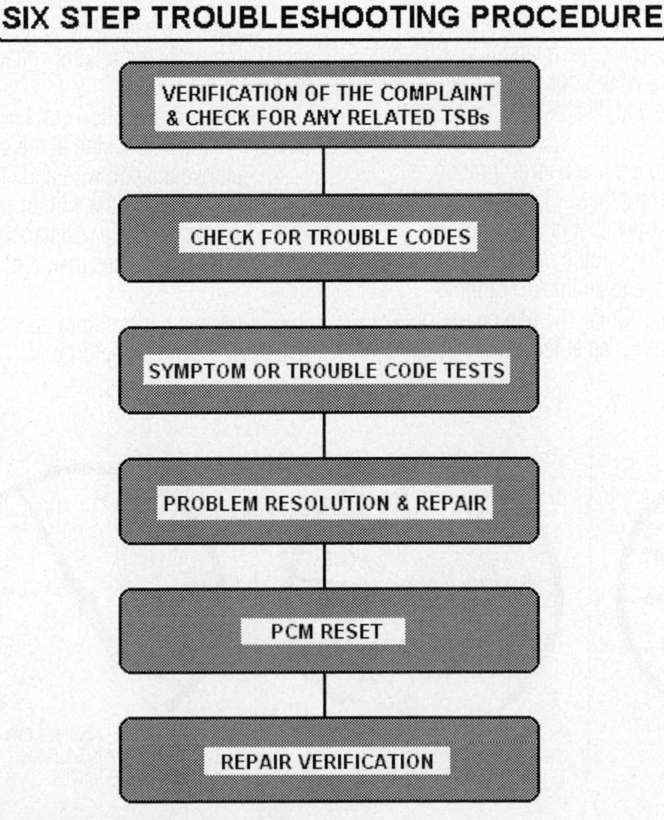

Fig. 3 Six-step diagnostic procedure

Base Engine Tests

To determine that an engine is mechanically sound, certain tests need to be performed to verify that the correct A/F mixture enters the engine, is compressed, ignited, burnt, and then discharged out of the exhaust system. These tests can be used to help determine the mechanical condition of the engine.

To diagnose an engine-related complaint, compare the results of the Compression, Cylinder Balance, Engine Cylinder Leakage (not included) and Engine Vacuum Tests.

Engine Compression Test

The Engine Compression Test is used to determine if each cylinder is contributing its equal share of power. The compression readings of all the cylinders are recorded and then compared to each other and to the manufacturer's specification (if available).

Cylinders that have low compression readings have lost their ability to seal. It this type of problem exists, the location of the compression leak must be identified. The leak can be in any of these areas: piston, head gasket, spark plugs, and exhaust or intake valves.

The results of this test can be used to determine the overall condition of the engine and to identify any problem cylinders as well as the most likely cause of the problem.

**** CAUTION**

Prior to starting this procedure, set the parking brake, place the gear selector in P/N and block the drive wheels for safety. The battery must be fully charged.

COMPRESSION TEST PROCEDURE

1. Allow the engine to run until it is fully warmed up.
2. Remove the spark plugs and disable the Ignition system and the Fuel system for safety. Disconnecting the CKP sensor harness connector will disable both fuel and ignition (except on NGC vehicles).
3. Carefully block the throttle to the wide-open position.
4. Insert the compression gauge into the cylinder and tighten it firmly by hand.
5. Use a remote starter switch or ignition key and crank the engine for 3-5 complete engine cycles. If the test is interrupted for any reason, release the gauge pressure and retest. Repeat this test procedure on all cylinders and record the readings.

The lowest cylinder compression reading should not be less than 70% of the highest cylinder compression reading and no cylinder should read less than 100 psi.

EVALUATING THE TEST RESULTS

To determine why an individual cylinder has a low compression reading, insert a small amount of engine oil (3 squirts) into the suspect cylinder. Reinstall the compression gauge and retest the cylinder and record the reading. Review the explanations that follow.

Reading is higher - If the reading is higher at this point, oil inserted into the cylinder helped to seal the piston rings against the cylinder walls. Look for worn piston rings.

Reading did not change - If the reading didn't change, the most likely cause of the low cylinder compression reading is the head gasket or valves.

Low readings on companion cylinders - If low compression readings were recorded from cylinders located next to each other, the most likely cause is a blown head gasket.

Readings are higher than normal - If the compression readings are higher than normal, excessive carbon may have collected on the pistons and in the exhaust areas. One way to remove the carbon is with an approved brand of "Top Engine Cleaner."

➡ **Always clean spark plug threads and seat with a spark plug thread chaser and seat cleaning tool prior to reinstallation. Use anti-seize compound on aluminum heads.**

Engine Vacuum Tests

An engine vacuum test can be used to determine if each cylinder is contributing an equal share of power. Engine vacuum, defined as any pressure lower than atmospheric pressure, is produced in each cylinder during the intake stroke. If each cylinder produces an equal amount of vacuum, the measured vacuum in the intake manifold will be even during engine cranking, at idle speed, and at off-idle speeds.

Engine vacuum is measured with a vacuum gauge calibrated to show the difference between engine vacuum (the lack of pressure in the intake manifold) and atmospheric pressure. Vacuum gauge measurements are usually shown in inches of Mercury (in. Hg).

➡ **In the tests described in this article, connect the vacuum gauge to an intake manifold vacuum source at a point below the throttle plate on the throttle body.**

ENGINE CRANKING VACUUM TEST PROCEDURE

The Engine Cranking Vacuum Test can be used to verify that low engine vacuum is not the cause of a No Start, Hard Start, Starts and Dies or Rough Idle condition (symptom).

The vacuum gauge needle fluctuations that occur during engine cranking are indications of individual cylinder problems. If a cylinder produces less than normal engine vacuum, the needle will respond by fluctuating between a steady high reading (from normal cylinders) and a lower reading (from the faulty cylinder). If more than one cylinder has a low vacuum reading, the needle will fluctuate very rapidly.

1. Prior to starting this test, set the parking brake, place the gearshift in P/N and block the drive wheels for safety. Then block the PCV valve and disable the idle air control device.
2. Disable the fuel and/or ignition system to prevent the vehicle from starting during the test (while it is cranking).
3. Close the throttle plate and connect a vacuum gauge to an intake manifold vacuum source. Crank the engine for three seconds (do this step at least twice).

The test results will vary due to engine design characteristics, the type of PCV valve and the position of the AIS or IAC motor and throttle plate. However, the engine vacuum should be steady between 1.0–4.0 in. Hg during normal cranking.

ENGINE RUNNING VACUUM TEST PROCEDURE

See Figure 4.

1. Allow the engine to run until fully warmed up. Connect a vacuum gauge to a clean intake manifold source. Connect a tachometer or Scan Tool to read engine speed.
2. Start the engine and let the idle speed stabilize. Raise the engine speed rapidly to just over 2000 rpm. Repeat the test (3) times. Compare the idle and cruise readings.

EVALUATING THE TEST RESULTS

If the engine wear is even, the gauge should read over 16 in. Hg and be steady. Test results can vary due to engine design and the altitude above or below sea level.

Ignition System Tests–Distributor

This next section provides an overview of ignition tests with examples of Engine Analyzer patterns for a Distributor Ignition System.

PRELIMINARY INSPECTION

1. Perform these checks prior to connecting the Engine Analyzer:
2. Check the battery condition (verify that it can sustain a cranking voltage of 9.6v).
3. Inspect the ignition coil for signs of damage or carbon tracking at the coil tower.
4. Remove the coil wire and check for signs of corrosion on the wire or tower.
5. Test the coil wire resistance with a DVOM (it should be less than 7 k/ohm per foot).
6. Connect a low output spark tester to the coil wire and engine ground. Verify that

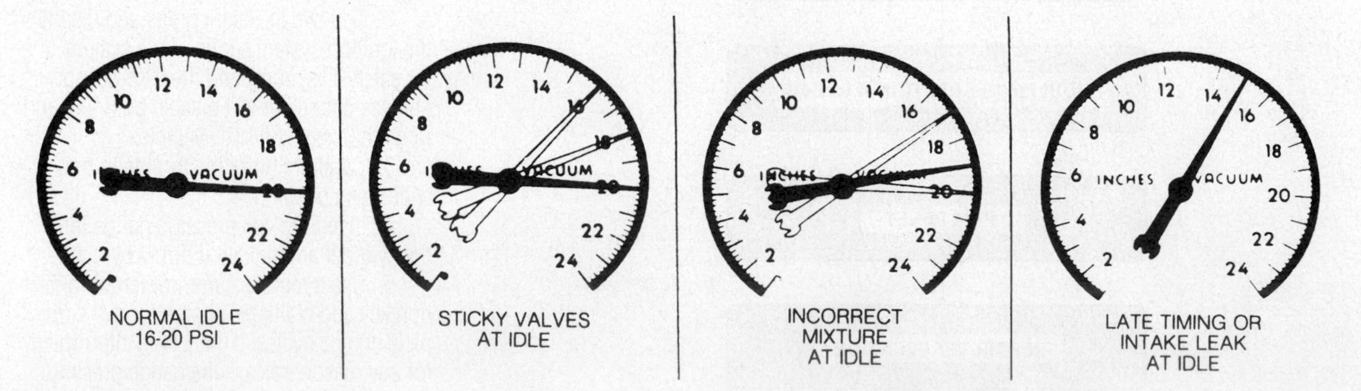

Fig. 4 Engine running vacuum test

the ignition coil can sustain adequate spark output while cranking for 3-6 seconds.

7. Connect the Engine Analyzer to the Ignition System, and choose Parade display. Run the engine at 2000 RPM, and note the display patterns, looking for any abnormalities.

Ignition System Tests–Distributorless

Perform the following checks prior to connecting the Engine Analyzer:

1. Check the battery condition (verify that it can sustain a cranking voltage of 9.6v).
2. Inspect the ignition coils for signs of damage or carbon tracking at the coil towers.
3. Remove the secondary ignition wires and check for signs of corrosion.
4. Test the plug wire resistance with a DVOM (specification varies from 15-30 k/ohm).
5. Connect a low output spark tester to a plug wire and to engine ground. Verify that the ignition coil can sustain adequate spark output for 3-6 seconds.

SECONDARY IGNITION SYSTEM SCOPE PATTERNS (V6 ENGINE)

See Figure 5.

1. Connect the Engine Analyzer to the ignition system.
2. Turn the scope selector to view the "Parade Display" of the ignition secondary.
3. Start the engine in Park or Neutral and slowly increase the engine speed from idle to 2000 rpm.
4. Compare actual display to the examples in the illustration.

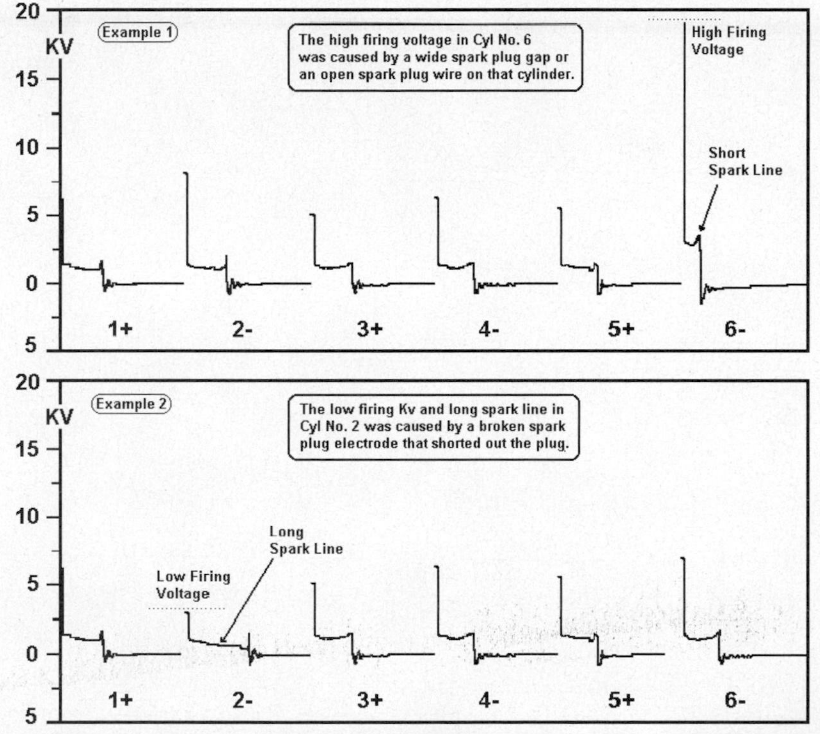

Fig. 5 Secondary ignition system (V6 engine)

21199_FDIA_G009

Symptom Diagnosis

To determine whether vehicle problems are identified by a set Diagnostic Trouble Code, you will first have to connect a proper scan tool to the Data Link Connector and retrieve any set codes. See DIAGNOSTIC TROUBLE CODES for information on retrieving and reading codes.

If no codes are set, the problem must be diagnosed using only vehicle operating symptoms. A complete set of "No Code" symptoms is found in the SYMPTOM DIAGNOSIS (NO CODES).

DO NOT attempt to diagnose driveability symptoms without having a logical plan to use to determine which engine control system is the cause of the symptom - this plan should include a way to determine which systems do NOT have a problem! Remember, there are 2 kinds of NO CODE conditions:

- Symptom diagnosis, in which a continuous problem exists, but no DTC is set as a result. Therefore, only the operating symptoms of the vehicle can be used to pinpoint the root cause of the problem.
- Intermittent problem diagnosis, in which the problem does not occur all the time and does not set any DTCs.
- Both of these NO CODE conditions are covered in the SYMPTOM DIAGNOSIS.

Accessing Components & Circuits

See Figures 6 and 7.

Every vehicle and every diagnostic situation is different. It is a good idea to first determine the best diagnostic path to follow using flow charts, wiring diagrams, TSBs, etc. Part of choosing steps is to determine how time-consuming and effective each step will be. It may be easy to access a component or circuit in one vehicle, but difficult in

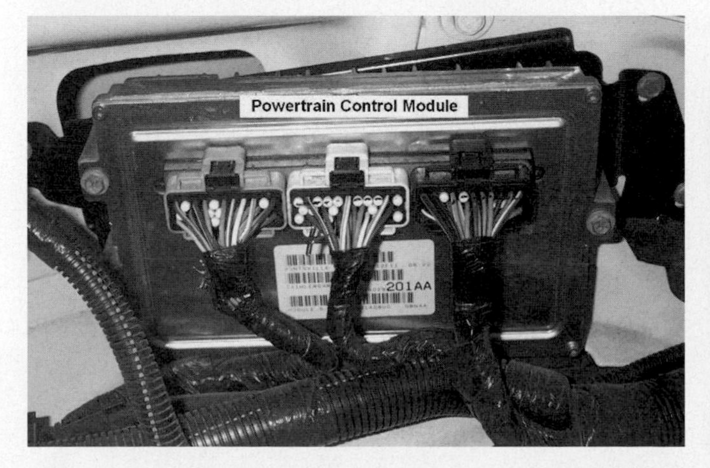

21199_FDIA_G010

Fig. 6 Circuits located at the back of the PCM connector

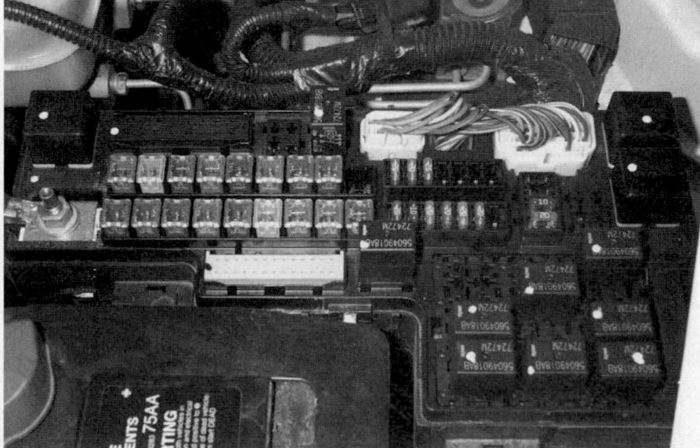

21199_FDIA_G010

Fig. 7 Typical underhood fuse block

another. Many circuits are integrated into a large harness and are difficult to test. Many components are inaccessible without disassembly of unrelated systems.

In the graphic, you will note that the protective covers have been removed from the PCM connectors, and any circuit can be easily identified and back probed. In other cases, PCM access is difficult, and it may be easier to access circuits at the component side of the harness.

Another important point to remember is that any circuit or component controlled by a relay or fused circuit can be monitored from the appropriate fuse box.

There is generally more than one of each type of relay or fuse. Therefore, swapping a suspect relay from another system may be more efficient than testing the relay itself. Relays and fuses may also be removed and replaced with fused jumper wires for testing circuits. Jumper wires can also provide a loop for inductive amperage tests.

Choosing the easiest way has its limitations, however. Remember that an appropriate signal on a PCM controlled circuit at an actuator means that the signal at the PCM is also good. However, a sensor signal at the sensor does not necessarily mean that the PCM is receiving the same signal. Think about the direction flow through a circuit, and not just what signal is appropriate, to save time without making costly assumptions.

INTRODUCTION TO OBD DIAGNOSTIC SYSTEMS 3

Table of Contents

OBD Systems

The California Air Resources Board (CARB) began regulating On-Board Diagnostic (OBD) systems for vehicles sold in California beginning with the 1988 model year. The initial requirements, known as OBD I, required the identification of the likely area of a fault with regard to the fuel metering system, EGR system, emission-related components and the PCM. Implementation of this new vehicle emission control monitoring regulation was done in several phases.

OBD I SYSTEMS

A Malfunction Indicator Lamp (MIL) labeled Check Engine Lamp or Service Engine Soon was required to illuminate and alert the driver of a fault, and the need to service the emission controls. A Diagnostic Trouble Code (DTC) was required to assist in identifying the system or component associated with the fault. If the fault that caused the MIL goes away, the MIL will go out and the code associated with the fault will disappear after a predetermined number of ignition cycles.

Following extensive research, CARB determined that by the time an Emission System component failed and caused the MIL to illuminate, that the vehicle could have emitted excess emissions over a long period of time. CARB also concluded that semi-annual or annual tailpipe tests were not catching enough of the vehicles with Emission Control systems operating at less than normal efficiency.

To take advantage of improvements in vehicle manufacturer adaptive and failsafe strategies, CARB developed new requirements designed to monitor the performance of Emission Control components, as well as to detect circuit and component hard faults. The new diagnostics were designed to operate under normal driving conditions, and the results of its tests would be viewable without any special equipment.

OBD II SYSTEMS

Beginning in the 1994 model year, both CARB and the EPA mandated Enhanced OBD systems, commonly known as OBD II. The objectives of OBD II were to improve air quality by reducing high in-use emissions caused by emission-related faults, reduce the time between the occurrence of a fault and its detection and repair, and assist in the diagnosis and repair of an emissions-related fault.

Differences Between OBD I & OBD II

As with OBD I, if an emission related problem is detected on a vehicle with OBD II, the MIL is activated and a code is set. However, that is the only real similarity between these systems. OBD II procedures that define emissions component and system tests, code clearing and drive cycles are more comprehensive than tests in the OBD I system.

Powertrain Control Module

The PCM in the OBD II system monitors almost all Emission Control systems that affect tailpipe or evaporative emissions. In most cases, the fault must be detected before tailpipe emissions exceed 1.5 times applicable 50K or 100K-mile FTP standards. If a component exceeds emission levels or fails to operate within the design specifications, the MIL is illuminated and a code is stored within two OBD II drive cycles.

The OBD II test runs continuously or once per trip (it depends on the driving mode requirement). Tests are run once per drive cycle during specific drive patterns called trips. Codes are stored in the PCM memory when a fault is first detected. In most cases, the MIL is turned on after two trips with a fault present. If the MIL is "on", it will go off after three consecutive trips if the same fault does not reappear. If the same fault is not detected after 40 engine warmup periods, the code will be erased (Fuel and Misfire faults require 80 warmup cycles).

OBD II Standardization

OBD II diagnostics require the use of a standardized Diagnostic Link Connector (DLC), standard communication protocol and messages, and standardized trouble codes and terminology. Examples of this standardization are Freeze Frame Data and I/M Readiness Monitors.

Changes in MIL Operation

An important change for OBD II involves when to activate the MIL. The MIL must be activated by at least the second trip if vehicle emissions could exceed 1.5 times the FTP standard. If any single component or system failure would allow the emissions to exceed this level, the MIL is activated and a related code is stored in the PCM.

1994 OBD II Phase-In Systems

Starting in 1994 some manufacturers began to "phase-in" the OBD II system on certain vehicles. The OBD II "phase-in" system on these vehicles included the use of a Misfire Monitor that operated with a "lower threshold" Misfire Detection system designed to monitor misfires without setting any codes. In addition, the EVAP Monitor was not operational on these vehicles.

1996 & Later OBD II Systems

By the 1996 model year, all California passenger cars and trucks up to 14,000 lb. GVWR, and all Federal passenger cars and trucks up to 8,600 lb. GWVR were required to comply with the CARB-OBD II or EPA OBD requirements. The requirements applied to diesel and gasoline vehicles, and were phased in on alternative-fuel vehicles.

Diagnostic Test Modes

The "test mode" messages available on a Scan Tool are listed below:

- Mode $01: Used to display Powertrain Data (PID data)
- Mode $02: Used to display any stored Freeze Frame data
- Mode $03: Used to request any trouble codes stored in memory
- Mode $04: Used to request that any trouble codes be cleared
- Mode $05: Used to monitor the Oxygen sensor test results
- Mode $06: Used to monitor Non-Continuous Monitor test results
- Mode $07: Used to monitor the Continuous Monitor test results
- Mode $08: Used to request control of a special test (EVAP Leak)
- Mode $09: Used to request vehicle information (INFO MENU)

Onboard Diagnostics

See Figure 1.

The Diagnostic Repair Chart should be used as follows:

- Trouble Code Diagnosis - Refer to the Code List or electronic media for a repair chart for a particular trouble code.
- Driveability Symptoms - Refer to the Driveability Symptom List in manuals or in electronic media.
- Intermittent Faults - Refer to the Intermittent Test Procedures.
- OBD II Drive Cycles - Refer to the Comprehensive Component Monitor or a Main Monitor drive cycle article.

OBD SYSTEM TERMINOLOGY

It is very important that service technicians understand terminology related to OBD II test procedures. Several of the essential OBD II terms and definitions are explained in the following text.

Two-Trip Detection

Frequently, an emission system or component must fail a Monitor test more

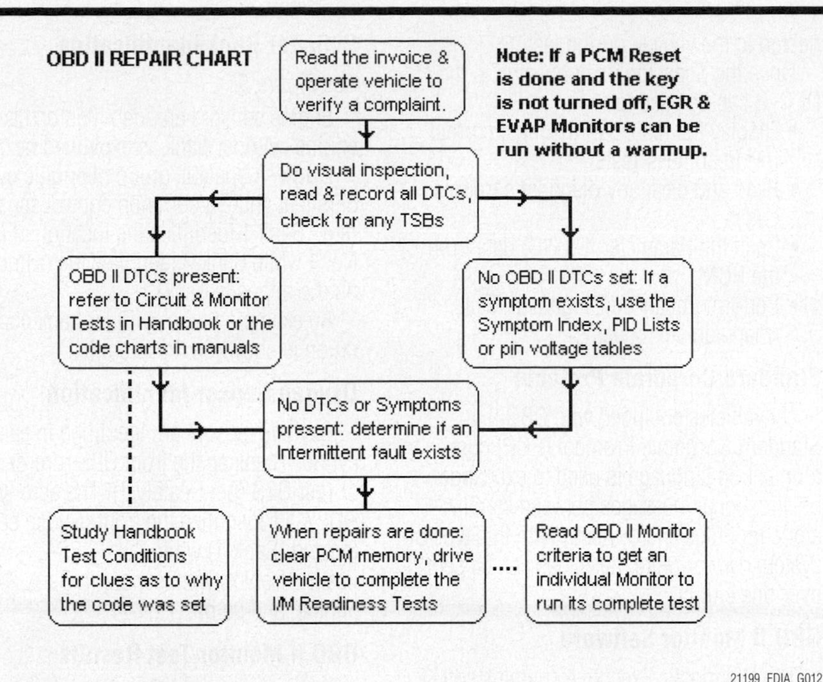

Fig. 1 OBD II repair chart

than once before the MIL is activated. In these cases, the first time an OBD II Monitor detects a fault during any drive cycle it sets a pending code in the PCM memory.

A pending code, which is read by selecting DDL from the Scan Tool menu, appears when Memory or Continuous codes are read. In order for a pending code to cause the MIL to activate, the original fault must be repeated under similar conditions.

This is a critical issue to understand as a pending code could remain in the PCM for a long time before the conditions that caused the code to set reappear. This type of OBD II trouble code logic is frequently referred to as the "Two-Trip Detection Logic".

➡ **Codes related to a Misfire fault and Fuel Trim can cause the PCM to activate the MIL after one trip because these codes are related to critical emission systems that could cause emissions to exceed the federally mandated limits.**

Similar Conditions

If a pending code is set because of a Misfire or Fuel System Monitor fault, the vehicle must meet similar conditions for a second trip before the code matures the PCM activates the MIL and stores the code in memory. Refer to prior Note for exceptions to this rule. The meaning of similar conditions is important when attempting to repair a fault detected by a Misfire or Fuel System Monitor.

To achieve similar conditions, the vehicle must reach the following engine running conditions simultaneously:

- Engine speed must be within 375 RPM of the speed when the trouble code set.
- Engine load must be within 10% of the engine load when the trouble code set.
- Engine warmup state must match a previous cold or warm state.

Summary–Similar conditions are defined as conditions that match the conditions recorded in Freeze Frame when the fault was first detected and the trouble code was set in the PCM memory.

OBD II Warmup Cycle

See Figure 2.

The meaning of the expression warmup cycle is important. Once the fault that caused an OBD II trouble code to set is gone and the MIL is turned off, the PCM will not erase that code until after 40 warmup cycles. This is the purpose of the warmup cycle: To help clear stored codes.

However, trouble codes related to a Fuel system or Misfire fault require that 80 warmup cycles occur without the fault reappearing before codes related to these monitors will be erased from the PCM memory.

➡ **A warmup cycle is defined as vehicle operation (after an engine off and cool-down period) when the engine temperature rises to at least 40°F and reaches at least 160°F.**

Malfunction Indicator Lamp

If the PCM detects an emission related component or system fault for two consecutive drive cycles on OBD II systems, the MIL is turned on and a trouble code is stored. The MIL is turned off if three consecutive drive cycles occur without the same fault being detected.

Most trouble codes related to a MIL are erased from memory after 40 warmup periods if the same fault is not repeated. The MIL can be turned off after a repair by using the Scan Tool PCM Reset function.

Freeze Frame Data

See Figure 3.

The term Freeze Frame is used to describe the engine conditions that are recorded in PCM memory at the time a Monitor detects an emissions related fault. These conditions include fuel control state, spark timing, engine speed and load.

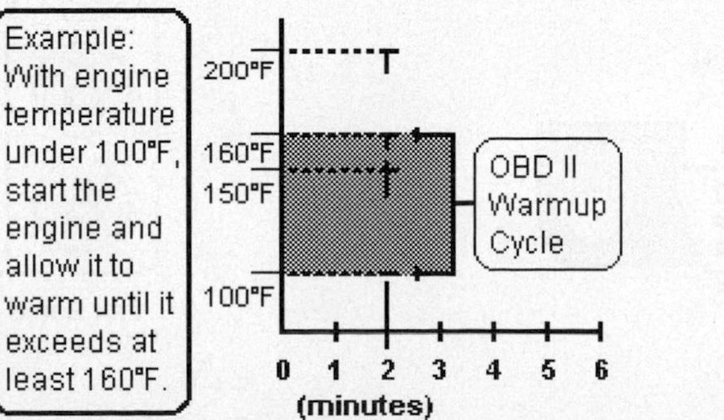

Fig. 2 OBD II warmup cycle

SCAN TOOL DISPLAY

Freeze Frame Data	
Fuel Sys Status	OL
Load Value	14%
ECT Deg F	+175°F
SHRTFT Adapt	+1.5%
MAP "Hg	18.1"
Engine RPM	750
DTC Priority	01

21199_FDIA_G014

Fig. 3 Scan tool freeze frame

Freeze Frame data is recorded when a system fails the first time for two-trip type faults. The Freeze Frame Data will only be overwritten by a different fault with a "higher emission priority."

Diagnostic Trouble Codes

The OBD II system uses a Diagnostic Trouble Code (DTC) identification system established by the Society of Automotive Engineers (SAE) and the EPA. The first letter of a DTC is used to identify the type of computer system that has failed as shown below:

- The letter 'P' indicates a Powertrain related device
- The letter 'C' indicates a Chassis related device
- The letter 'B' indicates a Body related device
- The letter 'U' indicates a Data Link or Network device code.

The first DTC number indicates a generic (P0xxx) or manufacturer (P1xxx) type code. A list of trouble codes is included.

The number in the hundreds position indicates the specific vehicle system or subgroup that failed (i.e., P0300 for a Misfire code, P0400 for an emission system code, etc.).

Data Link Connector

See Figure 4.

Vehicles equipped with OBD II use a standardized Data Link Connector (DLC). It is typically located between the left end of the instrument panel and 12 inches past vehicle centerline. The connector is mounted out of sight from vehicle passengers, but should be easy to see from outside by a technician in a kneeling position (door open). However, not all of the connectors are located in this exact area.

The DLC is rectangular in design and capable of accommodating up to 16 terminals. It has keying features to allow easy connection to the Scan Tool. Both the DLC and Scan Tool have latching features used to ensure that the Scan Tool will remain connected to the vehicle during testing.

Once the Scan Tool is connected to the DLC, it can be used to:

- Display the results of the most current I/M Readiness Tests
- Read and clear any diagnostic trouble codes
- Read the Parameter ID (PID) data from the PCM
- Perform Enhanced Diagnostic Tests (manufacturer specific)

Standard Corporate Protocol

On vehicles equipped with OBD II, a Standard Corporate Protocol (SCP) communication language is used to exchange bi-directional messages between standalone modules and devices. With this type of system, two or more messages can be sent over one circuit.

OBD II Monitor Software

The Diagnostic Executive contains software designed to allow the PCM to organize and prioritize the Main Monitor tests and procedures, and to record and display test results and diagnostic trouble codes.

The functions controlled by this software include:

- To control the diagnostic system so the vehicle continues to operate in a normal manner during testing.
- To ensure the OBD II Monitors run during the first two sample periods of the Federal Test Procedure.
- To ensure that all OBD II Monitors and their related tests are sequenced so that required inputs (enable criteria) for a particular Monitor are present prior to running that particular Monitor.
- To sequence the running of the Monitors to eliminate the possibility of different Monitor tests interfering with each other or upsetting normal vehicle operation.
- To provide a Scan Tool interface by coordinating the operation of special tests or data requests.

Cylinder Bank Identification

See Figure 5.

Engine sensors are identified on each engine cylinder bank as explained next.

Bank–A specific group of engine cylinders that share a common control sensor (e.g., Bank 1 identifies the location of Cyl. No. 1 while Bank 2 identifies the cylinders on the opposite bank).

An example of the cylinder bank configuration is shown in the Graphic.

Oxygen Sensor Identification

Oxygen sensors are identified in each cylinder bank as the front O2S (pre-catalyst) or rear O2S (post-catalyst). The acronym HO2S-11 identifies the front oxygen sensor located (Bank 1) while the HO2S-21 identifies the front oxygen sensor in Bank 2 of the engine, and so on.

OBD II Monitor Test Results

Generally, when an OBD II Monitor runs and fails a particular test during a trip, a pending code is set. If the same Monitor detects a fault for two consecutive trips, the MIL is activated and a code is set in PCM memory. The results of a particular Monitor test indicate that an emission system or component failed: NOT the circuit that failed!

To determine where the fault is located; follow the correct code repair chart, symptom diagnosis or intermittent test. The code and symptom repair charts are the most efficient way to repair an OBD II system.

➡ **Two important pieces of information that can help speed up a diagnosis are code conditions (including all enable criteria), and the parameter information (PID) stored in the Freeze Frame at the time a trouble code is set and stored in memory.**

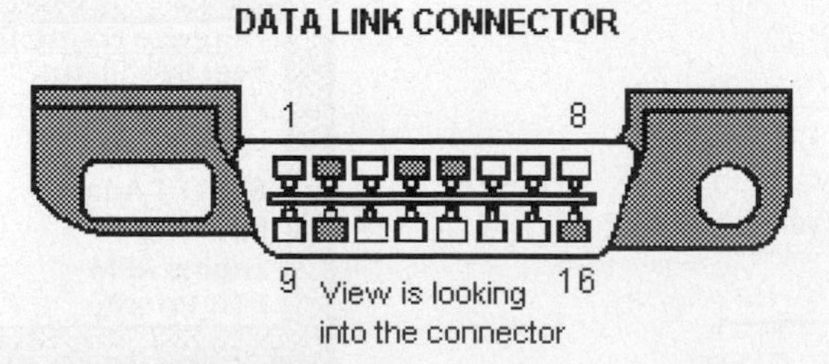

Fig. 4 Typical data link connector

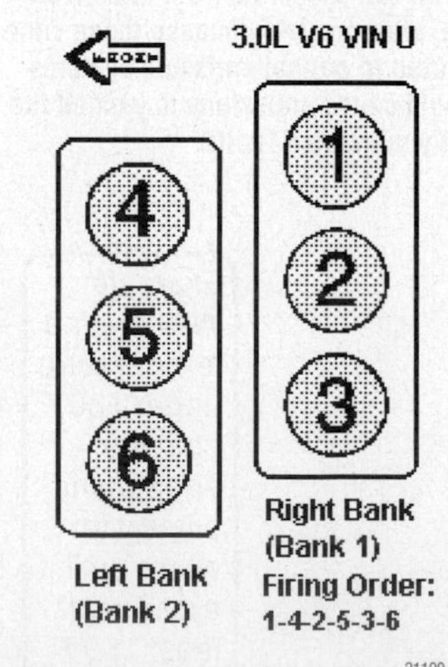

Fig. 5 Typical cylinder bank identification (V6 engine)

Adaptive Fuel Control Strategy

The PCM incorporates an Adaptive Fuel Control Strategy that includes an adaptive fuel control table stored to compensate for normal changes in fuel system devices due to age or engine wear.

During closed loop operation, the Fuel System Monitor has two methods of attempting to maintain an ideal A/F ratio of 14:7 to 1 (they are referred to as short term fuel trim and long term fuel trim).

➡ **If a fuel injector, fuel pressure regulator or oxygen sensor is replaced the, memory in the PCM should be cleared by a PCM Reset step so that the PCM will not use a previously learned strategy.**

Short Term Fuel Trim

Short term fuel trim (SHRTFT) is an engine operating parameter that indicates the amount of short term fuel adjustment made by the PCM to compensate for operating conditions that vary from the ideal A/F ratio condition. A SHRTFT number that is negative (-15%) means that the HO2S is indicating a richer than normal condition to the PCM, and that the PCM is attempting to lean the A/F mixture. If the A/F ratio conditions are near ideal, the SHRTFT number will be close to 0%.

Long Term Fuel Trim

Long term fuel trim (LONGFT) is an engine parameter that indicates the amount of long term fuel adjustment made by the PCM to correct for operating conditions that vary from ideal A/F ratios. A LONGFT number that is positive (+15%) means that the HO2S is indicating a leaner than normal condition, and that it is attempting to add more fuel to the A/F mixture. If A/F ratio conditions are near ideal, the LONGFT number will be close to 0%. The PCM adjusts the LONGFT in a range from -35 to +35%. The values are in percentage on a Scan Tool.

Enable Criteria

The term enable criteria describe the conditions necessary for any of the OBD II Monitors to run their diagnostic tests. Each Monitor has specific conditions that must be met before it will run its test.

Enable criteria information can be different for each vehicle and engine type. Examples of trouble code conditions for DTC P0460 and P1168 are shown below:

Code information includes any of the following examples:

- Air Conditioning Status
- BARO, ECT, IAT, TFT, TP and Vehicle Speed sensors
- Camshaft (CMP) and Crankshaft (CKP) sensors
- Canister Purge (duty cycle) and Ignition Control Module Signals
- Short (SHRTFT) and Long Term (LONGFT) Fuel Trim Values
- Transmission Shift Solenoid On/Off Status

Drive Cycle

The term drive cycle has been used to describe a drive pattern used to verify that a trouble code, driveability symptom or intermittent fault had been fixed. With OBD II systems, this term is used to describe a vehicle drive pattern that would allow all the OBD II Monitors to initiate and run their diagnostic tests. For OBD II purposes, a minimum drive cycle includes an engine startup with continued vehicle operation that exceeds the amount of time required to enter closed loop fuel control.

OBD II Trip

The term OBD II Trip describes a method of driving the vehicle so that one or more of the following OBD II Monitors complete their tests:

- Comprehensive Component Monitor (completes anytime in a trip)
- Fuel System Monitor (completes anytime during a trip)
- EGR System Monitor (completes after accomplishing a specific idle and acceleration period)
- Oxygen Sensor Monitor (completes after accomplishing a specific steady state cruise speed for a certain amount of time)

OBD II Drive Cycle

The ambient or inlet air temperature must be from 40-100°F to initiate the OBD II drive cycle. Allow the engine to warm to 130°F prior to starting the test.

Connect the Scan Tool prior to beginning the drive cycle. Some tools are designed to emit a three-pulse beep when all of the OBD II Monitors complete their tests.

➡ **The IAT PID must be from 50-100°F to start the drive cycle. If it is less than 50°F at any time during the highway part of the drive cycle, the EVAP Monitor may not complete. The engine should reach 130°F before starting before attempting to verify an EVAP system fault. Disengage the PTO before proceeding (PTO PID will show OFF) if applicable. For the EVAP Running Loss system, verify FLI PID is at 15-85%. Some Monitors require very specific idle and acceleration steps.**

Drive Cycle Procedure

The primary intention of the OBD II drive cycle is to clear a specific DTC. The drive cycle can also be used to assist in identifying any OBD II concerns present through total Monitor testing. Perform all of the Vehicle Preparation steps.

Connect a Scan Tool and have an assistant watch the Scan Tool I/M Readiness Status to determine when the Catalyst, EGR, EVAP, Fuel System, O2 Sensor, Secondary AIR and Misfire Monitors complete.

OBD II SYSTEM MONITORS

Comprehensive Component Monitor

OBD II regulations require that all emission related circuits and components controlled by the PCM that could affect emissions are monitored for circuit continuity and out-of-range faults. The Comprehensive Component Monitor (CCM) consists of four different monitoring strategies: two for inputs and two for output signals. The CCM is a two trip Monitor for emission faults on most vehicles.

Input Strategies

One input strategy is used to check devices with analog inputs for opens, shorts, or out-of-range values. The CCM accomplishes this task by monitoring A/D converter input voltages. The analog inputs monitored include the ECT, IAT, MAF, TP and Transmission Range Sensors signals.

DTC	Trouble Code Title & Conditions
	EVAP System Small Leak Conditions: Cold startup, engine running at off-idle conditions, then the PCM detected a small leak (a leak of more than 0.040") in the EVAP system.
	FRP Sensor in Range but Low Conditions: Engine running, then the PCM detected that the FRP sensor signal was out-of-range low. Scan Tool Tip: Monitor the FRP PID for a value below 80 psi (551 kPa).

A second input strategy is used to check devices with digital and frequency inputs by performing rationality checks. The PCM uses other sensor readings and calculations to determine if a sensor or switch reading is correct under existing conditions. Some tests run continuously, some only after actuation.

Output Strategies

An Output State Monitor in the PCM checks outputs for opens or shorts by observing the control voltage level of the related device. The control voltage is low with it on, and high with the device off.

IAC Motor Test

The PCM monitors the IAC system in order to "learn" the closed loop correlation it needs to reposition the IAC solenoid (a rationality check).

Catalyst Efficiency Monitor

The Catalyst Monitor is a PCM diagnostic run once per drive cycle that uses the downstream heated Oxygen Sensor (HO2S-12) to determine if a catalyst falls below a minimum level of effectiveness in its ability to control exhaust emissions. The PCM uses a program to determine the catalyst efficiency based on the oxygen storage capacity of the catalytic converter.

Catalyst Monitor Operation

See Figure 6.

The Catalyst Monitor is a diagnostic that tests the oxygen storage capacity of the catalyst. The PCM determines the capacity by comparing the switching frequency of the rear oxygen sensor to the switching frequency of the front oxygen sensor. If the catalyst is okay, the switching frequency of the rear oxygen sensor will be much slower than the frequency of the front oxygen sensor.

However, as the catalyst efficiency deteriorates its ability to store oxygen declines. This deterioration causes the rear oxygen sensor to switch more rapidly. If the PCM detects the switching frequency of the rear oxygen sensor is approaching the frequency of the front oxygen sensor, the test fails and a pending code is set. If the PCM detects a fault on consecutive trips (from two to six consecutive trips) the MIL is activated, and a trouble code is stored in the PCM memory.

The Catalyst Monitor runs after startup once a specified time has elapsed and the vehicle is in closed loop. The amount of time is subject to each PCM calibration. Certain inputs (enable criteria) from various engine sensors (i.e., CKP, ECT, IAT, TPS and VSS) are required before the Catalyst Monitor can run.

Once the Catalyst Monitor is activated, closed loop fuel control is temporarily transferred from the front oxygen sensor to the rear oxygen sensor. During the test, the Monitor analyzes the switching frequency of both sensors to determine if a catalyst has degraded.

Catalyst Efficiency Monitor

CATALYST TEST–STEADY STATE CATALYST EFFICIENCY TEST

The PCM transfers the input for closed loop fuel control from the front HO2S-11 to the rear HO2S-21 during this test. The PCM measures the output frequency of the rear HO2S. This "test frequency" indicates the current oxygen storage capacity of the converter. The slower the frequency of the test result, the higher the efficiency of the converter.

CATALYST TEST–CALIBRATED FREQUENCY TEST

In Part 2 of the test a second frequency is calculated based on engine speed and load. This frequency serves as a high limit threshold for the test frequency. If the PCM detects the test frequency is less than the calibrated frequency the catalyst passes the test. If the frequency is too high, the converter or system has failed (a pending code is set).

The sequence of counting the front and rear O2S switches continues until the drive cycle completes. The ratio of total HO2S-21 switches to the total of the HO2S-11 switches is calculated. If the switch ratio is over the stored threshold, the catalyst has failed and a code is set.

CATALYTIC MONITOR REPAIR VERIFICATION TRIP

See Figure 7.

Start the engine, and drive in stop and go traffic for over 20 minutes. (Ambient air temperature must be over 50°F to run this test). Drive at speeds from 25-40 mph (6 times) and then at cruise for five minutes.

POSSIBLE CAUSES OF A CATALYST EFFICIENCY FAULT

- Base Engine faults (engine mechanical)
- Exhaust leaks or contaminated fuel

EGR System Monitor

The EGR System Monitor is a PCM diagnostic run once per trip that monitors EGR system component functionality and components for faults that could cause vehicle tailpipe levels to exceed 1.5 times the FTP Standard. A series of sequenced tests is used to test the system.

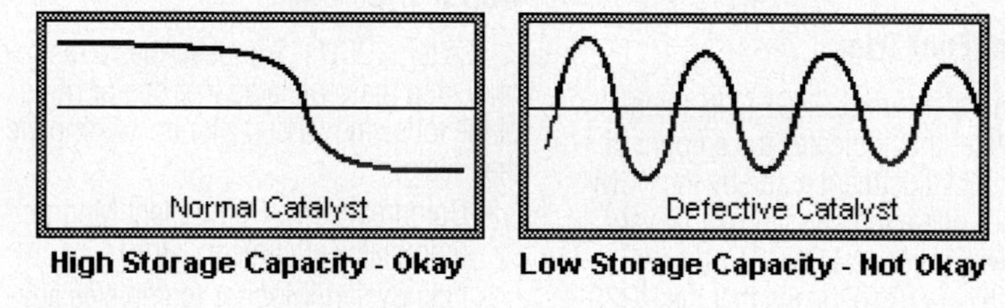

Fig. 6 Typical rear oxygen sensor waveform

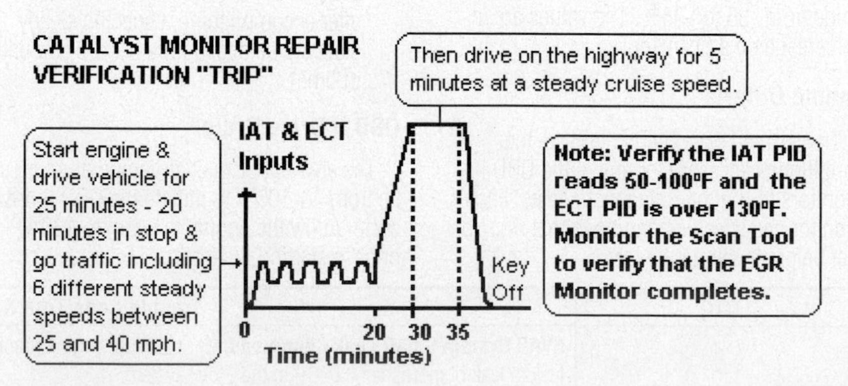

Fig. 7 Typical catalyst monitor trip

Possible Causes of an EGR System Failure

See Figure 8.

- Leaks or disconnects in upstream or downstream vacuum hoses
- Damaged DPFE or EGR EVP sensor
- Plugged or restricted DPFE or EGR VP sensor or orifice assembly

Evap System Monitor

The EVAP System Monitor is a PCM diagnostic run once per trip that monitors the EVAP system in order to detect a loss of system integrity or leaks in the system (anywhere from 0.020" to 0.040" in diameter).

Possible Causes of an EVAP System Failure

- Cracks, leaks or disconnected hoses in the fuel vapor lines, components or plastic connectors or lines
- Backed-out or loose connectors to the Canister Purge solenoid
- Fuel filler cap (gas cap) loose or missing
- PCM has failed

On-Board Refueling Vapor Recovery System

An On-Board Refueling Vapor Recovery (ORVR) system is used on late model vehicles to recover fuel vapors during vehicle refueling.

SYSTEM OPERATION

The operation of the ORVR system during refueling is described next:

- The fuel filler pipe forms a seal to stop vapors from escaping the fuel tank while liquid is entering the tank (liquid in the 1" diameter tube blocks fuel vapor from rushing back up the fuel filler pipe).
- The fuel vapor control valve controls the flow of vapors out of the tank (it closes when the liquid level reaches a height associated with the fuel tank usable capacity). The fuel vapor control valve:

  a. Limits the total amount of fuel dispensed into the fuel tank.

  b. Prevents liquid gasoline from exiting the fuel tank when submerged (and also when tipped well beyond a horizontal plane as part of the vehicle rollover protection in an accident).

  c. Minimizes vapor flow resistance in a refueling condition.

- Fuel vapor tubing connects the fuel vapor control valve to the EVAP canister. This routes the fuel tank vapors (that are displaced by the incoming fuel) to the canister.

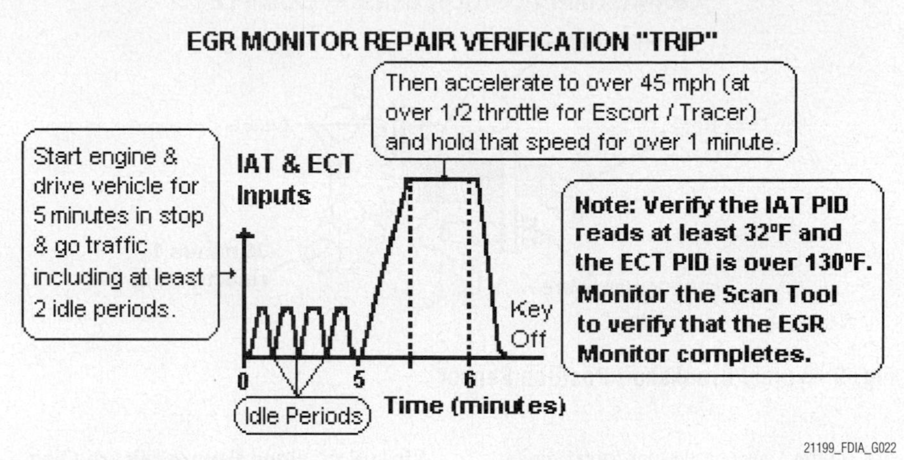

Fig. 8 Typical EGR monitor

- A check valve in the bottom of the pipe prevents any liquid from rushing back up the fuel filler pipe during liquid flow variations associated with the filler nozzle shut-off.
- Between refueling events, the charcoal canister is purged with fresh air so that it may be used again to store vapors accumulated during engine soak periods or subsequent refueling events. The vapors drawn from the canister are consumed in the engine.

Evap Monitor Test Conditions

The PCM allows canister purge to occur when the engine is warm, at wide open or part throttle (as long as the engine is not overheated). The engine can be in open or closed loop fuel control during purging.

Fuel System Monitor

The Fuel System Monitor is a PCM diagnostic that monitors the Adaptive Fuel Control system. The PCM uses adaptive fuel tables that are updated constantly and stored in long term memory (KAM) to compensate for wear and aging in the fuel system components.

FUEL SYSTEM MONITOR OPERATION

Once the PCM determines all the enable criteria has been are met (ECT, IAT and MAF PIDs in range and closed loop enabled), the PCM uses its adaptive strategy to "learn" changes needed to correct a Fuel system that is biased either rich or lean. The PCM accomplishes this task by monitoring Short Term and Long Term fuel trim in closed loop mode.

LONG AND SHORT TERM FUEL TRIM

Short Term fuel trim is a PCM parameter identification (PID) used to indicate Short Term fuel adjustments. This parameter is expressed as a percentage and its range of authority is from -10% to +10%. Once the engine enters closed loop, if the PCM receives a HO2S signal that indicates the A/F mixture is richer than desired, it moves the SHRTFT command to a more negative range to correct for the rich condition.

If the PCM detects the SHRTFT is adjusting for a rich condition for too long a time, the PCM will "learn" this fact, and move LONGFT into a negative range to compensate so that SHRTFT can return to a value close to 0%. Once a change occurs to LONGFT or SHRTFT, the PCM adds a correction factor to the injector pulsewidth calculation to adjust for variations. If the change is too large, the PCM will detect a fault.

➡ **If a fuel injector, fuel pressure regulator, etc. is replaced, clear the KAM and then drive the vehicle through the Fuel System Monitor drive pattern to reset the fuel control table in the PCM.**

Misfire Detection Monitor

The Misfire Monitor is a PCM diagnostic that continuously monitors for engine misfires under all engine positive load and speed conditions (accelerating, cruising and idling). The Misfire Monitor detects misfires caused by fuel, ignition or mechanical misfire conditions. If a misfire is detected, engine conditions present at the time of the fault are written to the Freeze Frame Data. These conditions overwrite existing data.

Misfire Monitor Operation

See Figure 9.

The Misfire Monitor is designed to measure the amount of power that each cylinder contributes to the engine. The amount of contribution is calculated based upon measurements determined by crankshaft acceleration (TDC of compression stroke to

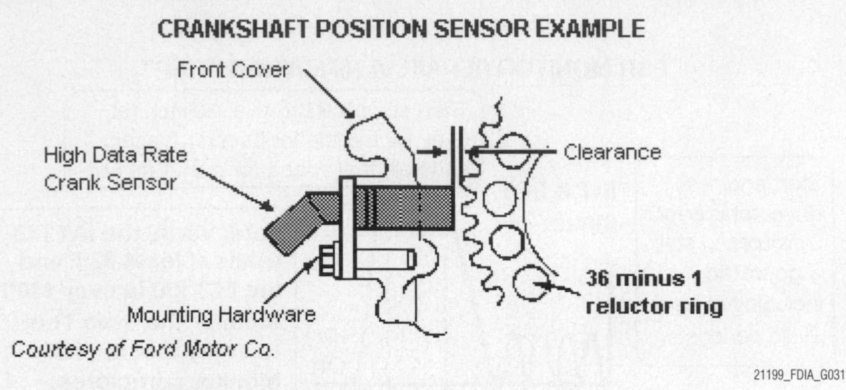

Fig. 9 Typical Crankshaft Position Sensor

BDC of the power stroke) for each cylinder. This calculation requires accurate measurement of the crankshaft angle. Crankshaft angle measurement is determined using a low data rate system on 4-Cyl engines. The high data rate system is used to determine crankshaft angle on all other engines.

Catalyst Damaging Misfire (One-Trip Detection)

If the PCM detects a Catalyst Damaging Misfire, the MIL will flash once per second within 200 engine revolutions from the point where misfire is detected. The MIL will stop flashing and remain on if the engine stops misfiring in a manner that could damage the catalyst.

High Emissions Misfire (Two-Trip Detection)

A High Emissions Misfire is set if a misfire condition is present that could cause the tailpipe emissions to exceed the FTP emissions standard by 1.5 times. If this fault is detected for two consecutive trips under similar engine speed, load and temperature conditions, the MIL is activated. It is also activated if a misfire is detected under similar conditions for two non-consecutive trips that are not 80 trips apart.

State Emissions Failure Misfire (Two-Trip Detection)

A State Emissions Failure Misfire is set if the misfire is sufficient to cause the vehicle to fail a State Inspection or Maintenance (I/M) Test. This fault is determined by identifying misfire percentages that would cause a "durability demonstration vehicle" to fail an Inspection Maintenance (I/M) Test. If the Misfire Monitor detects the fault for two consecutive trips with the engine at similar engine speed, load and temperature conditions, the MIL is activated and a code is set. The MIL is also activated if this type of misfire is detected under similar conditions for two non-consecutive trips of not more than 80 trips apart.

➡ **Some vehicles set Misfire codes because of an early version of OBD II hardware and software. If a misfire code is set and the cause of the fault is not found, clear the code and retest. Search the TSB list for possible answers or contact the dealer.**

Misfire Detection

See Figure 10.

The Misfire Monitor uses the CKP sensor signals to detect an engine misfire. The amount of contribution is calculated based upon measurements determined by crankshaft acceleration from each cylinder's power stroke.

The PCM performs various calculations to detect individual cylinder acceleration rates. If acceleration for a cylinder deviates beyond the average variation of acceleration for all cylinders, a misfire is detected.

Faults detected by the Misfire Monitor:

- Engine mechanical faults, restricted intake or exhaust system
- Dirty or faulty fuel injectors, loose or damaged injector connectors
- The vehicle has been run low on fuel or run until it ran out of fuel

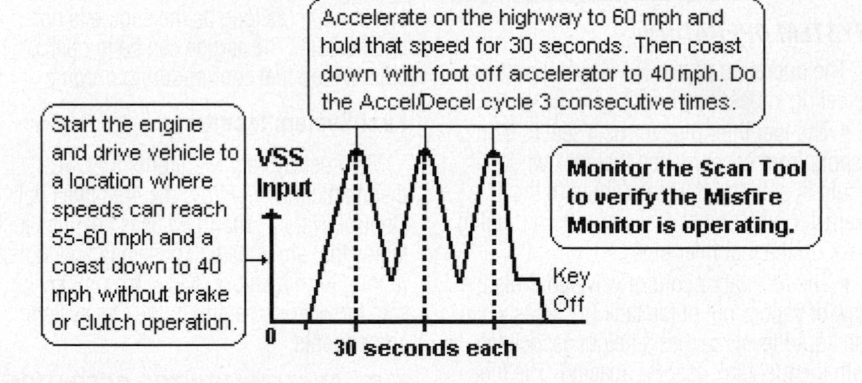

Fig. 10 Misfire Detection Monitor

Oxygen Sensor Monitor

The Oxygen Sensor Monitor is a PCM diagnostic designed to monitor the front and rear oxygen sensor for faults or deterioration that could cause tailpipe emissions to exceed 1.5 times the FTP standard. The front oxygen sensor voltage and response time are also monitored.

HO2S Monitor Operation

Fuel System and Misfire Monitors must be run and complete before the PCM will start the HO2S Monitor. Additionally, parts of the HO2S Sensor Monitor are enabled during the KOER Self-Test. The HO2S Monitor is run during each drive cycle after the CKP, ECT, IAT and MAF sensor signals are within a predetermined range.

Fixed Frequency Closed Loop Test

See Figure 11.

The HO2S Monitor constantly monitors the sensor voltage and frequency. The PCM detects a high voltage condition by comparing the HO2S signal to a preset level.

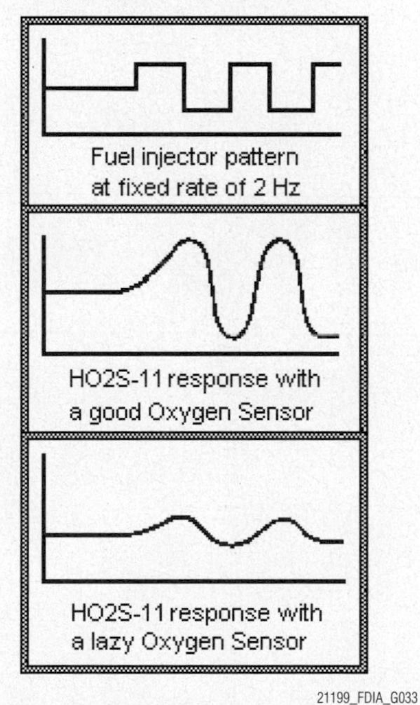

Fig. 11 Fixed Frequency Test

A Fixed Frequency Closed Loop Test is used to check the HO2S voltage and frequency. A sample of the HO2S signal is checked to determine if the sensor is capable of switching properly or has a slow response time (referred to as a lazy sensor).

Oxygen Sensor Heater Monitor

The Oxygen Sensor Heater Monitor is a PCM diagnostic designed to monitor the Oxygen Sensor Heater and its related circuits for faults.

OXYGEN SENSOR HEATER MONITOR OPERATION

The Oxygen Sensor Heater Monitor performs its task by detecting whether the proper amount of O2 sensor voltage change occurred as the HO2S Heater is turned from "on" to "off" with the engine in closed loop. The time it takes for the HO2S-11 and HO2S-12 signal to switch (the response time) is constantly monitored by the Oxygen Sensor Monitor. Once the Oxygen Sensor Heater Monitor is enabled, if the switch time for the HO2S-11 or HO2S-12 signal is too long, the PCM fails the test, the MIL is activated and a trouble code is set.

➡ **Response time is defined as the amount of time it takes for a HO2S signal to switch from Rich to Lean, and then Lean to Rich.**

FRONT AND REAR OXYGEN SENSOR HEATER OPERATION

Both upstream and downstream Oxygen sensors are used on the OBD II system. These sensors are designed with additional protection around the ceramic core to protect them from condensation that could crack them if the heater is turned on with condensation present.

The HO2S heaters are not turned on until the ECT sensor signal indicates that the engine is warm. The delay period can last for as long as 5 minutes from startup. The delay allows any condensation in the Exhaust system to evaporate.

Faults detected by the HO2S or HO2S Heater Monitor:

- A fault in the HO2S, the HO2S heater or its related circuits
- A fault in the HO2S connectors (look for moisture tracking)
- A defective Power Control Module

Air Injection System Monitor

The Air Injection System Monitor is an OBD diagnostic controlled by the PCM that monitors the Air Injection (AIR) system. The Oxygen Sensor Monitor must run and complete before the PCM will run this test. The PCM enables this test during AIR system operation after certain engine conditions are met and these enable criteria are met:

- Crankshaft Position sensor signal must be present
- ECT and IAT sensor input signals must be within limits

AIR MONITOR–ELECTRIC PUMP DESIGN

The AIR Monitor consists of these Solid State Monitor tests:

- A check of the Solid State relay for electrical faults.
- A check of the secondary side of the relay for electrical faults.
- A test to determine if the AIR system can inject additional air.

AIR MONITOR–MECHANICAL PUMP DESIGN

The AIR Monitor for the mechanical (belt-driven air pump) design uses two Output State Monitor configurations to perform two different circuit tests. One test is used to check for faults in the Secondary Air Bypass (AIRB) solenoid circuit. The normal function of the AIRB solenoid and valve assembly is to dump air into the atmosphere.

A second test is used to check for electrical faults in the Secondary Air Divert (AIRD) solenoid. The normal function of the AIRD solenoid and valve assembly is to direct the air either upstream or downstream.

FUNCTIONAL CHECK

See Figure 12.

An AIR system functional check is done at startup with the AIR pump on or during a hot idle period if the startup part of the test was not performed. A flow test is included that uses the HO2S signal to indicate the presence of extra air injected into the exhaust stream.

Diagnostic Trouble Codes

In the Diagnostic Trouble Code charts for the specific manufacturers you will see the following terms in the left column of the chart:

1. 1T–This means the code was activated when the PCM recognized the problem the first time it occurred.
2. 2T–This means the code was activated when the PCM recognized the problem and set the code after it occurred two times.
3. CCM–This means that the code and system affected is an emission related device and has a Comprehensive Component Monitor (CCM) tracking it.
4. MIL: Yes–This means that the Malfunction Indicator Light will be displayed.

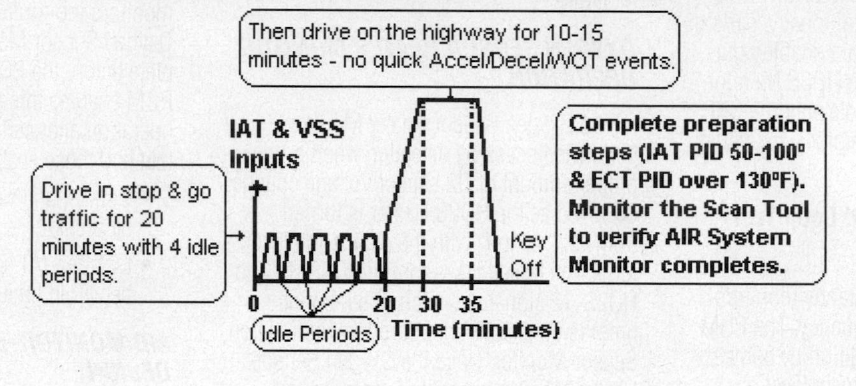

Fig. 12 Secondary AIR monitor

SYMPTOM DIAGNOSIS (NO CODES)

4

Table of Contents

WHAT TO DO WHEN THERE ARE NO DTCS

INTERMITTENT TESTS

What To Do When There Are No DTCs

Do not attempt to diagnose a Drivability Symptoms without having a logical plan to use to determine which Engine Control system is the cause of the symptom - this plan should include a way to determine which systems do not have a problem! Drivability symptom diagnosis is a part of an organized approach to problem solving and repair.

DRIVABILITY SYMPTOM INDEX TABLE

To use this list, locate the symptom that matches a particular problem and refer to the areas to test. The items listed under each symptom may not apply to all models, engines or vehicle systems. The repair steps indicate what vehicle component or system to test.

➡ **The Drivability Symptoms in this list are intended to be generic. While they apply to most vehicles, some vehicles may not have all of the components listed. Refer to other Chilton repair information and electronic media for specific tests.**

Symptom Test Table

Symptom Description	Suggested Areas to Test
Test 1 - No Start, Hard Start Condition • No Crank • Hard Start, Long Crank, Erratic Crank • Stall After Start • No Start, Normal Crank • No Start, MIL is off (if the VREF shorts to ground)	- Check battery, battery circuits to starter - Check for a damaged flywheel, engine compression, base timing and minimum air rate - Check for a failed fuel pump relay - Check for distributor rotor "punch-through" - Check for a faulty ignition control module (ICM) - Check for a VREF circuit shorted to ground - Check SKIM (security system) with a Scan Tool
Test 2 - Rough Idle or Stalls Condition • Low or slow idle speed • Fast idle speed • Hunting or rolling idle speed • Slow return to idle speed • Stalls or almost stalls	- Check for engine vacuum leaks - Check the condition of the PCV valve and lines - Check for excessive carbon buildup - Check for a restricted exhaust - Check base idle speed, check for low fuel pressure - Check the throttle linkage for sticking or binding
Test 3 - Runs Rough Condition • At idle speed • During acceleration • At cruise speed • During deceleration	- Check for engine vacuum leaks at intake manifold - Check condition of ignition secondary components - Check base timing and idle speed settings - Check for low or high fuel pressure - Check for dirty, leaking or shorted fuel injectors - Check for excessive carbon buildup on valves
Test 4 - Cuts-out, Misses Condition • At idle speed • During acceleration • At cruise speed • During deceleration	- Check for engine vacuum leaks at intake manifold - Check condition of ignition secondary components - Check that spark timing advance is available - Check for low or high fuel pressure - Check for dirty, leaking or shorted fuel injectors - Check for excessive carbon buildup on valves
Test 5 - Bucks, Jerks Condition • During acceleration • At cruise speed • During deceleration	- Check for engine vacuum leaks at intake manifold - Check condition of ignition secondary components - Check that spark timing advance is available - Check for low or high fuel pressure - Check for dirty, leaking or shorted fuel injectors - Check operation of the TCC solenoid, brake switch

Symptom Diagnosis Test 1 — No Start, Hard Start Condition

➡ **If there is no spark output or fuel pressure available, check for a failed fuel pump relay, no power to the PCM, or loss of the ignition reference signal to the PCM.**

PRELIMINARY CHECKS

Prior to starting this symptom test routine, inspect these underhood items:

- Check battery charge and condition, starter current draw.
- Verify the starter relay operation and that the engine cranks (turns over).
- Verify the check engine light (MIL) operation - if it does not activate, check the PCM power and ground circuits, and check for 5v supply at the MAP or TP sensor.
- Check Air Intake system for restrictions (inspect air inlet tubes, air filter for dirt, etc.).
- Check the status of the Smart Key Immobilizer System (SKIM) with the Scan Tool.

Test 1 Chart

Step	Action	Yes	No
1	**Step Description: No Start Condition Only** » Check battery cables, state of charge. » If the engine does not rotate, inspect for a locked engine (hydrostatic lockup condition). » Does the engine crank normally?	Go to Step 2.	Repair the fault in the battery, starter, or Base Engine. Retest for the symptom when all repairs are done.
2	**Step Description: Check the Fuel System** » Verify that the pump operates at key on. » Check the fuel pump relay operation. If the relay does not operate, check for blown fuse. » Inspect pump for a leak-down condition » Test fuel pressure, volume and quality. » Test the operation of the fuel regulator. » Are there any faults in the Fuel system?	Make needed repairs.	Go to Step 3.
3	**Step Description: Check the Ignition System** » Inspect ignition secondary components for damage (look for rotor "punch-through"). » Inspect the coils for signs of spark leakage at coil towers or primary connections. » Check the spark output with a spark tester. » Test Ignition system with an engine analyzer. » Are there any faults in the Ignition system?	Make repairs to the Ignition system. Then retest the symptom.	Go to Step 4.
4	**Step Description: Check the Exhaust System** » Check Exhaust system for leaks or damage. » Check the Exhaust system for a restriction using the Vacuum or Pressure Gauge Test (e.g., exhaust backpressure reading should not exceed 1.5 psi at cruise speeds). » Are there any faults in the Exhaust system?	Make repairs to the Exhaust system. Then retest the symptom.	Go to Step 5.
5	**Step Description: Check the MAP Sensor** » Disconnect the MAP sensor and attempt to start the engine. » Does the engine start and run normally?	Replace the MAP sensor. Retest for the symptom when repairs are completed.	Go to Step 6.
6	**Step Description: Check for a Hot Engine** » Check for signs of an engine overheating condition related to a Hard Start Symptom. » Does the engine appear to be overheated?	Make the repairs to correct the hot engine and then retest for the symptom when done.	Go to Step 7.
7	**Step Description: Check ECT Sensor PID** » Connect a Scan Tool and turn the key to on. » Read the ECT sensor (compare to chart). » Has the ECT sensor shifted out of range?	Replace the ECT sensor. Then retest for the symptom when all repairs are completed.	Go to Step 8.
8	**Step Description: Check the PCV System** » Inspect the PCV system components for broken parts or loose connections. » Test the operation of the PCV valve. » Are there any faults in the PCV system?	Repair the PCV system. Refer to the PCV system tests. Retest the symptom when all repairs are done.	Go to Step 9.
9	**Step Description: Check the EVAP System** » nspect for damaged or disconnected EVAP system components. » Inspect for a fuel saturated charcoal canister. » Are there any faults in the EVAP system?	Refer to the EVAP system tests. Retest for the symptom when all repairs are completed.	Go to Step 10.
10	**Step Description: Test the Base Engine** » Check the engine compression. » Test valve timing and timing chain condition. » Check for a worn camshaft or valve train. » Check for any large intake manifold leaks. » Are there any faults in the Base Engine?	Repair the Base Engine. Refer to the Base Engine Tests. Retest symptom when done.	Return to Step 2 to repeat the test steps in this series to locate and repair the "No Start, Hard Start" condition.

Symptom Diagnosis Test 2 — Rough, Low or High Idle Speed Condition

➡ If the vehicle has a rough idle and the base timing, idle speed and the IAC (or AIS) motor operates properly, check the engine for excessive carbon buildup.

PRELIMINARY CHECKS

Prior to starting this symptom test routine, inspect these underhood items:

- All related vacuum lines for proper routing and integrity.
- All related electrical connectors and wiring harnesses for faults (Wiggle Test).
- Check the throttle linkage for a sticking or binding condition.
- Air Intake system for restrictions (air inlet tubes, dirty air filter, etc.).
- Search for any technical service bulletins related to this symptom.
- Turn the key to off. Unplug the MAP sensor connection and restart the engine to recheck for the idle concern. If the condition is gone, replace the MAP sensor.

Test 2 Chart

Step	Action	Yes	No
1	**Step Description: Verify the rough idle or stall** » Does the engine have a warm engine rough idle, low idle or high idle condition in P or N?	Go to Step 2.	Fault is intermittent. Return to the Symptom List and select another fault.
2	**Step Description: Verify idle speed & timing** » Verify the base timing is within specifications » Verify that the base idle speed is set properly » Are the timing and idle speed set properly?	Go to Step 3.	Set the base idle speed and timing to the specifications and then retest for the symptom.
3	**Step Description: Check AIS / IAC Operation** » Check the AIS or IAC motor operation » Inspect the AIS/IAC housing in throttle body for restricted passages. Clean as needed. » Set the parking brake, block the drive wheels and turn the A/C off. Install the Scan Tool. » IAC Motor Tester - Turn the key off and then connect the IAC tester to the IAC valve. » Start the engine and use the IAC tester to extend and retract the IAC valve. » ATM Test - Start the engine. Use the tool to change the speed from min-idle to 1500 rpm. » Did the idle speed change as commanded?	Install an Aftermarket Noid light and check the operation of the PCM and AIS or IAC motor circuits. Check the motor for signs of open or shorted circuits. Replace the IAC motor or PCM as needed or make repairs to the IAC motor wiring. If all are okay, go to Step 4.	If the AIS/IAC motor passages are clean and engine speed did not change as described when the AIS/IAC motor was extended and retracted, replace the AIS/IAC motor. Then retest for the condition.
4	**Step Description: Check/compare PID values** » Connect Scan Tool & turn off all accessories. » Start the engine and allow it to fully warmup. » Monitor all related PIDs on the Scan Tool. » Verify the P/N switch input in gear and Park. » Check the O2S operation with a Lab Scope. » Are all PIDs within normal range?	Go to Step 5. Note: An IAC motor count of over 80 indicates the pintle is extended and an IAC count of (0) indicates the pintle is retracted.	One or more of the PIDs are out of range when compared to "known good" values. Make repairs to the system that is out of range, then retest for the symptom.

5	**Step Description: Check the Ignition System** » Inspect the coils for signs of spark leakage at coil towers or primary connections. » Check the spark output with a spark tester. » Test Ignition system with an engine analyzer. » Were any faults found in the Ignition system?	Make repairs as needed	Go to Step 6.
6	**Step Description: Check the Fuel System** » Inspect the Fuel delivery system for leaks. » Test the fuel pressure, quality and volume. » Test the operation of the pressure regulator. » Were any faults found in the Fuel system?	Make repairs as needed	Go to Step 7.
7	**Step Description: Check the Exhaust System** » Check Exhaust system for leaks or damage. » Check the Exhaust system for a restriction using the Vacuum or Pressure Gauge Test (e.g., exhaust backpressure reading should not exceed 1.5 psi at cruise speeds). » Were any faults found in Exhaust System?	Make repairs to the Exhaust system. Then retest the symptom.	Go to Step 8.
8	**Step Description: Check the PCV System** » Inspect the PCV system components for broken parts or loose connections. » Test the operation of the PCV valve. » Were any faults found in the PCV system?	Make repairs to the PCV system. Refer to the PCV system tests. Then retest for the condition.	Go to Step 9.
9	**Step Description: Check the EVAP System** » Inspect for damaged or disconnected EVAP system components or a saturated canister. » Were any faults found in the EVAP system?	Make repairs to EVAP system. Retest for the condition.	Go to Step 10.
10	**Step Description: Check the Base Engine** » Test the engine compression. » Test valve timing and timing chain condition. » Check for a worn camshaft or valve train. » Check for any large intake manifold leaks. » Were any faults found in the Base Engine?	Make repairs as needed to the Base Engine. Refer to the Base Engine tests. Then retest for the condition when repairs are completed.	Go to Step 2 and repeat the tests from the beginning to locate and repair the cause of the "Rough, Low or High Idle Speed" condition.

Symptom Diagnosis Test 3 — Runs Rough Condition

PRELIMINARY CHECKS

Prior to starting this symptom test routine, inspect these underhood items:

- All related vacuum lines for proper routing and integrity
- Air Intake system for restrictions (air inlet tubes, dirty air filter, etc.)
- Search for any technical service bulletins related to this symptom.

Test 3 Chart

Step	Action	Yes	No
1	**Step Description: Verify engine runs rough** » Start the engine and allow it to idle in P or N. » Does the engine run rough when warm in Park or Neutral position?	Check for any stored codes. If codes are set, repair codes and retest. If no codes are set, go to Step 3.	Go to Step 2.
2	**Step Description: Condition does not exist!** » Inspect various underhood items that could cause an intermittent Runs Rough condition (i.e., dirt in the throttle body, vacuum leaks, IAC motor connections, etc.). » Were any problems located in this step?	Correct the problems. Do a PCM reset and engine "idle relearn" procedure. Then verify the "runs rough" condition is repaired.	The problem is not present at this time. It may be an intermittent problem.
3	**Step Description: Check/compare PID values** » Connect a Scan Tool to the test connector. » Turn off all accessories. » Start the engine and allow it to fully warmup. » Monitor all related PIDs on the Scan Tool. » Were all PIDs within their normal range?	Go to Step 4. Note: The IAC motor should read from 5-50 counts. Check the LONGFT reading for a large shift into the negative range (due to a rich condition).	One or more of the PIDs are out of range when compared to "known good" values. Make repairs to the system that is out of range, then retest for the symptom.
4	**Step Description: Check the Ignition System** » Inspect the coils for signs of spark leakage at coil towers or primary connections. » Check the spark output with a spark tester. » Test Ignition system with an engine analyzer. » Were any faults found in the Ignition system?	Make repairs as needed	Go to Step 5.
5	**Step Description: Check the Fuel System** » Inspect the Fuel delivery system for leaks. » Test the fuel pressure, quality and volume. » Test the operation of the pressure regulator. » Were any faults found in the Fuel system?	Make repairs as needed	Go to Step 6.
6	**Step Description: Check the Exhaust System** » Check Exhaust system for leaks or damage. » Check the Exhaust system for a restriction using the Vacuum or Pressure Gauge Test (e.g., exhaust backpressure reading should not exceed 1.5 psi at cruise speeds). » Were any faults found in Exhaust System?	Make repairs to the Exhaust system. Then retest the symptom.	Go to Step 7.
7	**Step Description: Check the PCV System** » Inspect the PCV system components for broken parts or loose connections. » Test the operation of the PCV valve. » Were any faults found in the PCV system?	Make repairs to the PCV system. Refer to the PCV system tests. Then retest for the condition.	Go to Step 9.
8	**Step Description: Check the EVAP System** » Inspect for damaged or disconnected EVAP system components or a saturated canister. » Were any faults found in the EVAP system?	Make repairs to EVAP system. Retest for the condition.	Go to Step 10.
9	**Step Description: Check Engine Condition** » Test the engine compression. » Test valve timing and timing chain condition. » Check for a worn camshaft or valve train. » Check for any large intake manifold leaks. » Were any faults found in the Base Engine?	Make repairs as needed to the Base Engine. Refer to the Base Engine tests. Then retest for the condition when repairs are completed.	Return to Step 2 and repeat the tests from the beginning to locate and repair the cause of the "Runs Rough" condition.

Symptom Diagnosis Test 4 — Cuts-out or Misses Condition

PRELIMINARY CHECKS

Prior to starting this symptom test routine, inspect these underhood items:

- All related vacuum lines for proper routing and integrity
- Search for any technical service bulletins related to this symptom.

Test 4 Chart

Step	Action	Yes	No
1	**Step Description: Verify Cuts-out condition** » Start the engine and attempt to verify the Cuts-out or misses condition. » Does the engine have a cuts-out condition?	Check for any stored codes. If codes are set, repair codes and retest. If no codes are set, go to Step 3.	Go to Step 2.
2	**Step Description: Condition does not exist!** » Inspect various underhood items that could cause an intermittent Cuts-out condition (i.e., EVAP, Fuel or Ignition system components). » Were any problems located in this step?	Correct the problems. Do a PCM reset and "Fuel Trim Relearn" procedure. Then verify condition is repaired.	The problem is not present at this time. It may be an intermittent problem.
3	**Step Description: Check/compare PID values** » Connect a Scan Tool to the test connector. » Turn off all accessories. » Start the engine and allow it to fully warmup. » Monitor all related PIDs on the Scan Tool (i.e., ECT IAC Counts and LONGFT at idle). » Were all PIDs within their normal range?	Go to Step 4. Note: The IAC motor should be from 5-50 counts. Watch fuel trim (%) for a large shift into the negative (-) range (due to a rich condition).	One or more of the PIDs are out of range when compared to "known good" values. Make repairs to the system that is out of range, then retest for the symptom.
4	**Step Description: Check the Ignition System** » Inspect the coils for signs of spark leakage at coil towers or primary connections. » Check the spark output with a spark tester. » Test Ignition system with an engine analyzer. » Were any faults found in the Ignition system?	Make repairs as needed	Go to Step 5.
5	**Step Description: Check the Fuel System** » Inspect the Fuel delivery system for leaks. » Test the fuel pressure, quality and volume. » Test the operation of the pressure regulator. » Were any faults found in the Fuel system?	Make repairs as needed	Go to Step 6.
6	**Step Description: Check the Exhaust System** » Check Exhaust system for leaks or damage. » Check the Exhaust system for a restriction using the Vacuum or Pressure Gauge Test (e.g., exhaust backpressure reading should not exceed 1.5 psi at cruise speeds). » Were any faults found in Exhaust System?	Make repairs to the Exhaust system. Then retest the symptom.	Go to Step 7.
7	**Step Description: Check the PCV System** » Inspect the PCV system components for broken parts or loose connections. » Test the operation of the PCV valve. » Were any faults found in the PCV system?	Make repairs to the PCV system. Then retest for the condition.	Go to Step 8.
8	**Step Description: Check the EVAP System** » Inspect for damaged or disconnected EVAP system components » Check for a saturated EVAP canister. » Were any faults found in the EVAP system?	Make repairs to EVAP system. Retest for the condition.	Go to Step 9.
9	**Step Description: Check the AIR system** » Inspect AIR system for broken parts, leaking valves or disconnected hoses. » Test the operation of Secondary AIR system. » Were any faults found in the AIR system?	Make repairs as needed. Refer to the Secondary AIR system tests. Retest for the condition.	Go to Step 10.
10	**Step Description: Check Engine Condition** » Test the engine compression. » Test valve timing and timing chain condition. » Check for a worn camshaft or valve train. » Check for any large intake manifold leaks. » Were any faults found in the Base Engine?	Make repairs as needed to the Base Engine. Refer to the Base Engine tests. Then retest for the condition when repairs are completed.	Go to Step 2 and repeat the tests from the beginning to locate and repair the cause of the "Cuts Out or Misses" condition.

Symptom Diagnosis Test 5 — Surge Condition

PRELIMINARY CHECKS

1. Discuss how the operation of the torque converter clutch (TCC) or air conditioning compressor can affect the "feel" of the vehicle during normal operation. Refer to the information in the Owner's Manual to explain how these devices normally operate.
2. Search for any technical service bulletins related to this symptom.

Test 5 Chart

Step	Action	Yes	No
1	**Step Description: Verify the surge condition** » Drive the vehicle and attempt to verify that the vehicle surges at cruise speeds. » Does the engine have a surge condition?	Check for any stored codes. If codes are set, repair codes and retest. If no codes are set, go to Step 3.	Go to Step 2.
2	**Step Description: Condition does not exist!** » Inspect various underhood items that could cause an intermittent surge condition (check for leaks in the MAP sensor vacuum lines). » Were any problems located in this step?	Correct the problems. Do a PCM reset and "Fuel Trim Relearn" procedure. Then verify condition is repaired.	The problem is not present at this time. It may be an intermittent problem.
3	**Step Description: Check/compare PID values** » Connect a Scan Tool to the test connector. » lStart the engine and allow it to fully warmup. » Monitor all related PIDs on Scan Tool (HO2S switching, LONGFT, and the TCC operation) » Compare VSS PID reading to speedometer. » Were all PIDs within their normal range?	Go to Step 4. Note: Verify that the front HO2S responds quickly to throttle changes. Check for silicon contamination on the front HO2S (this can cause a rich A/F signal).	One or more of the PIDs are out of range when compared to "known good" values. Make repairs to the system that is out of range, then retest for the symptom.
4	**Step Description: Check the Ignition System** » Inspect the coils for signs of spark leakage at coil towers or primary connections. » Check the spark output with a spark tester. » Test Ignition system with an engine analyzer. » Were any faults found in the Ignition system?	Make repairs as needed	Go to Step 5.
5	**Step Description: Check the Fuel System** » Inspect the Fuel delivery system for leaks. » Test the fuel pressure, quality and volume. » Test the operation of the pressure regulator. » Were any faults found in the Fuel system?	Make repairs as needed	Go to Step 6.
6	**Step Description: Check the Exhaust System** » Check Exhaust system for leaks or damage. » Check the Exhaust system for a restriction using the Vacuum or Pressure Gauge Test (e.g., exhaust backpressure reading should not exceed 1.5 psi at cruise speeds). » Were any faults found in Exhaust System?	Make repairs to the Exhaust system. Then retest the symptom.	Return to Step 2 and repeat the tests from the beginning to locate and repair the cause of the "Surge" condition.

INTERMITTENT TESTS

Many trouble code repair charts end with a result that reads "Fault Not Present at this Time." What this expression means is that the conditions that were present when a code set or drivability symptom occurred are no longer there or were not met. In effect, the problem was present at least once, but is not present at this time. However, it is likely to return in the future, so it should be diagnosed and repaired if at all possible.

One way to find an intermittent problem is to gather the information that was present when the problem occurred. In the case of a Code Fault, this can be done in two ways: by capturing the data in Snapshot or Movie mode or by driver observations.

The PCM has to detect the fault for a specific period of time before a trouble code will set. While intermittent problems may appear to be occasional in nature, they usually occur under specific conditions. Therefore, you should identify and duplicate these conditions. Since intermittent faults are difficult to duplicate, a logical routine (checklist) must be followed when attempting to find the faulty component, system or circuit. The tests on the next page can be used to help find the cause of an intermittent fault.

Some intermittent faults occur due to a loose connection, wiring problem or warped circuit board. An intermittent fault can also be caused by poor test techniques that cause damage to the male or female ends of a connector.

Test for Loose Connectors

To test for a loose or damaged connection, take the male end of a connector from another wiring harness and carefully push it into the "suspect" female terminal to verify that the opening is tight. There should be some resistance felt as the male connector is inserted in the terminal connection.

The Wiggle Test

See Figures 1 and 2.

A wiggle test can be used to locate the cause of some intermittent faults. The sensor, switch or the PCM wiring can be back-probed, as shown, while the test is done.

During testing, move or wiggle the suspect device, connector or wiring while watching for a change.

If the DVOM has a Min/Max record mode, use this mode during the test.

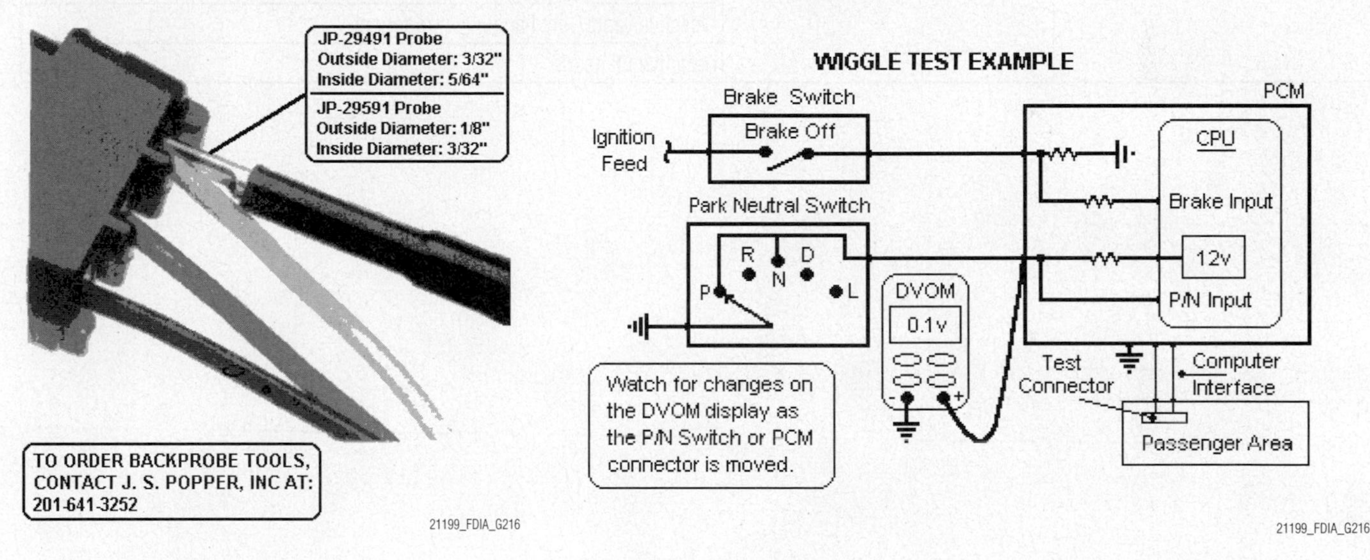

Fig. 1 Backprobing a connector

Fig. 2 Wiggle Test Example

Diagnosis And Testing - Vehicle Does Not Fill

CONDITION	POSSIBLE CAUSES	CORRECTION
Pre-Mature Nozzle Shut-Off	Defective fuel tank assembly components.	Fill tube improperly installed (sump)
		Fill tube hose pinched.
		Check valve stuck shut.
		Control valve stuck shut.
	Defective vapor/vent components.	Vent line from control valve to canister pinched.
		Vent line from canister to vent filter pinched.
		Canister vent valve failure (requires double failure, plugged to NVLD and atmosphere).
		Leak detection pump failed closed.
		Leak detection pump filter plugged.
	On-Board diagnostics evaporative system leak test just conducted.	Canister vent valve vent plugged to atmosphere.
		Engine still running when attempting to fill (System designed not to fill).
	Defective fill nozzle.	Try another nozzle.
Fuel Spits Out Of Filler Tube.	During fill.	See Pre-Mature Shut-Off.
	At conclusion of fill.	Defective fuel handling component. (Check valve stuck open).
		Defective vapor/vent handling component.
		Defective fill nozzle.

AUDI
DIAGNOSTIC TROUBLE CODES

5

TABLE OF CONTENTS

DIAGNOSTIC TROUBLE CODES

OBD II Vehicle Applications

AUDI

TT

2000–2006

1.8L L4 MPI . . . . .Engine Codes: ATC, AWP, AMU, BEA
3.2L VR6 MPI . . . . Engine Code: BHE

A4

1996–2006

1.8L L4 MPI . . . . .Engine Codes: AEB, AMB, ATW, AWM
2.7L V6 MPI . . . . Engine Codes: APB
2.8L V6 MPI . . . . Engine Codes: AFC, AHA, ATQ
3.0L V6 MPI . . . . Engine Codes: AVK

A4 Avant

1998–2006

1.8L L4 MPI . . . . .Engine Codes: AEB, AMB, ATW, AWM
2.7L V6 MPI . . . . Engine Codes: APB
2.8L V6 MPI . . . . .Engine Codes: AHA, ATQ
3.0L V6 MPI . . . . Engine Codes: AVK

A4 Cabriolet

2003–2006

1.8L L4 MPI . . . . .Engine Codes: AMB
3.0L V6 MPI . . . . .Engine Codes: AVK, BGN

A6

1995–2006

2.7L V6 MPI . . . . Engine Codes: APB, BEL
2.8L V6 MPI . . . . Engine Codes: AFC, AHA, ATQ
3.0L V6 MPI . . . . Engine Codes: AVK
3.2L V6 MPI . . . . Engine Codes: BKH
4.2L V8 MPI . . . . Engine Codes: ART, AWN, BNK

A6 Avant

1995–2004, 2006

2.8L V6 MPI . . . . Engine Codes: AAH, AFC, AHA, ATQ
3.0L V6 MPI . . . . Engine Codes: AVK
3.2L V6 MPI . . . . Engine Codes: BKH

Allroad Quattro

2001–2005

2.7L V6 MPI . . . . Engine Codes: APB, BEL
4.2L V8 MPI . . . . Engine Codes: BAS

A8L

2004–2005

4.2L V8 MPI . . . . .Engine Codes: BFM

A8

1997–2003, 2005

3.7L V8 MPI . . . . .Engine Codes: AEW
4.2L V8 MPI . . . . Engine Codes: ABZ, AKB, AUX, AYS, BFM

GAS ENGINE TROUBLE CODE LIST

Introduction

To use this information, first read and record all codes in memory along with any Freeze Frame data. If the ECM reset function is done prior to recording any data, all codes and freeze frame data will be lost! Look up the desired code by DTC number, Code Title and Conditions (enable criteria) that indicate why a code set, and how to drive the vehicle. 1T and 2T indicate a 1-trip or 2-trip fault and the Monitor type.

Gas Engine OBD II Trouble Code List (P0xxx Codes)

DTC	Trouble Code Title, Conditions & Possible Causes:
DTC: P0010 **1T CCM, MIL: Yes** **1996, 1997, 1998, 1999, 2000, 2001, 2002, 2003, 2004, 2005, 2006** **Models:** TT, A4, A6, Allroad Quattro **Engines:** 1.8L, 2.7L, 2.8L, 3.0L, 3.2L, 4.2L **Transmissions:** All	**"A" Camshaft Position Actuator Circuit (Bank 1) Conditions:** Key on or engine running; and the ECM detected an unexpected high voltage or low voltage condition on the camshaft position sensor. The relative position between the camshaft and crankshaft needs to be optimal so the engine has better torque, fuel economy and emissions. **Note: The camshaft adjustment is load- and RPM dependant. The electrical camshaft adjustment valve 1 switches oil pressure onto camshaft adjuster (mechanical adjustment mechanism), which adjusts the camshaft.** **Possible Causes:** • Fuel pump has failed • Actuator circuit is open • ECM has failed • Battery voltage below 11.5 volts • Position actuator circuit may short to B+ or Ground
DTC: P0011 **1T CCM, MIL: Yes** **1996, 1997, 1998, 1999, 2000, 2001, 2002, 2003, 2004, 2005, 2006** **Models:** TT, A4, A6, Allroad Quattro, A8, A8L **Engines:** 1.8L, 2.7L, 2.8L, 3.0L, 3.2L, 4.2L **Transmissions:** All	**"A" Camshaft Position Timing Over-Advanced (Bank 1) Conditions:** Engine started and driven at an engine speed of more than 400rpm; and the ECM detected the camshaft timing exceeded the maximum calibrated advance value, or the camshaft remained in an advanced position during the CCM test. The valve timing did not change from the current valve timing or it remained fixed during the testing. **Note: The camshaft adjustment is load- and RPM dependant. The electrical camshaft adjustment valve 1 switches oil pressure onto camshaft adjuster (mechanical adjustment mechanism), which adjusts the camshaft.** **Possible Causes:** • Fuel pump has failed • CPS circuit is open, shorted to ground or shorted to power • ECM has failed • Battery voltage below 11.5 volts • Position actuator circuit may short to B+ or Ground • Camshaft timing improperly set, or continuous oil flow to the VCT piston chamber • Camshaft advance mechanism (the VCT unit) is sticking or binding mechanically • VCT solenoid valve is stuck in open position
DTC: P0012 **1T CCM, MIL: Yes** **1996, 1997, 1998, 1999, 2000, 2001, 2002, 2003, 2004, 2005, 2006** **Models:** TT, A4, A6, Allroad Quattro, A8, A8L **Engines:** 1.8L, 2.7L, 2.8L, 3.0L, 3.2L, 4.2L **Transmissions:** All	**"A" Camshaft Position Over-Retarded (Bank 1) Conditions:** Engine started and driven at an engine speed of more than 400rpm; and the ECM detected the camshaft timing exceeded the minimu calibrated retarded value, or the camshaft remained in an retarted position during the CCM test. The valve timing did not change from the current valve timing or it remained fixed during the testing. **Note: The camshaft adjustment is load- and RPM dependant. The electrical camshaft adjustment valve 1 switches oil pressure onto camshaft adjuster (mechanical adjustment mechanism), which adjusts the camshaft.** **Possible Causes:** • Fuel pump has failed • CPS circuit is open, shorted to ground or shorted to power • ECM has failed • Battery voltage below 11.5 volts • Position actuator circuit may short to B+ or Ground • Camshaft timing improperly set, or continuous oil flow to the VCT piston chamber • Camshaft advance mechanism (the VCT unit) is sticking or binding mechanically • VCT solenoid valve is stuck in open position
DTC: P0013 **1T CCM, MIL: Yes** **1996, 1997, 1998, 1999, 2000, 2001, 2002, 2003, 2004, 2005, 2006** **Models:** A4, A6, Allroad Quattro **Engines:** 2.7L, 2.8L, 3.0L, 3.2L **Transmissions:** All	**"B" Camshaft Position Actuator Circuit (Bank 1) Conditions:** Key on or engine running; and the ECM detected an unexpected high voltage or low voltage condition on the camshaft position sensor. The relative position between the camshaft and crankshaft needs to be optimal so the engine has better torque, fuel economy and emissions. **Note: The camshaft adjustment is load- and RPM dependant. The electrical camshaft adjustment valve 1 switches oil pressure onto camshaft adjuster (mechanical adjustment mechanism), which adjusts the camshaft.** **Possible Causes:** • Fuel pump has failed • ECM has failed • Battery voltage below 11.5 volts • Position actuator circuit may short to B+ or Ground

DTC	Trouble Code Title, Conditions & Possible Causes
DTC: P0014 **1T CCM, MIL: Yes** **1996, 1997, 1998, 1999, 2000, 2001, 2002, 2003, 2004, 2005, 2006** **Models:** TT, A4, A6, Allroad Quattro **Engines:** 2.7L, 2.8L, 3.0L, 3.2L **Transmissions:** All	**"B" Camshaft Position Timing Over-Advanced (Bank 1) Conditions:** Engine started and driven at an engine speed of more than 400rpm; and the ECM detected the camshaft timing exceeded the maximum calibrated advance value, or the camshaft remained in an advanced position during the CCM test. The valve timing did not change from the current valve timing or it remained fixed during the testing. **Note: The camshaft adjustment is load- and RPM dependant. The electrical camshaft adjustment valve 1 switches oil pressure onto camshaft adjuster (mechanical adjustment mechanism), which adjusts the camshaft.** **Possible Causes:** • Fuel pump has failed • CPS circuit is open, shorted to ground or shorted to power • ECM has failed • Battery voltage below 11.5 volts • Position actuator circuit may short to B+ or Ground • Camshaft timing improperly set, or continuous oil flow to the VCT piston chamber • Camshaft advance mechanism (the VCT unit) is sticking or binding mechanically • VCT solenoid valve is stuck in open position
DTC: P0015 **1T CCM, MIL: Yes** **1996, 1997, 1998, 1999, 2000, 2001, 2002, 2003, 2004, 2005, 2006** **Models:** TT, A4, A6, Allroad Quattro **Engines:** 2.7L, 2.8L, 3.0L, 3.2L **Transmissions:** All	**"B" Camshaft Position Over-Retarded (Bank 1) Conditions:** Engine started and driven at an engine speed of more than 400rpm; and the ECM detected the camshaft timing exceeded the minimu calibrated retarded value, or the camshaft remained in an retarted position during the CCM test. The valve timing did not change from the current valve timing or it remained fixed during the testing. **Note: The camshaft adjustment is load- and RPM dependant. The electrical camshaft adjustment valve 1 switches oil pressure onto camshaft adjuster (mechanical adjustment mechanism), which adjusts the camshaft.** **Possible Causes:** • Fuel pump has failed • CPS circuit is open, shorted to ground or shorted to power • ECM has failed • Battery voltage below 11.5 volts • Position actuator circuit may short to B+ or Ground • Camshaft timing improperly set, or continuous oil flow to the VCT piston chamber • Camshaft advance mechanism (the VCT unit) is sticking or binding mechanically • VCT solenoid valve is stuck in open position
DTC: P0016 **2T CCM, MIL: Yes** **2004, 2005, 2006** **Models:** TT: **Engines:** 3.2L (BHE); **Transmissions:** All	**Crankshaft Position - Camshaft Position Correlation Bank 1 Sensor A Conditions:** Engine started, engine running, and the ECM detected a diviation between the crankshaft position sensor signal and the camshaft position sensor. A rationality error has been detected for camshaft position out of phase with crankshaft. **Possible Causes:** • Camshaft Position (CMP) sensor is faulty • CMP circuit short to ground, power or open • Engine Speed (RPM) sensor is faulty • ECM has failed
DTC: P0018 **1T CCM, MIL: Yes** **2004, 2005, 2006** **Models:** TT **Engines:** 3.2L **Transmissions:** All	**Crankshaft Position - Camshaft Position Correlation Bank 2 Sensor A Conditions:** Engine started, engine running, and the ECM detected a diviation between the crankshaft position sensor signal and the camshaft position sensor. A rationality error has been detected for camshaft position out of phase with crankshaft. **Possible Causes:** • Camshaft Position (CMP) sensor is faulty • CMP circuit short to ground, power or open • Engine Speed (RPM) sensor is faulty • ECM has failed
DTC: P0020 **2T CCM, MIL: Yes** **1996, 1997, 1998, 1999, 2000, 2001, 2002, 2003, 2004, 2005, 2006** **Models:** A4, A6, Allroad Quattro **Engines:** 2.7L, 2.8L, 3.0L, 3.2L **Transmissions:** All	**"A" Camshaft Position Timing Over-Advanced (Bank 2) Conditions:** Engine started and driven at an engine speed of more than 400rpm; and the ECM detected the camshaft timing exceeded the maximum calibrated advance value, or the camshaft remained in an advanced position during the CCM test. The valve timing did not change from the current valve timing or it remained fixed during the testing. **Possible Causes:** • Fuel pump has failed • CPS circuit is open, shorted to ground or shorted to power • ECM has failed • Battery voltage below 11.5 volts • Position actuator circuit may short to B+ or Ground • Camshaft timing improperly set, or continuous oil flow to the VCT piston chamber • Camshaft advance mechanism (the VCT unit) is sticking or binding mechanically • VCT solenoid valve is stuck in open position

DTC	Trouble Code Title, Conditions & Possible Causes
DTC: P0021 **2T CCM, MIL: Yes** **1996, 1997, 1998, 1999, 2000, 2001, 2002, 2003, 2004, 2005, 2006** **Models:** A4, A6, Allroad Quattro, A8, A8L **Engines:** 2.7L, 2.8L, 3.0L, 3.2L, 4.2L **Transmissions:** All	**"A" Camshaft Position Actuator Circuit (Bank 2) Conditions:** Key on or engine running; and the ECM detected an unexpected high voltage or low voltage condition on the camshaft position sensor. The relative position between the camshaft and crankshaft needs to be optimal so the engine has better torque, fuel economy and emissions. **Possible Causes:** • Fuel pump has failed • Actuator circuit is open, shorted to ground or shorted to power • ECM has failed • Battery voltage below 11.5 volts • Position actuator circuit may short to B+ or Ground
DTC: P0022 **2T CCM, MIL: Yes** **1996, 1997, 1998, 1999, 2000, 2001, 2002, 2003, 2004, 2005, 2006** **Models:** A4, A6, Allroad Quattro **Engines:** 2.7L, 2.8L, 3.0L, 3.2L **Transmissions:** All	**"A" Camshaft Position Over-Retarded (Bank 2) Conditions:** Engine started and driven at an engine speed of more than 400rpm; and the ECM detected the camshaft timing exceeded the minimu calibrated retarded value, or the camshaft remained in an retarted position during the CCM test. The valve timing did not change from the current valve timing or it remained fixed during the testing. **Possible Causes:** • Fuel pump has failed • CPS circuit is open, shorted to ground or shorted to power • ECM has failed • Battery voltage below 11.5 volts • Position actuator circuit may short to B+ or Ground • Camshaft timing improperly set, or continuous oil flow to the VCT piston chamber • Camshaft advance mechanism (the VCT unit) is sticking or binding mechanically • VCT solenoid valve is stuck in open position
DTC: P0023 **2T CCM, MIL: Yes** **1996, 1997, 1998, 1999, 2000, 2001, 2002, 2003, 2004, 2005, 2006** **Models:** A4, A6, Allroad Quattro **Engines:** 2.7L, 2.8L, 3.0L, 3.2l **Transmissions:** All **Transmissions:** All	**"B" Camshaft Position Actuator Circuit (Bank 2) Conditions:** Key on or engine running; and the ECM detected an unexpected high voltage or low voltage condition on the camshaft position sensor. The relative position between the camshaft and crankshaft needs to be optimal so the engine has better torque, fuel economy and emissions. **Possible Causes:** • Fuel pump has failed • Actuator circuit is open, shorted to ground or shorted to power • ECM has failed • Battery voltage below 11.5 volts • Position actuator circuit may short to B+ or Ground
DTC: P0024 **2T CCM, MIL: Yes** **1996, 1997, 1998, 1999, 2000, 2001, 2002, 2003, 2004, 2005, 2006** **Models:** A4, A6, Allroad Quattro **Engines:** 2.7L, 2.8L, 3.0L, 3.2L **Transmissions:** All **Transmissions:** All	**"B" Camshaft Position Timing Over-Advanced (Bank 2) Conditions:** Engine started and driven at an engine speed of more than 400rpm; and the ECM detected the camshaft timing exceeded the maximum calibrated advance value, or the camshaft remained in an advanced position during the CCM test. The valve timing did not change from the current valve timing or it remained fixed during the testing. **Possible Causes:** • Fuel pump has failed • CPS circuit is open, shorted to ground or shorted to power • ECM has failed • Battery voltage below 11.5 volts • Position actuator circuit may short to B+ or Ground • Camshaft timing improperly set, or continuous oil flow to the VCT piston chamber • Camshaft advance mechanism (the VCT unit) is sticking or binding mechanically • VCT solenoid valve is stuck in open position
DTC: P0025 **2T CCM, MIL: Yes** **1996, 1997, 1998, 1999, 2000, 2001, 2002, 2003, 2004, 2005, 2006** **Models:** A4, A6, Allroad Quattro **Engines:** 2.7L, 2.8L, 3.0L **Transmissions:** All	**"B" Camshaft Position Over-Retarded (Bank 2) Conditions:** Engine started and driven at an engine speed of more than 400rpm; and the ECM detected the camshaft timing exceeded the minimu calibrated retarded value, or the camshaft remained in an retarted position during the CCM test. The valve timing did not change from the current valve timing or it remained fixed during the testing. **Possible Causes:** • Fuel pump has failed • CPS circuit is open, shorted to ground or shorted to power • ECM has failed • Battery voltage below 11.5 volts • Position actuator circuit may short to B+ or Ground • Camshaft timing improperly set, or continuous oil flow to the VCT piston chamber • Camshaft advance mechanism (the VCT unit) is sticking or binding mechanically • VCT solenoid valve is stuck in open position

DTC	Trouble Code Title, Conditions & Possible Causes
DTC: P0030 **2T CCM, MIL: Yes** **1996, 1997, 1998, 1999, 2000, 2001, 2002, 2003, 2004, 2005, 2006** **Models:** TT, A4, A6, Allroad Quattro, A8, A8L **Engines:** 1.8L, 2.7L, 2.8L, 3.0L, 3.2L, 4.2L **Transmissions:** All	**HO2S Heater (Bank 1 Sensor 1) Control Circuit Malfunction Conditions:** Engine started, battery voltage must be at least 11.5v, all electrical components must be off, the ground between the engine and the chassis must be well connected, the exhaust system must be properly sealed between the catalytic converter and the cylinder head, the coolant temperature must be 80 degrees Celsius, and the oxygen sensor heater for oxygen sensor before the catalytic converter must be properly functioning. The ECM detected the HO2S signal was in a negative voltage range referred to as "character shift downward". This code sets when the HO2S signal remains in a low state (usually less than 156 mv). In effect, it does not switch properly between 0.1v and 1.1v in closed loop operation. **Possible Causes:** • HO2S is contaminated (due to presence of silicone in fuel) • HO2S signal and ground circuit wires crossed in wiring harness • HO2S signal circuit is shorted to sensor or chassis ground • HO2S element has failed (internal short condition) • ECM has failed
DTC: P0031 **2T CCM, MIL: Yes** **1996, 1997, 1998, 1999, 2000, 2001, 2002, 2003, 2004, 2005, 2006** **Models:** TT, A4, A6, Allroad Quattro, A8, A8L **Engines:** 1.8L, 2.7L, 2.8L, 3.0L, 3.2L, 4.2L **Transmissions:** All	**HO2S Heater (Bank 1 Sensor 1) Circuit Low Input Conditions:** Engine started, battery voltage must be at least 11.5v, all electrical components must be off, the ground between the engine and the chassis must be well connected, the exhaust system must be properly sealed between the catalytic converter and the cylinder head, the coolant temperature must be 80 degrees Celsius, and the oxygen sensor heater for oxygen sensor before the catalytic converter must be properly functioning. The ECM detected the HO2S signal was in a negative voltage range referred to as "character shift downward". This code sets when the HO2S signal remains in a low state. In effect, it does not switch properly in the closed loop operation. The HO2S (before the three-way catalytic converter) has a short circuit to ground that has lasted longer than 200 seconds **Possible Causes:** • HO2S is contaminated (due to presence of silicone in fuel) • HO2S signal and ground circuit wires crossed in wiring harness • HO2S signal circuit is shorted to sensor or chassis ground • HO2S element has failed (internal short condition) • ECM has failed
DTC: P0032 **2T CCM, MIL: Yes** **1996, 1997, 1998, 1999, 2000, 2001, 2002, 2003, 2004, 2005, 2006** **Models:** TT, A4, A6, Allroad Quattro, A8, A8L **Engines:** 1.8L, 2.7L, 2.8L, 3.0L, 3.2L, 4.2L **Transmissions:** All	**HO2S Heater (Bank 1 Sensor 1) Circuit High Input Conditions:** Engine started, battery voltage must be at least 11.5v, all electrical components must be off, the ground between the engine and the chassis must be well connected, the exhaust system must be properly sealed between the catalytic converter and the cylinder head, the coolant temperature must be 80 degrees Celsius, and the oxygen sensor heater for oxygen sensor before the catalytic converter must be properly functioning. The ECM detected the HO2S signal remained in a high state. **Note: The HO2S signal circuit may be shorted to the heater power circuit due to tracking inside of the HO2S connector. Remove the connector and visually inspect the connector for signs of oil or water.** **Possible Causes:** • HO2S signal shorted to heater power circuit inside connector • HO2S signal circuit shorted to ground or to system voltage • ECM has failed
DTC: P0036 **2T CCM, MIL: Yes** **1996, 1997, 1998, 1999, 2000, 2001, 2002, 2003, 2004, 2005, 2006** **Models:** TT, A4, A6, Allroad Quattro, A8, A8L **Engines:** 1.8L, 2.7L, 2.8L, 3.0L, 3.2L, 4.2L **Transmissions:** All	**HO2S Heater (Bank 1 Sensor 2) Control Circuit Malfunction Conditions:** Engine started, battery voltage must be at least 11.5v, all electrical components must be off, the ground between the engine and the chassis must be well connected, the exhaust system must be properly sealed between the catalytic converter and the cylinder head, the coolant temperature must be 80 degrees Celsius, and the oxygen sensor heater for oxygen sensor before the catalytic converter must be properly functioning. The ECM detected the HO2S signal was in a negative voltage range referred to as "character shift downward". This code sets when the HO2S signal remains in a low state. **Possible Causes:** • HO2S is contaminated (due to presence of silicone in fuel) • HO2S signal and ground circuit wires crossed in wiring harness • HO2S signal circuit is shorted to sensor or chassis ground • HO2S element has failed (internal short condition) • ECM has failed
DTC: P0037 **2T CCM, MIL: Yes** **1996, 1997, 1998, 1999, 2000, 2001, 2002, 2003, 2004, 2005, 2006** **Models:** TT, A4, A6, Allroad Quattro, A8, A8L **Engines:** 1.8L, 2.7L, 2.8L, 3.0L, 3.2L, 4.2L **Transmissions:** All	**HO2S Heater (Bank 1 Sensor 2) Circuit Low Input Conditions:** Engine started, battery voltage must be at least 11.5v, all electrical components must be off, the ground between the engine and the chassis must be well connected, the exhaust system must be properly sealed between the catalytic converter and the cylinder head, the coolant temperature must be 80 degrees Celsius, and the oxygen sensor heater for oxygen sensor before the catalytic converter must be properly functioning. The ECM detected the HO2S signal was in a negative voltage range referred to as "character shift downward". This code sets when the HO2S signal remains in a low state. In effect, it does not switch properly in the closed loop operation. The HO2S (before the three-way catalytic converter) has a short circuit to ground that has lasted longer than 200 seconds **Possible Causes:** • HO2S is contaminated (due to presence of silicone in fuel) • HO2S signal and ground circuit wires crossed in wiring harness • HO2S signal circuit is shorted to sensor or chassis ground • HO2S element has failed (internal short condition) • ECM has failed

DTC	Trouble Code Title, Conditions & Possible Causes
DTC: P0038 **2T CCM, MIL: Yes** **1996, 1997, 1998, 1999, 2000, 2001, 2002, 2003, 2004, 2005, 2006** **Models:** TT, A4, A6, Allroad Quattro, A8, A8L **Engines:** 1.8L, 2.7L, 2.8L, 3.0L, 3.2L, 4.2L **Transmissions:** All	**HO2S Heater (Bank 1 Sensor 2) Circuit High Input Conditions:** Engine started, battery voltage must be at least 11.5v, all electrical components must be off, the ground between the engine and the chassis must be well connected, the exhaust system must be properly sealed between the catalytic converter and the cylinder head, the coolant temperature must be 80 degrees Celsius, and the oxygen sensor heater for oxygen sensor before the catalytic converter must be properly functioning. The ECM detected the HO2S signal remained in a high state. **Note: The HO2S signal circuit may be shorted to the heater power circuit due to tracking inside of the HO2S connector. Remove the connector and visually inspect the connector for signs of oil or water.** **Possible Causes:** • HO2S signal shorted to heater power circuit inside connector • HO2S signal circuit shorted to ground or to system voltage • ECM has failed
DTC: P0040 **2T CCM, MIL: Yes** **1996, 1997, 1998, 1999, 2000, 2001, 2002, 2003, 2004, 2005, 2006** **Models:** TT, A4, A6, Allroad Quattro **Engines:** 2.7L, 2.8L, 3.0L, 3.2L, 4.2L **Transmissions:** All	**O2 Sensor Signals Swapped (Bank 1 Sensor 1/Bank 2 Sensor 1) Conditions:** Engine started, battery voltage must be at least 11.5v, all electrical components must be off, the ground between the engine and the chassis must be well connected, the exhaust system must be properly sealed between the catalytic converter and the cylinder head, and the coolant temperature must be 80 degrees Celsius. The ECM detected the O2 signals were mixed and reading implausible results from both. **Possible Causes:** • HO2S-11 and HO2S-21 harness connectors are swapped • HO2S-11 and HO2S-21 wiring is crossed inside the harness • HO2S-11 and HO2S-21 wires are crossed at 104-pin connector • Connector coding and color mixed with correct catalytic converter
DTC: P0041 **2T CCM, MIL: Yes** **1996, 1997, 1998, 1999, 2000, 2001, 2002, 2003, 2004, 2005, 2006** **Models:** A4, A6, Allroad Quattro, A8, A8L **Engines:** 2.7L, 2.8L, 3.0L, 3.2L, 4.2L **Transmissions:** All	**O2 Sensor Signals Swapped (Bank 1 Sensor 2/Bank 2 Sensor 2) Conditions:** Engine started, battery voltage must be at least 11.5v, all electrical components must be off, the ground between the engine and the chassis must be well connected, the exhaust system must be properly sealed between the catalytic converter and the cylinder head, and the coolant temperature must be 80 degrees Celsius. The ECM detected the O2 signals were mixed and reading implausible results from both. **Possible Causes:** • HO2S-12 and HO2S-22 harness connectors are swapped • HO2S-12 and HO2S-22 wiring is crossed inside the harness • HO2S-12 and HO2S-22 wires are crossed at 104-pin connector • Connector coding and color mixed with correct catalytic converter
DTC: P0050 **2T CCM, MIL: Yes** **1996, 1997, 1998, 1999, 2000, 2001, 2002, 2003, 2004, 2005, 2006** **Models:** A4, A6, Allroad Quattro, A8, A8L **Engines:** 2.7L, 2.8L, 3.0L, 3.2L, 4.2L **Transmissions:** All	**HO2S Heater (Bank 2 Sensor 1) Control Circuit Malfunction Conditions:** Engine started, battery voltage must be at least 11.5v, all electrical components must be off, the ground between the engine and the chassis must be well connected, the exhaust system must be properly sealed between the catalytic converter and the cylinder head, and the coolant temperature must be 80 degrees Celsius. The ECM detected the HO2S signal was in a negative voltage range referred to as "character shift downward". **Possible Causes:** • HO2S is contaminated (due to presence of silicone in fuel) • HO2S signal and ground circuit wires crossed in wiring harness • HO2S signal circuit is shorted to sensor or chassis ground • HO2S element has failed (internal short condition) • ECM has failed
DTC: P0051 **2T CCM, MIL: Yes** **1996, 1997, 1998, 1999, 2000, 2001, 2002, 2003, 2004, 2005, 2006** **Models:** A4, A6, Allroad Quattro, A8, A8L **Engines:** 2.7L, 2.8L, 3.0L, 3.2L, 4.2L **Transmissions:** All	**HO2S Heater (Bank 2 Sensor 1) Circuit Low Input Conditions:** Engine started, battery voltage must be at least 11.5v, all electrical components must be off, the ground between the engine and the chassis must be well connected, the exhaust system must be properly sealed between the catalytic converter and the cylinder head, and the coolant temperature must be 80 degrees Celsius. The ECM detected the HO2S signal was in a negative voltage range referred to as "character shift downward". This code sets when the HO2S signal remains in a low state. In effect, it does not switch properly in the closed loop operation. The HO2S (before the three-way catalytic converter) has a short circuit to ground that has lasted longer than a specified time. **Possible Causes:** • HO2S is contaminated (due to presence of silicone in fuel) • HO2S signal and ground circuit wires crossed in wiring harness • HO2S signal circuit is shorted to sensor or chassis ground • HO2S element has failed (internal short condition) • ECM has failed

DTC	Trouble Code Title, Conditions & Possible Causes
DTC: P0052 **2T CCM, MIL: Yes** **1996, 1997, 1998, 1999, 2000, 2001, 2002, 2003, 2004, 2005, 2006** **Models:** A4, A6, Allroad Quattro, A8, A8L **Engines:** 2.7L, 2.8L, 3.0L, 3.2L, 4.2L **Transmissions:** All	**HO2S Heater (Bank 2 Sensor 1) Circuit High Input Conditions:** Engine started, battery voltage must be at least 11.5v, all electrical components must be off, the ground between the engine and the chassis must be well connected, the exhaust system must be properly sealed between the catalytic converter and the cylinder head, and the coolant temperature must be 80 degrees Celsius. The ECM detected the HO2S signal was in a negative voltage range referred to as "character shift downward". This code sets when the HO2S signal remains in a low state. In effect, it does not switch properly in the closed loop operation. The HO2S (before the three-way catalytic converter) has a short circuit to ground that has lasted longer than a specified time. **Possible Causes:** • HO2S is contaminated (due to presence of silicone in fuel) • HO2S signal and ground circuit wires crossed in wiring harness • HO2S signal circuit is shorted to sensor or chassis ground • HO2S element has failed (internal short condition) • ECM has failed
DTC: P0056 **2T CCM, MIL: Yes** **1996, 1997, 1998, 1999, 2000, 2001, 2002, 2003, 2004, 2005, 2006** **Models:** A4, A6, Allroad Quattro, A8, A8L **Engines:** 2.7L, 2.8L, 3.0L, 3.2L, 4.2L **Transmissions:** All	**HO2S Heater (Bank 2 Sensor 2) Circuit High Input Conditions:** Engine started, battery voltage must be at least 11.5v, all electrical components must be off, the ground between the engine and the chassis must be well connected, the exhaust system must be properly sealed between the catalytic converter and the cylinder head, and the coolant temperature must be 80 degrees Celsius. The ECM detected the HO2S signal remained in a high state. **Note: The HO2S signal circuit may be shorted to the heater power circuit due to tracking inside of the HO2S connector. Remove the connector and visually inspect the connector for signs of oil or water.** **Possible Causes:** • HO2S signal shorted to heater power circuit inside connector • HO2S signal circuit shorted to ground or to system voltage • ECM has failed
DTC: P0057 **2T CCM, MIL: Yes** **1996, 1997, 1998, 1999, 2000, 2001, 2002, 2003, 2004, 2005, 2006** **Models:** A4, A6, Allroad Quattro, A8, A8L **Engines:** 2.7L, 2.8L, 3.0L, 3.2L, 4.2L **Transmissions:** All	**HO2S Heater (Bank 2 Sensor 2) Control Circuit Malfunction Conditions:** Engine started, battery voltage must be at least 11.5v, all electrical components must be off, the ground between the engine and the chassis must be well connected, the exhaust system must be properly sealed between the catalytic converter and the cylinder head, and the coolant temperature must be 80 degrees Celsius. The ECM detected the HO2S signal was in a negative voltage range referred to as "character shift downward". **Possible Causes:** • HO2S is contaminated (due to presence of silicone in fuel) • HO2S signal and ground circuit wires crossed in wiring harness • HO2S signal circuit is shorted to sensor or chassis ground • HO2S element has failed (internal short condition) • ECM has failed
DTC: P0058 **2T CCM, MIL: Yes** **1996, 1997, 1998, 1999, 2000, 2001, 2002, 2003, 2004, 2005, 2006** **Models:** A4, A6, Allroad Quattro, A8, A8L **Engines:** 2.7L, 2.8L, 3.0L, 3.2L, 4.2L **Transmissions:** All	**HO2S Heater (Bank 2 Sensor 2) Circuit Low Input Conditions:** Engine started, battery voltage must be at least 11.5v, all electrical components must be off, the ground between the engine and the chassis must be well connected, the exhaust system must be properly sealed between the catalytic converter and the cylinder head, and the coolant temperature must be 80 degrees Celsius. The ECM detected the HO2S signal was in a negative voltage range referred to as "character shift downward". This code sets when the HO2S signal remains in a low state. In effect, it does not switch properly in the closed loop operation. The HO2S (before the three-way catalytic converter) has a short circuit to ground that has lasted longer than a specified time. **Possible Causes:** • HO2S is contaminated (due to presence of silicone in fuel) • HO2S signal and ground circuit wires crossed in wiring harness • HO2S signal circuit is shorted to sensor or chassis ground • HO2S element has failed (internal short condition) • ECM has failed
DTC: P0087 **2T CCM, MIL: Yes** **2005, 2006** **Models:** A6 **Engines:** 3.2L **Transmissions:** All	**Fuel Rail/System Pressure Too Low Conditions** Engine started, battery voltage must be at least 11.5v, all electrical components must be off, the ground between the engine and the chassis must be well connected, the exhaust system must be properly sealed between the catalytic converter and the cylinder head, and the coolant temperature must be 80 degrees Celsius. The ECM detected that the system's fuel pressure has fallen below the accepted normal calibrated value. **Possible Causes:** • Fuel Pressure Regulator Valve faulty • Fuel Pressure Sensor faulty • Fuel Pump (FP) Control Module faulty • Fuel pump faulty • Low fuel

DTC	Trouble Code Title, Conditions & Possible Causes
DTC: P0088 **2T CCM, MIL: Yes** **2005, 2006** **Models:** A6 **Engines:** 3.2L **Transmissions:** All	**Fuel Rail/System Pressure Too High Conditions** Engine started, battery voltage must be at least 11.5v, all electrical components must be off, the ground between the engine and the chassis must be well connected, the exhaust system must be properly sealed between the catalytic converter and the cylinder head, and the coolant temperature must be 80 degrees Celsius. The ECM detected that the system's fuel pressure has risen above the accepted normal calibrated value. **Possible Causes:** • Fuel Pressure Regulator Valve faulty • Fuel Pressure Sensor faulty • Fuel Pump (FP) Control Module faulty • Fuel pump faulty
DTC: P0089 **2T CCM, MIL: Yes** **2005, 2006** **Models:** A6 **Engines:** 3.2L **Transmissions:** All	**Fuel Pressure Regulator Range/Performance Conditions** Engine started, battery voltage must be at least 11.5v, all electrical components must be off, the ground between the engine and the chassis must be well connected, the exhaust system must be properly sealed between the catalytic converter and the cylinder head, and the coolant temperature must be 80 degrees Celsius. The ECM detected that the system's fuel pressure sensor is providing a signal that is either outside the accepted normal values or is not receiving a signal at all. **Possible Causes:** • Fuel Pressure Regulator Valve faulty • Fuel Pressure Sensor faulty • Fuel Pump (FP) Control Module faulty • Fuel pump faulty
DTC: P0101 **2T CCM, MIL: Yes** **1996, 1997, 1998, 1999, 2000, 2001, 2002, 2003, 2004, 2005, 2006** **Models:** TT, A4, A6, Allroad Quattro **Engines:** 1.8L, 2.7L, 2.8L, 3.0L, 3.2L, 4.2L **Transmissions:** All	**Mass or Volume Air Flow Circuit Range/Performance Conditions** Engine running, with the system voltage more than 11.0v, and the temperature must be at least 185-degrees (F) and all electrical equipment (A/C, lights, etc) must be off. The ECM has detected that the MAF signal was out of a calculated range with the engine (or undetectable) for a certain period of time. **Possible Causes:** • Mass air flow (MAF) sensor has failed or is damaged • ECM has failed • Signal and ground wires of Mass Air Flow (MAF) sensor has short circuited
DTC: P0102 **1T CCM, MIL: Yes** **1996, 1997, 1998, 1999, 2000, 2001, 2002, 2003, 2004, 2005, 2006** **Models:** TT, A4, A6, A6 Avant, Allroad Quattro, A8, A8L **Engines:** 1.8L, 2.7L, 2.8L, 3.0L, 3.2L, 3.7L, 4.2L **Transmissions:** All	**MAF Sensor Circuit Low Input Conditions:** Key on, engine started, and the ECM detected the MAF sensor signal was less than the minimum calibrated value. The engine temperature must beat least 185-degrees (F) and all electrical equipment (A/C, lights, etc) must be off. The ECM has detected that the MAF signal was less than the required minimum. **Possible Causes:** • Check for leaks between MAF sensor and throttle valve control module • Voltage supply faulty. • Sensor power circuit open from fuel pump relay to MAF sensor • Sensor signal circuit open (may be disconnected) from ECM and MAF • Faulty ground cable resistance between connector terminal 1 and Ground • MAF Sensor malfunction
DTC: P0103 **1T CCM, MIL: Yes** **1996, 1997, 1998, 1999, 2000, 2001, 2002, 2003, 2004, 2005, 2006** **Models:** TT, A4, A6, A6 Avant, Allroad Quattro, A8, A8L **Engines:** 1.8L, 2.7L, 2.8L, 3.0L, 3.2L, 3.7L, 4.2L **Transmissions:** All	**MAF Sensor Circuit High Input Conditions:** Key on, engine started, and the ECM detected the MAF sensor signal was more than the minimum calibrated value. The engine temperature must beat least 185-degrees (F) and all electrical equipment (A/C, lights, etc) must be off. The ECM has detected that the MAF signal was more than the required minimum. **Possible Causes:** • Check for leaks between MAF sensor and throttle valve control module • Voltage supply faulty. • Sensor power circuit open from fuel pump relay to MAF sensor • Sensor signal circuit open (may be disconnected) from ECM and MAF • Faulty ground cable resistance between connector terminal 1 and Ground • MAF Sensor malfunction

DTC	Trouble Code Title, Conditions & Possible Causes
DTC: P0106 **2T CCM, MIL: Yes** **1996, 1997, 1998, 1999, 2000, 2001, 2002, 2003, 2004, 2005, 2006** **Models:** TT, A4, A6, Allroad Quattro **Engines:** 1.8L, 2.7L, 2.8L, 3.0L, 3.2L, 4.2L **Transmissions:** All	**Manifold Absolute Pressure/Barometric Pressure Sensor Circuit Performance Conditions:** Engine started, the temperature must beat least 185-degrees (F) and all electrical equipment (A/C, lights, etc) must be off. The ECM detected the BARO sensor was out of range during the CCM test. The BARO sensor signal should be in 4.5v. **Possible Causes:** • Sensor has deteriorated (response time too slow) or has failed • MAP sensor signal circuit is shorted to ground • MAP sensor circuit (5v) is open • MAP sensor is damaged or it has failed • BARO sensor signal circuit is shorted to ground • BARO sensor circuit (5v) is open • BARO sensor is damaged or it has failed • ECM is not connected properly • ECM has failed
DTC: P0107 **1T CCM, MIL: Yes** **1996, 1997, 1998, 1999, 2000, 2001, 2002, 2003, 2004, 2005, 2006** **Models:** TT, A4, A6 **Engines:** 1.8L, 3.2L **Transmissions:** All	**Manifold Absolute Pressure/Barometric Pressure Sensor Circuit Low Input Conditions:** Engine started, the temperature must beat least 185-degrees (F) and all electrical equipment (A/C, lights, etc) must be off. The ECM detected the BARO sensor was out of range during the CCM test. The BARO sensor signal should be in 4.5v. The BARO sensor is a variable capacitance unit used to detect altitude. **Possible Causes:** • Sensor has deteriorated (response time too slow) or has failed • MAP sensor signal circuit is shorted to ground • MAP sensor circuit (5v) is open • MAP sensor is damaged or it has failed • BARO sensor signal circuit is shorted to ground • BARO sensor circuit (5v) is open • BARO sensor is damaged or it has failed • ECM is not connected properly • ECM has failed
DTC: P0108 **1T CCM, MIL: Yes** **1996, 1997, 1998, 1999, 2000, 2001, 2002, 2003, 2004, 2005, 2006** **Models:** TT, A4, A6 **Engines:** 1.8L, 3.2L **Transmissions:** All	**Manifold Absolute Pressure/Barometric Sensor Circuit High Input Conditions:** Engine started, the temperature must beat least 185-degrees (F) and all electrical equipment (A/C, lights, etc) must be off. The ECM detected the BARO sensor was out of range during the CCM test. The BARO sensor signal should be in 4.5v. The BARO sensor is a variable capacitance unit used to detect altitude. **Possible Causes:** • Sensor has deteriorated (response time too slow) or has failed • MAP sensor signal circuit is shorted to ground • MAP sensor circuit (5v) is open • MAP sensor is damaged or it has failed • BARO sensor signal circuit is shorted to ground • BARO sensor circuit (5v) is open • BARO sensor is damaged or it has failed • ECM is not connected properly • ECM has failed
DTC: P0111 **2T CCM, MIL: Yes** **1996, 1997, 1998, 1999, 2000, 2001, 2002, 2003, 2004, 2005, 2006** **Models:** TT, A4, A6, Allroad Quattro **Engines:** 1.8L, 3.2L **Transmissions:** All: All	**Intake Air Temperature Sensor Circuit Low Input Conditions:** Key on or engine running, the temperature must beat least 185-degrees (F) and all electrical equipment (A/C, lights, etc) must be off; and the ECM detected the IAT sensor signal was less than the self-test minimum. This is a thermistor-type sensor with a variable resistance that changes when exposed to different temperatures. This means: the higher the temperature, the lower the resistance value. **Possible Causes:** • IAT sensor signal circuit is grounded (check wiring & connector) • Resistance value between sockets 33 and 36 out of range • IAT sensor has an open circuit • IAT sensor is damaged or it has failed • ECM has failed
DTC: P0112 **1T CCM, MIL: Yes** **1996, 1997, 1998, 1999, 2000, 2001, 2002, 2003, 2004, 2005, 2006** **Models:** TT, A4, A6, Allroad Quattro, A8, A8L **Engines:** 1.8L, 2.7L, 2.8L, 3.0L, 3.2L, 3.7L, 4.2L **Transmissions:** All	**Intake Air Temperature Sensor Circuit Low Input Conditions:** Key on or Engine running, the temperature must beat least 185-degrees (F) and all electrical equipment (A/C, lights, etc) must be off; and the ECM detected the IAT sensor signal was less than the self-test minimum. This is a thermistor-type sensor with a variable resistance that changes when exposed to different temperatures. This means: the higher the temperature, the lower the resistance value. **Possible Causes:** • IAT sensor signal circuit is grounded (check wiring & connector) • Resistance value between sockets 33 and 36 out of range • IAT sensor has an open circuit • IAT sensor is damaged or it has failed • ECM has failed

DTC	Trouble Code Title, Conditions & Possible Causes
DTC: P0113 **1T CCM, MIL: Yes** **1996, 1997, 1998, 1999, 2000, 2001, 2002, 2003, 2004, 2005, 2006** **Models:** TT, A4, A6, Allroad Quattro, A8, A8L **Engines:** 1.8L, 2.7L, 2.8L, 3.0L, 3.2L, 3.7L, 4.2L **Transmissions:** All	**Intake Air Temperature Sensor Circuit High Input Conditions:** Key on or engine running, the temperature must beat least 185-degrees (F) and all electrical equipment (A/C, lights, etc) must be off; and the ECM detected the IAT sensor signal was more than the self-test maximum. This is a thermistor-type sensor with a variable resistance that changes when exposed to different temperatures. This means: the higher the temperature, the lower the resistance value. **Possible Causes:** • IAT sensor signal circuit is open (inspect wiring & connector) • IAT sensor signal circuit is shorted • Resistance value between sockets 33 and 36 out of range • IAT sensor is damaged or it has failed • ECM has failed
DTC: P0116 **2T CCM, MIL: Yes** **1996, 1997, 1998, 1999, 2000, 2001, 2002, 2003, 2004, 2005, 2006** **Models:** TT, A4, A6, Allroad Quattro, A8, A8L **Engines:** 1.8L, 2.7L, 2.8L, 3.0L, 3.2L, 3.7L, 4.2L **Transmissions:** All	**ECT Sensor / CHT Sensor Signal Range/Performance Conditions:** Engine started (cold), battery voltage must be 11.5, and all equipment must be off. The ECM detected the ECT sensor exceeded the required calibrated value, or the engine is at idle and doesn't reach operating temperature quickly enough; the Catalyst, Fuel System, HO2S and Misfire Monitor did not complete, or the timer expired. Testing completion of procedure, the engine's temperature must rise uniformly during idle. **Possible Causes:** • Check for low coolant level or incorrect coolant mixture • ECM detects a short circuit wiring in the ECT • CHT sensor is out-of-calibration or it has failed • ECT sensor is out-of-calibration or it has failed
DTC: P0117 **1T CCM, MIL: Yes** **1996, 1997, 1998, 1999, 2000, 2001, 2002, 2003, 2004, 2005, 2006** **Models:** TT, A4, A6, Allroad Quattro, A8, A8L **Engines:** 1.8L, 2.7L, 2.8L, 3.0L, 3.2L, 3.7L, 4.2L **Transmissions:** All	**ECT Sensor Circuit Low Input Conditions:** Engine started (cold), battery voltage must be 11.5, and all equipment must be off. The ECM detected the ECT sensor signal was less than the self-test minimum. This is a thermistor-type sensor with a variable resistance that changes when exposed to different temperatures **Possible Causes:** • ECT sensor signal circuit is grounded in the wiring harness • ECT sensor doesn't react to changes in temperature • ECT sensor is damaged or the ECM has failed
DTC: P0118 **1T CCM, MIL: Yes** **1996, 1997, 1998, 1999, 2000, 2001, 2002, 2003, 2004, 2005, 2006** **Models:** TT, A4, A6, Allroad Quattro, A8, A8L **Engines:** 1.8L, 2.7L, 2.8L, 3.0L, 3.2L, 3.7L, 4.2L **Transmissions:** All	**ECT Sensor Circuit High Input Conditions:** Engine started (cold), battery voltage must be 11.5, and all equipment must be off. The ECM detected the ECT sensor signal was more than the self-test maximum. This is a thermistor-type sensor with a variable resistance that changes when exposed to different temperatures **Possible Causes:** • ECT sensor signal circuit is open (inspect wiring & connector) • ECT sensor signal circuit is shorted to ground • ECT sensor is damaged or it has failed • ECM has failed
DTC: P0120 **1T CCM, MIL: Yes** **1995, 1996, 1997, 1998, 1999, 2000, 2001** **Models:** A4, A6, A6 Avant **Engines:** 2.8L **Transmissions:** All	**Throttle/Pedal Position Sensor (A) Circuit Malfunction Conditions:** Engine started, at idle, the temperature must be 80 degrees Celsius. The throttle position sensor supplies implausible signal to the ECM. The throttle valve activation occurs via an electric motor (throttle drive) in the throttle valve control module. It is activated by the Engine Control Module (ECM) according to specifications of the two sensors, Throttle Position (TP) Sensor and Accelerator Pedal Position Sensor 2. **Possible Causes:** • TP sensor signal circuit is open (inspect wiring & connector) • TP sensor signal circuit is shorted to ground • TP sensor or module is damaged or it has failed • Throttle valve is damaged or dirty • Throttle valve control module is faulty • ECM has failed

DTC	Trouble Code Title, Conditions & Possible Causes
DTC: P0121 **1T CCM, MIL: Yes** **1996, 1997, 1998, 1999, 2000, 2001, 2002, 2003, 2004, 2005, 2006** **Models:** TT, A4, A6, A6 Avant, Allroad Quattro, A8, A8L **Engines:** 1.8L, 2.7L, 2.8L, 3.0L, 3.2L, 3.7L, 4.2L **Transmissions:** All	**Throttle/Pedal Position Sensor Signal Range/Performance Conditions:** Engine started; then immediately following a condition where the engine was running under at off-idle, the ECM detected the TP sensor signal indicated the throttle did not return to its previous closed position during the Rationality test. **Possible Causes:** • Throttle plate is binding, dirty or sticking • Throttle valve is damaged or dirty • Throttle valve control module is faulty • TP sensor signal circuit open (inspect wiring & connector) • TP sensor ground circuit open (inspect wiring & connector) • TP sensor and/or control module is damaged or has failed • MAF sensor signal is damaged, has failed or a short is present
DTC: P0122 **1T CCM, MIL: Yes** **1996, 1997, 1998, 1999, 2000, 2001, 2002, 2003, 2004, 2005, 2006** **Models:** TT, A4, A6, A6 Avant, Allroad Quattro, A8, A8L **Engines:** 1.8L, 2.7L, 2.8L, 3.0L, 3.2L, 3.7L, 4.2L **Transmissions:** All	**Throttle/Pedal Position Sensor Circuit Low Input Conditions:** Engine started, at idle, the temperature must be at least 80 degrees Celsius. The throttle position sensor supplies implausible signal to the ECM. **Possible Causes:** • TP sensor signal circuit open (inspect wiring & connector) • TP sensor signal shorted to ground (inspect wiring & connector) • TP sensor is damaged or has failed • Throttle control module's voltage supply is shorted or open • ECM has failed
DTC: P0123 **1T CCM, MIL: Yes** **1996, 1997, 1998, 1999, 2000, 2001, 2002, 2003, 2004, 2005, 2006** **Models:** TT, A4, A6, A6 Avant, Allroad Quattro, A8, A8L **Engines:** 1.8L, 2.7L, 2.8L, 3.0L, 3.2L, 3.7L, 4.2L **Transmissions:** All	**TP Sensor Circuit High Input Conditions:** Engine started, at idle, the temperature must be at least 80 degrees Celsius. The ECM detected the TP sensor signal was more than the self-test maximum during testing. **Possible Causes:** • TP sensor not seated correctly in housing (may be damaged) • TP sensor signal is circuit shorted to ground or system voltage • TP sensor ground circuit is open (check the wiring harness) • TP sensor and/or ECM has failed
DTC: P0130 **2T OBD/O2S1, MIL: Yes** **1996, 1997, 1998, 1999, 2000, 2001, 2002, 2003, 2004, 2005, 2006** **Models:** TT, A4, A6, A6 Avant, Allroad Quattro, A8, A8L **Engines:** 1.8L, 2.7L, 2.8L, 3.0L, 3.2L, 3.7L, 4.2L **Transmissions:** All	**O2 Sensor Circuit Bank 1 Sensor 1 Conditions:** Engine running, battery voltage 11.5, all electrical components off, ground between engine and chassis well connected and the exhaust system must be properly sealed between catalytic converter and the cylinder head. The ECM detected the HO2S signal was implausible or not detected. **Possible Causes:** • Oxygen sensor heater for oxygen sensor (HO2S) before catalytic converter is faulty • HO2S is contaminated (due to presence of silicone in fuel) • HO2S signal and ground circuit wires crossed in wiring harness • HO2S signal circuit is shorted to sensor or chassis ground • HO2S element before the catalytic converter has failed (internal short condition) • Leaks present in the exhaust manifold or exhaust pipes • ECM has failed
DTC: P0131 **2T CCM, MIL: Yes** **1996, 1997, 1998, 1999, 2000, 2001, 2002, 2003, 2004, 2005, 2006** **Models:** TT, A4, A6, A6 Avant, Allroad Quattro, A8, A8L **Engines:** 1.8L, 2.7L, 2.8L, 3.0L, 3.2L, 3.7L, 4.2L **Transmissions:** All	**HO2S (Bank 1 Sensor 1) Circuit Low Input Conditions:** Engine running, battery voltage 11.5, all electrical components off, ground between engine and chassis well connected and the exhaust system must be properly sealed between catalytic converter and the cylinder head. The ECM detected the HO2S signal was in a negative voltage range referred to as "character shift downward". This code sets when the HO2S signal remains in a low state for a measured period of time. In effect, it does not switch properly in the closed loop operation. **Possible Causes:** • HO2S is contaminated (due to presence of silicone in fuel) • HO2S signal and ground circuit wires crossed in wiring harness • HO2S signal circuit is shorted to sensor or chassis ground • HO2S element has failed (internal short condition) • Leaks present in the exhaust manifold or exhaust pipes • ECM has failed

DTC	Trouble Code Title, Conditions & Possible Causes
DTC: P0132 **2T CCM, MIL: Yes** **1996, 1997, 1998, 1999, 2000, 2001, 2002, 2003, 2004, 2005, 2006** **Models:** TT, A4, A6, A6 Avant, Allroad Quattro, A8, A8L **Engines:** 1.8L, 2.7L, 2.8L, 3.0L, 3.2L, 3.7L, 4.2L **Transmissions:** All	**HO2S (Bank 1 Sensor 1) Circuit High Input Conditions:** Engine running, battery voltage 11.5, all electrical components off, ground between engine and chassis well connected and the exhaust system must be properly sealed between catalytic converter and the cylinder head. The ECM detected the HO2S signal was in a high state. This code sets when the HO2S signal remains in a high state for a measured period of time. In effect, it does not switch properly in the closed loop operation. **Note: The HO2S signal circuit may be shorted to the heater power circuit due to tracking inside of the HO2S connector. Remove the connector and visually inspect the connector for signs of oil or water.** **Possible Causes:** • HO2S is contaminated (due to presence of silicone in fuel) • HO2S signal and ground circuit wires crossed in wiring harness • HO2S signal circuit is shorted to sensor or chassis ground • HO2S element has failed (internal short condition) • Leaks present in the exhaust manifold or exhaust pipes • ECM has failed
DTC: P0133 **2T OBD/O2S1, MIL: Yes** **1996, 1997, 1998, 1999, 2000, 2001, 2002, 2003, 2004, 2005, 2006** **Models:** TT, A4, A6, A6 Avant, Allroad Quattro, A8, A8L **Engines:** 1.8L, 2.7L, 2.8L, 3.0L, 3.2L, 3.7L, 4.2L **Transmissions:** All	**HO2S (Bank 1 Sensor 1) Circuit Slow Response Conditions:** Engine running, battery voltage 11.5, all electrical components off, ground between engine and chassis well connected and the exhaust system must be properly sealed between catalytic converter and the cylinder head. The ECM detected the HO2S amplitude and frequency were out of the normal range (e.g., the HO2S rich to lean switch) during the HO2S Monitor test. **Possible Causes:** • HO2S before the three-way catalytic converter is contaminated (due to presence of silicone in fuel); Run the engine for three minutes at 3500rpm as a self-cleaning effect • HO2S signal circuit open • Leaks present in the exhaust manifold or exhaust pipes • HO2S is damaged or has failed • ECM has failed
DTC: P0134 **2T OBD/O2S1, MIL: Yes** **1996, 1997, 1998, 1999, 2000, 2001, 2002, 2003, 2004, 2005, 2006** **Models:** TT, A4, A6, A6 Avant, Allroad Quattro, A8 **Engines:** 1.8L, 2.7L, 2.8L, 3.0L, 3.2L, 3.7L, 4.2L **Transmissions:** All	**HO2S (Bank 1 Sensor 1) Circuit No Activity Conditions:** Engine running, battery voltage 11.5, all electrical components off, ground between engine and chassis well connected and the exhaust system must be properly sealed between catalytic converter and the cylinder head. The ECM detected the HO2S signal failed to meet the maximum or minimum voltage levels (i.e., it failed the voltage range check). **Possible Causes:** • Leaks present in the exhaust manifold or exhaust pipes • HO2S signal wire and ground wire crossed in connector (voltage jumps) • HO2S element is fuel contaminated or has failed • ECM has failed
DTC: P0135 **2T OBD/O2S1, MIL: Yes** **1996, 1997, 1998, 1999, 2000, 2001, 2002, 2003, 2004, 2005, 2006** **Models:** TT, A4, A6, A6 Avant, Allroad Quattro, A8, A8L **Engines:** 1.8L, 2.7L, 2.8L, 3.0L, 3.2L, 3.7L, 4.2L **Transmissions:** All	**HO2S (Bank 1 Sensor 1) Heater Circuit Malfunction Conditions:** Engine running, battery voltage 11.5, all electrical components off, ground between engine and chassis well connected and the exhaust system must be properly sealed between catalytic converter and the cylinder head. The ECM detected an unexpected voltage condition, or it detected excessive current draw in the heater circuit during the CCM test. **Possible Causes:** • HO2S heater power circuit is open or heater ground circuit open • HO2S signal tracking (due to oil or moisture in the connector) • HO2S is damaged or has failed • ECM has failed
DTC: P0136 **2T OBD/O2S1, MIL: Yes** **1996, 1997, 1998, 1999, 2000, 2001, 2002, 2003, 2004, 2005, 2006** **Models:** TT, A4, A6, A6 Avant, Allroad Quattro, A8 **Engines:** 1.8L, 2.7L, 2.8L, 3.0L, 3.2L, 3.7L, 4.2L **Transmissions:** All	**HO2S (Bank 1 Sensor 2) Circuit Malfunction Conditions:** Engine running, battery voltage 11.5, all electrical components off, ground between engine and chassis well connected and the exhaust system must be properly sealed between catalytic converter and the cylinder head. The ECM detected the HO2S signal failed to meet the maximum or minimum voltage levels (i.e., it failed the voltage range check). **Possible Causes:** • Leaks present in the exhaust manifold or exhaust pipes • HO2S signal wire and ground wire crossed in connector • HO2S element is fuel contaminated or has failed • ECM has failed

DTC	Trouble Code Title, Conditions & Possible Causes
DTC: P0137 **2T CCM, MIL: Yes** **1996, 1997, 1998, 1999, 2000, 2001, 2002, 2003, 2004, 2005, 2006** **Models:** TT, A4, A6, A6 Avant, Allroad Quattro, A8, A8L **Engines:** 1.8L, 2.7L, 2.8L, 3.0L, 3.2L, 3.7L, 4.2L **Transmissions:** All	**HO2S (Bank 1 Sensor 2) Circuit Low Input Conditions:** Engine running, battery voltage 11.5, all electrical components off, ground between engine and chassis well connected and the exhaust system must be properly sealed between catalytic converter and the cylinder head. The ECM detected the HO2S signal remained in a high state. **Note: The HO2S signal circuit may be shorted to the heater power circuit due to "tracking inside of the HO2S connector. Remove the connector and visually inspect the connector for signs of oil or water.** **Possible Causes:** • HO2S signal shorted to heater power circuit in the connector • HO2S signal circuit shorted to ground (for more than 200 seconds) or to system voltage • ECM has failed
DTC: P0138 **2T CCM, MIL: Yes** **1996, 1997, 1998, 1999, 2000, 2001, 2002, 2003, 2004, 2005, 2006** **Models:** TT, A4, A6, A6 Avant, Allroad Quattro, A8, A8L **Engines:** 1.8L, 2.7L, 2.8L, 3.0L, 3.2L, 3.7L, 4.2L **Transmissions:** All	**HO2S (Bank 1 Sensor 2) Circuit High Input Conditions:** Engine running, battery voltage 11.5, all electrical components off, ground between engine and chassis well connected and the exhaust system must be properly sealed between catalytic converter and the cylinder head. The ECM detected the HO2S signal remained in a high state. **Note: The HO2S signal circuit may be shorted to the heater power circuit due to "tracking inside of the HO2S connector. Remove the connector and visually inspect the connector for signs of oil or water.** **Possible Causes:** • HO2S signal shorted to heater power circuit in the positive connector • HO2S signal circuit shorted to ground or to system voltage • HO2S has failed • ECM has failed
DTC: P0139 **2T CCM, MIL: Yes** **1996, 1997, 1998, 1999, 2000, 2001, 2002, 2003, 2004, 2005, 2006** **Models:** TT, A4, A6, Allroad Quattro, A8, A8L **Engines:** 1.8L, 2.7L, 2.8L, 3.0L, 3.2L, 3.7L, 4.2L **Transmissions:** All	**HO2S (Bank 1 Sensor 2) Slow Response Conditions:** Engine running, battery voltage 11.5, all electrical components off, ground between engine and chassis well connected and the exhaust system must be properly sealed between catalytic converter and the cylinder head. The ECM detected the HO2S amplitude and frequency were out of the normal range during the HO2S Monitor test. **Possible Causes:** • HO2S signal shorted to heater power circuit in the connector • HO2S signal circuit shorted to VREF or to system voltage • ECM has failed
DTC: P0140 **2T OBD/O2S1, MIL: Yes** **1996, 1997, 1998, 1999, 2000, 2001, 2002, 2003, 2004, 2005, 2006** **Models:** TT, A4, A6, Allroad Quattro, A8, A8L **Engines:** 1.8L, 2.7L, 2.8L, 3.0L, 3.2L, 3.7L, 4.2L **Transmissions:** All	**HO2S (Bank 1 Sensor 2) No Activity Conditions:** Engine running, battery voltage 11.5, all electrical components off, ground between engine and chassis well connected and the exhaust system must be properly sealed between catalytic converter and the cylinder head. The ECM detected the HO2S signal failed to meet the maximum or minimum voltage levels (i.e., it failed the voltage range check). **Possible Causes:** • HO2S before the three-way catalytic converter is contaminated (due to presence of silicone in fuel); Run the engine for three minutes at 3500rpm as a self-cleaning effect • Leaks present in the exhaust manifold or exhaust pipes • HO2S signal wire and ground wire crossed in connector (voltage jumps) • HO2S element is contaminated or has failed • ECM has failed
DTC: P0141 **2T OBD/O2S1, MIL: Yes** **1996, 1997, 1998, 1999, 2000, 2001, 2002, 2003, 2004, 2005, 2006** **Models:** TT, A4, A6, A6 Avant, Allroad Quattro, A8, A8L **Engines:** 1.8L, 2.7L, 2.8L, 3.0L, 3.2L, 3.7L, 4.2L **Transmissions:** All	**HO2S (Bank 1 Sensor 2) Malfunction Conditions:** Engine running, battery voltage 11.5, all electrical components off, ground between engine and chassis well connected and the exhaust system must be properly sealed between catalytic converter and the cylinder head. The ECM detected the HO2S signal failed to meet the maximum or minimum voltage levels (i.e., it failed the voltage range check). **Possible Causes:** • Leaks present in the exhaust manifold or exhaust pipes • HO2S signal wire and ground wire crossed in connector • HO2S element is fuel contaminated or has failed • ECM has failed

DTC	Trouble Code Title, Conditions & Possible Causes
DTC: P0150 **2T CCM, MIL: Yes** **1996, 1997, 1998, 1999, 2000, 2001, 2002, 2003, 2004, 2005, 2006** **Models:** TT, A4, A6, A6 Avant, Allroad Quattro, A8, A8L **Engines:** 1.8L, 2.7L, 2.8L, 3.0L, 3.2L, 3.7L, 4.2L **Transmissions:** All	**HO2S (Bank 2 Sensor 1) Circuit Malfunction Conditions:** Engine running, battery voltage 11.5, all electrical components off, ground between engine and chassis well connected and the exhaust system must be properly sealed between catalytic converter and the cylinder head. The ECM detected the HO2S signal failed to meet the maximum or minimum voltage levels (i.e., it failed the voltage range check). **Possible Causes:** • Leaks present in the exhaust manifold or exhaust pipes • HO2S signal wire and ground wire crossed in connector • HO2S element is fuel contaminated or has failed • ECM has failed
DTC: P0151 **2T CCM, MIL: Yes** 1996, 1997, 1998, 1999, 2000, 2001, 2002, 2003, 2004 **Models:** A4, A6, Allroad Quattro, A8 **Engines:** 2.7L, 2.8L, 3.2L, 3.7L, 4.2L **Transmissions:** All	**HO2S (Bank 2 Sensor 1) Low Input Conditions:** Engine running, battery voltage 11.5, all electrical components off, ground between engine and chassis well connected and the exhaust system must be properly sealed between catalytic converter and the cylinder head. The ECM detected the HO2S signal remained in a high state. **Note: The HO2S signal circuit may be shorted to the heater power circuit due to "tracking inside of the HO2S connector. Remove the connector and visually inspect the connector for signs of oil or water.** **Possible Causes:** • HO2S is contaminated (due to presence of silicone in fuel) • HO2S signal tracking (due to oil or moisture in the connector) • HO2S signal circuit is open or shorted to VREF • ECM has failed
DTC: P0152 **2T CCM, MIL: Yes** **1996, 1997, 1998, 1999, 2000, 2001, 2002, 2003, 2004** **Models:** A4, A6, A6 Avant, Allroad Quattro, A8 **Engines:** 2.7L, 2.8L, 3.2L, 3.7L, 4.2L **Transmissions:** All	**HO2S (Bank 2 Sensor 1) Circuit High Input Conditions:** Engine running, battery voltage 11.5, all electrical components off, ground between engine and chassis well connected and the exhaust system must be properly sealed between catalytic converter and the cylinder head. The ECM detected the HO2S signal remained in a high state (more than 1.5v). **Note: The HO2S signal circuit may be shorted to the heater power circuit due to "tracking inside of the HO2S connector. Remove the connector and visually inspect the connector for signs of oil or water.** **Possible Causes:** • HO2S is contaminated (due to presence of silicone in fuel) • HO2S signal tracking (due to oil or moisture in the connector) • HO2S signal circuit is open or shorted to VREF • ECM has failed
DTC: P0153 **2T CCM, MIL: Yes** **1996, 1997, 1998, 1999, 2000, 2001, 2002, 2003, 2004, 2005, 2006** **Models:** TT, A4, A6, A6 Avant, Allroad Quattro, A8, A8L **Engines:** 1.8L, 2.7L, 2.8L, 3.0L, 3.2L, 3.7L, 4.2L **Transmissions:** All	**HO2S (Bank 2 Sensor 1) Circuit Slow Response Conditions:** Engine running, battery voltage 11.5, all electrical components off, ground between engine and chassis well connected and the exhaust system must be properly sealed between catalytic converter and the cylinder head. The the ECM detected the HO2S amplitude and frequency were out of the normal range during the HO2S Monitor test. **Possible Causes:** • HO2S is contaminated (due to presence of silicone in fuel) • Leaks present in the exhaust manifold or exhaust pipes • HO2S is damaged or has failed • ECM has failed
DTC: P0154 **2T OBD/O2S1, MIL: Yes** **1996, 1997, 1998, 1999, 2000, 2001, 2002, 2003, 2004, 2005, 2006** **Models:** TT, A4, A6, A6 Avant, Allroad Quattro, A8, A8L **Engines:** 1.8L, 2.7L, 2.8L, 3.0L, 3.2L, 3.7L, 4.2L **Transmissions:** All	**HO2S (Bank 2 Sensor 1) Circuit No Activity Conditions:** Engine running, battery voltage 11.5, all electrical components off, ground between engine and chassis well connected and the exhaust system must be properly sealed between catalytic converter and the cylinder head. The ECM detected the HO2S signal failed to meet the maximum or minimum voltage (i.e., it failed the voltage check). **Possible Causes:** • Leaks present in the exhaust manifold or exhaust pipes • HO2S signal wire and ground wire crossed in connector • HO2S element is fuel contaminated or has failed • ECM has failed

DTC	Trouble Code Title, Conditions & Possible Causes
DTC: P0155 **2T OBD/O2S1, MIL: Yes** **1996, 1997, 1998, 1999, 2000, 2001, 2002, 2003, 2004, 2005, 2006** **Models:** TT, A4, A6, A6 Avant, Allroad Quattro, A8, A8L **Engines:** 1.8L, 2.7L, 2.8L, 3.0L, 3.2L, 3.7L, 4.2L **Transmissions:** All	**HO2S (Bank 2 Sensor 1) Heater Circuit Malfunction Conditions:** Engine running, battery voltage 11.5, all electrical components off, ground between engine and chassis well connected and the exhaust system must be properly sealed between catalytic converter and the cylinder head. The ECM detected an open or shorted condition, or excessive current draw in the heater circuit. **Possible Causes:** • HO2S heater power circuit is open • HO2S heater ground circuit is open • HO2S signal tracking (due to oil or moisture in the connector) • HO2S is damaged or has failed • ECM has failed
DTC: P0156 **2T OBD/O2S1, MIL: Yes** **1996, 1997, 1998, 1999, 2000, 2001, 2002, 2003, 2004, 2005, 2006** **Models:** TT, A4, A6, A6 Avant, Allroad Quattro, A8 **Engines:** 1.8L, 2.7L, 2.8L, 3.0L, 3.2L, 3.7L, 4.2L **Transmissions:** All	**HO2S (Bank 2 Sensor 2) Circuit No Activity Conditions:** Engine running, battery voltage 11.5, all electrical components off, ground between engine and chassis well connected and the exhaust system must be properly sealed between catalytic converter and the cylinder head. The ECM detected the HO2S signal failed to meet the maximum or minimum voltage (i.e., it failed the voltage check). **Possible Causes:** • Leaks present in the exhaust manifold or exhaust pipes • HO2S signal wire and ground wire crossed in connector • HO2S element is fuel contaminated or has failed • ECM has failed
DTC: P0157 **2T OBD/O2S1, MIL: Yes** **1996, 1997, 1998, 1999, 2000, 2001, 2002, 2003, 2004, 2005, 2006** **Models:** TT, A4, A6, A6 Avant, Allroad Quattro, A8, A8L **Engines:** 1.8L, 2.7L, 2.8L, 3.0L, 3.2L, 3.7L, 4.2L **Transmissions:** All	**HO2S (Bank 2 Sensor 2) Circuit Low Voltage Conditions:** Engine running, battery voltage 11.5, all electrical components off, ground between engine and chassis well connected and the exhaust system must be properly sealed between catalytic converter and the cylinder head. The ECM detected the HO2S signal remained in a high state. **Note: The HO2S signal circuit may be shorted to the heater power circuit due to "tracking inside of the HO2S connector. Remove the connector and visually inspect the connector for signs of oil or water** **Possible Causes:** • HO2S is contaminated (due to presence of silicone in fuel) • HO2S signal tracking (due to oil or moisture in the connector) • HO2S signal circuit is open or shorted to VREF • ECM has failed
DTC: P0158 **2T OBD/O2S1, MIL: Yes** **1996, 1997, 1998, 1999, 2000, 2001, 2002, 2003, 2004, 2005, 2006** **Models:** TT, A4, A6, A6 Avant, Allroad Quattro, A8, A8L **Engines:** 1.8L, 2.7L, 2.8L, 3.0L, 3.2L, 3.7L, 4.2L **Transmissions:** All	**HO2S (Bank 2 Sensor 2) Circuit High Input Conditions:** Engine running, battery voltage 11.5, all electrical components off, ground between engine and chassis well connected and the exhaust system must be properly sealed between catalytic converter and the cylinder head. The ECM detected the HO2S signal remained in a high state (i.e., more than 1.5v). **Note: The HO2S signal circuit may be shorted to the heater power circuit due to "tracking inside of the HO2S connector. Remove the connector and visually inspect the connector for signs of oil or water.** **Possible Causes:** • HO2S signal shorted to the heater power circuit (due to oil or moisture in the connector) • HO2S signal circuit shorted to VREF or to system voltage • ECM has failed
DTC: P0159 **2T OBD/O2S1, MIL: Yes** **1996, 1997, 1998, 1999, 2000, 2001, 2002, 2003, 2004, 2005, 2006** **Models:** TT, A4, A6, Allroad Quattro, A8, A8L **Engines:** 1.8L, 2.7L, 2.8L, 3.0L, 3.2L, 3.7L, 4.2L **Transmissions:** All	**HO2S (Bank 2 Sensor 2) Circuit Slow Response Conditions:** Engine running, battery voltage 11.5, all electrical components off, ground between engine and chassis well connected and the exhaust system must be properly sealed between catalytic converter and the cylinder head. The ECM detected the HO2S amplitude and frequency were out of the normal range during the HO2S Monitor test. **Possible Causes:** • HO2S is contaminated (due to presence of silicone in fuel) • Leaks present in the exhaust manifold or exhaust pipes • HO2S is damaged or has failed • ECM has failed

DTC	Trouble Code Title, Conditions & Possible Causes
DTC: P0160 **2T OBD/O2S1, MIL: Yes** **1996, 1997, 1998, 1999, 2000, 2001, 2002, 2003, 2004, 2005, 2006** **Models:** TT, A4, A6, Allroad Quattro, A8, A8L **Engines:** 1.8L, 2.7L, 2.8L, 3.0L, 3.2L, 3.7L, 4.2L **Transmissions:** All	**HO2S (Bank 2 Sensor 2) Circuit No Activity Detected Conditions:** Engine running, battery voltage 11.5, all electrical components off, ground between engine and chassis well connected and the exhaust system must be properly sealed between catalytic converter and the cylinder head. The ECM detected the HO2S signal failed to meet the maximum or minimum voltage (i.e., it failed the voltage check). **Possible Causes:** • Leaks present in the exhaust manifold or exhaust pipes • HO2S signal wire and ground wire crossed in connector • HO2S element is fuel contaminated or has failed • ECM has failed
DTC: P0161 **2T OBD/O2S1, MIL: Yes** **1996, 1997, 1998, 1999, 2000, 2001, 2002, 2003, 2004, 2005, 2006** **Models:** TT, A4, A6, A6 Avant, Allroad Quattro, A8, A8L **Engines:** 1.8L, 2.7L, 2.8L, 3.0L, 3.2L, 3.7L, 4.2L **Transmissions:** All	**HO2S (Bank 2 Sensor 2) Heater Circuit Malfunction Conditions:** Engine running, battery voltage 11.5, all electrical components off, ground between engine and chassis well connected and the exhaust system must be properly sealed between catalytic converter and the cylinder head. The the ECM detected an open or shorted condition, or excessive current draw in the heater circuit. **Possible Causes:** • HO2S heater power circuit or the heater ground circuit is open • HO2S signal tracking (due to oil or moisture in the connector) • HO2S has failed, or the ECM has failed
DTC: P0170 **2T CCM, MIL: Yes** **1995, 1996, 2000, 2001, 2002, 2003, 2004, 2005, 2006** **Models:** TT, A4, A6, A6 Avant **Engines:** 1.8L, 2.8L **Transmissions:** All	**Fuel System Malfunction (Cylinder Bank 1) Conditions:** The engine is running in a closed loop at a stable engine speed, and the ECM detected the lean or rich fuel trim correction valve was more than or less than a calibrated limit. **Possible Causes:** • Air leaks after the MAF sensor, or leaks in the PCV system • Exhaust leaks before or near where the HO2S is mounted • Fuel injector(s) restricted or not supplying enough fuel • Fuel system not supplying enough fuel during high fuel demand conditions (e.g., the fuel pump may not supply enough fuel) • Leaking EGR gasket, or leaking EGR valve diaphragm • MAF sensor dirty (causes ECM to underestimate airflow) • Vehicle running out of fuel or engine oil dip stick not seated
DTC: P0171 **2T CCM, MIL: Yes** **1996, 1997, 1998, 1999, 2000, 2001, 2002, 2003, 2004, 2005, 2006** **Models:** TT, A4, A6, A6 Avant, Allroad Quattro, A8, A8L **Engines:** 1.8L, 2.7L, 2.8L, 3.0L, 3.2L, 3.7L, 4.2L **Transmissions:** All	**Fuel System Too Lean (Cylinder Bank 1) Conditions:** Key on or engine running, all electrical components off and coolant temperature at least 80 degrees Celsius; and the ECM detected the Bank 1 Adaptive Fuel Control System reached its rich correction limit (a lean A/F condition). **Possible Causes:** • Air leaks after the MAF sensor, or leaks in the PCV system • Exhaust leaks before or near where the HO2S is mounted • Fuel injector(s) restricted or not supplying enough fuel • Fuel pump not supplying enough fuel during high fuel demand conditions • Leaking EGR gasket, or leaking EGR valve diaphragm • MAF sensor dirty (causes ECM to underestimate airflow) • Vehicle running out of fuel or engine oil dip stick not seated
DTC: P0172 **2T CCM, MIL: Yes** **1996, 1997, 1998, 1999, 2000, 2001, 2002, 2003, 2004, 2005, 2006** **Models:** TT, A4, A6, A6 Avant, Allroad Quattro, A8, A8L **Engines:** 1.8L, 2.7L, 2.8L, 3.0L, 3.2L, 3.7L, 4.2L **Transmissions:** All	**Fuel System Too Rich (Cylinder Bank 1) Conditions:** Key on or engine running, all electrical components off and coolant temperature at least 80 degrees Celsius; and the ECM detected the Bank 1 Adaptive Fuel Control System reached its rich correction limit (a rich A/F condition). **Possible Causes:** • Camshaft timing is incorrect, or the engine has an oil overfill condition • EVAP vapor recovery system failure (may be pulling vacuum) • Fuel pressure regulator is damaged or leaking • HO2S element is contaminated with alcohol or water • MAF or MAP sensor values are incorrect or out-of-range • One of more fuel injectors is leaking

DTC	Trouble Code Title, Conditions & Possible Causes
DTC: P0173 **2T CCM, MIL: Yes** **1995, 1996** **Models:** A4, A6, A6 Avant **Engines:** 2.8L **Transmissions:** All	**Fuel System Malfunction (Cylinder Bank 1) Conditions:** Key on or engine running, all electrical components off and coolant temperature at least 80 degrees Celsius; and the ECM detected the Bank 1 Fuel Control System experienced a implausible signal **Possible Causes:** • Air leaks after the MAF sensor, or leaks in the PCV system • Exhaust leaks before or near where the HO2S is mounted • Fuel injector(s) restricted or not supplying enough fuel • Fuel system not supplying enough fuel during high fuel demand conditions (e.g., the fuel pump may not supply enough fuel) • Leaking EGR gasket, or leaking EGR valve diaphragm • MAF sensor dirty (causes ECM to underestimate airflow) • Vehicle running out of fuel or engine oil dip stick not seated
DTC: P0174 **2T CCM, MIL: Yes** **1996, 1997, 1998, 1999, 2000, 2001, 2002, 2003, 2004, 2005, 2006** **Models:** A4, A6, A6 Avant, Allroad Quattro, A8, A8L **Engines:** 2.7L, 2.8L, 3.0L, 3.2L, 3.7L, 4.2L **Transmissions:** All	**Fuel System Too Lean (Cylinder Bank 2) Conditions:** Key on or engine running, all electrical components off and coolant temperature at least 80 degrees Celsius; and the ECM detected the Bank 2 Fuel Control System reached its lean correction limit **Possible Causes:** • Air leaks after the MAF sensor, or leaks in the PCV system • Exhaust leaks before or near where the HO2S is mounted • Fuel injector(s) restricted or not supplying enough fuel • Fuel pump not supplying enough fuel during high fuel demand conditions • Leaking EGR gasket, or leaking EGR valve diaphragm • MAF sensor dirty (causes ECM to underestimate airflow) • Vehicle running out of fuel or engine oil dip stick not seated
DTC: P0175 **2T CCM, MIL: Yes** **1996, 1997, 1998, 1999, 2000, 2001, 2002, 2003, 2004, 2005, 2006** **Models:** A4, A6, A6 Avant, Allroad Quattro, A8, A8L **Engines:** 2.7L, 2.8L, 3.0L, 3.2L, 3.7L, 4.2L **Transmissions:** All	**Fuel System Too Rich (Cylinder Bank 2) Conditions:** Key on or engine running, all electrical components off and coolant temperature at least 80 degrees Celsius; and the ECM detected the Bank 2 Adaptive Fuel Control System reached its rich correction limit (a rich A/F condition). **Possible Causes:** • Air leaks after the MAF sensor, or leaks in the PCV system • Exhaust leaks before or near where the HO2S is mounted • Fuel injector(s) restricted or not supplying enough fuel • Fuel pump not supplying enough fuel during high fuel demand conditions • Leaking EGR gasket, or leaking EGR valve diaphragm • MAF sensor dirty (causes ECM to underestimate airflow) • Vehicle running out of fuel or engine oil dip stick not seated
DTC: P0190 **1T CCM, MIL: Yes** **2005, 2006** **Models:** A6 **Engines:** 3.2L **Transmissions:** All	**Fuel Rail Pressure Sensor Circuit Conditions** Key on or engine running, all electrical components off and coolant temperature at least 80 degrees Celsius; and the ECM detected the fuel rail pressure sensor signal was outside the required voltage parameters in the self-test. **Possible Causes:** • Fuel Pressure Regulator Valve faulty • Fuel Pressure Sensor faulty • Fuel Pump (FP) Control Module faulty • Fuel pump faulty
DTC: P0192 **1T CCM, MIL: Yes** **2005, 2006** **Models:** A6 **Engines:** 3.2L **Transmissions:** All	**Fuel Rail Pressure Sensor Circuit Low Conditions** Key on or engine running, all electrical components off and coolant temperature at least 80 degrees Celsius; and the ECM detected the fuel rail pressure sensor signal was below the required voltage in the self-test. **Possible Causes:** • Fuel Pressure Regulator Valve faulty • Fuel Pressure Sensor faulty • Fuel Pump (FP) Control Module faulty • Fuel pump faulty
DTC: P0201 **1T CCM, MIL: Yes** **1996, 1997, 1998, 1999, 2000, 2001, 2002, 2003, 2004, 2005, 2006** **Models:** TT, A4, A6, Allroad Quattro, A8, A8L **Engines:** 1.8L, 2.7L, 2.8L, 3.0L, 3.2L, 3.7L, 4.2L **Transmissions:** All	**Cylinder 1 Injector Circuit Malfunction Conditions:** Engine started, and the ECM detected the fuel injector "1" control circuit was in a high state when it should have been low, or in a low state when it should have been high (wiring harness & injector okay). **Possible Causes:** • Injector 1 connector is damaged, open or shorted • Injector 1 control circuit is open, shorted to ground or to power • ECM has failed (the injector driver circuit may be damaged)

DTC	Trouble Code Title, Conditions & Possible Causes
DTC: P0202 **1T CCM, MIL: Yes** **1996, 1997, 1998, 1999, 2000, 2001, 2002, 2003, 2004, 2005, 2006** **Models:** TT, A4, A6, Allroad Quattro, A8, A8L **Engines:** 1.8L, 2.7L, 2.8L, 3.0L, 3.2L, 3.7L, 4.2L **Transmissions:** All	**Cylinder 2 Injector Circuit Malfunction Conditions:** Engine started, and the ECM detected the fuel injector "2" control circuit was in a high state when it should have been low, or in a low state when it should have been high (wiring harness & injector okay). **Possible Causes:** • Injector 2 connector is damaged, open or shorted • Injector 2 control circuit is open, shorted to ground or to power • ECM has failed (the injector driver circuit may be damaged)
DTC: P0203 **1T CCM, MIL: Yes** **1996, 1997, 1998, 1999, 2000, 2001, 2002, 2003, 2004, 2005, 2006** **Models:** TT, A4, A6, Allroad Quattro, A8, A8L **Engines:** 1.8L, 2.7L, 2.8L, 3.0L, 3.2L, 3.7L, 4.2L **Transmissions:** All **Transmissions:** All	**Cylinder 3 Injector Circuit Malfunction Conditions:** Engine started, and the ECM detected the fuel injector "3" control circuit was in a high state when it should have been low, or in a low state when it should have been high (wiring harness & injector okay). **Possible Causes:** • Injector 3 connector is damaged, open or shorted • Injector 3 control circuit is open, shorted to ground or to power • ECM has failed (the injector driver circuit may be damaged)
DTC: P0204 **1T CCM, MIL: Yes** **1996, 1997, 1998, 1999, 2000, 2001, 2002, 2003, 2004, 2005, 2006** **Models:** TT, A4, A6, Allroad Quattro, A8, A8L **Engines:** 1.8L, 2.7L, 2.8L, 3.0L, 3.2L, 3.7L, 4.2L **Transmissions:** All	**Cylinder 4 Injector Circuit Malfunction Conditions:** Engine started, and the ECM detected the fuel injector "4" control circuit was in a high state when it should have been low, or in a low state when it should have been high (wiring harness & injector okay). **Possible Causes:** • Injector 4 connector is damaged, open or shorted • Injector 4 control circuit is open, shorted to ground or to power • ECM has failed (the injector driver circuit may be damaged)
DTC: P0205 **1T CCM, MIL: Yes** **1996, 1997, 1998, 1999, 2000, 2001, 2002, 2003, 2004, 2005, 2006** **Models:** A4, A6, Allroad Quattro, A8, A8L **Engines:** 2.7L, 2.8L, 3.0L, 3.2L, 4.2L **Transmissions:** All	**Cylinder 5 Injector Circuit Malfunction Conditions:** Engine started, and the ECM detected the fuel injector "5" control circuit was in a high state when it should have been low, or in a low state when it should have been high (wiring harness & injector okay). **Possible Causes:** • Injector 5 connector is damaged, open or shorted • Injector 5 control circuit is open, shorted to ground or to power • ECM has failed (the injector driver circuit may be damaged)
DTC: P0206 **1T CCM, MIL: Yes** **1996, 1997, 1998, 1999, 2000, 2001, 2002, 2003, 2004, 2005, 2006** **Models:** A4, A6, Allroad Quattro, A8, A8L **Engines:** 2.7L, 2.8L, 3.0L, 3.2L, 4.2L **Transmissions:** All	**Cylinder 6 Injector Circuit Malfunction Conditions:** Engine started, and the ECM detected the fuel injector control circuit was in a high state when it should have been low, or in a low state when it should have been high (wiring harness & injector okay). **Possible Causes:** • Injector 6 connector is damaged, open or shorted • Injector 6 control circuit is open, shorted to ground or to power • ECM has failed (the injector driver circuit may be damaged)

DTC	Trouble Code Title, Conditions & Possible Causes
DTC: P0207 **1T CCM, MIL: Yes** **2003, 2004, 2005, 2006** **Models:** A6, Allroad Quattro, A8, A8L **Engines:** 4.2L **Transmissions:** All	**Cylinder 7 Injector Circuit Malfunction Conditions:** Engine started, and the ECM detected the fuel injector "7" control circuit was in a high state when it should have been low, or in a low state when it should have been high (wiring harness & injector okay). **Note: Monitor the INJIF PID Fault "flags" with the Scan Tool. The appropriate INJF PID "flag" will read Yes when this code is set.** **Possible Causes:** • Injector 7 connector is damaged, open or shorted • Injector 7 control circuit is open, shorted to ground or to power • ECM has failed (the injector driver circuit may be damaged)
DTC: P0208 **1T CCM, MIL: Yes** **2003, 2004, 2005, 2006** **Models:** A6, Allroad Quattro, A8, A8L **Engines:** 4.2L **Transmissions:** All	**Cylinder 8 Injector Circuit Malfunction Conditions:** Engine started, and the ECM detected the fuel injector "8" control circuit was in a high state when it should have been low, or in a low state when it should have been high (wiring harness & injector okay). **Note: Monitor the INJIF PID Fault "flags" with the Scan Tool. The appropriate INJF PID "flag" will read Yes when this code is set.** **Possible Causes:** • Injector 8 connector is damaged, open or shorted • Injector 8 control circuit is open, shorted to ground or to power • ECM has failed (the injector driver circuit may be damaged)
DTC: P0219 **1T CCM, MIL: Yes** **2004, 2005, 2006** **Models:** TT, A6 **Engines:** 3.2L, 4.2L **Transmissions:** All	**Engine Over-Speed Condition Conditions:** Engine started, and the ECM determined the vehicle had been driven in a manner that caused the engine to over-speed, and to exceed the engine speed calibration limit stored in memory. **Possible Causes:** • Engine operated in the wrong transmission gear position • Excessive engine speed with gear selector in Neutral position • Wheel slippage due to wet, muddy or snowing conditions
DTC: P0221 **1T CCM, MIL: Yes** **1996, 1997, 1998, 1999, 2000, 2001, 2002, 2003, 2004, 2005, 2006** **Models:** TT, A4, A6, Allroad Quattro, A8, A8L **Engines:** 1.8L, 2.7L, 2.8L, 3.0L, 3.2L, 4.2L **Transmissions:** All	**Throttle Position Sensor 'B' Signal Performance Conditions:** Engine started, battery voltage at least 11.5v, all electrical components off, ground connections between engine and chassis well connected, coolant temperature at least 80-degrees Celicius and the throttle valve must not be damaged or dirty; and the ECM detected the TP Sensor 'B' circuit was out of its normal operating range during a condition with the throttle wide open, or with it completely closed. The throttle valve activation occurs via an electric motor (throttle drive) in the throttle valve control module. It is activated by the ECM according to specifications of the two sensors, Throttle Position Sensor and Accelerator Pedal Position Sensor 2. Slowly depress accelerator pedal up to Wide Open Throttle (WOT) stop while observing the percentage display on the PID data function of the scan tool. The percentage display must increase uniformly. **Possible Causes:** • Throttle body is damaged • Throttle linkage is binding or sticking • ETC TP Sensor 'B' signal circuit to the ECM is open • ETC TP Sensor 'B' ground circuit is open • ETC TP Sensor 'B' is damaged or it has failed
DTC: P0222 **1T CCM, MIL: Yes** **1996, 1997, 1998, 1999, 2000, 2001, 2002, 2003, 2004, 2005, 2006** **Models:** TT, A4, A6, Allroad Quattro, A8, A8L **Engines:** 1.8L, 2.7L, 2.8L, 3.0L, 3.2L, 4.2L **Transmissions:** All	**Throttle Position Sensor 'B' Circuit Low Input Conditions:** Engine started, battery voltage at least 11.5v, all electrical components off, ground connections between engine and chassis well connected, coolant temperature at least 80-degrees Celicius and the throttle valve must not be damaged or dirty; and the ECM detected the TP Sensor 'B' circuit was out of its normal operating range during a condition with the throttle wide open, or with it completely closed. The throttle valve activation occurs via an electric motor (throttle drive) in the throttle valve control module. It is activated by the ECM according to specifications of the two sensors, Throttle Position Sensor and Accelerator Pedal Position Sensor 2. Slowly depress accelerator pedal up to Wide Open Throttle (WOT) stop while observing the percentage display on the PID data function of the scan tool. The percentage display must increase uniformly. **Possible Causes:** • ETC TP Sensor 'B' connector is damaged or shorted • ETC TP Sensor 'B' signal circuit is shorted to ground • ETC TP Sensor 'B' is damaged or it has failed • ECM has failed

DTC	Trouble Code Title, Conditions & Possible Causes
DTC: P0223 **1T CCM, MIL: Yes** **1996, 1997, 1998, 1999, 2000, 2001, 2002, 2003, 2004, 2005, 2006** **Models:** TT, A4, A6, Allroad Quattro, A8, A8L **Engines:** 1.8L, 2.7L, 2.8L, 3.0L, 3.2L, 4.2L **Transmissions:** All	**Throttle Position Sensor 'B' Circuit High Input Conditions:** Engine started, battery voltage at least 11.5v, all electrical components off, ground connections between engine and chassis well connected, coolant temperature at least 80-degrees Celicius and the throttle valve must not be damaged or dirty; and the ECM detected the TP Sensor 'B' circuit was out of its normal operating range during a condition with the throttle wide open, or with it completely closed. The throttle valve activation occurs via an electric motor (throttle drive) in the throttle valve control module. It is activated by the ECM according to specifications of the two sensors, Throttle Position Sensor and Accelerator Pedal Position Sensor 2. Slowly depress accelerator pedal up to Wide Open Throttle (WOT) stop while observing the percentage display on the PID data function of the scan tool. The percentage display must increase uniformly. **Possible Causes:** • ETC TP Sensor 'B' connector is damaged or open • ETC TP Sensor 'B' signal circuit is open • ETC TP Sensor 'B' signal circuit is shorted to VREF (5v) • ETC TP Sensor 'B' is damaged or it has failed
DTC: P0230 **1T CCM, MIL: Yes** **1996, 1997, 1998, 1999, 2000, 2001, 2002, 2003, 2004, 2005, 2006** **Models:** TT, A4, A6, Allroad Quattro, A8, A8L **Engines:** 1.8L, 2.7L, 2.8L, 3.0L, 3.2L, 4.2L **Transmissions:** All	**Fuel Pump Primary Circuit Malfunction Conditions:** Engine started, battery voltage at least 11.5v, all electrical components off, ground connections between engine and chassis well connected, coolant temperature at least 80-degrees Celicius. The ECM detected high current in fuel pump or fuel shutoff valve (FSV) circuit, or it detected voltage with the valve off, or it did not detect voltage on the circuit. The circuit is used to energize the fuel pump relay at key on or while running. Fuel pressure value should be 3000 to 5000 kPa at idle. **Possible Causes:** • FP or FSV circuit is open or shorted • Fuel pump relay VPWR circuit open • Fuel pump relay is damaged or has failed • Fuel pressure sensor has failed • Fuel pump control module is faulty • ECM has failed
DTC: P0234 **1T CCM, MIL: Yes** **1997, 1998, 1999, 2000, 2001, 2002, 2003, 2004, 2005, 2006** **Models:** TT, A4 **Engines:** 1.8L **Transmissions:** All	**Turbo/Supercharger Overboost Condition Conditions:** Engine started, battery voltage at least 11.5v, all electrical components off, ground connections between engine and chassis well connected, coolant temperature at least 80-degrees Celicius. The ECM detected an operating condition that could harm the engine or automatic transmission. **Possible Causes:** • Ignition misfire condition exceeds the calibrated threshold • Knock sensor circuit has failed, or excessive knock detected • Low speed fuel pump relay not switching properly • Transmission oil temperature beyond the calibrated threshold • Shaft bearing of charge pressure regulator valve in turbocharger is blocked
DTC: P0235 **1T CCM, MIL: Yes** **1997, 1998, 1999, 2000, 2001, 2002, 2003, 2004, 2005, 2006** **Models:** TT, A4 **Engines:** 1.8L **Transmissions:** All	**Turbocharger Boost Sensor (A) Circ Control Limit Not Reached Conditions:** Engine started, battery voltage at least 11.5v, all electrical components off, ground connections between engine and chassis well connected, coolant temperature at least 80-degrees Celicius. The ECM detected an operating condition that could harm the engine or automatic transmission. **Possible Causes:** • Charge air pressure sensor is faulty • Voltage supply to the charge air pressure sensor is open or shorted • Check the charge air system for leaks • Recirculating valve for turbocharger is faulty • Turbocharging system is damaged or not functioning properly • Turbocharger recirculating valve is faulty • Vacuum diaphragm for turbocharger is out of adjustment • Wastegate bypass regulator valve is faulty • Boost sensor has failed • ECM has failed

DTC	Trouble Code Title, Conditions & Possible Causes
DTC: P0236 **1T CCM, MIL: Yes** **1997, 1998, 1999, 2000, 2001, 2002, 2003, 2004, 2005, 2006** **Models:** TT, A4 **Engines:** 1.8L **Transmissions:** All	**Turbocharger Boost Sensor (A) Circ Control Range/Performance Conditions:** Engine started, battery voltage at least 11.5v, all electrical components off, ground connections between engine and chassis well connected, coolant temperature at least 80-degrees Celicius. The ECM detected an operating condition that could harm the engine or automatic transmission. **Possible Causes:** • Charge air pressure sensor is faulty • Voltage supply to the charge air pressure sensor is open or shorted • Check the charge air system for leaks • Recirculating valve for turbocharger is faulty • Turbocharging system is damaged or not functioning properly • Turbocharger recirculating valve is faulty • Vacuum diaphragm for turbocharger is out of adjustment • Wastegate bypass regulator valve is faulty • Boost sensor has failed • ECM has failed
DTC: P0237 **1T CCM, MIL: Yes** **1997, 1998, 1999, 2000, 2001, 2002, 2003, 2004, 2005, 2006** **Models:** TT, A4 **Engines:** 1.8L **Transmissions:** All	**Turbocharger Boost Sensor (A) Circ Low Input Conditions:** Engine started, battery voltage at least 11.5v, all electrical components off, ground connections between engine and chassis well connected, coolant temperature at least 80-degrees Celicius. The ECM detected an operating condition that could harm the engine or automatic transmission. **Possible Causes:** • Charge air pressure sensor is faulty • Voltage supply to the charge air pressure sensor is open or shorted • Check the charge air system for leaks • Recirculating valve for turbocharger is faulty • Turbocharging system is damaged or not functioning properly • Turbocharger recirculating valve is faulty • Vacuum diaphragm for turbocharger is out of adjustment • Wastegate bypass regulator valve is faulty • Boost sensor has failed • ECM has failed
DTC: P0238 **1T CCM, MIL: Yes** **1997, 1998, 1999, 2000, 2001, 2002, 2003, 2004, 2005, 2006** **Models:** TT, A4 **Engines:** 1.8L **Transmissions:** All	**Turbocharger Boost Sensor (A) Circ High Input Conditions:** Engine started, battery voltage at least 11.5v, all electrical components off, ground connections between engine and chassis well connected, coolant temperature at least 80-degrees Celicius. The ECM detected an operating condition that could harm the engine or automatic transmission. **Possible Causes:** • Charge air pressure sensor is faulty • Voltage supply to the charge air pressure sensor is open or shorted • Check the charge air system for leaks • Recirculating valve for turbocharger is faulty • Turbocharging system is damaged or not functioning properly • Turbocharger recirculating valve is faulty • Vacuum diaphragm for turbocharger is out of adjustment • Wastegate bypass regulator valve is faulty • Boost sensor has failed • ECM has failed
DTC: P0243 **1T CCM, MIL: Yes** **1997, 1998, 1999, 2000, 2001, 2002, 2003, 2004, 2005, 2006** **Models:** TT, A4 **Engines:** 1.8L **Transmissions:** All	**Turbocharger Boost Bypass Solenoid (A) Circuit Open/Short Circuit Conditions:** Engine started, battery voltage at least 11.5v, all electrical components off, ground connections between engine and chassis well connected, coolant temperature at least 80-degrees Celicius. The ECM detected an unexpected voltage condition on the Bypass Solenoid control circuit **Possible Causes:** • Bypass solenoid power supply circuit is open • Bypass solenoid control circuit is open, shorted to ground or system power • Bypass solenoid assembly is damaged or has failed • Charge air pressure sensor is faulty • Voltage supply to the charge air pressure sensor is open or shorted • Check the charge air system for leaks • Recirculating valve for turbocharger is faulty • Turbocharging system is damaged or not functioning properly • Turbocharger recirculating valve is faulty • Vacuum diaphragm for turbocharger is out of adjustment • Wastegate bypass regulator valve is faulty • Boost sensor has failed • ECM has failed

DTC	Trouble Code Title, Conditions & Possible Causes
DTC: P0245 **1T CCM, MIL: Yes** **1997, 1998, 1999, 2000, 2001, 2002, 2003, 2004, 2005, 2006** **Models:** TT, A4 **Engines:** 1.8L **Transmissions:** All	**Turbocharger Boost Bypass Solenoid (A) Circuit Low Input/Short to Ground Conditions:** Engine started, battery voltage at least 11.5v, all electrical components off, ground connections between engine and chassis well connected, coolant temperature at least 80-degrees Celicius. The ECM detected an unexpected voltage condition on the Bypass Solenoid control circuit **Possible Causes:** • Bypass solenoid power supply circuit is open • Bypass solenoid control circuit is open, shorted to ground or system power • Bypass solenoid assembly is damaged or has failed • Charge air pressure sensor is faulty • Voltage supply to the charge air pressure sensor is open or shorted • Check the charge air system for leaks • Recirculating valve for turbocharger is faulty • Turbocharging system is damaged or not functioning properly • Turbocharger recirculating valve is faulty • Vacuum diaphragm for turbocharger is out of adjustment • Wastegate bypass regulator valve is faulty • Boost sensor has failed • ECM has failed
DTC: P0246 **1T CCM, MIL: Yes** **1997, 1998, 1999, 2000, 2001, 2002, 2003, 2004, 2005, 2006** **Models:** TT, A4 **Engines:** 1.8L **Transmissions:** All	**Turbocharger Boost Bypass Solenoid (A) Circuit High Input/Short to B+ Conditions:** Engine started, battery voltage at least 11.5v, all electrical components off, ground connections between engine and chassis well connected, coolant temperature at least 80-degrees Celicius. The ECM detected an unexpected voltage condition on the Bypass Solenoid control circuit **Possible Causes:** • Bypass solenoid power supply circuit is open • Bypass solenoid control circuit is open, shorted to ground or system power • Bypass solenoid assembly is damaged or has failed • Charge air pressure sensor is faulty • Voltage supply to the charge air pressure sensor is open or shorted • Check the charge air system for leaks • Recirculating valve for turbocharger is faulty • Turbocharging system is damaged or not functioning properly • Turbocharger recirculating valve is faulty • Vacuum diaphragm for turbocharger is out of adjustment • Wastegate bypass regulator valve is faulty • Boost sensor has failed • ECM has failed
DTC: P0261 **2T CCM, MIL: Yes** **1996, 1997, 1998, 1999, 2000, 2001, 2002, 2003, 2004, 2005, 2006** **Models:** TT, A4, A6, Allroad Quattro, A8, A8L **Engines:** 1.8L, 2.7L, 2.8L, 3.0L, 3.2L, 4.2L **Transmissions:** All	**Cylinder 1 Injector Circuit Low Input/Short to Ground Conditions:** Key on or engine running, fuses in the instrument panel and the E-box in the engine compartment must be functioning, and the ground connections between the engine ad the chassis must be well connected; and the ECM detected an unexpected voltage condition on the injector circuit **Possible Causes:** • Injector 1 control circuit is open • Injector 1 power circuit (B+) is open • Injector 1 control circuit is shorted to chassis ground • Injector 1 is damaged or has failed • ECM is not connected or has failed
DTC: P0262 **2T CCM, MIL: Yes** **1996, 1997, 1998, 1999, 2000, 2001, 2002, 2003, 2004, 2005, 2006** **Models:** TT, A4, A6, Allroad Quattro, A8, A8L **Engines:** 1.8L, 2.7L, 2.8L, 3.0L, 3.2L, 4.2L **Transmissions:** All	**Cylinder 1 Injector Circuit Low Input/Short to B+ Conditions:** Key on or engine running, fuses in the instrument panel and the E-box in the engine compartment must be functioning, and the ground connections between the engine ad the chassis must be well connected; and the ECM detected an unexpected voltage condition on the injector circuit **Possible Causes:** • Injector control circuit is open • Injector power circuit (B+) is open • Injector control circuit is shorted to chassis ground • Injector is damaged or has failed • ECM is not connected or has failed • Fuel pump relay has failed • Fuel injectors may have malfunctioned • Faulty engine speed sensor

DTC	Trouble Code Title, Conditions & Possible Causes
DTC: P0264 **2T CCM, MIL: Yes** **1996, 1997, 1998, 1999, 2000, 2001, 2002, 2003, 2004, 2005, 2006** **Models:** TT, A4, A6, Allroad Quattro, A8, A8L **Engines:** 1.8L, 2.7L, 2.8L, 3.0L, 3.2L, 4.2L **Transmissions:** All	**Cylinder 2 Injector Circuit Low Input/Short to Ground Conditions:** Key on or engine running, fuses in the instrument panel and the E-box in the engine compartment must be functioning, and the ground connections between the engine ad the chassis must be well connected; and the ECM detected an unexpected voltage condition on the injector circuit **Possible Causes:** • Injector control circuit is open • Injector power circuit (B+) is open • Injector control circuit is shorted to chassis ground • Injector is damaged or has failed • ECM is not connected or has failed • Fuel pump relay has failed • Fuel injectors may have malfunctioned • Faulty engine speed sensor
DTC: P0265 **2T CCM, MIL: Yes** **1996, 1997, 1998, 1999, 2000, 2001, 2002, 2003, 2004, 2005, 2006** **Models:** TT, A4, A6, Allroad Quattro, A8, A8L **Engines:** 1.8L, 2.7L, 2.8L, 3.0L, 3.2L, 4.2L **Transmissions:** All	**Cylinder 2 Injector Circuit Low Input/Short to B+ Conditions:** Key on or engine running, fuses in the instrument panel and the E-box in the engine compartment must be functioning, and the ground connections between the engine ad the chassis must be well connected; and the ECM detected an unexpected voltage condition on the injector circuit **Possible Causes:** • Injector control circuit is open • Injector power circuit (B+) is open • Injector control circuit is shorted to chassis ground • Injector is damaged or has failed • ECM is not connected or has failed • Fuel pump relay has failed • Fuel injectors may have malfunctioned • Faulty engine speed sensor
DTC: P0267 **2T CCM, MIL: Yes** **1996, 1997, 1998, 1999, 2000, 2001, 2002, 2003, 2004, 2005, 2006** **Models:** TT, A4, A6, Allroad Quattro, A8, A8L **Engines:** 1.8L, 2.7L, 2.8L, 3.0L, 3.2L, 4.2L **Transmissions:** All	**Cylinder 3 Injector Circuit Low Input/Short to Ground Conditions:** Key on or engine running, fuses in the instrument panel and the E-box in the engine compartment must be functioning, and the ground connections between the engine ad the chassis must be well connected; and the ECM detected an unexpected voltage condition on the injector circuit **Possible Causes:** • Injector control circuit is open • Injector power circuit (B+) is open • Injector control circuit is shorted to chassis ground • Injector is damaged or has failed • ECM is not connected or has failed • Fuel pump relay has failed • Fuel injectors may have malfunctioned • Faulty engine speed sensor
DTC: P0268 **2T CCM, MIL: Yes** **1996, 1997, 1998, 1999, 2000, 2001, 2002, 2003, 2004, 2005, 2006** **Models:** TT, A4, A6, Allroad Quattro, A8, A8L **Engines:** 1.8L, 2.7L, 2.8L, 3.0L, 3.2L, 4.2L **Transmissions:** All	**Cylinder 3 Injector Circuit Low Input/Short to B+ Conditions:** Key on or engine running, fuses in the instrument panel and the E-box in the engine compartment must be functioning, and the ground connections between the engine ad the chassis must be well connected; and the ECM detected an unexpected voltage condition on the injector circuit **Possible Causes:** • Injector control circuit is open • Injector power circuit (B+) is open • Injector control circuit is shorted to chassis ground • Injector is damaged or has failed • ECM is not connected or has failed • Fuel pump relay has failed • Fuel injectors may have malfunctioned • Faulty engine speed sensor

DTC	Trouble Code Title, Conditions & Possible Causes
DTC: P0270 **2T CCM, MIL: Yes** **1996, 1997, 1998, 1999, 2000, 2001, 2002, 2003, 2004, 2005, 2006** **Models:** TT, A4, A6, Allroad Quattro, A8, A8L **Engines:** 1.8L, 2.7L, 2.8L, 3.0L, 3.2L, 4.2L **Transmissions:** All	**Cylinder 4 Injector Circuit Low Input/Short to Ground Conditions:** Key on or engine running, fuses in the instrument panel and the E-box in the engine compartment must be functioning, and the ground connections between the engine ad the chassis must be well connected; and the ECM detected an unexpected voltage condition on the injector circuit **Possible Causes:** • Injector control circuit is open • Injector power circuit (B+) is open • Injector control circuit is shorted to chassis ground • Injector is damaged or has failed • ECM is not connected or has failed • Fuel pump relay has failed • Fuel injectors may have malfunctioned • Faulty engine speed sensor
DTC: P0271 **2T CCM, MIL: Yes** **1996, 1997, 1998, 1999, 2000, 2001, 2002, 2003, 2004, 2005, 2006** **Models:** TT, A4, A6, Allroad Quattro, A8, A8L **Engines:** 1.8L, 2.7L, 2.8L, 3.0L, 3.2L, 4.2L **Transmissions:** All	**Cylinder 4 Injector Circuit Low Input/Short to B+ Conditions:** Key on or engine running, fuses in the instrument panel and the E-box in the engine compartment must be functioning, and the ground connections between the engine ad the chassis must be well connected; and the ECM detected an unexpected voltage condition on the injector circuit **Possible Causes:** • Injector control circuit is open • Injector power circuit (B+) is open • Injector control circuit is shorted to chassis ground • Injector is damaged or has failed • ECM is not connected or has failed • Fuel pump relay has failed • Fuel injectors may have malfunctioned • Faulty engine speed sensor
DTC: P0274 **2T CCM, MIL: Yes** **1998, 1999, 2000, 2001, 2002, 2003, 2004, 2005, 2006** **Models:** A4, A6, Allroad Quattro, A8, A8L **Engines:** 2.7L, 2.8L, 3.0L, 3.2L, 4.2L **Transmissions:** All	**Cylinder 5 Injector Circuit Low Input/Short to Ground Conditions:** Key on or engine running, fuses in the instrument panel and the E-box in the engine compartment must be functioning, and the ground connections between the engine ad the chassis must be well connected; and the ECM detected an unexpected voltage condition on the injector circuit **Possible Causes:** • Injector control circuit is open • Injector power circuit (B+) is open • Injector control circuit is shorted to chassis ground • Injector is damaged or has failed • ECM is not connected or has failed • Fuel pump relay has failed • Fuel injectors may have malfunctioned • Faulty engine speed sensor
DTC: P0274 **2T CCM, MIL: Yes** **1998, 1999, 2000, 2001, 2002, 2003, 2004, 2005, 2006** **Models:** A4, A6, Allroad Quattro, A8, A8L **Engines:** 2.7L, 2.8L, 3.0L, 3.2L, 4.2L **Transmissions:** All	**Cylinder 5 Injector Circuit Low Input/Short to B+ Conditions:** Key on or engine running, fuses in the instrument panel and the E-box in the engine compartment must be functioning, and the ground connections between the engine ad the chassis must be well connected; and the ECM detected an unexpected voltage condition on the injector circuit **Possible Causes:** • Injector control circuit is open • Injector power circuit (B+) is open • Injector control circuit is shorted to chassis ground • Injector is damaged or has failed • ECM is not connected or has failed • Fuel pump relay has failed • Fuel injectors may have malfunctioned • Faulty engine speed sensor

DTC	Trouble Code Title, Conditions & Possible Causes
DTC: P0276 **2T CCM, MIL: Yes** **1998, 1999, 2000, 2001, 2002, 2003, 2004, 2005, 2006** **Models:** A4, A6, Allroad Quattro, A8, A8L **Engines:** 2.7L, 2.8L, 3.0L, 3.2L, 4.2L **Transmissions:** All	**Cylinder 6 Injector Circuit Low Input/Short to Ground Conditions:** Key on or engine running, fuses in the instrument panel and the E-box in the engine compartment must be functioning, and the ground connections between the engine ad the chassis must be well connected; and the ECM detected an unexpected voltage condition on the injector circuit **Possible Causes:** • Injector control circuit is open • Injector power circuit (B+) is open • Injector control circuit is shorted to chassis ground • Injector is damaged or has failed • ECM is not connected or has failed • Fuel pump relay has failed • Fuel injectors may have malfunctioned • Faulty engine speed sensor
DTC: P0277 **2T CCM, MIL: Yes** **1998, 1999, 2000, 2001, 2002, 2003, 2004, 2005, 2006** **Models:** A4, A6, Allroad Quattro, A8, A8L **Engines:** 2.7L, 2.8L, 3.0L, 3.2L, 4.2L **Transmissions:** All	**Cylinder 6 Injector Circuit Low Input/Short to B+ Conditions:** Key on or engine running, fuses in the instrument panel and the E-box in the engine compartment must be functioning, and the ground connections between the engine ad the chassis must be well connected; and the ECM detected an unexpected voltage condition on the injector circuit **Possible Causes:** • Injector control circuit is open • Injector power circuit (B+) is open • Injector control circuit is shorted to chassis ground • Injector is damaged or has failed • ECM is not connected or has failed • Fuel pump relay has failed • Fuel injectors may have malfunctioned • Faulty engine speed sensor
DTC: P0279 **2T CCM, MIL: Yes** **2003, 2004, 2005, 2006** **Models:** A6, Allroad Quattro, A8, A8L **Engines:** 4.2L **Transmissions:** All	**Cylinder 7 Injector Circuit Low Input/Short to Ground Conditions:** Key on or engine running, fuses in the instrument panel and the E-box in the engine compartment must be functioning, and the ground connections between the engine ad the chassis must be well connected; and the ECM detected an unexpected voltage condition on the injector circuit **Possible Causes:** • Injector control circuit is open • Injector power circuit (B+) is open • Injector control circuit is shorted to chassis ground • Injector is damaged or has failed • ECM is not connected or has failed • Fuel pump relay has failed • Fuel injectors may have malfunctioned • Faulty engine speed sensor
DTC: P0280 **2T CCM, MIL: Yes** **2003, 2004, 2005, 2006** **Models:** A6, Allroad Quattro, A8, A8L **Engines:** 4.2L **Transmissions:** All	**Cylinder 7 Injector Circuit Low Input/Short to B+ Conditions:** Key on or engine running, fuses in the instrument panel and the E-box in the engine compartment must be functioning, and the ground connections between the engine ad the chassis must be well connected; and the ECM detected an unexpected voltage condition on the injector circuit **Possible Causes:** • Injector control circuit is open • Injector power circuit (B+) is open • Injector control circuit is shorted to chassis ground • Injector is damaged or has failed • ECM is not connected or has failed • Fuel pump relay has failed • Fuel injectors may have malfunctioned • Faulty engine speed sensor

DTC	Trouble Code Title, Conditions & Possible Causes
DTC: P0282 **2T CCM, MIL: Yes** **2003, 2004, 2005, 2006** **Models:** A6, Allroad Quattro, A8, A8L **Engines:** 4.2L **Transmissions:** All	**Cylinder 8 Injector Circuit Low Input/Short to Ground Conditions:** Key on or engine running, fuses in the instrument panel and the E-box in the engine compartment must be functioning, and the ground connections between the engine ad the chassis must be well connected; and the ECM detected an unexpected voltage condition on the injector circuit **Possible Causes:** • Injector control circuit is open • Injector power circuit (B+) is open • Injector control circuit is shorted to chassis ground • Injector is damaged or has failed • ECM is not connected or has failed • Fuel pump relay has failed • Fuel injectors may have malfunctioned • Faulty engine speed sensor
DTC: P0283 **2T CCM, MIL: Yes** **2003, 2004, 2005, 2006** **Models:** A6, Allroad Quattro, A8, A8L **Engines:** 4.2L **Transmissions:** All	**Cylinder 8 Injector Circuit Low Input/Short to B+ Conditions:** Key on or engine running, fuses in the instrument panel and the E-box in the engine compartment must be functioning, and the ground connections between the engine ad the chassis must be well connected; and the ECM detected an unexpected voltage condition on the injector circuit **Possible Causes:** • Injector control circuit is open • Injector power circuit (B+) is open • Injector control circuit is shorted to chassis ground • Injector is damaged or has failed • ECM is not connected or has failed • Fuel pump relay has failed • Fuel injectors may have malfunctioned • Faulty engine speed sensor
DTC: P0300 **2T CCM, MIL: Yes** **1996, 1997, 1998, 1999, 2000, 2001, 2002, 2003, 2004, 2005, 2006** **Models:** TT, A4, A6, A6 Avant, Allroad Quattro, A8, A8L **Engines:** 1.8L, 2.7L, 2.8L, 3.0L, 3.2L, 3.7L, 4.2L **Transmissions:** All	**Random/Multiple Misfire Detected Conditions:** Engine running under positive torque conditions, and the ECM detected a misfire or uneven engine running in two or more cylinders. **Note: If the misfire is severe, the MIL will flash on/off on the first trip!** **Possible Causes:** • Base engine mechanical fault that affects two or more cylinders • Fuel metering fault that affects two or more cylinders • Fuel pressure too low or too high, fuel supply contaminated • EVAP system problem or the EVAP canister is fuel saturated • EGR valve is stuck open or the PCV system has a vacuum leak • Ignition system fault (coil, plugs) affecting two or more cylinders • MAF sensor contamination (it can cause a very lean condition) • Vehicle driven while very low on fuel (less than 1/8 of a tank)
DTC: P0301 **2T CCM, MIL: Yes** **1996, 1997, 1998, 1999, 2000, 2001, 2002, 2003, 2004, 2005, 2006** **Models:** TT, A4, A6, A6 Avant, Allroad Quattro, A8, A8L **Engines:** 1.8L, 2.7L, 2.8L, 3.0L, 3.2L, 3.7L, 4.2L **Transmissions:** All	**Cylinder Number 1 Misfire Detected Conditions:** Engine running under positive torque conditions, and the ECM detected a misfire or uneven engine function. **Note: If the misfire is severe, the MIL will flash on/off on the 1st trip!** **Possible Causes:** • Air leak in the intake manifold, or in the EGR or ECM system • Base engine mechanical problem • Fuel delivery component problem (i.e., a contaminated, dirty or sticking fuel injector) • Fuel pump relay defective • Ignition coil fuses have failed • Ignition system problem (dirty damaged coil or plug) • Engine speed (RPM) sensor has failed • Camshaft position sensors have failed • Ignition coil is faulty • Spark plugs are not working properly or are not gapped properly

DTC	Trouble Code Title, Conditions & Possible Causes
DTC: P0302 **2T CCM, MIL: Yes** **1996, 1997, 1998, 1999, 2000, 2001, 2002, 2003, 2004, 2005, 2006** **Models:** TT, A4, A6, A6 Avant, Allroad Quattro, A8, A8L **Engines:** 1.8L, 2.7L, 2.8L, 3.0L, 3.2L, 3.7L, 4.2L **Transmissions:** All	**Cylinder Number 2 Misfire Detected Conditions:** Engine running under positive torque conditions, and the ECM detected a misfire or uneven engine function. **Note: If the misfire is severe, the MIL will flash on/off on the 1st trip!** **Possible Causes:** • Air leak in the intake manifold, or in the EGR or ECM system • Base engine mechanical problem • Fuel delivery component problem (i.e., a contaminated, dirty or sticking fuel injector) • Fuel pump relay defective • Ignition coil fuses have failed • Ignition system problem (dirty damaged coil or plug) • Engine speed (RPM) sensor has failed • Camshaft position sensors have failed • Ignition coil is faulty • Spark plugs are not working properly or are not gapped properly
DTC: P0303 **2T CCM, MIL: Yes** **1996, 1997, 1998, 1999, 2000, 2001, 2002, 2003, 2004, 2005, 2006** **Models:** TT, A4, A6, A6 Avant, Allroad Quattro, A8, A8L **Engines:** 1.8L, 2.7L, 2.8L, 3.0L, 3.2L, 3.7L, 4.2L **Transmissions:** All	**Cylinder Number 3 Misfire Detected Conditions:** Engine running under positive torque conditions, and the ECM detected a misfire or uneven engine function. **Note: If the misfire is severe, the MIL will flash on/off on the 1st trip!** **Possible Causes:** • Air leak in the intake manifold, or in the EGR or ECM system • Base engine mechanical problem • Fuel delivery component problem (i.e., a contaminated, dirty or sticking fuel injector) • Fuel pump relay defective • Ignition coil fuses have failed • Ignition system problem (dirty damaged coil or plug) • Engine speed (RPM) sensor has failed • Camshaft position sensors have failed • Ignition coil is faulty • Spark plugs are not working properly or are not gapped properly
DTC: P0304 **2T CCM, MIL: Yes** **1996, 1997, 1998, 1999, 2000, 2001, 2002, 2003, 2004, 2005, 2006** **Models:** TT, A4, A6, A6 Avant, Allroad Quattro, A8, A8L **Engines:** 1.8L, 2.7L, 2.8L, 3.0L, 3.2L, 3.7L, 4.2L **Transmissions:** All	**Cylinder Number 4 Misfire Detected Conditions:** Engine running under positive torque conditions, and the ECM detected a misfire or uneven engine function. **Note: If the misfire is severe, the MIL will flash on/off on the 1st trip!** **Possible Causes:** • Air leak in the intake manifold, or in the EGR or ECM system • Base engine mechanical problem • Fuel delivery component problem (i.e., a contaminated, dirty or sticking fuel injector) • Fuel pump relay defective • Ignition coil fuses have failed • Ignition system problem (dirty damaged coil or plug) • Engine speed (RPM) sensor has failed • Camshaft position sensors have failed • Ignition coil is faulty • Spark plugs are not working properly or are not gapped properly
DTC: P0305 **2T CCM, MIL: Yes** **1996, 1997, 1998, 1999, 2000, 2001, 2002, 2003, 2004, 2005, 2006** **Models:** TT, A4, A6, A6 Avant, Allroad Quattro, A8, A8L **Engines:** 1.8L, 2.7L, 2.8L, 3.0L, 3.2L, 3.7L, 4.2L **Transmissions:** All	**Cylinder Number 5 Misfire Detected Conditions:** Engine running under positive torque conditions, and the ECM detected a misfire or uneven engine function. **Note: If the misfire is severe, the MIL will flash on/off on the 1st trip!** **Possible Causes:** • Air leak in the intake manifold, or in the EGR or ECM system • Base engine mechanical problem • Fuel delivery component problem (i.e., a contaminated, dirty or sticking fuel injector) • Fuel pump relay defective • Ignition coil fuses have failed • Ignition system problem (dirty damaged coil or plug) • Engine speed (RPM) sensor has failed • Camshaft position sensors have failed • Ignition coil is faulty • Spark plugs are not working properly or are not gapped properly

DTC	Trouble Code Title, Conditions & Possible Causes
DTC: P0306 **2T CCM, MIL: Yes** **1996, 1997, 1998, 1999, 2000, 2001, 2002, 2003, 2004, 2005, 2006** **Models:** TT, A4, A6, A6 Avant, Allroad Quattro, A8, A8L **Engines:** 1.8L, 2.7L, 2.8L, 3.0L, 3.2L, 3.7L, 4.2L **Transmissions:** All	**Cylinder Number 6 Misfire Detected Conditions:** Engine running under positive torque conditions, and the ECM detected a misfire or uneven engine function. **Note: If the misfire is severe, the MIL will flash on/off on the 1st trip!** **Possible Causes:** • Air leak in the intake manifold, or in the EGR or ECM system • Base engine mechanical problem • Fuel delivery component problem (i.e., a contaminated, dirty or sticking fuel injector) • Fuel pump relay defective • Ignition coil fuses have failed • Ignition system problem (dirty damaged coil or plug) • Engine speed (RPM) sensor has failed • Camshaft position sensors have failed • Ignition coil is faulty • Spark plugs are not working properly or are not gapped properly
DTC: P0307 **2T CCM, MIL: Yes** **1997, 1998, 1999, 2000, 2001, 2002, 2003, 2004, 2005, 2006** **Models:** A6, Allroad Quattro, A8, A8L **Engines:** 3.7L, 4.2L **Transmissions:** All	**Cylinder Number 7 Misfire Detected Conditions:** Engine running under positive torque conditions, and the ECM detected a misfire or uneven engine function. **Note: If the misfire is severe, the MIL will flash on/off on the 1st trip!** **Possible Causes:** • Air leak in the intake manifold, or in the EGR or ECM system • Base engine mechanical problem • Fuel delivery component problem (i.e., a contaminated, dirty or sticking fuel injector) • Fuel pump relay defective • Ignition coil fuses have failed • Ignition system problem (dirty damaged coil or plug) • Engine speed (RPM) sensor has failed • Camshaft position sensors have failed • Ignition coil is faulty • Spark plugs are not working properly or are not gapped properly
DTC: P0308 **2T CCM, MIL: Yes** **1997, 1998, 1999, 2000, 2001, 2002, 2003, 2004, 2005, 2006** **Models:** A6, Allroad Quattro, A8, A8L **Engines:** 3.7L, 4.2L **Transmissions:** All	**Cylinder Number 8 Misfire Detected Conditions:** Engine running under positive torque conditions, and the ECM detected a misfire or uneven engine function. **Note: If the misfire is severe, the MIL will flash on/off on the 1st trip!** **Possible Causes:** • Air leak in the intake manifold, or in the EGR or ECM system • Base engine mechanical problem • Fuel delivery component problem (i.e., a contaminated, dirty or sticking fuel injector) • Fuel pump relay defective • Ignition coil fuses have failed • Ignition system problem (dirty damaged coil or plug) • Engine speed (RPM) sensor has failed • Camshaft position sensors have failed • Ignition coil is faulty • Spark plugs are not working properly or are not gapped properly
DTC: P0321 **1T CCM, MIL: Yes** **1996, 1997, 1998, 1999, 2000, 2001, 2002, 2003, 2004, 2005, 2006** **Models:** TT, A4, A6, Allroad Quattro, A8, A8L **Engines:** 1.8L, 2.7L, 2.8L, 3.0L, 3.2L, 3.7L, 4.2L **Transmissions:** All	**Ignition/Distributor Engine Speed Input Circuit Range/Performance Conditions:** Engine started, vehicle driven, and the ECM detected the engine speed signal was more than the calibrated value. **Note: The engine will not start if there is no speed signal. If the speed signal fails when the engine is running, it will cause the engine to stall immediately.** **Possible Causes:** • Engine speed sensor has failed or is damaged • ECM has failed • Sensor wheel is damaged or doesn't fit properly • Sensor wheel spacer isn't seated properly

DTC	Trouble Code Title, Conditions & Possible Causes
DTC: P0322 **1T CCM, MIL: Yes** **1996, 1997, 1998, 1999, 2000, 2001, 2002, 2003, 2004, 2005, 2006** **Models:** TT, A4, A6, Allroad Quattro, A8, A8L **Engines:** 1.8L, 2.7L, 2.8L, 3.0L, 3.2L, 3.7L, 4.2L **Transmissions:** All	**Ignition/Distributor Engine Input Circuit No Signal Conditions:** Key on, and the ECM could not detect the engine speed signal or the signal was erratic. **Note: The engine will not start if there is no speed signal. If the speed signal fails when the engine is running, it will cause the engine to stall immediately.** **Possible Causes:** • Engine speed sensor has failed or is damaged • ECM has failed • Sensor wheel is damaged or doesn't fit properly • Sensor wheel spacer isn't seated properly
DTC: P0324 **1T CCM, MIL: Yes** **2000, 2001, 2002, 2003, 2004, 2005, 2006** **Models:** TT, A4 **Engines:** 1.8L, 3.2L **Transmissions:** All	**Knock Control System Error Conditions:** Engine started, vehicle driven, and the ECM detected the Knock Sensor 1 (KS1) signal was too low or not recognized by the ECM **Possible Causes:** • Knock sensor circuit is open • Knock sensor is loose (tighten to 20 NM) • Contact between the knock sensor and cylinder block is dirty, corroded or greasy • Knock sensor circuit is shorted to ground, or shorted to power • Knock sensor is damaged or it has failed • Wrong kind of fuel used • A component in the engine compartment is loose or not properly secured • ECM has failed
DTC: P0327 **1T CCM, MIL: Yes** **1996, 1997, 1998, 1999, 2000, 2001, 2002, 2003, 2004, 2005, 2006** **Models:** TT, A4, A6, A6 Avant, A8 **Engines:** 1.8L, 2.7L, 2.8L, 3.2L, 4.2L **Transmissions:** All	**Knock Sensor 1 Signal Low Input Conditions:** Engine started, vehicle driven, and the ECM detected the Knock Sensor 1 (KS1) signal was too low or not recognized by the ECM **Possible Causes:** • Knock sensor circuit is open • Knock sensor is loose (tighten to 20 NM) • Contact between the knock sensor and cylinder block is dirty, corroded or greasy • Knock sensor circuit is shorted to ground, or shorted to power • Knock sensor is damaged or it has failed • Wrong kind of fuel used • A component in the engine compartment is loose or not properly secured • ECM has failed
DTC: P0328 **1T CCM, MIL: Yes** **1996, 1997, 1998, 1999, 2000, 2001, 2002, 2003, 2004, 2005, 2006** **Models:** TT, A4, A6, A8 **Engines:** 1.8L, 2.7L, 2.8L, 3.2L, 4.2L **Transmissions:** All	**Knock Sensor 1 Signal High Input Conditions:** Engine started, vehicle driven, and the ECM detected the Knock Sensor 1 (KS1) signal was too high **Possible Causes:** • Knock sensor circuit is open • Knock sensor is loose (tighten to 20 NM) • Contact between the knock sensor and cylinder block is dirty, corroded or greasy • Knock sensor circuit is shorted to ground, or shorted to power • Knock sensor is damaged or it has failed • Wrong kind of fuel used • A component in the engine compartment is loose or not properly secured • ECM has failed
DTC: P0332 **1T CCM, MIL: Yes** **1996, 1997, 1998, 1999, 2000, 2001, 2002, 2003, 2004, 2005, 2006** **Models:** TT, A4, A6, A6 Avant, A8 **Engines:** 1.8L, 2.7L, 2.8L, 3.2L, 4.2L **Transmissions:** All	**Knock Sensor 2 Signal Low Input Conditions:** Engine started, vehicle driven, and the ECM detected the Knock Sensor 1 (KS1) signal was too low or not recognized by the ECM **Possible Causes:** • Knock sensor circuit is open • Knock sensor is loose (tighten to 20 NM) • Contact between the knock sensor and cylinder block is dirty, corroded or greasy • Knock sensor circuit is shorted to ground, or shorted to power • Knock sensor is damaged or it has failed • Wrong kind of fuel used • A component in the engine compartment is loose or not properly secured • ECM has failed

DTC	Trouble Code Title, Conditions & Possible Causes
DTC: P0333 **1T CCM, MIL: Yes** **1996, 1997, 1998, 1999, 2000, 2001, 2002, 2003, 2004, 2005, 2006** **Models:** TT, A4, A6, A8 **Engines:** 1.8L, 2.7L, 2.8L, 3.2L, 4.2L **Transmissions:** All	**Knock Sensor 2 Signal High Input Conditions:** Engine started, vehicle driven, and the ECM detected the Knock Sensor 1 (KS1) signal was too high **Possible Causes:** • Knock sensor circuit is open • Knock sensor is loose (tighten to 20 NM) • Contact between the knock sensor and cylinder block is dirty, corroded or greasy • Knock sensor circuit is shorted to ground, or shorted to power • Knock sensor is damaged or it has failed • Wrong kind of fuel used • A component in the engine compartment is loose or not properly secured • ECM has failed
DTC: P0340 **1T CCM, MIL: Yes** **1996, 1997, 1998, 1999, 2000, 2001, 2002, 2003, 2004, 2005, 2006** **Models:** TT, A4, A6, A6 Avant, Allroad Quattro, A8, A8L **Engines:** 1.8L, 2.7L, 2.8L, 3.0L, 3.2L, 4.2L **Transmissions:** All	**Camshaft Position Sensor Circuit Malfunction Conditions:** Engine started, battery voltage must be at least 11.5v, all electrical components must be off, parking brake must be engaged (to keep daytime driving lights off), automatic transmission selector must be in park and the ground between the engine and the chassis must be well connected. The ECM detected the CMP sensor signal was missing or it was erratic. **Possible Causes:** • CMP sensor circuit is open or shorted to ground • CMP sensor circuit is shorted to power • CMP sensor ground (return) circuit is open • CMP sensor installation incorrect (Hall-effect type) • CMP sensor is damaged or CMP sensor shielding damaged • CMP sensor has failed • ECM has failed
DTC: P0341 **1T CCM, MIL: Yes** **1996, 1997, 1998, 1999, 2000, 2001, 2002, 2003, 2004, 2005, 2006** **Models:** TT, A4, A6, A6 Avant, Allroad Quattro, A8, A8L **Engines:** 1.8L, 2.7L, 2.8L, 3.0L, 3.2L, 4.2L **Transmissions:** All	**Camshaft Position Sensor Circ Range/Performance Conditions:** Engine started, battery voltage must be at least 11.5v, all electrical components must be off, parking brake must be engaged (to keep daytime driving lights off), automatic transmission selector must be in park and the ground between the engine and the chassis must be well connected. The ECM detected the CMP sensor signal was implausible. **Possible Causes:** • CMP sensor circuit is open or shorted to ground • CMP sensor circuit is shorted to power • CMP sensor ground (return) circuit is open • CMP sensor installation incorrect (Hall-effect type) • CMP sensor is damaged or CMP sensor shielding damaged • ECM has failed
DTC: P0342 **1T CCM, MIL: Yes** **1996, 1997, 1998, 1999, 2000, 2001, 2002, 2003, 2004, 2005, 2006** **Models:** TT, A4, A6, A6 Avant, Allroad Quattro, A8, A8L **Engines:** 1.8L, 2.7L, 2.8L, 3.0L, 3.2L, 4.2L **Transmissions:** All	**Camshaft Position Sensor "A" Circuit (Bank 1 or Single Sensor) Low Input Conditions:** Engine started, battery voltage must be at least 11.5v, all electrical components must be off, parking brake must be engaged (to keep daytime driving lights off), automatic transmission selector must be in park and the ground between the engine and the chassis must be well connected. The ECM detected the CMP sensor signal exceeded the bounds of the specified maximum limit. **Possible Causes:** • CMP sensor circuit is open or shorted to ground • CMP sensor circuit is shorted to power • CMP sensor ground (return) circuit is open • CMP sensor installation incorrect (Hall-effect type) • CMP sensor is damaged or CMP sensor shielding damaged • ECM has failed
DTC: P0343 **1T CCM, MIL: Yes** **1996, 1997, 1998, 1999, 2000, 2001, 2002, 2003, 2004, 2005, 2006** **Models:** A4, A6, Allroad Quattro, A8, A8L **Engines:** 2.7L, 2.8L, 3.0L, 3.2L, 4.2L **Transmissions:** All	**Camshaft Position Sensor "A" Circuit (Bank 1 or Single Sensor) High Input Conditions:** Engine started, battery voltage must be at least 11.5v, all electrical components must be off, parking brake must be engaged (to keep daytime driving lights off), automatic transmission selector must be in park and the ground between the engine and the chassis must be well connected. The ECM detected the CMP sensor signal did not reach the specified minimum limit. **Possible Causes:** • CMP sensor circuit is open or shorted to ground • CMP sensor circuit is shorted to power • CMP sensor ground (return) circuit is open • CMP sensor installation incorrect (Hall-effect type) • CMP sensor is damaged or CMP sensor shielding damaged • ECM has failed

DTC	Trouble Code Title, Conditions & Possible Causes
DTC: P0345 **1T CCM, MIL: Yes** **1996, 1997, 1998, 1999, 2000, 2001, 2002, 2003, 2004, 2005, 2006** **Models:** A4, A6, Allroad Quattro, A8, A8L **Engines:** 2.7L, 2.8L, 3.0L, 3.2L, 4.2L **Transmissions:** All	**Camshaft Position Sensor "A" Circuit (Bank 2) Malfunction Conditions:** Engine started, battery voltage must be at least 11.5v, all electrical components must be off, parking brake must be engaged (to keep daytime driving lights off), automatic transmission selector must be in park and the ground between the engine and the chassis must be well connected. The ECM detected the CMP sensor signal was missing or it was erratic. **Possible Causes:** • CMP sensor circuit is open or shorted to ground • CMP sensor circuit is shorted to power • CMP sensor ground (return) circuit is open • CMP sensor installation incorrect (Hall-effect type) • CMP sensor is damaged or CMP sensor shielding damaged • ECM has failed
DTC: P0346 **1T CCM, MIL: Yes** **1996, 1997, 1998, 1999, 2000, 2001, 2002, 2003, 2004, 2005, 2006** **Models:** A4, A6, Allroad Quattro, A8, A8L **Engines:** 2.7L, 2.8L, 3.0L, 3.2L, 4.2L **Transmissions:** All	**Camshaft Position Sensor "A" Circuit (Bank 2) Range/Performance Conditions:** Engine started, battery voltage must be at least 11.5v, all electrical components must be off, parking brake must be engaged (to keep daytime driving lights off), automatic transmission selector must be in park and the ground between the engine and the chassis must be well connected. The ECM detected the CMP sensor signal was implausible. **Possible Causes:** • CMP sensor circuit is open or shorted to ground • CMP sensor circuit is shorted to power • CMP sensor ground (return) circuit is open • CMP sensor installation incorrect (Hall-effect type) • CMP sensor is damaged or CMP sensor shielding damaged • ECM has failed
DTC: P0347 **1T CCM, MIL: Yes** **1996, 1997, 1998, 1999, 2000, 2001, 2002, 2003, 2004, 2005, 2006** **Models:** TT, A4, A6, Allroad Quattro, A8, A8L **Engines:** 2.7L, 2.8L, 3.0L, 4.2L **Transmissions:** All	**Camshaft Position Sensor "A" Circuit (Bank 2) Low Input Conditions:** Engine started, battery voltage must be at least 11.5v, all electrical components must be off, parking brake must be engaged (to keep daytime driving lights off), automatic transmission selector must be in park and the ground between the engine and the chassis must be well connected. The ECM detected the CMP sensor signal exceeded the bounds of the specified maximum limit. **Possible Causes:** • CMP sensor circuit is open or shorted to ground • CMP sensor circuit is shorted to power • CMP sensor ground (return) circuit is open • CMP sensor installation incorrect (Hall-effect type) • CMP sensor is damaged or CMP sensor shielding damaged • ECM has failed
DTC: P0348 **1T CCM, MIL: Yes** **1996, 1997, 1998, 1999, 2000, 2001, 2002, 2003, 2004, 2005, 2006** **Models:** TT, A4, A6, Allroad Quattro, A8, A8L **Engines:** 2.7L, 2.8L, 3.0L, 4.2L **Transmissions:** All	**Camshaft Position Sensor "A" Circuit "A" Circuit (Bank 2) High Input Conditions:** Engine started, battery voltage must be at least 11.5v, all electrical components must be off, parking brake must be engaged (to keep daytime driving lights off), automatic transmission selector must be in park and the ground between the engine and the chassis must be well connected. The ECM detected the CMP sensor signal did not reach the specified minimum limit. **Possible Causes:** • CMP sensor circuit is open or shorted to ground • CMP sensor circuit is shorted to power • CMP sensor ground (return) circuit is open • CMP sensor installation incorrect (Hall-effect type) • CMP sensor is damaged or CMP sensor shielding damaged • ECM has failed
DTC: P0351 **2T CCM, MIL: Yes** **1996, 1997, 1998, 1999, 2000, 2001, 2002, 2003, 2004, 2005, 2006** **Models:** TT, A4, A6, Allroad Quattro, A8, A8L **Engines:** 1.8L, 2.7L, 2.8L, 3.0L, 3.2L, 4.2L **Transmissions:** All	**Ignition Coilpack A Primary/Secondary Circuit Malfunction Conditions:** Engine started, battery voltage must be at least 11.5v, all electrical components must be off, parking brake must be engaged (to keep daytime driving lights off), automatic transmission selector must be in park and the ground between the engine and the chassis must be well connected. The ECM did not receive any valid pulses from the ignition module for the Ignition Coilpack A primary circuit. **Note: Ignition coils and power output stages are one component and cannot be replaced individually.** **Possible Causes:** • Engine speed (RPM) sensor has failed • Camshaft Position (CMP) sensor has failed • Power Supply Relay is shorted to an open circuit • There is a malfunction in voltage supply • Ignition coilpack is damaged or it has failed • Cylinder 1 to 4 Fuel Injector(s) have failed • ECM has failed •

DTC	Trouble Code Title, Conditions & Possible Causes
DTC: P0352 **2T CCM, MIL: Yes** **1996, 1997, 1998, 1999, 2000, 2001, 2002, 2003, 2004, 2005, 2006** **Models:** TT, A4, A6, Allroad Quattro, A8, A8L **Engines:** 1.8L, 2.7L, 2.8L, 3.0L, 3.2L, 4.2L **Transmissions:** All	**Ignition Coilpack B Primary/Secondary Circuit Malfunction Conditions:** Engine started, battery voltage must be at least 11.5v, all electrical components must be off, parking brake must be engaged (to keep daytime driving lights off), automatic transmission selector must be in park and the ground between the engine and the chassis must be well connected. The ECM did not receive any valid pulses from the ignition module for the Ignition Coilpack B primary circuit. **Note: Ignition coils and power output stages are one component and cannot be replaced individually.** **Possible Causes:** • Engine speed (RPM) sensor has failed • Camshaft Position (CMP) sensor has failed • Power Supply Relay is shorted to an open circuit • There is a malfunction in voltage supply • Ignition coilpack is damaged or it has failed • Cylinder 1 to 4 Fuel Injector(s) have failed • ECM has failed •
DTC: P0353 **2T CCM, MIL: Yes** **1996, 1997, 1998, 1999, 2000, 2001, 2002, 2003, 2004, 2005, 2006** **Models:** TT, A4, A6, Allroad Quattro, A8, A8L **Engines:** 1.8L, 2.7L, 2.8L, 3.0L, 3.2L, 4.2L **Transmissions:** All	**Ignition Coilpack C Primary/Secondary Circuit Malfunction Conditions:** Engine started, battery voltage must be at least 11.5v, all electrical components must be off, parking brake must be engaged (to keep daytime driving lights off), automatic transmission selector must be in park and the ground between the engine and the chassis must be well connected. The ECM did not receive any valid pulses from the ignition module for the Ignition Coilpack C primary circuit. **Note: Ignition coils and power output stages are one component and cannot be replaced individually.** **Possible Causes:** • Engine speed (RPM) sensor has failed • Camshaft Position (CMP) sensor has failed • Power Supply Relay is shorted to an open circuit • There is a malfunction in voltage supply • Ignition coilpack is damaged or it has failed • Cylinder 1 to 4 Fuel Injector(s) have failed • ECM has failed •
DTC: P0354 **2T CCM, MIL: Yes** **1996, 1997, 1998, 1999, 2000, 2001, 2002, 2003, 2004, 2005, 2006** **Models:** TT, A4, A6, Allroad Quattro, A8, A8L **Engines:** 1.8L, 2.7L, 2.8L, 3.0L, 3.2L, 4.2L **Transmissions:** All	**Ignition Coilpack D Primary/Secondary Circuit Malfunction Conditions:** Engine started, battery voltage must be at least 11.5v, all electrical components must be off, parking brake must be engaged (to keep daytime driving lights off), automatic transmission selector must be in park and the ground between the engine and the chassis must be well connected. The ECM did not receive any valid pulses from the ignition module for the Ignition Coilpack D primary circuit. **Note: Ignition coils and power output stages are one component and cannot be replaced individually.** **Possible Causes:** • Engine speed (RPM) sensor has failed • Camshaft Position (CMP) sensor has failed • Power Supply Relay is shorted to an open circuit • There is a malfunction in voltage supply • Ignition coilpack is damaged or it has failed • Cylinder 1 to 4 Fuel Injector(s) have failed • ECM has failed •
DTC: P0355 **2T CCM, MIL: Yes** **1996, 1997, 1998, 1999, 2000, 2001, 2002, 2003, 2004, 2005, 2006** **Models:** A4, A6, Allroad Quattro, A8, A8L **Engines:** 2.7L, 2.8L, 3.0L, 3.2L, 4.2L **Transmissions:** All	**Ignition Coilpack E Primary/Secondary Circuit Malfunction Conditions:** Engine started, battery voltage must be at least 11.5v, all electrical components must be off, parking brake must be engaged (to keep daytime driving lights off), automatic transmission selector must be in park and the ground between the engine and the chassis must be well connected. The ECM did not receive any valid pulses from the ignition module for the Ignition Coilpack E primary circuit. **Note: Ignition coils and power output stages are one component and cannot be replaced individually.** **Possible Causes:** • Engine speed (RPM) sensor has failed • Camshaft Position (CMP) sensor has failed • Power Supply Relay is shorted to an open circuit • There is a malfunction in voltage supply • Ignition coilpack is damaged or it has failed • Cylinder 1 to 4 Fuel Injector(s) have failed • ECM has failed •

DTC	Trouble Code Title, Conditions & Possible Causes
DTC: P0356 **2T CCM, MIL: Yes** **1996, 1997, 1998, 1999, 2000, 2001, 2002, 2003, 2004, 2005, 2006** **Models:** A4, A6, Allroad Quattro, A8, A8L **Engines:** 2.7L, 2.8L, 3.0L, 3.2L, 4.2L **Transmissions:** All	**Ignition Coilpack F Primary/Secondary Circuit Malfunction Conditions:** Engine started, battery voltage must be at least 11.5v, all electrical components must be off, parking brake must be engaged (to keep daytime driving lights off), automatic transmission selector must be in park and the ground between the engine and the chassis must be well connected. The ECM did not receive any valid pulses from the ignition module for the Ignition Coilpack F primary circuit. **Note: Ignition coils and power output stages are one component and cannot be replaced individually.** **Possible Causes:** • Engine speed (RPM) sensor has failed • Camshaft Position (CMP) sensor has failed • Power Supply Relay is shorted to an open circuit • There is a malfunction in voltage supply • Ignition coilpack is damaged or it has failed • Cylinder 1 to 4 Fuel Injector(s) have failed • ECM has failed •
DTC: P0357 **2T CCM, MIL: Yes** **2003, 2004, 2005, 2006** **Models:** A6, Allroad Quattro, A8, A8L **Engines:** 4.2L **Transmissions:** All	**Ignition Coilpack G Primary/Secondary Circuit Malfunction Conditions:** Engine started, battery voltage must be at least 11.5v, all electrical components must be off, parking brake must be engaged (to keep daytime driving lights off), automatic transmission selector must be in park and the ground between the engine and the chassis must be well connected. The ECM did not receive any valid pulses from the ignition module for the Ignition Coilpack G primary circuit. **Note: Ignition coils and power output stages are one component and cannot be replaced individually.** **Possible Causes:** • Engine speed (RPM) sensor has failed • Camshaft Position (CMP) sensor has failed • Power Supply Relay is shorted to an open circuit • There is a malfunction in voltage supply • Ignition coilpack is damaged or it has failed • Cylinder 1 to 4 Fuel Injector(s) have failed • ECM has failed •
DTC: P0366 **2T CCM, MIL: Yes** **2002, 2003, 2004, 2005, 2006** **Models:** A4, A6 **Engines:** 3.0L, 3.2L, 4.2L **Transmissions:** All	**Camshaft Position Sensor "B" Circuit (Bank 1) Range/Performance Conditions:** Engine started, battery voltage must be at least 11.5v, all electrical components must be off, parking brake must be engaged (to keep daytime driving lights off), automatic transmission selector must be in park and the ground between the engine and the chassis must be well connected. The ECM detected the CMP sensor signal exceeded the bounds of the specified maximum limit. **Possible Causes:** • CMP sensor circuit is open or shorted to ground • CMP sensor circuit is shorted to power • CMP sensor ground (return) circuit is open • CMP sensor installation incorrect (Hall-effect type) • CMP sensor is damaged or CMP sensor shielding damaged • ECM has failed
DTC: P0367 **1T CCM, MIL: Yes** **2002, 2003, 2004, 2005, 2006** **Models:** A4, A6 **Engines:** 3.0L, 3.2L, 4.2L **Transmissions:** All	**Camshaft Position Sensor "B" Circuit (Bank 1) Low Input Conditions:** Engine started, battery voltage must be at least 11.5v, all electrical components must be off, parking brake must be engaged (to keep daytime driving lights off), automatic transmission selector must be in park and the ground between the engine and the chassis must be well connected. The ECM detected the CMP sensor signal exceeded the bounds of the specified maximum limit. **Possible Causes:** • CMP sensor circuit is open or shorted to ground • CMP sensor circuit is shorted to power • CMP sensor ground (return) circuit is open • CMP sensor installation incorrect (Hall-effect type) • CMP sensor is damaged or CMP sensor shielding damaged • ECM has failed
DTC: P0368 **1T CCM, MIL: Yes** **2002, 2003, 2004, 2005, 2006** **Models:** A4, A6 **Engines:** 3.0L, 3.2L, 4.2L **Transmissions:** All	**Camshaft Position Sensor "B" Circuit (Bank 1) High Input Conditions:** Engine started, battery voltage must be at least 11.5v, all electrical components must be off, parking brake must be engaged (to keep daytime driving lights off), automatic transmission selector must be in park and the ground between the engine and the chassis must be well connected. The ECM detected the CMP sensor signal did not reach the specified minimum limit. **Possible Causes:** • CMP sensor circuit is open or shorted to ground • CMP sensor circuit is shorted to power • CMP sensor ground (return) circuit is open • CMP sensor installation incorrect (Hall-effect type) • CMP sensor is damaged or CMP sensor shielding damaged • ECM has failed

DTC	Trouble Code Title, Conditions & Possible Causes
DTC: P0391 **1T CCM, MIL: Yes** **2002, 2003, 2004, 2005, 2006** **Models:** A4, A6 **Engines:** 3.0L, 3.2L, 4.2L **Transmissions:** All	**Camshaft Position Sensor "B" Circuit (Bank 2) Range/Performance Conditions:** Engine started, battery voltage must be at least 11.5v, all electrical components must be off, parking brake must be engaged (to keep daytime driving lights off), automatic transmission selector must be in park and the ground between the engine and the chassis must be well connected. The ECM detected the CMP sensor signal exceeded the bounds of the specified maximum limit. **Possible Causes:** • CMP sensor circuit is open or shorted to ground • CMP sensor circuit is shorted to power • CMP sensor ground (return) circuit is open • CMP sensor installation incorrect (Hall-effect type) • CMP sensor is damaged or CMP sensor shielding damaged • ECM has failed
DTC: P0392 **1T CCM, MIL: Yes** **2002, 2003, 2004, 2005, 2006** **Models:** A4, A6 **Engines:** 3.0L, 3.2L, 4.2L **Transmissions:** All	**Camshaft Position Sensor "B" Circuit (Bank 2) Low Input Conditions:** Engine started, battery voltage must be at least 11.5v, all electrical components must be off, parking brake must be engaged (to keep daytime driving lights off), automatic transmission selector must be in park and the ground between the engine and the chassis must be well connected. The ECM detected the CMP sensor signal exceeded the bounds of the specified maximum limit. **Possible Causes:** • CMP sensor circuit is open or shorted to ground • CMP sensor circuit is shorted to power • CMP sensor ground (return) circuit is open • CMP sensor installation incorrect (Hall-effect type) • CMP sensor is damaged or CMP sensor shielding damaged • ECM has failed
DTC: P0393 **1T CCM, MIL: Yes** **2002, 2003, 2004, 2005, 2006** **Models:** A4, A6 **Engines:** 3.0L, 3.2L, 4.2L **Transmissions:** All	**Camshaft Position Sensor "B" Circuit (Bank 2) High Input Conditions:** Engine started, battery voltage must be at least 11.5v, all electrical components must be off, parking brake must be engaged (to keep daytime driving lights off), automatic transmission selector must be in park and the ground between the engine and the chassis must be well connected. The ECM detected the CMP sensor signal did not reach the specified minimum limit. **Possible Causes:** • CMP sensor circuit is open or shorted to ground • CMP sensor circuit is shorted to power • CMP sensor ground (return) circuit is open • CMP sensor installation incorrect (Hall-effect type) • CMP sensor is damaged or CMP sensor shielding damaged • ECM has failed
DTC: P0411 **2T CCM, MIL: Yes** **1997, 1998, 1999, 2000, 2001, 2002, 2003, 2004, 2005, 2006** **Models:** TT, A4 **Engines:** 1.8L, 3.2L **Transmissions:** All	**Secondary Air Injection System Upstream Flow Detected Conditions:** Engine started, battery voltage must be at least 11.5v, all electrical components must be off, parking brake must be engaged (to keep daytime driving lights off), automatic transmission selector must be in park and the ground between the engine and the chassis must be well connected. The ECM detected the Secondary AIR pump airflow was not diverted correctly when requested during the self-test. The pump is functioning but the quantity of air is recognized as insufficient by HO2S. **Note: The solenoid valve is closed when no voltage is present.** **Possible Causes:** • Air pump output is blocked or restricted • AIR bypass solenoid is leaking or it is restricted • AIR bypass solenoid is stuck open or stuck closed • Check valve (one or more) is damaged or leaking • Electric air injection pump hose(s) leaking • Electric air injection pump is damaged or faulty • ECM has failed
DTC: P0412 **2T CCM, MIL: Yes** **1996, 1997, 1998, 1999, 2000, 2001, 2002, 2003, 2004, 2005, 2006** **Models:** TT, A4, A6, Allroad Quattro, A8L **Engines:** 1.8L, 2.7L, 2.8L, 4.2L **Transmissions:** All	**Secondary Air Injection Solenoid Circuit Malfunction Conditions:** Engine started, battery voltage must be at least 11.5v, all electrical components must be off, parking brake must be engaged (to keep daytime driving lights off), automatic transmission selector must be in park and the ground between the engine and the chassis must be well connected. The ECM detected an unexpected low or high voltage condition on the AIR solenoid control circuit during testing. **Possible Causes:** • AIR solenoid power circuit (B+) is open (check dedicated fuse) • AIR bypass solenoid control circuit is open or shorted to ground • AIR diverter solenoid control circuit open or shorted to ground • AIR pump control circuit is open or shorted to ground • Check valve (one or more) is damaged or leaking • Solid State relay is damaged or it has failed • ECM has failed

DTC	Trouble Code Title, Conditions & Possible Causes
DTC: P0413 **2T CCM, MIL: Yes** **1996, 1997, 1998, 1999, 2000, 2001, 2002, 2003, 2004, 2005, 2006** **Models:** TT, A4, A6, Allroad Quattro, A8L, A8 **Engines:** 1.8L, 2.7L, 2.8L, 4.2L **Transmissions:** All	**Secondary Air Injection Solenoid Circuit Open Conditions:** Engine started, battery voltage must be at least 11.5v, all electrical components must be off, parking brake must be engaged (to keep daytime driving lights off), automatic transmission selector must be in park and the ground between the engine and the chassis must be well connected. The ECM detected an unexpected low or high voltage condition on the AIR solenoid control circuit during testing. **Possible Causes:** • AIR solenoid power circuit (B+) is open (check dedicated fuse) • AIR bypass solenoid control circuit is open or shorted to ground • AIR diverter solenoid control circuit open or shorted to ground • AIR pump control circuit is open or shorted to ground • Check valve (one or more) is damaged or leaking • Solid State relay is damaged or it has failed • ECM has failed
DTC: P0414 **2T CCM, MIL: Yes** **1996, 1997, 1998, 1999, 2000, 2001, 2002, 2003, 2004, 2005, 2006** **Models:** TT, A4, A6, Allroad Quattro, A8L, A8 **Engines:** 1.8L, 2.7L, 2.8L, 4.2L **Transmissions:** All	**Secondary Air Injection Solenoid Circuit Short Conditions:** Engine started, battery voltage must be at least 11.5v, all electrical components must be off, parking brake must be engaged (to keep daytime driving lights off), automatic transmission selector must be in park and the ground between the engine and the chassis must be well connected. The ECM detected an unexpected low or high voltage condition on the AIR solenoid control circuit during testing. **Possible Causes:** • AIR solenoid power circuit (B+) is open (check dedicated fuse) • AIR bypass solenoid control circuit is open or shorted to ground • AIR diverter solenoid control circuit open or shorted to ground • AIR pump control circuit is open or shorted to ground • Check valve (one or more) is damaged or leaking • Solid State relay is damaged or it has failed • ECM has failed
DTC: P0418 **2T CCM, MIL: Yes** **1996, 1997, 1998, 1999, 2000, 2001, 2002, 2003, 2004, 2005, 2006** **Models:** TT, A4, A6, Allroad Quattro, A8L, A8 **Engines:** 1.8L, 2.7L, 2.8L, 4.2L **Transmissions:** All	**Secondary Air Injection Relay (A) Circuit Malfunction Conditions:** Engine started, battery voltage must be at least 11.5v, all electrical components must be off, parking brake must be engaged (to keep daytime driving lights off), automatic transmission selector must be in park and the ground between the engine and the chassis must be well connected. The ECM detected an unexpected low or high voltage condition on the AIR solenoid control circuit during testing. **Possible Causes:** • AIR solenoid power circuit (B+) is open (check dedicated fuse) • AIR bypass solenoid control circuit is open or shorted to ground • AIR diverter solenoid control circuit open or shorted to ground • AIR pump control circuit is open or shorted to ground • Check valve (one or more) is damaged or leaking • Solid State relay is damaged or it has failed • ECM has failed
DTC: P0420 **2T OBD/CAT1, MIL: Yes** **1996, 1997, 1998, 1999, 2000, 2001, 2002, 2003, 2004, 2005, 2006** **Models:** TT, A4, A6, Allroad Quattro, A8L, A8 **Engines:** 1.8L, 2.7L, 2.8L, 4.2L **Transmissions:** All	**Catalyst System Efficiency (Bank 1) Below Threshold Conditions:** Engine started, battery voltage must be at least 11.5v, all electrical components must be off, parking brake must be engaged (to keep daytime driving lights off), automatic transmission selector must be in park, the exhaust system must be properly sealed between the catalytic converter and the cylinder head, coolant temperature must be at least 80 degrees Celsius and oxygen sensor heaters for oxygen sensors before the catalytic converter must be functioning properly and the ground between the engine and the chassis must be well connected. The ECM detected the switch rate of the rear HO2S-12 was close to the switch rate of front HO2S (it should be much slower). **Possible Causes:** • Air leaks at the exhaust manifold or in the exhaust pipes • Catalytic converter is damaged, contaminated or it has failed • ECT/CHT sensor has lost its calibration (the signal is incorrect) • Engine cylinders misfiring, or the ignition timing is over retarded • Engine oil is contaminated • Front HO2S or rear HO2S is contaminated with fuel or moisture • Front HO2S and/or the rear HO2S is loose in the mounting hole • Front HO2S much older than the rear HO2S (HO2S-11 is lazy) • Fuel system pressure is too high (check the pressure regulator) • Rear HO2S wires improperly connected or the HO2S has failed

DTC	Trouble Code Title, Conditions & Possible Causes
DTC: P0421 **2T OBD/CAT1, MIL: Yes** **1998, 1999, 2000, 2001, 2002, 2003, 2004, 2005, 2006** **Models:** A4, A8L, A8 **Engines:** 2.7L, 2.8L, 3.0L, 4.2L **Transmissions:** All	**Warm Up Catalyst System Efficiency (Bank 1) Below Threshold Conditions:** Engine started, battery voltage must be at least 11.5v, all electrical components must be off, parking brake must be engaged (to keep daytime driving lights off), automatic transmission selector must be in park, the exhaust system must be properly sealed between the catalytic converter and the cylinder head, coolant temperature must be at least 80 degrees Celsius and oxygen sensor heaters for oxygen sensors before the catalytic converter must be functioning properly and the ground between the engine and the chassis must be well connected. The ECM detected the switch rate of the rear HO2S-12 was close to the switch rate of front HO2S (it should be much slower). **Possible Causes:** • Air leaks at the exhaust manifold or in the exhaust pipes • Catalytic converter is damaged, contaminated or it has failed • ECT/CHT sensor has lost its calibration (the signal is incorrect) • Engine cylinders misfiring, or the ignition timing is over retarded • Engine oil is contaminated • Front HO2S or rear HO2S is contaminated with fuel or moisture • Front HO2S and/or the rear HO2S is loose in the mounting hole • Front HO2S much older than the rear HO2S (HO2S-11 is lazy) • Fuel system pressure is too high (check the pressure regulator) • Rear HO2S wires improperly connected or the HO2S has failed
DTC: P0422 **2T OBD/CAT1, MIL: Yes** **1996, 1997, 1998, 1999, 2000, 2001, 2002, 2003, 2004, 2005, 2006** **Models:** TT, A4, A6, A6 Avant, A8 **Engines:** 1.8L, 2.8L, 3.7L, 4.2L **Transmissions:** All	**Main Catalyst (Bank 1) Efficiency Below Threshold Conditions:** Engine started, battery voltage must be at least 11.5v, all electrical components must be off, parking brakc must be engaged (to keep daytime driving lights off), automatic transmission selector must be in park, the exhaust system must be properly sealed between the catalytic converter and the cylinder head, coolant temperature must be at least 80 degrees Celsius and oxygen sensor heaters for oxygen sensors before the catalytic converter must be functioning properly and the ground between the engine and the chassis must be well connected. The ECM detected the switch rate of the rear HO2S-12 was close to the switch rate of front HO2S (it should be much slower). **Possible Causes:** • Air leaks at the exhaust manifold or in the exhaust pipes • Catalytic converter is damaged, contaminated or it has failed • ECT/CHT sensor has lost its calibration (the signal is incorrect) • Engine cylinders misfiring, or the ignition timing is over retarded • Engine oil is contaminated • Front HO2S or rear HO2S is contaminated with fuel or moisture • Front HO2S and/or the rear HO2S is loose in the mounting hole • Front HO2S much older than the rear HO2S • Fuel system pressure is too high (check the pressure regulator) • Rear HO2S wires improperly connected or the HO2S has failed
DTC: P0430 **2T OBD/CAT1, MIL: Yes** **1998, 1999, 2000, 2001, 2002, 2003, 2004, 2005, 2006** **Models:** TT, A4, A6, Allroad Quatro, A8, A8L **Engines:** 1.8L, 2.7L, 2.8L, 3.2L, 3.7L, 4.2L **Transmissions:** All	**Catalyst System Efficiency (Bank 2) Below Threshold Conditions:** Engine started, battery voltage must be at least 11.5v, all electrical components must be off, parking brake must be engaged (to keep daytime driving lights off), automatic transmission selector must be in park, the exhaust system must be properly sealed between the catalytic converter and the cylinder head, coolant temperature must be at least 80 degrees Celsius and oxygen sensor heaters for oxygen sensors before the catalytic converter must be functioning properly and the ground between the engine and the chassis must be well connected. The ECM detected the switch rate of the rear HO2S-12 was close to the switch rate of front HO2S (it should be much slower). **Possible Causes:** • Air leaks at the exhaust manifold or in the exhaust pipes • Catalytic converter is damaged, contaminated or it has failed • ECT/CHT sensor has lost its calibration (the signal is incorrect) • Engine cylinders misfiring, or the ignition timing is over retarded • Engine oil is contaminated • Front HO2S or rear HO2S is contaminated with fuel or moisture • Front HO2S and/or the rear HO2S is loose in the mounting hole • Front HO2S much older than the rear HO2S (HO2S-11 is lazy) • Fuel system pressure is too high (check the pressure regulator) • Rear HO2S wires improperly connected or the HO2S has failed

DTC	Trouble Code Title, Conditions & Possible Causes
DTC: P0431 **2T OBD/CAT1, MIL: Yes** **1998, 1999, 2000, 2001, 2002, 2003, 2004, 2005, 2006** **Models:** A4, A6, Allroad Quatro, A8, A8L **Engines:** 2.7L, 2.8L, 3.0L, 3.2L, 3.7L, 4.2L **Transmissions:** All	**Warm Up Catalyst System Efficiency (Bank 2) Below Threshold Conditions:** Engine started, battery voltage must be at least 11.5v, all electrical components must be off, parking brake must be engaged (to keep daytime driving lights off), automatic transmission selector must be in park, the exhaust system must be properly sealed between the catalytic converter and the cylinder head, coolant temperature must be at least 80 degrees Celsius and oxygen sensor heaters for oxygen sensors before the catalytic converter must be functioning properly and the ground between the engine and the chassis must be well connected. The ECM detected the switch rate of the rear HO2S-12 was close to the switch rate of front HO2S (it should be much slower). **Possible Causes:** • Air leaks at the exhaust manifold or in the exhaust pipes • Catalytic converter is damaged, contaminated or it has failed • ECT/CHT sensor has lost its calibration (the signal is incorrect) • Engine cylinders misfiring, or the ignition timing is over retarded • Engine oil is contaminated • Front HO2S or rear HO2S is contaminated with fuel or moisture • Front HO2S and/or the rear HO2S is loose in the mounting hole • Front HO2S much older than the rear HO2S (HO2S-11 is lazy) • Fuel system pressure is too high (check the pressure regulator) • Rear HO2S wires improperly connected or the HO2S has failed
DTC: P0440 **2T CCM, MIL: Yes** **1995, 1996, 1997, 1998, 1999, 2000, 2001, 2002, 2003, 2004** **Models:** TT, A4, A6, A6 Avant **Engines:** 1.8L, 2.8L, 4.2L **Transmissions:** All	**EVAP System Malfunction Conditions:** ECT sensor is cold during startup, engine started, battery voltage must be at least 11.5v, all electrical components must be off, parking brake must be engaged (to keep daytime driving lights off), automatic transmission selector must be in park, the exhaust system must be properly sealed between the catalytic converter and the cylinder head, coolant temperature must be at least 80 degrees Celsius and oxygen sensor heaters for oxygen sensors before the catalytic converter must be functioning properly and the ground between the engine and the chassis must be well connected. The ECM detected the switch rate of the rear HO2S-12 was close to the switch rate of front HO2S (it should be much slower). ECM detected a problem in the EVAP system during the EVAP System Monitor test. **Possible Causes:** • EVAP canister purge valve is damaged • EVAP canister has an improper seal • Vapor line between purge solenoid and intake manifold vacuum reservoir is damaged, or vapor line between EVAP canister purge solenoid and charcoal canister is damaged • Vapor line between charcoal canister and check valve, or vapor line between check valve and fuel vapor valves is damaged • ECM has failed
DTC: P0441 **2T CCM, MIL: Yes** **1997, 1998, 1999, 2000, 2001, 2002, 2003, 2004, 2005, 2006** **Models:** TT, A4, A6, Allroad Quatro, A8, A8L **Engines:** 1.8L, 2.7L, 2.8L, 3.0L, 3.2L, 3.7L, 4.2L **Transmissions:** All	**EVAP Control System Incorrect Purge Flow Conditions:** ECT sensor is cold during startup, engine started, battery voltage must be at least 11.5v, all electrical components must be off, parking brake must be engaged (to keep daytime driving lights off), automatic transmission selector must be in park, the exhaust system must be properly sealed between the catalytic converter and the cylinder head, coolant temperature must be at least 80 degrees Celsius and oxygen sensor heaters for oxygen sensors before the catalytic converter must be functioning properly and the ground between the engine and the chassis must be well connected. The ECM detected the switch rate of the rear HO2S-12 was close to the switch rate of front HO2S (it should be much slower). ECM detected a problem in the EVAP system during the EVAP System Monitor test. **Possible Causes:** • EVAP canister purge valve is damaged • EVAP canister has an improper seal • Vapor line between purge solenoid and intake manifold vacuum reservoir is damaged, or vapor line between EVAP canister purge solenoid and charcoal canister is damaged • Vapor line between charcoal canister and check valve, or vapor line between check valve and fuel vapor valves is damaged • ECM has failed
DTC: P0442 **2T CCM, MIL: Yes** **1996, 1997, 1998, 1999, 2000, 2001, 2002, 2003, 2004, 2005, 2006** **Models:** TT, A4, A6, Allroad Quatro, A8, A8L **Engines:** 1.8L, 2.7L, 2.8L, 3.0L, 3.2L, 3.7L, 4.2L **Transmissions:** All	**EVAP Control System Small Leak Detected Conditions:** Engine started, battery voltage must be at least 11.5v, all electrical components must be off, parking brake must be engaged (to keep daytime driving lights off), automatic transmission selector must be in park, the exhaust system must be properly sealed between the catalytic converter and the cylinder head, coolant temperature must be at least 80 degrees Celsius and oxygen sensor heaters for oxygen sensors before the catalytic converter must be functioning properly and the ground between the engine and the chassis must be well connected. The ECM detected a leak in the EVAP system as small as 0.040" during the EVAP Monitor Test. **Possible Causes:** • Aftermarket EVAP parts that do not conform to specifications • CV solenoid remains partially open when commanded to close • EVAP component seals leaking (i.e., leaks in the Purge valve, fuel tank pressure sensor, canister vent solenoid, fuel vapor control valve tube assembly or fuel vapor vent valve). • Fuel filler cap damaged, cross-threaded or loosely installed • Loose fuel vapor hose/tube connections to EVAP components • Small holes or cuts in fuel vapor hoses or EVAP canister tubes

DTC	Trouble Code Title, Conditions & Possible Causes
DTC: P0443 **2T CCM, MIL: Yes** **1996, 1997, 1998, 1999, 2000, 2001, 2002, 2003, 2004, 2005, 2006** **Models:** TT, A4, Allroad Quatro **Engines:** 1.8L, 2.8L, 4.2L **Transmissions:** All	**EVAP Vapor Management Valve Circuit Malfunction Conditions:** Engine started, battery voltage must be at least 11.5v, all electrical components must be off, parking brake must be engaged (to keep daytime driving lights off), automatic transmission selector must be in park, the exhaust system must be properly sealed between the catalytic converter and the cylinder head, coolant temperature must be at least 80 degrees Celsius and oxygen sensor heaters for oxygen sensors before the catalytic converter must be functioning properly and the ground between the engine and the chassis must be well connected. The ECM detected an unexpected high or low voltage condition on the Vapor Management Valve (VMV) circuit when the device was cycled On/Off during testing. **Possible Causes:** • EVAP power supply circuit is open • EVAP solenoid control circuit is open or shorted to ground • EVAP solenoid control circuit is shorted to power (B+) • EVAP solenoid valve is damaged or it has failed • ECM has failed
DTC: P0444 **2T CCM, MIL: Yes** **1997, 1998, 1999, 2000, 2001, 2002, 2003, 2004, 2005, 2006** **Models:** TT, A4, A6, Allroad Quatro, A8, A8L **Engines:** 1.8L, 2.7L, 2.8L, 3.0L, 3.2L, 3.7L, 4.2L **Transmissions:** All	**Evaporative Emission System Purge Control Valve Circuit Open Conditions:** Engine started, battery voltage must be at least 11.5v, all electrical components must be off, parking brake must be engaged (to keep daytime driving lights off), automatic transmission selector must be in park, the exhaust system must be properly sealed between the catalytic converter and the cylinder head, coolant temperature must be at least 80 degrees Celsius and oxygen sensor heaters for oxygen sensors before the catalytic converter must be functioning properly and the ground between the engine and the chassis must be well connected. The ECM detected an unexpected voltage condition on the EVAP circuit when the device was cycled On/Off during testing. **Possible Causes:** • EVAP power supply circuit is open • EVAP solenoid control circuit is open or shorted to ground • EVAP solenoid control circuit is shorted to power (B+) • EVAP solenoid valve is damaged or it has failed • EVAP canister has a leak or a poor seal • ECM has failed
DTC: P0445 **2T CCM, MIL: Yes** **1996, 1997, 1998, 1999, 2000, 2001, 2002, 2003, 2004, 2005, 2006** **Models:** TT, A4, A6, Allroad Quatro, A8, A8L **Engines:** 1.8L, 2.7L, 2.8L, 3.0L, 3.2L, 3.7L, 4.2L **Transmissions:** All	**Evaporative Emission System Purge Control Valve Circuit Shorted Conditions:** Engine started, battery voltage must be at least 11.5v, all electrical components must be off, parking brake must be engaged (to keep daytime driving lights off), automatic transmission selector must be in park, the exhaust system must be properly sealed between the catalytic converter and the cylinder head, coolant temperature must be at least 80 degrees Celsius and oxygen sensor heaters for oxygen sensors before the catalytic converter must be functioning properly and the ground between the engine and the chassis must be well connected. The ECM detected an unexpected voltage condition on the EVAP circuit when the device was cycled On/Off during testing. **Possible Causes:** • EVAP power supply circuit is open • EVAP solenoid control circuit is open or shorted to ground • EVAP solenoid control circuit is shorted to power (B+) • EVAP solenoid valve is damaged or it has failed • EVAP canister has a leak or a poor seal • ECM has failed
DTC: P0449 **2T CCM, MIL: Yes** **2000, 2001, 2002, 2003, 2004, 2005, 2006** **Models:** TT, A8, A8L **Engines:** 1.8L, 4.2L **Transmissions:** All	**Evaporative Emission System Vent Valve/Solenoid Circuit Conditions:** Engine started, battery voltage must be at least 11.5v, all electrical components must be off, parking brake must be engaged (to keep daytime driving lights off), automatic transmission selector must be in park, the exhaust system must be properly sealed between the catalytic converter and the cylinder head, coolant temperature must be at least 80 degrees Celsius and oxygen sensor heaters for oxygen sensors before the catalytic converter must be functioning properly and the ground between the engine and the chassis must be well connected. The ECM detected an unexpected voltage condition on the EVAP circuit when the device was cycled On/Off during testing. **Possible Causes:** • EVAP power supply circuit is open • EVAP solenoid control circuit is open or shorted to ground • EVAP solenoid control circuit is shorted to power (B+) • EVAP solenoid valve is damaged or it has failed • EVAP canister has a leak or a poor seal • ECM has failed

DTC	Trouble Code Title, Conditions & Possible Causes
DTC: P0455 **2T CCM, MIL: Yes** **1996, 1997, 1998, 1999, 2000, 2001, 2002, 2003, 2004, 2005, 2006** **Models:** TT, A4, A6, Allroad Quatro, A8, A8L **Engines:** 1.8L, 2.7L, 2.8L, 3.0L, 3.2L, 3.7L, 4.2L **Transmissions:** All	**EVAP Control System Large Leak Detected Conditions:** Engine started, battery voltage must be at least 11.5v, all electrical components must be off, parking brake must be engaged (to keep daytime driving lights off), automatic transmission selector must be in park, the exhaust system must be properly sealed between the catalytic converter and the cylinder head, coolant temperature must be at least 80 degrees Celsius and oxygen sensor heaters for oxygen sensors before the catalytic converter must be functioning properly and the ground between the engine and the chassis must be well connected. The ECM detected multiple small fuel vapor leaks; or it detected a large leak in the system during the leak test. **Possible Causes:** • Aftermarket EVAP hardware non-conforming to specifications • EVAP canister tube, EVAP canister purge outlet tube or EVAP return tube disconnected or cracked, or canister is damaged • EVAP canister purge valve stuck closed, or canister damaged • Fuel filler cap missing, loose (not tightened) or the wrong part • Loose fuel vapor hose/tube connections to EVAP components • Canister vent (CV) solenoid stuck open • Fuel tank pressure (FTP) sensor has failed mechanically
DTC: P0456 **2T CCM, MIL: Yes** **1996, 1997, 1998, 1999, 2000, 2001, 2002, 2003, 2004, 2005, 2006** **Models:** TT, A4, A6, Allroad Quatro, A8, A8L **Engines:** 1.8L, 2.7L, 2.8L, 3.0L, 3.2L, 3.7L, 4.2L **Transmissions:** All	**EVAP Control System Small Leak Detected Conditions:** Engine started, battery voltage must be at least 11.5v, all electrical components must be off, parking brake must be engaged (to keep daytime driving lights off), automatic transmission selector must be in park, the exhaust system must be properly sealed between the catalytic converter and the cylinder head, coolant temperature must be at least 80 degrees Celsius and oxygen sensor heaters for oxygen sensors before the catalytic converter must be functioning properly and the ground between the engine and the chassis must be well connected. The ECM detected multiple small fuel vapor leaks; or it detected a large leak in the system during the leak test. **Possible Causes:** • Aftermarket EVAP hardware non-conforming to specifications • EVAP canister tube, EVAP canister purge outlet tube or EVAP return tube disconnected or cracked, or canister is damaged • EVAP canister purge valve stuck closed, or canister damaged • Fuel filler cap missing, loose (not tightened) or the wrong part • Loose fuel vapor hose/tube connections to EVAP components • Canister vent (CV) solenoid stuck open • Fuel tank pressure (FTP) sensor has failed mechanically
DTC: P0458 **2T CCM, MIL: Yes** **2005, 2006** **Models:** A6 **Engines:** 3.2L, 4.2L **Transmissions:** All	**Evaporative Emission System Purge Control Valve Circuit Low Conditions:** Engine started, battery voltage must be at least 11.5v, all electrical components must be off, parking brake must be engaged (to keep daytime driving lights off), automatic transmission selector must be in park, the exhaust system must be properly sealed between the catalytic converter and the cylinder head, coolant temperature must be at least 80 degrees Celsius and oxygen sensor heaters for oxygen sensors before the catalytic converter must be functioning properly and the ground between the engine and the chassis must be well connected. The ECM detected an unexpected voltage condition on the EVAP circuit when the device was cycled On/Off during testing. **Possible Causes:** • EVAP power supply circuit is open • EVAP solenoid control circuit is open or shorted to ground • EVAP solenoid control circuit is shorted to power (B+) • EVAP solenoid valve is damaged or it has failed • EVAP canister has a leak or a poor seal • ECM has failed
DTC: P0459 **2T CCM, MIL: Yes** **2005, 2006** **Models:** A6 **Engines:** 3.2L, 4.2L **Transmissions:** All	**Evaporative Emission System Purge Control Valve Circuit High Conditions:** Engine started, battery voltage must be at least 11.5v, all electrical components must be off, parking brake must be engaged (to keep daytime driving lights off), automatic transmission selector must be in park, the exhaust system must be properly sealed between the catalytic converter and the cylinder head, coolant temperature must be at least 80 degrees Celsius and oxygen sensor heaters for oxygen sensors before the catalytic converter must be functioning properly and the ground between the engine and the chassis must be well connected. The ECM detected an unexpected voltage condition on the EVAP circuit when the device was cycled On/Off during testing. **Possible Causes:** • EVAP power supply circuit is open • EVAP solenoid control circuit is open or shorted to ground • EVAP solenoid control circuit is shorted to power (B+) • EVAP solenoid valve is damaged or it has failed • EVAP canister has a leak or a poor seal • ECM has failed

DTC	Trouble Code Title, Conditions & Possible Causes
DTC: P0491 **2T CCM, MIL: Yes** **1998, 1999, 2000, 2001, 2002, 2003, 2004, 2005, 2006** **Models:** TT, A4, A6, Allroad Quatro, A8, A8L **Engines:** 1.8L, 2.7L, 2.8L, 3.0L, 4.2L **Transmissions:** All	**Secondary Air Injection System Insufficient Flow (Bank 1) Conditions:** Engine started, battery voltage must be at least 11.5v, all electrical components must be off, parking brake must be engaged (to keep daytime driving lights off), automatic transmission selector must be in park and the ground between the engine and the chassis must be well connected. The ECM detected the Secondary AIR pump airflow was not diverted correctly when requested during the self-test. The pump is functioning but the quantity of air is recognized as insufficient by HO2S **Possible Causes:** • Air pump output is blocked or restricted • AIR bypass solenoid is leaking or it is restricted • AIR bypass solenoid is stuck open or stuck closed • Check valve (one or more) is damaged or leaking • Electric air injection pump hose(s) leaking • Electric air injection pump is damaged or faulty • ECM has failed
DTC: P0492 **2T CCM, MIL: Yes** **1998, 1999, 2000, 2001, 2002, 2003, 2004, 2005, 2006** **Models:** TT, A4, A6, Allroad Quatro, A8, A8L **Engines:** 1.8L, 2.7L, 2.8L, 3.0L, 4.2L **Transmissions:** All	**Secondary Air Injection System Insufficient Flow (Bank 2) Conditions:** Engine started, battery voltage must be at least 11.5v, all electrical components must be off, parking brake must be engaged (to keep daytime driving lights off), automatic transmission selector must be in park and the ground between the engine and the chassis must be well connected. The ECM detected the Secondary AIR pump airflow was not diverted correctly when requested during the self-test. The pump is functioning but the quantity of air is recognized as insufficient by HO2S **Possible Causes:** • Air pump output is blocked or restricted • AIR bypass solenoid is leaking or it is restricted • AIR bypass solenoid is stuck open or stuck closed • Check valve (one or more) is damaged or leaking • Electric air injection pump hose(s) leaking • Electric air injection pump is damaged or faulty • ECM has failed
DTC: P0498 **2T CCM, MIL: Yes** **2005, 2006** **Models:** A8L **Engines:** 4.2L **Transmissions:** All	**Evaporative Emission System Vent Valve Control Circuit Low Conditions:** Engine started, battery voltage must be at least 11.5v, all electrical components must be off, parking brake must be engaged (to keep daytime driving lights off), automatic transmission selector must be in park, the exhaust system must be properly sealed between the catalytic converter and the cylinder head, coolant temperature must be at least 80 degrees Celsius and oxygen sensor heaters for oxygen sensors before the catalytic converter must be functioning properly and the ground between the engine and the chassis must be well connected. The ECM detected an unexpected voltage condition on the EVAP circuit when the device was cycled On/Off during testing. **Possible Causes:** • EVAP power supply circuit is open • EVAP solenoid control circuit is open or shorted to ground • EVAP solenoid control circuit is shorted to power (B+) • EVAP solenoid valve is damaged or it has failed • EVAP canister has a leak or a poor seal • ECM has failed
DTC: P0501 **2T CCM, MIL: Yes** **1996, 1997, 1998, 1999, 2000, 2001, 2002, 2003, 2004, 2005, 2006** **Models:** TT, A4, A6, A6 Avant, Allroad Quatro, A8, A8L **Engines:** 1.8L, 2.7L, 2.8L, 3.0L, 3.2L, 4.2L **Transmissions:** All	**Vehicle Speed Sensor or PSOM Range/Performance Conditions:** Engine started; engine speed above the TCC stall speed, and the ECM detected a loss of the VSS signal over a period of time or the signal is not usable. **Note: The ECM receives vehicle speed data from the VSS, TCSS, ABS module, CTM or GEM controller, depending up the application.** **Possible Causes:** • VSS signal circuit is open or shorted to ground • VSS harness circuit is shorted to ground • VSS harness circuit is shorted to power • VSS circuit open between the ECM and related control module • VSS or wheel speed sensors circuits are damaged • Modules connected to VSC/VSS harness circuits are damaged • Mechanical drive mechanism for the VSS is damaged

DTC	Trouble Code Title, Conditions & Possible Causes
DTC: P0506 **2T CCM, MIL: Yes** **1996, 1997, 1998, 1999, 2000, 2001, 2002, 2003, 2004, 2005, 2006** **Models:** TT, A4, A6, A6 Avant, Allroad Quatro, A8, A8L **Engines:** 1.8L, 2.7L, 2.8L, 3.0L, 3.2L, 4.2L **Transmissions:** All	**Idle Air Control System RPM Lower Than Expected Conditions:** Engine started, battery voltage must be at least 11.5v, all electrical components must be off, parking brake must be engaged (to keep daytime driving lights off), automatic transmission selector must be in park, the exhaust system must be properly sealed between the catalytic converter and the cylinder head, coolant temperature must be at least 80 degrees Celsius and oxygen sensor heaters for oxygen sensors before the catalytic converter must be functioning properly and the ground between the engine and the chassis must be well connected. The ECM detected it could not control the idle speed correctly, as it is constantly more than 100 rpm less than specification. **Possible Causes:** • Air inlet is plugged or the air filter element is severely clogged • IAC circuit is open or shorted • IAC circuit VPWR circuit is open • IAC solenoid is damaged or has failed • ECM has failed • The VSS has failed
DTC: P0507 **2T CCM, MIL: Yes** **1996, 1997, 1998, 1999, 2000, 2001, 2002, 2003, 2004, 2005, 2006** **Models:** TT, A4, A6, Allroad Quatro, A8, A8L **Engines:** 1.8L, 2.7L, 2.8L, 3.0L, 3.2L, 4.2L **Transmissions:** All	**Idle Air Control System RPM Higher Than Expected Conditions:** Engine started, battery voltage must be at least 11.5v, all electrical components must be off, parking brake must be engaged (to keep daytime driving lights off), automatic transmission selector must be in park, the exhaust system must be properly sealed between the catalytic converter and the cylinder head, coolant temperature must be at least 80 degrees Celsius and oxygen sensor heaters for oxygen sensors before the catalytic converter must be functioning properly and the ground between the engine and the chassis must be well connected. The ECM detected it could not control the idle speed correctly, as it is constantly more than 200 rpm more than specification. **Possible Causes:** • Air intake leak located somewhere after the throttle body • IAC control circuit is shorted to chassis ground • IAC solenoid is damaged or has failed • Throttle Valve Control module has failed or is clogged with carbon • ECM has failed • The VSS has failed
DTC: P0560 **2T CCM, MIL: Yes** **1997, 1998, 1999, 2000, 2001, 2002, 2003, 2004, 2005** **Models:** A4 **Engines:** 1.8L **Transmissions:** All	**System Voltage Malfunction Conditions:** Engine started, battery voltage must be at least 11.5v, all electrical components must be off, parking brake must be engaged (to keep daytime driving lights off), automatic transmission selector must be in park, and the ground between the engine and the chassis must be well connected. The ECM has detected a voltage value that is implausible or erratic. **Possible Causes:** • Alternator damaged or faulty • Battery voltage low or insufficient • Fuses blown or circuits open • Battery connection to terminal not clean • Voltage regulator has failed
DTC: P0562 **2T CCM, MIL: Yes** **1997, 1998, 1999, 2000, 2001, 2002, 2003, 2004, 2005, 2006** **Models:** A4, A6, **Engines:** 1.8L, 3.2L, 4.2L **Transmissions:** All	**System Voltage Low Conditions:** Engine started, battery voltage must be at least 11.5v, all electrical components must be off, parking brake must be engaged (to keep daytime driving lights off), automatic transmission selector must be in park, and the ground between the engine and the chassis must be well connected. The ECM has detected a voltage value that is below the specified minimum limit for the system to function properly. **Possible Causes:** • Alternator damaged or faulty • Battery voltage low or insufficient • Fuses blown or circuits open • Battery connection to terminal not clean • Voltage regulator has failed
DTC: P0563 **2T CCM, MIL: Yes** **1997, 1998, 1999, 2000, 2001, 2002, 2003, 2004, 2005, 2006** **Models:** A4, A6, **Engines:** 1.8L, 3.2L, 4.2L **Transmissions:** All	**System Voltage High Conditions:** Engine started, battery voltage must be at least 11.5v, all electrical components must be off, parking brake must be engaged (to keep daytime driving lights off), automatic transmission selector must be in park, and the ground between the engine and the chassis must be well connected. The ECM has detected a voltage value that has exceeded the specified maximum limit for the system to function properly. **Possible Causes:** • Alternator damaged or faulty • Battery voltage low or insufficient • Fuses blown or circuits open • Battery connection to terminal not clean • Voltage regulator has failed

DTC	Trouble Code Title, Conditions & Possible Causes
DTC: P0571 **2T CCM, MIL: Yes** **2002, 2003, 2004, 2005** **Models:** A4 **Engines:** 1.8L **Transmissions:** All	**Cruise/Brake Switch (A) Circuit Malfunction Conditions:** Engine started, battery voltage must be at least 11.5v, all electrical components must be off, parking brake must be engaged (to keep daytime driving lights off), automatic transmission selector must be in park, and the ground between the engine and the chassis must be well connected. The ECM has detected a voltage value that is implausible or erratic. **Possible Causes:** • Brake light switch is faulty • Control circuit is shorted to chassis ground
DTC: P0600 **1T CCM, MIL: Yes** **1996, 1997, 1998, 1999, 2000, 2001, 2002, 2003, 2004, 2005, 2006** **Models:** TT, A4, A6, Allroad Quatro **Engines:** 1.8L, 2.7L, 2.8L, 3.2L, 4.2L **Transmissions:** All	**Serial Communication Link (Data BUS) Message Missing Conditions:** The Engine Control Module (ECM) communicates with all databus-capable control modules via a CAN databus. These databus-capable control modules are connected via two data bus wires which are twisted together (CAN_High and CAN_Low), and exchange information (messages). Missing information on the databus is recognized as a malfunction and stored. Trouble-free operation of the CAN-Bus requires that it have a terminal resistance. This central terminal resistor is located in the Engine Control Module (ECM). **Possible Causes:** • ECM has failed • CAN data bus wires have short circuited to each other
DTC: P0601 **1T CCM, MIL: Yes** **1996, 1997, 1998, 1999, 2000, 2001, 2002, 2003, 2004, 2005, 2006** **Models:** TT, A4, A6, Allroad Quatro, A8, A8L **Engines:** 1.8L, 2.7L, 2.8L, 3.0L, 3.2L, 3.7L, 4.2L **Transmissions:** All	**Internal Control Module Memory Check Sum Error Conditions:** Key on, the ECM has detected a programming error **Possible Causes:** • Battery terminal corrosion, or loose battery connection • Connection to the ECM interrupted, or the circuit has been opened • Reprogramming error has occurred • ECM has failed and needs replacement. Remember to check for Aftermarket Performance Products before replacing a ECM.
DTC: P0602 **1T CCM, MIL: Yes** **1998, 1999, 2000, 2001, 2002, 2003, 2004, 2005** **Models:** A4, A6, Allroad Quatro, **Engines:** 1.8L, 2.7L, 2.8L, 4.2L **Transmissions:** All	**Control Module Programming Error Conditions:** Key on, and the ECM detected a programming error in the VID block. This fault requires that the VID Block be reprogrammed, or that the EEPROM be re-flashed. **Possible Causes:** • During the VID reprogramming function, the Vehicle ID (VID) data block failed during reprogramming wit the Scan Tool. • Battery terminal corrosion, or loose battery connection • Connection to the ECM interrupted, or the circuit has been opened • Reprogramming error has occurred • ECM has failed and needs replacement. Remember to check for Aftermarket Performance Products before replacing a ECM.
DTC: P0603 **1T CCM, MIL: Yes** **1999, 2000, 2001, 2002, 2003, 2004** **Models:** A6, A8 **Engines:** 3.2L, 4.2L **Transmissions:** All	**ECM Keep Alive Memory Test Error Conditions:** Key on, and the ECM detected an internal memory fault. This code will set if KAPWR to the ECM is interrupted (at the initial key on). **Possible Causes:** • Battery terminal corrosion, or loose battery connection • KAPWR to ECM interrupted, or the circuit has been opened • Reprogramming error has occurred • ECM has failed and needs replacement. Remember to check for Aftermarket Performance Products before replacing a ECM.
DTC: P0604 **1T CCM, MIL: Yes** **1996, 1997, 1998, 1999, 2000, 2001, 2002, 2003, 2004, 2005, 2006** **Models:** TT, A4, A6, Allroad Quatro, A8, A8L **Engines:** 1.8L, 2.7L, 2.8L, 3.0L, 3.2L, 3.7L, 4.2L **Transmissions:** All	**Internal Control Module Random Access Memory (RAM) Error Conditions:** Key on, and the ECM detected an internal memory fault. This code will set if KAPWR to the ECM is interrupted (at the initial key on). **Possible Causes:** • Battery terminal corrosion, or loose battery connection • Connection to the ECM interrupted, or the circuit has been opened • Reprogramming error has occurred • ECM has failed and needs replacement. Remember to check for Aftermarket Performance Products before replacing a ECM.

DTC	Trouble Code Title, Conditions & Possible Causes
DTC: P0605 **1T CCM, MIL: Yes** **1996, 1997, 1998, 1999, 2000, 2001, 2002, 2003, 2004, 2005, 2006** **Models:** TT, A4, A6, Allroad Quatro, A8, A8L **Engines:** 1.8L, 2.7L, 2.8L, 3.0L, 3.2L, 3.7L, 4.2L **Transmissions:** All	**ECM Read Only Memory (ROM) Test Error Conditions:** Key on, and the ECM detected a ROM test error (ROM inside ECM is corrupted). The ECM is normally replaced if this code has set. **Possible Causes:** • An attempt was made to change the module calibration, or a module programming error may have occurred • Clear the trouble codes and then check for this trouble code. If it resets, the ECM has failed and needs replacement. • Aftermarket performance products may have been installed. • The Transmission Control Module (TCM) has failed.
DTC: P0606 **1T CCM, MIL: Yes** **1996, 1997, 1998, 1999, 2000, 2001, 2002, 2003, 2004, 2005, 2006** **Models:** TT, A4, A6, Allroad Quatro, A8, A8L **Engines:** 1.8L, 2.7L, 2.8L, 3.0L, 3.2L, 3.7L, 4.2L **Transmissions:** All	**ECM Internal Communication Error Conditions:** Key on, and the ECM detected an internal communications register read back error during the initial key on check period. **Possible Causes:** • Clear the trouble codes and then check for this trouble code. If it resets, the ECM has failed and needs replacement. • Remember to check for signs of Aftermarket Performance Products installation before replacing the ECM.
DTC: P0614 **1T CCM, MIL: Yes** **1999, 2000, 2001, 2002, 2003, 2004** **Models:** A6 **Engines:** 3.2L, 4.2L **Transmissions:** All	**ECM / TCM Incompatible Conditions:** Key on, and the ECM detected a communication error between the Transmission control module and the ECM **Possible Causes:** • TCM failed • ECM failed • Circuit shorting between ECM and TCM • Replacement control module ID doesn't match old control module ID
DTC: P0627 **1T CCM, MIL: Yes** **2001, 2002, 2003, 2004** **Models:** A6 **Engines:** 4.2L **Transmissions:** All	**Fuel Pump "A" Control Circuit Open Conditions:** Engine started, battery voltage must be at least 11.5v, all electrical components must be off, parking brake must be engaged (to keep daytime driving lights off), automatic transmission selector must be in park, and the ground between the engine and the chassis must be well connected. The ECM has detected a voltage value across the fuel pump control circuit that is out of the specified limits for the system to function properly. **Possible Causes:** • Fuel Pressure Regulator Valve is faulty • Fuel Pressure Sensor is faulty • Fuel Pump (FP) Control Module is faulty • Fuel pump is faulty
DTC: P0629 **1T CCM, MIL: Yes** **2001, 2002, 2003, 2004** **Models:** A6 **Engines:** 4.2L **Transmissions:** All	**Fuel Pump "A" Control Circuit High Conditions:** Engine started, battery voltage must be at least 11.5v, all electrical components must be off, parking brake must be engaged (to keep daytime driving lights off), automatic transmission selector must be in park, and the ground between the engine and the chassis must be well connected. The ECM has detected a voltage value across the fuel pump control circuit that is above the specified limit for the system to function properly. **Possible Causes:** • Fuel Pressure Regulator Valve is faulty • Fuel Pressure Sensor is faulty • Fuel Pump (FP) Control Module is faulty • Fuel pump is faulty
DTC: P0638 **1T CCM, MIL: Yes** **1998, 1999, 2000, 2001, 2002, 2003, 2004, 2005, 2006** **Models:** TT, A4, A6, Allroad Quatro, A8, A8L **Engines:** 1.8L, 2.7L, 2.8L, 3.0L, 4.2L **Transmissions:** All	**Throttle Actuator Control Range/Performance Bank 1 Conditions:** Engine started, battery voltage must be at least 11.5v, all electrical components must be off, parking brake must be engaged (to keep daytime driving lights off), automatic transmission selector must be in park, and the ground between the engine and the chassis must be well connected. The ECM has detected a voltage value across the throttle actuator control circuit that is out of the specified limit for the system to function properly. Both Throttle Position (TP) Sensor / Accelerator Pedal Position Sensor 2 are located at the accelerator pedal and communicate the driver's intentions to the Motronic engine control module (ECM) completely independently of each other. Both sensors are integrated into one housing. **Possible Causes:** • Throttle Position (TP) sensor is faulty • Throttle valve control module is faulty • ECM is faulty • Circuit wires have short circuited to each other, to vehicle Ground (GND) or to B+. • Accelerator pedal module is faulty

DTC	Trouble Code Title, Conditions & Possible Causes
DTC: P0641 **2T CCM, MIL: Yes** **1999, 2000, 2001, 2002, 2003, 2004** **Models:** A6 **Engines:** 3.2L, 4.2L **Transmissions:** All	**Sensor Reference Voltage "A" Circuit Open Conditions:** Engine started, battery voltage must be at least 11.5v, all electrical components must be off, parking brake must be engaged (to keep daytime driving lights off), automatic transmission selector must be in park, and the ground between the engine and the chassis must be well connected. **Possible Causes:** • Circuit harness connector contacts are corroded or ingressed of water • Circuit wires have shorted to each other, to battery or ground • Automatic Transmission Hydraulic Pressure Sensor 1 has failed • Solenoid valves in valve body are faulty • Transmission Control Module (TCM) needs replacing • Transmission Input Speed (RPM) Sensor has failed • Transmission Output Speed (RPM) Sensor has failed
DTC: P0642 **2T CCM, MIL: Yes** **2005, 2006** **Models:** A6 **Engines:** 3.2L **Transmissions:** All	**Sensor Reference Voltage "A" Circuit Low Conditions:** Engine started, battery voltage must be at least 11.5v, all electrical components must be off, parking brake must be engaged (to keep daytime driving lights off), automatic transmission selector must be in park, and the ground between the engine and the chassis must be well connected. **Possible Causes:** • Circuit harness connector contacts are corroded or ingressed of water • Circuit wires have shorted to each other, to battery or ground • Automatic Transmission Hydraulic Pressure Sensor 1 has failed • Solenoid valves in valve body are faulty • Transmission Control Module (TCM) needs replacing • Transmission Input Speed (RPM) Sensor has failed • Transmission Output Speed (RPM) Sensor has failed
DTC: P0643 **2T CCM, MIL: Yes** **2005, 2006** **Models:** A6 **Engines:** 3.2L **Transmissions:** All	**Sensor Reference Voltage "A" Circuit High Conditions:** Engine started, battery voltage must be at least 11.5v, all electrical components must be off, parking brake must be engaged (to keep daytime driving lights off), automatic transmission selector must be in park, and the ground between the engine and the chassis must be well connected. **Possible Causes:** • Circuit harness connector contacts are corroded or ingressed of water • Circuit wires have shorted to each other, to battery or ground • Automatic Transmission Hydraulic Pressure Sensor 1 has failed • Solenoid valves in valve body are faulty • Transmission Control Module (TCM) needs replacing • Transmission Input Speed (RPM) Sensor has failed • Transmission Output Speed (RPM) Sensor has failed
DTC: P0652 **2T CCM, MIL: Yes** **2005, 2006** **Models:** A6 **Engines:** 3.2L **Transmissions:** All	**Sensor Reference Voltage "B" Circuit Low Conditions:** Engine started, battery voltage must be at least 11.5v, all electrical components must be off, parking brake must be engaged (to keep daytime driving lights off), automatic transmission selector must be in park, and the ground between the engine and the chassis must be well connected. **Possible Causes:** • Circuit harness connector contacts are corroded or ingressed of water • Circuit wires have shorted to each other, to battery or ground • Automatic Transmission Hydraulic Pressure Sensor 1 has failed • Solenoid valves in valve body are faulty • Transmission Control Module (TCM) needs replacing • Transmission Input Speed (RPM) Sensor has failed • Transmission Output Speed (RPM) Sensor has failed
DTC: P0653 **2T CCM, MIL: Yes** **2005, 2006** **Models:** A6 **Engines:** 3.2L **Transmissions:** All	**Sensor Reference Voltage "B" Circuit High Conditions:** Engine started, battery voltage must be at least 11.5v, all electrical components must be off, parking brake must be engaged (to keep daytime driving lights off), automatic transmission selector must be in park, and the ground between the engine and the chassis must be well connected. **Possible Causes:** • Circuit harness connector contacts are corroded or ingressed of water • Circuit wires have shorted to each other, to battery or ground • Automatic Transmission Hydraulic Pressure Sensor 1 has failed • Solenoid valves in valve body are faulty • Transmission Control Module (TCM) needs replacing • Transmission Input Speed (RPM) Sensor has failed • Transmission Output Speed (RPM) Sensor has failed

DTC	Trouble Code Title, Conditions & Possible Causes
DTC: P0657 **2T CCM, MIL: Yes** **2005, 2006** **Models:** A6 **Engines:** 3.2L **Transmissions:** All	**Actuator Supply Voltage "A" Circuit Open Conditions:** Engine started, battery voltage must be at least 11.5v, all electrical components must be off, parking brake must be engaged (to keep daytime driving lights off), automatic transmission selector must be in park, and the ground between the engine and the chassis must be well connected. **Possible Causes:** • Circuit harness connector contacts are corroded or ingressed of water • Circuit wires have shorted to each other, to battery or ground • Automatic Transmission Hydraulic Pressure Sensor 1 has failed • Solenoid valves in valve body are faulty • Transmission Control Module (TCM) needs replacing • Transmission Input Speed (RPM) Sensor has failed • Transmission Output Speed (RPM) Sensor has failed
DTC: P0658 **2T CCM, MIL: Yes** **2005, 2006** **Models: A6** **Engines:** 3.2L **Transmissions:** All	**Actuator Supply Voltage "A" Circuit Low Conditions:** Engine started, battery voltage must be at least 11.5v, all electrical components must be off, parking brake must be engaged (to keep daytime driving lights off), automatic transmission selector must be in park, and the ground between the engine and the chassis must be well connected. **Possible Causes:** • Circuit harness connector contacts are corroded or ingressed of water • Circuit wires have shorted to each other, to battery or ground • Automatic Transmission Hydraulic Pressure Sensor 1 has failed • Solenoid valves in valve body are faulty • Transmission Control Module (TCM) needs replacing • Transmission Input Speed (RPM) Sensor has failed • Transmission Output Speed (RPM) Sensor has failed
DTC: P0659 **2T CCM, MIL: Yes** **2005, 2006** **Models:** A6 **Engines:** 3.2L **Transmissions:** All	**Actuator Supply Voltage "A" Circuit High Conditions:** Engine started, battery voltage must be at least 11.5v, all electrical components must be off, parking brake must be engaged (to keep daytime driving lights off), automatic transmission selector must be in park, and the ground between the engine and the chassis must be well connected. **Possible Causes:** • Circuit harness connector contacts are corroded or ingressed of water • Circuit wires have shorted to each other, to battery or ground • Automatic Transmission Hydraulic Pressure Sensor 1 has failed • Solenoid valves in valve body are faulty • Transmission Control Module (TCM) needs replacing • Transmission Input Speed (RPM) Sensor has failed • Transmission Output Speed (RPM) Sensor has failed
DTC: P0685 **1T CCM, MIL: Yes** **1997, 1998, 1999, 2000, 2001, 2002, 2003, 2004, 2005, 2006** **Models:** TT, A4, A6, Allroad Quatro, A8, A8L **Engines:** 1.8L, 2.7L, 2.8L, 3.0L, 4.2L **Transmissions:** All	**ECM Power Relay Control Circuit Open Conditions:** Engine started, battery voltage must be at least 11.5v, all electrical components must be off, parking brake must be engaged (to keep daytime driving lights off), automatic transmission selector must be in park and the ground between the engine and the chassis must be well connected. The ECM detected the ECM power relay control circuit has a voltage outside requirement for proper function. **Possible Causes:** • Generator has failed or is damaged • Fuel pump relay is faulty • Circuit is grounded to power or chassis • ECM has failed
DTC: P0686 **1T CCM, MIL: Yes** **1997, 1998, 1999, 2000, 2001, 2002, 2003, 2004, 2005, 2006** **Models:** TT, A4, A6, Allroad Quatro, A8, A8L **Engines:** 1.8L, 2.7L, 2.8L, 3.0L, 4.2L **Transmissions:** All	**ECM/PCM Power Relay Control Circuit Low Conditions:** Engine started, battery voltage must be at least 11.5v, all electrical components must be off, parking brake must be engaged (to keep daytime driving lights off), automatic transmission selector must be in park and the ground between the engine and the chassis must be well connected. The ECM detected the ECM power relay control circuit has a voltage outside requirement for proper function. **Possible Causes:** • Generator has failed or is damaged • Fuel pump relay is faulty • Circuit is grounded to power or chassis • ECM has failed

DTC	Trouble Code Title, Conditions & Possible Causes
DTC: P0687 **1T CCM, MIL: Yes** **1997, 1998, 1999, 2000, 2001, 2002, 2003, 2004, 2005, 2006** **Models:** TT, A4, A6, Allroad Quatro, A8, A8L **Engines:** 1.8L, 2.7L, 2.8L, 3.0L, 4.2L **Transmissions:** All	**ECM/PCM Power Relay Control Circuit High Conditions:** Engine started, battery voltage must be at least 11.5v, all electrical components must be off, parking brake must be engaged (to keep daytime driving lights off), automatic transmission selector must be in park and the ground between the engine and the chassis must be well connected. The ECM detected the ECM power relay control circuit has a voltage outside requirement for proper function. **Possible Causes:** • Generator has failed or is damaged • Fuel pump relay is faulty • Circuit is grounded to power or chassis • ECM has failed
DTC: P0688 **1T CCM, MIL: Yes** **1997, 1998, 1999, 2000, 2001, 2002, 2003, 2004, 2005, 2006** **Models:** TT, A4, A6, Allroad Quatro, A8, A8L **Engines:** 1.8L, 2.7L, 2.8L, 3.0L, 4.2L **Transmissions:** All	**ECM/PCM Power Relay Control Sense Circuit Open Conditions:** Engine started, battery voltage must be at least 11.5v, all electrical components must be off, parking brake must be engaged (to keep daytime driving lights off), automatic transmission selector must be in park and the ground between the engine and the chassis must be well connected. The ECM detected the ECM power relay control circuit has a voltage outside requirement for proper function. **Possible Causes:** • Generator has failed or is damaged • Fuel pump relay is faulty • Circuit is grounded to power or chassis • ECM has failed
DTC: P0700 **2T CCM, MIL: Yes** **1999, 2000, 2001, 2002, 2003, 2004** **Models:** A6 **Engines:** 3.2L, 4.2L **Transmissions:** A/T	**Transmission Control System Malfunction Conditions:** Engine started, battery voltage must be at least 11.5v, all electrical components must be off, parking brake must be engaged (to keep daytime driving lights off), automatic transmission selector must be in park, and the ground between the engine and the chassis must be well connected. The ECM detected a malfunction int the transmission control system. **Possible Causes:** • Circuit harness connector contacts are corroded or ingressed of water • Circuit wires have shorted to each other, to battery or ground • Automatic Transmission Hydraulic Pressure Sensor 1 has failed • Solenoid valves in valve body are faulty • Transmission Input Speed (RPM) Sensor has failed • Transmission Output Speed (RPM) Sensor has failed • Engine Control Module (ECM) is faulty • Voltage supply for Engine Control Module (ECM) is faulty • Transmission Control Module (TCM) is faulty
DTC: P0701 **2T CCM, MIL: Yes** **2004, 2005, 2006** **Models:** TT **Engines:** 3.2L **Transmissions:** A/T	**Transmission Control System Range/Performance Conditions:** Engine started, battery voltage must be at least 11.5v, all electrical components must be off, parking brake must be engaged (to keep daytime driving lights off), automatic transmission selector must be in park, and the ground between the engine and the chassis must be well connected. The ECM detected a voltage outside the normal performance range to allow the system to properly function. **Possible Causes:** • Circuit harness connector contacts are corroded or ingressed of water • Circuit wires have shorted to each other, to battery or ground • Automatic Transmission Hydraulic Pressure Sensor 1 has failed • Solenoid valves in valve body are faulty • Transmission Input Speed (RPM) Sensor has failed • Transmission Output Speed (RPM) Sensor has failed • Engine Control Module (ECM) is faulty • Voltage supply for Engine Control Module (ECM) is faulty • Transmission Control Module (TCM) is faulty
DTC: P0702 **2T CCM, MIL: Yes** **2004, 2005, 2006** **Models:** TT **Engines:** 3.2L **Transmissions:** A/T	**Transmission Control System Electrical Conditions:** Engine started, battery voltage must be at least 11.5v, all electrical components must be off, parking brake must be engaged (to keep daytime driving lights off), automatic transmission selector must be in park, and the ground between the engine and the chassis must be well connected. The ECM detected a voltage outside the normal performance range to allow the system to properly function. **Possible Causes:** • Circuit harness connector contacts are corroded or ingressed of water • Circuit wires have shorted to each other, to battery or ground • Automatic Transmission Hydraulic Pressure Sensor 1 has failed • Solenoid valves in valve body are faulty • Transmission Input Speed (RPM) Sensor has failed • Transmission Output Speed (RPM) Sensor has failed • Engine Control Module (ECM) is faulty • Voltage supply for Engine Control Module (ECM) is faulty • Transmission Control Module (TCM) is faulty

DTC	Trouble Code Title, Conditions & Possible Causes
DTC: P0704 **2T CCM, MIL: Yes** **2000, 2001, 2002, 2003, 2004, 2005, 2006** **Models:** A4, A6, Allroad Quatro **Engines:** 1.8L, 2.7L, 2.8L, 3.0L **Transmissions:** A/T	**Clutch Switch Input Circuit Malfunction Conditions:** Engine started, battery voltage must be at least 11.5v, all electrical components must be off, parking brake must be engaged (to keep daytime driving lights off), automatic transmission selector must be in park, and the ground between the engine and the chassis must be well connected. The ECM detected a voltage outside the normal performance range to allow the system to properly function. **Possible Causes:** • Circuit harness connector contacts are corroded or ingressed of water • Circuit wires have shorted to each other, to battery or ground • Automatic Transmission Hydraulic Pressure Sensor 1 has failed • Solenoid valves in valve body are faulty • Transmission Input Speed (RPM) Sensor has failed • Transmission Output Speed (RPM) Sensor has failed • Engine Control Module (ECM) is faulty • Voltage supply for Engine Control Module (ECM) is faulty • Transmission Control Module (TCM) is faulty
DTC: P0705 **2T CCM, MIL: Yes** **1996, 1997, 1998, 1999** **Models:** A4, A8 **Engines:** 2.8L, 3.7L, 4.2L **Transmissions:** A/T	**TR Sensor Circuit Malfunction Conditions:** Engine started, battery voltage must be at least 11.5v, all electrical components must be off, parking brake must be engaged (to keep daytime driving lights off), automatic transmission selector must be in park, and the ground between the engine and the chassis must be well connected. The ECM detected a voltage outside the normal performance range to allow the system to properly function. **Possible Causes:** • Circuit harness connector contacts are corroded or ingressed of water • Circuit wires have shorted to each other, to battery or ground • Automatic Transmission Hydraulic Pressure Sensor 1 has failed • Solenoid valves in valve body are faulty • Transmission Input Speed (RPM) Sensor has failed • Transmission Output Speed (RPM) Sensor has failed • Engine Control Module (ECM) is faulty • Voltage supply for Engine Control Module (ECM) is faulty • Transmission Control Module (TCM) is faulty
DTC: P0706 **2T CCM, MIL: Yes** **1996, 1997, 1998, 1999, 2000, 2001, 2002, 2003, 2004** **Models:** A4, A6, A8 **Engines:** 2.8L, 3.2L, 3.7L, 4.2L **Transmissions:** A/T	**TR Sensor Circuit Range/Performance Conditions:** Engine started, battery voltage must be at least 11.5v, all electrical components must be off, parking brake must be engaged (to keep daytime driving lights off), automatic transmission selector must be in park, and the ground between the engine and the chassis must be well connected. The ECM detected a voltage outside the normal performance range to allow the system to properly function. **Possible Causes:** • Circuit harness connector contacts are corroded or ingressed of water • Circuit wires have shorted to each other, to battery or ground • Automatic Transmission Hydraulic Pressure Sensor 1 has failed • Solenoid valves in valve body are faulty • Transmission Input Speed (RPM) Sensor has failed • Transmission Output Speed (RPM) Sensor has failed • Engine Control Module (ECM) is faulty • Voltage supply for Engine Control Module (ECM) is faulty • Transmission Control Module (TCM) is faulty
DTC: P0707 **2T CCM, MIL: Yes** **1996, 1997, 1998, 1999, 2000, 2001, 2002, 2003, 2004, 2005, 2006** **Models:** A4, A6 **Engines:** 1.8L, 2.8L **Transmissions:** A/T	**Transmission Range Sensor Circuit Low Input Conditions:** Engine started, battery voltage must be at least 11.5v, all electrical components must be off, parking brake must be engaged (to keep daytime driving lights off), automatic transmission selector must be in park, and the ground between the engine and the chassis must be well connected. The ECM detected the Transmission Range sensor (TR) signal was less than the self-test minimum value in the test. **Possible Causes:** • Circuit harness connector contacts are corroded or ingressed of water • Circuit wires have shorted to each other, to battery or ground • Automatic Transmission Hydraulic Pressure Sensor 1 has failed • Solenoid valves in valve body are faulty • Transmission Input Speed (RPM) Sensor has failed • Transmission Output Speed (RPM) Sensor has failed • Engine Control Module (ECM) is faulty • Voltage supply for Engine Control Module (ECM) is faulty • Transmission Control Module (TCM) is faulty

DTC	Trouble Code Title, Conditions & Possible Causes
DTC: P0708 **2T CCM, MIL: Yes** **1997, 1998, 1999, 2000, 2001, 2002, 2003, 2004, 2005** **Models:** A4 **Engines:** 1.8L **Transmissions:** A/T	**Transmission Range Sensor Circuit High Input Conditions:** Engine started, battery voltage must be at least 11.5v, all electrical components must be off, parking brake must be engaged (to keep daytime driving lights off), automatic transmission selector must be in park, and the ground between the engine and the chassis must be well connected. The ECM detected the Transmission Range sensor (TR) input was more than the self-test maximum range in the test. **Possible Causes:** • Circuit harness connector contacts are corroded or ingressed of water • Circuit wires have shorted to each other, to battery or ground • Automatic Transmission Hydraulic Pressure Sensor 1 has failed • Solenoid valves in valve body are faulty • Transmission Input Speed (RPM) Sensor has failed • Transmission Output Speed (RPM) Sensor has failed • Engine Control Module (ECM) is faulty • Voltage supply for Engine Control Module (ECM) is faulty • Transmission Control Module (TCM) is faulty
DTC: P0710 **2T CCM, MIL: Yes** **1999, 2000, 2001, 2002, 2003, 2004** **Models:** A6 **Engines:** 3.2L, 4.2L **Transmissions:** A/T	**Transmission Fluid Temperature Sensor Circuit Malfunction Conditions:** Engine started, battery voltage must be at least 11.5v, all electrical components must be off, parking brake must be engaged (to keep daytime driving lights off), automatic transmission selector must be in park, and the ground between the engine and the chassis must be well connected. The ECM detected the Transmission fluid temperature sensor circuit was outside the normal range in the test to allow proper function. **Possible Causes:** • ATF is low, contaminated, dirty or burnt • Circuit harness connector contacts are corroded or ingressed of water • Circuit wires have shorted to each other, to battery or ground • Automatic Transmission Hydraulic Pressure Sensor 1 has failed • Solenoid valves in valve body are faulty • Transmission Input Speed (RPM) Sensor has failed • Transmission Output Speed (RPM) Sensor has failed • Engine Control Module (ECM) is faulty • Voltage supply for Engine Control Module (ECM) is faulty • Transmission Control Module (TCM) is faulty
DTC: P0711 **2T CCM, MIL: Yes** **1999, 2000, 2001, 2002, 2003, 2004** **Models:** A6 **Engines:** 3.2L, 4.2L **Transmissions:** A/T	**Transmission Fluid Temperature Sensor Signal Range/Performance Conditions:** Engine started, battery voltage must be at least 11.5v, all electrical components must be off, parking brake must be engaged (to keep daytime driving lights off), automatic transmission selector must be in park, and the ground between the engine and the chassis must be well connected. The ECM detected the Transmission Fluid Temperature (TFT) sensor value was not close its normal operating temperature. **Possible Causes:** • ATF is low, contaminated, dirty or burnt • TFT sensor signal circuit has a high resistance condition • TFT sensor is out-of-calibration ("skewed") or it has failed • ECM has failed
DTC: P0712 **2T CCM, MIL: Yes** **1999, 2000, 2001, 2002, 2003, 2004** **Models:** A6 **Engines:** 3.2L, 4.2L **Transmissions:** A/T	**Transmission Fluid Temperature Sensor Circuit Low Input Conditions:** Engine started, battery voltage must be at least 11.5v, all electrical components must be off, parking brake must be engaged (to keep daytime driving lights off), automatic transmission selector must be in park, and the ground between the engine and the chassis must be well connected. The ECM detected the Transmission Fluid Temperature (TFT) sensor was less than its minimum self-test range in the test. **Possible Causes:** • TFT sensor signal circuit is shorted to chassis ground • TFT sensor signal circuit is shorted to sensor ground • TFT sensor is damaged, or out-of-calibration, or has failed • ECM has failed
DTC: P0713 **1999, 2000, 2001, 2002, 2003, 2004** **Models:** A6 **Engines:** 3.2L, 4.2L **Transmissions:** A/T	**Transmission Fluid Temperature Sensor Circuit High Input Conditions:** Engine started, battery voltage must be at least 11.5v, all electrical components must be off, parking brake must be engaged (to keep daytime driving lights off), automatic transmission selector must be in park, and the ground between the engine and the chassis must be well connected. The ECM detected the Transmission Fluid Temperature (TFT) sensor was more than its maximum self-test range in the test. **Possible Causes:** • TFT sensor signal circuit is open between the sensor and ECM • TFT sensor ground circuit is open between sensor and ECM • TFT sensor is damaged or has failed • ECM has failed

DTC	Trouble Code Title, Conditions & Possible Causes
DTC: P0714 **2T CCM, MIL: Yes** **1999, 2000, 2001, 2002, 2003, 2004** **Models:** A6 **Engines:** 3.2L, 4.2L **Transmissions:** A/T	**Transmission Fluid Temperature Sensor Circuit Intermittent Conditions:** Engine started, battery voltage must be at least 11.5v, all electrical components must be off, parking brake must be engaged (to keep daytime driving lights off), automatic transmission selector must be in park, and the ground between the engine and the chassis must be well connected. The ECM detected the Transmission Fluid Temperature (TFT) sensor was giving a false reading or was not reading at all. **Possible Causes:** • TFT sensor signal circuit is open between the sensor and ECM • TFT sensor ground circuit is open between sensor and ECM • TFT sensor is damaged or has failed • ECM has failed
DTC: P0715 **1T CCM, MIL: Yes** **1997, 1998, 1999, 2000, 2001, 2002, 2003, 2004, 2005, 2006** **Models:** A4, A6, A8 **Engines:** 2.8L, 3.2L, 3.7L, 4.2L **Transmissions:** A/T	**Input/Turbine Speed Sensor Circuit Malfunction Conditions:** Engine started, vehicle driven with the vehicle speed sensor indicating more than 1 mph, and the ECM detected the Transmission Vehicle Speed Sensor signals were erratic, or that they were missing for a period of time. **Possible Causes:** • TVSS signal circuit is open • TVSS signal is shorted to chassis ground • TVSS signal is shorted to sensor ground • TVSS assembly is damaged or it has failed • ECM has failed
DTC: P0716 **1T CCM, MIL: Yes** **1997, 1998, 1999, 2000, 2001, 2002, 2003, 2004, 2005, 2006** **Models:** A4, A6, A8 **Engines:** 2.8L, 3.2L, 3.7L, 4.2L **Transmissions:** A/T	**Input Turbine/Speed Sensor Circuit Range/Performance Conditions:** Engine started, vehicle driven with the vehicle speed sensor indicating more than 1 mph, and the ECM detected the Transmission Vehicle Speed Sensor signals were erratic, or that they were missing for a period of time. **Possible Causes:** • TVSS signal circuit is open • TVSS signal is shorted to chassis ground • TVSS signal is shorted to sensor ground • TVSS assembly is damaged or it has failed • ECM has failed
DTC: P0717 **1T CCM, MIL: Yes** **1997, 1998, 1999, 2000, 2001, 2002, 2003, 2004, 2005, 2006** **Models:** A4, A6, A8 **Engines:** 2.8L, 3.2L, 3.7L, 4.2L **Transmissions:** A/T	**Transmission Speed Shaft Sensor Signal Intermittent Conditions:** Engine started, vehicle speed sensor indicating over 1 mph, and the ECM detected an intermittent loss of TSS signals (i.e., the TSS signals were erratic, irregular or missing). **Possible Causes:** • TSS connector is damaged, loose or shorted • TSS signal circuit has an intermittent open condition • TSS signal circuit has an intermittent short to ground condition • TSS assembly is damaged or is has failed • ECM has failed
DTC: P0721 **1996, 1997, 1998, 1999, 2000, 2001, 2002, 2003, 2004, 2005, 2006** **Models:** A4, A6 **Engines:** 2.8L, 3.2L, 4.2L **Transmissions:** A/T	**A/T Output Shaft Speed Sensor Noise Interference Conditions:** Engine started, VSS signal more than 1 mph, and the ECM detected "noise" interference on the Output Shaft Speed (OSS) sensor circuit. **Possible Causes:** • After market add-on devices interfering with the OSS signal • OSS connector is damaged, loose or shorted, or the wiring is misrouted or it is damaged • OSS assembly is damaged or it has failed • ECM has failed
DTC: P0722 **1T CCM, MIL: Yes** **1997, 1998, 1999, 2000, 2001, 2002, 2003, 2004, 2005, 2006** **Models:** A4, A6, A8 **Engines:** 2.8L, 3.2L, 3.7L, 4.2L **Transmissions:** A/T	**A/T Output Speed Sensor No Signal Conditions:** Engine started, and the ECM did not detect any Vehicle Speed Sensor (VSS) sensor signals upon initial vehicle movement. **Possible Causes:** • After market add-on devices interfering with the VSS signal • VSS sensor wiring is misrouted, damaged or shorting • ECM and/or TCM has failed
DTC: P0725 **1T CCM, MIL: Yes** **1996** **Models:** A4 **Engines:** 2.8L **Transmissions:** A/T	**Engine Speed Input Circuit Malfunction Conditions:** The Transmission Control Module (TCM) does not receive a signal from the Engine Control Module (ECM). **Possible Causes:** • The TCM circuit is shorting to ground, B+ or is open • TCM has failed • ECM has failed

DTC	Trouble Code Title, Conditions & Possible Causes
DTC: P0726 **1T CCM, MIL: Yes** **1996** **Models:** A4 **Engines:** 2.8L **Transmissions:** A/T	**Engine Speed Input Circuit Range/Performance Conditions:** The Engine Speed (RPM) Sensor detects engine speed and reference marks. Without an engine speed signal, the engine will not start. If the engine speed signal fails while the engine is running, the engine will stop immediately. **Note: There is a larger-sized gap on the sensor wheel. This gap is the reference mark and does not mean that the sensor wheel is damaged.** **Possible Causes:** • Engine speed sensor has failed • Circuit is shorting to ground, B+ or is open • Sensor wheel is damaged, run out or not properly secured • ECM has failed
DTC: P0727 **1T CCM, MIL: Yes** **1997, 1998, 1999, 2000, 2001, 2002, 2003, 2004, 2005, 2006** **Models:** A4, A6, A8 **Engines:** 2.8L, 3.2L, 3.7L, 4.2L **Transmissions:** A/T	**Engine Speed Input Circuit No Signal Conditions:** The Engine Speed (RPM) Sensor detects engine speed and reference marks. Without an engine speed signal, the engine will not start. If the engine speed signal fails while the engine is running, the engine will stop immediately. **Note: There is a larger-sized gap on the sensor wheel. This gap is the reference mark and does not mean that the sensor wheel is damaged.** **Possible Causes:** • Engine speed sensor has failed • Circuit is shorting to ground, B+ or is open • Sensor wheel is damaged, run out or not properly secured • ECM has failed
DTC: P0729 **1T CCM, MIL: Yes** **1999, 2000, 2001, 2002, 2003, 2004, 2005, 2006** **Models:** TT, A6 **Engines:** 3.2L, 4.2L **Transmissions:** A/T	**Gear 6 Incorrect Ratio Conditions:** Engine started, battery voltage must be at least 11.5v, all electrical components must be off, and the ground between the engine and the chassis must be well connected. The ECM detected an incorrect ratio within the sixth gear. **Possible Causes:** • ATF level is low • Circuit harness connector contacts are corroded or ingressed of water • Circuit wires have shorted to each other, to battery or ground • Automatic Transmission Hydraulic Pressure Sensor 1 has failed • Solenoid valves in valve body are faulty • Transmission Control Module (TCM) needs replacing • Transmission Input Speed (RPM) Sensor has failed • Transmission Output Speed (RPM) Sensor has failed
DTC: P0730 **1T CCM, MIL: Yes** **1996, 1997, 1998, 1999** **Models:** A4, A8 **Engines:** 2.8L, 3.7L, 4.2L **Transmissions:** A/T	**Gear Incorrect Ratio Conditions:** Engine started, battery voltage must be at least 11.5v, all electrical components must be off, and the ground between the engine and the chassis must be well connected. The ECM detected an incorrect gear ratio. **Possible Causes:** • ATF level is low • Circuit harness connector contacts are corroded or ingressed of water • Circuit wires have shorted to each other, to battery or ground • Automatic Transmission Hydraulic Pressure Sensor 1 has failed • Solenoid valves in valve body are faulty • Transmission Control Module (TCM) needs replacing • Transmission Input Speed (RPM) Sensor has failed • Transmission Output Speed (RPM) Sensor has failed
DTC: P0731 **2T CCM, MIL: Yes** **1997, 1998, 1999, 2000, 2001, 2002, 2003, 2004, 2005, 2006** **Models:** TT, A4, A6, A8 **Engines:** 2.8L, 3.2L, 3.7L, 4.2L **Transmissions:** A/T	**Incorrect First Gear Ratio Conditions:** Engine started, vehicle operating with 1st gear commanded "on", and the ECM detected an incorrect 1st gear ratio during the test. **Possible Causes:** • 1st Gear solenoid harness connector not properly seated • 1st Gear solenoid signal shorted to ground, or open • 1st Gear solenoid wiring harness connector is damaged • 1st Gear solenoid is damaged or not properly installed • ATF level is low • Circuit harness connector contacts are corroded or ingressed of water • Circuit wires have shorted to each other, to battery or ground • Automatic Transmission Hydraulic Pressure Sensor 1 has failed • Transmission Control Module (TCM) needs replacing • Transmission Input Speed (RPM) Sensor has failed • Transmission Output Speed (RPM) Sensor has failed

DTC	Trouble Code Title, Conditions & Possible Causes
DTC: P0732 **2T CCM, MIL: Yes** **1997, 1998, 1999, 2000, 2001, 2002, 2003, 2004, 2005, 2006** **Models:** TT, A4, A6, A8 **Engines:** 2.8L, 3.2L, 3.7L, 4.2L **Transmissions:** A/T	**Incorrect Second Gear Ratio Conditions:** Engine started, vehicle operating with 2nd Gear commanded "on", and the ECM detected an incorrect 2nd gear ratio during the test. **Possible Causes:** • 2nd Gear solenoid harness connector not properly seated • 2nd Gear solenoid signal shorted to ground, or open • 2nd Gear solenoid wring harness connector is damaged • 2nd Gear solenoid is damaged or not properly installed • ATF level is low • Circuit harness connector contacts are corroded or ingressed of water • Circuit wires have shorted to each other, to battery or ground • Automatic Transmission Hydraulic Pressure Sensor 1 has failed • Transmission Control Module (TCM) needs replacing • Transmission Input Speed (RPM) Sensor has failed • Transmission Output Speed (RPM) Sensor has failed
DTC: P0733 **2T CCM, MIL: Yes** **1997, 1998, 1999, 2000, 2001, 2002, 2003, 2004, 2005, 2006** **Models:** TT, A4, A6, A8 **Engines:** 2.8L, 3.2L, 3.7L, 4.2L **Transmissions:** A/T	**Incorrect Third Gear Ratio Conditions:** Engine started, vehicle operating with 3rd Gear commanded "on", and the ECM detected an incorrect 3rd gear ratio during the test. **Possible Causes:** • 3rd Gear solenoid harness connector not properly seated • 3rd Gear solenoid signal shorted to ground, or open • 3rd Gear solenoid wiring harness connector is damaged • 3rd Gear solenoid is damaged or not properly installed • ATF level is low • Circuit harness connector contacts are corroded or ingressed of water • Circuit wires have shorted to each other, to battery or ground • Automatic Transmission Hydraulic Pressure Sensor 1 has failed • Transmission Control Module (TCM) needs replacing • Transmission Input Speed (RPM) Sensor has failed • Transmission Output Speed (RPM) Sensor has failed
DTC: P0734 **2T CCM, MIL: Yes** **1997, 1998, 1999, 2000, 2001, 2002, 2003, 2004, 2005, 2006** **Models:** TT, A4, A6, A8 **Engines:** 2.8L, 3.2L, 3.7L, 4.2L **Transmissions:** A/T	**Incorrect Fourth Gear Ratio Conditions:** Engine started, vehicle operating with 4th Gear commanded "on", and the ECM detected an incorrect 4th gear ratio during the test. **Possible Causes:** • 4th Gear solenoid harness connector not properly seated • 4th Gear solenoid signal shorted to ground, or open • 4th Gear solenoid wiring harness connector is damaged • 4th Gear solenoid is damaged or not properly installed • ATF level is low • Circuit harness connector contacts are corroded or ingressed of water • Circuit wires have shorted to each other, to battery or ground • Automatic Transmission Hydraulic Pressure Sensor 1 has failed • Transmission Control Module (TCM) needs replacing • Transmission Input Speed (RPM) Sensor has failed • Transmission Output Speed (RPM) Sensor has failed
DTC: P0735 **2T CCM, MIL: Yes** **1997, 1998, 1999, 2000, 2001, 2002, 2003, 2004, 2005, 2006** **Models:** TT, A4, A6, A8 **Engines:** 2.8L, 3.2L, 3.7L, 4.2L **Transmissions:** A/T	**Incorrect Fifth Gear Ratio Conditions:** Engine started, vehicle operating with 5th Gear commanded "on", and the ECM detected an incorrect 5th gear ratio during the test. **Possible Causes:** • 5th Gear solenoid harness connector not properly seated • 5th Gear solenoid signal shorted to ground, or open • 5th Gear solenoid wiring harness connector is damaged • 5th Gear solenoid is damaged or not properly installed • ATF level is low • Circuit harness connector contacts are corroded or ingressed of water • Circuit wires have shorted to each other, to battery or ground • Automatic Transmission Hydraulic Pressure Sensor 1 has failed • Transmission Control Module (TCM) needs replacing • Transmission Input Speed (RPM) Sensor has failed • Transmission Output Speed (RPM) Sensor has failed

DTC	Trouble Code Title, Conditions & Possible Causes
DTC: P0740 **2T CCM, MIL: Yes** **1996** **Models:** A4 **Engines:** 2.8L **Transmissions:** A/T	**TCC Solenoid Circuit Malfunction Conditions:** Engine started, KOER Self-Test enabled, vehicle driven at cruise speed, and the ECM did not detect any voltage drop across the TCC solenoid circuit during the test period. **Possible Causes:** • TCC solenoid control circuit is open or shorted to ground • TCC solenoid wiring harness connector is damaged • TCC solenoid is damaged or has failed • ECM has failed
DTC: P0741 **2T CCM, MIL: Yes** **1997, 1998, 1999, 2000, 2001, 2002, 2003, 2004, 2005, 2006** **Models:** A4, A6, A8 **Engines:** 2.8L, 3.2L, 3.7L, 4.2L **Transmissions:** A/T	**TCC Mechanical System Range/Performance Conditions:** Engine started, vehicle driven in gear with VSS signals received, and the ECM detected excessive slippage while in normal operation. **Possible Causes:** • TCC solenoid has a mechanical failure • TCC solenoid has a hydraulic failure • ECM has failed
DTC: P0746 **1T CCM, MIL: Yes** **2004, 2005, 2006** **Models:** TT, A6 **Engines:** 3.2L, 4.2L **Transmissions:** A/T	**Pressure Control Solenoid "A" Performance or Stuck Off Conditions:** Engine started, battery voltage must be at least 11.5v, all electrical components must be off, and the ground between the engine and the chassis must be well connected. The ECM detected the pressure control solenoid was in the "stuck off" position. **Possible Causes:** • ATF level is low • Circuit harness connector contacts are corroded or ingressed of water • Circuit wires have shorted to each other, to battery or ground • Automatic Transmission Hydraulic Pressure Sensor 1 has failed • Solenoid valves in valve body are faulty • Transmission Control Module (TCM) needs replacing • Transmission Input Speed (RPM) Sensor has failed • Transmission Output Speed (RPM) Sensor has failed
DTC: P0747 **1T CCM, MIL: Yes** **2004, 2005, 2006** **Models:** TT, A6 **Engines:** 3.2L, 4.2L **Transmissions:** A/T	**Pressure Control Solenoid "A" Performance or Stuck On Conditions:** Engine started, battery voltage must be at least 11.5v, all electrical components must be off, and the ground between the engine and the chassis must be well connected. The ECM detected the pressure control solenoid was in the "stuck on" position. **Possible Causes:** • ATF level is low • Circuit harness connector contacts are corroded or ingressed of water • Circuit wires have shorted to each other, to battery or ground • Automatic Transmission Hydraulic Pressure Sensor 1 has failed • Solenoid valves in valve body are faulty • Transmission Control Module (TCM) needs replacing • Transmission Input Speed (RPM) Sensor has failed • Transmission Output Speed (RPM) Sensor has failed
DTC: P0748 **1T CCM, MIL: Yes** **1996, 2004, 2005, 2006** **Models:** A4, A6, A8 **Engines:** 2.8L, 3.2L, 4.2L **Transmissions:** A/T **Transmissions:** A/T	**Pressure Control Solenoid Electrical Conditions:** The valve body solenoid valve is not receiving a signal. **Possible Causes:** • Pressure control solenoid circuit is shorting to ground • Pressure control solenoid circuit is open • Valve has failed • TCM has failed • ECM has failed
DTC: P0751 **1T CCM, MIL: Yes** **1997, 1998, 1999, 2000, 2001, 2002, 2003, 2004, 2005, 2006** **Models:** TT, A4, A6, A8 **Engines:** 2.8L, 3.2L, 3.7L, 4.2L **Transmissions:** A/T **Transmissions:** A/T	**Shift Solenoid "A" Performance or Stuck Off Conditions:** Engine started, vehicle driven with the solenoid applied, and the ECM detected an unexpected voltage condition on the SS1/A solenoid circuit was incorrect during the test. **Possible Causes:** • Solenoid valves in valve body are faulty • Solenoid circuit is shorting to ground • Solenoid circuit is open • TCM has failed or wiring is shorting • ECM has failed

DTC	Trouble Code Title, Conditions & Possible Causes
DTC: P0752 **2T CCM, MIL: Yes** **1997, 1998, 1999, 2000, 2001, 2002, 2003, 2004, 2005, 2006** **Models:** A4, A6, A8 **Engines:** 2.8L, 3.2L, 3.7L, 4.2L **Transmissions:** A/T **Transmissions:** A/T	**A/T Shift Solenoid 1/A Function Range/Performance Conditions:** Engine started, vehicle driven with the solenoid applied, and the ECM detected a mechanical failure while operating the Shift Solenoid 1/A during the CCM test period. **Possible Causes:** • SS1/A solenoid is stuck in the "on" position • SS1/A solenoid has a mechanical failure • SS1/A solenoid has a hydraulic failure • ECM has failed
DTC: P0753 **2T CCM, MIL: Yes** **1996, 1997, 1998, 1999** **Models:** A4, A8 **Engines:** 2.8L, 3.7L, 4.2L **Transmissions:** A/T	**A/T Shift Solenoid 1/A Circuit Malfunction Conditions:** Engine started, vehicle driven with the solenoid applied, and the ECM detected an unexpected voltage condition on the SS1/A solenoid circuit was incorrect during the test. **Possible Causes:** • SS1/A solenoid control circuit is open • SS1/A solenoid control circuit is shorted to ground • SS1/A solenoid wiring harness connector is damaged • SS1/A solenoid is damaged or has failed • ECM has failed
DTC: P0756 **2T CCM, MIL: Yes** **2004, 2005, 2006** **Models:** TT, A6, A8 **Engines:** 3.2L, 3.7L, 4.2L **Transmissions:** A/T	**A/T Shift Solenoid 2/B Function Range/Performance Conditions:** Engine started, vehicle driven with the solenoid applied, and the ECM detected a mechanical failure while operating the Shift Solenoid 2/B during the CCM test period. **Possible Causes:** • SS2/B solenoid is stuck in the "on" position • SS2/B solenoid has a mechanical failure • SS2/B solenoid has a hydraulic failure • ECM has failed
DTC: P0757 **2T CCM, MIL: Yes** **1996, 1997, 1998, 1999** **Models:** A4, A8 **Engines:** 2.8L, 3.7L, 4.2L **Transmissions:** A/T	**A/T Shift Solenoid 2/B Function Range/Performance Conditions:** Engine started, vehicle driven with the solenoid applied, and the ECM detected a mechanical failure while operating the Shift Solenoid 2/B during the CCM test period. **Possible Causes:** • SS2/B solenoid is stuck in the "on" position • SS2/B solenoid has a mechanical failure • SS2/B solenoid has a hydraulic failure • ECM has failed
DTC: P0758 **2T CCM, MIL: Yes** **1996, 1997, 1998, 1999** **Models:** A4, A8 **Engines:** 2.8L, 3.7L, 4.2L **Transmissions:** A/T	**A/T Shift Solenoid 2/B Circuit Malfunction Conditions:** Engine started, vehicle driven with the solenoid applied, and the ECM detected an unexpected voltage condition on the SS1/A solenoid circuit was incorrect during the test.. **Possible Causes:** • Shift Solenoid 2/B connector is damaged, open or shorted • Shift Solenoid 2/B control circuit is open • Shift Solenoid 2/B control circuit is shorted to ground • Shift Solenoid 2/B is damaged or it has failed • ECM has failed
DTC: P0761 **2T CCM, MIL: Yes** **1996, 1997, 1998, 1999** **Models:** TT, A4, A8 **Engines:** 2.8L, 3.7L, 4.2L **Transmissions:** A/T	**A/T Shift Solenoid 3/C Function Range/Performance Conditions:** Engine started, vehicle driven with Shift Solenoid 3/C applied, and the ECM detected a mechanical failure occurred (stuck "off") while operating Shift Solenoid 3/C during the test. **Possible Causes:** • SS3/C solenoid may be stuck "off" • SS3/C solenoid has a mechanical failure • SS3/C solenoid has a hydraulic failure • ECM has failed
DTC: P0762 **2T CCM, MIL: Yes** **1996, 1997, 1998, 1999** **Models:** A4, A8 **Engines:** 2.8L, 3.7L, 4.2L **Transmissions:** A/T	**A/T Shift Solenoid 3/C Function Range/Performance Conditions:** Engine started, vehicle driven with Shift Solenoid 3/C applied, and the ECM detected a mechanical failure occurred (stuck "on") while operating Shift Solenoid 3/C during the test. **Possible Causes:** • SS3/C solenoid may be stuck "on" • SS3/C solenoid has a mechanical failure • SS3/C solenoid has a hydraulic failure • ECM has failed

DTC	Trouble Code Title, Conditions & Possible Causes
DTC: P0763 **2T CCM, MIL: Yes** **1996, 1997, 1998, 1999** **Models:** A4, A8 **Engines:** 2.8L, 3.7L, 4.2L **Transmissions:** A/T	**A/T Shift Solenoid 3/C Electrical Conditions:** Engine started, vehicle driven with the solenoid applied, and the ECM detected an unexpected voltage condition on the SS3/C solenoid circuit was incorrect during the test.. **Possible Causes:** • Shift Solenoid 3/C connector is damaged, open or shorted • Shift Solenoid 3/C control circuit is open • Shift Solenoid 3/C control circuit is shorted to ground • Shift Solenoid 3/C is damaged or it has failed • ECM has failed
DTC: P0766 **2T CCM, MIL: Yes** **2004, 2005, 2006** **Models:** TT **Engines:** 3.2L **Transmissions:** A/T	**A/T Shift Solenoid D Performance Conditions:** Engine started, vehicle driven with the solenoid applied, and the ECM detected an unexpected voltage condition on the SS3/C solenoid circuit was incorrect during the test.. **Possible Causes:** • Shift Solenoid D connector is damaged, open or shorted • Shift Solenoid D control circuit is open • Shift Solenoid D control circuit is shorted to ground • Shift Solenoid D is damaged or it has failed • ECM has failed
DTC: P0771 **2T CCM, MIL: Yes** **2004, 2005, 2006** **Models:** TT **Engines:** 3.2L **Transmissions:** A/T	**A/T Shift Solenoid E Performance Conditions:** Engine started, vehicle driven with the solenoid applied, and the ECM detected an unexpected voltage condition on the SS3/C solenoid circuit was incorrect during the test.. **Possible Causes:** • Shift Solenoid D connector is damaged, open or shorted • Shift Solenoid D control circuit is open • Shift Solenoid D control circuit is shorted to ground • Shift Solenoid D is damaged or it has failed • ECM has failed
DTC: P0773 **2T CCM, MIL: Yes** **1996** **Models:** A4 **Engines:** 2.8L **Transmissions:** A/T **Transmissions:** A/T	**A/T Shift Solenoid E Electrical Conditions:** Engine started, vehicle driven with the solenoid applied, and the ECM detected an unexpected voltage condition on the SS3/D solenoid circuit was incorrect during the test.. **Possible Causes:** • Shift Solenoid connector is damaged, open or shorted • Shift Solenoid control circuit is open • Shift Solenoid control circuit is shorted to ground • Shift Solenoid is damaged or it has failed • ECM has failed
DTC: P0776 **1T CCM, MIL: Yes** **2004, 2005, 2006** **Models:** TT, A6 **Engines:** 3.2L, 4.2L **Transmissions:** A/T	**Pressure Control Solenoid "B" Performance or Stuck Off Conditions:** Engine started, vehicle driven with Shift Solenoid 3/C applied, and the ECM detected a mechanical failure occurred (stuck "off") while operating Shift Solenoid 3/C during the test. **Possible Causes:** • SS3/C solenoid may be stuck "off" • SS3/C solenoid has a mechanical failure • SS3/C solenoid has a hydraulic failure • ECM has failed
DTC: P0777 **1T CCM, MIL: Yes** **2004, 2005, 2006** **Models:** TT, A6 **Engines:** 3.2L **Transmissions:** A/T	**Pressure Control Solenoid "B" Stuck On Conditions:** Engine started, vehicle driven with Shift Solenoid 3/C applied, and the ECM detected a mechanical failure occurred (stuck "on") while operating Shift Solenoid 3/C during the test. **Possible Causes:** • SS3/C solenoid may be stuck "on" • SS3/C solenoid has a mechanical failure • SS3/C solenoid has a hydraulic failure • ECM has failed
DTC: P0778 **1T CCM, MIL: Yes** **2004, 2005, 2006** **Models:** A6 **Engines:** 3.2L, 4.2L **Transmissions:** A/T	**Pressure Control Solenoid "B" Electrical Conditions:** Engine started, vehicle driven with the solenoid applied, and the ECM detected an unexpected voltage condition on the SS3/C solenoid circuit was incorrect during the test.. **Possible Causes:** • Shift Solenoid connector is damaged, open or shorted • Shift Solenoid control circuit is open • Shift Solenoid control circuit is shorted to ground • Shift Solenoid is damaged or it has failed • ECM has failed

DTC	Trouble Code Title, Conditions & Possible Causes
DTC: P0781 **2T CCM, MIL: Yes** **2004, 2005, 2006** **Models:** TT, A6 **Engines:** 3.2L, 4.2L **Transmissions:** A/T	**1-2 Shift Conditions:** Engine running and vehicle driven, the ECM detected a mechanical malfunction within the transmission **Possible Causes:** • Solenoid valves in valve body are faulty • Solenoid circuit is shorting to ground • Solenoid circuit is open • TCM has failed or wiring is shorting • ECM has failed • Mechanical malfunction in transmission
DTC: P0782 **2T CCM, MIL: Yes** **2004, 2005, 2006** **Models:** TT, A6 **Engines:** 3.2L, 4.2L **Transmissions:** A/T	**2-3 Shift Conditions:** Engine running and vehicle driven, the ECM detected a mechanical malfunction within the transmission **Possible Causes:** • Solenoid valves in valve body are faulty • Solenoid circuit is shorting to ground • Solenoid circuit is open • TCM has failed or wiring is shorting • ECM has failed • Mechanical malfunction in transmission
DTC: P0783 **2T CCM, MIL: Yes** **2004, 2005, 2006** **Models:** TT, A6 **Engines:** 3.2L, 4.2L **Transmissions:** A/T	**3-4 Shift Conditions:** Engine running and vehicle driven, the ECM detected a mechanical malfunction within the transmission **Possible Causes:** • Solenoid valves in valve body are faulty • Solenoid circuit is shorting to ground • Solenoid circuit is open • TCM has failed or wiring is shorting • ECM has failed • Mechanical malfunction in transmission
DTC: P0784 **2T CCM, MIL: Yes** **2004, 2005, 2006** **Models:** TT, A6 **Engines:** 3.2L, 4.2L **Transmissions:** A/T	**4-5 Shift Conditions:** Engine running and vehicle driven, the ECM detected a mechanical malfunction within the transmission **Possible Causes:** • Solenoid valves in valve body are faulty • Solenoid circuit is shorting to ground • Solenoid circuit is open • TCM has failed or wiring is shorting • ECM has failed • Mechanical malfunction in transmission
DTC: P0791 **2T CCM, MIL: Yes** **2004, 2005, 2006** **Models:** TT **Engines:** 3.2L **Transmissions:** A/T	**Intermediate Shaft Speed Sensor "A" Circuit Conditions:** Engine running and vehicle driven, the ECM detected a mechanical malfunction within the transmission **Possible Causes:** • Solenoid valves in valve body are faulty • Solenoid circuit is shorting to ground • Solenoid circuit is open • TCM has failed or wiring is shorting • ECM has failed • Mechanical malfunction in transmission
DTC: P0796 **2T CCM, MIL: Yes** **2004, 2005, 2006** **Models:** A6 **Engines:** 3.2L, 4.2L **Transmissions:** A/T	**Pressure Solenoid "C" Performance or Stuck Off Conditions:** Engine started, vehicle driven with the solenoid applied, and the ECM detected an unexpected voltage condition on the SS1/C solenoid circuit was incorrect during the test. **Possible Causes:** • Solenoid valves in valve body are faulty • Solenoid circuit is shorting to ground • Solenoid circuit is open • TCM has failed or wiring is shorting • ECM has failed

DTC	Trouble Code Title, Conditions & Possible Causes
DTC: P0797 **2T CCM, MIL: Yes** **2004, 2005, 2006** **Models:** TT, A6 **Engines:** 3.2L, 4.2L **Transmissions:** A/T	**Pressure Solenoid "C" Performance or Stuck On Conditions:** Engine started, vehicle driven with the solenoid applied, and the ECM detected an unexpected voltage condition on the SS1/C solenoid circuit was incorrect during the test. **Possible Causes:** • Solenoid valves in valve body are faulty • Solenoid circuit is shorting to ground • Solenoid circuit is open • TCM has failed or wiring is shorting • ECM has failed
DTC: P0798 **2T CCM, MIL: Yes** **2004, 2005, 2006** **Models:** A6 **Engines:** 3.2L, 4.2L **Transmissions:** A/T	**Pressure Solenoid "C" Electrical Conditions:** Engine started, vehicle driven with the solenoid applied, and the ECM detected an unexpected voltage condition on the SS1/C solenoid circuit was incorrect during the test. **Possible Causes:** • Solenoid valves in valve body are faulty • Solenoid circuit is shorting to ground • Solenoid circuit is open • TCM has failed or wiring is shorting • ECM has failed
DTC: P0840 **2T CCM, MIL: Yes** **2004, 2005, 2006** **Models:** TT **Engines:** 3.2L **Transmissions:** A/T	**Transmission Fluid Pressure Sensor/Switch "A" Circuit Conditions:** Engine started, vehicle driven with the solenoid applied, and the ECM detected an unexpected voltage condition on the pressure sensor/switch was incorrect during the test. **Possible Causes:** • Solenoid valves in valve body are faulty • Solenoid circuit is shorting to ground • Solenoid circuit is open • TCM has failed or wiring is shorting • Transmission Input Speed (RPM) Sensor has failed • Transmission Output Speed (RPM) Sensor has failed • ECM has failed
DTC: P0841 **2T CCM, MIL: Yes** **2004, 2005, 2006** **Models:** TT **Engines:** 3.2L **Transmissions:** A/T	**Transmission Fluid Pressure Sensor/Switch "A" Circuit Range/Performance Conditions:** Engine started, vehicle driven with the solenoid applied, and the ECM detected an unexpected voltage condition on the pressure sensor/switch was incorrect during the test. **Possible Causes:** • Solenoid valves in valve body are faulty • Solenoid circuit is shorting to ground • Solenoid circuit is open • TCM has failed or wiring is shorting • Transmission Input Speed (RPM) Sensor has failed • Transmission Output Speed (RPM) Sensor has failed • ECM has failed
DTC: P0845 **2T CCM, MIL: Yes** **2004, 2005, 2006** **Models:** TT **Engines:** 3.2L **Transmissions:** A/T	**Transmission Fluid Pressure Sensor/Switch "B" Circuit Conditions:** Engine started, vehicle driven with the solenoid applied, and the ECM detected an unexpected voltage condition on the pressure sensor/switch was incorrect during the test. **Possible Causes:** • Solenoid valves in valve body are faulty • Solenoid circuit is shorting to ground • Solenoid circuit is open • TCM has failed or wiring is shorting • Transmission Input Speed (RPM) Sensor has failed • Transmission Output Speed (RPM) Sensor has failed • ECM has failed
DTC: P0846 **2T CCM, MIL: Yes** **2004, 2005, 2006** **Models:** TT **Engines:** 3.2L **Transmissions:** A/T	**Transmission Fluid Pressure Sensor/Switch "B" Circuit Range/Performance Conditions:** Engine started, vehicle driven with the solenoid applied, and the ECM detected an unexpected voltage condition on the pressure sensor/switch was incorrect during the test. **Possible Causes:** • Solenoid valves in valve body are faulty • Solenoid circuit is shorting to ground • Solenoid circuit is open • TCM has failed or wiring is shorting • Transmission Input Speed (RPM) Sensor has failed • Transmission Output Speed (RPM) Sensor has failed • ECM has failed

DTC	Trouble Code Title, Conditions & Possible Causes
DTC: P0864 **2T CCM, MIL: Yes** **2004, 2005, 2006** **Models:** TT **Engines:** 3.2L **Transmissions:** A/T	**TCM Communication Circuit Range/Performance Conditions:** The Transmission Control Module (ECM) communicates with all databus-capable control modules via a CAN databus. These databus-capable control modules are connected via two data bus wires which are twisted together (CAN_High and CAN_Low), and exchange information (messages). Missing information on the databus is recognized as a malfunction and stored. Trouble-free operation of the CAN-Bus requires that it have a terminal resistance. **Possible Causes:** • ECM has failed • Terminal resistance for CAN-bus are faulty • Can data bus wires have short circuited to each other • TCM has failed
DTC: P0889 **2T CCM, MIL: Yes** **2004, 2005, 2006** **Models:** A6 **Engines:** 3.2L, 4.2L **Transmissions:** A/T	**TCM Power Relay Sense Circuit Range/Performance Conditions:** The Transmission Control Module (ECM) communicates with all databus-capable control modules via a CAN databus. These databus-capable control modules are connected via two data bus wires which are twisted together (CAN_High and CAN_Low), and exchange information (messages). Missing information on the databus is recognized as a malfunction and stored. Trouble-free operation of the CAN-Bus requires that it have a terminal resistance. **Possible Causes:** • Solenoid valves in valve body are faulty • Solenoid circuit is shorting to ground • Solenoid circuit is open • TCM has failed or wiring is shorting • ECM has failed
DTC: P0890 **2T CCM, MIL: Yes** **2004, 2005, 2006** **Models:** TT, A6 **Engines:** 3.2L, 4.2L **Transmissions:** A/T	**TCM Power Relay Sense Circuit Low Conditions:** The Transmission Control Module (ECM) communicates with all databus-capable control modules via a CAN databus. These databus-capable control modules are connected via two data bus wires which are twisted together (CAN_High and CAN_Low), and exchange information (messages). Missing information on the databus is recognized as a malfunction and stored. Trouble-free operation of the CAN-Bus requires that it have a terminal resistance. **Possible Causes:** • Solenoid valves in valve body are faulty • Solenoid circuit is shorting to ground • Solenoid circuit is open • TCM has failed or wiring is shorting • ECM has failed
DTC: P0891 **2T CCM, MIL: Yes** **2004, 2005, 2006** **Models:** A6 **Engines:** 3.2L, 4.2L **Transmissions:** A/T	**TCM Power Relay Sense Circuit High Conditions:** The Transmission Control Module (ECM) communicates with all databus-capable control modules via a CAN databus. These databus-capable control modules are connected via two data bus wires which are twisted together (CAN_High and CAN_Low), and exchange information (messages). Missing information on the databus is recognized as a malfunction and stored. Trouble-free operation of the CAN-Bus requires that it have a terminal resistance. **Possible Causes:** • Solenoid valves in valve body are faulty • Solenoid circuit is shorting to ground • Solenoid circuit is open • TCM has failed or wiring is shorting • ECM has failed
DTC: P0892 **2T CCM, MIL: Yes** **2004, 2005, 2006** **Models:** A6 **Engines:** 3.2L, 4.2L **Transmissions:** A/T	**TCM Power Relay Sense Circuit Intermittent Conditions:** The Transmission Control Module (ECM) communicates with all databus-capable control modules via a CAN databus. These databus-capable control modules are connected via two data bus wires which are twisted together (CAN_High and CAN_Low), and exchange information (messages). Missing information on the databus is recognized as a malfunction and stored. Trouble-free operation of the CAN-Bus requires that it have a terminal resistance. **Possible Causes:** • Solenoid valves in valve body are faulty • Solenoid circuit is shorting to ground • Solenoid circuit is open • TCM has failed or wiring is shorting • ECM has failed

DTC	Trouble Code Title, Conditions & Possible Causes:
DTC: P1102 **1T CCM, MIL: No** **1997, 1998, 1999, 2000, 2001, 2002, 2003, 2004, 2005, 2006** **Models:** TT, A4, A6,A6 Avant, A8 **Engines:** 1.8L, 2.8L, 3.7L, 4.2L **Transmissions:** A/T	**O2 Sensor Circuit (Bank 1-Sensor 1) Short to B+ Conditions:** Engine started, battery voltage must be at least 11.5v, all electrical components must be off, the ground between the engine and the chassis must be well connected, the exhaust system must be properly sealed between the catalytic converter and the cylinder head, and the oxygen sensor heater for oxygen sensor before the catalytic converter must be properly functioning. The ECM detected a voltage on the O2 sensor circuit that was outside the parameters to function properly. **Note: For resistance testing of sensor heating, oxygen sensor should be cooled to ambient temperature. High temperatures at oxygen sensor may lead to inaccurate measurements.** **Possible Causes:** • Oxygen sensor (before catalytic converter) is faulty • Oxygen sensor (behind catalytic converter) is faulty • Oxygen sensor heater (before catalytic converter) is faulty • Oxygen sensor heater (behind catalytic converter) is faulty • Circuit wiring has a short to power or ground • Engine Component Power Supply Relay is faulty • E-box fuses for oxygen sensor are faulty • Leaks present in the exhaust manifold or exhaust pipes • HO2S signal wire and ground wire crossed in connector • HO2S element is fuel contaminated or has failed • ECM has failed
DTC: P1103 **1T CCM, MIL: No** **1997, 1998, 1999, 2000, 2001, 2002, 2003, 2004, 2005, 2006** **Models:** TT, A4, A6,A6 Avant **Engines:** 1.8L, 2.8L **Transmissions:** A/T	**O2 Sensor Circuit (Bank 1-Sensor 1) Output Too Low Conditions:** Engine started, battery voltage must be at least 11.5v, all electrical components must be off, the ground between the engine and the chassis must be well connected, the exhaust system must be properly sealed between the catalytic converter and the cylinder head, and the oxygen sensor heater for oxygen sensor before the catalytic converter must be properly functioning. The ECM detected a voltage on the O2 sensor circuit that was outside the parameters to function properly. **Note: For resistance testing of sensor heating, oxygen sensor should be cooled to ambient temperature. High temperatures at oxygen sensor may lead to inaccurate measurements.** **Possible Causes:** • Oxygen sensor (before catalytic converter) is faulty • Oxygen sensor (behind catalytic converter) is faulty • Oxygen sensor heater (before catalytic converter) is faulty • Oxygen sensor heater (behind catalytic converter) is faulty • Circuit wiring has a short to power or ground • Engine Component Power Supply Relay is faulty • E-box fuses for oxygen sensor are faulty • Leaks present in the exhaust manifold or exhaust pipes • HO2S signal wire and ground wire crossed in connector • HO2S element is fuel contaminated or has failed • ECM has failed
DTC: P1105 **1T CCM, MIL: No** **1997, 1998, 1999, 2000, 2001, 2002, 2003, 2004, 2005, 2006** **Models:** TT, A4, A6,A6 Avant, A8 **Engines:** 1.8L, 2.8L, 3.7L, 4.2L **Transmissions:** A/T	**O2 Sensor Circuit (Bank 1-Sensor 2) Short to B+ Conditions:** Engine started, battery voltage must be at least 11.5v, all electrical components must be off, the ground between the engine and the chassis must be well connected, the exhaust system must be properly sealed between the catalytic converter and the cylinder head, and the oxygen sensor heater for oxygen sensor before the catalytic converter must be properly functioning. The ECM detected a voltage on the O2 sensor circuit that was outside the parameters to function properly. **Note: For resistance testing of sensor heating, oxygen sensor should be cooled to ambient temperature. High temperatures at oxygen sensor may lead to inaccurate measurements.** **Possible Causes:** • Oxygen sensor (before catalytic converter) is faulty • Oxygen sensor (behind catalytic converter) is faulty • Oxygen sensor heater (before catalytic converter) is faulty • Oxygen sensor heater (behind catalytic converter) is faulty • Circuit wiring has a short to power or ground • Engine Component Power Supply Relay is faulty • E-box fuses for oxygen sensor are faulty • Leaks present in the exhaust manifold or exhaust pipes • HO2S signal wire and ground wire crossed in connector • HO2S element is fuel contaminated or has failed • ECM has failed

DTC	Trouble Code Title, Conditions & Possible Causes
DTC: P1107 **1T CCM, MIL: No** **1997, 1998, 1999, 2000, 2001, 2002, 2003, 2004, 2005, 2006** **Models:** A4, A6,A6 Avant, A8 **Engines:** 2.8L, 3.7L, 4.2L **Transmissions:** A/T	**O2 Sensor Circuit (Bank 2-Sensor 1) Voltage Too Low Conditions:** Engine started, battery voltage must be at least 11.5v, all electrical components must be off, the ground between the engine and the chassis must be well connected, the exhaust system must be properly sealed between the catalytic converter and the cylinder head, and the oxygen sensor heater for oxygen sensor before the catalytic converter must be properly functioning. The ECM detected a voltage on the O2 sensor circuit that was outside the parameters to function properly. **Note: For resistance testing of sensor heating, oxygen sensor should be cooled to ambient temperature. High temperatures at oxygen sensor may lead to inaccurate measurements.** **Possible Causes:** • Oxygen sensor (before catalytic converter) is faulty • Oxygen sensor (behind catalytic converter) is faulty • Oxygen sensor heater (before catalytic converter) is faulty • Oxygen sensor heater (behind catalytic converter) is faulty • Circuit wiring has a short to power or ground • Engine Component Power Supply Relay is faulty • E-box fuses for oxygen sensor are faulty • Leaks present in the exhaust manifold or exhaust pipes • HO2S signal wire and ground wire crossed in connector • HO2S element is fuel contaminated or has failed • ECM has failed
DTC: P1110 **1T CCM, MIL: No** **1997, 1998, 1999, 2000, 2001, 2002, 2003, 2004, 2005, 2006** **Models:** A4, A6,A6 Avant, A8 **Engines:** 2.8L, 3.7L, 4.2L **Transmissions:** A/T	**O2 Sensor Circuit (Bank 2-Sensor 2) Short to B+ Conditions:** Engine started, battery voltage must be at least 11.5v, all electrical components must be off, the ground between the engine and the chassis must be well connected, the exhaust system must be properly sealed between the catalytic converter and the cylinder head, and the oxygen sensor heater for oxygen sensor before the catalytic converter must be properly functioning. The ECM detected a voltage on the O2 sensor circuit that was outside the parameters to function properly. **Note: For resistance testing of sensor heating, oxygen sensor should be cooled to ambient temperature. High temperatures at oxygen sensor may lead to inaccurate measurements.** **Possible Causes:** • Oxygen sensor (before catalytic converter) is faulty • Oxygen sensor (behind catalytic converter) is faulty • Oxygen sensor heater (before catalytic converter) is faulty • Oxygen sensor heater (behind catalytic converter) is faulty • Circuit wiring has a short to power or ground • Engine Component Power Supply Relay is faulty • E-box fuses for oxygen sensor are faulty • Leaks present in the exhaust manifold or exhaust pipes • HO2S signal wire and ground wire crossed in connector • HO2S element is fuel contaminated or has failed • ECM has failed
DTC: P1111 **1T CCM, MIL: No** **1997, 1998, 1999, 2000, 2001, 2002, 2003, 2004, 2005, 2006** **Models:** TT, A4 **Engines:** 1.8L **Transmissions:** All	**O2 Control (Bank 1) System Too Lean Conditions:** Engine started, battery voltage must be at least 11.5v, all electrical components must be off, the ground between the engine and the chassis must be well connected, the exhaust system must be properly sealed between the catalytic converter and the cylinder head, and the oxygen sensor heater for oxygen sensor before the catalytic converter must be properly functioning. The ECM detected a measurement on the O2 sensor circuit that was outside the parameters to function properly. **Note: For resistance testing of sensor heating, oxygen sensor should be cooled to ambient temperature. High temperatures at oxygen sensor may lead to inaccurate measurements.** **Note: When an O2S malfunction (P0131 to P0414) is also stored with this malfunction, the O2S malfunction(s) should be repaired first.** **Possible Causes:** • Oxygen sensor (before catalytic converter) is faulty • Oxygen sensor (behind catalytic converter) is faulty • Oxygen sensor heater (before catalytic converter) is faulty • Oxygen sensor heater (behind catalytic converter) is faulty • Circuit wiring has a short to power or ground • Engine Component Power Supply Relay is faulty • E-box fuses for oxygen sensor are faulty • Leaks present in the exhaust manifold or exhaust pipes • HO2S signal wire and ground wire crossed in connector • HO2S element is fuel contaminated or has failed • ECM has failed

DTC	Trouble Code Title, Conditions & Possible Causes
DTC: P1112 **1T CCM, MIL: No** **1997, 1998, 1999, 2000, 2001, 2002, 2003, 2004, 2005, 2006** **Models:** TT, A4 **Engines:** 1.8L **Transmissions:** All	**O2 Control (Bank 1) System Too Rich Conditions:** Engine started, battery voltage must be at least 11.5v, all electrical components must be off, the ground between the engine and the chassis must be well connected, the exhaust system must be properly sealed between the catalytic converter and the cylinder head, and the oxygen sensor heater for oxygen sensor before the catalytic converter must be properly functioning. The ECM detected a measurement on the O2 sensor circuit that was outside the parameters to function properly. **Note: For resistance testing of sensor heating, oxygen sensor should be cooled to ambient temperature. High temperatures at oxygen sensor may lead to inaccurate measurements.** **Note: When an O2S malfunction (P0131 to P0414) is also stored with this malfunction, the O2S malfunction(s) should be repaired first.** **Possible Causes:** • Oxygen sensor (before catalytic converter) is faulty • Oxygen sensor (behind catalytic converter) is faulty • Oxygen sensor heater (before catalytic converter) is faulty • Oxygen sensor heater (behind catalytic converter) is faulty • Circuit wiring has a short to power or ground • Engine Component Power Supply Relay is faulty • E-box fuses for oxygen sensor are faulty • Leaks present in the exhaust manifold or exhaust pipes • HO2S signal wire and ground wire crossed in connector • HO2S element is fuel contaminated or has failed • ECM has failed
DTC: P1113 **1T CCM, MIL: No** **1997, 1998, 1999, 2000, 2001, 2002, 2003, 2004, 2005, 2006** **Models:** TT, A4, A6, A8 **Engines:** 1.8L, 4.2L **Transmissions:** All	**O2 Control (Bank 1 Sensor 1) Internal Resistance Too High Conditions:** Engine started, battery voltage must be at least 11.5v, all electrical components must be off, the ground between the engine and the chassis must be well connected, the exhaust system must be properly sealed between the catalytic converter and the cylinder head, and the oxygen sensor heater for oxygen sensor before the catalytic converter must be properly functioning. The ECM detected a measurement on the O2 sensor circuit that was outside the parameters to function properly. **Note: For resistance testing of sensor heating, oxygen sensor should be cooled to ambient temperature. High temperatures at oxygen sensor may lead to inaccurate measurements.** **Possible Causes:** • Oxygen sensor (before catalytic converter) is faulty • Oxygen sensor (behind catalytic converter) is faulty • Oxygen sensor heater (before catalytic converter) is faulty • Oxygen sensor heater (behind catalytic converter) is faulty • Circuit wiring has a short to power or ground • Engine Component Power Supply Relay is faulty • E-box fuses for oxygen sensor are faulty • Leaks present in the exhaust manifold or exhaust pipes • HO2S signal wire and ground wire crossed in connector • HO2S element is fuel contaminated or has failed • ECM has failed
DTC: P1114 **1T CCM, MIL: No** **1997, 1998, 1999, 2000, 2001, 2002, 2003, 2004, 2005, 2006** **Models:** TT, A4, A6, A8 **Engines:** 1.8L, 4.2L **Transmissions:** All	**O2 Control (Bank 1 Sensor 2) Internal Resistance Too High Conditions:** Engine started, battery voltage must be at least 11.5v, all electrical components must be off, the ground between the engine and the chassis must be well connected, the exhaust system must be properly sealed between the catalytic converter and the cylinder head, and the oxygen sensor heater for oxygen sensor before the catalytic converter must be properly functioning. The ECM detected a measurement on the O2 sensor circuit that was outside the parameters to function properly. **Note: For resistance testing of sensor heating, oxygen sensor should be cooled to ambient temperature. High temperatures at oxygen sensor may lead to inaccurate measurements.** **Possible Causes:** • Oxygen sensor (before catalytic converter) is faulty • Oxygen sensor (behind catalytic converter) is faulty • Oxygen sensor heater (before catalytic converter) is faulty • Oxygen sensor heater (behind catalytic converter) is faulty • Circuit wiring has a short to power or ground • Engine Component Power Supply Relay is faulty • E-box fuses for oxygen sensor are faulty • Leaks present in the exhaust manifold or exhaust pipes • HO2S signal wire and ground wire crossed in connector • HO2S element is fuel contaminated or has failed • ECM has failed

DTC	Trouble Code Title, Conditions & Possible Causes
DTC: P1115 **1T CCM, MIL: No** **1997, 1998, 1999, 2000, 2001, 2002, 2003, 2004, 2005, 2006** **Models:** TT, A4, A6, A8 **Engines:** 1.8L, 4.2L **Transmissions:** All	**O2 Control (Bank 1 Sensor 1) Short to Ground Conditions:** Engine started, battery voltage must be at least 11.5v, all electrical components must be off, the ground between the engine and the chassis must be well connected, the exhaust system must be properly sealed between the catalytic converter and the cylinder head, and the oxygen sensor heater for oxygen sensor before the catalytic converter must be properly functioning. The ECM detected a measurement on the O2 sensor circuit that was outside the parameters to function properly. **Note: For resistance testing of sensor heating, oxygen sensor should be cooled to ambient temperature. High temperatures at oxygen sensor may lead to inaccurate measurements.** **Possible Causes:** • Oxygen sensor (before catalytic converter) is faulty • Oxygen sensor (behind catalytic converter) is faulty • Oxygen sensor heater (before catalytic converter) is faulty • Oxygen sensor heater (behind catalytic converter) is faulty • Circuit wiring has a short to power or ground • Engine Component Power Supply Relay is faulty • E-box fuses for oxygen sensor are faulty • Leaks present in the exhaust manifold or exhaust pipes • HO2S signal wire and ground wire crossed in connector • HO2S element is fuel contaminated or has failed • ECM has failed
DTC: P1116 **1T CCM, MIL: No** **1997, 1998, 1999, 2000, 2001, 2002, 2003, 2004, 2005, 2006** **Models:** TT, A4, A6, A8 **Engines:** 1.8L, 4.2L **Transmissions:** All	**O2 Control (Bank 1 Sensor 1) Open Conditions:** Engine started, battery voltage must be at least 11.5v, all electrical components must be off, the ground between the engine and the chassis must be well connected, the exhaust system must be properly sealed between the catalytic converter and the cylinder head, and the oxygen sensor heater for oxygen sensor before the catalytic converter must be properly functioning. The ECM detected a measurement on the O2 sensor circuit that was outside the parameters to function properly. **Note: For resistance testing of sensor heating, oxygen sensor should be cooled to ambient temperature. High temperatures at oxygen sensor may lead to inaccurate measurements.** **Possible Causes:** • Oxygen sensor (before catalytic converter) is faulty • Oxygen sensor (behind catalytic converter) is faulty • Oxygen sensor heater (before catalytic converter) is faulty • Oxygen sensor heater (behind catalytic converter) is faulty • Circuit wiring has a short to power or ground • Engine Component Power Supply Relay is faulty • E-box fuses for oxygen sensor are faulty • Leaks present in the exhaust manifold or exhaust pipes • HO2S signal wire and ground wire crossed in connector • HO2S element is fuel contaminated or has failed • ECM has failed
DTC: P1117 **1T CCM, MIL: No** **1997, 1998, 1999, 2000, 2001, 2002, 2003, 2004, 2005, 2006** **Models:** TT, A4, A6, A8 **Engines:** 1.8L, 4.2L **Transmissions:** All	**O2 Control (Bank 1 Sensor 2) Open Conditions:** Engine started, battery voltage must be at least 11.5v, all electrical components must be off, the ground between the engine and the chassis must be well connected, the exhaust system must be properly sealed between the catalytic converter and the cylinder head, and the oxygen sensor heater for oxygen sensor before the catalytic converter must be properly functioning. The ECM detected a measurement on the O2 sensor circuit that was outside the parameters to function properly. **Note: For resistance testing of sensor heating, oxygen sensor should be cooled to ambient temperature. High temperatures at oxygen sensor may lead to inaccurate measurements.** **Possible Causes:** • Oxygen sensor (before catalytic converter) is faulty • Oxygen sensor (behind catalytic converter) is faulty • Oxygen sensor heater (before catalytic converter) is faulty • Oxygen sensor heater (behind catalytic converter) is faulty • Circuit wiring has a short to power or ground • Engine Component Power Supply Relay is faulty • E-box fuses for oxygen sensor are faulty • Leaks present in the exhaust manifold or exhaust pipes • HO2S signal wire and ground wire crossed in connector • HO2S element is fuel contaminated or has failed • ECM has failed

DTC	Trouble Code Title, Conditions & Possible Causes
DTC: P1118 **1T CCM, MIL: No** **1997, 1998, 1999, 2000, 2001, 2002, 2003, 2004, 2005, 2006** **Models:** TT, A4, A6, A8 **Engines:** 1.8L, 4.2L **Transmissions:** All	**O2 Sensor Heater Circ. (Bank 1-Sensor2) Open Conditions:** Engine started, battery voltage must be at least 11.5v, all electrical components must be off, the ground between the engine and the chassis must be well connected, the exhaust system must be properly sealed between the catalytic converter and the cylinder head, and the oxygen sensor heater for oxygen sensor before the catalytic converter must be properly functioning. The ECM detected a measurement on the O2 sensor circuit that was outside the parameters to function properly. **Note: For resistance testing of sensor heating, oxygen sensor should be cooled to ambient temperature. High temperatures at oxygen sensor may lead to inaccurate measurements.** **Possible Causes:** • Oxygen sensor (before catalytic converter) is faulty • Oxygen sensor (behind catalytic converter) is faulty • Oxygen sensor heater (before catalytic converter) is faulty • Oxygen sensor heater (behind catalytic converter) is faulty • Circuit wiring has a short to power or ground • Engine Component Power Supply Relay is faulty • E-box fuses for oxygen sensor are faulty • Leaks present in the exhaust manifold or exhaust pipes • HO2S signal wire and ground wire crossed in connector • HO2S element is fuel contaminated or has failed • ECM has failed
DTC: P1127 **2T CCM, MIL: Yes** **1996, 1997, 1998, 1999, 2000, 2001, 2002, 2003, 2004, 2005, 2006** **Models:** TT, A4, A6, Allroad Quattro, A8 **Engines:** 1.8L, 2.7L, 2.8L, 3.7L, 4.2L **Transmissions:** All	**Long Term Fuel Trim Add. Air. Bank 1 System Too Rich Conditions:** Engine started, battery voltage must be at least 11.5v, all electrical components must be off, the ground between the engine and the chassis must be well connected, the exhaust system must be properly sealed between the catalytic converter and the cylinder head, and the oxygen sensor heater for oxygen sensor before the catalytic converter must be properly functioning. The fuel mixture is so rich that the O2S control is on lean limit. **Note: After exhaust system repairs, make sure exhaust system is not under stress and that it has sufficient clearance from the bodywork. If necessary, loosen double clamps and align exhaust pipe so that sufficient clearance is maintained to the bodywork and support rings carry uniform loads. Do not use any silicone sealant. Traces of silicone components which are sucked into the engine are not burned there, and they damage the oxygen sensor.** **Possible Causes:** • MAF sensor circuit open • MAF sensor circuit shorted to ground • Air leak in the manifold • Secondary air injection system combi-valve stuck open • Secondary air injection system electrical short • Fuel pressure too high, leaks in the vacuum hose to fuel pressure regulator • Fuel pressure regulator has failed • Fuel injectors are dirty, faulty or do not close properly • ECM has failed
DTC: P1128 **2T CCM, MIL: Yes** **1996, 1997, 1998, 1999, 2000, 2001, 2002, 2003, 2004, 2005, 2006** **Models:** TT, A4, A6, Allroad Quattro, A8 **Engines:** 1.8L, 2.7L, 2.8L, 3.7L, 4.2L **Transmissions:** All	**Long Term Fuel Trim Add. Air. Bank 1 System Too Lean Conditions:** Engine started, battery voltage must be at least 11.5v, all electrical components must be off, the ground between the engine and the chassis must be well connected, the exhaust system must be properly sealed between the catalytic converter and the cylinder head, and the oxygen sensor heater for oxygen sensor before the catalytic converter must be properly functioning. The fuel mixture is so rich that the O2S control is on lean limit. **Note: When an O2S malfunction (P0131 to P0414) is also stored with this malfunction, the O2S malfunction(s) should be repaired first.** **Note: After exhaust system repairs, make sure exhaust system is not under stress and that it has sufficient clearance from the bodywork. If necessary, loosen double clamps and align exhaust pipe so that sufficient clearance is maintained to the bodywork and support rings carry uniform loads. Do not use any silicone sealant. Traces of silicone components which are sucked into the engine are not burned there, and they damage the oxygen sensor.** **Possible Causes:** • Fuel pressure is too low or fuel quantity supplied is too low • Fuel filter faulty • Transfer fuel pump has failed • Fuel injector is faulty (sticking or not opening) • Engine speed (RPM) sensor is faulty • MAF sensor circuit open • MAF sensor circuit shorted to ground • Air leak in the manifold • Secondary air injection system combi-valve stuck open • Secondary air injection system electrical short • ECM has failed

DTC	Trouble Code Title, Conditions & Possible Causes
DTC: P1129 **2T CCM, MIL: Yes** **1997, 1998, 1999, 2000, 2001, 2002, 2003, 2004, 2005, 2006** **Models:** A4, A6,A6 Avant, A8 **Engines:** 2.8L, 3.7L, 4.2L **Transmissions:** A/T	**Long Term Fuel Trim at Rich Limit Conditions:** Engine started, battery voltage must be at least 11.5v, all electrical components must be off, the ground between the engine and the chassis must be well connected, the exhaust system must be properly sealed between the catalytic converter and the cylinder head, and the oxygen sensor heater for oxygen sensor before the catalytic converter must be properly functioning. The ECM detected the HO2S circuit was too rich, or that it could no longer change Fuel Trim because it was at its lean limit. **Possible Causes:** • Air intake system leaking, vacuum hoses leaking or damaged • Air leaks located after the MAF sensor mounting location • EGR valve sticking, EGR diaphragm leaking, or gasket leaking • EVAP vapor recovery system has failed • Excessive fuel pressure, leaking or contaminated fuel injectors • Exhaust leaks before or near the HO2S(s) mounting location • Fuel pressure regulator is leaking or damaged • HO2S circuits wet or oily, corroded, or poor terminal contact • HO2S is damaged or it has failed • HO2S signal circuit open, shorted to ground, shorted to power • Low fuel pressure or vehicle driven until it was out of fuel • Oil dipstick not seated or engine oil level too high (overfilled)
DTC: P1130 **2T CCM, MIL: Yes** **1997, 1998, 1999, 2000, 2001, 2002, 2003, 2004, 2005, 2006** **Models:** A4, A6,A6 Avant, A8 **Engines:** 2.8L, 3.7L, 4.2L **Transmissions:** A/T	**Long Term Fuel Trim at Lean Limit Conditions:** Engine started, battery voltage must be at least 11.5v, all electrical components must be off, the ground between the engine and the chassis must be well connected, the exhaust system must be properly sealed between the catalytic converter and the cylinder head, and the oxygen sensor heater for oxygen sensor before the catalytic converter must be properly functioning. The ECM detected the HO2S circuit was too lean, or that it could no longer change Fuel Trim because it was at its lean limit. **Possible Causes:** • Air intake system leaking, vacuum hoses leaking or damaged • Air leaks located after the MAF sensor mounting location • EGR valve sticking, EGR diaphragm leaking, or gasket leaking • EVAP vapor recovery system has failed • Excessive fuel pressure, leaking or contaminated fuel injectors • Exhaust leaks before or near the HO2S(s) mounting location • Fuel pressure regulator is leaking or damaged • HO2S circuits wet or oily, corroded, or poor terminal contact • HO2S is damaged or it has failed • HO2S signal circuit open, shorted to ground, shorted to power • Low fuel pressure or vehicle driven until it was out of fuel • Oil dipstick not seated or engine oil level too high (overfilled)
DTC: P1136 **2T CCM, MIL: Yes** **1996, 1997, 1998, 1999, 2000, 2001, 2002, 2003, 2004, 2005, 2006** **Models:** TT, A4, A6, Allroad Quattro, A8 **Engines:** 1.8L, 2.7L, 2.8L, 3.7L, 4.2L **Transmissions:** All	**Long Term Fuel Trim Add. Fuel, Bank 1 System Too Lean Conditions:** Engine started, battery voltage must be at least 11.5v, all electrical components must be off, the ground between the engine and the chassis must be well connected, the exhaust system must be properly sealed between the catalytic converter and the cylinder head, and the oxygen sensor heater for oxygen sensor before the catalytic converter must be properly functioning. The ECM detected the HO2S circuit was too lean, or that it could no longer change Fuel Trim because it was at its lean limit. **Possible Causes:** • Air intake system leaking, vacuum hoses leaking or damaged • Air leaks located after the MAF sensor mounting location • EGR valve sticking, EGR diaphragm leaking, or gasket leaking • EVAP vapor recovery system has failed • Excessive fuel pressure, leaking or contaminated fuel injectors • Exhaust leaks before or near the HO2S(s) mounting location • Fuel pressure regulator is leaking or damaged • HO2S circuits wet or oily, corroded, or poor terminal contact • HO2S is damaged or it has failed • HO2S signal circuit open, shorted to ground, shorted to power • Low fuel pressure or vehicle driven until it was out of fuel • Oil dipstick not seated or engine oil level too high (overfilled)

DTC	Trouble Code Title, Conditions & Possible Causes
DTC: P1137 **2T CCM, MIL: Yes** **1997, 1998, 1999, 2000, 2001, 2002, 2003, 2004, 2005, 2006** **Models:** TT, A4, A6,Allroad Quattro, A8 **Engines:** 1.8L, 2.7L, 2.8L, 3.7L, 4.2L **Transmissions:** All	**Long Term Fuel Trim Add. Fuel, Bank 1 System Too Rich Conditions:** Engine started, battery voltage must be at least 11.5v, all electrical components must be off, the ground between the engine and the chassis must be well connected, the exhaust system must be properly sealed between the catalytic converter and the cylinder head, and the oxygen sensor heater for oxygen sensor before the catalytic converter must be properly functioning. The ECM detected the HO2S circuit was too rich, or that it could no longer change Fuel Trim because it was at its lean limit. **Possible Causes:** • Air intake system leaking, vacuum hoses leaking or damaged • Air leaks located after the MAF sensor mounting location • EGR valve sticking, EGR diaphragm leaking, or gasket leaking • EVAP vapor recovery system has failed • Excessive fuel pressure, leaking or contaminated fuel injectors • Exhaust leaks before or near the HO2S(s) mounting location • Fuel pressure regulator is leaking or damaged • HO2S circuits wet or oily, corroded, or poor terminal contact • HO2S is damaged or it has failed • HO2S signal circuit open, shorted to ground, shorted to power • Low fuel pressure or vehicle driven until it was out of fuel • Oil dipstick not seated or engine oil level too high (overfilled)
DTC: P1138 **2T CCM, MIL: Yes** **1997, 1998, 1999, 2000, 2001, 2002, 2003, 2004, 2005, 2006** **Models:** A4, A6,A6 Avant, A8 **Engines:** 2.8L, 3.7L, 4.2L **Transmissions:** A/T	**Long Term Fuel Trim Add. Fuel, Bank 2 System Too Lean Conditions:** Engine started, battery voltage must be at least 11.5v, all electrical components must be off, the ground between the engine and the chassis must be well connected, the exhaust system must be properly sealed between the catalytic converter and the cylinder head, and the oxygen sensor heater for oxygen sensor before the catalytic converter must be properly functioning. The ECM detected the HO2S circuit was too lean, or that it could no longer change Fuel Trim because it was at its lean limit. **Possible Causes:** • Air intake system leaking, vacuum hoses leaking or damaged • Air leaks located after the MAF sensor mounting location • EGR valve sticking, EGR diaphragm leaking, or gasket leaking • EVAP vapor recovery system has failed • Excessive fuel pressure, leaking or contaminated fuel injectors • Exhaust leaks before or near the HO2S(s) mounting location • Fuel pressure regulator is leaking or damaged • HO2S circuits wet or oily, corroded, or poor terminal contact • HO2S is damaged or it has failed • HO2S signal circuit open, shorted to ground, shorted to power • Low fuel pressure or vehicle driven until it was out of fuel • Oil dipstick not seated or engine oil level too high (overfilled)
DTC: P1139 **2T CCM, MIL: Yes** **1997, 1998, 1999, 2000, 2001, 2002, 2003, 2004, 2005, 2006** **Models:** A4, A6,A6 Avant, A8 **Engines:** 2.8L, 3.7L, 4.2L **Transmissions:** A/T	**Long Term Fuel Trim Add. Fuel, Bank 2 System Too Rich Conditions:** Engine started, battery voltage must be at least 11.5v, all electrical components must be off, the ground between the engine and the chassis must be well connected, the exhaust system must be properly sealed between the catalytic converter and the cylinder head, and the oxygen sensor heater for oxygen sensor before the catalytic converter must be properly functioning. The ECM detected the HO2S circuit was too rich, or that it could no longer change Fuel Trim because it was at its lean limit. **Possible Causes:** • Air intake system leaking, vacuum hoses leaking or damaged • Air leaks located after the MAF sensor mounting location • EGR valve sticking, EGR diaphragm leaking, or gasket leaking • EVAP vapor recovery system has failed • Excessive fuel pressure, leaking or contaminated fuel injectors • Exhaust leaks before or near the HO2S(s) mounting location • Fuel pressure regulator is leaking or damaged • HO2S circuits wet or oily, corroded, or poor terminal contact • HO2S is damaged or it has failed • HO2S signal circuit open, shorted to ground, shorted to power • Low fuel pressure or vehicle driven until it was out of fuel • Oil dipstick not seated or engine oil level too high (overfilled)

DTC	Trouble Code Title, Conditions & Possible Causes
DTC: P1141 **2T CCM, MIL: Yes** **1997, 1998, 1999, 2000, 2001, 2002, 2003, 2004, 2005, 2006** **Models:** TT, A4, A6, A8 **Engines:** 1.8L, 4.2L **Transmissions:** All	**Load Calculation Cross Check Range/Performance Conditions:** Engine started, battery voltage must be at least 11.5v, all electrical components must be off, the ground between the engine and the chassis must be well connected, the exhaust system must be properly sealed between the catalytic converter and the cylinder head, and the oxygen sensor heater for oxygen sensor before the catalytic converter must be properly functioning. **Note: Vacuum in the intake system sucks in the leak detection spray with false air. Leak detection spray decreases ignition quality of the fuel mixture. This causes a drop in engine speed and changes the value produced by the Heated Oxygen Sensor.** **Note: Both Throttle Position (TP) sensor and Sender 2 for accelerator pedal position are located at the accelerator pedal and communicate the driver's intentions to the ECM completely independently of each other. Both sensors are stored in one housing.** **Possible Causes:** • Intake system is leaking • Signal is grounding • ECM has failed • Intake Manifold Runner Position Sensor is faulty • Intake system for leaks (false air) is faulty • Motor for intake flap is faulty • Mass Air Flow (MAF) sensor is faulty • Throttle Position (TP) sensor is faulty • Throttle valve control module is faulty
DTC: P1143 **2T CCM, MIL: Yes** **1997, 1998, 1999, 2000, 2001, 2002, 2003, 2004, 2005, 2006** **Models:** TT, A4, A6, A8 **Engines:** 1.8L, 4.2L **Transmissions:** All	**Load Calculation Cross Check Upper Limit Conditions:** Engine started, battery voltage must be at least 11.5v, all electrical components must be off, the ground between the engine and the chassis must be well connected, the exhaust system must be properly sealed between the catalytic converter and the cylinder head, and the oxygen sensor heater for oxygen sensor before the catalytic converter must be properly functioning. **Note: Vacuum in the intake system sucks in the leak detection spray with false air. Leak detection spray decreases ignition quality of the fuel mixture. This causes a drop in engine speed and changes the value produced by the Heated Oxygen Sensor.** **Note: Both Throttle Position (TP) sensor and Sender 2 for accelerator pedal position are located at the accelerator pedal and communicate the driver's intentions to the ECM completely independently of each other. Both sensors are stored in one housing.** **Possible Causes:** • Intake Manifold Runner Position Sensor is faulty • Intake system for leaks (false air) is faulty • Motor for intake flap is faulty • Mass Air Flow (MAF) sensor is faulty • Throttle Position (TP) sensor is faulty • Throttle valve control module is faulty • Intake system is leaking • Signal is grounding • ECM has failed
DTC: P1149 **2T CCM, MIL: Yes** **1997, 1998, 1999, 2000, 2001, 2002, 2003, 2004, 2005, 2006** **Models:** TT, A4 **Engines:** 1.8L **Transmissions:** All	**O2 Control (Bank 1) Out of Range Conditions:** Engine started, battery voltage must be at least 11.5v, all electrical components must be off, the ground between the engine and the chassis must be well connected, the exhaust system must be properly sealed between the catalytic converter and the cylinder head, and the oxygen sensor heater for oxygen sensor before the catalytic converter must be properly functioning. The ECM detected a voltage on the O2 sensor circuit that was outside the parameters to function properly. **Note: For resistance testing of sensor heating, oxygen sensor should be cooled to ambient temperature. High temperatures at oxygen sensor may lead to inaccurate measurements.** **Possible Causes:** • Oxygen sensor (before catalytic converter) is faulty • Oxygen sensor (behind catalytic converter) is faulty • Oxygen sensor heater (before catalytic converter) is faulty • Oxygen sensor heater (behind catalytic converter) is faulty • Circuit wiring has a short to power or ground • Engine Component Power Supply Relay is faulty • E-box fuses for oxygen sensor are faulty • Leaks present in the exhaust manifold or exhaust pipes • HO2S signal wire and ground wire crossed in connector • HO2S element is fuel contaminated or has failed • ECM has failed

DTC	Trouble Code Title, Conditions & Possible Causes
DTC: P1171 **1997, 1998, 1999, 2000, 2001, 2002, 2003, 2004, 2005, 2006** **Models:** TT, A4, A6, A8 **Engines:** 1.8L, 4.2L **Transmissions:** All	**Throttle Actuation Potentiometer Sign.2 Range/Performance Conditions:** Engine started, battery voltage must be at least 11.5v, all electrical components must be off, the ground between the engine and the chassis must be well connected, coolant temperature must be at least 80 degrees Celsius and the accelerator pedal must be properly adjusted. The ECM detected an incorrect singal from the throttle potentiometer. **Note: If the complete throttle valve control module is current-less (e.g. connector disconnected) the throttle valve moves into a particular, specified mechanical position, which signals an increased idle speed with an engine at operating temperature. If only the Throttle Position (TP) actuator is current-less, the throttle valve also moves into the specified mechanical position (emergency running gap), however, since Closed Throttle Position (CTP) switch can still be recognized, an "almost normal idle RPM" is reached via the respective ignition angle retardation.** **Note: Terminal assignment at throttle control module is different in vehicles with and without cruise control.** **Characteristic: Steering column switch with operating module for cruise control.** **Possible Causes:** • Throttle valve control module has failed • Throttle valve is dirty or damaged • Throttle valve is not in a closed position • Voltage supply of throttle valve control module is shorted or open • ECM has failed
DTC: P1172 **1999, 2000, 2001, 2002, 2003, 2004, 2005, 2006** **Models:** TT, A6, A8 **Engines:** 1.8L, 4.2L **Transmissions:** All	**Throttle Actuation Potentiometer Sign.2 Signal Too Low Conditions:** Engine started, battery voltage must be at least 11.5v, all electrical components must be off, the ground between the engine and the chassis must be well connected, coolant temperature must be at least 80 degrees Celsius and the accelerator pedal must be properly adjusted. The ECM detected an incorrect singal from the throttle potentiometer. **Note: If the complete throttle valve control module is current-less (e.g. connector disconnected) the throttle valve moves into a particular, specified mechanical position, which signals an increased idle speed with an engine at operating temperature. If only the Throttle Position (TP) actuator is current-less, the throttle valve also moves into the specified mechanical position (emergency running gap), however, since Closed Throttle Position (CTP) switch can still be recognized, an "almost normal idle RPM" is reached via the respective ignition angle retardation.** **Note: Terminal assignment at throttle control module is different in vehicles with and without cruise control.** **Characteristic: Steering column switch with operating module for cruise control.** **Possible Causes:** • Throttle valve control module has failed • Throttle valve is dirty or damaged • Throttle valve is not in a closed position • Voltage supply of throttle valve control module is shorted or open • ECM has failed
DTC: P1173 **1999, 2000, 2001, 2002, 2003, 2004, 2005, 2006** **Models:** TT, A6, A8 **Engines:** 1.8L, 4.2L **Transmissions:** All	**Throttle Actuation Potentiometer Sign.2 Signal Too High Conditions:** Engine started, battery voltage must be at least 11.5v, all electrical components must be off, the ground between the engine and the chassis must be well connected, coolant temperature must be at least 80 degrees Celsius and the accelerator pedal must be properly adjusted. The ECM detected an incorrect singal from the throttle potentiometer. **Note: If the complete throttle valve control module is current-less (e.g. connector disconnected) the throttle valve moves into a particular, specified mechanical position, which signals an increased idle speed with an engine at operating temperature. If only the Throttle Position (TP) actuator is current-less, the throttle valve also moves into the specified mechanical position (emergency running gap), however, since Closed Throttle Position (CTP) switch can still be recognized, an "almost normal idle RPM" is reached via the respective ignition angle retardation.** **Note: Terminal assignment at throttle control module is different in vehicles with and without cruise control.** **Characteristic: Steering column switch with operating module for cruise control.** **Possible Causes:** • Throttle valve control module has failed • Throttle valve is dirty or damaged • Throttle valve is not in a closed position • Voltage supply of throttle valve control module is shorted or open • ECM has failed

DTC	Trouble Code Title, Conditions & Possible Causes
DTC: P1176 **2T CCM, MIL: Yes** **1996, 1997, 1998, 1999, 2000, 2001, 2002, 2003, 2004, 2005, 2006** **Models:** TT, A4, A6, Allroad Quattro, A8 **Engines:** 1.8L, 2.7L, 2.8L, 3.7L, 4.2L **Transmissions:** All	**O2 Correction Behind Catalyst B1 Limit Attained Conditions:** Engine started, battery voltage must be at least 11.5v, all electrical components must be off, the ground between the engine and the chassis must be well connected, the exhaust system must be properly sealed between the catalytic converter and the cylinder head, the coolant temperature must be at least 80 degrees Celsius, and the oxygen sensor heater for oxygen sensor before the catalytic converter must be properly functioning. The ECM has detected a malfunction of the oxygen sensor. **Note: Vacuum in the intake system sucks in the leak detection spray with false air. Leak detection spray decreases ignition quality of the fuel mixture. This causes a drop in engine speed and changes the value produced by the Heated Oxygen Sensor (HO2S).** **Note: Vehicle must be raised before connector for oxygen sensor is accessible.** **Note: The oxygen sensor before catalytic converter has a static regulation and can be differentiated from the oxygen sensor behind catalytic converter via a 6-pin connector.** **Possible Causes:** • O2 sensor circuit has shorted to ground or B+ • O2 sensor circuit is open • ECM has failed • O2 sensor has failed • Intake Manifold Runner Position Sensor is faulty • Intake system for leaks (false air) is faulty • Motor for intake flap is faulty
DTC: P1177 **2T CCM, MIL: Yes** **1996, 1997, 1998, 1999, 2000, 2001, 2002, 2003, 2004, 2005, 2006** **Models:** TT, A4, A6, Allroad Quattro, A8 **Engines:** 1.8L, 2.7L, 2.8L, 3.7L, 4.2L **Transmissions:** All	**O2 Correction Behind Catalyst B2 Limit Attained Conditions:** Engine started, battery voltage must be at least 11.5v, all electrical components must be off, the ground between the engine and the chassis must be well connected, the exhaust system must be properly sealed between the catalytic converter and the cylinder head, the coolant temperature must be at least 80 degrees Celsius, and the oxygen sensor heater for oxygen sensor before the catalytic converter must be properly functioning. The ECM has detected a malfunction of the oxygen sensor. **Note: Vacuum in the intake system sucks in the leak detection spray with false air. Leak detection spray decreases ignition quality of the fuel mixture. This causes a drop in engine speed and changes the value produced by the Heated Oxygen Sensor (HO2S).** **Note: Vehicle must be raised before connector for oxygen sensor is accessible.** **Note: The oxygen sensor before catalytic converter has a static regulation and can be differentiated from the oxygen sensor behind catalytic converter via a 6-pin connector.** **Possible Causes:** • O2 sensor circuit has shorted to ground or B+ • O2 sensor circuit is open • ECM has failed • O2 sensor has failed • Intake Manifold Runner Position Sensor is faulty • Intake system for leaks (false air) is faulty • Motor for intake flap is faulty
DTC: P1196 **2T CCM, MIL: Yes** **1997, 1998, 1999, 2000, 2001, 2002, 2003, 2004, 2005, 2006** **Models:** TT, A4, A6, A8 **Engines:** 1.8L, 2.8L, 4.2L **Transmissions:** All	**O2 Sensor Heater Circuit (Bank 1-Sensor 1) Electrical Malfunction Conditions:** Engine started, battery voltage must be at least 11.5v, all electrical components must be off, the ground between the engine and the chassis must be well connected, the exhaust system must be properly sealed between the catalytic converter and the cylinder head, and the oxygen sensor heater for oxygen sensor before the catalytic converter must be properly functioning. **Note: For resistance testing of sensor heating, oxygen sensor should be cooled to ambient temperature. High temperatures at oxygen sensor may lead to inaccurate measurements. The ECM detected an open or shorted condition, or excessive current draw in the heater circuit.** **Possible Causes:** • HO2S heater power circuit is open • HO2S heater ground circuit is open • HO2S signal tracking (due to oil or moisture in the connector) • HO2S is damaged or has failed • ECM has failed • Oxygen sensor (before catalytic converter) is faulty • Oxygen sensor (behind catalytic converter) is faulty • Oxygen sensor heater (before catalytic converter) is faulty • Oxygen sensor heater (behind catalytic converter) is faulty

DTC	Trouble Code Title, Conditions & Possible Causes
DTC: P1197 **2T CCM, MIL: Yes** **1996, 1997, 1998, 1999, 2000, 2001** **Models:** A4, A8 **Engines:** 2.8L, 3.7L, 4.2L **Transmissions:** All	**O2 Sensor Heater Circuit (Bank 2-Sensor 1) Electrical Malfunction Conditions:** Engine started, battery voltage must be at least 11.5v, all electrical components must be off, the ground between the engine and the chassis must be well connected, the exhaust system must be properly sealed between the catalytic converter and the cylinder head, and the oxygen sensor heater for oxygen sensor before the catalytic converter must be properly functioning. **Note: For resistance testing of sensor heating, oxygen sensor should be cooled to ambient temperature. High temperatures at oxygen sensor may lead to inaccurate measurements. The ECM detected an open or shorted condition, or excessive current draw in the heater circuit.** **Possible Causes:** • HO2S heater power circuit is open • HO2S heater ground circuit is open • HO2S signal tracking (due to oil or moisture in the connector) • HO2S is damaged or has failed • ECM has failed • Oxygen sensor (before catalytic converter) is faulty • Oxygen sensor (behind catalytic converter) is faulty • Oxygen sensor heater (before catalytic converter) is faulty • Oxygen sensor heater (behind catalytic converter) is faulty
DTC: P1198 **2T CCM, MIL: Yes** **1997, 1998, 1999, 2000, 2001, 2002, 2003, 2004, 2005, 2006** **Models:** TT, A4, A6, A8 **Engines:** 1.8L, 4.2L **Transmissions:** All	**O2 Sensor Heater Circuit (Bank 1-Sensor 2) Electrical Malfunction Conditions:** Engine started, battery voltage must be at least 11.5v, all electrical components must be off, the ground between the engine and the chassis must be well connected, the exhaust system must be properly sealed between the catalytic converter and the cylinder head, and the oxygen sensor heater for oxygen sensor before the catalytic converter must be properly functioning. **Note: For resistance testing of sensor heating, oxygen sensor should be cooled to ambient temperature. High temperatures at oxygen sensor may lead to inaccurate measurements. The ECM detected an open or shorted condition, or excessive current draw in the heater circuit.** **Possible Causes:** • HO2S heater power circuit is open • HO2S heater ground circuit is open • HO2S signal tracking (due to oil or moisture in the connector) • HO2S is damaged or has failed • ECM has failed • Oxygen sensor (before catalytic converter) is faulty • Oxygen sensor (behind catalytic converter) is faulty • Oxygen sensor heater (before catalytic converter) is faulty • Oxygen sensor heater (behind catalytic converter) is faulty
DTC: P1199 **2T CCM, MIL: Yes** **1996, 1997, 1998, 1999, 2000, 2001, 2002, 2003** **Models:** A4, A6, A8 **Engines:** 2.8L, 3.7L, 4.2L **Transmissions:** All	**O2 Sensor Heater Circuit (Bank 2-Sensor 2) Electrical Malfunction Conditions:** Engine started, battery voltage must be at least 11.5v, all electrical components must be off, the ground between the engine and the chassis must be well connected, the exhaust system must be properly sealed between the catalytic converter and the cylinder head, and the oxygen sensor heater for oxygen sensor before the catalytic converter must be properly functioning. **Note: For resistance testing of sensor heating, oxygen sensor should be cooled to ambient temperature. High temperatures at oxygen sensor may lead to inaccurate measurements. The ECM detected an open or shorted condition, or excessive current draw in the heater circuit.** **Possible Causes:** • HO2S heater power circuit is open • HO2S heater ground circuit is open • HO2S signal tracking (due to oil or moisture in the connector) • HO2S is damaged or has failed • ECM has failed • Oxygen sensor (before catalytic converter) is faulty • Oxygen sensor (behind catalytic converter) is faulty • Oxygen sensor heater (before catalytic converter) is faulty • Oxygen sensor heater (behind catalytic converter) is faulty
DTC: P1201 **1T CCM, MIL: Yes** **1997, 1998, 1999, 2000, 2001, 2002, 2003, 2004, 2005, 2006** **Models:** TT, A4, A6, A6 Avant, A8 **Engines:** 1.8L, 2.8L,3.7L, 4.2L **Transmissions:** All	**Cylinder 1 Fuel Injection Circuit Electrical Malfunction Conditions:** Key on or engine running, fuses in the instrument panel and the E-box in the engine compartment must be functioning, and the ground connections between the engine ad the chassis must be well connected; and the ECM detected an unexpected voltage condition on the injector circuit **Possible Causes:** • Injector control circuit is open • Injector power circuit (B+) is open • Injector control circuit is shorted to chassis ground • Injector is damaged or has failed • ECM is not connected or has failed • Fuel pump relay has failed • Fuel injectors may have malfunctioned • Faulty engine speed sensor

DTC	Trouble Code Title, Conditions & Possible Causes
DTC: P1202 **1T CCM, MIL: Yes** **1997, 1998, 1999, 2000, 2001, 2002, 2003, 2004, 2005, 2006** **Models:** TT, A4, A6, A6 Avant, A8 **Engines:** 1.8L, 2.8L,3.7L, 4.2L **Transmissions:** All	**Cylinder 2 Fuel Injection Circuit Electrical Malfunction Conditions:** Key on or engine running, fuses in the instrument panel and the E-box in the engine compartment must be functioning, and the ground connections between the engine ad the chassis must be well connected; and the ECM detected an unexpected voltage condition on the injector circuit **Possible Causes:** • Injector control circuit is open • Injector power circuit (B+) is open • Injector control circuit is shorted to chassis ground • Injector is damaged or has failed • ECM is not connected or has failed • Fuel pump relay has failed • Fuel injectors may have malfunctioned • Faulty engine speed sensor
DTC: P1203 **1T CCM, MIL: Yes** **1997, 1998, 1999, 2000, 2001, 2002, 2003, 2004, 2005, 2006** **Models:** TT, A4, A6, A6 Avant, A8 **Engines:** 1.8L, 2.8L,3.7L, 4.2L **Transmissions:** All	**Cylinder 3 Fuel Injection Circuit Electrical Malfunction Conditions:** Key on or engine running, fuses in the instrument panel and the E-box in the engine compartment must be functioning, and the ground connections between the engine ad the chassis must be well connected; and the ECM detected an unexpected voltage condition on the injector circuit **Possible Causes:** • Injector control circuit is open • Injector power circuit (B+) is open • Injector control circuit is shorted to chassis ground • Injector is damaged or has failed • ECM is not connected or has failed • Fuel pump relay has failed • Fuel injectors may have malfunctioned • Faulty engine speed sensor
DTC: P1204 **1997, 1998, 1999, 2000, 2001, 2002, 2003, 2004, 2005, 2006** **Models:** TT, A4, A6, A6 Avant, A8 **Engines:** 1.8L, 2.8L,3.7L, 4.2L **Transmissions:** All	**Cylinder 4 Fuel Injection Circuit Electrical Malfunction Conditions:** Key on or engine running, fuses in the instrument panel and the E-box in the engine compartment must be functioning, and the ground connections between the engine ad the chassis must be well connected; and the ECM detected an unexpected voltage condition on the injector circuit **Possible Causes:** • Injector control circuit is open • Injector power circuit (B+) is open • Injector control circuit is shorted to chassis ground • Injector is damaged or has failed • ECM is not connected or has failed • Fuel pump relay has failed • Fuel injectors may have malfunctioned • Faulty engine speed sensor
DTC: P1213 **1T CCM, MIL: Yes** **1997, 1998, 1999, 2000, 2001, 2002, 2003, 2004, 2005, 2006** **Models:** TT, A4, A6, A6 Avant, A8 **Engines:** 1.8L, 2.8L,3.7L, 4.2L **Transmissions:** All	**Cylinder 1 Fuel Injection Circuit Short to B+ Conditions:** Key on or engine running, fuses in the instrument panel and the E-box in the engine compartment must be functioning, and the ground connections between the engine ad the chassis must be well connected; and the ECM detected an unexpected voltage condition on the injector circuit. Wiring or fuel injector has a short circuit to positive supply. **Possible Causes:** • Injector control circuit is open • Injector power circuit (B+) is open • Injector control circuit is shorted to chassis ground • Injector is damaged or has failed • ECM is not connected or has failed • Fuel pump relay has failed • Engine speed sensor has failed
DTC: P1214 **1997, 1998, 1999, 2000, 2001, 2002, 2003, 2004, 2005, 2006** **Models:** TT, A4, A6, A6 Avant, A8 **Engines:** 1.8L, 2.8L,3.7L, 4.2L **Transmissions:** All	**Cylinder 2 Fuel Injection Circuit Short to B+ Conditions:** Key on or engine running, fuses in the instrument panel and the E-box in the engine compartment must be functioning, and the ground connections between the engine ad the chassis must be well connected; and the ECM detected an unexpected voltage condition on the injector circuit. Wiring or fuel injector has a short circuit to positive supply. **Possible Causes:** • Injector control circuit is open • Injector power circuit (B+) is open • Injector control circuit is shorted to chassis ground • Injector is damaged or has failed • ECM is not connected or has failed • Fuel pump relay has failed • Engine speed sensor has failed

DTC	Trouble Code Title, Conditions & Possible Causes
DTC: P1215 **1T CCM, MIL: Yes** **1997, 1998, 1999, 2000, 2001, 2002, 2003, 2004, 2005, 2006** **Models:** TT, A4, A6, A6 Avant, A8 **Engines:** 1.8L, 2.8L,3.7L, 4.2L **Transmissions:** All	**Cylinder 3 Fuel Injection Circuit Short to B+ Conditions:** Key on or engine running, fuses in the instrument panel and the E-box in the engine compartment must be functioning, and the ground connections between the engine ad the chassis must be well connected; and the ECM detected an unexpected voltage condition on the injector circuit. Wiring or fuel injector has a short circuit to positive supply. **Possible Causes:** • Injector control circuit is open • Injector power circuit (B+) is open • Injector control circuit is shorted to chassis ground • Injector is damaged or has failed • ECM is not connected or has failed • Fuel pump relay has failed • Engine speed sensor has failed
DTC: P1216 **1T CCM, MIL: Yes** **1997, 1998, 1999, 2000, 2001, 2002, 2003, 2004, 2005, 2006** **Models:** TT, A4, A6, A6 Avant, A8 **Engines:** 1.8L, 2.8L,3.7L, 4.2L **Transmissions:** All	**Cylinder 4 Fuel Injection Circuit Short to B+ Conditions:** Key on or engine running, fuses in the instrument panel and the E-box in the engine compartment must be functioning, and the ground connections between the engine ad the chassis must be well connected; and the ECM detected an unexpected voltage condition on the injector circuit. Wiring or fuel injector has a short circuit to positive supply. **Possible Causes:** • Injector control circuit is open • Injector power circuit (B+) is open • Injector control circuit is shorted to chassis ground • Injector is damaged or has failed • ECM is not connected or has failed • Fuel pump relay has failed • Engine speed sensor has failed
DTC: P1217 **1T CCM, MIL: Yes** **1996, 1997, 1998, 1999, 2000, 2001, 2002, 2003** **Models:** A4, A6, A6 Avant, A8 **Engines:** 2.8L, 3.7L, 4.2L **Transmissions:** All	**Cylinder 5 Fuel Injection Circuit Short to B+ Conditions:** Key on or engine running, fuses in the instrument panel and the E-box in the engine compartment must be functioning, and the ground connections between the engine ad the chassis must be well connected; and the ECM detected an unexpected voltage condition on the injector circuit. Wiring or fuel injector has a short circuit to positive supply. **Possible Causes:** • Injector control circuit is open • Injector power circuit (B+) is open • Injector control circuit is shorted to chassis ground • Injector is damaged or has failed • ECM is not connected or has failed • Fuel pump relay has failed • Engine speed sensor has failed
DTC: P1218 **1T CCM, MIL: Yes** **1996, 1997, 1998, 1999, 2000, 2001, 2002, 2003** **Models:** A4, A6, A6 Avant, A8 **Engines:** 2.8L, 3.7L, 4.2L **Transmissions:** All	**Cylinder 6 Fuel Injection Circuit Short to B+ Conditions:** Key on or engine running, fuses in the instrument panel and the E-box in the engine compartment must be functioning, and the ground connections between the engine ad the chassis must be well connected; and the ECM detected an unexpected voltage condition on the injector circuit. Wiring or fuel injector has a short circuit to positive supply. **Possible Causes:** • Injector control circuit is open • Injector power circuit (B+) is open • Injector control circuit is shorted to chassis ground • Injector is damaged or has failed • ECM is not connected or has failed • Fuel pump relay has failed • Engine speed sensor has failed
DTC: P1225 **1T CCM, MIL: Yes** **1997, 1998, 1999, 2000, 2001, 2002, 2003, 2004, 2005, 2006** **Models:** TT, A4, A6, A8 **Engines:** 1.8L, 4.2L **Transmissions:** All	**Cylinder 1 Fuel Injection Circuit Short to Ground Conditions:** Key on or engine running, fuses in the instrument panel and the E-box in the engine compartment must be functioning, and the ground connections between the engine ad the chassis must be well connected; and the ECM detected an unexpected voltage condition on the injector circuit. Wiring or fuel injector has a short circuit to ground. **Possible Causes:** • Injector control circuit is open • Injector power circuit (B+) is open • Injector control circuit is shorted to chassis ground • Injector is damaged or has failed • ECM is not connected or has failed • Fuel pump relay has failed • Engine speed sensor has failed

DTC	Trouble Code Title, Conditions & Possible Causes
DTC: P1226 **1T CCM, MIL: Yes** **1997, 1998, 1999, 2000, 2001, 2002, 2003, 2004, 2005, 2006** **Models:** TT, A4, A6, A8 **Engines:** 1.8L, 4.2L **Transmissions:** All	**Cylinder 2 Fuel Injection Circuit Short to Ground Conditions:** Key on or engine running, fuses in the instrument panel and the E-box in the engine compartment must be functioning, and the ground connections between the engine ad the chassis must be well connected; and the ECM detected an unexpected voltage condition on the injector circuit. Wiring or fuel injector has a short circuit to ground. **Possible Causes:** • Injector control circuit is open • Injector power circuit (B+) is open • Injector control circuit is shorted to chassis ground • Injector is damaged or has failed • ECM is not connected or has failed • Fuel pump relay has failed • Engine speed sensor has failed
DTC: P1227 **1T CCM, MIL: Yes** **1997, 1998, 1999, 2000, 2001, 2002, 2003, 2004, 2005, 2006** **Models:** TT, A4, A6, A8 **Engines:** 1.8L, 4.2L **Transmissions:** All	**Cylinder 3 Fuel Injection Circuit Short to Ground Conditions:** Key on or engine running, fuses in the instrument panel and the E-box in the engine compartment must be functioning, and the ground connections between the engine ad the chassis must be well connected; and the ECM detected an unexpected voltage condition on the injector circuit. Wiring or fuel injector has a short circuit to ground. **Possible Causes:** • Injector control circuit is open • Injector power circuit (B+) is open • Injector control circuit is shorted to chassis ground • Injector is damaged or has failed • ECM is not connected or has failed • Fuel pump relay has failed • Engine speed sensor has failed
DTC: P1228 **1T CCM, MIL: Yes** **1997, 1998, 1999, 2000, 2001, 2002, 2003, 2004, 2005, 2006** **Models:** TT, A4, A6, A8 **Engines:** 1.8L, 4.2L **Transmissions:** All	**Cylinder 4 Fuel Injection Circuit Short to Ground Conditions:** Key on or engine running, fuses in the instrument panel and the E-box in the engine compartment must be functioning, and the ground connections between the engine ad the chassis must be well connected; and the ECM detected an unexpected voltage condition on the injector circuit. Wiring or fuel injector has a short circuit to ground. **Possible Causes:** • Injector control circuit is open • Injector power circuit (B+) is open • Injector control circuit is shorted to chassis ground • Injector is damaged or has failed • ECM is not connected or has failed • Fuel pump relay has failed • Engine speed sensor has failed
DTC: P1229 **1T CCM, MIL: Yes** **1996, 1997, 1998, 1999, 2000, 2001, 2002, 2003** **Models:** A4, A6, A8 **Engines:** 2.8L, 4.2L **Transmissions:** All	**Cylinder 5 Fuel Injection Circuit Short to Ground Conditions:** Key on or engine running, fuses in the instrument panel and the E-box in the engine compartment must be functioning, and the ground connections between the engine ad the chassis must be well connected; and the ECM detected an unexpected voltage condition on the injector circuit. Wiring or fuel injector has a short circuit to ground. **Possible Causes:** • Injector control circuit is open • Injector power circuit (B+) is open • Injector control circuit is shorted to chassis ground • Injector is damaged or has failed • ECM is not connected or has failed • Fuel pump relay has failed • Engine speed sensor has failed
DTC: P1230 **1T CCM, MIL: Yes** **1996, 1997, 1998, 1999, 2000, 2001, 2002, 2003** **Models:** A4, A6, A8 **Engines:** 2.8L, 4.2L **Transmissions:** All	**Cylinder 6 Fuel Injection Circuit Short to Ground Conditions:** Key on or engine running, fuses in the instrument panel and the E-box in the engine compartment must be functioning, and the ground connections between the engine ad the chassis must be well connected; and the ECM detected an unexpected voltage condition on the injector circuit. Wiring or fuel injector has a short circuit to ground. **Possible Causes:** • Injector control circuit is open • Injector power circuit (B+) is open • Injector control circuit is shorted to chassis ground • Injector is damaged or has failed • ECM is not connected or has failed • Fuel pump relay has failed • Engine speed sensor has failed

DTC	Trouble Code Title, Conditions & Possible Causes
DTC: P1237 **1T CCM, MIL: Yes** **1997, 1998, 1999, 2000, 2001, 2002, 2003, 2004, 2005, 2006** **Models:** TT, A4, A6, A8 **Engines:** 1.8L, 4.2L **Transmissions:** All	**Cylinder 1 Fuel Injection Circuit Open Circuit Conditions:** Key on or engine running, fuses in the instrument panel and the E-box in the engine compartment must be functioning, and the ground connections between the engine ad the chassis must be well connected; and the ECM detected an unexpected voltage condition on the injector circuit. Wiring or fuel injector has a short circuit that is open. **Possible Causes:** • Injector control circuit is open • Injector power circuit (B+) is open • Injector control circuit is shorted to chassis ground • Injector is damaged or has failed • ECM is not connected or has failed • Fuel pump relay has failed • Engine speed sensor has failed
DTC: P1238 **1T CCM, MIL: Yes** **1997, 1998, 1999, 2000, 2001, 2002, 2003, 2004, 2005, 2006** **Models:** TT, A4, A6, A8 **Engines:** 1.8L, 4.2L **Transmissions:** All	**Cylinder 2 Fuel Injection Circuit Open Circuit Conditions:** Key on or engine running, fuses in the instrument panel and the E-box in the engine compartment must be functioning, and the ground connections between the engine ad the chassis must be well connected; and the ECM detected an unexpected voltage condition on the injector circuit. Wiring or fuel injector has a short circuit that is open. **Possible Causes:** • Injector control circuit is open • Injector power circuit (B+) is open • Injector control circuit is shorted to chassis ground • Injector is damaged or has failed • ECM is not connected or has failed • Fuel pump relay has failed • Engine speed sensor has failed
DTC: P1239 **1T CCM, MIL: Yes** **1997, 1998, 1999, 2000, 2001, 2002, 2003, 2004, 2005, 2006** **Models:** TT, A4, A6, A8 **Engines:** 1.8L, 4.2L **Transmissions:** All	**Cylinder 3 Fuel Injection Circuit Open Circuit Conditions:** Key on or engine running, fuses in the instrument panel and the E-box in the engine compartment must be functioning, and the ground connections between the engine ad the chassis must be well connected; and the ECM detected an unexpected voltage condition on the injector circuit. Wiring or fuel injector has a short circuit that is open. **Possible Causes:** • Injector control circuit is open • Injector power circuit (B+) is open • Injector control circuit is shorted to chassis ground • Injector is damaged or has failed • ECM is not connected or has failed • Fuel pump relay has failed • Engine speed sensor has failed
DTC: P1240 **1T CCM, MIL: Yes** **1997, 1998, 1999, 2000, 2001, 2002, 2003, 2004, 2005, 2006** **Models:** TT, A4, A6, A8 **Engines:** 1.8L, 4.2L **Transmissions:** All	**Cylinder 4 Fuel Injection Circuit Open Circuit Conditions:** Key on or engine running, fuses in the instrument panel and the E-box in the engine compartment must be functioning, and the ground connections between the engine ad the chassis must be well connected; and the ECM detected an unexpected voltage condition on the injector circuit. Wiring or fuel injector has a short circuit that is open. **Possible Causes:** • Injector control circuit is open • Injector power circuit (B+) is open • Injector control circuit is shorted to chassis ground • Injector is damaged or has failed • ECM is not connected or has failed • Fuel pump relay has failed • Engine speed sensor has failed
DTC: P1241 **1T CCM, MIL: Yes** **1996, 1997, 1998, 1999, 2000, 2001, 2002, 2003** **Models:** A4, A6, A8 **Engines:** 2.8L, 4.2L **Transmissions:** All	**Cylinder 5 Fuel Injection Circuit Open Circuit Conditions:** Key on or engine running, fuses in the instrument panel and the E-box in the engine compartment must be functioning, and the ground connections between the engine ad the chassis must be well connected; and the ECM detected an unexpected voltage condition on the injector circuit. Wiring or fuel injector has a short circuit that is open. **Possible Causes:** • Injector control circuit is open • Injector power circuit (B+) is open • Injector control circuit is shorted to chassis ground • Injector is damaged or has failed • ECM is not connected or has failed • Fuel pump relay has failed • Engine speed sensor has failed

DTC	Trouble Code Title, Conditions & Possible Causes
DTC: P1242 **1T CCM, MIL: Yes** **1996, 1997, 1998, 1999, 2000, 2001, 2002, 2003** **Models:** A4, A6, A8 **Engines:** 2.8L, 4.2L **Transmissions:** All	**Cylinder 6 Fuel Injection Circuit Open Circuit Conditions:** Key on or engine running, fuses in the instrument panel and the E-box in the engine compartment must be functioning, and the ground connections between the engine ad the chassis must be well connected; and the ECM detected an unexpected voltage condition on the injector circuit. Wiring or fuel injector has a short circuit that is open. **Possible Causes:** • Injector control circuit is open • Injector power circuit (B+) is open • Injector control circuit is shorted to chassis ground • Injector is damaged or has failed • ECM is not connected or has failed • Fuel pump relay has failed • Engine speed sensor has failed
DTC: P1250 **1T CCM, MIL: Yes** **1997, 1998, 1999, 2000, 2001, 2002, 2003, 2004, 2005** **Models:** A4, A6, A6 Avant, Allroad Quattro, A8 **Engines:** 1.8L, 2.7L, 2.8L, 3.7L, 4.2L **Transmissions:** All	**Fuel Pressure Regulator Control Circuit Malfunction (Fuel Level too Low) Conditions:** KOEO or KOER Self-Test enabled, and the ECM detected a lack of power (VPWR) to the Fuel Pressure Regulator Control (FPRC) solenoid circuit. **Possible Causes:** • FPRC solenoid valve harness circuits are open or shorted • FPRC input port or output port vacuum lines are damaged • FRPC solenoid is damaged • Fuel level is too low • ECM has failed
DTC: P1287 **2T CCM, MIL: Yes** **1997, 1998, 1999, 2000, 2001** **Models:** A4 **Engines:** 1.8L **Transmissions:** All	**Turbocharger Bypass Valve Open Conditions:** Engine started, battery voltage at least 11.5v, all electrical components off, ground connections between engine and chassis well connected, coolant temperature at least 80-degrees Celicius. The ECM detected an unexpected voltage condition on the bypass valve control circuit **Possible Causes:** • Charge air system check for leaks • Recirculating valve for turbocharger is faulty • Turbocharging system may be damaged • Vacuum diaphragm for turbocharger needs adjusting • Wastegate bypass regulator valve is faulty • Bypass solenoid power supply circuit is open • Bypass solenoid control circuit is open, shorted to ground or system power • Bypass solenoid assembly is damaged or has failed • Charge air pressure sensor is faulty • Voltage supply to the charge air pressure sensor is open or shorted • Check the charge air system for leaks • Recirculating valve for turbocharger is faulty • Turbocharging system is damaged or not functioning properly • Turbocharger recirculating valve is faulty • Vacuum diaphragm for turbocharger is out of adjustment • Wastegate bypass regulator valve is faulty • ECM has failed

DTC	Trouble Code Title, Conditions & Possible Causes
DTC: P1288 **2T CCM, MIL: Yes** **1997, 1998, 1999, 2000, 2001** **Models:** A4 **Engines:** 1.8L **Transmissions:** All	**Turbocharger Bypass Valve Short to B+ Conditions:** Engine started, battery voltage at least 11.5v, all electrical components off, ground connections between engine and chassis well connected, coolant temperature at least 80-degrees Celicius. The ECM detected an unexpected voltage condition on the bypass valve control circuit **Possible Causes:** • Charge air system check for leaks • Recirculating valve for turbocharger is faulty • Turbocharging system may be damaged • Vacuum diaphragm for turbocharger needs adjusting • Wastegate bypass regulator valve is faulty • Bypass solenoid power supply circuit is open • Bypass solenoid control circuit is open, shorted to ground or system power • Bypass solenoid assembly is damaged or has failed • Charge air pressure sensor is faulty • Voltage supply to the charge air pressure sensor is open or shorted • Check the charge air system for leaks • Recirculating valve for turbocharger is faulty • Turbocharging system is damaged or not functioning properly • Turbocharger recirculating valve is faulty • Vacuum diaphragm for turbocharger is out of adjustment • Wastegate bypass regulator valve is faulty • ECM has failedvalve is faulty
DTC: P1289 **2T CCM, MIL: Yes** **1997, 1998, 1999, 2000, 2001** **Models:** A4 **Engines:** 1.8L **Transmissions:** All	**Turbocharger Bypass Valve Short to Ground Conditions:** Engine started, battery voltage at least 11.5v, all electrical components off, ground connections between engine and chassis well connected, coolant temperature at least 80-degrees Celicius. The ECM detected an unexpected voltage condition on the bypass valve control circuit **Possible Causes:** • Charge air system check for leaks • Recirculating valve for turbocharger is faulty • Turbocharging system may be damaged • Vacuum diaphragm for turbocharger needs adjusting • Wastegate bypass regulator valve is faulty • Bypass solenoid power supply circuit is open • Bypass solenoid control circuit is open, shorted to ground or system power • Bypass solenoid assembly is damaged or has failed • Charge air pressure sensor is faulty • Voltage supply to the charge air pressure sensor is open or shorted • Check the charge air system for leaks • Recirculating valve for turbocharger is faulty • Turbocharging system is damaged or not functioning properly • Turbocharger recirculating valve is faulty • Vacuum diaphragm for turbocharger is out of adjustment • Wastegate bypass regulator valve is faulty • ECM has failed

DTC	Trouble Code Title, Conditions & Possible Causes
DTC: P1295 **2T CCM, MIL: Yes** **1997, 1998, 1999, 2000, 2001** **Models:** A4 **Engines:** 1.8L **Transmissions:** All	**Turbocharger Bypass Valve Throughput Faulty Conditions:** Engine started, battery voltage at least 11.5v, all electrical components off, ground connections between engine and chassis well connected, coolant temperature at least 80-degrees Celicius. The ECM detected an unexpected voltage condition on the bypass valve control circuit **Possible Causes:** • Charge air system check for leaks • Recirculating valve for turbocharger is faulty • Turbocharging system may be damaged • Vacuum diaphragm for turbocharger needs adjusting • Wastegate bypass regulator valve is faulty • Bypass solenoid power supply circuit is open • Bypass solenoid control circuit is open, shorted to ground or system power • Bypass solenoid assembly is damaged or has failed • Charge air pressure sensor is faulty • Voltage supply to the charge air pressure sensor is open or shorted • Check the charge air system for leaks • Recirculating valve for turbocharger is faulty • Turbocharging system is damaged or not functioning properly • Turbocharger recirculating valve is faulty • Vacuum diaphragm for turbocharger is out of adjustment • Wastegate bypass regulator valve is faulty • ECM has failed
DTC: P1296 **1T CCM, MIL: Yes** **1996, 1997, 1998, 1999, 2000, 2001, 2002, 2003, 2004, 2005, 2006** **Models:** TT, A4, A6, Allroad Quattro, A8 **Engines:** 1.8L, 2.7L, 2.8L, 3.7L, 4.2L **Transmissions:** All	**Cooling System Malfunction Conditions:** Key on, engine not running, the Engine Control Module (ECM) will use the intake air temperature as a replacement value for an engine start (start temperature replacement value) as soon as there is a Diagnostic Trouble Code (DTC) stored in DTC memory for the Engine Coolant Temperature (ECT) sensor. The temperature then rises according to a program stored in the ECM. When the engine has reached normal operating temperature a fixed replacement value will be displayed. This fixed value is also dependent upon the intake air temperature. **Possible Causes:** • Engine coolant temperature sensor has failed • An open circuit or a short to B+ is present • Sensor circuit is short to ground • ECM has failed
DTC: P1297 **2T CCM, MIL: Yes** **1997, 1998, 1999, 2000, 2001** **Models:** A4 **Engines:** 1.8L **Transmissions:** All	**Connection Turbocharger/Throttle Valve Pressure Hose Conditions:** Engine started, battery voltage at least 11.5v, all electrical components off, ground connections between engine and chassis well connected, coolant temperature at least 80-degrees Celicius. The ECM detected an unexpected voltage condition on the turbo valve pressure hose. **Possible Causes:** • Charge air system check for leaks • Recirculating valve for turbocharger is faulty • Turbocharging system may be damaged • Vacuum diaphragm for turbocharger needs adjusting • Wastegate bypass regulator valve is faulty • Bypass solenoid power supply circuit is open • Bypass solenoid control circuit is open, shorted to ground or system power • Bypass solenoid assembly is damaged or has failed • Charge air pressure sensor is faulty • Voltage supply to the charge air pressure sensor is open or shorted • Check the charge air system for leaks • Recirculating valve for turbocharger is faulty • Turbocharging system is damaged or not functioning properly • Turbocharger recirculating valve is faulty • Vacuum diaphragm for turbocharger is out of adjustment • Wastegate bypass regulator valve is faulty • ECM has failed

DTC	Trouble Code Title, Conditions & Possible Causes
DTC: P1325 **2T CCM, MIL: Yes** **1996, 1997, 1998, 1999, 2000, 2001, 2002, 2003, 2004, 2005, 2006** **Models:** TT, A4, A6, Allroad Quattro, A8 **Engines:** 1.8L, 2.7L, 2.8L, 3.7L, 4.2L **Transmissions:** All	**Cylinder 1-Knock Control Limit Attained Conditions:** Engine started, battery voltage at least 11.5v, all electrical components off, ground connections between engine and chassis well connected, and the ECM detected the Knock Sensor signal was more than the calibrated value. **Possible Causes:** • Knock sensor circuit is open • Knock sensor circuit is shorted to ground, or shorted to power • Knock sensor is damaged or it has failed • Poor fuel quality • Loosen knock sensors and tighten again to 20 Nm • ECM has failed
DTC: P1326 **2T CCM, MIL: Yes** **1996, 1997, 1998, 1999, 2000, 2001, 2002, 2003, 2004, 2005, 2006** **Models:** TT, A4, A6, Allroad Quattro, A8 **Engines:** 1.8L, 2.7L, 2.8L, 3.7L, 4.2L **Transmissions:** All	**Cylinder 2-Knock Control Limit Attained Conditions:** Engine started, battery voltage at least 11.5v, all electrical components off, ground connections between engine and chassis well connected, and the ECM detected the Knock Sensor signal was more than the calibrated value. **Possible Causes:** • Knock sensor circuit is open • Knock sensor circuit is shorted to ground, or shorted to power • Knock sensor is damaged or it has failed • Poor fuel quality • Loosen knock sensors and tighten again to 20 Nm • ECM has failed
DTC: P1327 **2T CCM, MIL: Yes** **1996, 1997, 1998, 1999, 2000, 2001, 2002, 2003, 2004, 2005, 2006** **Models:** TT, A4, A6, Allroad Quattro, A8 **Engines:** 1.8L, 2.7L, 2.8L, 3.7L, 4.2L **Transmissions:** All	**Cylinder 3-Knock Control Limit Attained Conditions:** Engine started, battery voltage at least 11.5v, all electrical components off, ground connections between engine and chassis well connected, and the ECM detected the Knock Sensor signal was more than the calibrated value. **Possible Causes:** • Knock sensor circuit is open • Knock sensor circuit is shorted to ground, or shorted to power • Knock sensor is damaged or it has failed • Poor fuel quality • Loosen knock sensors and tighten again to 20 Nm • ECM has failed
DTC: P1328 **2T CCM, MIL: Yes** **1995, 1996, 1997, 1998, 1999, 2000, 2001, 2002, 2003, 2004** **Models:** A4, A6, A6 Avant, A8 **Engines:** 1.8L, 2.8L, 4.2L **Transmissions:** All A4 (all): 1.8L (AEB, ATW, AWM), 2.8L (AFC, AHA, ATQ); A6 (all): 2.8L (AFC), 4.2L (ART, AWN); A6 Avant: 2.8L (AAH); A8: 4.2L (AKB, AUX, AYS) **Transmissions:** All	**Cylinder 4-Knock Control Limit Attained Conditions:** Engine started, battery voltage at least 11.5v, all electrical components off, ground connections between engine and chassis well connected, and the ECM detected the Knock Sensor signal was more than the calibrated value. **Possible Causes:** • Knock sensor circuit is open • Knock sensor circuit is shorted to ground, or shorted to power • Knock sensor is damaged or it has failed • ECM has failed
DTC: P1329 **2T CCM, MIL: Yes** **1998, 1999, 2000, 2001, 2002, 2003, 2004** **Models:** A4, A6, A6 Avant, A8 **Engines:** 1.8L, 2.8L, 4.2L **Transmissions:** All	**Cylinder 5-Knock Control Limit Attained Conditions:** Engine started, battery voltage at least 11.5v, all electrical components off, ground connections between engine and chassis well connected, and the ECM detected the Knock Sensor signal was more than the calibrated value. **Possible Causes:** • Knock sensor circuit is open • Knock sensor circuit is shorted to ground, or shorted to power • Knock sensor is damaged or it has failed • Poor fuel quality • Loosen knock sensors and tighten again to 20 Nm • ECM has failed

DTC	Trouble Code Title, Conditions & Possible Causes
DTC: P1330 **2T CCM, MIL: Yes** **1998, 1999, 2000, 2001, 2002, 2003, 2004** **Models:** A4, A6, A6 Avant, A8 **Engines:** 1.8L, 2.8L, 4.2L **Transmissions:** All	**Cylinder 6-Knock Control Limit Attained Conditions:** Engine started, battery voltage at least 11.5v, all electrical components off, ground connections between engine and chassis well connected, and the ECM detected the Knock Sensor signal was more than the calibrated value. **Possible Causes:** • Knock sensor circuit is open • Knock sensor circuit is shorted to ground, or shorted to power • Knock sensor is damaged or it has failed • Poor fuel quality • Loosen knock sensors and tighten again to 20 Nm • ECM has failed
DTC: P1335 **1997, 1998, 1999, 2000, 2001, 2002, 2003, 2004, 2005, 2006** **Models:** TT, A4, A6, A8 **Engines:** 1.8L, 4.2L **Transmissions:** All	**Engine Torque Monitoring 2 Control Limit Exceeded Conditions:** Engine cold, battery voltage at least 11.5v, all electrical components off, ground connections between engine and chassis well connected, the ECM detected a signal beyond the required limit. **Possible Causes:** • Engine Control Module (ECM) has failed • Voltage supply for Engine Control Module (ECM) is shorted • Engine Coolant Temperature (ECT) sensor is faulty • Intake Air Temperature (IAT) sensor is faulty • Intake Manifold Runner Position Sensor is faulty • Intake system for leaks (false air) is faulty • Motor for intake flap is faulty • Mass Air Flow (MAF) sensor is faulty
DTC: P1336 **1997, 1998, 1999, 2000, 2001, 2002, 2003, 2004, 2005, 2006** **Models:** TT, A4 **Engines:** 1.8L **Transmissions:** All	**Engine Torque Monitoring Control Limit Exceeded Conditions:** Engine cold, battery voltage at least 11.5v, all electrical components off, ground connections between engine and chassis well connected, the ECM detected a signal beyond the required limit. **Possible Causes:** • Engine Control Module (ECM) has failed • Voltage supply for Engine Control Module (ECM) is shorted • Engine Coolant Temperature (ECT) sensor is faulty • Intake Air Temperature (IAT) sensor is faulty • Intake Manifold Runner Position Sensor is faulty • Intake system for leaks (false air) is faulty • Motor for intake flap is faulty • Mass Air Flow (MAF) sensor is faulty
DTC: P1337 **2T CCM, MIL: Yes** **1997, 1998, 1999, 2000, 2001, 2002, 2003, 2004, 2005, 2006** **Models:** TT, A4, A6, A8 **Engines:** 1.8L, 4.2L **Transmissions:** All	**Camshaft Position Sensor (Bank 1) Short to Ground Conditions:** Engine started, battery voltage at least 11.5v, all electrical components off, ground connections between engine and chassis well connected, and the ECM detected an unexpected low or high voltage condition on the camshaft position sensor circuit **Possible Causes:** • Faulty CPM sensor • ECM has failed
DTC: P1338 **2T CCM, MIL: Yes** **1997, 1998, 1999, 2000, 2001, 2002, 2003, 2004, 2005, 2006** **Models:** TT, A4, A6, A8 **Engines:** 1.8L, 4.2L **Transmissions:** All	**Camshaft Position Sensor (Bank 1) Open/Short to B+ Conditions:** Engine started, battery voltage at least 11.5v, all electrical components off, ground connections between engine and chassis well connected, and the ECM detected an unexpected low or high voltage condition on the camshaft position sensor circuit **Possible Causes:** • Faulty CPM sensor • ECM has failed
DTC: P1340 **2T CCM, MIL: Yes** **1996, 1997, 1998, 1999, 2000, 2001, 2002, 2003, 2004, 2005, 2006** **Models:** TT, A4, A6, A6 Avant, A8 **Engines:** 1.8L, 2.8L, 4.2L **Transmissions:** All	**Crankshaft Position/Camshaft Sensor Signal Out of Sequence Conditions:** Engine started, battery voltage at least 11.5v, all electrical components off, ground connections between engine and chassis well connected, and the ECM detected the crankshaft position sensor and the camshaft sensor were out of sequence with each other. **Note: The Engine Speed (RPM) Sensor detects engine speed and reference marks. Without an engine speed signal, the engine will not start. If the engine speed signal fails while the engine is running, the engine will stop immediately.** **Possible Causes:** • Engine speed sensor has failed or is contaminated (metal filings) • Engine speed sensor's wheel is damaged • Engine speed sensor circuit is shorted to the cable shield • Engine speed sensor circuit is open • ECM is faulty • Canshaft position sensor is faulty

DTC	Trouble Code Title, Conditions & Possible Causes
DTC: P1341 **1995, 1996** **Models:** A4, A6, A6 Avant **Engines:** 2.8L **Transmissions:** All	**Ignition Coil Power Output Stage 1 Short to Ground Conditions:** Key on or Engine started, battery voltage at least 11.5v, all electrical components off, ground connections between engine and chassis well connected, and the ECM detected the voltage of the ignition coil was outside the designed parameters. **Possible Causes:** • Fuel pump relay faulty • Canshaft position sensor has failed • Engine speed sensor has failed • Circuit wires have shorted to ground or are open • Ignition coils with power output stages are faulty • ECM has failed
DTC: P1343 **1995, 1996** **Models:** A4, A6, A6 Avant **Engines:** 2.8L **Transmissions:** All	**Ignition Coil Power Output Stage 2 Short to Ground Conditions:** Key on or Engine started, battery voltage at least 11.5v, all electrical components off, ground connections between engine and chassis well connected, and the ECM detected the voltage of the ignition coil was outside the designed parameters. **Possible Causes:** • Fuel pump relay faulty • Canshaft position sensor has failed • Engine speed sensor has failed • Circuit wires have shorted to ground or are open • Ignition coils with power output stages are faulty • ECM has failed
DTC: P1345 **1995, 1996** **Models:** A4, A6, A6 Avant **Engines:** 2.8L **Transmissions:** All	**Ignition Coil Power Output Stage 3 Short to Ground Conditions:** Key on or Engine started, battery voltage at least 11.5v, all electrical components off, ground connections between engine and chassis well connected, and the ECM detected the voltage of the ignition coil was outside the designed parameters. **Possible Causes:** • Fuel pump relay faulty • Canshaft position sensor has failed • Engine speed sensor has failed • Circuit wires have shorted to ground or are open • Ignition coils with power output stages are faulty • ECM has failed
DTC: P1355 **1997, 1998, 1999, 2000, 2001, 2002, 2003, 2004, 2005, 2006** **Models:** TT, A4 **Engines:** 1.8L **Transmissions:** All	**Cylinder 1 Ignition Circuit Open Circuit Conditions:** Key on or Engine started, battery voltage at least 11.5v, all electrical components off, ground connections between engine and chassis well connected, and the ECM detected the voltage of the ignition was outside the designed parameters. **Possible Causes:** • Fuel pump relay faulty • Canshaft position sensor has failed • Engine speed sensor has failed • Circuit wires have shorted to ground or are open • Ignition coils with power output stages are faulty • ECM has failed
DTC: P1356 **1997, 1998, 1999, 2000, 2001, 2002, 2003, 2004, 2005, 2006** **Models:** TT, A4 **Engines:** 1.8L **Transmissions:** All	**Cylinder 1 Ignition Circuit Short to B+ Conditions:** Key on or Engine started, battery voltage at least 11.5v, all electrical components off, ground connections between engine and chassis well connected, and the ECM detected the voltage of the ignition was outside the designed parameters. **Possible Causes:** • Fuel pump relay faulty • Canshaft position sensor has failed • Engine speed sensor has failed • Circuit wires have shorted to ground or are open • Ignition coils with power output stages are faulty • ECM has failed
DTC: P1357 **1997, 1998, 1999, 2000, 2001, 2002, 2003, 2004, 2005, 2006** **Models:** TT, A4 **Engines:** 1.8L **Transmissions:** All	**Cylinder 1 Ignition Circuit Short to Ground Conditions:** Key on or Engine started, battery voltage at least 11.5v, all electrical components off, ground connections between engine and chassis well connected, and the ECM detected the voltage of the ignition was outside the designed parameters. **Possible Causes:** • Fuel pump relay faulty • Canshaft position sensor has failed • Engine speed sensor has failed • Circuit wires have shorted to ground or are open • Ignition coils with power output stages are faulty • ECM has failed

DTC	Trouble Code Title, Conditions & Possible Causes
DTC: P1358 **1997, 1998, 1999, 2000, 2001, 2002, 2003, 2004, 2005, 2006** **Models:** TT, A4 **Engines:** 1.8L **Transmissions:** All	**Cylinder 2 Ignition Circuit Open Circuit Conditions:** Key on or Engine started, battery voltage at least 11.5v, all electrical components off, ground connections between engine and chassis well connected, and the ECM detected the voltage of the ignition was outside the designed parameters. **Possible Causes:** • Fuel pump relay faulty • Canshaft position sensor has failed • Engine speed sensor has failed • Circuit wires have shorted to ground or are open • Ignition coils with power output stages are faulty • ECM has failed
DTC: P1359 **1997, 1998, 1999, 2000, 2001, 2002, 2003, 2004, 2005, 2006** **Models:** TT, A4 **Engines:** 1.8L **Transmissions:** All	**Cylinder 2 Ignition Circuit Short to B+ Conditions:** Key on or Engine started, battery voltage at least 11.5v, all electrical components off, ground connections between engine and chassis well connected, and the ECM detected the voltage of the ignition was outside the designed parameters. **Possible Causes:** • Fuel pump relay faulty • Canshaft position sensor has failed • Engine speed sensor has failed • Circuit wires have shorted to ground or are open • Ignition coils with power output stages are faulty • ECM has failed
DTC: P1360 **1997, 1998, 1999, 2000, 2001, 2002, 2003, 2004, 2005, 2006** **Models:** TT, A4 **Engines:** 1.8L **Transmissions:** All	**Cylinder 2 Ignition Circuit Short to Ground Conditions:** Key on or Engine started, battery voltage at least 11.5v, all electrical components off, ground connections between engine and chassis well connected, and the ECM detected the voltage of the ignition was outside the designed parameters. **Possible Causes:** • Fuel pump relay faulty • Canshaft position sensor has failed • Engine speed sensor has failed • Circuit wires have shorted to ground or are open • Ignition coils with power output stages are faulty • ECM has failed
DTC: P1361 **1997, 1998, 1999, 2000, 2001, 2002, 2003, 2004, 2005, 2006** **Models:** TT, A4 **Engines:** 1.8L **Transmissions:** All	**Cylinder 3 Ignition Circuit Open Circuit Conditions:** Key on or Engine started, battery voltage at least 11.5v, all electrical components off, ground connections between engine and chassis well connected, and the ECM detected the voltage of the ignition was outside the designed parameters. **Possible Causes:** • Fuel pump relay faulty • Canshaft position sensor has failed • Engine speed sensor has failed • Circuit wires have shorted to ground or are open • Ignition coils with power output stages are faulty • ECM has failed
DTC: P1362 **1997, 1998, 1999, 2000, 2001, 2002, 2003, 2004, 2005, 2006** **Models:** TT, A4 **Engines:** 1.8L **Transmissions:** All	**Cylinder 3 Ignition Circuit Short to B+ Conditions:** Key on or Engine started, battery voltage at least 11.5v, all electrical components off, ground connections between engine and chassis well connected, and the ECM detected the voltage of the ignition was outside the designed parameters. **Possible Causes:** • Fuel pump relay faulty • Canshaft position sensor has failed • Engine speed sensor has failed • Circuit wires have shorted to ground or are open • Ignition coils with power output stages are faulty • ECM has failed
DTC: P1363 **1997, 1998, 1999, 2000, 2001, 2002, 2003, 2004, 2005, 2006** **Models:** TT, A4 **Engines:** 1.8L **Transmissions:** All	**Cylinder 3 Ignition Circuit Short to Ground Conditions:** Key on or Engine started, battery voltage at least 11.5v, all electrical components off, ground connections between engine and chassis well connected, and the ECM detected the voltage of the ignition was outside the designed parameters. **Possible Causes:** • Fuel pump relay faulty • Canshaft position sensor has failed • Engine speed sensor has failed • Circuit wires have shorted to ground or are open • Ignition coils with power output stages are faulty • ECM has failed

DTC	Trouble Code Title, Conditions & Possible Causes
DTC: P1364 **1997, 1998, 1999, 2000, 2001, 2002, 2003, 2004, 2005, 2006** **Models:** TT, A4 **Engines:** 1.8l **Transmissions:** All	**Cylinder 4 Ignition Circuit Open Circuit Conditions:** Key on or Engine started, battery voltage at least 11.5v, all electrical components off, ground connections between engine and chassis well connected, and the ECM detected the voltage of the ignition was outside the designed parameters. **Possible Causes:** • Fuel pump relay faulty • Canshaft position sensor has failed • Engine speed sensor has failed • Circuit wires have shorted to ground or are open • Ignition coils with power output stages are faulty • ECM has failed
DTC: P1365 **1997, 1998, 1999, 2000, 2001, 2002, 2003, 2004, 2005, 2006** **Models:** TT, A4 **Engines:** 1.8L **Transmissions:** All	**Cylinder 4 Ignition Circuit Short to B+ Conditions:** Key on or Engine started, battery voltage at least 11.5v, all electrical components off, ground connections between engine and chassis well connected, and the ECM detected the voltage of the ignition was outside the designed parameters. **Possible Causes:** • Fuel pump relay faulty • Canshaft position sensor has failed • Engine speed sensor has failed • Circuit wires have shorted to ground or are open • Ignition coils with power output stages are faulty • ECM has failed
DTC: P1366 **1997, 1998, 1999, 2000, 2001, 2002, 2003, 2004, 2005, 2006** **Models:** TT, A4 **Engines:** 1.8L **Transmissions:** All	**Cylinder 4 Ignition Circuit Short to Ground Conditions:** Key on or Engine started, battery voltage at least 11.5v, all electrical components off, ground connections between engine and chassis well connected, and the ECM detected the voltage of the ignition was outside the designed parameters. **Possible Causes:** • Fuel pump relay faulty • Canshaft position sensor has failed • Engine speed sensor has failed • Circuit wires have shorted to ground or are open • Ignition coils with power output stages are faulty • ECM has failed
DTC: P1386 **2T CCM, MIL: Yes** **1997, 1998, 1999, 2000, 2001, 2002, 2003, 2004, 2005, 2006** **Models:** TT, A4, A6, A8 **Engines:** 1.8L, 4.2L **Transmissions:** All	**Internal Control Module, Knock Control Circuit Error Conditions:** Engine started, and the ECM detected a too high or too low voltage condition on the knock control circuits, or a miscommunication between the knock control and the ECM. **Possible Causes:** • ECM has failed
DTC: P1387 **2T CCM, MIL: Yes** **1997, 1998, 1999, 2000, 2001, 2002, 2003, 2004, 2005, 2006** **Models:** TT, A4, A6, A8 **Engines:** 1.8L, 4.2L **Transmissions:** All	**Internal Control Module Altitude Sensor Error Conditions:** Ignition on, the ECM detected and altitude sensor error. To achieve optimal anti-theft protection for the vehicle, an anti-theft immobilizer is installed. The anti-theft immobilizer is a system for enabling and locking the Engine Control Module (ECM). So that this system cannot be circumvented, it is necessary to perform adaptation of the anti-theft immobilizer using the Vehicle Diagnostic and Information System VAS 5052 in the On Board Diagnostic (OBD) function. The great availability of equipment options makes it necessary to adapt the Engine Control Module (ECM) to the vehicle (e.g. throttle valve control module or cruise control system). This "writing" function is not possible with the generic scan tool. **Possible Causes:** • (If ECM was replaced) ECM ID not the same as the replaced unit • ECM has failed • Voltage supply for Engine Control Module (ECM) has shorted
DTC: P1388 **2T CCM, MIL: Yes** **1997, 1998, 1999, 2000, 2001, 2002, 2003, 2004, 2005, 2006** **Models:** TT, A4, A6, A8 **Engines:** 1.8L, 4.2L **Transmissions:** All	**Internal Control Module Drive By Wire Error Conditions:** Ignition on, the ECM detected and drive by wire error. To achieve optimal anti-theft protection for the vehicle, an anti-theft immobilizer is installed. The anti-theft immobilizer is a system for enabling and locking the Engine Control Module (ECM). So that this system cannot be circumvented, it is necessary to perform adaptation of the anti-theft immobilizer using the Vehicle Diagnostic and Information System VAS 5052 in the On Board Diagnostic (OBD) function. The great availability of equipment options makes it necessary to adapt the Engine Control Module (ECM) to the vehicle (e.g. throttle valve control module or cruise control system). This "writing" function is not possible with the generic scan tool. **Possible Causes:** • Engine Control Module (ECM) has failed • Voltage supply for Engine Control Module (ECM) has shorted

DTC	Trouble Code Title, Conditions & Possible Causes
DTC: P1391 **2T CCM, MIL: Yes** **1996, 1997, 1998, 1999, 2000, 2001, 2002, 2003** **Models:** A4, A6, A6 Avant, A8 **Engines:** 2.8L, 3.7L, 4.2L **Transmissions:** All	**Camshaft Position Sensor (Bank 2) Short to Ground Conditions:** Key on or Engine started, battery voltage must be at least 11.5v, all electrical components must be off, parking brake must be engaged (to keep daytime driving lights off), automatic transmission selector must be in park and the ground between the engine and the chassis must be well connected. The ECM detected an unexpected low or high voltage condition on the camshaft position sensor circuit. **Possible Causes:** • CMP sensor circuit is open or shorted to ground • CMP sensor circuit is shorted to power • CMP sensor ground (return) circuit is open • CMP sensor installation incorrect (Hall-effect type) • CMP sensor is damaged or CMP sensor shielding damaged • ECM has failed
DTC: P1392 **2T CCM, MIL: Yes** **1996, 1997, 1998, 1999, 2000, 2001, 2002, 2003** **Models:** A4, A6, A6 Avant, A8 **Engines:** 2.8L, 3.7L, 4.2L **Transmissions:** All	**Camshaft Position Sensor (Bank 2) Open/Short to B+ Conditions:** Key on or Engine started, battery voltage must be at least 11.5v, all electrical components must be off, parking brake must be engaged (to keep daytime driving lights off), automatic transmission selector must be in park and the ground between the engine and the chassis must be well connected. The ECM detected an unexpected low or high voltage condition on the camshaft position sensor circuit. **Possible Causes:** • CMP sensor circuit is open or shorted to ground • CMP sensor circuit is shorted to power • CMP sensor ground (return) circuit is open • CMP sensor installation incorrect (Hall-effect type) • CMP sensor is damaged or CMP sensor shielding damaged • ECM has failed
DTC: P1393 **1T CCM, MIL: Yes** **1996, 1997** **Models:** A4, A6, A6 Avant **Engines:** 2.8L **Transmissions:** All	**Ignition Coil Power Output Stage 1 Electrical Malfunction Conditions:** Key on or Engine started, battery voltage at least 11.5v, all electrical components off, ground connections between engine and chassis well connected, parking brake must be engaged (to keep daytime driving lights off), automatic transmission selector must be in park, and the ECM detected the an electrical malfunction of the ignition coil so that it won't properly function. **Possible Causes:** • Fuel pump relay faulty • Canshaft position sensor has failed • Engine speed sensor has failed • Circuit wires have shorted to ground or are open • Ignition coils with power output stages are faulty • ECM has failed
DTC: P1394 **1T CCM, MIL: Yes** **1996, 1997** **Models:** A4, A6, A6 Avant **Engines:** 2.8L **Transmissions:** All	**Ignition Coil Power Output Stage 2 Electrical Malfunction Conditions:** Key on or Engine started, battery voltage at least 11.5v, all electrical components off, ground connections between engine and chassis well connected, parking brake must be engaged (to keep daytime driving lights off), automatic transmission selector must be in park, and the ECM detected the an electrical malfunction of the ignition coil so that it won't properly function. **Possible Causes:** • Fuel pump relay faulty • Canshaft position sensor has failed • Engine speed sensor has failed • Circuit wires have shorted to ground or are open • Ignition coils with power output stages are faulty • ECM has failed
DTC: P1395 **1T CCM, MIL: Yes** **1996, 1997** **Models:** A4, A6, A6 Avant **Engines:** 2.8L **Transmissions:** All	**Ignition Coil Power Output Stage 3 Electrical Malfunction Conditions:** Key on or Engine started, battery voltage at least 11.5v, all electrical components off, ground connections between engine and chassis well connected, parking brake must be engaged (to keep daytime driving lights off), automatic transmission selector must be in park, and the ECM detected the an electrical malfunction of the ignition coil so that it won't properly function. **Possible Causes:** • Fuel pump relay faulty • Canshaft position sensor has failed • Engine speed sensor has failed • Circuit wires have shorted to ground or are open • Ignition coils with power output stages are faulty • ECM has failed
DTC: P1398 **2T CCM, MIL: Yes** **1996** **Models:** A4 **Engines:** 2.8L **Transmissions:** All	**Engine RPM Signal, TD Short to Ground Conditions:** Key on or Engine started, battery voltage at least 11.5v, all electrical components off, ground connections between engine and chassis well connected, parking brake must be engaged (to keep daytime driving lights off), automatic transmission selector must be in park, and the ECM detected the speed sensor signal short to ground **Possible Causes:** • Can-bus signal faulty • Speed sensor has failed

DTC	Trouble Code Title, Conditions & Possible Causes
DTC: P1409 **2T CCM, MIL: Yes** **1996, 1997, 1998, 1999, 2000, 2001, 2002, 2003, 2004, 2005, 2006** **Models:** TT, A4, A6, A6 Avant, A8 **Engines:** 1.8L, 2.8L, 4.2L **Transmissions:** All	**Tank Ventilation Valve Circuit Malfunction Conditions** Key on or engine running; and the ECM detected a too high or too low voltage level in the tank ventilation valve circuit. **Possible Causes:** • EVAP canister purge regulator valve has failed • Activation wire is shorting to positive • EVAP canister system has an improper or broken seal • Evaporative Emission (EVAP) canister purge regulator valve 1 is faulty • Leak Detection Pump (LDP) is faulty • Fuel filler cap is not properly closed • Lock ring on fuel pump not tightened • Hoses between EVAP canister and purge regulator valve have failed • ECM has failed
DTC: P1410 **2T CCM, MIL: Yes** **1996, 1997, 1998, 1999, 2000, 2001, 2002, 2003, 2004, 2005, 2006** **Models:** TT, A4, A6, A6 Avant, A8 **Engines:** 1.8L, 2.8L, 3.7L, 4.2L **Transmissions:** All	**Tank Ventilation Valve Circuit Short to B+:** Key on or engine running; and the ECM detected a too high or too low voltage level in the tank ventilation valve circuit. **Possible Causes:** • EVAP canister purge regulator valve has failed • Activation wire is shorting to positive • EVAP canister system has an improper or broken seal • Evaporative Emission (EVAP) canister purge regulator valve 1 is faulty • Leak Detection Pump (LDP) is faulty • Fuel filler cap is not properly closed • Lock ring on fuel pump not tightened • Hoses between EVAP canister and purge regulator valve have failed • ECM has failed
DTC: P1420 **2T CCM, MIL: Yes** **1997, 1998, 1999, 2000, 2001, 2002, 2003, 2004, 2005, 2006** **Models:** TT, A4, A6, A8 **Engines:** 1.8L, 2.8L, 4.2L **Transmissions:** All	**Secondary Air Injector Valve Circuit Electrical Malfunction Conditions:** The Engine Control Module activates the secondary air injection solenoid valve, but the Heated Oxygen Sensor (HO2S) does not detect secondary air injection. **Note: Solenoid valve is closed when no voltage is present.** **Possible Causes:** • Connector to the secondary air injection valve is loose or disconnected • Secondary air injector valve circuit short • Secondary air injector valve circuit is open • Faulty secondary air injector valve • ECM has failed
DTC: P1421 **2T CCM, MIL: Yes** **1997, 1998, 1999, 2000, 2001, 2002, 2003, 2004, 2005, 2006** **Models:** TT, A4, A6, A8 **Engines:** 1.8L, 2.8L, 4.2L **Transmissions:** All	**Secondary Air Injector Valve Circuit Short to Ground Conditions:** The Engine Control Module detects a short circuit to ground when activating the secondary air injection solenoid valve. **Note: Solenoid valve is closed when no voltage is present.** **Possible Causes:** • Connector to the secondary air injection valve is loose or disconnected • Secondary air injector valve circuit short • Secondary air injector valve circuit is open • Faulty secondary air injector valve • ECM has failed
DTC: P1422 **2T CCM, MIL: Yes** **1997, 1998, 1999, 2000, 2001, 2002, 2003, 2004, 2005, 2006** **Models:** TT, A4, A6, A8 **Engines:** 1.8L, 2.8L, 4.2L **Transmissions:** All	**Secondary Air Injector Valve Circuit Short to B+ Conditions:** The Engine Control Module detects a short circuit to B+ when activating the secondary air injection solenoid valve. **Note: Solenoid valve is closed when no voltage is present.** **Possible Causes:** • Connector to the secondary air injection valve is loose or disconnected • Secondary air injector valve circuit short • Secondary air injector valve circuit is open • Faulty secondary air injector valve • ECM has failed
DTC: P1424 **2T CCM, MIL: Yes** **1997, 1998, 1999, 2000, 2001, 2002, 2003, 2004, 2005, 2006** **Models:** TT, A4, A6, A8 **Engines:** 1.8L, 2.8L, 4.2L **Transmissions:** All	**Secondary Air Injector System (Bank 1) Leak Detected Conditions:** Ignition on or vehicle running, and the ECM detected a leak in the secondary air injector system. **Possible Causes:** • Poor hose/pipe connections between the secondary air injector pump motor and valve • Faulty hoses or pipes • Mechanical faults in the secondary air injector system •

DTC	Trouble Code Title, Conditions & Possible Causes
DTC: P1425 **1997, 1998, 1999, 2000, 2001, 2002, 2003, 2004, 2005, 2006** **Models:** TT, A4, A6, A8 **Engines:** 1.8L, 2.8L, 4.2L **Transmissions:** All	**Tank Ventilation Valve Short to Ground Conditions:** Ignition off. The Evaporative Emission (EVAP) canister purge regulator valve in the tank venting system or activation wire has a short circuit to ground. Engine started, engine running at a steady cruise speed, canister vent solenoid enabled, and the ECM detected an unexpected voltage condition on the Canister Vent solenoid circuit. **Note: Solenoid valve is closed when no voltage is present.** **Possible Causes:** • Activation wire has a short to ground • ECM has failed • EVAP canister has failed • EVAP canister system has an improper or broken seal • Evaporative Emission (EVAP) canister purge regulator valve is faulty • Leak Detection Pump (LDP) is faulty
DTC: P1426 **1997, 1998, 1999, 2000, 2001, 2002, 2003, 2004, 2005, 2006** **Models:** TT, A4, A6, A8 **Engines:** 1.8L, 2.8L, 4.2L **Transmissions:** All	**Tank Ventilation Valve Open Conditions:** Ignition off. The Evaporative Emission (EVAP) canister purge regulator valve in the tank venting system or activation wire has a short circuit to ground. Engine started, engine running at a steady cruise speed, canister vent solenoid enabled, and the ECM detected an unexpected voltage condition on the Canister Vent solenoid circuit. **Possible Causes:** • Activation wire has a short to ground • ECM has failed • EVAP canister has failed • EVAP canister system has an improper or broken seal • Evaporative Emission (EVAP) canister purge regulator valve 1 is faulty • Leak Detection Pump (LDP) is faulty
DTC: P1432 **2T CCM, MIL: Yes** **1996, 1997, 1998, 1999, 2000, 2001, 2002, 2003, 2004, 2005, 2006** **Models:** TT, A4, A6 **Engines:** 1.8L, 2.8L, 4.2L **Transmissions:** All	**Secondary Air Injection Valve Open Conditions:** The output Diagnostic Test Mode (DTM) can be activated only with the ignition switched on and the engine not running. The output DTM is interrupted if the engine is started, or if a rotary pulse from the ignition system is recognized.. **Possible Causes:** • Fuel pump relays have failed • Fuel injector has failed • Hoses on the EVAP canister may be clogged • EVAP canister purge regulator valve may be faulty • ECM may have failed • Manifold Tuning Valve (IMT) may have failed
DTC: P1433 **2T CCM, MIL: Yes** **1997, 1998, 1999, 2000, 2001, 2002, 2003, 2004, 2005, 2006** **Models:** TT, A4, A6, A8 **Engines:** 1.8L, 2.8L, 4.2L **Transmissions:** All	**Secondary Air Injection System Pump Relay Circuit Open Conditions:** The output Diagnostic Test Mode (DTM) can be activated only with the ignition switched on and the engine not running. The output DTM is interrupted if the engine is started, or if a rotary pulse from the ignition system is recognized.. **Possible Causes:** • Fuel pump relays have failed • Fuel injector has failed • Hoses on the EVAP canister may be clogged • EVAP canister purge regulator valve may be faulty • ECM may have failed • Manifold Tuning Valve (IMT) may have failed
DTC: P1434 **2T CCM, MIL: Yes** **1997, 1998, 1999, 2000, 2001, 2002, 2003, 2004, 2005, 2006** **Models:** TT, A4, A6, A8 **Engines:** 1.8L, 2.8L, 4.2L **Transmissions:** All	**Secondary Air Injection System Pump Relay Circuit Short to B+ Conditions:** The output Diagnostic Test Mode (DTM) can be activated only with the ignition switched on and the engine not running. The output DTM is interrupted if the engine is started, or if a rotary pulse from the ignition system is recognized.. **Possible Causes:** • Fuel pump relays have failed • Fuel injector has failed • Hoses on the EVAP canister may be clogged • EVAP canister purge regulator valve may be faulty • ECM may have failed • Manifold Tuning Valve (IMT) may have failed

DTC	Trouble Code Title, Conditions & Possible Causes
DTC: P1435 **2T CCM, MIL: Yes** **1997, 1998, 1999, 2000, 2001, 2002, 2003, 2004, 2005, 2006** **Models:** TT, A4, A6, A8 **Engines:** 1.8L, 2.8L, 4.2L **Transmissions:** All	**Secondary Air Injection System Pump Relay Circuit Short to Ground Conditions:** The output Diagnostic Test Mode (DTM) can be activated only with the ignition switched on and the engine not running. The output DTM is interrupted if the engine is started, or if a rotary pulse from the ignition system is recognized.. **Possible Causes:** • Fuel pump relays have failed • Fuel injector has failed • Hoses on the EVAP canister may be clogged • EVAP canister purge regulator valve may be faulty • ECM may have failed • Manifold Tuning Valve (IMT) may have failed
DTC: P1436 **1997, 1998, 1999, 2000, 2001, 2002, 2003, 2004, 2005, 2006** **Models:** TT, A4, A6, A8 **Engines:** 1.8L, 2.8L, 4.2L **Transmissions:** All	**A/C Evaporator Temperature (ACET) Circuit Low Input Conditions:** Key on or engine running; and the ECM detected the ACET signal was less than the self-test minimum amount of in the self-test. **Possible Causes:** • ACET signal circuit shorted to sensor ground (return) • ACET signal circuit shorted to chassis ground • ACET sensor is damaged or has failed • Check activation of Secondary Air Injection (AIR) Pump Relay • ECM has failed
DTC: P1470 **2T CCM, MIL: Yes** **1996, 1997, 1998, 1999** **Models:** A4 **Engines:** 1.8L, 2.8L **Transmissions:** All	**EVAP Emission Control LDP Circuit Electrical Malfunction Conditions:** Key on, KOEO Self-Test enabled, and the ECM detected an unexpected voltage condition on the EVAP emission control leak detection pump circuit. **Possible Causes:** • EVAP canister system has an improper or broken seal • Evaporative Emission (EVAP) canister purge regulator valve 1 is faulty • Leak Detection Pump (LDP) is faulty • ECM has failed
DTC: P1471 **2T CCM, MIL: Yes** **1997, 1998, 1999, 2000, 2001, 2002, 2003, 2004, 2005, 2006** **Models:** TT, A4, A6, A8 **Engines:** 1.8L, 2.8L, 4.2L **Transmissions:** All	**EVAP Emission Control Leak Detection Pump Circuit Short to B+ Conditions:** Key on, KOEO Self-Test enabled, and the ECM detected an unexpected voltage condition on the EVAP emission control leak detection pump circuit. **Possible Causes:** • Leak Detection Pump has failed • EVAP canister system has an improper or broken seal • Evaporative Emission (EVAP) canister purge regulator valve 1 is faulty • Hoses between the fuel pump and the EVAP canister are faulty • Fuel filler cap is loose • Fuel pump seal is defective, faulty or otherwise leaking • Hoses between the EVAP canister and the fuel flap unit are faulty • Hoses between the EVAP canister and the evaporative emission canister purge regulator valve are faulty • ECM has failed
DTC: P1472 **2T CCM, MIL: Yes** **1997, 1998, 1999, 2000, 2001, 2002, 2003, 2004, 2005, 2006** **Models:** TT, A4, A6, A8 **Engines:** 1.8L, 2.8L, 4.2L **Transmissions:** All	**EVAP Emission Control Leak Detection Pump Circuit Short to Ground Conditions:** Key on, KOEO Self-Test enabled, and the ECM detected an unexpected voltage condition on the EVAP emission control leak detection pump circuit. **Possible Causes:** • Leak Detection Pump has failed • EVAP canister system has an improper or broken seal • Evaporative Emission (EVAP) canister purge regulator valve 1 is faulty • Hoses between the fuel pump and the EVAP canister are faulty • Fuel filler cap is loose • Fuel pump seal is defective, faulty or otherwise leaking • Hoses between the EVAP canister and the fuel flap unit are faulty • Hoses between the EVAP canister and the evaporative emission canister purge regulator valve are faulty • ECM has failed

DTC	Trouble Code Title, Conditions & Possible Causes
DTC: P1473 **2T CCM, MIL: Yes** **1997, 1998, 1999, 2000, 2001, 2002, 2003, 2004, 2005, 2006** **Models:** TT, A4, A6, A8 **Engines:** 1.8L, 2.8L, 4.2L **Transmissions:** All	**EVAP Emission Control Leak Detection Pump Circuit Open Conditions:** Key on, KOEO Self-Test enabled, and the ECM detected an unexpected voltage condition on the EVAP emission control leak detection pump circuit. **Possible Causes:** • Leak Detection Pump has failed • EVAP canister system has an improper or broken seal • Evaporative Emission (EVAP) canister purge regulator valve 1 is faulty • Hoses between the fuel pump and the EVAP canister are faulty • Fuel filler cap is loose • Fuel pump seal is defective, faulty or otherwise leaking • Hoses between the EVAP canister and the fuel flap unit are faulty • Hoses between the EVAP canister and the evaporative emission canister purge regulator valve are faulty • ECM has failed
DTC: P1475 **2T CCM, MIL: Yes** **1997, 1998, 1999, 2000, 2001, 2002, 2003, 2004, 2005, 2006** **Models:** TT, A4, A6, A8 **Engines:** 1.8L, 2.8L, 4.2L **Transmissions:** All	**EVAP Emission Control LDP Circuit Malfunction/Signal Circuit Open Conditions:** Key on, KOEO Self-Test enabled, and the ECM detected an unexpected voltage condition on the EVAP emission control leak detection pump circuit. **Possible Causes:** • Leak Detection Pump has failed • EVAP canister system has an improper or broken seal • Evaporative Emission (EVAP) canister purge regulator valve 1 is faulty • Hoses between the fuel pump and the EVAP canister are faulty • Fuel filler cap is loose • Fuel pump seal is defective, faulty or otherwise leaking • Hoses between the EVAP canister and the fuel flap unit are faulty • Hoses between the EVAP canister and the evaporative emission canister purge regulator valve are faulty • ECM has failed
DTC: P1476 **2T CCM, MIL: Yes** **1997, 1998, 1999, 2000, 2001, 2002, 2003, 2004, 2005, 2006** **Models:** TT, A4, A6, A8 **Engines:** 1.8L, 2.8L, 4.2L **Transmissions:** All	**EVAP Emission Control LDP Circuit Malfunction/Insufficient Vacuum Conditions:** Key on, KOEO Self-Test enabled, and the ECM detected an unexpected voltage condition on the EVAP emission control leak detection pump circuit. **Possible Causes:** • Leak Detection Pump has failed • EVAP canister system has an improper or broken seal • Evaporative Emission (EVAP) canister purge regulator valve 1 is faulty • Hoses between the fuel pump and the EVAP canister are faulty • Fuel filler cap is loose • Fuel pump seal is defective, faulty or otherwise leaking • Hoses between the EVAP canister and the fuel flap unit are faulty • Hoses between the EVAP canister and the evaporative emission canister purge regulator valve are faulty • ECM has failed
DTC: P1477 **2T CCM, MIL: Yes** **1997, 1998, 1999, 2000, 2001, 2002, 2003, 2004, 2005, 2006** **Models:** TT, A4, A6, A8 **Engines:** 1.8L, 2.8L, 4.2L **Transmissions:** All	**EVAP Emission Control LDP Circuit Malfunction Conditions:** Key on, KOEO Self-Test enabled, and the ECM detected an unexpected voltage condition on the EVAP emission control leak detection pump circuit. **Possible Causes:** • Leak Detection Pump has failed • EVAP canister system has an improper or broken seal • Evaporative Emission (EVAP) canister purge regulator valve 1 is faulty • Hoses between the fuel pump and the EVAP canister are faulty • Fuel filler cap is loose • Fuel pump seal is defective, faulty or otherwise leaking • Hoses between the EVAP canister and the fuel flap unit are faulty • Hoses between the EVAP canister and the evaporative emission canister purge regulator valve are faulty • ECM has failed

DTC	Trouble Code Title, Conditions & Possible Causes
DTC: P1478 **2T CCM, MIL: Yes** **1997, 1998, 1999, 2000, 2001, 2002, 2003, 2004, 2005, 2006** **Models:** TT, A4, A6, A8 **Engines:** 1.8L, 2.8L, 4.2L **Transmissions:** All	**EVAP Emission Control LDP Circuit Clamped Tube Detected Conditions:** Key on, KOEO Self-Test enabled, and the ECM detected an unexpected voltage condition on the EVAP emission control leak detection pump circuit. **Possible Causes:** • Leak Detection Pump has failed • EVAP canister system has an improper or broken seal • Evaporative Emission (EVAP) canister purge regulator valve 1 is faulty • Hoses between the fuel pump and the EVAP canister are faulty • Fuel filler cap is loose • Fuel pump seal is defective, faulty or otherwise leaking • Hoses between the EVAP canister and the fuel flap unit are faulty • Hoses between the EVAP canister and the evaporative emission canister purge regulator valve are faulty • ECM has failed
DTC: P1500 **1995, 1996, 1997, 1998, 1999, 2000, 2001, 2002, 2003, 2004, 2005, 2006** **Models:** TT, A4, A6, A6 Avant, A8 **Engines:** 1.8L, 2.8L, 3.7L, 4.2L **Transmissions:** All	**Fuel Pump Relay Circuit Electrical Malfunction Conditions:** Engine running the ECM detected that the fuel pump relay signal was intermittent **Possible Causes:** • Fuel delivery unit connector is loose or not attached • Fuse 18 cause a short to the transfer fuel pump or the O2S • Fuel pump has failed • Fuel pump relay circuit is shorted to ground, B+ or is open • Fuel Pump (FP) Relay not activated • ECM has failed
DTC: P1501 **1996, 1997, 1998, 1999, 2000, 2001, 2002, 2003, 2004, 2005, 2006** **Models:** TT, A4 **Engines:** 1.8L, 2.8L **Transmissions:** All	**Fuel Pump Relay Circuit Electrical Short to Ground Conditions:** Engine running the ECM detected that the fuel pump relay signal was intermittent **Possible Causes:** • Fuel delivery unit connector is loose or not attached • Fuse 18 cause a short to the transfer fuel pump or the O2S • Fuel pump has failed • Fuel pump relay circuit is shorted to ground, B+ or is open • Fuel Pump (FP) Relay not activated • ECM has failed
DTC: P1502 **1996, 1997, 1998, 1999, 2000, 2001, 2002, 2003, 2004, 2005, 2006** **Models:** TT, A4, A6, A8 **Engines:** 1.8L, 2.8L, 3.7L, 4.2L **Transmissions:** All	**Fuel Pump Relay Circuit Short to B+ Conditions:** Engine running the ECM detected that the fuel pump relay signal was intermittent **Possible Causes:** • Fuel delivery unit connector is loose or not attached • Fuse 18 cause a short to the transfer fuel pump or the O2S • Fuel pump has failed • Fuel pump relay circuit is shorted to ground, B+ or is open • Fuel Pump (FP) Relay not activated • ECM has failed
DTC: P1512 **1998, 1999, 2000, 2001, 2002, 2003, 2004** **Models:** A4, A6 **Engines:** 2.8L, 4.2L **Transmissions:** All	**Intake Manifold Changeover Valve Circuit Short to B+ Conditions:** Engine started, and the ECM detected the changeover valve circuit was shorting to positive during the continuous self test. **Possible Causes:** • Leaky vacuum reservoir, vacuum lines loose or damaged • Vacuum solenoid or vacuum actuator is damaged • IMRC actuator cable/gears are seized, or the cables are improperly routed or seized • IMRC housing return springs are damaged or disconnected • Lever/shaft return stop may be obstructed or bent, or the lever/shaft wide open stop may be obstructed or bent, or the IMRC lever/shaft may be sticking, binding or disconnected • IMRC control circuit open, shorted or the VPWR circuit is open • ECM has failed
DTC: P1515 **1998, 1999, 2000, 2001** **Models:** A4 **Engines:** 2.8L **Transmissions:** All	**Intake Manifold Changeover Valve Circuit Short to Ground Conditions:** Engine started, and the ECM detected the changeover valve circuit was shorting to ground during the continuous self test. **Possible Causes:** • Leaky vacuum reservoir, vacuum lines loose or damaged • Vacuum solenoid or vacuum actuator is damaged • IMRC actuator cable/gears are seized, or the cables are improperly routed or seized • IMRC housing return springs are damaged or disconnected • Lever/shaft return stop may be obstructed or bent, or the lever/shaft wide open stop may be obstructed or bent, or the IMRC lever/shaft may be sticking, binding or disconnected • IMRC control circuit open, shorted or the VPWR circuit is open • ECM has failed

DTC	Trouble Code Title, Conditions & Possible Causes
DTC: P1516 **1998, 1999, 2000, 2001** **Models:** A4 **Engines:** 2.8L **Transmissions:** All	**Intake Manifold Runner Control Input Error (Bank 1) Conditions:** Key on or engine running; and the ECM detected the IMRC Monitor signal for Bank 1 was outside of its expected calibrated range during the Continuous self test. **Possible Causes:** • IMRC mechanical fault – the linkage may be bound or seized • Inspect for binding or improper routing. The cable core wire at the IMRC/IMSC housing attachment must have slack and lever must contact close plate stop screw
DTC: P1517 **1997, 1998, 1999, 2000, 2001** **Models:** A4, A8 **Engines:** 2.8L, 4.2L **Transmissions:** All	**Main Relay Circuit Electrical Malfunction Conditions:** The ECM detected an electrical malfunction on the main relay circuit **Possible Causes:** • Engine Control Module (ECM) has failed • Voltage supply for Engine Control Module (ECM) is faulty • Check activation of Motronic Engine Control Module (ECM) Power Supply Relay
DTC: P1519 **1998, 1999, 2000, 2001, 2002, 2003, 2004** **Models:** A4, A6, A8 **Engines:** 2.8L, 4.2L **Transmissions:** All	**Intake Manifold Runner Control Stuck Closed Conditions:** Key on, and the ECM detected the IMRC Monitor was more than the expected calibrated range at closed throttle. **Possible Causes:** • IMRC monitor signal circuit shorted to power ground • IMRC Monitor signal circuit shorted to signal ground (return) • IMRC actuator is damaged or has failed (e.g., there may be a small leak in the vacuum diaphragm of the actuator) • ECM has failed
DTC: P1522 **1998, 1999, 2000, 2001, 2002, 2003, 2004** **Models:** A4, A6, A8 **Engines:** 2.8L, 4.2L **Transmissions:** All	**Intake Camshaft Control (Bank 2) Malfunction Conditions:** Key on or engine running; and the ECM detected the intake manifold control signal for was outside of its expected calibrated range. **Possible Causes:** • Camshaft control circuit is open or shorted to ground • Camshaft sensor is damaged or the ECM has failed • Camshaft out of adjustment
DTC: P1529 **1996, 1997, 1998, 1999, 2000, 2001, 2002, 2003, 2004** **Models:** TT, A6, A8 **Engines:** 1.8L, 4.2L **Transmissions:** All	**Camshaft Control Circuit Short to B+ Conditions:** Engine started and driven at an engine speed of more than 400rpm; and the ECM detected the camshaft timing exceeded the calibrated voltage levels. The valve timing did not change from the current valve timing or it remained fixed during the testing. **Note: The camshaft adjustment is load- and RPM dependant. The electrical camshaft adjustment valve 1 switches oil pressure onto camshaft adjuster (mechanical adjustment mechanism), which adjusts the camshaft.** **Possible Causes:** • Fuel pump has failed • CPS circuit is open, shorted to ground or shorted to power • ECM has failed • Battery voltage below 11.5 volts • Position actuator circuit may short to B+ or Ground • Camshaft timing improperly set, or continuous oil flow to the VCT piston chamber • Camshaft advance mechanism (the VCT unit) is sticking or binding mechanically • VCT solenoid valve is stuck in open position
DTC: P1530 **2T CCM, MIL: Yes** **1996, 1997, 1998, 1999, 2000, 2001, 2002, 2003, 2004** **Models:** TT, A6, A8 **Engines:** 1.8L, 4.2L **Transmissions:** All	**Camshaft Control Circuit Short to Ground Conditions:** Engine started and driven at an engine speed of more than 400rpm; and the ECM detected the camshaft timing exceeded the calibrated levels. The valve timing did not change from the current valve timing or it remained fixed during the testing. **Note: The camshaft adjustment is load- and RPM dependant. The electrical camshaft adjustment valve 1 switches oil pressure onto camshaft adjuster (mechanical adjustment mechanism), which adjusts the camshaft.** **Possible Causes:** • Fuel pump has failed • CPS circuit is open, shorted to ground or shorted to power • ECM has failed • Battery voltage below 11.5 volts • Position actuator circuit may short to B+ or Ground • Camshaft timing improperly set, or continuous oil flow to the VCT piston chamber • Camshaft advance mechanism (the VCT unit) is sticking or binding mechanically • VCT solenoid valve is stuck in open position

DTC	Trouble Code Title, Conditions & Possible Causes
DTC: P1531 **2T CCM, MIL: Yes** **1996, 1997, 1998, 1999, 2000, 2001, 2002, 2003, 2004** **Models:** TT, A6, A8 **Engines:** 1.8L, 4.2L **Transmissions:** All	**Camshaft Control Circuit Open Conditions:** Engine started and driven at an engine speed of more than 400rpm; and the ECM detected the camshaft timing exceeded the calibrated levels. The valve timing did not change from the current valve timing or it remained fixed during the testing. **Note: The camshaft adjustment is load- and RPM dependant. The electrical camshaft adjustment valve 1 switches oil pressure onto camshaft adjuster (mechanical adjustment mechanism), which adjusts the camshaft.** **Possible Causes:** • Fuel pump has failed • CPS circuit is open, shorted to ground or shorted to power • ECM has failed • Battery voltage below 11.5 volts • Position actuator circuit may short to B+ or Ground • Camshaft timing improperly set, or continuous oil flow to the VCT piston chamber • Camshaft advance mechanism (the VCT unit) is sticking or binding mechanically • VCT solenoid valve is stuck in open position
DTC: P1541 **1996, 1997, 1998, 1999, 2000, 2001, 2002, 2003, 2004** **Models:** TT, A4 **Engines:** 1.8L **Transmissions:** All	**Fuel Pump Relay Circuit Open Conditions:** The ECM detected an electrical malfunction on the fuel pump relay circuit **Possible Causes:** • Fuel pump relay not activiated
DTC: P1542 **1997, 1998, 1999, 2000, 2001, 2002, 2003, 2004, 2005, 2006** **Models:** TT, A4, A6, A8 **Engines:** 1.8L, 2.8L, 4.2L **Transmissions:** All	**Throttle Actuation Potentiometer Range/Performance Conditions:** Engine started, battery voltage must be at least 11.5v, all electrical components must be off, parking brake must be engaged (to keep daytime driving lights off), automatic transmission selector must be in park, the exhaust system must be properly sealed between the catalytic converter and the cylinder head, coolant temperature must be at least 80 degrees Celsius, and the ground between the engine and the chassis must be well connected. The signal from the Throttle Position Valve Module to the ECM detected was erratic, non existent or unreliable. **Note: If the complete throttle valve control module is current-less (e.g. connector disconnected) the throttle valve moves into a particular, specified mechanical position, which signals an increased idle speed with an engine at operating temperature. If only the Throttle Position (TP) actuator –V60- is current-less, the throttle valve also moves into the specified mechanical position (emergency running gap), however, since Closed Throttle Position (CTP) switch –F60- can still be recognized, an "almost normal idle RPM" is reached via the respective ignition angle retardation. If the Engine Control Module (ECM) detects a malfunction at Throttle Position (TP) sensor –G69-, Throttle Position (TP) actuator –V60- is switched current-less by the Engine Control Module (ECM) and the throttle valve moves into the specified mechanical position (emergency running gap) again.** **Note: Terminal assignment at throttle control module is different in vehicles with and without cruise control.** **Characteristic: Steering column switch with operating module for cruise control.** **Possible Causes:** • Throttle valve control module is faulty • Throttle valve is damaged or dirty • Throttle valve must be in closed throttle position • Accelerator pedal is out of adjustment (AEG engines only) • Throttle position actuator is shorting to ground or power

DTC	Trouble Code Title, Conditions & Possible Causes
DTC: P1543 **1997, 1998, 1999, 2000, 2001, 2002, 2003, 2004, 2005, 2006** **Models:** TT, A4, A6, A8 **Engines:** 1.8L, 2.8L, 4.2L **Transmissions:** All	**Throttle Actuation Potentiometer Signal Too Low Conditions:** Engine started, battery voltage must be at least 11.5v, all electrical components must be off, parking brake must be engaged (to keep daytime driving lights off), automatic transmission selector must be in park, the exhaust system must be properly sealed between the catalytic converter and the cylinder head, coolant temperature must be at least 80 degrees Celsius, and the ground between the engine and the chassis must be well connected. The signal from the Throttle Position Valve Module to the ECM detected was erratic, non existent or unreliable. **Note: If the complete throttle valve control module is current-less (e.g. connector disconnected) the throttle valve moves into a particular, specified mechanical position, which signals an increased idle speed with an engine at operating temperature. If only the Throttle Position (TP) actuator –V60- is current-less, the throttle valve also moves into the specified mechanical position (emergency running gap), however, since Closed Throttle Position (CTP) switch –F60- can still be recognized, an "almost normal idle RPM" is reached via the respective ignition angle retardation. If the Engine Control Module (ECM) detects a malfunction at Throttle Position (TP) sensor –G69-, Throttle Position (TP) actuator –V60- is switched current-less by the Engine Control Module (ECM) and the throttle valve moves into the specified mechanical position (emergency running gap) again.** **Note: Terminal assignment at throttle control module is different in vehicles with and without cruise control.** **Characteristic: Steering column switch with operating module for cruise control.** **Possible Causes:** • Throttle valve control module is faulty • Throttle valve is damaged or dirty • Throttle valve must be in closed throttle position • Accelerator pedal is out of adjustment (AEG engines only) • Throttle position actuator is shorting to ground or power
DTC: P1544 **1997, 1998, 1999, 2000, 2001, 2002, 2003, 2004, 2005, 2006** **Models:** TT, A4, A6, A8 **Engines:** 1.8L, 2.8L, 4.2L **Transmissions:** All	**Throttle Actuation Potentiometer Signal Too High Conditions:** Engine started, battery voltage must be at least 11.5v, all electrical components must be off, parking brake must be engaged (to keep daytime driving lights off), automatic transmission selector must be in park, the exhaust system must be properly sealed between the catalytic converter and the cylinder head, coolant temperature must be at least 80 degrees Celsius, and the ground between the engine and the chassis must be well connected. The signal from the Throttle Position Valve Module to the ECM detected was erratic, non existent or unreliable. **Note: If the complete throttle valve control module is current-less (e.g. connector disconnected) the throttle valve moves into a particular, specified mechanical position, which signals an increased idle speed with an engine at operating temperature. If only the Throttle Position (TP) actuator –V60- is current-less, the throttle valve also moves into the specified mechanical position (emergency running gap), however, since Closed Throttle Position (CTP) switch –F60- can still be recognized, an "almost normal idle RPM" is reached via the respective ignition angle retardation. If the Engine Control Module (ECM) detects a malfunction at Throttle Position (TP) sensor –G69-, Throttle Position (TP) actuator –V60- is switched current-less by the Engine Control Module (ECM) and the throttle valve moves into the specified mechanical position (emergency running gap) again.** **Note: Terminal assignment at throttle control module is different in vehicles with and without cruise control.** **Characteristic: Steering column switch with operating module for cruise control.** **Possible Causes:** • Throttle valve control module is faulty • Throttle valve is damaged or dirty • Throttle valve must be in closed throttle position • Accelerator pedal is out of adjustment (AEG engines only) • Throttle position actuator is shorting to ground or power

DTC	Trouble Code Title, Conditions & Possible Causes
DTC: P1545 **1997, 1998, 1999, 2000, 2001, 2002, 2003, 2004, 2005, 2006** **Models:** TT, A4, A6, A8 **Engines:** 1.8L, 2.8L, 4.2L **Transmissions:** All	**Throttle Position Control Malfunction Conditions:** Engine started, battery voltage must be at least 11.5v, all electrical components must be off, parking brake must be engaged (to keep daytime driving lights off), automatic transmission selector must be in park, the exhaust system must be properly sealed between the catalytic converter and the cylinder head, coolant temperature must be at least 80 degrees Celsius, and the ground between the engine and the chassis must be well connected. The signal from the Throttle Position Valve Module to the ECM detected was erratic, non existent or unreliable. **Note: If the complete throttle valve control module is current-less (e.g. connector disconnected) the throttle valve moves into a particular, specified mechanical position, which signals an increased idle speed with an engine at operating temperature. If only the Throttle Position (TP) actuator is current-less, the throttle valve also moves into the specified mechanical position (emergency running gap), however, since Closed Throttle Position (CTP) switch – can still be recognized, an "almost normal idle RPM" is reached via the respective ignition angle retardation. If the Engine Control Module (ECM) detects a malfunction at Throttle Position (TP) sensor – Throttle Position (TP) actuator is switched current-less by the Engine Control Module (ECM) and the throttle valve moves into the specified mechanical position (emergency running gap) again.** **Note: Terminal assignment at throttle control module is different in vehicles with and without cruise control.** **Characteristic: Steering column switch with operating module for cruise control.** **Possible Causes:** • Throttle valve control module is faulty • Throttle valve is damaged or dirty • Throttle valve must be in closed throttle position • Accelerator pedal is out of adjustment (AEG engines only) • Throttle position actuator is shorting to ground or power
DTC: P1546 **1997, 1998, 1999, 2000, 2001, 2002, 2003, 2004, 2005, 2006** **Models:** TT, A4 **Engines:** 1.8L **Transmissions:** All	**Boost Pressure Control Valve Short to B+ Conditions:** Engine started, battery voltage at least 11.5v, all electrical components off, ground connections between engine and chassis well connected, coolant temperature at least 80-degrees Celicius. The ECM detected an short in the boost pressure control valve. **Possible Causes:** • Charge air pressure sensor is faulty • Voltage supply to the charge air pressure sensor is open or shorted • Check the charge air system for leaks • Recirculating valve for turbocharger is faulty • Turbocharging system is damaged or not functioning properly • Turbocharger recirculating valve is faulty • Vacuum diaphragm for turbocharger is out of adjustment • Wastegate bypass regulator valve is faulty • Boost sensor has failed • ECM has failed
DTC: P1547 **1997, 1998, 1999, 2000, 2001, 2002, 2003, 2004, 2005, 2006** **Models:** TT, A4 **Engines:** 1.8L **Transmissions:** All	**Boost Pressure Control Valve Short to Ground Conditions:** Engine started, battery voltage at least 11.5v, all electrical components off, ground connections between engine and chassis well connected, coolant temperature at least 80-degrees Celicius. The ECM detected an short in the boost pressure control valve. **Possible Causes:** • Charge air pressure sensor is faulty • Voltage supply to the charge air pressure sensor is open or shorted • Check the charge air system for leaks • Recirculating valve for turbocharger is faulty • Turbocharging system is damaged or not functioning properly • Turbocharger recirculating valve is faulty • Vacuum diaphragm for turbocharger is out of adjustment • Wastegate bypass regulator valve is faulty • Boost sensor has failed • ECM has failed

DTC	Trouble Code Title, Conditions & Possible Causes
DTC: P1548 **1997, 1998, 1999, 2000, 2001, 2002, 2003, 2004, 2005, 2006** **Models:** TT, A4 **Engines:** 1.8L **Transmissions:** All	**Boost Pressure Control Valve Open Conditions:** Engine started, battery voltage at least 11.5v, all electrical components off, ground connections between engine and chassis well connected, coolant temperature at least 80-degrees Celicius. The ECM detected an short in the boost pressure control valve. **Possible Causes:** • Charge air pressure sensor is faulty • Voltage supply to the charge air pressure sensor is open or shorted • Check the charge air system for leaks • Recirculating valve for turbocharger is faulty • Turbocharging system is damaged or not functioning properly • Turbocharger recirculating valve is faulty • Vacuum diaphragm for turbocharger is out of adjustment • Wastegate bypass regulator valve is faulty • Boost sensor has failed • ECM has failed
DTC: P1550 **1997, 1998, 1999, 2000, 2001, 2002, 2003, 2004, 2005, 2006** **Models:** TT, A4 **Engines:** 1.8L **Transmissions:** All	**Charge Pressure Deviation Conditions:** Engine started, battery voltage at least 11.5v, all electrical components off, ground connections between engine and chassis well connected, coolant temperature at least 80-degrees Celicius. The ECM detected deviation from the normal operating parameters of the charge pressure sensor. **Possible Causes:** • Charge air system leaks • Recirculating valve for turbocharger is faulty • Turbocharging system is damaged • Vacuum diaphragm for turbocharger needs adjusting • Wastegate bypass regulator valve is faulty • ECM has failed
DTC: P1555 **1997, 1998, 1999, 2000, 2001, 2002, 2003, 2004, 2005, 2006** **Models:** TT, A4 **Engines:** 1.8L **Transmissions:** All	**Charge Pressure Upper Limit Exceeded Conditions:** Engine started, battery voltage at least 11.5v, all electrical components off, ground connections between engine and chassis well connected, coolant temperature at least 80-degrees Celicius. The ECM detected deviation from the normal operating parameters of the charge pressure sensor. **Possible Causes:** • Charge air system leaks • Recirculating valve for turbocharger is faulty • Turbocharging system is damaged • Vacuum diaphragm for turbocharger needs adjusting • Wastegate bypass regulator valve is faulty • ECM has failed
DTC: P1556 **1997, 1998, 1999, 2000, 2001, 2002, 2003, 2004, 2005, 2006** **Models:** TT, A4 **Engines:** 1.8L **Transmissions:** All	**Charge Pressure Control Negative Deviation Conditions:** Engine started, battery voltage at least 11.5v, all electrical components off, ground connections between engine and chassis well connected, coolant temperature at least 80-degrees Celicius. The ECM detected deviation from the normal operating parameters of the charge pressure sensor. **Possible Causes:** • Charge air system leaks • Recirculating valve for turbocharger is faulty • Turbocharging system is damaged • Vacuum diaphragm for turbocharger needs adjusting • Wastegate bypass regulator valve is faulty • ECM has failed
DTC: P1557 **1997, 1998, 1999, 2000, 2001, 2002, 2003, 2004, 2005, 2006** **Models:** TT, A4 **Engines:** 1.8L **Transmissions:** All	**Charge Pressure Control Positive Deviation Conditions:** Engine started, battery voltage at least 11.5v, all electrical components off, ground connections between engine and chassis well connected, coolant temperature at least 80-degrees Celicius. The ECM detected deviation from the normal operating parameters of the charge pressure sensor. **Possible Causes:** • Charge air system leaks • Recirculating valve for turbocharger is faulty • Turbocharging system is damaged • Vacuum diaphragm for turbocharger needs adjusting • Wastegate bypass regulator valve is faulty • ECM has failed

DTC	Trouble Code Title, Conditions & Possible Causes
DTC: P1558 **1997, 1998, 1999, 2000, 2001, 2002, 2003, 2004, 2005, 2006** **Models:** TT, A4, A6, A8 **Engines:** 1.8L, 2.8L, 4.2L **Transmissions:** All	**Throttle Actuator Electrical Malfunction Conditions:** Engine started, battery voltage at least 11.5v, all electrical components off, ground connections between engine and chassis well connected, coolant temperature at least 80-degrees Celicius and the throttle valve must not be damaged or dirty; and the ECM detected the signal from the Throttle Position Valve Module to the ECM detected was erratic, non existent or unreliable (too high or too low). **Possible Causes:** • Throttle valve control module has failed • Throttle valve control module's circuit has shorted or is open • The ECM has failed
DTC: P1559 **1997, 1998, 1999, 2000, 2001, 2002, 2003, 2004, 2005, 2006** **Models:** TT, A6, A8 **Engines:** 1.8L, 2.8L, 4.2L **Transmissions:** All	**Idle Speed Control Throttle Position Adaptation Malfunction Conditions:** Engine started, battery voltage at least 11.5v, all electrical components off, ground connections between engine and chassis well connected, coolant temperature at least 80-degrees Celicius and the throttle valve must not be damaged or dirty; and the ECM detected the signal from the Throttle Position Valve Module to the ECM detected was erratic, non existent or unreliable (too high or too low). **Possible Causes:** • Throttle valve control module has failed • Throttle valve control module's circuit has shorted or is open • The ECM has failed
DTC: P1560 **1T CCM, MIL: Yes** **1995, 1996, 1997, 1998, 1999, 2000, 2001** **Models:** A4, A6, A8 **Engines:** 2.8L, 3.7L, 4.2L **Transmissions:** All	**Maximum Engine Speed Exceeded Conditions:** Engine running, the ECM has detected that the maximum engine speed had been attained. **Possible Causes:** • Throttle valve control module has failed • Throttle valve control module's circuit has shorted or is open • The ECM has failed • General engine damage
DTC: P1564 **1998, 1999, 2000, 2001** **Models:** A4 **Engines:** 2.8L **Transmissions:** All	**Idle Speed Control Throttle Position Low Voltage During Adaptation Conditions:** Engine started, battery voltage at least 11.5v, all electrical components off, ground connections between engine and chassis well connected, coolant temperature at least 80-degrees Celicius and the throttle valve must not be damaged or dirty; and the ECM detected the signal from the Throttle Position Valve Module to the ECM detected was erratic, non existent or unreliable (too high or too low). **Possible Causes:** • Alternator failed • ECM failed • Fuses blown or open circuits • Clean Throttle Valve Control Module • Faulty battery • Idle speed control throttle failed • Wire connections to relay carrier and ground connection of ECM may have shorted
DTC: P1565 **1997, 1998, 1999, 2000, 2001, 2002, 2003, 2004, 2005, 2006** **Models:** TT, A4, A6, A8 **Engines:** 1.8L, 2.8L, 4.2L **Transmissions:** All	**Idle Speed Control Throttle Position Lower Limit Not Attainted Conditions:** Engine started, battery voltage at least 11.5v, all electrical components off, ground connections between engine and chassis well connected, coolant temperature at least 80-degrees Celicius and the throttle valve must not be damaged or dirty; and the ECM detected the signal from the Throttle Position Valve Module to the ECM detected was erratic, non existent or unreliable (too high or too low). **Possible Causes:** • Alternator failed • ECM failed • Fuses blown or open circuits • Clean Throttle Valve Control Module • Accelerator cable not adjusted properly • Idle speed control throttle failed • Wire connections to relay carrier and ground connection of ECM may have shorted
DTC: P1568 **1999, 2000, 2001, 2002, 2003, 2004, 2005, 2006** **Models:** TT, A6, A8 **Engines:** 1.8L, 4.2L **Transmissions:** All	**Idle Speed Control Throttle Position Mechanical Malfunction Conditions:** Engine started, battery voltage at least 11.5v, all electrical components off, ground connections between engine and chassis well connected, coolant temperature at least 80-degrees Celicius and the throttle valve must not be damaged or dirty; and the ECM detected the signal from the Throttle Position Valve Module to the ECM detected was erratic, non existent or unreliable (too high or too low) suggesting a mechanicl malfunction. **Possible Causes:** • Alternator failed • ECM failed • Fuses blown or open circuits • Clean Throttle Valve Control Module • Accelerator cable not adjusted properly • Idle speed control throttle failed • Wire connections to relay carrier and ground connection of ECM may have shorted

DTC	Trouble Code Title, Conditions & Possible Causes
DTC: P1602 **1997, 1998, 1999, 2000, 2001, 2002, 2003, 2004, 2005, 2006** **Models:** TT, A4, A6, A6 Avant **Engines:** 1.8L, 2.8L **Transmissions:** All	**Power Supply (B+) Terminal 15 Low Voltage Conditions:** Ignition on, the ECM detected a low voltage condition on the power supply terminal (15). To achieve optimal anti-theft protection for the vehicle, an anti-theft immobilizer is installed. The anti-theft immobilizer is a system for enabling and locking the Engine Control Module (ECM). So that this system cannot be circumvented, it is necessary to perform adaptation of the anti-theft immobilizer using the Vehicle Diagnostic and Information System VAS 5052 in the On Board Diagnostic (OBD) function. The great availability of equipment options makes it necessary to adapt the Engine Control Module (ECM) to the vehicle (e.g. throttle valve control module or cruise control system). This "writing" function is not possible with the generic scan tool. **Possible Causes:** • (If ECM was replaced) ECM ID not the same as the replaced unit • ECM has failed • Voltage supply for Engine Control Module (ECM) has shorted
DTC: P1603 **1997, 1998, 1999, 2000, 2001, 2002, 2003, 2004, 2005, 2006** **Models:** TT, A4, A6, A6 Avant **Engines:** 1.8L, 2.8L **Transmissions:** All	**Internal Control Module Malfunction Conditions:** Ignition on, the ECM detected a control module malfunction. To achieve optimal anti-theft protection for the vehicle, an anti-theft immobilizer is installed. The anti-theft immobilizer is a system for enabling and locking the Engine Control Module (ECM). So that this system cannot be circumvented, it is necessary to perform adaptation of the anti-theft immobilizer using the Vehicle Diagnostic and Information System VAS 5052 in the On Board Diagnostic (OBD) function. The great availability of equipment options makes it necessary to adapt the Engine Control Module (ECM) to the vehicle (e.g. throttle valve control module or cruise control system). This "writing" function is not possible with the generic scan tool. **Possible Causes:** • (If ECM was replaced) ECM ID not the same as the replaced unit • ECM has failed • Voltage supply for Engine Control Module (ECM) has shorted
DTC: P1604 **1997, 1998, 1999, 2000, 2001, 2002, 2003, 2004, 2005, 2006** **Models:** TT, A4, A6, A6 Avant **Engines:** 1.8L, 2.8L **Transmissions:** All	**Internal Control Module Driver Error Conditions:** Ignition on, the ECM detected a control module malfunction. To achieve optimal anti-theft protection for the vehicle, an anti-theft immobilizer is installed. The anti-theft immobilizer is a system for enabling and locking the Engine Control Module (ECM). So that this system cannot be circumvented, it is necessary to perform adaptation of the anti-theft immobilizer using the Vehicle Diagnostic and Information System VAS 5052 in the On Board Diagnostic (OBD) function. The great availability of equipment options makes it necessary to adapt the Engine Control Module (ECM) to the vehicle (e.g. throttle valve control module or cruise control system). This "writing" function is not possible with the generic scan tool. **Possible Causes:** • (If ECM was replaced) ECM ID not the same as the replaced unit • ECM has failed • Voltage supply for Engine Control Module (ECM) has shorted
DTC: P1606 **1997, 1998, 1999, 2000, 2001, 2002, 2003, 2004, 2005, 2006** **Models:** A4, A6, A6 Avant **Engines:** 1.8L, 2.8L **Transmissions:** All	**Rough Road Spec Engine Torque ABS-ECU Electrical Malfunction Conditions:** Ignition on, the ECM detected an electrical malfunction. **Possible Causes:** • Check wire connection between Engine Control Module (ECM) and ABS Control Module
DTC: P1609 **1997, 1998, 1999, 2000, 2001** **Models:** A4 **Engines:** 1.8L **Transmissions:** All	**Crash Shut-Down Activated Conditions:** The ECM detected that the car has been in an accident. **Possible Causes:** • Check the vehicle for damage • Reset the ECU •
DTC: P1610 **1997, 1998, 1999, 2000, 2001, 2002, 2003, 2004, 2005, 2006** **Models:** TT, A4 **Engines:** 1.8L **Transmissions:** All	**ECU Defective Conditions:** To achieve optimal anti-theft protection for the vehicle, an anti-theft immobilizer is installed. The anti-theft immobilizer is a system for enabling and locking the Engine Control Module (ECM). So that this system cannot be circumvented, it is necessary to perform adaptation of the anti-theft immobilizer using the Vehicle Diagnostic and Information System VAS 5052 in the On Board Diagnostic (OBD) function. The great availability of equipment options makes it necessary to adapt the Engine Control Module (ECM) to the vehicle (e.g. throttle valve control module or cruise control system). This "writing" function is not possible with the generic scan tool. **Possible Causes:** • (If ECM was replaced) ECM ID not the same as the replaced unit • ECM has failed • Voltage supply for Engine Control Module (ECM) has shorted

DTC	Trouble Code Title, Conditions & Possible Causes
DTC: P1611 **1995, 1996, 1997** **Models:** A4, A6 **Engines:** 2.8L **Transmissions:** All	**MIL Call-Up Circuit, Transmission Control Module Short to Ground Conditions:** Engine started, VSS over 1 mph, and the ECM detected a problem in the Transmission Control system during the self-test. **Possible Causes:** • Open/short circuit to ground in the communication wire from the transmission to the ECM. • The ECM has failed
DTC: P1612 **1997, 1998, 1999, 2000, 2001, 2002, 2003, 2004, 2005, 2006** **Models:** TT, A4, A6, A8 **Engines:** 1.8L, 2.8L, 4.2L **Transmissions:** All	**Electronic Control Module Incorrect Coding Conditions:** Ignition on, the ECM detected a control module malfunction. To achieve optimal anti-theft protection for the vehicle, an anti-theft immobilizer is installed. The anti-theft immobilizer is a system for enabling and locking the Engine Control Module (ECM). So that this system cannot be circumvented, it is necessary to perform adaptation of the anti-theft immobilizer using the Vehicle Diagnostic and Information System VAS 5052 in the On Board Diagnostic (OBD) function. The great availability of equipment options makes it necessary to adapt the Engine Control Module (ECM) to the vehicle (e.g. throttle valve control module or cruise control system). This "writing" function is not possible with the generic scan tool. **Possible Causes:** • (If ECM was replaced) ECM ID not the same as the replaced unit • ECM has failed • Voltage supply for Engine Control Module (ECM) has shorted
DTC: P1613 **1995, 1996, 1997** **Models:** A4, A6 **Engines:** 2.8L **Transmissions:** All	**MIL Call-up Circuit Open/Short to B+ Conditions:** Engine started, VSS over 1 mph, and the ECM detected a problem in the Transmission Control system during the self-test. **Possible Causes:** • Open/short circuit to ground from the MIL to the ECM. • The ECM has failed • The MIL light has failed (check bulb)
DTC: P1624 **1997, 1998, 1999, 2000, 2001** **Models:** A4, A8 **Engines:** 1.8L, 2.8L, 3.7L, 4.2L **Transmissions:** All	**MIL Requested Signature Active Conditions:** Ignition on, the ECM detected a control module malfunction. To achieve optimal anti-theft protection for the vehicle, an anti-theft immobilizer is installed. The anti-theft immobilizer is a system for enabling and locking the Engine Control Module (ECM). So that this system cannot be circumvented, it is necessary to perform adaptation of the anti-theft immobilizer using the Vehicle Diagnostic and Information System VAS 5052 in the On Board Diagnostic (OBD) function. The great availability of equipment options makes it necessary to adapt the Engine Control Module (ECM) to the vehicle (e.g. throttle valve control module or cruise control system). This "writing" function is not possible with the generic scan tool. **Possible Causes:** • (If ECM was replaced) ECM ID not the same as the replaced unit • ECM has failed • Voltage supply for Engine Control Module (ECM) has shorted
DTC: P1626 **1997, 1998, 1999, 2000, 2001, 2002, 2003** **Models:** A4, A6, A8 **Engines:** 1.8L, 2.8L, 4.2L **Transmissions:** All	**Data BUS Powertrain Missing Message From Transmission Control Conditions:** Ignition on, the ECM detected a control module malfunction (Transmission). To achieve optimal anti-theft protection for the vehicle, an anti-theft immobilizer is installed. The anti-theft immobilizer is a system for enabling and locking the Engine Control Module (ECM). So that this system cannot be circumvented, it is necessary to perform adaptation of the anti-theft immobilizer using the Vehicle Diagnostic and Information System VAS 5052 in the On Board Diagnostic (OBD) function. The great availability of equipment options makes it necessary to adapt the Engine Control Module (ECM) to the vehicle (e.g. throttle valve control module or cruise control system). This "writing" function is not possible with the generic scan tool. **Possible Causes:** • (If ECM was replaced) ECM ID not the same as the replaced unit • ECM has failed • Voltage supply for Engine Control Module (ECM) has shorted
DTC: P1630 **1997, 1998, 1999, 2000, 2001, 2002, 2003, 2004, 2005, 2006** **Models:** TT, A4, A6, A8 **Engines:** 1.8L, 4.2L **Transmissions:** All	**Acceleration Pedal Position Sensor 1 Signal Too Low Conditions:** Engine started, battery voltage at least 11.5v, all electrical components off, ground connections between engine and chassis well connected, the ECM detected that the accelerator pedal position sensor signal was too low. **Note: Both the Throttle Position (TP) Sensor and Accelerator Pedal Position Sensor 2 are located at the accelerator pedal module and communicate the driver's intentions to the ECM completely independently of each other. Both sensors are stored in one housing.** **Possible Causes:** • Ground between engine and chassis may be broken • Throttle position sensor may have failed • Accelerator Pedal Position Sensor 2 has failed • Throttle position sensor wiring may have shorted • Faulty voltage supply • ECM has failed

DTC	Trouble Code Title, Conditions & Possible Causes
DTC: P1631 **1997, 1998, 1999, 2000, 2001, 2002, 2003, 2004, 2005, 2006** **Models:** TT, A4, A6, A8 **Engines:** 1.8L, 4.2L **Transmissions:** All	**Acceleration Pedal Position Sensor 1 Signal Too High Conditions:** Engine started, battery voltage at least 11.5v, all electrical components off, ground connections between engine and chassis well connected, the ECM detected that the accelerator pedal position sensor signal was too high. **Note: Both the Throttle Position (TP) Sensor and Accelerator Pedal Position Sensor 2 are located at the accelerator pedal module and communicate the driver's intentions to the ECM completely independently of each other. Both sensors are stored in one housing.** **Possible Causes:** • Ground between engine and chassis may be broken • Throttle position sensor may have failed • Accelerator Pedal Position Sensor 2 has failed • Throttle position sensor wiring may have shorted • Faulty voltage supply • ECM has failed
DTC: P1633 **1997, 1998, 1999, 2000, 2001, 2002, 2003, 2004, 2005, 2006** **Models:** TT, A4, A6, A8 **Engines:** 1.8L, 4.2L **Transmissions:** All	**Acceleration Pedal Position Sensor 2 Signal Too Low Conditions:** Engine started, battery voltage at least 11.5v, all electrical components off, ground connections between engine and chassis well connected, the ECM detected that the accelerator pedal position sensor signal was too low. **Note: Both the Throttle Position (TP) Sensor and Accelerator Pedal Position Sensor 2 are located at the accelerator pedal module and communicate the driver's intentions to the ECM completely independently of each other. Both sensors are stored in one housing.** **Possible Causes:** • Ground between engine and chassis may be broken • Throttle position sensor may have failed • Accelerator Pedal Position Sensor 2 has failed • Throttle position sensor wiring may have shorted • Faulty voltage supply • ECM has failed
DTC: P1634 **1997, 1998, 1999, 2000, 2001, 2002, 2003, 2004, 2005, 2006** **Models:** TT, A4, A6, A8 **Engines:** 1.8L, 4.2L **Transmissions:** All	**Acceleration Pedal Position Sensor 2 Signal Too High Conditions:** Engine started, battery voltage at least 11.5v, all electrical components off, ground connections between engine and chassis well connected, the ECM detected that the accelerator pedal position sensor signal was too high. **Note: Both the Throttle Position (TP) Sensor and Accelerator Pedal Position Sensor 2 are located at the accelerator pedal module and communicate the driver's intentions to the ECM completely independently of each other. Both sensors are stored in one housing.** **Possible Causes:** • Ground between engine and chassis may be broken • Throttle position sensor may have failed • Accelerator Pedal Position Sensor 2 has failed • Throttle position sensor wiring may have shorted • Faulty voltage supply • ECM has failed
DTC: P1639 **1997, 1998, 1999, 2000, 2001, 2002, 2003, 2004, 2005, 2006** **Models:** TT, A4, A6, A8 **Engines:** 1.8L, 4.2L **Transmissions:** All	**Accelerator Pedal Position Sensor 1+2 Range/Performance Conditions:** Engine started, battery voltage at least 11.5v, all electrical components off, ground connections between engine and chassis well connected, the ECM detected that the accelerator pedal position sensor signal was too high. **Note: Both the Throttle Position (TP) Sensor and Accelerator Pedal Position Sensor 2 are located at the accelerator pedal module and communicate the driver's intentions to the ECM completely independently of each other. Both sensors are stored in one housing.** **Possible Causes:** • Ground between engine and chassis may be broken • Throttle position sensor may have failed • Accelerator Pedal Position Sensor 2 has failed • Throttle position sensor wiring may have shorted • Faulty voltage supply • ECM has failed
DTC: P1640 **1995, 1996, 1997, 1998, 1999, 2000, 2001, 2002, 2003, 2004, 2005, 2006** **Models:** TT, A4, A6, A8 **Engines:** 1.8L, 2.8L, 3.7L, 4.2L **Transmissions:** All	**Internal Control Module (EEPROM) Error Conditions:** Ignition on, the ECM detected a control module malfunction (software). To achieve optimal anti-theft protection for the vehicle, an anti-theft immobilizer is installed. The anti-theft immobilizer is a system for enabling and locking the Engine Control Module (ECM). So that this system cannot be circumvented, it is necessary to perform adaptation of the anti-theft immobilizer using the Vehicle Diagnostic and Information System VAS 5052 in the On Board Diagnostic (OBD) function. The great availability of equipment options makes it necessary to adapt the Engine Control Module (ECM) to the vehicle (e.g. throttle valve control module or cruise control system). This "writing" function is not possible with the generic scan tool. **Possible Causes:** • Engine Control Module (ECM) has failed • Voltage supply for Engine Control Module (ECM) has shorted

DTC	Trouble Code Title, Conditions & Possible Causes
DTC: P1647 **2004, 2005, 2006** **Models:** TT **Engines:** 3.2L **Transmissions:** All	**Please Check Coding of the ECUs in the Data Bus Powertrain Conditions:** Ignition on, the ECM detected a control module malfunction (software). To achieve optimal anti-theft protection for the vehicle, an anti-theft immobilizer is installed. The anti-theft immobilizer is a system for enabling and locking the Engine Control Module (ECM). So that this system cannot be circumvented, it is necessary to perform adaptation of the anti-theft immobilizer using the Vehicle Diagnostic and Information System VAS 5052 in the On Board Diagnostic (OBD) function. The great availability of equipment options makes it necessary to adapt the Engine Control Module (ECM) to the vehicle (e.g. throttle valve control module or cruise control system). This "writing" function is not possible with the generic scan tool. **Possible Causes:** • Engine Control Module (ECM) has failed • Voltage supply for Engine Control Module (ECM) has shorted
DTC: P1648 **1997, 1998, 1999, 2000, 2001, 2002, 2003, 2004, 2005, 2006** **Models:** TT, A4, A6, A8 **Engines:** 1.8L, 4.2L **Transmissions:** All	**Data Bus Powertrain Malfunction Conditions:** Ignition on, the ECM detected a data bus malfunction (software). To achieve optimal anti-theft protection for the vehicle, an anti-theft immobilizer is installed. The anti-theft immobilizer is a system for enabling and locking the Engine Control Module (ECM). So that this system cannot be circumvented, it is necessary to perform adaptation of the anti-theft immobilizer using the Vehicle Diagnostic and Information System VAS 5052 in the On Board Diagnostic (OBD) function. The great availability of equipment options makes it necessary to adapt the Engine Control Module (ECM) to the vehicle (e.g. throttle valve control module or cruise control system). This "writing" function is not possible with the generic scan tool. **Possible Causes:** • Ground between engine and chassis may be broken • Throttle position sensor may have failed • Accelerator Pedal Position Sensor 2 has failed • Throttle position sensor wiring may have shorted • Faulty voltage supply • ECM has failed
DTC: P1649 **1997, 1998, 1999, 2000, 2001, 2002, 2003, 2004, 2005, 2006** **Models:** TT, A4 **Engines:** 1.8L **Transmissions:** All	**Data Bus Powertrain Missing Message from ABS Control Module Conditions:** Ignition off, the ECU is missing general Data BUS information from the central electrical control. The Engine Control Module (ECM) communicates with all databus-capable control modules via a CAN databus. These databus-capable control modules are connected via two data bus wires which are twisted together (CAN_High and CAN_Low), and exchange information (messages). Missing information on the databus is recognized as a malfunction and stored. Trouble-free operation of the CAN-bus requires that it have a terminal resistance. This central terminal resistor is located in the Engine Control Module (ECM). **Possible Causes:** • Ground between engine and chassis may be broken • Throttle position sensor may have failed • Accelerator Pedal Position Sensor 2 has failed • Throttle position sensor wiring may have shorted • Faulty voltage supply • Check the Terminal resistance for CAN-bus • Data-Bus wires have short • Data-Bus components are malfunctioning • ECM has failed
DTC: P1654 **1997, 1998, 1999, 2000, 2001** **Models:** A4 **Engines:** 1.8L **Transmissions:** All	**Please Check DTC Memory of the Control Panel ECU Conditions:** The Engine Control Module (ECM) communicates with all databus-capable control modules via a CAN databus. These databus-capable control modules are connected via two data bus wires which are twisted together (CAN_High and CAN_Low), and exchange information (messages). Missing information on the databus is recognized as a malfunction and stored. Trouble-free operation of the CAN-bus requires that it have a terminal resistance. This central terminal resistor is located in the Engine Control Module (ECM). **Possible Causes:** • Ground between engine and chassis may be broken • Faulty voltage supply • ECM has failed
DTC: P1676 **1997, 1998, 1999, 2000, 2001, 2002, 2003, 2004, 2005, 2006** **Models:** TT, A4 **Engines:** 1.8L **Transmissions:** All	**Drive by Wire-MIL Circuit Electrical Malfunction Conditions:** Key on or engine running, the ECM detected an electrical malfunction regarding the drive-by-wire circuit. **Note: EPC" is an abbreviation and stands for Electronic Power Control and means "electronic engine load control". If malfunctions are recognized in the EPC system during operation of the engine, the Engine Control Module (ECM) switches on the EPC warning lamp. An entry is made in DTC memory at the same time. After a few seconds of the engine at idle, the EPC should extinguish itself.** **Possible Causes:** • Circuit from the MIL to the ECM • ECM has failed • Circuit from the EPC to the ECM

DTC	Trouble Code Title, Conditions & Possible Causes
DTC: P1677 **1997, 1998, 1999, 2000, 2001, 2002, 2003, 2004, 2005, 2006** **Models:** TT, A4 **Engines:** 1.8L **Transmissions:** All	**Drive by Wire-MIL Circuit Short to B+ Conditions:** Key on or engine running, the ECM detected an electrical malfunction regarding the drive-by-wire circuit. **Note: EPC" is an abbreviation and stands for Electronic Power Control and means "electronic engine load control". If malfunctions are recognized in the EPC system during operation of the engine, the Engine Control Module (ECM) switches on the EPC warning lamp. An entry is made in DTC memory at the same time. After a few seconds of the engine at idle, the EPC should extinguish itself.** **Possible Causes:** • Circuit from the MIL to the ECM • ECM has failed • Circuit from the EPC to the ECM
DTC: P1681 **1997, 1998, 1999, 2000, 2001** **Models:** A4 **Engines:** 1.8L **Transmissions:** All	**Control Unit Programming, Programming Not Finished Conditions:** The Engine Control Module (ECM) communicates with all databus-capable control modules via a CAN databus. These databus-capable control modules are connected via two data bus wires which are twisted together (CAN_High and CAN_Low), and exchange information (messages). Missing information on the databus is recognized as a malfunction and stored. Trouble-free operation of the CAN-bus requires that it have a terminal resistance. This central terminal resistor is located in the Engine Control Module (ECM). **Possible Causes:** • ECM has failed
DTC: P1690 **1997, 1998, 1999, 2000, 2001, 2002, 2003, 2004, 2005, 2006** **Models:** TT, A4, A6, A8 **Engines:** 1.8L, 2.8L, 4.2L **Transmissions:** All	**Malfunction Indication Light Malfunction Conditions:** The exhaust Malfunction Indicator Lamp (MIL) lights up when exhaust relevant malfunctions are recognized by the Engine Control Module (ECM). The Malfunction Indicator Lamp (MIL) can blink or remain lit continuously. Blinking: There is a malfunction that causes damage to the catalytic converter in this driving condition. In this case, vehicle must only be driven at reduced power! Continuously lit: There is a malfunction that causes increased emissions. Check DTC memory for Motronic control module. DTC memory must still be checked if there are driveability problems or customer complaints and the MIL is not lit, since malfunctions can be stored without causing the MIL to light immediately. **Possible Causes:** • Wire from ECM to MIL is shorted or grounded • ECM has failed • MIL has failed
DTC: P1691 **1997, 1998, 1999, 2000, 2001, 2002, 2003, 2004, 2005, 2006** **Models:** TT, A4 **Engines:** 1.8L **Transmissions:** All	**Malfunction Indication Light Open Conditions:** The exhaust Malfunction Indicator Lamp (MIL) lights up when exhaust relevant malfunctions are recognized by the Engine Control Module (ECM). The Malfunction Indicator Lamp (MIL) can blink or remain lit continuously. Blinking: There is a malfunction that causes damage to the catalytic converter in this driving condition. In this case, vehicle must only be driven at reduced power! Continuously lit: There is a malfunction that causes increased emissions. Check DTC memory for Motronic control module. DTC memory must still be checked if there are driveability problems or customer complaints and the MIL is not lit, since malfunctions can be stored without causing the MIL to light immediately. **Possible Causes:** • Wire from ECM to MIL is shorted or grounded • ECM has failed • MIL has failed
DTC: P1692 **1997, 1998, 1999, 2000, 2001, 2002, 2003, 2004, 2005, 2006** **Models:** TT, A4 **Engines:** 1.8L **Transmissions:** All	**Malfunction Indication Light Short to Ground Conditions:** The exhaust Malfunction Indicator Lamp (MIL) lights up when exhaust relevant malfunctions are recognized by the Engine Control Module (ECM). The Malfunction Indicator Lamp (MIL) can blink or remain lit continuously. Blinking: There is a malfunction that causes damage to the catalytic converter in this driving condition. In this case, vehicle must only be driven at reduced power! Continuously lit: There is a malfunction that causes increased emissions. Check DTC memory for Motronic control module. DTC memory must still be checked if there are driveability problems or customer complaints and the MIL is not lit, since malfunctions can be stored without causing the MIL to light immediately. **Possible Causes:** • Wire from ECM to MIL is shorted or grounded • ECM has failed • MIL has failed
DTC: P1693 **1997, 1998, 1999, 2000, 2001, 2002, 2003, 2004, 2005, 2006** **Models:** TT, A4, A6, A8 **Engines:** 1.8L, 4.2L **Transmissions:** All	**Malfunction Indication Light Short to B+ Conditions:** The exhaust Malfunction Indicator Lamp (MIL) lights up when exhaust relevant malfunctions are recognized by the Engine Control Module (ECM). The Malfunction Indicator Lamp (MIL) can blink or remain lit continuously. Blinking: There is a malfunction that causes damage to the catalytic converter in this driving condition. In this case, vehicle must only be driven at reduced power! Continuously lit: There is a malfunction that causes increased emissions. Check DTC memory for Motronic control module. DTC memory must still be checked if there are driveability problems or customer complaints and the MIL is not lit, since malfunctions can be stored without causing the MIL to light immediately. **Possible Causes:** • Wire from ECM to MIL is shorted or grounded • ECM has failed • MIL has failed

DTC	Trouble Code Title, Conditions & Possible Causes
DTC: P1702 **2005, 2006** **Models:** A6 **Engines:** 3.2L, 4.2L **Transmissions:** All	**TR Sensor Signal Intermittent Conditions:** Key on or engine running; and the ECM detected the failure Trouble Code Conditions for DTC P0705 or P0708 were met intermittently. **Possible Causes:** • Refer to the appropriate Transmission Repair Manual or information in electronic media to perform a complete diagnosis of the automatic transmission when this code is set
DTC: P1823 **2005, 2006** **Models:** A8 **Engines:** 3.2L, 4.2L **Transmissions:** All	**Pressure Control Solenoid 3 Electrical Conditions:** Engine started, vehicle driven with the solenoid applied, and the ECM detected an unexpected voltage condition on the SS1/C solenoid circuit was incorrect during the test. **Possible Causes:** • Solenoid valves in valve body are faulty • Solenoid circuit is shorting to ground • Solenoid circuit is open • TCM has failed or wiring is shorting • Check harness connector for contact corrosion or water damage • Check resistance of solenoid valves may not be up to specification • Wires to Transmission Control Module (TCM) may have ground out or open • Transmission Control Module (TCM) has failed • ECM has failed
DTC: P1828 **2005, 2006** **Models:** A8 **Engines:** 3.2L, 4.2L **Transmissions:** All	**Pressure Control Solenoid 4 Electrical Conditions:** Engine started, vehicle driven with the solenoid applied, and the ECM detected an unexpected voltage condition on the SS1/C solenoid circuit was incorrect during the test. **Possible Causes:** • Solenoid valves in valve body are faulty • Solenoid circuit is shorting to ground • Solenoid circuit is open • TCM has failed or wiring is shorting • Check harness connector for contact corrosion or water damage • Check resistance of solenoid valves may not be up to specification • Wires to Transmission Control Module (TCM) may have ground out or open • Transmission Control Module (TCM) has failed • ECM has failed
DTC: P1850 **2005, 2006** **Models:** A8 **Engines:** 3.2L, 4.2L **Transmissions:** All	**Data BUS Powertrain Missing Message from Engine Control Conditions:** The Engine Control Module (ECM) communicates with all databus-capable control modules via a CAN databus. These databus-capable control modules are connected via two data bus wires which are twisted together (CAN_High and CAN_Low), and exchange information (messages). Missing information on the databus is recognized as a malfunction and stored. Trouble-free operation of the CAN-bus requires that it have a terminal resistance. This central terminal resistor is located in the Engine Control Module (ECM). **Possible Causes:** • Check the Terminal resistance for CAN-bus • Data-Bus wires have short • Data-Bus components are malfunctioning • ECM has failed
DTC: P1854 **1996, 1997, 1998, 1999** **Models:** A4, A8 **Engines:** 2.8L, 3.2L, 4.2L **Transmissions:** All	**Data BUS Powertrain Hardware Defective Conditions:** The Engine Control Module (ECM) communicates with all databus-capable control modules via a CAN databus. These databus-capable control modules are connected via two data bus wires which are twisted together (CAN_High and CAN_Low), and exchange information (messages). Missing information on the databus is recognized as a malfunction and stored. Trouble-free operation of the CAN-bus requires that it have a terminal resistance. This central terminal resistor is located in the Engine Control Module (ECM). **Possible Causes:** • Check the Terminal resistance for CAN-bus • Data-Bus wires have short • Data-Bus components are malfunctioning • ECM has failed

DTC	Trouble Code Title, Conditions & Possible Causes
DTC: P1855 **1996, 1997, 1998, 1999** **Models:** A4, A8 **Engines:** 2.8L, 3.2L, 4.2L **Transmissions:** All	**Data BUS Powertrain Software Version Control Conditions:** The Engine Control Module (ECM) communicates with all databus-capable control modules via a CAN databus. These databus-capable control modules are connected via two data bus wires which are twisted together (CAN_High and CAN_Low), and exchange information (messages). Missing information on the databus is recognized as a malfunction and stored. Trouble-free operation of the CAN-bus requires that it have a terminal resistance. This central terminal resistor is located in the Engine Control Module (ECM). **Possible Causes:** • Check the Terminal resistance for CAN-bus • Data-Bus wires have short • Data-Bus components are malfunctioning • ECM has failed
DTC: P1857 **1996** **Models:** A4 **Engines:** 2.8L **Transmissions:** All	**Load Signal Error Message From Engine Control Conditions:** The Engine Control Module (ECM) communicates with all databus-capable control modules via a CAN databus. These databus-capable control modules are connected via two data bus wires which are twisted together (CAN_High and CAN_Low), and exchange information (messages). Missing information on the databus is recognized as a malfunction and stored. Trouble-free operation of the CAN-bus requires that it have a terminal resistance. This central terminal resistor is located in the Engine Control Module (ECM). **Possible Causes:** • Intake Manifold Runner Position Sensor is faulty • Intake system has leaks (false air) • Motor for intake flap is faulty • Mass Air Flow (MAF) sensor has failed • ECM has failed
DTC: P1861 **1996** **Models:** A4 **Engines:** 2.8L **Transmissions:** All	**Throttle Position Sensor Message from ECM Conditions:** The Engine Control Module (ECM) communicates with all databus-capable control modules via a CAN databus. These databus-capable control modules are connected via two data bus wires which are twisted together (CAN_High and CAN_Low), and exchange information (messages). Missing information on the databus is recognized as a malfunction and stored. Trouble-free operation of the CAN-bus requires that it have a terminal resistance. This central terminal resistor is located in the Engine Control Module (ECM). **Note: Both the Throttle Position (TP) Sensor and Accelerator Pedal Position Sensor 2 are located at the accelerator pedal module and communicate the driver's intentions to the ECM completely independently of each other. Both sensors are stored in one housing.** **Possible Causes:** • Throttle Position (TP) Sensor has failed • Accelerator Pedal Position Sensor 2 has failed • ECM has failed • Ground (GND) connections between engine and chassis must be OK • Engine Control Module (ECM) may not connected
DTC: P1912 **1997, 1998, 1999, 2000, 2001** **Models:** A4 **Engines:** 2.8L **Transmissions:** All	**Brake Booster Pressure Sensor Short Circuit to B+ Conditions:** Key on or engine running, the ECM detected an error with the brake booster pressure sensor signal. **Possible Causes:** • Circuit short to ground or open • Brake booster pressure sensor has failed or is dirty • ECM has failed
DTC: P1913 **1997, 1998, 1999, 2000, 2001** **Models:** A4 **Engines:** 2.8L **Transmissions:** All	**Brake Booster Pressure Sensor Short Circuit to Ground Conditions:** Key on or engine running, the ECM detected an error with the brake booster pressure sensor signal. **Possible Causes:** • Circuit short to ground or open • Brake booster pressure sensor has failed or is dirty • ECM has failed

DTC	Trouble Code Title, Conditions & Possible Causes:
DTC: P2004 **2T CCM, MIL: Yes** **2005, 2006** **Models:** A6 **Engines:** 3.2L **Transmissions:** All	**Intake Manifold Runner Control Stuck Open Bank 1 Conditions:** Engine started, battery voltage must be at least 11.5v, all electrical components must be off, the ground between the engine and the chassis must be well connected. The ECM detected an unexpected voltage condition on the Intake Manifold Runner Control circuit during the CCM test period (i.e., the valve may be stuck open). **Note: Intake Flap Motor and Intake Manifold Runner Position Sensor are one component and cannot be replaced individually.** **Possible Causes:** • Test for a sticking Accelerator or speed control cable condition: Turn the key off and disconnect accelerator and speed control cable from the throttle body. Rotate the throttle body linkage to determine if it rotates freely (the throttle body may have failed). • Check the air cleaner and air inlet assembly for restrictions • Check the IAC motor response (it may be damaged or sticking) • Check the PCV system (valve and hoses) for leaks or plugging • Check for signs of vacuum leaks in the engine or components • Test TP sensor signal (due a sweep test at key on, engine off)
DTC: P2008 **2T CCM, MIL: Yes** **2005, 2006** **Models:** A6 **Engines:** 3.2L **Transmissions:** All	**Intake Manifold Runner Control Circuit/Open Bank 1 Conditions:** Engine started, battery voltage must be at least 11.5v, all electrical components must be off, the ground between the engine and the chassis must be well connected. The ECM detected an unexpected voltage condition on the Intake Manifold Runner Control circuit during the CCM test period (i.e., the valve may be stuck open). **Note: Intake Flap Motor and Intake Manifold Runner Position Sensor are one component and cannot be replaced individually.** **Possible Causes:** • Accelerator or speed control cable sticking or binding. To test for this condition, turn the key off. Then disconnect the accelerator and speed control cable from the throttle body. Then rotate the throttle body linkage to determine if it rotates freely. If it is sticking, the throttle body may need replacement. • Check the air cleaner and air inlet assembly for restrictions • Check the IAC motor response (it may be damaged or sticking) • Check the PCV system (valve and hoses) for leaks or plugging • Check for signs of vacuum leaks in the engine or components • Test TP sensor signal
DTC: P2009 **2T CCM, MIL: Yes** **2005, 2006** **Models:** A6 **Engines:** 3.2L **Transmissions:** All	**Intake Manifold Runner Control Circuit Low Bank 1 Conditions:** Engine started, battery voltage must be at least 11.5v, all electrical components must be off, the ground between the engine and the chassis must be well connected. The ECM detected an unexpected voltage condition on the Intake Manifold Runner Control circuit during the CCM test period (i.e., the valve may be stuck open). **Note: Intake Flap Motor and Intake Manifold Runner Position Sensor are one component and cannot be replaced individually.** **Possible Causes:** • Accelerator or speed control cable sticking or binding. To test for this condition, turn the key off. Then disconnect the accelerator and speed control cable from the throttle body. Then rotate the throttle body linkage to determine if it rotates freely. If it is sticking, the throttle body may need replacement. • Check the air cleaner and air inlet assembly for restrictions • Check the IAC motor response (it may be damaged or sticking) • Check the PCV system (valve and hoses) for leaks or plugging • Check for signs of vacuum leaks in the engine or components • Test TP sensor signal
DTC: P2014 **2T CCM, MIL: Yes** **2005, 2006** **Models:** A6 **Engines:** 3.2L **Transmissions:** All	**Intake Manifold Runner Position Sensor/Switch Circuit Bank 1 Conditions:** Engine started, battery voltage must be at least 11.5v, all electrical components must be off, the ground between the engine and the chassis must be well connected. The ECM detected an unexpected voltage condition on the Intake Manifold Runner Control circuit during the CCM test period (i.e., the valve may be stuck open). **Note: Intake Flap Motor and Intake Manifold Runner Position Sensor are one component and cannot be replaced individually.** **Possible Causes:** • Accelerator or speed control cable sticking or binding. To test for this condition, turn the key off. Then disconnect the accelerator and speed control cable from the throttle body. Then rotate the throttle body linkage to determine if it rotates freely. If it is sticking, the throttle body may need replacement. • Check the air cleaner and air inlet assembly for restrictions • Check the IAC motor response (it may be damaged or sticking) • Check the PCV system (valve and hoses) for leaks or plugging • Check for signs of vacuum leaks in the engine or components • Test TP sensor signal

DTC	Trouble Code Title, Conditions & Possible Causes
DTC: P2015 **2T CCM, MIL: Yes** **2005, 2006** **Models:** A6 **Engines:** 3.2L **Transmissions:**: All	**Intake Manifold Runner Position Sensor/Switch Circuit Range/Performance Bank 1 Conditions:** Engine started, battery voltage must be at least 11.5v, all electrical components must be off, the ground between the engine and the chassis must be well connected. The ECM detected an unexpected voltage condition on the Intake Manifold Runner Control circuit during the CCM test period (i.e., the valve may be stuck open). **Note: Intake Flap Motor and Intake Manifold Runner Position Sensor are one component and cannot be replaced individually.** **Possible Causes:** • Accelerator or speed control cable sticking or binding. To test for this condition, turn the key off. Then disconnect the accelerator and speed control cable from the throttle body. Then rotate the throttle body linkage to determine if it rotates freely. If it is sticking, the throttle body may need replacement. • Check the air cleaner and air inlet assembly for restrictions • Check the IAC motor response (it may be damaged or sticking) • Check the PCV system (valve and hoses) for leaks or plugging • Check for signs of vacuum leaks in the engine or components • Test TP sensor signal
DTC: P2017 **2T CCM, MIL: Yes** **2005, 2006** **Models:** A6 **Engines:** 3.2L **Transmissions:** All	**Intake Manifold Runner Position Sensor/Switch Circuit High Bank 1 Conditions:** Engine started, battery voltage must be at least 11.5v, all electrical components must be off, the ground between the engine and the chassis must be well connected. The ECM detected an unexpected voltage condition on the Intake Manifold Runner Control circuit during the CCM test period (i.e., the valve may be stuck open). **Note: Intake Flap Motor and Intake Manifold Runner Position Sensor are one component and cannot be replaced individually.** **Possible Causes:** • Accelerator or speed control cable sticking or binding. To test for this condition, turn the key off. Then disconnect the accelerator and speed control cable from the throttle body. Then rotate the throttle body linkage to determine if it rotates freely. If it is sticking, the throttle body may need replacement. • Check the air cleaner and air inlet assembly for restrictions • Check the IAC motor response (it may be damaged or sticking) • Check the PCV system (valve and hoses) for leaks or plugging • Check for signs of vacuum leaks in the engine or components • Test TP sensor signal
DTC: P2088 **2T CCM, MIL: Yes** **2005, 2006** **Models:** A6 **Engines:** 3.2L, 4.2L **Transmissions:** All	**"A" Camshaft Position Control Circuit Low Bank 1 Conditions:** Key on or engine running; and the ECM detected an unexpected voltage condition on the Camshaft Position Control circuit during the CCM test period. The relative position between the camshaft and crankshaft needs to be optimal so the engine has better torque, fuel economy and emissions. **Note: camshaft adjustment is load- and RPM dependant. The electrical camshaft adjustment valve 1 switches oil pressure onto camshaft adjuster (mechanical adjustment mechanism), which adjusts the camshaft.** **Possible Causes:** • Camshaft position control wiring harness connector is damaged or open • Camshaft adjustment valve has failed • Circuit is open or grounded • Assembly is damaged or it has failed (an open circuit) • ECM power supply relay has failed • ECM has failed
DTC: P2089 **2T CCM, MIL: Yes** **2005, 2006** **Models:** A6 **Engines:** 3.2L, 4.2L **Transmissions:** All	**"A" Camshaft Position Control Circuit High Bank 1 Conditions:** Key on or engine running; and the ECM detected an unexpected voltage condition on the Camshaft Position Control circuit during the CCM test period. The relative position between the camshaft and crankshaft needs to be optimal so the engine has better torque, fuel economy and emissions. **Note: camshaft adjustment is load- and RPM dependant. The electrical camshaft adjustment valve 1 switches oil pressure onto camshaft adjuster (mechanical adjustment mechanism), which adjusts the camshaft.** **Possible Causes:** • Camshaft position control wiring harness connector is damaged or open • Camshaft adjustment valve has failed • Circuit is open or grounded • Assembly is damaged or it has failed (an open circuit) • ECM power supply relay has failed • ECM has failed

DTC	Trouble Code Title, Conditions & Possible Causes
DTC: P2090 **2T CCM, MIL: Yes** **2005, 2006** **Models:** A6 **Engines:** 3.2L **Transmissions:** All	**"B" Camshaft Position Control Circuit Low Bank 1 Conditions:** Key on or engine running; and the ECM detected an unexpected voltage condition on the Camshaft Position Control circuit during the CCM test period. The relative position between the camshaft and crankshaft needs to be optimal so the engine has better torque, fuel economy and emissions. **Note: camshaft adjustment is load- and RPM dependant. The electrical camshaft adjustment valve 1 switches oil pressure onto camshaft adjuster (mechanical adjustment mechanism), which adjusts the camshaft.** **Possible Causes:** • Camshaft position control wiring harness connector is damaged or open • Camshaft adjustment valve has failed • Circuit is open or grounded • Assembly is damaged or it has failed (an open circuit) • ECM power supply relay has failed • ECM has failed
DTC: P2091 **2T CCM, MIL: Yes** **2005, 2006** **Models:** A6 **Engines:** 3.2L **Transmissions:** All	**"B" Camshaft Position Control Circuit High Bank 1 Conditions:** Key on or engine running; and the ECM detected an unexpected voltage condition on the Camshaft Position Control circuit during the CCM test period. The relative position between the camshaft and crankshaft needs to be optimal so the engine has better torque, fuel economy and emissions. **Note: camshaft adjustment is load- and RPM dependant. The electrical camshaft adjustment valve 1 switches oil pressure onto camshaft adjuster (mechanical adjustment mechanism), which adjusts the camshaft.** **Possible Causes:** • Camshaft position control wiring harness connector is damaged or open • Camshaft adjustment valve has failed • Circuit is open or grounded • Assembly is damaged or it has failed (an open circuit) • ECM power supply relay has failed • ECM has failed
DTC: P2094 **2T CCM, MIL: Yes** **2005, 2006** **Models:** A6 **Engines:** 3.2L **Transmissions:** All	**"B" Camshaft Position Control Circuit Low Bank 2 Conditions:** Key on or engine running; and the ECM detected an unexpected voltage condition on the Camshaft Position Control circuit during the CCM test period. The relative position between the camshaft and crankshaft needs to be optimal so the engine has better torque, fuel economy and emissions. **Note: camshaft adjustment is load- and RPM dependant. The electrical camshaft adjustment valve 1 switches oil pressure onto camshaft adjuster (mechanical adjustment mechanism), which adjusts the camshaft.** **Possible Causes:** • Camshaft position control wiring harness connector is damaged or open • Camshaft adjustment valve has failed • Circuit is open or grounded • Assembly is damaged or it has failed (an open circuit) • ECM power supply relay has failed • ECM has failed
DTC: P2095 **2T CCM, MIL: Yes** **2005, 2006** **Models:** A6 **Engines:** 3.2L **Transmissions:** All	**"B" Camshaft Position Control Circuit High Bank 2 Conditions:** Key on or engine running; and the ECM detected an unexpected voltage condition on the Camshaft Position Control circuit during the CCM test period. The relative position between the camshaft and crankshaft needs to be optimal so the engine has better torque, fuel economy and emissions. **Note: camshaft adjustment is load- and RPM dependant. The electrical camshaft adjustment valve 1 switches oil pressure onto camshaft adjuster (mechanical adjustment mechanism), which adjusts the camshaft.** **Possible Causes:** • Camshaft position control wiring harness connector is damaged or open • Camshaft adjustment valve has failed • Circuit is open or grounded • Assembly is damaged or it has failed (an open circuit) • ECM power supply relay has failed • ECM has failed

DTC	Trouble Code Title, Conditions & Possible Causes
DTC: P2096 **2T CCM, MIL: Yes** **1996, 1997, 1998, 1999, 2000, 2001, 2002, 2003, 2004, 2005, 2006** **Models:** TT, A4, A6, Allroad Quattro, A8, A8L **Engines:** 1.8L, 2.7L, 2.8L, 3.2L, 4.2L **Transmissions:** All	**Post Catalyst Fuel Trim System Too Lean (Bank 1) Conditions:** Engine started, battery voltage must be at least 11.5v, all electrical components must be off, the ground between the engine and the chassis must be well connected, the exhaust system must be properly sealed between the catalytic converter and the cylinder head, and the oxygen sensor heater for oxygen sensor before the catalytic converter must be properly functioning. The ECM detected a problem with the fuel mixture. **Note: For resistance testing of sensor heating, oxygen sensor should be cooled to ambient temperature. High temperatures at oxygen sensor may lead to inaccurate measurements.** **Possible Causes:** • Oxygen sensor (before catalytic converter) is faulty • Oxygen sensor (behind catalytic converter) is faulty • Oxygen sensor heater (before catalytic converter) is faulty • Oxygen sensor heater (behind catalytic converter) is faulty • Check circuits for shorts to each other, ground or power • ECM has failed
DTC: P2097 **2T CCM, MIL: Yes** **1996, 1997, 1998, 1999, 2000, 2001, 2002, 2003, 2004, 2005, 2006** **Models:** TT, A4, A6, Allroad Quattro, A8, A8L **Engines:** 1.8L, 2.7L, 2.8L, 3.2L, 4.2L **Transmissions:** All	**Post Catalyst Fuel Trim System Too Rich (Bank 1) Conditions:** Engine started, battery voltage must be at least 11.5v, all electrical components must be off, the ground between the engine and the chassis must be well connected, the exhaust system must be properly sealed between the catalytic converter and the cylinder head, and the oxygen sensor heater for oxygen sensor before the catalytic converter must be properly functioning. The ECM detected a problem with the fuel mixture. **Note: For resistance testing of sensor heating, oxygen sensor should be cooled to ambient temperature. High temperatures at oxygen sensor may lead to inaccurate measurements.** **Possible Causes:** • Oxygen sensor (before catalytic converter) is faulty • Oxygen sensor (behind catalytic converter) is faulty • Oxygen sensor heater (before catalytic converter) is faulty • Oxygen sensor heater (behind catalytic converter) is faulty • Check circuits for shorts to each other, ground or power • ECM has failed
DTC: P2098 **2T CCM, MIL: Yes** **1996, 1997, 1998, 1999, 2000, 2001, 2002, 2003, 2004, 2005, 2006** **Models:** A4, A6, Allroad Quattro, A8, A8L **Engines:** 1.8L, 2.7L, 2.8L, 3.0L, 3.2L, 4.2L **Transmissions:** All	**Post Catalyst Fuel Trim System Too Lean (Bank 2) Conditions:** Engine started, battery voltage must be at least 11.5v, all electrical components must be off, the ground between the engine and the chassis must be well connected, the exhaust system must be properly sealed between the catalytic converter and the cylinder head, and the oxygen sensor heater for oxygen sensor before the catalytic converter must be properly functioning. The ECM detected a problem with the fuel mixture. **Note: For resistance testing of sensor heating, oxygen sensor should be cooled to ambient temperature. High temperatures at oxygen sensor may lead to inaccurate measurements.** **Possible Causes:** • Oxygen sensor (before catalytic converter) is faulty • Oxygen sensor (behind catalytic converter) is faulty • Oxygen sensor heater (before catalytic converter) is faulty • Oxygen sensor heater (behind catalytic converter) is faulty • Check circuits for shorts to each other, ground or power • ECM has failed
DTC: P2099 **2T CCM, MIL: Yes** **1996, 1997, 1998, 1999, 2000, 2001, 2002, 2003, 2004, 2005, 2006** **Models:** A4, A6, Allroad Quattro, A8, A8L **Engines:** 1.8L, 2.7L, 2.8L, 3.0L, 3.2L, 4.2L **Transmissions:** All	**Post Catalyst Fuel Trim System Too Rich (Bank 2) Conditions:** Engine started, battery voltage must be at least 11.5v, all electrical components must be off, the ground between the engine and the chassis must be well connected, the exhaust system must be properly sealed between the catalytic converter and the cylinder head, and the oxygen sensor heater for oxygen sensor before the catalytic converter must be properly functioning. The ECM detected a problem with the fuel mixture. **Note: For resistance testing of sensor heating, oxygen sensor should be cooled to ambient temperature. High temperatures at oxygen sensor may lead to inaccurate measurements.** **Possible Causes:** • Oxygen sensor (before catalytic converter) is faulty • Oxygen sensor (behind catalytic converter) is faulty • Oxygen sensor heater (before catalytic converter) is faulty • Oxygen sensor heater (behind catalytic converter) is faulty • Check circuits for shorts to each other, ground or power • ECM has failed

DTC	Trouble Code Title, Conditions & Possible Causes
DTC: P2101 **1T CCM, MIL: Yes** **1996, 1997, 1998, 1999, 2000, 2001, 2002, 2003, 2004, 2005, 2006** **Models:** TT, A4, A6, Allroad Quattro, A8, A8L **Engines:** 1.8L, 2.7L, 2.8L, 3.0L, 3.2L, 4.2L **Transmissions:** All	**Throttle Actuator Control Motor Range/Performance Conditions:** Engine started, battery voltage must be at least 11.5v, all electrical components must be off, parking brake must be engaged (to keep daytime driving lights off), automatic transmission selector must be in park, the exhaust system must be properly sealed between the catalytic converter and the cylinder head, coolant temperature must be at least 80 degrees Celsius. The ECM detected an unexpected low or high voltage condition on the Throttle Actuator Control Motor (TACM) circuit during the CCM test. **Note: The throttle valve activation occurs via an electric motor (throttle drive) in the throttle valve control module. It is activated by the Engine Control Module (ECM) according to specifications of the two sensors, Throttle Position (TP) Sensor and Sender 2 for accelerator pedal position.** **Possible Causes:** • TACM wiring harness connector is damaged or open • TACM wiring may be crossed in the wire harness assembly • TACM (motor) circuit is open, or TACM assembly is damaged (possible open circuit) • TACM or the Throttle Valve is dirty • Throttle Position sensor has failed • ECM has failed
DTC: P2106 **1T CCM, MIL: Yes** **2000, 2001, 2002, 2003, 2004, 2005, 2006** **Models:** TT, A4, A6, Allroad Quattro, A8, A8L **Engines:** 2.7L, 2.8L, 3.0L, 3.2L, 4.2L **Transmissions:** All	**Throttle Actuator Control System – Forced Limited Power Conditions** Engine started, battery voltage must be at least 11.5v, all electrical components must be off, parking brake must be engaged (to keep daytime driving lights off), automatic transmission selector must be in park, the exhaust system must be properly sealed between the catalytic converter and the cylinder head, coolant temperature must be at least 80 degrees Celsius. The ECM detected an unexpected low or high voltage condition on the Throttle Actuator Control Motor (TACM) circuit during the CCM test. **Note: The throttle valve activation occurs via an electric motor (throttle drive) in the throttle valve control module. It is activated by the Engine Control Module (ECM) according to specifications of the two sensors, Throttle Position (TP) Sensor and Sender 2 for accelerator pedal position.** **Possible Causes:** • TACM wiring harness connector is damaged or open • TACM wiring may be crossed in the wire harness assembly • TACM (motor) circuit is open, or TACM assembly is damaged (possible open circuit) • TACM or the Throttle Valve is dirty • Throttle Position sensor has failed • ECM has failed
DTC: P2122 **1T CCM, MIL: Yes** **2000, 2001, 2002, 2003, 2004, 2005, 2006** **Models:** TT, A4, A6, Allroad Quattro, A8, A8L **Engines:** 2.7L, 2.8L, 3.0L, 3.2L, 4.2L **Transmissions:** All	**Accelerator Pedal Position Sensor 'D' Circuit Low Input Conditions:** Engine started, battery voltage at least 11.5v, all electrical components off, ground connections between engine and chassis well connected, the ECM detected that the accelerator pedal position sensor signal was outside the parameters to function normally. **Note: Both the Throttle Position (TP) Sensor and Accelerator Pedal Position Sensor are located at the accelerator pedal module and communicate the driver's intentions to the ECM completely independently of each other. Both sensors are stored in one housing.** **Possible Causes:** • Ground between engine and chassis may be broken • Throttle position sensor may have failed • Accelerator Pedal Position Sensor has failed • Throttle position sensor wiring may have shorted • Throttle position sensor has failed • Faulty voltage supply • ECM has failed
DTC: P2123 **1T CCM, MIL: Yes** **2000, 2001, 2002, 2003, 2004, 2005, 2006** **Models:** TT, A4, A6, Allroad Quattro, A8, A8L **Engines:** 1.8L, 2.7L, 2.8L, 3.0L, 3.2L, 4.2L **Transmissions:** All	**Accelerator Pedal Position Sensor 'D' Circuit High Input Conditions:** Engine started, battery voltage at least 11.5v, all electrical components off, ground connections between engine and chassis well connected, the ECM detected that the accelerator pedal position sensor signal was outside the parameters to function normally. **Note: Both the Throttle Position (TP) Sensor and Accelerator Pedal Position Sensor are located at the accelerator pedal module and communicate the driver's intentions to the ECM completely independently of each other. Both sensors are stored in one housing.** **Possible Causes:** • Ground between engine and chassis may be broken • Throttle position sensor may have failed • Accelerator Pedal Position Sensor has failed • Throttle position sensor wiring may have shorted • Throttle position sensor has failed • Faulty voltage supply • ECM has failed

DTC	Trouble Code Title, Conditions & Possible Causes
DTC: P2127 **1T CCM, MIL: Yes** **2000, 2001, 2002, 2003, 2004, 2005, 2006** **Models:** TT, A4, A6, Allroad Quattro, A8, A8L **Engines:** 1.8L, 2.7L, 2.8L, 3.0L, 3.2L, 4.2L **Transmissions:** All	**Accelerator Pedal Position Sensor 'E' Circuit Low Input Conditions:** Engine started, battery voltage at least 11.5v, all electrical components off, ground connections between engine and chassis well connected, the ECM detected that the accelerator pedal position sensor signal was outside the parameters to function normally. **Note: Both the Throttle Position (TP) Sensor and Accelerator Pedal Position Sensor are located at the accelerator pedal module and communicate the driver's intentions to the ECM completely independently of each other. Both sensors are stored in one housing.** **Possible Causes:** • Ground between engine and chassis may be broken • Throttle position sensor may have failed • Accelerator Pedal Position Sensor has failed • Throttle position sensor wiring may have shorted • Throttle position sensor has failed • Faulty voltage supply • ECM has failed
DTC: P2128 **1T CCM, MIL: Yes** **2000, 2001, 2002, 2003, 2004, 2005, 2006** **Models:** TT, A4, A6, Allroad Quattro, A8, A8L **Engines:** 1.8L, 2.7L, 2.8L, 3.0L, 3.2L, 4.2L **Transmissions:** All	**Accelerator Pedal Position Sensor 'E' Circuit High Input Conditions:** Engine started, battery voltage at least 11.5v, all electrical components off, ground connections between engine and chassis well connected, the ECM detected that the accelerator pedal position sensor signal was outside the parameters to function normally. **Note: Both the Throttle Position (TP) Sensor and Accelerator Pedal Position Sensor are located at the accelerator pedal module and communicate the driver's intentions to the ECM completely independently of each other. Both sensors are stored in one housing.** **Possible Causes:** • Ground between engine and chassis may be broken • Throttle position sensor may have failed • Accelerator Pedal Position Sensor has failed • Throttle position sensor wiring may have shorted • Throttle position sensor has failed • Faulty voltage supply • ECM has failed
DTC: P2138 **1T CCM, MIL: Yes** **2000, 2001, 2002, 2003, 2004, 2005, 2006** **Models:** TT, A4, A6, Allroad Quattro, A8, A8L **Engines:** 1.8L, 2.7L, 2.8L, 3.0L, 3.2L, 4.2L **Transmissions:** All	**Throttle Position Sensor D/E Voltage Correlation Conditions:** Engine started, battery voltage must be at least 11.5v, all electrical components must be off, parking brake must be engaged (to keep daytime driving lights off), automatic transmission selector must be in park; and the ECM detected the Throttle Position 'D' (TPD) and Throttle Position 'B' (TPE) sensors disagreed, or that the TPD sensor should not be in its detected position, or that the TPE sensor should not be in its detected position during testing. **Note: Both the Throttle Position (TP) Sensor and Accelerator Pedal Position Sensor are located at the accelerator pedal module and communicate the driver's intentions to the ECM completely independently of each other. Both sensors are stored in one housing.** **Possible Causes:** • ETC TP sensor connector is damaged or shorted • ETC TP sensor circuits shorted together in the wire harness • ETC TP sensor signal circuit is shorted to VREF (5v) • ETC TP sensor is damaged or the ECM has failed
DTC: P2177 **1T CCM, MIL: Yes** **2004, 2005, 2006** **Models:** TT **Engines:** 3.2L **Transmissions:** All	**System Too Lean Off Idle Bank 1 Conditions:** Engine started, battery voltage must be at least 11.5v, all electrical components must be off, the ground between the engine and the chassis must be well connected, the exhaust system must be properly sealed between the catalytic converter and the cylinder head, and the oxygen sensor heater for oxygen sensor before the catalytic converter must be properly functioning. The ECM detected the system indicated a lean signal, or it could no longer control bank 1 because it was at its lean limit. **Possible Causes:** • Intake Manifold Runner Position Sensor has failed • Intake system has leaks (false air) • Motor for intake flap is faulty • Oxygen sensor (before catalytic converter) is faulty • Oxygen sensor (behind catalytic converter) is faulty • Oxygen sensor heater (before catalytic converter) is faulty • Oxygen sensor heater (behind catalytic converter) is faulty • Check circuits for shorts to each other, ground or power • Fuel Injector(s) may have failed • ECM has failed

DTC	Trouble Code Title, Conditions & Possible Causes
DTC: P2178 **1T CCM, MIL: Yes** **2004, 2005, 2006** **Models:** TT **Engines:** 3.2L **Transmissions:** All	**System Too Rich Off Idle Bank 1 Conditions:** Engine started, battery voltage must be at least 11.5v, all electrical components must be off, the ground between the engine and the chassis must be well connected, the exhaust system must be properly sealed between the catalytic converter and the cylinder head, and the oxygen sensor heater for oxygen sensor before the catalytic converter must be properly functioning. The ECM detected the system indicated a rich signal, or it could no longer control bank 1 because it was at its rich limit. **Possible Causes:** • Intake Manifold Runner Position Sensor has failed • Intake system has leaks (false air) • Motor for intake flap is faulty • Oxygen sensor (before catalytic converter) is faulty • Oxygen sensor (behind catalytic converter) is faulty • Oxygen sensor heater (before catalytic converter) is faulty • Oxygen sensor heater (behind catalytic converter) is faulty • Check circuits for shorts to each other, ground or power • Fuel Injector(s) may have failed • ECM has failed
DTC: P2179 **1T CCM, MIL: Yes** **2004, 2005, 2006** **Models:** TT **Engines:** 3.2L **Transmissions:** All	**System Too Lean Off Idle Bank 2 Conditions:** Engine started, battery voltage must be at least 11.5v, all electrical components must be off, the ground between the engine and the chassis must be well connected, the exhaust system must be properly sealed between the catalytic converter and the cylinder head, and the oxygen sensor heater for oxygen sensor before the catalytic converter must be properly functioning. The ECM detected the system indicated a lean signal, or it could no longer control bank 2 because it was at its lean limit. **Possible Causes:** • Intake Manifold Runner Position Sensor has failed • Intake system has leaks (false air) • Motor for intake flap is faulty • Oxygen sensor (before catalytic converter) is faulty • Oxygen sensor (behind catalytic converter) is faulty • Oxygen sensor heater (before catalytic converter) is faulty • Oxygen sensor heater (behind catalytic converter) is faulty • Check circuits for shorts to each other, ground or power • Fuel Injector(s) may have failed • ECM has failed
DTC: P2180 **1T CCM, MIL: Yes** **2004, 2005, 2006** **Models:** TT **Engines:** 3.2L **Transmissions:** All	**System Too Rich Off Idle Bank 2 Conditions:** Engine started, battery voltage must be at least 11.5v, all electrical components must be off, the ground between the engine and the chassis must be well connected, the exhaust system must be properly sealed between the catalytic converter and the cylinder head, and the oxygen sensor heater for oxygen sensor before the catalytic converter must be properly functioning. The ECM detected the system indicated a rich signal, or it could no longer control bank 2 because it was at its rich limit. **Possible Causes:** • Intake Manifold Runner Position Sensor has failed • Intake system has leaks (false air) • Motor for intake flap is faulty • Oxygen sensor (before catalytic converter) is faulty • Oxygen sensor (behind catalytic converter) is faulty • Oxygen sensor heater (before catalytic converter) is faulty • Oxygen sensor heater (behind catalytic converter) is faulty • Check circuits for shorts to each other, ground or power • Fuel Injector(s) may have failed • ECM has failed
DTC: P2181 **1T CCM, MIL: Yes** **2000, 2001, 2002, 2003, 2004, 2005, 2006** **Models:** TT, A4, A6, Allroad Quattro, A8, A8L **Engines:** 1.8L, 2.7L, 2.8L, 3.0L, 3.2L, 4.2L **Transmissions:** All	**Cooling System Performance Malfunction Conditions:** Key on, engine cold; and the Engine Coolant Temperature (ECM) detected the ECT sensor signal was more or less than the self-test limits or has failed to gain a signal. This is a thermistor-type sensor with a variable resistance that changes when exposed to different temperatures **Possible Causes:** • ECT sensor has failed • ECT Sensor (on Radiator) has failed • ECT sensor signal circuit is open (inspect wiring & connector) • ECT sensor signal circuit is shorted • Cooling system malfunction, or the thermostat is stuck • Engine not operating at normal operating temperature • EOT sensor is damaged or it has failed

DTC	Trouble Code Title, Conditions & Possible Causes
DTC: P2191 **1T CCM, MIL: Yes** **2004, 2005, 2006** **Models:** TT **Engines:** 3.2L **Transmissions:** All	**System Too Lean at Higher Load Bank 1 Conditions:** Engine started, battery voltage must be at least 11.5v, all electrical components must be off, the ground between the engine and the chassis must be well connected, the exhaust system must be properly sealed between the catalytic converter and the cylinder head, and the oxygen sensor heater for oxygen sensor before the catalytic converter must be properly functioning. ECM detected the system indicated a lean signal, or it could no longer control bank 1 because it was at its lean limit. **Possible Causes:** • Evaporative Emission (EVAP) canister purge regulator valve is faulty • Exhaust system components are damaged • Fuel injectors are faulty • Fuel pressure regulator and residual pressure have failed • Fuel Pump (FP) in fuel tank is faulty • Intake system has leaks (false air) • Secondary Air Injection (AIR) system has an improper seal • Intake Manifold Runner Position Sensor has failed • Motor for intake flap is faulty • Oxygen sensor (before catalytic converter) is faulty • Oxygen sensor (behind catalytic converter) is faulty • Oxygen sensor heater (before catalytic converter) is faulty • Oxygen sensor heater (behind catalytic converter) is faulty • Check circuits for shorts to each other, ground or power • ECM has failed
DTC: P2192 **1T CCM, MIL: Yes** **2004, 2005, 2006** **Models:** TT **Engines:** 3.2L **Transmissions:** All	**System Too Rich at Higher Load Bank 1 Conditions:** Engine started, battery voltage must be at least 11.5v, all electrical components must be off, the ground between the engine and the chassis must be well connected, the exhaust system must be properly sealed between the catalytic converter and the cylinder head, and the oxygen sensor heater for oxygen sensor before the catalytic converter must be properly functioning. ECM detected the system indicated a rich signal, or it could no longer control bank 1 because it was at its rich limit. **Possible Causes:** • Evaporative Emission (EVAP) canister purge regulator valve is faulty • Exhaust system components are damaged • Fuel injectors are faulty • Fuel pressure regulator and residual pressure have failed • Fuel Pump (FP) in fuel tank is faulty • Intake system has leaks (false air) • Secondary Air Injection (AIR) system has an improper seal • Intake Manifold Runner Position Sensor has failed • Motor for intake flap is faulty • Oxygen sensor (before catalytic converter) is faulty • Oxygen sensor (behind catalytic converter) is faulty • Oxygen sensor heater (before catalytic converter) is faulty • Oxygen sensor heater (behind catalytic converter) is faulty • Check circuits for shorts to each other, ground or power • ECM has failed
DTC: P2193 **1T CCM, MIL: Yes** **2004, 2005, 2006** **Models:** TT **Engines:** 3.2L **Transmissions:** All	**System Too Lean at Higher Load Bank 2 Conditions:** Engine started, battery voltage must be at least 11.5v, all electrical components must be off, the ground between the engine and the chassis must be well connected, the exhaust system must be properly sealed between the catalytic converter and the cylinder head, and the oxygen sensor heater for oxygen sensor before the catalytic converter must be properly functioning. ECM detected the system indicated a lean signal, or it could no longer control bank 2 because it was at its lean limit. **Possible Causes:** • Evaporative Emission (EVAP) canister purge regulator valve is faulty • Exhaust system components are damaged • Fuel injectors are faulty • Fuel pressure regulator and residual pressure have failed • Fuel Pump (FP) in fuel tank is faulty • Intake system has leaks (false air) • Secondary Air Injection (AIR) system has an improper seal • Intake Manifold Runner Position Sensor has failed • Motor for intake flap is faulty • Oxygen sensor (before catalytic converter) is faulty • Oxygen sensor (behind catalytic converter) is faulty • Oxygen sensor heater (before catalytic converter) is faulty • Oxygen sensor heater (behind catalytic converter) is faulty • Check circuits for shorts to each other, ground or power • ECM has failed

DTC	Trouble Code Title, Conditions & Possible Causes
DTC: P2194 **1T CCM, MIL: Yes** **2004, 2005, 2006** **Models:** TT **Engines:** 3.2L **Transmissions:** All	**System Too Rich at Higher Load Bank 2 Conditions:** Engine started, battery voltage must be at least 11.5v, all electrical components must be off, the ground between the engine and the chassis must be well connected, the exhaust system must be properly sealed between the catalytic converter and the cylinder head, and the oxygen sensor heater for oxygen sensor before the catalytic converter must be properly functioning. ECM detected the system indicated a rich signal, or it could no longer control bank 2 because it was at its rich limit. **Possible Causes:** • Evaporative Emission (EVAP) canister purge regulator valve is faulty • Exhaust system components are damaged • Fuel injectors are faulty • Fuel pressure regulator and residual pressure have failed • Fuel Pump (FP) in fuel tank is faulty • Intake system has leaks (false air) • Secondary Air Injection (AIR) system has an improper seal • Intake Manifold Runner Position Sensor has failed • Motor for intake flap is faulty • Oxygen sensor (before catalytic converter) is faulty • Oxygen sensor (behind catalytic converter) is faulty • Oxygen sensor heater (before catalytic converter) is faulty • Oxygen sensor heater (behind catalytic converter) is faulty • Check circuits for shorts to each other, ground or power • ECM has failed
DTC: P2195 **1T CCM, MIL: Yes** **2003, 2004, 2005, 2006** **Models:** A6, Allroad Quattro, A8, A8L **Engines:** 3.2L, 4.2L **Transmissions:** All	**O2 Sensor Signal Stuck Lean Bank 1 Sensor 1 Conditions:** Engine running in closed loop, and the ECM detected the O2S indicated a lean signal, or it could no longer control Fuel Trim because it was at lean limit. **Possible Causes:** • Engine oil level high • Camshaft timing error • Cylinder compression low • Exhaust leaks in front of O2S • EGR valve is stuck open • EGR gasket is leaking • EVR diaphragm is leaking • Damaged fuel pressure regulator or extremely low fuel pressure • O2S circuit is open or shorted in the wiring harness • Oxygen sensor (before catalytic converter) is faulty • Oxygen sensor (behind catalytic converter) is faulty • Oxygen sensor heater (before catalytic converter) is faulty • Oxygen sensor heater (behind catalytic converter) is faulty • Air leaks after the MAF sensor • PCV system leaks • Dip stick not seated properly
DTC: P2196 **1T CCM, MIL: Yes** **2003, 2004, 2005, 2006** **Models:** A6, Allroad Quattro, A8, A8L **Engines:** 3.2L, 4.2L **Transmissions:** All	**O2 Sensor Signal Stuck Rich Bank 1 Sensor 1 Conditions:** Engine running in closed loop, and the ECM detected the O2S indicated a rich signal, or it could no longer control Fuel Trim because it was at its rich limit. **Possible Causes:** • Engine oil level high • Camshaft timing error • Cylinder compression low • Exhaust leaks in front of O2S • EGR valve is stuck open • EGR gasket is leaking • EVR diaphragm is leaking • Damaged fuel pressure regulator or extremely low fuel pressure • O2S circuit is open or shorted in the wiring harness • Oxygen sensor (before catalytic converter) is faulty • Oxygen sensor (behind catalytic converter) is faulty • Oxygen sensor heater (before catalytic converter) is faulty • Oxygen sensor heater (behind catalytic converter) is faulty • Air leaks after the MAF sensor • PCV system leaks • Dip stick not seated properly

DTC	Trouble Code Title, Conditions & Possible Causes
DTC: P2231 **1T CCM, MIL: Yes** **2003, 2004, 2005, 2006** **Models:** A6, Allroad Quattro, A8, A8L **Engines:** 3.2L, 4.2L **Transmissions:** All	**O2 Sensor Signal Circuit Shorted to Heater Circuit Bank 1 Sensor 1 Conditions:** Engine started, battery voltage must be at least 11.5v, all electrical components must be off, parking brake must be engaged (to keep daytime driving lights off), automatic transmission selector must be in park. The ECM detected an unexpected voltage condition, or it detected an unexpected current draw in the sensor circuit during the CCM test. **Note: Vehicle must be raised before connector for oxygen sensors is accessible.** **Possible Causes:** • Oxygen sensor (before catalytic converter) is faulty • Oxygen sensor heater (before catalytic converter) is faulty • Oxygen sensor heater (before catalytic converter) is faulty • Oxygen sensor heater (behind catalytic converter) is faulty • O2S circuit is open or shorted in the wiring harness • ECM has failed
DTC: P2234 **1T CCM, MIL: Yes** **2003, 2004, 2005, 2006** **Models:** A6, Allroad Quattro, A8, A8L **Engines:** 3.2L, 4.2L **Transmissions:** All	**O2 Sensor Signal Circuit Shorted to Heater Circuit Bank 2 Sensor 1 Conditions:** Engine started, battery voltage must be at least 11.5v, all electrical components must be off, parking brake must be engaged (to keep daytime driving lights off), automatic transmission selector must be in park. The ECM detected an unexpected voltage condition, or it detected an unexpected current draw in the sensor circuit during the CCM test. **Note: Vehicle must be raised before connector for oxygen sensors is accessible.** **Possible Causes:** • Oxygen sensor (before catalytic converter) is faulty • Oxygen sensor heater (before catalytic converter) is faulty • Oxygen sensor heater (before catalytic converter) is faulty • Oxygen sensor heater (behind catalytic converter) is faulty • O2S circuit is open or shorted in the wiring harness • ECM has failed
DTC: P2237 **1T CCM, MIL: Yes** **2003, 2004, 2005, 2006** **Models:** TT, A6, Allroad Quattro, A8, A8L **Engines:** 3.2L, 4.2L **Transmissions:** All	**O2 Sensor Positive Current Control Circuit/Open Bank 1 Sensor 1 Conditions:** Engine started, battery voltage must be at least 11.5v, all electrical components must be off, parking brake must be engaged (to keep daytime driving lights off), automatic transmission selector must be in park. The ECM detected an unexpected voltage condition, or it detected an unexpected current draw in the sensor circuit during the CCM test. **Note: Vehicle must be raised before connector for oxygen sensors is accessible.** **Possible Causes:** • Oxygen sensor (before catalytic converter) is faulty • Oxygen sensor heater (before catalytic converter) is faulty • Oxygen sensor heater (before catalytic converter) is faulty • Oxygen sensor heater (behind catalytic converter) is faulty • O2S circuit is open or shorted in the wiring harness • ECM has failed
DTC: P2240 **1T CCM, MIL: Yes** **2003, 2004, 2005, 2006** **Models:** TT, A6, Allroad Quattro, A8, A8L **Engines:** 3.2L, 4.2L **Transmissions:** All	**O2 Sensor Positive Current Control Circuit/Open Bank 2 Sensor 1 Conditions:** Engine started, battery voltage must be at least 11.5v, all electrical components must be off, parking brake must be engaged (to keep daytime driving lights off), automatic transmission selector must be in park. The ECM detected an unexpected voltage condition, or it detected an unexpected current draw in the sensor circuit during the CCM test. **Note: Vehicle must be raised before connector for oxygen sensors is accessible.** **Possible Causes:** • Oxygen sensor (before catalytic converter) is faulty • Oxygen sensor heater (before catalytic converter) is faulty • Oxygen sensor heater (before catalytic converter) is faulty • Oxygen sensor heater (behind catalytic converter) is faulty • O2S circuit is open or shorted in the wiring harness • ECM has failed
DTC: P2243 **1T CCM, MIL: Yes** **2003, 2004, 2005, 2006** **Models:** A6, Allroad Quattro, A8, A8L **Engines:** 3.2L, 4.2L **Transmissions:** All	**O2 Sensor Reference Voltage Circuit/Open Bank 1 Sensor 1 Conditions:** Engine started, battery voltage must be at least 11.5v, all electrical components must be off, parking brake must be engaged (to keep daytime driving lights off), automatic transmission selector must be in park. The ECM detected an unexpected voltage condition, or it detected an unexpected current draw in the sensor circuit during the CCM test. **Note: Vehicle must be raised before connector for oxygen sensors is accessible.** **Possible Causes:** • Oxygen sensor (before catalytic converter) is faulty • Oxygen sensor heater (before catalytic converter) is faulty • Oxygen sensor heater (before catalytic converter) is faulty • Oxygen sensor heater (behind catalytic converter) is faulty • O2S circuit is open or shorted in the wiring harness • ECM has failed

DTC	Trouble Code Title, Conditions & Possible Causes
DTC: P2247 **1T CCM, MIL: Yes** **2003, 2004, 2005, 2006** **Models:** A6, Allroad Quattro, A8, A8L **Engines:** 3.2L, 4.2L **Transmissions:** All	**O2 Sensor Reference Voltage Circuit/Open Bank 2 Sensor 1 Conditions:** Engine started, battery voltage must be at least 11.5v, all electrical components must be off, parking brake must be engaged (to keep daytime driving lights off), automatic transmission selector must be in park. The ECM detected an unexpected voltage condition, or it detected an unexpected current draw in the sensor circuit during the CCM test. **Note: Vehicle must be raised before connector for oxygen sensors is accessible.** **Possible Causes:** • Oxygen sensor (before catalytic converter) is faulty • Oxygen sensor heater (before catalytic converter) is faulty • Oxygen sensor heater (before catalytic converter) is faulty • Oxygen sensor heater (behind catalytic converter) is faulty • O2S circuit is open or shorted in the wiring harness • ECM has failed
DTC: P2251 **1T CCM, MIL: Yes** **2003, 2004, 2005, 2006** **Models:** A6, Allroad Quattro, A8, A8L **Engines:** 3.2L, 4.2L **Transmissions:** All	**O2 Sensor Negative Voltage Circuit/Open Bank 1 Sensor 1 Conditions:** Engine started, battery voltage must be at least 11.5v, all electrical components must be off, parking brake must be engaged (to keep daytime driving lights off), automatic transmission selector must be in park. The ECM detected an unexpected voltage condition, or it detected an unexpected current draw in the sensor circuit during the CCM test. **Note: Vehicle must be raised before connector for oxygen sensors is accessible.** **Possible Causes:** • Oxygen sensor (before catalytic converter) is faulty • Oxygen sensor heater (before catalytic converter) is faulty • Oxygen sensor heater (before catalytic converter) is faulty • Oxygen sensor heater (behind catalytic converter) is faulty • O2S circuit is open or shorted in the wiring harness • ECM has failed
DTC: P2254 **1T CCM, MIL: Yes** **2003, 2004, 2005, 2006** **Models:** A6, Allroad Quattro, A8, A8L **Engines:** 3.2L, 4.2L **Transmissions:** All	**O2 Sensor Negative Voltage Circuit/Open Bank 2 Sensor 1 Conditions:** Engine started, battery voltage must be at least 11.5v, all electrical components must be off, parking brake must be engaged (to keep daytime driving lights off), automatic transmission selector must be in park. The ECM detected an unexpected voltage condition, or it detected an unexpected current draw in the sensor circuit during the CCM test. **Note: Vehicle must be raised before connector for oxygen sensors is accessible.** **Possible Causes:** • Oxygen sensor (before catalytic converter) is faulty • Oxygen sensor heater (before catalytic converter) is faulty • Oxygen sensor heater (before catalytic converter) is faulty • Oxygen sensor heater (behind catalytic converter) is faulty • O2S circuit is open or shorted in the wiring harness • ECM has failed
DTC: P2257 **1T CCM, MIL: Yes** **2000, 2001, 2002, 2003, 2004, 2005, 2006** **Models:** TT, A4, A6, Allroad Quattro, A8, A8L **Engines:** 2.7L, 4.2L **Transmissions:** All	**Secondary Air Injection System Control "A" Circuit Low Conditions:** Engine started, battery voltage must be at least 11.5v, all electrical components must be off, parking brake must be engaged (to keep daytime driving lights off), automatic transmission selector must be in park and the ground between the engine and the chassis must be well connected. The ECM detected an unexpected voltage condition on the AIR system control circuit during testing. **Possible Causes:** • AIR solenoid power circuit (B+) is open (check dedicated fuse) • AIR bypass solenoid control circuit is open or shorted to ground • AIR diverter solenoid control circuit open or shorted to ground • AIR pump control circuit is open or shorted to ground • Check valve (one or more) is damaged or leaking • Solid State relay is damaged or it has failed • Check activation of Secondary Air Injection (AIR) Pump Relay • ECM has failed
DTC: P2258 **1T CCM, MIL: Yes** **2000, 2001, 2002, 2003, 2004, 2005, 2006** **Models:** TT, A4, A6, Allroad Quattro, A8, A8L **Engines:** 2.7L, 4.2L **Transmissions:** All	**Secondary Air Injection System Control "A" Circuit High Conditions:** Engine started, battery voltage must be at least 11.5v, all electrical components must be off, parking brake must be engaged (to keep daytime driving lights off), automatic transmission selector must be in park and the ground between the engine and the chassis must be well connected. The ECM detected an unexpected voltage condition on the AIR system control circuit during testing. **Possible Causes:** • AIR solenoid power circuit (B+) is open (check dedicated fuse) • AIR bypass solenoid control circuit is open or shorted to ground • AIR diverter solenoid control circuit open or shorted to ground • AIR pump control circuit is open or shorted to ground • Check valve (one or more) is damaged or leaking • Solid State relay is damaged or it has failed • Check activation of Secondary Air Injection (AIR) Pump Relay • ECM has failed

DTC	Trouble Code Title, Conditions & Possible Causes
DTC: P2270 **1T CCM, MIL: Yes** **2004, 2005, 2006** **Models:** TT, A6 **Engines:** 3.2L **Transmissions:** All	**O2 Sensor Signal Stuck Lean Bank 1 Sensor 2 Conditions:** Engine started, battery voltage must be at least 11.5v, all electrical components must be off, parking brake must be engaged (to keep daytime driving lights off), automatic transmission selector must be in park. The ECM detected an unexpected voltage condition, or it detected an unexpected current draw in the heater circuit during the CCM test. **Note: Vehicle must be raised before connector for oxygen sensors is accessible.** **Possible Causes:** • Oxygen sensor (before catalytic converter) is faulty • Oxygen sensor heater (before catalytic converter) is faulty • Oxygen sensor heater (before catalytic converter) is faulty • Oxygen sensor heater (behind catalytic converter) is faulty • O2S circuit is open or shorted in the wiring harness • ECM has failed
DTC: P2271 **1T CCM, MIL: Yes** **2004, 2005, 2006** **Models:** TT, A6 **Engines:** 3.2L **Transmissions:** All	**O2 Sensor Signal Stuck Rich Bank 1 Sensor 2 Conditions:** Engine started, battery voltage must be at least 11.5v, all electrical components must be off, parking brake must be engaged (to keep daytime driving lights off), automatic transmission selector must be in park. The ECM detected an unexpected voltage condition, or it detected an unexpected current draw in the heater circuit during the CCM test. **Note: Vehicle must be raised before connector for oxygen sensors is accessible.** **Possible Causes:** • Oxygen sensor (before catalytic converter) is faulty • Oxygen sensor heater (before catalytic converter) is faulty • Oxygen sensor heater (before catalytic converter) is faulty • Oxygen sensor heater (behind catalytic converter) is faulty • O2S circuit is open or shorted in the wiring harness • ECM has failed
DTC: P2272 **1T CCM, MIL: Yes** **2004, 2005, 2006** **Models:** TT, A6 **Engines:** 3.2L **Transmissions:** All	**O2 Sensor Signal Stuck Lean Bank 2 Sensor 2 Conditions:** Engine started, battery voltage must be at least 11.5v, all electrical components must be off, parking brake must be engaged (to keep daytime driving lights off), automatic transmission selector must be in park. The ECM detected an unexpected voltage condition, or it detected an unexpected current draw in the heater circuit during the CCM test. **Note: Vehicle must be raised before connector for oxygen sensors is accessible.** **Possible Causes:** • Oxygen sensor (before catalytic converter) is faulty • Oxygen sensor heater (before catalytic converter) is faulty • Oxygen sensor heater (before catalytic converter) is faulty • Oxygen sensor heater (behind catalytic converter) is faulty • O2S circuit is open or shorted in the wiring harness • ECM has failed
DTC: P2273 **1T CCM, MIL: Yes** **2004, 2005, 2006** **Models:** TT, A6 **Engines:** 3.2L **Transmissions:** All	**O2 Sensor Signal Stuck Rich Bank 2 Sensor 2 Conditions:** Engine started, battery voltage must be at least 11.5v, all electrical components must be off, parking brake must be engaged (to keep daytime driving lights off), automatic transmission selector must be in park. The ECM detected an unexpected voltage condition, or it detected an unexpected current draw in the heater circuit during the CCM test. **Note: Vehicle must be raised before connector for oxygen sensors is accessible.** **Possible Causes:** • Oxygen sensor (before catalytic converter) is faulty • Oxygen sensor heater (before catalytic converter) is faulty • Oxygen sensor heater (before catalytic converter) is faulty • Oxygen sensor heater (behind catalytic converter) is faulty • O2S circuit is open or shorted in the wiring harness • ECM has failed
DTC: P2279 **2004, 2005, 2006** **Models:** TT **Engines:** 3.2L **Transmissions:** All	**Intake Air System Leak Conditions:** Engine running and the vehicle speed more than 25mph, the ECM detected the intake air system has a potential leak. The IAT sensor is a variable resistor that includes an IAT signal circuit and a low reference circuit to measure the temperature of the air entering the engine. The ECM supplies the sensor with a low voltage singal circuit and a low reference ground circuit. When the IAT sensor is cold, its resistence is high. When the air temperature increases, its resistence decreases. With high sensor resisteance, the IAT sensor signal voltage is high. With lower sensor resistance, the IAT sensor signal voltage should be lower. **Possible Causes:** • Intake Manifold Runner Position Sensor is damaged or has failed • Intake system has leaks (false air) • Motor for intake flap is faulty • ECM has failed • IAT sensor signal circuit is shorted to sensor or chassis ground • IAT sensor is damaged or has failed • ECM has failed.

DTC	Trouble Code Title, Conditions & Possible Causes
DTC: P2294 **2005, 2006** **Models:** A6 **Engines:** 3.2L **Transmissions:** All	**Fuel Pressure Regulator 2 Control Circuit Conditions:** Engine started, battery voltage at least 11.5v, all electrical components off, ground connections between engine and chassis well connected, coolant temperature at least 80-degrees Celicius. The ECM detected a voltage condition that affected the performance of the fule pressure regulator. **Possible Causes:** • Fuel Pressure Regulator Valve has failed • Fuel Pressure Sensor has failed • Fuel Pump (FP) Control Module has failed • Fuel pump has failed • ECM has failed
DTC: P2295 **2005, 2006** **Models:** A6 **Engines:** 3.2L **Transmissions:** All	**Fuel Pressure Regulator 2 Control Circuit Low Conditions:** Engine started, battery voltage at least 11.5v, all electrical components off, ground connections between engine and chassis well connected, coolant temperature at least 80-degrees Celicius. The ECM detected a voltage condition that affected the performance of the fule pressure regulator. **Possible Causes:** • Fuel Pressure Regulator Valve has failed • Fuel Pressure Sensor has failed • Fuel Pump (FP) Control Module has failed • Fuel pump has failed • ECM has failed
DTC: P2296 **2005, 2006** **Models:** A6 **Engines:** 3.2L **Transmissions:** All	**Fuel Pressure Regulator 2 Control Circuit High Conditions:** Engine started, battery voltage at least 11.5v, all electrical components off, ground connections between engine and chassis well connected, coolant temperature at least 80-degrees Celicius. The ECM detected a voltage condition that affected the performance of the fule pressure regulator. **Possible Causes:** • Fuel Pressure Regulator Valve has failed • Fuel Pressure Sensor has failed • Fuel Pump (FP) Control Module has failed • Fuel pump has failed • ECM has failed
DTC: P2400 **2T CCM, MIL: Yes** **1998, 1999, 2000, 2001, 2002, 2003, 2004, 2005, 2006** **Models:** TT, A4, A6, Allroad Quattro, A8, A8L **Engines:** 1.8L, 2.7L, 2.8L, 3.0L, 4.2L **Transmissions:** All	**EVAP Leak Detection Pump (LDP) Control Circuit Open Conditions:** Engine started, battery voltage must be at least 11.5v, all electrical components must be off, parking brake must be engaged (to keep daytime driving lights off), automatic transmission selector must be in park, the exhaust system must be properly sealed between the catalytic converter and the cylinder head, coolant temperature must be at least 80 degrees Celsius and oxygen sensor heaters for oxygen sensors before the catalytic converter must be functioning properly and the ground between the engine and the chassis must be well connected. The ECM detected voltage irregularity in the leak detection pump control circuit. **Possible Causes:** • EVAP LDP power supply circuit is open • EVAP LDP solenoid valve is damaged or it has failed • EVAP LDP canister has a leak or a poor seal • ECM has failed • EVAP canister system has an improper seal • Evaporative Emission (EVAP) canister purge regulator valve 1 has failed • Leak Detection Pump (LDP) is faulty • Aftermarket EVAP parts that do not conform to specifications • EVAP component seals leaking (i.e., leaks in the Purge valve, fuel tank pressure sensor, canister vent solenoid, fuel vapor control valve tube assembly or fuel vapor vent valve).

DTC	Trouble Code Title, Conditions & Possible Causes
DTC: P2401 **2T CCM, MIL: Yes** **1998, 1999, 2000, 2001, 2002, 2003, 2004, 2005, 2006** **Models:** TT, A4, A6, Allroad Quattro, A8, A8L **Engines:** 1.8L, 2.7L, 2.8L, 3.0L, 4.2L **Transmissions:** All	**EVAP Leak Detection Pump Control Circuit Low Conditions:** Engine started, battery voltage must be at least 11.5v, all electrical components must be off, parking brake must be engaged (to keep daytime driving lights off), automatic transmission selector must be in park, the exhaust system must be properly sealed between the catalytic converter and the cylinder head, coolant temperature must be at least 80 degrees Celsius and oxygen sensor heaters for oxygen sensors before the catalytic converter must be functioning properly and the ground between the engine and the chassis must be well connected. The ECM detected voltage irregularity in the leak detection pump control circuit. **Possible Causes:** • EVAP LDP power supply circuit is open • EVAP LDP solenoid valve is damaged or it has failed • EVAP LDP canister has a leak or a poor seal • ECM has failed • EVAP canister system has an improper seal • Evaporative Emission (EVAP) canister purge regulator valve 1 has failed • Leak Detection Pump (LDP) is faulty • Aftermarket EVAP parts that do not conform to specifications • EVAP component seals leaking (i.e., leaks in the Purge valve, fuel tank pressure sensor, canister vent solenoid, fuel vapor control valve tube assembly or fuel vapor vent valve).
DTC: P2402 **2T CCM, MIL: Yes** **1998, 1999, 2000, 2001, 2002, 2003, 2004, 2005, 2006** **Models:** TT, A4, A6, Allroad Quattro, A8, A8L **Engines:** 1.8L, 2.7L, 2.8L, 3.0L, 4.2L **Transmissions:** All	**EVAP Leak Detection Pump Control Circuit High Conditions:** Engine started, battery voltage must be at least 11.5v, all electrical components must be off, parking brake must be engaged (to keep daytime driving lights off), automatic transmission selector must be in park, the exhaust system must be properly sealed between the catalytic converter and the cylinder head, coolant temperature must be at least 80 degrees Celsius and oxygen sensor heaters for oxygen sensors before the catalytic converter must be functioning properly and the ground between the engine and the chassis must be well connected. The ECM detected voltage irregularity in the leak detection pump control circuit. **Possible Causes:** • EVAP LDP power supply circuit is open • EVAP LDP solenoid valve is damaged or it has failed • EVAP LDP canister has a leak or a poor seal • ECM has failed • EVAP canister system has an improper seal • Evaporative Emission (EVAP) canister purge regulator valve 1 has failed • Leak Detection Pump (LDP) is faulty • Aftermarket EVAP parts that do not conform to specifications • EVAP component seals leaking (i.e., leaks in the Purge valve, fuel tank pressure sensor, canister vent solenoid, fuel vapor control valve tube assembly or fuel vapor vent valve).
DTC: P2403 **2T CCM, MIL: Yes** **1998, 1999, 2000, 2001, 2002, 2003, 2004, 2005, 2006** **Models:** TT, A4, A6, Allroad Quattro, A8, A8L **Engines:** 1.8L, 2.7L, 2.8L, 3.0L, 4.2L **Transmissions:** All	**EVAP Leak Detection Pump Sense Circuit Open Conditions:** Engine started, battery voltage must be at least 11.5v, all electrical components must be off, parking brake must be engaged (to keep daytime driving lights off), automatic transmission selector must be in park, the exhaust system must be properly sealed between the catalytic converter and the cylinder head, coolant temperature must be at least 80 degrees Celsius and oxygen sensor heaters for oxygen sensors before the catalytic converter must be functioning properly and the ground between the engine and the chassis must be well connected. The ECM detected voltage irregularity in the leak detection pump control circuit. **Possible Causes:** • EVAP LDP power supply circuit is open • EVAP LDP solenoid valve is damaged or it has failed • EVAP LDP canister has a leak or a poor seal • ECM has failed • EVAP canister system has an improper seal • Evaporative Emission (EVAP) canister purge regulator valve 1 has failed • Leak Detection Pump (LDP) is faulty • Aftermarket EVAP parts that do not conform to specifications • EVAP component seals leaking (i.e., leaks in the Purge valve, fuel tank pressure sensor, canister vent solenoid, fuel vapor control valve tube assembly or fuel vapor vent valve).

DTC	Trouble Code Title, Conditions & Possible Causes
DTC: P2404 **2T CCM, MIL: Yes** **1998, 1999, 2000, 2001, 2002, 2003, 2004, 2005, 2006** **Models:** TT, A4, A6, Allroad Quattro, A8, A8L **Engines:** 1.8L, 2.7L, 2.8L, 3.0L, 4.2L **Transmissions:** All	**EVAP Leak Detection Pump Sense Circuit Range/Performance Conditions:** Engine started, battery voltage must be at least 11.5v, all electrical components must be off, parking brake must be engaged (to keep daytime driving lights off), automatic transmission selector must be in park, the exhaust system must be properly sealed between the catalytic converter and the cylinder head, coolant temperature must be at least 80 degrees Celsius and oxygen sensor heaters for oxygen sensors before the catalytic converter must be functioning properly and the ground between the engine and the chassis must be well connected. The ECM detected voltage irregularity in the leak detection pump control circuit. **Possible Causes:** • EVAP LDP power supply circuit is open • EVAP LDP solenoid valve is damaged or it has failed • EVAP LDP canister has a leak or a poor seal • ECM has failed • EVAP canister system has an improper seal • Evaporative Emission (EVAP) canister purge regulator valve 1 has failed • Leak Detection Pump (LDP) is faulty • Aftermarket EVAP parts that do not conform to specifications • EVAP component seals leaking (i.e., leaks in the Purge valve, fuel tank pressure sensor, canister vent solenoid, fuel vapor control valve tube assembly or fuel vapor vent valve).
DTC: P2414 **1T CCM, MIL: Yes** **2003, 2004, 2005, 2006** **Models:** A6, Allroad Quattro, A8, A8L **Engines:** 3.2L, 4.2L **Transmissions:** All	**O2 Sensor Exhaust Sample Error Bank 1 Sensor 1 Conditions:** Engine running (ground connections between the engine and the chassis must be well connected), and the ECM detected an error on the OS Sensor. **Note: Intake Flap Motor and Intake Manifold Runner Position Sensor are one component and cannot be replaced individually.** **Note: Vacuum in the intake system sucks in the leak detection spray with false air. Leak detection spray decreases ignition quality of the fuel mixture. This causes a drop in engine speed and changes the value produced by the Heated Oxygen Sensor (HO2S).** **Possible Causes:** • Intake Manifold Runner Position Sensor is damaged or has failed • Intake system has leaks (false air) • Motor for intake flap is faulty • ECM has failed • Oxygen sensor (before catalytic converter) is faulty • Oxygen sensor heater (before catalytic converter) is faulty • Oxygen sensor heater (before catalytic converter) is faulty • Oxygen sensor heater (behind catalytic converter) is faulty • O2S circuit is open or shorted in the wiring harness
DTC: P2422 **2T CCM, MIL: Yes** **2004, 2005, 2006** **Models:** TT, A6, A8 **Engines:** 3.2L, 4.2L **Transmissions:** All	**Evaporative Emission System Vent Valve Stuck Closed Conditions:** Engine started, battery voltage must be at least 11.5v, all electrical components must be off, parking brake must be engaged (to keep daytime driving lights off), automatic transmission selector must be in park, the exhaust system must be properly sealed between the catalytic converter and the cylinder head, coolant temperature must be at least 80 degrees Celsius and oxygen sensor heaters for oxygen sensors before the catalytic converter must be functioning properly and the ground between the engine and the chassis must be well connected. The ECM detected an unexpected EVAP malfunction. **Note: Solenoid valve is closed when no voltage is present.** **Possible Causes:** • EVAP power supply circuit is open • EVAP solenoid control circuit is open or shorted to ground • EVAP solenoid control circuit is shorted to power (B+) • EVAP solenoid valve is damaged or it has failed • EVAP canister purge solenoid valve is faulty • ECM has failed
DTC: P2539 **2005, 2006** **Models:** A6 **Engines:** 3.2L **Transmissions:** All	**Low Pressure Fuel System Sensor Circuit Conditions:** Engine started, battery voltage must be at least 11.5v, all electrical components must be off, parking brake must be engaged (to keep daytime driving lights off), automatic transmission selector must be in park, the exhaust system must be properly sealed between the catalytic converter and the cylinder head, coolant temperature must be at least 80 degrees Celsius. The ECM detected an error on the fuel system sensor circuit. **Note: The specified fuel pressure should be between 3000 to 5000 kPA** **Possible Causes:** • Fuel Pressure Regulator Valve has failed • Fuel Pressure Sensor has failed • Fuel Pump (FP) Control Module has failed • Fuel pump has failed • ECM has failed

DTC	Trouble Code Title, Conditions & Possible Causes
DTC: P2540 **2005, 2006** **Models:** A6 **Engines:** 3.2L **Transmissions:** All	**Low Pressure Fuel System Sensor Circuit Range/Performance Conditions:** Engine started, battery voltage must be at least 11.5v, all electrical components must be off, parking brake must be engaged (to keep daytime driving lights off), automatic transmission selector must be in park, the exhaust system must be properly sealed between the catalytic converter and the cylinder head, coolant temperature must be at least 80 degrees Celsius. The ECM detected an error on the fuel system sensor circuit. **Note: The specified fuel pressure should be between 3000 to 5000 kPA** **Possible Causes:** • Fuel Pressure Regulator Valve has failed • Fuel Pressure Sensor has failed • Fuel Pump (FP) Control Module has failed • Fuel pump has failed • ECM has failed
DTC: P2541 **2005, 2006** **Models:** A6 **Engines:** 3.2L **Transmissions:** All	**Low Pressure Fuel System Sensor Circuit Low Conditions:** Engine started, battery voltage must be at least 11.5v, all electrical components must be off, parking brake must be engaged (to keep daytime driving lights off), automatic transmission selector must be in park, the exhaust system must be properly sealed between the catalytic converter and the cylinder head, coolant temperature must be at least 80 degrees Celsius. The ECM detected an error on the fuel system sensor circuit. **Note: The specified fuel pressure should be between 3000 to 5000 kPA** **Possible Causes:** • Fuel Pressure Regulator Valve has failed • Fuel Pressure Sensor has failed • Fuel Pump (FP) Control Module has failed • Fuel pump has failed • ECM has failed
DTC: P2626 **2003, 2004, 2005, 2006** **Models:** A6, Allroad Quattro, A8, A8L **Engines:** 3.2L, 4.2L **Transmissions:** All	**O2 Sensor Pumping Current Trim Circuit/Open Bank 1 Sensor 1 Conditions:** Engine started, battery voltage must be at least 11.5v, all electrical components must be off, parking brake must be engaged (to keep daytime driving lights off), automatic transmission selector must be in park, the exhaust system must be properly sealed between the catalytic converter and the cylinder head, coolant temperature must be at least 80 degrees Celsius and oxygen sensor heaters for oxygen sensors before the catalytic converter must be functioning properly and the ground between the engine and the chassis must be well connected. The ECM detected a voltage value that doesn't fall within the desired parameters for a properly functioning O2 system. **Possible Causes:** • Check activation of Recirculation Pump Relay • Oxygen sensor (before catalytic converter) is faulty • Oxygen sensor (behind catalytic converter) is faulty • Oxygen sensor heater (before catalytic converter) is faulty • Oxygen sensor heater (behind catalytic converter) is faulty
DTC: P2629 **1T CCM, MIL: Yes** **2003, 2004, 2005, 2006** **Models:** A6, Allroad Quattro, A8, A8L **Engines:** 3.2L, 4.2L **Transmissions:** All	**O2 Sensor Pumping Current Trim Circuit/Open Bank 2 Sensor 1 Conditions:** Engine started, battery voltage must be at least 11.5v, all electrical components must be off, parking brake must be engaged (to keep daytime driving lights off), automatic transmission selector must be in park, the exhaust system must be properly sealed between the catalytic converter and the cylinder head, coolant temperature must be at least 80 degrees Celsius and oxygen sensor heaters for oxygen sensors before the catalytic converter must be functioning properly and the ground between the engine and the chassis must be well connected. The ECM detected a voltage value that doesn't fall within the desired parameters for a properly functioning O2 system. **Possible Causes:** • Check activation of Recirculation Pump Relay • Oxygen sensor (before catalytic converter) is faulty • Oxygen sensor (behind catalytic converter) is faulty • Oxygen sensor heater (before catalytic converter) is faulty • Oxygen sensor heater (behind catalytic converter) is faulty
DTC: P2637 **1T CCM, MIL: Yes** **2005, 2006** **Models:** A6 **Engines:** 3.2L, 4.2L **Transmissions:** All	**Torque Management Feedback Signal "A" Conditions:** Engine started, battery voltage must be at least 11.5v, all electrical components must be off, parking brake must be engaged (to keep daytime driving lights off), automatic transmission selector must be in park, the exhaust system must be properly sealed between the catalytic converter and the cylinder head, coolant temperature must be at least 80 degrees Celsius and oxygen sensor heaters for oxygen sensors before the catalytic converter must be functioning properly and the ground between the engine and the chassis must be well connected. The ECM detected a voltage value on the torque management circuits that doesn't fall within the desired parameters **Possible Causes:** • Engine Control Module (ECM) has failed • Voltage supply for Engine Control Module (ECM) is damaged • Engine Coolant Temperature (ECT) sensor has failed • Intake Air Temperature (IAT) sensor has failed • Intake Manifold Runner Position Sensor has failed • Intake system has leaks (false air) • Motor for intake flap has failed • Mass Air Flow (MAF) sensor has failed

DTC	Trouble Code Title, Conditions & Possible Causes
DTC: P2714 **2T CCM, MIL: Yes** **2005, 2006** **Models:** A6 **Engines:** 3.2L, 4.2L **Transmissions:** All	**Pressure Control Solenoid "D" Performance or Stuck Off Conditions:** Engine started, battery voltage must be at least 11.5v, all electrical components must be off, and the ground between the engine and the chassis must be well connected. The ECM detected the pressure control solenoid was in the "stuck off" position. **Possible Causes:** • ATF level is low • Circuit harness connector contacts are corroded or ingressed of water • Circuit wires have shorted to each other, to battery or ground • Automatic Transmission Hydraulic Pressure Sensor 1 has failed • Solenoid valves in valve body are faulty • Transmission Control Module (TCM) needs replacing • Transmission Input Speed (RPM) Sensor has failed • Transmission Output Speed (RPM) Sensor has failed
DTC: P2715 **2T CCM, MIL: Yes** **2005, 2006** **Models:** A6 **Engines:** 3.2L, 4.2L **Transmissions:** All	**Pressure Control Solenoid "D" Performance or Stuck On Conditions:** Engine started, battery voltage must be at least 11.5v, all electrical components must be off, and the ground between the engine and the chassis must be well connected. The ECM detected the pressure control solenoid was in the "stuck on" position. **Possible Causes:** • ATF level is low • Circuit harness connector contacts are corroded or ingressed of water • Circuit wires have shorted to each other, to battery or ground • Automatic Transmission Hydraulic Pressure Sensor 1 has failed • Solenoid valves in valve body are faulty • Transmission Control Module (TCM) needs replacing • Transmission Input Speed (RPM) Sensor has failed • Transmission Output Speed (RPM) Sensor has failed
DTC: P2716 **2T CCM, MIL: Yes** **2005, 2006** **Models:** A6 **Engines:** 3.2L, 4.2L **Transmissions:** All	**Pressure Control Solenoid "D" Electrical Malfunction Conditions:** Engine started, battery voltage must be at least 11.5v, all electrical components must be off, and the ground between the engine and the chassis must be well connected. The ECM detected the pressure control solenoid was experiencing electrical malfunctions. **Possible Causes:** • ATF level is low • Circuit harness connector contacts are corroded or ingressed of water • Circuit wires have shorted to each other, to battery or ground • Automatic Transmission Hydraulic Pressure Sensor 1 has failed • Solenoid valves in valve body are faulty • Transmission Control Module (TCM) needs replacing • Transmission Input Speed (RPM) Sensor has failed • Transmission Output Speed (RPM) Sensor has failed
DTC: P2723 **2T CCM, MIL: Yes** **2004, 2005, 2006** **Models:** TT, A6 **Engines:** 3.2L **Transmissions:** All	**Pressure Control Solenoid "E" Performance or Stuck Off Conditions:** Engine started, battery voltage must be at least 11.5v, all electrical components must be off, and the ground between the engine and the chassis must be well connected. The ECM detected the pressure control solenoid was in the "stuck off" position. **Possible Causes:** • ATF level is low • Circuit harness connector contacts are corroded or ingressed of water • Circuit wires have shorted to each other, to battery or ground • Automatic Transmission Hydraulic Pressure Sensor 1 has failed • Solenoid valves in valve body are faulty • Transmission Control Module (TCM) needs replacing • Transmission Input Speed (RPM) Sensor has failed • Transmission Output Speed (RPM) Sensor has failed
DTC: P2724 **2T CCM, MIL: Yes** **2005, 2006** **Models:** A6 **Engines:** 3.2L, 4.2L **Transmissions:** All	**Pressure Control Solenoid "E" Performance or Stuck On Conditions:** Engine started, battery voltage must be at least 11.5v, all electrical components must be off, and the ground between the engine and the chassis must be well connected. The ECM detected the pressure control solenoid was in the "stuck on" position. **Possible Causes:** • ATF level is low • Circuit harness connector contacts are corroded or ingressed of water • Circuit wires have shorted to each other, to battery or ground • Automatic Transmission Hydraulic Pressure Sensor 1 has failed • Solenoid valves in valve body are faulty • Transmission Control Module (TCM) needs replacing • Transmission Input Speed (RPM) Sensor has failed • Transmission Output Speed (RPM) Sensor has failed

DTC	Trouble Code Title, Conditions & Possible Causes
DTC: P2732 **2T CCM, MIL: Yes** **2004, 2005, 2006** **Models:** TT, A6 **Engines:** 3.2L, 4.2L **Transmissions:** All	**Pressure Control Solenoid "F" Performance or Stuck Off Conditions:** Engine started, battery voltage must be at least 11.5v, all electrical components must be off, and the ground between the engine and the chassis must be well connected. The ECM detected the pressure control solenoid was in the "stuck off" position. **Possible Causes:** • ATF level is low • Circuit harness connector contacts are corroded or ingressed of water • Circuit wires have shorted to each other, to battery or ground • Automatic Transmission Hydraulic Pressure Sensor 1 has failed • Solenoid valves in valve body are faulty • Transmission Control Module (TCM) needs replacing • Transmission Input Speed (RPM) Sensor has failed • Transmission Output Speed (RPM) Sensor has failed
DTC: P2733 **2T CCM, MIL: Yes** **2005, 2006** **Models:** A6 **Engines:** 3.2L, 4.2L **Transmissions:** All	**Pressure Control Solenoid "F" Performance or Stuck On Conditions:** Engine started, battery voltage must be at least 11.5v, all electrical components must be off, and the ground between the engine and the chassis must be well connected. The ECM detected the pressure control solenoid was in the "stuck on" position. **Possible Causes:** • ATF level is low • Circuit harness connector contacts are corroded or ingressed of water • Circuit wires have shorted to each other, to battery or ground • Automatic Transmission Hydraulic Pressure Sensor 1 has failed • Solenoid valves in valve body are faulty • Transmission Control Module (TCM) needs replacing • Transmission Input Speed (RPM) Sensor has failed • Transmission Output Speed (RPM) Sensor has failed

DTC	Trouble Code Title, Conditions & Possible Causes:
DTC: P3081 **1998, 1999, 2000, 2001, 2002, 2003, 2004, 2005, 2006** **Models:** TT, A4, A6, Allroad Quattro, A8, A8L **Engines:** 1.8L, 2.7L, 2.8L, 3.0L, 4.2L **Transmissions:** All	**Engine Temperature Too Low Conditions:** Engine running and the ECM has detected that the engine temperature is too low. **Possible Causes:** • Engine hasn't completely warmed up • Radiator malfunction • Thermostat malfunction • ECM failure
DTC: P3096 **2004, 2005, 2006** **Models:** TT **Engines:** 3.2L **Transmissions:** All	**Internal Control Module Memory, Check Sum Error Conditions:** Key on, the ECM has detected a programming error **Possible Causes:** • Battery terminal corrosion, or loose battery connection • Connection to the ECM interrupted, or the circuit has been opened • Reprogramming error has occurred • ECM has failed and needs replacement. • Voltage supply for Engine Control Module (ECM) is faulty
DTC: P3097 **2004, 2005, 2006** **Models:** TT **Engines:** 3.2L **Transmissions:** All	**Internal Control Module Memory, Check Sum Error Conditions:** Key on, the ECM has detected a programming error **Possible Causes:** • Battery terminal corrosion, or loose battery connection • Connection to the ECM interrupted, or the circuit has been opened • Reprogramming error has occurred • ECM has failed and needs replacement. • Voltage supply for Engine Control Module (ECM) is faulty
DTC: P3262 **2000, 2001** **Models:** A4 **Engines:** 2.7L **Transmissions:** All A4 (all): 2.7L (APB) **Transmissions:** All	**Exhaust (Banks 1 and 2) Oxygen Sensors Behind Catalytic Converter Swpped Conditions:** Engine started, battery voltage must be at least 11.5v, all electrical components must be off, the ground between the engine and the chassis must be well connected, the exhaust system must be properly sealed between the catalytic converter and the cylinder head, and the coolant temperature must be 80 degrees Celsius. The ECM detected the O2 signals were mixed and reading implausible results from both. **Possible Causes:** • O2 harness connectors are swapped • O2 wiring is crossed inside the harness • O2 wires are crossed at 104-pin connector • Connector coding and color mixed with correct catalytic converter

Gas Engine OBD II Trouble Code List (U1xxx Codes)

DTC	Trouble Code Title, Conditions & Possible Causes:
DTC: U0001 **1T CCM, MIL: No** **2003, 2004, 2005, 2006** **Models:** TT, A4, A6, Allroad Quattro, A8, A8L **Engines:** 2.7L, 3.2L, 4.2L **Transmissions:** All	**High Speed CAN Communication Bus Conditions:** The Engine Control Module (ECM) communicates with all databus-capable control modules via a CAN databus. These databus-capable control modules are connected via two data bus wires which are twisted together (CAN_High and CAN_Low), and exchange information (messages). Missing information on the databus is recognized as a malfunction and stored. Trouble-free operation of the CAN-Bus requires that it have a terminal resistance. This central terminal resistor is located in the Engine Control Module (ECM). **Possible Causes:** • ECM has failed • CAN data bus wires have short circuited to each other
DTC: U0100 **1T CCM, MIL: No** **2004, 2005, 2006** **Models:** TT **Engines:** 3.2L **Transmissions:** All	**Lost Communication With ECM "A" Conditions:** Key on, and the ECM detected that it has lost communication during its initial startup. The Engine Control Module (ECM) communicates with all databus-capable control modules via a CAN databus. These databus-capable control modules are connected via two data bus wires which are twisted together (CAN_High and CAN_Low), and exchange information (messages). Missing information on the databus is recognized as a malfunction and stored. Trouble-free operation of the CAN-Bus requires that it have a terminal resistance. **Possible Causes:** • ECM has failed • Terminal resistance for CAN-bus are faulty • Can data bus wires have short circuited to each other
DTC: U0101 **2000, 2001, 2002, 2003, 2004, 2005, 2006** **Models:** A4, A6, A6 Avant, Allroad Quattro, A8, A8L **Engines:** 2.7L, 3.2L, 4.2L **Transmissions:** All	**Lost Communication With TCM Conditions:** Key on, and the ECM detected that it has lost communication with the Transmission Control Module (TCM) during its initial startup. The Engine Control Module (ECM) communicates with all databus-capable control modules via a CAN databus. These databus-capable control modules are connected via two data bus wires which are twisted together (CAN_High and CAN_Low), and exchange information (messages). Missing information on the databus is recognized as a malfunction and stored. Trouble-free operation of the CAN-Bus requires that it have a terminal resistance. **Possible Causes:** • ECM has failed • Terminal resistance for CAN-bus are faulty • Can data bus wires have short circuited to each other • TCM has failed
DTC: U0103 **1T CCM, MIL: No** **2004, 2005, 2006** **Models:** TT **Engines:** 3.2L **Transmissions:** All	**Lost Communication With Gear Shift Module Conditions:** Key on, and the ECM detected that it has lost communication with the gear shift module during its initial startup. The Engine Control Module (ECM) communicates with all databus-capable control modules via a CAN databus. These databus-capable control modules are connected via two data bus wires which are twisted together (CAN_High and CAN_Low), and exchange information (messages). Missing information on the databus is recognized as a malfunction and stored. Trouble-free operation of the CAN-Bus requires that it have a terminal resistance. **Possible Causes:** • ECM has failed • Terminal resistance for CAN-bus are faulty • Can data bus wires have short circuited to each other • The gear shift module has failed
DTC: U0121 **1T CCM, MIL: No** **2004, 2005, 2006** **Models:** TT **Engines:** 3.2L **Transmissions:** All	**Lost Communication With Anti-Lock Brake System (ABS) Control Module Conditions:** Key on, and the ECM detected that it has lost communication with the ABS Control Module during its initial startup. The Engine Control Module (ECM) communicates with all databus-capable control modules via a CAN databus. These databus-capable control modules are connected via two data bus wires which are twisted together (CAN_High and CAN_Low), and exchange information (messages). Missing information on the databus is recognized as a malfunction and stored. Trouble-free operation of the CAN-Bus requires that it have a terminal resistance. **Possible Causes:** • ECM has failed • Terminal resistance for CAN-bus are faulty • Can data bus wires have short circuited to each other • There is a fault with the ABS control module
DTC: U0155 **1T CCM, MIL: No** **2001, 2002, 2003, 2004, 2005, 2006** **Models:** TT, A4, A6, Allroad Quattro, A8, A8L **Engines:** 2.7L, 3.2L, 4.2L **Transmissions:** All	**Lost Communication With Instrument Cluster Conditions:** Key on, and the ECM detected that it has lost communication with the Instrument Cluster Panel (I/P) during its initial startup. The Engine Control Module (ECM) communicates with all databus-capable control modules via a CAN databus. These databus-capable control modules are connected via two data bus wires which are twisted together (CAN_High and CAN_Low), and exchange information (messages). Missing information on the databus is recognized as a malfunction and stored. Trouble-free operation of the CAN-Bus requires that it have a terminal resistance. **Possible Causes:** • ECM has failed • Terminal resistance for CAN-bus are faulty • Can data bus wires have short circuited to each other

DTC	Trouble Code Title, Conditions & Possible Causes
DTC: U0302 **1T CCM, MIL: No** **2003, 2004, 2005, 2006** **Models:** TT, A6, Allroad Quattro **Engines:** 3.2L, 4.2L **Transmissions:** All	**Software Incompatibility with Transmission Control Module Conditions:** Key on, and the ECM detected a software incompatibility condition with the Transmission Control Module during its initial startup. The Engine Control Module (ECM) communicates with all databus-capable control modules via a CAN databus. These databus-capable control modules are connected via two data bus wires which are twisted together (CAN_High and CAN_Low), and exchange information (messages). Missing information on the databus is recognized as a malfunction and stored. Trouble-free operation of the CAN-Bus requires that it have a terminal resistance. **Possible Causes:** • ECM or TCM has failed or is not properly coded • Terminal resistance for CAN-bus are faulty • Can data bus wires have short circuited to each other
DTC: U0315 **1T CCM, MIL: No** **2004, 2005, 2006** **Models:** TT **Engines:** 3.2L **Transmissions:** All	**Software Incompatibility with Anti-Lock Brake System Control Module Conditions:** Key on, and the ECM detected a software incompatibility condition with the Anti-Lock Brake System Control Module during its initial startup. The Engine Control Module (ECM) communicates with all databus-capable control modules via a CAN databus. These databus-capable control modules are connected via two data bus wires which are twisted together (CAN_High and CAN_Low), and exchange information (messages). Missing information on the databus is recognized as a malfunction and stored. Trouble-free operation of the CAN-Bus requires that it have a terminal resistance. **Possible Causes:** • ECM has failed • Terminal resistance for CAN-bus are faulty • Can data bus wires have short circuited to each other • The AB S control module has failed
DTC: U0402 **1T CCM, MIL: No** **2005, 2006** **Models:** A6, A8, A8L **Engines:** 3.2L, 4.2L **Transmissions:** All	**Invalid Data Received From Transmission Control Module Conditions:** Key on, and the ECM detected a software invalid data from the Cruise Control Module during its initial startup. The Engine Control Module (ECM) communicates with all databus-capable control modules via a CAN databus. These databus-capable control modules are connected via two data bus wires which are twisted together (CAN_High and CAN_Low), and exchange information (messages). Missing information on the databus is recognized as a malfunction and stored. Trouble-free operation of the CAN-Bus requires that it have a terminal resistance. **Possible Causes:** • ECM or TCM has failed • Terminal resistance for CAN-bus are faulty • Can data bus wires have short circuited to each other
DTC: U0404 **1T CCM, MIL: No** **2004, 2005, 2006** **Models:** TT **Engines:** 3.2L **Transmissions:** All	**Invalid Data Received From Gear Shift Control Module Conditions:** Key on, and the PCM detected a software invalid data from the Gear Shift Control Module during its initial startup. The Engine Control Module (ECM) communicates with all databus-capable control modules via a CAN databus. These databus-capable control modules are connected via two data bus wires which are twisted together (CAN_High and CAN_Low), and exchange information (messages). Missing information on the databus is recognized as a malfunction and stored. Trouble-free operation of the CAN-Bus requires that it have a terminal resistance. **Possible Causes:** • ECM has failed • Terminal resistance for CAN-bus are faulty • Can data bus wires have short circuited to each other • Gear shift control module has failed

AUDI
COMPONENT TESTING

6

TABLE OF CONTENTS

Barometric Pressure (BARO) Sensor

LOCATION

1.8L Engines

See Figure 1.

The Barometric Pressure (BARO) Sensor is positioned in a protective box within the fresh air plenum, on the left side, near the Engine Control Module.

TESTING

See Figure 2.

1. Disconnect the three-pin connector at sensor.
2. Switch on the ignition.
3. Connect the multimeter to the following connector terminals for voltage measurement:
   - Terminal 1 and Terminal 3
   - Terminal 2 and Terminal 3
4. Specified value: 4.5 to 5.5 V
5. If the specified values are not obtained:

   a. Connect the test box to the control module wiring harness.

   b. Check the wiring between the test box and the 3-pin connector for open circuit.
   - Terminal 1 and socket 61
   - Terminal 2 and socket 62
   - Terminal 3 and socket 67
6. The resistance between any of the wires should be 1.5 Kohms maximum.
7. Check the wiring for a short circuit. There should be no resistance.
8. Check the wiring for a short circuit to battery power or ground. There should be no resistance.

Camshaft Position (CMP) Sensor

LOCATION

1.8L Engines

See Figure 3.

The Camshaft Position (CMP) Sensor is mounted on the cylinder head, front, beneath timing belt cover.

2.7L Engines

See Figure 4.

The Camshaft Position (CMP) Sensor 2 is mounted on the cylinder head, in the rear.

3.0L Engines

See Figure 5.

The Camshaft Position (CMP) Sensor is located in the right side of the engine compartment at the rear of the valve cover, near the intake manifold.

3.2L Engines

See Figures 6 and 7.

The Camshaft Position (CMP) Sensor is mounted in the left side of the engine compartment to the left of the oil fill cap.

The Camshaft Position (CMP) Sensor 2 is mounted in the left side of the engine compartment with a black connector.

3.7L Engines

See Figure 8.

The Camshaft Position (CMP) Sensor is located on the rear of the left cylinder head.

29246_AUDI_G0034

Fig. 1 Barometric Pressure sensor location on the 1.8L engine

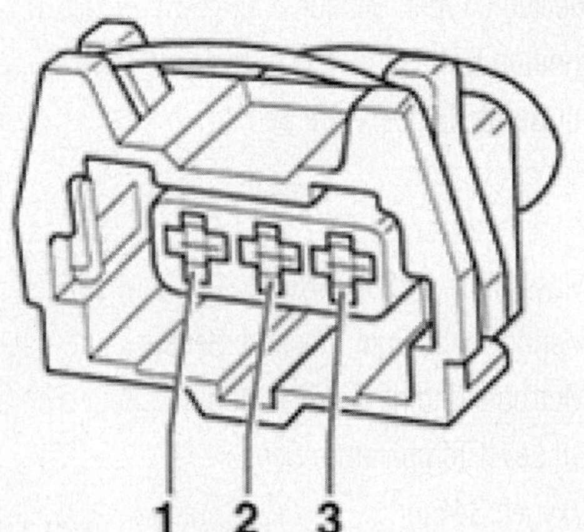

29246_VOLK_G0084

Fig. 2 The three-pin connector at the sensor is connected to the multimeter for testing

29246_AUDI_G0001

Fig. 3 Camshaft Position sensor location for the 1.8L engine

29246_AUDI_G0035

Fig. 4 Camshaft Position sensor 2 location for the 2.7L engine

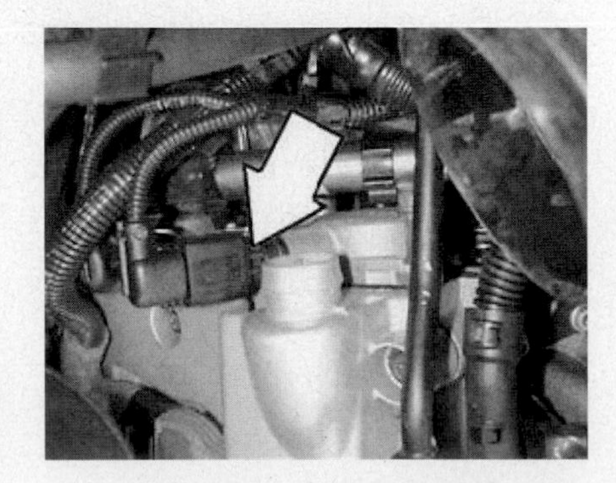

29246_AUDI_G0058

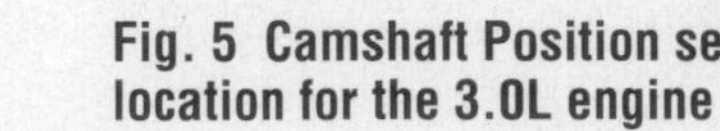

Fig. 5 Camshaft Position sensor location for the 3.0L engine

29246_AUDI_G0002

Fig. 6 Camshaft Position sensor location for the 3.2L engine

29246_AUDI_G0003

Fig. 7 Camshaft Position sensor 2 location for the 3.2L engine

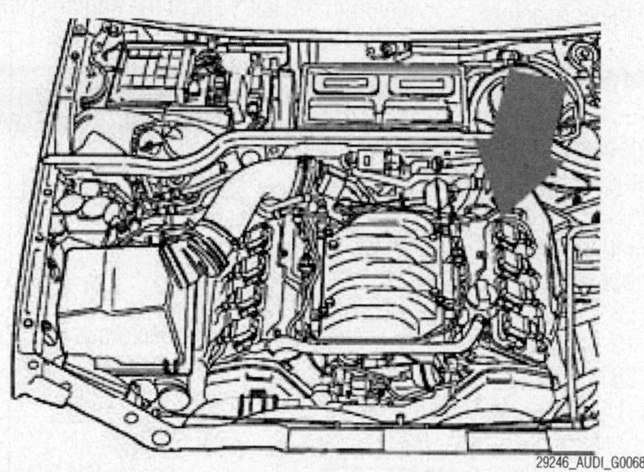

29246_AUDI_G0068

Fig. 8 Camshaft Position sensor location for the 3.7L engine

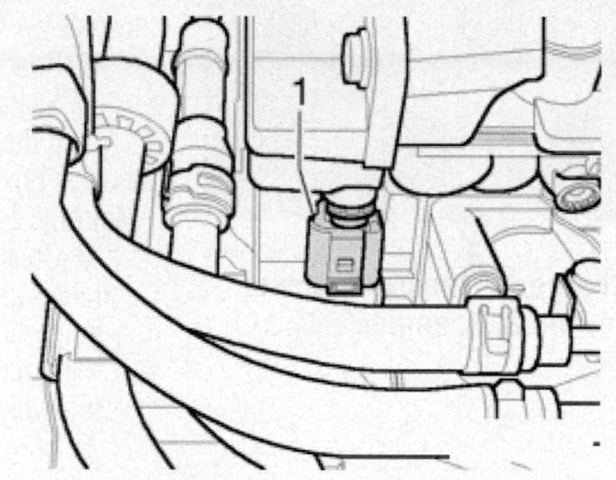

29246_VOLK_G0088

Fig. 9 Disconnect the 3-pin harness connector (1) at the Camshaft Position (CMP) Sensor

TESTING

See Figures 9, 10 and 11.

➡ **Use only gold-plated terminals when servicing terminals in harness connector of Camshaft Position (CMP) sensor.**

1. Remove the engine cover with the air filter.
2. Disconnect the 3-pin harness connector (1) at the Camshaft Position (CMP) Sensor
3. Switch the ignition on.
4. Measure voltage between terminals 1 and 3 of the connector using a multimeter and the adapter cables from special tool connector test kit V.A.G 1594 C.
5. Specified value: 4.5 to 5.5 V
6. Switch the ignition off.
7. If the voltage is not okay, connect the test box to wiring harness of Motronic Engine Control Module (ECM).
8. Check the wires between test box and 3-pin connector for open circuit.
   - Terminal 1 and socket 26
   - Terminal 2 and socket 44
   - Terminal 3 and socket 14
9. Wire resistance: max. 1.5 ohms
10. Check the wires for short circuit to each other, to Battery (+) and Ground (GND).
11. Specified value: infinite ohms
12. If the wires and voltage are okay, replace the Camshaft Position (CMP) Sensor.

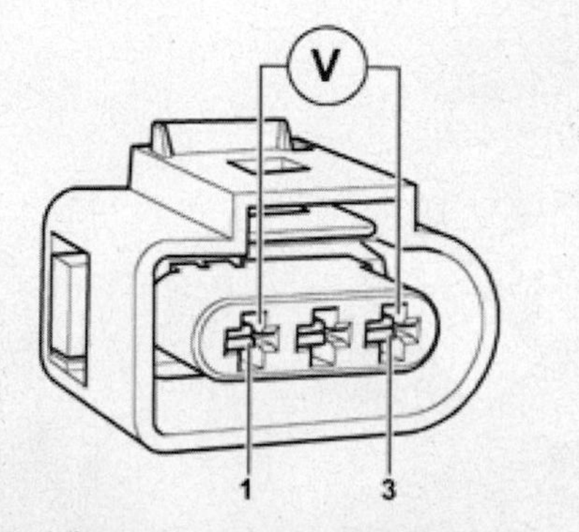

29246_VOLK_G0089

Fig. 10 Measure the voltage between terminals 1 and 3 of the connector using a multimeter

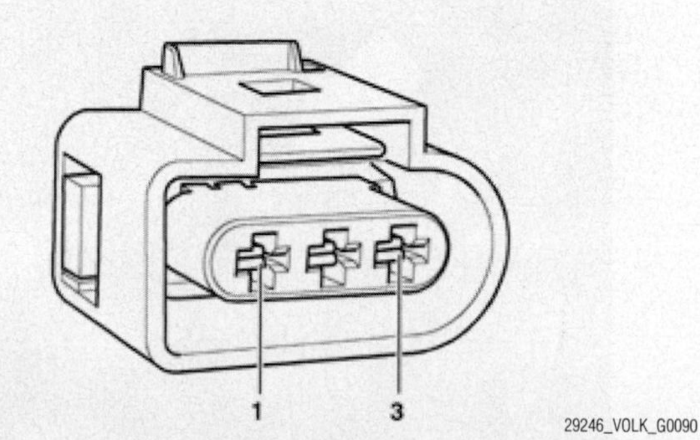

29246_VOLK_G0090

Fig. 11 Check wires between test box and 3-pin connector for open circuit

Charge Air Pressure Sensor

LOCATION

2.7L Engines

See Figure 12.

The Charge Air Pressure Sensor is located in the air intake boot, on the front of the intake manifold. It is attached to the throttle valve housing.

29246_AUDI_G0036

Fig. 12 Charge Air Pressure sensor location for the 2.7L engine

TESTING

See Figures 13 and 14.

➡ **Use only gold-plated terminals when servicing terminals in harness connector of Charge Air Pressure Sensor.**

1. Remove the engine cover with the air filter.
2. Disconnect the 4-pin harness connector at the Charge Air Pressure Sensor (arrow).
3. Start the engine and let it run at idle.
4. Measure the voltage between terminals 1and3 on connector to the wiring harness using a multimeter V.A.G. 1526 B and the adapter cables from the connector test kit V.A.G 1594 C.
5. Specified value: min. 4.5 V
6. Switch the ignition off.
7. If the the voltage is not okay, connect the test box to the wiring harness of the Motronic Engine Control Module (ECM).
8. Check the wires between the test box and the 4-pin connector for open circuit.
   - Terminal 1 and socket 53
   - Terminal 2 and socket 19
   - Terminal 3 and socket 381
9. Wire resistance: max. 1.5 ohms
10. Check the wires for short circuit to each other, to Battery (+) and Ground (GND).
11. Specified value: infinite ohms
12. If the voltage and wires are okay, replace Charge Air Pressure sensor.
13. If the voltage is not okay and the wires are okay, replace Motronic Engine Control Module (ECM).

Coolant Fan Control Module

LOCATION

All Engines

See Figure 15.

The Coolant Fan Control Module is located in the left front of the engine compartment next to the radiator.

Cylinder Fuel Injectors

LOCATION

3.2L Engines

See Figure 16.

The Cylinder Fuel Injectors are located under the intake manifold in the fuel rail (Cylinder One shown).

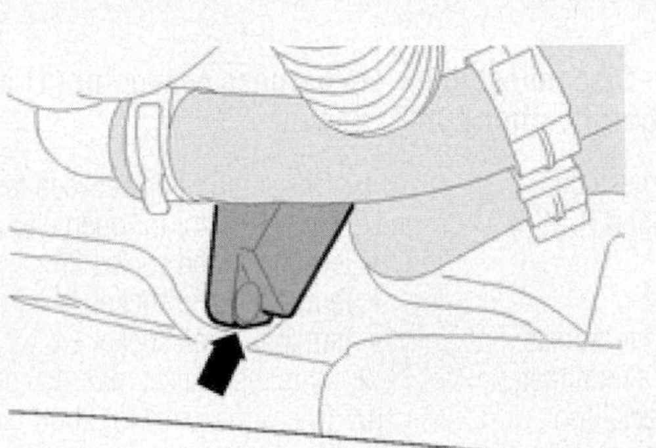

29246_VOLKS_G0107

Fig. 13 Disconnect 4-pin harness connector at Charge Air Pressure Sensor

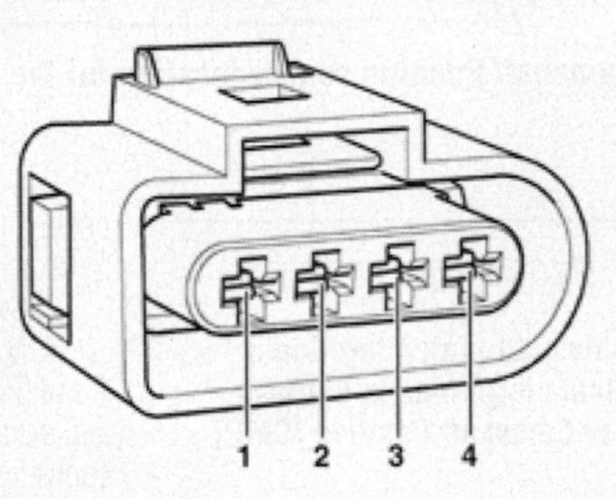

29246_VOLKS_G0108

Fig. 14 Check wires between test box and 4-pin connector for open circuit

29246_AUDI_G0004

Fig. 15 Coolant Fan Control Module location for all engines

29246_AUDI_G0005

Fig.16 Cylinder Fuel Injector location for the 3.2L engine

Engine Coolant Temperature (ECT) Sensor

LOCATION

1.8L and 3.2L Engines

See Figure 17.

The Engine Coolant Temperature Sensor is located in the coolant flange to the left of the cylinder head and is integral with the ECT sensor.

2.7L Engines

See Figure 18.

The Engine Coolant Temperature Sensor is located in the coolant pipe, behind the right cylinder head and is integral with the ECT sensor.

3.0L Engines

See Figure 19.

The Engine Coolant Temperature Sensor is located in the coolant pipe, below the air intake pipe and behind the right cylinder head and is integral with the ECT sensor.

3.7L and 4.2L Engines

See Figure 20.

The Engine Coolant Temperature Sensor is located on the back of the right cylinder head and is integral with the ECT sensor.

OPERATION

1. The engine cooling system consists of five main parts.
   - The engine cooler
   - The Engine Coolant Temperature Sensor
   - The thermostat valve
   - The small cooling circuit
   - The large cooling circuit

During the heating up of the engine, the coolant flows first inside the small cooling circuit. After the coolant reaches a sufficient temperature, the thermostat valve will open the large cooling circuit to integrate the engine cooler. The engine coolant temperature sensor measures a mixed temperature between the coolant coming from the small and large cooling circuit.

TESTING

**** WARNING**

The cooling system is under pressure. There is a danger of scalding when opening!

➡ **Use only gold-plated terminals when servicing terminals in harness connector of Engine Coolant Temperature (ECT) Sensor.**

1. Connect the vehicle to the diagnosis and service system.
2. Switch the ignition on.
3. Under address word "33 - On Board Diagnostic (OBD)", select "Diagnostic mode 1: Check measured values".
4. Select measured value "PID 05: Coolant temperature".
5. Check the specified value of the coolant temperature:

29246_AUDI_G0006

Fig. 17 Engine Coolant Temperature sensor location for the 1.8L and 3.2L engines

29246_AUDI_G0037

Fig. 18 Engine Coolant Temperature sensor location for the 2.7L engines

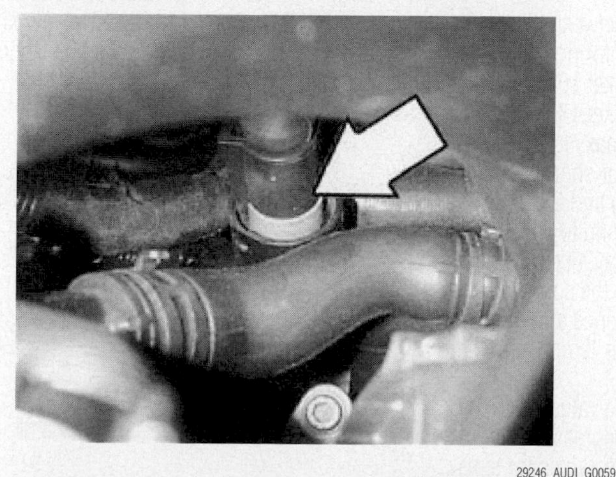

29246_AUDI_G0059

Fig. 19 Engine Coolant Temperature sensor location for the 3.0L engines

29246_AUDI_G0069

Fig. 20 Engine Coolant Temperature sensor location for the 3.7L and the 4.2L engines

6. If specified value is not obtained, continue test according to the following results:

- If the temperature indicated is approximately minus 40 degrees Celsius, there is an open circuit or a short to the battery.
- If the temperature indicated is approximately 140 degrees Celsius, there is a short circuit to the ground.

7. If the specified value is obtained, start the engine and let it run at idle.

8. Temperature value must increase uniformly in increments of 1.0 degrees C.

9. If the engine shows problems in certain temperature ranges and if the temperature does not climb uniformly, the temperature signal is intermittent. Replace Engine Coolant Temperature (ECT) Sensor.

TESTING IF DISPLAY APPROX. -40.0 DEGREES C:

See Figures 21 and 22.

1. Bridge terminals 1 and 2 of connector using adapter cables from connector test kit and observe indication on display.

2. If the indication jumps to approx. 140.0 degrees C, end diagnosis and switch ignition off. Replace Engine Coolant Temperature (ECT) Sensor.

⁂ WARNING

Cooling system is under pressure. Danger of scalding when opening!

3. If the indication remains at approx. -40.0 degrees C, end diagnosis, switch the ignition off and check the wires.

TESTING IF DISPLAY APPROX. 140.0 DEGREES C:

See Figures 23, 24 and 25.

1. Disconnect the 2-pin harness connector at Engine Coolant Temperature (ECT) Sensor (arrow).

2. If the indication jumps to approx. -40.0 degrees C, end diagnosis, switch the ignition off, and replace the Engine Coolant Temperature (ECT) sensor.

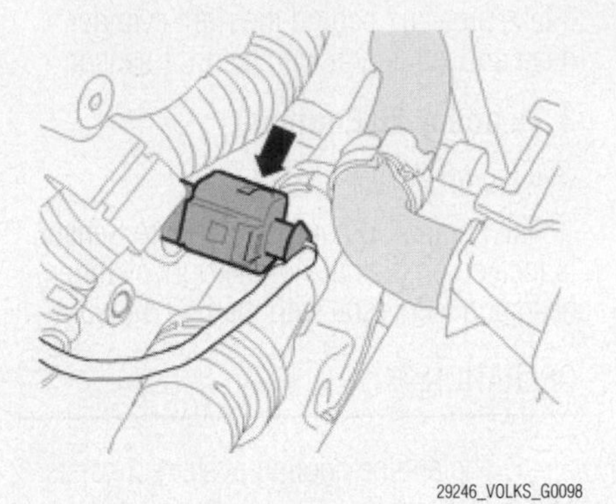

Fig. 21 Disconnect the 2-pin harness connector at Engine Coolant Temperature (ECT) Sensor (arrow)

29246_VOLKS_G0099

Fig. 22 Bridge terminals 1 and 2 of the connector using the adapter cables from the connector test kit and observe indication on display

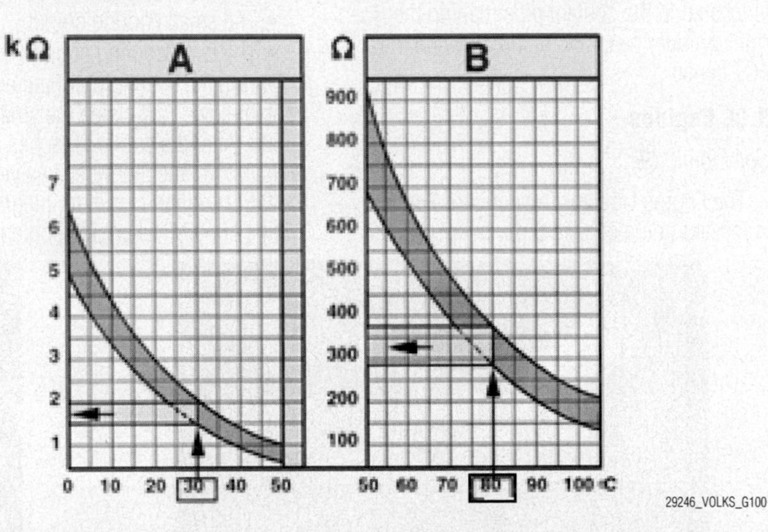

Fig. 24 Check wires between test box and 2-pin connector for open circuit according to wiring diagram.

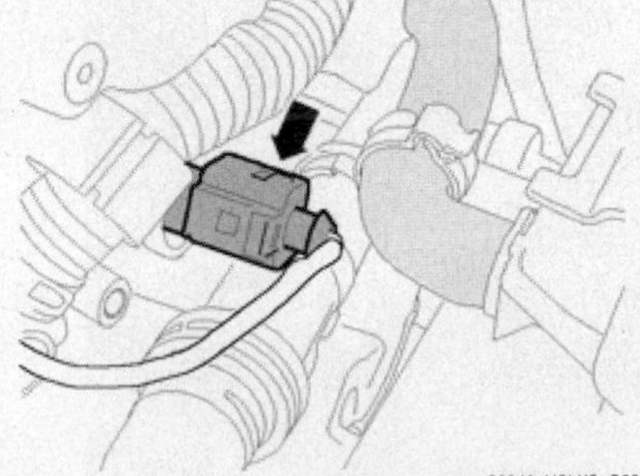

Fig. 23 Disconnect 2-pin harness connector at Engine Coolant Temperature (ECT) Sensor (arrow)

3. If the indication remains at approx. 140.0 degrees C, end diagnosis, switch the ignition off and check the wiring.

4. Connect the test box to the wiring harness of the Motronic Engine Control Module (ECM).

5. Check the wires between the test box and the 2-pin connector for open circuits.

- Terminal 1 and socket 10
- Terminal 2 and socket 14

6. Wire resistance: max. 1.5 ohms

7. Check the wires for short circuit to each other, to Battery (+) and Ground (GND).

8. Specified value: infinite ohms

9. If the wires are okay, use a multimeter and adapter cables from the connector test kit to measure the resistance at the connector to the Engine Coolant Temperature (ECT) Sensor terminals 1 and 2.

10. Area A: Resistance values for temperature range 0 to 50 degrees C.

11. Area B: Resistance values for temperature range 50 to 100 degrees C.

12. Read-out examples:

a. 30 degrees C is in range A and corresponds to a resistance of 1.5 to 2.0 kohms

b. 80 degrees C is in range B and corresponds to a resistance of 275 to 375ohms

13. If the resistance is not okay, replace Engine Coolant Temperature (ECT) Sensor.

14. If the wires and resistance values are okay, replace Motronic Engine Control Module (ECM).

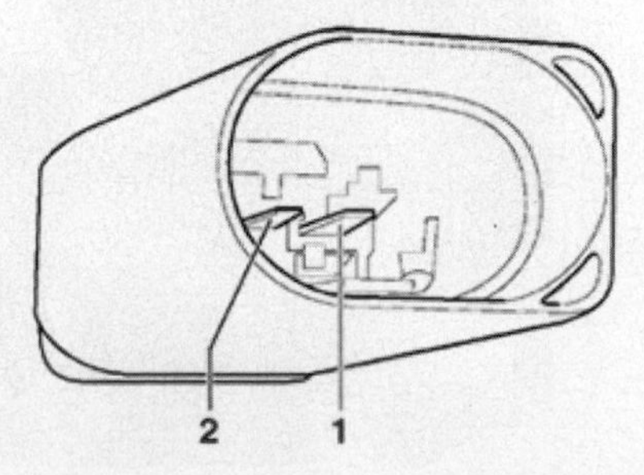

Fig. 25 Resistance read-out examples

Engine Oil Temperature Sensor

LOCATION

1.8L Engines

See Figure 26.

The Engine Oil Temperature Sensor is located on the driver's side of the cylinder block, in the center, in the oil filter housing.

2.7L and 2.8L Engines

See Figure 27.

The Engine Oil Temperature Sensor is located on the front of the engine, threaded into the oil pump housing, on the left side.

29246_AUDI_G0038

Fig. 26 Engine Oil Temperature sensor location for the 1.8L engine

Engine Speed (RPM) Sensor

LOCATION

1.8L Engines

See Figure 28.

The Engine Speed (RPM) Sensor is located on the front of the engine block, lower left, near the transmission flange.

2.7L and 2.8L Engines

See Figure 29.

The Engine Speed (RPM) Sensor is located in the front of the transmission bell-housing on the driver's side. (shown without the sensor).

3.7L Engines

See Figure 30.

The Engine Speed (RPM) Sensor is located on left side of the transmission to the rear of the flange for the engine.

29246_AUDI_G0039

Fig. 27 Engine Oil Temperature sensor location for the 2.7L and 2.8L engines

29246_AUDI_G0007

Fig. 28 Engine Speed sensor location for the 1.8L engine

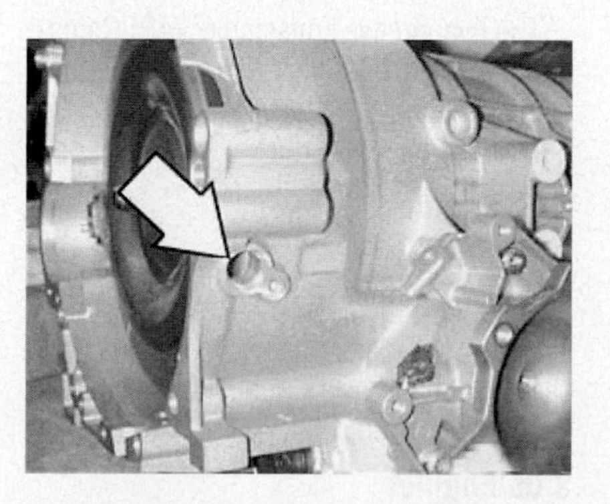

29246_AUDI_G0040

Fig. 29 Engine Speed sensor location for the 2.7L and 2.8L engines

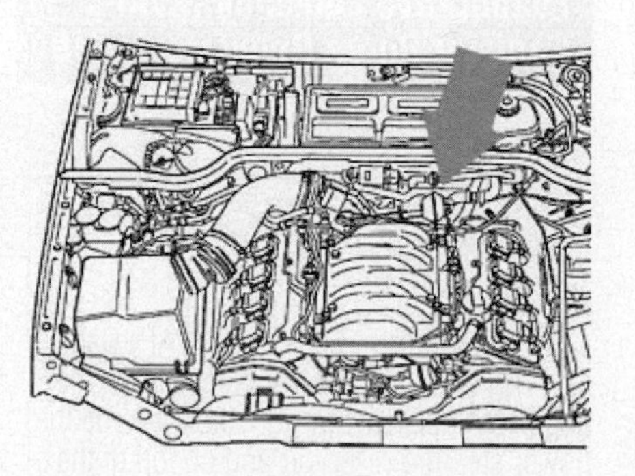

29246_AUDI_G0070

Fig. 30 Engine Speed sensor location for the 3.7L engine

OPERATION

The Engine Speed (RPM) Sensor detects the RPM and the reference marks. Without an engine speed signal, the engine will not start. If the engine speed signal fails while the engine is running, the engine will stop immediately.

TESTING

See Figures 31 and 32.

➡ **Use only gold-plated terminals when servicing terminals in harness connector of Camshaft Engine Speed (RPM) Sensor.**

1. Disconnect the gray 3-pin harness connector (5) to the Engine Speed (RPM) Sensor.
2. Measure the resistance between terminals 2 and 3 on connector (B) to Engine Speed (RPM) Sensor using multimeter and adapter cables from a connector test kit.
3. Specified value: 730 to 1000 ohms (at approx. 20 degrees C)
4. Check the Engine Speed (RPM) Sensor for short circuit between terminals 1 and 2 as well as 1 and 3.
5. Specified value: infinite ohms
6. If the resistance is not okay, replace the Engine Speed (RPM) Sensor.
7. If the Engine Speed (RPM) Sensor is okay, connect test box to wiring harness of Motronic Engine Control Module (ECM).
8. Check the wires between the test box and the 3-pin connector for an open circuit.
   - Terminal 1 and socket 52
   - Terminal 2 and socket 51
   - Terminal 3 and socket 36
9. Wire resistance: max. 1.5 ohms
10. Check the wires for short circuit to each other, to Battery (+) and Ground (GND).
11. Specified value: infinite ohms

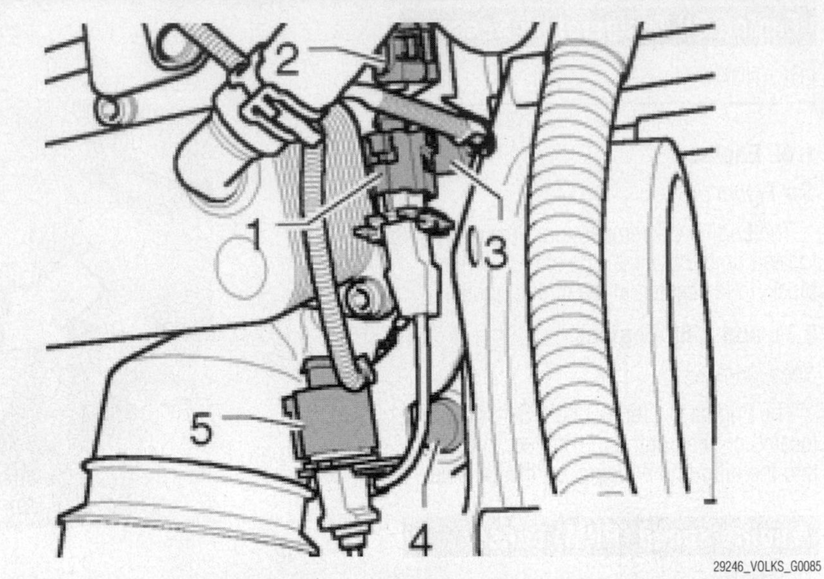

Fig. 31 Disconnect gray 3-pin harness connector (5) to Engine Speed (RPM) Sensor

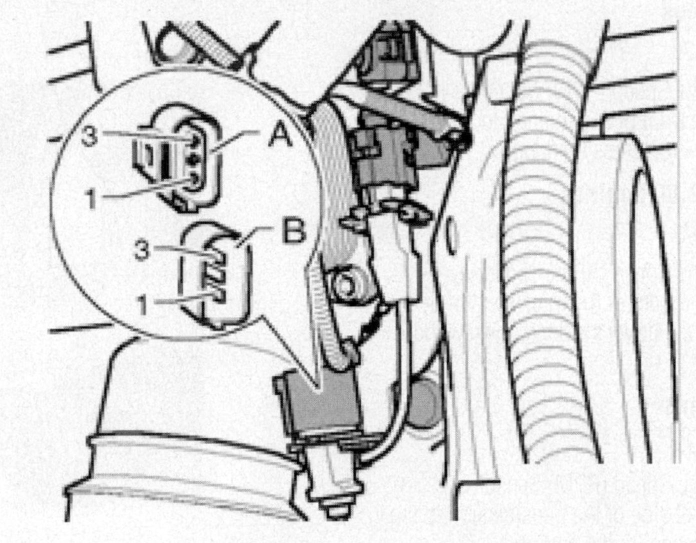

Fig. 32 The connector's terminals needed to be measured via a multimeter

Evaporative Emission (EVAP) Canister Purge Regulator Valve

LOCATION

1.8L Engines

See Figure 33.

The Evaporative Emission (EVAP) Canister Purge Regulator Valve is located in the engine compartment's right side, beside the power steering reservoir and on top of the air cleaner housing.

2.7L Engines

See Figure 34.

The Evaporative Emission (EVAP) Canister Purge Regulator Valve is located on top of the center of the intake manifold to the right of the air intake duct.

2.8L Engines

See Figure 35.

The Evaporative Emission (EVAP) Canister Purge Regulator Valve is located in the engine compartment's right side on top of the air cleaner housing.

3.0L Engines

See Figure 36.

The Evaporative Emission (EVAP) Canister Purge Regulator Valve is located in the engine compartment's right side, above the air box.

3.2L Engines

See Figure 37.

The Evaporative Emission (EVAP) Canister Purge Regulator Valve is located in the engine compartment's right side, in front of the power steering reservoir.

3.7L Engines

See Figure 38.

The Evaporative Emission (EVAP) Canister Purge Regulator Valve is located to the rear of the upper air cleaner housing.

29246_AUDI_G0008

Fig. 33 Evaporative Emission Canister Purge Regulator Valve location for the 1.8L engine

29246_AUDI_G0041

Fig. 34 Evaporative Emission Canister Purge Regulator Valve location for the 2.7L engine

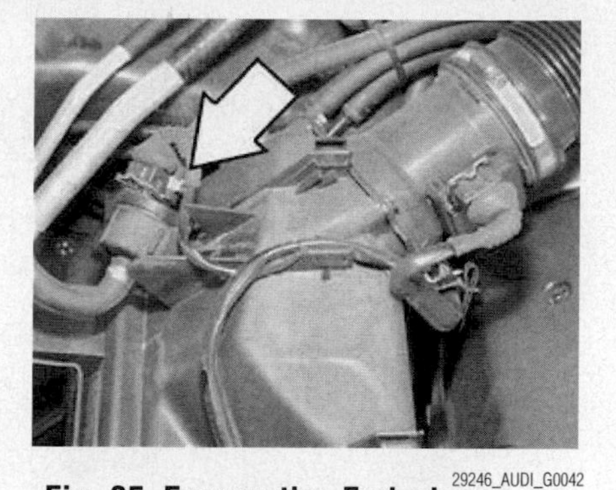

29246_AUDI_G0042

Fig. 35 Evaporative Emission Canister Purge Regulator Valve location for the 2.8L engine

29246_AUDI_G0060

Fig. 36 Evaporative Emission Canister Purge Regulator Valve location for the 3.0L engine

29246_AUDI_G0009

Fig. 37 Evaporative Emission Canister Purge Regulator Valve location for the 3.2L engine

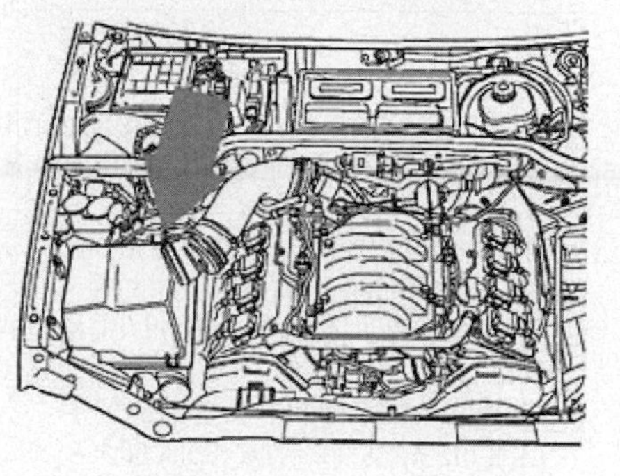

29246_AUDI_G0071

Fig. 38 Evaporative Emission Canister Purge Regulator Valve location for the 3.7L engine

TESTING

See Figures 39, 40 and 41.

1. Remove the engine cover with the air filter.
2. Disconnect the 2-pin harness connector (arrow) to the Evaporative Emission (EVAP) Canister Purge Regulator Valve.
3. Using a multimeter and the adapter cables from the connector test kit, measure the resistance between the terminals of the Evaporative Emission (EVAP) Canister Purge Regulator Valve.
4. Specified value: 22.0 to 30.0 ohms (at approx. 20 degrees C)
5. If the resistance is not okay, replace the Evaporative Emission (EVAP) Canister Purge Regulator Valve.
6. If the resistance is okay, connect the test box to the wiring harness of the Motronic Engine Control Module (ECM).
7. Check the wire between the test box socket and the 2-pin connector for a open circuit according.
   - Terminal 2 and socket 64
8. Wire resistance: max. 1.5 ohms
9. Check the wires for short circuit to Battery (+) and Ground (GND).
10. Specified value: infinite ohms

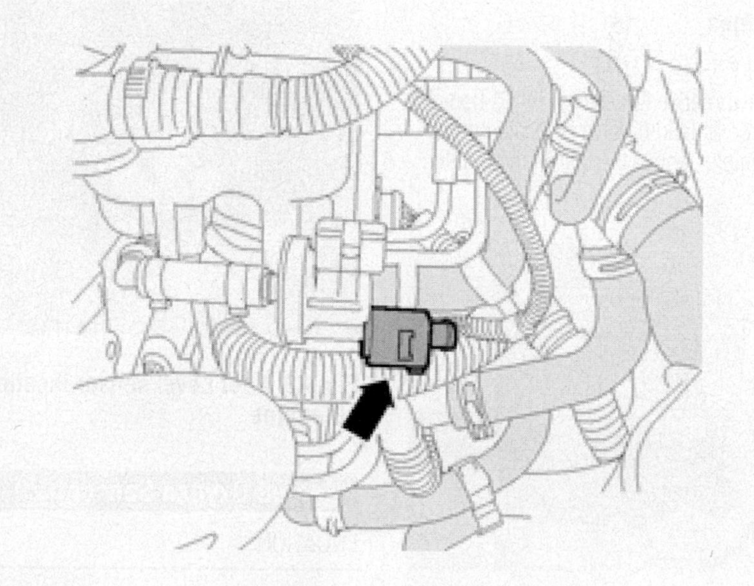

29246_VOLKS_G0082

Fig. 39 Disconnect 2-pin harness connector (arrow).

Fig. 40 Measure resistance between terminals of Evaporative Emission (EVAP) Canister Purge Regulator Valve

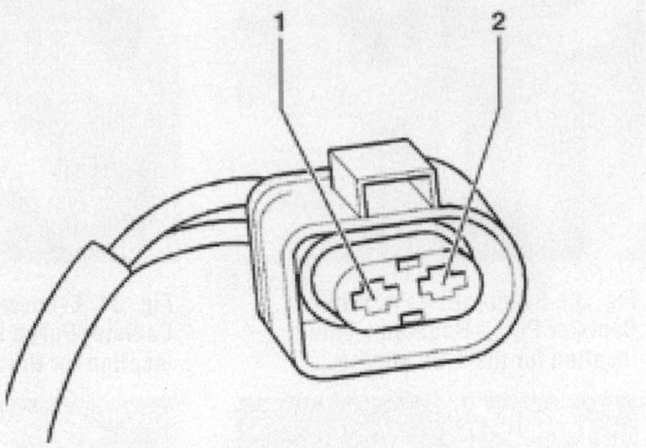

29246_VOLKS_G0104

Fig. 41 Check wire between test box socket and 2-pin connector for open circuit

11. Check the wire between the 2-pin connector terminal 1 and the Engine Component Power Supply Relay J757 for an open circuit.
12. Wire resistance: max. 1.5 ohms
13. If the wires are okay, replace Motronic Engine Control Module (ECM)

Evaporative Emission (EVAP) Canister Purge Solenoid Valve

LOCATION

All Engines

See Figure 42.

The Evaporative Emission (EVAP) Canister Purge Solenoid Valve is located in the EVAP canister, above the rear of the muffler.

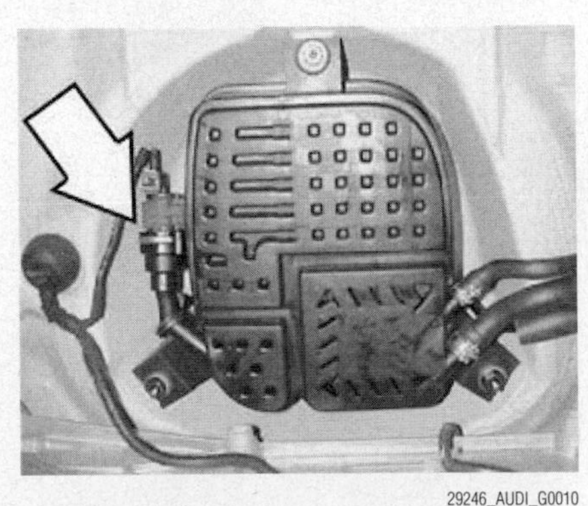

Fig. 42 Evaporative Emission Canister Purge Solenoid Valve location for all engines

Fuel Level Sensor

LOCATION

All Engines

See Figure 43.

The Fuel Level Sensor is located in the fuel tank, which is accessible under the rear seat on the passenger side (or in the trunk on some models). The sensor is a component of the fuel pump module.

Fig. 43 Fuel Level sensor location for all engine

Heated Oxygen Sensor

LOCATION

1.8L Engines

See Figure 44.

The Heated Oxygen Sensor is located on the exhaust system, in front of the three-way catalytic converter. It is integral with the Oxygen Sensor Heater.

2.7L and 2.8L Engines

See Figure 45.

The Heated Oxygen Sensor is located in the engine compartment on the right side, behind the cylinder head. It is integral with the Oxygen Sensor Heater.

3.0L Engines

See Figure 46.

The Heated Oxygen Sensor is located behind the engine in on the right side in the exhaust pipe.

3.2L Engines

See Figure 47

The Heated Oxygen Sensor is located behind the engine in the left front exhaust pipe. It is integral with the Oxygen Sensor Heater.

3.7L Engines

The Heated Oxygen Sensor is located behind the exhaust manifold in the right exhaust pipe. It is integral with the Oxygen Sensor Heater. (no image available)

TESTING

See Figure 48.

➡ **When servicing the terminals in the harness connector of Heated Oxygen Sensor (HO2S), use only gold-plated terminals.**

1. Connect to the vehicle diagnosis and service system (VAS).
2. Start the engine and let it run at idle.
3. Under address word "33 - On Board Diagnostic (OBD)", select "Diagnostic mode 6: Check test results of components that are not continuously monitored".

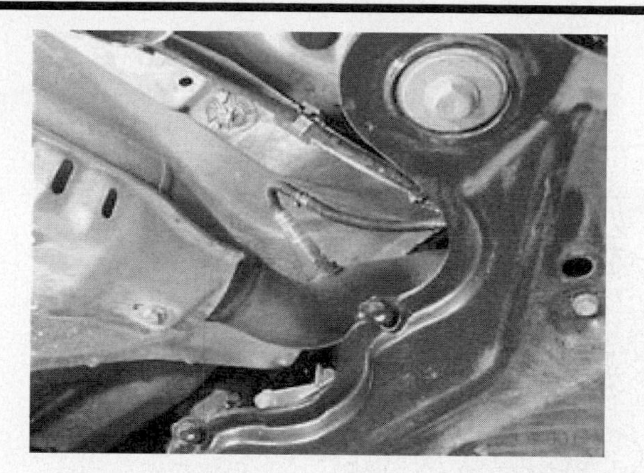

29246_AUDI_G0012

Fig. 44 Heated Oxygen sensor location for the 1.8L engine

29246_AUDI_G0043

Fig. 45 Heated Oxygen sensor location for the 2.7L and 2.8L engine

29246_AUDI_G0061

Fig. 46 Heated Oxygen sensor location for the 3.0L engine

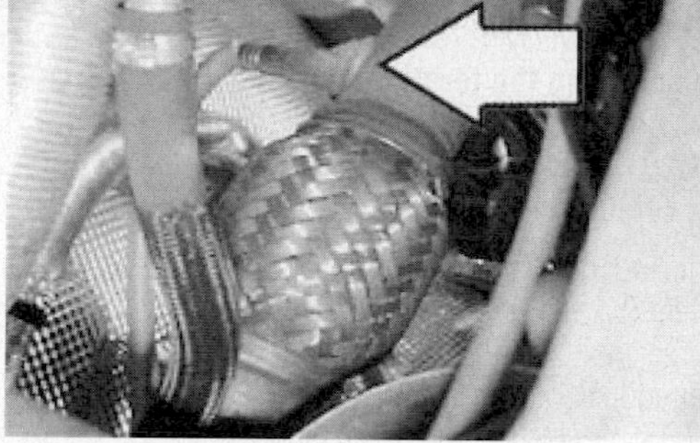

29246_AUDI_G0013

Fig. 47 Heated Oxygen sensor location for the 3.2L engine

4. Select "Test-ID 01: Oxygen sensor monitoring before catalytic converter".
5. Select "Test-ID 131: Dynamic test" and check whether specified values are obtained.
6. If the specified values are not obtained:
7. Check the oxygen sensor wiring.
8. Connect the test box to the wiring harness of the Motronic Engine Control Module (ECM)
9. Disconnect the black 6-pin harness connector to the Heated Oxygen Sensor (HO2S).
10. Check the wires between the test box and the 6-pin connector for open circuit.
   - Terminal 1 and socket 52
   - Terminal 2 and socket 51
   - Terminal 3 and socket 36
   - Terminal 5 and socket 71
   - Terminal 6 and socket 70
11. Wire resistance: max. 1.5 ohms
12. Check the wires for short circuit to each other, to Battery (+) and Ground (GND).
13. Specified value: infinite ohms
14. If the wires are okay, replace the Heated Oxygen Sensor (HO2S) G39 before catalytic converter.

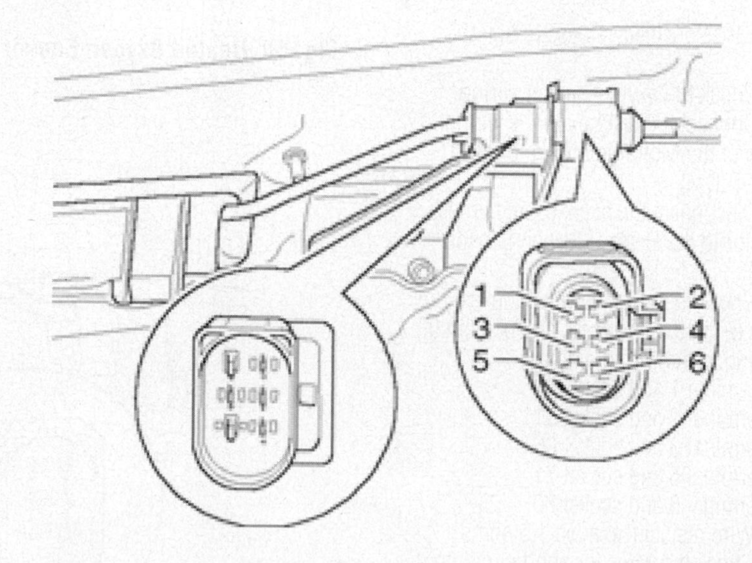

29246_VOLKS_G0091

Fig. 48 Check wires between test box and 6-pin connector for open circuit according to wiring diagram.

Heated Oxygen Sensor 2

LOCATION

2.7L and 2.8L Engines

See Figure 49.

The Heated Oxygen Sensor 2 is located in the engine compartment on the left side, behind the cylinder head. It is integral with the Oxygen Sensor Heater.

3.2L Engines

See Figure 50.

The Heated Oxygen Sensor 2 is located behind the engine in the right front exhaust pipe. It is integral with the Oxygen Sensor Heater 2.

TESTING

See Figure 51.

➡ **When servicing the terminals in the harness connector of Heated Oxygen Sensor (HO2S), use only gold-plated terminals.**

1. Connect to the vehicle diagnosis and service system (VAS).
2. Start the engine and let it run at idle.
3. Under address word "33 - On Board Diagnostic (OBD)", select "Diagnostic mode 6: Check test results of components that are not continuously monitored".
4. Select "Test-ID 01: Oxygen sensor monitoring before catalytic converter" .
5. Select "Test-ID 131: Dynamic test" and check whether specified values are obtained.
6. If the specified values are not obtained:
7. Check the oxygen sensor wiring.
8. Connect the test box to the wiring harness of the Motronic Engine Control Module (ECM)
9. Disconnect the black 6-pin harness connector to the Heated Oxygen Sensor (HO2S).
10. Check the wires between the test box and the 6-pin connector for an open circuit according to the wiring diagram.
    - Terminal 1 and socket 52
    - Terminal 2 and socket 51
    - Terminal 3 and socket 36
    - Terminal 5 and socket 71
    - Terminal 6 and socket 70
11. Wire resistance: max. 1.5 ohms
12. Check the wires for short circuit to each other, to Battery (+) and Ground (GND).
13. Specified value: infinite ohms
14. If the wires are okay, replace the Heated Oxygen Sensor (HO2S) G39 before catalytic converter.

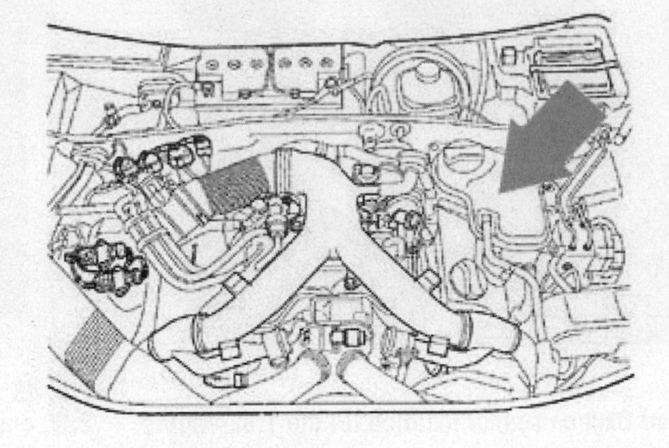

29246_AUDI_G0044

Fig. 49 Heated Oxygen Sensor 2 location for the 2.7L and 2.8L engines

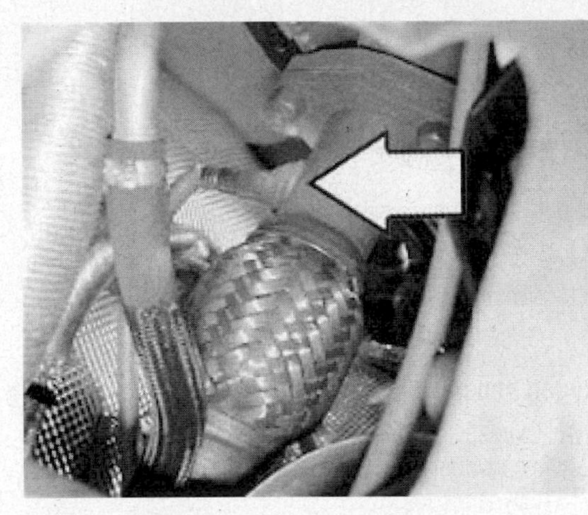

29246_AUDI_G0013

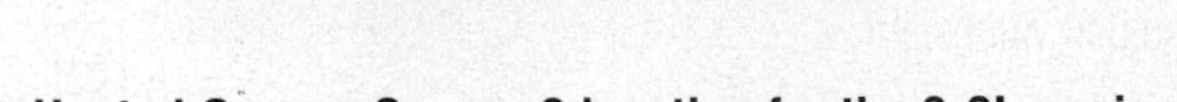

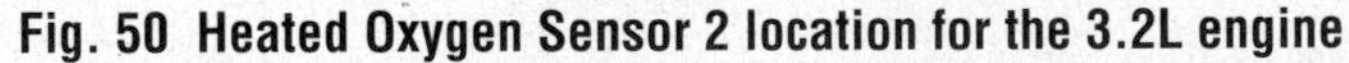

Fig. 50 Heated Oxygen Sensor 2 location for the 3.2L engine

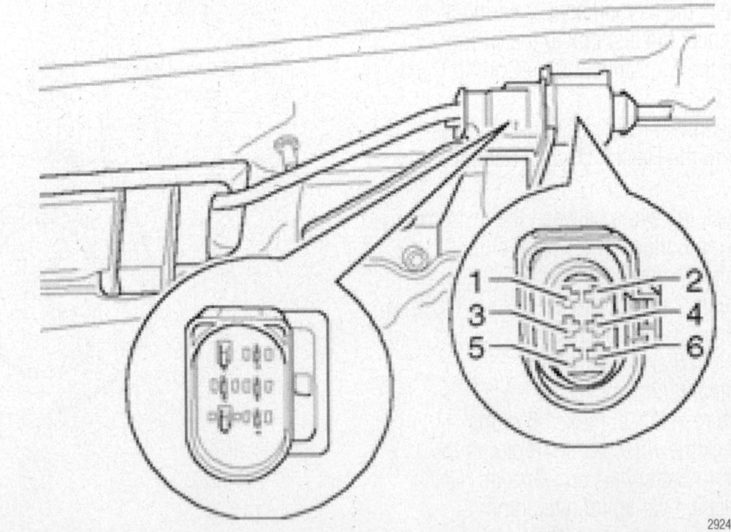

29246_VOLKS_G0091

Fig. 51 Check wires between test box and 6-pin connector for open circuit according to wiring diagram.

Ignition Coil

LOCATION

1.8L Engine

See Figure 52.

The Ignition Coils are located on the cylinder head above their respective cylinders (shown, Cylinder One).

2.7L Engine

See Figure 53.

The Ignition Coils are located on the cylinder head above their respective cylinders (shown, Cylinder One).

2.8L Engine

See Figure 54.

The Ignition Coils are located on the cylinder head above their respective cylinders (shown, Cylinder One).

3.0L Engines

See Figure 55.

The Ignition Coils (with power output stages) are located on the right side of the engine on the valve cover (shown, Cylinder One).

3.2L Engines

See Figure 56.

The Ignition Coils (with power output stages) are located on the cylinder head above their respective cylinders (shown, Cylinder One).

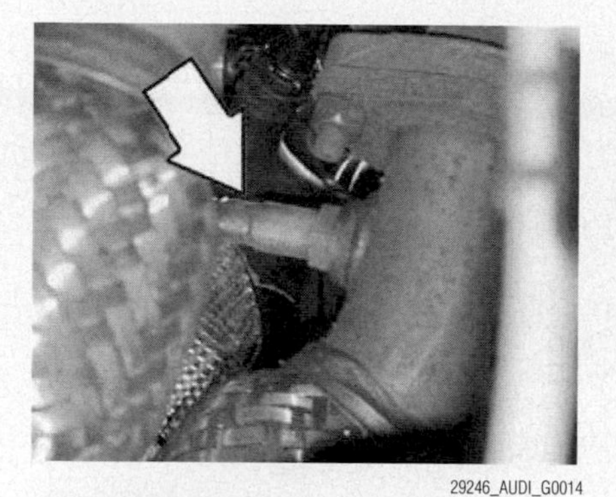

29246_AUDI_G0014

Fig. 52 Ignition Coil location for the 1.8L engine

29246_AUDI_G0045

Fig. 53 Ignition Coil location for the 2.7L engine

29246_AUDI_G0046

Fig. 54 Ignition Coil location for the 2.8L engine

29246_AUDI_G0062

Fig. 55 Ignition Coil location for the 3.0L engine

29246_AUDI_G0015

Fig. 56 Ignition Coil location for the 3.2L engine

Intake Air Temperature (IAT) Sensor

LOCATION

1.8L Engines

ATC AND AWP CODED ENGINES

See Figure 57.

The Intake Air Temperature (IAT) Sensor is located in the intake manifold, on the right side, near the throttle valve housing.

AMU CODED ENGINES

See Figure 58.

The Intake Air Temperature (IAT) Sensor is located on the intake manifold, in front of the engine compartment.

2.7L Engines

See Figure 59.

The Intake Air Temperature (IAT) Sensor is located on the bottom of the intake manifold, directly behind the throttle valve housing.

2.8L Engines

See Figure 60.

The Intake Air Temperature (IAT) Sensor is located in the air duct at the rear of the intake mainfold.

29246_AUDI_G0026

Fig. 57 Intake Air Temperature sensor location for the 1.8L engine (ATC and AWP codes)

29246_AUDI_G0017

Fig. 58 Intake Air Temperature sensor location for the 1.8L engine (AMU code)

29246_AUDI_G0047

Fig. 59 Intake Air Temperature sensor location for the 2.7L engine

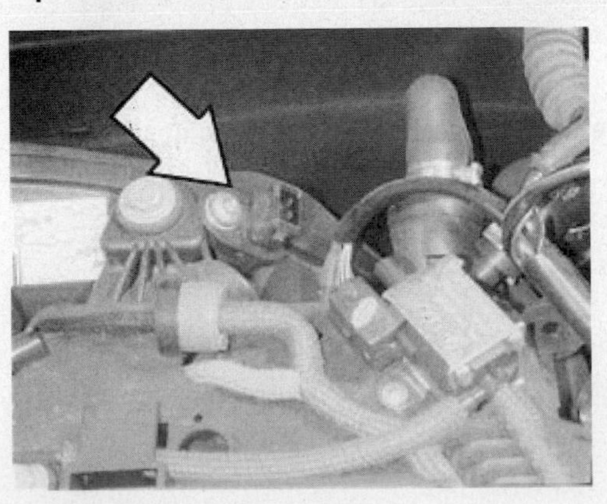

29246_AUDI_G0048

Fig. 60 Intake Air Temperature sensor location for the 2.8L engine

Knock Sensor (KS) 1

LOCATION

1.8L Engines

See Figure 61.

The Knock Sensor (KS) 1 is located in front of the engine block on the passenger's side.

2.7L and 2.8L Engines

See Figure 62.

The Knock Sensor (KS) is located below the intake manifold on the right rear cylinder head (shown with the intake manifold removed).

TESTING

See Figures 63, 64 and 65.

➡ **Contact surfaces between the Knock Sensor (KS) 1 and the cylinder block must be free of corrosion, dirt and grease.**

➡ **To ensure a problem-free function of the knock sensors, it is important that the torque specification of 20 Nm be adhered to exactly.**

➡ **Use only gold-plated terminals when servicing the terminals in the harness connector of the Knock Sensor (KS) 1.**

1. Disconnect the 3-pin connector from the Knock Sensor (KS) 1 / Knock Sensor (KS) 2: (1 - Green, for Knock Sensor (KS) 1 and 2 - Grey, for Knock sensor (KS) 2)

➡ **Before disconnecting the harness connectors, mark which component goes where.**

2. Measure the resistance between the terminals 1 and 2 at the connectors to the Knock Sensor (KS) 1.
3. Specification: infinite ohms
4. Connect the test box to the Motronic Engine Control Module (ECM).
5. Check the wires between the test box and the 3-pin connector for an open circuit according to wiring diagram.
   - Contact 1 and socket 106
   - Contact 2 and socket 99
6. Wiring resistance: max. 1.5 ohms
7. Check wires for short circuit to each other.
8. Specification: infinite ohms
9. If resistances and wires are okay., replace respective Knock Sensor (KS) 1.

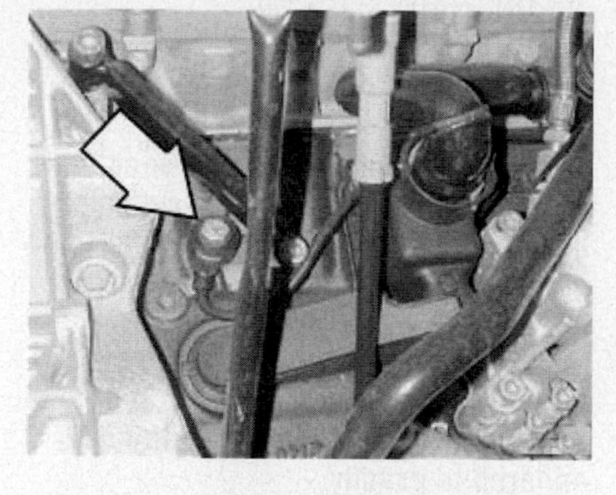

29246_AUDI_G0018

Fig. 61 Knock Sensor 1 location for the 1.8L engine

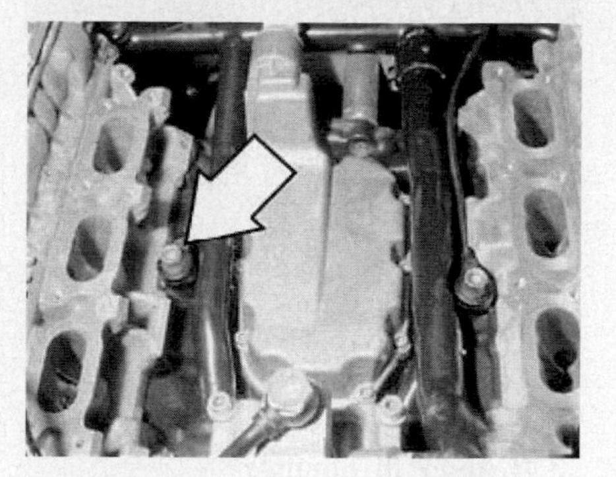

29246_AUDI_G0049

Fig. 62 Knock Sensor 1 location for the 2.7L and 2.8L engines

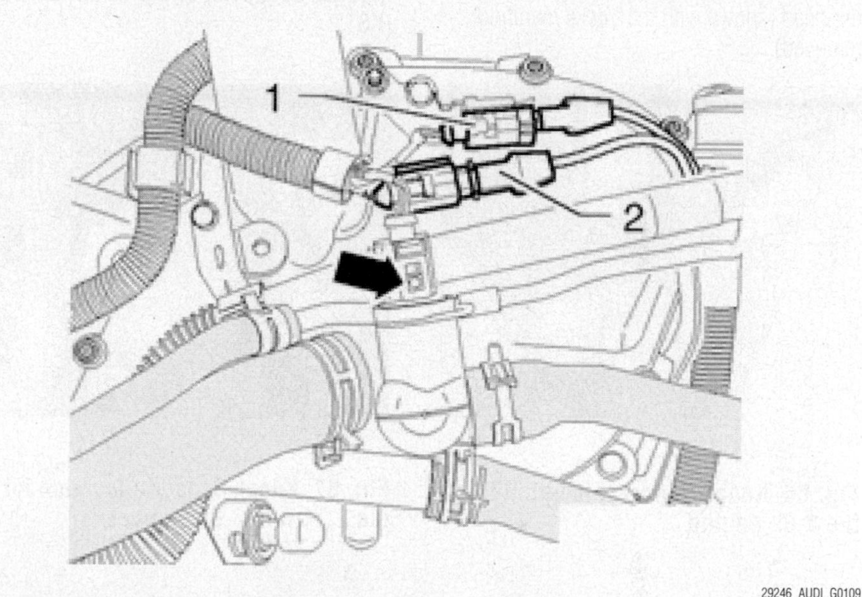

29246_AUDI_G0109

Fig. 63 Disconnect 3-pin connector from Knock Sensor (KS) 1 G61 / Knock Sensor (KS) 2 G66: (1 - Green, for Knock Sensor (KS) 1 and 2 - Grey, for Knock sensor (KS) 2)

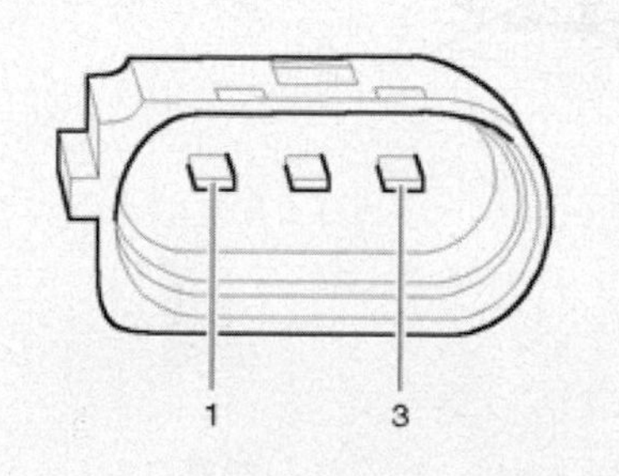

29246_AUDI_G0110

Fig. 64 Measure resistance between terminals 1 and 2 at connectors to Knock Sensor (KS) 1.

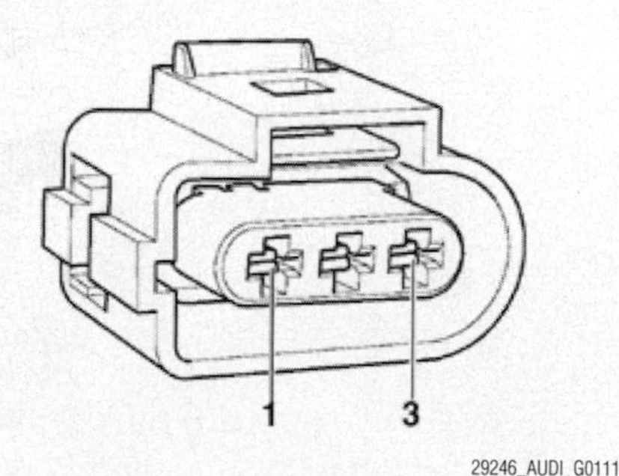

29246_AUDI_G0111

Fig. 65 Check wires between test box and 3-pin connector for open circuit.

Knock Sensor (KS) 2

LOCATION

1.8L Engines

See Figure 66.

The Knock Sensor (KS) 2 is located in front of the engine block on the driver's side behind the oil filter housing (Knock Sensor 1 shown).

2.7L and 2.8L Engines

See Figure 67.

The Knock Sensor (KS) 2 is located below the intake manifold on the left rear cylinder head (shown with the intake manifold removed).

29246_AUDI_G0019

Fig. 66 Knock Sensor 2 location for the 1.8L engine

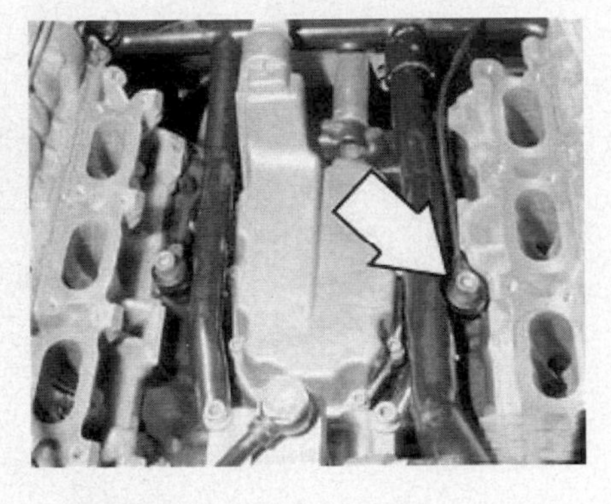

29246_AUDI_G0050

Fig. 67 Knock Sensor 2 location for the 2.7L and 2.8L engines

TESTING

See Figure 68, 69 and 70.

➡ **The contact surfaces between the Knock Sensor (KS) 2 and the cylinder block must be free of corrosion, dirt and grease.**

➡ **To ensure a problem-free function of the knock sensors, it is important that the torque specification of 20 Nm be adhered to exactly.**

➡ **Use only gold-plated terminals when servicing the terminals in the harness connector of the Knock Sensor (KS) 2.**

1. Disconnect the 3-pin connector from the Knock Sensor (KS) 1/ Knock Sensor (KS) 2: (1 - Green, for Knock Sensor (KS) 1 and 2 - Grey, for Knock sensor (KS) 2)

➡ **Before disconnecting the harness connectors, mark which component goes where.**

2. Measure the resistance between terminals 1 and 2 at the connectors to the Knock Sensor (KS) 2.
3. Specification: infinite ohms
4. Connect the test box to the Motronic Engine Control Module (ECM).
5. Check the wires between the test box and the 3-pin connector for an open circuit.
   - Contact 1 and socket 106
   - Contact 2 and socket 99
6. Wiring resistance: max. 1.5 ohms
7. Check the wires for short circuit to each other.
8. Specification: infinite ohms
9. If resistances and wires are okay, replace the respective Knock Sensor (KS) 2.

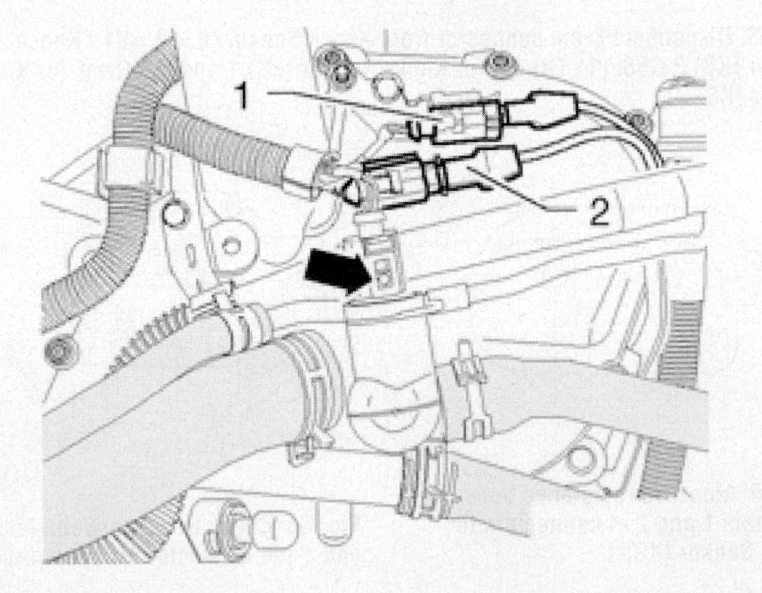

29246_VOLKS_G0109

Fig. 68 Disconnect 3-pin connector from Knock Sensor (KS) 1 G61 / Knock Sensor (KS) 2 G66: (1 - Green, for Knock Sensor (KS) 1 and 2 - Grey, for Knock sensor (KS) 2))

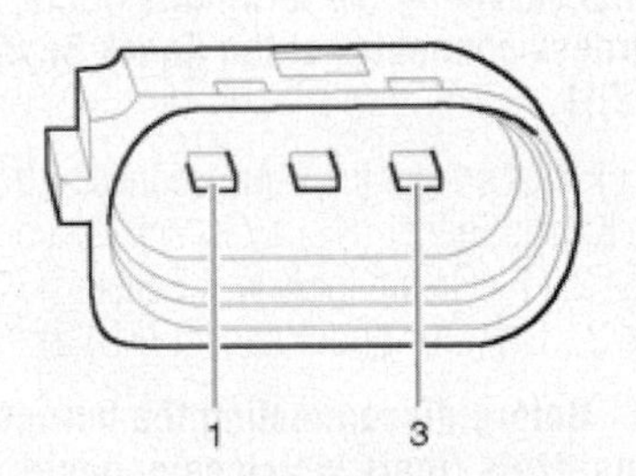

29246_VOLKS_G0110

Fig. 69 Measure resistance between terminals 1 and 2 at connectors to Knock Sensor (KS) 2.

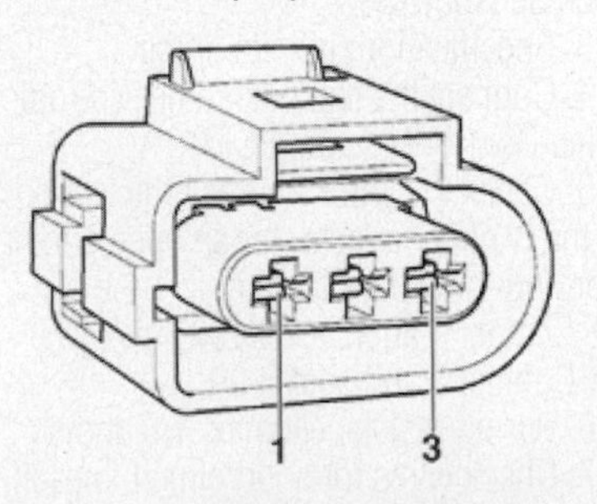

29246_VOLKS_G0111

Fig. 70 Check wires between test box and 3-pin connector for open circuit according to wiring diagram.

Manifold Absolute Pressure (MAP) Sensor

LOCATION

All Engines

See Figure 71.

The Manifold Absolute Pressure Sensor is located below the passenger side of the headlight assembly in the outlet flange of the intercooler.

TESTING

See Figures 72 and 73.

➡ **Use only gold-plated terminals when servicing the terminals in the harness connector of the Manifold Absolute Pressure (MAP) Sensor .**

1. Disconnect the 4-pin connector from the Manifold Absolute Pressure (MAP) Sensor (arrow).
2. Start the engine and run at idle speed.
3. Measure the voltage between terminals 1 and 3 on the harness connector.
4. Specification: min. 4.5 V
5. Switch off the ignition.
6. If no voltage is present, connect the test box to the Motronic Engine Control Module (ECM).
7. Check the wires between the test box and the 4-pin connector for an open circuit.
   - Contact 1 and socket 108
   - Contact 2 and socket 98
   - Contact 4 and socket 101
8. Wiring resistance: max. 1.5 ohms
9. Check the wires for short circuit to each other, to Band and Ground (GND).
10. Specification: infinite ohms
11. If no malfunction is detected in wiring, replace the Manifold Absolute Pressure (MAP) Sensor.
12. If no malfunction is detected in wiring, replace the Motronic Engine Control Module (ECM).

Mass Air Flow (MAF) Sensor

LOCATION

1.8L Engines

See Figure 74.

The Mass Air Flow Sensor is located in the left rear of the engine compartment, in the intake air duct, to the right of the air cleaner housing.

2.7L Engines

See Figure 75.

The Mass Air Flow Sensor is located on the left side of the engine compartment, attached to the air cleaner housing.

2.8L Engines

See Figure 76.

The Mass Air Flow Sensor is located on the left side of the engine compartment, attached to the air cleaner housing.

3.0L Engines

See Figure 77.

The Mass Air Flow Sensor is located in the right side of the engine compartment, near the intake air duct.

3.2L Engines

See Figure 78.

The Mass Air Flow Sensor is located in the left rear of the engine compartment, near the intake air duct.

29246_AUDI_G0020

Fig. 71 Manifold Absolute Pressure sensor location for all engines

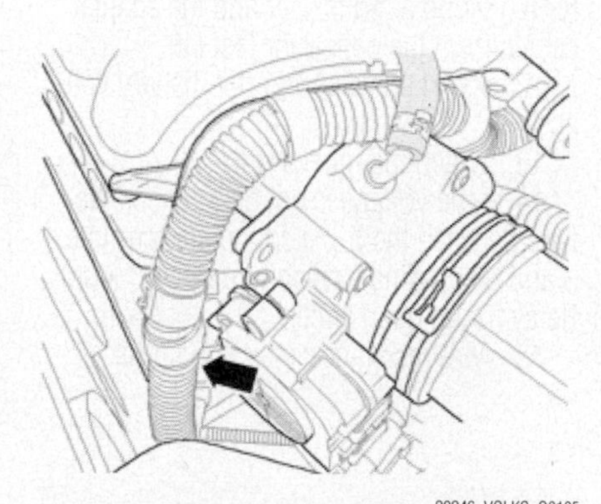

29246_VOLKS_G0105

Fig. 72 Disconnect 4-pin connector from Manifold Absolute Pressure (MAP) Sensor

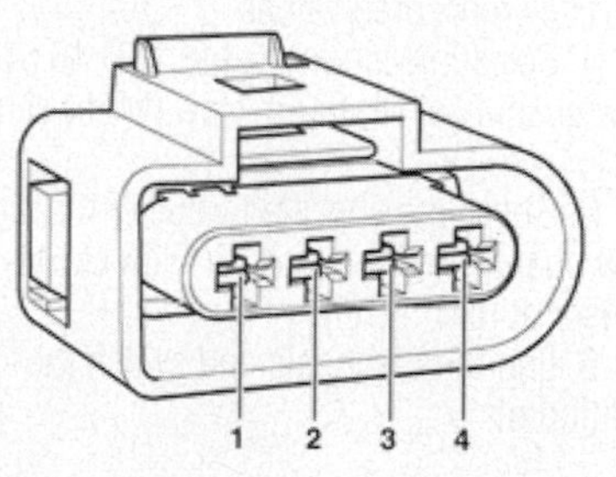

29246_VOLKS_G0106

Fig. 73 Check wires between test box and 4-pin connector for open circuit according to wiring diagram.

29246_AUDI_G0021

Fig. 74 Mass Air Flow sensor location for the 1.8L engine

29246_AUDI_G0051

Fig. 75 Mass Air Flow sensor location for the 2.7L engine

29246_AUDI_G0052

Fig. 76 Mass Air Flow sensor location for the 2.8L engine

29246_AUDI_G0063

Fig. 77 **Mass Air Flow sensor location for the 3.0L engine**

29246_AUDI_G0022

Fig. 78 **Mass Air Flow sensor location for the 3.2L engine**

TESTING

See Figures 79 and 80.

➡ **Use only gold-plated terminals when servicing the terminals in the harness connector of the Mass Air Flow (MAF) Sensor.**

1. Connect the vehicle to the diagnosis and service system.
2. Start the engine and let it run at idle.
3. Under address word "33 - On Board Diagnostic (OBD)", select "Diagnostic mode 1: Check measured values" .
4. Select measuring value "PID 16: Air flow quantity at Mass Air Flow (MAF) sensor" .
5. Check the specified value of the air flow quantity at the Mass Air Flow (MAF) Sensor at idle.
6. End the diagnosis and switch the ignition off.
7. If the specified value is obtained but DTC memory has a DTC concerning Mass Air Flow (MAF) Sensor, check the voltage supply of the Mass Air Flow (MAF) Sensor.
8. If the specified value is not obtained, check the wires of the Mass Air Flow (MAF) Sensor.
9. Disconnect the 5-pin harness connector at the Mass Air Flow (MAF) Sensor.
10. Start the engine and let it run at idle.
11. Measure the voltage between terminal 3 of connector and the engine ground (GND) using a multimeter and the adapter cables from the connector test kit.
12. Specified value: approx. battery voltage
13. Switch the ignition off.
14. If the voltage is not okay, check the wire between the 5-pin connector terminal 3 and the Engine Component Power Supply Relay for an open circuit.
15. Wire resistance: max. 1.5 ohms
16. If the voltage supply is okay, checking the wiring.
17. Connect the test box to the wiring harness of the Motronic Engine Control Module (ECM).
18. Check the wires between the test box and the 5-pin connector for an open circuit.
   - Terminal 1 and socket 22
   - Terminal 2 and socket 1
   - Terminal 2 and socket 2
   - Terminal 2 and socket 4
19. Wire resistance: max. 1.5 ohms
20. Check the wires for short circuit to each other, to Battery (+) and Ground (GND).
21. Specified value: infinite ohms
22. If the wires are okay, replace Mass Air Flow (MAF) Sensor.

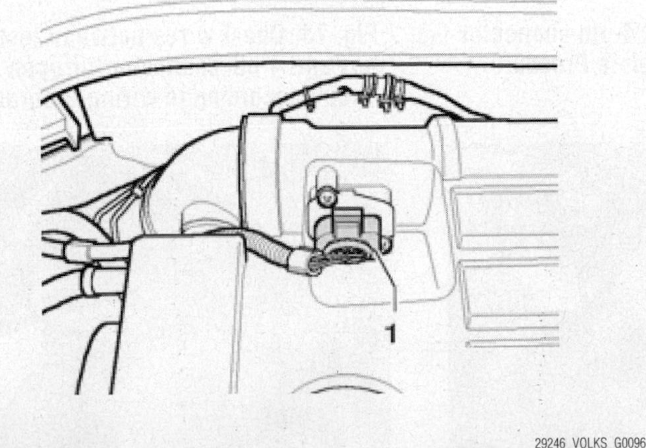

29246_VOLKS_G0096

Fig. 79 Disconnect the 5-pin harness connector (1) at Mass Air Flow (MAF) Sensor

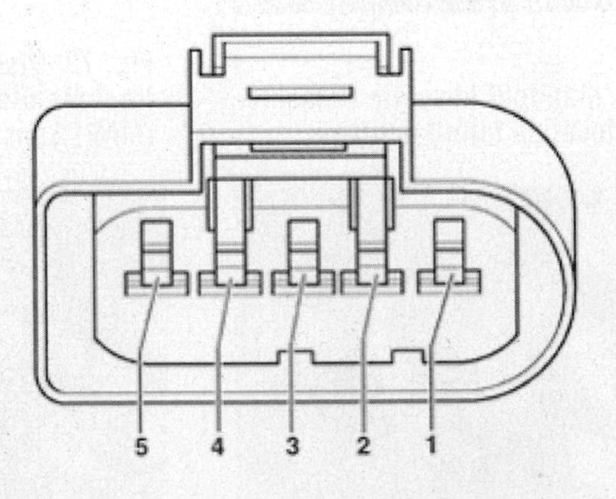

29246_VOLKS_G0097

Fig. 80 Check wires between test box and 5-pin connector for open circuit.

Motronic Engine Control Module

LOCATION

All Engines

See Figure 81.

The Motronic Engine Control Module is located in the center of the plenum.

29246_AUDI_G0023

Fig. 81 Motronic Engine Control Module location for all engines

REMOVAL & INSTALLATION

1. Remove the windshield wiper arms, the plenum chamber cover and the plenum chamber bulkhead.

See Figures 82, 83, 84, 85 and 86.

2. Cut the heads of shear bolts so that two parallel surfaces remain (arrows).
3. Remove the bolts with locking pliers.
4. Insert a screwdriver between the protective housing and the retaining plate.
5. Use screwdriver to pry the protective housing upward, and then pull if off sideways from the retaining plate.
6. Release and pull off the front connector from the Motronic Engine Control Module (ECM).
7. Pry the retaining tab slightly open.
8. Then, push the Motronic Engine Control Module (ECM) out of retainer.
9. Release the rear connector from the Motronic Engine Control Module (ECM) and pull it off.

To install:

10. Connect the rear connector to the Motronic Engine Control Module (ECM) and lock it into place.
11. Push the Motronic Engine Control Module (ECM) onto retaining plate.
12. Connect the front connector to the Motronic Engine Control Module (ECM) and lock it into place.
13. Check whether the new control module identification matches the old control module identification.
14. Perform function "Replacing control module" using the vehicle diagnosis and service information system.
15. Push the protective housing onto retaining plate.
16. Screw in the shear bolts uniformly until the bolt heads begin to shear.
17. Install the plenum chamber bulkhead, the plenum chamber cover and the windshield wiper arms.

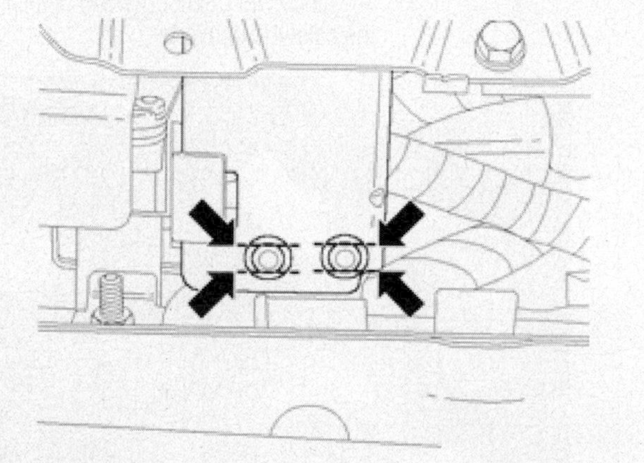
29246_VOLKS_G0115

Fig. 82 Cut the heads of shear bolts so that two parallel surfaces remain (arrows).

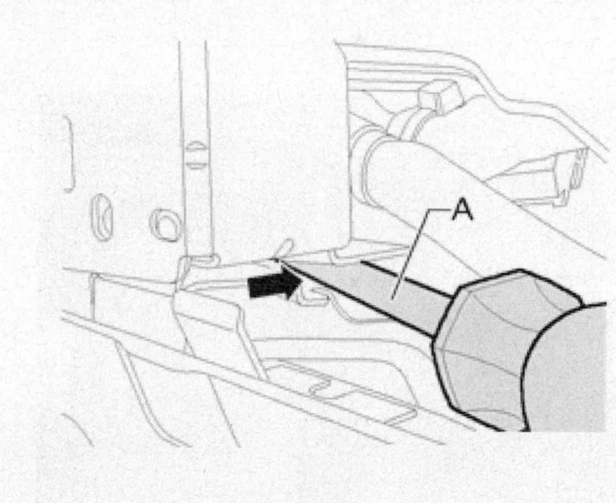

29246_VOLKS_G0116

Fig. 83 Insert a screwdriver (A) between protective housing and retaining plate (arrow).

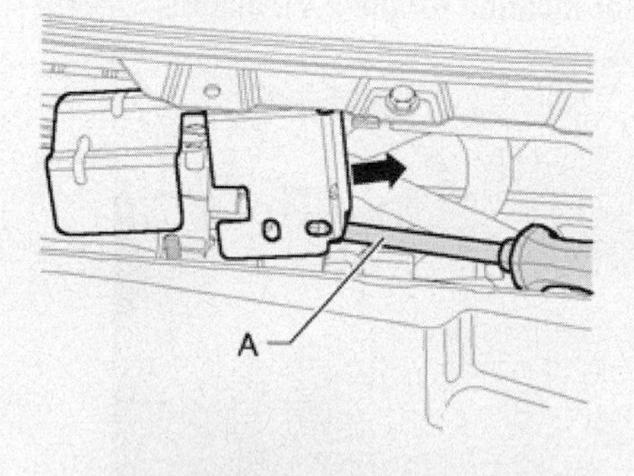

29246_VOLKS_G0117

Fig. 84 Use screwdriver (A) to pry protective housing upward, and then pull if off sideways from retaining plate (arrow)

29246_VOLKS_G0118

Fig. 85 Release and pull off front connector (1) from ECM and pry retaining tab (2) slightly open.

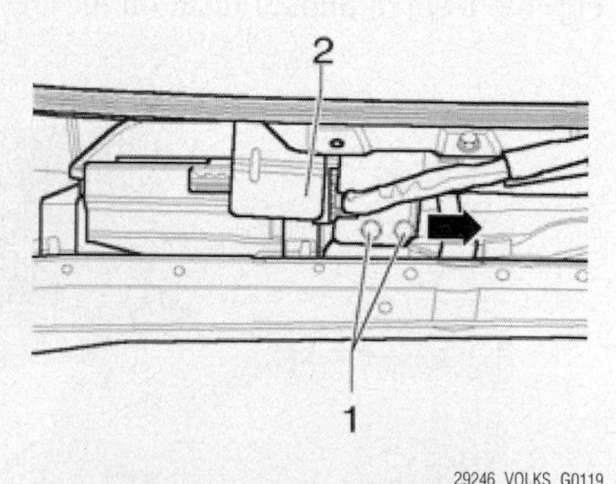

29246_VOLKS_G0119

Fig. 86 Push protective housing (2) onto retaining plate, and screw in shear bolts (1) uniformly until the bolt heads begin to shear.

Oil Level Temperature Sensor

LOCATION

1.8L Engines

See Figure 87.

The Oil Level Temperature Sensor is located on the bottom of the oil pan.

3.2L Engines

See Figure 88.

The Oil Level Temperature Sensor is located on the bottom of the oil pan.

29246_AUDI_G0024

Fig. 87 Oil Level Temperature sensor location for the 1.8L engine

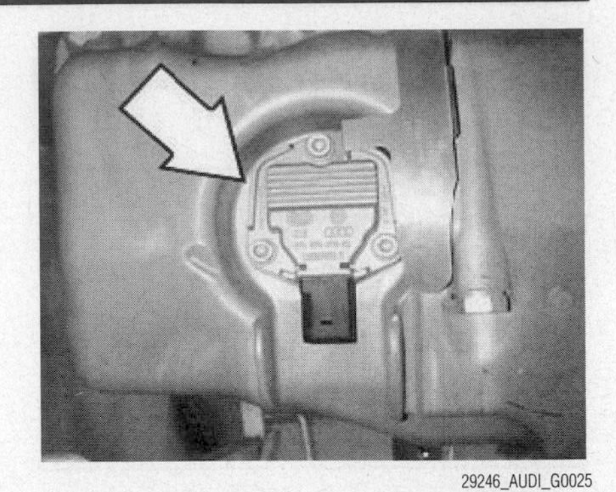

29246_AUDI_G0025

Fig. 88 Oil Level Temperature sensor location for the 3.2L engine

Oxygen Sensor (OS2)

LOCATION

1.8L Engines

See Figure 89.

The Oxygen Sensor (OS2) is located in the exhaust system, behind the three-way catalytic converter. It is integral with the heater .

2.7L Engines

See Figure 90.

The Oxygen Sensor (OS2) is located in the front exhaust piping, behind the three-way catalytic converter. It is integral with the heater .

2.8L Engines

See Figure 91.

The Oxygen Sensor (OS2) is located in the front exhaust piping, behind the three-way catalytic converter. It is integral with the heater .

3.2L Engines

See Figure 92.

The Oxygen Sensor (OS2) is located in the catalytic converter.

29246_AUDI_G0026

Fig. 89 Oxygen Sensor location for the 1.8L engine

29246_AUDI_G0054

Fig. 90 Oxygen Sensor location for the 2.7L engine

29246_AUDI_G0055

Fig. 91 Oxygen Sensor location for the 2.8L engine

29246_AUDI_G0027

Fig. 92 Oxygen Sensor location for the 3.2L engine

OPERATION

The static values of the oxygen sensors are indicated via this diagnostic mode. The values of the individual oxygen sensor signal outputs must reach the specified value or must be within the min.- and max.- limits.

TESTING

See Figures 93, 94, 95 and 96.

➡ **The vehicle must be raised before the connector for the Oxygen Sensor (O2S) Behind Three Way Catalytic Converter (TWC) is accessible.**

➡ **When servicing terminals 3 and 4 in the harness connector of the Oxygen Sensor (O2S) Behind Three Way Catalytic Converter (TWC), use only gold-plated terminals.**

1. Connect the vehicle to the diagnosis and service system.
2. Start the engine and let it run at idle.
3. Under the address word "33 - On Board Diagnostic (OBD)", select "Diagnostic mode 6: Check test results of components that are not continuously monitored" .
4. Select "Test-ID 02: Oxygen sensor monitoring behind catalytic converter" .
5. Select following Test-ID.
6. Check the specified values at idle.

Test-ID		Specified value	
		min.	max.
1	Rich to lean sensor barrier voltage	-	0.6241 V
2	Lean to rich sensor barrier voltage	-	0.6241 V
7	Minimum voltage at sensor for test cycle	-	0.450 V
8	Maximum voltage at sensor for test cycle	0.450 V	-
129	Sensor voltage lean	0	0.6241 V
130	Sensor voltage rich	0.6241 V	1.2998 V
131	Deceleration test	0 V	0.1599 V

7. If the specified values are obtained, end the diagnosis and switch the ignition off.
8. If the specified values are not obtained, end the diagnosis and let the engine continue to run at idle.
9. Check the primary voltage.
10. Remove the right vehicle floor cover.
11. Disconnect the brown 4-pin harness connector to the Oxygen Sensor (O2S) Behind Three Way Catalytic Converter (TWC).
12. Measure the voltage at terminals 3and 4 of the connector to the Motronic Engine Control Module (ECM) using a multimeter and the adapter cables from the connector test kit.
13. Specified value: 0.400 to 0.500 Volts
14. Switch the ignition off.
15. If the voltage is okay, replace the Oxygen Sensor (O2S) Behind Three Way Catalytic Converter (TWC).
16. If the voltage not okay, check the oxygen sensor wiring.
17. Connect the test box to the wiring harness of the Motronic Engine Control Module (ECM).
18. Check the wires between the test box and the 4-pin connector for open circuit.
    - Terminal 3 and socket 76
    - Terminal 4 and socket 77
19. Wire resistance: max. 1.5 ohms
20. Check the wires for short circuit to each other, to Battery (+) and Ground (GND).
21. Specified value: infinite ohms
22. If the wires are okay, replace Motronic Engine Control Module (ECM).

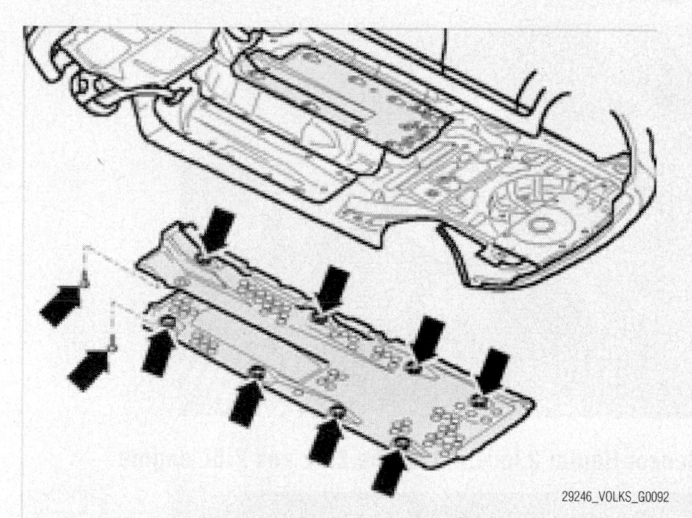

Fig. 93 Remove right vehicle floor cover (see arrows)

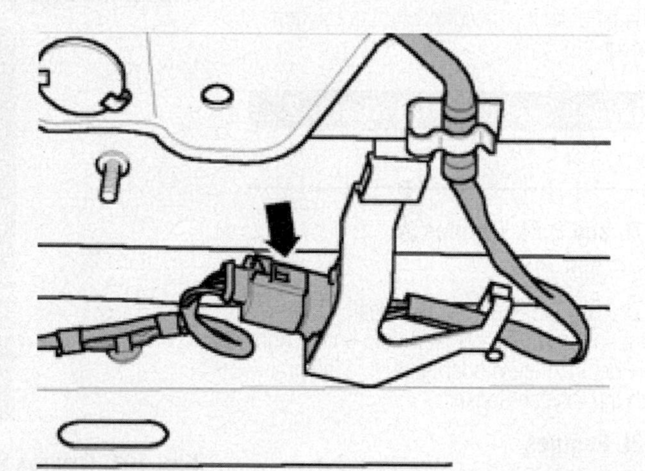

Fig. 94 Disconnect brown 4-pin harness connector (arrow) to Oxygen Sensor

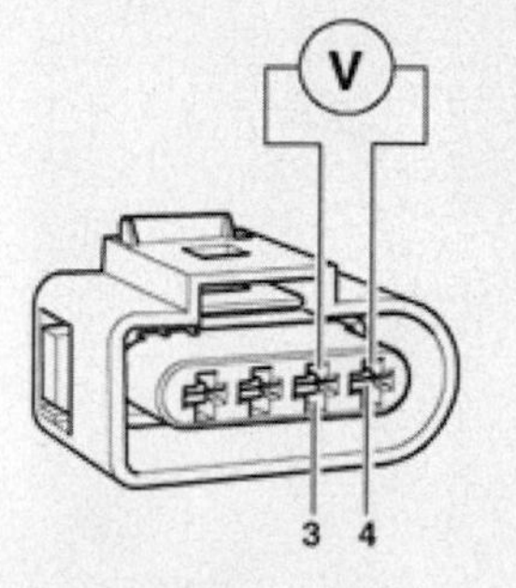

Fig. 95 Measure voltage at terminals 3and4 of connector to the ECM

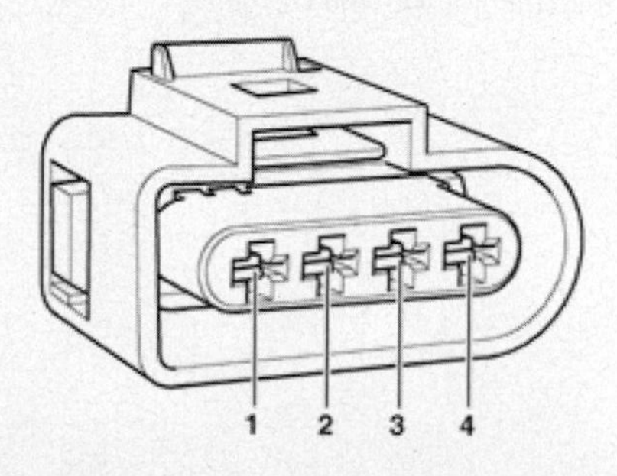

Fig. 96 Check wires between test box and 4-pin connector for open circuit

Oxygen Sensor (OS2) Heater

LOCATION

1.8L Engines

See Figure 97.

The Oxygen Sensor (OS2) Heater is located in the exhaust system, in front of the three-way catalytic converter. It is integral with the Oxygen Sensor .

2.7L and 2.8L Engines

See Figure 98.

The Oxygen Sensor (OS2) Heater is located in the engine compartment on the right side behind the cylinder head. It is integral with the Oxygen Sensor .

3.0L Engines

See Figure 99.

The Oxygen Sensor (OS2) Heater is located behind the engine in the right exhaust pipe. It is integral with the Oxygen Sensor .

3.2L Engines

See Figure 100.

The Oxygen Sensor (OS2) Heater is located behind the engine in the left front exhaust pipe. It is integral with the Oxygen Sensor .

Oxygen Sensor (OS2) Heater 2

LOCATION

2.7L and 2.8L Engines

See Figure 101.

The Oxygen Sensor (OS2) Heater 2 is located in the engine compartment on the right side behind the cylinder head. It is integral with the Oxygen Sensor .

3.2L Engines

See Figure 102.

The Oxygen Sensor (OS2) Heater is located behind the engine in the left front exhaust pipe. It is integral with the Oxygen Sensor .

29246_AUDI_G0028

Fig. 97 Oxygen Sensor Heater location for the 1.8L engine

29246_AUDI_G0053

Fig. 98 Oxygen Sensor Heater location for the 1.8L and 2.8L engine

29246_AUDI_G0064

Fig. 99 Oxygen Sensor Heater location for the 3.0L engine

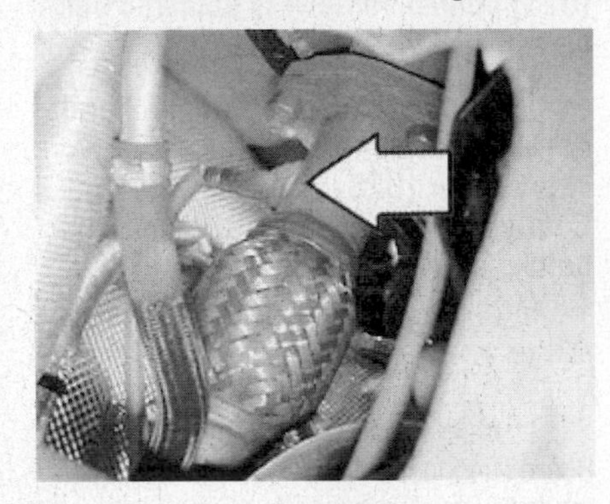

29246_AUDI_G0029

Fig. 100 Oxygen Sensor Heater location for the 3.2L engine

29246_AUDI_G0053

Fig. 101 Oxygen Sensor Heater 2 location for the 2.7L and 2.8L engine

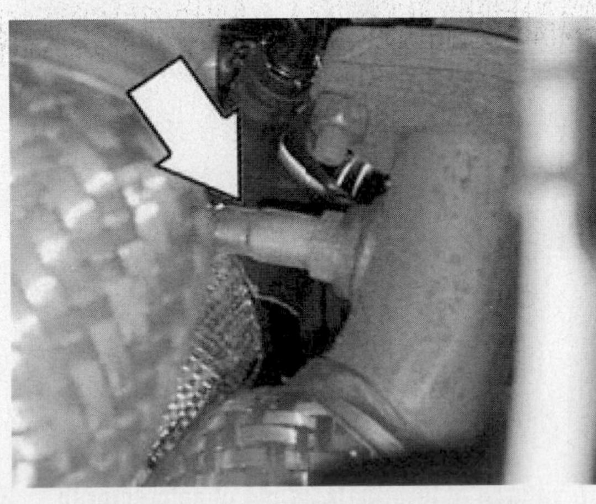

29246_AUDI_G0030

Fig. 102 Oxygen Sensor Heater 2 location for the 3.2L engine

Speedometer Vehicle Speed Sensor (VSS)

LOCATION

All Engines

The Speedometer Vehicle Speed Sensor (VSS) is located in the transmission housing on the passenger side output flange.

TESTING

See Figures 104, 105 and 106.

1. Disconnect the 3-pin harness connector to the Vehicle Speed Sensor (VSS) (arrow).
2. Check harness connector for contact corrosion and ingress of water.
3. If the harness connector is not okay, repair the malfunction.
4. If the harness connector is okay, measure the resistance at the 3-pin connector of the Vehicle Speed Sensor (VSS).
5. Measure the sensor resistance between terminals 1 and 2 at the connector to the sensor.
6. Specified value: 800 to 900 ohms
7. Check the sensor for short circuit between terminals 2 and 3 and 1 and 3.
8. Specified value: infinite ohms
9. If the specified values are not obtained, replace the Vehicle Speed Sensor (VSS).
10. If no malfunction is found on the sensor, check the wires to Transmission Control Module (TCM).
11. To check the wires to the Transmission Control Module (TCM), connect the test box to the control module wiring harness.
12. Check the wires between the test box and the 3-pin connector to the Transmission Control Module (TCM) for open circuit.
    - Terminal 1 and socket 65
    - Terminal 2 and socket 20
    - Terminal 3 and socket 43
13. Wire resistance: max. 1.5 ohms
14. Check the wires for short circuit to each other, to vehicle Ground (GND) and to Band.
15. Specified value: infinite ohms
16. If no malfunctions are found in the wires, replace the Transmission Control Module (TCM).

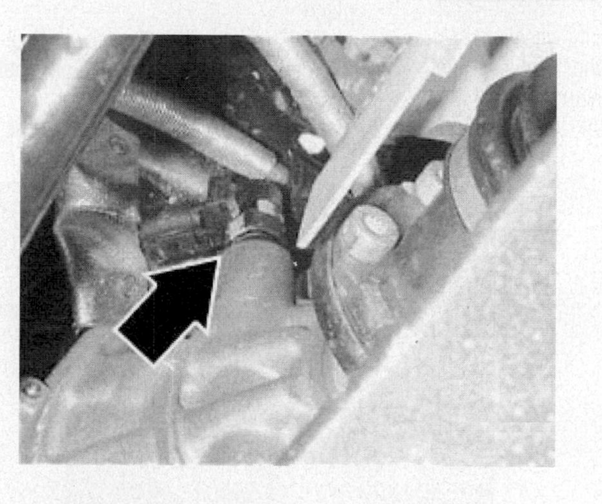

29246_AUDI_G0031

Fig. 103 Speedometer Vehicle Speed Sensor location for all engines

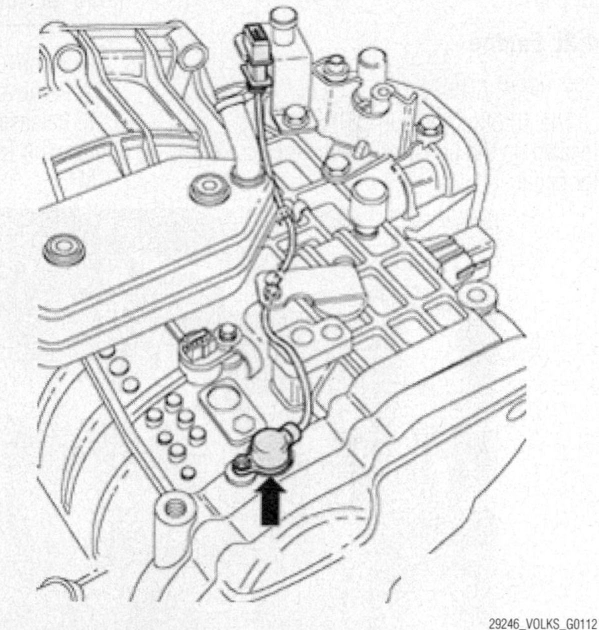

29246_VOLKS_G0112

Fig. 104 Disconnect 3-pin harness connector to Vehicle Speed Sensor (VSS) (arrow)

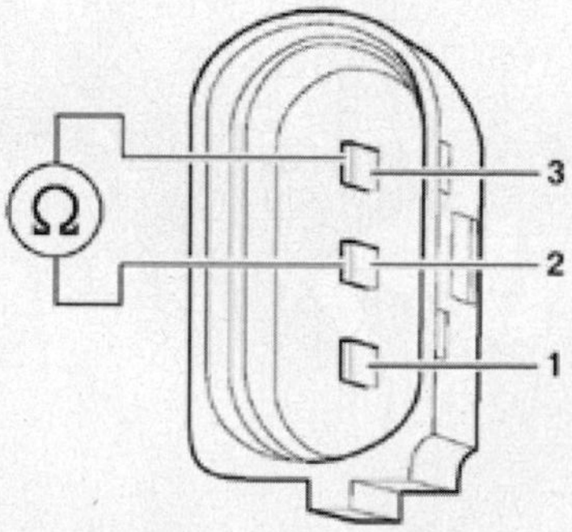

29246_VOLKS_G0113

Fig. 105 Measure sensor resistance between terminals 1 and 2 at connector to sensor.

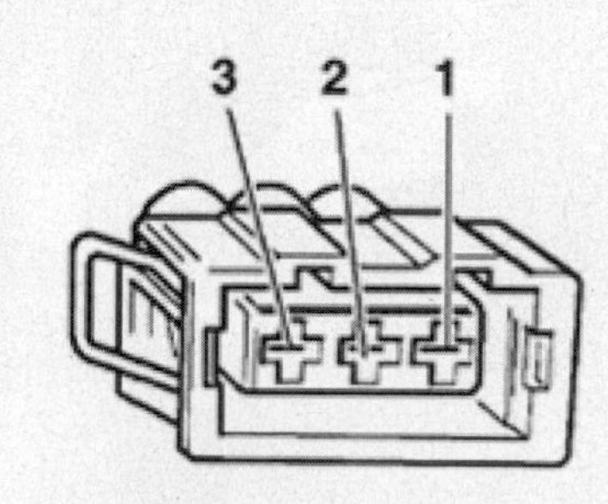

29246_VOLKS_G0114

Fig. 106 Check wires between test box and 3-pin connector to Transmission Control Module (TCM) for open circuit

Throttle Position (TP) Sensor

LOCATION

1.8L Engine

See Figure 107.

The Throttle Position (TP) Sensor is located on the bulkhead, above the accelerator pedal.

2.7L Engine

See Figure 108.

The Throttle Position (TP) Sensor is located on the rear of the intake manifold.

3.0L Engine

See Figure 109.

The Throttle Position (TP) Sensor is located on the bulkhead, above the accelerator pedal

4.2L Engine

See Figure 110.

The Throttle Position (TP) Sensor is located on the bulkhead, above the accelerator pedal.

Transmission Control Module (TCM)

LOCATION

All Engines

FIVE SPEED TRANSMISSIONS

See Figure 111.

The Transmission Control Module is located in an E-Box in the plenum chamber on the driver's side below the Engine Control Module (ECM).

AUTOMATIC TRANSMISSIONS

See Figure 112.

The Transmission Control Module is located inside the transmission at the rear of the case underneath the end cover.

REMOVAL & INSTALLTION

1. Remove the left front wheel.
2. Remove the left wheel housing liner.
3. Release the connector and pull it off from the Transmission Control Module (TCM).
4. If equipped, loosen the side turn signal wiring from the bracket on the Transmission Control Module (TCM).
5. If equipped, remove the left side turn signal from the fender toward the outside, via the bolts.

➡ **The Transmission Control Module (TCM) is still hooked in on the backside.**

6. Push the Transmission Control Module (TCM) upward together with the bracket and then remove them.
7. Remove the old Transmission Control Module (TCM) from the bracket via the bolts.

To install:

8. Set the new Transmission Control Module (TCM) in place and fasten it.
9. Set the connector onto the Transmission Control Module (TCM) and lock it into place.
10. Check whether the new control module identification matches the old control module identification.
11. When the harness connector is

29246_AUDI_G0032

Fig. 107 Throttle Position Sensor location for the 1.8L engine

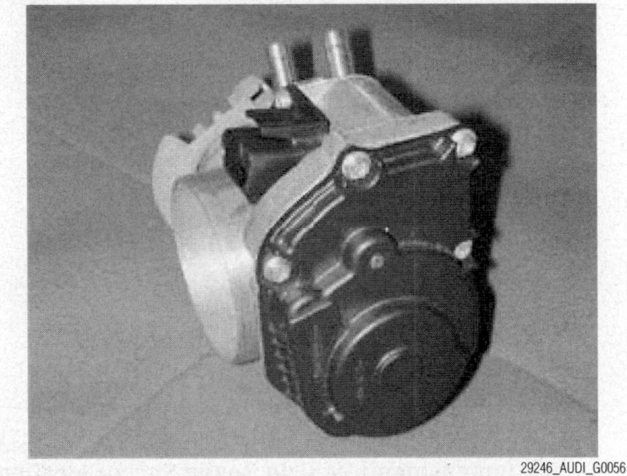

29246_AUDI_G0056

Fig. 108 Throttle Position Sensor location for the 2.7L engine

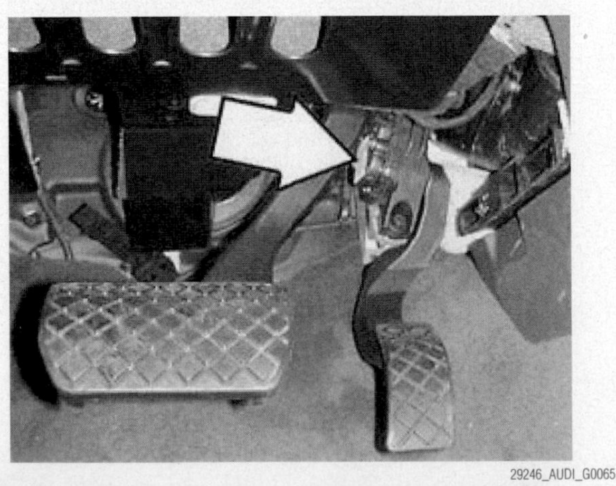

29246_AUDI_G0065

Fig. 109 Throttle Position Sensor location for the 3.0L engine

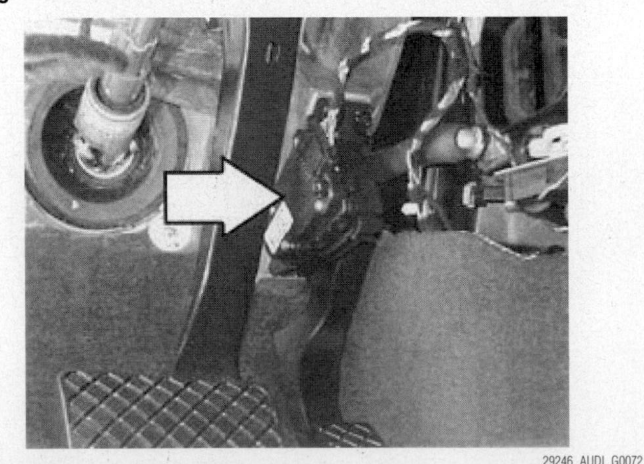

29246_AUDI_G0072

Fig. 110 Throttle Position Sensor location for the 4.2L engine

29246_AUDI_G0066

Fig. 111 Transmission Control Module location for engines with a five-speed transmission

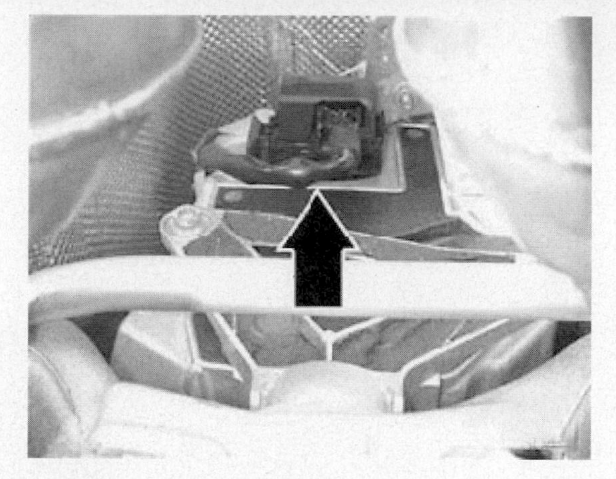
29246_AUDI_G0067

Fig. 112 Transmission Control Module location for engines with an automatic transmission

disconnected from the Transmission Control Module (TCM) or the battery is disconnected, all of the adaptation values in Transmission Control Module (TCM) are erased. However, the DTC memory content will remain intact. If the engine is started after this, the idle may be rough for a short period. In this case, the readiness code must be generated again.

Wastegate Bypass Regulator Valve

LOCATION

1.8L Engine

See Figure 113.

The Wastegate Bypass Regulator Valve is located on the intake hose, to the left of the brake fluid reservoir.

2.7L Engine

See Figure 114.

The Wastegate Bypass Regulator Valve is located on top of the intake manifold, to the right of the air intake duct.

29246_AUDI_G0033

Fig. 113 Wastegate Bypass Regulator Valve location for the 1.8L engine

29246_AUDI_G0033

Fig. 114 Wastegate Bypass Regulator Valve location for the 2.7L engine

AUDI
PIN CHARTS

7

TABLE OF CONTENTS

PIN CHARTS

Introduction

A Pin Voltage Table is a term used to describe a table that identifies PCM pins, wire colors of the PCM circuits, circuit descriptions and "known good" values for devices that connect to the PCM. These tables include the following information:

- Signals from various sensors (ECT, IAT, MAP, TPS, etc.)
- Signals from various switches (PNP, PSP, WOT, etc.)
- Signals from oxygen sensors (O2S, HO2S)
- Signals from output devices (IAC, INJ, TCC, etc.)
- Power & ground signals

Pin Voltage Tables

Information contained within the Pin Voltage Tables can be used to:

- Test circuits for open, short to power or short to ground faults
- Check the operation of a component before or after a repair
- Check the operation of a component or system by viewing signals on PCM input/output circuits with a DVOM or Lab Scope

Using a Breakout Box

There are several Breakout Box (BOB) designs available for use to test the PCM and its input and output circuits. However, all of them require removal of the wire harness to the PCM so that the BOB can be installed between the PCM and wire harness connector. Several breakout boxes require the use of overlays in order to allow the tool to be used on more than one year or engine type. Always verify that the correct adapter and overlays are used to prevent connection to the wrong circuits and a misdiagnosis.

Power and Ground Circuit Checks

Measurements made at the BOB are accomplished via test leads and probes from the DVOM or a Lab Scope. If any of the terminals on the PCM or BOB are damaged or loose, test measurements made at the Breakout Box will be inaccurate. To verify the PCM battery power and ground circuits are normal (correct) at the BOB, test the condition of the circuit between the battery negative (-) post and these circuits prior to starting a test sequence.

Diagnosis with Pin Voltage Tables

See Figure 1.

Once an actual PCM pin voltage reading is recorded, it can be compared to an example from a vehicle with "known good" values. In the example shown the Value at Hot Idle for the EVP sensor signal (0.4v) is the "known good" value.

Wire Color Changes

Every effort has been made to obtain and list the correct circuit wire colors for all vehicles. However, running changes from the vehicle manufacturer can cause the wrong colors to be listed.

PCM Pin #	W/Color	Circuit Description (60-Pin)	Value at Hot Idle
27	BN/LG	EVP Sensor Signal	0.4v

Fig. 1 Example

AUDI PIN CHARTS

Standard Colors and Abbreviations

Abbreviation	Color	Abbreviation	Color
BLK	Black	ORN	Orange
BLU	Blue	RED	Red
BRN	Brown	LIL	Lilac
GRY	Gray	WHT	White
GRN	Green	YEL	Yellow

TT

2000–2006 Sedan 1.8L MFI, ATC, AWP, AMU, BEA, ECM Connector

ECM Pin #	Wire Color	Circuit Description (121-Pin)	Value at Hot Idle
1	BRN/RED	Ground Connection	N/A
2	BRN/RED	Ground Connection	N/A
3	BLK	Connection 3 in Instrument Panel Wiring Harness	N/A
4	--	Not Used	--
5	BRN/GRN	Oxygen Sensor Heater	N/A
6	--	Not Used	--
7	--	Not Used	--

2000–2006 Sedan 1.8L MFI, ATC, AWP, AMU, BEA, ECM Connector, *continued*

8	--	Not Used	--
9	BRN/YEL	Secondary Air Injection Solenoid Valve	N/A
10	--	Not Used	--
11	--	Not Used	--
12	--	Not Used	--
13	--	Not Used	--
14	--	Not Used	--
15	--	Not Used	--
16	--	Not Used	--
17	--	Not Used	--
18	--	Not Used	--
19	--	Not Used	--
20	--	Not Used	--
21	LIL	ECM Power Supply Relay	N/A
22	--	Not Used	--
23	--	Not Used	--
24	--	Not Used	--
25	YEL/BRN	Leak Detection Pump	N/A
26	--	Not Used	--
27	GRN	Mass Air Flow Sensor	N/A
28	BLU/LIL	T-6 Connector Brown, in plenum chamber	N/A
29	BLK	Mass Air Flow Sensor	N/A
30	WHT/GRN	T-10 Connection, White, in plenum chamber	N/A
31	--	Not Used	--
32	--	Not Used	--
33	GRY/BLU	Sensor 2 for Accelerator Pedal Position	N/A
34	BRN/WHT	Sensor 2 for Accelerator Pedal Position	N/A
35	WHT/BLU	Throttle Position Sensor	N/A
36	GRY/RED	Throttle Position Sensor	N/A
37	GRN/BRN	Wire Connection (RPM signal) in instrument panel	N/A
38	BLK/WHT	Cruise Control Switch	N/A
39	WHT/RED	Clutch Vacuum Vent Valve Switch	N/A
40	BLU/RED	A/C Control Head	N/A
41	GRN/GRY	A/C Control Head	N/A
42	--	Not Used	--
43	GRY/WHT	T-10 Connection, Orange, in plenum chamber	N/A
44	--	Not Used	--
45	--	Not Used	--
46	--	Not Used	--
47	LIL/YEL	T-10 Connection, White, in plenum chamber	N/A
48	WHT/RED	T-10 Connection, White, in plenum chamber	N/A
49	WHT/LIL	Power Steering Pressure Switch	N/A
50	WHT/GRY	Power Steering Pressure Switch	N/A
51	GRN-GRY/WHT	Heated Oxygen Sensor	N/A
52	GRY/RED	Heated Oxygen Sensor	N/A
53	RED/LIL	Mass Air Flow Sensor	N/A
54	BLU/WHT	T-10 Connection, Orange, in plenum chamber	N/A

2000–2006 Sedan 1.8L MFI, ATC, AWP, AMU, BEA, ECM Connector, *continued*

55	WHT/YEL	Brake Vacuum Vent Valve Switch for Cruise Control	N/A
56	BLK/RED	Suppressor Filter	N/A
57	RED/YEL	Cruise Control Switch	N/A
58	ORN/BRN	CAN-Bus (Data-Bus) Connection	N/A
59	--	Not Used	--
60	ORN/BLK	CAN-Bus (Data-Bus) Connection	N/A
61	GRY/LIL	Sensor 1 for Exhaust Temperature	N/A
62	RED/GRN	Plus Connection in engine compartment wiring	N/A
63	BRN/GRN-GRY/YEL	Oxygen Sensor Heater and/or Oxygen Sensor Heater (behind TWC)	N/A
64	LIL/RED	Evap Emission Canister Purge Regulator Valve	N/A
65	LIL/WHT	Fuel Pump Relay	N/A
66	GRY/BRN	Secondary Air Injection Pump Relay	N/A
67	YEL/LIL	T-10 Connection (crash signal) on instrument panel	N/A
68	GRY/WHT	Heated Oxygen Sensor (behind TWC)	N/A
69	GRY/RED	Heated Oxygen Sensor (behind TWC)	N/A
70	BLK-GRN	Heated Oxygen Sensor	N/A
71	BLK	Heated Oxygen Sensor	N/A
72	GRN/WHT	Sensor 2 for Accelerator Pedal Position	N/A
73	YEL/GRN	Throttle Position Sensor	N/A
74	--	Not Used	--
75	BLU/GRY	Cruise Control Switch	N/A
76	WHT	Cruise Control Switch	N/A
77	--	Not Used	--
78	--	Not Used	--
79	--	Not Used	--
80	GRN/LIL	Leak Detection Pump	N/A
81	GRN/RED	T-10 Connection, Orange, in plenum chamber	N/A
82	WHT	Engine Speed Sensor	N/A
83	LIL/WHT	Angel Sensor 1 for Throttle Drive	N/A
84	BLU/GRY	Angel Sensor 2 for Throttle Drive	N/A
85	BLU/GRN	Intake Air Temperature Sensor	N/A
86	LIL/YEL	Camshaft Position Sensor 2	N/A
87	--	Not Used	--
88	LIL/BLU	Cylinder Four Fuel Injector	N/A
89	LIL/GRN	Cylinder Two Fuel Injector	N/A
90	BRN	Engine Speed Sensor	N/A
91	WHT/GRY	Angel Sensor 2 for Throttle Drive	N/A
92	BLU/WHT	Angel Sensor 1 for Throttle Drive	N/A
93	GRY/YEL	Engine Coolant Temperature Sensor	N/A
94	BLK/LIL	Ignition Coil 4 with Power Output Stage	N/A
95	BLK/YEL	Ignition Coil 2 with Power Output Stage	N/A
96	LIL	Cylinder One Fuel Injector	N/A
97	LIL/RED	Cylinder Three Fuel Injector	N/A
98	LIL/GRN	Connector (5V) in engine wiring harness	N/A
99	BLU	Wire Connection 1 in engine compartment	N/A
100	--	Not Used	--
101	LIL/GRY	Charge Air Pressure Sensor	N/A

2000–2006 Sedan 1.8L MFI, ATC, AWP, AMU, BEA, ECM Connector, *continued*

102	LIL/BLK	Ignition Coil 1 with Power Output Stage	N/A
103	BLK/BRN	Ignition Coil 3 with Power Output Stage	N/A
104	GRN/BRN	Wastegate Bypass Regulator Valve	N/A
105	GRY/GRN	Recirculating Valve for Turbocharger	N/A
106	GRY	Knock Sensor 1	N/A
107	GRN	Knock Sensor 2	N/A
108	BRN/BLU	Engine Speed Sensor	N/A
109	--	Not Used	--
110	--	Not Used	--
111	--	Not Used	--
112	--	Not Used	--
113	--	Not Used	--
114	--	Not Used	--
115	GRN/WHT	Valve 1 for Camshaft Adjustment (only ATC, AWP)	N/A
116	LIL/GRY	Evap Emission Canister Purge Solenoid Valve	N/A
117	WHT	Throttle Drive (power accelerator actuation)	N/A
118	LIL/BLK	Throttle Drive (power accelerator actuation)	N/A
119	--	Not Used	--
120	--	Not Used	--
121	BLK/LIL	Ignition Coils with Power Output Stage Ground	N/A

A4

1996 Sedan 2.8L MFI AFC Connector

ECM Pin #	Wire Color	Circuit Description (68-Pin)	Value at Hot Idle
A2	RED/BLK	Camshaft Position Sensor power	N/A
A3	BLK/BRN	Throttle Position Sensor power	N/A
A4	BLK/WHT	Closed Throttle Position Switch ground	N/A
A5	RED/BLU	Fuel Pump Relay	N/A
A7	GRN/WHT	Data Link Connector Rapid Data Transfer	N/A
A9	LIL	Crankshaft Position Sensor power	N/A
A10	RED	Crankshaft Position Sensor signal	N/A
A11	GRY	Engine Speed Sensor	N/A
A12	GRN	Mass Air Flow Sensor	N/A
A13	BRN	Knock Sensor 2 signal	N/A
A14	WHT	Knock Sensors 1 and 2 power	N/A
A15	YEL	Knock Sensor 1 signal	N/A
A16	GRY/RED	A/C Engine Coolant Temperature Sensor	N/A
A19	BRN/RED	Ground Connection 1	N/A
A20	GRN	Heated Oxygen Sensor	N/A
A21	GRN	Heated Oxygen Sensor 2	N/A
A23	WHT/BLK	Outside Air Temperature Sensor	N/A
A24	BLU/BRN	Closed Throttle Position Switch signal	N/A
B1	YEL/BLK	Transmission Control Module	N/A
B2	GRN/LIL	Camshaft Position Sensor signal	N/A
B3	GRN/WHT	Transmission Control Module	N/A

1996 Sedan 2.8L MFI AFC Connector, *continued*

B4	GRN/WHT	Intake Manifold Change-Over Valve	N/A
B5	WHT/GRN	Power Output Stage	N/A
B6	GRN/GRY	Power Output Stage	N/A
B7	GRY/YEL	Power Output Stage	N/A
B8	WHT/RED	Transmission Control Module	N/A
B9	LIL/BLK	Transmission Control Module	N/A
B10	RED	Power	N/A
B11	BRN/BLU	Central Lock, Alarm, Interior Light Dimmer	N/A
B16	YEL/LIL	Sensor for Evap Canister	N/A
C2	RED/LIL	Sensor for Evap Canister	N/A
C5	GRN/BRN	ABS Control Module (w/EDL)	N/A
C6	WHT/LIL	Sensor for Evap Canister	N/A
C10	GRN/BLU	Mini Check System Control Module	N/A
C11	BLK/GRY	A/C Head	N/A
C12	YEL	Auto Check System	N/A
C13	BRN/RED	Speedometer	N/A
C14	BLK/YEL	T-10 Connector, Brown, in Plenum Chamber	N/A
C15	GRN/WHT	ABS Control Module (w/EDL)	N/A
C17	RED/GRN	Heated Oxygen Sensor 2 (behind TWC)	N/A
C20	YEL	Heated Oxygen Sensor 2 (behind TWC)	N/A
C21	GRN	Heated Oxygen Sensor 2 (behind TWC)	N/A
C22	BLK/LIL	T-10 Connector, Yellow, in the Plenum Chamber	N/A
C23	YEL/BRN	Malfunction Indicator Light	N/A
D1	BRN/WHT	Cylinder One Fuel Injector signal	N/A
D2	BRN/BLU	Cylinder Two Fuel Injector signal	N/A
D3	BRN/WHT	Ground Connection 2	N/A
D4	BRN/LIL	Cylinder Three Fuel Injector signal	N/A
D5	BRN/GRN	Cylinder Four Fuel Injector signal	N/A
D6	BLK/GRY	Cylinder Five Fuel Injector signal	N/A
D7	BRN/WHT	Ground Connection 2	N/A
D8	BLK/YEL	Cylinder Six Fuel Injector signal	N/A
D9	BLU/GRN	Idle Air Control Valve ground	N/A
D10	BLK/BLU	Wire connector above circuit breaker	N/A
D11	BLK/RED	Idle Air Control Valve signal	N/A
D12	BRN/RED	Ground Connection 1	N/A
E1	BLU/LIL	EGR Vacuum Regulator Solenoid Valve	N/A
E3	BRN/WHT	Ground Connection 2	N/A
E4	YEL/WHT	Oxygen Sensor Heater (behind TWC) 2 power	N/A
E7	BRN/WHT	Ground Connection 2	N/A
E8	YEL/WHT	Oxygen Sensor Heater (behind TWC) power	N/A
E9	YEL/BLU	Oxygen Sensor Heater 2 power	N/A
E10	YEL/GRN	Oxygen Sensor Heater power	N/A
E11	BRN/WHT	Ground Connection 2	N/A
E12	LIL/WHT	Evap Emission Canister Purge Regulator Valve	N/A

A4

1997–1999 Sedan 1.8L MFI AEB Connector

ECM Pin #	Wire Color	Circuit Description (80-Pin)	Value at Hot Idle
1	BLK/BLU	Fuse Connection	N/A
2	BRN	Ground Connection	N/A
3	RED	Power	N/A
4	RED/BLU	Fuel Pump	N/A
5	GRN/BLK	T-10 Connector, Yellow, in Plenum Chamber	N/A
6	GRN/BLU	T-10 Connector, Black, in Plenum Chamber	N/A
7	YEL/BLK	Transmission Control Module	N/A
8	BLK/GRY	Transmission Control Module	N/A
10	BLK/YEL	T-10 Connector, Brown, in Plenum Chamber	N/A
11	GRN/LIL	Camshaft Position Sensor	N/A
12	YEL	Mass Air Flow Sensor	N/A
13	GRN	Mass Air Flow Sensor	N/A
15	LIL	Evap Emission Canister Purge Regulator Valve	N/A
16	YEL/RED	Diagnosis Pump for Fuel System	N/A
17	YEL/BRN	Malfunction Indicator Light	N/A
18	YEL	T-10 Connector, Brown, in Plenum Chamber	N/A
19	GRN/BLK	T-6 Connector, Red, in Plenum Chamber	N/A
20	WHT/BLU	T-10 Connector, Brown, in Plenum Chamber	N/A
21	BLK/LIL	T-10 Connector, Yellow, in Plenum Chamber	N/A
22	BRN/WHT	T-10 Connector, Blue, in Plenum Chamber	N/A
23	LIL/BLK	Transmission Control Module	N/A
25	GRY	Heated Oxygen Sensor	N/A
26	BLU	Heated Oxygen Sensor	N/A
27	RED/WHT	Oxygen Sensor Heater	N/A
28	RED/WHT	Heater for Lambda Probe	N/A
32	WHT/RED	Transmission Control Module	N/A
37	WHT/YEL	Diagnosis Pump for Fuel System	N/A
43	WHT/BLK	T-6 Connector, Red, in Plenum Chamber	N/A
45	GRN/BRN	T-10 Connector, Yellow, in Plenum Chamber	N/A
49	GRN/WHT	Transmission Control Module	N/A
51	LIL	Heated Oxygen Sensor (behind TWC)	N/A
52	RED	Heated Oxygen Sensor (behind TWC)	N/A
53	BRN/GRY	Engine Coolant Temperature Sensor	N/A
54	BLK/RED	Intake Air Temperature Sensor	N/A
56	GRY	Engine Speed Sensor	N/A
58	BRN/GRN	Cylinder Three Fuel Injector signal	N/A
59	RED/LIL	Throttle Valve Control Module	N/A
60	GRN	Knock Sensor 2	N/A
61	WHT	Barometric Pressure Sensor	N/A
62	LIL/GRY	Barometric Pressure Sensor	N/A
63	BLU	Engine Speed Sensor	N/A
64	YEL/WHT	Wastegate Bypass Regulator Valve signal	N/A
65	BLK/LIL	Cylinder Four Fuel Injector signal	N/A
66	BRN/LIL	Throttle Valve Control Module	N/A

1997–1999 Sedan 1.8L MFI AEB Connector, *continued*

67	GRY/WHT	Camshaft Position Sensor	N/A
68	BRN	Knock Sensor 1	N/A
69	LIL/YEL	Throttle Valve Control Module	N/A
70	GRY/YEL	Power Output Stage	N/A
71	GRN/WHT	Power Output Stage	N/A
73	BLK/GRN	Cylinder One Fuel Injector signal	N/A
74	LIL/RED	Throttle Valve Control Module	N/A
75	WHT/YEL	Throttle Valve Control Module	N/A
76	GRN/GRY	Camshaft Position Sensor	N/A
77	RED/GRY	Power Output Stage	N/A
78	YEL/GRY	Power Output Stage	N/A
80	BRN/BLU	Cylinder Two Fuel Injector signal	N/A

A4

1998–1999 Sedan 2.8L MFI AHA Connector

ECM Pin #	Wire Color	Circuit Description (80-Pin)	Value at Hot Idle
1	BLK/BLU	Fuse Connection	N/A
2	BRN	Ground Connection	N/A
3	RED	Power	N/A
4	RED/BLU	Fuel Pump Relay	N/A
5	GRN/BLK	ABS Control Module	N/A
6	GRN/BLU	T-10 Connector, Black, in Plenum Chamber	N/A
8	BLK/GRY	A/C Control Head	N/A
10	BLK/YEL	T-10 Connector, Brown, in Plenum Chamber	N/A
11	RED/BLK	Camshaft Position Sensor	N/A
12	YEL	Mass Air Flow Sensor	N/A
13	GRN	Mass Air Flow Sensor	N/A
14	BLK	Ground	N/A
15	LIL	Evap Emission Canister Purge Regulator Valve	N/A
16	YEL/RED	Diagnosis Pump for Fuel System	N/A
17	YEL/BRN	Malfunction Indicator Light	N/A
18	YEL	T-10 Connector, Brown, in Plenum Chamber	N/A
19	GRN/BLK	T-6 Connector, Red, in Plenum Chamber	N/A
20	WHT/BLU	T-10 Connector, Brown, in Plenum Chamber	N/A
21	BLK/LIL	T-10 Connector, Yellow, in Plenum Chamber	N/A
25	GRY	Heated Oxygen Sensor	N/A
26	BLU	Heated Oxygen Sensor	N/A
27	RED/WHT	Oxygen Sensor Heater	N/A
28	BRN/BLK	Heater for Lambda Probe	N/A
29	GRY/WHT	Transmission Control Module	N/A
30	BLU/GRY	Secondary Air Injection Relay	N/A

1998–1999 Sedan 2.8L MFI AHA Connector, ***continued***

33	YEL/GRN	Secondary Air Injection Solenoid Valve	N/A
37	WHT/YEL	Diagnosis Pump for Fuel System	N/A
38	LIL	Oxygen Sensor (behind TWC)	N/A
39	GRY	Heated Oxygen Sensor 2	N/A
40	BLU	Heated Oxygen Sensor 2	N/A
41	GRY/RED	Transmission Control Module	N/A
43	WHT/BLK	T-6 Connector, Red, in Plenum Chamber	N/A
44	GRN/BLK	Camshaft Position Sensor	N/A
45	GRN/BRN	ABS Control Module	N/A
50	RED	Oxygen Sensor (behind TWC)	N/A
51	LIL	Heated Oxygen Sensor (behind TWC)	N/A
52	RED	Heated Oxygen Sensor (behind TWC)	N/A
53	BRN/GRY	Engine Coolant Temperature Sensor	N/A
54	GRN/RED	Temperature Sensor 2	N/A
55	BRN/GRY	Valve 1 for Camshaft Adjustment Valve	N/A
56	GRY	Engine Speed Sensor	N/A
58	BRN/GRN	Cylinder Three Fuel Injector signal	N/A
59	RED/LIL	Throttle Valve Control Module	N/A
60	GRN	Knock Sensor 2	N/A
62	GRN/LIL	Throttle Valve Control Module	N/A
63	BLU	Engine Speed Sensor	N/A
64	YEL/WHT	Wastegate Bypass Regulator Valve signal	N/A
65	BRN/WHT	Cylinder Four Fuel Injector signal	N/A
66	BRN/LIL	Throttle Valve Control Module	N/A
67	GRY/WHT	Knock Sensor 2	N/A
68	BRN	Knock Sensor 1	N/A
69	LIL/YEL	Throttle Valve Control Module	N/A
70	GRN/GRY	Power Output Stage	N/A
71	GRY/YEL	Power Output Stage	N/A
72	BLK/RED	Cylinder Five Fuel Injector signal	N/A
73	BLK/GRN	Cylinder One Fuel Injector signal	N/A
74	LIL/RED	Throttle Valve Control Module	N/A
75	WHT/YEL	Throttle Valve Control Module	N/A
76	GRN/BRN	Camshaft Position Sensor	N/A
79	BLK/WHT	Cylinder Six Fuel Injector signal	N/A
80	BRN/BLU	Cylinder Two Fuel Injector signal	N/A

A4

2000–2001 Sedan 2.8L MFI, ATQ, ECM Connector

ECM Pin #	Wire Color	Circuit Description (121-Pin)	Value at Hot Idle
1	BRN	Ground Connection	N/A
2	BRN	Ground Connection	N/A
3	BLK/BLU	Wire Connector (above fuse) in Engine Compart.	N/A
4	RED/WHT	Oxygen Sensor Heater 2	N/A
5	RED/WHT	Oxygen Sensor Heater	N/A
6	BRN/BLK	Heater for Lambda Probe 2	N/A
7	--	Not Used	--
8	--	Not Used	--
9	--	Not Used	--
10	LIL	Heated Oxygen Sensor 2 (behind TWC)	N/A
11	RED	Heated Oxygen Sensor 2 (behind TWC)	N/A
12	LIL	Heated Oxygen Sensor 2	N/A
13	RED	Heated Oxygen Sensor 2	N/A
14	--	Not Used	--
15	--	Not Used	--
16	--	Not Used	--
17	--	Not Used	--
18	--	Not Used	--
19	--	Not Used	--
20	--	Not Used	--
21	--	Not Used	--
22	--	Not Used	--
23	--	Not Used	--
24	--	Not Used	--
25	YEL/RED	Leak Detection Pump	N/A
26	--	Not Used	--
27	YEL	Mass Air Flow Sensor	N/A
28	--	Not Used	--
29	GRN	Mass Air Flow Sensor	N/A
30	BLK/LIL	Instrument Cluster Combination Processor	N/A
31	--	Not Used	--
32	BLK	Sensor Ground	N/A
33	GRY/YEL	Sensor 2 for Accelerator Pedal Position	N/A
34	BRN/GRN	Sensor 2 for Accelerator Pedal Position	N/A
35	YEL/BLU	Throttle Position Sensor	N/A
36	BRN/RED	Throttle Position Sensor	N/A
37	GRN/BLU	Wire Connection (RPM signal)	N/A
38	BLK/WHT	Cruise Control Switch	N/A
39	RED/GRN	Clutch Vacuum Vent Valve Switch (manual only)	N/A
40	BLK/YEL	A/C Control Head	N/A
41	BLK/GRY	Connector (C15, A/C) in Instrument Panel Wiring	N/A
42	--	Not Used	--
43	GRN/BLK	T-10 Connection, Brown, in plenum chamber	N/A
44	YEL/GRN	Secondary Air Injection Solenoid Valve	N/A

2000–2001 Sedan 2.8L MFI, ATQ, ECM Connector, *continued*

45	--	Not Used	--
46	BLU/GRY	Secondary Air Injection Pump Relay	N/A
47	WHT/RED	T-10 Connection, Brown, in plenum chamber	N/A
48	RED/BRN	T-10 Connection, Orange, in plenum chamber	N/A
49	--	Not Used	--
50	--	Not Used	--
51	GRN	Heated Oxygen Sensor	N/A
52	--	Not Used	--
53	BLU/GRN	Mass Air Flow Sensor	N/A
54	WHT/BLU	Wire Connection (vehicle speed signal)	N/A
55	WHT/RED	Brake Light Switch	N/A
56	RED/BLK	Brake Light Switch	N/A
57	RED/YEL	Cruise Control Switch	N/A
58	RED/BRN	CAN-Bus (Data-Bus) Connection	N/A
59	--	Not Used	--
60	RED/BLK	CAN-Bus (Data-Bus) Connection	N/A
61	--	Not Used	--
62	--	Not Used	--
63	BRN/BLK	Heater for Lambda Probe 1	N/A
64	LIL	Evap Emission Canister Purge Regulator Valve	N/A
65	--	Not Used	--
66	--	Not Used	--
67	LIL/BLU	Emergency Flasher Switch	N/A
68	GRN	Heated Oxygen Sensor (behind TWC)	N/A
69	BLU	Heated Oxygen Sensor (behind TWC)	N/A
70	YEL	Heated Oxygen Sensor	N/A
71	--	Not Used	--
72	GRY	Sensor 2 for Accelerator Pedal Position	N/A
73	YEL/LIL	Throttle Position Sensor	N/A
74	GRN/BRN	ABS Control Module (w/EDL)	N/A
75	BLU	Cruise Control Switch	N/A
76	RED/GRY	Cruise Control Switch	N/A
77	--	Not Used	--
78	--	Not Used	--
79	--	Not Used	--
80	WHT/YEL	Leak Detection Pump	N/A
81	YEL	T-10 Connection, Brown, in plenum chamber	N/A
82	GRY	Engine Speed Sensor	N/A
83	GRN/LIL	Throttle Valve Control Module	N/A
84	WHT/YEL	Throttle Valve Control Module	N/A
85	GRN/RED	Intake Air Temperature Sensor	N/A
86	GRN/BRN	Camshaft Position Sensor	N/A
87	GRY/BLK	Camshaft Position Sensor 2	N/A
88	BRN/GRN	Cylinder Three Fuel Injector	N/A
89	BLK/WHT	Cylinder Six Fuel Injector	N/A
90	BLU	Engine Speed Sensor	N/A
91	GRN/WHT	Throttle Valve Control Module	N/A

2000–2001 Sedan 2.8L MFI, ATQ, ECM Connector, *continued*

92	LIL/RED	Throttle Valve Control Module	N/A
93	BRN/GRY	Engine Coolant Temperature Sensor	N/A
94	GRN/GRY	Power Output Stage	N/A
95	--	Not Used	--
96	BLK/GRN	Cylinder One Fuel Injector	N/A
97	BRN/WHT	Cylinder Four Fuel Injector	N/A
98	RED/BLK	Camshaft Position Sensor 1 and 2	N/A
99	GRN/BRN	Knock Sensor 1	N/A
100	--	Not Used	--
101	--	Not Used	--
102	GRY/YEL	Power Output Stage	N/A
103	GRN/WHT	Power Output Stage 2	N/A
104	YEL/WHT	Intake Manifold Change-Over Valve	N/A
105	--	Not Used	--
106	WHT	Knock Sensor 1	N/A
107	YEL	Knock Sensor 2	N/A
108	GRY/WHT	Sensor Ground	N/A
109	--	Not Used	--
110	--	Not Used	--
111	--	Not Used	--
112	BRN/BLU	Cylinder Two Fuel Injector	N/A
113	BLK/RED	Cylinder Five Fuel Injector	N/A
114	--	Not Used	--
115	BRN/GRY	Valve 1 for Camshaft Adjustment	N/A
116	--	Not Used	--
117	RED/LIL	Throttle Valve Control Module	N/A
118	BRN/LIL	Throttle Valve Control Module	N/A
119	--	Not Used	--
120	--	Not Used	--
121	--	Not Used	--

A4

2000 Sedan 1.8L MFI, ATW, ECM Connector

ECM Pin #	Wire Color	Circuit Description (121-Pin)	Value at Hot Idle
1	BRN	Ground Connection	N/A
2	BRN	Ground Connection	N/A
3	BLK/BLU	Wire Connector (above fuse) in Engine Compart.	N/A
4	--	Not Used	--
5	RED/WHT	Oxygen Sensor Heater	N/A
6	--	Not Used	--
7	--	Not Used	--
8	--	Not Used	--
9	YEL/GRN	Secondary Air Injection Solenoid Valve	N/A
10	--	Not Used	--
11	--	Not Used	--

2000 Sedan 1.8L MFI, ATW, ECM Connector, *continued*

12	--	Not Used	--
13	--	Not Used	--
14	--	Not Used	--
15	--	Not Used	--
16	--	Not Used	--
17	--	Not Used	--
18	--	Not Used	--
19	--	Not Used	--
20	--	Not Used	--
21	WHT/BLK	ECM Power Supply	N/A
22	--	Not Used	--
23	--	Not Used	--
24	--	Not Used	--
25	YEL/RED	Leak Detection Pump	N/A
26	--	Not Used	--
27	YEL	Mass Air Flow Sensor	N/A
28	--	Not Used	--
29	GRN	Mass Air Flow Sensor	N/A
30	BLK/LIL	T-15 Connector, Red, in plenum chamber	N/A
31	--	Not Used	--
32	BLK	Sensor Ground	N/A
33	GRY/YEL	Sensor 2 for Accelerator Pedal Position	N/A
34	BRN/GRN	Sensor 2 for Accelerator Pedal Position	N/A
35	YEL/BLU	Throttle Position Sensor	N/A
36	BRN/RED	Throttle Position Sensor	N/A
37	GRN/BLU	Wire Connection (RPM signal)	N/A
38	BLK/WHT	Cruise Control Switch	N/A
39	RED/GRN	Clutch Vacuum Vent Valve Switch (manual only)	N/A
40	BLK/YEL	A/C Control Head	N/A
41	BLK/GRY	A/C Control Head	N/A
42	--	Not Used	--
43	GRN/BLK	T-10 Connection, Brown, in plenum chamber	N/A
44	--	Not Used	--
45	--	Not Used	--
46	--	Not Used	--
47	YEL/BRN	T-10 Connection, Brown, in plenum chamber	N/A
48	RED/BRN	Fault Light for Power Accelerator Activation	N/A
49	--	Not Used	--
50	--	Not Used	--
51	BRN	Heated Oxygen Sensor	N/A
52	--	Not Used	--
53	BLK/GRN	Mass Air Flow Sensor	N/A
54	WHT/BLU	Wire Connection (vehicle speed signal)	N/A
55	WHT/RED	Brake Light Switch	N/A
56	RED/BLK	Brake Light Switch	N/A
57	RED/YEL	Cruise Control Switch	N/A
58	RED/BRN	CAN-Bus (Data-Bus) Connection	N/A

2000 Sedan 1.8L MFI, ATW, ECM Connector, *continued*

59	--	Not Used	--
60	RED/BLK	CAN-Bus (Data-Bus) Connection	N/A
61	--	Not Used	--
62	RED	Plus Connection (30) in engine compartment wiring	N/A
63	WHT/GRN	Heater for Lambda Probe 2	N/A
64	LIL	Evap Emission Canister Purge Regulator Valve	N/A
65	RED/BLU	Fuel Pump Relay	N/A
66	BLU/GRY	Secondary Air Injection Pump Relay	N/A
67	LIL/BLU	Emergency Flasher Switch	N/A
68	LIL	Heated Oxygen Sensor (behind TWC)	N/A
69	RED	Heated Oxygen Sensor (behind TWC)	N/A
70	WHT	Heated Oxygen Sensor	N/A
71	--	Not Used	--
72	GRY	Sensor 2 for Accelerator Pedal Position	N/A
73	YEL/LIL	Throttle Position Sensor	N/A
74	GRN/BRN	ABS Control Module (w/EDL)	N/A
75	BLU	Cruise Control Switch	N/A
76	RED/GRY	Cruise Control Switch	N/A
77	--	Not Used	--
78	--	Not Used	--
79	--	Not Used	--
80	WHT/YEL	Leak Detection Pump	N/A
81	YEL	T-10 Connection, Brown, in plenum chamber	N/A
82	GRY	Engine Speed Sensor	N/A
83	GRN/LIL	Throttle Valve Control Module	N/A
84	LIL/YEL	Throttle Valve Control Module	N/A
85	BLK/RED	Intake Air Temperature Sensor	N/A
86	GRN/GRY	Camshaft Position Sensor	N/A
87	--	Not Used	--
88	GRY/LIL	Cylinder Four Fuel Injector	N/A
89	BRN/BLU	Cylinder Two Fuel Injector	N/A
90	BLU	Engine Speed Sensor	N/A
91	GRN/BRN	Throttle Valve Control Module	N/A
92	LIL/RED	Throttle Valve Control Module	N/A
93	BRN/GRY	Engine Coolant Temperature Sensor	N/A
94	BRN/WHT	Ignition Coil 4	N/A
95	GRY/BRN	Ignition Coil 2	N/A
96	BLK/GRN	Cylinder One Fuel Injector	N/A
97	GRY/GRN	Cylinder Three Fuel Injector	N/A
98	WHT/BLK	Charge Air Pressure and Camshaft Position Sensors	N/A
99	BRN	Knock Sensor 2	N/A
100	--	Not Used	--
101	GRY/BLU	Charge Air Pressure Sensor	N/A
102	GRY	Ignition Coil	N/A
103	GRY/BLK	Ignition Coil 3	N/A
104	YEL/WHT	Intake Manifold Change-Over Valve	N/A
105	BRN/BLK	Not Used	N/A

2000 Sedan 1.8L MFI, ATW, ECM Connector, *continued*

106	WHT	Knock Sensor 1	N/A
107	YEL	Knock Sensor 2	N/A
108	BLK	Sensor Ground	N/A
109	--	Not Used	--
110	--	Not Used	--
111	--	Not Used	--
112	--	Not Used	--
113	--	Not Used	--
114	--	Not Used	--
115	--	Not Used	--
116	--	Not Used	--
117	RED/LIL	Throttle Valve Control Module	N/A
118	BRN/LIL	Throttle Valve Control Module	N/A
119	--	Not Used	--
120	--	Not Used	--
121	RED/GRN	Ignition Coils	N/A

A4

2001 Sedan 1.8L MFI, AWM, ECM Connector

ECM Pin #	Wire Color	Circuit Description (121-Pin)	Value at Hot Idle
1	BRN	Ground Connection	N/A
2	BRN	Ground Connection	N/A
3	BLK/BLU	Wire Connector (above fuse) in Engine Compart.	N/A
4	--	Not Used	--
5	RED/WHT	Oxygen Sensor Heater	N/A
6	--	Not Used	--
7	--	Not Used	--
8	--	Not Used	--
9	YEL/GRN	Secondary Air Injection Solenoid Valve	N/A
10	--	Not Used	--
11	--	Not Used	--
12	--	Not Used	--
13	--	Not Used	--
14	--	Not Used	--
15	--	Not Used	--
16	--	Not Used	--
17	--	Not Used	--
18	GRY/YEL	Brake Booster Pressure Sensor (auto only)	N/A
19	--	Not Used	--
20	--	Not Used	--
21	WHT/BLK	ECM Power Supply	N/A
22	WHT/YEL	Brake Booster Relay	N/A
23	--	Not Used	--
24	--	Not Used	--
25	YEL/RED	Leak Detection Pump	N/A

2001 Sedan 1.8L MFI, AWM, ECM Connector, *continued*

26	--	Not Used	--
27	YEL	Mass Air Flow Sensor	N/A
28	--	Not Used	--
29	GRN	Mass Air Flow Sensor	N/A
30	BLK/LIL	T-15 Connector, Red, in plenum chamber	N/A
31	--	Not Used	--
32	BLK	Sensor Ground	N/A
33	GRY/YEL	Sensor 2 for Accelerator Pedal Position	N/A
34	BRN/GRN	Sensor 2 for Accelerator Pedal Position	N/A
35	YEL/BLU	Throttle Position Sensor	N/A
36	BRN/RED	Throttle Position Sensor	N/A
37	GRN/BLU	Wire Connection (RPM signal)	N/A
38	BLK/WHT	Cruise Control Switch	N/A
39	RED/GRN	Clutch Vacuum Vent Valve Switch (manual only)	N/A
40	BLK/YEL	A/C Control Head	N/A
41	BLK/GRY	A/C Control Head	N/A
42	--	Not Used	--
43	GRN/BLK	T-10 Connection, Brown, in plenum chamber	N/A
44	--	Not Used	--
45	--	Not Used	--
46	--	Not Used	--
47	YEL/BRN	T-10 Connection, Brown, in plenum chamber	N/A
48	RED/BRN	Fault Light for Power Accelerator Activation	N/A
49	--	Not Used	--
50	--	Not Used	--
51	BRN	Heated Oxygen Sensor	N/A
52	YEL/GRY	Heated Oxygen Sensor	N/A
53	BLK/GRN	Mass Air Flow Sensor	N/A
54	WHT/BLU	Wire Connection (vehicle speed signal)	N/A
55	WHT/RED	Brake Light Switch	N/A
56	RED/BLK	Brake Light Switch	N/A
57	RED/YEL	Cruise Control Switch	N/A
58	ORN/BRN	CAN-Bus (Data-Bus) Connection	N/A
59	--	Not Used	--
60	ORN/BLK	CAN-Bus (Data-Bus) Connection	N/A
61	--	Not Used	--
62	RED	Plus Connection (30) in engine compartment wiring	N/A
63	WHT/GRN	Heater for Lambda Probe 2	N/A
64	LIL	Evap Emission Canister Purge Regulator Valve	N/A
65	RED/BLU	Fuel Pump Relay	N/A
66	BLU/GRY	Secondary Air Injection Pump Relay	N/A
67	LIL/BLU	Emergency Flasher Switch	N/A
68	LIL	Heated Oxygen Sensor (behind TWC)	N/A
69	RED	Heated Oxygen Sensor (behind TWC)	N/A

2001 Sedan 1.8L MFI, AWM, ECM Connector, *continued*

70	WHT	Heated Oxygen Sensor	N/A
71	--	Not Used	--
72	GRY	Sensor 2 for Accelerator Pedal Position	N/A
73	YEL/LIL	Throttle Position Sensor	N/A
74	GRN/BRN	ABS Control Module (w/EDL)	N/A
75	BLU	Cruise Control Switch	N/A
76	RED/GRY	Cruise Control Switch	N/A
77	--	Not Used	--
78	--	Not Used	--
79	--	Not Used	--
80	WHT/YEL	Leak Detection Pump	N/A
81	YEL	T-10 Connection, Brown, in plenum chamber	N/A
82	GRY	Engine Speed Sensor	N/A
83	GRN/LIL	Throttle Valve Control Module	N/A
84	LIL/YEL	Throttle Valve Control Module	N/A
85	BLK/RED	Intake Air Temperature Sensor	N/A
86	GRN/GRY	Camshaft Position Sensor	N/A
87	--	Not Used	--
88	GRY/LIL	Cylinder Four Fuel Injector	N/A
89	BRN/BLU	Cylinder Two Fuel Injector	N/A
90	BLU	Engine Speed Sensor	N/A
91	GRN/BRN	Throttle Valve Control Module	N/A
92	LIL/RED	Throttle Valve Control Module	N/A
93	BRN/GRY	Engine Coolant Temperature Sensor	N/A
94	BRN/GRN	Ignition Coil 4	N/A
95	BLU/YEL	Ignition Coil 2	N/A
96	BLK/GRN	Cylinder One Fuel Injector	N/A
97	GRY/GRN	Cylinder Three Fuel Injector	N/A
98	WHT/BLK	Charge Air Pressure and Camshaft Position Sensors	N/A
99	BRN	Knock Sensor 2	N/A
100	--	Not Used	--
101	GRY/BLU	Charge Air Pressure Sensor	N/A
102	BLU/RED	Ignition Coil	N/A
103	YEL/GRN	Ignition Coil 3	N/A
104	YEL/WHT	Intake Manifold Change-Over Valve	N/A
105	BRN/BLK	Not Used	--
106	WHT	Knock Sensor 1	N/A
107	YEL	Knock Sensor 2	N/A
108	BLK	Sensor Ground	N/A
109	--	Not Used	--
110	--	Not Used	--
111	--	Not Used	--
112	--	Not Used	--
113	--	Not Used	--

2001 Sedan 1.8L MFI, AWM, ECM Connector, *continued*

114	--	Not Used	--
115	GRN/YEL	Valve 1 for Camshaft Adjustment	N/A
116	--	Not Used	--
117	RED/LIL	Throttle Valve Control Module	N/A
118	BRN/LIL	Throttle Valve Control Module	N/A
119	--	Not Used	--
120	--	Not Used	--
121	RED/GRN	Ignition Coils	N/A

A4

2002–2005 Sedan 1.8L MFI, AMB, ECM Connector

ECM Pin #	Wire Color	Circuit Description (121-Pin)	Value at Hot Idle
1	BRN	Ground Connection	N/A
2	BRN	Ground Connection	N/A
3	BLK/LIL	Connection 3 in Engine Wiring Harness	N/A
4	--	Not Used	--
5	RED/WHT	Oxygen Sensor Heater	N/A
6	--	Not Used	--
7	--	Not Used	--
8	--	Not Used	--
9	YEL/GRN	Secondary Air Injection Solenoid Valve	N/A
10	--	Not Used	--
11	--	Not Used	--
12	--	Not Used	--
13	--	Not Used	--
14	--	Not Used	--
15	--	Not Used	--
16	--	Not Used	--
17	YEL/BLK	Engine Coolant Temperature Sensor (on radiator)	N/A
18	GRY/YEL	Brake Booster Pressure Sensor	N/A
19	--	Not Used	--
20	--	Not Used	--
21	--	Not Used	--
22	WHT/YEL	Brake Booster Relay	N/A
23	--	Not Used	--
24	RED/WHT	Coolant Fan Control Module	N/A
25	GRY/RED	Leak Detection Pump	N/A
26	--	Not Used	--
27	YEL	Mass Air Flow Sensor	N/A
28	BLU/BLK	Voltage Regulator	N/A
29	GRN	Mass Air Flow Sensor	N/A
30	--	Not Used	--
31	--	Not Used	--
32	BLK	Sensor Ground	N/A
33	GRY/YEL	Sensor 2 for Accelerator Pedal Position	N/A

2002–2005 Sedan 1.8L MFI, AMB, ECM Connector, *continued*

34	BRN	Sensor 2 for Accelerator Pedal Position	N/A
35	YEL/BLU	Throttle Position Sensor	N/A
36	BRN/RED	Throttle Position Sensor	N/A
37	GRN/BLU	Transmission Control Module	N/A
38	BLU/GRN	Steering Column Electronic Control Module	N/A
39	RED/GRN	Clutch Vacuum Vent Valve Switch (manual only)	N/A
40	--	Not Used	--
41	--	Not Used	--
42	--	Not Used	--
43	GRN/RED	T-10 Connection, Orange, in plenum chamber	N/A
44	--	Not Used	--
45	--	Not Used	--
46	--	Not Used	--
47	--	Not Used	--
48	--	Not Used	--
49	--	Not Used	--
50	BLU/BLK	Engine Coolant Temperature Sensor (on radiator)	N/A
51	BRN	Heated Oxygen Sensor	N/A
52	GRY/GRN	Heated Oxygen Sensor	N/A
53	WHT/GRN	Mass Air Flow Sensor	N/A
54	--	Not Used	--
55	WHT/RED	Brake Light Switch	N/A
56	RED/BLK	Brake Light Switch	N/A
57	--	Not Used	--
58	ORN/BRN	CAN-Bus (Data-Bus) Connection	N/A
59	--	Not Used	--
60	ORN/BLK	CAN-Bus (Data-Bus) Connection	N/A
61	--	Not Used	--
62	RED	Plus Connection in engine compartment wiring	N/A
63	--	Not Used	--
64	LIL	Evap Emission Canister Purge Regulator Valve	N/A
65	RED/BLU	Fuel Pump Relay	N/A
66	BLU/GRY	Secondary Air Injection Pump Relay	N/A
67	LIL/BLU	Airbag Control Module	N/A
68	LIL	Heated Oxygen Sensor (behind TWC)	N/A
69	RED	Heated Oxygen Sensor (behind TWC)	N/A
70	WHT	Heated Oxygen Sensor	N/A
71	BLK/WHT	Heated Oxygen Sensor	N/A
72	BRN/GRN	Sensor 2 for Accelerator Pedal Position	N/A
73	YEL/LIL	Throttle Position Sensor	N/A
74	--	Not Used	--
75	--	Not Used	--
76	--	Not Used	--
77	ORN/BRN	Transmission Control Module	N/A
78	--	Not Used	--
79	ORN/BLK	Transmission Control Module	N/A
80	WHT/YEL	Leak Detection Pump	N/A

2002–2005 Sedan 1.8L MFI, AMB, ECM Connector, *continued*

81	--	Not Used	--
82	GRY	Engine Speed Sensor	N/A
83	GRN/LIL	Throttle Valve Control Module	N/A
84	LIL/YEL	Throttle Valve Control Module	N/A
85	BLK/RED	Intake Air Temperature Sensor	N/A
86	GRN/GRY	Camshaft Position Sensor	N/A
87	--	Not Used	--
88	GRY/LIL	Cylinder Four Fuel Injector	N/A
89	BRN/BLU	Cylinder Two Fuel Injector	N/A
90	BLU	Engine Speed Sensor	N/A
91	LIL/RED	Throttle Valve Control Module	N/A
92	GRN/BRN	Throttle Valve Control Module	N/A
93	BRN/GRY	Engine Coolant Temperature Sensor	N/A
94	BRN/BLK	Ignition Coil 4 with Power Output Stage	N/A
95	GRY/BRN	Ignition Coil 2 with Power Output Stage	N/A
96	BLK/GRN	Cylinder One Fuel Injector	N/A
97	GRY/GRN	Cylinder Three Fuel Injector	N/A
98	WHT/BLK	Camshaft Position Sensor	N/A
99	BRN/GRN	Knock Sensor 1	N/A
100	--	Not Used	--
101	GRY/BLU	Charge Air Pressure Sensor	N/A
102	RED/GRN	Ignition Coil 1 with Power Output Stage	N/A
103	GRN/BLK	Ignition Coil 3 with Power Output Stage	N/A
104	YEL/WHT	Wastegate Bypass Regulator Valve	N/A
105	BRN/BLK	Recirculating Valve for Turbocharger	N/A
106	WHT	Knock Sensor 1	N/A
107	YEL	Knock Sensor 2	N/A
108	BLK	Sensor Ground	N/A
109	--	Not Used	--
110	--	Not Used	--
111	--	Not Used	--
112	--	Not Used	--
113	--	Not Used	--
114	--	Not Used	--
115	YEL/GRY	Valve 1 for Camshaft Adjustment	N/A
116	GRY/GRN	Map Controlled Engine Cooling Thermostat	N/A
117	RED/LIL	Throttle Valve Control Module	N/A
118	BRN/LIL	Throttle Valve Control Module	N/A
119	--	Not Used	--
120	--	Not Used	--
121	RED/GRN	Fuse Connection in Fuse Holder	N/A

A4

1998–1999 Sedan 2.8L MFI AHA Connector

ECM Pin #	Wire Color	Circuit Description (80-Pin)	Value at Hot Idle
1	BLK/BLU	Fuse Connection	N/A
2	BRN	Ground Connection	N/A
3	RED	Power	N/A
4	RED/BLU	Fuel Pump Relay	N/A
5	GRN/BLK	ABS Control Module	N/A
6	GRN/BLU	T-10 Connector, Black, in Plenum Chamber	N/A
8	BLK/GRY	A/C Control Head	N/A
10	BLK/YEL	T-10 Connector, Brown, in Plenum Chamber	N/A
11	RED/BLK	Camshaft Position Sensor	N/A
12	YEL	Mass Air Flow Sensor	N/A
13	GRN	Mass Air Flow Sensor	N/A
14	BLK	Ground	N/A
15	LIL	Evap Emission Canister Purge Regulator Valve	N/A
16	YEL/RED	Diagnosis Pump for Fuel System	N/A
17	YEL/BRN	Malfunction Indicator Light	N/A
18	YEL	T-10 Connector, Brown, in Plenum Chamber	N/A
19	GRN/BLK	T-6 Connector, Red, in Plenum Chamber	N/A
20	WHT/BLU	T-10 Connector, Brown, in Plenum Chamber	N/A
21	BLK/LIL	T-10 Connector, Yellow, in Plenum Chamber	N/A
25	GRY	Heated Oxygen Sensor	N/A
26	BLU	Heated Oxygen Sensor	N/A
27	RED/WHT	Oxygen Sensor Heater	N/A
28	BRN/BLK	Heater for Lambda Probe	N/A
29	GRY/WHT	Transmission Control Module	N/A
33	YEL/GRN	Secondary Air Injection Solenoid Valve	N/A
37	WHT/YEL	Diagnosis Pump for Fuel System	N/A
38	LIL	Oxygen Sensor (behind TWC)	N/A
39	GRY	Heated Oxygen Sensor 2	N/A
40	BLU	Heated Oxygen Sensor 2	N/A
41	ORN/BLK	CAN-Bus Connection	N/A
43	WHT/BLK	T-6 Connector, Red, in Plenum Chamber	N/A
44	GRN/BLK	Camshaft Position Sensor	N/A
45	GRN/BRN	ABS Control Module	N/A
50	RED	Oxygen Sensor (behind TWC)	N/A
51	LIL	Heated Oxygen Sensor (behind TWC)	N/A
52	RED	Heated Oxygen Sensor (behind TWC)	N/A
53	BRN/GRY	Engine Coolant Temperature Sensor	N/A
55	BRN/GRY	Valve 1 for Camshaft Adjustment Valve	N/A
56	GRY	Engine Speed Sensor	N/A
58	BRN/GRN	Cylinder Three Fuel Injector signal	N/A
59	RED/LIL	Throttle Valve Control Module	N/A

1998–1999 Sedan 2.8L MFI AHA Connector, *continued*

60	GRN	Knock Sensor 2	N/A
62	GRN/LIL	Throttle Valve Control Module	N/A
63	BLU	Engine Speed Sensor	N/A
64	YEL/WHT	Wastegate Bypass Regulator Valve signal	N/A
65	BRN/WHT	Cylinder Four Fuel Injector signal	N/A
66	BRN/LIL	Throttle Valve Control Module	N/A
67	GRY/WHT	Knock Sensor 2	N/A
68	BRN	Knock Sensor 1	N/A
69	LIL/YEL	Throttle Valve Control Module	N/A
70	GRN/GRY	Power Output Stage	N/A
71	GRY/YEL	Power Output Stage	N/A
72	BLK/RED	Cylinder Five Fuel Injector signal	N/A
73	BLK/GRN	Cylinder One Fuel Injector signal	N/A
74	LIL/RED	Throttle Valve Control Module	N/A
75	WHT/YEL	Throttle Valve Control Module	N/A
76	GRN/BRN	Camshaft Position Sensor	N/A
78	GRN/WHT	Power Output Stage	N/A
79	BLK/WHT	Cylinder Six Fuel Injector signal	N/A
80	BRN/BLU	Cylinder Two Fuel Injector signal	N/A

A6

1995 Sedan 2.8L MFI AAH ECM Connector

ECM Pin #	Wire Color	Circuit Description (68-Pin)	Value at Hot Idle
A2	BLU	Mass Air Flow Sensor	N/A
A3	GRN	Heated Oxygen Sensor 2	N/A
A4	BLU	Mass Air Flow Sensor	N/A
A6	BLK/BRN	Throttle Position Sensor power	N/A
A7	BLU/BRN	Closed Throttle Position Switch signal	N/A
A8	GRN	Heated Oxygen Sensor 2	N/A
A9	BLK/GRY	Closed Throttle Position Switch ground	N/A
A10	GRN	Heated Oxygen Sensor	N/A
A11	GRN	Heated Oxygen Sensor	N/A
B1	WHT	Knock Sensor 2 power	N/A
B2	BRN	Knock Sensor 2 signal	N/A
B3	N/A	Knock Sensor 2 ground	N/A
B4	YEL	Knock Sensor 1 power	N/A
B5	GRN	Knock Sensor 1 signal	N/A
B6	N/A	Knock Sensor 1 ground	N/A
B7	GRY/BLK	EGR Temperature Sensor	N/A
B8	YEL/GRY	Transmission Control Module	N/A
B9	WHT/BLU	A/C Head	N/A
B10	GRN	T-6 Connector, Green, on the Connector Station 2	N/A
B11	BLU/BLK	T-6 Connector, Green, on the Connector Station 2	N/A
B12	BRN/BLU	Central Lock, Alarm, Interior Light Dimmer	N/A
B13	YEL	Transmission Control Module	N/A
B15	GRY/YEL	Power Output Stage	N/A
B17	GRN/WHT	Power Output Stage	N/A
B18	GRN/GRY	Power Output Stage	N/A
B20	RED	Power	N/A
C1	GRY	Engine Speed Sensor power	N/A
C2	BLU	Engine Speed Sensor signal	N/A
C3	YEL/WHT	Intake Manifold Change-Over Valve	N/A
C4	LIL	Crankshaft Position Sensor signal	N/A
C5	RED	Crankshaft Position Sensor power	N/A
C7	RED/BLK	Camshaft Position Sensor power	N/A
C8	GRN/LIL	Camshaft Position Sensor signal	N/A
C10	GRN/YEL	A/C Head	N/A
C12	GRN/RED	Data Link Connector Rapid Data Transfer	N/A
C13	WHT/RED	Data Link Connector Rapid Data Transfer	N/A
C14	WHT/BLU	Malfunction Indicator Light	N/A
C15	GRY/RED	Engine Coolant Temperature Sensor	N/A
D1	BRN/WHT	Ground	N/A
D2	BRN/WHT	Ground	N/A
D3	BRN/WHT	Ground	N/A
D4	BRN/BLK	Cylinder One Fuel Injector signal	N/A

1995 Sedan 2.8L MFI AAH ECM Connector, *continued*

D5	BRN/BLU	Cylinder Two Fuel Injector signal	N/A
D6	BRN/YEL	Cylinder Three Fuel Injector signal	N/A
D7	BLU/GRN	Idle Air Control Valve signal	N/A
D8	BLK/GRN	Wire connector above circuit breaker	N/A
D9	BRN/GRN	Cylinder Four Fuel Injector signal	N/A
D10	BRN/GRY	Cylinder Five Fuel Injector signal	N/A
D11	BLK/RED	Idle Air Control Valve ground	N/A
D12	LIL	Evap Emission Canister Purge Regulator Valve	N/A
D13	BLU/YEL	Cylinder Six Fuel Injector signal	N/A
D14	BLU/LIL	EGR Vacuum Regulator Solenoid Valve	N/A
D15	BLU/RED	Fuel Pump Relay	N/A
D16	BRN/RED	Ground	N/A

A6

1995 Sedan 2.8L MFI AFC Connector

ECM Pin #	Wire Color	Circuit Description	Value at Hot Idle
A2	RED/BLK	Camshaft Position Sensor power	N/A
A3	BLK/BRN	Throttle Position Sensor power	N/A
A4	BLK/GRY	Closed Throttle Position Switch ground	N/A
A5	BLU/RED	Fuel Pump	N/A
A7	GRN/RED	Data Link Connector Rapid Data Transfer	N/A
A9	LIL	Crankshaft Position Sensor power	N/A
A10	RED	Crankshaft Position Sensor signal	N/A
A11	GRY	Engine Speed Sensor	N/A
A12	GRN	Mass Air Flow Sensor	N/A
A13	BRN	Knock Sensor 2 signal	N/A
A14	WHT	Knock Sensors 1 and 2 power	N/A
A15	GRN	Knock Sensor 1 signal	N/A
A16	GRY/RED	Engine Coolant Temperature Sensor	N/A
A18	BLU/GRY	Secondary Air Injection Pump Relay	N/A
A19	BRN	Ground	N/A
A20	GRN	Heated Oxygen Sensor	N/A
A21	GRN	Heated Oxygen Sensor 2	N/A
A23	GRY/BLK	EGR Temperature Sensor	N/A
A24	BLU/BRN	Closed Throttle Position Switch signal	N/A
B1	YEL/GRY	Transmission Control Module	N/A
B2	GRN/LIL	Camshaft Position Sensor signal	N/A
B4	YEL/WHT	Intake Manifold Change-Over Valve	N/A
B5	GRN/WHT	Power Output Stage	N/A
B6	GRN/GRY	Power Output Stage	N/A
B7	GRY/YEL	Power Output Stage	N/A
B8	YEL/BLK	Transmission Control Module	N/A
B9	YEL	Transmission Control Module	N/A
B10	RED	Power	N/A
B11	BRN/BLU	Central Lock, Alarm, Interior Light Dimmer	N/A

1995 Sedan 2.8L MFI AFC Connector, *continued*

C1	GRN/LIL	Leak Detection Pump	N/A
C5	GRN/WHT	ABS Control Module (w/EDL)	N/A
C10	GRN	A/C Head	N/A
C11	GRN/YEL	A/C Head	N/A
C12	BLU/BLK	T-6 Connector, Red, on the Connector Station 2	N/A
C13	BRN/RED	Speedometer	N/A
C15	GRN/RED	ABS Control Module (w/EDL)	N/A
C17	YEL	Heated Oxygen Sensor 2 (behind TWC)	N/A
C20	GRN	Heated Oxygen Sensor 2 (behind TWC)	N/A
C21	BLU	Heated Oxygen Sensor 2 (behind TWC)	N/A
C22	LIL/BLK	T-6 Connector, Gray, on the Connector Station 2	N/A
C23	WHT/BLK	Malfunction Indicator Light	N/A
D1	BRN/WHT	Cylinder One Fuel Injector signal	N/A
D2	BRN/BLU	Cylinder Two Fuel Injector signal	N/A
D3	BRN/WHT	Ground Connection 2	N/A
D4	BRN/YEL	Cylinder Three Fuel Injector signal	N/A
D5	BRN/GRN	Cylinder Four Fuel Injector signal	N/A
D6	BRN/GRY	Cylinder Five Fuel Injector signal	N/A
D7	BRN/WHT	Ground Connection 2	N/A
D8	BLU/YEL	Cylinder Six Fuel Injector signal	N/A
D9	BLU/GRN	Idle Air Control Valve ground	N/A
D10	BLK/GRN	Wire connector above circuit breaker	N/A
D11	BLK/RED	Idle Air Control Valve signal	N/A
D12	BRN/RED	Ground Connection 1	N/A
E1	BLU/LIL	EGR Vacuum Regulator Solenoid Valve	N/A
E2	YEL/GRN	Secondary Air Injection Solenoid Valve	N/A
E3	BRN/WHT	Ground Connection 2	N/A
E4	YEL/WHT	Oxygen Sensor Heater (behind TWC) 2 power	N/A
E6	GRY/BLU	Leak Detection Pump	N/A
E7	BRN/RED	Ground Connection 2	N/A
E8	YEL/RED	Oxygen Sensor Heater (behind TWC) power	N/A
E9	YEL/BLU	Oxygen Sensor Heater 2 power	N/A
E10	YEL/GRN	Oxygen Sensor Heater power	N/A
E11	BRN/RED	Ground Connection 2	N/A
E12	LIL	Evap Emission Canister Purge Regulator Valve	N/A

A6

1999–2000 Sedan 4.2L MFI, ART, ECM Connector

ECM Pin #	Wire Color	Circuit Description (121-Pin)	Value at Hot Idle
1	BRN/WHT	Ground Connection	N/A
2	BRN/WHT	Ground Connection	N/A
3	BLK/BLU	Wire Connector (above fuse) in Engine Compart.	N/A
4	YEL/BLU	Oxygen Sensor Heater 2	N/A
5	YEL/GRN	Oxygen Sensor Heater	N/A
6	YEL/WHT	Heater for Lambda Probe 2	N/A

1999–2000 Sedan 4.2L MFI, ART, ECM Connector, *continued*

7	WHT/BLK	Ignition Coil 2	N/A
8	BLU/RED	Ignition Coil 7	N/A
9	--	Not Used	--
10	GRN	Heated Oxygen Sensor 2 (behind TWC)	N/A
11	YEL	Heated Oxygen Sensor 2 (behind TWC)	N/A
12	YEL	Heated Oxygen Sensor 2	N/A
13	GRN	Heated Oxygen Sensor 2	N/A
14	--	Not Used	--
15	--	Not Used	--
16	--	Not Used	--
17	--	Not Used	--
18	--	Not Used	--
19	BLU/LIL	Connector (engine overhead warning light)	N/A
20	--	Not Used	--
21	--	Not Used	--
22	--	Not Used	--
23	GRN	Cylinder Two Fuel Injector	N/A
24	GRN/WHT	Cylinder Seven Fuel Injector	N/A
25	YEL/RED	Leak Detection Pump	N/A
26	--	Not Used	--
27	BLK/GRN	Mass Air Flow Sensor	N/A
28	--	Not Used	--
29	GRN	Mass Air Flow Sensor	N/A
30	WHT/LIL	Instrument Cluster Combination Processor	N/A
31	--	Not Used	--
32	BRN/GRN	Sensor Ground	N/A
33	GRY/YEL	Sensor 2 for Accelerator Pedal Position	N/A
34	BRN/GRN	Sensor 2 for Accelerator Pedal Position	N/A
35	YEL/BLU	Throttle Position Sensor	N/A
36	BRN/RED	Throttle Position Sensor	N/A
37	GRN/BLU	Wire Connection (RPM signal)	N/A
38	BLK/WHT	Wire Connection (Cruise Control)	N/A
39	--	Not Used	--
40	BLK/YEL	A/C Control Head	N/A
41	BLK/GRY	A/C Control Head	N/A
42	--	Not Used	--
43	GRN/BLK	T-10 Connection, Brown, in plenum chamber	N/A
44	YEL/GRN	Secondary Air Injection Solenoid Valve	N/A
45	GRN/RED	Not Used	--
46	BLU/GRY	Secondary Air Injection Pump Relay	N/A
47	WHT/RED	T-10 Connection, Brown, in plenum chamber	N/A
48	YEL/BRN	T-10 Connection, Orange, in plenum chamber	N/A
49	--	Not Used	--
50	--	Not Used	--
51	RED	Heated Oxygen Sensor	N/A
52	--	Not Used	--
53	BLU/GRN	Mass Air Flow Sensor	N/A

1999–2000 Sedan 4.2L MFI, ART, ECM Connector, *continued*

54	WHT/BLU	Wire Connection (vehicle speed signal)	N/A
55	WHT/RED	Brake Light Switch	N/A
56	RED/BLK	Brake Light Switch	N/A
57	RED/YEL	Cruise Control Switch	N/A
58	ORN/BRN	CAN-Bus (Data-Bus) Connection	N/A
59	--	Not Used	--
60	ORN/BLK	CAN-Bus (Data-Bus) Connection	N/A
61	--	Not Used	--
62	RED	Power	N/A
63	YEL/RED	Heater for Lambda Probe 1	N/A
64	LIL/WHT	Evap Emission Canister Purge Regulator Valve	N/A
65	RED/BLU	Fuel Pump Relay	N/A
66	--	Not Used	--
67	LIL/BLU	Emergency Flasher Switch	N/A
68	GRY	Heated Oxygen Sensor (behind TWC)	N/A
69	BLU	Heated Oxygen Sensor (behind TWC)	N/A
70	LIL	Heated Oxygen Sensor	N/A
71	--	Not Used	--
72	GRY	Sensor 2 for Accelerator Pedal Position	N/A
73	YEL/LIL	Throttle Position Sensor	N/A
74	GRN/BRN	ABS Control Module (w/EDL)	N/A
75	BLU	Cruise Control Switch	N/A
76	RED/GRY	Cruise Control Switch	N/A
77	--	Not Used	--
78	--	Not Used	--
79	--	Not Used	--
80	WHT/YEL	Leak Detection Pump	N/A
81	YEL	T-10 Connection, Brown, in plenum chamber	N/A
82	GRY	Engine Speed Sensor	N/A
83	BRN/RED	Throttle Valve Control Module	N/A
84	BLK/BRN	Throttle Valve Control Module	N/A
85	GRN/GRY	Mass Air Flow Sensor	N/A
86	GRN/LIL	Camshaft Position Sensor	N/A
87	BLU/LIL	Camshaft Position Sensor 2	N/A
88	--	Not Used	--
89	WHT/YEL	Cylinder Eight Fuel Injector	N/A
90	BLU	Engine Speed Sensor	N/A
91	BLU/BRN	Throttle Valve Control Module	N/A
92	BLK/GRY	Throttle Valve Control Module	N/A
93	GRY/RED	Engine Coolant Temperature Sensor	N/A
94	GRY/YEL	Ignition Coil 3	N/A
95	GRY/BRN	Ignition Coil 8	N/A
96	BRN/BLK	Cylinder One Fuel Injector	N/A
97	BRN/GRN	Cylinder Five Fuel Injector	N/A
98	GRN/BLK	Camshaft Position Sensor	N/A
99	WHT	Knock Sensor 1	N/A
100	--	Not Used	--

1999–2000 Sedan 4.2L MFI, ART, ECM Connector, *continued*

101	--	Not Used	--
102	WHT/GRN	Ignition Coil 1	N/A
103	WHT/BRN	Ignition Coil 5	N/A
104	GRN/YEL	Wastegate Bypass Regulator Valve	N/A
105	BLU/RED	Not Used	--
106	BRN	Knock Sensor 1	N/A
107	GRN	Knock Sensor 2	N/A
108	BRN/GRY	Camshaft Position Sensor 2	N/A
109	--	Not Used	--
110	GRN/GRY	Ignition Coil 6	N/A
111	GRN/BRN	Ignition Coil 3	N/A
112	BRN/BLU	Cylinder Six Fuel Injector	N/A
113	BRN/LIL	Cylinder Four Fuel Injector	N/A
114	--	Not Used	--
115	BRN/GRY	Valve 1 for Camshaft Adjustment	N/A
116	BLU/RED	Not Used	--
117	RED/LIL	Throttle Valve Control Module	N/A
118	BRN/LIL	Throttle Valve Control Module	N/A
119	--	Not Used	--
120	--	Not Used	--
121	--	Not Used	--

A6

2000–2001 Sedan 2.8L MFI, ATQ, ECM Connector

ECM Pin #	Wire Color	Circuit Description (121-Pin)	Value at Hot Idle
1	BRN	Ground Connection	N/A
2	BRN	Ground Connection	N/A
3	BLK/BLU	Wire Connector (above fuse) in Engine Compart.	N/A
4	RED/WHT	Oxygen Sensor Heater 2	N/A
5	RED/WHT	Oxygen Sensor Heater	N/A
6	BRN/YEL	Heater for Lambda Probe 2	N/A
7	--	Not Used	--
8	--	Not Used	--
9	--	Not Used	--
10	WHT	Heated Oxygen Sensor 2 (behind TWC)	N/A
11	BRN	Heated Oxygen Sensor 2 (behind TWC)	N/A
12	LIL	Heated Oxygen Sensor 2	N/A
13	RED	Heated Oxygen Sensor 2	N/A
14	--	Not Used	--
15	--	Not Used	--
16	--	Not Used	--
17	--	Not Used	--
18	--	Not Used	--
19	--	Not Used	--
20	--	Not Used	--

2000–2001 Sedan 2.8L MFI, ATQ, ECM Connector, *continued*

21	--	Not Used	--
22	--	Not Used	--
23	--	Not Used	--
24	--	Not Used	--
25	YEL/RED	Leak Detection Pump	N/A
26	--	Not Used	--
27	YEL	Mass Air Flow Sensor	N/A
28	--	Not Used	--
29	GRN	Mass Air Flow Sensor	N/A
30	BLK/LIL	Instrument Cluster Combination Processor	N/A
31	--	Not Used	--
32	BLK	Sensor Ground	N/A
33	GRY/YEL	Sensor 2 for Accelerator Pedal Position	N/A
34	BRN/GRN	Sensor 2 for Accelerator Pedal Position	N/A
35	YEL/BLU	Throttle Position Sensor	N/A
36	BRN/RED	Throttle Position Sensor	N/A
37	GRN/BLU	Wire Connection (RPM signal)	N/A
38	BLK/WHT	Cruise Control Switch	N/A
39	RED/GRN	Clutch Vacuum Vent Valve Switch (manual only)	N/A
40	BLK/YEL	A/C Control Head	N/A
41	BLK/GRY	Connector (C15, A/C) in Instrument Panel Wiring	N/A
42	--	Not Used	--
43	GRN/BLK	T-10 Connection, Brown, in plenum chamber	N/A
44	YEL/GRN	Secondary Air Injection Solenoid Valve	N/A
45	--	Not Used	--
46	BLU/GRY	Secondary Air Injection Pump Relay	N/A
47	YEL/BRN	T-10 Connection, Brown, in plenum chamber	N/A
48	--	Not Used	--
49	--	Not Used	--
50	--	Not Used	--
51	GRN	Heated Oxygen Sensor	N/A
52	--	Not Used	--
53	BLU/GRN	Mass Air Flow Sensor	N/A
54	WHT/BLU	Wire Connection (vehicle speed signal)	N/A
55	WHT/RED	Brake Light Switch	N/A
56	RED/BLK	Brake Light Switch	N/A
57	RED/YEL	Cruise Control Switch	N/A
58	ORN/BRN	CAN-Bus (Data-Bus) Connection	N/A
59	--	Not Used	--
60	ORN/BLK	CAN-Bus (Data-Bus) Connection	N/A
61	--	Not Used	--
62	RED	Power	N/A
63	--	Not Used	--
64	LIL	Evap Emission Canister Purge Regulator Valve	N/A
65	RED/BLU	Fuel Pump Relay	N/A
66	--	Not Used	--
67	LIL/BLU	Emergency Flasher Switch	N/A

2000–2001 Sedan 2.8L MFI, ATQ, ECM Connector, *continued*

68	GRN	Heated Oxygen Sensor (behind TWC)	N/A
69	BLU	Heated Oxygen Sensor (behind TWC)	N/A
70	YEL	Heated Oxygen Sensor	N/A
71	--	Not Used	--
72	GRY	Sensor 2 for Accelerator Pedal Position	N/A
73	YEL/LIL	Throttle Position Sensor	N/A
74	GRN/BRN	ABS Control Module (w/EDL)	N/A
75	BLU	Cruise Control Switch	N/A
76	RED/GRY	Cruise Control Switch	N/A
77	--	Not Used	--
78	--	Not Used	--
79	--	Not Used	--
80	WHT/YEL	Leak Detection Pump	N/A
81	YEL	T-10 Connection, Brown, in plenum chamber	N/A
82	GRY	Engine Speed Sensor	N/A
83	GRN/LIL	Throttle Valve Control Module	N/A
84	WHT/YEL	Throttle Valve Control Module	N/A
85	GRN/RED	Intake Air Temperature Sensor	N/A
86	GRN/BRN	Camshaft Position Sensor	N/A
87	GRY/BLK	Camshaft Position Sensor 2	N/A
88	BRN/GRN	Cylinder Three Fuel Injector	N/A
89	BLK/WHT	Cylinder Six Fuel Injector	N/A
90	BLU	Engine Speed Sensor	N/A
91	GRN/WHT	Throttle Valve Control Module	N/A
92	LIL/RED	Throttle Valve Control Module	N/A
93	BRN/GRY	Engine Coolant Temperature Sensor	N/A
94	GRN/GRY	Power Output Stage	N/A
95	--	Not Used	--
96	BLK/GRN	Cylinder One Fuel Injector	N/A
97	BRN/WHT	Cylinder Four Fuel Injector	N/A
98	RED/BLK	Camshaft Position Sensor 1 and 2	N/A
99	GRN	Knock Sensor 2	N/A
100	--	Not Used	--
101	--	Not Used	--
102	GRY/YEL	Power Output Stage	N/A
103	GRN/WHT	Power Output Stage	N/A
104	YEL/WHT	Intake Manifold Change-Over Valve	N/A
105	--	Not Used	--
106	WHT	Knock Sensor 1	N/A
107	YEL	Knock Sensor 2	N/A
108	GRY/WHT	Sensor Ground	N/A
109	--	Not Used	--
110	--	Not Used	--
111	--	Not Used	--
112	BRN/BLU	Cylinder Two Fuel Injector	N/A
113	BLK/RED	Cylinder Five Fuel Injector	N/A
114	--	Not Used	--

2000–2001 Sedan 2.8L MFI, ATQ, ECM Connector, *continued*

115	BRN/GRY	Valve 1 for Camshaft Adjustment	N/A
116	--	Not Used	--
117	RED/LIL	Throttle Valve Control Module	N/A
118	BRN/LIL	Throttle Valve Control Module	N/A
119	--	Not Used	--
120	--	Not Used	--
121	--	Not Used	--

A6

2000–2002 Sedan 2.7L MFI, APB, ECM Connector

ECM Pin #	Wire Color	Circuit Description (121-Pin)	Value at Hot Idle
1	BRN/WHT	Ground Connection	N/A
2	BRN/WHT	Ground Connection	N/A
3	BLK/BLU	Wire Connector (above fuse) in Engine Compart.	N/A
4	YEL/BLU	Oxygen Sensor Heater 2	N/A
5	YEL/GRN	Oxygen Sensor Heater	N/A
6	YEL/WHT	Heater for Lambda Probe 2	N/A
7	--	Not Used	--
8	--	Not Used	--
9	GNR/YEL	Central Idle Air Control Cut-off Valve	N/A
10	BRN	Heated Oxygen Sensor 2 (behind TWC)	N/A
11	WHT	Heated Oxygen Sensor 2 (behind TWC)	N/A
12	YEL	Heated Oxygen Sensor 2	N/A
13	GRN	Heated Oxygen Sensor 2	N/A
14	--	Not Used	--
15	--	Not Used	--
16	--	Not Used	--
17	--	Not Used	--
18	--	Not Used	--
19	BLU/LIL	Connector (engine overhead warning light)	N/A
20	BLU/RED	Sensor 2 for Exhaust Temperature	N/A
21	--	Not Used	--
22	--	Not Used	--
23	GRN	Cylinder Two Fuel Injector	N/A
24	--	Not Used	--
25	YEL/RED	Leak Detection Pump	N/A
26	--	Not Used	--
27	BLK/GRN	Mass Air Flow Sensor	N/A
28	--	Not Used	--
29	GRN	Mass Air Flow Sensor	N/A
30	BLK/LIL	Instrument Cluster Combination Processor	N/A
31	--	Not Used	--
32	BRN/GRN	Sensor Ground	N/A
33	GRY/YEL	Sensor 2 for Accelerator Pedal Position	N/A
34	BRN/GRN	Sensor 2 for Accelerator Pedal Position	N/A

2000–2002 Sedan 2.7L MFI, APB, ECM Connector, *continued*

35	YEL/BLU	Throttle Position Sensor	N/A
36	BRN/RED	Throttle Position Sensor	N/A
37	GRN/BLU	Wire Connection (RPM signal)	N/A
38	BLK/WHT	Cruise Control Switch	N/A
39	RED/GRN	Clutch Vacuum Vent Valve Switch (manual only)	N/A
40	BLK/YEL	A/C Control Head	N/A
41	BLK/GRY	Connector (C15, A/C) in Instrument Panel Wiring	N/A
42	--	Not Used	--
43	GRN/BLK	T-10 Connection, Brown, in plenum chamber	N/A
44	YEL/GRN	Secondary Air Injection Solenoid Valve	N/A
45	--	Not Used	--
46	BLU/GRY	Secondary Air Injection Pump Relay	N/A
47	YEL/BRN	T-10 Connection, Brown, in plenum chamber	N/A
48	RED/BRN	T-10 Connection, Orange, in plenum chamber	N/A
49	--	Not Used	--
50	LIL/GRN	Sensor 2 for Exhaust Temperature	N/A
51	YEL	Heated Oxygen Sensor	N/A
52	--	Not Used	--
53	LIL/GRY	Mass Air Flow Sensor	N/A
54	WHT/BLU	Wire Connection (vehicle speed signal)	N/A
55	WHT/RED	Brake Light Switch	N/A
56	RED/BLK	Brake Light Switch	N/A
57	RED/YEL	Cruise Control Switch	N/A
58	ORN/BRN	CAN-Bus (Data-Bus) Connection	N/A
59	--	Not Used	--
60	ORN/BLK	CAN-Bus (Data-Bus) Connection	N/A
61	BLU/BRN	Sensor 1 for Exhaust Temperature	N/A
62	RED	Power	N/A
63	YEL/RED	Heater for Lambda Probe 1	N/A
64	LIL/WHT	Evap Emission Canister Purge Regulator Valve	N/A
65	RED/BLU	Fuel Pump Relay	N/A
66	--	Not Used	--
67	LIL/BLU	Emergency Flasher Switch	N/A
68	GRY	Heated Oxygen Sensor (behind TWC)	N/A
69	BLU	Heated Oxygen Sensor (behind TWC)	N/A
70	GRN	Heated Oxygen Sensor	N/A
71	--	Not Used	--
72	GRY	Sensor 2 for Accelerator Pedal Position	N/A
73	YEL/LIL	Throttle Position Sensor	N/A
74	GRN/BRN	ABS Control Module (w/EDL)	N/A
75	BLU	Cruise Control Switch	N/A
76	RED/GRY	Cruise Control Switch	N/A
77	--	Not Used	--
78	--	Not Used	--

2000–2002 Sedan 2.7L MFI, APB, ECM Connector, *continued*

79	--	Not Used	--
80	WHT/YEL	Leak Detection Pump	N/A
81	YEL	T-10 Connection, Brown, in plenum chamber	N/A
82	GRY	Engine Speed Sensor	N/A
83	BRN/RED	Throttle Valve Control Module	N/A
84	BLK/BRN	Throttle Valve Control Module	N/A
85	GRN/WHT	Intake Air Temperature Sensor	N/A
86	GRN/LIL	Camshaft Position Sensor	N/A
87	BLU/LIL	Camshaft Position Sensor 2	N/A
88	BRN/LIL	Cylinder Three Fuel Injector	N/A
89	WHT/YEL	Cylinder Six Fuel Injector	N/A
90	BLU	Engine Speed Sensor	N/A
91	GRN/RED	Throttle Valve Control Module	N/A
92	BLK/GRY	Throttle Valve Control Module	N/A
93	GRY/RED	Engine Coolant Temperature Sensor	N/A
94	GRY/YEL	Power Output Stage	N/A
95	GRY/BRN	Power Output Stage 2	N/A
96	BRN/BLK	Cylinder One Fuel Injector	N/A
97	BRN/GRN	Cylinder Four Fuel Injector	N/A
98	GRN/BLK	Camshaft Position Sensor 2	N/A
99	WHT	Knock Sensor 1	N/A
100	--	Not Used	--
101	BLU/GRY	Charge Air Pressure Sensor	N/A
102	WHT/GRN	Power Output Stage	N/A
103	WHT/BRN	Power Output Stage 2	N/A
104	BLU/GRN	Wastegate Bypass Regulator Valve	N/A
105	--	Not Used	--
106	BRN	Knock Sensor 1	N/A
107	GRN	Knock Sensor 2	N/A
108	BRN/GRY	Sensor Ground	N/A
109	--	Not Used	--
110	GRN/GRY	Power Output Stage	N/A
111	GRN/BRN	Power Output Stage 2	N/A
112	BRN/BLU	Cylinder Two Fuel Injector	N/A
113	BRN/GRN	Cylinder Four Fuel Injector	N/A
114	--	Not Used	--
115	BRN/GRY	Valve 2 for Camshaft Adjustment	N/A
116	--	Not Used	--
117	RED/LIL	Throttle Valve Control Module	N/A
118	BRN/LIL	Throttle Valve Control Module	N/A
119	--	Not Used	--
120	--	Not Used	--
121	--	Not Used	--

A6 and All-Road Quattro

2002–2005 Sedan 4.2L MFI, AWN and BAS, ECM Connector

ECM Pin #	Wire Color	Circuit Description (121-Pin)	Value at Hot Idle
1	BRN/WHT	Ground Connection	N/A
2	BRN/WHT	Ground Connection	N/A
3	RED/GRN	Sensor Ground	N/A
4	YEL/BLU	Oxygen Sensor Heater 2	N/A
5	YEL	Oxygen Sensor Heater	N/A
6	YEL/WHT	Oxygen Sensor Heater 2 (behind TWC)	N/A
7	WHT/BLU	Ignition Coil 2with Power Output Stage	N/A
8	BLU/YEL	Ignition Coil 7 with Power Output Stage	N/A
9	--	Not Used	--
10	GRN	Heated Oxygen Sensor 2 (behind TWC)	N/A
11	YEL	Heated Oxygen Sensor 2 (behind TWC)	N/A
12	WHT	Heated Oxygen Sensor 2	N/A
13	BRN	Heated Oxygen Sensor 2	N/A
14	--	Not Used	--
15	--	Not Used	--
16	--	Not Used	--
17	--	Not Used	--
18	--	Not Used	--
19	GRN	Cylinder Two Fuel Injector	N/A
20	--	Not Used	--
21	RED/BLU	Connection 1 (15a) in ABS wiring harness	N/A
22	--	Not Used	--
23	WHT/BLK	ECM Power Supply Relay	N/A
24	GRN/WHT	Cylinder Seven Fuel Injector	N/A
25	YEL/RED	Leak Detection Pump	N/A
26	GRY	Mass Air Flow Sensor	N/A
27	BLK/GRN	Mass Air Flow Sensor	N/A
28	--	Not Used	--
29	GRN	Mass Air Flow Sensor	N/A
30	--	Not Used	--
31	--	Not Used	--
32	--	Not Used	--
33	GRY/YEL	Sensor 2 for Accelerator Pedal Position	N/A
34	BLU	Sensor 2 for Accelerator Pedal Position	N/A
35	YEL/BLU	Throttle Position Sensor	N/A
36	BRN/RED	Throttle Position Sensor	N/A
37	GRN/BLU	Wire Connection (RPM signal)	N/A
38	BLK/WHT	Connection (cruise control) in instrument panel	N/A
39	RED/GRN	Clutch Pedal Switch (manual only)	N/A
40	BLK/YEL	A/C Control Head	N/A

2002–2005 Sedan 4.2L MFI, AWN and BAS, ECM Connector, *continued*

41	BLK/GRY	A/C Control Head	N/A
42	--	Not Used	--
43	GRN/RED	T-10 Connection, Brown, in plenum chamber	N/A
44	LIL/YEL	Secondary Air Injection Solenoid Valve	N/A
45	GRN/BLK	Intake Manifold Tuning Valve 2 (AWN only)	N/A
46	BLU/GRY	Secondary Air Injection Pump Relay	N/A
47	--	Not Used	--
48	BLU/BLK	Intake Change-Over Valve	N/A
49	--	Not Used	--
50	--	Not Used	--
51	RED	Heated Oxygen Sensor	N/A
52	--	Not Used	--
53	BLU/GRN	Mass Air Flow Sensor	N/A
54	--	Not Used	--
55	WHT/RED	Brake Light Switch	N/A
56	RED/BLK	Brake Light Switch	N/A
57	RED/YEL	Cruise Control Switch	N/A
58	ORN/BRN	CAN-Bus (Data-Bus) Connection	N/A
59	--	Not Used	--
60	ORN/BLK	CAN-Bus (Data-Bus) Connection	N/A
61	--	Not Used	--
62	RED	Plus Connection (30) in engine compartment wiring	N/A
63	YEL/RED	Oxygen Sensor Heater (behind TWC)	N/A
64	LIL/WHT	Evap Emission Canister Purge Regulator Valve	N/A
65	RED/BLU	Fuel Pump Relay	N/A
66	--	Not Used	--
67	LIL/BLU	Emergency Flasher Switch	N/A
68	GRY	Heated Oxygen Sensor (behind TWC)	N/A
69	BLU	Heated Oxygen Sensor (behind TWC)	N/A
70	LIL	Heated Oxygen Sensor	N/A
71	--	Not Used	--
72	BRN/GRN	Sensor 2 for Accelerator Pedal Position	N/A
73	YEL/LIL	Throttle Position Sensor	N/A
74	--	Not Used	--
75	BLU	Cruise Control Switch	N/A
76	RED/GRY	Cruise Control Switch	N/A
77	--	Not Used	--
78	--	Not Used	--
79	--	Not Used	--
80	WHT/YEL	Leak Detection Pump	N/A
81	--	Not Used	--
82	GRY	Engine Speed Sensor	N/A
83	BRN/WHT	Throttle Valve Control Module	N/A

2002–2005 Sedan 4.2L MFI, AWN and BAS, ECM Connector, *continued*

84	BLK/BRN	Throttle Valve Control Module	N/A
85	--	Not Used	--
86	GRN/LIL	Camshaft Position Sensor 2	N/A
87	BLU/LIL	Camshaft Position Sensor	N/A
88	LIL	Cylinder Four Fuel Injector	N/A
89	WHT/LIL	Cylinder Eight Fuel Injector	N/A
90	BLU	Engine Speed Sensor	N/A
91	YEL/BLK	Throttle Valve Control Module	N/A
92	BLK/BLU	Throttle Valve Control Module	N/A
93	GRY/BLU	Engine Coolant Temperature Sensor	N/A
94	GRN/WHT	Ignition Coil 4 with Power Output Stage	N/A
95	GRY/BLK	Ignition Coil 8 with Power Output Stage	N/A
96	BLK/RED	Cylinder One Fuel Injector	N/A
97	BRN/YEL	Cylinder Five Fuel Injector	N/A
98	GRN/BLK	Camshaft Position Sensor 2	N/A
99	WHT	Knock Sensor 1	N/A
100	--	Not Used	--
101	--	Not Used	--
102	WHT/GRN	Ignition Coil 1 with Power Output Stage	N/A
103	WHT/BRN	Ignition Coil 5 with Power Output Stage	N/A
104	--	Not Used	--
105	BLU/RED	Left Electro-Hydraulic Engine Mount Solenoid Valve	N/A
106	BRN	Knock Sensor 1	N/A
107	GRN	Knock Sensor 2	N/A
108	BRN/GRY	Sensor Ground	N/A
109	--	Not Used	--
110	GRN/GRY	Ignition Coil 5 with Power Output Stage	N/A
111	GRN/BRN	Ignition Coil 3 with Power Output Stage	N/A
112	BRN/BLU	Cylinder Six Fuel Injector	N/A
113	GRY/GRN	Cylinder Three Fuel Injector	N/A
114	--	Not Used	--
115	BRN/GRY	Valve 1 for Camshaft Adjustment	N/A
116	BLU/RED	Right Electro-Hydro Engine Mount Solenoid Valve	N/A
117	RED/LIL	Throttle Valve Control Module	N/A
118	BRN/LIL	Throttle Valve Control Module	N/A
119	--	Not Used	--
120	BRN/YEL	Valve 2 for Camshaft Adjustment	N/A
121	--	Not Used	--

A8

1997 Sedan 3.7L MFI AEW Connector

ECM Pin #	Wire Color	Circuit Description (68-Pin)	Value at Hot Idle
A2	RED/BLU	Transmission Control Module	N/A
A3	BLK/WHT	Throttle Position Sensor	N/A
A4	BRN/GRN	Fuel Pump Relay	N/A
A5	YEL/WHT	Intake Change Over Valve	N/A
A7	GRN/BLK	Thermotronic Control Module	N/A
A8	LIL	Evap Emission Canister Purge Regulator Valve	N/A
A9	BLU/GRN	Transmission Control Module	N/A
A10	YEL	Transmission Control Module	N/A
A12	GRN/LIL	ABS Control Module (w/EDL)	N/A
A17	BLU	Engine Speed Sensor	N/A
A18	GRY	Engine Speed Sensor	N/A
A19	BLK/BRN	Knock Sensor 1	N/A
A20	BRN	Knock Sensor 1	N/A
A21	GRN	Knock Sensor 2	N/A
A22	BLK/GRN	Intake Air Temperature Sensor	N/A
A23	LIL/YEL	Engine Coolant Temperature Sensor	N/A
A24	BLU/BRN	Throttle Position Sensor	N/A
B2	BLK/BLU	Camshaft Position Sensor	N/A
B3	BLK/GRY	Ground Connection	N/A
B4	BLK/YEL	Closed Throttle Position Switch	N/A
B5	GRY/LIL	Power Output Stages	N/A
B6	BLK/GRN	Power Output Stages	N/A
B7	GRY/BLU	Power Output Stages	N/A
B8	GRY/GRN	Power Output Stages	N/A
B11	GRN	Multi-Function Transmission Range Switch	N/A
B13	GRY/YEL	Power Output Stage 2	N/A
B14	GRY/RED	Power Output Stage 2	N/A
B15	BLK/LIL	Power Output Stage 2	N/A
B16	BLK/RED	Power Output Stage 2	N/A
B18	BRN/GRY	Mass Air Flow Sensor	N/A
B19	YEL/BRN	Mass Air Flow Sensor	N/A
B20	BRN	Heated Oxygen Sensor 1	N/A
B21	YEL	Heated Oxygen Sensor 2	N/A
B23	GRN	Heated Oxygen Sensor 2	N/A
B24	WHT	Heated Oxygen Sensor 1	N/A
C7	BLU/GRN	Leak Detection Pump	N/A
C8	GRY/GRN	Fuel Gauge	N/A
C9	BLU/WHT	Airbag Control Module	N/A
C10	YEL	Tachometer	N/A
C11	GRN/RED	A/C Control Head	N/A
C12	GRN/WHT	Thermotronic Control Module	N/A
C13	WHT/BLU	Instrument Cluster Combination Processor	N/A

1997 Sedan 3.7L MFI AEW Connector, *continued*

C14	RED/BLK	A/C Control Head	N/A
C16	GRN	Central Lock, Alarm, Interior Light Dimmer	N/A
C17	BRN	Oxygen Sensor (behind TWC)	N/A
C18	YEL	Heated Oxygen Sensor (behind TWC)	N/A
C20	GRN	Heated Oxygen Sensor (behind TWC)	N/A
C21	WHT	Oxygen Sensor (behind TWC)	N/A
C22	WHT	Rear Window Defogger Switch	N/A
C24	WHT/BLK	ABS Control Module (w/EDL)	N/A
D1	BRN/BLU	Idle Air Control Valve	N/A
D2	BLK/GRN	Cylinder Six Fuel Injector	N/A
D3	BLK	Cylinder Seven Fuel Injector	N/A
D4	BRN/YEL	Air Control Valve for Fuel Injectors	N/A
D5	BRN/RED	Idle Air Control Valve	N/A
D6	BLK/BLU	Cylinder Three Fuel Injector	N/A
D7	BLK/BRN	Cylinder Two Fuel Injector	N/A
D8	GRY/BLU	Leak Detection Pump	N/A
D9	RED	Transmission Control Module	N/A
D10	WHT	Plus Connection in the Fuel Injection Wiring Harness	N/A
D12	BRN	Ground Connection	N/A
D13	BRN	Ground Connection	N/A
E1	BLK/RED	Cylinder One Fuel Injector	N/A
E2	BLK/WHT	Cylinder Four Fuel Injector	N/A
E4	BLU/RED	Oxygen Sensor Heater	N/A
E5	BLK/GRY	Cylinder Five Fuel Injector	N/A
E6	BLK/YEL	Cylinder Eight Fuel Injector	N/A
E7	BRN	Ground Connection	N/A
E8	BLU/LIL	Oxygen Sensor Heater (behind TWC)	N/A
E10	BLU/RED	Oxygen Sensor Heater	N/A
E11	BRN	Ground Connection	N/A
E12	BRN	Ground Connection	N/A

A8

1997 Sedan 4.2L MFI AEW Connector

ECM Pin #	Wire Color	Circuit Description (68-Pin)	Value at Hot Idle
A2	RED/BLU	Transmission Control Module	N/A
A3	BLK/WHT	Throttle Position Sensor	N/A
A4	BRN/GRN	Fuel Pump Relay	N/A
A5	YEL/WHT	Intake Change Over Valve	N/A
A7	GRN/BLK	Thermotronic Control Module	N/A
A8	LIL	Evap Emission Canister Purge Regulator Valve	N/A
A9	BLU/GRN	Transmission Control Module	N/A
A10	YEL	Transmission Control Module	N/A
A17	BLU	Engine Speed Sensor	N/A
A18	GRY	Engine Speed Sensor	N/A
A19	BLK/BRN	Knock Sensor 1	N/A

1997 Sedan 4.2L MFI AEW Connector, *continued*

A20	BRN	Knock Sensor 1	N/A
A21	GRN	Knock Sensor 2	N/A
A22	BLK/GRN	Intake Air Temperature Sensor	N/A
A23	LIL/YEL	Engine Coolant Temperature Sensor	N/A
A24	BLU/BRN	Throttle Position Sensor	N/A
B2	BLK/BLU	Camshaft Position Sensor	N/A
B3	BLK/GRY	Ground Connection	N/A
B4	BLK/YEL	Closed Throttle Position Switch	N/A
B5	GRY/LIL	Power Output Stages	N/A
B6	BLK/GRN	Power Output Stages	N/A
B7	GRY/BLU	Power Output Stages	N/A
B8	GRY/GRN	Power Output Stages	N/A
B11	GRN	Multi-Function Transmission Range Switch	N/A
B13	GRY/YEL	Power Output Stage 2	N/A
B14	GRY/RED	Power Output Stage 2	N/A
B15	BLK/LIL	Power Output Stage 2	N/A
B16	BLK/RED	Power Output Stage 2	N/A
B18	BRN/GRY	Mass Air Flow Sensor	N/A
B19	YEL/BRN	Mass Air Flow Sensor	N/A
B20	BRN	Heated Oxygen Sensor 1	N/A
B21	YEL	Heated Oxygen Sensor 2	N/A
B23	GRN	Heated Oxygen Sensor 2	N/A
B24	WHT	Heated Oxygen Sensor 1	N/A
C8	GRY/GRN	Fuel Gauge	N/A
C9	BLU/WHT	Airbag Control Module	N/A
C11	GRN/RED	A/C Control Head	N/A
C14	RED/BLK	A/C Control Head	N/A
C16	GRN	Central Lock, Alarm, Interior Light Dimmer	N/A
C17	BRN	Oxygen Sensor (behind TWC)	N/A
C18	YEL	Heated Oxygen Sensor (behind TWC)	N/A
C20	GRN	Heated Oxygen Sensor (behind TWC)	N/A
C21	WHT	Oxygen Sensor (behind TWC)	N/A
C22	WHT	Rear Window Defogger Switch	N/A
D1	BRN/BLU	Idle Air Control Valve	N/A
D2	BLK/GRN	Cylinder Six Fuel Injector	N/A
D3	BLK	Cylinder Seven Fuel Injector	N/A
D4	BRN/YEL	Air Control Valve for Fuel Injectors	N/A
D5	BRN/RED	Idle Air Control Valve	N/A
D6	BLK/BLU	Cylinder Three Fuel Injector	N/A
D7	BLK/BRN	Cylinder Two Fuel Injector	N/A
D9	RED	Transmission Control Module	N/A
D10	WHT	Plus Connection in the Fuel Injection Wiring Harness	N/A
D11	BRN	Ground Connection	N/A
D12	BRN	Ground Connection	N/A
E1	BLK/RED	Cylinder One Fuel Injector	N/A
E2	BLK/WHT	Cylinder Four Fuel Injector	N/A
E4	BLU/RED	Oxygen Sensor Heater	N/A

1997 Sedan 4.2L MFI AEW Connector, *continued*

E5	BLK/GRY	Cylinder Five Fuel Injector	N/A
E6	BLK/YEL	Cylinder Eight Fuel Injector	N/A
E7	BRN	Ground Connection	N/A
E8	BLU/LIL	Oxygen Sensor Heater (behind TWC)	N/A
E9	YEL/BLU	Oxygen Sensor Heater 2	N/A
E10	BLU/RED	Oxygen Sensor Heater	N/A
E11	BRN	Ground Connection	N/A
E12	BRN	Ground Connection	N/A

A8

2000–2002 Sedan 4.2L MFI AKB, AUX, AYS ECM Connector

ECM Pin #	Wire Color	Circuit Description (121-Pin)	Value at Hot Idle
1	BRN	Ground Connection	N/A
2	BRN	Ground Connection	N/A
3	BLK/GRY	Wire Connectors for the Injectors	N/A
4	BRN/YEL	Oxygen Sensor Heater 2	N/A
5	BRN/WHT	Oxygen Sensor Heater	N/A
6	BRN/GRN	Oxygen Sensor 2 Heater	N/A
7	BLK/RED	Ignition Coil 2	N/A
8	GRY/GRN	Ignition Coil 7	N/A
9	--	Not Used	--
10	LIL	Heated Oxygen Sensor 2 (behind TWC)	N/A
11	RED	Heated Oxygen Sensor 2 (behind TWC)	N/A
12	WHT	Heated Oxygen Sensor 2	N/A
13	BRN	Heated Oxygen Sensor 2	N/A
14	--	Not Used	--
15	--	Not Used	--
16	--	Not Used	--
17	--	Not Used	--
18	--	Not Used	--
19	--	Not Used	--
20	--	Not Used	--
21	--	Not Used	--
22	--	Not Used	--
23	GRN	Cylinder Two Fuel Injector	N/A
24	GRN/WHT	Cylinder Seven Fuel Injector	N/A
25	GRY/BLU	Leak Detection Pump	N/A
26	--	Not Used	--
27	--	Not Used	--
28	--	Not Used	--
29	GRN	Mass Air Flow Sensor	N/A
30	--	Not Used	--
31	--	Not Used	--
32	BLK	Shielding Ground Connection	N/A
33	GRY/YEL	Sensor 2 for Accelerator Pedal Position	N/A

2000–2002 Sedan 4.2L MFI AKB, AUX, AYS ECM Connector, *continued*

34	BRN/GRN	Sensor 2 for Accelerator Pedal Position	N/A
35	YEL/BLU	Throttle Position Sensor	N/A
36	BRN/RED	Throttle Position Sensor	N/A
37	--	Not Used	--
38	BLK/WHT	Wire Connection (Cruise Control) in instrument panel	N/A
39	--	Not Used	--
40	RED/BLK	Air Control Head	N/A
41	GRN/RED	Air Control Head	N/A
42	--	Not Used	--
43	GRN/BLK	T-6 Connector Brown, in plenum chamber	N/A
44	YEL/GRN	Secondary Air Injection Solenoid Valve	N/A
45	GRN/RED	Intake Manifold Tuning Valve	N/A
46	BLU/GRY	Secondary Air Injection Pump Relay	N/A
47	--	Not Used	--
48	--	Not Used	--
49	--	Not Used	--
50	--	Not Used	--
51	GRN	Heated Oxygen Sensor	N/A
52	--	Not Used	--
53	LIL/GRY	Mass Air Flow Sensor	N/A
54	--	Not Used	--
55	WHT/RED	Brake Vacuum Vent Valve	N/A
56	RED/BLK	Brake Light Switch	N/A
57	RED/YEL	Cruise Control Switch	N/A
58	BRN/RED	Connection (low bus) in instrument panel	N/A
59	BLK	Connection (data bus shielding) in instrument panel	N/A
60	WHT/BRN	Connection (high bus) in instrument panel	N/A
61	--	Not Used	--
62	RED	Fuse Control for the Automatic Transmission	N/A
63	BRN/BLK	Oxygen Sensor Heater	N/A
64	LIL	Evap Emission Canister Purge Regulator Valve	N/A
65	BRN/GRN	Fuel Pump Relay	N/A
66	--	Not Used	--
67	WHT/BLU	Emergency Flasher Switch	N/A
68	GRY	Heated Oxygen Sensor (behind TWC)	N/A
69	BLU	Heated Oxygen Sensor (behind TWC)	N/A
70	YEL	Heated Oxygen Sensor	N/A
71	--	Not Used	--
72	GRY	Sensor 2 for Accelerator Pedal Position	N/A
73	YEL/LIL	Throttle Position Sensor	N/A
74	WHT/BLK	ABS Control Module	N/A
75	BLU	Cruise Control Switch	N/A
76	RED/GRY	Cruise Control Switch	N/A
77	--	Not Used	--
78	--	Not Used	--
79	--	Not Used	--
80	WHT/YEL	Leak Detection Pump	N/A

2000–2002 Sedan 4.2L MFI AKB, AUX, AYS ECM Connector, *continued*

81	--	Not Used	--
82	GRY	Engine Speed Sensor	N/A
83	GRN/LIL	Angle Sensor 2 for Throttle Drive	N/A
84	WHT/YEL	Angle Sensor 2 for Throttle Drive	N/A
85	GRN/RED	Mass Air Flow Sensor	N/A
86	GRN/BRN	Camshaft Position Sensor 1	N/A
87	GRN/BLK	Camshaft Position Sensor 2	N/A
88	BRN/GRN	Cylinder Four Fuel Injector	N/A
89	BLK/WHT	Cylinder Eight Fuel Injector	N/A
90	BLU	Engine Speed Sensor	N/A
91	GRY/WHT	Angle Sensor 1 for Throttle Drive	N/A
92	LIL/RED	Angle Sensor 1 for Throttle Drive	N/A
93	BRN/GRN	Engine Coolant Temperature Sensor	N/A
94	GRY/YEL	Ignition Coil 4	N/A
95	GRY/BRN	Ignition Coil 8	N/A
96	BLK/GRN	Cylinder One Fuel Injector	N/A
97	BRN/WHT	Cylinder Five Fuel Injector	N/A
98	RED/BLK	Camshaft Position Sensor 2	N/A
99	GRN/BRN	Knock Sensor 1	N/A
100	--	Not Used	--
101	--	Not Used	--
102	WHT/GRN	Ignition Coil	N/A
103	WHT/BRN	Ignition Coil 5	N/A
104	YEL/WHT	Intake Change-Over Valve	N/A
105	BLU	Right Electro-Hydro Engine Mount Solenoid Valve	N/A
106	BRN	Knock Sensor 1	N/A
107	GRN	Knock Sensor 2	N/A
108	GRY/WHT	Sensor Ground Connection	N/A
109	--	Not Used	--
110	GRN/GRY	Ignition Coil 6	N/A
111	GRN/BRN	Ignition Coil 3	N/A
112	BRN/BLU	Cylinder Six Fuel Injector	N/A
113	BLK/RED	Cylinder Three Fuel Injector	N/A
114	--	Not Used	--
115	BRN/GRY	Valve One for Camshaft Adjustment	N/A
116	BLU	Left Electro-Hydro Engine Mount Solenoid Valve	N/A
117	RED/LIL	Throttle Drive (power accelerator actuation)	N/A
118	BRN/LIL	Throttle Drive (power accelerator actuation)	N/A
119	--	Not Used	--
120	--	Not Used	--
121	--	Not Used	--

BMW
DIAGNOSTIC TROUBLE CODES

8

TABLE OF CONTENTS

DIAGNOSTIC TROUBLE CODES

OBD II Vehicle Applications

BMW AUTOMOBILES

318i
1996–1998
E36 Model, M44 Engine .... VIN CG73

318is, 318iC
1996–1998
E36 Model, M44 Engine .... VIN BE73, BH73

318ti
1996–1999
E36 Model, M44 Engine .... VIN CG73

323i, 323is
1998
E46 Model, M52TU Engine .... VIN AM33, BF73

325i, 325Ci
2000–2005
E46 Model, M54 Engine .... VIN ET37, BD33

325i, 325xi
2006
E46 Model, N52 Engine .... VIN ET37, EU33

325xi
2000–2004
E46 Model, M54 Engine .... VIN EU33

325xi, 330xi
2005
E46 Model, S43 Engine .... VIN EU33, EW53

325xiT
2000–2003
E46 Model, M54 Engine .... VIN EN33

328i, 328is, 328iC
1996–2000
E36 Model, M52 Engine .... VIN CD33, BG13, BK73

330i, 330Ci, 330xi
2000–2005
E46 Model, M54 Engine .... VIN EV53, BD53, EW53

330i, 330xi
2006
E46 Model, N52 Engine .... VIN EV53, EW53

525i, 530i
2000–2005
E46 Model, M54 Engine .... VIN DT33, DT53

525i, 525xi
2006
E60 Model, M54/N52 Engine .... VIN DT33, NF33

528i
1997–2000
E39 Model, M52 Engine .... VIN DM53

530i, 530xi
2006
E60 Model, N52 Engine .... VIN NA73, NF73

540i
1997–2003
E39 Model, M62 Engine .... VIN DN53

545i
2004–2005
E60 Model, N62 Engine .... VIN NB33

550i
2006
E60 Model, N52 Engine .... VIN NB53

740i, 740iL
1996–2001
E38 Model, M62 Engine .... VIN GG83, GJ83

745i, 745Li
2002–2005
E66 Model, N62 Engine .... VIN GL63, GN63

750iL
1996–2001
E38 Model, M73 Engine .... VIN GK23

750i, 750iL
2006
E65 Model, N62TU Engine .... VIN HL83, GK23

M3
1996–2005
E36 Model, S52/M52 Engines .... VIN BG93

M5
2000–2003
E39 Model, S62 Engine .... VIN DE93

Z3
1996–1998
E36 Model, M44 Engine .... VIN CD73

Z3 Coupe, Roadster
1999–2000
E36 Model, M52 Engine .... VIN CK53, CH03

Z3, Z3 Coupe
2001–2002
E36 Model, M54 Engine .... VIN CN33, CK73

Z4-2.5, Z4-3.0 Roadsters
2001–2006
E85 Model, M54 Engine .... VIN BT33, BT53

Z8
2000–2003
E39 Model, S62 Engine .... VIN DE93

GAS ENGINE TROUBLE CODE LIST

Introduction

To use this information, first read and record all codes in memory along with any Freeze Frame data. If the DME reset function is done prior to recording any data, all codes and freeze frame data will be lost! Look up the desired code by DTC number, Code Title and Conditions (enable criteria) that indicate why a code set, and how to drive the vehicle. 1T and 2T indicate a 1-trip or 2-trip fault and the Monitor type.

Gas Engine OBD II Trouble Code List (P0xxx Codes)

DTC	Trouble Code Title, Conditions & Possible Causes:
DTC: P0010 **2T MIL: Yes** **2001, 2002, 2003, 2004, 2005, 2006** **Models:** 320i, 325i, 330i, 525i, 540i, 545i, 550i, 740i, 745i, 745Li, 750i, Z4 **Engines:** 2.5L, 3.0L, 3.2L, 4.4L, 4.8L, **Transmissions:** All	**"A" Camshaft Position Actuator Circuit (Bank 1) Conditions:** Key on or engine running; and the DME detected an unexpected high voltage or low voltage condition on the camshaft position sensor. The relative position between the camshaft and crankshaft needs to be optimal so the engine has better torque, fuel economy and emissions. **Note: The camshaft adjustment is load- and RPM-dependant. The electrical camshaft adjustment valve 1 switches oil pressure onto camshaft adjuster (mechanical adjustment mechanism), which adjusts the camshaft.** **Possible Causes:** • Fuel pump has failed • Actuator circuit is open • Battery voltage below 11.5 volts • Position actuator circuit may short to B+ or Ground
DTC: P0011 **2T MIL: Yes** **2001, 2002, 2003, 2004, 2005, 2006** **Models:** 320i, 325i, 330i, 525i, 540i, 545i, 550i, 740i, 745i, 745Li, 750i, Z4 **Engines:** 2.5L, 3.0L, 3.2L, 4.4L, 4.8L, **Transmissions:** All	**"A" Camshaft Position Timing Over-Advanced (Bank 1) Conditions:** Engine started and driven at an engine speed of more than 400rpm; and the DME detected the camshaft timing exceeded the maximum calibrated advance value, or the camshaft remained in an advanced position during the CCM test. The valve timing did not change from the current valve timing or it remained fixed during the testing. **Note: The camshaft adjustment is load- and RPM-dependant. The electrical camshaft adjustment valve 1 switches oil pressure onto camshaft adjuster (mechanical adjustment mechanism), which adjusts the camshaft.** **Possible Causes:** • Fuel pump has failed • CPS circuit is open, shorted to ground or shorted to power • Battery voltage below 11.5 volts • Position actuator circuit may short to B+ or Ground • Camshaft timing improperly set, or continuous oil flow to the VCT piston chamber • Camshaft advance mechanism (the VCT unit) is sticking or binding mechanically • VCT solenoid valve is stuck in open position
DTC: P0012 **2T MIL: Yes** **2001, 2002, 2003, 2004, 2005, 2006** **Models:** 320i, 325i, 330i, 525i, 540i, 545i, 550i, 740i, 745i, 745Li, 750i, Z4 **Engines:** 2.5L, 3.0L, 3.2L, 4.4L, 4.8L, **Transmissions:** All	**"A" Camshaft Position Over-Retarded (Bank 1) Conditions:** Engine started and driven at an engine speed of more than 400rpm; and the DME detected the camshaft timing exceeded the minimum calibrated retarded value, or the camshaft remained in an retarded position during the CCM test. The valve timing did not change from the current valve timing or it remained fixed during the testing. Note: The camshaft adjustment is load- and RPM dependant. The electrical camshaft adjustment valve 1 switches oil pressure onto camshaft adjuster (mechanical adjustment mechanism), which adjusts the camshaft. **Possible Causes:** • Fuel pump has failed • CPS circuit is open, shorted to ground or shorted to power • Battery voltage below 11.5 volts • Position actuator circuit may short to B+ or Ground • Camshaft timing improperly set, or continuous oil flow to the VCT piston chamber • Camshaft advance mechanism (the VCT unit) is sticking or binding mechanically • VCT solenoid valve is stuck in open position
DTC: P0013 **2T MIL: Yes** **2001, 2002, 2003, 2004, 2005, 2006** **Models:** 320i, 325i, 330i, 525i, 540i, 545i, 550i, 740i, 745i, 745Li, 750i, M3, Z4 **Engines:** 2.5L, 3.0L, 3.2L, 4.4L, 4.8L, **Transmissions:** All	**"B" Camshaft Position Actuator Circuit (Bank 1) Conditions:** Key on or engine running; and the DME detected an unexpected high voltage or low voltage condition on the camshaft position sensor. The relative position between the camshaft and crankshaft needs to be optimal so the engine has better torque, fuel economy and emissions. Note: The camshaft adjustment is load- and RPM dependant. The electrical camshaft adjustment valve 1 switches oil pressure onto camshaft adjuster (mechanical adjustment mechanism), which adjusts the camshaft. **Possible Causes:** • Fuel pump has failed • Battery voltage below 11.5 volts • Position actuator circuit may short to B+ or Ground

DTC	Trouble Code Title, Conditions & Possible Causes
DTC: P0014 **2T MIL: Yes** **2001, 2002, 2003, 2004, 2005, 2006** **Models:** 320i, 325i, 330i, 525i, 540i, 545i, 550i, 740i, 745i, 745Li, 750i, Z3, Z4 **Engines:** 2.5L, 3.0L, 3.2L, 4.4L, 4.8L, 5.0L **Transmissions:** All	**"B" Camshaft Position Timing Over-Advanced (Bank 1) Conditions:** Engine started and driven at an engine speed of more than 400rpm; and the DME detected the camshaft timing exceeded the maximum calibrated advance value, or the camshaft remained in an advanced position during the CCM test. The valve timing did not change from the current valve timing or it remained fixed during the testing. The VANOS is in the end position. **Note: The camshaft adjustment is load- and RPM dependant. The electrical camshaft adjustment valve 1 switches oil pressure onto camshaft adjuster (mechanical adjustment mechanism), which adjusts the camshaft.** **Possible Causes:** • Fuel pump has failed • CPS circuit is open, shorted to ground or shorted to power • Battery voltage below 11.5 volts • Position actuator circuit may short to B+ or Ground • Camshaft timing improperly set, or continuous oil flow to the VCT piston chamber • Camshaft advance mechanism (the VCT unit) is sticking or binding mechanically • VCT solenoid valve is stuck in open position
DTC: P0015 **2T MIL: Yes** **2001, 2002, 2003, 2004, 2005, 2006** **Models:** 320i, 325i, 330i, 525i, 540i, 545i, 550i, 740i, 745i, 745Li, 750i, Z3, Z4 **Engines:** 2.5L, 3.0L, 3.2L, 4.4L, 4.8L, 5.0L **Transmissions:** All	**"B" Camshaft Position Over-Retarded (Bank 1) Conditions:** Engine started and driven at an engine speed of more than 400rpm; and the DME detected the camshaft timing exceeded the minimum calibrated retarded value, or the camshaft remained in an retarded position during the CCM test. The valve timing did not change from the current valve timing or it remained fixed during the testing. **Note: The camshaft adjustment is load- and RPM dependant. The electrical camshaft adjustment valve 1 switches oil pressure onto camshaft adjuster (mechanical adjustment mechanism), which adjusts the camshaft.** **Possible Causes:** • Fuel pump has failed • CPS circuit is open, shorted to ground or shorted to power • Battery voltage below 11.5 volts • Position actuator circuit may short to B+ or Ground • Camshaft timing improperly set, or continuous oil flow to the VCT piston chamber • Camshaft advance mechanism (the VCT unit) is sticking or binding mechanically • VCT solenoid valve is stuck in open position
DTC: P0016 **2T MIL: Yes** **2006** **Models:** 325i, 330i, 525i, 530i, 550i, 750i **Engines:** 3.0L, 4.8L **Transmissions:** All	**Crankshaft Position - Camshaft Position Correlation Bank 1 Sensor A Conditions:** Engine started, engine running, and the DME detected a deviation between the crankshaft position sensor signal and the camshaft position sensor. A rationality error has been detected for camshaft position out of phase with crankshaft. **Possible Causes:** • Camshaft Position (CMP) sensor is faulty • CMP circuit short to ground, power or open • Engine Speed (RPM) sensor is faulty
DTC: P0017 **2T MIL: Yes** **2003, 2005, 2006** **Models:** 325i, 330i, 525i, 530i, 550i, 750i, M3, M5, Z4, Z8 **Engines:** 3.0L, 3.2L, 4.8L, 5.0L **Transmissions:** All	**Crankshaft Position - Camshaft Position Correlation Bank 1 Sensor B Conditions:** Engine started, engine running, and the DME detected a deviation between the crankshaft position sensor signal and the camshaft position sensor. A rationality error has been detected for camshaft position out of phase with crankshaft. **Possible Causes:** • Camshaft Position (CMP) sensor is faulty • CMP circuit short to ground, power or open • Engine Speed (RPM) sensor is faulty
DTC: P0018 **2T MIL: Yes** **2006** **Models:** 325i, 330i, 525i, 530i, 550i, 750i **Engines:** 3.0L, 4.8L **Transmissions:** All	**Crankshaft Position - Camshaft Position Correlation Bank 2 Sensor A Conditions:** Engine started, engine running, and the DME detected a deviation between the crankshaft position sensor signal and the camshaft position sensor. A rationality error has been detected for camshaft position out of phase with crankshaft. **Possible Causes:** • Camshaft Position (CMP) sensor is faulty • CMP circuit short to ground, power or open • Engine Speed (RPM) sensor is faulty
DTC: P0019 **2T MIL: Yes** **2006** **Models:** 325i, 330i, 525i, 530i, 550i, 750i **Engines:** 3.0L, 4.8L **Transmissions:** All	**Crankshaft Position - Camshaft Position Correlation Bank 2 Sensor B Conditions:** Engine started, engine running, and the DME detected a deviation between the crankshaft position sensor signal and the camshaft position sensor. A rationality error has been detected for camshaft position out of phase with crankshaft. **Possible Causes:** • Camshaft Position (CMP) sensor is faulty • CMP circuit short to ground, power or open • Engine Speed (RPM) sensor is faulty

DTC	Trouble Code Title, Conditions & Possible Causes
DTC: P0020 **2T MIL: Yes** **2001, 2002, 2003, 2004, 2005, 2006** **Models:** 325i, 330i, 525i, 540i, 545i, 550i, 740i, 745i, 745Li, 750i, M5, Z4, Z8 **Engines:** 3.0L, 4.4L, 4.8L, 5.0L **Transmissions:** All	**"A" Camshaft Position Timing Over-Advanced (Bank 2) Conditions:** Engine started and driven at an engine speed of more than 400rpm; and the DME detected the camshaft timing exceeded the maximum calibrated advance value, or the camshaft remained in an advanced position during the CCM test. The valve timing did not change from the current valve timing or it remained fixed during the testing. **Possible Causes:** • Fuel pump has failed • CPS circuit is open, shorted to ground or shorted to power • Battery voltage below 11.5 volts • Position actuator circuit may short to B+ or Ground • Camshaft timing improperly set, or continuous oil flow to the VCT piston chamber • Camshaft advance mechanism (the VCT unit) is sticking or binding mechanically • VCT solenoid valve is stuck in open position
DTC: P0021 **2T MIL: Yes** **2001, 2002, 2003, 2004, 2005, 2006** **Models:** 320i, 325i, 330i, 525i, 540i, 545i, 550i, 740i, 745i, 745Li, 750i, Z3, Z4 **Engines:** 2.5L, 3.0L, 3.2L, 4.4L, 4.8L, 5.0L **Transmissions:** All	**"A" Camshaft Position Actuator Circuit (Bank 2) Conditions:** Key on or engine running; and the DME detected an unexpected high voltage or low voltage condition on the camshaft position sensor. The relative position between the camshaft and crankshaft needs to be optimal so the engine has better torque, fuel economy and emissions. **Possible Causes:** • Fuel pump has failed • Actuator circuit is open, shorted to ground or shorted to power • Battery voltage below 11.5 volts • Position actuator circuit may short to B+ or Ground
DTC: P0022 **2T MIL: Yes** **2001, 2002, 2003, 2004, 2005, 2006** **Models:** 320i, 325i, 330i, 525i, 540i, 545i, 550i, 740i, 745i, 745Li, 750i, Z3, Z4 **Engines:** 2.5L, 3.0L, 3.2L, 4.4L, 4.8L, 5.0L **Transmissions:** All	**"A" Camshaft Position Over-Retarded (Bank 2) Conditions:** Engine started and driven at an engine speed of more than 400rpm; and the DME detected the camshaft timing exceeded the minimum calibrated retarded value, or the camshaft remained in an retarded position during the CCM test. The valve timing did not change from the current valve timing or it remained fixed during the testing. **Possible Causes:** • Fuel pump has failed • CPS circuit is open, shorted to ground or shorted to power • Battery voltage below 11.5 volts • Position actuator circuit may short to B+ or Ground • Camshaft timing improperly set, or continuous oil flow to the VCT piston chamber • Camshaft advance mechanism (the VCT unit) is sticking or binding mechanically • VCT solenoid valve is stuck in open position
DTC: P0023 **2T MIL: Yes** **2005, 2006** **Models:** 325i, 330i, 525i, 530i, 545i, 550i, 745i, 750i **Engines:** 3.0L, 4.8L **Transmissions:** All	**"B" Camshaft Position Actuator Circuit (Bank 2) Conditions:** Key on or engine running; and the DME detected an unexpected high voltage or low voltage condition on the camshaft position sensor. The relative position between the camshaft and crankshaft needs to be optimal so the engine has better torque, fuel economy and emissions. **Possible Causes:** • Fuel pump has failed • Actuator circuit is open, shorted to ground or shorted to power • Battery voltage below 11.5 volts • Position actuator circuit may short to B+ or Ground
DTC: P0024 **2T MIL: Yes** **2001, 2002, 2003, 2004, 2005, 2006** **Models:** 325i, 330i, 525i, 545i, 550i, 745i, 745Li, 750i, M5, Z3, Z4, Z8 **Engines:** 2.5L, 3.0L, 3.2L, 4.4L, 4.8L, 5.0L **Transmissions:** All	**"B" Camshaft Position Timing Over-Advanced (Bank 2) Conditions:** Engine started and driven at an engine speed of more than 400rpm; and the DME detected the camshaft timing exceeded the maximum calibrated advance value, or the camshaft remained in an advanced position during the CCM test. The valve timing did not change from the current valve timing or it remained fixed during the testing. The engine speed should be more than 500rpm and the VANOS is in the end position. **Possible Causes:** • Fuel pump has failed • CPS circuit is open, shorted to ground or shorted to power • Battery voltage below 11.5 volts • Position actuator circuit may short to B+ or Ground • Camshaft timing improperly set, or continuous oil flow to the VCT piston chamber • Camshaft advance mechanism (the VCT unit) is sticking or binding mechanically • VCT solenoid valve is stuck in open position

DTC	Trouble Code Title, Conditions & Possible Causes
DTC: P0025 **2T MIL: Yes** **2001, 2002, 2003, 2004, 2005, 2006** **Models:** 325i, 330i, 525i, 545i, 550i, 745i, 745Li, 750i, M5, Z3, Z4, Z8 **Engines:** 2.5L, 3.0L, 3.2L, 4.4L, 4.8L, 5.0L **Transmissions:** All	**"B" Camshaft Position Over-Retarded (Bank 2) Conditions:** Engine started and driven at an engine speed of more than 600rpm; and the DME detected the camshaft timing exceeded the minimum calibrated retarded value, or the camshaft remained in an retarded position during the CCM test. The valve timing did not change from the current valve timing or it remained fixed during the testing. **Possible Causes:** • Fuel pump has failed • CPS circuit is open, shorted to ground or shorted to power • Battery voltage below 11.5 volts • Position actuator circuit may short to B+ or Ground • Camshaft timing improperly set, or continuous oil flow to the VCT piston chamber • Camshaft advance mechanism (the VCT unit) is sticking or binding mechanically • VCT solenoid valve is stuck in open position
DTC: P0030 **2T MIL: Yes** **2001, 2002, 2003, 2004, 2005, 2006** **Models:** 325i, 330i, 525i, 540i, 545i, 550i, 740i, 745i, 745Li, 750i, M5, Z4, Z8 **Engines:** 3.0L, 4.4L, 4.8L, 5.0L **Transmissions:** All	**HO2S Heater (Bank 1 Sensor 1) Control Circuit Malfunction Conditions:** Engine started, battery voltage must be at least 11.5v, all electrical components must be off, the ground between the engine and the chassis must be well connected, the exhaust system must be properly sealed between the catalytic converter and the cylinder head, the coolant temperature must be 80 degrees Celsius, and the oxygen sensor heater for oxygen sensor before the catalytic converter must be properly functioning. The DME detected the HO2S signal was in a negative voltage range referred to as "character shift downward". This code sets when the HO2S signal remains in a low state (usually less than 156 mv). In effect, it does not switch properly between 0.1v and 1.1v in closed loop operation. **Possible Causes:** • HO2S is contaminated (due to presence of silicone in fuel) • HO2S signal and ground circuit wires crossed in wiring harness • HO2S signal circuit is shorted to sensor or chassis ground • HO2S element has failed (internal short condition)
DTC: P0031 **2T MIL: Yes** **2001, 2002, 2003, 2004, 2005, 2006** **Models:** 325i, 330i, 525i, 540i, 545i, 550i, 740i, 745i, 745Li, 750i, M5, Z4, Z8 **Engines:** 3.0L, 4.4L, 4.8L, 5.0L **Transmissions:** All	**HO2S Heater (Bank 1 Sensor 1) Circuit Low Input Conditions:** Engine started, battery voltage must be at least 11.5v, all electrical components must be off, the ground between the engine and the chassis must be well connected, the exhaust system must be properly sealed between the catalytic converter and the cylinder head, the coolant temperature must be 80 degrees Celsius, and the oxygen sensor heater for oxygen sensor before the catalytic converter must be properly functioning. The DME detected the HO2S signal was in a negative voltage range referred to as "character shift downward". This code sets when the HO2S signal remains in a low state. In effect, it does not switch properly in the closed loop operation. The HO2S (before the three-way catalytic converter) has a short circuit to ground that has lasted longer than 200 seconds. **Possible Causes:** • HO2S is contaminated (due to presence of silicone in fuel) • HO2S signal and ground circuit wires crossed in wiring harness • HO2S signal circuit is shorted to sensor or chassis ground • HO2S element has failed (internal short condition)
DTC: P0032 **2T MIL: Yes** **2001, 2002, 2003, 2004, 2005, 2006** **Models:** 325i, 330i, 525i, 540i, 545i, 550i, 740i, 745i, 745Li, 750i, M5, Z4, Z8 **Engines:** 3.0L, 4.4L, 4.8L, 5.0L **Transmissions:** All	**HO2S Heater (Bank 1 Sensor 1) Circuit High Input Conditions:** Engine started, battery voltage must be at least 11.5v, all electrical components must be off, the ground between the engine and the chassis must be well connected, the exhaust system must be properly sealed between the catalytic converter and the cylinder head, the coolant temperature must be 80 degrees Celsius, and the oxygen sensor heater for oxygen sensor before the catalytic converter must be properly functioning. The DME detected the HO2S signal remained in a high state. Note: The HO2S signal circuit may be shorted to the heater power circuit due to tracking inside of the HO2S connector. Remove the connector and visually inspect the connector for signs of oil or water. **Possible Causes:** • HO2S signal shorted to heater power circuit inside connector • HO2S signal circuit shorted to ground or to system voltage
DTC: P0036 **2T MIL: Yes** **2002, 2003, 2004, 2005, 2006** **Models:** 320i, 325i, 330i, 525i, 530i, 540i, 545i, 550i, 740i, 745i, 745Li, 750i, M3, M5, Z3, Z4, Z8 **Engines:** 2.5L, 3.0L, 3.2L, 4.4L, 4.8L, 5.0L **Transmissions:** All	**HO2S Heater (Bank 1 Sensor 2) Control Circuit Malfunction Conditions:** Engine started, battery voltage must be at least 11.5v, all electrical components must be off, the ground between the engine and the chassis must be well connected, the exhaust system must be properly sealed between the catalytic converter and the cylinder head, the coolant temperature must be 80 degrees Celsius, and the oxygen sensor heater for oxygen sensor before the catalytic converter must be properly functioning. The DME detected the HO2S signal was in a negative voltage range referred to as "character shift downward". This code sets when the HO2S signal remains in a low state. **Possible Causes:** • HO2S is contaminated (due to presence of silicone in fuel) • HO2S signal and ground circuit wires crossed in wiring harness • HO2S signal circuit is shorted to sensor or chassis ground • HO2S element has failed (internal short condition)

DTC	Trouble Code Title, Conditions & Possible Causes
DTC: P0037 **2T MIL: Yes** **2001, 2002, 2003, 2004, 2005, 2006** **Models:** 325i, 330i, 525i, 540i, 545i, 550i, 740i, 745i, 745Li, 750i, M5, Z4, Z8 **Engines:** 3.0L, 4.4L, 4.8L, 5.0L **Transmissions:** All	**HO2S Heater (Bank 1 Sensor 2) Circuit Low Input Conditions:** Engine started, battery voltage must be at least 11.5v, all electrical components must be off, the ground between the engine and the chassis must be well connected, the exhaust system must be properly sealed between the catalytic converter and the cylinder head, the coolant temperature must be 80 degrees Celsius, and the oxygen sensor heater for oxygen sensor before the catalytic converter must be properly functioning. The DME detected the HO2S signal was in a negative voltage range referred to as "character shift downward". This code sets when the HO2S signal remains in a low state. In effect, it does not switch properly in the closed loop operation. The HO2S (before the three-way catalytic converter) has a short circuit to ground that has lasted longer than 200 seconds. **Possible Causes:** • HO2S is contaminated (due to presence of silicone in fuel) • HO2S signal and ground circuit wires crossed in wiring harness • HO2S signal circuit is shorted to sensor or chassis ground • HO2S element has failed (internal short condition)
DTC: P0038 **2T MIL: Yes** **2001, 2002, 2003, 2004, 2005, 2006** **Models:** 325i, 330i, 525i, 540i, 545i, 550i, 740i, 745i, 745Li, 750i, M5, Z4, Z8 **Engines:** 3.0L, 4.4L, 4.8L, 5.0L **Transmissions:** All	**HO2S Heater (Bank 1 Sensor 2) Circuit High Input Conditions:** Engine started, battery voltage must be at least 11.5v, all electrical components must be off, the ground between the engine and the chassis must be well connected, the exhaust system must be properly sealed between the catalytic converter and the cylinder head, the coolant temperature must be 80 degrees Celsius, and the oxygen sensor heater for oxygen sensor before the catalytic converter must be properly functioning. The DME detected the HO2S signal remained in a high state. Note: The HO2S signal circuit may be shorted to the heater power circuit due to tracking inside of the HO2S connector. Remove the connector and visually inspect the connector for signs of oil or water. **Possible Causes:** • HO2S signal shorted to heater power circuit inside connector • HO2S signal circuit shorted to ground or to system voltage
DTC: P0040 **2T MIL: Yes** **2002, 2005, 2006** **Models:** 325i, 330i, 525i, 530i, 545i, 550i, 745i, 745Li, 750i **Engines:** 3.0L, 4.8L **Transmissions:** All	**O2 Sensor Signals Swapped (Bank 1 Sensor 1/Bank 2 Sensor 1) Conditions:** Engine started, battery voltage must be at least 11.5v, all electrical components must be off, the ground between the engine and the chassis must be well connected, the exhaust system must be properly sealed between the catalytic converter and the cylinder head, and the coolant temperature must be 80 degrees Celsius. The DME detected the O2 signals were mixed and reading implausible results from both. The Lambda controllers for the two cylinder banks display mutual displays exceeding 20 percent. The conditions for monitoring faults must be present longer than 10 seconds from the point at which the lambda control assumes active operation. **Possible Causes:** • HO2S-11 and HO2S-21 harness connectors are swapped • HO2S-11 and HO2S-21 wiring is crossed inside the harness • HO2S-11 and HO2S-21 wires are crossed at 104-pin connector • Connector coding and color mixed with correct catalytic converter
DTC: P0050 **2T MIL: Yes** **2002, 2003, 2004, 2005, 2006** **Models:** 320i, 325i, 330i, 525i, 530i, 540i, 545i, 550i, 740i, 745i, 745Li, 750i, M3, M5, Z3, Z4, Z8 **Engines:** 2.5L, 3.0L, 3.2L, 4.4L, 4.8L, 5.0L **Transmissions:** All	**HO2S Heater (Bank 2 Sensor 1) Control Circuit Malfunction Conditions:** Engine started, battery voltage must be at least 11.5v, all electrical components must be off, the ground between the engine and the chassis must be well connected, the exhaust system must be properly sealed between the catalytic converter and the cylinder head, and the coolant temperature must be 80 degrees Celsius. The DME detected the HO2S signal was in a negative voltage range referred to as "character shift downward". **Possible Causes:** • HO2S is contaminated (due to presence of silicone in fuel) • HO2S signal and ground circuit wires crossed in wiring harness • HO2S signal circuit is shorted to sensor or chassis ground • HO2S element has failed (internal short condition)
DTC: P0051 **2T MIL: Yes** **2002, 2003, 2004, 2005, 2006** **Models:** 320i, 325i, 330i, 525i, 530i, 540i, 545i, 550i, 740i, 745i, 745Li, 750i, M3, M5, Z3, Z4, Z8 **Engines:** 2.5L, 3.0L, 3.2L, 4.4L, 4.8L, 5.0L **Transmissions:** All	**HO2S Heater (Bank 2 Sensor 1) Circuit Low Input Conditions:** Engine started, battery voltage must be at least 11.5v, all electrical components must be off, the ground between the engine and the chassis must be well connected, the exhaust system must be properly sealed between the catalytic converter and the cylinder head, and the coolant temperature must be 80 degrees Celsius. The DME detected the HO2S signal was in a negative voltage range referred to as "character shift downward". This code sets when the HO2S signal remains in a low state. In effect, it does not switch properly in the closed loop operation. The HO2S (before the three-way catalytic converter) has a short circuit to ground that has lasted longer than a specified time. **Possible Causes:** • HO2S is contaminated (due to presence of silicone in fuel) • HO2S signal and ground circuit wires crossed in wiring harness • HO2S signal circuit is shorted to sensor or chassis ground • HO2S element has failed (internal short condition)

DTC	Trouble Code Title, Conditions & Possible Causes
DTC: P0052 **2T MIL: Yes** **2002, 2003, 2004, 2005, 2006** **Models:** 320i, 325i, 330i, 525i, 530i, 540i, 545i, 550i, 740i, 745i, 745Li, 750i, M3, M5, Z3, Z4, Z8 **Engines:** 2.5L, 3.0L, 3.2L, 4.4L, 4.8L, 5.0L **Transmissions:** All	**HO2S Heater (Bank 2 Sensor 1) Circuit High Input Conditions:** Engine started, battery voltage must be at least 11.5v, all electrical components must be off, the ground between the engine and the chassis must be well connected, the exhaust system must be properly sealed between the catalytic converter and the cylinder head, and the coolant temperature must be 80 degrees Celsius. The DME detected the HO2S signal was in a negative voltage range referred to as "character shift downward". This code sets when the HO2S signal remains in a low state. In effect, it does not switch properly in the closed loop operation. The HO2S (before the three-way catalytic converter) has a short circuit to ground that has lasted longer than a specified time. **Possible Causes:** • HO2S is contaminated (due to presence of silicone in fuel) • HO2S signal and ground circuit wires crossed in wiring harness • HO2S signal circuit is shorted to sensor or chassis ground • HO2S element has failed (internal short condition)
DTC: P0056 **2T MIL: Yes** **2002, 2003, 2004, 2005, 2006** **Models:** 320i, 325i, 330i, 525i, 530i, 540i, 545i, 550i, 740i, 745i, 745Li, 750i, M3, M5, Z3, Z4, Z8 **Engines:** 2.5L, 3.0L, 3.2L, 4.4L, 4.8L, 5.0L **Transmissions:** All	**HO2S Heater (Bank 2 Sensor 2) Circuit High Input Conditions:** Engine started, battery voltage must be at least 11.5v, all electrical components must be off, the ground between the engine and the chassis must be well connected, the exhaust system must be properly sealed between the catalytic converter and the cylinder head, and the coolant temperature must be 80 degrees Celsius. The DME detected the HO2S signal remained in a high state. Note: The HO2S signal circuit may be shorted to the heater power circuit due to tracking inside of the HO2S connector. Remove the connector and visually inspect the connector for signs of oil or water. **Possible Causes:** • HO2S signal shorted to heater power circuit inside connector • HO2S signal circuit shorted to ground or to system voltage
DTC: P0057 **2T MIL: Yes** **2002, 2003, 2004, 2005, 2006** **Models:** 320i, 325i, 330i, 525i, 530i, 540i, 545i, 550i, 740i, 745i, 745Li, 750i, M3, M5, Z3, Z4, Z8 **Engines:** 2.5L, 3.0L, 3.2L, 4.4L, 4.8L, 5.0L **Transmissions:** All	**HO2S Heater (Bank 2 Sensor 2) Control Circuit Malfunction Conditions:** Engine started, battery voltage must be at least 11.5v, all electrical components must be off, the ground between the engine and the chassis must be well connected, the exhaust system must be properly sealed between the catalytic converter and the cylinder head, and the coolant temperature must be 80 degrees Celsius. The DME detected the HO2S signal was in a negative voltage range referred to as "character shift downward". **Possible Causes:** • HO2S is contaminated (due to presence of silicone in fuel) • HO2S signal and ground circuit wires crossed in wiring harness • HO2S signal circuit is shorted to sensor or chassis ground • HO2S element has failed (internal short condition)
DTC: P0058 **2T MIL: Yes** **2002, 2003, 2004, 2005, 2006** **Models:** 320i, 325i, 330i, 525i, 530i, 540i, 545i, 550i, 740i, 745i, 745Li, 750i, M3, M5, Z3, Z4, Z8 **Engines:** 2.5L, 3.0L, 3.2L, 4.4L, 4.8L, 5.0L **Transmissions:** All	**HO2S Heater (Bank 2 Sensor 2) Circuit Low Input Conditions:** Engine started, battery voltage must be at least 11.5v, all electrical components must be off, the ground between the engine and the chassis must be well connected, the exhaust system must be properly sealed between the catalytic converter and the cylinder head, and the coolant temperature must be 80 degrees Celsius. The DME detected the HO2S signal was in a negative voltage range referred to as "character shift downward". This code sets when the HO2S signal remains in a low state. In effect, it does not switch properly in the closed loop operation. The HO2S (before the three-way catalytic converter) has a short circuit to ground that has lasted longer than a specified time. The difference between the outside and coolant temperature is greater than 3 degrees Celsius. **Possible Causes:** • HO2S is contaminated (due to presence of silicone in fuel) • HO2S signal and ground circuit wires crossed in wiring harness • HO2S signal circuit is shorted to sensor or chassis ground • HO2S element has failed (internal short condition)
DTC: P0070 **2T MIL: Yes** 2005 **Models:** 325i, 330i, **Engines:** S43 **Transmissions:** All	**Ambient Air Temperature Sensor Malfunction Conditions:** Key on or engine running (at over 800rpm), the vehicle velocity is over 25mph for 26 seconds, the ambient temperature is 20 degrees above or below the model figure for four seconds. This is a thermistor-type sensor with a variable resistance that changes when exposed to different temperatures. This means: the higher the temperature, the lower the resistance value. **Possible Causes:** • IAT sensor signal circuit is grounded (check wiring & connector) • Resistance value between sockets 33 and 36 out of range • IAT sensor has an open circuit • IAT sensor is damaged or it has failed • Ambient temperature sensor at the cluster is defective

DTC	Trouble Code Title, Conditions & Possible Causes
DTC: P0071 **2T MIL: Yes** **2005, 2006** **Models:** 325i, 330i, 525i, 530i, 545i, 550i, 745i, 745Li, 750i, M3 **Engines:** 3.0L, 3.2L, 4.8L **Transmissions:** All	**Ambient Air Temperature Sensor Range/Performance Conditions:** Key on or engine running (at over 800rpm), the vehicle velocity is over 25mph, the ambient temperature is 20 degrees above or below the model figure for four seconds. This is a thermistor-type sensor with a variable resistance that changes when exposed to different temperatures. This means: the higher the temperature, the lower the resistance value. **Possible Causes:** • IAT sensor signal circuit is grounded (check wiring & connector) • Resistance value between sockets 33 and 36 out of range • IAT sensor has an open circuit • IAT sensor is damaged or it has failed • Ambient temperature sensor at the cluster is defective
DTC: P0100 **2T MIL: Yes** **1996, 1997, 1998, 1999, 2000, 2006** **Models:** 318i, 328i, 325i, 330i, 525i, 530i, 540i, 550i, 740i, 750i, 850i, M5, Z3, Z4, Z8 **Engines:** 1.9L, 3.0L, 4.4L, 4.89L, 5.0L, 5.4L **Transmissions:** All	**Mass or Volume Air Flow Circuit "A" Conditions** Engine running, with the system voltage more than 11.0v, and the temperature must be at least 185-degrees (F) and all electrical equipment (A/C, lights, etc) must be off. The DME has detected that the MAF signal was out of a calculated range with the engine (or undetectable) for a certain period of time. The engine speed is greater than 200rpm. **Possible Causes:** • Mass air flow (MAF) sensor has failed or is damaged • Signal and ground wires of Mass Air Flow (MAF) sensor has short circuited
DTC: P0101 **2T MIL: Yes** **1996, 1997, 1998, 1999, 2000, 2001, 2002, 2003, 2004, 2005, 2006** **Models:** 318i, 320i, 323i, 325i, 328i, 330i, 525i, 528i, 530i, 540i, 545i, 550i, 740i, 745i, 745Li, 750i, 840i, M3, M5, Z3, Z4, Z8 **Engines:** 1.9L, 2.5L, 2.8L, 3.0L, 3.2L, 4.4L, 4.8L, 5.0L **Transmissions:** All	**Mass or Volume Air Flow Circuit Range/Performance Conditions** Engine running, with the system voltage more than 11.0v, and the temperature must be at least 185-degrees (F) and all electrical equipment (A/C, lights, etc) must be off. The DME has detected that the MAF signal was out of a calculated range with the engine (or undetectable) for a certain period of time. **Possible Causes:** • Mass air flow (MAF) sensor has failed or is damaged • Signal and ground wires of Mass Air Flow (MAF) sensor has short circuited
DTC: P0102 **2T MIL: Yes** **2001, 2002, 2003, 2004, 2005, 2006** **Models:** 320i, 325i, 330i, 525i, 540i, 545i, 550i, 740i, 745i, 745Li, 750i, M3, M5, Z3 **Engines:** 2.5L, 3.0L, 3.2L, 4.4L, 4.8L **Transmissions:** All	**MAF Sensor Circuit Low Input Conditions:** Key on, engine started, and the DME detected the MAF sensor signal was less than the minimum calibrated value. The engine temperature must beat least 185-degrees (F) and all electrical equipment (A/C, lights, etc) must be off. The DME has detected that the MAF signal was less than the required minimum. The engine speed is greater than 150rpm and the battery voltage is greater than 6 volts. **Possible Causes:** • Check for leaks between MAF sensor and throttle valve control module • Voltage supply faulty. • Sensor power circuit open from fuel pump relay to MAF sensor • Sensor signal circuit open (may be disconnected) from DME and MAF • Faulty ground cable resistance between connector terminal 1 and Ground • MAF Sensor malfunction
DTC: P0103 **2T MIL: Yes** **2001, 2002, 2003, 2004, 2005, 2006** **Models:** 320i, 325i, 330i, 525i, 540i, 545i, 550i, 740i, 745i, 745Li, 750i, M3, M5, Z3 **Engines:** 2.5L, 3.0L, 3.2L, 4.4L, 4.8L **Transmissions:** All	**MAF Sensor Circuit High Input Conditions:** Key on, engine started, and the DME detected the MAF sensor signal was more than the minimum calibrated value. The engine temperature must beat least 185-degrees (F) and all electrical equipment (A/C, lights, etc) must be off. The DME has detected that the MAF signal was more than the required minimum. The engine speed is greater than 150rpm and the battery voltage is greater than 6 volts. **Possible Causes:** • Check for leaks between MAF sensor and throttle valve control module • Voltage supply faulty. • Sensor power circuit open from fuel pump relay to MAF sensor • Sensor signal circuit open (may be disconnected) from DME and MAF • Faulty ground cable resistance between connector terminal 1 and Ground • MAF Sensor malfunction

DTC	Trouble Code Title, Conditions & Possible Causes
DTC: P0111 **2T MIL: Yes** **1996, 1997, 1998, 1999, 2000, 2001, 2002, 2003, 2004, 2005, 2006** **Models:** 318i, 320i, 323i, 325i, 328i, 330i, 525i, 528i, 530i, 540i, 545i, 550i, 740i, 745i, 745Li, 750i, 840i, M3, M5, Z3, Z4, Z8 **Engines:** 1.9L, 2.5L, 2.8L, 3.0L, 3.2L, 4.4L, 4.8L, 5.0L **Transmissions:** All	**Intake Air Temperature Sensor Circuit Low Input Conditions:** Key on or engine running, the temperature must beat least 185-degrees (F) and all electrical equipment (A/C, lights, etc) must be off; and the DME detected the IAT sensor signal was less than the self-test minimum. This is a thermistor-type sensor with a variable resistance that changes when exposed to different temperatures. This means: the higher the temperature, the lower the resistance value. **Possible Causes:** • IAT sensor signal circuit is grounded (check wiring & connector) • Resistance value between sockets 33 and 36 out of range • IAT sensor has an open circuit • IAT sensor is damaged or it has failed
DTC: P0112 **2T MIL: Yes** **2001, 2002, 2003, 2004, 2005, 2006** **Models:** 320i, 325i, 330i, 525i, 540i, 545i, 550i, 740i, 745i, 745Li, 750i, M3, M5, Z3 **Engines:** 2.5L, 3.0L, 3.2L, 4.4L, 4.8L **Transmissions:** All	**Intake Air Temperature Sensor Circuit Low Input Conditions:** Key on or Engine running, the temperature must beat least 185-degrees (F) and all electrical equipment (A/C, lights, etc) must be off; and the DME detected the IAT sensor signal was less than the self-test minimum. This is a thermistor-type sensor with a variable resistance that changes when exposed to different temperatures. This means: the higher the temperature, the lower the resistance value. **Possible Causes:** • IAT sensor signal circuit is grounded (check wiring & connector) • Resistance value between sockets 33 and 36 out of range • IAT sensor has an open circuit • IAT sensor is damaged or it has failed
DTC: P0113 **2T MIL: Yes** **1996, 2001, 2002, 2003, 2004, 2005, 2006** **Models:** 320i, 325i, 330i, 525i, 540i, 545i, 550i, 740i, 745i, 745Li, 750i, M3, M5, Z3, Z4 **Engines:** 2.5L, 3.0L, 3.2L, 4.4L, 4.8L **Transmissions:** All	**Intake Air Temperature Sensor Circuit High Input Conditions:** Key on or engine running, the temperature must beat least 185-degrees (F) and all electrical equipment (A/C, lights, etc) must be off; and the DME detected the IAT sensor signal was more than the self-test maximum. This is a thermistor-type sensor with a variable resistance that changes when exposed to different temperatures. This means: the higher the temperature, the lower the resistance value. **Possible Causes:** • IAT sensor signal circuit is open (inspect wiring & connector) • IAT sensor signal circuit is shorted • Resistance value between sockets 33 and 36 out of range • IAT sensor is damaged or it has failed
DTC: P0115 **2T MIL: Yes** **2002, 2003** **Models:** 745i, 745Li **Engines:** 4.8L **Transmissions:** All	**ECT Sensor Signal Rationality Check Conditions:** Engine started (cold), battery voltage must be 11.5, and all equipment must be off. The DME detected the ECT sensor exceeded the required calibrated value, or the engine is at idle and doesn't reach operating temperature quickly enough; the Catalyst, Fuel System, HO2S and Misfire Monitor did not complete, or the timer expired. Testing completion of procedure, the engine's temperature must rise uniformly during idle. The time to reach a closed loop status enabled temperature was not obtained. **Possible Causes:** • Check for low coolant level or incorrect coolant mixture • DME detects a short circuit wiring in the ECT • CHT sensor is out-of-calibration or it has failed • ECT sensor is out-of-calibration or it has failed
DTC: P0116 **2T MIL: Yes** **1996, 1997, 1998, 1999, 2000, 2001, 2002, 2003, 2004, 2005, 2006** **Models:** 318i, 320i, 323i, 325i, 328i, 330i, 525i, 528i, 530i, 540i, 545i, 550i, 740i, 745i, 745Li, 750i, 840i, M3, M5, Z3, Z4, Z8 **Engines:** 1.9L, 2.5L, 2.8L, 3.0L, 3.2L, 4.4L, 4.8L, 5.0L **Transmissions:** All	**ECT Sensor Signal Range/Performance Conditions:** Engine started (cold), battery voltage must be 11.5, and all equipment must be off. The DME detected the ECT sensor exceeded the required calibrated value, or the engine is at idle and doesn't reach operating temperature quickly enough; the Catalyst, Fuel System, HO2S and Misfire Monitor did not complete, or the timer expired. Testing completion of procedure, the engine's temperature must rise uniformly during idle. **Possible Causes:** • Check for low coolant level or incorrect coolant mixture • DME detects a short circuit wiring in the ECT • CHT sensor is out-of-calibration or it has failed • ECT sensor is out-of-calibration or it has failed

DTC	Trouble Code Title, Conditions & Possible Causes
DTC: P0117 **2T MIL: Yes** **1996, 2001, 2002, 2003, 2004, 2005, 2006** **Models:** 320i, 325i, 330i, 525i, 540i, 545i, 550i, 740i, 745i, 745Li, 750i, 840i, M3, M5, Z3, Z4 **Engines:** 2.5L, 3.0L, 3.2L, 4.4L, 4.8L **Transmissions:** All	**ECT Sensor Circuit Low Input Conditions:** Engine started (cold) for 10 seconds, battery voltage must be 11.5, and all equipment must be off. The DME detected the ECT sensor signal was less than the self-test minimum. This is a thermistor-type sensor with a variable resistance that changes when exposed to different temperatures **Possible Causes:** • ECT sensor signal circuit is grounded in the wiring harness • ECT sensor doesn't react to changes in temperature • ECT sensor is damaged or the DME has failed
DTC: P0118 **2T MIL: Yes** **2001, 2002, 2003, 2004, 2005, 2006** **Models:** 320i, 325i, 330i, 525i, 540i, 545i, 550i, 740i, 745i, 745Li, 750i, 840i, M3, M5, Z3, Z4 **Engines:** 2.5L, 3.0L, 3.2L, 4.4L, 4.8L **Transmissions:** All	**ECT Sensor Circuit High Input Conditions:** Engine started (cold) for 10 seconds, battery voltage must be 11.5, and all equipment must be off. The DME detected the ECT sensor signal was more than the self-test maximum. This is a thermistor-type sensor with a variable resistance that changes when exposed to different temperatures **Possible Causes:** • ECT sensor signal circuit is open (inspect wiring & connector) • ECT sensor signal circuit is shorted to ground • ECT sensor is damaged or it has failed
DTC: P0119 **2T MIL: Yes** **2001, 2002** **Models:** 540i, 740i **Engines:** 4.4L **Transmissions:** All	**ECT Sensor Circuit Continuity Conditions:** Engine started (cold) for 10 seconds, battery voltage must be 11.5, and all equipment must be off. The DME detected the ECT sensor signal was out of the specified range. This is a thermistor-type sensor with a variable resistance that changes when exposed to different temperatures **Possible Causes:** • ECT sensor signal circuit is open (inspect wiring & connector) • ECT sensor signal circuit is shorted to ground • ECT sensor is damaged or it has failed
DTC: P0120 **2T MIL: Yes** **1996, 1997, 1998, 1999, 2000, 2001, 2002, 2003, 2004, 2005** **Models:** 318i, 320i, 323i, 325i, 328i, 330i, 525i, 528i, 530i, 540i, 545i, 550i, 740i, 745i, 745Li, 750i, 840i, M3, M5, Z3, Z4, Z8 **Engines:** 1.9L, 2.5L, 2.8L, 3.0L, 3.2L, 4.4L, 4.8L, 5.0L **Transmissions:** All	**Throttle/Pedal Position Sensor (A) Circuit Malfunction Conditions:** Engine started, at idle (to 1320rpm), the temperature must be 80 degrees Celsius. The throttle position sensor supplies implausible signal to the DME. The throttle valve activation occurs via an electric motor (throttle drive) in the throttle valve control module. It is activated by the Engine Control Module (DME) according to specifications of the two sensors, Throttle Position (TP) Sensor and Accelerator Pedal Position Sensor 2. **Possible Causes:** • TP sensor signal circuit is open (inspect wiring & connector) • TP sensor signal circuit is shorted to ground • TP sensor or module is damaged or it has failed • Throttle valve is damaged or dirty • Throttle valve control module is faulty
DTC: P0121 **2T MIL: Yes** **2001, 2002, 2003, 2004, 2005** **Models:** 320i, 323i, 325i, 328i, 330i, 525i, 530i, 540i, 545i, 550i, 740i, 745i, 745Li, 750i, M3, M5, Z3, Z4, Z8 **Engines:** 2.5L, 3.0L, 3.2L, 4.4L, 4.8L, 5.0L **Transmissions:** All	**Throttle/Pedal Position Sensor Signal Range/Performance Conditions:** Engine started; then immediately following a condition where the engine was running under at off-idle, the DME detected the TP sensor signal indicated the throttle did not return to its previous closed position during the Rationality test. The engine speed is greater than 1320rpm. **Possible Causes:** • Throttle plate is binding, dirty or sticking • Throttle valve is damaged or dirty • Throttle valve control module is faulty • TP sensor signal circuit open (inspect wiring & connector) • TP sensor ground circuit open (inspect wiring & connector) • TP sensor and/or control module is damaged or has failed • MAF sensor signal is damaged, has failed or a short is present

DTC	Trouble Code Title, Conditions & Possible Causes
DTC: P0122 **2T MIL: Yes** **2001, 2002, 2003, 2004, 2005** **Models:** 320i, 323i, 325i, 328i, 330i, 525i, 530i, 540i, 545i, 550i, 740i, 745i, 745Li, 750i, M3, M5, Z3, Z4, Z8 **Engines:** 2.5L, 3.0L, 3.2L, 4.4L, 4.8L, 5.0L **Transmissions:** All	**Throttle/Pedal Position Sensor Circuit Low Input Conditions:** Engine started, at idle, the temperature must be at least 80 degrees Celsius. The throttle position sensor supplies implausible signal to the DME. **Possible Causes:** • TP sensor signal circuit open (inspect wiring & connector) • TP sensor signal shorted to ground (inspect wiring & connector) • TP sensor is damaged or has failed • Throttle control module's voltage supply is shorted or open
DTC: P0123 **2T MIL: Yes** **2001, 2002, 2003, 2004, 2005** **Models:** 320i, 323i, 325i, 328i, 330i, 525i, 530i, 540i, 545i, 550i, 740i, 745i, 745Li, 750i, M3, M5, Z3, Z4, Z8 **Engines:** 2.5L, 3.0L, 3.2L, 4.4L, 4.8L, 5.0L **Transmissions:** All	**TP Sensor Circuit High Input Conditions:** Engine started, at idle, the temperature must be at least 80 degrees Celsius. The DME detected the TP sensor signal was more than the self-test maximum during testing. **Possible Causes:** • TP sensor not seated correctly in housing (may be damaged) • TP sensor signal is circuit shorted to ground or system voltage • TP sensor ground circuit is open (check the wiring harness) • TP sensor and/or DME has failed
DTC: P0125 **2T MIL: Yes** **1996, 1997, 1998, 1999, 2000, 2001, 2002, 2003, 2004, 2005** **Models:** 318i, 320i, 323i, 325i, 328i, 330i, 525i, 528i, 530i, 540i, 545i, 550i, 740i, 745i, 745Li, 750i, 840i, M3, M5, Z3, Z4, Z8 **Engines:** 1.9L, 2.5L, 2.8L, 3.0L, 3.2L, 4.4L, 4.8L, 5.0L **Transmissions:** All	**ECT Sensor Insufficient for Closed Loop Fuel Control Conditions:** Engine started (cold), battery voltage must be 11.5, and all equipment must be off. The DME detected the ECT sensor exceeded the required calibrated value, or the engine is at idle and doesn't reach operating temperature quickly enough; the Catalyst, Fuel System, HO2S and Misfire Monitor did not complete, or the timer expired. Testing completion of procedure, the engine's temperature must rise uniformly during idle. **Possible Causes:** • Check for low coolant level or incorrect coolant mixture • DME detects a short circuit wiring in the ECT • CHT sensor is out-of-calibration or it has failed • ECT sensor is out-of-calibration or it has failed
DTC: P0128 **2T MIL: Yes** **2002, 2003, 2004, 2005** **Models:** 320i, 323i, 325i, 328i, 330i, 525i, 530i, 540i, 545i, 550i, 740i, 745i, 745Li, 750i, M3, M5, Z3, Z4, Z8 **Engines:** 2.5L, 3.0L, 3.2L, 4.4L, 4.8L, 5.0L **Transmissions:** All	**Coolant Thermostat (Coolant Temperature Below Thermostat Regulating Temperature) Conditions:** The engine's warm up performance is monitored by comparing measured coolant temperature with the modeled coolant temperature to detect a defective coolant thermostat. The engine temperature must be less than 65 degrees Celsius, engine speed greater than 800rpm (with the vehicle speed greater than 10 but less than 90km/h) and the ambient temperature greater than -8 degrees Celsius. The thermostat should be wide open when cold, but is in error if it opens below desired control temperature. **Possible Causes:** • Check for low coolant level or incorrect coolant mixture • DME detects a short circuit wiring in the ECT • CHT sensor is out-of-calibration or it has failed • ECT sensor is out-of-calibration or it has failed • Replace the thermostat
DTC: P0130 **2T MIL: Yes** **1996, 1997, 1998, 1999, 2000, 2001, 2002, 2003, 2004, 2005** **Models:** 318i, 320i, 323i, 325i, 328i, 330i, 525i, 528i, 530i, 540i, 545i, 550i, 740i, 745i, 745Li, 750i, 840i, M3, M5, Z3, Z4, Z8 **Engines:** 1.9L, 2.5L, 2.8L, 3.0L, 3.2L, 4.4L, 4.8L, 5.0L **Transmissions:** All	**O2 Sensor Circuit Bank 1 Sensor 1 Conditions:** Engine running, battery voltage 11.5, all electrical components off, ground between engine and chassis well connected and the exhaust system must be properly sealed between catalytic converter and the cylinder head. The DME detected the HO2S signal was implausible or not detected. The response rate for the sensor signal period is greater than 3.8/second. The engine speed is 1280 to 2400rpm, the catalyst temperature is greater than 300 degrees Celsius and the heater has been on for less than 90 seconds. **Possible Causes:** • Oxygen sensor heater for oxygen sensor (HO2S) before catalytic converter is faulty • HO2S is contaminated (due to presence of silicone in fuel) • HO2S signal and ground circuit wires crossed in wiring harness • HO2S signal circuit is shorted to sensor or chassis ground • HO2S element before the catalytic converter has failed (internal short condition) • Leaks present in the exhaust manifold or exhaust pipes

DTC	Trouble Code Title, Conditions & Possible Causes
DTC: P0131 **2T MIL: Yes** **2001, 2002, 2003, 2004, 2005** **Models:** 320i, 323i, 325i, 328i, 330i, 525i, 530i, 540i, 545i, 550i, 740i, 745i, 745Li, 750i, M3, M5, Z3, Z4, Z8 **Engines:** 2.5L, 3.0L, 3.2L, 4.4L, 4.8L, 5.0L **Transmissions:** All	**HO2S (Bank 1 Sensor 1) Circuit Low Input Conditions:** Engine running, battery voltage 11.5, all electrical components off, ground between engine and chassis well connected and the exhaust system must be properly sealed between catalytic converter and the cylinder head. The DME detected the HO2S signal was in a negative voltage range referred to as "character shift downward". This code sets when the HO2S signal remains in a low state for a measured period of time. In effect, it does not switch properly in the closed loop operation. **Possible Causes:** • HO2S is contaminated (due to presence of silicone in fuel) • HO2S signal and ground circuit wires crossed in wiring harness • HO2S signal circuit is shorted to sensor or chassis ground • HO2S element has failed (internal short condition) • Leaks present in the exhaust manifold or exhaust pipes
DTC: P0132 **2T MIL: Yes** **2001, 2002, 2003, 2004, 2005** **Models:** 320i, 323i, 325i, 328i, 330i, 525i, 530i, 540i, 545i, 550i, 740i, 745i, 745Li, 750i, M3, M5, Z3, Z4, Z8 **Engines:** 2.5L, 3.0L, 3.2L, 4.4L, 4.8L, 5.0L **Transmissions:** All	**HO2S (Bank 1 Sensor 1) Circuit High Input Conditions:** Engine running, battery voltage 11.5, all electrical components off, ground between engine and chassis well connected and the exhaust system must be properly sealed between catalytic converter and the cylinder head. The DME detected the HO2S signal was in a high state. This code sets when the HO2S signal remains in a high state for a measured period of time. In effect, it does not switch properly in the closed loop operation. **Note: The HO2S signal circuit may be shorted to the heater power circuit due to tracking inside of the HO2S connector. Remove the connector and visually inspect the connector for signs of oil or water.** **Possible Causes:** • HO2S is contaminated (due to presence of silicone in fuel) • HO2S signal and ground circuit wires crossed in wiring harness • HO2S signal circuit is shorted to sensor or chassis ground • HO2S element has failed (internal short condition) • Leaks present in the exhaust manifold or exhaust pipes
DTC: P0133 **2T MIL: Yes** **1996, 1997, 1998, 1999, 2000, 2001, 2002, 2003, 2004, 2005** **Models:** 318i, 320i, 323i, 325i, 328i, 330i, 525i, 528i, 530i, 540i, 545i, 550i, 740i, 745i, 745Li, 750i, 840i, M3, M5, Z3, Z4, Z8 **Engines:** 1.9L, 2.5L, 2.8L, 3.0L, 3.2L, 4.4L, 4.8L, 5.0L **Transmissions:** All	**HO2S (Bank 1 Sensor 1) Circuit Slow Response Conditions:** **Engine running, battery voltage 11.5, all electrical components off, ground between engine and chassis well connected and the exhaust system must be properly sealed between catalytic converter and the cylinder head. The DME detected the HO2S amplitude and frequency were out of the normal range (e.g., the HO2S rich to lean switch) during the HO2S Monitor test. The response rate for the sensor signal period is greater than 3.8/second. The engine speed is 1280 to 2400rpm, the catalyst temperature is greater than 300 degrees Celsius and the heater has been on for less than 90 seconds. For the 1999 M62: The idle speed variation is between 1400 and 2600rpm, the engine load variation is between 20 and 54 while the catalyst temperature should be greater than 360 degrees Celsius.** **Possible Causes:** • HO2S before the three-way catalytic converter is contaminated (due to presence of silicone in fuel); Run the engine for three minutes at 3500rpm as a self-cleaning effect • HO2S signal circuit open • Leaks present in the exhaust manifold or exhaust pipes • HO2S is damaged or has failed
DTC: P0134 **2T MIL: Yes** **1996, 1997, 1998, 1999, 2000, 2001, 2002, 2003, 2004, 2005** **Models:** 318i, 320i, 323i, 325i, 328i, 330i, 525i, 528i, 530i, 540i, 545i, 550i, 740i, 745i, 745Li, 750i, 840i, M3, M5, Z3, Z4, Z8 **Engines:** 1.9L, 2.5L, 2.8L, 3.0L, 3.2L, 4.4L, 4.8L, 5.0L **Transmissions:** All	**HO2S (Bank 1 Sensor 1) Circuit No Activity Conditions:** Engine running, battery voltage 11.5, all electrical components off, ground between engine and chassis well connected and the exhaust system must be properly sealed between catalytic converter and the cylinder head. The DME detected the HO2S signal failed to meet the maximum or minimum voltage levels (i.e., it failed the voltage range check). **Possible Causes:** • Leaks present in the exhaust manifold or exhaust pipes • HO2S signal wire and ground wire crossed in connector (voltage jumps) • HO2S element is fuel contaminated or has failed
DTC: P0135 **2T MIL: Yes** **1996, 1997, 1998, 1999, 2000, 2001, 2002, 2003, 2004, 2005** **Models:** 318i, 320i, 323i, 325i, 328i, 330i, 525i, 528i, 530i, 540i, 545i, 550i, 740i, 745i, 745Li, 750i, 840i, M3, M5, Z3, Z4, Z8 **Engines:** 1.9L, 2.5L, 2.8L, 3.0L, 3.2L, 4.4L, 4.8L, 5.0L **Transmissions:** All	**HO2S (Bank 1 Sensor 1) Heater Circuit Malfunction Conditions:** Engine running, battery voltage 11.5, all electrical components off, ground between engine and chassis well connected and the exhaust system must be properly sealed between catalytic converter and the cylinder head. The DME detected an unexpected voltage condition, or it detected excessive current draw in the heater circuit during the CCM test. The response rate for the sensor signal period is greater than 3.8/second. The engine speed is 1280 to 2400rpm, the catalyst temperature is greater than 300 degrees Celsius and the heater has been on for less than 90 seconds. **Possible Causes:** • HO2S heater power circuit is open or heater ground circuit open • HO2S signal tracking (due to oil or moisture in the connector) • HO2S is damaged or has failed

DTC	Trouble Code Title, Conditions & Possible Causes
DTC: P0136 **2T MIL: Yes** **1996, 1997, 1998, 1999, 2000, 2001, 2002, 2003, 2004, 2005** **Models:** 318i, 320i, 323i, 325i, 328i, 330i, 525i, 528i, 530i, 540i, 545i, 550i, 740i, 745i, 745Li, 750i, 840i, M3, M5, Z3, Z4, Z8 **Engines:** 1.9L, 2.5L, 2.8L, 3.0L, 3.2L, 4.4L, 4.8L, 5.0L **Transmissions:** All	**HO2S (Bank 1 Sensor 2) Circuit Malfunction Conditions:** Engine running, battery voltage 11.5, all electrical components off, ground between engine and chassis well connected and the exhaust system must be properly sealed between catalytic converter and the cylinder head. The DME detected the HO2S signal failed to meet the maximum or minimum voltage levels (i.e., it failed the voltage range check). The heater has been on for less than 90 seconds, the fuel system status is in fuel cut-off, the output voltage is between 400mV and 500mV and it is 120 seconds after engine start up. **Possible Causes:** • Leaks present in the exhaust manifold or exhaust pipes • HO2S signal wire and ground wire crossed in connector • HO2S element is fuel contaminated or has failed
DTC: P0137 **2T MIL: Yes** **2001, 2002, 2003, 2004, 2005** **Models:** 320i, 323i, 325i, 328i, 330i, 525i, 530i, 540i, 545i, 550i, 740i, 745i, 745Li, 750i, M3, M5, Z3, Z4, Z8 **Engines:** 2.5L, 3.0L, 3.2L, 4.4L, 4.8L, 5.0L **Transmissions:** All	**HO2S (Bank 1 Sensor 2) Circuit Low Input Conditions:** Engine running, battery voltage 11.5, all electrical components off, ground between engine and chassis well connected and the exhaust system must be properly sealed between catalytic converter and the cylinder head. The DME detected the HO2S signal remained in a high state. **Note: The HO2S signal circuit may be shorted to the heater power circuit due to "tracking inside of the HO2S connector. Remove the connector and visually inspect the connector for signs of oil or water.** **Possible Causes:** • HO2S signal shorted to heater power circuit in the connector • HO2S signal circuit shorted to ground (for more than 200 seconds) or to system voltage
DTC: P0138 **2T MIL: Yes** **2001, 2002, 2003, 2004, 2005** **Models:** 320i, 323i, 325i, 328i, 330i, 525i, 530i, 540i, 545i, 550i, 740i, 745i, 745Li, 750i, M3, M5, Z3, Z4, Z8 **Engines:** 2.5L, 3.0L, 3.2L, 4.4L, 4.8L, 5.0L **Transmissions:** All	**HO2S (Bank 1 Sensor 2) Circuit High Input Conditions:** Engine running, battery voltage 11.5, all electrical components off, ground between engine and chassis well connected and the exhaust system must be properly sealed between catalytic converter and the cylinder head. The DME detected the HO2S signal remained in a high state. **Note: The HO2S signal circuit may be shorted to the heater power circuit due to "tracking inside of the HO2S connector. Remove the connector and visually inspect the connector for signs of oil or water.** **Possible Causes:** • HO2S signal shorted to heater power circuit in the positive connector • HO2S signal circuit shorted to ground or to system voltage • HO2S has failed
DTC: P0139 **2T MIL: Yes** **1999, 2000, 2001, 2002, 2003, 2004, 2005** **Models:** 320i, 323i, 325i, 328i, 330i, 525i, 530i, 540i, 545i, 550i, 740i, 745i, 745Li, 750i, M3, M5, Z3, Z8 **Engines:** 2.5L, 3.0L, 3.2L, 4.4L, 4.8L, 5.0L **Transmissions:** All	**HO2S (Bank 1 Sensor 2) Slow Response Conditions:** Engine running, battery voltage 11.5, all electrical components off, ground between engine and chassis well connected and the exhaust system must be properly sealed between catalytic converter and the cylinder head. The DME detected the HO2S amplitude and frequency were out of the normal range during the HO2S Monitor test. The heater has been on for less than 90 seconds, the fuel system status is in fuel cut-off, the output voltage is between 400mV and 500mV and it is 120 seconds after engine start up. **Possible Causes:** • HO2S signal shorted to heater power circuit in the connector • HO2S signal circuit shorted to VREF or to system voltage
DTC: P0140 **2T MIL: Yes** **2001, 2002, 2003, 2004, 2005** **Models:** 320i, 323i, 325i, 328i, 330i, 525i, 530i, 540i, 545i, 550i, 740i, 745i, 745Li, 750i, M3, M5, Z3, Z4, Z8 **Engines:** 2.5L, 3.0L, 3.2L, 4.4L, 4.8L, 5.0L **Transmissions:** All	**HO2S (Bank 1 Sensor 2) No Activity Conditions:** Engine running, battery voltage 11.5, all electrical components off, ground between engine and chassis well connected and the exhaust system must be properly sealed between catalytic converter and the cylinder head. The DME detected the HO2S signal failed to meet the maximum or minimum voltage levels (i.e., it failed the voltage range check). **Possible Causes:** • HO2S before the three-way catalytic converter is contaminated (due to presence of silicone in fuel); Run the engine for three minutes at 3500rpm as a self-cleaning effect • Leaks present in the exhaust manifold or exhaust pipes • HO2S signal wire and ground wire crossed in connector (voltage jumps) • HO2S element is contaminated or has failed

DTC	Trouble Code Title, Conditions & Possible Causes
DTC: P0141 **2T MIL: Yes** **1996, 1997, 1998, 1999, 2000, 2001, 2002, 2003, 2004, 2005** **Models:** 318i, 320i, 323i, 325i, 328i, 330i, 525i, 528i, 530i, 540i, 545i, 550i, 740i, 745i, 745Li, 750i, 840i, M3, M5, Z3, Z4, Z8 **Engines:** 1.9L, 2.5L, 2.8L, 3.0L, 3.2L, 4.4L, 4.8L, 5.0L **Transmissions:** All	**HO2S (Bank 1 Sensor 2) Malfunction Conditions:** Engine running, battery voltage 11.5, all electrical components off, ground between engine and chassis well connected and the exhaust system must be properly sealed between catalytic converter and the cylinder head. The DME detected the HO2S signal failed to meet the maximum or minimum voltage levels (i.e., it failed the voltage range check). The engine speed is greater than 40rpm, the battery voltage must be between 10.7 and 15.5 volts, and the fault occurs 200 seconds after engine start up. **Possible Causes:** • Leaks present in the exhaust manifold or exhaust pipes • HO2S signal wire and ground wire crossed in connector • HO2S element is fuel contaminated or has failed
DTC: P0150 **2T MIL: Yes** **1996, 1997, 1998, 1999, 2000, 2001, 2002, 2003, 2004, 2005** **Models:** 318i, 320i, 323i, 325i, 328i, 330i, 525i, 528i, 530i, 540i, 545i, 550i, 740i, 745i, 745Li, 750i, 840i, M3, M5, Z3, Z4, Z8 **Engines:** 1.9L, 2.5L, 2.8L, 3.0L, 3.2L, 4.4L, 4.8L, 5.0L **Transmissions:** All	**HO2S (Bank 2 Sensor 1) Circuit Malfunction Conditions:** Engine running, battery voltage 11.5, all electrical components off, ground between engine and chassis well connected and the exhaust system must be properly sealed between catalytic converter and the cylinder head. The DME detected the HO2S signal failed to meet the maximum or minimum voltage levels (i.e., it failed the voltage range check). The response rate for the sensor signal period is greater than 3.8/second. The engine speed is 1280 to 2400rpm, the catalyst temperature is greater than 300 degrees Celsius and the heater has been on for less than 90 seconds. **Possible Causes:** • Leaks present in the exhaust manifold or exhaust pipes • HO2S signal wire and ground wire crossed in connector • HO2S element is fuel contaminated or has failed
DTC: P0151 **2T MIL: Yes** **2001, 2002, 2003, 2004, 2005** **Models:** 320i, 323i, 325i, 328i, 330i, 525i, 530i, 540i, 545i, 550i, 740i, 745i, 745Li, 750i, M3, M5, Z3, Z4, Z8 **Engines:** 2.5L, 3.0L, 3.2L, 4.4L, 4.8L, 5.0L **Transmissions:** All	**HO2S (Bank 2 Sensor 1) Low Input Conditions:** Engine running, battery voltage 11.5, all electrical components off, ground between engine and chassis well connected and the exhaust system must be properly sealed between catalytic converter and the cylinder head. The DME detected the HO2S signal remained in a high state. **Note: The HO2S signal circuit may be shorted to the heater power circuit due to "tracking inside of the HO2S connector. Remove the connector and visually inspect the connector for signs of oil or water.** **Possible Causes:** • HO2S is contaminated (due to presence of silicone in fuel) • HO2S signal tracking (due to oil or moisture in the connector) • HO2S signal circuit is open or shorted to VREF
DTC: P0152 **2T MIL: Yes** **2001, 2002, 2003, 2004, 2005** **Models:** 320i, 323i, 325i, 328i, 330i, 525i, 530i, 540i, 545i, 550i, 740i, 745i, 745Li, 750i, M3, M5, Z3, Z4, Z8 **Engines:** 2.5L, 3.0L, 3.2L, 4.4L, 4.8L, 5.0L **Transmissions:** All	**HO2S (Bank 2 Sensor 1) Circuit High Input Conditions:** Engine running, battery voltage 11.5, all electrical components off, ground between engine and chassis well connected and the exhaust system must be properly sealed between catalytic converter and the cylinder head. The DME detected the HO2S signal remained in a high state (more than 1.5v). **Note: The HO2S signal circuit may be shorted to the heater power circuit due to "tracking inside of the HO2S connector. Remove the connector and visually inspect the connector for signs of oil or water.** **Possible Causes:** • HO2S is contaminated (due to presence of silicone in fuel) • HO2S signal tracking (due to oil or moisture in the connector) • HO2S signal circuit is open or shorted to VREF
DTC: P0153 **2T MIL: Yes** **1996, 1997, 1998, 1999, 2000, 2001, 2002, 2003, 2004, 2005** **Models:** 318i, 320i, 323i, 325i, 328i, 330i, 525i, 528i, 530i, 540i, 545i, 550i, 740i, 745i, 745Li, 750i, 840i, M3, M5, Z3, Z4, Z8 **Engines:** 1.9L, 2.5L, 2.8L, 3.0L, 3.2L, 4.4L, 4.8L, 5.0L **Transmissions:** All	**HO2S (Bank 2 Sensor 1) Circuit Slow Response Conditions:** **Engine running, battery voltage 11.5, all electrical components off, ground between engine and chassis well connected and the exhaust system must be properly sealed between catalytic converter and the cylinder head. The DME detected the HO2S amplitude and frequency were out of the normal range during the HO2S Monitor test. For the 1999 M62: The idle speed variation is between 1400 and 2600rpm, the engine load variation is between 20 and 54 while the catalyst temperature should be greater than 360 degrees Celsius.** **Possible Causes:** • HO2S is contaminated (due to presence of silicone in fuel) • Leaks present in the exhaust manifold or exhaust pipes • HO2S is damaged or has failed

DTC	Trouble Code Title, Conditions & Possible Causes
DTC: P0154 **2T MIL: Yes** **1996, 1997, 1998, 1999, 2000, 2001, 2002, 2003, 2004, 2005** **Models:** 318i, 320i, 323i, 325i, 328i, 330i, 525i, 528i, 530i, 540i, 545i, 550i, 740i, 745i, 745Li, 750i, 840i, M3, M5, Z3, Z4, Z8 **Engines:** 1.9L, 2.5L, 2.8L, 3.0L, 3.2L, 4.4L, 4.8L, 5.0L **Transmissions:** All	**HO2S (Bank 2 Sensor 1) Circuit No Activity Conditions:** Engine running, battery voltage 11.5, all electrical components off, ground between engine and chassis well connected and the exhaust system must be properly sealed between catalytic converter and the cylinder head. The DME detected the HO2S signal failed to meet the maximum or minimum voltage (i.e., it failed the voltage check). **Possible Causes:** • Leaks present in the exhaust manifold or exhaust pipes • HO2S signal wire and ground wire crossed in connector • HO2S element is fuel contaminated or has failed
DTC: P0155 **2T MIL: Yes** **1996, 1997, 1998, 1999, 2000, 2001, 2002, 2003, 2004, 2005, 2006** **Models:** 318i, 320i, 323i, 325i, 328i, 330i, 525i, 528i, 530i, 540i, 545i, 550i, 740i, 745i, 745Li, 750i, 840i, M3, M5, Z3, Z4, Z8 **Engines:** 1.9L, 2.5L, 2.8L, 3.0L, 3.2L, 4.4L, 4.8L, 5.0L **Transmissions:** All	**HO2S (Bank 2 Sensor 1) Heater Circuit Malfunction Conditions:** Engine running, battery voltage 11.5, all electrical components off, ground between engine and chassis well connected and the exhaust system must be properly sealed between catalytic converter and the cylinder head. The DME detected an open or shorted condition, or excessive current draw in the heater circuit. The response rate for the sensor signal period is greater than 3.8/second. The engine speed is 1280 to 2400rpm, the catalyst temperature is greater than 300 degrees Celsius and the heater has been on for less than 90 seconds. **Possible Causes:** • HO2S heater power circuit is open • HO2S heater ground circuit is open • HO2S signal tracking (due to oil or moisture in the connector) • HO2S is damaged or has failed
DTC: P0156 **2T MIL: Yes** **1996, 1997, 1998, 1999, 2000, 2001, 2002, 2003** **Models:** 318i, 320i, 323i, 325i, 328i, 330i, 525i, 528i, 530i, 540i, 545i, 550i, 740i, 745i, 745Li, 750i, 840i, M3, M5, Z3, Z4, Z8 **Engines:** 1.9L, 2.5L, 2.8L, 3.0L, 3.2L, 4.4L, 4.8L, 5.0L **Transmissions:** All	**HO2S (Bank 2 Sensor 2) Circuit No Activity Conditions:** Engine running, battery voltage 11.5, all electrical components off, ground between engine and chassis well connected and the exhaust system must be properly sealed between catalytic converter and the cylinder head. The DME detected the HO2S signal failed to meet the maximum or minimum voltage (i.e., it failed the voltage check). The heater has been on for less than 90 seconds, the fuel system status is in fuel cut-off, the output voltage is between 400mV and 500mV and it is 120 seconds after engine start up. **Possible Causes:** • Leaks present in the exhaust manifold or exhaust pipes • HO2S signal wire and ground wire crossed in connector • HO2S element is fuel contaminated or has failed
DTC: P0157 **2T MIL: Yes** **2001, 2002, 2003, 2004, 2005** **Models:** 320i, 323i, 325i, 328i, 330i, 525i, 530i, 540i, 545i, 550i, 740i, 745i, 745Li, 750i, M3, M5, Z3, Z4, Z8 **Engines:** 2.5L, 3.0L, 3.2L, 4.4L, 4.8L, 5.0L **Transmissions:** All	**HO2S (Bank 2 Sensor 2) Circuit Low Voltage Conditions:** Engine running, battery voltage 11.5, all electrical components off, ground between engine and chassis well connected and the exhaust system must be properly sealed between catalytic converter and the cylinder head. The DME detected the HO2S signal remained in a high state. **Note: The HO2S signal circuit may be shorted to the heater power circuit due to "tracking inside of the HO2S connector. Remove the connector and visually inspect the connector for signs of oil or water.** **Possible Causes:** • HO2S is contaminated (due to presence of silicone in fuel) • HO2S signal tracking (due to oil or moisture in the connector) • HO2S signal circuit is open or shorted to VREF
DTC: P0158 **2T MIL: Yes** **1996, 2001, 2002, 2003, 2004, 2005, 2006** **Models:** 318i, 320i, 323i, 325i, 328i, 330i, 525i, 528i, 530i, 540i, 545i, 550i, 740i, 745i, 745Li, 750i, 840i, M3, M5, Z3, Z4, Z8 **Engines:** 1.9L, 2.5L, 2.8L, 3.0L, 3.2L, 4.4L, 4.8L, 5.0L **Transmissions:** All	**HO2S (Bank 2 Sensor 2) Circuit High Input Conditions:** Engine running, battery voltage 11.5, all electrical components off, ground between engine and chassis well connected and the exhaust system must be properly sealed between catalytic converter and the cylinder head. The DME detected the HO2S signal remained in a high state (i.e., more than 1.5v). **Note: The HO2S signal circuit may be shorted to the heater power circuit due to "tracking inside of the HO2S connector. Remove the connector and visually inspect the connector for signs of oil or water.** **Possible Causes:** • HO2S signal shorted to the heater power circuit (due to oil or moisture in the connector) • HO2S signal circuit shorted to VREF or to system voltage

DTC	Trouble Code Title, Conditions & Possible Causes
DTC: P0159 **2T MIL: Yes** **1999, 2000, 2001, 2002, 2003, 2004, 2005, 2006** **Models:** 323i, 325i, 328i, 330i, 525i, 530i, 540i, 545i, 550i, 740i, 745i, 745Li, 750i, M3, M5, Z3, Z8 **Engines:** 2.5L, 2.8L, 3.0L, 3.2L, 4.4L, 4.8L, 5.0L **Transmissions:** All	**HO2S (Bank 2 Sensor 2) Circuit Slow Response Conditions:** Engine running, battery voltage 11.5, all electrical components off, ground between engine and chassis well connected and the exhaust system must be properly sealed between catalytic converter and the cylinder head. The DME detected the HO2S amplitude and frequency were out of the normal range during the HO2S Monitor test. The heater has been on for less than 90 seconds, the fuel system status is in fuel cut-off, the output voltage is between 400mV and 500mV and it is 120 seconds after engine start up. **Possible Causes:** • HO2S is contaminated (due to presence of silicone in fuel) • Leaks present in the exhaust manifold or exhaust pipes • HO2S is damaged or has failed
DTC: P0160 **2T MIL: Yes** **2001, 2002, 2003, 2004, 2005, 2006** **Models:** 323i, 325i, 328i, 330i, 525i, 530i, 540i, 545i, 550i, 740i, 745i, 745Li, 750i, M3, M5, Z3, Z8 **Engines:** 2.5L, 2.8L, 3.0L, 3.2L, 4.4L, 4.8L, 5.0L **Transmissions:** All	**HO2S (Bank 2 Sensor 2) Circuit No Activity Detected Conditions:** Engine running, battery voltage 11.5, all electrical components off, ground between engine and chassis well connected and the exhaust system must be properly sealed between catalytic converter and the cylinder head. The DME detected the HO2S signal failed to meet the maximum or minimum voltage (i.e., it failed the voltage check). **Possible Causes:** • Leaks present in the exhaust manifold or exhaust pipes • HO2S signal wire and ground wire crossed in connector • HO2S element is fuel contaminated or has failed
DTC: P0161 **2T MIL: Yes** **1996, 1997, 1998, 1999, 2000, 2001, 2002, 2003, 2004, 2005, 2006** **Models:** 318i, 320i, 323i, 325i, 328i, 330i, 525i, 528i, 530i, 540i, 545i, 550i, 740i, 745i, 745Li, 750i, 840i, M3, M5, Z3, Z4, Z8 **Engines:** 1.9L, 2.5L, 2.8L, 3.0L, 3.2L, 4.4L, 4.8L, 5.0L **Transmissions:** All	**HO2S (Bank 2 Sensor 2) Heater Circuit Malfunction Conditions:** Engine running, battery voltage 11.5, all electrical components off, ground between engine and chassis well connected and the exhaust system must be properly sealed between catalytic converter and the cylinder head. The DME detected an open or shorted condition, or excessive current draw in the heater circuit. The engine speed is greater than 40rpm, the battery voltage must be between 10.7 and 15.5 volts, and the fault occurs 200 seconds after engine start up. **Possible Causes:** • HO2S heater power circuit or the heater ground circuit is open • HO2S signal tracking (due to oil or moisture in the connector) • HO2S has failed, or the DME has failed
DTC: P0170 **2T MIL: Yes** **1996, 1997, 1998, 1999, 2000** **Models:** 318i, 320i, 323i, 325i, 328i, 330i, 525i, 528i, 530i, 540i, 545i, 550i, 740i, 745i, 745Li, 750i, 840i, M3, M5, Z3, Z4, Z8 **Engines:** 1.9L, 2.5L, 2.8L, 3.0L, 3.2L, 4.4L, 4.8L, 5.0L **Transmissions:** All	**Fuel System Malfunction (Cylinder Bank 1) Conditions:** The engine is running in a closed loop at a stable engine speed, and the DME detected the lean or rich fuel trim correction valve was more than or less than a calibrated limit. The fuel system status is in a closed loop. **Possible Causes:** • Air leaks after the MAF sensor, or leaks in the PCV system • Exhaust leaks before or near where the HO2S is mounted • Fuel injector(s) restricted or not supplying enough fuel • Fuel system not supplying enough fuel during high fuel demand conditions (e.g., the fuel pump may not supply enough fuel) • Leaking EGR gasket, or leaking EGR valve diaphragm • MAF sensor dirty (causes DME to underestimate airflow) • Vehicle running out of fuel or engine oil dip stick not seated
DTC: P0171 **2T MIL: Yes** **2001, 2002, 2003, 2004, 2005** **Models:** 320i, 323i, 325i, 328i, 330i, 525i, 530i, 540i, 545i, 550i, 740i, 745i, 745Li, 750i, M3, M5, Z3, Z4, Z8 **Engines:** 2.5L, 3.0L, 3.2L, 4.4L, 4.8L, 5.0L **Transmissions:** All	**Fuel System Too Lean (Cylinder Bank 1) Conditions:** Key on or engine running, all electrical components off and coolant temperature at least 80 degrees Celsius; and the DME detected the Bank 1 Adaptive Fuel Control System reached its rich correction limit (a lean A/F condition). The fuel status is in a closed loop pattern, the coolant temperature is between 69 and 100 degrees Celsius, and the engine speed is between 800 and 6000rpm. **Possible Causes:** • Air leaks after the MAF sensor, or leaks in the PCV system • Exhaust leaks before or near where the HO2S is mounted • Fuel injector(s) restricted or not supplying enough fuel • Fuel pump not supplying enough fuel during high fuel demand conditions • Leaking EGR gasket, or leaking EGR valve diaphragm • MAF sensor dirty (causes DME to underestimate airflow) • Vehicle running out of fuel or engine oil dip stick not seated

DTC	Trouble Code Title, Conditions & Possible Causes
DTC: P0172 **2T MIL: Yes** **2001, 2002, 2003, 2004, 2005** **Models:** 320i, 323i, 325i, 328i, 330i, 525i, 530i, 540i, 545i, 550i, 740i, 745i, 745Li, 750i, M3, M5, Z3, Z4, Z8 **Engines:** 2.5L, 3.0L, 3.2L, 4.4L, 4.8L, 5.0L **Transmissions:** All	**Fuel System Too Rich (Cylinder Bank 1) Conditions:** Key on or engine running, all electrical components off and coolant temperature at least 80 degrees Celsius; and the DME detected the Bank 1 Adaptive Fuel Control System reached its rich correction limit (a rich A/F condition). The fuel status is in a closed loop pattern, the coolant temperature is between 69 and 100 degrees Celsius, and the engine speed is between 800 and 6000rpm. **Possible Causes:** • Camshaft timing is incorrect, or the engine has an oil overfill condition • EVAP vapor recovery system failure (may be pulling vacuum) • Fuel pressure regulator is damaged or leaking • HO2S element is contaminated with alcohol or water • MAF or MAP sensor values are incorrect or out-of-range • One of more fuel injectors is leaking
DTC: P0173 **2T MIL: Yes** **1996, 1997, 1998, 1999, 2000** **Models:** 318i, 320i, 323i, 325i, 328i, 330i, 525i, 528i, 530i, 540i, 545i, 550i, 740i, 745i, 745Li, 750i, 840i, M3, M5, Z3, Z4, Z8 **Engines:** 1.9L, 2.5L, 2.8L, 3.0L, 3.2L, 4.4L, 4.8L, 5.0L **Transmissions:** All	**Fuel System Malfunction (Cylinder Bank 1) Conditions:** Key on or engine running, all electrical components off and coolant temperature at least 80 degrees Celsius; and the DME detected the Bank 1 Fuel Control System experienced a implausible signal. The fuel system status is in a closed loop. **Possible Causes:** • Air leaks after the MAF sensor, or leaks in the PCV system • Exhaust leaks before or near where the HO2S is mounted • Fuel injector(s) restricted or not supplying enough fuel • Fuel system not supplying enough fuel during high fuel demand conditions (e.g., the fuel pump may not supply enough fuel) • Leaking EGR gasket, or leaking EGR valve diaphragm • MAF sensor dirty (causes DME to underestimate airflow) • Vehicle running out of fuel or engine oil dip stick not seated
DTC: P0174 **2T MIL: Yes** **1996, 1997, 1998, 1999, 2000, 2001, 2002, 2003, 2004, 2005** **Models:** 320i, 323i, 325i, 328i, 330i, 525i, 528i, 530i, 540i, 545i, 550i, 740i, 745i, 745Li, 750i, 840i, M3, M5, Z3, Z4, Z8 **Engines:** 2.5L, 3.0L, 3.2L, 4.4L, 4.8L, 5.0L **Transmissions:** All	**Fuel System Too Lean (Cylinder Bank 2) Conditions:** Key on or engine running, all electrical components off and coolant temperature at least 80 degrees Celsius; and the DME detected the Bank 2 Fuel Control System reached its lean correction limit. The fuel status is in a closed loop pattern, the coolant temperature is between 69 and 100 degrees Celsius, and the engine speed is between 800 and 6000rpm. **Possible Causes:** • Air leaks after the MAF sensor, or leaks in the PCV system • Exhaust leaks before or near where the HO2S is mounted • Fuel injector(s) restricted or not supplying enough fuel • Fuel pump not supplying enough fuel during high fuel demand conditions • Leaking EGR gasket, or leaking EGR valve diaphragm • MAF sensor dirty (causes DME to underestimate airflow) • Vehicle running out of fuel or engine oil dip stick not seated
DTC: P0175 **2T MIL: Yes** **1996, 1997, 1998, 1999, 2000, 2001, 2002, 2003, 2004, 2005** **Models:** 320i, 323i, 325i, 328i, 330i, 525i, 528i, 530i, 540i, 545i, 550i, 745i, 745Li, M3, M5, Z3, Z4, Z8 **Engines:** 2.5L, 2.8L, 3.0L, 3.2L, 5.0L, 5.4L **Transmissions:** All	**Fuel System Too Rich (Cylinder Bank 2) Conditions:** Key on or engine running, all electrical components off and coolant temperature at least 80 degrees Celsius; and the DME detected the Bank 2 Adaptive Fuel Control System reached its rich correction limit (a rich A/F condition). The fuel status is in a closed loop pattern, the coolant temperature is between 69 and 100 degrees Celsius, and the engine speed is between 800 and 6000rpm. **Possible Causes:** • Air leaks after the MAF sensor, or leaks in the PCV system • Exhaust leaks before or near where the HO2S is mounted • Fuel injector(s) restricted or not supplying enough fuel • Fuel pump not supplying enough fuel during high fuel demand conditions • Leaking EGR gasket, or leaking EGR valve diaphragm • MAF sensor dirty (causes DME to underestimate airflow) • Vehicle running out of fuel or engine oil dip stick not seated
DTC: P0201 **2T MIL: Yes** **1996, 1997, 1998, 1999, 2000, 2001, 2002, 2003, 2004, 2005, 2006** **Models:** 318i, 320i, 323i, 325i, 328i, 330i, 525i, 528i, 530i, 540i, 545i, 550i, 740i, 745i, 745Li, 750i, 840i, M3, M5, Z3, Z4, Z8 **Engines:** 1.9L, 2.5L, 2.8L, 3.0L, 3.2L, 4.4L, 4.8L, 5.0L **Transmissions:** All	**Cylinder 1 Injector Circuit Malfunction Conditions:** Engine started, and the DME detected the fuel injector "1" control circuit was in a high state when it should have been low, or in a low state when it should have been high (wiring harness & injector okay). The battery voltage should be between 9.5 and 17 volts while the engine speed is less than 40rpm. **Possible Causes:** • Injector 1 connector is damaged, open or shorted • Injector 1 control circuit is open, shorted to ground or to power (the injector driver circuit may be damaged)

DTC	Trouble Code Title, Conditions & Possible Causes
DTC: P0202 **2T MIL: Yes** **1996, 1997, 1998, 1999, 2000, 2001, 2002, 2003, 2004, 2005, 2006** **Models:** 318i, 320i, 323i, 325i, 328i, 330i, 525i, 528i, 530i, 540i, 545i, 550i, 740i, 745i, 745Li, 750i, 840i, M3, M5, Z3, Z4, Z8 **Engines:** 1.9L, 2.5L, 2.8L, 3.0L, 3.2L, 4.4L, 4.8L, 5.0L **Transmissions:** All	**Cylinder 2 Injector Circuit Malfunction Conditions:** Engine started, and the DME detected the fuel injector "2" control circuit was in a high state when it should have been low, or in a low state when it should have been high (wiring harness & injector okay). The battery voltage should be between 9.5 and 17 volts while the engine speed is less than 40rpm. **Possible Causes:** • Injector 2 connector is damaged, open or shorted • Injector 2 control circuit is open, shorted to ground or to power (the injector driver circuit may be damaged)
DTC: P0203 **2T MIL: Yes** **1996, 1997, 1998, 1999, 2000, 2001, 2002, 2003, 2004, 2005, 2006** **Models:** 318i, 320i, 323i, 325i, 328i, 330i, 525i, 528i, 530i, 540i, 545i, 550i, 740i, 745i, 745Li, 750i, 840i, M3, M5, Z3, Z4, Z8 **Engines:** 1.9L, 2.5L, 2.8L, 3.0L, 3.2L, 4.4L, 4.8L, 5.0L **Transmissions:** All	**Cylinder 3 Injector Circuit Malfunction Conditions:** Engine started, and the DME detected the fuel injector "3" control circuit was in a high state when it should have been low, or in a low state when it should have been high (wiring harness & injector okay). The battery voltage should be between 9.5 and 17 volts while the engine speed is less than 40rpm. **Possible Causes:** • Injector 3 connector is damaged, open or shorted • Injector 3 control circuit is open, shorted to ground or to power (the injector driver circuit may be damaged)
DTC: P0204 **2T MIL: Yes** **1996, 1997, 1998, 1999, 2000, 2001, 2002, 2003, 2004, 2005, 2006** **Models:** 318i, 320i, 323i, 325i, 328i, 330i, 525i, 528i, 530i, 540i, 545i, 550i, 740i, 745i, 745Li, 750i, 840i, M3, M5, Z3, Z4, Z8 **Engines:** 1.9L, 2.5L, 2.8L, 3.0L, 3.2L, 4.4L, 4.8L, 5.0L **Transmissions:** All	**Cylinder 4 Injector Circuit Malfunction Conditions:** Engine started, and the DME detected the fuel injector "4" control circuit was in a high state when it should have been low, or in a low state when it should have been high (wiring harness & injector okay). The battery voltage should be between 9.5 and 17 volts while the engine speed is less than 40rpm. **Possible Causes:** • Injector 4 connector is damaged, open or shorted • Injector 4 control circuit is open, shorted to ground or to power (the injector driver circuit may be damaged)
DTC: P0205 **2T MIL: Yes** **1996, 1997, 1998, 1999, 2000, 2001, 2002, 2003, 2004, 2005, 2006** **Models:** 318i, 320i, 323i, 325i, 328i, 330i, 525i, 528i, 530i, 540i, 545i, 550i, 740i, 745i, 745Li, 750i, 840i, M3, M5, Z3, Z4, Z8 **Engines:** 1.9L, 2.5L, 2.8L, 3.0L, 3.2L, 4.4L, 4.8L, 5.0L **Transmissions:** All	**Cylinder 5 Injector Circuit Malfunction Conditions:** Engine started, and the DME detected the fuel injector "5" control circuit was in a high state when it should have been low, or in a low state when it should have been high (wiring harness & injector okay). The battery voltage should be between 9.5 and 17 volts while the engine speed is less than 40rpm. **Possible Causes:** • Injector 5 connector is damaged, open or shorted • Injector 5 control circuit is open, shorted to ground or to power (the injector driver circuit may be damaged)
DTC: P0206 **2T MIL: Yes** **1996, 1997, 1998, 1999, 2000, 2001, 2002, 2003, 2004, 2005, 2006** **Models:** 318i, 320i, 323i, 325i, 328i, 330i, 525i, 528i, 530i, 540i, 545i, 550i, 740i, 745i, 745Li, 750i, 840i, M3, M5, Z3, Z4, Z8 **Engines:** 1.9L, 2.5L, 2.8L, 3.0L, 3.2L, 4.4L, 4.8L, 5.0L **Transmissions:** All	**Cylinder 6 Injector Circuit Malfunction Conditions:** Engine started, and the DME detected the fuel injector "6" control circuit was in a high state when it should have been low, or in a low state when it should have been high (wiring harness & injector okay). The battery voltage should be between 9.5 and 17 volts while the engine speed is less than 40rpm. **Possible Causes:** • Injector 6 connector is damaged, open or shorted • Injector 6 control circuit is open, shorted to ground or to power (the injector driver circuit may be damaged)

DTC	Trouble Code Title, Conditions & Possible Causes
DTC: P0207 **2T MIL: Yes** **1996, 1997, 1998, 1999, 2000, 2001, 2002, 2003, 2004, 2005, 2006** **Models:** 320i, 323i, 325i, 328i, 330i, 525i, 528i, 530i, 540i, 545i, 550i, 745i, 745Li, M3, M5, Z3, Z4, Z8 **Engines:** 2.5L, 2.8L, 3.0L, 3.2L, 5.0L, 5.4L **Transmissions:** All	**Cylinder 7 Injector Circuit Malfunction Conditions:** Engine started, and the DME detected the fuel injector "7" control circuit was in a high state when it should have been low, or in a low state when it should have been high (wiring harness & injector okay). The battery voltage should be between 9.5 and 17 volts while the engine speed is less than 40rpm. **Note: Monitor the INJIF PID Fault "flags" with the Scan Tool. The appropriate INJF PID "flag" will read Yes when this code is set.** **Possible Causes:** • Injector 7 connector is damaged, open or shorted • Injector 7 control circuit is open, shorted to ground or to power (the injector driver circuit may be damaged)
DTC: P0208 **2T MIL: Yes** **1996, 1997, 1998, 1999, 2000, 2001, 2002, 2003, 2004, 2005** **Models:** 320i, 323i, 325i, 328i, 330i, 525i, 528i, 530i, 540i, 545i, 550i, 740i, 745i, 745Li, 750i, M3, M5, Z3, Z4, Z8 **Engines:** 3.0L, 3.2L, 4.4L, 4.8L, 5.0L, 5.4L **Transmissions:** All	**Cylinder 8 Injector Circuit Malfunction Conditions:** Engine started, and the DME detected the fuel injector "8" control circuit was in a high state when it should have been low, or in a low state when it should have been high (wiring harness & injector okay). The battery voltage should be between 9.5 and 17 volts while the engine speed is less than 40rpm. **Note: Monitor the INJIF PID Fault "flags" with the Scan Tool. The appropriate INJF PID "flag" will read Yes when this code is set.** **Possible Causes:** • Injector 8 connector is damaged, open or shorted • Injector 8 control circuit is open, shorted to ground or to power (the injector driver circuit may be damaged)
DTC: P0209 2T MIL: Yes **1996, 1997, 1998, 1999, 2000, 2001** **Models:** 750i **Engines:** 5.4L **Transmissions:** All	**Cylinder 9 Injector Circuit Malfunction Conditions:** Engine started, and the DME detected the fuel injector "9" control circuit was in a high state when it should have been low, or in a low state when it should have been high (wiring harness & injector okay). **Note: Monitor the INJIF PID Fault "flags" with the Scan Tool. The appropriate INJF PID "flag" will read Yes when this code is set.** **Possible Causes:** • Injector 8 connector is damaged, open or shorted • Injector 8 control circuit is open, shorted to ground or to power (the injector driver circuit may be damaged)
DTC: P0210 **2T MIL: Yes** **1996, 1997, 1998, 1999, 2000, 2001** **Models:** 750i **Engines:** 5.4L **Transmissions:** All	**Cylinder 10 Injector Circuit Malfunction Conditions:** Engine started, and the DME detected the fuel injector "10" control circuit was in a high state when it should have been low, or in a low state when it should have been high (wiring harness & injector okay). **Note: Monitor the INJIF PID Fault "flags" with the Scan Tool. The appropriate INJF PID "flag" will read Yes when this code is set.** **Possible Causes:** • Injector 8 connector is damaged, open or shorted • Injector 8 control circuit is open, shorted to ground or to power (the injector driver circuit may be damaged)
DTC: P0211 **2T MIL: Yes** **1996, 1997, 1998, 1999, 2000, 2001** **Models:** 750i **Engines:** 5.4L **Transmissions:** All	**Cylinder 11 Injector Circuit Malfunction Conditions:** Engine started, and the DME detected the fuel injector "11" control circuit was in a high state when it should have been low, or in a low state when it should have been high (wiring harness & injector okay). **Note: Monitor the INJIF PID Fault "flags" with the Scan Tool. The appropriate INJF PID "flag" will read Yes when this code is set.** **Possible Causes:** • Injector 8 connector is damaged, open or shorted • Injector 8 control circuit is open, shorted to ground or to power (the injector driver circuit may be damaged)
DTC: P0212 **2T MIL: Yes** **1996, 1997, 1998, 1999, 2000, 2001** **Models:** 750i **Engines:** 5.4L **Transmissions:** All	**Cylinder 12 Injector Circuit Malfunction Conditions:** Engine started, and the DME detected the fuel injector "12" control circuit was in a high state when it should have been low, or in a low state when it should have been high (wiring harness & injector okay). **Note: Monitor the INJIF PID Fault "flags" with the Scan Tool. The appropriate INJF PID "flag" will read Yes when this code is set.** **Possible Causes:** • Injector 8 connector is damaged, open or shorted • Injector 8 control circuit is open, shorted to ground or to power (the injector driver circuit may be damaged)

DTC	Trouble Code Title, Conditions & Possible Causes
DTC: P0221 **2T MIL: Yes** **2001, 2002, 2003, 2004, 2005, 2006** **Models:** 320i, 323i, 325i, 328i, 330i, 525i, 530i, 540i, 545i, 550i, 740i, 745i, 745Li, 750i, M3, M5, Z3, Z4, Z8 **Engines:** 2.5L, 3.0L, 3.2L, 4.4L, 4.8L, 5.0L **Transmissions:** All	**Throttle Position Sensor 'B' Signal Performance Conditions:** Engine started, battery voltage at least 11.5v, all electrical components off, ground connections between engine and chassis well connected, coolant temperature at least 80-degrees Celsius and the throttle valve must not be damaged or dirty; and the DME detected the TP Sensor 'B' circuit was out of its normal operating range during a condition with the throttle wide open, or with it completely closed. The throttle valve activation occurs via an electric motor (throttle drive) in the throttle valve control module. It is activated by the DME according to specifications of the two sensors, Throttle Position Sensor and Accelerator Pedal Position Sensor 2. Slowly depress accelerator pedal up to Wide Open Throttle (WOT) stop while observing the percentage display on the PID data function of the scan tool. The percentage display must increase uniformly. The engine speed is greater than 1320rpm. **Possible Causes:** • Throttle body is damaged • Throttle linkage is binding or sticking • ETC TP Sensor 'B' signal circuit to the DME is open • ETC TP Sensor 'B' ground circuit is open • ETC TP Sensor 'B' is damaged or it has failed
DTC: P0222 **2T MIL: Yes** **2001, 2002, 2003, 2004, 2005, 2006** **Models:** 320i, 323i, 325i, 328i, 330i, 525i, 530i, 540i, 545i, 550i, 740i, 745i, 745Li, 750i, M3, M5, Z3, Z4, Z8 **Engines:** 2.5L, 3.0L, 3.2L, 4.4L, 4.8L, 5.0L **Transmissions:** All	**Throttle Position Sensor 'B' Circuit Low Input Conditions:** Engine started, battery voltage at least 11.5v, all electrical components off, ground connections between engine and chassis well connected, coolant temperature at least 80-degrees Celsius and the throttle valve must not be damaged or dirty; and the DME detected the TP Sensor 'B' circuit was out of its normal operating range during a condition with the throttle wide open, or with it completely closed. The throttle valve activation occurs via an electric motor (throttle drive) in the throttle valve control module. It is activated by the DME according to specifications of the two sensors, Throttle Position Sensor and Accelerator Pedal Position Sensor 2. Slowly depress accelerator pedal up to Wide Open Throttle (WOT) stop while observing the percentage display on the PID data function of the scan tool. The percentage display must increase uniformly. **Possible Causes:** • ETC TP Sensor 'B' connector is damaged or shorted • ETC TP Sensor 'B' signal circuit is shorted to ground • ETC TP Sensor 'B' is damaged or it has failed
DTC: P0223 **2T MIL: Yes** **2001, 2002, 2003, 2004, 2005, 2006** **Models:** 320i, 323i, 325i, 328i, 330i, 525i, 530i, 540i, 545i, 550i, 740i, 745i, 745Li, 750i, M3, M5, Z3, Z4, Z8 **Engines:** 2.5L, 3.0L, 3.2L, 4.4L, 4.8L, 5.0L **Transmissions:** All	**Throttle Position Sensor 'B' Circuit High Input Conditions:** Engine started, battery voltage at least 11.5v, all electrical components off, ground connections between engine and chassis well connected, coolant temperature at least 80-degrees Celsius and the throttle valve must not be damaged or dirty; and the DME detected the TP Sensor 'B' circuit was out of its normal operating range during a condition with the throttle wide open, or with it completely closed. The throttle valve activation occurs via an electric motor (throttle drive) in the throttle valve control module. It is activated by the DME according to specifications of the two sensors, Throttle Position Sensor and Accelerator Pedal Position Sensor 2. Slowly depress accelerator pedal up to Wide Open Throttle (WOT) stop while observing the percentage display on the PID data function of the scan tool. The percentage display must increase uniformly. **Possible Causes:** • ETC TP Sensor 'B' connector is damaged or open • ETC TP Sensor 'B' signal circuit is open • ETC TP Sensor 'B' signal circuit is shorted to VREF (5v) • ETC TP Sensor 'B' is damaged or it has failed
DTC: P0261 **2T MIL: Yes** **2001, 2002, 2003, 2004, 2005, 2006** **Models:** 320i, 323i, 325i, 328i, 330i, 525i, 530i, 540i, 545i, 550i, 740i, 745i, 745Li, 750i, M3, M5, Z3, Z4, Z8 **Engines:** 2.5L, 3.0L, 3.2L, 4.4L, 4.8L, 5.0L **Transmissions:** All	**Cylinder 1 Injector Circuit Low Input/Short to Ground Conditions:** Key on or engine running, fuses in the instrument panel and the E-box in the engine compartment must be functioning, and the ground connections between the engine ad the chassis must be well connected; and the DME detected an unexpected voltage condition on the injector circuit. **Possible Causes:** • Injector 1 control circuit is open • Injector 1 power circuit (B+) is open • Injector 1 control circuit is shorted to chassis ground • Injector 1 is damaged or has failed • DME is not connected or has failed
DTC: P0262 **2T MIL: Yes** **2001, 2002, 2003, 2004, 2005, 2006** **Models:** 320i, 323i, 325i, 328i, 330i, 525i, 530i, 540i, 545i, 550i, 740i, 745i, 745Li, 750i, M3, M5, Z3, Z4, Z8 **Engines:** 2.5L, 3.0L, 3.2L, 4.4L, 4.8L, 5.0L **Transmissions:** All	**Cylinder 1 Injector Circuit Low Input/Short to B+ Conditions:** Key on or engine running, fuses in the instrument panel and the E-box in the engine compartment must be functioning, and the ground connections between the engine ad the chassis must be well connected; and the DME detected an unexpected voltage condition on the injector circuit. **Possible Causes:** • Injector control circuit is open • Injector power circuit (B+) is open • Injector control circuit is shorted to chassis ground • Injector is damaged or has failed • DME is not connected or has failed • Fuel pump relay has failed • Fuel injectors may have malfunctioned • Faulty engine speed sensor

DTC	Trouble Code Title, Conditions & Possible Causes
DTC: P0264 **2T MIL: Yes** **2001, 2002, 2003, 2004, 2005, 2006** **Models:** 320i, 323i, 325i, 328i, 330i, 525i, 530i, 540i, 545i, 550i, 740i, 745i, 745Li, 750i, M3, M5, Z3, Z4, Z8 **Engines:** 2.5L, 3.0L, 3.2L, 4.4L, 4.8L, 5.0L **Transmissions:** All	**Cylinder 2 Injector Circuit Low Input/Short to Ground Conditions:** Key on or engine running, fuses in the instrument panel and the E-box in the engine compartment must be functioning, and the ground connections between the engine ad the chassis must be well connected; and the DME detected an unexpected voltage condition on the injector circuit. **Possible Causes:** • Injector control circuit is open • Injector power circuit (B+) is open • Injector control circuit is shorted to chassis ground • Injector is damaged or has failed • DME is not connected or has failed • Fuel pump relay has failed • Fuel injectors may have malfunctioned • Faulty engine speed sensor
DTC: P0265 **2T MIL: Yes** **2001, 2002, 2003, 2004, 2005, 2006** **Models:** 320i, 323i, 325i, 328i, 330i, 525i, 530i, 540i, 545i, 550i, 740i, 745i, 745Li, 750i, M3, M5, Z3, Z4, Z8 **Engines:** 2.5L, 3.0L, 3.2L, 4.4L, 4.8L, 5.0L **Transmissions:** All	**Cylinder 2 Injector Circuit Low Input/Short to B+ Conditions:** Key on or engine running, fuses in the instrument panel and the E-box in the engine compartment must be functioning, and the ground connections between the engine ad the chassis must be well connected; and the DME detected an unexpected voltage condition on the injector circuit. **Possible Causes:** • Injector control circuit is open • Injector power circuit (B+) is open • Injector control circuit is shorted to chassis ground • Injector is damaged or has failed • DME is not connected or has failed • Fuel pump relay has failed • Fuel injectors may have malfunctioned • Faulty engine speed sensor
DTC: P0267 **2T MIL: Yes** **2001, 2002, 2003, 2004, 2005, 2006** **Models:** 320i, 323i, 325i, 328i, 330i, 525i, 530i, 540i, 545i, 550i, 740i, 745i, 745Li, 750i, M3, M5, Z3, Z4, Z8 **Engines:** 2.5L, 3.0L, 3.2L, 4.4L, 4.8L, 5.0L **Transmissions:** All	**Cylinder 3 Injector Circuit Low Input/Short to Ground Conditions:** Key on or engine running, fuses in the instrument panel and the E-box in the engine compartment must be functioning, and the ground connections between the engine ad the chassis must be well connected; and the DME detected an unexpected voltage condition on the injector circuit. **Possible Causes:** • Injector control circuit is open • Injector power circuit (B+) is open • Injector control circuit is shorted to chassis ground • Injector is damaged or has failed • DME is not connected or has failed • Fuel pump relay has failed • Fuel injectors may have malfunctioned • Faulty engine speed sensor
DTC: P0268 **2T MIL: Yes** **2001, 2002, 2003, 2004, 2005, 2006** **Models:** 320i, 323i, 325i, 328i, 330i, 525i, 530i, 540i, 545i, 550i, 740i, 745i, 745Li, 750i, M3, M5, Z3, Z4, Z8 **Engines:** 2.5L, 3.0L, 3.2L, 4.4L, 4.8L, 5.0L **Transmissions:** All	**Cylinder 3 Injector Circuit Low Input/Short to B+ Conditions:** Key on or engine running, fuses in the instrument panel and the E-box in the engine compartment must be functioning, and the ground connections between the engine ad the chassis must be well connected; and the DME detected an unexpected voltage condition on the injector circuit. **Possible Causes:** • Injector control circuit is open • Injector power circuit (B+) is open • Injector control circuit is shorted to chassis ground • Injector is damaged or has failed • DME is not connected or has failed • Fuel pump relay has failed • Fuel injectors may have malfunctioned • Faulty engine speed sensor

DTC	Trouble Code Title, Conditions & Possible Causes
DTC: P0270 **2T MIL: Yes** **2001, 2002, 2003, 2004, 2005, 2006** **Models:** 320i, 323i, 325i, 328i, 330i, 525i, 530i, 540i, 545i, 550i, 740i, 745i, 745Li, 750i, M3, M5, Z3, Z4, Z8 **Engines:** 2.5L, 3.0L, 3.2L, 4.4L, 4.8L, 5.0L **Transmissions:** All	**Cylinder 4 Injector Circuit Low Input/Short to Ground Conditions:** Key on or engine running, fuses in the instrument panel and the E-box in the engine compartment must be functioning, and the ground connections between the engine ad the chassis must be well connected; and the DME detected an unexpected voltage condition on the injector circuit. **Possible Causes:** • Injector control circuit is open • Injector power circuit (B+) is open • Injector control circuit is shorted to chassis ground • Injector is damaged or has failed • DME is not connected or has failed • Fuel pump relay has failed • Fuel injectors may have malfunctioned • Faulty engine speed sensor
DTC: P0271 **2T MIL: Yes** **2001, 2002, 2003, 2004, 2005, 2006** **Models:** 320i, 323i, 325i, 328i, 330i, 525i, 530i, 540i, 545i, 550i, 740i, 745i, 745Li, 750i, M3, M5, Z3, Z4, Z8 **Engines:** 2.5L, 3.0L, 3.2L, 4.4L, 4.8L, 5.0L **Transmissions:** All	**Cylinder 4 Injector Circuit Low Input/Short to B+ Conditions:** Key on or engine running, fuses in the instrument panel and the E-box in the engine compartment must be functioning, and the ground connections between the engine ad the chassis must be well connected; and the DME detected an unexpected voltage condition on the injector circuit. **Possible Causes:** • Injector control circuit is open • Injector power circuit (B+) is open • Injector control circuit is shorted to chassis ground • Injector is damaged or has failed • DME is not connected or has failed • Fuel pump relay has failed • Fuel injectors may have malfunctioned • Faulty engine speed sensor
DTC: P0273 **2T MIL: Yes** **2001, 2002, 2003, 2004, 2005, 2006** **Models:** 320i, 323i, 325i, 328i, 330i, 525i, 530i, 540i, 545i, 550i, 740i, 745i, 745Li, 750i, M3, M5, Z3, Z4, Z8 **Engines:** 2.5L, 3.0L, 3.2L, 4.4L, 4.8L, 5.0L **Transmissions:** All	**Cylinder 5 Injector Circuit Low Input/Short to Ground Conditions:** Key on or engine running, fuses in the instrument panel and the E-box in the engine compartment must be functioning, and the ground connections between the engine ad the chassis must be well connected; and the DME detected an unexpected voltage condition on the injector circuit. **Possible Causes:** • Injector control circuit is open • Injector power circuit (B+) is open • Injector control circuit is shorted to chassis ground • Injector is damaged or has failed • DME is not connected or has failed • Fuel pump relay has failed • Fuel injectors may have malfunctioned • Faulty engine speed sensor
DTC: P0274 **2T MIL: Yes** **2001, 2002, 2003, 2004, 2005** **Models:** 320i, 323i, 325i, 328i, 330i, 525i, 530i, 540i, 545i, 550i, 740i, 745i, 745Li, 750i, M3, M5, Z3, Z4, Z8 **Engines:** 2.5L, 3.0L, 3.2L, 4.4L, 4.8L, 5.0L **Transmissions:** All	**Cylinder 5 Injector Circuit Low Input/Short to B+ Conditions:** Key on or engine running, fuses in the instrument panel and the E-box in the engine compartment must be functioning, and the ground connections between the engine ad the chassis must be well connected; and the DME detected an unexpected voltage condition on the injector circuit. **Possible Causes:** • Injector control circuit is open • Injector power circuit (B+) is open • Injector control circuit is shorted to chassis ground • Injector is damaged or has failed • DME is not connected or has failed • Fuel pump relay has failed • Fuel injectors may have malfunctioned • Faulty engine speed sensor

DTC	Trouble Code Title, Conditions & Possible Causes
DTC: P0276 **2T MIL: Yes** **2001, 2002, 2003, 2004, 2005, 2006** **Models:** 320i, 323i, 325i, 328i, 330i, 525i, 530i, 540i, 545i, 550i, 740i, 745i, 745Li, 750i, M3, M5, Z3, Z4, Z8 **Engines:** 2.5L, 3.0L, 3.2L, 4.4L, 4.8L, 5.0L **Transmissions:** All	**Cylinder 6 Injector Circuit Low Input/Short to Ground Conditions:** Key on or engine running, fuses in the instrument panel and the E-box in the engine compartment must be functioning, and the ground connections between the engine ad the chassis must be well connected; and the DME detected an unexpected voltage condition on the injector circuit. **Possible Causes:** • Injector control circuit is open • Injector power circuit (B+) is open • Injector control circuit is shorted to chassis ground • Injector is damaged or has failed • DME is not connected or has failed • Fuel pump relay has failed • Fuel injectors may have malfunctioned • Faulty engine speed sensor
DTC: P0277 **2T MIL: Yes** **2001, 2002, 2003, 2004, 2005, 2006** **Models:** 320i, 323i, 325i, 328i, 330i, 525i, 530i, 540i, 545i, 550i, 740i, 745i, 745Li, 750i, M3, M5, Z3, Z4, Z8 **Engines:** 2.5L, 3.0L, 3.2L, 4.4L, 4.8L, 5.0L **Transmissions:** All	**Cylinder 6 Injector Circuit Low Input/Short to B+ Conditions:** Key on or engine running, fuses in the instrument panel and the E-box in the engine compartment must be functioning, and the ground connections between the engine ad the chassis must be well connected; and the DME detected an unexpected voltage condition on the injector circuit. **Possible Causes:** • Injector control circuit is open • Injector power circuit (B+) is open • Injector control circuit is shorted to chassis ground • Injector is damaged or has failed • DME is not connected or has failed • Fuel pump relay has failed • Fuel injectors may have malfunctioned • Faulty engine speed sensor
DTC: P0279 **2T MIL: Yes** **2001, 2002, 2003, 2004, 2005, 2006** **Models:** 325i, 330i, 525i, 530i, 540i, 740i, 745i, 745Li, 750i, M3, M5, Z4, Z8 **Engines:** 3.0L, 3.2L, 4.4L, 4.8L, 5.0L, 5.4L **Transmissions:** All	**Cylinder 7 Injector Circuit Low Input/Short to Ground Conditions:** Key on or engine running, fuses in the instrument panel and the E-box in the engine compartment must be functioning, and the ground connections between the engine ad the chassis must be well connected; and the DME detected an unexpected voltage condition on the injector circuit. **Possible Causes:** • Injector control circuit is open • Injector power circuit (B+) is open • Injector control circuit is shorted to chassis ground • Injector is damaged or has failed • DME is not connected or has failed • Fuel pump relay has failed • Fuel injectors may have malfunctioned • Faulty engine speed sensor
DTC: P0280 **2T MIL: Yes** **2001, 2002, 2003, 2004, 2005, 2006** **Models:** 325i, 330i, 525i, 530i, 540i, 740i, 745i, 745Li, 750i, M3, M5, Z4, Z8 **Engines:** 3.0L, 3.2L, 4.4L, 4.8L, 5.0L, 5.4L **Transmissions:** All	**Cylinder 7 Injector Circuit Low Input/Short to B+ Conditions:** Key on or engine running, fuses in the instrument panel and the E-box in the engine compartment must be functioning, and the ground connections between the engine ad the chassis must be well connected; and the DME detected an unexpected voltage condition on the injector circuit. **Possible Causes:** • Injector control circuit is open • Injector power circuit (B+) is open • Injector control circuit is shorted to chassis ground • Injector is damaged or has failed • DME is not connected or has failed • Fuel pump relay has failed • Fuel injectors may have malfunctioned • Faulty engine speed sensor

DTC	Trouble Code Title, Conditions & Possible Causes
DTC: P0282 **2T MIL: Yes** **2001, 2002, 2003, 2004, 2005, 2006** **Models:** 325i, 330i, 525i, 530i, 540i, 740i, 745i, 745Li, 750i, M3, M5, Z4, Z8 **Engines:** 3.0L, 3.2L, 4.4L, 4.8L, 5.0L, 5.4L **Transmissions:** All	**Cylinder 8 Injector Circuit Low Input/Short to Ground Conditions:** Key on or engine running, fuses in the instrument panel and the E-box in the engine compartment must be functioning, and the ground connections between the engine ad the chassis must be well connected; and the DME detected an unexpected voltage condition on the injector circuit. **Possible Causes:** • Injector control circuit is open • Injector power circuit (B+) is open • Injector control circuit is shorted to chassis ground • Injector is damaged or has failed • DME is not connected or has failed • Fuel pump relay has failed • Fuel injectors may have malfunctioned • Faulty engine speed sensor
DTC: P0283 **2T MIL: Yes** **2001, 2002, 2003, 2004, 2005, 2006** **Models:** 325i, 330i, 525i, 530i, 540i, 740i, 745i, 745Li, 750i, M3, M5, Z4, Z8 **Engines:** 3.0L, 3.2L, 4.4L, 4.8L, 5.0L, 5.4L **Transmissions:** All	**Cylinder 8 Injector Circuit Low Input/Short to B+ Conditions:** Key on or engine running, fuses in the instrument panel and the E-box in the engine compartment must be functioning, and the ground connections between the engine ad the chassis must be well connected; and the DME detected an unexpected voltage condition on the injector circuit. **Possible Causes:** • Injector control circuit is open • Injector power circuit (B+) is open • Injector control circuit is shorted to chassis ground • Injector is damaged or has failed • DME is not connected or has failed • Fuel pump relay has failed • Fuel injectors may have malfunctioned • Faulty engine speed sensor
DTC: P0285 **2T MIL: Yes** **2001** **Models:** 750i **Engines:** 5.4L **Transmissions:** All	**Cylinder 9 Injector Circuit Low Input/Short to Ground Conditions:** Key on or engine running, fuses in the instrument panel and the E-box in the engine compartment must be functioning, and the ground connections between the engine ad the chassis must be well connected; and the DME detected an unexpected voltage condition on the injector circuit. **Possible Causes:** • Injector control circuit is open • Injector power circuit (B+) is open • Injector control circuit is shorted to chassis ground • Injector is damaged or has failed • DME is not connected or has failed • Fuel pump relay has failed • Fuel injectors may have malfunctioned • Faulty engine speed sensor
DTC: P0286 **2T MIL: Yes** **2001** **Models:** 750i **Engines:** 5.4L **Transmissions:** All	**Cylinder 9 Injector Circuit Low Input/Short to B+ Conditions:** Key on or engine running, fuses in the instrument panel and the E-box in the engine compartment must be functioning, and the ground connections between the engine ad the chassis must be well connected; and the DME detected an unexpected voltage condition on the injector circuit. **Possible Causes:** • Injector control circuit is open • Injector power circuit (B+) is open • Injector control circuit is shorted to chassis ground • Injector is damaged or has failed • DME is not connected or has failed • Fuel pump relay has failed • Fuel injectors may have malfunctioned • Faulty engine speed sensor

DTC	Trouble Code Title, Conditions & Possible Causes
DTC: P0288 **2T MIL: Yes** **2001** **Models:** 750i **Engines:** 5.4L **Transmissions:** All	**Cylinder 10 Injector Circuit Low Input/Short to Ground Conditions:** Key on or engine running, fuses in the instrument panel and the E-box in the engine compartment must be functioning, and the ground connections between the engine ad the chassis must be well connected; and the DME detected an unexpected voltage condition on the injector circuit. **Possible Causes:** • Injector control circuit is open • Injector power circuit (B+) is open • Injector control circuit is shorted to chassis ground • Injector is damaged or has failed • DME is not connected or has failed • Fuel pump relay has failed • Fuel injectors may have malfunctioned • Faulty engine speed sensor
DTC: P0289 **2T MIL: Yes** **2001** **Models:** 750i **Engines:** 5.4L **Transmissions:** All	**Cylinder 10 Injector Circuit Low Input/Short to B+ Conditions:** Key on or engine running, fuses in the instrument panel and the E-box in the engine compartment must be functioning, and the ground connections between the engine ad the chassis must be well connected; and the DME detected an unexpected voltage condition on the injector circuit. **Possible Causes:** • Injector control circuit is open • Injector power circuit (B+) is open • Injector control circuit is shorted to chassis ground • Injector is damaged or has failed • DME is not connected or has failed • Fuel pump relay has failed • Fuel injectors may have malfunctioned • Faulty engine speed sensor
DTC: P0291 **2T MIL: Yes** **2001** **Models:** 750i **Engines:** 5.4L **Transmissions:** All	**Cylinder 11 Injector Circuit Low Input/Short to Ground Conditions:** Key on or engine running, fuses in the instrument panel and the E-box in the engine compartment must be functioning, and the ground connections between the engine ad the chassis must be well connected; and the DME detected an unexpected voltage condition on the injector circuit. **Possible Causes:** • Injector control circuit is open • Injector power circuit (B+) is open • Injector control circuit is shorted to chassis ground • Injector is damaged or has failed • DME is not connected or has failed • Fuel pump relay has failed • Fuel injectors may have malfunctioned • Faulty engine speed sensor
DTC: P0292 **2T MIL: Yes** **2001** **Models:** 750i **Engines:** 5.4L **Transmissions:** All	**Cylinder 11 Injector Circuit Low Input/Short to B+ Conditions:** Key on or engine running, fuses in the instrument panel and the E-box in the engine compartment must be functioning, and the ground connections between the engine ad the chassis must be well connected; and the DME detected an unexpected voltage condition on the injector circuit. **Possible Causes:** • Injector control circuit is open • Injector power circuit (B+) is open • Injector control circuit is shorted to chassis ground • Injector is damaged or has failed • DME is not connected or has failed • Fuel pump relay has failed • Fuel injectors may have malfunctioned • Faulty engine speed sensor

DTC	Trouble Code Title, Conditions & Possible Causes
DTC: P0294 **2T MIL: Yes** **2001** **Models:** 750i **Engines:** 5.4L **Transmissions:** All	**Cylinder 12 Injector Circuit Low Input/Short to Ground Conditions:** Key on or engine running, fuses in the instrument panel and the E-box in the engine compartment must be functioning, and the ground connections between the engine ad the chassis must be well connected; and the DME detected an unexpected voltage condition on the injector circuit. **Possible Causes:** • Injector control circuit is open • Injector power circuit (B+) is open • Injector control circuit is shorted to chassis ground • Injector is damaged or has failed • DME is not connected or has failed • Fuel pump relay has failed • Fuel injectors may have malfunctioned • Faulty engine speed sensor
DTC: P0295 **2T MIL: Yes** **2001** **Models:** 750i **Engines:** 5.4L **Transmissions:** All	**Cylinder 12 Injector Circuit Low Input/Short to B+ Conditions:** Key on or engine running, fuses in the instrument panel and the E-box in the engine compartment must be functioning, and the ground connections between the engine ad the chassis must be well connected; and the DME detected an unexpected voltage condition on the injector circuit. **Possible Causes:** • Injector control circuit is open • Injector power circuit (B+) is open • Injector control circuit is shorted to chassis ground • Injector is damaged or has failed • DME is not connected or has failed • Fuel pump relay has failed • Fuel injectors may have malfunctioned • Faulty engine speed sensor
DTC: P0298 **2006** **Models:** 325i, 330i, 525i, 530i, 550i, 750i, Z4 **Engines:** 3.0L, 4.8L, 5.4L **Transmissions:** All	**Engine Oil Over Temperature Conditions:** The oil temperature difference of greater than 100 degrees within one second. The ignition must be on. The DME detected an error in the Engine Oil Temperature sensor. This occurs during attempted start value calibration. **Possible Causes:** • Replace the oil temperature sensor • Engine Oil temperature is too high • Engine coolant temperature is too high • Highest possible gear engaged in transmission • Check coolant
DTC: P0300 **1T MISFIRE, MIL: Yes** **1996, 1997, 1998, 1999, 2000, 2001, 2002, 2003, 2004, 2005, 2006** **Models:** 318i, 320i, 323i, 325i, 328i, 330i, 525i, 528i, 530i, 540i, 545i, 550i, 740i, 745i, 745Li, 750i, 840i, M3, M5, Z3, Z4, Z8 **Engines:** 1.9L, 2.5L, 2.8L, 3.0L, 3.2L, 4.4L, 4.8L, 5.0L **Transmissions:** All	**Random/Multiple Misfire Detected Conditions:** Engine running at an RPM greater than 400 but less than 6400 the DME detected a misfire or uneven engine running in two or more cylinders within 200 crankshaft rotations. Engine speed is between 480 and 4500rpm, load change is 0.4ms at ignition with a speed change of 2800rpms and the ASC is not active. **Note: If the misfire is severe, the MIL will flash on/off on the first trip!** **Possible Causes:** • Fuel metering fault that affects two or more cylinders • Fuel pressure too low or too high, fuel supply contaminated • EVAP system problem or the EVAP canister is fuel saturated • EGR valve is stuck open or the PCV system has a vacuum leak • Ignition system fault (coil, plugs) affecting two or more cylinders • MAF sensor contamination (it can cause a very lean condition) • Vehicle driven while very low on fuel (less than 1/8 of a tank)

DTC	Trouble Code Title, Conditions & Possible Causes
DTC: P0301 **1T MISFIRE, MIL: Yes** **1996, 1997, 1998, 1999, 2000, 2001, 2002, 2003, 2004, 2005, 2006** **Models:** 318i, 320i, 323i, 325i, 328i, 330i, 525i, 528i, 530i, 540i, 545i, 550i, 740i, 745i, 745Li, 750i, 840i, M3, M5, Z3, Z4, Z8 **Engines:** 1.9L, 2.5L, 2.8L, 3.0L, 3.2L, 4.4L, 4.8L, 5.0L **Transmissions:** All	**Cylinder Number 1 Misfire Detected Conditions:** Engine running at an RPM greater than 400 but less than 6400 the DME detected a misfire or uneven engine running in two or more cylinders within 200 crankshaft rotations. Engine speed is between 480 and 4500rpm, load change is 0.4ms at ignition with a speed change of 2800rpms and the ASC is not active. **Note: If the misfire is severe, the MIL will flash on/off on the first trip!** **Possible Causes:** • Air leak in the intake manifold, or in the EGR or DME system • Base engine mechanical problem • Fuel delivery component problem (i.e., a contaminated, dirty or sticking fuel injector) • Fuel pump relay defective • Ignition coil fuses have failed • Ignition system problem (dirty damaged coil or plug) • Engine speed (RPM) sensor has failed • Camshaft position sensors have failed • Ignition coil is faulty • Spark plugs are not working properly or are not gapped properly
DTC: P0302 **1T MISFIRE, MIL: Yes** **1996, 1997, 1998, 1999, 2000, 2001, 2002, 2003, 2004, 2005, 2006** **Models:** 318i, 320i, 323i, 325i, 328i, 330i, 525i, 528i, 530i, 540i, 545i, 550i, 740i, 745i, 745Li, 750i, 840i, M3, M5, Z3, Z4, Z8 **Engines:** 1.9L, 2.5L, 2.8L, 3.0L, 3.2L, 4.4L, 4.8L, 5.0L **Transmissions:** All	**Cylinder Number 2 Misfire Detected Conditions:** Engine running at an RPM greater than 400 but less than 6400 the DME detected a misfire or uneven engine running in two or more cylinders within 200 crankshaft rotations. Engine speed is between 480 and 4500rpm, load change is 0.4ms at ignition with a speed change of 2800rpms and the ASC is not active. **Note: If the misfire is severe, the MIL will flash on/off on the 1st trip!** **Possible Causes:** • Air leak in the intake manifold, or in the EGR or DME system • Base engine mechanical problem • Fuel delivery component problem (i.e., a contaminated, dirty or sticking fuel injector) • Fuel pump relay defective • Ignition coil fuses have failed • Ignition system problem (dirty damaged coil or plug) • Engine speed (RPM) sensor has failed • Camshaft position sensors have failed • Ignition coil is faulty • Spark plugs are not working properly or are not gapped properly
DTC: P0303 **1T MISFIRE, MIL: Yes** **1996, 1997, 1998, 1999, 2000, 2001, 2002, 2003, 2004, 2005, 2006** **Models:** 318i, 320i, 323i, 325i, 328i, 330i, 525i, 528i, 530i, 540i, 545i, 550i, 740i, 745i, 745Li, 750i, 840i, M3, M5, Z3, Z4, Z8 **Engines:** 1.9L, 2.5L, 2.8L, 3.0L, 3.2L, 4.4L, 4.8L, 5.0L **Transmissions:** All	**Cylinder Number 3 Misfire Detected Conditions:** Engine running at an RPM greater than 400 but less than 6400 the DME detected a misfire or uneven engine running in two or more cylinders within 200 crankshaft rotations. Engine speed is between 480 and 4500rpm, load change is 0.4ms at ignition with a speed change of 2800rpms and the ASC is not active. **Note: If the misfire is severe, the MIL will flash on/off on the 1st trip!** **Possible Causes:** • Air leak in the intake manifold, or in the EGR or DME system • Base engine mechanical problem • Fuel delivery component problem (i.e., a contaminated, dirty or sticking fuel injector) • Fuel pump relay defective • Ignition coil fuses have failed • Ignition system problem (dirty damaged coil or plug) • Engine speed (RPM) sensor has failed • Camshaft position sensors have failed • Ignition coil is faulty • Spark plugs are not working properly or are not gapped properly

DTC	Trouble Code Title, Conditions & Possible Causes
DTC: P0304 **1T MISFIRE, MIL: Yes** **1996, 1997, 1998, 1999, 2000, 2001, 2002, 2003, 2004, 2005, 2006** **Models:** 318i, 320i, 323i, 325i, 328i, 330i, 525i, 528i, 530i, 540i, 545i, 550i, 740i, 745i, 745Li, 750i, 840i, M3, M5, Z3, Z4, Z8 **Engines:** 1.9L, 2.5L, 2.8L, 3.0L, 3.2L, 4.4L, 4.8L, 5.0L **Transmissions:** All	**Cylinder Number 4 Misfire Detected Conditions:** Engine running at an RPM greater than 400 but less than 6400 the DME detected a misfire or uneven engine running in two or more cylinders within 200 crankshaft rotations. Engine speed is between 480 and 4500rpm, load change is 0.4ms at ignition with a speed change of 2800rpms and the ASC is not active. **Note: If the misfire is severe, the MIL will flash on/off on the 1st trip!** **Possible Causes:** • Air leak in the intake manifold, or in the EGR or DME system • Base engine mechanical problem • Fuel delivery component problem (i.e., a contaminated, dirty or sticking fuel injector) • Fuel pump relay defective • Ignition coil fuses have failed • Ignition system problem (dirty damaged coil or plug) • Engine speed (RPM) sensor has failed • Camshaft position sensors have failed • Ignition coil is faulty • Spark plugs are not working properly or are not gapped properly
DTC: P0305 **1T MISFIRE, MIL: Yes** **1996, 1997, 1998, 1999, 2000, 2001, 2002, 2003, 2004, 2005, 2006** **Models:** 318i, 320i, 323i, 325i, 328i, 330i, 525i, 528i, 530i, 540i, 545i, 550i, 740i, 745i, 745Li, 750i, 840i, M3, M5, Z3, Z4, Z8 **Engines:** 1.9L, 2.5L, 2.8L, 3.0L, 3.2L, 4.4L, 4.8L, 5.0L **Transmissions:** All	**Cylinder Number 5 Misfire Detected Conditions:** Engine running under positive torque conditions, and the DME detected a misfire or uneven engine function. Engine speed is between 480 and 4500rpm, load change is 0.4ms at ignition with a speed change of 2800rpms and the ASC is not active. **Note: If the misfire is severe, the MIL will flash on/off on the 1st trip!** **Possible Causes:** • Air leak in the intake manifold, or in the EGR or DME system • Base engine mechanical problem • Fuel delivery component problem (i.e., a contaminated, dirty or sticking fuel injector) • Fuel pump relay defective • Ignition coil fuses have failed • Ignition system problem (dirty damaged coil or plug) • Engine speed (RPM) sensor has failed • Camshaft position sensors have failed • Ignition coil is faulty • Spark plugs are not working properly or are not gapped properly
DTC: P0306 **1T MISFIRE, MIL: Yes** **1996, 1997, 1998, 1999, 2000, 2001, 2002, 2003, 2004, 2005, 2006** **Models:** 318i, 320i, 323i, 325i, 328i, 330i, 525i, 528i, 530i, 540i, 545i, 550i, 740i, 745i, 745Li, 750i, 840i, M3, M5, Z3, Z4, Z8 **Engines:** 1.9L, 2.5L, 2.8L, 3.0L, 3.2L, 4.4L, 4.8L, 5.0L **Transmissions:** All	**Cylinder Number 6 Misfire Detected Conditions:** Engine running under positive torque conditions, and the DME detected a misfire or uneven engine function. Engine speed is between 480 and 4500rpm, load change is 0.4ms at ignition with a speed change of 2800rpms and the ASC is not active. **Note: If the misfire is severe, the MIL will flash on/off on the 1st trip!** **Possible Causes:** • Air leak in the intake manifold, or in the EGR or DME system • Base engine mechanical problem • Fuel delivery component problem (i.e., a contaminated, dirty or sticking fuel injector) • Fuel pump relay defective • Ignition coil fuses have failed • Ignition system problem (dirty damaged coil or plug) • Engine speed (RPM) sensor has failed • Camshaft position sensors have failed • Ignition coil is faulty • Spark plugs are not working properly or are not gapped properly
DTC: P0307 **1T MISFIRE, MIL: Yes** **1996, 1997, 1998, 1999, 2000, 2001, 2002, 2003, 2004, 2005, 2006** **Models:** 318i, 320i, 323i, 325i, 328i, 330i, 525i, 528i, 530i, 540i, 545i, 550i, 740i, 745i, 745Li, 750i, 840i, M3, M5, Z3, Z4, Z8 **Engines:** 1.9L, 2.5L, 2.8L, 3.0L, 3.2L, 4.4L, 4.8L, 5.0L **Transmissions:** All	**Cylinder Number 7 Misfire Detected Conditions:** Engine running under positive torque conditions, and the DME detected a misfire or uneven engine function. Engine speed is between 480 and 4500rpm, load change is 0.4ms at ignition with a speed change of 2800rpms and the ASC is not active. **Note: If the misfire is severe, the MIL will flash on/off on the 1st trip!** **Possible Causes:** • Air leak in the intake manifold, or in the EGR or DME system • Base engine mechanical problem • Fuel delivery component problem (i.e., a contaminated, dirty or sticking fuel injector) • Fuel pump relay defective • Ignition coil fuses have failed • Ignition system problem (dirty damaged coil or plug) • Engine speed (RPM) sensor has failed • Camshaft position sensors have failed • Ignition coil is faulty • Spark plugs are not working properly or are not gapped properly

DTC	Trouble Code Title, Conditions & Possible Causes
DTC: P0308 **1T MISFIRE, MIL: Yes** **1996, 1997, 1998, 1999, 2000, 2001, 2002, 2003, 2004, 2005, 2006** **Models:** 318i, 320i, 323i, 325i, 328i, 330i, 525i, 528i, 530i, 540i, 545i, 550i, 740i, 745i, 745Li, 750i, 840i, M3, M5, Z3, Z4, Z8 **Engines:** 1.9L, 2.5L, 2.8L, 3.0L, 3.2L, 4.4L, 4.8L, 5.0L **Transmissions:** All	**Cylinder Number 8 Misfire Detected Conditions:** Engine running under positive torque conditions, and the DME detected a misfire or uneven engine function. Engine speed is between 480 and 4500rpm, load change is 0.4ms at ignition with a speed change of 2800rpms and the ASC is not active. **Note: If the misfire is severe, the MIL will flash on/off on the 1st trip!** **Possible Causes:** • Air leak in the intake manifold, or in the EGR or DME system • Base engine mechanical problem • Fuel delivery component problem (i.e., a contaminated, dirty or sticking fuel injector) • Fuel pump relay defective • Ignition coil fuses have failed • Ignition system problem (dirty damaged coil or plug) • Engine speed (RPM) sensor has failed • Camshaft position sensors have failed • Ignition coil is faulty • Spark plugs are not working properly or are not gapped properly
DTC: P0309 **1T MISFIRE, MIL: Yes** **1996, 1997, 1998** **Models:** 750i **Engines:** 5.4L **Transmissions:** All	**Cylinder Number 9 Misfire Detected Conditions:** Engine running under positive torque conditions, and the DME detected a misfire or uneven engine function. **Note: If the misfire is severe, the MIL will flash on/off on the 1st trip!** **Possible Causes:** • Air leak in the intake manifold, or in the EGR or DME system • Base engine mechanical problem • Fuel delivery component problem (i.e., a contaminated, dirty or sticking fuel injector) • Fuel pump relay defective • Ignition coil fuses have failed • Ignition system problem (dirty damaged coil or plug) • Engine speed (RPM) sensor has failed • Camshaft position sensors have failed • Ignition coil is faulty • Spark plugs are not working properly or are not gapped properly
DTC: P0310 **1T MISFIRE, MIL: Yes** **1996, 1997, 1998** **Models:** 750i **Engines:** 5.4L **Transmissions:** All	**Cylinder Number 10 Misfire Detected Conditions:** Engine running under positive torque conditions, and the DME detected a misfire or uneven engine function. **Note: If the misfire is severe, the MIL will flash on/off on the 1st trip!** **Possible Causes:** • Air leak in the intake manifold, or in the EGR or DME system • Base engine mechanical problem • Fuel delivery component problem (i.e., a contaminated, dirty or sticking fuel injector) • Fuel pump relay defective • Ignition coil fuses have failed • Ignition system problem (dirty damaged coil or plug) • Engine speed (RPM) sensor has failed • Camshaft position sensors have failed • Ignition coil is faulty • Spark plugs are not working properly or are not gapped properly
DTC: P0311 **1T MISFIRE, MIL: Yes** **1996, 1997, 1998** **Models:** 750i **Engines:** 5.4L **Transmissions:** All	**Cylinder Number 11 Misfire Detected Conditions:** Engine running under positive torque conditions, and the DME detected a misfire or uneven engine function. **Note: If the misfire is severe, the MIL will flash on/off on the 1st trip!** **Possible Causes:** • Air leak in the intake manifold, or in the EGR or DME system • Base engine mechanical problem • Fuel delivery component problem (i.e., a contaminated, dirty or sticking fuel injector) • Fuel pump relay defective • Ignition coil fuses have failed • Ignition system problem (dirty damaged coil or plug) • Engine speed (RPM) sensor has failed • Camshaft position sensors have failed • Ignition coil is faulty • Spark plugs are not working properly or are not gapped properly

DTC	Trouble Code Title, Conditions & Possible Causes
DTC: P0312 **1T MISFIRE, MIL: Yes** **1996, 1997, 1998** **Models:** 750i **Engines:** 5.4L **Transmissions:** All	**Cylinder Number 12 Misfire Detected Conditions:** Engine running under positive torque conditions, and the DME detected a misfire or uneven engine function. **Note: If the misfire is severe, the MIL will flash on/off on the 1st trip!** **Possible Causes:** • Air leak in the intake manifold, or in the EGR or DME system • Base engine mechanical problem • Fuel delivery component problem (i.e., a contaminated, dirty or sticking fuel injector) • Fuel pump relay defective • Ignition coil fuses have failed • Ignition system problem (dirty damaged coil or plug) • Engine speed (RPM) sensor has failed • Camshaft position sensors have failed • Ignition coil is faulty • Spark plugs are not working properly or are not gapped properly
DTC: P0313 **1T MISFIRE, MIL: Yes** **2005, 2006** **Models:** 325i, 330i, 352i, 530i, 750i, Z4 **Engines:** 2.5L, 3.0L, **Transmissions:** All	**Misfire Detected with Low Fuel Conditions:** Engine running under positive torque conditions, and the DME detected a misfire or uneven engine function. **Note: If the misfire is severe, the MIL will flash on/off on the 1st trip!** **Possible Causes:** • Air leak in the intake manifold, or in the EGR or DME system • Base engine mechanical problem • Fuel delivery component problem (i.e., a contaminated, dirty or sticking fuel injector) • Fuel pump relay defective • Ignition coil fuses have failed • Ignition system problem (dirty damaged coil or plug) • Engine speed (RPM) sensor has failed • Camshaft position sensors have failed • Ignition coil is faulty • Spark plugs are not working properly or are not gapped properly
DTC: P0324 **2T MIL: Yes** **2002** **Models:** 540i **Engines:** 4.4L **Transmissions:** All	**Knock Control System Error Conditions:** Engine started, vehicle driven, and the DME detected the Knock Sensor 1 (KS1) signal was too low or not recognized by the DME **Possible Causes:** • Knock sensor circuit is open • Knock sensor is loose (tighten to 20 NM) • Contact between the knock sensor and cylinder block is dirty, corroded or greasy • Knock sensor circuit is shorted to ground, or shorted to power • Knock sensor is damaged or it has failed • Wrong kind of fuel used • A component in the engine compartment is loose or not properly secured
DTC: P0325 **2T MIL: Yes** **1996, 1997, 1998, 1999, 2000, 2001, 2002, 2003, 2004, 2005, 2006** **Models:** 318i, 320i, 323i, 325i, 328i, 330i, 525i, 528i, 530i, 540i, 545i, 550i, 740i, 745i, 745Li, 750i, 840i, M3, M5, Z3, Z4, Z8 **Engines:** 1.9L, 2.5L, 2.8L, 3.0L, 3.2L, 4.4L, 4.8L, 5.0L, 5.4L **Transmissions:** All	**Knock Sensor 1 Circuit Malfunction Conditions:** Engine started, vehicle driven at 1520rpm for 3 seconds or to a temperature of 40 degrees Celsius, and the DME detected the Knock Sensor 1 (KS1) signal was not recognized. The engine speed is greater than 2080rpm but less than 6000rpm and the coolant temperature is greater than 40.5 degrees Celsius. **Possible Causes:** • Knock sensor circuit is open • Knock sensor is loose (tighten to 20 NM) • Contact between the knock sensor and cylinder block is dirty, corroded or greasy • Knock sensor circuit is shorted to ground, or shorted to power • Knock sensor is damaged or it has failed • Wrong kind of fuel used • A component in the engine compartment is loose or not properly secured
DTC: P0327 **2T MIL: Yes** **1996, 2001, 2002, 2003, 2004, 2005, 2006** **Models:** 318i, 320i, 323i, 325i, 328i, 330i, 525i, 528i, 530i, 540i, 545i, 550i, 740i, 745i, 745Li, 750i, 840i, M3, M5, Z3, Z4, Z8 **Engines:** 1.9L, 2.5L, 2.8L, 3.0L, 3.2L, 4.4L, 4.8L, 5.0L, 5.4L **Transmissions:** All	**Knock Sensor 1 Signal Low Input Conditions:** **Engine started, vehicle driven at 2000rpm for 3 seconds or to a temperature of 40 degrees Celsius, and the DME detected the Knock Sensor 1 (KS1) signal was too low or not recognized by the DME** **Possible Causes:** • Knock sensor circuit is open • Knock sensor is loose (tighten to 20 NM) • Contact between the knock sensor and cylinder block is dirty, corroded or greasy • Knock sensor circuit is shorted to ground, or shorted to power • Knock sensor is damaged or it has failed • Wrong kind of fuel used • A component in the engine compartment is loose or not properly secured

DTC	Trouble Code Title, Conditions & Possible Causes
DTC: P0328 **2T MIL: Yes** **2001, 2002, 2003, 2004, 2005, 2006** **Models:** 320i, 325i, 330i, 525i, 530i, 540i, 545i, 550i, 740i, 745i, 750i, M3, Z4 **Engines:** 2.5L, 3.0L, 3.2L, 4.4L, 4.8L, 5.4L **Transmissions:** All	**Knock Sensor 1 Signal High Input Conditions:** Engine started, vehicle driven at 1600rpm for 3 seconds or to a temperature of 40 degrees Celsius, and the DME detected the Knock Sensor 1 (KS1) signal was too high **Possible Causes:** • Knock sensor circuit is open • Knock sensor is loose (tighten to 20 NM) • Contact between the knock sensor and cylinder block is dirty, corroded or greasy • Knock sensor circuit is shorted to ground, or shorted to power • Knock sensor is damaged or it has failed • Wrong kind of fuel used • A component in the engine compartment is loose or not properly secured
DTC: P0330 **2T MIL: Yes** **1996, 2001, 2002, 2003, 2004, 2005, 2006** **Models:** 318i, 320i, 323i, 325i, 328i, 330i, 525i, 528i, 530i, 540i, 545i, 550i, 740i, 745i, 745Li, 750i, 840i, M3, M5, Z3, Z4, Z8 **Engines:** 1.9L, 2.5L, 2.8L, 3.0L, 3.2L, 4.4L, 4.8L, 5.0L, 5.4L **Transmissions:** All	**Knock Sensor 1 Circuit Malfunction Conditions:** Engine started, vehicle driven at 1520rpm for 3 seconds or to a temperature of 40 degrees Celsius, and the DME detected the Knock Sensor 1 (KS1) signal was not recognized. The engine speed is greater than 2080rpm but less than 6000rpm and the coolant temperature is greater than 40.5 degrees Celsius. **Possible Causes:** • Knock sensor circuit is open • Knock sensor is loose (tighten to 20 NM) • Contact between the knock sensor and cylinder block is dirty, corroded or greasy • Knock sensor circuit is shorted to ground, or shorted to power • Knock sensor is damaged or it has failed • Wrong kind of fuel used • A component in the engine compartment is loose or not properly secured
DTC: P0332 **2T MIL: Yes** **1996, 2001, 2002, 2003, 2004, 2005, 2006** **Models:** 318i, 320i, 323i, 325i, 328i, 330i, 525i, 528i, 530i, 540i, 545i, 550i, 740i, 745i, 745Li, 750i, 840i, M3, M5, Z3, Z4, Z8 **Engines:** 1.9L, 2.5L, 2.8L, 3.0L, 3.2L, 4.4L, 4.8L, 5.0L, 5.4L **Transmissions:** All	**Knock Sensor 2 Signal Low Input Conditions:** Engine started, vehicle driven, and the DME detected the Knock Sensor 1 (KS1) signal was too low or not recognized by the DME **Possible Causes:** • Knock sensor circuit is open • Knock sensor is loose (tighten to 20 NM) • Contact between the knock sensor and cylinder block is dirty, corroded or greasy • Knock sensor circuit is shorted to ground, or shorted to power • Knock sensor is damaged or it has failed • Wrong kind of fuel used • A component in the engine compartment is loose or not properly secured
DTC: P0333 **2T MIL: Yes** **2001, 2002, 2003** **Models:** 540i, 740i, 750i **Engines:** 4.4L, 4.8L, 5.4L **Transmissions:** All	**Knock Sensor 2 Signal High Input Conditions:** Engine started, vehicle driven, and the DME detected the Knock Sensor 1 (KS1) signal was too high **Possible Causes:** • Knock sensor circuit is open • Knock sensor is loose (tighten to 20 NM) • Contact between the knock sensor and cylinder block is dirty, corroded or greasy • Knock sensor circuit is shorted to ground, or shorted to power • Knock sensor is damaged or it has failed • Wrong kind of fuel used • A component in the engine compartment is loose or not properly secured
DTC: P0335 **2T MIL: Yes** **1996, 1997, 1998, 1999, 2000, 2001, 2002, 2003, 2004, 2005, 2006** **Models:** 318i, 320i, 323i, 325i, 328i, 330i, 525i, 528i, 530i, 540i, 545i, 550i, 740i, 745i, 745Li, 750i, 840i, M3, M5, Z3, Z4, Z8 **Engines:** 1.9L, 2.5L, 2.8L, 3.0L, 3.2L, 4.4L, 4.8L, 5.0L, 5.4L **Transmissions:** All	**Camshaft Position Sensor "A" Circ Malfunction Conditions:** Engine started, battery voltage must be at least 11.5v, all electrical components must be off, parking brake must be engaged (to keep daytime driving lights off), automatic transmission selector must be in park and the ground between the engine and the chassis must be well connected. The DME detected the CMP sensor signal was implausible. Engine speed is greater than 500rpm, and the fault is tolerable as long as there are no misfired occurring at the same time. **Possible Causes:** • CMP sensor circuit is open or shorted to ground • CMP sensor circuit is shorted to power • CMP sensor ground (return) circuit is open • CMP sensor installation incorrect (Hall-effect type) • CMP sensor is damaged or CMP sensor shielding damaged

DTC	Trouble Code Title, Conditions & Possible Causes
DTC: P0336 **2T MIL: Yes** **2005, 2006** **Models:** 325i, 330i, 525i, 530i, 545i, 550i, 745i, 750i, Z4 **Engines:** 3.0L, 4.8L **Transmissions:** All	**Camshaft Position Sensor "A" Circ Range/Performance Conditions:** Engine started (and engine speed is less than 25rpm), battery voltage must be at least 11.5v, all electrical components must be off, parking brake must be engaged (to keep daytime driving lights off), automatic transmission selector must be in park and the ground between the engine and the chassis must be well connected. The DME detected the CMP sensor signal was implausible. **Possible Causes:** • CMP sensor circuit is open or shorted to ground • CMP sensor circuit is shorted to power • CMP sensor ground (return) circuit is open • CMP sensor installation incorrect (Hall-effect type) • CMP sensor is damaged or CMP sensor shielding damaged
DTC: P0339 **2T MIL: Yes** **2001, 2002, 2003, 2004, 2005** **Models:** 320i, 325i, 330i, 525i, 530i, 540i, 745i, 750i, M3, Z3, Z4 **Engines:** 2.5L, 3.0L, 3.2L, 4.8L **Transmissions:** All	**Camshaft Position Sensor Circuit Malfunction Conditions:** Engine started, battery voltage must be at least 11.5v, all electrical components must be off, parking brake must be engaged (to keep daytime driving lights off), automatic transmission selector must be in park and the ground between the engine and the chassis must be well connected. The DME detected the CMP sensor signal was missing or it was erratic. There is no signal or an invalid one, and the engine speed is greater than 200rpm for two cycles. **Possible Causes:** • CMP sensor circuit is open or shorted to ground • CMP sensor circuit is shorted to power • CMP sensor ground (return) circuit is open • CMP sensor installation incorrect (Hall-effect type) • CMP sensor is damaged or CMP sensor shielding damaged • CMP sensor has failed
DTC: P0340 **2T MIL: Yes** **1996, 1997, 1998, 1999, 2000, 2001, 2002, 2003, 2004, 2005, 2006** **Models:** 318i, 320i, 323i, 325i, 328i, 330i, 525i, 528i, 530i, 540i, 545i, 550i, 740i, 745i, 745Li, 750i, 840i, M3, M5, Z3, Z4, Z8 **Engines:** 1.9L, 2.5L, 2.8L, 3.0L, 3.2L, 4.4L, 4.8L, 5.0L, 5.4L **Transmissions:** All	**Camshaft Position Sensor Circuit Malfunction Conditions:** Engine started, battery voltage must be at least 11.5v, all electrical components must be off, parking brake must be engaged (to keep daytime driving lights off), automatic transmission selector must be in park and the ground between the engine and the chassis must be well connected. The DME detected the CMP sensor signal was missing or it was erratic. There is no signal or an invalid one, and the engine speed is greater than 200rpm for two cycles. **Possible Causes:** • CMP sensor circuit is open or shorted to ground • CMP sensor circuit is shorted to power • CMP sensor ground (return) circuit is open • CMP sensor installation incorrect (Hall-effect type) • CMP sensor is damaged or CMP sensor shielding damaged • CMP sensor has failed
DTC: P0341 **2T MIL: Yes** **2002, 2003, 2004, 2005, 2006** **Models:** 325i, 330i, 525i, 530i, 545i, 550i, 745i, 750i, Z4 **Engines:** 3.0L, 4.8L **Transmissions:** All	**Camshaft Position Sensor Circ Range/Performance Conditions:** Engine started, battery voltage must be at least 11.5v, all electrical components must be off, parking brake must be engaged (to keep daytime driving lights off), automatic transmission selector must be in park and the ground between the engine and the chassis must be well connected. The DME detected the CMP sensor signal was implausible. **Possible Causes:** • CMP sensor circuit is open or shorted to ground • CMP sensor circuit is shorted to power • CMP sensor ground (return) circuit is open • CMP sensor installation incorrect (Hall-effect type) • CMP sensor is damaged or CMP sensor shielding damaged
DTC: P0342 **2T MIL: Yes** **2002, 2003, 2004, 2005, 2006** **Models:** 325i, 330i, 525i, 530i, 545i, 550i, 745i, 750i, Z4 **Engines:** 3.0L, 4.8L **Transmissions:** All	**Camshaft Position Sensor "A" Circuit (Bank 1 or Single Sensor) Low Input Conditions:** Engine started, battery voltage must be at least 11.5v, all electrical components must be off, parking brake must be engaged (to keep daytime driving lights off), automatic transmission selector must be in park and the ground between the engine and the chassis must be well connected. The DME detected the CMP sensor signal exceeded the bounds of the specified maximum limit. **Possible Causes:** • CMP sensor circuit is open or shorted to ground • CMP sensor circuit is shorted to power • CMP sensor ground (return) circuit is open • CMP sensor installation incorrect (Hall-effect type) • CMP sensor is damaged or CMP sensor shielding damaged

DTC	Trouble Code Title, Conditions & Possible Causes
DTC: P0343 **2T MIL: Yes** **2002, 2003, 2004, 2005, 2006** **Models:** 325i, 330i, 525i, 530i, 545i, 550i, 745i, 750i, Z4 **Engines:** 3.0L, 4.8L **Transmissions:** All	**Camshaft Position Sensor "A" Circuit (Bank 1 or Single Sensor) High Input Conditions:** Engine started, battery voltage must be at least 11.5v, all electrical components must be off, parking brake must be engaged (to keep daytime driving lights off), automatic transmission selector must be in park and the ground between the engine and the chassis must be well connected. The DME detected the CMP sensor signal did not reach the specified minimum limit. **Possible Causes:** • CMP sensor circuit is open or shorted to ground • CMP sensor circuit is shorted to power • CMP sensor ground (return) circuit is open • CMP sensor installation incorrect (Hall-effect type) • CMP sensor is damaged or CMP sensor shielding damaged
DTC: P0344 **2T MIL: Yes** **2001, 2002, 2003** **Models:** 325i, 330i, 525i, 530i, 540i, 745i, 750i, Z3, Z4 **Engines:** 2.5L, 3.0L **Transmissions:** All	**Camshaft Position Sensor Circuit Malfunction Conditions:** Engine started, battery voltage must be at least 11.5v, all electrical components must be off, parking brake must be engaged (to keep daytime driving lights off), automatic transmission selector must be in park and the ground between the engine and the chassis must be well connected. The DME detected the CMP sensor signal was missing or it was erratic. There is no signal or an invalid one, and the engine speed is greater than 200rpm for two cycles. **Possible Causes:** • CMP sensor circuit is open or shorted to ground • CMP sensor circuit is shorted to power • CMP sensor ground (return) circuit is open • CMP sensor installation incorrect (Hall-effect type) • CMP sensor is damaged or CMP sensor shielding damaged • CMP sensor has failed
DTC: P0345 **2T MIL: Yes** **2000, 2001, 2002, 2003** **Models:** 325i, 330i, 525i, 530i, 540i, 550i, 740i, 745i, 750i, M5, Z3, Z4, Z8 **Engines:** 2.5L, 3.0L, 4.4L, 4.8L, 5.0L **Transmissions:** All	**Camshaft Position Sensor "A" Circuit (Bank 2) Conditions:** Engine started, battery voltage must be at least 11.5v, all electrical components must be off, parking brake must be engaged (to keep daytime driving lights off), automatic transmission selector must be in park and the ground between the engine and the chassis must be well connected. The DME detected the CMP sensor signal was missing or it was erratic. **Possible Causes:** • CMP sensor circuit is open or shorted to ground • CMP sensor circuit is shorted to power • CMP sensor ground (return) circuit is open • CMP sensor installation incorrect (Hall-effect type) • CMP sensor is damaged or CMP sensor shielding damaged
DTC: P0346 **2T MIL: Yes** **2002, 2003, 2004, 2005, 2006** **Models:** 325i, 330i, 525i, 530i, 545i, 550i, 745i, 750i, Z4 **Engines:** 3.0L, 4.8L **Transmissions:** All	**Camshaft Position Sensor "A" Circuit (Bank 2) Range/Performance Conditions:** Engine started, battery voltage must be at least 11.5v, all electrical components must be off, parking brake must be engaged (to keep daytime driving lights off), automatic transmission selector must be in park and the ground between the engine and the chassis must be well connected. The DME detected the CMP sensor signal was implausible. **Possible Causes:** • CMP sensor circuit is open or shorted to ground • CMP sensor circuit is shorted to power • CMP sensor ground (return) circuit is open • CMP sensor installation incorrect (Hall-effect type) • CMP sensor is damaged or CMP sensor shielding damaged
DTC: P0347 **2T MIL: Yes** **2002, 2003, 2004, 2005, 2006** **Models:** 325i, 330i, 525i, 530i, 545i, 550i, 745i, 750i, Z4 **Engines:** 3.0L, 4.8L **Transmissions:** All	**Camshaft Position Sensor "A" Circuit (Bank 2) Low Input Conditions:** Engine started, battery voltage must be at least 11.5v, all electrical components must be off, parking brake must be engaged (to keep daytime driving lights off), automatic transmission selector must be in park and the ground between the engine and the chassis must be well connected. The DME detected the CMP sensor signal exceeded the bounds of the specified maximum limit. **Possible Causes:** • CMP sensor circuit is open or shorted to ground • CMP sensor circuit is shorted to power • CMP sensor ground (return) circuit is open • CMP sensor installation incorrect (Hall-effect type) • CMP sensor is damaged or CMP sensor shielding damaged

DTC	Trouble Code Title, Conditions & Possible Causes
DTC: P0348 **2T MIL: Yes** **2002, 2003, 2004, 2005, 2006** **Models:** 325i, 330i, 525i, 530i, 545i, 550i, 745i, 750i, Z4 **Engines:** 3.0L, 4.8L **Transmissions:** All	**Camshaft Position Sensor "A" Circuit (Bank 2) High Input Conditions:** Engine started, battery voltage must be at least 11.5v, all electrical components must be off, parking brake must be engaged (to keep daytime driving lights off), automatic transmission selector must be in park and the ground between the engine and the chassis must be well connected. The DME detected the CMP sensor signal did not reach the specified minimum limit. **Possible Causes:** • CMP sensor circuit is open or shorted to ground • CMP sensor circuit is shorted to power • CMP sensor ground (return) circuit is open • CMP sensor installation incorrect (Hall-effect type) • CMP sensor is damaged or CMP sensor shielding damaged
DTC: P0351 **2006** **Models:** 325i, 330i, 525i, 530i, 550i, 750i, Z4 **Engines:** 3.0L, 4.8L **Transmissions:** All	**Ignition Coilpack A Primary/Secondary Circuit Malfunction Conditions:** Engine started, battery voltage must be at least 11.5v, all electrical components must be off, parking brake must be engaged (to keep daytime driving lights off), automatic transmission selector must be in park and the ground between the engine and the chassis must be well connected. The DME did not receive any valid pulses from the ignition module for the Ignition Coilpack A primary circuit. **Note: Ignition coils and power output stages are one component and cannot be replaced individually.** **Possible Causes:** • Engine speed (RPM) sensor has failed • Camshaft Position (CMP) sensor has failed • Power Supply Relay is shorted to an open circuit • There is a malfunction in voltage supply • Ignition coilpack is damaged or it has failed • Cylinder 1 to 4 Fuel Injector(s) have failed
DTC: P0353 **2006** **Models:** 325i, 330i, 525i, 530i, 550i, 750i, Z4 **Engines:** 3.0L, 4.8L **Transmissions:** All	**Ignition Coilpack C Primary/Secondary Circuit Malfunction Conditions:** Engine started, battery voltage must be between 9 and 17 volts. The DME did not receive any valid pulses from the ignition module for the Ignition Coilpack C primary circuit. Voltage supplied and ground must be connected for ignition system spark plugs and coils. Check wiring harness, ground connection and plug-in contacts. Visual inspection of spark plug, ignition coil (replace if damaged). After excluding all of these faults, replace the control module. The injection is deactivated with a combustion miss and supplementary recognition of a rough running diagnosis. **Note: Ignition coils and power output stages are one component and cannot be replaced individually.** **Possible Causes:** • Engine speed (RPM) sensor has failed • Camshaft Position (CMP) sensor has failed • Power Supply Relay is shorted to an open circuit • There is a malfunction in voltage supply • Ignition coilpack is damaged or it has failed • Cylinder 1 to 4 Fuel Injector(s) have failed
DTC: P0354 **2006** **Models:** 325i, 330i, 525i, 530i, 550i, 750i, Z4 **Engines:** 3.0L, 4.8L **Transmissions:** All	**Ignition Coilpack D Primary/Secondary Circuit Malfunction Conditions:** Engine started, battery voltage must be between 9 and 17 volts. The DME did not receive any valid pulses from the ignition module for the Ignition Coilpack C primary circuit. Voltage supplied and ground must be connected for ignition system spark plugs and coils. Check wiring harness, ground connection and plug-in contacts. Visual inspection of spark plug, ignition coil (replace if damaged). After excluding all of these faults, replace the control module. The injection is deactivated with a combustion miss and supplementary recognition of a rough running diagnosis. **Note: Ignition coils and power output stages are one component and cannot be replaced individually.** **Possible Causes:** • Engine speed (RPM) sensor has failed • Camshaft Position (CMP) sensor has failed • Power Supply Relay is shorted to an open circuit • There is a malfunction in voltage supply • Ignition coilpack is damaged or it has failed • Cylinder 1 to 4 Fuel Injector(s) have failed

DTC	Trouble Code Title, Conditions & Possible Causes
DTC: P0355 **2006** **Models:** 325i, 330i, 525i, 530i, 550i, 750i, Z4 **Engines:** 3.0L, 4.8L **Transmissions:** All	**Ignition Coilpack E Primary/Secondary Circuit Malfunction Conditions:** Engine started, battery voltage must be between 9 and 17 volts. The DME did not receive any valid pulses from the ignition module for the Ignition Coilpack C primary circuit. Voltage supplied and ground must be connected for ignition system spark plugs and coils. Check wiring harness, ground connection and plug-in contacts. Visual inspection of spark plug, ignition coil (replace if damaged). After excluding all of these faults, replace the control module. The injection is deactivated with a combustion miss and supplementary recognition of a rough running diagnosis. **Note: Ignition coils and power output stages are one component and cannot be replaced individually.** **Possible Causes:** • Engine speed (RPM) sensor has failed • Camshaft Position (CMP) sensor has failed • Power Supply Relay is shorted to an open circuit • There is a malfunction in voltage supply • Ignition coilpack is damaged or it has failed • Cylinder 1 to 4 Fuel Injector(s) have failed
DTC: P0356 **2006** **Models:** 325i, 330i, 525i, 530i, 550i, 750i, Z4 **Engines:** 3.0L, 4.8L **Transmissions:** All	**Ignition Coilpack F Primary/Secondary Circuit Malfunction Conditions:** Engine started, battery voltage must be between 9 and 17 volts. The DME did not receive any valid pulses from the ignition module for the Ignition Coilpack C primary circuit. Voltage supplied and ground must be connected for ignition system spark plugs and coils. Check wiring harness, ground connection and plug-in contacts. Visual inspection of spark plug, ignition coil (replace if damaged). After excluding all of these faults, replace the control module. The injection is deactivated with a combustion miss and supplementary recognition of a rough running diagnosis. **Note: Ignition coils and power output stages are one component and cannot be replaced individually.** **Possible Causes:** • Engine speed (RPM) sensor has failed • Camshaft Position (CMP) sensor has failed • Power Supply Relay is shorted to an open circuit • There is a malfunction in voltage supply • Ignition coilpack is damaged or it has failed • Cylinder 1 to 4 Fuel Injector(s) have failed
DTC: P0357 **2006** **Models:** 325i, 330i, 525i, 530i, 550i, 750i, Z4 **Engines:** 3.0L, 4.8L **Transmissions:** All	**Ignition Coilpack G Primary/Secondary Circuit Malfunction Conditions:** Engine started, battery voltage must be between 9 and 17 volts. The DME did not receive any valid pulses from the ignition module for the Ignition Coilpack C primary circuit. Voltage supplied and ground must be connected for ignition system spark plugs and coils. Check wiring harness, ground connection and plug-in contacts. Visual inspection of spark plug, ignition coil (replace if damaged). After excluding all of these faults, replace the control module. The injection is deactivated with a combustion miss and supplementary recognition of a rough running diagnosis. **Note: Ignition coils and power output stages are one component and cannot be replaced individually.** **Possible Causes:** • Engine speed (RPM) sensor has failed • Camshaft Position (CMP) sensor has failed • Power Supply Relay is shorted to an open circuit • There is a malfunction in voltage supply • Ignition coilpack is damaged or it has failed • Cylinder 1 to 4 Fuel Injector(s) have failed
DTC: P0358 **2006** **Models:** 325i, 330i, 525i, 530i, 550i, 750i, Z4 **Engines:** 3.0L, 4.8L **Transmissions:** All	**Ignition Coilpack H Primary/Secondary Circuit Malfunction Conditions:** Engine started, battery voltage must be between 9 and 17 volts. The DME did not receive any valid pulses from the ignition module for the Ignition Coilpack C primary circuit. Voltage supplied and ground must be connected for ignition system spark plugs and coils. Check wiring harness, ground connection and plug-in contacts. Visual inspection of spark plug, ignition coil (replace if damaged). After excluding all of these faults, replace the control module. The injection is deactivated with a combustion miss and supplementary recognition of a rough running diagnosis. **Note: Ignition coils and power output stages are one component and cannot be replaced individually.** **Possible Causes:** • Engine speed (RPM) sensor has failed • Camshaft Position (CMP) sensor has failed • Power Supply Relay is shorted to an open circuit • There is a malfunction in voltage supply • Ignition coilpack is damaged or it has failed • Cylinder 1 to 4 Fuel Injector(s) have failed

DTC	Trouble Code Title, Conditions & Possible Causes
DTC: P0365 **2T MIL: Yes** **2001, 2002, 2003, 2004, 2005, 2006** **Models:** All **Engines:** All **Transmissions:** All	**Camshaft Position Sensor "B" Circuit (Bank 1) Conditions:** Engine started, battery voltage must be at least 11.5v, all electrical components must be off, parking brake must be engaged (to keep daytime driving lights off), automatic transmission selector must be in park and the ground between the engine and the chassis must be well connected. The DME detected the CMP sensor signal exceeded the bounds of the specified maximum limit. Flank number within three camshaft revolutions not 0, 1, 11, 12, 13. The number of phase flanks per cycle is implausible. **Possible Causes:** • CMP sensor circuit is open or shorted to ground • CMP sensor circuit is shorted to power • CMP sensor ground (return) circuit is open • CMP sensor installation incorrect (Hall-effect type) • CMP sensor is damaged or CMP sensor shielding damaged
DTC: P0366 **2T MIL: Yes** **2002, 2003, 2004, 2005, 2006** **Models:** 325i, 330i, 525i, 530i, 545i, 550i, 745i, 750i, Z4 **Engines:** 3.0L, 4.8L **Transmissions:** All	**Camshaft Position Sensor "B" Circuit (Bank 1) Range/Performance Conditions:** Engine started, battery voltage must be at least 11.5v, all electrical components must be off, parking brake must be engaged (to keep daytime driving lights off), automatic transmission selector must be in park and the ground between the engine and the chassis must be well connected. The DME detected the CMP sensor signal exceeded the bounds of the specified maximum limit. **Possible Causes:** • CMP sensor circuit is open or shorted to ground • CMP sensor circuit is shorted to power • CMP sensor ground (return) circuit is open • CMP sensor installation incorrect (Hall-effect type) • CMP sensor is damaged or CMP sensor shielding damaged
DTC: P0367 **2T MIL: Yes** **2002, 2003, 2004, 2005, 2006** **Models:** 325i, 330i, 525i, 530i, 545i, 550i, 745i, 750i, Z4 **Engines:** 3.0L, 4.8L **Transmissions:** All	**Camshaft Position Sensor "B" Circuit (Bank 1) Low Input Conditions:** Engine started, battery voltage must be at least 11.5v, all electrical components must be off, parking brake must be engaged (to keep daytime driving lights off), automatic transmission selector must be in park and the ground between the engine and the chassis must be well connected. The DME detected the CMP sensor signal exceeded the bounds of the specified maximum limit. **Possible Causes:** • CMP sensor circuit is open or shorted to ground • CMP sensor circuit is shorted to power • CMP sensor ground (return) circuit is open • CMP sensor installation incorrect (Hall-effect type) • CMP sensor is damaged or CMP sensor shielding damaged
DTC: P0368 **2T MIL: Yes** **2002, 2003, 2004, 2005, 2006** **Models:** 325i, 330i, 525i, 530i, 545i, 550i, 745i, 750i, Z4 **Engines:** 3.0L, 4.8L **Transmissions:** All	**Camshaft Position Sensor "B" Circuit (Bank 1) High Input Conditions:** Engine turning over for at least nine faults, battery voltage must be at least 11.5v, there must be multiple reference points lost, signal faults or intermittent contact on KWG signal wire. VVT emergency default mode (max stroke) active, VANOS emergency default mode (spec. 120 degrees) active, RPM sensor emergency default mode active. **Possible Causes:** • CMP sensor circuit is open or shorted to ground • CMP sensor circuit is shorted to power • CMP sensor ground (return) circuit is open • CMP sensor installation incorrect (Hall-effect type) • Defective KWG • Excessive gap between KWG and KW (or deformed KW)
DTC: P0369 **2T MIL: Yes** **2001, 2002, 2003, 2004, 2005, 2006** **Models:** 325i, 330i, 525i, 530i, M3, Z3, Z4 **Engines:** 2.5L, 3.0L, 3.2L, 5.0L **Transmissions:** All	**Crankshaft Position Sensor Rationality Check Conditions:** Engine started, battery voltage must be at least 11.5v, all electrical components must be off, parking brake must be engaged (to keep daytime driving lights off), automatic transmission selector must be in park and the ground between the engine and the chassis must be well connected. The DME detected the CMP sensor signal did not reach the specified minimum or maximum limit, or the difference between the actual and target position was incorrectly reported. This fault occurs 120 seconds after start up. **Possible Causes:** • CMP sensor circuit is open or shorted to ground • CMP sensor circuit is shorted to power • CMP sensor ground (return) circuit is open • CMP sensor installation incorrect (Hall-effect type) • CMP sensor is damaged or CMP sensor shielding damaged
DTC: P0370 **2T MIL: Yes** **2001, 2002, 2003, 2004, 2005, 2006** **Models:** 325i, 330i, 525i, 530i, 540i, 550i, 740i, 750i, Z4 **Engines:** 3.0L, 4.4L, 4.8L, 5.0L, 5.4L **Transmissions:** All	**Crankshaft Position Sensor Timing Reference High Conditions:** Engine started, battery voltage must be at least 11.5v, all electrical components must be off, parking brake must be engaged (to keep daytime driving lights off), automatic transmission selector must be in park and the ground between the engine and the chassis must be well connected. The DME detected the CMP sensor signal did not reach the specified minimum limit. **Possible Causes:** • CMP sensor circuit is open or shorted to ground • CMP sensor circuit is shorted to power • CMP sensor ground (return) circuit is open • CMP sensor installation incorrect (Hall-effect type) • CMP sensor is damaged or CMP sensor shielding damaged

DTC	Trouble Code Title, Conditions & Possible Causes
DTC: P0372 **2T MIL: Yes** **2006** **Models:** 325i, 330i, 525i, 530i, 550i, 750i, Z4 **Engines:** 3.0L, 4.8L **Transmissions:** All	**Crankshaft Position Sensor Timing Reference High Resolution Signal "A" Too Few Pulses Conditions:** Engine started, battery voltage must be at least 11.5v, all electrical components must be off, parking brake must be engaged (to keep daytime driving lights off), automatic transmission selector must be in park and the ground between the engine and the chassis must be well connected. The DME detected the CMP sensor signal did not reach the specified minimum limit. **Possible Causes:** • CMP sensor circuit is open or shorted to ground • CMP sensor circuit is shorted to power • CMP sensor ground (return) circuit is open • CMP sensor installation incorrect (Hall-effect type) • CMP sensor is damaged or CMP sensor shielding damaged
DTC: P0373 **2T MIL: Yes** **2006** **Models:** 325i, 330i, 525i, 530i, 550i, 750i, Z4 **Engines:** 3.0L, 4.8L **Transmissions:** All	**Crankshaft Position Sensor Timing Reference High Resolution Signal "A" Intermittent/Erratic Pulses Conditions:** Engine started, battery voltage must be at least 11.5v, all electrical components must be off, parking brake must be engaged (to keep daytime driving lights off), automatic transmission selector must be in park and the ground between the engine and the chassis must be well connected. The DME detected the CMP sensor signal did not reach the specified minimum limit. **Possible Causes:** • CMP sensor circuit is open or shorted to ground • CMP sensor circuit is shorted to power • CMP sensor ground (return) circuit is open • CMP sensor installation incorrect (Hall-effect type) • CMP sensor is damaged or CMP sensor shielding damaged
DTC: P0385 **2T MIL: Yes** **2000** **Models:** M5, Z8 **Engines:** 5.0L **Transmissions:** All	**Camshaft Position Sensor Circuit Conditions:** Engine started, battery voltage must be at least 11.5v, all electrical components must be off, parking brake must be engaged (to keep daytime driving lights off), automatic transmission selector must be in park and the ground between the engine and the chassis must be well connected. The DME detected the CMP sensor signal was missing or it was erratic. **Possible Causes:** • CMP sensor circuit is open or shorted to ground • CMP sensor circuit is shorted to power • CMP sensor ground (return) circuit is open • CMP sensor installation incorrect (Hall-effect type) • CMP sensor is damaged or CMP sensor shielding damaged
DTC: P0390 **2T MIL: Yes** **2000, 2005, 2006** **Models:** 325i, 330i, 525i, 530i, 550i, 745i, 750i, M5, Z4, Z8 **Engines:** 3.0L, 4.8L, 5.0L **Transmissions:** All	**Camshaft Position Sensor "B" Circuit (Bank 2) Conditions:** Engine started, battery voltage must be at least 11.5v, all electrical components must be off, parking brake must be engaged (to keep daytime driving lights off), automatic transmission selector must be in park and the ground between the engine and the chassis must be well connected. The DME detected the CMP sensor signal was missing or it was erratic. **Possible Causes:** • CMP sensor circuit is open or shorted to ground • CMP sensor circuit is shorted to power • CMP sensor ground (return) circuit is open • CMP sensor installation incorrect (Hall-effect type) • CMP sensor is damaged or CMP sensor shielding damaged
DTC: P0391 **2T MIL: Yes** **2002, 2003, 2004, 2005, 2006** **Models:** 325i, 330i, 525i, 530i, 545i, 550i, 745i, 750i, Z4 **Engines:** 3.0L, 4.8L **Transmissions:** All	**Camshaft Position Sensor "B" Circuit (Bank 2) Range/Performance Conditions:** Engine started, battery voltage must be at least 11.5v, all electrical components must be off, parking brake must be engaged (to keep daytime driving lights off), automatic transmission selector must be in park and the ground between the engine and the chassis must be well connected. The DME detected the CMP sensor signal exceeded the bounds of the specified maximum limit. **Possible Causes:** • CMP sensor circuit is open or shorted to ground • CMP sensor circuit is shorted to power • CMP sensor ground (return) circuit is open • CMP sensor installation incorrect (Hall-effect type) • CMP sensor is damaged or CMP sensor shielding damaged
DTC: P0392 **2T MIL: Yes** **2002, 2003, 2004, 2005, 2006** **Models:** 325i, 330i, 525i, 530i, 545i, 550i, 745i, 750i, Z4 **Engines:** 3.0L, 4.8L **Transmissions:** All	**Camshaft Position Sensor "B" Circuit (Bank 2) Low Input Conditions:** Engine started, battery voltage must be at least 11.5v, all electrical components must be off, parking brake must be engaged (to keep daytime driving lights off), automatic transmission selector must be in park and the ground between the engine and the chassis must be well connected. The DME detected the CMP sensor signal exceeded the bounds of the specified maximum limit. **Possible Causes:** • CMP sensor circuit is open or shorted to ground • CMP sensor circuit is shorted to power • CMP sensor ground (return) circuit is open • CMP sensor installation incorrect (Hall-effect type) • CMP sensor is damaged or CMP sensor shielding damaged

DTC	Trouble Code Title, Conditions & Possible Causes
DTC: P0393 **2T MIL: Yes** **2002, 2003, 2004, 2005, 2006** **Models:** 325i, 330i, 525i, 530i, 545i, 550i, 745i, 750i, Z4 **Engines:** 3.0L, 4.8L **Transmissions:** All	**Camshaft Position Sensor "B" Circuit (Bank 2) High Input Conditions:** Engine started, battery voltage must be at least 11.5v, all electrical components must be off, parking brake must be engaged (to keep daytime driving lights off), automatic transmission selector must be in park and the ground between the engine and the chassis must be well connected. The DME detected the CMP sensor signal did not reach the specified minimum limit. **Possible Causes:** • CMP sensor circuit is open or shorted to ground • CMP sensor circuit is shorted to power • CMP sensor ground (return) circuit is open • CMP sensor installation incorrect (Hall-effect type) • CMP sensor is damaged or CMP sensor shielding damaged
DTC: P0394 **2T MIL: Yes** **2001, 2002** **Models:** M5, Z8 **Engines:** 5.0L **Transmissions:** All	**Camshaft Position Sensor "B" Circuit (Bank 2) Conditions:** Engine started, battery voltage must be at least 11.5v, all electrical components must be off, parking brake must be engaged (to keep daytime driving lights off), automatic transmission selector must be in park and the ground between the engine and the chassis must be well connected. The DME detected the CMP sensor signal was missing or it was erratic. **Possible Causes:** • CMP sensor circuit is open or shorted to ground • CMP sensor circuit is shorted to power • CMP sensor ground (return) circuit is open • CMP sensor installation incorrect (Hall-effect type) • CMP sensor is damaged or CMP sensor shielding damaged
DTC: P0410 **2T MIL: Yes** **1999** **Models:** 750i **Engines:** 5.4L **Transmissions:** All	**Secondary Air Injection System Upstream Low Flow Detected Conditions:** Engine started, battery voltage must be at least 11.5v, all electrical components must be off, parking brake must be engaged (to keep daytime driving lights off), automatic transmission selector must be in park and the ground between the engine and the chassis must be well connected. The DME detected the Secondary AIR pump airflow was not diverted correctly when requested during the self-test. The pump is functioning but the quantity of air is recognized as insufficient by HO2S. The secondary pump is on, and the oxygen sensor is heated up. **Note: The solenoid valve is closed when no voltage is present.** **Possible Causes:** • Air pump output is blocked or restricted • AIR bypass solenoid is leaking or it is restricted • AIR bypass solenoid is stuck open or stuck closed • Check valve (one or more) is damaged or leaking • Electric air injection pump hose(s) leaking • Electric air injection pump is damaged or faulty
DTC: P0411 **2T MIL: Yes** **2000, 2001, 2002, 2003, 2004, 2005, 2006** **Models:** 325i, 330i, 525i, 530i, 540i, 550i, 740i, 750i, M5, Z3, Z4, Z8 **Engines:** 2.5L, 3.0L, 3.2L, 4.4L, 4.8L, 5.0L **Transmissions:** All	**Secondary Air Injection System Upstream Flow Detected Conditions:** Engine started, battery voltage must be at least 11.5v, all electrical components must be off, parking brake must be engaged (to keep daytime driving lights off), automatic transmission selector must be in park and the ground between the engine and the chassis must be well connected. The DME detected the Secondary AIR pump airflow was not diverted correctly when requested during the self-test. The pump is functioning but the quantity of air is recognized as insufficient by HO2S. **Note: The solenoid valve is closed when no voltage is present.** **Possible Causes:** • Air pump output is blocked or restricted • AIR bypass solenoid is leaking or it is restricted • AIR bypass solenoid is stuck open or stuck closed • Check valve (one or more) is damaged or leaking • Electric air injection pump hose(s) leaking • Electric air injection pump is damaged or faulty
DTC: P0412 **2T MIL: Yes** **1996, 1997, 1998, 1999, 2000, 2001, 2002, 2003, 2004, 2005** **Models:** 318i, 320i, 323i, 325i, 328i, 330i, 525i, 528i, 530i, 540i, 545i, 550i, 740i, 745i, 745Li, 750i, 840i, M3, M5, Z3, Z4, Z8 **Engines:** 1.9L, 2.5L, 2.8L, 3.0L, 3.2L, 4.4L, 4.8L, 5.0L, 5.4L **Transmissions:** All	**Secondary Air Injection Solenoid Circuit Malfunction Conditions:** Engine started, battery voltage must be at least 11.5v, all electrical components must be off, parking brake must be engaged (to keep daytime driving lights off), automatic transmission selector must be in park and the ground between the engine and the chassis must be well connected. The DME detected an unexpected low or high voltage condition on the AIR solenoid control circuit during testing. **Possible Causes:** • AIR solenoid power circuit (B+) is open (check dedicated fuse) • AIR bypass solenoid control circuit is open or shorted to ground • AIR diverter solenoid control circuit open or shorted to ground • AIR pump control circuit is open or shorted to ground • Check valve (one or more) is damaged or leaking • Solid State relay is damaged or it has failed

DTC	Trouble Code Title, Conditions & Possible Causes
DTC: P0413 **2T MIL: Yes** **2000, 2001, 2002, 2003, 2004, 2005** **Models:** 325i, 330i, 525i, 530i, 540i, 550i, 740i, 750i, M5, Z3, Z4, Z8 **Engines:** 2.5L, 3.0L, 3.2L, 4.4L, 4.8L, 5.0L **Transmissions:** All	**Secondary Air Injection Solenoid Circuit Open Conditions:** Engine started, battery voltage must be at least 11.5v, all electrical components must be off, parking brake must be engaged (to keep daytime driving lights off), automatic transmission selector must be in park and the ground between the engine and the chassis must be well connected. The DME detected an unexpected low or high voltage condition on the AIR solenoid control circuit during testing. **Possible Causes:** • AIR solenoid power circuit (B+) is open (check dedicated fuse) • AIR bypass solenoid control circuit is open or shorted to ground • AIR diverter solenoid control circuit open or shorted to ground • AIR pump control circuit is open or shorted to ground • Check valve (one or more) is damaged or leaking • Solid State relay is damaged or it has failed
DTC: P0414 **2T MIL: Yes** **2000, 2001, 2002, 2003, 2004, 2005** **Models:** 325i, 330i, 525i, 530i, 540i, 550i, 740i, 750i, M5, Z3, Z4, Z8 **Engines:** 2.5L, 3.0L, 3.2L, 4.4L, 4.8L, 5.0L **Transmissions:** All	**Secondary Air Injection Solenoid Circuit Short Conditions:** Engine started, battery voltage must be at least 11.5v, all electrical components must be off, parking brake must be engaged (to keep daytime driving lights off), automatic transmission selector must be in park and the ground between the engine and the chassis must be well connected. The DME detected an unexpected low or high voltage condition on the AIR solenoid control circuit during testing. **Possible Causes:** • AIR solenoid power circuit (B+) is open (check dedicated fuse) • AIR bypass solenoid control circuit is open or shorted to ground • AIR diverter solenoid control circuit open or shorted to ground • AIR pump control circuit is open or shorted to ground • Check valve (one or more) is damaged or leaking • Solid State relay is damaged or it has failed
DTC: P0418 **2T MIL: Yes** **2000, 2001, 2002, 2003, 2004, 2005, 2006** **Models:** 325i, 330i, 525i, 530i, 540i, 550i, 740i, 750i, M5, Z3, Z4, Z8 **Engines:** 2.5L, 3.0L, 3.2L, 4.4L, 4.8L, 5.0L **Transmissions:** All	**Secondary Air Injection Relay (A) Circuit Malfunction Conditions:** **Engine started, battery voltage must be at least 11.5v, all electrical components must be off, parking brake must be engaged (to keep daytime driving lights off), automatic transmission selector must be in park and the ground between the engine and the chassis must be well connected. The DME detected an unexpected low or high voltage condition on the AIR solenoid control circuit during testing. The fuel status is in a closed loop pattern, the coolant temperature is between 69 and 100 degrees Celsius, and the engine speed is between 800 and 6000rpm. 2001: 3.2L-S54 (M3, M-Coupe/Roadster);** **Possible Causes:** • AIR solenoid power circuit (B+) is open (check dedicated fuse) • AIR bypass solenoid control circuit is open or shorted to ground • AIR diverter solenoid control circuit open or shorted to ground • AIR pump control circuit is open or shorted to ground • Check valve (one or more) is damaged or leaking • Solid State relay is damaged or it has failed
DTC: P0420 **MIL: Yes** **1996, 1997, 1998, 1999, 2000, 2001, 2002, 2003, 2004, 2005, 2006** **Models:** 318i, 320i, 323i, 325i, 328i, 330i, 525i, 528i, 530i, 540i, 545i, 550i, 740i, 745i, 745Li, 750i, 840i, M3, M5, Z3, Z4, Z8 **Engines:** 1.9L, 2.5L, 2.8L, 3.0L, 3.2L, 4.4L, 4.8L, 5.0L, 5.4L **Transmissions:** All	**Catalyst System Efficiency (Bank 1) Below Threshold Conditions:** Engine started, battery voltage must be at least 11.5v, all electrical components must be off, parking brake must be engaged (to keep daytime driving lights off), automatic transmission selector must be in park, the exhaust system must be properly sealed between the catalytic converter and the cylinder head, coolant temperature must be at least 80 degrees Celsius and oxygen sensor heaters for oxygen sensors before the catalytic converter must be functioning properly and the ground between the engine and the chassis must be well connected. The DME detected the switch rate of the rear HO2S-12 was close to the switch rate of front HO2S (it should be much slower). The exhaust-gas mass airflow is less than 22g/sec. The engine speed is between 980 and 1920rpm, the catalyst temperature is greater than 300 degrees Celsius, the fuel system status is in a closed loop and the purge vapor factor is less than 3.5. **Possible Causes:** • Air leaks at the exhaust manifold or in the exhaust pipes • Catalytic converter is damaged, contaminated or it has failed • ECT/CHT sensor has lost its calibration (the signal is incorrect) • Engine cylinders misfiring, or the ignition timing is over retarded • Engine oil is contaminated • Front HO2S or rear HO2S is contaminated with fuel or moisture • Front HO2S and/or the rear HO2S is loose in the mounting hole • Front HO2S much older than the rear HO2S (HO2S-11 is lazy) • Fuel system pressure is too high (check the pressure regulator) • Rear HO2S wires improperly connected or the HO2S has failed

DTC	Trouble Code Title, Conditions & Possible Causes
DTC: P0422 **2T MIL: Yes** **1996, 1997, 1998** **Models:** 750i, 850i **Engines:** 5.4L **Transmissions:** All	**Main Catalyst (Bank 1) Efficiency Below Threshold Conditions:** Engine started, battery voltage must be at least 11.5v, all electrical components must be off, parking brake must be engaged (to keep daytime driving lights off), automatic transmission selector must be in park, the exhaust system must be properly sealed between the catalytic converter and the cylinder head, coolant temperature must be at least 80 degrees Celsius and oxygen sensor heaters for oxygen sensors before the catalytic converter must be functioning properly and the ground between the engine and the chassis must be well connected. The DME detected the switch rate of the rear HO2S-12 was close to the switch rate of front HO2S (it should be much slower). **Possible Causes:** • Air leaks at the exhaust manifold or in the exhaust pipes • Catalytic converter is damaged, contaminated or it has failed • ECT/CHT sensor has lost its calibration (the signal is incorrect) • Engine cylinders misfiring, or the ignition timing is over retarded • Engine oil is contaminated • Front HO2S or rear HO2S is contaminated with fuel or moisture • Front HO2S and/or the rear HO2S is loose in the mounting hole • Front HO2S much older than the rear HO2S • Fuel system pressure is too high (check the pressure regulator) • Rear HO2S wires improperly connected or the HO2S has failed
DTC: P0423 **2T MIL: Yes** **1998, 2000, 2001** **Models:** 750i **Engines:** 5.4L **Transmissions:** All	**Heated Catalyst System Efficiency Below Threshold Conditions:** Engine started, battery voltage must be at least 11.5v, all electrical components must be off, parking brake must be engaged (to keep daytime driving lights off), automatic transmission selector must be in park, the exhaust system must be properly sealed between the catalytic converter and the cylinder head, coolant temperature must be at least 80 degrees Celsius and oxygen sensor heaters for oxygen sensors before the catalytic converter must be functioning properly and the ground between the engine and the chassis must be well connected. The coolant temperature is less than 90 degrees Celsius, catalyst temperature is less than 300 degrees Celsius, engine crank time less than five seconds, vehicle speed is less than three mph and the engine speed is less than 200rpm. **Possible Causes:** • Air leaks at the exhaust manifold or in the exhaust pipes • Catalytic converter is damaged, contaminated or it has failed • ECT/CHT sensor has lost its calibration (the signal is incorrect) • Engine cylinders misfiring, or the ignition timing is over retarded • Engine oil is contaminated • Front HO2S or rcar HO2S is contaminated with fuel or moisture • Front HO2S and/or the rear HO2S is loose in the mounting hole • Front HO2S much older than the rear HO2S (HO2S-11 is lazy) • Fuel system pressure is too high (check the pressure regulator) • Rear HO2S wires improperly connected or the HO2S has failed
DTC: P0430 **MIL: Yes** **1996, 1997, 1998, 1999, 2000, 2001, 2002, 2003, 2004, 2005, 2006** **Models:** 318i, 320i, 323i, 325i, 328i, 330i, 525i, 528i, 530i, 540i, 545i, 550i, 740i, 745i, 745Li, 750i, 840i, M3, M5, Z3, Z4, Z8 **Engines:** 1.9L, 2.5L, 2.8L, 3.0L, 3.2L, 4.4L, 4.8L, 5.0L, 5.4L **Transmissions:** All	**Catalyst System Efficiency (Bank 2) Below Threshold Conditions:** Engine started, battery voltage must be at least 11.5v, all electrical components must be off, parking brake must be engaged (to keep daytime driving lights off), automatic transmission selector must be in park, the exhaust system must be properly sealed between the catalytic converter and the cylinder head, coolant temperature must be at least 80 degrees Celsius and oxygen sensor heaters for oxygen sensors before the catalytic converter must be functioning properly and the ground between the engine and the chassis must be well connected. The DME detected the switch rate of the rear HO2S-12 was close to the switch rate of front HO2S (it should be much slower). The engine speed is between 980 and 1920rpm, the catalyst temperature is greater than 300 degrees Celsius, the fuel system status is in a closed loop and the purge vapor factor is less than 3.5. **Possible Causes:** • Air leaks at the exhaust manifold or in the exhaust pipes • Catalytic converter is damaged, contaminated or it has failed • ECT/CHT sensor has lost its calibration (the signal is incorrect) • Engine cylinders misfiring, or the ignition timing is over retarded • Engine oil is contaminated • Front HO2S or rear HO2S is contaminated with fuel or moisture • Front HO2S and/or the rear HO2S is loose in the mounting hole • Front HO2S much older than the rear HO2S (HO2S-11 is lazy) • Fuel system pressure is too high (check the pressure regulator) • Rear HO2S wires improperly connected or the HO2S has failed

DTC	Trouble Code Title, Conditions & Possible Causes
DTC: P0432 **MIL: Yes** **1996, 1997, 1998** **Models:** 750i, 850i **Engines:** 5.4L **Transmissions:** All	**Main Catalyst (Bank 1) Efficiency Below Threshold Conditions:** Engine started, battery voltage must be at least 11.5v, all electrical components must be off, parking brake must be engaged (to keep daytime driving lights off), automatic transmission selector must be in park, the exhaust system must be properly sealed between the catalytic converter and the cylinder head, coolant temperature must be at least 80 degrees Celsius and oxygen sensor heaters for oxygen sensors before the catalytic converter must be functioning properly and the ground between the engine and the chassis must be well connected. The DME detected the switch rate of the rear HO2S-12 was close to the switch rate of front HO2S (it should be much slower). **Possible Causes:** • Air leaks at the exhaust manifold or in the exhaust pipes • Catalytic converter is damaged, contaminated or it has failed • ECT/CHT sensor has lost its calibration (the signal is incorrect) • Engine cylinders misfiring, or the ignition timing is over retarded • Engine oil is contaminated • Front HO2S or rear HO2S is contaminated with fuel or moisture • Front HO2S and/or the rear HO2S is loose in the mounting hole • Front HO2S much older than the rear HO2S • Fuel system pressure is too high (check the pressure regulator) • Rear HO2S wires improperly connected or the HO2S has failed
DTC: P0433 **2T MIL: Yes** **1998, 2000, 2001** **Models:** 750i **Engines:** 5.4L **Transmissions:** All	**Heated Catalyst System Efficiency Below Threshold Conditions:** Engine started, battery voltage must be at least 11.5v, all electrical components must be off, parking brake must be engaged (to keep daytime driving lights off), automatic transmission selector must be in park, the exhaust system must be properly sealed between the catalytic converter and the cylinder head, coolant temperature must be at least 80 degrees Celsius and oxygen sensor heaters for oxygen sensors before the catalytic converter must be functioning properly and the ground between the engine and the chassis must be well connected. The coolant temperature is less than 90 degrees Celsius, catalyst temperature is less than 300 degrees Celsius, engine crank time less than five seconds, vehicle speed is less than three mph and the engine speed is less than 200rpm. **Possible Causes:** • Air leaks at the exhaust manifold or in the exhaust pipes • Catalytic converter is damaged, contaminated or it has failed • ECT/CHT sensor has lost its calibration (the signal is incorrect) • Engine cylinders misfiring, or the ignition timing is over retarded • Engine oil is contaminated • Front HO2S or rear HO2S is contaminated with fuel or moisture • Front HO2S and/or the rear HO2S is loose in the mounting hole • Front HO2S much older than the rear HO2S (HO2S-11 is lazy) • Fuel system pressure is too high (check the pressure regulator) • Rear HO2S wires improperly connected or the HO2S has failed
DTC: P0440 **2T MIL: Yes** **1996, 1997, 1998, 1999, 2000, 2001, 2002, 2003, 2004, 2005, 2006** **Models:** 318i, 320i, 323i, 325i, 328i, 330i, 525i, 528i, 530i, 540i, 545i, 550i, 740i, 745i, 745Li, 750i, 840i, M3, M5, Z3, Z4, Z8 **Engines:** 1.9L, 2.5L, 2.8L, 3.0L, 3.2L, 4.4L, 4.8L, 5.0L, 5.4L **Transmissions:** All	**EVAP System Malfunction Conditions:** ECT sensor is cold during startup, engine started, battery voltage must be at least 11.5v, all electrical components must be off, parking brake must be engaged (to keep daytime driving lights off), automatic transmission selector must be in park, the exhaust system must be properly sealed between the catalytic converter and the cylinder head, coolant temperature must be at least 80 degrees Celsius and oxygen sensor heaters for oxygen sensors before the catalytic converter must be functioning properly and the ground between the engine and the chassis must be well connected. The DME detected the switch rate of the rear HO2S-12 was close to the switch rate of front HO2S (it should be much slower). DME detected a problem in the EVAP system during the EVAP System Monitor test. The fuel system adaptation has finished, the coolant temperature is greater than 60 degrees Celsius, normal purge is on, vehicle speed is zero, and engine is at idle. **Possible Causes:** • EVAP canister purge valve is damaged • EVAP canister has an improper seal • Vapor line between purge solenoid and intake manifold vacuum reservoir is damaged, or vapor line between EVAP canister purge solenoid and charcoal canister is damaged • Vapor line between charcoal canister and check valve, or vapor line between check valve and fuel vapor valves is damaged

DTC	Trouble Code Title, Conditions & Possible Causes
DTC: P0441 **2T MIL: Yes** **1996, 1997, 1998, 1999, 2000, 2001, 2002, 2003, 2004, 2005** **Models:** 318i, 320i, 323i, 325i, 328i, 330i, 525i, 528i, 530i, 540i, 545i, 550i, 740i, 745i, 745Li, 750i, 840i, M3, M5, Z3, Z4, Z8 **Engines:** 1.9L, 2.5L, 2.8L, 3.0L, 3.2L, 4.4L, 4.8L, 5.0L, 5.4L **Transmissions:** All	**EVAP Control System Incorrect Purge Flow Conditions:** ECT sensor is cold during startup, engine started, battery voltage must be at least 11.5v, all electrical components must be off, parking brake must be engaged (to keep daytime driving lights off), automatic transmission selector must be in park, the exhaust system must be properly sealed between the catalytic converter and the cylinder head, coolant temperature must be at least 80 degrees Celsius and oxygen sensor heaters for oxygen sensors before the catalytic converter must be functioning properly and the ground between the engine and the chassis must be well connected. The DME detected the switch rate of the rear HO2S-12 was close to the switch rate of front HO2S (it should be much slower). DME detected a problem in the EVAP system during the EVAP System Monitor test. **Possible Causes:** • EVAP canister purge valve is damaged • EVAP canister has an improper seal • Vapor line between purge solenoid and intake manifold vacuum reservoir is damaged, or vapor line between EVAP canister purge solenoid and charcoal canister is damaged • Vapor line between charcoal canister and check valve, or vapor line between check valve and fuel vapor valves is damaged
DTC: P0442 **2T MIL: Yes** **1996, 1997, 1998, 1999, 2000, 2001, 2002, 2003, 2004, 2005, 2006** **Models:** 318i, 320i, 323i, 325i, 328i, 330i, 525i, 528i, 530i, 540i, 545i, 550i, 740i, 745i, 745Li, 750i, 840i, M3, M5, Z3, Z4, Z8 **Engines:** 1.9L, 2.5L, 2.8L, 3.0L, 3.2L, 4.4L, 4.8L, 5.0L, 5.4L **Transmissions:** All	**EVAP Control System Small Leak Detected Conditions:** Engine started, battery voltage must be at least 11.5v, all electrical components must be off, parking brake must be engaged (to keep daytime driving lights off), automatic transmission selector must be in park, the exhaust system must be properly sealed between the catalytic converter and the cylinder head, coolant temperature must be at least 80 degrees Celsius and oxygen sensor heaters for oxygen sensors before the catalytic converter must be functioning properly and the ground between the engine and the chassis must be well connected. The DME detected a leak in the EVAP system as small as 0.040 inches during the EVAP Monitor Test. The fuel system adaptation has finished, the coolant temperature is greater than 60 degrees Celsius, normal purge is on, vehicle speed is zero, and engine is at idle. Engine start temperature must be greater than 2 degrees Celsius and the last driving cycle greater than 20 minutes. **Possible Causes:** • Aftermarket EVAP parts that do not conform to specifications • CV solenoid remains partially open when commanded to close • EVAP component seals leaking (i.e., leaks in the Purge valve, fuel tank pressure sensor, canister vent solenoid, fuel vapor control valve tube assembly or fuel vapor vent valve). • Fuel filler cap damaged, cross-threaded or loosely installed • Loose fuel vapor hose/tube connections to EVAP components • Small holes or cuts in fuel vapor hoses or EVAP canister tubes
DTC: P0443 **2T MIL: Yes** **1996, 1997, 1998, 1999, 2000, 2001, 2002, 2003, 2004, 2005, 2006** **Models:** 318i, 320i, 323i, 325i, 328i, 330i, 525i, 528i, 530i, 540i, 545i, 550i, 740i, 745i, 745Li, 750i, 840i, M3, M5, Z3, Z4, Z8 **Engines:** 1.9L, 2.5L, 2.8L, 3.0L, 3.2L, 4.4L, 4.8L, 5.0L, 5.4L **Transmissions:** All	**EVAP Vapor Management Valve Circuit Malfunction Conditions:** Engine started, battery voltage must be at least 11.5v, all electrical components must be off, parking brake must be engaged (to keep daytime driving lights off), automatic transmission selector must be in park, the exhaust system must be properly sealed between the catalytic converter and the cylinder head, coolant temperature must be at least 80 degrees Celsius and oxygen sensor heaters for oxygen sensors before the catalytic converter must be functioning properly and the ground between the engine and the chassis must be well connected. The DME detected an unexpected high or low voltage condition on the Vapor Management Valve (VMV) circuit when the device was cycled On/Off during testing. **Possible Causes:** • EVAP power supply circuit is open • EVAP solenoid control circuit is open or shorted to ground • EVAP solenoid control circuit is shorted to power (B+) • EVAP solenoid valve is damaged or it has failed
DTC: P0444 **2T MIL: Yes** **2001, 2002, 2003, 2004, 2005** **Models:** All **Engines:** All **Transmissions:** All	**Evaporative Emission System Purge Control Valve Circuit Open Conditions:** Engine started, battery voltage must be at least 11.5v, all electrical components must be off, parking brake must be engaged (to keep daytime driving lights off), automatic transmission selector must be in park, the exhaust system must be properly sealed between the catalytic converter and the cylinder head, coolant temperature must be at least 80 degrees Celsius and oxygen sensor heaters for oxygen sensors before the catalytic converter must be functioning properly and the ground between the engine and the chassis must be well connected. The DME detected an unexpected voltage condition on the EVAP circuit when the device was cycled On/Off during testing. **Possible Causes:** • EVAP power supply circuit is open • EVAP solenoid control circuit is open or shorted to ground • EVAP solenoid control circuit is shorted to power (B+) • EVAP solenoid valve is damaged or it has failed • EVAP canister has a leak or a poor seal

DTC	Trouble Code Title, Conditions & Possible Causes
DTC: P0445 **2T MIL: Yes** **2001, 2002, 2003, 2004, 2005** **Models:** All **Engines:** All **Transmissions:** All	**Evaporative Emission System Purge Control Valve Circuit Shorted Conditions:** Engine started, battery voltage must be at least 11.5v, all electrical components must be off, parking brake must be engaged (to keep daytime driving lights off), automatic transmission selector must be in park, the exhaust system must be properly sealed between the catalytic converter and the cylinder head, coolant temperature must be at least 80 degrees Celsius and oxygen sensor heaters for oxygen sensors before the catalytic converter must be functioning properly and the ground between the engine and the chassis must be well connected. The DME detected an unexpected voltage condition on the EVAP circuit when the device was cycled On/Off during testing. **Possible Causes:** • EVAP power supply circuit is open • EVAP solenoid control circuit is open or shorted to ground • EVAP solenoid control circuit is shorted to power (B+) • EVAP solenoid valve is damaged or it has failed • EVAP canister has a leak or a poor seal
DTC: P0446 **2T MIL: Yes** **1997, 1998, 1999, 2000, 2001, 2002** **Models:** 318i, 323i, 325i, 328i, 330i, 525i, M3, Z3 **Engines:** 1.9L, 2.5L, 2.8L, 3.0L, 3.2L **Transmissions:** All	**EVAP Control System Large Leak Detected Conditions:** Engine started, battery voltage must be at least 11.5v, all electrical components must be off, parking brake must be engaged (to keep daytime driving lights off), automatic transmission selector must be in park, the exhaust system must be properly sealed between the catalytic converter and the cylinder head, coolant temperature must be at least 80 degrees Celsius and oxygen sensor heaters for oxygen sensors before the catalytic converter must be functioning properly and the ground between the engine and the chassis must be well connected. The DME detected multiple small fuel vapor leaks; or it detected a large leak in the system during the leak test. **Possible Causes:** • Aftermarket EVAP hardware non-conforming to specifications • EVAP canister tube, EVAP canister purge outlet tube or EVAP return tube disconnected or cracked, or canister is damaged • EVAP canister purge valve stuck closed, or canister damaged • Fuel filler cap missing, loose (not tightened) or the wrong part • Loose fuel vapor hose/tube connections to EVAP components • Canister vent (CV) solenoid stuck open • Fuel tank pressure (FTP) sensor has failed mechanically
DTC: P0450 **2T MIL: Yes** **2001** **Models:** 540i, 740i **Engines:** 4.4L, 5.4L **Transmissions:** All	**EVAP Control System Leak Detection Pump Functionality Check Conditions:** Engine started, battery voltage must be at least 11.5v, all electrical components must be off, parking brake must be engaged (to keep daytime driving lights off), automatic transmission selector must be in park, the exhaust system must be properly sealed between the catalytic converter and the cylinder head, coolant temperature must be at least 80 degrees Celsius and oxygen sensor heaters for oxygen sensors before the catalytic converter must be functioning properly and the ground between the engine and the chassis must be well connected. The DME detected a functionality error in the leak detection pump. The secondary pump is on, the oxygen sensor is heated up, the coolant temperature is less than 15 degrees Celsius. **Possible Causes:** • Aftermarket EVAP hardware non-conforming to specifications • EVAP canister tube, EVAP canister purge outlet tube or EVAP return tube disconnected or cracked, or canister is damaged • EVAP canister purge valve stuck closed, or canister damaged • Fuel filler cap missing, loose (not tightened) or the wrong part • Loose fuel vapor hose/tube connections to EVAP components • Canister vent (CV) solenoid stuck open • Fuel tank pressure (FTP) sensor has failed mechanically
DTC: P0451 **2T MIL: Yes** **1996, 1997, 1998, 1999, 2000** **Models:** 318i, 323i, 325i, 328i, 330i, 525i, 528i, M3, Z3 **Engines:** 1.9L, 2.8L, 3.2L **Transmissions:** All	**Evaporative Emission System Pressure Sensor Conditions:** Engine started, battery voltage must be at least 11.5v, all electrical components must be off, parking brake must be engaged (to keep daytime driving lights off), automatic transmission selector must be in park, the exhaust system must be properly sealed between the catalytic converter and the cylinder head, coolant temperature must be at least 80 degrees Celsius and oxygen sensor heaters for oxygen sensors before the catalytic converter must be functioning properly and the ground between the engine and the chassis must be well connected. The DME detected an unexpected voltage condition on the EVAP circuit when the device was cycled On/Off during testing. **Possible Causes:** • EVAP power supply circuit is open • EVAP solenoid control circuit is open or shorted to ground • EVAP solenoid control circuit is shorted to power (B+) • EVAP solenoid valve is damaged or it has failed • EVAP canister has a leak or a poor seal

DTC	Trouble Code Title, Conditions & Possible Causes
DTC: P0452 **2T MIL: Yes** **2001** **Models:** 540i, 740i **Engines:** 4.4L, 5.4L **Transmissions:** All	**EVAP Control System Leak Detection Pump Functionality Check Conditions:** Engine started, battery voltage must be at least 11.5v, all electrical components must be off, parking brake must be engaged (to keep daytime driving lights off), automatic transmission selector must be in park, the exhaust system must be properly sealed between the catalytic converter and the cylinder head, coolant temperature must be at least 80 degrees Celsius and oxygen sensor heaters for oxygen sensors before the catalytic converter must be functioning properly and the ground between the engine and the chassis must be well connected. The DME detected a functionality error in the leak detection pump. The secondary pump is on, the oxygen sensor is heated up, the coolant temperature is less than 15 degrees Celsius. **Possible Causes:** • Aftermarket EVAP hardware non-conforming to specifications • EVAP canister tube, EVAP canister purge outlet tube or EVAP return tube disconnected or cracked, or canister is damaged • EVAP canister purge valve stuck closed, or canister damaged • Fuel filler cap missing, loose (not tightened) or the wrong part • Loose fuel vapor hose/tube connections to EVAP components • Canister vent (CV) solenoid stuck open • Fuel tank pressure (FTP) sensor has failed mechanically
DTC: P0453 **2T MIL: Yes** **2001** **Models:** 540i, 740i **Engines:** 4.4L, 5.4L **Transmissions:** All	**EVAP Control System Leak Detection Pump Functionality Check Conditions:** Engine started, battery voltage must be at least 11.5v, all electrical components must be off, parking brake must be engaged (to keep daytime driving lights off), automatic transmission selector must be in park, the exhaust system must be properly sealed between the catalytic converter and the cylinder head, coolant temperature must be at least 80 degrees Celsius and oxygen sensor heaters for oxygen sensors before the catalytic converter must be functioning properly and the ground between the engine and the chassis must be well connected. The DME detected a functionality error in the leak detection pump. The secondary pump is on, the oxygen sensor is heated up, the coolant temperature is less than 15 degrees Celsius. **Possible Causes:** • Aftermarket EVAP hardware non-conforming to specifications • EVAP canister tube, EVAP canister purge outlet tube or EVAP return tube disconnected or cracked, or canister is damaged • EVAP canister purge valve stuck closed, or canister damaged • Fuel filler cap missing, loose (not tightened) or the wrong part • Loose fuel vapor hose/tube connections to EVAP components • Canister vent (CV) solenoid stuck open • Fuel tank pressure (FTP) sensor has failed mechanically
DTC: P0455 **2T MIL: Yes** **1996, 1997, 1998, 1999, 2000, 2001, 2002, 2003, 2004, 2005, 2006** **Models:** 318i, 320i, 323i, 325i, 328i, 330i, 525i, 528i, 530i, 540i, 545i, 550i, 740i, 745i, 745Li, 750i, 840i, M3, M5, Z3, Z4, Z8 **Engines:** 1.9L, 2.5L, 2.8L, 3.0L, 3.2L, 4.4L, 4.8L, 5.0L, 5.4L **Transmissions:** All	**EVAP Control System Large Leak Detected Conditions:** Engine started, battery voltage must be at least 11.5v, all electrical components must be off, parking brake must be engaged (to keep daytime driving lights off), automatic transmission selector must be in park, the exhaust system must be properly sealed between the catalytic converter and the cylinder head, coolant temperature must be at least 80 degrees Celsius and oxygen sensor heaters for oxygen sensors before the catalytic converter must be functioning properly and the ground between the engine and the chassis must be well connected. The DME detected multiple small fuel vapor leaks; or it detected a large leak in the system during the leak test. **Possible Causes:** • Aftermarket EVAP hardware non-conforming to specifications • EVAP canister tube, EVAP canister purge outlet tube or EVAP return tube disconnected or cracked, or canister is damaged • EVAP canister purge valve stuck closed, or canister damaged • Fuel filler cap missing, loose (not tightened) or the wrong part • Loose fuel vapor hose/tube connections to EVAP components • Canister vent (CV) solenoid stuck open • Fuel tank pressure (FTP) sensor has failed mechanically
DTC: P0456 **2T MIL: Yes** **2001, 2002, 2003, 2004, 2005** **Models:** All **Engines:** All **Transmissions:** All	**EVAP Control System Small Leak Detected Conditions:** Engine started, battery voltage must be at least 11.5v, all electrical components must be off, parking brake must be engaged (to keep daytime driving lights off), automatic transmission selector must be in park, the exhaust system must be properly sealed between the catalytic converter and the cylinder head, coolant temperature must be at least 80 degrees Celsius and oxygen sensor heaters for oxygen sensors before the catalytic converter must be functioning properly and the ground between the engine and the chassis must be well connected. The DME detected multiple small fuel vapor leaks; or it detected a large leak in the system during the leak test. **Possible Causes:** • Aftermarket EVAP hardware non-conforming to specifications • EVAP canister tube, EVAP canister purge outlet tube or EVAP return tube disconnected or cracked, or canister is damaged • EVAP canister purge valve stuck closed, or canister damaged • Fuel filler cap missing, loose (not tightened) or the wrong part • Loose fuel vapor hose/tube connections to EVAP components • Canister vent (CV) solenoid stuck open • Fuel tank pressure (FTP) sensor has failed mechanically

DTC	Trouble Code Title, Conditions & Possible Causes
DTC: P0458 **2T MIL: Yes** **2005, 2006** **Models:** 325i, 330i, 525i, 530i, 545i, 550i, 745i, 750i, Z4 **Engines:** 3.0L, 4.8L **Transmissions:** All	**Evaporative Emission System Purge Control Valve Circuit Low Conditions:** Engine started, battery voltage must be at least 11.5v, all electrical components must be off, parking brake must be engaged (to keep daytime driving lights off), automatic transmission selector must be in park, the exhaust system must be properly sealed between the catalytic converter and the cylinder head, coolant temperature must be at least 80 degrees Celsius and oxygen sensor heaters for oxygen sensors before the catalytic converter must be functioning properly and the ground between the engine and the chassis must be well connected. The DME detected an unexpected voltage condition on the EVAP circuit when the device was cycled On/Off during testing. **Possible Causes:** • EVAP power supply circuit is open • EVAP solenoid control circuit is open or shorted to ground • EVAP solenoid control circuit is shorted to power (B+) • EVAP solenoid valve is damaged or it has failed • EVAP canister has a leak or a poor seal
DTC: P0459 **2T MIL: Yes** **2005, 2006** **Models:** 325i, 330i, 525i, 530i, 545i, 550i, 745i, 750i, Z4 **Engines:** 3.0L, 4.8L **Transmissions:** All	**Evaporative Emission System Purge Control Valve Circuit High Conditions:** Engine started, battery voltage must be at least 11.5v, all electrical components must be off, parking brake must be engaged (to keep daytime driving lights off), automatic transmission selector must be in park, the exhaust system must be properly sealed between the catalytic converter and the cylinder head, coolant temperature must be at least 80 degrees Celsius and oxygen sensor heaters for oxygen sensors before the catalytic converter must be functioning properly and the ground between the engine and the chassis must be well connected. The DME detected an unexpected voltage condition on the EVAP circuit when the device was cycled On/Off during testing. **Possible Causes:** • EVAP power supply circuit is open • EVAP solenoid control circuit is open or shorted to ground • EVAP solenoid control circuit is shorted to power (B+) • EVAP solenoid valve is damaged or it has failed • EVAP canister has a leak or a poor seal
DTC: P0460 **2005, 2006** **Models:** 325i, 330i, 525i, 530i, 545i, 550i, 745i, 750i, Z4 **Engines:** 3.0L, 3.2L, 4.8L **Transmissions:** All	**Fuel Level Sensor "A" Circuit Malfunction Conditions:** KOEO or KOER Self-Test enabled, and the DME detected a lack of power (VPWR) to the Fuel Pressure Regulator Control (FPRC) solenoid circuit. Cluster received incorrect fuel level from CAN or no message at all, calculated consumption does not correspond to transmitted fuel quantity. **Possible Causes:** • FPRC solenoid valve harness circuits are open or shorted • FPRC input port or output port vacuum lines are damaged • FRPC solenoid is damaged • Fuel level is too low • Check fuel level sensor
DTC: P0461 **2005, 2006** **Models:** 325i, 330i, 525i, 530i, 545i, 550i, 745i, 750i, Z4 **Engines:** 3.0L, 3.2L, 4.8L **Transmissions:** All	**Fuel Level Sensor "A" Circuit Range/Performance Conditions:** KOEO or KOER Self-Test enabled, and the DME detected a lack of power (VPWR) to the Fuel Pressure Regulator Control (FPRC) solenoid circuit. Cluster received incorrect fuel level from CAN, calculated consumption does not correspond to transmitted fuel quantity. There is a stuck fuel level sensor, and the fault is recorded after driving roughly 50 miles (or 2.6 gallons of gas). **Possible Causes:** • FPRC solenoid valve harness circuits are open or shorted • FPRC input port or output port vacuum lines are damaged • FRPC solenoid is damaged • Fuel level is too low • Check fuel level sensor
DTC: P0462 2005 **Models:** M3 **Engines:** 3.2L **Transmissions:** All	**Fuel Level Sensor "A" Circuit Range/Performance Conditions:** KOEO or KOER Self-Test enabled, and the DME detected a lack of power (VPWR) to the Fuel Pressure Regulator Control (FPRC) solenoid circuit. Cluster received incorrect fuel level from CAN, calculated consumption does not correspond to transmitted fuel quantity. The fuel level sensor value is too small, and the fault is recorded after driving roughly 100 miles (or 5.2 gallons of gas). **Possible Causes:** • FPRC solenoid valve harness circuits are open or shorted • FPRC input port or output port vacuum lines are damaged • FRPC solenoid is damaged • Fuel level is too low • Check fuel level sensor

DTC	Trouble Code Title, Conditions & Possible Causes
DTC: P0463 2005 **Models:** M3 **Engines:** 3.2L **Transmissions:** All	**Fuel Level Sensor "A" Circuit Range/Performance Conditions:** KOEO or KOER Self-Test enabled, and the DME detected a lack of power (VPWR) to the Fuel Pressure Regulator Control (FPRC) solenoid circuit. Cluster received incorrect fuel level from CAN, calculated consumption does not correspond to transmitted fuel quantity. The fuel level sensor value is too big, and the fault is recorded after driving roughly 100 miles (or 5.2 gallons of gas). **Possible Causes:** • FPRC solenoid valve harness circuits are open or shorted • FPRC input port or output port vacuum lines are damaged • FRPC solenoid is damaged • Fuel level is too low • Check fuel level sensor
DTC: P0477 **2006** **Models:** 325i, 330i, 525i, 530i, 550i, 750i, Z4 **Engines:** 3.0L, 4.8L **Transmissions:** All	**Exhaust Pressure Control Valve Low Conditions:** Engine started, battery voltage must be at least 11.5v, all electrical components must be off, parking brake must be engaged (to keep daytime driving lights off), automatic transmission selector must be in park, the exhaust system must be properly sealed between the catalytic converter and the cylinder head, coolant temperature must be at least 80 degrees Celsius and oxygen sensor heaters for oxygen sensors before the catalytic converter must be functioning properly and the ground between the engine and the chassis must be well connected. The DME detected an unexpected voltage condition on the EVAP circuit when the device was cycled On/Off during testing. The driver circuit has detected a short to ground. **Possible Causes:** • EVAP power supply circuit is open • EVAP solenoid control circuit is open or shorted to ground • EVAP solenoid control circuit is shorted to power (B+) • EVAP solenoid valve is damaged or it has failed • EVAP canister has a leak or a poor seal • Check wiring harness, otherwise replace DME
DTC: P0478 **2006** **Models:** 325i, 330i, 525i, 530i, 550i, 750i, Z4 **Engines:** 3.0L, 4.8L **Transmissions:** All	**Exhaust Pressure Control Valve High Conditions:** Engine started, battery voltage must be at least 11.5v, all electrical components must be off, parking brake must be engaged (to keep daytime driving lights off), automatic transmission selector must be in park, the exhaust system must be properly sealed between the catalytic converter and the cylinder head, coolant temperature must be at least 80 degrees Celsius and oxygen sensor heaters for oxygen sensors before the catalytic converter must be functioning properly and the ground between the engine and the chassis must be well connected. The DME detected an unexpected voltage condition on the EVAP circuit when the device was cycled On/Off during testing. The driver circuit has detected a short to battery voltage **Possible Causes:** • EVAP power supply circuit is open • EVAP solenoid control circuit is open or shorted to ground • EVAP solenoid control circuit is shorted to power (B+) • EVAP solenoid valve is damaged or it has failed • EVAP canister has a leak or a poor seal • Check wiring harness, otherwise replace DME
DTC: P0479 **2006** **Models:** 325i, 330i, 525i, 530i, 550i, 750i, Z4 **Engines:** 3.0L, 4.8L **Transmissions:** All	**Exhaust Pressure Control Valve Intermittent Conditions:** Engine started, battery voltage must be at least 11.5v, all electrical components must be off, parking brake must be engaged (to keep daytime driving lights off), automatic transmission selector must be in park, the exhaust system must be properly sealed between the catalytic converter and the cylinder head, coolant temperature must be at least 80 degrees Celsius and oxygen sensor heaters for oxygen sensors before the catalytic converter must be functioning properly and the ground between the engine and the chassis must be well connected. The DME detected an unexpected voltage condition on the EVAP circuit when the device was cycled On/Off during testing. The driver circuit has detected a implausible signal. **Possible Causes:** • EVAP power supply circuit is open • EVAP solenoid control circuit is open or shorted to ground • EVAP solenoid control circuit is shorted to power (B+) • EVAP solenoid valve is damaged or it has failed • EVAP canister has a leak or a poor seal • Check wiring harness, otherwise replace DME

DTC	Trouble Code Title, Conditions & Possible Causes
DTC: P0491 **2T MIL: Yes** **2001, 2002, 2003, 2004, 2005, 2006** **Models:** 325i, 328i, 330i, 525i, 528i, 530i, 540i, 545i, 550i, 740i, 745i, 745Li, 750i, 840i, M3, M5, Z3, Z4, Z8 **Engines:** All **Transmissions:** All	**Secondary Air Injection System Insufficient Flow (Bank 1) Conditions:** Engine started, battery voltage must be at least 11.5v, all electrical components must be off, parking brake must be engaged (to keep daytime driving lights off), automatic transmission selector must be in park and the ground between the engine and the chassis must be well connected. The DME detected the Secondary AIR pump airflow was not diverted correctly when requested during the self-test. The pump is functioning but the quantity of air is recognized as insufficient by HO2S. The secondary air pump is on, the oxygen sensor is heated up the cold start enrichment is activated and the coolant temperature is between negative 12 and 30 degrees Celsius. **Possible Causes:** • Air pump output is blocked or restricted • AIR bypass solenoid is leaking or it is restricted • AIR bypass solenoid is stuck open or stuck closed • Check valve (one or more) is damaged or leaking • Electric air injection pump hose(s) leaking • Electric air injection pump is damaged or faulty
DTC: P0492 **2T MIL: Yes** **2001, 2002, 2003, 2004, 2005, 2006** **Models:** 325i, 328i, 330i, 525i, 528i, 530i, 540i, 545i, 550i, 740i, 745i, 745Li, 750i, 840i, M3, M5, Z3, Z4, Z8 **Engines:** All **Transmissions:** All	**Secondary Air Injection System Insufficient Flow (Bank 2) Conditions:** Engine started, battery voltage must be at least 11.5v, all electrical components must be off, parking brake must be engaged (to keep daytime driving lights off), automatic transmission selector must be in park and the ground between the engine and the chassis must be well connected. The DME detected the Secondary AIR pump airflow was not diverted correctly when requested during the self-test. The pump is functioning but the quantity of air is recognized as insufficient by HO2S. The secondary air pump is on, the oxygen sensor is heated up the cold start enrichment is activated and the coolant temperature is between negative 12 and 30 degrees Celsius. **Possible Causes:** • Air pump output is blocked or restricted • AIR bypass solenoid is leaking or it is restricted • AIR bypass solenoid is stuck open or stuck closed • Check valve (one or more) is damaged or leaking • Electric air injection pump hose(s) leaking • Electric air injection pump is damaged or faulty
DTC: P0500 **MIL: Yes** **1996, 1997, 1998, 1999, 2000, 2001, 2002, 2003, 2004, 2005, 2006** **Models:** 318i, 320i, 323i, 325i, 328i, 330i, 525i, 528i, 530i, 540i, 545i, 550i, 740i, 745i, 745Li, 750i, 840i, M3, M5, Z3, Z4, Z8 **Engines:** 1.9L, 2.5L, 2.8L, 3.0L, 3.2L, 4.4L, 4.8L, 5.0L, 5.4L **Transmissions:** All	**Vehicle Speed Sensor "A" Malfunction Conditions:** Engine started; engine speed above the TCC stall speed, and the DME detected a loss of the VSS signal over a period of time or the signal is not usable. **Note: The DME receives vehicle speed data from the VSS, TCSS, ABS module, CTM or GEM controller, depending up the application. Speed Signal from DSC too high because of possible tampering. Check DSC and wires. The engine speed is greater than 2000rpm, engine load greater than 3.5msec/rev., and the vehicle speed is more than 55mph. The fuel system status is in fuel cut-off mode.** **Possible Causes:** • VSS signal circuit is open or shorted to ground • VSS harness circuit is shorted to ground • VSS harness circuit is shorted to power • VSS circuit open between the DME and related control module • VSS or wheel speed sensors circuits are damaged • Modules connected to VSC/VSS harness circuits are damaged • Mechanical drive mechanism for the VSS is damaged
DTC: P0501 **MIL: Yes** **2001, 2005, 2006** **Models:** 325i, 330i, 525i, 528i, 530i, 545i, 550i, 745i, 750i, Z4 **Engines:** All **Transmissions:** All	**Vehicle Speed Sensor or PSOM Range/Performance Conditions:** Engine started; engine speed above the TCC stall speed, and the DME detected a loss of the VSS signal over a period of time or the signal is not usable. **Note: The DME receives vehicle speed data from the VSS, TCSS, ABS module, CTM or GEM controller, depending up the application. The engine speed is between 1000 and 1320rpm. The fuel system status is in the fuel cut-off mode. The coolant temperature is greater than 60 degrees Celsius. The vehicle speed is zero.** **Possible Causes:** • VSS signal circuit is open or shorted to ground • VSS harness circuit is shorted to ground • VSS harness circuit is shorted to power • VSS circuit open between the DME and related control module • VSS or wheel speed sensors circuits are damaged • Modules connected to VSC/VSS harness circuits are damaged • Mechanical drive mechanism for the VSS is damaged

DTC	Trouble Code Title, Conditions & Possible Causes
DTC: P0503 **MIL: Yes** **2005, 2006** **Models:** 325i, 330i, 525i, 530i, 545i, 550i, 745i, 750i, Z4 **Engines:** 3.0L, 4.8L **Transmissions:** All	**Vehicle Speed Sensor "A" Intermittent/Erratic/High Conditions:** Engine started; engine speed above the TCC stall speed, and the DME detected a loss of the VSS signal over a period of time or the signal is not usable. **Note: The DME receives vehicle speed data from the VSS, TCSS, ABS module, CTM or GEM controller, depending up the application. Speed Signal from DSC too high because of possible tampering. Check DSC and wires** **Possible Causes:** • VSS signal circuit is open or shorted to ground • VSS harness circuit is shorted to ground • VSS harness circuit is shorted to power • VSS circuit open between the DME and related control module • VSS or wheel speed sensors circuits are damaged • Modules connected to VSC/VSS harness circuits are damaged • Mechanical drive mechanism for the VSS is damaged
DTC: P0505 **2T MIL: Yes** **1996, 1997, 1998, 1999, 2000, 2001, 2002, 2003, 2004, 2005, 2006** **Models:** 318i, 320i, 323i, 325i, 328i, 330i, 525i, 528i, 530i, 540i, 545i, 550i, 740i, 745i, 745Li, 750i, 840i, M3, M5, Z3, Z4, Z8 **Engines:** 1.9L, 2.5L, 2.8L, 3.0L, 3.2L, 4.4L, 4.8L, 5.0L, 5.4L **Transmissions:** All	**Idle Air Control Valve Malfunction Conditions:** Engine started, battery voltage at least 11.5v, all electrical components off, ground connections between engine and chassis well connected, coolant temperature at least 80-degrees Celsius. The DME detected deviation from the normal operating parameters of the Idle Air Control Valve. The vehicle speed can be zero mph and the engine load must be less than 1.5ms. **Possible Causes:** • Charge air system leaks • Recirculating valve for turbocharger is faulty • Turbocharging system is damaged • Vacuum diaphragm for turbocharger needs adjusting • Wastegate bypass regulator valve is faulty
DTC: P0506 **2T MIL: Yes** **1996, 1997, 1998, 1999, 2000, 2001, 2002, 2003, 2004, 2005, 2006** **Models:** All **Engines:** All **Transmissions:** All	**Idle Air Control System RPM Lower Than Expected Conditions:** Engine started, battery voltage must be at least 11.5v, all electrical components must be off, parking brake must be engaged (to keep daytime driving lights off), automatic transmission selector must be in park, the exhaust system must be properly sealed between the catalytic converter and the cylinder head, coolant temperature must be at least 80 degrees Celsius and oxygen sensor heaters for oxygen sensors before the catalytic converter must be functioning properly and the ground between the engine and the chassis must be well connected. The DME detected it could not control the idle speed correctly, as it is constantly more than 100 rpm less than specification. **Possible Causes:** • Air inlet is plugged or the air filter element is severely clogged • IAC circuit is open or shorted • IAC circuit VPWR circuit is open • IAC solenoid is damaged or has failed • The VSS has failed
DTC: P0507 **2T MIL: Yes** **2001, 2002, 2003, 2004, 2005, 2006** **Models:** 325i, 328i, 330i, 525i, 528i, 530i, 540i, 545i, 550i, 740i, 745i, 745Li, 750i, M3, Z4 **Engines:** All **Transmissions:** All	**Idle Air Control System RPM Higher Than Expected Conditions:** Engine started, battery voltage must be at least 11.5v, all electrical components must be off, parking brake must be engaged (to keep daytime driving lights off), automatic transmission selector must be in park, the exhaust system must be properly sealed between the catalytic converter and the cylinder head, coolant temperature must be at least 80 degrees Celsius and oxygen sensor heaters for oxygen sensors before the catalytic converter must be functioning properly and the ground between the engine and the chassis must be well connected. The DME detected it could not control the idle speed correctly, as it is constantly more than 200 rpm more than specification. **Possible Causes:** • Air intake leak located somewhere after the throttle body • IAC control circuit is shorted to chassis ground • IAC solenoid is damaged or has failed • Throttle Valve Control module has failed or is clogged with carbon • The VSS has failed
DTC: P0512 **2006** **Models:** 325i, 330i, 525i, 530i, 550i, 750i, Z4 **Engines:** 3.0L, 4.8L **Transmissions:** All	**Starter Request Circuit Malfunction Conditions:** The engine is on for more than one second and the injection and ignition have not yet released. Engine rpm present before DME triggers the starter, or the starter relay sticks in IVM, crankshaft sensor intermittent contact, the starter is grounded. **Possible Causes:** • Check the starter and starter relay. • The DME has failed

DTC	Trouble Code Title, Conditions & Possible Causes
DTC: P0520 **2006** **Models:** 325i, 330i, 525i, 530i, 550i, 750i, Z4 **Engines:** 3.0L, 4.8L **Transmissions:** All	**Engine Oil Pressure Sensor/Switch Circuit Malfunction Conditions:** The ignition must be on. The DME detected an error in the Engine Oil Pressure sensor. There is a short to ground. The plug has fallen off of the oil pressure switch. There is an open circuit in the harness. The pressure switch is defective. **Possible Causes:** • Replace the oil pressure sensor • Check coolant
DTC: P0530 **2006** **Models:** 325i, 330i, 525i, 530i, 550i, 750i, Z4 **Engines:** 3.0L, 4.8L **Transmissions:** All	**A/C Refrigerant Pressure Sensor "A" Circuit Malfunction Conditions:** The DME detected an implausible condition on the sensor. **Possible Causes:** • The A/C Refrigerant Pressure Sensor "A" has failed. • The DME has failed
DTC: P0532 **2006** **Models:** 325i, 330i, 525i, 530i, 550i, 750i, Z4 **Engines:** 3.0L, 4.8L **Transmissions:** All	**A/C Refrigerant Pressure Sensor "A" Circuit Low Conditions:** The DME detected a low condition on the sensor. **Possible Causes:** • The A/C Refrigerant Pressure Sensor "A" has failed. • The DME has failed
DTC: P0533 **2006** **Models:** 325i, 330i, 525i, 530i, 550i, 750i, Z4 **Engines:** 3.0L, 4.8L **Transmissions:** All	**A/C Refrigerant Pressure Sensor "A" Circuit High Conditions:** The DME detected a high condition on the sensor. **Possible Causes:** • The A/C Refrigerant Pressure Sensor "A" has failed. • The DME has failed
DTC: P0560 **2001, 2006** **Models:** 325i, 330i, 525i, 528i, 530i, 540i, 550i, 740i, 745Li, 750i, Z4 **Engines:** 3.0L, 4.4L, 4.8L, 5.4L **Transmissions:** All	**System Voltage Malfunction Conditions:** Engine started, battery voltage must be at least 11.5v, all electrical components must be off, parking brake must be engaged (to keep daytime driving lights off), automatic transmission selector must be in park, and the ground between the engine and the chassis must be well connected. The DME has detected a voltage value that is implausible or erratic. Engine speed must be greater than 1400rpm. **Possible Causes:** • Alternator damaged or faulty • Battery voltage low or insufficient • Fuses blown or circuits open • Battery connection to terminal not clean • Voltage regulator has failed
DTC: P0561 **2006** **Models:** 325i, 330i, 525i, 530i, 550i, 750i, Z4 **Engines:** 3.0L, 4.8L **Transmissions:** All	**System Voltage Unstable Conditions:** Engine started, battery voltage must be at least 11.5v, all electrical components must be off, parking brake must be engaged (to keep daytime driving lights off), automatic transmission selector must be in park, and the ground between the engine and the chassis must be well connected. The DME has detected a voltage value that is too erratic for the system to function properly. **Possible Causes:** • Alternator damaged or faulty • Battery voltage low or insufficient • Fuses blown or circuits open • Battery connection to terminal not clean • Voltage regulator has failed
DTC: P0562 **2006** **Models:** 325i, 330i, 525i, 530i, 550i, 750i, Z4 **Engines:** 3.0L, 4.8L **Transmissions:** All	**System Voltage Low Conditions:** Engine started, battery voltage must be at least 11.5v, all electrical components must be off, parking brake must be engaged (to keep daytime driving lights off), automatic transmission selector must be in park, and the ground between the engine and the chassis must be well connected. The DME has detected a voltage value that is below the specified minimum limit for the system to function properly. **Possible Causes:** • Alternator damaged or faulty • Battery voltage low or insufficient • Fuses blown or circuits open • Battery connection to terminal not clean • Voltage regulator has failed

DTC	Trouble Code Title, Conditions & Possible Causes
DTC: P0563 **2006** **Models:** 325i, 330i, 525i, 530i, 550i, 750i, Z4 **Engines:** 3.0L, 4.8L **Transmissions:** All	**System Voltage High Conditions:** Engine started for 18 seconds, battery voltage must be at least 11.5v, all electrical components must be off, parking brake must be engaged (to keep daytime driving lights off), automatic transmission selector must be in park, and the ground between the engine and the chassis must be well connected. The DME has detected a voltage value that has exceeded the specified maximum limit for the system to function properly. The vehicle was connected to 24 volts for too long after a jump start. ADC in ECU is defective. Delete stored fault codes from log. If fault reoccurs replace the ECU. **Possible Causes:** • Alternator damaged or faulty • Battery voltage low or insufficient • Fuses blown or circuits open • Battery connection to terminal not clean • Voltage regulator has failed
DTC: P0571 **2006** **Models:** 325i, 330i, 525i, 530i, 550i, 750i, Z4 **Engines:** 3.0L, 4.8L **Transmissions:** All	**Cruise/Brake Switch (A) Circuit Malfunction Conditions:** Engine started, battery voltage must be at least 11.5v, all electrical components must be off, parking brake must be engaged (to keep daytime driving lights off), automatic transmission selector must be in park, and the ground between the engine and the chassis must be well connected. The DME has detected a voltage value that is implausible or erratic. **Possible Causes:** • Brake light switch is faulty • Control circuit is shorted to chassis ground
DTC: P0597 **2T MIL: Yes (U.S. only)** **2005, 2006** **Models:** 325i, 330i, 525i, 530i, 545i, 550i, 745i, 750i, Z4 **Engines:** 3.0L, 4.8L **Transmissions:** All	**Thermostat Heater Control Circuit Open Conditions:** The engine's warm up performance is monitored by comparing measured coolant temperature with the modeled coolant temperature to detect a defective coolant thermostat that is reading false. The engine temperature must be less than 65 degrees Celsius, engine speed greater than 800rpm (with the vehicle speed greater than 10 but less than 90km/h) and the ambient temperature greater than -8 degrees Celsius. The thermostat should be wide open when cold, but is in error if it opens below desired control temperature **Possible Causes:** • Check for low coolant level or incorrect coolant mixture • DME detects a short circuit wiring in the ECT • CHT sensor is out-of-calibration or it has failed • ECT sensor is out-of-calibration or it has failed • Replace the thermostat
DTC: P0598 **2T MIL: Yes (U.S. only)** **2005, 2006** **Models:** 325i, 330i, 525i, 530i, 545i, 550i, 745i, 750i, Z4 **Engines:** 3.0L, 4.8L **Transmissions:** All	**Thermostat Heater Control Circuit Low Conditions:** The engine's warm up performance is monitored by comparing measured coolant temperature with the modeled coolant temperature to detect a defective coolant thermostat that is reading false. The engine temperature must be less than 65 degrees Celsius, engine speed greater than 800rpm (with the vehicle speed greater than 10 but less than 90km/h) and the ambient temperature greater than -8 degrees Celsius. The thermostat should be wide open when cold, but is in error if it opens below desired control temperature **Possible Causes:** • Check for low coolant level or incorrect coolant mixture • DME detects a short circuit wiring in the ECT • CHT sensor is out-of-calibration or it has failed • ECT sensor is out-of-calibration or it has failed • Replace the thermostat
DTC: P0599 **2T MIL: Yes (U.S. only)** **2005, 2006** **Models:** 325i, 330i, 525i, 530i, 545i, 550i, 745i, 750i, Z4 **Engines:** 3.0L, 4.8L **Transmissions:** All	**Thermostat Heater Control Circuit High Conditions:** The engine's warm up performance is monitored by comparing measured coolant temperature with the modeled coolant temperature to detect a defective coolant thermostat that is reading false. The engine temperature must be less than 65 degrees Celsius, engine speed greater than 800rpm (with the vehicle speed greater than 10 but less than 90km/h) and the ambient temperature greater than -8 degrees Celsius. The thermostat should be wide open when cold, but is in error if it opens below desired control temperature **Possible Causes:** • Check for low coolant level or incorrect coolant mixture • DME detects a short circuit wiring in the ECT • CHT sensor is out-of-calibration or it has failed • ECT sensor is out-of-calibration or it has failed • Replace the thermostat

DTC	Trouble Code Title, Conditions & Possible Causes
DTC: P0600 **2T MIL: Yes** **1996, 1997, 1998, 1999, 2000, 2001, 2002, 2003, 2004, 2005, 2006** **Models:** 318i, 320i, 323i, 325i, 328i, 330i, 525i, 528i, 530i, 540i, 545i, 550i, 740i, 745i, 745Li, 750i, 840i, M3, M5, Z3, Z4, Z8 **Engines:** 1.9L, 2.5L, 2.8L, 3.0L, 3.2L, 4.4L, 4.8L, 5.0L, 5.4L **Transmissions:** All	**Serial Communication Link (Data BUS) Message Missing Conditions:** The Engine Control Module (DME) communicates with all databus-capable control modules via a CAN databus. These databus-capable control modules are connected via two data bus wires which are twisted together (CAN_High and CAN_Low), and exchange information (messages). Missing information on the databus is recognized as a malfunction and stored. Trouble-free operation of the CAN-Bus requires that it have a terminal resistance. This central terminal resistor is located in the Engine Control Module (DME). **Possible Causes:** • CAN data bus wires have short circuited to each other
DTC: P0601 **2T MIL: Yes** **1996, 1997, 1998, 1999, 2000, 2001, 2002, 2003, 2004, 2005, 2006** **Models:** 318i, 320i, 323i, 325i, 328i, 330i, 525i, 528i, 530i, 540i, 545i, 550i, 740i, 745i, 745Li, 750i, 840i, M3, M5, Z3, Z4, Z8 **Engines:** 1.9L, 2.5L, 2.8L, 3.0L, 3.2L, 4.4L, 4.8L, 5.0L, 5.4L **Transmissions:** All	**Internal Control Module Memory Check Sum Error Conditions:** Key on, the DME has detected a programming error. The RAM and ROM check displays an invalid check-sum at power up/down. **Possible Causes:** • Battery terminal corrosion, or loose battery connection • Connection to the DME interrupted, or the circuit has been opened • Reprogramming error has occurred and needs replacement. Remember to check for Aftermarket Performance Products before replacing a DME.
DTC: P0603 **2T MIL: Yes** **1996, 1997, 1998, 1999, 2000, 2001** **Models:** 318i, 540i, 740i, Z3 **Engines:** 1.9L, 4.4L, 5.4L **Transmissions:** All	**DME Keep Alive Memory Test Error Conditions:** Key on, and the DME detected an internal memory fault. This code will set if KAPWR to the DME is interrupted (at the initial key on). Watchdog on. **Possible Causes:** • Battery terminal corrosion, or loose battery connection • KAPWR to DME interrupted, or the circuit has been opened • Reprogramming error has occurred and needs replacement. Remember to check for Aftermarket Performance Products before replacing a DME.
DTC: P0604 **2T MIL: Yes** **1996, 1997, 1998, 1999, 2000, 2001, 2002, 2003, 2004, 2005, 2006** **Models:** 318i, 320i, 323i, 325i, 328i, 330i, 525i, 528i, 530i, 540i, 545i, 550i, 740i, 745i, 745Li, 750i, 840i, M3, M5, Z3, Z4, Z8 **Engines:** 1.9L, 2.5L, 2.8L, 3.0L, 3.2L, 4.4L, 4.8L, 5.0L, 5.4L **Transmissions:** All	**Internal Control Module Random Access Memory (RAM) Error Conditions:** Key on, and the DME detected an internal memory fault. This code will set if KAPWR to the DME is interrupted (at the initial key on). Watchdog on. **Possible Causes:** • Battery terminal corrosion, or loose battery connection • Connection to the DME interrupted, or the circuit has been opened • Reprogramming error has occurred and needs replacement. Remember to check for Aftermarket Performance Products before replacing a DME.
DTC: P0605 **2000, 2001, 2002, 2003, 2004, 2005** **Models:** 320i, 325i, 330i, 525i, 540i, 545i, 550i, 740i, 745i , M3, M5, Z4, Z8 **Engines:** 2.5L, 3.0L, 3.2L, 4.4L, 4.8L, 5.0L **Transmissions:** All	**DME Read Only Memory (ROM) Test Error Conditions:** Key on, and the DME detected a ROM test error (ROM inside DME is corrupted). The DME is normally replaced if this code has set. **Possible Causes:** • An attempt was made to change the module calibration, or a module programming error may have occurred • Clear the trouble codes and then check for this trouble code. If it resets, the DME has failed and needs replacement. • Aftermarket performance products may have been installed. • The Transmission Control Module (TCM) has failed.
DTC: P0606 **2T MIL: Yes** **2001, 2002, 2003, 2004, 2005** **Models:** 320i, 325i, 330i, 525i, 540i, 545i, 550i, 740i, 745i , M3, M5, Z4, Z8 **Engines:** 2.5L, 3.0L, 3.2L, 4.4L, 4.8L, 5.0L **Transmissions:** All	**DME Internal Communication Error Conditions:** Key on, and the DME detected an internal communications register read back error during the initial key on check period. **Possible Causes:** • Clear the trouble codes and then check for this trouble code. If it resets, the DME has failed and needs replacement. • Remember to check for signs of Aftermarket Performance Products installation before replacing the DME.

DTC	Trouble Code Title, Conditions & Possible Causes
DTC: P0620 **2006** **Models:** 325i, 330i, 525i, 530i, 550i, 750i, Z4 **Engines:** 3.0L, 4.8L **Transmissions:** All	**Generator Control Circuit Error Conditions:** The engine is running for at least 25 seconds, and there is no communication faults at the BSD Interface **Possible Causes:** • Wires have short circuited to each other • ECU has failed • Generator failure
DTC: P0701 **2002, 2003** **Models:** M3 **Engines:** 3.2L **Transmissions:** All	**Transmission Control System Range/Performance Conditions:** Engine started, battery voltage must be at least 11.5v, all electrical components must be off, parking brake must be engaged (to keep daytime driving lights off), automatic transmission selector must be in park, and the ground between the engine and the chassis must be well connected. The DME detected a voltage outside the normal performance range to allow the system to properly function. **Possible Causes:** • Circuit harness connector contacts are corroded or ingresses of water • Circuit wires have shorted to each other, to battery or ground • Automatic Transmission Hydraulic Pressure Sensor 1 has failed • Solenoid valves in valve body are faulty • Transmission Input Speed (RPM) Sensor has failed • Transmission Output Speed (RPM) Sensor has failed • Engine Control Module (DME) is faulty • Voltage supply for Engine Control Module (DME) is faulty • Transmission Control Module (TCM) is faulty
DTC: P0704 **2006** **Models:** 325i, 330i, 525i, 530i, 550i, 750i, Z4 **Engines:** 3.0L, 4.8L **Transmissions:** All	**Clutch Switch Input Circuit Malfunction Conditions:** Engine started, battery voltage must be at least 11.5v, all electrical components must be off, parking brake must be engaged (to keep daytime driving lights off), automatic transmission selector must be in park, and the ground between the engine and the chassis must be well connected. The DME detected a voltage outside the normal performance range to allow the system to properly function. **Possible Causes:** • Circuit harness connector contacts are corroded or ingresses of water • Circuit wires have shorted to each other, to battery or ground • Automatic Transmission Hydraulic Pressure Sensor 1 has failed • Solenoid valves in valve body are faulty • Transmission Input Speed (RPM) Sensor has failed • Transmission Output Speed (RPM) Sensor has failed • Engine Control Module (DME) is faulty • Voltage supply for Engine Control Module (DME) is faulty • Transmission Control Module (TCM) is faulty
DTC: P0705 **2T MIL: Yes** **1996, 1997, 1998, 1999, 2000, 2001, 2002, 2003, 2004, 2005, 2006** **Models:** 318i, 320i, 323i, 325i, 328i, 330i, 525i, 528i, 530i, 540i, 545i, 550i, 740i, 745i, 745Li, 750i, 840i, M3, M5, Z3, Z4, Z8 **Engines:** 1.9L, 2.5L, 2.8L, 3.0L, 3.2L, 4.4L, 4.8L, 5.0L, 5.4L **Transmissions:** A/T	**TR Sensor Circuit Malfunction Conditions:** Engine started, battery voltage must be at least 11.5v, all electrical components must be off, parking brake must be engaged (to keep daytime driving lights off), automatic transmission selector must be in park, and the ground between the engine and the chassis must be well connected. The DME detected a voltage or signal outside the normal performance range to allow the system to properly function. The engine speed is between 200 and 440rpm. **Possible Causes:** • Circuit harness connector contacts are corroded or ingresses of water • Circuit wires have shorted to each other, to battery or ground • Automatic Transmission Hydraulic Pressure Sensor 1 has failed • Solenoid valves in valve body are faulty • Transmission Input Speed (RPM) Sensor has failed • Transmission Output Speed (RPM) Sensor has failed • Engine Control Module (DME) is faulty • Voltage supply for Engine Control Module (DME) is faulty • Transmission Control Module (TCM) is faulty

DTC	Trouble Code Title, Conditions & Possible Causes
DTC: P0710 **2004, 2005, 2006** **Models:** 320i, 325i, 330i, 525i, 530i, 545i, 740i, 745i, M3, Z4 **Engines:** 2.5L, 3.0L **Transmissions:** A/T	**Transmission Fluid Temperature Sensor Circuit Malfunction Conditions:** Engine started, battery voltage must be at least 11.5v, all electrical components must be off, parking brake must be engaged (to keep daytime driving lights off), automatic transmission selector must be in park, and the ground between the engine and the chassis must be well connected. The DME detected the Transmission fluid temperature sensor circuit was outside the normal range in the test to allow proper function. **Possible Causes:** • ATF is low, contaminated, dirty or burnt • Circuit harness connector contacts are corroded or ingresses of water • Circuit wires have shorted to each other, to battery or ground • Automatic Transmission Hydraulic Pressure Sensor 1 has failed • Solenoid valves in valve body are faulty • Transmission Input Speed (RPM) Sensor has failed • Transmission Output Speed (RPM) Sensor has failed • Engine Control Module (DME) is faulty • Voltage supply for Engine Control Module (DME) is faulty • Transmission Control Module (TCM) is faulty
DTC: P0711 **2004, 2005, 2006** **Models:** 320i, 325i, 330i, 525i, 530i, 545i, 740i, 745i, M3, Z4 **Engines:** 2.5L, 3.0L **Transmissions:** A/T	**Transmission Fluid Temperature Sensor Signal Range/Performance Conditions:** Engine started, battery voltage must be at least 11.5v, all electrical components must be off, parking brake must be engaged (to keep daytime driving lights off), automatic transmission selector must be in park, and the ground between the engine and the chassis must be well connected. The DME detected the Transmission Fluid Temperature (TFT) sensor value was not close its normal operating temperature. **Possible Causes:** • ATF is low, contaminated, dirty or burnt • TFT sensor signal circuit has a high resistance condition • TFT sensor is out-of-calibration ("skewed") or it has failed
DTC: P0712 **2004, 2005, 2006** **Models:** 320i, 325i, 330i, 525i, 530i, 545i, 740i, 745i, M3, Z4 **Engines:** 2.5L, 3.0L **Transmissions:** A/T	**Transmission Fluid Temperature Sensor Circuit Low Input Conditions:** Engine started, battery voltage must be at least 11.5v, all electrical components must be off, parking brake must be engaged (to keep daytime driving lights off), automatic transmission selector must be in park, and the ground between the engine and the chassis must be well connected. The DME detected the Transmission Fluid Temperature (TFT) sensor was less than its minimum self-test range in the test. **Possible Causes:** • TFT sensor signal circuit is shorted to chassis ground • TFT sensor signal circuit is shorted to sensor ground • TFT sensor is damaged, or out-of-calibration, or has failed
DTC: P0713 **2004, 2005, 2006** **Models:** 320i, 325i, 330i, 525i, 530i, 545i, 740i, 745i, M3, Z4 **Engines:** 2.5L, 3.0L **Transmissions:** A/T	**Transmission Fluid Temperature Sensor Circuit High Input Conditions:** Engine started, battery voltage must be at least 11.5v, all electrical components must be off, parking brake must be engaged (to keep daytime driving lights off), automatic transmission selector must be in park, and the ground between the engine and the chassis must be well connected. The DME detected the Transmission Fluid Temperature (TFT) sensor was more than its maximum self-test range in the test. **Possible Causes:** • TFT sensor signal circuit is open between the sensor and DME • TFT sensor ground circuit is open between sensor and DME • TFT sensor is damaged or has failed
DTC: P0714 **2004, 2005, 2006** **Models:** 320i, 325i, 330i, 525i, 530i, 545i, 740i, 745i, M3, Z4 **Engines:** 2.5L, 3.0L **Transmissions:** A/T	**Transmission Fluid Temperature Sensor Circuit Intermittent Conditions:** Engine started, battery voltage must be at least 11.5v, all electrical components must be off, parking brake must be engaged (to keep daytime driving lights off), automatic transmission selector must be in park, and the ground between the engine and the chassis must be well connected. The DME detected the Transmission Fluid Temperature (TFT) sensor was giving a false reading or was not reading at all. **Possible Causes:** • TFT sensor signal circuit is open between the sensor and DME • TFT sensor ground circuit is open between sensor and DME • TFT sensor is damaged or has failed

DTC	Trouble Code Title, Conditions & Possible Causes
DTC: P0715 **2T MIL: Yes** **1996, 1997, 1998, 1999, 2000, 2001, 2002, 2003, 2004, 2005, 2006** **Models:** 318i, 320i, 323i, 325i, 328i, 330i, 525i, 528i, 530i, 540i, 545i, 550i, 740i, 745i, 745Li, 750i, 840i, M3, M5, Z3, Z4, Z8 **Engines:** 1.9L, 2.5L, 2.8L, 3.0L, 3.2L, 4.4L, 4.8L, 5.0L, 5.4L **Transmissions:** A/T	**Input/Turbine Speed Sensor Circuit Malfunction Conditions:** Engine started, vehicle driven with the vehicle speed sensor indicating more than 1 mph, and the DME detected the Transmission Vehicle Speed Sensor signals were erratic, or that they were missing for a period of time. The engine speed is greater than 600rpm. Any gear can be selected, output speed must be greater than 600rpm, and wheel speed greater than 400rpm. **Possible Causes:** • TVSS signal circuit is open • TVSS signal is shorted to chassis ground • TVSS signal is shorted to sensor ground • TVSS assembly is damaged or it has failed
DTC: P0716 **2T MIL: Yes** **2002, 2003, 2004** **Models:** 320i, 325i, 330i, 525i, 530i, 540i, 545i, M3, Z4 **Engines:** 2.5L, 3.0L, 3.2L, 4.4L, 4.8L **Transmissions:** A/T	**Input Turbine/Speed Sensor Circuit Range/Performance Conditions:** Engine started, vehicle driven with the vehicle speed sensor indicating more than 1 mph, and the DME detected the Transmission Vehicle Speed Sensor signals were erratic, or that they were missing for a period of time. **Possible Causes:** • TVSS signal circuit is open • TVSS signal is shorted to chassis ground • TVSS signal is shorted to sensor ground • TVSS assembly is damaged or it has failed
DTC: P0717 **2T MIL: Yes** **2002, 2003, 2004** **Models:** 320i, 325i, 330i, 525i, 530i, 540i, 545i, M3, Z4 **Engines:** 2.5L, 3.0L, 3.2L, 4.4L, 4.8L **Transmissions:** A/T	**Transmission Speed Shaft Sensor Signal Intermittent Conditions:** Engine started, vehicle speed sensor indicating over 1 mph, and the DME detected an intermittent loss of TSS signals (i.e., the TSS signals were erratic, irregular or missing). **Possible Causes:** • TSS connector is damaged, loose or shorted • TSS signal circuit has an intermittent open condition • TSS signal circuit has an intermittent short to ground condition • TSS assembly is damaged or is has failed
DTC: P0720 **2T MIL: Yes** **1996, 1997, 1998, 1999, 2000, 2001, 2002, 2003, 2004, 2005, 2006** **Models:** 318i, 320i, 323i, 325i, 328i, 330i, 525i, 528i, 530i, 540i, 545i, 550i, 740i, 745i, 745Li, 750i, 840i, M3, M5, Z3, Z4, Z8 **Engines:** 1.9L, 2.5L, 2.8L, 3.0L, 3.2L, 4.4L, 4.8L, 5.0L, 5.4L **Transmissions:** A/T	**Output/Turbine Speed Sensor Circuit Malfunction Conditions:** Engine started, vehicle driven with the vehicle speed sensor indicating more than 1 mph, and the DME detected the Transmission Vehicle Speed Sensor signals were erratic, or that they were missing for a period of time. The engine speed is greater than 600rpm. Any gear can be selected, output speed must be greater than 600rpm, and wheel speed greater than 400rpm. **Possible Causes:** • TVSS signal circuit is open • TVSS signal is shorted to chassis ground • TVSS signal is shorted to sensor ground • TVSS assembly is damaged or it has failed
DTC: P0721 **2002, 2003, 2004** **Models:** 320i, 325i, 330i, 525i, 530i, 540i, 545i, M3, Z4 **Engines:** 2.5L, 3.0L, 3.2L, 4.4L, 4.8L **Transmissions:** A/T	**A/T Output Shaft Speed Sensor Noise Interference Conditions:** Engine started, VSS signal more than 1 mph, and the DME detected "noise" interference on the Output Shaft Speed (OSS) sensor circuit. **Possible Causes:** • After market add-on devices interfering with the OSS signal • OSS connector is damaged, loose or shorted, or the wiring is misrouted or it is damaged • OSS assembly is damaged or it has failed
DTC: P0722 **2T MIL: Yes** **2002, 2003, 2004** **Models:** 320i, 325i, 330i, 525i, 530i, 540i, 545i, M3, Z4 **Engines:** 2.5L, 3.0L, 3.2L, 4.4L, 4.8L **Transmissions:** A/T	**A/T Output Speed Sensor No Signal Conditions:** Engine started, and the DME did not detect any Vehicle Speed Sensor (VSS) sensor signals upon initial vehicle movement. **Possible Causes:** • After market add-on devices interfering with the VSS signal • VSS sensor wiring is misrouted, damaged or shorting • DME and/or TCM has failed

DTC	Trouble Code Title, Conditions & Possible Causes
DTC: P0727 **1T MIL: Yes** **2001, 2002** **Models:** 325i, 330i, 525i, 530i, Z3 **Engines:** 2.5L, 3.0L **Transmissions:** A/T	**Engine Speed Input Circuit No Signal Conditions:** The Engine Speed (RPM) Sensor detects engine speed and reference marks. Without an engine speed signal, the engine will not start. If the engine speed signal fails while the engine is running, the engine will stop immediately. **Note: There is a larger-sized gap on the sensor wheel. This gap is the reference mark and does not mean that the sensor wheel is damaged.** **Possible Causes:** • Engine speed sensor has failed • Circuit is shorting to ground, B+ or is open • Sensor wheel is damaged, run out or not properly secured
DTC: P0729 **2T MIL: Yes** 2003 **Models:** 745i **Engines:** 4.8L **Transmissions:** A/T	**Gear 6 Incorrect Ratio Conditions:** Engine started, battery voltage must be at least 11.5v, all electrical components must be off, and the ground between the engine and the chassis must be well connected. The DME detected an incorrect ratio within the sixth gear. **Possible Causes:** • ATF level is low • Circuit harness connector contacts are corroded or ingresses of water • Circuit wires have shorted to each other, to battery or ground • Automatic Transmission Hydraulic Pressure Sensor 1 has failed • Solenoid valves in valve body are faulty • Transmission Control Module (TCM) needs replacing • Transmission Input Speed (RPM) Sensor has failed • Transmission Output Speed (RPM) Sensor has failed
DTC: P0730 **2T MIL: Yes** **1996, 1997** **Models:** 318i, 325i, 328i, 540i, 740i, 750i, 840i, Z3 **Engines:** 1.9L, 2.8L, 4.4L, 5.4L **Transmissions:** A/T	**Gear Incorrect Ratio Conditions:** Engine started, battery voltage must be at least 11.5v, all electrical components must be off, and the ground between the engine and the chassis must be well connected. The DME detected an incorrect gear ratio. **Possible Causes:** • ATF level is low • Circuit harness connector contacts are corroded or ingresses of water • Circuit wires have shorted to each other, to battery or ground • Automatic Transmission Hydraulic Pressure Sensor 1 has failed • Solenoid valves in valve body are faulty • Transmission Control Module (TCM) needs replacing • Transmission Input Speed (RPM) Sensor has failed • Transmission Output Speed (RPM) Sensor has failed
DTC: P0731 **2T MIL: Yes** **1997, 1998, 1999, 2000, 2001, 2002, 2003, 2004, 2005, 2006** **Models:** 318i, 320i, 323i, 325i, 328i, 330i, 525i, 528i, 530i, 540i, 545i, 550i, 740i, 745i, 745Li, 750i, 840i, M3, M5, Z3, Z4, Z8 **Engines:** 1.9L, 2.5L, 2.8L, 3.0L, 3.2L, 4.4L, 4.8L, 5.0L, 5.4L **Transmissions:** A/T	**Incorrect First Gear Ratio Conditions:** Engine started, vehicle operating with 1st gear commanded "on", and the DME detected an incorrect 1st gear ratio during the test. **Possible Causes:** • 1st Gear solenoid harness connector not properly seated • 1st Gear solenoid signal shorted to ground, or open • 1st Gear solenoid wiring harness connector is damaged • 1st Gear solenoid is damaged or not properly installed • ATF level is low • Circuit harness connector contacts are corroded or ingresses of water • Circuit wires have shorted to each other, to battery or ground • Automatic Transmission Hydraulic Pressure Sensor 1 has failed • Transmission Control Module (TCM) needs replacing • Transmission Input Speed (RPM) Sensor has failed • Transmission Output Speed (RPM) Sensor has failed
DTC: P0732 **2T MIL: Yes** **1997, 1998, 1999, 2000, 2001, 2002, 2003, 2004, 2005, 2006** **Models:** 318i, 320i, 323i, 325i, 328i, 330i, 525i, 528i, 530i, 540i, 545i, 550i, 740i, 745i, 745Li, 750i, 840i, M3, M5, Z3, Z4, Z8 **Engines:** 1.9L, 2.5L, 2.8L, 3.0L, 3.2L, 4.4L, 4.8L, 5.0L, 5.4L **Transmissions:** A/T	**Incorrect Second Gear Ratio Conditions:** Engine started, vehicle operating with 2nd Gear commanded "on", and the DME detected an incorrect 2nd gear ratio during the test. Input speed must be greater than 400rpm, and output speed must be greater than 250rpm for 10ms of continuous time. **Possible Causes:** • 2nd Gear solenoid harness connector not properly seated • 2nd Gear solenoid signal shorted to ground, or open • 2nd Gear solenoid wring harness connector is damaged • 2nd Gear solenoid is damaged or not properly installed • ATF level is low • Circuit harness connector contacts are corroded or ingresses of water • Circuit wires have shorted to each other, to battery or ground • Automatic Transmission Hydraulic Pressure Sensor 1 has failed • Transmission Control Module (TCM) needs replacing • Transmission Input Speed (RPM) Sensor has failed • Transmission Output Speed (RPM) Sensor has failed

DTC	Trouble Code Title, Conditions & Possible Causes
DTC: P0733 **2T MIL: Yes** **1997, 1998, 1999, 2000, 2001, 2002, 2003, 2004, 2005, 2006** **Models:** 318i, 320i, 323i, 325i, 328i, 330i, 525i, 528i, 530i, 540i, 545i, 550i, 740i, 745i, 745Li, 750i, 840i, M3, M5, Z3, Z4, Z8 **Engines:** 1.9L, 2.5L, 2.8L, 3.0L, 3.2L, 4.4L, 4.8L, 5.0L, 5.4L **Transmissions:** A/T	**Incorrect Third Gear Ratio Conditions:** Engine started, vehicle operating with 3rd Gear commanded "on", and the DME detected an incorrect 3rd gear ratio during the test. Input speed must be greater than 400rpm, and output speed must be greater than 250rpm for 10ms of continuous time. **Possible Causes:** • 3rd Gear solenoid harness connector not properly seated • 3rd Gear solenoid signal shorted to ground, or open • 3rd Gear solenoid wiring harness connector is damaged • 3rd Gear solenoid is damaged or not properly installed • ATF level is low • Circuit harness connector contacts are corroded or ingresses of water • Circuit wires have shorted to each other, to battery or ground • Automatic Transmission Hydraulic Pressure Sensor 1 has failed • Transmission Control Module (TCM) needs replacing • Transmission Input Speed (RPM) Sensor has failed • Transmission Output Speed (RPM) Sensor has failed
DTC: P0734 **2T MIL: Yes** **1997, 1998, 1999, 2000, 2001, 2002, 2003, 2004, 2005, 2006** **Models:** 318i, 320i, 323i, 325i, 328i, 330i, 525i, 528i, 530i, 540i, 545i, 550i, 740i, 745i, 745Li, 750i, 840i, M3, M5, Z3, Z4, Z8 **Engines:** 1.9L, 2.5L, 2.8L, 3.0L, 3.2L, 4.4L, 4.8L, 5.0L, 5.4L **Transmissions:** A/T	**Incorrect Fourth Gear Ratio Conditions:** Engine started, vehicle operating with 4th Gear commanded "on", and the DME detected an incorrect 4th gear ratio during the test. Input speed must be greater than 400rpm, and output speed must be greater than 250rpm for 10ms of continuous time. **Possible Causes:** • 4th Gear solenoid harness connector not properly seated • 4th Gear solenoid signal shorted to ground, or open • 4th Gear solenoid wiring harness connector is damaged • 4th Gear solenoid is damaged or not properly installed • ATF level is low • Circuit harness connector contacts are corroded or ingresses of water • Circuit wires have shorted to each other, to battery or ground • Automatic Transmission Hydraulic Pressure Sensor 1 has failed • Transmission Control Module (TCM) needs replacing • Transmission Input Speed (RPM) Sensor has failed • Transmission Output Speed (RPM) Sensor has failed
DTC: P0735 **2T MIL: Yes** **1996, 1997, 1998, 1999, 2000, 2001, 2002, 2003, 2004, 2005, 2006** **Models:** 318i, 320i, 323i, 325i, 328i, 330i, 525i, 528i, 530i, 540i, 545i, 550i, 740i, 745i, 745Li, 750i, 840i, M3, M5, Z3, Z4, Z8 **Engines:** 1.9L, 2.5L, 2.8L, 3.0L, 3.2L, 4.4L, 4.8L, 5.0L, 5.4L **Transmissions:** A/T	**Incorrect Fifth Gear Ratio Conditions:** Engine started, vehicle operating with 5th Gear commanded "on", and the DME detected an incorrect 5th gear ratio during the test. Input speed must be greater than 400rpm, and output speed must be greater than 250rpm for 10ms of continuous time. **Possible Causes:** • 5th Gear solenoid harness connector not properly seated • 5th Gear solenoid signal shorted to ground, or open • 5th Gear solenoid wiring harness connector is damaged • 5th Gear solenoid is damaged or not properly installed • ATF level is low • Circuit harness connector contacts are corroded or ingresses of water • Circuit wires have shorted to each other, to battery or ground • Automatic Transmission Hydraulic Pressure Sensor 1 has failed • Transmission Control Module (TCM) needs replacing • Transmission Input Speed (RPM) Sensor has failed • Transmission Output Speed (RPM) Sensor has failed
DTC: P0740 **2T MIL: Yes** **1996, 1997, 1998, 1999, 2000, 2001, 2002** **Models:** 318i, 320i, 323i, 325i, 328i, 330i, 525i, 528i, 530i, 540i, 545i, 550i, 740i, 745i, 745Li, 750i, 840i, M3, M5, Z3, Z4, Z8 **Engines:** 1.9L, 2.5L, 2.8L, 3.0L, 3.2L, 4.4L, 4.8L, 5.0L, 5.4L **Transmissions:** A/T	**TCC Solenoid Circuit Malfunction Conditions:** Engine started, KOER Self-Test enabled, vehicle driven at cruise speed, and the DME did not detect any voltage drop across the TCC solenoid circuit during the test period. **Possible Causes:** • TCC solenoid control circuit is open or shorted to ground • TCC solenoid wiring harness connector is damaged • TCC solenoid is damaged or has failed

DTC	Trouble Code Title, Conditions & Possible Causes
DTC: P0741 **2T MIL: Yes** **2001, 2002, 2003, 2004** **Models:** All **Engines:** All **Transmissions:** A/T	**TCC Mechanical System Range/Performance Conditions:** Engine started, vehicle driven in gear with VSS signals received, and the DME detected excessive slippage while in normal operation. The TCC is stuck off. **Possible Causes:** • TCC solenoid has a mechanical failure • TCC solenoid has a hydraulic failure
DTC: P0743 **2T MIL: Yes** **1996, 1997, 1998, 1999, 2000, 2001, 2002, 2003, 2004, 2005, 2006** **Models:** 318i, 320i, 323i, 325i, 328i, 330i, 525i, 528i, 530i, 540i, 545i, 550i, 740i, 745i, 745Li, 750i, 840i, M3, M5, Z3, Z4, Z8 **Engines:** 1.9L, 2.5L, 2.8L, 3.0L, 3.2L, 4.4L, 4.8L, 5.0L, 5.4L **Transmissions:** A/T	**TCC Solenoid Circuit Malfunction Conditions:** Engine started, KOER Self-Test enabled, vehicle driven at cruise speed, and the DME did not detect any voltage drop across the TCC solenoid circuit during the test period. **Possible Causes:** • TCC solenoid control circuit is open or shorted to ground • TCC solenoid wiring harness connector is damaged • TCC solenoid is damaged or has failed
DTC: P0745 **2T MIL: Yes** **2002, 2003, 2004** **Models:** 320i, 325i, 330i, 525i, 540i, 745i, M3, Z3, Z4 **Engines:** 2.5L, 3.0L, 3.2L, 4.4L, 4.8L **Transmissions:** A/T	**Pressure Regulator Valve 1 Plausibility Conditions:** The current to/from the pressure regulator valve is either higher or lower than the threshold value. **Possible Causes:** • Pressure control solenoid circuit is shorting to ground • Pressure control solenoid circuit is open • Valve has failed • TCM has failed
DTC: P0748 **2T MIL: Yes** **1996, 1997, 1998, 1999, 2000, 2001, 2002, 2003, 2004, 2005, 2006** **Models:** 318i, 320i, 323i, 325i, 328i, 330i, 525i, 528i, 530i, 540i, 545i, 550i, 740i, 745i, 745Li, 750i, 840i, M3, M5, Z3, Z4, Z8 **Engines:** 1.9L, 2.5L, 2.8L, 3.0L, 3.2L, 4.4L, 4.8L, 5.0L, 5.4L **Transmissions:** A/T	**Pressure Regulator Valve 2 Upper Threshold Conditions:** The signal to/from the pressure regulator valve has been interrupted. **Possible Causes:** • Pressure control solenoid circuit is shorting to ground • Pressure control solenoid circuit is open • Valve has failed • TCM has failed
DTC: P0750 **2002, 2003, 2004** **Models:** 320i, 325i, 330i, 525i, 540i, 745i, M3, Z3, Z4 **Engines:** 2.5L, 3.0L, 3.2L, 4.4L, 4.8L **Transmissions:** A/T	**Shift Solenoid "A" Circuit Continuity Short to Battery Conditions:** Engine started, vehicle driven with the solenoid applied, and the DME detected an unexpected voltage condition on the SS1/A solenoid circuit was incorrect during the test. **Possible Causes:** • Solenoid valves in valve body are faulty • Solenoid circuit is shorting to ground • Solenoid circuit is open • TCM has failed or wiring is shorting
DTC: P0751 **2T MIL: Yes** **2002, 2003, 2004** **Models:** 320i, 325i, 330i, 525i, 540i, 745i, M3, Z3, Z4 **Engines:** 2.5L, 3.0L, 3.2L, 4.4L, 4.8L **Transmissions:** A/T	**Solenoid Valve 1 Plausibility Conditions:** The signal to/from the pressure regulator valve is interrupted or does not exist. **Possible Causes:** • Pressure control solenoid circuit is shorting to ground • Pressure control solenoid circuit is open • Valve has failed • TCM has failed

DTC	Trouble Code Title, Conditions & Possible Causes
DTC: P0752 **2T MIL: Yes** **2002, 2003, 2004** **Models:** 320i, 325i, 330i, 525i, 540i, 745i, M3, Z3, Z4 **Engines:** 2.5L, 3.0L, 3.2L, 4.4L, 4.8L **Transmissions:** A/T	**A/T Shift Solenoid 1/A Function Range/Performance Conditions:** Engine started, vehicle driven with the solenoid applied, and the DME detected a mechanical failure while operating the Shift Solenoid 1/A during the CCM test period. **Possible Causes:** • SS1/A solenoid is stuck in the "on" position • SS1/A solenoid has a mechanical failure • SS1/A solenoid has a hydraulic failure
DTC: P0753 **2T MIL: Yes** **1996, 1997, 1998, 1999, 2000, 2001, 2002, 2003, 2004, 2005, 2006** **Models:** 318i, 320i, 323i, 325i, 328i, 330i, 525i, 528i, 530i, 540i, 545i, 550i, 740i, 745i, 745Li, 750i, 840i, M3, M5, Z3, Z4, Z8 **Engines:** 1.9L, 2.5L, 2.8L, 3.0L, 3.2L, 4.4L, 4.8L, 5.0L, 5.4L **Transmissions:** A/T	**Solenoid Valve 1 Upper Threshold Conditions:** The signal to/from the pressure regulator valve is interrupted or short circuited to supply. **Possible Causes:** • Pressure control solenoid circuit is shorting to ground • Pressure control solenoid circuit is open • Valve has failed • TCM has failed
DTC: P0755 **2T MIL: Yes** **2002, 2003, 2004** **Models:** 320i, 325i, 330i, 525i, 540i, 745i, M3, Z3, Z4 **Engines:** 2.5L, 3.0L, 3.2L, 4.4L, 4.8L **Transmissions:** A/T	**Shift Solenoid "B" Circuit Continuity Short to Battery Conditions:** Engine started, vehicle driven with the solenoid applied, and the DME detected an unexpected voltage condition on the SS1/B solenoid circuit was incorrect during the test. **Possible Causes:** • Solenoid valves in valve body are faulty • Solenoid circuit is shorting to ground • Solenoid circuit is open • TCM has failed or wiring is shorting
DTC: P0756 **2T MIL: Yes** **2002, 2003, 2004** **Models:** 320i, 325i, 330i, 525i, 540i, 745i, M3, Z3, Z4 **Engines:** 2.5L, 3.0L, 3.2L, 4.4L, 4.8L **Transmissions:** A/T	**Solenoid Valve 2 Plausibility Conditions:** The signal to/from the pressure regulator valve is interrupted or does not exist. **Possible Causes:** • Pressure control solenoid circuit is shorting to ground • Pressure control solenoid circuit is open • Valve has failed • TCM has failed
DTC: P0757 **2T MIL: Yes** **2002, 2003, 2004** **Models:** 320i, 325i, 330i, 525i, 540i, 745i, M3, Z3, Z4 **Engines:** 2.5L, 3.0L, 3.2L, 4.4L, 4.8L **Transmissions:** A/T	**A/T Shift Solenoid 2/B Function Range/Performance Conditions:** Engine started, vehicle driven with the solenoid applied, and the DME detected a mechanical failure while operating the Shift Solenoid 2/B during the CCM test period. **Possible Causes:** • SS2/B solenoid is stuck in the "on" position • SS2/B solenoid has a mechanical failure • SS2/B solenoid has a hydraulic failure
DTC: P0758 **2T MIL: Yes** **1996, 1997, 1998, 1999, 2000, 2001, 2002, 2003, 2004, 2005, 2006** **Models:** 318i, 320i, 323i, 325i, 328i, 330i, 525i, 528i, 530i, 540i, 545i, 550i, 740i, 745i, 745Li, 750i, 840i, M3, M5, Z3, Z4, Z8 **Engines:** 1.9L, 2.5L, 2.8L, 3.0L, 3.2L, 4.4L, 4.8L, 5.0L, 5.4L **Transmissions:** A/T	**Solenoid Valve 2 Upper Threshold Conditions:** The signal to/from the pressure regulator valve is interrupted or short circuited to supply. **Possible Causes:** • Pressure control solenoid circuit is shorting to ground • Pressure control solenoid circuit is open • Valve has failed • TCM has failed

DTC	Trouble Code Title, Conditions & Possible Causes
DTC: P0760 **2T MIL: Yes** **2002, 2003, 2004** **Models:** 320i, 325i, 330i, 525i, 540i, 745i, M3, Z3, Z4 **Engines:** 2.5L, 3.0L, 3.2L, 4.4L, 4.8L **Transmissions:** A/T	**Shift Solenoid "C" Circuit Continuity Short to Battery Conditions:** Engine started, vehicle driven with the solenoid applied, and the DME detected an unexpected voltage condition on the SS1/C solenoid circuit was incorrect during the test. **Possible Causes:** • Solenoid valves in valve body are faulty • Solenoid circuit is shorting to ground • Solenoid circuit is open • TCM has failed or wiring is shorting
DTC: P0761 **2T MIL: Yes** **2002, 2003, 2004** **Models:** 320i, 325i, 330i, 525i, 540i, 745i, M3, Z3, Z4 **Engines:** 2.5L, 3.0L, 3.2L, 4.4L, 4.8L **Transmissions:** A/T	**Solenoid Valve 3 Plausibility Conditions:** The signal to/from the pressure regulator valve is interrupted or does not exist. **Possible Causes:** • Pressure control solenoid circuit is shorting to ground • Pressure control solenoid circuit is open • Valve has failed • TCM has failed
DTC: P0762 **2T MIL: Yes** **2002, 2003, 2004** **Models:** 320i, 325i, 330i, 525i, 540i, 745i, M3, Z3, Z4 **Engines:** 2.5L, 3.0L, 3.2L, 4.4L, 4.8L **Transmissions:** A/T	**A/T Shift Solenoid 3/C Function Range/Performance Conditions:** Engine started, vehicle driven with Shift Solenoid 3/C applied, and the DME detected a mechanical failure occurred (stuck "on") while operating Shift Solenoid 3/C during the test. **Possible Causes:** • SS3/C solenoid may be stuck "on" • SS3/C solenoid has a mechanical failure • SS3/C solenoid has a hydraulic failure
DTC: P0763 **2T MIL: Yes** **1996, 1997, 1998, 1999, 2000, 2001, 2002, 2003, 2004, 2005** **Models:** 318i, 320i, 323i, 325i, 328i, 330i, 525i, 528i, 530i, 540i, 545i, 550i, 740i, 745i, 745Li, 750i, 840i, M3, M5, Z3, Z4, Z8 **Engines:** 1.9L, 2.5L, 2.8L, 3.0L, 3.2L, 4.4L, 4.8L, 5.0L, 5.4L **Transmissions:** A/T	**Solenoid Valve 3 Upper Threshold Conditions:** The signal to/from the pressure regulator valve is interrupted or short circuited to supply. **Possible Causes:** • Pressure control solenoid circuit is shorting to ground • Pressure control solenoid circuit is open • Valve has failed • TCM has failed
DTC: P0768 **2T MIL: Yes** 1998 **Models:** 328i, 528i, M3 **Engines:** 2.8L **Transmissions:** A/T	**A/T Shift Solenoid 4/D Electrical Conditions:** Engine started, vehicle driven with the solenoid applied, and the DME detected an unexpected voltage condition on the SS3/C solenoid circuit was incorrect during the test.. **Possible Causes:** • Shift Solenoid 3/C connector is damaged, open or shorted • Shift Solenoid 3/C control circuit is open • Shift Solenoid 3/C control circuit is shorted to ground • Shift Solenoid 3/C is damaged or it has failed
DTC: P0773 **2T MIL: Yes** 1998 **Models:** 328i, 528i, M3 **Engines:** 2.8L **Transmissions:** A/T	**A/T Shift Solenoid 5/E Electrical Conditions:** Engine started, vehicle driven with the solenoid applied, and the DME detected an unexpected voltage condition on the SS3/D solenoid circuit was incorrect during the test.. **Possible Causes:** • Shift Solenoid connector is damaged, open or shorted • Shift Solenoid control circuit is open • Shift Solenoid control circuit is shorted to ground • Shift Solenoid is damaged or it has failed

DTC	Trouble Code Title, Conditions & Possible Causes
DTC: P0775 **2T MIL: Yes** **2002, 2003, 2004** **Models:** 320i, 325i, 330i, 525i, 540i, 745i, M3, Z3, Z4 **Engines:** 2.5L, 3.0L, 3.2L, 4.4L, 4.8L **Transmissions:** A/T	**Pressure Regulator Valve 2 Plausibility Conditions:** The current to/from the pressure regulator valve is either higher or lower than the threshold value. **Possible Causes:** • Pressure control solenoid circuit is shorting to ground • Pressure control solenoid circuit is open • Valve has failed • TCM has failed
DTC: P0778 **2T MIL: Yes** **2001, 2002, 2003, 2004, 2005, 2006** **Models:** 320i, 325i, 330i, 525i, 530i, 540i, 545i, 550i, 740i, 745i, 750i, M3, Z4 **Engines:** 2.5L, 3.0L, 3.2L, 4.4L, 4.8L, 5.4L **Transmissions:** A/T	**Pressure Control Solenoid "B" Electrical Conditions:** Engine started, vehicle driven with the solenoid applied, and the DME detected an unexpected voltage condition on the SS3/C solenoid circuit was incorrect during the test.. **Possible Causes:** • Shift Solenoid connector is damaged, open or shorted • Shift Solenoid control circuit is open • Shift Solenoid control circuit is shorted to ground • Shift Solenoid is damaged or it has failed
DTC: P0781 **2T MIL: Yes** **2002, 2003, 2004, 2005, 2006** **Models:** 320i, 325i, 330i, 525i, 530i, 545i, 745i, M3, Z4 **Engines:** 2.5L, 3.0L, 3.2L, 4.8L, 5.4L **Transmissions:** A/T	**1-2 Shift Range Monitoring Conditions:** Engine running and vehicle driven, the DME detected a mechanical malfunction within the transmission. The output speed is greater than 300rpm, the transmission oil temperature is greater than 0 degrees Celsius, the engine speed is greater or equal to 600rpm and the range position is P, R, or N. **Possible Causes:** • Solenoid valves in valve body are faulty • Solenoid circuit is shorting to ground • Solenoid circuit is open • TCM has failed or wiring is shorting • Mechanical malfunction in transmission
DTC: P0782 **2T MIL: Yes** **2002, 2003, 2004** **Models:** 320i, 325i, 330i, 525i, 540i, 745i, M3, Z3, Z4 **Engines:** 2.5L, 3.0L, 3.2L, 4.4L, 4.8L **Transmissions:** A/T	**2-3 Shift Range Monitoring Conditions:** Engine running and vehicle driven, the DME detected a mechanical malfunction within the transmission. The output speed is greater than 300rpm, the transmission oil temperature is greater than 0 degrees Celsius, the engine speed is greater or equal to 600rpm and the range position is P, R, or N. **Possible Causes:** • Solenoid valves in valve body are faulty • Solenoid circuit is shorting to ground • Solenoid circuit is open • TCM has failed or wiring is shorting • Mechanical malfunction in transmission
DTC: P0783 **2T MIL: Yes** **2002, 2003, 2004** **Models:** 320i, 325i, 330i, 525i, 540i, 745i, M3, Z3, Z4 **Engines:** 2.5L, 3.0L, 3.2L, 4.4L, 4.8L **Transmissions:** A/T	**3-4 Shift Range Monitoring Conditions:** Engine running and vehicle driven, the DME detected a mechanical malfunction within the transmission. The output speed is greater than 300rpm, the transmission oil temperature is greater than 0 degrees Celsius, the engine speed is greater or equal to 600rpm and the range position is P, R, or N. **Possible Causes:** • Solenoid valves in valve body are faulty • Solenoid circuit is shorting to ground • Solenoid circuit is open • TCM has failed or wiring is shorting • Mechanical malfunction in transmission
DTC: P0784 **2T MIL: Yes** **2002, 2003, 2004, 2005, 2006** **Models:** 320i, 325i, 330i, 525i, 530i, 545i, 745i, M3, Z4 **Engines:** 2.5L, 3.0L, 3.2L, 4.8L, 5.4L **Transmissions:** A/T	**4-5 Shift Range Monitoring Conditions:** Engine running and vehicle driven, the DME detected a mechanical malfunction within the transmission. The output speed is greater than 300rpm, the transmission oil temperature is greater than 0 degrees Celsius, the engine speed is greater or equal to 600rpm and the range position is P, R, or N. **Possible Causes:** • Solenoid valves in valve body are faulty • Solenoid circuit is shorting to ground • Solenoid circuit is open • TCM has failed or wiring is shorting • Mechanical malfunction in transmission

DTC	Trouble Code Title, Conditions & Possible Causes
DTC: P0795 **2T MIL: Yes** **2002, 2003, 2004** **Models:** 320i, 325i, 330i, 525i, 540i, 745i, M3, Z3, Z4 **Engines:** 2.5L, 3.0L, 3.2L, 4.4L, 4.8L **Transmissions:** A/T	**Pressure Regulator Valve 3 Plausibility Conditions:** The current to/from the pressure regulator valve is either higher or lower than the threshold value. **Possible Causes:** • Pressure control solenoid circuit is shorting to ground • Pressure control solenoid circuit is open • Valve has failed • TCM has failed
DTC: P0798 **2T MIL: Yes** **2001, 2002, 2003, 2004, 2005, 2006** **Models:** 320i, 325i, 330i, 525i, 530i, 540i, 545i, 740i, 745i, 750i, M3, Z4 **Engines:** 2.5L, 3.0L, 3.2L, 4.8L, 5.4L **Transmissions:** A/T	**Pressure Regulator Valve 2 Upper Threshold Conditions:** The signal to/from the pressure regulator valve has been interrupted. **Possible Causes:** • Pressure control solenoid circuit is shorting to ground • Pressure control solenoid circuit is open • Valve has failed • TCM has failed
DTC: P0829 **2T MIL: Yes** 2003, 2004 **Models:** 320i, 325i, 330i, 525i, 530i, 545i, 745i, M3, Z4 **Engines:** 2.5L, 3.0L, 3.2L, 4.8L, 5.4L **Transmissions:** A/T	**5-6 Shift Range Monitoring Conditions:** Engine running and vehicle driven, the DME detected a mechanical malfunction within the transmission. The output speed is greater than 300rpm, the transmission oil temperature is greater than 0 degrees Celsius, the engine speed is greater or equal to 600rpm and the range position is P, R, or N. **Possible Causes:** • Solenoid valves in valve body are faulty • Solenoid circuit is shorting to ground • Solenoid circuit is open • TCM has failed or wiring is shorting • Mechanical malfunction in transmission
DTC: P0900 **2T MIL: Yes** 2005 **Models:** M3 **Engines:** 3.2L **Transmissions:** A/T	**Clutch Solenoid Circuit Range/Performance-Open Load Conditions:** The clutch solenoid circuit continuity is in error. Ignition key is placed in radio position. The battery voltage is greater than 10 volts. Clutch position is normal. **Possible Causes:** • Short to battery • Short to ground • Open circuit
DTC: P0901 **2T MIL: Yes** 2005 **Models:** M3 **Engines:** 3.2L **Transmissions:** A/T	**Clutch Solenoid Rationality Check Conditions:** The clutch solenoid circuit continuity is in error. Ignition key is placed in radio position. The battery voltage is greater than 10 volts. Clutch position is normal. **Possible Causes:** • Short to battery • Short to ground • Open circuit
DTC: P0902 **2T MIL: Yes** 2005 **Models:** M3 **Engines:** 3.2L **Transmissions:** A/T	**Clutch Solenoid Circuit Range/Performance-Short to Ground Conditions:** The clutch solenoid circuit continuity is in error. Ignition key is placed in radio position. The battery voltage is greater than 10 volts. Clutch position is normal. **Possible Causes:** • Short to battery • Short to ground • Open circuit
DTC: P0903 **2T MIL: Yes** 2005 **Models:** M3 **Engines:** 3.2L **Transmissions:** A/T	**Clutch Solenoid Circuit Range/Performance-Short to Power Conditions:** The clutch solenoid circuit continuity is in error. Ignition key is placed in radio position. The battery voltage is greater than 10 volts. Clutch position is normal. **Possible Causes:** • Short to battery • Short to ground • Open circuit

DTC	Trouble Code Title, Conditions & Possible Causes
DTC: P0962 **2T MIL: Yes** 2003, 2004 **Models:** 320i, 325i, 330i, 525i, 530i, 540i, 745i, M3, Z3, Z4 **Engines:** 2.5L, 3.0L, 3.2L, 4.4L, 4.8L **Transmissions:** A/T	**Pressure Regulator Valve 1 Lower Threshold Conditions:** The signal to/from the pressure regulator valve is interrupted or short circuited to ground. **Possible Causes:** • Pressure control solenoid circuit is shorting to ground • Pressure control solenoid circuit is open • Valve has failed • TCM has failed
DTC: P0963 **2T MIL: Yes** 2003, 2004 **Models:** 320i, 325i, 330i, 525i, 530i, 540i, 745i, M3, Z3, Z4 **Engines:** 2.5L, 3.0L, 3.2L, 4.4L, 4.8L **Transmissions:** A/T	**Pressure Regulator Valve 1 Upper Threshold Conditions:** The signal to/from the pressure regulator valve is interrupted or short circuited to supply. **Possible Causes:** • Pressure control solenoid circuit is shorting to ground • Pressure control solenoid circuit is open • Valve has failed • TCM has failed
DTC: P0966 **2T MIL: Yes** 2003, 2004 **Models:** 320i, 325i, 330i, 525i, 530i, 540i, 745i, M3, Z3, Z4 **Engines:** 2.5L, 3.0L, 3.2L, 4.4L, 4.8L **Transmissions:** A/T	**Pressure Regulator Valve 2 Lower Threshold Conditions:** The signal to/from the pressure regulator valve is interrupted or short circuited to ground. **Possible Causes:** • Pressure control solenoid circuit is shorting to ground • Pressure control solenoid circuit is open • Valve has failed • TCM has failed
DTC: P0967 **2T MIL: Yes** 2003, 2004 **Models:** 320i, 325i, 330i, 525i, 530i, 540i, 745i, M3, Z3, Z4 **Engines:** 2.5L, 3.0L, 3.2L, 4.4L, 4.8L **Transmissions:** A/T	**Pressure Regulator Valve 2 Upper Threshold Conditions:** The signal to/from the pressure regulator valve is interrupted or short circuited to supply. **Possible Causes:** • Pressure control solenoid circuit is shorting to ground • Pressure control solenoid circuit is open • Valve has failed • TCM has failed
DTC: P0970 **2T MIL: Yes** 2003, 2004 **Models:** 320i, 325i, 330i, 525i, 530i, 540i, 745i, M3, Z3, Z4 **Engines:** 2.5L, 3.0L, 3.2L, 4.4L, 4.8L **Transmissions:** A/T	**Pressure Regulator Valve 3 Lower Threshold Conditions:** The signal to/from the pressure regulator valve is interrupted or short circuited to ground. **Possible Causes:** • Pressure control solenoid circuit is shorting to ground • Pressure control solenoid circuit is open • Valve has failed • TCM has failed
DTC: P0971 **2T MIL: Yes** 2003, 2004 **Models:** 320i, 325i, 330i, 525i, 530i, 540i, 745i, M3, Z3, Z4 **Engines:** 2.5L, 3.0L, 3.2L, 4.4L, 4.8L **Transmissions:** A/T	**Pressure Regulator Valve 3 Upper Threshold Conditions:** The signal to/from the pressure regulator valve is interrupted or short circuited to supply. **Possible Causes:** • Pressure control solenoid circuit is shorting to ground • Pressure control solenoid circuit is open • Valve has failed • TCM has failed
DTC: P0973 **2T MIL: Yes** 2003, 2004 **Models:** 320i, 325i, 330i, 525i, 530i, 540i, 745i, M3, Z3, Z4 **Engines:** 2.5L, 3.0L, 3.2L, 4.4L, 4.8L **Transmissions:** A/T	**Solenoid Valve 1 Lower Threshold Conditions:** The signal to/from the pressure regulator valve is interrupted or short circuited to ground. **Possible Causes:** • Pressure control solenoid circuit is shorting to ground • Pressure control solenoid circuit is open • Valve has failed • TCM has failed

DTC	Trouble Code Title, Conditions & Possible Causes
DTC: P0976 **2T MIL: Yes** 2003, 2005 **Models:** 745i, M3 **Engines:** 3.2L, 4.8L **Transmissions:** A/T	**Solenoid Valve 2 Lower Threshold Conditions:** The signal to/from the pressure regulator valve is interrupted or short circuited to ground. **Possible Causes:** • Pressure control solenoid circuit is shorting to ground • Pressure control solenoid circuit is open • Valve has failed • TCM has failed
DTC: P0979 **2T MIL: Yes** 2003, 2004 **Models:** 320i, 325i, 330i, 525i, 530i, 540i, 745i, M3, Z3, Z4 **Engines:** 2.5L, 3.0L, 3.2L, 4.4L, 4.8L **Transmissions:** A/T	**Solenoid Valve 3 Lower Threshold Conditions:** The signal to/from the pressure regulator valve is interrupted or short circuited to ground. **Possible Causes:** • Pressure control solenoid circuit is shorting to ground • Pressure control solenoid circuit is open • Valve has failed • TCM has failed

DTC: P1000 **MIL: No** **2006** **Models:** 325i, 330i, 525i, 530i, 550i, 750i, Z4 **Engines:** 3.0L, 4.8L **Transmissions:** A/T	**Valvetronic (VVT) System Minimum Stroke Adaptation Number of Stops Exceeded Conditions:** After the ignition is on for 500ms, the minimum number of stroke adaptations was exceeded. It is a distribution balance issue, and raising the minimum stroke at idle doesn't produce the desired results. **Possible Causes:** • Mechanical components have worn
DTC: P1001 **2006** **Models:** 325i, 330i, 525i, 530i, 550i, 750i, Z4 **Engines:** 3.0L, 4.8L **Transmissions:** A/T	**Valvetronic (VVT) Limp Home Request High Input Conditions:** After 500ms the there is detected a short to battery voltage. If there are simultaneous CAN faults detected, Terminal 15 is probably open at VVT-SG, otherwise check Kb-B. **Possible Causes:** • Resolve the primary fault.
DTC: P1002 **2006** **Models:** 325i, 330i, 525i, 530i, 550i, 750i, Z4 **Engines:** 3.0L, 4.8L **Transmissions:** A/T	**Valvetronic (VVT) Limp Home Request Low Input Conditions:** After 500ms the there is detected a short to ground. If there are simultaneous CAN faults detected, Terminal 15 is probably open at VVT-SG, otherwise check Kb-B. **Possible Causes:** • Resolve the primary fault.
DTC: P1003 **2006** **Models:** 325i, 330i, 525i, 530i, 550i, 750i, Z4 **Engines:** 3.0L, 4.8L **Transmissions:** A/T	**Valvetronic (VVT) Limp Home Request Open Circuit Conditions:** After 500ms the there is detected a short. If there are simultaneous CAN faults detected, Terminal 15 is probably open at VVT-SG, otherwise check Kb-B. **Possible Causes:** • Resolve the primary fault.
DTC: P1004 **2006** **Models:** 325i, 330i, 525i, 530i, 550i, 750i, Z4 **Engines:** 3.0L, 4.8L **Transmissions:** A/T	**Valvetronic (VVT) Guiding Sensor Solenoid Loss (Bank 1) Conditions:** After 3ms and the DME and VVT are active, it is detected that the sensor is missing a magnet. **Possible Causes:** • Defective sensor.
DTC: P1005 **2006** **Models:** 325i, 330i, 525i, 530i, 550i, 750i, Z4 **Engines:** 3.0L, 4.8L **Transmissions:** A/T	**Valvetronic (VVT) Guiding Sensor Reset Error (Bank 1) Conditions:** After 1.5ms and the DME and VVT are active, it is detected that the sensor is not properly resetting. **Possible Causes:** • Plug contact problem at Pin 6 on sensor • Oil in plug at sensor • Replace sensor • Repair gaskets
DTC: P1006 **2006** **Models:** 325i, 330i, 525i, 530i, 550i, 750i, Z4 **Engines:** 3.0L, 4.8L **Transmissions:** A/T	**Valvetronic (VVT) Guiding Sensor Parity Error (Bank 1) Conditions:** After 12ms and the DME and VVT are active, it is detected that the sensor is not properly communicating. The plug is defective on the sensor or there is an open circuit **Possible Causes:** • Plug contact problem at Pin 6 on sensor • Oil in plug at sensor • Replace sensor • Repair gaskets
DTC: P1007 **2006** **Models:** 325i, 330i, 525i, 530i, 550i, 750i, Z4 **Engines:** 3.0L, 4.8L **Transmissions:** A/T	**Valvetronic (VVT) Guiding Sensor Gradient Error (Bank 1) Conditions:** After 9ms and the DME and VVT are active, it is detected that the sensor has a gradient violation/identity, causing a reading of implausible sensor data. The plug is defective on the sensor or there is an open circuit **Possible Causes:** • Replace sensor
DTC: P1008 **2006** **Models:** 325i, 330i, 525i, 530i, 550i, 750i, Z4 **Engines:** 3.0L, 4.8L **Transmissions:** A/T	**Valvetronic (VVT) Guiding Sensor Solenoid Loss (Bank 2) Conditions:** After 3ms and the DME and VVT are active, it is detected that the sensor is missing a magnet. **Possible Causes:** • Defective sensor.

DTC	Trouble Code Title, Conditions & Possible Causes
DTC: P1009 **2006** **Models:** 325i, 330i, 525i, 530i, 550i, 750i, Z4 **Engines:** 3.0L, 4.8L **Transmissions:** A/T	**Valvetronic (VVT) Guiding Sensor Reset Error (Bank 2) Conditions:** After 1.5ms and the DME and VVT are active, it is detected that the sensor is not properly resetting. **Possible Causes:** • Plug contact problem at Pin 6 on sensor • Oil in plug at sensor • Replace sensor • Repair gaskets
DTC: P1010 **2006** **Models:** 325i, 330i, 525i, 530i, 550i, 750i, Z4 **Engines:** 3.0L, 4.8L **Transmissions:** A/T	**Valvetronic (VVT) Guiding Sensor Parity Error (Bank 2) Conditions:** After 12ms and the DME and VVT are active, it is detected that the sensor is not properly communicating. The plug is defective on the sensor or there is an open circuit **Possible Causes:** • Plug contact problem at Pin 6 on sensor • Oil in plug at sensor • Replace sensor • Repair gaskets
DTC: P1011 **2006** **Models:** 325i, 330i, 525i, 530i, 550i, 750i, Z4 **Engines:** 3.0L, 4.8L **Transmissions:** A/T	**Valvetronic (VVT) Guiding Sensor Gradient Error (Bank 2) Conditions:** After 9ms and the DME and VVT are active, it is detected that the sensor has a gradient violation/identity, causing a reading of implausible sensor data. The plug is defective on the sensor or there is an open circuit **Possible Causes:** • Replace sensor
DTC: P1012 **2006** **Models:** 325i, 330i, 525i, 530i, 550i, 750i, Z4 **Engines:** 3.0L, 4.8L **Transmissions:** A/T	**Valvetronic (VVT) Reference Sensor Solenoid Loss (Bank 1) Conditions:** After 3ms and the DME and VVT are active, it is detected that the sensor is missing a magnet. **Possible Causes:** • Defective sensor.
DTC: P1013 **2006** **Models:** 325i, 330i, 525i, 530i, 550i, 750i, Z4 **Engines:** 3.0L, 4.8L **Transmissions:** A/T	**Valvetronic (VVT) Reference Sensor Reset Error (Bank 1) Conditions:** After 1.5ms and the DME and VVT are active, it is detected that the sensor is not properly resetting. **Possible Causes:** • Plug contact problem at Pin 6 on sensor • Oil in plug at sensor • Replace sensor • Repair gaskets
DTC: P1014 **2006** **Models:** 325i, 330i, 525i, 530i, 550i, 750i, Z4 **Engines:** 3.0L, 4.8L **Transmissions:** A/T	**Valvetronic (VVT) Reference Sensor Parity Error (Bank 1) Conditions:** After 12ms and the DME and VVT are active, it is detected that the sensor is not properly communicating. The plug is defective on the sensor or there is an open circuit **Possible Causes:** • Plug contact problem at Pin 6 on sensor • Oil in plug at sensor • Replace sensor • Repair gaskets
DTC: P1015 **2006** **Models:** 325i, 330i, 525i, 530i, 550i, 750i, Z4 **Engines:** 3.0L, 4.8L **Transmissions:** A/T	**Valvetronic (VVT) Reference Sensor Gradient Error (Bank 1) Conditions:** After 9ms and the DME and VVT are active, it is detected that the sensor has a gradient violation/identity, causing a reading of implausible sensor data. The plug is defective on the sensor or there is an open circuit **Possible Causes:** • Replace sensor
DTC: P1022 **MIL: No** **2006** **Models:** 325i, 330i, 525i, 530i, 550i, 750i, Z4 **Engines:** 3.0L, 4.8L **Transmissions:** A/T	**Valvetronic (VVT), Eccentric Shaft Sensor 2 Circuit Low Input Conditions:** With the engine running, the fault is a low voltage supply to the sensor after 3ms. The DME, VVT are active at 4.5 to 5.5 volts **Possible Causes:** • Short to ground in wiring harness • VVT-SG defective • Short circuit within the sensor • Check plug and wiring harness for sensor defect

DTC	Trouble Code Title, Conditions & Possible Causes
DTC: P1023 **MIL: No** **2006** **Models:** 325i, 330i, 525i, 530i, 550i, 750i, Z4 **Engines:** 3.0L, 4.8L **Transmissions:** A/T	**Valvetronic (VVT) Self-Learning Function Faulty Adjustment Range (Bank 1) Conditions:** With the engine running, the fault is an out of range adjustment for the self-learning function of the VVT. Check the balance spring installation and the mechanical components for wear. **Possible Causes:** • Stuck at upper travel limit • Travel limit worn or deformed • Wear in idler lever • Torque compensation spring missing or not connected
DTC: P1024 **MIL: No** **2006** **Models:** 325i, 330i, 525i, 530i, 550i, 750i, Z4 **Engines:** 3.0L, 4.8L **Transmissions:** A/T	**Valvetronic (VVT) Self-Learning Function Faulty Lower Learning Range (Bank 1) Conditions:** With the engine running, the fault is an out of range adjustment for the self-learning function of the VVT at the lower range. Check the installation and the mechanical components for wear. **Possible Causes:** • Stuck at upper travel limit • Defective sensor magnet missing) • Travel limit worn or deformed • Wear in idler lever • Torque compensation spring missing or not connected
DTC: P1025 **MIL: No** **2006** **Models:** 325i, 330i, 525i, 530i, 550i, 750i, Z4 **Engines:** 3.0L, 4.8L **Transmissions:** A/T	**Valvetronic (VVT) Self-Learning Function No Positions Stored (Bank 1) Conditions:** With the engine running, no travel limit has been initialized. Check the installation and the mechanical components for wear. This fault is because the system is operated for the first time with a new VVT-SG as there is no automatic limit initialization. **Possible Causes:** • Conduct travel limit initialization.
DTC: P1026 **MIL: No** **2006** **Models:** 325i, 330i, 525i, 530i, 550i, 750i, Z4 **Engines:** 3.0L, 4.8L **Transmissions:** A/T	**Valvetronic (VVT) Self-Learning Function Faulty Adjustment Range (Bank 2) Conditions:** With the engine running, the fault is an out of range adjustment for the self-learning function of the VVT. Check the balance spring installation and the mechanical components for wear. **Possible Causes:** • Stuck at upper travel limit • Travel limit worn or deformed • Wear in idler lever • Torque compensation spring missing or not connected
DTC: P1027 **MIL: No** **2006** **Models:** 325i, 330i, 525i, 530i, 550i, 750i, Z4 **Engines:** 3.0L, 4.8L **Transmissions:** A/T	**Valvetronic (VVT) Self-Learning Function Faulty Lower Learning Range (Bank 2) Conditions:** With the engine running, the fault is an out of range adjustment for the self-learning function of the VVT at the lower range. Check the installation and the mechanical components for wear. **Possible Causes:** • Stuck at upper travel limit • Defective sensor (magnet missing) • Travel limit worn or deformed • Wear in idler lever • Torque compensation spring missing or not connected
DTC: P1028 **MIL: No** **2006** **Models:** 325i, 330i, 525i, 530i, 550i, 750i, Z4 **Engines:** 3.0L, 4.8L **Transmissions:** A/T	**Valvetronic (VVT) Self-Learning Function No Positions Stored (Bank 2) Conditions:** With the engine running, no travel limit has been initialized. Check the sensor installation and the mechanical components for wear. This fault is because the system is operated for the first time with a new VVT-SG as there is no automatic limit initialization. **Possible Causes:** • Conduct travel limit initialization.
DTC: P1030 **MIL: No** **2006** **Models:** 325i, 330i, 525i, 530i, 550i, 750i, Z4 **Engines:** 3.0L, 4.8L **Transmissions:** A/T	**Valvetronic (VVT) Actuator Monitoring Position Control, Control Deviation (Bank 1) Conditions:** The engine is on and running for 45ms at 9.6 to 15.5 volts and a sluggish monitoring movement, direction or rotation was detected. This function monitors the VVT system for resistance to motion and is always active when the driver circuits are released for operation, there's no control with pulse-duty factor, no relay/enable fault, no under voltage, the travel limits are initialized, and there is no reference sensor faults after sensor switching. Check the wiring harness for shorts, the sensor installation, and for wear and mechanical sticking. **Possible Causes:** • Low battery charge • Open motor control circuit • Motor shorted to ground • Loose sensor or it is operating at the limit

DTC	Trouble Code Title, Conditions & Possible Causes
DTC: P1031 **MIL: No** **2006** **Models:** 325i, 330i, 525i, 530i, 550i, 750i, Z4 **Engines:** 3.0L, 4.8L **Transmissions:** A/T	**Valvetronic (VVT) Actuator Monitoring Recognition of Direction of Rotation Plausibility (Bank 1) Conditions:** The engine is on and running for 63 to 498ms at 9.6 to 15.5 volts. Check to determine whether adjustment can be approved. Check VVT system once before each power application within one driving cycle for correct servo motor polarity and sticking in the system. **Possible Causes:** • Low battery charge • Open motor control circuit • Motor shorted to ground • Loose sensor or it is operating at the limit
DTC: P1033 **MIL: No** **2006** **Models:** 325i, 330i, 525i, 530i, 550i, 750i, Z4 **Engines:** 3.0L, 4.8L **Transmissions:** A/T	**Valvetronic (VVT) Actuator Monitoring Position Control, Control Deviation (Bank 2) Conditions:** The engine is on and running for 45ms at 9.6 to 15.5 volts and a sluggish monitoring movement, direction or rotation was detected. This function monitors the VVT system for resistance to motion and is always active when the driver circuits are released for operation, there's no control with pulse-duty factor, no relay/enable fault, no under voltage, the travel limits are initialized, and there is no reference sensor faults after sensor switching. Check the wiring harness for shorts, the sensor installation, and for wear and mechanical sticking. **Possible Causes:** • Low battery charge • Open motor control circuit • Motor shorted to ground • Loose sensor or it is operating at the limit
DTC: P1034 **MIL: No** **2006** **Models:** 325i, 330i, 525i, 530i, 550i, 750i, Z4 **Engines:** 3.0L, 4.8L **Transmissions:** A/T	**Valvetronic (VVT) Actuator Monitoring Recognition of Direction of Rotation Plausibility (Bank 1) Conditions:** The engine is on and running for 63 to 498ms at 9.6 to 15.5 volts. Check to determine whether adjustment can be approved. Check VVT system once before each power application within one driving cycle for correct servo motor polarity and sticking in the system. **Possible Causes:** • Low battery charge • Open motor control circuit • Motor shorted to ground • Loose sensor or it is operating at the limit
DTC: P1035 **MIL: No** **2006** **Models:** 325i, 330i, 525i, 530i, 550i, 750i, Z4 **Engines:** 3.0L, 4.8L **Transmissions:** A/T	**Valvetronic (VVT) CAN Message Monitoring Faulty Desired Message (Bank 1) Conditions:** After the DME is active or the engine running for 500ms, with a voltage of 7 on Terminal 87, the monitoring system displayed a faulty message. **Possible Causes:** • ECU failure (or SZL or ZGM) • Bus system failure • Defective bus controller (SZL or ZGM) • Short circuit in CAN wire or open circuit. • Defective DME • Defective VVT-SG
DTC: P1036 **MIL: No** **2006** **Models:** 325i, 330i, 525i, 530i, 550i, 750i, Z4 **Engines:** 3.0L, 4.8L **Transmissions:** A/T	**Valvetronic (VVT) CAN Timeout VVT-Desired Message (Bank 1) Conditions:** After the key has been on for 800ms (within two messages) or the engine running for 400ms, with a battery voltage at 10 volts, the difference between the deactivation and the starting positions exceeds specification. No suspension of BUS activity. **Possible Causes:** • ECU failure (or SZL or ZGM) • Bus system failure • Defective bus controller (SZL or ZGM) • Short circuit in CAN wire or open circuit. • Defective DME • Defective VVT-SG
DTC: P1037 **MIL: No** **2006** **Models:** 325i, 330i, 525i, 530i, 550i, 750i, Z4 **Engines:** 3.0L, 4.8L **Transmissions:** A/T	**Valvetronic (VVT) CAN Timeout Message (Bank 1) Conditions:** After the key has been on for 800ms (within two messages) or the engine running for 400ms, with a battery voltage at 10 volts, the difference between the deactivation and the starting positions exceeds specification. No suspension of BUS activity. **Possible Causes:** • ECU failure (or SZL or ZGM) • Bus system failure • Defective bus controller (SZL or ZGM) • Short circuit in CAN wire or open circuit. • Defective DME • Defective VVT-SG

DTC	Trouble Code Title, Conditions & Possible Causes
DTC: P1038 **MIL: No** **2006** **Models:** 325i, 330i, 525i, 530i, 550i, 750i, Z4 **Engines:** 3.0L, 4.8L **Transmissions:** A/T	**Valvetronic (VVT) CAN Message Monitoring Faulty Desired Message (Bank 2) Conditions:** After the DME is active or the engine running for 500ms, with a voltage of 7 on Terminal 87, the monitoring system displayed a faulty message. **Possible Causes:** • ECU failure (or SZL or ZGM) • Bus system failure • Defective bus controller (SZL or ZGM) • Short circuit in CAN wire or open circuit. • Defective DME • Defective VVT-SG
DTC: P1039 **MIL: No** **2006** **Models:** 325i, 330i, 525i, 530i, 550i, 750i, Z4 **Engines:** 3.0L, 4.8L **Transmissions:** A/T	**Valvetronic (VVT) CAN Timeout VVT-Desired Message (Bank 2) Conditions:** After the key has been on for 800ms (within two messages) or the engine running for 400ms, with a battery voltage at 10 volts, the difference between the deactivation and the starting positions exceeds specification. No suspension of BUS activity. **Possible Causes:** • ECU failure (or SZL or ZGM) • Bus system failure • Defective bus controller (SZL or ZGM) • Short circuit in CAN wire or open circuit. • Defective DME • Defective VVT-SG
DTC: P1040 **MIL: No** **2006** **Models:** 325i, 330i, 525i, 530i, 550i, 750i, Z4 **Engines:** 3.0L, 4.8L **Transmissions:** A/T	**Valvetronic (VVT) CAN Timeout Message (Bank 2) Conditions:** After the key has been on for 800ms (within two messages) or the engine running for 400ms, with a battery voltage at 10 volts, the difference between the deactivation and the starting positions exceeds specification. No suspension of BUS activity. **Possible Causes:** • ECU failure (or SZL or ZGM) • Bus system failure • Defective bus controller (SZL or ZGM) • Short circuit in CAN wire or open circuit. • Defective DME • Defective VVT-SG
DTC: P1041 **2T MIL: Yes** **2002, 2003, 2004, 2005, 2006** **Models:** 325i, 330i, 525i, 545i, 550i, 745i, 750i, Z4 **Engines:** 3.0L, 4.8L **Transmissions:** A/T	**Valvetronic (VVT) Actuator Control Module EEPROM Error (Bank 1) Conditions:** Ignition on for 50ms, the DME detected a control module malfunction (software). To achieve optimal anti-theft protection for the vehicle, an anti-theft immobilizer is installed. The anti-theft immobilizer is a system for enabling and locking the Engine Control Module (DME). So that this system cannot be circumvented, it is necessary to perform adaptation of the anti-theft immobilizer using the Vehicle Diagnostic and Information System VAS 5052 in the On Board Diagnostic (OBD) function. The great availability of equipment options makes it necessary to adapt the Engine Control Module (DME) to the vehicle (e.g. throttle valve control module or cruise control system). This "writing" function is not possible with the generic scan tool. **Possible Causes:** • Engine Control Module (DME) has failed • Voltage supply for Engine Control Module (DME) has shorted
DTC: P1042 **2T MIL: Yes** **2002, 2003, 2004, 2005, 2006** **Models:** 325i, 330i, 525i, 545i, 550i, 745i, 750i, Z4 **Engines:** 3.0L, 4.8L **Transmissions:** A/T	**Valvetronic (VVT) Actuator Control Module Random Access Memory Error (Bank 1) Conditions:** Key on for 50ms, and the DME detected an internal memory fault. This code will set if KAPWR to the DME is interrupted (at the initial key on). **Possible Causes:** • Battery terminal corrosion, or loose battery connection • Connection to the DME interrupted, or the circuit has been opened • Reprogramming error has occurred and needs replacement. Remember to check for Aftermarket Performance Products before replacing a DME.
DTC: P1043 **2T MIL: Yes** **2002, 2003, 2004, 2005, 2006** **Models:** 325i, 330i, 525i, 545i, 550i, 745i, 750i, Z4 **Engines:** 3.0L, 4.8L **Transmissions:** A/T	**Valvetronic (VVT) Actuator Control Module Read Only Memory Error (Bank 1) Conditions:** Key on for 50ms, and the DME detected an internal memory fault. This code will set if KAPWR to the DME is interrupted (at the initial key on). **Possible Causes:** • Battery terminal corrosion, or loose battery connection • Connection to the DME interrupted, or the circuit has been opened • Reprogramming error has occurred and needs replacement. Remember to check for Aftermarket Performance Products before replacing a DME.

DTC	Trouble Code Title, Conditions & Possible Causes
DTC: P1044 **2T MIL: Yes** **2002, 2003, 2004, 2005, 2006** **Models:** 325i, 330i, 525i, 545i, 550i, 745i, 750i, Z4 **Engines:** 3.0L, 4.8L **Transmissions:** A/T	**Valvetronic (VVT) Actuator Control Module EEPROM Error (Bank 2) Conditions:** Ignition on for 50ms, the DME detected a control module malfunction (software). To achieve optimal anti-theft protection for the vehicle, an anti-theft immobilizer is installed. The anti-theft immobilizer is a system for enabling and locking the Engine Control Module (DME). So that this system cannot be circumvented, it is necessary to perform adaptation of the anti-theft immobilizer using the Vehicle Diagnostic and Information System VAS 5052 in the On Board Diagnostic (OBD) function. The great availability of equipment options makes it necessary to adapt the Engine Control Module (DME) to the vehicle (e.g. throttle valve control module or cruise control system). This "writing" function is not possible with the generic scan tool. **Possible Causes:** • Engine Control Module (DME) has failed • Voltage supply for Engine Control Module (DME) has shorted
DTC: P1045 **2T MIL: Yes** **2002, 2003, 2004, 2005, 2006** **Models:** 325i, 330i, 525i, 545i, 550i, 745i, 750i, Z4 **Engines:** 3.0L, 4.8L **Transmissions:** A/T	**Valvetronic (VVT) Actuator Control Module Random Access Memory Error (Bank 2) Conditions:** Key on for 50ms, and the DME detected an internal memory fault. This code will set if KAPWR to the DME is interrupted (at the initial key on). **Possible Causes:** • Battery terminal corrosion, or loose battery connection • Connection to the DME interrupted, or the circuit has been opened • Reprogramming error has occurred and needs replacement. Remember to check for Aftermarket Performance Products before replacing a DME.
DTC: P1046 **2T MIL: Yes** **2002, 2003, 2004, 2005, 2006** **Models:** 325i, 330i, 525i, 545i, 550i, 745i, 750i, Z4 **Engines:** 3.0L, 4.8L **Transmissions:** A/T	**Valvetronic (VVT) Actuator Control Module Read Only Memory Error (Bank 2) Conditions:** Key on for 50ms, and the DME detected an internal memory fault. This code will set if KAPWR to the DME is interrupted (at the initial key on). **Possible Causes:** • Battery terminal corrosion, or loose battery connection • Connection to the DME interrupted, or the circuit has been opened • Reprogramming error has occurred and needs replacement. Remember to check for Aftermarket Performance Products before replacing a DME.
DTC: P1047 **2T MIL: Yes** **2002, 2003, 2004, 2005, 2006** **Models:** 325i, 330i, 525i, 545i, 550i, 745i, 750i, Z4 **Engines:** 3.0L, 4.8L **Transmissions:** A/T	**Valvetronic (VVT) Actuator Control Circuit High Input (Bank 1) Conditions:** Key on for 3ms, and the DME detected a short to positive. **Possible Causes:** • Short to battery voltage in Kb-B • KS ground in Kb-B • KS sensor motor to ground • KS motor winding
DTC: P1048 **2T MIL: Yes** **2002, 2003, 2004, 2005, 2006** **Models:** 325i, 330i, 525i, 545i, 550i, 745i, 750i, Z4 **Engines:** 3.0L, 4.8L **Transmissions:** A/T	**Valvetronic (VVT) Actuator Control Circuit Low Input (Bank 1) Conditions:** Key on for 3ms, and the DME detected a short to ground. **Possible Causes:** • Short to ground in Kb-B • KS ground in Kb-B • KS motor to ground • KS motor winding
DTC: P1049 **2T MIL: Yes** **2002, 2003, 2005** **Models:** 545i, 745i **Engines:** 4.8L **Transmissions:** A/T	**Valvetronic (VVT) Control Circuit Short with Each Other Conditions:** Key on for 3ms, and the DME detected a short to ground. Check wiring harness or otherwise replace the servo. **Possible Causes:** • Short to ground in Kb-B • KS ground in Kb-B • KS motor to ground • KS motor winding
DTC: P1050 **2T MIL: Yes** **2002, 2003, 2004, 2005, 2006** **Models:** 325i, 330i, 525i, 545i, 550i, 745i, 750i, Z4 **Engines:** 3.0L, 4.8L **Transmissions:** A/T	**Valvetronic (VVT) Control Circuit (Bank 1) Conditions:** Key on for 3ms, and the DME detected that the Control Circuit triggered a general fault. **Possible Causes:** • This fault is usually overwritten by three other faults before the user recognizes it.

DTC	Trouble Code Title, Conditions & Possible Causes
DTC: P1051 **2T MIL: Yes** **2002, 2003, 2004, 2005, 2006** **Models:** 325i, 330i, 525i, 545i, 550i, 745i, 750i, Z4 **Engines:** 3.0L, 4.8L **Transmissions:** A/T	**Valvetronic (VVT) Control Circuit High Input (Bank 1) Conditions:** Key on for 3ms, and the DME detected a short to positive. Check wiring harness or otherwise replace the servo. **Possible Causes:** • Short to battery voltage in Kb-B • KS ground in Kb-B • KS motor to ground • KS motor winding
DTC: P1052 **2T MIL: Yes** **2002, 2003, 2004, 2005, 2006** **Models:** 325i, 330i, 525i, 545i, 550i, 745i, 750i, Z4 **Engines:** 3.0L, 4.8L **Transmissions:** A/T	**Valvetronic (VVT) Control Circuit Low Input (Bank 1) Conditions:** Key on for 3ms, and the DME detected a short to ground. Check wiring harness or otherwise replace the servo. **Possible Causes:** • Short to ground in Kb-B • KS ground in Kb-B • KS motor to ground • KS motor winding
DTC: P1053 **2T MIL: Yes** **2002, 2003, 2005** **Models:** 545i, 745i **Engines:** 4.8L **Transmissions:** A/T	**Valvetronic (VVT) Control Circuit Short with Each Other Conditions:** Key on for 3ms, and the DME detected a short to ground. Check wiring harness or otherwise replace the servo. **Possible Causes:** • Short to ground in Kb-B • KS ground in Kb-B • KS motor to ground • KS motor winding
DTC: P1054 **2T MIL: Yes** **2002, 2003, 2004, 2005, 2006** **Models:** 325i, 330i, 525i, 545i, 550i, 745i, 750i, Z4 **Engines:** 3.0L, 4.8L **Transmissions:** A/T	**Valvetronic (VVT) Control Circuit (Bank 2) Conditions:** Key on for 3ms, and the DME detected that the Control Circuit triggered a general fault. **Possible Causes:** • This fault is usually overwritten by three other faults before the user recognizes it.
DTC: P1055 **2T MIL: Yes** **2002, 2003, 2004, 2005, 2006** **Models:** 325i, 330i, 525i, 545i, 550i, 745i, 750i, Z4 **Engines:** 3.0L, 4.8L **Transmissions:** A/T	**Valvetronic (VVT) Supply Voltage Control Motor High Input (Bank 1) Conditions:** Key on for 200ms, the DME detected that the supply voltage was too high (more than 17 volts). This is a jump-start detection that throws a fault if there is 24 volts of power for longer than 21 seconds. **Possible Causes:** • Check electrical system for faults.
DTC: P1056 **2T MIL: Yes** **2002, 2003, 2004, 2005, 2006** **Models:** 325i, 330i, 525i, 545i, 550i, 745i, 750i, Z4 **Engines:** 3.0L, 4.8L **Transmissions:** A/T	**Valvetronic (VVT) Supply Voltage Control Motor Low Input (Bank 1) Conditions:** Key on for 200ms, the DME detected that the supply voltage was too low (less than 5 volts). **Possible Causes:** • VVT fuse is faulty • Defective load reduction relay • Plug/Kb-B open circuit in power supply circuit • Plug/Kb-B open circuit in relay supply circuit
DTC: P1057 **2T MIL: Yes** **2002, 2003, 2004, 2005, 2006** **Models:** 325i, 330i, 525i, 545i, 550i, 745i, 750i, Z4 **Engines:** 3.0L, 4.8L **Transmissions:** A/T	**Valvetronic (VVT) Supply Voltage Control Motor Electrical (Bank 1) Conditions:** Key on for 50ms, the DME detected that the supply voltage was irregular. **Possible Causes:** • Short to ground at power input or defective ECU (capacitor preload).
DTC: P1058 **2T MIL: Yes** **2002, 2003, 2004, 2005, 2006** **Models:** 325i, 330i, 525i, 545i, 550i, 745i, 750i, Z4 **Engines:** 3.0L, 4.8L **Transmissions:** A/T	**Valvetronic (VVT) Supply Voltage Control Motor High Input (Bank 2) Conditions:** Key on for 200ms, the DME detected that the supply voltage was too high (more than 17 volts). This is a jump-start detection that throws a fault if there is 24 volts of power for longer than 21 seconds. **Possible Causes:** • Check electrical system for faults.

DTC	Trouble Code Title, Conditions & Possible Causes
DTC: P1059 **2T MIL: Yes** **2002, 2003, 2004, 2005, 2006** **Models:** 325i, 330i, 525i, 545i, 550i, 745i, 750i, Z4 **Engines:** 3.0L, 4.8L **Transmissions:** A/T	**Valvetronic (VVT) Supply Voltage Control Motor Low Input (Bank 2) Conditions:** Key on for 200ms, the DME detected that the supply voltage was too low (less than 5 volts). **Possible Causes:** • VVT fuse is faulty • Defective load reduction relay • Plug/Kb-B open circuit in power supply circuit • Plug/Kb-B open circuit in relay supply circuit
DTC: P1060 **2T MIL: Yes** **2002, 2003, 2004, 2005, 2006** **Models:** 325i, 330i, 525i, 545i, 550i, 745i, 750i, Z4 **Engines:** 3.0L, 4.8L **Transmissions:** A/T	**Valvetronic (VVT) Supply Voltage Control Motor Electrical (Bank 2) Conditions:** Key on for 50ms, the DME detected that the supply voltage was irregular. **Possible Causes:** • Short to ground at power input or defective ECU (capacitor preload).
DTC: P1061 **2T MIL: Yes** **2002, 2003, 2004, 2005, 2006** **Models:** 325i, 330i, 525i, 545i, 550i, 745i, 750i, Z4 **Engines:** 3.0L, 4.8L **Transmissions:** A/T	**Valvetronic (VVT) Limp Home Request RPM and Charge Limitation (Bank 1) Conditions:** After 3000ms the charge difference between the two banks (caused by other VVT faults) lead to an rpm charge limit. **Possible Causes:** • Resolve the primary fault.
DTC: P1062 **2T MIL: Yes** **2002, 2003, 2006** **Models:** 325i, 330i, 525i, 545i, 550i, 745i, 750i, Z4 **Engines:** 3.0L, 4.8L **Transmissions:** A/T	**Valvetronic (VVT) Limp Home Request Full Stroke Position Not Reached (Bank 1) Conditions:** After 3000ms the eccentric angle fails to close at the full stroke position. Other VVT fault issues maximum stroke command but the position is not reached. **Possible Causes:** • Resolve the primary fault.
DTC: P1063 **2T MIL: Yes** **2006** **Models:** 325i, 330i, 525i, 530i, 550i, 750i, Z4 **Engines:** 3.0L, 4.8L **Transmissions:** A/T	**Valvetronic (VVT) Limp Home Request Air Mass Plausibility (Bank 1) Conditions:** After 4000ms the eccentric overload angle detected a fault with the mass airflow plausibility. Other VVT fault issues maximum stroke command but the position is not reached. **Possible Causes:** • Resolve the primary fault.
DTC: P1064 **MIL: No** **2006** **Models:** 325i, 330i, 525i, 530i, 550i, 750i, Z4 **Engines:** 3.0L, 4.8L **Transmissions:** A/T	**Valvetronic (VVT) Value Comparison Starting Position/Parking Position Plausibility (Bank 1) Conditions:** After 500ms the difference between the deactivation and the starting positions exceeds specification. Usually occurs after repairs. **Possible Causes:** • Turn off ignition and wait until the HR releases. Turn on the ignition and the VVT will initialize automatically.
DTC: P1065 **MIL: No** **2002, 2003, 2006** **Models:** 325i, 330i, 525i, 545i, 550i, 745i, 750i, Z4 **Engines:** 3.0L, 4.8L **Transmissions:** A/T	**Valvetronic (VVT) CAN Timeout No Signal Conditions:** After the key has been on for 800ms (within two messages) or the engine running for 400ms, with a battery voltage at 10 volts, the difference between the deactivation and the starting positions exceeds specification. No suspension of BUS activity. CAN signal is missing, therefore considered in a time out. **Possible Causes:** • ECU failure (or SZL or ZGM) • Bus system failure • Defective bus controller (SZL or ZGM) • Short circuit in CAN wire or open circuit.
DTC: P1066 **MIL: No** **2002, 2003, 2006** **Models:** 325i, 330i, 525i, 545i, 550i, 745i, 750i, Z4 **Engines:** 3.0L, 4.8L **Transmissions:** A/T	**Valvetronic (VVT) CAN Message Monitoring Faulty Actual Message Conditions:** After the DME is active or the engine running for 500ms, with a voltage of 7 on Terminal 87, the monitoring system displayed a faulty message. **Possible Causes:** • ECU failure (or SZL or ZGM) • Bus system failure • Defective bus controller (SZL or ZGM) • Short circuit in CAN wire or open circuit. • Defective DME • Defective VVT-SG

DTC	Trouble Code Title, Conditions & Possible Causes
DTC: P1067 **2006** **Models:** 325i, 330i, 525i, 530i, 550i, 750i, Z4 **Engines:** 3.0L, 4.8L **Transmissions:** A/T	**Valvetronic (VVT) Reference Sensor Solenoid Loss (Bank 2) Conditions:** After 3ms and the DME and VVT are active, it is detected that the sensor is missing a magnet. **Possible Causes:** • Defective sensor.
DTC: P1068 **2006** **Models:** 325i, 330i, 525i, 530i, 550i, 750i, Z4 **Engines:** 3.0L, 4.8L **Transmissions:** A/T	**Valvetronic (VVT) Reference Sensor Reset Error (Bank 2) Conditions:** After 1.5ms and the DME and VVT are active, it is detected that the sensor is not properly resetting. **Possible Causes:** • Plug contact problem at Pin 6 on sensor • Oil in plug at sensor • Replace sensor • Repair gaskets
DTC: P1069 **2006** **Models:** 325i, 330i, 525i, 530i, 550i, 750i, Z4 **Engines:** 3.0L, 4.8L **Transmissions:** A/T	**Valvetronic (VVT) Reference Sensor Parity Error (Bank 2) Conditions:** After 12ms and the DME and VVT are active, it is detected that the sensor is not properly communicating. The plug is defective on the sensor or there is an open circuit. **Possible Causes:** • Plug contact problem at Pin 6 on sensor • Oil in plug at sensor • Replace sensor • Repair gaskets
DTC: P1070 **2006** **Models:** 325i, 330i, 525i, 530i, 550i, 750i, Z4 **Engines:** 3.0L, 4.8L **Transmissions:** A/T	**Valvetronic (VVT) Reference Sensor Gradient Error (Bank 2) Conditions:** After 9ms and the DME and VVT are active, it is detected that the sensor has a gradient violation/identity, causing a reading of implausible sensor data. The plug is defective on the sensor or there is an open circuit. **Possible Causes:** • Replace sensor
DTC: P1071 **2T MIL: Yes** **2002, 2006** **Models:** 325i, 330i, 525i, 530i, 550i, 745i, 750i, Z4 **Engines:** 3.0L, 4.8L **Transmissions:** A/T	**Valvetronic (VVT) Control Module Watchdog or Temperature Sensor Error (Bank 1) Conditions:** Key on for 10ms, and the DME detected an internal fault relating to an internal temperature sensor. Ignore single isolated appearances and only respond to repeated occurrences by replacing the VVT-SG. **Possible Causes:** • Battery terminal corrosion, or loose battery connection • Connection to the DME interrupted, or the circuit has been opened • Reprogramming error has occurred and needs replacement. Remember to check for Aftermarket Performance Products before replacing a DME.
DTC: P1072 **2T MIL: Yes** **2002, 2006** **Models:** 325i, 330i, 525i, 530i, 550i, 745i, 750i, Z4 **Engines:** 3.0L, 4.8L **Transmissions:** A/T	**Valvetronic (VVT) Control Module Watchdog or Temperature Sensor Error (Bank 2) Conditions:** Key on for 10ms, and the DME detected an internal fault relating to an internal temperature sensor. Ignore single isolated appearances and only respond to repeated occurrences by replacing the VVT-SG. **Possible Causes:** • Battery terminal corrosion, or loose battery connection • Connection to the DME interrupted, or the circuit has been opened • Reprogramming error has occurred and needs replacement. Remember to check for Aftermarket Performance Products before replacing a DME.
DTC: P1075 **MIL: No** **2006** **Models:** 325i, 330i, 525i, 530i, 550i, 750i, Z4 **Engines:** 3.0L, 4.8L **Transmissions:** A/T	**Valvetronic (VVT) Overload Protection (Bank 1) Conditions:** After the ignition is on for 45ms, the temperature was recorded as too high. **Possible Causes:** • Sticking in VVT mechanicals, pinion gear, etc. • Loose sensor • Sensor servo motor has short circuit
DTC: P1076 **MIL: No** **2006** **Models:** 325i, 330i, 525i, 530i, 550i, 750i, Z4 **Engines:** 3.0L, 4.8L **Transmissions:** A/T	**Valvetronic (VVT) Overload Protection ECU Temperature High Input (Bank 1) Conditions:** After the ignition is on for 45ms, the ECU temperature was recorded as too high. **Possible Causes:** • Sticking in VVT mechanicals, pinion gear, etc. • Loose sensor • Sensor servo motor has short circuit

DTC	Trouble Code Title, Conditions & Possible Causes
DTC: P1077 **MIL: No** **2006** **Models:** 325i, 330i, 525i, 530i, 550i, 750i, Z4 **Engines:** 3.0L, 4.8L **Transmissions:** A/T	**Valvetronic (VVT) Overload Protection Control Motor Temperature High Input (Bank 1) Conditions:** After the ignition is on for 45ms, the E motor temperature was recorded as too high. **Possible Causes:** • Sticking in VVT mechanicals, pinion gear, etc. • Loose sensor • Sensor servo motor has short circuit
DTC: P1078 **MIL: No** **2006** **Models:** 325i, 330i, 525i, 530i, 550i, 750i, Z4 **Engines:** 3.0L, 4.8L **Transmissions:** A/T	**Valvetronic (VVT) Overload Protection Control Motor Current High Input (Bank 1) Conditions:** After the ignition is on for 45ms, the E motor activation current is too high. **Possible Causes:** • Sticking in VVT mechanicals, pinion gear, etc. • Loose sensor • Sensor servo motor has short circuit
DTC: P1079 **MIL: No** **2006** **Models:** 325i, 330i, 525i, 530i, 550i, 750i, Z4 **Engines:** 3.0L, 4.8L **Transmissions:** A/T	**Valvetronic (VVT) Overload Protection (Bank 1) Conditions:** After the ignition is on for 45ms, the temperature was recorded as too high. **Possible Causes:** • Sticking in VVT mechanicals, pinion gear, etc. • Loose sensor • Sensor servo motor has short circuit
DTC: P1080 **MIL: No** **2006** **Models:** 325i, 330i, 525i, 530i, 550i, 750i, Z4 **Engines:** 3.0L, 4.8L **Transmissions:** A/T	**Valvetronic (VVT) Overload Protection ECU Temperature High Input (Bank 2) Conditions:** After the ignition is on for 45ms, the ECU temperature was recorded as too high. **Possible Causes:** • Sticking in VVT mechanicals, pinion gear, etc. • Loose sensor • Sensor servo motor has short circuit
DTC: P1081 **MIL: No** **2006** **Models:** 325i, 330i, 525i, 530i, 550i, 750i, Z4 **Engines:** 3.0L, 4.8L **Transmissions:** A/T	**Valvetronic (VVT) Overload Protection Control Motor Temperature High Input (Bank 2) Conditions:** After the ignition is on for 45ms, the E motor temperature was recorded as too high. **Possible Causes:** • Sticking in VVT mechanicals, pinion gear, etc. • Loose sensor • Sensor servo motor has short circuit
DTC: P1082 **MIL: No** **2006** **Models:** 325i, 330i, 525i, 530i, 550i, 750i, Z4 **Engines:** 3.0L, 4.8L **Transmissions:** A/T	**Valvetronic (VVT) Overload Protection Control Motor Current High Input (Bank 2) Conditions:** After the ignition is on for 45ms, the E motor activation current is too high. **Possible Causes:** • Sticking in VVT mechanicals, pinion gear, etc. • Loose sensor • Sensor servo motor has short circuit
DTC: P1083 **2T MIL: Yes** 2003 **Models:** 325i, 330i, 525i, 530i, Z4 **Engines:** 3.0L **Transmissions:** All	**Fuel System Too Rich Conditions:** Key on or engine running, all electrical components off and coolant temperature at least 80 degrees Celsius; and the DME detected the Bank 1 Adaptive Fuel Control System reached its rich correction limit (a rich A/F condition). The fuel status is in a closed loop pattern, the coolant temperature is between 69 and 100 degrees Celsius, and the engine speed is between 800 and 6000rpm. **Possible Causes:** • Camshaft timing is incorrect, or the engine has an oil overfill condition • EVAP vapor recovery system failure (may be pulling vacuum) • Fuel pressure regulator is damaged or leaking • HO2S element is contaminated with alcohol or water • MAF or MAP sensor values are incorrect or out-of-range • One of more fuel injectors is leaking

DTC	Trouble Code Title, Conditions & Possible Causes
DTC: P1084 **2T MIL: Yes** **2002, 2003, 2006** **Models:** 325i, 330i, 525i, 545i, 550i, 745i, 750i, Z4 **Engines:** 3.0L, 4.8L **Transmissions:** A/T	**Fuel System Too Lean Conditions:** Key on or engine running, all electrical components off and coolant temperature at least 80 degrees Celsius; and the DME detected the Bank 1 Adaptive Fuel Control System reached its rich correction limit (a lean A/F condition). The fuel status is in a closed loop pattern, the coolant temperature is between 69 and 100 degrees Celsius, and the engine speed is between 800 and 6000rpm. **Possible Causes:** • Air leaks after the MAF sensor, or leaks in the PCV system • Exhaust leaks before or near where the HO2S is mounted • Fuel injector(s) restricted or not supplying enough fuel • Fuel pump not supplying enough fuel during high fuel demand conditions • Leaking EGR gasket, or leaking EGR valve diaphragm • MAF sensor dirty (causes DME to underestimate airflow) • Vehicle running out of fuel or engine oil dip stick not seated
DTC: P1085 **2T MIL: Yes** **2002, 2003, 2006** **Models:** 325i, 330i, 525i, 545i, 550i, 745i, 750i, Z4 **Engines:** 3.0L, 4.8L **Transmissions:** A/T	**Fuel System Too Rich Conditions:** Key on or engine running, all electrical components off and coolant temperature at least 80 degrees Celsius; and the DME detected the Bank 2 Adaptive Fuel Control System reached its rich correction limit (a rich A/F condition). The fuel status is in a closed loop pattern, the coolant temperature is between 69 and 100 degrees Celsius, and the engine speed is between 800 and 6000rpm. **Possible Causes:** • Air leaks after the MAF sensor, or leaks in the PCV system • Exhaust leaks before or near where the HO2S is mounted • Fuel injector(s) restricted or not supplying enough fuel • Fuel pump not supplying enough fuel during high fuel demand conditions • Leaking EGR gasket, or leaking EGR valve diaphragm • MAF sensor dirty (causes DME to underestimate airflow) • Vehicle running out of fuel or engine oil dip stick not seated
DTC: P1086 **2T MIL: Yes** **2002, 2003, 2006** **Models:** 325i, 330i, 525i, 545i, 550i, 745i, 750i, Z4 **Engines:** 3.0L, 4.8L **Transmissions:** A/T	**Fuel System Too Lean Conditions:** Key on or engine running, all electrical components off and coolant temperature at least 80 degrees Celsius; and the DME detected the Bank 2 Fuel Control System reached its lean correction limit. The fuel status is in a closed loop pattern, the coolant temperature is between 69 and 100 degrees Celsius, and the engine speed is between 800 and 6000rpm. **Possible Causes:** • Air leaks after the MAF sensor, or leaks in the PCV system • Exhaust leaks before or near where the HO2S is mounted • Fuel injector(s) restricted or not supplying enough fuel • Fuel pump not supplying enough fuel during high fuel demand conditions • Leaking EGR gasket, or leaking EGR valve diaphragm • MAF sensor dirty (causes DME to underestimate airflow) • Vehicle running out of fuel or engine oil dip stick not seated
DTC: P1100 **2T MIL: Yes** **2000** **Models:** 745i, M5, Z8 **Engines:** 5.0L **Transmissions:** All	**Mass Air Flow Circuit Conditions** Engine running, with the system voltage more than 11.0v, and the temperature must be at least 185-degrees (F) and all electrical equipment (A/C, lights, etc) must be off. The DME has detected that the MAF signal was out of a calculated range with the engine (or undetectable) for a certain period of time. The engine speed is greater than 200rpm. **Possible Causes:** • Mass air flow (MAF) sensor has failed or is damaged • Signal and ground wires of Mass Air Flow (MAF) sensor has short circuited
DTC: P1101 **2T MIL: Yes** **2001** **Models:** 540i, 740i, 745i **Engines:** 4.4L, 4.8L **Transmissions:** All	**Mass Air Flow Circuit Rationality Check Conditions** Engine running, with the system voltage more than 11.0v, and the temperature must be at least 185-degrees (F) and all electrical equipment (A/C, lights, etc) must be off. The DME has detected that the MAF signal was out of a calculated range with the engine (or undetectable) for a certain period of time. The engine speed is greater than 200rpm. **Possible Causes:** • Mass air flow (MAF) sensor has failed or is damaged • Signal and ground wires of Mass Air Flow (MAF) sensor has short circuited
DTC: P1103 **2T MIL: Yes** **2001** **Models:** 540i, 740i, 745i **Engines:** 4.4L, 4.8L **Transmissions:** All	**Mass Air Flow Circuit Rationality Check Conditions** Engine running, with the system voltage more than 11.0v, and the temperature must be at least 185-degrees (F) and all electrical equipment (A/C, lights, etc) must be off. The DME has detected that the MAF signal was out of a calculated range with the engine (or undetectable) for a certain period of time. The engine speed is greater than 200rpm. **Possible Causes:** • Mass air flow (MAF) sensor has failed or is damaged • Signal and ground wires of Mass Air Flow (MAF) sensor has short circuited

DTC	Trouble Code Title, Conditions & Possible Causes
DTC: P1104 **2006** **Models:** 325i, 330i, 525i, 530i, 550i, 750i, Z4 **Engines:** 3.0L, 4.8L **Transmissions:** A/T	**Differential Pressure Sensor Intake Manifold Pressure Too Low (Bank 1) Conditions:** Engine started, battery voltage must be at least 11v, and the differential pressure sensor detected a control deviation at the minimum limit. The closed loop control of the differential pressure in the intake manifold is suspended and replaced by a direct specification. **Possible Causes:** • Sensor's voltage supply on Terminal 87 • Sensor's ground connection faulty • Signal wire to DME faulty • Replace sensor
DTC: P1105 **2006** **Models:** 325i, 330i, 525i, 530i, 550i, 750i, Z4 **Engines:** 3.0L, 4.8L **Transmissions:** A/T	**Differential Pressure Sensor Intake Manifold Pressure Too High (Bank 1) Conditions:** Engine started, battery voltage must be at least 11v, and the differential pressure sensor detected a control deviation at the maximum limit. The closed loop control of the differential pressure in the intake manifold is suspended and replaced by a direct specification. **Possible Causes:** • Sensor's voltage supply on Terminal 87 • Sensor's ground connection faulty • Signal wire to DME faulty • Replace sensor
DTC: P1111 **2001, 2002, 2003, 2004** **Models:** 325i, 330i, 525i, 530i, 545i, 550i, 745i, 750i, M3, Z3, Z4 **Engines:** 2.5L, 3.0L, 3.2L, 4.8L, 5.0L **Transmissions:** All	**O2 Control (Bank 1) System Too Lean Conditions:** Engine started, battery voltage must be at least 11.5v, all electrical components must be off, the ground between the engine and the chassis must be well connected, the exhaust system must be properly sealed between the catalytic converter and the cylinder head, and the oxygen sensor heater for oxygen sensor before the catalytic converter must be properly functioning. The DME detected a measurement on the O2 sensor circuit that was outside the parameters to function properly. **Note: For resistance testing of sensor heating, oxygen sensor should be cooled to ambient temperature. High temperatures at oxygen sensor may lead to inaccurate measurements.** **Note: When an O2S malfunction (P0131 to P0414) is also stored with this malfunction, the O2S malfunction(s) should be repaired first.** **Possible Causes:** • Oxygen sensor (before catalytic converter) is faulty • Oxygen sensor (behind catalytic converter) is faulty • Oxygen sensor heater (before catalytic converter) is faulty • Oxygen sensor heater (behind catalytic converter) is faulty • Circuit wiring has a short to power or ground • Engine Component Power Supply Relay is faulty • E-box fuses for oxygen sensor are faulty • Leaks present in the exhaust manifold or exhaust pipes • HO2S signal wire and ground wire crossed in connector • HO2S element is fuel contaminated or has failed
DTC: P1112 **2001, 2002, 2003, 2004** **Models:** 325i, 330i, 525i, 530i, 545i, 550i, 745i, 750i, M3, Z3, Z4 **Engines:** 2.5L, 3.0L, 3.2L, 4.8L, 5.0L **Transmissions:** All	**O2 Control (Bank 1) System Too Rich Conditions:** Engine started, battery voltage must be at least 11.5v, all electrical components must be off, the ground between the engine and the chassis must be well connected, the exhaust system must be properly sealed between the catalytic converter and the cylinder head, and the oxygen sensor heater for oxygen sensor before the catalytic converter must be properly functioning. The DME detected a measurement on the O2 sensor circuit that was outside the parameters to function properly. **Note: For resistance testing of sensor heating, oxygen sensor should be cooled to ambient temperature. High temperatures at oxygen sensor may lead to inaccurate measurements.** **Note: When an O2S malfunction (P0131 to P0414) is also stored with this malfunction, the O2S malfunction(s) should be repaired first.** **Possible Causes:** • Oxygen sensor (before catalytic converter) is faulty • Oxygen sensor (behind catalytic converter) is faulty • Oxygen sensor heater (before catalytic converter) is faulty • Oxygen sensor heater (behind catalytic converter) is faulty • Circuit wiring has a short to power or ground • Engine Component Power Supply Relay is faulty • E-box fuses for oxygen sensor are faulty • Leaks present in the exhaust manifold or exhaust pipes • HO2S signal wire and ground wire crossed in connector • HO2S element is fuel contaminated or has failed

DTC	Trouble Code Title, Conditions & Possible Causes
DTC: P1129 **MIL: No (Oil warning lamp)** **2006** **Models:** 325i, 330i, 525i, 530i, 550i, 750i, Z4 **Engines:** 3.0L, 4.8L **Transmissions:** A/T	**Engine Oil Level Sensor Signal Oil Level Too Low Conditions:** Engine started, and the oil sensor has detected that the level is too low. **Possible Causes:** • Top off the oil
DTC: P1130 **2T MIL: Yes** **2006** **Models:** 325i, 330i, 525i, 530i, 550i, 750i, Z4 **Engines:** 3.0L, 4.8L **Transmissions:** A/T	**Long Term Fuel Trim at Lean Limit Conditions:** Engine started, battery voltage must be at least 11.5v, all electrical components must be off, the ground between the engine and the chassis must be well connected, the exhaust system must be properly sealed between the catalytic converter and the cylinder head, and the oxygen sensor heater for oxygen sensor before the catalytic converter must be properly functioning. The DME detected the HO2S circuit was too lean, or that it could no longer change Fuel Trim because it was at its lean limit. **Possible Causes:** • Air intake system leaking, vacuum hoses leaking or damaged • Air leaks located after the MAF sensor mounting location • EGR valve sticking, EGR diaphragm leaking, or gasket leaking • EVAP vapor recovery system has failed • Excessive fuel pressure, leaking or contaminated fuel injectors • Exhaust leaks before or near the HO2S(s) mounting location • Fuel pressure regulator is leaking or damaged • HO2S circuits wet or oily, corroded, or poor terminal contact • HO2S is damaged or it has failed • HO2S signal circuit open, shorted to ground, shorted to power • Low fuel pressure or vehicle driven until it was out of fuel • Oil dipstick not seated or engine oil level too high (overfilled)
DTC: P1132 **2T MIL: Yes** **1999** **Models:** 323i, 328i, 528i, Z3 **Engines:** 2.8L **Transmissions:** All	**HO2S (Bank 2 Sensor 1) Heater Circuit Malfunction Conditions:** Engine running, battery voltage 11.5, all electrical components off, ground between engine and chassis well connected and the exhaust system must be properly sealed between catalytic converter and the cylinder head. The DME detected an open or shorted condition, or excessive current draw in the heater circuit. The response rate for the sensor signal period is greater than 3.8/second. The engine speed is 1280 to 2400rpm, the catalyst temperature is greater than 300 degrees Celsius and the heater has been on for less than 90 seconds. **Possible Causes:** • HO2S heater power circuit is open • HO2S heater ground circuit is open • HO2S signal tracking (due to oil or moisture in the connector) • HO2S is damaged or has failed
DTC: P1133 **2T MIL: Yes** **1999** **Models:** 323i, 328i, 528i, Z3 **Engines:** 2.8L **Transmissions:** All	**HO2S (Bank 2 Sensor 1) Heater Circuit Malfunction Conditions:** Engine running, battery voltage 11.5, all electrical components off, ground between engine and chassis well connected and the exhaust system must be properly sealed between catalytic converter and the cylinder head. The DME detected an open or shorted condition, or excessive current draw in the heater circuit. The response rate for the sensor signal period is greater than 3.8/second. The engine speed is 1280 to 2400rpm, the catalyst temperature is greater than 300 degrees Celsius and the heater has been on for less than 90 seconds. **Possible Causes:** • HO2S heater power circuit is open • HO2S heater ground circuit is open • HO2S signal tracking (due to oil or moisture in the connector) • HO2S is damaged or has failed
DTC: P1134 **2T MIL: Yes** **2001** **Models:** All **Engines:** All **Transmissions:** All	**HO2S Heater Circuit Current Malfunction Conditions:** Engine running, battery voltage 11.5, all electrical components off, ground between engine and chassis well connected and the exhaust system must be properly sealed between catalytic converter and the cylinder head. The DME detected an open or shorted condition, or excessive current draw in the heater circuit. The response rate for the sensor signal period is greater than 3.8/second. The engine speed is 1280 to 2400rpm, the catalyst temperature is greater than 300 degrees Celsius and the heater has been on for less than 90 seconds. **Possible Causes:** • HO2S heater power circuit is open • HO2S heater ground circuit is open • HO2S signal tracking (due to oil or moisture in the connector) • HO2S is damaged or has failed

DTC	Trouble Code Title, Conditions & Possible Causes
DTC: P1135 **2T MIL: Yes** **2001** **Models:** All **Engines:** All **Transmissions:** All	**HO2S Heater Circuit Current Malfunction Conditions:** Engine running, battery voltage 11.5, all electrical components off, ground between engine and chassis well connected and the exhaust system must be properly sealed between catalytic converter and the cylinder head. The DME detected an open or shorted condition, or excessive current draw in the heater circuit. The response rate for the sensor signal period is greater than 3.8/second. The engine speed is 1280 to 2400rpm, the catalyst temperature is greater than 300 degrees Celsius and the heater has been on for less than 90 seconds. **Possible Causes:** • HO2S heater power circuit is open • HO2S heater ground circuit is open • HO2S signal tracking (due to oil or moisture in the connector) • HO2S is damaged or has failed
DTC: P1136 **2T MIL: Yes** **2001** **Models:** All **Engines:** All **Transmissions:** All	**HO2S Heater Circuit Heater Resistance Conditions:** Engine running, battery voltage 11.5, all electrical components off, ground between engine and chassis well connected and the exhaust system must be properly sealed between catalytic converter and the cylinder head. The DME detected an open or shorted condition, or excessive current draw in the heater circuit. The response rate for the sensor signal period is greater than 3.8/second. The engine speed is 1280 to 2400rpm, the catalyst temperature is greater than 300 degrees Celsius and the heater has been on for less than 90 seconds. **Possible Causes:** • HO2S heater power circuit is open • HO2S heater ground circuit is open • HO2S signal tracking (due to oil or moisture in the connector) • HO2S is damaged or has failed
DTC: P1137 **2T MIL: Yes** **2001** **Models:** 325i, 330i, 525i, 530i, M3, M5, Z8 **Engines:** 2.5L, 3.0L, 3.2L, 5.0L **Transmissions:** All	**Long Term Fuel Trim Add. Fuel, Bank 1 System Too Rich Conditions:** Engine started, battery voltage must be at least 11.5v, all electrical components must be off, the ground between the engine and the chassis must be well connected, the exhaust system must be properly sealed between the catalytic converter and the cylinder head, and the oxygen sensor heater for oxygen sensor before the catalytic converter must be properly functioning. The DME detected the HO2S circuit was too rich, or that it could no longer change Fuel Trim because it was at its lean limit. **Possible Causes:** • Air intake system leaking, vacuum hoses leaking or damaged • Air leaks located after the MAF sensor mounting location • EGR valve sticking, EGR diaphragm leaking, or gasket leaking • EVAP vapor recovery system has failed • Excessive fuel pressure, leaking or contaminated fuel injectors • Exhaust leaks before or near the HO2S(s) mounting location • Fuel pressure regulator is leaking or damaged • HO2S circuits wet or oily, corroded, or poor terminal contact • HO2S is damaged or it has failed • HO2S signal circuit open, shorted to ground, shorted to power • Low fuel pressure or vehicle driven until it was out of fuel • Oil dipstick not seated or engine oil level too high (overfilled)
DTC: P1138 **2T MIL: Yes** **2001** **Models:** 325i, 330i, 525i, 530i, 740i, 750i, Z3 **Engines:** 2.5L, 3.0L, 3.2L, 4.4L, 5.4L **Transmissions:** All	**HO2S Circuit Malfunction Conditions:** Engine running, battery voltage 11.5, all electrical components off, ground between engine and chassis well connected and the exhaust system must be properly sealed between catalytic converter and the cylinder head. The DME detected the HO2S signal failed to meet the maximum or minimum voltage levels (i.e., it failed the voltage range check). The heater has been on for less than 90 seconds, the fuel system status is in fuel cut-off, the output voltage is between 400mV and 500mV and it is 120 seconds after engine start up. **Possible Causes:** • Leaks present in the exhaust manifold or exhaust pipes • HO2S signal wire and ground wire crossed in connector • HO2S element is fuel contaminated or has failed
DTC: P1139 **2T MIL: Yes** **2001** **Models:** 325i, 330i, 525i, 530i, 540i, 740i, 750i, M3, M5, Z3, Z8 **Engines:** 2.5L, 3.0L, 3.2L, 4.4L, 5.0L, 5.4L **Transmissions:** All	**HO2S Circuit Malfunction Conditions:** Engine running, battery voltage 11.5, all electrical components off, ground between engine and chassis well connected and the exhaust system must be properly sealed between catalytic converter and the cylinder head. The DME detected the HO2S signal failed to meet the maximum or minimum voltage levels (i.e., it failed the voltage range check). The heater has been on for less than 90 seconds, the fuel system status is in fuel cut-off, the output voltage is between 400mV and 500mV and it is 120 seconds after engine start up. **Possible Causes:** • Leaks present in the exhaust manifold or exhaust pipes • HO2S signal wire and ground wire crossed in connector • HO2S element is fuel contaminated or has failed

DTC	Trouble Code Title, Conditions & Possible Causes
DTC: P1140 **2T MIL: Yes** **1997, 1998** **Models:** 318i, 540i, 740i, 840Ci, Z3 **Engines:** 1.9L, 4.4L **Transmissions:** All	**Throttle/Pedal Position Sensor Malfunction Conditions:** Engine started, at idle, the temperature must be 80 degrees Celsius. The throttle position sensor supplies implausible signal to the DME. The throttle valve activation occurs via an electric motor (throttle drive) in the throttle valve control module. It is activated by the Engine Control Module (DME) according to specifications of the two sensors, Throttle Position (TP) Sensor and Accelerator Pedal Position Sensor 2. **Possible Causes:** • TP sensor signal circuit is open (inspect wiring & connector) • TP sensor signal circuit is shorted to ground • TP sensor or module is damaged or it has failed • Throttle valve is damaged or dirty • Throttle valve control module is faulty
DTC: P1149 **2T MIL: Yes** **2002, 2003** **Models:** 325i, 330i, 525i, 530i, 745i, 840Ci, Z3, Z4 **Engines:** 2.5L, 3.0L, 4.8L **Transmissions:** All	**O2 Control (Bank 1) Out of Range (Sensor Aging) Conditions:** Engine started, battery voltage must be at least 11.5v, all electrical components must be off, the ground between the engine and the chassis must be well connected, the exhaust system must be properly sealed between the catalytic converter and the cylinder head, and the oxygen sensor heater for oxygen sensor before the catalytic converter must be properly functioning. The DME detected a voltage on the O2 sensor circuit that was outside the parameters to function properly. The voltage remains above or below threshold at normal operation. **Note: For resistance testing of sensor heating, oxygen sensor should be cooled to ambient temperature. High temperatures at oxygen sensor may lead to inaccurate measurements.** **Possible Causes:** • Oxygen sensor (before catalytic converter) is faulty • Oxygen sensor (behind catalytic converter) is faulty • Oxygen sensor heater (before catalytic converter) is faulty • Oxygen sensor heater (behind catalytic converter) is faulty • Circuit wiring has a short to power or ground • Engine Component Power Supply Relay is faulty • E-box fuses for oxygen sensor are faulty • Leaks present in the exhaust manifold or exhaust pipes • HO2S signal wire and ground wire crossed in connector • HO2S element is fuel contaminated or has failed
DTC: P1151 **2T MIL: Yes** **2001** **Models:** 540i, 740i, M3, M5, Z8 **Engines:** 3.2L, 4.4L, 5.0L **Transmissions:** All	**HO2S Heater Circuit Current Malfunction Conditions:** Engine running, battery voltage 11.5, all electrical components off, ground between engine and chassis well connected and the exhaust system must be properly sealed between catalytic converter and the cylinder head. The DME detected an open or shorted condition, or excessive current draw in the heater circuit. The response rate for the sensor signal period is greater than 3.8/second. The engine speed is 1280 to 2400rpm, the catalyst temperature is greater than 300 degrees Celsius and the heater has been on for less than 90 seconds. **Possible Causes:** • HO2S heater power circuit is open • HO2S heater ground circuit is open • HO2S signal tracking (due to oil or moisture in the connector) • HO2S is damaged or has failed
DTC: P1152 **2T MIL: Yes** **2001** **Models:** 540i, 740i, M3, M5, Z8 **Engines:** 3.2L, 4.4L, 5.0L **Transmissions:** All	**HO2S Heater Circuit Current Malfunction Conditions:** Engine running, battery voltage 11.5, all electrical components off, ground between engine and chassis well connected and the exhaust system must be properly sealed between catalytic converter and the cylinder head. The DME detected an open or shorted condition, or excessive current draw in the heater circuit. The response rate for the sensor signal period is greater than 3.8/second. The engine speed is 1280 to 2400rpm, the catalyst temperature is greater than 300 degrees Celsius and the heater has been on for less than 90 seconds. **Possible Causes:** • HO2S heater power circuit is open • HO2S heater ground circuit is open • HO2S signal tracking (due to oil or moisture in the connector) • HO2S is damaged or has failed
DTC: P1153 **2T MIL: Yes** **2001** **Models:** 540i, 740i, M3, M5, Z8 **Engines:** 3.2L, 4.4L, 5.0L **Transmissions:** All	**HO2S Heater Circuit Current Malfunction Conditions:** Engine running, battery voltage 11.5, all electrical components off, ground between engine and chassis well connected and the exhaust system must be properly sealed between catalytic converter and the cylinder head. The DME detected an open or shorted condition, or excessive current draw in the heater circuit. The response rate for the sensor signal period is greater than 3.8/second. The engine speed is 1280 to 2400rpm, the catalyst temperature is greater than 300 degrees Celsius and the heater has been on for less than 90 seconds. **Possible Causes:** • HO2S heater power circuit is open • HO2S heater ground circuit is open • HO2S signal tracking (due to oil or moisture in the connector) • HO2S is damaged or has failed

DTC	Trouble Code Title, Conditions & Possible Causes
DTC: P1155 **2T MIL: Yes** **2001** **Models:** 325i, 330i, 525i, 530i, 540i, 740i, 750i, M3, Z3, Z4 **Engines:** 2.5L, 3.0L, 3.2L, 4.4L, 5.4L **Transmissions:** All	**HO2S Heater Circuit Current Malfunction Conditions:** Engine running, battery voltage 11.5, all electrical components off, ground between engine and chassis well connected and the exhaust system must be properly sealed between catalytic converter and the cylinder head. The DME detected an open or shorted condition, or excessive current draw in the heater circuit. The response rate for the sensor signal period is greater than 3.8/second. The engine speed is 1280 to 2400rpm, the catalyst temperature is greater than 300 degrees Celsius and the heater has been on for less than 90 seconds. **Possible Causes:** • HO2S heater power circuit is open • HO2S heater ground circuit is open • HO2S signal tracking (due to oil or moisture in the connector) • HO2S is damaged or has failed
DTC: P1156 **2T MIL: Yes** **2001** **Models:** 540i, 740i, M3, M5, Z8 **Engines:** 3.2L, 4.4L, 5.0L **Transmissions:** All	**HO2S Heater Circuit Current Malfunction-Circuit Continuity Conditions:** Engine running, battery voltage 11.5, all electrical components off, ground between engine and chassis well connected and the exhaust system must be properly sealed between catalytic converter and the cylinder head. The DME detected an open or shorted condition, or excessive current draw in the heater circuit. The response rate for the sensor signal period is greater than 3.8/second. The engine speed is 1280 to 2400rpm, the catalyst temperature is greater than 300 degrees Celsius and the heater has been on for less than 90 seconds. **Possible Causes:** • HO2S heater power circuit is open • HO2S heater ground circuit is open • HO2S signal tracking (due to oil or moisture in the connector) • HO2S is damaged or has failed
DTC: P1157 **2T MIL: Yes** **2001** **Models:** 325i, 330i, 525i, 530i, 540i, 740i, 750i, M3, Z3, Z4 **Engines:** 2.5L, 3.0L, 3.2L, 4.4L, 5.4L **Transmissions:** All	**HO2S Heater Circuit Current Malfunction-Heater Resistance Conditions:** Engine running, battery voltage 11.5, all electrical components off, ground between engine and chassis well connected and the exhaust system must be properly sealed between catalytic converter and the cylinder head. The DME detected an open or shorted condition, or excessive current draw in the heater circuit. The response rate for the sensor signal period is greater than 3.8/second. The engine speed is 1280 to 2400rpm, the catalyst temperature is greater than 300 degrees Celsius and the heater has been on for less than 90 seconds. **Possible Causes:** • HO2S heater power circuit is open • HO2S heater ground circuit is open • HO2S signal tracking (due to oil or moisture in the connector) • HO2S is damaged or has failed
DTC: P1158 **2T MIL: Yes** **2002** **Models:** 540i, 745i **Engines:** 4.4L, 4.8L **Transmissions:** All	**Fuel Trim System Too Rich (Bank 1) Conditions:** Engine started, battery voltage must be at least 11.5v, all electrical components must be off, the ground between the engine and the chassis must be well connected, the exhaust system must be properly sealed between the catalytic converter and the cylinder head, and the oxygen sensor heater for oxygen sensor before the catalytic converter must be properly functioning. The DME detected a problem with the fuel mixture. Trim control 1 segment (precision controller with oxygen sensor behind cat.) below delta lambda threshold of less than -0.03. Fault monitoring criterion must remain present for over one second. **Note: For resistance testing of sensor heating, oxygen sensor should be cooled to ambient temperature. High temperatures at oxygen sensor may lead to inaccurate measurements.** **Possible Causes:** • Oxygen sensor (before catalytic converter) is faulty • Oxygen sensor (behind catalytic converter) is faulty • Oxygen sensor heater (before catalytic converter) is faulty • Oxygen sensor heater (behind catalytic converter) is faulty • Check circuits for shorts to each other, ground or power
DTC: P1159 **2T MIL: Yes** **2002** **Models:** 540i, 745i **Engines:** 4.4L, 4.8L **Transmissions:** All	**Fuel Trim System Too Lean (Bank 1) Conditions:** Engine started, battery voltage must be at least 11.5v, all electrical components must be off, the ground between the engine and the chassis must be well connected, the exhaust system must be properly sealed between the catalytic converter and the cylinder head, and the oxygen sensor heater for oxygen sensor before the catalytic converter must be properly functioning. The DME detected a problem with the fuel mixture. Trim control 1 segment (precision controller with oxygen sensor behind cat.) below delta lambda threshold of less than -0.03. Fault monitoring criterion must remain present for over one second. **Note: For resistance testing of sensor heating, oxygen sensor should be cooled to ambient temperature. High temperatures at oxygen sensor may lead to inaccurate measurements.** **Possible Causes:** • Oxygen sensor (before catalytic converter) is faulty • Oxygen sensor (behind catalytic converter) is faulty • Oxygen sensor heater (before catalytic converter) is faulty • Oxygen sensor heater (behind catalytic converter) is faulty • Check circuits for shorts to each other, ground or power

DTC	Trouble Code Title, Conditions & Possible Causes
DTC: P1160 **2T MIL: Yes** **2002** **Models:** 540i, 745i **Engines:** 4.4L, 4.8L **Transmissions:** All	**Fuel Trim System Too Rich (Bank 1) Conditions:** Engine started, battery voltage must be at least 11.5v, all electrical components must be off, the ground between the engine and the chassis must be well connected, the exhaust system must be properly sealed between the catalytic converter and the cylinder head, and the oxygen sensor heater for oxygen sensor before the catalytic converter must be properly functioning. The DME detected a problem with the fuel mixture. Trim control 1 segment (precision controller with oxygen sensor behind cat.) below delta lambda threshold of less than -0.03. Fault monitoring criterion must remain present for over one second. **Note: For resistance testing of sensor heating, oxygen sensor should be cooled to ambient temperature. High temperatures at oxygen sensor may lead to inaccurate measurements.** **Possible Causes:** • Oxygen sensor (before catalytic converter) is faulty • Oxygen sensor (behind catalytic converter) is faulty • Oxygen sensor heater (before catalytic converter) is faulty • Oxygen sensor heater (behind catalytic converter) is faulty • Check circuits for shorts to each other, ground or power
DTC: P1161 **2T MIL: Yes** **2002** **Models:** 540i, 745i **Engines:** 4.4L, 4.8L **Transmissions:** All	**Fuel Trim System Too Lean (Bank 1) Conditions:** Engine started, battery voltage must be at least 11.5v, all electrical components must be off, the ground between the engine and the chassis must be well connected, the exhaust system must be properly sealed between the catalytic converter and the cylinder head, and the oxygen sensor heater for oxygen sensor before the catalytic converter must be properly functioning. The DME detected a problem with the fuel mixture. Trim control 1 segment (precision controller with oxygen sensor behind cat.) below delta lambda threshold of less than -0.03. Fault monitoring criterion must remain present for over one second. **Note: For resistance testing of sensor heating, oxygen sensor should be cooled to ambient temperature. High temperatures at oxygen sensor may lead to inaccurate measurements.** **Possible Causes:** • Oxygen sensor (before catalytic converter) is faulty • Oxygen sensor (behind catalytic converter) is faulty • Oxygen sensor heater (before catalytic converter) is faulty • Oxygen sensor heater (behind catalytic converter) is faulty • Check circuits for shorts to each other, ground or power
DTC: P1163 **2T MIL: Yes** **2001** **Models:** M3 **Engines:** 3.2L **Transmissions:** All	**HO2S Heater Circuit Current Malfunction Conditions:** Engine running, battery voltage 11.5, all electrical components off, ground between engine and chassis well connected and the exhaust system must be properly sealed between catalytic converter and the cylinder head. The DME detected an open or shorted condition, or excessive current draw in the heater circuit. The response rate for the sensor signal period is greater than 3.8/second. The engine speed is 1280 to 2400rpm, the catalyst temperature is greater than 300 degrees Celsius and the heater has been on for less than 90 seconds. **Possible Causes:** • HO2S heater power circuit is open • HO2S heater ground circuit is open • HO2S signal tracking (due to oil or moisture in the connector) • HO2S is damaged or has failed
DTC: P1174 **2T MIL: Yes** **1996, 1997, 1998, 1999, 2000** **Models:** 318i, 540i, 740i, 750i, Z3 **Engines:** 1.9L, 4.4L, 5.4L **Transmissions:** All	**Fuel System Malfunction (Cylinder Bank 1) Conditions:** The engine is running in a closed loop at a stable engine speed, and the DME detected the lean or rich fuel trim correction valve was more than or less than a calibrated limit. The fuel system status is in a closed loop. **Possible Causes:** • Air leaks after the MAF sensor, or leaks in the PCV system • Exhaust leaks before or near where the HO2S is mounted • Fuel injector(s) restricted or not supplying enough fuel • Fuel system not supplying enough fuel during high fuel demand conditions (e.g., the fuel pump may not supply enough fuel) • Leaking EGR gasket, or leaking EGR valve diaphragm • MAF sensor dirty (causes DME to underestimate airflow) • Vehicle running out of fuel or engine oil dip stick not seated
DTC: P1175 **2T MIL: Yes** **1997, 1998, 1999, 2000** **Models:** 540i, 740i, 750i, 840Ci **Engines:** 4.4L, 5.4L **Transmissions:** All	**Fuel System Malfunction (Cylinder Bank 1) Conditions:** The engine is running in a closed loop at a stable engine speed, and the DME detected the lean or rich fuel trim correction valve was more than or less than a calibrated limit. The fuel system status is in a closed loop. **Possible Causes:** • Air leaks after the MAF sensor, or leaks in the PCV system • Exhaust leaks before or near where the HO2S is mounted • Fuel injector(s) restricted or not supplying enough fuel • Fuel system not supplying enough fuel during high fuel demand conditions (e.g., the fuel pump may not supply enough fuel) • Leaking EGR gasket, or leaking EGR valve diaphragm • MAF sensor dirty (causes DME to underestimate airflow) • Vehicle running out of fuel or engine oil dip stick not seated

DTC	Trouble Code Title, Conditions & Possible Causes
DTC: P1178 **2T MIL: Yes** **1998, 2000, 2001, 2002, 2003** **Models:** 328i, 528i, M3, M5, Z8 **Engines:** 2.8L, 3.2L, 5.0L **Transmissions:** All	**O2 Sensor Switching Time Conditions:** **Engine running, battery voltage 11.5, all electrical components off, ground between engine and chassis well connected and the exhaust system must be properly sealed between catalytic converter and the cylinder head. The DME detected the O2S signal was implausible or not detected, the switching time range from lean to rich and vies versa was too slow. The exhaust temperature is greater than 380 degrees Celsius, the fuel system is in a closed loop, and the engine speed is between 2000 and 3200rpm.** **Possible Causes:** • Oxygen sensor heater for oxygen sensor (HO2S) before catalytic converter is faulty • O2S is contaminated (due to presence of silicone in fuel) • O2S signal and ground circuit wires crossed in wiring harness • O2S signal circuit is shorted to sensor or chassis ground • O2S element before the catalytic converter has failed (internal short condition) • Leaks present in the exhaust manifold or exhaust pipes
DTC: P1179 **2T MIL: Yes** **1998, 2000, 2001, 2002, 2003** **Models:** 328i, 528i, M3, M5, Z8 **Engines:** 2.8L, 3.2L, 5.0L **Transmissions:** All	**O2 Sensor Switching Time Conditions:** **Engine running, battery voltage 11.5, all electrical components off, ground between engine and chassis well connected and the exhaust system must be properly sealed between catalytic converter and the cylinder head. The DME detected the O2S signal was implausible or not detected, the switching time range from lean to rich and vies versa was too slow. The exhaust temperature is greater than 380 degrees Celsius, the fuel system is in a closed loop, and the engine speed is between 2000 and 3200rpm.** **Possible Causes:** • Oxygen sensor heater for oxygen sensor (HO2S) before catalytic converter is faulty • O2S is contaminated (due to presence of silicone in fuel) • O2S signal and ground circuit wires crossed in wiring harness • O2S signal circuit is shorted to sensor or chassis ground • O2S element before the catalytic converter has failed (internal short condition) • Leaks present in the exhaust manifold or exhaust pipes
DTC: P1186 **2T MIL: Yes** **1996, 1997, 1998, 1999, 2000** **Models:** 318i, 323i, 328i, 528i, 540i, 740i, 750i, 840Ci, Z3 **Engines:** 1.9L, 2.8L, 3.2L, 4.4L, 5.4L **Transmissions:** All	**O2 Sensor Heater Circuit (Bank 1-Sensor 1) Electrical Malfunction Conditions:** Engine started, battery voltage must be at least 11.5v, all electrical components must be off, the ground between the engine and the chassis must be well connected, the exhaust system must be properly sealed between the catalytic converter and the cylinder head, and the oxygen sensor heater for oxygen sensor before the catalytic converter must be properly functioning. Note: For resistance testing of sensor heating, oxygen sensor should be cooled to ambient temperature. High temperatures at oxygen sensor may lead to inaccurate measurements. The DME detected an open or shorted condition, or excessive current draw in the heater circuit. The heater has been on for less than 90 seconds, the fuel system status is in fuel cut-off, the output voltage is between 400mV and 500mV and it is 120 seconds after engine start up. **Possible Causes:** • HO2S heater power circuit is open • HO2S heater ground circuit is open • HO2S signal tracking (due to oil or moisture in the connector) • HO2S is damaged or has failed • Oxygen sensor (before catalytic converter) is faulty • Oxygen sensor (behind catalytic converter) is faulty • Oxygen sensor heater (before catalytic converter) is faulty • Oxygen sensor heater (behind catalytic converter) is faulty
DTC: P1187 **2T MIL: Yes** **1996, 1997, 1998, 1999, 2000** **Models:** 318i, 323i, 328i, 528i, 540i, 740i, 750i, 840Ci, Z3 **Engines:** 1.9L, 2.8L, 3.2L, 4.4L, 5.4L **Transmissions:** All	**O2 Sensor Heater Circuit (Bank 1-Sensor 1) Electrical Malfunction Conditions:** Engine started, battery voltage must be at least 11.5v, all electrical components must be off, the ground between the engine and the chassis must be well connected, the exhaust system must be properly sealed between the catalytic converter and the cylinder head, and the oxygen sensor heater for oxygen sensor before the catalytic converter must be properly functioning. Note: For resistance testing of sensor heating, oxygen sensor should be cooled to ambient temperature. High temperatures at oxygen sensor may lead to inaccurate measurements. The DME detected an open or shorted condition, or excessive current draw in the heater circuit. The heater has been on for less than 90 seconds, the fuel system status is in fuel cut-off, the output voltage is between 400mV and 500mV and it is 120 seconds after engine start up. **Possible Causes:** • HO2S heater power circuit is open • HO2S heater ground circuit is open • HO2S signal tracking (due to oil or moisture in the connector) • HO2S is damaged or has failed • Oxygen sensor (before catalytic converter) is faulty • Oxygen sensor (behind catalytic converter) is faulty • Oxygen sensor heater (before catalytic converter) is faulty • Oxygen sensor heater (behind catalytic converter) is faulty

DTC	Trouble Code Title, Conditions & Possible Causes
DTC: P1188 **2T MIL: Yes** **1996, 1997, 1998, 1999, 2000** **Models:** 318i, 323i, 328i, 528i, 540i, 740i, 750i, 840Ci, Z3 **Engines:** 1.9L, 2.8L, 3.2L, 4.4L, 5.4L **Transmissions:** All	**Fuel System Malfunction (Cylinder Bank 1) Conditions:** The engine is running in a closed loop at a stable engine speed, and the DME detected the lean or rich fuel trim correction valve was more than or less than a calibrated limit. **Possible Causes:** • Air leaks after the MAF sensor, or leaks in the PCV system • Exhaust leaks before or near where the HO2S is mounted • Fuel injector(s) restricted or not supplying enough fuel • Fuel system not supplying enough fuel during high fuel demand conditions (e.g., the fuel pump may not supply enough fuel) • Leaking EGR gasket, or leaking EGR valve diaphragm • MAF sensor dirty (causes DME to underestimate airflow) • Vehicle running out of fuel or engine oil dip stick not seated
DTC: P1189 **2T MIL: Yes** **1996, 1997, 1998, 1999, 2000** **Models:** 318i, 323i, 328i, 528i, 540i, 740i, 750i, 840Ci, Z3 **Engines:** 1.9L, 2.8L, 3.2L, 4.4L, 5.4L **Transmissions:** All	**Fuel System Malfunction (Cylinder Bank 1) Conditions:** Key on or engine running, all electrical components off and coolant temperature at least 80 degrees Celsius; and the DME detected the Bank 1 Fuel Control System experienced a implausible signal. **Possible Causes:** • Air leaks after the MAF sensor, or leaks in the PCV system • Exhaust leaks before or near where the HO2S is mounted • Fuel injector(s) restricted or not supplying enough fuel • Fuel system not supplying enough fuel during high fuel demand conditions (e.g., the fuel pump may not supply enough fuel) • Leaking EGR gasket, or leaking EGR valve diaphragm • MAF sensor dirty (causes DME to underestimate airflow) • Vehicle running out of fuel or engine oil dip stick not seated
DTC: P1197 **2006** **Models:** 325i, 330i, 525i, 530i, 550i, 750i, Z4 **Engines:** 3.0L, 4.8L **Transmissions:** All	**Differential Pressure Sensor Intake Manifold High Input (Bank 1) Conditions:** **Engine started, battery voltage must be at least 11v, and the differential pressure sensor wiring shorted to battery voltage. The closed loop control of the differential pressure in the intake manifold is suspended and replaced by a direct specification.** **Possible Causes:** • Sensor's voltage supply on Terminal 87 • Sensor's ground connection faulty • Signal wire to DME faulty • Replace sensor
DTC: P1198 **2006** **Models:** 325i, 330i, 525i, 530i, 550i, 750i, Z4 **Engines:** 3.0L, 4.8L **Transmissions:** All	**Differential Pressure Sensor Intake Manifold Low Input (Bank 1) Conditions:** **Engine started, battery voltage must be at least 11v, and the differential pressure sensor wiring shorted to ground. The closed loop control of the differential pressure in the intake manifold is suspended and replaced by a direct specification.** **Possible Causes:** • Sensor's voltage supply on Terminal 87 • Sensor's ground connection faulty • Signal wire to DME faulty • Replace sensor
DTC: P1199 **2006** **Models:** 325i, 330i, 525i, 530i, 550i, 750i, Z4 **Engines:** 3.0L, 4.8L **Transmissions:** All	**Differential Pressure Sensor Intake Manifold Pressure Plausibility (Bank 1) Conditions:** **Engine started, battery voltage must be at least 11v, and the differential pressure sensor signal is malfunction or is not present. The closed loop control of the differential pressure in the intake manifold is suspended and replaced by a direct specification.** **Possible Causes:** • Sensor's voltage supply on Terminal 87 • Sensor's ground connection faulty • Signal wire to DME faulty • Replace sensor
DTC: P1200 **2T MIL: Yes** **2002** **Models:** 540i **Engines:** 4.4L **Transmissions:** All	**Fuel System Too Lean Conditions:** Key on or engine running, all electrical components off and coolant temperature at least 80 degrees Celsius; and the DME detected the Bank 1 Adaptive Fuel Control System reached its rich correction limit (a lean A/F condition). The fuel status is in a closed loop pattern, the coolant temperature is between 69 and 100 degrees Celsius, and the engine speed is between 800 and 6000rpm. **Possible Causes:** • Air leaks after the MAF sensor, or leaks in the PCV system • Exhaust leaks before or near where the HO2S is mounted • Fuel injector(s) restricted or not supplying enough fuel • Fuel pump not supplying enough fuel during high fuel demand conditions • Leaking EGR gasket, or leaking EGR valve diaphragm • MAF sensor dirty (causes DME to underestimate airflow) • Vehicle running out of fuel or engine oil dip stick not seated

DTC	Trouble Code Title, Conditions & Possible Causes
DTC: P1201 **2T MIL: Yes** **2002** **Models:** 540i **Engines:** 4.4L **Transmissions:** All	**Fuel System Too Rich Conditions:** Key on or engine running, all electrical components off and coolant temperature at least 80 degrees Celsius; and the DME detected the Bank 2 Adaptive Fuel Control System reached its rich correction limit (a rich A/F condition). The fuel status is in a closed loop pattern, the coolant temperature is between 69 and 100 degrees Celsius, and the engine speed is between 800 and 6000rpm. **Possible Causes:** • Air leaks after the MAF sensor, or leaks in the PCV system • Exhaust leaks before or near where the HO2S is mounted • Fuel injector(s) restricted or not supplying enough fuel • Fuel pump not supplying enough fuel during high fuel demand conditions • Leaking EGR gasket, or leaking EGR valve diaphragm • MAF sensor dirty (causes DME to underestimate airflow) • Vehicle running out of fuel or engine oil dip stick not seated
DTC: P1202 **2T MIL: Yes** **2002** **Models:** 540i **Engines:** 4.4L **Transmissions:** All	**Fuel System Too Lean Conditions:** Key on or engine running, all electrical components off and coolant temperature at least 80 degrees Celsius; and the DME detected the Bank 1 Adaptive Fuel Control System reached its rich correction limit (a lean A/F condition). The fuel status is in a closed loop pattern, the coolant temperature is between 69 and 100 degrees Celsius, and the engine speed is between 800 and 6000rpm. **Possible Causes:** • Air leaks after the MAF sensor, or leaks in the PCV system • Exhaust leaks before or near where the HO2S is mounted • Fuel injector(s) restricted or not supplying enough fuel • Fuel pump not supplying enough fuel during high fuel demand conditions • Leaking EGR gasket, or leaking EGR valve diaphragm • MAF sensor dirty (causes DME to underestimate airflow) • Vehicle running out of fuel or engine oil dip stick not seated
DTC: P1203 **2T MIL: Yes** **2002** **Models:** 540i **Engines:** 4.4L **Transmissions:** All	**Fuel System Too Rich Conditions:** Key on or engine running, all electrical components off and coolant temperature at least 80 degrees Celsius; and the DME detected the Bank 2 Adaptive Fuel Control System reached its rich correction limit (a rich A/F condition). The fuel status is in a closed loop pattern, the coolant temperature is between 69 and 100 degrees Celsius, and the engine speed is between 800 and 6000rpm. **Possible Causes:** • Air leaks after the MAF sensor, or leaks in the PCV system • Exhaust leaks before or near where the HO2S is mounted • Fuel injector(s) restricted or not supplying enough fuel • Fuel pump not supplying enough fuel during high fuel demand conditions • Leaking EGR gasket, or leaking EGR valve diaphragm • MAF sensor dirty (causes DME to underestimate airflow) • Vehicle running out of fuel or engine oil dip stick not seated
DTC: P1270 **2T MIL: Yes** **1996, 1997, 1998, 1999, 2000** **Models:** 325i, 525i, 530i, 550i, 750i, M5, Z4, Z8 **Engines:** 3.0L, 5.0L, 5.4L **Transmissions:** All	**Mass or Volume Air Flow Circuit "A" Conditions** Engine running, with the system voltage more than 11.0v, and the temperature must be at least 185-degrees (F) and all electrical equipment (A/C, lights, etc) must be off. The DME has detected that the MAF signal was out of a calculated range with the engine (or undetectable) for a certain period of time. The engine speed is greater than 200rpm. **Possible Causes:** • Mass air flow (MAF) sensor has failed or is damaged • Signal and ground wires of Mass Air Flow (MAF) sensor has short circuited
DTC: P1327 **MIL: Yes** **2001, 2002, 2003, 2005, 2006** **Models:** 325i, 330i, 525i, 530i, 540i, 550i, 740i, 745i, 750i, Z4 **Engines:** 3.0L, 4.8L, 5.4L **Transmissions:** All	**Knock Sensor 2 Signal Low Input Conditions:** **Engine started, vehicle driven at 2000rpm for 3 seconds or to a temperature of 40 degrees Celsius, and the DME detected the Knock Sensor 1 (KS1) signal was too low or not recognized by the DME.** **Possible Causes:** • Knock sensor circuit is open • Knock sensor is loose (tighten to 20 NM) • Contact between the knock sensor and cylinder block is dirty, corroded or greasy • Knock sensor circuit is shorted to ground, or shorted to power • Knock sensor is damaged or it has failed • Wrong kind of fuel used • A component in the engine compartment is loose or not properly secured

DTC	Trouble Code Title, Conditions & Possible Causes
DTC: P1328 **MIL: Yes** **2001, 2002, 2003, 2005, 2006** **Models:** 325i, 330i, 525i, 530i, 540i, 550i, 740i, 745i, 750i, Z4 **Engines:** 3.0L, 4.4L, 4.8L, 5.4L **Transmissions:** All	**Knock Sensor 2 Signal High Input Conditions:** Engine started, vehicle driven at 1600rpm for 3 seconds or to a temperature of 40 degrees Celsius, and the DME detected the Knock Sensor 1 (KS1) signal was too high. **Possible Causes:** • Knock sensor circuit is open • Knock sensor is loose (tighten to 20 NM) • Contact between the knock sensor and cylinder block is dirty, corroded or greasy • Knock sensor circuit is shorted to ground, or shorted to power • Knock sensor is damaged or it has failed • Wrong kind of fuel used • A component in the engine compartment is loose or not properly secured
DTC: P1329 **2T MIL: Yes** **2005, 2006** **Models:** 325i, 330i, 525i, 530i, 545i, 550i, 750i, Z4 **Engines:** 3.0L, 4.8L **Transmissions:** All	**Knock Sensor 3 Signal Low Input Conditions:** **Engine started, vehicle driven at 2000rpm for 3 seconds or to a temperature of 40 degrees Celsius, and the DME detected the Knock Sensor 1 (KS1) signal was too low or not recognized by the DME.** **Possible Causes:** • Knock sensor circuit is open • Knock sensor is loose (tighten to 20 NM) • Contact between the knock sensor and cylinder block is dirty, corroded or greasy • Knock sensor circuit is shorted to ground, or shorted to power • Knock sensor is damaged or it has failed • Wrong kind of fuel used • A component in the engine compartment is loose or not properly secured
DTC: P1330 **MIL: Yes** **2005, 2006** **Models:** 325i, 330i, 525i, 530i, 545i, 550i, 750i, Z4 **Engines:** 3.0L, 4.8L **Transmissions:** All	**Knock Sensor 3 Signal High Input Conditions:** Engine started, vehicle driven at 1600rpm for 3 seconds or to a temperature of 40 degrees Celsius, and the DME detected the Knock Sensor 1 (KS1) signal was too high. **Possible Causes:** • Knock sensor circuit is open • Knock sensor is loose (tighten to 20 NM) • Contact between the knock sensor and cylinder block is dirty, corroded or greasy • Knock sensor circuit is shorted to ground, or shorted to power • Knock sensor is damaged or it has failed • Wrong kind of fuel used • A component in the engine compartment is loose or not properly secured
DTC: P1332 **2T MIL: Yes** **2001, 2002, 2003, 2005, 2006** **Models:** 325i, 330i, 525i, 530i, 545i, 550i, 745i, 750i, Z4 **Engines:** 3.0L, 4.4L, 4.8L, 5.4L **Transmissions:** All	**Knock Sensor 4 Signal Low Input Conditions:** **Engine started, vehicle driven at 2000rpm for 3 seconds or to a temperature of 40 degrees Celsius, and the DME detected the Knock Sensor 1 (KS1) signal was too low or not recognized by the DME.** **Possible Causes:** • Knock sensor circuit is open • Knock sensor is loose (tighten to 20 NM) • Contact between the knock sensor and cylinder block is dirty, corroded or greasy • Knock sensor circuit is shorted to ground, or shorted to power • Knock sensor is damaged or it has failed • Wrong kind of fuel used • A component in the engine compartment is loose or not properly secured
DTC: P1333 **2T MIL: Yes** **2001, 2002, 2003, 2005, 2006** **Models:** 325i, 330i, 525i, 530i, 545i, 550i, 745i, 750i, Z4 **Engines:** 3.0L, 4.4L, 4.8L, 5.4L **Transmissions:** All	**Knock Sensor 4 Signal High Input Conditions:** Engine started, vehicle driven at 1600rpm for 3 seconds or to a temperature of 40 degrees Celsius, and the DME detected the Knock Sensor 1 (KS1) signal was too high. **Possible Causes:** • Knock sensor circuit is open • Knock sensor is loose (tighten to 20 NM) • Contact between the knock sensor and cylinder block is dirty, corroded or greasy • Knock sensor circuit is shorted to ground, or shorted to power • Knock sensor is damaged or it has failed • Wrong kind of fuel used • A component in the engine compartment is loose or not properly secured
DTC: P1338 **2T MIL: Yes** 2004 **Models:** 320i, 325i, 330i, 525i, 530i, 545i, 745i, M3, Z4 **Engines:** 2.5L, 3.2L **Transmissions:** All	**Camshaft Position Sensor (Bank 1) Open/Short to B+ Conditions:** Engine started, battery voltage at least 11.5v, all electrical components off, ground connections between engine and chassis well connected, and the DME detected an unexpected low or high voltage condition on the camshaft position sensor circuit. **Possible Causes:** • Faulty CPM sensor

DTC	Trouble Code Title, Conditions & Possible Causes
DTC: P1339 **2T MIL: Yes** 2004 **Models:** 320i, 325i, 330i, 525i, 530i, 545i, 745i, M3, Z4 **Engines:** 2.5L, 3.2L **Transmissions:** All	**Camshaft Position Sensor (Bank 1) Open/Short to B+ Conditions:** Engine started, battery voltage at least 11.5v, all electrical components off, ground connections between engine and chassis well connected, and the DME detected an unexpected low or high voltage condition on the camshaft position sensor circuit. **Possible Causes:** • Faulty CPM sensor
DTC: P1340 **2T MIL: Yes** **1999, 2000, 2001, 2002** **Models:** 325i, 330i, 525i, 530i, 540i, 550i, 740i, M3, M5, Z3, Z8 **Engines:** 3.0L, 4.4L, 4.8L, 5.4L **Transmissions:** All	**Crankshaft Position/Camshaft Sensor Signal Out of Sequence Conditions:** Engine started, battery voltage at least 11.5v, all electrical components off, ground connections between engine and chassis well connected, and the DME detected the crankshaft position sensor and the camshaft sensor were out of sequence with each other. **Note: The Engine Speed (RPM) Sensor detects engine speed and reference marks. Without an engine speed signal, the engine will not start. If the engine speed signal fails while the engine is running, the engine will stop immediately.** **Possible Causes:** • Engine speed sensor has failed or is contaminated (metal filings) • Engine speed sensor's wheel is damaged • Engine speed sensor circuit is shorted to the cable shield • Engine speed sensor circuit is open • DME is faulty • Camshaft position sensor is faulty
DTC: P1342 **1T MISFIRE, MIL: Yes** **2001, 2002, 2003, 2005, 2006** **Models:** 325i, 330i, 525i, 530i, 545i, 550i, 745i, 750i, Z4 **Engines:** 3.0L, 4.4L, 4.8L, 5.4L **Transmissions:** All	**Random/Multiple Misfire Detected Conditions:** Engine running at an RPM greater than 400 but less than 6400 the DME detected a misfire or uneven engine running in two or more cylinders within 200 crankshaft rotations. Engine speed is between 480 and 4500rpm, load change is 0.4ms at ignition with a speed change of 2800rpms and the ASC is not active. **Note: If the misfire is severe, the MIL will flash on/off on the first trip!** **Possible Causes:** • Fuel metering fault that affects two or more cylinders • Fuel pressure too low or too high, fuel supply contaminated • EVAP system problem or the EVAP canister is fuel saturated • EGR valve is stuck open or the PCV system has a vacuum leak • Ignition system fault (coil, plugs) affecting two or more cylinders • MAF sensor contamination (it can cause a very lean condition) • Vehicle driven while very low on fuel (less than 1/8 of a tank)
DTC: P1344 **1T MISFIRE, MIL: Yes** **2001, 2002, 2003, 2005, 2006** **Models:** 325i, 330i, 525i, 530i, 545i, 550i, 745i, 750i, Z4 **Engines:** 3.0L, 4.4L, 4.8L, 5.4L **Transmissions:** All	**Random/Multiple Misfire Detected Conditions:** Engine running at an RPM greater than 400 but less than 6400 the DME detected a misfire or uneven engine running in two or more cylinders within 200 crankshaft rotations. Engine speed is between 480 and 4500rpm, load change is 0.4ms at ignition with a speed change of 2800rpms and the ASC is not active. **Note: If the misfire is severe, the MIL will flash on/off on the first trip!** **Possible Causes:** • Fuel metering fault that affects two or more cylinders • Fuel pressure too low or too high, fuel supply contaminated • EVAP system problem or the EVAP canister is fuel saturated • EGR valve is stuck open or the PCV system has a vacuum leak • Ignition system fault (coil, plugs) affecting two or more cylinders • MAF sensor contamination (it can cause a very lean condition) • Vehicle driven while very low on fuel (less than 1/8 of a tank)
DTC: P1346 **1T MISFIRE, MIL: Yes** **2001, 2002, 2003, 2005, 2006** **Models:** 325i, 330i, 525i, 530i, 545i, 550i, 745i, 750i, Z4 **Engines:** 3.0L, 4.4L, 4.8L, 5.4L **Transmissions:** All	**Random/Multiple Misfire Detected Conditions:** Engine running at an RPM greater than 400 but less than 6400 the DME detected a misfire or uneven engine running in two or more cylinders within 200 crankshaft rotations. Engine speed is between 480 and 4500rpm, load change is 0.4ms at ignition with a speed change of 2800rpms and the ASC is not active. **Note: If the misfire is severe, the MIL will flash on/off on the first trip!** **Possible Causes:** • Fuel metering fault that affects two or more cylinders • Fuel pressure too low or too high, fuel supply contaminated • EVAP system problem or the EVAP canister is fuel saturated • EGR valve is stuck open or the PCV system has a vacuum leak • Ignition system fault (coil, plugs) affecting two or more cylinders • MAF sensor contamination (it can cause a very lean condition) • Vehicle driven while very low on fuel (less than 1/8 of a tank)

DTC	Trouble Code Title, Conditions & Possible Causes
DTC: P1348 **1T MISFIRE, MIL: Yes** **2001, 2002, 2003, 2005, 2006** **Models:** 325i, 330i, 525i, 530i, 545i, 550i, 745i, 750i, Z4 **Engines:** 3.0L, 4.4L, 4.8L, 5.4L **Transmissions:** All	**Random/Multiple Misfire Detected Conditions:** Engine running at an RPM greater than 400 but less than 6400 the DME detected a misfire or uneven engine running in two or more cylinders within 200 crankshaft rotations. Engine speed is between 480 and 4500rpm, load change is 0.4ms at ignition with a speed change of 2800rpms and the ASC is not active. **Note: If the misfire is severe, the MIL will flash on/off on the first trip!** **Possible Causes:** • Fuel metering fault that affects two or more cylinders • Fuel pressure too low or too high, fuel supply contaminated • EVAP system problem or the EVAP canister is fuel saturated • EGR valve is stuck open or the PCV system has a vacuum leak • Ignition system fault (coil, plugs) affecting two or more cylinders • MAF sensor contamination (it can cause a very lean condition) • Vehicle driven while very low on fuel (less than 1/8 of a tank)
DTC: P1350 **1T MISFIRE, MIL: Yes** **2001, 2002, 2003, 2005, 2006** **Models:** 325i, 330i, 525i, 530i, 545i, 550i, 745i, 750i, Z4 **Engines:** 3.0L, 4.4L, 4.8L, 5.4L **Transmissions:** All	**Random/Multiple Misfire Detected Conditions:** Engine running at an RPM greater than 400 but less than 6400 the DME detected a misfire or uneven engine running in two or more cylinders within 200 crankshaft rotations. Engine speed is between 480 and 4500rpm, load change is 0.4ms at ignition with a speed change of 2800rpms and the ASC is not active. **Note: If the misfire is severe, the MIL will flash on/off on the first trip!** **Possible Causes:** • Fuel metering fault that affects two or more cylinders • Fuel pressure too low or too high, fuel supply contaminated • EVAP system problem or the EVAP canister is fuel saturated • EGR valve is stuck open or the PCV system has a vacuum leak • Ignition system fault (coil, plugs) affecting two or more cylinders • MAF sensor contamination (it can cause a very lean condition) • Vehicle driven while very low on fuel (less than 1/8 of a tank)
DTC: P1352 **1T MISFIRE, MIL: Yes** **2001, 2002, 2003** **Models:** All **Engines:** All **Transmissions:** All	**Random/Multiple Misfire Detected Conditions:** Engine running at an RPM greater than 400 but less than 6400 the DME detected a misfire or uneven engine running in two or more cylinders within 200 crankshaft rotations. Engine speed is between 480 and 4500rpm, load change is 0.4ms at ignition with a speed change of 2800rpms and the ASC is not active. **Note: If the misfire is severe, the MIL will flash on/off on the first trip!** **Possible Causes:** • Fuel metering fault that affects two or more cylinders • Fuel pressure too low or too high, fuel supply contaminated • EVAP system problem or the EVAP canister is fuel saturated • EGR valve is stuck open or the PCV system has a vacuum leak • Ignition system fault (coil, plugs) affecting two or more cylinders • MAF sensor contamination (it can cause a very lean condition) • Vehicle driven while very low on fuel (less than 1/8 of a tank)
DTC: P1354 **1T MISFIRE, MIL: Yes** **2001, 2002, 2003** **Models:** 540i, 550i, 740i, 745i, M3, M5, Z8 **Engines:** 3.2L, 4.4L, 4.8L, 5.0L, 5.4L **Transmissions:** All	**Random/Multiple Misfire Detected Conditions:** Engine running at an RPM greater than 400 but less than 6400 the DME detected a misfire or uneven engine running in two or more cylinders within 200 crankshaft rotations. Engine speed is between 480 and 4500rpm, load change is 0.4ms at ignition with a speed change of 2800rpms and the ASC is not active. **Note: If the misfire is severe, the MIL will flash on/off on the first trip!** **Possible Causes:** • Fuel metering fault that affects two or more cylinders • Fuel pressure too low or too high, fuel supply contaminated • EVAP system problem or the EVAP canister is fuel saturated • EGR valve is stuck open or the PCV system has a vacuum leak • Ignition system fault (coil, plugs) affecting two or more cylinders • MAF sensor contamination (it can cause a very lean condition) • Vehicle driven while very low on fuel (less than 1/8 of a tank)

DTC	Trouble Code Title, Conditions & Possible Causes
DTC: P1355 **1T MISFIRE, MIL: Yes** **2001, 2002, 2003** **Models:** 540i, 550i, 740i, 745i, M3, M5, Z8 **Engines:** 3.2L, 4.4L, 4.8L, 5.0L, 5.4L **Transmissions:** All	**Random/Multiple Misfire Detected Conditions:** Engine running at an RPM greater than 400 but less than 6400 the DME detected a misfire or uneven engine running in two or more cylinders within 200 crankshaft rotations. Engine speed is between 480 and 4500rpm, load change is 0.4ms at ignition with a speed change of 2800rpms and the ASC is not active. **Note: If the misfire is severe, the MIL will flash on/off on the first trip!** **Possible Causes:** • Fuel metering fault that affects two or more cylinders • Fuel pressure too low or too high, fuel supply contaminated • EVAP system problem or the EVAP canister is fuel saturated • EGR valve is stuck open or the PCV system has a vacuum leak • Ignition system fault (coil, plugs) affecting two or more cylinders • MAF sensor contamination (it can cause a very lean condition) • Vehicle driven while very low on fuel (less than 1/8 of a tank)
DTC: P1356 **1T MISFIRE, MIL: Yes** **2001, 2002, 2003** **Models:** 540i, 550i, 740i, 745i, M3, M5, Z8 **Engines:** 3.2L, 4.4L, 4.8L, 5.0L, 5.4L **Transmissions:** All	**Random/Multiple Misfire Detected Conditions:** Engine running at an RPM greater than 400 but less than 6400 the DME detected a misfire or uneven engine running in two or more cylinders within 200 crankshaft rotations. Engine speed is between 480 and 4500rpm, load change is 0.4ms at ignition with a speed change of 2800rpms and the ASC is not active. **Note: If the misfire is severe, the MIL will flash on/off on the first trip!** **Possible Causes:** • Fuel metering fault that affects two or more cylinders • Fuel pressure too low or too high, fuel supply contaminated • EVAP system problem or the EVAP canister is fuel saturated • EGR valve is stuck open or the PCV system has a vacuum leak • Ignition system fault (coil, plugs) affecting two or more cylinders • MAF sensor contamination (it can cause a very lean condition) • Vehicle driven while very low on fuel (less than 1/8 of a tank)
DTC: P1357 **1T MISFIRE, MIL: Yes** **2001, 2002, 2003** **Models:** 540i, 550i, 740i, 745i, M3, M5, Z8 **Engines:** 3.2L, 4.4L, 4.8L, 5.0L, 5.4L **Transmissions:** All	**Random/Multiple Misfire Detected Conditions:** Engine running at an RPM greater than 400 but less than 6400 the DME detected a misfire or uneven engine running in two or more cylinders within 200 crankshaft rotations. Engine speed is between 480 and 4500rpm, load change is 0.4ms at ignition with a speed change of 2800rpms and the ASC is not active. **Note: If the misfire is severe, the MIL will flash on/off on the first trip!** **Possible Causes:** • Fuel metering fault that affects two or more cylinders • Fuel pressure too low or too high, fuel supply contaminated • EVAP system problem or the EVAP canister is fuel saturated • EGR valve is stuck open or the PCV system has a vacuum leak • Ignition system fault (coil, plugs) affecting two or more cylinders • MAF sensor contamination (it can cause a very lean condition) • Vehicle driven while very low on fuel (less than 1/8 of a tank)
DTC: P1377 **MIL: Yes** **2006** **Models:** 325i, 330i, 525i, 530i, 550i, 750i, Z4 **Engines:** 3.0L, 4.8L **Transmissions:** All	**Camshaft Position Sensor Master Camshaft Not Defined Conditions:** Engine started, battery voltage must be at least 11.5v, after five camshaft revolutions, the DME detected the CMP sensor signal was implausible. Perhaps a defect in the power supply. Reduced power occurs and once the engine is turned off, it is impossible to restart it. **Possible Causes:** • CMP sensor circuit is open or shorted to ground • CMP sensor circuit is shorted to power • CMP sensor ground (return) circuit is open • CMP sensor installation incorrect (Hall-effect type) • CMP sensor is damaged or CMP sensor shielding damaged
DTC: P1378 **MIL: Yes** 2005 **Models:** 545i, 745i **Engines:** 4.8L **Transmissions:** All	**Control Module Self-Test, Knock Control Offset (Bank 1) Conditions:** Engine started, vehicle driven at 1520rpm for 10 seconds or to a temperature of 40 degrees Celsius, and the DME detected the Knock chip is defective. **Possible Causes:** • Knock sensor circuit is open • Knock sensor is loose (tighten to 20 NM) • Contact between the knock sensor and cylinder block is dirty, corroded or greasy • Knock sensor circuit is shorted to ground, or shorted to power • Knock sensor is damaged or it has failed • Wrong kind of fuel used • A component in the engine compartment is loose or not properly secured

DTC	Trouble Code Title, Conditions & Possible Causes
DTC: P1381 **MIL: Yes** **2002, 2003, 2005, 2006** **Models:** 530i, 540i, 550i, 745i, Z4 **Engines:** 3.0L, 4.4L, 4.8L **Transmissions:** All	**Control Module Self-Test, Knock Control Offset (Bank 1) Conditions:** Engine started, vehicle driven at 1520rpm for 10 seconds or to a temperature of 40 degrees Celsius, and the DME detected the Knock chip is defective. **Possible Causes:** • Knock sensor circuit is open • Knock sensor is loose (tighten to 20 NM) • Contact between the knock sensor and cylinder block is dirty, corroded or greasy • Knock sensor circuit is shorted to ground, or shorted to power • Knock sensor is damaged or it has failed • Wrong kind of fuel used • A component in the engine compartment is loose or not properly secured
DTC: P1382 **MIL: Yes** **2002, 2003, 2005, 2006** **Models:** 530i, 540i, 550i, 745i, Z4 **Engines:** 3.0L, 4.4L, 4.8L **Transmissions:** All	**Control Module Self-Test, Knock Control Test Pulse (Bank 1) Conditions:** Engine started, vehicle driven at 1520rpm for 10 seconds or to a temperature of 40 degrees Celsius, and the DME detected the Knock chip is defective. **Possible Causes:** • Knock sensor circuit is open • Knock sensor is loose (tighten to 20 NM) • Contact between the knock sensor and cylinder block is dirty, corroded or greasy • Knock sensor circuit is shorted to ground, or shorted to power • Knock sensor is damaged or it has failed • Wrong kind of fuel used • A component in the engine compartment is loose or not properly secured
DTC: P1383 **1T MIL: Yes** **1997, 1998** **Models:** 750i, 850Ci **Engines:** 5.4L **Transmissions:** All	**Secondary Ignition Circuit Range Check Voltage Malfunction Conditions:** Engine started with a threshold value voltage of less than 20 or greater than 40, the DME detected the circuit range malfunction at 240 Ohms. **Possible Causes:** • Ignition circuit has short or is open • DME is defective
DTC: P1384 **2T MIL: Yes** **1996, 1997, 1998, 1999, 2000, 2006** **Models:** 325i, 330i, 525i, 530i, 540i, 550i, 740i, 750i, Z4 **Engines:** 3.0L, 4.4L, 5.4L **Transmissions:** All	**Knock Sensor 3 Circuit Malfunction Conditions:** Engine started, vehicle driven at 1520rpm for 3 seconds or to a temperature of 40 degrees Celsius, and the DME detected the Knock Sensor 1 (KS1) signal was not recognized. The engine speed is greater than 2080rpm but less than 6000rpm and the coolant temperature is greater than 40.5 degrees Celsius. **Possible Causes:** • Knock sensor circuit is open • Knock sensor is loose (tighten to 20 NM) • Contact between the knock sensor and cylinder block is dirty, corroded or greasy • Knock sensor circuit is shorted to ground, or shorted to power • Knock sensor is damaged or it has failed • Wrong kind of fuel used • A component in the engine compartment is loose or not properly secured
DTC: P1385 **2T MIL: Yes** **1996, 1997, 1998, 1999, 2000, 2006** **Models:** 325i, 330i, 525i, 530i, 540i, 550i, 740i, 750i, Z4 **Engines:** 3.0L, 4.4L, 5.4L **Transmissions:** All	**Knock Sensor 4 Circuit Malfunction Conditions:** Engine started, vehicle driven at 1520rpm for 3 seconds or to a temperature of 40 degrees Celsius, and the DME detected the Knock Sensor 1 (KS1) signal was not recognized. The engine speed is greater than 2080rpm but less than 6000rpm and the coolant temperature is greater than 40.5 degrees Celsius. **Possible Causes:** • Knock sensor circuit is open • Knock sensor is loose (tighten to 20 NM) • Contact between the knock sensor and cylinder block is dirty, corroded or greasy • Knock sensor circuit is shorted to ground, or shorted to power • Knock sensor is damaged or it has failed • Wrong kind of fuel used • A component in the engine compartment is loose or not properly secured

DTC	Trouble Code Title, Conditions & Possible Causes
DTC: P1386 **MIL: Yes** **1999, 2000, 2002, 2003, 2005, 2006** **Models:** 325i, 330i, 525i, 530i, 540i, 550i, 740i, 745i, 750i, Z4 **Engines:** 3.0L, 4.4L, 4.8L, 5.4L **Transmissions:** All	**Control Module Self-Test, Knock Control Circuit Baseline Test (Bank 1) Conditions:** Engine started, vehicle driven at 1520rpm for 10 seconds or to a temperature of 40 degrees Celsius, and the DME detected the Knock chip is defective. The engine speed is greater than 2080rpm but less than 6000rpm and the coolant temperature is greater than 40.5 degrees Celsius. **Possible Causes:** • Knock sensor circuit is open • Knock sensor is loose (tighten to 20 NM) • Contact between the knock sensor and cylinder block is dirty, corroded or greasy • Knock sensor circuit is shorted to ground, or shorted to power • Knock sensor is damaged or it has failed • Wrong kind of fuel used • A component in the engine compartment is loose or not properly secured
DTC: P1396 **2T MIL: Yes** **1996, 1997, 1998, 1999, 2000, 2001, 2002, 2003, 2004, 2005, 2006** **Models:** All **Engines:** All **Transmissions:** All	**Camshaft Position Sensor "A" Circ Malfunction Conditions:** Engine started, battery voltage must be at least 11.5v, all electrical components must be off, parking brake must be engaged (to keep daytime driving lights off), automatic transmission selector must be in park and the ground between the engine and the chassis must be well connected. The DME detected the CMP sensor signal was implausible. Engine speed is greater than 500rpm, and the fault is tolerable as long as there are no misfired occurring at the same time. **Possible Causes:** • CMP sensor circuit is open or shorted to ground • CMP sensor circuit is shorted to power • CMP sensor ground (return) circuit is open • CMP sensor installation incorrect (Hall-effect type) • CMP sensor is damaged or CMP sensor shielding damaged
DTC: P1397 **2T MIL: Yes** **1999, 2000** **Models:** 323i, 328i, 528i, 530i, Z3 **Engines:** 2.5L, 2.8L **Transmissions:** All	**Camshaft Position Sensor Circ Malfunction Conditions:** Engine started, battery voltage must be at least 11.5v, all electrical components must be off, parking brake must be engaged (to keep daytime driving lights off), automatic transmission selector must be in park and the ground between the engine and the chassis must be well connected. The DME detected the CMP sensor signal was implausible. Engine speed is greater than 500rpm, and the fault is tolerable as long as there are no misfired occurring at the same time. **Possible Causes:** • CMP sensor circuit is open or shorted to ground • CMP sensor circuit is shorted to power • CMP sensor ground (return) circuit is open • CMP sensor installation incorrect (Hall-effect type) • CMP sensor is damaged or CMP sensor shielding damaged
DTC: P1400 **2T MIL: Yes** **2001** **Models:** 750i **Engines:** 5.4L **Transmissions:** All	**Heated Catalyst System Minimum Battery Voltage Conditions:** Engine started, battery voltage must be at least 11.5v, all electrical components must be off, parking brake must be engaged (to keep daytime driving lights off), automatic transmission selector must be in park, the exhaust system must be properly sealed between the catalytic converter and the cylinder head, coolant temperature must be at least 80 degrees Celsius and oxygen sensor heaters for oxygen sensors before the catalytic converter must be functioning properly and the ground between the engine and the chassis must be well connected. The coolant temperature is less than 90 degrees Celsius, catalyst temperature is less than 300 degrees Celsius, engine crank time less than five seconds, vehicle speed is less than three mph and the engine speed is less than 200rpm. **Possible Causes:** • Air leaks at the exhaust manifold or in the exhaust pipes • Catalytic converter is damaged, contaminated or it has failed • ECT/CHT sensor has lost its calibration (the signal is incorrect) • Engine cylinders misfiring, or the ignition timing is over retarded • Engine oil is contaminated • Front HO2S or rear HO2S is contaminated with fuel or moisture • Front HO2S and/or the rear HO2S is loose in the mounting hole • Front HO2S much older than the rear HO2S (HO2S-11 is lazy) • Fuel system pressure is too high (check the pressure regulator) • Rear HO2S wires improperly connected or the HO2S has failed

DTC	Trouble Code Title, Conditions & Possible Causes
DTC: P1401 **2T MIL: Yes** **2001** **Models:** 750i **Engines:** 5.4L **Transmissions:** All	**Heated Catalyst System Minimum Battery Voltage Conditions:** Engine started, battery voltage must be at least 11.5v, all electrical components must be off, parking brake must be engaged (to keep daytime driving lights off), automatic transmission selector must be in park, the exhaust system must be properly sealed between the catalytic converter and the cylinder head, coolant temperature must be at least 80 degrees Celsius and oxygen sensor heaters for oxygen sensors before the catalytic converter must be functioning properly and the ground between the engine and the chassis must be well connected. The coolant temperature is less than 90 degrees Celsius, catalyst temperature is less than 300 degrees Celsius, engine crank time less than five seconds, vehicle speed is less than three mph and the engine speed is less than 200rpm. **Possible Causes:** • Air leaks at the exhaust manifold or in the exhaust pipes • Catalytic converter is damaged, contaminated or it has failed • ECT/CHT sensor has lost its calibration (the signal is incorrect) • Engine cylinders misfiring, or the ignition timing is over retarded • Engine oil is contaminated • Front HO2S or rear HO2S is contaminated with fuel or moisture • Front HO2S and/or the rear HO2S is loose in the mounting hole • Front HO2S much older than the rear HO2S (HO2S-11 is lazy) • Fuel system pressure is too high (check the pressure regulator) • Rear HO2S wires improperly connected or the HO2S has failed
DTC: P1403 **2T MIL: Yes** **2001** **Models:** 750i **Engines:** 5.4L **Transmissions:** All	**Heated Catalyst System Minimum Battery Voltage Conditions:** Engine started, battery voltage must be at least 11.5v, all electrical components must be off, parking brake must be engaged (to keep daytime driving lights off), automatic transmission selector must be in park, the exhaust system must be properly sealed between the catalytic converter and the cylinder head, coolant temperature must be at least 80 degrees Celsius and oxygen sensor heaters for oxygen sensors before the catalytic converter must be functioning properly and the ground between the engine and the chassis must be well connected. The coolant temperature is less than 90 degrees Celsius, catalyst temperature is less than 300 degrees Celsius, engine crank time less than five seconds, vehicle speed is less than three mph and the engine speed is less than 200rpm. **Possible Causes:** • Air leaks at the exhaust manifold or in the exhaust pipes • Catalytic converter is damaged, contaminated or it has failed • ECT/CHT sensor has lost its calibration (the signal is incorrect) • Engine cylinders misfiring, or the ignition timing is over retarded • Engine oil is contaminated • Front HO2S or rear HO2S is contaminated with fuel or moisture • Front HO2S and/or the rear HO2S is loose in the mounting hole • Front HO2S much older than the rear HO2S (HO2S-11 is lazy) • Fuel system pressure is too high (check the pressure regulator) • Rear HO2S wires improperly connected or the HO2S has failed
DTC: P1403 **2T MIL: Yes** **1996, 1997, 1998, 1999, 2000, 2006** **Models:** 318i, 323i, 328i, 528i, M3, Z3 **Engines:** 1.9L, 2.8L, 3.2L **Transmissions:** All	**Evaporative Emission System Shut Off Valve Conditions:** Engine started, battery voltage must be at least 11.5v, all electrical components must be off, parking brake must be engaged (to keep daytime driving lights off), automatic transmission selector must be in park, the exhaust system must be properly sealed between the catalytic converter and the cylinder head, coolant temperature must be at least 80 degrees Celsius and oxygen sensor heaters for oxygen sensors before the catalytic converter must be functioning properly and the ground between the engine and the chassis must be well connected. The DME detected an unexpected condition on the EVAP shut off valve when the device was cycled On/Off during testing. **Possible Causes:** • EVAP power supply circuit is open • EVAP solenoid control circuit is open or shorted to ground • EVAP solenoid control circuit is shorted to power (B+) • EVAP solenoid valve is damaged or it has failed • EVAP canister has a leak or a poor seal

DTC	Trouble Code Title, Conditions & Possible Causes
DTC: P1404 **2T MIL: Yes** **2001** **Models:** 750i **Engines:** 5.4L **Transmissions:** All	**Heated Catalyst System Minimum Battery Voltage Conditions:** Engine started, battery voltage must be at least 11.5v, all electrical components must be off, parking brake must be engaged (to keep daytime driving lights off), automatic transmission selector must be in park, the exhaust system must be properly sealed between the catalytic converter and the cylinder head, coolant temperature must be at least 80 degrees Celsius and oxygen sensor heaters for oxygen sensors before the catalytic converter must be functioning properly and the ground between the engine and the chassis must be well connected. The coolant temperature is less than 90 degrees Celsius, catalyst temperature is less than 300 degrees Celsius, engine crank time less than five seconds, vehicle speed is less than three mph and the engine speed is less than 200rpm. **Possible Causes:** • Air leaks at the exhaust manifold or in the exhaust pipes • Catalytic converter is damaged, contaminated or it has failed • ECT/CHT sensor has lost its calibration (the signal is incorrect) • Engine cylinders misfiring, or the ignition timing is over retarded • Engine oil is contaminated • Front HO2S or rear HO2S is contaminated with fuel or moisture • Front HO2S and/or the rear HO2S is loose in the mounting hole • Front HO2S much older than the rear HO2S (HO2S-11 is lazy) • Fuel system pressure is too high (check the pressure regulator) • Rear HO2S wires improperly connected or the HO2S has failed
DTC: P1411 **1996, 1997, 1998, 1999, 2000, 2005** **Models:** 325i, 330i, 525i, 530i, 540i, 550i, 740i, 750i, Z4 **Engines:** 3.0L, 4.4L, 5.4L **Transmissions:** All	**Secondary Air Pump Valve Plausibility Conditions:** The Engine Control Module detects an implausible signal when activating the secondary air injection solenoid valve. **Note: Solenoid valve is closed when no voltage is present.** **Possible Causes:** • Connector to the secondary air injection valve is loose or disconnected • Secondary air injector valve circuit short • Secondary air injector valve circuit is open • Faulty secondary air injector valve
DTC: P1412 2004, 2005 **Models:** 320i, 325i, 330i, 525i, 530i, 545i, 745i, M3 **Engines:** 3.0L, 3.2L **Transmissions:** All	**Secondary Air Pump Valve Plausibility Conditions:** The Engine Control Module detects an implausible signal when activating the secondary air injection solenoid valve. The max flow limit check detected leakage between the air pump and the valve. **Note: Solenoid valve is closed when no voltage is present.** **Possible Causes:** • Connector to the secondary air injection valve is loose or disconnected • Secondary air injector valve circuit short • Secondary air injector valve circuit is open • Faulty secondary air injector valve
DTC: P1413 **2001, 2002, 2003, 2004, 2005, 2006** **Models:** 320i, 325i, 330i, 525i, 530i, 545i, 550i, 745i, 750i, M3, Z3, Z4 **Engines:** 3.0L, 3.2L, 4.8L **Transmissions:** All	**Secondary Air Injector Pump Relay Control Circuit Signal Low Conditions:** The Engine Control Module detects a short circuit when activating the secondary air injection solenoid valve. **Note: Solenoid valve is closed when no voltage is present.** **Possible Causes:** • Connector to the secondary air injection valve is loose or disconnected • Secondary air injector valve circuit short • Secondary air injector valve circuit is open • Faulty secondary air injector valve
DTC: P1414 **2001, 2002, 2003, 2004, 2005, 2006** **Models:** 320i, 325i, 330i, 525i, 530i, 545i, 550i, 745i, 750i, M3, Z3, Z4 **Engines:** 3.0L, 3.2L, 4.8L **Transmissions:** All	**Secondary Air Injector Pump Relay Control Circuit Signal High Conditions:** The Engine Control Module detects a short circuit when activating the secondary air injection solenoid valve. **Note: Solenoid valve is closed when no voltage is present.** **Possible Causes:** • Connector to the secondary air injection valve is loose or disconnected • Secondary air injector valve circuit short • Secondary air injector valve circuit is open • Faulty secondary air injector valve
DTC: P1418 2004, 2005 **Models:** 320i, 325i, 330i, 525i, 530i, 545i, 745i, M3 **Engines:** 3.0L, 3.2L **Transmissions:** All	**Secondary Air Pump Valve Plausibility Conditions:** The Engine Control Module detects an implausible signal when activating the secondary air injection solenoid valve. The secondary air valve or tube is blocked. **Note: Solenoid valve is closed when no voltage is present.** **Possible Causes:** • Connector to the secondary air injection valve is loose or disconnected • Secondary air injector valve circuit short • Secondary air injector valve circuit is open • Faulty secondary air injector valve

DTC	Trouble Code Title, Conditions & Possible Causes
DTC: P1420 **2T MIL: Yes** **1999, 2000** **Models:** 540i, 740i, 750i **Engines:** 4.4L, 5.4L **Transmissions:** All	**Secondary Air Injector Valve Circuit Electrical Malfunction Conditions:** The Engine Control Module activates the secondary air injection solenoid valve, but the Heated Oxygen Sensor (HO2S) does not detect secondary air injection. **Note: Solenoid valve is closed when no voltage is present.** **Possible Causes:** • Connector to the secondary air injection valve is loose or disconnected • Secondary air injector valve circuit short • Secondary air injector valve circuit is open • Faulty secondary air injector valve
DTC: P1421 **2T MIL: Yes** **1996, 1997, 1998, 1999, 2000** **Models:** 325i, 328i, 330i, 525i, 528i, 540i, 740i, Z3 **Engines:** 2.8L, 3.0L, 4.4L, 5.4L **Transmissions:** All	**Secondary Air Injector Valve Circuit Short to Ground Conditions:** The Engine Control Module detects a short circuit to ground when activating the secondary air injection solenoid valve. The air mass is greater than 12 but less than 152 m kg/h, coolant temperature is greater than 3 but less than 75 degrees Celsius, intake temperature is greater than 3 but less than 50 degrees Celsius, the calculated air density is greater than 0.68, the O2 Sensor has heated up and the batter voltage is greater than 10.8 volts. **Note: Solenoid valve is closed when no voltage is present.** **Possible Causes:** • Connector to the secondary air injection valve is loose or disconnected • Secondary air injector valve circuit short • Secondary air injector valve circuit is open • Faulty secondary air injector valve
DTC: P1423 **2T MIL: Yes** **1996, 1997, 1998, 1999, 2000** **Models:** 325i, 328i, 330i, 525i, 528i, 540i, 740i, Z3 **Engines:** 2.8L, 3.0L, 4.4L, 5.4L **Transmissions:** All	**Secondary Air Pump Valve Plausibility Conditions:** The Engine Control Module detects an implausible signal when activating the secondary air injection solenoid valve. The air mass is greater than 12 but less than 152 m kg/h, coolant temperature is greater than 3 but less than 75 degrees Celsius, intake temperature is greater than 3 but less than 50 degrees Celsius, the calculated air density is greater than 0.68, the O2 Sensor has heated up and the batter voltage is greater than 10.8 volts. **Note: Solenoid valve is closed when no voltage is present.** **Possible Causes:** • Connector to the secondary air injection valve is loose or disconnected • Secondary air injector valve circuit short • Secondary air injector valve circuit is open • Faulty secondary air injector valve
DTC: P1432 **2T MIL: Yes** **1996, 1997, 1998, 1999, 2000** **Models:** 325i, 328i, 330i, 525i, 528i, 540i, 740i, Z3 **Engines:** 2.8L, 3.0L, 4.4L, 5.4L **Transmissions:** All	**Secondary Air Injection Valve Open Conditions:** The output Diagnostic Test Mode (DTM) can be activated only with the ignition switched on and the engine not running. The output DTM is interrupted if the engine is started, or if a rotary pulse from the ignition system is recognized. **Possible Causes:** • Fuel pump relays have failed • Fuel injector has failed • Hoses on the EVAP canister may be clogged • EVAP canister purge regulator valve may be faulty • DME may have failed • Manifold Tuning Valve (IMT) may have failed
DTC: P1453 **2T MIL: Yes** **1996, 1997, 1998, 1999, 2000** **Models:** 325i, 328i, 330i, 525i, 528i, 540i, 740i, Z3 **Engines:** 2.8L, 3.0L, 4.4L, 5.4L **Transmissions:** All	**Secondary Air Injector Valve Circuit Electrical Malfunction (Disconnection) Conditions:** The Engine Control Module activates the secondary air injection solenoid valve, but the Heated Oxygen Sensor (HO2S) does not detect secondary air injection. **Note: Solenoid valve is closed when no voltage is present.** **Possible Causes:** • Connector to the secondary air injection valve is loose or disconnected • Secondary air injector valve circuit short • Secondary air injector valve circuit is open • Faulty secondary air injector valve
DTC: P1454 **2T MIL: Yes** **1999, 2000, 2001** **Models:** 750i **Engines:** 5.4L **Transmissions:** All	**Secondary Air Injector Valve Circuit Electrical Malfunction (Open) Conditions:** The Engine Control Module activates the secondary air injection solenoid valve, but the Heated Oxygen Sensor (HO2S) does not detect secondary air injection. **Note: Solenoid valve is closed when no voltage is present.** **Possible Causes:** • Connector to the secondary air injection valve is loose or disconnected • Secondary air injector valve circuit short • Secondary air injector valve circuit is open • Faulty secondary air injector valve

DTC	Trouble Code Title, Conditions & Possible Causes
DTC: P1456 **2T MIL: Yes** **1999, 2000, 2001** **Models:** 750i **Engines:** 5.4L **Transmissions:** All	**Heated Catalyst System Efficiency Above Threshold Conditions:** Engine started, battery voltage must be at least 11.5v, all electrical components must be off, parking brake must be engaged (to keep daytime driving lights off), automatic transmission selector must be in park, the exhaust system must be properly sealed between the catalytic converter and the cylinder head, coolant temperature must be at least 80 degrees Celsius and oxygen sensor heaters for oxygen sensors before the catalytic converter must be functioning properly and the ground between the engine and the chassis must be well connected. The coolant temperature is less than 90 degrees Celsius, catalyst temperature is less than 300 degrees Celsius, engine crank time less than five seconds, vehicle speed is less than three mph and the engine speed is less than 200rpm. **Possible Causes:** • Air leaks at the exhaust manifold or in the exhaust pipes • Catalytic converter is damaged, contaminated or it has failed • ECT/CHT sensor has lost its calibration (the signal is incorrect) • Engine cylinders misfiring, or the ignition timing is over retarded • Engine oil is contaminated • Front HO2S or rear HO2S is contaminated with fuel or moisture • Front HO2S and/or the rear HO2S is loose in the mounting hole • Front HO2S much older than the rear HO2S (HO2S-11 is lazy) • Fuel system pressure is too high (check the pressure regulator) • Rear HO2S wires improperly connected or the HO2S has failed
DTC: P1459 **2T MIL: Yes** **1999, 2000, 2001** **Models:** 750i **Engines:** 5.4L **Transmissions:** All	**Heated Catalyst System Efficiency Below Threshold Conditions:** Engine started, battery voltage must be at least 11.5v, all electrical components must be off, parking brake must be engaged (to keep daytime driving lights off), automatic transmission selector must be in park, the exhaust system must be properly sealed between the catalytic converter and the cylinder head, coolant temperature must be at least 80 degrees Celsius and oxygen sensor heaters for oxygen sensors before the catalytic converter must be functioning properly and the ground between the engine and the chassis must be well connected. The coolant temperature is less than 90 degrees Celsius, catalyst temperature is less than 300 degrees Celsius, engine crank time less than five seconds, vehicle speed is less than three mph and the engine speed is less than 200rpm. **Possible Causes:** • Air leaks at the exhaust manifold or in the exhaust pipes • Catalytic converter is damaged, contaminated or it has failed • ECT/CHT sensor has lost its calibration (the signal is incorrect) • Engine cylinders misfiring, or the ignition timing is over retarded • Engine oil is contaminated • Front HO2S or rear HO2S is contaminated with fuel or moisture • Front HO2S and/or the rear HO2S is loose in the mounting hole • Front HO2S much older than the rear HO2S (HO2S-11 is lazy) • Fuel system pressure is too high (check the pressure regulator) • Rear HO2S wires improperly connected or the HO2S has failed
DTC: P1561 **2T MIL: Yes** **1999, 2000** **Models:** 750i **Engines:** 5.4L **Transmissions:** All	**Heated Catalyst System Self Rationality Check Conditions:** Engine started, battery voltage must be at least 11.5v, all electrical components must be off, parking brake must be engaged (to keep daytime driving lights off), automatic transmission selector must be in park, the exhaust system must be properly sealed between the catalytic converter and the cylinder head, coolant temperature must be at least 80 degrees Celsius and oxygen sensor heaters for oxygen sensors before the catalytic converter must be functioning properly and the ground between the engine and the chassis must be well connected. The coolant temperature is less than 90 degrees Celsius, catalyst temperature is less than 300 degrees Celsius, engine crank time less than five seconds, vehicle speed is less than three mph and the engine speed is less than 200rpm. The self/rationality check EEPROM resulted in an invalid checksum at power up. **Possible Causes:** • Air leaks at the exhaust manifold or in the exhaust pipes • Catalytic converter is damaged, contaminated or it has failed • ECT/CHT sensor has lost its calibration (the signal is incorrect) • Engine cylinders misfiring, or the ignition timing is over retarded • Engine oil is contaminated • Front HO2S or rear HO2S is contaminated with fuel or moisture • Front HO2S and/or the rear HO2S is loose in the mounting hole • Front HO2S much older than the rear HO2S (HO2S-11 is lazy) • Fuel system pressure is too high (check the pressure regulator) • Rear HO2S wires improperly connected or the HO2S has failed

DTC	Trouble Code Title, Conditions & Possible Causes
DTC: P1463 **2T MIL: Yes** **2001** **Models:** 750i **Engines:** 5.4L **Transmissions:** All	**Heated Catalyst System Rationality Check Conditions:** Engine started, battery voltage must be at least 11.5v, all electrical components must be off, parking brake must be engaged (to keep daytime driving lights off), automatic transmission selector must be in park, the exhaust system must be properly sealed between the catalytic converter and the cylinder head, coolant temperature must be at least 80 degrees Celsius and oxygen sensor heaters for oxygen sensors before the catalytic converter must be functioning properly and the ground between the engine and the chassis must be well connected. The coolant temperature is less than 90 degrees Celsius, catalyst temperature is less than 300 degrees Celsius, engine crank time less than five seconds, vehicle speed is less than three mph and the engine speed is less than 200rpm. **Possible Causes:** • Air leaks at the exhaust manifold or in the exhaust pipes • Catalytic converter is damaged, contaminated or it has failed • ECT/CHT sensor has lost its calibration (the signal is incorrect) • Engine cylinders misfiring, or the ignition timing is over retarded • Engine oil is contaminated • Front HO2S or rear HO2S is contaminated with fuel or moisture • Front HO2S and/or the rear HO2S is loose in the mounting hole • Front HO2S much older than the rear HO2S (HO2S-11 is lazy) • Fuel system pressure is too high (check the pressure regulator) • Rear HO2S wires improperly connected or the HO2S has failed
DTC: P1464 **2T MIL: Yes** **2001** **Models:** 750i **Engines:** 5.4L **Transmissions:** All	**Heated Catalyst System Rationality Check Conditions:** Engine started, battery voltage must be at least 11.5v, all electrical components must be off, parking brake must be engaged (to keep daytime driving lights off), automatic transmission selector must be in park, the exhaust system must be properly sealed between the catalytic converter and the cylinder head, coolant temperature must be at least 80 degrees Celsius and oxygen sensor heaters for oxygen sensors before the catalytic converter must be functioning properly and the ground between the engine and the chassis must be well connected. The coolant temperature is less than 90 degrees Celsius, catalyst temperature is less than 300 degrees Celsius, engine crank time less than five seconds, vehicle speed is less than three mph and the engine speed is less than 200rpm. **Possible Causes:** • Air leaks at the exhaust manifold or in the exhaust pipes • Catalytic converter is damaged, contaminated or it has failed • ECT/CHT sensor has lost its calibration (the signal is incorrect) • Engine cylinders misfiring, or the ignition timing is over retarded • Engine oil is contaminated • Front HO2S or rear HO2S is contaminated with fuel or moisture • Front HO2S and/or the rear HO2S is loose in the mounting hole • Front HO2S much older than the rear HO2S (HO2S-11 is lazy) • Fuel system pressure is too high (check the pressure regulator) • Rear HO2S wires improperly connected or the HO2S has failed
DTC: P1465 **2T MIL: Yes** 1996 **Models:** 750i **Engines:** 5.4L **Transmissions:** All	**Heated Catalyst System Rationality Check Conditions:** Engine started, battery voltage must be at least 11.5v, all electrical components must be off, parking brake must be engaged (to keep daytime driving lights off), automatic transmission selector must be in park, the exhaust system must be properly sealed between the catalytic converter and the cylinder head, coolant temperature must be at least 80 degrees Celsius and oxygen sensor heaters for oxygen sensors before the catalytic converter must be functioning properly and the ground between the engine and the chassis must be well connected. The coolant temperature is less than 90 degrees Celsius, catalyst temperature is less than 300 degrees Celsius, engine crank time less than five seconds, vehicle speed is less than three mph and the engine speed is less than 200rpm. **Possible Causes:** • Air leaks at the exhaust manifold or in the exhaust pipes • Catalytic converter is damaged, contaminated or it has failed • ECT/CHT sensor has lost its calibration (the signal is incorrect) • Engine cylinders misfiring, or the ignition timing is over retarded • Engine oil is contaminated • Front HO2S or rear HO2S is contaminated with fuel or moisture • Front HO2S and/or the rear HO2S is loose in the mounting hole • Front HO2S much older than the rear HO2S (HO2S-11 is lazy) • Fuel system pressure is too high (check the pressure regulator) • Rear HO2S wires improperly connected or the HO2S has failed

DTC	Trouble Code Title, Conditions & Possible Causes
DTC: P1466 **2T MIL: Yes** **1996, 2000, 2001** **Models:** 750i **Engines:** 5.4L **Transmissions:** All	**Heated Catalyst System Rationality Check Conditions:** Engine started, battery voltage must be at least 11.5v, all electrical components must be off, parking brake must be engaged (to keep daytime driving lights off), automatic transmission selector must be in park, the exhaust system must be properly sealed between the catalytic converter and the cylinder head, coolant temperature must be at least 80 degrees Celsius and oxygen sensor heaters for oxygen sensors before the catalytic converter must be functioning properly and the ground between the engine and the chassis must be well connected. The coolant temperature is less than 90 degrees Celsius, catalyst temperature is less than 300 degrees Celsius, engine crank time less than five seconds, vehicle speed is less than three mph and the engine speed is less than 200rpm. **Possible Causes:** • Air leaks at the exhaust manifold or in the exhaust pipes • Catalytic converter is damaged, contaminated or it has failed • ECT/CHT sensor has lost its calibration (the signal is incorrect) • Engine cylinders misfiring, or the ignition timing is over retarded • Engine oil is contaminated • Front HO2S or rear HO2S is contaminated with fuel or moisture • Front HO2S and/or the rear HO2S is loose in the mounting hole • Front HO2S much older than the rear HO2S (HO2S-11 is lazy) • Fuel system pressure is too high (check the pressure regulator) • Rear HO2S wires improperly connected or the HO2S has failed
DTC: P1467 **2T MIL: Yes** **2001** **Models:** 750i **Engines:** 5.4L **Transmissions:** All	**Heated Catalyst System Rationality Check Conditions:** Engine started, battery voltage must be at least 11.5v, all electrical components must be off, parking brake must be engaged (to keep daytime driving lights off), automatic transmission selector must be in park, the exhaust system must be properly sealed between the catalytic converter and the cylinder head, coolant temperature must be at least 80 degrees Celsius and oxygen sensor heaters for oxygen sensors before the catalytic converter must be functioning properly and the ground between the engine and the chassis must be well connected. The coolant temperature is less than 90 degrees Celsius, catalyst temperature is less than 300 degrees Celsius, engine crank time less than five seconds, vehicle speed is less than three mph and the engine speed is less than 200rpm. **Possible Causes:** • Air leaks at the exhaust manifold or in the exhaust pipes • Catalytic converter is damaged, contaminated or it has failed • ECT/CHT sensor has lost its calibration (the signal is incorrect) • Engine cylinders misfiring, or the ignition timing is over retarded • Engine oil is contaminated • Front HO2S or rear HO2S is contaminated with fuel or moisture • Front HO2S and/or the rear HO2S is loose in the mounting hole • Front HO2S much older than the rear HO2S (HO2S-11 is lazy) • Fuel system pressure is too high (check the pressure regulator) • Rear HO2S wires improperly connected or the HO2S has failed
DTC: P1470 **2T MIL: Yes** **1998, 1999, 2000** **Models:** 325i, 328i, 528i, 540i, 740i, 750i, Z3 **Engines:** 2.8L, 4.4L, 5.4L **Transmissions:** All	**EVAP Emission Control LDP Circuit Electrical Malfunction Conditions:** Key on, KOEO Self-Test enabled, and the DME detected an unexpected voltage condition on the EVAP emission control leak detection pump circuit. The engine speed is greater than 40rpm and the battery voltage is between 9.5 and 17 volts. **Possible Causes:** • EVAP canister system has an improper or broken seal • Evaporative Emission (EVAP) canister purge regulator valve 1 is faulty • Leak Detection Pump (LDP) is faulty
DTC: P1472 **2T MIL: Yes** **2000** **Models:** M3, Z8 **Engines:** 3.2L, 5.0L **Transmissions:** All	**EVAP Emission Control Leak Detection Pump Circuit Short to Ground Conditions:** Key on, KOEO Self-Test enabled, and the DME detected an unexpected voltage condition on the EVAP emission control leak detection pump circuit. **Possible Causes:** • Leak Detection Pump has failed • EVAP canister system has an improper or broken seal • Evaporative Emission (EVAP) canister purge regulator valve 1 is faulty • Hoses between the fuel pump and the EVAP canister are faulty • Fuel filler cap is loose • Fuel pump seal is defective, faulty or otherwise leaking • Hoses between the EVAP canister and the fuel flap unit are faulty • Hoses between the EVAP canister and the evaporative emission canister purge regulator valve are faulty

DTC	Trouble Code Title, Conditions & Possible Causes
DTC: P1473 **2T MIL: Yes** **2000** **Models:** M3, Z8 **Engines:** 3.2L, 5.0L **Transmissions:** All	**EVAP Emission Control Leak Detection Pump Circuit Open Conditions:** Key on, KOEO Self-Test enabled, and the DME detected an unexpected voltage condition on the EVAP emission control leak detection pump circuit. **Possible Causes:** • Leak Detection Pump has failed • EVAP canister system has an improper or broken seal • Evaporative Emission (EVAP) canister purge regulator valve 1 is faulty • Hoses between the fuel pump and the EVAP canister are faulty • Fuel filler cap is loose • Fuel pump seal is defective, faulty or otherwise leaking • Hoses between the EVAP canister and the fuel flap unit are faulty • Hoses between the EVAP canister and the evaporative emission canister purge regulator valve are faulty
DTC: P1475 **2T MIL: Yes** **1998, 1999, 2000** **Models:** 325i, 328i, 528i, 540i, 740i, 750i, Z3 **Engines:** 2.8L, 4.4L, 5.4L **Transmissions:** All	**EVAP Emission Control LDP Circuit Malfunction/Signal Circuit Open Conditions:** Key on, KOEO Self-Test enabled, and the DME detected an unexpected voltage condition on the EVAP emission control leak detection pump circuit. The engine speed is greater than 40rpm and the battery voltage is between 9.5 and 17 volts. **Possible Causes:** • Leak Detection Pump has failed • EVAP canister system has an improper or broken seal • Evaporative Emission (EVAP) canister purge regulator valve 1 is faulty • Hoses between the fuel pump and the EVAP canister are faulty • Fuel filler cap is loose • Fuel pump seal is defective, faulty or otherwise leaking • Hoses between the EVAP canister and the fuel flap unit are faulty • Hoses between the EVAP canister and the evaporative emission canister purge regulator valve are faulty
DTC: P1476 **2T MIL: Yes** **1998, 1999, 2000** **Models:** 325i, 328i, 528i, 540i, 740i, 750i, Z3 **Engines:** 2.8L, 4.4L, 5.4L **Transmissions:** All	**EVAP Emission Control LDP Circuit Malfunction/Insufficient Vacuum Conditions:** Key on, KOEO Self-Test enabled, and the DME detected an unexpected voltage condition on the EVAP emission control leak detection pump circuit. The engine speed is greater than 40rpm and the battery voltage is between 9.5 and 17 volts. **Possible Causes:** • Leak Detection Pump has failed • EVAP canister system has an improper or broken seal • Evaporative Emission (EVAP) canister purge regulator valve 1 is faulty • Hoses between the fuel pump and the EVAP canister are faulty • Fuel filler cap is loose • Fuel pump seal is defective, faulty or otherwise leaking • Hoses between the EVAP canister and the fuel flap unit are faulty • Hoses between the EVAP canister and the evaporative emission canister purge regulator valve are faulty
DTC: P1500 **2T MIL: Yes** **2001, 2002, 2003, 2004, 2005** **Models:** 325i, 330i, 525i, 530i, 545i, 745i, M5, Z3, Z4 **Engines:** 3.0L, 3.2L, 4.4L, 5.0L **Transmissions:** All	**Idle Air Control Valve Malfunction Conditions:** Engine started, battery voltage at least 11.5v, all electrical components off, ground connections between engine and chassis well connected, coolant temperature at least 80-degrees Celsius. The DME detected deviation from the normal operating parameters of the Idle Air Control Valve. The vehicle speed can be zero mph and the engine load must be less than 1.5ms. **Possible Causes:** • Charge air system leaks • Recirculating valve for turbocharger is faulty • Turbocharging system is damaged • Vacuum diaphragm for turbocharger needs adjusting • Wastegate bypass regulator valve is faulty
DTC: P1501 **2T MIL: Yes** **2001, 2002, 2003, 2004, 2005** **Models:** 325i, 330i, 525i, 530i, 545i, 745i, M5, Z3, Z4 **Engines:** 3.0L, 3.2L, 4.4L, 5.0L **Transmissions:** All	**Idle Air Control Valve Malfunction Conditions:** Engine started, battery voltage at least 11.5v, all electrical components off, ground connections between engine and chassis well connected, coolant temperature at least 80-degrees Celsius. The DME detected deviation from the normal operating parameters of the Idle Air Control Valve. The vehicle speed can be zero mph and the engine load must be less than 1.5ms. **Possible Causes:** • Charge air system leaks • Recirculating valve for turbocharger is faulty • Turbocharging system is damaged • Vacuum diaphragm for turbocharger needs adjusting • Wastegate bypass regulator valve is faulty

DTC	Trouble Code Title, Conditions & Possible Causes
DTC: P1502 **2T MIL: Yes** **2001, 2002, 2003, 2004, 2005** **Models:** 325i, 330i, 525i, 530i, 545i, 745i, M5, Z3, Z4 **Engines:** 3.0L, 3.2L, 4.4L, 5.0L **Transmissions:** All	**Idle Air Control Valve Circuit Short to B+ Conditions:** Engine running the DME detected that the idle air control valve was intermittent. **Possible Causes:** • Fuel delivery unit connector is loose or not attached • Fuse 18 cause a short to the transfer fuel pump or the O2S • Fuel pump has failed • Fuel pump relay circuit is shorted to ground, B+ or is open • Fuel Pump (FP) Relay not activated
DTC: P1503 **2T MIL: Yes** **2001, 2002, 2003, 2004, 2005** **Models:** 325i, 330i, 525i, 530i, 545i, 745i, M5, Z3, Z4 **Engines:** 3.0L, 3.2L, 4.4L, 5.0L **Transmissions:** All	**Idle Air Control Valve Circuit Short to Ground Conditions:** Engine running the DME detected that the idle air control valve was intermittent. **Possible Causes:** • Fuel delivery unit connector is loose or not attached • Fuse 18 cause a short to the transfer fuel pump or the O2S • Fuel pump has failed • Fuel pump relay circuit is shorted to ground, B+ or is open • Fuel Pump (FP) Relay not activated
DTC: P1504 **2T MIL: Yes** **2001, 2002, 2003, 2004, 2005** **Models:** 325i, 330i, 525i, 530i, 545i, 745i, M5, Z3, Z4 **Engines:** 3.0L, 3.2L, 4.4L, 5.0L **Transmissions:** All	**Idle Air Control Valve Circuit Continuity-Open Load Conditions:** Engine running the DME detected that the idle air control valve signal was intermittent. **Possible Causes:** • Fuel delivery unit connector is loose or not attached • Fuse 18 cause a short to the transfer fuel pump or the O2S • Fuel pump has failed • Fuel pump relay circuit is shorted to ground, B+ or is open • Fuel Pump (FP) Relay not activated
DTC: P1506 **2T MIL: Yes** **2001, 2002, 2003, 2004, 2005** **Models:** 325i, 330i, 525i, 530i, 545i, 745i, M5, Z3, Z4 **Engines:** 3.0L, 3.2L, 4.4L, 5.0L **Transmissions:** All	**Idle Air Control Valve Circuit Short to B+ Conditions:** Engine running the DME detected that the idle air control valve signal was intermittent. **Possible Causes:** • Fuel delivery unit connector is loose or not attached • Fuse 18 cause a short to the transfer fuel pump or the O2S • Fuel pump has failed • Fuel pump relay circuit is shorted to ground, B+ or is open• Fuel Pump (FP) Relay not activated
DTC: P1507 **2T MIL: Yes** **2001, 2002, 2003, 2004, 2005** **Models:** 325i, 330i, 525i, 530i, 545i, 745i, M5, Z3, Z4 **Engines:** 3.0L, 3.2L, 4.4L, 5.0L **Transmissions:** All	**Idle Air Control Valve Circuit Short to Ground Conditions:** Engine running the DME detected that the idle air control valve signal was intermittent. **Possible Causes:** • Fuel delivery unit connector is loose or not attached • Fuse 18 cause a short to the transfer fuel pump or the O2S • Fuel pump has failed • Fuel pump relay circuit is shorted to ground, B+ or is open • Fuel Pump (FP) Relay not activated
DTC: P1508 **2T MIL: Yes** **2001, 2002, 2003, 2004, 2005** **Models:** 325i, 330i, 525i, 530i, 545i, 745i, M5, Z3, Z4 **Engines:** 3.0L, 3.2L, 4.4L, 5.0L **Transmissions:** All	**Idle Air Control Valve Circuit Continuity-Open Load Conditions:** Engine running the DME detected that the idle air control valve signal was intermittent. **Possible Causes:** • Fuel delivery unit connector is loose or not attached • Fuse 18 cause a short to the transfer fuel pump or the O2S • Fuel pump has failed • Fuel pump relay circuit is shorted to ground, B+ or is open • Fuel Pump (FP) Relay not activated

DTC	Trouble Code Title, Conditions & Possible Causes
DTC: P1509 **2T MIL: Yes** **1996, 1997, 1998, 1999, 2000** **Models:** 318i, 323i, 325i, 328i, 330i, 525i, 528i, 540i, 740i, 840Ci, M5, Z3, Z8 **Engines:** 1.9L, 2.5L, 2.8L, 3.2L, 4.4L, 5.4L **Transmissions:** All	**Idle Air Control Valve Conditions:** Engine started, battery voltage at least 11.5v, all electrical components off, ground connections between engine and chassis well connected, coolant temperature at least 80-degrees Celsius. The DME detected deviation from the normal operating parameters of the idle air control valve. **Possible Causes:** • Charge air system leaks • Recirculating valve for turbocharger is faulty • Turbocharging system is damaged • Vacuum diaphragm for turbocharger needs adjusting • Wastegate bypass regulator valve is faulty
DTC: P1511 **2T MIL: Yes** **1996, 1997, 1998, 1999, 2000, 2006** **Models:** 318i, 323i, 328i, 528i, M3, Z3 **Engines:** 1.9L, 2.8L, 3.2L **Transmissions:** All	**Differentiated Intake Manifold Control Circuit Electrical Conditions:** Engine started, and the DME detected the changeover valve circuit was faulting during the continuous self test. **Possible Causes:** • Leaky vacuum reservoir, vacuum lines loose or damaged • Vacuum solenoid or vacuum actuator is damaged • IMRC actuator cable/gears are seized, or the cables are improperly routed or seized • IMRC housing return springs are damaged or disconnected • Lever/shaft return stop may be obstructed or bent, or the lever/shaft wide open stop may be obstructed or bent, or the IMRC lever/shaft may be sticking, binding or disconnected • IMRC control circuit open, shorted or the VPWR circuit is open
DTC: P1512 **2T MIL: Yes** **2005, 2006** **Models:** 325i, 330i, 525i, 530i, 550i, 750i, Z4 **Engines:** 3.0L, 4.8L **Transmissions:** All	**Differentiated Intake Manifold Control Circuit Signal Low Conditions:** Engine started, and the DME detected the changeover valve circuit was shorting to negative during the continuous self test. **Possible Causes:** • Leaky vacuum reservoir, vacuum lines loose or damaged • Vacuum solenoid or vacuum actuator is damaged • IMRC actuator cable/gears are seized, or the cables are improperly routed or seized • IMRC housing return springs are damaged or disconnected • Lever/shaft return stop may be obstructed or bent, or the lever/shaft wide open stop may be obstructed or bent, or the IMRC lever/shaft may be sticking, binding or disconnected • IMRC control circuit open, shorted or the VPWR circuit is open
DTC: P1513 **2T MIL: Yes** **2005, 2006** **Models:** 325i, 330i, 525i, 530i, 550i, 750i, Z4 **Engines:** 3.0L, 4.8L **Transmissions:** All	**Differentiated Intake Manifold Control Circuit Signal High Conditions:** Engine started, and the DME detected the changeover valve circuit was shorting to positive during the continuous self test. **Possible Causes:** • Leaky vacuum reservoir, vacuum lines loose or damaged • Vacuum solenoid or vacuum actuator is damaged • IMRC actuator cable/gears are seized, or the cables are improperly routed or seized • IMRC housing return springs are damaged or disconnected • Lever/shaft return stop may be obstructed or bent, or the lever/shaft wide open stop may be obstructed or bent, or the IMRC lever/shaft may be sticking, binding or disconnected • IMRC control circuit open, shorted or the VPWR circuit is open
DTC: P1515 **MIL: Yes** **2005, 2006** **Models:** 325i, 330i, 525i, 530i, 550i, 750i, Z4 **Engines:** 3.0L, 4.8L **Transmissions:** All	**Engine Off Timer Plausibility Conditions:** The DME detected an implausible instrument cluster and/or power module signal. CAN signal failure while the DME was operating at a range between 6 and 16 volts. CAN Bus lost communications. The system time (time pulse) is implausible relative to the DME's internal counter. **Possible Causes:** • Check the CAN bus • Check instrument cluster/power module/Can Signal if no other CAN faults are detected
DTC: P1517 **2006** **Models:** 325i, 330i, 525i, 530i, 550i, 750i, Z4 **Engines:** 3.0L, 4.8L **Transmissions:** All	**Rough Road Detection, No Wheel Speed Signal Conditions:** The DME detected an electrical malfunction on the main relay circuit **Possible Causes:** • Possible failure of wheel speed sensor on drive axle
DTC: P1518 **2006** **Models:** 325i, 330i, 525i, 530i, 550i, 750i, Z4 **Engines:** 3.0L, 4.8L **Transmissions:** All	**Rough Road Detection, Wheel Speed Too High Conditions:** The DME detected an electrical malfunction on the main relay circuit **Possible Causes:** • Possible failure of wheel speed sensor on drive axle

DTC	Trouble Code Title, Conditions & Possible Causes
DTC: P1519 **2T MIL: Yes** **1996, 1997, 1998, 1999, 2000** **Models:** 325i, 328i, 330i, 525i, 528i, 540i, 740i, Z3 **Engines:** 2.8L, 3.0L, 4.4L, 5.4L **Transmissions:** All	**Engine Oil Quality Sensor Temperature Measurement Conditions:** The oil temperature difference of greater than 100 degrees within one second. The ignition must be on. The DME detected an error in the Engine Oil Temperature sensor. **Possible Causes:** • Replace the oil temperature sensor
DTC: P1520 **2006** **Models:** 325i, 330i, 525i, 530i, 550i, 750i, Z4 **Engines:** 3.0L, 4.8L **Transmissions:** All	**Engine Oil Quality Sensor Level Measurement Error Conditions:** After 60 seconds with the ignition on, a miscommunication between the oil sensor and the DME was detected. **Possible Causes:** • Replace the oil temperature sensor • Check BSD wire, voltage supply and oil level sensor ground for an open circuit. If okay, replace the oil level sensor.
DTC: P1521 **2T MIL: Yes** **1999** **Models:** 540i, 740i **Engines:** 4.4L **Transmissions:** All	**Intake Camshaft Control (Bank 2) Malfunction Conditions:** Key on or engine running; and the DME detected the intake manifold control signal for was outside of its expected calibrated range. **Possible Causes:** • Camshaft control circuit is open or shorted to ground • Camshaft sensor is damaged or the DME has failed • Camshaft out of adjustment
DTC: P1521 **2006** **Models:** 325i, 330i, 525i, 530i, 550i, 750i, Z4 **Engines:** 3.0L, 4.8L **Transmissions:** All	**Engine Oil Quality Sensor Communication Error Conditions:** After 60 seconds with the ignition on, a miscommunication between the oil sensor and the DME was detected. **Possible Causes:** • Replace the oil temperature sensor • Check BSD wire, voltage supply and oil level sensor ground for an open circuit. If okay, replace the oil level sensor.
DTC: P1522 **2T MIL: Yes** **1999, 2000** **Models:** 323i, 328i, 528i, 540i, 740i, Z3 **Engines:** 2.8L, 4.4L **Transmissions:** All	**Intake Camshaft Control (Bank 2) Malfunction Conditions:** Key on or engine running; and the DME detected the intake manifold control signal for was outside of its expected calibrated range. **Possible Causes:** • Camshaft control circuit is open or shorted to ground • Camshaft sensor is damaged or the DME has failed • Camshaft out of adjustment
DTC: P1526 **2T MIL: Yes** **1999, 2001, 2002, 2003** **Models:** 528i, 540i, 740i, 745i, M5, Z3, Z8 **Engines:** 2.8L, 3.2L, 4.4L, 4.8L, 5.0L **Transmissions:** All	**Camshaft Control Circuit Ground Conditions:** Engine started and driven at an engine speed of more than 400rpm; and the DME detected the camshaft timing exceeded the calibrated levels. The valve timing did not change from the current valve timing or it remained fixed during the testing. **Note: The camshaft adjustment is load- and RPM dependant. The electrical camshaft adjustment valve 1 switches oil pressure onto camshaft adjuster (mechanical adjustment mechanism), which adjusts the camshaft.** **Possible Causes:** • Fuel pump has failed • CPS circuit is open, shorted to ground or shorted to power • Battery voltage below 11.5 volts • Position actuator circuit may short to B+ or Ground • Camshaft timing improperly set, or continuous oil flow to the VCT piston chamber • Camshaft advance mechanism (the VCT unit) is sticking or binding mechanically • VCT solenoid valve is stuck in open position
DTC: P1527 **2T MIL: Yes** **1996, 1997, 1998, 1999, 2000** **Models:** 325i, 328i, 330i, 525i, 528i, 540i, 740i, Z3 **Engines:** 2.8L, 3.0L, 4.4L, 5.4L **Transmissions:** All	**Variable Camshaft Timing (VANOS) Circuit Malfunction Conditions:** The DME detected an error in the camshaft sensor. The battery voltage is between 9.5 and 17 volts, while the engine speed is greater than 40rpm. **Possible Causes:** • Replace the sensor

DTC	Trouble Code Title, Conditions & Possible Causes
DTC: P1529 **2T MIL: Yes** **1999, 2000, 2001, 2002, 2003** **Models:** 323i, 325i, 328i, 330i, 525i, 528i, 530i, 540i, 740i, 745i, M5, Z3, Z8 **Engines:** 2.5L, 2.8L, 3.0L, 4.8L **Transmissions:** All	**Camshaft Control Circuit Short to B+ Conditions:** Engine started and driven at an engine speed of more than 400rpm; and the DME detected the camshaft timing exceeded the calibrated voltage levels. The valve timing did not change from the current valve timing or it remained fixed during the testing. **Note: The camshaft adjustment is load- and RPM dependant. The electrical camshaft adjustment valve 1 switches oil pressure onto camshaft adjuster (mechanical adjustment mechanism), which adjusts the camshaft.** **Possible Causes:** • Fuel pump has failed • CPS circuit is open, shorted to ground or shorted to power • Battery voltage below 11.5 volts • Position actuator circuit may short to B+ or Ground • Camshaft timing improperly set, or continuous oil flow to the VCT piston chamber • Camshaft advance mechanism (the VCT unit) is sticking or binding mechanically • VCT solenoid valve is stuck in open position
DTC: P1530 **2T MIL: Yes** **1999, 2000, 2001, 2002, 2003** **Models:** 323i, 325i, 328i, 330i, 525i, 528i, 530i, 540i, 740i, 745i, M5, Z3, Z8 **Engines:** 2.5L, 2.8L, 3.0L, 4.8L **Transmissions:** All	**Camshaft Control Circuit Short to Ground Conditions:** Engine started and driven at an engine speed of more than 400rpm; and the DME detected the camshaft timing exceeded the calibrated levels. The valve timing did not change from the current valve timing or it remained fixed during the testing. **Note: The camshaft adjustment is load- and RPM dependant. The electrical camshaft adjustment valve 1 switches oil pressure onto camshaft adjuster (mechanical adjustment mechanism), which adjusts the camshaft.** **Possible Causes:** • Fuel pump has failed • CPS circuit is open, shorted to ground or shorted to power • Battery voltage below 11.5 volts • Position actuator circuit may short to B+ or Ground • Camshaft timing improperly set, or continuous oil flow to the VCT piston chamber • Camshaft advance mechanism (the VCT unit) is sticking or binding mechanically • VCT solenoid valve is stuck in open position
DTC: P1532 **2T MIL: Yes** **1999, 2000, 2001, 2002, 2003** **Models:** 323i, 325i, 328i, 330i, 525i, 528i, 530i, 540i, 740i, 745i, M5, Z3, Z8 **Engines:** 2.5L, 2.8L, 3.0L, 4.8L **Transmissions:** All	**Camshaft Control Circuit Open Conditions:** Engine started and driven at an engine speed of more than 400rpm; and the DME detected the camshaft timing exceeded the calibrated levels. The valve timing did not change from the current valve timing or it remained fixed during the testing. **Note: The camshaft adjustment is load- and RPM dependant. The electrical camshaft adjustment valve 1 switches oil pressure onto camshaft adjuster (mechanical adjustment mechanism), which adjusts the camshaft.** **Possible Causes:** • Fuel pump has failed • CPS circuit is open, shorted to ground or shorted to power • Battery voltage below 11.5 volts • Position actuator circuit may short to B+ or Ground • Camshaft timing improperly set, or continuous oil flow to the VCT piston chamber • Camshaft advance mechanism (the VCT unit) is sticking or binding mechanically • VCT solenoid valve is stuck in open position
DTC: P1535 **2T MIL: No** **2006** **Models:** 325i, 330i, 525i, 530i, 550i, 750i, Z4 **Engines:** 3.0L, 4.8L **Transmissions:** All	**Differentiated Intake Manifold Coil Temperature Limit Value Exceeded Conditions:** This fault stems from the 4.8L-N62 equipped with an infinitely adjustable control of induction system. The temperature of the coil has exceeded the predetermined limits. **Possible Causes:** • Faulty controller • Faulty coil
DTC: P1536 **2T MIL: No** **2006** **Models:** 325i, 330i, 525i, 530i, 550i, 750i, Z4 **Engines:** 3.0L, 4.8L **Transmissions:** All	**Differentiated Intake Manifold Controller Monitoring, Control Deviation Conditions:** This fault stems from the 4.8L-N62 equipped with an infinitely adjustable control of induction system. **Possible Causes:** • Faulty controller
DTC: P1537 **2T MIL: No** **2006** **Models:** 325i, 330i, 525i, 530i, 550i, 750i, Z4 **Engines:** 3.0L, 4.8L **Transmissions:** All	**Differentiated Intake Manifold Potentiometer Voltage in Lower Diagnosis Range Conditions:** This fault stems from the 4.8L-N62 equipped with an infinitely adjustable control of induction system. **Possible Causes:** • Faulty potentiometer

DTC	Trouble Code Title, Conditions & Possible Causes
DTC: P1538 **2T MIL: No** **2006** **Models:** 325i, 330i, 525i, 530i, 550i, 750i, Z4 **Engines:** 3.0L, 4.8L **Transmissions:** All	**Differentiated Intake Manifold Potentiometer Voltage in Upper Diagnosis Range Conditions:** This fault stems from the 4.8L-N62 equipped with an infinitely adjustable control of induction system. **Possible Causes:** • Faulty potentiometer
DTC: P1539 **2T MIL: No** **2006** **Models:** 325i, 330i, 525i, 530i, 550i, 750i, Z4 **Engines:** 3.0L, 4.8L **Transmissions:** All	**Differentiated Intake Manifold Coil Temperature Threshold Exceeded Conditions:** This fault stems from the 4.8L-N62 equipped with an infinitely adjustable control of induction system. The temperature of the coil has exceeded the predetermined limits. **Possible Causes:** • Faulty controller • Faulty coil
DTC: P1542 **1T MIL: Yes** **1999, 2000** **Models:** 323i, 325i, 328i, 528i, Z3 **Engines:** 2.5L, 2.8L **Transmissions:** All	**Throttle Actuation Potentiometer Range/Performance Conditions:** Engine started, battery voltage must be at least 11.5v, all electrical components must be off, parking brake must be engaged (to keep daytime driving lights off), automatic transmission selector must be in park, the exhaust system must be properly sealed between the catalytic converter and the cylinder head, coolant temperature must be at least 80 degrees Celsius, and the ground between the engine and the chassis must be well connected. The signal from the Throttle Position Valve Module to the DME detected was erratic, non existent or unreliable. **Note: If the complete throttle valve control module is current-less (e.g. connector disconnected) the throttle valve moves into a particular, specified mechanical position, which signals an increased idle speed with an engine at operating temperature. If only the Throttle Position (TP) actuator –V60- is current-less, the throttle valve also moves into the specified mechanical position (emergency running gap), however, since Closed Throttle Position (CTP) switch –F60- can still be recognized, an "almost normal idle RPM" is reached via the respective ignition angle retardation. If the Engine Control Module (DME) detects a malfunction at Throttle Position (TP) sensor –G69-, Throttle Position (TP) actuator –V60- is switched current-less by the Engine Control Module (DME) and the throttle valve moves into the specified mechanical position (emergency running gap) again.** **Note: Terminal assignment at throttle control module is different in vehicles with and without cruise control.** **Characteristic: Steering column switch with operating module for cruise control.** **Possible Causes:** • Throttle valve control module is faulty • Throttle valve is damaged or dirty • Throttle valve must be in closed throttle position • Accelerator pedal is out of adjustment (AEG engines only) • Throttle position actuator is shorting to ground or power
DTC: P1543 **2T MIL: Yes** **1996, 1997, 1998, 1999** **Models:** 750i **Engines:** 5.4L **Transmissions:** All	**Throttle Actuation Potentiometer Signal Too Low Conditions:** Engine started, battery voltage must be at least 11.5v, all electrical components must be off, parking brake must be engaged (to keep daytime driving lights off), automatic transmission selector must be in park, the exhaust system must be properly sealed between the catalytic converter and the cylinder head, coolant temperature must be at least 80 degrees Celsius, and the ground between the engine and the chassis must be well connected. The signal from the Throttle Position Valve Module to the DME detected was erratic, non existent or unreliable. **Note: If the complete throttle valve control module is current-less (e.g. connector disconnected) the throttle valve moves into a particular, specified mechanical position, which signals an increased idle speed with an engine at operating temperature. If only the Throttle Position (TP) actuator –V60- is current-less, the throttle valve also moves into the specified mechanical position (emergency running gap), however, since Closed Throttle Position (CTP) switch –F60- can still be recognized, an "almost normal idle RPM" is reached via the respective ignition angle retardation. If the Engine Control Module (DME) detects a malfunction at Throttle Position (TP) sensor –G69-, Throttle Position (TP) actuator –V60- is switched current-less by the Engine Control Module (DME) and the throttle valve moves into the specified mechanical position (emergency running gap) again.** **Note: Terminal assignment at throttle control module is different in vehicles with and without cruise control.** **Characteristic: Steering column switch with operating module for cruise control.** **Possible Causes:** • Throttle valve control module is faulty • Throttle valve is damaged or dirty • Throttle valve must be in closed throttle position • Accelerator pedal is out of adjustment (AEG engines only) • Throttle position actuator is shorting to ground or power

DTC	Trouble Code Title, Conditions & Possible Causes
DTC: P1544 **2T MIL: Yes** **1996, 1997, 1998, 1999** **Models:** 750i **Engines:** 5.4L **Transmissions:** All	**Throttle Actuation Potentiometer Signal Too High Conditions:** Engine started, battery voltage must be at least 11.5v, all electrical components must be off, parking brake must be engaged (to keep daytime driving lights off), automatic transmission selector must be in park, the exhaust system must be properly sealed between the catalytic converter and the cylinder head, coolant temperature must be at least 80 degrees Celsius, and the ground between the engine and the chassis must be well connected. The signal from the Throttle Position Valve Module to the DME detected was erratic, non existent or unreliable. **Note: If the complete throttle valve control module is current-less (e.g. connector disconnected) the throttle valve moves into a particular, specified mechanical position, which signals an increased idle speed with an engine at operating temperature. If only the Throttle Position (TP) actuator –V60- is current-less, the throttle valve also moves into the specified mechanical position (emergency running gap), however, since Closed Throttle Position (CTP) switch –F60- can still be recognized, an "almost normal idle RPM" is reached via the respective ignition angle retardation. If the Engine Control Module (DME) detects a malfunction at Throttle Position (TP) sensor –G69-, Throttle Position (TP) actuator –V60- is switched current-less by the Engine Control Module (DME) and the throttle valve moves into the specified mechanical position (emergency running gap) again.** **Note: Terminal assignment at throttle control module is different in vehicles with and without cruise control.** **Characteristic: Steering column switch with operating module for cruise control.** **Possible Causes:** • Throttle valve control module is faulty • Throttle valve is damaged or dirty • Throttle valve must be in closed throttle position • Accelerator pedal is out of adjustment (AEG engines only) • Throttle position actuator is shorting to ground or power
DTC: P1545 **2T MIL: Yes** **1996, 1997, 1998, 1999** **Models:** 750i **Engines:** 5.4L **Transmissions:** All	**Throttle Position Control Malfunction Conditions:** Engine started, battery voltage must be at least 11.5v, all electrical components must be off, parking brake must be engaged (to keep daytime driving lights off), automatic transmission selector must be in park, the exhaust system must be properly sealed between the catalytic converter and the cylinder head, coolant temperature must be at least 80 degrees Celsius, and the ground between the engine and the chassis must be well connected. The signal from the Throttle Position Valve Module to the DME detected was erratic, non existent or unreliable. **Note: If the complete throttle valve control module is current-less (e.g. connector disconnected) the throttle valve moves into a particular, specified mechanical position, which signals an increased idle speed with an engine at operating temperature. If only the Throttle Position (TP) actuator is current-less, the throttle valve also moves into the specified mechanical position (emergency running gap), however, since Closed Throttle Position (CTP) switch – can still be recognized, an "almost normal idle RPM" is reached via the respective ignition angle retardation. If the Engine Control Module (DME) detects a malfunction at Throttle Position (TP) sensor – Throttle Position (TP) actuator is switched current-less by the Engine Control Module (DME) and the throttle valve moves into the specified mechanical position (emergency running gap) again.** **Note: Terminal assignment at throttle control module is different in vehicles with and without cruise control.** **Characteristic: Steering column switch with operating module for cruise control.** **Possible Causes:** • Throttle valve control module is faulty • Throttle valve is damaged or dirty • Throttle valve must be in closed throttle position • Accelerator pedal is out of adjustment (AEG engines only) • Throttle position actuator is shorting to ground or power
DTC: P1550 **1996, 1997, 1998, 1999, 2000** **Models:** All **Engines:** All **Transmissions:** All	**Idle Air Control Valve Conditions:** Engine started, battery voltage at least 11.5v, all electrical components off, ground connections between engine and chassis well connected, coolant temperature at least 80-degrees Celsius. The DME detected deviation from the normal operating parameters of the charge pressure sensor. **Possible Causes:** • Charge air system leaks • Recirculating valve for turbocharger is faulty • Turbocharging system is damaged • Vacuum diaphragm for turbocharger needs adjusting • Wastegate bypass regulator valve is faulty
DTC: P1551 **MIL: Yes** **2006** **Models:** 325i, 330i, 525i, 530i, 550i, 750i, Z4 **Engines:** 3.0L, 4.8L **Transmissions:** All	**Engine Off Timer Timeout Conditions:** The DME detected a CAN signal failure while the DME was operating at a range between 6 and 16 volts. CAN Bus lost communications. **Possible Causes:** • Check the CAN bus • Check instrument cluster/power module/Can Signal

DTC	Trouble Code Title, Conditions & Possible Causes
DTC: P1557 **2T MIL: Yes** 1998 **Models:** 750i **Engines:** 5.4L **Transmissions:** All	**Heated Catalyst System Exceeded Maximum Temperature Conditions:** Engine started, battery voltage must be at least 11.5v, all electrical components must be off, parking brake must be engaged (to keep daytime driving lights off), automatic transmission selector must be in park, the exhaust system must be properly sealed between the catalytic converter and the cylinder head, coolant temperature must be at least 80 degrees Celsius and oxygen sensor heaters for oxygen sensors before the catalytic converter must be functioning properly and the ground between the engine and the chassis must be well connected. The coolant temperature is less than 90 degrees Celsius, catalyst temperature is less than 300 degrees Celsius, engine crank time less than five seconds, vehicle speed is less than three mph and the engine speed is less than 200rpm. **Possible Causes:** • Air leaks at the exhaust manifold or in the exhaust pipes • Catalytic converter is damaged, contaminated or it has failed • ECT/CHT sensor has lost its calibration (the signal is incorrect) • Engine cylinders misfiring, or the ignition timing is over retarded • Engine oil is contaminated • Front HO2S or rear HO2S is contaminated with fuel or moisture • Front HO2S and/or the rear HO2S is loose in the mounting hole • Front HO2S much older than the rear HO2S (HO2S-11 is lazy) • Fuel system pressure is too high (check the pressure regulator) • Rear HO2S wires improperly connected or the HO2S has failed
DTC: P1560 **2T MIL: Yes** **1999, 2000** **Models:** 750i **Engines:** 5.4L **Transmissions:** All	**Heated Catalyst System Exceeded Maximum Temperature Conditions:** Engine started, battery voltage must be at least 11.5v, all electrical components must be off, parking brake must be engaged (to keep daytime driving lights off), automatic transmission selector must be in park, the exhaust system must be properly sealed between the catalytic converter and the cylinder head, coolant temperature must be at least 80 degrees Celsius and oxygen sensor heaters for oxygen sensors before the catalytic converter must be functioning properly and the ground between the engine and the chassis must be well connected. The coolant temperature is less than 90 degrees Celsius, catalyst temperature is less than 300 degrees Celsius, engine crank time less than five seconds, vehicle speed is less than three mph and the engine speed is less than 200rpm. **Possible Causes:** • Air leaks at the exhaust manifold or in the exhaust pipes • Catalytic converter is damaged, contaminated or it has failed • ECT/CHT sensor has lost its calibration (the signal is incorrect) • Engine cylinders misfiring, or the ignition timing is over retarded • Engine oil is contaminated • Front HO2S or rear HO2S is contaminated with fuel or moisture • Front HO2S and/or the rear HO2S is loose in the mounting hole • Front HO2S much older than the rear HO2S (HO2S-11 is lazy) • Fuel system pressure is too high (check the pressure regulator) • Rear HO2S wires improperly connected or the HO2S has failed
DTC: P1590 **2T MIL: Yes** **1996, 1997, 1998, 1999, 2000** **Models:** 750i, 850Ci **Engines:** 5.4L **Transmissions:** All	**Throttle Actuation Potentiometer Signal Too Low (right side) Conditions:** Engine started, battery voltage must be at least 11.5v, all electrical components must be off, parking brake must be engaged (to keep daytime driving lights off), automatic transmission selector must be in park, the exhaust system must be properly sealed between the catalytic converter and the cylinder head, coolant temperature must be at least 80 degrees Celsius, and the ground between the engine and the chassis must be well connected. The signal from the Throttle Position Valve Module to the DME detected was erratic, non existent or unreliable. **Note: If the complete throttle valve control module is current-less (e.g. connector disconnected) the throttle valve moves into a particular, specified mechanical position, which signals an increased idle speed with an engine at operating temperature. If only the Throttle Position (TP) actuator –V60- is current-less, the throttle valve also moves into the specified mechanical position (emergency running gap), however, since Closed Throttle Position (CTP) switch –F60- can still be recognized, an "almost normal idle RPM" is reached via the respective ignition angle retardation. If the Engine Control Module (DME) detects a malfunction at Throttle Position (TP) sensor –G69-, Throttle Position (TP) actuator –V60- is switched current-less by the Engine Control Module (DME) and the throttle valve moves into the specified mechanical position (emergency running gap) again.** **Note: Terminal assignment at throttle control module is different in vehicles with and without cruise control.** **Characteristic: Steering column switch with operating module for cruise control.** **Possible Causes:** • Throttle valve control module is faulty • Throttle valve is damaged or dirty • Throttle valve must be in closed throttle position • Accelerator pedal is out of adjustment (AEG engines only) • Throttle position actuator is shorting to ground or power

DTC	Trouble Code Title, Conditions & Possible Causes
DTC: P1591 **2T MIL: Yes** **1996, 1997, 1998, 1999, 2000** **Models:** 750i, 850Ci **Engines:** 5.4L **Transmissions:** All	**Throttle Actuation Potentiometer Signal Too High (Right Side) Conditions:** Engine started, battery voltage must be at least 11.5v, all electrical components must be off, parking brake must be engaged (to keep daytime driving lights off), automatic transmission selector must be in park, the exhaust system must be properly sealed between the catalytic converter and the cylinder head, coolant temperature must be at least 80 degrees Celsius, and the ground between the engine and the chassis must be well connected. The signal from the Throttle Position Valve Module to the DME detected was erratic, non existent or unreliable. **Note: If the complete throttle valve control module is current-less (e.g. connector disconnected) the throttle valve moves into a particular, specified mechanical position, which signals an increased idle speed with an engine at operating temperature. If only the Throttle Position (TP) actuator –V60- is current-less, the throttle valve also moves into the specified mechanical position (emergency running gap), however, since Closed Throttle Position (CTP) switch –F60- can still be recognized, an "almost normal idle RPM" is reached via the respective ignition angle retardation. If the Engine Control Module (DME) detects a malfunction at Throttle Position (TP) sensor –G69-, Throttle Position (TP) actuator –V60- is switched current-less by the Engine Control Module (DME) and the throttle valve moves into the specified mechanical position (emergency running gap) again.** **Note: Terminal assignment at throttle control module is different in vehicles with and without cruise control. Characteristic: Steering column switch with operating module for cruise control.** **Possible Causes:** • Throttle valve control module is faulty • Throttle valve is damaged or dirty • Throttle valve must be in closed throttle position • Accelerator pedal is out of adjustment (AEG engines only) • Throttle position actuator is shorting to ground or power
DTC: P1592 **2T MIL: Yes** **1996, 1997, 1998, 1999, 2000** **Models:** 750i, 850Ci **Engines:** 5.4L **Transmissions:** All	**Throttle Position Control Malfunction (Right Side) Conditions:** Engine started, battery voltage must be at least 11.5v, all electrical components must be off, parking brake must be engaged (to keep daytime driving lights off), automatic transmission selector must be in park, the exhaust system must be properly sealed between the catalytic converter and the cylinder head, coolant temperature must be at least 80 degrees Celsius, and the ground between the engine and the chassis must be well connected. The signal from the Throttle Position Valve Module to the DME detected was erratic, non existent or unreliable. **Note: If the complete throttle valve control module is current-less (e.g. connector disconnected) the throttle valve moves into a particular, specified mechanical position, which signals an increased idle speed with an engine at operating temperature. If only the Throttle Position (TP) actuator is current-less, the throttle valve also moves into the specified mechanical position (emergency running gap), however, since Closed Throttle Position (CTP) switch – can still be recognized, an "almost normal idle RPM" is reached via the respective ignition angle retardation. If the Engine Control Module (DME) detects a malfunction at Throttle Position (TP) sensor – Throttle Position (TP) actuator is switched current-less by the Engine Control Module (DME) and the throttle valve moves into the specified mechanical position (emergency running gap) again.** **Note: Terminal assignment at throttle control module is different in vehicles with and without cruise control. Characteristic: Steering column switch with operating module for cruise control.** **Possible Causes:** • Throttle valve control module is faulty • Throttle valve is damaged or dirty • Throttle valve must be in closed throttle position • Accelerator pedal is out of adjustment (AEG engines only) • Throttle position actuator is shorting to ground or power
DTC: P1603 **2T MIL: Yes** **2001, 2002, 2003, 2005** **Models:** 325i, 330i, 750i , M5, Z8 **Engines:** 3.2L, 5.0L **Transmissions:** All	**Control Module Self Test, Torque Monitoring Conditions:** Ignition on, the DME detected a control module malfunction. The torque monitoring feature compares the torque demand (from accelerator pedal, FGR, electrical equipment and transmission) with the torque provided (calculated from HFH, injector valves, ignition angle, throttle valve angle, differential pressure and lambda). Deviations trigger fuel-supply safety shutdown to prevent vehicle from autonomous acceleration. **Possible Causes:** • Check HFM for contamination and replace if necessary • Replace the DME
DTC: P1604 **2T MIL: Yes** 2005 **Models:** 325i, 330i **Engines:** All **Transmissions:** All	**Internal Control Module Driver Error Conditions:** Ignition on, the DME detected a control module malfunction. To achieve optimal anti-theft protection for the vehicle, an anti-theft immobilizer is installed. The anti-theft immobilizer is a system for enabling and locking the Engine Control Module (DME). So that this system cannot be circumvented, it is necessary to perform adaptation of the anti-theft immobilizer using the Vehicle Diagnostic and Information System VAS 5052 in the On Board Diagnostic (OBD) function. The great availability of equipment options makes it necessary to adapt the Engine Control Module (DME) to the vehicle (e.g. throttle valve control module or cruise control system). This "writing" function is not possible with the generic scan tool. **Possible Causes:** • (If DME was replaced) DME ID not the same as the replaced unit • Voltage supply for Engine Control Module (DME) has shorted

DTC	Trouble Code Title, Conditions & Possible Causes
DTC: P1611 **2T MIL: Yes** **2002, 2003** **Models:** 745i **Engines:** 4.8L **Transmissions:** All	**MIL Call-Up Circuit, Transmission Control Module Short to Ground Conditions:** Engine started, VSS over 1 mph, and the DME detected a problem in the Transmission Control system during the self-test. **Possible Causes:** • Open/short circuit to ground in the communication wire from the transmission to the DME. • The DME has failed
DTC: P1614 **MIL: No** **2006** **Models:** 325i, 330i, 525i, 530i, 550i, 750i, Z4 **Engines:** 3.0L, 4.8L **Transmissions:** All	**Serial Communication Link ACC Malfunction Conditions:** Key on after 800ms or engine running for 100ms, the DME detected an electrical malfunction regarding the Adaptive Cruise Control circuit. Check the ACC fuse, the ECU or CAN for fault, measure the resistance on the BUS, check the wiring harness. If this fault occurs when the ACC is not installed, the DME has been incorrectly initialized for the ACC version. **Possible Causes:** • Circuit from the MIL to the DME • ECU Failure, BUS system failure • Circuit from the EPC to the DME
DTC: P1619 **2001** **Models:** 750i **Engines:** 5.4L **Transmissions:** All	**Thermostat Control Circuit Ground Conditions:** The engine's warm up performance is monitored by comparing measured coolant temperature with the modeled coolant temperature to detect a defective coolant thermostat that is reading false. The engine temperature must be less than 65 degrees Celsius, engine speed greater than 800rpm (with the vehicle speed greater than 10 but less than 90km/h) and the ambient temperature greater than -8 degrees Celsius. The thermostat should be wide open when cold, but is in error if it opens below desired control temperature. There is a difference between the coolant temperature at the engine and the radiator outlet of less than 5 degrees Celsius. **Possible Causes:** • Check for low coolant level or incorrect coolant mixture • DME detects a short circuit wiring in the ECT • CHT sensor is out-of-calibration or it has failed • ECT sensor is out-of-calibration or it has failed • Replace the thermostat
DTC: P1620 **2001** **Models:** 750i **Engines:** 5.4L **Transmissions:** All	**Thermostat Control Circuit Open Conditions:** The engine's warm up performance is monitored by comparing measured coolant temperature with the modeled coolant temperature to detect a defective coolant thermostat that is reading false. The engine temperature must be less than 65 degrees Celsius, engine speed greater than 800rpm (with the vehicle speed greater than 10 but less than 90km/h) and the ambient temperature greater than -8 degrees Celsius. The thermostat should be wide open when cold, but is in error if it opens below desired control temperature. There is a difference between the coolant temperature at the engine and the radiator outlet of less than 5 degrees Celsius. **Possible Causes:** • Check for low coolant level or incorrect coolant mixture • DME detects a short circuit wiring in the ECT • CHT sensor is out-of-calibration or it has failed • ECT sensor is out-of-calibration or it has failed • Replace the thermostat
DTC: P1624 **2T MIL: Yes (U.S. only)** **2000** **Models:** 323i, 328i, 528i, Z3 **Engines:** 2.8L **Transmissions:** All	**Thermostat Control Circuit Malfunction Conditions:** The engine's warm up performance is monitored by comparing measured coolant temperature with the modeled coolant temperature to detect a defective coolant thermostat that is reading false. The engine temperature must be less than 65 degrees Celsius, engine speed greater than 800rpm (with the vehicle speed greater than 10 but less than 90km/h) and the ambient temperature greater than -8 degrees Celsius. The thermostat should be wide open when cold, but is in error if it opens below desired control temperature. There is a difference between the coolant temperature at the engine and the radiator outlet of less than 5 degrees Celsius. **Possible Causes:** • Check for low coolant level or incorrect coolant mixture • DME detects a short circuit wiring in the ECT • CHT sensor is out-of-calibration or it has failed • ECT sensor is out-of-calibration or it has failed • Replace the thermostat
DTC: P1626 **2T MIL: Yes** **1999** **Models:** 323i, 328i, 528i, Z3 **Engines:** 2.8L **Transmissions:** All	**Throttle/Pedal Position Sensor Rationality Check Conditions:** Engine started, at idle (to 1320rpm), the temperature must be 80 degrees Celsius. The throttle position sensor supplies implausible signal to the DME. The throttle valve activation occurs via an electric motor (throttle drive) in the throttle valve control module. It is activated by the Engine Control Module (DME) according to specifications of the two sensors, Throttle Position (TP) Sensor and Accelerator Pedal Position Sensor 2. **Possible Causes:** • TP sensor signal circuit is open (inspect wiring & connector) • TP sensor signal circuit is shorted to ground • TP sensor or module is damaged or it has failed • Throttle valve is damaged or dirty • Throttle valve control module is faulty

DTC	Trouble Code Title, Conditions & Possible Causes
DTC: P1628 **2T MIL: Yes** **2002, 2003, 2004, 2005, 2006** **Models:** 325i, 330i, 525i, 530i, 545i, 550i, 745i, 750i, Z4 **Engines:** 3.0L, 4.8L **Transmissions:** All	**Throttle Valve Adaptation Spring Test Malfunction During Opening (Bank 1) Conditions:** **Engine started, and the battery voltage is greater than 7 volts, this fault is only diagnosed during the throttle valve's adaptation phase. DVE fails to close from emergency air position. Vehicle speed is zero, throttle pedal is less than 14.9 percent, coolant temperature is between 5.3 and 100.5 degrees Celsius and the intake air temperature is 5.3 degrees Celsius.** **Possible Causes:** • Plate is sticking • Defective return spring • DVE electrical activation defective • DVE motor wiring may have shorted • Throttle valve is contaminated with foreign objects
DTC: P1629 **2T MIL: Yes** **2002, 2003, 2004, 2005, 2006** **Models:** 325i, 330i, 525i, 530i, 545i, 550i, 745i, 750i, Z4 **Engines:** 3.0L, 4.8L **Transmissions:** All	**Throttle Valve Adaptation Spring Test Stop, Spring Does Not Open (Bank 1) Conditions:** **Engine started, and the battery voltage is greater than 7 volts, this fault is only diagnosed during the throttle valve's adaptation phase. Emergency air position not achieved from closed valve. Vehicle speed is zero, throttle pedal is less than 14.9 percent, coolant temperature is between 5.3 and 100.5 degrees Celsius and the intake air temperature is 5.3 degrees Celsius.** **Possible Causes:** • Plate is sticking • Defective return spring • DVE electrical activation defective. • DVE motor wiring may have shorted • Throttle valve is contaminated with foreign objects
DTC: P1631 **2T MIL: Yes** **2002, 2003, 2004, 2005, 2006** **Models:** 325i, 330i, 525i, 530i, 545i, 550i, 745i, 750i, Z4 **Engines:** 3.0L, 4.8L **Transmissions:** All	**Throttle Valve Adaptation Spring Test (Bank 1) Conditions:** **Engine started, and the battery voltage is greater than 7 volts, this fault is only diagnosed during the throttle valve's adaptation phase. It is not possible to move DVE from emergency air position. Vehicle speed is zero, throttle pedal is less than 14.9 percent, coolant temperature is between 5.3 and 100.5 degrees Celsius and the intake air temperature is 5.3 degrees Celsius.** **Possible Causes:** • Plate is sticking • Defective return spring • DVE electrical activation defective. • Throttle position sensor wiring may have shorted • Throttle valve is contaminated with foreign objects
DTC: P1633 **2T MIL: Yes** **2005, 2006** **Models:** 323i, 330i, 525i, 530i, 550i, 750i, Z4 **Engines:** 3.0L, 4.8L **Transmissions:** All	**Throttle Valve Adaptation Limp-Home Position Unknown Conditions:** **Engine started, and the battery voltage is greater than 7 volts, this fault is only diagnosed during the throttle valve's adaptation phase. Check on throttle valve's emergency default position as determined during throttle valve adaptation. There is a failure to reach emergency air position with DVE switched off. Vehicle speed is zero, throttle pedal is less than 14.9 percent, coolant temperature is between 5.3 and 100.5 degrees Celsius and the intake air temperature is 5.3 degrees Celsius.** **Possible Causes:** • Plate is sticking • Defective return spring • DVE electrical activation defective. • DVE motor wiring may have shorted • Throttle valve is contaminated with foreign objects
DTC: P1634 **2T MIL: Yes** **2001, 2002, 2003, 2004, 2005, 2006** **Models:** 325i, 330i, 525i, 530i, 545i, 550i, 745i, 750i, M3, M5, Z3, Z4, Z8 **Engines:** 3.0L, 3.2L, 4.8L, 5.0L **Transmissions:** All	**Throttle Valve Adaptation Spring Test Failed (Bank 1) Conditions:** **Engine started, and the battery voltage is greater than 7 volts, this fault is only diagnosed during the throttle valve's adaptation phase. Vehicle speed is zero, throttle pedal is less than 14.9 percent, coolant temperature is between 5.3 and 100.5 degrees Celsius and the intake air temperature is 5.3 degrees Celsius.** **Possible Causes:** • Plate is sticking • Defective return spring • DVE electrical activation defective. • Throttle position sensor wiring may have shorted • Throttle valve is contaminated with foreign objects
DTC: P1635 **2T MIL: Yes** **2005, 2006** **Models:** 323i, 330i, 525i, 530i, 550i, 750i, Z4 **Engines:** 3.0L, 4.8L **Transmissions:** All	**Throttle Valve Adaptation Lower Mechanical Stop Not Adapted (Bank 1) Conditions:** **Engine started, and the battery voltage is greater than 7 volts, there was no throttle valve adaptation conducted yet and the fuel-supply safety was shutdown until adaptation is successful. There was a failure to reach the lower mechanical travel limit** **Possible Causes:** • Travel stop contaminated. • Defected DVE • Throttle valve contaminated

DTC	Trouble Code Title, Conditions & Possible Causes
DTC: P1636 **2T MIL: Yes** **2001, 2002, 2003, 2004, 2005, 2006** **Models:** 325i, 330i, 525i, 530i, 545i, 550i, 745i, 750i, M3, M5, Z3, Z4, Z8 **Engines:** 3.0L, 3.2L, 4.8L, 5.0L **Transmissions:** All	**Throttle Valve Position Control, Range Check (Bank 1) Conditions:** Engine started, and the battery voltage is greater than 7 volts, and the comparison of the throttle valve's actual angle to its specified angle is great than 0.2 seconds and less than 0.5 seconds greater than a valve calculated from rpm and temperature readings. The throttle valve activation occurs via an electric motor (throttle drive) in the throttle valve control module. It is activated by the Engine Control Module (DME) according to specifications of the two sensors, Throttle Position (TP) Sensor and Accelerator Pedal Position Sensor 2. **Possible Causes:** • TP sensor signal circuit is open (inspect wiring & connector) • TP sensor signal circuit is shorted to ground • TP sensor or module is damaged or it has failed • Throttle valve is damaged or dirty • Throttle valve control module is faulty
DTC: P1637 **2T MIL: Yes** **2001, 2002, 2003, 2004, 2005, 2006** **Models:** 325i, 330i, 525i, 530i, 545i, 550i, 745i, 750i, M3, M5, Z3, Z4, Z8 **Engines:** 3.0L, 3.2L, 4.8L, 5.0L **Transmissions:** All	**Throttle Valve Position Control, Control Deviation (Bank 1) Conditions:** Engine started, and the battery voltage is greater than 7 volts, and the comparison of the throttle valve's actual angle to its specified angle is great than 0.2 seconds and less than 0.5 seconds greater than a valve calculated from rpm and temperature readings. The throttle valve activation occurs via an electric motor (throttle drive) in the throttle valve control module. It is activated by the Engine Control Module (DME) according to specifications of the two sensors, Throttle Position (TP) Sensor and Accelerator Pedal Position Sensor 2. **Possible Causes:** • TP sensor signal circuit is open (inspect wiring & connector) • TP sensor signal circuit is shorted to ground • TP sensor or module is damaged or it has failed • Throttle valve is damaged or dirty • Throttle valve control module is faulty
DTC: P1638 **2T MIL: Yes** **2002, 2003, 2004, 2005, 2006** **Models:** 325i, 330i, 525i, 530i, 545i, 550i, 745i, 750i, M3, M5, Z3, Z4, Z8 **Engines:** 3.0L, 3.2L, 4.8L, 5.0L **Transmissions:** All	**Throttle Valve Position Control Throttle Stuck Temporarily (Bank 1) Conditions:** Engine started, and the battery voltage is greater than 7 volts, and despite control signal to throttle valve, no position change was detected in 0.6 seconds. The throttle valve activation occurs via an electric motor (throttle drive) in the throttle valve control module. It is activated by the Engine Control Module (DME) according to specifications of the two sensors, Throttle Position (TP) Sensor and Accelerator Pedal Position Sensor 2. **Possible Causes:** • TP sensor signal circuit is open (inspect wiring & connector) • TP sensor signal circuit is shorted to ground • TP sensor or module is damaged or it has failed • Throttle valve is damaged or dirty • Throttle valve control module is faulty
DTC: P1639 **2T MIL: Yes** **2001, 2002, 2003, 2004, 2005, 2006** **Models:** 325i, 330i, 525i, 530i, 545i, 550i, 745i, 750i, M3, M5, Z3, Z4, Z8 **Engines:** 3.0L, 3.2L, 4.8L, 5.0L **Transmissions:** All	**Accelerator Pedal Position Sensor 1+2 Range/Performance Conditions:** Engine started, battery voltage at least 11.5v, all electrical components off, ground connections between engine and chassis well connected, the DME detected that the accelerator pedal position sensor signal was too high. **Note: Both the Throttle Position (TP) Sensor and Accelerator Pedal Position Sensor 2 are located at the accelerator pedal module and communicate the driver's intentions to the DME completely independently of each other. Both sensors are stored in one housing.** **Possible Causes:** • Ground between engine and chassis may be broken • Throttle position sensor may have failed • Accelerator Pedal Position Sensor 2 has failed • Throttle position sensor wiring may have shorted • Faulty voltage supply
DTC: P1640 **2T MIL: Yes** **2000, 2001, 2002, 2003, 2005** **Models:** 325i, 330i, 750i , M5, Z8 **Engines:** 3.2L, 5.0L **Transmissions:** All	**Internal Control Module (EEPROM) Error Conditions:** Ignition on, the DME detected a control module malfunction (software). To achieve optimal anti-theft protection for the vehicle, an anti-theft immobilizer is installed. The anti-theft immobilizer is a system for enabling and locking the Engine Control Module (DME). So that this system cannot be circumvented, it is necessary to perform adaptation of the anti-theft immobilizer using the Vehicle Diagnostic and Information System VAS 5052 in the On Board Diagnostic (OBD) function. The great availability of equipment options makes it necessary to adapt the Engine Control Module (DME) to the vehicle (e.g. throttle valve control module or cruise control system). This "writing" function is not possible with the generic scan tool. **Possible Causes:** • Engine Control Module (DME) has failed • Voltage supply for Engine Control Module (DME) has shorted

DTC	Trouble Code Title, Conditions & Possible Causes
DTC: P1641 **2006** **Models:** 325i, 330i, 525i, 530i, 550i, 750i, Z4 **Engines:** 3.0L, 4.8L **Transmissions:** All	**Throttle Valve Adaptation Stop Due to Environmental Conditions Conditions:** **Engine started, and the battery voltage is greater than 7 volts, this fault is only diagnosed during the throttle valve's adaptation phase. Environmental conditions for throttle valve adaptation were not present and adaptation aborted and failed to satisfy conditions. If the previous adaptation was valid, the fault code entry is for information only.** **Possible Causes:** • There is no action required
DTC: P1642 **2006** **Models:** 325i, 330i, 525i, 530i, 550i, 750i, Z4 **Engines:** 3.0L, 4.8L **Transmissions:** All	**Throttle Valve Adaptation Stop Due to Environmental Values Conditions:** **Engine started, and the battery voltage is greater than 7 volts, this fault is only diagnosed during the throttle valve's adaptation phase. Environmental conditions for throttle valve adaptation were not present and adaptation aborted and failed to satisfy conditions. If the previous adaptation was valid, the fault code entry is for information only.** **Possible Causes:** • There is no action required
DTC: P1643 **2006** **Models:** 325i, 330i, 525i, 530i, 550i, 750i, Z4 **Engines:** 3.0L, 4.8L **Transmissions:** All	**Throttle Valve Actuator Start Test Amplifier Balancing Plausibility Conditions:** Engine started, and the battery voltage is greater than 7 volts, this fault is only diagnosed during the throttle valve's adaptation phase. The fault during amplifier calibration leads to operation with unamplified signal from potentiometer 1. The throttle valve activation occurs via an electric motor (throttle drive) in the throttle valve control module. It is activated by the Engine Control Module (DME) according to specifications of the two sensors, Throttle Position (TP) Sensor and Accelerator Pedal Position Sensor 2. **Possible Causes:** • TP sensor signal circuit is open (inspect wiring & connector) • TP sensor signal circuit is shorted to ground • TP sensor or module is damaged or it has failed • Throttle valve is damaged or dirty • Throttle valve control module is faulty
DTC: P1644 **2006** **Models:** 325i, 330i, 525i, 530i, 550i, 750i, Z4 **Engines:** 3.0L, 4.8L **Transmissions:** All	**Throttle Valve Adaptation Stop Relearning Lower Mechanical Stop Conditions:** **Engine started, and the battery voltage is greater than 7 volts, this fault is only diagnosed during the throttle valve's adaptation phase. There was no throttle valve adaptation conducted yet and the fuel-supply safety was shutdown until adaptation is successful. The fault was triggered during an attempt to repeat the initialization and there was a failure to reach the lower mechanical travel limit.** **Possible Causes:** • Travel stop contaminated. • Defected DVE • Throttle valve contaminated
DTC: P1645 **2006** **Models:** 325i, 330i, 525i, 530i, 550i, 750i, Z4 **Engines:** 3.0L, 4.8L **Transmissions:** All	**Internal Control Module Random Access Memory (RAM) Reading Error Conditions:** Key on, and the DME detected an internal memory fault. This code will set if KAPWR to the DME is interrupted (at the initial key on). **Possible Causes:** • Battery terminal corrosion, or loose battery connection • Connection to the DME interrupted, or the circuit has been opened • Reprogramming error has occurred and needs replacement. Remember to check for Aftermarket Performance Products before replacing a DME.
DTC: P1649 **2006** **Models:** 325i, 330i, 525i, 530i, 550i, 750i, Z4 **Engines:** 3.0L, 4.8L **Transmissions:** All	**Internal Control Module Random Access Memory (RAM) Writing Error Conditions:** Key on, and the DME detected an internal memory fault. This code will set if KAPWR to the DME is interrupted (at the initial key on). **Possible Causes:** • Battery terminal corrosion, or loose battery connection • Connection to the DME interrupted, or the circuit has been opened • Reprogramming error has occurred and needs replacement. Remember to check for Aftermarket Performance Products before replacing a DME.
DTC: P1650 **2006** **Models:** 325i, 330i, 525i, 530i, 550i, 750i, Z4 **Engines:** 3.0L, 4.8L **Transmissions:** All	**Start While Engine is Running Conditions:** Engine speed must be at least 1200rpm, the starter is engaged with the engine running and the DME detected an internal memory fault. This code will set if KAPWR to the DME is interrupted (at the initial key on). Check the starter and the starter relay and inspect CAS-SG. **Possible Causes:** • Battery terminal corrosion, or loose battery connection • Connection to the DME interrupted, or the circuit has been opened • Reprogramming error has occurred and needs replacement. Remember to check for Aftermarket Performance Products before replacing a DME.

DTC	Trouble Code Title, Conditions & Possible Causes
DTC: P1660 **2006** **Models:** 325i, 330i, 525i, 530i, 550i, 750i, Z4 **Engines:** 3.0L, 4.8L **Transmissions:** All	**EWS (Electronic Immobilizer) Telegram Error Conditions:** Key on the DME detected an electrical malfunction regarding the EWS. Check the EWS fuse, the ECU or CAN for fault, turn the ignition off and then on to repeat start calibration. **Possible Causes:** • Circuit from the MIL to the DME • ECU Failure, BUS system failure • Circuit from the EPC to the DME • Defective CAS
DTC: P1661 **2006** **Models:** 325i, 330i, 525i, 530i, 550i, 750i, Z4 **Engines:** 3.0L, 4.8L **Transmissions:** All	**Timeout EWS (Electronic Immobilizer) Telegram Conditions:** Key on the DME detected an electrical malfunction regarding the EWS. Check the EWS fuse, the ECU or CAN for fault, turn the ignition off and then on to repeat start calibration. **Possible Causes:** • Circuit from the MIL to the DME • ECU Failure, BUS system failure • Circuit from the EPC to the DME • Defective CAS
DTC: P1662 **2006** **Models:** 325i, 330i, 525i, 530i, 550i, 750i, Z4 **Engines:** 3.0L, 4.8L **Transmissions:** All	**EWS (Electronic Immobilizer) Telegram Parity Error Conditions:** Key on the DME detected an electrical malfunction regarding the EWS. Check the EWS fuse, the ECU or CAN for fault, turn the ignition off and then on to repeat start calibration. **Possible Causes:** • Circuit from the MIL to the DME • ECU Failure, BUS system failure • Circuit from the EPC to the DME • Defective CAS
DTC: P1663 **2006** **Models:** 325i, 330i, 525i, 530i, 550i, 750i, Z4 **Engines:** 3.0L, 4.8L **Transmissions:** All	**EWS (Electronic Immobilizer) Rolling Code Faulty Storage in EEPROM Conditions:** Key on the DME detected an electrical malfunction regarding the EWS. Check the EWS fuse, the ECU or CAN for fault, turn the ignition off and then on to repeat start calibration. **Possible Causes:** • Circuit from the MIL to the DME • ECU Failure, BUS system failure • Circuit from the EPC to the DME • Defective CAS
DTC: P1664 **2006** **Models:** 325i, 330i, 525i, 530i, 550i, 750i, Z4 **Engines:** 3.0L, 4.8L **Transmissions:** All	**EWS (Electronic Immobilizer) Writing/Reading Error in EEPROM Conditions:** Key on the DME detected an electrical malfunction regarding the EWS. Check the EWS fuse, the ECU or CAN for fault, turn the ignition off and then on to repeat start calibration. **Possible Causes:** • Circuit from the MIL to the DME • ECU Failure, BUS system failure • Circuit from the EPC to the DME • Defective CAS
DTC: P1665 **2006** **Models:** 325i, 330i, 525i, 530i, 550i, 750i, Z4 **Engines:** 3.0L, 4.8L **Transmissions:** All	**EWS (Electronic Immobilizer) Tampering Via Rolling Code Conditions:** Key on the DME detected an electrical malfunction regarding the EWS. Check the EWS fuse, the ECU or CAN for fault, turn the ignition off and then on to repeat start calibration. **Possible Causes:** • Circuit from the MIL to the DME • ECU Failure, BUS system failure • Circuit from the EPC to the DME • Defective CAS
DTC: P1666 **2006** **Models:** 325i, 330i, 525i, 530i, 550i, 750i, Z4 **Engines:** 3.0L, 4.8L **Transmissions:** All	**EWS (Electronic Immobilizer) Tampering/Start Value Not Yet Programmed Conditions:** Key on the DME detected an electrical malfunction regarding the EWS. Check the EWS fuse, the ECU or CAN for fault, turn the ignition off and then on to repeat start calibration. **Possible Causes:** • Circuit from the MIL to the DME • ECU Failure, BUS system failure • Circuit from the EPC to the DME • Defective CAS

DTC	Trouble Code Title, Conditions & Possible Causes
DTC: P1667 **2006** **Models:** 325i, 330i, 525i, 530i, 550i, 750i, Z4 **Engines:** 3.0L, 4.8L **Transmissions:** All	**EWS (Electronic Immobilizer) Start Value Not Yet Programmed Conditions:** Key on the DME detected an electrical malfunction regarding the EWS. Check the EWS fuse, the ECU or CAN for fault, turn the ignition off and then on to repeat start calibration. **Possible Causes:** • Circuit from the MIL to the DME • ECU Failure, BUS system failure • Circuit from the EPC to the DME • Defective CAS
DTC: P1668 **2006** **Models:** 325i, 330i, 525i, 530i, 550i, 750i, Z4 **Engines:** 3.0L, 4.8L **Transmissions:** All	**EWS (Electronic Immobilizer) Start Value Destroyed Conditions:** Key on the DME detected an electrical malfunction regarding the EWS. Check the EWS fuse, the ECU or CAN for fault, turn the ignition off and then on to repeat start calibration. **Possible Causes:** • Circuit from the MIL to the DME • ECU Failure, BUS system failure • Circuit from the EPC to the DME
DTC: P1677 **MIL: No** **2006** **Models:** 325i, 330i, 525i, 530i, 550i, 750i, Z4 **Engines:** 3.0L, 4.8L **Transmissions:** All	**Adaptive Cruise Control No Activity Detected Conditions:** Key on after 800ms or engine running for 100ms, the DME detected an electrical malfunction regarding the Adaptive Cruise Control circuit. Check the ACC fuse, the ECU or CAN for fault, measure the resistance on the BUS, check the wiring harness. If this fault occurs when the ACC is not installed, the DME has been incorrectly initialized for the ACC version. **Possible Causes:** • Circuit from the MIL to the DME • ECU Failure, BUS system failure • Circuit from the EPC to the DME
DTC: P1680 **MIL: No** **2006** **Models:** 325i, 330i, 525i, 530i, 550i, 750i, Z4 **Engines:** 3.0L, 4.8L **Transmissions:** All	**Electronic Throttle Control Monitor Level 2/3 ADC Processor Fault Conditions:** Key on, engine running to at least 1200rpm, the DME has detected an internal fault in the computer or internal fault in the control modules. The Torque monitoring feature compares the torque demand (from accelerator pedal, FGR, electrical equipment, transmission) with the torque provided (calculated from HFM, injector valves, ignition angle, throttle valve angle, differential pressure, lambda). Deviations trigger fuel-supply safety shutdown to prevent vehicle from autonomous acceleration. Internal fault in computer or in electronic control modules (check whether all ADC channels have been converted). If additional fault codes are entered in the DME, resolve these issues. If fault remains, replace the DME. **Possible Causes:** • Battery terminal corrosion, loose battery connection, or faulty • Connection to the DME interrupted, or the circuit has been opened • Reprogramming error has occurred and needs replacement. • Voltage supply for Engine Control Module (DME) is faulty
DTC: P1702 **2T MIL: Yes** 2003 **Models:** 745i **Engines:** 4.8L **Transmissions:** A/T	**TR Sensor Signal Intermittent Conditions:** Key on or engine running; and the DME detected the failure Trouble Code Conditions for DTC P0705 or P0708 were met intermittently. **Possible Causes:** • Refer to the appropriate Transmission Repair Manual or information in electronic media to perform a complete diagnosis of the automatic transmission when this code is set
DTC: P1719 **MIL: Yes (U.S. only)** **2002, 2003, 2004** **Models:** 325i, 330i, 525i, 530i, 545i, 550i, 745i, 750i, M3, M5, Z3, Z4, Z8 **Engines:** 3.0L, 3.2L, 4.8L, 5.0L **Transmissions:** All	**CAN Level Wrong Value Conditions:** The Engine Control Module (DME) communicates with all databus-capable control modules via a CAN databus. These databus-capable control modules are connected via two data bus wires which are twisted together (CAN_High and CAN_Low), and exchange information (messages). Missing information on the databus is recognized as a malfunction and stored. Trouble-free operation of the CAN-Bus requires that it have a terminal resistance. This central terminal resistor is located in the Engine Control Module (DME). The ignition is on for 800ms or the engine running for 15 seconds, the voltage is greater than 10 volts. This applies to vehicles with MIL only and does not affect bus activity. **Possible Causes:** • CAN data bus wires have short circuited to each other • ECU has failed • BUS system failure • Defective bus controller (EGS)

DTC	Trouble Code Title, Conditions & Possible Causes
DTC: P1720 **MIL: Yes (U.S. only)** **2002, 2003, 2004** **Models:** 325i, 330i, 525i, 530i, 545i, 550i, 745i, 750i, M3, M5, Z3, Z4, Z8 **Engines:** 3.0L, 3.2L, 4.8L, 5.0L **Transmissions:** All	**CAN Message Timeout Conditions:** The Engine Control Module (DME) communicates with all databus-capable control modules via a CAN databus. These databus-capable control modules are connected via two data bus wires which are twisted together (CAN_High and CAN_Low), and exchange information (messages). Missing information on the databus is recognized as a malfunction and stored. Trouble-free operation of the CAN-Bus requires that it have a terminal resistance. This central terminal resistor is located in the Engine Control Module (DME). The ignition is on for 800ms or the engine running for 15 seconds, the voltage is greater than 10 volts. This applies to vehicles with MIL only and does not affect bus activity. **Possible Causes:** • CAN data bus wires have short circuited to each other • ECU has failed • BUS system failure • Defective bus controller (EGS)
DTC: P1721 **MIL: Yes (U.S. only)** **2002, 2003, 2004** **Models:** 325i, 330i, 525i, 530i, 545i, 550i, 745i, 750i, M3, M5, Z3, Z4, Z8 **Engines:** 3.0L, 3.2L, 4.8L, 5.0L **Transmissions:** A/T	**CAN Timeout ASC/DSC Conditions:** The Engine Control Module (DME) communicates with all databus-capable control modules via a CAN databus. These databus-capable control modules are connected via two data bus wires which are twisted together (CAN_High and CAN_Low), and exchange information (messages). Missing information on the databus is recognized as a malfunction and stored. Trouble-free operation of the CAN-Bus requires that it have a terminal resistance. This central terminal resistor is located in the Engine Control Module (DME). The ignition is on for 800ms or the engine running for 15 seconds, the voltage is greater than 10 volts. This applies to vehicles with MIL only and does not affect bus activity. **Possible Causes:** • CAN data bus wires have short circuited to each other • ECU has failed • BUS system failure • Defective bus controller (EGS)
DTC: P1727 **2T MIL: Yes** **2002, 2003, 2004** **Models:** 325i, 330i, 525i, 530i, 545i, 550i, 745i, 750i, M3, M5, Z3, Z4, Z8 **Engines:** 3.0L, 3.2L, 4.8L, 5.0L **Transmissions:** A/T	**Engine Speed Signal Plausibility Conditions:** The CAN Message signal error flag alive counter or check sum is sending/receiving an incorrect signal. There is no alteration of the alive counter and no wrong check sum. The DME-CAN Connection is okay. The CAN-Bus is okay and the ignition is on. **Possible Causes:** • Pressure control solenoid circuit is shorting to ground • Pressure control solenoid circuit is open • Valve has failed • TCM has failed
DTC: P1728 **2T MIL: Yes** **2002, 2003, 2004** **Models:** 325i, 330i, 525i, 530i, 545i, 550i, 745i, 750i, M3, M5, Z3, Z4, Z8 **Engines:** 3.0L, 3.2L, 4.8L, 5.0L **Transmissions:** A/T	**Engine Overspeed Plausibility Conditions:** The engine speed is over 10,000rpm. The range position is D, and the park-lock sensor is okay. **Possible Causes:** • Pressure control solenoid circuit is shorting to ground • Pressure control solenoid circuit is open • Valve has failed • TCM has failed
DTC: P1734 **2T MIL: Yes** **1996, 1997, 1998, 1999, 2000** **Models:** 325i, 330i, 540i, 740i , 840Ci, 850Ci **Engines:** 4.4L, 5.4L **Transmissions:** A/T	**Pressure Control Valve 2 Electrical Conditions:** The pressure control valve is not receiving nor sending a signal. **Possible Causes:** • Pressure control solenoid circuit is shorting to ground • Pressure control solenoid circuit is open • Valve has failed • TCM has failed
DTC: P1738 **2T MIL: Yes** **1996, 1997, 1998, 1999, 2000** **Models:** 325i, 330i, 540i, 740i , 840Ci, 850Ci **Engines:** 4.4L, 5.4L **Transmissions:** A/T	**Pressure Control Valve 3 Electrical Conditions:** The pressure control valve is not receiving nor sending a signal. **Possible Causes:** • Pressure control solenoid circuit is shorting to ground • Pressure control solenoid circuit is open • Valve has failed • TCM has failed
DTC: P1740 **2T MIL: Yes** **2002** **Models:** 540i **Engines:** 4.4L **Transmissions:**A/T	**Clutch Solenoid Circuit Range/Performance Conditions:** The clutch solenoid circuit continuity is in error. **Possible Causes:** • Short to battery • Short to ground • Open circuit

DTC	Trouble Code Title, Conditions & Possible Causes
DTC: P1743 **2T MIL: Yes** **1996, 1997, 1998, 1999, 2000** **Models:** All **Engines:** All **Transmissions:** A/T	**Pressure Regulator Valve 5 Upper Threshold Conditions:** The signal to/from the pressure regulator valve has been interrupted. **Possible Causes:** • Pressure control solenoid circuit is shorting to ground • Pressure control solenoid circuit is open • Valve has failed • TCM has failed
DTC: P1745 **2T MIL: Yes** **2002** **Models:** 540i, 745i **Engines:** 4.4L, 4.8L **Transmissions:**A/T	**Pressure Regulator Valve 5 Plausibility Conditions:** The current to/from the pressure regulator valve is either higher or lower than the threshold value. **Possible Causes:** • Pressure control solenoid circuit is shorting to ground • Pressure control solenoid circuit is open • Valve has failed • TCM has failed
DTC: P1747 **MIL: Yes (U.S. only)** **2002, 2003, 2004** **Models:** 325i, 330i, 525i, 530i, 545i, 550i, 745i, 750i, M3, M5, Z3, Z4, Z8 **Engines:** 3.0L, 3.2L, 4.8L, 5.0L **Transmissions:** All	**CAN-Bus Plausibility-Disabled Conditions:** The Engine Control Module (DME) communicates with all databus-capable control modules via a CAN databus. These databus-capable control modules are connected via two data bus wires which are twisted together (CAN_High and CAN_Low), and exchange information (messages). Missing information on the databus is recognized as a malfunction and stored. Trouble-free operation of the CAN-Bus requires that it have a terminal resistance. This central terminal resistor is located in the Engine Control Module (DME). The ignition is on for 800ms or the engine running for 15 seconds, the voltage is greater than 10 volts. This applies to vehicles with MIL only and does not affect bus activity. **Possible Causes:** • CAN data bus wires have short circuited to each other • ECU has failed • BUS system failure • Defective bus controller (EGS)
DTC: P1750 **2T MIL: Yes** **1998, 1999, 2000** **Models:** 328i, 528i, 540i, 740i, 750i, M3 **Engines:** 2.8L, 4.4L, 5.4L **Transmissions:** A/T	**Shift Solenoid Power Supply Malfunction Conditions:** The shift solenoid valve is not receiving a proper power supply after 300ms of operation. **Possible Causes:** • Pressure control solenoid circuit is shorting to ground • Pressure control solenoid circuit is open • Valve has failed • TCM has failed
DTC: P1753 **2T MIL: Yes** **1998, 1999, 2000, 2001, 2002** **Models:** 328i, 528i, 540i, 740i, 750i, M3 **Engines:** 2.8L, 4.4L, 5.4L **Transmissions:** A/T	**Pressure Regulator Valve 4 Upper Threshold Conditions:** The signal to/from the pressure regulator valve has been interrupted. **Possible Causes:** • Pressure control solenoid circuit is shorting to ground • Pressure control solenoid circuit is open • Valve has failed • TCM has failed
DTC: P1755 **2T MIL: Yes** **2002** **Models:** 745i **Engines:** 4.8L **Transmissions:** A/T	**Pressure Regulator Valve 4 Plausibility Conditions:** The current to/from the pressure regulator valve is either higher or lower than the threshold value. **Possible Causes:** • Pressure control solenoid circuit is shorting to ground • Pressure control solenoid circuit is open • Valve has failed • TCM has failed
DTC: P1756 **2T MIL: Yes** 2003, 2005 **Models:** M3 **Engines:** 3.2L **Transmissions:** A/T	**Shift Monitoring Functionality Check Malfunction Conditions:** X-Position gear cannot be engaged. **Possible Causes:** • Pressure control solenoid circuit is shorting to ground • Pressure control solenoid circuit is open • Mechanical failure in the transmission • TCM has failed
DTC: P1758 **2T MIL: Yes** 2003 **Models:** M3 **Engines:** 3.2L **Transmissions:** A/T	**Shift Monitoring Functionality Check Malfunction Conditions:** Y-Position gear cannot be adjusted. **Possible Causes:** • Pressure control solenoid circuit is shorting to ground • Pressure control solenoid circuit is open • Mechanical failure in the transmission • TCM has failed

DTC	Trouble Code Title, Conditions & Possible Causes
DTC: P1758 **2T MIL: Yes** **1998, 1999, 2000** **Models:** 328i, 528i, 540i, 740i, 750i, M3 **Engines:** 2.8L, 4.4L, 5.4L **Transmissions:** A/T	**Shift Solenoid B Conditions:** The shift solenoid valve is not receiving a signal after 50ms of operation. **Possible Causes:** • Pressure control solenoid circuit is shorting to ground • Pressure control solenoid circuit is open • Valve has failed • TCM has failed
DTC: P1759 **2T MIL: Yes** 2003, 2005 **Models:** M3 **Engines:** 3.2L **Transmissions:** A/T	**Shift Monitoring Functionality Check Malfunction Conditions:** X-Position gear cannot be disengaged. **Possible Causes:** • Pressure control solenoid circuit is shorting to ground • Pressure control solenoid circuit is open • Mechanical failure in the transmission • TCM has failed
DTC: P1762 **2T MIL: Yes** **2002, 2003, 2004** **Models:** 325i, 330i, 525i, 530i, 545i, 550i, 745i, 750i, M3, M5, Z3, Z4, Z8 **Engines:** 3.0L, 3.2L, 4.8L, 5.0L **Transmissions:** A/T	**Shift Solenoid C Short to Power Conditions:** The shift solenoid valve is shorting to power after 50ms of operation. **Possible Causes:** • Pressure control solenoid circuit is shorting to ground • Pressure control solenoid circuit is open • Valve has failed • TCM has failed
DTC: P1763 **2T MIL: Yes** **1998, 1999, 2000, 2002, 2003, 2004** **Models:** All **Engines:** 2.5L, 3.0L, 3.2L, 4.4L, 5.4L **Transmissions:** A/T	**Shift Solenoid C Short to Ground Conditions:** The shift solenoid valve is shorting to ground after 50ms of operation. **Possible Causes:** • Pressure control solenoid circuit is shorting to ground • Pressure control solenoid circuit is open • Valve has failed • TCM has failed
DTC: P1764 **2T MIL: Yes** **2002, 2003, 2004** **Models:** 325i, 330i, 525i, 530i, 545i, 550i, 745i, 750i, M3, M5, Z3, Z4, Z8 **Engines:** 3.0L, 3.2L, 4.8L, 5.0L **Transmissions:** A/T	**Shift Solenoid C Short Circuit Continuity-Disconnection Conditions:** The shift solenoid valve has a disconnection continuity after 50ms of operation. **Possible Causes:** • Pressure control solenoid circuit is shorting to ground • Pressure control solenoid circuit is open • Valve has failed • TCM has failed
DTC: P1765 **2T MIL: Yes** 2003, 2004 **Models:** 325i, 330i, 525i, 530i, 545i, 550i, 745i, 750i, M3, M5, Z3, Z4, Z8 **Engines:** 3.0L, 3.2L, 4.8L, 5.0L **Transmissions:** A/T	**Throttle Valve Signal Plausibility Conditions:** The CAN Message signal error flag alive counter or check sum is sending/receiving an incorrect signal. There is no alteration of the alive counter and no wrong check sum. The DME-CAN Connection is okay. The CAN-Bus is okay and the ignition is on. **Possible Causes:** • Throttle solenoid circuit is shorting to ground • Throttle valve sensor has failed • Valve has failed • DME/TCM has failed
DTC: P1766 **2T MIL: Yes** 2003, 2004 **Models:** 325i, 330i, 525i, 530i, 545i, 550i, 745i, 750i, M3, M5, Z3, Z4, Z8 **Engines:** 3.0L, 3.2L, 4.8L, 5.0L **Transmissions:** A/T	**Engine Speed Plausibility Conditions:** The CAN Message signal error flag alive counter or check sum is sending/receiving an incorrect signal. There is no alteration of the alive counter and no wrong check sum: the sensor is disconnected or there is a multiple fault recorded. The DME-CAN Connection is okay. The CAN-Bus is okay and the ignition is on. **Possible Causes:** • Throttle solenoid circuit is shorting to ground • Throttle valve sensor has failed • Valve has failed • DME/TCM has failed

DTC	Trouble Code Title, Conditions & Possible Causes
DTC: P1771 **2T MIL: Yes** **2002, 2003, 2004** **Models:** 325i, 330i, 525i, 530i, 545i, 550i, 745i, 750i, M3, M5, Z3, Z4, Z8 **Engines:** 3.0L, 3.2L, 4.8L, 5.0L **Transmissions:** A/T	**Engine Torque Plausibility Conditions:** The CAN Message signal error flag alive counter or check sum is sending/receiving an incorrect signal. There is no alteration of the alive counter and no wrong check sum. The DME-CAN Connection is okay. The CAN-Bus is okay and the ignition is on. **Possible Causes:** • Throttle solenoid circuit is shorting to ground • Throttle valve sensor has failed • Valve has failed • DME/TCM has failed
DTC: P1782 **2T MIL: Yes** 2003, 2004 **Models:** 325i, 330i, 525i, 530i, 545i, 550i, 745i, 750i, M3, M5, Z3, Z4, Z8 **Engines:** 3.0L, 3.2L, 4.8L, 5.0L **Transmissions:** A/T	**Brake Pedal Signal Plausibility Conditions:** The CAN Message signal error flag alive counter or check sum is sending/receiving an incorrect signal. There is no alteration of the alive counter and no wrong check sum. The DME-CAN Connection is okay. The CAN-Bus is okay and the ignition is on. **Possible Causes:** • Mechanical malfunction in the brake system • Brake solenoid circuits are open or grounded • DME/TCM has failed
DTC: P1790 **2T MIL: Yes** 2003, 2004 **Models:** 325i, 330i, 525i, 530i, 545i, 550i, 745i, 750i, M3, M5, Z3, Z4, Z8 **Engines:** 3.0L, 3.2L, 4.8L, 5.0L **Transmissions:** A/T	**Internal Control Module Memory Checksum Error Conditions:** During the self check, the EEPROM check detected an invalid checksum. The self/rationality check EEPROM resulted in an invalid checksum at power up. **Possible Causes:** • Battery terminal corrosion, or loose battery connection • Connection to the DME interrupted, or the circuit has been opened • Reprogramming error has occurred and needs replacement. Remember to check for Aftermarket Performance Products before replacing an DME. • Connection to the DME interrupted, or the circuit has been opened • Voltage supply for Engine Control Module (DME) is faulty
DTC: P1791 **2T MIL: Yes** 2004 **Models:** 325i, 330i, 525i, 530i, 545i, 745i, M3, Z4 **Engines:** 2.5L, 3.0L, 3.2L **Transmissions:** A/T	**EEPROM Failure Conditions:** During the self check, the EEPROM check detected an invalid checksum. The self/rationality check EEPROM resulted in an invalid checksum at power up. **Possible Causes:** • Battery terminal corrosion, or loose battery connection • Connection to the DME interrupted, or the circuit has been opened • Reprogramming error has occurred and needs replacement. Remember to check for Aftermarket Performance Products before replacing an DME. • Connection to the DME interrupted, or the circuit has been opened • Voltage supply for Engine Control Module (DME) is faulty
DTC: P1801 **2T MIL: Yes** **2002** **Models:** 325i, 330i, 525i, 745i, Z3 **Engines:** 3.0L, 4.8L **Transmissions:** A/T	**Solenoid Valve 1 Lower Threshold Conditions:** The signal to/from the pressure regulator valve is interrupted or short circuited to ground. **Possible Causes:** • Pressure control solenoid circuit is shorting to ground • Pressure control solenoid circuit is open • Valve has failed • TCM has failed
DTC: P1802 **2T MIL: Yes** **2002** **Models:** 325i, 330i, 525i, 745i, Z3 **Engines:** 3.0L, 4.8L **Transmissions:** A/T	**Solenoid Valve 2 Lower Threshold Conditions:** The signal to/from the pressure regulator valve is interrupted or short circuited to ground. **Possible Causes:** • Pressure control solenoid circuit is shorting to ground • Pressure control solenoid circuit is open • Valve has failed • TCM has failed
DTC: P1803 **2T MIL: Yes** **2002** **Models:** 325i, 330i, 525i, 745i, Z3 **Engines:** 3.0L, 4.8L **Transmissions:** A/T	**Solenoid Valve 3 Lower Threshold Conditions:** The signal to/from the pressure regulator valve is interrupted or short circuited to ground. **Possible Causes:** • Pressure control solenoid circuit is shorting to ground • Pressure control solenoid circuit is open • Valve has failed • TCM has failed

DTC	Trouble Code Title, Conditions & Possible Causes
DTC: P1806 **2T MIL: Yes** **2002, 2003** **Models:** 745i **Engines:** 4.8L **Transmissions:** A/T	**Solenoid Valve 1 or 2 Plausibility Conditions:** The signal to/from the pressure regulator valve is out of range. The engine speed greater than 100rpm. **Possible Causes:** • Pressure control solenoid circuit is shorting to ground • Pressure control solenoid circuit is open • Valve has failed • TCM has failed
DTC: P1810 **2T MIL: Yes** **2002, 2003, 2004** **Models:** 320i, 330i, 525i, 530i, 545i, 745i, M3, Z4 **Engines:** 2.5L, 2.8L, 3.0L, 4.8L **Transmissions:** A/T	**Input/Turbine Speed Sensor Circuit Malfunction Upper Threshold Conditions:** Engine started, vehicle driven with the vehicle speed sensor indicating more than 1 mph, and the DME detected the Transmission Vehicle Speed Sensor signals were erratic, or that they were missing for a period of time. The engine speed is greater than 600rpm. Any gear can be selected, output speed must be greater than 600rpm, and wheel speed greater than 400rpm. **Possible Causes:** • TVSS signal circuit is open • TVSS signal is shorted to chassis ground • TVSS signal is shorted to sensor ground • TVSS assembly is damaged or it has failed
DTC: P1811 **2T MIL: Yes** **2002, 2003, 2004** **Models:** 320i, 330i, 525i, 530i, 545i, 745i, M3, Z4 **Engines:** 2.5L, 2.8L, 3.0L, 4.8L **Transmissions:** A/T	**Input/Turbine Speed Sensor Circuit Malfunction Lower Threshold Conditions:** Engine started, vehicle driven with the vehicle speed sensor indicating more than 1 mph, and the DME detected the Transmission Vehicle Speed Sensor signals were erratic, or that they were missing for a period of time. The engine speed is greater than 600rpm. Any gear can be selected, output speed must be greater than 600rpm, and wheel speed greater than 400rpm. **Possible Causes:** • TVSS signal circuit is open • TVSS signal is shorted to chassis ground • TVSS signal is shorted to sensor ground • TVSS assembly is damaged or it has failed
DTC: P1812 **2T MIL: Yes** **2002, 2003, 2004** **Models:** 320i, 330i, 525i, 530i, 545i, 745i, M3, Z4 **Engines:** 2.5L, 2.8L, 3.0L, 4.8L **Transmissions:** A/T	**Output Speed Sensor Circuit Malfunction Upper Threshold Conditions:** Engine started, vehicle driven with the vehicle speed sensor indicating more than 1 mph, and the DME detected the Transmission Vehicle Speed Sensor signals were erratic, or that they were missing for a period of time. The engine speed is greater than 600rpm. Any gear can be selected, output speed must be greater than 600rpm, and wheel speed greater than 400rpm. The sensor supply status is okay. **Possible Causes:** • TVSS signal circuit is open • TVSS signal is shorted to chassis ground • TVSS signal is shorted to sensor ground • TVSS assembly is damaged or it has failed
DTC: P1813 **2T MIL: Yes** **2002, 2003, 2004** **Models:** 320i, 330i, 525i, 530i, 545i, 745i, M3, Z4 **Engines:** 2.5L, 2.8L, 3.0L, 4.8L **Transmissions:** A/T	**Output Speed Sensor Circuit Malfunction Lower Threshold Conditions:** Engine started, vehicle driven with the vehicle speed sensor indicating more than 1 mph, and the DME detected the Transmission Vehicle Speed Sensor signals were erratic, or that they were missing for a period of time. The engine speed is greater than 600rpm. Any gear can be selected, output speed must be greater than 600rpm, and wheel speed greater than 400rpm. The sensor supply status is okay. **Possible Causes:** • TVSS signal circuit is open • TVSS signal is shorted to chassis ground • TVSS signal is shorted to sensor ground • TVSS assembly is damaged or it has failed
DTC: P1814 **2T MIL: Yes** 2003, 2004 **Models:** 320i, 330i, 525i, 530i, 545i, 745i, M3, Z4 **Engines:** 2.5L, 2.8L, 3.0L, 4.8L **Transmissions:** A/T	**Output Speed Sensor Circuit Malfunction Plausibility Conditions:** Engine started, vehicle driven with the vehicle speed sensor indicating more than 1 mph, and the DME detected the Transmission Vehicle Speed Sensor signals were erratic, or that they were missing for a period of time. There was a negative gradient of the signal which was greater than the threshold at 1000rpm or 10 seconds after start. The engine speed is greater than 600rpm. Any gear can be selected, output speed must be greater than 600rpm, and wheel speed greater than 400rpm. The sensor supply status is okay. **Possible Causes:** • TVSS signal circuit is open • TVSS signal is shorted to chassis ground • TVSS signal is shorted to sensor ground • TVSS assembly is damaged or it has failed

DTC	Trouble Code Title, Conditions & Possible Causes
DTC: P1830 **2T MIL: Yes** 2003 **Models:** 745i **Engines:** 4.8L **Transmissions:** A/T	**Pressure Regulator Valve 1 Plausibility Conditions:** The current to/from the pressure regulator valve is either higher or lower than the threshold value. **Possible Causes:** • Pressure control solenoid circuit is shorting to ground • Pressure control solenoid circuit is open • Valve has failed • TCM has failed
DTC: P1831 **2T MIL: Yes** **2002** **Models:** 325i, 525i, 540i, 745i, Z3 **Engines:** 3.0L, 4.4L **Transmissions:** A/T	**Pressure Regulator Valve 1 Upper Threshold Conditions:** The signal to/from the pressure regulator valve is interrupted or short circuited to supply. **Possible Causes:** • Pressure control solenoid circuit is shorting to ground • Pressure control solenoid circuit is open • Valve has failed • TCM has failed
DTC: P1832 **2T MIL: Yes** **2002** **Models:** 325i, 525i, 540i, 745i, Z3 **Engines:** 3.0L, 4.4L **Transmissions:** A/T	**Pressure Regulator Valve 2 Upper Threshold Conditions:** The signal to/from the pressure regulator valve is interrupted or short circuited to supply. **Possible Causes:** • Pressure control solenoid circuit is shorting to ground • Pressure control solenoid circuit is open • Valve has failed • TCM has failed
DTC: P1833 **2T MIL: Yes** **2002** **Models:** 325i, 525i, 540i, 745i, Z3 **Engines:** 3.0L, 4.4L **Transmissions:** A/T	**Pressure Regulator Valve 3 Upper Threshold Conditions:** The signal to/from the pressure regulator valve is interrupted or short circuited to supply. **Possible Causes:** • Pressure control solenoid circuit is shorting to ground • Pressure control solenoid circuit is open • Valve has failed • TCM has failed
DTC: P1834 **2T MIL: Yes** **2002** **Models:** 325i, 525i, 540i, 745i, Z3 **Engines:** 3.0L, 4.4L **Transmissions:** A/T	**Pressure Regulator Valve 4 Upper Threshold Conditions:** The signal to/from the pressure regulator valve is interrupted or short circuited to supply. **Possible Causes:** • Pressure control solenoid circuit is shorting to ground • Pressure control solenoid circuit is open • Valve has failed • TCM has failed
DTC: P1835 **2T MIL: Yes** **2002** **Models:** 540i, 745i **Engines:** 4.4L, 4.8L **Transmissions:** A/T	**Pressure Regulator Valve 5 Upper Threshold Conditions:** The signal to/from the pressure regulator valve is interrupted or short circuited to supply. **Possible Causes:** • Pressure control solenoid circuit is shorting to ground • Pressure control solenoid circuit is open • Valve has failed • TCM has failed
DTC: P1836 **2T MIL: Yes** **2002, 2003** **Models:** 745i **Engines:** 4.8L **Transmissions:** A/T	**TCC Valve Upper Threshold Conditions:** Engine started to 400rpm, FET enabled, vehicle driven in gear with VSS signals received, and the DME detected a short to supply while in normal operation. **Possible Causes:** • TCC solenoid has a mechanical failure • TCC solenoid has a hydraulic failure
DTC: P1841 **2T MIL: Yes** **2002** **Models:** 325i, 525i, 540i, 745i, Z3 **Engines:** 3.0L, 4.4L **Transmissions:** A/T	**Pressure Regulator Valve 1 Lower Threshold Conditions:** The signal to/from the pressure regulator valve is interrupted or short circuited to ground. **Possible Causes:** • Pressure control solenoid circuit is shorting to ground • Pressure control solenoid circuit is open • Valve has failed • TCM has failed

DTC	Trouble Code Title, Conditions & Possible Causes
DTC: P1842 **2T MIL: Yes** **2002** **Models:** 325i, 525i, 540i, 745i, Z3 **Engines:** 3.0L, 4.4L **Transmissions:** A/T	**Pressure Regulator Valve 2 Lower Threshold Conditions:** The signal to/from the pressure regulator valve is interrupted or short circuited to ground. **Possible Causes:** • Pressure control solenoid circuit is shorting to ground • Pressure control solenoid circuit is open • Valve has failed • TCM has failed
DTC: P1843 **2T MIL: Yes** **2002** **Models:** 325i, 525i, 540i, 745i, Z3 **Engines:** 3.0L, 4.4L **Transmissions:** A/T	**Pressure Regulator Valve 3 Lower Threshold Conditions:** The signal to/from the pressure regulator valve is interrupted or short circuited to ground. **Possible Causes:** • Pressure control solenoid circuit is shorting to ground • Pressure control solenoid circuit is open • Valve has failed • TCM has failed
DTC: P1844 **2T MIL: Yes** **2002** **Models:** 325i, 525i, 540i, 745i, Z3 **Engines:** 3.0L, 4.4L **Transmissions:** A/T	**Pressure Regulator Valve 4 Lower Threshold Conditions:** The signal to/from the pressure regulator valve is interrupted or short circuited to ground. **Possible Causes:** • Pressure control solenoid circuit is shorting to ground • Pressure control solenoid circuit is open • Valve has failed • TCM has failed
DTC: P1845 **2T MIL: Yes** **2002** **Models:** 540i, 745i **Engines:** 4.4L 4.8L **Transmissions:** A/T	**Pressure Regulator Valve 5 Lower Threshold Conditions:** The signal to/from the pressure regulator valve is interrupted or short circuited to ground. **Possible Causes:** • Pressure control solenoid circuit is shorting to ground • Pressure control solenoid circuit is open • Valve has failed • TCM has failed
DTC: P1846 **2T MIL: Yes** **2002, 2003** **Models:** 745i **Engines:** 4.8L **Transmissions:** A/T	**TCC Valve Lower Threshold Conditions:** Engine started to 400rpm, FET enabled, vehicle driven in gear with VSS signals received, and the DME detected an interruption or short to ground while in normal operation. **Possible Causes:** • TCC solenoid has a mechanical failure • TCC solenoid has a hydraulic failure
DTC: P1861 **2T MIL: Yes** 2003, 2004 **Models:** 320i, 330i, 525i, 530i, 545i, 745i, M3, Z4 **Engines:** 2.5L, 3.0L, 3.2L, 4.8L **Transmissions:** A/T	**2-1 Shift Range Monitoring General Malfunction Conditions:** Engine running and vehicle driven, the DME detected a mechanical malfunction within the transmission. The output speed is greater than 300rpm, the transmission oil temperature is greater than 0 degrees Celsius, the engine speed is greater or equal to 600rpm and the range position is P, R, or N. **Possible Causes:** • Solenoid valves in valve body are faulty • Solenoid circuit is shorting to ground • Solenoid circuit is open • TCM has failed or wiring is shorting • Mechanical malfunction in transmission
DTC: P1862 **2T MIL: Yes** 2003, 2004 **Models:** 320i, 330i, 525i, 530i, 545i, 745i, M3, Z4 **Engines:** 2.5L, 3.0L, 3.2L, 4.8L **Transmissions:** A/T	**3-2 Shift Range Monitoring Conditions:** Engine running and vehicle driven, the DME detected a mechanical malfunction within the transmission. The output speed is greater than 300rpm, the transmission oil temperature is greater than 0 degrees Celsius, the engine speed is greater or equal to 600rpm and the range position is P, R, or N. **Possible Causes:** • Solenoid valves in valve body are faulty • Solenoid circuit is shorting to ground • Solenoid circuit is open • TCM has failed or wiring is shorting • Mechanical malfunction in transmission

DTC	Trouble Code Title, Conditions & Possible Causes
DTC: P1863 **2T MIL: Yes** 2003, 2004 **Models:** 320i, 330i, 525i, 530i, 545i, 745i, M3, Z4 **Engines:** 2.5L, 3.0L, 3.2L, 4.8L **Transmissions:** A/T	**4-3 Shift Range Monitoring Conditions:** Engine running and vehicle driven, the DME detected a mechanical malfunction within the transmission. The output speed is greater than 300rpm, the transmission oil temperature is greater than 0 degrees Celsius, the engine speed is greater or equal to 600rpm and the range position is P, R, or N. **Possible Causes:** • Solenoid valves in valve body are faulty • Solenoid circuit is shorting to ground • Solenoid circuit is open • TCM has failed or wiring is shorting • Mechanical malfunction in transmission
DTC: P1864 **2T MIL: Yes** 2003, 2004 **Models:** 320i, 330i, 525i, 530i, 545i, 745i, M3, Z4 **Engines:** 2.5L, 3.0L, 3.2L, 4.8L **Transmissions:** A/T	**5-4 Shift Range Monitoring Conditions:** Engine running and vehicle driven, the DME detected a mechanical malfunction within the transmission. The output speed is greater than 300rpm, the transmission oil temperature is greater than 0 degrees Celsius, the engine speed is greater or equal to 600rpm and the range position is P, R, or N. **Possible Causes:** • Solenoid valves in valve body are faulty • Solenoid circuit is shorting to ground • Solenoid circuit is open • TCM has failed or wiring is shorting • Mechanical malfunction in transmission
DTC: P1865 **2T MIL: Yes** 2003, 2004 **Models:** 320i, 330i, 525i, 530i, 545i, 745i, M3, Z4 **Engines:** 2.5L, 3.0L, 3.2L, 4.8L **Transmissions:** A/T	**6-5 Shift Range Monitoring Conditions:** Engine running and vehicle driven, the DME detected a mechanical malfunction within the transmission. The output speed is greater than 300rpm, the transmission oil temperature is greater than 0 degrees Celsius, the engine speed is greater or equal to 600rpm and the range position is P, R, or N. **Possible Causes:** • Solenoid valves in valve body are faulty • Solenoid circuit is shorting to ground • Solenoid circuit is open • TCM has failed or wiring is shorting • Mechanical malfunction in transmission
DTC: P1871 **2T MIL: Yes** 2003 **Models:** 745i **Engines:** 4.8L **Transmissions:** A/T	**2-1 Shift Range Monitoring Upper Threshold Conditions:** Engine running and vehicle driven, the DME detected a mechanical malfunction within the transmission. The output speed is greater than 300rpm, the transmission oil temperature is greater than 0 degrees Celsius, the engine speed is greater or equal to 600rpm and the range position is P, R, or N. **Possible Causes:** • Solenoid valves in valve body are faulty • Solenoid circuit is shorting to ground • Solenoid circuit is open • TCM has failed or wiring is shorting • Mechanical malfunction in transmission
DTC: P1872 **2T MIL: Yes** 2003 **Models:** 745i **Engines:** 4.8L **Transmissions:** A/T	**3-2 Shift Range Monitoring Upper Threshold Conditions:** Engine running and vehicle driven, the DME detected a mechanical malfunction within the transmission. The output speed is greater than 300rpm, the transmission oil temperature is greater than 0 degrees Celsius, the engine speed is greater or equal to 600rpm and the range position is P, R, or N. **Possible Causes:** • Solenoid valves in valve body are faulty • Solenoid circuit is shorting to ground • Solenoid circuit is open • TCM has failed or wiring is shorting • Mechanical malfunction in transmission
DTC: P1873 **2T MIL: Yes** 2003 **Models:** 745i **Engines:** 4.8L **Transmissions:** A/T	**4-3 Shift Range Monitoring Upper Threshold Conditions:** Engine running and vehicle driven, the DME detected a mechanical malfunction within the transmission. The output speed is greater than 300rpm, the transmission oil temperature is greater than 0 degrees Celsius, the engine speed is greater or equal to 600rpm and the range position is P, R, or N. **Possible Causes:** • Solenoid valves in valve body are faulty • Solenoid circuit is shorting to ground • Solenoid circuit is open • TCM has failed or wiring is shorting • Mechanical malfunction in transmission

DTC	Trouble Code Title, Conditions & Possible Causes
DTC: P1874 **2T MIL: Yes** 2003 **Models:** 745i **Engines:** 4.8L **Transmissions:** A/T	**5-4 Shift Range Monitoring Upper Threshold Conditions:** Engine running and vehicle driven, the DME detected a mechanical malfunction within the transmission. The output speed is greater than 300rpm, the transmission oil temperature is greater than 0 degrees Celsius, the engine speed is greater or equal to 600rpm and the range position is P, R, or N. **Possible Causes:** • Solenoid valves in valve body are faulty • Solenoid circuit is shorting to ground • Solenoid circuit is open • TCM has failed or wiring is shorting • Mechanical malfunction in transmission
DTC: P1875 **2T MIL: Yes** 2003 **Models:** 745i **Engines:** 4.8L **Transmissions:** A/T	**6-5 Shift Range Monitoring Upper Threshold Conditions:** Engine running and vehicle driven, the DME detected a mechanical malfunction within the transmission. The output speed is greater than 300rpm, the transmission oil temperature is greater than 0 degrees Celsius, the engine speed is greater or equal to 600rpm and the range position is P, R, or N. **Possible Causes:** • Solenoid valves in valve body are faulty • Solenoid circuit is shorting to ground • Solenoid circuit is open • TCM has failed or wiring is shorting • Mechanical malfunction in transmission
DTC: P1881 **2T MIL: Yes** **2002, 2003, 2004** **Models:** 320i, 330i, 525i, 530i, 545i, 745i, M3, Z4 **Engines:** 2.5L, 3.0L, 3.2L, 4.8L **Transmissions:** A/T	**1-2 Shift Range Monitoring Upper Threshold Conditions:** Engine running and vehicle driven, the DME detected a mechanical malfunction within the transmission. The output speed is greater than 300rpm, the transmission oil temperature is greater than 0 degrees Celsius, the engine speed is greater or equal to 600rpm and the range position is P, R, or N. **Possible Causes:** • Solenoid valves in valve body are faulty • Solenoid circuit is shorting to ground • Solenoid circuit is open • TCM has failed or wiring is shorting • Mechanical malfunction in transmission
DTC: P1882 **2T MIL: Yes** **2002, 2003, 2004** **Models:** 320i, 325i, 330i, 525i, 530i, 545i, 745i, M3, Z3, Z4 **Engines:** 2.5L, 3.0L, 3.2L, 4.4L, 4.8L **Transmissions:** A/T	**2-3 Shift Range Monitoring Upper Threshold Conditions:** Engine running and vehicle driven, the DME detected a mechanical malfunction within the transmission. The output speed is greater than 300rpm, the transmission oil temperature is greater than 0 degrees Celsius, the engine speed is greater or equal to 600rpm and the range position is P, R, or N. **Possible Causes:** • Solenoid valves in valve body are faulty • Solenoid circuit is shorting to ground • Solenoid circuit is open • TCM has failed or wiring is shorting • Mechanical malfunction in transmission
DTC: P1883 **2T MIL: Yes** **2002, 2003, 2004** **Models:** 320i, 325i, 330i, 525i, 530i, 545i, 745i, M3, Z3, Z4 **Engines:** 2.5L, 3.0L, 3.2L, 4.4L, 4.8L **Transmissions:** A/T	**3-4 Shift Upper Threshold Conditions:** Engine running and vehicle driven, the DME detected a mechanical malfunction within the transmission. The ratio of the input speed minus the output speed is greater than the threshold value. The output speed is greater than 300rpm, the transmission oil temperature is greater than 0 degrees Celsius, the engine speed is greater or equal to 600rpm and the range position is P, R, or N. **Possible Causes:** • Solenoid valves in valve body are faulty • Solenoid circuit is shorting to ground • Solenoid circuit is open • TCM has failed or wiring is shorting • Mechanical malfunction in transmission
DTC: P1884 **2T MIL: Yes** **2002, 2003, 2004** **Models:** 320i, 330i, 525i, 530i, 545i, 745i, M3, Z4 **Engines:** 2.5L, 3.0L, 3.2L, 4.8L **Transmissions:** A/T	**4-5 Shift Upper Threshold Conditions:** Engine running and vehicle driven, the DME detected a mechanical malfunction within the transmission. The ratio of the input speed minus the output speed is greater than the threshold value. The output speed is greater than 300rpm, the transmission oil temperature is greater than 0 degrees Celsius, the engine speed is greater or equal to 600rpm and the range position is P, R, or N. **Possible Causes:** • Solenoid valves in valve body are faulty • Solenoid circuit is shorting to ground • Solenoid circuit is open • TCM has failed or wiring is shorting • Mechanical malfunction in transmission

DTC	Trouble Code Title, Conditions & Possible Causes
DTC: P1885 **2T MIL: Yes** **2002, 2003, 2004** **Models:** 320i, 330i, 525i, 530i, 545i, 745i, M3, Z4 **Engines:** 2.5L, 3.0L, 3.2L, 4.8L **Transmissions:** A/T	**5-6 Shift Upper Threshold Conditions:** Engine running and vehicle driven, the DME detected a mechanical malfunction within the transmission. The ratio of the input speed minus the output speed is greater than the threshold value. The output speed is greater than 300rpm, the transmission oil temperature is greater than 0 degrees Celsius, the engine speed is greater or equal to 600rpm and the range position is P, R, or N. **Possible Causes:** • Solenoid valves in valve body are faulty • Solenoid circuit is shorting to ground • Solenoid circuit is open • TCM has failed or wiring is shorting • Mechanical malfunction in transmission
DTC: P1886 **2T MIL: Yes** 2003 **Models:** 745i **Engines:** 4.8L **Transmissions:** A/T	**5-6 Shift Range Monitoring Conditions:** Engine running and vehicle driven, the DME detected a mechanical malfunction within the transmission. The output speed is greater than 300rpm, the transmission oil temperature is greater than 0 degrees Celsius, the engine speed is greater or equal to 600rpm and the range position is P, R, or N. **Possible Causes:** • Solenoid valves in valve body are faulty • Solenoid circuit is shorting to ground • Solenoid circuit is open • TCM has failed or wiring is shorting • Mechanical malfunction in transmission
DTC: P1889 **2T MIL: Yes** **2002, 2003, 2004** **Models:** 320i, 325i, 330i, 525i, 530i, 545i, 745i, M3, Z3, Z4 **Engines:** 2.5L, 3.0L, 3.2L, 4.4L, 4.8L **Transmissions:** A/T	**System Power Supply (B+) Terminal 15 Malfunction Conditions:** Ignition on, the DME detected a low voltage condition on the power supply terminal (15). To achieve optimal anti-theft protection for the vehicle, an anti-theft immobilizer is installed. The anti-theft immobilizer is a system for enabling and locking the Engine Control Module (DME). So that this system cannot be circumvented, it is necessary to perform adaptation of the anti-theft immobilizer using the Vehicle Diagnostic and Information System VAS 5052 in the On Board Diagnostic (OBD) function. The great availability of equipment options makes it necessary to adapt the Engine Control Module (DME) to the vehicle (e.g. throttle valve control module or cruise control system). This "writing" function is not possible with the generic scan tool. **Possible Causes:** • (If DME was replaced) DME ID not the same as the replaced unit • Voltage supply for Engine Control Module (DME) has shorted
DTC: P1890 **2T MIL: Yes** **2002, 2003, 2004** **Models:** 320i, 330i, 525i, 530i, 545i, 745i, M3, Z4 **Engines:** 2.5L, 3.0L, 3.2L, 4.8L **Transmissions:** A/T	**TCC Power Supply Upper Threshold Conditions:** Engine started to 400rpm, FET enabled, vehicle driven in gear with VSS signals received, and the DME detected a the battery voltage was less than the threshold value (less than 9 volts) while in normal operation. **Possible Causes:** • TCC solenoid has a mechanical failure • TCC solenoid has a hydraulic failure
DTC: P1891 **2T MIL: Yes** **2002, 2003, 2004** **Models:** 320i, 325i, 330i, 525i, 530i, 545i, 745i, M3, Z3, Z4 **Engines:** 2.5L, 3.0L, 3.2L, 4.4L, 4.8L **Transmissions:** A/T	**TCC Power Supply Upper Threshold Conditions:** Engine started to 400rpm, FET enabled, vehicle driven in gear with VSS signals received, and the DME detected a the battery voltage was greater than the threshold value (greater than 16 volts) while in normal operation. **Possible Causes:** • TCC solenoid has a mechanical failure • TCC solenoid has a hydraulic failure
DTC: P1892 **2T MIL: Yes** **2002, 2003, 2004** **Models:** 320i, 325i, 330i, 525i, 530i, 545i, 745i, M3, Z3, Z4 **Engines:** 2.5L, 3.0L, 3.2L, 4.4L, 4.8L **Transmissions:** A/T	**TCC Power Supply Lower Threshold Conditions:** Engine started to 400rpm, FET enabled, vehicle driven in gear with VSS signals received, and the DME detected a the battery voltage was less than the threshold value (less than 7 volts) while in normal operation. **Possible Causes:** • TCC solenoid has a mechanical failure • TCC solenoid has a hydraulic failure
DTC: P1893 **2T MIL: Yes** **2002, 2003, 2004** **Models:** 320i, 325i, 330i, 525i, 530i, 545i, 745i, M3, Z3, Z4 **Engines:** 2.5L, 3.0L, 3.2L, 4.4L, 4.8L **Transmissions:** A/T	**TCC Power Supply Circuit Continuity Power Short Conditions:** Engine started to 400rpm, FET enabled, vehicle driven in gear with VSS signals received, and the DME detected a the battery voltage shorting to power. **Possible Causes:** • TCC solenoid has a mechanical failure • TCC solenoid has a hydraulic failure

DTC	Trouble Code Title, Conditions & Possible Causes
DTC: P1894 **2T MIL: Yes** **2002, 2003, 2004** **Models:** 320i, 325i, 330i, 525i, 530i, 545i, 745i, M3, Z3, Z4 **Engines:** 2.5L, 3.0L, 3.2L, 4.4L, 4.8L **Transmissions:** A/T	**TCC Power Supply Circuit Continuity Ground Short Conditions:** Engine started to 400rpm, FET enabled, vehicle driven in gear with VSS signals received, and the DME detected a the battery voltage shorting to ground. **Possible Causes:** • TCC solenoid has a mechanical failure • TCC solenoid has a hydraulic failure
DTC: P1895 **2T MIL: Yes** **2002, 2003, 2004** **Models:** 320i, 325i, 330i, 525i, 530i, 545i, 745i, M3, Z3, Z4 **Engines:** 2.5L, 3.0L, 3.2L, 4.4L, 4.8L **Transmissions:** A/T	**TCC Power Supply Circuit Continuity Disconnection Conditions:** Engine started to 400rpm, FET enabled, vehicle driven in gear with VSS signals received, and the DME detected a the battery voltage seems to be disconnected. **Possible Causes:** • TCC solenoid has a mechanical failure • TCC solenoid has a hydraulic failure

DTC: P2088 **2T MIL: Yes** 2003, 2004, 2005 **Models:** 320i, 325i, 330i, 525i, 530i, 540i, 545i, 550i, 745i, 750i, M3, Z3, Z4 **Engines:** 2.5L, 3.0L, 3.2L, 4.4L, 4.8L **Transmissions:** All	**Inlet "A" Camshaft Position Control Circuit Low Bank 1 Conditions:** Key on or engine running; and the DME detected an unexpected voltage condition on the Camshaft Position Control circuit during the CCM test period. The relative position between the camshaft and crankshaft needs to be optimal so the engine has better torque, fuel economy and emissions. **Note: camshaft adjustment is load- and RPM dependant. The electrical camshaft adjustment valve 1 switches oil pressure onto camshaft adjuster (mechanical adjustment mechanism), which adjusts the camshaft.** **Possible Causes:** • Camshaft position control wiring harness connector is damaged or open • Camshaft adjustment valve has failed • Circuit is open or grounded • Assembly is damaged or it has failed (an open circuit) • DME power supply relay has failed
DTC: P2089 **2T MIL: Yes** 2003, 2005 **Models:** 320i, 325i, 330i, 525i, 530i, 540i, 545i, 550i, 745i, 750i, M3, Z3, Z4 **Engines:** 2.5L, 3.0L, 3.2L, 4.4L, 4.8L **Transmissions:** All	**Inlet "A" Camshaft Position Control Circuit High Bank 1 Conditions:** Key on or engine running; and the DME detected an unexpected voltage condition on the Camshaft Position Control circuit during the CCM test period. The relative position between the camshaft and crankshaft needs to be optimal so the engine has better torque, fuel economy and emissions. **Note: camshaft adjustment is load- and RPM dependant. The electrical camshaft adjustment valve 1 switches oil pressure onto camshaft adjuster (mechanical adjustment mechanism), which adjusts the camshaft.** **Possible Causes:** • Camshaft position control wiring harness connector is damaged or open • Camshaft adjustment valve has failed • Circuit is open or grounded • Assembly is damaged or it has failed (an open circuit) • DME power supply relay has failed
DTC: P2090 **2T MIL: Yes** 2003, 2004, 2005 **Models:** 320i, 325i, 330i, 525i, 530i, 540i, 545i, 550i, 745i, 750i, M3, Z3, Z4 **Engines:** 2.5L, 3.0L, 3.2L, 4.4L, 4.8L **Transmissions:** All	**Outlet "B" Camshaft Position Control Circuit Low Bank 1 Conditions:** Key on or engine running; and the DME detected an unexpected voltage condition on the Camshaft Position Control circuit during the CCM test period. The relative position between the camshaft and crankshaft needs to be optimal so the engine has better torque, fuel economy and emissions. **Note: camshaft adjustment is load- and RPM dependant. The electrical camshaft adjustment valve 1 switches oil pressure onto camshaft adjuster (mechanical adjustment mechanism), which adjusts the camshaft.** **Possible Causes:** • Camshaft position control wiring harness connector is damaged or open • Camshaft adjustment valve has failed • Circuit is open or grounded • Assembly is damaged or it has failed (an open circuit) • DME power supply relay has failed
DTC: P2091 **2T MIL: Yes** 2003, 2004, 2005 **Models:** 320i, 325i, 330i, 525i, 530i, 540i, 545i, 550i, 745i, 750i, M3, Z3, Z4 **Engines:** 2.5L, 3.0L, 3.2L, 4.4L, 4.8L **Transmissions:** All	**Outlet "B" Camshaft Position Control Circuit High Bank 1 Conditions:** Key on or engine running; and the DME detected an unexpected voltage condition on the Camshaft Position Control circuit during the CCM test period. The relative position between the camshaft and crankshaft needs to be optimal so the engine has better torque, fuel economy and emissions. **Note: camshaft adjustment is load- and RPM dependant. The electrical camshaft adjustment valve 1 switches oil pressure onto camshaft adjuster (mechanical adjustment mechanism), which adjusts the camshaft.** **Possible Causes:** • Camshaft position control wiring harness connector is damaged or open • Camshaft adjustment valve has failed • Circuit is open or grounded • Assembly is damaged or it has failed (an open circuit) • DME power supply relay has failed
DTC: P2092 **2T MIL: Yes** 2003, 2004, 2005 **Models:** 320i, 325i, 330i, 525i, 530i, 540i, 545i, 550i, 745i, 750i, M3, Z3, Z4 **Engines:** 2.5L, 3.0L, 3.2L, 4.4L, 4.8L **Transmissions:** All	**Inlet "A" Camshaft Position Control Circuit Low Bank 1 Conditions:** Key on or engine running; and the DME detected an unexpected voltage condition on the Camshaft Position Control circuit during the CCM test period. The relative position between the camshaft and crankshaft needs to be optimal so the engine has better torque, fuel economy and emissions. **Note: camshaft adjustment is load- and RPM dependant. The electrical camshaft adjustment valve 1 switches oil pressure onto camshaft adjuster (mechanical adjustment mechanism), which adjusts the camshaft.** **Possible Causes:** • Camshaft position control wiring harness connector is damaged or open • Camshaft adjustment valve has failed • Circuit is open or grounded • Assembly is damaged or it has failed (an open circuit) • DME power supply relay has failed

DTC	Trouble Code Title, Conditions & Possible Causes
DTC: P2093 **2T MIL: Yes** **2003, 2006** **Models:** 325i, 330i, 525i, 530i, 540i, 545i, 550i, 745i, 750i **Engines:** 3.0L, 4.4L, 4.8L **Transmissions:** All	**Inlet "A" Camshaft Position Control Circuit Low Bank 2 Conditions:** Key on or engine running; and the DME detected an unexpected voltage condition on the Camshaft Position Control circuit during the CCM test period. The relative position between the camshaft and crankshaft needs to be optimal so the engine has better torque, fuel economy and emissions. **Note: camshaft adjustment is load- and RPM dependant. The electrical camshaft adjustment valve 1 switches oil pressure onto camshaft adjuster (mechanical adjustment mechanism), which adjusts the camshaft.** **Possible Causes:** • Camshaft position control wiring harness connector is damaged or open • Camshaft adjustment valve has failed • Circuit is open or grounded • Assembly is damaged or it has failed (an open circuit) • DME power supply relay has failed
DTC: P2094 **2T MIL: Yes** **2003, 2006** **Models:** 325i, 330i, 525i, 530i, 540i, 545i, 550i, 745i, 750i **Engines:** 3.0L, 4.4L, 4.8L **Transmissions:** All	**Outlet "B" Camshaft Position Control Circuit Low Bank 2 Conditions:** Key on or engine running; and the DME detected an unexpected voltage condition on the Camshaft Position Control circuit during the CCM test period. The relative position between the camshaft and crankshaft needs to be optimal so the engine has better torque, fuel economy and emissions. **Note: camshaft adjustment is load- and RPM dependant. The electrical camshaft adjustment valve 1 switches oil pressure onto camshaft adjuster (mechanical adjustment mechanism), which adjusts the camshaft.** **Possible Causes:** • Camshaft position control wiring harness connector is damaged or open • Camshaft adjustment valve has failed • Circuit is open or grounded • Assembly is damaged or it has failed (an open circuit) • DME power supply relay has failed
DTC: P2095 **2T MIL: Yes** **2003, 2005, 2006** **Models:** 325i, 330i, 525i, 530i, 540i, 545i, 550i, 745i, 750i **Engines:** 3.0L, 4.4L, 4.8L **Transmissions:** All	**Outlet "B" Camshaft Position Control Circuit High Bank 2 Conditions:** Key on or engine running; and the DME detected an unexpected voltage condition on the Camshaft Position Control circuit during the CCM test period. The relative position between the camshaft and crankshaft needs to be optimal so the engine has better torque, fuel economy and emissions. **Note: camshaft adjustment is load- and RPM dependant. The electrical camshaft adjustment valve 1 switches oil pressure onto camshaft adjuster (mechanical adjustment mechanism), which adjusts the camshaft.** **Possible Causes:** • Camshaft position control wiring harness connector is damaged or open • Camshaft adjustment valve has failed • Circuit is open or grounded • Assembly is damaged or it has failed (an open circuit) • DME power supply relay has failed
DTC: P2096 **2T MIL: Yes** **2002, 2003, 2004, 2005, 2006** **Models:** 325i, 330i, 525i, 530i, 545i, 550i, 745i, 750i, M3, Z3, Z4 **Engines:** 2.5L, 3.0L, 3.2L, 4.8L **Transmissions:** All	**Post Catalyst Fuel Trim System Too Lean (Bank 1) Conditions:** Engine started, battery voltage must be at least 11.5v, all electrical components must be off, the ground between the engine and the chassis must be well connected, the exhaust system must be properly sealed between the catalytic converter and the cylinder head, and the oxygen sensor heater for oxygen sensor before the catalytic converter must be properly functioning. The DME detected a problem with the fuel mixture. Trim control 1 segment (precision controller with oxygen sensor behind cat.) below delta lambda threshold of less than -0.03. Fault monitoring criterion must remain present for over one second. The engine speed must be between 1060 and 3000 rpm and the catalytic converter temperature must be 280 degrees Celsius. **Note: For resistance testing of sensor heating, oxygen sensor should be cooled to ambient temperature. High temperatures at oxygen sensor may lead to inaccurate measurements.** **Possible Causes:** • Oxygen sensor (before catalytic converter) is faulty • Oxygen sensor (behind catalytic converter) is faulty • Oxygen sensor heater (before catalytic converter) is faulty • Oxygen sensor heater (behind catalytic converter) is faulty • Check circuits for shorts to each other, ground or power

DTC	Trouble Code Title, Conditions & Possible Causes
DTC: P2097 **2T MIL: Yes** **2004, 2005, 2006** **Models:** 325i, 330i, 525i, 530i, 545i, 550i, 745i, 750i, M3, Z3, Z4 **Engines:** 2.5L, 3.0L, 3.2L, 4.8L **Transmissions:** All	**Post Catalyst Fuel Trim System Too Rich (Bank 1) Conditions:** Engine started, battery voltage must be at least 11.5v, all electrical components must be off, the ground between the engine and the chassis must be well connected, the exhaust system must be properly sealed between the catalytic converter and the cylinder head, and the oxygen sensor heater for oxygen sensor before the catalytic converter must be properly functioning. The DME detected a problem with the fuel mixture. Trim control 1 segment (precision controller with oxygen sensor behind cat.) below delta lambda threshold of less than -0.03. Fault monitoring criterion must remain present for over one second. The engine speed must be between 1060 and 3000 rpm and the catalytic converter temperature must be 280 degrees Celsius. **Note: For resistance testing of sensor heating, oxygen sensor should be cooled to ambient temperature. High temperatures at oxygen sensor may lead to inaccurate measurements.** **Possible Causes:** • Oxygen sensor (before catalytic converter) is faulty • Oxygen sensor (behind catalytic converter) is faulty • Oxygen sensor heater (before catalytic converter) is faulty • Oxygen sensor heater (behind catalytic converter) is faulty • Check circuits for shorts to each other, ground or power
DTC: P2098 **2T MIL: Yes** **2002, 2003, 2004, 2005, 2006** **Models:** 325i, 330i, 525i, 530i, 545i, 550i, 745i, 750i, M3, Z3, Z4 **Engines:** 2.5L, 3.0L, 3.2L, 4.8L **Transmissions:** All	**Post Catalyst Fuel Trim System Too Lean (Bank 2) Conditions:** Engine started, battery voltage must be at least 11.5v, all electrical components must be off, the ground between the engine and the chassis must be well connected, the exhaust system must be properly sealed between the catalytic converter and the cylinder head, and the oxygen sensor heater for oxygen sensor before the catalytic converter must be properly functioning. The DME detected a problem with the fuel mixture. Trim control 1 segment (precision controller with oxygen sensor behind cat.) below delta lambda threshold of less than -0.03. Fault monitoring criterion must remain present for over one second. The engine speed must be between 1060 and 3000 rpm and the catalytic converter temperature must be 280 degrees Celsius. **Note: For resistance testing of sensor heating, oxygen sensor should be cooled to ambient temperature. High temperatures at oxygen sensor may lead to inaccurate measurements.** **Possible Causes:** • Oxygen sensor (before catalytic converter) is faulty • Oxygen sensor (behind catalytic converter) is faulty • Oxygen sensor heater (before catalytic converter) is faulty • Oxygen sensor heater (behind catalytic converter) is faulty • Check circuits for shorts to each other, ground or power
DTC: P2099 **2T MIL: Yes** **2005, 2006** **Models:** 325i, 330i, 525i, 530i, 545i, 550i, 745i, 750i, Z4 **Engines:** 3.0L, 4.8L **Transmissions:** All	**Post Catalyst Fuel Trim System Too Rich (Bank 2) Conditions:** Engine started, battery voltage must be at least 11.5v, all electrical components must be off, the ground between the engine and the chassis must be well connected, the exhaust system must be properly sealed between the catalytic converter and the cylinder head, and the oxygen sensor heater for oxygen sensor before the catalytic converter must be properly functioning. The DME detected a problem with the fuel mixture. Trim control 1 segment (precision controller with oxygen sensor behind cat.) below delta lambda threshold of less than -0.03. Fault monitoring criterion must remain present for over one second. The engine speed must be between 1060 and 3000 rpm and the catalytic converter temperature must be 400 degrees Celsius. **Note: For resistance testing of sensor heating, oxygen sensor should be cooled to ambient temperature. High temperatures at oxygen sensor may lead to inaccurate measurements.** **Possible Causes:** • Oxygen sensor (before catalytic converter) is faulty • Oxygen sensor (behind catalytic converter) is faulty • Oxygen sensor heater (before catalytic converter) is faulty • Oxygen sensor heater (behind catalytic converter) is faulty • Check circuits for shorts to each other, ground or power
DTC: P2100 **2T MIL: Yes** **2005, 2006** **Models:** 325i, 330i, 525i, 530i, 545i, 550i, 745i, 750i, Z4 **Engines:** 3.0L, 4.8L **Transmissions:** All	**Throttle Actuator "A" Control Motor Circuit Open Conditions:** Engine started, battery voltage must be at least 7v, coolant temperature must be at least 80 degrees Celsius. The DME detected an unexpected low or high voltage condition on the Throttle Actuator Control Motor (TACM) circuit during the CCM test. **Note: The throttle valve activation occurs via an electric motor (throttle drive) in the throttle valve control module. It is activated by the Engine Control Module (DME) according to specifications of the two sensors, Throttle Position (TP) Sensor and Sender 2 for accelerator pedal position.** **Possible Causes:** • TACM wiring harness connector is damaged or open • TACM wiring may be crossed in the wire harness assembly • TACM (motor) circuit is open, or TACM assembly is damaged (possible open circuit) • TACM or the Throttle Valve is dirty • Throttle Position sensor has failed • Heater defective

DTC	Trouble Code Title, Conditions & Possible Causes
DTC: P2102 **2T MIL: Yes** **2005, 2006** **Models:** 325i, 330i, 525i, 530i, 545i, 550i, 745i, 750i, Z4 **Engines:** 3.0L, 4.8L **Transmissions:** All	**Throttle Actuator "A" Control Motor Circuit Low Conditions:** Engine started, battery voltage must be at least 7v, coolant temperature must be at least 80 degrees Celsius. The DME detected an unexpected low or high voltage condition on the Throttle Actuator Control Motor (TACM) circuit during the CCM test. **Note: The throttle valve activation occurs via an electric motor (throttle drive) in the throttle valve control module. It is activated by the Engine Control Module (DME) according to specifications of the two sensors, Throttle Position (TP) Sensor and Sender 2 for accelerator pedal position.** **Possible Causes:** • TACM wiring harness connector is damaged or open • TACM wiring may be crossed in the wire harness assembly • TACM (motor) circuit is open, or TACM assembly is damaged (possible open circuit) • TACM or the Throttle Valve is dirty • Throttle Position sensor has failed • Heater defective
DTC: P2103 **2T MIL: Yes** **2005, 2006** **Models:** 325i, 330i, 525i, 530i, 545i, 550i, 745i, 750i, Z4 **Engines:** 3.0L, 4.8L **Transmissions:** All	**Throttle Actuator "A" Control Motor Circuit High Conditions:** Engine started, battery voltage must be at least 7v, coolant temperature must be at least 80 degrees Celsius. The DME detected an unexpected low or high voltage condition on the Throttle Actuator Control Motor (TACM) circuit during the CCM test. **Note: The throttle valve activation occurs via an electric motor (throttle drive) in the throttle valve control module. It is activated by the Engine Control Module (DME) according to specifications of the two sensors, Throttle Position (TP) Sensor and Sender 2 for accelerator pedal position.** **Possible Causes:** • TACM wiring harness connector is damaged or open • TACM wiring may be crossed in the wire harness assembly • TACM (motor) circuit is open, or TACM assembly is damaged (possible open circuit) • TACM or the Throttle Valve is dirty • Throttle Position sensor has failed • Heater defective
DTC: P2122 **2T MIL: Yes** **2004, 2005, 2006** **Models:** 325i, 330i, 525i, 530i, 545i, 550i, 745i, 750i, M3, Z3, Z4 **Engines:** 2.5L, 3.0L, 3.2L, 4.8L **Transmissions:** All	**Accelerator Pedal Position Sensor 'D' Circuit Low Input Conditions:** Engine started, battery voltage at least 11.5v, all electrical components off, ground connections between engine and chassis well connected, the DME detected that the accelerator pedal position sensor signal was outside the parameters to function normally. **Note: Both the Throttle Position (TP) Sensor and Accelerator Pedal Position Sensor are located at the accelerator pedal module and communicate the driver's intentions to the DME completely independently of each other. Both sensors are stored in one housing.** **Possible Causes:** • Ground between engine and chassis may be broken • Throttle position sensor may have failed • Accelerator Pedal Position Sensor has failed • Throttle position sensor wiring may have shorted • Throttle position sensor has failed • Faulty voltage supply
DTC: P2123 **2T MIL: Yes** **2004, 2005, 2006** **Models:** 325i, 330i, 525i, 530i, 545i, 550i, 745i, 750i, M3, Z3, Z4 **Engines:** 2.5L, 3.0L, 3.2L, 4.8L **Transmissions:** All	**Accelerator Pedal Position Sensor 'D' Circuit High Input Conditions:** Engine started, battery voltage at least 11.5v, all electrical components off, ground connections between engine and chassis well connected, the DME detected that the accelerator pedal position sensor signal was outside the parameters to function normally. **Note: Both the Throttle Position (TP) Sensor and Accelerator Pedal Position Sensor are located at the accelerator pedal module and communicate the driver's intentions to the DME completely independently of each other. Both sensors are stored in one housing.** **Possible Causes:** • Ground between engine and chassis may be broken • Throttle position sensor may have failed • Accelerator Pedal Position Sensor has failed • Throttle position sensor wiring may have shorted • Throttle position sensor has failed • Faulty voltage supply

DTC	Trouble Code Title, Conditions & Possible Causes
DTC: P2127 **2T MIL: Yes** **2004, 2005, 2006** **Models:** 325i, 330i, 525i, 530i, 545i, 550i, 745i, 750i, M3, Z3, Z4 **Engines:** 2.5L, 3.0L, 3.2L, 4.8L **Transmissions:** All	**Accelerator Pedal Position Sensor 'E' Circuit Low Input Conditions:** Engine started, battery voltage at least 11.5v, all electrical components off, ground connections between engine and chassis well connected, the DME detected that the accelerator pedal position sensor signal was outside the parameters to function normally. **Note: Both the Throttle Position (TP) Sensor and Accelerator Pedal Position Sensor are located at the accelerator pedal module and communicate the driver's intentions to the DME completely independently of each other. Both sensors are stored in one housing.** **Possible Causes:** • Ground between engine and chassis may be broken • Throttle position sensor may have failed • Accelerator Pedal Position Sensor has failed • Throttle position sensor wiring may have shorted • Throttle position sensor has failed • Faulty voltage supply
DTC: P2128 **2T MIL: Yes** **2004, 2005, 2006** **Models:** 325i, 330i, 525i, 530i, 545i, 550i, 745i, 750i, M3, Z3, Z4 **Engines:** 2.5L, 3.0L, 3.2L, 4.8L **Transmissions:** All	**Accelerator Pedal Position Sensor 'E' Circuit High Input Conditions:** Engine started, battery voltage at least 11.5v, all electrical components off, ground connections between engine and chassis well connected, the DME detected that the accelerator pedal position sensor signal was outside the parameters to function normally. **Note: Both the Throttle Position (TP) Sensor and Accelerator Pedal Position Sensor are located at the accelerator pedal module and communicate the driver's intentions to the DME completely independently of each other. Both sensors are stored in one housing.** **Possible Causes:** • Ground between engine and chassis may be broken • Throttle position sensor may have failed • Accelerator Pedal Position Sensor has failed • Throttle position sensor wiring may have shorted • Throttle position sensor has failed • Faulty voltage supply
DTC: P2138 **2T MIL: Yes** **2004, 2005, 2006** **Models:** 325i, 330i, 525i, 530i, 545i, 550i, 745i, 750i, M3, Z3, Z4 **Engines:** 2.5L, 3.0L, 3.2L, 4.8L **Transmissions:** All	**Throttle Position Sensor D/E Voltage Correlation Conditions:** Engine started, battery voltage must be at least 11.5v, all electrical components must be off, parking brake must be engaged (to keep daytime driving lights off), automatic transmission selector must be in park; and the DME detected the Throttle Position 'D' (TPD) and Throttle Position 'B' (TPE) sensors disagreed, or that the TPD sensor should not be in its detected position, or that the TPE sensor should not be in its detected position during testing. **Note: Both the Throttle Position (TP) Sensor and Accelerator Pedal Position Sensor are located at the accelerator pedal module and communicate the driver's intentions to the DME completely independently of each other. Both sensors are stored in one housing.** **Possible Causes:** • ETC TP sensor connector is damaged or shorted • ETC TP sensor circuits shorted together in the wire harness • ETC TP sensor signal circuit is shorted to VREF (5v) • ETC TP sensor is damaged or the DME has failed
DTC: P2177 **2T MIL: Yes** **2003, 2006** **Models:** 325i, 330i, 525i, 530i, 540i, 550i, 745i, 750i, Z4 **Engines:** 3.0L, 4.4L, 4.8L **Transmissions:** All	**System Too Lean Off Idle Bank 1 Conditions:** Engine started, battery voltage must be at least 11.5v, all electrical components must be off, the ground between the engine and the chassis must be well connected, the exhaust system must be properly sealed between the catalytic converter and the cylinder head, and the oxygen sensor heater for oxygen sensor before the catalytic converter must be properly functioning. The DME detected the system indicated a lean signal, or it could no longer control bank 1 because it was at its lean limit. **Possible Causes:** • Intake Manifold Runner Position Sensor has failed • Intake system has leaks (false air) • Motor for intake flap is faulty • Oxygen sensor (before catalytic converter) is faulty • Oxygen sensor (behind catalytic converter) is faulty • Oxygen sensor heater (before catalytic converter) is faulty • Oxygen sensor heater (behind catalytic converter) is faulty • Check circuits for shorts to each other, ground or power • Fuel Injector(s) may have failed

DTC	Trouble Code Title, Conditions & Possible Causes
DTC: P2178 **2T MIL: Yes** **2003, 2006** **Models:** 325i, 330i, 525i, 530i, 540i, 550i, 745i, 750i, Z4 **Engines:** 3.0L, 4.4L, 4.8L **Transmissions:** All	**System Too Rich Off Idle Bank 1 Conditions:** Engine started, battery voltage must be at least 11.5v, all electrical components must be off, the ground between the engine and the chassis must be well connected, the exhaust system must be properly sealed between the catalytic converter and the cylinder head, and the oxygen sensor heater for oxygen sensor before the catalytic converter must be properly functioning. The DME detected the system indicated a rich signal, or it could no longer control bank 1 because it was at its rich limit. **Possible Causes:** • Intake Manifold Runner Position Sensor has failed • Intake system has leaks (false air) • Motor for intake flap is faulty • Oxygen sensor (before catalytic converter) is faulty • Oxygen sensor (behind catalytic converter) is faulty • Oxygen sensor heater (before catalytic converter) is faulty • Oxygen sensor heater (behind catalytic converter) is faulty • Check circuits for shorts to each other, ground or power • Fuel Injector(s) may have failed
DTC: P2179 **2T MIL: Yes** **2003, 2006** **Models:** 325i, 330i, 525i, 530i, 540i, 550i, 745i, 750i, Z4 **Engines:** 3.0L, 4.4L, 4.8L **Transmissions:** All	**System Too Lean Off Idle Bank 2 Conditions:** Engine started, battery voltage must be at least 11.5v, all electrical components must be off, the ground between the engine and the chassis must be well connected, the exhaust system must be properly sealed between the catalytic converter and the cylinder head, and the oxygen sensor heater for oxygen sensor before the catalytic converter must be properly functioning. The DME detected the system indicated a lean signal, or it could no longer control bank 2 because it was at its lean limit. **Possible Causes:** • Intake Manifold Runner Position Sensor has failed • Intake system has leaks (false air) • Motor for intake flap is faulty • Oxygen sensor (before catalytic converter) is faulty • Oxygen sensor (behind catalytic converter) is faulty • Oxygen sensor heater (before catalytic converter) is faulty • Oxygen sensor heater (behind catalytic converter) is faulty • Check circuits for shorts to each other, ground or power • Fuel Injector(s) may have failed
DTC: P2180 **2T MIL: Yes** **2003, 2006** **Models:** 325i, 330i, 525i, 530i, 540i, 550i, 745i, 750i, Z4 **Engines:** 3.0L, 4.4L, 4.8L **Transmissions:** All	**System Too Rich Off Idle Bank 2 Conditions:** Engine started, battery voltage must be at least 11.5v, all electrical components must be off, the ground between the engine and the chassis must be well connected, the exhaust system must be properly sealed between the catalytic converter and the cylinder head, and the oxygen sensor heater for oxygen sensor before the catalytic converter must be properly functioning. The DME detected the system indicated a rich signal, or it could no longer control bank 2 because it was at its rich limit. **Possible Causes:** • Intake Manifold Runner Position Sensor has failed • Intake system has leaks (false air) • Motor for intake flap is faulty • Oxygen sensor (before catalytic converter) is faulty • Oxygen sensor (behind catalytic converter) is faulty • Oxygen sensor heater (before catalytic converter) is faulty • Oxygen sensor heater (behind catalytic converter) is faulty • Check circuits for shorts to each other, ground or power • Fuel Injector(s) may have failed
DTC: P2183 **2T MIL: Yes** **2005, 2006** **Models:** 325i, 330i, 525i, 530i, 545i, 550i, 745i, 750i, M3, Z4 **Engines:** 3.0L, 3.2L, 4.8L **Transmissions:** All	**ECT Sensor Signal Range/Performance Rationality Conditions:** Engine started (cold), battery voltage must be 11.5, and all equipment must be off. The DME detected the ECT sensor exceeded the required calibrated value, or the engine is at idle and doesn't reach operating temperature quickly enough; the Catalyst, Fuel System, HO2S and Misfire Monitor did not complete, or the timer expired. Testing completion of procedure, the engine's temperature must rise uniformly during idle. The ECT is greater than 101.3 degrees Celsius for more than 60 seconds. The engine speed is greater than 1100rpm. The ambient temperature is greater than negative 7 degrees Celsius. The vehicle speed less than 62.5mph. **Possible Causes:** • Check for low coolant level or incorrect coolant mixture • DME detects a short circuit wiring in the ECT • CHT sensor is out-of-calibration or it has failed • ECT sensor is out-of-calibration or it has failed
DTC: P2184 **2T MIL: Yes** **2005, 2006** **Models:** 325i, 330i, 525i, 530i, 545i, 550i, 745i, 750i, M3, Z4 **Engines:** 3.0L, 3.2L, 4.8L **Transmissions:** All	**ECT Sensor 2 Circuit Range Check (Minimum) Conditions:** Engine started (cold) for 10 seconds, battery voltage must be 11.5, and all equipment must be off. The DME detected the ECT sensor signal was less than the self-test minimum. This is a thermistor-type sensor with a variable resistance that changes when exposed to different temperatures **Possible Causes:** • ECT sensor signal circuit is grounded in the wiring harness • ECT sensor doesn't react to changes in temperature • ECT sensor is damaged or the DME has failed

DTC	Trouble Code Title, Conditions & Possible Causes
DTC: P2185 **2T MIL: Yes** **2005, 2006** **Models:** 325i, 330i, 525i, 530i, 545i, 550i, 745i, 750i, M3, Z4 **Engines:** 3.0L, 3.2L, 4.8L **Transmissions:** All	**ECT Sensor 2 Circuit Range Check (Maximum) Conditions:** Engine started (cold) for 10 seconds, battery voltage must be 11.5, and all equipment must be off. The DME detected the ECT sensor signal was less than the self-test minimum. This is a thermistor-type sensor with a variable resistance that changes when exposed to different temperatures **Possible Causes:** • ECT sensor signal circuit is grounded in the wiring harness • ECT sensor doesn't react to changes in temperature • ECT sensor is damaged or the DME has failed
DTC: P2186 **2T MIL: Yes** **2005, 2006** **Models:** 325i, 330i, 525i, 530i, 545i, 550i, 745i, 750i, M3, Z4 **Engines:** 3.0L, 3.2L, 4.8L **Transmissions:** All	**ECT Sensor 2 Circuit High Input Conditions:** Engine started (cold) for 10 seconds, battery voltage must be 11.5, and all equipment must be off. The DME detected the ECT sensor signal was more than the self-test maximum. This is a thermistor-type sensor with a variable resistance that changes when exposed to different temperatures **Possible Causes:** • ECT sensor signal circuit is open (inspect wiring & connector) • ECT sensor signal circuit is shorted to ground • ECT sensor is damaged or it has failed
DTC: P2187 **2T MIL: Yes** **2003, 2005, 2006** **Models:** 325i, 330i, 525i, 530i, 540i, 550i, 745i, 750i, Z4 **Engines:** 3.0L, 4.4L, 4.8L **Transmissions:** All	**System Too Lean at Idle Bank 1 Conditions:** Engine started, battery voltage must be at least 11v, all electrical components must be off, the ground between the engine and the chassis must be well connected, the exhaust system must be properly sealed between the catalytic converter and the cylinder head, and the oxygen sensor heater for oxygen sensor before the catalytic converter must be properly functioning. The engine temperature must be greater than 63 degrees Celsius for approximately 10 to 20 minutes. The air intake temperature must be less than or equal to 80 degrees Celsius, and the engine speed must be less than or equal to 800rpm. **Possible Causes:** • Evaporative Emission (EVAP) canister purge regulator valve is faulty • Exhaust system components are damaged • Fuel injectors are faulty • Fuel pressure regulator and residual pressure have failed • Fuel Pump (FP) in fuel tank is faulty • Intake system has leaks (false air) • Secondary Air Injection (AIR) system has an improper seal • Intake Manifold Runner Position Sensor has failed • Motor for intake flap is faulty • Oxygen sensor (before catalytic converter) is faulty • Oxygen sensor (behind catalytic converter) is faulty • Oxygen sensor heater (before catalytic converter) is faulty • Oxygen sensor heater (behind catalytic converter) is faulty • Check circuits for shorts to each other, ground or power
DTC: P2188 **2T MIL: Yes** **2003, 2005, 2006** **Models:** 325i, 330i, 525i, 530i, 540i, 550i, 745i, 750i, Z4 **Engines:** 3.0L, 4.4L, 4.8L **Transmissions:** All	**System Too Rich at Idle Bank 1 Conditions:** Engine started, battery voltage must be at least 11v, all electrical components must be off, the ground between the engine and the chassis must be well connected, the exhaust system must be properly sealed between the catalytic converter and the cylinder head, and the oxygen sensor heater for oxygen sensor before the catalytic converter must be properly functioning. he engine temperature must be greater than 63 degrees Celsius for approximately 10 to 20 minutes. The air intake temperature must be less than or equal to 80 degrees Celsius, and the engine speed must be less than or equal to 800rpm. **Possible Causes:** • Evaporative Emission (EVAP) canister purge regulator valve is faulty • Exhaust system components are damaged • Fuel injectors are faulty • Fuel pressure regulator and residual pressure have failed • Fuel Pump (FP) in fuel tank is faulty • Intake system has leaks (false air) • Secondary Air Injection (AIR) system has an improper seal • Intake Manifold Runner Position Sensor has failed • Motor for intake flap is faulty • Oxygen sensor (before catalytic converter) is faulty • Oxygen sensor (behind catalytic converter) is faulty • Oxygen sensor heater (before catalytic converter) is faulty • Oxygen sensor heater (behind catalytic converter) is faulty • Check circuits for shorts to each other, ground or power

DTC	Trouble Code Title, Conditions & Possible Causes
DTC: P2189 **2T MIL: Yes** **2003, 2005, 2006** **Models:** 325i, 330i, 525i, 530i, 540i, 550i, 745i, 750i, Z4 **Engines:** 3.0L, 4.4L, 4.8L **Transmissions:** All	**System Too Lean at Idle Bank 2 Conditions:** Engine started, battery voltage must be at least 11v, all electrical components must be off, the ground between the engine and the chassis must be well connected, the exhaust system must be properly sealed between the catalytic converter and the cylinder head, and the oxygen sensor heater for oxygen sensor before the catalytic converter must be properly functioning. he engine temperature must be greater than 63 degrees Celsius for approximately 10 to 20 minutes. The air intake temperature must be less than or equal to 80 degrees Celsius, and the engine speed must be less than or equal to 800rpm. **Possible Causes:** • Evaporative Emission (EVAP) canister purge regulator valve is faulty • Exhaust system components are damaged • Fuel injectors are faulty • Fuel pressure regulator and residual pressure have failed • Fuel Pump (FP) in fuel tank is faulty • Intake system has leaks (false air) • Secondary Air Injection (AIR) system has an improper seal • Intake Manifold Runner Position Sensor has failed • Motor for intake flap is faulty • Oxygen sensor (before catalytic converter) is faulty • Oxygen sensor (behind catalytic converter) is faulty • Oxygen sensor heater (before catalytic converter) is faulty • Oxygen sensor heater (behind catalytic converter) is faulty • Check circuits for shorts to each other, ground or power
DTC: P2190 **2T MIL: Yes** **2003, 2005, 2006** **Models:** 325i, 330i, 525i, 530i, 540i, 550i, 745i, 750i, Z4 **Engines:** 3.0L, 4.4L, 4.8L **Transmissions:** All	**System Too Rich at Idle Bank 2 Conditions:** Engine started, battery voltage must be at least 11v, all electrical components must be off, the ground between the engine and the chassis must be well connected, the exhaust system must be properly sealed between the catalytic converter and the cylinder head, and the oxygen sensor heater for oxygen sensor before the catalytic converter must be properly functioning. he engine temperature must be greater than 63 degrees Celsius for approximately 10 to 20 minutes. The air intake temperature must be less than or equal to 80 degrees Celsius, and the engine speed must be less than or equal to 800rpm. **Possible Causes:** • Evaporative Emission (EVAP) canister purge regulator valve is faulty • Exhaust system components are damaged • Fuel injectors are faulty • Fuel pressure regulator and residual pressure have failed • Fuel Pump (FP) in fuel tank is faulty • Intake system has leaks (false air) • Secondary Air Injection (AIR) system has an improper seal • Intake Manifold Runner Position Sensor has failed • Motor for intake flap is faulty • Oxygen sensor (before catalytic converter) is faulty • Oxygen sensor (behind catalytic converter) is faulty • Oxygen sensor heater (before catalytic converter) is faulty • Oxygen sensor heater (behind catalytic converter) is faulty • Check circuits for shorts to each other, ground or power
DTC: P2191 **2T MIL: Yes** **2003, 2005, 2006** **Models:** 325i, 330i, 525i, 530i, 540i, 550i, 745i, 750i, Z4 **Engines:** 3.0L, 4.4L, 4.8L **Transmissions:** All	**System Too Lean at Higher Load Bank 1 Conditions:** Engine started, battery voltage must be at least 11v, all electrical components must be off, the ground between the engine and the chassis must be well connected, the exhaust system must be properly sealed between the catalytic converter and the cylinder head, and the oxygen sensor heater for oxygen sensor before the catalytic converter must be properly functioning. The engine temperature must be greater than 63 degrees Celsius for approximately 10 to 20 minutes. The air intake temperature must be less than or equal to 80 degrees Celsius. **Possible Causes:** • Evaporative Emission (EVAP) canister purge regulator valve is faulty • Exhaust system components are damaged • Fuel injectors are faulty • Fuel pressure regulator and residual pressure have failed • Fuel Pump (FP) in fuel tank is faulty • Intake system has leaks (false air) • Secondary Air Injection (AIR) system has an improper seal • Intake Manifold Runner Position Sensor has failed • Motor for intake flap is faulty • Oxygen sensor (before catalytic converter) is faulty • Oxygen sensor (behind catalytic converter) is faulty • Oxygen sensor heater (before catalytic converter) is faulty • Oxygen sensor heater (behind catalytic converter) is faulty • Check circuits for shorts to each other, ground or power

DTC	Trouble Code Title, Conditions & Possible Causes
DTC: P2192 **2T MIL: Yes** **2003, 2005, 2006** **Models:** 325i, 330i, 525i, 530i, 540i, 550i, 745i, 750i, Z4 **Engines:** 3.0L, 4.4L, 4.8L **Transmissions:** All	**System Too Rich at Higher Load Bank 1 Conditions:** Engine started, battery voltage must be at least 11v, all electrical components must be off, the ground between the engine and the chassis must be well connected, the exhaust system must be properly sealed between the catalytic converter and the cylinder head, and the oxygen sensor heater for oxygen sensor before the catalytic converter must be properly functioning. The engine temperature must be greater than 63 degrees Celsius for approximately 10 to 20 minutes. The air intake temperature must be less than or equal to 80 degrees Celsius. **Possible Causes:** • Evaporative Emission (EVAP) canister purge regulator valve is faulty • Exhaust system components are damaged • Fuel injectors are faulty • Fuel pressure regulator and residual pressure have failed • Fuel Pump (FP) in fuel tank is faulty • Intake system has leaks (false air) • Secondary Air Injection (AIR) system has an improper seal • Intake Manifold Runner Position Sensor has failed • Motor for intake flap is faulty • Oxygen sensor (before catalytic converter) is faulty • Oxygen sensor (behind catalytic converter) is faulty • Oxygen sensor heater (before catalytic converter) is faulty • Oxygen sensor heater (behind catalytic converter) is faulty • Check circuits for shorts to each other, ground or power
DTC: P2193 **2T MIL: Yes** **2003, 2005, 2006** **Models:** 325i, 330i, 525i, 530i, 540i, 550i, 745i, 750i, Z4 **Engines:** 3.0L, 4.4L, 4.8L **Transmissions:** All	**System Too Lean at Higher Load Bank 2 Conditions:** Engine started, battery voltage must be at least 11.5v, all electrical components must be off, the ground between the engine and the chassis must be well connected, the exhaust system must be properly sealed between the catalytic converter and the cylinder head, and the oxygen sensor heater for oxygen sensor before the catalytic converter must be properly functioning. DME detected the system indicated a lean signal, or it could no longer control bank 2 because it was at its lean limit. The engine temperature must be greater than 63 degrees Celsius for approximately 10 to 20 minutes. The air intake temperature must be less than or equal to 80 degrees Celsius. **Possible Causes:** • Evaporative Emission (EVAP) canister purge regulator valve is faulty • Exhaust system components are damaged • Fuel injectors are faulty • Fuel pressure regulator and residual pressure have failed • Fuel Pump (FP) in fuel tank is faulty • Intake system has leaks (false air) • Secondary Air Injection (AIR) system has an improper seal • Intake Manifold Runner Position Sensor has failed • Motor for intake flap is faulty • Oxygen sensor (before catalytic converter) is faulty • Oxygen sensor (behind catalytic converter) is faulty • Oxygen sensor heater (before catalytic converter) is faulty • Oxygen sensor heater (behind catalytic converter) is faulty • Check circuits for shorts to each other, ground or power
DTC: P2194 **2T MIL: Yes** **2003, 2005, 2006** **Models:** 325i, 330i, 525i, 530i, 540i, 550i, 745i, 750i, Z4 **Engines:** 3.0L, 4.4L, 4.8L **Transmissions:** All	**System Too Rich at Higher Load Bank 2 Conditions:** Engine started, battery voltage must be at least 11v, all electrical components must be off, the ground between the engine and the chassis must be well connected, the exhaust system must be properly sealed between the catalytic converter and the cylinder head, and the oxygen sensor heater for oxygen sensor before the catalytic converter must be properly functioning. The engine temperature must be greater than 63 degrees Celsius for approximately 10 to 20 minutes. The air intake temperature must be less than or equal to 80 degrees Celsius. **Possible Causes:** • Evaporative Emission (EVAP) canister purge regulator valve is faulty • Exhaust system components are damaged • Fuel injectors are faulty • Fuel pressure regulator and residual pressure have failed • Fuel Pump (FP) in fuel tank is faulty • Intake system has leaks (false air) • Secondary Air Injection (AIR) system has an improper seal • Intake Manifold Runner Position Sensor has failed • Motor for intake flap is faulty • Oxygen sensor (before catalytic converter) is faulty • Oxygen sensor (behind catalytic converter) is faulty • Oxygen sensor heater (before catalytic converter) is faulty • Oxygen sensor heater (behind catalytic converter) is faulty • Check circuits for shorts to each other, ground or power

DTC	Trouble Code Title, Conditions & Possible Causes
DTC: P2195 **2T MIL: Yes** **2004, 2006** **Models:** 320i, 325i, 330i, 525i, 530i, 545i, 550i, 745i, 750i, M3, Z4 **Engines:** 2.5L, 3.0L, 3.2L, 4.8L **Transmissions:** All	**O2 Sensor Signal Stuck Lean Bank 1 Sensor 1 Conditions:** Engine running in closed loop, and the DME detected the O2S indicated a lean signal, or it could no longer control Fuel Trim because it was at lean limit. **Possible Causes:** • Engine oil level high • Camshaft timing error • Cylinder compression low • Exhaust leaks in front of O2S • EGR valve is stuck open • EGR gasket is leaking • EVR diaphragm is leaking • Damaged fuel pressure regulator or extremely low fuel pressure • O2S circuit is open or shorted in the wiring harness • Oxygen sensor (before catalytic converter) is faulty • Oxygen sensor (behind catalytic converter) is faulty • Oxygen sensor heater (before catalytic converter) is faulty • Oxygen sensor heater (behind catalytic converter) is faulty • Air leaks after the MAF sensor • PCV system leaks • Dip stick not seated properly
DTC: P2196 **2T MIL: Yes** **2004, 2006** **Models:** 320i, 325i, 330i, 525i, 530i, 545i, 550i, 745i, 750i, M3, Z4 **Engines:** 2.5L, 3.0L, 3.2L, 4.8L **Transmissions:** All	**O2 Sensor Signal Stuck Rich Bank 1 Sensor 1 Conditions:** Engine running in closed loop, and the DME detected the O2S indicated a rich signal, or it could no longer control Fuel Trim because it was at its rich limit. The sensor temperature is heated up. The relative engine load change is less than or equal to 3 percent per camshaft revolution. **Possible Causes:** • Engine oil level high • Camshaft timing error • Cylinder compression low • Exhaust leaks in front of O2S • EGR valve is stuck open • EGR gasket is leaking • EVR diaphragm is leaking • Damaged fuel pressure regulator or extremely low fuel pressure • O2S circuit is open or shorted in the wiring harness • Oxygen sensor (before catalytic converter) is faulty • Oxygen sensor (behind catalytic converter) is faulty • Oxygen sensor heater (before catalytic converter) is faulty • Oxygen sensor heater (behind catalytic converter) is faulty • Air leaks after the MAF sensor • PCV system leaks • Dip stick not seated properly
DTC: P2197 **2T MIL: Yes** **2004, 2006** **Models:** 320i, 325i, 330i, 525i, 530i, 545i, 550i, 745i, 750i, M3, Z4 **Engines:** 2.5L, 3.0L, 3.2L, 4.8L **Transmissions:** All	**O2 Sensor Signal Stuck Lean Bank 2 Sensor 1 Conditions:** Engine running in closed loop, and the DME detected the O2S indicated a lean signal, or it could no longer control Fuel Trim because it was at lean limit. The sensor temperature is heated up. The relative engine load change is less than or equal to 3 percent per camshaft revolution. **Possible Causes:** • Engine oil level high • Camshaft timing error • Cylinder compression low • Exhaust leaks in front of O2S • EGR valve is stuck open • EGR gasket is leaking • EVR diaphragm is leaking • Damaged fuel pressure regulator or extremely low fuel pressure • O2S circuit is open or shorted in the wiring harness • Oxygen sensor (before catalytic converter) is faulty • Oxygen sensor (behind catalytic converter) is faulty • Oxygen sensor heater (before catalytic converter) is faulty • Oxygen sensor heater (behind catalytic converter) is faulty • Air leaks after the MAF sensor • PCV system leaks • Dip stick not seated properly

DTC	Trouble Code Title, Conditions & Possible Causes
DTC: P2198 **2T MIL: Yes** **2004, 2006** **Models:** 320i, 325i, 330i, 525i, 530i, 545i, 550i, 745i, 750i, M3, Z4 **Engines:** 2.5L, 3.0L, 3.2L, 4.8L **Transmissions:** All	**O2 Sensor Signal Stuck Rich Bank 2 Sensor 1 Conditions:** Engine running in closed loop, and the DME detected the O2S indicated a rich signal, or it could no longer control Fuel Trim because it was at its rich limit. The sensor temperature is heated up. The relative engine load change is less than or equal to 3 percent per camshaft revolution. **Possible Causes:** • Engine oil level high • Camshaft timing error • Cylinder compression low • Exhaust leaks in front of O2S • EGR valve is stuck open • EGR gasket is leaking • EVR diaphragm is leaking • Damaged fuel pressure regulator or extremely low fuel pressure • O2S circuit is open or shorted in the wiring harness • Oxygen sensor (before catalytic converter) is faulty • Oxygen sensor (behind catalytic converter) is faulty • Oxygen sensor heater (before catalytic converter) is faulty • Oxygen sensor heater (behind catalytic converter) is faulty • Air leaks after the MAF sensor • PCV system leaks • Dip stick not seated properly
DTC: P2227 **2T MIL: Yes** 2004 **Models:** 325i, 330i, 525i, 530i, 545i, 745i, M3, Z4 **Engines:** 2.5L, 3.0L, 3.2L, 4.8L **Transmissions:** All	**Barometric Circuit Rationality Check Conditions:** Engine started, the temperature must beat least 185-degrees (F) and all electrical equipment (A/C, lights, etc) must be off. The DME detected the BARO sensor was out of range during the CCM test. The BARO sensor signal should be in 4.5v. The BARO sensor is a variable capacitance unit used to detect altitude. There is a circuit malfunction and the internal voltage measurement in ambient pressure sensor is greater than 4.7998 and the ambient pressure is greater than 1150 hPa. **Possible Causes:** • Sensor has deteriorated (response time too slow) or has failed • MAP sensor signal circuit is shorted to ground • MAP sensor circuit (5v) is open • MAP sensor is damaged or it has failed • BARO sensor signal circuit is shorted to ground • BARO sensor circuit (5v) is open • BARO sensor is damaged or it has failed • Replace the DME
DTC: P2228 **2T MIL: Yes** **2004, 2006** **Models:** 320i, 325i, 330i, 525i, 530i, 545i, 550i, 745i, 750i, M3, Z4 **Engines:** 2.5L, 3.0L, 3.2L, 4.8L **Transmissions:** All	**Barometric Circuit Low Conditions:** Engine started, the temperature must beat least 185-degrees (F) and all electrical equipment (A/C, lights, etc) must be off. The DME detected the BARO sensor was out of range during the CCM test. The BARO sensor signal should be in 4.5v. The BARO sensor is a variable capacitance unit used to detect altitude. There is a short to ground and the internal voltage measurement in ambient pressure sensor is greater than 4.7998 and the ambient pressure is greater than 1150 hPa. **Possible Causes:** • Sensor has deteriorated (response time too slow) or has failed • MAP sensor signal circuit is shorted to ground • MAP sensor circuit (5v) is open • MAP sensor is damaged or it has failed • BARO sensor signal circuit is shorted to ground • BARO sensor circuit (5v) is open • BARO sensor is damaged or it has failed • Replace the DME
DTC: P2229 **2T MIL: Yes** **2004, 2006** **Models:** 320i, 325i, 330i, 525i, 530i, 545i, 550i, 745i, 750i, M3, Z4 **Engines:** 2.5L, 3.0L, 3.2L, 4.8L **Transmissions:** All	**Barometric Circuit High Conditions:** Engine started, the temperature must beat least 185-degrees (F) and all electrical equipment (A/C, lights, etc) must be off. The DME detected the BARO sensor was out of range during the CCM test. The BARO sensor signal should be in 4.5v. The BARO sensor is a variable capacitance unit used to detect altitude. There is a short to battery voltage and the internal voltage measurement in ambient pressure sensor is greater than 4.7998 and the ambient pressure is greater than 1150 hPa. **Possible Causes:** • Sensor has deteriorated (response time too slow) or has failed • MAP sensor signal circuit is shorted to ground • MAP sensor circuit (5v) is open • MAP sensor is damaged or it has failed • BARO sensor signal circuit is shorted to ground • BARO sensor circuit (5v) is open • BARO sensor is damaged or it has failed • Replace the DME

DTC	Trouble Code Title, Conditions & Possible Causes
DTC: P2231 **2T MIL: Yes** **2003, 2004, 2006** **Models:** 320i, 325i, 330i, 525i, 530i, 545i, 550i, 745i, 750i, M3, Z4 **Engines:** 2.5L, 3.0L, 3.2L, 4.8L **Transmissions:** All	**O2 Sensor Signal Circuit Shorted to Heater Circuit Bank 1 Sensor 1 Conditions:** Engine started, battery voltage must be at least 11.5v, all electrical components must be off, parking brake must be engaged (to keep daytime driving lights off), automatic transmission selector must be in park. The DME detected an unexpected voltage condition, or it detected an unexpected current draw in the sensor circuit during the CCM test. **Note: Vehicle must be raised before connector for oxygen sensors is accessible.** **Possible Causes:** • Oxygen sensor (before catalytic converter) is faulty • Oxygen sensor heater (before catalytic converter) is faulty • Oxygen sensor heater (before catalytic converter) is faulty • Oxygen sensor heater (behind catalytic converter) is faulty • O2S circuit is open or shorted in the wiring harness
DTC: P2234 **2T MIL: Yes** **2003, 2004, 2006** **Models:** 320i, 325i, 330i, 525i, 530i, 545i, 550i, 745i, 750i, M3, Z4 **Engines:** 2.5L, 3.0L, 3.2L, 4.8L **Transmissions:** All	**O2 Sensor Signal Circuit Shorted to Heater Circuit Bank 2 Sensor 1 Conditions:** Engine started, battery voltage must be at least 11.5v, all electrical components must be off, parking brake must be engaged (to keep daytime driving lights off), automatic transmission selector must be in park. The DME detected an unexpected voltage condition, or it detected an unexpected current draw in the sensor circuit during the CCM test. **Note: Vehicle must be raised before connector for oxygen sensors is accessible.** **Possible Causes:** • Oxygen sensor (before catalytic converter) is faulty • Oxygen sensor heater (before catalytic converter) is faulty • Oxygen sensor heater (before catalytic converter) is faulty • Oxygen sensor heater (behind catalytic converter) is faulty • O2S circuit is open or shorted in the wiring harness
DTC: P2237 **2T MIL: Yes** **2004, 2006** **Models:** 320i, 325i, 330i, 525i, 530i, 545i, 550i, 745i, 750i, M3, Z4 **Engines:** 2.5L, 3.0L, 3.2L, 4.8L **Transmissions:** All	**O2S Sensor Positive Current Control Circuit/Open Circuit (Bank 1 Sensor 1) Conditions:** Engine started, the fault criterion must remain present for over 2 seconds, the Lambda specification is outside three percent relative to Lambda = 1, and the closed loop lambda control is active. **Possible Causes:** • Check plugs • Check wiring harness • Replace sensor
DTC: P2240 **2T MIL: Yes** **2004, 2006** **Models:** 320i, 325i, 330i, 525i, 530i, 545i, 550i, 745i, 750i, M3, Z4 **Engines:** 2.5L, 3.0L, 3.2L, 4.8L **Transmissions:** All	**O2S Sensor Positive Current Control Circuit/Open Circuit (Bank 2 Sensor 1) Conditions:** Engine started, the fault criterion must remain present for over 2 seconds, the Lambda specification is outside three percent relative to Lambda = 1, and the closed loop lambda control is active. **Possible Causes:** • Check plugs • Check wiring harness • Replace sensor
DTC: P2243 **2T MIL: Yes** **2004, 2006** **Models:** 320i, 325i, 330i, 525i, 530i, 545i, 550i, 745i, 750i, M3, Z4 **Engines:** 2.5L, 3.0L, 3.2L, 4.8L **Transmissions:** All	**O2 Sensor Reference Voltage Circuit/Open Bank 1 Sensor 1 Conditions:** Engine started, battery voltage must be at least 11.5v, all electrical components must be off, parking brake must be engaged (to keep daytime driving lights off), automatic transmission selector must be in park. The DME detected an unexpected voltage condition, or it detected an unexpected current draw in the sensor circuit during the CCM test. The voltage is out of range. The sensor temperature is heated up and the battery voltage is between 11 and 16 volts. **Note: Vehicle must be raised before connector for oxygen sensors is accessible.** **Possible Causes:** • Oxygen sensor (before catalytic converter) is faulty • Oxygen sensor heater (before catalytic converter) is faulty • Oxygen sensor heater (before catalytic converter) is faulty • Oxygen sensor heater (behind catalytic converter) is faulty • O2S circuit is open or shorted in the wiring harness
DTC: P2247 **2T MIL: Yes** **2004, 2006** **Models:** 320i, 325i, 330i, 525i, 530i, 545i, 550i, 745i, 750i, M3, Z4 **Engines:** 2.5L, 3.0L, 3.2L, 4.8L **Transmissions:** All	**O2 Sensor Reference Voltage Circuit/Open Bank 2 Sensor 1 Conditions:** Engine started, battery voltage must be at least 11.5v, all electrical components must be off, parking brake must be engaged (to keep daytime driving lights off), automatic transmission selector must be in park. The DME detected an unexpected voltage condition, or it detected an unexpected current draw in the sensor circuit during the CCM test. The voltage is out of range. The sensor temperature is heated up and the battery voltage is between 11 and 16 volts. **Note: Vehicle must be raised before connector for oxygen sensors is accessible.** **Possible Causes:** • Oxygen sensor (before catalytic converter) is faulty • Oxygen sensor heater (before catalytic converter) is faulty • Oxygen sensor heater (before catalytic converter) is faulty • Oxygen sensor heater (behind catalytic converter) is faulty • O2S circuit is open or shorted in the wiring harness

DTC	Trouble Code Title, Conditions & Possible Causes
DTC: P2251 **2T MIL: Yes** **2004, 2006** **Models:** 320i, 325i, 330i, 525i, 530i, 545i, 550i, 745i, 750i, M3, Z4 **Engines:** 2.5L, 3.0L, 3.2L, 4.8L **Transmissions:** All	**O2 Sensor Negative Voltage Circuit/Open Bank 1 Sensor 1 Conditions:** Engine started, battery voltage must be at least 11.5v, all electrical components must be off, parking brake must be engaged (to keep daytime driving lights off), automatic transmission selector must be in park. The DME detected an unexpected voltage condition, or it detected an unexpected current draw in the sensor circuit during the CCM test. Fault monitoring criterion must remain present for over five seconds. The voltage is within critical range. The sensor temperature is heated up and the battery voltage is between 11 and 16 volts. **Note: Vehicle must be raised before connector for oxygen sensors is accessible.** **Possible Causes:** • Oxygen sensor (before catalytic converter) is faulty • Oxygen sensor heater (before catalytic converter) is faulty • Oxygen sensor heater (before catalytic converter) is faulty • Oxygen sensor heater (behind catalytic converter) is faulty • O2S circuit is open or shorted in the wiring harness
DTC: P2254 **2T MIL: Yes** **2004, 2006** **Models:** 320i, 325i, 330i, 525i, 530i, 545i, 550i, 745i, 750i, M3, Z4 **Engines:** 2.5L, 3.0L, 3.2L, 4.8L **Transmissions:** All	**O2 Sensor Negative Voltage Circuit/Open Bank 2 Sensor 1 Conditions:** Engine started, battery voltage must be at least 11.5v, all electrical components must be off, parking brake must be engaged (to keep daytime driving lights off), automatic transmission selector must be in park. The DME detected an unexpected voltage condition, or it detected an unexpected current draw in the sensor circuit during the CCM test. Fault monitoring criterion must remain present for over five seconds. The voltage is within critical range. The sensor temperature is heated up and the battery voltage is between 11 and 16 volts. **Note: Vehicle must be raised before connector for oxygen sensors is accessible.** **Possible Causes:** • Oxygen sensor (before catalytic converter) is faulty • Oxygen sensor heater (before catalytic converter) is faulty • Oxygen sensor heater (before catalytic converter) is faulty • Oxygen sensor heater (behind catalytic converter) is faulty • O2S circuit is open or shorted in the wiring harness
DTC: P2270 **2T MIL: Yes** **2003, 2005, 2006** **Models:** 325i, 330i, 525i, 530i, 540i, 550i, 745i, 750i, Z4 **Engines:** 3.0L, 4.4L, 4.8L **Transmissions:** All	**O2 Sensor Signal Stuck Lean Bank 1 Sensor 2 Conditions:** Engine started, battery voltage must be at least 11.5v, all electrical components must be off, parking brake must be engaged (to keep daytime driving lights off), automatic transmission selector must be in park. The DME detected an unexpected voltage condition, or it detected an unexpected current draw in the heater circuit during the CCM test. **Note: Vehicle must be raised before connector for oxygen sensors is accessible.** **Possible Causes:** • Oxygen sensor (before catalytic converter) is faulty • Oxygen sensor heater (before catalytic converter) is faulty • Oxygen sensor heater (before catalytic converter) is faulty • Oxygen sensor heater (behind catalytic converter) is faulty • O2S circuit is open or shorted in the wiring harness
DTC: P2271 **2T MIL: Yes** **2003, 2005, 2006** **Models:** 325i, 330i, 525i, 530i, 540i, 550i, 745i, 750i, Z4 **Engines:** 3.0L, 4.4L, 4.8L **Transmissions:** All	**O2 Sensor Signal Stuck Rich Bank 1 Sensor 2 Conditions:** Engine started, battery voltage must be at least 11.5v, all electrical components must be off, parking brake must be engaged (to keep daytime driving lights off), automatic transmission selector must be in park. The DME detected an unexpected voltage condition, or it detected an unexpected current draw in the heater circuit during the CCM test. **Note: Vehicle must be raised before connector for oxygen sensors is accessible.** **Possible Causes:** • Oxygen sensor (before catalytic converter) is faulty • Oxygen sensor heater (before catalytic converter) is faulty • Oxygen sensor heater (before catalytic converter) is faulty • Oxygen sensor heater (behind catalytic converter) is faulty • O2S circuit is open or shorted in the wiring harness
DTC: P2272 **2T MIL: Yes** **2003, 2005, 2006** **Models:** 325i, 330i, 525i, 530i, 540i, 550i, 745i, 750i, Z4 **Engines:** 3.0L, 4.4L, 4.8L **Transmissions:** All	**O2 Sensor Signal Stuck Lean Bank 2 Sensor 2 Conditions:** Engine started, battery voltage must be at least 11.5v, all electrical components must be off, parking brake must be engaged (to keep daytime driving lights off), automatic transmission selector must be in park. The DME detected an unexpected voltage condition, or it detected an unexpected current draw in the heater circuit during the CCM test. **Note: Vehicle must be raised before connector for oxygen sensors is accessible.** **Possible Causes:** • Oxygen sensor (before catalytic converter) is faulty • Oxygen sensor heater (before catalytic converter) is faulty • Oxygen sensor heater (before catalytic converter) is faulty • Oxygen sensor heater (behind catalytic converter) is faulty • O2S circuit is open or shorted in the wiring harness

DTC	Trouble Code Title, Conditions & Possible Causes
DTC: P2273 **2T MIL: Yes** **2003, 2005, 2006** **Models:** 325i, 330i, 525i, 530i, 540i, 550i, 745i, 750i, Z4 **Engines:** 3.0L, 4.4L, 4.8L **Transmissions:** All	**O2 Sensor Signal Stuck Rich Bank 2 Sensor 2 Conditions:** Engine started, battery voltage must be at least 11.5v, all electrical components must be off, parking brake must be engaged (to keep daytime driving lights off), automatic transmission selector must be in park. The DME detected an unexpected voltage condition, or it detected an unexpected current draw in the heater circuit during the CCM test. **Note: Vehicle must be raised before connector for oxygen sensors is accessible.** **Possible Causes:** • Oxygen sensor (before catalytic converter) is faulty • Oxygen sensor heater (before catalytic converter) is faulty • Oxygen sensor heater (before catalytic converter) is faulty • Oxygen sensor heater (behind catalytic converter) is faulty • O2S circuit is open or shorted in the wiring harness
DTC: P2400 **2T MIL: Yes** **2004, 2006** **Models:** 320i, 325i, 330i, 525i, 530i, 545i, 550i, 745i, 750i, M3, Z4 **Engines:** 2.5L, 3.0L, 3.2L, 4.8L **Transmissions:** All	**EVAP Leak Detection Pump (LDP) Control Circuit Open Conditions:** Engine started, battery voltage must be at least 11.5v, all electrical components must be off, parking brake must be engaged (to keep daytime driving lights off), automatic transmission selector must be in park, the exhaust system must be properly sealed between the catalytic converter and the cylinder head, coolant temperature must be at least 80 degrees Celsius and oxygen sensor heaters for oxygen sensors before the catalytic converter must be functioning properly and the ground between the engine and the chassis must be well connected. The DME detected voltage irregularity in the leak detection pump control circuit. **Possible Causes:** • EVAP LDP power supply circuit is open • EVAP LDP solenoid valve is damaged or it has failed • EVAP LDP canister has a leak or a poor seal • EVAP canister system has an improper seal • Evaporative Emission (EVAP) canister purge regulator valve 1 has failed • Leak Detection Pump (LDP) is faulty • Aftermarket EVAP parts that do not conform to specifications • EVAP component seals leaking (i.e., leaks in the Purge valve, fuel tank pressure sensor, canister vent solenoid, fuel vapor control valve tube assembly or fuel vapor vent valve).
DTC: P2401 **2T MIL: Yes** **2004, 2006** **Models:** 320i, 325i, 330i, 525i, 530i, 545i, 550i, 745i, 750i, M3, Z4 **Engines:** 2.5L, 3.0L, 3.2L, 4.8L **Transmissions:** All	**EVAP Leak Detection Pump Control Circuit Low Conditions:** Engine started, battery voltage must be at least 11.5v, all electrical components must be off, parking brake must be engaged (to keep daytime driving lights off), automatic transmission selector must be in park, the exhaust system must be properly sealed between the catalytic converter and the cylinder head, coolant temperature must be at least 80 degrees Celsius and oxygen sensor heaters for oxygen sensors before the catalytic converter must be functioning properly and the ground between the engine and the chassis must be well connected. The DME detected voltage irregularity in the leak detection pump control circuit. **Possible Causes:** • EVAP LDP power supply circuit is open • EVAP LDP solenoid valve is damaged or it has failed • EVAP LDP canister has a leak or a poor seal • EVAP canister system has an improper seal • Evaporative Emission (EVAP) canister purge regulator valve 1 has failed • Leak Detection Pump (LDP) is faulty • Aftermarket EVAP parts that do not conform to specifications • EVAP component seals leaking (i.e., leaks in the Purge valve, fuel tank pressure sensor, canister vent solenoid, fuel vapor control valve tube assembly or fuel vapor vent valve).
DTC: P2402 **2T MIL: Yes** **2004, 2006** **Models:** 320i, 325i, 330i, 525i, 530i, 545i, 550i, 745i, 750i, M3, Z4 **Engines:** 2.5L, 3.0L, 3.2L, 4.8L **Transmissions:** All	**EVAP Leak Detection Pump Control Circuit High Conditions:** Engine started, battery voltage must be at least 11.5v, all electrical components must be off, parking brake must be engaged (to keep daytime driving lights off), automatic transmission selector must be in park, the exhaust system must be properly sealed between the catalytic converter and the cylinder head, coolant temperature must be at least 80 degrees Celsius and oxygen sensor heaters for oxygen sensors before the catalytic converter must be functioning properly and the ground between the engine and the chassis must be well connected. The DME detected voltage irregularity in the leak detection pump control circuit. **Possible Causes:** • EVAP LDP power supply circuit is open • EVAP LDP solenoid valve is damaged or it has failed • EVAP LDP canister has a leak or a poor seal • EVAP canister system has an improper seal • Evaporative Emission (EVAP) canister purge regulator valve 1 has failed • Leak Detection Pump (LDP) is faulty • Aftermarket EVAP parts that do not conform to specifications • EVAP component seals leaking (i.e., leaks in the Purge valve, fuel tank pressure sensor, canister vent solenoid, fuel vapor control valve tube assembly or fuel vapor vent valve).

DTC	Trouble Code Title, Conditions & Possible Causes
DTC: P2414 **2T MIL: Yes** **2004, 2006** **Models:** 320i, 325i, 330i, 525i, 530i, 545i, 550i, 745i, 750i, M3, Z4 **Engines:** 2.5L, 3.0L, 3.2L, 4.8L **Transmissions:** All	**O2 Sensor Exhaust Sample Error Bank 1 Sensor 1 Conditions:** Engine running (ground connections between the engine and the chassis must be well connected), and the DME detected an error on the OS Sensor. **Note: Intake Flap Motor and Intake Manifold Runner Position Sensor are one component and cannot be replaced individually.** **Note: Vacuum in the intake system sucks in the leak detection spray with false air. Leak detection spray decreases ignition quality of the fuel mixture. This causes a drop in engine speed and changes the value produced by the Heated Oxygen Sensor (HO2S). The voltage is out of range. The sensor temperature is heated up and the battery voltage is between 11 and 16 volts.** **Possible Causes:** • Intake Manifold Runner Position Sensor is damaged or has failed • Intake system has leaks (false air) • Motor for intake flap is faulty • Oxygen sensor (before catalytic converter) is faulty • Oxygen sensor heater (before catalytic converter) is faulty • Oxygen sensor heater (before catalytic converter) is faulty • Oxygen sensor heater (behind catalytic converter) is faulty • O2S circuit is open or shorted in the wiring harness
DTC: P2415 **2T MIL: Yes** **2004, 2006** **Models:** 320i, 325i, 330i, 525i, 530i, 545i, 550i, 745i, 750i, M3, Z4 **Engines:** 2.5L, 3.0L, 3.2L, 4.8L **Transmissions:** All	**O2 Sensor Exhaust Sample Error Bank 2 Sensor 1 Conditions:** Engine running (ground connections between the engine and the chassis must be well connected), and the DME detected an error on the OS Sensor. **Note: Intake Flap Motor and Intake Manifold Runner Position Sensor are one component and cannot be replaced individually. The voltage is out of range. The sensor temperature is heated up and the battery voltage is between 11 and 16 volts.** **Note: Vacuum in the intake system sucks in the leak detection spray with false air. Leak detection spray decreases ignition quality of the fuel mixture. This causes a drop in engine speed and changes the value produced by the Heated Oxygen Sensor (HO2S).** **Possible Causes:** • Intake Manifold Runner Position Sensor is damaged or has failed • Intake system has leaks (false air) • Motor for intake flap is faulty • Oxygen sensor (before catalytic converter) is faulty • Oxygen sensor heater (before catalytic converter) is faulty • Oxygen sensor heater (before catalytic converter) is faulty • Oxygen sensor heater (behind catalytic converter) is faulty • O2S circuit is open or shorted in the wiring harness
DTC: P2418 **2T MIL: Yes** **2004, 2005, 2006** **Models:** 320i, 325i, 330i, 525i, 530i, 545i, 550i, 745i, 750i, M3, Z4 **Engines:** 2.5L, 3.0L, 3.2L, 4.8L **Transmissions:** All	**Evaporative Emission System Switching Valve Control Circuit Open Conditions:** Engine started, battery voltage must be at least 11.5v, all electrical components must be off, parking brake must be engaged (to keep daytime driving lights off), automatic transmission selector must be in park, the exhaust system must be properly sealed between the catalytic converter and the cylinder head, coolant temperature must be at least 80 degrees Celsius and oxygen sensor heaters for oxygen sensors before the catalytic converter must be functioning properly and the ground between the engine and the chassis must be well connected. The DME detected an unexpected EVAP malfunction. **Note: Solenoid valve is closed when no voltage is present.** **Possible Causes:** • EVAP power supply circuit is open • EVAP solenoid control circuit is open or shorted to ground • EVAP solenoid control circuit is shorted to power (B+) • EVAP solenoid valve is damaged or it has failed • EVAP canister purge solenoid valve is faulty
DTC: P2419 **2T MIL: Yes** **2004, 2005, 2006** **Models:** 320i, 325i, 330i, 525i, 530i, 545i, 550i, 745i, 750i, M3, Z4 **Engines:** 2.5L, 3.0L, 3.2L, 4.8L **Transmissions:** All	**Evaporative Emission System Switching Valve Control Circuit Low Conditions:** Engine started, battery voltage must be at least 11.5v, all electrical components must be off, parking brake must be engaged (to keep daytime driving lights off), automatic transmission selector must be in park, the exhaust system must be properly sealed between the catalytic converter and the cylinder head, coolant temperature must be at least 80 degrees Celsius and oxygen sensor heaters for oxygen sensors before the catalytic converter must be functioning properly and the ground between the engine and the chassis must be well connected. The DME detected an unexpected EVAP malfunction. **Note: Solenoid valve is closed when no voltage is present.** **Possible Causes:** • EVAP power supply circuit is open • EVAP solenoid control circuit is open or shorted to ground • EVAP solenoid control circuit is shorted to power (B+) • EVAP solenoid valve is damaged or it has failed • EVAP canister purge solenoid valve is faulty

DTC	Trouble Code Title, Conditions & Possible Causes
DTC: P2420 **2T MIL: Yes** **2004, 2005, 2006** **Models:** 320i, 325i, 330i, 525i, 530i, 545i, 550i, 745i, 750i, M3, Z4 **Engines:** 2.5L, 3.0L, 3.2L, 4.8L **Transmissions:** All	**Evaporative Emission System Switching Valve Control Circuit High Conditions:** Engine started, battery voltage must be at least 11.5v, all electrical components must be off, parking brake must be engaged (to keep daytime driving lights off), automatic transmission selector must be in park, the exhaust system must be properly sealed between the catalytic converter and the cylinder head, coolant temperature must be at least 80 degrees Celsius and oxygen sensor heaters for oxygen sensors before the catalytic converter must be functioning properly and the ground between the engine and the chassis must be well connected. The DME detected an unexpected EVAP malfunction. **Note: Solenoid valve is closed when no voltage is present.** **Possible Causes:** • EVAP power supply circuit is open • EVAP solenoid control circuit is open or shorted to ground • EVAP solenoid control circuit is shorted to power (B+) • EVAP solenoid valve is damaged or it has failed • EVAP canister purge solenoid valve is faulty
DTC: P2430 2004, 2005 **Models:** 320i, 325i, 330i, 525i, 530i, 545i, 550i, 745i, 750i, M3, Z4 **Engines:** 2.5L, 3.0L, 3.2L, 4.8L **Transmissions:** All	**Secondary Air Mass Flow Sensor Rationality Check Conditions:** Engine started, battery voltage must be at least 11.5v, all electrical components must be off, parking brake must be engaged (to keep daytime driving lights off), automatic transmission selector must be in park, the exhaust system must be properly sealed between the catalytic converter and the cylinder head, coolant temperature must be at least 80 degrees Celsius and oxygen sensor heaters for oxygen sensors before the catalytic converter must be functioning properly and the ground between the engine and the chassis must be well connected. The DME detected an unexpected secondary air system malfunction. It is disconnected or stuck. **Note: Solenoid valve is closed when no voltage is present.** **Possible Causes:** • EVAP power supply circuit is open • EVAP solenoid control circuit is open or shorted to ground • EVAP solenoid control circuit is shorted to power (B+) • EVAP solenoid valve is damaged or it has failed • EVAP canister purge solenoid valve is faulty
DTC: P2540 2004, 2005 **Models:** 320i, 325i, 330i, 525i, 530i, 545i, 550i, 745i, 750i, M3, Z4 **Engines:** 2.5L, 3.0L, 3.2L, 4.8L **Transmissions:** All	**Secondary Air System Vent Valve Stuck Closed Conditions:** Engine started, battery voltage must be at least 11.5v, all electrical components must be off, parking brake must be engaged (to keep daytime driving lights off), automatic transmission selector must be in park, the exhaust system must be properly sealed between the catalytic converter and the cylinder head, coolant temperature must be at least 80 degrees Celsius and oxygen sensor heaters for oxygen sensors before the catalytic converter must be functioning properly and the ground between the engine and the chassis must be well connected. The DME detected an unexpected secondary air system malfunction. **Note: Solenoid valve is closed when no voltage is present.** **Possible Causes:** • EVAP power supply circuit is open • EVAP solenoid control circuit is open or shorted to ground • EVAP solenoid control circuit is shorted to power (B+) • EVAP solenoid valve is damaged or it has failed • EVAP canister purge solenoid valve is faulty
DTC: P2626 **2T MIL: Yes** **2004, 2005, 2006** **Models:** 320i, 325i, 330i, 525i, 530i, 545i, 550i, 745i, 750i, M3, Z4 **Engines:** 2.5L, 3.0L, 3.2L, 4.8L **Transmissions:** All	**O2 Sensor Pumping Current Trim Circuit/Open Bank 1 Sensor 1 Conditions:** Engine started and the fault entry trips after the criterion remains present for 1 second, battery voltage must be at least 11v, the trailing throttle overrun abort is present for at least two seconds, the O2 sensor is heated to an adequate temperature and the exhaust gas temperature before the catalytic converter is less than 750 degrees Celsius. **Possible Causes:** • Check activation of Recirculation Pump Relay • Oxygen sensor (before catalytic converter) is faulty • Oxygen sensor (behind catalytic converter) is faulty • Oxygen sensor heater (before catalytic converter) is faulty • Oxygen sensor heater (behind catalytic converter) is faulty
DTC: P2629 **2T MIL: Yes** **2004, 2005, 2006** **Models:** 320i, 325i, 330i, 525i, 530i, 545i, 550i, 745i, 750i, M3, Z4 **Engines:** 2.5L, 3.0L, 3.2L, 4.8L **Transmissions:** All	**O2 Sensor Pumping Current Trim Circuit/Open Bank 2 Sensor 1 Conditions:** Engine started, battery voltage must be at least 11.5v, all electrical components must be off, parking brake must be engaged (to keep daytime driving lights off), automatic transmission selector must be in park, the exhaust system must be properly sealed between the catalytic converter and the cylinder head, coolant temperature must be at least 80 degrees Celsius and oxygen sensor heaters for oxygen sensors before the catalytic converter must be functioning properly and the ground between the engine and the chassis must be well connected. The DME detected a voltage value that doesn't fall within the desired parameters for a properly functioning O2 system. **Possible Causes:** • Check activation of Recirculation Pump Relay • Oxygen sensor (before catalytic converter) is faulty • Oxygen sensor (behind catalytic converter) is faulty • Oxygen sensor heater (before catalytic converter) is faulty • Oxygen sensor heater (behind catalytic converter) is faulty

DTC	Trouble Code Title, Conditions & Possible Causes
DTC: P2713 **2T MIL: Yes** 2003, 2004 **Models:** 320i, 325i, 330i, 525i, 530i, 545i, 745i, 750i, M3, Z4 **Engines:** 2.5L, 3.0L, 3.2L, 4.8L **Transmissions:** A/T	**Pressure Regulator Valve 4 Plausibility Conditions:** The current to/from the pressure regulator valve is either higher or lower than the threshold value. **Possible Causes:** • Pressure control solenoid circuit is shorting to ground • Pressure control solenoid circuit is open • Valve has failed • TCM has failed
DTC: P2716 **2T MIL: Yes** 2003, 2004 **Models:** 320i, 325i, 330i, 525i, 530i, 545i, 745i, 750i, M3, Z4 **Engines:** 2.5L, 3.0L, 3.2L, 4.8L **Transmissions:** A/T	**Pressure Control Solenoid 4 Electrical Malfunction Conditions:** Engine started, battery voltage must be at least 11.5v, all electrical components must be off, and the ground between the engine and the chassis must be well connected. The DME detected the pressure control solenoid was experiencing electrical malfunctions. **Possible Causes:** • ATF level is low • Circuit harness connector contacts are corroded or ingresses of water • Circuit wires have shorted to each other, to battery or ground • Automatic Transmission Hydraulic Pressure Sensor 1 has failed • Solenoid valves in valve body are faulty • Transmission Control Module (TCM) needs replacing • Transmission Input Speed (RPM) Sensor has failed • Transmission Output Speed (RPM) Sensor has failed
DTC: P2720 **2T MIL: Yes** 2003, 2004 **Models:** 320i, 325i, 330i, 525i, 530i, 545i, 745i, 750i, M3, Z4 **Engines:** 2.5L, 3.0L, 3.2L, 4.8L **Transmissions:** A/T	**Pressure Regulator Valve 4 Lower Threshold Conditions:** The signal to/from the pressure regulator valve is interrupted or short circuited to ground. **Possible Causes:** • Pressure control solenoid circuit is shorting to ground • Pressure control solenoid circuit is open • Valve has failed • TCM has failed
DTC: P2721 2T MIL: Yes 2003, 2004 **Models:** 320i, 325i, 330i, 525i, 530i, 545i, 745i, 750i, M3, Z4 **Engines:** 2.5L, 3.0L, 3.2L, 4.8L **Transmissions:** A/T	**Pressure Regulator Valve 4 Upper Threshold Conditions:** The signal to/from the pressure regulator valve is interrupted or short circuited to supply. **Possible Causes:** • Pressure control solenoid circuit is shorting to ground • Pressure control solenoid circuit is open • Valve has failed • TCM has failed
DTC: P2722 **2T MIL: Yes** 2003, 2004 **Models:** 320i, 325i, 330i, 525i, 530i, 545i, 745i, 750i, M3, Z4 **Engines:** 2.5L, 3.0L, 3.2L, 4.8L **Transmissions:** A/T	**Pressure Regulator Valve 5 Plausibility Conditions:** The current to/from the pressure regulator valve is either higher or lower than the threshold value. **Possible Causes:** • Pressure control solenoid circuit is shorting to ground • Pressure control solenoid circuit is open • Valve has failed • TCM has failed
DTC: P2725 **2T MIL: Yes** 2003, 2004 **Models:** 320i, 325i, 330i, 525i, 530i, 545i, 745i, 750i, M3, Z4 **Engines:** 2.5L, 3.0L, 3.2L, 4.8L **Transmissions:** A/T	**Pressure Regulator Valve 5 No Signal Conditions:** The signal to/from the pressure regulator valve is interrupted and there is no signal. **Possible Causes:** • Pressure control solenoid circuit is shorting to ground • Pressure control solenoid circuit is open • Valve has failed • TCM has failed
DTC: P2729 **2T MIL: Yes** 2003, 2004 **Models:** 320i, 325i, 330i, 525i, 530i, 540i, 545i, 745i, 750i, M3, Z4 **Engines:** 2.5L, 3.0L, 3.2L, 4.4L, 4.8L **Transmissions:** A/T	**Pressure Regulator Valve 5 Lower Threshold Conditions:** The signal to/from the pressure regulator valve is interrupted or short circuited to ground. **Possible Causes:** • Pressure control solenoid circuit is shorting to ground • Pressure control solenoid circuit is open • Valve has failed • TCM has failed

DTC	Trouble Code Title, Conditions & Possible Causes
DTC: P2730 **2T MIL: Yes** 2003, 2004 **Models:** 320i, 325i, 330i, 525i, 530i, 540i, 545i, 745i, 750i, M3, Z4 **Engines:** 2.5L, 3.0L, 3.2L, 4.4L, 4.8L **Transmissions:** A/T	**Pressure Regulator Valve 5 Upper Threshold Conditions:** The signal to/from the pressure regulator valve is interrupted or short circuited to supply. **Possible Causes:** • Pressure control solenoid circuit is shorting to ground • Pressure control solenoid circuit is open • Valve has failed • TCM has failed
DTC: P2761 **2T MIL: Yes** 2003, 2004 **Models:** 320i, 325i, 330i, 525i, 530i, 540i, 545i, 745i, 750i, M3, Z4 **Engines:** 2.5L, 3.0L, 3.2L, 4.4L, 4.8L **Transmissions:** A/T	**Pressure Regulator Valve 4 Plausibility Conditions:** The current to/from the pressure regulator valve is either higher or lower than the threshold value. **Possible Causes:** • Pressure control solenoid circuit is shorting to ground • Pressure control solenoid circuit is open • Valve has failed • TCM has failed
DTC: P2763 **2T MIL: Yes** 2003, 2004 **Models:** 320i, 325i, 330i, 525i, 530i, 540i, 545i, 745i, 750i, M3, Z4 **Engines:** 2.5L, 3.0L, 3.2L, 4.4L, 4.8L **Transmissions:** A/T	**Pressure Regulator Valve 4 Upper Threshold Conditions:** The signal to/from the pressure regulator valve is interrupted or short circuited to power. **Possible Causes:** • Pressure control solenoid circuit is shorting to ground • Pressure control solenoid circuit is open • Valve has failed • TCM has failed
DTC: P2764 **2T MIL: Yes** 2003, 2004 **Models:** 320i, 325i, 330i, 525i, 530i, 540i, 545i, 745i, 750i, M3, Z4 **Engines:** 2.5L, 3.0L, 3.2L, 4.4L, 4.8L **Transmissions:** A/T	**Pressure Regulator Valve 4 Lower Threshold Conditions:** The signal to/from the pressure regulator valve is interrupted or short circuited to ground. **Possible Causes:** • Pressure control solenoid circuit is shorting to ground • Pressure control solenoid circuit is open • Valve has failed • TCM has failed

DTC: P3012 **2T MIL: Yes** **2005, 2006** **Models:** 325i, 330i, 525i, 530i, 540i, 550i, 745i, 750i, Z4 **Engines:** 3.0L, 4.8L **Transmissions:** All	**O2S Sensor Circuit Adaptation Value to High (Bank 1 Sensor 1) Conditions:** Engine started and the fault entry trips after 10 seconds, after the calibration runs after engine is in idle phases, battery voltage must be at least 11v, and an offset correction value was found to be above the maximum approved threshold value. The battery voltage must be between 11 and 16 volts. **Possible Causes:** • O2S signal shorted to power circuit inside connector • O2S signal circuit shorted to ground or to system voltage • Check wiring harness • Replace control module
DTC: P3013 **2T MIL: Yes** **2005, 2006** **Models:** 325i, 330i, 525i, 530i, 540i, 550i, 745i, 750i, Z4 **Engines:** 3.0L, 4.8L **Transmissions:** All	**O2S Sensor Circuit Adaptation Value to High (Bank 2 Sensor 1) Conditions:** Engine started and the fault entry trips after 10 seconds, after the calibration runs after engine is in idle phases, battery voltage must be at least 11v, and an offset correction value was found to be above the maximum approved threshold value. The battery voltage must be between 11 and 16 volts. **Possible Causes:** • O2S signal shorted to power circuit inside connector • O2S signal circuit shorted to ground or to system voltage • Check wiring harness • Replace control module
DTC: P3014 **2T MIL: Yes** **2005, 2006** **Models:** 325i, 330i, 525i, 530i, 540i, 550i, 745i, 750i, Z4 **Engines:** 3.0L, 4.8L **Transmissions:** All	**O2S Sensor WRAF-IC Supply Voltage Too Low (Bank 1 Sensor 1) Conditions:** Engine started and the fault entry trips after the criterion remains present for 2 seconds, battery voltage must be at least 11v, and an the monitor for the CJ125 chip power supply was found that the voltage was too low. The battery voltage must be between 11 and 16 volts. **Possible Causes:** • Internal HW fault in control module • Replace control module
DTC: P3015 **2T MIL: Yes** **2005, 2006** **Models:** 325i, 330i, 525i, 530i, 540i, 550i, 745i, 750i, Z4 **Engines:** 3.0L, 4.8L **Transmissions:** All	**O2S Sensor WRAF-IC Supply Voltage Too Low (Bank 2 Sensor 1) Conditions:** Engine started and the fault entry trips after the criterion remains present for 2 seconds, battery voltage must be at least 11v, and an the monitor for the CJ125 chip power supply was found that the voltage was too low. The battery voltage must be between 11 and 16 volts. **Possible Causes:** • Internal HW fault in control module • Replace control module
DTC: P3016 **2T MIL: Yes** **2004, 2006** **Models:** 320i, 325i, 330i, 525i, 530i, 545i, 550i, 745i, 750i, M3, Z4 **Engines:** 2.5L, 3.0L, 3.2L, 4.8L **Transmissions:** All	**O2 Sensor Calibration Resistance at WRAF-IC Plausibility Bank 1 Sensor 1 Conditions:** Engine running for 25 seconds, battery voltage 11.5, ground between engine and chassis well connected and the exhaust system must be properly sealed between catalytic converter and the cylinder head. The DME detected the O2S signal was implausible or not detected. There is no overrun cutoff and the exhaust gas temperature is less than 400 degrees Celsius. The battery voltage must be between 11 and 16 volts. **Possible Causes:** • Oxygen sensor for oxygen sensor (O2S) before catalytic converter is faulty • O2S is contaminated (due to presence of silicone in fuel) • O2S signal and ground circuit wires crossed in wiring harness • O2S signal circuit is shorted to sensor or chassis ground • O2S element before the catalytic converter has failed (internal short condition) • Leaks present in the exhaust manifold or exhaust pipes
DTC: P3017 **2T MIL: Yes** **2004, 2006** **Models:** 320i, 325i, 330i, 525i, 530i, 545i, 550i, 745i, 750i, M3, Z4 **Engines:** 2.5L, 3.0L, 3.2L, 4.8L **Transmissions:** All	**O2 Sensor Calibration Resistance at WRAF-IC Plausibility Bank 2 Sensor 1 Conditions:** Engine running for 25 seconds, battery voltage 11.5, ground between engine and chassis well connected and the exhaust system must be properly sealed between catalytic converter and the cylinder head. The DME detected the O2S signal was implausible or not detected. There is no overrun cutoff and the exhaust gas temperature is less than 400 degrees Celsius. The battery voltage must be between 11 and 16 volts. **Possible Causes:** • Oxygen sensor for oxygen sensor (O2S) before catalytic converter is faulty • O2S is contaminated (due to presence of silicone in fuel) • O2S signal and ground circuit wires crossed in wiring harness • O2S signal circuit is shorted to sensor or chassis ground • O2S element before the catalytic converter has failed (internal short condition) • Leaks present in the exhaust manifold or exhaust pipes

DTC	Trouble Code Title, Conditions & Possible Causes
DTC: P3018 **2T MIL: Yes** **2005, 2006** **Models:** 325i, 330i, 525i, 530i, 540i, 550i, 745i, 750i, Z4 **Engines:** 3.0L, 4.8L **Transmissions:** All	**O2S Sensor Lambda Controller Value Above Threshold due to Open Pumping Current Circuit (Bank 1 Sensor 1)** **Conditions:** Engine started and the fault entry trips after the criterion remains present for 2 seconds, battery voltage must be at least 11v, and there is a communications fault between the ECU and the CJ125 (SPI bus) chip. The sensor temperature status is heated up and the fuel system status is closed loop. **Possible Causes:** • Internal HW fault in control module • Replace control module
DTC: P3019 **2T MIL: Yes** **2005, 2006** **Models:** 325i, 330i, 525i, 530i, 540i, 550i, 745i, 750i, Z4 **Engines:** 3.0L, 4.8L **Transmissions:** All	**O2S Sensor Lambda Controller Value Above Threshold due to Open Pumping Current Circuit (Bank 2 Sensor 1)** **Conditions:** Engine started and the fault entry trips after the criterion remains present for 2 seconds, battery voltage must be at least 11v, and there is a communications fault between the ECU and the CJ125 (SPI bus) chip. The sensor temperature status is heated up and the fuel system status is closed loop. **Possible Causes:** • Internal HW fault in control module • Replace control module
DTC: P3020 **2T MIL: Yes** **2005, 2006** **Models:** 325i, 330i, 525i, 530i, 540i, 550i, 745i, 750i, Z4 **Engines:** 3.0L, 4.8L **Transmissions:** All	**O2S Sensor Signal Voltage Too Low During Coast Down Fuel Cut-Off Due to Open Pumping Current Circuit (Bank 1 Sensor 1) Conditions:** Engine started and the fault monitoring criterion must remain present for over 2 seconds, battery voltage must be at least 11v, and the voltage at CJ125 is at 1.5v so the sensor is heated adequately, there is an open pump current wire. **Possible Causes:** • Check plugs • Check wiring harness • Replace sensor
DTC: P3021 **2T MIL: Yes** **2005, 2006** **Models:** 325i, 330i, 525i, 530i, 540i, 550i, 745i, 750i, Z4 **Engines:** 3.0L, 4.8L **Transmissions:** All	**O2S Sensor Signal Voltage Too Low During Coast Down Fuel Cut-Off Due to Open Pumping Current Circuit (Bank 2 Sensor 1) Conditions:** Engine started and the fault monitoring criterion must remain present for over 2 seconds, battery voltage must be at least 11v, and the voltage at CJ125 is at 1.5v so the sensor is heated adequately, there is an open pump current wire. **Possible Causes:** • Check plugs • Check wiring harness • Replace sensor
DTC: P3022 **2T MIL: Yes** **2004, 2005, 2006** **Models:** 320i, 325i, 330i, 525i, 530i, 545i, 550i, 745i, 750i, M3, Z4 **Engines:** 2.5L, 3.0L, 3.2L, 4.8L **Transmissions:** All	**O2S Sensor Disturbed SPI Communication to WRAF-IC (Bank 1 Sensor 1) Conditions:** Engine started and the fault entry trips after the criterion remains present for 2 seconds, battery voltage must be at least 11v, and there is a communications fault between the ECU and the CJ125 (SPI bus) chip. **Possible Causes:** • Internal HW fault in control module • Replace control module
DTC: P3023 **2T MIL: Yes** **2004, 2005, 2006** **Models:** 320i, 325i, 330i, 525i, 530i, 545i, 550i, 745i, 750i, M3, Z4 **Engines:** 2.5L, 3.0L, 3.2L, 4.8L **Transmissions:** All	**O2S Sensor Disturbed SPI Communication to WRAF-IC (Bank 2 Sensor 1) Conditions:** Engine started and the fault entry trips after the criterion remains present for 2 seconds, battery voltage must be at least 11v, and there is a communications fault between the ECU and the CJ125 (SPI bus) chip. **Possible Causes:** • Internal HW fault in control module • Replace control module
DTC: P3024 **2T MIL: Yes** **2004, 2005, 2006** **Models:** 320i, 325i, 330i, 525i, 530i, 545i, 550i, 745i, 750i, M3, Z4 **Engines:** 2.5L, 3.0L, 3.2L, 4.8L **Transmissions:** All	**O2S Sensor Initialization Error WRAF-IC (Bank 1 Sensor 1) Conditions:** Engine started and the fault entry trips after the criterion remains present for 2 seconds, battery voltage must be at least 11v, and there is a communications fault between the ECU and the CJ125 (SPI bus) chip. **Possible Causes:** • Internal HW fault in control module • Replace control module
DTC: P3025 **2T MIL: Yes** **2004, 2005, 2006** **Models:** 320i, 325i, 330i, 525i, 530i, 545i, 550i, 745i, 750i, M3, Z4 **Engines:** 2.5L, 3.0L, 3.2L, 4.8L **Transmissions:** All	**O2S Sensor Initialization Error WRAF-IC (Bank 2 Sensor 1) Conditions:** Engine started and the fault entry trips after the criterion remains present for 2 seconds, battery voltage must be at least 11v, and there is a communications fault between the ECU and the CJ125 (SPI bus) chip. **Possible Causes:** • Internal HW fault in control module • Replace control module

DTC	Trouble Code Title, Conditions & Possible Causes
DTC: P3026 **2T MIL: Yes** **2004, 2005, 2006** **Models:** 320i, 325i, 330i, 525i, 530i, 545i, 550i, 745i, 750i, M3, Z4 **Engines:** 2.5L, 3.0L, 3.2L, 4.8L **Transmissions:** All	**O2 Sensor Operating Temperature Not Reached Bank 1 Sensor 1 Conditions:** Engine running for 40 seconds, battery voltage 11.5, ground between engine and chassis well connected and the exhaust system must be properly sealed between catalytic converter and the cylinder head. The DME detected the O2S signal was implausible or not detected. Monitoring of the sensor's ceramic temperature and control single pulse-duty factor detects reduced heater performance at the oxygen sensor before the catalytic converter. The heater controller value stays above the threshold (97 percent). There is no overrun cutoff and the exhaust gas temperature is less than 400 degrees Celsius. **Possible Causes:** • Oxygen sensor heater for oxygen sensor (HO2S) before catalytic converter is faulty • O2S is contaminated (due to presence of silicone in fuel) • O2S signal and ground circuit wires crossed in wiring harness • O2S signal circuit is shorted to sensor or chassis ground • O2S element before the catalytic converter has failed (internal short condition) • Leaks present in the exhaust manifold or exhaust pipes
DTC: P3027 **2T MIL: Yes** **2004, 2005, 2006** **Models:** 320i, 325i, 330i, 525i, 530i, 545i, 550i, 745i, 750i, M3, Z4 **Engines:** 2.5L, 3.0L, 3.2L, 4.8L **Transmissions:** All	**O2 Sensor Operating Temperature Not Reached Bank 2 Sensor 1 Conditions:** Engine running for 40 seconds, battery voltage 11.5, ground between engine and chassis well connected and the exhaust system must be properly sealed between catalytic converter and the cylinder head. The DME detected the O2S signal was implausible or not detected. Monitoring of the sensor's ceramic temperature and control single pulse-duty factor detects reduced heater performance at the oxygen sensor before the catalytic converter. The heater controller value stays above the threshold (97 percent). There is no overrun cutoff and the exhaust gas temperature is less than 400 degrees Celsius. **Possible Causes:** • Oxygen sensor heater for oxygen sensor (HO2S) before catalytic converter is faulty • O2S is contaminated (due to presence of silicone in fuel) • O2S signal and ground circuit wires crossed in wiring harness • O2S signal circuit is shorted to sensor or chassis ground • O2S element before the catalytic converter has failed (internal short condition) • Leaks present in the exhaust manifold or exhaust pipes
DTC: P3028 **2T MIL: Yes** **2006** **Models:** 325i, 330i, 525i, 530i, 550i, 750i, Z4 **Engines:** 3.0L, 4.8L **Transmissions:** All	**O2 Sensor Heater Control No Activity Detected Bank 1 Sensor 1 Conditions:** Engine running for 40 seconds, battery voltage 11.5, ground between engine and chassis well connected and the exhaust system must be properly sealed between catalytic converter and the cylinder head. The DME detected the O2S signal was implausible or not detected. Monitoring of the sensor's ceramic temperature and control single pulse-duty factor detects reduced heater performance at the oxygen sensor before the catalytic converter. There is no overrun cutoff and the exhaust gas temperature is less than 400 degrees Celsius. **Possible Causes:** • Oxygen sensor heater for oxygen sensor (HO2S) before catalytic converter is faulty • O2S is contaminated (due to presence of silicone in fuel) • O2S signal and ground circuit wires crossed in wiring harness • O2S signal circuit is shorted to sensor or chassis ground • O2S element before the catalytic converter has failed (internal short condition) • Leaks present in the exhaust manifold or exhaust pipes
DTC: P3029 **2T MIL: Yes** **2006** **Models:** 325i, 330i, 525i, 530i, 550i, 750i, Z4 **Engines:** 3.0L, 4.8L **Transmissions:** All	**O2 Sensor Heater Control No Activity Detected Bank 2 Sensor 1 Conditions:** Engine running for 40 seconds, battery voltage 11.5, ground between engine and chassis well connected and the exhaust system must be properly sealed between catalytic converter and the cylinder head. The DME detected the O2S signal was implausible or not detected. Monitoring of the sensor's ceramic temperature and control single pulse-duty factor detects reduced heater performance at the oxygen sensor before the catalytic converter. There is no overrun cutoff and the exhaust gas temperature is less than 400 degrees Celsius. **Possible Causes:** • Oxygen sensor heater for oxygen sensor (HO2S) before catalytic converter is faulty • O2S is contaminated (due to presence of silicone in fuel) • O2S signal and ground circuit wires crossed in wiring harness • O2S signal circuit is shorted to sensor or chassis ground • O2S element before the catalytic converter has failed (internal short condition) • Leaks present in the exhaust manifold or exhaust pipes
DTC: P3037 **2T MIL: Yes** **2006** **Models:** 325i, 330i, 525i, 530i, 550i, 750i, Z4 **Engines:** 3.0L, 4.8L **Transmissions:** All	**O2S Sensor Positive Current Control Circuit/Open Circuit (Bank 1 Sensor 1) Conditions:** Engine started, the fault criterion must remain present for over 2 seconds, the Lambda specification is outside three percent relative to Lambda = 1, and the closed loop lambda control is active. **Possible Causes:** • Check plugs • Check wiring harness • Replace sensor

DTC	Trouble Code Title, Conditions & Possible Causes
DTC: P3200 **2T MIL: Yes** **2005, 2006** **Models:** 325i, 330i, 525i, 530i, 540i, 550i, 745i, 750i, Z4 **Engines:** 3.0L, 4.8L **Transmissions:** All	**Power CAN, CAN Chip Defective Conditions:** The Engine Control Module (DME) communicates with all databus-capable control modules via a CAN databus. These databus-capable control modules are connected via two data bus wires which are twisted together (CAN_High and CAN_Low), and exchange information (messages). Missing information on the databus is recognized as a malfunction and stored. Trouble-free operation of the CAN-Bus requires that it have a terminal resistance. This central terminal resistor is located in the Engine Control Module (DME). The ignition is on for 800ms or the engine running for 360ms, the voltage is greater than 10 volts. This applies to vehicles with ARS only and does not affect bus activity. **Possible Causes:** • CAN data bus wires have short circuited to each other • ECU has failed • BUS system failure • Defective bus controller (EGS)
DTC: P3201 **2T MIL: Yes** **2006** **Models:** 325i, 330i, 525i, 530i, 550i, 750i, Z4 **Engines:** 3.0L, 4.8L **Transmissions:** All	**Power CAN, DPRAM-CAN Chip Defective Conditions:** The Engine Control Module (DME) communicates with all databus-capable control modules via a CAN databus. These databus-capable control modules are connected via two data bus wires which are twisted together (CAN_High and CAN_Low), and exchange information (messages). Missing information on the databus is recognized as a malfunction and stored. Trouble-free operation of the CAN-Bus requires that it have a terminal resistance. This central terminal resistor is located in the Engine Control Module (DME). The ignition is on for 800ms or the engine running for 360ms, the voltage is greater than 10 volts. This applies to vehicles with ARS only and does not affect bus activity. **Possible Causes:** • CAN data bus wires have short circuited to each other • ECU has failed • BUS system failure • Defective bus controller (EGS)
DTC: P3202 **2T MIL: Yes** **2005, 2006** **Models:** 325i, 330i, 525i, 530i, 540i, 550i, 745i, 750i, Z4 **Engines:** 3.0L, 4.8L **Transmissions:** All	**Powertrain CAN, CAN Chip Cut-Off Conditions:** The Engine Control Module (DME) communicates with all databus-capable control modules via a CAN databus. These databus-capable control modules are connected via two data bus wires which are twisted together (CAN_High and CAN_Low), and exchange information (messages). Missing information on the databus is recognized as a malfunction and stored. Trouble-free operation of the CAN-Bus requires that it have a terminal resistance. This central terminal resistor is located in the Engine Control Module (DME). The ignition is on for 800ms or the engine running for 360ms, the voltage is greater than 10 volts. This applies to vehicles with ARS only and does not affect bus activity. **Possible Causes:** • CAN data bus wires have short circuited to each other • ECU has failed • BUS system failure • Defective bus controller (EGS)
DTC: P3203 **2T MIL: Yes** **2005, 2006** **Models:** 325i, 330i, 525i, 530i, 540i, 550i, 745i, 750i, Z4 **Engines:** 3.0L, 4.8L **Transmissions:** All	**Local CAN, LoCAN Chip Defective Conditions:** The Engine Control Module (DME) communicates with all databus-capable control modules via a CAN databus. These databus-capable control modules are connected via two data bus wires which are twisted together (CAN_High and CAN_Low), and exchange information (messages). Missing information on the databus is recognized as a malfunction and stored. Trouble-free operation of the CAN-Bus requires that it have a terminal resistance. This central terminal resistor is located in the Engine Control Module (DME). The ignition is on for 800ms or the engine running for 360ms, the voltage is greater than 10 volts. This applies to vehicles with ARS only and does not affect bus activity. **Possible Causes:** • CAN data bus wires have short circuited to each other • ECU has failed • BUS system failure • Defective bus controller (EGS)
DTC: P3204 **2T MIL: Yes** **2006** **Models:** 325i, 330i, 525i, 530i, 550i, 750i, Z4 **Engines:** 3.0L, 4.8L **Transmissions:** All	**Local CAN, DPRAM-LoCAN Chip Defective Conditions:** The Engine Control Module (DME) communicates with all databus-capable control modules via a CAN databus. These databus-capable control modules are connected via two data bus wires which are twisted together (CAN_High and CAN_Low), and exchange information (messages). Missing information on the databus is recognized as a malfunction and stored. Trouble-free operation of the CAN-Bus requires that it have a terminal resistance. This central terminal resistor is located in the Engine Control Module (DME). The ignition is on for 800ms or the engine running for 360ms, the voltage is greater than 10 volts. This applies to vehicles with ARS only and does not affect bus activity. **Possible Causes:** • CAN data bus wires have short circuited to each other • ECU has failed • BUS system failure • Defective bus controller (EGS)

DTC	Trouble Code Title, Conditions & Possible Causes
DTC: P3205 **2T MIL: Yes** **2005, 2006** **Models:** 325i, 330i, 525i, 530i, 540i, 550i, 745i, 750i, Z4 **Engines:** 3.0L, 4.8L **Transmissions:** All	**Local CAN, DPRAM-LoCAN Chip Cut-Off Conditions:** The Engine Control Module (DME) communicates with all databus-capable control modules via a CAN databus. These databus-capable control modules are connected via two data bus wires which are twisted together (CAN_High and CAN_Low), and exchange information (messages). Missing information on the databus is recognized as a malfunction and stored. Trouble-free operation of the CAN-Bus requires that it have a terminal resistance. This central terminal resistor is located in the Engine Control Module (DME). The ignition is on for 800ms or the engine running for 360ms, the voltage is greater than 10 volts. This applies to vehicles with ARS only and does not affect bus activity. **Possible Causes:** • CAN data bus wires have short circuited to each other • ECU has failed • BUS system failure • Defective bus controller (EGS)
DTC: P3206 **MIL: Yes (U.S. only)** **2006** **Models:** 325i, 330i, 525i, 530i, 550i, 750i, Z4 **Engines:** 3.0L, 4.8L **Transmissions:** All	**CAN Timeout ARS Conditions:** The Engine Control Module (DME) communicates with all databus-capable control modules via a CAN databus. These databus-capable control modules are connected via two data bus wires which are twisted together (CAN_High and CAN_Low), and exchange information (messages). Missing information on the databus is recognized as a malfunction and stored. Trouble-free operation of the CAN-Bus requires that it have a terminal resistance. This central terminal resistor is located in the Engine Control Module (DME). The ignition is on for 800ms or the engine running for 360ms, the voltage is greater than 10 volts. This applies to vehicles with ARS only and does not affect bus activity. **Possible Causes:** • CAN data bus wires have short circuited to each other • ECU has failed • BUS system failure • Defective bus controller (EGS)
DTC: P3207 **MIL: Yes (U.S. only)** **2006** **Models:** 325i, 330i, 525i, 530i, 550i, 750i, Z4 **Engines:** 3.0L, 4.8L **Transmissions:** All	**CAN Message Monitoring ARS No Signal Conditions:** The Engine Control Module (DME) communicates with all databus-capable control modules via a CAN databus. These databus-capable control modules are connected via two data bus wires which are twisted together (CAN_High and CAN_Low), and exchange information (messages). Missing information on the databus is recognized as a malfunction and stored. Trouble-free operation of the CAN-Bus requires that it have a terminal resistance. This central terminal resistor is located in the Engine Control Module (DME). The ignition is on for 800ms or the engine running for 360ms, the voltage is greater than 10 volts. This applies to vehicles with ARS only and does not affect bus activity. After 40ms, no signal was received. **Possible Causes:** • CAN data bus wires have short circuited to each other • ECU has failed • BUS system failure • Defective bus controller (EGS)
DTC: P3208 **MIL: Yes (U.S. only)** **2006** **Models:** 325i, 330i, 525i, 530i, 550i, 750i, Z4 **Engines:** 3.0L, 4.8L **Transmissions:** All	**CAN Message Monitoring ARS Plausibility Conditions:** The Engine Control Module (DME) communicates with all databus-capable control modules via a CAN databus. These databus-capable control modules are connected via two data bus wires which are twisted together (CAN_High and CAN_Low), and exchange information (messages). Missing information on the databus is recognized as a malfunction and stored. Trouble-free operation of the CAN-Bus requires that it have a terminal resistance. This central terminal resistor is located in the Engine Control Module (DME). The ignition is on for 800ms or the engine running for 360ms, the voltage is greater than 10 volts. This applies to vehicles with ARS only and does not affect bus activity. **Possible Causes:** • CAN data bus wires have short circuited to each other • ECU has failed • BUS system failure • Defective bus controller (EGS)
DTC: P3209 **MIL: Yes (U.S. only)** **2006** **Models:** 325i, 330i, 525i, 530i, 550i, 750i, Z4 **Engines:** 3.0L, 4.8L **Transmissions:** All	**CAN Message Monitoring ASC/DSC Alive Check Malfunction Conditions:** The Engine Control Module (DME) communicates with all databus-capable control modules via a CAN databus. These databus-capable control modules are connected via two data bus wires which are twisted together (CAN_High and CAN_Low), and exchange information (messages). Missing information on the databus is recognized as a malfunction and stored. Trouble-free operation of the CAN-Bus requires that it have a terminal resistance. This central terminal resistor is located in the Engine Control Module (DME). The ignition is on for 800ms or the engine running for 20ms, the voltage is greater than 10 volts **Possible Causes:** • CAN data bus wires have short circuited to each other • ECU has failed • BUS system failure • Defective bus controller (EGS)

DTC	Trouble Code Title, Conditions & Possible Causes
DTC: P3210 **MIL: Yes (U.S. only)** **2006** **Models:** 325i, 330i, 525i, 530i, 550i, 750i, Z4 **Engines:** 3.0L, 4.8L **Transmissions:** All	**CAN Message Monitoring ASC/DSC Plausibility Conditions:** The Engine Control Module (DME) communicates with all databus-capable control modules via a CAN databus. These databus-capable control modules are connected via two data bus wires which are twisted together (CAN_High and CAN_Low), and exchange information (messages). Missing information on the databus is recognized as a malfunction and stored. Trouble-free operation of the CAN-Bus requires that it have a terminal resistance. This central terminal resistor is located in the Engine Control Module (DME). The ignition is on for 800ms or the engine running for 20ms, the voltage is greater than 10 volts **Possible Causes:** • CAN data bus wires have short circuited to each other • ECU has failed • BUS system failure • Defective bus controller (EGS)
DTC: P3211 **MIL: Yes (U.S. only)** **2006** **Models:** 325i, 330i, 525i, 530i, 550i, 750i, Z4 **Engines:** 3.0L, 4.8L **Transmissions:** All	**CAN Message Monitoring CAS No Signal Conditions:** The Engine Control Module (DME) communicates with all databus-capable control modules via a CAN databus. These databus-capable control modules are connected via two data bus wires which are twisted together (CAN_High and CAN_Low), and exchange information (messages). Missing information on the databus is recognized as a malfunction and stored. Trouble-free operation of the CAN-Bus requires that it have a terminal resistance. This central terminal resistor is located in the Engine Control Module (DME). The ignition is on for 800ms or the engine running for 400ms, the voltage is greater than 10 volts. This applies to vehicles with CAS only and does not affect bus activity. After 40ms, no signal was received. **Possible Causes:** • CAN data bus wires have short circuited to each other • ECU has failed • BUS system failure • Defective bus controller (EGS)
DTC: P3212 **MIL: Yes (U.S. only)** **2006** **Models:** 325i, 330i, 525i, 530i, 550i, 750i, Z4 **Engines:** 3.0L, 4.8L **Transmissions:** All	**CAN Message Monitoring CAS Plausibility Conditions:** The Engine Control Module (DME) communicates with all databus-capable control modules via a CAN databus. These databus-capable control modules are connected via two data bus wires which are twisted together (CAN_High and CAN_Low), and exchange information (messages). Missing information on the databus is recognized as a malfunction and stored. Trouble-free operation of the CAN-Bus requires that it have a terminal resistance. This central terminal resistor is located in the Engine Control Module (DME). The ignition is on for 800ms or the engine running for 400ms, the voltage is greater than 10 volts. This applies to vehicles with CAS only and does not affect bus activity. **Possible Causes:** • CAN data bus wires have short circuited to each other • ECU has failed • BUS system failure • Defective bus controller (EGS)
DTC: P3213 **MIL: Yes (U.S. only)** **2002, 2003, 2004, 2005, 2006** **Models:** 325i, 330i, 525i, 530i, 545i, 550i, 745i, 750i, Z4 **Engines:** 3.0L, 4.8L **Transmissions:** All	**CAN Message Monitoring ETC Alive Check Malfunction Conditions:** The Engine Control Module (DME) communicates with all databus-capable control modules via a CAN databus. These databus-capable control modules are connected via two data bus wires which are twisted together (CAN_High and CAN_Low), and exchange information (messages). Missing information on the databus is recognized as a malfunction and stored. Trouble-free operation of the CAN-Bus requires that it have a terminal resistance. This central terminal resistor is located in the Engine Control Module (DME). The ignition is on for 800ms or the engine running for 20ms, the voltage is greater than 10 volts **Possible Causes:** • CAN data bus wires have short circuited to each other • ECU has failed • BUS system failure • Defective bus controller (EGS)
DTC: P3214 **MIL: Yes (U.S. only)** **2002, 2003, 2004, 2005, 2006** **Models:** 325i, 330i, 525i, 530i, 545i, 550i, 745i, 750i, Z4 **Engines:** 3.0L, 4.8L **Transmissions:** All	**CAN Message Monitoring ETC Plausibility Conditions:** The Engine Control Module (DME) communicates with all databus-capable control modules via a CAN databus. These databus-capable control modules are connected via two data bus wires which are twisted together (CAN_High and CAN_Low), and exchange information (messages). Missing information on the databus is recognized as a malfunction and stored. Trouble-free operation of the CAN-Bus requires that it have a terminal resistance. This central terminal resistor is located in the Engine Control Module (DME). The ignition is on for 800ms or the engine running for 20ms, the voltage is greater than 10 volts **Possible Causes:** • CAN data bus wires have short circuited to each other • ECU has failed • BUS system failure • Defective bus controller (EGS)

DTC	Trouble Code Title, Conditions & Possible Causes
DTC: P3215 **MIL: Yes (U.S. only)** **2006** **Models:** 325i, 330i, 525i, 530i, 550i, 750i, Z4 **Engines:** 3.0L, 4.8L **Transmissions:** All	**CAN Message Monitoring IHKA (Automatic Heating and Air Conditioning) No Signal Conditions:** The Engine Control Module (DME) communicates with all databus-capable control modules via a CAN databus. These databus-capable control modules are connected via two data bus wires which are twisted together (CAN_High and CAN_Low), and exchange information (messages). Missing information on the databus is recognized as a malfunction and stored. Trouble-free operation of the CAN-Bus requires that it have a terminal resistance. This central terminal resistor is located in the Engine Control Module (DME). The ignition is on for 800ms or the engine running for 360ms, the voltage is greater than 10 volts. This applies to vehicles with ARS only and does not affect bus activity. After 40ms, no signal was received. **Possible Causes:** • CAN data bus wires have short circuited to each other • ECU has failed • BUS system failure • Defective bus controller (EGS)
DTC: P3216 **MIL: Yes (U.S. only)** **2005, 2006** **Models:** 325i, 330i, 525i, 530i, 540i, 550i, 745i, 750i, Z4 **Engines:** 3.0L, 4.8L **Transmissions:** All	**CAN Timeout Instrument Pack Conditions:** The Engine Control Module (DME) communicates with all databus-capable control modules via a CAN databus. These databus-capable control modules are connected via two data bus wires which are twisted together (CAN_High and CAN_Low), and exchange information (messages). Missing information on the databus is recognized as a malfunction and stored. Trouble-free operation of the CAN-Bus requires that it have a terminal resistance. This central terminal resistor is located in the Engine Control Module (DME). The ignition is on for 800ms or the engine running for 15 seconds, the voltage is greater than 10 volts. This applies to vehicles with MIL only and does not affect bus activity. **Possible Causes:** • CAN data bus wires have short circuited to each other • ECU has failed • BUS system failure • Defective bus controller (EGS)
DTC: P3217 **MIL: Yes (U.S. only)** **2005, 2006** **Models:** 325i, 330i, 525i, 530i, 540i, 550i, 745i, 750i, Z4 **Engines:** 3.0L, 4.8L **Transmissions:** All	**CAN Message Monitoring Instrument Pack Plausibility Conditions:** The Engine Control Module (DME) communicates with all databus-capable control modules via a CAN databus. These databus-capable control modules are connected via two data bus wires which are twisted together (CAN_High and CAN_Low), and exchange information (messages). Missing information on the databus is recognized as a malfunction and stored. Trouble-free operation of the CAN-Bus requires that it have a terminal resistance. This central terminal resistor is located in the Engine Control Module (DME). The ignition is on for 800ms or the engine running for 15 seconds, the voltage is greater than 10 volts. This applies to vehicles with MIL only and does not affect bus activity. **Possible Causes:** • CAN data bus wires have short circuited to each other • ECU has failed • BUS system failure • Defective bus controller (EGS)
DTC: P3219 **MIL: Yes (U.S. only)** **2006** **Models:** 325i, 330i, 525i, 530i, 550i, 750i, Z4 **Engines:** 3.0L, 4.8L **Transmissions:** All	**CAN Message Monitoring SZL (Switch Cluster Steering Column) Alive Check Malfunction Conditions:** The Engine Control Module (DME) communicates with all databus-capable control modules via a CAN databus. These databus-capable control modules are connected via two data bus wires which are twisted together (CAN_High and CAN_Low), and exchange information (messages). Missing information on the databus is recognized as a malfunction and stored. Trouble-free operation of the CAN-Bus requires that it have a terminal resistance. This central terminal resistor is located in the Engine Control Module (DME). The ignition is on for 800ms or the engine running for 20ms, the voltage is greater than 10 volts **Possible Causes:** • CAN data bus wires have short circuited to each other • ECU has failed • BUS system failure • Defective bus controller (EGS)
DTC: P3220 **MIL: Yes (U.S. only)** **2006** **Models:** 325i, 330i, 525i, 530i, 550i, 750i, Z4 **Engines:** 3.0L, 4.8L **Transmissions:** All	**CAN Message Monitoring SZL (Switch Cluster Steering Column) No Signal Conditions:** The Engine Control Module (DME) communicates with all databus-capable control modules via a CAN databus. These databus-capable control modules are connected via two data bus wires which are twisted together (CAN_High and CAN_Low), and exchange information (messages). Missing information on the databus is recognized as a malfunction and stored. Trouble-free operation of the CAN-Bus requires that it have a terminal resistance. This central terminal resistor is located in the Engine Control Module (DME). The ignition is on for 800ms or the engine running for 360ms, the voltage is greater than 10 volts. This applies to vehicles with SZL only and does not affect bus activity. After 40ms, no signal was received. **Possible Causes:** • CAN data bus wires have short circuited to each other • ECU has failed • BUS system failure • Defective bus controller (EGS)

DTC	Trouble Code Title, Conditions & Possible Causes
DTC: P3221 **MIL: Yes (U.S. only)** **2006** **Models:** 325i, 330i, 525i, 530i, 550i, 750i, Z4 **Engines:** 3.0L, 4.8L **Transmissions:** All	**CAN Message Monitoring SZL (Switch Cluster Steering Column) Plausibility Conditions:** The Engine Control Module (DME) communicates with all databus-capable control modules via a CAN databus. These databus-capable control modules are connected via two data bus wires which are twisted together (CAN_High and CAN_Low), and exchange information (messages). Missing information on the databus is recognized as a malfunction and stored. Trouble-free operation of the CAN-Bus requires that it have a terminal resistance. This central terminal resistor is located in the Engine Control Module (DME). The ignition is on for 800ms or the engine running for 15 seconds, the voltage is greater than 10 volts. This applies to vehicles with SZL only and does not affect bus activity. **Possible Causes:** • CAN data bus wires have short circuited to each other • ECU has failed • BUS system failure • Defective bus controller (EGS)
DTC: P3223 **2006** **Models:** 325i, 330i, 525i, 530i, 550i, 750i, Z4 **Engines:** 3.0L, 4.8L **Transmissions:** All	**Generator Mechanical Error Conditions:** The engine is running for at least 25 seconds, and there is no communication faults at the BSD Interface. Mechanical problems exist within the generator. An alternator is not approved for E60, E65 and E66 models specifically. **Possible Causes:** • Wires have short circuited to each other • ECU has failed • Generator failure or installed incorrectly
DTC: P3225 **2006** **Models:** 325i, 330i, 525i, 530i, 550i, 750i, Z4 **Engines:** 3.0L, 4.8L **Transmissions:** All	**Generator Communication Error Conditions:** The engine is running for at least 25 seconds, and there is no communication faults at the BSD Interface **Possible Causes:** • Wires have short circuited to each other • ECU has failed • Generator failure
DTC: P3226 **2006** **Models:** 325i, 330i, 525i, 530i, 550i, 750i, Z4 **Engines:** 3.0L, 4.8L **Transmissions:** All	**E-Box Control Circuit Fan High Conditions:** The engine is running for at least 800ms and the voltage is between 9.5 and 17.8 volts, and there is a short to the battery voltage. **Possible Causes:** • Wires have short circuited to each other • ECU has failed • Generator failure
DTC: P3227 **2006** **Models:** 325i, 330i, 525i, 530i, 550i, 750i, Z4 **Engines:** 3.0L, 4.8L **Transmissions:** All	**E-Box Control Circuit Fan Low Conditions:** The engine is running for at least 800ms and the voltage is between 9.5 and 17.8 volts, and there is a short to the ground. **Possible Causes:** • Wires have short circuited to each other • ECU has failed • Generator failure
DTC: P3228 **2006** **Models:** 325i, 330i, 525i, 530i, 550i, 750i, Z4 **Engines:** 3.0L, 4.8L **Transmissions:** All	**E-Box Control Circuit Open Circuit Conditions:** The engine is running for at least 800ms and the voltage is between 9.5 and 17.8 volts, and there is a short. **Possible Causes:** • Wires have short circuited to each other • ECU has failed • Generator failure

DTC	Trouble Code Title, Conditions & Possible Causes
DTC: P3231 **2006** **Models:** 325i, 330i, 525i, 530i, 550i, 750i, Z4 **Engines:** 3.0L, 4.8L **Transmissions:** All	**Control Module Monitoring Error Response Plausibility Conditions:** Key on, engine running to at least 1200rpm, the DME has detected an internal fault in the computer or internal fault in the control modules. The fault is deactivated and cannot be entered. If additional fault codes are entered in the DME, resolve these issues. If fault remains, replace the DME. **Possible Causes:** • Battery terminal corrosion, loose battery connection, or faulty • Connection to the DME interrupted, or the circuit has been opened • Reprogramming error has occurred and needs replacement. • Voltage supply for Engi ne Control Module (DME) is faulty
DTC: P3232 **2006** **Models:** 325i, 330i, 525i, 530i, 550i, 750i, Z4 **Engines:** 3.0L, 4.8L **Transmissions:** All	**Control Module Monitoring Ignition Timing Plausibility Conditions:** Key on, engine running to at least 1200rpm, the DME has detected an internal fault in the computer or internal fault in the control modules. The Torque monitoring feature compares the torque demand (from accelerator pedal, FGR, electrical equipment, transmission) with the torque provided (calculated from HFM, injector valves, ignition angle, throttle valve angle, differential pressure, lambda). Deviations trigger fuel-supply safety shutdown to prevent vehicle from autonomous acceleration. Internal fault in computer or in electronic control modules (check whether all ADC channels have been converted). If additional fault codes are entered in the DME, resolve these issues. If fault remains, replace the DME. **Possible Causes:** • Battery terminal corrosion, loose battery connection, or faulty • Connection to the DME interrupted, or the circuit has been opened • Reprogramming error has occurred and needs replacement. • Voltage supply for Engine Control Module (DME) is faulty
DTC: P3233 **2006** **Models:** 325i, 330i, 525i, 530i, 550i, 750i, Z4 **Engines:** 3.0L, 4.8L **Transmissions:** All	**Control Module Monitoring Relative Charge Plausibility Conditions:** Key on, engine running to at least 1200rpm, the DME has detected an internal fault in the computer or internal fault in the control modules. The Torque monitoring feature compares the torque demand (from accelerator pedal, FGR, electrical equipment, transmission) with the torque provided (calculated from HFM, injector valves, ignition angle, throttle valve angle, differential pressure, lambda). Deviations trigger fuel-supply safety shutdown to prevent vehicle from autonomous acceleration. Internal fault in computer or in electronic control modules (check whether all ADC channels have been converted). If additional fault codes are entered in the DME, resolve these issues. If fault remains, replace the DME. **Possible Causes:** • Battery terminal corrosion, loose battery connection, or faulty • Connection to the DME interrupted, or the circuit has been opened • Reprogramming error has occurred and needs replacement. • Voltage supply for Engine Control Module (DME) is faulty
DTC: P3236 **2006** **Models:** 325i, 330i, 525i, 530i, 550i, 750i, Z4 **Engines:** 3.0L, 4.8L **Transmissions:** All	**Control Module Monitoring Injection Time Relative Fuel Quantity Plausibility Conditions:** Key on, engine running to at least 1200rpm for 4.1 seconds, the DME has detected an internal fault in the computer or internal fault in the control modules. The Torque monitoring feature compares the torque demand (from accelerator pedal, FGR, electrical equipment, transmission) with the torque provided (calculated from HFM, injector valves, ignition angle, throttle valve angle, differential pressure, lambda). Deviations trigger fuel-supply safety shutdown to prevent vehicle from autonomous acceleration. Internal fault in computer or in electronic control modules (check whether all ADC channels have been converted). If additional fault codes are entered in the DME, resolve these issues. If fault remains, replace the DME. **Possible Causes:** • Battery terminal corrosion, loose battery connection, or faulty • Connection to the DME interrupted, or the circuit has been opened • Reprogramming error has occurred and needs replacement. • Voltage supply for Engine Control Module (DME) is faulty
DTC: P3237 **2006** **Models:** 325i, 330i, 525i, 530i, 550i, 750i, Z4 **Engines:** 3.0L, 4.8L **Transmissions:** All	**Control Module Monitoring Fuel Correction Error Conditions:** Key on, engine running to at least 1200rpm, the DME has detected an internal fault in the computer or internal fault in the control modules. The Torque monitoring feature compares the torque demand (from accelerator pedal, FGR, electrical equipment, transmission) with the torque provided (calculated from HFM, injector valves, ignition angle, throttle valve angle, differential pressure, lambda). Deviations trigger fuel-supply safety shutdown to prevent vehicle from autonomous acceleration. Internal fault in computer or in electronic control modules (check whether all ADC channels have been converted). If additional fault codes are entered in the DME, resolve these issues. If fault remains, replace the DME. **Possible Causes:** • Battery terminal corrosion, loose battery connection, or faulty • Connection to the DME interrupted, or the circuit has been opened • Reprogramming error has occurred and needs replacement. • Voltage supply for Engine Control Module (DME) is faulty

DTC	Trouble Code Title, Conditions & Possible Causes
DTC: P3238 **2006** **Models:** 325i, 330i, 525i, 530i, 550i, 750i, Z4 **Engines:** 3.0L, 4.8L **Transmissions:** All	**Control Module Monitoring TPU Chip Defective Conditions:** Key on, engine running to at least 1200rpm, the DME has detected a programming error. The Torque monitoring feature compares the torque demand (from accelerator pedal, FGR, electrical equipment, transmission) with the torque provided (calculated from HFM, injector valves, ignition angle, throttle valve angle, differential pressure, lambda). Deviations trigger fuel-supply safety shutdown to prevent vehicle from autonomous acceleration. Internal fault in computer or in electronic control modules (check whether all ADC channels have been converted). If additional fault codes are entered in the DME, resolve these issues. If fault remains, replace the DME. **Possible Causes:** • Battery terminal corrosion, loose battery connection, or faulty • Connection to the DME interrupted, or the circuit has been opened • Reprogramming error has occurred and needs replacement. • Voltage supply for Engine Control Module (DME) is faulty
DTC: P3247 **2006** **Models:** 325i, 330i, 525i, 530i, 550i, 750i, Z4 **Engines:** 3.0L, 4.8L **Transmissions:** All	**Internal Control Module NVRAM Backup Error Conditions:** Key on, the DME has detected a programming error in that the counters in NVRAM and the RAM backup not identical. The RAM test is initiated each time the DME starts. **Possible Causes:** • Battery terminal corrosion, loose battery connection, or faulty • Connection to the DME interrupted, or the circuit has been opened • Reprogramming error has occurred and needs replacement. • Voltage supply for Engine Control Module (DME) is faulty
DTC: P3300 **2006** **Models:** 325i, 330i, 525i, 530i, 550i, 750i, Z4 **Engines:** 3.0L, 4.8L **Transmissions:** All	**Ignition Coil Cylinder 1 High Input or None-Impedance Conditions:** Engine started, battery voltage must be at least 11.5v, all electrical components must be off, parking brake must be engaged (to keep daytime driving lights off), automatic transmission selector must be in park and the ground between the engine and the chassis must be well connected. The DME received high pulses from the ignition module for the Ignition Coil 1 primary circuit. **Possible Causes:** • Engine speed (RPM) sensor has failed • Camshaft Position (CMP) sensor has failed • Power Supply Relay is shorted to an open circuit • There is a malfunction in voltage supply • Ignition coilpack is damaged or it has failed • Cylinder 1 to 4 Fuel Injector(s) have failed
DTC: P3301 **2006** **Models:** 325i, 330i, 525i, 530i, 550i, 750i, Z4 **Engines:** 3.0L, 4.8L **Transmissions:** All	**Ignition Coil Cylinder 1 Contact Resistance or High-Impedance Conditions:** Engine started, battery voltage must be at least 11.5v, all electrical components must be off, parking brake must be engaged (to keep daytime driving lights off), automatic transmission selector must be in park and the ground between the engine and the chassis must be well connected. The DME received high pulses from the ignition module for the Ignition Coil 1 primary circuit. **Possible Causes:** • Engine speed (RPM) sensor has failed • Camshaft Position (CMP) sensor has failed • Power Supply Relay is shorted to an open circuit • There is a malfunction in voltage supply • Ignition coilpack is damaged or it has failed • Cylinder 1 to 4 Fuel Injector(s) have failed
DTC: P3302 **2006** **Models:** 325i, 330i, 525i, 530i, 550i, 750i, Z4 **Engines:** 3.0L, 4.8L **Transmissions:** All	**Ignition Coil Cylinder 1 Cut-Off Due to Over-temperature Condition or No Signal Conditions:** Engine started, battery voltage must be at least 11.5v, all electrical components must be off, parking brake must be engaged (to keep daytime driving lights off), automatic transmission selector must be in park and the ground between the engine and the chassis must be well connected. The DME received no pulses from the ignition module for the Ignition Coil 1 primary circuit due to an elevated temperature reading. **Possible Causes:** • Engine speed (RPM) sensor has failed • Camshaft Position (CMP) sensor has failed • Power Supply Relay is shorted to an open circuit • There is a malfunction in voltage supply • Ignition coilpack is damaged or it has failed • Cylinder 1 to 4 Fuel Injector(s) have failed

DTC	Trouble Code Title, Conditions & Possible Causes
DTC: P3303 **2006** **Models:** 325i, 330i, 525i, 530i, 550i, 750i, Z4 **Engines:** 3.0L, 4.8L **Transmissions:** All	**Ignition Coil Cylinder 5 High Input or None-Impedance Conditions:** Engine started, battery voltage must be at least 11.5v, all electrical components must be off, parking brake must be engaged (to keep daytime driving lights off), automatic transmission selector must be in park and the ground between the engine and the chassis must be well connected. The DME received high pulses from the ignition module for the Ignition Coil 1 primary circuit. **Possible Causes:** • Engine speed (RPM) sensor has failed • Camshaft Position (CMP) sensor has failed • Power Supply Relay is shorted to an open circuit • There is a malfunction in voltage supply • Ignition coilpack is damaged or it has failed • Cylinder 1 to 4 Fuel Injector(s) have failed
DTC: P3304 **2006** **Models:** 325i, 330i, 525i, 530i, 550i, 750i, Z4 **Engines:** 3.0L, 4.8L **Transmissions:** All	**Ignition Coil Cylinder 5 Contact Resistance or High-Impedance Conditions:** Engine started, battery voltage must be at least 11.5v, all electrical components must be off, parking brake must be engaged (to keep daytime driving lights off), automatic transmission selector must be in park and the ground between the engine and the chassis must be well connected. The DME received high pulses from the ignition module for the Ignition Coil 1 primary circuit. **Possible Causes:** • Engine speed (RPM) sensor has failed • Camshaft Position (CMP) sensor has failed • Power Supply Relay is shorted to an open circuit • There is a malfunction in voltage supply • Ignition coilpack is damaged or it has failed • Cylinder 1 to 4 Fuel Injector(s) have failed
DTC: P3305 **2006** **Models:** 325i, 330i, 525i, 530i, 550i, 750i, Z4 **Engines:** 3.0L, 4.8L **Transmissions:** All	**Ignition Coil Cylinder 5 Cut-Off Due to Over-temperature Condition or No Signal Conditions:** Engine started, battery voltage must be at least 11.5v, all electrical components must be off, parking brake must be engaged (to keep daytime driving lights off), automatic transmission selector must be in park and the ground between the engine and the chassis must be well connected. The DME received no pulses from the ignition module for the Ignition Coil 1 primary circuit due to an elevated temperature reading. **Possible Causes:** • Engine speed (RPM) sensor has failed • Camshaft Position (CMP) sensor has failed • Power Supply Relay is shorted to an open circuit • There is a malfunction in voltage supply • Ignition coilpack is damaged or it has failed • Cylinder 1 to 4 Fuel Injector(s) have failed
DTC: P3306 **2006** **Models:** 325i, 330i, 525i, 530i, 550i, 750i, Z4 **Engines:** 3.0L, 4.8L **Transmissions:** All	**Ignition Coil Cylinder 4 High Input or None-Impedance Conditions:** Engine started, battery voltage must be at least 11.5v, all electrical components must be off, parking brake must be engaged (to keep daytime driving lights off), automatic transmission selector must be in park and the ground between the engine and the chassis must be well connected. The DME received high pulses from the ignition module for the Ignition Coil 1 primary circuit. **Possible Causes:** • Engine speed (RPM) sensor has failed • Camshaft Position (CMP) sensor has failed • Power Supply Relay is shorted to an open circuit • There is a malfunction in voltage supply • Ignition coilpack is damaged or it has failed • Cylinder 1 to 4 Fuel Injector(s) have failed
DTC: P3307 **2006** **Models:** 325i, 330i, 525i, 530i, 550i, 750i, Z4 **Engines:** 3.0L, 4.8L **Transmissions:** All	**Ignition Coil Cylinder 4 Contact Resistance or High-Impedance Conditions:** Engine started, battery voltage must be at least 11.5v, all electrical components must be off, parking brake must be engaged (to keep daytime driving lights off), automatic transmission selector must be in park and the ground between the engine and the chassis must be well connected. The DME received high pulses from the ignition module for the Ignition Coil 1 primary circuit. **Possible Causes:** • Engine speed (RPM) sensor has failed • Camshaft Position (CMP) sensor has failed • Power Supply Relay is shorted to an open circuit • There is a malfunction in voltage supply • Ignition coilpack is damaged or it has failed • Cylinder 1 to 4 Fuel Injector(s) have failed

DTC	Trouble Code Title, Conditions & Possible Causes
DTC: P3308 **2006** **Models:** 325i, 330i, 525i, 530i, 550i, 750i, Z4 **Engines:** 3.0L, 4.8L **Transmissions:** All	**Ignition Coil Cylinder 4 Cut-Off Due to Over-temperature Condition or No Signal Conditions:** Engine started, battery voltage must be at least 11.5v, all electrical components must be off, parking brake must be engaged (to keep daytime driving lights off), automatic transmission selector must be in park and the ground between the engine and the chassis must be well connected. The DME received no pulses from the ignition module for the Ignition Coil 1 primary circuit due to an elevated temperature reading. **Possible Causes:** • Engine speed (RPM) sensor has failed • Camshaft Position (CMP) sensor has failed • Power Supply Relay is shorted to an open circuit • There is a malfunction in voltage supply • Ignition coilpack is damaged or it has failed • Cylinder 1 to 4 Fuel Injector(s) have failed
DTC: P3309 **2006** **Models:** 325i, 330i, 525i, 530i, 550i, 750i, Z4 **Engines:** 3.0L, 4.8L **Transmissions:** All	**Ignition Coil Cylinder 8 High Input or None-Impedance Conditions:** Engine started, battery voltage must be at least 11.5v, all electrical components must be off, parking brake must be engaged (to keep daytime driving lights off), automatic transmission selector must be in park and the ground between the engine and the chassis must be well connected. The DME received high pulses from the ignition module for the Ignition Coil 1 primary circuit. **Possible Causes:** • Engine speed (RPM) sensor has failed • Camshaft Position (CMP) sensor has failed • Power Supply Relay is shorted to an open circuit • There is a malfunction in voltage supply • Ignition coilpack is damaged or it has failed • Cylinder 1 to 4 Fuel Injector(s) have failed
DTC: P3310 **2006** **Models:** 325i, 330i, 525i, 530i, 550i, 750i, Z4 **Engines:** 3.0L, 4.8L **Transmissions:** All	**Ignition Coil Cylinder 8 Contact Resistance or High-Impedance Conditions:** Engine started, battery voltage must be at least 11.5v, all electrical components must be off, parking brake must be engaged (to keep daytime driving lights off), automatic transmission selector must be in park and the ground between the engine and the chassis must be well connected. The DME received high pulses from the ignition module for the Ignition Coil 1 primary circuit. **Possible Causes:** • Engine speed (RPM) sensor has failed • Camshaft Position (CMP) sensor has failed • Power Supply Relay is shorted to an open circuit • There is a malfunction in voltage supply • Ignition coilpack is damaged or it has failed • Cylinder 1 to 4 Fuel Injector(s) have failed
DTC: P3311 **2006** **Models:** 325i, 330i, 525i, 530i, 550i, 750i, Z4 **Engines:** 3.0L, 4.8L **Transmissions:** All	**Ignition Coil Cylinder 8 Cut-Off Due to Over-temperature Condition or No Signal Conditions:** Engine started, battery voltage must be at least 11.5v, all electrical components must be off, parking brake must be engaged (to keep daytime driving lights off), automatic transmission selector must be in park and the ground between the engine and the chassis must be well connected. The DME received no pulses from the ignition module for the Ignition Coil 1 primary circuit due to an elevated temperature reading. **Possible Causes:** • Engine speed (RPM) sensor has failed • Camshaft Position (CMP) sensor has failed • Power Supply Relay is shorted to an open circuit • There is a malfunction in voltage supply • Ignition coilpack is damaged or it has failed • Cylinder 1 to 4 Fuel Injector(s) have failed
DTC: P3312 **2006** **Models:** 325i, 330i, 525i, 530i, 550i, 750i, Z4 **Engines:** 3.0L, 4.8L **Transmissions:** All	**Ignition Coil Cylinder 6 High Input or None-Impedance Conditions:** Engine started, battery voltage must be at least 11.5v, all electrical components must be off, parking brake must be engaged (to keep daytime driving lights off), automatic transmission selector must be in park and the ground between the engine and the chassis must be well connected. The DME received high pulses from the ignition module for the Ignition Coil 1 primary circuit. **Possible Causes:** • Engine speed (RPM) sensor has failed • Camshaft Position (CMP) sensor has failed • Power Supply Relay is shorted to an open circuit • There is a malfunction in voltage supply • Ignition coilpack is damaged or it has failed • Cylinder 1 to 4 Fuel Injector(s) have failed

DTC	Trouble Code Title, Conditions & Possible Causes
DTC: P3313 **2006** **Models:** 325i, 330i, 525i, 530i, 550i, 750i, Z4 **Engines:** 3.0L, 4.8L **Transmissions:** All	**Ignition Coil Cylinder 6 Contact Resistance or High-Impedance Conditions:** Engine started, battery voltage must be at least 11.5v, all electrical components must be off, parking brake must be engaged (to keep daytime driving lights off), automatic transmission selector must be in park and the ground between the engine and the chassis must be well connected. The DME received high pulses from the ignition module for the Ignition Coil 1 primary circuit. **Possible Causes:** • Engine speed (RPM) sensor has failed • Camshaft Position (CMP) sensor has failed • Power Supply Relay is shorted to an open circuit • There is a malfunction in voltage supply • Ignition coilpack is damaged or it has failed • Cylinder 1 to 4 Fuel Injector(s) have failed
DTC: P3314 **2006** **Models:** 325i, 330i, 525i, 530i, 550i, 750i, Z4 **Engines:** 3.0L, 4.8L **Transmissions:** All	**Ignition Coil Cylinder 6 Cut-Off Due to Over-temperature Condition or No Signal Conditions:** Engine started, battery voltage must be at least 11.5v, all electrical components must be off, parking brake must be engaged (to keep daytime driving lights off), automatic transmission selector must be in park and the ground between the engine and the chassis must be well connected. The DME received no pulses from the ignition module for the Ignition Coil 1 primary circuit due to an elevated temperature reading. **Possible Causes:** • Engine speed (RPM) sensor has failed • Camshaft Position (CMP) sensor has failed • Power Supply Relay is shorted to an open circuit • There is a malfunction in voltage supply • Ignition coilpack is damaged or it has failed • Cylinder 1 to 4 Fuel Injector(s) have failed
DTC: P3315 **2006** **Models:** 325i, 330i, 525i, 530i, 550i, 750i, Z4 **Engines:** 3.0L, 4.8L **Transmissions:** All	**Ignition Coil Cylinder 31 High Input or None-Impedance Conditions:** Engine started, battery voltage must be at least 11.5v, all electrical components must be off, parking brake must be engaged (to keep daytime driving lights off), automatic transmission selector must be in park and the ground between the engine and the chassis must be well connected. The DME received high pulses from the ignition module for the Ignition Coil 3 primary circuit. **Possible Causes:** • Engine speed (RPM) sensor has failed • Camshaft Position (CMP) sensor has failed • Power Supply Relay is shorted to an open circuit • There is a malfunction in voltage supply • Ignition coilpack is damaged or it has failed • Cylinder 1 to 4 Fuel Injector(s) have failed
DTC: P3316 **2006** **Models:** 325i, 330i, 525i, 530i, 550i, 750i, Z4 **Engines:** 3.0L, 4.8L **Transmissions:** All	**Ignition Coil Cylinder 3 Contact Resistance or High-Impedance Conditions:** Engine started, battery voltage must be at least 11.5v, all electrical components must be off, parking brake must be engaged (to keep daytime driving lights off), automatic transmission selector must be in park and the ground between the engine and the chassis must be well connected. The DME received high pulses from the ignition module for the Ignition Coil 3 primary circuit. **Possible Causes:** • Engine speed (RPM) sensor has failed • Camshaft Position (CMP) sensor has failed • Power Supply Relay is shorted to an open circuit • There is a malfunction in voltage supply • Ignition coilpack is damaged or it has failed • Cylinder 1 to 4 Fuel Injector(s) have failed
DTC: P3317 **2006** **Models:** 325i, 330i, 525i, 530i, 550i, 750i, Z4 **Engines:** 3.0L, 4.8L **Transmissions:** All	**Ignition Coil Cylinder 3 Cut-Off Due to Over-temperature Condition or No Signal Conditions:** Engine started, battery voltage must be at least 11.5v, all electrical components must be off, parking brake must be engaged (to keep daytime driving lights off), automatic transmission selector must be in park and the ground between the engine and the chassis must be well connected. The DME received no pulses from the ignition module for the Ignition Coil 3 primary circuit due to an elevated temperature reading. **Possible Causes:** • Engine speed (RPM) sensor has failed • Camshaft Position (CMP) sensor has failed • Power Supply Relay is shorted to an open circuit • There is a malfunction in voltage supply • Ignition coilpack is damaged or it has failed • Cylinder 1 to 4 Fuel Injector(s) have failed

DTC	Trouble Code Title, Conditions & Possible Causes
DTC: P3318 **2006** **Models:** 325i, 330i, 525i, 530i, 550i, 750i, Z4 **Engines:** 3.0L, 4.8L **Transmissions:** All	**Ignition Coil Cylinder 7 High Input or None-Impedance Conditions:** Engine started, battery voltage must be at least 11.5v, all electrical components must be off, parking brake must be engaged (to keep daytime driving lights off), automatic transmission selector must be in park and the ground between the engine and the chassis must be well connected. The DME received high pulses from the ignition module for the Ignition Coil 1 primary circuit. **Possible Causes:** • Engine speed (RPM) sensor has failed • Camshaft Position (CMP) sensor has failed • Power Supply Relay is shorted to an open circuit • There is a malfunction in voltage supply • Ignition coilpack is damaged or it has failed • Cylinder 1 to 4 Fuel Injector(s) have failed
DTC: P3319 **2006** **Models:** 325i, 330i, 525i, 530i, 550i, 750i, Z4 **Engines:** 3.0L, 4.8L **Transmissions:** All	**Ignition Coil Cylinder 7 Contact Resistance or High-Impedance Conditions:** Engine started, battery voltage must be at least 11.5v, all electrical components must be off, parking brake must be engaged (to keep daytime driving lights off), automatic transmission selector must be in park and the ground between the engine and the chassis must be well connected. The DME received high pulses from the ignition module for the Ignition Coil 1 primary circuit. **Possible Causes:** • Engine speed (RPM) sensor has failed • Camshaft Position (CMP) sensor has failed • Power Supply Relay is shorted to an open circuit • There is a malfunction in voltage supply • Ignition coilpack is damaged or it has failed • Cylinder 1 to 4 Fuel Injector(s) have failed
DTC: P3320 **2006** **Models:** 325i, 330i, 525i, 530i, 550i, 750i, Z4 **Engines:** 3.0L, 4.8L **Transmissions:** All	**Ignition Coil Cylinder 7 Cut-Off Due to Over-temperature Condition or No Signal Conditions:** Engine started, battery voltage must be at least 11.5v, all electrical components must be off, parking brake must be engaged (to keep daytime driving lights off), automatic transmission selector must be in park and the ground between the engine and the chassis must be well connected. The DME received no pulses from the ignition module for the Ignition Coil 1 primary circuit due to an elevated temperature reading. **Possible Causes:** • Engine speed (RPM) sensor has failed • Camshaft Position (CMP) sensor has failed • Power Supply Relay is shorted to an open circuit • There is a malfunction in voltage supply • Ignition coilpack is damaged or it has failed • Cylinder 1 to 4 Fuel Injector(s) have failed

Gas Engine OBD II Trouble Code List (U1xxx Codes)

DTC	Trouble Code Title, Conditions & Possible Causes:
DTC: U1115 **2006** **Models:** 325i, 330i, 525i, 530i, 550i, 750i, Z4 **Engines:** 3.0L, 4.8L **Transmissions:** All	**Lost Communication With Vehicle Mode Status Conditions:** Key on for 800ms, and the DME detected that it has lost communication with the Vehicle Mode Status during its initial startup for 40ms (2 message cycles). The Engine Control Module (DME) communicates with all databus-capable control modules via a CAN databus. These databus-capable control modules are connected via two data bus wires which are twisted together (CAN_High and CAN_Low), and exchange information (messages). Missing information on the databus is recognized as a malfunction and stored. Trouble-free operation of the CAN-Bus requires that it have a terminal resistance. **Possible Causes:** • DME has failed • Terminal resistance for CAN-bus are faulty • Can data bus wires have short circuited to each other
DTC: U1116 **2006** **Models:** 325i, 330i, 525i, 530i, 550i, 750i, Z4 **Engines:** 3.0L, 4.8L **Transmissions:** All	**Lost Communication With Vehicle Mode Status Check Sum Error Conditions:** Key on for 800ms, and the DME detected that it has lost communication with the Vehicle Mode Status during its initial startup for 40ms (2 message cycles). The Engine Control Module (DME) communicates with all databus-capable control modules via a CAN databus. These databus-capable control modules are connected via two data bus wires which are twisted together (CAN_High and CAN_Low), and exchange information (messages). Missing information on the databus is recognized as a malfunction and stored. Trouble-free operation of the CAN-Bus requires that it have a terminal resistance. **Possible Causes:** • DME has failed • Terminal resistance for CAN-bus are faulty • Can data bus wires have short circuited to each other
DTC: U1120 **2006** **Models:** 325i, 330i, 525i, 530i, 550i, 750i, Z4 **Engines:** 3.0L, 4.8L **Transmissions:** All	**Lost Communication With Steering Angle Sensor Module Conditions:** Key on for 800ms, and the DME detected that it has lost communication with the Steering Angle Sensor during its initial startup for 40ms (2 message cycles). The Engine Control Module (DME) communicates with all databus-capable control modules via a CAN databus. These databus-capable control modules are connected via two data bus wires which are twisted together (CAN_High and CAN_Low), and exchange information (messages). Missing information on the databus is recognized as a malfunction and stored. Trouble-free operation of the CAN-Bus requires that it have a terminal resistance. **Possible Causes:** • DME has failed • Terminal resistance for CAN-bus are faulty • Can data bus wires have short circuited to each other • The steering angle sensor module is faulty
DTC: U1121 **2006** **Models:** 325i, 330i, 525i, 530i, 550i, 750i, Z4 **Engines:** 3.0L, 4.8L **Transmissions:** All	**Lost Communication With Power Management Battery Voltage Conditions:** Key on for 800ms, and the DME detected that it has lost communication with the power management battery voltage during its initial startup for 40ms (2 message cycles). The Engine Control Module (DME) communicates with all databus-capable control modules via a CAN databus. These databus-capable control modules are connected via two data bus wires which are twisted together (CAN_High and CAN_Low), and exchange information (messages). Missing information on the databus is recognized as a malfunction and stored. Trouble-free operation of the CAN-Bus requires that it have a terminal resistance. **Possible Causes:** • DME has failed • Terminal resistance for CAN-bus are faulty • Can data bus wires have short circuited to each other
DTC: U1121 **2006** **Models:** 325i, 330i, 525i, 530i, 550i, 750i, Z4 **Engines:** 3.0L, 4.8L **Transmissions:** All	**Lost Communication With Power Management Charge Voltage Conditions:** Key on for 800ms, and the DME detected that it has lost communication with the power management charge voltage during its initial startup for 40ms (2 message cycles). The Engine Control Module (DME) communicates with all databus-capable control modules via a CAN databus. These databus-capable control modules are connected via two data bus wires which are twisted together (CAN_High and CAN_Low), and exchange information (messages). Missing information on the databus is recognized as a malfunction and stored. Trouble-free operation of the CAN-Bus requires that it have a terminal resistance. **Possible Causes:** • DME has failed • Terminal resistance for CAN-bus are faulty • Can data bus wires have short circuited to each other

DTC	Trouble Code Title, Conditions & Possible Causes
DTC: U1129 **2006** **Models:** 325i, 330i, 525i, 530i, 550i, 750i, Z4 **Engines:** 3.0L, 4.8L **Transmissions:** All	**Lost Communication With Reverse Status Conditions:** Key on for 800ms, and the DME detected that it has lost communication with the power management charge voltage during its initial startup for 40ms (2 message cycles). The Engine Control Module (DME) communicates with all databus-capable control modules via a CAN databus. These databus-capable control modules are connected via two data bus wires which are twisted together (CAN_High and CAN_Low), and exchange information (messages). Missing information on the databus is recognized as a malfunction and stored. Trouble-free operation of the CAN-Bus requires that it have a terminal resistance. **Possible Causes:** • DME has failed • Terminal resistance for CAN-bus are faulty • Can data bus wires have short circuited to each other
DTC: U1134 **2006** **Models:** 325i, 330i, 525i, 530i, 550i, 750i, Z4 **Engines:** 3.0L, 4.8L **Transmissions:** All	**Lost Communication With Lamp Status Conditions:** Key on for 800ms, and the DME detected that it has lost communication with the lamp status during its initial startup for 40ms (2 message cycles). The Engine Control Module (DME) communicates with all databus-capable control modules via a CAN databus. These databus-capable control modules are connected via two data bus wires which are twisted together (CAN_High and CAN_Low), and exchange information (messages). Missing information on the databus is recognized as a malfunction and stored. Trouble-free operation of the CAN-Bus requires that it have a terminal resistance. **Possible Causes:** • DME has failed • Terminal resistance for CAN-bus are faulty • Can data bus wires have short circuited to each other
DTC: U1135 **2006** **Models:** 325i, 330i, 525i, 530i, 550i, 750i, Z4 **Engines:** 3.0L, 4.8L **Transmissions:** All	**Lost Communication With Status Water Valve Conditions:** Key on for 800ms, and the DME detected that it has lost communication with the status water valve during its initial startup for 40ms (2 message cycles). The Engine Control Module (DME) communicates with all databus-capable control modules via a CAN databus. These databus-capable control modules are connected via two data bus wires which are twisted together (CAN_High and CAN_Low), and exchange information (messages). Missing information on the databus is recognized as a malfunction and stored. Trouble-free operation of the CAN-Bus requires that it have a terminal resistance. **Possible Causes:** • DME has failed • Terminal resistance for CAN-bus are faulty • Can data bus wires have short circuited to each other

BMW
COMPONENT TESTING

9

TABLE OF CONTENTS

Accelerator Pedal Position (APP) Sensor

LOCATION

4.4L (M62) and 5.4L (M73) Engines

750IL (1996–01); 540I (1997–03); 740I, 740IL (1996–05);

See Figure 1.

The accelerator pedal position sensor (1) is attached to the pedal cluster under the driver's side dashboard.

OPERATION

The accelerator pedal position sensor is integrated into the accelerator pedal housing. Two hall sensors are used to provide the driver's input request for acceleration. The hall sensors receive power (5 volts) and ground from the control module and produce linear voltage signals as the pedal is pressed from one level to another.

REMOVAL & INSTALLATION

4.4L (M62) and 5.4L (M73) Engines

750IL (1996–01); 540I (1997–03); 740I, 740IL (1996–05);

See Figures 2 through 5.

**** CAUTION**

Make sure that the ignition is switched off and no electrical equipment is on or running.

➡ **Whenever the accelerator pedal position sensor is disconnected, basic adaptation of the accelerator pedal position sensor is required. To properly do so, you will need access to a DIS tester machine.**

1. Remove lower trim from instrument panel. It is located under the steering wheel and above the pedal cluster.

➡ **Press the accelerator pedal by hand (and not your foot) to avoid bending the pedal tip and falsifying the measured values.**

2. The accelerator sensor is attached to the pedal cluster under the driver's side dashboard.
3. Unscrew the three screws and remove the control unit from the bracket that holds it in position. The bracket and unit will come out together.
4. Remove the retaining ring and unhook the lever from the pedal assembly.
5. Unfasten the electrical plug connection and the screws holding the accelerator pedal position sensor (arrows).
6. When removing the sensor, note and remember its location and the position of the travel sensor's arm and lever (1 and 2) in relation to the body of the sensor. It is a 71 degree angle between the body of the sensor and the arm. This angle must be precisely maintained.

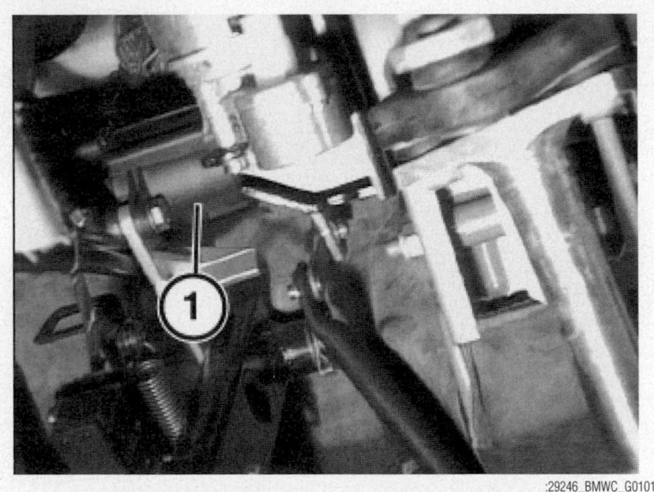

29246_BMWC_G0101

Fig. 1 The location of the accelerator pedal position sensor

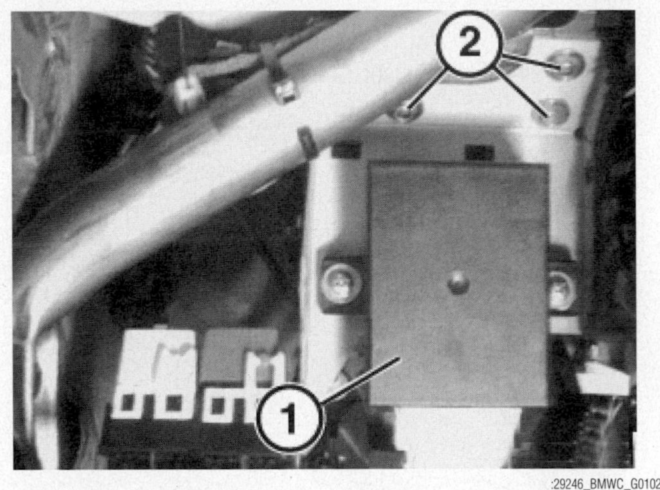

29246_BMWC_G0102

Fig. 2 The bracket screws (2) and cover will come out together

29246_BMWC_G0103

Fig. 3 Unfasten the electrical plug connection (1) and the screws holding the accelerator pedal position sensor (arrows)

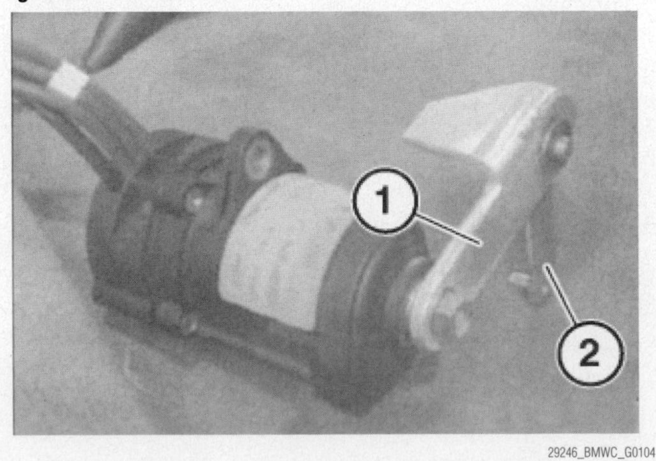

29246_BMWC_G0104

Fig. 4 Note the position of the levers

To install:

✻✻ WARNING

The lever of the sensor is set at the factory and must not be released. If the lever has been released (via the spring), the entire pedal travel sensor must be replaced.

7. Replace the bracket with the captured nuts.
8. Reconnect the electrical plug connection and the tighten the screws that hold the PWG.
9. Replace the retaining ring and hook the lever onto the pedal assembly.
10. Check the adjustment of the kickdown switch.

➡ Whenever the accelerator pedal position sensor is disconnected, basic adaptation of the accelerator sensor is required. Connect to the DIS machine and select diagnosis of EML IIIs control unit. To proceed, follow DIS (Diagnostic Information System) instructions. Measure the voltage of the accelerator pedal position sensor's potentiometer with the DIS tester. The specified values should be:

- Pedal travel sensor in idle position: 0.6 to 0.85 volts.
- Pedal travel sensor in full throttle position: 3.7 volts to 4.15 volts.
- Kickdown: 4.35 V to 4.55 volts

➡ If the idle measurement value is outside the above-mentioned limit value, the 71-degree angle of the lever must be checked and a new lever installed if that angle is not maintained.

11. Operate accelerator pedal for full throttle position by hand to prevent a distortion of the measured values.

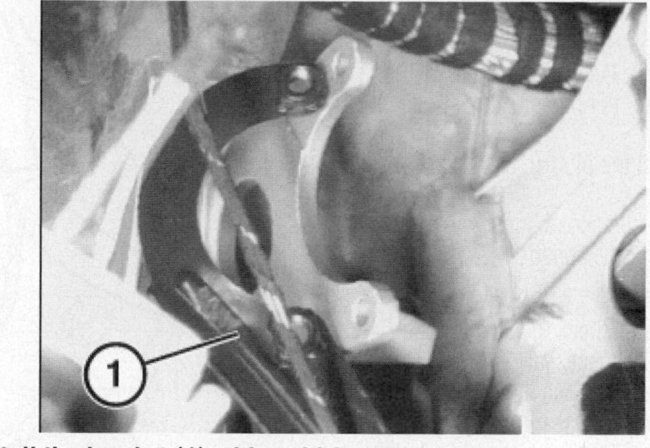

Fig. 5 Install the bracket (1) with weld-in nuts

12. Readjust the full throttle stop on the accelerator pedal if the voltage values in the full throttle position do not attain the values mentioned above.

Camshaft Position (CMP) Sensor

OPERATION

1. The static hall sensors are used on this sensor so that the camshaft positions are recognized once ignition is on and even before the engine is started.
2. There are four functions of the intake camshaft position sensor:
   - Cylinder bank detection for preliminary injection
   - Synchronization
   - Engine speed sensor (if crankshaft speed sensor fails)
   - Position control of the intake cam (VANOS)
3. The exhaust camshaft sensor is used for position control of the exhaust camshaft as part of the VANOS (variable camshaft control) system.
4. If these sensors fail, there will be no substitute values and the system will operate in the failsafe mode with no VANOS adjustment. The engine will still operate, but a significant reduction torque reduction will be quite noticeable.

➡ Use caution when removing or repairing the camshaft sensor and related mechanicals, as you should not want to bend the impulse wheels.

REMOVAL & INSTALLATION

4.4L (M62) and 5.0L (S62) Engines

540I (1997–03); 740I, 740IL (1996–05); M5 (2000–05)

See Figures 6, 7 and 8.

✻✻ CAUTION

Make sure that the ignition is switched off and no electrical equipment is on or running.

1. For cars equipped with an acoustic cover that is screwed down, remove the seal caps and unfasten the nuts.
2. For cars equipped with an acoustic cover that is held in place with press stud

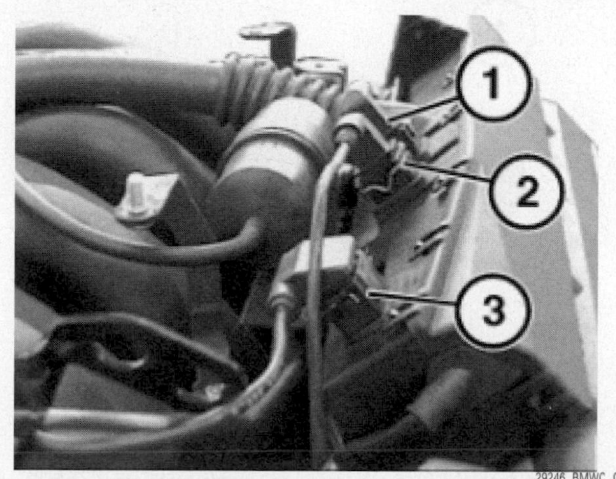

Fig. 6 Disconnect the plug-in electrical connection (3)

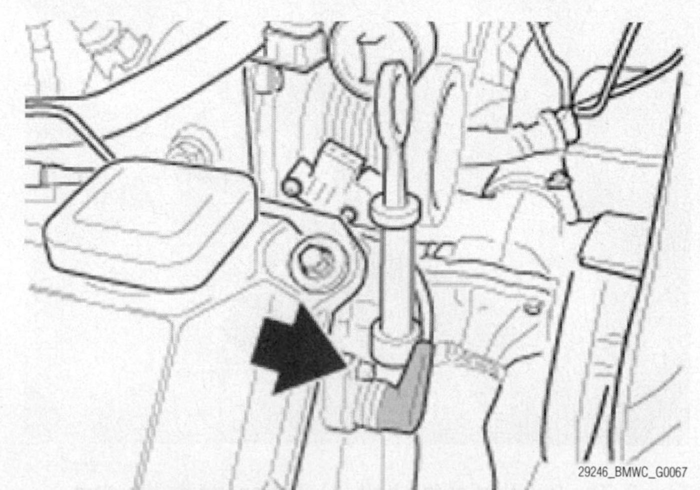

Fig. 7 Unscrew the screw and pull out the sensor

fasteners, open the press-studs one at a time, while lifting the acoustic cover slightly on all four corners. Once all four press-studs have been opened, remove the acoustic cover.

3. Disconnect the plug-in electrical connection (3).

4. To remove the sensor, unscrew the screw and pull out the sensor.

To install:

5. Make sure to check the seal for damage and replace it if necessary.

6. Make sure to replace the line routing in the same manner it was removed, by following the original route and securing it with a cable connector.

7. Reposition the acoustic cover and press downwards until the four press-studs fasteners click audibly into place.

8. Access the fault memory of control unit of Digital Motor Electronics (DME), check for any stored fault messages, rectify those faults and clear the fault memory.

5.4L (M73) Engine

750IL (1996–01)

See Figures 9 and 10.

✻✻ CAUTION

Make sure that the ignition is switched off and no electrical equipment is on or running.

1. Access the fault memory of control unit of Digital Motor Electronics (DME), check for any stored fault messages, rectify those faults and clear the fault memory.

2. Remove distributor cap with left ignition wiring harness,

3. Unscrew the bolt that retains the sensor (1) and remove the unit.

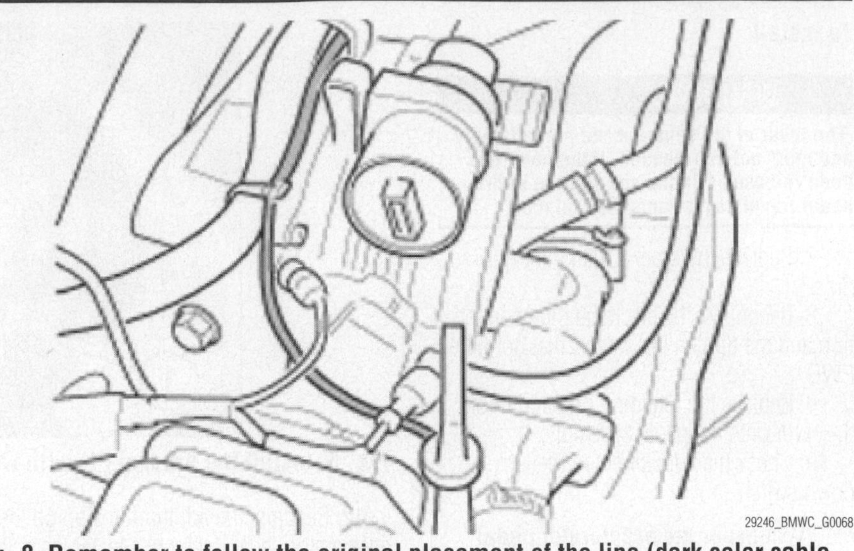

29246_BMWC_G0068

Fig. 8 Remember to follow the original placement of the line (dark color cable in the photo)

To install:

4. Check the seal of the sensor for damage and replace it if necessary.

5. To facilitate assembly coat seal with anti-friction agent.

Crankshaft Position (CKP) Sensor

OPERATION

The crankshaft sensor is a dynamic hall-effect sensor (mounted through a access hole in the engine block). A signal is sent to the DME the moment the crankshaft begins to rotate. The pulse wheel is mounted directly to the crankshaft.

Digital Motor Electronics (DME)

REMOVAL & INSTALLATION

1.9L (M44) Engines

Z3, 318I, 318IS, 318IC (1996–98); 318TI (1996–99)

See Figures 11 through 14.

✻✻ CAUTION

Before you begin, always make sure to communicate with the fault memory with a BMW DIS (or equivalent OBD-II scan tool) for existing faults. It may be helpful to print out the results.

✻✻ CAUTION

Make sure that the ignition is switched off and no electrical equipment is on or running.

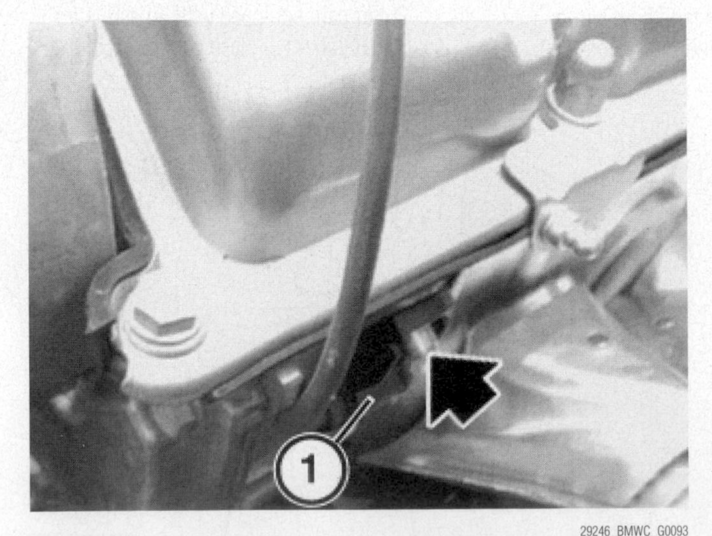

29246_BMWC_G0093

Fig. 9 The location of the bolt (1) that connects the camshaft position sensor to the engine block

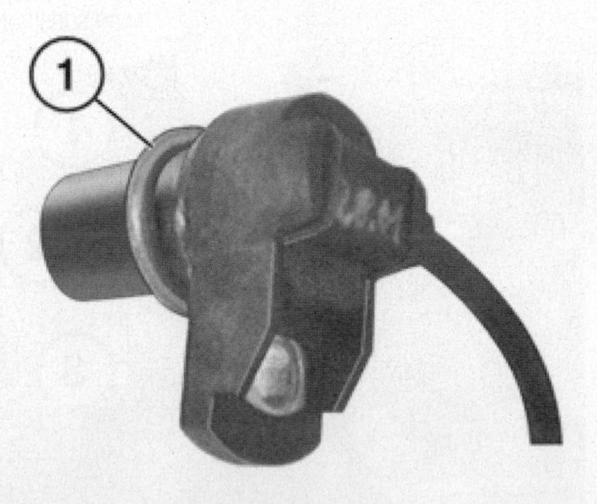

29246_BMWC_G0094

Fig. 10 Check the seal of the sensor (1) for damage and replace it if necessary

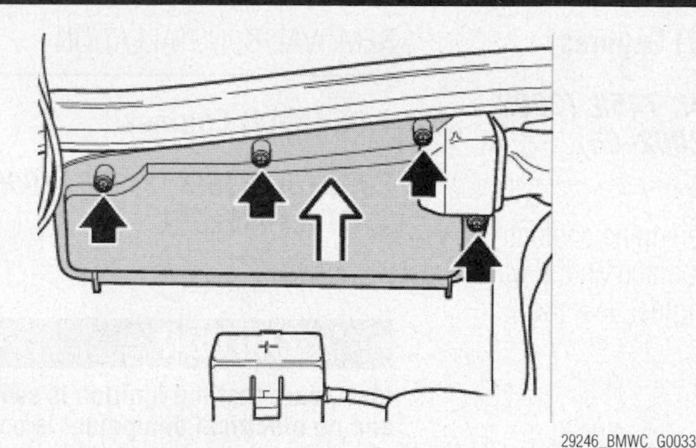

Fig. 11 Unscrew the four bolts on the wire cover and fold it to one side

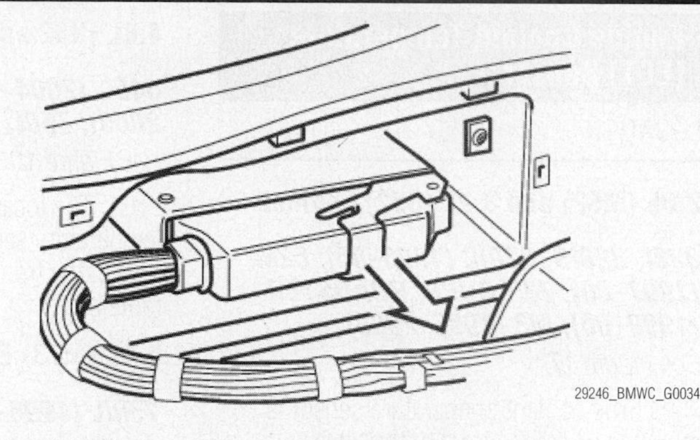

Fig. 12 Pull the connector in an upwards direction

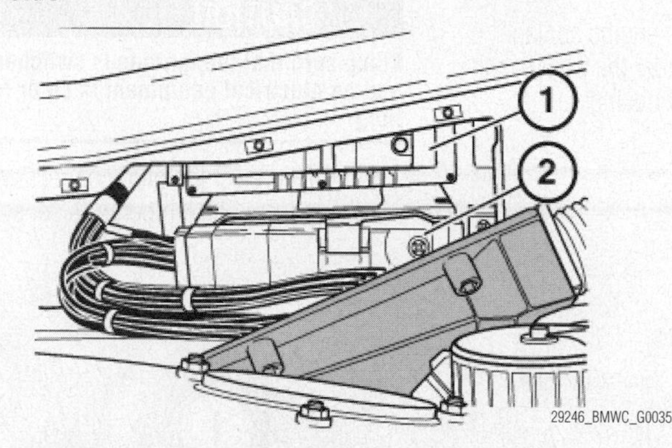

Fig. 13 In this area are both the control unit for the automatic transmission (1) and the control unit for the Digital Motor Electronics (DME)

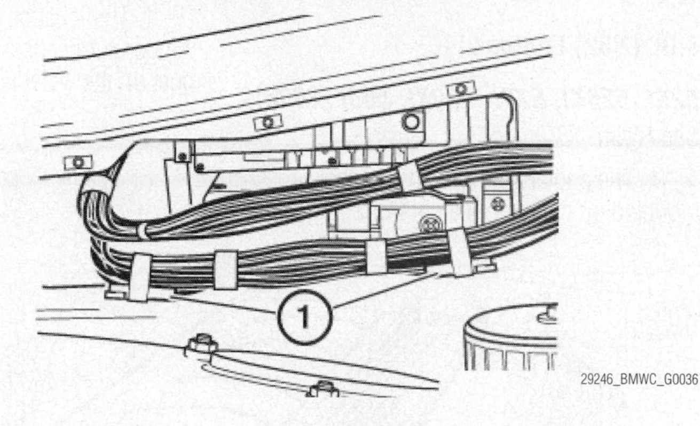

Fig. 14 The plug-in connection (1) must be inserted when the catalytic converter is removed

1. Unscrew the four bolts on the wire cover and fold it to one side. With it should come the wiring harness.

➡ **The control units are located behind the battery holder in the engine compartment.**

2. To disconnect the connector, lift the locking clamp on the plug-in connection. Pull it in an upwards direction.
3. Disconnect the connector.
4. Pull out the control unit from the clips. There are retaining screws on the clips and they do not need to be unscrewed).

To install:

5. When replacing a control unit, make sure that you have the exact part number and coding before making the connections.
6. Access the fault memory of control unit of Digital Motor Electronics (DME), check for any stored fault messages, rectify those faults and clear the fault memory.
7. The plug-in connection (1) must be inserted when the catalytic converter is removed. After which, the DME control unit then recognizes the output value of the idle speed potentiometer.

4.8L (N62 and N62TU) Engine

545I, (2004–05); 745I, 745IL (2002–2005); 750I, 750LI (2002–05)

See Figures 15 and 16.

**** CAUTION**

Make sure that the ignition is switched off and no electrical equipment is on or running.

1. Disconnect the battery's negative cable.
2. Remove the right fresh-air duct.
3. There is a sliding "stopper" cover that needs to first be removed. Do so by sliding it the direction of the white arrow in the photo.
4. Unscrew the retaining screws to the cover and lift it out. Some models are equipped with an armor plating that needs to be unscrewed as well.
5. Unlock (via small tabs) and detach all of the plug connections on the DME control unit (1).
6. Unlock the DME control unit (1) itself and remove out towards top.

To install:

7. Slide the DME control unit back into the box.
8. Lock the small tabs and reconnect all of the plug connectors
9. Screw on the retaining screws of the cover
10. Slide the "stopper" cover back over the DME box
11. Replace the fresh air duct.

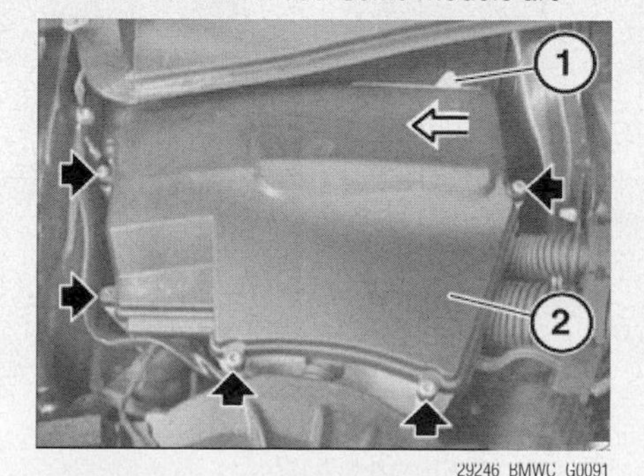

Fig. 15 Slide the locking stopper the direction of the white arrow

Fig. 16 Unlock the DME control unit (1) itself and remove out towards top

Engine Coolant Temperature (ECT) Sensor

LOCATION

2.8L (M52) and 3.2L (S52) Engines

328I, 328IS, 328IC (1996–00); 528I (1997–00); Z3 COUPE, ROADSTER (1999–00); M3 (1996–2005)

See Figure 17.

1. The coolant temperature sensor is mounted under the front of the intake air manifold.

3.0L (N52) Engines

525I, 525XI, 530I, 530XI, 550I (2006)

See Figure 18.

1. The coolant temperature sensor is mounted on cylinder head at front.

4.8L (N62 and N62TU) Engines

545I, (2004–05); 745I, 745IL (2002–2005); 750I, 750LI (2002–05)

See Figure 19.

1. The location of the engine coolant temperature sensor is mounted on the water pipe at the front of the engine, near the radiator.

5.4L (M73) Engine

750IL (1996–01)

See Figure 20.

1. The location of the engine coolant temperature sensor is under the front cable duct on the water pump housing.

REMOVAL & INSTALLATION

1.9L (M44) Engines

Z3, 318I, 318IS, 318IC (1996–98); 318TI (1996–99)

See Figure 21.

**** CAUTION**

Make sure that the ignition is switched off and no electrical equipment is on or running.

**** CAUTION**

Make sure that the ignition is switched off and no electrical equipment is on or running.

1. Disconnect the plug connection (1).
2. Unscrew the temperature sensor.

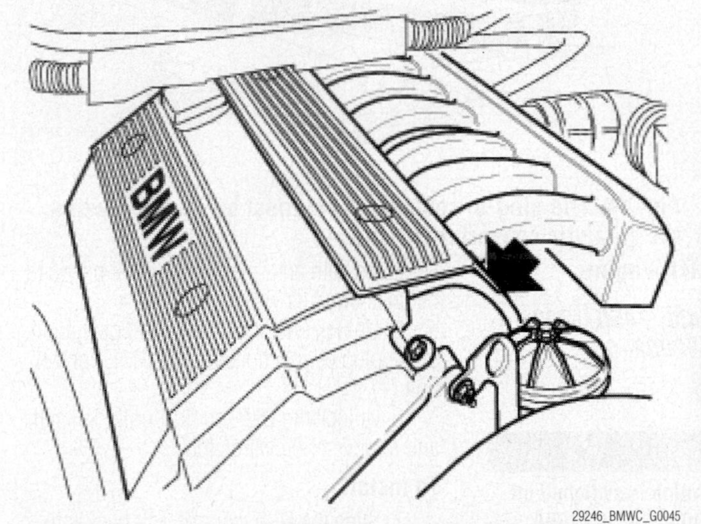

Fig. 17 The location of the engine coolant temperature sensor

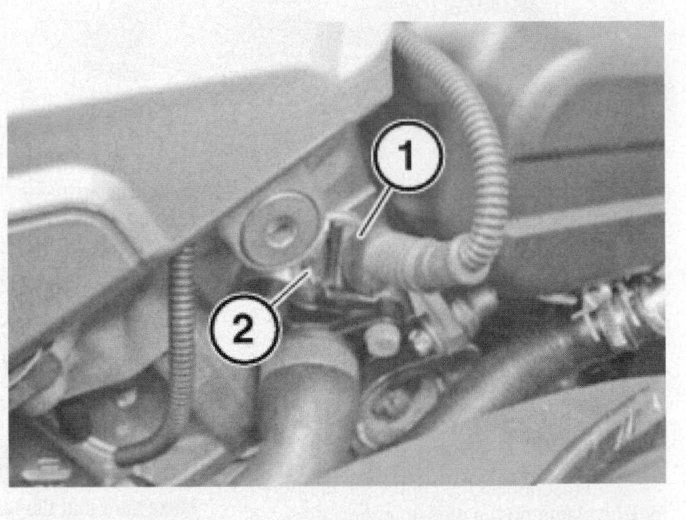

Fig. 18 The location of the engine coolant temperature sensor

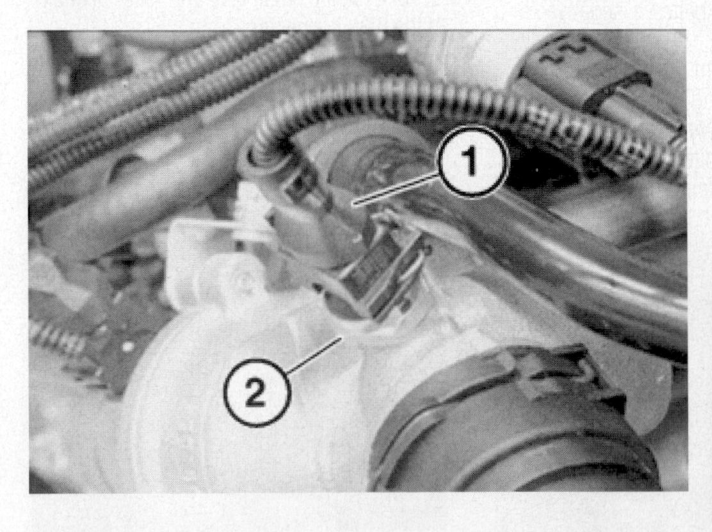

Fig. 19 The location of the engine coolant temperature sensor

Fig. 20 The location of the engine coolant temperature sensor

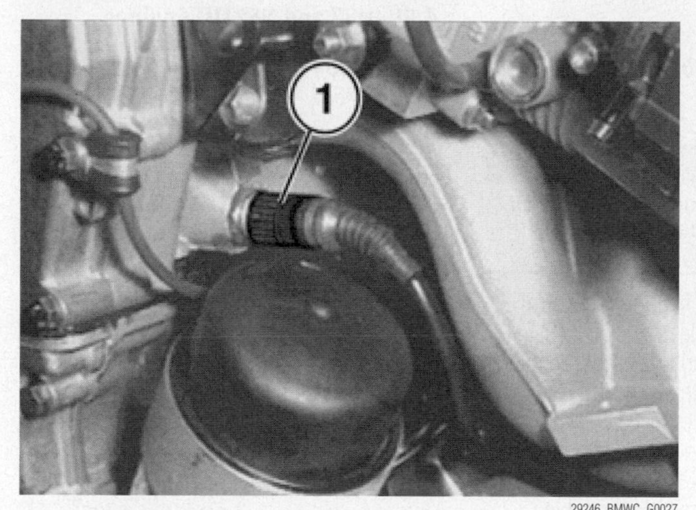

29246_BMWC_G0027

Fig. 21 The removal of the temperature sensor and the plug connector (1)

29246_BMWC_G0010

Fig. 22 The removal of the temperature sensor (1)

To install:

3. Regardless of its condition, always replace sealing ring.
4. Reclip the connection for the crankcase venting.
5. If necessary, top off the coolant in the reservoir.
6. End by clearing the fault memory.

2.5L (M52TU) and 3.0L (M54) Engines

325I, 325XI, 330I, 330CI, 330XI (2000 TO 2005); 325XIT; (2000–03); 525I, 530I (2000–05); Z3, Z3 COUPE (2001–02); Z4 2.5L, Z4 3.0L (2001–06)

See Figure 22.

✻✻ CAUTION

Make sure that the ignition is switched off and no electrical equipment is on or running.

1. Disconnect the plug connection (1).
2. Unscrew the temperature sensor.

To install:

3. Reclip the connection for the crankcase venting.
4. If necessary, top off the coolant in the reservoir.
5. End by clearing the fault memory.

2.8L (M52) and 3.2L (S52) Engines

328I, 328IS, 328IC (1996–00); 528I (1997–00); Z3 COUPE, ROADSTER (1999–00); M3 (1996–2005)

See Figures 23 and 24.

✻✻ CAUTION

Make sure that the ignition is switched off and no electrical equipment is on or running.

1. Access the fault memory of control unit of Digital Motor Electronics (DME), check for any stored fault messages, rectify those faults and clear the fault memory.
2. Remove the cover from injection valves.
3. Unclip the connection for the crankcase venting.
4. Unfasten the plug connection.
5. Unscrew and remove the double temperature sensor for Digital Motor Electronics (DME) and coolant indicator.

To install:

6. After returning the temperature sensor back to its position, make sure to vent the cooling system and check for leaks.
7. Reconnect the plug connection.
8. Reclip the connection for the crankcase venting.
9. If necessary, top off the coolant in the reservoir.
10. End by clearing the fault memory.

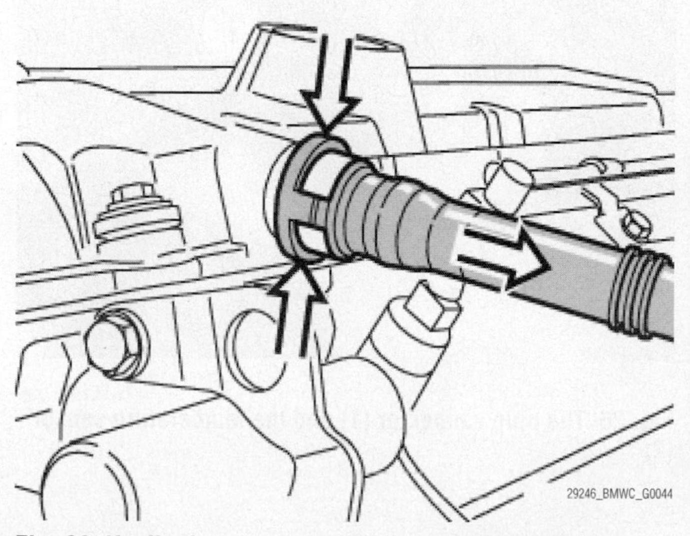
29246_BMWC_G0044

Fig. 23 Unclip the connection for the crankcase venting

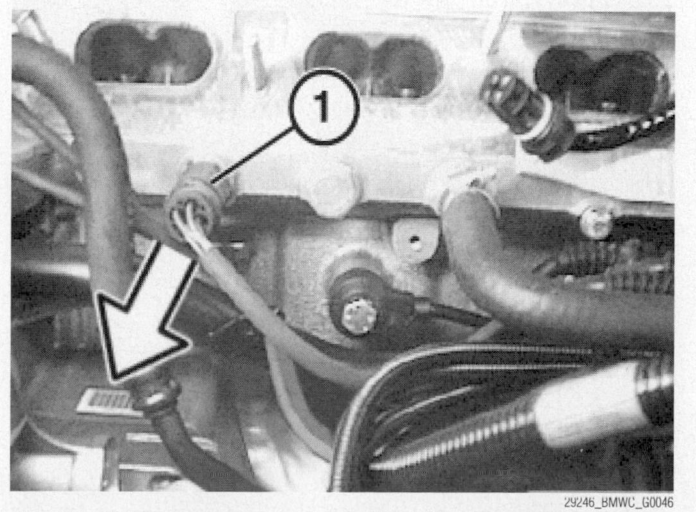

29246_BMWC_G0046

Fig. 24 Unfasten the plug connection (1) (illustrated here without intake air manifold)

3.0L (N52) Engines

525I, 525XI, 530I, 530XI, 550I (2006)

See Figure 25.

✻✻ WARNING

There is a danger of scalding so only perform this task on an engine that has completely cooled down.

✻✻ CAUTION

Make sure that the ignition is switched off and no electrical equipment is on or running.

1. Access the fault memory of control unit of Digital Motor Electronics (DME), check for any stored fault messages, rectify those faults and clear the fault memory.
2. Though you can avoid it, for easier access, remove the intake duct.
3. The coolant temperature sensor is mounted on cylinder head at the front of the engine
4. Unlock the plug on the sensor (1) and remove it.
5. Remove the temperature sensor (2).

To install:

6. Reposition the sensor and lock the plug.
7. After returning the temperature sensor back to its position, make sure to vent the cooling system and check for leaks.
8. If necessary, top off the coolant in the reservoir.
9. End by clearing the fault memory.

4.4L (M62) Engine

540I (1997–03); 740I, 740IL (1996–05); M5 (2000–05)

✻✻ CAUTION

Make sure that the ignition is switched off and no electrical equipment is on or running.

1. Access the fault memory of control unit of Digital Motor Electronics (DME), check for any stored fault messages, rectify those faults and clear the fault memory.
2. For cars with idle-speed control valves, simply unfasten the plug connection to remove the sensor
3. Unscrew and remove the double temperature sensor (1) for Digital Motor Electronics (DME) and remove coolant indicator.
4. For cars without idle-speed control valves, disconnect the vacuum lines and detach the plug from the sensor.
5. Unscrew the sensor.

To install:

6. Make sure to replace sealing ring before installing either type of sensor.
7. After returning the temperature sensor back to its position, make sure to vent the cooling system and check for leaks.
8. If necessary, top off the coolant in the reservoir.
9. End by clearing the fault memory.

4.8L (N62 and N62TU) Engines

545I, (2004–05); 745I, 745IL (2002–2005); 750I, 750LI (2002–05)

See Figure 26.

✻✻ WARNING

There is a danger of scalding so only perform this task on an engine that has completely cooled down.

✻✻ CAUTION

Make sure that the ignition is switched off and no electrical equipment is on or running.

1. Access the fault memory of control unit of Digital Motor Electronics (DME), check for any stored fault messages, rectify those faults and clear the fault memory.

➡ For the good of the environment, catch and dispose of any escaping coolant. Research and observe the waste-disposal regulations or laws in your area.

2. Unlock the plug connector (1) and remove it.
3. Release the temperature sensor (2) from its position.

To install:

4. After returning the temperature sensor back to its position, make sure to vent the cooling system and check for leaks.
5. If necessary, top off the coolant in the reservoir.
6. End by clearing the fault memory.

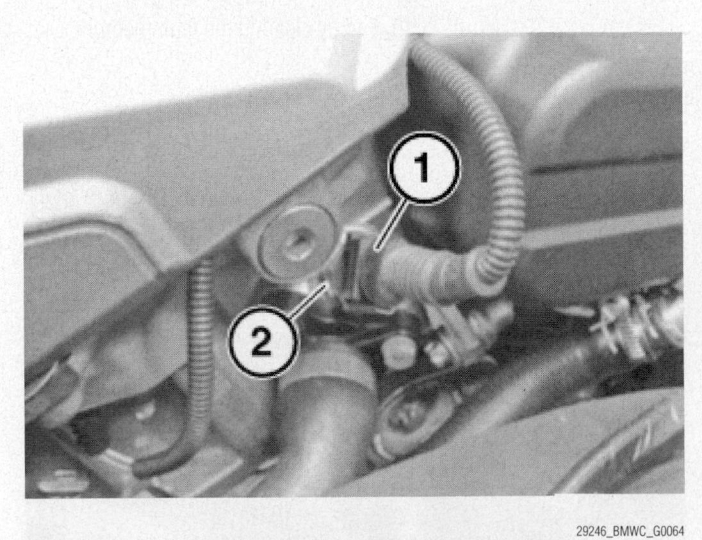

Fig. 25 The removal of the engine coolant temperature sensor

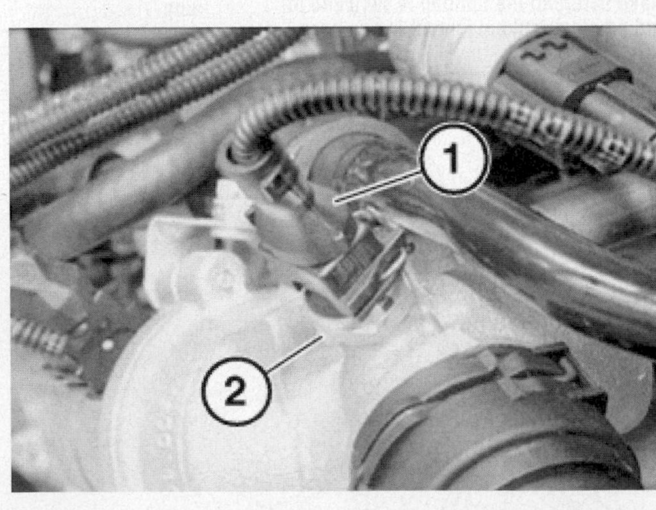

Fig. 26 The plug connector (1) and the temperature sensor (2)

5.4L (M73) Engine

750IL (1996–01)
See Figures 27 and 28.

**** CAUTION**
Make sure that the ignition is switched off and no electrical equipment is on or running.

1. Remove the non-return valve on Cylinders One through Six
2. Disconnect the connector by carefully pulling up on the unit.
3. Remove the temperature sensor.

To install:

4. Make sure to inspect the seal around the sensor, and if it is suspect replace it.
5. Check the coolant level and top it off with coolant, as necessary.
6. Reconnect the sensor by carefully pushing the unit back into its hole.
7. Access the fault memory of control unit of Digital Motor Electronics (DME), check for any stored fault messages, rectify those faults and clear the fault memory.

Engine Oil Pressure Sensor

REMOVAL & INSTALLATION

3.0L (N52) Engines

525I, 525XI, 530I, 530XI, 550I (2006)
See Figure 29.

**** WARNING**
There is a danger of scalding so only perform this task on an engine that has completely cooled down.

**** CAUTION**
Make sure that the ignition is switched off and no electrical equipment is on or running.

➡ Engine oil may emerge when the oil pressure sensor is removed and replaced, so make sure to have a cleaning cloth ready and wipe down any oil that has spilled on the engine.

1. Remove the intake filter housing
2. Unlock the plug connector and remove it.
3. Release the oil pressure sensor.

To install:

4. Reposition the oil pressure sensor.
5. Replace the plug connector and lock it into place
6. Reinstall the intake filter housing.
7. Check that the engine oil level is at the proper level (especially if some spilled), and top off if necessary.

29246_BMWC_G0109

Fig. 27 The non-return valve

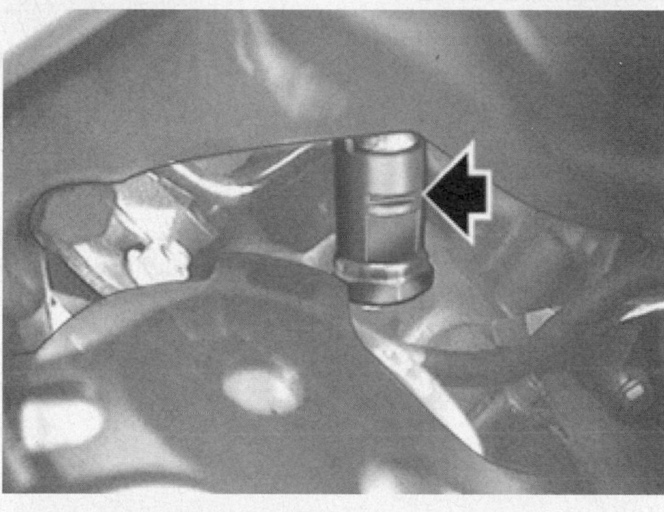

Fig. 28 Reconnect the sensor by carefully pushing the unit back into its hole

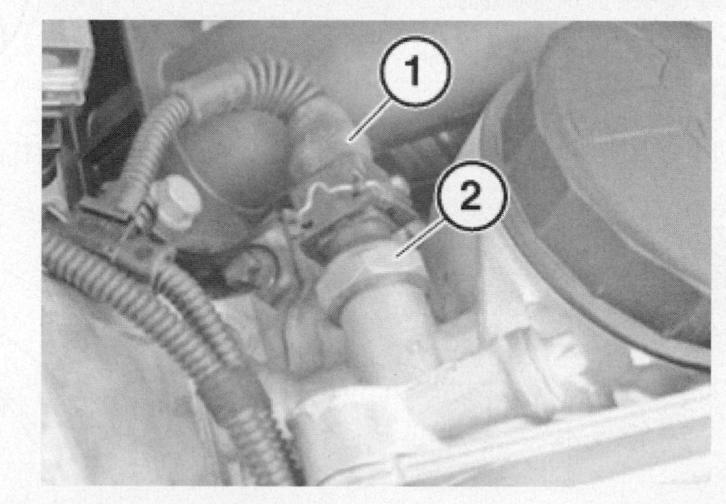

29246_BMWC_G0062

Fig. 29 The removal of the pressure sensor (2) and the plug connector (1)

Engine Oil Level Sensor

REMOVAL & INSTALLATION

3.0L (N52) Engines

525I, 525XI, 530I, 530XI, 550I (2006)
See Figure 30.

**** WARNING**

There is a danger of scalding so only perform this task on an engine that has completely cooled down.

**** CAUTION**

Since there is concerns with damage to the sensor from electrostatic charges, make sure you are properly grounded before handling the sensor.

**** CAUTION**

Make sure that the ignition is switched off and no electrical equipment is on or running.

1. Remove the underbody protection plate (aka reinforcement plate).
2. Drain the engine oil.

➡ **For the good of the environment, catch and dispose of the oil properly. Research and observe the waste-disposal regulations and laws in your area.**

3. Unlock plug connection from the sensor and remove it.
4. Unscrew the three nuts (black arrows in the illustration) and remove the sensor.

To install:

5. Clean the face of the oil sump, where the sensor covers, and make sure it is free from any oil or debris.
6. Replace the seal on oil level sensor. For best results, always replace the seal.
7. Remember, an excessively low torque value will result in oil leaks, while on the contrary, an excessively high torque value will result in damage to the oil level sensor, perhaps the threads and maybe the oil sump.
8. Lastly, refill the engine oil to the specifications of the vehicle.

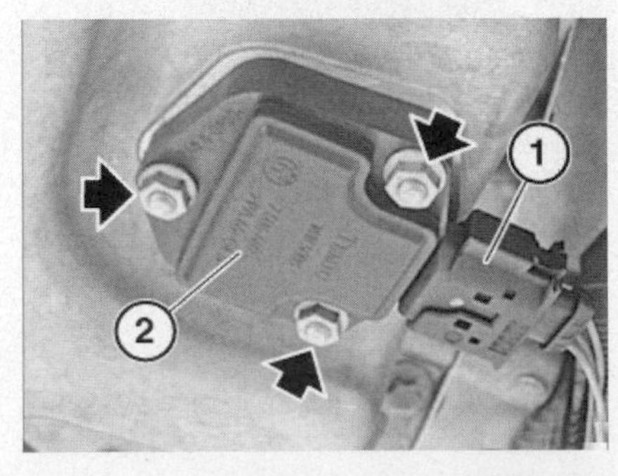

Fig. 30 The removal of the oil level sensor (2) and the plug connector (1)

Fuel Level Sensor

REMOVAL & INSTALLATION

2.8L (M52) and 3.2L (S52) Engines

328I, 328IS, 328IC (1996–00); 528I (1997–00); Z3 COUPE, ROADSTER (1999–00); M3 (1996–2005)
See Figures 31, 32, 33 and 34.

1. Remove the fuel from the tank. The easiest and safest method is via a suction pump and canister.

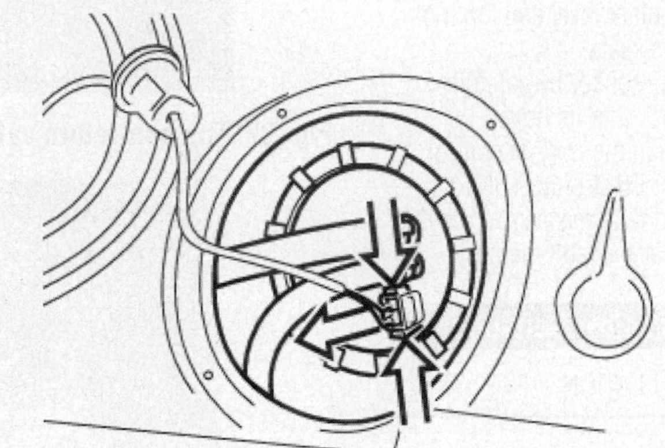

Fig. 31 The location of the plug-in connector (arrow)

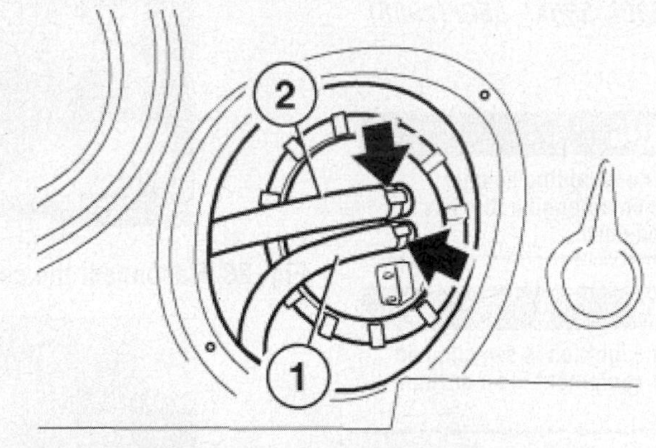

Fig. 32 The fuel return hose (1) and the tank expansion line (2)

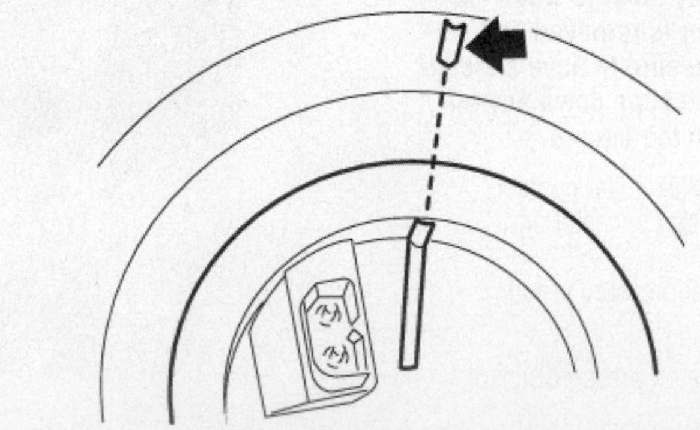

Fig. 33 The marker rib on the fuel level sensor must match the mark on the fuel tank

2. Disconnect the connector by pulling it free from the plug.

➡ **Ensure that area around fuel tank screw connection is clean.**

3. Remove the fuel return hose and tank expansion hose. Tank expansion lead is fitted inside fuel tank.
4. Covering the sensor, there is a rotary ring that must be unfastened.
5. Carefully withdraw the fuel level sensor by pressing the height button slightly towards the housing.

To install:

6. Replace the sealing ring. Check the rotary connection for damage and replace it if necessary.

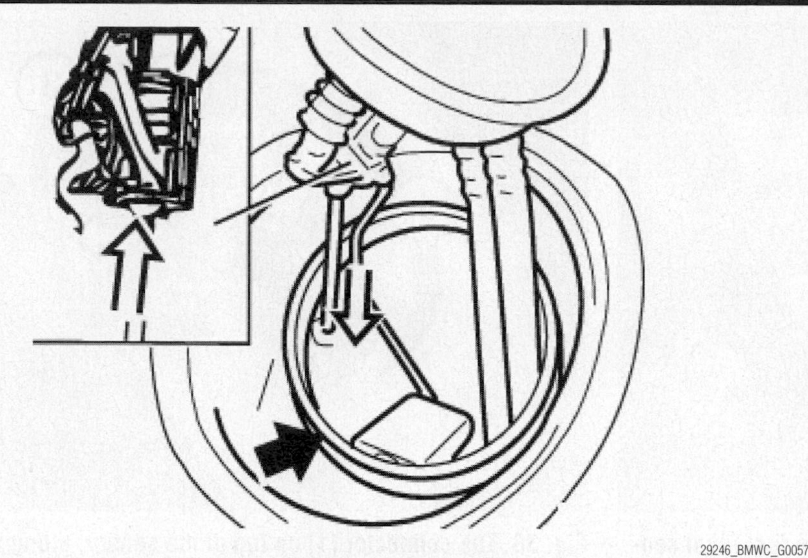

29246_BMWC_G0057

Fig. 34 The installation of the fuel level sensor unit

7. The marker rib on the fuel level sensor must match the mark on the fuel tank.

✻✻ CAUTION

The fuel return hose and the tank expansion line are pressed with preload to the base of fuel tank.

The swing-arm and the float must still move easily.

8. Swivel the air valve and float into the fuel tank.

9. Press the fuel level sensor vertically into its correct position with the height sensor perceptibly in contact with the base of the fuel tank.

3.0L (M54) Engines—Right Side Sensor

325I, 325XI, 330I, 330CI, 330XI (2000 TO 2005); 325XIT; (2000–03); 525I, 530I (2000–05); Z3, Z3 COUPE (2001–02); Z4 2.5L, Z4 3.0L (2001–06)

See Figures 35 and 36.

1. Remove the fuel from the tank. The easiest and safest method is via a suction pump and canister.
2. Remove the rear bench seat.
3. Fold the floor trim panel forwards and cut the insulating mat at the perforation and fold back and out of the way. This will expose the rubber grommet.
4. Unscrew the hexagon nuts, remove the cover to the fuel level sensor.
5. Unfasten plug connection on the sensor and disconnect it from the unit.
6. Release the hose clamps to disconnect the hose.
7. Covering the sensor, there is a rotary ring that must be unfastened.
8. Remove the sensor.

To install:

9. Replace the rubber seal that goes between the tank and the sensor
10. Tighten down the rotary ring
11. When installing the fuel level sensor, ensure that the lug is fitted properly into the recess on to of the tank.
12. When tightening the notch on the screw cap, it can be heard clicking and felt engaging the toothed portion of the tank.

3.0L (M54) Engines—Left Side Sensor

325I, 325XI, 330I, 330CI, 330XI (2000 TO 2005); 325XIT; (2000–03); 525I, 530I (2000–05); Z3, Z3 COUPE (2001–02); Z4 2.5L, Z4 3.0L (2001–06)

See Figures 37, 38, 39 and 40.

1. Remove the fuel from the tank. The easiest and safest method is via a suction pump and canister.
2. Remove the rear bench seat.
3. Fold the floor trim panel forwards and cut the insulating mat at the perforation and fold back and out of the way. This will expose the rubber grommet.
4. Unscrew the hexagon nuts, remove the cover to the fuel level sensor.
5. Unfasten plug connection on the sensor and disconnect it from the unit.

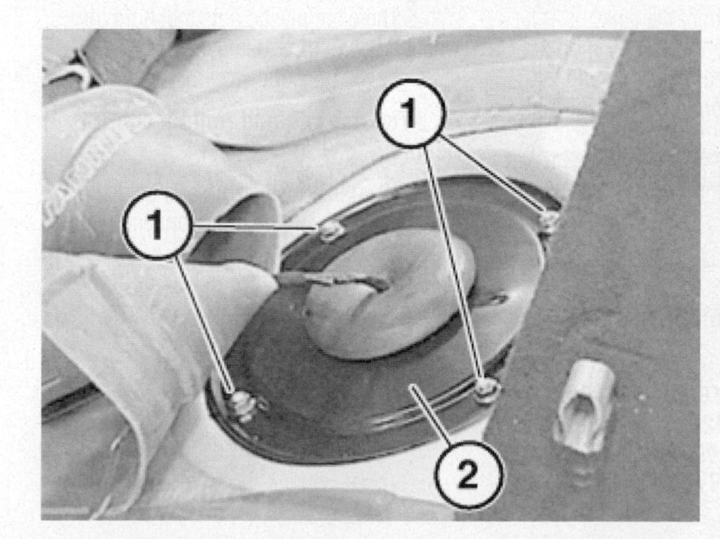

29246_BMWC_G0013

Fig. 35 The nuts (1) and the cover (2) to the fuel level sensor

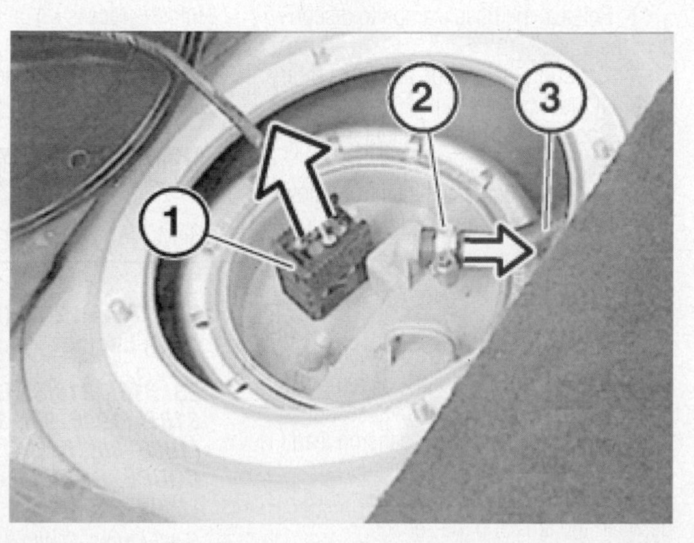

29246_BMWC_G0014

Fig. 36 The connector (1) on top of the sensor, along with the hose clamp (2) and the hose (3)

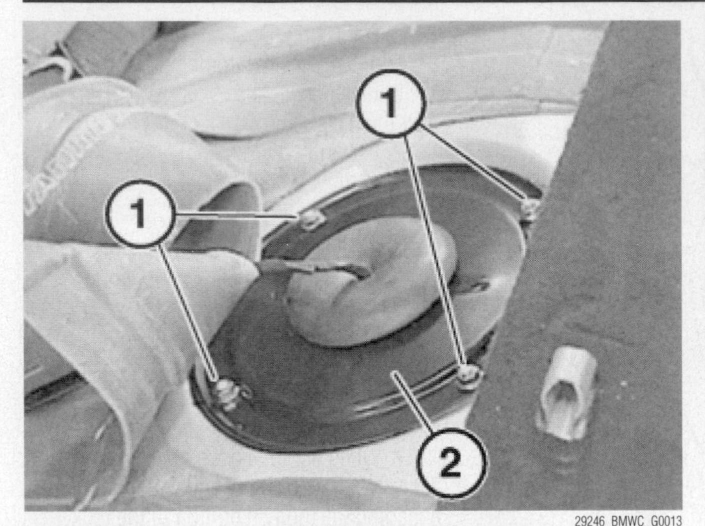

29246_BMWC_G0013

Fig. 37 The nuts (1) and the cover (2) to the fuel level sensor

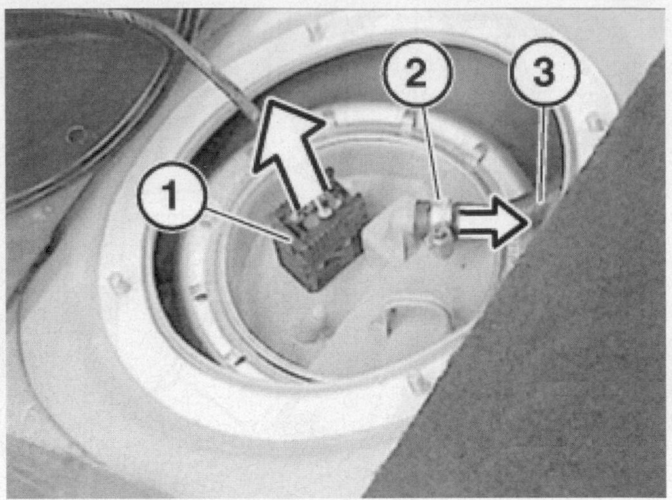

29246_BMWC_G0014

Fig. 38 The connector (1) on top of the sensor, along with the hose clamp (2) and the hose (3)

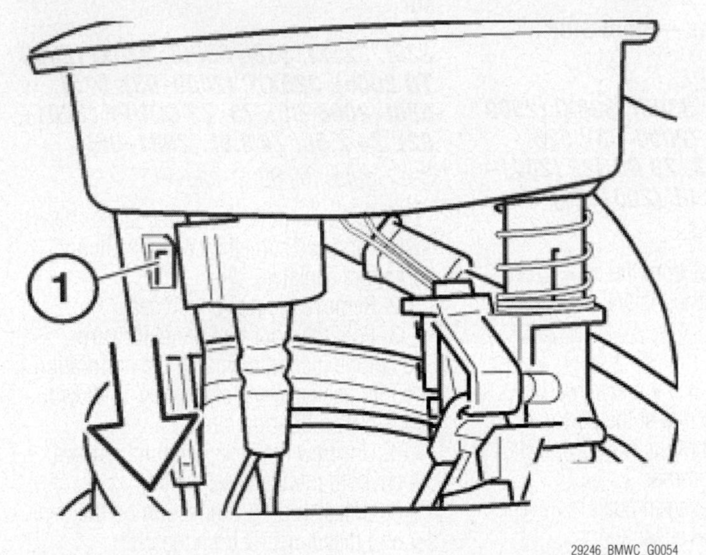

29246_BMWC_G0054

Fig. 39 Depress the tab (1) to remove the tank expansion line

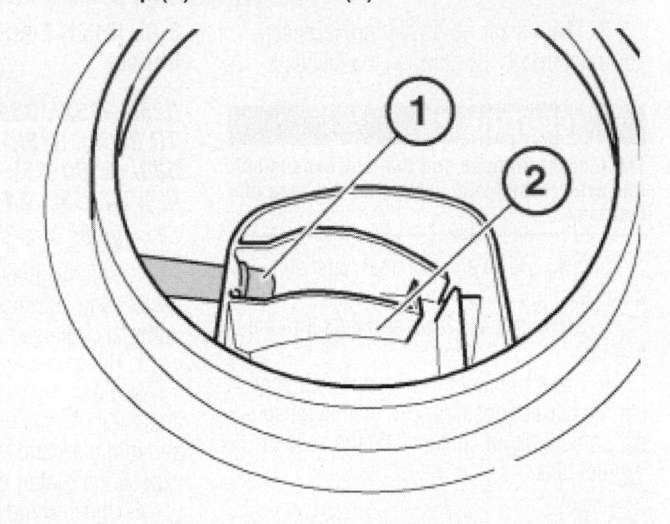

29246_BMWC_G0055

Fig. 40 Check whether the tank expansion lead (1) has engaged in surge tank (2)

6. Release the hose clamps to disconnect the hose.

7. Covering the sensor, there is a rotary ring that must be unfastened.

8. Carefully withdraw the fuel level sensor.

9. Once out, depress the tab to remove the tank expansion line.

To install:

10. Ensure that non-return valve is fitted in tank expansion line.

11. If the air valve does not function, the fuel delivery unit (right) must be removed.

12. After removal of fuel level sensor (left), check whether tank expansion lead (1) has engaged in surge tank (2).

13. Replace the rubber seal that goes between the tank and the sensor

14. Tighten down the rotary ring

15. When installing the fuel level sensor, ensure that the lug (1) is fitted properly into the recess (2) on to of the tank.

16. When tightening the notch (3) on the screw cap, it can be heard clicking and felt engaging the toothed portion (4) of the tank.

Fuel Tank Pressure Sensor

REMOVAL & INSTALLATION

1.9L (M44). 2.8L (M52) and 3.2L (S52) Engines

Z3, 318I, 318IS, 318IC (1996–98); 318TI (1996–99); 328I, 328IS, 328IC (1996–00); 528I (1997–00); Z3 COUPE, ROADSTER (1999–00); M3 (1996–2005)

See Figures 41 and 42.

1. Remove the rear right wheel to access the wheel housing trim.

2. There are plastic nuts that hold the wheel housing trim. Remove them and tilt the expansion tank downwards.

➡ **The lines should remain connected.**

3. Open the hose clamp and detach the hose. Unfasten the plug connection and disconnect it from the unit.

4. Detach tank pressure sensor from both retainers on the bottom of the sensor.

To install:

5. Replace the hose clip and ensure the proper locking of the plug connector.

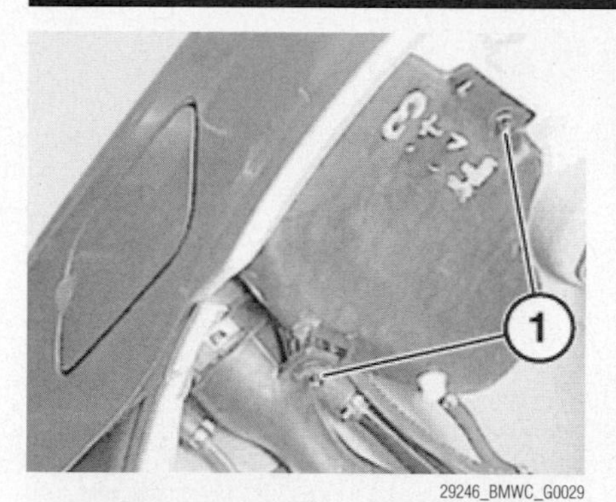

29246_BMWC_G0029

Fig. 41 Release plastic nuts (1) and tilt expansion tank downwards

Intake Air Temperature (IAT) Sensor

REMOVAL & INSTALLATION

1.9L (M44) Engines

Z3, 318I, 318IS, 318IC (1996–98); 318TI (1996–99)

See Figures 43 and 44.

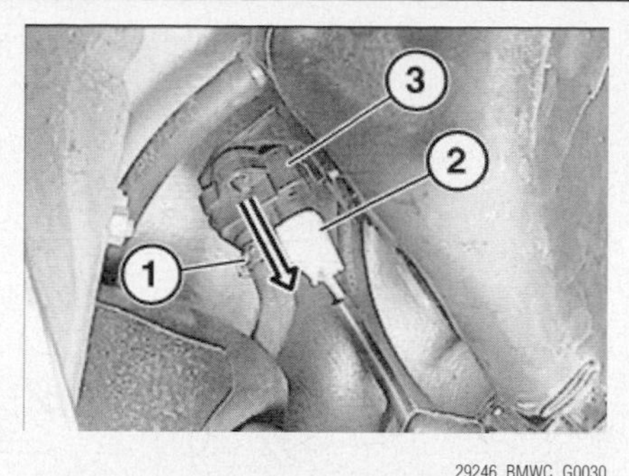

29246_BMWC_G0030

Fig. 42 The hose clamp (1), the plug connector (2) and the pressure sensor (3)

CAUTION

Make sure that the ignition is switched off and no electrical equipment is on or running.

1. Access the fault memory of control unit of Digital Motor Electronics (DME), check for any stored fault messages, rectify those faults and clear the fault memory.
2. Disconnect the plug connection on the temperature sensor.
3. Depress interlock button (1), unclip temperature sensor and remove from intake air manifold.

To install:

4. Depress interlock button, and return the temperature sensor to the intake air manifold.
5. Connect the plug connection on the temperature sensor.

2.5L (M52TU) and 3.0L (M54) Engines

325I, 325XI, 330I, 330CI, 330XI (2000 TO 2005); 325XIT; (2000–03); 525I, 530I (2000–05); Z3, Z3 COUPE (2001–02); Z4 2.5L, Z4 3.0L (2001–06)

See Figures 45 and 46.

CAUTION

Make sure that the ignition is switched off and no electrical equipment is on or running.

1. Access the fault memory of control unit of Digital Motor Electronics (DME), check for any stored fault messages, rectify those faults and clear the fault memory.

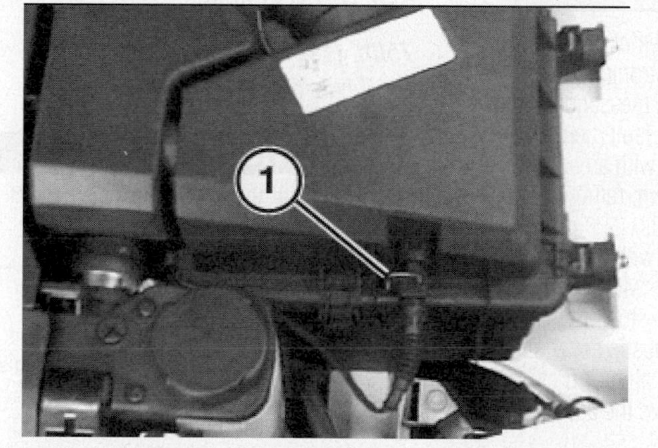

29246_BMWC_G0025

Fig. 43 Disconnect the plug connection on temperature sensor (1)

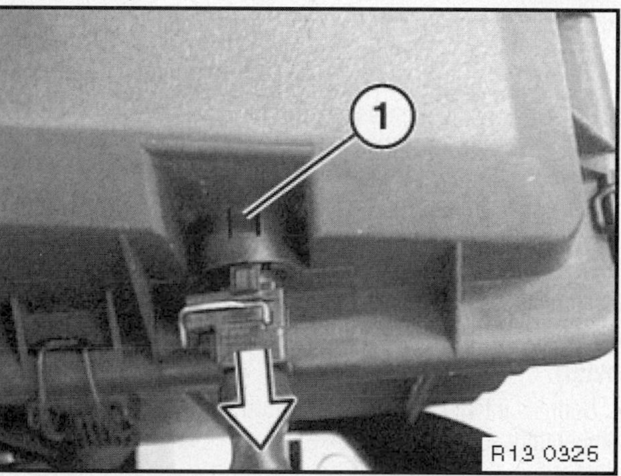

29246_BMWC_G0026

Fig. 44 Depress the interlock button (1) and attach temperature sensor

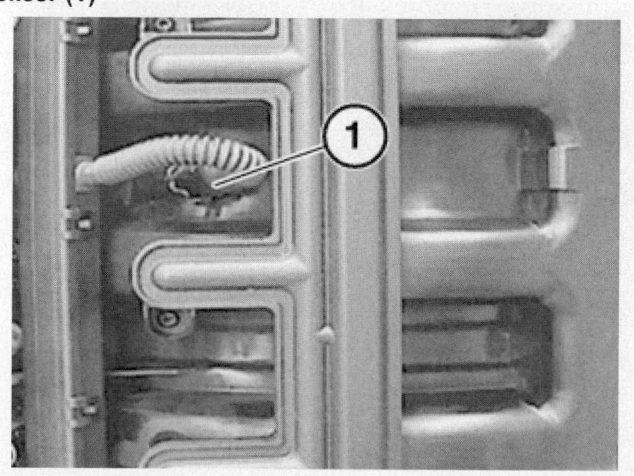

29246_BMWC_G0007

Fig. 45 Disconnect the plug connection (1) for temperature sensor

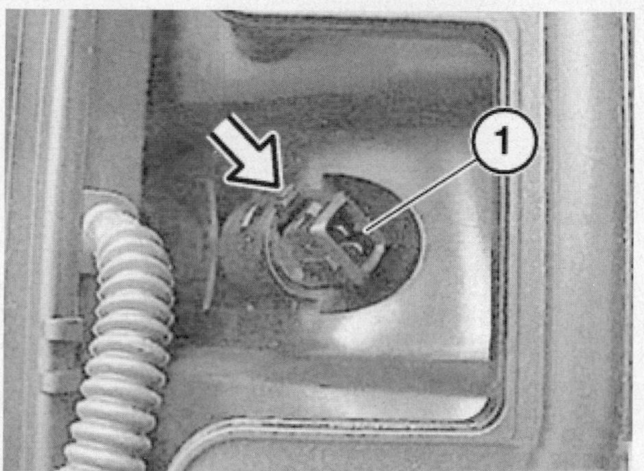

29246_BMWC_G0008

Fig. 46 Press down on the catch spring and replace the temperature sensor

2. Remove the cover from the fuel injectors.

3. Disconnect the plug connection for temperature sensor.

4. Press down on the catch spring and remove the temperature sensor from intake air manifold.

To install:

5. Check the sealing ring on temperature sensor for damage and replace it if necessary.

6. Before installing the sensor, coat the sealing ring with an anti-seize material.

7. Press down on the catch spring and replace the temperature sensor

2.8L (M52) and 3.2L (S52) Engines

328I, 328IS, 328IC (1996–00); 528I (1997–00); Z3 COUPE, ROADSTER (1999–00); M3 (1996–2005)

See Figures 47 and 48.

✱✱ CAUTION

Make sure that the ignition is switched off and no electrical equipment is on or running.

1. Access the fault memory of control unit of Digital Motor Electronics (DME), check for any stored fault messages, rectify those faults and clear the fault memory.

2. Remove the throttle valve neck, but the cables and throttle valve preheating unit remains connected.

3. Depress interlock button (1), unclip temperature sensor and remove from intake air manifold.

To install:

4. Depress interlock button, and return the temperature sensor to the intake air manifold.

5. Connect the plug connection on the temperature sensor.

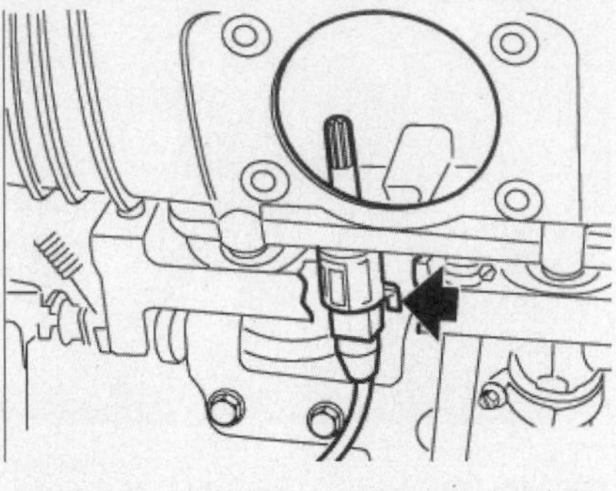

29246_BMWC_G0042

Fig. 47 Disconnect the plug connection on temperature sensor (arrow)

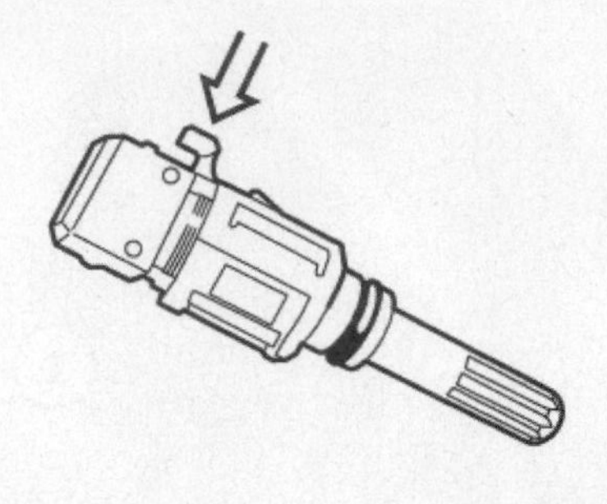

29246_BMWC_G0043

Fig. 48 Depress the interlock button and attach temperature sensor

4.4L (M62) Engine

540I (1997–03); 740I, 740IL (1996–05);

See Figure 49.

✱✱ CAUTION

Make sure that the ignition is switched off and no electrical equipment is on or running.

1. Access the fault memory of control unit of Digital Motor Electronics (DME), check for any stored fault messages, rectify those faults and clear the fault memory.

2. For cars equipped with an acoustic cover that is screwed down, remove the seal caps and unfasten the nuts.

3. For cars equipped with an acoustic cover that is held in place with press stud fasteners, open the press-studs one at a time, while lifting the acoustic cover slightly on all four corners. Once all four press-studs have been opened, remove the acoustic cover.

4. The temperature sensor is clipped into the upper section of intake filter housing.

5. Disconnect connector.

6. Press locking unit and remove temperature sensor.

To install:

7. Replace the temperature sensor and lock the clip into place.

8. Plug in the connector.

5.4L (M73) Engine

750IL (1996–01)

See Figure 50.

✱✱ CAUTION

Make sure that the ignition is switched off and no electrical equipment is on or running.

1. Access the fault memory of control unit of Digital Motor Electronics (DME), check for any stored fault messages, rectify those faults and clear the fault memory.

2. Press down on the retaining spring.

3. Remove the connector.

4. Unclip and remove the temperature sensor.

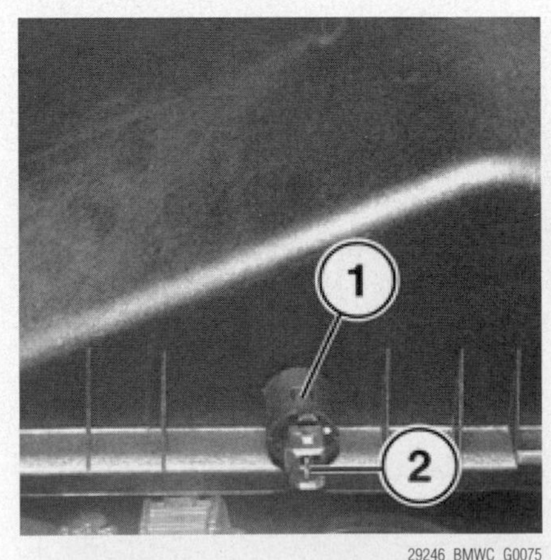

29246_BMWC_G0075

Fig. 49 Press down on the locking unit (1) and remove the temperature sensor (2)

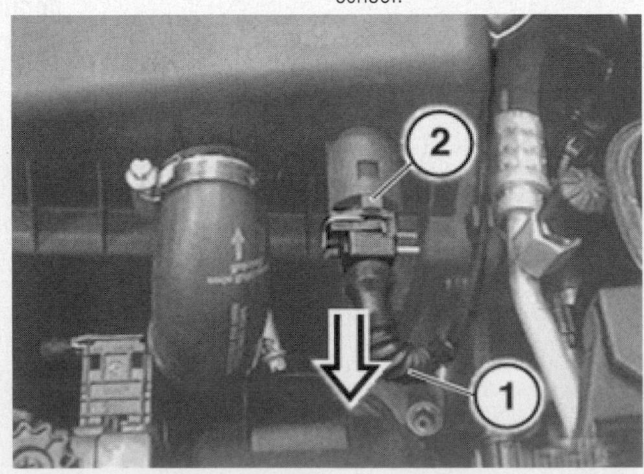

29246_BMWC_G0107

Fig. 50 The connector and the retaining spring for the sensor

To install:

5. Replace the temperature sensor and clip it into place
6. Plug in the connector .
7. Attach the retaining spring.

Knock Sensor (KS)

REMOVAL & INSTALLATION

1.9L (M44) Engines

Z3, 318I, 318IS, 318IC (1996–98); 318TI (1996–99)

See Figures 51 and 52.

**** CAUTION**

Make sure that the ignition is switched off and no electrical equipment is on or running.

1. Access the fault memory of control unit of Digital Motor Electronics (DME), check for any stored fault messages, rectify those faults and clear the fault memory.
2. Remove the upper section of the manifold.
3. For cylinder group 1–2, the knock sensor can be seen between the oil filter and the intake system.
4. To gain access, the air intake hose for the fuel injectors on the neck (1) of fuel rail must be removed.
5. To remove the knock sensor, slightly unscrew the screw that mounts the coolant pipe on the engine block, gently pull the pipe forward and fish out the sensor from behind the coolant pipe.
6. For the knock sensor on cylinder group 3–4, remove the screw and the sensor is released.

To install:

7. To install the sensors, follow the reverse of the removal procedure.

2.8L (M52) and 3.2L (S52) Engines

328I, 328IS, 328IC (1996–00); 528I (1997–00); Z3 COUPE, ROADSTER (1999–00); M3 (1996–2005)

See Figure 53.

**** CAUTION**

Make sure that the ignition is switched off and no electrical equipment is on or running.

1. Access the fault memory of control unit of Digital Motor Electronics (DME), check for any stored fault messages, rectify those faults and clear the fault memory.

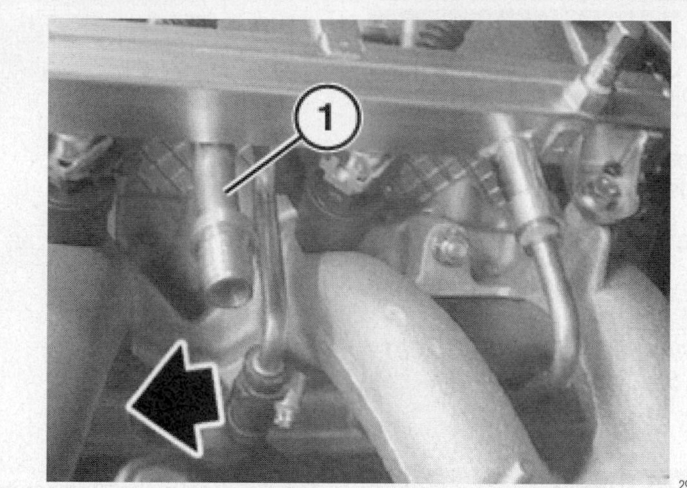

Fig. 51 The air intake hose for the fuel injectors

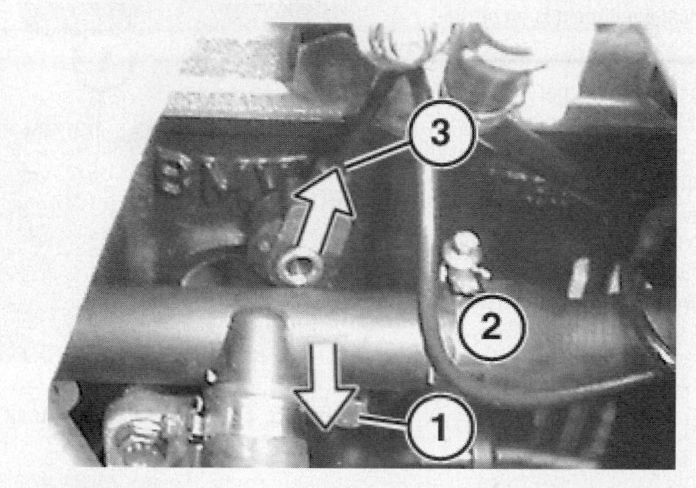

Fig. 52 The screw (1) on the coolant pipe (2) is loosened. Pull the coolant pipe (2) forward gently and fish out the knock sensor (3) from behind the coolant pipe

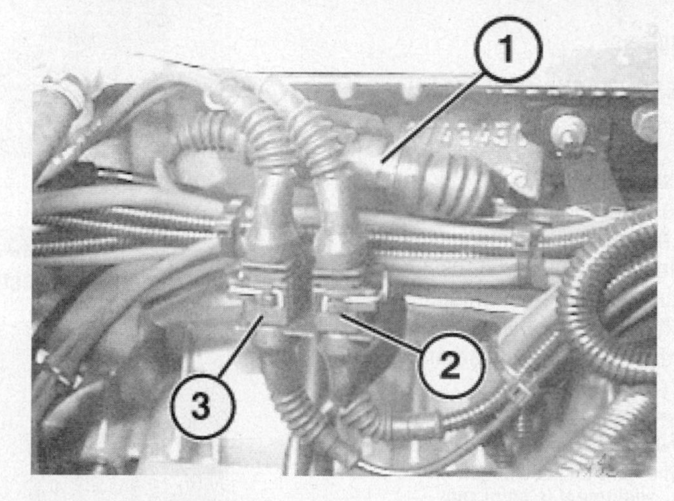

Fig. 53 The plug connection (1) for the knock sensor for cylinders 4 to 6

2. Remove the air intake manifold.
3. Unclip and disconnect the plug connection (1) for the knock sensor for cylinders 4 to 6.
4. The knock sensor for cylinders 1 to 3 is removed via a screw.

To install:

5. Clean the support face of the knock sensors on engine block.
6. Attach the knock sensors with the retaining screws.

✱✱ WARNING

Fitting the knock sensors incorrectly during the assembly will cause engine damage!

➡ The knock sensor with the short cable leads to cylinders 4 to 6.

3.0L (M54) Engines

325I, 325XI, 330I, 330CI, 330XI (2000 TO 2005); 325XIT; (2000–03); 525I, 530I (2000–05); Z3, Z3 COUPE (2001–02); Z4 2.5L, Z4 3.0L (2001–06)

See Figures 54, 55 and 56.

✱✱ CAUTION

Make sure that the ignition is switched off and no electrical equipment is on or running.

1. Access the fault memory of control unit of Digital Motor Electronics (DME), check for any stored fault messages, rectify those faults and clear the fault memory.
2. Remove the intake air manifold
3. Unclip plug connection from holder of cable duct and disconnect.
4. Unscrew the screw and remove the knock sensor for Cylinder Bank 1–3.
5. Unscrew the screw and remove the knock sensor for Cylinder Bank 4–6.

To install:

6. Clean support face of knock sensors on engine block.
7. Replace each sensor via the retaining screws.

3.0L (N52) Engines

525I, 525XI, 530I, 530XI, 550I (2006)

See Figures 57 and 58.

➡ No steel screws or bolts may be used due to the threat of electrochemical corrosion. A magnesium crankcase requires aluminum screws and bolts exclusively. The end of the aluminum screws and bolts are painted blue for the purposes of reliable identification.

✱✱ CAUTION

Make sure that the ignition is switched off and no electrical equipment is on or running.

1. Access the fault memory of control unit of Digital Motor Electronics (DME), check for any stored fault messages, rectify those faults and clear the fault memory.
2. Disconnect the battery.
3. Remove the air intake manifold.

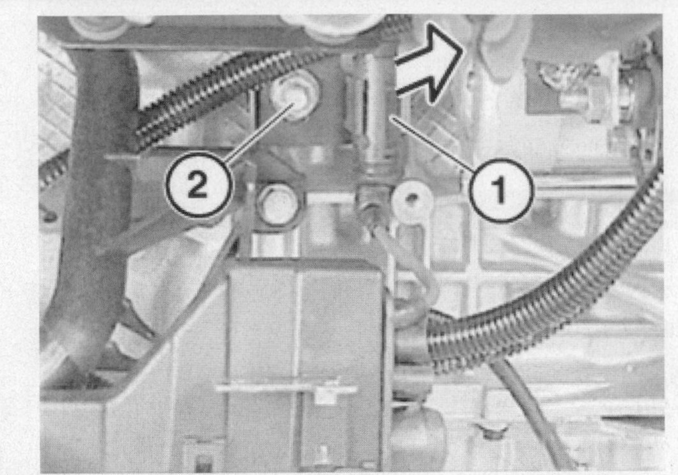

Fig. 54 Unclip the plug connection (1) from the holder of the cable duct (2) and disconnect it

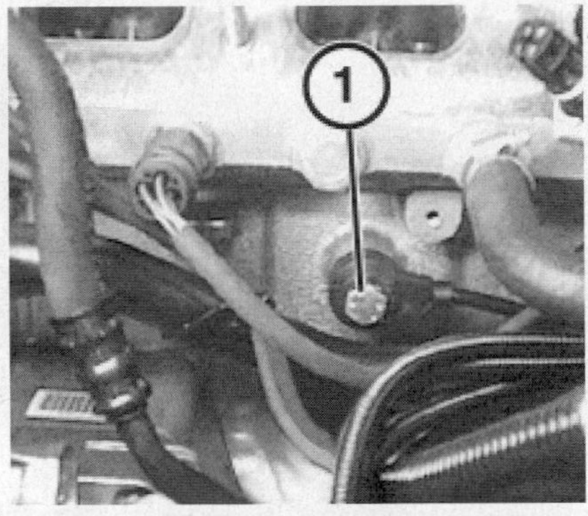

Fig. 55 The screw (1) for knock sensor on Cylinder Bank 1–3

Fig. 56 The screw (1) for knock sensor on Cylinder Bank 4–6

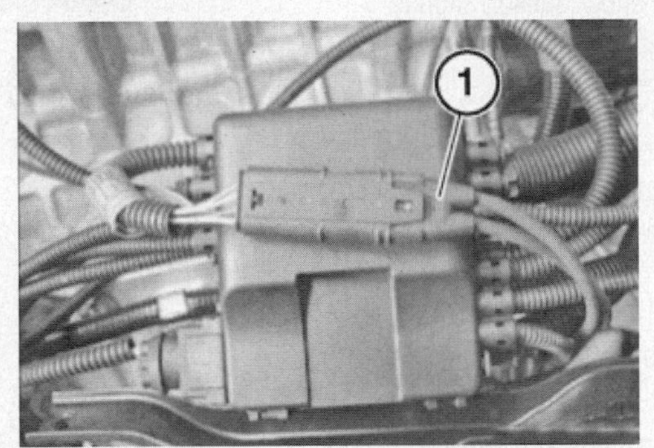

Fig. 57 Unlock the plug connection (1) and remove it

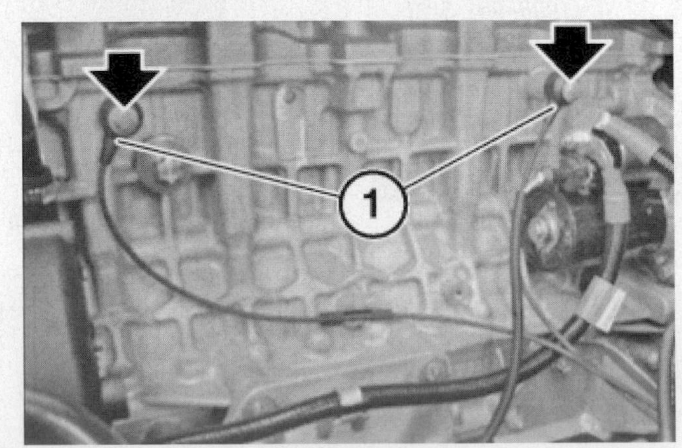

Fig. 58 The screw locations for both knock sensors

4. Unlock the plug connection and remove it.

5. Unscrew the screws on both knock sensors and remove the sensors.

To install:

6. Clean the support face of the knock sensors on engine block.

7. Install the screws on both knock sensors to replace the sensors.

4.4L (M62) Engine—Both Sides

540I (1997–03); 740I, 740IL (1996–05);

See Figure 59.

**** CAUTION**

Make sure that the ignition is switched off and no electrical equipment is on or running.

1. Access the fault memory of control unit of Digital Motor Electronics (DME), check for any stored fault messages, rectify those faults and clear the fault memory.

2. Disconnect the battery.

3. Remove the air intake manifold.

**** WARNING**

Fitting the knock sensors incorrectly during the assembly will cause engine damage!

4. Disconnect the plug-in wires to the sensors

5. Remove the screws that retain the sensors.

6. Remove the sensors

To install:

7. Replace the sensors

8. Tighten the screws and attach the plug-in connections.

9. Clear any fault messages.

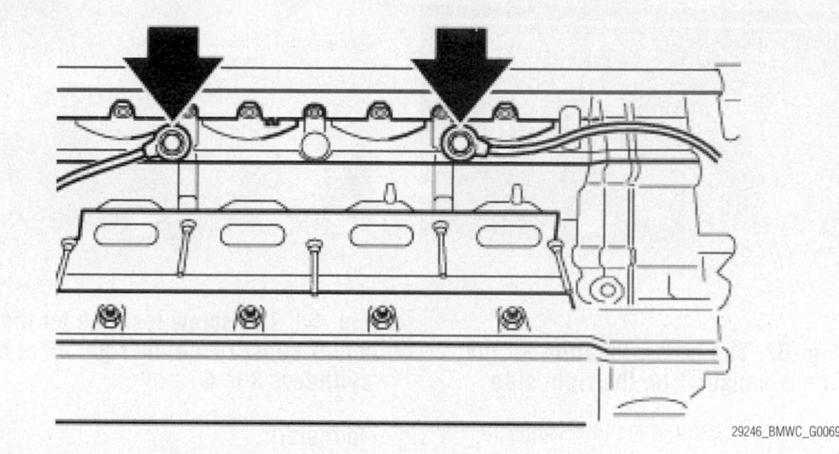

Fig. 59 The screw locations for both knock sensors

4.8L (N62 and N62TU) Engines—Both Sides

545I, (2004–05); 745I, 745IL (2002–2005); 750I, 750LI (2002–05)

See Figures 60 and 61.

**** CAUTION**

Make sure that the ignition is switched off and no electrical equipment is on or running.

1. Access the fault memory of control unit of Digital Motor Electronics (DME), check for any stored fault messages, rectify those faults and clear the fault memory.

2. Disconnect the battery.

3. Remove the air intake manifold.

**** WARNING**

Fitting the knock sensors incorrectly during the assembly will cause engine damage!

4. Disconnect the plug-in wires to the sensors

5. Remove the screws that retain the sensors.

6. Remove the sensors

To install:

7. Replace the sensors

8. Tighten the screws and attach the plug-in connections.

9. Clear any fault messages.

5.4L (M73) Engine—Right Sensor

750IL (1996–01)

See Figures 62 and 63.

**** CAUTION**

Make sure that the ignition is switched off and no electrical equipment is on or running.

1. Access the fault memory of control unit of Digital Motor Electronics (DME),

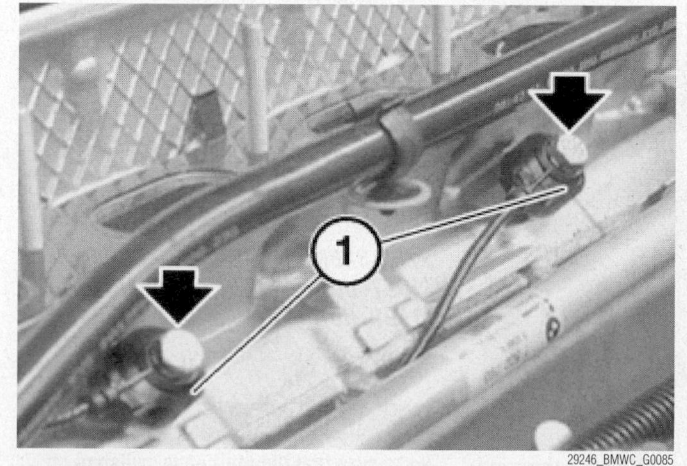

Fig. 60 The screw locations for both knock sensors on the left side

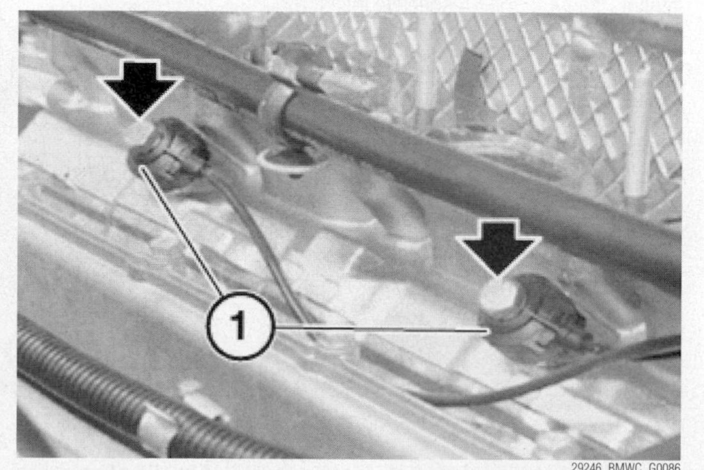

Fig. 61 The screw locations for both knock sensors on the right side

Fig. 62 The screw location for the knock sensor 1 on the right side

check for any stored fault messages, rectify those faults and clear the fault memory.

2. Disconnect the battery.
3. Remove the air intake manifold.
4. Unfasten the screws for the knock sensor 1 and 2 in their respective locations.

To install:

5. Return the sensors in their proper places (making sure not to mix them) and screw them into place.

5.4L (M73) Engine—Left Sensor

750IL (1996–01)

See Figures 64 and 65.

✱✱ CAUTION

Make sure that the ignition is switched off and no electrical equipment is on or running.

1. Access the fault memory of control unit of Digital Motor Electronics (DME), check for any stored fault messages, rectify those faults and clear the fault memory.
2. Disconnect the battery.
3. Remove the air intake manifold.
4. Unfasten the screws for the knock sensor 3and 4 in their respective locations.

Fig. 64 The screw location for the knock sensor 3 on the left side, for cylinders 7 to 9

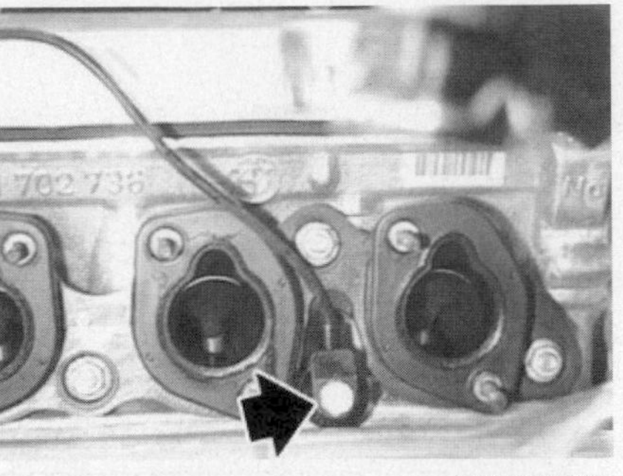

Fig. 63 The screw location for the knock sensor 2 on the right side, for cylinders 3 to 6

To install:

5. Return the sensors in their proper places (making sure not to mix them) and screw them into place.

Mass Air Flow (MAF) Sensor

REMOVAL & INSTALLATION

1.9L (M44) Engines

Z3, 318I, 318IS, 318IC (1996–98); 318TI (1996–99)

See Figure 66.

✱✱ CAUTION

Make sure that the ignition is switched off and no electrical equipment is on or running.

1. Access the fault memory of control unit of Digital Motor Electronics (DME), check for any stored fault messages, rectify those faults and clear the fault memory.
2. Unfasten the retaining clips for the connector.
3. Turn and pull off the multiple pin connector .

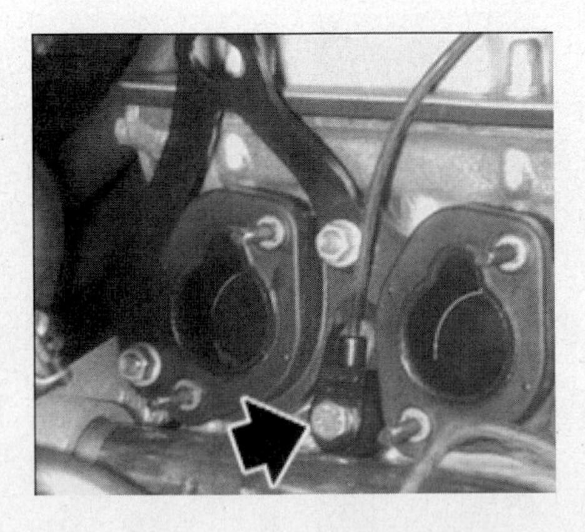

Fig. 65 The screw location for the knock sensor 4 on the left side, for cylinders 10 to 12

4. Unfasten the hose clip.
5. Remove the mass airflow sensor.

To install:

6. Install the mass airflow sensor.
7. Secure the hose clip.
8. Slide on the multiple pin connector .
9. Fasten the retaining clips for the connector.
10. Check stored fault messages.
11. Rectify faults.
12. Clear the fault memory.

2.5L (M52TU) and 3.0L (M54) Engines

325I, 325XI, 330I, 330CI, 330XI (2000 TO 2005); 325XIT; (2000–03); 525I, 530I (2000–05); Z3, Z3 COUPE (2001–02); Z4 2.5L, Z4 3.0L (2001–06)

See Figures 67, 68 and 69.

✱✱ CAUTION

Make sure that the ignition is switched off and no electrical equipment is on or running.

1. Access the fault memory of control unit of Digital Motor Electronics (DME), check for any stored fault messages, rectify those faults and clear the fault memory.
2. Unlock the retaining clips on the intake filter housing.
3. Unlock the plug and detach it from the mass airflow sensor.
4. Release the hose clip.
5. Detach the intake hose from the mass airflow sensor.
6. Remove the mass airflow sensor.

To install:

7. Check the seal between the mass airflow sensor and air cleaner upper section for damage and replace if necessary.
8. To facilitate the assembly, coat the seal with an acid-free grease.

➡ **Only reinstall mass air flow sensor with undamaged lattice frame.**

9. Check stored fault messages.
10. Rectify faults.
11. Clear the fault memory.

2.8L (M52) and 3.2L (S52) Engines

328I, 328IS, 328IC (1996–00); 528I (1997–00); Z3 COUPE, ROADSTER (1999–00); M3 (1996–2005)

See Figures 70, 71, 72 and 73.

✱✱ CAUTION

Make sure that the ignition is switched off and no electrical equipment is on or running.

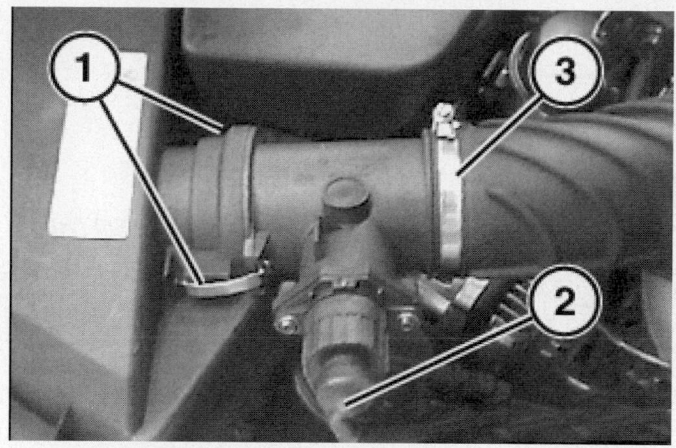

Fig. 66 The location of the retaining clips (1), the multiple pin connector (2), and the hose clip (3)

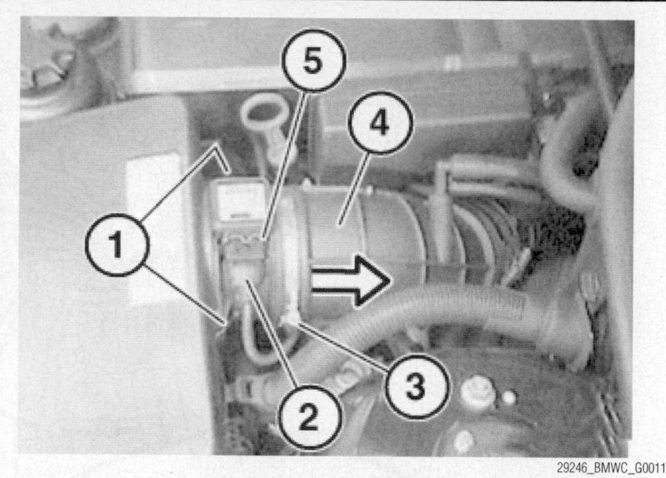

Fig. 67 The location of the retaining clips (1), the plug (2), the hose clip (3), the intake hose (4) and the sensor (5)

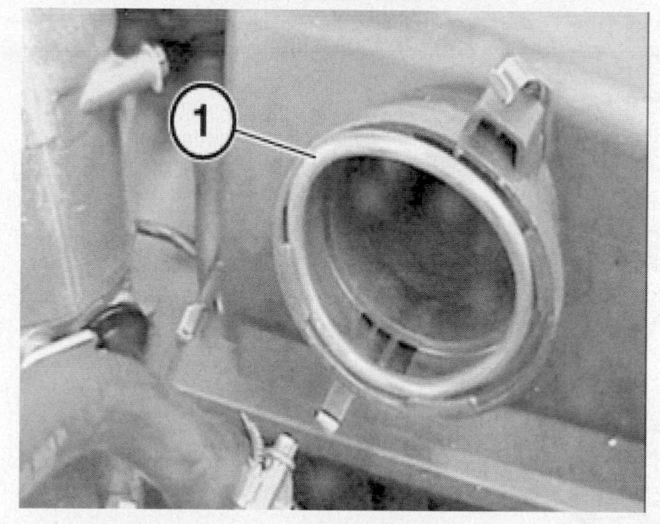

Fig. 68 Check the seal (1) between air-mass flow sensor and air cleaner

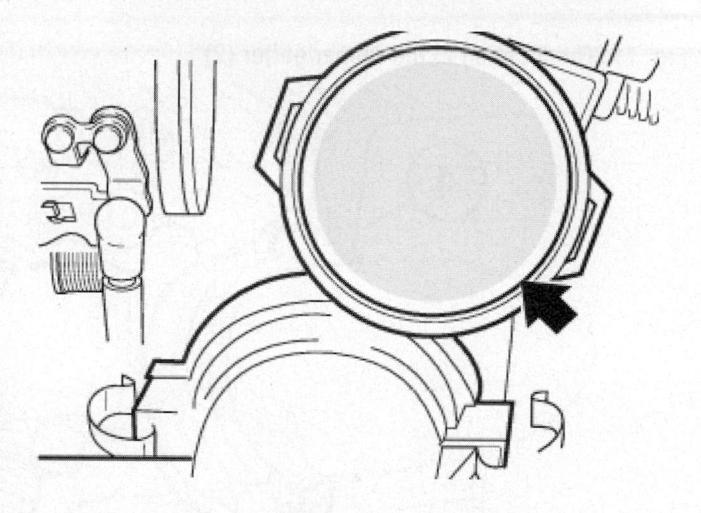

Fig. 69 The location of the lattice frame on the mass airflow sensor

1. Access the fault memory of control unit of Digital Motor Electronics (DME), check for any stored fault messages, rectify those faults and clear the fault memory.
2. Twist and remove connector on the throttle body.
3. Unfasten the hose clip and remove the rubber gaiter from throttle body.
4. Loosen clamps on the bellows.
5. Remove air-mass sensor with bellows
6. Remove mass airflow sensor from the rubber gaiter.

To install:

➡ **Only reinstall mass air flow sensor with undamaged lattice frame.**

7. Replace the mass airflow sensor into the rubber gaiter.
8. Tighten the clamps.
9. Replace the hose clip and the rubber gaiter onto the throttle body.

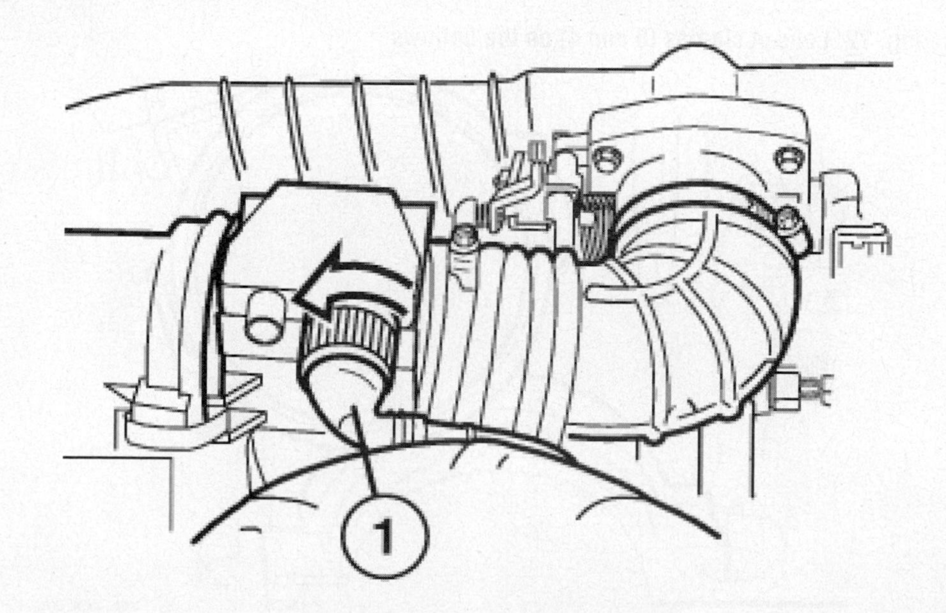

Fig. 70 The location of the connector (1) on the throttle body.

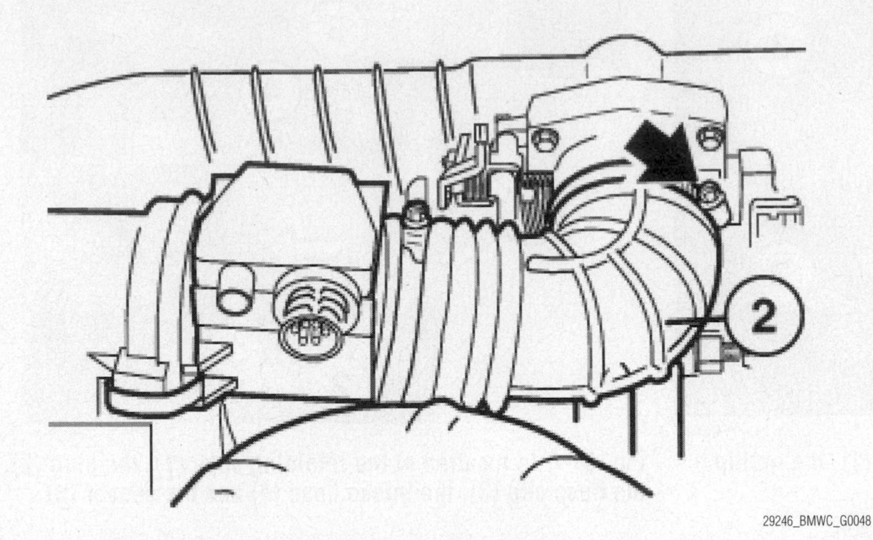

Fig. 71 The location of the rubber gaiter (2)

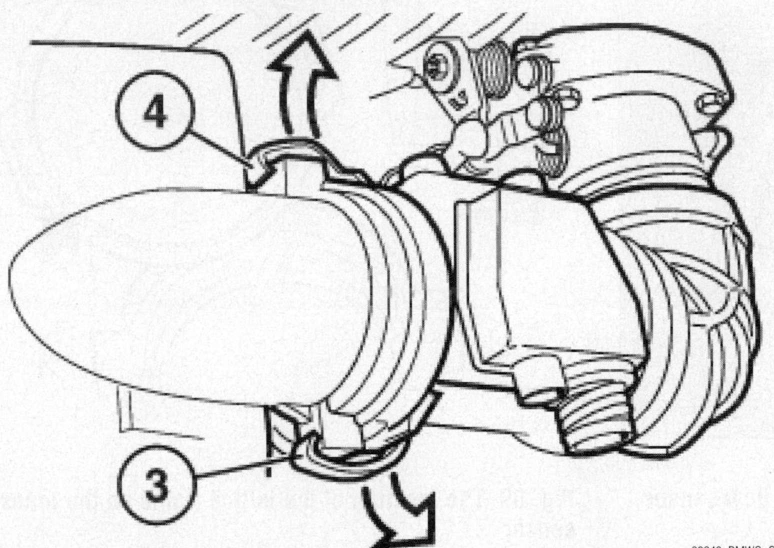

Fig. 72 Loosen clamps (3 and 4) on the bellows

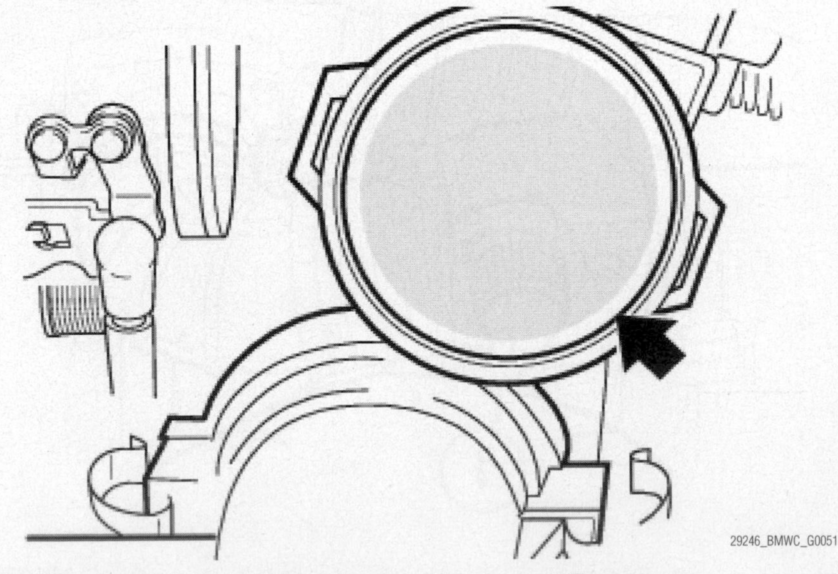

Fig. 73 The location of the lattice frame on the mass airflow sensor

10. Install the connector on the throttle body.
11. Check stored fault messages.
12. Rectify faults.
13. Clear the fault memory.

3.0L (N52) Engines

525I, 525XI, 530I, 530XI, 550I (2006)

See Figure 74.

**** CAUTION**

Make sure that the ignition is switched off and no electrical equipment is on or running.

1. Access the fault memory of control unit of Digital Motor Electronics (DME), check for any stored fault messages, rectify those faults and clear the fault memory.
2. Unfasten the screws that hold the sensor in place.
3. Unlock the plug and remove it.
4. Pull the mass airflow sensor out of the upper section of the intake filter housing.

To install:

5. Replace the sensor, screw it down and connect the plug-in connector.
6. Check stored fault messages.
7. Rectify faults.
8. Clear the fault memory.

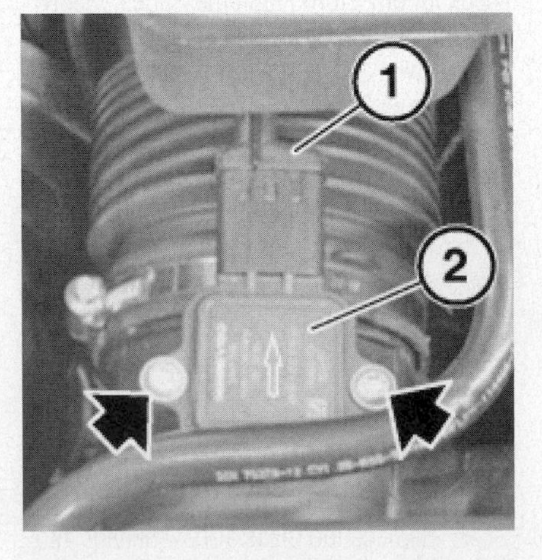

Fig. 74 The location of the plug (1), the sensor (2) and the screws (arrows)

4.4L (M62) and 5.4L (M73) Engines

540I (1997–03); 740I, 740IL (1996–05); 750IL (1996–01)

See Figures 75, 76 and 77.

✻✻ CAUTION

Make sure that the ignition is switched off and no electrical equipment is on or running.

1. Access the fault memory of control unit of Digital Motor Electronics (DME), check for any stored fault messages, rectify those faults and clear the fault memory.
2. Disconnect the plug connection from air-mass flow sensor.
3. Release the hose clip.
4. Detach the intake hose from air-mass flow sensor.
5. Open clips (1) on the air cleaner upper section and detach the mass airflow sensor.

To install:

6. Check the seal between the mass airflow sensor and air cleaner upper section for damage and replace if necessary.
7. To facilitate the assembly, coat the seal with an acid-free grease.
8. Check stored fault messages.
9. Rectify faults.
10. Clear the fault memory.

4.8L (N62 and N62TU) Engines

545I, (2004–05); 745I, 745IL (2002–2005); 750I, 750LI (2002–05)

See Figures 78 and 79.

✻✻ CAUTION

Make sure that the ignition is switched off and no electrical equipment is on or running.

1. Access the fault memory of control unit of Digital Motor Electronics (DME),

Fig. 75 Disconnect the plug connection (1) from air-mass flow sensor and Release the hose clip (2)

Fig. 76 Install the sensor into the upper section of the air cleaner and close the clips (1)

Fig. 77 Check the seal (1) between air-mass flow sensor and air cleaner

Fig. 78 The location of the clips (1), the intake filter housing (3) and the plug-in connector

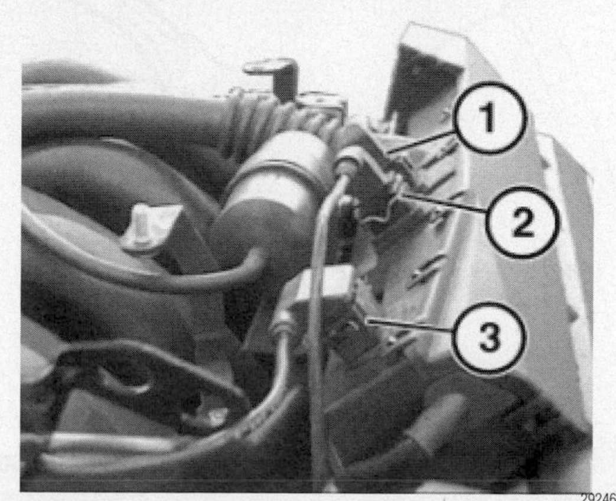

Fig. 79 Unscrew the screws for the sensor from the intake filter housing

check for any stored fault messages, rectify those faults and clear the fault memory.

2. Open the clips to the top section of the intake filter housing.

3. Remove the connector by unlocking it and pulling it aside.

4. Unscrew the two screws holding the mass airflow sensor to the upper section of the intake filter housing.

To install:

5. Install the sensor and tighten down the two screws.

6. Replace the connector.

7. Replace the cover to the intake filter housing and fasten the clips.

8. Check for stored fault memory and clear the codes.

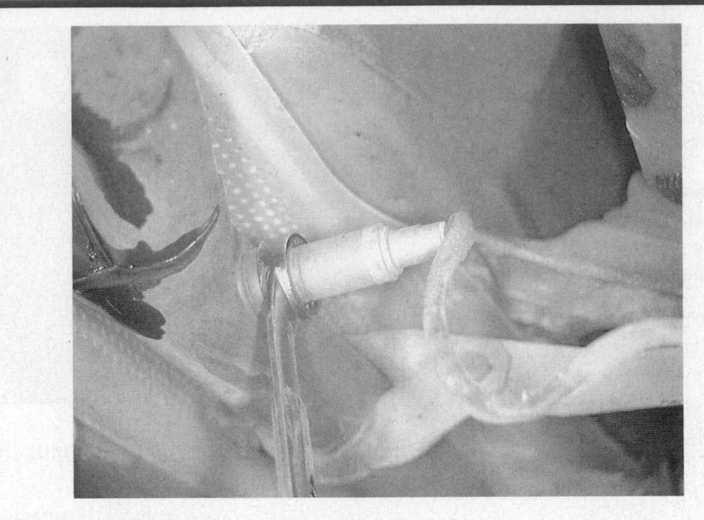

90964P77

Fig. 80 A typical location of an oxygen sensor

Oxygen (O2) Sensor

LOCATION

See Figure 80.

The typical location of an oxygen sensor

REMOVAL & INSTALLATION

1.9L (M44), M52, 4.4L (M62) and 5.0L (S62) Engines

Z3, 318I, 318IS, 318IC (1996–98); 318TI (1996–99); 328I, 328IS, 328IC (1996–00); 528I (1997–00); Z3 COUPE, ROADSTER (1999–00); 540I (1997–03); 740I, 740IL (1996–05); M5 (2000–05)

See Figures 81and 82.

**** WARNING**

There is a danger of scalding so only perform this task on an engine that has completely cooled down.

**** CAUTION**

Make sure that the ignition is switched off and no electrical equipment is on or running.

1. Access the fault memory of control unit of Digital Motor Electronics (DME), check for any stored fault messages, rectify those faults and clear the fault memory.

2. Disconnect the connector for the oxygen sensor.

3. Unclip the oxygen sensor cable from its bracket.

4. Remove the oxygen sensor.

To install:

➡ **The threads of a new oxygen sensors are already coated with an anti-seize compound. If an oxygen sensor is to be used again, apply a thin and even coat of an anti-seize compound to the thread only.**

➡ **Do not clean the oxygen sensor section which protrudes into the exhaust line and ensure that it avoids all contact with any lubricants.**

➡ **Observe cable routing of the oxygen sensor so it doesn't interfere with any other system or the exhaust pipes.**

5. Install the oxygen sensor.

6. Clip the oxygen sensor cable back into its bracket.

7. Connect the connector for the oxygen sensor.

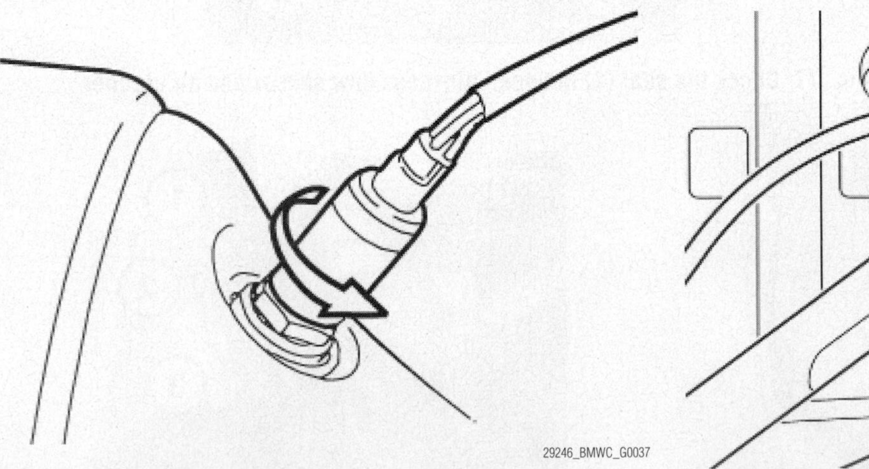

29246_BMWC_G0037

Fig. 81 Disconnect the connector for the oxygen sensor

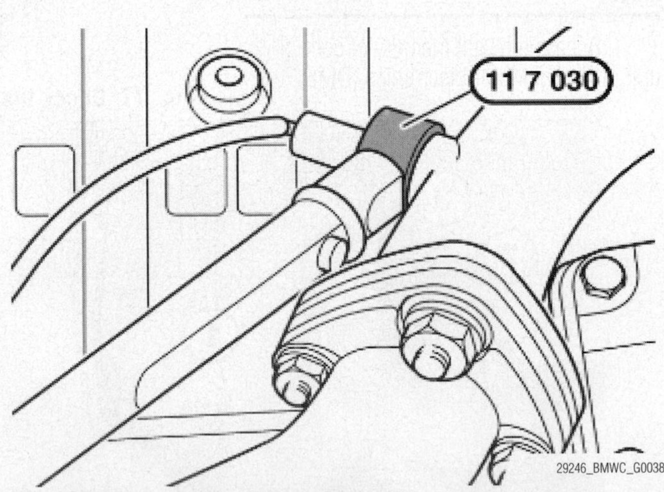

29246_BMWC_G0038

Fig. 82 Remove the oxygen sensor

2.5L (M52TU) and 3.0L (M54) Engines

325I, 325XI, 330I, 330CI, 330XI (2000 TO 2005); 325XIT; (2000–03); 525I, 530I (2000–05); Z3, Z3 COUPE (2001–02); Z4 2.5L, Z4 3.0L (2001–06)

See Figure 83.

**** WARNING**

There is a danger of scalding so only perform this task on an engine that has completely cooled down.

**** CAUTION**

Make sure that the ignition is switched off and no electrical equipment is on or running.

1. Access the fault memory of control unit of Digital Motor Electronics (DME), check for any stored fault messages, rectify those faults and clear the fault memory.

➡ The plug connection must be disconnected before the oxygen control sensor is removed.

If the plug connection separation point is below the cover for the fuel injectors:

- Remove the lower section of microfilter housing.
- Remove the cover from fuel injectors.
- Unclip the oxygen control sensor cable from retainer.
- Disconnect the plug connection for the oxygen sensor.
- Remove the oxygen control sensor.

To install:

➡ The threads of a new oxygen sensors are already coated with an anti-seize compound. If an oxygen sensor is to be used again, apply a thin and even coat of an anti-seize compound to the thread only.

➡ Do not clean the oxygen sensor section which protrudes into the exhaust line and ensure that it avoids all contact with any lubricants.

➡ Observe cable routing of the oxygen sensor so it doesn't interfere with any other system or the exhaust pipes.

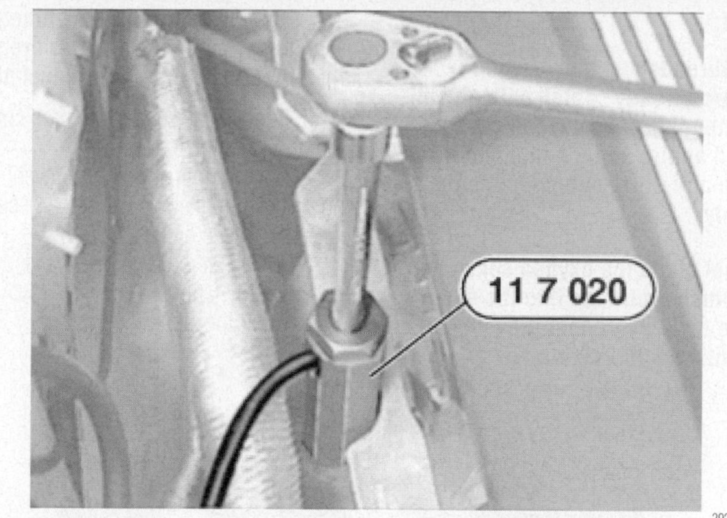

Fig. 83 The removal and installation of the oxygen sensor

3.0L (N52) Engines

525I, 525XI, 530I, 530XI, 550I (2006)

**** WARNING**

There is a danger of scalding so only perform this task on an engine that has completely cooled down.

**** CAUTION**

Make sure that the ignition is switched off and no electrical equipment is on or running.

1. Access the fault memory of control unit of Digital Motor Electronics (DME), check for any stored fault messages, rectify those faults and clear the fault memory.
2. Remove the exhaust system.
3. Disconnect the plug connection from the sensors.
4. Detach the oxygen sensor from cylinders 4 to 6.
5. The oxygen sensor at cylinders 1 to 3 is accessible from above without the exhaust system having to be removed.

To install:

➡ The threads of a new oxygen sensors are already coated with an anti-seize compound. If an oxygen sensor is to be used again, apply a thin and even coat of an anti-seize compound to the thread only.

➡ Do not clean the oxygen sensor section which protrudes into the exhaust line and ensure that it avoids all contact with any lubricants.

➡ Observe cable routing of the oxygen sensor so it doesn't interfere with any other system or the exhaust pipes.

6. Attach the oxygen sensors and their respective cables.
7. The cable color for the sensor that leads to cylinders 1 to 3 is black.
8. The cable color for the sensor that leads to cylinders 4 to 6 is gray.
9. Connect the plug connection to the sensors.

4.8L (N62 and N62TU) Engines—Both Sensors

545I, (2004–05); 745I, 745IL (2002–2005); 750I, 750LI (2002–05)

See Figures 84, 85 and 86.

**** WARNING**

There is a danger of scalding so only perform this task on an engine that has completely cooled down.

**** CAUTION**

Make sure that the ignition is switched off and no electrical equipment is on or running.

1. Access the fault memory of control unit of Digital Motor Electronics (DME), check for any stored fault messages, rectify those faults and clear the fault memory.
2. Remove the reinforcement plate.
3. Unlock and detach the plug connection.
4. Unclip the control sensor cable.
5. Remove the sensor.

To install:

➡ The threads of a new oxygen sensors are already coated with an anti-seize compound. If an oxygen sensor is to be used again, apply a thin and even coat of an anti-seize compound to the thread only.

➡ Do not clean the oxygen sensor section which protrudes into the exhaust line and ensure that it avoids all contact with any lubricants.

➡ Observe cable routing of the oxygen sensor so it doesn't interfere with any other system or the exhaust pipes.

6. Replace the sensor, the plug connection and the sensor cable.
7. For the right sensor, to prevent fraying, make sure that the wiring harness is correctly placed on heat shield.
8. Check stored fault messages.
9. Rectify faults.
10. Clear the fault memory.

Fig. 84 The location of the plug connection (3) and the sensor cable (1)

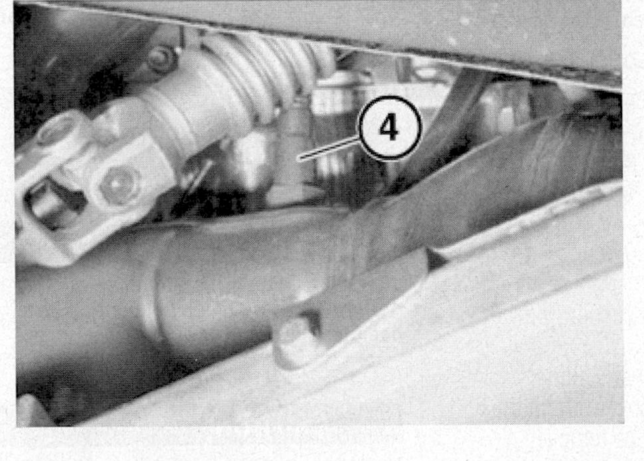

Fig. 85 The location of the sensor (4)

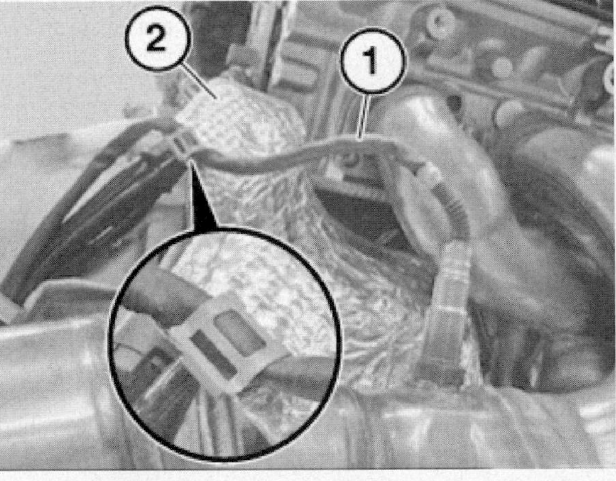

Fig. 86 For the right sensor, to prevent fraying, make sure that the wiring harness (1) is correctly placed on heat shield (2)

5.4L (M73) Engine

750IL (1996–01)

See Figures 87 and 88.

✱✱ CAUTION

Make sure that the ignition is switched off and no electrical equipment is on or running.

1. Access the fault memory of control unit of Digital Motor Electronics (DME), check for any stored fault messages, rectify those faults and clear the fault memory.
2. Unfasten the nuts that retain the cover.
3. Disconnect the connector on the sensor.
4. Remove the oxygen sensor.

To install:

➡ The threads of a new oxygen sensors are already coated with an anti-seize compound. If an oxygen sensor is to be used again, apply a thin and even coat of an anti-seize compound to the thread only.

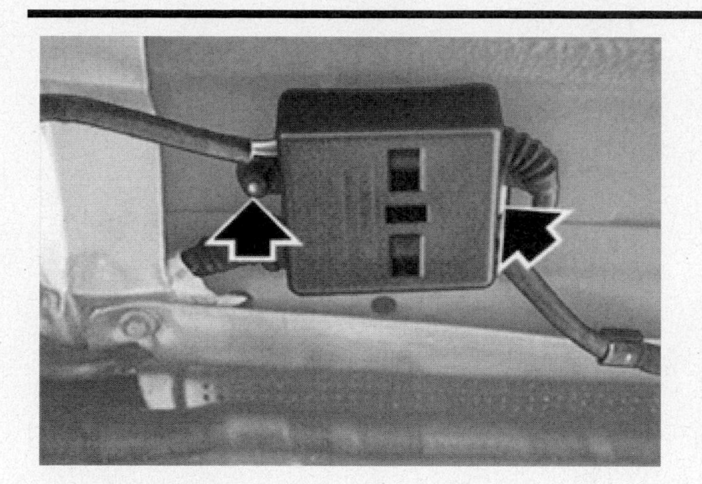

29246_BMWC_G0095

Fig. 87 Unfasten the nuts that retain the cover

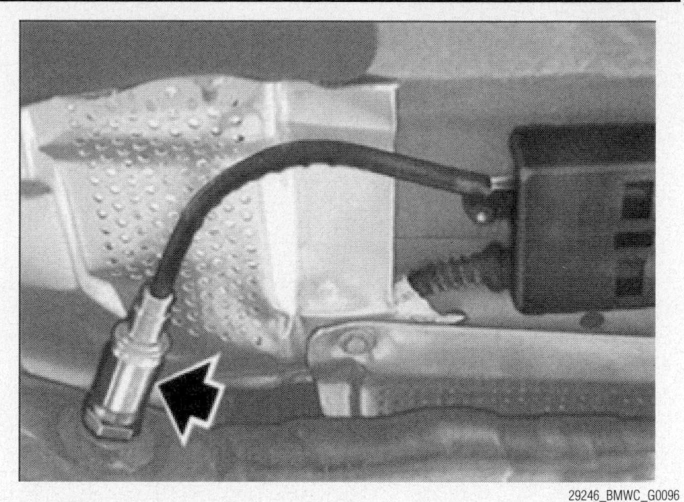

29246_BMWC_G0096

Fig. 88 Removing the oxygen sensor

➡ Do not clean the oxygen sensor section which protrudes into the exhaust line and ensure that it avoids all contact with any lubricants.

➡ Observe cable routing of the oxygen sensor so it doesn't interfere with any other system or the exhaust pipes.

5. Check stored fault messages.
6. Rectify faults.
7. Clear the fault memory.

Vehicle Speed Sensor (VSS)

REMOVAL & INSTALLATION

1.9L (M44), 2.8L (M52) and 3.2L (S52) Engines

Z3, 318I, 318IS, 318IC (1996–98); 318TI (1996–99); 328I, 328IS, 328IC (1996–00); 528I (1997–00); Z3 COUPE, ROADSTER (1999–00); M3 (1996–2005)

See Figures 89 and 90.

1. Remove the transmission.
2. Remove the screw that holds the speed sensor in place.
3. Pull out the sensor.
4. Disconnect the wire harness from the holder

To install:

5. Lubricate the o-ring lightly with grease.
6. Replace the sensor and tighten down the screw.

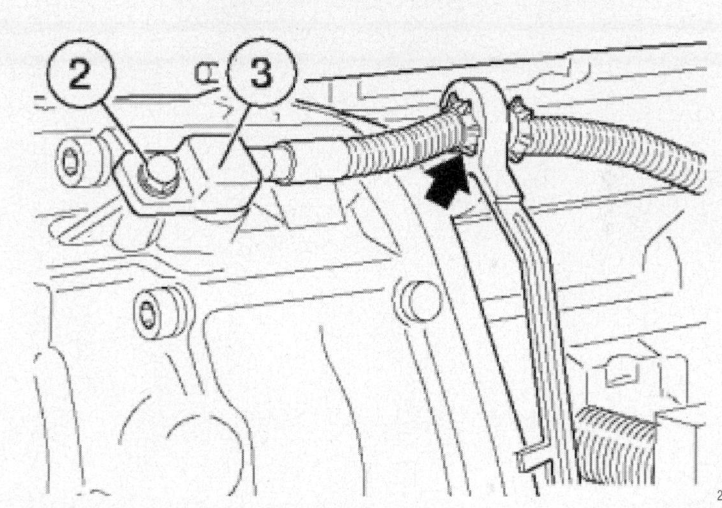

29246_BMWC_G0031

Fig. 89 The screw (2) location for speed sensor (3)

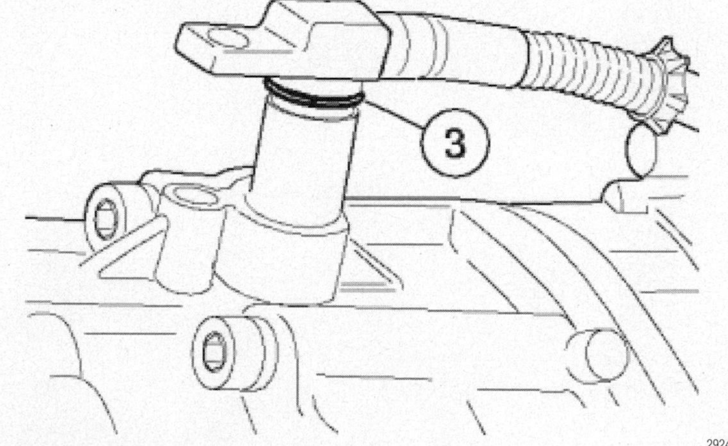

29246_BMWC_G0032

Fig. 90 Lubricate the o-ring (3) lightly with grease

MINI
DIAGNOSTIC TROUBLE CODES

10

TABLE OF CONTENTS

DIAGNOSTIC TROUBLE CODES

OBD II Vehicle Applications

MINI AUTOMOBILES

GAS ENGINE TROUBLE CODE LIST

Introduction

To use this information, first read and record all codes in memory along with any Freeze Frame data. If the DME reset function is done prior to recording any data, all codes and freeze frame data will be lost! Look up the desired code by DTC number, Code Title and Conditions (enable criteria) that indicate why a code set, and how to drive the vehicle. 1T and 2T indicate a 1-trip or 2-trip fault and the Monitor type.

Gas Engine OBD II Trouble Code List (P0xxx Codes)

DTC	Trouble Code Title, Conditions & Possible Causes
DTC: P0030 **2T CCM, MIL: Yes** **2002-05** **Models:** Cooper, Cooper S **Engines:** All **Transmissions:** All	**HO2S Heater (Bank 1 Sensor 1) Control Circuit Malfunction Conditions:** Engine started, battery voltage must be at least 11.5v, all electrical components must be off, the ground between the engine and the chassis must be well connected, the exhaust system must be properly sealed between the catalytic converter and the cylinder head, the coolant temperature must be 80 degrees Celsius, and the oxygen sensor heater for oxygen sensor before the catalytic converter must be properly functioning. The DME detected the HO2S signal was in a negative voltage range referred to as "character shift downward". This code sets when the HO2S signal remains in a low state (usually less than 156 mv). In effect, it does not switch properly between 0.1v and 1.1v in closed loop operation. **Possible Causes:** • HO2S is contaminated (due to presence of silicone in fuel) • HO2S signal and ground circuit wires crossed in wiring harness • HO2S signal circuit is shorted to sensor or chassis ground • HO2S element has failed (internal short condition)
DTC: P0032 **2T CCM, MIL: Yes** **2002-05** **Models:** Cooper, Cooper S **Engines:** All **Transmissions:** All	**HO2S Heater (Bank 1 Sensor 1) Circuit High Input Conditions:** Engine started, battery voltage must be at least 11.5v, all electrical components must be off, the ground between the engine and the chassis must be well connected, the exhaust system must be properly sealed between the catalytic converter and the cylinder head, the coolant temperature must be 80 degrees Celsius, and the oxygen sensor heater for oxygen sensor before the catalytic converter must be properly functioning. The DME detected the HO2S signal remained in a high state. **Note: The HO2S signal circuit may be shorted to the heater power circuit due to tracking inside of the HO2S connector. Remove the connector and visually inspect the connector for signs of oil or water.** **Possible Causes:** • HO2S signal shorted to heater power circuit inside connector • HO2S signal circuit shorted to ground or to system voltage
DTC: P0036 **2T CCM, MIL: Yes** **2002-05** **Models:** Cooper, Cooper S **Engines:** All **Transmissions:** All	**HO2S Heater (Bank 1 Sensor 2) Control Circuit Malfunction Conditions:** Engine started, battery voltage must be at least 11.5v, all electrical components must be off, the ground between the engine and the chassis must be well connected, the exhaust system must be properly sealed between the catalytic converter and the cylinder head, the coolant temperature must be 80 degrees Celsius, and the oxygen sensor heater for oxygen sensor before the catalytic converter must be properly functioning. The DME detected the HO2S signal was in a negative voltage range referred to as "character shift downward". This code sets when the HO2S signal remains in a low state. **Possible Causes:** • HO2S is contaminated (due to presence of silicone in fuel) • HO2S signal and ground circuit wires crossed in wiring harness • HO2S signal circuit is shorted to sensor or chassis ground • HO2S element has failed (internal short condition)
DTC: P0037 **2T CCM, MIL: Yes** **2002-05** **Models:** Cooper, Cooper S **Engines:** All **Transmissions:** All	**HO2S Heater (Bank 1 Sensor 2) Circuit Low Input Conditions:** Engine started, battery voltage must be at least 11.5v, all electrical components must be off, the ground between the engine and the chassis must be well connected, the exhaust system must be properly sealed between the catalytic converter and the cylinder head, the coolant temperature must be 80 degrees Celsius, and the oxygen sensor heater for oxygen sensor before the catalytic converter must be properly functioning. The DME detected the HO2S signal was in a negative voltage range referred to as "character shift downward". This code sets when the HO2S signal remains in a low state. In effect, it does not switch properly in the closed loop operation. The HO2S (before the three-way catalytic converter) has a short circuit to ground that has lasted longer than 200 seconds. **Possible Causes:** • HO2S is contaminated (due to presence of silicone in fuel) • HO2S signal and ground circuit wires crossed in wiring harness • HO2S signal circuit is shorted to sensor or chassis ground • HO2S element has failed (internal short condition)
DTC: P0038 **2T CCM, MIL: Yes** **2002-05** **Models:** Cooper, Cooper S **Engines:** All **Transmissions:** All	**HO2S Heater (Bank 1 Sensor 2) Circuit High Input Conditions:** Engine started, battery voltage must be at least 11.5v, all electrical components must be off, the ground between the engine and the chassis must be well connected, the exhaust system must be properly sealed between the catalytic converter and the cylinder head, the coolant temperature must be 80 degrees Celsius, and the oxygen sensor heater for oxygen sensor before the catalytic converter must be properly functioning. The DME detected the HO2S signal remained in a high state. **Note: The HO2S signal circuit may be shorted to the heater power circuit due to tracking inside of the HO2S connector. Remove the connector and visually inspect the connector for signs of oil or water.** **Possible Causes:** • HO2S signal shorted to heater power circuit inside connector • HO2S signal circuit shorted to ground or to system voltage

DTC	Trouble Code Title, Conditions & Possible Causes
DTC: P0053 **2T CCM, MIL: Yes** **2003-05** **Models:** Cooper, Cooper S **Engines:** All **Transmissions:** All	**HO2S Heater (Bank 1 Sensor 1) Control Circuit Malfunction Conditions:** Engine started, battery voltage must be at least 10.96v, all electrical components must be off, the ground between the engine and the chassis must be well connected, the exhaust system must be properly sealed between the catalytic converter and the cylinder head, and the coolant temperature must be 80 degrees Celsius. The DME detected the HO2S signal was in a negative voltage range referred to as "character shift downward". The resistance is out of limits. The engine speed is less than 7008rpm (6208 for A/T) and the exhaust temperatures are between 350.006 and 649.995 degrees Celsius. **Possible Causes:** • HO2S is contaminated (due to presence of silicone in fuel) • HO2S signal and ground circuit wires crossed in wiring harness • HO2S signal circuit is shorted to sensor or chassis ground • HO2S element has failed (internal short condition)
DTC: P0054 **2T CCM, MIL: Yes** **2003-05** **Models:** Cooper, Cooper S **Engines:** All **Transmissions:** All	**HO2S Heater (Bank 1 Sensor 2) Circuit High Input Conditions:** Engine started, battery voltage must be at least 11.5v, all electrical components must be off, the ground between the engine and the chassis must be well connected, the exhaust system must be properly sealed between the catalytic converter and the cylinder head, the coolant temperature must be 80 degrees Celsius, and the oxygen sensor heater for oxygen sensor before the catalytic converter must be properly functioning. The DME detected the HO2S signal remained in a high state. The resistance is out of limits. The engine speed is less than 7008rpm (6208 for A/T) and the exhaust temperatures are between 350.006 and 649.995 degrees Celsius. **Note: The HO2S signal circuit may be shorted to the heater power circuit due to tracking inside of the HO2S connector. Remove the connector and visually inspect the connector for signs of oil or water.** **Possible Causes:** • HO2S signal shorted to heater power circuit inside connector • HO2S signal circuit shorted to ground or to system voltage
DTC: P0070 **2T CCM, MIL: Yes** **2005** **Models:** Cooper, Cooper S **Engines:** All **Transmissions:** All	**Ambient Air Temperature Sensor Malfunction Conditions:** Key on or engine running (at over 800rpm), the vehicle velocity is over 25mph for 26 seconds, the ambient temperature is 20 degrees above or below the model figure for four seconds. This is a thermistor-type sensor with a variable resistance that changes when exposed to different temperatures. This means: the higher the temperature, the lower the resistance value. **Possible Causes:** • IAT sensor signal circuit is grounded (check wiring & connector) • Resistance value between sockets 33 and 36 out of range • IAT sensor has an open circuit • IAT sensor is damaged or it has failed • Ambient temperature sensor at the cluster is defective
DTC: P0106 **2T MIL: Yes** **2005** **Models:** Cooper, Cooper S **Engines:** All **Transmissions:** All	**Manifold Pressure Sensor Circuit High Conditions:** Engine started, battery voltage must be at least 11v, and the differential pressure sensor detected a control deviation at the minimum limit. The closed loop control of the differential pressure in the intake manifold is suspended and replaced by a direct specification. **Possible Causes:** • Sensor's voltage supply on Terminal 87 • Sensor's ground connection faulty • Signal wire to DME faulty • Replace sensor
DTC: P0107 **2T MIL: Yes** **2002-05** **Models:** Cooper, Cooper S **Engines:** All **Transmissions:** All	**Manifold Pressure Sensor Circuit Low Conditions:** Engine started, battery voltage must be at least 11v, and the differential pressure sensor detected a control deviation at the minimum limit. The closed loop control of the differential pressure in the intake manifold is suspended and replaced by a direct specification. The MAP was too low (less than 105.0016kPa); engine stopped. **Possible Causes:** • Sensor's voltage supply on Terminal 87 • Sensor's ground connection faulty • Signal wire to DME faulty • Replace sensor
DTC: P0108 **2T MIL: Yes** **2002-04** **Models:** Cooper, Cooper S **Engines:** All **Transmissions:** All	**Manifold Pressure Sensor Circuit Short to Battery Conditions:** Engine started, battery voltage must be at least 11v, and the differential pressure sensor detected a control deviation at the minimum limit. The closed loop control of the differential pressure in the intake manifold is suspended and replaced by a direct specification. The MAP was too low (less than 105.0016kPa); engine stopped. **Possible Causes:** • Sensor's voltage supply on Terminal 87 • Sensor's ground connection faulty • Signal wire to DME faulty • Replace sensor

DTC	Trouble Code Title, Conditions & Possible Causes
DTC: P0112 **2T MIL: Yes** **2002-05** **Models:** Cooper, Cooper S **Engines:** All **Transmissions:** All	**Intake Air Temperature Sensor Circuit Low Input Conditions:** Key on or Engine running, the temperature must beat least 185-degrees (F) and all electrical equipment (A/C, lights, etc) must be off; and the DME detected the IAT sensor signal was less than the self-test minimum. This is a thermistor-type sensor with a variable resistance that changes when exposed to different temperatures. This means: the higher the temperature, the lower the resistance value. **Possible Causes:** • IAT sensor signal circuit is grounded (check wiring & connector) • Resistance value between sockets 33 and 36 out of range • IAT sensor has an open circuit • IAT sensor is damaged or it has failed
DTC: P0113 **2T MIL: Yes** **2002-05** **Models:** Cooper, Cooper S **Engines:** All **Transmissions:** All	**Intake Air Temperature Sensor Circuit High Input Conditions:** Key on or engine running, the temperature must beat least 185-degrees (F) and all electrical equipment (A/C, lights, etc) must be off; and the DME detected the IAT sensor signal was more than the self-test maximum. This is a thermistor-type sensor with a variable resistance that changes when exposed to different temperatures. This means: the higher the temperature, the lower the resistance value. **Possible Causes:** • IAT sensor signal circuit is open (inspect wiring & connector) • IAT sensor signal circuit is shorted • Resistance value between sockets 33 and 36 out of range • IAT sensor is damaged or it has failed
DTC: P0114 **2T MIL: Yes** 2002, 2005 **Models:** Cooper, Cooper S **Engines:** All **Transmissions:** All	**Intake Air Temperature Sensor Circuit Intermittent Failure Conditions:** Key on or engine running, the temperature must beat least 185-degrees (F) and all electrical equipment (A/C, lights, etc) must be off; and the DME detected the IAT sensor signal was more than the self-test maximum. This is a thermistor-type sensor with a variable resistance that changes when exposed to different temperatures. This means: the higher the temperature, the lower the resistance value. The gradient between filtered and current intake air sensor values exceeds 9.75 degrees Celsius. **Possible Causes:** • IAT sensor signal circuit is open (inspect wiring & connector) • IAT sensor signal circuit is shorted • Resistance value between sockets 33 and 36 out of range • IAT sensor is damaged or it has failed
DTC: P0117 **2T MIL: Yes** **2002-05** **Models:** Cooper, Cooper S **Engines:** All **Transmissions:** All	**ECT Sensor Circuit Low Input Conditions:** Engine started (cold) for 10 seconds, battery voltage must be 11.5, and all equipment must be off. The DME detected the ECT sensor signal was less than the self-test minimum. This is a thermistor-type sensor with a variable resistance that changes when exposed to different temperatures **Possible Causes:** • ECT sensor signal circuit is grounded in the wiring harness • ECT sensor doesn't react to changes in temperature • ECT sensor is damaged or the DME has failed
DTC: P0118 **2T MIL: Yes** **2002-05** **Models:** Cooper, Cooper S **Engines:** All **Transmissions:** All	**ECT Sensor Circuit High Input Conditions:** Engine started (cold) for 10 seconds, battery voltage must be 11.5, and all equipment must be off. The DME detected the ECT sensor signal was more than the self-test maximum. This is a thermistor-type sensor with a variable resistance that changes when exposed to different temperatures **Possible Causes:** • ECT sensor signal circuit is open (inspect wiring & connector) • ECT sensor signal circuit is shorted to ground • ECT sensor is damaged or it has failed
DTC: P0119 **2T MIL: Yes** 2002, 2005 **Models:** Cooper, Cooper S **Engines:** All **Transmissions:** All	**ECT Sensor Circuit Continuity Conditions:** Engine started (cold) for 10 seconds, battery voltage must be 11.5, and all equipment must be off. The DME detected the ECT sensor signal was out of the specified range. This is a thermistor-type sensor with a variable resistance that changes when exposed to different temperatures **Possible Causes:** • ECT sensor signal circuit is open (inspect wiring & connector) • ECT sensor signal circuit is shorted to ground • ECT sensor is damaged or it has failed
DTC: P0122 **2T MIL: Yes** **2002-05** **Models:** Cooper, Cooper S **Engines:** All **Transmissions:** All	**Throttle/Pedal Position Sensor Circuit Low Input Conditions:** Engine started, at idle, the temperature must be at least 80 degrees Celsius. The throttle position sensor supplies implausible signal to the DME. **Possible Causes:** • TP sensor signal circuit open (inspect wiring & connector) • TP sensor signal shorted to ground (inspect wiring & connector) • TP sensor is damaged or has failed • Throttle control module's voltage supply is shorted or open

DTC	Trouble Code Title, Conditions & Possible Causes
DTC: P0123 **2T MIL: Yes** **2002-05** **Models:** Cooper, Cooper S **Engines:** All **Transmissions:** All	**TP Sensor Circuit High Input Conditions:** Engine started, at idle, the temperature must be at least 80 degrees Celsius. The DME detected the TP sensor signal was more than the self-test maximum during testing. **Possible Causes:** • TP sensor not seated correctly in housing (may be damaged) • TP sensor signal is circuit shorted to ground or system voltage • TP sensor ground circuit is open (check the wiring harness) • TP sensor and/or DME has failed
DTC: P0125 **2T MIL: Yes** **2002-05** **Models:** Cooper, Cooper S **Engines:** All **Transmissions:** All	**ECT Sensor Insufficient for Closed Loop Fuel Control Conditions:** Engine started (cold), battery voltage must be 11.5, and all equipment must be off. The DME detected the ECT sensor exceeded the required calibrated value, or the engine is at idle and doesn't reach operating temperature quickly enough; the Catalyst, Fuel System, HO2S and Misfire Monitor did not complete, or the timer expired. Testing completion of procedure, the engine's temperature must rise uniformly during idle. **Possible Causes:** • Check for low coolant level or incorrect coolant mixture • DME detects a short circuit wiring in the ECT • CHT sensor is out-of-calibration or it has failed • ECT sensor is out-of-calibration or it has failed
DTC: P0128 **2T MIL: Yes** **2002-05** **Models:** Cooper, Cooper S **Engines:** All **Transmissions:** All	**Coolant Thermostat (Coolant Temperature Below Thermostat Regulating Temperature) Conditions:** The engine's warm up performance is monitored by comparing measured coolant temperature with the modeled coolant temperature to detect a defective coolant thermostat. The engine temperature must be less than 65 degrees Celsius, engine speed greater than 800rpm (with the vehicle speed greater than 10 but less than 90km/h) and the ambient temperature greater than -8 degrees Celsius. The thermostat should be wide open when cold, but is in error if it opens below desired control temperature. **Possible Causes:** • Check for low coolant level or incorrect coolant mixture • DME detects a short circuit wiring in the ECT • CHT sensor is out-of-calibration or it has failed • ECT sensor is out-of-calibration or it has failed • Replace the thermostat
DTC: P0130 **2T MIL: Yes** **2002-05** **Models:** Cooper, Cooper S **Engines:** All **Transmissions:** All	**O2 Sensor Circuit Bank 1 Sensor 1 Conditions:** Engine running, battery voltage 11.5, all electrical components off, ground between engine and chassis well connected and the exhaust system must be properly sealed between catalytic converter and the cylinder head. The DME detected the HO2S signal was implausible or not detected. The engine speed is less than 8000 rpm. **Possible Causes:** • Oxygen sensor heater for oxygen sensor (HO2S) before catalytic converter is faulty • HO2S is contaminated (due to presence of silicone in fuel) • HO2S signal and ground circuit wires crossed in wiring harness • HO2S signal circuit is shorted to sensor or chassis ground • HO2S element before the catalytic converter has failed (internal short condition) • Leaks present in the exhaust manifold or exhaust pipes
DTC: P0131 **2T MIL: Yes** **2002-05** **Models:** Cooper, Cooper S **Engines:** All **Transmissions:** All	**HO2S (Bank 1 Sensor 1) Circuit Low Input Conditions:** Engine running, battery voltage 11.5, all electrical components off, ground between engine and chassis well connected and the exhaust system must be properly sealed between catalytic converter and the cylinder head. The DME detected the HO2S signal was in a negative voltage range referred to as "character shift downward". This code sets when the HO2S signal remains in a low state for a measured period of time. In effect, it does not switch properly in the closed loop operation. Engine speed is less than 8000rpm. **Possible Causes:** • HO2S is contaminated (due to presence of silicone in fuel) • HO2S signal and ground circuit wires crossed in wiring harness • HO2S signal circuit is shorted to sensor or chassis ground • HO2S element has failed (internal short condition) • Leaks present in the exhaust manifold or exhaust pipes

DTC	Trouble Code Title, Conditions & Possible Causes
DTC: P0132 **2T MIL: Yes** **2002-05** **Models:** Cooper, Cooper S **Engines:** All **Transmissions:** All	**HO2S (Bank 1 Sensor 1) Circuit High Input Conditions:** Engine running, battery voltage 11.5, all electrical components off, ground between engine and chassis well connected and the exhaust system must be properly sealed between catalytic converter and the cylinder head. The DME detected the HO2S signal was in a high state. This code sets when the HO2S signal remains in a high state for a measured period of time. In effect, it does not switch properly in the closed loop operation. **Note: The HO2S signal circuit may be shorted to the heater power circuit due to tracking inside of the HO2S connector. Remove the connector and visually inspect the connector for signs of oil or water.** **Possible Causes:** • HO2S is contaminated (due to presence of silicone in fuel) • HO2S signal and ground circuit wires crossed in wiring harness • HO2S signal circuit is shorted to sensor or chassis ground • HO2S element has failed (internal short condition) • Leaks present in the exhaust manifold or exhaust pipes
DTC: P0133 **2T MIL: Yes** **2002-05** **Models:** Cooper, Cooper S **Engines:** All **Transmissions:** All	**HO2S (Bank 1 Sensor 1) Circuit Slow Response Conditions:** Engine running, battery voltage 11.5, all electrical components off, ground between engine and chassis well connected and the exhaust system must be properly sealed between catalytic converter and the cylinder head. The DME detected the HO2S amplitude and frequency were out of the normal range (e.g., the HO2S rich to lean switch) during the HO2S Monitor test. The engine speed is 1984 to 3488rpm (1888 to 3296 for A/T), the coolant temperature is greater than 80.25 degrees Celsius and the vehicle speed is between 24.85 and 68.35mph. The ambient pressure is greater than 75.00114kPa. **Possible Causes:** • HO2S before the three-way catalytic converter is contaminated (due to presence of silicone in fuel); Run the engine for three minutes at 3500rpm as a self-cleaning effect • HO2S signal circuit open • Leaks present in the exhaust manifold or exhaust pipes • HO2S is damaged or has failed
DTC: P0135 **2T MIL: Yes** **2002-04** **Models:** Cooper, Cooper S **Engines:** All **Transmissions:** All	**HO2S (Bank 1 Sensor 1) Heater Circuit Malfunction Conditions:** Engine running, battery voltage is between 11 and 16 volts, all electrical components off, ground between engine and chassis well connected and the exhaust system must be properly sealed between catalytic converter and the cylinder head. The DME detected an unexpected voltage condition, or it detected excessive current draw in the heater circuit during the CCM test. The engine load is 25 to 160kg/h. The exhaust gas temperature is between 450 and 700 degrees Celsius. **Possible Causes:** • HO2S heater power circuit is open or heater ground circuit open • HO2S signal tracking (due to oil or moisture in the connector) • HO2S is damaged or has failed
DTC: P0136 **2T MIL: Yes** **2002-05** **Models:** Cooper, Cooper S **Engines:** All **Transmissions:** All	**HO2S (Bank 1 Sensor 2) Circuit Malfunction Conditions:** Engine running, battery voltage 11.5, all electrical components off, ground between engine and chassis well connected and the exhaust system must be properly sealed between catalytic converter and the cylinder head. The DME detected the HO2S signal failed to meet the maximum or minimum voltage levels (i.e., it failed the voltage range check). The heater has been on for less than 90 seconds, the fuel system status is in fuel cut-off, the output voltage is between 400mV and 500mV and it is 120 seconds after engine start up. The engine speed is less than 8000rpm. **Possible Causes:** • Leaks present in the exhaust manifold or exhaust pipes • HO2S signal wire and ground wire crossed in connector • HO2S element is fuel contaminated or has failed
DTC: P0137 **2T MIL: Yes** **2002-05** **Models:** Cooper, Cooper S **Engines:** All **Transmissions:** All	**HO2S (Bank 1 Sensor 2) Circuit Low Input Conditions:** Engine running, battery voltage 11.5, all electrical components off, ground between engine and chassis well connected and the exhaust system must be properly sealed between catalytic converter and the cylinder head. The DME detected the HO2S signal remained in a high state. **Note: The HO2S signal circuit may be shorted to the heater power circuit due to "tracking inside of the HO2S connector. Remove the connector and visually inspect the connector for signs of oil or water.** **Possible Causes:** • HO2S signal shorted to heater power circuit in the connector • HO2S signal circuit shorted to ground (for more than 200 seconds) or to system voltage
DTC: P0138 **2T MIL: Yes** **2002-05** **Models:** Cooper, Cooper S **Engines:** All **Transmissions:** All	**HO2S (Bank 1 Sensor 2) Circuit High Input Conditions:** Engine running, battery voltage 11.5, all electrical components off, ground between engine and chassis well connected and the exhaust system must be properly sealed between catalytic converter and the cylinder head. The DME detected the HO2S signal remained in a high state. **Note: The HO2S signal circuit may be shorted to the heater power circuit due to "tracking inside of the HO2S connector. Remove the connector and visually inspect the connector for signs of oil or water.** **Possible Causes:** • HO2S signal shorted to heater power circuit in the positive connector • HO2S signal circuit shorted to ground or to system voltage• HO2S has failed

DTC	Trouble Code Title, Conditions & Possible Causes
DTC: P0141 **2T MIL: Yes** **2002-04** **Models:** Cooper, Cooper S **Engines:** All **Transmissions:** All	**HO2S (Bank 1 Sensor 2) Malfunction Conditions:** Engine running, battery voltage 11.5, all electrical components off, ground between engine and chassis well connected and the exhaust system must be properly sealed between catalytic converter and the cylinder head. The DME detected the HO2S signal failed to meet the maximum or minimum voltage levels (i.e., it failed the voltage range check). The engine speed is greater than 40rpm, the battery voltage must be between 10.7 and 15.5 volts, and the fault occurs 200 seconds after engine start up. **Possible Causes:** • Leaks present in the exhaust manifold or exhaust pipes • HO2S signal wire and ground wire crossed in connector • HO2S element is fuel contaminated or has failed
DTC: P0153 **2T MIL: Yes** **2003-04** **Models:** Cooper, Cooper S **Engines:** All **Transmissions:** All	**HO2S (Bank 2 Sensor 1) Circuit Slow Response Conditions:** Engine running, battery voltage 11.5, all electrical components off, ground between engine and chassis well connected and the exhaust system must be properly sealed between catalytic converter and the cylinder head. The DME detected the HO2S amplitude and frequency were out of the normal range during the HO2S Monitor test. For the 1999 M62: The idle speed variation is between 1400 and 2600rpm, the engine load variation is between 20 and 54 while the catalyst temperature should be greater than 360 degrees Celsius. **Possible Causes:** • HO2S is contaminated (due to presence of silicone in fuel) • Leaks present in the exhaust manifold or exhaust pipes • HO2S is damaged or has failed
DTC: P0154 **2T MIL: Yes** **2003-04** **Models:** Cooper, Cooper S **Engines:** All **Transmissions:** All	**HO2S (Bank 2 Sensor 1) Circuit No Activity Conditions:** Engine running, battery voltage 11.5, all electrical components off, ground between engine and chassis well connected and the exhaust system must be properly sealed between catalytic converter and the cylinder head. The DME detected the HO2S signal failed to meet the maximum or minimum voltage (i.e., it failed the voltage check). **Possible Causes:** • Leaks present in the exhaust manifold or exhaust pipes • HO2S signal wire and ground wire crossed in connector • HO2S element is fuel contaminated or has failed
DTC: P0171 **2T MIL: Yes** **2002-05** **Models:** Cooper, Cooper S **Engines:** All **Transmissions:** All	**Fuel System Too Lean (Cylinder Bank 1) Conditions:** Key on or engine running, all electrical components off and coolant temperature at least 80 degrees Celsius; and the DME detected the Bank 1 Adaptive Fuel Control System reached its rich correction limit (a lean A/F condition). The fuel status is in a closed loop pattern, the coolant temperature is greater than 7 degrees Celsius, and the engine speed is less than 1400rpm. **Possible Causes:** • Air leaks after the MAF sensor, or leaks in the PCV system • Exhaust leaks before or near where the HO2S is mounted • Fuel injector(s) restricted or not supplying enough fuel • Fuel pump not supplying enough fuel during high fuel demand conditions • Leaking EGR gasket, or leaking EGR valve diaphragm • MAF sensor dirty (causes DME to underestimate airflow) • Vehicle running out of fuel or engine oil dip stick not seated
DTC: P0172 **2T MIL: Yes** **2002-05** **Models:** Cooper, Cooper S **Engines:** All **Transmissions:** All	**Fuel System Too Rich (Cylinder Bank 1) Conditions:** Key on or engine running, all electrical components off and coolant temperature at least 80 degrees Celsius; and the DME detected the Bank 1 Adaptive Fuel Control System reached its rich correction limit (a rich A/F condition). The fuel status is in a closed loop pattern, the coolant temperature is greater than 7 degrees Celsius, and the engine speed is less than 1400rpm. **Possible Causes:** • Camshaft timing is incorrect, or the engine has an oil overfill condition • EVAP vapor recovery system failure (may be pulling vacuum) • Fuel pressure regulator is damaged or leaking • HO2S element is contaminated with alcohol or water • MAF or MAP sensor values are incorrect or out-of-range • One of more fuel injectors is leaking
DTC: P0201 **2T MIL: Yes** **2002-05** **Models:** Cooper, Cooper S **Engines:** All **Transmissions:** All	**Cylinder 1 Injector Circuit Malfunction Conditions:** Engine started, and the DME detected the fuel injector "1" control circuit was in a high state when it should have been low, or in a low state when it should have been high (wiring harness & injector okay). The battery voltage should be between 9.5 and 17 volts while the engine speed is less than 40rpm. **Possible Causes:** • Injector 1 connector is damaged, open or shorted • Injector 1 control circuit is open, shorted to ground or to power (the injector driver circuit may be damaged)

DTC	Trouble Code Title, Conditions & Possible Causes
DTC: P0202 **2T MIL: Yes** **2002-05** **Models:** Cooper, Cooper S **Engines:** All **Transmissions:** All	**Cylinder 2 Injector Circuit Malfunction Conditions:** Engine started, and the DME detected the fuel injector "2" control circuit was in a high state when it should have been low, or in a low state when it should have been high (wiring harness & injector okay). The battery voltage should be between 9.5 and 17 volts while the engine speed is less than 40rpm. **Possible Causes:** • Injector 2 connector is damaged, open or shorted • Injector 2 control circuit is open, shorted to ground or to power (the injector driver circuit may be damaged)
DTC: P0203 **2T MIL: Yes** **2002-05** **Models:** Cooper, Cooper S **Engines:** All **Transmissions:** All	**Cylinder 3 Injector Circuit Malfunction Conditions:** Engine started, and the DME detected the fuel injector "3" control circuit was in a high state when it should have been low, or in a low state when it should have been high (wiring harness & injector okay). The battery voltage should be between 9.5 and 17 volts while the engine speed is less than 40rpm. **Possible Causes:** • Injector 3 connector is damaged, open or shorted • Injector 3 control circuit is open, shorted to ground or to power (the injector driver circuit may be damaged)
DTC: P0204 **2T MIL: Yes** **2002-05** **Models:** Cooper, Cooper S **Engines:** All **Transmissions:** All	**Cylinder 4 Injector Circuit Malfunction Conditions:** Engine started, and the DME detected the fuel injector "4" control circuit was in a high state when it should have been low, or in a low state when it should have been high (wiring harness & injector okay). The battery voltage should be between 9.5 and 17 volts while the engine speed is less than 40rpm. **Possible Causes:** • Injector 4 connector is damaged, open or shorted • Injector 4 control circuit is open, shorted to ground or to power (the injector driver circuit may be damaged)
DTC: P0218 **MIL: No** **2003-04** **Models:** Cooper, Cooper S **Engines:** All **Transmissions:** All	**Engine Oil Over Temperature Conditions:** The oil temperature difference of greater than 100 degrees within one second. The ignition must be on. The DME detected an error in the Engine Oil Temperature sensor. This occurs during attempted start value calibration. **Possible Causes:** • Replace the oil temperature sensor • Engine Oil temperature is too high • Engine coolant temperature is too high • Highest possible gear engaged in transmission • Check coolant
DTC: P0222 **2T MIL: Yes** **2002-05** **Models:** Cooper, Cooper S **Engines:** All **Transmissions:** All	**Throttle Position Sensor 'B' Circuit Low Input Conditions:** Engine started, battery voltage at least 11.5v, all electrical components off, ground connections between engine and chassis well connected, coolant temperature at least 80-degrees Celsius and the throttle valve must not be damaged or dirty; and the DME detected the TP Sensor 'B' circuit was out of its normal operating range during a condition with the throttle wide open, or with it completely closed. The throttle valve activation occurs via an electric motor (throttle drive) in the throttle valve control module. It is activated by the DME according to specifications of the two sensors, Throttle Position Sensor and Accelerator Pedal Position Sensor 2. Slowly depress accelerator pedal up to Wide Open Throttle (WOT) stop while observing the percentage display on the PID data function of the scan tool. The percentage display must increase uniformly. **Possible Causes:** • ETC TP Sensor 'B' connector is damaged or shorted • ETC TP Sensor 'B' signal circuit is shorted to ground • ETC TP Sensor 'B' is damaged or it has failed
DTC: P0223 **2T MIL: Yes** **2002-05** **Models:** Cooper, Cooper S **Engines:** All **Transmissions:** All	**Throttle Position Sensor 'B' Circuit High Input Conditions:** Engine started, battery voltage at least 11.5v, all electrical components off, ground connections between engine and chassis well connected, coolant temperature at least 80-degrees Celsius and the throttle valve must not be damaged or dirty; and the DME detected the TP Sensor 'B' circuit was out of its normal operating range during a condition with the throttle wide open, or with it completely closed. The throttle valve activation occurs via an electric motor (throttle drive) in the throttle valve control module. It is activated by the DME according to specifications of the two sensors, Throttle Position Sensor and Accelerator Pedal Position Sensor 2. Slowly depress accelerator pedal up to Wide Open Throttle (WOT) stop while observing the percentage display on the PID data function of the scan tool. The percentage display must increase uniformly. **Possible Causes:** • ETC TP Sensor 'B' connector is damaged or open • ETC TP Sensor 'B' signal circuit is open • ETC TP Sensor 'B' signal circuit is shorted to VREF (5v) • ETC TP Sensor 'B' is damaged or it has failed

DTC	Trouble Code Title, Conditions & Possible Causes
DTC: P0261 **2T MIL: Yes** **2002-05** **Models:** Cooper, Cooper S **Engines:** All **Transmissions:** All	**Cylinder 1 Injector Circuit Low Input/Short to Ground Conditions:** Key on or engine running, fuses in the instrument panel and the E-box in the engine compartment must be functioning, and the ground connections between the engine ad the chassis must be well connected; and the DME detected an unexpected voltage condition on the injector circuit. **Possible Causes:** • Injector 1 control circuit is open • Injector 1 power circuit (B+) is open • Injector 1 control circuit is shorted to chassis ground • Injector 1 is damaged or has failed • DME is not connected or has failed
DTC: P0262 **2T MIL: Yes** **2002-05** **Models:** Cooper, Cooper S **Engines:** All **Transmissions:** All	**Cylinder 1 Injector Circuit Low Input/Short to B+ Conditions:** Key on or engine running, fuses in the instrument panel and the E-box in the engine compartment must be functioning, and the ground connections between the engine ad the chassis must be well connected; and the DME detected an unexpected voltage condition on the injector circuit. **Possible Causes:** • Injector control circuit is open • Injector power circuit (B+) is open • Injector control circuit is shorted to chassis ground • Injector is damaged or has failed • DME is not connected or has failed • Fuel pump relay has failed • Fuel injectors may have malfunctioned • Faulty engine speed sensor
DTC: P0264 **2T MIL: Yes** **2002-05** **Models:** Cooper, Cooper S **Engines:** All **Transmissions:** All	**Cylinder 2 Injector Circuit Low Input/Short to Ground Conditions:** Key on or engine running, fuses in the instrument panel and the E-box in the engine compartment must be functioning, and the ground connections between the engine ad the chassis must be well connected; and the DME detected an unexpected voltage condition on the injector circuit. **Possible Causes:** • Injector control circuit is open • Injector power circuit (B+) is open • Injector control circuit is shorted to chassis ground • Injector is damaged or has failed • DME is not connected or has failed • Fuel pump relay has failed • Fuel injectors may have malfunctioned • Faulty engine speed sensor
DTC: P0265 **2T MIL: Yes** **2002-05** **Models:** Cooper, Cooper S **Engines:** All **Transmissions:** All	**Cylinder 2 Injector Circuit Low Input/Short to B+ Conditions:** Key on or engine running, fuses in the instrument panel and the E-box in the engine compartment must be functioning, and the ground connections between the engine ad the chassis must be well connected; and the DME detected an unexpected voltage condition on the injector circuit. **Possible Causes:** • Injector control circuit is open • Injector power circuit (B+) is open • Injector control circuit is shorted to chassis ground • Injector is damaged or has failed • DME is not connected or has failed • Fuel pump relay has failed • Fuel injectors may have malfunctioned • Faulty engine speed sensor
DTC: P0267 **2T MIL: Yes** **2002-05** **Models:** Cooper, Cooper S **Engines:** All **Transmissions:** All	**Cylinder 3 Injector Circuit Low Input/Short to Ground Conditions:** Key on or engine running, fuses in the instrument panel and the E-box in the engine compartment must be functioning, and the ground connections between the engine ad the chassis must be well connected; and the DME detected an unexpected voltage condition on the injector circuit. **Possible Causes:** • Injector control circuit is open • Injector power circuit (B+) is open • Injector control circuit is shorted to chassis ground • Injector is damaged or has failed • DME is not connected or has failed • Fuel pump relay has failed • Fuel injectors may have malfunctioned • Faulty engine speed sensor

DTC	Trouble Code Title, Conditions & Possible Causes
DTC: P0268 **2T MIL: Yes** **2002-05** **Models:** Cooper, Cooper S **Engines:** All **Transmissions:** All	**Cylinder 3 Injector Circuit Low Input/Short to B+ Conditions:** Key on or engine running, fuses in the instrument panel and the E-box in the engine compartment must be functioning, and the ground connections between the engine ad the chassis must be well connected; and the DME detected an unexpected voltage condition on the injector circuit. **Possible Causes:** • Injector control circuit is open • Injector power circuit (B+) is open • Injector control circuit is shorted to chassis ground • Injector is damaged or has failed • DME is not connected or has failed • Fuel pump relay has failed • Fuel injectors may have malfunctioned • Faulty engine speed sensor
DTC: P0270 **2T MIL: Yes** **2002-05** **Models:** Cooper, Cooper S **Engines:** All **Transmissions:** All	**Cylinder 4 Injector Circuit Low Input/Short to Ground Conditions:** Key on or engine running, fuses in the instrument panel and the E-box in the engine compartment must be functioning, and the ground connections between the engine ad the chassis must be well connected; and the DME detected an unexpected voltage condition on the injector circuit. **Possible Causes:** • Injector control circuit is open • Injector power circuit (B+) is open • Injector control circuit is shorted to chassis ground • Injector is damaged or has failed • DME is not connected or has failed • Fuel pump relay has failed • Fuel injectors may have malfunctioned • Faulty engine speed sensor
DTC: P0271 **2T MIL: Yes** **2002-05** **Models:** Cooper, Cooper S **Engines:** All **Transmissions:** All	**Cylinder 4 Injector Circuit Low Input/Short to B+ Conditions:** Key on or engine running, fuses in the instrument panel and the E-box in the engine compartment must be functioning, and the ground connections between the engine ad the chassis must be well connected; and the DME detected an unexpected voltage condition on the injector circuit. **Possible Causes:** • Injector control circuit is open • Injector power circuit (B+) is open • Injector control circuit is shorted to chassis ground • Injector is damaged or has failed • DME is not connected or has failed • Fuel pump relay has failed • Fuel injectors may have malfunctioned • Faulty engine speed sensor
DTC: P0300 **2T MISFIRE** **MIL: Yes** **2002-05** **Models:** Cooper, Cooper S **Engines:** All **Transmissions:** All	**Random/Multiple Misfire Detected Conditions:** Engine running at an RPM greater than 600 but less than 7000 the DME detected a misfire or uneven engine running in two or more cylinders within 1000 engine revolutions. The sum of misfires caused an increase in emissions for the first 1000 revolutions after start up, or the sum of misfires caused catalyst damage after the first 200 engine revolutions. Time after start less than one second. **Note: If the misfire is severe, the MIL will flash on/off on the first trip!** **Possible Causes:** • Fuel metering fault that affects two or more cylinders • Fuel pressure too low or too high, fuel supply contaminated • EVAP system problem or the EVAP canister is fuel saturated • EGR valve is stuck open or the PCV system has a vacuum leak • Ignition system fault (coil, plugs) affecting two or more cylinders • MAF sensor contamination (it can cause a very lean condition) • Vehicle driven while very low on fuel (less than 1/8 of a tank)

DTC	Trouble Code Title, Conditions & Possible Causes
DTC: P0301 **2T MISFIRE** **MIL: Yes** **2002-05** **Models:** Cooper, Cooper S **Engines:** All **Transmissions:** All	**Cylinder Number 1 Misfire Detected Conditions:** Engine running at an RPM greater than 600 but less than 7000 the DME detected a misfire or uneven engine running in two or more cylinders within 1000 engine revolutions. The sum of misfires caused an increase in emissions for the first 1000 revolutions after start up, or the sum of misfires caused catalyst damage after the first 200 engine revolutions. Time after start less than one second. **Note: If the misfire is severe, the MIL will flash on/off on the first trip!** **Possible Causes:** • Air leak in the intake manifold, or in the EGR or DME system • Base engine mechanical problem • Fuel delivery component problem (i.e., a contaminated, dirty or sticking fuel injector) • Fuel pump relay defective • Ignition coil fuses have failed • Ignition system problem (dirty damaged coil or plug) • Engine speed (RPM) sensor has failed • Camshaft position sensors have failed • Ignition coil is faulty • Spark plugs are not working properly or are not gapped properly
DTC: P0302 **2T MISFIRE** **MIL: Yes** **2002-05** **Models:** Cooper, Cooper S **Engines:** All **Transmissions:** All	**Cylinder Number 2 Misfire Detected Conditions:** Engine running at an RPM greater than 600 but less than 7000 the DME detected a misfire or uneven engine running in two or more cylinders within 1000 engine revolutions. The sum of misfires caused an increase in emissions for the first 1000 revolutions after start up, or the sum of misfires caused catalyst damage after the first 200 engine revolutions. Time after start less than one second. **Note: If the misfire is severe, the MIL will flash on/off on the 1st trip!** **Possible Causes:** • Air leak in the intake manifold, or in the EGR or DME system • Base engine mechanical problem • Fuel delivery component problem (i.e., a contaminated, dirty or sticking fuel injector) • Fuel pump relay defective • Ignition coil fuses have failed • Ignition system problem (dirty damaged coil or plug) • Engine speed (RPM) sensor has failed • Camshaft position sensors have failed • Ignition coil is faulty • Spark plugs are not working properly or are not gapped properly
DTC: P0303 **2T MISFIRE** **MIL: Yes** **2002-05** **Models:** Cooper, Cooper S **Engines:** All **Transmissions:** All	**Cylinder Number 3 Misfire Detected Conditions:** Engine running at an RPM greater than 600 but less than 7000 the DME detected a misfire or uneven engine running in two or more cylinders within 1000 engine revolutions. The sum of misfires caused an increase in emissions for the first 1000 revolutions after start up, or the sum of misfires caused catalyst damage after the first 200 engine revolutions. Time after start less than one second. **Note: If the misfire is severe, the MIL will flash on/off on the 1st trip!** **Possible Causes:** • Air leak in the intake manifold, or in the EGR or DME system • Base engine mechanical problem • Fuel delivery component problem (i.e., a contaminated, dirty or sticking fuel injector) • Fuel pump relay defective • Ignition coil fuses have failed • Ignition system problem (dirty damaged coil or plug) • Engine speed (RPM) sensor has failed • Camshaft position sensors have failed • Ignition coil is faulty • Spark plugs are not working properly or are not gapped properly

DTC	Trouble Code Title, Conditions & Possible Causes
DTC: P0304 **2T MISFIRE** **MIL: Yes** **2002-05** **Models:** Cooper, Cooper S **Engines:** All **Transmissions:** All	**Cylinder Number 4 Misfire Detected Conditions:** Engine running at an RPM greater than 600 but less than 7000 the DME detected a misfire or uneven engine running in two or more cylinders within 1000 engine revolutions. The sum of misfires caused an increase in emissions for the first 1000 revolutions after start up, or the sum of misfires caused catalyst damage after the first 200 engine revolutions. Time after start less than one second. **Note: If the misfire is severe, the MIL will flash on/off on the 1st trip!** **Possible Causes:** • Air leak in the intake manifold, or in the EGR or DME system • Base engine mechanical problem • Fuel delivery component problem (i.e., a contaminated, dirty or sticking fuel injector) • Fuel pump relay defective • Ignition coil fuses have failed • Ignition system problem (dirty damaged coil or plug) • Engine speed (RPM) sensor has failed • Camshaft position sensors have failed • Ignition coil is faulty • Spark plugs are not working properly or are not gapped properly
DTC: P0313 **MISFIRE** **2T MIL: Yes** **2002-05** **Models:** Cooper, Cooper S **Engines:** All **Transmissions:** All	**Misfire Detected with Low Fuel Conditions:** Engine running under positive torque conditions, and the DME detected a misfire or uneven engine function as well as an indication of low fuel level when another misfire was detected. **Note: If the misfire is severe, the MIL will flash on/off on the 1st trip!** **Possible Causes:** • Air leak in the intake manifold, or in the EGR or DME system • Base engine mechanical problem • Fuel delivery component problem (i.e., a contaminated, dirty or sticking fuel injector) • Fuel pump relay defective • Ignition coil fuses have failed • Ignition system problem (dirty damaged coil or plug) • Engine speed (RPM) sensor has failed • Camshaft position sensors have failed • Ignition coil is faulty • Spark plugs are not working properly or are not gapped properly
DTC: P0324 **2T MIL: Yes** **2002-05** **Models:** Cooper, Cooper S **Engines:** All **Transmissions:** All	**Knock Control System Error Conditions:** Engine started, vehicle driven, and the DME detected the Knock Sensor 1 (KS1) signal was too low or not recognized by the DME. **Possible Causes:** • Knock sensor circuit is open • Knock sensor is loose (tighten to 20 NM) • Contact between the knock sensor and cylinder block is dirty, corroded or greasy • Knock sensor circuit is shorted to ground, or shorted to power • Knock sensor is damaged or it has failed • Wrong kind of fuel used • A component in the engine compartment is loose or not properly secured
DTC: P0326 **2T MIL: Yes** **2002-05** **Models:** Cooper, Cooper S **Engines:** All **Transmissions:** All	**Knock Sensor Circuit Malfunction Conditions:** Engine started, vehicle driven at 1520rpm for 3 seconds or to a temperature of 40 degrees Celsius, and the DME detected the Knock Sensor 1 (KS1) signal was not recognized. The engine speed is greater than 2016rpm and the coolant temperature is greater than 50.25 degrees Celsius. The difference between raw and filtered knock sensor signal is less than 0.0499 to 0.0698 volts. **Possible Causes:** • Knock sensor circuit is open • Knock sensor is loose (tighten to 20 NM) • Contact between the knock sensor and cylinder block is dirty, corroded or greasy • Knock sensor circuit is shorted to ground, or shorted to power • Knock sensor is damaged or it has failed • Wrong kind of fuel used • A component in the engine compartment is loose or not properly secured

DTC	Trouble Code Title, Conditions & Possible Causes
DTC: P0335 **2T MIL: Yes** **2002-05** **Models:** Cooper, Cooper S **Engines:** All **Transmissions:** All	**Camshaft Position Sensor "A" Circ Malfunction Conditions:** Engine started, battery voltage must be at least 11.5v, all electrical components must be off, parking brake must be engaged (to keep daytime driving lights off), automatic transmission selector must be in park and the ground between the engine and the chassis must be well connected. The DME detected the CMP sensor signal was implausible or missing. Engine speed is greater than 500rpm, and the fault is tolerable as long as there are no misfired occurring at the same time. **Possible Causes:** • CMP sensor circuit is open or shorted to ground • CMP sensor circuit is shorted to power • CMP sensor ground (return) circuit is open • CMP sensor installation incorrect (Hall-effect type) • CMP sensor is damaged or CMP sensor shielding damaged
DTC: P0336 **2T MIL: Yes** **2002-05** **Models:** Cooper, Cooper S **Engines:** All **Transmissions:** All	**Camshaft Position Sensor "A" Circ Range/Performance Conditions:** Engine started (and engine speed is less than 25rpm), battery voltage must be at least 11.5v, all electrical components must be off, parking brake must be engaged (to keep daytime driving lights off), automatic transmission selector must be in park and the ground between the engine and the chassis must be well connected. The DME detected the CMP sensor signal was implausible. **Possible Causes:** • CMP sensor circuit is open or shorted to ground • CMP sensor circuit is shorted to power • CMP sensor ground (return) circuit is open • CMP sensor installation incorrect (Hall-effect type) • CMP sensor is damaged or CMP sensor shielding damaged
DTC: P0340 **2T MIL: Yes** **2002-05** **Models:** Cooper, Cooper S **Engines:** All **Transmissions:** All	**Camshaft Position Sensor Circuit Malfunction Conditions:** Engine started, battery voltage must be at least 11.5v, all electrical components must be off, parking brake must be engaged (to keep daytime driving lights off), automatic transmission selector must be in park and the ground between the engine and the chassis must be well connected. The DME detected the CMP sensor signal was missing or it was erratic. There is no signal or an invalid one, and the engine speed is greater than 200rpm for two cycles. **Possible Causes:** • CMP sensor circuit is open or shorted to ground • CMP sensor circuit is shorted to power • CMP sensor ground (return) circuit is open • CMP sensor installation incorrect (Hall-effect type) • CMP sensor is damaged or CMP sensor shielding damaged • CMP sensor has failed
DTC: P0341 **2T MIL: Yes** **2002-05** **Models:** Cooper, Cooper S **Engines:** All **Transmissions:** All	**Camshaft Position Sensor Circ Range/Performance Conditions:** Engine started, battery voltage must be at least 11.5v, all electrical components must be off, parking brake must be engaged (to keep daytime driving lights off), automatic transmission selector must be in park and the ground between the engine and the chassis must be well connected. The DME detected the CMP sensor signal was implausible. **Possible Causes:** • CMP sensor circuit is open or shorted to ground • CMP sensor circuit is shorted to power • CMP sensor ground (return) circuit is open • CMP sensor installation incorrect (Hall-effect type) • CMP sensor is damaged or CMP sensor shielding damaged
DTC: P0351 **2T MIL: Yes** **2002-04** **Models:** Cooper, Cooper S **Engines:** All **Transmissions:** All	**Ignition Coilpack A Primary/Secondary Circuit Malfunction Conditions:** Engine started, battery voltage must be at least 11.5v, all electrical components must be off, parking brake must be engaged (to keep daytime driving lights off), automatic transmission selector must be in park and the ground between the engine and the chassis must be well connected. The DME did not receive any valid pulses from the ignition module for the Ignition Coilpack A primary circuit. **Note: Ignition coils and power output stages are one component and cannot be replaced individually.** **Possible Causes:** • Engine speed (RPM) sensor has failed • Camshaft Position (CMP) sensor has failed • Power Supply Relay is shorted to an open circuit • There is a malfunction in voltage supply • Ignition coilpack is damaged or it has failed • Cylinder 1 to 4 Fuel Injector(s) have failed•

DTC	Trouble Code Title, Conditions & Possible Causes
DTC: P0352 **2T MIL: Yes** **2002-04** **Models:** Cooper, Cooper S **Engines:** All **Transmissions:** All	**Ignition Coilpack A Primary/Secondary Circuit Malfunction Conditions:** Engine started, battery voltage must be at least 11.5v, all electrical components must be off, parking brake must be engaged (to keep daytime driving lights off), automatic transmission selector must be in park and the ground between the engine and the chassis must be well connected. The DME did not receive any valid pulses from the ignition module for the Ignition Coilpack A primary circuit. **Note: Ignition coils and power output stages are one component and cannot be replaced individually.** **Possible Causes:** • Engine speed (RPM) sensor has failed • Camshaft Position (CMP) sensor has failed • Power Supply Relay is shorted to an open circuit • There is a malfunction in voltage supply • Ignition coilpack is damaged or it has failed • Cylinder 1 to 4 Fuel Injector(s) have failed
DTC: P0420 **MIL: Yes** **2002-05** **Models:** Cooper, Cooper S **Engines:** All **Transmissions:** All	**Catalyst System Efficiency (Bank 1) Below Threshold Conditions:** Engine started for longer than one second, battery voltage must be at least 11.5v, all electrical components must be off, parking brake must be engaged (to keep daytime driving lights off), automatic transmission selector must be in park, the exhaust system must be properly sealed between the catalytic converter and the cylinder head, coolant temperature must be at least 80 degrees Celsius and oxygen sensor heaters for oxygen sensors before the catalytic converter must be functioning properly and the ground between the engine and the chassis must be well connected. The DME detected the switch rate of the rear HO2S-12 was close to the switch rate of front HO2S (it should be much slower). The coolant temperature is greater than 80.25 degrees Celsius. The fuel system is in closed loop. The vehicle speed is between 28 and 80.8mph. The engine speed is between 1984 and 3648rpm. Exhaust gas temperature is between 450 and 700 degrees Celsius. Ambient pressure is 75.001kPa. **Possible Causes:** • Air leaks at the exhaust manifold or in the exhaust pipes • Catalytic converter is damaged, contaminated or it has failed • ECT/CHT sensor has lost its calibration (the signal is incorrect) • Engine cylinders misfiring, or the ignition timing is over retarded • Engine oil is contaminated • Front HO2S or rear HO2S is contaminated with fuel or moisture • Front HO2S and/or the rear HO2S is loose in the mounting hole • Front HO2S much older than the rear HO2S (HO2S-11 is lazy) • Fuel system pressure is too high (check the pressure regulator) • Rear HO2S wires improperly connected or the HO2S has failed
DTC: P0441 **2T MIL: Yes** **2002-05** **Models:** Cooper, Cooper S **Engines:** All **Transmissions:** All	**EVAP Control System Incorrect Purge Flow Conditions:** ECT sensor is cold during startup, engine started, battery voltage must be at least 11.5v, all electrical components must be off. The coolant temperature is less than 60 degrees Celsius, and the ambient pressure is greater than 76.2994kPa. The air intake temperature at start is between 9.04 and 16.04 degrees Celsius. The change in barometric pressure since engine start is less than 0.9998kPa. The vehicle speed is less than 74.56mph, and the purge valve has opened enough on previous driving cycle. The DME detected the switch rate of the rear HO2S-12 was close to the switch rate of front HO2S (it should be much slower). DME detected a problem in the EVAP system during the EVAP System Monitor test. **Possible Causes:** • EVAP canister purge valve is damaged • EVAP canister has an improper seal • Vapor line between purge solenoid and intake manifold vacuum reservoir is damaged, or vapor line between EVAP canister purge solenoid and charcoal canister is damaged • Vapor line between charcoal canister and check valve, or vapor line between check valve and fuel vapor valves is damaged
DTC: P0442 **2T MIL: Yes** **2002-05** **Models:** Cooper, Cooper S **Engines:** All **Transmissions:** All	**EVAP Control System Small Leak Detected Conditions:** Engine started, battery voltage must be at least 11.5v, all electrical components must be off. The DME detected a leak in the EVAP system as small as 0.040 inches during the EVAP Monitor Test. The coolant temperature is less than 60 degrees Celsius, and the ambient pressure is greater than 76.2994kPa. The air intake temperature at start is between 9.04 and 16.04 degrees Celsius. The change in barometric pressure since engine start is less than 0.9998kPa. The vehicle speed is less than 74.56mph, and the purge valve has opened enough on previous driving cycle. **Possible Causes:** • Aftermarket EVAP parts that do not conform to specifications • CV solenoid remains partially open when commanded to close • EVAP component seals leaking (i.e., leaks in the Purge valve, fuel tank pressure sensor, canister vent solenoid, fuel vapor control valve tube assembly or fuel vapor vent valve). • Fuel filler cap damaged, cross-threaded or loosely installed • Loose fuel vapor hose/tube connections to EVAP components • Small holes or cuts in fuel vapor hoses or EVAP canister tubes

DTC	Trouble Code Title, Conditions & Possible Causes
DTC: P0443 **2T MIL: Yes** **2002-05** **Models:** Cooper, Cooper S **Engines:** All **Transmissions:** All	**EVAP Vapor Management Valve Circuit Malfunction Conditions:** Engine started, battery voltage must be at least 11.5v, all electrical components must be off, parking brake must be engaged (to keep daytime driving lights off), automatic transmission selector must be in park, the exhaust system must be properly sealed between the catalytic converter and the cylinder head, coolant temperature must be at least 80 degrees Celsius and oxygen sensor heaters for oxygen sensors before the catalytic converter must be functioning properly and the ground between the engine and the chassis must be well connected. The DME detected an unexpected high or low voltage condition on the Vapor Management Valve (VMV) circuit when the device was cycled On/Off during testing. **Possible Causes:** • EVAP power supply circuit is open • EVAP solenoid control circuit is open or shorted to ground • EVAP solenoid control circuit is shorted to power (B+) • EVAP solenoid valve is damaged or it has failed
DTC: P0444 **2T MIL: Yes** **2002-05** **Models:** Cooper, Cooper S **Engines:** All **Transmissions:** All	**Evaporative Emission System Purge Control Valve Circuit Open Conditions:** Engine started, battery voltage must be at least 11.5v, all electrical components must be off, parking brake must be engaged (to keep daytime driving lights off), automatic transmission selector must be in park, the exhaust system must be properly sealed between the catalytic converter and the cylinder head, coolant temperature must be at least 80 degrees Celsius and oxygen sensor heaters for oxygen sensors before the catalytic converter must be functioning properly and the ground between the engine and the chassis must be well connected. The DME detected an unexpected voltage condition on the EVAP circuit when the device was cycled On/Off during testing. **Possible Causes:** • EVAP power supply circuit is open • EVAP solenoid control circuit is open or shorted to ground • EVAP solenoid control circuit is shorted to power (B+) • EVAP solenoid valve is damaged or it has failed • EVAP canister has a leak or a poor seal
DTC: P0445 **2T MIL: Yes** **2002-05** **Models:** Cooper, Cooper S **Engines:** All **Transmissions:** All	**Evaporative Emission System Purge Control Valve Circuit Shorted Conditions:** Engine started, battery voltage must be at least 11.5v, all electrical components must be off, parking brake must be engaged (to keep daytime driving lights off), automatic transmission selector must be in park, the exhaust system must be properly sealed between the catalytic converter and the cylinder head, coolant temperature must be at least 80 degrees Celsius and oxygen sensor heaters for oxygen sensors before the catalytic converter must be functioning properly and the ground between the engine and the chassis must be well connected. The DME detected an unexpected voltage condition on the EVAP circuit when the device was cycled On/Off during testing. **Possible Causes:** • EVAP power supply circuit is open • EVAP solenoid control circuit is open or shorted to ground • EVAP solenoid control circuit is shorted to power (B+) • EVAP solenoid valve is damaged or it has failed • EVAP canister has a leak or a poor seal
DTC: P0455 **2T MIL: Yes** **2002-05** **Models:** Cooper, Cooper S **Engines:** All **Transmissions:** All	**EVAP Control System Large Leak Detected Conditions:** Engine started, battery voltage must be at least 11.5v, all electrical components must be off. The coolant temperature is less than 60 degrees Celsius, and the ambient pressure is greater than 76.2994kPa. The air intake temperature at start is between 9.04 and 16.04 degrees Celsius. The change in barometric pressure since engine start is less than 0.9998kPa. The vehicle speed is less than 74.56mph, and the purge valve has opened enough on previous driving cycle. The DME detected multiple small fuel vapor leaks; or it detected a large leak in the system during the leak test. **Possible Causes:** • Aftermarket EVAP hardware non-conforming to specifications • EVAP canister tube, EVAP canister purge outlet tube or EVAP return tube disconnected or cracked, or canister is damaged • EVAP canister purge valve stuck closed, or canister damaged • Fuel filler cap missing, loose (not tightened) or the wrong part • Loose fuel vapor hose/tube connections to EVAP components • Canister vent (CV) solenoid stuck open • Fuel tank pressure (FTP) sensor has failed mechanically

DTC	Trouble Code Title, Conditions & Possible Causes
DTC: P0456 **2T MIL: Yes** **2002-05** **Models:** Cooper, Cooper S **Engines:** All **Transmissions:** All	**EVAP Control System Small Leak Detected Conditions:** Engine started, battery voltage must be at least 11.5v, all electrical components must be off. The coolant temperature is less than 60 degrees Celsius, and the ambient pressure is greater than 76.2994kPa. The air intake temperature at start is between 9.04 and 16.04 degrees Celsius. The change in barometric pressure since engine start is less than 0.9998kPa. The vehicle speed is less than 74.56mph, and the purge valve has opened enough on previous driving cycle. The DME detected multiple small fuel vapor leaks; or it detected a large leak in the system during the leak test. **Possible Causes:** • Aftermarket EVAP hardware non-conforming to specifications • EVAP canister tube, EVAP canister purge outlet tube or EVAP return tube disconnected or cracked, or canister is damaged • EVAP canister purge valve stuck closed, or canister damaged • Fuel filler cap missing, loose (not tightened) or the wrong part • Loose fuel vapor hose/tube connections to EVAP components • Canister vent (CV) solenoid stuck open • Fuel tank pressure (FTP) sensor has failed mechanically
DTC: P0500 **T2 MIL: Yes** **2002-05** **Models:** Cooper, Cooper S **Engines:** All **Transmissions:** All	**Vehicle Speed Sensor "A" Malfunction Conditions:** Engine started; engine speed above the TCC stall speed, and the DME detected a loss of the VSS signal over a period of time or the signal is not usable. **Note: The DME receives vehicle speed data from the VSS, TCSS, ABS module, CTM or GEM controller, depending up the application. Speed Signal from DSC too high because of possible tampering. Check DSC and wires.** **Possible Causes:** • VSS signal circuit is open or shorted to ground • VSS harness circuit is shorted to ground • VSS harness circuit is shorted to power • VSS circuit open between the DME and related control module • VSS or wheel speed sensors circuits are damaged • Modules connected to VSC/VSS harness circuits are damaged • Mechanical drive mechanism for the VSS is damaged
DTC: P0506 **2T MIL: Yes** **2002-05** **Models:** Cooper, Cooper S **Engines:** All **Transmissions:** All	**Idle Air Control System RPM Lower Than Expected Conditions:** Engine started, battery voltage must be at least 10.96v, all electrical components must be off, parking brake must be engaged (to keep daytime driving lights off), automatic transmission selector must be in park, the exhaust system must be properly sealed between the catalytic converter and the cylinder head, coolant temperature must be between 80.25 and 110.25 degrees Celsius and oxygen sensor heaters for oxygen sensors before the catalytic converter must be functioning properly and the ground between the engine and the chassis must be well connected. The DME detected it could not control the idle speed correctly, as it is constantly more than 100 rpm less than specification. **Possible Causes:** • Air inlet is plugged or the air filter element is severely clogged • IAC circuit is open or shorted • IAC circuit VPWR circuit is open • IAC solenoid is damaged or has failed • The VSS has failed
DTC: P0507 **2T MIL: Yes** **2002-05** **Models:** Cooper, Cooper S **Engines:** All **Transmissions:** All	**Idle Air Control System RPM Higher Than Expected Conditions:** Engine started, battery voltage must be at least 10.96v, all electrical components must be off, parking brake must be engaged (to keep daytime driving lights off), automatic transmission selector must be in park, the exhaust system must be properly sealed between the catalytic converter and the cylinder head, coolant temperature must be between 80.25 and 110.25 degrees Celsius and oxygen sensor heaters for oxygen sensors before the catalytic converter must be functioning properly and the ground between the engine and the chassis must be well connected. The DME detected it could not control the idle speed correctly, as it is constantly more than 200 rpm more than specification. **Possible Causes:** • Air intake leak located somewhere after the throttle body • IAC control circuit is shorted to chassis ground • IAC solenoid is damaged or has failed • Throttle Valve Control module has failed or is clogged with carbon • The VSS has failed
DTC: P0532 **MIL: No** **2002** **Models:** Cooper, Cooper S **Engines:** All **Transmissions:** All	**A/C Refrigerant Pressure Sensor "A" Circuit Low Conditions:** The DME detected a low condition on the sensor. **Possible Causes:** • The A/C Refrigerant Pressure Sensor "A" has failed. • The DME has failed

DTC	Trouble Code Title, Conditions & Possible Causes
DTC: P0533 **MIL: No** **2002** **Models:** Cooper, Cooper S **Engines:** All **Transmissions:** All	**A/C Refrigerant Pressure Sensor "A" Circuit High Conditions:** The DME detected a high condition on the sensor. **Possible Causes:** • The A/C Refrigerant Pressure Sensor "A" has failed. • The DME has failed
DTC: P0562 **MIL: No** **2002** **Models:** Cooper, Cooper S **Engines:** All **Transmissions:** All	**System Voltage Low Conditions:** Engine started, battery voltage must be at least 11.5v, all electrical components must be off, parking brake must be engaged (to keep daytime driving lights off), automatic transmission selector must be in park, and the ground between the engine and the chassis must be well connected. The DME has detected a voltage value that is below the specified minimum limit for the system to function properly. **Possible Causes:** • Alternator damaged or faulty • Battery voltage low or insufficient • Fuses blown or circuits open • Battery connection to terminal not clean • Voltage regulator has failed
DTC: P0563 **MIL: No** **2002** **Models:** Cooper, Cooper S **Engines:** All **Transmissions:** All	**System Voltage High Conditions:** Engine started for 18 seconds, battery voltage must be at least 11.5v, all electrical components must be off, parking brake must be engaged (to keep daytime driving lights off), automatic transmission selector must be in park, and the ground between the engine and the chassis must be well connected. The DME has detected a voltage value that has exceeded the specified maximum limit for the system to function properly. The vehicle was connected to 24 volts for too long after a jump start. ADC in ECU is defective. Delete stored fault codes from log. If fault reoccurs replace the ECU. **Possible Causes:** • Alternator damaged or faulty • Battery voltage low or insufficient • Fuses blown or circuits open • Battery connection to terminal not clean • Voltage regulator has failed
DTC: P0571 **MIL: No** **2002-04** **Models:** Cooper, Cooper S **Engines:** All **Transmissions:** All	**Cruise/Brake Switch (A) Circuit Malfunction Conditions:** Engine started, battery voltage must be at least 11.5v, all electrical components must be off, parking brake must be engaged (to keep daytime driving lights off), automatic transmission selector must be in park, and the ground between the engine and the chassis must be well connected. The DME has detected a voltage value that is implausible or erratic. **Possible Causes:** • Brake light switch is faulty • Control circuit is shorted to chassis ground
DTC: P0600 **2T MIL: Yes** **2005** **Models:** Cooper, Cooper S **Engines:** All **Transmissions:** All	**Serial Communication Link (Data BUS) Message Missing Conditions:** The Engine Control Module (DME) communicates with all databus-capable control modules via a CAN databus. These databus-capable control modules are connected via two data bus wires which are twisted together (CAN_High and CAN_Low), and exchange information (messages). Missing information on the databus is recognized as a malfunction and stored. Trouble-free operation of the CAN-Bus requires that it have a terminal resistance. This central terminal resistor is located in the Engine Control Module (DME). **Possible Causes:** • CAN data bus wires have short circuited to each other
DTC: P0601 **2T MIL: Yes** **2002-05** **Models:** Cooper, Cooper S **Engines:** All **Transmissions:** All	**Internal Control Module Memory Check Sum Error Conditions:** Key on, the DME has detected a programming error. The RAM and ROM check displays an invalid check-sum at power up/down. **Possible Causes:** • Battery terminal corrosion, or loose battery connection • Connection to the DME interrupted, or the circuit has been opened • Reprogramming error has occurred and needs replacement. Remember to check for Aftermarket Performance Products before replacing a DME.
DTC: P0603 **2T MIL: Yes** **2002-05** **Models:** Cooper, Cooper S **Engines:** All **Transmissions:** All	**DME Keep Alive Memory Test Error Conditions:** Key on, and the DME detected an internal memory fault. This code will set if KAPWR to the DME is interrupted (at the initial key on). Watchdog on. **Possible Causes:** • Battery terminal corrosion, or loose battery connection • KAPWR to DME interrupted, or the circuit has been opened • Reprogramming error has occurred and needs replacement. Remember to check for Aftermarket Performance Products before replacing a DME.

DTC	Trouble Code Title, Conditions & Possible Causes
DTC: P0604 **2T MIL: Yes** **2002-05** **Models:** Cooper, Cooper S **Engines:** All **Transmissions:** All	**Internal Control Module Random Access Memory (RAM) Error Conditions:** Key on, and the DME detected an internal memory fault. This code will set if KAPWR to the DME is interrupted (at the initial key on). Watchdog on. **Possible Causes:** • Battery terminal corrosion, or loose battery connection • Connection to the DME interrupted, or the circuit has been opened • Reprogramming error has occurred and needs replacement. Remember to check for Aftermarket Performance Products before replacing a DME.
DTC: P0646 **MIL: No** **2002** **Models:** Cooper, Cooper S **Engines:** All **Transmissions:** All	**A/C Compressor Circuit High Conditions:** The DME detected a high condition on the sensor. **Possible Causes:** • The A/C Compressor has failed. • The DME has failed • Check wiring
DTC: P0647 **MIL: No** **2002** **Models:** Cooper, Cooper S **Engines:** All **Transmissions:** All	**A/C Compressor Circuit Low Conditions:** The DME detected a low condition on the sensor. **Possible Causes:** • The A/C Compressor has failed. • The DME has failed • Check wiring
DTC: P0704 **MIL: No** **2002-04** **Models:** Cooper, Cooper S **Engines:** All **Transmissions:** A/T	**Clutch Switch Input Circuit Malfunction Conditions:** Engine started, battery voltage must be at least 11.5v, all electrical components must be off, parking brake must be engaged (to keep daytime driving lights off), automatic transmission selector must be in park, and the ground between the engine and the chassis must be well connected. The DME detected a voltage outside the normal performance range to allow the system to properly function. **Possible Causes:** • Circuit harness connector contacts are corroded or ingresses of water • Circuit wires have shorted to each other, to battery or ground • Automatic Transmission Hydraulic Pressure Sensor 1 has failed • Solenoid valves in valve body are faulty • Transmission Input Speed (RPM) Sensor has failed • Transmission Output Speed (RPM) Sensor has failed • Engine Control Module (DME) is faulty • Voltage supply for Engine Control Module (DME) is faulty • Transmission Control Module (TCM) is faulty
DTC: P0705 **2T MIL: Yes** **2003-04** **Models:** Cooper, Cooper S **Engines:** All **Transmissions:** A/T	**TR Sensor Circuit Malfunction Conditions:** Engine started, battery voltage must be at least 11.5v, all electrical components must be off, parking brake must be engaged (to keep daytime driving lights off), automatic transmission selector must be in park, and the ground between the engine and the chassis must be well connected. The DME detected a voltage or signal outside the normal performance range to allow the system to properly function. The engine speed is between 200 and 440rpm. **Possible Causes:** • Circuit harness connector contacts are corroded or ingresses of water • Circuit wires have shorted to each other, to battery or ground • Automatic Transmission Hydraulic Pressure Sensor 1 has failed • Solenoid valves in valve body are faulty • Transmission Input Speed (RPM) Sensor has failed • Transmission Output Speed (RPM) Sensor has failed • Engine Control Module (DME) is faulty • Voltage supply for Engine Control Module (DME) is faulty • Transmission Control Module (TCM) is faulty
DTC: P0712 **MIL: No** **2002-04** **Models:** Cooper, Cooper S **Engines:** All **Transmissions:** A/T	**Oil Temperature Sensor Circuit Low Input Conditions:** Engine started, battery voltage must be at least 11.5v, all electrical components must be off, parking brake must be engaged (to keep daytime driving lights off), automatic transmission selector must be in park, and the ground between the engine and the chassis must be well connected. The DME detected the oil temperature sensor was less than its minimum self-test range in the test. **Possible Causes:** • Sensor signal circuit is open between the sensor and DME • Sensor ground circuit is open between sensor and DME • Sensor is damaged or has failed

DTC	Trouble Code Title, Conditions & Possible Causes
DTC: P0713 **MIL: No** **2002-04** **Models:** Cooper, Cooper S **Engines:** All **Transmissions:** A/T	**Oil Temperature Sensor Circuit High Input Conditions:** Engine started, battery voltage must be at least 11.5v, all electrical components must be off, parking brake must be engaged (to keep daytime driving lights off), automatic transmission selector must be in park, and the ground between the engine and the chassis must be well connected. The DME detected the oil temperature sensor was more than its maximum self-test range in the test. **Possible Causes:** • Sensor signal circuit is open between the sensor and DME • Sensor ground circuit is open between sensor and DME • Sensor is damaged or has failed
DTC: P0721 **MIL: NO** **2002-04** **Models:** Cooper, Cooper S **Engines:** All **Transmissions:** A/T	**A/T Output Shaft Speed Sensor Noise Interference Conditions:** Engine started, VSS signal more than 1 mph, and the DME detected "noise" interference on the Output Shaft Speed (OSS) sensor circuit. The calculation of the road speed impossible, as the indicated speed is less than the minimum road speed value and the timer expired. **Possible Causes:** • After market add-on devices interfering with the OSS signal • OSS connector is damaged, loose or shorted, or the wiring is misrouted or it is damaged • OSS assembly is damaged or it has failed • Failure of the ABS CAN vehicle speed sensor

Gas Engine OBD II Trouble Code List (P1XXX Codes)

DTC	Trouble Code Title, Conditions & Possible Causes
DTC: P1104 **2T MIL: Yes** **2005** **Models:** Cooper, Cooper S **Engines:** All **Transmissions:** All	**Manifold Pressure Sensor Plausibility Conditions:** Engine started, battery voltage must be at least 11v, and the differential pressure sensor detected a control deviation at the minimum limit. The closed loop control of the pressure in the intake manifold is suspended and replaced by a direct specification. The MAP was too low (less than 105.0016kPa); engine stopped. **Possible Causes:** • Sensor's voltage supply on Terminal 87 • Sensor's ground connection faulty • Signal wire to DME faulty • Replace sensor
DTC: P1106 **2T MIL: Yes** **2002-05** **Models:** Cooper, Cooper S **Engines:** All **Transmissions:** All	**Manifold Pressure Too Low at Full Load for Low Engine Speed Conditions:** Engine started, battery voltage must be at least 11v, and the differential pressure sensor detected a control deviation at the minimum limit. The closed loop control of the differential pressure in the intake manifold is suspended and replaced by a direct specification. The engine speed is less than 4000rpm. The manifold pressure is less than 600hPa. **Possible Causes:** • Sensor's voltage supply on Terminal 87 • Sensor's ground connection faulty • Signal wire to DME faulty • Replace sensor
DTC: P1107 **2T MIL: Yes** **2002-05** **Models:** Cooper, Cooper S **Engines:** All **Transmissions:** All	**Manifold Pressure Too Low at Idle Conditions:** **Engine started, battery voltage must be at least 11v, and the differential pressure sensor detected a control deviation at the minimum limit. The closed loop control of the differential pressure in the intake manifold is suspended and replaced by a direct specification. The engine speed is less than 1504rpm. The manifold pressure is less than 120hPa.** **Possible Causes:** • Sensor's voltage supply on Terminal 87 • Sensor's ground connection faulty • Signal wire to DME faulty • Replace sensor
DTC: P1108 **2T MIL: Yes** **2002-05** **Models:** Cooper, Cooper S **Engines:** All **Transmissions:** All	**Manifold Pressure Too Low at Stable and in Full Load for Low Engine Speed Conditions:** Engine started, battery voltage must be at least 11v, and the differential pressure sensor detected a control deviation at the minimum limit. The closed loop control of the differential pressure in the intake manifold is suspended and replaced by a direct specification. The engine speed is less than 4000rpm. The manifold pressure is less than 600hPa. **Possible Causes:** • Sensor's voltage supply on Terminal 87 • Sensor's ground connection faulty • Signal wire to DME faulty • Replace sensor
DTC: P1109 **2T MIL: Yes** **2002-05** **Models:** Cooper, Cooper S **Engines:** All **Transmissions:** All	**Manifold Pressure Too High During Deceleration Conditions:** Engine started, battery voltage must be at least 11v, and the differential pressure sensor detected a control deviation at the minimum limit. The closed loop control of the differential pressure in the intake manifold is suspended and replaced by a direct specification. The engine speed is greater than 1696rpm. The manifold pressure is greater than 600hPa. **Possible Causes:** • Sensor's voltage supply on Terminal 87 • Sensor's ground connection faulty • Signal wire to DME faulty • Replace sensor
DTC: P1122 **2T MIL: Yes** **2002-05** **Models:** Cooper, Cooper S **Engines:** All **Transmissions:** All	**Accelerator Pedal Position Sensor 'D' Circuit Low Input Conditions:** Engine started, battery voltage at least 11.5v, all electrical components off, ground connections between engine and chassis well connected, the DME detected that the accelerator pedal position sensor signal was outside the parameters to function normally. **Note: Both the Throttle Position (TP) Sensor and Accelerator Pedal Position Sensor are located at the accelerator pedal module and communicate the driver's intentions to the DME completely independently of each other. Both sensors are stored in one housing.** **Possible Causes:** • Ground between engine and chassis may be broken • Throttle position sensor may have failed • Accelerator Pedal Position Sensor has failed • Throttle position sensor wiring may have shorted • Throttle position sensor has failed • Faulty voltage supply

DTC	Trouble Code Title, Conditions & Possible Causes
DTC: P1123 **2T MIL: Yes** **2002-05** **Models:** Cooper, Cooper S **Engines:** All **Transmissions:** All	**Accelerator Pedal Position Sensor 'D' Circuit High Input Conditions:** Engine started, battery voltage at least 11.5v, all electrical components off, ground connections between engine and chassis well connected, the DME detected that the accelerator pedal position sensor signal was outside the parameters to function normally. **Note: Both the Throttle Position (TP) Sensor and Accelerator Pedal Position Sensor are located at the accelerator pedal module and communicate the driver's intentions to the DME completely independently of each other. Both sensors are stored in one housing.** **Possible Causes:** • Ground between engine and chassis may be broken • Throttle position sensor may have failed • Accelerator Pedal Position Sensor has failed • Throttle position sensor wiring may have shorted • Throttle position sensor has failed • Faulty voltage supply
DTC: P1125 **2T MIL: Yes** **2002-05** **Models:** Cooper, Cooper S **Engines:** All **Transmissions:** All	**Throttle/Pedal Position Sensor Circuit Plausibility Error Conditions:** Engine started, at idle, the temperature must be at least 80 degrees Celsius. The throttle position sensor supplies implausible signal to the DME. The difference between the TPS1 and the TPS2 is greater than five percent. **Possible Causes:** • TP sensor signal circuit open (inspect wiring & connector) • TP sensor signal shorted to ground (inspect wiring & connector) • TP sensor is damaged or has failed • Throttle control module's voltage supply is shorted or open
DTC: P1126 **2T MIL: Yes** **2002-05** **Models:** Cooper, Cooper S **Engines:** All **Transmissions:** All	**Throttle/Pedal Position Sensor Circuit Large Plausibility Error Conditions:** Engine started, at idle, the temperature must be at least 80 degrees Celsius. The throttle position sensor supplies implausible signal to the DME. The difference between the TPS1 and the TPS2 is greater than five percent. **Possible Causes:** • TP sensor signal circuit open (inspect wiring & connector) • TP sensor signal shorted to ground (inspect wiring & connector) • TP sensor is damaged or has failed • Throttle control module's voltage supply is shorted or open
DTC: P1126 **2T MIL: Yes** **2005:** All **Engines,** All **Models;** **Transmissions:** All	**TP Sensor Circuit High Input Conditions:** Engine started, at idle, the temperature must be at least 80 degrees Celsius. The DME detected the TP sensor signal was more than the self-test maximum during testing. **Possible Causes:** • TP sensor not seated correctly in housing (may be damaged) • TP sensor signal is circuit shorted to ground or system voltage • TP sensor ground circuit is open (check the wiring harness) • TP sensor and/or DME has failed
DTC: P1143 **2T MIL: Yes** **2002-04** **Models:** Cooper, Cooper S **Engines:** All **Transmissions:** All	**O2 Sensor Signal Stuck Lean Bank 1 Sensor 2 Conditions:** Engine started, battery voltage must be at least 11.5v, all electrical components must be off, parking brake must be engaged (to keep daytime driving lights off), automatic transmission selector must be in park. The DME detected an unexpected voltage condition, or it detected an unexpected current draw in the heater circuit during the CCM test. Coolant temperature must been at least 80.25 degrees Celsius. The vehicle speed is greater than 27.96 and less than 80.76. The engine speed is between 1984 and 3647rpm. Ambient pressure is greater than 75.001kPa and the engine stability load is 6.94g/s. **Note: Vehicle must be raised before connector for oxygen sensors is accessible.** **Possible Causes:** • Oxygen sensor (before catalytic converter) is faulty • Oxygen sensor heater (before catalytic converter) is faulty • Oxygen sensor heater (before catalytic converter) is faulty • Oxygen sensor heater (behind catalytic converter) is faulty • O2S circuit is open or shorted in the wiring harness

DTC	Trouble Code Title, Conditions & Possible Causes
DTC: P1144 **2002-04** **Models:** Cooper, Cooper S **Engines:** All **Transmissions:** All	**O2 Sensor Signal Stuck Rich Bank 1 Sensor 2 Conditions:** Engine started, battery voltage must be at least 11.5v, all electrical components must be off, parking brake must be engaged (to keep daytime driving lights off), automatic transmission selector must be in park. The DME detected an unexpected voltage condition, or it detected an unexpected current draw in the heater circuit during the CCM test. Coolant temperature must been at least 80.25 degrees Celsius. The vehicle speed is greater than 27.96 and less than 80.76. The engine speed is between 1984 and 3647rpm. Ambient pressure is greater than 75.001kPa and the engine stability load is 6.94g/s. **Note: Vehicle must be raised before connector for oxygen sensors is accessible.** **Possible Causes:** • Oxygen sensor (before catalytic converter) is faulty • Oxygen sensor heater (before catalytic converter) is faulty • Oxygen sensor heater (before catalytic converter) is faulty • Oxygen sensor heater (behind catalytic converter) is faulty • O2S circuit is open or shorted in the wiring harness
DTC: P1222 **2T MIL: Yes** **2002-05** **Models:** Cooper, Cooper S **Engines:** All **Transmissions:** All	**Accelerator Pedal Position Sensor 'E' Circuit Low Input Conditions:** Engine started, battery voltage at least 11.5v, all electrical components off, ground connections between engine and chassis well connected, the DME detected that the accelerator pedal position sensor signal was outside the parameters to function normally. **Note: Both the Throttle Position (TP) Sensor and Accelerator Pedal Position Sensor are located at the accelerator pedal module and communicate the driver's intentions to the DME completely independently of each other. Both sensors are stored in one housing.** **Possible Causes:** • Ground between engine and chassis may be broken • Throttle position sensor may have failed • Accelerator Pedal Position Sensor has failed • Throttle position sensor wiring may have shorted • Throttle position sensor has failed • Faulty voltage supply
DTC: P1223 **2T MIL: Yes** **2002-05** **Models:** Cooper, Cooper S **Engines:** All **Transmissions:** All	**Accelerator Pedal Position Sensor 'E' Circuit High Input Conditions:** Engine started, battery voltage at least 11.5v, all electrical components off, ground connections between engine and chassis well connected, the DME detected that the accelerator pedal position sensor signal was outside the parameters to function normally. **Note: Both the Throttle Position (TP) Sensor and Accelerator Pedal Position Sensor are located at the accelerator pedal module and communicate the driver's intentions to the DME completely independently of each other. Both sensors are stored in one housing.** **Possible Causes:** • Ground between engine and chassis may be broken • Throttle position sensor may have failed • Accelerator Pedal Position Sensor has failed • Throttle position sensor wiring may have shorted • Throttle position sensor has failed • Faulty voltage supply
DTC: P1224 **2T MIL: Yes** **2002-05** **Models:** Cooper, Cooper S **Engines:** All **Transmissions:** All	**Throttle Position Sensor D/E Voltage Correlation Conditions:** Engine started, battery voltage must be at least 11.5v, all electrical components must be off, parking brake must be engaged (to keep daytime driving lights off), automatic transmission selector must be in park; and the DME detected the Throttle Position 'D' (TPD) and Throttle Position 'B' (TPE) sensors disagreed, or that the TPD sensor should not be in its detected position, or that the TPE sensor should not be in its detected position during testing. **Note: Both the Throttle Position (TP) Sensor and Accelerator Pedal Position Sensor are located at the accelerator pedal module and communicate the driver's intentions to the DME completely independently of each other. Both sensors are stored in one housing.** **Possible Causes:** • ETC TP sensor connector is damaged or shorted • ETC TP sensor circuits shorted together in the wire harness • ETC TP sensor signal circuit is shorted to VREF (5v) • ETC TP sensor is damaged or the DME has failed
DTC: P1229 **2T MIL: Yes** **2002-05** **Models:** Cooper, Cooper S **Engines:** All **Transmissions:** All	**Throttle/Pedal Position Sensor Adaptation Outside Tolerance Conditions:** Engine started, at idle, the temperature must be at least 80 degrees Celsius. The throttle position sensor supplies implausible signal to the DME and is outside the specified tolerance. The measured max/min TPS values within the limits is greater than 0.0244 volts. **Possible Causes:** • TP sensor signal circuit open (inspect wiring & connector) • TP sensor signal shorted to ground (inspect wiring & connector) • TP sensor is damaged or has failed • Throttle control module's voltage supply is shorted or open

DTC	Trouble Code Title, Conditions & Possible Causes
DTC: P1234 **MIL: No** **2002** **Models:** Cooper, Cooper S **Engines:** All **Transmissions:** All	**Electrical Fuel Pump Circuit Short to Battery Conditions:** The DME detected a low condition on the pump. **Possible Causes:** • The fuel pump has failed. • The DME has failed • Check wiring
DTC: P1236 **MIL: No** **2002** **Models:** Cooper, Cooper S **Engines:** All **Transmissions:** All	**Electrical Fuel Pump Short to Ground or Open Circuit Conditions:** The DME detected a high condition on the pump. **Possible Causes:** • The fuel pump has failed. • The DME has failed • Check wiring
DTC: P1320 **2T MISFIRE** **MIL: Yes** **2002-05** **Models:** Cooper, Cooper S **Engines:** All **Transmissions:** All	**Misfire Detected Crankshaft Segment Adaptation Conditions:** Engine running under positive torque conditions, and the DME detected a misfire or uneven engine function as well as the crankshaft adaptation at its limit. **Note: If the misfire is severe, the MIL will flash on/off on the 1st trip!** **Possible Causes:** • Air leak in the intake manifold, or in the EGR or DME system • Base engine mechanical problem • Fuel delivery component problem (i.e., a contaminated, dirty or sticking fuel injector) • Fuel pump relay defective • Ignition coil fuses have failed • Ignition system problem (dirty damaged coil or plug) • Engine speed (RPM) sensor has failed • Camshaft position sensors have failed • Ignition coil is faulty • Spark plugs are not working properly or are not gapped properly
DTC: P1321 **2T MISFIRE** **MIL: Yes** **2002-05** **Models:** Cooper, Cooper S **Engines:** All **Transmissions:** All	**Misfire Crank Wheel Tooth Count Conditions:** Engine running under positive torque conditions, and the DME detected a misfire or uneven engine function as well as a tooth error of plus or minus one or two teeth during the count. **Note: If the misfire is severe, the MIL will flash on/off on the 1st trip!** **Possible Causes:** • Air leak in the intake manifold, or in the EGR or DME system • Base engine mechanical problem • Fuel delivery component problem (i.e., a contaminated, dirty or sticking fuel injector) • Fuel pump relay defective • Ignition coil fuses have failed • Ignition system problem (dirty damaged coil or plug) • Engine speed (RPM) sensor has failed • Camshaft position sensors have failed • Ignition coil is faulty • Spark plugs are not working properly or are not gapped properly
DTC: P1366 **2T MIL: Yes** **2002-04** **Models:** Cooper, Cooper S **Engines:** All **Transmissions:** All	**Ignition Coilpack A Primary/Secondary Circuit Malfunction Open Circuit/Short to Ground Conditions:** Engine started, battery voltage must be at least 11.5v, all electrical components must be off, parking brake must be engaged (to keep daytime driving lights off), automatic transmission selector must be in park and the ground between the engine and the chassis must be well connected. The DME did not receive any valid pulses from the ignition module for the Ignition Coilpack A primary circuit. **Note: Ignition coils and power output stages are one component and cannot be replaced individually.** **Possible Causes:** • Engine speed (RPM) sensor has failed • Camshaft Position (CMP) sensor has failed • Power Supply Relay is shorted to an open circuit • There is a malfunction in voltage supply • Ignition coilpack is damaged or it has failed • Cylinder 1 to 4 Fuel Injector(s) have failed•

DTC	Trouble Code Title, Conditions & Possible Causes
DTC: P1367 **2T MIL: Yes** **2002-04** **Models:** Cooper, Cooper S **Engines:** All **Transmissions:** All	**Ignition Coilpack A Primary/Secondary Circuit Malfunction Open Circuit/Short to Ground Conditions:** Engine started, battery voltage must be at least 11.5v, all electrical components must be off, parking brake must be engaged (to keep daytime driving lights off), automatic transmission selector must be in park and the ground between the engine and the chassis must be well connected. The DME did not receive any valid pulses from the ignition module for the Ignition Coilpack A primary circuit. **Note: Ignition coils and power output stages are one component and cannot be replaced individually.** **Possible Causes:** • Engine speed (RPM) sensor has failed • Camshaft Position (CMP) sensor has failed • Power Supply Relay is shorted to an open circuit • There is a malfunction in voltage supply • Ignition coilpack is damaged or it has failed • Cylinder 1 to 4 Fuel Injector(s) have failed•
DTC: P1436 **2T MIL: Yes** **2003-04** **Models:** Cooper, Cooper S **Engines:** All **Transmissions:** All	**EVAP Leak Detection Pump (LDP) Control Circuit Open Conditions:** Engine started, battery voltage must be at least 11.5v, all electrical components must be off, parking brake must be engaged (to keep daytime driving lights off), automatic transmission selector must be in park, the exhaust system must be properly sealed between the catalytic converter and the cylinder head, coolant temperature must be at least 80 degrees Celsius and oxygen sensor heaters for oxygen sensors before the catalytic converter must be functioning properly and the ground between the engine and the chassis must bc well connected. The DME detected voltage irregularity in the leak detection pump control circuit. **Possible Causes:** • EVAP LDP power supply circuit is open • EVAP LDP solenoid valve is damaged or it has failed • EVAP LDP canister has a leak or a poor seal • EVAP canister system has an improper seal • Evaporative Emission (EVAP) canister purge regulator valve 1 has failed • Leak Detection Pump (LDP) is faulty • Aftermarket EVAP parts that do not conform to specifications • EVAP component seals leaking (i.e., leaks in the Purge valve, fuel tank pressure sensor, canister vent solenoid, fuel vapor control valve tube assembly or fuel vapor vent valve).
DTC: P1437 **2T MIL: Yes** **2002-04** **Models:** Cooper, Cooper S **Engines:** All **Transmissions:** All	**EVAP Emission Control LDP Circuit Malfunction Pump Problem Conditions:** Key on, KOEO Self-Test enabled, and the DME detected an unexpected voltage condition on the EVAP emission control leak detection pump circuit. The reed switch level stays low after activation of solenoids within the time threshold of more than 1 second. The coolant temperature is less than 60 degrees Celsius, and the ambient pressure is greater than 76.2994kPa. The air intake temperature at start is between 9.04 and 16.04 degrees Celsius. The change in barometric pressure since engine start is less than 0.9998kPa. The vehicle speed is less than 74.56mph, and the purge valve has opened enough on previous driving cycle. **Possible Causes:** • Leak Detection Pump has failed • EVAP canister system has an improper or broken seal • Evaporative Emission (EVAP) canister purge regulator valve 1 is faulty • Hoses between the fuel pump and the EVAP canister are faulty • Fuel filler cap is loose • Fuel pump seal is defective, faulty or otherwise leaking • Hoses between the EVAP canister and the fuel flap unit are faulty • Hoses between the EVAP canister and the evaporative emission canister purge regulator valve are faulty
DTC: P1442 **2T MIL: Yes** **2002-04** **Models:** Cooper, Cooper S **Engines:** All **Transmissions:** All	**EVAP Leak Detection Pump Control Circuit Low Conditions:** Engine started, battery voltage must be at least 11.5v, all electrical components must be off, parking brake must be engaged (to keep daytime driving lights off), automatic transmission selector must be in park, the exhaust system must be properly sealed between the catalytic converter and the cylinder head, coolant temperature must be at least 80 degrees Celsius and oxygen sensor heaters for oxygen sensors before the catalytic converter must be functioning properly and the ground between the engine and the chassis must be well connected. The DME detected voltage irregularity in the leak detection pump control circuit. **Possible Causes:** • EVAP LDP power supply circuit is open • EVAP LDP solenoid valve is damaged or it has failed • EVAP LDP canister has a leak or a poor seal • EVAP canister system has an improper seal • Evaporative Emission (EVAP) canister purge regulator valve 1 has failed • Leak Detection Pump (LDP) is faulty • Aftermarket EVAP parts that do not conform to specifications • EVAP component seals leaking (i.e., leaks in the Purge valve, fuel tank pressure sensor, canister vent solenoid, fuel vapor control valve tube assembly or fuel vapor vent valve).

DTC	Trouble Code Title, Conditions & Possible Causes
DTC: P1443 **2T MIL: Yes** **2002-04** **Models:** Cooper, Cooper S **Engines:** All **Transmissions:** All	**EVAP Leak Detection Pump Control Circuit High Conditions:** Engine started, battery voltage must be at least 11.5v, all electrical components must be off, parking brake must be engaged (to keep daytime driving lights off), automatic transmission selector must be in park, the exhaust system must be properly sealed between the catalytic converter and the cylinder head, coolant temperature must be at least 80 degrees Celsius and oxygen sensor heaters for oxygen sensors before the catalytic converter must be functioning properly and the ground between the engine and the chassis must be well connected. The DME detected voltage irregularity in the leak detection pump control circuit. **Possible Causes:** • EVAP LDP power supply circuit is open • EVAP LDP solenoid valve is damaged or it has failed • EVAP LDP canister has a leak or a poor seal • EVAP canister system has an improper seal • Evaporative Emission (EVAP) canister purge regulator valve 1 has failed • Leak Detection Pump (LDP) is faulty • Aftermarket EVAP parts that do not conform to specifications • EVAP component seals leaking (i.e., leaks in the Purge valve, fuel tank pressure sensor, canister vent solenoid, fuel vapor control valve tube assembly or fuel vapor vent valve).
DTC: P1475 **2T MIL: Yes** **2002-05** **Models:** Cooper, Cooper S **Engines:** All **Transmissions:** All	**EVAP Emission Control LDP Circuit Malfunction Conditions:** Key on, KOEO Self-Test enabled, and the DME detected an unexpected voltage condition on the EVAP emission control leak detection pump circuit. The reed switch level stays high after activation of solenoids within the time threshold of more than 0.5 seconds. The coolant temperature is less than 60 degrees Celsius, and the ambient pressure is greater than 76.2994kPa. The air intake temperature at start is between 9.04 and 16.04 degrees Celsius. The change in barometric pressure since engine start is less than 0.9998kPa. The vehicle speed is less than 74.56mph, and the purge valve has opened enough on previous driving cycle. **Possible Causes:** • Leak Detection Pump has failed • EVAP canister system has an improper or broken seal • Evaporative Emission (EVAP) canister purge regulator valve 1 is faulty • Hoses between the fuel pump and the EVAP canister are faulty • Fuel filler cap is loose • Fuel pump seal is defective, faulty or otherwise leaking • Hoses between the EVAP canister and the fuel flap unit are faulty • Hoses between the EVAP canister and the evaporative emission canister purge regulator valve are faulty
DTC: P1476 **2T MIL: Yes** **2002-05** **Models:** Cooper, Cooper S **Engines:** All **Transmissions:** All	**EVAP Emission Control LDP Circuit Malfunction/Insufficient Vacuum Conditions:** Key on, KOEO Self-Test enabled, and the DME detected an unexpected voltage condition on the EVAP emission control leak detection pump circuit. There is a clamped tube during the time period of any of the five first pump cycles. The coolant temperature is less than 60 degrees Celsius, and the ambient pressure is greater than 76.2994kPa. The air intake temperature at start is between 9.04 and 16.04 degrees Celsius. The change in barometric pressure since engine start is less than 0.9998kPa. The vehicle speed is less than 74.56mph, and the purge valve has opened enough on previous driving cycle. **Possible Causes:** • Leak Detection Pump has failed • EVAP canister system has an improper or broken seal • Evaporative Emission (EVAP) canister purge regulator valve 1 is faulty • Hoses between the fuel pump and the EVAP canister are faulty • Fuel filler cap is loose • Fuel pump seal is defective, faulty or otherwise leaking • Hoses between the EVAP canister and the fuel flap unit are faulty • Hoses between the EVAP canister and the evaporative emission canister purge regulator valve are faulty
DTC: P1477 **2T MIL: Yes** **2002-05** **Models:** Cooper, Cooper S **Engines:** All **Transmissions:** All	**EVAP Emission Control LDP Circuit Malfunction Conditions:** Key on, KOEO Self-Test enabled, and the DME detected an unexpected voltage condition on the EVAP emission control leak detection pump circuit. The reed switch level stays continuously low after activation of solenoids within the time threshold of more than 1 second. The coolant temperature is less than 60 degrees Celsius, and the ambient pressure is greater than 76.2994kPa. The air intake temperature at start is between 9.04 and 16.04 degrees Celsius. The change in barometric pressure since engine start is less than 0.9998kPa. The vehicle speed is less than 74.56mph, and the purge valve has opened enough on previous driving cycle. **Possible Causes:** • Leak Detection Pump has failed • EVAP canister system has an improper or broken seal • Evaporative Emission (EVAP) canister purge regulator valve 1 is faulty • Hoses between the fuel pump and the EVAP canister are faulty • Fuel filler cap is loose • Fuel pump seal is defective, faulty or otherwise leaking • Hoses between the EVAP canister and the fuel flap unit are faulty • Hoses between the EVAP canister and the evaporative emission canister purge regulator valve are faulty

DTC	Trouble Code Title, Conditions & Possible Causes
DTC: P1481 **MIL: No** **2002** **Models:** Cooper, Cooper S **Engines:** All **Transmissions:** All	**Cooling Fans Circuit Short to Ground or Open Circuit Conditions:** The DME detected a high condition on the sensor. **Possible Causes:** • The cooling fan has failed. • The DME has failed • Check wiring
DTC: P1482 **MIL: No** **2002** **Models:** Cooper, Cooper S **Engines:** All **Transmissions:** All	**Cooling Fans Circuit Short to Battery Conditions:** The DME detected a Low condition on the sensor. **Possible Causes:** • The cooling fan has failed. • The DME has failed • Check wiring
DTC: P1484 **MIL: No** **2002** **Models:** Cooper, Cooper S **Engines:** All **Transmissions:** All	**Cooling Fans Circuit Short to Ground or Open Circuit Conditions:** The DME detected a high condition on the sensor. **Possible Causes:** • The cooling fan has failed. • The DME has failed • Check wiring
DTC: P1485 **MIL: No** **2002** **Models:** Cooper, Cooper S **Engines:** All **Transmissions:** All	**Cooling Fans Circuit Short to Battery Conditions:** The DME detected a Low condition on the sensor. **Possible Causes:** • The cooling fan has failed. • The DME has failed • Check wiring
DTC: P1496 **1T MIL: Yes** **2005** **Models:** Cooper, Cooper S **Engines:** All **Transmissions:** All	**Air Intake System Leak (Block 3) Conditions:** Engine speed greater than 704rpm. The manifold pressure is greater than 15.002kPa. The throttle position is less than 89.98 percent. The DME detected that a comparison of the modeled mass airflow at the cylinder and the mass airflow at the throttle exceeds the threshold relative to the throttle opening by more than 1.3. **Possible Causes:** • Charge air system leaks • Recirculation valve for turbocharger is faulty • Turbocharging system is damaged • Vacuum diaphragm for turbocharger needs adjusting • Wastegate bypass regulator valve is faulty
DTC: P1600 **2T MIL: Yes** **2002-05** **Models:** Cooper, Cooper S **Engines:** All **Transmissions:** All	**Internal Control Module Random Access Memory (RAM) Error Conditions:** Key on, and the DME detected an internal memory fault. This code will set if KAPWR to the DME is interrupted (at the initial key on). Watchdog on. **Possible Causes:** • Battery terminal corrosion, or loose battery connection • Connection to the DME interrupted, or the circuit has been opened • Reprogramming error has occurred and needs replacement. Remember to check for Aftermarket Performance Products before replacing a DME.
DTC: P1607 **2T MIL: Yes** **2002-05** **Models:** Cooper, Cooper S **Engines:** All **Transmissions:** All	**CAN Bus Error Conditions:** Engine started, VSS over 1 mph, and the DME detected a problem in the CAN Bus system during the self-test. **Possible Causes:** • Open/short circuit to ground in the communication wire from the transmission to the DME. • The DME has failed
DTC: P1611 **2T MIL: Yes** **2002-05** **Models:** Cooper, Cooper S **Engines:** All **Transmissions:** All	**MIL Call-Up Circuit, Transmission Control Module Short to Ground Conditions:** Engine started, VSS over 1 mph, and the DME detected a problem in the Transmission Control system during the self-test. **Possible Causes:** • Open/short circuit to ground in the communication wire from the transmission to the DME. • The DME has failed
DTC: P1612 **2T MIL: Yes** **2002-05** **Models:** Cooper, Cooper S **Engines:** All **Transmissions:** All	**INSTR Module Error Conditions:** Engine started, VSS over 1 mph, and the DME detected a problem in the INSTR Module system during the self-test. **Possible Causes:** • Open/short circuit to ground in the communication wire from the transmission to the DME. • The DME has failed

DTC	Trouble Code Title, Conditions & Possible Causes
DTC: P1613 **2T MIL: Yes** **2002-05** **Models:** Cooper, Cooper S **Engines:** All **Transmissions:** All	**ASC Error Conditions:** Engine started, VSS over 1 mph, and the DME detected a problem in the ASC system during the self-test. **Possible Causes:** • Open/short circuit to ground in the communication wire from the transmission to the DME. • The DME has failed
DTC: P1615 **2T MIL: Yes** **2002-05** **Models:** Cooper, Cooper S **Engines:** All **Transmissions:** All	**SPI-Bus Error Conditions:** Engine started, VSS over 1 mph, and the DME detected a problem in the SPI Bus system during the self-test. **Possible Causes:** • Open/short circuit to ground in the communication wire from the transmission to the DME. • The DME has failed
DTC: U1116 **1T MIL: Yes** **2002** **Models:** Cooper, Cooper S **Engines:** All **Transmissions:** All	**Check Sum Error Conditions:** Key on, and the DME detected unofficial calibration modifications (flagged during ignition on). The Engine Control Module (DME) communicates with all databus-capable control modules via a CAN databus. These databus-capable control modules are connected via two data bus wires which are twisted together (CAN_High and CAN_Low), and exchange information (messages). Missing information on the databus is recognized as a malfunction and stored. Trouble-free operation of the CAN-Bus requires that it have a terminal resistance. **Possible Causes:** • DME has failed • Terminal resistance for CAN-bus are faulty • Can data bus wires have short circuited to each other
DTC: P1656 **MIL: No** **2002** **Models:** Cooper, Cooper S **Engines:** All **Transmissions:** All	**Timeout EWS (Electronic Immobilizer) Telegram Conditions:** Key on the DME detected an electrical malfunction regarding the EWS. The wrong message was received. Check the EWS fuse, the ECU or CAN for fault, turn the ignition off and then on to repeat start calibration. **Possible Causes:** • Circuit from the MIL to the DME • ECU Failure, BUS system failure • Circuit from the EPC to the DME • Defective CAS
DTC: P1661 **MIL: No** **2002** **Models:** Cooper, Cooper S **Engines:** All **Transmissions:** All	**Timeout EWS (Electronic Immobilizer) Telegram Conditions:** Key on the DME detected an electrical malfunction regarding the EWS. Check the EWS fuse, the ECU or CAN for fault, turn the ignition off and then on to repeat start calibration. **Possible Causes:** • Circuit from the MIL to the DME • ECU Failure, BUS system failure • Circuit from the EPC to the DME • Defective CAS
DTC: P1679 **1T MIL: Yes** **2002-04** **Models:** Cooper, Cooper S **Engines:** All **Transmissions:** All	**Monitoring of Torque Losses Conditions:** Key on, engine running, the DME has detected that there is an error in the torque loss calculation. The limit was exceeded in the threshold map during the first 360 ms of operation. **Possible Causes:** • Battery terminal corrosion, loose battery connection, or faulty • Connection to the DME interrupted, or the circuit has been opened • Reprogramming error has occurred and needs replacement. • Voltage supply for Engine Control Module (DME) is faulty
DTC: P1680 **1T MIL: Yes** **2002-05** **Models:** Cooper, Cooper S **Engines:** All **Transmissions:** All	**Monitoring of A to D Conversion Conditions:** Key on, engine running to at least 1200rpm, the DME has detected that the PVS ratio differences exceeds the threshold greater than 0.273 volts. **Possible Causes:** • Battery terminal corrosion, loose battery connection, or faulty • Connection to the DME interrupted, or the circuit has been opened • Reprogramming error has occurred and needs replacement. • Voltage supply for Engine Control Module (DME) is faulty

DTC	Trouble Code Title, Conditions & Possible Causes
DTC: P1681 **1T MIL: Yes** **2002-05** **Models:** Cooper, Cooper S **Engines:** All **Transmissions:** All	**Monitoring of Engine Speed Conditions:** Key on, engine running to at least 1200rpm, the DME has detected that the engine speed difference exceeds the threshold of 576rpm **Possible Causes:** • Battery terminal corrosion, loose battery connection, or faulty • Connection to the DME interrupted, or the circuit has been opened • Reprogramming error has occurred and needs replacement. • Voltage supply for Engine Control Module (DME) is faulty
DTC: P1682 **1T MIL: Yes** **2002-05** **Models:** Cooper, Cooper S **Engines:** All **Transmissions:** All	**Idle Speed Control, Monitoring of the Proportional Derivative Conditions:** Key on, engine running to at least 1200rpm, the DME has detected that there is an error in the torque demand from the proportional derivative part. The maximum limit has been exceeded. **Possible Causes:** • Battery terminal corrosion, loose battery connection, or faulty • Connection to the DME interrupted, or the circuit has been opened • Reprogramming error has occurred and needs replacement. • Voltage supply for Engine Control Module (DME) is faulty
DTC: P1683 **1T MIL: Yes** **2002-05** **Models:** Cooper, Cooper S **Engines:** All **Transmissions:** All	**Idle Speed Control, Monitoring of the Integral Part Conditions:** Key on, engine running to at least 1200rpm, the DME has detected that there is an error in the torque demand from the integral part is greater than 25NM. **Possible Causes:** • Battery terminal corrosion, loose battery connection, or faulty • Connection to the DME interrupted, or the circuit has been opened • Reprogramming error has occurred and needs replacement. • Voltage supply for Engine Control Module (DME) is faulty
DTC: P1684 **1T MIL: Yes** **2002-05** **Models:** Cooper, Cooper S **Engines:** All **Transmissions:** All	**Monitoring of Minimum Torque at Clutch Conditions:** Key on, engine running, the DME has detected that there is an error in the minimum torque at the clutch calculation. The limit was exceeded in the threshold map. **Possible Causes:** • Battery terminal corrosion, loose battery connection, or faulty • Connection to the DME interrupted, or the circuit has been opened • Reprogramming error has occurred and needs replacement. • Voltage supply for Engine Control Module (DME) is faulty
DTC: P1685 **1T MIL: Yes** **2002-05** **Models:** Cooper, Cooper S **Engines:** All **Transmissions:** All	**Monitoring of Maximum Torque at Clutch Conditions:** Key on, engine running, the DME has detected that there is an error in the maximum torque at the clutch calculation. The limit was exceeded in the threshold map. **Possible Causes:** • Battery terminal corrosion, loose battery connection, or faulty • Connection to the DME interrupted, or the circuit has been opened • Reprogramming error has occurred and needs replacement. • Voltage supply for Engine Control Module (DME) is faulty
DTC: P1686 **1T MIL: Yes** **2002-05** **Models:** Cooper, Cooper S **Engines:** All **Transmissions:** All	**Monitoring of Pedal Values Conditions:** Key on, engine running, the DME has detected that there is an error in pedal value checks. The difference exceeds the threshold map by 15.23 to 28.91 percent. **Possible Causes:** • Battery terminal corrosion, loose battery connection, or faulty • Connection to the DME interrupted, or the circuit has been opened • Reprogramming error has occurred and needs replacement. • Voltage supply for Engine Control Module (DME) is faulty
DTC: P1687 **1T MIL: Yes** **2002-05** **Models:** Cooper, Cooper S **Engines:** All **Transmissions:** All	**Monitoring of Throttle Position Conditions:** Key on, engine running, the DME has detected that there is an error in the throttle position sensor ratio calculation by greater than 0.313 volts. **Possible Causes:** • Battery terminal corrosion, loose battery connection, or faulty • Connection to the DME interrupted, or the circuit has been opened • Reprogramming error has occurred and needs replacement. • Voltage supply for Engine Control Module (DME) is faulty

DTC	Trouble Code Title, Conditions & Possible Causes
DTC: P1688 **1T MIL: Yes** **2002-05** **Models:** Cooper, Cooper S **Engines:** All **Transmissions:** All	**Monitoring of Mass Airflow Conditions:** Key on, engine running, the DME has detected that there is an error in the MAF calculation. The limit was exceeded in the threshold map by 0.044 to 0.218g/rev. **Possible Causes:** • Battery terminal corrosion, loose battery connection, or faulty • Connection to the DME interrupted, or the circuit has been opened • Reprogramming error has occurred and needs replacement. • Voltage supply for Engine Control Module (DME) is faulty
DTC: P1689 **1T MIL: Yes** **2002-05** **Models:** Cooper, Cooper S **Engines:** All **Transmissions:** All	**Monitoring of Actual Indicated Engine Torque Conditions:** Key on, engine running, the DME has detected that there is an error in the maximum torque at the clutch calculation. The limit was exceeded in the threshold map by 30 to 38NM. **Possible Causes:** • Battery terminal corrosion, loose battery connection, or faulty • Connection to the DME interrupted, or the circuit has been opened • Reprogramming error has occurred and needs replacement. • Voltage supply for Engine Control Module (DME) is faulty
DTC: P1691 **1T MIL: Yes** **2002-05** **Models:** Cooper, Cooper S **Engines:** All **Transmissions:** All	**Monitoring of Engine Speed Limit in Limp Home Conditions:** Key on, engine running, the DME has detected that monitoring of the engine speed limit in limp home condition exceeds the threshold map by greater than 2656rpm. **Possible Causes:** • Battery terminal corrosion, loose battery connection, or faulty • Connection to the DME interrupted, or the circuit has been opened • Reprogramming error has occurred and needs replacement. • Voltage supply for Engine Control Module (DME) is faulty
DTC: P1692 **1T MIL: Yes** **2002-05** **Models:** Cooper, Cooper S **Engines:** All **Transmissions:** All	**Monitoring of Processor Calculations Conditions:** Key on, engine running, the DME has detected that there is an error in the for the final request for disabled power stages of MTC and IV. **Possible Causes:** • Battery terminal corrosion, loose battery connection, or faulty • Connection to the DME interrupted, or the circuit has been opened • Reprogramming error has occurred and needs replacement. • Voltage supply for Engine Control Module (DME) is faulty
DTC: P1693 **1T MIL: Yes** **2002-05** **Models:** Cooper, Cooper S **Engines:** All **Transmissions:** All	**Monitoring of Processor Calculations Conditions:** Key on, engine running, the DME has detected that there is an error in the for the temporary request for disabled power stages of MTC and IV. **Possible Causes:** • Battery terminal corrosion, loose battery connection, or faulty • Connection to the DME interrupted, or the circuit has been opened • Reprogramming error has occurred and needs replacement. • Voltage supply for Engine Control Module (DME) is faulty
DTC: P1698 **1T MIL: Yes** **2002-05** **Models:** Cooper, Cooper S **Engines:** All **Transmissions:** A/T	**ECU Functionality Incorrect Conditions:** The ECU Functionality is in error as there are internal errors. This test is performed by the GIB (Gearbox Interface Box), a system dedicated to low level control of the transmission control unit. **Possible Causes:** • Short to battery • Short to ground • Open circuit • CAN data bus wires have short circuited to each other • ECU has failed • BUS system failure • Defective bus controller (EGS)

DTC	Trouble Code Title, Conditions & Possible Causes
DTC: P1699 **1T MIL: Yes** **2002-05** **Models:** Cooper, Cooper S **Engines:** All **Transmissions:** A/T	**EPROM Checksum Incorrect Conditions:** The EPROM Checksum is incorrect. This test is performed by the GIB (Gearbox Interface Box), a system dedicated to low level control of the transmission control unit. **Possible Causes:** • Short to battery • Short to ground • Open circuit • CAN data bus wires have short circuited to each other • ECU has failed • BUS system failure • Defective bus controller (EGS)
DTC: P1705 **2T MIL: Yes** **2002-04** **Models:** Cooper, Cooper S **Engines:** All **Transmissions:** A/T	**LED Drives Plausibility Conditions:** Key on or engine running; and the DME detected an implausible signal (fault performed by the Gearbox Interface Box). The battery voltage is greater than 9 volts and the CAN Bus is operational. **Possible Causes:** • Refer to the appropriate Transmission Repair Manual or information in electronic media to perform a complete diagnosis of the automatic transmission when this code is set
DTC: P1706 **2T MIL: Yes** **2002-04** **Models:** Cooper, Cooper S **Engines:** All **Transmissions:** A/T	**LED Drives Short Circuit Conditions:** Key on or engine running; and the DME detected short circuit (fault performed by the Gearbox Interface Box). The battery voltage is greater than 9 volts and the CAN Bus is operational. **Possible Causes:** • Refer to the appropriate Transmission Repair Manual or information in electronic media to perform a complete diagnosis of the automatic transmission when this code is set
DTC: P1739 **2T MIL: Yes** **2002-05** **Models:** Cooper, Cooper S **Engines:** All All **Models;** **Transmissions:** A/T	**Clutch Solenoid Circuit Communication Error Conditions:** The clutch solenoid circuit signal is implausible or missing. This test is performed by the GIB (Gearbox Interface Box), a system dedicated to low level control of the transmission control unit. **Possible Causes:** • Short to battery • Short to ground • Open circuit
DTC: P1741 **2T MIL: Yes** **2002-05** **Models:** Cooper, Cooper S **Engines:** All **Transmissions:** A/T	**Clutch Solenoid Circuit Open Circuit Conditions:** The clutch solenoid circuit continuity is in error. This test is performed by the GIB (Gearbox Interface Box), a system dedicated to low level control of the transmission control unit. **Possible Causes:** • Short to battery • Short to ground • Open circuit
DTC: P1742 **2T MIL: Yes** **2002-04** **Models:** Cooper, Cooper S **Engines:** All **Transmissions:** A/T	**Clutch Solenoid Circuit Short Circuit Conditions:** The clutch solenoid circuit continuity is in error. This test is performed by the GIB (Gearbox Interface Box), a system dedicated to low level control of the transmission control unit. **Possible Causes:** • Short to battery • Short to ground • Open circuit
DTC: P1749 **2T MIL: Yes** **2002, 2005** **Models:** Cooper, Cooper S **Engines:** All **Transmissions:** A/T	**Secondary Pressure Solenoid Circuit Communication Error Conditions:** The Secondary Pressure circuit signal is implausible or missing. This test is performed by the GIB (Gearbox Interface Box), a system dedicated to low level control of the transmission control unit. **Possible Causes:** • Short to battery • Short to ground • Open circuit

DTC	Trouble Code Title, Conditions & Possible Causes
DTC: P1751 **2T MIL: Yes** **2002, 2005** **Models:** Cooper, Cooper S **Engines:** All **Transmissions:** A/T	**Secondary Pressure Circuit Open Circuit Conditions:** The Secondary Pressure circuit continuity is in error. This test is performed by the GIB (Gearbox Interface Box), a system dedicated to low level control of the transmission control unit. **Possible Causes:** • Short to battery • Short to ground • Open circuit
DTC: P1752 **2T MIL: Yes** **2002, 2005** **Models:** Cooper, Cooper S **Engines:** All **Transmissions:** A/T	**Secondary Pressure Solenoid Circuit Short Circuit Conditions:** The Secondary Pressure circuit continuity is in error. This test is performed by the GIB (Gearbox Interface Box), a system dedicated to low level control of the transmission control unit. **Possible Causes:** • Short to battery • Short to ground • Open circuit

Gas Engine OBD II Trouble Code List (P2XXX Codes)

DTC	Trouble Code Title, Conditions & Possible Causes
DTC: P2096 **2T MIL: Yes** **2002-05** **Models:** Cooper, Cooper S **Engines:** All **Transmissions:** All	**Post Catalyst Fuel Trim System Too Lean (Bank 1) Conditions:** Engine started, battery voltage must be at least 11.5v, all electrical components must be off, the ground between the engine and the chassis must be well connected, the exhaust system must be properly sealed between the catalytic converter and the cylinder head, and the oxygen sensor heater for oxygen sensor before the catalytic converter must be properly functioning. The DME detected a problem with the fuel mixture. Trim control 1 segment (precision controller with oxygen sensor behind cat.) below delta lambda threshold of less than -1.56. Coolant temperature greater than 45 degrees Celsius. O2 heaters ready, fuel system in a closed loop, but the rear O2 sensor is in voltage outside the parameters. **Note: For resistance testing of sensor heating, oxygen sensor should be cooled to ambient temperature. High temperatures at oxygen sensor may lead to inaccurate measurements.** **Possible Causes:** • Oxygen sensor (before catalytic converter) is faulty • Oxygen sensor (behind catalytic converter) is faulty • Oxygen sensor heater (before catalytic converter) is faulty • Oxygen sensor heater (behind catalytic converter) is faulty • Check circuits for shorts to each other, ground or power
DTC: P2097 **2T MIL: Yes** **2002-05** **Models:** Cooper, Cooper S **Engines:** All **Transmissions:** All	**Post Catalyst Fuel Trim System Too Rich (Bank 1) Conditions:** Engine started, battery voltage must be at least 11.5v, all electrical components must be off, the ground between the engine and the chassis must be well connected, the exhaust system must be properly sealed between the catalytic converter and the cylinder head, and the oxygen sensor heater for oxygen sensor before the catalytic converter must be properly functioning. The DME detected a problem with the fuel mixture. Trim control 1 segment (precision controller with oxygen sensor behind cat.) below delta lambda threshold of less than -1.56. Coolant temperature greater than 45 degrees Celsius. O2 heaters ready, fuel system in a closed loop, but the rear O2 sensor is in voltage outside the parameters. **Note: For resistance testing of sensor heating, oxygen sensor should be cooled to ambient temperature. High temperatures at oxygen sensor may lead to inaccurate measurements.** **Possible Causes:** • Oxygen sensor (before catalytic converter) is faulty • Oxygen sensor (behind catalytic converter) is faulty • Oxygen sensor heater (before catalytic converter) is faulty • Oxygen sensor heater (behind catalytic converter) is faulty • Check circuits for shorts to each other, ground or power
DTC: P2122 **2T MIL: Yes** **2003-05** **Models:** Cooper, Cooper S **Engines:** All **Transmissions:** All	**Accelerator Pedal Position Sensor 'D' Circuit Low Input Conditions:** Engine started, battery voltage at least 11.5v, all electrical components off, ground connections between engine and chassis well connected, the DME detected that the accelerator pedal position sensor signal was outside the parameters to function normally. **Note: Both the Throttle Position (TP) Sensor and Accelerator Pedal Position Sensor are located at the accelerator pedal module and communicate the driver's intentions to the DME completely independently of each other. Both sensors are stored in one housing.** **Possible Causes:** • Ground between engine and chassis may be broken • Throttle position sensor may have failed • Accelerator Pedal Position Sensor has failed • Throttle position sensor wiring may have shorted • Throttle position sensor has failed • Faulty voltage supply
DTC: P2123 **2T MIL: Yes** **2003-05** **Models:** Cooper, Cooper S **Engines:** All **Transmissions:** All	**Accelerator Pedal Position Sensor 'D' Circuit High Input Conditions:** Engine started, battery voltage at least 11.5v, all electrical components off, ground connections between engine and chassis well connected, the DME detected that the accelerator pedal position sensor signal was outside the parameters to function normally. **Note: Both the Throttle Position (TP) Sensor and Accelerator Pedal Position Sensor are located at the accelerator pedal module and communicate the driver's intentions to the DME completely independently of each other. Both sensors are stored in one housing.** **Possible Causes:** • Ground between engine and chassis may be broken • Throttle position sensor may have failed • Accelerator Pedal Position Sensor has failed • Throttle position sensor wiring may have shorted • Throttle position sensor has failed • Faulty voltage supply

DTC	Trouble Code Title, Conditions & Possible Causes
DTC: P2127 **2T MIL: Yes** **2003-05** **Models:** Cooper, Cooper S **Engines:** All **Transmissions:** All	**Accelerator Pedal Position Sensor 'E' Circuit Low Input Conditions:** Engine started, battery voltage at least 11.5v, all electrical components off, ground connections between engine and chassis well connected, the DME detected that the accelerator pedal position sensor signal was outside the parameters to function normally. **Note: Both the Throttle Position (TP) Sensor and Accelerator Pedal Position Sensor are located at the accelerator pedal module and communicate the driver's intentions to the DME completely independently of each other. Both sensors are stored in one housing.** **Possible Causes:** • Ground between engine and chassis may be broken • Throttle position sensor may have failed • Accelerator Pedal Position Sensor has failed • Throttle position sensor wiring may have shorted • Throttle position sensor has failed • Faulty voltage supply
DTC: P2128 **2T MIL: Yes** **2003-05** **Models:** Cooper, Cooper S **Engines:** All **Transmissions:** All	**Accelerator Pedal Position Sensor 'E' Circuit High Input Conditions:** Engine started, battery voltage at least 11.5v, all electrical components off, ground connections between engine and chassis well connected, the DME detected that the accelerator pedal position sensor signal was outside the parameters to function normally. **Note: Both the Throttle Position (TP) Sensor and Accelerator Pedal Position Sensor are located at the accelerator pedal module and communicate the driver's intentions to the DME completely independently of each other. Both sensors are stored in one housing.** **Possible Causes:** • Ground between engine and chassis may be broken • Throttle position sensor may have failed • Accelerator Pedal Position Sensor has failed • Throttle position sensor wiring may have shorted • Throttle position sensor has failed • Faulty voltage supply
DTC: P2138 **2T MIL: Yes** **2003-05** **Models:** Cooper, Cooper S **Engines:** All **Transmissions:** All	**Throttle Position Sensor D/E Voltage Correlation Conditions:** Engine started, battery voltage must be at least 11.5v, all electrical components must be off, parking brake must be engaged (to keep daytime driving lights off), automatic transmission selector must be in park; and the DME detected the Throttle Position 'D' (TPD) and Throttle Position 'B' (TPE) sensors disagreed, or that the TPD sensor should not be in its detected position, or that the TPE sensor should not be in its detected position during testing. **Note: Both the Throttle Position (TP) Sensor and Accelerator Pedal Position Sensor are located at the accelerator pedal module and communicate the driver's intentions to the DME completely independently of each other. Both sensors are stored in one housing.** **Possible Causes:** • ETC TP sensor connector is damaged or shorted • ETC TP sensor circuits shorted together in the wire harness • ETC TP sensor signal circuit is shorted to VREF (5v) • ETC TP sensor is damaged or the DME has failed
DTC: P2270 **2T MIL: Yes** **2003-05** **Models:** Cooper, Cooper S **Engines:** All **Transmissions:** All	**O2 Sensor Signal Stuck Lean Bank 1 Sensor 2 Conditions:** Engine started, battery voltage must be at least 11.5v, all electrical components must be off, parking brake must be engaged (to keep daytime driving lights off), automatic transmission selector must be in park. The DME detected an unexpected voltage condition, or it detected an unexpected current draw in the heater circuit during the CCM test. Coolant temperature must been at least 80.25 degrees Celsius. The vehicle speed is greater than 27.96 and less than 80.76. The engine speed is between 1984 and 3647rpm. Ambient pressure is greater than 75.001kPa and the engine stability load is 6.94g/s. **Note: Vehicle must be raised before connector for oxygen sensors is accessible.** **Possible Causes:** • Oxygen sensor (before catalytic converter) is faulty • Oxygen sensor heater (before catalytic converter) is faulty • Oxygen sensor heater (before catalytic converter) is faulty • Oxygen sensor heater (behind catalytic converter) is faulty • O2S circuit is open or shorted in the wiring harness

DTC	Trouble Code Title, Conditions & Possible Causes
DTC: P2271 **2T MIL: Yes** **2003-05** **Models:** Cooper, Cooper S **Engines:** All **Transmissions:** All	**O2 Sensor Signal Stuck Rich Bank 1 Sensor 2 Conditions:** Engine started, battery voltage must be at least 11.5v, all electrical components must be off, parking brake must be engaged (to keep daytime driving lights off), automatic transmission selector must be in park. The DME detected an unexpected voltage condition, or it detected an unexpected current draw in the heater circuit during the CCM test. Coolant temperature must been at least 80.25 degrees Celsius. The vehicle speed is greater than 27.96 and less than 80.76. The engine speed is between 1984 and 3647rpm. Ambient pressure is greater than 75.001kPa and the engine stability load is 6.94g/s. **Note: Vehicle must be raised before connector for oxygen sensors is accessible.** **Possible Causes:** • Oxygen sensor (before catalytic converter) is faulty • Oxygen sensor heater (before catalytic converter) is faulty • Oxygen sensor heater (before catalytic converter) is faulty • Oxygen sensor heater (behind catalytic converter) is faulty • O2S circuit is open or shorted in the wiring harness
DTC: P2300 **2T MIL: Yes** **2003-04** **Models:** Cooper, Cooper S **Engines:** All **Transmissions:** All	**Ignition Coilpack A Primary/Secondary Circuit Malfunction Open Circuit/Short to Ground Conditions:** Engine started, battery voltage must be at least 11.5v, all electrical components must be off, parking brake must be engaged (to keep daytime driving lights off), automatic transmission selector must be in park and the ground between the engine and the chassis must be well connected. The DME did not receive any valid pulses from the ignition module for the Ignition Coilpack A primary circuit. **Note: Ignition coils and power output stages are one component and cannot be replaced individually.** **Possible Causes:** • Engine speed (RPM) sensor has failed • Camshaft Position (CMP) sensor has failed • Power Supply Relay is shorted to an open circuit • There is a malfunction in voltage supply • Ignition coilpack is damaged or it has failed • Cylinder 1 to 4 Fuel Injector(s) have failed•
DTC: P2301 **2T MIL: Yes** **2003-04** **Models:** Cooper, Cooper S **Engines:** All **Transmissions:** All	**Ignition Coilpack A Primary/Secondary Circuit Malfunction Short to Battery Conditions:** Engine started, battery voltage must be at least 11.5v, all electrical components must be off, parking brake must be engaged (to keep daytime driving lights off), automatic transmission selector must be in park and the ground between the engine and the chassis must be well connected. The DME did not receive any valid pulses from the ignition module for the Ignition Coilpack A primary circuit. **Note: Ignition coils and power output stages are one component and cannot be replaced individually.** **Possible Causes:** • Engine speed (RPM) sensor has failed • Camshaft Position (CMP) sensor has failed • Power Supply Relay is shorted to an open circuit • There is a malfunction in voltage supply • Ignition coilpack is damaged or it has failed • Cylinder 1 to 4 Fuel Injector(s) have failed
DTC: P2303 **2T MIL: Yes** **2003-04** **Models:** Cooper, Cooper S **Engines:** All **Transmissions:** All	**Ignition Coilpack A Primary/Secondary Circuit Malfunction Open Circuit/Short to Ground Conditions:** Engine started, battery voltage must be at least 11.5v, all electrical components must be off, parking brake must be engaged (to keep daytime driving lights off), automatic transmission selector must be in park and the ground between the engine and the chassis must be well connected. The DME did not receive any valid pulses from the ignition module for the Ignition Coilpack A primary circuit. **Note: Ignition coils and power output stages are one component and cannot be replaced individually.** **Possible Causes:** • Engine speed (RPM) sensor has failed • Camshaft Position (CMP) sensor has failed • Power Supply Relay is shorted to an open circuit • There is a malfunction in voltage supply • Ignition coilpack is damaged or it has failed • Cylinder 1 to 4 Fuel Injector(s) have failed
DTC: P2304 **2T MIL: Yes** **2003-04** **Models:** Cooper, Cooper S **Engines:** All **Transmissions:** All	**Ignition Coilpack A Primary/Secondary Circuit Malfunction Short to Battery Conditions:** Engine started, battery voltage must be at least 11.5v, all electrical components must be off, parking brake must be engaged (to keep daytime driving lights off), automatic transmission selector must be in park and the ground between the engine and the chassis must be well connected. The DME did not receive any valid pulses from the ignition module for the Ignition Coilpack A primary circuit. **Note: Ignition coils and power output stages are one component and cannot be replaced individually.** **Possible Causes:** • Engine speed (RPM) sensor has failed • Camshaft Position (CMP) sensor has failed • Power Supply Relay is shorted to an open circuit • There is a malfunction in voltage supply • Ignition coilpack is damaged or it has failed • Cylinder 1 to 4 Fuel Injector(s) have failed•

DTC	Trouble Code Title, Conditions & Possible Causes
DTC: P2400 **2T MIL: Yes** **2003-05** **Models:** Cooper, Cooper S **Engines:** All **Transmissions:** All	**EVAP Leak Detection Pump (LDP) Control Circuit Open Conditions:** Engine started, battery voltage must be at least 11.5v, all electrical components must be off, parking brake must be engaged (to keep daytime driving lights off), automatic transmission selector must be in park, the exhaust system must be properly sealed between the catalytic converter and the cylinder head, coolant temperature must be at least 80 degrees Celsius and oxygen sensor heaters for oxygen sensors before the catalytic converter must be functioning properly and the ground between the engine and the chassis must be well connected. The DME detected voltage irregularity in the leak detection pump control circuit. **Possible Causes:** • EVAP LDP power supply circuit is open • EVAP LDP solenoid valve is damaged or it has failed • EVAP LDP canister has a leak or a poor seal • EVAP canister system has an improper seal • Evaporative Emission (EVAP) canister purge regulator valve 1 has failed • Leak Detection Pump (LDP) is faulty • Aftermarket EVAP parts that do not conform to specifications • EVAP component seals leaking (i.e., leaks in the Purge valve, fuel tank pressure sensor, canister vent solenoid, fuel vapor control valve tube assembly or fuel vapor vent valve).
DTC: P2401 **2T MIL: Yes** **2003-05** **Models:** Cooper, Cooper S **Engines:** All **Transmissions:** All	**EVAP Leak Detection Pump Control Circuit Low Conditions:** Engine started, battery voltage must be at least 11.5v, all electrical components must be off, parking brake must be engaged (to keep daytime driving lights off), automatic transmission selector must be in park, the exhaust system must be properly sealed between the catalytic converter and the cylinder head, coolant temperature must be at least 80 degrees Celsius and oxygen sensor heaters for oxygen sensors before the catalytic converter must be functioning properly and the ground between the engine and the chassis must be well connected. The DME detected voltage irregularity in the leak detection pump control circuit. **Possible Causes:** • EVAP LDP power supply circuit is open • EVAP LDP solenoid valve is damaged or it has failed • EVAP LDP canister has a leak or a poor seal • EVAP canister system has an improper seal • Evaporative Emission (EVAP) canister purge regulator valve 1 has failed • Leak Detection Pump (LDP) is faulty • Aftermarket EVAP parts that do not conform to specifications • EVAP component seals leaking (i.e., leaks in the Purge valve, fuel tank pressure sensor, canister vent solenoid, fuel vapor control valve tube assembly or fuel vapor vent valve).
DTC: P2402 **2T MIL: Yes** **2003-05** **Models:** Cooper, Cooper S **Engines:** All **Transmissions:** All	**EVAP Leak Detection Pump Control Circuit High Conditions:** Engine started, battery voltage must be at least 11.5v, all electrical components must be off, parking brake must be engaged (to keep daytime driving lights off), automatic transmission selector must be in park, the exhaust system must be properly sealed between the catalytic converter and the cylinder head, coolant temperature must be at least 80 degrees Celsius and oxygen sensor heaters for oxygen sensors before the catalytic converter must be functioning properly and the ground between the engine and the chassis must be well connected. The DME detected voltage irregularity in the leak detection pump control circuit. **Possible Causes:** • EVAP LDP power supply circuit is open • EVAP LDP solenoid valve is damaged or it has failed • EVAP LDP canister has a leak or a poor seal • EVAP canister system has an improper seal • Evaporative Emission (EVAP) canister purge regulator valve 1 has failed • Leak Detection Pump (LDP) is faulty • Aftermarket EVAP parts that do not conform to specifications • EVAP component seals leaking (i.e., leaks in the Purge valve, fuel tank pressure sensor, canister vent solenoid, fuel vapor control valve tube assembly or fuel vapor vent valve).
DTC: P2404 **2T MIL: Yes** **2003-05** **Models:** Cooper, Cooper S **Engines:** All **Transmissions:** All	**EVAP Emission Control LDP Circuit Malfunction Pump Problem Conditions:** Key on, KOEO Self-Test enabled, and the DME detected an unexpected voltage condition on the EVAP emission control leak detection pump circuit. The reed switch level stays low after activation of solenoids within the time threshold of more than 1 second. The coolant temperature is less than 60 degrees Celsius, and the ambient pressure is greater than 76.2994kPa. The air intake temperature at start is between 9.04 and 16.04 degrees Celsius. The change in barometric pressure since engine start is less than 0.9998kPa. The vehicle speed is less than 74.56mph, and the purge valve has opened enough on previous driving cycle. **Possible Causes:** • Leak Detection Pump has failed • EVAP canister system has an improper or broken seal • Evaporative Emission (EVAP) canister purge regulator valve 1 is faulty • Hoses between the fuel pump and the EVAP canister are faulty • Fuel filler cap is loose • Fuel pump seal is defective, faulty or otherwise leaking • Hoses between the EVAP canister and the fuel flap unit are faulty • Hoses between the EVAP canister and the evaporative emission canister purge regulator valve are faulty

MINI
COMPONENT TESTING

11

TABLE OF CONTENTS

COMPONENT TESTING

Digital Motor Electronics (DME)

REMOVAL & INSTALLATION

1.6L (W10 and W11) Engines

COOPER AND COOPER S

See figure 1.

**** CAUTION**

Before beginning, always make sure to communicate with the fault memory with a BMW DIS (or equivalent OBD-II scan tool) for existing faults. It may be helpful to print out the results. Once the installation is complete, rerun the scan and correct the remaining faults.

**** CAUTION**

Make sure that all electrical accessories are off and the ignition is switched off.

1. Remove the battery cover and disconnect the battery.
2. Release the retaining clips and lift out the DME control unit.
3. Unlock the plug connections (pull outwards) and detach them in the direction of arrow.
4. Remove the control unit.

To install:

5. Replace the control unit.
6. Push in and lock the plug connections.
7. Replace the retaining clips and lift out the DME control unit.
8. Connect the battery and install the cover.
9. Complete the coding and programming of the replacement unit.

Engine Coolant Temperature (ECT) Sensor

REMOVAL & INSTALLATION

1.6L (W10 and W11) Engines

COOPER AND COOPER S

See figure 2.

**** WARNING**

There is a danger of scalding so only perform this task on an engine that has completely cooled down.

**** CAUTION**

Before beginning, always make sure to communicate with the fault memory with a BMW DIS (or equivalent OBD-II scan tool) for existing faults. It may be helpful to print out the results. Once the installation is complete, rerun the scan and correct the remaining faults.

**** CAUTION**

Make sure that all electrical accessories are off and the ignition is switched off.

1. Remove the intake filter housing.
2. Drain the coolant down to below the height of the thermostat housing.
3. For the R50 (W10 Cooper) model only: Remove the battery housing.
4. Unlock the plug connection and remove it.
5. Remove the coolant temperature sensor.

To install:

6. Replace the coolant temperature sensor.
7. Install the plug connection and lock down the tab.
8. Refill and vent the cooling system.
9. Clear the fault memory.

Fuel Level Sensor and Fuel Pump

REMOVAL & INSTALLATION

1.6L (W10 and W11) Engines

COOPER AND COOPER S

See figures 3 through 7.

1. Drain the fuel tank.
2. Remove the back seat.
3. Push the trim panel forward so it is clear of the sensor cover.
4. Unscrew the four nuts on the cover.
5. Remove the cover.
6. Disconnect the plug connection by pushing to the left and pulling up.
7. Release the screw cap using special tool 16 1 020.
8. Remove the locking clip from the line connection and set aside.
9. Detach the lines and plug from fuel level sensor unit.
10. Carefully lift the sensor unit out of the tank.

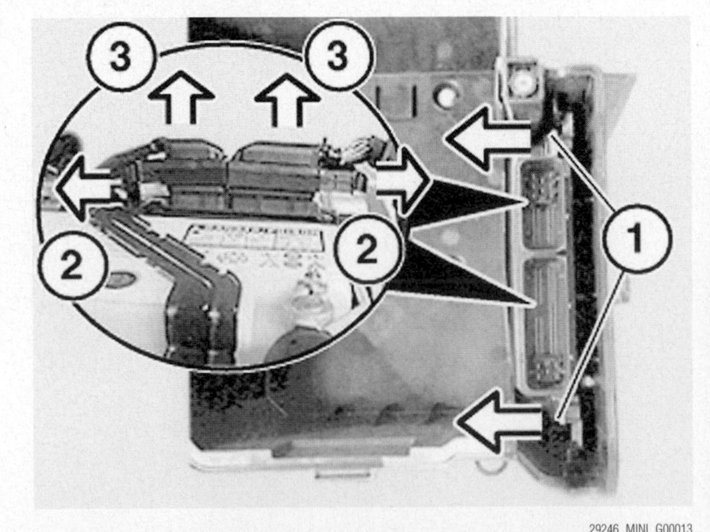

Fig. 1 The retaining clips (1), the plug connections (2) and the direction to remove them (3)

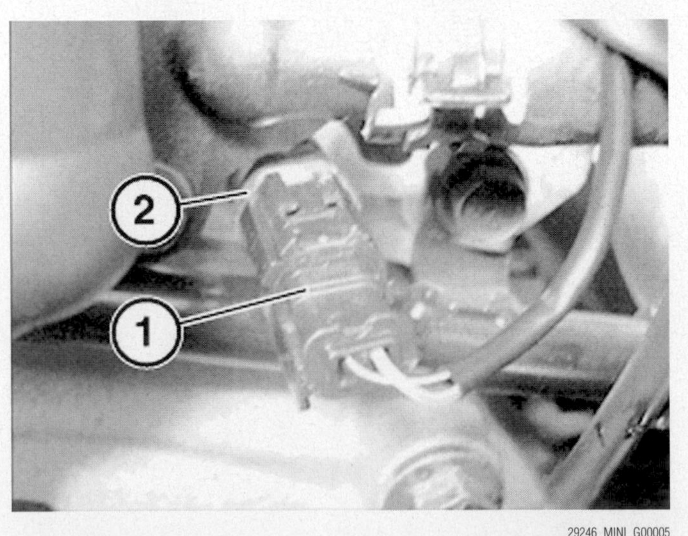

Fig. 2 The plug connection and the coolant temperature sensor

11. Catch any dripping fuel in a suitable container (and discard safely).

To install:

12. Before installing the sensor, always replace the rubber seal
13. When installing the fuel level sensor, make sure the lug on the sensor unit lines up with and engages in the recess on the tank.
14. Attach the lines and plug to fuel level sensor unit.
15. Replace the locking clip to the line connection.
16. Connect the plug connection by pushing down and to the right.
17. Replace the cover.
18. Replace the four nuts on the cover.
19. Replace the trim panel over the sensor cover.
20. Install the back seat.
21. Refuel the fuel tank.

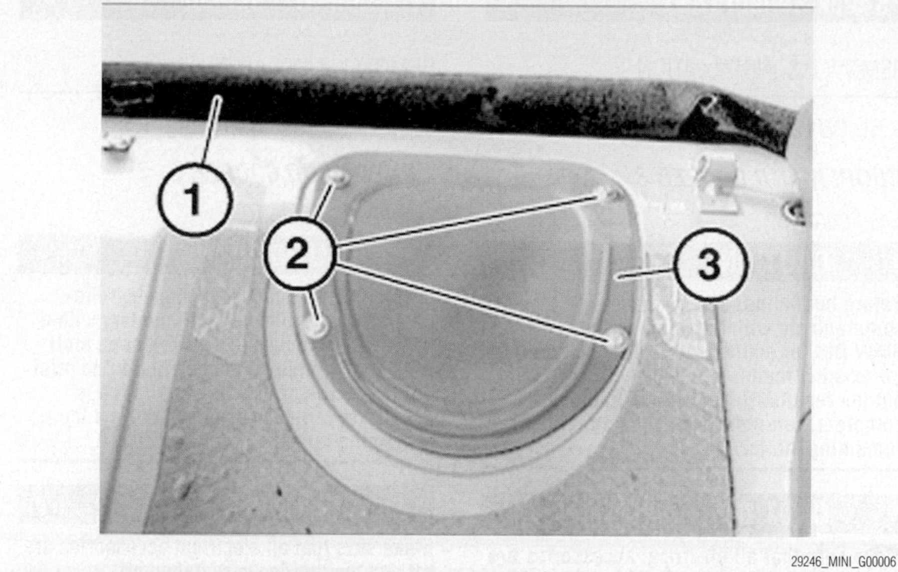

Fig. 3 The trim panel (1), the four retaining nuts (2) and the cover (3)

Fig. 4 The plug connection removal

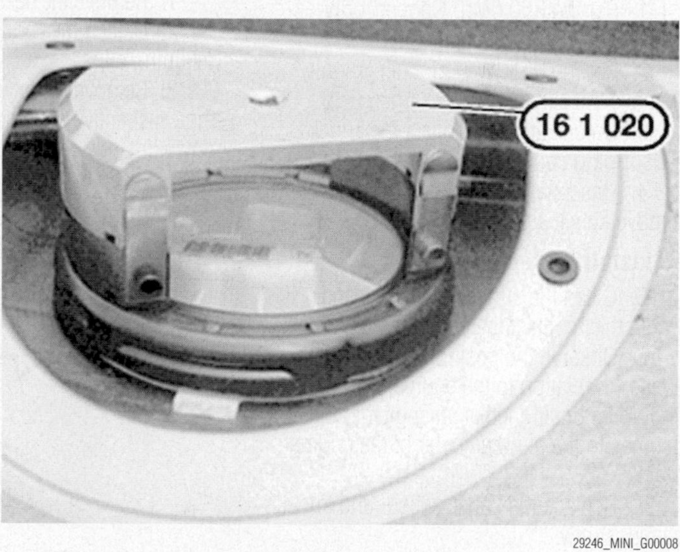

Fig. 5 Release the screw cap

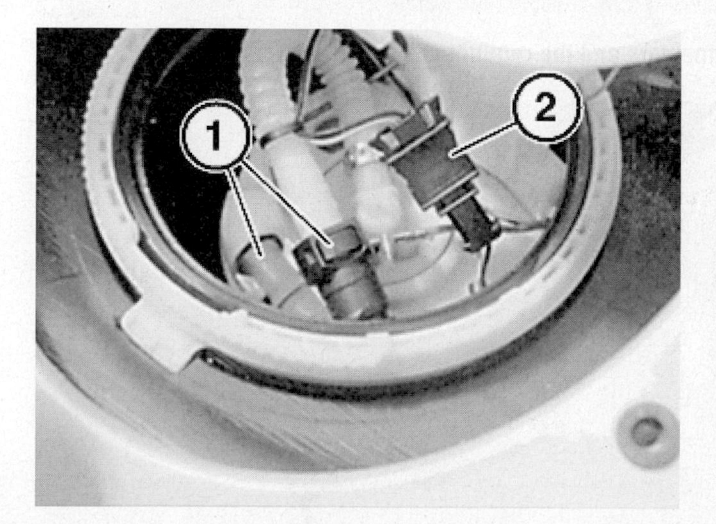

Fig. 6 The lines (1) and plugs (2) on the fuel level sensor

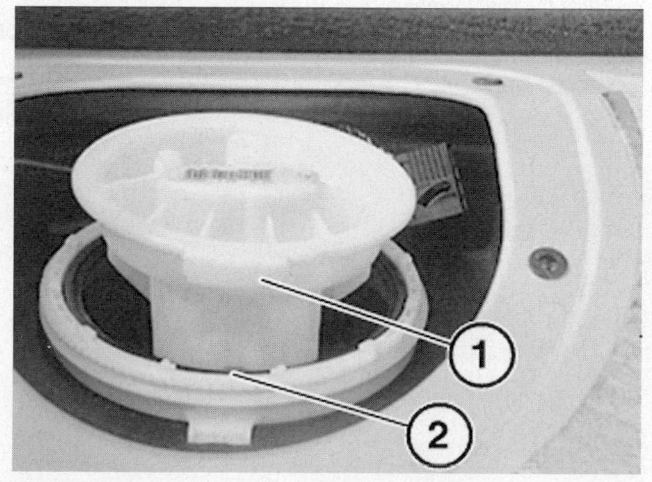

Fig. 7 The lug (1) on the sensor and the recess (2) on the tank

Knock Sensor (KS)

REMOVAL & INSTALLATION

1.6L (W10 and W11) Engines

COOPER AND COOPER S

See figure 8.

**** CAUTION**

Before beginning, always make sure to communicate with the fault memory with a BMW DIS (or equivalent OBD-II scan tool) for existing faults. It may be helpful to print out the results. Once the installation is complete, rerun the scan and correct the remaining faults.

**** CAUTION**

Make sure that all electrical accessories are off and the ignition is switched off.

1. For the R50 (W10 Cooper) only: Remove the intake air manifold.
2. For the R53 (W11 Cooper S) only: Remove the exhaust turbocharger.
3. Unlock and disconnect the knock sensor plug connection.
4. Unscrew the knock sensor screw and remove knock sensor.

To install:

5. Clean the surface of knock sensors where they contact the engine block.
6. Observe the position of the knock sensor in relation to the engine block. It should be positioned at an angle of 20 degrees to the perpendicular of the engine block.
7. Replace the knock sensor and knock sensor screw.
8. Connect and lock the knock sensor plug connection.
9. Clear the fault memory.

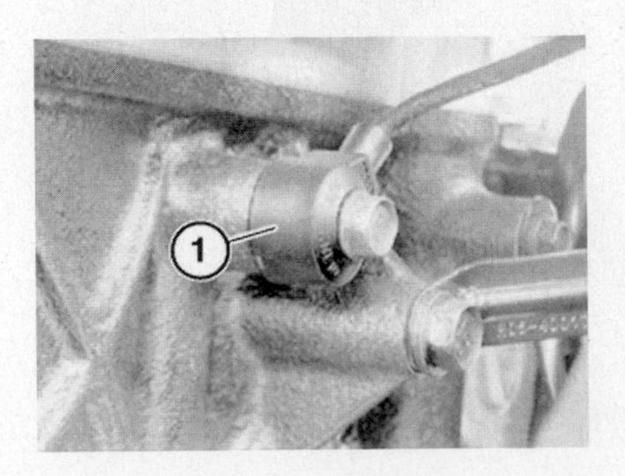

29246_MINI_G00004

Fig. 8 The knock sensor (1)

Oxygen (O2) Sensor

REMOVAL & INSTALLATION

1.6L (W10 and W11) Engines

COOPER AND COOPER S

See figures 9 and 10.

**** CAUTION**

Before beginning, always make sure to communicate with the fault memory with a BMW DIS (or equivalent OBD-II scan tool) for existing faults. It may be helpful to print out the results. Once the installation is complete, rerun the scan and correct the remaining faults.

**** CAUTION**

Make sure that all electrical accessories are off and the ignition is switched off.

1. Remove the heat shield.
2. Detach the connector from the retaining clips.
3. Remove the oxygen sensors from their respective positions upstream and downstream.

To install:

➡ **The threads of a new oxygen sensors are already coated with an anti-seize compound. If an oxygen sensor is to be used again, apply a thin and even coat of an anti-seize compound to the thread only.**

➡ **Do not clean the oxygen sensor section which protrudes into the exhaust line and ensure that it avoids all contact with any lubricants.**

➡ **Observe cable routing of the oxygen sensor so it doesn't interfere with any other system or the exhaust pipes.**

4. Replace the oxygen sensors in their respective positions upstream and downstream.
5. Attach the connector to the retaining clips.
6. Replace the heat shield.
7. Check for any remaining faults. Rectify them and clear the fault memory.

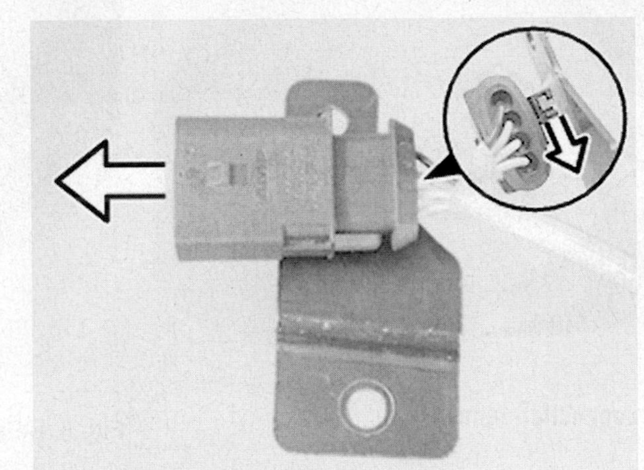

29246_MINI_G0001

Fig. 9 The plug connectors and the retaining clips

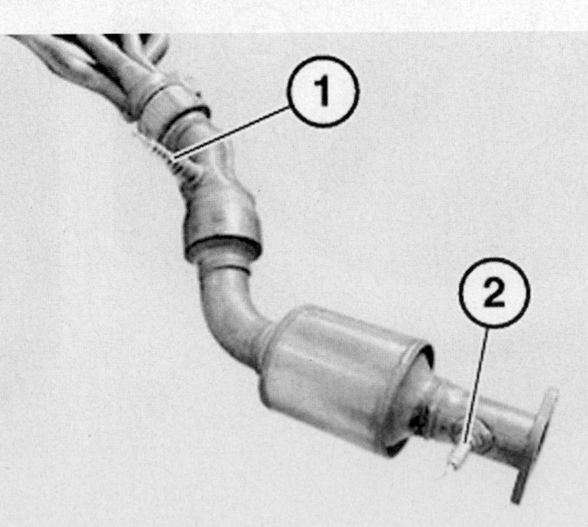

29246_MINI_G00002

Fig. 10 The upstream oxygen sensor (1) and the downstream oxygen sensor (2)

SAAB
DIAGNOSTIC TROUBLE CODES

12

TABLE OF CONTENTS

OBD II Vehicle Applications

SAAB

900
1996–1998

2.0L I4 B204I . . . . VIN J
2.0L I4 B204L. . . . . VIN N
2.3L I4 B234I . . . . VIN B
2.3L I4 B234R. . . . . VIN M
2.3L I4 B234R. . . . . VIN R

9000
1996–1998

2.0L I4 B204I . . . . VIN J
2.0L I4 B204L. . . . . VIN N
2.3L I4 B234I . . . . VIN B
2.3L I4 B234R. . . . . VIN M
2.3L I4 B234R. . . . . VIN R

9–3
1999–2000

2.0L I4 B204I . . . . VIN J
2.0L I4 B204L. . . . . VIN N
2.3L I4 B234R. . . . . VIN M
2.3L I4 B234R. . . . . VIN R

OBD II Trouble Code List (P0xxx Codes)

DTC	Trouble Code Title, Conditions & Possible Causes
DTC: P0105 **2T, MIL: Yes** **1996-2000** **Models:** 900, 9000, 9-3 **Engines:** 2.0L, 2.3L **Transmissions:** All	**Manifold Absolute Pressure (MAP) Sensor Malfunction Conditions:** All encompassing MAP sensor DTC. Poor driveability; other DTCs may be set; low turbocharger pressure; A/C compressor inoperative; Turbo/APC gauge readings faulty. **Possible Causes:** • MAP circuit open • MAP circuit shorted • MAP circuit connectors loose (an intermittent fault) • Vacuum hose between MAP sensor and intake manifold leaking or pinched • ECM failure (rare)
DTC: P0106 **2T, MIL: Yes** **1996-2000** **Models:** 900, 9000, 9-3 **Engines:** 2.0L, 2.3L **Transmissions:** All	**Manifold Absolute Pressure (MAP) Sensor Input Fault Conditions:** Engine is at least 3 minutes into closed loop operation; coolant temperature greater than 158°F (70°C); vehicle speed greater than 12 MPH; Poor driveability; P0171, P0172, P1171, P1172 or other DTCs may be set; low turbocharger pressure; A/C compressor inoperative; Turbo/APC gauge readings faulty; Pressure constant for 1 minute despite changes in throttle position. **Possible Causes:** • MAP sensor signal circuit open, shorted to ground • MAP sensor signal circuit is shorted to VREF • MAP sensor damaged or failed • ECM failure (rare)
DTC: P0107 **2T, MIL: Yes** **1996-2000** **Models:** 900, 9000, 9-3 **Engines:** 2.0L, 2.3L **Transmissions:** All	**Manifold Absolute Pressure (MAP) Sensor Circuit Low Input Conditions:** Poor driveability; other DTCs may be set; low turbocharger pressure; A/C compressor inoperative; Turbo/APC gauge readings faulty; Ignition ON or engine running; With wide throttle opening, ECM detected unexpected low voltage on MAP signal circuit during CCM test. **Possible Causes:** • MAP sensor signal circuit shorted to ground • MAP sensor damaged or failed • ECM failure (rare)
DTC: P0108 **2T, MIL: Yes** **1996-2000** **Models:** 900, 9000, 9-3 **Engines:** 2.0L, 2.3L **Transmissions:** All	**Manifold Absolute Pressure (MAP) Sensor Circuit High Input Conditions:** Poor driveability; other DTCs may be set; low turbocharger pressure; A/C compressor inoperative; Turbo/APC gauge readings faulty; Ignition ON or engine running; With slight throttle opening, ECM detected unexpected high voltage on MAP signal circuit during CCM test. **Possible Causes:** • MAP sensor connector is damaged (open circuit) • MAP sensor signal circuit open, shorted to VREF • MAP sensor ground circuit is open • MAP sensor damaged or failed • ECM failure (rare)
DTC: P0110 **2T, MIL: Yes** **1996-2000** **Models:** 900, 9000, 9-3 **Engines:** 2.0L, 2.3L **Transmissions:** All	**Intake Air Temperature (IAT) Sensor Signal Range/Performance Conditions:** All encompassing IAT sensor DTC. Ignition ON; ECM detects unexpected low or high voltage at IAT sensor circuit. **Possible Causes:** • IAT sensor signal circuit open, shorted to ground or to VREF • IAT sensor damaged or failed (out of calibration) • ECM failure (rare)
DTC: P0112 **2T, MIL: Yes** **1996-2000** **Models:** 900, 9000, 9-3 **Engines:** 2.0L, 2.3L **Transmissions:** All	**Intake Air Temperature (IAT) Sensor Circuit Low Input Conditions:** Ignition ON; ECM detects unexpected low voltage condition (less than 0.06 volts for 5 seconds) at IAT sensor signal circuit in CCM test. **Possible Causes:** • IAT sensor connector damaged (it may be shorted) • IAT sensor signal circuit shorted to ground • IAT sensor damaged or failed • ECM failure (rare)

DTC	Trouble Code Title, Conditions & Possible Causes
DTC: P0113 **2T, MIL: Yes** **1996-2000** **Models:** 900, 9000, 9-3 **Engines:** 2.0L, 2.3L **Transmissions:** All	**Intake Air Temperature (IAT) Sensor Circuit High Input Conditions:** Ignition ON; ECM detected an unexpected high voltage condition (greater than 4.96 volts for 5 seconds) at IAT sensor signal circuit in CCM test. **Possible Causes:** • IAT sensor signal circuit open or shorted to VREF • IAT sensor ground circuit open • IAT sensor damaged or failed • ECM failure (rare)
DTC: P0115 **2T, MIL: Yes** **1996-2000** **Models:** 900, 9000, 9-3 **Engines:** 2.0L, 2.3L **Transmissions:** All	**Engine Coolant Temperature (ECT) Sensor Circuit Malfunction Conditions:** All encompassing ECT sensor DTC. Impaired driveability, poor starting, misfiring, other DTCs set; Ignition ON; ECM detects ECT sensor input too high or too low during CCM test. **Possible Causes:** • ECT sensor connector damaged or loose (intermittent fault) • ECT sensor signal circuit open, shorted to ground or to VREF • ECT sensor damaged or failed (out of calibration) • ECM failure (rare)
DTC: P0117 **1T, MIL: Yes** **1996-2000** **Models:** 900, 9000, 9-3 **Engines:** 2.0L, 2.3L **Transmissions:** All	**Engine Coolant Temperature (ECT) Sensor Circuit Low Input Conditions:** Impaired driveability, poor starting, misfiring, other DTCs set; Ignition ON; ECM detects unexpected low voltage condition (less than 0.08 volts for 5 seconds) at ECT sensor signal circuit in CCM test. **Possible Causes:** • ECT sensor signal circuit shorted to ground • ECT sensor damaged or failed • ECM failure (rare)
DTC: P0118 **1T, MIL: Yes** **1996-2000** **Models:** 900, 9000, 9-3 **Engines:** 2.0L, 2.3L **Transmissions:** All	**Engine Coolant Temperature (ECT) Sensor Circuit High Input Conditions:** Impaired driveability, poor starting, misfiring, other DTCs set; Ignition ON; ECM detects unexpected high voltage condition (greater than 4.8 volts for 5 seconds) at ECT sensor signal circuit in CCM test. **Possible Causes:** • ECT sensor signal circuit open (connector may be loose) • ECT sensor ground circuit open (connector may be loose) • ECT sensor signal circuit shorted to sensor VREF • ECT sensor damaged or failed • ECM failure (rare)
DTC: P0120 **2T, MIL: Yes** **1996-2000** **Models:** 900, 9000, 9-3 **Engines:** 2.0L, 2.3L **Transmissions:** All	**Throttle Position Sensor (TPS) Circuit Malfunction Conditions:** All-encompassing TPS circuit DTC. Ignition ON; ECM detects unexpected TP sensor signal; High idle speed; A/C compressor inoperative. **Possible Causes:** • TP sensor signal circuit open or shorted to ground • TP sensor signal circuit shorted to VREF or to system power • TP sensor damaged or failed • ECM failure
DTC: P0121 **2T, MIL: Yes** **1996-2000** **Models:** 900, 9000, 9-3 **Engines:** 2.0L, 2.3L **Transmissions:** All	**Throttle Position Sensor (TPS) Signal Range/Performance Conditions:** No fuel system DTCs (P0171, P0172, P1171, P1172) are set. ECM detects unexpected TP sensor signal (greater than 1.0 volt or less than 0.24 volts); Engine RPM 1500-2500; vehicle speed about 12 MPH; coolant temperature above 158°F (70°C); A/C compressor inoperative; intake manifold pressure 11-24 in. Hg (38-80 kPa). **Possible Causes:** • TP sensor connector damaged or loose (intermittent fault) • TP sensor signal circuit open, shorted to ground or to VREF • TP sensor damaged • ECM failure
DTC: P0122 **1T, MIL: Yes** **1996-2000** **Models:** 900, 9000, 9-3 **Engines:** 2.0L, 2.3L **Transmissions:** All	**Throttle Position Sensor (TPS) Circuit Low Input Conditions:** Engine running; ECM detects unexpected TP sensor signal; High idle speed; A/C compressor inoperative; TP sensor signal voltage less than 0.2 volts for 5 seconds. **Possible Causes:** • TP sensor connector damaged (intermittent short) • TP sensor signal circuit shorted to ground • TP sensor damaged • ECM failure

DTC	Trouble Code Title, Conditions & Possible Causes
DTC: P0123 **1T, MIL: Yes** **1996-2000** **Models:** 900, 9000, 9-3 **Engines:** 2.0L, 2.3L **Transmissions:** All	**Throttle Position Sensor (TPS) Circuit High Input Conditions:** Engine running; ECM detects unexpected TP sensor signal; High idle speed; A/C compressor inoperative; TP sensor signal voltage greater than 4.96 volts for 5 seconds. **Possible Causes:** • TP sensor connector damaged or loose (intermittent open) • TP sensor signal circuit open or shorted to VREF • TP sensor damaged • ECM failure
DTC: P0125 **1T, MIL: Yes** **1996-2000** **Models:** 900, 9000, 9-3 **Engines:** 2.0L, 2.3L **Transmissions:** All	**Insufficient Coolant Temperature For Closed Loop Control Conditions:** Engine running; vehicle speed greater than 12 MPH; ECT sensor signal does not rise by 36°F (20°C) from starting temperature, or reach 104°F (40°C) within 7 minutes. **Possible Causes:** • Defective thermostat • Unapproved external block heater in use • Ice formed in cooling system • ECT sensor signal circuit fault or loose (intermittent) • ECT sensor damaged or failed (out of calibration)
DTC: P0130 **2T, MIL: Yes** **1996-2000** **Models:** 900, 9000, 9-3 **Engines:** 2.0L, 2.3L **Transmissions:** All	**Oxygen Sensor, Bank 1, Sensor 1 (HO2S) Circuit Conditions:** All-encompassing DTC for oxygen sensor operation. Other DTCs present for engine misfire, closed loop adaptation. **Possible Causes:** • First, diagnose other DTCs, such as P0105, P0106, P0107 and P0108 • Manifold Absolute Pressure (MAP) sensor damaged or circuit failure • HO2S damaged or failed • Fuel tank empty • Air leaks at MAP sensor or fuel pressure sensor and intake manifold
DTC: P0131 **2T, MIL: Yes** **1996-2000** **Models:** 900, 9000, 9-3 **Engines:** 2.0L, 2.3L **Transmissions:** All	**Oxygen Sensor, Bank 1, Sensor 1 (HO2S-11) Circuit Low Voltage Conditions:** Engine running in closed loop; short fuel trim greater than 25% below for more than 20 seconds while HO2S indicates rich fuel mixture. **Possible Causes:** • Low fuel pressure, fuel filter restricted or fuel injectors plugged • HO2S heater damaged or failed • HO2S contaminated or failed • Fuel tank empty
DTC: P0132 **2T, MIL: Yes** **1996-2000** **Models:** 900, 9000, 9-3 **Engines:** 2.0L, 2.3L **Transmissions:** All	**Oxygen Sensor, Bank 1, Sensor 1 (HO2S-11) High Voltage Conditions:** Engine running in closed loop; short fuel trim less than -25% below for more than 20 seconds while HO2S indicates lean fuel mixture. ECM detects minimum HO2S voltage greater than 1500 millivolts for more than 5 seconds. **Possible Causes:** • Fuel pressure regulator leaking or fuel injectors leaking • HO2S connector damaged or loose (intermittent open circuit) • HO2S signal circuit open or shorted to power • HO2S contaminated or failed • Fuel tank empty
DTC: P0133 **2T, MIL: Yes** **1996-2000** **Models:** 900, 9000, 9-3 **Engines:** 2.0L, 2.3L **Transmissions:** All	**Oxygen Sensor, Bank 1, Sensor 1 (HO2S-11) Circuit High Voltage Conditions:** Engine at idle; more than 3 minutes since engine started; closed loop active; engine coolant temperature greater than 140°F (60°C); HO2S preheating current less than 1500 milliamps; oxygen sensor voltage oscillates below standard specification of 0.1 to 0.9 volts. **Possible Causes:** • HO2S signal circuit open or shorted to ground • HO2S element contaminated, or HO2S heater failed • Intake air leaks, exhaust manifold leaks or PCV system leaks • MAP sensor out of calibration (it may be dirty or contaminated)
DTC: P0135 **2T, MIL: Yes** **1996-2000** **Models:** 900, 9000, 9-3 **Engines:** 2.0L, 2.3L **Transmissions:** All	**Oxygen Sensor, Bank 1, Sensor 1 (HO2S-11) Heater Circuit Conditions:** Engine running in closed loop, heating circuit active; ECM detects either too low or too high HO2S-11 circuit current (either below 500 milliamps or above 2300 milliamps) for 5 seconds. **Possible Causes:** • If DTC P0141 is also set, fuse no. 38 may be blown • If DTC P0141 is also set, check for wiring fault between H33-4 connector (behind glove box) pin 8 and fuel pump relay pin 87 • DTC HO2S heater control circuit open or shorted to ground • HO2S heater control circuit shorted to power • HO2S heater damaged or failed

DTC	Trouble Code Title, Conditions & Possible Causes
DTC: P0136 **2T, MIL: Yes** **1996-2000** **Models:** 900, 9000, 9-3 **Engines:** 2.0L, 2.3L **Transmissions:** All	**Oxygen Sensor, Bank 1, Sensor 2 (HO2S-12) Circuit Conditions:** All-encompassing DTC for oxygen sensor operation. Vehicle in motion; engine coolant temperature above 158°F (70°C); ECM detects HO2S voltage has not climbed above 0.4 volts, or it has not changed more than 0.5 volts in at least 8 minutes. **Possible Causes:** • HO2S signal circuit open or shorted to ground or power • HO2S deteriorated or has fuel contamination • HO2S connector contains moisture or is damaged • HO2S damaged
DTC: P0138 **2T, MIL: Yes** **1996-2000** **Models:** 900, 9000, 9-3 **Engines:** 2.0L, 2.3L **Transmissions:** All	**Oxygen Sensor, Bank 1, Sensor 2 (HO2S-12) High Voltage Conditions:** Ignition ON; HO2S voltage greater than 1500 millivolts for at least 5 seconds. DTC P0132 may also be set if reference ground or sensor voltage is shorted to power. **Possible Causes:** • Fuel injectors leaking • HO2S connector contains moisture or is damaged • HO2S signal circuit open or shorted to power • HO2S contaminated or failed
DTC: P0140 **2T, MIL: Yes** **1996-2000** **Models:** 900, 9000, 9-3 **Engines:** 2.0L, 2.3L **Transmissions:** All	**Oxygen Sensor, Bank 1, Sensor 2 (HO2S-12) No Activity Detected Conditions:** Ignition ON; HO2S signal is not greater than 0.7 volts at full throttle; HO2S signal is not approximately 0 volts during closed throttle. **Possible Causes:** • HO2S connector contains moisture or is damaged • HO2S signal circuit is open or shorted to ground • HO2S heater element contaminated, damaged or failed
DTC: P0141 **2T, MIL: Yes** **1996-2000** **Models:** 900, 9000, 9-3 **Engines:** 2.0L, 2.3L **Transmissions:** All	**Oxygen Sensor Heater Circuit, Bank 1, Sensor 2 (HO2S-12) Conditions:** Engine coolant temperature above 122°F (50°C); Heating circuit active; Heating current below 500 milliamps or above 2300 milliamps for 5 seconds. DTC P0135 may also be set. **Possible Causes:** • If DTC P0135 is also set, fuse no. 38 may be blown • If DTC P0135 is also set, check for wiring fault between H33-4 connector (behind glove box) pin 8 and fuel pump relay pin 87 • HO2S heater control circuit open or shorted to ground • HO2S heater control circuit shorted to power • HO2S heater element contaminated, damaged or failed
DTC: P0170 **2T, MIL: Yes** **1996-2000** **Models:** 900, 9000, 9-3 **Engines:** 2.0L, 2.3L **Transmissions:** All	**Fuel Trim, Bank 1 Conditions:** All-encompassing Fuel Trim or Adaptation DTC. Engine RPM oscillation at idle; Poor driveability; ECM detects lean or rich fuel mixture conditions; other DTCs are set (low fuel, MAP sensor, closed loop, misfiring). **Possible Causes:** • Low fuel level, indicated by DTC P1416 also being set • Intermittent short circuit in wiring leads and connectors • Air leak between MAP sensor and intake manifold • Air leak between fuel pressure regulator and intake manifold • Faulty MAP sensor • Exhaust system leaks before or near front HO2S • EVAP canister purge valve stuck closed • Fuel system not supplying enough fuel under high fuel demand conditions (i.e., fuel pump weak, fuel filter or fuel injectors dirty) • Vehicle running out of fuel, or driven until it runs out of fuel
DTC: P0171 **2T, MIL: Yes** **1996-2000** **Models:** 900, 9000, 9-3 **Engines:** 2.0L, 2.3L **Transmissions:** All	**System Too Lean, Bank 1 Conditions:** Engine started, engine running at cruise speed in closed loop, ECM detects a lean fuel mixture condition. **Possible Causes:** • Low fuel level, indicated by DTC P1416 also being set • Intermittent short circuit in wiring leads and connectors • Air leak between MAP sensor and intake manifold • Air leak between fuel pressure regulator and intake manifold • EVAP canister purge valve does not open when no current is applied • Faulty MAP sensor • Exhaust system leaks before or near front HO2S • Fuel system not supplying enough fuel under high fuel demand conditions (i.e., fuel pump weak, fuel filter or fuel injectors dirty) • Vehicle running out of fuel, or driven until it runs out of fuel

DTC	Trouble Code Title, Conditions & Possible Causes
DTC: P0172 **2T, MIL: Yes** **1996-2000** **Models:** 900, 9000, 9-3 **Engines:** 2.0L, 2.3L **Transmissions:** All	**System Too Rich, Bank 1 Conditions:** Engine started, engine running at cruise speed in closed loop, ECM detects a rich fuel mixture condition. **Possible Causes:** • Intermittent short circuit in wiring leads and connectors • Engine oil overfilled • EVAP canister purge valve does not close when current is applied • HO2S element contaminated with water or alcohol • Leaking fuel pressure regulator or leaking fuel injector(s) • MAP sensor signals invalid (it may be dirty or out of calibration)
DTC: P0300 **2T, MIL: Yes** **1996-2000** **Models:** 900, 9000, 9-3 **Engines:** 2.0L, 2.3L **Transmissions:** All	**Random, Multiple Cylinder Misfire Detected Conditions:** Engine running; ECM detects irregular, multiple misfires in two or more cylinders. Note: Diagnose any other DTCs set, then diagnose DTC P0300. Incorrectly detected misfire without actual misfire could be caused by ECM pin 17 or 18 shorted to ground. **Possible Causes:** • Low fuel level in tank; DTC P1416 also set • EVAP purge valve continuously open • Poor engine compression affecting two or more cylinders • Low fuel system pressure • Defective fuel injectors • Incorrect voltage from heated oxygen sensor (HO2S) • Defective ignition discharge module • MAP sensor faulty • Main relay faulty • ECT faulty
DTC: P0301 **2T, MIL: Yes** **1996-2000** **Models:** 900, 9000, 9-3 **Engines:** 2.0L, 2.3L **Transmissions:** All	**Cylinder 1 Misfire Detected Conditions:** Engine running from idle to high RPM; ECM detects misfire present in Cylinder 1. If misfiring occurs in more than one cylinder, cylinder 1 accounts for 95 percent of the misfires. **Possible Causes:** • Intermittent short circuit and breaks in wiring or loose connectors • Ignition coil for cylinder 1 inoperative • Ignition discharge module 10-pin connector trigger lead of cylinder 1 ignition coil has poor continuity • Ignition discharge module faulty, or has wet, or corroded contacts • Cylinder 1 spark plug electrode gap incorrect or in poor condition • Poor engine compression affecting only cylinder 1 • Fuel injector power supply faulty • Fuel injector resistance across pins 1 and 2 not about 12 ohms • Fuel injector frequency signal output not at 10 Hz for 10 seconds • Fuel injector fuel flow irregular
DTC: P0302 **2T, MIL: Yes** **1996-2000** **Models:** 900, 9000, 9-3 **Engines:** 2.0L, 2.3L **Transmissions:** All	**Cylinder 2 Misfire Detected Conditions:** Engine running from idle to high RPM; ECM detects misfire present in Cylinder 2. If misfiring occurs in more than one cylinder, cylinder 2 accounts for 95 percent of the misfires. **Possible Causes:** • Intermittent short circuit and breaks in wiring or loose connectors • Ignition coil for cylinder 2 inoperative • Ignition discharge module 10-pin connector trigger lead of cylinder 2 ignition coil has poor continuity • Ignition discharge module faulty, or has wet, or corroded contacts • Cylinder 2 spark plug electrode gap incorrect or in poor condition • Poor engine compression affecting only cylinder 2 • Fuel injector power supply faulty • Fuel injector resistance across pins 1 and 2 not about 12 ohms • Fuel injector frequency signal output not at 10 Hz for 10 seconds • Fuel injector fuel flow irregular

DTC	Trouble Code Title, Conditions & Possible Causes
DTC: P0303 **2T, MIL: Yes** **1996-2000** **Models:** 900, 9000, 9-3 **Engines:** 2.0L, 2.3L **Transmissions:** All	**Cylinder 3 Misfire Detected Conditions:** Engine running from idle to high RPM; ECM detects misfire present in Cylinder 3. If misfiring occurs in more than one cylinder, cylinder 3 accounts for 95 percent of the misfires. **Possible Causes:** • Intermittent short circuit and breaks in wiring or loose connectors • Ignition coil for cylinder 3 inoperative • Ignition discharge module 10-pin connector trigger lead of cylinder 3 ignition coil has poor continuity • Ignition discharge module faulty, or has wet, or corroded contacts • Cylinder 3 spark plug electrode gap incorrect or in poor condition • Poor engine compression affecting only cylinder 3 • Fuel injector power supply faulty • Fuel injector resistance across pins 1 and 2 not about 12 ohms • Fuel injector frequency signal output not at 10 Hz for 10 seconds • Fuel injector fuel flow irregular
DTC: P0304 **2T, MIL: Yes** **1996-2000** **Models:** 900, 9000, 9-3 **Engines:** 2.0L, 2.3L **Transmissions:** All	**Cylinder 4 Misfire Detected Conditions:** Engine running from idle to high RPM; ECM detects misfire present in Cylinder 4. If misfiring occurs in more than one cylinder, cylinder 4 accounts for 95 percent of the misfires. **Possible Causes:** • Intermittent short circuit and breaks in wiring or loose connectors • Ignition coil for cylinder 4 inoperative • Ignition discharge module 10-pin connector trigger lead of cylinder 4 ignition coil has poor continuity • Ignition discharge module faulty, or has wet, or corroded contacts • Cylinder 4 spark plug electrode gap incorrect or in poor condition • Poor engine compression affecting only cylinder 4 • Fuel injector power supply faulty • Fuel injector resistance across pins 1 and 2 not about 12 ohms • Fuel injector frequency signal output not at 10 Hz for 10 seconds • Fuel injector fuel flow irregular
DTC: P0327 **2T, MIL: No** **1996-2000** **Models:** 900, 9000, 9-3 **Engines:** 2.0L, 2.3L **Transmissions:** All	**Knock Sensor Circuit Low Input:** Engine running on basic turbocharger boost pressure; continuous ignition retard; No knock signal for 10 seconds. **Note: Saab Trionic system does not have a conventional knock sensor. Signal comes from ignition discharge module.** **Possible Causes:** • Open circuit between ignition discharge module 10-pin wiring connector pin 7 and battery power or ground. To test, connect multimeter alternately between pin 7 and power (voltage should be 6-12 volts, then drop rapidly to 0 volts); and pin 7 and ground (voltage should be -6 to -12 volts, then rise rapidly to 0 volts). • Short to ground (could be intermittent) • Break in signal lead from ECM pin 44 (indicated by 0.06-0.08 volts at various throttle positions) • Poor wiring continuity between ignition discharge module pin 7 and ECM pin 44 • Ignition discharge module faulty
DTC: P0335 **2T, MIL: Yes** **1996-2000** **Models:** 900, 9000, 9-3 **Engines:** 2.0L, 2.3L **Transmissions:** All	**Crankshaft Position (CKP) Sensor Circuit Malfunction Conditions:** Often an intermittent fault; engine misfires when vehicle speed is greater than 19 MPH, engine RPM is greater than 1500 RPM, and brakes are not applied; CKP reads incorrect number of ribs or teeth per crankshaft revolution; other DTCs may be set for misfire or camshaft position, especially if ECM pin 67 is not properly grounded. Engine may not start and DTC may not be set if CKP is completely inoperative. **Possible Causes:** • Crankshaft perforated ring damaged or loose • CKP sensor tip dirty • Voltage between CKP sensor connector pin 2 and B+ power not 12 volts • Voltage between CKP sensor connector pin 3 and B+ power not 12 volts • CKP sensor resistance between pins 1 and 2 is not 485 to 595 ohms • CKP sensor resistance between pins 1 and 3 is not infinite • Wiring between ECM pin 41 and CKP sensor pin 1 open or shorted • ECM pin 67 not grounded

DTC	Trouble Code Title, Conditions & Possible Causes
DTC: P0340 **2T, MIL: Yes** **1996-2000** **Models:** 900, 9000, 9-3 **Engines:** 2.0L, 2.3L **Transmissions:** All	**Camshaft Position (CMP) Sensor Circuit Malfunction Conditions:** Engine running for about 5 to 6 minutes (5000 engine revolutions since starting); ECM does not detect a CMP signal; additional DTCs (P0105-P0135, P0170-P0172, P0300-P0304, P0335, P0441-P0445, P1170-P1172) may be set. **Note: Diagnose other DTCs before diagnosing P0340.** **Possible Causes:** • Severe misfiring • Incorrect type, defective, or dirty spark plugs • Open circuit in wiring leads to ECM pins 17 or 18 • Voltage between ignition discharge module connector pin 8 and B+ power is not about 10 volts • Voltage between ignition discharge module connector pin 9 and ground is not about 10 volts • Ignition discharge module defective
DTC: P0420 **1T, MIL: Yes** **1996-2000** **Models:** 900, 9000, 9-3 **Engines:** 2.0L, 2.3L **Transmissions:** All	**Catalyst Efficiency Below Normal (Bank 1) Conditions:** To heat catalytic converter: Engine running under load for at least 2 minutes, obtaining at least 15 seconds of aggregated acceleration. To evaluate catalytic converter: Engine running under load between 19 and 63 MPH; coolant temperature greater than 140°F (60°C); voltage from rear HO2S greater than 0.4 volts at any time in drive cycle. **Possible Causes:** • Fuel additives • Using leaded fuel or fuel with sulfur • Oil additives if oil consumption is high • Exhaust system leaks between manifold and rear HO2S • Loose rear HO2S wiring or connector (intermittent fault) • Rear HO2S is loose (check it for a leak) • Fuel pressure and fuel pump volume
DTC: P0440 **1T, MIL: Yes** **1996-2000** **Models:** 900, 9000, 9-3 **Engines:** 2.0L, 2.3L **Transmissions:** All	**Evaporative Emission (EVAP) System Malfunction Conditions:** Engine idling, vehicle stationary, intake manifold pressure less than -0.18 in. Hg (-0.6 kPa) for 5 seconds. **Possible Causes:** • Intermittent faults usually caused by wiring short and open circuits • EVAP canister purge valve remains open when not supplied with current • EVAP canister shutoff valve is faulty
DTC: P0441 **2T, MIL: Yes** **1996-2000** **Models:** 900, 9000, 9-3 **Engines:** 2.0L, 2.3L **Transmissions:** All	**EVAP System Incorrect Purge Flow Conditions:** Other DTCs may be set for misfiring during closed loop operation, but front HO2S is OK; ECM detects pressure drop or low flow in EVAP system during the EVAP Monitor test period. Engine idling steadily, vehicle stationary, closed loop operation, coolant temperature greater than 140°F (60°C), no load changes. **Possible Causes:** • EVAP canister purge volume control solenoid stuck closed • EVAP canister purge volume control solenoid circuit is shorted • EVAP canister purge valve circuit resistance is not to specification; check for resistance of 23 to 29 ohms with ohmmeter connected between male wiring connector pins 5 and 6. • Open or shorted EVAP canister purge valve power circuit; check with engine at idle and test lamp between female connector pin 6 and ground • EVAP control system pressure sensor is damaged or has failed • EVAP emission canister or rubber tube is cracked or restricted • EVAP purge port is blocked or restricted • EVAP canister vent control valve is damaged or has failed
DTC: P0442 **2T, MIL: Yes** **1996-2000** **Models:** 900, 9000, 9-3 **Engines:** 2.0L, 2.3L **Transmissions:** All	**EVAP System Small Leak Detected Conditions:** Engine at idle; front HO2S is OK, possible fuel odor present, vehicle stationary for 7 seconds. ECM shuts off EVAP canister supply air, purge valve is active, and pressure drops. When purging is stopped, pressure rises too fast. ECM detects a small leak in the system. **Possible Causes:** • EVAP control system pressure sensor is damaged or has failed • EVAP purge solenoid or canister vacuum line loose or off • EVAP purge solenoid is damaged or has failed • EVAP emission canister leaks, is dirty, restricted or full of water • EVAP canister purge valve is damaged or has failed • Vacuum hose leak between EVAP canister and purge valve • Fuel tank cap loose, missing, incorrect type, or has a damaged seal • Vacuum leak at fuel tank above surface level, at filler pipe or breather pipe

DTC	Trouble Code Title, Conditions & Possible Causes
DTC: P0443 **2T, MIL: Yes** **1996-2000** **Models:** 900, 9000, 9-3 **Engines:** 2.0L, 2.3L **Transmissions:** All	**EVAP System Purge Control Valve Circuit Malfunction Conditions:** All-encompassing DTC. Engine running, purge valve active; ECM detects unexpected voltage in EVAP purge circuit, or it detects invalid EVAP signal during CCM test period. Other DTCs may be present. **Possible Causes:** • Purge connector is damaged or loose (intermittent fault) • Purge control circuit open, shorted to ground or power • Purge valve resistance not to specification; check with ohmmeter for 23 to 29 ohms across purge valve connector pins 1 and 2 • Purge valve power supply fault; check with test lamp between purge valve female connector pin 1 and ground • Purge valve ground connection fault; check for less than 1 ohm with ohmmeter connected between purge valve connector pin 2 and ECM pins 21 and 27 • Purge valve is damaged or has failed; with scan tool, activate valve for 10 seconds with 8 Hz, 50 percent; check for clicking sound, indicating activation
DTC: P0444 **2T, MIL: Yes** **1996-2000** **Models:** 900, 9000, 9-3 **Engines:** 2.0L, 2.3L **Transmissions:** All	**EVAP System Purge Control Valve Circuit Open Conditions:** Open circuit for more than 2 seconds. Engine running, purge valve active; ECM detects unexpected voltage in EVAP purge circuit, or it detects invalid EVAP signal during CCM test period. Other DTCs may be present. **Possible Causes:** • Purge connector is damaged or loose (intermittent fault) • Purge control circuit open, shorted to ground or power • Purge valve resistance not to specification; check with ohmmeter for 23 to 29 ohms across purge valve connector pins 1 and 2 • Purge valve power supply fault; check with test lamp between purge valve female connector pin 1 and ground • Purge valve ground connection fault; check for less than 1 ohm with ohmmeter connected between purge valve connector pin 2 and ECM pins 21 and 27 • Purge valve is damaged or has failed; activate valve for 10 seconds with 8 Hz, 50 percent; check for clicking sound, indicating activation
DTC: P0445 **2T, MIL: Yes** **1996-2000** **Models:** 900, 9000, 9-3 **Engines:** 2.0L, 2.3L **Transmissions:** All	**EVAP System Purge Control Valve Circuit Shorted Conditions:** Short circuit to ground for more than 2 seconds. Engine running, purge valve active; ECM detects unexpected voltage in EVAP purge circuit, or it detects invalid EVAP signal during CCM test period. Other DTCs may be present. **Possible Causes:** • Purge connector is damaged or loose (intermittent fault) • Purge control circuit open, shorted to ground or power • Purge valve resistance not to specification; check with ohmmeter for 23 to 29 ohms across purge valve connector pins 1 and 2 • Purge valve power supply fault; check with test lamp between purge valve female connector pin 1 and ground • Purge valve ground connection fault; check for less than 1 ohm with ohmmeter connected between purge valve connector pin 2 and ECM pins 21 and 27 • Purge valve is damaged or has failed; activate valve for 10 seconds with 8 Hz, 50 percent; check for clicking sound, indicating activation
DTC: P0451 **2T, MIL: Yes** **1996-2000** **Models:** 900, 9000, 9-3 **Engines:** 2.0L, 2.3L **Transmissions:** All	**EVAP System Pressure Sensor Circuit Range/Performance Conditions:** Ignition ON, vehicle stationary; EVAP pressure sensor signal oscillates more than 0.15 in. Hg (0.5 kPa) with more than 25 oscillations in 5 seconds. **Possible Causes:** • EVAP pressure sensor circuit connector is damaged or loose • EVAP pressure sensor circuit open, shorted to ground or 5 VREF; to test, turn Ignition ON then OFF, unplug fuel tank pressure sensor 4-pin connector near filler pipe, under car; check for 5 volts across connector pins 1 and 2, and between pin 1 and ground • EVAP pressure sensor is damaged or has failed
DTC: P0452 **2T, MIL: Yes** **1996-2000** **Models:** 900, 9000, 9-3 **Engines:** 2.0L, 2.3L **Transmissions:** All	**EVAP System Pressure Sensor Circuit Low Conditions:** Ignition ON or engine running; ECM detects unexpected low voltage (less than 0.1 volt for more than 5 seconds) at EVAP pressure sensor signal circuit. **Possible Causes:** • EVAP pressure sensor circuit connector is damaged or loose • EVAP pressure sensor circuit is shorted to ground; to test, turn Ignition ON then OFF, unplug fuel tank pressure sensor 4-pin connector near filler pipe, under car; then check for 5 volts across connector pins 1 and 2, and between pin 1 and ground • EVAP pressure sensor is damaged or has failed

DTC	Trouble Code Title, Conditions & Possible Causes
DTC: P0453 **2T, MIL: Yes** **1996-2000** **Models:** 900, 9000, 9-3 **Engines:** 2.0L, 2.3L **Transmissions:** All	**EVAP System Pressure Sensor Circuit High Conditions:** Ignition ON or engine running; ECM detects an unexpected high voltage (at least 5 volts for more than 5 seconds) at EVAP pressure sensor signal circuit. **Possible Causes:** • EVAP pressure sensor circuit connector is damaged (shorted) • EVAP pressure sensor circuit is open or shorted to 5 VREF; to test, turn Ignition ON then OFF, unplug fuel tank pressure sensor 4-pin connector near filler pipe, under car; then check for 5 volts across connector pins 1 and 2, and between pin 1 and ground • EVAP pressure sensor is damaged or has failed
DTC: P0455 **2T, MIL: Yes** **1996-2000** **Models:** 900, 9000, 9-3 **Engines:** 2.0L, 2.3L **Transmissions:** All	**EVAP System Large Leak Detected Conditions:** Engine at idle; front HO2S is OK, possible fuel odor present, vehicle stationary for 7 seconds. ECM shuts off EVAP canister supply air, no pressure drop occurs. ECM detects a large leak in the system. **Possible Causes:** • EVAP control system pressure sensor is damaged or has failed • EVAP purge solenoid or canister vacuum line loose or off • EVAP purge solenoid is damaged or has failed • EVAP emission canister leaks, is dirty, restricted or full of water • EVAP canister purge valve is damaged or has failed • Vacuum hose leak between EVAP canister and purge valve • Fuel tank cap loose, missing, incorrect type, or has a damaged seal • Vacuum leak at fuel tank above surface level, at filler pipe or breather pipe
DTC: P0500 **2T, MIL: Yes** **1996-2000** **Models:** 900, 9000, 9-3 **Engines:** 2.0L, 2.3L **Transmissions:** All	**Vehicle Speed Sensor (VSS) Circuit Malfunction Conditions:** All-encompassing DTC; engine speed greater than 1800 RPM, upshift lamp inoperative (manual transmission vehicles), fuel shut off in all gears, turbocharger pressure below 8.9 in. Hg (30 kPa) in 3rd, 4th and 5th gears (manual transmission vehicles), brake pedal not depressed, ABS warning lamp may be on, ECM did not detect a VSS signal for 4 seconds. **Possible Causes:** • Faulty ABS circuit wiring • ABS control unit damaged or failed • Speedometer damaged or failed • VSS signal circuit open, or shorted to ground or to power (B+) • VSS damaged or failed
DTC: P0501 **2T, MIL: Yes** **1996-2000** **Models:** 900, 9000, 9-3 **Engines:** 2.0L, 2.3L **Transmissions:** All	**Vehicle Speed Sensor (VSS) Circuit Range/Performance Conditions:** All-encompassing DTC; engine speed greater than 1800 RPM, upshift lamp inoperative (manual transmission vehicles), fuel shut off in all gears, turbocharger pressure below 8.9 in. Hg (30 kPa) in 3rd, 4th and 5th gears (manual transmission vehicles), indicated vehicle speed greater than 167 MPH (270 KPH). **Possible Causes:** • Faulty ABS circuit wiring • ABS control unit damaged or failed • Speedometer damaged or failed • VSS signal circuit open, or shorted to ground or to power (B+) • VSS damaged or failed
DTC: P0502 **2T, MIL: Yes** **1996-2000** **Models:** 900, 9000, 9-3 **Engines:** 2.0L, 2.3L **Transmissions:** All	**Vehicle Speed Sensor (VSS) Circuit Low Input Conditions:** All-encompassing DTC; ignition ON, engine speed greater than 1800 RPM, upshift lamp inoperative (manual transmission vehicles), fuel shut off in all gears, intake manifold pressure greater than 17.8 in. Hg (60 kPa), brake pedal not depressed, ECM did not detect a VSS signal for 4 seconds. **Possible Causes:** • Faulty ABS circuit wiring • ABS control unit damaged or failed • Speedometer damaged or failed • VSS signal circuit open, or shorted to ground or to power (B+) • VSS damaged or failed

DTC	Trouble Code Title, Conditions & Possible Causes
DTC: P0505 **2T, MIL: Yes** **1996-2000** **Models:** 900, 9000, 9-3 **Engines:** 2.0L, 2.3L **Transmissions:** All	**Idle Air Control System, Idle Air Control Valve Conditions:** Engine running, vehicle speed exceeded 12 MPH (20 KPH) during driving cycle, vehicle stationary now, engine coolant temperature greater than 140°F (60°C), TPS indicates idle speed, engine RPM is actually above 1200 RPM or below 700 RPM, normal idle cannot be achieved. **Possible Causes:** • IAC valve wiring circuit open or shorted to ground • IAC valve power circuit fault between wiring connector pin 2 and ground • Wiring continuity fault between female connector pin 1 and power (B+) • Air leakage after throttle valve • Throttle needs adjustment • EVAP purge valve stuck open • IAC valve damaged or failed • Brake servo damaged or failed
DTC: P0506 **2T, MIL: Yes** **1996-2000** **Models:** 900, 9000, 9-3 **Engines:** 2.0L, 2.3L **Transmissions:** All	**Idle Speed Control System RPM Lower Than Expected Conditions:** Engine running, vehicle speed exceeded 12 MPH (20 KPH) during driving cycle, vehicle stationary now, engine coolant temperature greater than 140°F (60°C), TPS indicates idle speed, engine RPM is actually below 700 RPM, normal idle cannot be achieved. **Possible Causes:** • IAC valve wiring circuit open or shorted to ground • IAC valve power circuit fault between wiring connector pin 2 and ground • Wiring continuity fault between female connector pin 1 and power (B+) • Air leakage after throttle valve • Throttle needs adjustment • EVAP purge valve stuck open • IAC valve damaged or failed • Brake servo damaged or failed
DTC: P0507 **2T, MIL: Yes** **1996-2000** **Models:** 900, 9000, 9-3 **Engines:** 2.0L, 2.3L **Transmissions:** All	**Idle Speed Control System RPM Higher Than Expected Conditions:** Engine running, vehicle speed exceeded 12 MPH (20 KPH) during driving cycle, vehicle stationary now, engine coolant temperature greater than 140°F (60°C), TPS indicates idle speed, engine RPM is actually above 1200 RPM, normal idle cannot be achieved. **Possible Causes:** • IAC valve wiring circuit open or shorted to ground • IAC valve power circuit fault between wiring connector pin 2 and ground • Wiring continuity fault between female connector pin 1 and power (B+) • Air leakage after throttle valve • Throttle needs adjustment • EVAP purge valve stuck open • IAC valve damaged or failed • Brake servo damaged or failed
DTC: P0605 **2T, MIL: Yes** **1996-2000** **Models:** 900, 9000, 9-3 **Engines:** 2.0L, 2.3L **Transmissions:** All	**ECM Internal Read Only Memory (ROM) Error Conditions:** Vehicle driven for 5 minutes over varying loads and engine speeds to complete a driving cycle, ignition switch ON, the ECM detected an internal calculation function problem or a problem in self-shutoff function. **Possible Causes:** • ECM connector damaged or not plugged in correctly • ECM ground or power supply pins 1, 24, 25, 48 and 50 shorted • ECM faulty. Before replacing ECM, double check ground and power points, clear DTCs, perform 5 minute driving cycle, turn ignition ON and OFF 10 times, and retest for DTC P0605. If same DTC resets, substitute a known good ECM and retest. If same DTC does not reset, original ECM has failed.

OBD II Trouble Code List (P1XXX Codes)

DTC	Trouble Code Title, Conditions & Possible Causes
DTC: P1170 **MIL: Yes** **1996-2000** **Models:** 900, 9000, 9-3 **Engines:** 2.0L, 2.3L **Transmissions:** All	**Closed Loop Malfunction Conditions:** All encompassing DTC; other DTCs may be set for low fuel level (P1416), misfiring, oxygen sensor, manifold pressure sensor; Engine running in closed loop mode; coolant temperature above 140°F (60°C), ECM determines engine is not operating in closed loop mode after driving cycle elapsed. **Possible Causes:** • If other DTCs are set, such as P1416, P0105-P0135, and P0441-P0445, fix those issues before proceeding. • Engine misfiring, usually indicated by DTC P1171 • Air leakage between manifold pressure sensor and intake manifold • Air leakage between fuel pressure regulator and intake manifold • Manifold pressure sensor faulty • HO2S signal circuit open or shorted to ground • HO2S damaged, contaminated or failed • Poor fuel pressure or inadequate fuel volume • EVAP canister purge valve stuck open • Exhaust system air leak ahead of HO2S
DTC: P1171 **MIL: Yes** **1996-2000** **Models:** 900, 9000, 9-3 **Engines:** 2.0L, 2.3L **Transmissions:** All	**Closed Loop Lean Mixture Conditions:** Other DTCs may be set for low fuel level (P1416), misfiring, oxygen sensor, manifold pressure sensor, Engine running in closed loop mode; coolant temperature above 140°F (60°C), ECM determines engine is not operating in closed loop mode after driving cycle elapsed. **Possible Causes:** • If other DTCs are set, such as P1416, P0105-P0135, and P0441-P0445, fix those issues before proceeding. • Engine misfiring • Air leakage between manifold pressure sensor and intake manifold • Air leakage between fuel pressure regulator and intake manifold • Manifold pressure sensor faulty • HO2S signal circuit open or shorted to ground • HO2S damaged, contaminated or failed • Poor fuel pressure or inadequate fuel volume • EVAP canister purge valve stuck open • Exhaust system air leak ahead of HO2S
DTC: P1172 **MIL: Yes** **1996-2000** **Models:** 900, 9000, 9-3 **Engines:** 2.0L, 2.3L **Transmissions:** All	**Closed Loop Rich Mixture Conditions:** Other DTCs may be set for low fuel level (P1416), misfiring, oxygen sensor, manifold pressure sensor; Engine running in closed loop mode; coolant temperature above 140°F (60°C), ECM determines engine is not operating in closed loop mode after driving cycle elapsed. **Possible Causes:** • If other DTCs are set, such as P1416, P0105-P0135, and P0441-P0445, fix those issues before proceeding. • Engine misfiring • Air leakage between manifold pressure sensor and intake manifold • Air leakage between fuel pressure regulator and intake manifold • Manifold pressure sensor faulty • HO2S signal circuit open or shorted to ground • HO2S damaged, contaminated or failed • Poor fuel pressure or inadequate fuel volume • EVAP canister purge valve stuck open • Exhaust system air leak ahead of HO2S
DTC: P1416 **MIL: No** **1996-2000** **Models:** 900, 9000, 9-3 **Engines:** 2.0L, 2.3L **Transmissions:** All	**Fuel Level Low Conditions:** Occurs when any of the following DTCs are set: P0170, P0171, P0172, P0300, P1170, P1171, P1172 and there is less than 2.64 gallons (10.0L) of fuel in the tank. Prevents unnecessary fault diagnosis. **Possible Causes:** • Low fuel level (below 2.64 gallons (10.0L) • DTCs may be set: P0170, P0171, P0172, P0300, P1170, P1171, P1172 • Intermittent wiring short or open circuit

DTC	Trouble Code Title, Conditions & Possible Causes
DTC: P1549 **MIL: No** **1996-2000** **Models:** 900, 9000, 9-3 **Engines:** 2.0L, 2.3L **Transmissions:** All	**Boost Pressure Control Valve (BPCV) Conditions:** ECM reduces boost pressure as much as possible, pushing Turbo/APC gauge to red zone. Boost pressure control switch is open. DTCs P0105, P0106, P0107, P0108 may be present; if so, resolve them first. **Possible Causes:** • Basic engine or turbocharger fault; To check, disconnect BPCV hoses W and C, connect them together with a pipe, and drive the car. Only basic boost pressure should be available during wide open throttle or sudden acceleration. If more than basic boost pressure is available, check for the following 4 issues: • Binding wastegate • Defective diaphragm • Incorrect or poorly adjusted basic boost pressure • Defective or incorrectly connected control hoses • Faulty fuse 24 • Faulty or loose BPCV air control hoses (C, W or R) • Inconsistent power supply to BPCV; To check, with ignition ON, connect test lamp between BPCV connector pin 2 and ground. • BPCV winding short circuit; Check resistance readings between BPCV pins 1 and 2, or between pins 2 and 3; correct resistance for either is 2-4 ohms. **Note: If any BPCV windings have a short circuit, the ECM is also likely to be damaged and will need to be replaced with the BPCV.** • BPCV wiring lead to pin 3 faulty; To check, with ignition ON, plug in BPCV connector and confirm that a soft buzzing noise can be heard. • BPCV wiring lead to pin 1 faulty; To check, with ignition ON, plug in BPCV connector, activate BPCV using scan tool, and confirm that a loud buzzing noise can be heard.
DTC: P1576 **MIL: Yes** **1996-2000** **Models:** 900, 9000, 9-3 **Engines:** 2.0L, 2.3L **Transmissions:** All	**Brake Light Switch Circuit Malfunction Conditions:** Engine running; basic boost pressure only; braking occurs 6 times from 22 MPH (35 KPH) to 0, with each braking session being shorter than 10 seconds, brake light switch is inoperative. **Possible Causes:** • Short circuit to power (B+) • If brake lights are inoperative, check for brake light wiring fault • Wiring fault between ECM pin 15 and wiring crimped connector J32
DTC: P1577 **MIL: Yes** **1996-2000** **Models:** 900, 9000, 9-3 **Engines:** 2.0L, 2.3L **Transmissions:** All	**Brake Light Switch Circuit Malfunction Conditions:** Engine running; basic boost pressure only; braking occurs 6 times from 22 MPH (35 KPH) to 0, with each braking session being shorter than 10 seconds, brake light switch is inoperative. **Possible Causes:** • Short circuit to ground, or open circuit • If brake lights are inoperative, check for brake light wiring fault • Wiring fault between ECM pin 15 and wiring crimped connector J32
DTC: P1611 **MIL: Yes** **1996-2000** **Models:** 900, 9000, 9-3 **Engines:** 2.0L, 2.3L **Transmissions:** A/T	**Check Engine Request Shorting to Ground Conditions:** Ignition ON; short circuit to ground for more than 2.5 seconds. **Possible Causes:** • Wiring continuity fault between Trionic ECM pin 61 and TCM pin 5. • Faulty TCM output stage; Check with ignition OFF, unplug Trionic ECM 70-pin connector, and connect test lamp between female connector pins 1 and 61; test lamp should go on for about 1 second then go out when ignition switch is turned ON. If TCM detects emission fault, test lamp will flash twice per second.

SAAB
COMPONENT TESTING

13

TABLE OF CONTENTS

Component Locations

See Figures 1 and 2.

The Saab Trionic OBD II engine management system consists of the following main components:

1. Electronic Control Module (ECM)
2. Crankshaft Position (CKP) sensor
3. Manifold Absolute Pressure (MAP) sensor
4. Intake Air Temperature (IAT) sensor
5. Engine Coolant Temperature (ECT) sensor
6. Throttle Position Sensor (TPS)
7. Heated Oxygen Sensor, front (HO2S1)
8. Heated Oxygen Sensor, rear (HO2S2)
9. Ignition Discharge Module (IDM)
10. Fuel injectors
11. Boost Pressure Control (BPC) valve
12. Idle Air Control (IAC) valve
13. Evaporative emission (EVAP) canister purge valve
14. Fuel tank pressure sensor
15. EVAP canister shutoff solenoid valve
16. Spark plugs

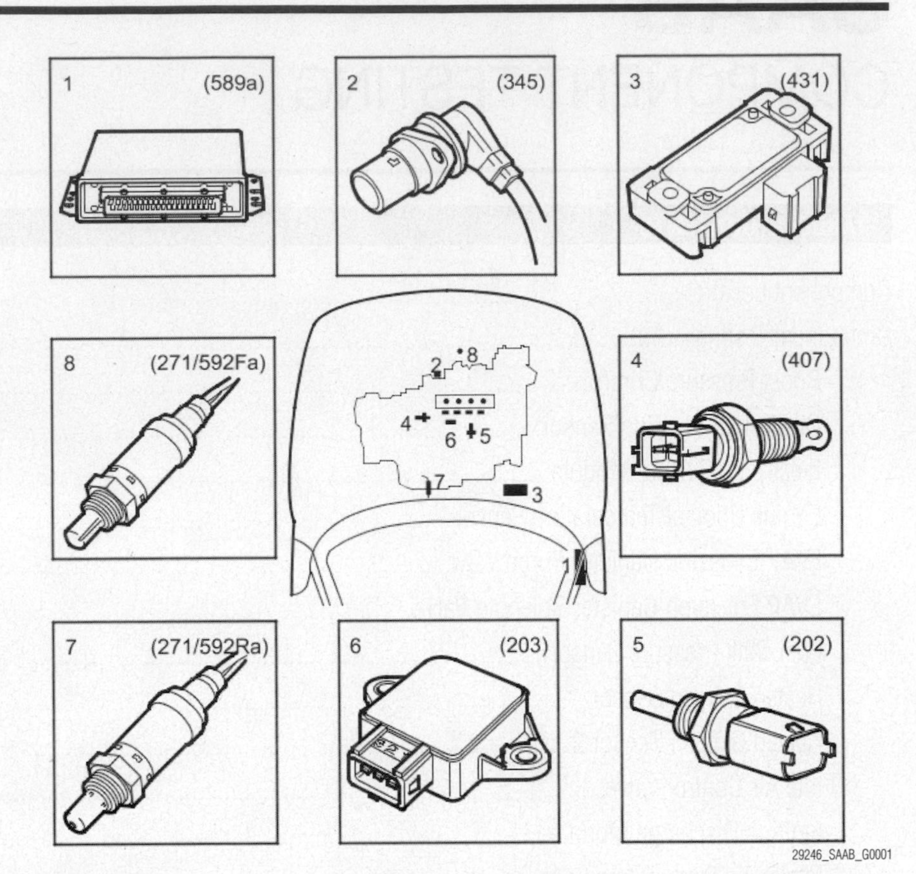

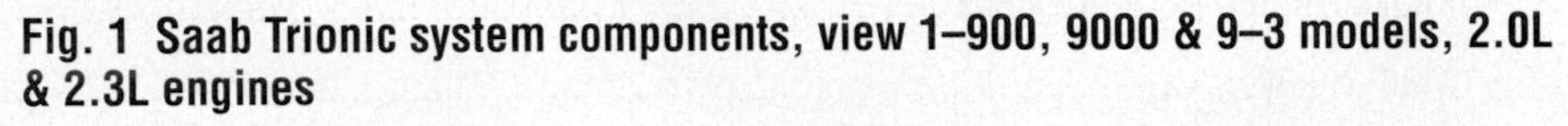

Fig. 1 Saab Trionic system components, view 1–900, 9000 & 9–3 models, 2.0L & 2.3L engines

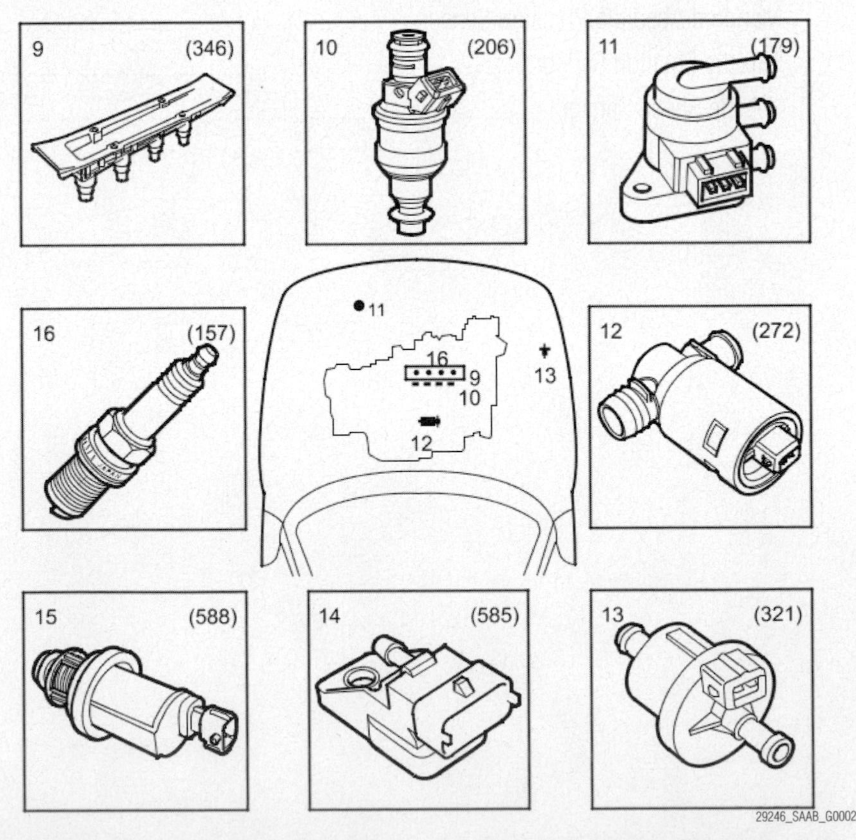

Fig. 2 Saab Trionic system components, view 2–900, 9000 & 9–3 models, 2.0L & 2.3L engines

Boost Pressure Control (BPC) Valve

LOCATION

See Figure 2, component 11. BPC valve is located at front of engine compartment.

OPERATION

See Figure 3.

Saab Trionic system regulates boost pressure by controlling a 2-coil, 3-way solenoid valve connected via hoses to turbocharger wastegate, compressor outlet and inlet. MAP sensor sends signal to ECM, which calculates correct boost pressure and sends signal to BPC valve, which controls the solenoid valve.

ECM pins 26 and 2 control the solenoid valve and send it power through circuit 54, via fuse 24. The control current is a pulse width modulated frequency at 90 Hz below 2500 rpm and 70 Hz above 2500 rpm. The change in frequency at 2500 rpm is to avoid resonance in the air hoses.

If pin 2 is grounded for longer than pin 26, boost pressure will drop. If pin 26 is grounded for longer than pin 2, the boost pressure will rise.

To regulate boost pressure, ECM first calculates the desired value based on programmed boost pressures for every engine speed and throttle position. Boost pressure for each engine speed is chosen so that

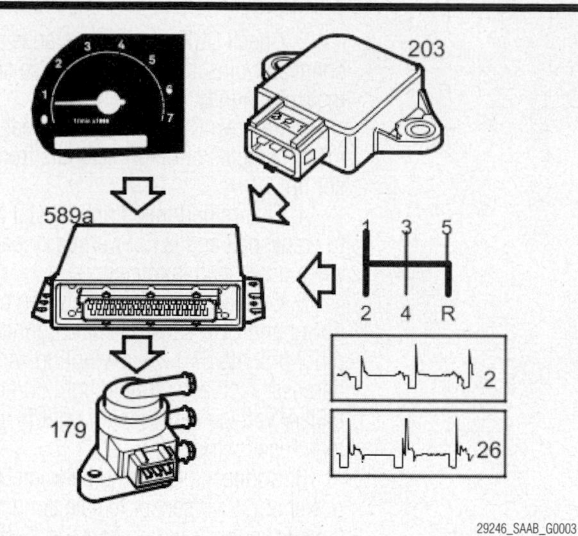

Fig. 3 **BCP valve operation–900, 9000 & 9–3 models, 2.0L & 2.3L engines**

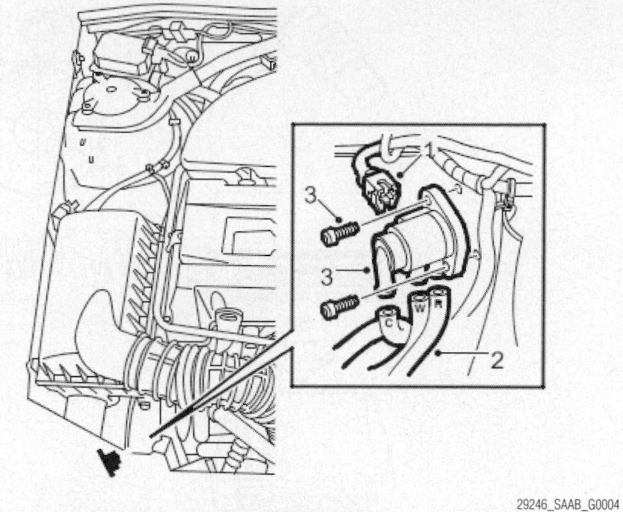

Fig. 4 **Removing BCP valve–900, 9000 & 9–3 models, 2.0L & 2.3L engines**

desired torque curve is applied to the engine at wide open throttle. Specific limits are set for certain situations:

- first, second or reverse gear position
- engine knock is detected
- vehicle top speed is attained

Basic boost pressure occurs when:

- brake pedal is depressed and ECM receives battery power on pin 15
- certain DTCs are set
- cruise control is active and ECM pin 36 is grounded
- battery power is low

REMOVAL & INSTALLATION

See Figure 4.

1. Disconnect BPC valve wiring connector.
2. Disconnect BPC valve vacuum hoses.

➡ **Note location of each hose to valve nipples.**

3. Remove BPC valve screws and remove valve.

To install:

Reverse removal procedure, reconnecting hoses to correct valve nipples.

TESTING

BPC valve can be activated with an appropriate diagnostic scan tool.

1. Check BPC valve resistance with the ignition OFF. Connect an ohmmeter across BPC valve's two terminals. Resistance should be 2 to 4 ohms at 68°F (20°C).
2. DTC 1549 will be set if intake pressure exceeds limits.
3. Check for intermittent wiring faults with a voltmeter while wiggling wiring harness and connectors.
4. Check fuse 24 and change it if necessary.

Crankshaft Position (CKP) Sensor

LOCATION

See Figure 5.

Crankshaft Position (CKP) sensor is mounted in engine crankcase wall at rear of engine. It reads a ring, perforated with 58 ribs, which is mounted to crankshaft.

OPERATION

See Figure 6.

The inductive Crankshaft Position (CKP) sensor sends crankshaft position and speed data via a perforated ring with 58 ribs to the Engine Control Module (ECM) pin 41. The distance between sensor and perforated ring is 0.016-0.051 in. (0.4-1.3 mm) and is not adjustable. CKP sensor acts as a generator, producing an alternating current with a sine wave frequency that can be viewed on an oscilloscope.

When rib number 1 passes the sensor, ECM knows crankshaft is 117° BTDC. CKP sensor voltage and frequency vary with engine RPM. At idle, voltage is 7-10 volts and AC frequency is about 30 Hz. At 2500 RPM, voltage output is 15 to 20 volts and AC frequency is about 85 Hz. Using AC frequency readings, ECM measures engine speed and crankshaft position primarily to monitor ignition timing, idle speed, fuel injection timing and duration, and turbocharger boost pressure. As ECM receives pulses from CKP sensor, it grounds the main and fuel pump relays. ECM cuts off fuel injection when engine speed exceeds 6200 RPM.

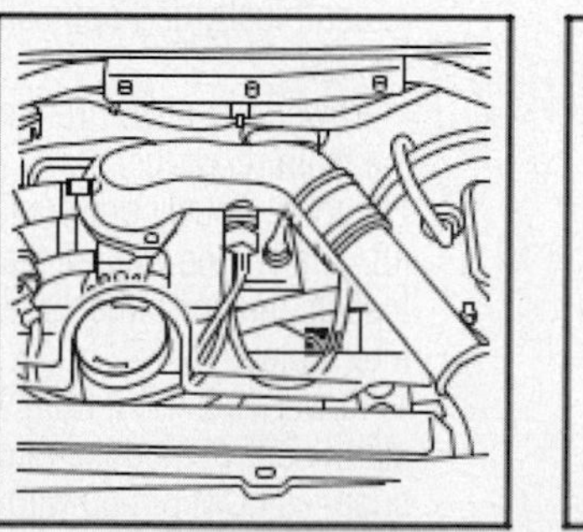

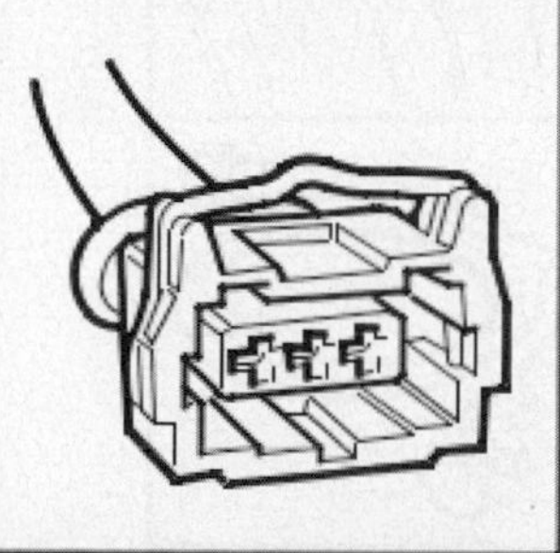

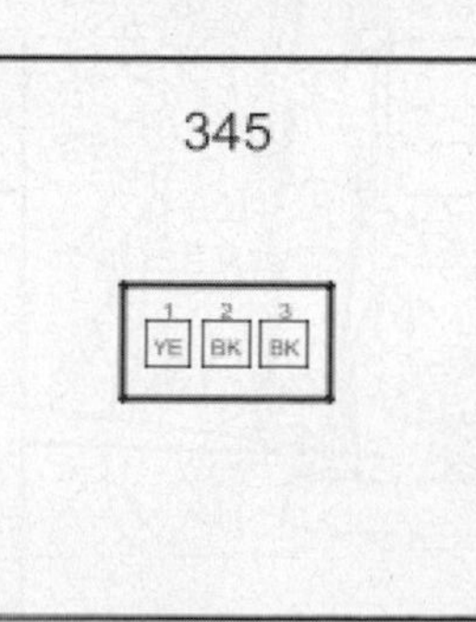

Fig. 5 **CKP sensor location–900, 9000 & 9–3 models, 2.0L & 2.3L engines**

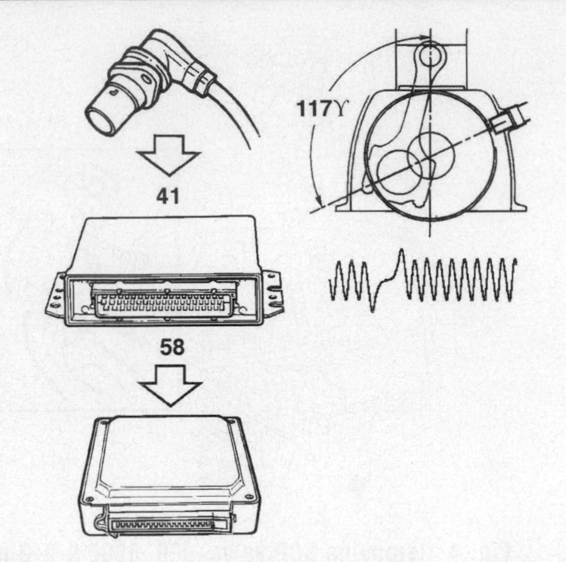

Fig. 6 CKP sensor operation–900, 9000 & 9-3 models, 2.0L & 2.3L engines

REMOVAL & INSTALLATION

See Figure 7.

1. Remove engine top cover.
2. Remove boost pressure delivery pipe and plug the delivery hose.
3. Disconnect CKP sensor connector wiring and remove clips and cable ties.

➡ **Remember wire positions or mark wires for correct installation.**

4. Remove CKP wiring.
5. Remove sensor heat shield, if equipped.
6. Remove CKP sensor retaining bolt and remove sensor while noting sealing ring position for correct installation.

To install:

7. Clean CKP sensor seating area and install sealing ring in correct position.
8. Tighten sensor retaining bolt to 6 ft. lbs. (8 Nm).
9. Reverse removal procedure for remaining components.

TESTING

See Figures 8 and 9.

Engine will usually not start and DTC P0335 will be set if there is a fault in Crankshaft Position (CKP) sensor circuit.

1. Using an ohmmeter, check CKP sensor resistance across connector pins 1 and 2 (middle pin). Resistance should measure 485-595 ohms.

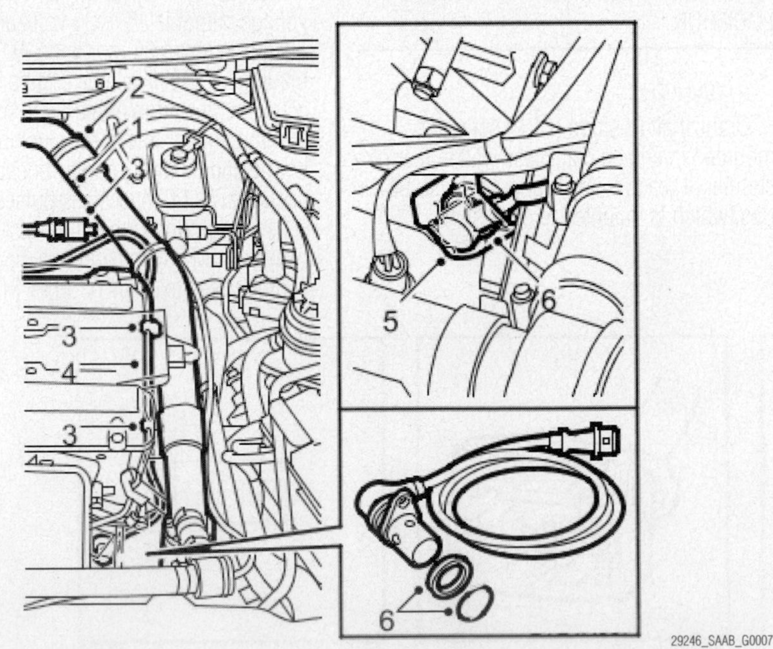

Fig. 7 Removing CKP sensor–900, 9000 & 9-3 models, 2.0L & 2.3L engines

2. Check CKP sensor resistance across connector pins 1 and 3. Resistance should measure infinity.
3. Remove CKP sensor and clean any metal particles or other materials from sensor tip.
4. Ensure perforated ring is well affixed to crankshaft and is not warped or damaged when crankshaft is rotated.
5. Connect a test lamp between battery power and CKP sensor female connector pin 2 (middle pin) while wiggling wiring harness. Test lamp should light, indicating battery voltage is present. If light is intermittent, repair wiring.
6. Connect a test lamp between battery power and CKP sensor female connector pin 3 while wiggling wiring harness. Test lamp should light, indicating battery voltage is present. If light is intermittent, repair wiring.
7. Check wiring between CKP pin 1 and ECM pin 41 for an open circuit, a short to ground or a short to power.
8. If all tests pass, exchange CKP with a known good unit, clear all DTCs and perform a drive cycle at varying engine loads and speeds for five minutes.
9. Evaluate drive cycle and check that DTCs have not reset.

Electronic Control Module (ECM)

LOCATION

See Figure 10.

Saab Trionic Electronic Control Module (ECM) is behind glove box, at right side of vehicle, below "A" pillar.

OPERATION

See Figures 11 and 12.

Saab Trionic Electronic Control Module controls ignition timing, fuel injection and turbocharger boost pressure via a 70-pin wiring connector that connects ECM to its various input and output components. It contains a 32-bit processor to receive inputs, process data and provide outputs to continually and optimally operate the vehicle.

Power circuit 30, from maxi fuse 2 to the main relay, fuel pump relay and fuse 28, continuously powers ECM pins 1 and 48. The engine will not start and ECM will lose its memory when power circuit 30 is interrupted.

Power circuit 15, from vehicle anti-theft alarm control module, through fuse 17, supplies ECM pin 60 with power. The engine will not start unless power is supplied to ECM via circuit 15.

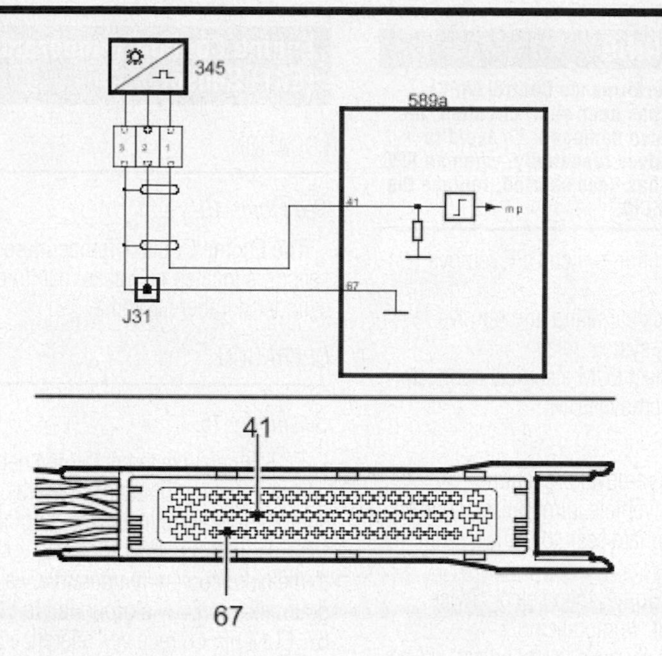

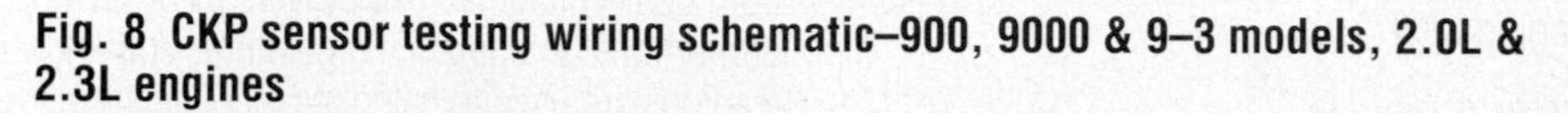

29246_SAAB_G0008

Fig. 8 CKP sensor testing wiring schematic–900, 9000 & 9-3 models, 2.0L & 2.3L engines

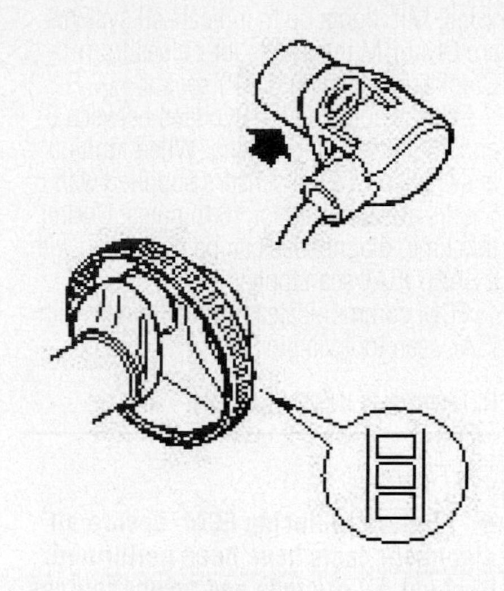

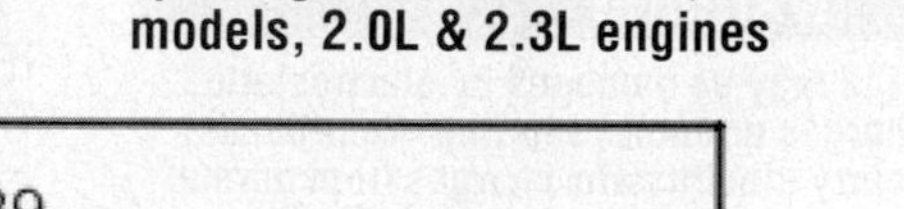

29246_SAAB_G0009

Fig. 9 Removing, cleaning and inspecting CKP sensor–900, 9000 & 9–3 models, 2.0L & 2.3L engines

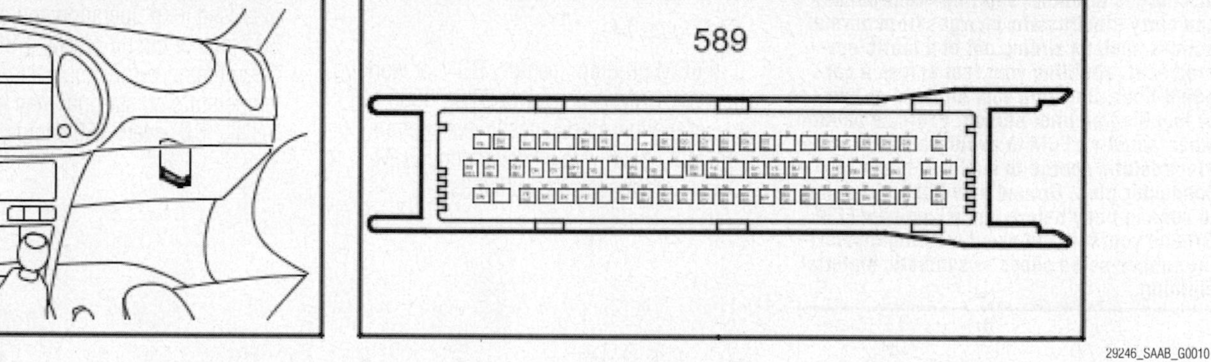

29246_SAAB_G0010

Fig. 10 ECM location below right side "A" pillar–900, 9000 & 9-3 models, 2.0L & 2.3L engines

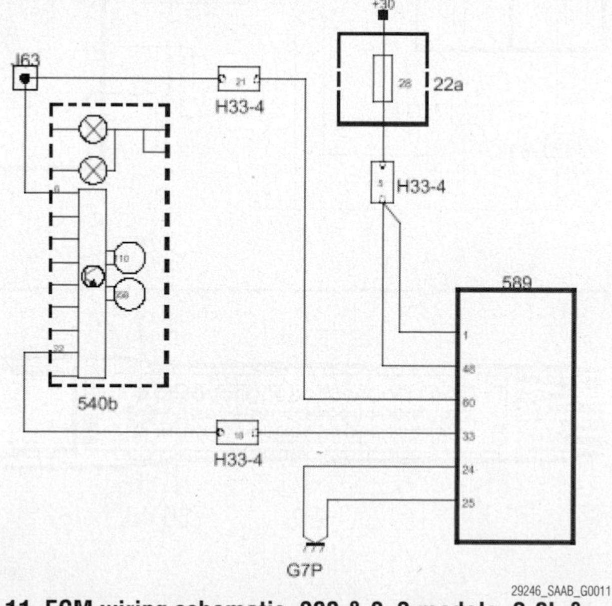

29246_SAAB_G0011

Fig. 11 ECM wiring schematic–900 & 9-3 models, 2.0L & 2.3L engines

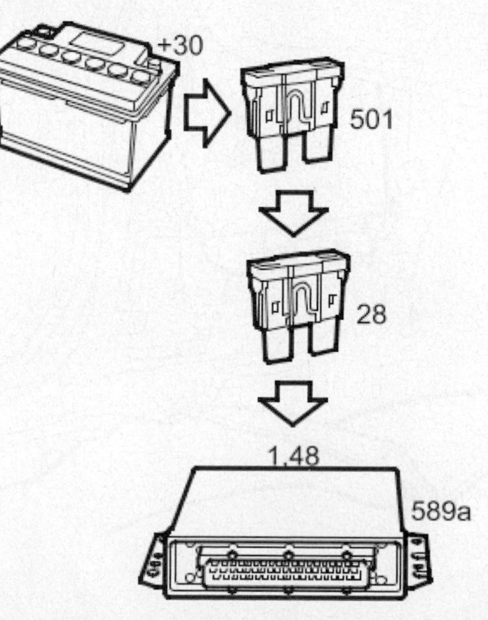

29246_SAAB_G0012

Fig. 12 ECM power supply–900 & 9-3 models, 2.0L & 2.3L engines

With ignition switch ON, ECM is activated, MIL lights up to indicate all systems are OK. ECM then waits for pulses from Crankshaft Position (CKP) sensor.

ECM is adjusted for a voltage between 8 and 16 volts during driving. When ignition is switched OFF, all sensors supplied with 5 volts are activated for 15 minutes. During this time, diagnostics can be performed with a SAAB ISAT scan tool.

ECM communicates bi-directionally with ISAT scan tool via pin 33.

REMOVAL & INSTALLATION

See Figure 13.

➡ **Before replacing ECM, ensure all diagnostic tests have been performed. Recheck all grounds and power sources to ECM. Recheck all electronic components.**

✱✱ WARNING

An ECM may be damaged by electrostatic discharges or output shorting. Your hands can carry electrostatic charges from normal actions such as sliding out of a fabric-covered seat, shuffling your feet across a carpeted floor, touching your clothing or hair, or touching another person. Exercise caution when handling ECM to avoid transmitting an electrostatic charge to it. Never touch ECM connector pins. Ground new ECM packaging to vehicle body before unpacking new ECM. Ground yourself and avoid wearing insulating rubber-soled shoes or synthetic material clothing.

✱✱ WARNING

If Automatic Performance Control (APC) valve winding has been short circuited, the ECM is likely also damaged. To avoid replacing APC valves repeatedly, when an APC valve winding has been shorted, replace the valve and the ECM.

1. With ignition switch OFF, remove glove box.
2. Fold back carpeting and remove central locking system relay.
3. Disconnect ECM electrical connector and carefully remove ECM.

To install:

4. Reverse removal procedure.
5. To reset vehicle immobilizer, connect ISAT or appropriate scan tool. Turn ignition switch ON.
6. Select "Body", then select "Twice", and then select "Immobilizing".
7. After 10 seconds, immobilizer will be programmed.

TESTING

See Figure 14.

If ECM program memory ROM or working memory RAM is defective, DTC P0605 will be set. Follow all diagnostic tests before replacing ECM with a known good ECM. Check for reoccurrence of faults.

Engine Coolant Temperature (ECT) Sensor

LOCATION

See Figure 15.

The Engine Coolant Temperature (ECT) sensor is located on intake manifold between cylinder numbers 2 and 3.

OPERATION

See Figure 16.

ECT is negative temperature coefficient device, meaning ohms decrease as measured temperature increases. Engine Control Module (ECM) pin 68 applies 5 volts to ECT to measure coolant temperature via an integral resistor. ECT is grounded to ECM pin 67. ECM pin 66 receives reference ground from ECT. Voltage across ECM resistor is proportional to coolant temperature. Voltage readings are used to calculate fuel injection duration for cold starts, warm-up, activation of closed loop operation and idle speed. If ECT fails or circuit breaks, ECM will set default temperature equivalent to intake air temperature on starting, then increase it by 1.8°F (1°C) every 150 engine revolutions.

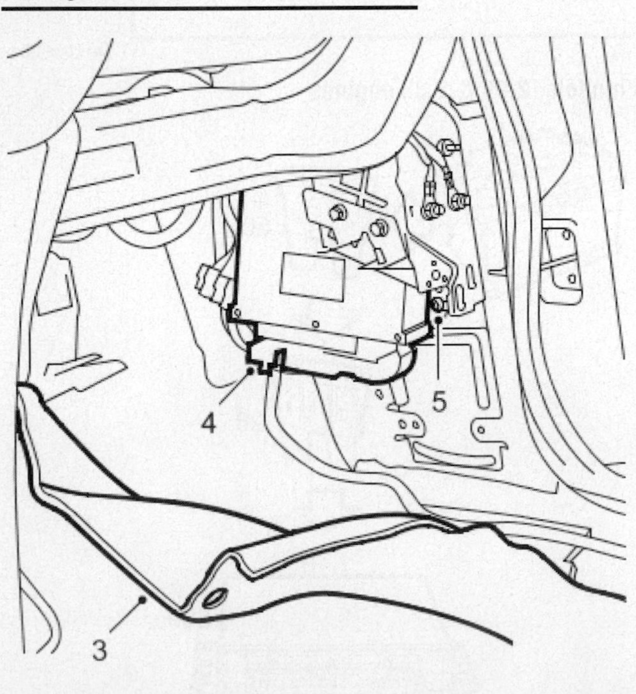

Fig. 13 Removing ECM after removing glove box–900 & 9–3 models, 2.0L & 2.3L engines

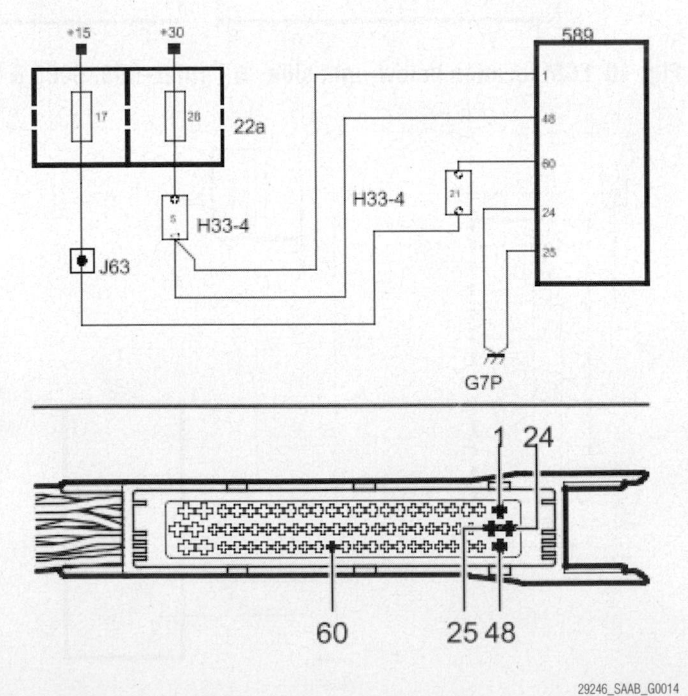

Fig. 14 ECM wiring schematic with connector pin identification–900 & 9–3 models, 2.0L & 2.3L engines

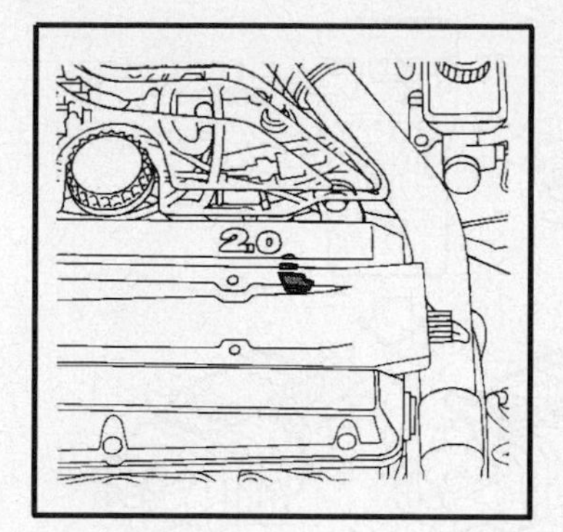

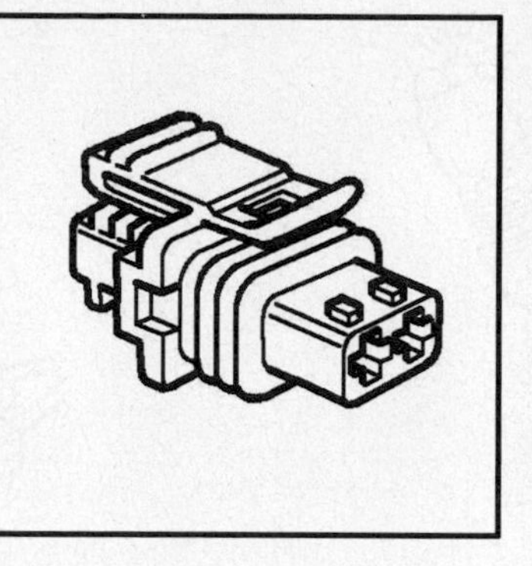

202
1 2
YE WH BK

Fig. 15 ECT sensor location–900, 9000 & 9–3 models, 2.0L & 2.3L engines

29246_SAAB_G0015

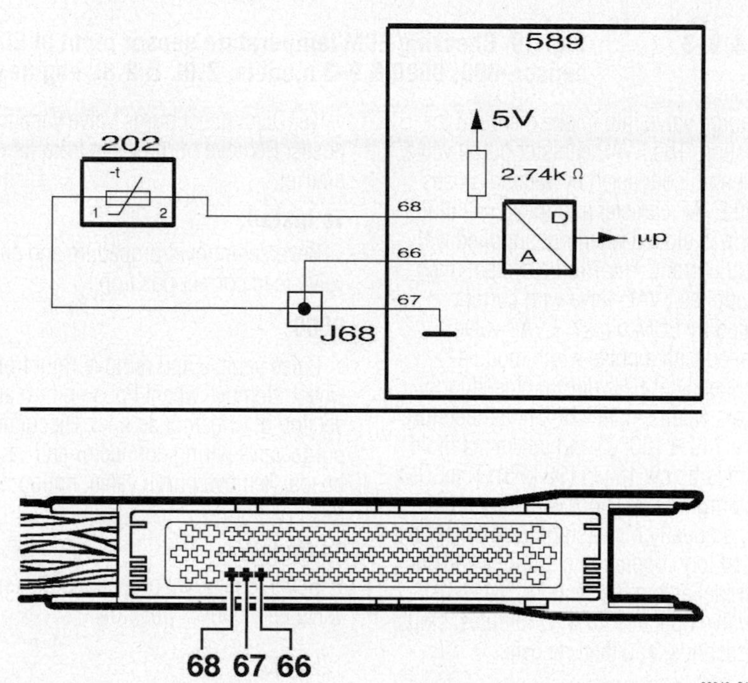

29246_SAAB_G0016

Fig. 16 ECT sensor wiring & ECM connector pins–900, 9000 & 9–3 models, 2.0L & 2.3L engines

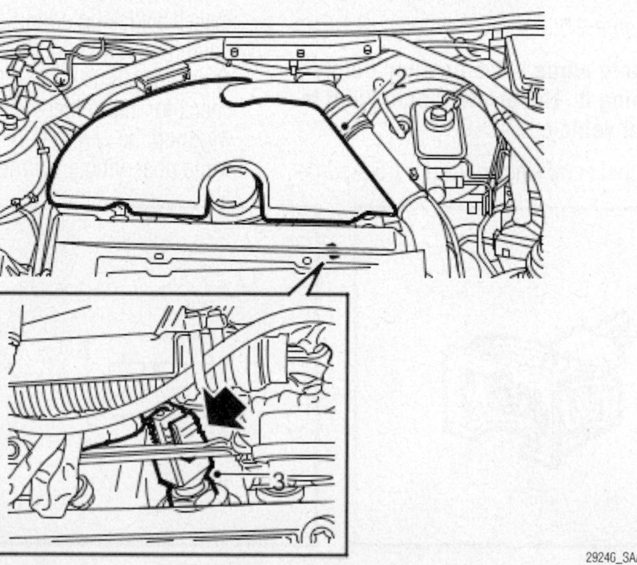

29246_SAAB_G0017

Fig. 17 Removing ECT sensor–900, 9000 & 9–3 models, 2.0L & 2.3L engines

REMOVAL & INSTALLATION

See Figure 17.

**** CAUTION**

Coolant escaping under pressure can cause burns.

1. Carefully unscrew coolant expansion tank filler cap to release system pressure. Screw cap back on.
2. Remove engine top cover.
3. Remove ECT sensor wiring connector and remove sensor.

To install:

Reverse removal procedure and tighten sensor mounting bolts to 10 ft. lbs. (13 Nm).

TESTING

See Figures 18 and 19.

1. Turn ignition ON then back OFF.
2. Connect SAAB ISAT or equivalent scan tool. Select “Read Functions,” then “Coolant Temp.”
3. Scan tool should display current coolant temperature. If not, check wiring for intermittent short circuits and open conditions. Wiggle connectors at several places in circuit while observing scan tool or test lamp for changes.
4. Check ECT sensor ground by disconnecting sensor 2-pin connector and connecting voltmeter or test lamp to battery positive terminal and ECT sensor pin 2. Test lamp should light or voltmeter should display 12 volts.
5. Check ECM's ECT sensor input as follows.
6. Turn ignition ON then back OFF.
7. Disconnect ECT sensor 2-pin connector.
8. Connect SAAB ISAT or equivalent scan tool, select “read Functions,” then “Coolant Temp.”
9. Scan tool should read -40°F (-40°C).

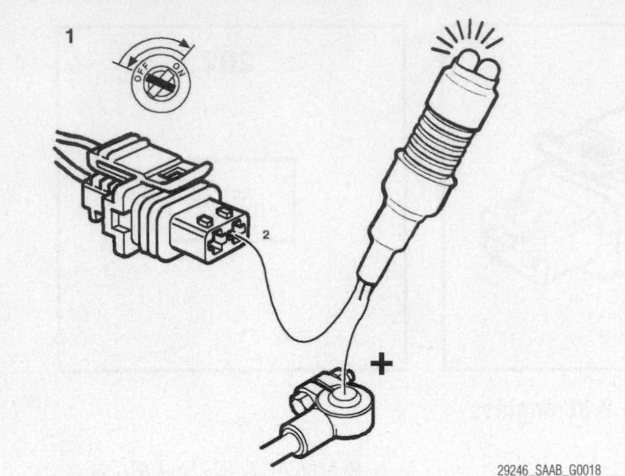

Fig. 18 Checking ECT sensor ground–900, 9000 & 9–3 models, 2.0L & 2.3L engines

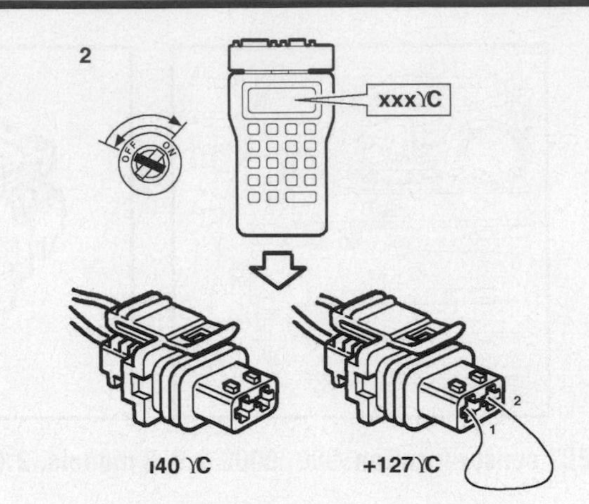

Fig. 19 Checking ECM temperature sensor input at ECT sensor–900, 9000 & 9–3 models, 2.0L & 2.3L engines

10. Connect jumper wire between ECT sensor wiring connector pins 1 and 2.

11. Scan tool should read 261°F (127°C).

12. If ECT sensor wiring connector readings match specifications and DTCs P0115, P0117, or P0118 are present, replace ECT sensor.

13. After replacing ECT sensor, clear DTCs by driving vehicle at varying loads and engine speeds for 5 minutes. Check that DTCs are cleared.

Evaporative Emission (EVAP) Canister Purge Valve

LOCATION

900 & 9–3

See Figure 20.

EVAP canister purge valve is located at the right side of the engine compartment.

OPERATION

The EVAP canister, which is filled with activated charcoal, temporarily stores vapors from fuel tank and releases them through EVAP purge valve into intake manifold for combustion. The EVAP canister purge valve is a solenoid connected by vacuum hoses between EVAP canister and intake manifold. The valve is closed when not supplied with electrical current. The Trionic system main relay supplies EVAP valve with current controlled by ECM pin 27. EVAP valve is supplied with a pulse width modulation frequency of 8 Hz during closed loop operation, when engine coolant temperature is above 140°F (60°C) and intake manifold pressure is below 13 psi (90 kPa). The ECM begins purging fuel vapor with very short pulses, gradually increasing pulse duration as closed loop operation continues and additional fuel enters the system. The canister purge valve operates for 4 ½ minutes, stops for 30 seconds, and then repeats.

REMOVAL & INSTALLATION

900 & 9–3

See Figure 21.

➡ **Note purge valve position before removing it. The arrow should point to front of vehicle.**

Disconnect wiring connector from purge valve. Disconnect purge valve vacuum hoses. Remove purge valve from its round bracket.

To install:

Reverse removal procedure and ensure valve is in correct position.

9000

Raise vehicle and remove right front wheel. Remove wheel housing trim and front section of wheelhouse liner. Disconnect purge valve wiring connector and vacuum hoses. Remove purge valve, noting its correct position.

To install:

Reverse removal procedure and ensure valve is in correct position.

TESTING

See Figure 22.

1. Check that the canister purge valve makes a clicking sound during closed loop operation. If no sound is heard and DTC codes P0443, P0444 or P0445 are set, check for intermittent wiring faults while wiggling the wiring harness and connectors while observing a voltmeter.

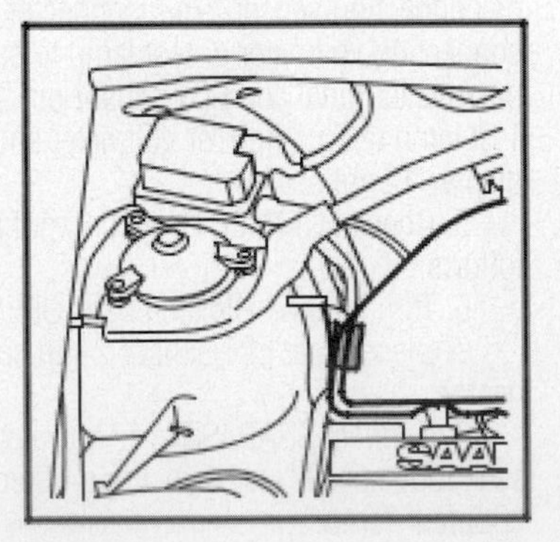

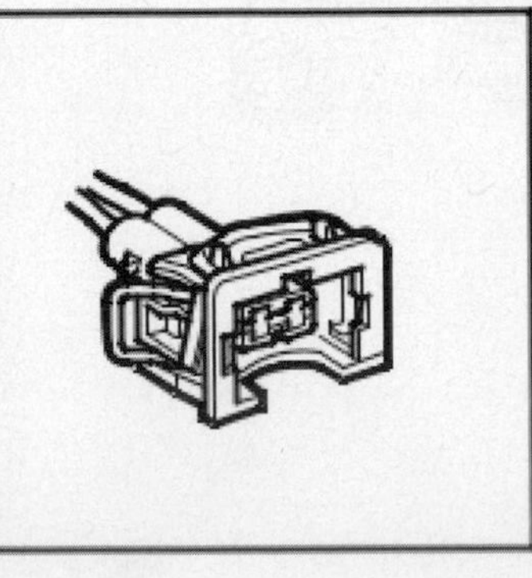

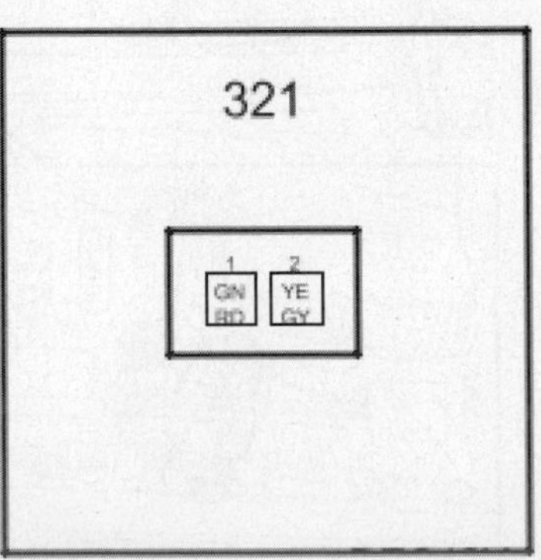

Fig. 20 EVAP canister purge valve location–900 & 9–3 models, 2.0L & 2.3L engines

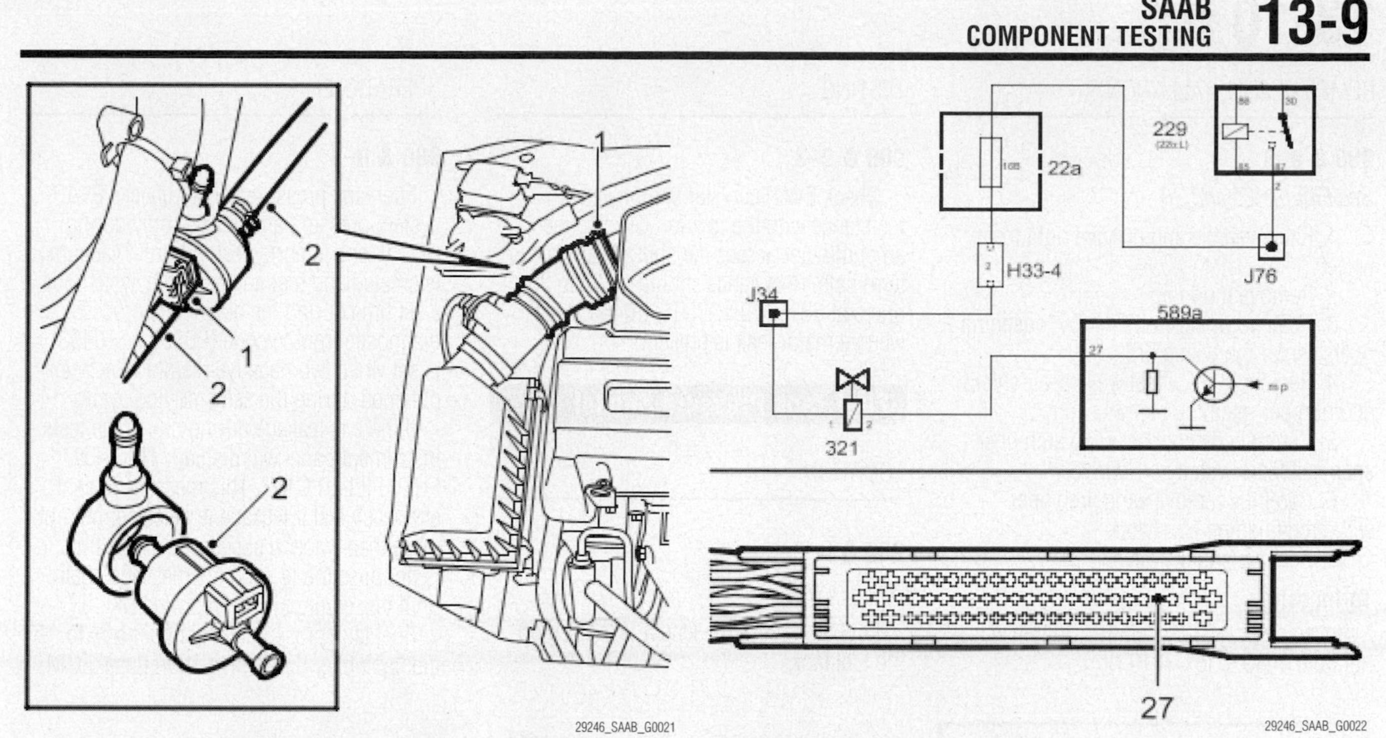

29246_SAAB_G0021

Fig. 21 Removing EVAP canister purge valve–900 & 9–3 models, 2.0L & 2.3L engines

29246_SAAB_G0022

Fig. 22 EVAP canister purge valve wiring schematic–900 & 9–3 models, 2.0L & 2.3L engines

2. Check the resistance of the EVAP canister purge valve. With the ignition OFF, disconnect the valve's two-pin wiring connector. Using an ohmmeter, measure across the valve's connector pins. The correct reading is 23 to 29 ohms. If it is not, replace the valve.

3. Check the purge valve's power supply with the engine idling. Connect a test lamp between purge valve female connector pin 1 and ground. Test lamp should light up. If not, repair or replace the wiring lead between connector pin 1 and crimped connection J67.

4. Check for purge valve ground connection fault. With ohmmeter connected between purge valve connector pin 2 and ECM pins 21 and 27, reading should be less than 1 ohm.

5. Use an appropriate scan tool to activate purge valve for 10 seconds with 8 Hz at 50 percent. Check for clicking sound, indicating activation.

Evaporative Emission Canister (EVAP) Shutoff Solenoid Valve

LOCATION

900 & 9–3

See Figure 23.

EVAP canister shutoff solenoid valve is located at right front side of engine compartment, accessed through the wheel well.

OPERATION

900 & 9–3

See Figure 24.

EVAP canister shutoff solenoid valve connects to EVAP canister air inlet hose and is powered by the main relay. The valve is grounded via ECM pin 28. The valve is normally open when engine is in closed loop operation. The valve is closed only during fuel tank integrity diagnostic check.

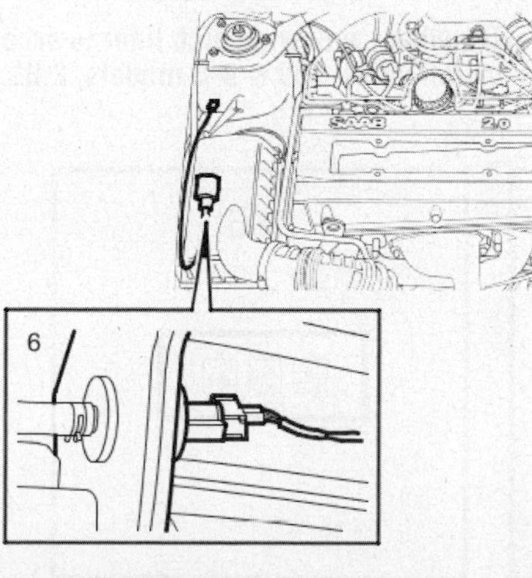

29246_SAAB_G0023

Fig. 23 EVAP canister shutoff solenoid valve location–900 & 9–3 models, 2.0L & 2.3L engines

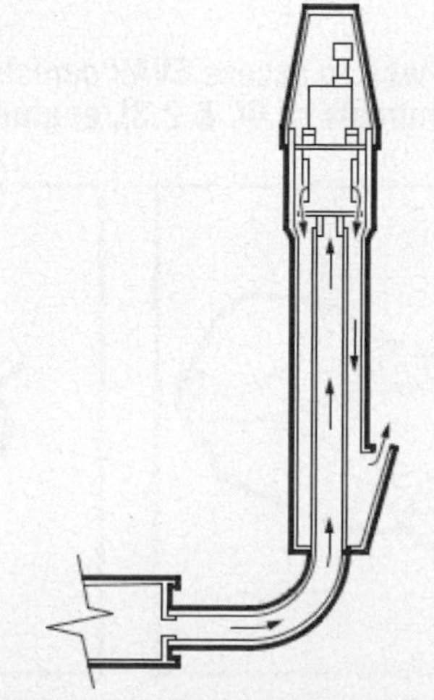

29246_SAAB_G0024

Fig. 24 EVAP canister shutoff solenoid valve operation–900 & 9–3 models, 2.0L & 2.3L engines

REMOVAL & INSTALLATION

900 & 9–3

See Figures 25 and 26.

1. Raise vehicle and remove right front wheel.
2. Remove mud flap.
3. Remove wheel well liner by loosening rivets center pin with a drift.
4. Remove front wheel well-to-air shield and bumper retaining screws.
5. Carefully disconnect wing arch liner snap fasteners with a screwdriver.
6. Carefully remove wing arch liner without damaging paintwork.
7. Remove shutoff valve.

To install:

Reverse removal procedure. Tighten wheel lug nuts to 80 ft. lbs. (110 Nm).

TESTING

900 & 9–3

Check EVAP canister shutoff valve resistance with the ignition OFF. Connect an ohmmeter across the shutoff valve's two terminals. Resistance should be 22 to 28 ohms at 68°F (20°C). DTC P0455 is set when a major leak is present.

Fuel Tank Pressure Sensor

LOCATION

900 & 9–3

See Figure 27.

Fuel tank pressure sensor is located on the fuel tank.

OPERATION

900 & 9–3

Fuel tank pressure sensor checks EVAP system pressure and is used only during OBD II tank integrity test. A partial vacuum is created and maintained in the system for a set time during the diagnostic process. A diagnostic trouble code (P0452 or P0453) is set when two negative results have been obtained during the same driving cycle.

If there is leakage during one of the tests, instrument panel will display "TIGHTEN FUEL FILLER CAP". This message is sent after each test if leakage is detected, but not more than twice in succession. After this, it is not possible to send the message again until test diagnosis has reported OK.

Fuel tank pressure sensor connects to fuel tank filler pipe with a short hose. ECM

29246_SAAB_G0025

Fig. 25 Removing wheel well to access EVAP canister shutoff valve–900 & 9–3 models, 2.0L & 2.3L engines

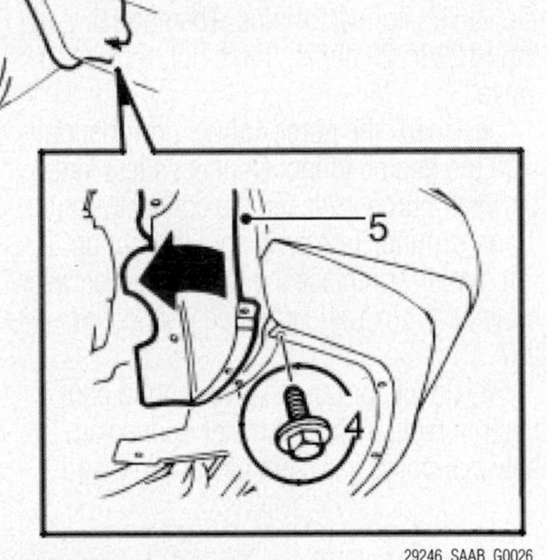

29246_SAAB_G0026

Fig. 26 Removing wheel well wing arch liner to access EVAP canister shutoff valve–900 & 9–3 models, 2.0L & 2.3L engines

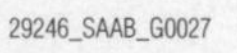
29246_SAAB_G0027

Fig. 27 EVAP system or fuel tank pressure sensor location–900 & 9–3 models, 2.0L & 2.3L engines

pin 42 delivers 5 volts of power to fuel tank pressure sensor. ECM pin 67 grounds the pressure sensor.

Pressure sensor delivers a voltage to ECM pin 21 that is proportional to difference between fuel tank pressure and atmospheric pressure. When there is no difference in pressure between fuel tank and atmosphere, pressure sensor voltage is 2.5 volts.

REMOVAL & INSTALLATION

900 & 9–3

See Figures 28 and 29.

1. Raise vehicle and remove right rear wheel.
2. Clean around filler pipe to prevent dirt from entering fuel tank.
3. Remove plastic screw securing filler pipe to vehicle body.
4. Remove hose clips and plastic hose connecting filler pipe to fixed tank pipe.
5. Disconnect fuel tank pressure sensor wiring connector.
6. Lower vehicle to floor.
7. Grip filler pipe with one hand under fender and use a screwdriver to remove filler pipe from its collar at filler cap.
8. Remove fuel tank pressure sensor.

To install:

9. Install pressure sensor to tank.
10. Connect but do not fully install plastic filler pipe.
11. Raise vehicle and connect pressure sensor wiring connector.
12. Connect plastic hose between filler pipe and fuel tank.
13. Install filler pipe to vehicle body using a new plastic screw. Tighten screw to 1.5 ft. lbs. (2 Nm).

∗∗ WARNING

Use a new plastic screw to secure filler pipe to vehicle body. In case of collision, the plastic screw must give way to avoid damaging plastic filler pipe.

14. Install right rear wheel and tighten lug nuts to 80 ft. lbs. (110 Nm).

TESTING

900 & 9–3

See Figure 30.

If DTCs P0451, P0452 or P0453 are set, EVAP system pressure is incorrect.

1. To test system power supply, turn ignition switch ON, then OFF.
2. Disconnect fuel tank pressure sensor 4-pin wiring connector near filler pipe, under vehicle.

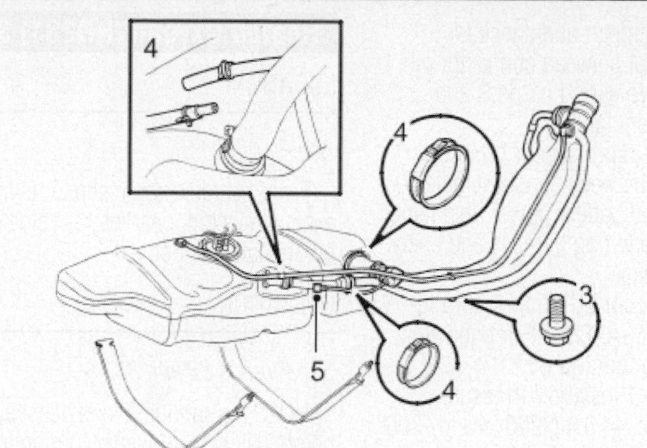

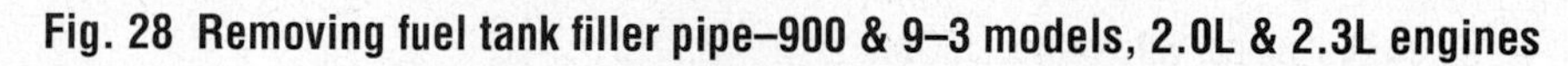

Fig. 28 Removing fuel tank filler pipe–900 & 9–3 models, 2.0L & 2.3L engines

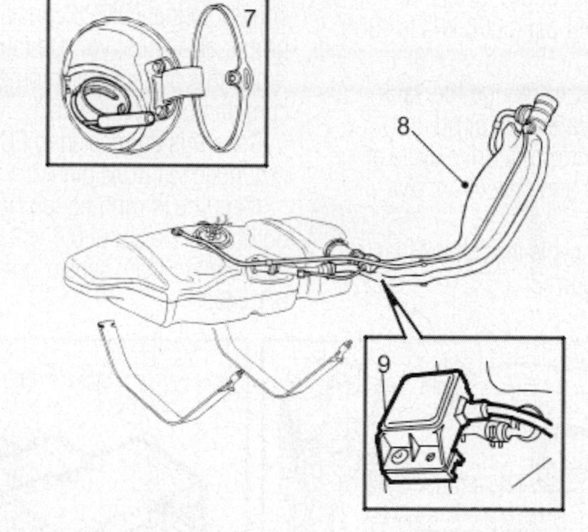

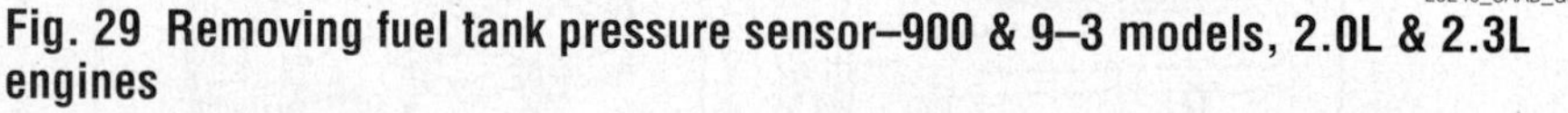

Fig. 29 Removing fuel tank pressure sensor–900 & 9–3 models, 2.0L & 2.3L engines

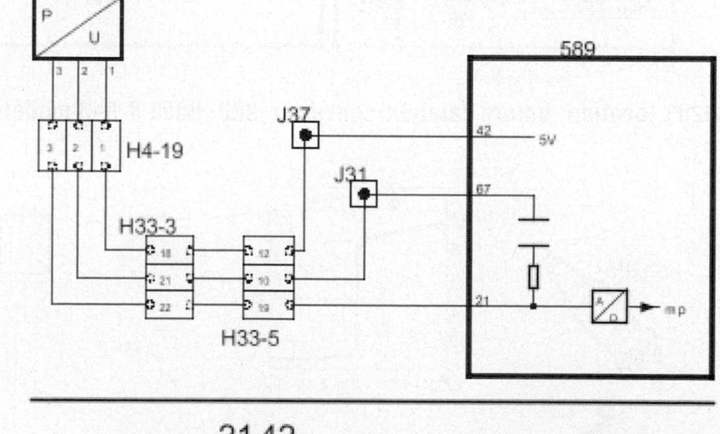

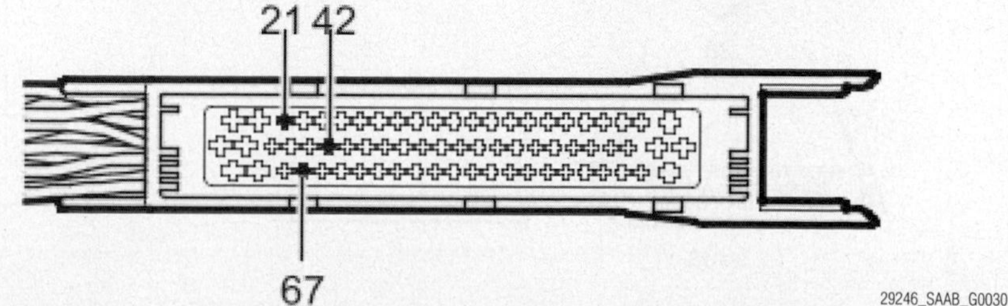

Fig. 30 Fuel tank pressure sensor wiring schematic–900 & 9–3 models, 2.0L & 2.3L engines

3. Connect voltmeter and check for 5 volt reference signal between connector pin 1 and ground. Move wiring harness and check for intermittent faults.

4. To check pressure sensor ground, connect test lamp between sensor female connector pin 2 and battery power. Test lamp should light. Move wiring harness and check for intermittent faults.

5. To check sensor ECM input, turn ignition ON, connect SAAB ISAT or equivalent scan tool, and turn ignition OFF.

6. Select "Tank Pressure." Pressure should read -29 to -44 psi (-200 kPa to -300 kPa).

7. Connect a jumper wire between ECM female connector pin 1 and 3. Pressure should read -29 to -44 psi (-200 kPa to -300 kPa).

8. If all tests pass, exchange fuel tank pressure sensor with a known good unit, clear all DTCs and perform a drive cycle at varying engine loads and speeds for five minutes.

9. Evaluate drive cycle and check that DTCs have not reoccurred.

Heated Oxygen (HO2S1) Sensor

LOCATION

See Figure 31.

Front heated oxygen sensor is mounted in exhaust system, near left rear side of engine, before catalytic converter.

OPERATION

See Figures 32 and 33.

HO2S1 is mounted in exhaust system before catalytic converter to monitor oxygen content of exhaust gases. Sensor sends a voltage signal proportional to oxygen content to the ECM to constantly maintain proper air-to-fuel volume ratio of 14.7 to 1. When this perfect stoichiometric ratio occurs, lambda is equal to 1.

Sensor is connected to ECM pin 23 and grounded via ECM pin 47.

If engine is running too rich, sensor output will be about 0.9 volts. If engine is running too lean, sensor output will be about 0.1 volts.

For sensor to begin supplying voltage signals quickly after engine is started, ECM powers sensor preheating circuit from the main relay, through fuse 38. ECM grounds sensor preheating circuit via pin 50. If estimated exhaust temperature is high, preheating stops.

REMOVAL & INSTALLATION

See Figure 34.

✱✱ WARNING

Oxygen sensors receive reference oxygen from surrounding air via connecting cables. Therefore do not use contact spray or grease on oxygen sensor connectors.

1. Remove engine cover. Disconnect turbo delivery pipe.

➡ **Note position of sensor wiring, clips and ties before disconnecting wiring. Ensure wires are not twisted together.**

2. Disconnect sensor wiring.

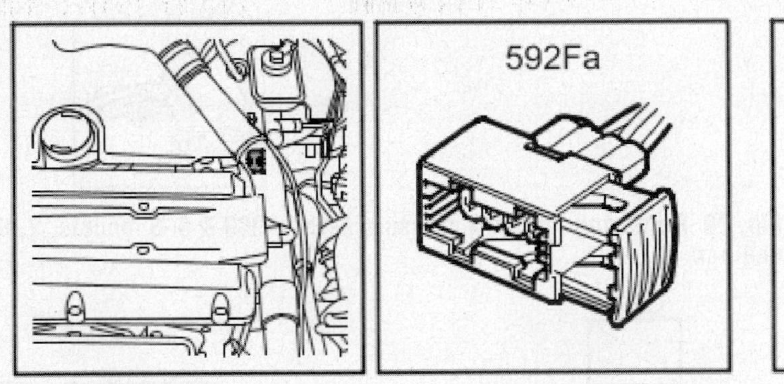

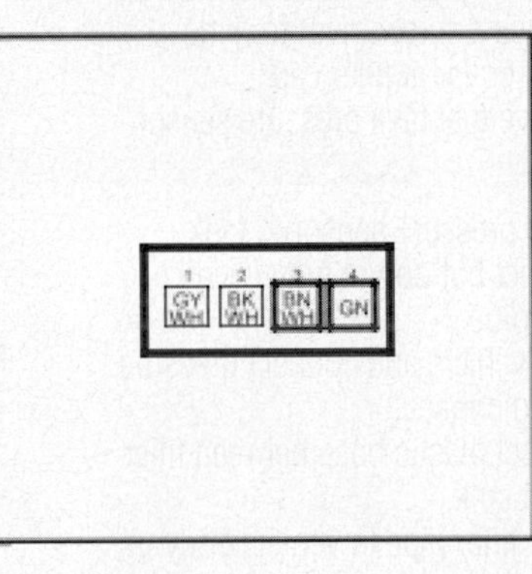

29246_SAAB_G0031

Fig. 31 HO2S1 location, before catalytic converter–900, 9000 & 9–3 models, 2.0L & 2.3L engines

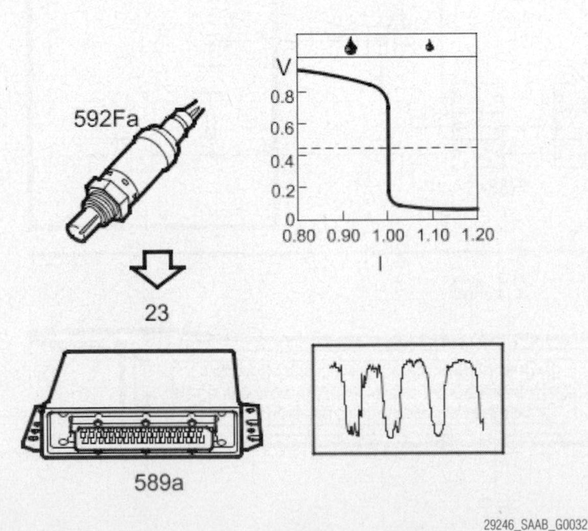

29246_SAAB_G0032

Fig. 32 HO2S1 operation–900, 9000 & 9–3 models, 2.0L & 2.3L engines

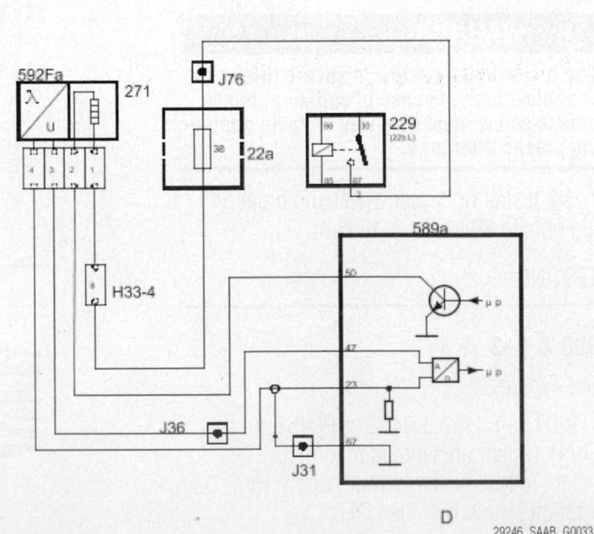

29246_SAAB_G0033

Fig. 33 HO2S1 wiring schematic–900 & 9–3 models, 2.0L & 2.3L engines

3. Raise vehicle. Remove sensor.

To install:

➡ **Coat sensor threads with anti-seize compound before installing sensor.**

4. With vehicle raised, install oxygen sensor to exhaust system. Tighten sensor to 40 ft. lbs. (55 Nm).
5. Lower vehicle. Attach sensor wiring, using new clips and ties.

**** WARNING**

Ensure sensor wiring does not touch exhaust system or turbocharger.

6. Reconnect turbo delivery pipe and install engine cover.

TESTING

See Figure 35.

1. With ignition OFF and using an ohmmeter, check resistance across sensor connector pins 1 and 2. For reference, pins 3 and 4 are gold plated. At 68°F (20°C), resistance across pins 1 and 2 should be 1.5 to 2.5 ohms.
2. If DTC P0135 is set, HO2S1 preheating circuit current is too low or too high.
3. If DTC P0131 is set, HO2S1 wiring is shorted to ground or there is an open wiring circuit.
4. If DTC P0132 is set, HO2S1 wiring is shorted to power.
5. If DTC P0133 is set, HO2S1 reacts too slowly to changes in air-fuel ratio, or sensor transmits a voltage representing a significant deviation from optimum air-to-fuel ratio.
6. If any of the DTCs above are set, closed loop operation will not function.
7. Clear all DTCs after completing their diagnoses. Perform final drive cycle under varying engine loads and speeds for 5 minutes. Evaluate drive cycle and confirm that DTCs have not reset.

Heated Oxygen (HO2S2) Sensor

LOCATION

See Figure 36.

Rear heated oxygen sensor is mounted in exhaust system, near left rear side of engine, after catalytic converter.

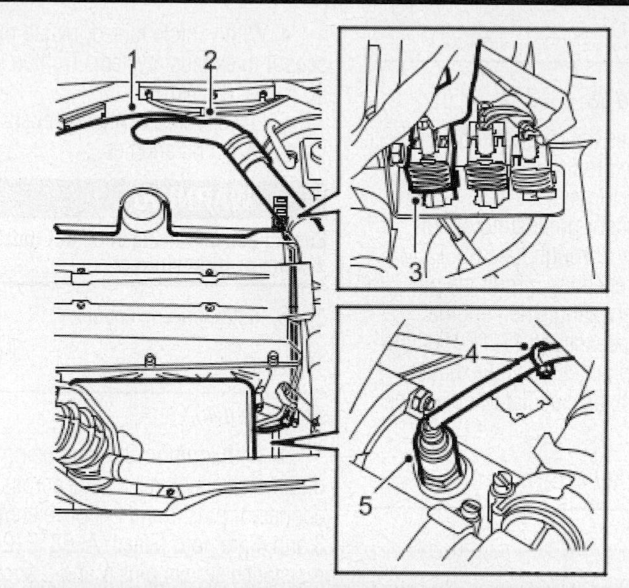

Fig. 34 Removing HO2S1–900 & 9–3 models, 2.0L & 2.3L engines

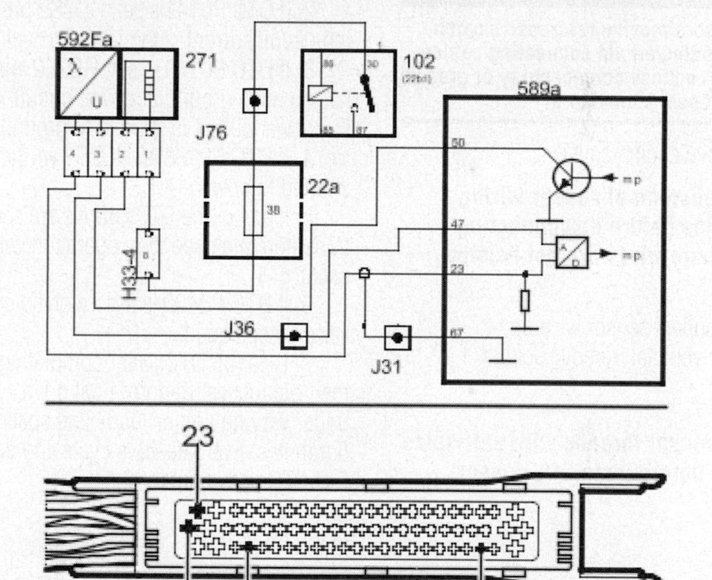

Fig. 35 HO2S1 wiring schematic–900 & 9–3 models, 2.0L & 2.3L engines

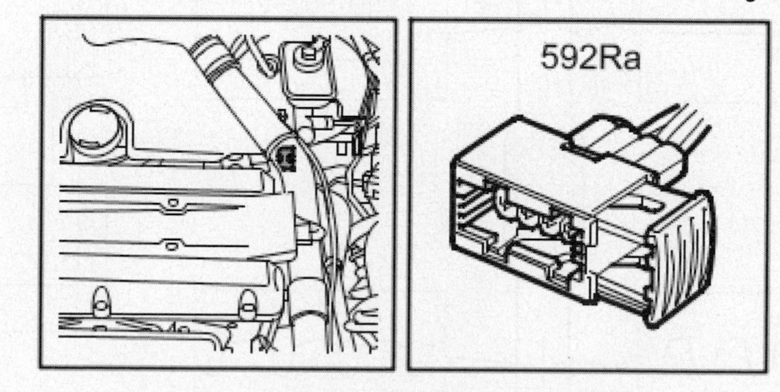

Fig. 36 HO2S2 location, after catalytic converter–900, 9000 & 9–3 models, 2.0L & 2.3L engines

OPERATION

See Figures 37 and 38.

HO2S2 diagnoses the catalytic converter. Sensor is connected to ECM pin 70 and grounded via ECM pin 47.

ECM powers sensor preheating circuit from the main relay, through fuse 38. ECM grounds sensor preheating circuit via pin 51. ECM starts preheating when engine coolant temperature exceeds 122°F (50°C). Preheating stops when estimated exhaust temperature is high, as calculated by engine load and speed.

REMOVAL & INSTALLATION

See Figure 39.

**** WARNING**

Oxygen sensors receive reference oxygen from surrounding air via connecting cables. Therefore do not use contact spray or grease on oxygen sensor connectors.

1. Remove engine cover.

➡ **Note position of sensor wiring, clips and ties before disconnecting wiring. Ensure wires are not twisted together.**

2. Disconnect sensor wiring.
3. Raise vehicle. Remove sensor.

To install:

➡ **Coat sensor threads with anti-seize compound before installing sensor.**

4. With vehicle raised, install oxygen sensor to exhaust system. Tighten sensor to 40 ft. lbs. (55 Nm).
5. Lower vehicle. Attach sensor wiring, using new clips and ties.

**** WARNING**

Ensure sensor wiring does not touch exhaust system or turbocharger.

6. Install engine cover.

TESTING

See Figure 40.

1. With ignition OFF and using an ohmmeter, check resistance across sensor connector pins 1 and 2. For reference, pins 3 and 4 are gold plated. At 68°F (20°C), resistance across pins 1 and 2 should be 1.5 to 2.5 ohms.
2. If DTC P0141is set, HO2S2 preheating circuit current is too low or too high.
3. If DTC P0140 is set, HO2S2 wiring has an open ground or power circuit, or sensor power circuit is shorted to ground.
4. If DTC P0138 is, HO2S2 wiring is shorted to power.
5. If any of the DTCs above are set, transition to closed loop operation will not occur.
6. If DTC P0420 is set, catalytic converter is damaged.
7. Clear all DTCs after completing their diagnoses. Perform final drive cycle under varying engine loads and speeds for 5 minutes. Evaluate drive cycle and confirm that DTCs have not reset.

Idle Air Control (IAC) Valve

LOCATION

900 & 9-3

See Figure 41.

Idle Air Control (IAC) valve is located at right side of engine compartment, under engine top cover.

9000

IAC valve is located on left rear side of engine, below throttle body.

OPERATION

See Figure 42.

IAC valve maintains idle speed at approximately 900 RPM. IAC valve winding resistance is 6.7 to 8.7 ohms at 68°F (20°C).

At idle, the engine receives air only via the idle air control valve. Engine Control Module (ECM) controls the degree of valve opening to keep idle speed constant. The valve opens slightly more when the A/C compressor is switched on, or if DRIVE has been selected on an automatic transmission, in order to compensate idle speed, which would otherwise fall.

IAC valve is a single-coil type and supplied with power from the main relay. It is controlled from ECM pin 49 by pulse width modulation frequency of 500 Hz. IAC valve opens more the longer the control module keeps pin 49 grounded. ECM is programmed to maintain engine idling speed

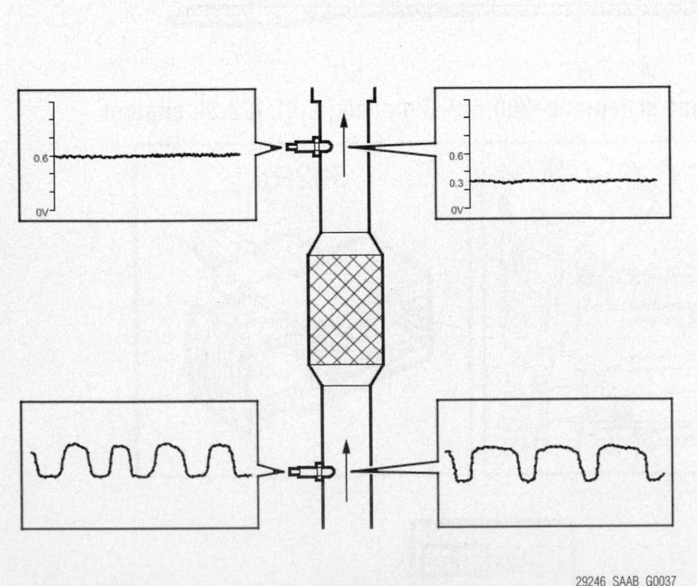

Fig. 37 HO2S2 operation–900, 9000 & 9-3 models, 2.0L & 2.3L engines

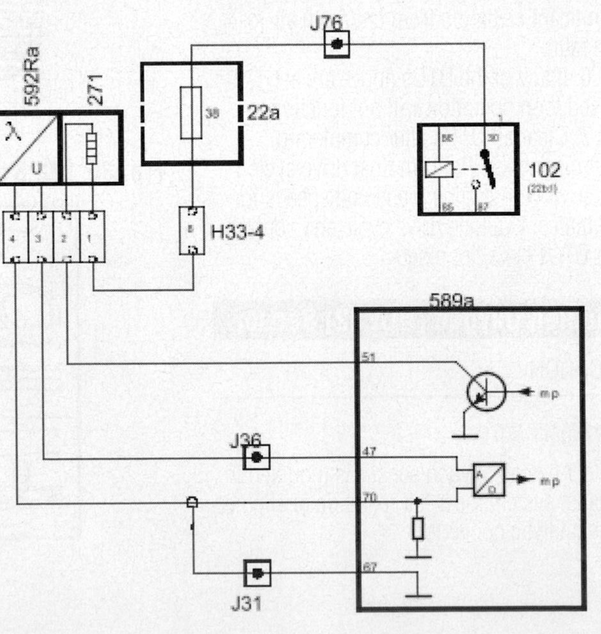

Fig. 38 HO2S2 wiring schematic–900 & 9-3 models, 2.0L & 2.3L engines

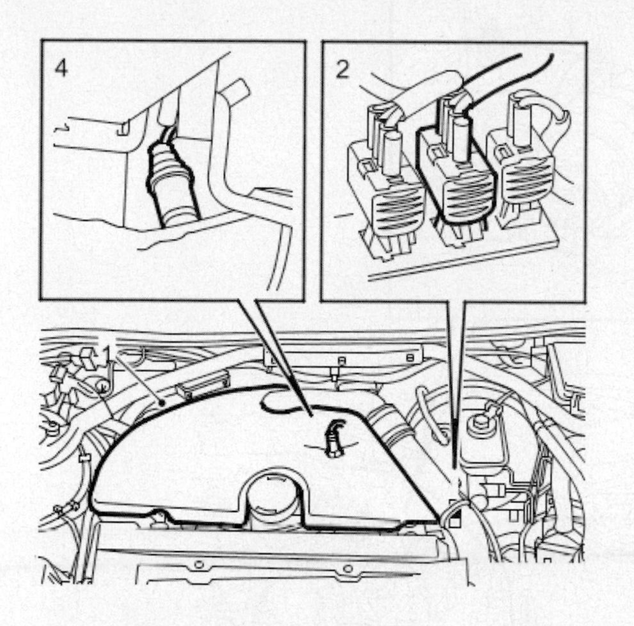

29246_SAAB_G0039

Fig. 39 Removing HO2S12–900 & 9–3 models, 2.0L & 2.3L engines

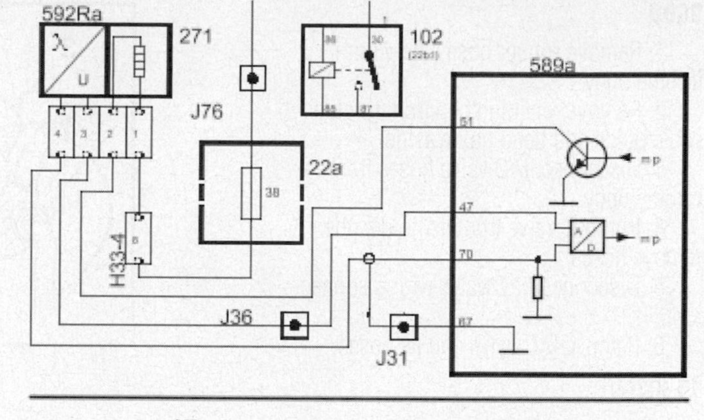

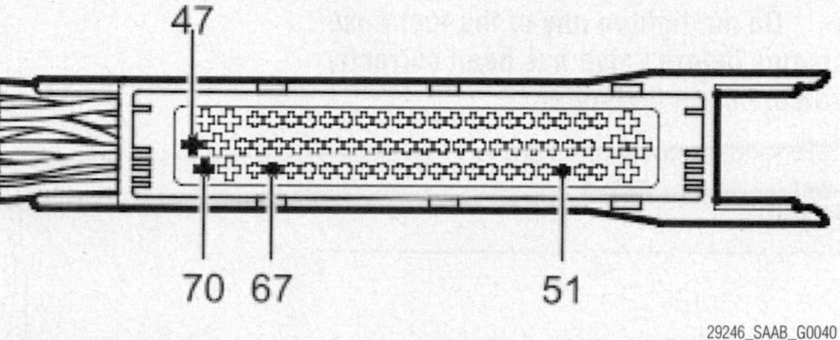

29246_SAAB_G0040

Fig. 40 HO2S2 wiring schematic–900 & 9–3 models, 2.0L & 2.3L engines

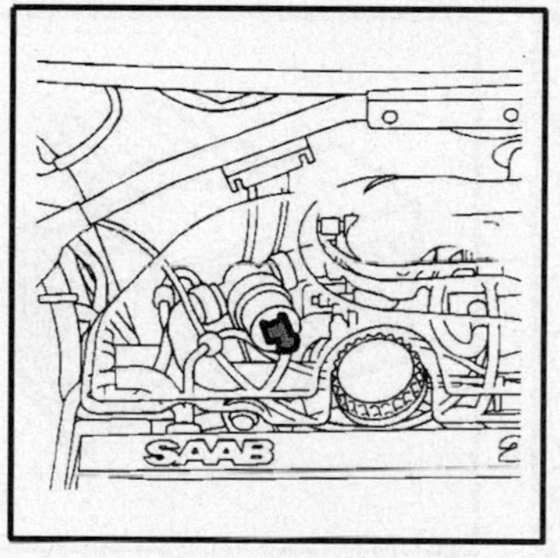

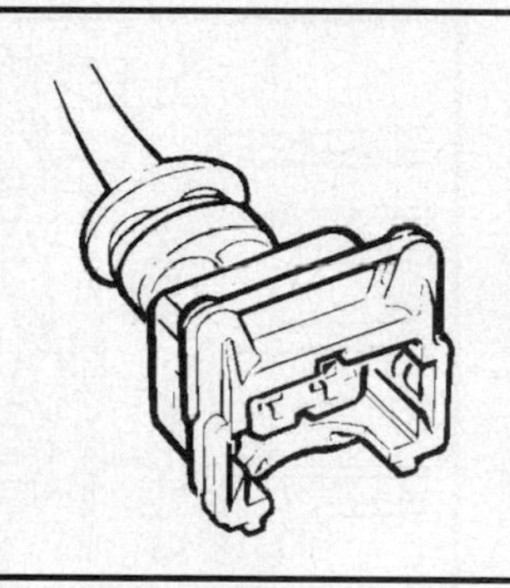

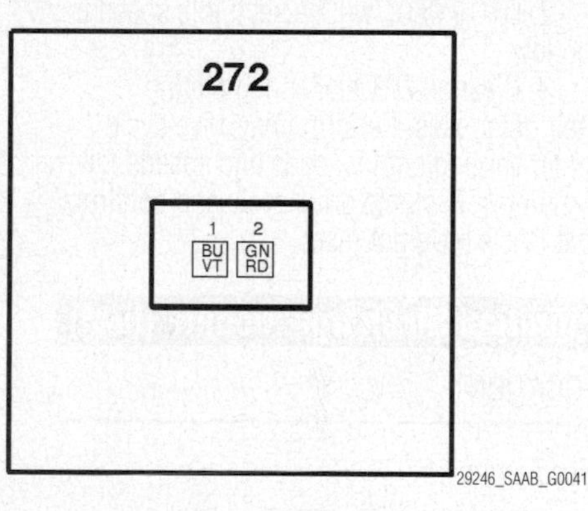

29246_SAAB_G0041

Fig. 41 IAC valve location–900 & 9–3 models, 2.0L & 2.3L engines

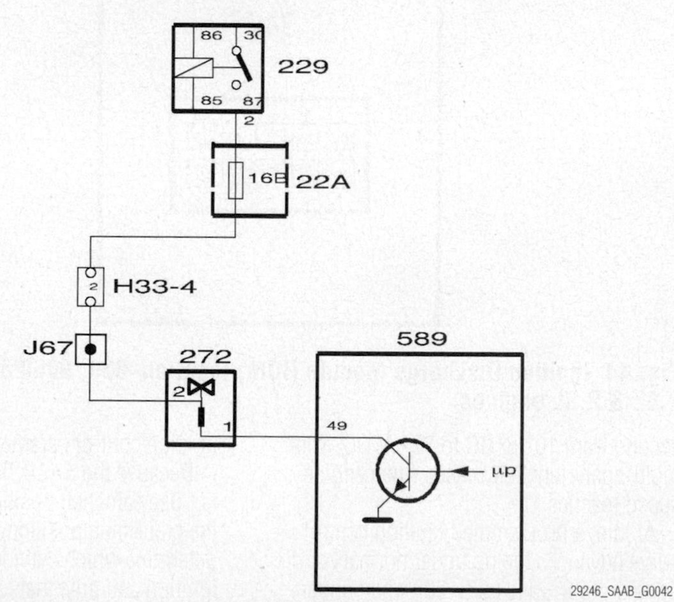

29246_SAAB_G0042

Fig. 42 IAC valve wiring schematic–900, 9000 & 9–3 models, 2.0L & 2.3L engines

at 850 to 950 RPM. Rapid changes in idling speed cannot be compensated for by changing the flow of air. Instead, ignition timing is modulated so that engine idle speed is maintained constant. If there is an open circuit to the idle air control valve, a spring sets valve opening so that idle speed is at least 1000 RPM when engine is warmed up.

REMOVAL & INSTALLATION

900 & 9–3

See Figure 43.

Remove engine top cover and disconnect IAC valve electrical connector and vacuum hoses. Remove valve from brackets.

To install, reverse removal procedure and ensure IAC valve is positioned correctly.

➡ **Arrow on IAC valve should point away from throttle body on 900 and 9–3 models.**

9000

1. Remove rubber hose elbow from throttle body.
2. Remove retaining clip from throttle cable guide and bend guide aside.
3. Disconnect IAC valve hoses from throttle body.
4. Pull IAC valve from its holder, then remove hoses.
5. Disconnect IAC valve wiring connector.
6. Remove IAC valve and its hoses.

To install:

➡ **Do not tighten any of the four hose clamps before valve has been correctly positioned in its holder.**

Reverse removal procedure.

TESTING

See Figure 42.

1. DTC P0506 will be set if idle speed is too high.
2. DTC P0507 will be set if idle speed is too low.
3. Clear all DTCs after completing their diagnoses. Perform final drive cycle under varying engine loads and speeds for 5 minutes. Evaluate drive cycle and confirm that DTCs have not reset.

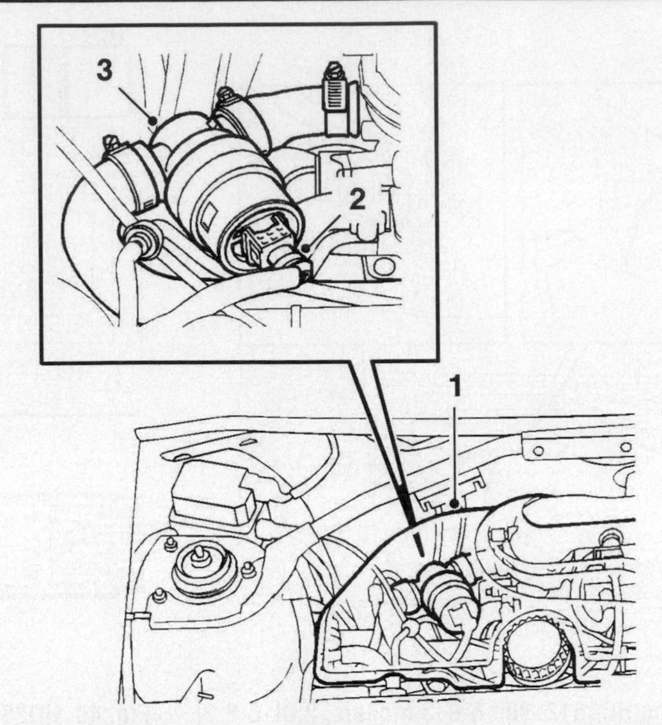

Fig. 43 Removing IAC valve–900 & 9–3 models, 2.0L & 2.3L engines

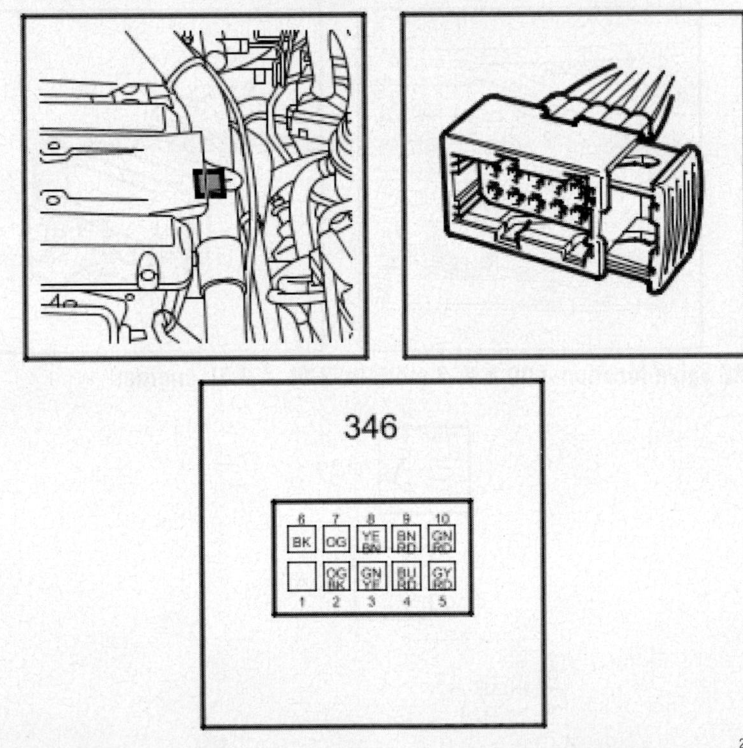

Fig. 44 Ignition Discharge Module (IDM) location–900, 9000 & 9–3 models, 2.0L & 2.3L engines

Ignition Discharge Module (IDM)

LOCATION

See Figure 44.

Ignition Discharge Module (IDM) is mounted on the camshaft cover over the spark plugs. It has a 10-pin connector.

OPERATION

See Figures 45 and 46.

The main relay supplies 12 volts of power to IDM, which converts it to 400 volts DC and charges a capacitor with the voltage. The 400 volts are supplied to one pole of the four ignition coils' primary windings.

Four trigger wires from ECM connect to IDM via four pins: 9 through 12, which trigger cylinders 1 through 4, respectively. When ECM grounds a pin, the other pole of the ignition coil's primary winding will be grounded via IDM power input. The 400 volts transforms to 40,000 volts, igniting the cylinder.

To facilitate engine starting when coolant temperature is below 32°F (0 °C), ECM will ground each trigger wire 210 times per second from 10° BTDC to 20° ATDC. This multi-spark function ceases when engine speed reaches idle.

At idle, a programmed ignition control curve advances timing under normal conditions up to 8°, and up to 20° when under sudden, auxiliary electrical load. When engine speed increases above idle, normal ignition control parameters resume.

Because the SAAB Trionic system does not use camshaft position sensor, it uses the crankshaft position (CKP) sensor to determine which cylinders are firing. The ignition coil pole that is not connected to the spark plug is connected to 80 volts, not ground. 80 volts is always present across

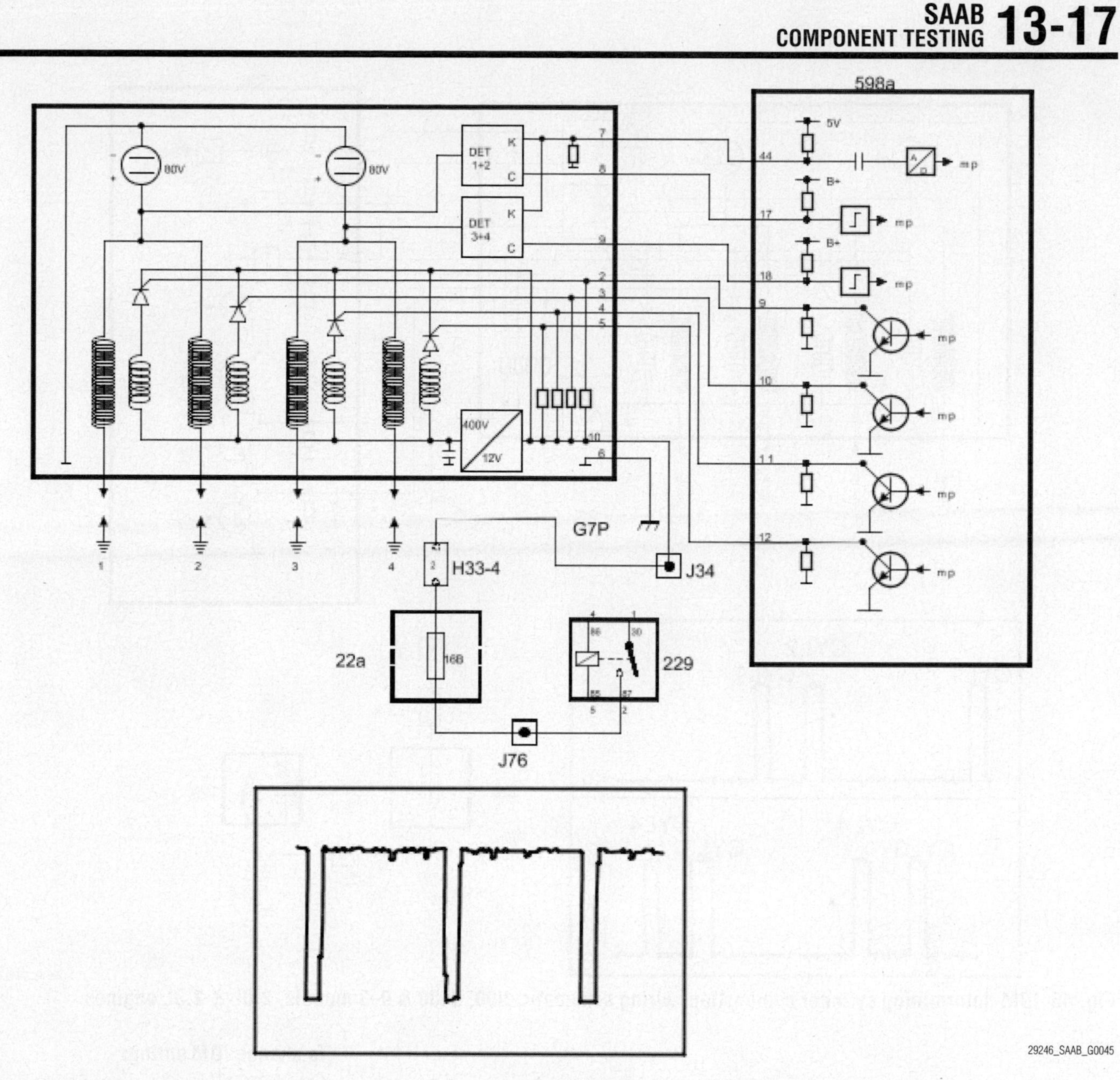

Fig. 45 Ignition Discharge Module (IDM) wiring schematic–900, 9000 & 9–3 models, 2.0L & 2.3L engines

the spark plug except when a spark is produced. When cylinder combustion occurs, gases ionize and conduct current across the spark plug gap without creating a spark. The ionization current is measured in cylinder pairs 1 and 2, or 3 and 4. The IDM sends a power pulse to ECM pin 17 if it measures ionization current from cylinders 1 or 2, and to ECM pin 18 if it measures ionization current from cylinders 3 or 4. If CKP indicates cylinders 1 and 4 are at TDC, and ECM pin 17 receives a power pulse from IDM, the ECM knows cylinder 1 has fired. The same process occurs for each cylinder.

For knock control, the SAAB Trionic system does not use a standard knock sensor. Instead, IDM analyzes cylinder combustion ionization currents and sends signals to ECM pin 44. If ECM detects sufficient knocking at any cylinder, it instructs IDM to retard ignition timing 1.5° at the knocking cylinder. If cylinder knocking continues, IDM retards ignition timing incrementally, up to 12° maximum, until knocking stops.

If cylinder knocking occurs when intake manifold pressure exceeds about 20 psi (140kPa), knock control measures change. Initially, ECM will adjust fuel injection and ignition parameters. If those adjustments are insufficient, ECM will reduce turbocharger boost pressure.

The main relay remains active for 6 seconds after ignition has been switched OFF. ECM then grounds all trigger wires to IDM at 210 times per second for 5 seconds, creating 1,050 sparks per cylinder, which allow spark plugs to burn off combustion residues.

REMOVAL & INSTALLATION

See Figure 47.

Disconnect IDM electrical connector. Remove 4 attachment screws or bolts. Lift up IDM.

To install:

Reverse removal procedure. Tighten attachment screws or bolts to 8 ft. lbs. (11 Nm). Spray electrical connector with Kontakt 61 or other anti-corrosive lubricant. Spray rubber seals with synthetic lubricant.

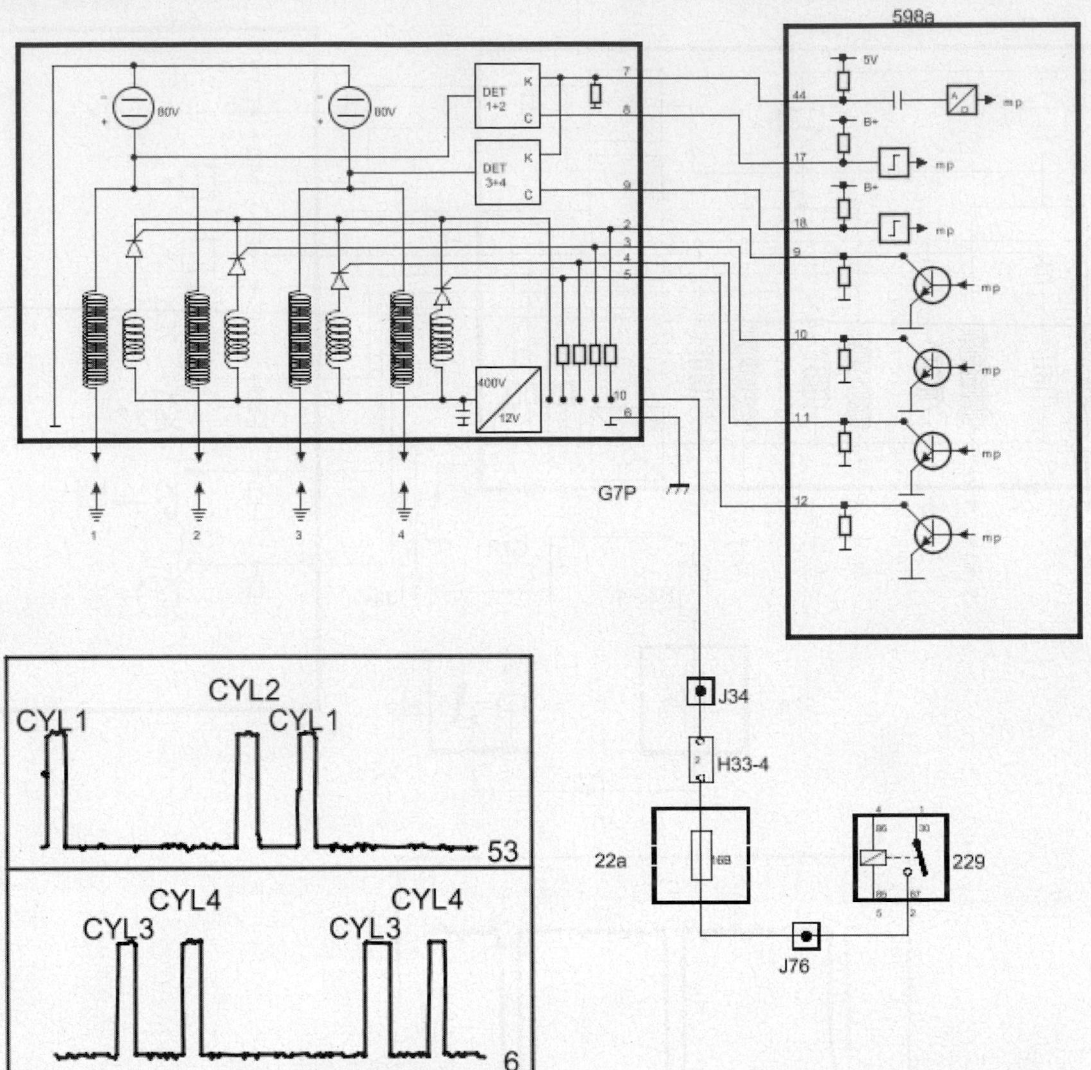

29246_SAAB_G0046

Fig. 46 IDM determining cylinder combustion wiring schematic–900, 9000 & 9–3 models, 2.0L & 2.3L engines

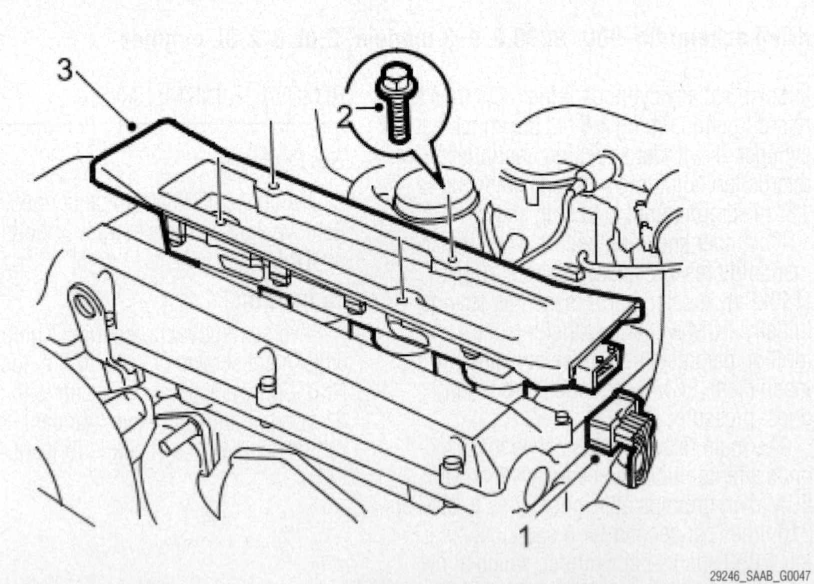

29246_SAAB_G0047

Fig. 47 Removing Ignition Discharge Module (IDM)–900, 9000 & 9–3 models, 2.0L & 2.3L engines

To change IDM spring:

See Figure 48.

1. Remove 8 screws from underneath IDM and remove black, lower section of IDM.
2. Use screwdriver to pry out spring.
3. Replace old spring with a new one.
4. Install black, lower section of IDM and install 8 screws.

TESTING

CAUTION

Ignition system generates 40,000 volts. People with pacemakers or heart problems should exercise extreme caution when testing ignition system.

No knock signal from IDM can be caused by an open circuit to ECM pin 44. DTC P0327 will be set.

Engine will not start if there is a break in IDM power supply or grounding.

A misfire DTC will set if there is a break in ignition wiring.

If ignition and fuel injection synchronization does not occur, knock control measures will be performed in parallel on cylinders 1 and 4, and cylinders 2 and 3.

If there are open circuits in wiring to ECM pin 17 and 18, ignition and fuel injection will not be synchronized and DTC P0340 will be set.

If a short circuit to ground occurs at wiring to ECM pins 17 and 18, ignition and fuel injection will be synchronized, but DTC P0300 will be set to indicate misfiring in more than one cylinder.

Misfiring in one cylinder will set DTC P0301 to P0304, depending on which cylinder is misfiring.

Fig. 48 Changing Ignition Discharge Module (IDM) spring–900, 9000 & 9-3 models, 2.0L & 2.3L engines

Intake Air Temperature (IAT) Sensor

LOCATION

See Figure 49.

Intake Air Temperature (IAT) sensor is located in the intake manifold, before throttle valve.

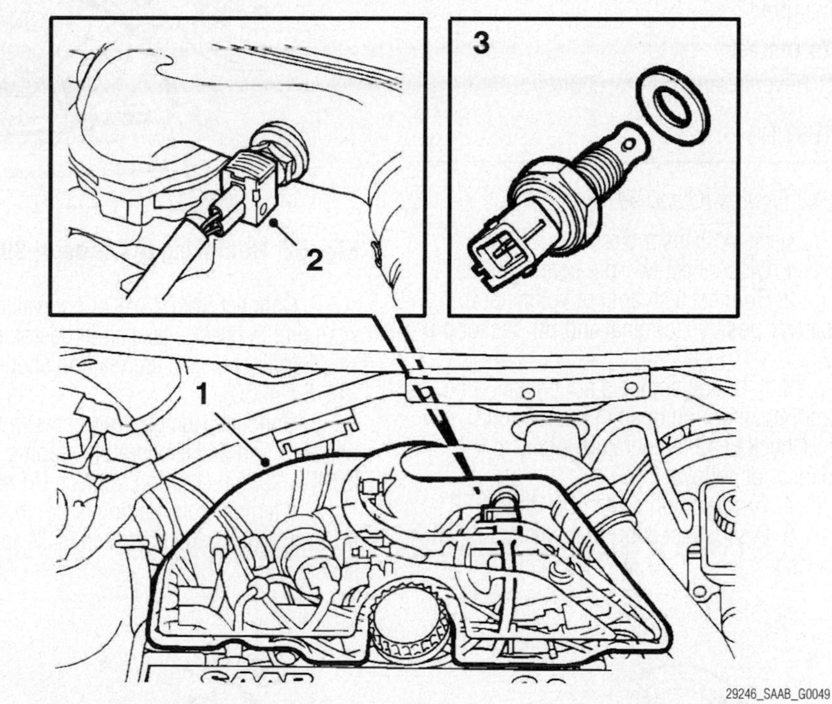

Fig. 49 IAT sensor location–900, 9000 & 9-3 models, 2.0L & 2.3L engines

OPERATION

See Figures 50 and 51.

Engine Control Module (ECM) measures air temperature in intake manifold by sending 5 volts to IAT sensor from ECM pin 46 via an integral resistor. IAT is a negative temperature coefficient device, meaning ohms decrease as measured temperature rises. IAT sensor is grounded via ECM pin 67. The voltage across ECM resistor is proportional

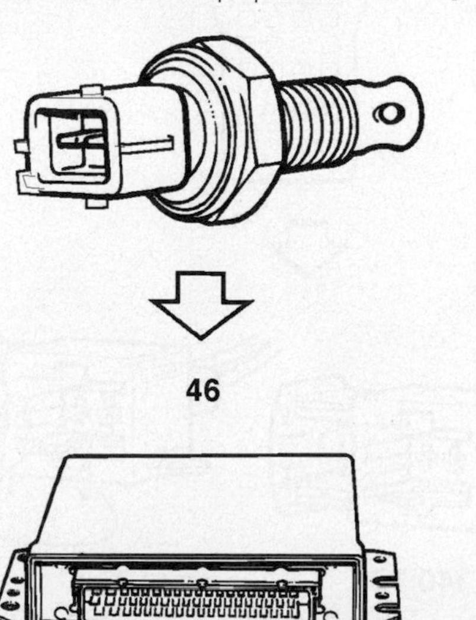

Fig. 50 IAT sensor powered by ECM pin 46–900, 9000 & 9-3 models, 2.0L & 2.3L engines

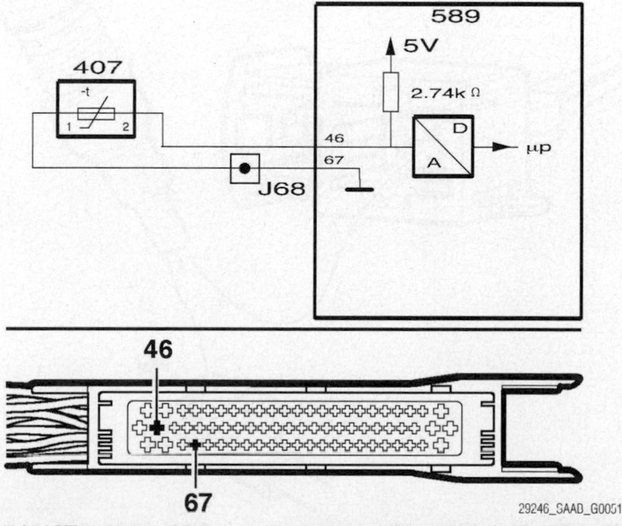

Fig. 51 IAT sensor wiring schematic–900, 9000 & 9-3 models, 2.0L & 2.3L engines

to intake air temperature. In combination with intake manifold pressure, the voltage is used to determine the intake air density and engine load. Fuel injection time is proportional to engine load. If sensor fails or the circuit breaks, the electronic control module will set a default temperature of 46°F (8°C) and fuel adaptation will not function.

REMOVAL & INSTALLATION

See Figure 52.

1. Remove engine top cover.
2. Disconnect IAT sensor electrical connector.
3. Remove IAT sensor from intake manifold.

To install:

Reverse removal procedure.

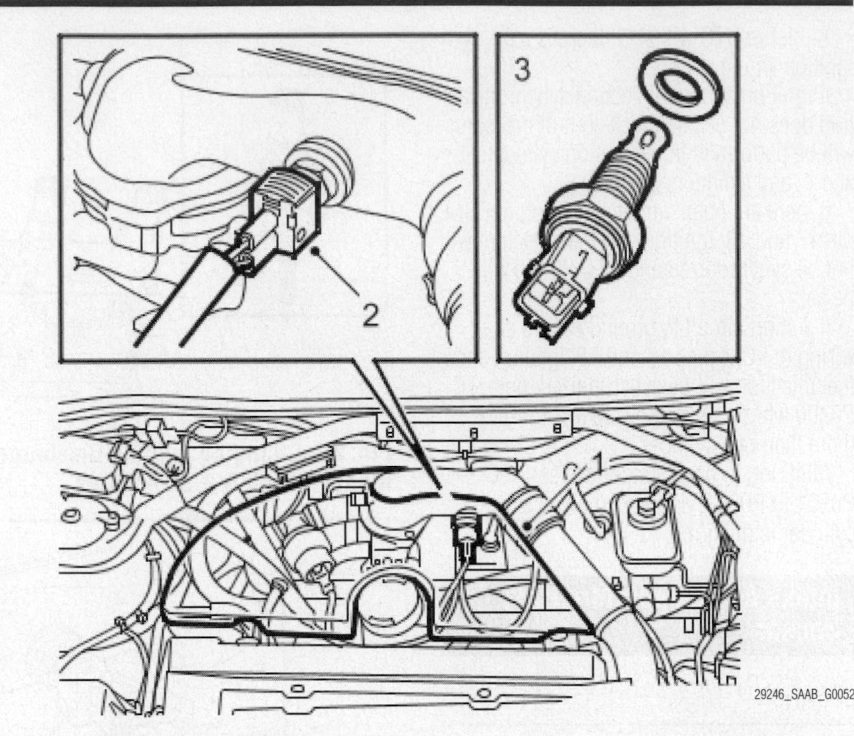

Fig. 52 Removing IAT sensor–900, 9000 & 9–3 models, 2.0L & 2.3L engines

TESTING

See Figures 53 and 54.

Check IAT sensor circuit as follows.

1. Disconnect two-pin connector.
2. Connect test lamp or voltmeter to battery positive terminal and IAT sensor pin 2.
3. If 12 volt signal is not present, find and repair wiring open or short circuit.

Check ECM temperature signal at IAT sensor as follows.

4. Turn ignition switch ON, then OFF.
5. Disconnect two-pin connector at IAT sensor.
6. Connect SAAB ISAT or equivalent scan tool to check intake air temperature.
7. Initial temperature reading should be -40°F (-40°C).
8. Connect a jumper lead between terminals 1 and 2. If temperature reading from ECM is 253° F (123° C), replace IAT sensor.
9. If temperature reading is not to specification, repair wiring between ECM and IAT sensor.

Perform final check and clear trouble code as follows.

10. Drive the vehicle for 5 minutes under varying engine RPM and loads.
11. Ensure DTCs P0110, P0112, or P0113 are cleared.

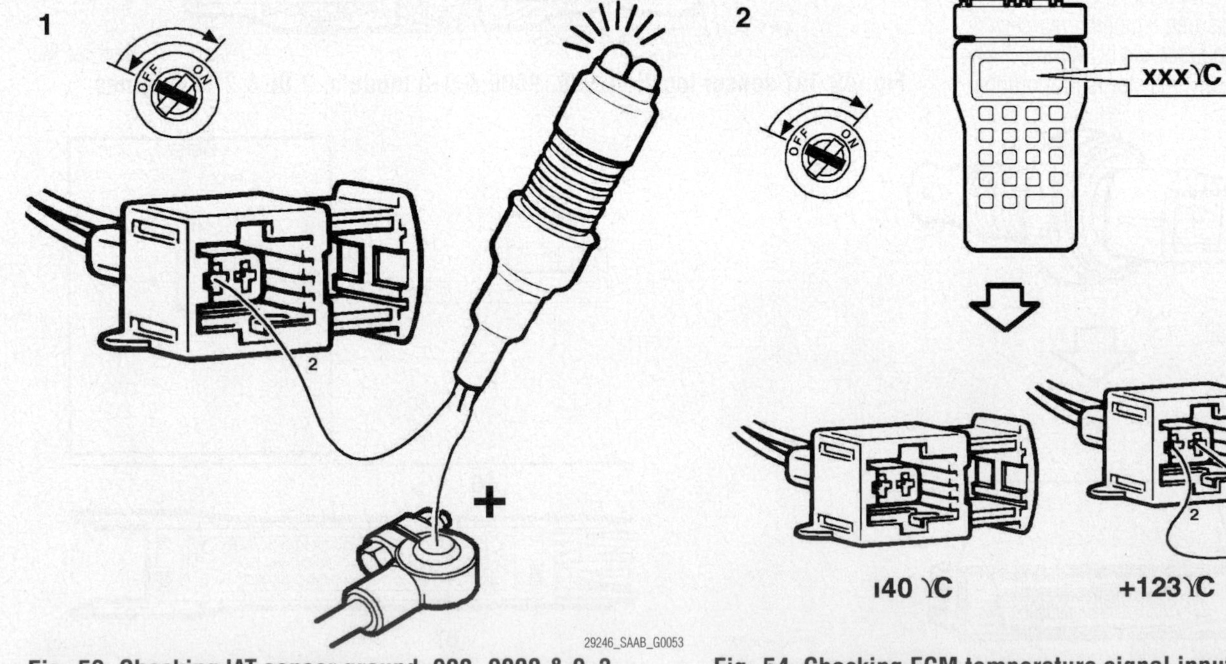

Fig. 53 Checking IAT sensor ground–900, 9000 & 9–3 models, 2.0L & 2.3L engines

Fig. 54 Checking ECM temperature signal input at IAT sensor –900, 9000 & 9–3 models, 2.0L & 2.3L engines

Main Relay

LOCATION

See Figure 55.

Engine management main relay is located in relay holder, behind left side of instrument panel.

OPERATION

See Figure 56.

Main relay supplies power to fuel pump relay, fuel injectors, Ignition Discharge Module (IDM), Idle Air Control (IAC) valve, EVAP canister purge valve and EVAP canister shutoff solenoid valve. ECM pin 31, which is grounded, controls the main relay. As soon as ECM receives impulses from Crankshaft Position (CKP) sensor, ECM operates main relay. It will remain active for about 6 seconds after ignition has been turned OFF, to supply power to IDM for spark plug burn off process. If ignition has been turned OFF for more than 15 minutes, main relay will operate for 6 seconds after ignition is turned ON.

REMOVAL & INSTALLATION

See Figure 57.

Remove instrument panel lower left cover. Unscrew relay holder screws and tilt relay holder down. Unplug main relay from relay holder.

To install:

Reverse removal procedure.

TESTING

See Figure 58, component 229.

There is no specific diagnosis for main relay wiring circuit.

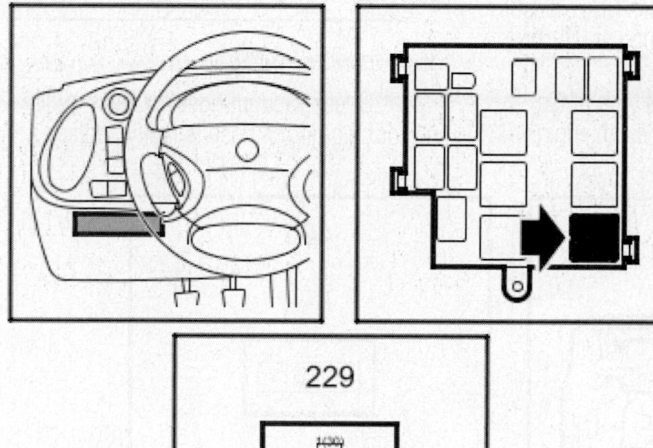

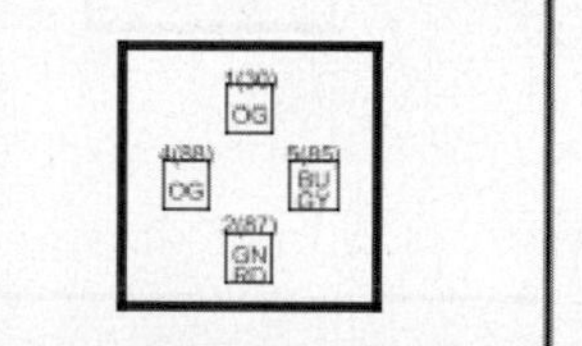

Fig. 55 Main relay location–900, 9000 & 9–3 models, 2.0L & 2.3L engines

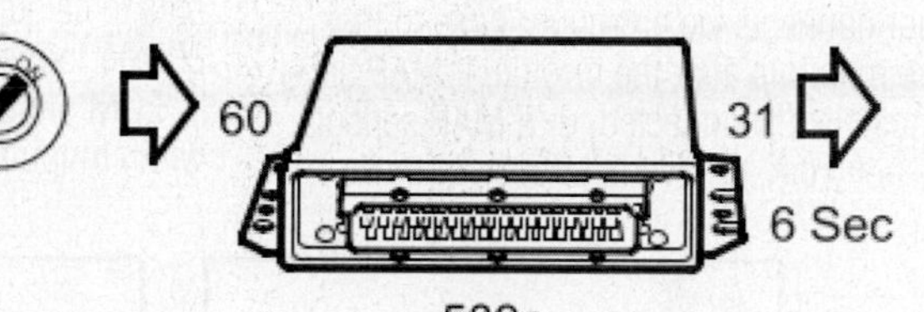

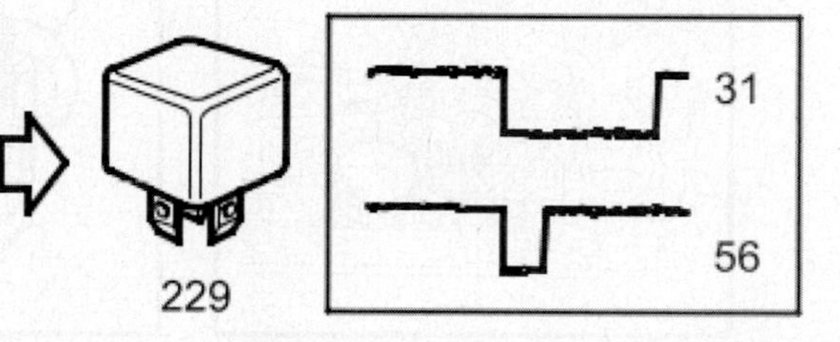

Fig. 56 Main relay operation–900, 9000 & 9–3 models, 2.0L & 2.3L engines

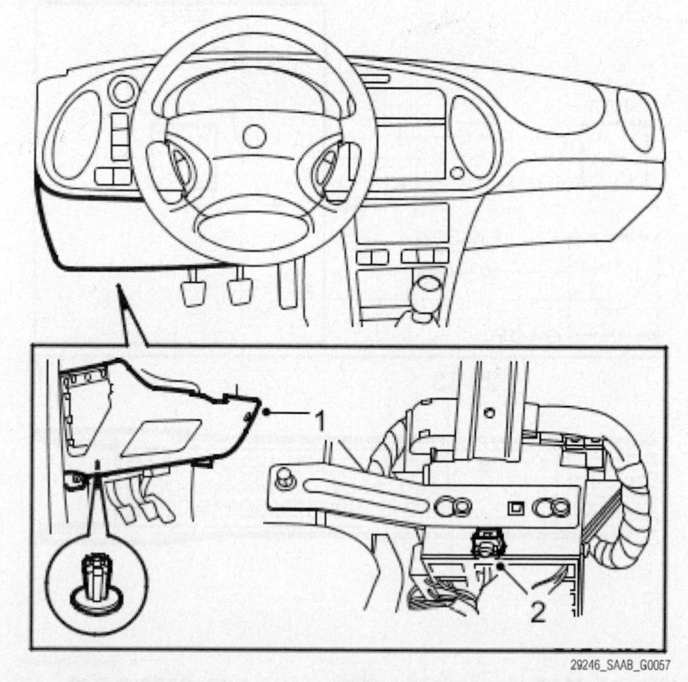

Fig. 57 Removing main relay–900, 9000 & 9–3 models, 2.0L & 2.3L engines

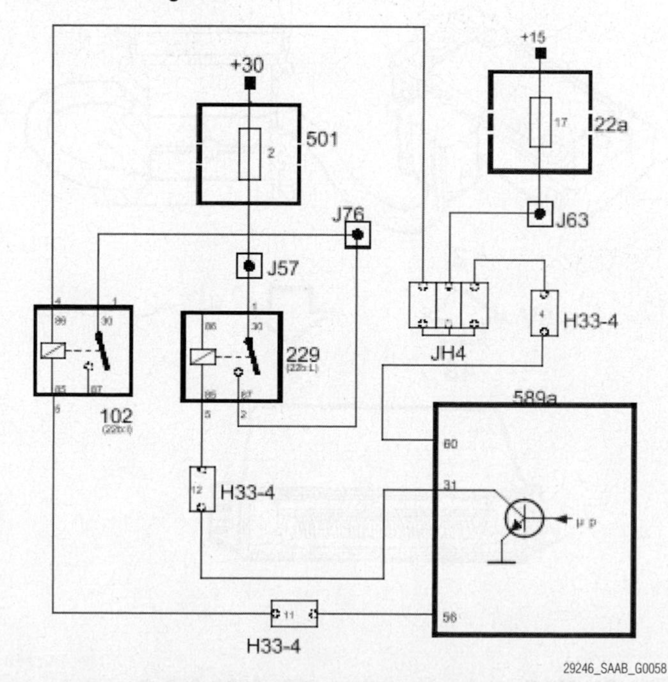

Fig. 58 Main relay wiring schematic–900, 9000 & 9–3 models, 2.0L & 2.3L engines

Manifold Absolute Pressure (MAP) Sensor

LOCATION

See Figure 59.

Manifold Absolute Pressure (MAP) sensor is located at right side of engine compartment, near the firewall.

OPERATION

See Figures 60 and 61.

MAP sensor contains a pressure-sensitive ceramic sensing device, an amplifier and a circuit for temperature compensation. MAP sensor connects via a vacuum hose to the intake manifold after the throttle. If MAP sensor hose is disconnected, or if MAP sensor does not work, or reverts to open circuit, ECM uses throttle position as a substitute value and the Malfunction Indicator Lamp (MIL) will come on.

Engine Control Module (ECM) pin 43 powers MAP sensor with 5 volts. MAP sensor is grounded to ECM pin 67. Depending on intake manifold pressure, MAP sensor supplies a proportional voltage to ECM pin 22. ECM pin 20 supplies MAP sensor with intake manifold pressure voltage. At sea level, the voltage is approximately 1.9 volts.

ECM uses intake manifold pressure and temperature to calculate engine load. Fuel injection time is proportional to air mass drawn in. With engine idling, MAP sensor operating pressure should be 5.8 to 8.7 psi (40 to 60 kPa). ECM also uses MAP sensor data to calculate ignition timing and turbocharger boost pressure control.

ECM pin 35 provides a varying pulse width modulation frequency as an engine load signal. The frequency is the same as engine speed signal, (2 pulses per crankshaft rotation), but varying pulse width carries engine load data.

REMOVAL & INSTALLATION

See Figure 62.

Remove engine top cover. Disconnect MAP sensor electrical connector. Disconnect vacuum hose from MAP sensor and remove sensor.

To install:

Reverse removal procedure.

TESTING

See Figures 63 and 64.

1. Check vacuum hose for leaks, pinching or clogging. If hose is faulty, replace it.

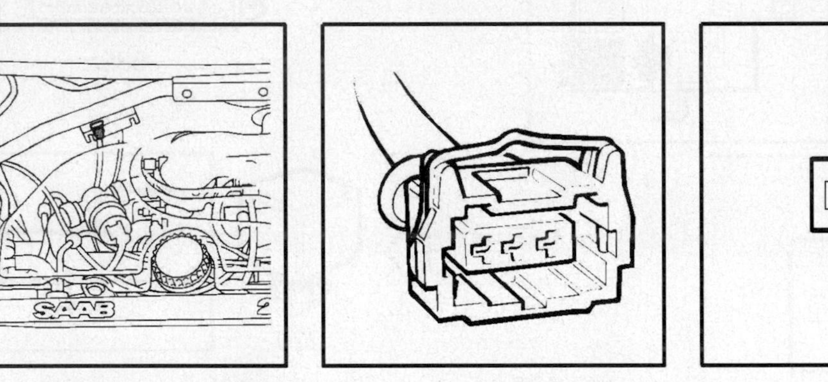

Fig. 59 MAP sensor location–900, 9000 & 9-3 models, 2.0L & 2.3L engines

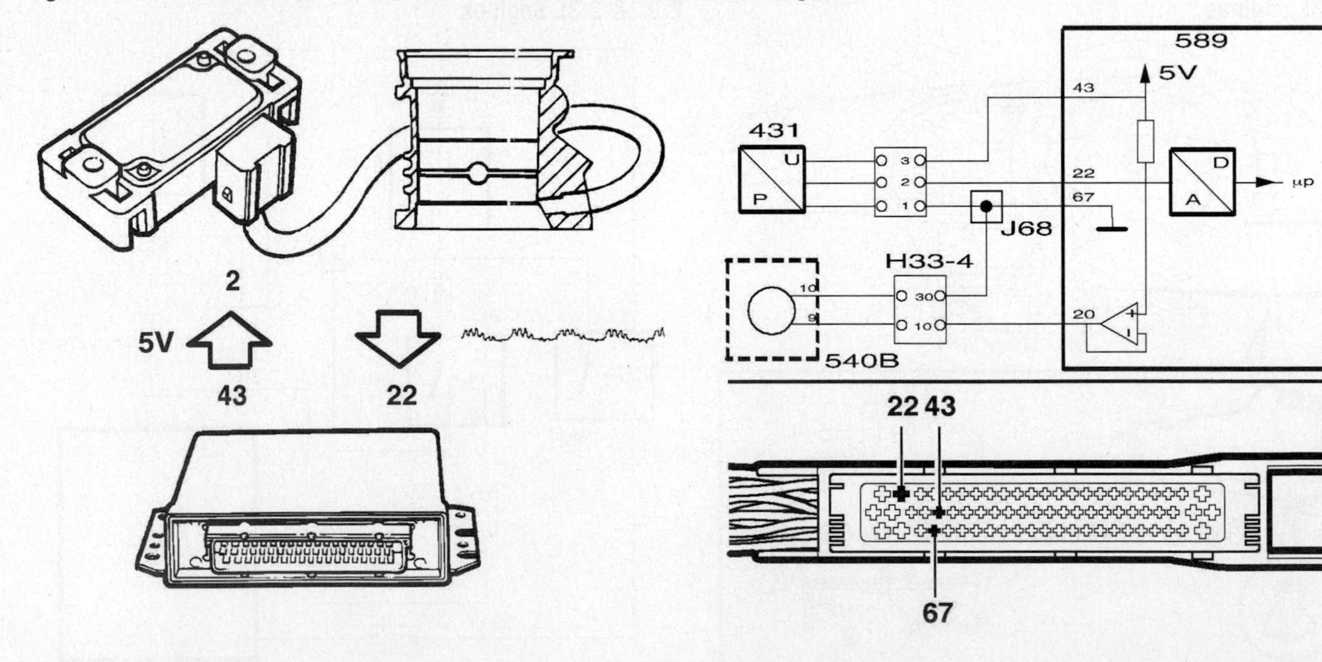

Fig. 60 MAP sensor operation–900, 9000 & 9-3 models, 2.0L & 2.3L engines

Fig. 61 MAP sensor wiring schematic–900, 9000 & 9-3 models, 2.0L & 2.3L engines

2. Turn ignition switch ON, and then OFF. Disconnect MAP sensor wiring connector.

3. Check voltage between MAP sensor wiring connector pin 3 and ground. Voltage should be 4.5 to 5.5 volts. Wiggle wiring at several locations to check for intermittent faults. Check switches, wiring connectors and crimps for signs of oxidation, retracted pins and loose pin connections. Correct any wiring faults between MAP sensor connector pin 3 and ECM pin 43.

4. Check voltage between MAP sensor wiring connector pin 1 and battery power. Voltage should be 11 to 13 volts. Wiggle wiring at several locations to check for intermittent faults. Check switches, wiring connectors and crimps for signs of oxidation, retracted pins and loose pin connections. Correct any wiring faults between MAP sensor connector pin 1 and crimped connection J31.

1. To test intake pressure, turn ignition switch ON, and then OFF. Connect scan tool.

2. Read intake pressure, which should be about 37 psi (255 kPa). If it is not to specification, replace MAP sensor.

3. Check ECM sensor input using scan tool and connecting a jumper wire between MAP sensor connector pins 1 and 2. Pressure reading should be 0 psi (0 kPa).

4. When a fault occurs, basic turbocharger pressure will result, ECM will use throttle position as a substitute value instead of MAP signal, EVAP and A/C will be inoperative.

5. If MAP sensor voltage is too low, DTC P0107 will be set. If MAP sensor voltage is too high, DTC P0108 will be set.

6. If pressure reading does not make sense, for example when a vacuum hose comes loose, DTC P0106 will be set.

7. As a final check, clear all DTCs, perform a drive cycle at varying engine speeds and loads for at least five minutes, evaluate the drive cycle and check for DTCs.

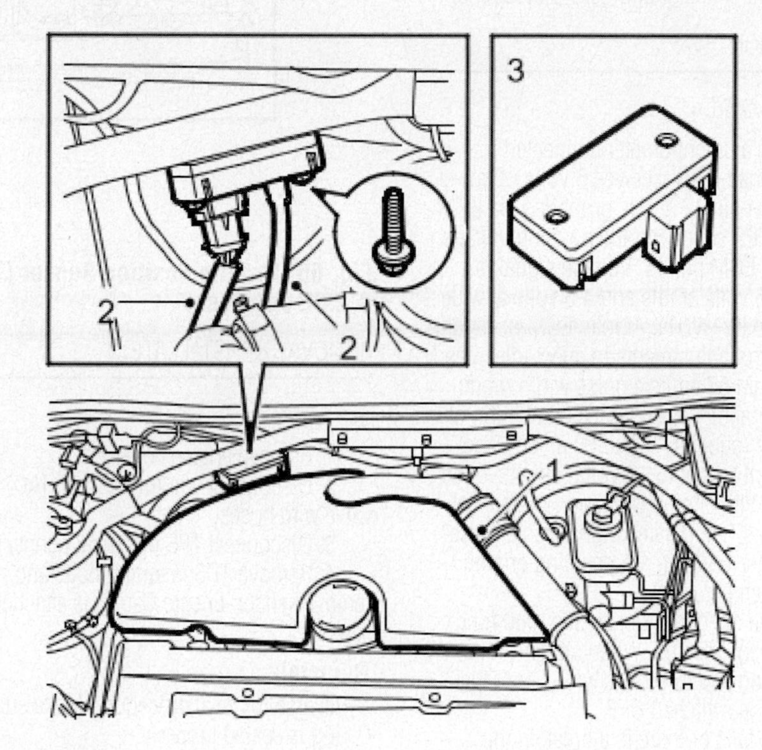

Fig. 62 Removing and installing MAP sensor–900, 9000 & 9–3 models, 2.0L & 2.3L engines

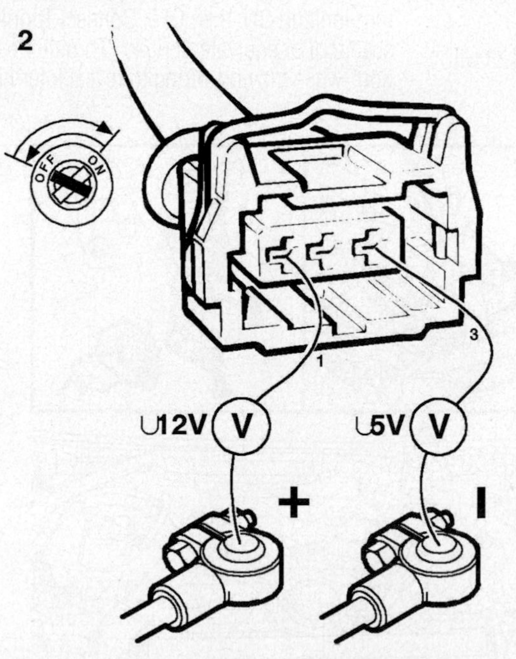

Fig. 63 Testing MAP sensor ground and power connections–900, 9000 & 9–3 models, 2.0L & 2.3L engines.

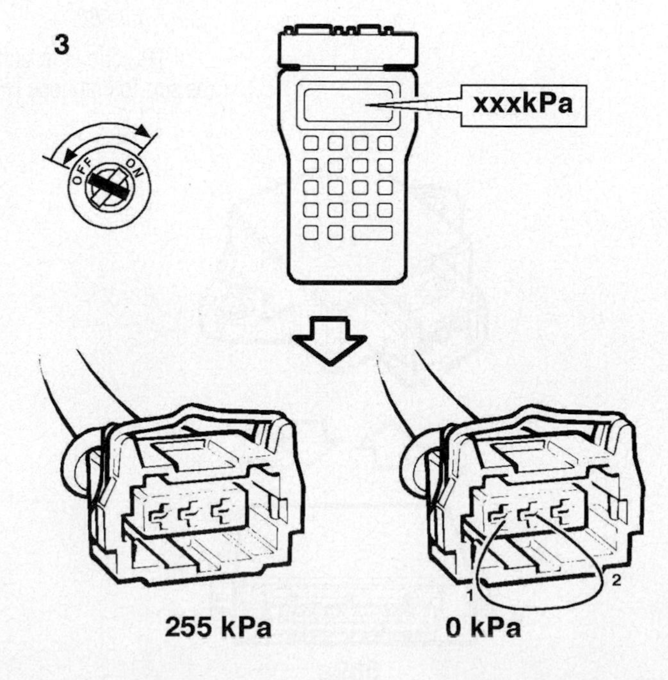

Fig. 64 Testing MAP sensor input to ECM–900, 9000 & 9–3 models, 2.0L & 2.3L engines.

Throttle Position Sensor (TPS)

LOCATION

See Figure 65.

Throttle Position Sensor (TPS) is located at throttle valve, at intake manifold. It has a three-pin connector and gold-plated pins.

OPERATION

See Figure 66.

TPS is a potentiometer connected to throttle shaft. TPS receives 5 volts of power from ECM pin 42 and is grounded via ECM pin 67. TPS sends throttle position voltage signal to ECM pin 45. Voltage signal is about 0.5 volts at idle and 4.5 volts at wide open throttle (WOT). At idle, ECM uses ignition control adjustments to affect idle speed.

ECM pin 57 emits a pulse width modulated signal of 100 Hz with a pulse ratio of 9 percent (equal to 1.2 volts) at idle, and 90 percent (equal to 12 volts) at wide open throttle. With engine OFF and ignition ON, ECM pin 57 modulated signal has a pulse ratio of 2-7 percent, representing engine coolant temperature.

At wide open throttle, closed loop function is disconnected. When starter motor is cranking and throttle is wide open, fuel injection is switched OFF.

If TPS fails or there is a break in the wiring circuit, ECM sets wide open throttle position as default value.

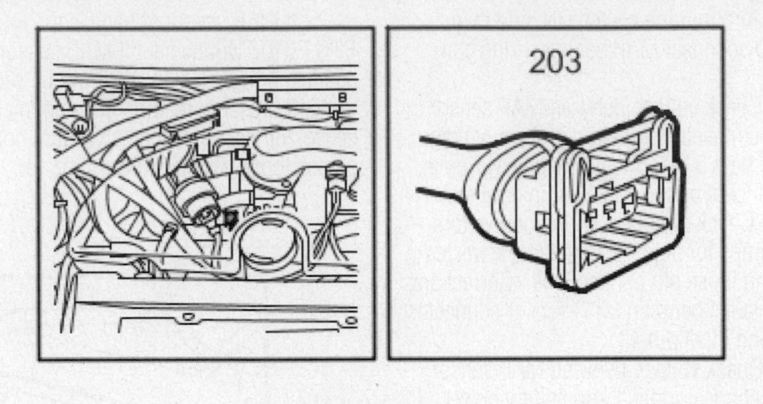

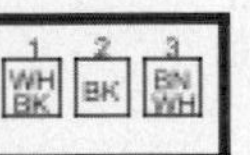

29246_SAAB_G0065

Fig. 65 Throttle Position Sensor (TPS) location–900, 9000 & 9–3 models, 2.0L & 2.3L engines.

REMOVAL & INSTALLATION

See Figure 67.

1. Remove engine cover.
2. Disconnect Idle Air Control (IAC) valve with hoses.
3. Disconnect TPS electrical connector.
4. Remove TPS retaining bolts and remove sensor. Ensure O-ring is attached to sensor.

To install:

Reverse removal procedure. Make sure O-ring is seated properly.

TESTING

See Figure 68.

If TPS circuit is faulty, DTCs 120-123 will be set. To diagnose TPS circuit:

1. Turn ignition switch ON, then OFF.
2. Disconnect TPS electrical connector.
3. To check TPS power supply, connect voltmeter between TPS pin 1 and ground. Take voltage reading while moving wiring to reveal intermittent faults. Reading should be 4.5-5.5 volts. If voltage is not to specification, repair wiring between TPS pin 1 and ECM pin 42.
4. To check TPS ground, connect voltmeter between TPS pin 2 and battery power. Take voltage reading while moving wiring to reveal intermittent faults. Reading should be 11-13 volts. If voltage is not to specification, repair wiring between TPS pin 2 and crimped connection J31. See Figure 6X.
5. To check TPS input signal to ECM, turn ignition ON, then OFF. Connect Trionic scan tool or equivalent. Read "Throttle Position" while moving wiring to reveal intermit-

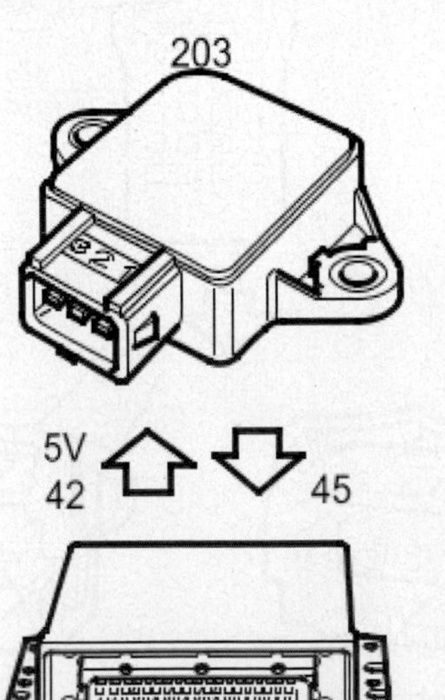

29246_SAAB_G0066

Fig. 66 Throttle Position Sensor (TPS) operation–900, 9000 & 9–3 models, 2.0L & 2.3L engines.

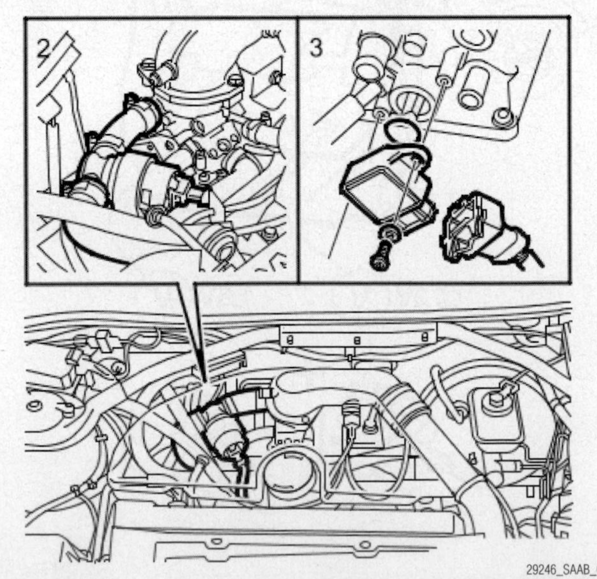

29246_SAAB_G0068

Fig. 67 Removing Throttle Position Sensor (TPS)–900, 9000 & 9–3 models, 2.0L & 2.3L engines.

tent faults. Reading should be 99 percent. If not to specification, repair wiring between TPS pin 3 and ECM pin 45.

6. Connect a jumper wire between pins 2 and 3 of TPS female connector. Read "Throttle Position" on scan tool while moving wiring to reveal intermittent faults. Reading should be 0 percent. If not to specification, repair wiring between TPS pin 3 and ECM pin 45.

7. If wiring and scan tool tests are OK, replace TPS.

8. Clear all DTCs. Perform drive cycle for 5 minutes at varying engine loads and speeds. Evaluate drive cycle and check that DTCs have not returned.

TPS Tests	Ohms	Volts
Pin 1 - ground		4.5-5.5
Pin 2 - power		11.0-13.0
Pins 1 - 2	1.6-2.4	4.9-5.1
Pins 2 - 3 at idle	0.8-1.2	0.1-0.9
Pins 2 - 3 at WOT	2.0-3.0	4.1-4.9

Vehicle Speed Sensor (VSS)

LOCATION

900 & 9-3

See Figure 69.

Vehicle speed signal is sent by ABS control module, which is located at hydraulic unit.

9000

Speedometer sends vehicle speed signal to ECM.

OPERATION

900 & 9-3

See Figures 70 and 71.

Speed sensors at all wheels send signals to ABS control unit. Right front wheel speed sensor sends signal to ABS control unit for engine management purposes. ABS control unit pin 26 sends a vehicle speed signal to ECM pin 39. Speed signal is a square wave that varies between 0 and battery positive voltage. A transistor in ABS control module grounds the output signal 29 times per wheel revolution. Vehicle speed signal is used to:

- determine which gear is currently engaged
- activate SHIFT UP lamp (manual transaxle vehicles, except 5th gear)
- limit boost pressure in Reverse, 1st and 2nd gear
- activate fuel shut-off in 3rd, 4th and 5th gear
- limit vehicle top speed by limiting boost pressure at 143 mph (230 km per hour)
- prevent idle speed control adjustment when vehicle is moving

If vehicle speed signal stops, the functions above will cease, and boost pressure will be limited on vehicles with manual transaxles.

9000

Speedometer sends vehicle speed signal to ECM pin 39. Speed signal is a square wave that varies between 0 and battery positive voltage. At 19 mph (30 km per hour) signal frequency is about 20 Hz. Vehicle speed signal is used to:

- determine which gear is currently engaged
- activate SHIFT UP lamp (manual transaxle vehicles, except 5th gear)
- limit boost pressure in 1st gear
- limit vehicle top speed by limiting boost pressure when maximum speed is reached
- prevent idle speed control adjustment when vehicle is moving

If vehicle speed signal stops, the functions above will cease, and boost pressure

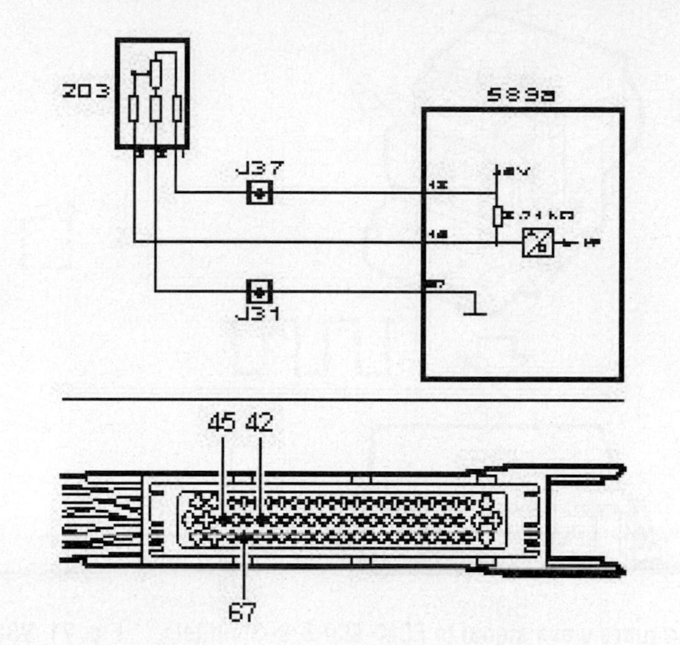

29246_SAAB_G0067

Fig. 68 Throttle Position Sensor (TPS) wiring schematic–900, 9000 & 9-3 models, 2.0L & 2.3L engines.

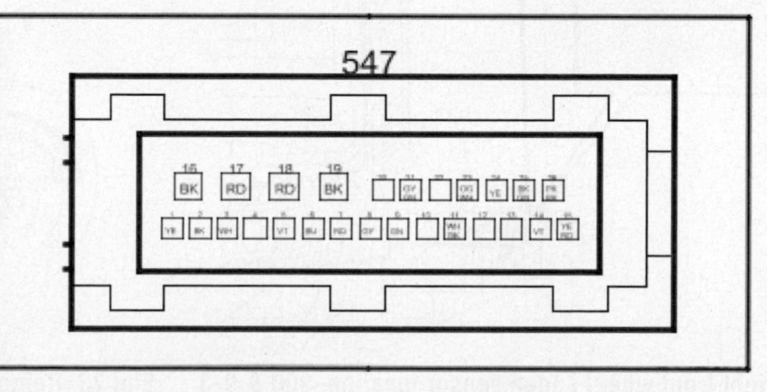

29246_SAAB_G0069

Fig. 69 ABS control module wiring connector location–900 & 9-3 models, 2.0L & 2.3L engines.

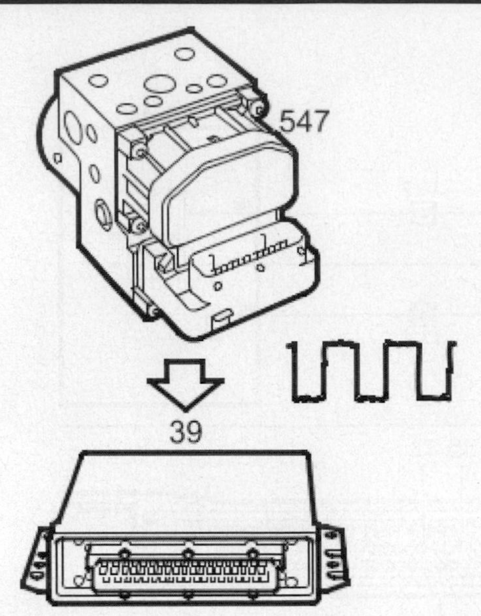

Fig. 70 VSS square wave signal to ECM–900 & 9–3 models, 2.0L & 2.3L engines.

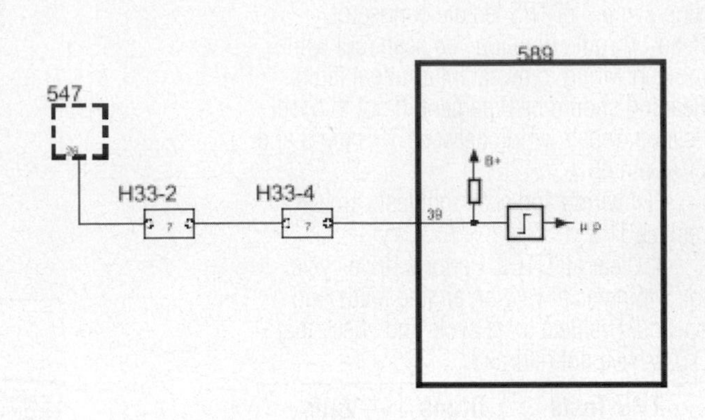

Fig. 71 VSS to ECM wiring schematic–900 & 9–3 models, 2.0L & 2.3L engines.

will be limited on vehicles with manual transaxles.

REMOVAL & INSTALLATION

900 & 9–3

RIGHT FRONT WHEEL SPEED SENSOR

See Figures 72 and 73.

1. Remove air filter top housing and MAP sensor, if needed, and move them aside.
2. Remove the connector's cable tie. Separate the connector and snip the cable tie by air filter housing attachment point.
3. Raise vehicle and remove right front wheel.
4. Clean area around wheel speed sensor with soft wire brush.
5. Remove sensor retaining bolt and sensor.
6. Remove rubber grommet and disconnect sensor wiring connector.
7. Disconnect wiring from retaining clips and remove sensor.

To install:

8. Insert wiring and connector into sensor.
9. Install sensor and tighten retaining bolt.
10. Spray rubber grommet with silicone or synthetic lubricant.
11. Insert rubber grommet into place.
12. Insert wiring into retaining clips.
13. Install wheel. Tighten wheel lug nuts to 80 ft. lbs. (110 Nm). Lower vehicle.
14. Insert wiring connector.
15. Install new cable tie to air filter housing attachment point.
16. Install air filter top housing and MAP sensor, if removed.

TESTING

If vehicle speed signal is absent for an extended period of time, DTC P0502 will be set. If the speed value is unrealistic, DTC P0501 will be set.

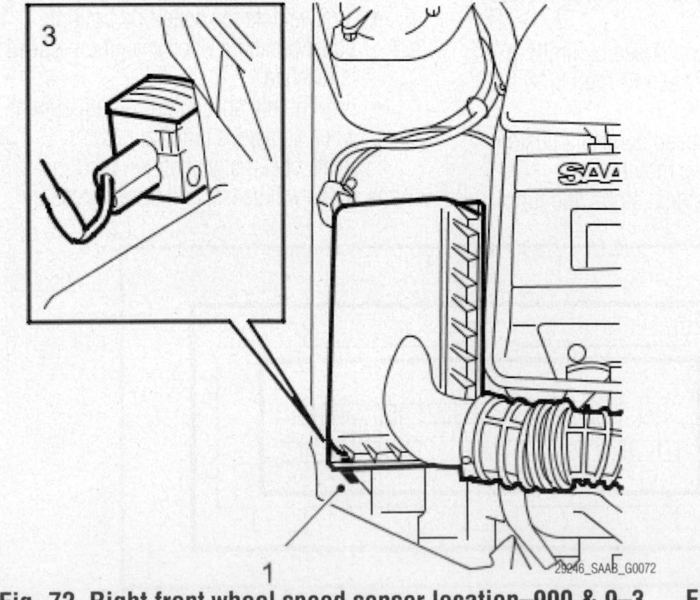

Fig. 72 Right front wheel speed sensor location–900 & 9–3 models, 2.0L & 2.3L engines.

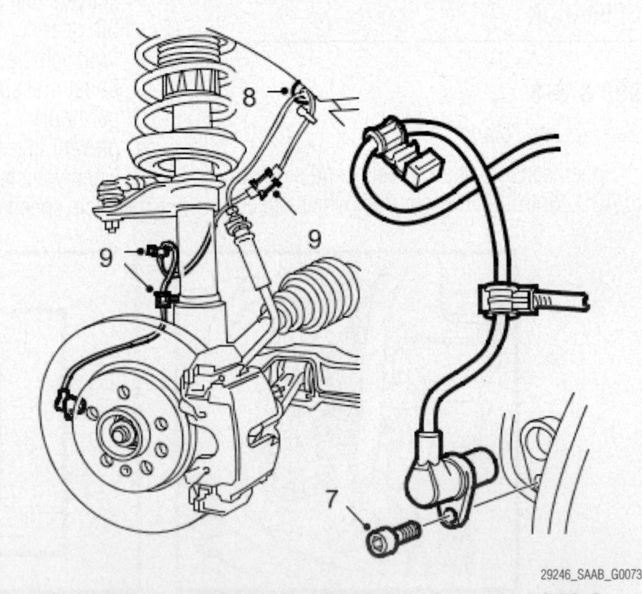

Fig. 73 Removing right front wheel speed sensor –900 & 9–3 models, 2.0L & 2.3L engines.

VOLKSWAGEN
DIAGNOSTIC TROUBLE CODES

14

TABLE OF CONTENTS

DIAGNOSTIC TROUBLE CODES

OBD II Vehicle Applications

VOLKSWAGEN AUTOMOBILES

Cabrio
1995–2002
2.0L V4 MPI . . . . . Engine Code: ABA

Golf
1995–2006
1.8L MPI . . . . . Engine Codes: AWD, AWW
1.9L DFI . . . . . Engine Codes: AHU, ALH, BEW
2.0L MPI . . . . . Engine Codes: ABA, AEG, AVH, AZG, BEV

GTI
1995–2007
1.8L MPI . . . . . Engine Codes: AWD, AWW, AWP
2.0L MPI . . . . . Engine Codes: ABVA, AEG, BPY
2.8L MPI . . . . . Engine Codes: AAA, AFP, BDF

Jetta
1995–2006
1.8L MPI . . . . . Engine Codes: AWD, AWW, AWP
1.9L DFI . . . . . Engine Codes: ALH, BEW, BRM
2.0L MPI . . . . . Engine Codes: ABA, AEG, AVH, AZG, BEV, BBW, BPY
2.5L MPI . . . . . Engine Codes: BGP, BGQ
2.8L MPI . . . . . Engine Codes: AAA, AFP, BDF

Jetta Wagon
2001–2005
1.8L MPI . . . . . Engine Codes: AWP
1.9L DFI . . . . . Engine Codes: ALH, BEW
2.0L MPI . . . . . Engine Codes: AVH, AZG, BEV
2.8L MPI . . . . . Engine Codes: AFP

New Beetle
1998–2006
1.8L MPI . . . . . Engine Codes: APH, AWV, AWP, BNU
1.9L DFI . . . . . Engine Codes: ALH, BEW
2.0L MPI . . . . . Engine Codes: AEG, AVH, AZG, BEV
2.5L MPI . . . . . Engine Codes: BPR, BPS

New Beetle Convertible
2003–2006
1.8L MPI . . . . . Engine Codes: AWV, BKF
2.0L MPI . . . . . Engine Codes: BDC, BEV, BGD
2.5L MPI . . . . . Engine Codes: BPR, BPS

Passat
1995–2007
1.8L MPI . . . . . Engine Codes: AEB, AUG, ATW, AWM
1.9L DFI . . . . . Engine Codes: AAZ, 1Z
2.0L MPI . . . . . Engine Codes: ABA, BHW, BPY
2.8L MPI . . . . . Engine Codes: AAA, AHA, ATQ
3.6L MPI . . . . . Engine Codes: BLV
4.0L MPI . . . . . Engine Codes: BDP

Passat Wagon
1995–2006
1.8L MPI . . . . . Engine Codes: AEB, AUG, ATW, AWM
2.0L MPI . . . . . Engine Codes: BHW, BPY
2.8L MPI . . . . . Engine Codes: AHA, ATQ
3.6L MPI . . . . . Engine Codes: BLV
4.0L MPI . . . . . Engine Codes: BDP

GAS ENGINE TROUBLE CODE LIST

Introduction

To use this information, first read and record all codes in memory along with any Freeze Frame data. *If the PCM reset function is done prior to recording any data, all codes and freeze frame data will be lost!* Look up the desired code by DTC number, Code Title and Conditions (enable criteria) that indicate why a code set, and how to drive the vehicle. **1T and 2T** (where available) indicate a 1-trip or 2-trip fault and the Monitor type.

Gas Engine OBD II Trouble Code List (P0xxx Codes)

DTC	Trouble Code Title, Conditions & Possible Causes:
DTC: P0010 **2T, MIL: Yes** **Passat**: 1.8L (AEB, AUG, AWM), 2.0 (BPY), 2.8L (AHA, ATQ), 4.0L (BDP); **Jetta:** 1.8L (AWP, AWW), 2.0L (BBW), 2.5L (BGP, BGQ), 2.8 (BDF); **Golf:** 1.8L (AWW); **GTI:** 1.8 (AWP, AWW), 2.8L (BDF); **NB:** 1.8L (AWP, AWV, BNU, BKF) **Transmissions:** All	**"A" Camshaft Position Actuator Circuit (Bank 1) Conditions:** Key on or engine running; and the ECM detected an unexpected high voltage or low voltage condition on the camshaft position sensor. The relative position between the camshaft and crankshaft needs to be optimal so the engine has better torque, fuel economy and emissions. **Note: The camshaft adjustment is load- and RPM dependant. The electrical camshaft adjustment valve 1 switches oil pressure onto camshaft adjuster (mechanical adjustment mechanism), which adjusts the camshaft.** **Possible Causes:** • Fuel pump has failed • Actuator circuit is open • ECM has failed • Battery voltage below 11.5 volts • Position actuator circuit may short to B+ or Ground
DTC: P0011 **2T, MIL: Yes** **Passat:** 1.8L (AEB, AUG, AWM), 2.0 (BPY), 2.8L (AHA, ATQ), 4.0L (BDP); **Jetta:** 1.8L (AWP, AWW), 2.0L (BBW, BPY), 2.5L (BGP, BGQ), 2.8 (BDF); **Golf:** 1.8L (AWW); **GTI:** 1.8 (AWP, AWW), 2.8L (BDF); **NB:** 1.8L (AWP, AWV, BNU, BKF) **Transmissions:** All	**"A" Camshaft Position Timing Over-Advanced (Bank 1) Conditions:** Engine started and driven at an engine speed of more than 400rpm; and the ECM detected the camshaft timing exceeded the maximum calibrated advance value, or the camshaft remained in an advanced position during the CCM test. The valve timing did not change from the current valve timing or it remained fixed during the testing. **Note: The camshaft adjustment is load- and RPM dependant. The electrical camshaft adjustment valve 1 switches oil pressure onto camshaft adjuster (mechanical adjustment mechanism), which adjusts the camshaft.** **Possible Causes:** • Fuel pump has failed • CPS circuit is open, shorted to ground or shorted to power • ECM has failed • Battery voltage below 11.5 volts • Position actuator circuit may short to B+ or Ground • Camshaft timing improperly set, or continuous oil flow to the VCT piston chamber • Camshaft advance mechanism (the VCT unit) is sticking or binding mechanically • VCT solenoid valve is stuck in open position
DTC: P0012 **2T, MIL: Yes** **Passat:** 1.8L (AEB, AUG, AWM), 2.0 (BPY), 2.8L (AHA, ATQ), 4.0L (BDP); **Jetta:** 1.8L (AWP, AWW), 2.0L (BBW), 2.5L (BGP, BGQ), 2.8 (BDF); **Golf:** 1.8L (AWW); **GTI:** 1.8 (AWP, AWW), 2.8L (BDF) **Transmissions:** All	**"A" Camshaft Position Over-Retarded (Bank 1) Conditions:** Engine started and driven at an engine speed of more than 400rpm; and the ECM detected the camshaft timing exceeded the minimu calibrated retarded value, or the camshaft remained in an retarted position during the CCM test. The valve timing did not change from the current valve timing or it remained fixed during the testing. **Note: The camshaft adjustment is load- and RPM dependant. The electrical camshaft adjustment valve 1 switches oil pressure onto camshaft adjuster (mechanical adjustment mechanism), which adjusts the camshaft.** **Possible Causes:** • Fuel pump has failed • CPS circuit is open, shorted to ground or shorted to power • ECM has failed • Battery voltage below 11.5 volts • Position actuator circuit may short to B+ or Ground • Camshaft timing improperly set, or continuous oil flow to the VCT piston chamber • Camshaft advance mechanism (the VCT unit) is sticking or binding mechanically • VCT solenoid valve is stuck in open position
DTC: P0013 **2T, MIL: Yes** **Passat:** 2.8L (AHA and ATQ), 4.0L (BDP); **Jetta:** 2.8 (BDF); **Golf:** 2.8L (BDF); **NB:** 1.8L (AWP, AWV, BNU, BKF) **Transmissions:** All	**"B" Camshaft Position Actuator Circuit (Bank 1) Conditions:** Key on or engine running; and the ECM detected an unexpected high voltage or low voltage condition on the camshaft position sensor. The relative position between the camshaft and crankshaft needs to be optimal so the engine has better torque, fuel economy and emissions. **Note: The camshaft adjustment is load- and RPM dependant. The electrical camshaft adjustment valve 1 switches oil pressure onto camshaft adjuster (mechanical adjustment mechanism), which adjusts the camshaft.** **Possible Causes:** • Fuel pump has failed • ECM has failed • Battery voltage below 11.5 volts • Position actuator circuit may short to B+ or Ground

DTC	Trouble Code Title, Conditions & Possible Causes
DTC: P0014 **2T, MIL: Yes** **Passat:** 2.8L (AHA and ATQ), 4.0L (BDP); **Jetta:** 2.8 (BDF); **Golf:** 2.8L (BDF) **Transmissions:** All	**"B" Camshaft Position Timing Over-Advanced (Bank 1) Conditions:** Engine started and driven at an engine speed of more than 400rpm; and the ECM detected the camshaft timing exceeded the maximum calibrated advance value, or the camshaft remained in an advanced position during the CCM test. The valve timing did not change from the current valve timing or it remained fixed during the testing. **Note: The camshaft adjustment is load- and RPM dependant. The electrical camshaft adjustment valve 1 switches oil pressure onto camshaft adjuster (mechanical adjustment mechanism), which adjusts the camshaft.** **Possible Causes:** • Fuel pump has failed • CPS circuit is open, shorted to ground or shorted to power • ECM has failed • Battery voltage below 11.5 volts • Position actuator circuit may short to B+ or Ground • Camshaft timing improperly set, or continuous oil flow to the VCT piston chamber • Camshaft advance mechanism (the VCT unit) is sticking or binding mechanically • VCT solenoid valve is stuck in open position
DTC: P0015 **2T, MIL: Yes** **Passat:** 2.8L (AHA and ATQ), 4.0L (BDP); **Jetta:** 2.8 (BDF); **Golf:** 2.8L (BDF) **Transmissions:** All	**"B" Camshaft Position Over-Retarded (Bank 1) Conditions:** Engine started and driven at an engine speed of more than 400rpm; and the ECM detected the camshaft timing exceeded the minimu calibrated retarded value, or the camshaft remained in an retarted position during the CCM test. The valve timing did not change from the current valve timing or it remained fixed during the testing. **Note: The camshaft adjustment is load- and RPM dependant. The electrical camshaft adjustment valve 1 switches oil pressure onto camshaft adjuster (mechanical adjustment mechanism), which adjusts the camshaft.** **Possible Causes:** • Fuel pump has failed • CPS circuit is open, shorted to ground or shorted to power • ECM has failed • Battery voltage below 11.5 volts • Position actuator circuit may short to B+ or Ground • Camshaft timing improperly set, or continuous oil flow to the VCT piston chamber • Camshaft advance mechanism (the VCT unit) is sticking or binding mechanically • VCT solenoid valve is stuck in open position
DTC: P0016 **2T, MIL: Yes** **Passat:** 2.0L (BPY), 4.0L (BDP); **Jetta:** 1.8L (AWD, AWP, AWW), 2.0L (AEG, AVH, AZG, BBW, BEV, BPY), 2.5L (BGP, BGQ), 2.8 (AFP, BDF); **Golf:** 1.8L (AWD, AWW), 2.0L (AEG, AVH, AZG, BEV); **GTI:** 1.8L (AWD, AWP, AWW), 2.0L (AEG), 2.8 (AFP, BDF); **NB:** 1.8L (APH, AWP, AWV, BNU, BKF), 2.0L (AEG, AVH, AZG, BEV); **NBvert:** (BDG) **Transmissions:** All	**Crankshaft Position - Camshaft Position Correlation Bank 1 Sensor A Conditions:** Engine started, engine running, and the ECM detected a diviation between the crankshaft position sensor signal and the camshaft position sensor. A rationality error has been detected for camshaft position out of phase with crankshaft. **Possible Causes:** • Camshaft Position (CMP) sensor is faulty • CMP circuit short to ground, power or open • Engine Speed (RPM) sensor is faulty • ECM has failed
DTC: P0017 **2T, MIL: Yes** **Passat:** 4.0L (BDP) **Transmissions:** All	**Crankshaft Position - Camshaft Position Correlation Bank 1 Sensor B Conditions:** Engine started, engine running, and the ECM detected a diviation between the crankshaft position sensor signal and the camshaft position sensor. A rationality error has been detected for camshaft position out of phase with crankshaft. **Possible Causes:** • Camshaft Position (CMP) sensor is faulty • CMP circuit short to ground, power or open • Engine Speed (RPM) sensor is faulty • ECM has failed
DTC: P0018 **2T, MIL: Yes** **Passat:** 4.0L (BDP); **GTI:** 2.8 (AFP, BDF) **Transmissions:** All	**Crankshaft Position - Camshaft Position Correlation Bank 2 Sensor A Conditions:** Engine started, engine running, and the ECM detected a diviation between the crankshaft position sensor signal and the camshaft position sensor. A rationality error has been detected for camshaft position out of phase with crankshaft. **Possible Causes:** • Camshaft Position (CMP) sensor is faulty • CMP circuit short to ground, power or open • Engine Speed (RPM) sensor is faulty • ECM has failed

DTC	Trouble Code Title, Conditions & Possible Causes
DTC: P0019 **2T, MIL: Yes** **Passat:** 4.0L (BDP) **Transmissions:** All	**Crankshaft Position - Camshaft Position Correlation Bank 2 Sensor B Conditions:** Engine started, engine running, and the ECM detected a diviation between the crankshaft position sensor signal and the camshaft position sensor. A rationality error has been detected for camshaft position out of phase with crankshaft. **Possible Causes:** • Camshaft Position (CMP) sensor is faulty • CMP circuit short to ground, power or open • Engine Speed (RPM) sensor is faulty • ECM has failed
DTC: P0020 **2T, MIL: Yes** **Passat:** 2.8L (AHA and ATQ), 4.0L (BDP) **Transmissions:** All	**"A" Camshaft Position Timing Over-Advanced (Bank 2) Conditions:** Engine started and driven at an engine speed of more than 400rpm; and the ECM detected the camshaft timing exceeded the maximum calibrated advance value, or the camshaft remained in an advanced position during the CCM test. The valve timing did not change from the current valve timing or it remained fixed during the testing. **Possible Causes:** • Fuel pump has failed • CPS circuit is open, shorted to ground or shorted to power • ECM has failed • Battery voltage below 11.5 volts • Position actuator circuit may short to B+ or Ground • Camshaft timing improperly set, or continuous oil flow to the VCT piston chamber • Camshaft advance mechanism (the VCT unit) is sticking or binding mechanically • VCT solenoid valve is stuck in open position
DTC: P0021 **2T, MIL: Yes** **Passat:** 2.8L (AHA and ATQ), 4.0L (BDP) **Transmissions:** All	**"A" Camshaft Position Actuator Circuit (Bank 2) Conditions:** Key on or engine running; and the ECM detected an unexpected high voltage or low voltage condition on the camshaft position sensor. The relative position between the camshaft and crankshaft needs to be optimal so the engine has better torque, fuel economy and emissions. **Possible Causes:** • Fuel pump has failed • Actuator circuit is open, shorted to ground or shorted to power • ECM has failed • Battery voltage below 11.5 volts • Position actuator circuit may short to B+ or Ground
DTC: P0022 **2T, MIL: Yes** **Passat:** 2.8L (AHA and ATQ) **Transmissions:** All	**"A" Camshaft Position Over-Retarded (Bank 2) Conditions:** Engine started and driven at an engine speed of more than 400rpm; and the ECM detected the camshaft timing exceeded the minimu calibrated retarded value, or the camshaft remained in an retarted position during the CCM test. The valve timing did not change from the current valve timing or it remained fixed during the testing. **Possible Causes:** • Fuel pump has failed • CPS circuit is open, shorted to ground or shorted to power • ECM has failed • Battery voltage below 11.5 volts • Position actuator circuit may short to B+ or Ground • Camshaft timing improperly set, or continuous oil flow to the VCT piston chamber • Camshaft advance mechanism (the VCT unit) is sticking or binding mechanically • VCT solenoid valve is stuck in open position
DTC: P0023 **2T, MIL: Yes** **Passat:** 2.8L (AHA and ATQ), 4.0L (BDP) **Transmissions:** All	**"B" Camshaft Position Actuator Circuit (Bank 2) Conditions:** Key on or engine running; and the ECM detected an unexpected high voltage or low voltage condition on the camshaft position sensor. The relative position between the camshaft and crankshaft needs to be optimal so the engine has better torque, fuel economy and emissions. **Possible Causes:** • Fuel pump has failed • Actuator circuit is open, shorted to ground or shorted to power • ECM has failed • Battery voltage below 11.5 volts • Position actuator circuit may short to B+ or Ground

DTC	Trouble Code Title, Conditions & Possible Causes
DTC: P0024 **2T, MIL: Yes** **Passat:** 2.8L (AHA and ATQ), 4.0L (BDP) **Transmissions:** All	**"B" Camshaft Position Timing Over-Advanced (Bank 2) Conditions:** Engine started and driven at an engine speed of more than 400rpm; and the ECM detected the camshaft timing exceeded the maximum calibrated advance value, or the camshaft remained in an advanced position during the CCM test. The valve timing did not change from the current valve timing or it remained fixed during the testing. **Possible Causes:** • Fuel pump has failed • CPS circuit is open, shorted to ground or shorted to power • ECM has failed • Battery voltage below 11.5 volts • Position actuator circuit may short to B+ or Ground • Camshaft timing improperly set, or continuous oil flow to the VCT piston chamber • Camshaft advance mechanism (the VCT unit) is sticking or binding mechanically • VCT solenoid valve is stuck in open position
DTC: P0025 **2T, MIL: Yes** **Passat:** 2.8L (AHA and ATQ), 4.0L (BDP) **Transmissions:** All	**"B" Camshaft Position Over-Retarded (Bank 2) Conditions:** Engine started and driven at an engine speed of more than 400rpm; and the ECM detected the camshaft timing exceeded the minimu calibrated retarded value, or the camshaft remained in an retarted position during the CCM test. The valve timing did not change from the current valve timing or it remained fixed during the testing. **Possible Causes:** • Fuel pump has failed • CPS circuit is open, shorted to ground or shorted to power • ECM has failed • Battery voltage below 11.5 volts • Position actuator circuit may short to B+ or Ground • Camshaft timing improperly set, or continuous oil flow to the VCT piston chamber • Camshaft advance mechanism (the VCT unit) is sticking or binding mechanically • VCT solenoid valve is stuck in open position
DTC: P0030 **2T, MIL: Yes** **Passat:** 1.8L (AWM, ATW, AUG), 2.0L (BPY), 4.0L (BDP); **Jetta:** 1.8L (AWP), 2.0 (ABA, AEG, AVH, AZG, BBW, BEV, BPY), 2.5L (BGP, BGQ), 2.8 (AFP, BDF); **Golf:** 1.8L (AWD, AWW), 2.0L (ABA, AEG, AVH, AZG, BEV); **GTI:** 1.8L (AWD, AWP, AWW), 2.0L (ABA, AEG), 2.8 (AFP, BDF); **NB:** 1.8L (APH, AWP, AWV, BNU, BKF), 2.0L (AEG, AVH, AZG, BEV); **NBvert:** (BDG); **Cabrio:** 2.0L (ABA) **Transmissions:** All	**HO2S Heater (Bank 1 Sensor 1) Control Circuit Malfunction Conditions:** Engine started, battery voltage must be at least 11.5v, all electrical components must be off, the ground between the engine and the chassis must be well connected, the exhaust system must be properly sealed between the catalytic converter and the cylinder head, the coolant temperature must be 80 degrees Celsius, and the oxygen sensor heater for oxygen sensor before the catalytic converter must be properly functioning. The ECM detected the HO2S signal was in a negative voltage range referred to as "character shift downward". This code sets when the HO2S signal remains in a low state (usually less than 156 mv). In effect, it does not switch properly between 0.1v and 1.1v in closed loop operation. **Possible Causes:** • HO2S is contaminated (due to presence of silicone in fuel) • HO2S signal and ground circuit wires crossed in wiring harness • HO2S signal circuit is shorted to sensor or chassis ground • HO2S element has failed (internal short condition) • ECM has failed
DTC: P0031 **2T, MIL: Yes** **Passat:** 1.8L (AWM, ATW, AUG), 2.0L (BPY), 2.8L (AHA), 4.0L (BDP); **Jetta:** 1.8L (AWD, AWP, AWW), 2.0 (ABA, AEG, AVH, BPY), 2.5L (BGP, BGQ), 2.8 (AFP, BDF); **Golf:** 1.8L (AWD, AWW), 2.0L (ABA, AEG, AVH, AZG, BEV); **GTI:** 1.8L (AWD, AWP, AWW), 2.0L (ABA, AEG), 2.8 (AFP, BDF); **NB:** 1.8L (APH, AWP, AWV, BNU, BKF), 2.0L (AEG, AVH, AZG, BEV); **NBvert:** (BDG); **Cabrio:** 2.0L (ABA) **Transmissions:** All	**HO2S Heater (Bank 1 Sensor 1) Circuit Low Input Conditions:** Engine started, battery voltage must be at least 11.5v, all electrical components must be off, the ground between the engine and the chassis must be well connected, the exhaust system must be properly sealed between the catalytic converter and the cylinder head, the coolant temperature must be 80 degrees Celsius, and the oxygen sensor heater for oxygen sensor before the catalytic converter must be properly functioning. The ECM detected the HO2S signal was in a negative voltage range referred to as "character shift downward". This code sets when the HO2S signal remains in a low state. In effect, it does not switch properly in the closed loop operation. The HO2S (before the three-way catalytic converter) has a short circuit to ground that has lasted longer than 200 seconds **Possible Causes:** • HO2S is contaminated (due to presence of silicone in fuel) • HO2S signal and ground circuit wires crossed in wiring harness • HO2S signal circuit is shorted to sensor or chassis ground • HO2S element has failed (internal short condition) • ECM has failed

DTC	Trouble Code Title, Conditions & Possible Causes
DTC: P0032 **2T, MIL: Yes** **Passat:** 1.8L (AWM, ATW, AUG), 2.0L (BPY), 2.8L (AHA, ATQ), 4.0L (BDP); **Jetta:** 1.8L (AWD, AWP, AWW), 2.0 (ABA, AEG, AVH, AZG, BBW, BEV, BPY), 2.5L (BGP, BGQ), 2.8 (AFP, BDF); **Golf:** 1.8L (AWD, AWW), 2.0L (ABA, AEG, AVH, AZG, BEV); **GTI:** 1.8L (AWD, AWP, AWW), 2.0L (ABA, AEG), 2.8 (AFP, BDF); **NB:** 1.8L (APH, AWP, AWV, BNU, BKF), 2.0L (AEG, AVH, AZG, BEV); **NBvert:** (BDG); **Cabrio:** 2.0L (ABA) **Transmissions:** All	**HO2S Heater (Bank 1 Sensor 1) Circuit High Input Conditions:** Engine started, battery voltage must be at least 11.5v, all electrical components must be off, the ground between the engine and the chassis must be well connected, the exhaust system must be properly sealed between the catalytic converter and the cylinder head, the coolant temperature must be 80 degrees Celsius, and the oxygen sensor heater for oxygen sensor before the catalytic converter must be properly functioning. The ECM detected the HO2S signal remained in a high state. **Note: The HO2S signal circuit may be shorted to the heater power circuit due to tracking inside of the HO2S connector. Remove the connector and visually inspect the connector for signs of oil or water.** **Possible Causes:** • HO2S signal shorted to heater power circuit inside connector • HO2S signal circuit shorted to ground or to system voltage • ECM has failed
DTC: P0033 **Passat:** 2.0L (BPY) **Transmissions:** All	**Turbocharger Bypass Valve Control Circuit Conditions:** The ECM detected the turbocharger bypass valve control circuit signal was providing an invalid input. **Possible Causes:** • Leaks in the air charger system • Turbocharger recirculating valve faulty • Turbocharging system damaged • Vacuum diaphragm out of adjustment • Wastegate bypass valve regulator valve faulty • ECM has failed
DTC: P0034 **Passat:** 2.0L (BPY) **Transmissions:** All	**Turbocharger Bypass Valve Control Circuit Low Conditions:** The ECM detected the turbocharger bypass valve control circuit signal was exceeding the minimum threshold. **Possible Causes:** • Leaks in the air charger system • Turbocharger recirculating valve faulty • Turbocharging system damaged • Vacuum diaphragm out of adjustment • Wastegate bypass valve regulator valve faulty • ECM has failed
DTC: P0035 **Passat:** 2.0L (BPY) **Transmissions:** All	**Turbocharger Bypass Valve Control Circuit High Conditions:** The ECM detected the turbocharger bypass valve control circuit signal was exceeding the maximum threshold. **Possible Causes:** • Leaks in the air charger system • Turbocharger recirculating valve faulty • Turbocharging system damaged • Vacuum diaphragm out of adjustment • Wastegate bypass valve regulator valve faulty • ECM has failed
DTC: P0036 **2T, MIL: Yes** **Passat:** 1.8L (AWM, ATW, AUG), 2.0L (BPY), 4.0L (BDP); **Jetta:** 1.8L (AWD, AWP, AWW), 2.0 (ABA, AEG, AVH, AZG, BBW, BEV, BPY), 2.5L (BGP, BGQ), 2.8 (AFP, BDF); **Golf:** 1.8L (AWD, AWW), 2.0L (ABA, AEG, AVH, AZG, BEV); **GTI:** 1.8L (AWD, AWP, AWW), 2.0L (ABA, AEG), 2.8 (AFP, BDF); **NB:** 1.8L (APH, AWP, AWV, BNU, BKF), 2.0L (AEG, AVH, AZG, BEV); **NBvert:** (BDG); **Cabrio:** 2.0L (ABA) **Transmissions:** All	**HO2S Heater (Bank 1 Sensor 2) Control Circuit Malfunction Conditions:** Engine started, battery voltage must be at least 11.5v, all electrical components must be off, the ground between the engine and the chassis must be well connected, the exhaust system must be properly sealed between the catalytic converter and the cylinder head, the coolant temperature must be 80 degrees Celsius, and the oxygen sensor heater for oxygen sensor before the catalytic converter must be properly functioning. The ECM detected the HO2S signal was in a negative voltage range referred to as "character shift downward". This code sets when the HO2S signal remains in a low state. **Possible Causes:** • HO2S is contaminated (due to presence of silicone in fuel) • HO2S signal and ground circuit wires crossed in wiring harness • HO2S signal circuit is shorted to sensor or chassis ground • HO2S element has failed (internal short condition) • ECM has failed

DTC	Trouble Code Title, Conditions & Possible Causes
DTC: P0037 **2T, MIL: Yes** **Passat:** 1.8L (AWM, ATW, AUG), 2.0L (BPY), 2.8L (AHA, ATQ), 4.0L (BDP); **Jetta:** 1.8L (AWD, AWP, AWW), 2.0L (ABA, AEG, AVH, AZG, BBW, BEV, BPY), 2.5L (BGP, BGQ), 2.8 (AFP, BDF); **Golf:** 1.8L (AWD, AWW), 2.0L (ABA, AEG, AVH, AZG, BEV); **GTI:** 1.8L (AWD, AWP, AWW), 2.0L (ABA, AEG), 2.8 (AFP, BDF); **NB:** 1.8L (APH, AWP, AWV, BNU, BKF), 2.0L (AEG, AVH, AZG, BEV); **NBvert:** (BDG); **Cabrio:** 2.0L (ABA) **Transmissions:** All	**HO2S Heater (Bank 1 Sensor 2) Circuit Low Input Conditions:** Engine started, battery voltage must be at least 11.5v, all electrical components must be off, the ground between the engine and the chassis must be well connected, the exhaust system must be properly sealed between the catalytic converter and the cylinder head, the coolant temperature must be 80 degrees Celsius, and the oxygen sensor heater for oxygen sensor before the catalytic converter must be properly functioning. The ECM detected the HO2S signal was in a negative voltage range referred to as "character shift downward". This code sets when the HO2S signal remains in a low state. In effect, it does not switch properly in the closed loop operation. The HO2S (before the three-way catalytic converter) has a short circuit to ground that has lasted longer than 200 seconds **Possible Causes:** • HO2S is contaminated (due to presence of silicone in fuel) • HO2S signal and ground circuit wires crossed in wiring harness • HO2S signal circuit is shorted to sensor or chassis ground • HO2S element has failed (internal short condition) • ECM has failed
DTC: P0038 **2T, MIL: Yes** **Passat:** 1.8L (AWM, ATW, AUG), 2.0L (BPY), 2.8L (AHA, ATQ), 4.0L (BDP); **Jetta:** 1.8L (AWD, AWP, AWW), 2.0 (ABA, AEG, AVH, AZG, BBW, BEV, BPY), 2.5L (BGP, BGQ), 2.8 (AFP, BDF); **Golf:** 1.8L (AWD, AWW), 2.0L (ABA, AEG, AVH, AZG, BEV); **GTI:** 1.8L (AWD, AWP, AWW), 2.0L (ABA, AEG), 2.8 (AFP, BDF); **NB:** 1.8L (APH, AWP, AWV, BNU, BKF), 2.0L (AEG, AVH, AZG, BEV); **NBvert:** (BDG); **Cabrio:** 2.0L (ABA) **Transmissions:** All	**HO2S Heater (Bank 1 Sensor 2) Circuit High Input Conditions:** Engine started, battery voltage must be at least 11.5v, all electrical components must be off, the ground between the engine and the chassis must be well connected, the exhaust system must be properly sealed between the catalytic converter and the cylinder head, the coolant temperature must be 80 degrees Celsius, and the oxygen sensor heater for oxygen sensor before the catalytic converter must be properly functioning. The ECM detected the HO2S signal remained in a high state. **Note: The HO2S signal circuit may be shorted to the heater power circuit due to tracking inside of the HO2S connector. Remove the connector and visually inspect the connector for signs of oil or water.** **Possible Causes:** • HO2S signal shorted to heater power circuit inside connector • HO2S signal circuit shorted to ground or to system voltage • ECM has failed
DTC: P0040 **2T, MIL: Yes** **Passat:** 4.0L (BDP); **Jetta:** 2.8 (BDF); **Golf:** 2.8L (BDF) **Transmissions:** All	**O2 Sensor Signals Swapped (Bank 1 Sensor 1/Bank 2 Sensor 1) Conditions:** Engine started, battery voltage must be at least 11.5v, all electrical components must be off, the ground between the engine and the chassis must be well connected, the exhaust system must be properly sealed between the catalytic converter and the cylinder head, and the coolant temperature must be 80 degrees Celsius. The ECM detected the O2 signals were mixed and reading implausible results from both. **Possible Causes:** • HO2S-11 and HO2S-21 harness connectors are swapped • HO2S-11 and HO2S-21 wiring is crossed inside the harness • HO2S-11 and HO2S-21 wires are crossed at 104-pin connector • Connector coding and color mixed with correct catalytic converter
DTC: P0041 **2T, MIL: Yes** **Passat:** 2.8L (AHA, ATQ), 4.0L (BDP); **Jetta:** 2.8 (BDF); **Golf:** 2.8L (BDF) **Transmissions:** All	**O2 Sensor Signals Swapped (Bank 1 Sensor 2/Bank 2 Sensor 2) Conditions:** Engine started, battery voltage must be at least 11.5v, all electrical components must be off, the ground between the engine and the chassis must be well connected, the exhaust system must be properly sealed between the catalytic converter and the cylinder head, and the coolant temperature must be 80 degrees Celsius. The ECM detected the O2 signals were mixed and reading implausible results from both. **Possible Causes:** • HO2S-12 and HO2S-22 harness connectors are swapped • HO2S-12 and HO2S-22 wiring is crossed inside the harness • HO2S-12 and HO2S-22 wires are crossed at 104-pin connector • Connector coding and color mixed with correct catalytic converter

DTC	Trouble Code Title, Conditions & Possible Causes
DTC: P0042 **Jetta:** 2.0L (BBW), 2.5L (BGP, BGQ); **NB:** 1.8L (BNU, BKF) **Transmissions:** All	**HO2S Heater Control Circuit Bank 1 Sensor 3 Conditions:** Engine started, battery voltage must be at least 11.5v, all electrical components must be off, the ground between the engine and the chassis must be well connected, the exhaust system must be properly sealed between the catalytic converter and the cylinder head, and the coolant temperature must be 80 degrees Celsius. The ECM detected the HO2S signal was in a negative voltage range referred to as "character shift downward". This code sets when the HO2S signal remains in a low state. In effect, it does not switch properly in the closed loop operation. The HO2S (before the three-way catalytic converter) has a short circuit to ground that has lasted longer than 200 seconds. **Possible Causes:** • HO2S is contaminated (due to presence of silicone in fuel) • HO2S signal and ground circuit wires crossed in wiring harness • HO2S signal circuit is shorted to sensor or chassis ground • HO2S element has failed (internal short condition) • ECM has failed
DTC: P0043 **Jetta:** 2.0L (BBW), 2.5L (BGP, BGQ); **NB:** 1.8L (BNU, BKF) **Transmissions:** All	**HO2S Heater Control Circuit Low Bank 1 Sensor 3 Conditions:** Engine started, battery voltage must be at least 11.5v, all electrical components must be off, the ground between the engine and the chassis must be well connected, the exhaust system must be properly sealed between the catalytic converter and the cylinder head, and the coolant temperature must be 80 degrees Celsius. The ECM detected the HO2S signal was in a negative voltage range referred to as "character shift downward". This code sets when the HO2S signal remains in a low state. In effect, it does not switch properly in the closed loop operation. The HO2S (before the three-way catalytic converter) has a short circuit to ground that has lasted longer than 200 seconds. **Possible Causes:** • HO2S is contaminated (due to presence of silicone in fuel) • HO2S signal and ground circuit wires crossed in wiring harness • HO2S signal circuit is shorted to sensor or chassis ground • HO2S element has failed (internal short condition) • ECM has failed
DTC: P0044 **Jetta:** 2.0L (BBW), 2.5L (BGP, BGQ); **NB:** 1.8L (BNU, BKF) **Transmissions:** All	**HO2S Heater Control Circuit High Bank 1 Sensor 3 Conditions:** Engine started, battery voltage must be at least 11.5v, all electrical components must be off, the ground between the engine and the chassis must be well connected, the exhaust system must be properly sealed between the catalytic converter and the cylinder head, and the coolant temperature must be 80 degrees Celsius. The ECM detected the HO2S signal remained in a high state. **Note: The HO2S signal circuit may be shorted to the heater power circuit due to tracking inside of the HO2S connector. Remove the connector and visually inspect the connector for signs of oil or water.** **Possible Causes:** • HO2S signal shorted to heater power circuit inside connector • HO2S signal circuit shorted to ground or to system voltage • ECM has failed
DTC: P0050 **Passat:** 2.8L (AHA, ATQ), 4.0L (BDP); **Jetta:** 2.8 (AFP, BDF); **GTI:** 2.8 (AFP, BDF) **Transmissions:** All	**HO2S Heater (Bank 2 Sensor 1) Control Circuit Malfunction Conditions:** Engine started, battery voltage must be at least 11.5v, all electrical components must be off, the ground between the engine and the chassis must be well connected, the exhaust system must be properly sealed between the catalytic converter and the cylinder head, and the coolant temperature must be 80 degrees Celsius. The ECM detected the HO2S signal was in a negative voltage range referred to as "character shift downward". **Possible Causes:** • HO2S is contaminated (due to presence of silicone in fuel) • HO2S signal and ground circuit wires crossed in wiring harness • HO2S signal circuit is shorted to sensor or chassis ground • HO2S element has failed (internal short condition) • ECM has failed
DTC: P0051 **Passat:** 2.8L (AHA, ATQ), 4.0L (BDP); **Jetta:** 2.8 (AFP, BDF); **GTI:** 2.8 (AFP, BDF) **Transmissions:** All	**HO2S Heater (Bank 2 Sensor 1) Circuit Low Input Conditions:** Engine started, battery voltage must be at least 11.5v, all electrical components must be off, the ground between the engine and the chassis must be well connected, the exhaust system must be properly sealed between the catalytic converter and the cylinder head, and the coolant temperature must be 80 degrees Celsius. The ECM detected the HO2S signal was in a negative voltage range referred to as "character shift downward". This code sets when the HO2S signal remains in a low state. In effect, it does not switch properly in the closed loop operation. The HO2S (before the three-way catalytic converter) has a short circuit to ground that has lasted longer than a specified time. **Possible Causes:** • HO2S is contaminated (due to presence of silicone in fuel) • HO2S signal and ground circuit wires crossed in wiring harness • HO2S signal circuit is shorted to sensor or chassis ground • HO2S element has failed (internal short condition) • ECM has failed

DTC	Trouble Code Title, Conditions & Possible Causes
DTC: P0052 **Passat:** 4.0L (BDP); **Jetta:** 2.8 (AFP, BDF); **GTI:** 2.8 (AFP, BDF) **Transmissions:** All	**HO2S Heater (Bank 2 Sensor 1) Circuit High Input Conditions:** Engine started, battery voltage must be at least 11.5v, all electrical components must be off, the ground between the engine and the chassis must be well connected, the exhaust system must be properly sealed between the catalytic converter and the cylinder head, and the coolant temperature must be 80 degrees Celsius. The ECM detected the HO2S signal was in a negative voltage range referred to as "character shift downward". This code sets when the HO2S signal remains in a low state. In effect, it does not switch properly in the closed loop operation. The HO2S (before the three-way catalytic converter) has a short circuit to ground that has lasted longer than a specified time. **Possible Causes:** • HO2S is contaminated (due to presence of silicone in fuel) • HO2S signal and ground circuit wires crossed in wiring harness • HO2S signal circuit is shorted to sensor or chassis ground • HO2S element has failed (internal short condition) • ECM has failed
DTC: P0056 **Passat:** 2.8L (AHA, ATQ), 4.0L (BDP); **Jetta:** 2.8 (AFP, BDF); **GTI:** 2.8 (AFP, BDF) **Transmissions:** All	**HO2S Heater (Bank 2 Sensor 2) Circuit High Input Conditions:** Engine started, battery voltage must be at least 11.5v, all electrical components must be off, the ground between the engine and the chassis must be well connected, the exhaust system must be properly sealed between the catalytic converter and the cylinder head, and the coolant temperature must be 80 degrees Celsius. The ECM detected the HO2S signal remained in a high state. **Note: The HO2S signal circuit may be shorted to the heater power circuit due to tracking inside of the HO2S connector. Remove the connector and visually inspect the connector for signs of oil or water.** **Possible Causes:** • HO2S signal shorted to heater power circuit inside connector • HO2S signal circuit shorted to ground or to system voltage • ECM has failed
DTC: P0057 **Passat:** 2.8L (AHA, ATQ), 4.0L (BDP); **Jetta:** 2.8 (AFP, BDF); **GTI:** 2.8 (AFP, BDF) **Transmissions:** All	**HO2S Heater (Bank 2 Sensor 2) Control Circuit Malfunction Conditions:** Engine started, battery voltage must be at least 11.5v, all electrical components must be off, the ground between the engine and the chassis must be well connected, the exhaust system must be properly sealed between the catalytic converter and the cylinder head, and the coolant temperature must be 80 degrees Celsius. The ECM detected the HO2S signal was in a negative voltage range referred to as "character shift downward". **Possible Causes:** • HO2S is contaminated (due to presence of silicone in fuel) • HO2S signal and ground circuit wires crossed in wiring harness • HO2S signal circuit is shorted to sensor or chassis ground • HO2S element has failed (internal short condition) • ECM has failed
DTC: P0058 **Passat:** 2.8L (AHA, ATQ), 4.0L (BDP); **Jetta:** 2.8 (AFP, BDF); **GTI:** 2.8 (AFP, BDF) **Transmissions:** All	**HO2S Heater (Bank 2 Sensor 2) Circuit Low Input Conditions:** Engine started, battery voltage must be at least 11.5v, all electrical components must be off, the ground between the engine and the chassis must be well connected, the exhaust system must be properly sealed between the catalytic converter and the cylinder head, and the coolant temperature must be 80 degrees Celsius. The ECM detected the HO2S signal was in a negative voltage range referred to as "character shift downward". This code sets when the HO2S signal remains in a low state. In effect, it does not switch properly in the closed loop operation. The HO2S (before the three-way catalytic converter) has a short circuit to ground that has lasted longer than a specified time. **Possible Causes:** • HO2S is contaminated (due to presence of silicone in fuel) • HO2S signal and ground circuit wires crossed in wiring harness • HO2S signal circuit is shorted to sensor or chassis ground • HO2S element has failed (internal short condition) • ECM has failed
DTC: P0052 **Passat:** 2.8L (AHA, ATQ) **Transmissions:** All	**HO2S Heater (Bank 2 Sensor 1) Circuit High Input Conditions:** Engine started, battery voltage must be at least 11.5v, all electrical components must be off, the ground between the engine and the chassis must be well connected, the exhaust system must be properly sealed between the catalytic converter and the cylinder head, and the coolant temperature must be 80 degrees Celsius. The ECM detected the HO2S signal remained in a high state. **Note: The HO2S signal circuit may be shorted to the heater power circuit due to tracking inside of the HO2S connector. Remove the connector and visually inspect the connector for signs of oil or water.** **Possible Causes:** • HO2S signal shorted to heater power circuit inside connector • HO2S signal circuit shorted to ground or to system voltage • ECM has failed

DTC	Trouble Code Title, Conditions & Possible Causes
DTC: P0068 **Passat:** 2.0L (BPY); **Jetta:** 2.5L (BGP, BGQ); **NB:** 1.8L (APH, AWP, AWV, BNU, BKF) **Transmissions:** All	**Mass Air Pressure and Mass Air Flow Miscommunication Conditions** Engine running and the temperature must be at least 185-degrees (F) and all electrical equipment (A/C, lights, etc) must be off. The ECM detected a miscommunication between the mass air pressure and the mass air flow sensors **Possible Causes:** • Intake motor flap faulty • Intake system is leaking • Intake manifold runner position sensor is faulty • Mass Air Flow sensor has failed • Throttle valve control module has failed
DTC: P0087 **Passat:** 2.0L (BPY) **Transmissions:** All	**Fuel Rail/System Pressure Too Low Conditions** Engine started, battery voltage must be at least 11.5v, all electrical components must be off, the ground between the engine and the chassis must be well connected, the exhaust system must be properly sealed between the catalytic converter and the cylinder head, and the coolant temperature must be 80 degrees Celsius. The ECM detected that the system's fuel pressure has fallen below the accepted normal calibrated value. **Possible Causes:** • Fuel Pressure Regulator Valve faulty • Fuel Pressure Sensor faulty • Fuel Pump (FP) Control Module faulty • Fuel pump faulty • Low fuel
DTC: P0088 **Passat:** 2.0L (BPY) **Transmissions:** All	**Fuel Rail/System Pressure Too High Conditions** Engine started, battery voltage must be at least 11.5v, all electrical components must be off, the ground between the engine and the chassis must be well connected, the exhaust system must be properly sealed between the catalytic converter and the cylinder head, and the coolant temperature must be 80 degrees Celsius. The ECM detected that the system's fuel pressure has risen above the accepted normal calibrated value. **Possible Causes:** • Fuel Pressure Regulator Valve faulty • Fuel Pressure Sensor faulty • Fuel Pump (FP) Control Module faulty • Fuel pump faulty
DTC: P0089 **Passat:** 2.0L (BPY) **Transmissions:** All	**Fuel Pressure Regulator Range/Performance Conditions** Engine started, battery voltage must be at least 11.5v, all electrical components must be off, the ground between the engine and the chassis must be well connected, the exhaust system must be properly sealed between the catalytic converter and the cylinder head, and the coolant temperature must be 80 degrees Celsius. The ECM detected that the system's fuel pressure sensor is providing a signal that is either outside the accepted normal values or is not receiving a signal at all. **Possible Causes:** • Fuel Pressure Regulator Valve faulty • Fuel Pressure Sensor faulty • Fuel Pump (FP) Control Module faulty • Fuel pump faulty
DTC: P0097 **NB:** 1.8L (BNU, BKF) **Transmissions:** All	**Intake Air Temperature Sensor 2 Circuit Low Conditions** Engine running and the vehicle speed more than 25mph, the ECM detected the IAT sensor was less than the minimum required value (temperature) over a certain period of time. The IAT sensor is a variable resistor that includes an IAT signal circuit and a low reference circuit to measure the temperature of the air entering the engine. The ECM supplies the sensor with a low voltage singal circuit and a low reference ground circuit. When the IAT sensor is cold, its resistence is high. When the air temperature increases, its resistence decreases. With high sensor resisteance, the IAT sensor signal voltage is high. With lower sensor resistance, the IAT sensor signal voltage should be lower. **Possible Causes:** • IAT sensor signal circuit is shorted to sensor or chassis ground • IAT sensor is damaged or has failed • ECM has failed.
DTC: P0098 **NB:** 1.8L (BNU, BKF) **Transmissions:** All	**Intake Air Temperature Sensor 2 Circuit High Conditions** Engine running and the vehicle speed more than 25mph, the ECM detected the IAT sensor was more than the maximum required value (temperature) over a certain period of time. The IAT sensor is a variable resistor that includes an IAT signal circuit and a low reference circuit to measure the temperature of the air entering the engine. The ECM supplies the sensor with a low voltage singal circuit and a low reference ground circuit. When the IAT sensor is cold, its resistence is high. When the air temperature increases, its resistence decreases. With high sensor resisteance, the IAT sensor signal voltage is high. With lower sensor resistance, the IAT sensor signal voltage should be lower. **Possible Causes:** • IAT sensor signal circuit is shorted to sensor or chassis ground • IAT sensor is damaged or has failed • ECM has failed.

DTC	Trouble Code Title, Conditions & Possible Causes
DTC: P0100 **Passat:** 2.0L (BPY); **Jetta:** **Transmissions:** All	**Mass Air Flow Circuit Range/Performance Conditions** Engine running, with the system voltage more than 11.0v, and the temperature must be at least 185-degrees (F) and all electrical equipment (A/C, lights, etc) must be off. The ECM has detected that the MAF signal was less than the required minimum or out of a calculated range with the engine (or undetectable) for a certain period of time. **Possible Causes:** • Mass air flow (MAF) sensor has failed or is damaged • MAF sensor signal circuit is open, shorted to ground or to B+ • ECM has failed
DTC: P0101 **2T, MIL: Yes** **Passat:** 1.8 (AEB, ATW, AUG, AWM), 2.0L (BPY), 2.8L (AHA, ATQ), 4.0L (BDP); **Jetta:** 1.8L (AWD, AWP, AWW), 2.0 (ABA, AEG, AVH, AZG, BBW, BEV, BPY), 2.5L (BGP, BGQ); **Jetta:** 2.8 (AFP, BDF); **Golf:** 1.8L (AWD, AWW), 2.0L (ABA, AEG, AVH, AZG, BEV); **GTI:** 1.8L (AWD, AWP, AWW), 2.0L (ABA, AEG), 2.8 (AFP, BDF); **NB:** 1.8L (APH, AWP, AWV, BNU, BKF), 2.0L (AEG, AVH, AZG, BEV); **NBvert:** (BDG); **Cabrio:** 2.0L (ABA) **Transmissions:** All	**Mass or Volume Air Flow Circuit Range/Performance Conditions** Engine running, with the system voltage more than 11.0v, and the temperature must be at least 185-degrees (F) and all electrical equipment (A/C, lights, etc) must be off. The ECM has detected that the MAF signal was out of a calculated range with the engine (or undetectable) for a certain period of time. **Possible Causes:** • Mass air flow (MAF) sensor has failed or is damaged • ECM has failed • Signal and ground wires of Mass Air Flow (MAF) sensor has short circuited
DTC: P0102 **2T, MIL: Yes** **Passat:** 1.8 (AEB, ATW, AUG, AWM), 2.0L (ABA, BPY), 2.8L (AAA, AHA, ATQ), 4.0L (BDP); **Jetta:** 1.8L (AWD, AWP, AWW), 2.0 (ABA, AEG, AVH, AZG, BBW, BEV, BPY), 2.5L (BGP, BGQ), 2.8L (AAA, AFP, BDF); **Golf:** 1.8L (AWD, AWW), 2.0L (ABA, AEG, AVH, AZG, BEV); **GTI:** 1.8L (AWD, AWP, AWW), 2.0L (ABA, AEG), 2.8L (AAA, AFP, BDF); **NB:** 1.8L (APH, AWP, AWV, BNU, BKF), 2.0L (AEG, AVH, AZG, BEV); **NBvert:** (BDG); **Cabrio:** 2.0L (ABA) **Transmissions:** All	**MAF Sensor Circuit Low Input Conditions:** Key on, engine started, and the ECM detected the MAF sensor signal was less than the minimum calibrated value. The engine temperature must beat least 185-degrees (F) and all electrical equipment (A/C, lights, etc) must be off. The ECM has detected that the MAF signal was less than the required minimum. **Possible Causes:** • Check for leaks between MAF sensor and throttle valve control module • Voltage supply faulty. • Sensor power circuit open from fuel pump relay to MAF sensor • Sensor signal circuit open (may be disconnected) from ECM and MAF • Faulty ground cable resistance between connector terminal 1 and Ground • MAF Sensor malfunction
DTC: P0103 **2T, MIL: Yes** **Passat:** 1.8 (AEB, ATW, AUG, AWM), 2.0L (ABA, BPY), 2.8L (AAA, AHA, ATQ), 4.0L (BDP); **Jetta:** 1.8L (AWD, AWP, AWW), 2.0 (ABA, AEG, AVH, AZG, BBW, BEV, BPY), 2.5L (BGP, BGQ), 2.8L (AAA, AFP, BDF); **Golf:** 1.8L (AWD, AWW), 2.0L (ABA, AEG, AVH, AZG, BEV); **GTI:** 1.8L (AWD, AWP, AWW), 2.0L (ABA, AEG), 2.8L (AAA, AFP); **NB:** 1.8L (APH, AWP, AWV, BNU, BKF), 2.0L (AEG, AVH, AZG, BEV); **NBvert:** (BDG); **Cabrio:** 2.0L (ABA) **Transmissions:** All	**MAF Sensor Circuit High Input Conditions:** Key on, engine started, and the ECM detected the MAF sensor signal was more than the minimum calibrated value. The engine temperature must beat least 185-degrees (F) and all electrical equipment (A/C, lights, etc) must be off. The ECM has detected that the MAF signal was more than the required minimum. **Possible Causes:** • Check for leaks between MAF sensor and throttle valve control module • Voltage supply faulty. • Sensor power circuit open from fuel pump relay to MAF sensor • Sensor signal circuit open (may be disconnected) from ECM and MAF • Faulty ground cable resistance between connector terminal 1 and Ground • MAF Sensor malfunction

DTC	Trouble Code Title, Conditions & Possible Causes
DTC: P0106 **2T, MIL: Yes** **Passat:** 1.8 (AEB, ATW, AUG, AWM), 2.0L (BPY); **Jetta:** 1.8L (AWD, AWP, AWW), 2.0 (ABA, AEG, AVH, AZG, BBW, BEV), 2.5L (BGP, BGQ), 2.8L (AFP, BDF); **Golf:** 1.8L (AWD, AWW), 2.0L (ABA, AEG, AVH, AZG, BEV); **GTI:** 1.8L (AWD, AWP, AWW), 2.0L (ABA, AEG), 2.8L (AFP, BDF); **NB:** 1.8L (APH, AWP, AWV, BNU, BKF), 2.0L (AEG, AVH, AZG, BEV); **NBvert:** (BDG); **Cabrio:** 2.0L (ABA) **Transmissions:** All	**Manifold Absolute Pressure/Barometric Pressure Sensor Circuit Performance Conditions:** Engine started, the temperature must beat least 185-degrees (F) and all electrical equipment (A/C, lights, etc) must be off. The ECM detected the BARO sensor was out of range during the CCM test. The BARO sensor signal should be in 4.5v. **Possible Causes:** • Sensor has deteriorated (response time too slow) or has failed • MAP sensor signal circuit is shorted to ground • MAP sensor circuit (5v) is open • MAP sensor is damaged or it has failed • BARO sensor signal circuit is shorted to ground • BARO sensor circuit (5v) is open • BARO sensor is damaged or it has failed • ECM is not connected properly • ECM has failed
DTC: P0107 **Passat:** 1.8 (AEB, ATW, AUG, AWM), 2.0L (BPY); **Jetta:** 2.0L (ABA), 2.5L (BGP, BGQ); **Golf:** 2.0L (ABA, AEG, AVH, AZG, BEV); **GTI:** 2.0L (ABA); **NB:** 1.8L (APH, AWP, AWV, BNU, BKF); **Cabrio:** 2.0L (ABA) **Transmissions:** All	**Manifold Absolute Pressure/Barometric Pressure Sensor Circuit Low Input Conditions:** Engine started, the temperature must beat least 185-degrees (F) and all electrical equipment (A/C, lights, etc) must be off. The ECM detected the BARO sensor was out of range during the CCM test. The BARO sensor signal should be in 4.5v. The BARO sensor is a variable capacitance unit used to detect altitude. **Possible Causes:** • Sensor has deteriorated (response time too slow) or has failed • MAP sensor signal circuit is shorted to ground • MAP sensor circuit (5v) is open • MAP sensor is damaged or it has failed • BARO sensor signal circuit is shorted to ground • BARO sensor circuit (5v) is open • BARO sensor is damaged or it has failed • ECM is not connected properly • ECM has failed
DTC: P0108 **Passat:** 1.8 (AEB, ATW, AUG, AWM), 2.0L (BPY); **Jetta:** 2.0L (ABA), 2.5L (BGP, BGQ); **Golf:** 2.0L (ABA, AEG, AVH, AZG, BEV); **GTI:** 2.0L (ABA); **NB:** 1.8L (APH, AWP, AWV, BNU, BKF); **Cabrio:** 2.0L (ABA) **Transmissions:** All	**Manifold Absolute Pressure/Barometric Sensor Circuit High Input Conditions:** Engine started, the temperature must beat least 185-degrees (F) and all electrical equipment (A/C, lights, etc) must be off. The ECM detected the BARO sensor was out of range during the CCM test. The BARO sensor signal should be in 4.5v. The BARO sensor is a variable capacitance unit used to detect altitude. **Possible Causes:** • Sensor has deteriorated (response time too slow) or has failed • MAP sensor signal circuit is shorted to ground • MAP sensor circuit (5v) is open • MAP sensor is damaged or it has failed • BARO sensor signal circuit is shorted to ground • BARO sensor circuit (5v) is open • BARO sensor is damaged or it has failed • ECM is not connected properly • ECM has failed
DTC: P0111 **Passat:** 1.8 (AEB, ATW, AUG, AWM), 2.0L (ABA, BPY), 2.8L (AHA, ATQ), 4.0L (BDP); Jetta 2.8L (AFP, BDF); **GTI:** 2.8L (AFP, BDF) **Transmissions:** All	**Intake Air Temperature Sensor Circuit Low Input Conditions:** Key on or engine running, the temperature must beat least 185-degrees (F) and all electrical equipment (A/C, lights, etc) must be off; and the ECM detected the IAT sensor signal was less than the self-test minimum. This is a thermistor-type sensor with a variable resistance that changes when exposed to different temperatures. This means: the higher the temperature, the lower the resistance value. **Possible Causes:** • IAT sensor signal circuit is grounded (check wiring & connector) • Resistance value between sockets 33 and 36 out of range • IAT sensor has an open circuit • IAT sensor is damaged or it has failed • ECM has failed

DTC	Trouble Code Title, Conditions & Possible Causes
DTC: P0112 **2T, MIL: Yes** **Passat:** 1.8 (AEB, ATW, AUG, AWM), 2.0L (ABA, BPY), 2.8L (AAA, AHA, ATQ), 4.0L (BDP); **Jetta:** 1.8L (AWD, AWP, AWW), 2.0 (ABA, AEG, AVH, AZG, BBW, BEV, BPY), 2.5L (BGP, BGQ), 2.8L (AAA, AFP, BDF); **Golf:** 1.8L (AWD, AWW), 2.0L (ABA, AEG, AVH, AZG, BEV); **GTI:** 1.8L (AWD, AWP, AWW), 2.0L (ABA, AEG), 2.8L (AAA, AFP, BDF); **NB:** 1.8L (APH, AWP, AWV, BNU, BKF), 2.0L (AEG, AVH, AZG, BEV); **NBvert:** (BDG); **Cabrio:** 2.0L (ABA) **Transmissions:** All	**Intake Air Temperature Sensor Circuit Low Input Conditions:** Key on or Engine running, the temperature must beat least 185-degrees (F) and all electrical equipment (A/C, lights, etc) must be off; and the ECM detected the IAT sensor signal was less than the self-test minimum. This is a thermistor-type sensor with a variable resistance that changes when exposed to different temperatures. This means: the higher the temperature, the lower the resistance value. **Possible Causes:** • IAT sensor signal circuit is grounded (check wiring & connector) • Resistance value between sockets 33 and 36 out of range • IAT sensor has an open circuit • IAT sensor is damaged or it has failed • ECM has failed
DTC: P0113 **2T, MIL: Yes** **Passat:** 1.8 (AEB, ATW, AUG, AWM), 2.0L (ABA, BPY), 2.8L (AAA, AHA, ATQ), 4.0L (BDP); **Jetta:** 1.8L (AWD, AWP, AWW), 2.0 (ABA, AEG, AVH, AZG, BBW, BEV, BPY), 2.5L (BGP, BGQ), 2.8L (AAA, AFP, BDF); **Golf:** 2.0L (ABA, AEG, AVH, AZG, BEV); **GTI:** 1.8L (AWD, AWP, AWW), 2.0L (ABA, AEG), 2.8L (AAA, AFP, BDF); **NB:** 1.8L (APH, AWP, AWV, BNU, BKF), 2.0L (AEG, AVH, AZG, BEV); **NBvert:** (BDG); **Cabrio:** 2.0L (ABA) **Transmissions:** All	**Intake Air Temperature Sensor Circuit High Input Conditions:** Key on or engine running, the temperature must beat least 185-degrees (F) and all electrical equipment (A/C, lights, etc) must be off; and the ECM detected the IAT sensor signal was more than the self-test maximum. This is a thermistor-type sensor with a variable resistance that changes when exposed to different temperatures. This means: the higher the temperature, the lower the resistance value. **Possible Causes:** • IAT sensor signal circuit is open (inspect wiring & connector) • IAT sensor signal circuit is shorted • Resistance value between sockets 33 and 36 out of range • IAT sensor is damaged or it has failed • ECM has failed
DTC: P0116 **2T, MIL: Yes** **Passat:** 1.8 (AEB, ATW, AUG, AWM), 2.0L (ABA, BPY), 2.8L (AAA, AHA, ATQ), 4.0L (BDP); **Jetta:** 1.8L (AWD, AWP, AWW), 2.0 (ABA, AEG, AVH, AZG, BBW, BEV, BPY), 2.5L (BGP, BGQ), 2.8L (AAA, AFP, BDF); **Golf:** 1.8L (AWD, AWW), 2.0L (ABA, AEG, AVH, AZG, BEV); **GTI:** 1.8L (AWD, AWP, AWW), 2.0L (ABA, AEG), 2.8L (AAA, AFP, BDF); **NB:** 1.8L (APH, AWP, AWV, BNU, BKF), 2.0L (AEG, AVH, AZG, BEV); **NBvert:** (BDG); **Cabrio:** 2.0L (ABA) **Transmissions:** All	**ECT Sensor / CHT Sensor Signal Range/Performance Conditions:** Engine started (cold), battery voltage must be 11.5, and all equipment must be off. The ECM detected the ECT sensor exceeded the required calibrated value, or the engine is at idle and doesn't reach operating temperature quickly enough; the Catalyst, Fuel System, HO2S and Misfire Monitor did not complete, or the timer expired. Testing completion of procedure, the engine's temperature must rise uniformly during idle. **Possible Causes:** • Check for low coolant level or incorrect coolant mixture • ECM detects a short circuit wiring in the ECT • CHT sensor is out-of-calibration or it has failed • ECT sensor is out-of-calibration or it has failed

DTC	Trouble Code Title, Conditions & Possible Causes
DTC: P0117 **2T, MIL: Yes** **Passat:** 1.8 (AEB, ATW, AUG, AWM), 2.0L (ABA, BPY), 2.8L (AAA, AHA, ATQ), 4.0L (BDP); **Jetta:** 1.8L (AWD, AWP, AWW), 2.0 (ABA, AEG, AVH, AZG, BPY), 2.5L (BGP, BGQ), 2.8L (AAA, AFP, BDF); **Golf:** 1.8L (AWD, AWW), 2.0L (ABA, AEG, AVH, AZG); **GTI:** 1.8L (AWD, AWP, AWW), 2.0L (ABA, AEG), 2.8L (AAA, AFP, BDF); **NB:** 1.8L (APH, AWP, AWV, BNU, BKF), 2.0L (AEG, AVH, AZG, BEV); **NBvert:** (BDG); **Cabrio:** 2.0L (ABA) **Transmissions:** All	**ECT Sensor Circuit Low Input Conditions:** Engine started (cold), battery voltage must be 11.5, and all equipment must be off. The ECM detected the ECT sensor signal was less than the self-test minimum. This is a thermistor-type sensor with a variable resistance that changes when exposed to different temperatures **Possible Causes:** • ECT sensor signal circuit is grounded in the wiring harness • ECT sensor doesn't react to changes in temperature • ECT sensor is damaged or the ECM has failed
DTC: P0118 **2T, MIL: Yes** **Passat:** 1.8 (AEB, ATW, AUG, AWM), 2.0L (ABA, BPY), 2.8L (AAA, AHA, ATQ), 4.0L (BDP); **Jetta:** 1.8L (AWD, AWP, AWW), 2.0 (ABA, AEG, AVH, AZG, BPY), 2.5L (BGP, BGQ), 2.8L (AAA, AFP, BDF); **Golf:** 1.8L (AWD, AWW), 2.0L (ABA, AEG, AVH, AZG, BEV); **GTI:** 1.8L (AWD, AWP, AWW), 2.0L (ABA, AEG, AEG), 2.8L (AAA, AFP, BDF); **NB:** 1.8L (APH, AWP, AWV, BNU, BKF), 2.0L (AEG, AVH, AZG, BEV); **NBvert:** (BDG); **Cabrio:** 2.0L (ABA) **Transmissions:** All	**ECT Sensor Circuit High Input Conditions:** Engine started (cold), battery voltage must be 11.5, and all equipment must be off. The ECM detected the ECT sensor signal was more than the self-test maximum. This is a thermistor-type sensor with a variable resistance that changes when exposed to different temperatures **Possible Causes:** • ECT sensor signal circuit is open (inspect wiring & connector) • ECT sensor signal circuit is shorted to ground • ECT sensor is damaged or it has failed • ECM has failed
DTC: P0120 **Passat:** 2.0L (ABA, BPY), 2.8L (AHA, ATQ); **Jetta:** 2.0 (ABA, AEG, AVH, AZG, BBW, BEV), 2.8L (AAA), 2.0L (ABA, AEG, AVH, AZG, BEV), 2.0L (ABA), 2.8L (AAA); **NB:** 1.8L (APH, AWP, AWV, BNU, BKF), 2.0L (AEG, AVH, AZG, BEV); **NBvert:** (BDG); **Cabrio:** 2.0L (ABA) **Transmissions:** All	**Throttle/Pedal Position Sensor (A) Circuit Malfunction Conditions:** Engine started, at idle, the temperature must be 80 degrees Celsius. The throttle position sensor supplies implausible signal to the ECM. The throttle valve activation occurs via an electric motor (throttle drive) in the throttle valve control module. It is activated by the Engine Control Module (ECM) according to specifications of the two sensors, Throttle Position (TP) Sensor and Accelerator Pedal Position Sensor 2. **Possible Causes:** • TP sensor signal circuit is open (inspect wiring & connector) • TP sensor signal circuit is shorted to ground • TP sensor or module is damaged or it has failed • Throttle valve is damaged or dirty • Throttle valve control module is faulty • ECM has failed

DTC	Trouble Code Title, Conditions & Possible Causes
DTC: P0121 **2T, MIL: Yes** **Passat:** 1.8 (AEB, ATW, AUG, AUG, AWM), 2.0L (ABA, BPY), 2.8L (AAA, AHA, ATQ), 4.0L (BDP); **Jetta:** 1.8L (AWD, AWP, AWW), 2.0 (ABA, AEG, AVH, AZG, BBW, BEV, BPY), 2.5L (BGP, BGQ), 2.8L (AAA, AFP, BDF); **Golf:** 1.8L (AWD, AWW), 2.0L (ABA, AEG, AVH, AZG, BEV); **GTI:** 1.8L (AWD, AWP, AWW), 2.0L (ABA, AEG), 2.8L (AAA, AFP, BDF); **NB:** 1.8L (APH, AWP, AWV, BNU, BKF), 2.0L (AEG, AVH, AZG, BEV); **NBvert:** (BDG); **Cabrio:** 2.0L (ABA) **Transmissions:** All	**Throttle/Pedal Position Sensor Signal Range/Performance Conditions:** Engine started; then immediately following a condition where the engine was running under at off-idle, the ECM detected the TP sensor signal indicated the throttle did not return to its previous closed position during the Rationality test. **Possible Causes:** • Throttle plate is binding, dirty or sticking • Throttle valve is damaged or dirty • Throttle valve control module is faulty • TP sensor signal circuit open (inspect wiring & connector) • TP sensor ground circuit open (inspect wiring & connector) • TP sensor and/or control module is damaged or has failed • MAF sensor signal is damaged, has failed or a short is present
DTC: P0122 **2T, MIL: Yes** **Passat:** 1.8 (AEB, ATW, AUG, AWM), 2.0L (ABA, BPY), 2.8L (AAA, AHA, ATQ), 4.0L (BDP); **Jetta:** 1.8L (AWD, AWP, AWW), 2.0 (ABA, AEG, AVH, AZG, BBW, BEV, BPY), 2.5L (BGP, BGQ), 2.8L (AAA, AFP, BDF); **Golf:** 1.8L (AWD, AWW), 2.0L (ABA, AEG, AVH, AZG, BEV); **GTI:** 1.8L (AWD, AWP, AWW), 2.0L (ABA, AEG), 2.8L (AAA, AFP, BDF), 2.0L (AEG, AVH, AZG, BEV); **NBvert:** (BDG); **Cabrio:** 2.0L (ABA) **Transmissions:** All	**Throttle/Pedal Position Sensor Circuit Low Input Conditions:** Engine started, at idle, the temperature must be at least 80 degrees Celsius. The throttle position sensor supplies implausible signal to the ECM. **Possible Causes:** • TP sensor signal circuit open (inspect wiring & connector) • TP sensor signal shorted to ground (inspect wiring & connector) • TP sensor is damaged or has failed • Throttle control module's voltage supply is shorted or open • ECM has failed
DTC: P0123 **2T, MIL: Yes** **Passat:** 1.8 (AEB, ATW, AUG, AWM), 2.0 (BPY), 2.8L (AAA, AHA, ATQ); **Jetta:** 1.8L (AWD, AWP, AWW), 2.0L (AEG, AVH, AZG, BBW, BEV, BPY), 2.5L (BGP, BGQ), 2.8L (AFP, BDF); **Golf:** 1.8L (AWD, AWW), 2.0L (AEG, AVH, AZG, BEV); **GTI:** 1.8L (AWD, AWP, AWW), 2.0L (AEG), 2.8L (AFP, BDF); **NB:** 1.8L (APH, AWP, AWV, BNU, BKF), 2.0L (AEG, AVH, AZG, BEV); **NBvert:** (BDG) **Transmissions:** All	**TP Sensor Circuit High Input Conditions:** Engine started, at idle, the temperature must be at least 80 degrees Celsius. The ECM detected the TP sensor signal was more than the self-test maximum during testing. **Possible Causes:** • TP sensor not seated correctly in housing (may be damaged) • TP sensor signal is circuit shorted to ground or system voltage • TP sensor ground circuit is open (check the wiring harness) • TP sensor and/or ECM has failed
DTC: P0125 **Passat:** 2.0L (ABA, BPY), 2.8L (AAA, AHA, ATQ), 2.0L (ABA), 2.8L (AAA); **Golf:** 2.0L (ABA); **GTI:** 2.0L (ABA), 2.8L (AAA); **NB:** 1.8L (APH, AWP, AWV, BNU, BKF); **Cabrio:** 2.0L (ABA) **Transmissions:** All	**Insufficient Coolant Temperature For Closed Loop Conditions:** Let engine run at idle until coolant temperature for oxygen sensor control has been reached. The ECM detected that the ECT sensor signal did not indicate the required engine temperature value to enter closed loop within a specified amount of time. The amount of time is calculated from the point at which the engine is started, and depends upon the ECT sensor signal value at startup. **Possible Causes:** • Check the coolant mixture for an incorrect mixture • Check the operation of the thermostat (it may be stuck open or has failed) • ECT sensor has failed or is disconnected • Inspect for low coolant level

DTC	Trouble Code Title, Conditions & Possible Causes
DTC: P0130 **2T, MIL: Yes** **Passat:** 1.8 (AEB, ATW, AUG, AWM), 2.0L (BPY), 2.8L (AHA), 4.0L (BDP); **Jetta:** 1.8L (AWD, AWP, AWW), 2.0 (ABA, AEG, AVH, AZG, BBW, BEV, BPY), 2.8L (AFP, BDF); **Golf:** 1.8L (AWD, AWW), 2.0L (ABA, AEG, AVH, AZG, BEV); **GTI:** 1.8L (AWD, AWP, AWW), 2.0L (ABA, AEG), 2.8L (AFP, BDF); **NB:** 1.8L (APH, AWP, AWV, BNU, BKF), 2.0L (AEG, AVH, AZG, BEV); **NBvert:** (BDG); **Cabrio:** 2.0L (ABA) **Transmissions:** All	**O2 Sensor Circuit Bank 1 Sensor 1 Conditions:** Engine running, battery voltage 11.5, all electrical components off, ground between engine and chassis well connected and the exhaust system must be properly sealed between catalytic converter and the cylinder head. The ECM detected the HO2S signal was implausible or not detected. **Possible Causes:** • Oxygen sensor heater for oxygen sensor (HO2S) before catalytic converter is faulty • HO2S is contaminated (due to presence of silicone in fuel) • HO2S signal and ground circuit wires crossed in wiring harness • HO2S signal circuit is shorted to sensor or chassis ground • HO2S element before the catalytic converter has failed (internal short condition) • Leaks present in the exhaust manifold or exhaust pipes • ECM has failed
DTC: P0131 **Passat:** 1.8 (AEB, ATW, AUG, AWM), 2.0L (ABA, BPY), 2.8L (AAA, AHA, ATQ), 4.0L (BDP); **Jetta:** 1.8L (AWD, AWP, AWW), 2.0 (ABA, AEG, AVH, AZG, BBW, BEV, BPY), 2.5L (BGP, BGQ), 2.8L (AAA, AFP, BDF); **Golf:** 1.8L (AWD, AWW), 2.0L (ABA, AEG, AVH, AZG, BEV); **GTI:** 1.8L (AWD, AWP, AWW), 2.0L (ABA, AEG), 2.8L (AAA, AFP, BDF); **NB:** 1.8L (APH, AWP, AWV, BNU, BKF), 2.0L (AEG, AVH, AZG, BEV); **NBvert:** (BDG); **Cabrio:** 2.0L (ABA) **Transmissions:** All	**HO2S (Bank 1 Sensor 1) Circuit Low Input Conditions:** Engine running, battery voltage 11.5, all electrical components off, ground between engine and chassis well connected and the exhaust system must be properly sealed between catalytic converter and the cylinder head. The ECM detected the HO2S signal was in a negative voltage range referred to as "character shift downward". This code sets when the HO2S signal remains in a low state for a measured period of time. In effect, it does not switch properly in the closed loop operation. **Possible Causes:** • HO2S is contaminated (due to presence of silicone in fuel) • HO2S signal and ground circuit wires crossed in wiring harness • HO2S signal circuit is shorted to sensor or chassis ground • HO2S element has failed (internal short condition) • Leaks present in the exhaust manifold or exhaust pipes • ECM has failed
DTC: P0132 **2T, MIL: Yes** **Passat:** 1.8 (AEB, ATW, AUG, AWM), 2.0L (ABA, BPY), 2.8L (AAA, AHA, ATQ), 4.0L (BDP); **Jetta:** 1.8L (AWD, AWP, AWW), 2.0 (ABA, AEG, AVH, AZG, BBW, BEV, BPY), 2.5L (BGP, BGQ), 2.8L (AAA, AFP, BDF); **Golf:** 1.8L (AWD, AWW), 2.0L (ABA, AEG, AVH, AZG, BEV); **GTI:** 1.8L (AWD, AWP, AWW), 2.0L (ABA, AEG), 2.8L (AAA, AFP, BDF); **NB:** 1.8L (APH, AWP, AWV, BNU, BKF), 2.0L (AEG, AVH, AZG, BEV); **NBvert:** (BDG); **Cabrio:** 2.0L (ABA) **Transmissions:** All	**HO2S (Bank 1 Sensor 1) Circuit High Input Conditions:** Engine running, battery voltage 11.5, all electrical components off, ground between engine and chassis well connected and the exhaust system must be properly sealed between catalytic converter and the cylinder head. The ECM detected the HO2S signal was in a high state. This code sets when the HO2S signal remains in a high state for a measured period of time. In effect, it does not switch properly in the closed loop operation. **Note: The HO2S signal circuit may be shorted to the heater power circuit due to tracking inside of the HO2S connector. Remove the connector and visually inspect the connector for signs of oil or water.** **Possible Causes:** • HO2S is contaminated (due to presence of silicone in fuel) • HO2S signal and ground circuit wires crossed in wiring harness • HO2S signal circuit is shorted to sensor or chassis ground • HO2S element has failed (internal short condition) • Leaks present in the exhaust manifold or exhaust pipes • ECM has failed

DTC	Trouble Code Title, Conditions & Possible Causes
DTC: P0133 **2T, MIL: Yes** **Passat:** 1.8 (AEB, ATW, AUG, AWM), 2.0L (ABA, BPY), 2.8L (AAA, AHA, ATQ), 4.0L (BDP); **Jetta:** 1.8L (AWD, AWP, AWW), 2.0 (ABA, AEG, AVH, AZG, BBW, BEV), 2.5L (BGP, BGQ), 2.8L (AAA, AFP, BDF); **Golf:** 1.8L (AWD, AWW), 2.0L (ABA, AEG, AVH, AZG, BEV); **GTI:** 1.8L (AWD, AWP, AWW), 2.0L (ABA, AEG), 2.8L (AAA, AFP, BDF); **NB:** 1.8L (APH, AWP, AWV, BNU, BKF), 2.0L (AEG, AVH, AZG, BEV); **NBvert:** (BDG); **Cabrio:** 2.0L (ABA) **Transmissions:** All	**HO2S (Bank 1 Sensor 1) Circuit Slow Response Conditions:** Engine running, battery voltage 11.5, all electrical components off, ground between engine and chassis well connected and the exhaust system must be properly sealed between catalytic converter and the cylinder head. The ECM detected the HO2S amplitude and frequency were out of the normal range (e.g., the HO2S rich to lean switch) during the HO2S Monitor test. **Possible Causes:** • HO2S before the three-way catalytic converter is contaminated (due to presence of silicone in fuel); Run the engine for three minutes at 3500rpm as a self-cleaning effect • HO2S signal circuit open • Leaks present in the exhaust manifold or exhaust pipes • HO2S is damaged or has failed • ECM has failed
DTC: P0134 **2T, MIL: Yes** **Passat:** 1.8 (AEB, ATW, AUG, AWM), 2.0L (ABA, BPY), 2.8L (AAA, AHA, ATQ), 4.0L (BDP); **Jetta:** 1.8L (AWD, AWP, AWW), 2.0 (ABA, AEG, AVH, AZG, BBW, BEV, BPY), 2.5L (BGP, BGQ), 2.8L (AAA, AFP, BDF); **Golf:** 1.8L (AWD, AWW), 2.0L (ABA, AEG, AVH, AZG, BEV); **GTI:** 1.8L (AWD, AWP, AWW), 2.0L (ABA, AEG), 2.8L (AAA, AFP, BDF); **NB:** 1.8L (APH, AWP, AWV, BNU, BKF), 2.0L (AEG, AVH, AZG, BEV); **NBvert:** (BDG); **Cabrio:** 2.0L (ABA) **Transmissions:** All	**HO2S (Bank 1 Sensor 1) Circuit No Activity Conditions:** Engine running, battery voltage 11.5, all electrical components off, ground between engine and chassis well connected and the exhaust system must be properly sealed between catalytic converter and the cylinder head. The ECM detected the HO2S signal failed to meet the maximum or minimum voltage levels (i.e., it failed the voltage range check). **Possible Causes:** • Leaks present in the exhaust manifold or exhaust pipes • HO2S signal wire and ground wire crossed in connector (voltage jumps) • HO2S element is fuel contaminated or has failed • ECM has failed
DTC: P0135 **2T, MIL: Yes** **Passat:** 1.8 (AEB, AUG, AUG, AWM), 2.0L (ABA, BPY), 2.8L (AHA, ATQ), 4.0L (BDP); **Jetta:** 1.8L (AWD, AWP, AWW), 2.0 (ABA, AEG, AVH, AZG, BBW, BEV), 2.5L (BGP, BGQ), 2.8L (AAA, AFP, BDF); **Golf:** 1.8L (AWD, AWW), 2.0L (ABA, AEG, AVH, AZG, BEV); **GTI:** 1.8L (AWD, AWP, AWW), 2.0L (ABA, AEG), 2.8L (AAA, AFP, BDF); **NB:** 1.8L (APH, AWP, AWV, BNU, BKF), 2.0L (AEG, AVH, AZG, BEV); **NBvert:** (BDG); **Cabrio:** 2.0L (ABA) **Transmissions:** All	**HO2S (Bank 1 Sensor 1) Heater Circuit Malfunction Conditions:** Engine running, battery voltage 11.5, all electrical components off, ground between engine and chassis well connected and the exhaust system must be properly sealed between catalytic converter and the cylinder head. The ECM detected an unexpected voltage condition, or it detected excessive current draw in the heater circuit during the CCM test. **Possible Causes:** • HO2S heater power circuit is open or heater ground circuit open • HO2S signal tracking (due to oil or moisture in the connector) • HO2S is damaged or has failed • ECM has failed

DTC	Trouble Code Title, Conditions & Possible Causes
DTC: P0136 **2T, MIL: Yes** **Passat:** 1.8 (AEB, ATW, AUG, AUG, AWM), 2.0L (ABA, BPY), 2.8L (AHA, ATQ), 4.0L (BDP); **Jetta:** 1.8L (AWD, AWP, AWW), 2.0 (ABA, AEG, AVH, AZG, BBW, BEV, BPY), 2.5L (BGP, BGQ), 2.8L (AFP, BDF); **Golf:** 1.8L (AWD, AWW), 2.0L (ABA, AEG, AVH, AZG, BEV); **GTI:** 1.8L (AWD, AWP, AWW), 2.0L (ABA, AEG), 2.8L (AFP, BDF); **NB:** 1.8L (APH, AWP, AWV, BNU, BKF), 2.0L (AEG, AVH, AZG, BEV); **NBvert:** (BDG); **Cabrio:** 2.0L (ABA) **Transmissions:** All	**HO2S (Bank 1 Sensor 2) Circuit Malfunction Conditions:** Engine running, battery voltage 11.5, all electrical components off, ground between engine and chassis well connected and the exhaust system must be properly sealed between catalytic converter and the cylinder head. The ECM detected the HO2S signal failed to meet the maximum or minimum voltage levels (i.e., it failed the voltage range check). **Possible Causes:** • Leaks present in the exhaust manifold or exhaust pipes • HO2S signal wire and ground wire crossed in connector • HO2S element is fuel contaminated or has failed • ECM has failed
DTC: P0137 **2T, MIL: Yes** **Passat:** 1.8 (AEB, ATW, AUG, AWM), 2.0L (ABA, BPY), 2.8L (AAA, AHA, ATQ), 4.0L (BDP); **Jetta:** 1.8L (AWD, AWP, AWW), 2.0 (ABA, AEG, AVH, AZG, BBW, BEV, BPY), 2.5L (BGP, BGQ), 2.8L (AAA, AFP, BDF); **Golf:** 1.8L (AWD, AWW), 2.0L (ABA, AEG, AVH, AZG, BEV); **GTI:** 1.8L (AWD, AWP, AWW), 2.0L (ABA, AEG), 2.8L (AAA, AFP, BDF); **NB:** 1.8L (APH, AWP, AWV, BNU, BKF), 2.0L (AEG, AVH, AZG, BEV); **NBvert:** (BDG); **Cabrio:** 2.0L (ABA) **Transmissions:** All	**HO2S (Bank 1 Sensor 2) Circuit Low Input Conditions:** Engine running, battery voltage 11.5, all electrical components off, ground between engine and chassis well connected and the exhaust system must be properly sealed between catalytic converter and the cylinder head. The ECM detected the HO2S signal remained in a high state. **Note: The HO2S signal circuit may be shorted to the heater power circuit due to "tracking inside of the HO2S connector. Remove the connector and visually inspect the connector for signs of oil or water.** **Possible Causes:** • HO2S signal shorted to heater power circuit in the connector • HO2S signal circuit shorted to ground (for more than 200 seconds) or to system voltage • ECM has failed
DTC: P0138 **2T, MIL: Yes** **Passat:** 1.8 (AEB, ATW, AUG, AWM), 2.0L (ABA, BPY), 2.8L (AAA, AHA, ATQ), 4.0L (BDP); **Jetta:** 1.8L (AWD, AWP, AWW), 2.0 (ABA, AEG, AVH, AZG, BBW, BEV, BPY), 2.5L (BGP, BGQ), 2.8L (AAA, AFP, BDF); **Golf:** 1.8L (AWD, AWW), 2.0L (ABA, AEG, AVH, AZG, BEV); **GTI:** 1.8L (AWD, AWP, AWW), 2.0L (ABA, AEG), 2.8L (AAA, AFP, BDF); **NB:** 1.8L (APH, AWP, AWV, BNU, BKF), 2.0L (AEG, AVH, AZG, BEV); **NBvert:** (BDG); **Cabrio:** 2.0L (ABA) **Transmissions:** All	**HO2S (Bank 1 Sensor 2) Circuit High Input Conditions:** Engine running, battery voltage 11.5, all electrical components off, ground between engine and chassis well connected and the exhaust system must be properly sealed between catalytic converter and the cylinder head. The ECM detected the HO2S signal remained in a high state. **Note: The HO2S signal circuit may be shorted to the heater power circuit due to "tracking inside of the HO2S connector. Remove the connector and visually inspect the connector for signs of oil or water.** **Possible Causes:** • HO2S signal shorted to heater power circuit in the positive connector • HO2S signal circuit shorted to ground or to system voltage • HO2S has failed • ECM has failed

DTC	Trouble Code Title, Conditions & Possible Causes
DTC: P0139 **2T, MIL: Yes** **Passat:** 1.8 (AEB, AUG, AWM), 2.0L (BPY), 4.0L (BDP); **Jetta:** 1.8L (AWD, AWP, AWW), 2.0 (ABA, AEG, AVH, AZG, BBW, BEV), 2.5L (BGP, BGQ), 2.8L (AFP, BDF); **Golf:** 1.8L (AWD, AWW), 2.0L (ABA, AEG, AVH, AZG, BEV); **GTI:** 1.8L (AWD, AWP, AWW), 2.0L (AEG), 2.8L (AFP, BDF); **NB:** 1.8L (APH, AWP, AWV, BNU, BKF), 2.0L (AEG, AVH, AZG, BEV); **NBvert:** (BDG); **Cabrio:** 2.0L (ABA) **Transmissions:** All	**HO2S (Bank 1 Sensor 2) Slow Response Conditions:** Engine running, battery voltage 11.5, all electrical components off, ground between engine and chassis well connected and the exhaust system must be properly sealed between catalytic converter and the cylinder head. The ECM detected the HO2S amplitude and frequency were out of the normal range during the HO2S Monitor test. **Possible Causes:** • HO2S signal shorted to heater power circuit in the connector • HO2S signal circuit shorted to VREF or to system voltage • ECM has failed
DTC: P0140 **2T, MIL: Yes** **Passat:** 1.8 (AEB, ATW, AUG, AWM), 2.0L (ABA, BPY), 2.8L (AAA, AHA, ATQ), 4.0L (BDP); **Jetta:** 1.8L (AWD, AWP, AWW), 2.0 (ABA, AEG, AVH, AZG, BBW, BEV, BPY), 2.5L (BGP, BGQ), 2.8L (AAA, AFP, BDF); **Golf:** 1.8L (AWD, AWW), 2.0L (ABA, AEG, AVH, AZG, BEV); **GTI:** 1.8L (AWD, AWP, AWW), 2.0L (ABA, AEG), 2.8L (AAA, AFP, BDF); **NB:** 1.8L (APH, AWP, AWV, BNU, BKF), 2.0L (AEG, AVH, AZG, BEV); **NBvert:** (BDG); **Cabrio:** 2.0L (ABA) **Transmissions:** All	**HO2S (Bank 1 Sensor 2) No Activity Conditions:** Engine running, battery voltage 11.5, all electrical components off, ground between engine and chassis well connected and the exhaust system must be properly sealed between catalytic converter and the cylinder head. The ECM detected the HO2S signal failed to meet the maximum or minimum voltage levels (i.e., it failed the voltage range check). **Possible Causes:** • HO2S before the three-way catalytic converter is contaminated (due to presence of silicone in fuel); Run the engine for three minutes at 3500rpm as a self-cleaning effect • Leaks present in the exhaust manifold or exhaust pipes • HO2S signal wire and ground wire crossed in connector (voltage jumps) • HO2S element is contaminated or has failed • ECM has failed
DTC: P0141 **2T, MIL: Yes** **Passat:** 1.8L (AUG, ATW), 2.0L (ABA, BPY), 2.8L (AAA, AHA, ATQ), 4.0L (BDP); **Jetta:** 1.8L (AWD, AWP, AWW), 2.0 (ABA, AEG, AVH, AZG, BBW, BEV, BPY), 2.5L (BGP, BGQ), 2.8L (AAA, AFP, BDF); **Golf:** 1.8L (AWD, AWW), 2.0L (ABA, AEG, AVH, AZG, BEV); **GTI:** 1.8L (AWD, AWP, AWW), 2.0L (ABA, AEG), 2.8L (AAA, AFP, BDF); **NB:** 1.8L (APH, AWP, AWV, BNU, BKF), 2.0L (AEG, AVH, AZG, BEV); **NBvert:** (BDG); **Cabrio:** 2.0L (ABA) **Transmissions:** All	**HO2S (Bank 1 Sensor 2) Malfunction Conditions:** Engine running, battery voltage 11.5, all electrical components off, ground between engine and chassis well connected and the exhaust system must be properly sealed between catalytic converter and the cylinder head. The ECM detected the HO2S signal failed to meet the maximum or minimum voltage levels (i.e., it failed the voltage range check). **Possible Causes:** • Leaks present in the exhaust manifold or exhaust pipes • HO2S signal wire and ground wire crossed in connector • HO2S element is fuel contaminated or has failed • ECM has failed
DTC: P0142 **Jetta:** 2.0 (BBW), 2.5L (BGP, BGQ) **Transmissions:** All	**O2 Sensor Circuit Bank 1 Sensor 3 Conditions:** Engine running, battery voltage 11.5, all electrical components off, ground between engine and chassis well connected and the exhaust system must be properly sealed between catalytic converter and the cylinder head. The ECM detected the HO2S amplitude and frequency were out of the normal range during the HO2S Monitor test. **Possible Causes:** • HO2S signal shorted to heater power circuit in the connector • HO2S signal circuit shorted to VREF or to system voltage • ECM has failed

DTC	Trouble Code Title, Conditions & Possible Causes
DTC: P0143 **Jetta:** 2.0 (BBW), 2.5L (BGP, BGQ) **Transmissions:** All	**02 Sensor Circuit Low Voltage Bank 1 Sensor 3 Conditions:** Engine running, battery voltage 11.5, all electrical components off, ground between engine and chassis well connected and the exhaust system must be properly sealed between catalytic converter and the cylinder head. The ECM detected the HO2S amplitude and frequency were out of the normal range during the HO2S Monitor test. **Possible Causes:** • HO2S signal shorted to heater power circuit in the connector • HO2S signal circuit shorted to VREF or to system voltage • ECM has failed
DTC: P0144 **Jetta:** 2.0 (BBW), 2.5L (BGP, BGQ) **Transmissions:** All	**02 Sensor Circuit High Voltage Bank 1 Sensor 3 Conditions:** Engine running, battery voltage 11.5, all electrical components off, ground between engine and chassis well connected and the exhaust system must be properly sealed between catalytic converter and the cylinder head. The ECM detected the HO2S amplitude and frequency were out of the normal range during the HO2S Monitor test. **Possible Causes:** • HO2S signal shorted to heater power circuit in the connector • HO2S signal circuit shorted to VREF or to system voltage • ECM has failed
DTC: P0145 **Jetta:** 2.0 (BBW), 2.5L (BGP, BGQ) **Transmissions:** All	**02 Sensor Circuit Slow Response Bank 1 Sensor 3 Conditions:** Engine running, battery voltage 11.5, all electrical components off, ground between engine and chassis well connected and the exhaust system must be properly sealed between catalytic converter and the cylinder head. The ECM detected the HO2S amplitude and frequency were out of the normal range during the HO2S Monitor test. **Possible Causes:** • HO2S signal shorted to heater power circuit in the connector • HO2S signal circuit shorted to VREF or to system voltage • ECM has failed
DTC: P0146 **Jetta:** 2.0 (BBW), 2.5L (BGP, BGQ) **Transmissions:** All	**02 Sensor Circuit No Activity Detected Bank 1 Sensor 3 Conditions:** Engine running, battery voltage 11.5, all electrical components off, ground between engine and chassis well connected and the exhaust system must be properly sealed between catalytic converter and the cylinder head. The ECM detected the HO2S amplitude and frequency were not detected during the HO2S Monitor test. **Possible Causes:** • HO2S signal shorted to heater power circuit in the connector • HO2S signal circuit shorted to VREF or to system voltage • ECM has failed
DTC: P0147 **Jetta:** 2.0 (BBW), 2.5L (BGP, BGQ) **Transmissions:** All	**02 Sensor Heater Circuit Bank 1 Sensor 3 Conditions:** Engine running, battery voltage 11.5, all electrical components off, ground between engine and chassis well connected and the exhaust system must be properly sealed between catalytic converter and the cylinder head. The ECM detected the HO2S amplitude and frequency were out of the normal range during the HO2S Monitor test. **Possible Causes:** • HO2S signal shorted to heater power circuit in the connector • HO2S signal circuit shorted to VREF or to system voltage • ECM has failed
DTC: P0150 **Passat:** 2.8L (AHA, ATQ), 4.0L (BDP); **Jetta:** 2.8L (AFP, BDF); **GTI:** 2.8L (AFP, BDF) **Transmissions:** All	**HO2S (Bank 2 Sensor 1) Circuit Malfunction Conditions:** Engine running, battery voltage 11.5, all electrical components off, ground between engine and chassis well connected and the exhaust system must be properly sealed between catalytic converter and the cylinder head. The ECM detected the HO2S signal failed to meet the maximum or minimum voltage levels (i.e., it failed the voltage range check). **Possible Causes:** • Leaks present in the exhaust manifold or exhaust pipes • HO2S signal wire and ground wire crossed in connector • HO2S element is fuel contaminated or has failed • ECM has failed
DTC: P0151 **Passat:** 2.8L (AHA, ATQ), 4.0L (BDP); **Jetta:** 2.8L (AFP, BDF); **GTI:** 2.8L (AFP, BDF) **Transmissions:** All	**HO2S (Bank 2 Sensor 1) Low Input Conditions:** Engine running, battery voltage 11.5, all electrical components off, ground between engine and chassis well connected and the exhaust system must be properly sealed between catalytic converter and the cylinder head. The ECM detected the HO2S signal remained in a high state. **Note: The HO2S signal circuit may be shorted to the heater power circuit due to "tracking inside of the HO2S connector. Remove the connector and visually inspect the connector for signs of oil or water.** **Possible Causes:** • HO2S is contaminated (due to presence of silicone in fuel) • HO2S signal tracking (due to oil or moisture in the connector) • HO2S signal circuit is open or shorted to VREF • ECM has failed

DTC	Trouble Code Title, Conditions & Possible Causes
DTC: P0152 **Passat:** 2.8L (AHA, ATQ), 4.0L (BDP); **Jetta:** 2.8L (AFP, BDF); **GTI:** 2.8L (AFP, BDF) **Transmissions:** All	**HO2S (Bank 2 Sensor 1) Circuit High Input Conditions:** Engine running, battery voltage 11.5, all electrical components off, ground between engine and chassis well connected and the exhaust system must be properly sealed between catalytic converter and the cylinder head. The ECM detected the HO2S signal remained in a high state (more than 1.5v). **Note: The HO2S signal circuit may be shorted to the heater power circuit due to "tracking inside of the HO2S connector. Remove the connector and visually inspect the connector for signs of oil or water.** **Possible Causes:** • HO2S is contaminated (due to presence of silicone in fuel) • HO2S signal tracking (due to oil or moisture in the connector) • HO2S signal circuit is open or shorted to VREF • ECM has failed
DTC: P0153 **2T, MIL: Yes** **Passat:** 2.8L (AHA, ATQ), 4.0L (BDP); **Jetta:** 2.8L (AFP, BDF); **GTI:** 2.8L (AFP, BDF) **Transmissions:** All	**HO2S (Bank 2 Sensor 1) Circuit Slow Response Conditions:** Engine running, battery voltage 11.5, all electrical components off, ground between engine and chassis well connected and the exhaust system must be properly sealed between catalytic converter and the cylinder head. The the ECM detected the HO2S amplitude and frequency were out of the normal range during the HO2S Monitor test. **Possible Causes:** • HO2S is contaminated (due to presence of silicone in fuel) • Leaks present in the exhaust manifold or exhaust pipes • HO2S is damaged or has failed • ECM has failed
DTC: P0154 **2T, MIL: Yes** **Passat:** 2.8L (AHA, ATQ), 4.0L (BDP); **Jetta:** 2.8L (AFP, BDF); **GTI:** 2.8L (AFP, BDF) **Transmissions:** All	**HO2S (Bank 2 Sensor 1) Circuit No Activity Conditions:** Engine running, battery voltage 11.5, all electrical components off, ground between engine and chassis well connected and the exhaust system must be properly sealed between catalytic converter and the cylinder head. The ECM detected the HO2S signal failed to meet the maximum or minimum voltage (i.e., it failed the voltage check). **Possible Causes:** • Leaks present in the exhaust manifold or exhaust pipes • HO2S signal wire and ground wire crossed in connector • HO2S element is fuel contaminated or has failed • ECM has failed
DTC: P0155 **2T, MIL: Yes** **Passat:** 2.8L (AHA, ATQ), 4.0L (BDP); **Jetta:** 2.8L (AFP, BDF); **GTI:** 2.8L (AFP, BDF) **Transmissions:** All	**HO2S (Bank 2 Sensor 1) Heater Circuit Malfunction Conditions:** Engine running, battery voltage 11.5, all electrical components off, ground between engine and chassis well connected and the exhaust system must be properly sealed between catalytic converter and the cylinder head. The ECM detected an open or shorted condition, or excessive current draw in the heater circuit. **Possible Causes:** • HO2S heater power circuit is open • HO2S heater ground circuit is open • HO2S signal tracking (due to oil or moisture in the connector) • HO2S is damaged or has failed • ECM has failed
DTC: P0156 **2T, MIL: Yes** **Passat:** 2.8L (AHA, ATQ), 4.0L (BDP); **Jetta:** 2.5L (BGP, BGQ), 2.8 (AFP, BDF); **GTI:** 2.8L (AFP, BDF) **Transmissions:** All	**HO2S (Bank 2 Sensor 2) Circuit No Activity Conditions:** Engine running, battery voltage 11.5, all electrical components off, ground between engine and chassis well connected and the exhaust system must be properly sealed between catalytic converter and the cylinder head. The ECM detected the HO2S signal failed to meet the maximum or minimum voltage (i.e., it failed the voltage check). **Possible Causes:** • Leaks present in the exhaust manifold or exhaust pipes • HO2S signal wire and ground wire crossed in connector • HO2S element is fuel contaminated or has failed • ECM has failed
DTC: P0157 **2T, MIL: Yes** **Passat:** 2.8L (AHA, ATQ); **Jetta:** 2.8L (AFP, BDF); **GTI:** 2.8L (AFP)**Transmissions:** All	**HO2S (Bank 2 Sensor 2) Circuit Low Voltage Conditions:** Engine running, battery voltage 11.5, all electrical components off, ground between engine and chassis well connected and the exhaust system must be properly sealed between catalytic converter and the cylinder head. The ECM detected the HO2S signal remained in a high state. **Note: The HO2S signal circuit may be shorted to the heater power circuit due to "tracking inside of the HO2S connector. Remove the connector and visually inspect the connector for signs of oil or water** **Possible Causes:** • HO2S is contaminated (due to presence of silicone in fuel) • HO2S signal tracking (due to oil or moisture in the connector) • HO2S signal circuit is open or shorted to VREF • ECM has failed

DTC	Trouble Code Title, Conditions & Possible Causes
DTC: P0158 **2T, MIL: Yes** **Passat:** 2.8L (AHA, ATQ), 4.0L (BDP); **Jetta:** 2.8L (AFP, BDF); **GTI:** 2.8L (AFP, BDF) **Transmissions:** All	**HO2S (Bank 2 Sensor 2) Circuit High Input Conditions:** Engine running, battery voltage 11.5, all electrical components off, ground between engine and chassis well connected and the exhaust system must be properly sealed between catalytic converter and the cylinder head. The ECM detected the HO2S signal remained in a high state (i.e., more than 1.5v). **Note: The HO2S signal circuit may be shorted to the heater power circuit due to "tracking inside of the HO2S connector. Remove the connector and visually inspect the connector for signs of oil or water.** **Possible Causes:** • HO2S signal shorted to the heater power circuit (due to oil or moisture in the connector) • HO2S signal circuit shorted to VREF or to system voltage • ECM has failed
DTC: P0159 **2T, MIL: Yes** **Passat:** 2.8L (AHA, ATQ), 4.0L (BDP); **Jetta:** 2.8L (AFP, BDF); **GTI:** 2.8L (AFP, BDF) **Transmissions:** All	**HO2S (Bank 2 Sensor 2) Circuit Slow Response Conditions:** Engine running, battery voltage 11.5, all electrical components off, ground between engine and chassis well connected and the exhaust system must be properly sealed between catalytic converter and the cylinder head. The ECM detected the HO2S amplitude and frequency were out of the normal range during the HO2S Monitor test. **Possible Causes:** • HO2S is contaminated (due to presence of silicone in fuel) • Leaks present in the exhaust manifold or exhaust pipes • HO2S is damaged or has failed • ECM has failed
DTC: P0160 **2T, MIL: Yes** **Passat:** 2.8L (AHA, ATQ), 4.0L (BDP); **Jetta:** 2.8L (AFP, BDF); **GTI:** 2.8L (AFP, BDF) **Transmissions:** All	**HO2S (Bank 2 Sensor 2) Circuit No Activity Detected Conditions:** Engine running, battery voltage 11.5, all electrical components off, ground between engine and chassis well connected and the exhaust system must be properly sealed between catalytic converter and the cylinder head. The ECM detected the HO2S signal failed to meet the maximum or minimum voltage (i.e., it failed the voltage check). **Possible Causes:** • Leaks present in the exhaust manifold or exhaust pipes • HO2S signal wire and ground wire crossed in connector • HO2S element is fuel contaminated or has failed • ECM has failed
DTC: P0161 **2T, MIL: Yes** **Passat:** 2.8L (AHA, ATQ), 4.0L (BDP); **Jetta:** 2.8L (AFP, BDF); **GTI:** 2.8L (AFP, BDF) **Transmissions:** All	**HO2S (Bank 2 Sensor 2) Heater Circuit Malfunction Conditions:** Engine running, battery voltage 11.5, all electrical components off, ground between engine and chassis well connected and the exhaust system must be properly sealed between catalytic converter and the cylinder head. The the ECM detected an open or shorted condition, or excessive current draw in the heater circuit. **Possible Causes:** • HO2S heater power circuit or the heater ground circuit is open • HO2S signal tracking (due to oil or moisture in the connector) • HO2S has failed, or the ECM has failed
DTC: P0169 **Passat:** 2.0L (BPY) **Transmissions:** All	**Incorrect Fuel Composition Conditions:** The engine is running in a closed loop at a stable engine speed, and the ECM detected the lean or rich fuel trim correction valve was more than or less than a calibrated limit. **Possible Causes:** • One or more fuel injectors have failed
DTC: P0170 **Passat:** 1.8L (AUG, ATW), 2.0L (BPY); **Jetta:** 1.8L (AWD, AWP, AWW), 2.0L (AEG, AVH, AZG, BBW, BEV), 2.8L (AFP, BDF); **Golf:** 1.8L (AWD, AWW), 2.0L (AEG, AVH, AZG, BEV); **GTI:** 1.8L (AWD, AWP, AWW), 2.0L (AEG), 2.8L (AFP, BDF); **NB:** 1.8L (APH, AWP, AWV, BNU, BKF), 2.0L (AEG, AVH, AZG, BEV); **NBvert:** (BDG) **Transmissions:** All	**Fuel System Malfunction (Cylinder Bank 1) Conditions:** The engine is running in a closed loop at a stable engine speed, and the ECM detected the lean or rich fuel trim correction valve was more than or less than a calibrated limit. **Possible Causes:** • Air leaks after the MAF sensor, or leaks in the PCV system • Exhaust leaks before or near where the HO2S is mounted • Fuel injector(s) restricted or not supplying enough fuel • Fuel system not supplying enough fuel during high fuel demand conditions (e.g., the fuel pump may not supply enough fuel) • Leaking EGR gasket, or leaking EGR valve diaphragm • MAF sensor dirty (causes ECM to underestimate airflow) • Vehicle running out of fuel or engine oil dip stick not seated

DTC	Trouble Code Title, Conditions & Possible Causes
DTC: P0171 **2T, MIL: Yes** **Passat:** 1.8 (AEB, ATW, AUG, AWM), 2.0L (ABA, BPY), 2.8L (AAA, AHA, ATQ); **Jetta:** 1.8L (AWD, AWP, AWW), 2.0 (ABA, AEG, AVH, AZG, BBW, BEV, BPY), 2.5L (BGP, BGQ), 2.8L (AAA, AFP, BDF); **Golf:** 1.8L (AWD, AWW), 2.0L (ABA, AEG, AVH, AZG, BEV); **GTI:** 1.8L (AWD, AWP, AWW), 2.0L (ABA, AEG), 2.8L (AAA, AFP, BDF); **NB:** 1.8L (APH, AWP, AWV, BNU, BKF), 2.0L (AEG, AVH, AZG, BEV); **NBvert:** (BDG); **Cabrio:** 2.0L (ABA) **Transmissions:** All	**Fuel System Too Lean (Cylinder Bank 1) Conditions:** Key on or engine running, all electrical components off and coolant temperature at least 80 degrees Celsius; and the ECM detected the Bank 1 Adaptive Fuel Control System reached its rich correction limit (a lean A/F condition). **Possible Causes:** • Air leaks after the MAF sensor, or leaks in the PCV system • Exhaust leaks before or near where the HO2S is mounted • Fuel injector(s) restricted or not supplying enough fuel • Fuel pump not supplying enough fuel during high fuel demand conditions • Leaking EGR gasket, or leaking EGR valve diaphragm • MAF sensor dirty (causes ECM to underestimate airflow) • Vehicle running out of fuel or engine oil dip stick not seated
DTC: P0172 **2T, MIL: Yes** **Passat:** 1.8 (AEB, ATW, AUG, AWM), 2.0L (ABA, BPY), 2.8L (AAA, AHA, ATQ); **Jetta:** 1.8L (AWD, AWP, AWW), 2.0 (ABA, AEG, AVH, AZG, BBW, BEV), 2.5L (BGP, BGQ), 2.8L (AAA, AFP, BDF); **Golf:** 1.8L (AWD, AWW), 2.0L (ABA); **GTI:** 1.8L (AWD, AWP, AWW), 2.0L (ABA, AEG), 2.8L (AAA, AFP, BDF); **NB:** 1.8L (APH, AWP, AWV, BNU, BKF); **Cabrio:** 2.0L (ABA) **Transmissions:** All	**Fuel System Too Rich (Cylinder Bank 1) Conditions:** Key on or engine running, all electrical components off and coolant temperature at least 80 degrees Celsius; and the ECM detected the Bank 1 Adaptive Fuel Control System reached its rich correction limit (a rich A/F condition). **Possible Causes:** • Camshaft timing is incorrect, or the engine has an oil overfill condition • EVAP vapor recovery system failure (may be pulling vacuum) • Fuel pressure regulator is damaged or leaking • HO2S element is contaminated with alcohol or water • MAF or MAP sensor values are incorrect or out-of-range • One of more fuel injectors is leaking
DTC: P0173 **Jetta:** 2.8L (AFP, BDF); **GTI:** 2.8L (AFP, BDF) **Transmissions:** All	**Fuel System Malfunction (Cylinder Bank 1) Conditions:** Key on or engine running, all electrical components off and coolant temperature at least 80 degrees Celsius; and the ECM detected the Bank 1 Fuel Control System experienced a implausible signal **Possible Causes:** • Air leaks after the MAF sensor, or leaks in the PCV system • Exhaust leaks before or near where the HO2S is mounted • Fuel injector(s) restricted or not supplying enough fuel • Fuel system not supplying enough fuel during high fuel demand conditions (e.g., the fuel pump may not supply enough fuel) • Leaking EGR gasket, or leaking EGR valve diaphragm • MAF sensor dirty (causes ECM to underestimate airflow) • Vehicle running out of fuel or engine oil dip stick not seated
DTC: P0174 **2T, MIL: Yes** **Passat:** 2.8L (AHA, ATQ); **Jetta:** 2.8L (AFP, BDF); **GTI:** 2.8L (AFP, BDF) **Transmissions:** All	**Fuel System Too Lean (Cylinder Bank 2) Conditions:** Key on or engine running, all electrical components off and coolant temperature at least 80 degrees Celsius; and the ECM detected the Bank 2 Fuel Control System reached its lean correction limit **Possible Causes:** • Air leaks after the MAF sensor, or leaks in the PCV system • Exhaust leaks before or near where the HO2S is mounted • Fuel injector(s) restricted or not supplying enough fuel • Fuel pump not supplying enough fuel during high fuel demand conditions • Leaking EGR gasket, or leaking EGR valve diaphragm • MAF sensor dirty (causes ECM to underestimate airflow) • Vehicle running out of fuel or engine oil dip stick not seated

DTC	Trouble Code Title, Conditions & Possible Causes
DTC: P0175 **2T, MIL: Yes** **Passat:** 2.8L (AHA, ATQ); **Jetta:** 2.8L (AFP, BDF); **GTI:** 2.8L (AFP, BDF) **Transmissions:** All	**Fuel System Too Rich (Cylinder Bank 2) Conditions:** Key on or engine running, all electrical components off and coolant temperature at least 80 degrees Celsius; and the ECM detected the Bank 2 Adaptive Fuel Control System reached its rich correction limit (a rich A/F condition). **Possible Causes:** • Air leaks after the MAF sensor, or leaks in the PCV system • Exhaust leaks before or near where the HO2S is mounted • Fuel injector(s) restricted or not supplying enough fuel • Fuel pump not supplying enough fuel during high fuel demand conditions • Leaking EGR gasket, or leaking EGR valve diaphragm • MAF sensor dirty (causes ECM to underestimate airflow) • Vehicle running out of fuel or engine oil dip stick not seated
DTC: P0190 **Passat:** 2.0L (BPY) **Transmissions:** All	**Fuel Rail Pressure Sensor Circuit Conditions** Key on or engine running, all electrical components off and coolant temperature at least 80 degrees Celsius; and the ECM detected the fuel rail pressure sensor signal was outside the required voltage parameters in the self-test. **Possible Causes:** • Fuel Pressure Regulator Valve faulty • Fuel Pressure Sensor faulty • Fuel Pump (FP) Control Module faulty • Fuel pump faulty
DTC: P0191 **Passat:** 2.0L (BPY) **Transmissions:** All	**Fuel Rail Pressure Sensor Circuit Range/Performance Conditions** Key on or engine running; and the ECM detected the fuel rail pressure sensor signal was outside the required voltage parameters in the self-test. **Possible Causes:** • Fuel Pressure Regulator Valve faulty • Fuel Pressure Sensor faulty • Fuel Pump (FP) Control Module faulty • Fuel pump faulty
DTC: P0192 **Passat:** 2.0L (BPY); **Jetta:** 2.0L (BPY) **Transmissions:** All	**Fuel Rail Pressure Sensor Circuit Low Conditions** Key on or engine running, all electrical components off and coolant temperature at least 80 degrees Celsius; and the ECM detected the fuel rail pressure sensor signal was below the required voltage in the self-test. **Possible Causes:** • Fuel Pressure Regulator Valve faulty • Fuel Pressure Sensor faulty • Fuel Pump (FP) Control Module faulty • Fuel pump faulty
DTC: P0193 **Passat:** 2.0L (BPY) **Transmissions:** All	**Fuel Rail Pressure Sensor Circuit High Conditions** Key on or engine running, all electrical components off and coolant temperature at least 80 degrees Celsius; and the ECM detected the fuel rail pressure sensor signal was above the required voltage in the self-test. **Possible Causes:** • Fuel Pressure Regulator Valve faulty • Fuel Pressure Sensor faulty • Fuel Pump (FP) Control Module faulty • Fuel pump faulty
DTC: P0194 **Passat:** 2.0L (BPY) **Transmissions:** All	**Fuel Rail Pressure Sensor Circuit Intermittent Conditions** Key on or engine running, all electrical components off and coolant temperature at least 80 degrees Celsius; and the ECM detected the fuel rail pressure sensor signal was implausible. **Possible Causes:** • Fuel Pressure Regulator Valve faulty • Fuel Pressure Sensor faulty • Fuel Pump (FP) Control Module faulty • Fuel pump faulty

DTC	Trouble Code Title, Conditions & Possible Causes
DTC: P0201 **2T, MIL: Yes** **Passat:** 1.8 (AEB, ATW, AUG, AWM), 2.0L (BPY), 2.8L (AHA, ATQ), 4.0L (BDP); **Jetta:** 1.8L (AWD, AWP, AWW), 2.0 (ABA, AEG, AVH, AZG, BBW, BEV, BPY), 2.5L (BGP, BGQ), 2.8L (AFP, BDF); **Golf:** 1.8L (AWD, AWW), 2.0L (ABA, AEG, AVH, AZG, BEV); **GTI:** 1.8L (AWD, AWP, AWW), 2.0L (ABA, AEG), 2.8L (AFP, BDF); **NB:** 1.8L (APH, AWP, AWV, BNU, BKF), 2.0L (AEG, AVH, AZG, BEV); **NBvert:** (BDG); **Cabrio:** 2.0L (ABA) **Transmissions:** All	**Cylinder 1 Injector Circuit Malfunction Conditions:** Engine started, and the ECM detected the fuel injector "1" control circuit was in a high state when it should have been low, or in a low state when it should have been high (wiring harness & injector okay). **Possible Causes:** • Injector 1 connector is damaged, open or shorted • Injector 1 control circuit is open, shorted to ground or to power • ECM has failed (the injector driver circuit may be damaged)
DTC: P0202 **2T, MIL: Yes** **Passat:** 1.8 (AEB, ATW, AUG, AWM), 2.0L (BPY), 2.8L (AHA, ATQ), 4.0L (BDP); **Jetta:** 1.8L (AWD, AWP, AWW), 2.0 (ABA, AEG, AVH, AZG, BBW, BEV, BPY), 2.5L (BGP, BGQ), 2.8L (AFP, BDF); **Golf:** 1.8L (AWD, AWW), 2.0L (ABA, AEG, AVH, AZG, BEV); **GTI:** 1.8L (AWD, AWP, AWW), 2.0L (ABA, AEG), 2.8L (AFP, BDF); **NB:** 1.8L (APH, AWP, AWV, BNU, BKF), 2.0L (AEG, AVH, AZG, BEV); **NBvert:** (BDG); **Cabrio:** 2.0L (ABA) **Transmissions:** All	**Cylinder 2 Injector Circuit Malfunction Conditions:** Engine started, and the ECM detected the fuel injector "2" control circuit was in a high state when it should have been low, or in a low state when it should have been high (wiring harness & injector okay). **Possible Causes:** • Injector 2 connector is damaged, open or shorted • Injector 2 control circuit is open, shorted to ground or to power • ECM has failed (the injector driver circuit may be damaged)
DTC: P0203 **2T, MIL: Yes** **Passat:** 1.8 (AEB, ATW, AUG, AWM), 2.0L (BPY), 2.8L (AHA, ATQ), 4.0L (BDP); **Jetta:** 1.8L (AWD, AWP, AWW), 2.0 (ABA, AEG, AVH, AZG, BBW, BEV, BPY), 2.5L (BGP, BGQ), 2.8L (AFP, BDF); **Golf:** 1.8L (AWD, AWW), 2.0L (ABA, AEG, AVH, AZG, BEV); **GTI:** 1.8L (AWD, AWP, AWW), 2.0L (ABA, AEG), 2.8L (AFP, BDF); **NB:** 1.8L (APH, AWP, AWV, BNU, BKF), 2.0L (AEG, AVH, AZG, BEV); **NBvert:** (BDG); **Cabrio:** 2.0L (ABA) **Transmissions:** All	**Cylinder 3 Injector Circuit Malfunction Conditions:** Engine started, and the ECM detected the fuel injector "3" control circuit was in a high state when it should have been low, or in a low state when it should have been high (wiring harness & injector okay). **Possible Causes:** • Injector 3 connector is damaged, open or shorted • Injector 3 control circuit is open, shorted to ground or to power • ECM has failed (the injector driver circuit may be damaged)

DTC	Trouble Code Title, Conditions & Possible Causes
DTC: P0204 **2T, MIL: Yes** **Passat:** 1.8 (AEB, ATW, AUG, AWM), 2.0L (BPY), 2.8L (AHA, ATQ), 4.0L (BDP); **Jetta:** 1.8L (AWD, AWP, AWW), 2.0 (ABA, AEG, AVH, AZG, BBW, BEV, BPY), 2.5L (BGP, BGQ), 2.8L (AFP, BDF); **Golf:** 1.8L (AWD, AWW), 2.0L (ABA, AEG, AVH, AZG, BEV); **GTI:** 1.8L (AWD, AWP, AWW), 2.0L (ABA, AEG), 2.8L (AFP, BDF); **NB:** 1.8L (APH, AWP, AWV, BNU, BKF), 2.0L (AEG, AVH, AZG, BEV); **NBvert:** (BDG); **Cabrio:** 2.0L (ABA) **Transmissions:** All	**Cylinder 4 Injector Circuit Malfunction Conditions:** Engine started, and the ECM detected the fuel injector "4" control circuit was in a high state when it should have been low, or in a low state when it should have been high (wiring harness & injector okay). **Possible Causes:** • Injector 4 connector is damaged, open or shorted • Injector 4 control circuit is open, shorted to ground or to power • ECM has failed (the injector driver circuit may be damaged)
DTC: P0205 **2T, MIL: Yes** **Passat:** 2.8L (AHA, ATQ), 4.0L (BDP); **Jetta:** 2.5L (BGP, BGQ), 2.8L (AFP, BDF); **GTI:** 2.8L (AFP, BDF) **Transmissions:** All	**Cylinder 5 Injector Circuit Malfunction Conditions:** Engine started, and the ECM detected the fuel injector "5" control circuit was in a high state when it should have been low, or in a low state when it should have been high (wiring harness & injector okay). **Possible Causes:** • Injector 5 connector is damaged, open or shorted • Injector 5 control circuit is open, shorted to ground or to power • ECM has failed (the injector driver circuit may be damaged)
DTC: P0206 **2T, MIL: Yes** **Passat:** 2.8L (AHA, ATQ), 4.0L (BDP); **Jetta:** 2.8L (AFP, BDF); **GTI:** 2.8L (AFP, BDF) **Transmissions:** All	**Cylinder 6 Injector Circuit Malfunction Conditions:** Engine started, and the ECM detected the fuel injector control circuit was in a high state when it should have been low, or in a low state when it should have been high (wiring harness & injector okay). **Possible Causes:** • Injector 6 connector is damaged, open or shorted • Injector 6 control circuit is open, shorted to ground or to power • ECM has failed (the injector driver circuit may be damaged)
DTC: P0207 **2T, MIL: Yes** **Passat:** 4.0L (BDP) **Transmissions:** All	**Cylinder 7 Injector Circuit Malfunction Conditions:** Engine started, and the ECM detected the fuel injector "7" control circuit was in a high state when it should have been low, or in a low state when it should have been high (wiring harness & injector okay). **Note: Monitor the INJIF PID Fault "flags" with the Scan Tool. The appropriate INJF PID "flag" will read Yes when this code is set.** **Possible Causes:** • Injector 7 connector is damaged, open or shorted • Injector 7 control circuit is open, shorted to ground or to power • ECM has failed (the injector driver circuit may be damaged)
DTC: P0208 **2T, MIL: Yes** **Passat:** 4.0L (BDP) **Transmissions:** All	**Cylinder 8 Injector Circuit Malfunction Conditions:** Engine started, and the ECM detected the fuel injector "8" control circuit was in a high state when it should have been low, or in a low state when it should have been high (wiring harness & injector okay). **Note: Monitor the INJIF PID Fault "flags" with the Scan Tool. The appropriate INJF PID "flag" will read Yes when this code is set.** **Possible Causes:** • Injector 8 connector is damaged, open or shorted • Injector 8 control circuit is open, shorted to ground or to power • ECM has failed (the injector driver circuit may be damaged)

DTC	Trouble Code Title, Conditions & Possible Causes
DTC: P0219 **Passat:** 1.8L (AUG, ATW), 2.0L (BPY); **Jetta:** 1.8L (AWD, AWP, AWW), 2.0 (ABA, BPY), 2.8L (AFP, BDF); **Golf:** 1.8L (AWD, AWW), 2.0L (ABA); **GTI:** 1.8L (AWD, AWP, AWW), 2.0L (ABA), 2.8L (AFP, BDF); **NB:** 1.8L (APH, AWP, AWV, BNU, BKF), 2.0L (AEG, AVH, AZG, BEV); **NBvert:** (BDG); **Cabrio:** 2.0L (ABA) **Transmissions:** All	**Engine Over-Speed Condition Conditions:** Engine started, and the ECM determined the vehicle had been driven in a manner that caused the engine to over-speed, and to exceed the engine speed calibration limit stored in memory. **Possible Causes:** • Engine operated in the wrong transmission gear position • Excessive engine speed with gear selector in Neutral position • Wheel slippage due to wet, muddy or snowing conditions
DTC: P0221 **2T, MIL: Yes** **Passat:** 1.8 (AEB, ATW, AUG, AWM), 2.0L (BPY), 2.8L (AHA, ATQ), 4.0L (BDP); **Jetta:** 1.8L (AWD, AWP, AWW), 2.0L (AEG, AVH, AZG, BBW, BEV, BPY), 2.5L (BGP, BGQ), 2.8L (AFP, BDF); **Golf:** 1.8L (AWD, AWW), 2.0L (AEG, AVH, AZG, BEV); **GTI:** 1.8L (AWD, AWP, AWW), 2.0L (AEG), 2.8L (AFP, BDF); **NB:** 1.8L (APH, AWP, AWV, BNU, BKF), 2.0L (AEG, AVH, AZG, BEV); **NBvert:** (BDG) **Transmissions:** All	**Throttle Position Sensor 'B' Signal Performance Conditions:** Engine started, battery voltage at least 11.5v, all electrical components off, ground connections between engine and chassis well connected, coolant temperature at least 80-degrees Celicius and the throttle valve must not be damaged or dirty; and the ECM detected the TP Sensor 'B' circuit was out of its normal operating range during a condition with the throttle wide open, or with it completely closed. The throttle valve activation occurs via an electric motor (throttle drive) in the throttle valve control module. It is activated by the ECM according to specifications of the two sensors, Throttle Position Sensor and Accelerator Pedal Position Sensor 2. Slowly depress accelerator pedal up to Wide Open Throttle (WOT) stop while observing the percentage display on the PID data function of the scan tool. The percentage display must increase uniformly. **Possible Causes:** • Throttle body is damaged • Throttle linkage is binding or sticking • ETC TP Sensor 'B' signal circuit to the ECM is open • ETC TP Sensor 'B' ground circuit is open • ETC TP Sensor 'B' is damaged or it has failed
DTC: P0222 **2T, MIL: Yes** **Passat:** 1.8 (AEB, ATW, AUG, AWM), 2.0L (BPY), 2.8L (AHA, ATQ), 4.0L (BDP); **Jetta:** 1.8L (AWD, AWP, AWW), 2.0L (AEG, AVH, AZG, BBW, BEV, BPY), 2.5L (BGP, BGQ), 2.8L (AFP, BDF); **Golf:** 1.8L (AWD, AWW), 2.0L (AEG, AVH, AZG, BEV); **GTI:** 1.8L (AWD, AWP, AWW), 2.0L (AEG), 2.8L (AFP, BDF); **NB:** 1.8L (APH, AWP, AWV, BNU, BKF), 2.0L (AEG, AVH, AZG, BEV); **NBvert:** (BDG) **Transmissions:** All	**Throttle Position Sensor 'B' Circuit Low Input Conditions:** Engine started, battery voltage at least 11.5v, all electrical components off, ground connections between engine and chassis well connected, coolant temperature at least 80-degrees Celicius and the throttle valve must not be damaged or dirty; and the ECM detected the TP Sensor 'B' circuit was out of its normal operating range during a condition with the throttle wide open, or with it completely closed. The throttle valve activation occurs via an electric motor (throttle drive) in the throttle valve control module. It is activated by the ECM according to specifications of the two sensors, Throttle Position Sensor and Accelerator Pedal Position Sensor 2. Slowly depress accelerator pedal up to Wide Open Throttle (WOT) stop while observing the percentage display on the PID data function of the scan tool. The percentage display must increase uniformly. **Possible Causes:** • ETC TP Sensor 'B' connector is damaged or shorted • ETC TP Sensor 'B' signal circuit is shorted to ground • ETC TP Sensor 'B' is damaged or it has failed • ECM has failed
DTC: P0223 **2T, MIL: Yes** **Passat:** 1.8 (AEB, ATW, AUG, AWM), 2.0L (BPY), 2.8L (AHA, ATQ), 4.0L (BDP); **Jetta:** 1.8L (AWD, AWP, AWW), 2.0L (AEG, AVH, AZG, BBW, BEV, BPY), 2.5L (BGP, BGQ), 2.8L (AFP, BDF); **Golf:** 1.8L (AWD, AWW), 2.0L (AEG, AVH, AZG, BEV); **GTI:** 1.8L (AWD, AWP, AWW), 2.0L (AEG), 2.8L (AFP, BDF); **NB:** 1.8L (APH, AWP, AWV, BNU, BKF), 2.0L (AEG, AVH, AZG, BEV); **NBvert:** (BDG) **Transmissions:** All	**Throttle Position Sensor 'B' Circuit High Input Conditions:** Engine started, battery voltage at least 11.5v, all electrical components off, ground connections between engine and chassis well connected, coolant temperature at least 80-degrees Celicius and the throttle valve must not be damaged or dirty; and the ECM detected the TP Sensor 'B' circuit was out of its normal operating range during a condition with the throttle wide open, or with it completely closed. The throttle valve activation occurs via an electric motor (throttle drive) in the throttle valve control module. It is activated by the ECM according to specifications of the two sensors, Throttle Position Sensor and Accelerator Pedal Position Sensor 2. Slowly depress accelerator pedal up to Wide Open Throttle (WOT) stop while observing the percentage display on the PID data function of the scan tool. The percentage display must increase uniformly. **Possible Causes:** • ETC TP Sensor 'B' connector is damaged or open • ETC TP Sensor 'B' signal circuit is open • ETC TP Sensor 'B' signal circuit is shorted to VREF (5v) • ETC TP Sensor 'B' is damaged or it has failed

DTC	Trouble Code Title, Conditions & Possible Causes
DTC: P0225 **NB:** 2.0L (AEG, AVH, AZG, BEV); **NBvert:** (BDG) **Transmissions:** All	**Throttle Position Sensor 'C' Signal Voltage Supply Conditions:** Engine started, battery voltage at least 11.5v, all electrical components off, ground connections between engine and chassis well connected, coolant temperature at least 80-degrees Celicius and the throttle valve must not be damaged or dirty; and the ECM detected the TP Sensor 'B' circuit was out of its normal operating range during a condition with the throttle wide open, or with it completely closed. The throttle valve activation occurs via an electric motor (throttle drive) in the throttle valve control module. It is activated by the ECM according to specifications of the two sensors, Throttle Position Sensor and Accelerator Pedal Position Sensor 2. Slowly depress accelerator pedal up to Wide Open Throttle (WOT) stop while observing the percentage display on the PID data function of the scan tool. The percentage display must increase uniformly. **Possible Causes:** • Throttle body is damaged • Throttle linkage is binding or sticking • ETC TP Sensor 'C' signal circuit to the ECM is open • ETC TP Sensor 'C' ground circuit is open • ETC TP Sensor 'C' is damaged or it has failed
DTC: P0226 **Passat:** 1.8 (AEB, ATW, AUG, AWM), 2.0L (BPY); **Jetta:** 1.8L (AWD, AWP, AWW), 2.0L (AEG, AVH, AZG, BBW, BEV), 2.8L (AFP, BDF); **Golf:** 1.8L (AWD, AWW), 2.0L (AEG, AVH, AZG, BEV); **GTI:** 1.8L (AWD, AWP, AWW), 2.0L (AEG), 2.8L (AFP, BDF); **NB:** 1.8L (APH, AWP, AWV, BNU, BKF), 2.0L (AEG, AVH, AZG, BEV); **NBvert:** (BDG) **Transmissions:** All	**Throttle Position Sensor 'C' Signal Performance Conditions:** Engine started, battery voltage at least 11.5v, all electrical components off, ground connections between engine and chassis well connected, coolant temperature at least 80-degrees Celicius and the throttle valve must not be damaged or dirty; and the ECM detected the TP Sensor 'B' circuit was out of its normal operating range during a condition with the throttle wide open, or with it completely closed. The throttle valve activation occurs via an electric motor (throttle drive) in the throttle valve control module. It is activated by the ECM according to specifications of the two sensors, Throttle Position Sensor and Accelerator Pedal Position Sensor 2. Slowly depress accelerator pedal up to Wide Open Throttle (WOT) stop while observing the percentage display on the PID data function of the scan tool. The percentage display must increase uniformly.. **Possible Causes:** • Throttle body is damaged • Throttle linkage is binding or sticking • ETC TP Sensor 'C' signal circuit to the ECM is open • ETC TP Sensor 'C' ground circuit is open • ETC TP Sensor 'C' is damaged or it has failed
DTC: P0227 **Passat:** 2.0L (BPY); **Jetta:** 2.0L (AEG, AVH, AZG, BBW, BEV), 2.8L (AFP, BDF); **Golf:** 2.0L (AEG, AVH, AZG, BEV); **GTI:** 2.0L (AEG), 2.8L (AFP, BDF); **NB:** 1.8L (APH, AWP, AWV, BNU, BKF), 2.0L (AEG, AVH, AZG, BEV); **NBvert:** (BDG) **Transmissions:** All	**Throttle Position Sensor 'C' Circuit Low Input Conditions:** Engine started, battery voltage at least 11.5v, all electrical components off, ground connections between engine and chassis well connected, coolant temperature at least 80-degrees Celicius and the throttle valve must not be damaged or dirty; and the ECM detected the TP Sensor 'B' circuit was out of its normal operating range during a condition with the throttle wide open, or with it completely closed. The throttle valve activation occurs via an electric motor (throttle drive) in the throttle valve control module. It is activated by the ECM according to specifications of the two sensors, Throttle Position Sensor and Accelerator Pedal Position Sensor 2. Slowly depress accelerator pedal up to Wide Open Throttle (WOT) stop while observing the percentage display on the PID data function of the scan tool. The percentage display must increase uniformly. **Possible Causes:** • ETC TP Sensor 'C' connector is damaged or shorted • ETC TP Sensor 'C' signal circuit is shorted to ground • ETC TP Sensor 'C' is damaged or it has failed • ECM has failed
DTC: P0228 **Passat:** 2.0L (BPY); **Jetta:** 2.0L (AEG, AVH, AZG, BBW, BEV), 2.8L (AFP, BDF); **Golf:** 2.0L (AEG, AVH, AZG, BEV); **GTI:** 2.0L (AEG), 2.8L (AFP, BDF); **NB:** 1.8L (APH, AWP, AWV, BNU, BKF), 2.0L (AEG, AVH, AZG, BEV); **NBvert:** (BDG) **Transmissions:** All	**Throttle Position Sensor 'C' Circuit High Input Conditions:** Engine started, battery voltage at least 11.5v, all electrical components off, ground connections between engine and chassis well connected, coolant temperature at least 80-degrees Celicius and the throttle valve must not be damaged or dirty; and the ECM detected the TP Sensor 'B' circuit was out of its normal operating range during a condition with the throttle wide open, or with it completely closed. The throttle valve activation occurs via an electric motor (throttle drive) in the throttle valve control module. It is activated by the ECM according to specifications of the two sensors, Throttle Position Sensor and Accelerator Pedal Position Sensor 2. Slowly depress accelerator pedal up to Wide Open Throttle (WOT) stop while observing the percentage display on the PID data function of the scan tool. The percentage display must increase uniformly. **Possible Causes:** • ETC TP Sensor 'C' connector is damaged or open • ETC TP Sensor 'C' signal circuit is open • ETC TP Sensor 'C' signal circuit is shorted to VREF (5v) • ETC TP Sensor 'C' is damaged or it has failed

DTC	Trouble Code Title, Conditions & Possible Causes
DTC: P0230 **2T, MIL: Yes** **Passat:** 1.8 (AEB, ATW, AUG, AWM), 2.0L (BPY), 2.8L (AHA, ATQ), 4.0L (BDP); **Jetta:** 1.8L (AWD, AWP, AWW), 2.0 (ABA, AEG, AVH, AZG, BBW, BEV), 2.8L (AFP, BDF); **Golf:** 1.8L (AWD, AWW), 2.0L (ABA, AEG, AVH, AZG, BEV); **GTI:** 1.8L (AWD, AWP, AWW), 2.0L (ABA, AEG), 2.8L (AFP, BDF); **NB:** 1.8L (APH, AWP, AWV, BNU, BKF), 2.0L (AEG, AVH, AZG, BEV); **NBvert:** (BDG); **Cabrio:** 2.0L (ABA) **Transmissions:** All	**Fuel Pump Primary Circuit Malfunction Conditions:** Engine started, battery voltage at least 11.5v, all electrical components off, ground connections between engine and chassis well connected, coolant temperature at least 80-degrees Celicius. The ECM detected high current in fuel pump or fuel shutoff valve (FSV) circuit, or it detected voltage with the valve off, or it did not detect voltage on the circuit. The circuit is used to energize the fuel pump relay at key on or while running. Fuel pressure value should be 3000 to 5000 kPa at idle. **Possible Causes:** • FP or FSV circuit is open or shorted • Fuel pump relay VPWR circuit open • Fuel pump relay is damaged or has failed • Fuel pressure sensor has failed • Fuel pump control module is faulty • ECM has failed
DTC: P0234 **Passat:** 1.8 (AEB, ATW, AUG, AWM), 2.0L (BPY); **Jetta:** 1.8L (AWD, AWP, AWW); 2.0L (BPY); **Golf:** 1.8L (AWD, AWW); **GTI:** 1.8L (AWD, AWP, AWW); **NB:** 1.8L (APH, AWP, AWV, BNU, BKF) **Transmissions:** All	**Turbo/Supercharger Overboost Condition Conditions:** Engine started, battery voltage at least 11.5v, all electrical components off, ground connections between engine and chassis well connected, coolant temperature at least 80-degrees Celicius. The ECM detected an operating condition that could harm the engine or automatic transmission. **Possible Causes:** • Ignition misfire condition exceeds the calibrated threshold • Knock sensor circuit has failed, or excessive knock detected • Low speed fuel pump relay not switching properly • Transmission oil temperature beyond the calibrated threshold • Shaft bearing of charge pressure regulator valve in turbocharger is blocked
DTC: P0235 **Passat:** 1.8 (AEB, ATW, AUG, AWM), 2.0L (BPY); **Jetta:** 1.8L (AWD, AWP, AWW); **Golf:** 1.8L (AWD, AWW); **GTI:** 1.8L (AWD, AWP, AWW); **NB:** 1.8L (APH, AWP, AWV, BNU, BKF) **Transmissions:** All	**Turbocharger Boost Sensor (A) Circ Control Limit Not Reached Conditions:** Engine started, battery voltage at least 11.5v, all electrical components off, ground connections between engine and chassis well connected, coolant temperature at least 80-degrees Celicius. The ECM detected an operating condition that could harm the engine or automatic transmission. **Possible Causes:** • Charge air pressure sensor is faulty • Voltage supply to the charge air pressure sensor is open or shorted • Check the charge air system for leaks • Recirculating valve for turbocharger is faulty • Turbocharging system is damaged or not functioning properly • Turbocharger recirculating valve is faulty • Vacuum diaphragm for turbocharger is out of adjustment • Wastegate bypass regulator valve is faulty • Boost sensor has failed • ECM has failed
DTC: P0236 **Passat:** 1.8 (AEB, ATW, AUG, AWM), 2.0L (BPY); **Jetta:** 1.8L (AWD, AWP, AWW); 2.0L (BPY); **Golf:** 1.8L (AWD, AWW); **GTI:** 1.8L (AWD, AWP, AWW); **NB:** 1.8L (APH, AWP, AWV, BNU, BKF) **Transmissions:** All	**Turbocharger Boost Sensor (A) Circ Control Range/Performance Conditions:** Engine started, battery voltage at least 11.5v, all electrical components off, ground connections between engine and chassis well connected, coolant temperature at least 80-degrees Celicius. The ECM detected an operating condition that could harm the engine or automatic transmission. **Possible Causes:** • Charge air pressure sensor is faulty • Voltage supply to the charge air pressure sensor is open or shorted • Check the charge air system for leaks • Recirculating valve for turbocharger is faulty • Turbocharging system is damaged or not functioning properly • Turbocharger recirculating valve is faulty • Vacuum diaphragm for turbocharger is out of adjustment • Wastegate bypass regulator valve is faulty • Boost sensor has failed • ECM has failed

DTC	Trouble Code Title, Conditions & Possible Causes
DTC: P0237 **2T, MIL: Yes** **Passat:** 1.8 (AEB, ATW, AUG, AWM), 2.0L (BPY); **Jetta:** 1.8L (AWD, AWP, AWW); 2.0L (BPY); **Golf:** 1.8L (AWD, AWW); **GTI:** 1.8L (AWD, AWP, AWW); **NB:** 1.8L (APH, AWP, AWV, BNU, BKF) **Transmissions:** All	**Turbocharger Boost Sensor (A) Circ Low Input Conditions:** Engine started, battery voltage at least 11.5v, all electrical components off, ground connections between engine and chassis well connected, coolant temperature at least 80-degrees Celicius. The ECM detected an operating condition that could harm the engine or automatic transmission. **Possible Causes:** • Charge air pressure sensor is faulty • Voltage supply to the charge air pressure sensor is open or shorted • Check the charge air system for leaks • Recirculating valve for turbocharger is faulty • Turbocharging system is damaged or not functioning properly • Turbocharger recirculating valve is faulty • Vacuum diaphragm for turbocharger is out of adjustment • Wastegate bypass regulator valve is faulty • Boost sensor has failed • ECM has failed
DTC: P0238 **2T, MIL: Yes** **Passat:** 1.8 (AEB, AUG, AWM), 2.0L (BPY); **Jetta:** 1.8L (AWD, AWP, AWW); 2.0L (BPY); **Golf:** 1.8L (AWD, AWW); **GTI:** 1.8L (AWD, AWP, AWW); **NB:** 1.8L (APH, AWP, AWV, BNU, BKF) **Transmissions:** All	**Turbocharger Boost Sensor (A) Circ High Input Conditions:** Engine started, battery voltage at least 11.5v, all electrical components off, ground connections between engine and chassis well connected, coolant temperature at least 80-degrees Celicius. The ECM detected an operating condition that could harm the engine or automatic transmission. **Possible Causes:** • Charge air pressure sensor is faulty • Voltage supply to the charge air pressure sensor is open or shorted • Check the charge air system for leaks • Recirculating valve for turbocharger is faulty • Turbocharging system is damaged or not functioning properly • Turbocharger recirculating valve is faulty • Vacuum diaphragm for turbocharger is out of adjustment • Wastegate bypass regulator valve is faulty • Boost sensor has failed • ECM has failed
DTC: P0243 **Passat:** 1.8 (AEB, ATW, AUG, AWM), 2.0L (BPY); **Jetta:** 1.8L (AWD, AWP, AWW); 2.0L (BPY); **Golf:** 1.8L (AWD, AWW); **GTI:** 1.8L (AWD, AWP, AWW); **NB:** 1.8L (APH, AWP, AWV, BNU, BKF) **Transmissions:** All	**Turbocharger Boost Bypass Solenoid (A) Circuit Open/Short Circuit Conditions:** Engine started, battery voltage at least 11.5v, all electrical components off, ground connections between engine and chassis well connected, coolant temperature at least 80-degrees Celicius. The ECM detected an unexpected voltage condition on the Bypass Solenoid control circuit **Possible Causes:** • Bypass solenoid power supply circuit is open • Bypass solenoid control circuit is open, shorted to ground or system power • Bypass solenoid assembly is damaged or has failed • Charge air pressure sensor is faulty • Voltage supply to the charge air pressure sensor is open or shorted • Check the charge air system for leaks • Recirculating valve for turbocharger is faulty • Turbocharging system is damaged or not functioning properly • Turbocharger recirculating valve is faulty • Vacuum diaphragm for turbocharger is out of adjustment • Wastegate bypass regulator valve is faulty • Boost sensor has failed • ECM has failed

DTC	Trouble Code Title, Conditions & Possible Causes
DTC: P0245 **Passat:** 1.8 (AEB, ATW, AUG, AWM), 2.0L (BPY); **Jetta:** 1.8L (AWD, AWP, AWW); 2.0L (BPY); **Golf:** 1.8L (AWD, AWW); **GTI:** 1.8L (AWD, AWP, AWW); **NB:** 1.8L (APH, AWP, AWV, BNU, BKF) **Transmissions:** All	**Turbocharger Boost Bypass Solenoid (A) Circuit Low Input/Short to Ground Conditions:** Engine started, battery voltage at least 11.5v, all electrical components off, ground connections between engine and chassis well connected, coolant temperature at least 80-degrees Celicius. The ECM detected an unexpected voltage condition on the Bypass Solenoid control circuit **Possible Causes:** • Bypass solenoid power supply circuit is open • Bypass solenoid control circuit is open, shorted to ground or system power • Bypass solenoid assembly is damaged or has failed • Charge air pressure sensor is faulty • Voltage supply to the charge air pressure sensor is open or shorted • Check the charge air system for leaks • Recirculating valve for turbocharger is faulty • Turbocharging system is damaged or not functioning properly • Turbocharger recirculating valve is faulty • Vacuum diaphragm for turbocharger is out of adjustment • Wastegate bypass regulator valve is faulty • Boost sensor has failed • ECM has failed
DTC: P0246 **Passat:** 1.8 (AEB, ATW, AUG, AWM), 2.0L (BPY); **Jetta:** 1.8L (AWD, AWP, AWW); 2.0L (BPY); **Golf:** 1.8L (AWD, AWW); **GTI:** 1.8L (AWD, AWP, AWW); **NB:** 1.8L (APH, AWP, AWV, BNU, BKF) **Transmissions:** All	**Turbocharger Boost Bypass Solenoid (A) Circuit High Input/Short to B+ Conditions:** Engine started, battery voltage at least 11.5v, all electrical components off, ground connections between engine and chassis well connected, coolant temperature at least 80-degrees Celicius. The ECM detected an unexpected voltage condition on the Bypass Solenoid control circuit **Possible Causes:** • Bypass solenoid power supply circuit is open • Bypass solenoid control circuit is open, shorted to ground or system power • Bypass solenoid assembly is damaged or has failed • Charge air pressure sensor is faulty • Voltage supply to the charge air pressure sensor is open or shorted • Check the charge air system for leaks • Recirculating valve for turbocharger is faulty • Turbocharging system is damaged or not functioning properly • Turbocharger recirculating valve is faulty • Vacuum diaphragm for turbocharger is out of adjustment • Wastegate bypass regulator valve is faulty • Boost sensor has failed • ECM has failed
DTC: P0261 **Passat:** 1.8 (AEB, ATW, AUG, AWM), 2.0L (BPY), 2.8L (AHA, ATQ), 4.0L (BDP); **Jetta:** 1.8L (AWD, AWP, AWW), 2.0 (ABA, AEG, AVH, AZG, BBW, BEV, BPY), 2.5L (BGP, BGQ), 2.8L (AFP, BDF); **Golf:** 1.8L (AWD, AWW), 2.0L (ABA, AEG, AVH, AZG, BEV); **GTI:** 1.8L (AWD, AWP, AWW), 2.0L (ABA, AEG), 2.8L (AFP, BDF); **NB:** 1.8L (APH, AWP, AWV, BNU, BKF), 2.0L (AEG, AVH, AZG, BEV); **NBvert:** (BDG); **Cabrio:** 2.0L (ABA) **Transmissions:** All	**Cylinder 1 Injector Circuit Low Input/Short to Ground Conditions:** Key on or engine running, fuses in the instrument panel and the E-box in the engine compartment must be functioning, and the ground connections between the engine ad the chassis must be well connected; and the ECM detected an unexpected voltage condition on the injector circuit **Possible Causes:** • Injector 1 control circuit is open • Injector 1 power circuit (B+) is open • Injector 1 control circuit is shorted to chassis ground • Injector 1 is damaged or has failed • ECM is not connected or has failed

DTC	Trouble Code Title, Conditions & Possible Causes
DTC: P0262 **2T, MIL: Yes** **Passat:** 1.8 (AEB, ATW, AUG, AWM), 2.0L (BPY), 2.8L (AHA, ATQ), 4.0L (BDP); **Jetta:** 1.8L (AWD, AWP, AWW), 2.0 (ABA, AEG, AVH, AZG, BBW, BEV, BPY), 2.5L (BGP, BGQ), 2.8L (AFP, BDF); **Golf:** 1.8L (AWD, AWW), 2.0L (ABA, AEG, AVH, AZG, BEV); **GTI:** 1.8L (AWD, AWP, AWW), 2.0L (ABA, AEG), 2.8L (AFP, BDF); **NB:** 1.8L (APH, AWP, AWV, BNU, BKF), 2.0L (AEG, AVH, AZG, BEV); **NBvert:** (BDG); **Cabrio:** 2.0L (ABA) **Transmissions:** All	**Cylinder 1 Injector Circuit Low Input/Short to B+ Conditions:** Key on or engine running, fuses in the instrument panel and the E-box in the engine compartment must be functioning, and the ground connections between the engine ad the chassis must be well connected; and the ECM detected an unexpected voltage condition on the injector circuit **Possible Causes:** • Injector control circuit is open • Injector power circuit (B+) is open • Injector control circuit is shorted to chassis ground • Injector is damaged or has failed • ECM is not connected or has failed • Fuel pump relay has failed • Fuel injectors may have malfunctioned • Faulty engine speed sensor
DTC: P0264 **2T, MIL: Yes** **Passat:** 1.8 (AEB, ATW, AUG, AWM), 2.0L (BPY), 2.8L (AHA, ATQ), 4.0L (BDP); **Jetta:** 1.8L (AWD, AWP, AWW), 2.0 (ABA, AEG, AVH, AZG, BBW, BEV, BPY), 2.5L (BGP, BGQ), 2.8L (AFP, BDF); **Golf:** 1.8L (AWD, AWW), 2.0L (ABA, AEG, AVH, AZG, BEV); **GTI:** 1.8L (AWD, AWP, AWW), 2.0L (ABA, AEG), 2.8L (AFP, BDF); **NB:** 1.8L (APH, AWP, AWV, BNU, BKF), 2.0L (AEG, AVH, AZG, BEV); **NBvert:** (BDG); **Cabrio:** 2.0L (ABA) **Transmissions:** All	**Cylinder 2 Injector Circuit Low Input/Short to Ground Conditions:** Key on or engine running, fuses in the instrument panel and the E-box in the engine compartment must be functioning, and the ground connections between the engine ad the chassis must be well connected; and the ECM detected an unexpected voltage condition on the injector circuit **Possible Causes:** • Injector control circuit is open • Injector power circuit (B+) is open • Injector control circuit is shorted to chassis ground • Injector is damaged or has failed • ECM is not connected or has failed • Fuel pump relay has failed • Fuel injectors may have malfunctioned • Faulty engine speed sensor
DTC: P0265 **2T, MIL: Yes** **Passat:** 1.8 (AEB, ATW, AUG, AWM), 2.0L (BPY), 2.8L (AHA, ATQ), 4.0L (BDP); **Jetta:** 1.8L (AWD, AWP, AWW), 2.0 (ABA, AEG, AVH, AZG, BBW, BEV, BPY), 2.5L (BGP, BGQ), 2.8L (AFP, BDF); **Golf:** 1.8L (AWD, AWW), 2.0L (ABA, AEG, AVH, AZG, BEV); **GTI:** 1.8L (AWD, AWP, AWW), 2.0L (ABA, AEG), 2.8L (AFP, BDF); **NB:** 1.8L (APH, AWP, AWV, BNU, BKF), 2.0L (AEG, AVH, AZG, BEV); **NBvert:** (BDG); **Cabrio:** 2.0L (ABA) **Transmissions:** All	**Cylinder 2 Injector Circuit Low Input/Short to B+ Conditions:** Key on or engine running, fuses in the instrument panel and the E-box in the engine compartment must be functioning, and the ground connections between the engine ad the chassis must be well connected; and the ECM detected an unexpected voltage condition on the injector circuit **Possible Causes:** • Injector control circuit is open • Injector power circuit (B+) is open • Injector control circuit is shorted to chassis ground • Injector is damaged or has failed • ECM is not connected or has failed • Fuel pump relay has failed • Fuel injectors may have malfunctioned • Faulty engine speed sensor

DTC	Trouble Code Title, Conditions & Possible Causes
DTC: P0267 **2T, MIL: Yes** **Passat:** 1.8 (AEB, ATW, AUG, AWM), 2.0L (BPY), 2.8L (AHA, ATQ), 4.0L (BDP); **Jetta:** 1.8L (AWD, AWP, AWW), 2.0 (ABA, AEG, AVH, AZG, BBW, BEV, BPY), 2.5L (BGP, BGQ), 2.8L (AFP, BDF); **Golf:** 1.8L (AWD, AWW), 2.0L (ABA, AEG, AVH, AZG, BEV); **GTI:** 1.8L (AWD, AWP, AWW), 2.0L (ABA, AEG), 2.8L (AFP, BDF); **NB:** 1.8L (APH, AWP, AWV, BNU, BKF), 2.0L (AEG, AVH, AZG, BEV); **NBvert:** (BDG); **Cabrio:** 2.0L (ABA) **Transmissions:** All	**Cylinder 3 Injector Circuit Low Input/Short to Ground Conditions:** Key on or engine running, fuses in the instrument panel and the E-box in the engine compartment must be functioning, and the ground connections between the engine ad the chassis must be well connected; and the ECM detected an unexpected voltage condition on the injector circuit **Possible Causes:** • Injector control circuit is open • Injector power circuit (B+) is open • Injector control circuit is shorted to chassis ground • Injector is damaged or has failed • ECM is not connected or has failed • Fuel pump relay has failed • Fuel injectors may have malfunctioned • Faulty engine speed sensor
DTC: P0268 **2T, MIL: Yes** **Passat:** 1.8 (AEB, ATW, AUG, AWM), 2.0L (BPY), 2.8L (AHA, ATQ), 4.0L (BDP); **Jetta:** 1.8L (AWD, AWP, AWW), 2.0 (ABA, AEG, AVH, AZG, BBW, BEV, BPY), 2.5L (BGP, BGQ), 2.8L (AFP, BDF); **Golf:** 1.8L (AWD, AWW), 2.0L (ABA, AEG, AVH, AZG, BEV); **GTI:** 1.8L (AWD, AWP, AWW), 2.0L (ABA, AEG), 2.8L (AFP, BDF); **NB:** 1.8L (APH, AWP, AWV, BNU, BKF), 2.0L (AEG, AVH, AZG, BEV); **NBvert:** (BDG); **Cabrio:** 2.0L (ABA) **Transmissions:** All	**Cylinder 3 Injector Circuit Low Input/Short to B+ Conditions:** Key on or engine running, fuses in the instrument panel and the E-box in the engine compartment must be functioning, and the ground connections between the engine ad the chassis must be well connected; and the ECM detected an unexpected voltage condition on the injector circuit **Possible Causes:** • Injector control circuit is open • Injector power circuit (B+) is open • Injector control circuit is shorted to chassis ground • Injector is damaged or has failed • ECM is not connected or has failed • Fuel pump relay has failed • Fuel injectors may have malfunctioned • Faulty engine speed sensor
DTC: P0270 **2T, MIL: Yes** **Passat:** 1.8 (AEB, ATW, AUG, AWM), 2.0L (BPY), 2.8L (AHA, ATQ), 4.0L (BDP); **Jetta:** 1.8L (AWD, AWP, AWW), 2.0 (ABA, AEG, AVH, AZG, BBW, BEV, BPY), 2.5L (BGP, BGQ), 2.8L (AFP, BDF); **Golf:** 1.8L (AWD, AWW), 2.0L (ABA, AEG, AVH, AZG, BEV); **GTI:** 1.8L (AWD, AWP, AWW), 2.0L (ABA, AEG), 2.8L (AFP, BDF); **NB:** 1.8L (APH, AWP, AWV, BNU, BKF), 2.0L (AEG, AVH, AZG, BEV); **NBvert:** (BDG); **Cabrio:** 2.0L (ABA) **Transmissions:** All	**Cylinder 4 Injector Circuit Low Input/Short to Ground Conditions:** Key on or engine running, fuses in the instrument panel and the E-box in the engine compartment must be functioning, and the ground connections between the engine ad the chassis must be well connected; and the ECM detected an unexpected voltage condition on the injector circuit **Possible Causes:** • Injector control circuit is open • Injector power circuit (B+) is open • Injector control circuit is shorted to chassis ground • Injector is damaged or has failed • ECM is not connected or has failed • Fuel pump relay has failed • Fuel injectors may have malfunctioned • Faulty engine speed sensor

DTC	Trouble Code Title, Conditions & Possible Causes
DTC: P0271 **2T, MIL: Yes** **Passat:** 1.8 (AEB, ATW, AUG, AWM), 2.0L (BPY), 2.8L (AHA, ATQ), 4.0L (BDP); **Jetta:** 1.8L (AWD, AWP, AWW), 2.0 (ABA, AEG, AVH, AZG, BBW, BEV, BPY), 2.5L (BGP, BGQ), 2.8L (AFP, BDF); **Golf:** 1.8L (AWD, AWW), 2.0L (ABA, AEG, AVH, AZG, BEV); **GTI:** 1.8L (AWD, AWP, AWW), 2.0L (ABA, AEG), 2.8L (AFP, BDF); **NB:** 1.8L (APH, AWP, AWV, BNU, BKF), 2.0L (AEG, AVH, AZG, BEV); **NBvert:** (BDG); **Cabrio:** 2.0L (ABA) **Transmissions:** All	**Cylinder 4 Injector Circuit Low Input/Short to B+ Conditions:** Key on or engine running, fuses in the instrument panel and the E-box in the engine compartment must be functioning, and the ground connections between the engine ad the chassis must be well connected; and the ECM detected an unexpected voltage condition on the injector circuit **Possible Causes:** • Injector control circuit is open • Injector power circuit (B+) is open • Injector control circuit is shorted to chassis ground • Injector is damaged or has failed • ECM is not connected or has failed • Fuel pump relay has failed • Fuel injectors may have malfunctioned • Faulty engine speed sensor
DTC: P0273 **Passat:** 2.8L (AHA, ATQ), 4.0L (BDP): **Jetta:** 2.5L (BGP, BGQ), 2.8L (AFP, BDF); **GTI:** 2.8L (AFP, BDF) **Transmissions:** All	**Cylinder 5 Injector Circuit Low Input/Short to Ground Conditions:** Key on or engine running, fuses in the instrument panel and the E-box in the engine compartment must be functioning, and the ground connections between the engine ad the chassis must be well connected; and the ECM detected an unexpected voltage condition on the injector circuit **Possible Causes:** • Injector control circuit is open • Injector power circuit (B+) is open • Injector control circuit is shorted to chassis ground • Injector is damaged or has failed • ECM is not connected or has failed • Fuel pump relay has failed • Fuel injectors may have malfunctioned • Faulty engine speed sensor
DTC: P0274 **Passat:** 2.8L (AHA, ATQ), 4.0L (BDP); **Jetta:** 2.5L (BGP, BGQ), 2.8L (AFP, BDF); **GTI:** 2.8L (AFP, BDF) **Transmissions:** All	**Cylinder 5 Injector Circuit Low Input/Short to B+ Conditions:** Key on or engine running, fuses in the instrument panel and the E-box in the engine compartment must be functioning, and the ground connections between the engine ad the chassis must be well connected; and the ECM detected an unexpected voltage condition on the injector circuit **Possible Causes:** • Injector control circuit is open • Injector power circuit (B+) is open • Injector control circuit is shorted to chassis ground • Injector is damaged or has failed • ECM is not connected or has failed • Fuel pump relay has failed • Fuel injectors may have malfunctioned • Faulty engine speed sensor
DTC: P0276 **Passat:** 2.8L (AHA, ATQ), 4.0L (BDP); **Jetta:** 2.8L (AFP, BDF); **GTI:** 2.8L (AFP, BDF) **Transmissions:** All	**Cylinder 6 Injector Circuit Low Input/Short to Ground Conditions:** Key on or engine running, fuses in the instrument panel and the E-box in the engine compartment must be functioning, and the ground connections between the engine ad the chassis must be well connected; and the ECM detected an unexpected voltage condition on the injector circuit **Possible Causes:** • Injector control circuit is open • Injector power circuit (B+) is open • Injector control circuit is shorted to chassis ground • Injector is damaged or has failed • ECM is not connected or has failed • Fuel pump relay has failed • Fuel injectors may have malfunctioned • Faulty engine speed sensor

DTC	Trouble Code Title, Conditions & Possible Causes
DTC: P0277 **Passat:** 2.8L (AHA, ATQ), 4.0L (BDP); **Jetta:** 2.8L (AFP, BDF); **GTI:** 2.8L (AFP, BDF) **Transmissions:** All	**Cylinder 6 Injector Circuit Low Input/Short to B+ Conditions:** Key on or engine running, fuses in the instrument panel and the E-box in the engine compartment must be functioning, and the ground connections between the engine ad the chassis must be well connected; and the ECM detected an unexpected voltage condition on the injector circuit **Possible Causes:** • Injector control circuit is open • Injector power circuit (B+) is open • Injector control circuit is shorted to chassis ground • Injector is damaged or has failed • ECM is not connected or has failed • Fuel pump relay has failed • Fuel injectors may have malfunctioned • Faulty engine speed sensor
DTC: P0279 **Passat:** 4.0L (BDP) **Transmissions:** All	**Cylinder 7 Injector Circuit Low Input/Short to Ground Conditions:** Key on or engine running, fuses in the instrument panel and the E-box in the engine compartment must be functioning, and the ground connections between the engine ad the chassis must be well connected; and the ECM detected an unexpected voltage condition on the injector circuit **Possible Causes:** • Injector control circuit is open • Injector power circuit (B+) is open • Injector control circuit is shorted to chassis ground • Injector is damaged or has failed • ECM is not connected or has failed • Fuel pump relay has failed • Fuel injectors may have malfunctioned • Faulty engine speed sensor
DTC: P0280 **Passat:** 4.0L (BDP) **Transmissions:** All	**Cylinder 7 Injector Circuit Low Input/Short to B+ Conditions:** Key on or engine running, fuses in the instrument panel and the E-box in the engine compartment must be functioning, and the ground connections between the engine ad the chassis must be well connected; and the ECM detected an unexpected voltage condition on the injector circuit **Possible Causes:** • Injector control circuit is open • Injector power circuit (B+) is open • Injector control circuit is shorted to chassis ground • Injector is damaged or has failed • ECM is not connected or has failed • Fuel pump relay has failed • Fuel injectors may have malfunctioned • Faulty engine speed sensor
DTC: P0282 **Passat:** 4.0L (BDP) **Transmissions:** All	**Cylinder 8 Injector Circuit Low Input/Short to Ground Conditions:** Key on or engine running, fuses in the instrument panel and the E-box in the engine compartment must be functioning, and the ground connections between the engine ad the chassis must be well connected; and the ECM detected an unexpected voltage condition on the injector circuit **Possible Causes:** • Injector control circuit is open • Injector power circuit (B+) is open • Injector control circuit is shorted to chassis ground • Injector is damaged or has failed • ECM is not connected or has failed • Fuel pump relay has failed • Fuel injectors may have malfunctioned • Faulty engine speed sensor

<table>
<tr><th>DTC</th><th>Trouble Code Title, Conditions & Possible Causes</th></tr>
<tr><td>DTC: P0283

Passat: 4.0L (BDP)

Transmissions: All</td><td>Cylinder 8 Injector Circuit Low Input/Short to B+ Conditions:
Key on or engine running, fuses in the instrument panel and the E-box in the engine compartment must be functioning, and the ground connections between the engine ad the chassis must be well connected; and the ECM detected an unexpected voltage condition on the injector circuit
Possible Causes:
• Injector control circuit is open
• Injector power circuit (B+) is open
• Injector control circuit is shorted to chassis ground
• Injector is damaged or has failed
• ECM is not connected or has failed
• Fuel pump relay has failed
• Fuel injectors may have malfunctioned
• Faulty engine speed sensor</td></tr>
<tr><td>DTC: P0299

Passat: 1.8L (AUG, ATW), 2.0L (BPY); Jetta: 1.8L (AWD, AWP, AWW), 2.0L (BPY); Golf: 1.8L (AWD, AWW); GTI: 1.8L (AWD, AWP, AWW); NB: 1.8L (APH, AWP, AWV, BNU, BKF)

Transmissions: All</td><td>Turbocharger Underboost Conditions:
Engine started, battery voltage at least 11.5v, all electrical components off, ground connections between engine and chassis well connected, coolant temperature at least 80-degrees Celicius. The ECM detected an operating condition that could harm the engine or automatic transmission.
Possible Causes:
• Charge air pressure sensor has failed
• Charge air system has leaks
• Recirculating valve for turbocharger is faulty
• Turbocharging system is faulty
• Vacuum diaphragm for turbocharger needs adjusting
• Wastegate bypass regulator valve is faulty</td></tr>
<tr><td>DTC: P0300
2T, MIL: Yes

Passat: 1.8 (AEB, ATW, AUG, AWM), 2.0L (ABA, BPY), 2.8L (AAA, AHA, ATQ), 2.8L (AHA, ATQ), 4.0L (BDP); Jetta: 1.8L (AWD, AWP, AWW), 2.0 (ABA, AEG, AVH, AZG, BBW, BEV, BPY), 2.5L (BGP, BGQ), 2.8L (AAA, AFP, BDF); Golf: 1.8L (AWD, AWW), 2.0L (ABA, AEG, AVH, AZG, BEV); GTI: 1.8L (AWD, AWP, AWW), 2.0L (ABA, AEG), 2.8L (AAA, AFP, BDF); NB: 1.8L (APH, AWP, AWV, BNU, BKF), 2.0L (AEG, AVH, AZG, BEV); NBvert: (BDG); Cabrio: 2.0L (ABA)

Transmissions: All</td><td>Random/Multiple Misfire Detected Conditions:
Engine running under positive torque conditions, and the ECM detected a misfire or uneven engine running in two or more cylinders.
Note: If the misfire is severe, the MIL will flash on/off on the first trip!
Possible Causes:
• Base engine mechanical fault that affects two or more cylinders
• Fuel metering fault that affects two or more cylinders
• Fuel pressure too low or too high, fuel supply contaminated
• EVAP system problem or the EVAP canister is fuel saturated
• EGR valve is stuck open or the PCV system has a vacuum leak
• Ignition system fault (coil, plugs) affecting two or more cylinders
• MAF sensor contamination (it can cause a very lean condition)
• Vehicle driven while very low on fuel (less than 1/8 of a tank)</td></tr>
<tr><td>DTC: P0301
2T, MIL: Yes

Passat: 1.8 (AEB, ATW, AUG, AWM), 2.0L (ABA, BPY), 2.8L (AAA, AHA, ATQ), 2.8L (AHA, ATQ), 4.0L (BDP); Jetta: 1.8L (AWD, AWP, AWW), 2.0 (ABA, AEG, AVH, AZG, BBW, BEV, BPY), 2.5L (BGP, BGQ), 2.8L (AAA, AFP, BDF); Golf: 1.8L (AWD, AWW), 2.0L (ABA, AEG, AVH, AZG, BEV); GTI: 1.8L (AWD, AWP, AWW), 2.0L (ABA, AEG), 2.8L (AAA, AFP, BDF); NB: 1.8L (APH, AWP, AWV, BNU, BKF), 2.0L (AEG, AVH, AZG, BEV); NBvert: (BDG); Cabrio: 2.0L (ABA)

Transmissions: All</td><td>Cylinder Number 1 Misfire Detected Conditions:
Engine running under positive torque conditions, and the ECM detected a misfire or uneven engine function.
Note: If the misfire is severe, the MIL will flash on/off on the 1st trip!
Possible Causes:
• Air leak in the intake manifold, or in the EGR or ECM system
• Base engine mechanical problem
• Fuel delivery component problem (i.e., a contaminated, dirty or sticking fuel injector)
• Fuel pump relay defective
• Ignition coil fuses have failed
• Ignition system problem (dirty damaged coil or plug)
• Engine speed (RPM) sensor has failed
• Camshaft position sensors have failed
• Ignition coil is faulty
• Spark plugs are not working properly or are not gapped properly</td></tr>
</table>

DTC	Trouble Code Title, Conditions & Possible Causes
DTC: P0302 **2T, MIL: Yes** **Passat:** 1.8 (AEB, ATW, AUG, AWM), 2.0L (ABA, BPY), 2.8L (AAA, AHA, ATQ), 2.8L (AHA, ATQ), 4.0L (BDP); **Jetta:** 1.8L (AWD, AWP, AWW), 2.0 (ABA, AEG, AVH, AZG, BBW, BEV), 2.5L (BGP, BGQ), 2.8L (AAA); **Golf:** 1.8L (AWD, AWW), 2.0L (ABA, AEG, AVH, AZG, BEV); **GTI:** 1.8L (AWD, AWP, AWW), 2.0L (ABA, AEG), 2.8L (AAA, AFP, BDF); **NB:** 1.8L (APH, AWP, AWV, BNU, BKF), 2.0L (AEG, AVH, AZG, BEV); **NBvert:** (BDG); **Cabrio:** 2.0L (ABA) **Transmissions:** All	**Cylinder Number 2 Misfire Detected Conditions:** Engine running under positive torque conditions, and the ECM detected a misfire or uneven engine function. **Note: If the misfire is severe, the MIL will flash on/off on the 1st trip!** **Possible Causes:** • Air leak in the intake manifold, or in the EGR or ECM system • Base engine mechanical problem • Fuel delivery component problem (i.e., a contaminated, dirty or sticking fuel injector) • Fuel pump relay defective • Ignition coil fuses have failed • Ignition system problem (dirty damaged coil or plug) • Engine speed (RPM) sensor has failed • Camshaft position sensors have failed • Ignition coil is faulty • Spark plugs are not working properly or are not gapped properly
DTC: P0303 **2T, MIL: Yes** **Passat:** 1.8 (AEB, ATW, AUG, AWM), 2.0L (ABA, BPY), 2.8L (AAA, AHA, ATQ), 2.8L (AHA, ATQ), 4.0L (BDP); **Jetta:** 1.8L (AWD, AWP, AWW), 2.0 (ABA, AEG, AVH, AZG, BBW, BEV, BPY), 2.5L (BGP, BGQ), 2.8L (AAA, AFP, BDF); **Golf:** 1.8L (AWD, AWW), 2.0L (ABA, AEG, AVH, AZG, BEV); **GTI:** 1.8L (AWD, AWP, AWW), 2.0L (ABA, AEG), 2.8L (AAA, AFP, BDF); **NB:** 1.8L (APH, AWP, AWV, BNU, BKF), 2.0L (AEG, AVH, AZG, BEV); **NBvert:** (BDG); **Cabrio:** 2.0L (ABA) **Transmissions:** All	**Cylinder Number 3 Misfire Detected Conditions:** Engine running under positive torque conditions, and the ECM detected a misfire or uneven engine function. **Note: If the misfire is severe, the MIL will flash on/off on the 1st trip!** **Possible Causes:** • Air leak in the intake manifold, or in the EGR or ECM system • Base engine mechanical problem • Fuel delivery component problem (i.e., a contaminated, dirty or sticking fuel injector) • Fuel pump relay defective • Ignition coil fuses have failed • Ignition system problem (dirty damaged coil or plug) • Engine speed (RPM) sensor has failed • Camshaft position sensors have failed • Ignition coil is faulty • Spark plugs are not working properly or are not gapped properly
DTC: P0304 **2T, MIL: Yes** **Passat:** 1.8 (AEB, ATW, AUG, AWM), 2.0L (ABA, BPY), 2.8L (AAA, AHA, ATQ), 2.8L (AHA, ATQ), 4.0L (BDP); **Jetta:** 1.8L (AWD, AWP, AWW), 2.0 (ABA, AEG, AVH, AZG, BBW, BEV, BPY), 2.5L (BGP, BGQ), 2.8L (AAA, AFP, BDF); **Golf:** 1.8L (AWD, AWW), 2.0L (ABA, AEG, AVH, AZG, BEV); **GTI:** 1.8L (AWD, AWP, AWW), 2.0L (ABA, AEG), 2.8L (AAA, AFP, BDF); **NB:** 1.8L (APH, AWP, AWV, BNU, BKF), 2.0L (AEG, AVH, AZG, BEV); **NBvert:** (BDG); **Cabrio:** 2.0L (ABA) **Transmissions:** All	**Cylinder Number 4 Misfire Detected Conditions:** Engine running under positive torque conditions, and the ECM detected a misfire or uneven engine function. **Note: If the misfire is severe, the MIL will flash on/off on the 1st trip!** **Possible Causes:** • Air leak in the intake manifold, or in the EGR or ECM system • Base engine mechanical problem • Fuel delivery component problem (i.e., a contaminated, dirty or sticking fuel injector) • Fuel pump relay defective • Ignition coil fuses have failed • Ignition system problem (dirty damaged coil or plug) • Engine speed (RPM) sensor has failed • Camshaft position sensors have failed • Ignition coil is faulty • Spark plugs are not working properly or are not gapped properly

DTC	Trouble Code Title, Conditions & Possible Causes
DTC: P0305 **Passat:** 2.8L (AAA, AHA, ATQ), 4.0L (BDP); **Jetta:** 2.5L (BGP, BGQ), 2.8L (AAA, AFP, BDF); **GTI:** 2.8L (AAA, AFP, BDF) **Transmissions:** All	**Cylinder Number 5 Misfire Detected Conditions:** Engine running under positive torque conditions, and the ECM detected a misfire or uneven engine function. **Note: If the misfire is severe, the MIL will flash on/off on the 1st trip!** **Possible Causes:** • Air leak in the intake manifold, or in the EGR or ECM system • Base engine mechanical problem • Fuel delivery component problem (i.e., a contaminated, dirty or sticking fuel injector) • Fuel pump relay defective • Ignition coil fuses have failed • Ignition system problem (dirty damaged coil or plug) • Engine speed (RPM) sensor has failed • Camshaft position sensors have failed • Ignition coil is faulty • Spark plugs are not working properly or are not gapped properly
DTC: P0306 **Passat:** 2.8L (AAA, AHA, ATQ), 4.0L (BDP); **Jetta:** 2.8L (AAA, AFP, BDF); **GTI:** 2.8L (AAA, AFP, BDF) **Transmissions:** All	**Cylinder Number 6 Misfire Detected Conditions:** Engine running under positive torque conditions, and the ECM detected a misfire or uneven engine function. **Note: If the misfire is severe, the MIL will flash on/off on the 1st trip!** **Possible Causes:** • Air leak in the intake manifold, or in the EGR or ECM system • Base engine mechanical problem • Fuel delivery component problem (i.e., a contaminated, dirty or sticking fuel injector) • Fuel pump relay defective • Ignition coil fuses have failed • Ignition system problem (dirty damaged coil or plug) • Engine speed (RPM) sensor has failed • Camshaft position sensors have failed • Ignition coil is faulty • Spark plugs are not working properly or are not gapped properly
DTC: P0307 **Passat:** 4.0L (BDP) **Transmissions:** All	**Cylinder Number 7 Misfire Detected Conditions:** Engine running under positive torque conditions, and the ECM detected a misfire or uneven engine function. **Note: If the misfire is severe, the MIL will flash on/off on the 1st trip!** **Possible Causes:** • Air leak in the intake manifold, or in the EGR or ECM system • Base engine mechanical problem • Fuel delivery component problem (i.e., a contaminated, dirty or sticking fuel injector) • Fuel pump relay defective • Ignition coil fuses have failed • Ignition system problem (dirty damaged coil or plug) • Engine speed (RPM) sensor has failed • Camshaft position sensors have failed • Ignition coil is faulty • Spark plugs are not working properly or are not gapped properly
DTC: P0308 **Passat:** 4.0L (BDP) **Transmissions:** All	**Cylinder Number 8 Misfire Detected Conditions:** Engine running under positive torque conditions, and the ECM detected a misfire or uneven engine function. **Note: If the misfire is severe, the MIL will flash on/off on the 1st trip!** **Possible Causes:** • Air leak in the intake manifold, or in the EGR or ECM system • Base engine mechanical problem • Fuel delivery component problem (i.e., a contaminated, dirty or sticking fuel injector) • Fuel pump relay defective • Ignition coil fuses have failed • Ignition system problem (dirty damaged coil or plug) • Engine speed (RPM) sensor has failed • Camshaft position sensors have failed • Ignition coil is faulty • Spark plugs are not working properly or are not gapped properly

DTC	Trouble Code Title, Conditions & Possible Causes
DTC: P0318 **Passat:** 2.0L (BPY); **Jetta:** 1.8L (AWD, AWP, AWW); **Golf:** 1.8L (AWD, AWW); **GTI:** 1.8L (AWD, AWP, AWW); **NB:** 1.8L (APH, AWP, AWV, BNU, BKF) **Transmissions:** All	**Rough Road Sensor Conditions:** Engine running, and the ECM detected an implausible signal from the rough road sensor. **Possible Causes:** • Wire connection between Engine Control Module (ECM) and ABS Control Module
DTC: P0321 **2T, MIL: Yes** **Passat:** 1.8 (AEB, ATW, AUG, AWM), 2.0L (BPY), 2.8L (AAA, AHA, ATQ), 4.0L (BDP); **Jetta:** 1.8L (AWD, AWP, AWW), 2.0 (ABA, AEG, AVH, AZG, BBW, BEV, BPY), 2.5L (BGP, BGQ), 2.8L (AAA, AFP, BDF); **Golf:** 1.8L (AWD, AWW), 2.0L (ABA, AEG, AVH, AZG, BEV); **GTI:** 1.8L (AWD, AWP, AWW), 2.0L (ABA, AEG), 2.8L (AAA, AFP, BDF); **NB:** 1.8L (APH, AWP, AWV, BNU, BKF), 2.0L (AEG, AVH, AZG, BEV); **NBvert:** (BDG); **Cabrio:** 2.0L (ABA) **Transmissions:** All	**Ignition/Distributor Engine Speed Input Circuit Range/Performance Conditions:** Engine started, vehicle driven, and the ECM detected the engine speed signal was more than the calibrated value. **Note: The engine will not start if there is no speed signal. If the speed signal fails when the engine is running, it will cause the engine to stall immediately.** **Possible Causes:** • Engine speed sensor has failed or is damaged • ECM has failed • Sensor wheel is damaged or doesn't fit properly • Sensor wheel spacer isn't seated properly
DTC: P0322 **2T, MIL: Yes** **Passat:** 1.8 (AEB, ATW, AUG, AWM), 2.0L (BPY), 2.8L (AAA, AHA, ATQ), 4.0L (BDP); **Jetta:** 1.8L (AWD, AWP, AWW), 2.0 (ABA, AEG, AVH, AZG, BBW, BEV, BPY), 2.5L (BGP, BGQ), 2.8L (AAA, AFP, BDF); **Golf:** 1.8L (AWD, AWW), 2.0L (ABA, AEG, AVH, AZG, BEV); **GTI:** 1.8L (AWD, AWP, AWW), 2.0L (ABA, AEG), 2.8L (AAA, AFP, BDF); **NB:** 1.8L (APH, AWP, AWV, BNU, BKF), 2.0L (AEG, AVH, AZG, BEV); **NBvert:** (BDG); **Cabrio:** 2.0L (ABA) **Transmissions:** All	**Ignition/Distributor Engine Input Circuit No Signal Conditions:** Key on, and the ECM could not detect the engine speed signal or the signal was erratic. **Note: The engine will not start if there is no speed signal. If the speed signal fails when the engine is running, it will cause the engine to stall immediately.** **Possible Causes:** • Engine speed sensor has failed or is damaged • ECM has failed • Sensor wheel is damaged or doesn't fit properly • Sensor wheel spacer isn't seated properly
DTC: P0324 **Passat:** 1.8L (AUG, ATW), 2.0L (BPY), 4.0L (BDP); **Jetta:** 1.8L (AWD, AWP, AWW), 2.0 (ABA, AEG, AVH, AZG, BBW, BEV), 2.5L (BGP, BGQ), 2.8L (AFP, BDF); **Golf:** 1.8L (AWD, AWW), 2.0L (ABA, AEG, AVH, AZG, BEV); **GTI:** 1.8L (AWD, AWP, AWW), 2.0L (ABA, AEG), 2.8L (AFP, BDF); **NB:** 1.8L (APH, AWP, AWV, BNU, BKF), 2.0L (AEG, AVH, AZG, BEV); **NBvert:** (BDG); **Cabrio:** 2.0L (ABA) **Transmissions:** All	**Knock Control System Error Conditions:** Engine started, vehicle driven, and the ECM detected the Knock Sensor 1 (KS1) signal was too low or not recognized by the ECM **Possible Causes:** • Knock sensor circuit is open • Knock sensor is loose (tighten to 20 NM) • Contact between the knock sensor and cylinder block is dirty, corroded or greasy • Knock sensor circuit is shorted to ground, or shorted to power • Knock sensor is damaged or it has failed • Wrong kind of fuel used • A component in the engine compartment is loose or not properly secured • ECM has failed

DTC	Trouble Code Title, Conditions & Possible Causes
DTC: P0327 **2T, MIL: Yes** **Passat:** 1.8 (AEB, ATW, AUG, AWM), 2.0L (ABA, BPY), 2.8L (AAA, AHA, ATQ), 4.0L (BDP); **Jetta:** 1.8L (AWD, AWP, AWW), 2.0 (ABA, AEG, AVH, AZG, BBW, BEV, BPY), 2.5L (BGP, BGQ), 2.8L (AAA, AFP, BDF); **Golf:** 1.8L (AWD, AWW), 2.0L (ABA, AEG, AVH, AZG, BEV); **GTI:** 1.8L (AWD, AWP, AWW), 2.0L (ABA, AEG), 2.8L (AAA, AFP, BDF); **NB:** 1.8L (APH, AWP, AWV, BNU, BKF), 2.0L (AEG, AVH, AZG, BEV); **NBvert:** (BDG); **Cabrio:** 2.0L (ABA) **Transmissions:** All	**Knock Sensor 1 Signal Low Input Conditions:** Engine started, vehicle driven, and the ECM detected the Knock Sensor 1 (KS1) signal was too low or not recognized by the ECM **Possible Causes:** • Knock sensor circuit is open • Knock sensor is loose (tighten to 20 NM) • Contact between the knock sensor and cylinder block is dirty, corroded or greasy • Knock sensor circuit is shorted to ground, or shorted to power • Knock sensor is damaged or it has failed • Wrong kind of fuel used • A component in the engine compartment is loose or not properly secured • ECM has failed
DTC: P0328 **2T, MIL: Yes** **Passat:** 1.8L (AUG, ATW), 2.8L (AHA, ATQ), 2.0L (BPY), 4.0L (BDP); **Jetta:** 1.8L (AWD, AWP, AWW), 2.0 (ABA, AEG, AVH, AZG, BBW, BEV, BPY), 2.5L (BGP, BGQ), 2.8L (AFP, BDF); **Golf:** 1.8L (AWD, AWW), 2.0L (ABA, AEG, AVH, AZG, BEV); **GTI:** 1.8L (AWD, AWP, AWW), 2.0L (ABA, AEG), 2.8L (AFP, BDF); **NB:** 1.8L (APH, AWP, AWV, BNU, BKF), 2.0L (AEG, AVH, AZG, BEV); **NBvert:** (BDG); **Cabrio:** 2.0L (ABA) **Transmissions:** All	**Knock Sensor 1 Signal High Input Conditions:** Engine started, vehicle driven, and the ECM detected the Knock Sensor 1 (KS1) signal was too high **Possible Causes:** • Knock sensor circuit is open • Knock sensor is loose (tighten to 20 NM) • Contact between the knock sensor and cylinder block is dirty, corroded or greasy • Knock sensor circuit is shorted to ground, or shorted to power • Knock sensor is damaged or it has failed • Wrong kind of fuel used • A component in the engine compartment is loose or not properly secured • ECM has failed
DTC: P0332 **2T, MIL: Yes** **Passat:** 1.8 (AEB, ATW, AUG, AWM), 2.0L (BPY), 2.8L (AAA, AHA, ATQ), 4.0L (BDP); **Jetta:** 1.8L (AWD, AWP, AWW), 2.0L (AEG, AVH, AZG, BBW, BEV, BPY), 2.5L (BGP, BGQ), 2.8L (AAA, AFP, BDF); **Golf:** 1.8L (AWD, AWW), 2.0L (AEG, AVH, AZG, BEV); **GTI:** 1.8L (AWD, AWP, AWW), 2.0L (AEG), 2.8L (AAA, AFP, BDF); **NB:** 1.8L (APH, AWP, AWV, BNU, BKF), 2.0L (AEG, AVH, AZG, BEV); **NBvert:** (BDG) **Transmissions:** All	**Knock Sensor 2 Signal Low Input Conditions:** Engine started, vehicle driven, and the ECM detected the Knock Sensor 1 (KS1) signal was too low or not recognized by the ECM **Possible Causes:** • Knock sensor circuit is open • Knock sensor is loose (tighten to 20 NM) • Contact between the knock sensor and cylinder block is dirty, corroded or greasy • Knock sensor circuit is shorted to ground, or shorted to power • Knock sensor is damaged or it has failed • Wrong kind of fuel used • A component in the engine compartment is loose or not properly secured • ECM has failed

DTC	Trouble Code Title, Conditions & Possible Causes
DTC: P0333 **2T, MIL: Yes** **Passat:** 2.8L (AHA, ATW, AUG, ATQ), 2.0L (BPY), 4.0L (BDP); **Jetta:** 1.8L (AWD, AWP, AWW), 2.0L (AEG, AVH, AZG, BBW, BEV, BPY), 2.5L (BGP, BGQ), 2.8L (AFP, BDF); **Golf:** 1.8L (AWD, AWW), 2.0L (AEG, AVH, AZG, BEV); **GTI:** 1.8L (AWD, AWP, AWW), 2.0L (AEG), 2.8L (AFP, BDF); **NB:** 1.8L (APH, AWP, AWV, BNU, BKF), 2.0L (AEG, AVH, AZG, BEV); **NBvert:** (BDG) **Transmissions:** All	**Knock Sensor 2 Signal High Input Conditions:** Engine started, vehicle driven, and the ECM detected the Knock Sensor 1 (KS1) signal was too high **Possible Causes:** • Knock sensor circuit is open • Knock sensor is loose (tighten to 20 NM) • Contact between the knock sensor and cylinder block is dirty, corroded or greasy • Knock sensor circuit is shorted to ground, or shorted to power • Knock sensor is damaged or it has failed • Wrong kind of fuel used • A component in the engine compartment is loose or not properly secured • ECM has failed
DTC: P0340 **2T, MIL: Yes** **Passat:** 1.8 (AEB, ATW, AUG, AWM), 2.8L (AHA, ATQ), 4.0L (BDP); **Jetta:** 1.8L (AWD, AWP, AWW), 2.0 (ABA, AEG, AVH, AZG, BBW, BEV), 2.8L (AFP, BDF); **Golf:** 1.8L (AWD, AWW), 2.0L (ABA, AEG, AVH, AZG, BEV); **GTI:** 1.8L (AWD, AWP, AWW), 2.0L (ABA, AEG), 2.8L (AFP, BDF); **NB:** 1.8L (APH, AWP, AWV, BNU, BKF), 2.0L (AEG, AVH, AZG, BEV); **NBvert:** (BDG); **Cabrio:** 2.0L (ABA) **Transmissions:** All	**Camshaft Position Sensor Circuit Malfunction Conditions:** Engine started, battery voltage must be at least 11.5v, all electrical components must be off, parking brake must be engaged (to keep daytime driving lights off), automatic transmission selector must be in park and the ground between the engine and the chassis must be well connected. The ECM detected the CMP sensor signal was missing or it was erratic. **Possible Causes:** • CMP sensor circuit is open or shorted to ground • CMP sensor circuit is shorted to power • CMP sensor ground (return) circuit is open • CMP sensor installation incorrect (Hall-effect type) • CMP sensor is damaged or CMP sensor shielding damaged • CMP sensor has failed • ECM has failed
DTC: P0341 **2T, MIL: Yes** **Passat:** 1.8 (AEB, ATW, AUG, AWM), 2.0L (ABA, BPY), 2.8L (AAA, AHA, ATQ), 4.0L (BDP); **Jetta:** 1.8L (AWD, AWP, AWW), 2.0 (ABA, AEG, AVH, AZG, BBW, BEV, BPY), 2.5L (BGP, BGQ), 2.8L (AAA, AFP, BDF); **Golf:** 1.8L (AWD, AWW), 2.0L (ABA, AEG, AVH, AZG, BEV); **GTI:** 1.8L (AWD, AWP, AWW), 2.0L (ABA, AEG), 2.8L (AAA, AFP, BDF); **NB:** 1.8L (APH, AWP, AWV, BNU, BKF), 2.0L (AEG, AVH, AZG, BEV); **NBvert:** (BDG); **Cabrio:** 2.0L (ABA) **Transmissions:** All	**Camshaft Position Sensor Circ Range/Performance Conditions:** Engine started, battery voltage must be at least 11.5v, all electrical components must be off, parking brake must be engaged (to keep daytime driving lights off), automatic transmission selector must be in park and the ground between the engine and the chassis must be well connected. The ECM detected the CMP sensor signal was implausible. **Possible Causes:** • CMP sensor circuit is open or shorted to ground • CMP sensor circuit is shorted to power • CMP sensor ground (return) circuit is open • CMP sensor installation incorrect (Hall-effect type) • CMP sensor is damaged or CMP sensor shielding damaged • ECM has failed

DTC	Trouble Code Title, Conditions & Possible Causes
DTC: P0342 **2T, MIL: Yes** **Passat:** 1.8 (AEB, ATW, AUG, AWM), 2.0L (BPY), 2.8L (AHA, ATQ), 4.0L (BDP); **Jetta:** 1.8L (AWD, AWP, AWW), 2.0 (ABA, AEG, AVH, AZG, BBW, BEV, BPY), 2.5L (BGP, BGQ), 2.8L (AFP, BDF); **Golf:** 1.8L (AWD, AWW), 2.0L (ABA, AEG, AVH, AZG, BEV); **GTI:** 1.8L (AWD, AWP, AWW), 2.0L (ABA, AEG), 2.8L (AFP, BDF); **NB:** 1.8L (APH, AWP, AWV, BNU, BKF), 2.0L (AEG, AVH, AZG, BEV); **NBvert:** (BDG); **Cabrio:** 2.0L (ABA) **Transmissions:** All	**Camshaft Position Sensor "A" Circuit (Bank 1 or Single Sensor) Low Input Conditions:** Engine started, battery voltage must be at least 11.5v, all electrical components must be off, parking brake must be engaged (to keep daytime driving lights off), automatic transmission selector must be in park and the ground between the engine and the chassis must be well connected. The ECM detected the CMP sensor signal exceeded the bounds of the specified maximum limit. **Possible Causes:** • CMP sensor circuit is open or shorted to ground • CMP sensor circuit is shorted to power • CMP sensor ground (return) circuit is open • CMP sensor installation incorrect (Hall-effect type) • CMP sensor is damaged or CMP sensor shielding damaged • ECM has failed
DTC: P0343 **2T, MIL: Yes** **Passat:** 1.8 (AEB, ATW, AUG, AWM), 2.0L (BPY), 2.8L (AHA, ATQ), 4.0L (BDP); **Jetta:** 1.8L (AWD, AWP, AWW), 2.0 (ABA, AEG, AVH, AZG, BBW, BEV, BPY), 2.5L (BGP, BGQ), 2.8L (AFP, BDF); **Golf:** 1.8L (AWD, AWW), 2.0L (ABA, AEG, AVH, AZG, BEV); **GTI:** 1.8L (AWD, AWP, AWW), 2.0L (ABA, AEG), 2.8L (AFP, BDF); **NB:** 1.8L (APH, AWP, AWV, BNU, BKF), 2.0L (AEG, AVH, AZG, BEV); **NBvert:** (BDG); **Cabrio:** 2.0L (ABA) **Transmissions:** All	**Camshaft Position Sensor "A" Circuit (Bank 1 or Single Sensor) High Input Conditions:** Engine started, battery voltage must be at least 11.5v, all electrical components must be off, parking brake must be engaged (to keep daytime driving lights off), automatic transmission selector must be in park and the ground between the engine and the chassis must be well connected. The ECM detected the CMP sensor signal did not reach the specified minimum limit. **Possible Causes:** • CMP sensor circuit is open or shorted to ground • CMP sensor circuit is shorted to power • CMP sensor ground (return) circuit is open • CMP sensor installation incorrect (Hall-effect type) • CMP sensor is damaged or CMP sensor shielding damaged • ECM has failed
DTC: P0345 **Passat:** 2.8L (AHA, ATQ), 4.0L (BDP); **Jetta:** 2.8L (AFP, BDF); **GTI:** 2.8L (AFP, BDF) **Transmissions:** All	**Camshaft Position Sensor "A" Circuit (Bank 2) Malfunction Conditions:** Engine started, battery voltage must be at least 11.5v, all electrical components must be off, parking brake must be engaged (to keep daytime driving lights off), automatic transmission selector must be in park and the ground between the engine and the chassis must be well connected. The ECM detected the CMP sensor signal was missing or it was erratic. **Possible Causes:** • CMP sensor circuit is open or shorted to ground • CMP sensor circuit is shorted to power • CMP sensor ground (return) circuit is open • CMP sensor installation incorrect (Hall-effect type) • CMP sensor is damaged or CMP sensor shielding damaged • ECM has failed
DTC: P0346 **Passat:** 2.8L (AHA, ATQ), 4.0L (BDP); **Jetta:** 2.8L (AFP, BDF); **GTI:** 2.8L (AFP, BDF) **Transmissions:** All	**Camshaft Position Sensor "A" Circuit (Bank 2) Range/Performance Conditions:** Engine started, battery voltage must be at least 11.5v, all electrical components must be off, parking brake must be engaged (to keep daytime driving lights off), automatic transmission selector must be in park and the ground between the engine and the chassis must be well connected. The ECM detected the CMP sensor signal was implausible. **Possible Causes:** • CMP sensor circuit is open or shorted to ground • CMP sensor circuit is shorted to power • CMP sensor ground (return) circuit is open • CMP sensor installation incorrect (Hall-effect type) • CMP sensor is damaged or CMP sensor shielding damaged • ECM has failed

DTC	Trouble Code Title, Conditions & Possible Causes
DTC: P0347 **Passat:** 2.8L (AHA, ATQ), 4.0L (BDP); **Jetta:** 2.8L (AFP, BDF); **GTI:** 2.8L (AFP, BDF) **Transmissions:** All	**Camshaft Position Sensor "A" Circuit (Bank 2) Low Input Conditions:** Engine started, battery voltage must be at least 11.5v, all electrical components must be off, parking brake must be engaged (to keep daytime driving lights off), automatic transmission selector must be in park and the ground between the engine and the chassis must be well connected. The ECM detected the CMP sensor signal exceeded the bounds of the specified maximum limit. **Possible Causes:** • CMP sensor circuit is open or shorted to ground • CMP sensor circuit is shorted to power • CMP sensor ground (return) circuit is open • CMP sensor installation incorrect (Hall-effect type) • CMP sensor is damaged or CMP sensor shielding damaged • ECM has failed
DTC: P0348 **Passat:** 2.8L (AHA, ATQ), 4.0L (BDP); **Jetta:** 2.8L (AFP, BDF); **GTI:** 2.8L (AFP, BDF) **Transmissions:** All	**Camshaft Position Sensor "A" Circuit "A" Circuit (Bank 2) High Input Conditions:** Engine started, battery voltage must be at least 11.5v, all electrical components must be off, parking brake must be engaged (to keep daytime driving lights off), automatic transmission selector must be in park and the ground between the engine and the chassis must be well connected. The ECM detected the CMP sensor signal did not reach the specified minimum limit. **Possible Causes:** • CMP sensor circuit is open or shorted to ground • CMP sensor circuit is shorted to power • CMP sensor ground (return) circuit is open • CMP sensor installation incorrect (Hall-effect type) • CMP sensor is damaged or CMP sensor shielding damaged • ECM has failed
DTC: P0351 **2T, MIL: Yes** **Passat:** 1.8 (AEB, ATW, AUG, AWM), 2.0L (BPY), 2.8L (AHA, ATQ), 4.0L (BDP); **Jetta:** 1.8L (AWD, AWP, AWW), 2.0 (ABA, AEG, AVH, AZG, BBW, BEV, BPY), 2.5L (BGP, BGQ), 2.8L (AFP); **Golf:** 1.8L (AWD, AWW), 2.0L (ABA, AEG, AVH, AZG, BEV); **GTI:** 1.8L (AWD, AWP, AWW), 2.0L (ABA, AEG), 2.8L (AFP); **NB:** 1.8L (APH, AWP, AWV, BNU, BKF), 2.0L (AEG, AVH, AZG, BEV); **NBvert:** (BDG); **Cabrio:** 2.0L (ABA) **Transmissions:** All	**Ignition Coilpack A Primary/Secondary Circuit Malfunction Conditions:** Engine started, battery voltage must be at least 11.5v, all electrical components must be off, parking brake must be engaged (to keep daytime driving lights off), automatic transmission selector must be in park and the ground between the engine and the chassis must be well connected. The ECM did not receive any valid pulses from the ignition module for the Ignition Coilpack A primary circuit. **Note: Ignition coils and power output stages are one component and cannot be replaced individually.** **Possible Causes:** • Engine speed (RPM) sensor has failed • Camshaft Position (CMP) sensor has failed • Power Supply Relay is shorted to an open circuit • There is a malfunction in voltage supply • Ignition coilpack is damaged or it has failed • Cylinder 1 to 4 Fuel Injector(s) have failed • ECM has failed •
DTC: P0352 **2T, MIL: Yes** **Passat:** 1.8 (AEB, ATW, AUG, AWM), 2.0L (BPY), 2.8L (AHA, ATQ), 4.0L (BDP); **Jetta:** 1.8L (AWD, AWP, AWW), 2.0 (ABA, AEG, AVH, AZG, BBW, BEV, BPY), 2.5L (BGP, BGQ), 2.8L (AFP, BDF); **Golf:** 1.8L (AWD, AWW), 2.0L (ABA, AEG, AVH, AZG, BEV); **GTI:** 1.8L (AWD, AWP, AWW), 2.0L (ABA, AEG), 2.8L (AFP, BDF); **NB:** 1.8L (APH, AWP, AWV, BNU, BKF), 2.0L (AEG, AVH, AZG, BEV); **NBvert:** (BDG); **Cabrio:** 2.0L (ABA) **Transmissions:** All	**Ignition Coilpack B Primary/Secondary Circuit Malfunction Conditions:** Engine started, battery voltage must be at least 11.5v, all electrical components must be off, parking brake must be engaged (to keep daytime driving lights off), automatic transmission selector must be in park and the ground between the engine and the chassis must be well connected. The ECM did not receive any valid pulses from the ignition module for the Ignition Coilpack B primary circuit. **Note: Ignition coils and power output stages are one component and cannot be replaced individually.** **Possible Causes:** • Engine speed (RPM) sensor has failed • Camshaft Position (CMP) sensor has failed • Power Supply Relay is shorted to an open circuit • There is a malfunction in voltage supply • Ignition coilpack is damaged or it has failed • Cylinder 1 to 4 Fuel Injector(s) have failed • ECM has failed •

DTC	Trouble Code Title, Conditions & Possible Causes
DTC: P0353 **2T, MIL: Yes** **Passat:** 1.8 (AEB, ATW, AUG, AWM), 2.0L (BPY), 2.8L (AHA, ATQ), 4.0L (BDP); **Jetta:** 1.8L (AWD, AWP, AWW), 2.0 (ABA, AEG, AVH, AZG, BBW, BEV, BPY), 2.5L (BGP, BGQ), 2.8L (AFP, BDF); **Golf:** 1.8L (AWD, AWW), 2.0L (ABA, AEG, AVH, AZG, BEV); **GTI:** 1.8L (AWD, AWP, AWW), 2.0L (ABA, AEG), 2.8L (AFP, BDF); **NB:** 1.8L (APH, AWP, AWV, BNU, BKF), 2.0L (AEG, AVH, AZG, BEV); **NBvert:** (BDG); **Cabrio:** 2.0L (ABA) **Transmissions:** All	**Ignition Coilpack C Primary/Secondary Circuit Malfunction Conditions:** Engine started, battery voltage must be at least 11.5v, all electrical components must be off, parking brake must be engaged (to keep daytime driving lights off), automatic transmission selector must be in park and the ground between the engine and the chassis must be well connected. The ECM did not receive any valid pulses from the ignition module for the Ignition Coilpack C primary circuit. **Note: Ignition coils and power output stages are one component and cannot be replaced individually.** **Possible Causes:** • Engine speed (RPM) sensor has failed • Camshaft Position (CMP) sensor has failed • Power Supply Relay is shorted to an open circuit • There is a malfunction in voltage supply • Ignition coilpack is damaged or it has failed • Cylinder 1 to 4 Fuel Injector(s) have failed • ECM has failed •
DTC: P0354 **2T, MIL: Yes** **Passat:** 1.8 (AEB, ATW, AUG, AWM), 2.0L (BPY), 2.8L (AHA, ATQ), 4.0L (BDP); **Jetta:** 1.8L (AWD, AWP, AWW), 2.0 (ABA, AEG, AVH, AZG, BBW, BEV, BPY), 2.5L (BGP, BGQ), 2.8L (AFP, BDF); **Golf:** 1.8L (AWD, AWW), 2.0L (ABA, AEG, AVH, AZG, BEV); **GTI:** 1.8L (AWD, AWP, AWW), 2.0L (ABA, AEG), 2.8L (AFP, BDF); **NB:** 1.8L (APH, AWP, AWV, BNU, BKF), 2.0L (AEG, AVH, AZG, BEV); **NBvert:** (BDG); **Cabrio:** 2.0L (ABA) **Transmissions:** All	**Ignition Coilpack D Primary/Secondary Circuit Malfunction Conditions:** Engine started, battery voltage must be at least 11.5v, all electrical components must be off, parking brake must be engaged (to keep daytime driving lights off), automatic transmission selector must be in park and the ground between the engine and the chassis must be well connected. The ECM did not receive any valid pulses from the ignition module for the Ignition Coilpack D primary circuit. **Note: Ignition coils and power output stages are one component and cannot be replaced individually.** **Possible Causes:** • Engine speed (RPM) sensor has failed • Camshaft Position (CMP) sensor has failed • Power Supply Relay is shorted to an open circuit • There is a malfunction in voltage supply • Ignition coilpack is damaged or it has failed • Cylinder 1 to 4 Fuel Injector(s) have failed • ECM has failed •
DTC: P0355 **Passat:** 2.8L (AHA, ATQ), 4.0L (BDP); **Jetta:** 2.5L (BGP, BGQ), 2.8L (AFP, BDF); **GTI:** 2.8L (AFP, BDF) **Transmissions:** All	**Ignition Coilpack E Primary/Secondary Circuit Malfunction Conditions:** Engine started, battery voltage must be at least 11.5v, all electrical components must be off, parking brake must be engaged (to keep daytime driving lights off), automatic transmission selector must be in park and the ground between the engine and the chassis must be well connected. The ECM did not receive any valid pulses from the ignition module for the Ignition Coilpack E primary circuit. **Note: Ignition coils and power output stages are one component and cannot be replaced individually.** **Possible Causes:** • Engine speed (RPM) sensor has failed • Camshaft Position (CMP) sensor has failed • Power Supply Relay is shorted to an open circuit • There is a malfunction in voltage supply • Ignition coilpack is damaged or it has failed • Cylinder 1 to 4 Fuel Injector(s) have failed • ECM has failed •
DTC: P0356 **Passat:** 2.8L (AHA, ATQ), 4.0L (BDP); **Jetta:** 2.8L (AFP, BDF); **GTI:** 2.8L (AFP, BDF) **Transmissions:** All	**Ignition Coilpack F Primary/Secondary Circuit Malfunction Conditions:** Engine started, battery voltage must be at least 11.5v, all electrical components must be off, parking brake must be engaged (to keep daytime driving lights off), automatic transmission selector must be in park and the ground between the engine and the chassis must be well connected. The ECM did not receive any valid pulses from the ignition module for the Ignition Coilpack F primary circuit. **Note: Ignition coils and power output stages are one component and cannot be replaced individually.** **Possible Causes:** • Engine speed (RPM) sensor has failed • Camshaft Position (CMP) sensor has failed • Power Supply Relay is shorted to an open circuit • There is a malfunction in voltage supply • Ignition coilpack is damaged or it has failed • Cylinder 1 to 4 Fuel Injector(s) have failed • ECM has failed •

DTC	Trouble Code Title, Conditions & Possible Causes
DTC: P0357 **Passat:**4.0L (BDP) **Transmissions:** All	**Ignition Coilpack G Primary/Secondary Circuit Malfunction Conditions:** Engine started, battery voltage must be at least 11.5v, all electrical components must be off, parking brake must be engaged (to keep daytime driving lights off), automatic transmission selector must be in park and the ground between the engine and the chassis must be well connected. The ECM did not receive any valid pulses from the ignition module for the Ignition Coilpack G primary circuit. **Note: Ignition coils and power output stages are one component and cannot be replaced individually.** **Possible Causes:** • Engine speed (RPM) sensor has failed • Camshaft Position (CMP) sensor has failed • Power Supply Relay is shorted to an open circuit • There is a malfunction in voltage supply • Ignition coilpack is damaged or it has failed • Cylinder 1 to 4 Fuel Injector(s) have failed • ECM has failed •
DTC: P0366 **Passat:** 4.0L (BDP) **Transmissions:** All	**Camshaft Position Sensor "B" Circuit (Bank 1) Range/Performance Conditions:** Engine started, battery voltage must be at least 11.5v, all electrical components must be off, parking brake must be engaged (to keep daytime driving lights off), automatic transmission selector must be in park and the ground between the engine and the chassis must be well connected. The ECM detected the CMP sensor signal exceeded the bounds of the specified maximum limit. **Possible Causes:** • CMP sensor circuit is open or shorted to ground • CMP sensor circuit is shorted to power • CMP sensor ground (return) circuit is open • CMP sensor installation incorrect (Hall-effect type) • CMP sensor is damaged or CMP sensor shielding damaged • ECM has failed
DTC: P0367 **Passat:** 4.0L (BDP) **Transmissions:** All	**Camshaft Position Sensor "B" Circuit (Bank 1) Low Input Conditions:** Engine started, battery voltage must be at least 11.5v, all electrical components must be off, parking brake must be engaged (to keep daytime driving lights off), automatic transmission selector must be in park and the ground between the engine and the chassis must be well connected. The ECM detected the CMP sensor signal exceeded the bounds of the specified maximum limit. **Possible Causes:** • CMP sensor circuit is open or shorted to ground • CMP sensor circuit is shorted to power • CMP sensor ground (return) circuit is open • CMP sensor installation incorrect (Hall-effect type) • CMP sensor is damaged or CMP sensor shielding damaged • ECM has failed
DTC: P0368 **Passat:** 4.0L (BDP) **Transmissions:** All	**Camshaft Position Sensor "B" Circuit (Bank 1) High Input Conditions:** Engine started, battery voltage must be at least 11.5v, all electrical components must be off, parking brake must be engaged (to keep daytime driving lights off), automatic transmission selector must be in park and the ground between the engine and the chassis must be well connected. The ECM detected the CMP sensor signal did not reach the specified minimum limit. **Possible Causes:** • CMP sensor circuit is open or shorted to ground • CMP sensor circuit is shorted to power • CMP sensor ground (return) circuit is open • CMP sensor installation incorrect (Hall-effect type) • CMP sensor is damaged or CMP sensor shielding damaged • ECM has failed
DTC: P0391 **Passat:** 4.0L (BDP) **Transmissions:** All	**Camshaft Position Sensor "B" Circuit (Bank 2) Range/Performance Conditions:** Engine started, battery voltage must be at least 11.5v, all electrical components must be off, parking brake must be engaged (to keep daytime driving lights off), automatic transmission selector must be in park and the ground between the engine and the chassis must be well connected. The ECM detected the CMP sensor signal exceeded the bounds of the specified maximum limit. **Possible Causes:** • CMP sensor circuit is open or shorted to ground • CMP sensor circuit is shorted to power • CMP sensor ground (return) circuit is open • CMP sensor installation incorrect (Hall-effect type) • CMP sensor is damaged or CMP sensor shielding damaged • ECM has failed

DTC	Trouble Code Title, Conditions & Possible Causes
DTC: P0392 **Passat:** 4.0L (BDP) **Transmissions:** All	**Camshaft Position Sensor "B" Circuit (Bank 2) Low Input Conditions:** Engine started, battery voltage must be at least 11.5v, all electrical components must be off, parking brake must be engaged (to keep daytime driving lights off), automatic transmission selector must be in park and the ground between the engine and the chassis must be well connected. The ECM detected the CMP sensor signal exceeded the bounds of the specified maximum limit. **Possible Causes:** • CMP sensor circuit is open or shorted to ground • CMP sensor circuit is shorted to power • CMP sensor ground (return) circuit is open • CMP sensor installation incorrect (Hall-effect type) • CMP sensor is damaged or CMP sensor shielding damaged • ECM has failed
DTC: P0393 **Passat:** 4.0L (BDP) **Transmissions:** All	**Camshaft Position Sensor "B" Circuit (Bank 2) High Input Conditions:** Engine started, battery voltage must be at least 11.5v, all electrical components must be off, parking brake must be engaged (to keep daytime driving lights off), automatic transmission selector must be in park and the ground between the engine and the chassis must be well connected. The ECM detected the CMP sensor signal did not reach the specified minimum limit. **Possible Causes:** • CMP sensor circuit is open or shorted to ground • CMP sensor circuit is shorted to power • CMP sensor ground (return) circuit is open • CMP sensor installation incorrect (Hall-effect type) • CMP sensor is damaged or CMP sensor shielding damaged • ECM has failed
DTC: P0411 **2T, MIL: Yes** **Passat:** 1.8 (AEB, ATW, AUG, AWM), 2.0L (ABA, BPY), 2.8L (AAA, AHA, ATQ); **Jetta:** 1.8L (AWD, AWP, AWW), 2.0 (ABA, AEG, AVH, AZG, BBW, BEV), 2.5L (BGP, BGQ), 2.8L (AAA, AFP, BDF); **Golf:** 1.8L (AWD, AWW), 2.0L (ABA, AEG, AVH, AZG, BEV); **GTI:** 1.8L (AWD, AWP, AWW), 2.0L (ABA, AEG), 2.8L (AAA, AFP, BDF); **NB:** 1.8L (APH, AWP, AWV, BNU, BKF), 2.0L (AEG, BEV); **NBvert:** (BDG); **Cabrio:** 2.0L (ABA) **Transmissions:** All	**Secondary Air Injection System Upstream Flow Detected Conditions:** Engine started, battery voltage must be at least 11.5v, all electrical components must be off, parking brake must be engaged (to keep daytime driving lights off), automatic transmission selector must be in park and the ground between the engine and the chassis must be well connected. The ECM detected the Secondary AIR pump airflow was not diverted correctly when requested during the self-test. The pump is functioning but the quantity of air is recognized as insufficient by HO2S. **Note: The solenoid valve is closed when no voltage is present.** **Possible Causes:** • Air pump output is blocked or restricted • AIR bypass solenoid is leaking or it is restricted • AIR bypass solenoid is stuck open or stuck closed • Check valve (one or more) is damaged or leaking • Electric air injection pump hose(s) leaking • Electric air injection pump is damaged or faulty • ECM has failed
DTC: P0412 **Passat:** 1.8 (AEB, ATW, AUG, AWM), 2.8L (AHA), 4.0L (BDP); **Jetta:** 1.8L (AWD, AWP, AWW), 2.0 (ABA, AEG, AVH, AZG, BEV), 2.8L (AFP, BDF); **Golf:** 1.8L (AWD, AWW), 2.0L (ABA, AEG); **GTI:** 1.8L (AWD, AWP, AWW), 2.0L (ABA, AEG), 2.8L (AFP, BDF); **NB:** 1.8L (APH, AWP, AWV, BNU, BKF), 2.0L (AEG, AVH); **NBvert:** (BDG); **Cabrio:** 2.0L (ABA) **Transmissions:** All	**Secondary Air Injection Solenoid Circuit Malfunction Conditions:** Engine started, battery voltage must be at least 11.5v, all electrical components must be off, parking brake must be engaged (to keep daytime driving lights off), automatic transmission selector must be in park and the ground between the engine and the chassis must be well connected. The ECM detected an unexpected low or high voltage condition on the AIR solenoid control circuit during testing. **Possible Causes:** • AIR solenoid power circuit (B+) is open (check dedicated fuse) • AIR bypass solenoid control circuit is open or shorted to ground • AIR diverter solenoid control circuit open or shorted to ground • AIR pump control circuit is open or shorted to ground • Check valve (one or more) is damaged or leaking • Solid State relay is damaged or it has failed • ECM has failed

DTC	Trouble Code Title, Conditions & Possible Causes
DTC: P0413 **Passat:** 1.8L (ATW, AUG), 2.8L (AHA), 4.0L (BDP); **Jetta:** 1.8L (AWD, AWP, AWW), 2.0 (ABA, AEG, AVH, AZG, BEV), 2.5L (BGP, BGQ), 2.8L (AFP, BDF); **Golf:** 1.8L (AWD, AWW), 2.0L (AEG); **GTI:** 1.8L (AWD, AWP, AWW), 2.0L (ABA, AEG), 2.8L (AFP, BDF); **NB:** 1.8L (APH, AWP, AWV, BNU, BKF), 2.0L (AEG); **NBvert:** (BDG); **Cabrio:** 2.0L (ABA) **Transmissions:** All	**Secondary Air Injection Solenoid Circuit Open Conditions:** Engine started, battery voltage must be at least 11.5v, all electrical components must be off, parking brake must be engaged (to keep daytime driving lights off), automatic transmission selector must be in park and the ground between the engine and the chassis must be well connected. The ECM detected an unexpected low or high voltage condition on the AIR solenoid control circuit during testing. **Possible Causes:** • AIR solenoid power circuit (B+) is open (check dedicated fuse) • AIR bypass solenoid control circuit is open or shorted to ground • AIR diverter solenoid control circuit open or shorted to ground • AIR pump control circuit is open or shorted to ground • Check valve (one or more) is damaged or leaking • Solid State relay is damaged or it has failed • ECM has failed
DTC: P0414 **Passat:** 1.8L (ATW, AUG), 2.8L (AHA), 4.0L (BDP); **Jetta:** 1.8L (AWD, AWP, AWW), 2.0 (ABA, AEG, AVH, AZG, BEV), 2.5L (BGP, BGQ), 2.8L (AFP, BDF); **Golf:** 1.8L (AWD, AWW), 2.0L (ABA, AEG); **GTI:** 1.8L (AWD, AWP, AWW), 2.0L (ABA, AEG), 2.8L (AFP, BDF); **NB:** 1.8L (APH, AWP, AWV, BNU, BKF), 2.0L (AEG, AVH); **NBvert:** (BDG); **Cabrio:** 2.0L (ABA) **Transmissions:** All	**Secondary Air Injection Solenoid Circuit Short Conditions:** Engine started, battery voltage must be at least 11.5v, all electrical components must be off, parking brake must be engaged (to keep daytime driving lights off), automatic transmission selector must be in park and the ground between the engine and the chassis must be well connected. The ECM detected an unexpected low or high voltage condition on the AIR solenoid control circuit during testing. **Possible Causes:** • AIR solenoid power circuit (B+) is open (check dedicated fuse) • AIR bypass solenoid control circuit is open or shorted to ground • AIR diverter solenoid control circuit open or shorted to ground • AIR pump control circuit is open or shorted to ground • Check valve (one or more) is damaged or leaking • Solid State relay is damaged or it has failed • ECM has failed
DTC: P0418 **2T, MIL: Yes** **Passat:** 1.8 (AEB, ATW, AUG, AWM), 2.8L (AHA), 4.0L (BDP); **Jetta:** 1.8L (AWD, AWP, AWW), 2.0 (ABA, AEG, AVH, AZG, BBW, BEV), 2.5L (BGP, BGQ), 2.8L (AFP, BDF); **Golf:** 1.8L (AWD, AWW), 2.0L (ABA, AEG, AVH, AZG, BEV); **GTI:** 1.8L (AWD, AWP, AWW), 2.0L (ABA, AEG), 2.8L (AFP, BDF); **NB:** 1.8L (APH, AWP, AWV, BNU, BKF), 2.0L (AEG, AZG, BEV); **NBvert:** (BDG); **Cabrio:** 2.0L (ABA) **Transmissions:** All	**Secondary Air Injection Relay (A) Circuit Malfunction Conditions:** Engine started, battery voltage must be at least 11.5v, all electrical components must be off, parking brake must be engaged (to keep daytime driving lights off), automatic transmission selector must be in park and the ground between the engine and the chassis must be well connected. The ECM detected an unexpected low or high voltage condition on the AIR solenoid control circuit during testing. **Possible Causes:** • AIR solenoid power circuit (B+) is open (check dedicated fuse) • AIR bypass solenoid control circuit is open or shorted to ground • AIR diverter solenoid control circuit open or shorted to ground • AIR pump control circuit is open or shorted to ground • Check valve (one or more) is damaged or leaking • Solid State relay is damaged or it has failed • ECM has failed

DTC	Trouble Code Title, Conditions & Possible Causes
DTC: P0420 **2T, MIL: Yes** **Passat:** 1.8 (AEB, ATW, AUG, AWM), 2.8L (AHA), 4.0L (BDP); **Jetta:** 1.8L (AWD, AWP, AWW), 2.0 (ABA, AEG, AVH, AZG, BBW, BEV, BPY), 2.5L (BGP, BGQ), 2.8L (AFP, BDF); **Golf:** 1.8L (AWD, AWW), 2.0L (ABA, AEG, AVH, AZG, BEV); **GTI:** 1.8L (AWD, AWP, AWW), 2.0L (ABA, AEG), 2.8L (AFP, BDF); **NB:** 1.8L (APH, AWP, AWV, BNU, BKF), 2.0L (AEG, AVH, AZG, BEV); **NBvert:** (BDG); **Cabrio:** 2.0L (ABA) **Transmissions:** All	**Catalyst System Efficiency (Bank 1) Below Threshold Conditions:** Engine started, battery voltage must be at least 11.5v, all electrical components must be off, parking brake must be engaged (to keep daytime driving lights off), automatic transmission selector must be in park, the exhaust system must be properly sealed between the catalytic converter and the cylinder head, coolant temperature must be at least 80 degrees Celsius and oxygen sensor heaters for oxygen sensors before the catalytic converter must be functioning properly and the ground between the engine and the chassis must be well connected. The ECM detected the switch rate of the rear HO2S-12 was close to the switch rate of front HO2S (it should be much slower). **Possible Causes:** • Air leaks at the exhaust manifold or in the exhaust pipes • Catalytic converter is damaged, contaminated or it has failed • ECT/CHT sensor has lost its calibration (the signal is incorrect) • Engine cylinders misfiring, or the ignition timing is over retarded • Engine oil is contaminated • Front HO2S or rear HO2S is contaminated with fuel or moisture • Front HO2S and/or the rear HO2S is loose in the mounting hole • Front HO2S much older than the rear HO2S (HO2S-11 is lazy) • Fuel system pressure is too high (check the pressure regulator) • Rear HO2S wires improperly connected or the HO2S has failed
DTC: P0421 **Passat:** 1.8 (AEB, AWM), 2.8L (AHA), 4.0L (BDP) **Transmissions:** All	**Warm Up Catalyst System Efficiency (Bank 1) Below Threshold Conditions:** Engine started, battery voltage must be at least 11.5v, all electrical components must be off, parking brake must be engaged (to keep daytime driving lights off), automatic transmission selector must be in park, the exhaust system must be properly sealed between the catalytic converter and the cylinder head, coolant temperature must be at least 80 degrees Celsius and oxygen sensor heaters for oxygen sensors before the catalytic converter must be functioning properly and the ground between the engine and the chassis must be well connected. The ECM detected the switch rate of the rear HO2S-12 was close to the switch rate of front HO2S (it should be much slower). **Possible Causes:** • Air leaks at the exhaust manifold or in the exhaust pipes • Catalytic converter is damaged, contaminated or it has failed • ECT/CHT sensor has lost its calibration (the signal is incorrect) • Engine cylinders misfiring, or the ignition timing is over retarded • Engine oil is contaminated • Front HO2S or rear HO2S is contaminated with fuel or moisture • Front HO2S and/or the rear HO2S is loose in the mounting hole • Front HO2S much older than the rear HO2S (HO2S-11 is lazy) • Fuel system pressure is too high (check the pressure regulator) • Rear HO2S wires improperly connected or the HO2S has failed
DTC: P0422 **2T, MIL: Yes** **Passat:** 1.8L (ATW, AUG), 2.0L (ABA, BPY), 2.8L (AAA, AHA, ATQ); **Jetta:** 1.8L (AWD, AWP, AWW), 2.0 (ABA, AEG, AVH, AZG, BBW, BEV), 2.8L (AAA), 2.0L (ABA, AEG, AVH, AZG, BEV); **GTI:** 1.8L (AWD, AWP, AWW), 2.0L (ABA, AEG), 2.8L (AAA); **NB:** 1.8L (APH, AWP, AWV, BNU, BKF), 2.0L (AEG, AVH, AZG, BEV); **NBvert:** (BDG); **Cabrio:** 2.0L (ABA) **Transmissions:** All	**Main Catalyst (Bank 1) Efficiency Below Threshold Conditions:** Engine started, battery voltage must be at least 11.5v, all electrical components must be off, parking brake must be engaged (to keep daytime driving lights off), automatic transmission selector must be in park, the exhaust system must be properly sealed between the catalytic converter and the cylinder head, coolant temperature must be at least 80 degrees Celsius and oxygen sensor heaters for oxygen sensors before the catalytic converter must be functioning properly and the ground between the engine and the chassis must be well connected. The ECM detected the switch rate of the rear HO2S-12 was close to the switch rate of front HO2S (it should be much slower). **Possible Causes:** • Air leaks at the exhaust manifold or in the exhaust pipes • Catalytic converter is damaged, contaminated or it has failed • ECT/CHT sensor has lost its calibration (the signal is incorrect) • Engine cylinders misfiring, or the ignition timing is over retarded • Engine oil is contaminated • Front HO2S or rear HO2S is contaminated with fuel or moisture • Front HO2S and/or the rear HO2S is loose in the mounting hole • Front HO2S much older than the rear HO2S • Fuel system pressure is too high (check the pressure regulator) • Rear HO2S wires improperly connected or the HO2S has failed

DTC	Trouble Code Title, Conditions & Possible Causes
DTC: P0430 **Passat:** 2.8L (AHA), 4.0L (BDP); **GTI:** 2.8L (AFP, BDF) **Transmissions:** All	**Catalyst System Efficiency (Bank 2) Below Threshold Conditions:** Engine started, battery voltage must be at least 11.5v, all electrical components must be off, parking brake must be engaged (to keep daytime driving lights off), automatic transmission selector must be in park, the exhaust system must be properly sealed between the catalytic converter and the cylinder head, coolant temperature must be at least 80 degrees Celsius and oxygen sensor heaters for oxygen sensors before the catalytic converter must be functioning properly and the ground between the engine and the chassis must be well connected. The ECM detected the switch rate of the rear HO2S-12 was close to the switch rate of front HO2S (it should be much slower). **Possible Causes:** • Air leaks at the exhaust manifold or in the exhaust pipes • Catalytic converter is damaged, contaminated or it has failed • ECT/CHT sensor has lost its calibration (the signal is incorrect) • Engine cylinders misfiring, or the ignition timing is over retarded • Engine oil is contaminated • Front HO2S or rear HO2S is contaminated with fuel or moisture • Front HO2S and/or the rear HO2S is loose in the mounting hole • Front HO2S much older than the rear HO2S (HO2S-11 is lazy) • Fuel system pressure is too high (check the pressure regulator) • Rear HO2S wires improperly connected or the HO2S has failed
DTC: P0431 **Passat:** 2.8L (AHA), 4.0L (BDP) **Transmissions:** All	**Warm Up Catalyst System Efficiency (Bank 2) Below Threshold Conditions:** Engine started, battery voltage must be at least 11.5v, all electrical components must be off, parking brake must be engaged (to keep daytime driving lights off), automatic transmission selector must be in park, the exhaust system must be properly sealed between the catalytic converter and the cylinder head, coolant temperature must be at least 80 degrees Celsius and oxygen sensor heaters for oxygen sensors before the catalytic converter must be functioning properly and the ground between the engine and the chassis must be well connected. The ECM detected the switch rate of the rear HO2S-12 was close to the switch rate of front HO2S (it should be much slower). **Possible Causes:** • Air leaks at the exhaust manifold or in the exhaust pipes • Catalytic converter is damaged, contaminated or it has failed • ECT/CHT sensor has lost its calibration (the signal is incorrect) • Engine cylinders misfiring, or the ignition timing is over retarded • Engine oil is contaminated • Front HO2S or rear HO2S is contaminated with fuel or moisture • Front HO2S and/or the rear HO2S is loose in the mounting hole • Front HO2S much older than the rear HO2S (HO2S-11 is lazy) • Fuel system pressure is too high (check the pressure regulator) • Rear HO2S wires improperly connected or the HO2S has failed
DTC: P0440 **Passat:** 1.8 (AEB, ATW, AUG, AWM), 2.0L (ABA, BPY), 2.8L (AAA, AHA, ATQ), 2.0 (ABA, AEG, AVH, AZG, BBW, BEV), 2.8L (AAA), 2.0L (ABA, AEG, AVH, AZG, BEV); **GTI:** 2.0L (ABA, AEG), 2.8L (AAA, AFP, BDF); **NB:** 1.8L (APH, AWP, AWV, BNU, BKF), 2.0L (AEG, AVH, AZG, BEV); **NBvert:** (BDG); **Cabrio:** 2.0L (ABA) **Transmissions:** All	**EVAP System Malfunction Conditions:** ECT sensor is cold during startup, engine started, battery voltage must be at least 11.5v, all electrical components must be off, parking brake must be engaged (to keep daytime driving lights off), automatic transmission selector must be in park, the exhaust system must be properly sealed between the catalytic converter and the cylinder head, coolant temperature must be at least 80 degrees Celsius and oxygen sensor heaters for oxygen sensors before the catalytic converter must be functioning properly and the ground between the engine and the chassis must be well connected. The ECM detected the switch rate of the rear HO2S-12 was close to the switch rate of front HO2S (it should be much slower). ECM detected a problem in the EVAP system during the EVAP System Monitor test. **Possible Causes:** • EVAP canister purge valve is damaged • EVAP canister has an improper seal • Vapor line between purge solenoid and intake manifold vacuum reservoir is damaged, or vapor line between EVAP canister purge solenoid and charcoal canister is damaged • Vapor line between charcoal canister and check valve, or vapor line between check valve and fuel vapor valves is damaged • ECM has failed

DTC	Trouble Code Title, Conditions & Possible Causes
DTC: P0441 **2T, MIL: Yes** **Passat:** 1.8 (AEB, ATW, AUG, AWM), 2.8L (AAA, AHA, ATQ), 4.0L (BDP); **Jetta:** 1.8L (AWD, AWP, AWW), 2.0 (ABA, AEG, AVH, AZG, BBW, BEV, BPY), 2.5L (BGP, BGQ), 2.8L (AAA, AFP, BDF); **Golf:** 1.8L (AWD, AWW), 2.0L (ABA, AEG, AVH, AZG, BEV); **GTI:** 1.8L (AWD, AWP, AWW), 2.0L (ABA, AEG), 2.8L (AAA, AFP, BDF); **NB:** 1.8L (APH, AWP, AWV, BNU, BKF), 2.0L (AEG, AVH, AZG, BEV); **NBvert:** (BDG); **Cabrio:** 2.0L (ABA) **Transmissions:** All	**EVAP Control System Incorrect Purge Flow Conditions:** ECT sensor is cold during startup, engine started, battery voltage must be at least 11.5v, all electrical components must be off, parking brake must be engaged (to keep daytime driving lights off), automatic transmission selector must be in park, the exhaust system must be properly sealed between the catalytic converter and the cylinder head, coolant temperature must be at least 80 degrees Celsius and oxygen sensor heaters for oxygen sensors before the catalytic converter must be functioning properly and the ground between the engine and the chassis must be well connected. The ECM detected the switch rate of the rear HO2S-12 was close to the switch rate of front HO2S (it should be much slower). ECM detected a problem in the EVAP system during the EVAP System Monitor test. **Possible Causes:** • EVAP canister purge valve is damaged • EVAP canister has an improper seal • Vapor line between purge solenoid and intake manifold vacuum reservoir is damaged, or vapor line between EVAP canister purge solenoid and charcoal canister is damaged • Vapor line between charcoal canister and check valve, or vapor line between check valve and fuel vapor valves is damaged • ECM has failed
DTC: P0442 **2T, MIL: Yes** **Passat:** 1.8 (AEB, ATW, AUG, AWM), 2.8L (AAA, AHA, ATQ), 4.0L (BDP); **Jetta:** 1.8L (AWD, AWP, AWW), 2.0 (ABA, AEG, AVH, AZG, BBW, BEV, BPY), 2.5L (BGP, BGQ), 2.8L (AAA, AFP, BDF); **Golf:** 1.8L (AWD, AWW), 2.0L (ABA, AEG, AVH, AZG, BEV); **GTI:** 1.8L (AWD, AWP, AWW), 2.0L (ABA, AEG), 2.8L (AAA); **NB:** 1.8L (APH, AWP, AWV, BNU, BKF), 2.0L (AEG, AVH, AZG, BEV); **NBvert:** (BDG); **Cabrio:** 2.0L (ABA) **Transmissions:** All	**EVAP Control System Small Leak Detected Conditions:** Engine started, battery voltage must be at least 11.5v, all electrical components must be off, parking brake must be engaged (to keep daytime driving lights off), automatic transmission selector must be in park, the exhaust system must be properly sealed between the catalytic converter and the cylinder head, coolant temperature must be at least 80 degrees Celsius and oxygen sensor heaters for oxygen sensors before the catalytic converter must be functioning properly and the ground between the engine and the chassis must be well connected. The ECM detected a leak in the EVAP system as small as 0.040" during the EVAP Monitor Test. **Possible Causes:** • Aftermarket EVAP parts that do not conform to specifications • CV solenoid remains partially open when commanded to close • EVAP component seals leaking (i.e., leaks in the Purge valve, fuel tank pressure sensor, canister vent solenoid, fuel vapor control valve tube assembly or fuel vapor vent valve). • Fuel filler cap damaged, cross-threaded or loosely installed • Loose fuel vapor hose/tube connections to EVAP components • Small holes or cuts in fuel vapor hoses or EVAP canister tubes
DTC: P0443 **Passat:** 1.8 (AEB, ATW, AUG, AWM); **Jetta:** 2.0L (ABA); Golf 2.0L (ABA); **GTI:** 2.0L (ABA, AEG), 2.0L (AEG, AVH, AZG, BEV); **NBvert:** (BDG); **Cabrio:** 2.0L (ABA) **Transmissions:** All	**EVAP Vapor Management Valve Circuit Malfunction Conditions:** Engine started, battery voltage must be at least 11.5v, all electrical components must be off, parking brake must be engaged (to keep daytime driving lights off), automatic transmission selector must be in park, the exhaust system must be properly sealed between the catalytic converter and the cylinder head, coolant temperature must be at least 80 degrees Celsius and oxygen sensor heaters for oxygen sensors before the catalytic converter must be functioning properly and the ground between the engine and the chassis must be well connected. The ECM detected an unexpected high or low voltage condition on the Vapor Management Valve (VMV) circuit when the device was cycled On/Off during testing. **Possible Causes:** • EVAP power supply circuit is open • EVAP solenoid control circuit is open or shorted to ground • EVAP solenoid control circuit is shorted to power (B+) • EVAP solenoid valve is damaged or it has failed • ECM has failed

DTC	Trouble Code Title, Conditions & Possible Causes
DTC: P0444 **2T, MIL: Yes** **Passat:** 1.8 (AEB ATW,, AUG, AWM), 2.8L (AHA, ATQ), 4.0L (BDP); **Jetta:** 1.8L (AWD, AWP, AWW), 2.0 (ABA, AEG, AVH, AZG, BBW, BEV, BPY), 2.5L (BGP, BGQ), 2.8L (AFP, BDF); **Golf:** 1.8L (AWD, AWW), 2.0L (ABA, AEG, AVH, AZG, BEV); **GTI:** 1.8L (AWD, AWP, AWW), 2.0L (ABA, AEG), 2.8L (AFP, BDF); **NB:** 1.8L (APH, AWP, AWV, BNU, BKF), 2.0L (AEG, AVH, AZG, BEV); **NBvert:** (BDG); **Cabrio:** 2.0L (ABA) **Transmissions:** All	**Evaporative Emission System Purge Control Valve Circuit Open Conditions:** Engine started, battery voltage must be at least 11.5v, all electrical components must be off, parking brake must be engaged (to keep daytime driving lights off), automatic transmission selector must be in park, the exhaust system must be properly sealed between the catalytic converter and the cylinder head, coolant temperature must be at least 80 degrees Celsius and oxygen sensor heaters for oxygen sensors before the catalytic converter must be functioning properly and the ground between the engine and the chassis must be well connected. The ECM detected an unexpected voltage condition on the EVAP circuit when the device was cycled On/Off during testing. **Possible Causes:** • EVAP power supply circuit is open • EVAP solenoid control circuit is open or shorted to ground • EVAP solenoid control circuit is shorted to power (B+) • EVAP solenoid valve is damaged or it has failed • EVAP canister has a leak or a poor seal • ECM has failed
DTC: P0445 **2T, MIL: Yes** **Passat:** 1.8 (AEB, ATW, AUG, AWM), 2.8L (AHA, ATQ), 4.0L (BDP); **Jetta:** 1.8L (AWD, AWP, AWW), 2.0 (ABA, AEG, AVH, AZG, BBW, BEV, BPY), 2.5L (BGP, BGQ), 2.8L (AAA, AFP, BDF); **Golf:** 1.8L (AWD, AWW), 2.0L (ABA, AEG, AVH, AZG, BEV); **GTI:** 1.8L (AWD, AWP, AWW), 2.0L (ABA, AEG), 2.8L (AFP, BDF); **NB:** 1.8L (APH, AWP, AWV, BNU, BKF), 2.0L (AEG, AVH, AZG, BEV); **NBvert:** (BDG); **Cabrio:** 2.0L (ABA) **Transmissions:** All	**Evaporative Emission System Purge Control Valve Circuit Shorted Conditions:** Engine started, battery voltage must be at least 11.5v, all electrical components must be off, parking brake must be engaged (to keep daytime driving lights off), automatic transmission selector must be in park, the exhaust system must be properly sealed between the catalytic converter and the cylinder head, coolant temperature must be at least 80 degrees Celsius and oxygen sensor heaters for oxygen sensors before the catalytic converter must be functioning properly and the ground between the engine and the chassis must be well connected. The ECM detected an unexpected voltage condition on the EVAP circuit when the device was cycled On/Off during testing. **Possible Causes:** • EVAP power supply circuit is open • EVAP solenoid control circuit is open or shorted to ground • EVAP solenoid control circuit is shorted to power (B+) • EVAP solenoid valve is damaged or it has failed • EVAP canister has a leak or a poor seal • ECM has failed
DTC: P0449 **Jetta:** 2.8L (BDF); **Golf:** 2.8L (BDF) **Transmissions:** All	**Evaporative Emission System Vent Valve/Solenoid Circuit Conditions:** Engine started, battery voltage must be at least 11.5v, all electrical components must be off, parking brake must be engaged (to keep daytime driving lights off), automatic transmission selector must be in park, the exhaust system must be properly sealed between the catalytic converter and the cylinder head, coolant temperature must be at least 80 degrees Celsius and oxygen sensor heaters for oxygen sensors before the catalytic converter must be functioning properly and the ground between the engine and the chassis must be well connected. The ECM detected an unexpected voltage condition on the EVAP circuit when the device was cycled On/Off during testing. **Possible Causes:** • EVAP power supply circuit is open • EVAP solenoid control circuit is open or shorted to ground • EVAP solenoid control circuit is shorted to power (B+) • EVAP solenoid valve is damaged or it has failed • EVAP canister has a leak or a poor seal • ECM has failed

DTC	Trouble Code Title, Conditions & Possible Causes
DTC: P0455 **2T, MIL: Yes** **Passat:** 1.8L (ATW, AUG), 2.8L (AAA, AHA, ATQ), 4.0L (BDP); **Jetta:** 2.0L AEG, AVH, AZG, BBW, BEV, BPY), 2.5L (BGP, BGQ), 2.8L (AFP, BDF); **Golf:** 1.8L (AWD, AWW), 2.0L (AEG, AVH, AZG, BEV); **GTI:** 1.8L (AWD, AWP, AWW), 2.0L (AEG), 2.8L (AAA, AFP, BDF); **NB:** 1.8L (APH, AWP, AWV, BNU, BKF), 2.0L (AEG, AVH, AZG, BEV); **NBvert:** (BDG) **Transmissions:** All	**EVAP Control System Large Leak Detected Conditions:** Engine started, battery voltage must be at least 11.5v, all electrical components must be off, parking brake must be engaged (to keep daytime driving lights off), automatic transmission selector must be in park, the exhaust system must be properly sealed between the catalytic converter and the cylinder head, coolant temperature must be at least 80 degrees Celsius and oxygen sensor heaters for oxygen sensors before the catalytic converter must be functioning properly and the ground between the engine and the chassis must be well connected. The ECM detected multiple small fuel vapor leaks; or it detected a large leak in the system during the leak test. **Possible Causes:** • Aftermarket EVAP hardware non-conforming to specifications • EVAP canister tube, EVAP canister purge outlet tube or EVAP return tube disconnected or cracked, or canister is damaged • EVAP canister purge valve stuck closed, or canister damaged • Fuel filler cap missing, loose (not tightened) or the wrong part • Loose fuel vapor hose/tube connections to EVAP components • Canister vent (CV) solenoid stuck open • Fuel tank pressure (FTP) sensor has failed mechanically
DTC: P0456 **2T, MIL: Yes** **Passat:** 1.8L (ATW, AUG), 4.0L (BDP); **Jetta:** 1.8L (AWD, AWP, AWW), 2.0L AEG, AVH, AZG, BBW, BEV, BPY), 2.5L (BGP, BGQ), 2.8L (AFP, BDF); **Golf:** 1.8L (AWD, AWW), 2.0L (AEG, AVH, AZG, BEV); **GTI:** 1.8L (AWD, AWP, AWW), 2.0L (AEG), 2.8L (AFP, BDF); **NB:** 1.8L (APH, AWP, AWV, BNU, BKF), 2.0L (AEG, AVH, AZG, BEV); **NBvert:** (BDG) **Transmissions:** All	**EVAP Control System Small Leak Detected Conditions:** Engine started, battery voltage must be at least 11.5v, all electrical components must be off, parking brake must be engaged (to keep daytime driving lights off), automatic transmission selector must be in park, the exhaust system must be properly sealed between the catalytic converter and the cylinder head, coolant temperature must be at least 80 degrees Celsius and oxygen sensor heaters for oxygen sensors before the catalytic converter must be functioning properly and the ground between the engine and the chassis must be well connected. The ECM detected multiple small fuel vapor leaks; or it detected a large leak in the system during the leak test. **Possible Causes:** • Aftermarket EVAP hardware non-conforming to specifications • EVAP canister tube, EVAP canister purge outlet tube or EVAP return tube disconnected or cracked, or canister is damaged • EVAP canister purge valve stuck closed, or canister damaged • Fuel filler cap missing, loose (not tightened) or the wrong part • Loose fuel vapor hose/tube connections to EVAP components • Canister vent (CV) solenoid stuck open • Fuel tank pressure (FTP) sensor has failed mechanically
DTC: P0458 **Passat:** 1.8 (ATW, AUG); **Jetta:** 2.0 (ABA, BPY), 2.5L (BGP, BGQ); Golf 2.0L (ABA); **GTI:** 2.0L (ABA); **NB:** 1.8L (APH, AWP, AWV, BNU, BKF); **Cabrio:** 2.0L (ABA) **Transmissions:** All	**Evaporative Emission System Purge Control Valve Circuit Low Conditions:** Engine started, battery voltage must be at least 11.5v, all electrical components must be off, parking brake must be engaged (to keep daytime driving lights off), automatic transmission selector must be in park, the exhaust system must be properly sealed between the catalytic converter and the cylinder head, coolant temperature must be at least 80 degrees Celsius and oxygen sensor heaters for oxygen sensors before the catalytic converter must be functioning properly and the ground between the engine and the chassis must be well connected. The ECM detected an unexpected voltage condition on the EVAP circuit when the device was cycled On/Off during testing. **Possible Causes:** • EVAP power supply circuit is open • EVAP solenoid control circuit is open or shorted to ground • EVAP solenoid control circuit is shorted to power (B+) • EVAP solenoid valve is damaged or it has failed • EVAP canister has a leak or a poor seal • ECM has failed
DTC: P0459 **Passat:** 1.8 (ATW, AUG); **Jetta:** 2.0 (ABA, BPY), 2.5L (BGP, BGQ); Golf 2.0L (ABA); **GTI:** 2.0L (ABA); **NB:** 1.8L (APH, AWP, AWV, BNU, BKF); **Cabrio:** 2.0L (ABA) **Transmissions:** All	**Evaporative Emission System Purge Control Valve Circuit High Conditions:** Engine started, battery voltage must be at least 11.5v, all electrical components must be off, parking brake must be engaged (to keep daytime driving lights off), automatic transmission selector must be in park, the exhaust system must be properly sealed between the catalytic converter and the cylinder head, coolant temperature must be at least 80 degrees Celsius and oxygen sensor heaters for oxygen sensors before the catalytic converter must be functioning properly and the ground between the engine and the chassis must be well connected. The ECM detected an unexpected voltage condition on the EVAP circuit when the device was cycled On/Off during testing. **Possible Causes:** • EVAP power supply circuit is open • EVAP solenoid control circuit is open or shorted to ground • EVAP solenoid control circuit is shorted to power (B+) • EVAP solenoid valve is damaged or it has failed • EVAP canister has a leak or a poor seal • ECM has failed

DTC	Trouble Code Title, Conditions & Possible Causes
DTC: P0480 **Jetta:** 2.0 (BPY), 2.5L (BGP, BGQ) **Transmissions:** All	**Fan 1 Control Circuit Conditions:** Engine running, battery voltage at least 11.5v, all electrical components off, and the ECM detected an problem with the fan control circuit **Possible Causes:** • Check connection of coolant fan control module according to wiring diagram • Short to B+ or ground in the circuit • Circuit open
DTC: P0491 **Passat:** 2.8L (AHA), 4.0L (BDP); **Jetta:** 2.8L (AFP, BDF); **GTI:** 2.8L (AFP, BDF) **Transmissions:** All	**Secondary Air Injection System Insufficient Flow (Bank 1) Conditions:** Engine started, battery voltage must be at least 11.5v, all electrical components must be off, parking brake must be engaged (to keep daytime driving lights off), automatic transmission selector must be in park and the ground between the engine and the chassis must be well connected. The ECM detected the Secondary AIR pump airflow was not diverted correctly when requested during the self-test. The pump is functioning but the quantity of air is recognized as insufficient by HO2S **Possible Causes:** • Air pump output is blocked or restricted • AIR bypass solenoid is leaking or it is restricted • AIR bypass solenoid is stuck open or stuck closed • Check valve (one or more) is damaged or leaking • Electric air injection pump hose(s) leaking • Electric air injection pump is damaged or faulty • ECM has failed
DTC: P0492 **Passat:** 2.8L (AHA), 4.0L (BDP); **Jetta:** 2.8L (AFP, BDF); **GTI:** 2.8L (AFP, BDF) **Transmissions:** All	**Secondary Air Injection System Insufficient Flow (Bank 2) Conditions:** Engine started, battery voltage must be at least 11.5v, all electrical components must be off, parking brake must be engaged (to keep daytime driving lights off), automatic transmission selector must be in park and the ground between the engine and the chassis must be well connected. The ECM detected the Secondary AIR pump airflow was not diverted correctly when requested during the self-test. The pump is functioning but the quantity of air is recognized as insufficient by HO2S **Possible Causes:** • Air pump output is blocked or restricted • AIR bypass solenoid is leaking or it is restricted • AIR bypass solenoid is stuck open or stuck closed • Check valve (one or more) is damaged or leaking • Electric air injection pump hose(s) leaking • Electric air injection pump is damaged or faulty • ECM has failed
DTC: P0498 **Jetta:** 2.8L (BDF); **Golf:** 2.8L (BDF) **Transmissions:** All	**Evaporative Emission System Vent Valve Control Circuit Low Conditions:** Engine started, battery voltage must be at least 11.5v, all electrical components must be off, parking brake must be engaged (to keep daytime driving lights off), automatic transmission selector must be in park, the exhaust system must be properly sealed between the catalytic converter and the cylinder head, coolant temperature must be at least 80 degrees Celsius and oxygen sensor heaters for oxygen sensors before the catalytic converter must be functioning properly and the ground between the engine and the chassis must be well connected. The ECM detected an unexpected voltage condition on the EVAP circuit when the device was cycled On/Off during testing. **Possible Causes:** • EVAP power supply circuit is open • EVAP solenoid control circuit is open or shorted to ground • EVAP solenoid control circuit is shorted to power (B+) • EVAP solenoid valve is damaged or it has failed • EVAP canister has a leak or a poor seal • ECM has failed

DTC	Trouble Code Title, Conditions & Possible Causes
DTC: P0501 **2T, MIL: Yes** **Passat:** 1.8 (AEB, ATW, AUG, AWM), 2.0L (ABA, BPY), 2.8L (AAA, AHA, ATQ), 4.0L (BDP); **Jetta:** 1.8L (AWD, AWP, AWW), 2.0 (ABA, AEG, AVH, AZG, BBW, BEV, BPY), 2.5L (BGP, BGQ), 2.8L (AAA, AFP, BDF); **Golf:** 1.8L (AWD, AWW), 2.0L (ABA, AEG, AVH, AZG, BEV); **GTI:** 1.8L (AWD, AWP, AWW), 2.0L (ABA, AEG), 2.8L (AAA, AFP, BDF); **NB:** 1.8L (APH, AWP, AWV, BNU, BKF), 2.0L (AEG, AVH, AZG, BEV); **NBvert:** (BDG); **Cabrio:** 2.0L (ABA) **Transmissions:** All	**Vehicle Speed Sensor or PSOM Range/Performance Conditions:** Engine started; engine speed above the TCC stall speed, and the ECM detected a loss of the VSS signal over a period of time or the signal is not usable. **Note: The ECM receives vehicle speed data from the VSS, TCSS, ABS module, CTM or GEM controller, depending up the application.** **Possible Causes:** • VSS signal circuit is open or shorted to ground • VSS harness circuit is shorted to ground • VSS harness circuit is shorted to power • VSS circuit open between the ECM and related control module • VSS or wheel speed sensors circuits are damaged • Modules connected to VSC/VSS harness circuits are damaged • Mechanical drive mechanism for the VSS is damaged
DTC: P0506 **2T, MIL: Yes** **Passat:** 1.8 (AEB, ATW, AUG, AWM), 2.0L (ABA, BPY), 2.8L (AAA, AHA, ATQ), 4.0L (BDP); **Jetta:** 1.8L (AWD, AWP, AWW), 2.0L (AEG, AVH, AZG, BBW, BEV, BPY), 2.5L (BGP, BGQ), 2.8L (AAA, AFP, BDF); **Golf:** 1.8L (AWD, AWW), 2.0L (AEG, AVH, AZG, BEV); **GTI:** 1.8L (AWD, AWP, AWW), 2.0L (AEG), 2.8L (AAA, AFP, BDF); **NB:** 1.8L (APH, AWP, AWV, BNU, BKF), 2.0L (AEG, AVH, AZG, BEV); **NBvert:** (BDG) **Transmissions:** All	**Idle Air Control System RPM Lower Than Expected Conditions:** Engine started, battery voltage must be at least 11.5v, all electrical components must be off, parking brake must be engaged (to keep daytime driving lights off), automatic transmission selector must be in park, the exhaust system must be properly sealed between the catalytic converter and the cylinder head, coolant temperature must be at least 80 degrees Celsius and oxygen sensor heaters for oxygen sensors before the catalytic converter must be functioning properly and the ground between the engine and the chassis must be well connected. The ECM detected it could not control the idle speed correctly, as it is constantly more than 100 rpm less than specification. **Possible Causes:** • Air inlet is plugged or the air filter element is severely clogged • IAC circuit is open or shorted • IAC circuit VPWR circuit is open • IAC solenoid is damaged or has failed • ECM has failed • The VSS has failed
DTC: P0507 **2T, MIL: Yes** **Passat:** 1.8 (AEB, ATW, AUG, AWM), 2.0L (ABA, BPY), 2.8L (AAA, AHA, ATQ), 4.0L (BDP); **Jetta:** 1.8L (AWD, AWP, AWW), 2.0 (ABA, BBW, BPY), 2.5L (BGP, BGQ), 2.8L (AFP, BDF); **Golf:** 1.8L (AWD, AWW), 2.0L (ABA, AEG, AVH, AZG, BEV); **GTI:** 1.8L (AWD, AWP, AWW), 2.0L (ABA, AEG), 2.8L (AAA, AFP, BDF); **NB:** 1.8L (APH, AWP, AWV, BNU, BKF), 2.0L (AEG, AVH, AZG, BEV); **NBvert:** (BDG); **Cabrio:** 2.0L (ABA) **Transmissions:** All	**Idle Air Control System RPM Higher Than Expected Conditions:** Engine started, battery voltage must be at least 11.5v, all electrical components must be off, parking brake must be engaged (to keep daytime driving lights off), automatic transmission selector must be in park, the exhaust system must be properly sealed between the catalytic converter and the cylinder head, coolant temperature must be at least 80 degrees Celsius and oxygen sensor heaters for oxygen sensors before the catalytic converter must be functioning properly and the ground between the engine and the chassis must be well connected. The ECM detected it could not control the idle speed correctly, as it is constantly more than 200 rpm more than specification. **Possible Causes:** • Air intake leak located somewhere after the throttle body • IAC control circuit is shorted to chassis ground • IAC solenoid is damaged or has failed • Throttle Valve Control module has failed or is clogged with carbon • ECM has failed • The VSS has failed

DTC	Trouble Code Title, Conditions & Possible Causes
DTC: P0510 **Passat:** 2.0L (ABA, BPY), 2.8L (AAA, AHA, ATQ); **Jetta:** 2.0 (ABA, AEG, AVH, AZG, BBW, BEV), 2.8L (AAA); Golf 2.0L (ABA, AEG, AVH, AZG, BEV); **GTI:** 2.0L (ABA, AEG), 2.8L (AAA); **NB:** 1.8L (APH, AWP, AWV, BNU, BKF), 2.0L (AEG, AVH, AZG, BEV); **NBvert:** (BDG); **Cabrio:** 2.0L (ABA) **Transmissions:** All	**Closed Throttle Position Switch Malfunction Conditions:** The Engine Control Module is not receiving a usable signal from the Closed Throttle Position switch. **Possible Causes:** • Throttle Valve Control module has failed • Throttle Valve Control module is shorted to ground • Throttle Valve Control module is open • ECM has failed
DTC: P0513 **Passat:** 2.0L (BPY); **Jetta:** 2.0L (BPY) **Transmissions:** All	**Incorrect Immobilizer Key Conditions:** The Immobilizer is a theft-deterrent device that adds a layer of security that is standard on every Volkswagen and Audi model. This is achieved with an electronic device embedded in the ignition key, an Immobilizer Control Module and an electronic link to the Engine Control Module. **Possible Causes:** • The key is incorrect or damaged • ECM has failed • Short within the circuit
DTC: P0544 **Jetta:** 2.0L (BBW) **Transmissions:** All	**Exhaust Gas Temperature Sensor Circuit Bank 1 Sensor 1 Conditions:** Engine started, battery voltage must be at least 11.5v, all electrical components must be off, parking brake must be engaged (to keep daytime driving lights off), automatic transmission selector must be in park, and the ground between the engine and the chassis must be well connected. The ECM has detected a voltage value that is implausible or erratic. **Possible Causes:** • Exhust gas temperature sensor is faulty • EGT control circuit is shorted to chassis ground • ECM has failed
DTC: P0545 **Jetta:** 2.0L (BBW) **Transmissions:** All	**Exhaust Gas Temperature Sensor Circuit Low Bank 1 Sensor 1 Conditions:** Engine started, battery voltage must be at least 11.5v, all electrical components must be off, parking brake must be engaged (to keep daytime driving lights off), automatic transmission selector must be in park, and the ground between the engine and the chassis must be well connected. The ECM has detected a voltage value below the minimum system requirement. **Possible Causes:** • Exhust gas temperature sensor is faulty • EGT control circuit is shorted to chassis ground • ECM has failed
DTC: P0546 **Jetta:** 2.0L (BBW) **Transmissions:** All	**Exhaust Gas Temperature Sensor Circuit High Bank 1 Sensor 1 Conditions:** Engine started, battery voltage must be at least 11.5v, all electrical components must be off, parking brake must be engaged (to keep daytime driving lights off), automatic transmission selector must be in park, and the ground between the engine and the chassis must be well connected. The ECM has detected a voltage value above the maximum system requirement. **Possible Causes:** • Exhust gas temperature sensor is faulty • EGT control circuit is shorted to chassis ground • ECM has failed
DTC: P0560 **Passat:** 1.8 (AEB, AWM), 2.0L (BPY), 2.8L (AHA); **Jetta:** 1.8L (AWD, AWP, AWW); 2.0L (BPY), 2.8L (AFP, BDF); **Golf:** 1.8L (AWD, AWW); **GTI:** 1.8L (AWD, AWP, AWW), 2.8L (AFP, BDF); **NB:** 1.8L (APH, AWP, AWV, BNU, BKF) **Transmissions:** All	**System Voltage Malfunction Conditions:** Engine started, battery voltage must be at least 11.5v, all electrical components must be off, parking brake must be engaged (to keep daytime driving lights off), automatic transmission selector must be in park, and the ground between the engine and the chassis must be well connected. The ECM has detected a voltage value that is implausible or erratic. **Possible Causes:** • Alternator damaged or faulty • Battery voltage low or insufficient • Fuses blown or circuits open • Battery connection to terminal not clean • Voltage regulator has failed

DTC	Trouble Code Title, Conditions & Possible Causes
DTC: P0562 **Passat:** 1.8 (AEB, ATW, AUG, AWM), 2.0L (BPY), 2.8L (AHA); **Jetta:** 1.8L (AWD, AWP, AWW), 2.0 (ABA, BPY), 2.8L (AFP, BDF); **Golf:** 1.8L (AWD, AWW), 2.0L (ABA); **GTI:** 1.8L (AWD, AWP, AWW), 2.0L (ABA), 2.8L (AFP, BDF); **NB:** 1.8L (APH, AWP, AWV, BNU, BKF); **Cabrio:** 2.0L (ABA) **Transmissions:** All	**System Voltage Low Conditions:** Engine started, battery voltage must be at least 11.5v, all electrical components must be off, parking brake must be engaged (to keep daytime driving lights off), automatic transmission selector must be in park, and the ground between the engine and the chassis must be well connected. The ECM has detected a voltage value that is below the specified minimum limit for the system to function properly. **Possible Causes:** • Alternator damaged or faulty • Battery voltage low or insufficient • Fuses blown or circuits open • Battery connection to terminal not clean • Voltage regulator has failed
DTC: P0563 **Passat:** 1.8 (AEB, ATW, AUG, AWM), 2.0L (BPY), 2.8L (AHA); **Jetta:** 1.8L (AWD, AWP, AWW), 2.0 (ABA, BPY), 2.8L (AFP, BDF); **Golf:** 1.8L (AWD, AWW), 2.0L (ABA); **GTI:** 1.8L (AWD, AWP, AWW), 2.0L (ABA), 2.8L (AFP, BDF); **NB:** 1.8L (APH, AWP, AWV, BNU, BKF); **Cabrio:** 2.0L (ABA) **Transmissions:** All	**System Voltage High Conditions:** Engine started, battery voltage must be at least 11.5v, all electrical components must be off, parking brake must be engaged (to keep daytime driving lights off), automatic transmission selector must be in park, and the ground between the engine and the chassis must be well connected. The ECM has detected a voltage value that has exceeded the specified maximum limit for the system to function properly. **Possible Causes:** • Alternator damaged or faulty • Battery voltage low or insufficient • Fuses blown or circuits open • Battery connection to terminal not clean • Voltage regulator has failed
DTC: P0568 **Passat:** 2.0L (BPY); **Jetta:** 1.8L (AWD, AWP, AWW). 2.0L (BPY), 2.5L (BGP, BGQ); **Golf:** 1.8L (AWD, AWW); **GTI:** 1.8L (AWD, AWP, AWW); **NB:** 1.8L (APH, AWP, AWV, BNU, BKF) **Transmissions:** All	**Cruise Control Set Signal Incorrect Signal Conditions:** Engine started, battery voltage must be at least 11.5v, all electrical components must be off, parking brake must be engaged (to keep daytime driving lights off), automatic transmission selector must be in park, and the ground between the engine and the chassis must be well connected. The ECM has detected a voltage value that has exceeded the specified maximum limit for the system to function properly. **Possible Causes:** • Cruise control system is damaged • Control circuit is shorted to chassis ground
DTC: P0571 **Passat:** 2.0L (BPY); **Jetta:** 1.8L (AWD, AWP, AWW), 2.0L (BPY), 2.8L (AFP, BDF); **Golf:** 1.8L (AWD, AWW); **GTI:** 1.8L (AWD, AWP, AWW), 2.8L (AFP, BDF); **NB:** 1.8L (APH, AWP, AWV, BNU, BKF) **Transmissions:** All	**Cruise/Brake Switch (A) Circuit Malfunction Conditions:** Engine started, battery voltage must be at least 11.5v, all electrical components must be off, parking brake must be engaged (to keep daytime driving lights off), automatic transmission selector must be in park, and the ground between the engine and the chassis must be well connected. The ECM has detected a voltage value that is implausible or erratic. **Possible Causes:** • Brake light switch is faulty • Control circuit is shorted to chassis ground
DTC: P0598 **Passat:** 4.0 (BDP) **Transmissions:** All	**Thermostat Heater Control Circuit Low Conditions:** Engine started, battery voltage must be at least 11.5v, all electrical components must be off, parking brake must be engaged (to keep daytime driving lights off), automatic transmission selector must be in park, and the ground between the engine and the chassis must be well connected. The ECM has detected a voltage value across the thermostat heater control circuit that is below the specified minimum limit for the system to function properly **Possible Causes:** • Battery voltage low or insufficient • Fuses blown or circuits open • Battery connection to terminal not clean • Voltage regulator has failed • Thermostat heater control faulty • ECM has failed

DTC	Trouble Code Title, Conditions & Possible Causes
DTC: P0599 **Passat:** 4.0 (BDP) **Transmissions:** All	**Thermostat Heater Control Circuit High Conditions:** Engine started, battery voltage must be at least 11.5v, all electrical components must be off, parking brake must be engaged (to keep daytime driving lights off), automatic transmission selector must be in park, and the ground between the engine and the chassis must be well connected. The ECM has detected a voltage value across the thermostat heater control circuit that is above the specified maximum limit for the system to function properly **Possible Causes:** • Battery voltage low or insufficient • Fuses blown or circuits open • Battery connection to terminal not clean • Voltage regulator has failed • Thermostat heater control faulty • ECM has failed
DTC: P0600 **2T, MIL: Yes** **Passat:** 1.8 (AEB, ATW, AUG, AWM), 2.0L (BPY), 4.0L (BDP); **Jetta:** 1.8L (AWD, AWP, AWW), 2.0L (AVH, AZG, BBW, BEV), 2.8L (AFP, BDF); **Golf:** 1.8L (AWD, AWW), 2.0L (AVH, AZG, BEV); **GTI:** 1.8L (AWD, AWP, AWW), 2.8L (AFP, BDF); **NB:** 1.8L (APH, AWP, AWV, BNU, BKF), 2.0L (AZG, BEV) **Transmissions:** All	**Serial Communication Link (Data BUS) Message Missing Conditions:** The Engine Control Module (ECM) communicates with all databus-capable control modules via a CAN databus. These databus-capable control modules are connected via two data bus wires which are twisted together (CAN_High and CAN_Low), and exchange information (messages). Missing information on the databus is recognized as a malfunction and stored. Trouble-free operation of the CAN-Bus requires that it have a terminal resistance. This central terminal resistor is located in the Engine Control Module (ECM). **Possible Causes:** • ECM has failed • CAN data bus wires have short circuited to each other
DTC: P0601 **2T, MIL: Yes** **Passat:** 1.8 (AEB, ATW, AUG, AWM), 2.0L (BPY), 2.8L (AHA, ATQ), 4.0L (BDP); **Jetta:** 1.8L (AWD, AWP, AWW), 2.0 (ABA, AEG, AVH, AZG, BBW, BEV, BPY), 2.5L (BGP, BGQ), 2.8L (AFP, BDF); **Golf:** 1.8L (AWD, AWW), 2.0L (ABA, AEG, AVH, AZG, BEV); **GTI:** 1.8L (AWD, AWP, AWW), 2.0L (ABA, AEG), 2.8L (AFP, BDF); **NB:** 1.8L (APH, AWP, AWV, BNU, BKF), 2.0L (AEG, AVH, AZG, BEV); **NBvert:** (BDG); **Cabrio:** 2.0L (ABA) **Transmissions:** All	**Internal Control Module Memory Check Sum Error Conditions:** Key on, the ECM has detected a programming error **Possible Causes:** • Battery terminal corrosion, or loose battery connection • Connection to the ECM interrupted, or the circuit has been opened • Reprogramming error has occurred • ECM has failed and needs replacement. Remember to check for Aftermarket Performance Products before replacing a ECM.
DTC: P0602 **Passat:** 1.8L (ATW, AUG), 2.0L (BPY), 4.0L (BDP); **Jetta:** 1.8L (AWD, AWP, AWW), 2.0 (ABA, AEG, AVH, AZG, BBW, BEV), 2.5L (BGP, BGQ), 2.8L (AFP, BDF); **Golf:** 1.8L (AWD, AWW), 2.0L (ABA, AEG, AVH, AZG, BEV); **GTI:** 1.8L (AWD, AWP, AWW), 2.0L (ABA, AEG), 2.8L (AFP, BDF); **NB:** 1.8L (APH, AWP, AWV, BNU, BKF), 2.0L (AEG, AVH, AZG, BEV); **NBvert:** (BDG); **Cabrio:** 2.0L (ABA) **Transmissions:** All	**Control Module Programming Error Conditions:** Key on, and the ECM detected a programming error in the VID block. This fault requires that the VID Block be reprogrammed, or that the EEPROM be re-flashed. **Possible Causes:** • During the VID reprogramming function, the Vehicle ID (VID) data block failed during reprogramming wit the Scan Tool. • Battery terminal corrosion, or loose battery connection • Connection to the ECM interrupted, or the circuit has been opened • Reprogramming error has occurred • ECM has failed and needs replacement. Remember to check for Aftermarket Performance Products before replacing a ECM.

DTC	Trouble Code Title, Conditions & Possible Causes
DTC: P0603 **Passat:** 1.8L (ATW, AUG) **Transmissions:** All	**ECM Keep Alive Memory Test Error Conditions:** Key on, and the ECM detected an internal memory fault. This code will set if KAPWR to the ECM is interrupted (at the initial key on). **Possible Causes:** • Battery terminal corrosion, or loose battery connection • KAPWR to ECM interrupted, or the circuit has been opened • Reprogramming error has occurred • ECM has failed and needs replacement. Remember to check for Aftermarket Performance Products before replacing a ECM.
DTC: P0604 **2T, MIL: Yes** **Passat:** 1.8 (AEB, ATW, AUG, AWM), 2.0L (BPY), 2.8L (AHA, ATQ), 4.0L (BDP); **Jetta:** 1.8L (AWD, AWP, AWW), 2.0L (AEG, AVH, AZG, BBW, BEV, BPY), 2.5L (BGP, BGQ), 2.8L (AFP, BDF); **Golf:** 1.8L (AWD, AWW), 2.0L (AEG, AVH, AZG, BEV); **GTI:** 1.8L (AWD, AWP, AWW), 2.0l (AEG), 2.8L (AFP, BDF); **NB:** 1.8L (APH, AWP, AWV, BNU, BKF), 2.0L (AEG, AVH, AZG, BEV); **NBvert:** (BDG) **Transmissions:** All	**Internal Control Module Random Access Memory (RAM) Error Conditions:** Key on, and the ECM detected an internal memory fault. This code will set if KAPWR to the ECM is interrupted (at the initial key on). **Possible Causes:** • Battery terminal corrosion, or loose battery connection • Connection to the ECM interrupted, or the circuit has been opened • Reprogramming error has occurred • ECM has failed and needs replacement. Remember to check for Aftermarket Performance Products before replacing a ECM.
DTC: P0605 **2T, MIL: Yes** **Passat:** 1.8 (AEB, ATW, AUG, AWM), 2.0L (ABA, BPY), 2.8L (AAA, AHA, ATQ), 4.0L (BDP); **Jetta:** 1.8L (AWD, AWP, AWW), 2.0 (ABA, AEG, AVH, AZG, BBW, BEV, BPY), 2.5L (BGP, BGQ), 2.8L (AAA, AFP, BDF); **Golf:** 1.8L (AWD, AWW), 2.0L (ABA, AEG, AVH, AZG, BEV); **GTI:** 1.8L (AWD, AWP, AWW), 2.0L (ABA, AEG), 2.8L (AAA, AFP, BDF); **NB:** 1.8L (APH, AWP, AWV, BNU, BKF), 2.0L (AEG, AVH, AZG, BEV); **NBvert:** (BDG); **Cabrio:** 2.0L (ABA) **Transmissions:** All	**ECM Read Only Memory (ROM) Test Error Conditions:** Key on, and the ECM detected a ROM test error (ROM inside ECM is corrupted). The ECM is normally replaced if this code has set. **Possible Causes:** • An attempt was made to change the module calibration, or a module programming error may have occurred • Clear the trouble codes and then check for this trouble code. If it resets, the ECM has failed and needs replacement. • Aftermarket performance products may have been installed. • The Transmission Control Module (TCM) has failed.
DTC: P0606 **2T, MIL: Yes** **Passat:** 1.8 (AEB, ATW, AUG, AWM), 2.0L (BPY), 4.0L (BDP); **Jetta:** 1.8L (AWD, AWP, AWW), 2.0 (ABA, AEG, AVH, AZG, BBW, BEV), 2.5L (BGP, BGQ), 2.8L (AFP, BDF); **Golf:** 1.8L (AWD, AWW), 2.0L (ABA, AEG, AVH, AZG, BEV); **GTI:** 1.8L (AWD, AWP, AWW), 2.0L (ABA, AEG), 2.8L (AFP, BDF); **NB:** 1.8L (APH, AWP, AWV, BNU, BKF), 2.0L (AEG, AVH, AZG, BEV); **NBvert:** (BDG); **Cabrio:** 2.0L (ABA) **Transmissions:** All	**ECM Internal Communication Error Conditions:** Key on, and the ECM detected an internal communications register read back error during the initial key on check period. **Possible Causes:** • Clear the trouble codes and then check for this trouble code. If it resets, the ECM has failed and needs replacement. • Remember to check for signs of Aftermarket Performance Products installation before replacing the ECM.

DTC	Trouble Code Title, Conditions & Possible Causes
DTC: P0613 **NB:** 1.8L (APH, AWP, AWV, BNU, BKF), 2.0L (AEG, AVH, AZG, BEV); **NBvert:** (BDG) **Transmissions:** All	**TCM Processor Conditions:** Key on, and the ECM detected an internal communication error with the Transmission control module **Possible Causes:** • TCM failed • ECM failed • Replacement control module ID doesn't match old control module ID
DTC: P0614 **Passat:** 2.8 (AAA); **Jetta:** 1.8L (AWD, AWP, AWW), 2.0 (ABA, AEG, AVH, AZG, BBW, BEV), 2.5L (BGP, BGQ), 2.8L (AAA, AFP, BDF); **Golf:** 1.8L (AWD, AWW), 2.0L (ABA, AEG, AVH, AZG, BEV); **GTI:** 1.8L (AWD, AWP, AWW), 2.0L (ABA, AEG), 2.8L (AAA, AFP, BDF); **NB:** 1.8L (APH, AWP, AWV, BNU, BKF), 2.0L (AEG, AVH, AZG, BEV); **NBvert:** (BDG); **Cabrio:** 2.0L (ABA) **Transmissions:** All	**ECM / TCM Incompatible Conditions:** Key on, and the ECM detected a communication error between the Transmission control module and the ECM **Possible Causes:** • TCM failed • ECM failed • Circuit shorting between ECM and TCM • Replacement control module ID doesn't match old control module ID
DTC: P0627 **Passat:** 1.8L (ATW, AUG), 2.0L (BPY); **Jetta:** 2.5L (BGP, BGQ); **NB:** 1.8L (APH, AWP, AWV, BNU, BKF) **Transmissions:** All	**Fuel Pump "A" Control Circuit Open Conditions:** Engine started, battery voltage must be at least 11.5v, all electrical components must be off, parking brake must be engaged (to keep daytime driving lights off), automatic transmission selector must be in park, and the ground between the engine and the chassis must be well connected. The ECM has detected a voltage value across the fuel pump control circuit that is out of the specified limits for the system to function properly. **Possible Causes:** • Fuel Pressure Regulator Valve is faulty • Fuel Pressure Sensor is faulty • Fuel Pump (FP) Control Module is faulty • Fuel pump is faulty
DTC: P0628 **Passat:** 2.0L (BPY); **NB:** 1.8L (APH, AWP, AWV, BNU, BKF) **Transmissions:** All	**Fuel Pump "A" Control Circuit Low Conditions:** Engine started, battery voltage must be at least 11.5v, all electrical components must be off, parking brake must be engaged (to keep daytime driving lights off), automatic transmission selector must be in park, and the ground between the engine and the chassis must be well connected. The ECM has detected a voltage value across the fuel pump control circuit that is below the specified limit for the system to function properly. **Possible Causes:** • Fuel Pressure Regulator Valve is faulty • Fuel Pressure Sensor is faulty • Fuel Pump (FP) Control Module is faulty • Fuel pump is faulty
DTC: P0629 **Passat:** 1.8L (ATW, AUG), 2.0L (BPY); **Jetta:** 2.5L (BGP, BGQ); **NB:** 1.8L (APH, AWP, AWV, BNU, BKF) **Transmissions:** All	**Fuel Pump "A" Control Circuit High Conditions:** Engine started, battery voltage must be at least 11.5v, all electrical components must be off, parking brake must be engaged (to keep daytime driving lights off), automatic transmission selector must be in park, and the ground between the engine and the chassis must be well connected. The ECM has detected a voltage value across the fuel pump control circuit that is above the specified limit for the system to function properly. **Possible Causes:** • Fuel Pressure Regulator Valve is faulty • Fuel Pressure Sensor is faulty • Fuel Pump (FP) Control Module is faulty • Fuel pump is faulty

DTC	Trouble Code Title, Conditions & Possible Causes
DTC: P0638 **2T, MIL: Yes** **Passat:** 1.8L (ATW, AUG), 2.0L (BPY), 4.0L (BDP); **Jetta:** 1.8L (AWD, AWP, AWW), 2.0 (ABA, AEG, AVH, AZG, BBW, BEV), 2.5L (BGP, BGQ), 2.8L (AFP, BDF); **Golf:** 1.8L (AWD, AWW), 2.0L (ABA, AEG, AVH, AZG, BEV); **GTI:** 1.8L (AWD, AWP, AWW), 2.0L (ABA, AEG), 2.8L (AFP, BDF); **NB:** 1.8L (APH, AWP, AWV, BNU, BKF), 2.0L (AEG, AVH, AZG, BEV); **NBvert:** (BDG); **Cabrio:** 2.0L (ABA) **Transmissions:** All	**Throttle Actuator Control Range/Performance Bank 1 Conditions:** Engine started, battery voltage must be at least 11.5v, all electrical components must be off, parking brake must be engaged (to keep daytime driving lights off), automatic transmission selector must be in park, and the ground between the engine and the chassis must be well connected. The ECM has detected a voltage value across the throttle actuator control circuit that is out of the specified limit for the system to function properly. Both Throttle Position (TP) Sensor / Accelerator Pedal Position Sensor 2 are located at the accelerator pedal and communicate the driver's intentions to the Motronic engine control module (ECM) completely independently of each other. Both sensors are integrated into one housing. **Possible Causes:** • Throttle Position (TP) sensor is faulty • Throttle valve control module is faulty • ECM is faulty • Circuit wires have short circuited to each other, to vehicle Ground (GND) or to B+. • Accelerator pedal module is faulty
DTC: P0639 **Jetta:** 2.5L (BGP, BGQ) **Transmissions:** All	**Throttle Actuator Control Range/Performance Bank 2 Conditions:** Engine started, battery voltage must be at least 11.5v, all electrical components must be off, parking brake must be engaged (to keep daytime driving lights off), automatic transmission selector must be in park, and the ground between the engine and the chassis must be well connected. The ECM has detected a voltage value across the throttle actuator control circuit that is out of the specified limit for the system to function properly. Both Throttle Position (TP) Sensor / Accelerator Pedal Position Sensor 2 are located at the accelerator pedal and communicate the driver's intentions to the Motronic engine control module (ECM) completely independently of each other. Both sensors are integrated into one housing. **Possible Causes:** • Throttle Position (TP) sensor is faulty • Throttle valve control module is faulty • ECM is faulty • Circuit wires have short circuited to each other, to vehicle Ground (GND) or to B+. • Accelerator pedal module is faulty
DTC: P0641 **Passat:** 2.0L (BPY); **Jetta:** 2.0L (BPY), 2.5L (BGP, BGQ) **Transmissions:** All	**Sensor Reference Voltage "A" Circuit Open Conditions:** Engine started, battery voltage must be at least 11.5v, all electrical components must be off, parking brake must be engaged (to keep daytime driving lights off), automatic transmission selector must be in park, and the ground between the engine and the chassis must be well connected. **Possible Causes:** • Circuit harness connector contacts are corroded or ingressed of water • Circuit wires have shorted to each other, to battery or ground • Automatic Transmission Hydraulic Pressure Sensor 1 has failed • Solenoid valves in valve body are faulty • Transmission Control Module (TCM) needs replacing • Transmission Input Speed (RPM) Sensor has failed • Transmission Output Speed (RPM) Sensor has failed
DTC: P0642 **Passat:** 2.0L (BPY); **Jetta:** 2.0 (ABA, AEG, AVH, AZG, BBW, BEV, BPY); **Golf:** 2.0L (ABA, AEG, AVH, AZG, BEV); **GTI:** 2.0L (ABA, AEG); **NB:** 2.0L (AEG, AVH, AZG, BEV); **NBvert:** (BDG); **Cabrio:** 2.0L (ABA) **Transmissions:** A/M	**Sensor Reference Voltage "A" Circuit Low Conditions:** Engine started, battery voltage must be at least 11.5v, all electrical components must be off, parking brake must be engaged (to keep daytime driving lights off), automatic transmission selector must be in park, and the ground between the engine and the chassis must be well connected. **Possible Causes:** • Circuit harness connector contacts are corroded or ingressed of water • Circuit wires have shorted to each other, to battery or ground • Automatic Transmission Hydraulic Pressure Sensor 1 has failed • Solenoid valves in valve body are faulty • Transmission Control Module (TCM) needs replacing • Transmission Input Speed (RPM) Sensor has failed • Transmission Output Speed (RPM) Sensor has failed

DTC	Trouble Code Title, Conditions & Possible Causes
DTC: P0643 **Passat:** 2.0L (BPY); **Jetta:** 2.0L (BPY) **Transmissions:** A/M	**Sensor Reference Voltage "A" Circuit High Conditions:** Engine started, battery voltage must be at least 11.5v, all electrical components must be off, parking brake must be engaged (to keep daytime driving lights off), automatic transmission selector must be in park, and the ground between the engine and the chassis must be well connected. **Possible Causes:** • Circuit harness connector contacts are corroded or ingressed of water • Circuit wires have shorted to each other, to battery or ground • Automatic Transmission Hydraulic Pressure Sensor 1 has failed • Solenoid valves in valve body are faulty • Transmission Control Module (TCM) needs replacing • Transmission Input Speed (RPM) Sensor has failed • Transmission Output Speed (RPM) Sensor has failed
DTC: P0650 **Jetta:** 1.8L (AWD, AWP, AWW), 2.8L (AFP, BDF); **Golf:** 1.8L (AWD, AWW); **GTI:** 1.8L (AWD, AWP, AWW), 2.8L (AFP, BDF) **Transmissions:** All	**Malfunction Indicator Lamp (MIL) Control Circuit Conditions:** The exhaust Malfunction Indicator Lamp (MIL) lights up when exhaust relevant malfunctions are recognized by the Engine Control Module (ECM). The Malfunction Indicator Lamp (MIL) can blink or remain lit continuously. Blinking: There is a malfunction that causes damage to the catalytic converter in this driving condition. In this case, vehicle must only be driven at reduced power! Continuously lit: There is a malfunction that causes increased emissions. Check DTC memory for Motronic control module. DTC memory must still be checked if there are driveability problems or customer complaints and the MIL is not lit, since malfunctions can be stored without causing the MIL to light immediately. **Possible Causes:** • Wire from ECM to MIL is shorted or grounded • ECM has failed • MIL has failed
DTC: P0651 **Passat:** 2.0L (BPY); **Jetta:** 2.0L (BPY) **Transmissions:** All	**Sensor Reference Voltage "B" Circuit Open Conditions:** Engine started, battery voltage must be at least 11.5v, all electrical components must be off, parking brake must be engaged (to keep daytime driving lights off), automatic transmission selector must be in park, and the ground between the engine and the chassis must be well connected. **Possible Causes:** • Circuit harness connector contacts are corroded or ingressed of water • Circuit wires have shorted to each other, to battery or ground • Automatic Transmission Hydraulic Pressure Sensor 1 has failed • Solenoid valves in valve body are faulty • Transmission Control Module (TCM) needs replacing • Transmission Input Speed (RPM) Sensor has failed • Transmission Output Speed (RPM) Sensor has failed
DTC: P0652 **Passat:** 2.0L (BPY); **Jetta:** 2.0L (BPY) **Transmissions:** All	**Sensor Reference Voltage "B" Circuit Low Conditions:** Engine started, battery voltage must be at least 11.5v, all electrical components must be off, parking brake must be engaged (to keep daytime driving lights off), automatic transmission selector must be in park, and the ground between the engine and the chassis must be well connected. **Possible Causes:** • Circuit harness connector contacts are corroded or ingressed of water • Circuit wires have shorted to each other, to battery or ground • Automatic Transmission Hydraulic Pressure Sensor 1 has failed • Solenoid valves in valve body are faulty • Transmission Control Module (TCM) needs replacing • Transmission Input Speed (RPM) Sensor has failed • Transmission Output Speed (RPM) Sensor has failed
DTC: P0653 **Passat:** 2.0L (BPY); **Jetta:** 2.0L (BPY) **Transmissions:** All	**Sensor Reference Voltage "B" Circuit High Conditions:** Engine started, battery voltage must be at least 11.5v, all electrical components must be off, parking brake must be engaged (to keep daytime driving lights off), automatic transmission selector must be in park, and the ground between the engine and the chassis must be well connected. **Possible Causes:** • Circuit harness connector contacts are corroded or ingressed of water • Circuit wires have shorted to each other, to battery or ground • Automatic Transmission Hydraulic Pressure Sensor 1 has failed • Solenoid valves in valve body are faulty • Transmission Control Module (TCM) needs replacing • Transmission Input Speed (RPM) Sensor has failed • Transmission Output Speed (RPM) Sensor has failed

DTC	Trouble Code Title, Conditions & Possible Causes
DTC: P0654 **Jetta:** 2.0L (BPY) **Transmissions:** All	**Engine RPM Output Circuit Conditions:** Engine started, battery voltage must be at least 11.5v, all electrical components must be off, parking brake must be engaged (to keep daytime driving lights off), automatic transmission selector must be in park, and the ground between the engine and the chassis must be well connected. **Possible Causes:** • Circuit harness connector contacts are corroded or ingressed of water • Circuit wires have shorted to each other, to battery or ground • Automatic Transmission Hydraulic Pressure Sensor 1 has failed • Solenoid valves in valve body are faulty • Transmission Control Module (TCM) needs replacing • Transmission Input Speed (RPM) Sensor has failed • Transmission Output Speed (RPM) Sensor has failed
DTC: P0657 **Passat:** 2.0L (BPY) **Transmissions:** All	**Actuator Supply Voltage "A" Circuit Open Conditions:** Engine started, battery voltage must be at least 11.5v, all electrical components must be off, parking brake must be engaged (to keep daytime driving lights off), automatic transmission selector must be in park, and the ground between the engine and the chassis must be well connected. **Possible Causes:** • Circuit harness connector contacts are corroded or ingressed of water • Circuit wires have shorted to each other, to battery or ground • Automatic Transmission Hydraulic Pressure Sensor 1 has failed • Solenoid valves in valve body are faulty • Transmission Control Module (TCM) needs replacing • Transmission Input Speed (RPM) Sensor has failed • Transmission Output Speed (RPM) Sensor has failed
DTC: P0658 **Passat:** 2.0L (BPY) **Transmissions:** All	**Actuator Supply Voltage "A" Circuit Low Conditions:** Engine started, battery voltage must be at least 11.5v, all electrical components must be off, parking brake must be engaged (to keep daytime driving lights off), automatic transmission selector must be in park, and the ground between the engine and the chassis must be well connected. **Possible Causes:** • Circuit harness connector contacts are corroded or ingressed of water • Circuit wires have shorted to each other, to battery or ground • Automatic Transmission Hydraulic Pressure Sensor 1 has failed • Solenoid valves in valve body are faulty • Transmission Control Module (TCM) needs replacing • Transmission Input Speed (RPM) Sensor has failed • Transmission Output Speed (RPM) Sensor has failed
DTC: P0659 **Passat:** 2.0L (BPY) **Transmissions:** All	**Actuator Supply Voltage "A" Circuit High Conditions:** Engine started, battery voltage must be at least 11.5v, all electrical components must be off, parking brake must be engaged (to keep daytime driving lights off), automatic transmission selector must be in park, and the ground between the engine and the chassis must be well connected. **Possible Causes:** • Circuit harness connector contacts are corroded or ingressed of water • Circuit wires have shorted to each other, to battery or ground • Automatic Transmission Hydraulic Pressure Sensor 1 has failed • Solenoid valves in valve body are faulty • Transmission Control Module (TCM) needs replacing • Transmission Input Speed (RPM) Sensor has failed • Transmission Output Speed (RPM) Sensor has failed
DTC: P0660 **Jetta:** 2.8L (AFP, BDF); **GTI:** 2.8L (AFP, BDF) **Transmissions:** All	**Intake Manifold Tuning Valve Control Circuit/Open Bank 1 Conditions:** Engine started, battery voltage must be at least 11.5v, all electrical components must be off, parking brake must be engaged (to keep daytime driving lights off), automatic transmission selector must be in park and the ground between the engine and the chassis must be well connected. The ECM detected the intake manifold tuning valve control circuit is open. **Note: The solenoid valve is closed when no voltage is present.** **Possible Causes:** • Fuel pump relay is faulty • ECM has failed • Intake manifold change-over valve needs replacing • Circuit wires have shorted to each other, to battery or ground

DTC	Trouble Code Title, Conditions & Possible Causes
DTC: P0661 **Jetta:** 2.8L (AFP, BDF); **GTI:** 2.8L (AFP, BDF) **Transmissions:** All	**Intake Manifold Tuning Valve Control Circuit Low Bank 1 Conditions:** Engine started, battery voltage must be at least 11.5v, all electrical components must be off, parking brake must be engaged (to keep daytime driving lights off), automatic transmission selector must be in park and the ground between the engine and the chassis must be well connected. The ECM detected the intake manifold tuning valve control circuit has a voltage less than the minimum requirement for proper function. **Note: The solenoid valve is closed when no voltage is present.** **Possible Causes:** • Fuel pump relay is faulty • ECM has failed • Intake manifold change-over valve needs replacing • Circuit wires have shorted to each other, to battery or ground
DTC: P0662 **Jetta:** 2.8L (AFP, BDF); **GTI:** 2.8L (AFP, BDF) **Transmissions:** All	**Intake Manifold Tuning Valve Control Circuit High Bank 1 Conditions:** Engine started, battery voltage must be at least 11.5v, all electrical components must be off, parking brake must be engaged (to keep daytime driving lights off), automatic transmission selector must be in park and the ground between the engine and the chassis must be well connected. The ECM detected the intake manifold tuning valve control circuit has a voltage more than the maximum requirement for proper function. **Note: The solenoid valve is closed when no voltage is present.** **Possible Causes:** • Fuel pump relay is faulty • ECM has failed • Intake manifold change-over valve needs replacing • Circuit wires have shorted to each other, to battery or ground
DTC: P0685 **Passat:** 1.8 (AEB, ATW, AUG, AWM), 2.0L (BPY), 4.0L (BDP); **Jetta:** 1.8L (AWD, AWP, AWW), 2.0L (AEG, AVH, AZG, BBW, BEV, BPY), 2.5L (BGP, BGQ), 2.8L (AFP, BDF); **Golf:** 1.8L (AWD, AWW), 2.0L (AEG, AVH, AZG, BEV); **GTI:** 1.8L (AWD, AWP, AWW), 2.0L (AEG), 2.8L (AFP, BDF); **NB:** 1.8L (APH, AWP, AWV, BNU, BKF), 2.0L (AEG, AVH, AZG, BEV); **NBvert:** (BDG) **Transmissions:** All	**ECM Power Relay Control Circuit Open Conditions:** Engine started, battery voltage must be at least 11.5v, all electrical components must be off, parking brake must be engaged (to keep daytime driving lights off), automatic transmission selector must be in park and the ground between the engine and the chassis must be well connected. The ECM detected the ECM power relay control circuit has a voltage outside requirement for proper function. **Possible Causes:** • Generator has failed or is damaged • Fuel pump relay is faulty • Circuit is grounded to power or chassis • ECM has failed
DTC: P0686 **Passat:** 1.8 (AEB, ATW, AUG, AWM), 2.0L (BPY), 4.0L (BDP); **Jetta:** 1.8L (AWD, AWP, AWW), 2.0L (AEG, AVH, AZG, BBW, BEV, BPY), 2.5L (BGP, BGQ), 2.8L (AFP, BDF); **Golf:** 1.8L (AWD, AWW), 2.0L (AEG, AVH, AZG, BEV); **GTI:** 1.8L (AWD, AWP, AWW), 2.0L (AEG), 2.8L (AFP, BDF); **NB:** 1.8L (APH, AWP, AWV, BNU, BKF), 2.0L (AEG, AVH, AZG, BEV); **NBvert:** (BDG) **Transmissions:** All	**ECM/PCM Power Relay Control Circuit Low Conditions:** Engine started, battery voltage must be at least 11.5v, all electrical components must be off, parking brake must be engaged (to keep daytime driving lights off), automatic transmission selector must be in park and the ground between the engine and the chassis must be well connected. The ECM detected the ECM power relay control circuit has a voltage outside requirement for proper function. **Possible Causes:** • Generator has failed or is damaged • Fuel pump relay is faulty • Circuit is grounded to power or chassis • ECM has failed

DTC	Trouble Code Title, Conditions & Possible Causes
DTC: P0687 **Passat:** 1.8 (AEB, ATW, AUG, AWM), 2.0L (BPY), 4.0L (BDP); **Jetta:** 1.8L (AWD, AWP, AWW), 2.0L (AEG, AVH, AZG, BBW, BEV, BPY), 2.5L (BGP, BGQ), 2.8L (AFP, BDF); **Golf:** 1.8L (AWD, AWW), 2.0L (AEG, AVH, AZG, BEV); **GTI:** 1.8L (AWD, AWP, AWW), 2.0L (AEG), 2.8L (AFP, BDF); **NB:** 1.8L (APH, AWP, AWV, BNU, BKF), 2.0L (AEG, AVH, AZG, BEV); **NBvert:** (BDG) **Transmissions:** All	**ECM/PCM Power Relay Control Circuit High Conditions:** Engine started, battery voltage must be at least 11.5v, all electrical components must be off, parking brake must be engaged (to keep daytime driving lights off), automatic transmission selector must be in park and the ground between the engine and the chassis must be well connected. The ECM detected the ECM power relay control circuit has a voltage outside requirement for proper function. **Possible Causes:** • Generator has failed or is damaged • Fuel pump relay is faulty • Circuit is grounded to power or chassis • ECM has failed
DTC: P0688 **Passat:** 1.8 (AEB, ATW, AUG, AWM), 2.0L (BPY), 4.0L (BDP); **Jetta:** 1.8L (AWD, AWP, AWW), 2.0L (BPY), 2.5L (BGP, BGQ), 2.8L (AFP, BDF); **Golf:** 1.8L (AWD, AWW); **GTI:** 1.8L (AWD, AWP, AWW); **GTI:** 2.8L (AFP, BDF); **NB:** 1.8L (APH, AWP, AWV, BNU, BKF) **Transmissions:** All	**ECM/PCM Power Relay Control Sense Circuit Open Conditions:** Engine started, battery voltage must be at least 11.5v, all electrical components must be off, parking brake must be engaged (to keep daytime driving lights off), automatic transmission selector must be in park and the ground between the engine and the chassis must be well connected. The ECM detected the ECM power relay control circuit has a voltage outside requirement for proper function. **Possible Causes:** • Generator has failed or is damaged • Fuel pump relay is faulty • Circuit is grounded to power or chassis • ECM has failed
DTC: P0691 **Passat:** 2.0L (BPY); **Jetta:** 2.0L (BPY), 2.5L (BGP, BGQ) **Transmissions:** All	**Fan 1 Control Circuit Low Conditions:** Engine running, battery voltage at least 11.5v, all electrical components off, and the ECM detected an problem with the fan control circuit **Possible Causes:** • Check connection of coolant fan control module according to wiring diagram • Short to B+ or ground in the circuit • Circuit open
DTC: P0692 **Passat:** 2.0L (BPY); **Jetta:** 2.0L (BPY), 2.5L (BGP, BGQ) **Transmissions:** All	**Fan 1 Control Circuit High Conditions:** Engine running, battery voltage at least 11.5v, all electrical components off, and the ECM detected an problem with the fan control circuit **Possible Causes:** • Check connection of coolant fan control module according to wiring diagram • Short to B+ or ground in the circuit • Circuit open
DTC: P0697 **Passat:** 2.0L (BPY); **Jetta:** 2.0L (BPY) **Transmissions:** All	**Sensor Reference Voltage "C" Circuit Open Conditions:** Engine started, battery voltage must be at least 11.5v, all electrical components must be off, parking brake must be engaged (to keep daytime driving lights off), automatic transmission selector must be in park, and the ground between the engine and the chassis must be well connected. **Possible Causes:** • Circuit harness connector contacts are corroded or ingressed of water • Circuit wires have shorted to each other, to battery or ground • Automatic Transmission Hydraulic Pressure Sensor 1 has failed • Solenoid valves in valve body are faulty • Transmission Control Module (TCM) needs replacing • Transmission Input Speed (RPM) Sensor has failed • Transmission Output Speed (RPM) Sensor has failed

DTC	Trouble Code Title, Conditions & Possible Causes
DTC: P0698 **Passat:** 2.0L (BPY); **Jetta:** 2.0L (BPY) **Transmissions:** All	**Sensor Reference Voltage "C" Circuit Low Conditions:** Engine started, battery voltage must be at least 11.5v, all electrical components must be off, parking brake must be engaged (to keep daytime driving lights off), automatic transmission selector must be in park, and the ground between the engine and the chassis must be well connected. **Possible Causes:** • Circuit harness connector contacts are corroded or ingressed of water • Circuit wires have shorted to each other, to battery or ground • Automatic Transmission Hydraulic Pressure Sensor 1 has failed • Solenoid valves in valve body are faulty • Transmission Control Module (TCM) needs replacing • Transmission Input Speed (RPM) Sensor has failed • Transmission Output Speed (RPM) Sensor has failed
DTC: P0699 **Passat:** 2.0L (BPY); **Jetta:** 2.0L (BPY) **Transmissions:** All	**Sensor Reference Voltage "C" Circuit High Conditions:** Engine started, battery voltage must be at least 11.5v, all electrical components must be off, parking brake must be engaged (to keep daytime driving lights off), automatic transmission selector must be in park, and the ground between the engine and the chassis must be well connected. **Possible Causes:** • Circuit harness connector contacts are corroded or ingressed of water • Circuit wires have shorted to each other, to battery or ground • Automatic Transmission Hydraulic Pressure Sensor 1 has failed • Solenoid valves in valve body are faulty • Transmission Control Module (TCM) needs replacing • Transmission Input Speed (RPM) Sensor has failed • Transmission Output Speed (RPM) Sensor has failed
DTC: P0700 **Passat:** 1.8L (ATW, AUG), 2.0L (BPY) 2.8L (AAA); **Jetta:** 1.8L (AWD, AWP, AWW), 2.0 (ABA, AEG, AVH, AZG, BBW, BEV, BPY), 2.5L (BGP, BGQ), 2.8L (AAA, AFP, BDF); **Golf:** 1.8L (AWD, AWW), 2.0L (ABA, AEG, AVH, AZG, BEV); **GTI:** 1.8L (AWD, AWP, AWW), 2.0L (ABA, AEG), 2.8L (AAA, AFP, BDF); **NB:** 1.8L (APH, AWP, AWV, BNU, BKF), 2.0L (AEG, AVH, AZG, BEV); **NBvert:** (BDG); **Cabrio:** 2.0L (ABA) **Transmissions:** All	**Transmission Control System Malfunction Conditions:** Engine started, battery voltage must be at least 11.5v, all electrical components must be off, parking brake must be engaged (to keep daytime driving lights off), automatic transmission selector must be in park, and the ground between the engine and the chassis must be well connected. The ECM detected a malfunction int the transmission control system. **Possible Causes:** • Circuit harness connector contacts are corroded or ingressed of water • Circuit wires have shorted to each other, to battery or ground • Automatic Transmission Hydraulic Pressure Sensor 1 has failed • Solenoid valves in valve body are faulty • Transmission Input Speed (RPM) Sensor has failed • Transmission Output Speed (RPM) Sensor has failed • Engine Control Module (ECM) is faulty • Voltage supply for Engine Control Module (ECM) is faulty • Transmission Control Module (TCM) is faulty
DTC: P0701 **Passat:** 2.0L (BPY) **Transmissions:** All	**Transmission Control System Range/Performance Conditions:** Engine started, battery voltage must be at least 11.5v, all electrical components must be off, parking brake must be engaged (to keep daytime driving lights off), automatic transmission selector must be in park, and the ground between the engine and the chassis must be well connected. The ECM detected a voltage outside the normal performance range to allow the system to properly function. **Possible Causes:** • Circuit harness connector contacts are corroded or ingressed of water • Circuit wires have shorted to each other, to battery or ground • Automatic Transmission Hydraulic Pressure Sensor 1 has failed • Solenoid valves in valve body are faulty • Transmission Input Speed (RPM) Sensor has failed • Transmission Output Speed (RPM) Sensor has failed • Engine Control Module (ECM) is faulty • Voltage supply for Engine Control Module (ECM) is faulty • Transmission Control Module (TCM) is faulty

DTC	Trouble Code Title, Conditions & Possible Causes
DTC: P0702 **Passat:** 1.8L (ATW, AUG), 2.0L (BPY); **Jetta:** 2.0 (ABA, BPY), 2.5L (BGP, BGQ), 2.8L (AFP, BDF); Golf 2.0L (ABA); **GTI:** 2.0L (ABA), 2.8L (AFP, BDF); **NB:** 1.8L (APH, AWP, AWV, BNU, BKF), 2.0L (AEG, AVH, AZG, BEV); **NBvert:** (BDG); **Cabrio:** 2.0L (ABA) **Transmissions:** All	**Transmission Control System Electrical Conditions:** Engine started, battery voltage must be at least 11.5v, all electrical components must be off, parking brake must be engaged (to keep daytime driving lights off), automatic transmission selector must be in park, and the ground between the engine and the chassis must be well connected. The ECM detected a voltage outside the normal performance range to allow the system to properly function. **Possible Causes:** • Circuit harness connector contacts are corroded or ingressed of water • Circuit wires have shorted to each other, to battery or ground • Automatic Transmission Hydraulic Pressure Sensor 1 has failed • Solenoid valves in valve body are faulty • Transmission Input Speed (RPM) Sensor has failed • Transmission Output Speed (RPM) Sensor has failed • Engine Control Module (ECM) is faulty • Voltage supply for Engine Control Module (ECM) is faulty • Transmission Control Module (TCM) is faulty
DTC: P0704 **Passat:** 1.8L (ATW, AUG), 2.0L (BPY); **Jetta:** 1.8L (AWD, AWP, AWW), 2.0L (AEG, AVH, AZG, BBW, BEV, BPY), 2.5L (BGP, BGQ), 2.8L (AFP, BDF); **Golf:** 1.8L (AWD, AWW), 2.0L (AEG, AVH, AZG, BEV); **GTI:** 1.8L (AWD, AWP, AWW), 2.0L (AEG), 2.8L (AFP, BDF); **NB:** 1.8L (APH, AWP, AWV, BNU, BKF), 2.0L (AEG, AVH, AZG, BEV); **NBvert:** (BDG) **Transmissions:** All	**Clutch Switch Input Circuit Malfunction Conditions:** Engine started, battery voltage must be at least 11.5v, all electrical components must be off, parking brake must be engaged (to keep daytime driving lights off), automatic transmission selector must be in park, and the ground between the engine and the chassis must be well connected. The ECM detected a voltage outside the normal performance range to allow the system to properly function. **Possible Causes:** • Circuit harness connector contacts are corroded or ingressed of water • Circuit wires have shorted to each other, to battery or ground • Automatic Transmission Hydraulic Pressure Sensor 1 has failed • Solenoid valves in valve body are faulty • Transmission Input Speed (RPM) Sensor has failed • Transmission Output Speed (RPM) Sensor has failed • Engine Control Module (ECM) is faulty • Voltage supply for Engine Control Module (ECM) is faulty • Transmission Control Module (TCM) is faulty
DTC: P0705 **Passat:** 1.8L (ATW, AUG), 2.0L (BPY), 2.8L (AAA); **Jetta:** 1.8L (AWD, AWP, AWW), 2.0 (ABA, AEG, AVH, AZG, BBW, BEV, BPY), 2.5L (BGP, BGQ), 2.8L (AAA); **Golf:** 1.8L (AWD, AWW), 2.0L (ABA, AEG, AVH, AZG, BEV); **GTI:** 1.8L (AWD, AWP, AWW), 2.0L (ABA, AEG), 2.8L (AAA, AFP, BDF); **NB:** 1.8L (APH, AWP, AWV, BNU, BKF), 2.0L (AEG, AVH, AZG, BEV); **NBvert:** (BDG); **Cabrio:** 2.0L (ABA) **Transmissions:** A/T	**TR Sensor Circuit Malfunction Conditions:** Engine started, battery voltage must be at least 11.5v, all electrical components must be off, parking brake must be engaged (to keep daytime driving lights off), automatic transmission selector must be in park, and the ground between the engine and the chassis must be well connected. The ECM detected a voltage outside the normal performance range to allow the system to properly function. **Possible Causes:** • Circuit harness connector contacts are corroded or ingressed of water • Circuit wires have shorted to each other, to battery or ground • Automatic Transmission Hydraulic Pressure Sensor 1 has failed • Solenoid valves in valve body are faulty • Transmission Input Speed (RPM) Sensor has failed • Transmission Output Speed (RPM) Sensor has failed • Engine Control Module (ECM) is faulty • Voltage supply for Engine Control Module (ECM) is faulty • Transmission Control Module (TCM) is faulty
DTC: P0706 **Passat:** 1.8L (ATW, AUG), 2.0L (BPY); **Jetta:** 2.0L (ABA); Golf 2.0L (ABA); **GTI:** 2.0L (ABA); **NB:** 1.8L (APH, AWP, AWV, BNU, BKF); **Cabrio:** 2.0L (ABA) **Transmissions:** A/T	**TR Sensor Circuit Range/Performance Conditions:** Engine started, battery voltage must be at least 11.5v, all electrical components must be off, parking brake must be engaged (to keep daytime driving lights off), automatic transmission selector must be in park, and the ground between the engine and the chassis must be well connected. The ECM detected a voltage outside the normal performance range to allow the system to properly function. **Possible Causes:** • Circuit harness connector contacts are corroded or ingressed of water • Circuit wires have shorted to each other, to battery or ground • Automatic Transmission Hydraulic Pressure Sensor 1 has failed • Solenoid valves in valve body are faulty • Transmission Input Speed (RPM) Sensor has failed • Transmission Output Speed (RPM) Sensor has failed • Engine Control Module (ECM) is faulty • Voltage supply for Engine Control Module (ECM) is faulty • Transmission Control Module (TCM) is faulty

DTC	Trouble Code Title, Conditions & Possible Causes
DTC: P0707 **Passat:** 1.8 (AEB, AWM), 2.8L (AHA, ATQ) **Transmissions:** A/T	**Transmission Range Sensor Circuit Low Input Conditions:** Engine started, battery voltage must be at least 11.5v, all electrical components must be off, parking brake must be engaged (to keep daytime driving lights off), automatic transmission selector must be in park, and the ground between the engine and the chassis must be well connected. The ECM detected the Transmission Range sensor (TR) signal was less than the self-test minimum value in the test. **Possible Causes:** • Circuit harness connector contacts are corroded or ingressed of water • Circuit wires have shorted to each other, to battery or ground • Automatic Transmission Hydraulic Pressure Sensor 1 has failed • Solenoid valves in valve body are faulty • Transmission Input Speed (RPM) Sensor has failed • Transmission Output Speed (RPM) Sensor has failed • Engine Control Module (ECM) is faulty • Voltage supply for Engine Control Module (ECM) is faulty • Transmission Control Module (TCM) is faulty
DTC: P0708 **Passat:** 1.8 (AEB, AWM), 2.8L (AHA, ATQ) **Transmissions:** A/T	**Transmission Range Sensor Circuit High Input Conditions:** Engine started, battery voltage must be at least 11.5v, all electrical components must be off, parking brake must be engaged (to keep daytime driving lights off), automatic transmission selector must be in park, and the ground between the engine and the chassis must be well connected. The ECM detected the Transmission Range sensor (TR) input was more than the self-test maximum range in the test. **Possible Causes:** • Circuit harness connector contacts are corroded or ingressed of water • Circuit wires have shorted to each other, to battery or ground • Automatic Transmission Hydraulic Pressure Sensor 1 has failed • Solenoid valves in valve body are faulty • Transmission Input Speed (RPM) Sensor has failed • Transmission Output Speed (RPM) Sensor has failed • Engine Control Module (ECM) is faulty • Voltage supply for Engine Control Module (ECM) is faulty • Transmission Control Module (TCM) is faulty
DTC: P0710 **Passat:** 1.8L (ATW, AUG), 2.0L (BPY); **Jetta:** 1.8L (AWD, AWP, AWW), 2.0L (ABA), 2.8L (AFP, BDF); **Golf:** 1.8L (AWD, AWW), 2.0L (ABA); **GTI:** 1.8L (AWD, AWP, AWW), 2.0L (ABA), 2.8L (AFP, BDF); **NB:** 1.8L (APH, AWP, AWV, BNU, BKF); **Cabrio:** 2.0L (ABA) **Transmissions:** A/T	**Transmission Fluid Temperature Sensor Circuit Malfunction Conditions:** Engine started, battery voltage must be at least 11.5v, all electrical components must be off, parking brake must be engaged (to keep daytime driving lights off), automatic transmission selector must be in park, and the ground between the engine and the chassis must be well connected. The ECM detected the Transmission fluid temperature sensor circuit was outside the normal range in the test to allow proper function. **Possible Causes:** • ATF is low, contaminated, dirty or burnt • Circuit harness connector contacts are corroded or ingressed of water • Circuit wires have shorted to each other, to battery or ground • Automatic Transmission Hydraulic Pressure Sensor 1 has failed • Solenoid valves in valve body are faulty • Transmission Input Speed (RPM) Sensor has failed • Transmission Output Speed (RPM) Sensor has failed • Engine Control Module (ECM) is faulty • Voltage supply for Engine Control Module (ECM) is faulty • Transmission Control Module (TCM) is faulty
DTC: P0711 **Passat:** 1.8L (ATW, AUG), 2.0L (BPY); **Jetta:** 2.0 (ABA, BPY), 2.5L (BGP, BGQ); **Golf:** 2.0L (ABA); **GTI:** 2.0L (ABA); **NB:** 1.8L (APH, AWP, AWV, BNU, BKF), 2.0L (AEG, AVH, AZG, BEV); **NBvert:** (BDG); **Cabrio:** 2.0L (ABA) **Transmissions:** A/T	**Transmission Fluid Temperature Sensor Signal Range/Performance Conditions:** Engine started, battery voltage must be at least 11.5v, all electrical components must be off, parking brake must be engaged (to keep daytime driving lights off), automatic transmission selector must be in park, and the ground between the engine and the chassis must be well connected. The ECM detected the Transmission Fluid Temperature (TFT) sensor value was not close its normal operating temperature. **Possible Causes:** • ATF is low, contaminated, dirty or burnt • TFT sensor signal circuit has a high resistance condition • TFT sensor is out-of-calibration ("skewed") or it has failed • ECM has failed

DTC	Trouble Code Title, Conditions & Possible Causes
DTC: P0712 **Passat:** 1.8L (ATW, AUG), 2.0L (BPY); **Jetta:** 2.0 (ABA, BPY), 2.5L (BGP, BGQ); **Golf:** 2.0L (ABA); **GTI:** 2.0L (ABA); **NB:** 1.8L (APH, AWP, AWV, BNU, BKF), 2.0L (AEG, AVH, AZG, BEV); **NBvert:** (BDG); **Cabrio:** 2.0L (ABA) **Transmissions:** A/T	**Transmission Fluid Temperature Sensor Circuit Low Input Conditions:** Engine started, battery voltage must be at least 11.5v, all electrical components must be off, parking brake must be engaged (to keep daytime driving lights off), automatic transmission selector must be in park, and the ground between the engine and the chassis must be well connected. The ECM detected the Transmission Fluid Temperature (TFT) sensor was less than its minimum self-test range in the test. **Possible Causes:** • TFT sensor signal circuit is shorted to chassis ground • TFT sensor signal circuit is shorted to sensor ground • TFT sensor is damaged, or out-of-calibration, or has failed • ECM has failed
DTC: P0713 **Passat:** 1.8L (ATW, AUG), 2.0L (BPY); **Jetta:** 2.0 (ABA, BPY), 2.5L (BGP, BGQ); **Golf:** 2.0L (ABA); **GTI:** 2.0L (ABA); **NB:** 1.8L (APH, AWP, AWV, BNU, BKF), 2.0L (AEG, AVH, AZG, BEV); **NBvert:** (BDG); **Cabrio:** 2.0L (ABA) **Transmissions:** A/T	**Transmission Fluid Temperature Sensor Circuit High Input Conditions:** Engine started, battery voltage must be at least 11.5v, all electrical components must be off, parking brake must be engaged (to keep daytime driving lights off), automatic transmission selector must be in park, and the ground between the engine and the chassis must be well connected. The ECM detected the Transmission Fluid Temperature (TFT) sensor was more than its maximum self-test range in the test. **Possible Causes:** • TFT sensor signal circuit is open between the sensor and ECM • TFT sensor ground circuit is open between sensor and ECM • TFT sensor is damaged or has failed • ECM has failed
DTC: P0714 **Passat:** 1.8L (ATW, AUG), 2.0L (BPY); **Jetta:** 2.0 (ABA, BPY), 2.5L (BGP, BGQ); **Golf:** 2.0L (ABA); **GTI:** 2.0L (ABA); **NB:** 1.8L (APH, AWP, AWV, BNU, BKF), 2.0L (AEG, AVH, AZG, BEV); **NBvert:** (BDG); **Cabrio:** 2.0L (ABA) **Transmissions:** A/T	**Transmission Fluid Temperature Sensor Circuit Intermittent Conditions:** Engine started, battery voltage must be at least 11.5v, all electrical components must be off, parking brake must be engaged (to keep daytime driving lights off), automatic transmission selector must be in park, and the ground between the engine and the chassis must be well connected. The ECM detected the Transmission Fluid Temperature (TFT) sensor was giving a false reading or was not reading at all. **Possible Causes:** • TFT sensor signal circuit is open between the sensor and ECM • TFT sensor ground circuit is open between sensor and ECM • TFT sensor is damaged or has failed • ECM has failed
DTC: P0715 **Passat:** 1.8L (ATW, AUG), 2.0L (ABA, BPY), 2.8L (AAA); **Jetta:** 1.8L (AWD, AWP, AWW), 2.0 (ABA, AEG, AVH, AZG, BBW, BEV, BPY), 2.5L (BGP, BGQ), 2.8L (AAA, AFP, BDF); **Golf:** 1.8L (AWD, AWW), 2.0L (ABA, AEG, AVH, AZG, BEV); **GTI:** 1.8L (AWD, AWP, AWW), 2.0L (ABA, AEG), 2.8L (AAA, AFP, BDF); **NB:** 1.8L (APH, AWP, AWV, BNU, BKF), 2.0L (AEG, AVH, AZG, BEV); **NBvert:** (BDG); **Cabrio:** 2.0L (ABA) **Transmissions:** A/T	**Input/Turbine Speed Sensor Circuit Malfunction Conditions:** Engine started, vehicle driven with the vehicle speed sensor indicating more than 1 mph, and the ECM detected the Transmission Vehicle Speed Sensor signals were erratic, or that they were missing for a period of time. **Possible Causes:** • TVSS signal circuit is open • TVSS signal is shorted to chassis ground • TVSS signal is shorted to sensor ground • TVSS assembly is damaged or it has failed • ECM has failed
DTC: P0716 **Passat:** 1.8L (ATW, AUG), 2.0L (BPY); **Jetta:** 2.0L (BPY), 2.5L (BGP, BGQ), 2.8L (AFP, BDF); **GTI:** 2.8L (AFP, BDF); **NB:** 1.8L (APH, AWP, AWV, BNU, BKF), 2.0L (AEG, AVH, AZG, BEV); **NBvert:** (BDG) **Transmissions:** A/T	**Input Turbine/Speed Sensor Circuit Range/Performance Conditions:** Engine started, vehicle driven with the vehicle speed sensor indicating more than 1 mph, and the ECM detected the Transmission Vehicle Speed Sensor signals were erratic, or that they were missing for a period of time. **Possible Causes:** • TVSS signal circuit is open • TVSS signal is shorted to chassis ground • TVSS signal is shorted to sensor ground • TVSS assembly is damaged or it has failed • ECM has failed

DTC	Trouble Code Title, Conditions & Possible Causes
DTC: P0717 **Passat:** 1.8L (ATW, AUG), 2.0L (BPY); **Jetta:** 1.8L (AWD, AWP, AWW), 2.0L (BPY), 2.5L (BGP, BGQ), 2.8L (AFP, BDF); **Golf:** 1.8L (AWD, AWW); **GTI:** 1.8L (AWD, AWP, AWW), 2.8L (AFP, BDF); **NB:** 1.8L (APH, AWP, AWV, BNU, BKF), 2.0L (AEG, AVH, AZG, BEV); **NBvert:** (BDG) **Transmissions:** A/T	**Transmission Speed Shaft Sensor Signal Intermittent Conditions:** Engine started, vehicle speed sensor indicating over 1 mph, and the ECM detected an intermittent loss of TSS signals (i.e., the TSS signals were erratic, irregular or missing). **Possible Causes:** • TSS connector is damaged, loose or shorted • TSS signal circuit has an intermittent open condition • TSS signal circuit has an intermittent short to ground condition • TSS assembly is damaged or is has failed • ECM has failed
DTC: P0720 **Passat:** 1.8L (ATW), 2.0L (BPY); **Jetta:** 2.0L (BPY), 2.5L (BGP, BGQ); **GTI:** 2.8L (AFP, BDF) **Transmissions:** A/T	**Output Shaft Speed Sensor Circuit Conditions:** Engine started, VSS signal more than 1 mph, and the ECM detected "noise" interference on the Output Shaft Speed (OSS) sensor circuit. **Possible Causes:** • After market add-on devices interfering with the OSS signal • OSS connector is damaged, loose or shorted, or the wiring is misrouted or it is damaged • OSS assembly is damaged or it has failed • ECM has failed
DTC: P0721 **Passat:** 1.8L (ATW, AUG), 2.0L (BPY); **Jetta:** 2.0L (BPY), 2.5L (BGP, BGQ); **NB:** 1.8L (APH, AWP, AWV, BNU, BKF), 2.0L (AEG, AVH, AZG, BEV); **NBvert:** (BDG) **Transmissions:** A/T	**A/T Output Shaft Speed Sensor Noise Interference Conditions:** Engine started, VSS signal more than 1 mph, and the ECM detected "noise" interference on the Output Shaft Speed (OSS) sensor circuit. **Possible Causes:** • After market add-on devices interfering with the OSS signal • OSS connector is damaged, loose or shorted, or the wiring is misrouted or it is damaged • OSS assembly is damaged or it has failed • ECM has failed
DTC: P0722 **Passat:** 1.8L (ATW), 2.0L (ABA, BPY), 2.8L (AAA); **Jetta:** 1.8L (AWD, AWP, AWW), 2.0 (ABA, BBW, BPY), 2.5L (BGP, BGQ), 2.8L (AFP, BDF); **Golf:** 1.8L (AWD, AWW), 2.0L (ABA, AEG, AVH, AZG, BEV); **GTI:** 1.8L (AWD, AWP, AWW), 2.0L (ABA, AEG), 2.8L (AAA, AFP, BDF); **NB:** 1.8L (APH, AWP, AWV, BNU, BKF), 2.0L (AEG, AVH, AZG, BEV); **NBvert:** (BDG); **Cabrio:** 2.0L (ABA) **Transmissions:** A/T	**A/T Output Speed Sensor No Signal Conditions:** Engine started, and the ECM did not detect any Vehicle Speed Sensor (VSS) sensor signals upon initial vehicle movement. **Possible Causes:** • After market add-on devices interfering with the VSS signal • VSS sensor wiring is misrouted, damaged or shorting • ECM and/or TCM has failed
DTC: P0725 **Passat:** 1.8L (ATW, AUG), 2.0L (ABA, BPY), 2.8L (AAA); **Jetta:** 1.8L (AWD, AWP, AWW), 2.0 (ABA, AEG, AVH, AZG, BBW, BEV, BPY), 2.5L (BGP, BGQ), 2.8L (AAA, AFP, BDF); **Golf:** 1.8L (AWD, AWW), 2.0L (ABA, AEG, AVH, AZG, BEV); **GTI:** 1.8L (AWD, AWP, AWW), 2.0L (ABA, AEG), 2.8L (AAA, AFP, BDF); **NB:** 1.8L (APH, AWP, AWV, BNU, BKF), 2.0L (AEG, AVH, AZG, BEV); **NBvert:** (BDG); **Cabrio:** 2.0L (ABA) **Transmissions:** A/T	**Engine Speed Input Circuit Malfunction Conditions:** The Transmission Control Module (TCM) does not receive a signal from the Engine Control Module (ECM). **Possible Causes:** • The TCM circuit is shorting to ground, B+ or is open • TCM has failed • ECM has failed

DTC	Trouble Code Title, Conditions & Possible Causes
DTC: P0726 **Passat:** 1.8L (ATW, AUG), 2.0L (BPY); **Jetta:** 2.0L (ABA); **Golf:** 2.0L (ABA); **GTI:** 2.0L (ABA); **NB:** 1.8L (APH, AWP, AWV, BNU, BKF); **Cabrio:** 2.0L (ABA) **Transmissions:** A/T	**Engine Speed Input Circuit Range/Performance Conditions:** The Engine Speed (RPM) Sensor detects engine speed and reference marks. Without an engine speed signal, the engine will not start. If the engine speed signal fails while the engine is running, the engine will stop immediately. **Note: There is a larger-sized gap on the sensor wheel. This gap is the reference mark and does not mean that the sensor wheel is damaged.** **Possible Causes:** • Engine speed sensor has failed • Circuit is shorting to ground, B+ or is open • Sensor wheel is damaged, run out or not properly secured • ECM has failed
DTC: P0727 **Passat:** 1.8L (ATW, AUG), 2.0L (BPY); **Jetta:** 2.0L (ABA); **Golf:** 2.0L (ABA); **GTI:** 2.0L (ABA); **NB:** 1.8L (APH, AWP, AWV, BNU, BKF); **Cabrio:** 2.0L (ABA) **Transmissions:** A/T	**Engine Speed Input Circuit No Signal Conditions:** The Engine Speed (RPM) Sensor detects engine speed and reference marks. Without an engine speed signal, the engine will not start. If the engine speed signal fails while the engine is running, the engine will stop immediately. **Note: There is a larger-sized gap on the sensor wheel. This gap is the reference mark and does not mean that the sensor wheel is damaged.** **Possible Causes:** • Engine speed sensor has failed • Circuit is shorting to ground, B+ or is open • Sensor wheel is damaged, run out or not properly secured • ECM has failed
DTC: P0729 **Passat:** 2.0L (BPY); **Jetta:** 2.0L (BPY), 2.5L (BGP, BGQ); **NB:** 1.8L (APH, AWP, AWV, BNU, BKF), 2.0L (AEG, AVH, AZG, BEV); **NBvert:** (BDG) **Transmissions:** A/T	**Gear 6 Incorrect Ratio Conditions:** Engine started, battery voltage must be at least 11.5v, all electrical components must be off, and the ground between the engine and the chassis must be well connected. The ECM detected an incorrect ratio within the sixth gear. **Possible Causes:** • ATF level is low • Circuit harness connector contacts are corroded or ingressed of water • Circuit wires have shorted to each other, to battery or ground • Automatic Transmission Hydraulic Pressure Sensor 1 has failed • Solenoid valves in valve body are faulty • Transmission Control Module (TCM) needs replacing • Transmission Input Speed (RPM) Sensor has failed • Transmission Output Speed (RPM) Sensor has failed
DTC: P0730 **Passat:** 1.8L (ATW, AUG), 2.0L (BPY), 2.8L (AAA); **Jetta:** 1.8L (AWD, AWP, AWW), 2.0 (ABA, AEG, AVH, AZG, BBW, BEV, BPY), 2.5L (BGP, BGQ), 2.8L (AAA, AFP, BDF); **Golf:** 1.8L (AWD, AWW), 2.0L (ABA, AEG, AVH, AZG, BEV); **GTI:** 1.8L (AWD, AWP, AWW), 2.0L (ABA, AEG), 2.8L (AAA, AFP, BDF); **NB:** 1.8L (APH, AWP, AWV, BNU, BKF), 2.0L (AEG, AVH, AZG, BEV); **NBvert:** (BDG); **Cabrio:** 2.0L (ABA) **Transmissions:** A/T	**Gear Incorrect Ratio Conditions:** Engine started, battery voltage must be at least 11.5v, all electrical components must be off, and the ground between the engine and the chassis must be well connected. The ECM detected an incorrect gear ratio. **Possible Causes:** • ATF level is low • Circuit harness connector contacts are corroded or ingressed of water • Circuit wires have shorted to each other, to battery or ground • Automatic Transmission Hydraulic Pressure Sensor 1 has failed • Solenoid valves in valve body are faulty • Transmission Control Module (TCM) needs replacing • Transmission Input Speed (RPM) Sensor has failed • Transmission Output Speed (RPM) Sensor has failed

DTC	Trouble Code Title, Conditions & Possible Causes
DTC: P0731 **Passat:** 1.8L (ATW, AUG), 2.0L (BPY), 2.8L (AAA); **Jetta:** 1.8L (AWD, AWP, AWW), 2.0 (ABA, AEG, AVH, AZG, BBW, BEV, BPY), 2.5L (BGP, BGQ), 2.8L (AAA, AFP, BDF); **Golf:** 1.8L (AWD, AWW), 2.0L (ABA, AEG, AVH, AZG, BEV); **GTI:** 1.8L (AWD, AWP, AWW), 2.0L (ABA, AEG), 2.8L (AAA, AFP, BDF); **NB:** 1.8L (APH, AWP, AWV, BNU, BKF), 2.0L (AEG, AVH, AZG, BEV); **NBvert:** (BDG); **Cabrio:** 2.0L (ABA) **Transmissions:** A/T	**Incorrect First Gear Ratio Conditions:** Engine started, vehicle operating with 1st gear commanded "on", and the ECM detected an incorrect 1st gear ratio during the test. **Possible Causes:** • 1st Gear solenoid harness connector not properly seated • 1st Gear solenoid signal shorted to ground, or open • 1st Gear solenoid wiring harness connector is damaged • 1st Gear solenoid is damaged or not properly installed • ATF level is low • Circuit harness connector contacts are corroded or ingressed of water • Circuit wires have shorted to each other, to battery or ground • Automatic Transmission Hydraulic Pressure Sensor 1 has failed • Transmission Control Module (TCM) needs replacing • Transmission Input Speed (RPM) Sensor has failed • Transmission Output Speed (RPM) Sensor has failed
DTC: P0732 **Passat:** 1.8L (ATW, AUG), 2.0L (BPY), 2.8L (AAA); **Jetta:** 1.8L (AWD, AWP, AWW), 2.0 (ABA, AEG, AVH, AZG, BBW, BEV, BPY), 2.5L (BGP, BGQ), 2.8L (AAA, AFP, BDF); **Golf:** 1.8L (AWD, AWW), 2.0L (ABA, AEG, AVH, AZG, BEV); **GTI:** 1.8L (AWD, AWP, AWW), 2.0L (ABA, AEG), 2.8L (AAA, AFP, BDF); **NB:** 1.8L (APH, AWP, AWV, BNU, BKF), 2.0L (AEG, AVH, AZG, BEV); **NBvert:** (BDG); **Cabrio:** 2.0L (ABA) **Transmissions:** A/T	**Incorrect Second Gear Ratio Conditions:** Engine started, vehicle operating with 2nd Gear commanded "on", and the ECM detected an incorrect 2nd gear ratio during the test. **Possible Causes:** • 2nd Gear solenoid harness connector not properly seated • 2nd Gear solenoid signal shorted to ground, or open • 2nd Gear solenoid wring harness connector is damaged • 2nd Gear solenoid is damaged or not properly installed • ATF level is low • Circuit harness connector contacts are corroded or ingressed of water • Circuit wires have shorted to each other, to battery or ground • Automatic Transmission Hydraulic Pressure Sensor 1 has failed • Transmission Control Module (TCM) needs replacing • Transmission Input Speed (RPM) Sensor has failed • Transmission Output Speed (RPM) Sensor has failed
DTC: P0733 **Passat:** 1.8L (ATW), 2.0L (BPY), 2.8L (AAA); **Jetta:** 1.8L (AWD, AWP, AWW), 2.0 (ABA, AEG, AVH, AZG, BBW, BEV, BPY), 2.5L (BGP, BGQ), 2.8L (AAA, AFP, BDF); **Golf:** 1.8L (AWD, AWW), 2.0L (ABA, AEG, AVH, AZG, BEV); **GTI:** 1.8L (AWD, AWP, AWW), 2.0L (ABA, AEG), 2.8L (AAA, AFP, BDF); **NB:** 1.8L (APH, AWP, AWV, BNU, BKF), 2.0L (AEG, AVH, AZG, BEV); **NBvert:** (BDG); **Cabrio:** 2.0L (ABA) **Transmissions:** A/T	**Incorrect Third Gear Ratio Conditions:** Engine started, vehicle operating with 3rd Gear commanded "on", and the ECM detected an incorrect 3rd gear ratio during the test. **Possible Causes:** • 3rd Gear solenoid harness connector not properly seated • 3rd Gear solenoid signal shorted to ground, or open • 3rd Gear solenoid wiring harness connector is damaged • 3rd Gear solenoid is damaged or not properly installed • ATF level is low • Circuit harness connector contacts are corroded or ingressed of water • Circuit wires have shorted to each other, to battery or ground • Automatic Transmission Hydraulic Pressure Sensor 1 has failed • Transmission Control Module (TCM) needs replacing • Transmission Input Speed (RPM) Sensor has failed • Transmission Output Speed (RPM) Sensor has failed

DTC	Trouble Code Title, Conditions & Possible Causes
DTC: P0734 **Passat:** 1.8L (ATW, AUG), 2.0L (BPY), 2.8L (AAA); **Jetta:** 1.8L (AWD, AWP, AWW), 2.0 (ABA, AEG, AVH, AZG, BBW, BEV, BPY), 2.5L (BGP, BGQ), 2.8L (AAA, AFP, BDF); **Golf:** 1.8L (AWD, AWW), 2.0L (ABA, AEG, AVH, AZG, BEV); **GTI:** 1.8L (AWD, AWP, AWW), 2.0L (ABA, AEG), 2.8L (AAA, AFP, BDF); **NB:** 1.8L (APH, AWP, AWV, BNU, BKF), 2.0L (AEG, AVH, AZG, BEV); **NBvert:** (BDG); **Cabrio:** 2.0L (ABA) **Transmissions:** A/T	**Incorrect Fourth Gear Ratio Conditions:** Engine started, vehicle operating with 4th Gear commanded "on", and the ECM detected an incorrect 4th gear ratio during the test. **Possible Causes:** • 4th Gear solenoid harness connector not properly seated • 4th Gear solenoid signal shorted to ground, or open • 4th Gear solenoid wiring harness connector is damaged • 4th Gear solenoid is damaged or not properly installed • ATF level is low • Circuit harness connector contacts are corroded or ingressed of water • Circuit wires have shorted to each other, to battery or ground • Automatic Transmission Hydraulic Pressure Sensor 1 has failed • Transmission Control Module (TCM) needs replacing • Transmission Input Speed (RPM) Sensor has failed • Transmission Output Speed (RPM) Sensor has failed
DTC: P0735 **Passat:** 1.8L (ATW, AUG), 2.0L (BPY); **Jetta:** 2.0L (BPY), 2.5L (BGP, BGQ); **NB:** 1.8L (APH, AWP, AWV, BNU, BKF) **Transmissions:** A/T	**Incorrect Fifth Gear Ratio Conditions:** Engine started, vehicle operating with 5th Gear commanded "on", and the ECM detected an incorrect 5th gear ratio during the test. **Possible Causes:** • 5th Gear solenoid harness connector not properly seated • 5th Gear solenoid signal shorted to ground, or open • 5th Gear solenoid wiring harness connector is damaged • 5th Gear solenoid is damaged or not properly installed • ATF level is low • Circuit harness connector contacts are corroded or ingressed of water • Circuit wires have shorted to each other, to battery or ground • Automatic Transmission Hydraulic Pressure Sensor 1 has failed • Transmission Control Module (TCM) needs replacing • Transmission Input Speed (RPM) Sensor has failed • Transmission Output Speed (RPM) Sensor has failed
DTC: P0740 **Passat:** 2.0L (BPY), 2.8L (AAA); **Jetta:** 1.8L (AWD, AWP, AWW), 2.0 (ABA, AEG, AVH, AZG, BBW, BEV), 2.8L (AAA, AFP, BDF); **Golf:** 1.8L (AWD, AWW), 2.0L (ABA, AEG, AVH, AZG, BEV); **GTI:** 1.8L (AWD, AWP, AWW), 2.0L (ABA, AEG), 2.8L (AAA, AFP, BDF); **NB:** 1.8L (APH, AWP, AWV, BNU, BKF), 2.0L (AEG, AVH, AZG, BEV); **NBvert:** (BDG); **Cabrio:** 2.0L (ABA) **Transmissions:** A/T	**TCC Solenoid Circuit Malfunction Conditions:** Engine started, KOER Self-Test enabled, vehicle driven at cruise speed, and the ECM did not detect any voltage drop across the TCC solenoid circuit during the test period. **Possible Causes:** • TCC solenoid control circuit is open or shorted to ground • TCC solenoid wiring harness connector is damaged • TCC solenoid is damaged or has failed • ECM has failed
DTC: P0741 **Passat:** 1.8L (ATW, AUG) **Transmissions:** A/T	**TCC Mechanical System Range/Performance Conditions:** Engine started, vehicle driven in gear with VSS signals received, and the ECM detected excessive slippage while in normal operation. **Possible Causes:** • TCC solenoid has a mechanical failure • TCC solenoid has a hydraulic failure • ECM has failed

DTC	Trouble Code Title, Conditions & Possible Causes
DTC: P0743 **Passat:** 1.8L (ATW, AUG), 2.0L (BPY); **Jetta:** 1.8L (AWD, AWP, AWW), 2.0L (BPY), 2.5L (BGP, BGQ), 2.8L (AFP, BDF); **Golf:** 1.8L (AWD, AWW); **GTI:** 1.8L (AWD, AWP, AWW), 2.8L (AFP, BDF); **NB:** 1.8L (APH, AWP, AWV, BNU, BKF), 2.0L (AEG, AVH, AZG, BEV); **NBvert:** (BDG) **Transmissions:** A/T	**Torque Converter Clutch Circuit Electrical Malfunction Conditions:** Engine started, KOER Self-Test enabled, vehicle driven at cruise speed, and the ECM did not detect any voltage drop across the TCC solenoid circuit during the test period. **Possible Causes:** • TCC solenoid control circuit is open or shorted to ground • TCC solenoid wiring harness connector is damaged • TCC solenoid is damaged or has failed • ECM has failed
DTC: P0746 **Passat:** 1.8L (ATW, AUG), 2.0L (BPY); **Jetta:** 1.8L (AWD, AWP, AWW), 2.0L (BPY), 2.5L (BGP, BGQ), 2.8L (AFP, BDF); **Golf:** 1.8L (AWD, AWW); **GTI:** 1.8L (AWD, AWP, AWW), 2.8L (AFP, BDF); **NB:** 1.8L (APH, AWP, AWV, BNU, BKF), 2.0L (AEG, AVH, AZG, BEV); **NBvert:** (BDG) **Transmissions:** A/T	**Pressure Control Solenoid "A" Performance or Stuck Off Conditions:** Engine started, battery voltage must be at least 11.5v, all electrical components must be off, and the ground between the engine and the chassis must be well connected. The ECM detected the pressure control solenoid was in the "stuck off" position. **Possible Causes:** • ATF level is low • Circuit harness connector contacts are corroded or ingressed of water • Circuit wires have shorted to each other, to battery or ground • Automatic Transmission Hydraulic Pressure Sensor 1 has failed • Solenoid valves in valve body are faulty • Transmission Control Module (TCM) needs replacing • Transmission Input Speed (RPM) Sensor has failed • Transmission Output Speed (RPM) Sensor has failed
DTC: P0747 **Passat:** 1.8L (ATW, AUG), 2.0L (BPY); **Jetta:** 2.0L (BPY), 2.5L (BGP, BGQ), 2.8L (AFP, BDF) **GTI:** 2.8L (AFP, BDF); **NB:** 1.8L (APH, AWP, AWV, BNU, BKF), 2.0L (AEG, AVH, AZG, BEV); **NBvert:** (BDG) **Transmissions:** A/T	**Pressure Control Solenoid "A" Performance or Stuck On Conditions:** Engine started, battery voltage must be at least 11.5v, all electrical components must be off, and the ground between the engine and the chassis must be well connected. The ECM detected the pressure control solenoid was in the "stuck on" position. **Possible Causes:** • ATF level is low • Circuit harness connector contacts are corroded or ingressed of water • Circuit wires have shorted to each other, to battery or ground • Automatic Transmission Hydraulic Pressure Sensor 1 has failed • Solenoid valves in valve body are faulty • Transmission Control Module (TCM) needs replacing • Transmission Input Speed (RPM) Sensor has failed • Transmission Output Speed (RPM) Sensor has failed
DTC: P0748 **Passat:** 1.8L (ATW, AUG), 2.0L (ABA, BPY), 2.8L (AAA); **Jetta:** 1.8L (AWD, AWP, AWW), 2.0 (ABA, AEG, AVH, AZG, BBW, BEV, BPY), 2.5L (BGP, BGQ), 2.8L (AAA, AFP, BDF); **Golf:** 1.8L (AWD, AWW), 2.0L (ABA, AEG, AVH, AZG, BEV); **GTI:** 1.8L (AWD, AWP, AWW), 2.0L (ABA, AEG), 2.8L (AAA, AFP, BDF); **NB:** 1.8L (APH, AWP, AWV, BNU, BKF), 2.0L (AEG, AVH, AZG, BEV); **NBvert:** (BDG); **Cabrio:** 2.0L (ABA) **Transmissions:** A/T	**Pressure Control Solenoid Electrical Conditions:** The valve body solenoid valve is not receiving a signal. **Possible Causes:** • Pressure control solenoid circuit is shorting to ground • Pressure control solenoid circuit is open • Valve has failed • TCM has failed • ECM has failed
DTC: P0749 **Passat:** 1.8L (ATW, AUG), 2.0L (ABA, BPY); **Jetta:** 2.0L (BPY), 2.5L (BGP, BGQ), 2.8L (AFP, BDF); **GTI:** 2.8L (AFP, BDF); **NB:** 1.8L (APH, AWP, AWV, BNU, BKF), 2.0L (AEG, AVH, AZG, BEV); **NBvert:** (BDG) **Transmissions:** A/T	**Pressure Control Solenoid "A" Intermittent Conditions:** The valve body solenoid valve is receiving an improper signal. **Possible Causes:** • Pressure control solenoid circuit is shorting to ground • Pressure control solenoid circuit is open • Valve has failed • TCM has failed • ECM has failed

DTC	Trouble Code Title, Conditions & Possible Causes
DTC: P0751 **Passat:** 1.8L (ATW, AUG), 2.0L (BPY); **Jetta:** 2.0L (BPY), 2.5L (BGP, BGQ), 2.8L (AFP, BDF); **GTI:** 2.8L (AFP, BDF); **NB:** 1.8L (APH, AWP, AWV, BNU, BKF), 2.0L (AEG, AVH, AZG, BEV); **NBvert:** (BDG) **Transmissions:** A/T	**Shift Solenoid "A" Performance or Stuck Off Conditions:** Engine started, vehicle driven with the solenoid applied, and the ECM detected an unexpected voltage condition on the SS1/A solenoid circuit was incorrect during the test. **Possible Causes:** • Solenoid valves in valve body are faulty • Solenoid circuit is shorting to ground • Solenoid circuit is open • TCM has failed or wiring is shorting • ECM has failed
DTC: P0752 **Passat:** 1.8L (ATW, AUG), 2.0L (BPY); **Jetta:** 2.0L (BPY), 2.5L (BGP, BGQ), 2.8L (AFP, BDF); **GTI:** 2.8L (AFP, BDF); **NB:** 1.8L (APH, AWP, AWV, BNU, BKF), 2.0L (AEG, AVH, AZG, BEV); **NBvert:** (BDG) **Transmissions:** A/T	**A/T Shift Solenoid 1/A Function Range/Performance Conditions:** Engine started, vehicle driven with the solenoid applied, and the ECM detected a mechanical failure while operating the Shift Solenoid 1/A during the CCM test period. **Possible Causes:** • SS1/A solenoid is stuck in the "on" position • SS1/A solenoid has a mechanical failure • SS1/A solenoid has a hydraulic failure • ECM has failed
DTC: P0753 **Passat:** 1.8L (ATW, AUG), 2.0L (ABA, BPY), 2.8L (AAA); **Jetta:** 1.8L (AWD, AWP, AWW), 2.0 (ABA, AEG, AVH, AZG, BBW, BEV, BPY), 2.5L (BGP, BGQ), 2.8L (AAA, AFP, BDF); **Golf:** 1.8L (AWD, AWW), 2.0L (AEG, AVH, AZG, BEV); **GTI:** 1.8L (AWD, AWP, AWW), 2.0L (ABA, AEG), 2.8L (AAA, AFP, BDF); **NB:** 1.8L (APH, AWP, AWV, BNU, BKF); **Cabrio:** 2.0L (ABA) **Transmissions:** A, 2.0L (AEG)/T	**A/T Shift Solenoid 1/A Circuit Malfunction Conditions:** Engine started, vehicle driven with the solenoid applied, and the ECM detected an unexpected voltage condition on the SS1/A solenoid circuit was incorrect during the test. **Possible Causes:** • SS1/A solenoid control circuit is open • SS1/A solenoid control circuit is shorted to ground • SS1/A solenoid wiring harness connector is damaged • SS1/A solenoid is damaged or has failed • ECM has failed
DTC: P0756 **Passat:** 1.8L (ATW, AUG), 2.0L (BPY); **Jetta:** 2.0L (BPY), 2.5L (BGP, BGQ), 2.8L (AFP, BDF); **GTI:** 2.8L (AFP, BDF); **NB:** 1.8L (APH, AWP, AWV, BNU, BKF), 2.0L (AEG, AVH, AZG, BEV); **NBvert:** (BDG) **Transmissions:** A/T	**A/T Shift Solenoid 2/B Function Range/Performance Conditions:** Engine started, vehicle driven with the solenoid applied, and the ECM detected a mechanical failure while operating the Shift Solenoid 2/B during the CCM test period. **Possible Causes:** • SS2/B solenoid is stuck in the "on" position • SS2/B solenoid has a mechanical failure • SS2/B solenoid has a hydraulic failure • ECM has failed
DTC: P0757 **Passat:** 1.8L (ATW, AUG), 2.0L (BPY); **Jetta:** 2.0L (BPY), 2.5L (BGP, BGQ), 2.8L (AFP, BDF); **GTI:** 2.8L (AFP, BDF); **NB:** 1.8L (APH, AWP, AWV, BNU, BKF), 2.0L (AEG, AVH, AZG, BEV); **NBvert:** (BDG) **Transmissions:** A/T	**A/T Shift Solenoid 2/B Function Range/Performance Conditions:** Engine started, vehicle driven with the solenoid applied, and the ECM detected a mechanical failure while operating the Shift Solenoid 2/B during the CCM test period. **Possible Causes:** • SS2/B solenoid is stuck in the "on" position • SS2/B solenoid has a mechanical failure • SS2/B solenoid has a hydraulic failure • ECM has failed

DTC	Trouble Code Title, Conditions & Possible Causes
DTC: P0758 **Passat:** 1.8L (ATW, AUG), 2.0L (ABA, BPY), 2.8L (AAA); **Jetta:** 1.8L (AWD, AWP, AWW), 2.0 (ABA, AEG, AVH, AZG, BBW, BEV, BPY), 2.5L (BGP, BGQ), 2.8L (AAA, AFP, BDF); **Golf:** 1.8L (AWD, AWW), 2.0L (AEG, AVH, AZG, BEV); **GTI:** 1.8L (AWD, AWP, AWW), 2.0L (ABA, AEG), 2.8L (AAA, AFP, BDF); **NB:** 1.8L (APH, AWP, AWV, BNU, BKF), 2.0L (AEG, AVH, AZG, BEV); **NBvert:** (BDG); **Cabrio:** 2.0L (ABA) **Transmissions:** A/T	**A/T Shift Solenoid 2/B Circuit Malfunction Conditions:** Engine started, vehicle driven with the solenoid applied, and the ECM detected an unexpected voltage condition on the SS1/A solenoid circuit was incorrect during the test.. **Possible Causes:** • Shift Solenoid 2/B connector is damaged, open or shorted • Shift Solenoid 2/B control circuit is open • Shift Solenoid 2/B control circuit is shorted to ground • Shift Solenoid 2/B is damaged or it has failed • ECM has failed
DTC: P0761 **Passat:** 1.8L (ATW, AUG), 2.0L (BPY); **Jetta:** 2.0L (BPY), 2.5L (BGP, BGQ), 2.8L (AFP, BDF); **GTI:** 2.8L (AFP, BDF); **NB:** 1.8L (APH, AWP, AWV, BNU, BKF), 2.0L (AEG, AVH, AZG, BEV); **NBvert:** (BDG) **Transmissions:** A/T	**A/T Shift Solenoid 3/C Function Range/Performance Conditions:** Engine started, vehicle driven with Shift Solenoid 3/C applied, and the ECM detected a mechanical failure occurred (stuck "off") while operating Shift Solenoid 3/C during the test. **Possible Causes:** • SS3/C solenoid may be stuck "off" • SS3/C solenoid has a mechanical failure • SS3/C solenoid has a hydraulic failure • ECM has failed
DTC: P0762 **Passat:** 1.8L (ATW, AUG), 2.0L (BPY); **Jetta:** 2.0L (BPY), 2.5L (BGP, BGQ), 2.8L (AFP, BDF); **GTI:** 2.8L (AFP, BDF); **NB:** 1.8L (APH, AWP, AWV, BNU, BKF), 2.0L (AEG, AVH, AZG, BEV); **NBvert:** (BDG) **Transmissions:** A/T	**A/T Shift Solenoid 3/C Function Range/Performance Conditions:** Engine started, vehicle driven with Shift Solenoid 3/C applied, and the ECM detected a mechanical failure occurred (stuck "on") while operating Shift Solenoid 3/C during the test. **Possible Causes:** • SS3/C solenoid may be stuck "on" • SS3/C solenoid has a mechanical failure • SS3/C solenoid has a hydraulic failure • ECM has failed
DTC: P0763 **Passat:** 1.8L (ATW, AUG), 2.0L (ABA, BPY), 2.8L (AAA); **Jetta:** 1.8L (AWD, AWP, AWW), 2.0 (ABA, AEG, AVH, AZG, BBW, BEV, BPY), 2.5L (BGP, BGQ), 2.8L (AAA, AFP, BDF); **Golf:** 1.8L (AWD, AWW), 2.0L (ABA, AEG, AVH, AZG, BEV); **GTI:** 1.8L (AWD, AWP, AWW), 2.0L (ABA, AEG), 2.8L (AAA, AFP, BDF); **NB:** 1.8L (APH, AWP, AWV, BNU, BKF), 2.0L (AEG, AVH, AZG, BEV); **NBvert:** (BDG); **Cabrio:** 2.0L (ABA) **Transmissions:** A/T	**A/T Shift Solenoid 3/C Electrical Conditions:** Engine started, vehicle driven with the solenoid applied, and the ECM detected an unexpected voltage condition on the SS3/C solenoid circuit was incorrect during the test.. **Possible Causes:** • Shift Solenoid 3/C connector is damaged, open or shorted • Shift Solenoid 3/C control circuit is open • Shift Solenoid 3/C control circuit is shorted to ground • Shift Solenoid 3/C is damaged or it has failed • ECM has failed
DTC: P0766 **Jetta:** 2.0L (BPY), 2.5L (BGP, BGQ), 2.8L (AFP, BDF); **GTI:** 2.8L (AFP, BDF); **NB:** 1.8L (APH, AWP, AWV, BNU, BKF), 2.0L (AEG, AVH, AZG, BEV); **NBvert:** (BDG) **Transmissions:** A/T	**A/T Shift Solenoid D Performance Conditions:** Engine started, vehicle driven with the solenoid applied, and the ECM detected an unexpected voltage condition on the SS3/C solenoid circuit was incorrect during the test.. **Possible Causes:** • Shift Solenoid D connector is damaged, open or shorted • Shift Solenoid D control circuit is open • Shift Solenoid D control circuit is shorted to ground • Shift Solenoid D is damaged or it has failed • ECM has failed

DTC	Trouble Code Title, Conditions & Possible Causes
DTC: P0768 **Passat:** 2.0L (ABA, BPY), 2.8L (AAA); **Jetta:** 1.8L (AWD, AWP, AWW), 2.0 (ABA, AEG, AVH, AZG, BBW, BEV, BPY), 2.5L (BGP, BGQ), 2.8L (AAA, AFP, BDF); **Golf:** 1.8L (AWD, AWW), 2.0L (ABA, AEG, AVH, AZG, BEV); **GTI:** 1.8L (AWD, AWP, AWW), 2.0L (ABA), 2.8L (AFP, BDF); **NB:** 1.8L (APH, AWP, AWV, BNU, BKF), 2.0L (AEG, AVH, AZG, BEV); **NBvert:** (BDG); **Cabrio:** 2.0L (ABA) **Transmissions:** A/T	**A/T Shift Solenoid 3/D Electrical Conditions:** Engine started, vehicle driven with the solenoid applied, and the ECM detected an unexpected voltage condition on the SS3/D solenoid circuit was incorrect during the test.. **Possible Causes:** • Shift Solenoid 3/D connector is damaged, open or shorted • Shift Solenoid 3/D control circuit is open • Shift Solenoid 3/D control circuit is shorted to ground • Shift Solenoid 3/D is damaged or it has failed • ECM has failed
DTC: P0771 **Jetta:** 2.0L (BPY), 2.5L (BGP, BGQ), 2.8L (AFP, BDF); **GTI:** 2.8L (AFP, BDF); **NB:** 1.8L (APH, AWP, AWV, BNU, BKF), 2.0L (AEG, AVH, AZG, BEV); **NBvert:** (BDG) **Transmissions:** A/T	**A/T Shift Solenoid E Performance Conditions:** Engine started, vehicle driven with the solenoid applied, and the ECM detected an unexpected voltage condition on the SS3/C solenoid circuit was incorrect during the test.. **Possible Causes:** • Shift Solenoid D connector is damaged, open or shorted • Shift Solenoid D control circuit is open • Shift Solenoid D control circuit is shorted to ground • Shift Solenoid D is damaged or it has failed • ECM has failed
DTC: P0773 **Passat:** 2.0L (ABA, BPY), 2.8L (AAA); **Jetta:** 1.8L (AWD, AWP, AWW), 2.0 (ABA, AEG, AVH, AZG, BBW, BEV, BPY), 2.5L (BGP, BGQ), 2.8L (AAA, AFP, BDF), 2.0L (ABA, AEG, AVH, AZG, BEV); **GTI:** 1.8L (AWD, AWP, AWW), 2.0L (ABA, AEG), 2.8L (AAA, AFP, BDF); **NB:** 1.8L (APH, AWP, AWV, BNU, BKF), 2.0L (AEG, AVH, AZG, BEV); **NBvert:** (BDG); **Cabrio:** 2.0L (ABA) **Transmissions:** A/T	**A/T Shift Solenoid E Electrical Conditions:** Engine started, vehicle driven with the solenoid applied, and the ECM detected an unexpected voltage condition on the SS3/D solenoid circuit was incorrect during the test.. **Possible Causes:** • Shift Solenoid connector is damaged, open or shorted • Shift Solenoid control circuit is open • Shift Solenoid control circuit is shorted to ground • Shift Solenoid is damaged or it has failed • ECM has failed
DTC: P0776 **Passat:** 1.8L (ATW, AUG), 2.0L (BPY); **Jetta:** 2.0L (BPY), 2.5L (BGP, BGQ), 2.8L (AFP, BDF); **GTI:** 2.8L (AFP, BDF); **NB:** 1.8L (APH, AWP, AWV, BNU, BKF), 2.0L (AEG, AVH, AZG, BEV); **NBvert:** (BDG) **Transmissions:** A/T	**Pressure Control Solenoid "B" Performance or Stuck Off Conditions:** Engine started, vehicle driven with Shift Solenoid 3/C applied, and the ECM detected a mechanical failure occurred (stuck "off") while operating Shift Solenoid 3/C during the test. **Possible Causes:** • SS3/C solenoid may be stuck "off" • SS3/C solenoid has a mechanical failure • SS3/C solenoid has a hydraulic failure • ECM has failed
DTC: P0777 **Passat:** 1.8L (ATW, AUG), 2.0L (BPY); **Jetta:** 2.0L (BPY), 2.5L (BGP, BGQ), 2.8L (AFP, BDF); **GTI:** 2.8L (AFP, BDF); **NB:** 1.8L (APH, AWP, AWV, BNU, BKF), 2.0L (AEG, AVH, AZG, BEV); **NBvert:** (BDG) **Transmissions:** A/T	**Pressure Control Solenoid "B" Stuck On Conditions:** Engine started, vehicle driven with Shift Solenoid 3/C applied, and the ECM detected a mechanical failure occurred (stuck "on") while operating Shift Solenoid 3/C during the test. **Possible Causes:** • SS3/C solenoid may be stuck "on" • SS3/C solenoid has a mechanical failure • SS3/C solenoid has a hydraulic failure • ECM has failed

DTC	Trouble Code Title, Conditions & Possible Causes
DTC: P0778 **Passat:** 1.8L (ATW, AUG), 2.0L (ABA, BPY); **Jetta:** 1.8L (AWD, AWP, AWW), 2.0L (BPY), 2.5L (BGP, BGQ), 2.8L (AFP, BDF); **Golf:** 1.8L (AWD, AWW); **GTI:** 2.8L (AFP, BDF); **NB:** 1.8L (APH, AWP, AWV, BNU, BKF), 2.0L (AEG, AVH, AZG, BEV); **NBvert:** (BDG) **Transmissions:** A/T	**Pressure Control Solenoid "B" Electrical Conditions:** Engine started, vehicle driven with the solenoid applied, and the ECM detected an unexpected voltage condition on the SS3/C solenoid circuit was incorrect during the test.. **Possible Causes:** • Shift Solenoid connector is damaged, open or shorted • Shift Solenoid control circuit is open • Shift Solenoid control circuit is shorted to ground • Shift Solenoid is damaged or it has failed • ECM has failed
DTC: P0779 **Passat:** 1.8L (ATW, AUG) **Transmissions:** A/T	**Pressure Control Solenoid "B" Intermittent Conditions:** Engine started, vehicle driven with the solenoid applied, and the ECM detected an unexpected voltage condition on the SS3/C solenoid circuit was incorrect during the test.. **Possible Causes:** • Shift Solenoid connector is damaged, open or shorted • Shift Solenoid control circuit is open • Shift Solenoid control circuit is shorted to ground • Shift Solenoid is damaged or it has failed • ECM has failed
DTC: P0781 **Passat:** 1.8L (ATW, AUG) **Transmissions:** A/T	**1-2 Shift Conditions:** Engine running and vehicle driven, the ECM detected a mechanical malfunction within the transmission **Possible Causes:** • Solenoid valves in valve body are faulty • Solenoid circuit is shorting to ground • Solenoid circuit is open • TCM has failed or wiring is shorting • ECM has failed • Mechanical malfunction in transmission
DTC: P0782 **Passat:** 1.8L (ATW, AUG) **Transmissions:** A/T	**2-3 Shift Conditions:** Engine running and vehicle driven, the ECM detected a mechanical malfunction within the transmission **Possible Causes:** • Solenoid valves in valve body are faulty • Solenoid circuit is shorting to ground • Solenoid circuit is open • TCM has failed or wiring is shorting • ECM has failed • Mechanical malfunction in transmission
DTC: P0783 **Passat:** 1.8L (ATW, AUG) **Transmissions:** A/T	**3-4 Shift Conditions:** xx **Possible Causes:** • Solenoid valves in valve body are faulty • Solenoid circuit is shorting to ground • Solenoid circuit is open • TCM has failed or wiring is shorting • ECM has failed • Mechanical malfunction in transmission
DTC: P0784 **Passat:** 1.8L (ATW, AUG) **Transmissions:** A/T	**4-5 Shift Conditions:** Engine running and vehicle driven, the ECM detected a mechanical malfunction within the transmission **Possible Causes:** • Solenoid valves in valve body are faulty • Solenoid circuit is shorting to ground • Solenoid circuit is open • TCM has failed or wiring is shorting • ECM has failed • Mechanical malfunction in transmission

DTC	Trouble Code Title, Conditions & Possible Causes
DTC: P0785 **Passat:** 2.8L (AAA); **Jetta:** 1.8L (AWD, AWP, AWW), 2.0 (ABA, AEG, AVH, AZG, BBW, BEV, BPY), 2.5L (BGP, BGQ), 2.8L (AAA, AFP, BDF); **Golf:** 1.8L (AWD, AWW), 2.0L (ABA, AEG, AVH, AZG, BEV); **GTI:** 1.8L (AWD, AWP, AWW), 2.0L (ABA, AEG), 2.8L (AAA, AFP, BDF); **NB:** 1.8L (APH, AWP, AWV, BNU, BKF), 2.0L (AEG, AVH, AZG, BEV); **NBvert:** (BDG); **Cabrio:** 2.0L (ABA) **Transmissions:** A/T	**Shift/Timing Solenoid Conditions:** Engine running and vehicle driven, the ECM detected a mechanical malfunction within the transmission **Possible Causes:** • Solenoid valves in valve body are faulty • Solenoid circuit is shorting to ground • Solenoid circuit is open • TCM has failed or wiring is shorting • ECM has failed • Mechanical malfunction in transmission
DTC: P0791 **Jetta:** 1.8L (AWD, AWP, AWW), 2.8L (AFP, BDF); **Golf:** 1.8L (AWD, AWW); **GTI:** 1.8L (AWD, AWP, AWW), 2.8L (AFP, BDF) **Transmissions:** A/T	**Intermediate Shaft Speed Sensor "A" Circuit Conditions:** Engine running and vehicle driven, the ECM detected a mechanical malfunction within the transmission **Possible Causes:** • Solenoid valves in valve body are faulty • Solenoid circuit is shorting to ground • Solenoid circuit is open • TCM has failed or wiring is shorting • ECM has failed • Mechanical malfunction in transmission
DTC: P0796 **Passat:** 1.8L (ATW, AUG), 2.0L (BPY); **Jetta:** 2.0L (BPY), 2.5L (BGP, BGQ), 2.8L (AFP, BDF); **GTI:** 2.8L (AFP, BDF); **NB:** 1.8L (APH, AWP, AWV, BNU, BKF), 2.0L (AEG, AVH, AZG, BEV); **NBvert:** (BDG) **Transmissions:** A/T	**Pressure Solenoid "C" Performance or Stuck Off Conditions:** Engine started, vehicle driven with the solenoid applied, and the ECM detected an unexpected voltage condition on the SS1/C solenoid circuit was incorrect during the test. **Possible Causes:** • Solenoid valves in valve body are faulty • Solenoid circuit is shorting to ground • Solenoid circuit is open • TCM has failed or wiring is shorting • ECM has failed
DTC: P0797 **Passat:** 1.8L (ATW, AUG), 2.0L (BPY); **Jetta:** 2.0L (BPY), 2.5L (BGP, BGQ), 2.8L (AFP, BDF); **GTI:** 2.8L (AFP, BDF); **NB:** 1.8L (APH, AWP, AWV, BNU, BKF), 2.0L (AEG, AVH, AZG, BEV); **NBvert:** (BDG) **Transmissions:** A/T	**Pressure Solenoid "C" Performance or Stuck On Conditions:** Engine started, vehicle driven with the solenoid applied, and the ECM detected an unexpected voltage condition on the SS1/C solenoid circuit was incorrect during the test. **Possible Causes:** • Solenoid valves in valve body are faulty • Solenoid circuit is shorting to ground • Solenoid circuit is open • TCM has failed or wiring is shorting • ECM has failed
DTC: P0798 **Passat:** 1.8L (ATW, AUG), 2.0L (BPY); **Jetta:** 2.0L (BPY), 2.5L (BGP, BGQ), 2.8L (AFP, BDF); **GTI:** 2.8L (AFP, BDF); **NB:** 1.8L (APH, AWP, AWV, BNU, BKF), 2.0L (AEG, AVH, AZG, BEV); **NBvert:** (BDG) **Transmissions:** A/T	**Pressure Solenoid "C" Electrical Conditions:** Engine started, vehicle driven with the solenoid applied, and the ECM detected an unexpected voltage condition on the SS1/C solenoid circuit was incorrect during the test. **Possible Causes:** • Solenoid valves in valve body are faulty • Solenoid circuit is shorting to ground • Solenoid circuit is open • TCM has failed or wiring is shorting • ECM has failed
DTC: P0799 **Passat:** 1.8L (ATW, AUG) **Transmissions:** A/T	**Pressure Solenoid "C" Intermittent Conditions:** Engine started, vehicle driven with the solenoid applied, and the ECM detected an unexpected voltage condition on the SS1/C solenoid circuit was incorrect during the test. **Possible Causes:** • Solenoid valves in valve body are faulty • Solenoid circuit is shorting to ground • Solenoid circuit is open • TCM has failed or wiring is shorting • ECM has failed

DTC	Trouble Code Title, Conditions & Possible Causes
DTC: P0811 **Passat:** 1.8L (ATW, AUG), 2.0L (BPY), 2.8L (AAA); **Jetta:** 1.8L (AWD, AWP, AWW), 2.0 (ABA, AEG, AVH, AZG, BBW, BEV, BPY), 2.5L (BGP, BGQ), 2.8L (AAA, AFP, BDF); **Golf:** 1.8L (AWD, AWW), 2.0L (ABA, AEG, AVH, AZG, BEV); **GTI:** 1.8L (AWD, AWP, AWW), 2.0L (ABA, AEG), 2.8L (AAA, AFP, BDF); **NB:** 1.8L (APH, AWP, AWV, BNU, BKF), 2.0L (AEG, AVH, AZG, BEV); **NBvert:** (BDG); **Cabrio:** 2.0L (ABA) **Transmissions:** A/T	**Excessive Clutch Slippage Conditions:** Engine started, vehicle driven and the ECM and/or TCM has detected that the clutch has slipped multiple times in a given time frame. **Possible Causes:** • Solenoid valves in valve body are faulty • Solenoid circuit is shorting to ground • Solenoid circuit is open • TCM has failed or wiring is shorting • ECM has failed
DTC: P0840 **Jetta:** 2.0L (BPY), 2.5L (BGP, BGQ); **NB:** 1.8L (APH, AWP, AWV, BNU, BKF), 2.0L (AEG, AVH, AZG, BEV); **NBvert:** (BDG) **Transmissions:** A/T	**Transmission Fluid Pressure Sensor/Switch "A" Circuit Conditions:** Engine started, vehicle driven with the solenoid applied, and the ECM detected an unexpected voltage condition on the pressure sensor/switch was incorrect during the test. **Possible Causes:** • Solenoid valves in valve body are faulty • Solenoid circuit is shorting to ground • Solenoid circuit is open • TCM has failed or wiring is shorting • Transmission Input Speed (RPM) Sensor has failed • Transmission Output Speed (RPM) Sensor has failed • ECM has failed
DTC: P0841 **Jetta:** 2.0L (BPY), 2.5L (BGP, BGQ); **NB:** 1.8L (APH, AWP, AWV, BNU, BKF), 2.0L (AEG, AVH, AZG, BEV); **NBvert:** (BDG) **Transmissions:** A/T	**Transmission Fluid Pressure Sensor/Switch "A" Circuit Range/Performance Conditions:** Engine started, vehicle driven with the solenoid applied, and the ECM detected an unexpected voltage condition on the pressure sensor/switch was incorrect during the test. **Possible Causes:** • Solenoid valves in valve body are faulty • Solenoid circuit is shorting to ground • Solenoid circuit is open • TCM has failed or wiring is shorting • Transmission Input Speed (RPM) Sensor has failed • Transmission Output Speed (RPM) Sensor has failed • ECM has failed
DTC: P0842 **NB:** 1.8L (APH, AWP, AWV, BNU, BKF) **Transmissions:** A/T	**Transmission Fluid Pressure Sensor/Switch "A" Circuit Low Conditions:** Engine started, vehicle driven with the solenoid applied, and the ECM detected an unexpected voltage condition on the pressure sensor/switch was incorrect during the test. **Possible Causes:** • Solenoid valves in valve body are faulty • Solenoid circuit is shorting to ground • Solenoid circuit is open • TCM has failed or wiring is shorting • Transmission Input Speed (RPM) Sensor has failed • Transmission Output Speed (RPM) Sensor has failed • ECM has failed
DTC: P0843 **NB:** 1.8L (APH, AWP, AWV, BNU, BKF) **Transmissions:** A/T	**Transmission Fluid Pressure Sensor/Switch "A" Circuit High Conditions:** Engine started, vehicle driven with the solenoid applied, and the ECM detected an unexpected voltage condition on the pressure sensor/switch was incorrect during the test. **Possible Causes:** • Solenoid valves in valve body are faulty • Solenoid circuit is shorting to ground • Solenoid circuit is open • TCM has failed or wiring is shorting • Transmission Input Speed (RPM) Sensor has failed • Transmission Output Speed (RPM) Sensor has failed • ECM has failed

DTC	Trouble Code Title, Conditions & Possible Causes
DTC: P0844 **NB:** 1.8L (APH, AWP, AWV, BNU, BKF) **Transmissions:** A/T	**Transmission Fluid Pressure Sensor/Switch "A" Circuit Intermittent Conditions:** Engine started, vehicle driven with the solenoid applied, and the ECM detected an unexpected voltage condition on the pressure sensor/switch was incorrect during the test. **Possible Causes:** • Solenoid valves in valve body are faulty • Solenoid circuit is shorting to ground • Solenoid circuit is open • TCM has failed or wiring is shorting • Transmission Input Speed (RPM) Sensor has failed • Transmission Output Speed (RPM) Sensor has failed • ECM has failed
DTC: P0845 **Jetta:** 2.0L (BPY), 2.5L (BGP, BGQ); **NB:** 1.8L (APH, AWP, AWV, BNU, BKF), 2.0L (AEG, AVH, AZG, BEV); **NBvert:** (BDG) **Transmissions:** A/T	**Transmission Fluid Pressure Sensor/Switch "B" Circuit Conditions:** Engine started, vehicle driven with the solenoid applied, and the ECM detected an unexpected voltage condition on the pressure sensor/switch was incorrect during the test. **Possible Causes:** • Solenoid valves in valve body are faulty • Solenoid circuit is shorting to ground • Solenoid circuit is open • TCM has failed or wiring is shorting • Transmission Input Speed (RPM) Sensor has failed • Transmission Output Speed (RPM) Sensor has failed • ECM has failed
DTC: P0846 **Jetta:** 2.0L (BPY), 2.5L (BGP, BGQ); **NB:** 1.8L (APH, AWP, AWV, BNU, BKF), 2.0L (AEG, AVH, AZG, BEV); **NBvert:** (BDG) **Transmissions:** A/T	**Transmission Fluid Pressure Sensor/Switch "B" Circuit Range/Performance Conditions:** Engine started, vehicle driven with the solenoid applied, and the ECM detected an unexpected voltage condition on the pressure sensor/switch was incorrect during the test. **Possible Causes:** • Solenoid valves in valve body are faulty • Solenoid circuit is shorting to ground • Solenoid circuit is open • TCM has failed or wiring is shorting • Transmission Input Speed (RPM) Sensor has failed • Transmission Output Speed (RPM) Sensor has failed • ECM has failed
DTC: P0847 **NB:** 1.8L (APH, AWP, AWV, BNU, BKF) **Transmissions:** A/T	**Transmission Fluid Pressure Sensor/Switch "B" Circuit Low Conditions:** Engine started, vehicle driven with the solenoid applied, and the ECM detected an unexpected voltage condition on the pressure sensor/switch was incorrect during the test. **Possible Causes:** • Solenoid valves in valve body are faulty • Solenoid circuit is shorting to ground • Solenoid circuit is open • TCM has failed or wiring is shorting • Transmission Input Speed (RPM) Sensor has failed • Transmission Output Speed (RPM) Sensor has failed • ECM has failed
DTC: P0848 **NB:** 1.8L (APH, AWP, AWV, BNU, BKF) **Transmissions:** A/T	**Transmission Fluid Pressure Sensor/Switch "B" Circuit High Conditions:** Engine started, vehicle driven with the solenoid applied, and the ECM detected an unexpected voltage condition on the pressure sensor/switch was incorrect during the test. **Possible Causes:** • Solenoid valves in valve body are faulty • Solenoid circuit is shorting to ground • Solenoid circuit is open • TCM has failed or wiring is shorting • Transmission Input Speed (RPM) Sensor has failed • Transmission Output Speed (RPM) Sensor has failed • ECM has failed

DTC	Trouble Code Title, Conditions & Possible Causes
DTC: P0849 **NB:** 1.8L (APH, AWP, AWV, BNU, BKF) **Transmissions:** A/T	**Transmission Fluid Pressure Sensor/Switch "B" Circuit Intermittent Conditions:** Engine started, vehicle driven with the solenoid applied, and the ECM detected an unexpected voltage condition on the pressure sensor/switch was incorrect during the test. **Possible Causes:** • Solenoid valves in valve body are faulty • Solenoid circuit is shorting to ground • Solenoid circuit is open • TCM has failed or wiring is shorting • Transmission Input Speed (RPM) Sensor has failed • Transmission Output Speed (RPM) Sensor has failed • ECM has failed
DTC: P0863 **Passat:** 2.0L (BPY); **Jetta:** 1.8L (AWD, AWP, AWW), 2.0L (AVH, AZG, BBW, BEV, BPY), 2.5L (BGP, BGQ), 2.8L (AFP, BDF); **Golf:** 1.8L (AWD, AWW), 2.0L (AVH, AZG, BEV); **GTI:** 1.8L (AWD, AWP, AWW), 2.8L (AFP, BDF); **NB:** 1.8L (APH, AWP, AWV, BNU, BKF), 2.0L (AEG, AVH, AZG, BEV); **NBvert:** (BDG) **Transmissions:** A/T	**TCM Communication Circuit Conditions:** The Transmission Control Module (ECM) communicates with all databus-capable control modules via a CAN databus. These databus-capable control modules are connected via two data bus wires which are twisted together (CAN_High and CAN_Low), and exchange information (messages). Missing information on the databus is recognized as a malfunction and stored. Trouble-free operation of the CAN-Bus requires that it have a terminal resistance. **Possible Causes:** • ECM has failed • Terminal resistance for CAN-bus are faulty • Can data bus wires have short circuited to each other • TCM has failed
DTC: P0864 **Passat:** 1.8L (ATW, AUG), 2.0L (BPY); **Jetta:** 1.8L (AWD, AWP, AWW), 2.0 (ABA, AVH, AZG, BBW, BEV, BPY), 2.5L (BGP, BGQ), 2.8L (AFP, BDF); **Golf:** 1.8L (AWD, AWW), 2.0L (ABA, AVH, AZG, BEV); **GTI:** 1.8L (AWD, AWP); **GTI:** 1.8L (AWD, AWP, AWW), 2.0L (ABA), 2.8L (AFP, BDF); **NB:** 1.8L (APH, AWP, AWV, BNU, BKF), 2.0L (AEG, AVH, AZG, BEV); **NBvert:** (BDG); **Cabrio:** 2.0L (ABA) **Transmissions:** A/T	**TCM Communication Circuit Range/Performance Conditions:** The Transmission Control Module (ECM) communicates with all databus-capable control modules via a CAN databus. These databus-capable control modules are connected via two data bus wires which are twisted together (CAN_High and CAN_Low), and exchange information (messages). Missing information on the databus is recognized as a malfunction and stored. Trouble-free operation of the CAN-Bus requires that it have a terminal resistance. **Possible Causes:** • ECM has failed • Terminal resistance for CAN-bus are faulty • Can data bus wires have short circuited to each other • TCM has failed
DTC: P0865 **Passat:** 1.8L (ATW, AUG), 2.0L (BPY); **Jetta:** 1.8L (AWD, AWP, AWW), 2.0 (ABA, AVH, AZG, BBW, BEV, BPY), 2.5L (BGP, BGQ), 2.8L (AFP, BDF); **Golf:** 1.8L (AWD, AWW), 2.0L (ABA, AVH, AZG, BEV); **GTI:** 1.8L (AWD, AWP, AWW), 2.0L (ABA), 2.8L (AFP, BDF); **NB:** 1.8L (APH, AWP, AWV, BNU, BKF), 2.0L (AEG, AVH, AZG, BEV); **NBvert:** (BDG); **Cabrio:** 2.0L (ABA) **Transmissions:** A/T	**TCM Communication Circuit Low Conditions:** The Transmission Control Module (ECM) communicates with all databus-capable control modules via a CAN databus. These databus-capable control modules are connected via two data bus wires which are twisted together (CAN_High and CAN_Low), and exchange information (messages). Missing information on the databus is recognized as a malfunction and stored. Trouble-free operation of the CAN-Bus requires that it have a terminal resistance. **Possible Causes:** • ECM has failed • Terminal resistance for CAN-bus are faulty • Can data bus wires have short circuited to each other • TCM has failed

DTC	Trouble Code Title, Conditions & Possible Causes
DTC: P0884 **Passat:** 1.8L (ATW, AUG), 2.0L (BPY); **Jetta:** 2.0L (ABA); **Golf:** 2.0L (ABA); **GTI:** 2.0L (ABA); **NB:** 1.8L (APH, AWP, AWV, BNU, BKF); **Cabrio:** 2.0L (ABA) **Transmissions:** A/T	**TCM Power Input Signal Intermittent Conditions:** The Transmission Control Module (ECM) communicates with all databus-capable control modules via a CAN databus. These databus-capable control modules are connected via two data bus wires which are twisted together (CAN_High and CAN_Low), and exchange information (messages). Missing information on the databus is recognized as a malfunction and stored. Trouble-free operation of the CAN-Bus requires that it have a terminal resistance. **Possible Causes:** • Solenoid valves in valve body are faulty • Solenoid circuit is shorting to ground • Solenoid circuit is open • TCM has failed or wiring is shorting • ECM has failed
DTC: P0886 **Passat:** 1.8L (ATW, AUG), 2.0L (BPY); **Jetta:** 2.0L (ABA); **Golf:** 2.0L (ABA); **GTI:** 2.0L (ABA); **NB:** 1.8L (APH, AWP, AWV, BNU, BKF); **Cabrio:** 2.0L (ABA) **Transmissions:** A/T	**TCM Power Relay Control Circuit Low Conditions:** The Transmission Control Module (ECM) communicates with all databus-capable control modules via a CAN databus. These databus-capable control modules are connected via two data bus wires which are twisted together (CAN_High and CAN_Low), and exchange information (messages). Missing information on the databus is recognized as a malfunction and stored. Trouble-free operation of the CAN-Bus requires that it have a terminal resistance. **Possible Causes:** • Solenoid valves in valve body are faulty • Solenoid circuit is shorting to ground • Solenoid circuit is open • TCM has failed or wiring is shorting • ECM has failed
DTC: P0887 **Passat:** 1.8L (ATW, AUG), 2.0L (BPY); **Jetta:** 2.0L (ABA); **GTI:** 2.0L (ABA); **NB:** 1.8L (APH, AWP, AWV, BNU, BKF); **Cabrio:** 2.0L (ABA) **Transmissions:** A/T	**TCM Power Relay Control Circuit High Conditions:** The Transmission Control Module (ECM) communicates with all databus-capable control modules via a CAN databus. These databus-capable control modules are connected via two data bus wires which are twisted together (CAN_High and CAN_Low), and exchange information (messages). Missing information on the databus is recognized as a malfunction and stored. Trouble-free operation of the CAN-Bus requires that it have a terminal resistance. **Possible Causes:** • Solenoid valves in valve body are faulty
DTC: P0889 **Passat:** 2.0L (BPY); **Jetta:** 2.0L (BPY), 2.5L (BGP, BGQ), 2.8L (AFP, BDF); **GTI:** 2.8L (AFP, BDF); **NB:** 1.8L (APH, AWP, AWV, BNU, BKF), 2.0L (AEG, AVH, AZG, BEV); **NBvert:** (BDG) **Transmissions:** A/T	**TCM Power Relay Sense Circuit Range/Performance Conditions:** The Transmission Control Module (ECM) communicates with all databus-capable control modules via a CAN databus. These databus-capable control modules are connected via two data bus wires which are twisted together (CAN_High and CAN_Low), and exchange information (messages). Missing information on the databus is recognized as a malfunction and stored. Trouble-free operation of the CAN-Bus requires that it have a terminal resistance. **Possible Causes:** • Solenoid valves in valve body are faulty • Solenoid circuit is shorting to ground • Solenoid circuit is open • TCM has failed or wiring is shorting • ECM has failed
DTC: P0890 **Passat:** 2.0L (BPY); **Jetta:** 2.0L (BPY), 2.8L (AFP, BDF); **GTI:** 2.8L (AFP, BDF); **NB:** 1.8L (APH, AWP, AWV, BNU, BKF), 2.0L (AEG, AVH, AZG, BEV); **NBvert:** (BDG) **Transmissions:** A/T	**TCM Power Relay Sense Circuit Low Conditions:** The Transmission Control Module (ECM) communicates with all databus-capable control modules via a CAN databus. These databus-capable control modules are connected via two data bus wires which are twisted together (CAN_High and CAN_Low), and exchange information (messages). Missing information on the databus is recognized as a malfunction and stored. Trouble-free operation of the CAN-Bus requires that it have a terminal resistance. **Possible Causes:** • Solenoid valves in valve body are faulty • Solenoid circuit is shorting to ground • Solenoid circuit is open • TCM has failed or wiring is shorting • ECM has failed
DTC: P0891 **Passat:** 2.0L (BPY); **Jetta:** 2.0L (BPY), 2.5L (BGP, BGQ), 2.8L (AFP, BDF); **GTI:** 2.8L (AFP, BDF); **NB:** 1.8L (APH, AWP, AWV, BNU, BKF), 2.0L (AEG, AVH, AZG, BEV); **NBvert:** (BDG) **Transmissions:** A/T	**TCM Power Relay Sense Circuit High Conditions:** The Transmission Control Module (ECM) communicates with all databus-capable control modules via a CAN databus. These databus-capable control modules are connected via two data bus wires which are twisted together (CAN_High and CAN_Low), and exchange information (messages). Missing information on the databus is recognized as a malfunction and stored. Trouble-free operation of the CAN-Bus requires that it have a terminal resistance. **Possible Causes:** • Solenoid valves in valve body are faulty • Solenoid circuit is shorting to ground • Solenoid circuit is open • TCM has failed or wiring is shorting • ECM has failed

DTC	Trouble Code Title, Conditions & Possible Causes
DTC: P0892 **Passat:** 2.0L (BPY); **Jetta:** 2.0L (BPY), 2.5L (BGP, BGQ), 2.8L (AFP, BDF); **GTI:** 2.8L (AFP, BDF); **NB:** 1.8L (APH, AWP, AWV, BNU, BKF), 2.0L (AEG, AVH, AZG, BEV); **NBvert:** (BDG) **Transmissions:** A/T	**TCM Power Relay Sense Circuit Intermittent Conditions:** The Transmission Control Module (ECM) communicates with all databus-capable control modules via a CAN databus. These databus-capable control modules are connected via two data bus wires which are twisted together (CAN_High and CAN_Low), and exchange information (messages). Missing information on the databus is recognized as a malfunction and stored. Trouble-free operation of the CAN-Bus requires that it have a terminal resistance. **Possible Causes:** • Solenoid valves in valve body are faulty • Solenoid circuit is shorting to ground • Solenoid circuit is open • TCM has failed or wiring is shorting • ECM has failed

Gas Engine OBD II Trouble Code List (P1xxx Codes)

DTC	Trouble Code Title, Conditions & Possible Causes:
DTC: P1025 **Jetta:** 2.0L (BPY) **Transmissions:** All	**Fuel Pressure Regulator Valve Mechanical Malfunction Conditions:** Engine started, battery voltage must be at least 11.5v, all electrical components must be off, the ground between the engine and the chassis must be well connected, the exhaust system must be properly sealed between the catalytic converter and the cylinder head, the coolant temperature must be 80 degrees Celsius, and the oxygen sensor heater for oxygen sensor before the catalytic converter must be properly functioning. The ECM detected a mechanical malfunction of the fuel pressure regulator valve. **Possible Causes:** • Fuel Pressure Regulator Valve is faulty • Fuel Pressure Sensor is faulty • Low Fuel Pressure Sensor is faulty
DTC: P1093 **Jetta:** 2.0L (BPY) **Transmissions:** All	**Fuel Trim 2, Bank 1 Malfunction Conditions:** Engine started, battery voltage must be at least 11.5v, all electrical components must be off, the ground between the engine and the chassis must be well connected, the exhaust system must be properly sealed between the catalytic converter and the cylinder head, the coolant temperature must be 80 degrees Celsius, and the oxygen sensor heater for oxygen sensor before the catalytic converter must be properly functioning. The ECM detected a mechanical malfunction of the fuel pressure regulator valve. **Possible Causes:** • Fuel Pressure Regulator Valve is faulty • Fuel Pressure Sensor is faulty • Low Fuel Pressure Sensor is faulty
DTC: P1100 **Passat:** 2.0L (BPY) **Transmissions:** All	**O2 Sensor Circuit (Bank 1-Sensor 2) Heating Circuit Voltage Too Low Conditions:** Engine started, battery voltage must be at least 11.5v, all electrical components must be off, the ground between the engine and the chassis must be well connected, the exhaust system must be properly sealed between the catalytic converter and the cylinder head, and the oxygen sensor heater for oxygen sensor before the catalytic converter must be properly functioning. The ECM detected a voltage on the O2 sensor circuit that was below the parameters to function properly. **Note: For resistance testing of sensor heating, oxygen sensor should be cooled to ambient temperature. High temperatures at oxygen sensor may lead to inaccurate measurements.** **Possible Causes:** • Oxygen sensor (before catalytic converter) is faulty • Oxygen sensor (behind catalytic converter) is faulty • Oxygen sensor heater (before catalytic converter) is faulty • Oxygen sensor heater (behind catalytic converter) is faulty • Circuit wiring has a short to power or ground • Engine Component Power Supply Relay is faulty • E-box fuses for oxygen sensor are faulty • Leaks present in the exhaust manifold or exhaust pipes • HO2S signal wire and ground wire crossed in connector • HO2S element is fuel contaminated or has failed • ECM has failed
DTC: P1102 **2T, MIL: Yes** **Passat:** 1.8 (AEB, ATW, AUG, AWM), 2.0L (BPY), 2.8L (AHA, ATQ); **Jetta:** 1.8L (AWD, AWP, AWW), 2.0 (ABA, AEG, AVH, AZG, BBW, BBW, BEV), 2.8L (AFP, BDF); **Golf:** 1.8L (AWD, AWW), 2.0L (ABA, AEG, AVH, AZG, BEV); **GTI:** 1.8L (AWD, AWP, AWW), 2.0L (ABA, AEG), 2.8L (AFP, BDF); **NB:** 1.8L (APH, AWP, AWV, BNU, BKF), 2.0L (AEG, AVH, AZG, BEV); **NBvert:** (BDG); **Cabrio:** 2.0L (ABA) **Transmissions:** A/T	**O2 Sensor Circuit (Bank 1-Sensor 1) Short to B+ Conditions:** Engine started, battery voltage must be at least 11.5v, all electrical components must be off, the ground between the engine and the chassis must be well connected, the exhaust system must be properly sealed between the catalytic converter and the cylinder head, and the oxygen sensor heater for oxygen sensor before the catalytic converter must be properly functioning. The ECM detected a voltage on the O2 sensor circuit that was outside the parameters to function properly. **Note: For resistance testing of sensor heating, oxygen sensor should be cooled to ambient temperature. High temperatures at oxygen sensor may lead to inaccurate measurements.** **Possible Causes:** • Oxygen sensor (before catalytic converter) is faulty • Oxygen sensor (behind catalytic converter) is faulty • Oxygen sensor heater (before catalytic converter) is faulty • Oxygen sensor heater (behind catalytic converter) is faulty • Circuit wiring has a short to power or ground • Engine Component Power Supply Relay is faulty • E-box fuses for oxygen sensor are faulty • Leaks present in the exhaust manifold or exhaust pipes • HO2S signal wire and ground wire crossed in connector • HO2S element is fuel contaminated or has failed • ECM has failed

DTC	Trouble Code Title, Conditions & Possible Causes
DTC: P1103 **Passat:** 1.8 (AEB, ATW, AUG, AWM), 2.0L (BPY); **Jetta:** 2.0 (ABA, AEG, AVH, AZG, BBW, BEV), 2.8L (AFP, BDF); **Golf:** 2.0L (ABA, AEG, AVH, AZG, BEV); **GTI:** 2.0L (ABA, AEG), 2.8L (AFP, BDF); **NB:** 1.8L (APH, AWP, AWV, BNU, BKF), 2.0L (AEG, AVH, AZG, BEV); **Nbvert:** (BDG); **Cabrio:** 2.0L (ABA) **Transmissions:** A/T	**O2 Sensor Circuit (Bank 1-Sensor 1) Output Too Low Conditions:** Engine started, battery voltage must be at least 11.5v, all electrical components must be off, the ground between the engine and the chassis must be well connected, the exhaust system must be properly sealed between the catalytic converter and the cylinder head, and the oxygen sensor heater for oxygen sensor before the catalytic converter must be properly functioning. The ECM detected a voltage on the O2 sensor circuit that was outside the parameters to function properly. **Note: For resistance testing of sensor heating, oxygen sensor should be cooled to ambient temperature. High temperatures at oxygen sensor may lead to inaccurate measurements.** **Possible Causes:** • Oxygen sensor (before catalytic converter) is faulty • Oxygen sensor (behind catalytic converter) is faulty • Oxygen sensor heater (before catalytic converter) is faulty • Oxygen sensor heater (behind catalytic converter) is faulty • Circuit wiring has a short to power or ground • Engine Component Power Supply Relay is faulty • E-box fuses for oxygen sensor are faulty • Leaks present in the exhaust manifold or exhaust pipes • HO2S signal wire and ground wire crossed in connector • HO2S element is fuel contaminated or has failed • ECM has failed
DTC: P1105 **2T, MIL: Yes** **Passat:** 1.8 (AEB, ATW, AUG, AWM), 2.0L (BPY), 2.8L (AHA, ATQ); **Jetta:** 1.8L (AWD, AWP, AWW), 2.0 (ABA, AEG, AVH, AZG, BBW, BEV), 2.8L (AFP, BDF); **Golf:** 1.8L (AWD, AWW), 2.0L (ABA, AEG, AVH, AZG, BEV); **GTI:** 1.8L (AWD, AWP, AWW), 2.0L (ABA, AEG), 2.8L (AFP, BDF); **NB:** 1.8L (APH, AWP, AWV, BNU, BKF), 2.0L (AEG, AVH, AZG, BEV); **Nbvert:** (BDG); **Cabrio:** 2.0L (ABA) **Transmissions:** A/T	**O2 Sensor Circuit (Bank 1-Sensor 2) Short to B+ Conditions:** Engine started, battery voltage must be at least 11.5v, all electrical components must be off, the ground between the engine and the chassis must be well connected, the exhaust system must be properly sealed between the catalytic converter and the cylinder head, and the oxygen sensor heater for oxygen sensor before the catalytic converter must be properly functioning. The ECM detected a voltage on the O2 sensor circuit that was outside the parameters to function properly. **Note: For resistance testing of sensor heating, oxygen sensor should be cooled to ambient temperature. High temperatures at oxygen sensor may lead to inaccurate measurements.** **Possible Causes:** • Oxygen sensor (before catalytic converter) is faulty • Oxygen sensor (behind catalytic converter) is faulty • Oxygen sensor heater (before catalytic converter) is faulty • Oxygen sensor heater (behind catalytic converter) is faulty • Circuit wiring has a short to power or ground • Engine Component Power Supply Relay is faulty • E-box fuses for oxygen sensor are faulty • Leaks present in the exhaust manifold or exhaust pipes • HO2S signal wire and ground wire crossed in connector • HO2S element is fuel contaminated or has failed • ECM has failed
DTC: P1107 **Passat:** 2.8L (AHA, ATQ) **Transmissions:** A/T	**O2 Sensor Circuit (Bank 2-Sensor 1) Voltage Too Low Conditions:** Engine started, battery voltage must be at least 11.5v, all electrical components must be off, the ground between the engine and the chassis must be well connected, the exhaust system must be properly sealed between the catalytic converter and the cylinder head, and the oxygen sensor heater for oxygen sensor before the catalytic converter must be properly functioning. The ECM detected a voltage on the O2 sensor circuit that was outside the parameters to function properly. **Note: For resistance testing of sensor heating, oxygen sensor should be cooled to ambient temperature. High temperatures at oxygen sensor may lead to inaccurate measurements.** **Possible Causes:** • Oxygen sensor (before catalytic converter) is faulty • Oxygen sensor (behind catalytic converter) is faulty • Oxygen sensor heater (before catalytic converter) is faulty • Oxygen sensor heater (behind catalytic converter) is faulty • Circuit wiring has a short to power or ground • Engine Component Power Supply Relay is faulty • E-box fuses for oxygen sensor are faulty • Leaks present in the exhaust manifold or exhaust pipes • HO2S signal wire and ground wire crossed in connector • HO2S element is fuel contaminated or has failed • ECM has failed

DTC	Trouble Code Title, Conditions & Possible Causes
DTC: P1110 **Passat:** 2.8L (AHA, ATQ) **Transmissions:** A/T	**O2 Sensor Circuit (Bank 2-Sensor 2) Short to B+ Conditions:** Engine started, battery voltage must be at least 11.5v, all electrical components must be off, the ground between the engine and the chassis must be well connected, the exhaust system must be properly sealed between the catalytic converter and the cylinder head, and the oxygen sensor heater for oxygen sensor before the catalytic converter must be properly functioning. The ECM detected a voltage on the O2 sensor circuit that was outside the parameters to function properly. **Note: For resistance testing of sensor heating, oxygen sensor should be cooled to ambient temperature. High temperatures at oxygen sensor may lead to inaccurate measurements.** **Possible Causes:** • Oxygen sensor (before catalytic converter) is faulty • Oxygen sensor (behind catalytic converter) is faulty • Oxygen sensor heater (before catalytic converter) is faulty • Oxygen sensor heater (behind catalytic converter) is faulty • Circuit wiring has a short to power or ground • Engine Component Power Supply Relay is faulty • E-box fuses for oxygen sensor are faulty • Leaks present in the exhaust manifold or exhaust pipes • HO2S signal wire and ground wire crossed in connector • HO2S element is fuel contaminated or has failed • ECM has failed
DTC: P1111 **Passat:** 1.8 (AEB, ATW, AUG, AWM), 2.0L (BPY); **Jetta:** 1.8L (AWD, AWP, AWW), 2.0 (ABA, BPY); **Golf:** 1.8L (AWD, AWW), 2.0L (ABA); **GTI:** 1.8L (AWD, AWP, AWW), 2.0L (ABA); **NB:** 1.8L (APH, AWP, AWV, BNU, BKF); **Cabrio:** 2.0L (ABA) **Transmissions:** All	**O2 Control (Bank 1) System Too Lean Conditions:** Engine started, battery voltage must be at least 11.5v, all electrical components must be off, the ground between the engine and the chassis must be well connected, the exhaust system must be properly sealed between the catalytic converter and the cylinder head, and the oxygen sensor heater for oxygen sensor before the catalytic converter must be properly functioning. The ECM detected a measurement on the O2 sensor circuit that was outside the parameters to function properly. **Note: For resistance testing of sensor heating, oxygen sensor should be cooled to ambient temperature. High temperatures at oxygen sensor may lead to inaccurate measurements.** **Note: When an O2S malfunction (P0131 to P0414) is also stored with this malfunction, the O2S malfunction(s) should be repaired first.** **Possible Causes:** • Oxygen sensor (before catalytic converter) is faulty • Oxygen sensor (behind catalytic converter) is faulty • Oxygen sensor heater (before catalytic converter) is faulty • Oxygen sensor heater (behind catalytic converter) is faulty • Circuit wiring has a short to power or ground • Engine Component Power Supply Relay is faulty • E-box fuses for oxygen sensor are faulty • Leaks present in the exhaust manifold or exhaust pipes • HO2S signal wire and ground wire crossed in connector • HO2S element is fuel contaminated or has failed • ECM has failed
DTC: P1112 **Passat:** 1.8 (AEB, ATW, AUG, AWM), 2.0L (BPY); **Jetta:** 1.8L (AWD, AWP, AWW), 2.0 (ABA, BPY); **Golf:** 1.8L (AWD, AWW), 2.0L (ABA); **GTI:** 1.8L (AWD, AWP, AWW), 2.0L (ABA); **NB:** 1.8L (APH, AWP, AWV, BNU, BKF); **Cabrio:** 2.0L (ABA) **Transmissions:** All	**O2 Control (Bank 1) System Too Rich Conditions:** Engine started, battery voltage must be at least 11.5v, all electrical components must be off, the ground between the engine and the chassis must be well connected, the exhaust system must be properly sealed between the catalytic converter and the cylinder head, and the oxygen sensor heater for oxygen sensor before the catalytic converter must be properly functioning. The ECM detected a measurement on the O2 sensor circuit that was outside the parameters to function properly. **Note: For resistance testing of sensor heating, oxygen sensor should be cooled to ambient temperature. High temperatures at oxygen sensor may lead to inaccurate measurements.** **Note: When an O2S malfunction (P0131 to P0414) is also stored with this malfunction, the O2S malfunction(s) should be repaired first.** **Possible Causes:** • Oxygen sensor (before catalytic converter) is faulty • Oxygen sensor (behind catalytic converter) is faulty • Oxygen sensor heater (before catalytic converter) is faulty • Oxygen sensor heater (behind catalytic converter) is faulty • Circuit wiring has a short to power or ground • Engine Component Power Supply Relay is faulty • E-box fuses for oxygen sensor are faulty • Leaks present in the exhaust manifold or exhaust pipes • HO2S signal wire and ground wire crossed in connector • HO2S element is fuel contaminated or has failed • ECM has failed

DTC	Trouble Code Title, Conditions & Possible Causes
DTC: P1113 **2T, MIL: Yes** **Passat:** 1.8 (AEB, ATW, AUG, AWM), 2.0L (BPY); **Jetta:** 1.8L (AWD, AWP, AWW), 2.0L (ABA); **Golf:** 1.8L (AWD, AWW), 2.0L (ABA); **GTI:** 1.8L (AWD, AWP, AWW), 2.0L (ABA); **NB:** 1.8L (APH, AWP, AWV, BNU, BKF); **Cabrio:** 2.0L (ABA) **Transmissions:** All	**O2 Control (Bank 1 Sensor 1) Internal Resistance Too High Conditions:** Engine started, battery voltage must be at least 11.5v, all electrical components must be off, the ground between the engine and the chassis must be well connected, the exhaust system must be properly sealed between the catalytic converter and the cylinder head, and the oxygen sensor heater for oxygen sensor before the catalytic converter must be properly functioning. The ECM detected a measurement on the O2 sensor circuit that was outside the parameters to function properly. **Note: For resistance testing of sensor heating, oxygen sensor should be cooled to ambient temperature. High temperatures at oxygen sensor may lead to inaccurate measurements.** **Possible Causes:** • Oxygen sensor (before catalytic converter) is faulty • Oxygen sensor (behind catalytic converter) is faulty • Oxygen sensor heater (before catalytic converter) is faulty • Oxygen sensor heater (behind catalytic converter) is faulty • Circuit wiring has a short to power or ground • Engine Component Power Supply Relay is faulty • E-box fuses for oxygen sensor are faulty • Leaks present in the exhaust manifold or exhaust pipes • HO2S signal wire and ground wire crossed in connector • HO2S element is fuel contaminated or has failed • ECM has failed
DTC: P1114 **2T, MIL: Yes** **Passat:** 1.8 (AEB, ATW, AUG, AWM), 2.0L (BPY), 2.8L (AHA); **Jetta:** 1.8L (AWD, AWP, AWW), 2.0L (ABA); **Golf:** 1.8L (AWD, AWW), 2.0L (ABA); **GTI:** 1.8L (AWD, AWP, AWW), 2.0L (ABA); **NB:** 1.8L (APH, AWP, AWV, BNU, BKF); **Cabrio:** 2.0L (ABA) **Transmissions:** All	**O2 Control (Bank 1 Sensor 2) Internal Resistance Too High Conditions:** Engine started, battery voltage must be at least 11.5v, all electrical components must be off, the ground between the engine and the chassis must be well connected, the exhaust system must be properly sealed between the catalytic converter and the cylinder head, and the oxygen sensor heater for oxygen sensor before the catalytic converter must be properly functioning. The ECM detected a measurement on the O2 sensor circuit that was outside the parameters to function properly. **Note: For resistance testing of sensor heating, oxygen sensor should be cooled to ambient temperature. High temperatures at oxygen sensor may lead to inaccurate measurements.** **Possible Causes:** • Oxygen sensor (before catalytic converter) is faulty • Oxygen sensor (behind catalytic converter) is faulty • Oxygen sensor heater (before catalytic converter) is faulty • Oxygen sensor heater (behind catalytic converter) is faulty • Circuit wiring has a short to power or ground • Engine Component Power Supply Relay is faulty • E-box fuses for oxygen sensor are faulty • Leaks present in the exhaust manifold or exhaust pipes • HO2S signal wire and ground wire crossed in connector • HO2S element is fuel contaminated or has failed • ECM has failed
DTC: P1115 **2T, MIL: Yes** **Passat:** 1.8 (AEB, ATW, AUG, AWM), 2.0L (BPY); **Jetta:** 1.8L (AWD, AWP, AWW), 2.0 (ABA, AEG, AVH, AZG, BBW, BEV), 2.8L (AFP, BDF); **Golf:** 1.8L (AWD, AWW), 2.0L (ABA, AEG, AVH, AZG, BEV); **GTI:** 1.8L (AWD, AWP, AWW), 2.0L (ABA, AEG), 2.8L (AFP, BDF); **NB:** 1.8L (APH, AWP, AWV, BNU, BKF), 2.0L (AEG, AVH, AZG, BEV); **Nbvert:** (BDG); **Cabrio:** 2.0L (ABA) **Transmissions:** All	**O2 Control (Bank 1 Sensor 1) Short to Ground Conditions:** Engine started, battery voltage must be at least 11.5v, all electrical components must be off, the ground between the engine and the chassis must be well connected, the exhaust system must be properly sealed between the catalytic converter and the cylinder head, and the oxygen sensor heater for oxygen sensor before the catalytic converter must be properly functioning. The ECM detected a measurement on the O2 sensor circuit that was outside the parameters to function properly. **Note: For resistance testing of sensor heating, oxygen sensor should be cooled to ambient temperature. High temperatures at oxygen sensor may lead to inaccurate measurements.** **Possible Causes:** • Oxygen sensor (before catalytic converter) is faulty • Oxygen sensor (behind catalytic converter) is faulty • Oxygen sensor heater (before catalytic converter) is faulty • Oxygen sensor heater (behind catalytic converter) is faulty • Circuit wiring has a short to power or ground • Engine Component Power Supply Relay is faulty • E-box fuses for oxygen sensor are faulty • Leaks present in the exhaust manifold or exhaust pipes • HO2S signal wire and ground wire crossed in connector • HO2S element is fuel contaminated or has failed • ECM has failed

DTC	Trouble Code Title, Conditions & Possible Causes
DTC: P1116 **2T, MIL: Yes** **Passat:** 1.8 (AEB, ATW, AUG, AWM), 2.0L (BPY); **Jetta:** 1.8L (AWD, AWP, AWW), 2.0 (ABA, AEG, AVH, AZG, BBW, BEV), 2.8L (AFP, BDF); **Golf:** 1.8L (AWD, AWW), 2.0L (ABA, AEG, AVH, AZG, BEV); **GTI:** 1.8L (AWD, AWP, AWW), 2.0L (ABA, AEG), 2.8L (AFP, BDF); **NB:** 1.8L (APH, AWP, AWV, BNU, BKF), 2.0L (AEG, AVH, AZG, BEV); **Nbvert:** (BDG); **Cabrio:** 2.0L (ABA) **Transmissions:** All	**O2 Control (Bank 1 Sensor 1) Open Conditions:** Engine started, battery voltage must be at least 11.5v, all electrical components must be off, the ground between the engine and the chassis must be well connected, the exhaust system must be properly sealed between the catalytic converter and the cylinder head, and the oxygen sensor heater for oxygen sensor before the catalytic converter must be properly functioning. The ECM detected a measurement on the O2 sensor circuit that was outside the parameters to function properly. **Note: For resistance testing of sensor heating, oxygen sensor should be cooled to ambient temperature. High temperatures at oxygen sensor may lead to inaccurate measurements.** **Possible Causes:** • Oxygen sensor (before catalytic converter) is faulty • Oxygen sensor (behind catalytic converter) is faulty • Oxygen sensor heater (before catalytic converter) is faulty • Oxygen sensor heater (behind catalytic converter) is faulty • Circuit wiring has a short to power or ground • Engine Component Power Supply Relay is faulty • E-box fuses for oxygen sensor are faulty • Leaks present in the exhaust manifold or exhaust pipes • HO2S signal wire and ground wire crossed in connector • HO2S element is fuel contaminated or has failed • ECM has failed
DTC: P1117 **2T, MIL: Yes** **Passat:** 1.8 (AEB, ATW, AUG, AWM), 2.0L (BPY); **Jetta:** 1.8L (AWD, AWP, AWW), 2.0 (ABA, AEG, AVH, AZG, BBW, BEV), 2.8L (AFP, BDF); **Golf:** 1.8L (AWD, AWW), 2.0L (ABA, AEG, AVH, AZG, BEV); **GTI:** 1.8L (AWD, AWP, AWW), 2.0L (ABA, AEG), 2.8L (AFP, BDF); **NB:** 1.8L (APH, AWP, AWV, BNU, BKF), 2.0L (AEG, AVH, AZG, BEV); **Nbvert:** (BDG); **Cabrio:** 2.0L (ABA) **Transmissions:** All	**O2 Control (Bank 1 Sensor 2) Open Conditions:** Engine started, battery voltage must be at least 11.5v, all electrical components must be off, the ground between the engine and the chassis must be well connected, the exhaust system must be properly sealed between the catalytic converter and the cylinder head, and the oxygen sensor heater for oxygen sensor before the catalytic converter must be properly functioning. The ECM detected a measurement on the O2 sensor circuit that was outside the parameters to function properly. **Note: For resistance testing of sensor heating, oxygen sensor should be cooled to ambient temperature. High temperatures at oxygen sensor may lead to inaccurate measurements.** **Possible Causes:** • Oxygen sensor (before catalytic converter) is faulty • Oxygen sensor (behind catalytic converter) is faulty • Oxygen sensor heater (before catalytic converter) is faulty • Oxygen sensor heater (behind catalytic converter) is faulty • Circuit wiring has a short to power or ground • Engine Component Power Supply Relay is faulty • E-box fuses for oxygen sensor are faulty • Leaks present in the exhaust manifold or exhaust pipes • HO2S signal wire and ground wire crossed in connector • HO2S element is fuel contaminated or has failed • ECM has failed
DTC: P1118 **2T, MIL: Yes** **Passat:** 2.8L (AHA, ATW, AUG, ATQ), 2.0L (BPY); **Jetta:** 1.8L (AWD, AWP, AWW), 2.0 (ABA, AEG, AVH, AZG, BBW, BEV), 2.8L (AFP, BDF); **Golf:** 1.8L (AWD, AWW), 2.0L (ABA, AEG, AVH, AZG, BEV); **GTI:** 1.8L (AWD, AWP, AWW), 2.0L (ABA, AEG), 2.8L (AFP, BDF); **NB:** 1.8L (APH, AWP, AWV, BNU, BKF), 2.0L (AEG, AVH, AZG, BEV); **Nbvert:** (BDG); **Cabrio:** 2.0L (ABA) **Transmissions:** All	**O2 Sensor Heater Circ. (Bank 1-Sensor2) Open Conditions:** Engine started, battery voltage must be at least 11.5v, all electrical components must be off, the ground between the engine and the chassis must be well connected, the exhaust system must be properly sealed between the catalytic converter and the cylinder head, and the oxygen sensor heater for oxygen sensor before the catalytic converter must be properly functioning. The ECM detected a measurement on the O2 sensor circuit that was outside the parameters to function properly. **Note: For resistance testing of sensor heating, oxygen sensor should be cooled to ambient temperature. High temperatures at oxygen sensor may lead to inaccurate measurements.** **Possible Causes:** • Oxygen sensor (before catalytic converter) is faulty • Oxygen sensor (behind catalytic converter) is faulty • Oxygen sensor heater (before catalytic converter) is faulty • Oxygen sensor heater (behind catalytic converter) is faulty • Circuit wiring has a short to power or ground • Engine Component Power Supply Relay is faulty • E-box fuses for oxygen sensor are faulty • Leaks present in the exhaust manifold or exhaust pipes • HO2S signal wire and ground wire crossed in connector • HO2S element is fuel contaminated or has failed • ECM has failed

DTC	Trouble Code Title, Conditions & Possible Causes
DTC: P1127 **2T, MIL: Yes** **Passat:** 1.8 (AEB, ATW, AUG, AWM), 2.0L (ABA, BPY), 2.8L (AAA, AHA, ATQ), 4.0L (BDP); **Jetta:** 1.8L (AWD, AWP, AWW), 2.0 (ABA, AEG, AVH, AZG, BBW, BEV), 2.8L (AAA, AFP, BDF); **Golf:** 1.8L (AWD, AWW), 2.0L (ABA, AEG, AVH, AZG, BEV); **GTI:** 1.8L (AWD, AWP, AWW), 2.0L (ABA, AEG), 2.8L (AAA, AFP, BDF); **NB:** 1.8L (APH, AWP, AWV, BNU, BKF), 2.0L (AEG, AVH, AZG, BEV); **Nbvert:** (BDG); **Cabrio:** 2.0L (ABA) **Transmissions:** All	**Long Term Fuel Trim Add. Air. Bank 1 System Too Rich Conditions:** Engine started, battery voltage must be at least 11.5v, all electrical components must be off, the ground between the engine and the chassis must be well connected, the exhaust system must be properly sealed between the catalytic converter and the cylinder head, and the oxygen sensor heater for oxygen sensor before the catalytic converter must be properly functioning. The fuel mixture is so rich that the O2S control is on lean limit. **Note: After exhaust system repairs, make sure exhaust system is not under stress and that it has sufficient clearance from the bodywork. If necessary, loosen double clamps and align exhaust pipe so that sufficient clearance is maintained to the bodywork and support rings carry uniform loads. Do not use any silicone sealant. Traces of silicone components which are sucked into the engine are not burned there, and they damage the oxygen sensor.** **Possible Causes:** • MAF sensor circuit open • MAF sensor circuit shorted to ground • Air leak in the manifold • Secondary air injection system combi-valve stuck open • Secondary air injection system electrical short • Fuel pressure too high, leaks in the vacuum hose to fuel pressure regulator • Fuel pressure regulator has failed • Fuel injectors are dirty, faulty or do not close properly • ECM has failed
DTC: P1128 **2T, MIL: Yes** **Passat:** 2.0L (ABA, ATW, AUG, BPY), 2.8L (AAA, AHA, ATQ), 4.0L (BDP); **Jetta:** 1.8L (AWD, AWP, AWW), 2.0 (ABA, AEG, AVH, AZG, BEV), 2.8L (AAA, AFP, BDF); **Golf:** 1.8L (AWD, AWW), 2.0L (ABA, AEG, AVH, AZG, BEV); **GTI:** 1.8L (AWD, AWP, AWW), 2.0L (ABA, AEG), 2.8L (AAA, AFP, BDF); **NB:** 1.8L (APH, AWP, AWV, BNU, BKF), 2.0L (AEG, AVH, AZG, BEV); **Nbvert:** (BDG); **Cabrio:** 2.0L (ABA) **Transmissions:** All	**Long Term Fuel Trim Add. Air. Bank 1 System Too Lean Conditions:** Engine started, battery voltage must be at least 11.5v, all electrical components must be off, the ground between the engine and the chassis must be well connected, the exhaust system must be properly sealed between the catalytic converter and the cylinder head, and the oxygen sensor heater for oxygen sensor before the catalytic converter must be properly functioning. The fuel mixture is so rich that the O2S control is on lean limit. **Note: When an O2S malfunction (P0131 to P0414) is also stored with this malfunction, the O2S malfunction(s) should be repaired first.** **Note: After exhaust system repairs, make sure exhaust system is not under stress and that it has sufficient clearance from the bodywork. If necessary, loosen double clamps and align exhaust pipe so that sufficient clearance is maintained to the bodywork and support rings carry uniform loads. Do not use any silicone sealant. Traces of silicone components which are sucked into the engine are not burned there, and they damage the oxygen sensor.** **Possible Causes:** • Fuel pressure is too low or fuel quantity supplied is too low • Fuel filter faulty • Transfer fuel pump has failed • Fuel injector is faulty (sticking or not opening) • Engine speed (RPM) sensor is faulty • MAF sensor circuit open • MAF sensor circuit shorted to ground • Air leak in the manifold • Secondary air injection system combi-valve stuck open • Secondary air injection system electrical short • ECM has failed
DTC: P1129 **Passat:** 2.8L (AHA, ATQ), 4.0L (BDP) **Transmissions:** All	**Long Term Fuel Trim at Rich Limit Conditions:** Engine started, battery voltage must be at least 11.5v, all electrical components must be off, the ground between the engine and the chassis must be well connected, the exhaust system must be properly sealed between the catalytic converter and the cylinder head, and the oxygen sensor heater for oxygen sensor before the catalytic converter must be properly functioning. The ECM detected the HO2S circuit was too rich, or that it could no longer change Fuel Trim because it was at its lean limit. **Possible Causes:** • Air intake system leaking, vacuum hoses leaking or damaged • Air leaks located after the MAF sensor mounting location • EGR valve sticking, EGR diaphragm leaking, or gasket leaking • EVAP vapor recovery system has failed • Excessive fuel pressure, leaking or contaminated fuel injectors • Exhaust leaks before or near the HO2S(s) mounting location • Fuel pressure regulator is leaking or damaged • HO2S circuits wet or oily, corroded, or poor terminal contact • HO2S is damaged or it has failed • HO2S signal circuit open, shorted to ground, shorted to power • Low fuel pressure or vehicle driven until it was out of fuel • Oil dipstick not seated or engine oil level too high (overfilled)

DTC	Trouble Code Title, Conditions & Possible Causes
DTC: P1130 **Passat:** 2.8L (AHA, ATQ), 4.0L (BDP) **Transmissions:** All	**Long Term Fuel Trim at Lean Limit Conditions:** Engine started, battery voltage must be at least 11.5v, all electrical components must be off, the ground between the engine and the chassis must be well connected, the exhaust system must be properly sealed between the catalytic converter and the cylinder head, and the oxygen sensor heater for oxygen sensor before the catalytic converter must be properly functioning. The ECM detected the HO2S circuit was too lean, or that it could no longer change Fuel Trim because it was at its lean limit. **Possible Causes:** • Air intake system leaking, vacuum hoses leaking or damaged • Air leaks located after the MAF sensor mounting location • EGR valve sticking, EGR diaphragm leaking, or gasket leaking • EVAP vapor recovery system has failed • Excessive fuel pressure, leaking or contaminated fuel injectors • Exhaust leaks before or near the HO2S(s) mounting location • Fuel pressure regulator is leaking or damaged • HO2S circuits wet or oily, corroded, or poor terminal contact • HO2S is damaged or it has failed • HO2S signal circuit open, shorted to ground, shorted to power • Low fuel pressure or vehicle driven until it was out of fuel • Oil dipstick not seated or engine oil level too high (overfilled)
DTC: P1136 **2T, MIL: Yes** **Passat:** 1.8 (AEB, ATW, AUG, AWM), 2.0L (BPY), 2.8L (AHA, ATQ), 4.0L (BDP); **Jetta:** 1.8L (AWD, AWP, AWW), 2.0 (ABA, AEG, AVH, AZG, BBW, BEV), 2.8L (AFP, BDF); **Golf:** 1.8L (AWD, AWW), 2.0L (ABA, AEG, AVH, AZG, BEV); **GTI:** 1.8L (AWD, AWP, AWW), 2.0L (ABA, AEG), 2.8L (AFP, BDF); **NB:** 1.8L (APH, AWP, AWV, BNU, BKF), 2.0L (AEG, AVH, AZG, BEV); **Nbvert:** (BDG); **Cabrio:** 2.0L (ABA) **Transmissions:** All	**Long Term Fuel Trim Add. Fuel, Bank 1 System Too Lean Conditions:** Engine started, battery voltage must be at least 11.5v, all electrical components must be off, the ground between the engine and the chassis must be well connected, the exhaust system must be properly sealed between the catalytic converter and the cylinder head, and the oxygen sensor heater for oxygen sensor before the catalytic converter must be properly functioning. The ECM detected the HO2S circuit was too lean, or that it could no longer change Fuel Trim because it was at its lean limit. **Possible Causes:** • Air intake system leaking, vacuum hoses leaking or damaged • Air leaks located after the MAF sensor mounting location • EGR valve sticking, EGR diaphragm leaking, or gasket leaking • EVAP vapor recovery system has failed • Excessive fuel pressure, leaking or contaminated fuel injectors • Exhaust leaks before or near the HO2S(s) mounting location • Fuel pressure regulator is leaking or damaged • HO2S circuits wet or oily, corroded, or poor terminal contact • HO2S is damaged or it has failed • HO2S signal circuit open, shorted to ground, shorted to power • Low fuel pressure or vehicle driven until it was out of fuel • Oil dipstick not seated or engine oil level too high (overfilled)
DTC: P1137 **2T, MIL: Yes** **Passat:** 1.8 (AEB, ATW, AUG, AWM), 2.0L (BPY), 2.8L (AHA, ATQ), 4.0L (BDP); **Jetta:** 1.8L (AWD, AWP, AWW), 2.0 (ABA, AEG, AVH, AZG, BBW, BEV), 2.8L (AFP, BDF); **Golf:** 1.8L (AWD, AWW), 2.0L (ABA, AEG, AVH, AZG, BEV); **GTI:** 1.8L (AWD, AWP, AWW), 2.0L (ABA, AEG), 2.8L (AFP, BDF); **NB:** 1.8L (APH, AWP, AWV, BNU, BKF), 2.0L (AEG, AVH, AZG, BEV); **Nbvert:** (BDG); **Cabrio:** 2.0L (ABA) **Transmissions:** All	**Long Term Fuel Trim Add. Fuel, Bank 1 System Too Rich Conditions:** Engine started, battery voltage must be at least 11.5v, all electrical components must be off, the ground between the engine and the chassis must be well connected, the exhaust system must be properly sealed between the catalytic converter and the cylinder head, and the oxygen sensor heater for oxygen sensor before the catalytic converter must be properly functioning. The ECM detected the HO2S circuit was too rich, or that it could no longer change Fuel Trim because it was at its lean limit. **Possible Causes:** • Air intake system leaking, vacuum hoses leaking or damaged • Air leaks located after the MAF sensor mounting location • EGR valve sticking, EGR diaphragm leaking, or gasket leaking • EVAP vapor recovery system has failed • Excessive fuel pressure, leaking or contaminated fuel injectors • Exhaust leaks before or near the HO2S(s) mounting location • Fuel pressure regulator is leaking or damaged • HO2S circuits wet or oily, corroded, or poor terminal contact • HO2S is damaged or it has failed • HO2S signal circuit open, shorted to ground, shorted to power • Low fuel pressure or vehicle driven until it was out of fuel • Oil dipstick not seated or engine oil level too high (overfilled)

DTC	Trouble Code Title, Conditions & Possible Causes
DTC: P1138 **Passat:** 2.8L (AHA, ATQ), 4.0L (BDP) **Transmissions:** All	**Long Term Fuel Trim Add. Fuel, Bank 2 System Too Lean Conditions:** Engine started, battery voltage must be at least 11.5v, all electrical components must be off, the ground between the engine and the chassis must be well connected, the exhaust system must be properly sealed between the catalytic converter and the cylinder head, and the oxygen sensor heater for oxygen sensor before the catalytic converter must be properly functioning. The ECM detected the HO2S circuit was too lean, or that it could no longer change Fuel Trim because it was at its lean limit. **Possible Causes:** • Air intake system leaking, vacuum hoses leaking or damaged • Air leaks located after the MAF sensor mounting location • EGR valve sticking, EGR diaphragm leaking, or gasket leaking • EVAP vapor recovery system has failed • Excessive fuel pressure, leaking or contaminated fuel injectors • Exhaust leaks before or near the HO2S(s) mounting location • Fuel pressure regulator is leaking or damaged • HO2S circuits wet or oily, corroded, or poor terminal contact • HO2S is damaged or it has failed • HO2S signal circuit open, shorted to ground, shorted to power • Low fuel pressure or vehicle driven until it was out of fuel • Oil dipstick not seated or engine oil level too high (overfilled)
DTC: P1139 **Passat:** 2.8L (AHA, ATQ), 4.0L (BDP) **Transmissions:** All	**Long Term Fuel Trim Add. Fuel, Bank 2 System Too Rich Conditions:** Engine started, battery voltage must be at least 11.5v, all electrical components must be off, the ground between the engine and the chassis must be well connected, the exhaust system must be properly sealed between the catalytic converter and the cylinder head, and the oxygen sensor heater for oxygen sensor before the catalytic converter must be properly functioning. The ECM detected the HO2S circuit was too rich, or that it could no longer change Fuel Trim because it was at its lean limit. **Possible Causes:** • Air intake system leaking, vacuum hoses leaking or damaged • Air leaks located after the MAF sensor mounting location • EGR valve sticking, EGR diaphragm leaking, or gasket leaking • EVAP vapor recovery system has failed • Excessive fuel pressure, leaking or contaminated fuel injectors • Exhaust leaks before or near the HO2S(s) mounting location • Fuel pressure regulator is leaking or damaged • HO2S circuits wet or oily, corroded, or poor terminal contact • HO2S is damaged or it has failed • HO2S signal circuit open, shorted to ground, shorted to power • Low fuel pressure or vehicle driven until it was out of fuel • Oil dipstick not seated or engine oil level too high (overfilled)
DTC: P1141 **Passat:** 1.8 (AEB, ATW, AUG, AWM), 2.0L (BPY); **Jetta:** 1.8L (AWD, AWP, AWW), 2.0 (ABA, AEG, AVH, AZG, BBW, BEV), 2.8L (AFP, BDF); **Golf:** 1.8L (AWD, AWW), 2.0L (ABA, AEG, AVH, AZG, BEV); **GTI:** 1.8L (AWD, AWP, AWW), 2.0L (ABA, AEG), 2.8L (AFP, BDF); **NB:** 1.8L (APH, AWP, AWV, BNU, BKF), 2.0L (AEG, AVH, AZG, BEV); **Nbvert:** (BDG); **Cabrio:** 2.0L (ABA) **Transmissions:** All	**Load Calculation Cross Check Range/Performance Conditions:** Engine started, battery voltage must be at least 11.5v, all electrical components must be off, the ground between the engine and the chassis must be well connected, the exhaust system must be properly sealed between the catalytic converter and the cylinder head, and the oxygen sensor heater for oxygen sensor before the catalytic converter must be properly functioning. **Note: Vacuum in the intake system sucks in the leak detection spray with false air. Leak detection spray decreases ignition quality of the fuel mixture. This causes a drop in engine speed and changes the value produced by the Heated Oxygen Sensor.** **Note: Both Throttle Position (TP) sensor and Sender 2 for accelerator pedal position are located at the accelerator pedal and communicate the driver's intentions to the ECM completely independently of each other. Both sensors are stored in one housing.** **Possible Causes:** • Intake system is leaking • Signal is grounding • ECM has failed • Intake Manifold Runner Position Sensor is faulty • Intake system for leaks (false air) is faulty • Motor for intake flap is faulty • Mass Air Flow (MAF) sensor is faulty • Throttle Position (TP) sensor is faulty • Throttle valve control module is faulty

DTC	Trouble Code Title, Conditions & Possible Causes
DTC: P1142 **Passat:** 2.0L (BPY); **Jetta:** 2.8L (AFP, BDF); **Golf:** (AEG, AVH, AZG, BEV); **GTI:** 2.0L (AEG), 2.8L (AFP, BDF); **NB:** 1.8L (APH, AWP, AWV, BNU, BKF), 2.0L (AEG, AVH, AZG, BEV); **Nbvert:** (BDG) **Transmissions:** All	**Load Calculation Cross Check Lower Limit Conditions:** Engine started, battery voltage must be at least 11.5v, all electrical components must be off, the ground between the engine and the chassis must be well connected, the exhaust system must be properly sealed between the catalytic converter and the cylinder head, and the oxygen sensor heater for oxygen sensor before the catalytic converter must be properly functioning. **Note: Vacuum in the intake system sucks in the leak detection spray with false air. Leak detection spray decreases ignition quality of the fuel mixture. This causes a drop in engine speed and changes the value produced by the Heated Oxygen Sensor.** **Note: Both Throttle Position (TP) sensor and Sender 2 for accelerator pedal position are located at the accelerator pedal and communicate the driver's intentions to the ECM completely independently of each other. Both sensors are stored in one housing.** **Possible Causes:** • Intake Manifold Runner Position Sensor is faulty • Intake system for leaks (false air) is faulty • Motor for intake flap is faulty • Mass Air Flow (MAF) sensor is faulty • Throttle Position (TP) sensor is faulty • Throttle valve control module is faulty • Intake system is leaking • Signal is grounding • ECM has failed
DTC: P1143 **Passat:** 1.8 (AEB, ATW, AUG, AWM), 2.0L (BPY); **Jetta:** 2.0 (ABA, AEG, AVH, AZG, BBW, BEV), 2.8L (AFP, BDF); **Golf:** 2.0L (ABA, AEG, AVH, AZG, BEV); **GTI:** 2.0L (ABA, AEG), 2.8L (AFP, BDF); **NB:** 1.8L (APH, AWP, AWV, BNU, BKF), 2.0L (AEG, AVH, AZG, BEV); **Nbvert:** (BDG); **Cabrio:** 2.0L (ABA) **Transmissions:** All	**Load Calculation Cross Check Upper Limit Conditions:** Engine started, battery voltage must be at least 11.5v, all electrical components must be off, the ground between the engine and the chassis must be well connected, the exhaust system must be properly sealed between the catalytic converter and the cylinder head, and the oxygen sensor heater for oxygen sensor before the catalytic converter must be properly functioning. **Note: Vacuum in the intake system sucks in the leak detection spray with false air. Leak detection spray decreases ignition quality of the fuel mixture. This causes a drop in engine speed and changes the value produced by the Heated Oxygen Sensor.** **Note: Both Throttle Position (TP) sensor and Sender 2 for accelerator pedal position are located at the accelerator pedal and communicate the driver's intentions to the ECM completely independently of each other. Both sensors are stored in one housing.** **Possible Causes:** • Intake Manifold Runner Position Sensor is faulty • Intake system for leaks (false air) is faulty • Motor for intake flap is faulty • Mass Air Flow (MAF) sensor is faulty • Throttle Position (TP) sensor is faulty • Throttle valve control module is faulty • Intake system is leaking • Signal is grounding • ECM has failed
DTC: P1149 **Passat:** 1.8 (AEB, ATW, AUG, AWM), 2.0L (BPY); **Jetta:** 1.8L (AWD, AWP, AWW), 2.0 (ABA, BPY); **Golf:** 1.8L (AWD, AWW), 2.0L (ABA); **GTI:** 1.8L (AWD, AWP, AWW), 2.0L (ABA); **NB:** 1.8L (APH, AWP, AWV, BNU, BKF); **Cabrio:** 2.0L (ABA) **Transmissions:** All	**O2 Control (Bank 1) Out of Range Conditions:** Engine started, battery voltage must be at least 11.5v, all electrical components must be off, the ground between the engine and the chassis must be well connected, the exhaust system must be properly sealed between the catalytic converter and the cylinder head, and the oxygen sensor heater for oxygen sensor before the catalytic converter must be properly functioning. The ECM detected a voltage on the O2 sensor circuit that was outside the parameters to function properly. **Note: For resistance testing of sensor heating, oxygen sensor should be cooled to ambient temperature. High temperatures at oxygen sensor may lead to inaccurate measurements.** **Possible Causes:** • Oxygen sensor (before catalytic converter) is faulty • Oxygen sensor (behind catalytic converter) is faulty • Oxygen sensor heater (before catalytic converter) is faulty • Oxygen sensor heater (behind catalytic converter) is faulty • Circuit wiring has a short to power or ground • Engine Component Power Supply Relay is faulty • E-box fuses for oxygen sensor are faulty • Leaks present in the exhaust manifold or exhaust pipes • HO2S signal wire and ground wire crossed in connector • HO2S element is fuel contaminated or has failed • ECM has failed

DTC	Trouble Code Title, Conditions & Possible Causes
DTC: P1150 All Models **Transmissions:** All	**Lack of HO2S-21 Switching, Fuel Trim At Rich/Lean Limit Conditions:** Engine running in closed loop, and the ECM detected the HO2S circuit was too lean or too rich, or that it could no longer correct Fuel Trim (i.e., the Fuel Trim was at its calibrated rich limit or its calibrated lean limit). **Possible Causes:** • Air intake system leaking, vacuum hoses leaking or damaged • Air leaks located after the MAF sensor mounting location • EGR valve sticking, EGR diaphragm leaking, or gasket leaking • EVAP vapor recovery system has failed • Excessive fuel pressure, leaking or contaminated fuel injectors • Exhaust leaks before or near the HO2S(s) mounting location • Fuel pressure regulator is leaking or damaged • HO2S circuits wet or oily, corroded, or poor terminal contact • HO2S signal circuit open, shorted to ground, shorted to power, or the sensor has failed • Low fuel pressure or vehicle driven until it was out of fuel • Oil dipstick not seated or engine oil level too high (overfilled)
DTC: P1151 **Passat:** 2.0L (BPY); **Jetta:** 2.0L (AEG, AVH, AZG, BBW, BEV), 2.8L (AFP, BDF); **Golf:** 2.0L (AEG, AVH, AZG, BEV); **GTI:** 2.0L (AEG), 2.8L (AFP, BDF); **NB:** 1.8L (APH, AWP, AWV, BNU, BKF), 2.0L (AEG, AVH, AZG, BEV); **Nbvert:** (BDG) **Transmissions:** All	**Long Term Fuel Trim (Bank1, Range 1) Leanness Lower Limit Exceeded Conditions:** Engine started, battery voltage must be at least 11.5v, all electrical components must be off, the ground between the engine and the chassis must be well connected, the exhaust system must be properly sealed between the catalytic converter and the cylinder head, and the oxygen sensor heater for oxygen sensor before the catalytic converter must be properly functioning. The fuel mixture is so lean that the O2S control is on rich limit. **Note: After exhaust system repairs, make sure exhaust system is not under stress and that it has sufficient clearance from the bodywork. If necessary, loosen double clamps and align exhaust pipe so that sufficient clearance is maintained to the bodywork and support rings carry uniform loads. Do not use any silicone sealant. Traces of silicone components which are sucked into the engine are not burned there, and they damage the oxygen sensor.** **Possible Causes:** • Exhaust system is damaged • Intake Manifold Runner Position Sensor is faulty • Intake system for leaks (false air) • Motor for intake flap has failed • MAF sensor circuit open • MAF sensor circuit shorted to ground • Air leak in the manifold • Secondary air injection system combi-valve stuck open • Secondary air injection system electrical short • Fuel pressure too high, leaks in the vacuum hose to fuel pressure regulator • Fuel pressure regulator has failed • Fuel injectors are dirty, faulty or do not close properly
DTC: P1152 **Passat:** 2.0L (BPY); **Jetta:** 2.0L (AEG, AVH, AZG, BBW, BEV), 2.8L (AFP, BDF); **Golf:** 2.0L (AEG, AVH, AZG, BEV); **GTI:** 2.0L (AEG), 2.8L (AFP, BDF); **NB:** 1.8L (APH, AWP, AWV, BNU, BKF), 2.0L (AEG, AVH, AZG, BEV); **Nbvert:** (BDG) **Transmissions:** All	**Long Term Fuel Trim (Bank1, Range 2) Leanness Lower Limit Exceeded Conditions:** Engine started, battery voltage must be at least 11.5v, all electrical components must be off, the ground between the engine and the chassis must be well connected, the exhaust system must be properly sealed between the catalytic converter and the cylinder head, and the oxygen sensor heater for oxygen sensor before the catalytic converter must be properly functioning. The fuel mixture is so lean that the O2S control is on rich limit. **Note: After exhaust system repairs, make sure exhaust system is not under stress and that it has sufficient clearance from the bodywork. If necessary, loosen double clamps and align exhaust pipe so that sufficient clearance is maintained to the bodywork and support rings carry uniform loads. Do not use any silicone sealant. Traces of silicone components which are sucked into the engine are not burned there, and they damage the oxygen sensor.** **Possible Causes:** • Exhaust system is damaged • Intake Manifold Runner Position Sensor is faulty • Intake system for leaks (false air) • Motor for intake flap has failed • MAF sensor circuit open • MAF sensor circuit shorted to ground • Air leak in the manifold • Secondary air injection system combi-valve stuck open • Secondary air injection system electrical short • Fuel pressure too high, leaks in the vacuum hose to fuel pressure regulator • Fuel pressure regulator has failed • Fuel injectors are dirty, faulty or do not close properly

DTC	Trouble Code Title, Conditions & Possible Causes
DTC: P1165 **Passat:** 2.0L (BPY); **Jetta:** 2.0L (AEG, AVH, AZG, BBW, BEV), 2.8L (AFP, BDF); **Golf:** 2.0 (AEG, AEG, AVH, AZG, BEV); **GTI:** 2.0L (AEG), 2.8L (AFP, BDF); **NB:** 1.8L (APH, AWP, AWV, BNU, BKF), 2.0L (AEG, AVH, AZG, BEV); **Nbvert:** (BDG) **Transmissions:** All	**Bank 1, Long Term Fuel Trim, Range 1 Rich Limit Exceeded Conditions:** Engine started, battery voltage must be at least 11.5v, all electrical components must be off, the ground between the engine and the chassis must be well connected, the exhaust system must be properly sealed between the catalytic converter and the cylinder head, and the oxygen sensor heater for oxygen sensor before the catalytic converter must be properly functioning. The fuel mixture is so rich that the O2S control is on lean limit. **Note: After exhaust system repairs, make sure exhaust system is not under stress and that it has sufficient clearance from the bodywork. If necessary, loosen double clamps and align exhaust pipe so that sufficient clearance is maintained to the bodywork and support rings carry uniform loads. Do not use any silicone sealant. Traces of silicone components which are sucked into the engine are not burned there, and they damage the oxygen sensor.** **Possible Causes:** • Exhaust system is damaged • Intake Manifold Runner Position Sensor is faulty • Intake system for leaks (false air) • Motor for intake flap has failed • MAF sensor circuit open • MAF sensor circuit shorted to ground • Air leak in the manifold • Secondary air injection system combi-valve stuck open • Secondary air injection system electrical short • Fuel pressure too high, leaks in the vacuum hose to fuel pressure regulator • Fuel pressure regulator has failed • Fuel injectors are dirty, faulty or do not close properly • EVAP canister system lacks a proper seal • Evaporative Emission (EVAP) canister purge regulator valve 1 is faulty • Leak Detection Pump (LDP) is faulty • Fuel injectors have failed • Oxygen sensor (before catalytic converter) is faulty • Oxygen sensor (behind catalytic converter) is faulty • Oxygen sensor heater (before catalytic converter) is faulty • Oxygen sensor heater (behind catalytic converter) is faulty
DTC: P1166 **Passat:** 2.0L (BPY); **Jetta:** 2.0L (AEG, AVH, AZG, BBW, BEV), 2.8L (AFP, BDF); **Golf:** 2.0L (AEG, AVH, AZG, BEV); **GTI:** 2.0L (AEG), 2.8L (AFP, BDF); **NB:** 1.8L (APH, AWP, AWV, BNU, BKF), 2.0L (AEG, AVH, AZG, BEV); **Nbvert:** (BDG) **Transmissions:** All	**Bank 1, Long Term Fuel Trim, Range 2 Rich Limit Exceeded Conditions:** Engine started, battery voltage must be at least 11.5v, all electrical components must be off, the ground between the engine and the chassis must be well connected, the exhaust system must be properly sealed between the catalytic converter and the cylinder head, and the oxygen sensor heater for oxygen sensor before the catalytic converter must be properly functioning. The fuel mixture is so rich that the O2S control is on lean limit. **Note: After exhaust system repairs, make sure exhaust system is not under stress and that it has sufficient clearance from the bodywork. If necessary, loosen double clamps and align exhaust pipe so that sufficient clearance is maintained to the bodywork and support rings carry uniform loads. Do not use any silicone sealant. Traces of silicone components which are sucked into the engine are not burned there, and they damage the oxygen sensor.** **Possible Causes:** • Exhaust system is damaged • Intake Manifold Runner Position Sensor is faulty • Intake system for leaks (false air) • Motor for intake flap has failed • MAF sensor circuit open • MAF sensor circuit shorted to ground • Air leak in the manifold • Secondary air injection system combi-valve stuck open • Secondary air injection system electrical short • Fuel pressure too high, leaks in the vacuum hose to fuel pressure regulator • Fuel pressure regulator has failed • Fuel injectors are dirty, faulty or do not close properly • EVAP canister system lacks a proper seal • Evaporative Emission (EVAP) canister purge regulator valve 1 is faulty • Leak Detection Pump (LDP) is faulty • Fuel injectors have failed • Oxygen sensor (before catalytic converter) is faulty • Oxygen sensor (behind catalytic converter) is faulty • Oxygen sensor heater (before catalytic converter) is faulty • Oxygen sensor heater (behind catalytic converter) is faulty

DTC	Trouble Code Title, Conditions & Possible Causes
DTC: P1171 **2T, MIL: Yes** **Passat:** 2.0L (BPY); **Jetta:** 1.8L (AWD, AWP, AWW), 2.0L (AEG, AVH, AZG, BBW, BEV), 2.8L (AFP, BDF); **Golf:** 1.8L (AWD, AWW), 2.0 (AEG, AVH, AZG, BEV); **GTI:** 1.8L (AWD, AWP, AWW), 2.0L (AEG), 2.8L (AFP, BDF); **NB:** 1.8L (APH, AWP, AWV, BNU, BKF), 2.0L (AEG, AVH, AZG, BEV); **Nbvert:** (BDG) **Transmissions:** All	**Throttle Actuation Potentiometer Sign.2 Range/Performance Conditions:** Engine started, battery voltage must be at least 11.5v, all electrical components must be off, the ground between the engine and the chassis must be well connected, coolant temperature must be at least 80 degrees Celsius and the accelerator pedal must be properly adjusted. The ECM detected an incorrect singal from the throttle potentiometer. **Note: If the complete throttle valve control module is current-less (e.g. connector disconnected) the throttle valve moves into a particular, specified mechanical position, which signals an increased idle speed with an engine at operating temperature. If only the Throttle Position (TP) actuator is current-less, the throttle valve also moves into the specified mechanical position (emergency running gap), however, since Closed Throttle Position (CTP) switch can still be recognized, an "almost normal idle RPM" is reached via the respective ignition angle retardation.** **Note: Terminal assignment at throttle control module is different in vehicles with and without cruise control. Characteristic: Steering column switch with operating module for cruise control.** **Possible Causes:** • Throttle valve control module has failed • Throttle valve is dirty or damaged • Throttle valve is not in a closed position • Voltage supply of throttle valve control module is shorted or open • ECM has failed
DTC: P1172 **2T, MIL: Yes** **Passat:** 2.0L (BPY); **Jetta:** 1.8L (AWD, AWP, AWW), 2.0L (AEG, AVH, AZG, BBW, BEV), 2.8L (AFP, BDF); **Golf:** 1.8L (AWD, AWW), 2.0 (AEG, AVH, AZG, BEV); **GTI:** 1.8L (AWD, AWP, AWW), 2.0L (AEG), 2.8L (AFP, BDF); **NB:** 1.8L (APH, AWP, AWV, BNU, BKF), 2.0L (AEG, AVH, AZG, BEV); **Nbvert:** (BDG) **Transmissions:** All	**Throttle Actuation Potentiometer Sign.2 Signal Too Low Conditions:** Engine started, battery voltage must be at least 11.5v, all electrical components must be off, the ground between the engine and the chassis must be well connected, coolant temperature must be at least 80 degrees Celsius and the accelerator pedal must be properly adjusted. The ECM detected an incorrect singal from the throttle potentiometer. **Note: If the complete throttle valve control module is current-less (e.g. connector disconnected) the throttle valve moves into a particular, specified mechanical position, which signals an increased idle speed with an engine at operating temperature. If only the Throttle Position (TP) actuator is current-less, the throttle valve also moves into the specified mechanical position (emergency running gap), however, since Closed Throttle Position (CTP) switch can still be recognized, an "almost normal idle RPM" is reached via the respective ignition angle retardation.** **Note: Terminal assignment at throttle control module is different in vehicles with and without cruise control. Characteristic: Steering column switch with operating module for cruise control.** **Possible Causes:** • Throttle valve control module has failed • Throttle valve is dirty or damaged • Throttle valve is not in a closed position • Voltage supply of throttle valve control module is shorted or open • ECM has failed
DTC: P1173 **2T, MIL: Yes** **Passat:** 2.0L (BPY); **Jetta:** 1.8L (AWD, AWP, AWW), 2.0L (AEG, AVH, AZG, BBW, BEV); **Golf:** 1.8L (AWD, AWW), 2.0 (AEG, AVH, AZG, BEV); **GTI:** 1.8L (AWD, AWP, AWW), 2.0L (AEG), 2.8L (AFP, BDF); **NB:** 1.8L (APH, AWP, AWV, BNU, BKF), 2.0L (AEG, AVH, AZG, BEV); **Nbvert:** (BDG) **Transmissions:** All	**Throttle Actuation Potentiometer Sign.2 Signal Too High Conditions:** Engine started, battery voltage must be at least 11.5v, all electrical components must be off, the ground between the engine and the chassis must be well connected, coolant temperature must be at least 80 degrees Celsius and the accelerator pedal must be properly adjusted. The ECM detected an incorrect singal from the throttle potentiometer. **Note: If the complete throttle valve control module is current-less (e.g. connector disconnected) the throttle valve moves into a particular, specified mechanical position, which signals an increased idle speed with an engine at operating temperature. If only the Throttle Position (TP) actuator is current-less, the throttle valve also moves into the specified mechanical position (emergency running gap), however, since Closed Throttle Position (CTP) switch can still be recognized, an "almost normal idle RPM" is reached via the respective ignition angle retardation.** **Note: Terminal assignment at throttle control module is different in vehicles with and without cruise control. Characteristic: Steering column switch with operating module for cruise control.** **Possible Causes:** • Throttle valve control module has failed • Throttle valve is dirty or damaged • Throttle valve is not in a closed position • Voltage supply of throttle valve control module is shorted or open • ECM has failed

DTC	Trouble Code Title, Conditions & Possible Causes
DTC: P1176 **2T, MIL: Yes** **Passat:** 1.8 (AEB, ATW, AWM), 2.0L (BPY), 2.8L (AAA, AHA, ATQ), 4.0L (BDP); **Jetta:** 1.8L (AWD, AWP, AWW), 2.0 (ABA, AEG, AVH, AZG, BBW, BEV), 2.5L (BGP, BGQ), 2.8L (AAA, AFP, BDF); **Golf:** 1.8L (AWD, AWW), 2.0L (ABA, AEG, AVH, AZG, BEV); **GTI:** 1.8L (AWD, AWP, AWW), 2.0L (ABA, AEG), 2.8L (AAA, AFP, BDF); **NB:** 1.8L (APH, AWP, AWV, BNU, BKF), 2.0L (AEG, AVH, AZG, BEV); **Nbvert:** (BDG); **Cabrio:** 2.0L (ABA) **Transmissions:** All	**O2 Correction Behind Catalyst B1 Limit Attained Conditions:** Engine started, battery voltage must be at least 11.5v, all electrical components must be off, the ground between the engine and the chassis must be well connected, the exhaust system must be properly sealed between the catalytic converter and the cylinder head, the coolant temperature must be at least 80 degrees Celsius, and the oxygen sensor heater for oxygen sensor before the catalytic converter must be properly functioning. The ECM has detected a malfunction of the oxygen sensor. **Note: Vacuum in the intake system sucks in the leak detection spray with false air. Leak detection spray decreases ignition quality of the fuel mixture. This causes a drop in engine speed and changes the value produced by the Heated Oxygen Sensor (HO2S).** **Note: Vehicle must be raised before connector for oxygen sensor is accessible.** **Note: The oxygen sensor before catalytic converter has a static regulation and can be differentiated from the oxygen sensor behind catalytic converter via a 6-pin connector.** **Possible Causes:** • O2 sensor circuit has shorted to ground or B+ • O2 sensor circuit is open • ECM has failed • O2 sensor has failed • Intake Manifold Runner Position Sensor is faulty • Intake system for leaks (false air) is faulty • Motor for intake flap is faulty
DTC: P1177 **Passat:** 2.8L (AHA, ATQ), 4.0L (BDP); **Jetta:** 2.8L (AFP, BDF); **GTI:** 2.8L (AFP, BDF) **Transmissions:** All	**O2 Correction Behind Catalyst B2 Limit Attained Conditions:** Engine started, battery voltage must be at least 11.5v, all electrical components must be off, the ground between the engine and the chassis must be well connected, the exhaust system must be properly sealed between the catalytic converter and the cylinder head, the coolant temperature must be at least 80 degrees Celsius, and the oxygen sensor heater for oxygen sensor before the catalytic converter must be properly functioning. The ECM has detected a malfunction of the oxygen sensor. **Note: Vacuum in the intake system sucks in the leak detection spray with false air. Leak detection spray decreases ignition quality of the fuel mixture. This causes a drop in engine speed and changes the value produced by the Heated Oxygen Sensor (HO2S).** **Note: Vehicle must be raised before connector for oxygen sensor is accessible.** **Note: The oxygen sensor before catalytic converter has a static regulation and can be differentiated from the oxygen sensor behind catalytic converter via a 6-pin connector.** **Possible Causes:** • O2 sensor circuit has shorted to ground or B+ • O2 sensor circuit is open • ECM has failed • O2 sensor has failed • Intake Manifold Runner Position Sensor is faulty • Intake system for leaks (false air) is faulty • Motor for intake flap is faulty
DTC: P1178 **Jetta:** 2.0L (AEG, AVH, AZG, BBW, BEV); **Golf:** 2.0 (AEG, AVH, AZG, BEV); **GTI:** 2.0L (AEG); **NB:** 2.0L (AEG, AVH, AZG, BEV); **Nbvert:** (BDG) **Transmissions:** All	**Linear O2 Sensor/Pump Current Open Circuit Conditions:** Engine started, battery voltage must be at least 11.5v, all electrical components must be off, the ground between the engine and the chassis must be well connected, the exhaust system must be properly sealed between the catalytic converter and the cylinder head, the coolant temperature must be at least 80 degrees Celsius, and the oxygen sensor heater for oxygen sensor before the catalytic converter must be properly functioning. The ECM has detected a malfunction of the oxygen sensor. **Note: Vacuum in the intake system sucks in the leak detection spray with false air. Leak detection spray decreases ignition quality of the fuel mixture. This causes a drop in engine speed and changes the value produced by the Heated Oxygen Sensor (HO2S).** **Note: Vehicle must be raised before connector for oxygen sensor is accessible.** **Possible Causes:** • O2 sensor circuit has shorted to ground or B+ • O2 sensor circuit is open • ECM has failed • O2 sensor has failed • Intake Manifold Runner Position Sensor is faulty • Intake system for leaks (false air) is faulty • Motor for intake flap is faulty

DTC	Trouble Code Title, Conditions & Possible Causes
DTC: P1179 **Jetta:** 2.0L (AEG, AVH, AZG, BBW, BEV); **Golf:** 2.0 (AEG, AVH, AZG, BEV); **GTI:** 2.0L (AEG); **NB:** 2.0L (AEG, AVH, AZG, BEV); **Nbvert:** (BDG) **Transmissions:** All	**Linear 02 Sensor/Pump Current Short to Ground Conditions:** Engine started, battery voltage must be at least 11.5v, all electrical components must be off, the ground between the engine and the chassis must be well connected, the exhaust system must be properly sealed between the catalytic converter and the cylinder head, the coolant temperature must be at least 80 degrees Celsius, and the oxygen sensor heater for oxygen sensor before the catalytic converter must be properly functioning. The ECM has detected a malfunction of the oxygen sensor. **Note: Vacuum in the intake system sucks in the leak detection spray with false air. Leak detection spray decreases ignition quality of the fuel mixture. This causes a drop in engine speed and changes the value produced by the Heated Oxygen Sensor (HO2S).** **Note: Vehicle must be raised before connector for oxygen sensor is accessible.** **Possible Causes:** • 02 sensor circuit has shorted to ground or B+ • 02 sensor circuit is open • ECM has failed • 02 sensor has failed • Intake Manifold Runner Position Sensor is faulty • Intake system for leaks (false air) is faulty • Motor for intake flap is faulty
DTC: P1180 **Jetta:** 2.0L (AEG, AVH, AZG, BBW, BEV); **Golf:** 2.0 (AEG, AVH, AZG, BEV); **GTI:** 2.0L (AEG); **NB:** 2.0L (AEG, AVH, AZG, BEV); **Nbvert:** (BDG) **Transmissions:** All	**Linear 02 Sensor / Pump Current Short to B+ Conditions:** Engine started, battery voltage must be at least 11.5v, all electrical components must be off, the ground between the engine and the chassis must be well connected, the exhaust system must be properly sealed between the catalytic converter and the cylinder head, the coolant temperature must be at least 80 degrees Celsius, and the oxygen sensor heater for oxygen sensor before the catalytic converter must be properly functioning. The ECM has detected a malfunction of the oxygen sensor. **Note: Vacuum in the intake system sucks in the leak detection spray with false air. Leak detection spray decreases ignition quality of the fuel mixture. This causes a drop in engine speed and changes the value produced by the Heated Oxygen Sensor (HO2S).** **Note: Vehicle must be raised before connector for oxygen sensor is accessible.** **Possible Causes:** • 02 sensor circuit has shorted to ground or B+ • 02 sensor circuit is open • ECM has failed • 02 sensor has failed • Intake Manifold Runner Position Sensor is faulty • Intake system for leaks (false air) is faulty • Motor for intake flap is faulty
DTC: P1181 **Jetta:** 2.0L (AEG, AVH, AZG, BBW, BEV); **Golf:** 2.0 (AEG, AVH, AZG, BEV); **GTI:** 2.0L (AEG); **NB:** 2.0L (AEG, AVH, AZG, BEV); **Nbvert:** (BDG) **Transmissions:** All	**Linear 02 Sensor / Reference Voltage Open Circuit Conditions:** Engine started, battery voltage must be at least 11.5v, all electrical components must be off, the ground between the engine and the chassis must be well connected, the exhaust system must be properly sealed between the catalytic converter and the cylinder head, the coolant temperature must be at least 80 degrees Celsius, and the oxygen sensor heater for oxygen sensor before the catalytic converter must be properly functioning. The ECM has detected a malfunction of the oxygen sensor. **Note: Vacuum in the intake system sucks in the leak detection spray with false air. Leak detection spray decreases ignition quality of the fuel mixture. This causes a drop in engine speed and changes the value produced by the Heated Oxygen Sensor (HO2S).** **Note: Vehicle must be raised before connector for oxygen sensor is accessible.** **Possible Causes:** • 02 sensor circuit has shorted to ground or B+ • 02 sensor circuit is open • ECM has failed • 02 sensor has failed • Intake Manifold Runner Position Sensor is faulty • Intake system for leaks (false air) is faulty • Motor for intake flap is faulty

DTC	Trouble Code Title, Conditions & Possible Causes
DTC: P1182 **Jetta:** 2.0L (AEG, AVH, AZG, BBW, BEV); **Golf:** 2.0 (AEG, AVH, AZG, BEV); **GTI:** 2.0L (AEG); **NB:** 2.0L (AEG, AVH, AZG, BEV); **Nbvert:** (BDG) **Transmissions:** All	**Linear O2 Sensor/Reference Voltage Short to Ground Conditions:** Engine started, battery voltage must be at least 11.5v, all electrical components must be off, the ground between the engine and the chassis must be well connected, the exhaust system must be properly sealed between the catalytic converter and the cylinder head, the coolant temperature must be at least 80 degrees Celsius, and the oxygen sensor heater for oxygen sensor before the catalytic converter must be properly functioning. The ECM has detected a malfunction of the oxygen sensor. **Note: Vacuum in the intake system sucks in the leak detection spray with false air. Leak detection spray decreases ignition quality of the fuel mixture. This causes a drop in engine speed and changes the value produced by the Heated Oxygen Sensor (HO2S).** **Note: Vehicle must be raised before connector for oxygen sensor is accessible.** **Possible Causes:** • O2 sensor circuit has shorted to ground or B+ • O2 sensor circuit is open • ECM has failed • O2 sensor has failed • Intake Manifold Runner Position Sensor is faulty • Intake system for leaks (false air) is faulty • Motor for intake flap is faulty
DTC: P1183 **Jetta:** 2.0L (AEG, AVH, AZG, BBW, BEV); **Golf:** 2.0 (AEG, AVH, AZG, BEV); **GTI:** 2.0L (AEG); **NB:** 2.0L (AEG, AVH, AZG, BEV); **Nbvert:** (BDG) **Transmissions:** All	**Linear O2 Sensor / Reference Voltage Short to B+ Conditions:** Engine started, battery voltage must be at least 11.5v, all electrical components must be off, the ground between the engine and the chassis must be well connected, the exhaust system must be properly sealed between the catalytic converter and the cylinder head, the coolant temperature must be at least 80 degrees Celsius, and the oxygen sensor heater for oxygen sensor before the catalytic converter must be properly functioning. The ECM has detected a malfunction of the oxygen sensor. **Note: Vacuum in the intake system sucks in the leak detection spray with false air. Leak detection spray decreases ignition quality of the fuel mixture. This causes a drop in engine speed and changes the value produced by the Heated Oxygen Sensor (HO2S).** **Note: Vehicle must be raised before connector for oxygen sensor is accessible.** **Possible Causes:** • O2 sensor circuit has shorted to ground or B+ • O2 sensor circuit is open • ECM has failed • O2 sensor has failed • Intake Manifold Runner Position Sensor is faulty • Intake system for leaks (false air) is faulty • Motor for intake flap is faulty
DTC: P1184 **Jetta:** 2.0L (AEG, AVH, AZG, BEV); **Golf:** 2.0 (AEG, AVH, AZG, BEV); **GTI:** 2.0L (AEG); **NB:** 2.0L (AEG, AVH, AZG, BEV); **Nbvert:** (BDG) **Transmissions:** All	**Linear O2 Sensor / Common Ground Wire Open Circuit Conditions:** Engine started, battery voltage must be at least 11.5v, all electrical components must be off, the ground between the engine and the chassis must be well connected, the exhaust system must be properly sealed between the catalytic converter and the cylinder head, the coolant temperature must be at least 80 degrees Celsius, and the oxygen sensor heater for oxygen sensor before the catalytic converter must be properly functioning. The ECM has detected a malfunction of the oxygen sensor. **Note: Vacuum in the intake system sucks in the leak detection spray with false air. Leak detection spray decreases ignition quality of the fuel mixture. This causes a drop in engine speed and changes the value produced by the Heated Oxygen Sensor (HO2S).** **Note: Vehicle must be raised before connector for oxygen sensor is accessible.** **Possible Causes:** • O2 sensor circuit has shorted to ground or B+ • O2 sensor circuit is open • ECM has failed • O2 sensor has failed • Intake Manifold Runner Position Sensor is faulty • Intake system for leaks (false air) is faulty • Motor for intake flap is faulty

DTC	Trouble Code Title, Conditions & Possible Causes
DTC: P1185 **Jetta:** 2.0L (AEG, AVH, AZG, BBW, BEV); **Golf:** 2.0 (AEG, AVH, AZG, BEV); **GTI:** 2.0L (AEG); **NB:** 2.0L (AEG, AVH, AZG, BEV); **Nbvert:** (BDG) **Transmissions:** All	**Linear O2 Sensor/Common Ground Wire Short to Ground Conditions:** Engine started, battery voltage must be at least 11.5v, all electrical components must be off, the ground between the engine and the chassis must be well connected, the exhaust system must be properly sealed between the catalytic converter and the cylinder head, the coolant temperature must be at least 80 degrees Celsius, and the oxygen sensor heater for oxygen sensor before the catalytic converter must be properly functioning. The ECM has detected a malfunction of the oxygen sensor. **Note: Vacuum in the intake system sucks in the leak detection spray with false air. Leak detection spray decreases ignition quality of the fuel mixture. This causes a drop in engine speed and changes the value produced by the Heated Oxygen Sensor (HO2S).** **Note: Vehicle must be raised before connector for oxygen sensor is accessible.** **Possible Causes:** • O2 sensor circuit has shorted to ground or B+ • O2 sensor circuit is open • ECM has failed • O2 sensor has failed • Intake Manifold Runner Position Sensor is faulty • Intake system for leaks (false air) is faulty • Motor for intake flap is faulty
DTC: P1186 **Jetta:** 2.0L (AEG, AVH, AZG, BBW, BEV); **Golf:** 2.0 (AEG, AVH, AZG, BEV); **GTI:** 2.0L (AEG); **NB:** 2.0L (AEG, AVH, AZG, BEV); **Nbvert:** (BDG) **Transmissions:** All	**Linear O2 Sensor/Common Ground Wire Short to B+ Conditions:** Engine started, battery voltage must be at least 11.5v, all electrical components must be off, the ground between the engine and the chassis must be well connected, the exhaust system must be properly sealed between the catalytic converter and the cylinder head, the coolant temperature must be at least 80 degrees Celsius, and the oxygen sensor heater for oxygen sensor before the catalytic converter must be properly functioning. The ECM has detected a malfunction of the oxygen sensor. **Note: Vacuum in the intake system sucks in the leak detection spray with false air. Leak detection spray decreases ignition quality of the fuel mixture. This causes a drop in engine speed and changes the value produced by the Heated Oxygen Sensor (HO2S).** **Note: Vehicle must be raised before connector for oxygen sensor is accessible.** **Possible Causes:** • O2 sensor circuit has shorted to ground or B+ • O2 sensor circuit is open • ECM has failed • O2 sensor has failed • Intake Manifold Runner Position Sensor is faulty • Intake system for leaks (false air) is faulty • Motor for intake flap is faulty
DTC: P1196 **Passat:** 1.8 (AEB, ATW, AUG, AWM), 2.0L (BPY), 2.8L (AHA, ATQ); **Jetta:** 2.0 (ABA, AEG, AVH, AZG, BBW, BEV), 2.8L (AFP, BDF); **Golf:** 2.0L (ABA, AEG, AVH, AZG, BEV); **GTI:** 2.0L (ABA, AEG), 2.8L (AFP, BDF); **NB:** 1.8L (APH, AWP, AWV, BNU, BKF), 2.0L (AEG, AVH, AZG, BEV); **Nbvert:** (BDG); **Cabrio:** 2.0L (ABA) **Transmissions:** All	**O2 Sensor Heater Circuit (Bank 1-Sensor 1) Electrical Malfunction Conditions:** Engine started, battery voltage must be at least 11.5v, all electrical components must be off, the ground between the engine and the chassis must be well connected, the exhaust system must be properly sealed between the catalytic converter and the cylinder head, and the oxygen sensor heater for oxygen sensor before the catalytic converter must be properly functioning. **Note: For resistance testing of sensor heating, oxygen sensor should be cooled to ambient temperature. High temperatures at oxygen sensor may lead to inaccurate measurements. The ECM detected an open or shorted condition, or excessive current draw in the heater circuit.** **Possible Causes:** • HO2S heater power circuit is open • HO2S heater ground circuit is open • HO2S signal tracking (due to oil or moisture in the connector) • HO2S is damaged or has failed • ECM has failed • Oxygen sensor (before catalytic converter) is faulty • Oxygen sensor (behind catalytic converter) is faulty • Oxygen sensor heater (before catalytic converter) is faulty • Oxygen sensor heater (behind catalytic converter) is faulty

DTC	Trouble Code Title, Conditions & Possible Causes
DTC: P1197 **Passat:** 2.8L (AHA, ATQ) **Transmissions:** All	**O2 Sensor Heater Circuit (Bank 2-Sensor 1) Electrical Malfunction Conditions:** Engine started, battery voltage must be at least 11.5v, all electrical components must be off, the ground between the engine and the chassis must be well connected, the exhaust system must be properly sealed between the catalytic converter and the cylinder head, and the oxygen sensor heater for oxygen sensor before the catalytic converter must be properly functioning. **Note: For resistance testing of sensor heating, oxygen sensor should be cooled to ambient temperature. High temperatures at oxygen sensor may lead to inaccurate measurements. The ECM detected an open or shorted condition, or excessive current draw in the heater circuit.** **Possible Causes:** • HO2S heater power circuit is open • HO2S heater ground circuit is open • HO2S signal tracking (due to oil or moisture in the connector) • HO2S is damaged or has failed • ECM has failed • Oxygen sensor (before catalytic converter) is faulty • Oxygen sensor (behind catalytic converter) is faulty • Oxygen sensor heater (before catalytic converter) is faulty • Oxygen sensor heater (behind catalytic converter) is faulty
DTC: P1198 **Passat:** 1.8 (AEB, ATW, AUG, AWM), 2.0L (BPY), 2.8L (AHA, ATQ); **Jetta:** 1.8L (AWD, AWP, AWW), 2.0 (ABA, AEG, AVH, AZG, BBW, BEV), 2.8L (AFP, BDF); **Golf:** 1.8L (AWD, AWW), 2.0L (ABA, AEG, AVH, AZG, BEV); **GTI:** 1.8L (AWD, AWP, AWW), 2.0L (ABA, AEG), 2.8L (AFP, BDF); **NB:** 1.8L (APH, AWP, AWV, BNU, BKF), 2.0L (AEG, AVH, AZG, BEV); **Nbvert:** (BDG); **Cabrio:** 2.0L (ABA) **Transmissions:** All	**O2 Sensor Heater Circuit (Bank 1-Sensor 2) Electrical Malfunction Conditions:** Engine started, battery voltage must be at least 11.5v, all electrical components must be off, the ground between the engine and the chassis must be well connected, the exhaust system must be properly sealed between the catalytic converter and the cylinder head, and the oxygen sensor heater for oxygen sensor before the catalytic converter must be properly functioning. **Note: For resistance testing of sensor heating, oxygen sensor should be cooled to ambient temperature. High temperatures at oxygen sensor may lead to inaccurate measurements. The ECM detected an open or shorted condition, or excessive current draw in the heater circuit.** **Possible Causes:** • HO2S heater power circuit is open • HO2S heater ground circuit is open • HO2S signal tracking (due to oil or moisture in the connector) • HO2S is damaged or has failed • ECM has failed • Oxygen sensor (before catalytic converter) is faulty • Oxygen sensor (behind catalytic converter) is faulty • Oxygen sensor heater (before catalytic converter) is faulty • Oxygen sensor heater (behind catalytic converter) is faulty
DTC: P1199 **Passat:** 2.8L (AHA, ATQ) **Transmissions:** All	**O2 Sensor Heater Circuit (Bank 2-Sensor 2) Electrical Malfunction Conditions:** Engine started, battery voltage must be at least 11.5v, all electrical components must be off, the ground between the engine and the chassis must be well connected, the exhaust system must be properly sealed between the catalytic converter and the cylinder head, and the oxygen sensor heater for oxygen sensor before the catalytic converter must be properly functioning. **Note: For resistance testing of sensor heating, oxygen sensor should be cooled to ambient temperature. High temperatures at oxygen sensor may lead to inaccurate measurements. The ECM detected an open or shorted condition, or excessive current draw in the heater circuit.** **Possible Causes:** • HO2S heater power circuit is open • HO2S heater ground circuit is open • HO2S signal tracking (due to oil or moisture in the connector) • HO2S is damaged or has failed • ECM has failed • Oxygen sensor (before catalytic converter) is faulty • Oxygen sensor (behind catalytic converter) is faulty • Oxygen sensor heater (before catalytic converter) is faulty • Oxygen sensor heater (behind catalytic converter) is faulty

DTC	Trouble Code Title, Conditions & Possible Causes
DTC: P1200 **Passat:** 1.8 (AEB, ATW, AUG, AWM) **Transmissions:** All	**Turbocharger Bypass Valve Mechanical Malfunction Conditions:** Engine started, battery voltage at least 11.5v, all electrical components off, ground connections between engine and chassis well connected, coolant temperature at least 80-degrees Celicius. The ECM detected an operating condition that could harm the engine or automatic transmission. **Possible Causes:** • Charge air pressure sensor is faulty • Voltage supply to the charge air pressure sensor is open or shorted • Check the charge air system for leaks • Recirculating valve for turbocharger is faulty • Turbocharging system is damaged or not functioning properly • Turbocharger recirculating valve is faulty • Vacuum diaphragm for turbocharger is out of adjustment • Wastegate bypass regulator valve is faulty • Boost sensor has failed • ECM has failed
DTC: P1201 **Passat:** 1.8 (AEB, ATW, AUG, AWM), 2.0L (BPY); **Jetta:** 1.8L (AWD, AWP, AWW), 2.0L (ABA); **Golf:** 1.8L (AWD, AWW), 2.0L (ABA); **GTI:** 1.8L (AWD, AWP, AWW), 2.0L (ABA); **NB:** 1.8L (APH, AWP, AWV, BNU, BKF); **Cabrio:** 2.0L (ABA) **Transmissions:** All	**Cylinder 1 Fuel Injection Circuit Electrical Malfunction Conditions:** Key on or engine running, fuses in the instrument panel and the E-box in the engine compartment must be functioning, and the ground connections between the engine ad the chassis must be well connected; and the ECM detected an unexpected voltage condition on the injector circuit **Possible Causes:** • Injector control circuit is open • Injector power circuit (B+) is open • Injector control circuit is shorted to chassis ground • Injector is damaged or has failed • ECM is not connected or has failed • Fuel pump relay has failed • Fuel injectors may have malfunctioned • Faulty engine speed sensor
DTC: P1202 **Passat:** 1.8 (AEB, ATW, AUG, AWM), 2.0L (BPY); **Jetta:** 1.8L (AWD, AWP, AWW), 2.0L (ABA); **Golf:** 1.8L (AWD, AWW), 2.0L (ABA); **GTI:** 1.8L (AWD, AWP, AWW), 2.0L (ABA); **NB:** 1.8L (APH, AWP, AWV, BNU, BKF); **Cabrio:** 2.0L (ABA) **Transmissions:** All	**Cylinder 2 Fuel Injection Circuit Electrical Malfunction Conditions:** Key on or engine running, fuses in the instrument panel and the E-box in the engine compartment must be functioning, and the ground connections between the engine ad the chassis must be well connected; and the ECM detected an unexpected voltage condition on the injector circuit **Possible Causes:** • Injector control circuit is open • Injector power circuit (B+) is open • Injector control circuit is shorted to chassis ground • Injector is damaged or has failed • ECM is not connected or has failed • Fuel pump relay has failed • Fuel injectors may have malfunctioned • Faulty engine speed sensor
DTC: P1203 **Passat:** 1.8 (AEB, ATW, AUG, AWM), 2.0L (BPY); **Jetta:** 1.8L (AWD, AWP, AWW), 2.0L (ABA); **Golf:** 1.8L (AWD, AWW), 2.0L (ABA); **GTI:** 1.8L (AWD, AWP, AWW), 2.0L (ABA); **NB:** 1.8L (APH, AWP, AWV, BNU, BKF); **Cabrio:** 2.0L (ABA) **Transmissions:** All	**Cylinder 3 Fuel Injection Circuit Electrical Malfunction Conditions:** Key on or engine running, fuses in the instrument panel and the E-box in the engine compartment must be functioning, and the ground connections between the engine ad the chassis must be well connected; and the ECM detected an unexpected voltage condition on the injector circuit **Possible Causes:** • Injector control circuit is open • Injector power circuit (B+) is open • Injector control circuit is shorted to chassis ground • Injector is damaged or has failed • ECM is not connected or has failed • Fuel pump relay has failed • Fuel injectors may have malfunctioned • Faulty engine speed sensor

DTC	Trouble Code Title, Conditions & Possible Causes
DTC: P1204 **Passat:** 1.8 (AEB, ATW, AUG, AWM), 2.0L (BPY); **Jetta:** 1.8L (AWD, AWP, AWW), 2.0L (ABA); **Golf:** 1.8L (AWD, AWW), 2.0L (ABA); **GTI:** 1.8L (AWD, AWP, AWW), 2.0L (ABA); **NB:** 1.8L (APH, AWP, AWV, BNU, BKF); **Cabrio:** 2.0L (ABA) **Transmissions:** All	**Cylinder 4 Fuel Injection Circuit Electrical Malfunction Conditions:** Key on or engine running, fuses in the instrument panel and the E-box in the engine compartment must be functioning, and the ground connections between the engine ad the chassis must be well connected; and the ECM detected an unexpected voltage condition on the injector circuit **Possible Causes:** • Injector control circuit is open • Injector power circuit (B+) is open • Injector control circuit is shorted to chassis ground • Injector is damaged or has failed • ECM is not connected or has failed • Fuel pump relay has failed • Fuel injectors may have malfunctioned • Faulty engine speed sensor
DTC: P1213 **2T, MIL: Yes** **Passat:** 1.8 (AEB, ATW, AUG, AWM), 2.0L (ABA, BPY), 2.8L (AAA, AHA, ATQ); **Jetta:** 1.8L (AWD, AWP, AWW), 2.0 (ABA, AEG, AVH, AZG, BBW, BEV), 2.8L (AAA, AFP, BDF); **Golf:** 1.8L (AWD, AWW), 2.0L (ABA, AEG, AVH, AZG, BEV); **GTI:** 1.8L (AWD, AWP, AWW), 2.0L (ABA, AEG), 2.8L (AAA, AFP, BDF); **NB:** 1.8L (APH, AWP, AWV, BNU, BKF), 2.0L (AEG, AVH, AZG, BEV); **Nbvert:** (BDG); **Cabrio:** 2.0L (ABA) **Transmissions:** All	**Cylinder 1 Fuel Injection Circuit Short to B+ Conditions:** Key on or engine running, fuses in the instrument panel and the E-box in the engine compartment must be functioning, and the ground connections between the engine ad the chassis must be well connected; and the ECM detected an unexpected voltage condition on the injector circuit. Wiring or fuel injector has a short circuit to positive supply. **Possible Causes:** • Injector control circuit is open • Injector power circuit (B+) is open • Injector control circuit is shorted to chassis ground • Injector is damaged or has failed • ECM is not connected or has failed • Fuel pump relay has failed • Engine speed sensor has failed
DTC: P1214 **2T, MIL: Yes** **Passat:** 1.8 (AEB, ATW, AUG, AWM), 2.0L (ABA, BPY), 2.8L (AAA, AHA, ATQ); **Jetta:** 1.8L (AWD, AWP, AWW), 2.0 (ABA, AEG, AVH, AZG, BBW, BEV), 2.8L (AAA, AFP, BDF); **Golf:** 1.8L (AWD, AWW), 2.0L (ABA, AEG, AVH, AZG, BEV); **GTI:** 1.8L (AWD, AWP, AWW), 2.0L (ABA, AEG), 2.8L (AAA, AFP, BDF); **NB:** 1.8L (APH, AWP, AWV, BNU, BKF), 2.0L (AEG, AVH, AZG, BEV); **Nbvert:** (BDG); **Cabrio:** 2.0L (ABA) **Transmissions:** All	**Cylinder 2 Fuel Injection Circuit Short to B+ Conditions:** Key on or engine running, fuses in the instrument panel and the E-box in the engine compartment must be functioning, and the ground connections between the engine ad the chassis must be well connected; and the ECM detected an unexpected voltage condition on the injector circuit. Wiring or fuel injector has a short circuit to positive supply. **Possible Causes:** • Injector control circuit is open • Injector power circuit (B+) is open • Injector control circuit is shorted to chassis ground • Injector is damaged or has failed • ECM is not connected or has failed • Fuel pump relay has failed • Engine speed sensor has failed
DTC: P1215 **2T, MIL: Yes** **Passat:** 1.8 (AEB, ATW, AUG, AWM), 2.0L (ABA, BPY), 2.8L (AAA, AHA, ATQ); **Jetta:** 1.8L (AWD, AWP, AWW), 2.0 (ABA, AEG, AVH, AZG, BEV), 2.8L (AAA, AFP, BDF); **Golf:** 1.8L (AWD, AWW), 2.0L (ABA, AEG, AVH, AZG, BEV); **GTI:** 1.8L (AWD, AWP, AWW), 2.0L (ABA, AEG), 2.8L (AAA, AFP, BDF); **NB:** 1.8L (APH, AWP, AWV, BNU, BKF), 2.0L (AEG, AVH, AZG, BEV); **Nbvert:** (BDG); **Cabrio:** 2.0L (ABA) **Transmissions:** All	**Cylinder 3 Fuel Injection Circuit Short to B+ Conditions:** Key on or engine running, fuses in the instrument panel and the E-box in the engine compartment must be functioning, and the ground connections between the engine ad the chassis must be well connected; and the ECM detected an unexpected voltage condition on the injector circuit. Wiring or fuel injector has a short circuit to positive supply. **Possible Causes:** • Injector control circuit is open • Injector power circuit (B+) is open • Injector control circuit is shorted to chassis ground • Injector is damaged or has failed • ECM is not connected or has failed • Fuel pump relay has failed • Engine speed sensor has failed

DTC	Trouble Code Title, Conditions & Possible Causes
DTC: P1216 **2T, MIL: Yes** **Passat:** 1.8 (AEB, ATW, AUG, AWM), 2.0L (ABA, BPY), 2.8L (AAA, AHA, ATQ); **Jetta:** 1.8L (AWD, AWP, AWW), 2.0 (ABA, AEG, AVH, AZG, BBW, BEV), 2.8L (AAA, APF); **Golf:** 1.8L (AWD, AWW), 2.0L (ABA, AEG, AVH, AZG, BEV); **GTI:** 1.8L (AWD, AWP, AWW), 2.0L (ABA, AEG), 2.8L (AAA, AFP, BDF); **NB:** 1.8L (APH, AWP, AWV, BNU, BKF), 2.0L (AEG, AVH, AZG, BEV); **Nbvert:** (BDG); **Cabrio:** 2.0L (ABA) **Transmissions:** All	**Cylinder 4 Fuel Injection Circuit Short to B+ Conditions:** Key on or engine running, fuses in the instrument panel and the E-box in the engine compartment must be functioning, and the ground connections between the engine ad the chassis must be well connected; and the ECM detected an unexpected voltage condition on the injector circuit. Wiring or fuel injector has a short circuit to positive supply. **Possible Causes:** • Injector control circuit is open • Injector power circuit (B+) is open • Injector control circuit is shorted to chassis ground • Injector is damaged or has failed • ECM is not connected or has failed • Fuel pump relay has failed • Engine speed sensor has failed
DTC: P1217 **Passat:** 2.8L (AAA, AHA, ATQ); **Jetta:** 2.8L (AAA, AFP, BDF); **GTI:** 2.8L (AAA, AFP, BDF) **Transmissions:** All	**Cylinder 5 Fuel Injection Circuit Short to B+ Conditions:** Key on or engine running, fuses in the instrument panel and the E-box in the engine compartment must be functioning, and the ground connections between the engine ad the chassis must be well connected; and the ECM detected an unexpected voltage condition on the injector circuit. Wiring or fuel injector has a short circuit to positive supply. **Possible Causes:** • Injector control circuit is open • Injector power circuit (B+) is open • Injector control circuit is shorted to chassis ground • Injector is damaged or has failed • ECM is not connected or has failed • Fuel pump relay has failed • Engine speed sensor has failed
DTC: P1218 **Passat:** 2.8L (AAA, AHA, ATQ); **Jetta:** 2.8L (AAA, AFP, BDF); **GTI:** 2.8L (AAA, AFP, BDF) **Transmissions:** All	**Cylinder 6 Fuel Injection Circuit Short to B+ Conditions:** Key on or engine running, fuses in the instrument panel and the E-box in the engine compartment must be functioning, and the ground connections between the engine ad the chassis must be well connected; and the ECM detected an unexpected voltage condition on the injector circuit. Wiring or fuel injector has a short circuit to positive supply. **Possible Causes:** • Injector control circuit is open • Injector power circuit (B+) is open • Injector control circuit is shorted to chassis ground • Injector is damaged or has failed • ECM is not connected or has failed • Fuel pump relay has failed • Engine speed sensor has failed
DTC: P1225 **2T, MIL: Yes** **Passat:** 1.8 (AEB, ATW, AUG, AWM), 2.0L (ABA, BPY), 2.8L (AAA, AHA, ATQ); **Jetta:** 1.8L (AWD, AWP, AWW), 2.0 (ABA, AEG, AVH, AZG, BBW, BEV), 2.8L (AAA, AFP, BDF); **Golf:** 1.8L (AWD, AWW), 2.0L (ABA, AEG, AVH, AZG, BEV); **GTI:** 1.8L (AWD, AWP, AWW), 2.0L (ABA, AEG), 2.8L (AAA, AFP, BDF); **NB:** 1.8L (APH, AWP, AWV, BNU, BKF), 2.0L (AEG, AVH, AZG, BEV); **Nbvert:** (BDG); **Cabrio:** 2.0L (ABA) **Transmissions:** All	**Cylinder 1 Fuel Injection Circuit Short to Ground Conditions:** Key on or engine running, fuses in the instrument panel and the E-box in the engine compartment must be functioning, and the ground connections between the engine ad the chassis must be well connected; and the ECM detected an unexpected voltage condition on the injector circuit. Wiring or fuel injector has a short circuit to ground. **Possible Causes:** • Injector control circuit is open • Injector power circuit (B+) is open • Injector control circuit is shorted to chassis ground • Injector is damaged or has failed • ECM is not connected or has failed • Fuel pump relay has failed • Engine speed sensor has failed

DTC	Trouble Code Title, Conditions & Possible Causes
DTC: P1226 **2T, MIL: Yes** **Passat:** 1.8 (AEB, ATW, AUG, AWM), 2.0L (ABA, BPY), 2.8L (AAA, AHA, ATQ); **Jetta:** 1.8L (AWD, AWP, AWW), 2.0 (ABA, AEG, AVH, AZG, BBW, BEV), 2.8L (AAA, AFP, BDF); **Golf:** 1.8L (AWD, AWW), 2.0L (ABA, AEG, AVH, AZG, BEV); **GTI:** 1.8L (AWD, AWP, AWW), 2.0L (ABA, AEG), 2.8L (AAA, AFP, BDF); **NB:** 1.8L (APH, AWP, AWV, BNU, BKF), 2.0L (AEG, AVH, AZG, BEV); **Nbvert:** (BDG); **Cabrio:** 2.0L (ABA) **Transmissions:** All	**Cylinder 2 Fuel Injection Circuit Short to Ground Conditions:** Key on or engine running, fuses in the instrument panel and the E-box in the engine compartment must be functioning, and the ground connections between the engine ad the chassis must be well connected; and the ECM detected an unexpected voltage condition on the injector circuit. Wiring or fuel injector has a short circuit to ground. **Possible Causes:** • Injector control circuit is open • Injector power circuit (B+) is open • Injector control circuit is shorted to chassis ground • Injector is damaged or has failed • ECM is not connected or has failed • Fuel pump relay has failed • Engine speed sensor has failed
DTC: P1227 **2T, MIL: Yes** **Passat:** 1.8 (AEB, ATW, AUG, AWM), 2.0L (ABA, BPY), 2.8L (AAA, AHA, ATQ); **Jetta:** 1.8L (AWD, AWP, AWW), 2.0 (ABA, AEG, AVH, AZG, BEV), 2.8L (AAA, AFP, BDF); **Golf:** 1.8L (AWD, AWW), 2.0L (ABA, AEG, AVH, AZG, BEV); **GTI:** 1.8L (AWD, AWP, AWW), 2.0L (ABA, AEG), 2.8L (AAA, AFP, BDF); **NB:** 1.8L (APH, AWP, AWV, BNU, BKF), 2.0L (AEG, AVH, AZG, BEV); **Nbvert:** (BDG); **Cabrio:** 2.0L (ABA) **Transmissions:** All	**Cylinder 3 Fuel Injection Circuit Short to Ground Conditions:** Key on or engine running, fuses in the instrument panel and the E-box in the engine compartment must be functioning, and the ground connections between the engine ad the chassis must be well connected; and the ECM detected an unexpected voltage condition on the injector circuit. Wiring or fuel injector has a short circuit to ground. **Possible Causes:** • Injector control circuit is open • Injector power circuit (B+) is open • Injector control circuit is shorted to chassis ground • Injector is damaged or has failed • ECM is not connected or has failed • Fuel pump relay has failed • Engine speed sensor has failed
DTC: P1228 **2T, MIL: Yes** **Passat:** 1.8 (AEB, ATW, AUG, AWM), 2.0L (ABA, BPY), 2.8L (AHA, ATQ); **Jetta:** 1.8L (AWD, AWP, AWW), 2.0 (ABA, AEG, AVH, AZG, BBW, BEV), 2.8L (AAA, AFP, BDF); **Golf:** 1.8L (AWD, AWW), 2.0L (ABA, AEG, AVH, AZG, BEV); **GTI:** 1.8L (AWD, AWP, AWW), 2.0L (ABA, AEG), 2.8L (AAA, AFP, BDF); **NB:** 1.8L (APH, AWP, AWV, BNU, BKF), 2.0L (AEG, AVH, AZG, BEV); **Nbvert:** (BDG); **Cabrio:** 2.0L (ABA) **Transmissions:** All	**Cylinder 4 Fuel Injection Circuit Short to Ground Conditions:** Key on or engine running, fuses in the instrument panel and the E-box in the engine compartment must be functioning, and the ground connections between the engine ad the chassis must be well connected; and the ECM detected an unexpected voltage condition on the injector circuit. Wiring or fuel injector has a short circuit to ground. **Possible Causes:** • Injector control circuit is open • Injector power circuit (B+) is open • Injector control circuit is shorted to chassis ground • Injector is damaged or has failed • ECM is not connected or has failed • Fuel pump relay has failed • Engine speed sensor has failed

DTC	Trouble Code Title, Conditions & Possible Causes
DTC: P1229 **Passat:** 2.8L (AAA, AHA, ATQ); **Jetta:** 2.8L (AAA, AFP, BDF); **GTI:** 2.8L (AAA, AFP, BDF) **Transmissions:** All	**Cylinder 5 Fuel Injection Circuit Short to Ground Conditions:** Key on or engine running, fuses in the instrument panel and the E-box in the engine compartment must be functioning, and the ground connections between the engine ad the chassis must be well connected; and the ECM detected an unexpected voltage condition on the injector circuit. Wiring or fuel injector has a short circuit to ground. **Possible Causes:** • Injector control circuit is open • Injector power circuit (B+) is open • Injector control circuit is shorted to chassis ground • Injector is damaged or has failed • ECM is not connected or has failed • Fuel pump relay has failed • Engine speed sensor has failed
DTC: P1230 **Passat:** 2.8L (AAA, AHA, ATQ); **Jetta:** 2.8L (AAA, AFP, BDF); **GTI:** 2.8L (AAA, AFP, BDF) **Transmissions:** All	**Cylinder 6 Fuel Injection Circuit Short to Ground Conditions:** Key on or engine running, fuses in the instrument panel and the E-box in the engine compartment must be functioning, and the ground connections between the engine ad the chassis must be well connected; and the ECM detected an unexpected voltage condition on the injector circuit. Wiring or fuel injector has a short circuit to ground. **Possible Causes:** • Injector control circuit is open • Injector power circuit (B+) is open • Injector control circuit is shorted to chassis ground • Injector is damaged or has failed • ECM is not connected or has failed • Fuel pump relay has failed • Engine speed sensor has failed
DTC: P1237 **2T, MIL: Yes** **Passat:** 1.8 (AEB, ATW, AUG, AWM), 2.0L (ABA, BPY), 2.8L (AAA, AHA, ATQ); **Jetta:** 1.8L (AWD, AWP, AWW), 2.0 (ABA, AEG, AVH, AZG, BBW, BEV), 2.8L (AAA, AFP, BDF); **Golf:** 1.8L (AWD, AWW), 2.0L (ABA); **GTI:** 1.8L (AWD, AWP, AWW), 2.0L (ABA, AEG), 2.8L (AAA, AFP, BDF); **NB:** 1.8L (APH, AWP, AWV, BNU, BKF), 2.0L (AEG, AVH, AZG, BEV); **Nbvert:** (BDG); **Cabrio:** 2.0L (ABA) **Transmissions:** All	**Cylinder 1 Fuel Injection Circuit Open Circuit Conditions:** Key on or engine running, fuses in the instrument panel and the E-box in the engine compartment must be functioning, and the ground connections between the engine ad the chassis must be well connected; and the ECM detected an unexpected voltage condition on the injector circuit. Wiring or fuel injector has a short circuit that is open. **Possible Causes:** • Injector control circuit is open • Injector power circuit (B+) is open • Injector control circuit is shorted to chassis ground • Injector is damaged or has failed • ECM is not connected or has failed • Fuel pump relay has failed • Engine speed sensor has failed
DTC: P1238 **2T, MIL: Yes** **Passat:** 1.8 (AEB, ATW, AUG, AWM), 2.0L (ABA, BPY), 2.8L (AAA, AHA, ATQ); **Jetta:** 1.8L (AWD, AWP, AWW), 2.0 (ABA, AEG, AVH, AZG, BBW, BEV), 2.8L (AAA, AFP, BDF); **Golf:** 1.8L (AWD, AWW), 2.0L (ABA, AEG, AVH, AZG, BEV); **GTI:** 1.8L (AWD, AWP, AWW), 2.0L (ABA, AEG), 2.8L (AAA, AFP, BDF); **NB:** 1.8L (APH, AWP, AWV, BNU, BKF), 2.0L (AEG, AVH, AZG, BEV); **Nbvert:** (BDG); **Cabrio:** 2.0L (ABA) **Transmissions:** All	**Cylinder 2 Fuel Injection Circuit Open Circuit Conditions:** Key on or engine running, fuses in the instrument panel and the E-box in the engine compartment must be functioning, and the ground connections between the engine ad the chassis must be well connected; and the ECM detected an unexpected voltage condition on the injector circuit. Wiring or fuel injector has a short circuit that is open. **Possible Causes:** • Injector control circuit is open • Injector power circuit (B+) is open • Injector control circuit is shorted to chassis ground • Injector is damaged or has failed • ECM is not connected or has failed • Fuel pump relay has failed • Engine speed sensor has failed

DTC	Trouble Code Title, Conditions & Possible Causes
DTC: P1239 **2T, MIL: Yes** **Passat:** 1.8 (AEB, ATW, AUG, AWM), 2.0L (ABA, BPY), 2.8L (AAA, AHA, ATQ); **Jetta:** 1.8L (AWD, AWP, AWW), 2.0 (ABA, AEG, AVH, AZG, BBW, BEV), 2.8L (AAA, AFP, BDF); **Golf:** 1.8L (AWD, AWW), 2.0L (ABA, AEG); **GTI:** 1.8L (AWD, AWP, AWW), 2.0L (ABA, AEG), 2.8L (AAA, AFP, BDF); **NB:** 1.8L (APH, AWP, AWV, BNU, BKF), 2.0L (AEG, AVH, AZG, BEV); **Nbvert:** (BDG); **Cabrio:** 2.0L (ABA) **Transmissions:** All	**Cylinder 3 Fuel Injection Circuit Open Circuit Conditions:** Key on or engine running, fuses in the instrument panel and the E-box in the engine compartment must be functioning, and the ground connections between the engine ad the chassis must be well connected; and the ECM detected an unexpected voltage condition on the injector circuit. Wiring or fuel injector has a short circuit that is open. **Possible Causes:** • Injector control circuit is open • Injector power circuit (B+) is open • Injector control circuit is shorted to chassis ground • Injector is damaged or has failed • ECM is not connected or has failed • Fuel pump relay has failed • Engine speed sensor has failed
DTC: P1240 **2T, MIL: Yes** **Passat:** 1.8 (AEB, ATW, AUG, AWM), 2.0L (ABA, BPY), 2.8L (AAA, AHA, ATQ); **Jetta:** 1.8L (AWD, AWP, AWW), 2.0 (ABA, AEG, AVH, AZG, BBW, BEV), 2.8L (AAA, AFP, BDF); **Golf:** 1.8L (AWD, AWW), 2.0L (ABA, AEG, AVH, AZG, BEV); **GTI:** 1.8L (AWD, AWP, AWW), 2.0L (ABA, AEG), 2.8L (AAA, AFP, BDF); **NB:** 1.8L (APH, AWP, AWV, BNU, BKF), 2.0L (AEG, AVH, AZG, BEV); **Nbvert:** (BDG); **Cabrio:** 2.0L (ABA) **Transmissions:** All	**Cylinder 4 Fuel Injection Circuit Open Circuit Conditions:** Key on or engine running, fuses in the instrument panel and the E-box in the engine compartment must be functioning, and the ground connections between the engine ad the chassis must be well connected; and the ECM detected an unexpected voltage condition on the injector circuit. Wiring or fuel injector has a short circuit that is open. **Possible Causes:** • Injector control circuit is open • Injector power circuit (B+) is open • Injector control circuit is shorted to chassis ground • Injector is damaged or has failed • ECM is not connected or has failed • Fuel pump relay has failed • Engine speed sensor has failed
DTC: P1241 **Passat:** 2.8L (AAA, AHA, ATQ); **Jetta:** 2.8L (AAA, AFP, BDF); **GTI:** 2.8L (AAA, AFP, BDF) **Transmissions:** All	**Cylinder 5 Fuel Injection Circuit Open Circuit Conditions:** Key on or engine running, fuses in the instrument panel and the E-box in the engine compartment must be functioning, and the ground connections between the engine ad the chassis must be well connected; and the ECM detected an unexpected voltage condition on the injector circuit. Wiring or fuel injector has a short circuit that is open. **Possible Causes:** • Injector control circuit is open • Injector power circuit (B+) is open • Injector control circuit is shorted to chassis ground • Injector is damaged or has failed • ECM is not connected or has failed • Fuel pump relay has failed • Engine speed sensor has failed
DTC: P1242 **Passat:** 2.8L (AAA, AHA, ATQ); **Jetta:** 2.8L (AAA, AFP, BDF); **GTI:** 2.8L (AAA, AFP, BDF) **Transmissions:** All	**Cylinder 6 Fuel Injection Circuit Open Circuit Conditions:** Key on or engine running, fuses in the instrument panel and the E-box in the engine compartment must be functioning, and the ground connections between the engine ad the chassis must be well connected; and the ECM detected an unexpected voltage condition on the injector circuit. Wiring or fuel injector has a short circuit that is open. **Possible Causes:** • Injector control circuit is open • Injector power circuit (B+) is open • Injector control circuit is shorted to chassis ground • Injector is damaged or has failed • ECM is not connected or has failed • Fuel pump relay has failed • Engine speed sensor has failed

DTC	Trouble Code Title, Conditions & Possible Causes
DTC: P1250 **Passat:** 1.8 (AEB, AWM), 2.0L (BPY), 2.8L (AHA, ATQ); **Jetta:** 1.8L (AWD, AWP, AWW), 2.0L (BPY); **Golf:** 1.8L (AWD, AWW); **GTI:** 1.8L (AWD, AWP, AWW) **Transmissions:** All	**Fuel Pressure Regulator Control Circuit Malfunction (Fuel Level too Low) Conditions:** KOEO or KOER Self-Test enabled, and the ECM detected a lack of power (VPWR) to the Fuel Pressure Regulator Control (FPRC) solenoid circuit. **Possible Causes:** • FPRC solenoid valve harness circuits are open or shorted • FPRC input port or output port vacuum lines are damaged • FRPC solenoid is damaged • Fuel level is too low • ECM has failed
DTC: P1255 **Passat:** 2.0L (BPY); **Jetta:** 2.0L (AEG, AVH, AZG, BBW, BEV); **Golf:** 2.0 (AEG, AVH, AZG, BEV); **GTI:** 2.0L (AEG); **NB:** 1.8L (APH, AWP, AWV, BNU, BKF), 2.0L (AEG, AVH, AZG, BEV); **Nbvert:** (BDG) **Transmissions:** All	**Engine Coolant Temperature Circuit Short to Ground Conditions:** Key on, engine started, the ECM detected an unexpected voltage condition on the ECT circuit. **Possible Causes:** • Coolant Temperature Sensor has failed • Circuit short to ground, open or other component • ECM has failed
DTC: P1256 **Passat:** 2.0L (BPY); **Jetta:** 2.0L (BBW); **Golf:** 2.L (AEG, AVH, AZG, BEV); **GTI:** 2.0L (AEG); **NB:** 1.8L (APH, AWP, AWV, BNU, BKF), 2.0L (AEG, AVH, AZG, BEV); **Nbvert:** (BDG) **Transmissions:** All	**Engine Coolant Temperature Circuit Open/Short to B+ Conditions:** Key on, engine started, the ECM detected an unexpected voltage condition on the ECT circuit. **Possible Causes:** • Coolant Temperature Sensor has failed • Circuit short to ground, open or other component • ECM has failed
DTC: P1287 **Passat:** 1.8 (AEB, ATW, AUG, AWM), 2.0L (BPY); **Jetta:** 1.8L (AWD, AWP, AWW); **Golf:** 1.8L (AWD, AWW); **GTI:** 1.8L (AWD, AWP, AWW); **NB:** 1.8L (APH, AWP, AWV, BNU, BKF) **Transmissions:** All	**Turbocharger Bypass Valve Open Conditions:** Engine started, battery voltage at least 11.5v, all electrical components off, ground connections between engine and chassis well connected, coolant temperature at least 80-degrees Celicius. The ECM detected an unexpected voltage condition on the bypass valve control circuit **Possible Causes:** • Charge air system check for leaks • Recirculating valve for turbocharger is faulty • Turbocharging system may be damaged • Vacuum diaphragm for turbocharger needs adjusting • Wastegate bypass regulator valve is faulty • Bypass solenoid power supply circuit is open • Bypass solenoid control circuit is open, shorted to ground or system power • Bypass solenoid assembly is damaged or has failed • Charge air pressure sensor is faulty • Voltage supply to the charge air pressure sensor is open or shorted • Check the charge air system for leaks • Recirculating valve for turbocharger is faulty • Turbocharging system is damaged or not functioning properly • Turbocharger recirculating valve is faulty • Vacuum diaphragm for turbocharger is out of adjustment • Wastegate bypass regulator valve is faulty • ECM has failed

DTC	Trouble Code Title, Conditions & Possible Causes
DTC: P1288 **Passat:** 1.8 (AEB, ATW, AUG, AWM), 2.0L (BPY); **Jetta:** 1.8L (AWD, AWP, AWW); **Golf:** 1.8L (AWD, AWW); **GTI:** 1.8L (AWD, AWP, AWW); **NB:** 1.8L (APH, AWP, AWV, BNU, BKF) **Transmissions:** All	**Turbocharger Bypass Valve Short to B+ Conditions:** Engine started, battery voltage at least 11.5v, all electrical components off, ground connections between engine and chassis well connected, coolant temperature at least 80-degrees Celicius. The ECM detected an unexpected voltage condition on the bypass valve control circuit **Possible Causes:** • Charge air system check for leaks • Recirculating valve for turbocharger is faulty • Turbocharging system may be damaged • Vacuum diaphragm for turbocharger needs adjusting • Wastegate bypass regulator valve is faulty • Bypass solenoid power supply circuit is open • Bypass solenoid control circuit is open, shorted to ground or system power • Bypass solenoid assembly is damaged or has failed • Charge air pressure sensor is faulty • Voltage supply to the charge air pressure sensor is open or shorted • Check the charge air system for leaks • Recirculating valve for turbocharger is faulty • Turbocharging system is damaged or not functioning properly • Turbocharger recirculating valve is faulty • Vacuum diaphragm for turbocharger is out of adjustment • Wastegate bypass regulator valve is faulty • ECM has failedvalve is faulty
DTC: P1289 **Passat:** 1.8 (AEB, ATW, AUG, AWM), 2.0L (BPY); **Jetta:** 1.8L (AWD, AWP, AWW); **Golf:** 1.8L (AWD, AWW); **GTI:** 1.8L (AWD, AWP, AWW); **NB:** 1.8L (APH, AWP, AWV, BNU, BKF) **Transmissions:** All	**Turbocharger Bypass Valve Short to Ground Conditions:** Engine started, battery voltage at least 11.5v, all electrical components off, ground connections between engine and chassis well connected, coolant temperature at least 80-degrees Celicius. The ECM detected an unexpected voltage condition on the bypass valve control circuit **Possible Causes:** • Charge air system check for leaks • Recirculating valve for turbocharger is faulty • Turbocharging system may be damaged • Vacuum diaphragm for turbocharger needs adjusting • Wastegate bypass regulator valve is faulty • Bypass solenoid power supply circuit is open • Bypass solenoid control circuit is open, shorted to ground or system power • Bypass solenoid assembly is damaged or has failed • Charge air pressure sensor is faulty • Voltage supply to the charge air pressure sensor is open or shorted • Check the charge air system for leaks • Recirculating valve for turbocharger is faulty • Turbocharging system is damaged or not functioning properly • Turbocharger recirculating valve is faulty • Vacuum diaphragm for turbocharger is out of adjustment • Wastegate bypass regulator valve is faulty • ECM has failed

DTC	Trouble Code Title, Conditions & Possible Causes
DTC: P1295 **Passat:** 1.8 (AEB, ATW, AUG, AWM), 2.0L (BPY); **Jetta:** 1.8L (AWD, AWP, AWW); **Golf:** 1.8L (AWD, AWW); **GTI:** 1.8L (AWD, AWP, AWW); **NB:** 1.8L (APH, AWP, AWV, BNU, BKF) **Transmissions:** All	**Turbocharger Bypass Valve Throughput Faulty Conditions:** Engine started, battery voltage at least 11.5v, all electrical components off, ground connections between engine and chassis well connected, coolant temperature at least 80-degrees Celicius. The ECM detected an unexpected voltage condition on the bypass valve control circuit **Possible Causes:** • Charge air system check for leaks • Recirculating valve for turbocharger is faulty • Turbocharging system may be damaged • Vacuum diaphragm for turbocharger needs adjusting • Wastegate bypass regulator valve is faulty • Bypass solenoid power supply circuit is open • Bypass solenoid control circuit is open, shorted to ground or system power • Bypass solenoid assembly is damaged or has failed • Charge air pressure sensor is faulty • Voltage supply to the charge air pressure sensor is open or shorted • Check the charge air system for leaks • Recirculating valve for turbocharger is faulty • Turbocharging system is damaged or not functioning properly • Turbocharger recirculating valve is faulty • Vacuum diaphragm for turbocharger is out of adjustment • Wastegate bypass regulator valve is faulty • ECM has failed
DTC: P1296 **Passat:** 2.8L (AHA, ATQ); **Jetta:** 2.0L (AEG, AVH, AZG, BBW, BEV), 2.8L (AFP, BDF); **Golf:** 2.0L (AEG, AVH, AZG, BEV); **GTI:** 2.0L (AEG), 2.8L (AFP, BDF); **NB:** 1.8L (APH, AWP, AWV, BNU, BKF), 2.0L (AEG, AVH, AZG, BEV); **Nbvert:** (BDG) **Transmissions:** All	**Cooling System Malfunction Conditions:** Key on, engine not running, the Engine Control Module (ECM) will use the intake air temperature as a replacement value for an engine start (start temperature replacement value) as soon as there is a Diagnostic Trouble Code (DTC) stored in DTC memory for the Engine Coolant Temperature (ECT) sensor. The temperature then rises according to a program stored in the ECM. When the engine has reached normal operating temperature a fixed replacement value will be displayed. This fixed value is also dependent upon the intake air temperature. **Possible Causes:** • Engine coolant temperature sensor has failed • An open circuit or a short to B+ is present • Sensor circuit is short to ground • ECM has failed
DTC: P1297 **Passat:** 1.8 (AEB, ATW, AUG, AWM), 2.0L (BPY); **Jetta:** 1.8L (AWD, AWP, AWW), 2.0L (BPY); **Golf:** 1.8L (AWD, AWW); **GTI:** 1.8L (AWD, AWP, AWW); **NB:** 1.8L (APH, AWP, AWV, BNU, BKF) **Transmissions:** All	**Connection Turbocharger/Throttle Valve Pressure Hose Conditions:** Engine started, battery voltage at least 11.5v, all electrical components off, ground connections between engine and chassis well connected, coolant temperature at least 80-degrees Celicius. The ECM detected an unexpected voltage condition on the turbo valve pressure hose. **Possible Causes:** • Charge air system check for leaks • Recirculating valve for turbocharger is faulty • Turbocharging system may be damaged • Vacuum diaphragm for turbocharger needs adjusting • Wastegate bypass regulator valve is faulty • Bypass solenoid power supply circuit is open • Bypass solenoid control circuit is open, shorted to ground or system power • Bypass solenoid assembly is damaged or has failed • Charge air pressure sensor is faulty • Voltage supply to the charge air pressure sensor is open or shorted • Check the charge air system for leaks • Recirculating valve for turbocharger is faulty • Turbocharging system is damaged or not functioning properly • Turbocharger recirculating valve is faulty • Vacuum diaphragm for turbocharger is out of adjustment • Wastegate bypass regulator valve is faulty • ECM has failed

DTC	Trouble Code Title, Conditions & Possible Causes
DTC: P1300 **Passat:** 2.0L (BPY), 2.8L (AAA); **Jetta:** 2.0 (ABA, AEG, AVH, AZG, BBW, BEV), 2.8L (AAA); **Golf:** 2.0L (ABA, AEG, AVH, AZG, BEV); **GTI:** 2.0L (ABA, AEG), 2.8L (AAA); **NB:** 1.8L (APH, AWP, AWV, BNU, BKF), 2.0L (AEG, AVH, AZG, BEV); **Nbvert:** (BDG); **Cabrio:** 2.0L (ABA) **Transmissions:** All	**Misfire Detected Reason: Fuel Level Too Low Conditions:** Engine running, the ECM detected a misfire because of lack of fuel **Possible Causes:** • Fuel level too low • Fuel leak • Fuel injector faulty
DTC: P1321 **Passat:** 4.0L (BDP) **Transmissions:** All	**Knock Sensor 3 Circuit Low Input Conditions:** Engine started, battery voltage at least 11.5v, all electrical components off, ground connections between engine and chassis well connected and the ECM detected the Knock Sensor signal was too low or not recognized by the ECM **Possible Causes:** • Knock sensor circuit is open • Knock sensor is loose (tighten to 20 NM) • Contact between the knock sensor and cylinder block is dirty, corroded or greasy • Knock sensor circuit is shorted to ground, or shorted to power • Knock sensor is damaged or it has failed • Wrong kind of fuel used • A component in the engine compartment is loose or not properly secured • ECM has failed
DTC: P1322 **Passat:** 4.0L (BDP) **Transmissions:** All	**Knock Sensor 3 Circuit High Input Conditions:** Engine started, battery voltage at least 11.5v, all electrical components off, ground connections between engine and chassis well connected and the ECM detected the Knock Sensor signal was too high or not recognized by the ECM **Possible Causes:** • Knock sensor circuit is open • Knock sensor is loose (tighten to 20 NM) • Contact between the knock sensor and cylinder block is dirty, corroded or greasy • Knock sensor circuit is shorted to ground, or shorted to power • Knock sensor is damaged or it has failed • Wrong kind of fuel used • A component in the engine compartment is loose or not properly secured • ECM has failed
DTC: P1323 **Passat:** 4.0L (BDP) **Transmissions:** All	**Knock Sensor 4 Circuit Low Input Conditions:** Engine started, battery voltage at least 11.5v, all electrical components off, ground connections between engine and chassis well connected, and the ECM detected the Knock Sensor signal was too low or not recognized by the ECM **Possible Causes:** • Knock sensor circuit is open • Knock sensor is loose (tighten to 20 NM) • Contact between the knock sensor and cylinder block is dirty, corroded or greasy • Knock sensor circuit is shorted to ground, or shorted to power • Knock sensor is damaged or it has failed • Wrong kind of fuel used • A component in the engine compartment is loose or not properly secured • ECM has failed
DTC: P1324 **Passat:** 4.0L (BDP) **Transmissions:** All	**Knock Sensor 4 Circuit High Input Conditions:** Engine started, battery voltage at least 11.5v, all electrical components off, ground connections between engine and chassis well connected, and the ECM detected the Knock Sensor signal was too high or not recognized by the ECM **Possible Causes:** • Knock sensor circuit is open • Knock sensor is loose (tighten to 20 NM) • Contact between the knock sensor and cylinder block is dirty, corroded or greasy • Knock sensor circuit is shorted to ground, or shorted to power • Knock sensor is damaged or it has failed • Wrong kind of fuel used • A component in the engine compartment is loose or not properly secured • ECM has failed

DTC	Trouble Code Title, Conditions & Possible Causes
DTC: P1325 **Passat:** 1.8 (AEB, ATW, AUG, AWM), 2.0L (BPY), 2.8L (AHA, ATQ); **Jetta:** 1.8L (AWD, AWP, AWW), 2.0L (ABA); **Golf:** 1.8L (AWD, AWW), 2.0L (ABA); **GTI:** 1.8L (AWD, AWP, AWW), 2.0L (ABA); **NB:** 1.8L (APH, AWP, AWV, BNU, BKF); **Cabrio:** 2.0L (ABA) **Transmissions:** All	**Cylinder 1-Knock Control Limit Attained Conditions:** Engine started, battery voltage at least 11.5v, all electrical components off, ground connections between engine and chassis well connected, and the ECM detected the Knock Sensor signal was more than the calibrated value. **Possible Causes:** • Knock sensor circuit is open • Knock sensor circuit is shorted to ground, or shorted to power • Knock sensor is damaged or it has failed • Poor fuel quality • Loosen knock sensors and tighten again to 20 Nm • ECM has failed
DTC: P1326 **Passat:** 1.8 (AEB, ATW, AUG, AWM), 2.0L (BPY), 2.8L (AHA, ATQ); **Jetta:** 1.8L (AWD, AWP, AWW), 2.0L (ABA); **Golf:** 1.8L (AWD, AWW), 2.0L (ABA); **GTI:** 1.8L (AWD, AWP, AWW), 2.0L (ABA); **NB:** 1.8L (APH, AWP, AWV, BNU, BKF); **Cabrio:** 2.0L (ABA) **Transmissions:** All	**Cylinder 2-Knock Control Limit Attained Conditions:** Engine started, battery voltage at least 11.5v, all electrical components off, ground connections between engine and chassis well connected, and the ECM detected the Knock Sensor signal was more than the calibrated value. **Possible Causes:** • Knock sensor circuit is open • Knock sensor circuit is shorted to ground, or shorted to power • Knock sensor is damaged or it has failed • Poor fuel quality • Loosen knock sensors and tighten again to 20 Nm • ECM has failed
DTC: P1327 **Passat:** 1.8 (AEB, ATW, AUG, AWM), 2.0L (BPY), 2.8L (AHA, ATQ); **Jetta:** 1.8L (AWD, AWP, AWW), 2.0L (ABA); **Golf:** 1.8L (AWD, AWW), 2.0L (ABA); **GTI:** 1.8L (AWD, AWP, AWW), 2.0L (ABA); **NB:** 1.8L (APH, AWP, AWV, BNU, BKF); **Cabrio:** 2.0L (ABA) **Transmissions:** All	**Cylinder 3-Knock Control Limit Attained Conditions:** Engine started, battery voltage at least 11.5v, all electrical components off, ground connections between engine and chassis well connected, and the ECM detected the Knock Sensor signal was more than the calibrated value. **Possible Causes:** • Knock sensor circuit is open • Knock sensor circuit is shorted to ground, or shorted to power • Knock sensor is damaged or it has failed • Poor fuel quality • Loosen knock sensors and tighten again to 20 Nm • ECM has failed
DTC: P1328 **Passat:** 1.8 (AEB, ATW, AUG, AWM), 2.0L (BPY), 2.8L (AHA, ATQ); **Jetta:** 1.8L (AWD, AWP, AWW), 2.0L (ABA); **Golf:** 1.8L (AWD, AWW), 2.0L (ABA); **GTI:** 1.8L (AWD, AWP, AWW), 2.0L (ABA); **NB:** 1.8L (APH, AWP, AWV, BNU, BKF); **Cabrio:** 2.0L (ABA) **Transmissions:** All	**Cylinder 4-Knock Control Limit Attained Conditions:** Engine started, battery voltage at least 11.5v, all electrical components off, ground connections between engine and chassis well connected, and the ECM detected the Knock Sensor signal was more than the calibrated value. **Possible Causes:** • Knock sensor circuit is open • Knock sensor circuit is shorted to ground, or shorted to power • Knock sensor is damaged or it has failed • ECM has failed
DTC: P1329 **Passat:** 2.8L (AHA, ATQ) **Transmissions:** All	**Cylinder 5-Knock Control Limit Attained Conditions:** Engine started, battery voltage at least 11.5v, all electrical components off, ground connections between engine and chassis well connected, and the ECM detected the Knock Sensor signal was more than the calibrated value. **Possible Causes:** • Knock sensor circuit is open • Knock sensor circuit is shorted to ground, or shorted to power • Knock sensor is damaged or it has failed • Poor fuel quality • Loosen knock sensors and tighten again to 20 Nm • ECM has failed

DTC	Trouble Code Title, Conditions & Possible Causes
DTC: P1330 **Passat:** 2.8L (AHA, ATQ) **Transmissions:** All	**Cylinder 6-Knock Control Limit Attained Conditions:** Engine started, battery voltage at least 11.5v, all electrical components off, ground connections between engine and chassis well connected, and the ECM detected the Knock Sensor signal was more than the calibrated value. **Possible Causes:** • Knock sensor circuit is open • Knock sensor circuit is shorted to ground, or shorted to power • Knock sensor is damaged or it has failed • Poor fuel quality • Loosen knock sensors and tighten again to 20 Nm • ECM has failed
DTC: P1335 **Passat:** 1.8 (AEB, ATW, AUG, AWM), 2.0L (BPY); **Jetta:** 1.8L (AWD, AWP, AWW), 2.0L (AEG, AVH, AZG, BBW, BEV, BPY), 2.8L (AFP, BDF); **Golf:** 1.8L (AWD, AWW), 2.0L (AEG, AVH, AZG, BEV); **GTI:** 1.8L (AWD, AWP, AWW), 2.0L (AEG), 2.8L (AFP, BDF); **NB:** 1.8L (APH, AWP, AWV, BNU, BKF), 2.0L (AEG, AVH, AZG, BEV); **Nbvert:** (BDG) **Transmissions:** All	**Engine Torque Monitoring 2 Control Limit Exceeded Conditions:** Engine cold, battery voltage at least 11.5v, all electrical components off, ground connections between engine and chassis well connected, the ECM detected a signal beyond the required limit. **Possible Causes:** • Engine Control Module (ECM) has failed • Voltage supply for Engine Control Module (ECM) is shorted • Engine Coolant Temperature (ECT) sensor is faulty • Intake Air Temperature (IAT) sensor is faulty • Intake Manifold Runner Position Sensor is faulty • Intake system for leaks (false air) is faulty • Motor for intake flap is faulty • Mass Air Flow (MAF) sensor is faulty
DTC: P1336 **Passat:** 1.8 (AEB, ATW, AUG, AWM), 2.0L (BPY); **Jetta:** 1.8L (AWD, AWP, AWW), 2.0L (AEG, AVH, AZG, BBW, BEV, BPY), 2.8L (AFP, BDF); **Golf:** 1.8L (AWD, AWW), 2.0L (AEG, AVH, AZG, BEV); **GTI:** 1.8L (AWD, AWP, AWW), 2.0L (AEG), 2.8L (AFP, BDF); **NB:** 1.8L (APH, AWP, AWV, BNU, BKF), 2.0L (AEG, AVH, AZG, BEV); **Nbvert:** (BDG) **Transmissions:** All	**Engine Torque Monitoring Control Limit Exceeded Conditions:** Engine cold, battery voltage at least 11.5v, all electrical components off, ground connections between engine and chassis well connected, the ECM detected a signal beyond the required limit. **Possible Causes:** • Engine Control Module (ECM) has failed • Voltage supply for Engine Control Module (ECM) is shorted • Engine Coolant Temperature (ECT) sensor is faulty • Intake Air Temperature (IAT) sensor is faulty • Intake Manifold Runner Position Sensor is faulty • Intake system for leaks (false air) is faulty • Motor for intake flap is faulty • Mass Air Flow (MAF) sensor is faulty
DTC: P1337 **2T, MIL: Yes** **Passat:** 1.8 (AEB, ATW, AUG, AWM), 2.0L (BPY), 2.8L (AHA, ATQ); **Jetta:** 1.8L (AWD, AWP, AWW); **Golf:** 1.8L (AWD, AWW); **GTI:** 1.8L (AWD, AWP, AWW); **NB:** 1.8L (APH, AWP, AWV, BNU, BKF) **Transmissions:** All	**Camshaft Position Sensor (Bank 1) Short to Ground Conditions:** Engine started, battery voltage at least 11.5v, all electrical components off, ground connections between engine and chassis well connected, and the ECM detected an unexpected low or high voltage condition on the camshaft position sensor circuit **Possible Causes:** • Faulty CPM sensor • ECM has failed
DTC: P1338 **2T, MIL: Yes** **Passat:** 1.8 (AEB, ATW, AUG, AWM), 2.0L (BPY), 2.8L (AHA, ATQ); **Jetta:** 1.8L (AWD, AWP, AWW); **Golf:** 1.8L (AWD, AWW); **GTI:** 1.8L (AWD, AWP, AWW); **NB:** 1.8L (APH, AWP, AWV, BNU, BKF) **Transmissions:** All	**Camshaft Position Sensor (Bank 1) Open/Short to B+ Conditions:** Engine started, battery voltage at least 11.5v, all electrical components off, ground connections between engine and chassis well connected, and the ECM detected an unexpected low or high voltage condition on the camshaft position sensor circuit **Possible Causes:** • Faulty CPM sensor • ECM has failed

DTC	Trouble Code Title, Conditions & Possible Causes
DTC: P1340 **2T, MIL: Yes** **Passat:** 1.8 (AEB, ATW, AUG, AWM), 2.0L (ABA, BPY), 2.8L (AAA, AHA, ATQ); **Jetta:** 1.8L (AWD, AWP, AWW), 2.0L (AEG, AVH, AZG, BBW, BEV), 2.8L (AAA, AFP, BDF); **Golf:** 1.8L (AWD, AWW), 2.0L (AEG, AVH, AZG, BEV); **GTI:** 1.8L (AWD, AWP, AWW), 2.0L (AEG), 2.8L (AAA, AFP, BDF); **NB:** 1.8L (APH, AWP, AWV, BNU, BKF), 2.0L (AEG, AVH, AZG, BEV); **Nbvert:** (BDG) **Transmissions:** All	Crankshaft Position/Camshaft Sensor Signal Out of Sequence Conditions: Engine started, battery voltage at least 11.5v, all electrical components off, ground connections between engine and chassis well connected, and the ECM detected the crankshaft position sensor and the camshaft sensor were out of sequence with each other. **Note: The Engine Speed (RPM) Sensor detects engine speed and reference marks. Without an engine speed signal, the engine will not start. If the engine speed signal fails while the engine is running, the engine will stop immediately.** **Possible Causes:** • Engine speed sensor has failed or is contaminated (metal filings) • Engine speed sensor's wheel is damaged • Engine speed sensor circuit is shorted to the cable shield • Engine speed sensor circuit is open • ECM is faulty • Canshaft position sensor is faulty
DTC: P1341 **2T, MIL: Yes** **Jetta:** 2.8L (AFP, BDF); **GTI:** 2.8L (AFP) **Transmissions:** All	**Ignition Coil Power Output Stage 1 Short to Ground Conditions:** Key on or Engine started, battery voltage at least 11.5v, all electrical components off, ground connections between engine and chassis well connected, and the ECM detected the voltage of the ignition coil was outside the designed parameters. **Possible Causes:** • Fuel pump relay faulty • Canshaft position sensor has failed • Engine speed sensor has failed • Circuit wires have shorted to ground or are open • Ignition coils with power output stages are faulty • ECM has failed
DTC: P1343 **Jetta:** 2.8L (AFP); **GTI:** 2.8L (AFP) **Transmissions:** All	**Ignition Coil Power Output Stage 2 Short to Ground Conditions:** Key on or Engine started, battery voltage at least 11.5v, all electrical components off, ground connections between engine and chassis well connected, and the ECM detected the voltage of the ignition coil was outside the designed parameters. **Possible Causes:** • Fuel pump relay faulty • Canshaft position sensor has failed • Engine speed sensor has failed • Circuit wires have shorted to ground or are open • Ignition coils with power output stages are faulty • ECM has failed
DTC: P1345 **Jetta:** 2.8L (AFP); **GTI:** 2.8L (AFP) **Transmissions:** All	**Ignition Coil Power Output Stage 3 Short to Ground Conditions:** Key on or Engine started, battery voltage at least 11.5v, all electrical components off, ground connections between engine and chassis well connected, and the ECM detected the voltage of the ignition coil was outside the designed parameters. **Possible Causes:** • Fuel pump relay faulty • Canshaft position sensor has failed • Engine speed sensor has failed • Circuit wires have shorted to ground or are open • Ignition coils with power output stages are faulty • ECM has failed
DTC: P1355 **Passat:** 1.8 (AEB, ATW, AUG, AWM), 2.0L (BPY); **Jetta:** 1.8L (AWD, AWP, AWW), 2.0 (ABA, AEG, AVH, AZG, BBW, BEV, BPY); **Golf:** 1.8L (AWD, AWW), 2.0L (ABA, AEG, AVH, AZG, BEV); **GTI:** 1.8L (AWD, AWP, AWW), 2.0L (ABA, AEG); **NB:** 1.8L (APH, AWP, AWV, BNU, BKF), 2.0L (AEG, AVH, AZG, BEV); **Nbvert:** (BDG); **Cabrio:** 2.0L (ABA) **Transmissions:** All	**Cylinder 1 Ignition Circuit Open Circuit Conditions:** Key on or Engine started, battery voltage at least 11.5v, all electrical components off, ground connections between engine and chassis well connected, and the ECM detected the voltage of the ignition was outside the designed parameters. **Possible Causes:** • Fuel pump relay faulty • Canshaft position sensor has failed • Engine speed sensor has failed • Circuit wires have shorted to ground or are open • Ignition coils with power output stages are faulty • ECM has failed

DTC	Trouble Code Title, Conditions & Possible Causes
DTC: P1356 **Passat:** 1.8 (AEB, ATW, AUG, AWM), 2.0L (BPY); **Jetta:** 1.8L (AWD, AWP, AWW), 2.0L (ABA); **Golf:** 1.8L (AWD, AWW), 2.0L (ABA); **GTI:** 1.8L (AWD, AWP, AWW), 2.0L (ABA, AEG); **NB:** 1.8L (APH, AWP, AWV, BNU, BKF); **Cabrio:** 2.0L (ABA) **Transmissions:** All	**Cylinder 1 Ignition Circuit Short to B+ Conditions:** Key on or Engine started, battery voltage at least 11.5v, all electrical components off, ground connections between engine and chassis well connected, and the ECM detected the voltage of the ignition was outside the designed parameters. **Possible Causes:** • Fuel pump relay faulty • Canshaft position sensor has failed • Engine speed sensor has failed • Circuit wires have shorted to ground or are open • Ignition coils with power output stages are faulty • ECM has failed
DTC: P1357 **Passat:** 1.8 (AEB, ATW, AUG, AWM), 2.0L (BPY); **Jetta:** 1.8L (AWD, AWP, AWW), 2.0L (ABA); **Golf:** 1.8L (AWD, AWW), 2.0L (ABA); **GTI:** 1.8L (AWD, AWP, AWW), 2.0L (ABA); **NB:** 1.8L (APH, AWP, AWV, BNU, BKF); **Cabrio:** 2.0L (ABA) **Transmissions:** All	**Cylinder 1 Ignition Circuit Short to Ground Conditions:** Key on or Engine started, battery voltage at least 11.5v, all electrical components off, ground connections between engine and chassis well connected, and the ECM detected the voltage of the ignition was outside the designed parameters. **Possible Causes:** • Fuel pump relay faulty • Canshaft position sensor has failed • Engine speed sensor has failed • Circuit wires have shorted to ground or are open • Ignition coils with power output stages are faulty • ECM has failed
DTC: P1358 **Passat:** 1.8 (AEB, ATW, AUG, AWM), 2.0L (BPY); **Jetta:** 1.8L (AWD, AWP, AWW), 2.0 (ABA, AEG, AVH, AZG, BBW, BEV, BPY); **Golf:** 1.8L (AWD, AWW), 2.0L (ABA, AEG, AVH, AZG, BEV); **GTI:** 1.8L (AWD, AWP, AWW), 2.0L (ABA, AEG); **NB:** 1.8L (APH, AWP, AWV, BNU, BKF), 2.0L (AEG, AVH, AZG, BEV); **Nbvert:** (BDG); **Cabrio:** 2.0L (ABA) **Transmissions:** All	**Cylinder 2 Ignition Circuit Open Circuit Conditions:** Key on or Engine started, battery voltage at least 11.5v, all electrical components off, ground connections between engine and chassis well connected, and the ECM detected the voltage of the ignition was outside the designed parameters. **Possible Causes:** • Fuel pump relay faulty • Canshaft position sensor has failed • Engine speed sensor has failed • Circuit wires have shorted to ground or are open • Ignition coils with power output stages are faulty • ECM has failed
DTC: P1359 **2T, MIL: Yes** **Passat:** 1.8 (AEB, ATW, AUG, AWM), 2.0L (BPY); **Jetta:** 1.8L (AWD, AWP, AWW), 2.0L (ABA); **Golf:** 1.8L (AWD, AWW), 2.0L (ABA); **GTI:** 1.8L (AWD, AWP, AWW), 2.0L (ABA); **NB:** 1.8L (APH, AWP, AWV, BNU, BKF); **Cabrio:** 2.0L (ABA) **Transmissions:** All	**Cylinder 2 Ignition Circuit Short to B+ Conditions:** Key on or Engine started, battery voltage at least 11.5v, all electrical components off, ground connections between engine and chassis well connected, and the ECM detected the voltage of the ignition was outside the designed parameters. **Possible Causes:** • Fuel pump relay faulty • Canshaft position sensor has failed • Engine speed sensor has failed • Circuit wires have shorted to ground or are open • Ignition coils with power output stages are faulty • ECM has failed
DTC: P1360 **Passat:** 1.8 (AEB, ATW, AUG, AWM), 2.0L (BPY); **Jetta:** 1.8L (AWD, AWP, AWW), 2.0L (ABA); **Golf:** 1.8L (AWD, AWW), 2.0L (ABA); **GTI:** 1.8L (AWD, AWP, AWW), 2.0L (ABA); **NB:** 1.8L (APH, AWP, AWV, BNU, BKF); **Cabrio:** 2.0L (ABA) **Transmissions:** All	**Cylinder 2 Ignition Circuit Short to Ground Conditions:** Key on or Engine started, battery voltage at least 11.5v, all electrical components off, ground connections between engine and chassis well connected, and the ECM detected the voltage of the ignition was outside the designed parameters. **Possible Causes:** • Fuel pump relay faulty • Canshaft position sensor has failed • Engine speed sensor has failed • Circuit wires have shorted to ground or are open • Ignition coils with power output stages are faulty • ECM has failed

DTC	Trouble Code Title, Conditions & Possible Causes
DTC: P1361 **Passat:** 1.8 (AEB, ATW, AUG, AWM); **Jetta:** 1.8L (AWD, AWP, AWW), 2.0 (ABA, AEG, AVH, AZG, BBW, BEV, BPY); **Golf:** 1.8L (AWD, AWW), 2.0L (ABA, AEG, AVH, AZG, BEV); **GTI:** 1.8L (AWD, AWP, AWW), 2.0L (ABA, AEG); **NB:** 1.8L (APH, AWP, AWV, BNU, BKF), 2.0L (AEG, AVH, AZG, BEV); **Nbvert:** (BDG); **Cabrio:** 2.0L (ABA) **Transmissions:** All	**Cylinder 3 Ignition Circuit Open Circuit Conditions:** Key on or Engine started, battery voltage at least 11.5v, all electrical components off, ground connections between engine and chassis well connected, and the ECM detected the voltage of the ignition was outside the designed parameters. **Possible Causes:** • Fuel pump relay faulty • Canshaft position sensor has failed • Engine speed sensor has failed • Circuit wires have shorted to ground or are open • Ignition coils with power output stages are faulty • ECM has failed
DTC: P1362 **Passat:** 1.8 (AEB, ATW, AWM), 2.0L (BPY); **Jetta:** 1.8L (AWD, AWP, AWW), 2.0L (ABA); **Golf:** 1.8L (AWD, AWW), 2.0L (ABA); **GTI:** 1.8L (AWD, AWP, AWW), 2.0L (ABA); **NB:** 1.8L (APH, AWP, AWV, BNU, BKF); **Cabrio:** 2.0L (ABA) **Transmissions:** All	**Cylinder 3 Ignition Circuit Short to B+ Conditions:** Key on or Engine started, battery voltage at least 11.5v, all electrical components off, ground connections between engine and chassis well connected, and the ECM detected the voltage of the ignition was outside the designed parameters. **Possible Causes:** • Fuel pump relay faulty • Canshaft position sensor has failed • Engine speed sensor has failed • Circuit wires have shorted to ground or are open • Ignition coils with power output stages are faulty • ECM has failed
DTC: P1363 **Passat:** 1.8 (AEB, ATW, AUG, AWM), 2.0L (BPY); **Jetta:** 1.8L (AWD, AWP, AWW), 2.0L (ABA); **Golf:** 1.8L (AWD, AWW), 2.0L (ABA); **GTI:** 1.8L (AWD, AWP, AWW), 2.0L (ABA); **NB:** 1.8L (APH, AWP, AWV, BNU, BKF); **Cabrio:** 2.0L (ABA) **Transmissions:** All	**Cylinder 3 Ignition Circuit Short to Ground Conditions:** Key on or Engine started, battery voltage at least 11.5v, all electrical components off, ground connections between engine and chassis well connected, and the ECM detected the voltage of the ignition was outside the designed parameters. **Possible Causes:** • Fuel pump relay faulty • Canshaft position sensor has failed • Engine speed sensor has failed • Circuit wires have shorted to ground or are open • Ignition coils with power output stages are faulty • ECM has failed
DTC: P1364 **Passat:** 1.8 (AEB, ATW, AUG, AWM), 2.0L (BPY); **Jetta:** 1.8L (AWD, AWP, AWW), 2.0 (ABA, AEG, AVH, AZG, BBW, BEV, BPY); **Golf:** 1.8L (AWD, AWW), 2.0L (ABA, AEG, AVH, AZG, BEV); **GTI:** 1.8L (AWD, AWP, AWW), 2.0L (ABA, AEG); **NB:** 1.8L (APH, AWP, AWV, BNU, BKF), 2.0L (AEG, AVH, AZG, BEV); **Nbvert:** (BDG); **Cabrio:** 2.0L (ABA) **Transmissions:** All	**Cylinder 4 Ignition Circuit Open Circuit Conditions:** Key on or Engine started, battery voltage at least 11.5v, all electrical components off, ground connections between engine and chassis well connected, and the ECM detected the voltage of the ignition was outside the designed parameters. **Possible Causes:** • Fuel pump relay faulty • Canshaft position sensor has failed • Engine speed sensor has failed • Circuit wires have shorted to ground or are open • Ignition coils with power output stages are faulty • ECM has failed
DTC: P1365 **Passat:** 1.8 (AEB, ATW, AUG, AWM), 2.0L (BPY); **Jetta:** 1.8L (AWD, AWP, AWW), 2.0L (ABA); **Golf:** 1.8L (AWD, AWW), 2.0L (ABA); **GTI:** 1.8L (AWD, AWP, AWW), 2.0L (ABA); **NB:** 1.8L (APH, AWP, AWV, BNU, BKF); **Cabrio:** 2.0L (ABA) **Transmissions:** All	**Cylinder 4 Ignition Circuit Short to B+ Conditions:** Key on or Engine started, battery voltage at least 11.5v, all electrical components off, ground connections between engine and chassis well connected, and the ECM detected the voltage of the ignition was outside the designed parameters. **Possible Causes:** • Fuel pump relay faulty • Canshaft position sensor has failed • Engine speed sensor has failed • Circuit wires have shorted to ground or are open • Ignition coils with power output stages are faulty • ECM has failed

DTC	Trouble Code Title, Conditions & Possible Causes
DTC: P1366 **Passat:** 1.8 (AEB, ATW, AUG, AWM), 2.0L (BPY); **Jetta:** 1.8L (AWD, AWP, AWW), 2.0L (ABA); **Golf:** 1.8L (AWD, AWW), 2.0L (ABA); **GTI:** 1.8L (AWD, AWP, AWW), 2.0L (ABA); **NB:** 1.8L (APH, AWP, AWV, BNU, BKF); **Cabrio:** 2.0L (ABA) **Transmissions:** All	**Cylinder 4 Ignition Circuit Short to Ground Conditions:** Key on or Engine started, battery voltage at least 11.5v, all electrical components off, ground connections between engine and chassis well connected, and the ECM detected the voltage of the ignition was outside the designed parameters. **Possible Causes:** • Fuel pump relay faulty • Canshaft position sensor has failed • Engine speed sensor has failed • Circuit wires have shorted to ground or are open • Ignition coils with power output stages are faulty • ECM has failed
DTC: P1386 **Passat:** 1.8 (AEB, ATW, AUG, AWM), 2.8L (AHA, ATQ); **Jetta:** 1.8L (AWD, AWP, AWW), 2.0 (ABA, BPY); **Golf:** 1.8L (AWD, AWW), 2.0L (ABA); **GTI:** 1.8L (AWD, AWP, AWW), 2.0L (ABA); **NB:** 1.8L (APH, AWP, AWV, BNU, BKF) **Transmissions:** All	**Internal Control Module, Knock Control Circuit Error Conditions:** Engine started, and the ECM detected a too high or too low voltage condition on the knock control circuits, or a miscommunication between the knock control and the ECM. **Possible Causes:** • ECM has failed
DTC: P1387 **2T, MIL: Yes** **Passat:** 1.8L (ATW, AUG); **Jetta:** 1.8L (AWD, AWP, AWW), 2.0 (ABA, BPY), 2.8L (AFP, BDF); **Golf:** 1.8L (AWD, AWW), 2.0L (ABA); **GTI:** 1.8L (AWD, AWP, AWW), 2.0L (ABA), 2.8L (AFP, BDF); **NB:** 1.8L (APH, AWP, AWV, BNU, BKF); **Cabrio:** 2.0L (ABA) **Transmissions:** All	**Internal Control Module Altitude Sensor Error Conditions:** Ignition on, the ECM detected and altitude sensor error. To achieve optimal anti-theft protection for the vehicle, an anti-theft immobilizer is installed. The anti-theft immobilizer is a system for enabling and locking the Engine Control Module (ECM). So that this system cannot be circumvented, it is necessary to perform adaptation of the anti-theft immobilizer using the Vehicle Diagnostic and Information System VAS 5052 in the On Board Diagnostic (OBD) function. The great availability of equipment options makes it necessary to adapt the Engine Control Module (ECM) to the vehicle (e.g. throttle valve control module or cruise control system). This "writing" function is not possible with the generic scan tool. **Possible Causes:** • (If ECM was replaced) ECM ID not the same as the replaced unit • ECM has failed • Voltage supply for Engine Control Module (ECM) has shorted
DTC: P1388 **Passat:** 1.8L (ATW, AUG), 2.0L (BPY); **Jetta:** 1.8L (AWD, AWP, AWW), 2.0 (ABA, BPY), 2.8L (AFP, BDF); **Golf:** 1.8L (AWD, AWW), 2.0L (ABA); **GTI:** 1.8L (AWD, AWP, AWW), 2.0L (ABA), 2.8L (AFP, BDF); **NB:** 1.8L (APH, AWP, AWV, BNU, BKF); **Cabrio:** 2.0L (ABA) **Transmissions:** All	**Internal Control Module Drive By Wire Error Conditions:** Ignition on, the ECM detected and drive by wire error. To achieve optimal anti-theft protection for the vehicle, an anti-theft immobilizer is installed. The anti-theft immobilizer is a system for enabling and locking the Engine Control Module (ECM). So that this system cannot be circumvented, it is necessary to perform adaptation of the anti-theft immobilizer using the Vehicle Diagnostic and Information System VAS 5052 in the On Board Diagnostic (OBD) function. The great availability of equipment options makes it necessary to adapt the Engine Control Module (ECM) to the vehicle (e.g. throttle valve control module or cruise control system). This "writing" function is not possible with the generic scan tool. **Possible Causes:** • Engine Control Module (ECM) has failed • Voltage supply for Engine Control Module (ECM) has shorted
DTC: P1391 **Passat:** 2.8L (AHA, ATQ) **Transmissions:** All	**Camshaft Position Sensor (Bank 2) Short to Ground Conditions:** Key on or Engine started, battery voltage must be at least 11.5v, all electrical components must be off, parking brake must be engaged (to keep daytime driving lights off), automatic transmission selector must be in park and the ground between the engine and the chassis must be well connected. The ECM detected an unexpected low or high voltage condition on the camshaft position sensor circuit. **Possible Causes:** • CMP sensor circuit is open or shorted to ground • CMP sensor circuit is shorted to power • CMP sensor ground (return) circuit is open • CMP sensor installation incorrect (Hall-effect type) • CMP sensor is damaged or CMP sensor shielding damaged • ECM has failed

DTC	Trouble Code Title, Conditions & Possible Causes
DTC: P1392 **Passat:** 2.8L (AHA, ATQ) **Transmissions:** All	**Camshaft Position Sensor (Bank 2) Open/Short to B+ Conditions:** Key on or Engine started, battery voltage must be at least 11.5v, all electrical components must be off, parking brake must be engaged (to keep daytime driving lights off), automatic transmission selector must be in park and the ground between the engine and the chassis must be well connected. The ECM detected an unexpected low or high voltage condition on the camshaft position sensor circuit. **Possible Causes:** • CMP sensor circuit is open or shorted to ground • CMP sensor circuit is shorted to power • CMP sensor ground (return) circuit is open • CMP sensor installation incorrect (Hall-effect type) • CMP sensor is damaged or CMP sensor shielding damaged • ECM has failed
DTC: P1393 **Jetta:** 2.8L (AFP); **GTI:** 2.8L (AFP) **Transmissions:** All	**Ignition Coil Power Output Stage 1 Electrical Malfunction Conditions:** Key on or Engine started, battery voltage at least 11.5v, all electrical components off, ground connections between engine and chassis well connected, parking brake must be engaged (to keep daytime driving lights off), automatic transmission selector must be in park, and the ECM detected the an electrical malfunction of the ignition coil so that it won't properly function. **Possible Causes:** • Fuel pump relay faulty • Canshaft position sensor has failed • Engine speed sensor has failed • Circuit wires have shorted to ground or are open • Ignition coils with power output stages are faulty • ECM has failed
DTC: P1394 **Jetta:** 2.8L (AFP); **GTI:** 2.8L (AFP) **Transmissions:** All	**Ignition Coil Power Output Stage 2 Electrical Malfunction Conditions:** Key on or Engine started, battery voltage at least 11.5v, all electrical components off, ground connections between engine and chassis well connected, parking brake must be engaged (to keep daytime driving lights off), automatic transmission selector must be in park, and the ECM detected the an electrical malfunction of the ignition coil so that it won't properly function. **Possible Causes:** • Fuel pump relay faulty • Canshaft position sensor has failed • Engine speed sensor has failed • Circuit wires have shorted to ground or are open • Ignition coils with power output stages are faulty • ECM has failed
DTC: P1395 **Jetta:** 2.8L (AFP); **GTI:** 2.8L (AFP) **Transmissions:** All	**Ignition Coil Power Output Stage 3 Electrical Malfunction Conditions:** Key on or Engine started, battery voltage at least 11.5v, all electrical components off, ground connections between engine and chassis well connected, parking brake must be engaged (to keep daytime driving lights off), automatic transmission selector must be in park, and the ECM detected the an electrical malfunction of the ignition coil so that it won't properly function. **Possible Causes:** • Fuel pump relay faulty • Canshaft position sensor has failed • Engine speed sensor has failed • Circuit wires have shorted to ground or are open • Ignition coils with power output stages are faulty • ECM has failed
DTC: P1398 **Jetta:** 2.0L (BPY) **Transmissions:** All	**Engine RPM Signal, TD Short to Ground Conditions:** Key on or Engine started, battery voltage at least 11.5v, all electrical components off, ground connections between engine and chassis well connected, parking brake must be engaged (to keep daytime driving lights off), automatic transmission selector must be in park, and the ECM detected the speed sensor signal short to ground **Possible Causes:** • Can-bus signal faulty • Speed sensor has failed
DTC: P1399 **Jetta:** 2.0L (BPY) **Transmissions:** All	**Engine RPM Signal, TD Short to B+ Conditions:** Key on or Engine started, battery voltage at least 11.5v, all electrical components off, ground connections between engine and chassis well connected, parking brake must be engaged (to keep daytime driving lights off), automatic transmission selector must be in park, and the ECM detected the speed sensor signal short to power **Possible Causes:** • Can-bus signal faulty • Speed sensor has failed

DTC	Trouble Code Title, Conditions & Possible Causes
DTC: P1409 **Passat:** 1.8 (AEB, ATW, AUG, AWM), 2.0L (BPY), 2.8L (AHA, ATQ); **Jetta:** 1.8L (AWD, AWP, AWW), 2.0L (ABA); **Golf:** 1.8L (AWD, AWW), 2.0L (ABA); **GTI:** 1.8L (AWD, AWP, AWW), 2.0L (ABA); **NB:** 1.8L (APH, AWP, AWV, BNU, BKF); **Cabrio:** 2.0L (ABA) **Transmissions:** All	**Tank Ventilation Valve Circuit Malfunction Conditions** Key on or engine running; and the ECM detected a too high or too low voltage level in the tank ventilation valve circuit. **Possible Causes:** • EVAP canister purge regulator valve has failed • Activation wire is shorting to positive • EVAP canister system has an improper or broken seal • Evaporative Emission (EVAP) canister purge regulator valve 1 is faulty • Leak Detection Pump (LDP) is faulty • Fuel filler cap is not properly closed • Lock ring on fuel pump not tightened • Hoses between EVAP canister and purge regulator valve have failed • ECM has failed
DTC: P1410 **2T, MIL: Yes** **Passat:** 1.8 (AEB, ATW, AUG, AWM), 2.0L (BPY), 2.8L (AAA, AHA, ATQ); **Jetta:** 1.8L (AWD, AWP, AWW), 2.0 (ABA, AEG, AVH, AZG, BBW, BEV), 2.8L (AAA, AFP, BDF); **Golf:** 1.8L (AWD, AWW), 2.0L (ABA, AEG, AVH, AZG, BEV); **GTI:** 1.8L (AWD, AWP, AWW), 2.0L (ABA, AEG), 2.8L (AAA, AFP, BDF); **NB:** 1.8L (APH, AWP, AWV, BNU, BKF), 2.0L (AEG, AVH, AZG, BEV); **Nbvert:** (BDG); **Cabrio:** 2.0L (ABA) **Transmissions:** All	Tank Ventilation Valve Circuit Short to B+: Key on or engine running; and the ECM detected a too high or too low voltage level in the tank ventilation valve circuit. **Possible Causes:** • EVAP canister purge regulator valve has failed • Activation wire is shorting to positive • EVAP canister system has an improper or broken seal • Evaporative Emission (EVAP) canister purge regulator valve 1 is faulty • Leak Detection Pump (LDP) is faulty • Fuel filler cap is not properly closed • Lock ring on fuel pump not tightened • Hoses between EVAP canister and purge regulator valve have failed • ECM has failed
DTC: P1420 **Passat:** 1.8L (ATW, AUG), 2.0L (ABA), 2.8L (AAA, AHA, ATQ); **Jetta:** 1.8L (AWD, AWP, AWW), 2.0 (ABA, AEG), 2.8L (AAA); **Golf:** 1.8L (AWD, AWW), 2.0L (ABA, AEG); **GTI:** 1.8L (AWD, AWP, AWW), 2.0L (ABA, AEG), 2.8L (AAA); **NB:** 1.8L (APH, AWP, AWV, BNU, BKF), 2.0L (AEG, AVH, AZG, BEV); **Nbvert:** (BDG); **Cabrio:** 2.0L (ABA) **Transmissions:** All	**Secondary Air Injector Valve Circuit Electrical Malfunction Conditions:** The Engine Control Module activates the secondary air injection solenoid valve, but the Heated Oxygen Sensor (HO2S) does not detect secondary air injection. **Note: Solenoid valve is closed when no voltage is present.** **Possible Causes:** • Connector to the secondary air injection valve is loose or disconnected • Secondary air injector valve circuit short • Secondary air injector valve circuit is open • Faulty secondary air injector valve • ECM has failed
DTC: P1421 **2T, MIL: Yes** **Passat:** 1.8L (ATW, AUG), 2.0L (ABA), 2.8L (AAA, AHA, ATQ); **Jetta:** 1.8L (AWD, AWP, AWW), 2.0 (ABA, AEG), 2.8L (AAA, AFP, BDF); **Golf:** 1.8L (AWD, AWW), 2.0L (ABA, AEG); **GTI:** 1.8L (AWD, AWP, AWW), 2.0L (ABA, AEG), 2.8L (AAA, AFP, BDF); **NB:** 1.8L (APH, AWP, AWV, BNU, BKF), 2.0L (AEG, AVH, AZG, BEV); **Nbvert:** (BDG); **Cabrio:** 2.0L (ABA) **Transmissions:** All	**Secondary Air Injector Valve Circuit Short to Ground Conditions:** The Engine Control Module detects a short circuit to ground when activating the secondary air injection solenoid valve. **Note: Solenoid valve is closed when no voltage is present.** **Possible Causes:** • Connector to the secondary air injection valve is loose or disconnected • Secondary air injector valve circuit short • Secondary air injector valve circuit is open • Faulty secondary air injector valve • ECM has failed

DTC	Trouble Code Title, Conditions & Possible Causes
DTC: P1422 **2T, MIL: Yes** **Passat:** 1.8L (ATW, AUG), 2.0L (ABA), 2.8L (AAA, AHA, ATQ); **Jetta:** 1.8L (AWD, AWP, AWW), 2.0 (ABA, AEG), 2.8L (AAA, AFP, BDF); **Golf:** 1.8L (AWD, AWW), 2.0L (ABA, AEG); **GTI:** 1.8L (AWD, AWP, AWW), 2.0L (ABA, AEG), 2.8L (AAA, AFP, BDF); **NB:** 1.8L (APH, AWP, AWV, BNU, BKF), 2.0L (AEG, AVH, AZG, BEV); **Nbvert:** (BDG); **Cabrio:** 2.0L (ABA) **Transmissions:** All	**Secondary Air Injector Valve Circuit Short to B+ Conditions:** The Engine Control Module detects a short circuit to B+ when activating the secondary air injection solenoid valve. **Note: Solenoid valve is closed when no voltage is present.** **Possible Causes:** • Connector to the secondary air injection valve is loose or disconnected • Secondary air injector valve circuit short • Secondary air injector valve circuit is open • Faulty secondary air injector valve • ECM has failed
DTC: P1424 **2T, MIL: Yes** **Passat:** 1.8L (ATW, AUG); **Jetta:** 1.8L (AWD, AWP, AWW); 2.0L (AEG, AVH, AZG, BBW, BEV), 2.8L (AFP, BDF); **Golf:** 1.8L (AWD, AWW), 2.0L (AEG, AVH, AZG, BEV); **GTI:** 1.8L (AWD, AWP, AWW), 2.0L (AEG), 2.8L (AFP, BDF); **NB:** 1.8L (APH, AWP, AWV, BNU, BKF), 2.0L (AEG, AVH, AZG, BEV); **Nbvert:** (BDG) **Transmissions:** All	**Secondary Air Injector System (Bank 1) Leak Detected Conditions:** Ignition on or vehicle running, and the ECM detected a leak in the secondary air injector system. **Possible Causes:** • Poor hose/pipe connections between the secondary air injector pump motor and valve • Faulty hoses or pipes • Mechanical faults in the secondary air injector system •
DTC: P1425 **2T, MIL: Yes** **Passat:** 1.8 (AEB, ATW, AUG, AWM), 2.0L (ABA, BPY), 2.8L (AAA, AHA, ATQ); **Jetta:** 1.8L (AWD, AWP, AWW), 2.0 (ABA, AEG, AVH, AZG, BBW, BEV), 2.8L (AAA, AFP, BDF); **Golf:** 1.8L (AWD, AWW), 2.0L (ABA, AEG, AVH, AZG, BEV); **GTI:** 1.8L (AWD, AWP, AWW), 2.0L (ABA, AEG), 2.8L (AAA, AFP, BDF); **NB:** 1.8L (APH, AWP, AWV, BNU, BKF), 2.0L (AEG, AVH, AZG, BEV); **Nbvert:** (BDG); **Cabrio:** 2.0L (ABA) **Transmissions:** All	**Tank Ventilation Valve Short to Ground Conditions:** Ignition off. The Evaporative Emission (EVAP) canister purge regulator valve in the tank venting system or activation wire has a short circuit to ground. Engine started, engine running at a steady cruise speed, canister vent solenoid enabled, and the ECM detected an unexpected voltage condition on the Canister Vent solenoid circuit. **Note: Solenoid valve is closed when no voltage is present.** **Possible Causes:** • Activation wire has a short to ground • ECM has failed • EVAP canister has failed • EVAP canister system has an improper or broken seal • Evaporative Emission (EVAP) canister purge regulator valve is faulty • Leak Detection Pump (LDP) is faulty
DTC: P1426 **2T, MIL: Yes** **Passat:** 1.8 (AEB, ATW, AUG, AWM), 2.0L (ABA, BPY), 2.8L (AAA, AHA, ATQ); **Jetta:** 1.8L (AWD, AWP, AWW), 2.0 (ABA, AEG, AVH, AZG, BBW, BEV), 2.8L (AAA, AFP, BDF); **Golf:** 1.8L (AWD, AWW), 2.0L (ABA, AEG, AVH, AZG, BEV); **GTI:** 1.8L (AWD, AWP, AWW), 2.0L (ABA, AEG), 2.8L (AAA, AFP, BDF); **NB:** 1.8L (APH, AWP, AWV, BNU, BKF), 2.0L (AEG, AVH, AZG, BEV); **Nbvert:** (BDG); **Cabrio:** 2.0L (ABA) **Transmissions:** All	**Tank Ventilation Valve Open Conditions:** Ignition off. The Evaporative Emission (EVAP) canister purge regulator valve in the tank venting system or activation wire has a short circuit to ground. Engine started, engine running at a steady cruise speed, canister vent solenoid enabled, and the ECM detected an unexpected voltage condition on the Canister Vent solenoid circuit. **Possible Causes:** • Activation wire has a short to ground • ECM has failed • EVAP canister has failed • EVAP canister system has an improper or broken seal • Evaporative Emission (EVAP) canister purge regulator valve 1 is faulty • Leak Detection Pump (LDP) is faulty

DTC	Trouble Code Title, Conditions & Possible Causes
DTC: P1432 **2T, MIL: Yes** **Passat:** 1.8L (ATW, AUG), 2.8L (AHA, ATQ); **Jetta:** 1.8L (AWD, AWP, AWW), 2.0 (ABA, AEG), 2.8L (AFP, BDF); **Golf:** 1.8L (AWD, AWW), 2.0L (ABA, AEG); **GTI:** 1.8L (AWD, AWP, AWW), 2.0L (ABA, AEG), 2.8L (AFP, BDF); **NB:** 1.8L (APH, AWP, AWV, BNU, BKF), 2.0L (AEG, AVH, AZG, BEV); **Nbvert:** (BDG); **Cabrio:** 2.0L (ABA) **Transmissions:** All	**Secondary Air Injection Valve Open Conditions:** The output Diagnostic Test Mode (DTM) can be activated only with the ignition switched on and the engine not running. The output DTM is interrupted if the engine is started, or if a rotary pulse from the ignition system is recognized.. **Possible Causes:** • Fuel pump relays have failed • Fuel injector has failed • Hoses on the EVAP canister may be clogged • EVAP canister purge regulator valve may be faulty • ECM may have failed • Manifold Tuning Valve (IMT) may have failed
DTC: P1433 **2T, MIL: Yes** **Passat:** 1.8L (ATW, AUG), 2.8L (AHA, ATQ); **Jetta:** 1.8L (AWD, AWP, AWW), 2.0 (ABA, AEG, AVH, AZG, BBW, BEV), 2.8L (AFP, BDF); **Golf:** 1.8L (AWD, AWW), 2.0L (ABA, AEG, AVH, AZG, BEV); **GTI:** 1.8L (AWD, AWP, AWW), 2.0L (ABA, AEG), 2.8L (AFP, BDF); **NB:** 1.8L (APH, AWP, AWV, BNU, BKF), 2.0L (AEG, AVH, AZG, BEV); **Nbvert:** (BDG); **Cabrio:** 2.0L (ABA) **Transmissions:** All	**Secondary Air Injection System Pump Relay Circuit Open Conditions:** The output Diagnostic Test Mode (DTM) can be activated only with the ignition switched on and the engine not running. The output DTM is interrupted if the engine is started, or if a rotary pulse from the ignition system is recognized.. **Possible Causes:** • Fuel pump relays have failed • Fuel injector has failed • Hoses on the EVAP canister may be clogged • EVAP canister purge regulator valve may be faulty • ECM may have failed • Manifold Tuning Valve (IMT) may have failed
DTC: P1434 **2T, MIL: Yes** **Passat:** 1.8L (ATW, AUG), 2.8L (AHA, ATQ); **Jetta:** 1.8L (AWD, AWP, AWW), 2.0 (ABA, AEG, AVH, AZG, BBW, BEV), 2.8L (AFP, BDF); **Golf:** 1.8L (AWD, AWW), 2.0L (ABA, AEG, AVH, AZG, BEV); **GTI:** 1.8L (AWD, AWP, AWW), 2.0L (ABA, AEG), 2.8L (AFP, BDF); **NB:** 1.8L (APH, AWP, AWV, BNU, BKF), 2.0L (AEG, AVH, AZG, BEV); **Nbvert:** (BDG); **Cabrio:** 2.0L (ABA) **Transmissions:** All	**Secondary Air Injection System Pump Relay Circuit Short to B+ Conditions:** The output Diagnostic Test Mode (DTM) can be activated only with the ignition switched on and the engine not running. The output DTM is interrupted if the engine is started, or if a rotary pulse from the ignition system is recognized.. **Possible Causes:** • Fuel pump relays have failed • Fuel injector has failed • Hoses on the EVAP canister may be clogged • EVAP canister purge regulator valve may be faulty • ECM may have failed • Manifold Tuning Valve (IMT) may have failed
DTC: P1435 **2T, MIL: Yes** **Passat:** 1.8L (ATW, AUG), 2.8L (AHA, ATQ); **Jetta:** 1.8L (AWD, AWP, AWW), 2.0 (ABA, AEG, AVH, AZG, BBW, BEV), 2.8L (AFP, BDF); **Golf:** 1.8L (AWD, AWW), 2.0L (ABA, AEG, AVH, AZG, BEV); **GTI:** 1.8L (AWD, AWP, AWW), 2.0L (ABA, AEG), 2.8L (AFP, BDF); **NB:** 1.8L (APH, AWP, AWV, BNU, BKF), 2.0L (AEG, AVH, AZG, BEV); **Nbvert:** (BDG); **Cabrio:** 2.0L (ABA) **Transmissions:** All	**Secondary Air Injection System Pump Relay Circuit Short to Ground Conditions:** The output Diagnostic Test Mode (DTM) can be activated only with the ignition switched on and the engine not running. The output DTM is interrupted if the engine is started, or if a rotary pulse from the ignition system is recognized.. **Possible Causes:** • Fuel pump relays have failed • Fuel injector has failed • Hoses on the EVAP canister may be clogged • EVAP canister purge regulator valve may be faulty • ECM may have failed • Manifold Tuning Valve (IMT) may have failed

DTC	Trouble Code Title, Conditions & Possible Causes
DTC: P1436 **Passat:** 1.8L (ATW, AUG), 2.8L (AHA, ATQ); **Jetta:** 1.8L (AWD, AWP, AWW), 2.0L (ABA); **Golf:** 1.8L (AWD, AWW), 2.0L (ABA); **GTI:** 1.8L (AWD, AWP, AWW), 2.0L (ABA); **NB:** 1.8L (APH, AWP, AWV, BNU, BKF); **Cabrio:** 2.0L (ABA) **Transmissions:** All	**A/C Evaporator Temperature (ACET) Circuit Low Input Conditions:** Key on or engine running; and the ECM detected the ACET signal was less than the self-test minimum amount of in the self-test. **Possible Causes:** • ACET signal circuit shorted to sensor ground (return) • ACET signal circuit shorted to chassis ground • ACET sensor is damaged or has failed • Check activation of Secondary Air Injection (AIR) Pump Relay • ECM has failed
DTC: P1450 **Passat:** 2.0L (ABA, BPY), 2.8L (AAA, AHA, ATQ); **Jetta:** 2.0 (ABA, AEG, AVH, AZG, BBW, BEV), 2.8L (AAA); **Golf:** 2.0L (ABA, AEG, AVH, AZG, BEV); **GTI:** 2.0L (ABA, AEG), 2.8L (AAA); **NB:** 1.8L (APH, AWP, AWV, BNU, BKF), 2.0L (AEG, AVH, AZG, BEV); **Nbvert:** (BDG); **Cabrio:** 2.0L (ABA) **Transmissions:** All	**Secondary Air Injector Valve Circuit Short to B+ Conditions:** The Engine Control Module detects a short circuit to positive (B+) when activating the secondary air injection solenoid valve. **Possible Causes:** • Connector to the secondary air injection valve is loose or disconnected • Secondary air injector valve circuit short • Secondary air injector valve circuit is open • Faulty secondary air injector valve • ECM has failed
DTC: P1451 **Passat:** 2.0L (ABA, BPY), 2.8L (AAA, AHA, ATQ); **Jetta:** 2.0 (ABA, AEG, AVH, AZG, BBW, BEV), 2.8L (AAA); **Golf:** 2.0L (ABA, AEG, AVH, AZG, BEV); **GTI:** 2.0L (ABA, AEG), 2.8L (AAA); **NB:** 1.8L (APH, AWP, AWV, BNU, BKF), 2.0L (AEG, AVH, AZG, BEV); **Nbvert:** (BDG); **Cabrio:** 2.0L (ABA) **Transmissions:** All	**Secondary Air Injector Valve Circuit Short to Ground Conditions:** Engine started, engine running at a steady cruise speed, the Engine Control Module detects a short circuit open when activating the secondary air injection solenoid valve. **Possible Causes:** • Connector to the secondary air injection valve is loose or disconnected • Secondary air injector valve circuit short • Secondary air injector valve circuit is open • Faulty secondary air injector valve • ECM has failed
DTC: P1452 **Passat:** 2.0L (ABA, BPY), 2.8L (AAA, AHA, ATQ); **Jetta:** 2.0 (ABA, AEG, AVH, AZG, BBW, BEV), 2.8L (AAA); **Golf:** 2.0L (ABA, AEG, AVH, AZG, BEV); **GTI:** 2.0L (ABA, AEG), 2.8L (AAA); **NB:** 1.8L (APH, AWP, AWV, BNU, BKF), 2.0L (AEG, AVH, AZG, BEV); **Nbvert:** (BDG); **Cabrio:** 2.0L (ABA) **Transmissions:** All	**Secondary Air Injector Valve Circuit Open Conditions:** Engine started, engine running at a steady cruise speed, the Engine Control Module detects a short circuit open when activating the secondary air injection solenoid valve. **Possible Causes:** • Connector to the secondary air injection valve is loose or disconnected • Secondary air injector valve circuit short • Secondary air injector valve circuit is open • Faulty secondary air injector valve • ECM has failed
DTC: P1470 **Passat:** 1.8L (ATW, AUG), 2.0 (BPY); **Jetta:** 1.8L (AWD, AWP, AWW); **Golf:** 1.8L (AWD, AWW); **GTI:** 1.8L (AWD, AWP, AWW); **NB:** 1.8L (APH, AWP, AWV, BNU, BKF) **Transmissions:** All	**EVAP Emission Control LDP Circuit Electrical Malfunction Conditions:** Key on, KOEO Self-Test enabled, and the ECM detected an unexpected voltage condition on the EVAP emission control leak detection pump circuit. **Possible Causes:** • EVAP canister system has an improper or broken seal • Evaporative Emission (EVAP) canister purge regulator valve 1 is faulty • Leak Detection Pump (LDP) is faulty • ECM has failed

DTC	Trouble Code Title, Conditions & Possible Causes
DTC: P1471 **2T, MIL: Yes** **Passat:** 1.8 (AEB, ATW, AUG, AWM), 2.0L (BPY), 2.8L (AAA, AHA, ATQ); **Jetta:** 1.8L (AWD, AWP, AWW), 2.0 (ABA, AEG, AVH, AZG, BBW, BEV), 2.8L (AAA, AFP, BDF); **Golf:** 1.8L (AWD, AWW), 2.0L (ABA, AEG, AVH, AZG, BEV); **GTI:** 2.0L (ABA, AEG), 2.8L (AAA, AFP, BDF); **NB:** 1.8L (APH, AWP, AWV, BNU, BKF), 2.0L (AEG, AVH, AZG, BEV); **Nbvert:** (BDG); **Cabrio:** 2.0L (ABA) **Transmissions:** All	**EVAP Emission Control Leak Detection Pump Circuit Short to B+ Conditions:** Key on, KOEO Self-Test enabled, and the ECM detected an unexpected voltage condition on the EVAP emission control leak detection pump circuit. **Possible Causes:** • Leak Detection Pump has failed • EVAP canister system has an improper or broken seal • Evaporative Emission (EVAP) canister purge regulator valve 1 is faulty • Hoses between the fuel pump and the EVAP canister are faulty • Fuel filler cap is loose • Fuel pump seal is defective, faulty or otherwise leaking • Hoses between the EVAP canister and the fuel flap unit are faulty • Hoses between the EVAP canister and the evaporative emission canister purge regulator valve are faulty • ECM has failed
DTC: P1472 **2T, MIL: Yes** **Passat:** 1.8 (AEB, ATW, AUG, AWM), 2.0L (BPY), 2.8L (AAA, AHA, ATQ); **Jetta:** 1.8L (AWD, AWP, AWW), 2.0 (ABA, AEG, AVH, AZG, BBW, BEV), 2.8L (AAA, AFP, BDF); **Golf:** 1.8L (AWD, AWW), 2.0L (ABA, AEG, AVH, AZG, BEV); **GTI:** 1.8L (AWD, AWP, AWW), 2.0L (ABA, AEG), 2.8L (AAA, AFP, BDF); **NB:** 1.8L (APH, AWP, AWV, BNU, BKF), 2.0L (AEG, AVH, AZG, BEV); **Nbvert:** (BDG); **Cabrio:** 2.0L (ABA) **Transmissions:** All	**EVAP Emission Control Leak Detection Pump Circuit Short to Ground Conditions:** Key on, KOEO Self-Test enabled, and the ECM detected an unexpected voltage condition on the EVAP emission control leak detection pump circuit. **Possible Causes:** • Leak Detection Pump has failed • EVAP canister system has an improper or broken seal • Evaporative Emission (EVAP) canister purge regulator valve 1 is faulty • Hoses between the fuel pump and the EVAP canister are faulty • Fuel filler cap is loose • Fuel pump seal is defective, faulty or otherwise leaking • Hoses between the EVAP canister and the fuel flap unit are faulty • Hoses between the EVAP canister and the evaporative emission canister purge regulator valve are faulty • ECM has failed
DTC: P1473 **2T, MIL: Yes** **Passat:** 1.8 (AEB, ATW, AUG, AWM), 2.0L (BPY), 2.8L (AAA, AHA, ATQ); **Jetta:** 1.8L (AWD, AWP, AWW), 2.0 (ABA, AEG, AVH, AZG, BEV), 2.8L (AAA); **Golf:** 1.8L (AWD, AWW), 2.0L (ABA, AEG); **GTI:** 1.8L (AWD, AWP, AWW), 2.0L (ABA, AEG), 2.8L (AAA, AFP, BDF); **NB:** 1.8L (APH, AWP, AWV, BNU, BKF), 2.0L (AEG, AVH, AZG, BEV); **Nbvert:** (BDG); **Cabrio:** 2.0L (ABA) **Transmissions:** All	**EVAP Emission Control Leak Detection Pump Circuit Open Conditions:** Key on, KOEO Self-Test enabled, and the ECM detected an unexpected voltage condition on the EVAP emission control leak detection pump circuit. **Possible Causes:** • Leak Detection Pump has failed • EVAP canister system has an improper or broken seal • Evaporative Emission (EVAP) canister purge regulator valve 1 is faulty • Hoses between the fuel pump and the EVAP canister are faulty • Fuel filler cap is loose • Fuel pump seal is defective, faulty or otherwise leaking • Hoses between the EVAP canister and the fuel flap unit are faulty • Hoses between the EVAP canister and the evaporative emission canister purge regulator valve are faulty • ECM has failed

DTC	Trouble Code Title, Conditions & Possible Causes
DTC: P1475 **2T, MIL: Yes** **Passat:** 1.8 (AEB, ATW, AUG, AWM), 2.0L (BPY), 2.8L (AAA); **Jetta:** 1.8L (AWD, AWP, AWW), 2.0 (ABA, AEG, AVH, AZG, BBW, BEV), 2.8L (AAA, AFP, BDF); **Golf:** 1.8L (AWD, AWW), 2.0L (ABA, AEG, AVH, AZG, BEV); **GTI:** 1.8L (AWD, AWP, AWW), 2.0L (ABA, AEG), 2.8L (AAA, AFP, BDF); **NB:** 1.8L (APH, AWP, AWV, BNU, BKF), 2.0L (AEG, AVH, AZG, BEV); **Nbvert:** (BDG); **Cabrio:** 2.0L (ABA) **Transmissions:** All	**EVAP Emission Control LDP Circuit Malfunction/Signal Circuit Open Conditions:** Key on, KOEO Self-Test enabled, and the ECM detected an unexpected voltage condition on the EVAP emission control leak detection pump circuit. **Possible Causes:** • Leak Detection Pump has failed • EVAP canister system has an improper or broken seal • Evaporative Emission (EVAP) canister purge regulator valve 1 is faulty • Hoses between the fuel pump and the EVAP canister are faulty • Fuel filler cap is loose • Fuel pump seal is defective, faulty or otherwise leaking • Hoses between the EVAP canister and the fuel flap unit are faulty • Hoses between the EVAP canister and the evaporative emission canister purge regulator valve are faulty • ECM has failed
DTC: P1476 **2T, MIL: Yes** **Passat:** 1.8 (AEB, ATW, AUG, AWM), 2.0L (BPY), 2.8L (AAA, AHA, ATQ); **Jetta:** 1.8L (AWD, AWP, AWW), 2.0 (ABA, AEG, AVH, AZG, BBW, BEV), 2.8L (AAA, AFP, BDF); **Golf:** 1.8L (AWD, AWW), 2.0L (ABA, AEG, AVH, AZG, BEV); **GTI:** 1.8L (AWD, AWP, AWW), 2.0L (ABA, AEG), 2.8L (AAA, AFP, BDF); **NB:** 1.8L (APH, AWP, AWV, BNU, BKF), 2.0L (AEG, AVH, AZG, BEV); **Nbvert:** (BDG); **Cabrio:** 2.0L (ABA) **Transmissions:** All	**EVAP Emission Control LDP Circuit Malfunction/Insufficient Vacuum Conditions:** Key on, KOEO Self-Test enabled, and the ECM detected an unexpected voltage condition on the EVAP emission control leak detection pump circuit. **Possible Causes:** • Leak Detection Pump has failed • EVAP canister system has an improper or broken seal • Evaporative Emission (EVAP) canister purge regulator valve 1 is faulty • Hoses between the fuel pump and the EVAP canister are faulty • Fuel filler cap is loose • Fuel pump seal is defective, faulty or otherwise leaking • Hoses between the EVAP canister and the fuel flap unit are faulty • Hoses between the EVAP canister and the evaporative emission canister purge regulator valve are faulty • ECM has failed
DTC: P1477 **Passat:** 1.8 (ATW, AUG), 2.8L (AAA); **Jetta:** 1.8L (AWD, AWP, AWW), 2.0L (ABA), 2.8L (AAA); **Golf:** 1.8L (AWD, AWW), 2.0L (ABA); **GTI:** 1.8L (AWD, AWP, AWW), 2.0L (ABA), 2.8L (AAA); **NB:** 1.8L (APH, AWP, AWV, BNU, BKF); **Cabrio:** 2.0L (ABA) **Transmissions:** All	**EVAP Emission Control LDP Circuit Malfunction Conditions:** Key on, KOEO Self-Test enabled, and the ECM detected an unexpected voltage condition on the EVAP emission control leak detection pump circuit. **Possible Causes:** • Leak Detection Pump has failed • EVAP canister system has an improper or broken seal • Evaporative Emission (EVAP) canister purge regulator valve 1 is faulty • Hoses between the fuel pump and the EVAP canister are faulty • Fuel filler cap is loose • Fuel pump seal is defective, faulty or otherwise leaking • Hoses between the EVAP canister and the fuel flap unit are faulty • Hoses between the EVAP canister and the evaporative emission canister purge regulator valve are faulty • ECM has failed
DTC: P1478 **Passat:** 1.8 (AEB, ATW, AUG, AWM), 2.0L (BPY), 2.8L (AAA, AHA, ATQ); **Jetta:** 1.8L (AWD, AWP, AWW), 2.8L (AAA); **Golf:** 1.8L (AWD, AWW); **GTI:** 1.8L (AWD, AWP, AWW), 2.8L (AAA); **NB:** 1.8L (APH, AWP, AWV, BNU, BKF) **Transmissions:** All	**EVAP Emission Control LDP Circuit Clamped Tube Detected Conditions:** Key on, KOEO Self-Test enabled, and the ECM detected an unexpected voltage condition on the EVAP emission control leak detection pump circuit. **Possible Causes:** • Leak Detection Pump has failed • EVAP canister system has an improper or broken seal • Evaporative Emission (EVAP) canister purge regulator valve 1 is faulty • Hoses between the fuel pump and the EVAP canister are faulty • Fuel filler cap is loose • Fuel pump seal is defective, faulty or otherwise leaking • Hoses between the EVAP canister and the fuel flap unit are faulty • Hoses between the EVAP canister and the evaporative emission canister purge regulator valve are faulty • ECM has failed

DTC	Trouble Code Title, Conditions & Possible Causes
DTC: P1500 **2T, MIL: Yes** **Passat:** 1.8 (AEB, ATW, AUG, AWM), 2.0L (ABA, BPY), 2.8L (AAA, AHA, ATQ); **Jetta:** 1.8L (AWD, AWP, AWW), 2.0 (ABA, AEG, AVH, AZG, BBW, BEV), 2.8L (AAA); **Golf:** 1.8L (AWD, AWW), 2.0L (ABA, AEG, AVH, AZG, BEV); **GTI:** 1.8L (AWD, AWP, AWW), 2.0L (AEG), 2.8L (AAA); **NB:** 1.8L (APH, AWP, AWV, BNU, BKF), 2.0L (AEG, AVH, AZG, BEV); **Nbvert:** (BDG); **Cabrio:** 2.0L (ABA) **Transmissions:** All	**Fuel Pump Relay Circuit Electrical Malfunction Conditions:** Engine running the ECM detected that the fuel pump relay signal was intermittent **Possible Causes:** • Fuel delivery unit connector is loose or not attached • Fuse 18 cause a short to the transfer fuel pump or the O2S • Fuel pump has failed • Fuel pump relay circuit is shorted to ground, B+ or is open • Fuel Pump (FP) Relay not activated • ECM has failed
DTC: P1501 **Passat:** 1.8 (AEB, ATW, AUG, AWM), 2.8L (AAA, AHA, ATQ); **Jetta:** 2.0 (ABA, AEG, AVH, AZG, BBW, BEV), 2.8L (AAA, AFP, BDF); **Golf:** 2.0L (ABA, AEG, AVH, AZG, BEV); **GTI:** 2.0L (ABA, AEG), 2.8L (AAA, AFP, BDF); **NB:** 1.8L (APH, AWP, AWV, BNU, BKF), 2.0L (AEG, AVH, AZG, BEV); **Nbvert:** (BDG); **Cabrio:** 2.0L (ABA) **Transmissions:** All	**Fuel Pump Relay Circuit Electrical Short to Ground Conditions:** Engine running the ECM detected that the fuel pump relay signal was intermittent **Possible Causes:** • Fuel delivery unit connector is loose or not attached • Fuse 18 cause a short to the transfer fuel pump or the O2S • Fuel pump has failed • Fuel pump relay circuit is shorted to ground, B+ or is open • Fuel Pump (FP) Relay not activated • ECM has failed
DTC: P1502 **2T, MIL: Yes** **Passat:** 1.8 (AEB, ATW, AUG, AWM), 2.0L (ABA, BPY), 2.8L (AAA, AHA, ATQ); **Jetta:** 1.8L (AWD, AWP, AWW), 2.0 (ABA, AEG, AVH, AZG, BBW, BEV), 2.8L (AAA, AFP, BDF); **Golf:** 1.8L (AWD, AWW), 2.0L (ABA, AEG, AVH, AZG, BEV); **GTI:** 1.8L (AWD, AWP, AWW), 2.0L (ABA, AEG), 2.8L (AAA, AFP, BDF); **NB:** 1.8L (APH, AWP, AWV, BNU, BKF), 2.0L (AEG, AVH, AZG, BEV); **Nbvert:** (BDG); **Cabrio:** 2.0L (ABA) **Transmissions:** All	**Fuel Pump Relay Circuit Short to B+ Conditions:** Engine running the ECM detected that the fuel pump relay signal was intermittent **Possible Causes:** • Fuel delivery unit connector is loose or not attached • Fuse 18 cause a short to the transfer fuel pump or the O2S • Fuel pump has failed • Fuel pump relay circuit is shorted to ground, B+ or is open • Fuel Pump (FP) Relay not activated • ECM has failed
DTC: P1512 **Passat:** 2.8L (AHA, ATQ) **Transmissions:** All	**Intake Manifold Changeover Valve Circuit Short to B+ Conditions:** Engine started, and the ECM detected the changeover valve circuit was shorting to positive during the continuous self test. **Possible Causes:** • Leaky vacuum reservoir, vacuum lines loose or damaged • Vacuum solenoid or vacuum actuator is damaged • IMRC actuator cable/gears are seized, or the cables are improperly routed or seized • IMRC housing return springs are damaged or disconnected • Lever/shaft return stop may be obstructed or bent, or the lever/shaft wide open stop may be obstructed or bent, or the IMRC lever/shaft may be sticking, binding or disconnected • IMRC control circuit open, shorted or the VPWR circuit is open • ECM has failed

DTC	Trouble Code Title, Conditions & Possible Causes
DTC: P1515 **Passat:** 2.8L (AHA, ATQ) **Transmissions:** All	**Intake Manifold Changeover Valve Circuit Short to Ground Conditions:** Engine started, and the ECM detected the changeover valve circuit was shorting to ground during the continuous self test. **Possible Causes:** • Leaky vacuum reservoir, vacuum lines loose or damaged • Vacuum solenoid or vacuum actuator is damaged • IMRC actuator cable/gears are seized, or the cables are improperly routed or seized • IMRC housing return springs are damaged or disconnected • Lever/shaft return stop may be obstructed or bent, or the lever/shaft wide open stop may be obstructed or bent, or the IMRC lever/shaft may be sticking, binding or disconnected • IMRC control circuit open, shorted or the VPWR circuit is open • ECM has failed
DTC: P1516 **Passat:** 2.8L (AHA, ATQ) **Transmissions:** All	**Intake Manifold Runner Control Input Error (Bank 1) Conditions:** Key on or engine running; and the ECM detected the IMRC Monitor signal for Bank 1 was outside of its expected calibrated range during the Continuous self test. **Possible Causes:** • IMRC mechanical fault – the linkage may be bound or seized • Inspect for binding or improper routing. The cable core wire at the IMRC/IMSC housing attachment must have slack and lever must contact close plate stop screw
DTC: P1517 **Passat:** 1.8L (ATW, AUG), 2.0L (BPY); **Jetta:** 1.8L (AWD, AWP, AWW), 2.0L (BPY); **Golf:** 1.8L (AWD, AWW); **GTI:** 1.8L (AWD, AWP, AWW); **NB:** 1.8L (APH, AWP, AWV, BNU, BKF) **Transmissions:** All	**Main Relay Circuit Electrical Malfunction Conditions:** The ECM detected an electrical malfunction on the main relay circuit **Possible Causes:** • Engine Control Module (ECM) has failed • Voltage supply for Engine Control Module (ECM) is faulty • Check activation of Motronic Engine Control Module (ECM) Power Supply Relay
DTC: P1519 **Passat:** 2.8L (AHA, ATQ) **Transmissions:** All	**Intake Manifold Runner Control Stuck Closed Conditions:** Key on, and the ECM detected the IMRC Monitor was more than the expected calibrated range at closed throttle. **Possible Causes:** • IMRC monitor signal circuit shorted to power ground • IMRC Monitor signal circuit shorted to signal ground (return) • IMRC actuator is damaged or has failed (e.g., there may be a small leak in the vacuum diaphragm of the actuator) • ECM has failed
DTC: P1522 **Passat:** 2.8L (AHA, ATQ) **Transmissions:** All	**Intake Camshaft Control (Bank 2) Malfunction Conditions:** Key on or engine running; and the ECM detected the intake manifold control signal for was outside of its expected calibrated range. **Possible Causes:** • Camshaft control circuit is open or shorted to ground • Camshaft sensor is damaged or the ECM has failed • Camshaft out of adjustment
DTC: P1529 **Passat:** 1.8L (AUG) **Transmissions:** All	**Camshaft Control Circuit Short to B+ Conditions:** Engine started and driven at an engine speed of more than 400rpm; and the ECM detected the camshaft timing exceeded the calibrated voltage levels. The valve timing did not change from the current valve timing or it remained fixed during the testing. **Note: The camshaft adjustment is load- and RPM dependant. The electrical camshaft adjustment valve 1 switches oil pressure onto camshaft adjuster (mechanical adjustment mechanism), which adjusts the camshaft.** **Possible Causes:** • Fuel pump has failed • CPS circuit is open, shorted to ground or shorted to power • ECM has failed • Battery voltage below 11.5 volts • Position actuator circuit may short to B+ or Ground • Camshaft timing improperly set, or continuous oil flow to the VCT piston chamber • Camshaft advance mechanism (the VCT unit) is sticking or binding mechanically • VCT solenoid valve is stuck in open position

DTC	Trouble Code Title, Conditions & Possible Causes
DTC: P1530 **Passat:** 1.8L (AUG); **Jetta:** 2.0L (BPY) **Transmissions:** All	**Camshaft Control Circuit Short to Ground Conditions:** Engine started and driven at an engine speed of more than 400rpm; and the ECM detected the camshaft timing exceeded the calibrated levels. The valve timing did not change from the current valve timing or it remained fixed during the testing. **Note: The camshaft adjustment is load- and RPM dependant. The electrical camshaft adjustment valve 1 switches oil pressure onto camshaft adjuster (mechanical adjustment mechanism), which adjusts the camshaft.** **Possible Causes:** • Fuel pump has failed • CPS circuit is open, shorted to ground or shorted to power • ECM has failed • Battery voltage below 11.5 volts • Position actuator circuit may short to B+ or Ground • Camshaft timing improperly set, or continuous oil flow to the VCT piston chamber • Camshaft advance mechanism (the VCT unit) is sticking or binding mechanically • VCT solenoid valve is stuck in open position
DTC: P1531 **Passat:** 1.8L (AUG); **Jetta:** 2.0L (BPY) **Transmissions:** All	**Camshaft Control Circuit Open Conditions:** Engine started and driven at an engine speed of more than 400rpm; and the ECM detected the camshaft timing exceeded the calibrated levels. The valve timing did not change from the current valve timing or it remained fixed during the testing. **Note: The camshaft adjustment is load- and RPM dependant. The electrical camshaft adjustment valve 1 switches oil pressure onto camshaft adjuster (mechanical adjustment mechanism), which adjusts the camshaft.** **Possible Causes:** • Fuel pump has failed • CPS circuit is open, shorted to ground or shorted to power • ECM has failed • Battery voltage below 11.5 volts • Position actuator circuit may short to B+ or Ground • Camshaft timing improperly set, or continuous oil flow to the VCT piston chamber • Camshaft advance mechanism (the VCT unit) is sticking or binding mechanically • VCT solenoid valve is stuck in open position
DTC: P1539 **Jetta:** 1.8 (AWD, AWP, AWW); **Golf:** 1.8L (AWD, AWW); **GTI:** 1.8L (AWD, AWP, AWW); **NB:** 1.8L (APH, AWP, AWV, BNU, BKF) **Transmissions:** All	**Clutch Vacuum Vent Valve Switch Incorrect Signal Conditions:** Engine started, battery voltage must be at least 11.5v, all electrical components must be off, parking brake must be engaged (to keep daytime driving lights off), automatic transmission selector must be in park, and the ground between the engine and the chassis must be well connected. The ECM detected an incorrect signal from the clutch vacuum vent valve switch. **Possible Causes:** • Signal from clutch vacuum vent valve switch is faulty • Clutch vacuum vent valve is faulty • Circuit wires are short circuiting to each other, to vehicle ground or to B+ • ECM has failed
DTC: P1541 **Passat:** 1.8L (ATW, AUG), 2.8L (AAA); **Jetta:** 2.0 (ABA, AEG, AVH, AZG, BBW, BEV), 2.8L (AAA, AFP, BDF); **Golf:** 2.0L (ABA, AEG, AVH, AZG, BEV); **GTI:** 2.0L (ABA, AEG), 2.8L (AAA, AFP, BDF); **NB:** 1.8L (APH, AWP, AWV, BNU, BKF), 2.0L (AEG, AVH, AZG, BEV); **Nbvert:** (BDG); **Cabrio:** 2.0L (ABA) **Transmissions:** All	**Fuel Pump Relay Circuit Open Conditions:** The ECM detected an electrical malfunction on the fuel pump relay circuit **Possible Causes:** • Fuel pump relay not activiated

DTC	Trouble Code Title, Conditions & Possible Causes
DTC: P1542 **2T, MIL: Yes** **Passat:** 1.8L (ATW, AUG), 2.0L (BPY); **Jetta:** 1.8L (AWD, AWP, AWW), 2.0 (ABA, AEG, AVH, AZG, BBW, BEV), 2.8L (AAA, AFP, BDF); **Golf:** 1.8L (AWD, AWW), 2.0L (ABA, AEG, AVH, AZG, BEV); **GTI:** 1.8L (AWD, AWP, AWW), 2.0L (AEG), 2.8L (AAA, AFP, BDF); **NB:** 1.8L (APH, AWP, AWV, BNU, BKF), 2.0L (AEG, AVH, AZG, BEV); **Nbvert:** (BDG); **Cabrio:** 2.0L (ABA) **Transmissions:** All	**Throttle Actuation Potentiometer Range/Performance Conditions:** Engine started, battery voltage must be at least 11.5v, all electrical components must be off, parking brake must be engaged (to keep daytime driving lights off), automatic transmission selector must be in park, the exhaust system must be properly sealed between the catalytic converter and the cylinder head, coolant temperature must be at least 80 degrees Celsius, and the ground between the engine and the chassis must be well connected. The signal from the Throttle Position Valve Module to the ECM detected was erratic, non existent or unreliable. **Note: If the complete throttle valve control module is current-less (e.g. connector disconnected) the throttle valve moves into a particular, specified mechanical position, which signals an increased idle speed with an engine at operating temperature. If only the Throttle Position (TP) actuator –V60- is current-less, the throttle valve also moves into the specified mechanical position (emergency running gap), however, since Closed Throttle Position (CTP) switch –F60- can still be recognized, an "almost normal idle RPM" is reached via the respective ignition angle retardation. If the Engine Control Module (ECM) detects a malfunction at Throttle Position (TP) sensor –G69-, Throttle Position (TP) actuator –V60- is switched current-less by the Engine Control Module (ECM) and the throttle valve moves into the specified mechanical position (emergency running gap) again.** **Note: Terminal assignment at throttle control module is different in vehicles with and without cruise control.** **Characteristic: Steering column switch with operating module for cruise control.** **Possible Causes:** • Throttle valve control module is faulty • Throttle valve is damaged or dirty • Throttle valve must be in closed throttle position • Accelerator pedal is out of adjustment (AEG engines only) • Throttle position actuator is shorting to ground or power
DTC: P1543 **2T, MIL: Yes** **Passat:** 1.8 (AEB, ATW, AUG, AWM), 2.0L (ABA, BPY), 2.8L (AAA, AHA, ATQ); **Jetta:** 1.8L (AWD, AWP, AWW), 2.0 (ABA, AEG, AVH, AZG, BBW, BEV), 2.8L (AAA, AFP, BDF); **Golf:** 1.8L (AWD, AWW), 2.0L (ABA, AEG, AVH, AZG, BEV); **GTI:** 1.8L (AWD, AWP, AWW), 2.0L (ABA, AEG), 2.8L (AAA, AFP, BDF); **NB:** 1.8L (APH, AWP, AWV, BNU, BKF), 2.0L (AEG, AVH, AZG, BEV); **Nbvert:** (BDG); **Cabrio:** 2.0L (ABA) **Transmissions:** All	**Throttle Actuation Potentiometer Signal Too Low Conditions:** Engine started, battery voltage must be at least 11.5v, all electrical components must be off, parking brake must be engaged (to keep daytime driving lights off), automatic transmission selector must be in park, the exhaust system must be properly sealed between the catalytic converter and the cylinder head, coolant temperature must be at least 80 degrees Celsius, and the ground between the engine and the chassis must be well connected. The signal from the Throttle Position Valve Module to the ECM detected was erratic, non existent or unreliable. **Note: If the complete throttle valve control module is current-less (e.g. connector disconnected) the throttle valve moves into a particular, specified mechanical position, which signals an increased idle speed with an engine at operating temperature. If only the Throttle Position (TP) actuator –V60- is current-less, the throttle valve also moves into the specified mechanical position (emergency running gap), however, since Closed Throttle Position (CTP) switch –F60- can still be recognized, an "almost normal idle RPM" is reached via the respective ignition angle retardation. If the Engine Control Module (ECM) detects a malfunction at Throttle Position (TP) sensor –G69-, Throttle Position (TP) actuator –V60- is switched current-less by the Engine Control Module (ECM) and the throttle valve moves into the specified mechanical position (emergency running gap) again.** **Note: Terminal assignment at throttle control module is different in vehicles with and without cruise control.** **Characteristic: Steering column switch with operating module for cruise control.** **Possible Causes:** • Throttle valve control module is faulty • Throttle valve is damaged or dirty • Throttle valve must be in closed throttle position • Accelerator pedal is out of adjustment (AEG engines only) • Throttle position actuator is shorting to ground or power
DTC: P1544 **2T, MIL: Yes** **Passat:** 1.8 (AEB, ATW, AUG, AWM), 2.0L (ABA, BPY), 2.8L (AAA, AHA, ATQ); **Jetta:** 1.8L (AWD, AWP, AWW), 2.0 (ABA, BBW), 2.8L (AAA, AFP, BDF); **Golf:** 1.8L (AWD, AWW), 2.0L (ABA, AEG, AVH, AZG, BEV); **GTI:** 1.8L (AWD, AWP, AWW), 2.0L (AEG), 2.8L (AAA, AFP, BDF); **NB:** 1.8L (APH, AWP, AWV, BNU, BKF), 2.0L (AEG, AVH, AZG, BEV); **Nbvert:** (BDG); **Cabrio:** 2.0L (ABA) **Transmissions:** All	**Throttle Actuation Potentiometer Signal Too High Conditions:** Engine started, battery voltage must be at least 11.5v, all electrical components must be off, parking brake must be engaged (to keep daytime driving lights off), automatic transmission selector must be in park, the exhaust system must be properly sealed between the catalytic converter and the cylinder head, coolant temperature must be at least 80 degrees Celsius, and the ground between the engine and the chassis must be well connected. The signal from the Throttle Position Valve Module to the ECM detected was erratic, non existent or unreliable. **Note: If the complete throttle valve control module is current-less (e.g. connector disconnected) the throttle valve moves into a particular, specified mechanical position, which signals an increased idle speed with an engine at operating temperature. If only the Throttle Position (TP) actuator –V60- is current-less, the throttle valve also moves into the specified mechanical position (emergency running gap), however, since Closed Throttle Position (CTP) switch –F60- can still be recognized, an "almost normal idle RPM" is reached via the respective ignition angle retardation. If the Engine Control Module (ECM) detects a malfunction at Throttle Position (TP) sensor –G69-, Throttle Position (TP) actuator –V60- is switched current-less by the Engine Control Module (ECM) and the throttle valve moves into the specified mechanical position (emergency running gap) again.** **Note: Terminal assignment at throttle control module is different in vehicles with and without cruise control.** **Characteristic: Steering column switch with operating module for cruise control.** **Possible Causes:** • Throttle valve control module is faulty • Throttle valve is damaged or dirty • Throttle valve must be in closed throttle position • Accelerator pedal is out of adjustment (AEG engines only) • Throttle position actuator is shorting to ground or power

DTC	Trouble Code Title, Conditions & Possible Causes
DTC: P1545 **2T, MIL: Yes** **Passat:** 1.8 (AEB, ATW, AUG, AWM), 2.0L (BPY), 2.8L (AHA, ATQ); **Jetta:** 1.8L (AWD, AWP, AWW), 2.0 (ABA, AEG, AVH, AZG, BBW, BEV, BPY), 2.8L (AFP, BDF); **Golf:** 1.8L (AWD, AWW), 2.0L (AEG, AVH, AZG, BEV); **GTI:** 1.8L (AWD, AWP, AWW), 2.0L (ABA, AEG), 2.8L (AFP, BDF); **NB:** 1.8L (APH, AWP, AWV, BNU, BKF), 2.0L (AEG, AVH, AZG, BEV); **Nbvert:** (BDG); **Cabrio:** 2.0L (ABA) **Transmissions:** All	**Throttle Position Control Malfunction Conditions:** Engine started, battery voltage must be at least 11.5v, all electrical components must be off, parking brake must be engaged (to keep daytime driving lights off), automatic transmission selector must be in park, the exhaust system must be properly sealed between the catalytic converter and the cylinder head, coolant temperature must be at least 80 degrees Celsius, and the ground between the engine and the chassis must be well connected. The signal from the Throttle Position Valve Module to the ECM detected was erratic, non existent or unreliable. **Note: If the complete throttle valve control module is current-less (e.g. connector disconnected) the throttle valve moves into a particular, specified mechanical position, which signals an increased idle speed with an engine at operating temperature. If only the Throttle Position (TP) actuator is current-less, the throttle valve also moves into the specified mechanical position (emergency running gap), however, since Closed Throttle Position (CTP) switch – can still be recognized, an "almost normal idle RPM" is reached via the respective ignition angle retardation. If the Engine Control Module (ECM) detects a malfunction at Throttle Position (TP) sensor – Throttle Position (TP) actuator is switched current-less by the Engine Control Module (ECM) and the throttle valve moves into the specified mechanical position (emergency running gap) again.** **Note: Terminal assignment at throttle control module is different in vehicles with and without cruise control. Characteristic: Steering column switch with operating module for cruise control.** **Possible Causes:** • Throttle valve control module is faulty • Throttle valve is damaged or dirty • Throttle valve must be in closed throttle position • Accelerator pedal is out of adjustment (AEG engines only) • Throttle position actuator is shorting to ground or power
DTC: P1546 **Passat:** 1.8 (AEB, ATW, AUG, AWM), 2.0L (BPY); **Jetta:** 1.8L (AWD, AWP, AWW); **Golf:** 1.8L (AWD, AWW); **GTI:** 1.8L (AWD, AWP, AWW); **NB:** 1.8L (APH, AWP, AWV, BNU, BKF) **Transmissions:** All	**Boost Pressure Control Valve Short to B+ Conditions:** Engine started, battery voltage at least 11.5v, all electrical components off, ground connections between engine and chassis well connected, coolant temperature at least 80-degrees Celicius. The ECM detected an short in the boost pressure control valve. **Possible Causes:** • Charge air pressure sensor is faulty • Voltage supply to the charge air pressure sensor is open or shorted • Check the charge air system for leaks • Recirculating valve for turbocharger is faulty • Turbocharging system is damaged or not functioning properly • Turbocharger recirculating valve is faulty • Vacuum diaphragm for turbocharger is out of adjustment • Wastegate bypass regulator valve is faulty • Boost sensor has failed • ECM has failed
DTC: P1547 **Passat:** 1.8 (AEB, ATW, AUG, AWM), 2.0L (BPY); **Jetta:** 1.8L (AWD, AWP, AWW); **Golf:** 1.8L (AWD, AWW); **GTI:** 1.8L (AWD, AWP, AWW); **NB:** 1.8L (APH, AWP, AWV, BNU, BKF) **Transmissions:** All	**Boost Pressure Control Valve Short to Ground Conditions:** Engine started, battery voltage at least 11.5v, all electrical components off, ground connections between engine and chassis well connected, coolant temperature at least 80-degrees Celicius. The ECM detected an short in the boost pressure control valve. **Possible Causes:** • Charge air pressure sensor is faulty • Voltage supply to the charge air pressure sensor is open or shorted • Check the charge air system for leaks • Recirculating valve for turbocharger is faulty • Turbocharging system is damaged or not functioning properly • Turbocharger recirculating valve is faulty • Vacuum diaphragm for turbocharger is out of adjustment • Wastegate bypass regulator valve is faulty • Boost sensor has failed • ECM has failed
DTC: P1548 **Passat:** 1.8 (AEB, ATW, AUG, AWM), 2.0L (BPY); **Jetta:** 1.8L (AWD, AWP, AWW); **Golf:** 1.8L (AWD, AWW); **GTI:** 1.8L (AWD, AWP, AWW), 2.0L (AEG); **NB:** 1.8L (APH, AWP, AWV, BNU, BKF) **Transmissions:** All	**Boost Pressure Control Valve Open Conditions:** Engine started, battery voltage at least 11.5v, all electrical components off, ground connections between engine and chassis well connected, coolant temperature at least 80-degrees Celicius. The ECM detected an short in the boost pressure control valve. **Possible Causes:** • Charge air pressure sensor is faulty • Voltage supply to the charge air pressure sensor is open or shorted • Check the charge air system for leaks • Recirculating valve for turbocharger is faulty • Turbocharging system is damaged or not functioning properly • Turbocharger recirculating valve is faulty • Vacuum diaphragm for turbocharger is out of adjustment • Wastegate bypass regulator valve is faulty • Boost sensor has failed • ECM has failed

DTC	Trouble Code Title, Conditions & Possible Causes
DTC: P1550 **Passat:** 1.8 (AEB, ATW, AUG, AWM), 2.0L (BPY) **Transmissions:** All	**Charge Pressure Deviation Conditions:** Engine started, battery voltage at least 11.5v, all electrical components off, ground connections between engine and chassis well connected, coolant temperature at least 80-degrees Celicius. The ECM detected deviation from the normal operating parameters of the charge pressure sensor. **Possible Causes:** • Charge air system leaks • Recirculating valve for turbocharger is faulty • Turbocharging system is damaged • Vacuum diaphragm for turbocharger needs adjusting • Wastegate bypass regulator valve is faulty • ECM has failed
DTC: P1555 **Passat:** 1.8 (AEB, ATW, AUG, AWM), 2.0L (BPY); **Jetta:** 1.8L (AWD, AWP, AWW); **Golf:** 1.8L (AWD, AWW); **GTI:** 1.8L (AWD, AWP, AWW); **NB:** 1.8L (APH, AWP, AWV, BNU, BKF) **Transmissions:** All	**Charge Pressure Upper Limit Exceeded Conditions:** Engine started, battery voltage at least 11.5v, all electrical components off, ground connections between engine and chassis well connected, coolant temperature at least 80-degrees Celicius. The ECM detected deviation from the normal operating parameters of the charge pressure sensor. **Possible Causes:** • Charge air system leaks • Recirculating valve for turbocharger is faulty • Turbocharging system is damaged • Vacuum diaphragm for turbocharger needs adjusting • Wastegate bypass regulator valve is faulty • ECM has failed
DTC: P1556 **Passat:** 1.8 (AEB, ATW, AUG, AWM), 2.0L (BPY); **Jetta:** 1.8L (AWD, AWP, AWW); **Golf:** 1.8L (AWD, AWW); **GTI:** 1.8L (AWD, AWP, AWW); **NB:** 1.8L (APH, AWP, AWV, BNU, BKF) **Transmissions:** All	**Charge Pressure Control Negative Deviation Conditions:** Engine started, battery voltage at least 11.5v, all electrical components off, ground connections between engine and chassis well connected, coolant temperature at least 80-degrees Celicius. The ECM detected deviation from the normal operating parameters of the charge pressure sensor. **Possible Causes:** • Charge air system leaks • Recirculating valve for turbocharger is faulty • Turbocharging system is damaged • Vacuum diaphragm for turbocharger needs adjusting • Wastegate bypass regulator valve is faulty • ECM has failed
DTC: P1557 **Passat:** 1.8 (AEB, ATW, AUG, AWM), 2.0L (BPY); **Jetta:** 1.8L (AWD, AWP, AWW); **Golf:** 1.8L (AWD, AWW); **GTI:** 1.8L (AWD, AWP, AWW); **NB:** 1.8L (APH, AWP, AWV, BNU, BKF) **Transmissions:** All	**Charge Pressure Control Positive Deviation Conditions:** Engine started, battery voltage at least 11.5v, all electrical components off, ground connections between engine and chassis well connected, coolant temperature at least 80-degrees Celicius. The ECM detected deviation from the normal operating parameters of the charge pressure sensor. **Possible Causes:** • Charge air system leaks • Recirculating valve for turbocharger is faulty • Turbocharging system is damaged • Vacuum diaphragm for turbocharger needs adjusting • Wastegate bypass regulator valve is faulty • ECM has failed
DTC: P1558 **2T, MIL: Yes** **Passat:** 1.8 (AEB, ATW, AUG, AWM), 2.0L (BPY), 2.8L (AHA, ATQ); **Jetta:** 1.8L (AWD, AWP, AWW), 2.0L (AEG, AVH, AZG, BBW, BEV, BPY), 2.8L (AFP, BDF); **Golf:** 1.8L (AWD, AWW), 2.0L (AEG, AVH, AZG, BEV); **GTI:** 1.8L (AWD, AWP, AWW), 2.0L (AEG), 2.8L (AFP, BDF); **NB:** 1.8L (APH, AWP, AWV, BNU, BKF), 2.0L (AEG, AVH, AZG, BEV); **Nbvert:** (BDG) **Transmissions:** All	**Throttle Actuator Electrical Malfunction Conditions:** Engine started, battery voltage at least 11.5v, all electrical components off, ground connections between engine and chassis well connected, coolant temperature at least 80-degrees Celicius and the throttle valve must not be damaged or dirty; and the ECM detected the signal from the Throttle Position Valve Module to the ECM detected was erratic, non existent or unreliable (too high or too low). **Possible Causes:** • Throttle valve control module has failed • Throttle valve control module's circuit has shorted or is open • The ECM has failed

DTC	Trouble Code Title, Conditions & Possible Causes
DTC: P1559 **2T, MIL: Yes** **Passat:** 1.8 (AEB, AWM), 2.0L (BPY), 2.8L (AHA, ATQ); **Jetta:** 1.8L (AWD, AWP, AWW), 2.0L (AEG, AVH, AZG, BBW, BEV, BPY), 2.8L (AFP, BDF); **Golf:** 1.8L (AWD, AWW), 2.0L (AEG); **GTI:** 1.8L (AWD, AWP, AWW), 2.0L (AEG), 2.8L (AFP, BDF); **NB:** 1.8L (APH, AWP, AWV, BNU, BKF), 2.0L (AEG, AVH, AZG, BEV); **Nbvert:** (BDG) **Transmissions:** All	**Idle Speed Control Throttle Position Adaptation Malfunction Conditions:** Engine started, battery voltage at least 11.5v, all electrical components off, ground connections between engine and chassis well connected, coolant temperature at least 80-degrees Celicius and the throttle valve must not be damaged or dirty; and the ECM detected the signal from the Throttle Position Valve Module to the ECM detected was erratic, non existent or unreliable (too high or too low). **Possible Causes:** • Throttle valve control module has failed • Throttle valve control module's circuit has shorted or is open • The ECM has failed
DTC: P1560 **Passat:** 1.8 (AEB, AWM), 2.0L (BPY), 2.8L (AHA, ATQ); **Jetta:** 1.8L (AWD, AWP, AWW); **Golf:** 1.8L (AWD, AWW); **GTI:** 1.8L (AWD, AWP, AWW); **NB:** 1.8L (APH, AWP, AWV, BNU, BKF) **Transmissions:** All	**Maximum Engine Speed Exceeded Conditions:** Engine running, the ECM has detected that the maximum engine speed had been attained. **Possible Causes:** • Throttle valve control module has failed • Throttle valve control module's circuit has shorted or is open • The ECM has failed • General engine damage
DTC: P1564 **Passat:** 1.8 (AEB, AWM), 2.0L (BPY), 2.8L (AHA, ATQ); **Jetta:** 1.8L (AWD, AWP, AWW), 2.0L (BPY); **Golf:** 1.8L (AWD, AWW); **GTI:** 1.8L (AWD, AWP, AWW); **NB:** 1.8L (APH, AWP, AWV, BNU, BKF) **Transmissions:** All	**Idle Speed Control Throttle Position Low Voltage During Adaptation Conditions:** Engine started, battery voltage at least 11.5v, all electrical components off, ground connections between engine and chassis well connected, coolant temperature at least 80-degrees Celicius and the throttle valve must not be damaged or dirty; and the ECM detected the signal from the Throttle Position Valve Module to the ECM detected was erratic, non existent or unreliable (too high or too low). **Possible Causes:** • Alternator failed • ECM failed • Fuses blown or open circuits • Clean Throttle Valve Control Module • Faulty battery • Idle speed control throttle failed • Wire connections to relay carrier and ground connection of ECM may have shorted
DTC: P1565 **2T, MIL: Yes** **Passat:** 2.8L (AHA, ATQ), 2.0L (BPY); **Jetta:** 1.8L (AWD, AWP, AWW), 2.0L (AEG, AVH, AZG, BEV, BPY), 2.8L (AFP, BDF); **Golf:** 1.8L (AWD, AWW), 2.0L (AEG); **GTI:** 1.8L (AWD, AWP, AWW), 2.0L (AEG), 2.8L (AFP, BDF); **NB:** 1.8L (APH, AWP, AWV, BNU, BKF) **Transmissions:** All	**Idle Speed Control Throttle Position Lower Limit Not Attainted Conditions:** Engine started, battery voltage at least 11.5v, all electrical components off, ground connections between engine and chassis well connected, coolant temperature at least 80-degrees Celicius and the throttle valve must not be damaged or dirty; and the ECM detected the signal from the Throttle Position Valve Module to the ECM detected was erratic, non existent or unreliable (too high or too low). **Possible Causes:** • Alternator failed • ECM failed • Fuses blown or open circuits • Clean Throttle Valve Control Module • Accelerator cable not adjusted properly • Idle speed control throttle failed • Wire connections to relay carrier and ground connection of ECM may have shorted
DTC: P1568 **2T, MIL: Yes** **Passat:** 2.0L (BPY); **Jetta:** 1.8L (AWD, AWP, AWW), 2.0L (BPY); **Golf:** 1.8L (AWD, AWW); **GTI:** 1.8L (AWD, AWP, AWW); **NB:** 1.8L (APH, AWP, AWV, BNU, BKF) **Transmissions:** All	**Idle Speed Control Throttle Position Mechanical Malfunction Conditions:** Engine started, battery voltage at least 11.5v, all electrical components off, ground connections between engine and chassis well connected, coolant temperature at least 80-degrees Celicius and the throttle valve must not be damaged or dirty; and the ECM detected the signal from the Throttle Position Valve Module to the ECM detected was erratic, non existent or unreliable (too high or too low) suggesting a mechanicl malfunction. **Possible Causes:** • Alternator failed • ECM failed • Fuses blown or open circuits • Clean Throttle Valve Control Module • Accelerator cable not adjusted properly • Idle speed control throttle failed • Wire connections to relay carrier and ground connection of ECM may have shorted

DTC	Trouble Code Title, Conditions & Possible Causes
DTC: P1569 **Passat:** 2.0L (BPY); **Jetta:** 1.8L (AWD, AWP, AWW); **Golf:** 1.8L (AWD, AWW); **GTI:** 1.8L (AWD, AWP, AWW); **NB:** 1.8L (APH, AWP, AWV, BNU, BKF) **Transmissions:** All	**Cruise Control Switch Incorrect Signal Conditions:** Key on or engine started and the ECM detected an incorrect signal from the cruise control switch **Possible Causes:** • Check Cruise Control System (CCS) wiring shorts to ground or B+
DTC: P1579 **Jetta:** 1.8L (AWD, AWP, AWW), 2.0L (BPY); **Golf:** 1.8L (AWD, AWW); **GTI:** 1.8L (AWD, AWP, AWW) **Transmissions:** All	**Idle Speed Control Throttle Position Adaptation Not Started Conditions:** Key on or engine started and the ECM detected an incorrect signal between the idle speed control and the throttle valve control module **Possible Causes:** • Adapt Engine Control Module (ECM) to throttle valve control module.
DTC: P1580 **Passat:** 2.0L (ABA, BPY), 2.8L (AAA, AHA, ATQ); **Jetta:** 2.0 (ABA, AEG, AVH, AZG, BBW, BEV), 2.8L (AAA); **Golf:** 2.0L (ABA, AEG, AVH, AZG, BEV); **GTI:** 2.0L (ABA, AEG), 2.8L (AAA); **NB:** 1.8L (APH, AWP, AWV, BNU, BKF), 2.0L (AEG, AVH, AZG, BEV); **Nbvert:** (BDG); **Cabrio:** 2.0L (ABA) **Transmissions:** All	**Throttle Actuator B1 Malfunction Conditions:** Engine started, battery voltage at least 11.5v, all electrical components off, ground connections between engine and chassis well connected, coolant temperature at least 80-degrees Celicius. The ECM detected the throttle actuator B1 input failed the rationality test (i.e., the input did not change as expected by the ECM). The throttle valve activation occurs via an electric motor (throttle drive) in the throttle valve control module. It is activated by the Engine Control Module (ECM) according to specifications of the two sensors, Throttle Position (TP) Sensor –G79- and Accelerator Pedal Position Sensor 2 **Possible Causes:** • The throttle position actuator B1 circuit is open • The ECM has failed • The throttle valve control module circuit is open • The throttle position actuator has failed
DTC: P1582 **Passat:** 2.0L (ABA, BPY), 2.8L (AAA, AHA, ATQ); **Jetta:** 2.0 (ABA, AEG, AVH, AZG, BBW, BEV), 2.8L (AAA); **Golf:** 2.0L (ABA, AEG, AVH, AZG, BEV); **GTI:** 2.0L (ABA, AEG), 2.8L (AAA); **NB:** 1.8L (APH, AWP, AWV, BNU, BKF), 2.0L (AEG, AVH, AZG, BEV); **Nbvert:** (BDG); **Cabrio:** 2.0L (ABA) **Transmissions:** All	**Idle Adaptation at Limit** Key on or engine running, the ECM detected that the idle adaptation reached its limit **Possible Causes:** • Crankcase oil is diluted (change the oil) • Fuel injectors are worn • Compression ratio is low • Throttle valve control module is faulty • Leak Detection Pump has failed • EVAP canister system has an improper or broken seal • Evaporative Emission (EVAP) canister purge regulator valve 1 is faulty • Hoses between the fuel pump and the EVAP canister are faulty • Fuel filler cap is loose • Fuel pump seal is defective, faulty or otherwise leaking • Hoses between the EVAP canister and the fuel flap unit are faulty • Hoses between the EVAP canister and the evaporative emission canister purge regulator valve are faulty • ECM has failed
DTC: P1586 **Passat:** 4.0L (BDP) **Transmissions:** All	**Engine Mount Valves Short to B+ Conditions:** Ignition off, the ECM detected a short in the engine mount valve **Possible Causes:** • Electro-hydraulic engine mount solenoid valve has failed • ECM has failed • Circuit wires have shorted to B+ or ground
DTC: P1587 **Passat:** 4.0L (BDP) **Transmissions:** All	**Engine Mount Valves Short to Ground Conditions:** Ignition off, the ECM detected a short in the engine mount valve **Possible Causes:** • Electro-hydraulic engine mount solenoid valve has failed • ECM has failed • Circuit wires have shorted to B+ or ground
DTC: P1588 **Passat:** 4.0L (BDP) **Transmissions:** All	**Engine Mount Valves Open Conditions:** Ignition off, the ECM detected a short in the engine mount valve **Possible Causes:** • Electro-hydraulic engine mount solenoid valve has failed • ECM has failed • Circuit wires have shorted to B+ or ground

DTC	Trouble Code Title, Conditions & Possible Causes
DTC: P1592 **Passat:** 2.0L (BPY); **NB:** 1.8L (APH, AWP, AWV, BNU, BKF) **Transmissions:** All	**Barometric/Boost Pressure Signal Ratio out of Range Conditions:** Engine started, battery voltage at least 11.5v, all electrical components off, ground connections between engine and chassis well connected, coolant temperature at least 80-degrees Celicius. The ECM detected that the boost pressure signal was faulty or out of range. **Possible Causes:** • Charge air pressure sensor is faulty • Voltage supply to the charge air pressure sensor is open or shorted • Check the charge air system for leaks • Recirculating valve for turbocharger is faulty • Turbocharging system is damaged or not functioning properly • Turbocharger recirculating valve is faulty • Vacuum diaphragm for turbocharger is out of adjustment • Wastegate bypass regulator valve is faulty • Boost sensor has failed • ECM has failed
DTC: P1602 **Passat:** 1.8 (AEB, ATW, AUG, AWM), 2.0L (BPY), 2.8L (AHA, ATQ); **Jetta:** 1.8L (AWD, AWP, AWW), 2.0L (ABA); **Golf:** 1.8L (AWD, AWW), 2.0L (ABA); **GTI:** 1.8L (AWD, AWP, AWW), 2.0L (ABA); **NB:** 1.8L (APH, AWP, AWV, BNU, BKF); **Cabrio:** 2.0L (ABA) **Transmissions:** All	**Power Supply (B+) Terminal 15 Low Voltage Conditions:** Ignition on, the ECM detected a low voltage condition on the power supply terminal (15). To achieve optimal anti-theft protection for the vehicle, an anti-theft immobilizer is installed. The anti-theft immobilizer is a system for enabling and locking the Engine Control Module (ECM). So that this system cannot be circumvented, it is necessary to perform adaptation of the anti-theft immobilizer using the Vehicle Diagnostic and Information System VAS 5052 in the On Board Diagnostic (OBD) function. The great availability of equipment options makes it necessary to adapt the Engine Control Module (ECM) to the vehicle (e.g. throttle valve control module or cruise control system). This "writing" function is not possible with the generic scan tool. **Possible Causes:** • (If ECM was replaced) ECM ID not the same as the replaced unit • ECM has failed • Voltage supply for Engine Control Module (ECM) has shorted
DTC: P1603 **Passat:** 1.8 (AEB, ATW, AUG, AWM), 2.0L (BPY); **Jetta:** 1.8L (AWD, AWP, AWW), 2.0 (ABA, BPY), 2.8L (AFP, BDF); **Golf:** 1.8L (AWD, AWW), 2.0L (ABA); **GTI:** 1.8L (AWD, AWP, AWW), 2.0L (ABA), 2.8L (AFP, BDF); **NB:** 1.8L (APH, AWP, AWV, BNU, BKF); **Cabrio:** 2.0L (ABA) **Transmissions:** All	**Internal Control Module Malfunction Conditions:** Ignition on, the ECM detected a control module malfunction. To achieve optimal anti-theft protection for the vehicle, an anti-theft immobilizer is installed. The anti-theft immobilizer is a system for enabling and locking the Engine Control Module (ECM). So that this system cannot be circumvented, it is necessary to perform adaptation of the anti-theft immobilizer using the Vehicle Diagnostic and Information System VAS 5052 in the On Board Diagnostic (OBD) function. The great availability of equipment options makes it necessary to adapt the Engine Control Module (ECM) to the vehicle (e.g. throttle valve control module or cruise control system). This "writing" function is not possible with the generic scan tool. **Possible Causes:** • (If ECM was replaced) ECM ID not the same as the replaced unit • ECM has failed • Voltage supply for Engine Control Module (ECM) has shorted
DTC: P1604 **2T, MIL: Yes** **Passat:** 1.8 (AEB, ATW, AUG, AWM), 2.0L (BPY); **Jetta:** 1.8L (AWD, AWP, AWW), 2.0L (ABA); **Golf:** 1.8L (AWD, AWW), 2.0L (ABA); **GTI:** 1.8L (AWD, AWP, AWW), 2.0L (ABA); **NB:** 1.8L (APH, AWP, AWV, BNU, BKF); **Cabrio:** 2.0L (ABA) **Transmissions:** All	**Internal Control Module Driver Error Conditions:** Ignition on, the ECM detected a control module malfunction. To achieve optimal anti-theft protection for the vehicle, an anti-theft immobilizer is installed. The anti-theft immobilizer is a system for enabling and locking the Engine Control Module (ECM). So that this system cannot be circumvented, it is necessary to perform adaptation of the anti-theft immobilizer using the Vehicle Diagnostic and Information System VAS 5052 in the On Board Diagnostic (OBD) function. The great availability of equipment options makes it necessary to adapt the Engine Control Module (ECM) to the vehicle (e.g. throttle valve control module or cruise control system). This "writing" function is not possible with the generic scan tool. **Possible Causes:** • (If ECM was replaced) ECM ID not the same as the replaced unit • ECM has failed • Voltage supply for Engine Control Module (ECM) has shorted
DTC: P1606 **Passat:** 1.8 (AEB, ATW, AUG, AWM), 2.0L (BPY), 2.8L (AHA, ATQ); **Jetta:** 1.8L (AWD, AWP, AWW); **Golf:** 1.8L (AWD, AWW); **GTI:** 1.8L (AWD, AWP, AWW); **NB:** 1.8L (APH, AWP, AWV, BNU, BKF) **Transmissions:** All	**Rough Road Spec Engine Torque ABS-ECU Electrical Malfunction Conditions:** Ignition on, the ECM detected an electrical malfunction. **Possible Causes:** • Check wire connection between Engine Control Module (ECM) and ABS Control Module

DTC	Trouble Code Title, Conditions & Possible Causes
DTC: P1609 **Passat:** 1.8 (AEB, ATW, AUG, AWM), 2.0L (BPY); **Jetta:** 1.8L (AWD, AWP, AWW), 2.0L (BPY); **Golf:** 1.8L (AWD, AWW); **GTI:** 1.8L (AWD, AWP, AWW); **NB:** 1.8L (APH, AWP, AWV, BNU, BKF) **Transmissions:** All	**Crash Shut-Down Activated Conditions:** The ECM detected that the car has been in an accident. **Possible Causes:** • Check the vehicle for damage • Reset the ECU •
DTC: P1610 **Passat:** 1.8L (ATW, AUG), 2.0L (BPY); **Jetta:** 2.0 (ABA, BPY); **Golf:** 2.0L (ABA); **GTI:** 2.0L (ABA); **NB:** 1.8L (APH, AWP, AWV, BNU, BKF); **Cabrio:** 2.0L (ABA) **Transmissions:** All	**ECU Defective Conditions:** To achieve optimal anti-theft protection for the vehicle, an anti-theft immobilizer is installed. The anti-theft immobilizer is a system for enabling and locking the Engine Control Module (ECM). So that this system cannot be circumvented, it is necessary to perform adaptation of the anti-theft immobilizer using the Vehicle Diagnostic and Information System VAS 5052 in the On Board Diagnostic (OBD) function. The great availability of equipment options makes it necessary to adapt the Engine Control Module (ECM) to the vehicle (e.g. throttle valve control module or cruise control system). This "writing" function is not possible with the generic scan tool. **Possible Causes:** • (If ECM was replaced) ECM ID not the same as the replaced unit • ECM has failed • Voltage supply for Engine Control Module (ECM) has shorted
DTC: P1611 MIL: Yes **Passat:** 2.0L (ABA, BPY), 2.8L (AAA, AHA, ATQ); **Jetta:** 2.0L (ABA), 2.8L (AAA); **Golf:** 2.0L (ABA); **GTI:** 2.0L (ABA), 2.8L (AAA); **Cabrio:** 2.0L (ABA) **Transmissions:** All	**MIL Call-Up Circuit, Transmission Control Module Short to Ground Conditions:** Engine started, VSS over 1 mph, and the ECM detected a problem in the Transmission Control system during the self-test. **Possible Causes:** • Open/short circuit to ground in the communication wire from the transmission to the ECM. • The ECM has failed
DTC: P1612 **2T, MIL: Yes** **Passat:** 1.8 (AEB, ATW, AUG, AWM), 2.0L (BPY), 2.8L (AHA, ATQ); **Jetta:** 1.8L (AWD, AWP, AWW), 2.0L (BPY), 2.8L (AFP, BDF); **Golf:** 1.8L (AWD, AWW); **GTI:** 1.8L (AWD, AWP, AWW), 2.8L (AFP, BDF); **NB:** 1.8L (APH, AWP, AWV, BNU, BKF) **Transmissions:** All	**Electronic Control Module Incorrect Coding Conditions:** Ignition on, the ECM detected a control module malfunction. To achieve optimal anti-theft protection for the vehicle, an anti-theft immobilizer is installed. The anti-theft immobilizer is a system for enabling and locking the Engine Control Module (ECM). So that this system cannot be circumvented, it is necessary to perform adaptation of the anti-theft immobilizer using the Vehicle Diagnostic and Information System VAS 5052 in the On Board Diagnostic (OBD) function. The great availability of equipment options makes it necessary to adapt the Engine Control Module (ECM) to the vehicle (e.g. throttle valve control module or cruise control system). This "writing" function is not possible with the generic scan tool. **Possible Causes:** • (If ECM was replaced) ECM ID not the same as the replaced unit • ECM has failed • Voltage supply for Engine Control Module (ECM) has shorted
DTC: P1613 MIL: Yes **Passat:** 1.8 (AEB, AWM), 2.0L (ABA, BPY), 2.8L (AAA, AHA, ATQ); **Jetta:** 2.0L (ABA), 2.8L (AAA); **Golf:** 2.0L (ABA); **GTI:** 2.0L (ABA), 2.8L (AAA); **Cabrio:** 2.0L (ABA) **Transmissions:** All	**MIL Call-up Circuit Open/Short to B+ Conditions:** Engine started, VSS over 1 mph, and the ECM detected a problem in the Transmission Control system during the self-test. **Possible Causes:** • Open/short circuit to ground from the MIL to the ECM. • The ECM has failed • The MIL light has failed (check bulb)

DTC	Trouble Code Title, Conditions & Possible Causes
DTC: P1624 **Passat:** 1.8 (AEB, ATW, AUG, AWM), 2.0L (BPY), 2.8L (AHA, ATQ); **Jetta:** 1.8L (AWD, AWP, AWW), 2.0L (BPY), 2.8L (AFP, BDF); **Golf:** 1.8L (AWD, AWW); **GTI:** 1.8L (AWD, AWP, AWW), 2.8L (AFP, BDF); **NB:** 1.8L (APH, AWP, AWV, BNU, BKF) **Transmissions:** All	**MIL Requested Signature Active Conditions:** Ignition on, the ECM detected a control module malfunction. To achieve optimal anti-theft protection for the vehicle, an anti-theft immobilizer is installed. The anti-theft immobilizer is a system for enabling and locking the Engine Control Module (ECM). So that this system cannot be circumvented, it is necessary to perform adaptation of the anti-theft immobilizer using the Vehicle Diagnostic and Information System VAS 5052 in the On Board Diagnostic (OBD) function. The great availability of equipment options makes it necessary to adapt the Engine Control Module (ECM) to the vehicle (e.g. throttle valve control module or cruise control system). This "writing" function is not possible with the generic scan tool. **Possible Causes:** • (If ECM was replaced) ECM ID not the same as the replaced unit • ECM has failed • Voltage supply for Engine Control Module (ECM) has shorted
DTC: P1626 **Passat:** 1.8 (AEB, ATW, AUG, AWM), 2.0L (BPY), 2.8L (AHA, ATQ); **Jetta:** 1.8L (AWD, AWP, AWW), 2.8L (AFP, BDF); **Golf:** 1.8L (AWD, AWW); **GTI:** 1.8L (AWD, AWP, AWW), 2.8L (AFP, BDF); **NB:** 1.8L (APH, AWP, AWV, BNU, BKF), 2.0L (AVH, AZG, BEV); **Nbvert:** (BGD) **Transmissions:** All	**Data BUS Powertrain Missing Message From Transmission Control Conditions:** Ignition on, the ECM detected a control module malfunction (Transmission). To achieve optimal anti-theft protection for the vehicle, an anti-theft immobilizer is installed. The anti-theft immobilizer is a system for enabling and locking the Engine Control Module (ECM). So that this system cannot be circumvented, it is necessary to perform adaptation of the anti-theft immobilizer using the Vehicle Diagnostic and Information System VAS 5052 in the On Board Diagnostic (OBD) function. The great availability of equipment options makes it necessary to adapt the Engine Control Module (ECM) to the vehicle (e.g. throttle valve control module or cruise control system). This "writing" function is not possible with the generic scan tool. **Possible Causes:** • (If ECM was replaced) ECM ID not the same as the replaced unit • ECM has failed • Voltage supply for Engine Control Module (ECM) has shorted
DTC: P1630 **2T, MIL: Yes** **Passat:** 1.8 (AEB, ATW, AUG, AWM), 2.0L (BPY); **Jetta:** 1.8L (AWD, AWP, AWW), 2.0L (AEG, AVH, AZG, BBW, BEV), 2.8L (AFP, BDF); **Golf:** 1.8L (AWD, AWW), 2.0L (AEG, AVH, AZG, BEV); **GTI:** 1.8L (AWD, AWP, AWW), 2.0L (AEG), 2.8L (AFP, BDF); **NB:** 1.8L (APH, AWP, AWV, BNU, BKF), 2.0L (AEG, AVH, AZG, BEV); **Nbvert:** (BDG) **Transmissions:** All	**Acceleration Pedal Position Sensor 1 Signal Too Low Conditions:** Engine started, battery voltage at least 11.5v, all electrical components off, ground connections between engine and chassis well connected, the ECM detected that the accelerator pedal position sensor signal was too low. **Note: Both the Throttle Position (TP) Sensor and Accelerator Pedal Position Sensor 2 are located at the accelerator pedal module and communicate the driver's intentions to the ECM completely independently of each other. Both sensors are stored in one housing.** **Possible Causes:** • Ground between engine and chassis may be broken • Throttle position sensor may have failed • Accelerator Pedal Position Sensor 2 has failed • Throttle position sensor wiring may have shorted • Faulty voltage supply • ECM has failed
DTC: P1631 **2T, MIL: Yes** **Passat:** 1.8 (AEB, ATW, AUG, AWM), 2.0L (BPY); **Jetta:** 1.8L (AWD, AWP, AWW), 2.0L (AEG, AVH, AZG, BBW, BEV), 2.8L (AFP, BDF); **Golf:** 1.8L (AWD, AWW), 2.0L (AEG, AVH, AZG, BEV); **GTI:** 1.8L (AWD, AWP, AWW), 2.0L (AEG), 2.8L (AFP, BDF); **NB:** 1.8L (APH, AWP, AWV, BNU, BKF), 2.0L (AEG, AVH, AZG, BEV); **Nbvert:** (BDG) **Transmissions:** All	**Acceleration Pedal Position Sensor 1 Signal Too High Conditions:** Engine started, battery voltage at least 11.5v, all electrical components off, ground connections between engine and chassis well connected, the ECM detected that the accelerator pedal position sensor signal was too high. **Note: Both the Throttle Position (TP) Sensor and Accelerator Pedal Position Sensor 2 are located at the accelerator pedal module and communicate the driver's intentions to the ECM completely independently of each other. Both sensors are stored in one housing.** **Possible Causes:** • Ground between engine and chassis may be broken • Throttle position sensor may have failed • Accelerator Pedal Position Sensor 2 has failed • Throttle position sensor wiring may have shorted • Faulty voltage supply • ECM has failed

DTC	Trouble Code Title, Conditions & Possible Causes
DTC: P1633 **2T, MIL: Yes** **Passat:** 1.8L (ATW, AUG), 2.0L (BPY); **Jetta:** 1.8L (AWD, AWP, AWW), 2.0L (AEG, AVH, AZG, BBW, BEV), 2.8L (AFP, BDF); **Golf:** 1.8L (AWD, AWW), 2.0L (AEG, AVH, AZG, BEV); **GTI:** 1.8L (AWD, AWP, AWW), 2.0L (AEG), 2.8L (AFP, BDF); **NB:** 1.8L (APH, AWP, AWV, BNU, BKF), 2.0L (AEG, AVH, AZG, BEV); **Nbvert:** (BDG) **Transmissions:** All	**Acceleration Pedal Position Sensor 2 Signal Too Low Conditions:** Engine started, battery voltage at least 11.5v, all electrical components off, ground connections between engine and chassis well connected, the ECM detected that the accelerator pedal position sensor signal was too low. **Note: Both the Throttle Position (TP) Sensor and Accelerator Pedal Position Sensor 2 are located at the accelerator pedal module and communicate the driver's intentions to the ECM completely independently of each other. Both sensors are stored in one housing.** **Possible Causes:** • Ground between engine and chassis may be broken • Throttle position sensor may have failed • Accelerator Pedal Position Sensor 2 has failed • Throttle position sensor wiring may have shorted • Faulty voltage supply • ECM has failed
DTC: P1634 **2T, MIL: Yes** **Passat:** 1.8 (AEB, ATW, AUG, AWM), 2.0L (BPY); **Jetta:** 1.8L (AWD, AWP, AWW), 2.0L (AEG, AVH, AZG, BBW, BEV), 2.8L (AFP, BDF); **Golf:** 1.8L (AWD, AWW), 2.0L (AEG, AVH, AZG, BEV); **GTI:** 1.8L (AWD, AWP, AWW), 2.0L (AEG), 2.8L (AFP, BDF); **NB:** 1.8L (APH, AWP, AWV, BNU, BKF), 2.0L (AEG, AVH, AZG, BEV); **Nbvert:** (BDG) **Transmissions:** All	**Acceleration Pedal Position Sensor 2 Signal Too High Conditions:** Engine started, battery voltage at least 11.5v, all electrical components off, ground connections between engine and chassis well connected, the ECM detected that the accelerator pedal position sensor signal was too high. **Note: Both the Throttle Position (TP) Sensor and Accelerator Pedal Position Sensor 2 are located at the accelerator pedal module and communicate the driver's intentions to the ECM completely independently of each other. Both sensors are stored in one housing.** **Possible Causes:** • Ground between engine and chassis may be broken • Throttle position sensor may have failed • Accelerator Pedal Position Sensor 2 has failed • Throttle position sensor wiring may have shorted • Faulty voltage supply • ECM has failed
DTC: P1635 **Passat:** 2.0L (BPY); **Jetta:** 2.0L (AVH, AZG, BBW, BEV), 2.8L (AFP, BDF); **Golf:** 2.0L (AVH, AZG, BEV); **GTI:** 2.8L (AFP, BDF); **NB:** 1.8L (APH, AWP, AWV, BNU, BKF), 2.0L (AVH, AZG, BEV); **Nbvert:** (BGD) **Transmissions:** All	**Data BUS Powertrain Missing Message From Central A/C Control Conditions:** Ignition off, the ECU is missing general Data BUS information from the A/C control. The Engine Control Module (ECM) communicates with all databus-capable control modules via a CAN databus. These databus-capable control modules are connected via two data bus wires which are twisted together (CAN_High and CAN_Low), and exchange information (messages). Missing information on the databus is recognized as a malfunction and stored. Trouble-free operation of the CAN-bus requires that it have a terminal resistance. This central terminal resistor is located in the Engine Control Module (ECM). **Possible Causes:** • Check the Terminal resistance for CAN-bus • Data-Bus wires have short • Data-Bus components are malfunctioning • ECM has failed
DTC: P1637 **Passat:** 2.0L (BPY); **Jetta:** 2.0L (BPY) **Transmissions:** All	**Data BUS Powertrain Missing Message From Central Electrical Control Conditions:** Ignition off, the ECU is missing general Data BUS information from the central electrical control. The Engine Control Module (ECM) communicates with all databus-capable control modules via a CAN databus. These databus-capable control modules are connected via two data bus wires which are twisted together (CAN_High and CAN_Low), and exchange information (messages). Missing information on the databus is recognized as a malfunction and stored. Trouble-free operation of the CAN-bus requires that it have a terminal resistance. This central terminal resistor is located in the Engine Control Module (ECM). **Possible Causes:** • Check the Terminal resistance for CAN-bus • Data-Bus wires have short • Data-Bus components are malfunctioning • ECM has failed

DTC	Trouble Code Title, Conditions & Possible Causes
DTC: P1639 **2T, MIL: Yes** **Passat:** 1.8 (AEB, ATW, AUG, AWM), 2.0L (BPY); **Jetta:** 1.8L (AWD, AWP, AWW), 2.0L (AEG, AVH, AZG, BBW, BEV), 2.8L (AFP, BDF); **Golf:** 1.8L (AWD, AWW), 2.0L (AEG, AVH, AZG, BEV); **GTI:** 1.8L (AWD, AWP, AWW), 2.0L (AEG), 2.8L (AFP, BDF); **NB:** 1.8L (APH, AWP, AWV, BNU, BKF), 2.0L (AEG, AVH, AZG, BEV); **Nbvert:** (BDG) **Transmissions:** All	**Accelerator Pedal Position Sensor 1+2 Range/Performance Conditions:** Engine started, battery voltage at least 11.5v, all electrical components off, ground connections between engine and chassis well connected, the ECM detected that the accelerator pedal position sensor signal was too high. **Note: Both the Throttle Position (TP) Sensor and Accelerator Pedal Position Sensor 2 are located at the accelerator pedal module and communicate the driver's intentions to the ECM completely independently of each other. Both sensors are stored in one housing.** **Possible Causes:** • Ground between engine and chassis may be broken • Throttle position sensor may have failed • Accelerator Pedal Position Sensor 2 has failed • Throttle position sensor wiring may have shorted • Faulty voltage supply • ECM has failed
DTC: P1640 **Passat:** 1.8 (AEB, ATW, AUG, AWM), 2.0L (BPY), 2.8L (AHA, ATQ); **Jetta:** 1.8L (AWD, AWP, AWW), 2.0 (ABA, AEG, AVH, AZG, BBW, BEV, BPY), 2.8L (AFP, BDF); **Golf:** 1.8L (AWD, AWW), 2.0L (ABA, AEG, AVH, AZG, BEV); **GTI:** 1.8L (AWD, AWP, AWW), 2.0L (ABA, AEG), 2.8L (AFP, BDF); **NB:** 1.8L (APH, AWP, AWV, BNU, BKF), 2.0L (AEG, AVH, AZG, BEV); **Nbvert:** (BDG); **Cabrio:** 2.0L (ABA) **Transmissions:** All	**Internal Control Module (EEPROM) Error Conditions:** Ignition on, the ECM detected a control module malfunction (software). To achieve optimal anti-theft protection for the vehicle, an anti-theft immobilizer is installed. The anti-theft immobilizer is a system for enabling and locking the Engine Control Module (ECM). So that this system cannot be circumvented, it is necessary to perform adaptation of the anti-theft immobilizer using the Vehicle Diagnostic and Information System VAS 5052 in the On Board Diagnostic (OBD) function. The great availability of equipment options makes it necessary to adapt the Engine Control Module (ECM) to the vehicle (e.g. throttle valve control module or cruise control system). This "writing" function is not possible with the generic scan tool. **Possible Causes:** • Engine Control Module (ECM) has failed • Voltage supply for Engine Control Module (ECM) has shorted
DTC: P1647 **Passat:** 2.0L (BPY); **Jetta:** 2.5L (BGP, BGQ), 2.8L (AFP, BDF); **GTI:** 2.8L (AFP, BDF) **Transmissions:** All	**Please Check Coding of the ECUs in the Data Bus Powertrain Conditions:** Ignition on, the ECM detected a control module malfunction (software). To achieve optimal anti-theft protection for the vehicle, an anti-theft immobilizer is installed. The anti-theft immobilizer is a system for enabling and locking the Engine Control Module (ECM). So that this system cannot be circumvented, it is necessary to perform adaptation of the anti-theft immobilizer using the Vehicle Diagnostic and Information System VAS 5052 in the On Board Diagnostic (OBD) function. The great availability of equipment options makes it necessary to adapt the Engine Control Module (ECM) to the vehicle (e.g. throttle valve control module or cruise control system). This "writing" function is not possible with the generic scan tool. **Possible Causes:** • Engine Control Module (ECM) has failed • Voltage supply for Engine Control Module (ECM) has shorted
DTC: P1648 **Passat:** 1.8 (AEB, ATW, AUG, AWM), 2.0L (BPY); **Jetta:** 1.8L (AWD, AWP, AWW), 2.0L (AVH, AZG, BBW, BEV), 2.8L (AFP, BDF); **Golf:** 1.8L (AWD, AWW), 2.0L (AVH, AZG, BEV); **GTI:** 1.8L (AWD, AWP, AWW), 2.8L (AFP, BDF); **NB:** 1.8L (APH, AWP, AWV, BNU, BKF), 2.0L (AVH, AZG, BEV); **Nbvert:** (BGD) **Transmissions:** All	**Data Bus Powertrain Malfunction Conditions:** Ignition on, the ECM detected a data bus malfunction (software). To achieve optimal anti-theft protection for the vehicle, an anti-theft immobilizer is installed. The anti-theft immobilizer is a system for enabling and locking the Engine Control Module (ECM). So that this system cannot be circumvented, it is necessary to perform adaptation of the anti-theft immobilizer using the Vehicle Diagnostic and Information System VAS 5052 in the On Board Diagnostic (OBD) function. The great availability of equipment options makes it necessary to adapt the Engine Control Module (ECM) to the vehicle (e.g. throttle valve control module or cruise control system). This "writing" function is not possible with the generic scan tool. **Possible Causes:** • Ground between engine and chassis may be broken • Throttle position sensor may have failed • Accelerator Pedal Position Sensor 2 has failed • Throttle position sensor wiring may have shorted • Faulty voltage supply • ECM has failed

DTC	Trouble Code Title, Conditions & Possible Causes
DTC: P1649 **2T, MIL: Yes** **Passat:** 1.8 (AEB, ATW, AUG, AWM), 2.0L (BPY); **Jetta:** 1.8L (AWD, AWP, AWW), 2.0L (AVH, AZG, BBW, BEV); **Golf:** 1.8L (AWD, AWW), 2.0L (AVH, AZG, BEV); **GTI:** 1.8L (AWD, AWP, AWW); **NB:** 1.8L (APH, AWP, AWV, BNU, BKF), 2.0L (AVH, AZG, BEV); **Nbvert:** (BGD) **Transmissions:** All	**Data Bus Powertrain Missing Message from ABS Control Module Conditions:** Ignition off, the ECU is missing general Data BUS information from the central electrical control. The Engine Control Module (ECM) communicates with all databus-capable control modules via a CAN databus. These databus-capable control modules are connected via two data bus wires which are twisted together (CAN_High and CAN_Low), and exchange information (messages). Missing information on the databus is recognized as a malfunction and stored. Trouble-free operation of the CAN-bus requires that it have a terminal resistance. This central terminal resistor is located in the Engine Control Module (ECM). **Possible Causes:** • Ground between engine and chassis may be broken • Throttle position sensor may have failed • Accelerator Pedal Position Sensor 2 has failed • Throttle position sensor wiring may have shorted • Faulty voltage supply • Check the Terminal resistance for CAN-bus • Data-Bus wires have short • Data-Bus components are malfunctioning • ECM has failed
DTC: P1650 **Passat:** 2.0L (BPY); **Jetta:** 1.8L (AWD, AWP, AWW), 2.0L (AVH, AZG, BBW, BEV, BPY), 2.8L (AFP, BDF); **Golf:** 1.8L (AWD, AWW), 2.0L (AVH, AZG, BEV); **GTI:** 1.8L (AWD, AWP, AWW), 2.8L (AFP, BDF); **NB:** 1.8L (APH, AWP, AWV, BNU, BKF), 2.0L (AVH, AZG, BEV); **Nbvert:** (BGD) **Transmissions:** All	**Data Bus Powertrain Missing Message from Instrument Panel ECU Conditions:** Ignition off, the ECU is missing general Data BUS information from the instrument panel. The Engine Control Module (ECM) communicates with all databus-capable control modules via a CAN databus. These databus-capable control modules are connected via two data bus wires which are twisted together (CAN_High and CAN_Low), and exchange information (messages). Missing information on the databus is recognized as a malfunction and stored. Trouble-free operation of the CAN-bus requires that it have a terminal resistance. This central terminal resistor is located in the Engine Control Module (ECM). **Possible Causes:** • Ground between engine and chassis may be broken • Faulty voltage supply • ECM has failed
DTC: P1653 **Passat:** 2.0L (BPY) **Transmissions:** All	**Please Check DTC Memory of the ABS Control Module Conditions:** Ignition off, the ECU detected a memory fault of the ABS control module. The Engine Control Module (ECM) communicates with all databus-capable control modules via a CAN databus. These databus-capable control modules are connected via two data bus wires which are twisted together (CAN_High and CAN_Low), and exchange information (messages). Missing information on the databus is recognized as a malfunction and stored. Trouble-free operation of the CAN-bus requires that it have a terminal resistance. This central terminal resistor is located in the Engine Control Module (ECM). **Possible Causes:** • Ground between engine and chassis may be broken • Throttle position sensor may have failed • Accelerator Pedal Position Sensor 2 has failed • Throttle position sensor wiring may have shorted • Faulty voltage supply • ECM has failed
DTC: P1654 **Passat:** 1.8 (AEB, ATW, AUG, AWM) **Transmissions:** All	**Please Check DTC Memory of the Control Panel ECU Conditions:** The Engine Control Module (ECM) communicates with all databus-capable control modules via a CAN databus. These databus-capable control modules are connected via two data bus wires which are twisted together (CAN_High and CAN_Low), and exchange information (messages). Missing information on the databus is recognized as a malfunction and stored. Trouble-free operation of the CAN-bus requires that it have a terminal resistance. This central terminal resistor is located in the Engine Control Module (ECM). **Possible Causes:** • Ground between engine and chassis may be broken • Faulty voltage supply • ECM has failed
DTC: P1655 **Passat:** 2.0L (BPY) **Transmissions:** All	**Please Check DTC Memory of the ADR Control Module Conditions:** The Engine Control Module (ECM) communicates with all databus-capable control modules via a CAN databus. These databus-capable control modules are connected via two data bus wires which are twisted together (CAN_High and CAN_Low), and exchange information (messages). Missing information on the databus is recognized as a malfunction and stored. Trouble-free operation of the CAN-bus requires that it have a terminal resistance. This central terminal resistor is located in the Engine Control Module (ECM). **Possible Causes:** • Ground between engine and chassis may be broken • Throttle position sensor may have failed • Accelerator Pedal Position Sensor 2 has failed • Throttle position sensor wiring may have shorted • Faulty voltage supply • ECM has failed

DTC	Trouble Code Title, Conditions & Possible Causes
DTC: P1676 **Passat:** 1.8 (AEB, ATW, AUG, AWM); **Jetta:** 1.8L (AWD, AWP, AWW); **Golf:** 1.8L (AWD, AWW); **GTI:** 1.8L (AWD, AWP, AWW); **NB:** 1.8L (APH, AWP, AWV, BNU, BKF) **Transmissions:** All	**Drive by Wire-MIL Circuit Electrical Malfunction Conditions:** Key on or engine running, the ECM detected an electrical malfunction regarding the drive-by-wire circuit. **Note: EPC" is an abbreviation and stands for Electronic Power Control and means "electronic engine load control". If malfunctions are recognized in the EPC system during operation of the engine, the Engine Control Module (ECM) switches on the EPC warning lamp. An entry is made in DTC memory at the same time. After a few seconds of the engine at idle, the EPC should extinguish itself.** **Possible Causes:** • Circuit from the MIL to the ECM • ECM has failed • Circuit from the EPC to the ECM
DTC: P1677 **Passat:** 1.8 (AEB, ATW, AUG, AWM); **Jetta:** 1.8L (AWD, AWP, AWW); **Golf:** 1.8L (AWD, AWW); **GTI:** 1.8L (AWD, AWP, AWW); **NB:** 1.8L (APH, AWP, AWV, BNU, BKF) **Transmissions:** All	**Drive by Wire-MIL Circuit Short to B+ Conditions:** Key on or engine running, the ECM detected an electrical malfunction regarding the drive-by-wire circuit. **Note: EPC" is an abbreviation and stands for Electronic Power Control and means "electronic engine load control". If malfunctions are recognized in the EPC system during operation of the engine, the Engine Control Module (ECM) switches on the EPC warning lamp. An entry is made in DTC memory at the same time. After a few seconds of the engine at idle, the EPC should extinguish itself.** **Possible Causes:** • Circuit from the MIL to the ECM • ECM has failed • Circuit from the EPC to the ECM
DTC: P1681 **2T, MIL: Yes** **Passat:** 1.8 (AEB, AWM), 2.8L (AHA, ATQ); **Jetta:** 2.0 (ABA, AEG, AVH, AZG, BBW, BEV); **Golf:** 2.0L (ABA, AEG, AVH, AZG, BEV); **GTI:** 2.0L (ABA, AEG), 2.0L (AEG, AVH, AZG, BEV); **Nbvert:** (BDG); **Cabrio:** 2.0L (ABA) **Transmissions:** All	**Control Unit Programming, Programming Not Finished Conditions:** The Engine Control Module (ECM) communicates with all databus-capable control modules via a CAN databus. These databus-capable control modules are connected via two data bus wires which are twisted together (CAN_High and CAN_Low), and exchange information (messages). Missing information on the databus is recognized as a malfunction and stored. Trouble-free operation of the CAN-bus requires that it have a terminal resistance. This central terminal resistor is located in the Engine Control Module (ECM). **Possible Causes:** • ECM has failed
DTC: P1682 **Jetta:** 2.0L (BPY) **Transmissions:** All	**Powertrain Data Bus Implausible Message from ABS Control Module Conditions:** Ignition off, the ECU detected a I fault of the ABS control module. The Engine Control Module (ECM) communicates with all databus-capable control modules via a CAN databus. These databus-capable control modules are connected via two data bus wires which are twisted together (CAN_High and CAN_Low), and exchange information (messages). Missing information on the databus is recognized as a malfunction and stored. Trouble-free operation of the CAN-bus requires that it have a terminal resistance. This central terminal resistor is located in the Engine Control Module (ECM). **Possible Causes:** • Ground between engine and chassis may be broken • Faulty voltage supply • ECM has failed
DTC: P1683 **Passat:** 2.0L (BPY); **Jetta:** 2.0L (BPY) **Transmissions:** All	**Data Bus Powertrain Implausible Message from Airbag Control Conditions:** Ignition off, the ECU detected a circuit fault of the airbag control module. The Engine Control Module (ECM) communicates with all databus-capable control modules via a CAN databus. These databus-capable control modules are connected via two data bus wires which are twisted together (CAN_High and CAN_Low), and exchange information (messages). Missing information on the databus is recognized as a malfunction and stored. Trouble-free operation of the CAN-bus requires that it have a terminal resistance. This central terminal resistor is located in the Engine Control Module (ECM). **Possible Causes:** • Ground between engine and chassis may be broken • Throttle position sensor may have failed • Accelerator Pedal Position Sensor 2 has failed • Throttle position sensor wiring may have shorted • Faulty voltage supply • ECM has failed

DTC	Trouble Code Title, Conditions & Possible Causes
DTC: P1690 **Passat:** 1.8 (AEB, ATW, AUG, AWM), 2.8L (AHA, ATQ); **Jetta:** 1.8L (AWD, AWP, AWW); **Golf:** 1.8L (AWD, AWW); **GTI:** 1.8L (AWD, AWP, AWW); **NB:** 1.8L (APH, AWP, AWV, BNU, BKF) **Transmissions:** All	**Malfunction Indication Light Malfunction Conditions:** The exhaust Malfunction Indicator Lamp (MIL) lights up when exhaust relevant malfunctions are recognized by the Engine Control Module (ECM). The Malfunction Indicator Lamp (MIL) can blink or remain lit continuously. Blinking: There is a malfunction that causes damage to the catalytic converter in this driving condition. In this case, vehicle must only be driven at reduced power! Continuously lit: There is a malfunction that causes increased emissions. Check DTC memory for Motronic control module. DTC memory must still be checked if there are driveability problems or customer complaints and the MIL is not lit, since malfunctions can be stored without causing the MIL to light immediately. **Possible Causes:** • Wire from ECM to MIL is shorted or grounded • ECM has failed • MIL has failed
DTC: P1691 **Passat:** 1.8 (AEB, ATW, AUG, AWM); **Jetta:** 2.8L (AFP, BDF); **GTI:** 2.8L (AFP, BDF) **Transmissions:** All	**Malfunction Indication Light Open Conditions:** The exhaust Malfunction Indicator Lamp (MIL) lights up when exhaust relevant malfunctions are recognized by the Engine Control Module (ECM). The Malfunction Indicator Lamp (MIL) can blink or remain lit continuously. Blinking: There is a malfunction that causes damage to the catalytic converter in this driving condition. In this case, vehicle must only be driven at reduced power! Continuously lit: There is a malfunction that causes increased emissions. Check DTC memory for Motronic control module. DTC memory must still be checked if there are driveability problems or customer complaints and the MIL is not lit, since malfunctions can be stored without causing the MIL to light immediately. **Possible Causes:** • Wire from ECM to MIL is shorted or grounded • ECM has failed • MIL has failed
DTC: P1692 **Passat:** 1.8 (AEB, ATW, AUG, AWM); **Jetta:** 2.8L (AFP, BDF); **GTI:** 2.8L (AFP, BDF) **Transmissions:** All	**Malfunction Indication Light Short to Ground Conditions:** The exhaust Malfunction Indicator Lamp (MIL) lights up when exhaust relevant malfunctions are recognized by the Engine Control Module (ECM). The Malfunction Indicator Lamp (MIL) can blink or remain lit continuously. Blinking: There is a malfunction that causes damage to the catalytic converter in this driving condition. In this case, vehicle must only be driven at reduced power! Continuously lit: There is a malfunction that causes increased emissions. Check DTC memory for Motronic control module. DTC memory must still be checked if there are driveability problems or customer complaints and the MIL is not lit, since malfunctions can be stored without causing the MIL to light immediately. **Possible Causes:** • Wire from ECM to MIL is shorted or grounded • ECM has failed • MIL has failed
DTC: P1693 **Passat:** 1.8 (AEB, ATW, AUG, AWM), 2.8L (AHA, ATQ); **Jetta:** 1.8L (AWD, AWP, AWW), 2.8L (AFP, BDF); **Golf:** 1.8L (AWD, AWW); **GTI:** 1.8L (AWD, AWP, AWW), 2.8L (AFP, BDF); **NB:** 1.8L (APH, AWP, AWV, BNU, BKF) **Transmissions:** All	**Malfunction Indication Light Short to B+ Conditions:** The exhaust Malfunction Indicator Lamp (MIL) lights up when exhaust relevant malfunctions are recognized by the Engine Control Module (ECM). The Malfunction Indicator Lamp (MIL) can blink or remain lit continuously. Blinking: There is a malfunction that causes damage to the catalytic converter in this driving condition. In this case, vehicle must only be driven at reduced power! Continuously lit: There is a malfunction that causes increased emissions. Check DTC memory for Motronic control module. DTC memory must still be checked if there are driveability problems or customer complaints and the MIL is not lit, since malfunctions can be stored without causing the MIL to light immediately. **Possible Causes:** • Wire from ECM to MIL is shorted or grounded • ECM has failed • MIL has failed
DTC: P1698 **Jetta:** 2.0L (BPY) **Transmissions:** All	**Check DTC Memory of Steering Column ECU Conditions:** The Engine Control Module (ECM) communicates with all databus-capable control modules via a CAN databus. These databus-capable control modules are connected via two data bus wires which are twisted together (CAN_High and CAN_Low), and exchange information (messages). Missing information on the databus is recognized as a malfunction and stored. Trouble-free operation of the CAN-bus requires that it have a terminal resistance. This central terminal resistor is located in the Engine Control Module (ECM). **Possible Causes:** • Ground between engine and chassis may be broken • Faulty voltage supply • ECM has failed
DTC: P1702 **Passat:** 1.8L (ATW, AUG) **Transmissions:** A/T	**TR Sensor Signal Intermittent Conditions:** Key on or engine running; and the ECM detected the failure Trouble Code Conditions for DTC P0705 or P0708 were met intermittently. **Possible Causes:** • Refer to the appropriate Transmission Repair Manual or information in electronic media to perform a complete diagnosis of the automatic transmission when this code is set

DTC	Trouble Code Title, Conditions & Possible Causes
DTC: P1778 **Passat:** 2.0L (ABA, BPY), 2.8L (AAA); **Jetta:** 1.8L (AWD, AWP, AWW), 2.0 (ABA, AEG, AVH, AZG, BBW, BEV), 2.8L (AAA); **Golf:** 1.8L (AWD, AWW), 2.0L (ABA, AEG, AVH, AZG, BEV); **GTI:** 1.8L (AWD, AWP, AWW), 2.0L (ABA, AEG), 2.8L (AAA, AFP, BDF); **NB:** 1.8L (APH, AWP, AWV, BNU, BKF), 2.0L (AEG, AVH, AZG, BEV); **Nbvert:** (BDG); **Cabrio:** 2.0L (ABA) **Transmissions:** All	**Solenoid EV7 Electrical Malfunction Conditions:** Engine started, battery voltage must be at least 11.5v, all electrical components must be off, the ground between the engine and the chassis must be well connected, the exhaust system must be properly sealed between the catalytic converter and the cylinder head, and the oxygen sensor heater for oxygen sensor before the catalytic converter must be properly functioning. The ECM detected a loss of communication between the TCM and the valve body solenoid valve. **Possible Causes:** • Valve body solenoid valve has a short • The TCM has failed • ECM has failed
DTC: P1780 **Passat:** 2.0L (ABA, BPY), 2.8L (AAA); **Jetta:** 1.8L (AWD, AWP, AWW), 2.0L (AEG, AVH, AZG, BBW, BEV), 2.8L (AAA); **Golf:** 1.8L (AWD, AWW), 2.0L (AEG, AVH, AZG, BEV); **GTI:** 1.8L (AWD, AWP, AWW), 2.0L (ABA, AEG), 2.8L (AAA, AFP, BDF); **NB:** 1.8L (APH, AWP, AWV, BNU, BKF), 2.0L (AEG, AVH, AZG, BEV); **Nbvert:** (BDG) **Transmissions:** A/T	**Engine Intervention Readable Conditions:** Key on or engine started, Self-Test enabled, and the ECM detected the Transmission Control Switch (TCS) was out of range during the test and a signal was not received. **Note: The seal on the ATF level plug must always be replaced if the ATF level is checked.** **Possible Causes:** • TCS circuit open or shorted in the wiring harness • TCS not cycled during the self-test • TCS is damaged or failed, or the ECM has failed • Check ATF level • Mechanical malfunction in transmission
DTC: P1781 **Jetta:** 2.0L (ABA); **Golf:** 2.0L (ABA); **Cabrio:** 2.0L (ABA) **Transmissions:** All	**Engine Torque Reduction Open/Short to Ground Conditions:** Key on or engine started, Self-Test enabled, and the ECM detected the Transmission Control Switch (TCS) was out of range during the test and a signal was not received. **Note: The seal on the ATF level plug must always be replaced if the ATF level is checked.** **Possible Causes:** • Check ATM fluid (if applicable) • The TCM has failed • ECM has failed
DTC: P1823 **Passat:** 2.0L (BPY); **Jetta:** 1.8L (AWD, AWP, AWW), 2.5L (BGP, BGQ), 2.8L (AFP, BDF); **Golf:** 1.8L (AWD, AWW); **GTI:** 1.8L (AWD, AWP, AWW), 2.8L (AFP, BDF); **NB:** 1.8L (APH, AWP, AWV, BNU, BKF), 2.0L (AEG, AVH, AZG, BEV); **Nbvert:** (BDG) **Transmissions:** All	**Pressure Control Solenoid 3 Electrical Conditions:** Engine started, vehicle driven with the solenoid applied, and the ECM detected an unexpected voltage condition on the SS1/C solenoid circuit was incorrect during the test. **Possible Causes:** • Solenoid valves in valve body are faulty • Solenoid circuit is shorting to ground • Solenoid circuit is open • TCM has failed or wiring is shorting • Check harness connector for contact corrosion or water damage • Check resistance of solenoid valves may not be up to specification • Wires to Transmission Control Module (TCM) may have ground out or open • Transmission Control Module (TCM) has failed • ECM has failed
DTC: P1828 **Passat:** 2.0L (BPY); **Jetta:** 1.8L (AWD, AWP, AWW), 2.5L (BGP, BGQ), 2.8L (AFP, BDF); **Golf:** 1.8L (AWD, AWW); **GTI:** 1.8L (AWD, AWP, AWW), 2.8L (AFP, BDF); **NB:** 1.8L (APH, AWP, AWV, BNU, BKF), 2.0L (AEG, AVH, AZG, BEV); **Nbvert:** (BDG) **Transmissions:** All	**Pressure Control Solenoid 4 Electrical Conditions:** Engine started, vehicle driven with the solenoid applied, and the ECM detected an unexpected voltage condition on the SS1/C solenoid circuit was incorrect during the test. **Possible Causes:** • Solenoid valves in valve body are faulty • Solenoid circuit is shorting to ground • Solenoid circuit is open • TCM has failed or wiring is shorting • Check harness connector for contact corrosion or water damage • Check resistance of solenoid valves may not be up to specification • Wires to Transmission Control Module (TCM) may have ground out or open • Transmission Control Module (TCM) has failed • ECM has failed

DTC	Trouble Code Title, Conditions & Possible Causes
DTC: P1847 **Passat:** 2.0L (BPY); **NB:** 1.8L (APH, AWP, AWV, BNU, BKF) **Transmissions:** All	**Please Check DTC Memory of Brake System ECU Conditions:** The ECU detected a memory fault of the brake system. The Engine Control Module (ECM) communicates with all databus-capable control modules via a CAN databus. These databus-capable control modules are connected via two data bus wires which are twisted together (CAN_High and CAN_Low), and exchange information (messages). Missing information on the databus is recognized as a malfunction and stored. Trouble-free operation of the CAN-bus requires that it have a terminal resistance. This central terminal resistor is located in the Engine Control Module (ECM). **Possible Causes:** • Fuses on E-box in engine compartment, left side, may be faulty • Check harness connector for contact corrosion or water damage • Check resistance of solenoid valves may not be up to specification • Solenoid valve may be faulty • Wires to Transmission Control Module (TCM) may have ground out or open • Transmission Control Module (TCM) has failed
DTC: P1850 **Passat:** 2.0L (BPY), 2.8L (AAA); **Jetta:** 1.8L (AWD, AWP, AWW), 2.0 (ABA, AVH, AZG, BBW, BEV), 2.5L (BGP, BGQ), 2.8L (AAA, AFP, BDF); **Golf:** 1.8L (AWD, AWW), 2.0L (ABA, AVH, AZG, BEV); **GTI:** 1.8L (AWD, AWP, AWW), 2.0L (ABA), 2.8L (AAA, AFP, BDF); **NB:** 1.8L (APH, AWP, AWV, BNU, BKF), 2.0L (AVH, AZG, BEV); **Nbvert:** (BGD); **Cabrio:** 2.0L (ABA) **Transmissions:** All	**Data BUS Powertrain Missing Message from Engine Control Conditions:** The Engine Control Module (ECM) communicates with all databus-capable control modules via a CAN databus. These databus-capable control modules are connected via two data bus wires which are twisted together (CAN_High and CAN_Low), and exchange information (messages). Missing information on the databus is recognized as a malfunction and stored. Trouble-free operation of the CAN-bus requires that it have a terminal resistance. This central terminal resistor is located in the Engine Control Module (ECM). **Possible Causes:** • Check the Terminal resistance for CAN-bus • Data-Bus wires have short • Data-Bus components are malfunctioning • ECM has failed
DTC: P1853 **Passat:** 2.0L (BPY); **Jetta:** 1.8L (AWD, AWP, AWW); **Golf:** 1.8L (AWD, AWW); **GTI:** 1.8L (AWD, AWP, AWW); **NB:** 1.8L (APH, AWP, AWV, BNU, BKF) **Transmissions:** All	**Data BUS Powertrain Implausible Message from Brake Control Conditions:** The Engine Control Module (ECM) communicates with all databus-capable control modules via a CAN databus. These databus-capable control modules are connected via two data bus wires which are twisted together (CAN_High and CAN_Low), and exchange information (messages). Missing information on the databus is recognized as a malfunction and stored. Trouble-free operation of the CAN-bus requires that it have a terminal resistance. This central terminal resistor is located in the Engine Control Module (ECM). **Possible Causes:** • Check the Terminal resistance for CAN-bus • Data-Bus wires have short • Data-Bus components are malfunctioning • ECM has failed
DTC: P1854 **Passat:** 2.0L (BPY), 2.8L (AAA); **Jetta:** 1.8L (AWD, AWP, AWW), 2.0 (ABA, AVH, AZG, BBW, BEV), 2.5L (BGP, BGQ), 2.8L (AAA, AFP, BDF); **Golf:** 1.8L (AWD, AWW), 2.0L (ABA, AVH, AZG, BEV); **GTI:** 1.8L (AWD, AWP, AWW), 2.0L (ABA), 2.8L (AAA, AFP, BDF); **NB:** 1.8L (APH, AWP, AWV, BNU, BKF), 2.0L (AVH, AZG, BEV); **Nbvert:** (BGD); **Cabrio:** 2.0L (ABA) **Transmissions:** All	**Data BUS Powertrain Hardware Defective Conditions:** The Engine Control Module (ECM) communicates with all databus-capable control modules via a CAN databus. These databus-capable control modules are connected via two data bus wires which are twisted together (CAN_High and CAN_Low), and exchange information (messages). Missing information on the databus is recognized as a malfunction and stored. Trouble-free operation of the CAN-bus requires that it have a terminal resistance. This central terminal resistor is located in the Engine Control Module (ECM). **Possible Causes:** • Check the Terminal resistance for CAN-bus • Data-Bus wires have short • Data-Bus components are malfunctioning • ECM has failed

DTC	Trouble Code Title, Conditions & Possible Causes
DTC: P1855 **Passat:** 2.0L (BPY), 2.8L (AAA); **Jetta:** 1.8L (AWD, AWP, AWW), 2.0 (ABA, AVH, AZG, BBW, BEV), 2.5L (BGP, BGQ), 2.8L (AAA, AFP, BDF); **Golf:** 1.8L (AWD, AWW), 2.0L (ABA, AVH, AZG, BEV); **GTI:** 1.8L (AWD, AWP, AWW), 2.0L (ABA), 2.8L (AAA, AFP, BDF); **NB:** 1.8L (APH, AWP, AWV, BNU, BKF), 2.0L (AVH, AZG, BEV); **Nbvert:** (BGD); **Cabrio:** 2.0L (ABA) **Transmissions:** All	**Data BUS Powertrain Software Version Control Conditions:** The Engine Control Module (ECM) communicates with all databus-capable control modules via a CAN databus. These databus-capable control modules are connected via two data bus wires which are twisted together (CAN_High and CAN_Low), and exchange information (messages). Missing information on the databus is recognized as a malfunction and stored. Trouble-free operation of the CAN-bus requires that it have a terminal resistance. This central terminal resistor is located in the Engine Control Module (ECM). **Possible Causes:** • Check the Terminal resistance for CAN-bus • Data-Bus wires have short • Data-Bus components are malfunctioning • ECM has failed
DTC: P1857 **Passat:** 1.8L (ATW, AUG), 2.0L (BPY), 2.8L (AAA); **Jetta:** 1.8L (AWD, AWP, AWW), 2.0 (ABA, AVH, AZG, BBW, BEV), 2.5L (BGP, BGQ), 2.8L (AAA, AFP, BDF); **Golf:** 1.8L (AWD, AWW), 2.0L (ABA, AVH, AZG, BEV); **GTI:** 1.8L (AWD, AWP, AWW), 2.0L (ABA), 2.8L (AAA, AFP, BDF); **NB:** 1.8L (APH, AWP, AWV, BNU, BKF), 2.0L (AVH, AZG, BEV); **Nbvert:** (BGD); **Cabrio:** 2.0L (ABA) **Transmissions:** All	**Load Signal Error Message From Engine Control Conditions:** The Engine Control Module (ECM) communicates with all databus-capable control modules via a CAN databus. These databus-capable control modules are connected via two data bus wires which are twisted together (CAN_High and CAN_Low), and exchange information (messages). Missing information on the databus is recognized as a malfunction and stored. Trouble-free operation of the CAN-bus requires that it have a terminal resistance. This central terminal resistor is located in the Engine Control Module (ECM). **Possible Causes:** • Intake Manifold Runner Position Sensor is faulty • Intake system has leaks (false air) • Motor for intake flap is faulty • Mass Air Flow (MAF) sensor has failed • ECM has failed
DTC: P1861 **Passat:** 2.0L (BPY); **Jetta:** 1.8L (AWD, AWP, AWW), 2.5L (BGP, BGQ), 2.8L (AFP, BDF); **Golf:** 1.8L (AWD, AWW); **GTI:** 1.8L (AWD, AWP, AWW), 2.8L (AFP, BDF); **NB:** 1.8L (APH, AWP, AWV, BNU, BKF), 2.0L (AEG, AVH, AZG, BEV); **Nbvert:** (BDG) **Transmissions:** All	**Throttle Position Sensor Message from ECM Conditions:** The Engine Control Module (ECM) communicates with all databus-capable control modules via a CAN databus. These databus-capable control modules are connected via two data bus wires which are twisted together (CAN_High and CAN_Low), and exchange information (messages). Missing information on the databus is recognized as a malfunction and stored. Trouble-free operation of the CAN-bus requires that it have a terminal resistance. This central terminal resistor is located in the Engine Control Module (ECM). **Note: Both the Throttle Position (TP) Sensor and Accelerator Pedal Position Sensor 2 are located at the accelerator pedal module and communicate the driver's intentions to the ECM completely independently of each other. Both sensors are stored in one housing.** **Possible Causes:** • Throttle Position (TP) Sensor has failed • Accelerator Pedal Position Sensor 2 has failed • ECM has failed • Ground (GND) connections between engine and chassis must be OK • Engine Control Module (ECM) may not connected
DTC: P1866 **Passat:** 2.0L (BPY), 2.8L (AAA); **Jetta:** 1.8L (AWD, AWP, AWW), 2.0 (ABA, AVH, AZG, BBW, BEV), 2.5L (BGP, BGQ), 2.8L (AAA, AFP, BDF); **Golf:** 1.8L (AWD, AWW), 2.0L (ABA, AVH, AZG, BEV); **GTI:** 1.8L (AWD, AWP, AWW), 2.0L (ABA), 2.8L (AAA, AFP, BDF); **NB:** 1.8L (APH, AWP, AWV, BNU, BKF), 2.0L (AVH, AZG, BEV); **Nbvert:** (BGD); **Cabrio:** 2.0L (ABA) **Transmissions:** All	**Data Bus Powertrain Missing Messages Conditions:** The Engine Control Module (ECM) communicates with all databus-capable control modules via a CAN databus. These databus-capable control modules are connected via two data bus wires which are twisted together (CAN_High and CAN_Low), and exchange information (messages). Missing information on the databus is recognized as a malfunction and stored. Trouble-free operation of the CAN-bus requires that it have a terminal resistance. This central terminal resistor is located in the Engine Control Module (ECM). **Possible Causes:** • Check the Terminal resistance for CAN-bus • Data-Bus wires have short • Data-Bus components are malfunctioning • ECM has failed

DTC	Trouble Code Title, Conditions & Possible Causes
DTC: P1912 **Passat:** 1.8 (AEB, AWM); **Jetta:** 2.8L (AFP, BDF); **GTI:** 2.8L (AFP, BDF) **Transmissions:** All	**Brake Booster Pressure Sensor Short Circuit to B+ Conditions:** Key on or engine running, the ECM detected an error with the brake booster pressure sensor signal. **Possible Causes:** • Circuit short to ground or open • Brake booster pressure sensor has failed or is dirty • ECM has failed
DTC: P1913 **Passat:** 1.8 (AEB, AWM); **Jetta:** 2.8L (AFP, BDF); **GTI:** 2.8L (AFP, BDF) **Transmissions:** All	**Brake Booster Pressure Sensor Short Circuit to Ground Conditions:** Key on or engine running, the ECM detected an error with the brake booster pressure sensor signal. **Possible Causes:** • Circuit short to ground or open • Brake booster pressure sensor has failed or is dirty • ECM has failed

DTC	Trouble Code Title, Conditions & Possible Causes:
DTC: P2004 **Passat:** 2.0L (BPY) **Transmissions:** All	**Intake Manifold Runner Control Stuck Open Bank 1 Conditions:** Engine started, battery voltage must be at least 11.5v, all electrical components must be off, the ground between the engine and the chassis must be well connected. The ECM detected an unexpected voltage condition on the Intake Manifold Runner Control circuit during the CCM test period (i.e., the valve may be stuck open). **Note: Intake Flap Motor and Intake Manifold Runner Position Sensor are one component and cannot be replaced individually.** **Possible Causes:** • Test for a sticking Accelerator or speed control cable condition: Turn the key off and disconnect accelerator and speed control cable from the throttle body. Rotate the throttle body linkage to determine if it rotates freely (the throttle body may have failed). • Check the air cleaner and air inlet assembly for restrictions • Check the IAC motor response (it may be damaged or sticking) • Check the PCV system (valve and hoses) for leaks or plugging • Check for signs of vacuum leaks in the engine or components • Test TP sensor signal (due a sweep test at key on, engine off)
DTC: P2008 **Passat:** 2.0L (BPY); **Jetta:** 2.0L (BPY), 2.5L (BGP, BGQ) **Transmissions:** All	**Intake Manifold Runner Control Circuit/Open Bank 1 Conditions:** Engine started, battery voltage must be at least 11.5v, all electrical components must be off, the ground between the engine and the chassis must be well connected. The ECM detected an unexpected voltage condition on the Intake Manifold Runner Control circuit during the CCM test period (i.e., the valve may be stuck open). **Note: Intake Flap Motor and Intake Manifold Runner Position Sensor are one component and cannot be replaced individually.** **Possible Causes:** • Accelerator or speed control cable sticking or binding. To test for this condition, turn the key off. Then disconnect the accelerator and speed control cable from the throttle body. Then rotate the throttle body linkage to determine if it rotates freely. If it is sticking, the throttle body may need replacement. • Check the air cleaner and air inlet assembly for restrictions • Check the IAC motor response (it may be damaged or sticking) • Check the PCV system (valve and hoses) for leaks or plugging • Check for signs of vacuum leaks in the engine or components • Test TP sensor signal
DTC:P2009 **Passat:** 2.0L (BPY) **Transmissions:** All	**Intake Manifold Runner Control Circuit Low Bank 1 Conditions:** Engine started, battery voltage must be at least 11.5v, all electrical components must be off, the ground between the engine and the chassis must be well connected. The ECM detected an unexpected voltage condition on the Intake Manifold Runner Control circuit during the CCM test period (i.e., the valve may be stuck open). **Note: Intake Flap Motor and Intake Manifold Runner Position Sensor are one component and cannot be replaced individually.** **Possible Causes:** • Accelerator or speed control cable sticking or binding. To test for this condition, turn the key off. Then disconnect the accelerator and speed control cable from the throttle body. Then rotate the throttle body linkage to determine if it rotates freely. If it is sticking, the throttle body may need replacement. • Check the air cleaner and air inlet assembly for restrictions • Check the IAC motor response (it may be damaged or sticking) • Check the PCV system (valve and hoses) for leaks or plugging • Check for signs of vacuum leaks in the engine or components • Test TP sensor signal
DTC:P2014 **Passat:** 2.0L (BPY) **Transmissions:** All	**Intake Manifold Runner Position Sensor/Switch Circuit Bank 1 Conditions:** Engine started, battery voltage must be at least 11.5v, all electrical components must be off, the ground between the engine and the chassis must be well connected. The ECM detected an unexpected voltage condition on the Intake Manifold Runner Control circuit during the CCM test period (i.e., the valve may be stuck open). **Note: Intake Flap Motor and Intake Manifold Runner Position Sensor are one component and cannot be replaced individually.** **Possible Causes:** • Accelerator or speed control cable sticking or binding. To test for this condition, turn the key off. Then disconnect the accelerator and speed control cable from the throttle body. Then rotate the throttle body linkage to determine if it rotates freely. If it is sticking, the throttle body may need replacement. • Check the air cleaner and air inlet assembly for restrictions • Check the IAC motor response (it may be damaged or sticking) • Check the PCV system (valve and hoses) for leaks or plugging • Check for signs of vacuum leaks in the engine or components • Test TP sensor signal

DTC	Trouble Code Title, Conditions & Possible Causes
DTC:P2015 **Passat:** 2.0L (BPY) **Transmissions:** All	**Intake Manifold Runner Position Sensor/Switch Circuit Range/Performance Bank 1 Conditions:** Engine started, battery voltage must be at least 11.5v, all electrical components must be off, the ground between the engine and the chassis must be well connected. The ECM detected an unexpected voltage condition on the Intake Manifold Runner Control circuit during the CCM test period (i.e., the valve may be stuck open). **Note: Intake Flap Motor and Intake Manifold Runner Position Sensor are one component and cannot be replaced individually.** **Possible Causes:** • Accelerator or speed control cable sticking or binding. To test for this condition, turn the key off. Then disconnect the accelerator and speed control cable from the throttle body. Then rotate the throttle body linkage to determine if it rotates freely. If it is sticking, the throttle body may need replacement. • Check the air cleaner and air inlet assembly for restrictions • Check the IAC motor response (it may be damaged or sticking) • Check the PCV system (valve and hoses) for leaks or plugging • Check for signs of vacuum leaks in the engine or components • Test TP sensor signal
DTC:P2016 **Passat:** 2.0L (BPY) **Transmissions:** All	**Intake Manifold Runner Position Sensor/Switch Circuit Low Bank 1 Conditions:** Engine started, battery voltage must be at least 11.5v, all electrical components must be off, the ground between the engine and the chassis must be well connected. The ECM detected an unexpected voltage condition on the Intake Manifold Runner Control circuit during the CCM test period (i.e., the valve may be stuck open). **Note: Intake Flap Motor and Intake Manifold Runner Position Sensor are one component and cannot be replaced individually.** Accelerator or speed control cable sticking or binding. To test for this condition, turn the key off. Then disconnect the accelerator and speed control cable from the throttle body. Then rotate the throttle body linkage to determine if it rotates freely. If it is sticking, the throttle body may need replacement. • Check the air cleaner and air inlet assembly for restrictions • Check the IAC motor response (it may be damaged or sticking) • Check the PCV system (valve and hoses) for leaks or plugging • Check for signs of vacuum leaks in the engine or components • Test TP sensor signal
DTC:P2017 **Passat:** 2.0L (BPY) **Transmissions:** All	**Intake Manifold Runner Position Sensor/Switch Circuit High Bank 1 Conditions:** Engine started, battery voltage must be at least 11.5v, all electrical components must be off, the ground between the engine and the chassis must be well connected. The ECM detected an unexpected voltage condition on the Intake Manifold Runner Control circuit during the CCM test period (i.e., the valve may be stuck open). **Note: Intake Flap Motor and Intake Manifold Runner Position Sensor are one component and cannot be replaced individually.** **Possible Causes:** • Accelerator or speed control cable sticking or binding. To test for this condition, turn the key off. Then disconnect the accelerator and speed control cable from the throttle body. Then rotate the throttle body linkage to determine if it rotates freely. If it is sticking, the throttle body may need replacement. • Check the air cleaner and air inlet assembly for restrictions • Check the IAC motor response (it may be damaged or sticking) • Check the PCV system (valve and hoses) for leaks or plugging • Check for signs of vacuum leaks in the engine or components • Test TP sensor signal
DTC:P2088 **2T ECM, MIL: No** **Passat:** 1.8L (AUG), 2.0L (BPY); **NB:** 1.8L (AWP, AWV, BNU, BKF) **Transmissions:** All	**"A" Camshaft Position Control Circuit Low Bank 1 Conditions:** Key on or engine running; and the ECM detected an unexpected voltage condition on the Camshaft Position Control circuit during the CCM test period. The relative position between the camshaft and crankshaft needs to be optimal so the engine has better torque, fuel economy and emissions. **Note: camshaft adjustment is load- and RPM dependant. The electrical camshaft adjustment valve 1 switches oil pressure onto camshaft adjuster (mechanical adjustment mechanism), which adjusts the camshaft.** **Possible Causes:** • Camshaft position control wiring harness connector is damaged or open • Camshaft adjustment valve has failed • Circuit is open or grounded • Assembly is damaged or it has failed (an open circuit) • ECM power supply relay has failed • ECM has failed

DTC	Trouble Code Title, Conditions & Possible Causes
DTC:P2089 **2T ECM, MIL: No** **Passat:** 1.8L (AUG), 2.0L (BPY); **Jetta:** 2.5L (BGP, BGQ); **NB:** 1.8L (AWP, AWV, BNU, BKF) **Transmissions:** All	**"A" Camshaft Position Control Circuit High Bank 1 Conditions:** Key on or engine running; and the ECM detected an unexpected voltage condition on the Camshaft Position Control circuit during the CCM test period. The relative position between the camshaft and crankshaft needs to be optimal so the engine has better torque, fuel economy and emissions. **Note: camshaft adjustment is load- and RPM dependant. The electrical camshaft adjustment valve 1 switches oil pressure onto camshaft adjuster (mechanical adjustment mechanism), which adjusts the camshaft.** **Possible Causes:** • Camshaft position control wiring harness connector is damaged or open • Camshaft adjustment valve has failed • Circuit is open or grounded • Assembly is damaged or it has failed (an open circuit) • ECM power supply relay has failed • ECM has failed
DTC:P2090 **2T ECM, MIL: No** **Passat:** 2.0L (BPY); **Jetta:** 2.8L (AFP, BDF); **GTI:** 2.8L (AFP, BDF) **Transmissions:** All	**"B" Camshaft Position Control Circuit Low Bank 1 Conditions:** Key on or engine running; and the ECM detected an unexpected voltage condition on the Camshaft Position Control circuit during the CCM test period. The relative position between the camshaft and crankshaft needs to be optimal so the engine has better torque, fuel economy and emissions. **Note: camshaft adjustment is load- and RPM dependant. The electrical camshaft adjustment valve 1 switches oil pressure onto camshaft adjuster (mechanical adjustment mechanism), which adjusts the camshaft.** **Possible Causes:** • Camshaft position control wiring harness connector is damaged or open • Camshaft adjustment valve has failed • Circuit is open or grounded • Assembly is damaged or it has failed (an open circuit) • ECM power supply relay has failed • ECM has failed
DTC:P2091 **2T ECM, MIL: No** **Passat:** 2.0L (BPY); **Jetta:** 2.8L (AFP, BDF); **GTI:** 2.8L (AFP, BDF) **Transmissions:** All	**"B" Camshaft Position Control Circuit High Bank 1 Conditions:** Key on or engine running; and the ECM detected an unexpected voltage condition on the Camshaft Position Control circuit during the CCM test period. The relative position between the camshaft and crankshaft needs to be optimal so the engine has better torque, fuel economy and emissions. **Note: camshaft adjustment is load- and RPM dependant. The electrical camshaft adjustment valve 1 switches oil pressure onto camshaft adjuster (mechanical adjustment mechanism), which adjusts the camshaft.** **Possible Causes:** • Camshaft position control wiring harness connector is damaged or open • Camshaft adjustment valve has failed • Circuit is open or grounded • Assembly is damaged or it has failed (an open circuit) • ECM power supply relay has failed • ECM has failed
DTC:P2094 **2T ECM, MIL: No** **Passat:** 2.0L (BPY) **Transmissions:** All	**"B" Camshaft Position Control Circuit Low Bank 2 Conditions:** Key on or engine running; and the ECM detected an unexpected voltage condition on the Camshaft Position Control circuit during the CCM test period. The relative position between the camshaft and crankshaft needs to be optimal so the engine has better torque, fuel economy and emissions. **Note: camshaft adjustment is load- and RPM dependant. The electrical camshaft adjustment valve 1 switches oil pressure onto camshaft adjuster (mechanical adjustment mechanism), which adjusts the camshaft.** **Possible Causes:** • Camshaft position control wiring harness connector is damaged or open • Camshaft adjustment valve has failed • Circuit is open or grounded • Assembly is damaged or it has failed (an open circuit) • ECM power supply relay has failed • ECM has failed

DTC	Trouble Code Title, Conditions & Possible Causes
DTC:P2095 **2T ECM, MIL: No** **Passat:** 2.0L (BPY) **Transmissions:** All	**"B" Camshaft Position Control Circuit High Bank 2 Conditions:** Key on or engine running; and the ECM detected an unexpected voltage condition on the Camshaft Position Control circuit during the CCM test period. The relative position between the camshaft and crankshaft needs to be optimal so the engine has better torque, fuel economy and emissions. **Note: camshaft adjustment is load- and RPM dependant. The electrical camshaft adjustment valve 1 switches oil pressure onto camshaft adjuster (mechanical adjustment mechanism), which adjusts the camshaft.** **Possible Causes:** • Camshaft position control wiring harness connector is damaged or open • Camshaft adjustment valve has failed • Circuit is open or grounded • Assembly is damaged or it has failed (an open circuit) • ECM power supply relay has failed • ECM has failed
DTC:P2096 **Passat:** 1.8 (AEB, ATW, AUG, AWM), 2.0L (BPY), 2.8L (AHA, ATQ); **Jetta:** 2.0 (ABA, AEG, AVH, AZG, BBW, BEV, BPY), 2.5L (BGP, BGQ), 2.8L (AFP, BDF); **Golf:** 2.0L (ABA); **GTI:** 2.0L (ABA, AEG), 2.8L (AFP, BDF); **NB:** 1.8L (APH, AWP, AWV, BNU, BKF); **Cabrio:** 2.0L (ABA) **Transmissions:** All	**Post Catalyst Fuel Trim System Too Lean (Bank 1) Conditions:** Engine started, battery voltage must be at least 11.5v, all electrical components must be off, the ground between the engine and the chassis must be well connected, the exhaust system must be properly sealed between the catalytic converter and the cylinder head, and the oxygen sensor heater for oxygen sensor before the catalytic converter must be properly functioning. The ECM detected a problem with the fuel mixture. **Note: For resistance testing of sensor heating, oxygen sensor should be cooled to ambient temperature. High temperatures at oxygen sensor may lead to inaccurate measurements.** **Possible Causes:** • Oxygen sensor (before catalytic converter) is faulty • Oxygen sensor (behind catalytic converter) is faulty • Oxygen sensor heater (before catalytic converter) is faulty • Oxygen sensor heater (behind catalytic converter) is faulty • Check circuits for shorts to each other, ground or power • ECM has failed
DTC:P2097 **Passat:** 1.8 (AEB, ATW, AUG, AWM), 2.0L (BPY), 2.8L (AHA, ATQ); **Jetta:** 2.0 (ABA, AEG, AVH, AZG, BBW, BEV, BPY), 2.5L (BGP, BGQ), 2.8L (AFP, BDF); **Golf:** 2.0L (ABA); **GTI:** 2.0L (ABA, AEG), 2.8L (AFP, BDF); **NB:** 1.8L (APH, AWP, AWV, BNU, BKF); **Cabrio:** 2.0L (ABA) **Transmissions:** All	**Post Catalyst Fuel Trim System Too Rich (Bank 1) Conditions:** Engine started, battery voltage must be at least 11.5v, all electrical components must be off, the ground between the engine and the chassis must be well connected, the exhaust system must be properly sealed between the catalytic converter and the cylinder head, and the oxygen sensor heater for oxygen sensor before the catalytic converter must be properly functioning. The ECM detected a problem with the fuel mixture. **Note: For resistance testing of sensor heating, oxygen sensor should be cooled to ambient temperature. High temperatures at oxygen sensor may lead to inaccurate measurements.** **Possible Causes:** • Oxygen sensor (before catalytic converter) is faulty • Oxygen sensor (behind catalytic converter) is faulty • Oxygen sensor heater (before catalytic converter) is faulty • Oxygen sensor heater (behind catalytic converter) is faulty • Check circuits for shorts to each other, ground or power • ECM has failed
DTC:P2098 **Passat:** 2.8L (AHA, ATQ) **Transmissions:** All	**Post Catalyst Fuel Trim System Too Lean (Bank 2) Conditions:** Engine started, battery voltage must be at least 11.5v, all electrical components must be off, the ground between the engine and the chassis must be well connected, the exhaust system must be properly sealed between the catalytic converter and the cylinder head, and the oxygen sensor heater for oxygen sensor before the catalytic converter must be properly functioning. The ECM detected a problem with the fuel mixture. **Note: For resistance testing of sensor heating, oxygen sensor should be cooled to ambient temperature. High temperatures at oxygen sensor may lead to inaccurate measurements.** **Possible Causes:** • Oxygen sensor (before catalytic converter) is faulty • Oxygen sensor (behind catalytic converter) is faulty • Oxygen sensor heater (before catalytic converter) is faulty • Oxygen sensor heater (behind catalytic converter) is faulty • Check circuits for shorts to each other, ground or power • ECM has failed

DTC	Trouble Code Title, Conditions & Possible Causes
DTC:P2099 **Passat:** 2.8L (AHA, ATQ) **Transmissions:** All	**Post Catalyst Fuel Trim System Too Rich (Bank 2) Conditions:** Engine started, battery voltage must be at least 11.5v, all electrical components must be off, the ground between the engine and the chassis must be well connected, the exhaust system must be properly sealed between the catalytic converter and the cylinder head, and the oxygen sensor heater for oxygen sensor before the catalytic converter must be properly functioning. The ECM detected a problem with the fuel mixture. **Note: For resistance testing of sensor heating, oxygen sensor should be cooled to ambient temperature. High temperatures at oxygen sensor may lead to inaccurate measurements.** **Possible Causes:** • Oxygen sensor (before catalytic converter) is faulty • Oxygen sensor (behind catalytic converter) is faulty • Oxygen sensor heater (before catalytic converter) is faulty • Oxygen sensor heater (behind catalytic converter) is faulty • Check circuits for shorts to each other, ground or power • ECM has failed
DTC:P2101 **2T, MIL: Yes** **Passat:** 1.8 (AEB, ATW, AUG, AWM), 2.0L (BPY), 2.8L (AHA, ATQ), 4.0L (BDP); **Jetta:** 1.8L (AWD, AWP, AWW), 2.0L (AEG, AVH, AZG, BBW, BEV), 2.5L (BGP, BGQ), 2.8L (AFP, BDF); **Golf:** 1.8L (AWD, AWW), 2.0L (AEG, AVH, AZG, BEV); **GTI:** 1.8L (AWD, AWP, AWW), 2.0L (AEG), 2.8L (AFP, BDF); **NB:** 1.8L (APH, AWP, AWV, BNU, BKF), 2.0L (AEG, AVH, AZG, BEV); **Nbvert:** (BDG) **Transmissions:** All	**Throttle Actuator Control Motor Range/Performance Conditions:** Engine started, battery voltage must be at least 11.5v, all electrical components must be off, parking brake must be engaged (to keep daytime driving lights off), automatic transmission selector must be in park, the exhaust system must be properly sealed between the catalytic converter and the cylinder head, coolant temperature must be at least 80 degrees Celsius. The ECM detected an unexpected low or high voltage condition on the Throttle Actuator Control Motor (TACM) circuit during the CCM test. **Note: The throttle valve activation occurs via an electric motor (throttle drive) in the throttle valve control module. It is activated by the Engine Control Module (ECM) according to specifications of the two sensors, Throttle Position (TP) Sensor and Sender 2 for accelerator pedal position.** **Possible Causes:** • TACM wiring harness connector is damaged or open • TACM wiring may be crossed in the wire harness assembly • TACM (motor) circuit is open, or TACM assembly is damaged (possible open circuit) • TACM or the Throttle Valve is dirty • Throttle Position sensor has failed • ECM has failed
DTC:P2106 **Passat:** 1.8L (ATW, AUG), 2.0L (BPY), 4.0L (BDP); **Jetta:** 1.8L (AWD, AWP, AWW), 2.0L (AEG, AVH, AZG, BBW, BEV, BPY), 2.5L (BGP, BGQ), 2.8L (AFP, BDF); **Golf:** 1.8L (AWD, AWW), 2.0L (AEG, AVH, AZG, BEV); **GTI:** 1.8L (AWD, AWP, AWW), 2.0L (AEG), 2.8L (AFP, BDF); **NB:** 1.8L (APH, AWP, AWV, BNU, BKF), 2.0L (AEG, AVH, AZG, BEV); **Nbvert:** (BDG) **Transmissions:** All	**Throttle Actuator Control System – Forced Limited Power Conditions** Engine started, battery voltage must be at least 11.5v, all electrical components must be off, parking brake must be engaged (to keep daytime driving lights off), automatic transmission selector must be in park, the exhaust system must be properly sealed between the catalytic converter and the cylinder head, coolant temperature must be at least 80 degrees Celsius. The ECM detected an unexpected low or high voltage condition on the Throttle Actuator Control Motor (TACM) circuit during the CCM test. **Note: The throttle valve activation occurs via an electric motor (throttle drive) in the throttle valve control module. It is activated by the Engine Control Module (ECM) according to specifications of the two sensors, Throttle Position (TP) Sensor and Sender 2 for accelerator pedal position.** **Possible Causes:** • TACM wiring harness connector is damaged or open • TACM wiring may be crossed in the wire harness assembly • TACM (motor) circuit is open, or TACM assembly is damaged (possible open circuit) • TACM or the Throttle Valve is dirty • Throttle Position sensor has failed • ECM has failed
DTC:P2108 2T ECM, MIL: No **Passat:** 2.0L (BPY) **Transmissions:** All	**Throttle Actuator Control Motor Performance Conditions:** Engine started, battery voltage must be at least 11.5v, all electrical components must be off, parking brake must be engaged (to keep daytime driving lights off), automatic transmission selector must be in park, the exhaust system must be properly sealed between the catalytic converter and the cylinder head, coolant temperature must be at least 80 degrees Celsius. The ECM detected an unexpected low or high voltage condition on the Throttle Actuator Control Motor (TACM) circuit during the CCM test. **Note: The throttle valve activation occurs via an electric motor (throttle drive) in the throttle valve control module. It is activated by the Engine Control Module (ECM) according to specifications of the two sensors, Throttle Position (TP) Sensor and Sender 2 for accelerator pedal position.** **Possible Causes:** • TACM wiring harness connector is damaged or open • TACM wiring may be crossed in the wire harness assembly • TACM (motor) circuit is open, or TACM assembly is damaged (possible open circuit) • TACM or the Throttle Valve is dirty • Throttle Position sensor has failed • ECM has failed

DTC	Trouble Code Title, Conditions & Possible Causes
DTC:P2110 2T ECM, MIL: No **Passat:** 2.0L (BPY) **Transmissions:** All	**Throttle Actuator Control System – Forced Limited RPM Conditions:** Engine started, battery voltage must be at least 11.5v, all electrical components must be off, parking brake must be engaged (to keep daytime driving lights off), automatic transmission selector must be in park, the exhaust system must be properly sealed between the catalytic converter and the cylinder head, coolant temperature must be at least 80 degrees Celsius. The ECM detected an unexpected low or high voltage condition on the Throttle Actuator Control Motor (TACM) circuit during the CCM test. **Note: The throttle valve activation occurs via an electric motor (throttle drive) in the throttle valve control module. It is activated by the Engine Control Module (ECM) according to specifications of the two sensors, Throttle Position (TP) Sensor and Sender 2 for accelerator pedal position.** **Possible Causes:** • TACM wiring harness connector is damaged or open • TACM wiring may be crossed in the wire harness assembly • TACM (motor) circuit is open, or TACM assembly is damaged (possible open circuit) • TACM or the Throttle Valve is dirty • Throttle Position sensor has failed • ECM has failed
DTC:P2122 **2T, MIL: Yes** **Passat:** 1.8 (AEB, ATW, AUG, AWM), 2.0 (BPY), 2.8L (AHA, ATQ), 4.0L (BDP); **Jetta:** 1.8L (AWD, AWP, AWW), 2.0L (AEG, AVH, AZG, BBW, BEV, BPY), 2.5L (BGP, BGQ), 2.8L (AFP, BDF); **Golf:** 1.8L (AWD, AWW), 2.0L (AEG, AVH, AZG, BEV); **GTI:** 1.8L (AWD, AWP, AWW), 2.0L (AEG), 2.8L (AFP, BDF); **NB:** 1.8L (APH, AWP, AWV, BNU, BKF), 2.0L (AEG, AVH, AZG, BEV); **Nbvert:** (BDG) **Transmissions:** All	**Accelerator Pedal Position Sensor 'D' Circuit Low Input Conditions:** Engine started, battery voltage at least 11.5v, all electrical components off, ground connections between engine and chassis well connected, the ECM detected that the accelerator pedal position sensor signal was outside the parameters to function normally. **Note: Both the Throttle Position (TP) Sensor and Accelerator Pedal Position Sensor are located at the accelerator pedal module and communicate the driver's intentions to the ECM completely independently of each other. Both sensors are stored in one housing.** **Possible Causes:** • Ground between engine and chassis may be broken • Throttle position sensor may have failed • Accelerator Pedal Position Sensor has failed • Throttle position sensor wiring may have shorted • Throttle position sensor has failed • Faulty voltage supply • ECM has failed
DTC:P2123 **2T, MIL: Yes** **Passat:** 1.8 (AEB, ATW, AUG, AWM), 2.0 (BPY), 2.8L (AHA, ATQ), 4.0L (BDP); **Jetta:** 1.8L (AWD, AWP, AWW), 2.0L (AEG, AVH, AZG, BBW, BEV), 2.5L (BGP, BGQ), 2.8L (AFP, BDF); **Golf:** 1.8L (AWD, AWW), 2.0L (AEG, AVH, AZG, BEV); **GTI:** 1.8L (AWD, AWP, AWW), 2.0L (AEG), 2.8L (AFP, BDF); **NB:** 1.8L (APH, AWP, AWV, BNU, BKF), 2.0L (AEG, AVH, AZG, BEV); **Nbvert:** (BDG) **Transmissions:** All	**Accelerator Pedal Position Sensor 'D' Circuit High Input Conditions:** Engine started, battery voltage at least 11.5v, all electrical components off, ground connections between engine and chassis well connected, the ECM detected that the accelerator pedal position sensor signal was outside the parameters to function normally. **Note: Both the Throttle Position (TP) Sensor and Accelerator Pedal Position Sensor are located at the accelerator pedal module and communicate the driver's intentions to the ECM completely independently of each other. Both sensors are stored in one housing.** **Possible Causes:** • Ground between engine and chassis may be broken • Throttle position sensor may have failed • Accelerator Pedal Position Sensor has failed • Throttle position sensor wiring may have shorted • Throttle position sensor has failed • Faulty voltage supply • ECM has failed

DTC	Trouble Code Title, Conditions & Possible Causes
DTC:P2127 **2T, MIL: Yes** **Passat:** 1.8 (AEB, ATW, AUG, AWM), 2.0 (BPY), 2.8L (AHA, ATQ), 4.0L (BDP); **Jetta:** 1.8L (AWD, AWP, AWW), 2.0L (AEG, AVH, AZG, BBW, BEV, BPY), 2.5L (BGP, BGQ), 2.8L (AFP, BDF); **Golf:** 1.8L (AWD, AWW), 2.0L (AEG, AVH, AZG, BEV); **GTI:** 1.8L (AWD, AWP, AWW), 2.0L (AEG), 2.8L (AFP, BDF); **NB:** 1.8L (APH, AWP, AWV, BNU, BKF), 2.0L (AEG, AVH, AZG, BEV); **Nbvert:** (BDG) **Transmissions:** All	**Accelerator Pedal Position Sensor 'E' Circuit Low Input Conditions:** Engine started, battery voltage at least 11.5v, all electrical components off, ground connections between engine and chassis well connected, the ECM detected that the accelerator pedal position sensor signal was outside the parameters to function normally. **Note: Both the Throttle Position (TP) Sensor and Accelerator Pedal Position Sensor are located at the accelerator pedal module and communicate the driver's intentions to the ECM completely independently of each other. Both sensors are stored in one housing.** **Possible Causes:** • Ground between engine and chassis may be broken • Throttle position sensor may have failed • Accelerator Pedal Position Sensor has failed • Throttle position sensor wiring may have shorted • Throttle position sensor has failed • Faulty voltage supply • ECM has failed
DTC:P2128 **2T, MIL: Yes** **Passat:** 1.8 (AEB, ATW, AUG, AWM), 2.0 (BPY), 2.8L (AHA, ATQ), 4.0L (BDP); **Jetta:** 1.8L (AWD, AWP, AWW), 2.0L (AEG, AVH, AZG, BBW, BEV, BPY), 2.5L (BGP, BGQ), 2.8L (AFP, BDF); **Golf:** 1.8L (AWD, AWW), 2.0L (AEG, AVH, AZG, BEV); **GTI:** 1.8L (AWD, AWP, AWW), 2.0L (AEG), 2.8L (AFP, BDF); **NB:** 1.8L (APH, AWP, AWV, BNU, BKF), 2.0L (AEG, AVH, AZG, BEV); **Nbvert:** (BDG) **Transmissions:** All	**Accelerator Pedal Position Sensor 'E' Circuit High Input Conditions:** Engine started, battery voltage at least 11.5v, all electrical components off, ground connections between engine and chassis well connected, the ECM detected that the accelerator pedal position sensor signal was outside the parameters to function normally. **Note: Both the Throttle Position (TP) Sensor and Accelerator Pedal Position Sensor are located at the accelerator pedal module and communicate the driver's intentions to the ECM completely independently of each other. Both sensors are stored in one housing.** **Possible Causes:** • Ground between engine and chassis may be broken • Throttle position sensor may have failed • Accelerator Pedal Position Sensor has failed • Throttle position sensor wiring may have shorted • Throttle position sensor has failed • Faulty voltage supply • ECM has failed
DTC:P2133 **2T, MIL: Yes** **Jetta:** 1.8L (AWD, AWP, AWW), 2.0L (AEG, AVH, AZG, BBW, BEV); **Golf:** 1.8L (AWD, AWW), 2.0L (AEG, AVH, AZG, BEV); **GTI:** 1.8L (AWD, AWP, AWW), 2.0L (AEG); **NB:** 2.0L (AEG, AVH, AZG, BEV); **Nbvert:** (BDG) **Transmissions:** All	**Accelerator Pedal Position Sensor 'F' Circuit High Input Conditions:** Engine started, battery voltage at least 11.5v, all electrical components off, ground connections between engine and chassis well connected, the ECM detected that the accelerator pedal position sensor signal was outside the parameters to function normally. **Note: Both the Throttle Position (TP) Sensor and Accelerator Pedal Position Sensor are located at the accelerator pedal module and communicate the driver's intentions to the ECM completely independently of each other. Both sensors are stored in one housing.** **Possible Causes:** • Ground between engine and chassis may be broken • Throttle position sensor may have failed • Accelerator Pedal Position Sensor has failed • Throttle position sensor wiring may have shorted • Throttle position sensor has failed • Faulty voltage supply • ECM has failed
DTC:P2138 **2T, MIL: Yes** **Passat:** 1.8 (AEB, ATW, AUG, AWM), 2.0 (BPY), 2.8L (AHA, ATQ), 4.0L (BDP); **Jetta:** 1.8L (AWD, AWP, AWW), 2.0L (AEG, AVH, AZG, BBW, BEV, BPY), 2.5L (BGP, BGQ), 2.8L (AFP, BDF); **Golf:** 1.8L (AWD, AWW), 2.0L (AEG, AVH, AZG, BEV); **GTI:** 1.8L (AWD, AWP, AWW), 2.0L (AEG), 2.8L (AFP, BDF); **NB:** 1.8L (APH, AWP, AWV, BNU, BKF), 2.0L (AEG, AVH, AZG, BEV); **Nbvert:** (BDG) **Transmissions:** All	**Throttle Position Sensor D/E Voltage Correlation Conditions:** Engine started, battery voltage must be at least 11.5v, all electrical components must be off, parking brake must be engaged (to keep daytime driving lights off), automatic transmission selector must be in park; and the ECM detected the Throttle Position 'D' (TPD) and Throttle Position 'B' (TPE) sensors disagreed, or that the TPD sensor should not be in its detected position, or that the TPE sensor should not be in its detected position during testing. **Note: Both the Throttle Position (TP) Sensor and Accelerator Pedal Position Sensor are located at the accelerator pedal module and communicate the driver's intentions to the ECM completely independently of each other. Both sensors are stored in one housing.** **Possible Causes:** • ETC TP sensor connector is damaged or shorted • ETC TP sensor circuits shorted together in the wire harness • ETC TP sensor signal circuit is shorted to VREF (5v) • ETC TP sensor is damaged or the ECM has failed

DTC	Trouble Code Title, Conditions & Possible Causes
DTC:P2146 **Passat:** 2.0 (BPY); **Jetta:** 2.0L (BPY) **Transmissions:** All	**Fuel Injector Group "A" Supply Voltage Circuit/Open Conditions:** Engine started, battery voltage must be at least 11.5v, all electrical components must be off, the ground between the engine and the chassis must be well connected, the exhaust system must be properly sealed between the catalytic converter and the cylinder head, and the oxygen sensor heater for oxygen sensor before the catalytic converter must be properly functioning. The ECM detected the fuel injector supply voltage circuit was outside the normal range during the test period. **Note: For resistance testing of sensor heating, oxygen sensor should be cooled to ambient temperature. High temperatures at oxygen sensor may lead to inaccurate measurements.** **Possible Causes:** • Oxygen sensor (before catalytic converter) is faulty • Oxygen sensor (behind catalytic converter) is faulty • Oxygen sensor heater (before catalytic converter) is faulty • Oxygen sensor heater (behind catalytic converter) is faulty • Check circuits for shorts to each other, ground or power • Fuel Injector(s) may have failed • ECM has failed
DTC:P2149 **Passat:** 2.0 (BPY); **Jetta:** 2.0L (BPY) **Transmissions:** All	**Fuel Injector Group "B" Supply Voltage Circuit/Open Conditions:** Engine started, battery voltage must be at least 11.5v, all electrical components must be off, the ground between the engine and the chassis must be well connected, the exhaust system must be properly sealed between the catalytic converter and the cylinder head, and the oxygen sensor heater for oxygen sensor before the catalytic converter must be properly functioning. The ECM detected the fuel injector supply voltage circuit was outside the normal range during the test period. **Note: For resistance testing of sensor heating, oxygen sensor should be cooled to ambient temperature. High temperatures at oxygen sensor may lead to inaccurate measurements.** **Possible Causes:** • Oxygen sensor (before catalytic converter) is faulty • Oxygen sensor (behind catalytic converter) is faulty • Oxygen sensor heater (before catalytic converter) is faulty • Oxygen sensor heater (behind catalytic converter) is faulty • Check circuits for shorts to each other, ground or power • Fuel Injector(s) may have failed • ECM has failed
DTC:P2177 **2T, MIL: Yes** **Passat:** 2.0 (BPY), 4.0L (BDP); **Jetta:** 2.0L (AEG, AVH, AZG, BBW, BEV), 2.8L (AFP, BDF); **Golf:** 2.0L (AEG, AVH, AZG, BEV); **GTI:** 2.0L (AEG), 2.8L (AFP, BDF); **NB:** 1.8L (APH, AWP, AWV, BNU, BKF), 2.0L (AEG, AVH, AZG, BEV); **Nbvert:** (BDG) **Transmissions:** All	**System Too Lean Off Idle Bank 1 Conditions:** Engine started, battery voltage must be at least 11.5v, all electrical components must be off, the ground between the engine and the chassis must be well connected, the exhaust system must be properly sealed between the catalytic converter and the cylinder head, and the oxygen sensor heater for oxygen sensor before the catalytic converter must be properly functioning. The ECM detected the system indicated a lean signal, or it could no longer control bank 1 because it was at its lean limit. **Possible Causes:** • Intake Manifold Runner Position Sensor has failed • Intake system has leaks (false air) • Motor for intake flap is faulty • Oxygen sensor (before catalytic converter) is faulty • Oxygen sensor (behind catalytic converter) is faulty • Oxygen sensor heater (before catalytic converter) is faulty • Oxygen sensor heater (behind catalytic converter) is faulty • Check circuits for shorts to each other, ground or power • Fuel Injector(s) may have failed • ECM has failed
DTC:P2178 **2T, MIL: Yes** **Passat:** 2.0 (BPY), 4.0L (BDP); **Jetta:** 2.0L (AEG, AVH, AZG, BBW, BEV), 2.8L (AFP, BDF); **Golf:** 2.0L (AEG, AVH, AZG, BEV); **GTI:** 2.0L (AEG), 2.8L (AFP, BDF); **NB:** 1.8L (APH, AWP, AWV, BNU, BKF), 2.0L (AEG, AVH, AZG, BEV); **Nbvert:** (BDG) **Transmissions:** All	**System Too Rich Off Idle Bank 1 Conditions:** Engine started, battery voltage must be at least 11.5v, all electrical components must be off, the ground between the engine and the chassis must be well connected, the exhaust system must be properly sealed between the catalytic converter and the cylinder head, and the oxygen sensor heater for oxygen sensor before the catalytic converter must be properly functioning. The ECM detected the system indicated a rich signal, or it could no longer control bank 1 because it was at its rich limit. **Possible Causes:** • Intake Manifold Runner Position Sensor has failed • Intake system has leaks (false air) • Motor for intake flap is faulty • Oxygen sensor (before catalytic converter) is faulty • Oxygen sensor (behind catalytic converter) is faulty • Oxygen sensor heater (before catalytic converter) is faulty • Oxygen sensor heater (behind catalytic converter) is faulty • Check circuits for shorts to each other, ground or power • Fuel Injector(s) may have failed • ECM has failed

DTC	Trouble Code Title, Conditions & Possible Causes
DTC:P2179 **Passat:** 4.0L (BDP); **Jetta:** 2.8L (AFP, BDF); **GTI:** 2.8L (AFP, BDF) **Transmissions:** All	**System Too Lean Off Idle Bank 2 Conditions:** Engine started, battery voltage must be at least 11.5v, all electrical components must be off, the ground between the engine and the chassis must be well connected, the exhaust system must be properly sealed between the catalytic converter and the cylinder head, and the oxygen sensor heater for oxygen sensor before the catalytic converter must be properly functioning. The ECM detected the system indicated a lean signal, or it could no longer control bank 2 because it was at its lean limit. **Possible Causes:** • Intake Manifold Runner Position Sensor has failed • Intake system has leaks (false air) • Motor for intake flap is faulty • Oxygen sensor (before catalytic converter) is faulty • Oxygen sensor (behind catalytic converter) is faulty • Oxygen sensor heater (before catalytic converter) is faulty • Oxygen sensor heater (behind catalytic converter) is faulty • Check circuits for shorts to each other, ground or power • Fuel Injector(s) may have failed • ECM has failed
DTC:P2180 **Passat:** 4.0L (BDP); **Jetta:** 2.8L (AFP, BDF); **GTI:** 2.8L (AFP, BDF) **Transmissions:** All	**System Too Rich Off Idle Bank 2 Conditions:** Engine started, battery voltage must be at least 11.5v, all electrical components must be off, the ground between the engine and the chassis must be well connected, the exhaust system must be properly sealed between the catalytic converter and the cylinder head, and the oxygen sensor heater for oxygen sensor before the catalytic converter must be properly functioning. The ECM detected the system indicated a rich signal, or it could no longer control bank 2 because it was at its rich limit. **Possible Causes:** • Intake Manifold Runner Position Sensor has failed • Intake system has leaks (false air) • Motor for intake flap is faulty • Oxygen sensor (before catalytic converter) is faulty • Oxygen sensor (behind catalytic converter) is faulty • Oxygen sensor heater (before catalytic converter) is faulty • Oxygen sensor heater (behind catalytic converter) is faulty • Check circuits for shorts to each other, ground or power • Fuel Injector(s) may have failed • ECM has failed
DTC:P2181 **2T, MIL: Yes** **Passat:** 1.8L (ATW, AUG), 2.8L (AHA, ATQ), 2.0 (BPY), 4.0L (BDP); **Jetta:** 1.8L (AWD, AWP, AWW), 2.0 (ABA, AEG, AVH, AZG, BBW, BEV, BPY), 2.5L (BGP, BGQ), 2.8L (AFP, BDF); **Golf:** 1.8L (AWD, AWW), 2.0L (ABA, AEG, AVH, AZG, BEV); **GTI:** 1.8L (AWD, AWP, AWW), 2.0L (ABA, AEG), 2.8L (AFP, BDF); **NB:** 1.8L (APH, AWP, AWV, BNU, BKF), 2.0L (AEG, AVH, AZG, BEV); **Nbvert:** (BDG); **Cabrio:** 2.0L (ABA) **Transmissions:** All	**Cooling System Performance Malfunction Conditions:** Key on, engine cold; and the Engine Coolant Temperature (ECM) detected the ECT sensor signal was more or less than the self-test limits or has failed to gain a signal. This is a thermistor-type sensor with a variable resistance that changes when exposed to different temperatures **Possible Causes:** • ECT sensor has failed • ECT Sensor (on Radiator) has failed • ECT sensor signal circuit is open (inspect wiring & connector) • ECT sensor signal circuit is shorted • Cooling system malfunction, or the thermostat is stuck • Engine not operating at normal operating temperature • EOT sensor is damaged or it has failed
DTC:P2184 **Passat:** 2.8L (AHA, ATQ), 2.0 (BPY), 4.0L (BDP); **Jetta:** 2.0L (BPY), 2.5L (BGP, BGQ) **Transmissions:** All	**Engine Coolant Temperature Sensor 2 Circuit Low Conditions:** Key on or engine running; and the Engine Coolant Temperature (ECM) detected the ECT sensor signal was less than the self-test minimum. This is a thermistor-type sensor with a variable resistance that changes when exposed to different temperatures **Possible Causes:** • ECT sensor has failed • ECT Sensor (on Radiator) has failed • ECT sensor signal circuit is open (inspect wiring & connector) • ECT sensor signal circuit is shorted • Cooling system malfunction, or the thermostat is stuck • Engine not operating at normal operating temperature • EOT sensor is damaged or it has failed

DTC	Trouble Code Title, Conditions & Possible Causes
DTC:P2185 **Passat:** 2.8L (AHA, ATQ), 2.0 (BPY), 4.0L (BDP); **Jetta:** 2.0L (BPY), 2.5L (BGP, BGQ) **Transmissions:** All	**Engine Coolant Temperature Sensor 2 Circuit High Conditions:** Key on or engine running; and the Engine Coolant Temperature (ECM) detected the ECT sensor signal was more than the self-test maximum. This is a thermistor-type sensor with a variable resistance that changes when exposed to different temperatures **Possible Causes:** • ECT sensor has failed • ECT Sensor (on Radiator) has failed • ECT sensor signal circuit is open (inspect wiring & connector) • ECT sensor signal circuit is shorted • Cooling system malfunction, or the thermostat is stuck • Engine not operating at normal operating temperature • EOT sensor is damaged or it has failed
DTC:P2187 **2T, MIL: Yes** **Passat:** 4.0L (BDP); **Jetta:** 2.0L (AEG, AVH, AZG, BBW, BEV), 2.8L (AFP, BDF); **Golf:** 2.0L (AEG, AVH, AZG, BEV); **GTI:** 2.0L (AEG), 2.8L (AFP, BDF); **NB:** 2.0L (AEG, AVH, AZG, BEV); **Nbvert:** (BDG) **Transmissions:** All	**System Too Lean On Idle Bank 1 Conditions:** Engine started, battery voltage must be at least 11.5v, all electrical components must be off, the ground between the engine and the chassis must be well connected, the exhaust system must be properly sealed between the catalytic converter and the cylinder head, and the oxygen sensor heater for oxygen sensor before the catalytic converter must be properly functioning. ECM detected the system indicated a lean signal, or it could no longer control bank 1 because it was at its lean limit. **Possible Causes:** • Evaporative Emission (EVAP) canister purge regulator valve is faulty • Exhaust system components are damaged • Fuel injectors are faulty • Fuel pressure regulator and residual pressure have failed • Fuel Pump (FP) in fuel tank is faulty • Intake system has leaks (false air) • Secondary Air Injection (AIR) system has an improper seal • Intake Manifold Runner Position Sensor has failed • Motor for intake flap is faulty • Oxygen sensor (before catalytic converter) is faulty • Oxygen sensor (behind catalytic converter) is faulty • Oxygen sensor heater (before catalytic converter) is faulty • Oxygen sensor heater (behind catalytic converter) is faulty • Check circuits for shorts to each other, ground or power • ECM has failed
DTC:P2188 **2T, MIL: Yes** **Passat:** 4.0L (BDP); **Jetta:** 2.0L (AEG, AVH, AZG, BBW, BEV), 2.8L (AFP, BDF); **Golf:** 2.0L (AEG, AVH, AZG, BEV); **GTI:** 2.0L (AEG), 2.8L (AFP, BDF); **NB:** 2.0L (AEG, AVH, AZG, BEV); **Nbvert:** (BDG) **Transmissions:** All	**System Too Rich On Idle Bank 1 Conditions:** Engine started, battery voltage must be at least 11.5v, all electrical components must be off, the ground between the engine and the chassis must be well connected, the exhaust system must be properly sealed between the catalytic converter and the cylinder head, and the oxygen sensor heater for oxygen sensor before the catalytic converter must be properly functioning. ECM detected the system indicated a rich signal, or it could no longer control bank 1 because it was at its rich limit. **Possible Causes:** • Evaporative Emission (EVAP) canister purge regulator valve is faulty • Exhaust system components are damaged • Fuel injectors are faulty • Fuel pressure regulator and residual pressure have failed • Fuel Pump (FP) in fuel tank is faulty • Intake system has leaks (false air) • Secondary Air Injection (AIR) system has an improper seal • Intake Manifold Runner Position Sensor has failed • Motor for intake flap is faulty • Oxygen sensor (before catalytic converter) is faulty • Oxygen sensor (behind catalytic converter) is faulty • Oxygen sensor heater (before catalytic converter) is faulty • Oxygen sensor heater (behind catalytic converter) is faulty • Check circuits for shorts to each other, ground or power • ECM has failed

DTC	Trouble Code Title, Conditions & Possible Causes
DTC:P2189 **Passat:** 4.0L (BDP); **Jetta:** 2.8L (AFP, BDF); **GTI:** 2.8L (AFP, BDF) **Transmissions:** All	**System Too Lean On Idle Bank 2 Conditions:** Engine started, battery voltage must be at least 11.5v, all electrical components must be off, the ground between the engine and the chassis must be well connected, the exhaust system must be properly sealed between the catalytic converter and the cylinder head, and the oxygen sensor heater for oxygen sensor before the catalytic converter must be properly functioning. ECM detected the system indicated a lean signal, or it could no longer control bank 2 because it was at its lean limit. **Possible Causes:** • Evaporative Emission (EVAP) canister purge regulator valve is faulty • Exhaust system components are damaged • Fuel injectors are faulty • Fuel pressure regulator and residual pressure have failed • Fuel Pump (FP) in fuel tank is faulty • Intake system has leaks (false air) • Secondary Air Injection (AIR) system has an improper seal • Intake Manifold Runner Position Sensor has failed • Motor for intake flap is faulty • Oxygen sensor (before catalytic converter) is faulty • Oxygen sensor (behind catalytic converter) is faulty • Oxygen sensor heater (before catalytic converter) is faulty • Oxygen sensor heater (behind catalytic converter) is faulty • Check circuits for shorts to each other, ground or power • ECM has failed
DTC:P2190 **Passat:** 4.0L (BDP); **Jetta:** 2.8L (AFP, BDF); **GTI:** 2.8L (AFP, BDF) **Transmissions:** All	**System Too Rich On Idle Bank 2 Conditions:** Engine started, battery voltage must be at least 11.5v, all electrical components must be off, the ground between the engine and the chassis must be well connected, the exhaust system must be properly sealed between the catalytic converter and the cylinder head, and the oxygen sensor heater for oxygen sensor before the catalytic converter must be properly functioning. ECM detected the system indicated a rich signal, or it could no longer control bank 1 because it was at its rich limit. **Possible Causes:** • Evaporative Emission (EVAP) canister purge regulator valve is faulty • Exhaust system components are damaged • Fuel injectors are faulty • Fuel pressure regulator and residual pressure have failed • Fuel Pump (FP) in fuel tank is faulty • Intake system has leaks (false air) • Secondary Air Injection (AIR) system has an improper seal • Intake Manifold Runner Position Sensor has failed • Motor for intake flap is faulty • Oxygen sensor (before catalytic converter) is faulty • Oxygen sensor (behind catalytic converter) is faulty • Oxygen sensor heater (before catalytic converter) is faulty • Oxygen sensor heater (behind catalytic converter) is faulty • Check circuits for shorts to each other, ground or power • ECM has failed
DTC:P2191 **Passat:** 4.0L (BDP); **Jetta:** 2.0L (AEG, AVH, AZG, BBW, BEV), 2.8L (AFP, BDF); **Golf:** 2.0L (AEG, AVH, AZG, BEV); **GTI:** 2.0L (AEG), 2.8L (AFP, BDF); **NB:** 2.0L (AEG, AVH, AZG, BEV); **Nbvert:** (BDG) **Transmissions:** All	**System Too Lean at Higher Load Bank 1 Conditions:** Engine started, battery voltage must be at least 11.5v, all electrical components must be off, the ground between the engine and the chassis must be well connected, the exhaust system must be properly sealed between the catalytic converter and the cylinder head, and the oxygen sensor heater for oxygen sensor before the catalytic converter must be properly functioning. ECM detected the system indicated a lean signal, or it could no longer control bank 1 because it was at its lean limit. **Possible Causes:** • Evaporative Emission (EVAP) canister purge regulator valve is faulty • Exhaust system components are damaged • Fuel injectors are faulty • Fuel pressure regulator and residual pressure have failed • Fuel Pump (FP) in fuel tank is faulty • Intake system has leaks (false air) • Secondary Air Injection (AIR) system has an improper seal • Intake Manifold Runner Position Sensor has failed • Motor for intake flap is faulty • Oxygen sensor (before catalytic converter) is faulty • Oxygen sensor (behind catalytic converter) is faulty • Oxygen sensor heater (before catalytic converter) is faulty • Oxygen sensor heater (behind catalytic converter) is faulty • Check circuits for shorts to each other, ground or power • ECM has failed

DTC	Trouble Code Title, Conditions & Possible Causes
DTC:P2192 **Passat:** 4.0L (BDP); **Jetta:** 2.0L (AEG, AVH, AZG, BBW, BEV), 2.8L (AFP, BDF); **Golf:** 2.0L (AEG, AVH, AZG, BEV); **GTI:** 2.0L (AEG, 2.8L (AFP, BDF); **NB:** 2.0L (AEG, AVH, AZG, BEV); **Nbvert:** (BDG) **Transmissions:** All	**System Too Rich at Higher Load Bank 1 Conditions:** Engine started, battery voltage must be at least 11.5v, all electrical components must be off, the ground between the engine and the chassis must be well connected, the exhaust system must be properly sealed between the catalytic converter and the cylinder head, and the oxygen sensor heater for oxygen sensor before the catalytic converter must be properly functioning. ECM detected the system indicated a rich signal, or it could no longer control bank 1 because it was at its rich limit. **Possible Causes:** • Evaporative Emission (EVAP) canister purge regulator valve is faulty • Exhaust system components are damaged • Fuel injectors are faulty • Fuel pressure regulator and residual pressure have failed • Fuel Pump (FP) in fuel tank is faulty • Intake system has leaks (false air) • Secondary Air Injection (AIR) system has an improper seal • Intake Manifold Runner Position Sensor has failed • Motor for intake flap is faulty • Oxygen sensor (before catalytic converter) is faulty • Oxygen sensor (behind catalytic converter) is faulty • Oxygen sensor heater (before catalytic converter) is faulty • Oxygen sensor heater (behind catalytic converter) is faulty • Check circuits for shorts to each other, ground or power • ECM has failed
DTC:P2193 **Passat:** 4.0L (BDP); **Jetta:** 2.8L (AFP, BDF) **Transmissions:** All	**System Too Lean at Higher Load Bank 2 Conditions:** Engine started, battery voltage must be at least 11.5v, all electrical components must be off, the ground between the engine and the chassis must be well connected, the exhaust system must be properly sealed between the catalytic converter and the cylinder head, and the oxygen sensor heater for oxygen sensor before the catalytic converter must be properly functioning. ECM detected the system indicated a lean signal, or it could no longer control bank 2 because it was at its lean limit. **Possible Causes:** • Evaporative Emission (EVAP) canister purge regulator valve is faulty • Exhaust system components are damaged • Fuel injectors are faulty • Fuel pressure regulator and residual pressure have failed • Fuel Pump (FP) in fuel tank is faulty • Intake system has leaks (false air) • Secondary Air Injection (AIR) system has an improper seal • Intake Manifold Runner Position Sensor has failed • Motor for intake flap is faulty • Oxygen sensor (before catalytic converter) is faulty • Oxygen sensor (behind catalytic converter) is faulty • Oxygen sensor heater (before catalytic converter) is faulty • Oxygen sensor heater (behind catalytic converter) is faulty • Check circuits for shorts to each other, ground or power • ECM has failed
DTC:P2194 **Passat:** 4.0L (BDP); **Jetta:** 2.8L (AFP, BDF); **GTI:** 2.8L (AFP, BDF) **Transmissions:** All	**System Too Rich at Higher Load Bank 2 Conditions:** Engine started, battery voltage must be at least 11.5v, all electrical components must be off, the ground between the engine and the chassis must be well connected, the exhaust system must be properly sealed between the catalytic converter and the cylinder head, and the oxygen sensor heater for oxygen sensor before the catalytic converter must be properly functioning. ECM detected the system indicated a rich signal, or it could no longer control bank 2 because it was at its rich limit. **Possible Causes:** • Evaporative Emission (EVAP) canister purge regulator valve is faulty • Exhaust system components are damaged • Fuel injectors are faulty • Fuel pressure regulator and residual pressure have failed • Fuel Pump (FP) in fuel tank is faulty • Intake system has leaks (false air) • Secondary Air Injection (AIR) system has an improper seal • Intake Manifold Runner Position Sensor has failed • Motor for intake flap is faulty • Oxygen sensor (before catalytic converter) is faulty • Oxygen sensor (behind catalytic converter) is faulty • Oxygen sensor heater (before catalytic converter) is faulty • Oxygen sensor heater (behind catalytic converter) is faulty • Check circuits for shorts to each other, ground or power • ECM has failed

DTC	Trouble Code Title, Conditions & Possible Causes
DTC:P2195 **Passat:** 2.8L (AHA, ATQ), 2.0 (BPY); **Jetta:** 2.0L (AEG, AVH, AZG, BBW, BEV, BPY), 2.5L (BGP, BGQ); **Golf:** 2.0L (AEG, AVH, AZG, BEV); **GTI:** 2.0L (AEG); **NB:** 1.8L (APH, AWP, AWV, BNU, BKF), 2.0L (AEG, AVH, AZG, BEV); **Nbvert:** (BDG) **Transmissions:** All	**O2 Sensor Signal Stuck Lean Bank 1 Sensor 1 Conditions:** Engine running in closed loop, and the ECM detected the O2S indicated a lean signal, or it could no longer control Fuel Trim because it was at lean limit. **Possible Causes:** • Engine oil level high • Camshaft timing error • Cylinder compression low • Exhaust leaks in front of O2S • EGR valve is stuck open • EGR gasket is leaking • EVR diaphragm is leaking • Damaged fuel pressure regulator or extremely low fuel pressure • O2S circuit is open or shorted in the wiring harness • Oxygen sensor (before catalytic converter) is faulty • Oxygen sensor (behind catalytic converter) is faulty • Oxygen sensor heater (before catalytic converter) is faulty • Oxygen sensor heater (behind catalytic converter) is faulty • Air leaks after the MAF sensor • PCV system leaks • Dip stick not seated properly
DTC:P2196 **Passat:** 2.0 (BPY); **Jetta:** 2.0L (AEG, AVH, AZG, BBW, BEV, BPY), 2.5L (BGP, BGQ); **Golf:** 2.0L (AEG, AVH, AZG, BEV); **GTI:** 2.0L (AEG); **NB:** 1.8L (APH, AWP, AWV, BNU, BKF), 2.0L (AEG, AVH, AZG, BEV); **Nbvert:** (BDG) **Transmissions:** All	**O2 Sensor Signal Stuck Rich Bank 1 Sensor 1 Conditions:** Engine running in closed loop, and the ECM detected the O2S indicated a rich signal, or it could no longer control Fuel Trim because it was at its rich limit. **Possible Causes:** • Engine oil level high • Camshaft timing error • Cylinder compression low • Exhaust leaks in front of O2S • EGR valve is stuck open • EGR gasket is leaking • EVR diaphragm is leaking • Damaged fuel pressure regulator or extremely low fuel pressure • O2S circuit is open or shorted in the wiring harness • Oxygen sensor (before catalytic converter) is faulty • Oxygen sensor (behind catalytic converter) is faulty • Oxygen sensor heater (before catalytic converter) is faulty • Oxygen sensor heater (behind catalytic converter) is faulty • Air leaks after the MAF sensor • PCV system leaks • Dip stick not seated properly
DTC:P2231 **Passat:** 2.0L (BPY), 4.0L (BDP); **Jetta:** 2.0 (ABA, AEG, AVH, AZG, BBW, BEV, BPY), 2.5L (BGP, BGQ), 2.8L (AFP, BDF); **Golf:** 2.0L (ABA, AEG, AVH, AZG, BEV); **GTI:** 2.0L (ABA, AEG), 2.8L (AFP, BDF); **NB:** 1.8L (APH, AWP, AWV, BNU, BKF), 2.0L (AEG, AVH, AZG, BEV); **Nbvert:** (BDG); **Cabrio:** 2.0L (ABA) **Transmissions:** All	**O2 Sensor Signal Circuit Shorted to Heater Circuit Bank 1 Sensor 1 Conditions:** Engine started, battery voltage must be at least 11.5v, all electrical components must be off, parking brake must be engaged (to keep daytime driving lights off), automatic transmission selector must be in park. The ECM detected an unexpected voltage condition, or it detected an unexpected current draw in the sensor circuit during the CCM test. **Note: Vehicle must be raised before connector for oxygen sensors is accessible.** **Possible Causes:** • Oxygen sensor (before catalytic converter) is faulty • Oxygen sensor heater (before catalytic converter) is faulty • Oxygen sensor heater (before catalytic converter) is faulty • Oxygen sensor heater (behind catalytic converter) is faulty • O2S circuit is open or shorted in the wiring harness • ECM has failed

DTC	Trouble Code Title, Conditions & Possible Causes
DTC:P2234 **Passat:** 4.0L (BDP); **Jetta:** 2.8L (AFP, BDF); **GTI:** 2.8L (AFP, BDF) **Transmissions:** All	**O2 Sensor Signal Circuit Shorted to Heater Circuit Bank 2 Sensor 1 Conditions:** Engine started, battery voltage must be at least 11.5v, all electrical components must be off, parking brake must be engaged (to keep daytime driving lights off), automatic transmission selector must be in park. The ECM detected an unexpected voltage condition, or it detected an unexpected current draw in the sensor circuit during the CCM test. **Note: Vehicle must be raised before connector for oxygen sensors is accessible.** **Possible Causes:** • Oxygen sensor (before catalytic converter) is faulty • Oxygen sensor heater (before catalytic converter) is faulty • Oxygen sensor heater (before catalytic converter) is faulty • Oxygen sensor heater (behind catalytic converter) is faulty • O2S circuit is open or shorted in the wiring harness • ECM has failed
DTC:P2237 **Passat:** 2.0L (BPY), 4.0L (BDP); **Jetta:** 2.0 (ABA, AEG, AVH, AZG, BBW, BEV, BPY), 2.5L (BGP, BGQ), 2.8L (AFP, BDF); **Golf:** 2.0L (ABA, AEG, AVH, AZG, BEV); **GTI:** 2.0L (ABA, AEG), 2.8L (AFP, BDF); **NB:** 1.8L (APH, AWP, AWV, BNU, BKF), 2.0L (AEG, AVH, AZG, BEV); **Nbvert:** (BDG); **Cabrio:** 2.0L (ABA) **Transmissions:** All	**O2 Sensor Positive Current Control Circuit/Open Bank 1 Sensor 1 Conditions:** Engine started, battery voltage must be at least 11.5v, all electrical components must be off, parking brake must be engaged (to keep daytime driving lights off), automatic transmission selector must be in park. The ECM detected an unexpected voltage condition, or it detected an unexpected current draw in the sensor circuit during the CCM test. **Note: Vehicle must be raised before connector for oxygen sensors is accessible.** **Possible Causes:** • Oxygen sensor (before catalytic converter) is faulty • Oxygen sensor heater (before catalytic converter) is faulty • Oxygen sensor heater (before catalytic converter) is faulty • Oxygen sensor heater (behind catalytic converter) is faulty • O2S circuit is open or shorted in the wiring harness • ECM has failed
DTC:P2240 **Passat:** 2.0L (BPY), 4.0L (BDP); **Jetta:** 2.8L (AFP, BDF); **GTI:** 2.8L (AFP, BDF) **Transmissions:** All	**O2 Sensor Positive Current Control Circuit/Open Bank 2 Sensor 1 Conditions:** Engine started, battery voltage must be at least 11.5v, all electrical components must be off, parking brake must be engaged (to keep daytime driving lights off), automatic transmission selector must be in park. The ECM detected an unexpected voltage condition, or it detected an unexpected current draw in the sensor circuit during the CCM test. **Note: Vehicle must be raised before connector for oxygen sensors is accessible.** **Possible Causes:** • Oxygen sensor (before catalytic converter) is faulty • Oxygen sensor heater (before catalytic converter) is faulty • Oxygen sensor heater (before catalytic converter) is faulty • Oxygen sensor heater (behind catalytic converter) is faulty • O2S circuit is open or shorted in the wiring harness • ECM has failed
DTC:P2243 **Passat:** 2.0L (BPY), 4.0L (BDP); **Jetta:** 2.0 (ABA, AEG, AVH, AZG, BBW, BEV, BPY), 2.5L (BGP, BGQ), 2.8L (AFP, BDF); **Golf:** 2.0L (ABA, AEG, AVH, AZG, BEV); **GTI:** 2.0L (ABA, AEG), 2.8L (AFP, BDF); **NB:** 1.8L (APH, AWP, AWV, BNU, BKF), 2.0L (AEG, AVH, AZG, BEV); **Nbvert:** (BDG); **Cabrio:** 2.0L (ABA) **Transmissions:** All	**O2 Sensor Reference Voltage Circuit/Open Bank 1 Sensor 1 Conditions:** Engine started, battery voltage must be at least 11.5v, all electrical components must be off, parking brake must be engaged (to keep daytime driving lights off), automatic transmission selector must be in park. The ECM detected an unexpected voltage condition, or it detected an unexpected current draw in the sensor circuit during the CCM test. **Note: Vehicle must be raised before connector for oxygen sensors is accessible.** **Possible Causes:** • Oxygen sensor (before catalytic converter) is faulty • Oxygen sensor heater (before catalytic converter) is faulty • Oxygen sensor heater (before catalytic converter) is faulty • Oxygen sensor heater (behind catalytic converter) is faulty • O2S circuit is open or shorted in the wiring harness • ECM has failed
DTC:P2247 **Passat:** 4.0L (BDP); **Jetta:** 2.8L (AFP, BDF); **GTI:** 2.8L (AFP, BDF) **Transmissions:** All	**O2 Sensor Reference Voltage Circuit/Open Bank 2 Sensor 1 Conditions:** Engine started, battery voltage must be at least 11.5v, all electrical components must be off, parking brake must be engaged (to keep daytime driving lights off), automatic transmission selector must be in park. The ECM detected an unexpected voltage condition, or it detected an unexpected current draw in the sensor circuit during the CCM test. **Note: Vehicle must be raised before connector for oxygen sensors is accessible.** **Possible Causes:** • Oxygen sensor (before catalytic converter) is faulty • Oxygen sensor heater (before catalytic converter) is faulty • Oxygen sensor heater (before catalytic converter) is faulty • Oxygen sensor heater (behind catalytic converter) is faulty • O2S circuit is open or shorted in the wiring harness • ECM has failed

DTC	Trouble Code Title, Conditions & Possible Causes
DTC:P2251 **Passat:** 2.0L (BPY), 4.0L (BDP); **Jetta:** 2.0 (ABA, AEG, AVH, AZG, BBW, BEV, BPY), 2.5L (BGP, BGQ), 2.8L (AFP, BDF); **Golf:** 2.0L (ABA, AEG, AVH, AZG, BEV); **GTI:** 2.0L (ABA, AEG), 2.8L (AFP, BDF); **NB:** 1.8L (APH, AWP, AWV, BNU, BKF), 2.0L (AEG, AVH, AZG, BEV); **Nbvert:** (BDG); **Cabrio:** 2.0L (ABA) **Transmissions:** All	**O2 Sensor Negative Voltage Circuit/Open Bank 1 Sensor 1 Conditions:** Engine started, battery voltage must be at least 11.5v, all electrical components must be off, parking brake must be engaged (to keep daytime driving lights off), automatic transmission selector must be in park. The ECM detected an unexpected voltage condition, or it detected an unexpected current draw in the sensor circuit during the CCM test. **Note: Vehicle must be raised before connector for oxygen sensors is accessible.** **Possible Causes:** • Oxygen sensor (before catalytic converter) is faulty • Oxygen sensor heater (before catalytic converter) is faulty • Oxygen sensor heater (before catalytic converter) is faulty • Oxygen sensor heater (behind catalytic converter) is faulty • O2S circuit is open or shorted in the wiring harness • ECM has failed
DTC:P2254 **Passat:** 4.0L (BDP); **Jetta:** 2.8L (AFP, BDF); **GTI:** 2.8L (AFP, BDF) **Transmissions:** All	**O2 Sensor Negative Voltage Circuit/Open Bank 2 Sensor 1 Conditions:** Engine started, battery voltage must be at least 11.5v, all electrical components must be off, parking brake must be engaged (to keep daytime driving lights off), automatic transmission selector must be in park. The ECM detected an unexpected voltage condition, or it detected an unexpected current draw in the sensor circuit during the CCM test. **Note: Vehicle must be raised before connector for oxygen sensors is accessible.** **Possible Causes:** • Oxygen sensor (before catalytic converter) is faulty • Oxygen sensor heater (before catalytic converter) is faulty • Oxygen sensor heater (before catalytic converter) is faulty • Oxygen sensor heater (behind catalytic converter) is faulty • O2S circuit is open or shorted in the wiring harness • ECM has failed
DTC:P2257 **Passat:** 1.8L (ATW, AUG), 4.0L (BDP); **Jetta:** 2.0 (ABA, AEG, AVH, AZG, BBW, BEV), 2.5L (BGP, BGQ), 2.8L (AFP, BDF); **Golf:** 2.0L (ABA, AEG, AVH, AZG, BEV); **GTI:** 2.0L (ABA, AEG), 2.8L (AFP, BDF); **NB:** 1.8L (APH, AWP, AWV, BNU, BKF), 2.0L (AEG, AVH, AZG, BEV); **Nbvert:** (BDG); **Cabrio:** 2.0L (ABA) **Transmissions:** All	**Secondary Air Injection System Control "A" Circuit Low Conditions:** Engine started, battery voltage must be at least 11.5v, all electrical components must be off, parking brake must be engaged (to keep daytime driving lights off), automatic transmission selector must be in park and the ground between the engine and the chassis must be well connected. The ECM detected an unexpected voltage condition on the AIR system control circuit during testing. **Possible Causes:** • AIR solenoid power circuit (B+) is open (check dedicated fuse) • AIR bypass solenoid control circuit is open or shorted to ground • AIR diverter solenoid control circuit open or shorted to ground • AIR pump control circuit is open or shorted to ground • Check valve (one or more) is damaged or leaking • Solid State relay is damaged or it has failed • Check activation of Secondary Air Injection (AIR) Pump Relay • ECM has failed
DTC:P2258 **Passat:** 1.8L (ATW, AUG), 4.0L (BDP); **Jetta:** 2.0 (ABA, AEG, AVH, AZG, BBW, BEV), 2.5L (BGP, BGQ), 2.8L (AFP, BDF); **Golf:** 2.0L (ABA, AEG, AVH, AZG, BEV); **GTI:** 2.0L (ABA, AEG), 2.8L (AFP, BDF); **NB:** 1.8L (APH, AWP, AWV, BNU, BKF), 2.0L (AEG, AVH, AZG, BEV); **Nbvert:** (BDG); **Cabrio:** 2.0L (ABA) **Transmissions:** All	**Secondary Air Injection System Control "A" Circuit High Conditions:** Engine started, battery voltage must be at least 11.5v, all electrical components must be off, parking brake must be engaged (to keep daytime driving lights off), automatic transmission selector must be in park and the ground between the engine and the chassis must be well connected. The ECM detected an unexpected voltage condition on the AIR system control circuit during testing. **Possible Causes:** • AIR solenoid power circuit (B+) is open (check dedicated fuse) • AIR bypass solenoid control circuit is open or shorted to ground • AIR diverter solenoid control circuit open or shorted to ground • AIR pump control circuit is open or shorted to ground • Check valve (one or more) is damaged or leaking • Solid State relay is damaged or it has failed • Check activation of Secondary Air Injection (AIR) Pump Relay • ECM has failed

DTC	Trouble Code Title, Conditions & Possible Causes
DTC:P2261 **Passat:** 2.0L (BPY); **Jetta:** 2.0L (BPY) **Transmissions:** All	**Turbo/Super Charger Bypass Valve – Mechanical Conditions:** Engine started, battery voltage at least 11.5v, all electrical components off, ground connections between engine and chassis well connected, coolant temperature at least 80-degrees Celicius. The ECM detected an operating condition of the turbo bypass valve that could harm the engine or automatic transmission. **Possible Causes:** • Charge air system has leaks • Recirculating valve for turbocharger is faulty • Turbocharging system is damaged • Vacuum diaphragm for turbocharger needs adjusting • Wastegate bypass regulator valve is faulty • Ignition misfire condition exceeds the calibrated threshold • Knock sensor circuit has failed, or excessive knock detected • Low speed fuel pump relay not switching properly • Transmission oil temperature beyond the calibrated threshold • Shaft bearing of charge pressure regulator valve in turbocharger is blocked
DTC:P2270 **Passat:** 2.0L (BPY), 4.0L (BDP); **Jetta:** 2.0 (ABA, AEG, AVH, AZG, BBW, BEV), 2.5L (BGP, BGQ), 2.8L (AFP, BDF); **Golf:** 2.0L (ABA, AEG, AVH, AZG, BEV); **GTI:** 2.0L (ABA, AEG), 2.8L (AFP, BDF); **NB:** 1.8L (APH, AWP, AWV, BNU, BKF), 2.0L (AEG, AVH, AZG, BEV); **Nbvert:** (BDG); **Cabrio:** 2.0L (ABA) **Transmissions:** All	**O2 Sensor Signal Stuck Lean Bank 1 Sensor 2 Conditions:** Engine started, battery voltage must be at least 11.5v, all electrical components must be off, parking brake must be engaged (to keep daytime driving lights off), automatic transmission selector must be in park. The ECM detected an unexpected voltage condition, or it detected an unexpected current draw in the heater circuit during the CCM test. **Note: Vehicle must be raised before connector for oxygen sensors is accessible.** **Possible Causes:** • Oxygen sensor (before catalytic converter) is faulty • Oxygen sensor heater (before catalytic converter) is faulty • Oxygen sensor heater (before catalytic converter) is faulty • Oxygen sensor heater (behind catalytic converter) is faulty • O2S circuit is open or shorted in the wiring harness • ECM has failed
DTC:P2271 **Passat:** 2.0L (BPY), 4.0L (BDP); **Jetta:** 2.0 (ABA, AEG, AVH, AZG, BBW, BEV), 2.5L (BGP, BGQ), 2.8L (AFP, BDF); **Golf:** 2.0L (ABA, AEG, AVH, AZG, BEV); **GTI:** 2.0L (ABA, AEG), 2.8L (AFP), 2.8L (AFP, BDF); **NB:** 1.8L (APH, AWP, AWV, BNU, BKF), 2.0L (AEG, AVH, AZG, BEV); **Nbvert:** (BDG); **Cabrio:** 2.0L (ABA) **Transmissions:** All	**O2 Sensor Signal Stuck Rich Bank 1 Sensor 2 Conditions:** Engine started, battery voltage must be at least 11.5v, all electrical components must be off, parking brake must be engaged (to keep daytime driving lights off), automatic transmission selector must be in park. The ECM detected an unexpected voltage condition, or it detected an unexpected current draw in the heater circuit during the CCM test. **Note: Vehicle must be raised before connector for oxygen sensors is accessible.** **Possible Causes:** • Oxygen sensor (before catalytic converter) is faulty • Oxygen sensor heater (before catalytic converter) is faulty • Oxygen sensor heater (before catalytic converter) is faulty • Oxygen sensor heater (behind catalytic converter) is faulty • O2S circuit is open or shorted in the wiring harness • ECM has failed
DTC:P2272 **Passat:** 4.0L (BDP); **Jetta:** 2.8L (AFP, BDF); **GTI:** 2.8L (AFP, BDF) **Transmissions:** All	**O2 Sensor Signal Stuck Lean Bank 2 Sensor 2 Conditions:** Engine started, battery voltage must be at least 11.5v, all electrical components must be off, parking brake must be engaged (to keep daytime driving lights off), automatic transmission selector must be in park. The ECM detected an unexpected voltage condition, or it detected an unexpected current draw in the heater circuit during the CCM test. **Note: Vehicle must be raised before connector for oxygen sensors is accessible.** **Possible Causes:** • Oxygen sensor (before catalytic converter) is faulty • Oxygen sensor heater (before catalytic converter) is faulty • Oxygen sensor heater (before catalytic converter) is faulty • Oxygen sensor heater (behind catalytic converter) is faulty • O2S circuit is open or shorted in the wiring harness • ECM has failed

DTC	Trouble Code Title, Conditions & Possible Causes
DTC:P2273 **Passat:** 4.0L (BDP); **Jetta:** 2.5L (BGP, BGQ), 2.8L (AFP, BDF); **GTI:** 2.8L (AFP, BDF) **Transmissions:** All	**O2 Sensor Signal Stuck Rich Bank 2 Sensor 2 Conditions:** Engine started, battery voltage must be at least 11.5v, all electrical components must be off, parking brake must be engaged (to keep daytime driving lights off), automatic transmission selector must be in park. The ECM detected an unexpected voltage condition, or it detected an unexpected current draw in the heater circuit during the CCM test. **Note: Vehicle must be raised before connector for oxygen sensors is accessible.** **Possible Causes:** • Oxygen sensor (before catalytic converter) is faulty • Oxygen sensor heater (before catalytic converter) is faulty • Oxygen sensor heater (before catalytic converter) is faulty • Oxygen sensor heater (behind catalytic converter) is faulty • O2S circuit is open or shorted in the wiring harness • ECM has failed
DTC:P2274 **Jetta:** 2.0L (BBW), 2.5L (BGP, BGQ); **NB:** 1.8L (BNU, BKF) **Transmissions:** All	**O2 Sensor Signal Stuck Lean Bank 2 Sensor 3 Conditions:** Engine started, battery voltage must be at least 11.5v, all electrical components must be off, parking brake must be engaged (to keep daytime driving lights off), automatic transmission selector must be in park. The ECM detected an unexpected voltage condition, or it detected an unexpected current draw in the heater circuit during the CCM test. **Note: Vehicle must be raised before connector for oxygen sensors is accessible.** **Possible Causes:** • Oxygen sensor (beforc catalytic converter) is faulty • Oxygen sensor heater (before catalytic converter) is faulty • Oxygen sensor heater (before catalytic converter) is faulty • Oxygen sensor heater (behind catalytic converter) is faulty • O2S circuit is open or shorted in the wiring harness • ECM has failed
DTC:P2275 **Jetta:** 2.0L (BBW), 2.5L (BGP, BGQ); **NB:** 1.8L (BNU, BKF) **Transmissions:** All	**O2 Sensor Signal Stuck Rich Bank 2 Sensor 3 Conditions:** Engine started, battery voltage must be at least 11.5v, all electrical components must be off, parking brake must be engaged (to keep daytime driving lights off), automatic transmission selector must be in park. The ECM detected an unexpected voltage condition, or it detected an unexpected current draw in the heater circuit during the CCM test. **Note: Vehicle must be raised before connector for oxygen sensors is accessible.** **Possible Causes:** • Oxygen sensor (beforc catalytic converter) is faulty • Oxygen sensor heater (before catalytic converter) is faulty • Oxygen sensor heater (before catalytic converter) is faulty • Oxygen sensor heater (behind catalytic converter) is faulty • O2S circuit is open or shorted in the wiring harness • ECM has failed
DTC:P2279 **Passat:** 2.0L (BPY), 4.0L (BDP); **Jetta:** 1.8L (AWD, AWP, AWW), 2.0 (ABA, BPY), 2.5L (BGP, BGQ), 2.8L (AFP, BDF); **Golf:** 1.8L (AWD, AWW), 2.0L (ABA); **GTI:** 1.8L (AWD, AWP, AWW), 2.0L (ABA), 2.8L (AFP, BDF); **NB:** 1.8L (APH, AWP, AWV, BNU, BKF); **Cabrio:** 2.0L (ABA) **Transmissions:** All	**Intake Air System Leak Conditions:** Engine running and thc vehicle speed more than 25mph, the ECM detected the intake air system has a potential leak. The IAT sensor is a variable resistor that includes an IAT signal circuit and a low reference circuit to measure the temperature of the air entering the engine. The ECM supplies the sensor with a low voltage singal circuit and a low reference ground circuit. When the IAT sensor is cold, its resistence is high. When the air temperature increases, its resistence decreases. With high sensor resisteance, the IAT sensor signal voltage is high. With lower sensor resistance, the IAT sensor signal voltage should be lower. **Possible Causes:** • Intake Manifold Runner Position Sensor is damaged or has failed • Intake system has leaks (false air) • Motor for intake flap is faulty • ECM has failed • IAT sensor signal circuit is shorted to sensor or chassis ground • IAT sensor is damaged or has failed • ECM has failed.
DTC:P2293 **Passat:** 2.0L (BPY); **Jetta:** 2.0L (BPY) **Transmissions:** All	**Fuel Pressure Regulator 2 Performance Conditions:** Engine started, battery voltage at least 11.5v, all electrical components off, ground connections between engine and chassis well connected, coolant temperature at least 80-degrees Celicius. The ECM detected a condition that affected the performance of the fule pressure regulator. **Possible Causes:** • Fuel Pressure Regulator Valve has failed • Fuel Pressure Sensor has failed • Fuel Pump (FP) Control Module has failed • Fuel pump has failed • ECM has failed

DTC	Trouble Code Title, Conditions & Possible Causes
DTC:P2294 **Passat:** 2.0L (BPY); **Jetta:** 2.0L (BPY) **Transmissions:** All	**Fuel Pressure Regulator 2 Control Circuit Conditions:** Engine started, battery voltage at least 11.5v, all electrical components off, ground connections between engine and chassis well connected, coolant temperature at least 80-degrees Celicius. The ECM detected a voltage condition that affected the performance of the fule pressure regulator. **Possible Causes:** • Fuel Pressure Regulator Valve has failed • Fuel Pressure Sensor has failed • Fuel Pump (FP) Control Module has failed • Fuel pump has failed • ECM has failed
DTC:P2295 **Passat:** 2.0L (BPY); **Jetta:** 2.0L (BPY) **Transmissions:** All	**Fuel Pressure Regulator 2 Control Circuit Low Conditions:** Engine started, battery voltage at least 11.5v, all electrical components off, ground connections between engine and chassis well connected, coolant temperature at least 80-degrees Celicius. The ECM detected a voltage condition that affected the performance of the fule pressure regulator. **Possible Causes:** • Fuel Pressure Regulator Valve has failed • Fuel Pressure Sensor has failed • Fuel Pump (FP) Control Module has failed • Fuel pump has failed • ECM has failed
DTC:P2296 **Passat:** 2.0L (BPY); **Jetta:** 2.0L (BPY) **Transmissions:** All	**Fuel Pressure Regulator 2 Control Circuit High Conditions:** Engine started, battery voltage at least 11.5v, all electrical components off, ground connections between engine and chassis well connected, coolant temperature at least 80-degrees Celicius. The ECM detected a voltage condition that affected the performance of the fule pressure regulator. **Possible Causes:** • Fuel Pressure Regulator Valve has failed • Fuel Pressure Sensor has failed • Fuel Pump (FP) Control Module has failed • Fuel pump has failed • ECM has failed
DTC:P2300 **Passat:** 2.0L (BPY) **Transmissions:** All	**Ignition Coil "A" Primary Control Circuit Low Conditions:** Engine started, battery voltage must be at least 11.5v, all electrical components must be off, parking brake must be engaged (to keep daytime driving lights off), automatic transmission selector must be in park and the ground between the engine and the chassis must be well connected. The ECM detected voltage values from the ignition module for the Ignition coilpack primary circuit that were outside the normal values required for proper function. **Note: Ignition coils and power output stages are one component and cannot be replaced individually.** **Possible Causes:** • Engine speed (RPM) sensor has failed • Camshaft Position (CMP) sensor has failed • Power Supply Relay is shorted to an open circuit • There is a malfunction in voltage supply • Ignition coilpack is damaged or it has failed • Cylinder 1 to 4 Fuel Injector(s) have failed • ECM has failed
DTC:P2301 **Passat:** 2.0L (BPY) **Transmissions:** All	**Ignition Coil "A" Primary Control Circuit High Conditions:** Engine started, battery voltage must be at least 11.5v, all electrical components must be off, parking brake must be engaged (to keep daytime driving lights off), automatic transmission selector must be in park and the ground between the engine and the chassis must be well connected. The ECM detected voltage values from the ignition module for the Ignition coilpack primary circuit that were outside the normal values required for proper function. **Note: Ignition coils and power output stages are one component and cannot be replaced individually.** **Possible Causes:** • Engine speed (RPM) sensor has failed • Camshaft Position (CMP) sensor has failed • Power Supply Relay is shorted to an open circuit • There is a malfunction in voltage supply • Ignition coilpack is damaged or it has failed • Cylinder 1 to 4 Fuel Injector(s) have failed • ECM has failed•

DTC	Trouble Code Title, Conditions & Possible Causes
DTC:P2303 **Passat:** 2.0L (BPY) **Transmissions:** All	**Ignition Coil "B" Primary Control Circuit Low Conditions:** Engine started, battery voltage must be at least 11.5v, all electrical components must be off, parking brake must be engaged (to keep daytime driving lights off), automatic transmission selector must be in park and the ground between the engine and the chassis must be well connected. The ECM detected voltage values from the ignition module for the Ignition coilpack primary circuit that were outside the normal values required for proper function. **Note: Ignition coils and power output stages are one component and cannot be replaced individually.** **Possible Causes:** • Engine speed (RPM) sensor has failed • Camshaft Position (CMP) sensor has failed • Power Supply Relay is shorted to an open circuit • There is a malfunction in voltage supply • Ignition coilpack is damaged or it has failed • Cylinder 1 to 4 Fuel Injector(s) have failed • ECM has failed•
DTC:P2304 **Passat:** 2.0L (BPY) **Transmissions:** All	**Ignition Coil "B" Primary Control Circuit High Conditions:** Engine started, battery voltage must be at least 11.5v, all electrical components must be off, parking brake must be engaged (to keep daytime driving lights off), automatic transmission selector must be in park and the ground between the engine and the chassis must be well connected. The ECM detected voltage values from the ignition module for the Ignition coilpack primary circuit that were outside the normal values required for proper function. **Note: Ignition coils and power output stages are one component and cannot be replaced individually.** **Possible Causes:** • Engine speed (RPM) sensor has failed • Camshaft Position (CMP) sensor has failed • Power Supply Relay is shorted to an open circuit • There is a malfunction in voltage supply • Ignition coilpack is damaged or it has failed • Cylinder 1 to 4 Fuel Injector(s) have failed • ECM has failed•
DTC:P2306 **Passat:** 2.0L (BPY) **Transmissions:** All	**Ignition Coil "C" Primary Control Circuit Low Conditions:** Engine started, battery voltage must be at least 11.5v, all electrical components must be off, parking brake must be engaged (to keep daytime driving lights off), automatic transmission selector must be in park and the ground between the engine and the chassis must be well connected. The ECM detected voltage values from the ignition module for the Ignition coilpack primary circuit that were outside the normal values required for proper function. **Note: Ignition coils and power output stages are one component and cannot be replaced individually.** **Possible Causes:** • Engine speed (RPM) sensor has failed • Camshaft Position (CMP) sensor has failed • Power Supply Relay is shorted to an open circuit • There is a malfunction in voltage supply • Ignition coilpack is damaged or it has failed • Cylinder 1 to 4 Fuel Injector(s) have failed • ECM has failed•
DTC:P2307 **Passat:** 2.0L (BPY) **Transmissions:** All	**Ignition Coil "C" Primary Control Circuit High Conditions:** Engine started, battery voltage must be at least 11.5v, all electrical components must be off, parking brake must be engaged (to keep daytime driving lights off), automatic transmission selector must be in park and the ground between the engine and the chassis must be well connected. The ECM detected voltage values from the ignition module for the Ignition coilpack primary circuit that were outside the normal values required for proper function. **Note: Ignition coils and power output stages are one component and cannot be replaced individually.** **Possible Causes:** • Engine speed (RPM) sensor has failed • Camshaft Position (CMP) sensor has failed • Power Supply Relay is shorted to open circuit • There is a malfunction in voltage supply • Cylinder 1 to 4 Fuel Injector(s) have failed • ECM has failed

DTC	Trouble Code Title, Conditions & Possible Causes
DTC:P2309 **Passat:** 2.8L (BPY) **Transmissions:** All	**Ignition Coil "D" Primary Control Circuit Low Conditions:** Engine started, battery voltage must be at least 11.5v, all electrical components must be off, parking brake must be engaged (to keep daytime driving lights off), automatic transmission selector must be in park and the ground between the engine and the chassis must be well connected. The ECM detected voltage values from the ignition module for the Ignition coilpack primary circuit that were outside the normal values required for proper function. **Note: Ignition coils and power output stages are one component and cannot be replaced individually.** **Possible Causes:** • Engine speed (RPM) sensor has failed • Camshaft Position (CMP) sensor has failed • Power Supply Relay is shorted to an open circuit • There is a malfunction in voltage supply • Ignition coilpack is damaged or it has failed • Cylinder 1 to 4 Fuel Injector(s) have failed • ECM has failed•
DTC:P2310 **Passat:** 2.0L (BPY) **Transmissions:** All	**Ignition Coil "D" Primary Control Circuit High Conditions:** Engine started, battery voltage must be at least 11.5v, all electrical components must be off, parking brake must be engaged (to keep daytime driving lights off), automatic transmission selector must be in park and the ground between the engine and the chassis must be well connected. The ECM detected voltage values from the ignition module for the Ignition coilpack primary circuit that were outside the normal values required for proper function. **Note: Ignition coils and power output stages are one component and cannot be replaced individually.** **Possible Causes:** • Engine speed (RPM) sensor has failed • Camshaft Position (CMP) sensor has failed • Power Supply Relay is shorted to an open circuit • There is a malfunction in voltage supply • Ignition coilpack is damaged or it has failed • Cylinder 1 to 4 Fuel Injector(s) have failed • ECM has failed•
DTC:P2336 **Passat:** 2.0L (BPY); **Jetta:** 2.8L (AFP, BDF); **GTI:** 2.8L (AFP, BDF) **Transmissions:** All	**Cylinder #1 Above Knock Threshold Conditions:** Engine running (ground connections between the engine and the chassis must be well connected), and the ECM detected that the knock sensor for this particular cylinder was above the threashold to properly function. **Note: Contact surfaces between knock sensor and cylinder block must be free of corrosion, dirt and grease. For the Knock Sensors to function properly, it is important for tightening torque to be exactly 20 Nm.** **Possible Causes:** • Contact surfaces between knock sensor and cylinder block are dirty • Knock Sensors not tightened to be exactly 20 Nm. • Knock sensor(s) have failed • Engine Control Module (ECM) is not connected or has failed • There is a fuel contamination • Engine equipment is lose or broken
DTC:P2337 **Passat:** 2.0L (BPY); **Jetta:** 2.8L (AFP, BDF); **GTI:** 2.8L (AFP, BDF) **Transmissions:** All	**Cylinder #2 Above Knock Threshold Conditions:** Engine running (ground connections between the engine and the chassis must be well connected), and the ECM detected that the knock sensor for this particular cylinder was above the threashold to properly function. **Note: Contact surfaces between knock sensor and cylinder block must be free of corrosion, dirt and grease. For the Knock Sensors to function properly, it is important for tightening torque to be exactly 20 Nm.** **Possible Causes:** • Contact surfaces between knock sensor and cylinder block are dirty • Knock Sensors not tightened to be exactly 20 Nm. • Knock sensor(s) have failed • Engine Control Module (ECM) is not connected or has failed • There is a fuel contamination • Engine equipment is lose or broken
DTC:P2338 **Passat:** 2.0L (BPY); **Jetta:** 2.8L (AFP, BDF); **GTI:** 2.8L (AFP, BDF) **Transmissions:** All	**Cylinder #3 Above Knock Threshold Conditions:** Engine running (ground connections between the engine and the chassis must be well connected), and the ECM detected that the knock sensor for this particular cylinder was above the threashold to properly function. **Note: Contact surfaces between knock sensor and cylinder block must be free of corrosion, dirt and grease. For the Knock Sensors to function properly, it is important for tightening torque to be exactly 20 Nm.** **Possible Causes:** • Contact surfaces between knock sensor and cylinder block are dirty • Knock Sensors not tightened to be exactly 20 Nm. • Knock sensor(s) have failed • Engine Control Module (ECM) is not connected or has failed • There is a fuel contamination • Engine equipment is lose or broken

DTC	Trouble Code Title, Conditions & Possible Causes
DTC:P2339 **Passat:** 2.0L (BPY); **Jetta:** 2.8L (AFP, BDF); **GTI:** 2.8L (AFP, BDF) **Transmissions:** All	**Cylinder #4 Above Knock Threshold Conditions:** Engine running (ground connections between the engine and the chassis must be well connected), and the ECM detected that the knock sensor for this particular cylinder was above the threashold to properly function. **Note: Contact surfaces between knock sensor and cylinder block must be free of corrosion, dirt and grease. For the Knock Sensors to function properly, it is important for tightening torque to be exactly 20 Nm.** **Possible Causes:** • Contact surfaces between knock sensor and cylinder block are dirty • Knock Sensors not tightened to be exactly 20 Nm. • Knock sensor(s) have failed • Engine Control Module (ECM) is not connected or has failed • There is a fuel contamination • Engine equipment is lose or broken
DTC:P2340 **Jetta:** 2.8L (AFP, BDF); **GTI:** 2.8L (AFP, BDF) **Transmissions:** All	**Cylinder #5 Above Knock Threshold Conditions:** Engine running (ground connections between the engine and the chassis must be well connected), and the ECM detected that the knock sensor for this particular cylinder was above the threashold to properly function. **Note: Contact surfaces between knock sensor and cylinder block must be free of corrosion, dirt and grease. For the Knock Sensors to function properly, it is important for tightening torque to be exactly 20 Nm.** **Possible Causes:** • Contact surfaces between knock sensor and cylinder block are dirty • Knock Sensors not tightened to be exactly 20 Nm. • Knock sensor(s) have failed • Engine Control Module (ECM) is not connected or has failed • There is a fuel contamination • Engine equipment is lose or broken
DTC:P2341 **Jetta:** 2.8L (AFP, BDF); **GTI:** 2.8L (AFP, BDF) **Transmissions:** All	**Cylinder #6 Above Knock Threshold Conditions:** Engine running (ground connections between the engine and the chassis must be well connected), and the ECM detected that the knock sensor for this particular cylinder was above the threashold to properly function. **Note: Contact surfaces between knock sensor and cylinder block must be free of corrosion, dirt and grease. For the Knock Sensors to function properly, it is important for tightening torque to be exactly 20 Nm.** **Possible Causes:** • Contact surfaces between knock sensor and cylinder block are dirty • Knock Sensors not tightened to be exactly 20 Nm. • Knock sensor(s) have failed • Engine Control Module (ECM) is not connected or has failed • There is a fuel contamination • Engine equipment is lose or broken
DTC:P2400 **2T, MIL: Yes** **Passat:** 1.8L (ATW, AUG), 2.8L (AHA, ATQ), 2.0L (BPY), 4.0L (BDP); **Jetta:** 1.8L (AWD, AWP, AWW), 2.0L (AEG, AVH, AZG, BBW, BEV), 2.5L (BGP, BGQ), 2.8L (AFP, BDF); **Golf:** 1.8L (AWD, AWW), 2.0L (AEG, AVH, AZG, BEV); **GTI:** 1.8L (AWD, AWP, AWW), 2.0L (AEG), 2.8L (AFP, BDF); **NB:** 1.8L (APH, AWP, AWV, BNU, BKF), 2.0L (AEG, AVH, AZG, BEV); **Nbvert:** (BDG) **Transmissions:** All	**EVAP Leak Detection Pump (LDP) Control Circuit Open Conditions:** Engine started, battery voltage must be at least 11.5v, all electrical components must be off, parking brake must be engaged (to keep daytime driving lights off), automatic transmission selector must be in park, the exhaust system must be properly sealed between the catalytic converter and the cylinder head, coolant temperature must be at least 80 degrees Celsius and oxygen sensor heaters for oxygen sensors before the catalytic converter must be functioning properly and the ground between the engine and the chassis must be well connected. The ECM detected voltage irregularity in the leak detection pump control circuit. **Possible Causes:** • EVAP LDP power supply circuit is open • EVAP LDP solenoid valve is damaged or it has failed • EVAP LDP canister has a leak or a poor seal • ECM has failed • EVAP canister system has an improper seal • Evaporative Emission (EVAP) canister purge regulator valve 1 has failed • Leak Detection Pump (LDP) is faulty • Aftermarket EVAP parts that do not conform to specifications • EVAP component seals leaking (i.e., leaks in the Purge valve, fuel tank pressure sensor, canister vent solenoid, fuel vapor control valve tube assembly or fuel vapor vent valve).

DTC	Trouble Code Title, Conditions & Possible Causes
DTC:P2401 **2T, MIL: Yes** **Passat:** 1.8L (ATW, AUG), 2.8L (AHA, ATQ), 2.0L (BPY), 4.0L (BDP); **Jetta:** 1.8L (AWD, AWP, AWW), 2.0L (AEG, AVH, AZG, BBW, BEV), 2.5L (BGP, BGQ), 2.8L (AFP, BDF); **Golf:** 1.8L (AWD, AWW), 2.0L (AEG, AVH, AZG, BEV); **GTI:** 1.8L (AWD, AWP, AWW), 2.0L (AEG), 2.8L (AFP, BDF); **NB:** 1.8L (APH, AWP, AWV, BNU, BKF), 2.0L (AEG, AVH, AZG, BEV); **Nbvert:** (BDG) **Transmissions:** All	**EVAP Leak Detection Pump Control Circuit Low Conditions:** Engine started, battery voltage must be at least 11.5v, all electrical components must be off, parking brake must be engaged (to keep daytime driving lights off), automatic transmission selector must be in park, the exhaust system must be properly sealed between the catalytic converter and the cylinder head, coolant temperature must be at least 80 degrees Celsius and oxygen sensor heaters for oxygen sensors before the catalytic converter must be functioning properly and the ground between the engine and the chassis must be well connected. The ECM detected voltage irregularity in the leak detection pump control circuit. **Possible Causes:** • EVAP LDP power supply circuit is open • EVAP LDP solenoid valve is damaged or it has failed • EVAP LDP canister has a leak or a poor seal • ECM has failed • EVAP canister system has an improper seal • Evaporative Emission (EVAP) canister purge regulator valve 1 has failed • Leak Detection Pump (LDP) is faulty • Aftermarket EVAP parts that do not conform to specifications • EVAP component seals leaking (i.e., leaks in the Purge valve, fuel tank pressure sensor, canister vent solenoid, fuel vapor control valve tube assembly or fuel vapor vent valve).
DTC:P2402 **2T, MIL: Yes** **Passat:** 1.8L (ATW, AUG), 2.8L (AHA, ATQ), 2.0L (BPY), 4.0L (BDP); **Jetta:** 1.8L (AWD, AWP, AWW), 2.0L (AEG, AVH, AZG, BBW, BEV), 2.5L (BGP, BGQ), 2.8L (AFP, BDF); **Golf:** 1.8L (AWD, AWW), 2.0L (AEG, AVH, AZG, BEV); **GTI:** 1.8L (AWD, AWP, AWW), 2.0L (AEG), 2.8L (AFP, BDF); **NB:** 1.8L (APH, AWP, AWV, BNU, BKF), 2.0L (AEG, AVH, AZG, BEV); **Nbvert:** (BDG) **Transmissions:** All	**EVAP Leak Detection Pump Control Circuit High Conditions:** Engine started, battery voltage must be at least 11.5v, all electrical components must be off, parking brake must be engaged (to keep daytime driving lights off), automatic transmission selector must be in park, the exhaust system must be properly sealed between the catalytic converter and the cylinder head, coolant temperature must be at least 80 degrees Celsius and oxygen sensor heaters for oxygen sensors before the catalytic converter must be functioning properly and the ground between the engine and the chassis must be well connected. The ECM detected voltage irregularity in the leak detection pump control circuit. **Possible Causes:** • EVAP LDP power supply circuit is open • EVAP LDP solenoid valve is damaged or it has failed • EVAP LDP canister has a leak or a poor seal • ECM has failed • EVAP canister system has an improper seal • Evaporative Emission (EVAP) canister purge regulator valve 1 has failed • Leak Detection Pump (LDP) is faulty • Aftermarket EVAP parts that do not conform to specifications • EVAP component seals leaking (i.e., leaks in the Purge valve, fuel tank pressure sensor, canister vent solenoid, fuel vapor control valve tube assembly or fuel vapor vent valve).
DTC:P2403 **2T, MIL: Yes** **Passat:** 1.8L (ATW, AUG), 2.8L (AHA, ATQ), 2.0L (BPY), 4.0L (BDP); **Jetta:** 1.8L (AWD, AWP, AWW), 2.0L (AEG, AVH, AZG, BBW, BEV), 2.5L (BGP, BGQ), 2.8L (AFP, BDF); **Golf:** 1.8L (AWD, AWW), 2.0L (AEG, AVH, AZG, BEV); **GTI:** 1.8L (AWD, AWP, AWW), 2.8L (AFP, BDF); **NB:** 1.8L (APH, AWP, AWV, BNU, BKF), 2.0L (AEG, AVH, AZG, BEV); **Nbvert:** (BDG) **Transmissions:** All	**EVAP Leak Detection Pump Sense Circuit Open Conditions:** Engine started, battery voltage must be at least 11.5v, all electrical components must be off, parking brake must be engaged (to keep daytime driving lights off), automatic transmission selector must be in park, the exhaust system must be properly sealed between the catalytic converter and the cylinder head, coolant temperature must be at least 80 degrees Celsius and oxygen sensor heaters for oxygen sensors before the catalytic converter must be functioning properly and the ground between the engine and the chassis must be well connected. The ECM detected voltage irregularity in the leak detection pump control circuit. **Possible Causes:** • EVAP LDP power supply circuit is open • EVAP LDP solenoid valve is damaged or it has failed • EVAP LDP canister has a leak or a poor seal • ECM has failed • EVAP canister system has an improper seal • Evaporative Emission (EVAP) canister purge regulator valve 1 has failed • Leak Detection Pump (LDP) is faulty • Aftermarket EVAP parts that do not conform to specifications • EVAP component seals leaking (i.e., leaks in the Purge valve, fuel tank pressure sensor, canister vent solenoid, fuel vapor control valve tube assembly or fuel vapor vent valve).

DTC	Trouble Code Title, Conditions & Possible Causes
DTC:P2404 **2T, MIL: Yes** **Passat:** 1.8L (ATW, AUG), 2.8L (AHA, ATQ), 2.0L (BPY), 4.0L (BDP); **Jetta:** 1.8L (AWD, AWP, AWW), 2.0L (AEG, AVH, AZG, BBW, BEV), 2.5L (BGP, BGQ), 2.8L (AFP, BDF); **Golf:** 1.8L (AWD, AWW), 2.0L (AEG, AVH, AZG, BEV); **GTI:** 1.8L (AWD, AWP, AWW), 2.0L (AEG), 2.8L (AFP, BDF); **NB:** 1.8L (APH, AWP, AWV, BNU, BKF), 2.0L (AEG, AVH, AZG, BEV); **Nbvert:** (BDG) **Transmissions:** All	**EVAP Leak Detection Pump Sense Circuit Range/Performance Conditions:** Engine started, battery voltage must be at least 11.5v, all electrical components must be off, parking brake must be engaged (to keep daytime driving lights off), automatic transmission selector must be in park, the exhaust system must be properly sealed between the catalytic converter and the cylinder head, coolant temperature must be at least 80 degrees Celsius and oxygen sensor heaters for oxygen sensors before the catalytic converter must be functioning properly and the ground between the engine and the chassis must be well connected. The ECM detected voltage irregularity in the leak detection pump control circuit. **Possible Causes:** • EVAP LDP power supply circuit is open • EVAP LDP solenoid valve is damaged or it has failed • EVAP LDP canister has a leak or a poor seal • ECM has failed • EVAP canister system has an improper seal • Evaporative Emission (EVAP) canister purge regulator valve 1 has failed • Leak Detection Pump (LDP) is faulty • Aftermarket EVAP parts that do not conform to specifications • EVAP component seals leaking (i.e., leaks in the Purge valve, fuel tank pressure sensor, canister vent solenoid, fuel vapor control valve tube assembly or fuel vapor vent valve).
DTC:P2414 **Passat:** 2.0L (BPY); **Jetta:** 2.0L (AEG, AVH, AZG, BBW, BEV, BPY), 2.5L (BGP, BGQ); **Golf:** 2.0L (AEG, AVH, AZG, BEV), 2.0L (AEG); **NB:** 1.8L (APH, AWP, AWV, BNU, BKF), 2.0L (AEG, AVH, AZG, BEV); **Nbvert:** (BDG) **Transmissions:** All	**O2 Sensor Exhaust Sample Error Bank 1 Sensor 1 Conditions:** Engine running (ground connections between the engine and the chassis must be well connected), and the ECM detected an error on the OS Sensor. **Note: Intake Flap Motor and Intake Manifold Runner Position Sensor are one component and cannot be replaced individually.** **Note: Vacuum in the intake system sucks in the leak detection spray with false air. Leak detection spray decreases ignition quality of the fuel mixture. This causes a drop in engine speed and changes the value produced by the Heated Oxygen Sensor (HO2S).** **Possible Causes:** • Intake Manifold Runner Position Sensor is damaged or has failed • Intake system has leaks (false air) • Motor for intake flap is faulty • ECM has failed • Oxygen sensor (before catalytic converter) is faulty • Oxygen sensor heater (before catalytic converter) is faulty • Oxygen sensor heater (before catalytic converter) is faulty • Oxygen sensor heater (behind catalytic converter) is faulty • O2S circuit is open or shorted in the wiring harness
DTC:P2416 **NB:** 1.8L (BNU, BKF) **Transmissions:** All	**O2 Sensor Signals Swapped Bank 1 Sensor 2 / Bank 1 Sensor 3 Conditions:** Engine started, battery voltage must be at least 11.5v, all electrical components must be off, the ground between the engine and the chassis must be well connected, the exhaust system must be properly sealed between the catalytic converter and the cylinder head, and the coolant temperature must be 80 degrees Celsius. The ECM detected the O2 signals were mixed and reading implausible results from both. **Possible Causes:** • HO2S-12 and HO2S-13 harness connectors are swapped • HO2S-12 and HO2S-13 wiring is crossed inside the harness • HO2S-12 and HO2S-13 wires are crossed at 104-pin connector • Connector coding and color mixed with correct catalytic converter
DTC:P2422 **Jetta:** 2.8L (AFP, BDF); **GTI:** 2.8L (AFP, BDF) **Transmissions:** All	**Evaporative Emission System Vent Valve Stuck Closed Conditions:** Engine started, battery voltage must be at least 11.5v, all electrical components must be off, parking brake must be engaged (to keep daytime driving lights off), automatic transmission selector must be in park, the exhaust system must be properly sealed between the catalytic converter and the cylinder head, coolant temperature must be at least 80 degrees Celsius and oxygen sensor heaters for oxygen sensors before the catalytic converter must be functioning properly and the ground between the engine and the chassis must be well connected. The ECM detected an unexpected EVAP malfunction. **Note: Solenoid valve is closed when no voltage is present.** **Possible Causes:** • EVAP power supply circuit is open • EVAP solenoid control circuit is open or shorted to ground • EVAP solenoid control circuit is shorted to power (B+) • EVAP solenoid valve is damaged or it has failed • EVAP canister purge solenoid valve is faulty • ECM has failed

DTC	Trouble Code Title, Conditions & Possible Causes
DTC:P2539 **Passat:** 2.0L (BPY) **Transmissions:** All	**Low Pressure Fuel System Sensor Circuit Conditions:** Engine started, battery voltage must be at least 11.5v, all electrical components must be off, parking brake must be engaged (to keep daytime driving lights off), automatic transmission selector must be in park, the exhaust system must be properly sealed between the catalytic converter and the cylinder head, coolant temperature must be at least 80 degrees Celsius. The ECM detected an error on the fuel system sensor circuit. **Note: The specified fuel pressure should be between 3000 to 5000 kPA** **Possible Causes:** • Fuel Pressure Regulator Valve has failed • Fuel Pressure Sensor has failed • Fuel Pump (FP) Control Module has failed • Fuel pump has failed • ECM has failed
DTC:P2540 **Passat:** 2.0L (BPY) **Transmissions:** All	**Low Pressure Fuel System Sensor Circuit Range/Performance Conditions:** Engine started, battery voltage must be at least 11.5v, all electrical components must be off, parking brake must be engaged (to keep daytime driving lights off), automatic transmission selector must be in park, the exhaust system must be properly sealed between the catalytic converter and the cylinder head, coolant temperature must be at least 80 degrees Celsius. The ECM detected an error on the fuel system sensor circuit. **Note: The specified fuel pressure should be between 3000 to 5000 kPA** **Possible Causes:** • Fuel Pressure Regulator Valve has failed • Fuel Pressure Sensor has failed • Fuel Pump (FP) Control Module has failed • Fuel pump has failed • ECM has failed
DTC:P2541 **Passat:** 2.0L (BPY); **Jetta:** 2.0L (BPY) **Transmissions:** All	**Low Pressure Fuel System Sensor Circuit Low Conditions:** Engine started, battery voltage must be at least 11.5v, all electrical components must be off, parking brake must be engaged (to keep daytime driving lights off), automatic transmission selector must be in park, the exhaust system must be properly sealed between the catalytic converter and the cylinder head, coolant temperature must be at least 80 degrees Celsius. The ECM detected an error on the fuel system sensor circuit. **Note: The specified fuel pressure should be between 3000 to 5000 kPA** **Possible Causes:** • Fuel Pressure Regulator Valve has failed • Fuel Pressure Sensor has failed • Fuel Pump (FP) Control Module has failed • Fuel pump has failed • ECM has failed
DTC:P2600 **Passat:** 2.0L (BPY), 4.0L (BDP) **Transmissions:** All	**Coolant Pump Control Circuit/Open Conditions:** Key on, engine started, the ECM detected an unexpected voltage condition on the ECT circuit. **Possible Causes:** • Check activation of Recirculation Pump Relay • Recirculation pump has failed • Coolant Temperature Sensor has failed • Circuit short to ground, open or other component • ECM has failed
DTC:P2602 **Passat:** 2.0L (BPY), 4.0L (BDP) **Transmissions:** All	**Coolant Pump Control Circuit Low Conditions:** Key on, engine started, the ECM detected an unexpected voltage condition on the ECT circuit. **Possible Causes:** • Check activation of Recirculation Pump Relay • Recirculation pump has failed • Coolant Temperature Sensor has failed • Circuit short to ground, open or other component • ECM has failed
DTC:P2603 **Passat:** 2.0L (BPY), 4.0L (BDP) **Transmissions:** All	**Coolant Pump Control Circuit High Conditions:** Key on, engine started, the ECM detected an unexpected voltage condition on the ECT circuit. **Possible Causes:** • Check activation of Recirculation Pump Relay • Recirculation pump has failed • Coolant Temperature Sensor has failed • Circuit short to ground, open or other component • ECM has failed

DTC	Trouble Code Title, Conditions & Possible Causes
DTC:P2626 **Passat:** 2.0L (BPY), 4.0L (BDP); **Jetta:** 2.0 (ABA, AEG, AVH, AZG, BBW, BEV, BPY), 2.5L (BGP, BGQ), 2.8L (AFP, BDF); **Golf:** 2.0L (ABA, AEG, AVH, AZG, BEV); **GTI:** 2.0L (ABA, AEG), 2.8L (AFP, BDF); **NB:** 1.8L (APH, AWP, AWV, BNU, BKF), 2.0L (AEG, AVH, AZG, BEV); **Nbvert:** (BDG); **Cabrio:** 2.0L (ABA) **Transmissions:** All	**O2 Sensor Pumping Current Trim Circuit/Open Bank 1 Sensor 1 Conditions:** Engine started, battery voltage must be at least 11.5v, all electrical components must be off, parking brake must be engaged (to keep daytime driving lights off), automatic transmission selector must be in park, the exhaust system must be properly sealed between the catalytic converter and the cylinder head, coolant temperature must be at least 80 degrees Celsius and oxygen sensor heaters for oxygen sensors before the catalytic converter must be functioning properly and the ground between the engine and the chassis must be well connected. The ECM detected a voltage value that doesn't fall within the desired parameters for a properly functioning O2 system. **Possible Causes:** • Check activation of Recirculation Pump Relay • Oxygen sensor (before catalytic converter) is faulty • Oxygen sensor (behind catalytic converter) is faulty • Oxygen sensor heater (before catalytic converter) is faulty • Oxygen sensor heater (behind catalytic converter) is faulty
DTC:P2629 **Passat:** 2.0L (BPY), 4.0L (BDP); **Jetta:** 2.8L (AFP, BDF); **GTI:** 2.8L (AFP, BDF) **Transmissions:** All	**O2 Sensor Pumping Current Trim Circuit/Open Bank 2 Sensor 1 Conditions:** Engine started, battery voltage must be at least 11.5v, all electrical components must be off, parking brake must be engaged (to keep daytime driving lights off), automatic transmission selector must be in park, the exhaust system must be properly sealed between the catalytic converter and the cylinder head, coolant temperature must be at least 80 degrees Celsius and oxygen sensor heaters for oxygen sensors before the catalytic converter must be functioning properly and the ground between the engine and the chassis must be well connected. The ECM detected a voltage value that doesn't fall within the desired parameters for a properly functioning O2 system. **Possible Causes:** • Check activation of Recirculation Pump Relay • Oxygen sensor (before catalytic converter) is faulty • Oxygen sensor (behind catalytic converter) is faulty • Oxygen sensor heater (before catalytic converter) is faulty • Oxygen sensor heater (behind catalytic converter) is faulty
DTC:P2637 **Passat:** 2.0L (BPY); **Jetta:** 1.8L (AWD, AWP, AWW), 2.5L (BGP, BGQ), 2.8L (AFP, BDF); **Golf:** 1.8L (AWD, AWW); **GTI:** 1.8L (AWD, AWP, AWW), 2.8L (AFP, BDF); **NB:** 1.8L (APH, AWP, AWV, BNU, BKF), 2.0L (AEG, AVH, AZG, BEV); **Nbvert:** (BDG) **Transmissions:** All	**Torque Management Feedback Signal "A" Conditions:** Engine started, battery voltage must be at least 11.5v, all electrical components must be off, parking brake must be engaged (to keep daytime driving lights off), automatic transmission selector must be in park, the exhaust system must be properly sealed between the catalytic converter and the cylinder head, coolant temperature must be at least 80 degrees Celsius and oxygen sensor heaters for oxygen sensors before the catalytic converter must be functioning properly and the ground between the engine and the chassis must be well connected. The ECM detected a voltage value on the torque management circuits that doesn't fall within the desired parameters **Possible Causes:** • Engine Control Module (ECM) has failed • Voltage supply for Engine Control Module (ECM) is damaged • Engine Coolant Temperature (ECT) sensor has failed • Intake Air Temperature (IAT) sensor has failed • Intake Manifold Runner Position Sensor has failed • Intake system has leaks (false air) • Motor for intake flap has failed • Mass Air Flow (MAF) sensor has failed
DTC:P2714 **Passat:** 1.8L (ATW, AUG), 2.0L (BPY); **NB:** 1.8L (APH, AWP, AWV, BNU, BKF) **Transmissions:** All	**Pressure Control Solenoid "D" Performance or Stuck Off Conditions:** Engine started, battery voltage must be at least 11.5v, all electrical components must be off, and the ground between the engine and the chassis must be well connected. The ECM detected the pressure control solenoid was in the "stuck off" position. **Possible Causes:** • ATF level is low • Circuit harness connector contacts are corroded or ingressed of water • Circuit wires have shorted to each other, to battery or ground • Automatic Transmission Hydraulic Pressure Sensor 1 has failed • Solenoid valves in valve body are faulty • Transmission Control Module (TCM) needs replacing • Transmission Input Speed (RPM) Sensor has failed • Transmission Output Speed (RPM) Sensor has failed

DTC	Trouble Code Title, Conditions & Possible Causes
DTC:P2715 **Passat:** 1.8L (ATW, AUG), 2.0L (BPY); **NB:** 1.8L (APH, AWP, AWV, BNU, BKF) **Transmissions:** All	**Pressure Control Solenoid "D" Performance or Stuck On Conditions:** Engine started, battery voltage must be at least 11.5v, all electrical components must be off, and the ground between the engine and the chassis must be well connected. The ECM detected the pressure control solenoid was in the "stuck on" position. **Possible Causes:** • ATF level is low • Circuit harness connector contacts are corroded or ingressed of water • Circuit wires have shorted to each other, to battery or ground • Automatic Transmission Hydraulic Pressure Sensor 1 has failed • Solenoid valves in valve body are faulty • Transmission Control Module (TCM) needs replacing • Transmission Input Speed (RPM) Sensor has failed • Transmission Output Speed (RPM) Sensor has failed
DTC:P2716 **Passat:** 1.8L (ATW, AUG), 2.0L (BPY); **Jetta:** 2.0L (BPY), 2.5L (BGP, BGQ); **NB:** 1.8L (APH, AWP, AWV, BNU, BKF), 2.0L (AEG, AVH, AZG, BEV); **Nbvert:** (BDG) **Transmissions:** All	**Pressure Control Solenoid "D" Electrical Malfunction Conditions:** Engine started, battery voltage must be at least 11.5v, all electrical components must be off, and the ground between the engine and the chassis must be well connected. The ECM detected the pressure control solenoid was experiencing electrical malfunctions. **Possible Causes:** • ATF level is low • Circuit harness connector contacts are corroded or ingressed of water • Circuit wires have shorted to each other, to battery or ground • Automatic Transmission Hydraulic Pressure Sensor 1 has failed • Solenoid valves in valve body are faulty • Transmission Control Module (TCM) needs replacing • Transmission Input Speed (RPM) Sensor has failed • Transmission Output Speed (RPM) Sensor has failed
DTC:P2717 **Passat:** 1.8L (ATW, AUG), 2.0L (BPY); **NB:** 1.8L (APH, AWP, AWV, BNU, BKF) **Transmissions:** All	**Pressure Control Solenoid "D" Intermittent Conditions:** Engine started, battery voltage must be at least 11.5v, all electrical components must be off, and the ground between the engine and the chassis must be well connected. The ECM detected the pressure control solenoid sending intermittent signals. **Possible Causes:** • ATF level is low • Circuit harness connector contacts are corroded or ingressed of water • Circuit wires have shorted to each other, to battery or ground • Automatic Transmission Hydraulic Pressure Sensor 1 has failed • Solenoid valves in valve body are faulty • Transmission Control Module (TCM) needs replacing • Transmission Input Speed (RPM) Sensor has failed • Transmission Output Speed (RPM) Sensor has failed
DTC:P2723 **Passat:** 2.0L (BPY); **Jetta:** 2.0L (BPY), 2.5L (BGP, BGQ); **NB:** 1.8L (APH, AWP, AWV, BNU, BKF), 2.0L (AEG, AVH, AZG, BEV); **Nbvert:** (BDG) **Transmissions:** All	**Pressure Control Solenoid "E" Performance or Stuck Off Conditions:** Engine started, battery voltage must be at least 11.5v, all electrical components must be off, and the ground between the engine and the chassis must be well connected. The ECM detected the pressure control solenoid was in the "stuck off" position. **Possible Causes:** • ATF level is low • Circuit harness connector contacts are corroded or ingressed of water • Circuit wires have shorted to each other, to battery or ground • Automatic Transmission Hydraulic Pressure Sensor 1 has failed • Solenoid valves in valve body are faulty • Transmission Control Module (TCM) needs replacing • Transmission Input Speed (RPM) Sensor has failed • Transmission Output Speed (RPM) Sensor has failed
DTC:P2724 **Passat:** 2.0L (BPY); **Jetta:** 2.0L (BPY), 2.5L (BGP, BGQ); **NB:** 1.8L (APH, AWP, AWV, BNU, BKF), 2.0L (AEG, AVH, AZG, BEV); **Nbvert:** (BDG) **Transmissions:** All	**Pressure Control Solenoid "E" Performance or Stuck On Conditions:** Engine started, battery voltage must be at least 11.5v, all electrical components must be off, and the ground between the engine and the chassis must be well connected. The ECM detected the pressure control solenoid was in the "stuck on" position. **Possible Causes:** • ATF level is low • Circuit harness connector contacts are corroded or ingressed of water • Circuit wires have shorted to each other, to battery or ground • Automatic Transmission Hydraulic Pressure Sensor 1 has failed • Solenoid valves in valve body are faulty • Transmission Control Module (TCM) needs replacing • Transmission Input Speed (RPM) Sensor has failed • Transmission Output Speed (RPM) Sensor has failed

DTC	Trouble Code Title, Conditions & Possible Causes
DTC:P2725 **Passat:** 2.0L (BPY); **Jetta:** 2.0L (BPY), 2.5L (BGP, BGQ); **NB:** 1.8L (APH, AWP, AWV, BNU, BKF), 2.0L (AEG, AVH, AZG, BEV); **Nbvert:** (BDG) **Transmissions:** All	**Pressure Control Solenoid "E" Electrical Malfunction Conditions:** Engine started, battery voltage must be at least 11.5v, all electrical components must be off, and the ground between the engine and the chassis must be well connected. The ECM detected the pressure control solenoid was experiencing electrical malfunctions. **Possible Causes:** • ATF level is low • Circuit harness connector contacts are corroded or ingressed of water • Circuit wires have shorted to each other, to battery or ground • Automatic Transmission Hydraulic Pressure Sensor 1 has failed • Solenoid valves in valve body are faulty • Transmission Control Module (TCM) needs replacing • Transmission Input Speed (RPM) Sensor has failed • Transmission Output Speed (RPM) Sensor has failed
DTC:P2726 **Passat:** 2.0L (BPY); **Jetta:** 2.0L (BPY), 2.5L (BGP, BGQ); **NB:** 1.8L (APH, AWP, AWV, BNU, BKF), 2.0L (AEG, AVH, AZG, BEV); **Nbvert:** (BDG) **Transmissions:** All	**Pressure Control Solenoid "E" Intermittent Conditions:** Engine started, battery voltage must be at least 11.5v, all electrical components must be off, and the ground between the engine and the chassis must be well connected. The ECM detected the pressure control solenoid sending intermittent signals. **Possible Causes:** • ATF level is low • Circuit harness connector contacts are corroded or ingressed of water • Circuit wires have shorted to each other, to battery or ground • Automatic Transmission Hydraulic Pressure Sensor 1 has failed • Solenoid valves in valve body are faulty • Transmission Control Module (TCM) needs replacing • Transmission Input Speed (RPM) Sensor has failed • Transmission Output Speed (RPM) Sensor has failed
DTC:P2732 **Passat:** 2.0L (BPY); **Jetta:** 2.0L (BPY), 2.5L (BGP, BGQ); **NB:** 1.8L (APH, AWP, AWV, BNU, BKF), 2.0L (AEG, AVH, AZG, BEV); **Nbvert:** (BDG) **Transmissions:** All	**Pressure Control Solenoid "F" Performance or Stuck Off Conditions:** Engine started, battery voltage must be at least 11.5v, all electrical components must be off, and the ground between the engine and the chassis must be well connected. The ECM detected the pressure control solenoid was in the "stuck off" position. **Possible Causes:** • ATF level is low • Circuit harness connector contacts are corroded or ingressed of water • Circuit wires have shorted to each other, to battery or ground • Automatic Transmission Hydraulic Pressure Sensor 1 has failed • Solenoid valves in valve body are faulty • Transmission Control Module (TCM) needs replacing • Transmission Input Speed (RPM) Sensor has failed • Transmission Output Speed (RPM) Sensor has failed
DTC:P2733 **Passat:** 2.0L (BPY); **Jetta:** 2.0L (BPY), 2.5L (BGP, BGQ); **NB:** 1.8L (APH, AWP, AWV, BNU, BKF), 2.0L (AEG, AVH, AZG, BEV); **Nbvert:** (BDG) **Transmissions:** All	**Pressure Control Solenoid "F" Performance or Stuck On Conditions:** Engine started, battery voltage must be at least 11.5v, all electrical components must be off, and the ground between the engine and the chassis must be well connected. The ECM detected the pressure control solenoid was in the "stuck on" position. **Possible Causes:** • ATF level is low • Circuit harness connector contacts are corroded or ingressed of water • Circuit wires have shorted to each other, to battery or ground • Automatic Transmission Hydraulic Pressure Sensor 1 has failed • Solenoid valves in valve body are faulty • Transmission Control Module (TCM) needs replacing • Transmission Input Speed (RPM) Sensor has failed • Transmission Output Speed (RPM) Sensor has failed
DTC:P2734 **Passat:** 2.0L (BPY); **Jetta:** 2.0L (BPY), 2.5L (BGP, BGQ); **NB:** 1.8L (APH, AWP, AWV, BNU, BKF), 2.0L (AEG, AVH, AZG, BEV); **Nbvert:** (BDG) **Transmissions:** All	**Pressure Control Solenoid "F" Electrical Malfunction Conditions:** Engine started, battery voltage must be at least 11.5v, all electrical components must be off, and the ground between the engine and the chassis must be well connected. The ECM detected the pressure control solenoid was experiencing electrical malfunctions. **Possible Causes:** • ATF level is low • Circuit harness connector contacts are corroded or ingressed of water • Circuit wires have shorted to each other, to battery or ground • Automatic Transmission Hydraulic Pressure Sensor 1 has failed • Solenoid valves in valve body are faulty • Transmission Control Module (TCM) needs replacing • Transmission Input Speed (RPM) Sensor has failed • Transmission Output Speed (RPM) Sensor has failed

DTC	Trouble Code Title, Conditions & Possible Causes
DTC:P2735 **Passat:** 2.0L (BPY); **Jetta:** 2.0L (BPY), 2.5L (BGP, BGQ); **NB:** 1.8L (APH, AWP, AWV, BNU, BKF), 2.0L (AEG, AVH, AZG, BEV); **Nbvert:** (BDG) **Transmissions:** All	**Pressure Control Solenoid "F" Intermittent Conditions:** Engine started, battery voltage must be at least 11.5v, all electrical components must be off, and the ground between the engine and the chassis must be well connected. The ECM detected the pressure control solenoid sending intermittent signals. **Possible Causes:** • ATF level is low • Circuit harness connector contacts are corroded or ingressed of water • Circuit wires have shorted to each other, to battery or ground • Automatic Transmission Hydraulic Pressure Sensor 1 has failed • Solenoid valves in valve body are faulty • Transmission Control Module (TCM) needs replacing • Transmission Input Speed (RPM) Sensor has failed • Transmission Output Speed (RPM) Sensor has failed

Gas Engine OBD II Trouble Code List (P3XXX Codes)

DTC	Trouble Code Title, Conditions & Possible Causes:
DTC: P3028 **Jetta:** 2.5L (BGP, BGQ) **Transmissions:** All	**Throttle Actuation 2 Potentiometer Sign.2 Range/Performance Conditions:** Engine started, battery voltage must be at least 11.5v, all electrical components must be off, the ground between the engine and the chassis must be well connected, coolant temperature must be at least 80 degrees Celsius and the accelerator pedal must be properly adjusted. The ECM detected an incorrect singal from the throttle potentiometer. **Note: If the complete throttle valve control module is current-less (e.g. connector disconnected) the throttle valve moves into a particular, specified mechanical position, which signals an increased idle speed with an engine at operating temperature. If only the Throttle Position (TP) actuator is current-less, the throttle valve also moves into the specified mechanical position (emergency running gap), however, since Closed Throttle Position (CTP) switch can still be recognized, an "almost normal idle RPM" is reached via the respective ignition angle retardation.** **Note: Terminal assignment at throttle control module is different in vehicles with and without cruise control. Characteristic: Steering column switch with operating module for cruise control.** **Possible Causes:** • Throttle Position (TP) sensor is faulty • Throttle valve control module is faulty • ECM is fault • Circuit wires have short circuited to each other, to vehicle Ground (GND) or to B+ • Accelerator pedal module is faulty
DTC: P3031 **Jetta:** 2.5L (BGP, BGQ) **Transmissions:** All	**Throttle Actuator 2 Electrical Malfunction Conditions:** Engine started, battery voltage must be at least 11.5v, all electrical components must be off, the ground between the engine and the chassis must be well connected, coolant temperature must be at least 80 degrees Celsius and the accelerator pedal must be properly adjusted. The ECM detected an incorrect singal from the throttle potentiometer. **Note: If the complete throttle valve control module is current-less (e.g. connector disconnected) the throttle valve moves into a particular, specified mechanical position, which signals an increased idle speed with an engine at operating temperature. If only the Throttle Position (TP) actuator is current-less, the throttle valve also moves into the specified mechanical position (emergency running gap), however, since Closed Throttle Position (CTP) switch can still be recognized, an "almost normal idle RPM" is reached via the respective ignition angle retardation.** **Note: Terminal assignment at throttle control module is different in vehicles with and without cruise control. Characteristic: Steering column switch with operating module for cruise control.** **Possible Causes:** • Throttle Position (TP) sensor is faulty • Throttle valve control module is faulty • ECM is fault • Circuit wires have short circuited to each other, to vehicle Ground (GND) or to B+ • Accelerator pedal module is faulty
DTC: P3032 **Jetta:** 2.5L (BGP, BGQ) **Transmissions:** All	**Idle Speed Control Throttle Position 2 Adaptation Malfunction Conditions:** Engine started, battery voltage at least 11.5v, all electrical components off, ground connections between engine and chassis well connected, coolant temperature at least 80-degrees Celicius and the throttle valve must not be damaged or dirty; and the ECM detected the signal from the Throttle Position Valve Module to the ECM detected was erratic, non existent or unreliable (too high or too low). **Possible Causes:** • Throttle valve control module has failed • Throttle valve control module's circuit has shorted or is open • The ECM has failed
DTC: P3035 **Jetta:** 2.5L (BGP, BGQ) **Transmissions:** All	**Idle Speed Control Throttle Position 2 Mechanical Malfunction Conditions:** Engine started, battery voltage at least 11.5v, all electrical components off, ground connections between engine and chassis well connected, coolant temperature at least 80-degrees Celicius and the throttle valve must not be damaged or dirty; and the ECM detected the signal from the Throttle Position Valve Module to the ECM detected was erratic, non existent or unreliable (too high or too low). **Possible Causes:** • Throttle valve control module has failed • Throttle valve control module's circuit has shorted or is open • The ECM has failed
DTC: P3047 **Jetta:** 2.0L (AEG, AVH, AZG, BBW, BEV); **Golf:** 2.0L (AEG, AVH, BEV); GIT: 2.0L (AEG); **NB:** 2.0L (AEG, AVH, AZG, BEV) **Transmissions:** All	**Activation Starter Relay 2 Short Circuit to B+ Conditions:** The ECM detected a short circuit on the starter relay circuit. **Possible Causes:** • Check activation of Starting interlock relay

DTC	Trouble Code Title, Conditions & Possible Causes
DTC: P3048 **Jetta:** 2.0L (AEG, AVH, AZG, BBW, BEV); **Golf:** 2.0L (AEG, AVH, BEV); GIT: 2.0L (AEG); **NB:** 2.0L (AEG, AVH, AZG, BEV) **Transmissions:** All	**Activation Starter Relay 2 Short Circuit to Ground Conditions:** The ECM detected a short circuit on the starter relay circuit. **Possible Causes:** • Check activation of Starting interlock relay
DTC: P3049 **Jetta:** 2.0L (AEG, AVH, AZG, BBW, BEV); **Golf:** 2.0L (AEG, AVH, BEV); GIT: 2.0L (AEG); **NB:** 2.0L (AEG, AVH, AZG, BEV) **Transmissions:** All	**Activation Starter Relay 2 Open Conditions:** The ECM detected a short circuit on the starter relay circuit. **Possible Causes:** • Check activation of Starting interlock relay
DTC: P3050 **Jetta:** 2.0L (AEG, AVH, AZG, BBW, BEV); **Golf:** 2.0L (AEG, AVH, BEV); GIT: 2.0L (AEG); **NB:** 2.0L (AEG, AVH, AZG, BEV) **Transmissions:** All	**Starter Relay 2 Electrical Malfunction in Circuit (Relay Stuck) Conditions:** The ECM detected a malfunction of the starter relay circuit. **Possible Causes:** • Check activation of Starting interlock relay
DTC: P3081 **2T, MIL: Yes** **Passat:** 1.8L (ATW, AUG), 2.8L (AHA, ATQ), 2.0L (BPY), 4.0L (BDP); **Jetta:** 1.8L (AWD, AWP, AWW), 2.0L (ABA), 2.5L (BGP, BGQ), 2.8L (AFP, BDF); **Golf:** 1.8L (AWD, AWW), 2.0L (ABA); **GTI:** 1.8L (AWD, AWP, AWW), 2.0L (ABA, AEG), 2.8L (AFP, BDF); **NB:** 1.8L (APH, AWP, AWV, BNU, BKF), 2.0L (AEG, AVH, AZG, BEV); **NBvert:** (BDG); **Cabrio:** 2.0L (ABA) **Transmissions:** All	**Engine Temperature Too Low Conditions:** Engine running and the ECM has detected that the engine temperature is too low. **Possible Causes:** • Engine hasn't completely warmed up • Radiator malfunction • Thermostat malfunction • ECM failure
DTC: P3096 **Passat:** 2.0L (BPY), 4.0L (BDP); **Jetta:** 2.0L (ABA); **Golf:** 2.0L (ABA); **GTI:** 2.0L (ABA) **Transmissions:** All	**Internal Control Module Memory, Check Sum Error Conditions:** Key on, the ECM has detected a programming error **Possible Causes:** • Battery terminal corrosion, or loose battery connection • Connection to the ECM interrupted, or the circuit has been opened • Reprogramming error has occurred • ECM has failed and needs replacement. • Voltage supply for Engine Control Module (ECM) is faulty
DTC: P3097 **Passat:** 2.0L (BPY), 4.0L (BDP); **Jetta:** 2.0L (ABA); **Golf:** 2.0L (ABA); **GTI:** 2.0L (ABA); **Cabrio:** 2.0L (ABA) **Transmissions:** All	**Internal Control Module Memory, Check Sum Error Conditions:** Key on, the ECM has detected a programming error **Possible Causes:** • Battery terminal corrosion, or loose battery connection • Connection to the ECM interrupted, or the circuit has been opened • Reprogramming error has occurred • ECM has failed and needs replacement. • Voltage supply for Engine Control Module (ECM) is faulty

DTC	Trouble Code Title, Conditions & Possible Causes
DTC: P310A **Passat:** 2.0L (BPY); **Jetta:** 2.0L (BPY) **Transmissions:** All	**Low Fuel Pressure Regulation Coolant Fuel Pressure Outside Specification Conditions:** Engine started, battery voltage must be at least 11.5v, all electrical components must be off, the ground between the engine and the chassis must be well connected, the exhaust system must be properly sealed between the catalytic converter and the cylinder head, and the coolant temperature must be 80 degrees Celsius. The ECM detected that the system's fuel pressure has fallen below the accepted normal calibrated value. **Possible Causes:** • Fuel Pressure Regulator Valve faulty • Fuel Pressure Sensor faulty • Fuel Pump (FP) Control Module faulty • Fuel pump faulty • Low fuel
DTC: P310B **Passat:** 2.0L (BPY); **Jetta:** 2.0L (BPY) **Transmissions:** All	**Low fuel Pressure Regulation Fuel Pressure Fluctuates Conditions:** Engine started, battery voltage must be at least 11.5v, all electrical components must be off, the ground between the engine and the chassis must be well connected, the exhaust system must be properly sealed between the catalytic converter and the cylinder head, and the coolant temperature must be 80 degrees Celsius. The ECM detected that the system's fuel pressure is irratic. **Possible Causes:** • Fuel Pressure Regulator Valve faulty • Fuel Pressure Sensor faulty • Fuel Pump (FP) Control Module faulty • Fuel pump faulty • Low fuel
DTC: P310C **Passat:** 2.0L (BPY); **Jetta:** 2.0L (BPY) **Transmissions:** All	**Low fuel Pressure Regulation Fuel Pressure Breaks Down Conditions:** Engine started, battery voltage must be at least 11.5v, all electrical components must be off, the ground between the engine and the chassis must be well connected, the exhaust system must be properly sealed between the catalytic converter and the cylinder head, and the coolant temperature must be 80 degrees Celsius. The ECM detected that the system's fuel pressure is irratic. **Possible Causes:** • Fuel Pressure Regulator Valve faulty • Fuel Pressure Sensor faulty • Fuel Pump (FP) Control Module faulty • Fuel pump faulty • Low fuel
DTC: P3137 **Passat:** 2.0L (BPY); **Jetta:** 2.0L (BPY) **Transmissions:** All	**Intake Manifold Runner Control (IMRC) Basic Setting Not Carried Out Conditions:** Engine started, battery voltage must be at least 11.5v, all electrical components must be off, the ground between the engine and the chassis must be well connected, the exhaust system must be properly sealed between the catalytic converter and the cylinder head, and the coolant temperature must be 80 degrees Celsius. **Note: The throttle valve activation occurs via an electric motor (throttle drive) in the throttle valve control module. It is activated by the Engine Control Module (ECM) according to specifications of the two sensors, Throttle Position (TP) Sensor and Accelerator Pedal Position Sensor.** **Possible Causes:** •
DTC: P3211 **2T, MIL: Yes** **Passat:** 2.0L (ABA); **Jetta:** 1.8L (AWD, AWP), 2.0L (ABA, AEG, AVH, AZG, BBW, BEV), 2.8L (AFP, BDF); **Golf:** 1.8L (AWD, AWP), 2.0L (ABA, AEG), 2.8L (AFP, BDF); **NB:** 1.8L APH, AWV, BNU, BKF); **NBvert:** 2.0L (BGD); Cabrio 2.0L (ABA) **Transmissions:** All	**Exhaust (Bank 1 Sensor 1) Heater Return Connection Conditions:** Engine started, battery voltage must be at least 11.5v, all electrical components must be off, parking brake must be engaged (to keep daytime driving lights off), automatic transmission selector must be in park, the exhaust system must be properly sealed between the catalytic converter and the cylinder head, coolant temperature must be at least 80 degrees Celsius and oxygen sensor heaters for oxygen sensors before the catalytic converter must be functioning properly and the ground between the engine and the chassis must be well connected. The ECM detected a voltage value that doesn't fall within the desired parameters for a properly functioning exhaust system. **Possible Causes:** • Check activation of Recirculation Pump Relay • Oxygen sensor (before catalytic converter) is faulty • Oxygen sensor (behind catalytic converter) is faulty • Oxygen sensor heater (before catalytic converter) is faulty • Oxygen sensor heater (behind catalytic converter) is faulty

DTC	Trouble Code Title, Conditions & Possible Causes
DTC: P3255 **Passat:** 2.0L (ABA, BPY); **Jetta:** 2.0L (ABA, BPY); **Golf:** 2.0L (ABA); **GTI:** 2.0L (ABA); **NB:** 1.8L (APH, AWV, BNU, BKF); **NBvert:** 2.0L (BGD) **Transmissions:** All	**O2 Sensor Before Catalytic Converter (Bank 1), Heating Circuit Regulation at Upper Impact Conditions:** Engine started, battery voltage must be at least 11.5v, all electrical components must be off, parking brake must be engaged (to keep daytime driving lights off), automatic transmission selector must be in park, the exhaust system must be properly sealed between the catalytic converter and the cylinder head, coolant temperature must be at least 80 degrees Celsius and oxygen sensor heaters for oxygen sensors before the catalytic converter must be functioning properly and the ground between the engine and the chassis must be well connected. The ECM detected a voltage value that doesn't fall within the desired parameters for a properly functioning O2 system. **Possible Causes:** • Check activation of Recirculation Pump Relay • Oxygen sensor (before catalytic converter) is faulty • Oxygen sensor (behind catalytic converter) is faulty • Oxygen sensor heater (before catalytic converter) is faulty • Oxygen sensor heater (behind catalytic converter) is faulty
DTC: P3056 **Passat:** 2.0L (ABA, BPY); **Jetta:** 2.0L (ABA, BPY); **Golf:** 2.0L (ABA); **GTI:** 2.0L (ABA); **NB:** 1.8L (APH, AWV, BNU, BKF); **NBvert:** 2.0L (BGD) **Transmissions:** All	**O2 Sensor Before Catalytic Converter (Bank 1), Heating Circuit Regulation at Lower Impact Conditions:** Engine started, battery voltage must be at least 11.5v, all electrical components must be off, parking brake must be engaged (to keep daytime driving lights off), automatic transmission selector must be in park, the exhaust system must be properly sealed between the catalytic converter and the cylinder head, coolant temperature must be at least 80 degrees Celsius and oxygen sensor heaters for oxygen sensors before the catalytic converter must be functioning properly and the ground between the engine and the chassis must be well connected. The ECM detected a voltage value that doesn't fall within the desired parameters for a properly functioning O2 system. **Possible Causes:** • Check activation of Recirculation Pump Relay • Oxygen sensor (before catalytic converter) is faulty • Oxygen sensor (behind catalytic converter) is faulty • Oxygen sensor heater (before catalytic converter) is faulty • Oxygen sensor heater (behind catalytic converter) is faulty
DTC: P3266 **Passat:** 2.0L (ABA, BPY); **Jetta:** 2.0L (ABA, BPY); **Golf:** 2.0L (ABA); **GTI:** 2.0L (ABA); **NB:** 1.8L (APH, AWV, BNU, BKF); **NBvert:** 2.0L (BGD) **Transmissions:** All	**Internal Resistance (Bank 1, Sensor 1) Too Large Conditions:** Engine started, battery voltage must be at least 11.5v, all electrical components must be off, the ground between the engine and the chassis must be well connected, the exhaust system must be properly sealed between the catalytic converter and the cylinder head, and the oxygen sensor heater for oxygen sensor before the catalytic converter must be properly functioning. The ECM detected a measurement on the O2 sensor circuit that was outside the parameters to function properly. **Note: For resistance testing of sensor heating, oxygen sensor should be cooled to ambient temperature. High temperatures at oxygen sensor may lead to inaccurate measurements.** **Possible Causes:** • Oxygen sensor (before catalytic converter) is faulty • Oxygen sensor (behind catalytic converter) is faulty • Oxygen sensor heater (before catalytic converter) is faulty • Oxygen sensor heater (behind catalytic converter) is faulty • Circuit wiring has a short to power or ground • Engine Component Power Supply Relay is faulty • E-box fuses for oxygen sensor are faulty • Leaks present in the exhaust manifold or exhaust pipes • HO2S signal wire and ground wire crossed in connector • HO2S element is fuel contaminated or has failed • ECM has failed
DTC: P3262 **Passat:** 2.8L (AHA, ATQ) **Transmissions:** All	**Exhaust (Banks 1 and 2) Oxygen Sensors Behind Catalytic Converter Swpped Conditions:** Engine started, battery voltage must be at least 11.5v, all electrical components must be off, the ground between the engine and the chassis must be well connected, the exhaust system must be properly sealed between the catalytic converter and the cylinder head, and the coolant temperature must be 80 degrees Celsius. The ECM detected the O2 signals were mixed and reading implausible results from both. **Possible Causes:** • O2 harness connectors are swapped • O2 wiring is crossed inside the harness • O2 wires are crossed at 104-pin connector • Connector coding and color mixed with correct catalytic converter

Gas Engine OBD II Trouble Code List (U1xxx Codes)

DTC	Trouble Code Title, Conditions & Possible Causes:
DTC: U0001 Passat 1.8L (AWM, AUG), 2.0L (BHW, BPY), 4.0L (BDP); **Jetta:** 2.0L (AVH, AZG, BBW, BEV), 2.8L (AFP, BDF); **Golf:** 2.0 (AVH, AZG, BEV); **GTI:** 2.8L (AFP, BDF); **NB:** 1.8L (APH, AWV, BNU, BKF), 2.0L (AEG, AVH, AZG, BEV); **NBvert:** 2.0L (BGD) **Transmissions:** All	**High Speed CAN Communication Bus Conditions:** The Engine Control Module (ECM) communicates with all databus-capable control modules via a CAN databus. These databus-capable control modules are connected via two data bus wires which are twisted together (CAN_High and CAN_Low), and exchange information (messages). Missing information on the databus is recognized as a malfunction and stored. Trouble-free operation of the CAN-Bus requires that it have a terminal resistance. This central terminal resistor is located in the Engine Control Module (ECM). **Possible Causes:** • ECM has failed • CAN data bus wires have short circuited to each other
DTC: U0100 **Jetta:** 2.8L (AFP, BDF) ; **GTI:** 2.8L (AFP, BDF); **NB:** 1.8L (APH, AWV, BNU, BKF); **NBvert:** 2.0L (BGD) **Transmissions:** All	**Lost Communication With ECM "A" Conditions:** Key on, and the ECM detected that it has lost communication during its initial startup. The Engine Control Module (ECM) communicates with all databus-capable control modules via a CAN databus. These databus-capable control modules are connected via two data bus wires which are twisted together (CAN_High and CAN_Low), and exchange information (messages). Missing information on the databus is recognized as a malfunction and stored. Trouble-free operation of the CAN-Bus requires that it have a terminal resistance. **Possible Causes:** • ECM has failed • Terminal resistance for CAN-bus are faulty • Can data bus wires have short circuited to each other
DTC: U0101 Passat 1.8L (ATW, AWM, AUG) , 2.0L (BHW, BPY) , 4.0L (BDP) ; **Jetta:** 2.0L (AVH, AZG, BBW, BEV), 2.8L (AFP, BDF); **Golf:** 2.0 (AVH, AZG, BEV); **GTI:** 2.8L (AFP, BDF); **NB:** 1.8L (APH, AWV, BNU, BKF), 2.0L (AEG, AVH, AZG, BEV); **NBvert:** 2.0L (BGD) **Transmissions:** All	**Lost Communication With TCM Conditions:** Key on, and the ECM detected that it has lost communication with the Transmission Control Module (TCM) during its initial startup. The Engine Control Module (ECM) communicates with all databus-capable control modules via a CAN databus. These databus-capable control modules are connected via two data bus wires which are twisted together (CAN_High and CAN_Low), and exchange information (messages). Missing information on the databus is recognized as a malfunction and stored. Trouble-free operation of the CAN-Bus requires that it have a terminal resistance. **Possible Causes:** • ECM has failed • Terminal resistance for CAN-bus are faulty • Can data bus wires have short circuited to each other • TCM has failed
DTC: U0103 **Jetta:** 2.8L (AFP, BDF) ; **GTI:** 2.8L (AFP, BDF); **NB:** 1.8L (APH, AWV, BNU, BKF), 2.0L (BGD) **Transmissions:** All	**Lost Communication With Gear Shift Module Conditions:** Key on, and the ECM detected that it has lost communication with the gear shift module during its initial startup. The Engine Control Module (ECM) communicates with all databus-capable control modules via a CAN databus. These databus-capable control modules are connected via two data bus wires which are twisted together (CAN_High and CAN_Low), and exchange information (messages). Missing information on the databus is recognized as a malfunction and stored. Trouble-free operation of the CAN-Bus requires that it have a terminal resistance. **Possible Causes:** • ECM has failed • Terminal resistance for CAN-bus are faulty • Can data bus wires have short circuited to each other • The gear shift module has failed
DTC: U0104 **Passat:** 2.0L (BPY) **Transmissions:** All	**Lost Communication With Cruise Control Module Conditions:** Key on, and the ECM detected that it has lost communication with the Cruise Control Module during its initial startup. The Engine Control Module (ECM) communicates with all databus-capable control modules via a CAN databus. These databus-capable control modules are connected via two data bus wires which are twisted together (CAN_High and CAN_Low), and exchange information (messages). Missing information on the databus is recognized as a malfunction and stored. Trouble-free operation of the CAN-Bus requires that it have a terminal resistance. **Possible Causes:** • ECM has failed • Terminal resistance for CAN-bus are faulty • Can data bus wires have short circuited to each other • The cruise control module has failed

DTC	Trouble Code Title, Conditions & Possible Causes
DTC: U0115 Passat 1.8L (AUG) **Transmissions:** All	**Lost Communication With ECM "B" Conditions:** Key on, and the ECM detected that it has lost communication during its initial startup. The Engine Control Module (ECM) communicates with all databus-capable control modules via a CAN databus. These databus-capable control modules are connected via two data bus wires which are twisted together (CAN_High and CAN_Low), and exchange information (messages). Missing information on the databus is recognized as a malfunction and stored. Trouble-free operation of the CAN-Bus requires that it have a terminal resistance. **Possible Causes:** • ECM has failed • Terminal resistance for CAN-bus are faulty • Can data bus wires have short circuited to each other
DTC: U0121 Passat 1.8L (ATW, AWM, AUG) , 2.0L (BHW, BPY) , 4.0L (BDP) ; **Jetta:** 2.0L (AVH, AZG, BBW, BEV), 2.8L (AFP, BDF) ; **Golf:** 2.0 (AVH, AZG, BEV); **GTI:** 2.8L (AFP, BDF); **NB:** 1.8L (APH, AWV, BNU, BKF), 2.0L (AEG, AVH, AZG, BEV); **NBvert:** 2.0L (BGD) **Transmissions:** All	**Lost Communication With Anti-Lock Brake System (ABS) Control Module Conditions:** Key on, and the ECM detected that it has lost communication with the ABS Control Module during its initial startup. The Engine Control Module (ECM) communicates with all databus-capable control modules via a CAN databus. These databus-capable control modules are connected via two data bus wires which are twisted together (CAN_High and CAN_Low), and exchange information (messages). Missing information on the databus is recognized as a malfunction and stored. Trouble-free operation of the CAN-Bus requires that it have a terminal resistance. **Possible Causes:** • ECM has failed • Terminal resistance for CAN-bus are faulty • Can data bus wires have short circuited to each other • There is a fault with the ABS control module
DTC: U0126 Passat 2.0L (BPY) **Transmissions:** All	**Lost Communication With Steering Angle Sensor Module Conditions:** Key on, and the ECM detected that it has lost communication with the Steering Angle Sensor during its initial startup. The Engine Control Module (ECM) communicates with all databus-capable control modules via a CAN databus. These databus-capable control modules are connected via two data bus wires which are twisted together (CAN_High and CAN_Low), and exchange information (messages). Missing information on the databus is recognized as a malfunction and stored. Trouble-free operation of the CAN-Bus requires that it have a terminal resistance. **Possible Causes:** • ECM has failed • Terminal resistance for CAN-bus are faulty • Can data bus wires have short circuited to each other • The steering angle sensor module is faulty
DTC: U0128 Passat 2.0L (BPY) **Transmissions:** All	**Lost Communication With Park Brake Control Module Conditions:** Key on, and the ECM detected that it has lost communication with the Parking Brake Control Module during its initial startup. The Engine Control Module (ECM) communicates with all databus-capable control modules via a CAN databus. These databus-capable control modules are connected via two data bus wires which are twisted together (CAN_High and CAN_Low), and exchange information (messages). Missing information on the databus is recognized as a malfunction and stored. Trouble-free operation of the CAN-Bus requires that it have a terminal resistance. **Possible Causes:** • ECM has failed • Terminal resistance for CAN-bus are faulty • Can data bus wires have short circuited to each other • The parking brake control module has failed
DTC: U0146 Passat 2.0L (BPY) **Transmissions:** All	**Lost Communication With Gateway "A" Conditions:** Key on, and the ECM detected that it has lost communication with the Gateway "A" during its initial startup. The Engine Control Module (ECM) communicates with all databus-capable control modules via a CAN databus. These databus-capable control modules are connected via two data bus wires which are twisted together (CAN_High and CAN_Low), and exchange information (messages). Missing information on the databus is recognized as a malfunction and stored. Trouble-free operation of the CAN-Bus requires that it have a terminal resistance. **Possible Causes:** • ECM has failed or isn't properly coded • Terminal resistance for CAN-bus are faulty • Can data bus wires have short circuited to each other

DTC	Trouble Code Title, Conditions & Possible Causes
DTC: U0151 Passat 2.0L (BPY) **Transmissions:** All	**Lost Communication With Restraints Control Module Conditions:** Key on, and the ECM detected that it has lost communication with the Restraints Control Module during its initial startup. The Engine Control Module (ECM) communicates with all databus-capable control modules via a CAN databus. These databus-capable control modules are connected via two data bus wires which are twisted together (CAN_High and CAN_Low), and exchange information (messages). Missing information on the databus is recognized as a malfunction and stored. Trouble-free operation of the CAN-Bus requires that it have a terminal resistance. **Possible Causes:** • ECM has failed or isn't properly coded • Terminal resistance for CAN-bus are faulty • Can data bus wires have short circuited to each other
DTC: U0155 Passat 1.8L (ATW, AWM, AUG) , 2.0L (BHW, BPY) , 4.0L (BDP) ; **Jetta:** 2.0L (AVH, AZG, BBW, BEV), 2.8L (AFP, BDF) ; **Golf:** 2.0 (AVH, AZG, BEV); **GTI:** 2.8L (AFP, BDF); **NB:** 1.8L (APH, AWV, BNU, BKF), 2.0L (AEG, AVH, AZG, BEV); **NBvert:** 2.0L (BGD) **Transmissions:** All	**Lost Communication With Instrument Cluster Conditions:** Key on, and the ECM detected that it has lost communication with the Instrument Cluster Panel (I/P) during its initial startup. The Engine Control Module (ECM) communicates with all databus-capable control modules via a CAN databus. These databus-capable control modules are connected via two data bus wires which are twisted together (CAN_High and CAN_Low), and exchange information (messages). Missing information on the databus is recognized as a malfunction and stored. Trouble-free operation of the CAN-Bus requires that it have a terminal resistance. **Possible Causes:** • ECM has failed • Terminal resistance for CAN-bus are faulty • Can data bus wires have short circuited to each other
DTC: U0164 Passat 2.0L (BPY) **Transmissions:** All	**Lost Communication With HVAC Control Module Conditions:** Key on, and the ECM detected that it has lost communication with the HVAC Control Module during its initial startup. The Engine Control Module (ECM) communicates with all databus-capable control modules via a CAN databus. These databus-capable control modules are connected via two data bus wires which are twisted together (CAN_High and CAN_Low), and exchange information (messages). Missing information on the databus is recognized as a malfunction and stored. Trouble-free operation of the CAN-Bus requires that it have a terminal resistance. **Possible Causes:** • ECM has failed • Terminal resistance for CAN-bus are faulty • Can data bus wires have short circuited to each other • The HVAC control module has failed
DTC: U0302 Passat 1.8L (AEB, ATW, AWM, AUG) , 2.0L (BHW, BPY); **Jetta:** 2.8L (AFP, BDF) ; **GTI:** 2.8L (AFP, BDF); **NB:** 1.8L (APH, AWV, BNU, BKF); **NBvert:** 2.0L (BGD) **Transmissions:** All	**Software Incompatibility with Transmission Control Module Conditions:** Key on, and the ECM detected a software incompatibility condition with the Transmission Control Module during its initial startup. The Engine Control Module (ECM) communicates with all databus-capable control modules via a CAN databus. These databus-capable control modules are connected via two data bus wires which are twisted together (CAN_High and CAN_Low), and exchange information (messages). Missing information on the databus is recognized as a malfunction and stored. Trouble-free operation of the CAN-Bus requires that it have a terminal resistance. **Possible Causes:** • ECM or TCM has failed or is not properly coded • Terminal resistance for CAN-bus are faulty • Can data bus wires have short circuited to each other
DTC: U0305 Passat 2.0L (BPY) **Transmissions:** All	**Software Incompatibility with Cruise Control Module Conditions:** Key on, and the ECM detected that it has lost communication with the Cruise Control Module during its initial startup. The Engine Control Module (ECM) communicates with all databus-capable control modules via a CAN databus. These databus-capable control modules are connected via two data bus wires which are twisted together (CAN_High and CAN_Low), and exchange information (messages). Missing information on the databus is recognized as a malfunction and stored. Trouble-free operation of the CAN-Bus requires that it have a terminal resistance. **Possible Causes:** • ECM has failed • Terminal resistance for CAN-bus are faulty • Can data bus wires have short circuited to each other • The cruise control module has failed

DTC	Trouble Code Title, Conditions & Possible Causes
DTC: U0315 Passat 1.8L (AEB, AWM, AUG) , 2.0L (BHW, BPY); **Jetta:** 2.8L (AFP, BDF) ; **GTI:** 2.8L (AFP, BDF); **NB:** 1.8L (APH, AWV, BNU, BKF); **NBvert:** 2.0L (BGD) **Transmissions:** All	**Software Incompatibility with Anti-Lock Brake System Control Module Conditions:** Key on, and the ECM detected a software incompatibility condition with the Anti-Lock Brake System Control Module during its initial startup. The Engine Control Module (ECM) communicates with all databus-capable control modules via a CAN databus. These databus-capable control modules are connected via two data bus wires which are twisted together (CAN_High and CAN_Low), and exchange information (messages). Missing information on the databus is recognized as a malfunction and stored. Trouble-free operation of the CAN-Bus requires that it have a terminal resistance. **Possible Causes:** • ECM has failed • Terminal resistance for CAN-bus are faulty • Can data bus wires have short circuited to each other • The AB S control module has failed
DTC: U0402 Passat 1.8L (ATW, AWM, AUG) , 2.0L (BHW, BPY); **Jetta:** 2.8L (AFP, BDF) ; **GTI:** 2.8L (AFP, BDF); **NB:** 1.8L (APH, AWV, BNU, BKF); **NBvert:** 2.0L (BGD) **Transmissions:** All	**Invalid Data Received From Transmission Control Module Conditions:** Key on, and the ECM detected a software invalid data from the Cruise Control Module during its initial startup. The Engine Control Module (ECM) communicates with all databus-capable control modules via a CAN databus. These databus-capable control modules are connected via two data bus wires which are twisted together (CAN_High and CAN_Low), and exchange information (messages). Missing information on the databus is recognized as a malfunction and stored. Trouble-free operation of the CAN-Bus requires that it have a terminal resistance. **Possible Causes:** • ECM or TCM has failed • Terminal resistance for CAN-bus are faulty • Can data bus wires have short circuited to each other
DTC: U0404 **Jetta:** 2.8L (AFP, BDF); **GTI:** 2.8L (AFP, BDF); **NB:** 1.8L (APH, AWV, BNU, BKF); **NBvert:** 2.0L (BGD) **Transmissions:** All	**Invalid Data Received From Gear Shift Control Module Conditions:** Key on, and the PCM detected a software invalid data from the Gear Shift Control Module during its initial startup. The Engine Control Module (ECM) communicates with all databus-capable control modules via a CAN databus. These databus-capable control modules are connected via two data bus wires which are twisted together (CAN_High and CAN_Low), and exchange information (messages). Missing information on the databus is recognized as a malfunction and stored. Trouble-free operation of the CAN-Bus requires that it have a terminal resistance. **Possible Causes:** • ECM has failed • Terminal resistance for CAN-bus are faulty • Can data bus wires have short circuited to each other • Gear shift control module has failed
DTC: U0405 Passat 2.0L (BPY) **Transmissions:** All	**Invalid Data Received From Cruise Control Module Conditions:** Key on, and the ECM detected that it is getting invalid data from the Cruise Control Module during its initial startup. The Engine Control Module (ECM) communicates with all databus-capable control modules via a CAN databus. These databus-capable control modules are connected via two data bus wires which are twisted together (CAN_High and CAN_Low), and exchange information (messages). Missing information on the databus is recognized as a malfunction and stored. Trouble-free operation of the CAN-Bus requires that it have a terminal resistance. **Possible Causes:** • ECM has failed • Terminal resistance for CAN-bus are faulty • Can data bus wires have short circuited to each other • The cruise control module has failed
DTC: U0415 Passat 1.8L (ATW, AWM, AUG) , 2.0L (BHW) **Jetta:** 2.8L (AFP, BDF); **GTI:** 2.8L (AFP, BDF); **NB:** 1.8L (APH, AWV, BNU, BKF) **Transmissions:** All	**Software Incompatibility with Anti-Lock Brake System Control Module Conditions:** Key on, and the ECM detected a software incompatibility condition with the Anti-Lock Brake System Control Module (FICM) during its initial startup. The Engine Control Module (ECM) communicates with all databus-capable control modules via a CAN databus. These databus-capable control modules are connected via two data bus wires which are twisted together (CAN_High and CAN_Low), and exchange information (messages). Missing information on the databus is recognized as a malfunction and stored. Trouble-free operation of the CAN-Bus requires that it have a terminal resistance. **Possible Causes:** • ECM has failed • Terminal resistance for CAN-bus are faulty • Can data bus wires have short circuited to each other • The ABS control module has failed

DTC	Trouble Code Title, Conditions & Possible Causes
DTC: U0417 Passat 2.0L (BPY) **Transmissions:** All	**Invalid Data Received From Park Brake Control Module Conditions:** Key on, and the ECM detected invalid data from the Park Brake Control Module during its initial startup. The Engine Control Module (ECM) communicates with all databus-capable control modules via a CAN databus. These databus-capable control modules are connected via two data bus wires which are twisted together (CAN_High and CAN_Low), and exchange information (messages). Missing information on the databus is recognized as a malfunction and stored. Trouble-free operation of the CAN-Bus requires that it have a terminal resistance. **Possible Causes:** • ECM has failed • Terminal resistance for CAN-bus are faulty • Can data bus wires have short circuited to each other • Park brake control module has failed
DTC: U0433 Passat 2.0L (BPY) **Transmissions:** All	**Invalid Data Received from Cruise Control Front Distance Range Sensor Conditions:** Key on, and the ECM detected invalid data from the Cruise Control Module during its initial startup. The Engine Control Module (ECM) communicates with all databus-capable control modules via a CAN databus. These databus-capable control modules are connected via two data bus wires which are twisted together (CAN_High and CAN_Low), and exchange information (messages). Missing information on the databus is recognized as a malfunction and stored. Trouble-free operation of the CAN-Bus requires that it have a terminal resistance. **Possible Causes:** • ECM has failed • Terminal resistance for CAN-bus are faulty • Can data bus wires have short circuited to each other • Cruise control distance range sensor has failed

DIESEL ENGINE TROUBLE CODE LIST

Introduction

To use this information, first read and record all codes in memory along with any Freeze Frame data. *If the PCM reset function is done prior to recording any data, all codes and freeze frame data will be lost!* Look up the desired code by DTC number, Code Title and Conditions (enable criteria) that indicate why a code set, and how to drive the vehicle. **1T and 2T** indicate a 1-trip or 2-trip fault and the Monitor type.

Diesel Engine OBD II Trouble Code List (P0xxx Codes)

DTC	Trouble Code Title, Conditions & Possible Causes:
DTC: P0030 **Passat:** 2.0L (BHW); **Jetta:** 1.9 (ALH, BEW); **Golf:** 1.9: (ALH, BEW); **NB:** 1.9L (ALH, BEW) **Transmissions:** All	**HO2S Heater (Bank 1 Sensor 1) Control Circuit Malfunction Conditions:** Engine started, battery voltage must be at least 11.5v, all electrical components must be off, the ground between the engine and the chassis must be well connected, the exhaust system must be properly sealed between the catalytic converter and the cylinder head, the coolant temperature must be 80 degrees Celsius, and the oxygen sensor heater for oxygen sensor before the catalytic converter must be properly functioning. The ECM detected the HO2S signal was in a negative voltage range referred to as "character shift downward". This code sets when the HO2S signal remains in a low state (usually less than 156 mv). In effect, it does not switch properly between 0.1v and 1.1v in closed loop operation. **Possible Causes:** • HO2S is contaminated (due to presence of silicone in fuel) • HO2S signal and ground circuit wires crossed in wiring harness • HO2S signal circuit is shorted to sensor or chassis ground • HO2S element has failed (internal short condition) • ECM has failed
DTC: P0031 **Passat:** 2.0L (BHW); **Jetta:** 1.9 (ALH, BEW); **Golf:** 1.9: (ALH, BEW); **NB:** 1.9L (ALH, BEW) **Transmissions:** All	**HO2S Heater (Bank 1 Sensor 1) Circuit Low Input Conditions:** Engine started, battery voltage must be at least 11.5v, all electrical components must be off, the ground between the engine and the chassis must be well connected, the exhaust system must be properly sealed between the catalytic converter and the cylinder head, the coolant temperature must be 80 degrees Celsius, and the oxygen sensor heater for oxygen sensor before the catalytic converter must be properly functioning. The ECM detected the HO2S signal was in a negative voltage range referred to as "character shift downward". This code sets when the HO2S signal remains in a low state. In effect, it does not switch properly in the closed loop operation. The HO2S (before the three-way catalytic converter) has a short circuit to ground that has lasted longer than 200 seconds **Possible Causes:** • HO2S is contaminated (due to presence of silicone in fuel) • HO2S signal and ground circuit wires crossed in wiring harness • HO2S signal circuit is shorted to sensor or chassis ground • HO2S element has failed (internal short condition) • ECM has failed
DTC: P0032 **Passat:** 2.0L (BHW); **Jetta:** 1.9 (ALH, BEW); **Golf:** 1.9: (ALH, BEW); **NB:** 1.9L (ALH, BEW) **Transmissions:** All	**HO2S Heater (Bank 1 Sensor 1) Circuit High Input Conditions:** Engine started, battery voltage must be at least 11.5v, all electrical components must be off, the ground between the engine and the chassis must be well connected, the exhaust system must be properly sealed between the catalytic converter and the cylinder head, the coolant temperature must be 80 degrees Celsius, and the oxygen sensor heater for oxygen sensor before the catalytic converter must be properly functioning. The ECM detected the HO2S signal remained in a high state. **Note: The HO2S signal circuit may be shorted to the heater power circuit due to tracking inside of the HO2S connector. Remove the connector and visually inspect the connector for signs of oil or water.** **Possible Causes:** • HO2S signal shorted to heater power circuit inside connector • HO2S signal circuit shorted to ground or to system voltage • ECM has failed
DTC: P0100 **Passat:** 2.0L (BHW); **Jetta:** 1.9 (ALH, BEW); **Golf:** 1.9: (ALH, BEW); **NB:** 1.9L (ALH, BEW) **Transmissions:** All	**Mass Air Flow Circuit Range/Performance Conditions** Engine running, with the system voltage more than 11.0v, and the temperature must be at least 185-degrees (F) and all electrical equipment (A/C, lights, etc) must be off. The ECM has detected that the MAF signal was less than the required minimum or out of a calculated range with the engine (or undetectable) for a certain period of time. **Possible Causes:** • Mass air flow (MAF) sensor has failed or is damaged • MAF sensor signal circuit is open, shorted to ground or to B+ • ECM has failed
DTC: P0101 **2T, MIL: Yes** **Passat:** 2.0L (BHW); **Jetta:** 1.9 (ALH, BEW); **Golf:** 1.9: (ALH, BEW); **NB:** 1.9L (ALH, BEW) **Transmissions:** All	**Mass or Volume Air Flow Circ. Range/Performance Conditions** Engine running, with the system voltage more than 11.0v, and the temperature must be at least 185-degrees (F) and all electrical equipment (A/C, lights, etc) must be off. The ECM has detected that the MAF signal was out of a calculated range with the engine (or undetectable) for a certain period of time. **Possible Causes:** • Mass air flow (MAF) sensor has failed or is damaged • ECM has failed • Signal and ground wires of Mass Air Flow (MAF) sensor has short circuited

DTC	Trouble Code Title, Conditions & Possible Causes
DTC: P0102 **2T, MIL: Yes** **Passat:** 2.0L (BHW); **Jetta:** 1.9 (ALH, BEW); **Golf:** 1.9: (ALH, BEW); **NB:** 1.9L (ALH, BEW) **Transmissions:** All	**MAF Sensor Circuit Low Input Conditions:** Key on, engine started, and the ECM detected the MAF sensor signal was less than the minimum calibrated value. The engine temperature must beat least 185-degrees (F) and all electrical equipment (A/C, lights, etc) must be off. The ECM has detected that the MAF signal was less than the required minimum. **Possible Causes:** • Check for leaks between MAF sensor and throttle valve control module • Voltage supply faulty. • Sensor power circuit open from fuel pump relay to MAF sensor • Sensor signal circuit open (may be disconnected) from ECM and MAF • Faulty ground cable resistance between connector terminal 1 and Ground • MAF Sensor malfunction
DTC: P0103 **2T, MIL: Yes** **Passat:** 2.0L (BHW); **Jetta:** 1.9 (ALH, BEW); **Golf:** 1.9: (ALH, BEW); **NB:** 1.9L (ALH, BEW) **Transmissions:** All	**MAF Sensor Circuit High Input Conditions:** Key on, engine started, and the ECM detected the MAF sensor signal was more than the minimum calibrated value. The engine temperature must beat least 185-degrees (F) and all electrical equipment (A/C, lights, etc) must be off. The ECM has detected that the MAF signal was more than the required minimum. **Possible Causes:** • Check for leaks between MAF sensor and throttle valve control module • Voltage supply faulty. • Sensor power circuit open from fuel pump relay to MAF sensor • Sensor signal circuit open (may be disconnected) from ECM and MAF • Faulty ground cable resistance between connector terminal 1 and Ground • MAF Sensor malfunction
DTC: P0105 **2T, MIL: Yes** **Jetta:** 1.9 (ALH, BEW); **Golf:** 1.9: (ALH, BEW); **NB:** 1.9L (ALH, BEW) **Transmissions:** All	**Barometric Pressure Sensor Circuit Conditions:** Engine started, the temperature must beat least 185-degrees (F) and all electrical equipment (A/C, lights, etc) must be off. The ECM detected the BARO sensor was out of range during the CCM test. The BARO sensor signal should be in 4.5v. **Possible Causes:** • Sensor has deteriorated (response time too slow) or has failed • MAP sensor signal circuit is shorted to ground • MAP sensor circuit (5v) is open • MAP sensor is damaged or it has failed • BARO sensor signal circuit is shorted to ground • BARO sensor circuit (5v) is open • BARO sensor is damaged or it has failed • ECM is not connected properly • ECM has failed
DTC: P0106 **2T, MIL: Yes** **Jetta:** 1.9 (ALH, BEW); **Golf:** 1.9: (ALH, BEW); **NB:** 1.9L (ALH, BEW) **Transmissions:** All	**Manifold Absolute Pressure/Barometric Pressure Sensor Circuit Performance Conditions:** Engine started, the temperature must beat least 185-degrees (F) and all electrical equipment (A/C, lights, etc) must be off. The ECM detected the BARO sensor was out of range during the CCM test. The BARO sensor signal should be in 4.5v. **Possible Causes:** • Sensor has deteriorated (response time too slow) or has failed • MAP sensor signal circuit is shorted to ground • MAP sensor circuit (5v) is open • MAP sensor is damaged or it has failed • BARO sensor signal circuit is shorted to ground • BARO sensor circuit (5v) is open • BARO sensor is damaged or it has failed • ECM is not connected properly • ECM has failed
DTC: P0107 **2T, MIL: Yes** **Jetta:** 1.9 (ALH, BEW); **Golf:** 1.9: (ALH, BEW); **NB:** 1.9L (ALH, BEW) **Transmissions:** All	**Manifold Absolute Pressure/Barometric Pressure Sensor Circuit Low Input Conditions:** Engine started, the temperature must beat least 185-degrees (F) and all electrical equipment (A/C, lights, etc) must be off. The ECM detected the BARO sensor was out of range during the CCM test. The BARO sensor signal should be in 4.5v. The BARO sensor is a variable capacitance unit used to detect altitude. **Possible Causes:** • Sensor has deteriorated (response time too slow) or has failed • MAP sensor signal circuit is shorted to ground • MAP sensor circuit (5v) is open • MAP sensor is damaged or it has failed • BARO sensor signal circuit is shorted to ground • BARO sensor circuit (5v) is open • BARO sensor is damaged or it has failed • ECM is not connected properly • ECM has failed

DTC	Trouble Code Title, Conditions & Possible Causes
DTC: P0108 **2T, MIL: Yes** **Jetta:** 1.9 (ALH, BEW); **Golf:** 1.9: (ALH, BEW); **NB:** 1.9L (ALH, BEW) **Transmissions:** All	**Manifold Absolute Pressure/Barometric Sensor Circuit High Input Conditions:** Engine started, the temperature must beat least 185-degrees (F) and all electrical equipment (A/C, lights, etc) must be off. The ECM detected the BARO sensor was out of range during the CCM test. The BARO sensor signal should be in 4.5v. The BARO sensor is a variable capacitance unit used to detect altitude. **Possible Causes:** • Sensor has deteriorated (response time too slow) or has failed • MAP sensor signal circuit is shorted to ground • MAP sensor circuit (5v) is open • MAP sensor is damaged or it has failed • BARO sensor signal circuit is shorted to ground • BARO sensor circuit (5v) is open • BARO sensor is damaged or it has failed • ECM is not connected properly • ECM has failed
DTC: P0112 **Passat:** 2.0L (BHW); **Jetta:** 1.9 (ALH, BEW); **Golf:** 1.9: (ALH, BEW); **NB:** 1.9L (ALH, BEW) **Transmissions:** All	**Intake Air Temperature Sensor Circuit Low Input Conditions:** Key on or engine running, the temperature must beat least 185-degrees (F) and all electrical equipment (A/C, lights, etc) must be off; and the ECM detected the IAT sensor signal was less than the self-test minimum. This is a thermistor-type sensor with a variable resistance that changes when exposed to different temperatures. This means: the higher the temperature, the lower the resistance value. **Possible Causes:** • IAT sensor signal circuit is grounded (check wiring & connector) • Resistance value between sockets 33 and 36 out of range • IAT sensor has an open circuit • IAT sensor is damaged or it has failed • ECM has failed
DTC: P0113 **Passat:** 2.0L (BHW); **Jetta:** 1.9 (ALH, BEW); **Golf:** 1.9: (ALH, BEW); **NB:** 1.9L (ALH, BEW) **Transmissions:** All	**Intake Air Temperature Sensor Circuit High Input Conditions:** Key on or engine running, the temperature must beat least 185-degrees (F) and all electrical equipment (A/C, lights, etc) must be off; and the ECM detected the IAT sensor signal was more than the self-test maximum. This is a thermistor-type sensor with a variable resistance that changes when exposed to different temperatures. This means: the higher the temperature, the lower the resistance value. **Possible Causes:** • IAT sensor signal circuit is open (inspect wiring & connector) • IAT sensor signal circuit is shorted • Resistance value between sockets 33 and 36 out of range • IAT sensor is damaged or it has failed • ECM has failed
DTC: P0116 **2T, MIL: Yes** **Passat:** 2.0L (BHW); **Jetta:** 1.9 (ALH, BEW); **Golf:** 1.9: (ALH, BEW); **NB:** 1.9L (ALH, BEW) **Transmissions:** All	**ECT Sensor / CHT Sensor Signal Range/Performance Conditions:** Engine started (cold), battery voltage must be 11.5, and all equipment must be off. The ECM detected the ECT sensor exceeded the required calibrated value, or the engine is at idle and doesn't reach operating temperature quickly enough; the Catalyst, Fuel System, HO2S and Misfire Monitor did not complete, or the timer expired. Testing completion of procedure, the engine's temperature must rise uniformily during idle. **Possible Causes:** • Check for low coolant level or incorrect coolant mixture • ECM detects a short circuit wiring in the ECT • CHT sensor is out-of-calibration or it has failed • ECT sensor is out-of-calibration or it has failed
DTC: P0117 **2T, MIL: Yes** **Passat:** 2.0L (BHW); **Jetta:** 1.9 (ALH, BEW); **Golf:** 1.9: (ALH, BEW); **NB:** 1.9L (ALH, BEW) **Transmissions:** All	**ECT Sensor Circuit Low Input Conditions:** Engine started (cold), battery voltage must be 11.5, and all equipment must be off. The ECM detected the ECT sensor signal was less than the self-test minimum. This is a thermistor-type sensor with a variable resistance that changes when exposed to different temperatures **Possible Causes:** • ECT sensor signal circuit is grounded in the wiring harness • ECT sensor doesn't react to changes in temperature • ECT sensor is damaged or the ECM has failed
DTC: P0118 **2T, MIL: Yes** **Passat:** 2.0L (BHW); **Jetta:** 1.9 (ALH, BEW); **Golf:** 1.9: (ALH, BEW); **NB:** 1.9L (ALH, BEW) **Transmissions:** All	**ECT Sensor Circuit High Input Conditions:** Engine started (cold), battery voltage must be 11.5, and all equipment must be off. The ECM detected the ECT sensor signal was more than the self-test maximum. This is a thermistor-type sensor with a variable resistance that changes when exposed to different temperatures **Possible Causes:** • ECT sensor signal circuit is open (inspect wiring & connector) • ECT sensor signal circuit is shorted to ground • ECT sensor is damaged or it has failed • ECM has failed

DTC	Trouble Code Title, Conditions & Possible Causes
DTC: P0121 **Passat:** 2.0L (BHW); **Jetta:** 1.9 (ALH, BEW); **Golf:** 1.9: (ALH, BEW); **NB:** 1.9L (ALH, BEW) **Transmissions:** All	**TP Sensor Signal Range/Performance Conditions:** Engine started; then immediately following a condition where the engine was running under at off-idle, the ECM detected the TP sensor signal indicated the throttle did not return to its previous closed position during the Rationality test. **Possible Causes:** • Throttle plate is binding, dirty or sticking • Throttle valve is damaged or dirty • Throttle valve control module is faulty • TP sensor signal circuit open (inspect wiring & connector) • TP sensor ground circuit open (inspect wiring & connector) • TP sensor and/or control module is damaged or has failed • MAF sensor signal is damaged, has failed or a short is present
DTC: P0123 **Passat:** 2.0L (BHW); **Jetta:** 1.9 (ALH, BEW); **Golf:** 1.9: (ALH, BEW); **NB:** 1.9L (ALH, BEW) **Transmissions:** All	**TP Sensor Circuit High Input Conditions:** Engine started, at idle, the temperature must be at least 80 degrees Celsius. The ECM detected the TP sensor signal was more than the self-test maximum during testing. **Possible Causes:** • TP sensor not seated correctly in housing (may be damaged) • TP sensor signal is circuit shorted to ground or system voltage • TP sensor ground circuit is open (check the wiring harness) • TP sensor and/or ECM has failed
DTC: P0128 **2T, MIL: Yes** **Passat:** 2.0L (BHW); **Jetta:** 1.9 (ALH, BEW); **Golf:** 1.9: (ALH, BEW); **NB:** 1.9L (ALH, BEW) **Transmissions:** All	**Intake Air Temperature Sensor 2 Circuit High Input Conditions:** Engine started, vehicle driven for over 10 minutes, and the ECM detected the engine did not reach an engine operating temperature of 160°F after an additional runtime of 2 minutes. **Possible Causes:** • Check the operation of the thermostat (it may be stuck open) • ECT sensor or CHT sensor is out-of-calibration, or has failed • Inspect for low coolant level or an incorrect coolant mixture
DTC: P0135 **Passat:** 2.0L (BHW); **Jetta:** 1.9 (ALH, BEW); **Golf:** 1.9: (ALH, BEW); **NB:** 1.9L (ALH, BEW) **Transmissions:** All	**HO2S (Bank 1 Sensor 1) Heater Circuit Malfunction Conditions:** Engine running, battery voltage 11.5, all electrical components off, ground between engine and chassis well connected and the exhaust system must be properly sealed between catalytic converter and the cylinder head. The ECM detected an unexpected voltage condition, or it detected excessive current draw in the heater circuit during the CCM test. **Possible Causes:** • HO2S heater power circuit is open or heater ground circuit open • HO2S signal tracking (due to oil or moisture in the connector) • HO2S is damaged or has failed • ECM has failed
DTC: P0181 **Passat:** 2.0L (BHW); **Jetta:** 1.9 (ALH, BEW); **Golf:** 1.9: (ALH, BEW); **NB:** 1.9L (ALH, BEW) **Transmissions:** All	**Engine Fuel Temperature Sensor 'A' Circuit Range/Performance Conditions:** Key on or engine running, all electrical components off and coolant temperature at least 80 degrees Celsius; and the ECM detected the Engine Fuel Temperature (EFT) Sensor 'A' signal was intermittent in the self-test. **Possible Causes:** • EFT sensor connector is damaged or shorted • EFT sensor VREF circuit is open or shorted to ground • EFT sensor circuit is shorted to chassis or to sensor ground • EFT sensor is damaged or it has failed • ECM has failed
DTC: P0182 **2T, MIL: Yes** **Passat:** 2.0L (BHW); **Jetta:** 1.9 (ALH, BEW); **Golf:** 1.9: (ALH, BEW); **NB:** 1.9L (ALH, BEW) **Transmissions:** All	**Engine Fuel Temperature Sensor 'A' Circuit Low Input Conditions:** Key on or engine running; and the ECM detected the Engine Fuel Temperature (EFT) Sensor 'A' signal was under the required voltage in the self-test. **Possible Causes:** • EFT sensor connector is damaged or shorted • EFT sensor VREF circuit is open or shorted to ground • EFT sensor circuit is shorted to chassis or to sensor ground • EFT sensor is damaged or it has failed • ECM has failed
DTC: P0183 **2T, MIL: Yes** **Passat:** 2.0L (BHW); **Jetta:** 1.9 (ALH, BEW); **Golf:** 1.9: (ALH, BEW); **NB:** 1.9L (ALH, BEW) **Transmissions:** All	**Engine Fuel Temperature Sensor 'A' Circuit High Input Conditions:** Key on or engine running, all electrical components off and coolant temperature at least 80 degrees Celsius; and the ECM detected the Engine Fuel Temperature (EFT) Sensor 'A' signal was intermittent in the self-test. **Possible Causes:** • EFT sensor connector is damaged, loose or open • EFT sensor signal circuit is open or it is shorted to VREF (5v) • EFT sensor is damaged or it has failed • ECM has failed

DTC	Trouble Code Title, Conditions & Possible Causes
DTC: P0200 **Passat:** 2.0L (BHW); **Jetta:** 1.9 (ALH, BEW); **Golf:** 1.9: (ALH, BEW); **NB:** 1.9L (ALH, BEW) **Transmissions:** All	**Injector Circuit Open Conditions:** Key on or engine running, all electrical components off and coolant temperature at least 80 degrees Celsius; and the ECM detected the fuel injector control circuit was not function within specs **Possible Causes:** • Injector connector is damaged, open or shorted • Injector control circuit is open, shorted to ground or to power • ECM has failed (the injector driver circuit may be damaged)
DTC: P0201 **Passat:** 2.0L (BHW); **Jetta:** 1.9 (ALH, BEW); **Golf:** 1.9: (ALH, BEW); **NB:** 1.9L (ALH, BEW) **Transmissions:** All	**Cylinder 1 Injector Circuit Malfunction Conditions:** Engine started, and the ECM detected the fuel injector "1" control circuit was in a high state when it should have been low, or in a low state when it should have been high (wiring harness & injector okay). **Possible Causes:** • Injector 1 connector is damaged, open or shorted • Injector 1 control circuit is open, shorted to ground or to power • ECM has failed (the injector driver circuit may be damaged)
DTC: P0202 **Passat:** 2.0L (BHW); **Jetta:** 1.9 (ALH, BEW); **Golf:** 1.9: (ALH, BEW); **NB:** 1.9L (ALH, BEW) **Transmissions:** All	**Cylinder 2 Injector Circuit Malfunction Conditions:** Engine started, and the ECM detected the fuel injector "2" control circuit was in a high state when it should have been low, or in a low state when it should have been high (wiring harness & injector okay). **Possible Causes:** • Injector 2 connector is damaged, open or shorted • Injector 2 control circuit is open, shorted to ground or to power • ECM has failed (the injector driver circuit may be damaged)
DTC: P0203 **Passat:** 2.0L (BHW); **Jetta:** 1.9 (ALH, BEW); **Golf:** 1.9: (ALH, BEW); **NB:** 1.9L (ALH, BEW) **Transmissions:** All	**Cylinder 3 Injector Circuit Malfunction Conditions:** Engine started, and the ECM detected the fuel injector "3" control circuit was in a high state when it should have been low, or in a low state when it should have been high (wiring harness & injector okay). **Possible Causes:** • Injector 3 connector is damaged, open or shorted • Injector 3 control circuit is open, shorted to ground or to power • ECM has failed (the injector driver circuit may be damaged)
DTC: P0204 **Passat:** 2.0L (BHW); **Jetta:** 1.9 (ALH, BEW); **Golf:** 1.9: (ALH, BEW); **NB:** 1.9L (ALH, BEW) **Transmissions:** All	**Cylinder 4 Injector Circuit Malfunction Conditions:** Engine started, and the ECM detected the fuel injector "4" control circuit was in a high state when it should have been low, or in a low state when it should have been high (wiring harness & injector okay). **Possible Causes:** • Injector 4 connector is damaged, open or shorted • Injector 4 control circuit is open, shorted to ground or to power • ECM has failed (the injector driver circuit may be damaged)
DTC: P0216 **2T, MIL: Yes** **Jetta:** 1.9 (ALH, BEW); **Golf:** 1.9: (ALH, BEW); **NB:** 1.9L (ALH, BEW) **Transmissions:** All	**Injector/Injection Timing Control Malfunction Conditions:** Engine started, and the ECM detected a malfunction in the injector timing control **Possible Causes:** • The cold start injector has failed • Circuit wires have short circuited to each other ground or power • The diesel direct fuel injection engine control module has failed
DTC: P0219 **Passat:** 2.0L (BHW **Transmissions:** All	**Engine Over-Speed Condition Conditions:** Engine started, and the ECM determined the vehicle had been driven in a manner that caused the engine to over-speed, and to exceed the engine speed calibration limit stored in memory. **Possible Causes:** • Engine operated in the wrong transmission gear position • Excessive engine speed with gear selector in Neutral position • Wheel slippage due to wet, muddy or snowing conditions
DTC: P0225 **2T, MIL: Yes** **Passat:** 2.0L (BHW); **Jetta:** 1.9 (ALH, BEW); **Golf:** 1.9: (ALH, BEW); **NB:** 1.9L (ALH, BEW) **Transmissions:** All	**Throttle Position Sensor 'C' Signal Voltage Supply Conditions:** Engine started, battery voltage at least 11.5v, all electrical components off, ground connections between engine and chassis well connected, coolant temperature at least 80-degrees Celicius and the throttle valve must not be damaged or dirty; and the ECM detected the TP Sensor 'B' circuit was out of its normal operating range during a condition with the throttle wide open, or with it completely closed. The throttle valve activation occurs via an electric motor (throttle drive) in the throttle valve control module. It is activated by the ECM according to specifications of the two sensors, Throttle Position Sensor and Accelerator Pedal Position Sensor 2. Slowly depress accelerator pedal up to Wide Open Throttle (WOT) stop while observing the percentage display on the PID data function of the scan tool. The percentage display must increase uniformly. **Possible Causes:** • Throttle body is damaged • Throttle linkage is binding or sticking • ETC TP Sensor 'C' signal circuit to the ECM is open • ETC TP Sensor 'C' ground circuit is open • ETC TP Sensor 'C' is damaged or it has failed

DTC	Trouble Code Title, Conditions & Possible Causes
DTC: P0226 **2T, MIL: Yes** **Passat:** 2.0L (BHW); **Jetta:** 1.9 (ALH, BEW); **Golf:** 1.9: (ALH, BEW); **NB:** 1.9L (ALH, BEW) **Transmissions:** All	**Throttle Position Sensor 'C' Signal Performance Conditions:** Engine started, battery voltage at least 11.5v, all electrical components off, ground connections between engine and chassis well connected, coolant temperature at least 80-degrees Celicius and the throttle valve must not be damaged or dirty; and the ECM detected the TP Sensor 'B' circuit was out of its normal operating range during a condition with the throttle wide open, or with it completely closed. The throttle valve activation occurs via an electric motor (throttle drive) in the throttle valve control module. It is activated by the ECM according to specifications of the two sensors, Throttle Position Sensor and Accelerator Pedal Position Sensor 2. Slowly depress accelerator pedal up to Wide Open Throttle (WOT) stop while observing the percentage display on the PID data function of the scan tool. The percentage display must increase uniformly.. **Possible Causes:** • Throttle body is damaged • Throttle linkage is binding or sticking • ETC TP Sensor 'C' signal circuit to the ECM is open • ETC TP Sensor 'C' ground circuit is open • ETC TP Sensor 'C' is damaged or it has failed
DTC: P0228 **2T, MIL: Yes** **Passat:** 2.0L (BHW); **Jetta:** 1.9 (ALH, BEW); **Golf:** 1.9: (ALH, BEW); **NB:** 1.9L (ALH, BEW) **Transmissions:** All	**Throttle Position Sensor 'C' Circuit High Input Conditions:** Engine started, battery voltage at least 11.5v, all electrical components off, ground connections between engine and chassis well connected, coolant temperature at least 80-degrees Celicius and the throttle valve must not be damaged or dirty; and the ECM detected the TP Sensor 'B' circuit was out of its normal operating range during a condition with the throttle wide open, or with it completely closed. The throttle valve activation occurs via an electric motor (throttle drive) in the throttle valve control module. It is activated by the ECM according to specifications of the two sensors, Throttle Position Sensor and Accelerator Pedal Position Sensor 2. Slowly depress accelerator pedal up to Wide Open Throttle (WOT) stop while observing the percentage display on the PID data function of the scan tool. The percentage display must increase uniformly. **Possible Causes:** • ETC TP Sensor 'C' connector is damaged or open • ETC TP Sensor 'C' signal circuit is open • ETC TP Sensor 'C' signal circuit is shorted to VREF (5v) • ETC TP Sensor 'C' is damaged or it has failed
DTC: P0230 **Passat:** 2.0L (BHW); **Jetta:** 1.9 (ALH, BEW); **Golf:** 1.9: (ALH, BEW); **NB:** 1.9L (ALH, BEW) **Transmissions:** All	**Fuel Pump Primary Circuit Malfunction Conditions:** Engine started, battery voltage at least 11.5v, all electrical components off, ground connections between engine and chassis well connected, coolant temperature at least 80-degrees Celicius. The ECM detected high current in fuel pump or fuel shutoff valve (FSV) circuit, or it detected voltage with the valve off, or it did not detect voltage on the circuit. The circuit is used to energize the fuel pump relay at key on or while running. Fuel pressure value should be 3000 to 5000 kPa at idle. **Possible Causes:** • FP or FSV circuit is open or shorted • Fuel pump relay VPWR circuit open • Fuel pump relay is damaged or has failed • Fuel pressure sensor has failed • Fuel pump control module is faulty • ECM has failed
DTC: P0231 **Passat:** 2.0L (BHW); **Jetta:** 1.9 (ALH, BEW); **Golf:** 1.9: (ALH, BEW); **NB:** 1.9L (ALH, BEW) **Transmissions:** All	**Fuel Pump Secondary Circuit Low Conditions:** Engine started, battery voltage at least 11.5v, all electrical components off, ground connections between engine and chassis well connected, coolant temperature at least 80-degrees Celicius. The ECM detected high current in fuel pump or fuel shutoff valve (FSV) circuit, or it detected voltage with the valve off, or it did not detect voltage on the circuit. The circuit is used to energize the fuel pump relay at key on or while running. Fuel pressure value should be 3000 to 5000 kPa at idle. **Possible Causes:** • FP or FSV circuit is open or shorted • Fuel pump relay VPWR circuit open • Fuel pump relay is damaged or has failed • Fuel pressure sensor has failed • Fuel pump control module is faulty • ECM has failed
DTC: P0232 **Passat:** 2.0L (BHW); **Jetta:** 1.9 (ALH, BEW); **Golf:** 1.9: (ALH, BEW); **NB:** 1.9L (ALH, BEW) **Transmissions:** All	**Fuel Pump Secondary Circuit High Conditions:** Engine started, battery voltage at least 11.5v, all electrical components off, ground connections between engine and chassis well connected, coolant temperature at least 80-degrees Celicius. The ECM detected high current in fuel pump or fuel shutoff valve (FSV) circuit, or it detected voltage with the valve off, or it did not detect voltage on the circuit. The circuit is used to energize the fuel pump relay at key on or while running. Fuel pressure value should be 3000 to 5000 kPa at idle. **Possible Causes:** • FP or FSV circuit is open or shorted • Fuel pump relay VPWR circuit open • Fuel pump relay is damaged or has failed • Fuel pressure sensor has failed • Fuel pump control module is faulty • ECM has failed

DTC	Trouble Code Title, Conditions & Possible Causes
DTC: P0234 **2T, MIL: Yes** **Passat:** 2.0L (BHW); **Jetta:** 1.9 (ALH, BEW); **Golf:** 1.9: (ALH, BEW); **NB:** 1.9L (ALH, BEW) **Transmissions:** All	**Turbo/Supercharger Overboost Condition Conditions:** Engine started, battery voltage at least 11.5v, all electrical components off, ground connections between engine and chassis well connected, coolant temperature at least 80-degrees Celicius. The ECM detected an operating condition that could harm the engine or automatic transmission. **Possible Causes:** • Ignition misfire condition exceeds the calibrated threshold • Knock sensor circuit has failed, or excessive knock detected • Low speed fuel pump relay not switching properly • Transmission oil temperature beyond the calibrated threshold • Shaft bearing of charge pressure regulator valve in turbocharger is blocked
DTC: P0236 **Passat:** 2.0L (BHW); **Jetta:** 1.9 (ALH, BEW); **Golf:** 1.9: (ALH, BEW); **NB:** 1.9L (ALH, BEW) **Transmissions:** All	**Turbocharger Boost Sensor (A) Circ Control Range/Performance Conditions:** Engine started, battery voltage at least 11.5v, all electrical components off, ground connections between engine and chassis well connected, coolant temperature at least 80-degrees Celicius. The ECM detected an operating condition that could harm the engine or automatic transmission. **Possible Causes:** • Charge air pressure sensor is faulty • Voltage supply to the charge air pressure sensor is open or shorted • Check the charge air system for leaks • Recirculating valve for turbocharger is faulty • Turbocharging system is damaged or not functioning properly • Turbocharger recirculating valve is faulty • Vacuum diaphragm for turbocharger is out of adjustment • Wastegate bypass regulator valve is faulty • Boost sensor has failed • ECM has failed
DTC: P0237 **Passat:** 2.0L (BHW); **Jetta:** 1.9 (ALH, BEW); **Golf:** 1.9: (ALH, BEW); **NB:** 1.9L (ALH, BEW) **Transmissions:** All	**Turbocharger Boost Sensor (A) Circ Low Input Conditions:** Engine started, battery voltage at least 11.5v, all electrical components off, ground connections between engine and chassis well connected, coolant temperature at least 80-degrees Celicius. The ECM detected an operating condition that could harm the engine or automatic transmission. **Possible Causes:** • Charge air pressure sensor is faulty • Voltage supply to the charge air pressure sensor is open or shorted • Check the charge air system for leaks • Recirculating valve for turbocharger is faulty • Turbocharging system is damaged or not functioning properly • Turbocharger recirculating valve is faulty • Vacuum diaphragm for turbocharger is out of adjustment • Wastegate bypass regulator valve is faulty • Boost sensor has failed • ECM has failed
DTC: P0238 **Passat:** 2.0L (BHW); **Jetta:** 1.9 (ALH, BEW); **Golf:** 1.9: (ALH, BEW); **NB:** 1.9L (ALH, BEW) **Transmissions:** All	**Turbocharger Boost Sensor (A) Circ High Input Conditions:** Engine started, battery voltage at least 11.5v, all electrical components off, ground connections between engine and chassis well connected, coolant temperature at least 80-degrees Celicius. The ECM detected an operating condition that could harm the engine or automatic transmission. **Possible Causes:** • Charge air pressure sensor is faulty • Voltage supply to the charge air pressure sensor is open or shorted • Check the charge air system for leaks • Recirculating valve for turbocharger is faulty • Turbocharging system is damaged or not functioning properly • Turbocharger recirculating valve is faulty • Vacuum diaphragm for turbocharger is out of adjustment • Wastegate bypass regulator valve is faulty • Boost sensor has failed • ECM has failed

DTC	Trouble Code Title, Conditions & Possible Causes
DTC: P0243 **Passat:** 2.0L (BHW); **Jetta:** 1.9 (ALH, BEW); **Golf:** 1.9: (ALH, BEW); **NB:** 1.9L (ALH, BEW) **Transmissions:** All	**Turbocharger Boost Bypass Solenoid (A) Circuit Open/Short Circuit Conditions:** Engine started, battery voltage at least 11.5v, all electrical components off, ground connections between engine and chassis well connected, coolant temperature at least 80-degrees Celicius. The ECM detected an unexpected voltage condition on the Bypass Solenoid control circuit **Possible Causes:** • Bypass solenoid power supply circuit is open • Bypass solenoid control circuit is open, shorted to ground or system power • Bypass solenoid assembly is damaged or has failed • Charge air pressure sensor is faulty • Voltage supply to the charge air pressure sensor is open or shorted • Check the charge air system for leaks • Recirculating valve for turbocharger is faulty • Turbocharging system is damaged or not functioning properly • Turbocharger recirculating valve is faulty • Vacuum diaphragm for turbocharger is out of adjustment • Wastegate bypass regulator valve is faulty • Boost sensor has failed • ECM has failed
DTC: P0245 **2T, MIL: Yes** **Passat:** 2.0L (BHW); **Jetta:** 1.9 (ALH, BEW); **Golf:** 1.9: (ALH, BEW); **NB:** 1.9L (ALH, BEW) **Transmissions:** All	**Turbocharger Boost Bypass Solenoid (A) Circuit Low Input/Short to Ground Conditions:** Engine started, battery voltage at least 11.5v, all electrical components off, ground connections between engine and chassis well connected, coolant temperature at least 80-degrees Celicius. The ECM detected an unexpected voltage condition on the Bypass Solenoid control circuit **Possible Causes:** • Bypass solenoid power supply circuit is open • Bypass solenoid control circuit is open, shorted to ground or system power • Bypass solenoid assembly is damaged or has failed • Charge air pressure sensor is faulty • Voltage supply to the charge air pressure sensor is open or shorted • Check the charge air system for leaks • Recirculating valve for turbocharger is faulty • Turbocharging system is damaged or not functioning properly • Turbocharger recirculating valve is faulty • Vacuum diaphragm for turbocharger is out of adjustment • Wastegate bypass regulator valve is faulty • Boost sensor has failed • ECM has failed
DTC: P0246 **2T, MIL: Yes** **Passat:** 2.0L (BHW); **Jetta:** 1.9 (ALH, BEW); **Golf:** 1.9: (ALH, BEW); **NB:** 1.9L (ALH, BEW) **Transmissions:** All	**Turbocharger Boost Bypass Solenoid (A) Circuit High Input/Short to B+ Conditions:** Engine started, battery voltage at least 11.5v, all electrical components off, ground connections between engine and chassis well connected, coolant temperature at least 80-degrees Celicius. The ECM detected an unexpected voltage condition on the Bypass Solenoid control circuit **Possible Causes:** • Bypass solenoid power supply circuit is open • Bypass solenoid control circuit is open, shorted to ground or system power • Bypass solenoid assembly is damaged or has failed • Charge air pressure sensor is faulty • Voltage supply to the charge air pressure sensor is open or shorted • Check the charge air system for leaks • Recirculating valve for turbocharger is faulty • Turbocharging system is damaged or not functioning properly • Turbocharger recirculating valve is faulty • Vacuum diaphragm for turbocharger is out of adjustment • Wastegate bypass regulator valve is faulty • Boost sensor has failed • ECM has failed
DTC: P0251 **Jetta:** 1.9 (ALH, BEW); **Golf:** 1.9: (ALH, BEW); **NB:** 1.9L (ALH, BEW) **Transmissions:** All	**Injection Pump Fuel Metering Control "A" (Cam/Rotor/Injector) Conditions:** Engine running and the battery voltage must be 11.5v. The ECM detected a injection pump metering control problem. **Note: The quantity adjuster is an electro-magnetic swiveling positioner which is controlled by the ECM via a directed duty cycle. The eccentric shaft on the quantity adjuster moves the modulating piston on the high pressure piston thereby regulating the quantity of fuel injected. The modulating piston displacement sensor informs the ECM of the position of Quantity Adjuster there determines the amount injected.** **Possible Causes:** • Circuit wires have short circuited to each other ground or power • The diesel direct fuel injection engine control module has failed

DTC	Trouble Code Title, Conditions & Possible Causes
DTC: P0252 **Jetta:** 1.9 (ALH, BEW); **Golf:** 1.9: (ALH, BEW); **NB:** 1.9L (ALH, BEW) **Transmissions:** All	**Injection Pump Metering Control (A) Range/Performance Conditions:** Engine running and the battery voltage must be 11.5v. The ECM detected a injection pump metering control problem. **Note: The quantity adjuster is an electro-magnetic swiveling positioner which is controlled by the ECM via a directed duty cycle. The eccentric shaft on the quantity adjuster moves the modulating piston on the high pressure piston thereby regulating the quantity of fuel injected. The modulating piston displacement sensor informs the ECM of the position of Quantity Adjuster there determines the amount injected.** **Possible Causes:** • Circuit wires have short circuited to each other ground or power • The diesel direct fuel injection engine control module has failed
DTC: P0263 **Passat:** 2.0L (BHW); **Jetta:** 1.9 (ALH, BEW); **Golf:** 1.9: (ALH, BEW); **NB:** 1.9L (ALH, BEW) **Transmissions:** All	**Cylinder 1 Contribution/Balance Conditions:** Key on or engine running; and the ECM detected an unexpected low or high voltage condition on the injector circuit **Possible Causes:** • Check activation of valve for pump/injector • ECM has failed • Fuel pump relay has failed • Fuel injectors may have malfunctioned
DTC: P0266 **Passat:** 2.0L (BHW); **Jetta:** 1.9 (ALH, BEW); **Golf:** 1.9: (ALH, BEW); **NB:** 1.9L (ALH, BEW) **Transmissions:** All	**Cylinder 2 Contribution/Balance Conditions:** Key on or engine running; and the ECM detected an unexpected low or high voltage condition on the injector circuit **Possible Causes:** • Check activation of valve for pump/injector • ECM has failed • Fuel pump relay has failed • Fuel injectors may have malfunctioned
DTC: P0269 **Passat:** 2.0L (BHW); **Jetta:** 1.9 (ALH, BEW); **Golf:** 1.9: (ALH, BEW); **NB:** 1.9L (ALH, BEW) **Transmissions:** All	**Cylinder 3 Contribution/Balance Conditions:** Key on or engine running; and the ECM detected an unexpected low or high voltage condition on the injector circuit **Possible Causes:** • Check activation of valve for pump/injector • ECM has failed • Fuel pump relay has failed • Fuel injectors may have malfunctioned
DTC: P0272 **Passat:** 2.0L (BHW); **Jetta:** 1.9 (ALH, BEW); **Golf:** 1.9: (ALH, BEW); **NB:** 1.9L (ALH, BEW) **Transmissions:** All	**Cylinder 4 Contribution/Balance Conditions:** Key on or engine running; and the ECM detected an unexpected low or high voltage condition on the injector circuit **Possible Causes:** • Check activation of valve for pump/injector • ECM has failed • Fuel pump relay has failed • Fuel injectors may have malfunctioned
DTC: P0299 **Passat:** 2.0L (BHW); **Jetta:** 1.9 (ALH, BEW); **Golf:** 1.9: (ALH, BEW); **NB:** 1.9L (ALH, BEW) **Transmissions:** All	**Turbocharger Underboost Conditions:** Engine started, battery voltage at least 11.5v, all electrical components off, ground connections between engine and chassis well connected, coolant temperature at least 80-degrees Celicius. The ECM detected an operating condition that could harm the engine or automatic transmission. **Possible Causes:** • Charge air pressure sensor has failed • Charge air system has leaks • Recirculating valve for turbocharger is faulty • Turbocharging system is faulty • Vacuum diaphragm for turbocharger needs adjusting • Wastegate bypass regulator valve is faulty
DTC: P0300 **2T, MIL: Yes** **Passat:** 2.0L (BHW); **Jetta:** 1.9 (ALH, BEW); **Golf:** 1.9: (ALH, BEW); **NB:** 1.9L (ALH, BEW) **Transmissions:** All	**Random Misfire Detected Conditions:** Engine running under positive torque conditions, and the ECM detected a misfire or uneven engine running in two or more cylinders. **Note: If the misfire is severe, the MIL will flash on/off on the 1st trip!** **Possible Causes:** • Base engine mechanical fault that affects two or more cylinders • Fuel metering fault that affects two or more cylinders • Fuel pressure too low or too high, fuel supply contaminated • EVAP system problem or the EVAP canister is fuel saturated • EGR valve is stuck open or the PCV system has a vacuum leak • Ignition system fault (coil, plugs) affecting two or more cylinders • MAF sensor contamination (it can cause a very lean condition) • Vehicle driven while very low on fuel (less than 1/8 of a tank)

DTC	Trouble Code Title, Conditions & Possible Causes
DTC: P0301 **2T, MIL: Yes** **Passat:** 2.0L (BHW); **Jetta:** 1.9 (ALH, BEW); **Golf:** 1.9: (ALH, BEW); **NB:** 1.9L (ALH, BEW) **Transmissions:** All	**Cylinder Number 1 Misfire Detected Conditions:** Engine running under positive torque conditions, and the ECM detected a misfire or uneven engine function. **Note: If the misfire is severe, the MIL will flash on/off on the 1st trip!** **Possible Causes:** • Air leak in the intake manifold, or in the EGR or ECM system • Base engine mechanical problem • Fuel delivery component problem (i.e., a contaminated, dirty or sticking fuel injector) • Fuel pump relay defective • Ignition coil fuses have failed • Ignition system problem (dirty damaged coil or plug) • Engine speed (RPM) sensor has failed • Camshaft position sensors have failed • Ignition coil is faulty • Spark plugs are not working properly or are not gapped properly
DTC: P0302 **2T, MIL: Yes** **Passat:** 2.0L (BHW); **Jetta:** 1.9 (ALH, BEW); **Golf:** 1.9: (ALH, BEW); **NB:** 1.9L (ALH, BEW) **Transmissions:** All	**Cylinder Number 2 Misfire Detected Conditions:** Engine running under positive torque conditions, and the ECM detected a misfire or uneven engine function. **Note: If the misfire is severe, the MIL will flash on/off on the 1st trip!** **Possible Causes:** • Air leak in the intake manifold, or in the EGR or ECM system • Base engine mechanical problem • Fuel delivery component problem (i.e., a contaminated, dirty or sticking fuel injector) • Fuel pump relay defective • Ignition coil fuses have failed • Ignition system problem (dirty damaged coil or plug) • Engine speed (RPM) sensor has failed • Camshaft position sensors have failed • Ignition coil is faulty • Spark plugs are not working properly or are not gapped properly
DTC: P0303 **2T, MIL: Yes** **Passat:** 2.0L (BHW); **Jetta:** 1.9 (ALH, BEW); **Golf:** 1.9: (ALH, BEW); **NB:** 1.9L (ALH, BEW) **Transmissions:** All	**Cylinder Number 3 Misfire Detected Conditions:** Engine running under positive torque conditions, and the ECM detected a misfire or uneven engine function. **Note: If the misfire is severe, the MIL will flash on/off on the 1st trip!** **Possible Causes:** • Air leak in the intake manifold, or in the EGR or ECM system • Base engine mechanical problem • Fuel delivery component problem (i.e., a contaminated, dirty or sticking fuel injector) • Fuel pump relay defective • Ignition coil fuses have failed • Ignition system problem (dirty damaged coil or plug) • Engine speed (RPM) sensor has failed • Camshaft position sensors have failed • Ignition coil is faulty • Spark plugs are not working properly or are not gapped properly
DTC: P0304 **2T, MIL: Yes** **Passat:** 2.0L (BHW); **Jetta:** 1.9 (ALH, BEW); **Golf:** 1.9: (ALH, BEW); **NB:** 1.9L (ALH, BEW) **Transmissions:** All	**Cylinder Number 4 Misfire Detected Conditions:** Engine running under positive torque conditions, and the ECM detected a misfire or uneven engine function. **Note: If the misfire is severe, the MIL will flash on/off on the 1st trip!** **Possible Causes:** • Air leak in the intake manifold, or in the EGR or ECM system • Base engine mechanical problem • Fuel delivery component problem (i.e., a contaminated, dirty or sticking fuel injector) • Fuel pump relay defective • Ignition coil fuses have failed • Ignition system problem (dirty damaged coil or plug) • Engine speed (RPM) sensor has failed • Camshaft position sensors have failed • Ignition coil is faulty • Spark plugs are not working properly or are not gapped properly

DTC	Trouble Code Title, Conditions & Possible Causes
DTC: P0321 **2T, MIL: Yes** **Passat:** 2.0L (BHW); **Jetta:** 1.9 (ALH, BEW); **Golf:** 1.9: (ALH, BEW); **NB:** 1.9L (ALH, BEW) **Transmissions:** All	**Ignition/Distributor Engine Speed Input Circuit Range/Performance Conditions:** Engine started, vehicle driven, and the ECM detected the engine speed signal was more than the calibrated value. **Note: The engine will not start if there is no speed signal. If the speed signal fails when the engine is running, it will cause the engine to stall immediately.** **Possible Causes:** • Engine speed sensor has failed or is damaged • ECM has failed • Sensor wheel is damaged or doesn't fit properly • Sensor wheel spacer isn't seated properly
DTC: P0322 **2T, MIL: Yes** **Passat:** 2.0L (BHW); **Transmissions:** All	**Ignition/Distributor Engine Input Circuit No Signal Conditions:** Key on, and the ECM could not detect the engine speed signal or the signal was erratic. **Note: The engine will not start if there is no speed signal. If the speed signal fails when the engine is running, it will cause the engine to stall immediately.** **Possible Causes:** • Engine speed sensor has failed or is damaged • ECM has failed • Sensor wheel is damaged or doesn't fit properly • Sensor wheel spacer isn't seated properly
DTC: P0381 **Passat:** 2.0L (BHW) **Transmissions:** All	**Glow Plug/Heater Indicator Circuit Conditions:** The ECM detected a problem with the glow plug/heater indicator circuit. **Possible Causes:** • Check activation of Glow Plug Indicator Lamp
DTC: P0401 **2T, MIL: Yes** **Passat:** 2.0L (BHW); **Jetta:** 1.9 (ALH, BEW); **Golf:** 1.9: (ALH, BEW); **NB:** 1.9L (ALH, BEW) **Transmissions:** All	**Exhaust Gas Recirculation Malfunction (ESM System) Conditions:** Engine started, battery voltage must be at least 11.5v, all electrical components must be off, parking brake must be engaged (to keep daytime driving lights off), automatic transmission selector must be in park and the ground between the engine and the chassis must be well connected. The ECM detected a problem in the EGR ESM system. Run the KOER self-test, make sure the EGR valve is set by switching the ignition on and then off, wait one minute, start the car and let it idle for at least one minute. **Possible Causes:** • EGR valve hoses are damaged, leaking or restricted • EGR vacuum regulator solenoid valve has failed or is dirty • EGR valve source hoses loose or connected wrong • Potentiometer for the EGR is faulty • EGR valve connector is damaged, loose or shorted • EGR valve is damaged or it has failed • ECM has failed
DTC: P0402 **2T, MIL: Yes** **Passat:** 2.0L (BHW); **Jetta:** 1.9 (ALH, BEW); **Golf:** 1.9: (ALH, BEW); **NB:** 1.9L (ALH, BEW) **Transmissions:** All	**EGR Flow At Idle Speed Detected (ESM System) Conditions:** Engine started, battery voltage must be at least 11.5v, all electrical components must be off, parking brake must be engaged (to keep daytime driving lights off), automatic transmission selector must be in park and the ground between the engine and the chassis must be well connected. The ECM detected a problem in the EGR ESM system. Run the KOER self-test, make sure the EGR valve is set by switching the ignition on and then off, wait one minute, start the car and let it idle for at least one minute. **Possible Causes:** • EGR valve hoses are damaged, leaking or restricted • EGR vacuum regulator solenoid valve has failed or is dirty • EGR valve source hoses loose or connected wrong • Potentiometer for the EGR is faulty • EGR valve connector is damaged, loose or shorted • EGR valve is damaged or it has failed • ECM has failed
DTC: P0403 **Passat:** 2.0L (BHW); **Jetta:** 1.9 (ALH, BEW); **Golf:** 1.9: (ALH, BEW); **NB:** 1.9L (ALH, BEW) **Transmissions:** All	**EGR Solenoid Circuit Malfunction Conditions:** Engine started, battery voltage must be at least 11.5v, all electrical components must be off, parking brake must be engaged (to keep daytime driving lights off), automatic transmission selector must be in park and the ground between the engine and the chassis must be well connected. The ECM detected a problem in the EGR ESM system. Run the KOER self-test, make sure the EGR valve is set by switching the ignition on and then off, wait one minute, start the car and let it idle for at least one minute. **Possible Causes:** • EGR valve hoses are damaged, leaking or restricted • EGR vacuum regulator solenoid valve has failed or is dirty • EGR valve source hoses loose or connected wrong • Potentiometer for the EGR is faulty • EGR valve connector is damaged, loose or shorted • EGR valve is damaged or it has failed • ECM has failed

DTC	Trouble Code Title, Conditions & Possible Causes
DTC: P0404 **Passat:** 2.0L (BHW); **Jetta:** 1.9 (ALH, BEW); **Golf:** 1.9: (ALH, BEW); **NB:** 1.9L (ALH, BEW) **Transmissions:** All	**EGR Control Circuit Range/Performance Conditions:** Engine started, battery voltage must be at least 11.5v, all electrical components must be off, parking brake must be engaged (to keep daytime driving lights off), automatic transmission selector must be in park and the ground between the engine and the chassis must be well connected. The ECM detected a problem in the EGR ESM system. Run the KOER self-test, make sure the EGR valve is set by switching the ignition on and then off, wait one minute, start the car and let it idle for at least one minute. **Possible Causes:** • EGR valve hoses are damaged, leaking or restricted • EGR vacuum regulator solenoid valve has failed or is dirty • EGR valve source hoses loose or connected wrong • Potentiometer for the EGR is faulty • EGR valve connector is damaged, loose or shorted • EGR valve is damaged or it has failed • ECM has failed
DTC: P0545 **Passat:** 2.0L (BHW) **Transmissions:** All	**Exhaust Gas Temperature Sensor Circuit Low Bank 1 Sensor 1 Conditions:** Engine started, battery voltage must be at least 11.5v, all electrical components must be off, parking brake must be engaged (to keep daytime driving lights off), automatic transmission selector must be in park, and the ground between the engine and the chassis must be well connected. The ECM has detected a voltage value below the minimum system requirement. **Possible Causes:** • Exhust gas temperature sensor is faulty • EGT control circuit is shorted to chassis ground • ECM has failed
DTC: P0546 **Passat:** 2.0L (BHW) **Transmissions:** All	**Exhaust Gas Temperature Sensor Circuit High Bank 1 Sensor 1 Conditions:** Engine started, battery voltage must be at least 11.5v, all electrical components must be off, parking brake must be engaged (to keep daytime driving lights off), automatic transmission selector must be in park, and the ground between the engine and the chassis must be well connected. The ECM has detected a voltage value above the maximum system requirement. **Possible Causes:** • Exhust gas temperature sensor is faulty • EGT control circuit is shorted to chassis ground • ECM has failed
DTC: P0562 **Passat:** 2.0L (BHW); **Transmissions:** All	**System Voltage Low Conditions:** Engine started, battery voltage must be at least 11.5v, all electrical components must be off, parking brake must be engaged (to keep daytime driving lights off), automatic transmission selector must be in park, and the ground between the engine and the chassis must be well connected. The ECM has detected a voltage value that is below the specified minimum limit for the system to function properly. **Possible Causes:** • Alternator damaged or faulty • Battery voltage low or insufficient • Fuses blown or circuits open • Battery connection to terminal not clean • Voltage regulator has failed
DTC: P0563 **Passat:** 2.0L (BHW); **Transmissions:** All	**System Voltage High Conditions:** Engine started, battery voltage must be at least 11.5v, all electrical components must be off, parking brake must be engaged (to keep daytime driving lights off), automatic transmission selector must be in park, and the ground between the engine and the chassis must be well connected. The ECM has detected a voltage value that has exceeded the specified maximum limit for the system to function properly. **Possible Causes:** • Alternator damaged or faulty • Battery voltage low or insufficient • Fuses blown or circuits open • Battery connection to terminal not clean • Voltage regulator has failed
DTC: P0600 **Passat:** 2.0L (BHW); **Jetta:** 1.9 (ALH, BEW); **Golf:** 1.9: (ALH, BEW); **NB:** 1.9L (ALH, BEW) **Transmissions:** All	**Serial Communication Link (Data BUS) Message Missing Conditions:** The Engine Control Module (ECM) communicates with all databus-capable control modules via a CAN databus. These databus-capable control modules are connected via two data bus wires which are twisted together (CAN_High and CAN_Low), and exchange information (messages). Missing information on the databus is recognized as a malfunction and stored. Trouble-free operation of the CAN-Bus requires that it have a terminal resistance. This central terminal resistor is located in the Engine Control Module (ECM). **Possible Causes:** • ECM has failed • CAN data bus wires have short circuited to each other

DTC	Trouble Code Title, Conditions & Possible Causes
DTC: P0601 **2T, MIL: Yes** **Passat:** 2.0L (BHW); **Jetta:** 1.9 (ALH, BEW); **Golf:** 1.9: (ALH, BEW); **NB:** 1.9L (ALH, BEW) **Transmissions:** All	**Internal Control Module Memory Check Sum Error Conditions:** Key on, the ECM has detected a programming error **Possible Causes:** • Battery terminal corrosion, or loose battery connection • Connection to the ECM interrupted, or the circuit has been opened • Reprogramming error has occurred • ECM has failed and needs replacement. Remember to check for Aftermarket Performance Products before replacing a ECM.
DTC: P0603 **Passat:** 2.0L (BHW) **Transmissions:** All	**ECM Keep Alive Memory Test Error Conditions:** Key on, and the ECM detected an internal memory fault. This code will set if KAPWR to the ECM is interrupted (at the initial key on). **Possible Causes:** • Battery terminal corrosion, or loose battery connection • KAPWR to ECM interrupted, or the circuit has been opened • Reprogramming error has occurred • ECM has failed and needs replacement. Remember to check for Aftermarket Performance Products before replacing a ECM.
DTC: P0605 **2T, MIL: Yes** **Passat:** 2.0L (BHW); **Jetta:** 1.9 (ALH, BEW); **Golf:** 1.9: (ALH, BEW); **NB:** 1.9L (ALH, BEW) **Transmissions:** All	**ECM Read Only Memory (ROM) Test Error Conditions:** Key on, and the ECM detected a ROM test error (ROM inside ECM is corrupted). The ECM is normally replaced if this code has set. **Possible Causes:** • An attempt was made to change the module calibration, or a module programming error may have occurred • Clear the trouble codes and then check for this trouble code. If it resets, the ECM has failed and needs replacement. • Aftermarket performance products may have been installed. • The Transmission Control Module (TCM) has failed.
DTC: P0606 **Passat:** 2.0L (BHW); **Jetta:** 1.9 (ALH, BEW); **Golf:** 1.9: (ALH, BEW); **NB:** 1.9L (ALH, BEW) **Transmissions:** All	**ECM Internal Communication Error Conditions:** Key on, and the ECM detected an internal communications register read back error during the initial key on check period. **Possible Causes:** • Clear the trouble codes and then check for this trouble code. If it resets, the ECM has failed and needs replacement. • Remember to check for signs of Aftermarket Performance Products installation before replacing the ECM.
DTC: P0607 **Passat:** 2.0L (BHW); **Jetta:** 1.9 (ALH, BEW); **Golf:** 1.9: (ALH, BEW); **NB:** 1.9L (ALH, BEW) **Transmissions:** All	**Control Module Performance Conditions:** Key on, and the ECM detected an internal communications register read back error during the initial key on check period. **Possible Causes:** • Clear the trouble codes and then check for this trouble code. If it resets, the ECM has failed and needs replacement. • Remember to check for signs of Aftermarket Performance Products installation before replacing the ECM.
DTC: P0614 **2T, MIL: Yes** **Jetta:** 1.9 (ALH, BEW); **Golf:** 1.9: (ALH, BEW); **NB:** 1.9L (ALH, BEW) **Transmissions:** All	**ECM / TCM Incompatible Conditions:** Key on, and the ECM detected a communication error between the Transmission control module and the ECM **Possible Causes:** • TCM failed • ECM failed • Circuit shorting between ECM and TCM • Replacement control module ID doesn't match old control module ID
DTC: P0642 **Passat:** 2.0L (BHW); **Jetta:** 1.9 (ALH, BEW); **Golf:** 1.9: (ALH, BEW); **NB:** 1.9L (ALH, BEW) **Transmissions:** All	**Sensor Reference Voltage "A" Circuit Low Conditions:** Engine started, battery voltage must be at least 11.5v, all electrical components must be off, parking brake must be engaged (to keep daytime driving lights off), automatic transmission selector must be in park, and the ground between the engine and the chassis must be well connected. **Possible Causes:** • Circuit harness connector contacts are corroded or ingressed of water • Circuit wires have shorted to each other, to battery or ground • Automatic Transmission Hydraulic Pressure Sensor 1 has failed • Solenoid valves in valve body are faulty • Transmission Control Module (TCM) needs replacing • Transmission Input Speed (RPM) Sensor has failed • Transmission Output Speed (RPM) Sensor has failed

DTC	Trouble Code Title, Conditions & Possible Causes
DTC: P0643 **Passat:** 2.0L (BHW); **Jetta:** 1.9 (ALH, BEW); **Golf:** 1.9: (ALH, BEW); **NB:** 1.9L (ALH, BEW) **Transmissions:** All	**Sensor Reference Voltage "A" Circuit High Conditions:** Engine started, battery voltage must be at least 11.5v, all electrical components must be off, parking brake must be engaged (to keep daytime driving lights off), automatic transmission selector must be in park, and the ground between the engine and the chassis must be well connected. **Possible Causes:** • Circuit harness connector contacts are corroded or ingressed of water • Circuit wires have shorted to each other, to battery or ground • Automatic Transmission Hydraulic Pressure Sensor 1 has failed • Solenoid valves in valve body are faulty • Transmission Control Module (TCM) needs replacing • Transmission Input Speed (RPM) Sensor has failed • Transmission Output Speed (RPM) Sensor has failed
DTC: P0652 **Passat:** 2.0L (BHW); **Jetta:** 1.9 (ALH, BEW); **Golf:** 1.9: (ALH, BEW); **NB:** 1.9L (ALH, BEW) **Transmissions:** All	**Sensor Reference Voltage "B" Circuit Low Conditions:** Engine started, battery voltage must be at least 11.5v, all electrical components must be off, parking brake must be engaged (to keep daytime driving lights off), automatic transmission selector must be in park, and the ground between the engine and the chassis must be well connected. **Possible Causes:** • Circuit harness connector contacts are corroded or ingressed of water • Circuit wires have shorted to each other, to battery or ground • Automatic Transmission Hydraulic Pressure Sensor 1 has failed • Solenoid valves in valve body are faulty • Transmission Control Module (TCM) needs replacing • Transmission Input Speed (RPM) Sensor has failed • Transmission Output Speed (RPM) Sensor has failed
DTC: P0653 **Passat:** 2.0L (BHW); **Jetta:** 1.9 (ALH, BEW); **Golf:** 1.9: (ALH, BEW); **NB:** 1.9L (ALH, BEW) **Transmissions:** All	**Sensor Reference Voltage "B" Circuit High Conditions:** Engine started, battery voltage must be at least 11.5v, all electrical components must be off, parking brake must be engaged (to keep daytime driving lights off), automatic transmission selector must be in park, and the ground between the engine and the chassis must be well connected. **Possible Causes:** • Circuit harness connector contacts are corroded or ingressed of water • Circuit wires have shorted to each other, to battery or ground • Automatic Transmission Hydraulic Pressure Sensor 1 has failed • Solenoid valves in valve body are faulty • Transmission Control Module (TCM) needs replacing • Transmission Input Speed (RPM) Sensor has failed • Transmission Output Speed (RPM) Sensor has failed
DTC: P0670 **2T, MIL: Yes** **Passat:** 2.0L (BHW); **Jetta:** 1.9 (ALH, BEW); **Golf:** 1.9: (ALH, BEW); **NB:** 1.9L (ALH, BEW) **Transmissions:** All	**Glow Plug Module Control Circuit Conditions:** Key on, and the ECM detected an unexpected voltage condition on the Glow Plug Lamp circuit during the CCM test. The Glow Plug Lamp remains "on" for 1-12 seconds (depending on the Glow Plug relay on-time which can vary from 1 and 120 seconds). **Possible Causes:** • Glow plug lamp circuit is open or shorted to ground • Glow plug relay control circuit is open or shorted to ground • Glow plug relay power circuit is open (test the 12GA fuse link) • Glow plug relay is damaged or it has failed
DTC: P0671 **2T, MIL: Yes** **Passat:** 2.0L (BHW); **Jetta:** 1.9 (ALH, BEW); **Golf:** 1.9: (ALH, BEW); **NB:** 1.9L (ALH, BEW) **Transmissions:** All	**Cylinder 1 Glow Plug Circuit Conditions:** Key on, and the ECM detected an unexpected voltage condition on the Glow Plug Lamp circuit during the CCM test. The Glow Plug Lamp remains "on" for 1-12 seconds (depending on the Glow Plug relay on-time which can vary from 1 and 120 seconds). **Possible Causes:** • Glow plug lamp circuit is open or shorted to ground • Glow plug relay control circuit is open or shorted to ground • Glow plug relay power circuit is open (test the 12GA fuse link) • Glow plug relay is damaged or it has failed
DTC: P0672 **2T, MIL: Yes** **Passat:** 2.0L (BHW); **Jetta:** 1.9 (ALH, BEW); **Golf:** 1.9: (ALH, BEW); **NB:** 1.9L (ALH, BEW) **Transmissions:** All	**Cylinder 2 Glow Plug Circuit Conditions:** Key on, and the ECM detected an unexpected voltage condition on the Glow Plug Lamp circuit during the CCM test. The Glow Plug Lamp remains "on" for 1-12 seconds (depending on the Glow Plug relay on-time which can vary from 1 and 120 seconds). **Possible Causes:** • Glow plug lamp circuit is open or shorted to ground • Glow plug relay control circuit is open or shorted to ground • Glow plug relay power circuit is open (test the 12GA fuse link) • Glow plug relay is damaged or it has failed

DTC	Trouble Code Title, Conditions & Possible Causes
DTC: P0673 **2T, MIL: Yes** **Passat:** 2.0L (BHW); **Jetta:** 1.9 (ALH, BEW); **Golf:** 1.9: (ALH, BEW); **NB:** 1.9L (ALH, BEW) **Transmissions:** All	**Cylinder 3 Glow Plug Circuit Conditions:** Key on, and the ECM detected an unexpected voltage condition on the Glow Plug Lamp circuit during the CCM test. The Glow Plug Lamp remains "on" for 1-12 seconds (depending on the Glow Plug relay on-time which can vary from 1 and 120 seconds). **Possible Causes:** • Glow plug lamp circuit is open or shorted to ground • Glow plug relay control circuit is open or shorted to ground • Glow plug relay power circuit is open (test the 12GA fuse link) • Glow plug relay is damaged or it has failed
DTC: P0674 **2T, MIL: Yes** **Passat:** 2.0L (BHW); **Jetta:** 1.9 (ALH, BEW); **Golf:** 1.9: (ALH, BEW); **NB:** 1.9L (ALH, BEW) **Transmissions:** All	**Cylinder 4 Glow Plug Circuit Conditions:** Key on, and the ECM detected an unexpected voltage condition on the Glow Plug Lamp circuit during the CCM test. The Glow Plug Lamp remains "on" for 1-12 seconds (depending on the Glow Plug relay on-time which can vary from 1 and 120 seconds). **Possible Causes:** • Glow plug lamp circuit is open or shorted to ground • Glow plug relay control circuit is open or shorted to ground • Glow plug relay power circuit is open (test the 12GA fuse link) • Glow plug relay is damaged or it has failed
DTC: P0684 **2T, MIL: Yes** **Passat:** 2.0L (BHW); **Jetta:** 1.9 (ALH, BEW); **Golf:** 1.9: (ALH, BEW); **NB:** 1.9L (ALH, BEW) **Transmissions:** All	**Glow Plug Control Module to PCM Communication Circuit Range/Performance Conditions:** Key on, and the ECM detected an unexpected voltage condition on the Glow Plug Lamp circuit during the CCM test. The Glow Plug Lamp remains "on" for 1-12 seconds (depending on the Glow Plug relay on-time which can vary from 1 and 120 seconds). **Possible Causes:** • Glow plug lamp circuit is open or shorted to ground • Glow plug relay control circuit is open or shorted to ground • Glow plug relay power circuit is open (test the 12GA fuse link) • Glow plug relay is damaged or it has failed
DTC: P0700 **2T, MIL: Yes** **Passat:** 2.0L (BHW); **Jetta:** 1.9 (ALH, BEW); **Golf:** 1.9: (ALH, BEW); **NB:** 1.9L (ALH, BEW) **Transmissions:** All	**Transmission Control System Malfunction Conditions:** Engine started, battery voltage must be at least 11.5v, all electrical components must be off, parking brake must be engaged (to keep daytime driving lights off), automatic transmission selector must be in park, and the ground between the engine and the chassis must be well connected. The ECM detected a malfunction int the transmission control system. **Possible Causes:** • Circuit harness connector contacts are corroded or ingressed of water • Circuit wires have shorted to each other, to battery or ground • Automatic Transmission Hydraulic Pressure Sensor 1 has failed • Solenoid valves in valve body are faulty • Transmission Input Speed (RPM) Sensor has failed • Transmission Output Speed (RPM) Sensor has failed • Engine Control Module (ECM) is faulty • Voltage supply for Engine Control Module (ECM) is faulty • Transmission Control Module (TCM) is faulty
DTC: P0702 **Passat:** 2.0L (BHW); **Transmissions:** All	**Transmission Control System Electrical Conditions:** Engine started, battery voltage must be at least 11.5v, all electrical components must be off, parking brake must be engaged (to keep daytime driving lights off), automatic transmission selector must be in park, and the ground between the engine and the chassis must be well connected. The ECM detected a voltage outside the normal performance range to allow the system to properly function. **Possible Causes:** • Circuit harness connector contacts are corroded or ingressed of water • Circuit wires have shorted to each other, to battery or ground • Automatic Transmission Hydraulic Pressure Sensor 1 has failed • Solenoid valves in valve body are faulty • Transmission Input Speed (RPM) Sensor has failed • Transmission Output Speed (RPM) Sensor has failed • Engine Control Module (ECM) is faulty • Voltage supply for Engine Control Module (ECM) is faulty • Transmission Control Module (TCM) is faulty

DTC	Trouble Code Title, Conditions & Possible Causes
DTC: P0705 **2T, MIL: Yes** **Passat:** 2.0L (BHW); **Jetta:** 1.9 (ALH, BEW); **Golf:** 1.9: (ALH, BEW); **NB:** 1.9L (ALH, BEW) **Transmissions:** A/T	**TR Sensor Circuit Malfunction Conditions:** Engine started, battery voltage must be at least 11.5v, all electrical components must be off, parking brake must be engaged (to keep daytime driving lights off), automatic transmission selector must be in park, and the ground between the engine and the chassis must be well connected. The ECM detected a voltage outside the normal performance range to allow the system to properly function. **Possible Causes:** • Circuit harness connector contacts are corroded or ingressed of water • Circuit wires have shorted to each other, to battery or ground • Automatic Transmission Hydraulic Pressure Sensor 1 has failed • Solenoid valves in valve body are faulty • Transmission Input Speed (RPM) Sensor has failed • Transmission Output Speed (RPM) Sensor has failed • Engine Control Module (ECM) is faulty • Voltage supply for Engine Control Module (ECM) is faulty • Transmission Control Module (TCM) is faulty
DTC: P0706 **Passat:** 2.0L (BHW); **Transmissions:** A/T	**TR Sensor Circuit Range/Performance Conditions:** Engine started, battery voltage must be at least 11.5v, all electrical components must be off, parking brake must be engaged (to keep daytime driving lights off), automatic transmission selector must be in park, and the ground between the engine and the chassis must be well connected. The ECM detected a voltage outside the normal performance range to allow the system to properly function. **Possible Causes:** • Circuit harness connector contacts are corroded or ingressed of water • Circuit wires have shorted to each other, to battery or ground • Automatic Transmission Hydraulic Pressure Sensor 1 has failed • Solenoid valves in valve body are faulty • Transmission Input Speed (RPM) Sensor has failed • Transmission Output Speed (RPM) Sensor has failed • Engine Control Module (ECM) is faulty • Voltage supply for Engine Control Module (ECM) is faulty • Transmission Control Module (TCM) is faulty
DTC: P0710 **Passat:** 2.0L (BHW); **Jetta:** 1.9 (ALH, BEW); **Golf:** 1.9: (ALH, BEW); **NB:** 1.9L (ALH, BEW) **Transmissions:** A/T	**TFT Sensor Circuit Malfunction Conditions:** Engine started, battery voltage must be at least 11.5v, all electrical components must be off, parking brake must be engaged (to keep daytime driving lights off), automatic transmission selector must be in park, and the ground between the engine and the chassis must be well connected. The ECM detected the Transmission Fluid Temperature (TFT) sensor value was not close its normal operating temperature. **Possible Causes:** • ATF is low, contaminated, dirty or burnt • TFT sensor signal circuit has a high resistance condition • TFT sensor is out-of-calibration ("skewed") or it has failed • ECM has failed
DTC: P0711 **Passat:** 2.0L (BHW); **Transmissions:** A/T	**Transmission Fluid Temperature Sensor Signal Range/Performance Conditions:** Engine started, battery voltage must be at least 11.5v, all electrical components must be off, parking brake must be engaged (to keep daytime driving lights off), automatic transmission selector must be in park, and the ground between the engine and the chassis must be well connected. The ECM detected the Transmission Fluid Temperature (TFT) sensor value was not close its normal operating temperature. **Possible Causes:** • ATF is low, contaminated, dirty or burnt • TFT sensor signal circuit has a high resistance condition • TFT sensor is out-of-calibration ("skewed") or it has failed • ECM has failed
DTC: P0712 **Passat:** 2.0L (BHW); **Transmissions:** A/T	**Transmission Fluid Temperature Sensor Circuit Low Input Conditions:** Engine started, battery voltage must be at least 11.5v, all electrical components must be off, parking brake must be engaged (to keep daytime driving lights off), automatic transmission selector must be in park, and the ground between the engine and the chassis must be well connected. The ECM detected the Transmission Fluid Temperature (TFT) sensor was less than its minimum self-test range in the test. **Possible Causes:** • TFT sensor signal circuit is shorted to chassis ground • TFT sensor signal circuit is shorted to sensor ground • TFT sensor is damaged, or out-of-calibration, or has failed • ECM has failed

DTC	Trouble Code Title, Conditions & Possible Causes
DTC: P0713 **Passat:** 2.0L (BHW); **Transmissions:** A/T	**Transmission Fluid Temperature Sensor Circuit High Input Conditions:** Engine started, battery voltage must be at least 11.5v, all electrical components must be off, parking brake must be engaged (to keep daytime driving lights off), automatic transmission selector must be in park, and the ground between the engine and the chassis must be well connected. The ECM detected the Transmission Fluid Temperature (TFT) sensor was more than its maximum self-test range in the test. **Possible Causes:** • TFT sensor signal circuit is open between the sensor and ECM • TFT sensor ground circuit is open between sensor and ECM • TFT sensor is damaged or has failed • ECM has failed
DTC: P0714 **Passat:** 2.0L (BHW); **Transmissions:** A/T	**Transmission Fluid Temperature Sensor Circuit Intermittent Conditions:** Engine started, battery voltage must be at least 11.5v, all electrical components must be off, parking brake must be engaged (to keep daytime driving lights off), automatic transmission selector must be in park, and the ground between the engine and the chassis must be well connected. The ECM detected the Transmission Fluid Temperature (TFT) sensor was giving a false reading or was not reading at all. **Possible Causes:** • TFT sensor signal circuit is open between the sensor and ECM • TFT sensor ground circuit is open between sensor and ECM • TFT sensor is damaged or has failed • ECM has failed
DTC: P0715 **2T, MIL: Yes** **Passat:** 2.0L (BHW); **Jetta:** 1.9 (ALH, BEW); **Golf:** 1.9: (ALH, BEW); **NB:** 1.9L (ALH, BEW) **Transmissions:** A/T	**Input/Turbine Speed Sensor Circuit Malfunction Conditions:** Engine started, vehicle driven with the vehicle speed sensor indicating more than 1 mph, and the ECM detected the Transmission Vehicle Speed Sensor signals were erratic, or that they were missing for a period of time. **Possible Causes:** • TVSS signal circuit is open • TVSS signal is shorted to chassis ground • TVSS signal is shorted to sensor ground • TVSS assembly is damaged or it has failed • ECM has failed
DTC: P0716 **Passat:** 2.0L (BHW); **Transmissions:** A/T	**Input Turbine/Speed Sensor Circuit Range/Performance Conditions:** Engine started, vehicle driven with the vehicle speed sensor indicating more than 1 mph, and the ECM detected the Transmission Vehicle Speed Sensor signals were erratic, or that they were missing for a period of time. **Possible Causes:** • TVSS signal circuit is open • TVSS signal is shorted to chassis ground • TVSS signal is shorted to sensor ground • TVSS assembly is damaged or it has failed • ECM has failed
DTC: P0717 **Passat:** 2.0L (BHW); **Jetta:** 1.9 (ALH, BEW); **Golf:** 1.9: (ALH, BEW); **NB:** 1.9L (ALH, BEW) **Transmissions:** A/T	**Transmission Speed Shaft Sensor Signal Intermittent Conditions:** Engine started, vehicle speed sensor indicating over 1 mph, and the ECM detected an intermittent loss of TSS signals (i.e., the TSS signals were erratic, irregular or missing). **Possible Causes:** • TSS connector is damaged, loose or shorted • TSS signal circuit has an intermittent open condition • TSS signal circuit has an intermittent short to ground condition • TSS assembly is damaged or is has failed • ECM has failed
DTC: P0720 **Passat:** 2.0L (BHW); **Transmissions:** A/T	**Output Shaft Speed Sensor Circuit Conditions:** Engine started, VSS signal more than 1 mph, and the ECM detected "noise" interference on the Output Shaft Speed (OSS) sensor circuit. **Possible Causes:** • After market add-on devices interfering with the OSS signal • OSS connector is damaged, loose or shorted, or the wiring is misrouted or it is damaged • OSS assembly is damaged or it has failed • ECM has failed
DTC: P0721 **Passat:** 2.0L (BHW); **Transmissions:** A/T	**A/T Output Shaft Speed Sensor Noise Interference Conditions:** Engine started, VSS signal more than 1 mph, and the ECM detected "noise" interference on the Output Shaft Speed (OSS) sensor circuit. **Possible Causes:** • After market add-on devices interfering with the OSS signal • OSS connector is damaged, loose or shorted, or the wiring is misrouted or it is damaged • OSS assembly is damaged or it has failed • ECM has failed

DTC	Trouble Code Title, Conditions & Possible Causes
DTC: P0722 **2T, MIL: Yes** **Passat:** 2.0L (BHW); **Jetta:** 1.9 (ALH, BEW); **Golf:** 1.9: (ALH, BEW); **NB:** 1.9L (ALH, BEW) **Transmissions:** A/T	**A/T Output Speed Sensor No Signal Conditions:** Engine started, and the ECM did not detect any Vehicle Speed Sensor (VSS) sensor signals upon initial vehicle movement. **Possible Causes:** • After market add-on devices interfering with the VSS signal • VSS sensor wiring is misrouted, damaged or shorting • ECM and/or TCM has failed
DTC: P0725 **2T, MIL: Yes** **Passat:** 2.0L (BHW); **Jetta:** 1.9 (ALH, BEW); **Golf:** 1.9: (ALH, BEW); **NB:** 1.9L (ALH, BEW) **Transmissions:** A/T	**Engine Speed Input Circuit Malfunction Conditions:** The Transmission Control Module (TCM) does not receive a signal from the Engine Control Module (ECM). **Possible Causes:** • The TCM circuit is shorting to ground, B+ or is open • TCM has failed • ECM has failed
DTC: P0726 **Passat:** 2.0L (BHW); **Jetta:** 1.9 (ALH, BEW); **Golf:** 1.9: (ALH, BEW); **NB:** 1.9L (ALH, BEW) **Transmissions:** A/T	**Engine Speed Input Circuit Range/Performance Conditions:** The Engine Speed (RPM) Sensor detects engine speed and reference marks. Without an engine speed signal, the engine will not start. If the engine speed signal fails while the engine is running, the engine will stop immediately. **Note: There is a larger-sized gap on the sensor wheel. This gap is the reference mark and does not mean that the sensor wheel is damaged.** **Possible Causes:** • Engine speed sensor has failed • Circuit is shorting to ground, B+ or is open • Sensor wheel is damaged, run out or not properly secured • ECM has failed
DTC: P0727 **Passat:** 2.0L (BHW); **Jetta:** 1.9 (ALH, BEW); **Golf:** 1.9: (ALH, BEW); **NB:** 1.9L (ALH, BEW) **Transmissions:** A/T	**Engine Speed Input Circuit No Signal Conditions:** The Engine Speed (RPM) Sensor detects engine speed and reference marks. Without an engine speed signal, the engine will not start. If the engine speed signal fails while the engine is running, the engine will stop immediately. **Note: There is a larger-sized gap on the sensor wheel. This gap is the reference mark and does not mean that the sensor wheel is damaged.** **Possible Causes:** • Engine speed sensor has failed • Circuit is shorting to ground, B+ or is open • Sensor wheel is damaged, run out or not properly secured • ECM has failed
DTC: P0730 **2T, MIL: Yes** **Passat:** 2.0L (BHW); **Jetta:** 1.9 (ALH, BEW); **Golf:** 1.9: (ALH, BEW); **NB:** 1.9L (ALH, BEW) **Transmissions:** A/T	**Gear Incorrect Ratio Conditions:** Engine started, battery voltage must be at least 11.5v, all electrical components must be off, and the ground between the engine and the chassis must be well connected. The ECM detected an incorrect gear ratio. **Possible Causes:** • ATF level is low • Circuit harness connector contacts are corroded or ingressed of water • Circuit wires have shorted to each other, to battery or ground • Automatic Transmission Hydraulic Pressure Sensor 1 has failed • Solenoid valves in valve body are faulty • Transmission Control Module (TCM) needs replacing • Transmission Input Speed (RPM) Sensor has failed • Transmission Output Speed (RPM) Sensor has failed
DTC: P0731 **2T, MIL: Yes** **Passat:** 2.0L (BHW); **Jetta:** 1.9 (ALH, BEW); **Golf:** 1.9: (ALH, BEW); **NB:** 1.9L (ALH, BEW) **Transmissions:** A/T	**Incorrect First Gear Ratio Conditions:** Engine started, vehicle operating with 1st gear commanded "on", and the ECM detected an incorrect 1st gear ratio during the test. **Possible Causes:** • 1st Gear solenoid harness connector not properly seated • 1st Gear solenoid signal shorted to ground, or open • 1st Gear solenoid wiring harness connector is damaged • 1st Gear solenoid is damaged or not properly installed • ATF level is low • Circuit harness connector contacts are corroded or ingressed of water • Circuit wires have shorted to each other, to battery or ground • Automatic Transmission Hydraulic Pressure Sensor 1 has failed • Transmission Control Module (TCM) needs replacing • Transmission Input Speed (RPM) Sensor has failed • Transmission Output Speed (RPM) Sensor has failed

DTC	Trouble Code Title, Conditions & Possible Causes
DTC: P0732 **2T, MIL: Yes** **Passat:** 2.0L (BHW); **Jetta:** 1.9 (ALH, BEW); **Golf:** 1.9: (ALH, BEW); **NB:** 1.9L (ALH, BEW) **Transmissions:** A/T	**Incorrect Second Gear Ratio Conditions:** Engine started, vehicle operating with 2nd Gear commanded "on", and the ECM detected an incorrect 2nd gear ratio during the test. **Possible Causes:** • 2nd Gear solenoid harness connector not properly seated • 2nd Gear solenoid signal shorted to ground, or open • 2nd Gear solenoid wiring harness connector is damaged • 2nd Gear solenoid is damaged or not properly installed • ATF level is low • Circuit harness connector contacts are corroded or ingressed of water • Circuit wires have shorted to each other, to battery or ground • Automatic Transmission Hydraulic Pressure Sensor 1 has failed • Transmission Control Module (TCM) needs replacing • Transmission Input Speed (RPM) Sensor has failed • Transmission Output Speed (RPM) Sensor has failed
DTC: P0733 **2T, MIL: Yes** **Passat:** 2.0L (BHW); **Jetta:** 1.9 (ALH, BEW); **Golf:** 1.9: (ALH, BEW); **NB:** 1.9L (ALH, BEW) **Transmissions:** A/T	**Incorrect Third Gear Ratio Conditions:** Engine started, vehicle operating with 3rd Gear commanded "on", and the ECM detected an incorrect 3rd gear ratio during the test. **Possible Causes:** • 3rd Gear solenoid harness connector not properly seated • 3rd Gear solenoid signal shorted to ground, or open • 3rd Gear solenoid wiring harness connector is damaged • 3rd Gear solenoid is damaged or not properly installed • ATF level is low • Circuit harness connector contacts are corroded or ingressed of water • Circuit wires have shorted to each other, to battery or ground • Automatic Transmission Hydraulic Pressure Sensor 1 has failed • Transmission Control Module (TCM) needs replacing • Transmission Input Speed (RPM) Sensor has failed • Transmission Output Speed (RPM) Sensor has failed
DTC: P0734 **2T, MIL: Yes** **Passat:** 2.0L (BHW); **Jetta:** 1.9 (ALH, BEW); **Golf:** 1.9: (ALH, BEW); **NB:** 1.9L (ALH, BEW) **Transmissions:** A/T	**Incorrect Fourth Gear Ratio Conditions:** Engine started, vehicle operating with 4th Gear commanded "on", and the ECM detected an incorrect 4th gear ratio during the test. **Possible Causes:** • 4th Gear solenoid harness connector not properly seated • 4th Gear solenoid signal shorted to ground, or open • 4th Gear solenoid wiring harness connector is damaged • 4th Gear solenoid is damaged or not properly installed • ATF level is low • Circuit harness connector contacts are corroded or ingressed of water • Circuit wires have shorted to each other, to battery or ground • Automatic Transmission Hydraulic Pressure Sensor 1 has failed • Transmission Control Module (TCM) needs replacing • Transmission Input Speed (RPM) Sensor has failed • Transmission Output Speed (RPM) Sensor has failed
DTC: P0735 **Passat:** 2.0L (BHW); **Transmissions:** A/T	**Incorrect Fifth Gear Ratio Conditions:** Engine started, vehicle operating with 5th Gear commanded "on", and the ECM detected an incorrect 5th gear ratio during the test. **Possible Causes:** • 5th Gear solenoid harness connector not properly seated • 5th Gear solenoid signal shorted to ground, or open • 5th Gear solenoid wiring harness connector is damaged • 5th Gear solenoid is damaged or not properly installed • ATF level is low • Circuit harness connector contacts are corroded or ingressed of water • Circuit wires have shorted to each other, to battery or ground • Automatic Transmission Hydraulic Pressure Sensor 1 has failed • Transmission Control Module (TCM) needs replacing • Transmission Input Speed (RPM) Sensor has failed • Transmission Output Speed (RPM) Sensor has failed
DTC: P0740 **2T, MIL: Yes** **Jetta:** 1.9 (ALH, BEW); **Golf:** 1.9: (ALH, BEW); **NB:** 1.9L (ALH, BEW) **Transmissions:** A/T	**TCC Solenoid Circuit Malfunction Conditions:** Engine started, KOER Self-Test enabled, vehicle driven at cruise speed, and the ECM did not detect any voltage drop across the TCC solenoid circuit during the test period. **Possible Causes:** • TCC solenoid control circuit is open or shorted to ground • TCC solenoid wiring harness connector is damaged • TCC solenoid is damaged or has failed • ECM has failed

DTC	Trouble Code Title, Conditions & Possible Causes
DTC: P0743 **2T, MIL: Yes** **Passat:** 2.0L (BHW); **Jetta:** 1.9 (ALH, BEW); **Golf:** 1.9: (ALH, BEW); **NB:** 1.9L (ALH, BEW) **Transmissions:** A/T	**Torque Converter Clutch Circuit Electrical Malfunction Conditions:** Engine started, KOER Self-Test enabled, vehicle driven at cruise speed, and the ECM did not detect any voltage drop across the TCC solenoid circuit during the test period. **Possible Causes:** • TCC solenoid control circuit is open or shorted to ground • TCC solenoid wiring harness connector is damaged • TCC solenoid is damaged or has failed • ECM has failed
DTC: P0746 **Passat:** 2.0L (BHW); **Jetta:** 1.9 (ALH, BEW); **Golf:** 1.9: (ALH, BEW); **NB:** 1.9L (ALH, BEW) **Transmissions:** A/T	**Pressure Control Solenoid "A" Performance or Stuck Off Conditions:** Engine started, battery voltage must be at least 11.5v, all electrical components must be off, and the ground between the engine and the chassis must be well connected. The ECM detected the pressure control solenoid was in the "stuck off" position. **Possible Causes:** • ATF level is low • Circuit harness connector contacts are corroded or ingressed of water • Circuit wires have shorted to each other, to battery or ground • Automatic Transmission Hydraulic Pressure Sensor 1 has failed • Solenoid valves in valve body are faulty • Transmission Control Module (TCM) needs replacing • Transmission Input Speed (RPM) Sensor has failed • Transmission Output Speed (RPM) Sensor has failed
DTC: P0747 **Passat:** 2.0L (BHW); **Transmissions:** A/T	**Pressure Control Solenoid "A" Performance or Stuck On Conditions:** Engine started, battery voltage must be at least 11.5v, all electrical components must be off, and the ground between the engine and the chassis must be well connected. The ECM detected the pressure control solenoid was in the "stuck on" position. **Possible Causes:** • ATF level is low • Circuit harness connector contacts are corroded or ingressed of water • Circuit wires have shorted to each other, to battery or ground • Automatic Transmission Hydraulic Pressure Sensor 1 has failed • Solenoid valves in valve body are faulty • Transmission Control Module (TCM) needs replacing • Transmission Input Speed (RPM) Sensor has failed • Transmission Output Speed (RPM) Sensor has failed
DTC: P0748 **2T, MIL: Yes** **Passat:** 2.0L (BHW); **Jetta:** 1.9 (ALH, BEW); **Golf:** 1.9: (ALH, BEW); **NB:** 1.9L (ALH, BEW) **Transmissions:** A/T	**Pressure Control Solenoid Electrical Conditions:** The valve body solenoid valve is not receiving a signal. **Possible Causes:** • Pressure control solenoid circuit is shorting to ground • Pressure control solenoid circuit is open • Valve has failed • TCM has failed • ECM has failed
DTC: P0751 **Passat:** 2.0L (BHW); **Transmissions:** A/T	**Shift Solenoid "A" Performance or Stuck Off Conditions:** Engine started, vehicle driven with the solenoid applied, and the ECM detected an unexpected voltage condition on the SS1/A solenoid circuit was incorrect during the test. **Possible Causes:** • Solenoid valves in valve body are faulty • Solenoid circuit is shorting to ground • Solenoid circuit is open • TCM has failed or wiring is shorting • ECM has failed
DTC: P0752 **Passat:** 2.0L (BHW); **Transmissions:** A/T	**A/T Shift Solenoid 1/A Function Range/Performance Conditions:** Engine started, vehicle driven with the solenoid applied, and the ECM detected a mechanical failure while operating the Shift Solenoid 1/A during the CCM test period. **Possible Causes:** • SS1/A solenoid is stuck in the "on" position • SS1/A solenoid has a mechanical failure • SS1/A solenoid has a hydraulic failure • ECM has failed

DTC	Trouble Code Title, Conditions & Possible Causes
DTC: P0753 **2T, MIL: Yes** **Jetta:** 1.9 (ALH, BEW); **Golf:** 1.9: (ALH, BEW); **NB:** 1.9L (ALH, BEW) **Transmissions:** A/T	**A/T Shift Solenoid 1/A Circuit Malfunction Conditions:** Engine started, vehicle driven with the solenoid applied, and the ECM detected an unexpected voltage condition on the SS1/A solenoid circuit was incorrect during the test. **Possible Causes:** • SS1/A solenoid control circuit is open • SS1/A solenoid control circuit is shorted to ground • SS1/A solenoid wiring harness connector is damaged • SS1/A solenoid is damaged or has failed • ECM has failed
DTC: P0756 **Passat:** 2.0L (BHW); **Transmissions:** A/T	**A/T Shift Solenoid 2/B Function Range/Performance Conditions:** Engine started, vehicle driven with the solenoid applied, and the ECM detected a mechanical failure while operating the Shift Solenoid 2/B during the CCM test period. **Possible Causes:** • SS2/B solenoid is stuck in the "on" position • SS2/B solenoid has a mechanical failure • SS2/B solenoid has a hydraulic failure • ECM has failed
DTC: P0757 **Passat:** 2.0L (BHW); **Transmissions:** A/T	**A/T Shift Solenoid 2/B Function Range/Performance Conditions:** Engine started, vehicle driven with the solenoid applied, and the ECM detected a mechanical failure while operating the Shift Solenoid 2/B during the CCM test period. **Possible Causes:** • SS2/B solenoid is stuck in the "on" position • SS2/B solenoid has a mechanical failure • SS2/B solenoid has a hydraulic failure • ECM has failed
DTC: P0758 **2T, MIL: Yes** **Passat:** 2.0L (BHW); **Jetta:** 1.9 (ALH, BEW); **Golf:** 1.9: (ALH, BEW); **NB:** 1.9L (ALH, BEW) **Transmissions:** A/T	**A/T Shift Solenoid 2/B Circuit Malfunction Conditions:** Engine started, vehicle driven with the solenoid applied, and the ECM detected an unexpected voltage condition on the SS1/A solenoid circuit was incorrect during the test.. **Possible Causes:** • Shift Solenoid 2/B connector is damaged, open or shorted • Shift Solenoid 2/B control circuit is open • Shift Solenoid 2/B control circuit is shorted to ground • Shift Solenoid 2/B is damaged or it has failed • ECM has failed
DTC: P0761 **Passat:** 2.0L (BHW) **Transmissions:** A/T	**A/T Shift Solenoid 3/C Function Range/Performance Conditions:** Engine started, vehicle driven with Shift Solenoid 3/C applied, and the ECM detected a mechanical failure occurred (stuck "off") while operating Shift Solenoid 3/C during the test. **Possible Causes:** • SS3/C solenoid may be stuck "off" • SS3/C solenoid has a mechanical failure • SS3/C solenoid has a hydraulic failure • ECM has failed
DTC: P0762 **Passat:** 2.0L (BHW) **Transmissions:** A/T	**A/T Shift Solenoid 3/C Function Range/Performance Conditions:** Engine started, vehicle driven with Shift Solenoid 3/C applied, and the ECM detected a mechanical failure occurred (stuck "on") while operating Shift Solenoid 3/C during the test. **Possible Causes:** • SS3/C solenoid may be stuck "on" • SS3/C solenoid has a mechanical failure • SS3/C solenoid has a hydraulic failure • ECM has failed
DTC: P0763 **2T, MIL: Yes** **Passat:** 2.0L (BHW); **Jetta:** 1.9 (ALH, BEW); **Golf:** 1.9: (ALH, BEW); **NB:** 1.9L (ALH, BEW) **Transmissions:** A/T	**A/T Shift Solenoid 3/C Electrical Conditions:** Engine started, vehicle driven with the solenoid applied, and the ECM detected an unexpected voltage condition on the SS3/C solenoid circuit was incorrect during the test.. **Possible Causes:** • Shift Solenoid 3/C connector is damaged, open or shorted • Shift Solenoid 3/C control circuit is open • Shift Solenoid 3/C control circuit is shorted to ground • Shift Solenoid 3/C is damaged or it has failed • ECM has failed

DTC	Trouble Code Title, Conditions & Possible Causes
DTC: P0773 **2T, MIL: Yes** **Jetta:** 1.9 (ALH, BEW); **Golf:** 1.9: (ALH, BEW); **NB:** 1.9L (ALH, BEW) **Transmissions:** A/T	**A/T Shift Solenoid E Electrical Conditions:** Engine started, vehicle driven with the solenoid applied, and the ECM detected an unexpected voltage condition on the SS3/D solenoid circuit was incorrect during the test.. **Possible Causes:** • Shift Solenoid connector is damaged, open or shorted • Shift Solenoid control circuit is open • Shift Solenoid control circuit is shorted to ground • Shift Solenoid is damaged or it has failed • ECM has failed
DTC: P0776 **Passat:** 2.0L (BHW); **Transmissions:** A/T	**Pressure Control Solenoid "B" Performance or Stuck Off Conditions:** Engine started, vehicle driven with Shift Solenoid 3/C applied, and the ECM detected a mechanical failure occurred (stuck "off") while operating Shift Solenoid 3/C during the test. **Possible Causes:** • SS3/C solenoid may be stuck "off" • SS3/C solenoid has a mechanical failure • SS3/C solenoid has a hydraulic failure • ECM has failed
DTC: P0777 **Passat:** 2.0L (BHW); **Transmissions:** A/T	**Pressure Control Solenoid "B" Stuck On Conditions:** Engine started, vehicle driven with Shift Solenoid 3/C applied, and the ECM detected a mechanical failure occurred (stuck "on") while operating Shift Solenoid 3/C during the test. **Possible Causes:** • SS3/C solenoid may be stuck "on" • SS3/C solenoid has a mechanical failure • SS3/C solenoid has a hydraulic failure • ECM has failed
DTC: P0778 **Passat:** 2.0L (BHW); **Jetta:** 1.9 (ALH, BEW); **Golf:** 1.9: (ALH, BEW); **NB:** 1.9L (ALH, BEW) **Transmissions:** A/T	**Pressure Control Solenoid "B" Electrical Conditions:** Engine started, vehicle driven with the solenoid applied, and the ECM detected an unexpected voltage condition on the SS3/C solenoid circuit was incorrect during the test.. **Possible Causes:** • Shift Solenoid connector is damaged, open or shorted • Shift Solenoid control circuit is open • Shift Solenoid control circuit is shorted to ground • Shift Solenoid is damaged or it has failed • ECM has failed
DTC: P0779 **Passat:** 2.0L (BHW); **Transmissions:** A/T	**Pressure Control Solenoid "B" Intermittent Conditions:** Engine started, vehicle driven with the solenoid applied, and the ECM detected an unexpected voltage condition on the SS3/C solenoid circuit was incorrect during the test.. **Possible Causes:** • Shift Solenoid connector is damaged, open or shorted • Shift Solenoid control circuit is open • Shift Solenoid control circuit is shorted to ground • Shift Solenoid is damaged or it has failed • ECM has failed
DTC: P0781 **Passat:** 2.0L (BHW); **Transmissions:** A/T	**1-2 Shift Conditions:** Engine running and vehicle driven, the ECM detected a mechanical malfunction within the transmission **Possible Causes:** • Solenoid valves in valve body are faulty • Solenoid circuit is shorting to ground • Solenoid circuit is open • TCM has failed or wiring is shorting • ECM has failed • Mechanical malfunction in transmission
DTC: P0782 **Passat:** 2.0L (BHW); **Transmissions:** A/T	**2-3 Shift Conditions:** Engine running and vehicle driven, the ECM detected a mechanical malfunction within the transmission **Possible Causes:** • Solenoid valves in valve body are faulty • Solenoid circuit is shorting to ground • Solenoid circuit is open • TCM has failed or wiring is shorting • ECM has failed • Mechanical malfunction in transmission

DTC	Trouble Code Title, Conditions & Possible Causes
DTC: P0783 **Passat:** 2.0L (BHW); **Transmissions:** A/T	**3-4 Shift Conditions:** xx **Possible Causes:** • Solenoid valves in valve body are faulty • Solenoid circuit is shorting to ground • Solenoid circuit is open • TCM has failed or wiring is shorting • ECM has failed • Mechanical malfunction in transmission
DTC: P0784 **Passat:** 2.0L (BHW); **Transmissions:** A/T	**4-5 Shift Conditions:** Engine running and vehicle driven, the ECM detected a mechanical malfunction within the transmission **Possible Causes:** • Solenoid valves in valve body are faulty • Solenoid circuit is shorting to ground • Solenoid circuit is open • TCM has failed or wiring is shorting • ECM has failed • Mechanical malfunction in transmission
DTC: P0785 **Jetta:** 1.9 (ALH, BEW); **Golf:** 1.9: (ALH, BEW); **NB:** 1.9L (ALH, BEW) **Transmissions:** A/T	**Shift/Timing Solenoid Conditions:** Engine running and vehicle driven, the ECM detected a mechanical malfunction within the transmission **Possible Causes:** • Solenoid valves in valve body are faulty • Solenoid circuit is shorting to ground • Solenoid circuit is open • TCM has failed or wiring is shorting • ECM has failed • Mechanical malfunction in transmission
DTC: P0791 **2T, MIL: Yes** **Jetta:** 1.9 (ALH, BEW); **Golf:** 1.9: (ALH, BEW); **NB:** 1.9L (ALH, BEW) **Transmissions:** A/T	**Intermediate Shaft Speed Sensor "A" Circuit Conditions:** Engine running and vehicle driven, the ECM detected a mechanical malfunction within the transmission **Possible Causes:** • Solenoid valves in valve body are faulty • Solenoid circuit is shorting to ground • Solenoid circuit is open • TCM has failed or wiring is shorting • ECM has failed • Mechanical malfunction in transmission
DTC: P0796 **Passat:** 2.0L (BHW); **Transmissions:** A/T	**Pressure Solenoid "C" Performance or Stuck Off Conditions:** Engine started, vehicle driven with the solenoid applied, and the ECM detected an unexpected voltage condition on the SS1/C solenoid circuit was incorrect during the test. **Possible Causes:** • Solenoid valves in valve body are faulty • Solenoid circuit is shorting to ground • Solenoid circuit is open • TCM has failed or wiring is shorting • ECM has failed
DTC: P0797 **Passat:** 2.0L (BHW); **Transmissions:** A/T	**Pressure Solenoid "C" Performance or Stuck On Conditions:** Engine started, vehicle driven with the solenoid applied, and the ECM detected an unexpected voltage condition on the SS1/C solenoid circuit was incorrect during the test. **Possible Causes:** • Solenoid valves in valve body are faulty • Solenoid circuit is shorting to ground • Solenoid circuit is open • TCM has failed or wiring is shorting • ECM has failed
DTC: P0798 **Passat:** 2.0L (BHW); **Transmissions:** A/T	**Pressure Solenoid "C" Electrical Conditions:** Engine started, vehicle driven with the solenoid applied, and the ECM detected an unexpected voltage condition on the SS1/C solenoid circuit was incorrect during the test. **Possible Causes:** • Solenoid valves in valve body are faulty • Solenoid circuit is shorting to ground • Solenoid circuit is open • TCM has failed or wiring is shorting • ECM has failed

DTC	Trouble Code Title, Conditions & Possible Causes
DTC: P0799 **Passat:** 2.0L (BHW); **Transmissions:** A/T	**Pressure Solenoid "C" Intermittent Conditions:** Engine started, vehicle driven with the solenoid applied, and the ECM detected an unexpected voltage condition on the SS1/C solenoid circuit was incorrect during the test. **Possible Causes:** • Solenoid valves in valve body are faulty • Solenoid circuit is shorting to ground • Solenoid circuit is open • TCM has failed or wiring is shorting • ECM has failed
DTC: P0811 **Passat:** 2.0L (BHW); **Jetta:** 1.9 (ALH, BEW); **Golf:** 1.9: (ALH, BEW); **NB:** 1.9L (ALH, BEW) **Transmissions:** A/T	**Excessive Clutch Slippage Conditions:** Engine started, vehicle driven and the ECM and/or TCM has detected that the clutch has slipped multiple times in a given time frame. **Possible Causes:** • Solenoid valves in valve body are faulty • Solenoid circuit is shorting to ground • Solenoid circuit is open • TCM has failed or wiring is shorting • ECM has failed
DTC: P0863 **2T, MIL: Yes** **Jetta:** 1.9 (ALH, BEW); **Golf:** 1.9: (ALH, BEW); **NB:** 1.9L (ALH, BEW) **Transmissions:** A/T	**TCM Communication Circuit Conditions:** The Transmission Control Module (ECM) communicates with all databus-capable control modules via a CAN databus. These databus-capable control modules are connected via two data bus wires which are twisted together (CAN_High and CAN_Low), and exchange information (messages). Missing information on the databus is recognized as a malfunction and stored. Trouble-free operation of the CAN-Bus requires that it have a terminal resistance. **Possible Causes:** • ECM has failed • Terminal resistance for CAN-bus are faulty • Can data bus wires have short circuited to each other • TCM has failed
DTC: P0864 **2T, MIL: Yes** **Passat:** 2.0L (BHW); **Jetta:** 1.9 (ALH, BEW); **Golf:** 1.9: (ALH, BEW); **NB:** 1.9L (ALH, BEW) **Transmissions:** A/T	**TCM Communication Circuit Range/Performance Conditions:** The Transmission Control Module (ECM) communicates with all databus-capable control modules via a CAN databus. These databus-capable control modules are connected via two data bus wires which are twisted together (CAN_High and CAN_Low), and exchange information (messages). Missing information on the databus is recognized as a malfunction and stored. Trouble-free operation of the CAN-Bus requires that it have a terminal resistance. **Possible Causes:** • ECM has failed • Terminal resistance for CAN-bus are faulty • Can data bus wires have short circuited to each other • TCM has failed
DTC: P0865 **2T, MIL: Yes** **Passat:** 2.0L (BHW); **Jetta:** 1.9 (ALH, BEW); **Golf:** 1.9: (ALH, BEW); **NB:** 1.9L (ALH, BEW) **Transmissions:** A/T	**TCM Communication Circuit Low Conditions:** The Transmission Control Module (ECM) communicates with all databus-capable control modules via a CAN databus. These databus-capable control modules are connected via two data bus wires which are twisted together (CAN_High and CAN_Low), and exchange information (messages). Missing information on the databus is recognized as a malfunction and stored. Trouble-free operation of the CAN-Bus requires that it have a terminal resistance. **Possible Causes:** • ECM has failed • Terminal resistance for CAN-bus are faulty • Can data bus wires have short circuited to each other • TCM has failed
DTC: P0884 **Passat:** 2.0L (BHW); **Transmissions:** A/T	**TCM Power Input Signal Intermittent Conditions:** The Transmission Control Module (ECM) communicates with all databus-capable control modules via a CAN databus. These databus-capable control modules are connected via two data bus wires which are twisted together (CAN_High and CAN_Low), and exchange information (messages). Missing information on the databus is recognized as a malfunction and stored. Trouble-free operation of the CAN-Bus requires that it have a terminal resistance. **Possible Causes:** • Solenoid valves in valve body are faulty • Solenoid circuit is shorting to ground • Solenoid circuit is open • TCM has failed or wiring is shorting • ECM has failed

DTC	Trouble Code Title, Conditions & Possible Causes
DTC: P0886 **Passat:** 2.0L (BHW); **Transmissions:** A/T	**TCM Power Relay Control Circuit Low Conditions:** The Transmission Control Module (ECM) communicates with all databus-capable control modules via a CAN databus. These databus-capable control modules are connected via two data bus wires which are twisted together (CAN_High and CAN_Low), and exchange information (messages). Missing information on the databus is recognized as a malfunction and stored. Trouble-free operation of the CAN-Bus requires that it have a terminal resistance. **Possible Causes:** • Solenoid valves in valve body are faulty • Solenoid circuit is shorting to ground • Solenoid circuit is open • TCM has failed or wiring is shorting • ECM has failed
DTC: P0887 **Passat:** 2.0L (BHW); **Transmissions:** A/T	**TCM Power Relay Control Circuit High Conditions:** The Transmission Control Module (ECM) communicates with all databus-capable control modules via a CAN databus. These databus-capable control modules are connected via two data bus wires which are twisted together (CAN_High and CAN_Low), and exchange information (messages). Missing information on the databus is recognized as a malfunction and stored. Trouble-free operation of the CAN-Bus requires that it have a terminal resistance. **Possible Causes:** • Solenoid valves in valve body are faulty

Diesel Engine OBD II Trouble Code List (P1xxx Codes)

DTC	Trouble Code Title, Conditions & Possible Causes:
DTC: P1026 **Jetta:** 1.9 (ALH, BEW); **Golf:** 1.9: (ALH, BEW); **NB:** 1.9L (ALH, BEW) **Transmissions:** All	**Activation Intake Manifold Flap for Air Stream Regulation Short Circuit to B+ Conditions:** The ECM detected a voltage irregularity on the intake manifold flap. **Possible Causes:** • Check activation of intake manifold flap for air stream regulation
DTC: P1027 **Jetta:** 1.9 (ALH, BEW); **Golf:** 1.9: (ALH, BEW); **NB:** 1.9L (ALH, BEW) **Transmissions:** All	**Activation Intake Manifold Flap for Air Stream Regulation Short Circuit to Ground Conditions:** The ECM detected a voltage irregularity on the intake manifold flap. **Possible Causes:** • Check activation of intake manifold flap for air stream regulation
DTC: P1028 **Jetta:** 1.9 (ALH, BEW); **Golf:** 1.9: (ALH, BEW); **NB:** 1.9L (ALH, BEW) **Transmissions:** All	**Activation Intake Manifold Flap for Air Stream Regulation Open Circuit Conditions:** The ECM detected a voltage irregularity on the intake manifold flap. **Possible Causes:** • Check activation of intake manifold flap for air stream regulation
DTC: P1702 Passat: 2.0L (BHW); **Transmissions:** A/T	**TR Sensor Signal Intermittent Conditions:** Key on or engine running; and the ECM detected the failure Trouble Code Conditions for DTC P0705 or P0708 were met intermittently. **Possible Causes:** • Refer to the appropriate Transmission Repair Manual or information in electronic media to perform a complete diagnosis of the automatic transmission when this code is set
DTC: P1778 **2T, MIL: Yes** **Jetta:** 1.9 (ALH, BEW); **Golf:** 1.9: (ALH, BEW); **NB:** 1.9L (ALH, BEW) **Transmissions:** All	**Solenoid EV7 Electrical Malfunction Conditions:** Engine started, battery voltage must be at least 11.5v, all electrical components must be off, the ground between the engine and the chassis must be well connected, the exhaust system must be properly sealed between the catalytic converter and the cylinder head, and the oxygen sensor heater for oxygen sensor before the catalytic converter must be properly functioning. The ECM detected a loss of communication between the TCM and the valve body solenoid valve. **Possible Causes:** • Valve body solenoid valve has a short • The TCM has failed • ECM has failed
DTC: P1780 **2T, MIL: Yes** **Jetta:** 1.9 (ALH, BEW); **Golf:** 1.9: (ALH, BEW); **NB:** 1.9L (ALH, BEW) **Transmissions:** A/T	**Engine Intervention Readable Conditions:** Key on or engine started, Self-Test enabled, and the ECM detected the Transmission Control Switch (TCS) was out of range during the test and a signal was not received. **Note: The seal on the ATF level plug must always be replaced if the ATF level is checked.** **Possible Causes:** • TCS circuit open or shorted in the wiring harness • TCS not cycled during the self-test • TCS is damaged or failed, or the ECM has failed • Check ATF level • Mechanical malfunction in transmission
DTC: P1850 **Jetta:** 1.9 (ALH, BEW); **Golf:** 1.9: (ALH, BEW); **NB:** 1.9L (ALH, BEW) **Transmissions:** All	**Data BUS Powertrain Missing Message from Engine Control Conditions:** The Engine Control Module (ECM) communicates with all databus-capable control modules via a CAN databus. These databus-capable control modules are connected via two data bus wires which are twisted together (CAN_High and CAN_Low), and exchange information (messages). Missing information on the databus is recognized as a malfunction and stored. Trouble-free operation of the CAN-bus requires that it have a terminal resistance. This central terminal resistor is located in the Engine Control Module (ECM). **Possible Causes:** • Check the Terminal resistance for CAN-bus • Data-Bus wires have short • Data-Bus components are malfunctioning • ECM has failed
DTC: P1854 **Jetta:** 1.9 (ALH, BEW); **Golf:** 1.9: (ALH, BEW); **NB:** 1.9L (ALH, BEW) **Transmissions:** All	**Data BUS Powertrain Hardware Defective Conditions:** The Engine Control Module (ECM) communicates with all databus-capable control modules via a CAN databus. These databus-capable control modules are connected via two data bus wires which are twisted together (CAN_High and CAN_Low), and exchange information (messages). Missing information on the databus is recognized as a malfunction and stored. Trouble-free operation of the CAN-bus requires that it have a terminal resistance. This central terminal resistor is located in the Engine Control Module (ECM). **Possible Causes:** • Check the Terminal resistance for CAN-bus • Data-Bus wires have short • Data-Bus components are malfunctioning • ECM has failed

DTC	Trouble Code Title, Conditions & Possible Causes
DTC: P1855 **Jetta:** 1.9 (ALH, BEW); **Golf:** 1.9: (ALH, BEW); **NB:** 1.9L (ALH, BEW) **Transmissions:** All	**Data BUS Powertrain Software Version Control Conditions:** The Engine Control Module (ECM) communicates with all databus-capable control modules via a CAN databus. These databus-capable control modules are connected via two data bus wires which are twisted together (CAN_High and CAN_Low), and exchange information (messages). Missing information on the databus is recognized as a malfunction and stored. Trouble-free operation of the CAN-bus requires that it have a terminal resistance. This central terminal resistor is located in the Engine Control Module (ECM). **Possible Causes:** • Check the Terminal resistance for CAN-bus • Data-Bus wires have short • Data-Bus components are malfunctioning • ECM has failed
DTC: P1857 **Passat:** 2.0L (BHW); **Jetta:** 1.9 (ALH, BEW); **Golf:** 1.9: (ALH, BEW); **NB:** 1.9L (ALH, BEW) **Transmissions:** All	**Load Signal Error Message From Engine Control Conditions:** The Engine Control Module (ECM) communicates with all databus-capable control modules via a CAN databus. These databus-capable control modules are connected via two data bus wires which are twisted together (CAN_High and CAN_Low), and exchange information (messages). Missing information on the databus is recognized as a malfunction and stored. Trouble-free operation of the CAN-bus requires that it have a terminal resistance. This central terminal resistor is located in the Engine Control Module (ECM). **Possible Causes:** • Intake Manifold Runner Position Sensor is faulty • Intake system has leaks (false air) • Motor for intake flap is faulty • Mass Air Flow (MAF) sensor has failed • ECM has failed
DTC: P1866 **Jetta:** 1.9 (ALH, BEW); **Golf:** 1.9: (ALH, BEW); **NB:** 1.9L (ALH, BEW) **Transmissions:** All	**Data Bus Powertrain Missing Messages Conditions:** The Engine Control Module (ECM) communicates with all databus-capable control modules via a CAN databus. These databus-capable control modules are connected via two data bus wires which are twisted together (CAN_High and CAN_Low), and exchange information (messages). Missing information on the databus is recognized as a malfunction and stored. Trouble-free operation of the CAN-bus requires that it have a terminal resistance. This central terminal resistor is located in the Engine Control Module (ECM). **Possible Causes:** • Check the Terminal resistance for CAN-bus • Data-Bus wires have short • Data-Bus components are malfunctioning • ECM has failed

Diesel Engine OBD II Trouble Code List (P2xxx Codes)

DTC	Trouble Code Title, Conditions & Possible Causes:
DTC: P2100 **Jetta:** 1.9 (ALH, BEW); **Golf:** 1.9: (ALH, BEW); **NB:** 1.9L (ALH, BEW) **Transmissions:** All	**Throttle Actuator Control Motor Circuit Open Conditions:** Engine started, battery voltage must be at least 11.5v, all electrical components must be off, parking brake must be engaged (to keep daytime driving lights off), automatic transmission selector must be in park, the exhaust system must be properly sealed between the catalytic converter and the cylinder head, coolant temperature must be at least 80 degrees Celsius. The ECM detected an unexpected low or high voltage condition on the Throttle Actuator Control Motor (TACM) circuit during the CCM test. **Note: The throttle valve activation occurs via an electric motor (throttle drive) in the throttle valve control module. It is activated by the Engine Control Module (ECM) according to specifications of the two sensors, Throttle Position (TP) Sensor and Sender 2 for accelerator pedal position.** **Possible Causes:** • TACM wiring harness connector is damaged or open • TACM wiring may be crossed in the wire harness assembly • TACM (motor) circuit is open, or TACM assembly is damaged (possible open circuit) • TACM or the Throttle Valve is dirty • Throttle Position sensor has failed • ECM has failed
DTC: P2102 **Jetta:** 1.9 (ALH, BEW); **Golf:** 1.9: (ALH, BEW); **NB:** 1.9L (ALH, BEW) **Transmissions:** All	**Throttle Actuator Control Motor Circuit Low Conditions:** Engine started, battery voltage must be at least 11.5v, all electrical components must be off, parking brake must be engaged (to keep daytime driving lights off), automatic transmission selector must be in park, the exhaust system must be properly sealed between the catalytic converter and the cylinder head, coolant temperature must be at least 80 degrees Celsius. The ECM detected an unexpected low or high voltage condition on the Throttle Actuator Control Motor (TACM) circuit during the CCM test. **Note: The throttle valve activation occurs via an electric motor (throttle drive) in the throttle valve control module. It is activated by the Engine Control Module (ECM) according to specifications of the two sensors, Throttle Position (TP) Sensor and Sender 2 for accelerator pedal position.** **Possible Causes:** • TACM wiring harness connector is damaged or open • TACM wiring may be crossed in the wire harness assembly • TACM (motor) circuit is open, or TACM assembly is damaged (possible open circuit) • TACM or the Throttle Valve is dirty • Throttle Position sensor has failed • ECM has failed
DTC: P2103 **Jetta:** 1.9 (ALH, BEW); **Golf:** 1.9: (ALH, BEW); **NB:** 1.9L (ALH, BEW) **Transmissions:** All	**Throttle Actuator Control Motor Circuit High Conditions:** Engine started, battery voltage must be at least 11.5v, all electrical components must be off, parking brake must be engaged (to keep daytime driving lights off), automatic transmission selector must be in park, the exhaust system must be properly sealed between the catalytic converter and the cylinder head, coolant temperature must be at least 80 degrees Celsius. The ECM detected an unexpected low or high voltage condition on the Throttle Actuator Control Motor (TACM) circuit during the CCM test. **Note: The throttle valve activation occurs via an electric motor (throttle drive) in the throttle valve control module. It is activated by the Engine Control Module (ECM) according to specifications of the two sensors, Throttle Position (TP) Sensor and Sender 2 for accelerator pedal position.** **Possible Causes:** • TACM wiring harness connector is damaged or open • TACM wiring may be crossed in the wire harness assembly • TACM (motor) circuit is open, or TACM assembly is damaged (possible open circuit) • TACM or the Throttle Valve is dirty • Throttle Position sensor has failed • ECM has failed
DTC: P2108 **Jetta:** 1.9 (ALH, BEW); **Golf:** 1.9: (ALH, BEW); **NB:** 1.9L (ALH, BEW) **Transmissions:** All	**Throttle Actuator Control Motor Performance Conditions:** Engine started, battery voltage must be at least 11.5v, all electrical components must be off, parking brake must be engaged (to keep daytime driving lights off), automatic transmission selector must be in park, the exhaust system must be properly sealed between the catalytic converter and the cylinder head, coolant temperature must be at least 80 degrees Celsius. The ECM detected an unexpected low or high voltage condition on the Throttle Actuator Control Motor (TACM) circuit during the CCM test. **Note: The throttle valve activation occurs via an electric motor (throttle drive) in the throttle valve control module. It is activated by the Engine Control Module (ECM) according to specifications of the two sensors, Throttle Position (TP) Sensor and Sender 2 for accelerator pedal position.** **Possible Causes:** • TACM wiring harness connector is damaged or open • TACM wiring may be crossed in the wire harness assembly • TACM (motor) circuit is open, or TACM assembly is damaged (possible open circuit) • TACM or the Throttle Valve is dirty • Throttle Position sensor has failed • ECM has failed

DTC	Trouble Code Title, Conditions & Possible Causes
DTC: P2122 **Jetta:** 1.9 (ALH, BEW); **Golf:** 1.9: (ALH, BEW); **NB:** 1.9L (ALH, BEW) **Transmissions:** All	**Accelerator Pedal Position Sensor 'D' Circuit Low Input Conditions:** Engine started, battery voltage at least 11.5v, all electrical components off, ground connections between engine and chassis well connected, the ECM detected that the accelerator pedal position sensor signal was outside the parameters to function normally. **Note: Both the Throttle Position (TP) Sensor and Accelerator Pedal Position Sensor are located at the accelerator pedal module and communicate the driver's intentions to the ECM completely independently of each other. Both sensors are stored in one housing.** **Possible Causes:** • Ground between engine and chassis may be broken • Throttle position sensor may have failed • Accelerator Pedal Position Sensor has failed • Throttle position sensor wiring may have shorted • Throttle position sensor has failed • Faulty voltage supply • ECM has failed
DTC: P2195 **Jetta:** 1.9 (ALH, BEW); **Golf:** 1.9: (ALH, BEW); **NB:** 1.9L (ALH, BEW) **Transmissions:** All	**O2 Sensor Signal Stuck Lean Bank 1 Sensor 1 Conditions:** Engine running in closed loop, and the ECM detected the O2S indicated a lean signal, or it could no longer control Fuel Trim because it was at lean limit. **Possible Causes:** • Engine oil level high • Camshaft timing error • Cylinder compression low • Exhaust leaks in front of O2S • EGR valve is stuck open • EGR gasket is leaking • EVR diaphragm is leaking • Damaged fuel pressure regulator or extremely low fuel pressure • O2S circuit is open or shorted in the wiring harness • Oxygen sensor (before catalytic converter) is faulty • Oxygen sensor (behind catalytic converter) is faulty • Oxygen sensor heater (before catalytic converter) is faulty • Oxygen sensor heater (behind catalytic converter) is faulty • Air leaks after the MAF sensor • PCV system leaks • Dip stick not seated properly
DTC: P2196 **Jetta:** 1.9 (ALH, BEW); **Golf:** 1.9: (ALH, BEW); **NB:** 1.9L (ALH, BEW) **Transmissions:** All	**O2 Sensor Signal Stuck Rich Bank 1 Sensor 1 Conditions:** Engine running in closed loop, and the ECM detected the O2S indicated a rich signal, or it could no longer control Fuel Trim because it was at its rich limit. **Possible Causes:** • Engine oil level high • Camshaft timing error • Cylinder compression low • Exhaust leaks in front of O2S • EGR valve is stuck open • EGR gasket is leaking • EVR diaphragm is leaking • Damaged fuel pressure regulator or extremely low fuel pressure • O2S circuit is open or shorted in the wiring harness • Oxygen sensor (before catalytic converter) is faulty • Oxygen sensor (behind catalytic converter) is faulty • Oxygen sensor heater (before catalytic converter) is faulty • Oxygen sensor heater (behind catalytic converter) is faulty • Air leaks after the MAF sensor • PCV system leaks • Dip stick not seated properly

DTC	Trouble Code Title, Conditions & Possible Causes
DTC: P2237 **Passat:** 2.0L (BHW); **Jetta:** 1.9 (ALH, BEW); **Golf:** 1.9: (ALH, BEW); **NB:** 1.9L (ALH, BEW) **Transmissions:** All	**O2 Sensor Positive Current Control Circuit/Open Bank 1 Sensor 1 Conditions:** Engine started, battery voltage must be at least 11.5v, all electrical components must be off, parking brake must be engaged (to keep daytime driving lights off), automatic transmission selector must be in park. The ECM detected an unexpected voltage condition, or it detected an unexpected current draw in the sensor circuit during the CCM test. **Note: Vehicle must be raised before connector for oxygen sensors is accessible.** **Possible Causes:** • Oxygen sensor (before catalytic converter) is faulty • Oxygen sensor heater (before catalytic converter) is faulty • Oxygen sensor heater (before catalytic converter) is faulty • Oxygen sensor heater (behind catalytic converter) is faulty • O2S circuit is open or shorted in the wiring harness • ECM has failed
DTC: P2238 **Passat:** 2.0L (BHW); **Jetta:** 1.9 (ALH, BEW); **Golf:** 1.9: (ALH, BEW); **NB:** 1.9L (ALH, BEW) **Transmissions:** All	**O2 Sensor Positive Current Control Circuit Low Bank 1 Sensor 1 Conditions:** Engine started, battery voltage must be at least 11.5v, all electrical components must be off, parking brake must be engaged (to keep daytime driving lights off), automatic transmission selector must be in park. The ECM detected an unexpected voltage condition, or it detected an unexpected current draw in the sensor circuit during the CCM test. **Note: Vehicle must be raised before connector for oxygen sensors is accessible.** **Possible Causes:** • Oxygen sensor (before catalytic converter) is faulty • Oxygen sensor heater (before catalytic converter) is faulty • Oxygen sensor heater (before catalytic converter) is faulty • Oxygen sensor heater (behind catalytic converter) is faulty • O2S circuit is open or shorted in the wiring harness • ECM has failed
DTC: P2239 **Passat:** 2.0L (BHW); **Jetta:** 1.9 (ALH, BEW); **Golf:** 1.9: (ALH, BEW); **NB:** 1.9L (ALH, BEW) **Transmissions:** All	**O2 Sensor Positive Current Control Circuit High Bank 1 Sensor 1 Conditions:** Engine started, battery voltage must be at least 11.5v, all electrical components must be off, parking brake must be engaged (to keep daytime driving lights off), automatic transmission selector must be in park. The ECM detected an unexpected voltage condition, or it detected an unexpected current draw in the sensor circuit during the CCM test. **Note: Vehicle must be raised before connector for oxygen sensors is accessible.** **Possible Causes:** • Oxygen sensor (before catalytic converter) is faulty • Oxygen sensor heater (before catalytic converter) is faulty • Oxygen sensor heater (before catalytic converter) is faulty • Oxygen sensor heater (behind catalytic converter) is faulty • O2S circuit is open or shorted in the wiring harness • ECM has failed
DTC: P2243 **Passat:** 2.0L (BHW); **Jetta:** 1.9 (ALH, BEW); **Golf:** 1.9: (ALH, BEW); **NB:** 1.9L (ALH, BEW) **Transmissions:** All	**O2 Sensor Reference Voltage Circuit/Open Bank 1 Sensor 1 Conditions:** Engine started, battery voltage must be at least 11.5v, all electrical components must be off, parking brake must be engaged (to keep daytime driving lights off), automatic transmission selector must be in park. The ECM detected an unexpected voltage condition, or it detected an unexpected current draw in the sensor circuit during the CCM test. **Note: Vehicle must be raised before connector for oxygen sensors is accessible.** **Possible Causes:** • Oxygen sensor (before catalytic converter) is faulty • Oxygen sensor heater (before catalytic converter) is faulty • Oxygen sensor heater (before catalytic converter) is faulty • Oxygen sensor heater (behind catalytic converter) is faulty • O2S circuit is open or shorted in the wiring harness • ECM has failed
DTC: P2245 **Passat:** 2.0L (BHW); **Jetta:** 1.9 (ALH, BEW); **Golf:** 1.9: (ALH, BEW); **NB:** 1.9L (ALH, BEW) **Transmissions:** All	**O2 Sensor Reference Voltage Circuit Low Bank 1 Sensor 1 Conditions:** Engine started, battery voltage must be at least 11.5v, all electrical components must be off, parking brake must be engaged (to keep daytime driving lights off), automatic transmission selector must be in park. The ECM detected an unexpected voltage condition, or it detected an unexpected current draw in the sensor circuit during the CCM test. **Note: Vehicle must be raised before connector for oxygen sensors is accessible.** **Possible Causes:** • Oxygen sensor (before catalytic converter) is faulty • Oxygen sensor heater (before catalytic converter) is faulty • Oxygen sensor heater (before catalytic converter) is faulty • Oxygen sensor heater (behind catalytic converter) is faulty • O2S circuit is open or shorted in the wiring harness • ECM has failed

DTC	Trouble Code Title, Conditions & Possible Causes
DTC: P2246 **Passat:** 2.0L (BHW); **Jetta:** 1.9 (ALH, BEW); **Golf:** 1.9: (ALH, BEW); **NB:** 1.9L (ALH, BEW) **Transmissions:** All	**O2 Sensor Reference Voltage Circuit High Bank 1 Sensor 1 Conditions:** Engine started, battery voltage must be at least 11.5v, all electrical components must be off, parking brake must be engaged (to keep daytime driving lights off), automatic transmission selector must be in park. The ECM detected an unexpected voltage condition, or it detected an unexpected current draw in the sensor circuit during the CCM test. **Note: Vehicle must be raised before connector for oxygen sensors is accessible.** **Possible Causes:** • Oxygen sensor (before catalytic converter) is faulty • Oxygen sensor heater (before catalytic converter) is faulty • Oxygen sensor heater (before catalytic converter) is faulty • Oxygen sensor heater (behind catalytic converter) is faulty • O2S circuit is open or shorted in the wiring harness • ECM has failed
DTC: P2251 **Passat:** 2.0L (BHW); **Jetta:** 1.9 (ALH, BEW); **Golf:** 1.9: (ALH, BEW); **NB:** 1.9L (ALH, BEW) **Transmissions:** All	**O2 Sensor Negative Voltage Circuit/Open Bank 1 Sensor 1 Conditions:** Engine started, battery voltage must be at least 11.5v, all electrical components must be off, parking brake must be engaged (to keep daytime driving lights off), automatic transmission selector must be in park. The ECM detected an unexpected voltage condition, or it detected an unexpected current draw in the sensor circuit during the CCM test. **Note: Vehicle must be raised before connector for oxygen sensors is accessible.** **Possible Causes:** • Oxygen sensor (before catalytic converter) is faulty • Oxygen sensor heater (before catalytic converter) is faulty • Oxygen sensor heater (before catalytic converter) is faulty • Oxygen sensor heater (behind catalytic converter) is faulty • O2S circuit is open or shorted in the wiring harness • ECM has failed
DTC: P2252 **Passat:** 2.0L (BHW); **Jetta:** 1.9 (ALH, BEW); **Golf:** 1.9: (ALH, BEW); **NB:** 1.9L (ALH, BEW) **Transmissions:** All	**O2 Sensor Negative Voltage Circuit Low Bank 1 Sensor 1 Conditions:** Engine started, battery voltage must be at least 11.5v, all electrical components must be off, parking brake must be engaged (to keep daytime driving lights off), automatic transmission selector must be in park. The ECM detected an unexpected voltage condition, or it detected an unexpected current draw in the sensor circuit during the CCM test. **Note: Vehicle must be raised before connector for oxygen sensors is accessible.** **Possible Causes:** • Oxygen sensor (before catalytic converter) is faulty • Oxygen sensor heater (before catalytic converter) is faulty • Oxygen sensor heater (before catalytic converter) is faulty • Oxygen sensor heater (behind catalytic converter) is faulty • O2S circuit is open or shorted in the wiring harness • ECM has failed
DTC: P2253 **Passat:** 2.0L (BHW); **Jetta:** 1.9 (ALH, BEW); **Golf:** 1.9: (ALH, BEW); **NB:** 1.9L (ALH, BEW) **Transmissions:** All	**O2 Sensor Negative Voltage Circuit High Bank 1 Sensor 1 Conditions:** Engine started, battery voltage must be at least 11.5v, all electrical components must be off, parking brake must be engaged (to keep daytime driving lights off), automatic transmission selector must be in park. The ECM detected an unexpected voltage condition, or it detected an unexpected current draw in the sensor circuit during the CCM test. **Note: Vehicle must be raised before connector for oxygen sensors is accessible.** **Possible Causes:** • Oxygen sensor (before catalytic converter) is faulty • Oxygen sensor heater (before catalytic converter) is faulty • Oxygen sensor heater (before catalytic converter) is faulty • Oxygen sensor heater (behind catalytic converter) is faulty • O2S circuit is open or shorted in the wiring harness • ECM has failed
DTC: P2413 **Passat:** 2.0L (BHW); **Jetta:** 1.9 (ALH, BEW); **Golf:** 1.9: (ALH, BEW); **NB:** 1.9L (ALH, BEW) **Transmissions:** All	**Exhaust Gas Recirculation System Performance Conditions:** Engine started, battery voltage must be at least 11.5v, all electrical components must be off, parking brake must be engaged (to keep daytime driving lights off), automatic transmission selector must be in park. The ECM detected a EGR fault that could affect the system's performance. **Note: If the EGR Vacuum Regulator Solenoid Valve with Potentiometer for EGR was replaced, a basic setting must be performed.** **Possible Causes:** • EGR Vacuum Regulator Solenoid Valve is faulty • EGR valve is malfunctioning • Circuit wires have short circuited to each other, to ground or to power • Potentiometer for EGR has failed

DTC	Trouble Code Title, Conditions & Possible Causes
DTC: P2425 **Jetta:** 1.9 (ALH, BEW); **Golf:** 1.9: (ALH, BEW); **NB:** 1.9L (ALH, BEW) **Transmissions:** All	**Exhaust Gas Recirculation Cooling Valve Control Circuit/Open Conditions:** Engine started, battery voltage must be at least 11.5v. The ECM detected a problem with the EGR cooling valve circuit that could affect the performance of the vehicle. **Possible Causes:** • The exhaust gas recirculation cooler switch-over valve has failed • Circuit wires have short circuited to each other, to ground or to power • ECM has failed
DTC: P2426 **Jetta:** 1.9 (ALH, BEW); **Golf:** 1.9: (ALH, BEW); **NB:** 1.9L (ALH, BEW) **Transmissions:** All	**Exhaust Gas Recirculation Cooling Valve Control Circuit Low Conditions:** Engine started, battery voltage must be at least 11.5v. The ECM detected a problem with the EGR cooling valve circuit that could affect the performance of the vehicle. **Possible Causes:** • The exhaust gas recirculation cooler switch-over valve has failed • Circuit wires have short circuited to each other, to ground or to power • ECM has failed
DTC: P2427 **Jetta:** 1.9 (ALH, BEW); **Golf:** 1.9: (ALH, BEW); **NB:** 1.9L (ALH, BEW) **Transmissions:** All	**Exhaust Gas Recirculation Cooling Valve Control Circuit High Conditions:** Engine started, battery voltage must be at least 11.5v. The ECM detected a problem with the EGR cooling valve circuit that could affect the performance of the vehicle. **Possible Causes:** • The exhaust gas recirculation cooler switch-over valve has failed • Circuit wires have short circuited to each other, to ground or to power • ECM has failed
DTC: P2637 **Jetta:** 1.9 (ALH, BEW); **Golf:** 1.9: (ALH, BEW); **NB:** 1.9L (ALH, BEW) **Transmissions:** All	**Torque Management Feedback Signal "A" Conditions:** Engine started, battery voltage must be at least 11.5v, all electrical components must be off, parking brake must be engaged (to keep daytime driving lights off), automatic transmission selector must be in park, the exhaust system must be properly sealed between the catalytic converter and the cylinder head, coolant temperature must be at least 80 degrees Celsius and oxygen sensor heaters for oxygen sensors before the catalytic converter must be functioning properly and the ground between the engine and the chassis must be well connected. The ECM detected a voltage value on the torque management circuits that doesn't fall within the desired parameters **Possible Causes:** • Engine Control Module (ECM) has failed • Voltage supply for Engine Control Module (ECM) is damaged • Engine Coolant Temperature (ECT) sensor has failed • Intake Air Temperature (IAT) sensor has failed • Intake Manifold Runner Position Sensor has failed • Intake system has leaks (false air) • Motor for intake flap has failed • Mass Air Flow (MAF) sensor has failed
DTC: P2714 **Passat:** 2.0L (BHW); **Transmissions:** All	**Pressure Control Solenoid "D" Performance or Stuck Off Conditions:** Engine started, battery voltage must be at least 11.5v, all electrical components must be off, and the ground between the engine and the chassis must be well connected. The ECM detected the pressure control solenoid was in the "stuck off" position. **Possible Causes:** • ATF level is low • Circuit harness connector contacts are corroded or ingressed of water • Circuit wires have shorted to each other, to battery or ground • Automatic Transmission Hydraulic Pressure Sensor 1 has failed • Solenoid valves in valve body are faulty • Transmission Control Module (TCM) needs replacing • Transmission Input Speed (RPM) Sensor has failed • Transmission Output Speed (RPM) Sensor has failed
DTC: P2715 **Passat:** 2.0L (BHW); **Transmissions:** All	**Pressure Control Solenoid "D" Performance or Stuck On Conditions:** Engine started, battery voltage must be at least 11.5v, all electrical components must be off, and the ground between the engine and the chassis must be well connected. The ECM detected the pressure control solenoid was in the "stuck on" position. **Possible Causes:** • ATF level is low • Circuit harness connector contacts are corroded or ingressed of water • Circuit wires have shorted to each other, to battery or ground • Automatic Transmission Hydraulic Pressure Sensor 1 has failed • Solenoid valves in valve body are faulty • Transmission Control Module (TCM) needs replacing • Transmission Input Speed (RPM) Sensor has failed • Transmission Output Speed (RPM) Sensor has failed

DTC	Trouble Code Title, Conditions & Possible Causes
DTC: P2716 **Passat:** 2.0L (BHW); **Transmissions:** All	**Pressure Control Solenoid "D" Electrical Malfunction Conditions:** Engine started, battery voltage must be at least 11.5v, all electrical components must be off, and the ground between the engine and the chassis must be well connected. The ECM detected the pressure control solenoid was experiencing electrical malfunctions. **Possible Causes:** • ATF level is low • Circuit harness connector contacts are corroded or ingressed of water • Circuit wires have shorted to each other, to battery or ground • Automatic Transmission Hydraulic Pressure Sensor 1 has failed • Solenoid valves in valve body are faulty • Transmission Control Module (TCM) needs replacing • Transmission Input Speed (RPM) Sensor has failed • Transmission Output Speed (RPM) Sensor has failed
DTC: P2717 **Passat:** 2.0L (BHW); **Transmissions:** All	**Pressure Control Solenoid "D" Intermittent Conditions:** Engine started, battery voltage must be at least 11.5v, all electrical components must be off, and the ground between the engine and the chassis must be well connected. The ECM detected the pressure control solenoid sending intermittent signals. **Possible Causes:** • ATF level is low • Circuit harness connector contacts are corroded or ingressed of water • Circuit wires have shorted to each other, to battery or ground • Automatic Transmission Hydraulic Pressure Sensor 1 has failed • Solenoid valves in valve body are faulty • Transmission Control Module (TCM) needs replacing • Transmission Input Speed (RPM) Sensor has failed • Transmission Output Speed (RPM) Sensor has failed

VOLKSWAGEN
COMPONENT TESTING

15

TABLE OF CONTENTS

Barometric Pressure (BARO) Sensor

LOCATION

See Figure 1.

1.8L Engines

The Barometric Pressure (BARO) Sensor is positioned in a protective box within the fresh air plenum, on the left side, near the Engine Control Module.

TESTING

See Figure 2.

1. Disconnect three-pin connector at sensor.
2. Switch ignition on.
3. Connect multimeter to the following connector terminals for voltage measurement:
   - Terminal 1 + 3
   - Terminal 2 + 3
4. Specified value: 4.5 to 5.5 V
5. If specified values are not obtained:
   a. Connect test box to control module wiring harness.
   b. Check wiring between test box and 3-pin connector for open circuit according to wiring diagram.
   - Terminal 1 + socket 61
   - Terminal 2 + socket 62
   - Terminal 3 + socket 67
6. Wire resistance: max: 1.5 OHMS
7. Check wires for short circuit to each other. Specified value: infinite ohms.
8. Check wire for short circuit to B+ or Ground (GND). Specified value: infinite ohms.

Camshaft Position (CMP) Sensor

LOCATION

1.8L Engines

See Figure 3.

The Camshaft Position (CMP) Sensor is mounted on right side of the cylinder head.

2.0L Engines

See Figure 4.

The Camshaft Position (CMP) Sensor is mounted on right side of the cylinder head, behind the camshaft sprocket (black arrow, while the white arrow is the harness connector).

2.5L Engines

See Figure 5.

The Camshaft Position (CMP) Sensor is mounted on the front edge of the cylinder head cover, on the right side

2.8L Engines

See Figure 6.

The Camshaft Position (CMP) Sensor is mounted on left side of the cylinder head, behind the ignition coil, in the rear.

3.2L Engines

See Figure 6.

The Camshaft Position (CMP) Sensor is mounted on left side of the cylinder head.

3.6L Engines

See Figure 7.

The Camshaft Position (CMP) Sensor is mounted on right side of the engine near the rubber hoses.

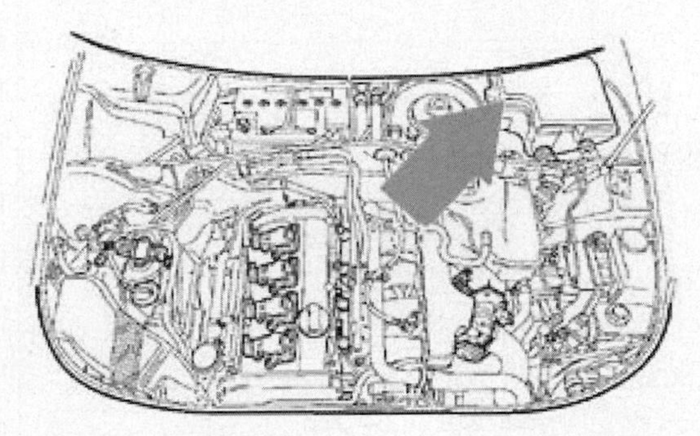

Fig. 1 Barometric Pressure sensor location for the 1.8L engine.

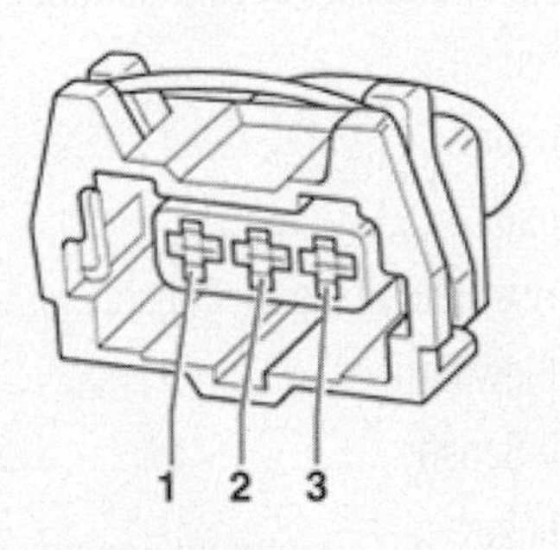

Fig. 2 The three-pin connector at the sensor is connected to the multimeter for testing.

Fig. 3 Camshaft Position sensor location for the 1.8L engine.

Fig. 4 Camshaft Position sensor location for the 2.0L engine.

4.0L Engines

CAMSHAFT POSITION (CMP) SENSOR 1

See Figure 8.

The Camshaft Position (CMP) Sensor 1 (intake camshaft sensor) is located on the engine, on the back of the left cylinder head.

CAMSHAFT POSITION (CMP) SENSOR 2

See Figure 9.

The Camshaft Position (CMP) Sensor 2 (exhaust camshaft sensor) is located on the engine, on the back of the right cylinder head.

CAMSHAFT POSITION (CMP) SENSOR 3

See Figure 10.

The Camshaft Position (CMP) Sensor 3 (intake camshaft sensor) is located on the engine, on the back of the left cylinder head.

CAMSHAFT POSITION (CMP) SENSOR 4

See Figure 11.

The Camshaft Position (CMP) Sensor 4 is mounted on back of the cylinder head.

TESTING

See Figures 12 through 14.

➡ Use only gold-plated terminals when servicing terminals in harness connector of Camshaft Position (CMP) sensor.

1. Remove engine cover with air filter.
2. Disconnect 3-pin harness connector (1) at Camshaft Position (CMP) Sensor.
3. Switch ignition on.
4. Measure voltage at terminals 1 + 3 of connector using hand multimeter and adapter cables from connector test kit.
5. Specified value: about 5.0 V
6. Switch ignition off.
7. If the voltage is not OK, connect test box to wiring harness of Motronic Engine Control Module (ECM).

Fig. 5 Camshaft Position sensor location for the 2.5L engine.

Fig. 6 Camshaft Position sensor location for the 2.8L and 3.2L engines.

Fig. 7 Camshaft Position sensor location for the 3.6 L engine.

Fig. 8 Camshaft Position sensor 1 location for the 4.0L engine.

Fig. 9 Camshaft Position sensor 2 location for the 4.0L engine.

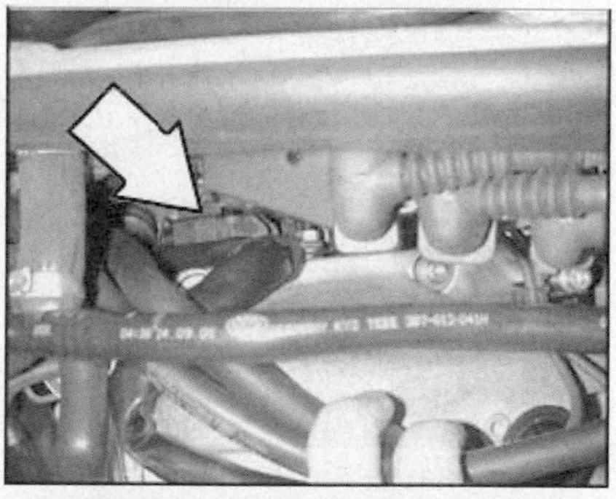

Fig. 10 Camshaft Position sensor 3 location for the 4.0L engine.

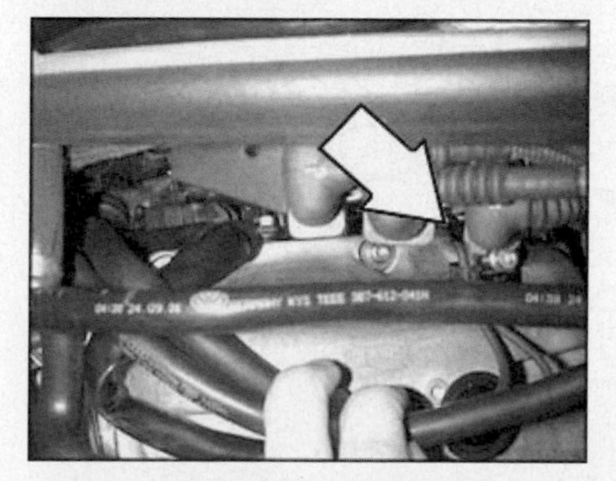

Fig. 11 Camshaft Position sensor 4 location for the 4.0L engine.

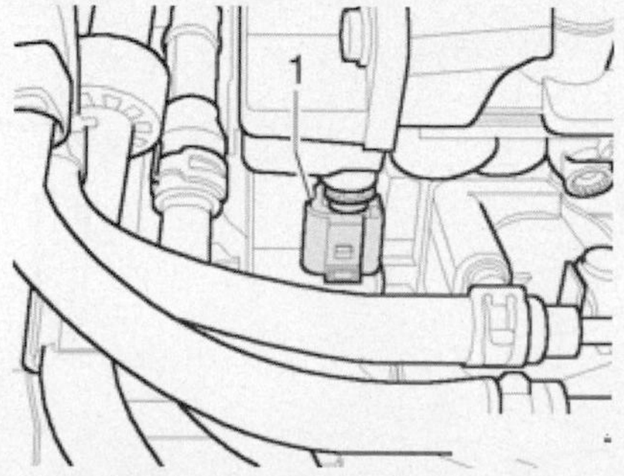

Fig. 12 Disconnect 3-pin harness connector (1) at Camshaft Position (CMP) Sensor

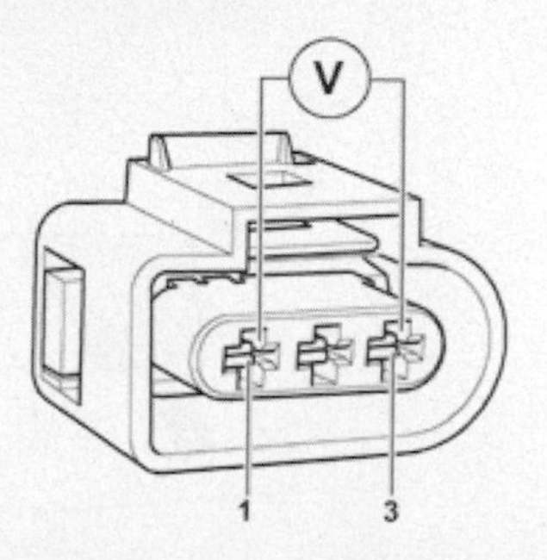

Fig. 13 Measure voltage at terminals 1 + 3 of connector using hand multimeter

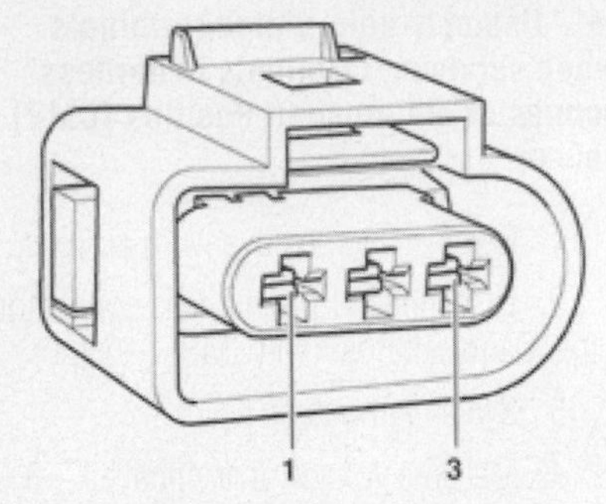

Fig.14 Check wires between test box and 3-pin connector for open circuit according to wiring diagram

8. Check wires between test box and 3-pin connector for open circuit according to the wiring diagram.

Terminal	Test box socket
1	26
2	44
3	14

9. Wire resistance: max. 1.5 OHMS
10. Check wires for short circuit to each other, to Battery A (+) and Ground (GND).
11. Specified value: infinite ohms.
12. If the wires and voltage are OK, replace Camshaft Position (CMP) Sensor.
13. If the wires and voltage are OK replace Motronic Engine Control Module (ECM).

Charge Air Pressure Sensor

LOCATION

See Figure 15.

All Engines Except 1.8L and 2.0L TDI Engines

The Charge Air Pressure Sensor is located in the lower right front part of the engine compartment, forward of the windshield washer fluid reservoir and on top of the intercooler.

1.8L Engines

See Figure 16.

The Charge Air Pressure Sensor is located in the engine compartment, on the driver's side, behind the headlight assembly (shown from above).

2.0L TDI Engines

See Figure 17.

The Charge Air Pressure Sensor is located in the engine compartment, on intake air duct behind the engine.

TESTING

See Figures 18 and 19.

➡ **Use only gold-plated terminals when servicing terminals in harness connector of Charge Air Pressure Sensor.**

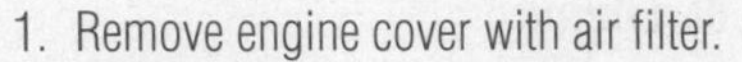

1. Remove engine cover with air filter.
2. Disconnect 4-pin harness connector at Charge Air Pressure Sensor (arrow).
3. Start engine and let it run at idle.
4. Measure voltage between terminals 1+3 on connector to wiring harness using hand multimeter V.A.G 1526 B and adapter cables from connector test kit V.A.G 1594 C .
5. Specified value: min. 4.5 V
6. Switch ignition off.
7. If the voltage is not OK, connect test box to wiring harness of Motronic Engine Control Module (ECM).
8. Check wires between test box and 4-pin connector for open circuit according to wiring diagram.

Terminal	Test box socket
1	53
3	19
4	381

9. Wire resistance: max.: 1.5 OHMS
10. Check wires for short circuit to each other, to Battery A (+) and Ground (GND).
11. Specified value: infinite ohms.
12. If the voltage and wires are OK, replace Charge Air Pressure.
13. If the voltage is not OK and the wires are OK, replace Motronic Engine Control Module (ECM).

Fig. 15 Charge Air Pressure sensor location for all engines except the 1.8L and 2.0L TDI engines.

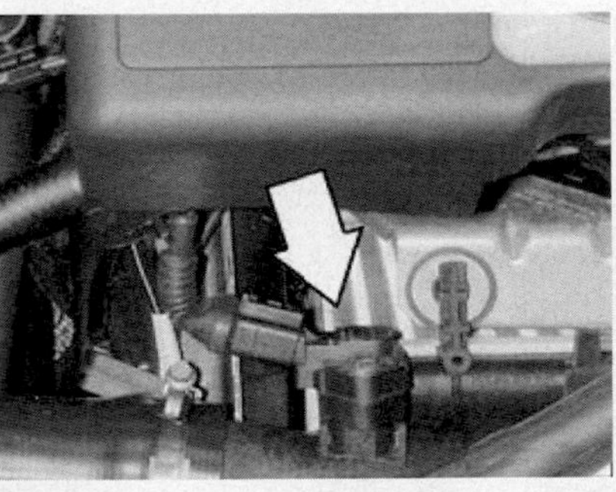

Fig. 16 Charge Air Pressure sensor location for the 1.8L engine.

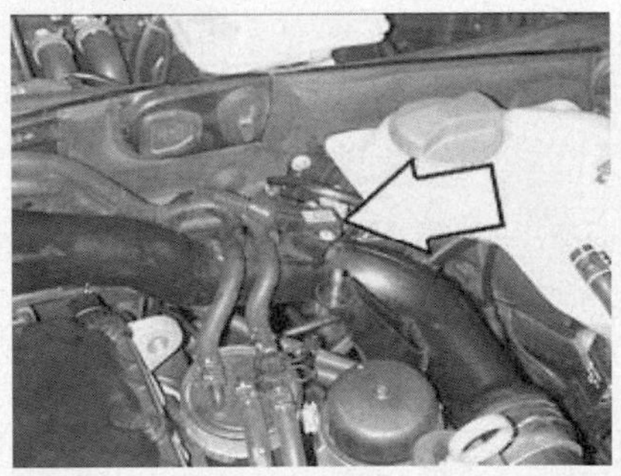

Fig. 17 Charge Air Pressure sensor location for the 2.0L TDI engine.

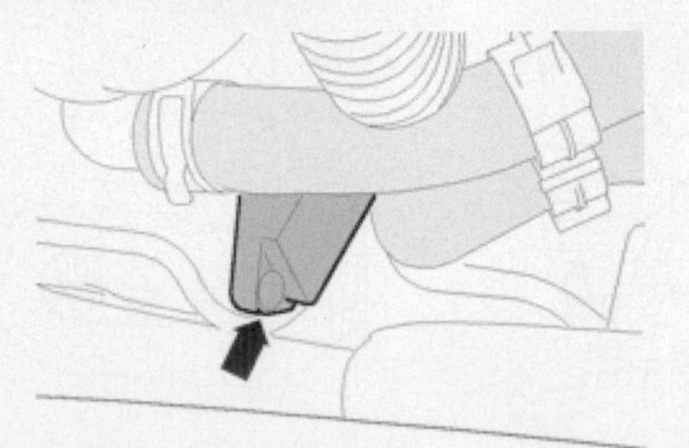

Fig.18 Disconnect 4-pin harness connector at Charge Air Pressure Sensor.

1 2 3 4

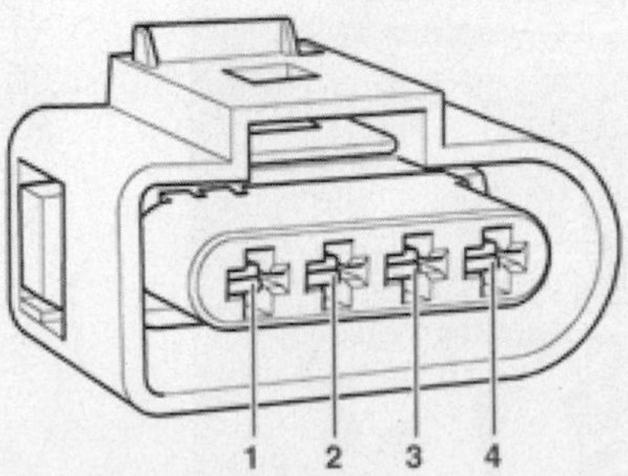

Fig.19 Check wires between test box and 4-pin connector for open circuit according to wiring diagram

Diesel Direct Fuel Injection (DFI) Engine Control Module (ECM)

LOCATION

See Figure 20.

The Diesel Direct Fuel Injection (DFI) Engine Control Module (ECM) is located in the center of the plenum.

REMOVAL & INSTALLATION

See Figures 21 through 25.

1. Remove windshield wiper arms, the plenum chamber cover and plenum chamber bulkhead.

2. Cut the heads of shear bolts so that two parallel surfaces remain (arrows).

3. Remove bolts with locking pliers.

4. Insert a screwdriver (A) between protective housing and retaining plate (arrow).

5. Use screwdriver (A) to pry protective housing upward, and then pull if off sideways from retaining plate (arrow).

6. Release and pull off front connector (1) from Motronic Engine Control Module (ECM).

7. Pry retaining tab (2) slightly open.

8. Then, push Diesel Direct Fuel Injection (DFI) Engine Control Module (ECM) out of retainer (arrow).

9. Release rear connector from Motronic Engine Control Module (ECM) and pull it off.

To install:

10. Connect rear connector to Diesel Direct Fuel Injection (DFI) Engine Control-Module (ECM) and lock it into place.

11. Push Diesel Direct Fuel Injection (DFI) Engine Control Module (ECM) onto retaining plate.

12. Connect front connector to Diesel Direct Fuel Injection (DFI) Engine Control Module (ECM) and lock it into place.

13. Check whether the new control module identification matches the old control module identification.

14. Perform function "Replacing control module" using Vehicle diagnosis and service information system.

15. Push protective housing (2) onto retaining plate.

16. Screw in shear bolts (1) uniformly until the bolt heads begin to shear.

17. Install plenum chamber bulkhead, plenum chamber cover and windshield wiper arms.

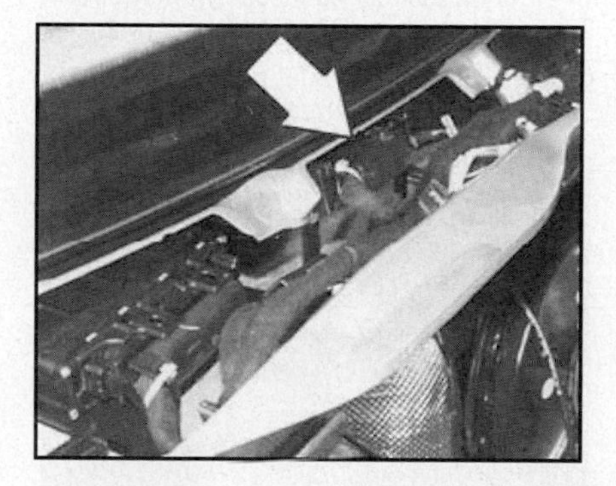

Fig. 20 The location of the Diesel Direct Fuel Injection Engine Control Module.

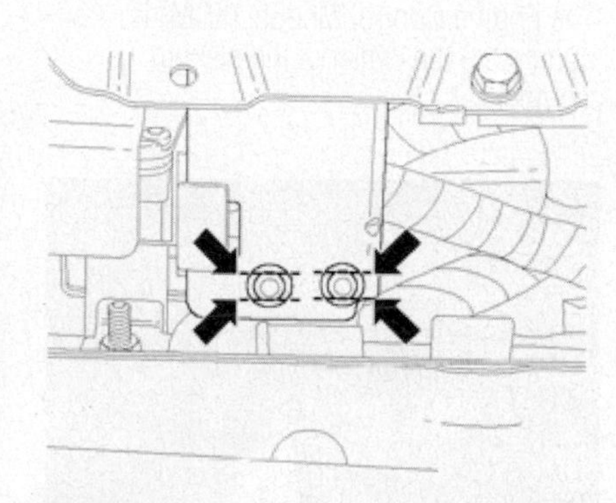

Fig. 21 Cut the heads of shear bolts so that two parallel surfaces remain (arrows).

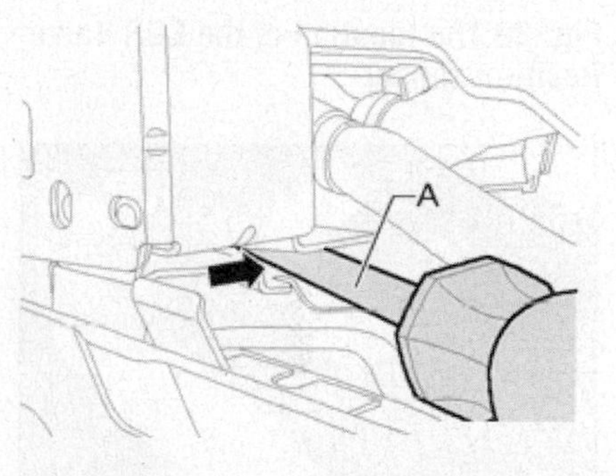

Fig. 22 Insert a screwdriver (A) between protective housing and retaining plate (arrow).

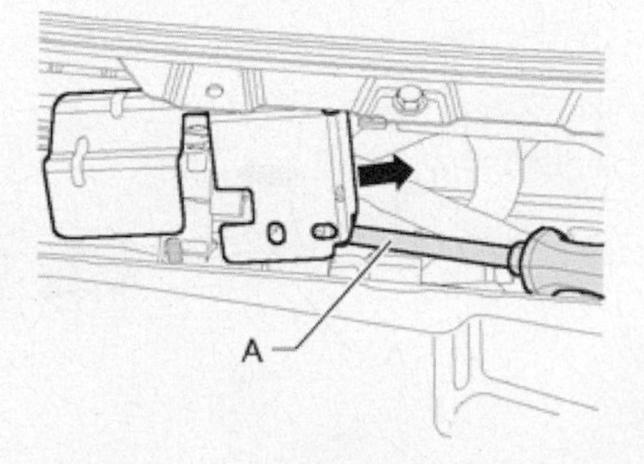

Fig. 23 Use screwdriver (A) to pry protective housing upward, and then pull if off sideways from retaining plate (arrow).

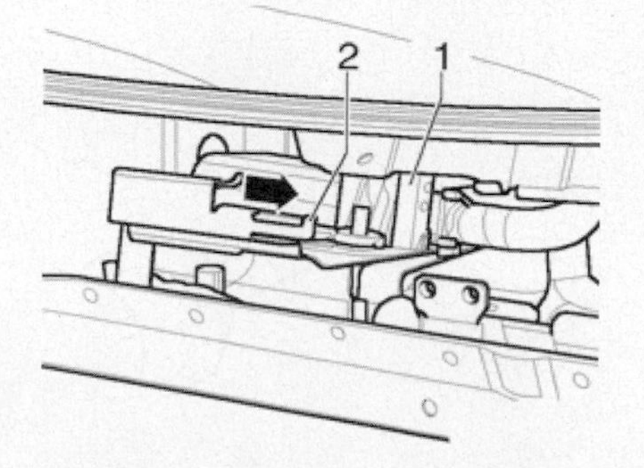

Fig.24 Release and pull off front con nector (1) from ECM and pry retaining tab (2) slightly open

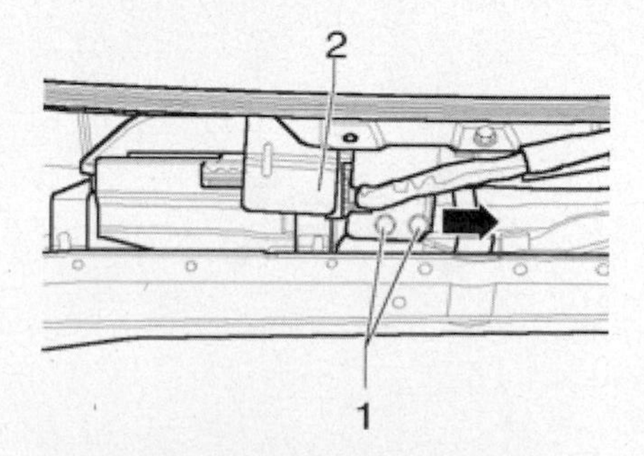

Fig. 25 Push protective housing (2) onto retaining plate, and screw in shear bolts (1) uniformly until the bolt heads begin to shear.

EGR Valve Position (EVP) Sensor

LOCATION

See Figure 26.

The EGR Valve Position (EVP) Sensor is mounted on the left side of the engine compartment in the rear and to the right of the wastegate bypass valve.

OPERATION

The Exhaust Gas Recirculation (EGR) system is activated by the Diesel Direct Fuel Injection (DFI) Engine Control Module (ECM) or the MFI ECM via EGR Vacuum Regulator Solenoid Valve with EGR Potentiometer to the EGR valve.

Fig. 26 The location of the EGR Valve Position sensor.

Engine Control Module (ECM)

LOCATION

2.0L Engines

See Figure 27.

The Engine Control Module (ECM) is mounted in the center of the plenum chamber.

2.5L Engines

See Figure 28.

The Engine Control Module (ECM) is mounted in the rear of the engine compartment, near the center of the bulkhead plenum.

3.2L Engines

See Figure 29.

The Engine Control Module (ECM) is mounted in the center of the plenum chamber.

REMOVAL & INSTALLATION

See Figures 30 through 34.

1. Remove windshield wiper arms, the plenum chamber cover and plenum chamber bulkhead.
2. Cut the heads of shear bolts so that two parallel surfaces remain (arrows).
3. Remove bolts with locking pliers.
4. Insert a screwdriver (A) between protective housing and retaining plate (arrow).
5. Use screwdriver (A) to pry protective housing upward, and then pull if off sideways from retaining plate (arrow).
6. Release and pull off front connector (1) from Motronic Engine Control Module (ECM).
7. Pry retaining tab (2) slightly open.
8. Then, push Motronic Engine Control Module (ECM) out of retainer (arrow).
9. Release rear connector from Motronic Engine Control Module (ECM) and pull it off.

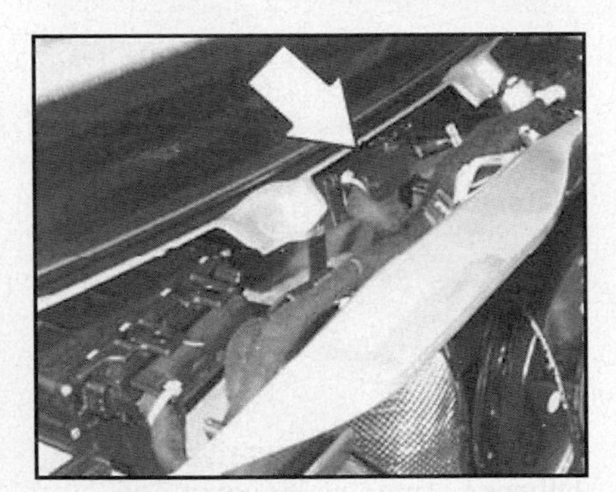

Fig. 27 Engine Control Module location for the 2.0L engine.

Fig. 28 Engine Control Module location for the 2.5L engine.

Fig. 29 Engine Control Module location for the 3.2L engine.

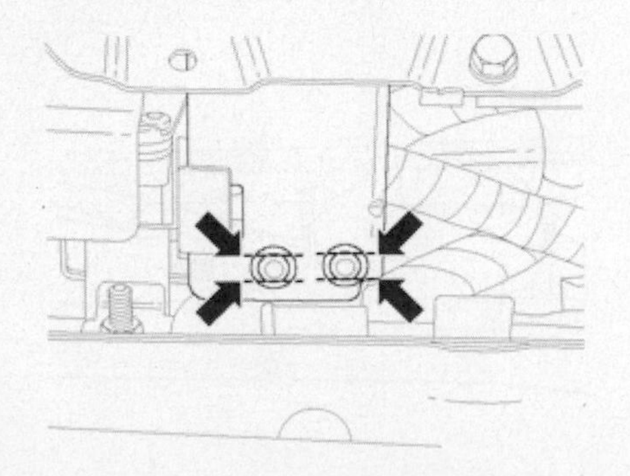

Fig. 30 Cut the heads of shear bolts so that two parallel surfaces remain (arrows).

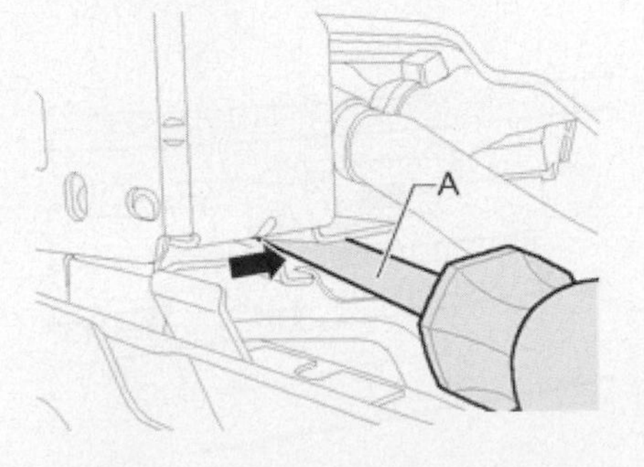

Fig. 31 Insert a screwdriver (A) between protective housing and retaining plate (arrow).

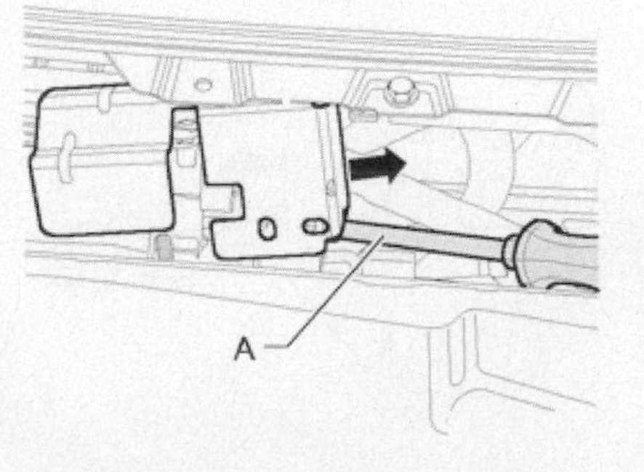

Fig. 32 Use screwdriver (A) to pry protective housing upward, and then pull if off sideways from retaining plate (arrow).

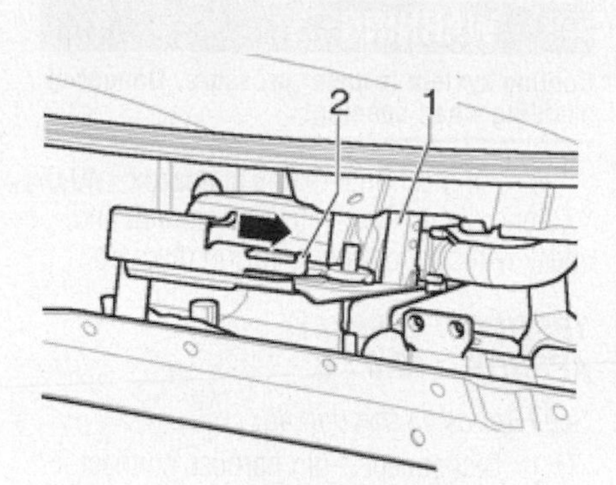

Fig. 33 Release and pull off front connector (1) from ECM and pry retaining tab (2) slightly open

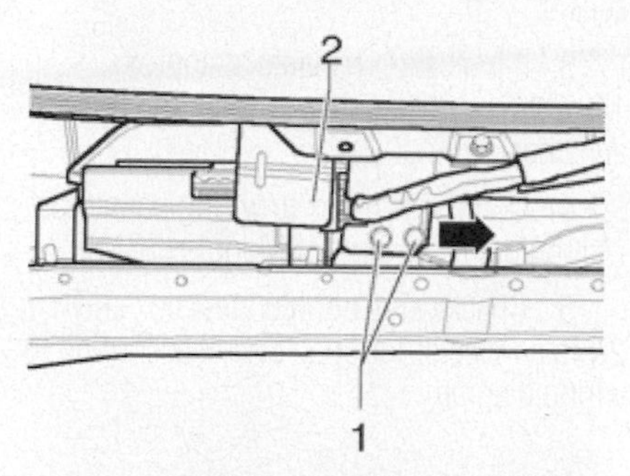

Fig. 34 Push protective housing (2) onto retaining plate, and screw in shear bolts (1) uniformly until the bolt heads begin to shear.

To install:

10. Connect rear connector to Motronic Engine Control Module (ECM) and lock it into place.
11. Push Motronic Engine Control Module (ECM) onto retaining plate.
12. Connect front connector to Motronic Engine Control Module (ECM) and lock it into place.
13. Check whether the new control module identification matches the old control module identification.
14. Perform function "Replacing control module" using Vehicle diagnosis and service information system.
15. Push protective housing (2) onto retaining plate.
16. Screw in shear bolts (1) uniformly until the bolt heads begin to shear.
17. Install plenum chamber bulkhead, plenum chamber cover and windshield wiper arms.

Engine Coolant Temperature (ECT) Sensor

LOCATION

1.8L Engines

See Figure 35.

The Engine Coolant Temperature (ECT) Sensor is mounted to the left of the cylinder head, in the engine coolant flange.

1.9L Engines

See Figure 36.

The Engine Coolant Temperature (ECT) Sensor is mounted to the left of the cylinder head, in the engine coolant flange.

2.0L Engines

See Figure 37.

The Engine Coolant Temperature (ECT) Sensor is mounted to the left of the cylinder head, in the engine coolant flange.

2.5L Engines

See Figure 38.

The Engine Coolant Temperature (ECT) Sensor is mounted to the left of the cylinder head, on top of the coolant manifold.

2.8L Engines

See Figure 39.

The Engine Coolant Temperature (ECT) Sensor is mounted to the left of the cylinder head, in the engine coolant flange.

3.2L Engines

See Figure 40.

The Engine Coolant Temperature (ECT) Sensor is mounted to the left of the cylinder head, in the engine coolant flange.

Fig. 35 Engine Coolant Temperature sensor for the 1.8L engine.

Fig. 36 Engine Coolant Temperature sensor for the 1.9L engine.

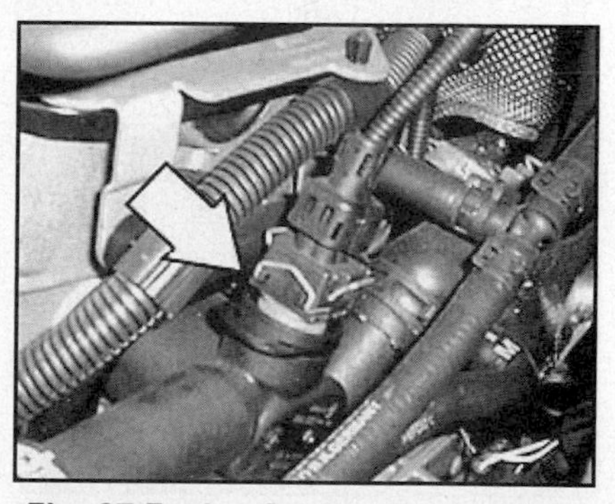

Fig. 37 Engine Coolant Temperature sensor for the 2.0L engine.

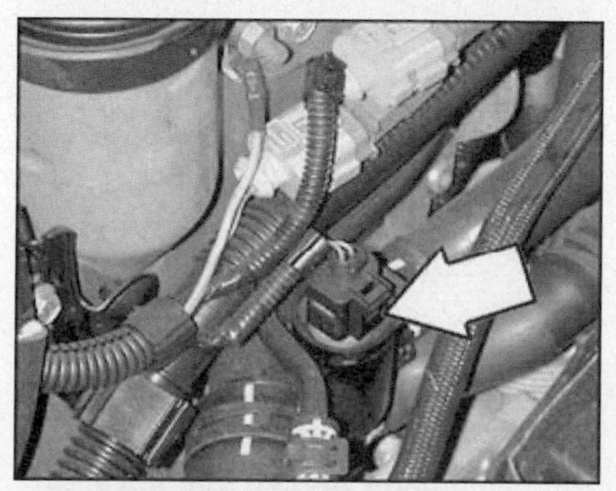

Fig. 38 Engine Coolant Temperature sensor for the 2.5L engine.

Fig. 39 Engine Coolant Temperature sensor for the 2.8L engine.

Fig. 40 Engine Coolant Temperature sensor for the 3.2L engine.

3.6L Engines

See Figure 41.

The Engine Coolant Temperature (ECT) Sensor is mounted on the left bottom of the engine, below the camshaft housing.

4.0L Engines

See Figure 42.

The Engine Coolant Temperature (ECT) Sensor is mounted on the radiator hose leading to the radiator.

OPERATION

1. The engine cooling system consists of five main parts.

- The Engine Cooler
- The Engine Coolant Temperature Sensor
- The Thermostat Valve
- The small Cooling Circuit
- The large Cooling Circuit

During heating up the Engine the coolant flows first inside the small cooling circuit. After the coolant reach a sufficient temperature the thermostat valve will open the large cooling circuit to integrate the engine cooler. The engine coolant temperature sensor measures a mixed temperature between the coolant coming from the small and large cooling circuit.

Fig. 41 Engine Coolant Temperature sensor for the 3.6L engine.

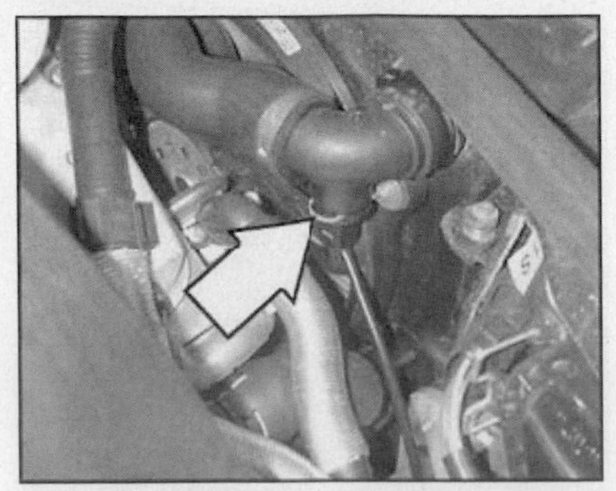

Fig. 42 Engine Coolant Temperature sensor for the 4.0L engine.

TESTING

**** WARNING**

Cooling system is under pressure. Danger of scalding when opening!

➡ Use only gold-plated terminals when servicing terminals in harness connector of Engine Coolant Temperature (ECT) Sensor.

1. Connect Vehicle diagnosis and service system.
2. Switch ignition on.
3. Under address word "33 - On Board Diagnostic (OBD)" , select "Diagnostic mode 1: Check measured values" .
4. Select measured value "PID 05: Coolant temperature" .
5. Check specified value of coolant temperature:

PID	Diagnostic text	Specified value
05	Coolant temperature	approx. coolant temperature

6. If specified value is not obtained, continue test according to the following table:

Indicated	Cause	Test
approx. -40.0 ° C	Open circuit or short circuit to B+	
approx. 140.0 ° C	Short circuit to Ground (GND)	

7. If specified value is obtained, start engine and let it run at idle.
8. Temperature value must increase uniformly in increments of 1.0 ° C.
9. If the engine shows problems in certain temperature ranges and if the temperature does not climb uniformly, the temperature signal is intermittent, replace Engine Coolant Temperature (ECT) Sensor.

TESTING IF DISPLAY APPROX. -40.0 ° C:

See Figures 43 and 44.

1. Bridge terminals 1 + 2 of connector using adapter cables from connector test kit and observe indication on display.
2. If indication jumps to approx. 140.0 ° C, end diagnosis and switch ignition off. Replace Engine Coolant Temperature (ECT) Sensor.

**** WARNING**

Cooling system is under pressure. Danger of scalding when opening!

3. If indication remains at approx. -40.0 ° C, end diagnosis, switch ignition off and check wires according to wiring diagram .

TESTING IF DISPLAY APPROX. 140.0 ° C:

See Figures 43 through 46.

1. Disconnect 2-pin harness connector at Engine Coolant Temperature (ECT) Sensor (arrow).
2. If indication jumps to approx. -40.0 ° C, end diagnosis, switch ignition off, and replace Engine Coolant Temperature (ECT) sensor.
3. If indication remains at approx. 140.0 ° C, end diagnosis, switch ignition off and check wiring.
4. Connect test box to wiring harness of Motronic Engine Control Module (ECM).
5. Check wires between test box and 2-pin connector for open circuit according to wiring diagram.

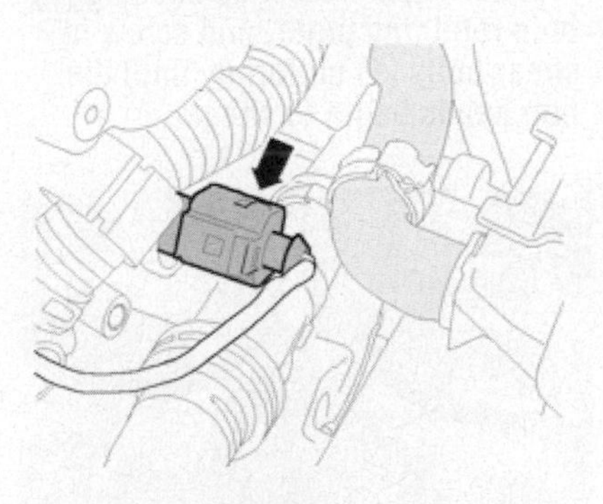

Fig. 43 Disconnect 2-pin harness connector at Engine Coolant Temperature (ECT) Sensor (arrow).

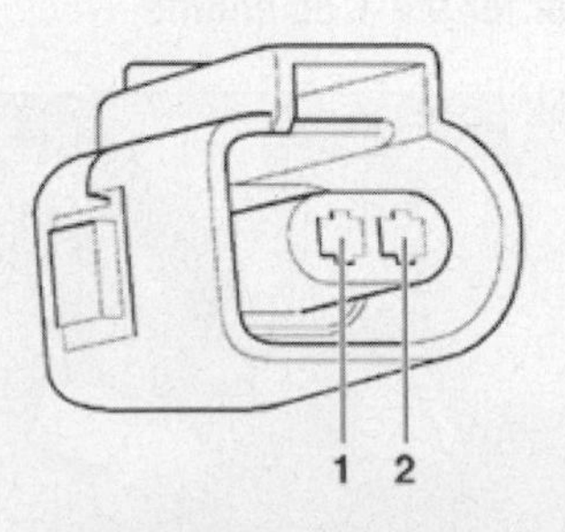

Fig. 44 Bridge terminals 1 + 2 of connector using adapter cables from connector test kit and observe indication on display

Terminal	Test box socket
1	10
2	14

9. Wire resistance: max.: 1.5 OHMS
10. Check wires for short circuit to each other, to Battery A (+) and Ground (GND).
11. Specified value: infinite ohms.
12. If the wires are OK, use a hand multimeter and adapter cables from connector test kit to measure resistance at connector to Engine Coolant Temperature (ECT) Sensor terminals 1+2.
13. Area A: Resistance values for temperature range 0 to 50 ° C.
14. Area B: Resistance values for temperature range 50 to 100 ° C.
15. Read-out examples:
    a. 30 ° C is in range A and corresponds to a resistance of 1.5 to 2.0 kohms
    b. 80 ° C is in range B and corresponds to a resistance of 275 to 375 OHMS,

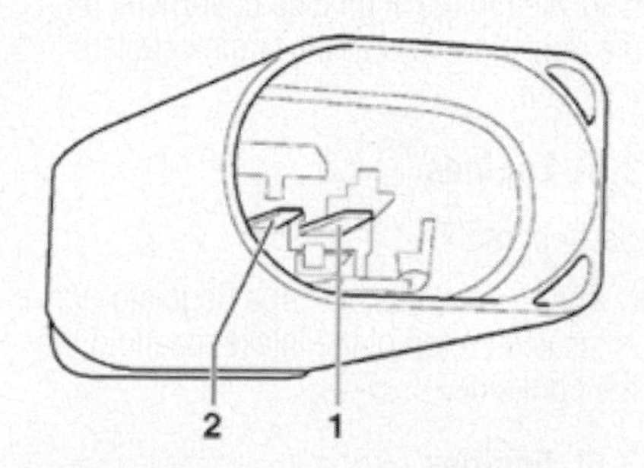

Fig. 45 Check wires between test box and 2-pin connector for open circuit according to wiring diagram.

16. If the resistance is not OK, replace Engine Coolant Temperature (ECT) Sensor.
17. If the wires and resistance values are OK, replace Motronic Engine Control Module (ECM).

Engine Oil Temperature (EOT) Sensor

LOCATION

1.8L Engines

See Figure 47.

The Engine Oil Temperature (EOT) Sensor is located on the driver's side of the cylinder block, in the center, in the oil filter housing.

2.8L Engines

See Figure 48.

The Engine Oil Temperature (EOT) Sensor is located on the front of the engine threaded into the oil pump housing, on the left side.

4.0L Engines

See Figure 49.

The Engine Oil Temperature (EOT) Sensor is located on the front of the engine threaded into the oil pump housing, on the left side.

Engine Speed (RPM) Sensor

LOCATION

See Figure 50.

1.8L, 1.9L and 2.0L Engines

The Engine Speed (RPM) Sensor is located on the front of the engine cylinder block, lower left, near flange for the transmission.

2.5L Engines

See Figure 51.

The Engine Speed (RPM) Sensor is located on the bottom of the engine, between the engine and the transmission.

2.8L Engines

See Figure 52.

The Engine Speed (RPM) Sensor is located on the front of the engine cylinder block, lower left, near flange for the transmission.

3.2L Engines

See Figure 53.

The Engine Speed (RPM) Sensor is located in front of the engine, left side, next to the engine oil filter housing.

Fig. 47 Engine Oil Temperature sensor location for the 1.8L engine.

Fig. 48 Engine Oil Temperature sensor location for the 2.8L engine.

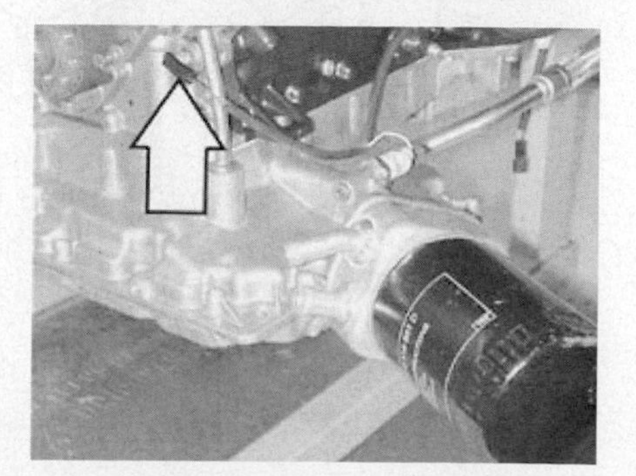

Fig. 49 Engine Oil Temperature sensor location for the 4.0L engine.

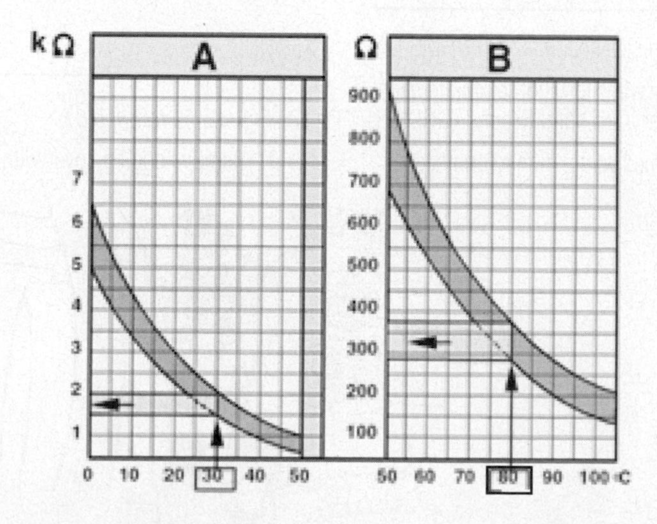

Fig. 46 Resistance read-out examples.

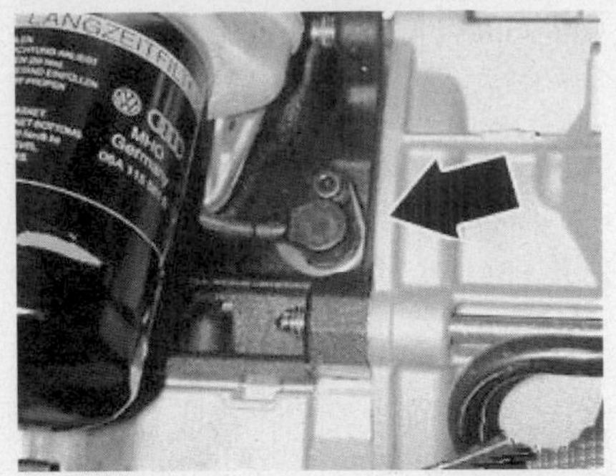

Fig. 50 Engine Speed sensor location for the 1.8L, 1.9L and 2.0L engines.

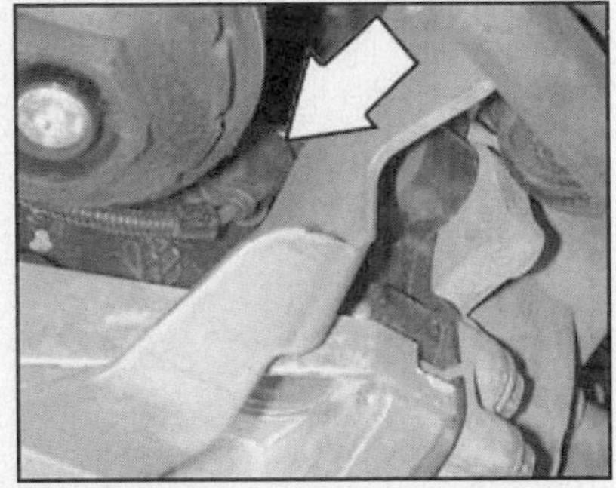

Fig. 53 Engine Speed sensor location for the 3.2L engines.

OPERATION

The Engine Speed (RPM) Sensor detects RPM and reference marks. Without an engine speed signal, the engine will not start. If the engine speed signal fails while the engine is running, the engine will stop immediately.

TESTING

See Figures 54 and 55.

➡ **Use only gold-plated terminals when servicing terminals in harness connector of Camshaft Engine Speed (RPM) Sensor.**

1. Disconnect gray 3-pin harness connector (5) to Engine Speed (RPM) Sensor.

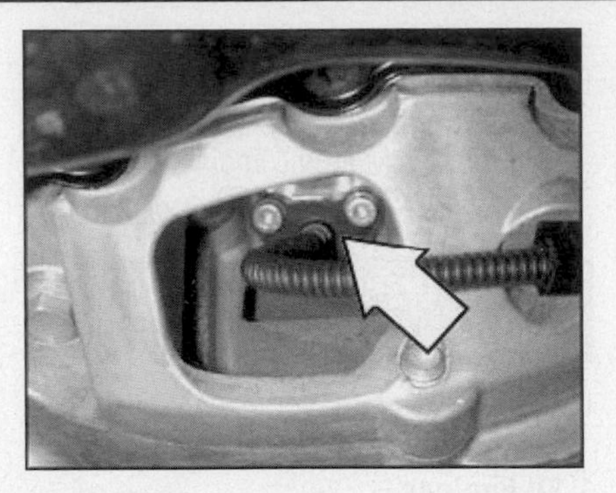

Fig. 51 Engine Speed sensor location for the 2.5L engines.

2. Measure resistance between terminals 2+3 on connector (B) to Engine Speed (RPM) Sensor using hand multimeter and adapter cables from a connector test kit.
3. Specified value: 730 to 1000 OHMS (at approx. 20 ° C)
4. Check Engine Speed (RPM) Sensor for short circuit between terminals 1 + 2 as well as 1 + 3.
5. Specified value: infinite ohms.
6. If the resistance is not OK, replace the Engine Speed (RPM) Sensor.
7. If the Engine Speed (RPM) Sensor is okay, connect test box to wiring harness of Motronic Engine Control Module (ECM).
8. Check wires between test box and 3-pin connector - A - for open circuit according to wiring diagram.

Terminal	Test box socket
1	52
2	51
3	36

9. Wire resistance: max. 1.5 OHMS
10. Check wires for short circuit to each other, to Battery A (+) and Ground (GND).
11. Specified value: infinite ohms.

Fig. 52 Engine Speed sensor location for the 2.8L engines.

EVAP Canister Purge Regulator Valve

LOCATION

See Figure 56.

All Engines Except 2.5L and 3.6L Engines

The EVAP Canister Purge Regulator Valve is located in the engine compartment, on the right side, next to the engine coolant reservoir.

2.5L Engines

See Figure 57.

The EVAP Canister Purge Regulator Valve is located on top of the intake manifold, on the right side.

3.6L Engines

See Figure 58.

The EVAP Canister Purge Regulator Valve is located on the front of the center of the enging.

TESTING

See Figures 59 through 61.

1. Remove engine cover with air filter.

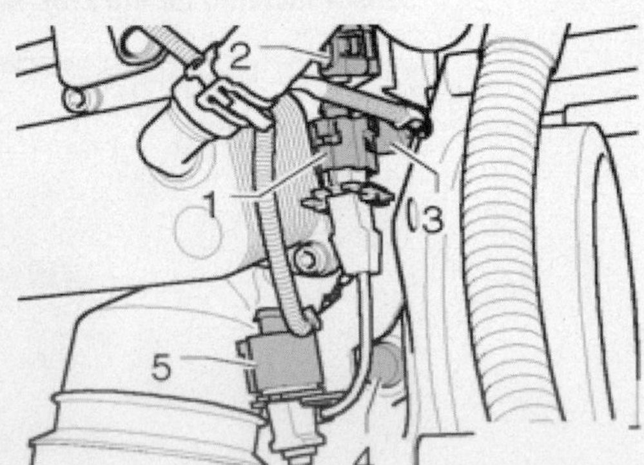

Fig. 54 Disconnect gray 3-pin harness connector (5) to Engine Speed (RPM) Sensor

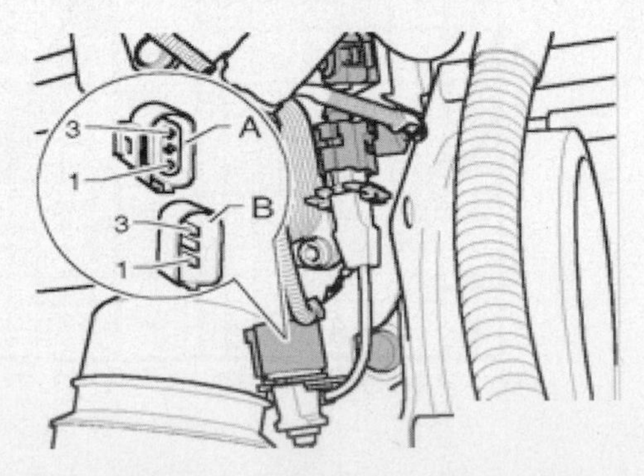

Fig. 55 The connector's terminals needed to be measured via a multimeter.

Fig. 56 EVAP Canister Purge Regulator Valve location for all engines except 2.5L and 3.6L engines.

2. Disconnect 2-pin harness connector (arrow) to Evaporative Emission (EVAP) Canister Purge Regulator Valve.

3. Using hand multimeter and adapter cables from connector test kit, measure resistance between terminals of Evaporative Emission (EVAP) Canister Purge Regulator Valve.

4. Specified value: 22.0 to 30.0 OHMS (at approx. 20 ° C)

5. If resistance is not OK, replace Evaporative Emission (EVAP) Canister Purge Regulator Valve.

6. If resistance is OK, connect test box to wiring harness of Motronic Engine Control Module (ECM).

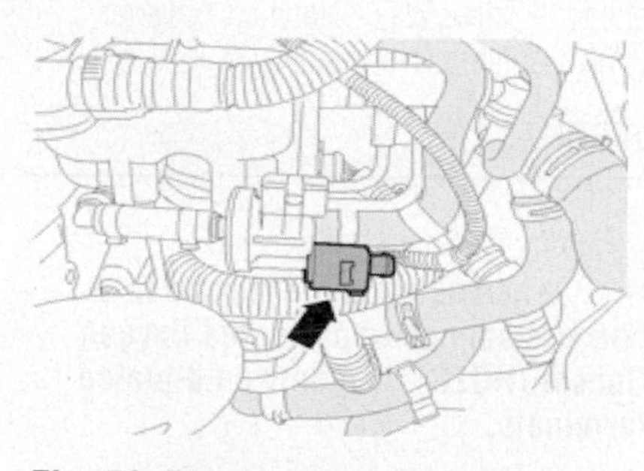

Fig. 59 Disconnect 2-pin harness connector (arrow).

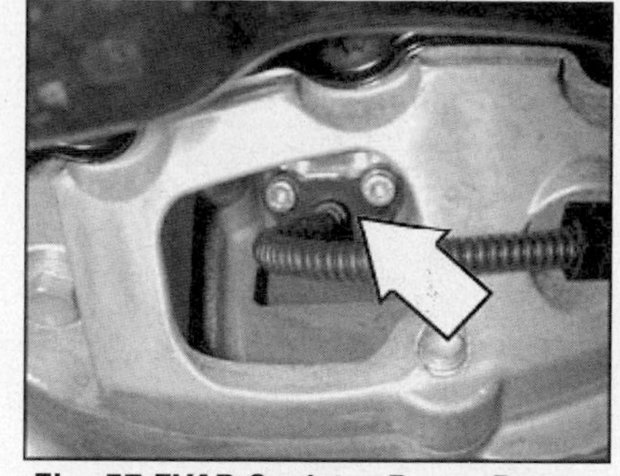

Fig. 57 EVAP Canister Purge Regulator Valve location for the 2.5L engine.

7. Check wire between test box socket and 2-pin connector for open circuit according to wiring diagram.

Terminal	Test box socket
2	64

8. Wire resistance: max.: 1.5 OHMS

9. Check wires for short circuit to Battery A (+) and Ground (GND).

10. Specified value: infinite ohms.

11. Check wire between 2-pin connector terminal 1 and Engine Component Power Supply Relay J757 for open circuit according to wiring diagram.

12. Wire resistance: max. 1.5 Ω

13. If wires are OK, replace Motronic Engine Control Module (ECM)

Fuel Temperature Sensor

LOCATION

DFI Engines

See Figure 62.

The Fuel Temperature Sensor is located in front of the Diesel DFI pump.

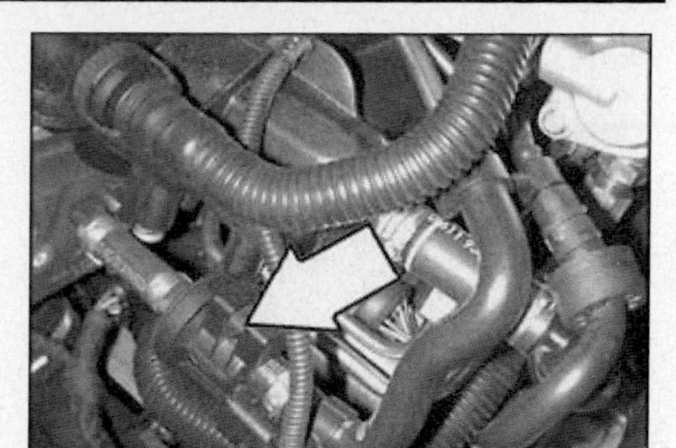

Fig. 58 EVAP Canister Purge Regulator Valve location for the 3.6L engine.

MFI Engines

See Figure 63.

The Fuel Temperature Sensor is located in the engine compartment, on the driver's side, attached to fuel line.

Heated Oxygen (HO2S) Sensor

LOCATION

1.8L Engines

See Figure 64.

The Heated Oxygen (HO2S) Sensor is located behind the engine, in the exhaust pipe and forward of the Three Way Catalytic Converter (TWC)..

2.0L Engines

See Figure 65.

The Heated Oxygen (HO2S) Sensor is located in exhaust manifold near flange, upstream of Three Way Catalytic Converter (TWC).

2.5L Engines

See Figure 66.

The Heated Oxygen (HO2S) Sensor is located at the exhaust manifold junction.

Fig. 60 Measure resistance between terminals of Evaporative Emission (EVAP) Canister Purge Regulator Valve

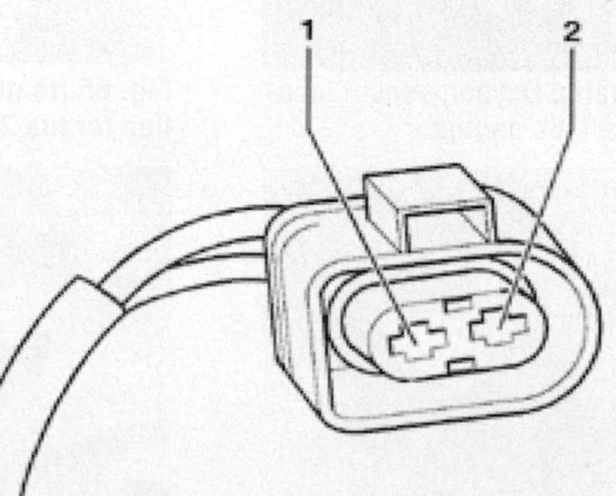

Fig. 61 Check wire between test box socket and 2-pin connector for open circuit.

Fig. 62 Fuel Temperature sensor location for all DFI engines.

Fig. 63 Fuel Temperature sensor location for all MFI engines.

2.8L Engines

See Figure 67.

The Heated Oxygen (HO2S) Sensor is located in exhaust manifold near flange, upstream of Three Way Catalytic Converter (TWC).

3.2L Engines

See Figure 68.

The Heated Oxygen (HO2S) Sensor is located in exhaust manifold near flange, upstream of Three Way Catalytic Converter (TWC) on the passenger side.

TESTING

See Figure 69.

➡ **When servicing terminals in harness connector of Heated Oxygen Sensor (HO2S), use only gold-plated terminals.**

1. Connect Vehicle diagnosis and service system VAS.
2. Start engine and let it run at idle.
3. Under address word "33 - On Board Diagnostic (OBD)", select "Diagnostic mode 6: Check test results of components that are not continuously monitored".
4. Select "Test-ID 01: Oxygen sensor monitoring before catalytic converter".
5. Select "Test-ID 131: Dynamic test" and check whether specified values are obtained.

Dynamic test	Specified value	
	min.	max.
Normal operation	0.35009	-
Reduced threshold upon suspicion of malfunction in catalytic converter	0.39990	-

6. If specified values are not obtained:
7. Check oxygen sensor wiring.
8. Connect test box to wiring harness of Motronic Engine Control Module (ECM)
9. Disconnect black 6-pin harness connector to Heated Oxygen Sensor (HO2S).
10. Check wires between test box and 6-pin connector for open circuit according to wiring diagram.

Terminal	Test box socket
1	52
2	51
5	71
6	70

11. Wire resistance: max. 1.5Ω
12. Check wires for short circuit to each other, to Battery A (+) and Ground (GND).
13. Specified value: infinite ohms.
14. If the wires are OK, replace Heated Oxygen Sensor (HO2S) G39 before catalytic converter.

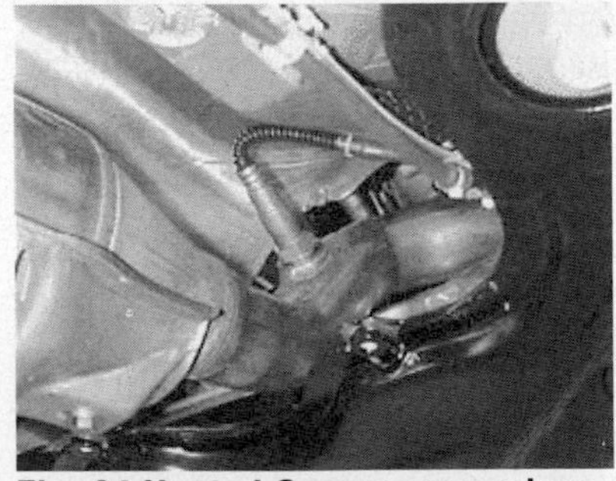

Fig. 64 Heated Oxygen sensor location for the 1.8L engine.

Fig. 65 Heated Oxygen sensor location for the 2.0L engine.

Heated Oxygen (HO2S) 2 Sensor

LOCATION

See Figure 70.

3.2L Engines

The Heated Oxygen (HO2S) 2 Sensor is located in exhaust manifold near flange, ahead of Three Way Catalytic Converter (TWC) on the driver's side.

TESTING

See Figure 71.

➡ **When servicing terminals in harness connector of Heated Oxygen Sensor (HO2S), use only gold-plated terminals.**

1. Connect Vehicle diagnosis and service system VAS.
2. Start engine and let it run at idle.
3. Under address word "33 - On Board Diagnostic (OBD)", select "Diagnostic mode 6: Check test results of components that are not continuously monitored".

Fig. 66 Heated Oxygen sensor location for the 2.5L engine.

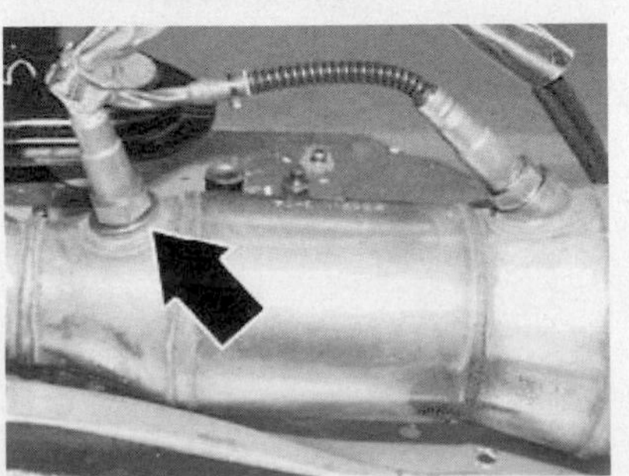

Fig. 67 Heated Oxygen sensor location for the 2.8L engine.

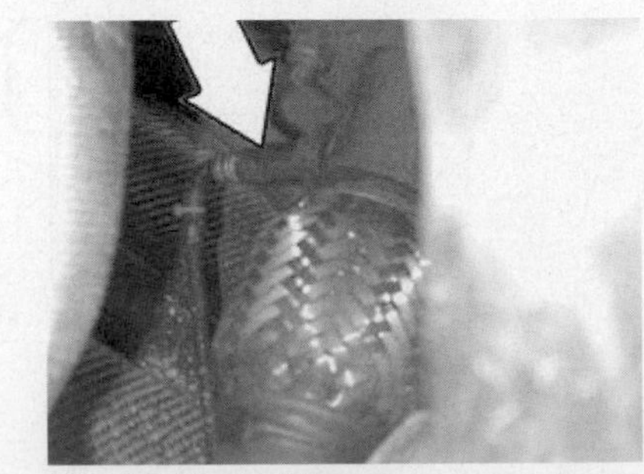

Fig. 68 Heated Oxygen sensor location for the 3.2L engine.

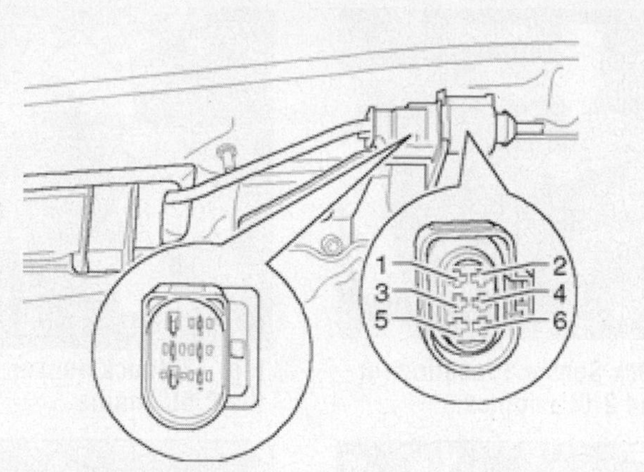

Fig. 69 Check wires between test box and 6-pin connector for open circuit according to wiring diagram.

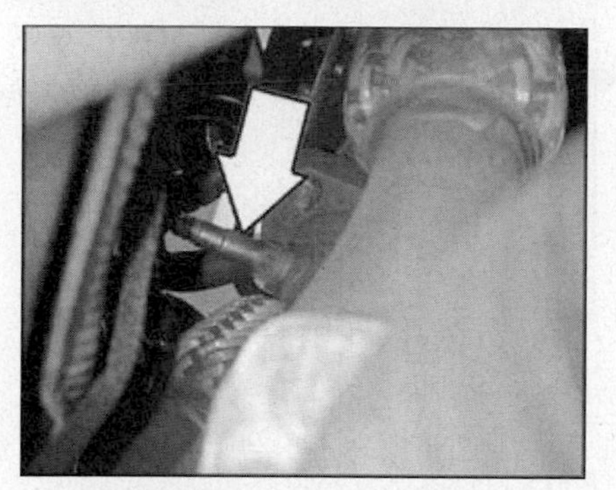

Fig. 70 Heated Oxygen sensor 2 location for the 3.2L engine.

4. Select "Test-ID 01: Oxygen sensor monitoring before catalytic converter".
5. Select "Test-ID 131: Dynamic test" and check whether specified values are obtained.

Dynamic test	Specified value	
	min.	max.
Normal operation	0.35009	-
Reduced threshold upon suspicion of malfunction in catalytic converter	0.39990	-

6. If specified values are not obtained:
7. Check oxygen sensor wiring.
8. Connect test box to wiring harness of Motronic Engine Control Module (ECM)
9. Disconnect black 6-pin harness connector to Heated Oxygen Sensor (HO2S).
10. Check wires between test box and 6-pin connector for open circuit according to wiring diagram.

Terminal	Test box socket
1	52
2	51
5	71
6	70

11. Wire resistance: max. 1.5 OHMS
12. Check wires for short circuit to each other, to Battery A (+) and Ground (GND).
13. Specified value: infinite ohms.
14. If the wires are OK, replace Heated Oxygen Sensor (HO2S) G39 before catalytic converter.

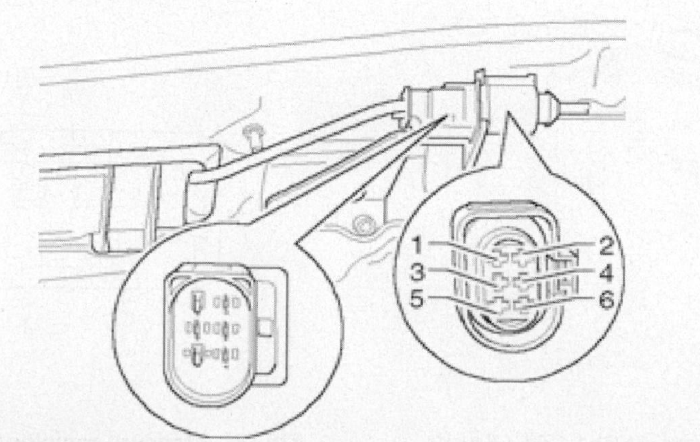

Fig. 71 Check wires between test box and 6-pin connector for open circuit according to wiring diagram.

Intake Air Temperature (IAT) Sensor

LOCATION

1.8L Engines

See Figure 72.

The Intake Air Temperature (IAT) Sensor is located in the intake manifold, on the right side, near the throttle valve.

1.9L Engines

See Figure 73.

The Intake Air Temperature (IAT) Sensor is located in the intake air duct between the charge air cooler and the intake manifold.

2.0L Engines

See Figure 74.

The Intake Air Temperature (IAT) Sensor is located in the intake duct, to the right of the air cleaner housing.

2.5L Engines

See Figure 75.

The Intake Air Temperature (IAT) Sensor is located on the left side of the engine cover.

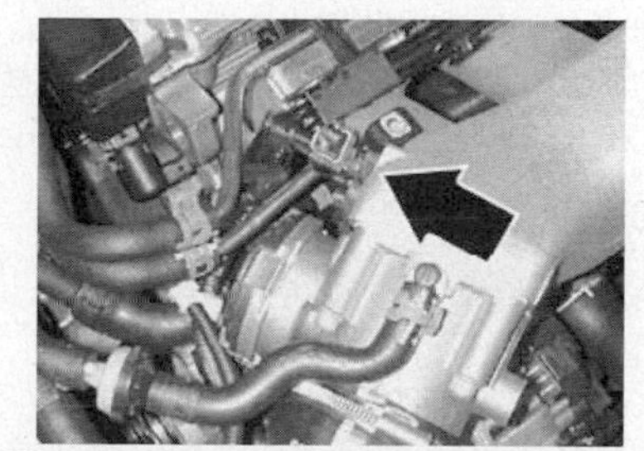

Fig. 72 Intake Air Temperature sensor location for the 1.8L engine.

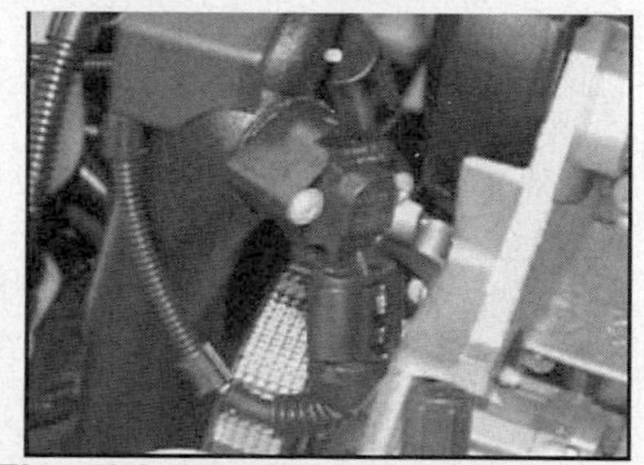

Fig. 73 Intake Air Temperature sensor location for the 1.9L engine.

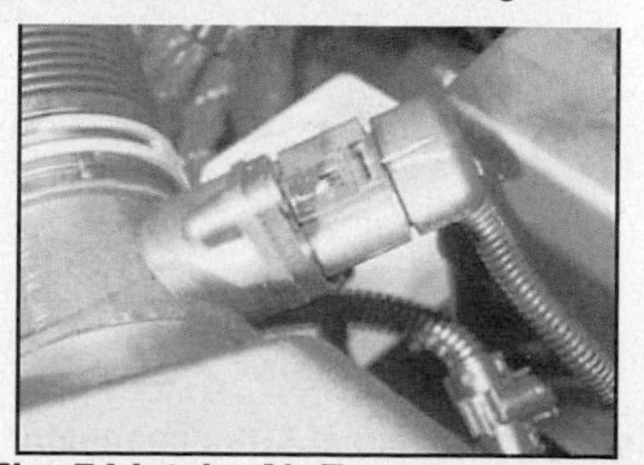

Fig. 74 Intake Air Temperature sensor location for the 2.0L engine.

Fig. 75 Intake Air Temperature sensor location for the 2.5L engine.

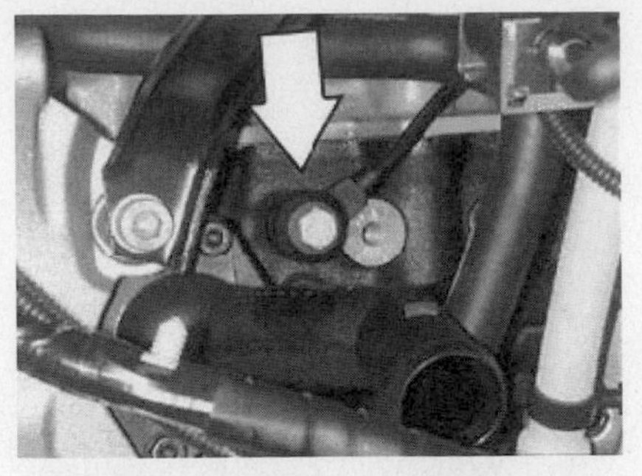

Fig. 76 Knock Sensor 1 location for the 1.8L and 2.0L engines.

Fig. 77 Knock Sensor 1 location for the 2.5L engine.

Knock Sensor (KS) 1

LOCATION

1.8L and 2.0L Engines

See Figure 76.

The Knock Sensor (KS) 1 is located on the front of the engine cylinder block, on the right side, above the engine coolant flange.

2.5L Engines

See Figure 77.

The Knock Sensor (KS) 1 is located below the exhaust manifold, on the left side (shown here with the heat shield removed).

2.8L Engines

See Figure 78.

The Knock Sensor (KS) 1 is located on the engine cylinder block, right rear, below the exhaust manifold.

3.2L Engines

See Figure 79.

The Knock Sensor (KS) 1 is located on the engine cylinder block, right rear, below the exhaust manifold.

Fig. 78 Knock Sensor 1 location for the 2.8L engine.

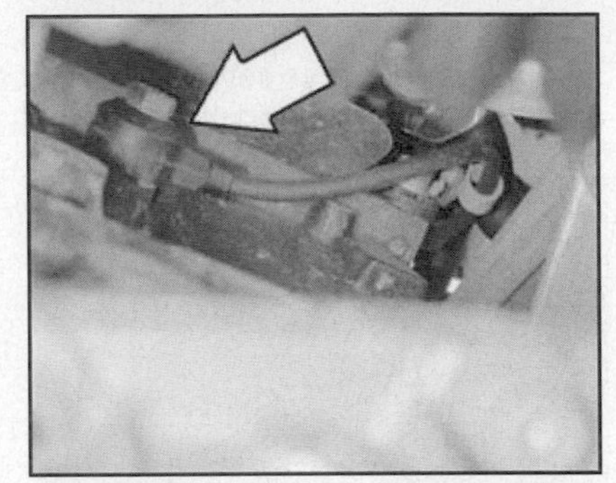

Fig. 79 Knock Sensor 1 location for the 3.2L engine.

TESTING

See Figures 80 through 82.

➡ **Contact surfaces between Knock Sensor (KS) 1 and the cylinder block must be free of corrosion, dirt and grease.**

➡ **To ensure problem-free function of knock sensors, it is important that torque specification of 20 Nm be adhered to exactly.**

➡ **Use only gold-plated terminals when servicing terminals in harness connector of Knock Sensor (KS) 1.**

1. Disconnect 3-pin connector from Knock Sensor (KS) 1 G61 / Knock Sensor (KS) 2 G66: (1 - Green, for Knock Sensor (KS) 1 and 2 - Grey, for Knock sensor (KS) 2)

➡ **Before disconnecting harness connectors, mark which component goes where.**

2. Measure resistance between terminals 1 + 2 at connectors to Knock Sensor (KS) 1.
3. Specification: infinite ohms
4. Connect test box to Motronic Engine Control Module (ECM).

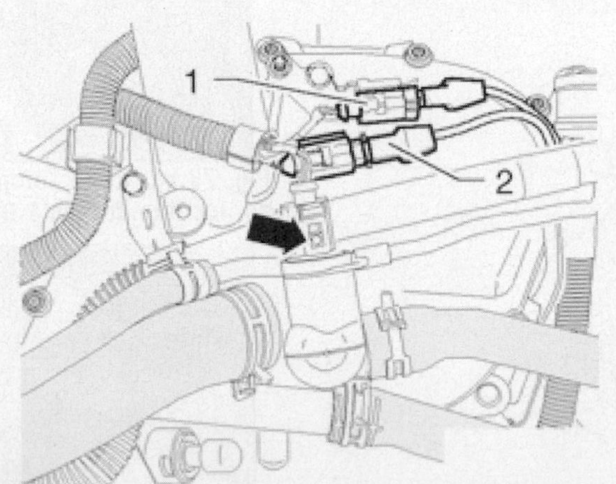

Fig. 80 Disconnect 3-pin connector from Knock Sensor (KS) 1 G61 / Knock Sensor (KS) 2 G66: (1 - Green, for Knock Sensor (KS) 1 and 2 - Grey, for Knock sensor (KS) 2)

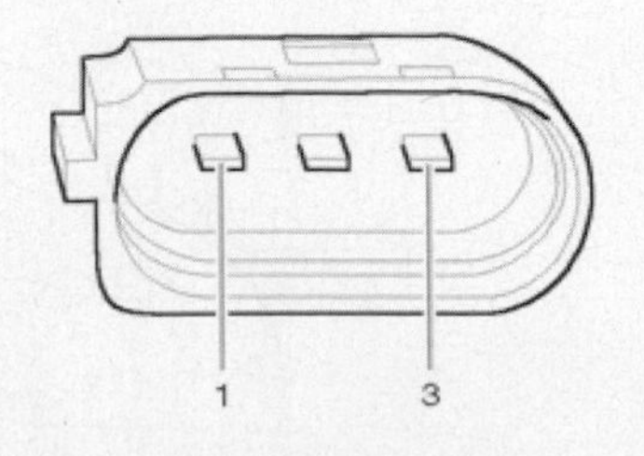

Fig. 81 Measure resistance between terminals 1 + 2 at connectors to Knock Sensor (KS) 1

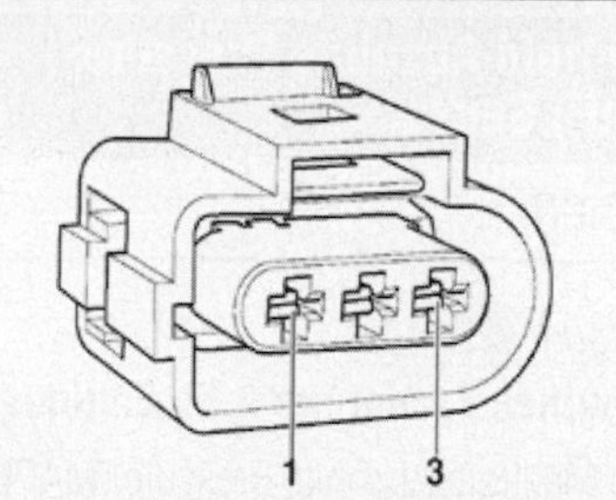

Fig. 82 Check wires between test box and 3-pin connector for open circuit according to wiring diagram.

5. Check wires between test box and 3-pin connector for open circuit according to wiring diagram.

Component	Contact	Test box socket
Knock Sensor (KS) 1 G61	1	106
	2	99

6. Wiring resistance: max. 1.5Ω

7. Check wires for short circuit to each other.

8. Specification: infinite ohms

9. If resistances and wires are OK., replace respective Knock Sensor (KS) 1.

Knock Sensor (KS) 2

LOCATION

1.8L Engines

See Figure 83.

The Knock Sensor (KS) 2 is located on the front of the engine cylinder block, on the left side, above the oil coolant flange.

2.0L Engines

See Figure 84.

The Knock Sensor (KS) 2 is located on the front of the engine cylinder block, behind the ignition coil.

2.5L Engines

See Figure 85.

The Knock Sensor (KS) 2 is located below the exhaust manifold, on the right side (shown here with the heat shield removed).

2.8L Engines

See Figure 86.

The Knock Sensor (KS) 2 is located on the engine cylinder block, left front, above the engine oil cooler.

3.2L Engines

See Figure 87.

The Knock Sensor (KS) 2 is located on the engine cylinder block, left front, above the engine oil filter housing.

TESTING

See Figures 88 through 90.

➡ **Contact surfaces between Knock Sensor (KS) 2 and the cylinder block must be free of corrosion, dirt and grease.**

➡ **To ensure problem-free function of knock sensors, it is important that torque specification of 20 Nm be adhered to exactly.**

➡ **Use only gold-plated terminals when servicing terminals in harness connector of Knock Sensor (KS) 2.**

1. Disconnect 3-pin connector from Knock Sensor (KS) 1 G61 / Knock Sensor (KS) 2 G66: (1 - Green, for Knock Sensor (KS) 1 and 2 - Grey, for Knock sensor (KS) 2)

➡ **Before disconnecting harness connectors, mark which component goes where.**

2. Measure resistance between terminals 1 + 2 at connectors to Knock Sensor (KS) 2.

3. Specification: infinite ohms

4. Connect test box to Motronic Engine Control Module (ECM).

5. Check wires between test box and 3-pin connector for open circuit according to wiring diagram.

Fig. 83 Knock Sensor 2 location for the 1.8L engine.

Fig. 84 Knock Sensor 2 location for the 2.0L engine.

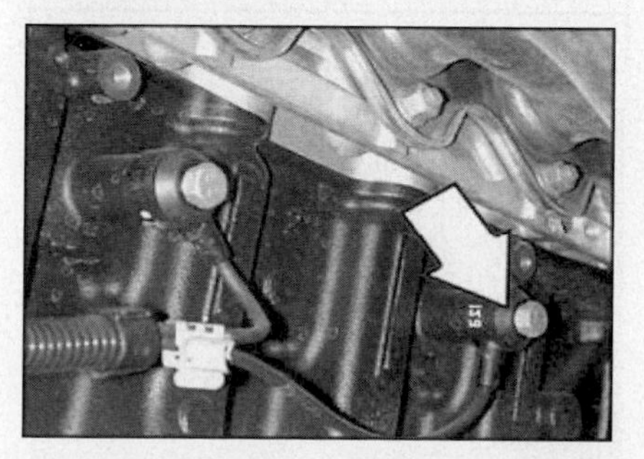

Fig. 85 Knock Sensor 2 location for the 2.5L engine.

Fig. 86 Knock Sensor 2 location for the 2.8L engine.

Fig. 87 Knock Sensor 2 location for the 3.2L engine.

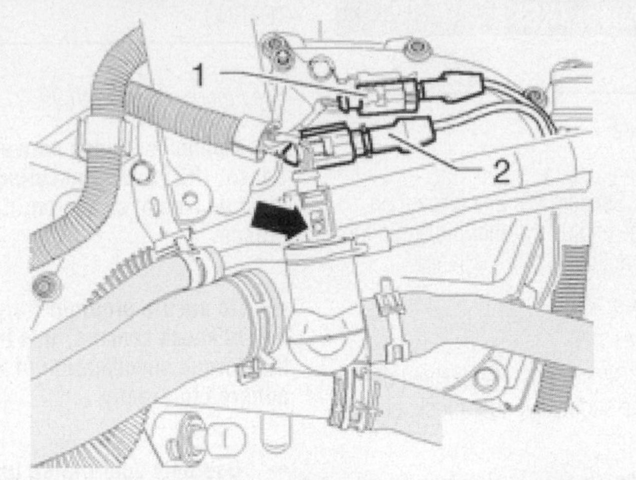

Fig. 88 Disconnect 3-pin connector from Knock Sensor (KS) 1 G61 / Knock Sensor (KS) 2 G66: (1 - Green, for Knock Sensor (KS) 1 and 2 - Grey, for Knock sensor (KS) 2)

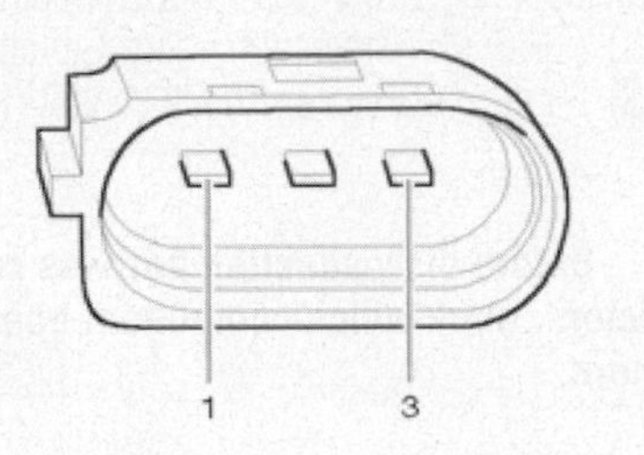

Fig. 89 Measure resistance between terminals 1 + 2 at connectors to Knock Sensor (KS) 2.

Component	Contact	Test box socket
Knock Sensor (KS) 2 G66	1	107
	2	99

6. Wiring resistance: max. 1.5Ω
7. Check wires for short circuit to each other.
8. Specification: infinite ohms
9. If resistances and wires are OK., replace respective Knock Sensor (KS) 2.

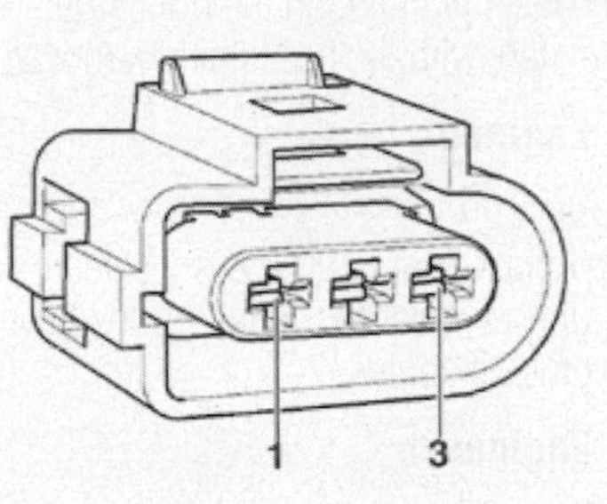

Fig. 90 Check wires between test box and 3-pin connector for open circuit according to wiring diagram.

Knock Sensor (KS) 3

LOCATION

See Figure 91.

4.0L Engines

The Knock Sensor (KS) 3 is located on the front of the engine cylinder block, on the left side, above the serpentine belt tensioner.

Manifold Absolute Pressure (MAP) Sensor

LOCATION

See Figure 92.

All Engines Except the 2.5L Engines

The Manifold Absolute Pressure (MAP) Sensor is located in the intake air duct between the charge air cooler and the intake manifold.

2.5L Engines

See Figure 93.

The Manifold Absolute Pressure (MAP) Sensor is located at the left end of the intake manifold, next to the throttle control valve.

TESTING

See Figures 94 and 95.

➡ **Use only gold-plated terminals when servicing terminals in harness connector of Manifold Absolute Pressure (MAP) Sensor .**

1. Disconnect 4-pin connector from Manifold Absolute Pressure (MAP) Sensor (arrow).
2. Start engine and run at idle speed.
3. Measure voltage between terminals 1 + 3 on harness connector.
4. Specification: min. 4.5 V
5. Switch off ignition.
6. If no voltage is present, connect test box to Motronic Engine Control Module (ECM).
7. Check wires between test box and 4-pin connector for open circuit according to wiring diagram.

Contact	Test box socket
1	108
3	98
4	101

Fig. 91 Knock Sensor 3 location for the 4.0L engine.

Fig. 92 Manifold Absolute Pressure sensor location for all engines except the 2.5L engine.

Fig. 93 Manifold Absolute Pressure sensor location for the 2.5L engine.

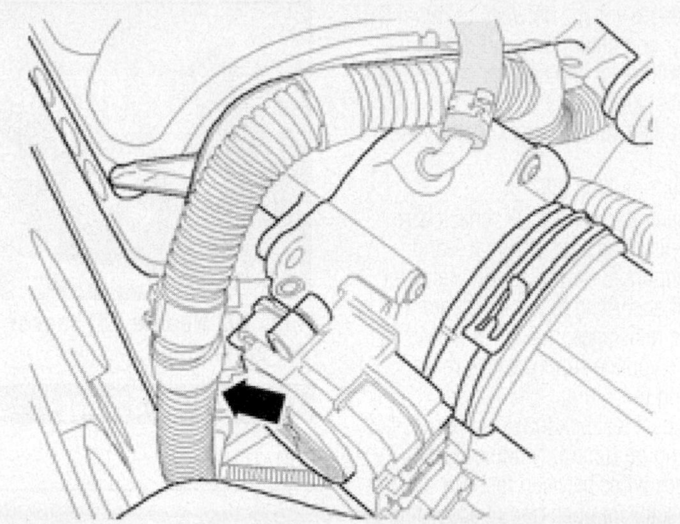

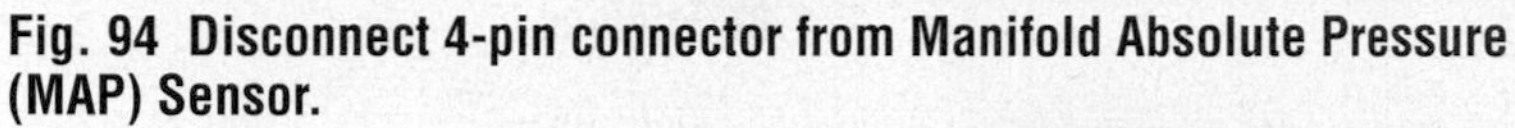

Fig. 94 Disconnect 4-pin connector from Manifold Absolute Pressure (MAP) Sensor.

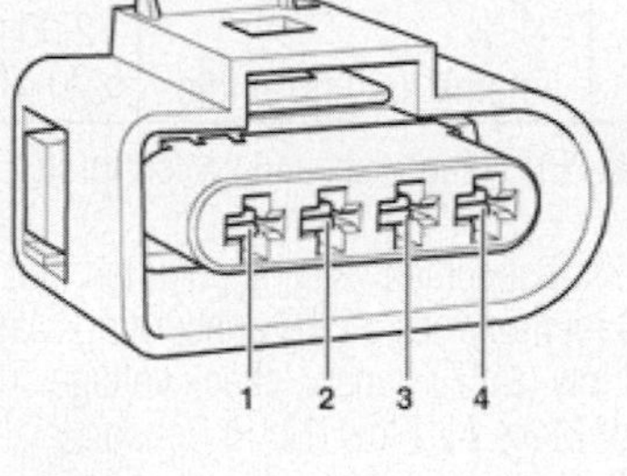

Fig. 95 Check wires between test box and 4-pin connector for open circuit according to wiring diagram.

8. Wiring resistance: max. 1.5Ω
9. Check wires for short circuit to each other, to B+ and Ground (GND).
10. Specification: infinite ohms
11. If no malfunction is detected in wiring, replace Manifold Absolute Pressure (MAP) Sensor.
12. If no malfunction is detected in wiring, replace Motronic Engine Control Module (ECM).

Mass Air Flow (MAF) Sensor

LOCATION

2.0L Engines

See Figure 96.

The Mass Air Flow (MAF) Sensor is located in the intake air duct, to the right of the air cleaner housing.

2.5L Engines

See Figure 97.

The Mass Air Flow (MAF) Sensor is located on the left side of the engine cover.

3.2L Engines

See Figure 98.

The Mass Air Flow (MAF) Sensor is located in the intake air duct, to the right of the air cleaner housing.

4.0L Engines

See Figure 99.

The Mass Air Flow (MAF) Sensor is located on the right side of the engine compartment next to the air cleaner assembly.

Fig. 96 Mass Air Flow sensor location for the 2.0L engine.

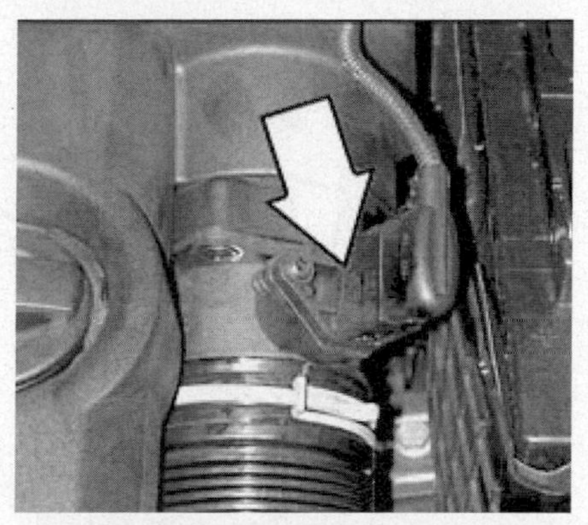

Fig. 97 Mass Air Flow sensor location for the 2.5L engine.

Fig. 98 Mass Air Flow sensor location for the 3.2L engine.

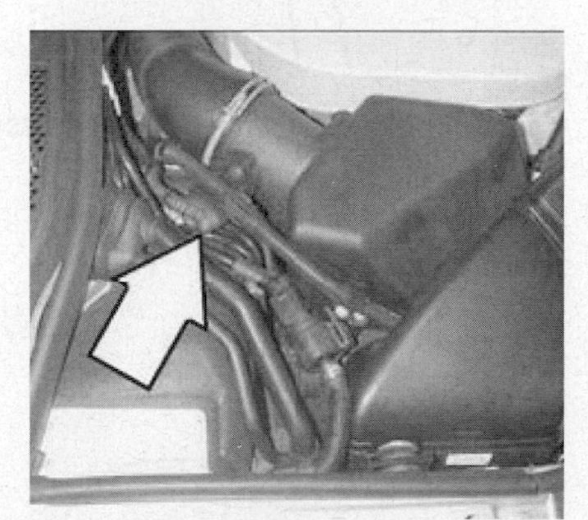

Fig. 99 Mass Air Flow sensor location for the 4.0L engine.

TESTING

See Figures 100 through 102.

➡ **Use only gold-plated terminals when servicing terminals in harness connector of Mass Air Flow (MAF) Sensor.**

1. Connect vehicle diagnosis and service system.
2. Start engine and let it run at idle.
3. Under address word "33 - On Board Diagnostic (OBD)", select "Diagnostic mode 1: Check measured values" .
4. Select measuring value "PID 16: Air flow quantity at Mass Air Flow (MAF) sensor" .
5. Check specified value of air flow quantity at Mass Air Flow (MAF) Sensor at idle:

PID	Diagnostic text	Specified value
16	Air flow quantity at Mass Air Flow (MAF) sensor	
	Engine running at idle	2.00 to 5.00 g/sec

6. End diagnosis and switch ignition off.
7. If specified value is obtained, but DTC memory has a DTC concerning Mass Air Flow (MAF) Sensor, check voltage supply of Mass Air Flow (MAF) Sensor.
8. If specified value is not obtained, check wires of Mass Air Flow (MAF) Sensor G70 .
9. Disconnect 5-pin harness connector (1) at Mass Air Flow (MAF) Sensor.
10. Start engine and let it run at idle.
11. Measure voltage between terminal 3 of connector and engine Ground (GND) using hand multimeter and adapter cables from connector test kit.
12. Specified value: approx. battery voltage
13. Switch ignition off.
14. If the voltage is not OK, check wire between 5-pin connector terminal 3 and Engine Component Power Supply Relay for open circuit according to wiring diagram.
15. Wire resistance: max.: 1.5 OHMS
16. If the voltage supply and wire are OK, checking the wiring.
17. Connect test box to wiring harness of Motronic Engine Control Module (ECM).
18. Check wires between test box and 5-pin connector for open circuit according to wiring diagram.

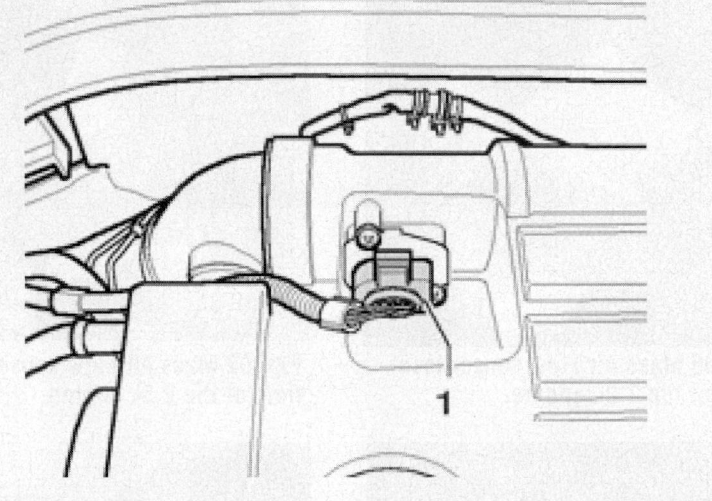

Fig. 100 Disconnect the 5-pin harness connector (1) at Mass Air Flow (MAF) Sensor.

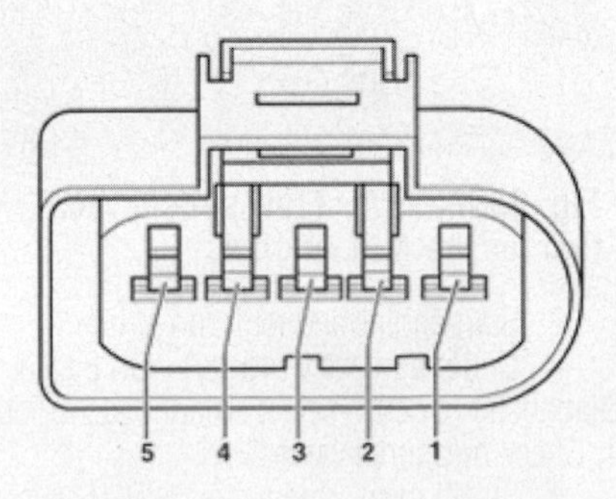

Fig. 101 Check wires between test box and 5-pin connector for open circuit.

Terminal	Test box socket
1	22
2	1
2	2
2	4

19. Wire resistance: max. 1.5 Ω
20. Check wires for short circuit to each other, to Battery A (+) and Ground (GND).
21. Specified value: infinite ohms.
22. If the wires are OK, replace Mass Air Flow (MAF) Sensor G70.

Needle Lift Sensor

LOCATION

The Needle Lift Sensor is located in front of the cylinder head and is a component of the Cylinder Three fuel injector.

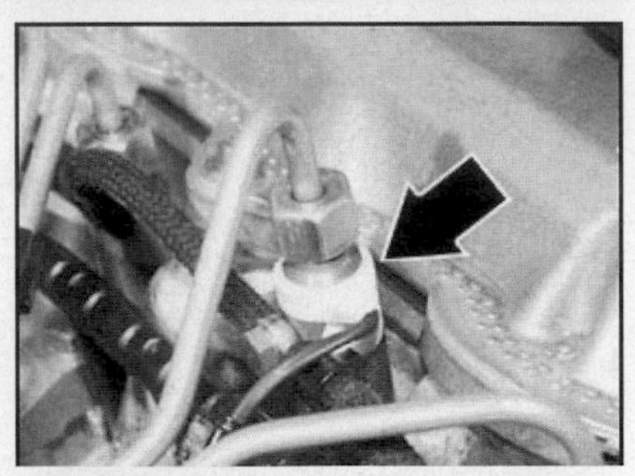

Fig. 102 Needle Lift sensor location for all engines.

Oxygen (O2) Sensor

LOCATION

1.8L Engines

See Figure 103.

The Oxygen Sensor (O2S) is located behind the Three-Way Catalytic Converter, beneath the floor and within the exhaust pipe.

2.0L Engines

See Figure 104.

The Oxygen Sensor (O2S) is located behind the Three-Way Catalytic Converter, within the exhaust pipe.

2.5L Engines

See Figure 105.

The Oxygen Sensor (O2S) is located in the exhaust tunnel downstream from the Catalytic Converter.

2.8L Engines

See Figure 106.

The Oxygen Sensor (O2S) is located behind the Three-Way Catalytic Converter.

3.2L Engines

See Figure 107.

The Oxygen Sensor (O2S) is located behind the Three-Way Catalytic Converter on the passenger side.

3.2L Engines

See Figure 108.

The Oxygen Sensor (O2S) is located behind the Three-Way Catalytic Converter on the drivers side.

4.0L Engines

See Figure 109.

The Oxygen Sensor (O2S) is located behind the Three-Way Catalytic Converter on the right exhaust pipe.

Fig. 103 Oxygen sensor location for the 1.8L engine.

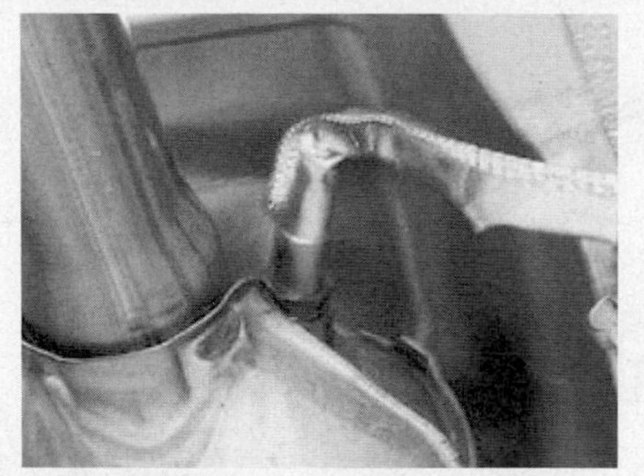

Fig. 104 Oxygen sensor location for the 2.0L engine.

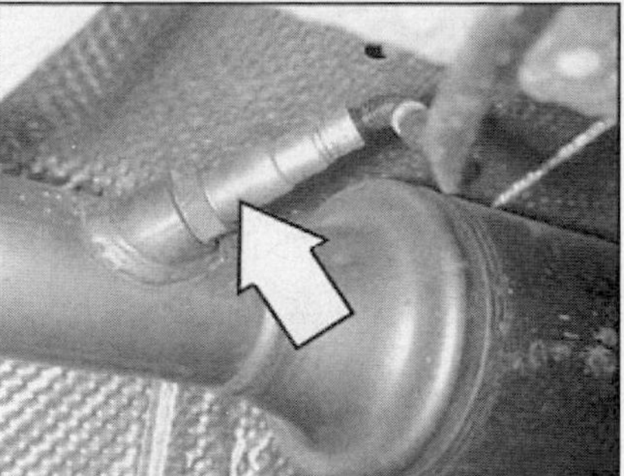

Fig. 105 Oxygen sensor location for the 2.5L engine.

Fig. 106 Oxygen sensor location for the 2.8L engine.

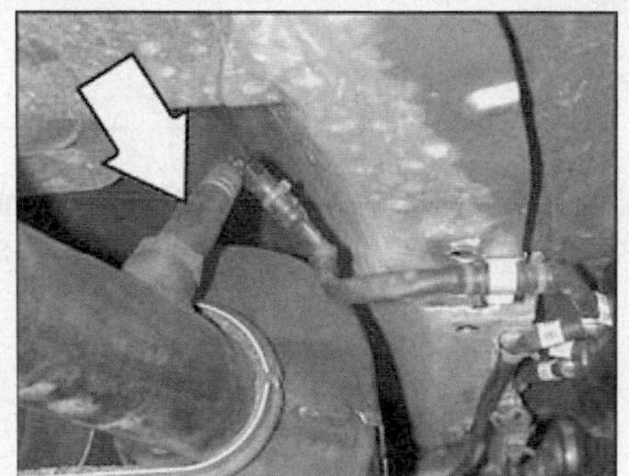

Fig. 107 Oxygen sensor location for the 3.2L engine.

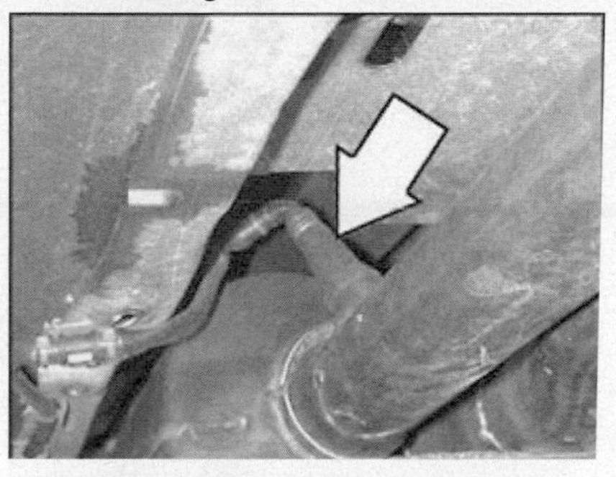

Fig. 108 Oxygen sensor location for the 3.2L engine.

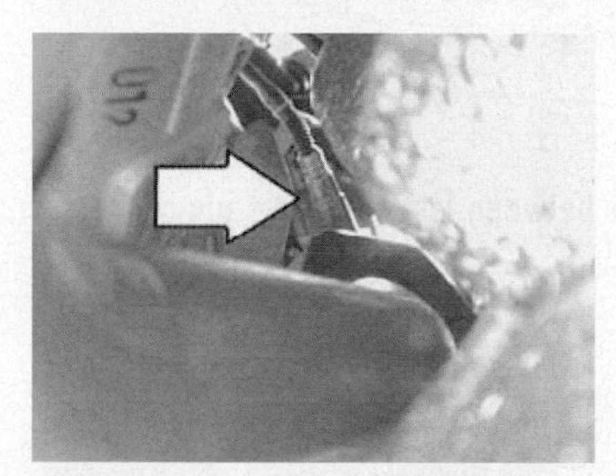

Fig. 109 Oxygen sensor location for the 4.0L engine.

OPERATION

The static values of the oxygen sensors are indicated via this diagnostic mode. The values of the individual oxygen sensor signal outputs must reach the specified value or must be within the min.- and max.- limits.

TESTING

See Figures 110 through 113.

➡ **Vehicle must be raised before connector for Oxygen Sensor (O2S) Behind Three Way Catalytic Converter (TWC) is accessible.**

➡ **When servicing terminals 3 and 4 in harness connector of Oxygen Sensor (O2S) Behind Three Way Catalytic Converter (TWC), use only gold-plated terminals.**

1. Connect Vehicle diagnosis and service system.
2. Start engine and let it run at idle.
3. Under address word "33 - On Board Diagnostic (OBD)" , select "Diagnostic mode 6: Check test results of components that are not continuously monitored" .
4. Select "Test-ID 02: Oxygen sensor monitoring behind catalytic converter" .
5. Select following Test-ID.
6. Check specified values at idle.**Test-ID**
7. If specified values are obtained, end diagnosis and switch ignition off.
8. If specified values are not obtained, end diagnosis and let engine continue to run at idle.
9. Check primary voltage.
10. Remove right vehicle floor cover (see arrows).
11. Disconnect brown 4-pin harness connector (arrow) to Oxygen Sensor (O2S) Behind Three Way Catalytic Converter (TWC).
12. Measure voltage at terminals 3+4 of connector to Motronic Engine Control Module (ECM) using hand multimeter and adapter cables from connector test kit.
13. Specified value: 0.400 to 0.500 Volts
14. Switch ignition off.
15. If the voltage is OK, replace Oxygen Sensor (O2S) Behind Three Way Catalytic Converter (TWC).
16. If the voltage not OK, check the oxygen sensor wiring.
17. Connect test box to wiring harness of Motronic Engine Control Module (ECM).

Test-ID		Specified value	
		min.	max.
1	Rich to lean sensor barrier voltage	-	0.6241 V
2	Lean to rich sensor barrier voltage	-	0.6241 V
7	Minimum voltage at sensor for test cycle	-	0.450 V
8	Maximum voltage at sensor for test cycle	0.450 V	-
129	Sensor voltage lean	0	0.6241 V
130	Sensor voltage rich	0.6241 V	1.2998 V
131	Deceleration test	0 V	0.1599 V

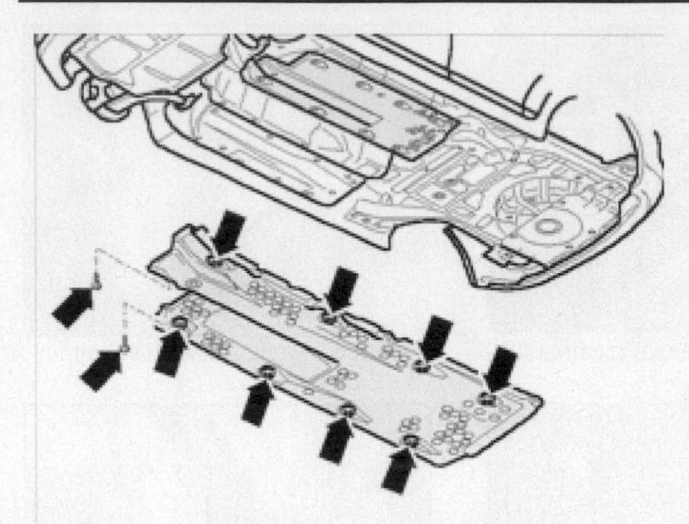

Fig. 110 Remove right vehicle floor cover (see arrows)

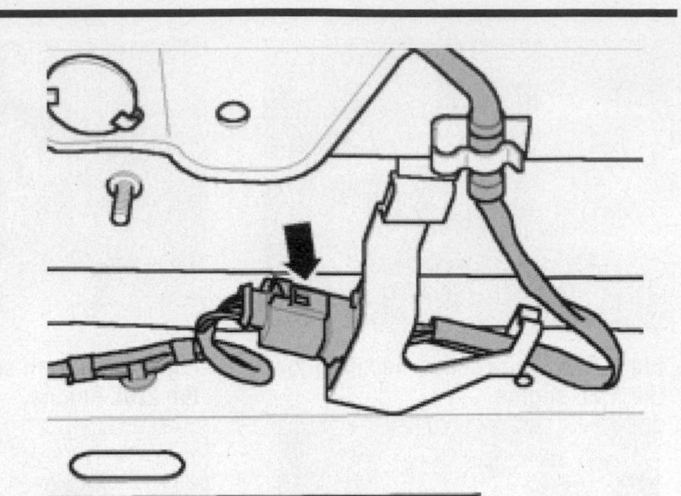

Fig. 111 Disconnect brown 4-pin harness connector (arrow) to Oxygen Sensor

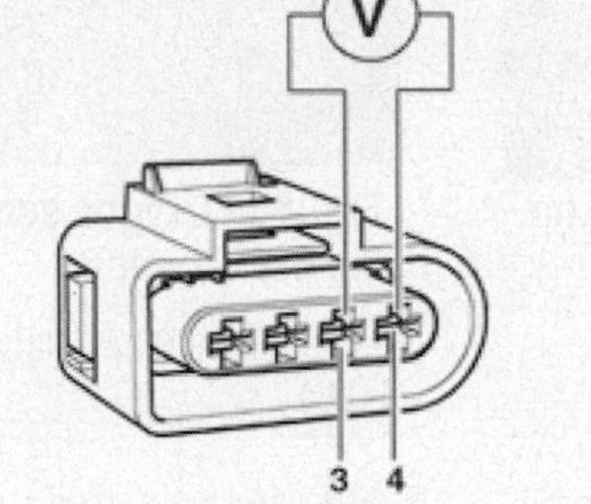

Fig. 112 Measure voltage at terminals 3+4 of connector to the ECM.

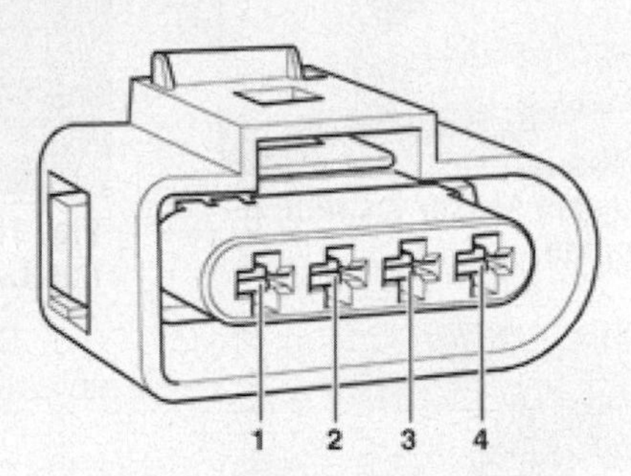

Fig. 113 Check wires between test box and 4-pin connector for open circuit

18. Check wires between test box and 4-pin connector for open circuit according to wiring diagram.

Terminal	Test box socket
3	76
4	77

19. Wire resistance: max. 1.5 Ω
20. Check wires for short circuit to each other, to Battery A (+) and Ground (GND).
21. Specified value: infinite ohms.
22. If the wires are OK, replace Motronic Engine Control Module (ECM).

Throttle Position Sensor (TPS)

LOCATION

1.8L and 1.9L Engines

See Figure 114.

The Throttle Position (TPS) Sensor is mounted on bulkhead, above accelerator pedal.

2.0L Engines

See Figure 115.

The Throttle Position (TPS) Sensor is mounted on the throttle body.

2.8L Engines

See Figure 116.

The Throttle Position (TPS) Sensor is a component of the throttle valve control module (note the ground wire to the cylinder head (A to B).

3.2L Engines

See Figure 117.

The Throttle Position (TPS) Sensor is mounted on bulkhead, above accelerator pedal.

Transmission Control Module (TCM)

LOCATION

All Engines Except the 2.5L

See Figure 118.

The Transmission Control Module is located in the plenum chamber on the right side.

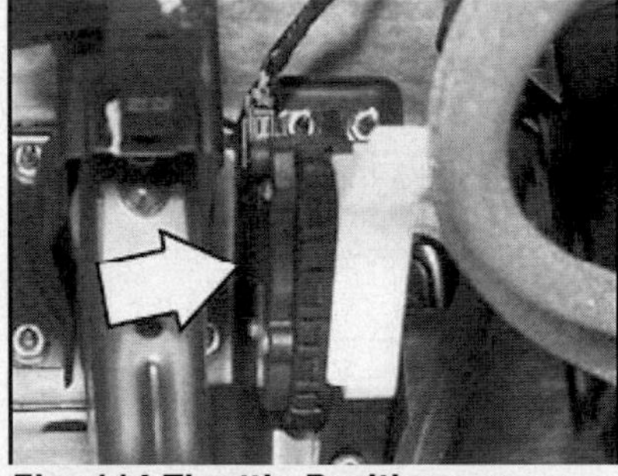

Fig. 114 Throttle Position sensor location for the 1.8L and 1.9L engines.

Fig. 115 Throttle Position sensor location for the 2.0L engine.

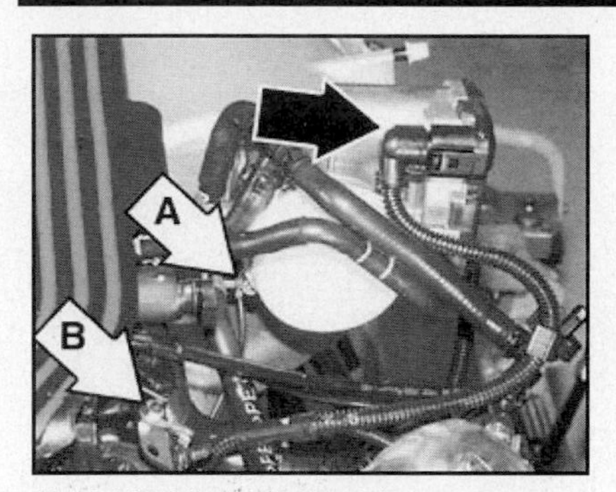

Fig. 116 Throttle Position sensor location for the 2.8L engine.

2.5L Engine

See Figure 119.

The Transmission Control Module is located behind the left front wheelhouse cover near the A-Pillar.

REMOVAL & INSTALLTION

1. Remove left front wheel.
2. Remove left wheel housing liner.
3. Release connector and pull it off from Transmission Control Module (TCM).
4. If equipped, loosen side turn signal wiring from bracket on Transmission Control Module (TCM).
5. If equipped, remove left side turn signal from fender toward the outside, via the bolts.

➡ **The Transmission Control Module (TCM) is still hooked in on the backside.**

Fig. 118 Transmission Control Module location for all engines except the 2.5L engine.

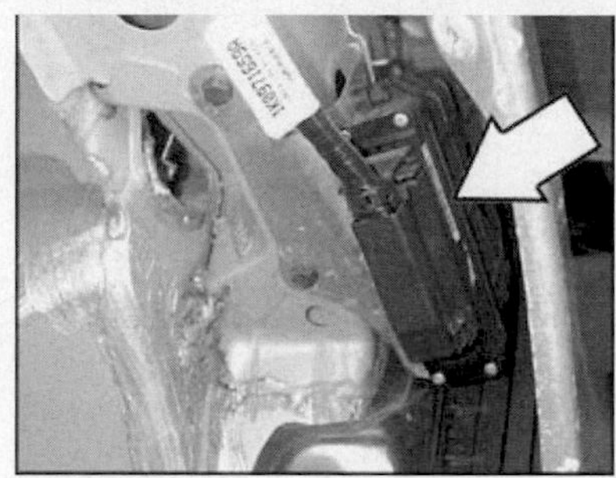

Fig. 119 Transmission Control Module location for the 2.5L engine.

Fig. 117 Throttle Position sensor location for the 3.2L engine.

6. Push Transmission Control Module (TCM) upward together with bracket and then remove it.
7. Remove old Transmission Control Module (TCM) from bracket via the bolts.

To install:

8. Set new Transmission Control Module (TCM) in place and fasten it.
9. Set connector onto Transmission Control Module (TCM) and lock it into place.
10. Further installation is performed in reverse order.
11. Check whether the new control module identification matches the old control module identification.
12. When the harness connector is disconnected from Transmission Control Module (TCM) or the battery is disconnected, all adaptation values in Transmission Control Module (TCM) are erased. However, DTC memory content will remain intact. If the engine is started after this, the idle may be rough for a short period. In this case, the readiness code must be generated again.

Vehicle Speed Sensor (VSS)

LOCATION

Five-Speed Manual Transmission

See Figure 120.

The Vehicle Speed Sensor (VSS) is located on top of the transmission final drive housing.

Six-Speed Manual Transmission

See Figure 121.

The Vehicle Speed Sensor (VSS) is located on top of the transmission final drive housing.

Automatic Transmission

See Figure 122.

The Vehicle Speed Sensor (VSS) is located on top of the transmission final drive housing.

TESTING

See Figures 123 through 125.

1. Disconnect 3-pin harness connector to Vehicle Speed Sensor (VSS) (arrow).
2. Check harness connector for contact corrosion and ingress of water.
3. If harness connector is not OK, repair the malfunction.
4. If harness connector is OK, measure resistance at 3-pin connector of Vehicle Speed Sensor (VSS).
5. Measure sensor resistance between terminals 1 + 2 at connector to sensor.
6. Specified value: 800 to 900 OHMS
7. Check sensor for short circuit between terminals 2 + 3 and 1 + 3.
8. Specified value: infinite ohms.
9. If specified values are not obtained, replace Vehicle Speed Sensor (VSS).

Fig. 120 Vehicle Speed sensor location for all engines equipped with a five-speed transmission.

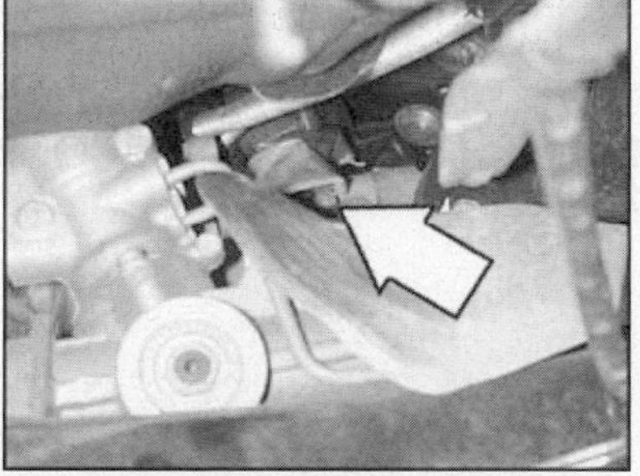

Fig. 121 Vehicle Speed sensor location for all engines equipped with a six-speed transmission.

Fig. 122 Vehicle Speed sensor location for all engines equipped with an automatic transmission.

10. If no malfunction is found on sensor, check wires to Transmission Control Module (TCM) according to wiring diagram
11. To check the wires to Transmission Control Module (TCM), connect test box to control module wiring harness.
12. Check wires between test box and 3-pin connector to Transmission Control Module (TCM) for open circuit according to wiring diagram.
    - Terminal 1 + socket 65
    - Terminal 2 + socket 20
    - Terminal 3 + socket 43
13. Wire resistance: max. 1.5 Ω
14. Check wires for short circuit to each other, to vehicle Ground (GND) and to B+.
15. Specified value: infinite ohms.
16. If no malfunctions are found in wires, replace Transmission Control Module (TCM).

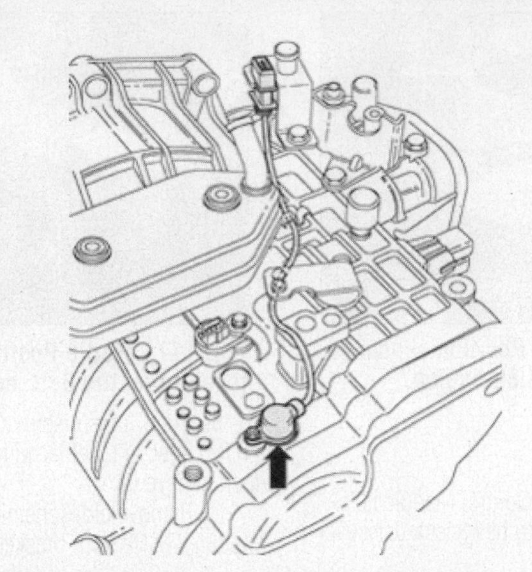

Fig. 123 Disconnect 3-pin harness connector to Vehicle Speed Sensor (VSS) (arrow)

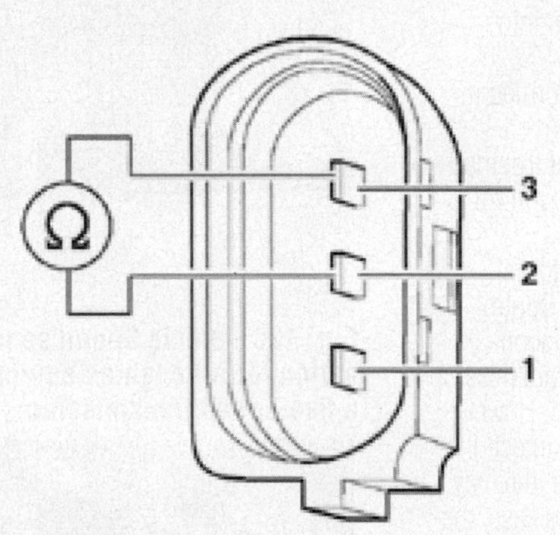

Fig. 124 Measure sensor resistance between terminals 1 + 2 at connector to sensor.

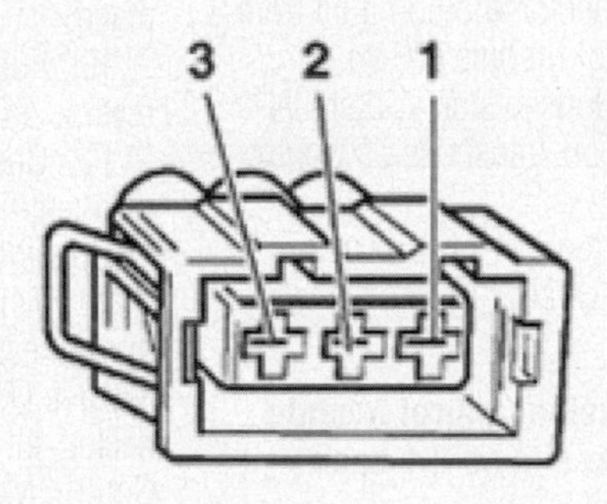

Fig. 125 Check wires between test box and 3-pin connector to Transmission Control Module (TCM) for open circuit.

VOLKSWAGEN
PIN CHARTS

16

TABLE OF CONTENTS

PIN CHARTS

Introduction

A Pin Voltage Table is a term used to describe a table that identifies PCM pins, wire colors of the PCM circuits, circuit descriptions and "known good" values for devices that connect to the PCM. These tables include the following information:

- Signals from various sensors (ECT, IAT, MAP, TPS, etc.)
- Signals from various switches (PNP, PSP, WOT, etc.)
- Signals from oxygen sensors (O2S, HO2S)
- Signals from output devices (IAC, INJ, TCC, etc.)
- Power & ground signals

Pin Voltage Tables

Information contained within the Pin Voltage Tables can be used to:

- Test circuits for open, short to power or short to ground faults
- Check the operation of a component before or after a repair
- Check the operation of a component or system by viewing signals on PCM input/output circuits with a DVOM or Lab Scope

Using a Breakout Box

There are several Breakout Box (BOB) designs available for use to test the PCM and its input and output circuits. However, all of them require removal of the wire harness to the PCM so that the BOB can be installed between the PCM and wire harness connector. Several breakout boxes require the use of overlays in order to allow the tool to be used on more than one year or engine type. Always verify that the correct adapter and overlays are used to prevent connection to the wrong circuits and a misdiagnosis.

Power and Ground Circuit Checks

Measurements made at the BOB are accomplished via test leads and probes from the DVOM or a Lab Scope. If any of the terminals on the PCM or BOB are damaged or loose, test measurements made at the Breakout Box will be inaccurate. To verify the PCM battery power and ground circuits are normal (correct) at the BOB, test the condition of the circuit between the battery negative (-) post and these circuits prior to starting a test sequence.

Diagnosis with Pin Voltage Tables

See Figure 1.

Once an actual PCM pin voltage reading is recorded, it can be compared to an example from a vehicle with "known good" values. In the example shown the Value at Hot Idle for the EVP sensor signal (0.4v) is the "known good" value.

Wire Color Changes

Every effort has been made to obtain and list the correct circuit wire colors for all vehicles. However, running changes from the vehicle manufacturer can cause the wrong colors to be listed.

PCM Pin #	W/Color	Circuit Description (60-Pin)	Value at Hot Idle
27	BN/LG	EVP Sensor Signal	0.4v

Fig. 1 Example

VOLKSWAGEN PIN CHARTS

Standard Colors and Abbreviations

Abbreviation	Color	Abbreviation	Color
BLK	Black	ORN	Orange
BLU	Blue	RED	Red
BRN	Brown	LIL	Lilac
GRY	Gray	WHT	White
GRN	Green	YEL	Yellow

Cabrio

1995–1999 Sedan 2.0L MFI ABA T68 Connector

ECM Pin #	Wire Color	Circuit Description (68-Pin)	Value at Hot Idle
1	BLU	Ground	N/A
2	GRY/BLU	Cylinder 4 Fuel Injector Control	N/A
3	YEL/BLK	Malfunction Indicator Lamp (MIL)	N/A
4	--	Not Used	--
5	--	Not Used	--
6	YEL/BLU	Fuel Pump Relay	N/A
7	BRN/BLK	Ground	N/A

1995–1999 Sedan 2.0L MFI ABA T68 Connector, *continued*

8	BLK/RED	Ignition Coil Power Output Stage	N/A
9		Not Used	--
10	WHT/GRN	Throttle Position Actuator Signal	N/A
11	BLU/YEL	T-10 Connector (behind fuse panel) (A/T only)	N/A
12	BRN	Heated Oxygen Sensor 1 Power	N/A
13	WHT/RED	Heated Oxygen Sensor 2 Signal	N/A
14	BLU	Engine Coolant Temperature Sensor	N/A
15	--	Not Used	--
16	BRN/BLU	Mass Airflow Sensor	N/A
17	BLU	Mass Airflow Sensor	N/A
18	YEL/RED	T-10 Connector (behind fuse panel)	N/A
19	--	Not Used	--
20	WHT	Heated Oxygen Sensor 1 Signal	N/A
21	YEL	T-10 Connector (behind fuse panel) (A/T only)	N/A
22	GRN/BLK	Relay Panel	N/A
23	BLK/YEL	Power Supply Relay	N/A
24	GRY	Cylinder 1 Fuel Injector Control	N/A
25	GRY/GRN	Cylinder 2 Fuel Injector Control	N/A
26	GRY/RED	Cylinder 3 Fuel Injector Control	N/A
27	BLK	Throttle Position Actuator Power	N/A
28	--	Not Used	--
29	--	Not Used	--
30	--	Not Used	--
31	GRN/YEL	EVAP Canister Purge Regulator Valve	N/A
32	--	Not Used	--
33	BRN/BLU	Knock Sensor 1 Ground	N/A
34	GRY	Knock Sensor 1 Power	N/A
35	--	Not Used	--
36	WHT/YEL	Intake Air Temperature Sensor Signal	N/A
37	--	Not Used	--
38	--	Not Used	--
39	BLU	T-2 Connector (A/C Pressure Switch)	N/A
40	RED/BLU	Throttle Position Sensor Signal	N/A
41	RED/WHT	Throttle Position Sensor Power	N/A
42	YEL	Heated Oxygen Sensor 1Ground	N/A
43	GRY/WHT	Data Link Connector (DLC)	N/A
44	GRN/WHT	Camshaft Position Sensor	N/A
45	--	Not Used	--
46	--	Not Used	--
47	--	Not Used	--
48	--	Not Used	--
49	GRN	Relay for the A/C Shutoff	N/A
50	--	Not Used	--
51	--	Not Used	--
52	--	Not Used	--
53	WHT	Throttle Position Actuator Ground	N/A

1995–1999 Sedan 2.0L MFI ABA T68 Connector, *continued*

54	RED	Fuel Pump Relay Signal	N/A
55	BLU/WHT	Vehicle Speed Signal Wire Connector	N/A
56	BRN/BLK	Ground	N/A
57	--	Not Used	--
58	BLK/YEL	Heated Oxygen Sensor 2 Ground	N/A
59	--	Not Used	--
60	--	Not Used	--
61	--	Not Used	--
62	RED/GRY	Throttle Position Sensor Power	N/A
63	--	Not Used	--
64	--	Not Used	--
65	BRN/WHT	Heated Oxygen Sensor 2 Power	N/A
66	--	Not Used	--
67	RED	Engine Speed Sensor	N/A
68	GRN	Engine Speed Sensor	N/A

Cabrio

2000 Sedan 2.0L MFI ABA T80 Connector

ECM Pin #	Wire Color	Circuit Description (80-Pin)	Value at Hot Idle
1	BLK/YEL	Terminal 15	N/A
2	BRN	Power Ground	N/A
3	RED	Battery Power	N/A
4	YEL/BLU	Fuel Pump Relay Control	N/A
5	--	Not Used	--
6	GRN/BLK	Relay Panel	N/A
7	--	Not Used	--
8	GRN	Relay Panel (A/C Connector)	N/A
9	--	Not Used	--
10	BLU/RED	A/C Pressure Switch	N/A
11	BRN/WHT	Mass Air Flow Sensor	N/A
12	BLU	Mass Air Flow Sensor	N/A
13	BRN/BLU	Mass Air Flow Sensor	N/A
14	--	Not Used	--
15	GRN/YEL	EVAP Canister Purge Regulator Valve Control	N/A
16	YEL/BRN	Leak Detection Pump	N/A
17	YEL/BLK	Malfunction Indicator Light	N/A
18	--	Not Used	--
19	GRY/WHT	Control Module for Anti-Theft Immobilizer	N/A
20	BLU/WHT	Vehicle Speed Sensor Wire Connector	N/A
21	--	Not Used	--
22	--	Not Used	--
23	--	Not Used	--
24	--	Not Used	--
25	BRN	Heated Oxygen Sensor Ground	N/A
26	BLK	Heated Oxygen Sensor Signal	N/A

2000 Sedan 2.0L MFI ABA T80 Connector, *continued*

27	WHT	Oxygen Sensor Heater Control	N/A
28	BRN/BLK	Oxygen Sensor 2 Heater Control	N/A
29	BLU/BRN	Transmission Control Module (A/T only)	N/A
30	GRY/BRN	Secondary Air Injection Pump Relay Control	N/A
31	--	Not Used	--
32	--	Not Used	--
33	RED/YEL	Secondary Air Injection Solenoid Valve Control	N/A
34	--	Not Used	--
35	--	Not Used	--
36	--	Not Used	--
37	GRN/LIL	Leak Detection Pump Switch Signal	N/A
38	--	Not Used	--
39	--	Not Used	--
40	--	Not Used	--
41	BLU/RED	Transmission Control Module (A/T only)	N/A
42	--	Not Used	--
43	GRY/YEL	T-16 Connection, Data Link Connector	N/A
44	--	Not Used	--
45	--	Not Used	--
46	--	Not Used	--
47	--	Not Used	--
48	--	Not Used	--
49	--	Not Used	--
50	--	Not Used	--
51	BLK/YEL	Heated Oxygen Sensor 2 Ground	N/A
52	WHT/RED	Heated Oxygen Sensor Signal	N/A
53	BLU	Engine Coolant Temperature Sensor Power	N/A
54	WHT/YEL	Intake Air Temperature Sensor	N/A
55	--	Not Used	--
56	RED	Engine Speed Sensor Signal	N/A
57	--	Not Used	--
58	GRY/RED	Cylinder 3 Fuel Injector Control	N/A
59	WHT	Throttle Position Actuator Power	N/A
60	--	Not Used	--
61	--	Not Used	--
62	RED/WHT	Throttle Position Sensor Signal	N/A
63	BLK	Engine Speed Sensor Signal	N/A
64	--	Not Used	--
65	GRY/BLU	Cylinder 4 Fuel Injector Control	N/A
66	BLK	Throttle Position Actuator Control	N/A
67	BRN/BLU	Sensor Ground	N/A
68	RED	Knock Sensor 1 Signal	N/A
69	WHT/GRN	Throttle Position Actuator Ground	N/A
70	--	Not Used	--
71	BLK/RED	Ignition Coil Power Output Stage	N/A
72	--	Not Used	--

2000 Sedan 2.0L MFI ABA T80 Connector, *continued*

73	GRY	Cylinder 1 Fuel Injector Control	N/A
74	RED/GRY	Throttle Position Sensor Signal	N/A
75	RED/BLU	Throttle Position Sensor Ground	N/A
76	GRN/WHT	Camshaft Position Sensor Signal	N/A
77	--	Not Used	--
78	--	Not Used	--
79	--	Not Used	--
80	GRY/GRN	Cylinder 2 Fuel Injector Control	N/A

Cabrio

2001–2002 Sedan 2.0L MFI ABA T80 Connector

ECM Pin #	Wire Color	Circuit Description (80-Pin)	Value at Hot Idle
1	BLK/YEL	Terminal 15	N/A
2	BRN	Power Ground	N/A
3	RED	Battery Power	N/A
4	YEL/BLU	Fuel Pump Relay Control	N/A
5	--	Not Used	--
6	GRN/BLK	Relay Panel	N/A
7	--	Not Used	--
8	GRN	Relay Panel (A/C Connector)	N/A
9	--	Not Used	--
10	BLU/RED	A/C Pressure Switch	N/A
11	BRN/WHT	Mass Air Flow Sensor	N/A
12	BLU	Mass Air Flow Sensor	N/A
13	BRN/BLU	Mass Air Flow Sensor	N/A
14	--	Not Used	--
15	GRN/YEL	EVAP Canister Purge Regulator Valve Control	N/A
16	YEL/BRN	Leak Detection Pump	N/A
17	YEL/BLK	Malfunction Indicator Light	N/A
18	--	Not Used	--
19	GRY/WHT	Control Module for Anti-Theft Immobilizer	N/A
20	BLU/WHT	Vehicle Speed Sensor Wire Connector	N/A
21	--	Not Used	--
22	--	Not Used	--
23	--	Not Used	--
24	--	Not Used	--
25	BRN	Heated Oxygen Sensor Ground	N/A
26	BLK	Heated Oxygen Sensor Signal	N/A
27	WHT	Oxygen Sensor Heater Control	N/A
28	BRN/BLK	Oxygen Sensor 2 Heater Control	N/A
29	BLU/BRN	Transmission Control Module (A/T only)	N/A
30	GRY/BRN	Secondary Air Injection Pump Relay Control	N/A
31	--	Not Used	--
32	--	Not Used	--
33	--	Not Used	--

2001–2002 Sedan 2.0L MFI ABA T80 Connector, *continued*

34	--	Not Used	--
35	--	Not Used	--
36	--	Not Used	--
37	GRN/LIL	Leak Detection Pump Switch Signal	N/A
38	--	Not Used	--
39	--	Not Used	--
40	--	Not Used	--
41	BLU/RED	Transmission Control Module (A/T only)	N/A
42	--	Not Used	--
43	GRY/YEL	T-16 Connection, Data Link Connector	N/A
44	--	Not Used	--
45	--	Not Used	--
46	--	Not Used	--
47	--	Not Used	--
48	--	Not Used	--
49	--	Not Used	--
50	--	Not Used	--
51	BLK/YEL	Heated Oxygen Sensor 2 Ground	N/A
52	WHT/RED	Heated Oxygen Sensor Signal	N/A
53	BLU	Engine Coolant Temperature Sensor Power	N/A
54	WHT/YEL	Intake Air Temperature Sensor	N/A
55	--	Not Used	--
56	RED	Engine Speed Sensor Signal	N/A
57	--	Not Used	--
58	GRY/RED	Cylinder 3 Fuel Injector Control	N/A
59	WHT	Throttle Position Actuator Power	N/A
60	--	Not Used	--
61	--	Not Used	--
62	RED/WHT	Throttle Position Sensor Signal	N/A
63	BLK	Engine Speed Sensor Signal	N/A
64	--	Not Used	--
65	GRY/BLU	Cylinder 4 Fuel Injector Control	N/A
66	BLK	Throttle Position Actuator Control	N/A
67	BRN/BLU	Sensor Ground	N/A
68	RED	Knock Sensor 1 Signal	N/A
69	WHT/GRN	Throttle Position Actuator Ground	N/A
70	--	Not Used	--
71	BLK/RED	Ignition Coil Power Output Stage	N/A
72	--	Not Used	--
73	GRY	Cylinder 1 Fuel Injector Control	N/A
74	RED/GRY	Throttle Position Sensor Signal	N/A
75	RED/BLU	Throttle Position Sensor Ground	N/A
76	GRN/WHT	Camshaft Position Sensor Signal	N/A
77	--	Not Used	--
78	--	Not Used	--
79	--	Not Used	--
80	GRY/GRN	Cylinder 2 Fuel Injector Control	N/A

Cabrio, Golf, Jetta

1995 Sedan 2.0L MFI ABA T68 Connector

ECM Pin #	Wire Color	Circuit Description (68-Pin)	Value at Hot Idle
1	BRN/RED	Ground	N/A
2	WHT/GRY	Cylinder 2 Fuel Injector Control	N/A
3	--	Not Used	--
4	--	Not Used	--
5	YEL/BLK	Multi-Function Indicator Light (MIL)	N/A
6	YEL/BLU	Fuel Pump Relay	N/A
7	BRN/RED	Ground	N/A
8	BLK/RED	Ignition Coil Power Output Stage	N/A
9	BRN/BLK	Power Supply Relay	N/A
10	BRN/WHT	Ground (CA and NY only)	N/A
11	BLU/YEL	Transmission Control Module	N/A
12	--	Not Used	--
13	--	Not Used	--
14	BLU	Engine Coolant Temperature Sensor	N/A
15	LIL	EGR Temperature Sensor Signal	N/A
16	BRN/RED	Mass Airflow Sensor	N/A
17	RED	Mass Airflow Sensor	N/A
18	--	Not Used	--
19	--	Not Used	--
20	WHT	Heated Oxygen Sensor Signal	N/A
21	YEL	T-44 Connector for OBD	N/A
22	GRN/BLK	Control Module Display Unit in Instrument Cluster	N/A
23	RED/BLU	Power Supply Relay	N/A
24	WHT/BLK	Cylinder 1 Fuel Injector Control	N/A
25	WHT/GRY	Cylinder 3 Fuel Injector Control	N/A
26	WHT/BLU	Cylinder 4 Fuel Injector Control	N/A
27	WHT	Idle Air Control Valve	N/A
28	WHT	Heated Oxygen Sensor Relay	N/A
29	--	Not Used	--
30	GRN/GRY	EGR Vacuum Regulator Solenoid Valve	N/A
31	GRN/YEL	EVAP Canister Purge Regulator Valve	N/A
32	--	Not Used	--
33	BRN	Ground	N/A
34	GRY	Knock Sensor 1	N/A
35	--	Not Used	--
36	BLU/GRN	Intake Air Temperature Sensor Signal	N/A
37	GRN	T-2 Connector (A/C Connection)	N/A
38	BLK	Plus Connection	N/A
39	BLU	T-2 Connector (A/C Connection)	N/A
40	GRN/GRY	Throttle Position Sensor Signal	N/A
41	BLK/YEL	Throttle Position Sensor Power	N/A
42	YEL	Heated Oxygen Sensor Power	N/A
43	GRY/WHT	T-44 Connector for OBD	N/A
44	WHT/RED	Camshaft Position Sensor	N/A

1995 Sedan 2.0L MFI ABA T68 Connector, *continued*

45	--	Not Used	--
46	--	Not Used	--
47	--	Not Used	--
48	--	Not Used	--
49	--	Not Used	--
50	--	Not Used	--
51	LIL/WHT	T-1 Connector (behind fuse panel)	N/A
52	--	Not Used	--
53	WHT	Idle Air Control Valve Ground	N/A
54	RED/YEL	Fuel Pump Relay Signal	N/A
55	BRN/BLK	Ground	N/A
56	GRY/WHT	Ground	N/A
57	RED	Engine Speed Sensor	N/A
58	WHT	Ground (A/T only)	N/A
59	--	Not Used	--
60	--	Not Used	--
61	--	Not Used	--
62	--	Not Used	--
63	--	Not Used	--
64	--	Not Used	--
65	BLU/WHT	T-43 Connector, Vehicle Speed Sensor	N/A
66	--	Not Used	--
67	--	Not Used	--
68	GRN	Engine Speed Sensor	N/A

Cabrio, Golf, Jetta

1996 Sedan 2.0L MFI ABA T68 Connector

ECM Pin #	Wire Color	Circuit Description (68-Pin)	Value at Hot Idle
1	BRN	Ground	N/A
2	GRY/RED	Cylinder 4 Fuel Injector Control	N/A
3	YEL/BLK	Malfunction Indicator Light (MIL)	N/A
4	--	Not Used	--
5	--	Not Used	--
6	YEL/BLU	Fuel Pump Relay	N/A
7	BRN/RED	Ground (Right Seatbelt Control Module)	N/A
8	BLK/RED	Ignition Coil Power Output Stage	N/A
9	--	Not Used	--
10	WHT/GRN	Throttle Valve Control Module	N/A
11	--	Not Used	--
12	RED/WHT	Heated Oxygen Sensor 1 Ground	N/A
13	WHT/RED	Heated Oxygen Sensor 2 Power	N/A
14	BLU	Engine Coolant Temperature Sensor	N/A
15	--	Not Used	--
16	BRN/BLU	Mass Airflow Sensor	N/A
17	BLU	Mass Airflow Sensor	N/A
18	YEL/RED	Transmission Control Module	N/A

1996 Sedan 2.0L MFI ABA T68 Connector, *continued*

19	--	Not Used	--
20	WHT	Heated Oxygen Sensor Signal	N/A
21	YEL	Transmission Control Module	N/A
22	GRY/BLK	Relay Panel	N/A
23	BLK/YEL	Power Supply Relay	N/A
24	GRY	Cylinder 1 Fuel Injector Control	N/A
25	GRY/GRN	Cylinder 2 Fuel Injector Control	N/A
26	GRY/BLU	Cylinder 3 Fuel Injector Control	N/A
27	BLK	Throttle Valve Control Module Motor Power	N/A
28	--	Not Used	--
29	--	Not Used	--
30	--	Not Used	--
31	GRN/YEL	EVAP Canister Purge Regulator Valve	N/A
32	--	Not Used	--
33	BRN/BLU	T-28 Connector (near distributor)	N/A
34	GRN	Knock Sensor 1 Power	N/A
35	YEL/BRN	Leak Detection Pump	N/A
36	WHT/YEL	Intake Air Temperature Sensor Signal	N/A
37	GRN	Transmission Control Module	N/A
38	--	Not Used	--
39	BLU/RED	A/C Cutoff Relay	N/A
40	RED/BLU	Throttle Valve Control Module	N/A
41	RED/WHT	Throttle Valve Control Module Signal	N/A
42	YEL	Heated Oxygen Sensor Power	N/A
43	GRY/WHT	T-44 Connector for OBD	N/A
44	GRN/WHT	Camshaft Position Sensor	N/A
45	--	Not Used	--
46	GRY/LIL	Leak Detection Pump	N/A
47	--	Not Used	--
48	--	Not Used	--
49	GRY/BRN	First or Third Speed Coolant Fan Cutoff Relay	N/A
50	RED/YEL	Secondary Air Injection Solenoid Valve (CA only)	N/A
51	--	Not Used	--
52	--	Not Used	--
53	WHT	Throttle Valve Control Module	N/A
54	RED	Fuel Pump Relay Signal	N/A
55	--	Not Used	--
56	BRN/YEL	Ground	N/A
57	--	Not Used	--
58	WHT/YEL	Heated Oxygen Sensor 2 Ground	N/A
59	--	Not Used	--
60	--	Not Used	--
61	--	Not Used	--
62	RED/GRY	Throttle Valve Control Module	N/A
63	--	Not Used	--
64	--	Not Used	--
65	BLU/WHT	Vehicle Speed Sensor Connector	N/A

1996 Sedan 2.0L MFI ABA T68 Connector, *continued*

66	BRN/WHT	Heated Oxygen Sensor 2	N/A
67	RED	Engine Speed Sensor	N/A
68	GRN	Engine Speed Sensor	N/A

Cabrio, GTI, Jetta

1997–1999 Sedan 2.0L MFI ABA T68 Connector

ECM Pin #	Wire Color	Circuit Description (68-Pin)	Value at Hot Idle
1	BRN	Ground	N/A
2	GRY/BLU	Cylinder 4 Fuel Injector Control	N/A
3	YEL/BLK	Malfunction Indicator Light (MIL)	N/A
4	--	Not Used	--
5	--	Not Used	--
6	YEL/BLU	Fuel Pump Relay	N/A
7	BRN/BLK	Ground	N/A
8	BLK/RED	Ignition Coil Power Output Stage	N/A
9	--	Not Used	--
10	WHT/GRN	Throttle Valve Control Module	N/A
11	BLU/WHT	T-10 Connector	N/A
12	BRN	Heated Oxygen Sensor 1 Ground	N/A
13	WHT/RED	Heated Oxygen Sensor Power	N/A
14	BLU	Engine Coolant Temperature Sensor	N/A
15	--	Not Used	--
16	BRN/BLU	Mass Airflow Sensor	N/A
17	BLU	Mass Airflow Sensor	N/A
18	YEL/RED	T-10 Connector	N/A
19	--	Not Used	--
20	WHT	Heated Oxygen Sensor 1 Signal	N/A
21	YEL	T-10 Connector	N/A
22	GRN/BLK	Relay Panel	N/A
23	BLK/YEL	Power Supply Relay	N/A
24	GRY	Cylinder 1 Fuel Injector Control	N/A
25	GRY/GRN	Cylinder 2 Fuel Injector Control	N/A
26	GRY/RED	Cylinder 3 Fuel Injector Control	N/A
27	BLK	Throttle Valve Control Module Motor Power	N/A
28	--	Not Used	--
29	--	Not Used	--
30	--	Not Used	--
31	GRN/YEL	EVAP Canister Purge Regulator Valve	N/A
32	--	Not Used	--
33	BRN/BLU	Sensor Ground	N/A
34	GRN	Knock Sensor 1 Power	N/A
35	YEL/BRN	Leak Detection Pump	N/A
36	WHT/YEL	Intake Air Temperature Sensor Signal	N/A
37	GRN	T-10 Connector	N/A
38	--	Not Used	--
39	BLU/RED	T-2 Connector	N/A

1997–1999 Sedan 2.0L MFI ABA T68 Connector, *continued*

40	RED/BLU	Throttle Valve Control Module	N/A
41	RED/WHT	Throttle Valve Control Module Signal	N/A
42	YEL	Heated Oxygen Sensor 1 Power	N/A
43	GRN/WHT	T-44 Connector for OBD	N/A
44	GRN/WHT	Camshaft Position Sensor	N/A
45	--	Not Used	--
46	GRN/LIL	Leak Detection Pump	N/A
47	--	Not Used	--
48	--	Not Used	--
49	GRY/BRN	Secondary Air Injection Pump Motor	N/A
50	RED/YEL	Secondary Air Injection Solenoid Valve	N/A
51	--	Not Used	--
52	--	Not Used	--
53	WHT	Throttle Valve Control Module	N/A
54	RED	Fuel Pump Relay Signal	N/A
55	--	Not Used	--
56	BRN/YEL	Ground	N/A
57	--	Not Used	--
58	WHT/YEL	Heated Oxygen Sensor 2 Ground	N/A
59	--	Not Used	--
60	YEL/WHT	T-1 Connector	N/A
61	--	Not Used	--
62	RED/GRY	Throttle Valve Control Module	N/A
63	--	Not Used	--
64	LIL/BLK	T-1 Connector	N/A
65	WHT/BLU	Vehicle Speed Sensor Connector	N/A
66	BRN/WHT	Heated Oxygen Sensor 2	N/A
67	RED	Engine Speed Sensor	N/A
68	GRN	Engine Speed Sensor	N/A

Cabrio, GTI, Jetta

1999–2001 Sedan 2.0L MFI AEG T80 Connector

ECM Pin #	Wire Color	Circuit Description (80-Pin)	Value at Hot Idle
1	BLK/LIL	Power Ground	N/A
2	BRN/RED	Power Ground	N/A
3	BLK/LIL	Fuse 10 Connection	N/A
4	LIL/WHT	T-6 Connector, brown on ECM (Fuel Pump Relay)	N/A
5	--	Not Used	--
6	GRN/BRN	T-10 Connector, orange on ECM	N/A
7	--	Not Used	--
8	GRN	A/C Coolant Fan Control	N/A
9	--	Not Used	--
10	BLU/RED	A/C Coolant Fan Control	N/A
11	LIL/RED	Mass Air Flow Sensor	N/A
12	BRN/BLU	Mass Air Flow Sensor	N/A
13	BLU	Mass Air Flow Sensor	N/A

1999–2001 Sedan 2.0L MFI AEG T80 Connector, *continued*

14	--	Not Used	--
15	LIL/RED	EVAP Canister Purge Regulator Valve Signal	N/A
16	YEL/BRN	Leak Detection Pump Motor Signal	N/A
17	WHT/BLU	T-10 Connector, orange on ECM	N/A
18	GRN/WHT	T-10 Connector, white on ECM	N/A
19	GRY/WHT	T-10 Connector, white on ECM	N/A
20	BLU/WHT	T-10 Connector, orange on ECM	N/A
21	WHT	T-10 Connector, orange on ECM	N/A
22	RED/GRY	T-10 Connector, gray on ECM	N/A
23	--	Not Used	--
24	--	Not Used	--
25	GRY	Heated Oxygen Sensor Ground	N/A
26	BLK	Heated Oxygen Sensor Signal	N/A
27	BRN/BLK	Heated Oxygen Sensor Power	N/A
28	BRN/BLK	Heated Oxygen Sensor 2 Power	N/A
29	ORN/BRN	ABS Control Module	N/A
30	GRY/BRN	Secondary Air Injection Pump Relay Control	N/A
31	--	Not Used	--
32	--	Not Used	--
33	LIL/RED	Secondary Air Injection Solenoid Valve Signal	N/A
34	--	Not Used	--
35	--	Not Used	--
36	--	Not Used	--
37	GRN/LIL	Leak Detection Pump Signal	N/A
38	--	Not Used	--
39	--	Not Used	--
40	BRN/BLU	Mass Air Flow Sensor	N/A
41	ORN/BLK	ABS Control Module	N/A
42	--	Not Used	--
43	LIL/YEL	T-10 Connector, white on ECM	N/A
44	--	Not Used	--
45	--	Not Used	--
46	--	Not Used	--
47	BLK/RED	T-10 Connector, black on ECM (1999 only)	N/A
48	--	Not Used	--
49	--	Not Used	--
50	--	Not Used	--
51	GRY	Heated Oxygen Sensor 2 Ground	N/A
52	BLK	Heated Oxygen Sensor 2 Signal	N/A
53	BLU	Engine Coolant Temperature Sensor	N/A
54	--	Not Used	--
55	--	Not Used	--
56	YEL	Engine Speed Sensor Signal	N/A
57	--	Not Used	--
58	LIL/RED	Cylinder 3 Fuel Injector Control	N/A
59	LIL/WHT	Throttle Position Actuator Motor Power	N/A
60	GRN/YEL	Knock Sensor 2 Signal	N/A

1999–2001 Sedan 2.0L MFI AEG T80 Connector, *continued*

61	--	Not Used	--
62	LIL/RED	Camshaft Position Sensor	N/A
63	WHT	Engine Speed Sensor Signal	N/A
64	--	Not Used	--
65	LIL/BLU	Cylinder 4 Fuel Injector Control	N/A
66	LIL	Throttle Position Actuator Motor Control	N/A
67	BRN/BLU	Sensor Ground	N/A
68	GRN	Knock Sensor 1 Signal	N/A
69	WHT	Throttle Position Actuator Motor Signal	N/A
70	--	Not Used	--
71	GRN/BRN	Ignition Coil Control	N/A
72	--	Not Used	--
73	LIL	Cylinder 1 Fuel Injector Control	N/A
74	LIL/YEL	Throttle Position Sensor Signal	N/A
75	LIL/BLK	Throttle Position Sensor Ground	N/A
76	LIL/YEL	Camshaft Position Sensor Signal	N/A
77	--	Not Used	--
78	GRN/YEL	Ignition Coil Control	N/A
79	--	Not Used	--
80	LIL/GRN	Cylinder 2 Fuel Injector Control	N/A

Golf, GTI, Jetta

2000 Sedan 1.8L MFI AWD T121 Connector

ECM Pin #	Wire Color	Circuit Description (121-Pin)	Value at Hot Idle
1	BRN/RED	PCV Heating Element Control	N/A
2	BRN/RED	Ground	N/A
3	BLK	Instrument Panel wiring harness	N/A
4	--	Not Used	--
5	WHT	Heated Oxygen Sensor Power	N/A
6	--	Not Used	--
7	--	Not Used	--
8	--	Not Used	--
9	BRN/YEL	Secondary Air Injection Solenoid Valve Signal	N/A
10	--	Not Used	--
11	--	Not Used	--
12	--	Not Used	--
13	--	Not Used	--
14	--	Not Used	--
15	--	Not Used	--
16	--	Not Used	--
17	--	Not Used	--
18	--	Not Used	--
19	--	Not Used	--
20	--	Not Used	--
21	LIL	ECM Power Supply Relay	N/A
22	--	Not Used	--

2000 Sedan 1.8L MFI AWD T121 Connector, *continued*

23	--	Not Used	--
24	--	Not Used	--
25	YEL/BRN	Leak Detection Pump Motor Power	N/A
26	--	Not Used	--
27	GRN	Mass Air Flow Sensor	N/A
28	BLU/LIL	T-10 Connector, white in plenum chamber	N/A
29	BLK	Mass Air Flow Sensor	N/A
30	WHT/GRN	T-10 Connector, white in plenum chamber	N/A
31	--	Not Used	--
32	--	Not Used	--
33	GRY/BLU	Accelerator Pedal Position Sender Ground	N/A
34	BRN/WHT	Accelerator Pedal Position Sender Signal	N/A
35	WHT/BLU	Throttle Position Sensor Signal	N/A
36	GRY/RED	Throttle Position Sensor Power	N/A
37	GRN/BRN	Instrument Panel Control Module (to 4/05)	N/A
38	BLK/WHT	Cruise Control Switch Ground	N/A
39	WHT/RED	Clutch Vacuum Vent Valve Switch Signal	N/A
40	BLU/RED	T-10 Connector, white in plenum (A/C connection)	N/A
41	GRN/GRY	T-10 Connector, orange in plenum (GRN for manual)	N/A
42	--	Not Used	N/A
43	GRY/WHT	T-10 Connector, white in plenum chamber	N/A
44	--	Not Used	--
45	--	Not Used	--
46	--	Not Used	--
47	LIL/YEL	T-10 Connector, white in plenum chamber	N/A
48	WHT/RED	T-10 Connector, white in plenum chamber	N/A
49	WHT/LIL	Power Steering Pressure Switch Control	N/A
50	WHT/GRY	Power Steering Pressure Switch Signal	N/A
51	GRN	Heated Oxygen Sensor Signal	N/A
52	--	Not Used	--
53	RED/LIL	Mass Air Flow Sensor	N/A
54	BLU/WHT	Vehicle Speed Signal (instrument panel)	N/A
55	WHT/YEL	Brake Vacuum Vent Valve Switch Signal	N/A
56	BLK/RED	T-10 Connector, black in plenum chamber	N/A
57	RED/YEL	Set Switch for Cruise Control Signal	N/A
58	--	Not Used	--
59	--	Not Used	--
60	--	Not Used	--
61	--	Not Used	--
62	RED/GRN	Fuse Box connection	N/A
63	GRY/YEL	Oxygen Sensor Power	N/A
64	LIL/RED	EVAP Canister Purge Regulator Valve Signal	N/A
65	LIL/WHT	Fuel Pump Relay	N/A
66	GRY/BRN	Secondary Air Injection Pump Relay Control	N/A
67	GRN/LIL	T-10 Connector, white in plenum chamber	N/A
68	GRN/WHT	Oxygen Sensor Ground	N/A
69	GRN/RED	Oxygen Sensor Signal	N/A

2000 Sedan 1.8L MFI AWD T121 Connector, *continued*

70	BLK	Heated Oxygen Sensor Signal	N/A
71	--	Not Used	--
72	GRN/WHT	Accelerator Pedal Position Sender Power	N/A
73	YEL/GRN	Throttle Position Sensor Ground	N/A
74	--	Not Used	--
75	BLU/GRY	Cruise Control Switch Signal	N/A
76	WHT	Cruise Control Switch Signal	N/A
77	--	Not Used	--
78	--	Not Used	--
79	--	Not Used	--
80	GRN/LIL	Leak Detection Pump Signal	N/A
81	GRN/RED	T-10 Connector, white in plenum chamber	N/A
82	WHT	Engine Speed Sensor Signal	N/A
83	LIL/WHT	Throttle Drive Angle Sensor Ground	N/A
84	BLU/GRY	Throttle Drive Angle Sensor Signal	N/A
85	BLU/GRN	Intake Air Temperature Sensor Control	N/A
86	LIL/YEL	Camshaft Position Sensor	N/A
87	--	Not Used	--
88	LIL/BLU	Cylinder 2 Fuel Injector Control	N/A
89	LIL/GRN	Cylinder 4 Fuel Injector Control	N/A
90	BRN	Engine Speed Sensor Signal	N/A
91	WHT/GRY	Throttle Drive Angle Sensor Power	N/A
92	BLU/WHT	Throttle Drive Angle Sensor Signal	N/A
93	GRY/YEL	Engine Coolant Temperature Sensor Control	N/A
94	BLK/LIL	Ignition Coil 4 with Power Output Stage	N/A
95	BLK/YEL	Ignition Coil 2 with Power Output Stage	N/A
96	LIL	Cylinder 1 Fuel Injector Control	N/A
97	LIL/RED	Cylinder 3 Fuel Injector Control	N/A
98	LIL/GRN	Camshaft Position Sensor	N/A
99	BLU	Knock Sensor 1 and 2 Control	N/A
100	--	Not Used	--
101	LIL/GRN	Charge Air Pressure Sensor	N/A
102	BLK/BLU	Ignition Coil 1 with Power Output Stage	N/A
103	BLK/BRN	Ignition Coil 3 with Power Output Stage	N/A
104	GRN/BRN	Wastegate Bypass Regulator Valve Signal	N/A
105	GRN/BRN	Turbocharger Recirculating Valve Signal	N/A
106	GRY	Knock Sensor 1 Signal	N/A
107	GRY	Knock Sensor 2	N/A
108	BRN/BLU	Sensor Ground	N/A
109	--	Not Used	--
110	--	Not Used	--
111	--	Not Used	--
112	--	Not Used	--
113	--	Not Used	--
114	--	Not Used	--
115	--	Not Used	--
116	--	Not Used	--

2000 Sedan 1.8L MFI AWD T121 Connector, *continued*

117	WHT	Throttle Drive Motor Power	N/A
118	LIL/BLK	Throttle Drive Motor Control	N/A
119	--	Not Used	--
120	--	Not Used	--
121	BLK/LIL	Ground	N/A

GTI, Jetta

2001 Jetta and GTI; 2002 Jetta 1.8L MFI AWW T121 Connector

ECM Pin #	Wire Color	Circuit Description (121-Pin)	Value at Hot Idle
1	BRN/RED	PCV Heating Element Control	N/A
2	BRN/RED	Ground	N/A
3	BLK	Instrument Panel wiring harness	N/A
4	--	Not Used	--
5	WHT	Heated Oxygen Sensor Power	N/A
6	--	Not Used	--
7	--	Not Used	--
8	--	Not Used	--
9	BRN/YEL	Secondary Air Injection Solenoid Valve Signal	N/A
10	--	Not Used	--
11	--	Not Used	--
12	--	Not Used	--
13	--	Not Used	--
14	--	Not Used	--
15	--	Not Used	--
16	--	Not Used	--
17	--	Not Used	--
18	--	Not Used	--
19	--	Not Used	--
20	--	Not Used	--
21	LIL	ECM Power Supply Relay	N/A
22	BLU/GRN	Brake Booster Control Module	N/A
23	--	Not Used	--
24	--	Not Used	--
25	YEL/BRN	Leak Detection Pump Motor Power	N/A
26	--	Not Used	--
27	GRN	Mass Air Flow Sensor	N/A
28	BLU/LIL	T-10 Connector, white in plenum chamber	N/A
29	BLK	Mass Air Flow Sensor	N/A
30	WHT/GRN	T-10 Connector, white in plenum chamber	N/A
31	--	Not Used	--
32	--	Not Used	--
33	GRY/BLU	Accelerator Pedal Position Sender Ground	N/A
34	BRN/WHT	Accelerator Pedal Position Sender Signal	N/A
35	WHT/BLU	Throttle Position Sensor Signal	N/A
36	GRY/RED	Throttle Position Sensor Power	N/A
37	GRN/BRN	Instrument Panel Control Module (to 4/05)	N/A

2001 Jetta and GTI; 2002 Jetta 1.8L MFI AWW T121 Connector, *continued*

38	BLK/WHT	Cruise Control Switch Ground	N/A
39	WHT/RED	Clutch Vacuum Vent Valve Switch Signal (man only)	N/A
40	BLU/RED	T-10 Connector, white in plenum (A/C connection)	N/A
41	GRN/GRY	T-10 Connector, orange in plenum (GRN for auto)	N/A
42	--	Not Used	--
43	GRY/WHT	T-10 Connector, white in plenum chamber	N/A
44	--	Not Used	--
45	--	Not Used	--
46	--	Not Used	--
47	LIL/YEL	T-10 Connector, white in plenum chamber	N/A
48	WHT/RED	T-10 Connector, white in plenum chamber	N/A
49	WHT/LIL	Power Steering Pressure Switch Control	N/A
50	WHT/GRY	Power Steering Pressure Switch Signal	N/A
51	GRN/BLK	Heated Oxygen Sensor Signal	N/A
52	GRY/RED	Heated Oxygen Sensor Ground	N/A
53	RED/LIL	Mass Air Flow Sensor	N/A
54	BLU/WHT	Vehicle Speed Signal (instrument panel)	N/A
55	WHT/YEL	Brake Vacuum Vent Valve Switch Signal	N/A
56	BLK/RED	T-10 Connector, black in plenum chamber	N/A
57	RED/YEL	Set Switch for Cruise Control Signal	N/A
58	--	Not Used	--
59	--	Not Used	--
60	--	Not Used	--
61	--	Not Used	--
62	RED/GRN	Fuse Box connection	N/A
63	GRY/YEL	Oxygen Sensor Power	N/A
64	LIL/RED	EVAP Canister Purge Regulator Valve Signal	N/A
65	LIL/WHT	Fuel Pump Relay	N/A
66	GRY/BRN	Secondary Air Injection Pump Relay Control	N/A
67	GRN/LIL	T-10 Connector, white in plenum chamber	N/A
68	GRN/WHT	Oxygen Sensor Ground	N/A
69	GRN/RED	Oxygen Sensor Signal	N/A
70	GRN	Heated Oxygen Sensor Signal	N/A
71	BLK	Heated Oxygen Sensor Signal	N/A
72	GRN/WHT	Accelerator Pedal Position Sender Power	N/A
73	YEL/GRN	Throttle Position Sensor Ground	N/A
74	--	Not Used	--
75	BLU/GRY	Cruise Control Switch Signal	N/A
76	WHT	Cruise Control Switch Signal	N/A
77	--	Not Used	--
78	--	Not Used	--
79	--	Not Used	--
80	GRN/LIL	Leak Detection Pump Signal	N/A
81	GRN/RED	T-10 Connector, white in plenum chamber	N/A
82	WHT	Engine Speed Sensor Signal	N/A
83	LIL/WHT	Throttle Drive Angle Sensor Ground	N/A
84	BLU/GRY	Throttle Drive Angle Sensor Signal	N/A

2001 Jetta and GTI; 2002 Jetta 1.8L MFI AWW T121 Connector, *continued*

85	BLU/GRN	Intake Air Temperature Sensor Control	N/A
86	LIL/YEL	Camshaft Position Sensor	N/A
87	--	Not Used	--
88	LIL/BLU	Cylinder 2 Fuel Injector Control	N/A
89	LIL/GRN	Cylinder 4 Fuel Injector Control	N/A
90	BRN	Engine Speed Sensor Signal	N/A
91	WHT/GRY	Throttle Drive Angle Sensor Power	N/A
92	BLU/WHT	Throttle Drive Angle Sensor Signal	N/A
93	GRY/YEL	Engine Coolant Temperature Sensor Control	N/A
94	BLK/LIL	Ignition Coil 4 with Power Output Stage	N/A
95	BLK/YEL	Ignition Coil 2 with Power Output Stage	N/A
96	LIL	Cylinder 1 Fuel Injector Control	N/A
97	LIL/RED	Cylinder 3 Fuel Injector Control	N/A
98	LIL/GRN	Camshaft Position Sensor	N/A
99	BLU	Knock Sensor 1 and 2 Control	N/A
100	--	Not Used	--
101	LIL/GRN	Charge Air Pressure Sensor	N/A
102	LIL/BLK	Ignition Coil 1 with Power Output Stage	N/A
103	BLK/BRN	Ignition Coil 3 with Power Output Stage	N/A
104	GRN/BRN	Wastegate Bypass Regulator Valve Signal	N/A
105	GRY/GRN	Turbocharger Recirculating Valve Signal	N/A
106	GRY	Knock Sensor 1 Signal	N/A
107	GRY	Knock Sensor 2	N/A
108	BRN/BLU	Sensor Ground	N/A
109	--	Not Used	--
110	--	Not Used	--
111	--	Not Used	--
112	--	Not Used	--
113	--	Not Used	--
114	--	Not Used	--
115	GRN/WHT	Camshaft Adjustment, Valve One	N/A
116	--	Not Used	--
117	WHT	Throttle Drive Motor Power	N/A
118	LIL/BLK	Throttle Drive Motor Control	N/A
119	--	Not Used	--
120	--	Not Used	--
121	BLK/LIL	Ground	N/A

Golf, Jetta

1999 Model 1.9L DFI ALH (manual) T80 Connector

ECM Pin #	Wire Color	Circuit Description (80-Pin)	Value at Hot Idle
1	BRN/RED	Power Ground	N/A
2	RED/LIL	Power Ground	N/A
3	LIL/GRY	Change-Over Valve for Intake Manifold Flap Control	N/A
4	BLU	Mass Air Flow Sensor	N/A
5	--	Not Used	--

1999 Model 1.9L DFI ALH (manual) T80 Connector, *continued*

6	GRN/BRN	Instrument Panel Control Module	N/A
7	--	Not Used	--
8	GRY/WHT	Kick Down Switch Power	N/A
9	WHT/YEL	Brake Vacuum Vent Valve Switch (for DFI)	N/A
10	RED	Cruise Control Set Switch Signal	N/A
11	YEL/GRN	Throttle Position Sensor Ground	N/A
12	GRY/BLU	Closed Throttle Position Switch Signal	N/A
13	GRY/GRN	Manifold Absolute Pressure Sensor	N/A
14	--	Not Used	--
15	BRN/YEL	Wastegate Bypass Regulator Valve Control	N/A
16	GRN	T-10 Connector, Orange for A/C Connection	N/A
17	--	Not Used	N/A
18	GRN/WHT	T-10 Connector, Orange in plenum chamber	N/A
19	BLK/WHT	Cruise Control Switch Ground	N/A
20	RED/BLK	Ground	N/A
21	BLU/GRY	Cruise Control Switch Signal	N/A
22	BRN/RED	T-10 Connector, Green in plenum chamber	N/A
23	GRY/RED	Throttle Position Sensor Power	N/A
24	WHT/BLU	Throttle Position Sensor Signal	N/A
25	BRN/BLU	Sensor Ground	N/A
26	--	Not Used	--
27	BRN/RED	Power Ground	N/A
28	BLK/WHT	Cruise Control Switch Ground	N/A
29	RED/BLU	EGR Vacuum Regulator Solenoid Valve Control	N/A
30	--	Not Used	--
31	BRN/WHT	T-10 Connector, Blue on ECM	N/A
32	--	Not Used	--
33	BLU/YEL	Power Supply Relay	N/A
34	--	Not Used	--
35	WHT	Cruise Control Switch Signal	N/A
36	--	Not Used	--
37	GRN	Glow Plug Relay	N/A
38	--	Not Used	--
39	LIL/RED	Manifold Absolute Pressure Sensor	N/A
40	YEL/BLK	Manifold Absolute Pressure Sensor	N/A
41	BLU/GRN	T-10 Connector, Green in plenum chamber	N/A
42	LIL/WHT	Glow Plug Relay	N/A
43	BLK/BLU	T-10 Connector, Blue on ECM	N/A
44	GRN	Transmission Control Module	N/A
45	GRY/WHT	T-10 Connector, Orange in plenum chamber	N/A
46	--	Not Used	N/A
47	BLK/LIL	T-6 Connector, brown on ECM	N/A
48	BLU/RED	T-10 Connector, Orange for A/C Connection	N/A
49	--	Not Used	--
50	RED/GRN	Mass Air Flow Sensor	N/A
51	BLU/WHT	Instrument Panel Control Module	N/A
52	BRN/BLU	Mass Air Flow Sensor	N/A

1999 Model 1.9L DFI ALH (manual) T80 Connector, *continued*

53	YEL/BLU	Fuel Temperature Sensor Control	N/A
54	BRU/BRN	Engine Coolant Temperature Sensor Ground	N/A
55	BLU	Needle Lift Sensor Signal	N/A
56	LIL/BLK	Modulating Piston Displacement Sensor Signal	N/A
57	GRY/GRN	Modulating Piston Displacement Sensor Ground	N/A
58	--	Not Used	N/A
59	BRN/RED	Connection to Diesel Direct Fuel Injection System	N/A
60	--	Not Used	--
61	--	Not Used	--
62	GRY	Needle Lift Sensor Ground	N/A
63	--	Not Used	--
64	BLK/GRN	Modulating Piston Displacement Sensor Signal	N/A
65	--	Not Used	--
66	BRN/RED	Connection to Diesel Direct Fuel Injection System	N/A
67	BLK	Engine Speed Sensor Signal	N/A
68	ORN/BLK	Double connector in engine compartment	N/A
69	YEL	Engine Speed Sensor Signal	N/A
70	BRN/GRN	Engine Coolant Temperature Sensor Control	N/A
71	BRN/YEL	Shielding Ground	N/A
72	--	Not Used	--
73	--	Not Used	--
74	--	Not Used	--
75	ORN/BRN	Double connector in engine compartment	N/A
76	BRN/BLU	Fuel Temperature Sensor Ground	N/A
77	BLK/WHT	Fuel Cut-Off Valve Control	N/A
78	--	Not Used	--
79	BRN/BLK	Cold Start Injector Control	N/A
80	BRN/RED	Connection to Diesel Direct Fuel Injection System	N/A

Golf, Jetta

1999 Model 1.9L DFI ALH (manual) T80 Connector

ECM Pin #	Wire Color	Circuit Description (80-Pin)	Value at Hot Idle
1	BRN/RED	Power Ground	N/A
2	RED/LIL	Power Ground	N/A
3	LIL/GRY	Change-Over Valve for Intake Manifold Flap Control	N/A
4	BLU	Mass Air Flow Sensor	N/A
5	--	Not Used	--
6	GRN/BRN	Instrument Panel Control Module	N/A
7	--	Not Used	--
8	GRY/WHT	Kick Down Switch Power	N/A
9	WHT/YEL	Brake Vacuum Vent Valve Switch (for DFI)	N/A
10	RED	Cruise Control Set Switch Signal	N/A
11	YEL/GRN	Throttle Position Sensor Ground	N/A
12	GRY/BLU	Closed Throttle Position Switch Signal	N/A
13	GRY/GRN	Manifold Absolute Pressure Sensor	N/A
14	--	Not Used	--

1999 Model 1.9L DFI ALH (manual) T80 Connector, *continued*

15	BRN/YEL	Wastegate Bypass Regulator Valve Control	N/A
16	GRN	T-10 Connector, Orange for A/C Connection	N/A
17	BLK/BLU	Relay for Pre-Heating Coolant (low heat)	N/A
18	GRN/WHT	T-10 Connector, Orange in plenum chamber	N/A
19	BLK/WHT	Cruise Control Switch Ground	N/A
20	RED/BLK	Ground	N/A
21	BLU/GRY	Cruise Control Switch Signal	N/A
22	BRN/RED	T-10 Connector, Green in plenum chamber	N/A
23	GRY/RED	Throttle Position Sensor Power	N/A
24	WHT/BLU	Throttle Position Sensor Signal	N/A
25	BRN/BLU	Sensor Ground	N/A
26	--	Not Used	--
27	BRN/RED	Power Ground	N/A
28	BLK/WHT	Cruise Control Switch Ground	N/A
29	RED/BLU	EGR Vacuum Regulator Solenoid Valve Control	N/A
30	--	Not Used	--
31	BRN/WHT	T-10 Connector, Blue on ECM	N/A
32	--	Not Used	--
33	BLU/YEL	Power Supply Relay	N/A
34	BLK/BRN	Relay for Pre-Heating Coolant (high heat)	N/A
35	WHT	Cruise Control Switch Signal	N/A
36	--	Not Used	--
37	GRN	Glow Plug Relay	N/A
38	--	Not Used	--
39	LIL/RED	Manifold Absolute Pressure Sensor	N/A
40	YEL/BLK	Manifold Absolute Pressure Sensor	N/A
41	BLU/GRN	T-10 Connector, Green in plenum chamber	N/A
42	LIL/WHT	Glow Plug Relay	N/A
43	BLK/BLU	T-10 Connector, Blue on ECM	N/A
44	--	Not Used	--
45	GRY/WHT	T-10 Connector, Orange in plenum chamber	N/A
46	WHT/RED	Clutch Vacuum Vent Valve Switch Signal	N/A
47	BLK/LIL	T-6 Connector, brown on ECM	N/A
48	BLU/RED	T-10 Connector, Orange for A/C Connection	N/A
49	--	Not Used	--
50	RED/GRN	Mass Air Flow Sensor	N/A
51	BLU/WHT	Instrument Panel Control Module	N/A
52	BRN/BLU	Mass Air Flow Sensor	N/A
53	YEL/BLU	Fuel Temperature Sensor Control	N/A
54	BRU/BRN	Engine Coolant Temperature Sensor Ground	N/A
55	BLU	Needle Lift Sensor Signal	N/A
56	LIL/BLK	Modulating Piston Displacement Sensor Signal	N/A
57	GRY/GRN	Modulating Piston Displacement Sensor Ground	N/A
58	--	Not Used	--
59	BRN/RED	Connection to Diesel Direct Fuel Injection System	N/A
60	--	Not Used	--
61	--	Not Used	--

1999 Model 1.9L DFI ALH (manual) T80 Connector, *continued*

62	GRY	Needle Lift Sensor Ground	N/A
63	--	Not Used	--
64	BLK/GRN	Modulating Piston Displacement Sensor Signal	N/A
65	--	Not Used	--
66	BRN/RED	Connection to Diesel Direct Fuel Injection System	N/A
67	BLK	Engine Speed Sensor Signal	N/A
68	ORN/BLK	Double connector in engine compartment	N/A
69	YEL	Engine Speed Sensor Signal	N/A
70	BRN/GRN	Engine Coolant Temperature Sensor Control	N/A
71	BRN/YEL	Shielding Ground	N/A
72	--	Not Used	--
73	--	Not Used	--
74	--	Not Used	--
75	ORN/BRN	Double connector in engine compartment	N/A
76	BRN/BLU	Fuel Temperature Sensor Ground	N/A
77	BLK/WHT	Fuel Cut-Off Valve Control	N/A
78	--	Not Used	--
79	BRN/BLK	Cold Start Injector Control	N/A
80	BRN/RED	Connection to Diesel Direct Fuel Injection System	N/A

Golf, Jetta

2000–2002 Sedan, 2002 Wagon 1.9L DFI ALH (automatic) T121 Connector

ECM Pin #	Wire Color	Circuit Description (121-Pin)	Value at Hot Idle
1	RED/LIL	Ground (wire connection in cngine compartment)	N/A
2	RED/LIL	Ground (wire connection in engine compartment)	N/A
3	--	Not Used	N/A
4	BRN/RED	PCV Heating Element Ground	N/A
5	BRN/RED	Ground	N/A
6	ORN/BRN	T-10 Connector, white in plenum chamber	N/A
7	ORN/BLK	T-10 Connector, white in plenum chamber	N/A
8	--	Not Used	--
9	GRY	Transmission Control Module	N/A
10	--	Not Used	--
11	BRN/WHT	T-10 Connector, white in plenum chamber	N/A
12	YEL/GRN	Throttle Position Sensor Ground	N/A
13	--	Not Used	--
14	BLK/WHT	Cruise Control Switch Ground	N/A
15	--	Not Used	--
16	GRY/WHT	T-10 Connector, orange in plenum chamber	N/A
17	--	Not Used	--
18	BLU/YEL	Power Supply Relay	N/A
19	YEL/RED	Transmission Control Module	N/A
20	BLU/WHT	Instrument Panel Control Module	N/A
21	--	Not Used	--
22	--	Not Used	--
23	--	Not Used	--

2000–2002 Sedan, 2002 Wagon 1.9L DFI ALH (automatic) T121 Connector, *continued*

24	WHT/BLU	T-10 Connector, white in plenum chamber	N/A
25	--	Not Used	--
26	--	Not Used	--
27	GRN/BRN	Instrument Panel Control Module	N/A
28	GRN/WHT	T-10 Connector, white in plenum chamber	N/A
29	GRN	Transmission Control Module	N/A
30	RED/LIL	Mass Air Flow Sensor	N/A
31	LIL/RED	Manifold Absolute Pressure Sensor	N/A
32	RED/BLK	Power Ground	N/A
33	GRN	Glow Plug Relay	N/A
34	BLU/RED	T-10 Connector, orange in plenum chamber	N/A
35	BLU/RED	Transmission Control Module	N/A
36	--	Not Used	--
37	BLK/LIL	Fuse Panel	N/A
38	BRN/RED	T-10 Connector, white in plenum chamber	N/A
39	--	Not Used	--
40	BLU/GRN	T-10 Connector, white in plenum chamber	N/A
41	--	Not Used	--
42	LIL/WHT	Glow Plug Relay	N/A
43	--	Not Used	--
44	RED	Cruise Control Set Switch Signal	N/A
45	BLU/GRY	Cruise Control Switch Signal	N/A
46	WHT	Cruise Control Switch Signal	N/A
47	GRN/LIL	T-10 Connector, white in plenum chamber	N/A
48	--	Not Used	--
49	GRN	Mass Air Flow Sensor	N/A
50	GRY/RED	Throttle Position Sensor Power	N/A
51	GRN/RED	Closed Throttle Position Switch Ground	N/A
52	BRN/BLU	Intake Air Temperature Sensor Signal	N/A
53	--	Not Used	--
54	--	Not Used	--
55	--	Not Used	--
56	--	Not Used	--
57	--	Not Used	--
58	--	Not Used	--
59	--	Not Used	--
60	--	Not Used	--
61	RED/BLU	EGR Vacuum Regulator Solenoid Valve Control	N/A
62	RED/BRN	Wastegate Bypass Regulator Valve Control	N/A
63	WHT/BLU	Kick Down Switch Power	N/A
64	--	Not Used	--
65	WHT/YEL	Brake Vacuum Vent Valve Switch (for cruise control)	N/A
66	--	Not Used	--
67	--	Not Used	--
68	BLK	Mass Air Flow Sensor	N/A
69	GRY/BLU	Throttle Position Sensor Signal	N/A
70	GRY/WHT	Throttle Position Sensor Ground	N/A

2000–2002 Sedan, 2002 Wagon 1.9L DFI ALH (automatic) T121 Connector, *continued*

71	YEL/BLK	Manifold Absolute Pressure Sensor	N/A
72	--	Not Used	--
73	GRY/GRN	Intake Air Temperature Sensor Control	N/A
74	--	Not Used	--
75	--	Not Used	--
76	--	Not Used	--
77	--	Not Used	--
78	--	Not Used	--
79	--	Not Used	--
80	--	Not Used	--
81	LIL/GRY	Change-Over Valve for Intake Manifold Flap Control	N/A
82	--	Not Used	--
83	--	Not Used	--
84	--	Not Used	--
85	--	Not Used	--
86	BRN/YEL	Ground	N/A
87	--	Not Used	--
88	BLK/GRN	Fuse Panel	N/A
89	--	Not Used	--
90	--	Not Used	--
91	--	Not Used	--
92	--	Not Used	--
93	--	Not Used	--
94	--	Not Used	--
95	--	Not Used	--
96	--	Not Used	--
97	--	Not Used	--
98	--	Not Used	--
99	WHT/GRN	Modulating Piston Displacement Sensor Control	N/A
100	--	Not Used	--
101	BLU	Needle Lift Sensor Signal	N/A
102	BRN	Engine Speed Sensor Signal	N/A
103	BRN/BLU	Fuel Temperature Sensor Ground	N/A
104	BRN/GRN	Engine Coolant Temperature Sensor Signal	N/A
105	--	Not Used	--
106	GRY/GRN	Modulating Piston Displacement Sensor Signal	N/A
107	--	Not Used	--
108	LIL/BLK	Modulating Piston Displacement Sensor Control	N/A
109	GRY	Needle Lift Sensor Ground	N/A
110	WHT	Engine Speed Sensor Signal	N/A
111	YEL/BLU	Fuel Temperature Sensor Power	N/A
112	BLU/BRN	Engine Coolant Temperature Sensor Ground	N/A
113	--	Not Used	--
114	BRN/BLK	Cold Start Injector Control	N/A
115	--	Not Used	--
116	BRN/RED	Quantity Adjuster Control	N/A
117	--	Not Used	--

2000–2002 Sedan, 2002 Wagon 1.9L DFI ALH (automatic) T121 Connector, *continued*

118	--	Not Used	--
119	--	Not Used	--
120	BLK/WHT	Fuel Cut-off Valve Control	N/A
121	BRN/RED	Ground	N/A

Golf, Jetta

2000–2002 Sedan, 2002 Wagon 1.9L DFI ALH (manual) T121 Connector

ECM Pin #	Wire Color	Circuit Description (121-Pin)	Value at Hot Idle
1	RED/LIL	Ground (wire connection in engine compartment)	N/A
2	RED/LIL	Ground (wire connection in engine compartment)	N/A
3	--	Not Used	--
4	BRN/RED	PCV Heating Element Ground	N/A
5	BRN/RED	Ground	N/A
6	ORN/BRN	T-10 Connector, white in plenum chamber	N/A
7	ORN/BLK	T-10 Connector, white in plenum chamber	N/A
8	--	Not Used	--
9	--	Not Used	--
10	--	Not Used	--
11	BRN/WHT	T-10 Connector, white in plenum chamber	N/A
12	YEL/GRN	Throttle Position Sensor Ground	N/A
13	--	Not Used	--
14	BLK/WHT	Cruise Control Switch Ground	N/A
15	--	Not Used	--
16	GRY/WHT	T-10 Connector, orange in plenum chamber	N/A
17	--	Not Used	--
18	BLU/YEL	Power Supply Relay	N/A
19	--	Not Used	--
20	BLU/WHT	Instrument Panel Control Module	N/A
21	BLK/BLU	Preheating Coolant (low heat)	N/A
22	BLK/BRN	Preheating Coolant (high heat)	N/A
23	--	Not Used	--
24	WHT/BLU	T-10 Connector, white in plenum chamber	N/A
25	--	Not Used	--
26	--	Not Used	--
27	GRN/BRN	Instrument Panel Control Module	N/A
28	GRN/WHT	T-10 Connector, white in plenum chamber	N/A
29	GRN	Transmission Control Module	N/A
30	RED/GRN	Mass Air Flow Sensor	N/A
31	LIL/RED	Manifold Absolute Pressure Sensor	N/A
32	RED/BLK	Power Ground	N/A
33	GRN	Glow Plug Relay	N/A
34	BLU/RED	T-10 Connector, orange in plenum chamber	N/A
35	--	Not Used	--
36	--	Not Used	--
37	BLK/LIL	Fuse Panel	N/A
38	BRN/RED	T-10 Connector, white in plenum chamber	N/A

2000–2002 Sedan, 2002 Wagon 1.9L DFI ALH (manual) T121 Connector, *continued*

39	--	Not Used	--
40	BLU/GRN	T-10 Connector, white in plenum chamber	N/A
41	--	Not Used	--
42	LIL/WHT	Glow Plug Relay	N/A
43	--	Not Used	--
44	RED	Cruise Control Set Switch Signal	N/A
45	BLU/GRY	Cruise Control Switch Signal	N/A
46	WHT	Cruise Control Switch Signal	N/A
47	GRN/LIL	T-10 Connector, white in plenum chamber	N/A
48	--	Not Used	--
49	GRN	Mass Air Flow Sensor	N/A
50	GRY/RED	Throttle Position Sensor Power	N/A
51	GRN/RED	Closed Throttle Position Switch Ground	N/A
52	BRN/BLU	Intake Air Temperature Sensor Signal	N/A
53	--	Not Used	--
54	--	Not Used	--
55	--	Not Used	--
56	--	Not Used	--
57	--	Not Used	--
58	--	Not Used	--
59	--	Not Used	--
60	--	Not Used	--
61	RED/BLU	EGR Vacuum Regulator Solenoid Valve Control	N/A
62	RED/BRN	Wastegate Bypass Regulator Valve Control	N/A
63	WHT/BLU	Kick Down Switch Power	N/A
64	--	Not Used	--
65	WHT/YEL	Brake Vacuum Vent Valve Switch (for cruise control)	N/A
66	WHT/RED	Clutch Vacuum Vent Valve Switch Signal	N/A
67	--	Not Used	--
68	BLK	Mass Air Flow Sensor	N/A
69	GRY/BLU	Throttle Position Sensor Signal	N/A
70	GRY/WHT	Throttle Position Sensor Ground	N/A
71	YEL/BLK	Manifold Absolute Pressure Sensor	N/A
72	--	Not Used	--
73	GRY/GRN	Intake Air Temperature Sensor Control	N/A
74	--	Not Used	--
75	--	Not Used	--
76	--	Not Used	--
77	--	Not Used	--
78	--	Not Used	--
79	--	Not Used	--
80	--	Not Used	--
81	LIL/GRY	Change-Over Valve for Intake Manifold Flap Control	N/A
82	--	Not Used	--
83	--	Not Used	--
84	--	Not Used	--
85	--	Not Used	--

2000–2002 Sedan, 2002 Wagon 1.9L DFI ALH (manual) T121 Connector, *continued*

86	BRN/YEL	Ground	N/A
87	--	Not Used	--
88	BLK/GRN	Fuse Panel	N/A
89	--	Not Used	--
90	--	Not Used	--
91	--	Not Used	--
92	--	Not Used	--
93	--	Not Used	--
94	--	Not Used	--
95	--	Not Used	--
96	--	Not Used	--
97	--	Not Used	--
98	--	Not Used	--
99	WHT/GRN	Modulating Piston Displacement Sensor Control	N/A
100	--	Not Used	--
101	BLU	Needle Lift Sensor Signal	N/A
102	BRN	Engine Speed Sensor Signal	N/A
103	BRN/BLU	Fuel Temperature Sensor Ground	N/A
104	BRN/GRN	Engine Coolant Temperature Sensor Signal	N/A
105	--	Not Used	--
106	GRY/GRN	Modulating Piston Displacement Sensor Signal	N/A
107	--	Not Used	--
108	LIL/BLK	Modulating Piston Displacement Sensor Control	N/A
109	GRY	Needle Lift Sensor Ground	N/A
110	WHT	Engine Speed Sensor Signal	N/A
111	YEL/BLU	Fuel Temperature Sensor Power	N/A
112	BLU/BRN	Engine Coolant Temperature Sensor Ground	N/A
113	--	Not Used	--
114	BRN/BLK	Cold Start Injector Control	N/A
115	--	Not Used	N/A
116	BRN/RED	Quantity Adjuster Control	N/A
117	--	Not Used	--
118	--	Not Used	--
119	--	Not Used	--
120	BLK/WHT	Fuel Cut-off Valve Control	N/A
121	BRN/RED	Ground	N/A

Golf, Jetta

2003–2004 Sedan, 2002 Wagon 1.9L DFI ALH (automatic) T121 Connector

ECM Pin #	Wire Color	Circuit Description (121-Pin)	Value at Hot Idle
1	RED/LIL	Ground (wire connection in engine compartment)	N/A
2	RED/LIL	Ground (wire connection in engine compartment)	N/A
3	--	Not Used	--
4	BRN/RED	PCV Heating Element Ground	N/A
5	BRN/RED	Ground	N/A
6	ORN/BRN	T-10 Connector, white in plenum chamber	N/A

2003–2004 Sedan, 2002 Wagon 1.9L DFI ALH (automatic) T121 Connector, *continued*

7	ORN/BLK	T-10 Connector, white in plenum chamber	N/A
8	--	Not Used	--
9	GRY	Transmission Control Module	N/A
10	--	Not Used	--
11	BRN/WHT	T-10 Connector, white in plenum chamber	N/A
12	YEL/GRN	Throttle Position Sensor Ground	N/A
13	--	Not Used	--
14	BLK/WHT	Cruise Control Switch Ground	N/A
15	--	Not Used	--
16	GRY/WHT	T-10 Connector, orange in plenum chamber	N/A
17	--	Not Used	--
18	BLU/YEL	Power Supply Relay	N/A
19	YEL/RED	Transmission Control Module	N/A
20	BLU/WHT	Instrument Panel Control Module	N/A
21	--	Not Used	--
22	--	Not Used	--
23	--	Not Used	--
24	WHT/BLU	T-10 Connector, white in plenum chamber	N/A
25	--	Not Used	--
26	--	Not Used	--
27	GRN/BRN	Instrument Panel Control Module	N/A
28	GRN/WHT	T-10 Connector, white in plenum chamber	N/A
29	GRN	Transmission Control Module	N/A
30	RED/LIL	Mass Air Flow Sensor	N/A
31	LIL/RED	Manifold Absolute Pressure Sensor	N/A
32	RED/BLK	Power Ground	N/A
33	GRN	Glow Plug Relay	N/A
34	BLU/RED	T-10 Connector, orange in plenum chamber	N/A
35	GRN/LIL	Transmission Control Module	N/A
36	--	Not Used	--
37	BLK/LIL	Fuse Panel	N/A
38	BRN/RED	T-10 Connector, white in plenum chamber	N/A
39	--	Not Used	--
40	BLU/GRN	T-10 Connector, white in plenum chamber	N/A
41	--	Not Used	N/A
42	LIL/WHT	Glow Plug Relay	N/A
43	--	Not Used	--
44	RED	Cruise Control Set Switch Signal	N/A
45	BLU/GRY	Cruise Control Switch Signal	N/A
46	WHT	Cruise Control Switch Signal	N/A
47	GRN/LIL	T-10 Connector, white in plenum chamber	N/A
48	--	Not Used	--
49	GRN	Mass Air Flow Sensor	N/A
50	GRY/RED	Throttle Position Sensor Power	N/A
51	GRN/RED	Closed Throttle Position Switch Ground	N/A
52	BRN/BLU	Intake Air Temperature Sensor Signal	N/A
53	--	Not Used	--

2003–2004 Sedan, 2002 Wagon 1.9L DFI ALH (automatic) T121 Connector, *continued*

54	--	Not Used	--
55	--	Not Used	--
56	--	Not Used	--
57	--	Not Used	--
58	--	Not Used	--
59	--	Not Used	--
60	--	Not Used	--
61	RED/BLU	EGR Vacuum Regulator Solenoid Valve Control	N/A
62	RED/BRN	Wastegate Bypass Regulator Valve Control	N/A
63	WHT/BLU	Kick Down Switch Power	N/A
64	--	Not Used	--
65	WHT/YEL	Brake Vacuum Vent Valve Switch (for cruise control)	N/A
66	--	Not Used	--
67	--	Not Used	--
68	BLK	Mass Air Flow Sensor	N/A
69	GRY/BLU	Throttle Position Sensor Signal	N/A
70	GRY/WHT	Throttle Position Sensor Ground	N/A
71	YEL/BLK	Manifold Absolute Pressure Sensor	N/A
72	--	Not Used	--
73	GRY/GRN	Intake Air Temperature Sensor Control	N/A
74	--	Not Used	--
75	--	Not Used	--
76	--	Not Used	--
77	--	Not Used	--
78	--	Not Used	--
79	--	Not Used	--
80	LIL/YEL	T-10 Connector, white in plenum chamber	N/A
81	LIL/GRY	Change-Over Valve for Intake Manifold Flap Control	N/A
82	--	Not Used	--
83	--	Not Used	--
84	--	Not Used	--
85	--	Not Used	--
86	BRN/YEL	Ground	N/A
87	--	Not Used	--
88	BLK/GRN	Fuse Panel	N/A
89	--	Not Used	--
90	--	Not Used	--
91	--	Not Used	--
92	--	Not Used	--
93	--	Not Used	--
94	--	Not Used	--
95	--	Not Used	--
96	--	Not Used	--
97	--	Not Used	--
98	--	Not Used	--
99	WHT/GRN	Modulating Piston Displacement Sensor Control	N/A
100	--	Not Used	--

2003–2004 Sedan, 2002 Wagon 1.9L DFI ALH (automatic) T121 Connector, *continued*

101	BLU	Needle Lift Sensor Signal	N/A
102	BRN	Engine Speed Sensor Signal	N/A
103	BRN/BLU	Fuel Temperature Sensor Ground	N/A
104	BRN/GRN	Engine Coolant Temperature Sensor Signal	N/A
105	--	Not Used	--
106	GRY/GRN	Modulating Piston Displacement Sensor Signal	N/A
107	--	Not Used	--
108	LIL/BLK	Modulating Piston Displacement Sensor Control	N/A
109	GRY	Needle Lift Sensor Ground	N/A
110	WHT	Engine Speed Sensor Signal	N/A
111	YEL/BLU	Fuel Temperature Sensor Power	N/A
112	BLU/BRN	Engine Coolant Temperature Sensor Ground	N/A
113	--	Not Used	--
114	BRN/BLK	Cold Start Injector Control	N/A
115	--	Not Used	--
116	BRN/RED	Quantity Adjuster Control	N/A
117	--	Not Used	--
118	--	Not Used	--
119	--	Not Used	--
120	BLK/WHT	Fuel Cut-off Valve Control	N/A
121	BRN/RED	Ground	N/A

Golf, Jetta

2003–2004 Sedan, 2003 Wagon 1.9L DFI ALH (manual) T121 Connector

ECM Pin #	Wire Color	Circuit Description (121-Pin)	Value at Hot Idle
1	RED/LIL	Ground (wire connection in engine compartment)	N/A
2	RED/LIL	Ground (wire connection in engine compartment)	N/A
3	--	Not Used	N/A
4	BRN/RED	PCV Heating Element Ground	N/A
5	BRN/RED	Ground	N/A
6	ORN/BRN	T-10 Connector, white in plenum chamber	N/A
7	ORN/BLK	T-10 Connector, white in plenum chamber	N/A
8	--	Not Used	--
9	--	Not Used	--
10	--	Not Used	--
11	BRN/WHT	T-10 Connector, white in plenum chamber	N/A
12	YEL/GRN	Throttle Position Sensor Ground	N/A
13	--	Not Used	--
14	BLK/WHT	Cruise Control Switch Ground	N/A
15	--	Not Used	--
16	GRY/WHT	T-10 Connector, orange in plenum chamber	N/A
17	--	Not Used	--
18	BLU/YEL	Power Supply Relay	N/A
19	--	Not Used	--
20	BLU/WHT	Instrument Panel Control Module	N/A
21	BLK/BLU	Preheating Coolant Output Relay (low heat)	N/A

2003–2004 Sedan, 2003 Wagon 1.9L DFI ALH (manual) T121 Connector, *continued*

22	BLK/BRN	Preheating Coolant Output Relay (high heat)	N/A
23	--	Not Used	--
24	WHT/BLU	T-10 Connector, white in plenum chamber	N/A
25	--	Not Used	--
26	--	Not Used	--
27	GRN/BRN	Instrument Panel Control Module	N/A
28	GRN/WHT	T-10 Connector, white in plenum chamber	N/A
29	GRN	Transmission Control Module	N/A
30	RED/GRN	Mass Air Flow Sensor	N/A
31	LIL/RED	Manifold Absolute Pressure Sensor	N/A
32	RED/BLK	Power Ground	N/A
33	GRN	Glow Plug Relay	N/A
34	BLU/RED	T-10 Connector, orange in plenum chamber	N/A
35	--	Not Used	--
36	--	Not Used	--
37	BLK/LIL	Fuse Panel	N/A
38	BRN/RED	T-10 Connector, white in plenum chamber	N/A
39	--	Not Used	--
40	BLU/GRN	T-10 Connector, white in plenum chamber	N/A
41	--	Not Used	--
42	LIL/WHT	Glow Plug Relay	N/A
43	--	Not Used	--
44	RED	Cruise Control Set Switch Signal	N/A
45	BLU/GRY	Cruise Control Switch Signal	N/A
46	WHT	Cruise Control Switch Signal	N/A
47	GRN/LIL	T-10 Connector, white in plenum chamber	N/A
48	--	Not Used	--
49	GRN	Mass Air Flow Sensor	N/A
50	GRY/RED	Throttle Position Sensor Power	N/A
51	GRN/RED	Closed Throttle Position Switch Ground	N/A
52	BRN/BLU	Intake Air Temperature Sensor Signal	N/A
53	--	Not Used	--
54	--	Not Used	--
55	--	Not Used	--
56	--	Not Used	--
57	--	Not Used	--
58	--	Not Used	--
59	--	Not Used	--
60	--	Not Used	--
61	RED/BLU	EGR Vacuum Regulator Solenoid Valve Control	N/A
62	RED/BRN	Wastegate Bypass Regulator Valve Control	N/A
63	WHT/BLU	Kick Down Switch Power	N/A
64	--	Not Used	--
65	WHT/YEL	Brake Vacuum Vent Valve Switch (for cruise control)	N/A
66	WHT/RED	Clutch Vacuum Vent Valve Switch Control	N/A
67	--	Not Used	--
68	BLK	Mass Air Flow Sensor	N/A

2003–2004 Sedan, 2003 Wagon 1.9L DFI ALH (manual) T121 Connector, *continued*

69	GRY/BLU	Throttle Position Sensor Signal	N/A
70	GRY/WHT	Throttle Position Sensor Ground	N/A
71	YEL/BLK	Manifold Absolute Pressure Sensor	N/A
72	--	Not Used	--
73	GRY/GRN	Intake Air Temperature Sensor Control	N/A
74	--	Not Used	--
75	--	Not Used	--
76	--	Not Used	--
77	--	Not Used	--
78	--	Not Used	--
79	--	Not Used	--
80	--	Not Used	--
81	LIL/GRY	Change-Over Valve for Intake Manifold Flap Control	N/A
82	--	Not Used	--
83	--	Not Used	--
84	--	Not Used	--
85	--	Not Used	--
86	BRN/YEL	Ground	N/A
87	--	Not Used	--
88	BLK/GRN	Fuse Panel	N/A
89	--	Not Used	--
90	--	Not Used	--
91	--	Not Used	--
92	--	Not Used	--
93	--	Not Used	--
94	--	Not Used	--
95	--	Not Used	--
96	--	Not Used	--
97	--	Not Used	--
98	--	Not Used	--
99	WHT/GRN	Modulating Piston Displacement Sensor Control	N/A
100	--	Not Used	--
101	BLU	Needle Lift Sensor Signal	N/A
102	BRN	Engine Speed Sensor Signal	N/A
103	BRN/BLU	Fuel Temperature Sensor Ground	N/A
104	BRN/GRN	Engine Coolant Temperature Sensor Signal	N/A
105	--	Not Used	--
106	GRY/GRN	Modulating Piston Displacement Sensor Signal	N/A
107	--	Not Used	--
108	LIL/BLK	Modulating Piston Displacement Sensor Control	N/A
109	GRY	Needle Lift Sensor Ground	N/A
110	WHT	Engine Speed Sensor Signal	N/A
111	YEL/BLU	Fuel Temperature Sensor Power	N/A
112	BLU/BRN	Engine Coolant Temperature Sensor Ground	N/A
113	--	Not Used	--
114	BRN/BLK	Cold Start Injector Control	N/A
115	--	Not Used	--

2003–2004 Sedan, 2003 Wagon 1.9L DFI ALH (manual) T121 Connector, *continued*

116	BRN/RED	Quantity Adjuster Control	N/A
117	--	Not Used	--
118	--	Not Used	--
119	--	Not Used	--
120	BLK/WHT	Fuel Cut-off Valve Control	N/A
121	BRN/RED	Ground	N/A

GTI, Jetta

1995 Sedan 2.8L AAA T68 Connector

ECM Pin #	Wire Color	Circuit Description (68-Pin)	Value at Hot Idle
1	BRN	Power Ground	N/A
2	GRY/YEL	Fuel Injector Cylinder 6	N/A
3	GRY/GRN	Fuel Injector Cylinder 2	N/A
4	GRY/BLU	Fuel Injector Cylinder 4	N/A
5	YEL/BLK	Relay Panel Multi Function Indicator	N/A
6	YEL/BLU	Fuel Pump Relay	N/A
7	BRN/RED	Sensor Ground (A/T only)	N/A
8	BLK/BRN	Ignition Coil	N/A
9	BLK/BLU	ECM Power Supply Relay	N/A
10	WHT	Throttle Valve Control Module Sensor	N/A
11	BLU/YEL	Transmission Control Module	N/A
12	BRN	Heated Oxygen Sensor Ground	N/A
13	--	Heated Oxygen Sensor 2 Power	N/A
14	BLU	Engine Coolant Temperature Sensor Ground	N/A
15	LIL	EGR Temperature Sensor	N/A
16	BRN/BLU	Mass Air Flow Sensor Signal	N/A
17	BLU	Mass Air Flow Sensor Ground	N/A
18	YEL/RED	Transmission Control Module	N/A
19	--	Not Used	--
20	WHT	Heated Oxygen Sensor	N/A
21	YEL	Transmission Control Module	N/A
22	GRN/BLK	Control Module for Instrument Panel	N/A
23	BLK/YEL	ECM Power Supply Relay	N/A
24	GRY	Fuel Injector Cylinder 1	N/A
25	GRY/BLK	Fuel Injector Cylinder 5	N/A
26	GRY/RED	Fuel Injector Cylinder 3	N/A
27	BLK	Throttle Valve Control Module	N/A
28	WHT	Heated Oxygen Sensor Relay	N/A
29	--	Not Used	--
30	GRN/GRY	EGR Vacuum Regulator Valve Power	N/A
31	GRN/YEL	EVAP Canister Purge Regulator Valve	N/A
32	--	Not Used	--
33	BRN	Knock Sensor 1 Signal	N/A
34	GRN	Knock Sensor 1 Power	N/A
35	YEL/BRN	Leak Detection Pump	N/A

1995 Sedan 2.8L AAA T68 Connector, *continued*

36	BLU/GRN	Intake Air Temperature Sensor Power	N/A
37	GRN	Relay Panel	N/A
38	BLK/GRN	E-6 Connector	N/A
39	BLU/RED	Double Connector (behind fuse panel)	N/A
40	GRN/WHT	Throttle Valve Control Module	N/A
41	BLK/GRY	Throttle Valve Control Module	N/A
42	YEL	Heated Oxygen Sensor Ground	N/A
43	GRY/WHT	Junction Box for OBD	N/A
44	WHT/RED	Camshaft Position Sensor	N/A
45	--	Not Used	--
46	GRN/LIL	Leak Detection Pump	N/A
47	--	Not Used	--
48	--	Not Used	--
49	RED/GRY	Secondary Air Injection Pump Relay	N/A
50	RED/YEL	Secondary Air Injection Solenoid Valve	N/A
51	LIL/WHT	Radio Power	N/A
52	BLK/LIL	Ignition Coil	N/A
53	RED/GRY	Throttle Valve Control Module	N/A
54	RED/BLU	Battery Power	N/A
55	BRN/BLK	Power Ground	N/A
56	BRN/WHT	Sensor Ground	N/A
57	GRN/YEL	Knock Sensor 2 Power	N/A
58	WHT/YEL	Heated Oxygen Sensor 2 Ground	N/A
59		Not Used	--
60	BLK/BLU	Ignition Coil	N/A
61	--	Not Used	--
62	BLK/WHT	Throttle Valve Control Module	N/A
63	--	Not Used	--
64	LIL/BLK	T-1 Connector	N/A
65	BLU/WHT	TV-13 Vehicle Speed Signal Wire Connector	N/A
66	BRN/WHT	Fuel Injector Ground	N/A
67	RED	Engine Speed Sensor Power	N/A
68	GRN	Engine Speed Sensor Signal	N/A

GTI, Jetta

1999–2002 Sedan, 2001–2002 Wagon 2.8L MFI AFP T121 Connector

ECM Pin #	Wire Color	Circuit Description (121-Pin)	Value at Hot Idle
1	BRN/RED	Power Ground	N/A
2	BRN/RED	Power Ground	N/A
3	BLK/LIL	Fuse Connection	N/A
4	--	Not Used	--
5	WHT	Heated Oxygen Sensor Power	N/A
6	--	Not Used	--
7	--	Not Used	--
8	--	Not Used	--

1999–2002 Sedan, 2001–2002 Wagon 2.8L MFI AFP T121 Connector, *continued*

9	--	Not Used	--
10	--	Not Used	--
11	--	Not Used	--
12	--	Not Used	--
13	--	Not Used	--
14	--	Not Used	--
15	--	Not Used	--
16	--	Not Used	--
17	--	Not Used	--
18	--	Not Used	--
19	--	Not Used	--
20	--	Not Used	--
21	--	Not Used	--
22	--	Not Used	--
23	--	Not Used	--
24	--	Not Used	--
25	YEL/BRN	Leak Detection Pump Motor Signal	N/A
26	LIL/GRN	Mass Air Flow Sensor	N/A
27	GRN	Mass Air Flow Sensor	N/A
28	--	Not Used	--
29	BLK	Mass Air Flow Sensor	N/A
30	WHT/GRN	T-10 Connection, Green on ECM	N/A
31	--	Not Used	--
32	--	Not Used	--
33	GRY/BLU	Throttle Position Sender Ground	N/A
34	BRN/WHT	Throttle Position Sender Signal	N/A
35	WHT/BLU	Throttle Position Sensor Power	N/A
36	GRY/RED	Throttle Position Sensor Signal	N/A
37	GRN/BRN	Malfunction Indicator Lamp (MIL)	N/A
38	BLK/WHT	Cruise Control Switch Ground	N/A
39	WHT/RED	Clutch Vacuum Vent Valve Switch Signal	N/A
40	BLU/RED	T-10 Connection, Orange on ECM (A/C Control)	N/A
41	GRN/GRY	T-10 Connection, Orange on ECM (A/C Control)	N/A
42	--	Not Used	--
43	GRY/WHT	T-10 Connection, Orange on ECM	N/A
44	BRN/YEL	Secondary Air Injection Solenoid Valve Control	N/A
45	--	Not Used	--
46	GRY/BRN	Secondary Air Injection Pump Relay	N/A
47	LIL/YEL	T-10 Connection, Blue on ECM	N/A
48	WHT/RED	T-10 Connection, Blue on ECM	N/A
49	--	Not Used	--
50	--	Not Used	--
51	YEL/BLK	Heated Oxygen Sensor Signal (also ORN/LIL)	N/A
52	RED	Heated Oxygen Sensor Ground (also ORN/BRN)	N/A
53	RED/LIL	Mass Air Flow Sensor	N/A
54	BLU/WHT	Vehicle Speed Sensor Signal	N/A
55	WHT/YEL	Brake Pedal Switch (cruise control) Signal	N/A

1999–2002 Sedan, 2001–2002 Wagon 2.8L MFI AFP T121 Connector, *continued*

56	BLK/RED	Sensor Ground	N/A
57	RED/YEL	Cruise Control Set Switch Signal	N/A
58	ORN/BRN	T-10 Connection, Green on ECM	N/A
59	--	Not Used	--
60	ORN/BLK	T-10 Connection, Green on ECM	N/A
61	WHT	T-10 Connection, Orange on ECM	N/A
62	RED/GRN	Fuse Connection	N/A
63	BRN/BLK	Heated Oxygen Sensor 2 Power	N/A
64	LIL/RED	EVAP Canister Purge Regulatory Valve Control	N/A
65	LIL/WHT	Fuel Pump Relay	N/A
66	YEL	T-10 Connection, Black on ECM	N/A
67	--	Not Used	--
68	GRY/WHT	Heated Oxygen Sensor 2 Ground	N/A
69	GRY/RED	Heated Oxygen Sensor 2 Signal	N/A
70	BLK	Heated Oxygen Sensor Signal (also GRN)	N/A
71	GRN/YEL	Heated Oxygen Sensor Ground (also BLK)	N/A
72	GRN/WHT	Throttle Position Sender Power	N/A
73	YEL/GRN	Throttle Position Sensor Ground	N/A
74	--	Not Used	--
75	BLU/GRY	Cruise Control Switch Signal	N/A
76	WHT	Cruise Control Switch Signal	N/A
77	ORN/BRN	Transmission Control Module (1999 auto only)	N/A
78	--	Not Used	--
79	ORN/BLK	Transmission Control Module (1999 auto only)	N/A
80	GRN/LIL	Leak Detection Pump Motor Signal	N/A
81	GRN/RED	T-10 Connection, Orange on ECM	N/A
82	WHT	Speed Sensor Signal	N/A
83	LIL/RED	Throttle Position Sensor Ground	N/A
84	LIL/YEL	Throttle Position Sensor 2 Signal	N/A
85	--	Not Used	--
86	LIL/BRN	Camshaft Position Sensor Signal	N/A
87	--	Not Used	--
88	LIL/BLU	Cylinder 3 Fuel Injector Control	N/A
89	GRN/YEL	Cylinder 6 Fuel Injector Control	N/A
90	BRN	Speed Sensor Signal	N/A
91	WHT	Throttle Position Sensor 1 Signal	N/A
92	LIL/BLK	Throttle Position Sensor Power	N/A
93	BLU	Engine Coolant Temperature Sensor	N/A
94	LIL/BLK	Ignition Coil Control	N/A
95	--	Not Used	--
96	LIL	Cylinder 1 Fuel Injector Control	N/A
97	LIL/GRY	Cylinder 5 Fuel Injector Control	N/A
98	LIL/RED	Camshaft Position Sensor Ground	N/A
99	BLK	Knock Sensor 2 Ground	N/A
100	--	Not Used	--
101	--	Not Used	--
102	LIL/GRN	Ignition Coil Control	N/A

1999–2002 Sedan, 2001–2002 Wagon 2.8L MFI AFP T121 Connector, *continued*

103	LIL/YEL	Ignition Coil Control	N/A
104	LIL/BRN	Intake Manifold Change-Over Valve Signal	N/A
105	GRY/YEL	Brake Booster Vacuum Valve Signal (2000-02 only)	N/A
106	GRY	Knock Sensor 1 Signal	N/A
107	GRN	Knock Sensor 2 Signal	N/A
108	BRN/BLU	Sensor Ground	N/A
109	--	Not Used	--
110	--	Not Used	--
111	--	Not Used	--
112	LIL/GRN	Cylinder 2 Fuel Injector Control	N/A
113	LIL/BLU	Cylinder 4 Fuel Injector Control	N/A
114	--	Not Used	--
115	--	Not Used	--
116	--	Not Used	--
117	LIL	Throttle Position Actuator Motor Power	N/A
118	LIL/WHT	Throttle Position Actuator Motor Control	N/A
119	--	Not Used	--
120	--	Not Used	--
121	--	Not Used	--

GTI, Jetta

2002–2005 Sedan 1.8L MFI AWP T121 Connector

ECM Pin #	Wire Color	Circuit Description (121-Pin)	Value at Hot Idle
1	BRN/RED	PCV Heating Element Control	N/A
2	BRN/RED	Ground	N/A
3	BLK	Instrument Panel wiring harness	N/A
4	--	Not Used	--
5	WHT	Heated Oxygen Sensor Power	N/A
6	--	Not Used	--
7	--	Not Used	--
8	--	Not Used	--
9	BRN/YEL	Secondary Air Injection Solenoid Valve Signal	N/A
10	--	Not Used	--
11	--	Not Used	--
12	--	Not Used	--
13	--	Not Used	--
14	--	Not Used	--
15	--	Not Used	--
16	--	Not Used	--
17	--	Not Used	--
18	--	Not Used	--
19	--	Not Used	--
20	--	Not Used	--
21	LIL	ECM Power Supply Relay	N/A
22	BLU/GRN	Brake Booster Control Module	N/A

2002–2005 Sedan 1.8L MFI AWP T121 Connector, *continued*

23	--	Not Used	--
24	--	Not Used	--
25	YEL/BRN	Leak Detection Pump Motor Power	N/A
26	--	Not Used	--
27	GRN	Mass Air Flow Sensor	N/A
28	BLU/LIL	T-10 Connector, white in plenum chamber	N/A
29	BLK	Mass Air Flow Sensor	N/A
30	WHT/GRN	T-10 Connector, white in plenum chamber	N/A
31	--	Not Used	--
32	--	Not Used	--
33	GRY/BLU	Accelerator Pedal Position Sender Ground	N/A
34	BRN/WHT	Accelerator Pedal Position Sender Signal	N/A
35	WHT/BLU	Throttle Position Sensor Signal	N/A
36	GRY/RED	Throttle Position Sensor Power	N/A
37	GRN/BRN	Instrument Panel Control Module (to 4/05)	N/A
38	BLK/WHT	Cruise Control Switch Ground	N/A
39	WHT/RED	Clutch Vacuum Vent Valve Switch Signal	N/A
40	BLU/RED	T-10 Connector, white in plenum (A/C connection)	N/A
41	GRN/GRY	T-10 Connector, orange in plenum (GRN for manual)	N/A
42	--	Not Used	N/A
43	GRY/WHT	T-10 Connector, white in plenum chamber	N/A
44	--	Not Used	--
45	--	Not Used	--
46	--	Not Used	--
47	LIL/YEL	T-10 Connector, white in plenum chamber	N/A
48	WHT/RED	T-10 Connector, white in plenum chamber	N/A
49	WHT/LIL	Power Steering Pressure Switch Control	N/A
50	WHT/GRY	Power Steering Pressure Switch Signal	N/A
51	GRN/BLK	Heated Oxygen Sensor Signal	N/A
52	GRY/RED	Heated Oxygen Sensor Ground	N/A
53	RED/LIL	Mass Air Flow Sensor	N/A
54	BLU/WHT	Vehicle Speed Signal (instrument panel)	N/A
55	WHT/YEL	Brake Vacuum Vent Valve Switch Signal	N/A
56	BLK/RED	T-10 Connector, black in plenum chamber	N/A
57	RED/YEL	Set Switch for Cruise Control Signal	N/A
58	--	Not Used	--
59	--	Not Used	--
60	--	Not Used	--
61	--	Not Used	--
62	RED/GRN	Fuse Box connection	N/A
63	GRY/YEL	Oxygen Sensor Power	N/A
64	LIL/RED	EVAP Canister Purge Regulator Valve Signal	N/A
65	LIL/WHT	Fuel Pump Relay	N/A
66	GRY/BRN	Secondary Air Injection Pump Relay Control	N/A
67	GRN/LIL	T-10 Connector, white in plenum chamber	N/A
68	GRN/WHT	Oxygen Sensor Ground	N/A
69	GRN/RED	Oxygen Sensor Signal	N/A

2002–2005 Sedan 1.8L MFI AWP T121 Connector, *continued*

70	GRN	Heated Oxygen Sensor Signal	N/A
71	BLK	Heated Oxygen Sensor Signal	N/A
72	GRN/WHT	Accelerator Pedal Position Sender Power	N/A
73	YEL/GRN	Throttle Position Sensor Ground	N/A
74	--	Not Used	--
75	BLU/GRY	Cruise Control Switch Signal	N/A
76	WHT	Cruise Control Switch Signal	N/A
77	--	Not Used	--
78	--	Not Used	--
79	--	Not Used	--
80	GRN/LIL	Leak Detection Pump Signal	N/A
81	GRN/RED	T-10 Connector, white in plenum chamber	N/A
82	WHT	Engine Speed Sensor Signal	N/A
83	LIL/WHT	Throttle Drive Angle Sensor Ground	N/A
84	BLU/GRY	Throttle Drive Angle Sensor Signal	N/A
85	BLU/GRN	Intake Air Temperature Sensor Control	N/A
86	LIL/YEL	Camshaft Position Sensor	N/A
87	--	Not Used	--
88	LIL/BLU	Cylinder 2 Fuel Injector Control	N/A
89	LIL/GRN	Cylinder 4 Fuel Injector Control	N/A
90	BRN	Engine Speed Sensor Signal	N/A
91	WHT/GRY	Throttle Drive Angle Sensor Power	N/A
92	BLU/WHT	Throttle Drive Angle Sensor Signal	N/A
93	GRY/YEL	Engine Coolant Temperature Sensor Control	N/A
94	BLK/LIL	Ignition Coil 4 with Power Output Stage	N/A
95	BLK/YEL	Ignition Coil 2 with Power Output Stage	N/A
96	LIL	Cylinder 1 Fuel Injector Control	N/A
97	LIL/RED	Cylinder 3 Fuel Injector Control	N/A
98	LIL/GRN	Camshaft Position Sensor	N/A
99	BLU	Knock Sensor 1 and 2 Control	N/A
100	--	Not Used	--
101	LIL/GRN	Charge Air Pressure Sensor	N/A
102	LIL/BLK	Ignition Coil 1 with Power Output Stage	N/A
103	BLK/BRN	Ignition Coil 3 with Power Output Stage	N/A
104	GRN/BRN	Wastegate Bypass Regulator Valve Signal	N/A
105	GRY/GRN	Turbocharger Recirculating Valve Signal	N/A
106	GRY	Knock Sensor 1 Signal	N/A
107	GRY	Knock Sensor 2	N/A
108	BRN/BLU	Sensor Ground	N/A
109	--	Not Used	--
110	--	Not Used	--
111	--	Not Used	--
112	--	Not Used	--
113	--	Not Used	--
114	--	Not Used	--
115	GRN/WHT	Camshaft Adjustment, Valve One	N/A
116	--	Not Used	--

2002–2005 Sedan 1.8L MFI AWP T121 Connector, *continued*

117	WHT	Throttle Drive Motor Power	N/A
118	LIL/BLK	Throttle Drive Motor Control	N/A
119	--	Not Used	--
120	--	Not Used	--
121	BLK/LIL	Ground	N/A

GTI, Jetta

2002–2005 Sedan 2.8L MFI BDF T121 Connector

ECM Pin #	Wire Color	Circuit Description (121-Pin)	Value at Hot Idle
1	BRN/RED	Ground	N/A
2	BRN/RED	Ground	N/A
3	BLK/LIL	Ignition Coils 1-4 with Power Output Stage Ground	N/A
4	--	Not Used	--
5	WHT	Heated Oxygen Sensor Power	N/A
6	--	Not Used	--
7	--	Not Used	--
8	--	Not Used	--
9	--	Not Used	--
10	--	Not Used	--
11	--	Not Used	--
12	--	Not Used	--
13	--	Not Used	--
14	--	Not Used	--
15	--	Not Used	--
16	--	Not Used	--
17	--	Not Used	--
18	--	Not Used	--
19	--	Not Used	--
20	--	Not Used	--
21	BLK	Fuse Box Connection	N/A
22	BLU/GRN	Brake Booster Control Module (A/T only)	N/A
23	LIL	ECM Power Supply Relay	N/A
24	--	Not Used	--
25	YEL/BRN	Leak Detection Pump Motor Power	N/A
26	LIL/GRN	Mass Air Flow Sensor	N/A
27	GRN	Mass Air Flow Sensor	N/A
28	BLU/LIL	T-10 Connector, White, in plenum chamber	N/A
29	BLK	Mass Air Flow Sensor	N/A
30	WHT/GRN	T-10 Connector, White, in plenum chamber	N/A
31	--	Not Used	--
32	--	Not Used	--
33	GRY/BLU	Accelerator Pedal Position Sender Ground	N/A
34	BRN/WHT	Accelerator Pedal Position Sender Signal	N/A
35	WHT/BLU	Throttle Position Sensor Signal	N/A
36	GRY/RED	Throttle Position Sensor Power	N/A

2002–2005 Sedan 2.8L MFI BDF T121 Connector, *continued*

37	GRN/BRN	Instrument Panel Control Module	N/A
38	BLK/WHT	Cruise Control Switch Ground	N/A
39	WHT/RED	Clutch Vacuum Vent Valve Switch Signal	N/A
40	BLU/RED	T-10 Connector, Orange, in plenum chamber	N/A
41	GRN/GRY	T-10 Connector, Orange, in plenum chamber	N/A
42	--	Not Used	N/A
43	GRY/WHT	T-10 Connector, Orange, in plenum chamber	N/A
44	BRN/YEL	Secondary Air Injection Solenoid Valve	N/A
45	--	Not Used	N/A
46	GRN/BRN	Secondary Air Injection Pump Relay	N/A
47	LIL/YEL	T-10 Connector, White, in plenum chamber	N/A
48	WHT/RED	T-10 Connector, White, in plenum chamber	N/A
49	--	Not Used	--
50	--	Not Used	--
51	WHT	Heated Oxygen Sensor Signal	N/A
52	GRN	Heated Oxygen Sensor Ground	N/A
53	RED/LIL	Mass Air Flow Sensor	N/A
54	BLU/WHT	Vehicle Speed Signal (instrument panel)	N/A
55	WHT/YEL	Brake Vacuum Vent Valve Switch Signal	N/A
56	BLK/RED	T-10 Connector, Black, in plenum chamber	N/A
57	RED/YEL	Set Switch for Cruise Control Signal	N/A
58	ORN/BRN	ABS control module (TCM for automatic)	N/A
59	--	Not Used	--
60	ORN/BLK	ABS control module (TCM for automatic)	N/A
61	--	Not Used	--
62	RED/GRN	Fuse Box connection	N/A
63	BRN	Heated Oxygen Sensor Power	N/A
64	LIL/RED	EVAP Canister Purge Regulator Valve Signal	N/A
65	LIL/WHT	T-6 Connector, Brown, in plenum chamber	N/A
66	--	Not Used	--
67	--	Not Used	--
68	BLK	Heated Oxygen Sensor Ground	N/A
69	BLU	Heated Oxygen Sensor Signal	N/A
70	GRN	Heated Oxygen Sensor Signal	N/A
71	BLK	Heated Oxygen Sensor Ground	N/A
72	GRN/WHT	Accelerator Pedal Position Sender Power	N/A
73	YEL/GRN	Throttle Position Sensor Ground	N/A
74	--	Not Used	--
75	BLU/GRY	Cruise Control Switch Signal	N/A
76	WHT	Cruise Control Switch Signal	N/A
77	--	Not Used	--
78	--	Not Used	--
79	--	Not Used	--
80	GRN/LIL	Leak Detection Pump Signal	N/A
81	--	Not Used	--
82	WHT	Engine Speed Sensor Signal	N/A
83	LIL/RED	Throttle Drive Angle Sensor Ground	N/A

2002–2005 Sedan 2.8L MFI BDF T121 Connector, *continued*

84	LIL/YEL	Throttle Drive Angle Sensor Signal	N/A
85	--	Not Used	--
86	LIL/BRN	Camshaft Position Sensor	N/A
87	BLU/RED	Camshaft Position Sensor 2	N/A
88	LIL/BLU	Cylinder 3 Fuel Injector Control	N/A
89	GRN/YEL	Cylinder 6 Fuel Injector Control	N/A
90	BRN	Engine Speed Sensor Signal	N/A
91	WHT	Throttle Drive Angle Sensor Power	N/A
92	LIL/BLK	Throttle Drive Angle Sensor Signal	N/A
93	BLU	Engine Coolant Temperature Sensor Signal	N/A
94	GRY/GRN	Ignition Coil 3 with Power Output Stage	N/A
95	BLU/YEL	Ignition Coil 6 with Power Output Stage	N/A
96	LIL	Cylinder 1 Fuel Injector Control	N/A
97	LIL/GRN	Cylinder 5 Fuel Injector Control	N/A
98	LIL/RED	Camshaft Position Sensor	N/A
99	BLK	Knock Sensor 1 and 2 Control	N/A
100	--	Not Used	--
101	--	Not Used	--
102	LIL/GRN	Ignition Coil 1 with Power Output Stage	N/A
103	BLU/GRY	Ignition Coil 5 with Power Output Stage	N/A
104	--	Not Used	--
105	--	Not Used	--
106	RED	Knock Sensor 1 Signal	N/A
107	GRN	Knock Sensor 2 Signal	N/A
108	BRN/BLU	Sensor Ground	N/A
109	--	Not Used	--
110	LIL/YEL	Ignition Coil 2 with Power Output Stage	N/A
111	GRY/RED	Ignition Coil 4 with Power Output Stage	N/A
112	LIL/GRN	Cylinder 2 Fuel Injector Control	N/A
113	LIL/BLU	Cylinder 4 Fuel Injector Control	N/A
114	--	Not Used	N/A
115	BLU/BRN	Camshaft Adjustment Valve 1 Control	N/A
116	--	Not Used	--
117	LIL	Throttle Drive Motor Power	N/A
118	LIL/WHT	Throttle Drive Motor Control	N/A
119	--	Not Used	--
120	GRY/BLU	Camshaft Adjustment Valve 1 (exhaust) Control	N/A
121	LIL/BRN	Intake Manifold Change Over Valve Control	N/A

GTI, Jetta

2006–2007 Sedan 2.0L MFI BPY T94 plus T60 Connectors

ECM Pin #	Wire Color	Circuit Description (94-Pin)	Value at Hot Idle
1	BRN	Ground Connection 2	N/A
2	BRN	Ground Connection 2	N/A
3	RED/GRN	Connection 3 in engine compartment wiring harness	N/A

2006–2007 Sedan 2.0L MFI BPY T94 plus T60 Connectors, *continued*

4	BRN	Ground Connection 2	N/A
5	RED/GRN	Connection 3 in engine compartment wiring harness	N/A
6	RED/GRN	Connection 3 in engine compartment wiring harness	N/A
7	BRN/GRN	Heated Oxygen Sensor Power (behind cat. conv.)	N/A
8	--	Not Used	--
9	--	Not Used	--
10	--	Not Used	--
11	--	Not Used	--
12	YEL	Engine Coolant Temperature Sensor (on radiator)	N/A
13	--	Not Used	--
14	--	Not Used	--
15	--	Not Used	--
16	--	Not Used	--
17	--	Not Used	--
18	BLK/WHT	Steering Column Electronic Systems Control Module	N/A
19	BLK	Charge Air Pressure Sensor	N/A
20	--	Not Used	--
21	--	Not Used	--
22	YEL	Mass Air Flow Sensor	N/A
23	--	Not Used	--
24	--	Not Used	--
25	BLK/RED	Connection (54) in interior wiring harness	N/A
26	--	Not Used	--
27	YEL/BLU	Fuel Pump Control Module	N/A
28	YEL/LIL	T-4 Connector on coolant fan, bottom left	N/A
29	--	Not Used	--
30	--	Not Used	--
31	--	Not Used	--
32	BLK/BRN	Engine Control Module Power Supply Relay	N/A
33	LIL/BRN	Engine Coolant Pump Relay	N/A
34	--	Not Used	--
35	--	Not Used	--
36	--	Not Used	--
37	--	Not Used	--
38	LIL/GRN	Charge Air Pressure Sensor	N/A
39	--	Not Used	--
40	LIL	Leak Detection Pump Ground	N/A
41	WHT/RED	Clutch Position Sensor	N/A
42	--	Not Used	--
43	--	Not Used	--
44	--	Not Used	--
45	--	Not Used	--
46	--	Not Used	--
47	WHT/GRN	Brake Pedal Position Sensor	N/A
48	--	Not Used	--
49	--	Not Used	--
50	BRN/LIL	Leak Detection Pump Motor	N/A

2006–2007 Sedan 2.0L MFI BPY T94 plus T60 Connectors, *continued*

51	WHT	Heated Oxygen Sensor Power	N/A
52	--	Not Used	--
53	BRN	Sensor Ground Connection	N/A
54	--	Not Used	--
55	--	Not Used	--
56	GRY/BLU	Accelerator Pedal Position Sensor 2	N/A
57	BRN/BLU	Accelerator Pedal Position Sensor 2	N/A
58	GRY/BLK	Accelerator Pedal Position Sensor 2	N/A
59	--	Not Used	--
60	GRN	Heated Oxygen Sensor Signal	N/A
61	BLK	Heated Oxygen Sensor Ground	N/A
62	--	Not Used	--
63	--	Not Used	--
64	--	Not Used	--
65	BLU/WHT	Connection 8 in engine compartment wiring harness	N/A
66	--	Not Used	--
67	ORN/BRN	Powertrain CAN-Bus Low Connection 1	N/A
68	ORN/BLK	Powertrain CAN-Bus High Connection 1	N/A
69	BLK/GRY	Power Supply Relay (Terminal 30)	N/A
70	--	Not Used	--
71	--	Not Used	--
72	--	Not Used	--
73	--	Not Used	--
74	--	Not Used	--
75	--	Not Used	--
76	BLK	Heated Oxygen Sensor Ground (behind cat conv.)	N/A
77	BLU	Heated Oxygen Sensor Signal (behind cat converter)	N/A
78	GRY/RED	Throttle Position Sensor	N/A
79	WHT/BLU	Throttle Position Sensor	N/A
80	YEL/GRN	Throttle Position Sensor	N/A
81	GRY/WHT	Heated Oxygen Sensor Signal	N/A
82	GRY/RED	Heated Oxygen Sensor Ground	N/A
83	--	Not Used	--
84	--	Not Used	--
85	--	Not Used	--
86	GRY/WHT	Diagnostic Wire K Connection	N/A
87	BLK/BLU	Fuse 1 on Fuse Panel C	N/A
88	--	Not Used	--
89	--	Not Used	--
90	--	Not Used	--
91	--	Not Used	--
92	RED/BLK	Plus Connection 1 in engine wiring harness	N/A
93	--	Not Used	--
94	--	Not Used	--
ECM Pin #	Wire Color	Circuit Description (60-Pin)	Value at Hot Idle
1	RED/WHT	Cylinder 2 Fuel Injector Control	N/A
2	RED/BLK	Cylinder 1 Fuel Injector Control	N/A

2006–2007 Sedan 2.0L MFI BPY T94 plus T60 Connectors, *continued*

3	--	Not Used	--
4	WHT/LIL	Wastegate Bypass Regulator Valve	N/A
5	BLU/LIL	EVAP Canister Purge Regulator Valve 1	N/A
6	--	Not Used	--
7	LIL/GRN	Low Fuel Pressure Sensor	N/A
8	--	Not Used	--
9	RED	Knock Sensor 2 Signal	N/A
10	GRN	Engine Speed Sensor Signal	N/A
11	--	Not Used	--
12	GRN/YEL	Throttle Valve Switch Signal	N/A
13	WHT	Intake Air Temperature Sensor	N/A
14	BRN/BLU	Ground Connection	N/A
15	LIL	Throttle Valve Motor Power	N/A
16	RED/GRY	Cylinder 4 Fuel Injector Control	N/A
17	RED/LIL	Cylinder 3 Fuel Injector Control	N/A
18	--	Not Used	--
19	LIL/BRN	Fuel Pressure Regulating Valve	N/A
20	LILI/WHT	Camshaft Adjustment Valve 1	N/A
21	LIL/BRN	Turbocharger Recirculating Valve	N/A
22	GRN	Intake Flap Motor Ground	N/A
23	--	Not Used	--
24	LIL	Knock Sensor 2 Signal	N/A
25	GRY/BLU	Fuel Pressure Sensor	N/A
26	BLK/GRY	Fuel Pressure, Camshaft Position Sensors Grounds	N/A
27	BLU/GRN	Throttle Valve Switch Ground	N/A
28	BLK/BLU	Throttle Valve Control Module Signal	N/A
29	LIL/BRN	Throttle Valve Switch Ground	N/A
30	BRN/LIL	Throttle Valve Motor Ground	N/A
31	--	Not Used	--
32	BRN/LIL	Cylinder 3 Fuel Injector Ground	N/A
33	BRN/BLK	Cylinder 1 Fuel Injector Ground	N/A
34	GRN/LIL	Intake Flap Motor Ground	N/A
35	--	Not Used	--
36	WHT	Engine Speed Sensor Signal	N/A
37	BLK/GRN	Intake Flap Motor Signal	N/A
38	--	Not Used	--
39	BRN/RED	Knock Sensor 1 Signal	N/A
40	--	Not Used	--
41	LIL	Ignition Coil 2 with Power Output Stage	N/A
42	--	Not Used	--
43	BLU	Ignition Coil 3 with Power Output Stage	N/A
44	GRN/LIL	Camshaft Position Sensor	N/A
45	--	Not Used	--
46	--	Not Used	--
47	BRN/WHT	Cylinder 2 Fuel Injector Ground	N/A
48	BRN/GRY	Cylinder 4 Fuel Injector Ground	N/A
49	BRN/GRN	Intake Flap Motor Power	N/A

2006–2007 Sedan 2.0L MFI BPY T94 plus T60 Connectors, *continued*

50	--	Not Used	--
51	BRN	Engine Speed Sensor Signal	N/A
52	BLK	Knock Sensor 1 and 2 Ground	N/A
53	--	Not Used	--
54	WHT/BRN	Knock Sensor 1 Signal	N/A
55	--	Not Used	--
56	LIL/YEL	Ignition Coil 4 with Power Output Stage	N/A
57	--	Not Used	--
58	LIL/GRY	Ignition Coil 1 with Power Output Stage	N/A
59	--	Not Used	--
60	--	Not Used	--

GTI, Jetta, Passat

1995 Sedan 2.8L AAA T68 Connector

ECM Pin #	Wire Color	Circuit Description (68-Pin)	Value at Hot Idle
1	BRN	Power Ground	N/A
2	GRY/YEL	Fuel Injector Cylinder 6	N/A
3	GRY/GRN	Fuel Injector Cylinder 2	N/A
4	GRY/BLU	Fuel Injector Cylinder 4	N/A
5	YEL/BLK	Relay Panel Multi Function Indicator	N/A
6	YEL/BLU	Fuel Pump Relay	N/A
7	BRN/RED	Sensor Ground	N/A
8	BLK/BRN	Ignition Coil	N/A
9	BLK/BLU	ECM Power Supply Relay	N/A
10	WHT	Sensor Ground	N/A
11	BLU/YEL	T-8 Connector (behind fuse panel) (A/T only)	N/A
12	--	Not Used	--
13	--	Not Used	--
14	BLU	Engine Coolant Temperature Sensor Power	N/A
15	LIL	EGR Temperature Sensor	N/A
16	BRN/BLU	Mass Air Flow Sensor Signal	N/A
17	RED	Mass Air Flow Sensor Ground	N/A
18	YEL/RED	T-8 Connector (behind fuse panel) (A/T only)	N/A
19	--	Not Used	--
20	WHT	Heated Oxygen Sensor	N/A
21	YEL	Double Connector (behind fuse panel)	N/A
22	GRN/BLK	Control Module for Instrument Panel	N/A
23	RED/BLU	ECM Power Supply Relay	N/A
24	GRY	Fuel Injector Cylinder 1	N/A
25	GRY/BLK	Fuel Injector Cylinder 5	N/A
26	GRY/RED	Fuel Injector Cylinder 3	N/A
27	RED/BLU	Idle Air Control Valve Power	N/A
28	WHT	Heated Oxygen Sensor Relay	N/A
29	--	Not Used	--
30	GRN/GRY	EGR Vacuum Regulator Valve Power	N/A
31	GRN/YEL	EVAP Canister Purge Regulator Valve	N/A

1995 Sedan 2.8L AAA T68 Connector, *continued*

32	--	Not Used	--
33	BRN	Knock Sensor Signal	N/A
34	GRN	Knock Sensor 1 Power	N/A
35	--	Not Used	--
36	BLU/GRN	Intake Air Temperature Sensor Power	N/A
37	GRN	Double Connector (behind fuse panel)	N/A
38	BLK/GRN	E-6 Connector	N/A
39	BLU/RED	Double Connector (behind fuse panel)	N/A
40	GRN/WHT	Throttle Position Sensor Power	N/A
41	BLK/GRY	Throttle Position Sensor Signal	N/A
42	YEL	T-4 Connector (near intake manifold)	N/A
43	GRY/WHT	Double Connector (behind fuse panel)	N/A
44	WHT/RED	Camshaft Position Sensor	N/A
45	--	Not Used	--
46	--	Not Used	--
47	--	Not Used	--
48	--	Not Used	--
49	RED/GRY	Secondary Air Injection Pump Relay	N/A
50	RED/YEL	Secondary Air Injection Solenoid Valve	N/A
51	LIL/WHT	Radio Power	N/A
52	BLK/LIL	Ignition Coil	N/A
53	BLK/WHT	Idle Air Control Valve Ground	N/A
54	RED/BLU	Battery Power	N/A
55	BRN/BLK	Power Ground	N/A
56	BRN/WHT	Sensor Ground	N/A
57	GRN/YEL	Knock Sensor 2 Power	N/A
58	WHT	Sensor Ground (A/T only)	N/A
59		Not Used	--
60	BLK/BLU	Ignition Coil	N/A
61	--	Not Used	--
62	--	Not Used	--
63	--	Not Used	--
64	--	Not Used	--
65	BLU/WHT	TV-13 Vehicle Speed Signal Wire Connector	N/A
66	--	Not Used	--
67	RED	Engine Speed Sensor Power	N/A
68	GRN	Engine Speed Sensor Signal	N/A

Golf, Jetta

2001–2002 Jetta Sedan, Wagon and Golf; 2003–2004 Jetta Sedan; 2003–2004 Jetta Wagon; and 2003–2006 Golf 2.0L MFI AVH and AZG T121 Connector

ECM Pin #	Wire Color	Circuit Description (121-Pin)	Value at Hot Idle
1	BRN/RED	PCV Heating Element Control	N/A
2	BRN/RED	Ground	N/A
3	BLK	Instrument Panel wiring harness	N/A

2.0L MFI AVH and AZG T121 Connector, *continued*

4	--	Not Used	--
5	WHT	Heated Oxygen Sensor Power	N/A
6	--	Not Used	--
7	--	Not Used	--
8	--	Not Used	--
9	GRY/BRN	Secondary Air Injection Pump Relay Control	N/A
10	--	Not Used	--
11	--	Not Used	--
12	--	Not Used	--
13	--	Not Used	--
14	--	Not Used	--
15	--	Not Used	--
16	--	Not Used	--
17	--	Not Used	--
18	--	Not Used	--
19	--	Not Used	--
20	--	Not Used	--
21	LIL	ECM Power Supply Relay	N/A
22	--	Not Used	--
23	--	Not Used	--
24	--	Not Used	--
25	YEL/BRN	Leak Detection Pump Motor Power	N/A
26	LIL/GRN	Mass Air Flow Sensor	N/A
27	GRN	Mass Air Flow Sensor	N/A
28	BLU/LIL	T-10 Connector, white in plenum chamber	N/A
29	BLK	Mass Air Flow Sensor	N/A
30	WHT/GRN	T-10 Connector, white in plenum chamber	N/A
31	--	Not Used	--
32	--	Not Used	--
33	GRY/BLU	Accelerator Pedal Position Sender Ground	N/A
34	BRN/WHT	Accelerator Pedal Position Sender Signal	N/A
35	WHT/BLU	Throttle Position Sensor Signal	N/A
36	GRY/RED	Throttle Position Sensor Power	N/A
37	GRN/BRN	Instrument Panel Control Module (to 4/05)	N/A
38	BLK/WHT	Cruise Control Switch Ground	N/A
39	WHT/RED	Clutch Vacuum Vent Valve Switch Signal	N/A
40	BLU/RED	T-10 Connector, white in plenum (A/C connection)	N/A
41	GRN/GRY	T-10 Connector, orange in plenum (GRN for manual)	N/A
42	--	Not Used	N/A
43	GRY/WHT	T-10 Connector, white in plenum chamber	N/A
44	--	Not Used	--
45	--	Not Used	--
46	--	Not Used	--
47	LIL/YEL	T-10 Connector, white in plenum chamber	N/A
48	WHT/RED	T-10 Connector, white in plenum chamber	N/A
49	--	Not Used	--
50	--	Not Used	--

2.0L MFI AVH and AZG T121 Connector, *continued*

51	WHT	Heated Oxygen Sensor Signal	N/A
52	GRN	Heated Oxygen Sensor Ground	N/A
53	RED/LIL	Mass Air Flow Sensor	N/A
54	BLU/WHT	Vehicle Speed Signal (instrument panel)	N/A
55	WHT/YEL	Brake Vacuum Vent Valve Switch Signal	N/A
56	BLK/RED	T-10 Connector, black in plenum chamber	N/A
57	RED/YEL	Set Switch for Cruise Control Signal	N/A
58	ORN/BRN	ABS control module (TCM for automatic)	N/A
59	--	Not Used	--
60	ORN/BLK	ABS control module (TCM for automatic)	N/A
61	--	Not Used	--
62	RED/GRN	Fuse Box connection	N/A
63	BRN	Oxygen Sensor Power	N/A
64	LIL/RED	EVAP Canister Purge Regulator Valve Signal	N/A
65	LIL/WHT	Fuel Pump Relay	N/A
66	--	Not Used	--
67	--	Not Used	--
68	GRN/WHT	Oxygen Sensor Ground	N/A
69	GRN/RED	Oxygen Sensor Signal	N/A
70	GRN	Heated Oxygen Sensor Signal	N/A
71	BLK	Heated Oxygen Sensor Ground	N/A
72	GRN/WHT	Accelerator Pedal Position Sender Power	N/A
73	YEL/GRN	Throttle Position Sensor Ground	N/A
74	--	Not Used	--
75	BLU/GRY	Cruise Control Switch Signal	N/A
76	WHT	Cruise Control Switch Signal	N/A
77	--	Not Used	--
78	--	Not Used	--
79	--	Not Used	--
80	GRN/LIL	Leak Detection Pump Signal	N/A
81	GRN/RED	T-10 Connector, white in plenum chamber	N/A
82	WHT	Engine Speed Sensor Signal	N/A
83	LIL/RED	Throttle Drive Angle Sensor Ground	N/A
84	LIL/YEL	Throttle Drive Angle Sensor Signal	N/A
85	--	Not Used	--
86	GRN/LIL	Camshaft Position Sensor	N/A
87	--	Not Used	--
88	LIL/BLU	Cylinder 2 Fuel Injector Control	N/A
89	LIL/GRN	Cylinder 4 Fuel Injector Control	N/A
90	BRN	Engine Speed Sensor Signal	N/A
91	WHT	Throttle Drive Angle Sensor Power	N/A
92	LIL/WHT	Throttle Drive Angle Sensor Signal	N/A
93	BLU	Engine Coolant Temperature Sensor Control	N/A
94	LIL/RED	Ignition Coil 4 with Power Output Stage	N/A
95	LIL/BLU	Ignition Coil 2 with Power Output Stage	N/A
96	LIL	Cylinder 1 Fuel Injector Control	N/A
97	LIL/RED	Cylinder 3 Fuel Injector Control	N/A

2.0L MFI AVH and AZG T121 Connector, *continued*

98	GRN/YEL	Camshaft Position Sensor	N/A
99	BLU	Knock Sensor 1 and 2 Control	N/A
100	--	Not Used	--
101	--	Not Used	--
102	BLU/GRN	Ignition Coil 1 with Power Output Stage	N/A
103	BLU/YEL	Ignition Coil 3 with Power Output Stage	N/A
104	--	Not Used	--
105	--	Not Used	--
106	GRN	Knock Sensor 1 Signal	N/A
107	GRY	Knock Sensor 2 Signal	N/A
108	BRN/BLU	Sensor Ground	N/A
109	--	Not Used	--
110	--	Not Used	--
111	--	Not Used	--
112	--	Not Used	--
113	--	Not Used	--
114	--	Not Used	--
115	--	Not Used	--
116	--	Not Used	--
117	LIL	Throttle Drive Motor Power	N/A
118	LIL/WHT	Throttle Drive Motor Control	N/A
119	--	Not Used	--
120	--	Not Used	--
121	BLK/LIL	Ground	N/A

Golf, GTI, Jetta

2004–2005 Sedan and Wagon, 2004–2006 Golf 2.0L MFI BEV T121 Connector

ECM Pin #	Wire Color	Circuit Description (121-Pin)	Value at Hot Idle
1	BRN/RED	PCV Heating Element Control	N/A
2	BRN/RED	Ground	N/A
3	BLK	Instrument Panel wiring harness	N/A
4	--	Not Used	--
5	WHT	Heated Oxygen Sensor Power	N/A
6	BLK	Oxygen Sensor Power	N/A
7	--	Not Used	--
8	--	Not Used	--
9	--	Not Used	--
10	BLU	Oxygen Sensor Ground	N/A
11	RED	Oxygen Sensor Signal	N/A
12	--	Not Used	--
13	--	Not Used	--
14	--	Not Used	--
15	--	Not Used	--
16	--	Not Used	--
17	--	Not Used	--
18	--	Not Used	--

2004–2005 Sedan and Wagon, 2004–2006 Golf 2.0L MFI BEV T121 Connector, *continued*

19	BLU	T-10 Connector, white in plenum chamber	N/A
20	--	Not Used	--
21	LIL	ECM Power Supply Relay	N/A
22	BLU/GRN	Brake Booster Control Module	N/A
23	LIL	ECM Power Supply	N/A
24	BLU/GRN	Brake System Vacuum Pump	N/A
25	YEL/BRN	Leak Detection Pump Motor Power	N/A
26	LIL/BLU	Mass Air Flow Sensor	N/A
27	LIL/BLK	Mass Air Flow Sensor	N/A
28	BLU/LIL	T-10 Connector, white in plenum chamber	N/A
29	BLK	Mass Air Flow Sensor	N/A
30	--	Not Used	--
31	--	Not Used	--
32	--	Not Used	--
33	GRY/BLU	Accelerator Pedal Position Sender Ground	N/A
34	BRN/WHT	Accelerator Pedal Position Sender Signal	N/A
35	WHT/BLU	Throttle Position Sensor Signal	N/A
36	GRY/RED	Throttle Position Sensor Power	N/A
37	GRN/BRN	Instrument Panel Control Module (to 4/05)	N/A
38	BLK/WHT	Cruise Control Switch Ground	N/A
39	WHT/RED	Clutch Vacuum Vent Valve Switch Signal (man only)	N/A
40	BLU/RED	T-10 Connector, white in plenum (A/C connection)	N/A
41	GRN/GRY	T-10 Connector, orange in plenum (GRN for auto)	N/A
42	--	Not Used	--
43	GRY/WHT	T-10 Connector, white in plenum chamber	N/A
44	--	Not Used	--
45	--	Not Used	--
46	BRN/YEL	Secondary Air Injection Solenoid Valve Signal	N/A
47	--	Not Used	--
48	--	Not Used	--
49	WHT/LIL	Power Steering Pressure Switch Control	N/A
50	WHT/GRY	Power Steering Pressure Switch Signal	N/A
51	GRN/BLK	Heated Oxygen Sensor Signal	N/A
52	GRY/RED	Heated Oxygen Sensor Ground	N/A
53	RED/LIL	Mass Air Flow Sensor	N/A
54	BLU/WHT	Vehicle Speed Signal (instrument panel)	N/A
55	WHT/YEL	Brake Vacuum Vent Valve Switch Signal	N/A
56	BLK/RED	T-10 Connector, black in plenum chamber	N/A
57	RED/YEL	Set Switch for Cruise Control Signal	N/A
58	--	Not Used	--
59	--	Not Used	--
60	--	Not Used	--
61	--	Not Used	--
62	RED/GRN	Fuse Box connection	N/A
63	--	Not Used	--
64	LIL/RED	EVAP Canister Purge Regulator Valve Signal	N/A
65	LIL/WHT	Fuel Pump Relay	N/A

2004–2005 Sedan and Wagon, 2004–2006 Golf 2.0L MFI BEV T121 Connector, *continued*

66	GRY/BRN	Secondary Air Injection Pump Relay Control	N/A
67	GRN/LIL	T-10 Connector, white in plenum chamber	N/A
68	--	Not Used	--
69	--	Not Used	--
70	GRN	Heated Oxygen Sensor Signal	N/A
71	BLK	Heated Oxygen Sensor Signal	N/A
72	GRN/WHT	Accelerator Pedal Position Sender Power	N/A
73	YEL/GRN	Throttle Position Sensor Ground	N/A
74	--	Not Used	--
75	BLU/GRY	Cruise Control Switch Signal	N/A
76	WHT	Cruise Control Switch Signal	N/A
77	--	Not Used	--
78	--	Not Used	--
79	--	Not Used	--
80	GRN/LIL	Leak Detection Pump Signal	N/A
81	--	Not Used	--
82	WHT	Engine Speed Sensor Signal	N/A
83	LIL/WHT	Throttle Drive Angle Sensor Ground	N/A
84	BLU/GRY	Throttle Drive Angle Sensor Signal	N/A
85	BLU/GRN	Intake Air Temperature Sensor Control	N/A
86	LIL/YEL	Camshaft Position Sensor	N/A
87	--	Not Used	--
88	LIL/BLU	Cylinder 2 Fuel Injector Control	N/A
89	LIL/GRN	Cylinder 4 Fuel Injector Control	N/A
90	BRN	Engine Speed Sensor Signal	N/A
91	WHT/GRY	Throttle Drive Angle Sensor Power	N/A
92	BLU/WHT	Throttle Drive Angle Sensor Signal	N/A
93	GRY/YEL	Engine Coolant Temperature Sensor Control	N/A
94	BLK/LIL	Ignition Coil 4 with Power Output Stage	N/A
95	BLK/YEL	Ignition Coil 2 with Power Output Stage	N/A
96	LIL	Cylinder 1 Fuel Injector Control	N/A
97	LIL/RED	Cylinder 3 Fuel Injector Control	N/A
98	LIL/GRN	Camshaft Position Sensor	N/A
99	BLU	Knock Sensor 1 and 2 Control	N/A
100	--	Not Used	--
101	LIL/GRN	Charge Air Pressure Sensor	N/A
102	LIL/BLK	Ignition Coil 1 with Power Output Stage	N/A
103	BLK/BRN	Ignition Coil 3 with Power Output Stage	N/A
104	GRN/BRN	Wastegate Bypass Regulator Valve Signal	N/A
105	GRY/GRN	Turbocharger Recirculating Valve Signal	N/A
106	GRY	Knock Sensor 1 Signal	N/A
107	GRY	Knock Sensor 2	N/A
108	BRN/BLU	Sensor Ground	N/A
109	--	Not Used	--
110	--	Not Used	--
111	--	Not Used	--
112	--	Not Used	--

2004–2005 Sedan and Wagon, 2004–2006 Golf 2.0L MFI BEV T121 Connector, *continued*

113	--	Not Used	--
114	--	Not Used	--
115	GRN/WHT	Camshaft Adjustment, Valve One	N/A
116	--	Not Used	--
117	WHT	Throttle Drive Motor Power	N/A
118	LIL/BLK	Throttle Drive Motor Control	N/A
119	--	Not Used	--
120	--	Not Used	--
121	BLK/LIL	Ground	N/A

Golf, Jetta

2004–2006 Sedan, 2004–2005 Wagon 1.9L DFI BEW T94 plus T60 Connectors

ECM Pin #	Wire Color	Circuit Description (94-Pin)	Value at Hot Idle
1	BRN/RED	PVC Heating Element Power	N/A
2	BRN/RED	Ground Connection	N/A
3	RED/LIL	Wire Connection in Engine Compartment	N/A
4	BRN/RED	Intake Flap Motor	N/A
5	RED/LIL	Wire Connection in Engine Compartment	N/A
6	RED/LIL	Wire Connection in Engine Compartment	N/A
7	--	Not Used	--
8	--	Not Used	--
9	--	Not Used	--
10	GRY/RED	Heated Oxygen Sensor Ground	N/A
11	--	Not Used	--
12	--	Not Used	--
13	--	Not Used	--
14	--	Not Used	--
15	--	Not Used	--
16	--	Not Used	--
17	GRN/RED	Closed Throttle Position Switch Ground	N/A
18	BLK/LIL	T-6 Connector, Brown, in plenum chamber	N/A
19	--	Not Used	--
20	BLK/WHT	Cruise Control Switch Ground	N/A
21	BLU/GRY	Cruise Control Switch Signal	N/A
22	--	Not Used	--
23	--	Not Used	--
24	GRY/RED	Intake Manifold Change-Over Valve Signal	N/A
25	--	Not Used	--
26	--	Not Used	--
27	--	Not Used	--
28	GRN/BRN	T-10 Connector, orange in plenum chamber	N/A
29	--	Not Used	--
30	BLU/BLK	Glow Plug Actuation Control Module	N/A
31	--	Not Used	--
32	GRY/WHT	Heated Oxygen Sensor Signal	N/A
33	--	Not Used	--

2004–2006 Sedan, 2004–2005 Wagon 1.9L DFI BEW T94 plus T60 Connectors, *continued*

34	--	Not Used	--
35	--	Not Used	--
36	--	Not Used	--
37	--	Not Used	--
38	GRN	Intake Air Temperature Sensor Control	N/A
39	GRY/RED	Throttle Position Sensor Power	N/A
40	RED/GRN	Mass Airflow Sensor	N/A
41	--	Not Used	--
42	--	Not Used	--
43	WHT/RED	Clutch Vacuum Vent Valve Switch Control	N/A
44	WHT	Cruise Control Switch Signal	N/A
45	--	Not Used	--
46	--	Not Used	--
47	BRN/WHT	Coolant Fan Control Module	N/A
48	--	Not Used	--
49	BLU/YEL	Power Supply Relay	N/A
50	GRN	A/C Connection	N/A
51	WHT	Heated Oxygen Sensor Ground	N/A
52	LIL/YEL	Fuel Pump Relay	N/A
53	--	Not Used	--
54	GRN	Heated Oxygen Sensor Ground	N/A
55	--	Not Used	--
56	--	Not Used	--
57	--	Not Used	--
58	--	Not Used	--
59	WHT/BLU	Kick Down Switch Power	N/A
60	GRN	Mass Airflow Sensor	N/A
61	GRY/WHT	Throttle Position Sensor Ground	N/A
62	LIL/RED	Charge Air Pressure Sensor	N/A
63	GRN/WHT	Glow Plug Actuation Control Module	N/A
64	BRN/RED	T-10 Connector, white in plenum chamber	N/A
65	WHT/YEL	Brake Pedal Switch Control	N/A
66	ORN/BRN	T-10 Connector, white in plenum chamber	N/A
67	--	Not Used	--
68	--	Not Used	--
69	--	Not Used	--
70	BLK/BLU	Preheating Coolant Output Relay (low heat)	N/A
71	BLK/BRN	Preheating Coolant Output Relay (high heat)	N/A
72	GRY/WHT	T-10 Connector, orange in plenum chamber	N/A
73	--	Not Used	--
74	--	Not Used	--
75	--	Not Used	--
76	BRN	Intake Air Temperature Sensor Signal	N/A
77	BLK	Heated Oxygen Sensor Signal	N/A
78	YEL/BLK	Charge Air Pressure Sensor	N/A
79	--	Not Used	--
80	--	Not Used	--

2004–2006 Sedan, 2004–2005 Wagon 1.9L DFI BEW T94 plus T60 Connectors, *continued*

81	--	Not Used	--
82	BLK	Mass Airflow Sensor	N/A
83	GRY/BLU	Throttle Position Sensor Signal	N/A
84	YEL/GRN	Throttle Position Sensor Ground	N/A
85	BLU/RED	A/C Connection	N/A
86	--	Not Used	--
87	RED/BLK	Power Ground	N/A
88	RED	Cruise Control Set Switch Signal	N/A
89	ORN/BLK	T-10 Connector, white in plenum chamber	N/A
90	BLU/WHT	Vehicle Speed Sensor in instrument panel	N/A
91	--	Not Used	--
92	--	Not Used	--
93	--	Not Used	--
94	--	Not Used	--
ECM Pin #	Wire Color	Circuit Description (60-Pin)	Value at Hot Idle
1	BRN/GRY	Pump/Injector Valve for Cylinder 2	N/A
2	--	Not Used	--
3	--	Not Used	--
4	--	Not Used	--
5	--	Not Used	--
6	--	Not Used	--
7	--	Not Used	--
8	--	Not Used	--
9	GRN/BLK	Turbocharger Vane Position Sensor	N/A
10	--	Not Used	--
11	--	Not Used	--
12	BLK/GRN	Camshaft Position Sensor	N/A
13	--	Not Used	--
14	--	Not Used	--
15	BRN/WHT	EGR Cooler Switch Over Valve Signal	N/A
16	--	Not Used	--
17	--	Not Used	--
18	--	Not Used	--
19	--	Not Used	--
20	--	Not Used	--
21	--	Not Used	--
22	--	Not Used	--
23	LIL/GRY	EGR Potentiometer Signal	N/A
24	LIL/BLU	EGR Potentiometer Ground	N/A
25	BLU/GRN	Intake Flap Motor	N/A
26	YEL/GRN	Turbocharger Vane Position Sensor	N/A
27	--	Not Used	--
28	BLK/BLU	Camshaft Position Sensor	N/A
29	RED/BRN	Wastegate Bypass Regulator Valve	N/A
30	--	Not Used	--
31	BLK/YEL	Ground Connection	N/A
32	BLK/YEL	Ground Connection	N/A

2004–2006 Sedan, 2004–2005 Wagon 1.9L DFI BEW T94 plus T60 Connectors, *continued*

33	--	Not Used	--
34	--	Not Used	--
35	BLK/GRN	Fuse Connection (#37)	N/A
36	--	Not Used	--
37	--	Not Used	--
38	--	Not Used	--
39	LIL	Fuel Temperature Sensor Signal	N/A
40	BLK/YEL	Fuel Temperature Sensor Ground	N/A
41	--	Not Used	--
42	--	Not Used	--
43	WHT	Engine Speed Sensor Signal	N/A
44	--	Not Used	--
45	--	Not Used	--
46	BRN/GRN	Pump/Injector Valve for Cylinder 1	N/A
47	BRN/WHT	Pump/Injector Valve for Cylinder 3	N/A
48	BRN/BLK	Pump/Injector Valve for Cylinder 4	N/A
49	RED/BLU	EGR Vacuum Regulator Solenoid Valve	N/A
50	YEL/BLK	EGR Vacuum Regulator Solenoid Motor	N/A
51	--	Not Used	--
52	BLU/BRN	Engine Coolant Temperature Sensor Ground	N/A
53	BRN/GRN	Engine Coolant Temperature Sensor Signal	N/A
54	--	Not Used	--
55	--	Not Used	--
56	--	Not Used	--
57	--	Not Used	--
58	BRN	Engine Speed Sensor Signal	N/A
59	--	Not Used	--
60	BLK/RED	Intake Flap Motor	N/A

Jetta

2005 Sedan 2.0L MFI BBW T121 Connector

ECM Pin #	Wire Color	Circuit Description (121-Pin)	Value at Hot Idle
1	BRN/RED	PCV Heating Element Control	N/A
2	BRN/RED	Ground	N/A
3	BLK	Instrument Panel wiring harness	N/A
4	--	Not Used	--
5	WHT	Heated Oxygen Sensor Power	N/A
6	BLK/BRN	Heated Oxygen Sensor 2 Power	N/A
7	--	Not Used	--
8	--	Not Used	--
9	--	Not Used	--
10	BLU	Heated Oxygen Sensor 2 Ground	N/A
11	RED	Heated Oxygen Sensor 2 Signal	N/A
12	--	Not Used	--
13	--	Not Used	--

2005 Sedan 2.0L MFI BBW T121 Connector, *continued*

14	--	Not Used	--
15	--	Not Used	--
16	GRN/YEL	Fuel Pump Relay	N/A
17	--	Not Used	--
18	--	Not Used	--
19	--	Not Used	--
20	--	Not Used	--
21	--	Not Used	--
22	BLU/GRN	Brake System Vacuum Pump (A/T only)	N/A
23	LIL	ECM Power Supply Relay	N/A
24	--	Not Used	--
25	YEL/BRN	Leak Detection Pump Motor Power	N/A
26	LIL/GRN	Mass Air Flow Sensor	N/A
27	GRN	Mass Air Flow Sensor	N/A
28	BLU/LIL	T-10 Connector, white in plenum chamber	N/A
29	BLK	Mass Air Flow Sensor	N/A
30	RED/BLK	T-10 Connector, white in plenum chamber	N/A
31	GRN	Exhaust Gas Temperature Sensor 1 Ground	N/A
32	YEL/RED	T-10 Connector, white in plenum (A/C connection)	N/A
33	GRY/BLU	Accelerator Pedal Position Sender Ground	N/A
34	BRN/WHT	Accelerator Pedal Position Sender Signal	N/A
35	WHT/BLU	Throttle Position Sensor Signal	N/A
36	GRY/RED	Throttle Position Sensor Power	N/A
37	GRN/BRN	T-10 Connector, Orange in plenum chamber	N/A
38	BLK/WHT	Cruise Control Switch Ground	N/A
39	WHT/RED	Clutch Vacuum Vent Valve Switch Signal	N/A
40	BLU/RED	T-10 Connector, white in plenum (A/C connection)	N/A
41	GRN/GRY	T-10 Connector, orange in plenum (GRN for manual)	N/A
42	RED/BRN	T-10 Connector, white in plenum chamber	N/A
43	GRY/WHT	T-10 Connector, white in plenum chamber	N/A
44	--	Not Used	--
45	--	Not Used	--
46	GRY/BRN	Secondary Air Injection Pump Relay Control	N/A
47	--	Not Used	--
48	WHT/RED	T-10 Connector, white in plenum chamber	N/A
49	--	Not Used	--
50	BLK	Exhaust Gas Temperature Sensor 1 Signal	N/A
51	WHT	Heated Oxygen Sensor Signal	N/A
52	GRN	Heated Oxygen Sensor Ground	N/A
53	RED/LIL	Mass Air Flow Sensor	N/A
54	BLU/WHT	Vehicle Speed Signal (instrument panel)	N/A
55	WHT/YEL	Brake Vacuum Vent Valve Switch Signal	N/A
56	BLK/RED	T-10 Connector, black in plenum chamber	N/A
57	RED/YEL	Set Switch for Cruise Control Signal	N/A
58	ORN/BRN	ABS control module (TCM for automatic)	N/A
59	--	Not Used	--
60	ORN/BLK	ABS control module (TCM for automatic)	N/A

2005 Sedan 2.0L MFI BBW T121 Connector, *continued*

61	--	Not Used	--
62	RED/GRN	Fuse Box connection	N/A
63	BRN	Oxygen Sensor Power	N/A
64	LIL/RED	EVAP Canister Purge Regulator Valve Signal	N/A
65	LIL/WHT	Fuel Pump Relay	N/A
66	GRN/RED	T-10 Connector, white in plenum (A/C connection)	N/A
67	--	Not Used	--
68	GRN/WHT	Oxygen Sensor Ground	N/A
69	GRN/RED	Oxygen Sensor Signal	N/A
70	GRN	Heated Oxygen Sensor Signal	N/A
71	BLK	Heated Oxygen Sensor Ground	N/A
72	GRN/WHT	Accelerator Pedal Position Sender Power	N/A
73	YEL/GRN	Throttle Position Sensor Ground	N/A
74	--	Not Used	--
75	BLU/GRY	Cruise Control Switch Signal	N/A
76	WHT	Cruise Control Switch Signal	N/A
77	--	Not Used	--
78	--	Not Used	--
79	--	Not Used	--
80	GRN/LIL	Leak Detection Pump Signal	N/A
81	--	Not Used	--
82	WHT	Engine Speed Sensor Signal	N/A
83	LIL/RED	Throttle Drive Angle Sensor Ground	N/A
84	LIL/YEL	Throttle Drive Angle Sensor Signal	N/A
85	--	Not Used	--
86	GRN/LIL	Camshaft Position Sensor	N/A
87	--	Not Used	--
88	LIL/BLU	Cylinder 2 Fuel Injector Control	N/A
89	LIL/GRN	Cylinder 4 Fuel Injector Control	N/A
90	BRN	Engine Speed Sensor Signal	N/A
91	WHT	Throttle Drive Angle Sensor Power	N/A
92	LIL/WHT	Throttle Drive Angle Sensor Signal	N/A
93	GRN	Engine Coolant Temperature Sensor Control	N/A
94	LIL/RED	Ignition Coil 4 with Power Output Stage	N/A
95	LIL/BLU	Ignition Coil 2 with Power Output Stage	N/A
96	LIL	Cylinder 1 Fuel Injector Control	N/A
97	LIL/RED	Cylinder 3 Fuel Injector Control	N/A
98	GRN/YEL	Camshaft Position Sensor	N/A
99	BLU	Knock Sensor 1 and 2 Control	N/A
100	--	Not Used	--
101	--	Not Used	--
102	BLU/GRN	Ignition Coil 1 with Power Output Stage	N/A
103	BLU/YEL	Ignition Coil 3 with Power Output Stage	N/A
104	--	Not Used	--
105	--	Not Used	--
106	GRN	Knock Sensor 1 Signal	N/A
107	GRY	Knock Sensor 2 Signal	N/A

2005 Sedan 2.0L MFI BBW T121 Connector, *continued*

108	BRN/BLU	Sensor Ground	N/A
109	--	Not Used	--
110	--	Not Used	--
111	--	Not Used	--
112	--	Not Used	--
113	--	Not Used	--
114	--	Not Used	--
115	GRN/YEL	Camshaft Adjustment Valve 1 Control	N/A
116	--	Not Used	--
117	LIL	Throttle Drive Motor Power	N/A
118	LIL/WHT	Throttle Drive Motor Control	N/A
119	--	Not Used	--
120	--	Not Used	--
121	BLK/LIL	Ground	N/A

Jetta

2005 Sedan 2.5L MFI BGP and BGQ T121 Connector

ECM Pin #	Wire Color	Circuit Description (121-Pin)	Value at Hot Idle
1	BRN	Ground	N/A
2	BRN	Ground	N/A
3	RED/GRN	Power Supply Relay	N/A
4	--	Not Used	--
5	WHT	Heated Oxygen Sensor Power	N/A
6	--	Not Used	--
7	--	Not Used	--
8	--	Not Used	--
9	--	Not Used	--
10	--	Not Used	--
11	--	Not Used	--
12	--	Not Used	--
13	--	Not Used	--
14	--	Not Used	--
15	--	Not Used	--
16	--	Not Used	--
17	BLU	Engine Coolant Temperature Sensor (on radiator)	N/A
18	--	Not Used	--
19	--	Not Used	--
20	--	Not Used	--
21	BLK/BLU	Fuse Panel Connection	N/A
22	--	Not Used	--
23	BLK/GRY	Power Supply Relay	N/A
24	--	Not Used	--
25	--	Not Used	--
26	LIL	Mass Air Flow Sensor	N/A
27	WHT/BRN	Mass Air Flow Sensor	N/A

2005 Sedan 2.5L MFI BGP and BGQ T121 Connector, *continued*

28	BLU/WHT	Voltage Regulator	N/A
29	GRN/BRN	Mass Air Flow Sensor	N/A
30	--	Not Used	--
31	--	Not Used	--
32	--	Not Used	--
33	GRY/BLU	Accelerator Pedal Position Sender Ground	N/A
34	BRN/BLU	Accelerator Pedal Position Sender Signal	N/A
35	WHT/BLU	Throttle Position Sensor Signal	N/A
36	GRY/RED	Throttle Position Sensor Power	N/A
37	--	Not Used	--
38	BLK/WHT	Ground on Left Headlight	N/A
39	WHT/RED	Clutch Position Switch Signal	N/A
40	--	Not Used	--
41	--	Not Used	--
42	--	Not Used	--
43	GRY/WHT	Connection (K-diagnosis wire), instrument panel	N/A
44	--	Not Used	--
45	--	Not Used	--
46	BRN/LIL	Secondary Air Injection Pump Relay	N/A
47	--	Not Used	--
48	--	Not Used	--
49	--	Not Used	--
50	BRN	Engine Coolant Temperature Sensor (on radiator)	N/A
51	BLK	Heated Oxygen Sensor Signal	N/A
52	GRN	Heated Oxygen Sensor Ground	N/A
53	LIL	Mass Air Flow Sensor	N/A
54	--	Not Used	--
55	WHT/GRN	Brake Pedal Switch Signal	N/A
56	BLK/RED	Connection in interior wiring harness	N/A
57	--	Not Used	--
58	ORN/BRN	Power Can-Bus Low Connection 1	N/A
59	--	Not Used	--
60	ORN/BLK	Power Can-Bus High Connection 1	N/A
61	--	Not Used	--
62	RED/BLK	Power Supply Relay	N/A
63	BRN/GRN	Heated Oxygen Sensor Power	N/A
64	BLU/LIL	EVAP Purge Regulator Valve Signal	N/A
65	BLK/BRN	Power Supply Relay	N/A
66	YEL/LIL	T-4 Connectio	N/A
67	--	Not Used	--
68	WHT/YEL	Heated Oxygen Sensor Ground	N/A
69	GRN/YEL	Heated Oxygen Sensor Signal	N/A
70	GRY/RED	Heated Oxygen Sensor Signal	N/A
71	GRY/WHT	Heated Oxygen Sensor Ground	N/A
72	GRY/BLK	Accelerator Pedal Position Sender Power	N/A
73	YEL/GRN	Throttle Position Sensor Ground	N/A
74	--	Not Used	--

2005 Sedan 2.5L MFI BGP and BGQ T121 Connector, *continued*

75	--	Not Used	--
76	--	Not Used	--
77	--	Not Used	--
78	--	Not Used	--
79	--	Not Used	--
80	BRN/BLK	Leak Detection Pump Motor Signal	N/A
81	--	Not Used	--
82	LIL	Engine Speed Sensor Signal	N/A
83	BLK	Throttle Drive Angle Sensor Ground	N/A
84	LIL/YEL	Throttle Drive Angle Sensor Signal	N/A
85	--	Not Used	--
86	YEL/WHT	Camshaft Position Sensor Signal	N/A
87	--	Not Used	--
88	RED/GRY	Cylinder 4 Fuel Injector Control	N/A
89	RED/BLU	Cylinder 5 Fuel Injector Control	N/A
90	BRN/BLU	Engine Speed Sensor Signal	N/A
91	GRN	Throttle Drive Angle Sensor Power	N/A
92	BLK/LIL	Throttle Drive Angle Sensor Signal	N/A
93	GRN	Engine Coolant Temperature Sensor Signal	N/A
94	LIL/GRN	Ignition Coil 4 with Power Output Stage	N/A
95	LIL/WHT	Ignition Coil 5 with Power Output Stage	N/A
96	RED/BLK	Cylinder 1 Fuel Injector Control	N/A
97	RED/WHT	Cylinder 2 Fuel Injector Control	N/A
98	BLK/GRY	Engine Speed Sensor Ground	N/A
99	WHT/BRN	Knock Sensor 1 and 2 Control	N/A
100	--	Not Used	--
101	BLU	Manifold Absolute Pressure Sensor Control	N/A
102	LIL/GRY	Ignition Coil 1 with Power Output Stage	N/A
103	LIL/BLU	Ignition Coil 2 with Power Output Stage	N/A
104	--	Not Used	--
105	--	Not Used	--
106	BRN/RED	Knock Sensor 1 Signal	N/A
107	RED	Knock Sensor 2 Signal	N/A
108	BRN	Camshaft Position Sensor Ground	N/A
109	--	Not Used	--
110	BLU/LIL	Ignition Coil 3 with Power Output Stage	N/A
111	--	Not Used	--
112	RED/LIL	Cylinder 3 Fuel Injector Control	N/A
113	--	Not Used	--
114	--	Not Used	--
115	LIL/WHT	Camshaft Adjustment Valve 1 Control	N/A
116	--	Not Used	--
117	LIL	Throttle Drive Motor Power	N/A
118	BRN/LIL	Throttle Drive Motor Control	N/A
119	--	Not Used	--
120	--	Not Used	--

2005 Sedan 2.5L MFI BGP and BGQ T121 Connector, *continued*

121	BRN/BLK	Secondary Air Injection Solenoid Valve	N/A

Jetta

2005–2006 Sedan 1.9L DFI BRM T94 plus T60 Connectors

ECM Pin #	Wire Color	Circuit Description (94-Pin)	Value at Hot Idle
1	BRN	Ground Connection 2	N/A
2	BRN	Ground Connection 2	N/A
3	RED/GRN	Connection 3 in engine compartment wiring harness	N/A
4	BRN	Ground Connection 2	N/A
5	RED/GRN	Connection 3 in engine compartment wiring harness	N/A
6	RED/GRN	Connection 3 in engine compartment wiring harness	N/A
7	--	Not Used	--
8	--	Not Used	--
9	--	Not Used	--
10	BLK	Heated Oxygen Sensor Ground	N/A
11	--	Not Used	--
12	--	Not Used	--
13	--	Not Used	--
14	--	Not Used	--
15	GRY/BLK	Accelerator Position Sensor 2	N/A
16	--	Not Used	--
17	GRY/BLU	Accelerator Position Sensor 2	N/A
18	BLK/BLU	Fuse 29 on Fuse Panel	N/A
19	--	Not Used	--
20	BLK/WHT	Steering Column Electronic Systems Control Module	N/A
21	--	Not Used	--
22	--	Not Used	--
23	--	Not Used	--
24	--	Not Used	--
25	--	Not Used	--
26	--	Not Used	--
27	--	Not Used	--
28	--	Not Used	--
29	--	Not Used	--
30	LIL/YEL	Automatic Glow Time Control Module	N/A
31	--	Not Used	--
32	GRN	Heated Oxygen Sensor Signal	N/A
33	--	Not Used	--
34	--	Not Used	--
35	--	Not Used	--
36	--	Not Used	--
37	--	Not Used	--
38	YEL	Intake Air Temperature Sensor Signal	N/A
39	GRY/RED	Throttle Position Sensor	N/A
40	BLK/LIL	Mass Air Flow Sensor	N/A
41	--	Not Used	--

2005–2006 Sedan 1.9L DFI BRM T94 plus T60 Connectors, *continued*

42	--	Not Used	--
43	WHT/RED	Clutch Position Sensor	N/A
44	--	Not Used	--
45	--	Not Used	--
46	--	Not Used	--
47	YEL/LIL	T-4 Connection (Coolant Fan)	N/A
48	--	Not Used	--
49	BLK/GRY	Engine Control Module Power Supply Relay	N/A
50	--	Not Used	--
51	WHT	Heated Oxygen Sensor Power	N/A
52	BLK/BRN	Fuel Pump Relay	N/A
53	--	Not Used	--
54	GRY/WHT	Heated Oxygen Sensor Signal	N/A
55	--	Not Used	--
56	--	Not Used	--
57	--	Not Used	--
58	--	Not Used	--
59	--	Not Used	--
60	BLK/BRN	Mass Air Flow Sensor	N/A
61	BRN/BLU	Accelerator Position Sensor 2	N/A
62	GRY/BLK	Charge Air Pressure Sensor	N/A
63	LIL/GRY	Automatic Glow Time Control Module	N/A
64	BLU/WHT	Generator	N/A
65	WHT/GRN	Brake Pedal Switch	N/A
66	ORN/BRN	Powertrain CAN-Bus Low Connection 1	N/A
67	--	Not Used	--
68	--	Not Used	--
69	--	Not Used	--
70	--	Not Used	--
71	--	Not Used	--
72	GRN/WHT	Connection K-Diagnosis wire) in instrument panel	N/A
73	--	Not Used	--
74	--	Not Used	--
75	--	Not Used	--
76	BRN	Intake Air Temperature Sensor Ground	N/A
77	GRY/RED	Heated Oxygen Sensor Ground	N/A
78	LIL	Charge Air Pressure Sensor	N/A
79	--	Not Used	--
80	--	Not Used	--
81	--	Not Used	--
82	GRN/BRN	Mass Air Flow Sensor	N/A
83	WHT/BLU	Throttle Position Sensor	N/A
84	YEL/GRN	Throttle Position Sensor	N/A
85	--	Not Used	--
86	RED	Power Supply Relay (Terminal 50)	N/A
87	BLK/RED	Connection (53) in engine wiring harness	N/A
88	--	Not Used	--

2005–2006 Sedan 1.9L DFI BRM T94 plus T60 Connectors, *continued*

89	ORN/BLK	Powertrain CAN-Bus High Connection 1	N/A
90	--	Not Used	--
91	--	Not Used	--
92	--	Not Used	--
93	--	Not Used	--
94	--	Not Used	--
ECM Pin #	Wire Color	Circuit Description (60-Pin)	Value at Hot Idle
1	BRN/GRN	Pump/Injector Valve Control Cylinder 2	N/A
2	--	Not Used	--
3	--	Not Used	--
4	BLU/WHT	Intake Flap Motor	N/A
5	--	Not Used	--
6	--	Not Used	--
7	--	Not Used	--
8	--	Not Used	--
9	LIL/BLU	Turbocharger Van Position Sensor	N/A
10	--	Not Used	--
11	--	Not Used	--
12	BRN/GRN	Camshaft Position Sensor	N/A
13	--	Not Used	--
14	--	Not Used	--
15	GRN/BLK	EGR Recirculation Cooler Switch-Over Valve	N/A
16	--	Not Used	--
17	--	Not Used	--
18	--	Not Used	--
19	BLK/RED	Intake Flap Motor	N/A
20	--	Not Used	--
21	--	Not Used	--
22	--	Not Used	--
23	BRN/GRN	EGR Potentiometer	N/A
24	BLU/GRN	EGR Potentiometer	N/A
25	--	Not Used	--
26	BLK/RED	EGR Potentiometer	N/A
27	BLK/WHT	Camshaft Position Sensor	N/A
28	GRN/LILL	Camshaft Position Sensor	N/A
29	BRN/YEL	Wastegate Bypass Regulator Valve	N/A
30	BLK/WHT	Intake Flap Motor Power	N/A
31	BRN/YEL	Connection (screening), MFI wiring harness	N/A
32	BRN/YEL	Connection (injectors) in engine wiring harness	N/A
33	--	Not Used	--
34	--	Not Used	--
35	--	Not Used	--
36	--	Not Used	--
37	YEL	Engine Coolant Temperature Sensor (on radiator)	N/A
38	BRN	Engine Coolant Temperature Sensor (on radiator)	N/A
39	GRY/WHT	Fuel Temperature Sensor	N/A
40	GRY/RED	Fuel Temperature Sensor	N/A

2005–2006 Sedan 1.9L DFI BRM T94 plus T60 Connectors, *continued*

41	BRN	Intake Flap Motor	N/A
42	BLK/RED	Engine Speed Sensor Ground	N/A
43	--	Not Used	--
44	--	Not Used	--
45	--	Not Used	--
46	BRN/RED	Pump/Injector Valve Control Cylinder 1	N/A
47	BRN/WHT	Pump/Injector Valve Control Cylinder 3	N/A
48	BRN/BLU	Pump/Injector Valve Control Cylinder 4	N/A
49	GRN	EGR Vacuum Regulator Solenoid Valve	N/A
50	BRN/LIL	EGR Vacuum Regulator Solenoid Valve	N/A
51	--	Not Used	--
52	RED/GRY	Engine Coolant Temperature Sensor	N/A
53	BRN/GRY	Engine Coolant Temperature Sensor	N/A
54	--	Not Used	--
55	--	Not Used	--
56	GRN/GRY	Intake Flap Motor	N/A
57	BRN/BLU	Engine Speed Sensor Signal	N/A
58	LIL	Engine Speed Sensor Signal	N/A
59	--	Not Used	--
60	--	Not Used	--

Jetta

2006 Sedan 2.5L MFI BGP and BGQ T121 Connector

ECM Pin #	Wire Color	Circuit Description (121-Pin)	Value at Hot Idle
1	BRN	Ground	N/A
2	BRN	Ground	N/A
3	RED/GRN	Power Supply Relay	N/A
4	--	Not Used	--
5	WHT	Heated Oxygen Sensor Power	N/A
6	--	Not Used	--
7	--	Not Used	--
8	--	Not Used	--
9	--	Not Used	--
10	--	Not Used	--
11	--	Not Used	--
12	--	Not Used	--
13	--	Not Used	--
14	--	Not Used	--
15	--	Not Used	--
16	--	Not Used	--
17	YEL	Engine Coolant Temperature Sensor (on radiator)	N/A
18	--	Not Used	--
19	--	Not Used	--
20	--	Not Used	--
21	BLK/BLU	Fuse Panel Connection	N/A
22	--	Not Used	--

2006 Sedan 2.5L MFI BGP and BGQ T121 Connector, *continued*

23	BLK/GRY	Power Supply Relay	N/A
24	--	Not Used	--
25	BRN/BLK	Leak Detection Pump Motor Control	N/A
26	LIL	Mass Air Flow Sensor	N/A
27	WHT/BRN	Mass Air Flow Sensor	N/A
28	BLU/WHT	Voltage Regulator	N/A
29	GRN/BRN	Mass Air Flow Sensor	N/A
30	--	Not Used	--
31	--	Not Used	--
32	--	Not Used	--
33	GRY/BLU	Accelerator Pedal Position Sender Ground	N/A
34	BRN/BLU	Accelerator Pedal Position Sender Signal	N/A
35	WHT/BLU	Throttle Position Sensor Signal	N/A
36	GRY/RED	Throttle Position Sensor Power	N/A
37	--	Not Used	--
38	BLK/WHT	Ground on Left Headlight	N/A
39	WHT/RED	Clutch Pedal Switch Signal	N/A
40	--	Not Used	--
41	--	Not Used	--
42	--	Not Used	--
43	GRY/WHT	Connection (K-diagnosis wire), instrument panel	N/A
44	--	Not Used	--
45	--	Not Used	--
46	BRN/LIL	Secondary Air Injection Pump Relay	N/A
47	--	Not Used	--
48	--	Not Used	--
49	RED	Intake Air Temperature Sensor 2 Signal	N/A
50	BRN	Engine Coolant Temperature Sensor (on radiator)	N/A
51	BLK	Heated Oxygen Sensor Signal	N/A
52	GRN	Heated Oxygen Sensor Ground	N/A
53	LIL	Mass Air Flow Sensor	N/A
54	--	Not Used	--
55	WHT/GRN	Brake Pedal Switch Signal	N/A
56	BLK/RED	Connection in interior wiring harness	N/A
57	--	Not Used	--
58	ORN/BRN	Power Can-Bus Low Connection 1	N/A
59	--	Not Used	--
60	ORN/BLK	Power Can-Bus High Connection 1	N/A
61	--	Not Used	--
62	RED/BLK	Power Supply Relay	N/A
63	BRN/GRN	Heated Oxygen Sensor Power	N/A
64	BLU/LIL	EVAP Canister Purge Regulator Valve Signal	N/A
65	BLK/BRN	Power Supply Relay	N/A
66	YEL/LIL	T-4 Connection	N/A
67	--	Not Used	--
68	WHT/YEL	Heated Oxygen Sensor Ground	N/A
69	GRN/YEL	Heated Oxygen Sensor Signal	N/A

2006 Sedan 2.5L MFI BGP and BGQ T121 Connector, *continued*

70	GRY/RED	Heated Oxygen Sensor Signal	N/A
71	GRY/WHT	Heated Oxygen Sensor Ground	N/A
72	GRY/BLK	Accelerator Pedal Position Sender Power	N/A
73	YEL/GRN	Throttle Position Sensor Ground	N/A
74	--	Not Used	--
75	--	Not Used	--
76	--	Not Used	--
77	--	Not Used	--
78	--	Not Used	--
79	--	Not Used	--
80	BRN/BLK	Leak Detection Pump Motor Signal	N/A
81	--	Not Used	--
82	LIL	Engine Speed Sensor Signal	N/A
83	BLK	Throttle Drive Angle Sensor Ground	N/A
84	LIL/YEL	Throttle Drive Angle Sensor Signal	N/A
85	--	Not Used	--
86	YEL/WHT	Camshaft Position Sensor Signal	N/A
87	--	Not Used	--
88	RED/GRY	Cylinder 4 Fuel Injector Control	N/A
89	RED/BLU	Cylinder 5 Fuel Injector Control	N/A
90	BRN/BLU	Engine Speed Sensor Signal	N/A
91	GRN	Throttle Drive Angle Sensor Power	N/A
92	BLK/LIL	Throttle Drive Angle Sensor Signal	N/A
93	GRN	Engine Coolant Temperature Sensor Signal	N/A
94	LIL/GRN	Ignition Coil 4 with Power Output Stage	N/A
95	LIL/WHT	Ignition Coil 5 with Power Output Stage	N/A
96	RED/BLK	Cylinder 1 Fuel Injector Control	N/A
97	RED/WHT	Cylinder 2 Fuel Injector Control	N/A
98	BLK/GRY	Engine Speed Sensor Ground	N/A
99	WHT/BRN	Knock Sensor 1 and 2 Control	N/A
100	--	Not Used	--
101	BLU	Manifold Absolute Pressure Sensor Control	N/A
102	LIL/GRY	Ignition Coil 1 with Power Output Stage	N/A
103	LIL/BLU	Ignition Coil 2 with Power Output Stage	N/A
104	--	Not Used	--
105	--	Not Used	--
106	BRN/RED	Knock Sensor 1 Signal	N/A
107	RED	Knock Sensor 2 Signal	N/A
108	BRN	Camshaft Position Sensor Ground	N/A
109	--	Not Used	--
110	BLU/LIL	Ignition Coil 3 with Power Output Stage	N/A
111	--	Not Used	--
112	RED/LIL	Cylinder 3 Fuel Injector Control	N/A
113	--	Not Used	--
114	--	Not Used	--
115	LIL/WHT	Camshaft Adjustment Valve 1 Control	N/A
116	--	Not Used	--

2006 Sedan 2.5L MFI BGP and BGQ T121 Connector, *continued*

117	LIL	Throttle Drive Motor Power	N/A
118	BRN/LIL	Throttle Drive Motor Control	N/A
119	--	Not Used	--
120	--	Not Used	--
121	BRN/BLK	Secondary Air Injection Solenoid Valve	N/A

New Beetle

1998 Sedan 1.9L DFI ALH T80 Connector

ECM Pin #	Wire Color	Circuit Description (80-Pin)	Value at Hot Idle
1	BRN/RED	Power Ground	N/A
2	RED/LIL	T-10 Connector, White on ECM	N/A
3	LIL/GRY	Change-Over Valve for Intake Manifold Flap Control	N/A
4	BRN/BLU	Mass Air Flow Sensor	N/A
5	--	Not Used	--
6	GRN/BRN	Instrument Panel Control Module	N/A
7	--	Not Used	--
8	GRY/WHT	Kick Down Switch Power	N/A
9	WHT/YEL	Brake Vacuum Vent Valve Switch (for DFI)	N/A
10	RED	Cruise Control Set Switch Signal	N/A
11	YEL/GRN	Throttle Position Sensor Ground	N/A
12	GRY/BLU	Closed Throttle Position Switch Signal	N/A
13	GRY/GRN	Intake Air Temperature Sensor Control	N/A
14	--	Not Used	--
15	RED/BLU	Wastegate Bypass Regulator Valve Control	N/A
16	GRN	Cooling Fan Control Module	N/A
17	BLK/BLU	Relay for Pre-Heating Coolant (low heat)	N/A
18	--	Not Used	--
19	BLK/WHT	Cruise Control Switch Ground	N/A
20	RED/BLK	Ground	N/A
21	BLU/GRY	Cruise Control Switch Signal	N/A
22	BRN/RED	T-10 Connector, Green in plenum chamber	N/A
23	GRY/RED	Throttle Position Sensor Power	N/A
24	WHT/BLU	Throttle Position Sensor Signal	N/A
25	BRN/BLU	Sensor Ground	N/A
26	--	Not Used	--
27	BRN/RED	Power Ground	N/A
28	RED/LIL	Ground	N/A
29	BRN/WHT	EGR Vacuum Regulator Solenoid Valve Control	N/A
30	--	Not Used	--
31	--	Not Used	--
32	--	Not Used	--
33	BLU/YEL	Power Supply Relay	N/A
34	BLK/BRN	Relay for Pre-Heating Coolant (high heat)	N/A
35	WHT	Cruise Control Switch Signal	N/A
36	--	Not Used	--

1998 Sedan 1.9L DFI ALH T80 Connector, *continued*

37	BLU/YEL	Glow Plug Relay	N/A
38	--	Not Used	--
39	LIL/RED	Manifold Absolute Pressure Sensor	N/A
40	YEL/BLK	Manifold Absolute Pressure Sensor	N/A
41	BLU/GRN	T-10 Connector, Green in plenum chamber	N/A
42	LIL/WHT	Glow Plug Relay	N/A
43	WHT/BLU	T-10 Connector, Blue on ECM	N/A
44	GRY	Transmission Control Module (A/T only)	N/A
45	GRY/WHT	T-10 Connector, Orange on ECM	N/A
46	WHT/RED	Clutch Vacuum Vent Valve Switch Signal	N/A
47	BLK/LIL	Fuse box	N/A
48	BLU/RED	Cooling Fan Control Module	N/A
49	--	Not Used	--
50	LIL/RED	Mass Air Flow Sensor	N/A
51	BLU/WHT	Instrument Panel Control Module	N/A
52	BLU	Mass Air Flow Sensor	N/A
53	YEL/BLK	Fuel Temperature Sensor Control	N/A
54	BLU/BRN	Engine Coolant Temperature Sensor Ground	N/A
55	BLU	Needle Lift Sensor Signal	N/A
56	LIL/BLK	Modulating Piston Displacement Sensor Signal	N/A
57	GRY/GRN	Modulating Piston Displacement Sensor Ground	N/A
58	--	Not Used	--
59	BRN/RED	Connection to Diesel Direct Fuel Injection System	N/A
60	--	Not Used	--
61	--	Not Used	--
62	GRY	Needle Lift Sensor Ground	N/A
63	--	Not Used	--
64	WHT/GRN	Modulating Piston Displacement Sensor Signal	N/A
65	--	Not Used	--
66	BRN/RED	Connection to Diesel Direct Fuel Injection System	N/A
67	WHT	Engine Speed Sensor Signal	N/A
68	ORN/BLK	T-10 Connector, Orange on ECM	N/A
69	YEL	Engine Speed Sensor Signal	N/A
70	BRN/YEL	Engine Coolant Temperature Sensor Control	N/A
71	BRN/YEL	Shielding Ground	N/A
72	--	Not Used	--
73	--	Not Used	--
74	--	Not Used	--
75	ORN/BRN	T-10 Connector, Orange on ECM	N/A
76	BRN/BLU	Fuel Temperature Sensor Ground	N/A
77	BLK/WHT	Fuel Cut-Off Valve Control	N/A
78	--	Not Used	--
79	BRN/BLK	Cold Start Injector Control	N/A
80	BRN/RED	Connection to Diesel Direct Fuel Injection System	N/A

New Beetle

1998–2000 Sedan 2.0L MFI AEG T80 Connector

ECM Pin #	Wire Color	Circuit Description (80-Pin)	Value at Hot Idle
1	BLK/LIL	Power Ground	N/A
2	BRN/RED	Power Ground	N/A
3	BLK/LIL	Fuse 10 Connection	N/A
4	LIL/WHT	T-10 Connector, orange on ECM (white on early yrs)	N/A
5	--	Not Used	N/A
6	GRN/BRN	T-10 Connector, orange on ECM (white on early yrs)	N/A
7	--	Not Used	--
8	GRN	Coolant Fan Control	N/A
9	GRN	Coolant Fan Control Module (2000 only)	N/A
10	BLU/RED	Coolant Fan Control	N/A
11	LIL/RED	Mass Air Flow Sensor	N/A
12	BRN/BLU	Mass Air Flow Sensor	N/A
13	BLU	Mass Air Flow Sensor	N/A
14	--	Not Used	--
15	LIL/RED	EVAP Canister Purge Regulator Valve Signal	N/A
16	BRN/YEL	Leak Detection Pump Motor Signal	N/A
17	LIL/GRN	T-10 Connector, blue on ECM	N/A
18	--	Not Used	--
19	GRY/WHT	T-10 Connector, white on ECM	N/A
20	BLU/WHT	T-10 Connector, orange on ECM (white on early yrs)	N/A
21	WHT	T-10 Connector, white on ECM (1998-1999 only)	N/A
22	--	Not Used	--
23	--	Not Used	--
24	--	Not Used	--
25	BRN	Heated Oxygen Sensor Ground	N/A
26	BLK	Heated Oxygen Sensor Signal	N/A
27	BRN/BLK	Heated Oxygen Sensor Power	N/A
28	BRN/BLK	Heated Oxygen Sensor 2 Power	N/A
29	ORN/BRN	ABS Control Module	N/A
30	GRY/BRN	Secondary Air Injection Pump Relay Control	N/A
31	n/a	Low BUS Connection, Instrument Panel (2000 only)	N/A
32	--	Not Used	--
33	LIL/RED	Secondary Air Injection Solenoid Valve Signal	N/A
34	--	Not Used	--
35	--	Not Used	--
36	--	Not Used	--
37	GRN/LIL	Leak Detection Pump Signal	N/A
38	--	Not Used	--
39	--	Not Used	--
40	BRN/BLU	Mass Air Flow Sensor	N/A
41	ORN/BLK	ABS Control Module	N/A
42	--	Not Used	--
43	GRN/GRY	T-10 Connector, white on ECM (orange on early yrs)	N/A
44	--	Not Used	--

1998–2000 Sedan 2.0L MFI AEG T80 Connector, *continued*

45	--	Not Used	--
46	BLK/YEL	T-10 Connector (2000 only)	N/A
47	RED/BLK	T-10 Connector (2000 only)	N/A
48	--	Not Used	--
49	--	Not Used	--
50	--	Not Used	--
51	BRN	Heated Oxygen Sensor 2 Ground	N/A
52	BLK	Heated Oxygen Sensor 2 Signal	N/A
53	BLU	Engine Coolant Temperature Sensor	N/A
54	--	Not Used	--
55	--	Not Used	--
56	YEL	Engine Speed Sensor Signal	N/A
57	--	Not Used	--
58	LIL/RED	Cylinder 3 Fuel Injector Control	N/A
59	LIL/WHT	Throttle Position Actuator Motor Power	N/A
60	GRN/YEL	Knock Sensor 2 Signal	N/A
61	--	Not Used	--
62	LIL/RED	Camshaft Position Sensor	N/A
63	WHT	Engine Speed Sensor Signal	N/A
64	--	Not Used	--
65	LIL/BLU	Cylinder 4 Fuel Injector Control	N/A
66	LIL	Throttle Position Actuator Motor Control	N/A
67	BRN/BLU	Sensor Ground	N/A
68	GRN	Knock Sensor 1 Signal	N/A
69	WHT	Throttle Position Actuator Motor Signal	N/A
70	--	Not Used	--
71	GRN/BRN	Ignition Coil Control	N/A
72	--	Not Used	--
73	LIL	Cylinder 1 Fuel Injector Control	N/A
74	LIL/GRN	Throttle Position Sensor Signal	N/A
75	LIL/BLK	Throttle Position Sensor Ground	N/A
76	LIL/YEL	Camshaft Position Sensor Signal	N/A
77	--	Not Used	--
78	GRN/YEL	Ignition Coil Control	N/A
79	--	Not Used	--
80	LIL/GRN	Cylinder 2 Fuel Injector Control	N/A

New Beetle

1999 Sedan 1.9L DFI ALH T80 Connector

ECM Pin #	Wire Color	Circuit Description (80-Pin)	Value at Hot Idle
1	BRN/RED	Power Ground	N/A
2	RED/LIL	T-10 Connector, White on ECM	N/A
3	LIL/GRY	Change-Over Valve for Intake Manifold Flap Control	N/A
4	BRN/BLU	Mass Air Flow Sensor	N/A
5	--	Not Used	--

1999 Sedan 1.9L DFI ALH T80 Connector, *continued*

6	GRN/BRN	Instrument Panel Control Module	N/A
7	--	Not Used	--
8	GRY/WHT	Throttle Position Sensor Signal	N/A
9	WHT/YEL	Brake Vacuum Vent Valve Switch (for DFI)	N/A
10	RED	Cruise Control Set Switch Signal	N/A
11	YEL/GRN	Throttle Position Sensor Ground	N/A
12	GRY/BLU	Kick Down Switch Power	N/A
13	GRY/GRN	Intake Air Temperature Sensor Control	N/A
14	--	Not Used	--
15	BRN/YEL	Wastegate Bypass Regulator Valve Control	N/A
16	GRN	Cooling Fan Control Module	N/A
17	BLK/BLU	Relay for Pre-Heating Coolant (low heat)	N/A
18	--	Not Used	--
19	BLK/WHT	Cruise Control Switch Ground	N/A
20	RED/BLK	Ground	N/A
21	BLU/GRY	Cruise Control Switch Signal	N/A
22	BRN/RED	T-10 Connector, Orange on ECM	N/A
23	GRY/RED	Throttle Position Sensor Power	N/A
24	WHT/BLU	Closed Throttle Position Switch Signal	N/A
25	BRN/BLU	Sensor Ground	N/A
26	--	Not Used	--
27	BRN/RED	Power Ground	N/A
28	RED/LIL	Ground	N/A
29	RED/BLU	EGR Vacuum Regulator Solenoid Valve Control	N/A
30	--	Not Used	--
31	--	Not Used	--
32	--	Not Used	--
33	BLU	Power Supply Relay	N/A
34	BLK/BRN	Relay for Pre-Heating Coolant (high heat)	N/A
35	WHT	Cruise Control Switch Signal	N/A
36	--	Not Used	--
37	GRN/WHT	Glow Plug Relay	N/A
38	--	Not Used	--
39	LIL/RED	Manifold Absolute Pressure Sensor	N/A
40	YEL/BLK	Manifold Absolute Pressure Sensor	N/A
41	BLU/GRN	T-10 Connector, Orange on ECM	N/A
42	LIL/WHT	Glow Plug Relay	N/A
43	WHT/BLU	T-10 Connector, Blue on ECM	N/A
44	--	Not Used	--
45	GRY/WHT	T-10 Connector, Orange on ECM	N/A
46	WHT/RED	Clutch Vacuum Vent Valve Switch Signal	N/A
47	BLK/LIL	Fuse box	N/A
48	BLU/RED	Cooling Fan Control Module	N/A
49	--	Not Used	--
50	LIL/RED	Mass Air Flow Sensor	N/A
51	BLU/WHT	Instrument Panel Control Module	N/A
52	BLU	Mass Air Flow Sensor	N/A

1999 Sedan 1.9L DFI ALH T80 Connector, *continued*

53	YEL/BLK	Fuel Temperature Sensor Control	N/A
54	BLU/BRN	Engine Coolant Temperature Sensor Ground	N/A
55	BLU	Needle Lift Sensor Signal	N/A
56	LIL/BLK	Modulating Piston Displacement Sensor Signal	N/A
57	GRY/GRN	Modulating Piston Displacement Sensor Ground	N/A
58	--	Not Used	--
59	BRN/RED	Connection to Diesel Direct Fuel Injection System	N/A
60	--	Not Used	--
61	--	Not Used	--
62	GRY	Needle Lift Sensor Ground	N/A
63	--	Not Used	--
64	WHT/GRN	Modulating Piston Displacement Sensor Signal	N/A
65	--	Not Used	--
66	BRN/RED	Connection to Diesel Direct Fuel Injection System	N/A
67	BLU	Engine Speed Sensor Signal	N/A
68	ORN/BLK	T-10 Connector, Orange on ECM	N/A
69	GRY	Engine Speed Sensor Signal	N/A
70	BRN/GRN	Engine Coolant Temperature Sensor Control	N/A
71	BRN/YEL	Shielding Ground	N/A
72	--	Not Used	--
73	--	Not Used	--
74	--	Not Used	--
75	ORN/BRN	T-10 Connector, Orange on ECM	N/A
76	BRN/BLU	Fuel Temperature Sensor Ground	N/A
77	BLK/WHT	Fuel Cut-Off Valve Control	N/A
78	--	Not Used	--
79	BRN/BLK	Cold Start Injector Control	N/A
80	BRN/RED	Connection to Diesel Direct Fuel Injection System	N/A

New Beetle

1999 Sedan 1.8L MFI APH T121 Connector

ECM Pin #	Wire Color	Circuit Description (121-Pin)	Value at Hot Idle
1	BRN/RED	Ground	N/A
2	BRN/RED	Ground	N/A
3	BLK	Instrument Panel wiring harness	N/A
4	--	Not Used	--
5	BRN/BLK	Heated Oxygen Sensor (black) Power	N/A
6	--	Not Used	--
7	--	Not Used	--
8	--	Not Used	--
9	BRN/YEL	Secondary Air Injection Solenoid Valve	N/A
10	--	Not Used	--
11	--	Not Used	--
12	--	Not Used	--
13	--	Not Used	--
14	--	Not Used	--

1999 Sedan 1.8L MFI APH T121 Connector, *continued*

15	--	Not Used	--
16	--	Not Used	--
17	--	Not Used	--
18	--	Not Used	--
19	--	Not Used	--
20	--	Not Used	--
21	--	Not Used	--
22	--	Not Used	--
23	--	Not Used	--
24	--	Not Used	--
25	YEL/BRN	Leak Detection Pump Motor Power	N/A
26	--	Not Used	--
27	GRN	Mass Air Flow Sensor	N/A
28	BLU/LIL	T-10 Connector, White, behind instrument panel	N/A
29	BLK	Mass Air Flow Sensor	N/A
30	WHT/GRN	T-10 Connector, Green, behind instrument panel	N/A
31	--	Not Used	--
32	--	Not Used	--
33	GRY/BLU	Accelerator Pedal Position Sender Ground	N/A
34	BRN/WHT	Accelerator Pedal Position Sender Signal	N/A
35	WHT/BLU	Throttle Position Sensor Signal	N/A
36	GRY/RED	Throttle Position Sensor Power	N/A
37	GRN/BRN	Instrument Panel Control Module	N/A
38	BLK/WHT	Cruise Control Switch Ground	N/A
39	WHT/RED	Clutch Vacuum Vent Valve Switch Signal	N/A
40	--	Not Used	--
41	GRN	Coolant Fan Control Module	N/A
42	--	Not Used	--
43	GRY/WHT	T-10 Connector, Orange, behind instrument panel	N/A
44	--	Not Used	--
45	--	Not Used	--
46	--	Not Used	--
47	WHT/BLU	Instrument Panel relay	N/A
48	WHT/RED	T-10 Connector, Blue, behind instrument panel	N/A
49	--	Not Used	--
50	--	Not Used	--
51	YEL/BLK	Heated Oxygen Sensor (black) Ground	N/A
52	YEL/BLK	Heated Oxygen Sensor (black) Signal	N/A
53	RED/LIL	Mass Air Flow Sensor	N/A
54	BLU/WHT	Vehicle Speed Signal (instrument panel)	N/A
55	WHT/YEL	Brake Vacuum Vent Valve Switch Signal	N/A
56	BLK/RED	T-10 Connector, Blue, behind instrument panel	N/A
57	RED/YEL	Set Switch for Cruise Control Signal	N/A
58	ORN/BRN	ABS control module (TCM for automatic)	N/A
59	--	Not Used	--
60	ORN/BLK	ABS control module (TCM for automatic)	N/A
61	WHT	Coolant Fan Control Module	N/A

1999 Sedan 1.8L MFI APH T121 Connector, *continued*

62	RED/GRN	Fuse Box connection	N/A
63	BRN/BLK	Heated Oxygen Sensor (brown) Power	N/A
64	LIL/RED	EVAP Canister Purge Regulator Valve Signal	N/A
65	LIL/WHT	Fuel Pump Relay	N/A
66	GRY/BRN	Secondary Air Injection Pump Relay Control	N/A
67	--	Not Used	--
68	GRN	Heated Oxygen Sensor (brown) Ground	N/A
69	BLK	Heated Oxygen Sensor (brown) Signal	N/A
70	BLK	Heated Oxygen Sensor (black)	N/A
71	GRN/YEL	Heated Oxygen Sensor (black)	N/A
72	GRN/WHT	Accelerator Pedal Position Sender Power	N/A
73	YEL/GRN	Throttle Position Sensor Ground	N/A
74	--	Not Used	--
75	BLU/GRY	Cruise Control Switch Signal	N/A
76	WHT	Cruise Control Switch Signal	N/A
77	ORN/BRN	Transmission Control Module (A/T only)	N/A
78	--	Not Used	--
79	ORN/BLK	Transmission Control Module (A/T only)	N/A
80	GRN/LIL	Leak Detection Pump Signal	N/A
81	GRN/WHT	T-10 Connector, Orange, for fuel gauge	N/A
82	WHT	Engine Speed Sensor Signal	N/A
83	LIL/WHT	Throttle Drive Angle Sensor Ground	N/A
84	LIL/YEL	Throttle Drive Angle Sensor Signal	N/A
85	BLU/GRN	Intake Air Temperature Sensor Control	N/A
86	LIL/YEL	Camshaft Position Sensor	N/A
87	--	Not Used	--
88	LIL/BLU	Cylinder 2 Fuel Injector Control	N/A
89	LIL/GRN	Cylinder 4 Fuel Injector Control	N/A
90	BRN	Engine Speed Sensor Signal	N/A
91	WHT	Throttle Drive Angle Sensor Power	N/A
92	LIL	Throttle Drive Angle Sensor Signal	N/A
93	BLU	Engine Coolant Temperature Sensor Control	N/A
94	BLK/LIL	Ignition Coil 4 with Power Output Stage	N/A
95	BLK/YEL	Ignition Coil 2 with Power Output Stage	N/A
96	LIL	Cylinder 1 Fuel Injector Control	N/A
97	LIL/RED	Cylinder 3 Fuel Injector Control	N/A
98	LIL/GRN	Camshaft Position Sensor	N/A
99	BLU	Knock Sensor 1 and 2 Control	N/A
100	--	Not Used	--
101	LIL/GRY	Charge Air Pressure Sensor Signal	N/A
102	BLK/BLU	Ignition Coil 1 with Power Output Stage	N/A
103	BLK/BRN	Ignition Coil 3 with Power Output Stage	N/A
104	GRN/BRN	Wastegate Bypass Regulator Valve Control	N/A
105	LIL/BLK	Recirculating Valve for the Turbocharger Control	N/A
106	GRN	Knock Sensor 1 Signal	N/A
107	GRY	Knock Sensor 2 Signal	N/A
108	BRN/BLU	Sensor Ground	N/A

1999 Sedan 1.8L MFI APH T121 Connector, *continued*

109	--	Not Used	--
110	--	Not Used	--
111	--	Not Used	--
112	--	Not Used	--
113	--	Not Used	--
114	--	Not Used	--
115	--	Not Used	--
116	--	Not Used	--
117	WHT	Throttle Drive Motor Power	N/A
118	LIL/BLK	Throttle Drive Motor Control	N/A
119	--	Not Used	--
120	--	Not Used	--
121	--	Not Used	--

New Beetle

2000 Sedan 1.8L MFI APH T121 Connector

ECM Pin #	Wire Color	Circuit Description (121-Pin)	Value at Hot Idle
1	BRN/RED	Ground	N/A
2	BRN/RED	Ground	N/A
3	BLK	Instrument Panel wiring harness	N/A
4	--	Not Used	--
5	BRN/BLK	Heated Oxygen Sensor (black) Power	N/A
6	--	Not Used	--
7	--	Not Used	--
8	--	Not Used	--
9	BRN/YEL	Secondary Air Injection Solenoid Valve	N/A
10	--	Not Used	--
11	--	Not Used	--
12	--	Not Used	--
13	--	Not Used	--
14	--	Not Used	--
15	--	Not Used	--
16	--	Not Used	--
17	--	Not Used	--
18	--	Not Used	--
19	--	Not Used	--
20	--	Not Used	--
21	--	Not Used	--
22	--	Not Used	--
23	--	Not Used	--
24	--	Not Used	--
25	YEL/BRN	Leak Detection Pump Motor Power	N/A
26	--	Not Used	--
27	GRN	Mass Air Flow Sensor	N/A
28	BLU/LIL	T-10 Connector, White, behind instrument panel	N/A

2000 Sedan 1.8L MFI APH T121 Connector, *continued*

29	BLK	Mass Air Flow Sensor	N/A
30	WHT/GRN	T-10 Connector, White, behind instrument panel	N/A
31	--	Not Used	--
32	--	Not Used	--
33	GRY/BLU	Accelerator Pedal Position Sender Ground	N/A
34	BRN/WHT	Accelerator Pedal Position Sender Signal	N/A
35	WHT/BLU	Throttle Position Sensor Signal	N/A
36	GRY/RED	Throttle Position Sensor Power	N/A
37	GRN/BRN	Instrument Panel Control Module	N/A
38	BLK/WHT	Cruise Control Switch Ground	N/A
39	WHT/RED	Clutch Vacuum Vent Valve Switch Signal	N/A
40	BLU/RED	T-10 Connector, Orange, behind instrument panel	N/A
41	GRN	Coolant Fan Control Module	N/A
42	--	Not Used	--
43	GRY/WHT	T-10 Connector, Orange, behind instrument panel	N/A
44	--	Not Used	--
45	--	Not Used	--
46	--	Not Used	--
47	--	Not Used	--
48	WHT/RED	T-10 Connector, White, behind instrument panel	N/A
49	--	Not Used	--
50	--	Not Used	--
51	YEL/BLK	Heated Oxygen Sensor (black) Ground	N/A
52	YEL/BLK	Heated Oxygen Sensor (black) Signal	N/A
53	RED/LIL	Mass Air Flow Sensor	N/A
54	BLU/WHT	Vehicle Speed Signal (instrument panel)	N/A
55	WHT/YEL	Brake Vacuum Vent Valve Switch Signal	N/A
56	BLK/RED	T-10 Connector, Black, behind instrument panel	N/A
57	RED/YEL	Set Switch for Cruise Control Signal	N/A
58	ORN/BRN	ABS control module (TCM for automatic)	N/A
59	--	Not Used	--
60	ORN/BLK	ABS control module (TCM for automatic)	N/A
61	WHT	Coolant Fan Control Module	N/A
62	RED/GRN	Fuse Box connection	N/A
63	BRN/BLK	Heated Oxygen Sensor (brown) Power	N/A
64	LIL/RED	EVAP Canister Purge Regulator Valve Signal	N/A
65	LIL/WHT	Fuel Pump Relay	N/A
66	GRY/BRN	Secondary Air Injection Pump Relay Control	N/A
67	--	Not Used	--
68	GRN	Heated Oxygen Sensor (brown) Ground	N/A
69	BLK	Heated Oxygen Sensor (brown) Signal	N/A
70	BLK	Heated Oxygen Sensor (black)	N/A
71	GRN/YEL	Heated Oxygen Sensor (black)	N/A
72	GRN/WHT	Accelerator Pedal Position Sender Power	N/A
73	YEL/GRN	Throttle Position Sensor Ground	N/A
74	--	Not Used	--
75	BLU/GRY	Cruise Control Switch Signal	N/A

2000 Sedan 1.8L MFI APH T121 Connector, *continued*

76	WHT	Cruise Control Switch Signal	N/A
77	--	Not Used	--
78	--	Not Used	--
79	--	Not Used	--
80	GRN/LIL	Leak Detection Pump Signal	N/A
81	GRN/WHT	T-10 Connector, Orange, for fuel gauge	N/A
82	WHT	Engine Speed Sensor Signal	N/A
83	LIL/WHT	Throttle Drive Angle Sensor Ground	N/A
84	LIL/YEL	Throttle Drive Angle Sensor Signal	N/A
85	BLU/GRN	Intake Air Temperature Sensor Control	N/A
86	LIL/YEL	Camshaft Position Sensor	N/A
87	--	Not Used	--
88	LIL/BLU	Cylinder 2 Fuel Injector Control	N/A
89	LIL/GRN	Cylinder 4 Fuel Injector Control	N/A
90	BRN	Engine Speed Sensor Signal	N/A
91	WHT	Throttle Drive Angle Sensor Power	N/A
92	LIL	Throttle Drive Angle Sensor Signal	N/A
93	BLU	Engine Coolant Temperature Sensor Control	N/A
94	BLK/LIL	Ignition Coil 4 with Power Output Stage	N/A
95	BLK/YEL	Ignition Coil 2 with Power Output Stage	N/A
96	LIL	Cylinder 1 Fuel Injector Control	N/A
97	LIL/RED	Cylinder 3 Fuel Injector Control	N/A
98	LIL/GRN	Camshaft Position Sensor	N/A
99	BLU	Knock Sensor 1 and 2 Control	N/A
100	--	Not Used	--
101	LIL/GRY	Charge Air Pressure Sensor Signal	N/A
102	BLK/BLU	Ignition Coil 1 with Power Output Stage	N/A
103	BLK/BRN	Ignition Coil 3 with Power Output Stage	N/A
104	GRN/BRN	Wastegate Bypass Regulator Valve Control	N/A
105	LIL/BLK	Recirculating Valve for the Turbocharger Control	N/A
106	GRN	Knock Sensor 1 Signal	N/A
107	GRY	Knock Sensor 2 Signal	N/A
108	BRN/BLU	Sensor Ground	N/A
109	--	Not Used	--
110	--	Not Used	--
111	--	Not Used	--
112	--	Not Used	--
113	--	Not Used	--
114	--	Not Used	--
115	--	Not Used	--
116	--	Not Used	--
117	WHT	Throttle Drive Motor Power	N/A
118	LIL/BLK	Throttle Drive Motor Control	N/A
119	--	Not Used	--
120	--	Not Used	--
121	--	Not Used	--

New Beetle

2000–2001 New Beetle 1.9L DFI ALH T121 Connector

ECM Pin #	Wire Color	Circuit Description (121-Pin)	Value at Hot Idle
1	RED/LIL	Ground (wire connection in engine compartment)	N/A
2	RED/LIL	Ground (wire connection in engine compartment)	N/A
3	--	Not Used	--
4	BRN/RED	Ground	N/A
5	BRN/RED	Ground	N/A
6	ORN/BRN	T-10 Connector, White, behind instrument panel	N/A
7	ORN/WHT	T-10 Connector, White, behind instrument panel	N/A
8	--	Not Used	--
9	GRY	Transmission Control Module (A/T only)	N/A
10	--	Not Used	--
11	--	Not Used	--
12	YEL/GRN	Throttle Position Sensor Ground	N/A
13	--	Not Used	--
14	BLK/WHT	Cruise Control Switch Ground	N/A
15	--	Not Used	--
16	GRY/WHT	T-10 Connector, Orange, behind instrument panel	N/A
17	--	Not Used	--
18	BLU/YEL	Power Supply Relay	N/A
19	--	Not Used	--
20	BLU/WHT	Instrument Panel Control Module	N/A
21	BLK/BLU	Relay for Preheating Coolant (low heat output)	N/A
22	BLK/BRN	Relay for Preheating Coolant (high heat output)	N/A
23	--	Not Used	N/A
24	WHT/BLU	T-10 Connector, White, behind instrument panel	N/A
25	--	Not Used	--
26	--	Not Used	--
27	--	Not Used	--
28	GRN/WHT	T-10 Connector, White, behind instrument panel	N/A
29	GRN	Coolant Fan Control Module	N/A
30	RED/LIL	Mass Air Flow Sensor	N/A
31	LIL/RED	Charge Air Pressure Sensor Signal	N/A
32	--	Not Used	N/A
33	GRN/WHT	T-10 Connector, White, behind instrument panel	N/A
34	BLU/RED	T-10 Connector, Orange, behind instrument panel	N/A
35	--	Not Used	--
36	--	Not Used	--
37	BLK/LIL	Fuse Panel	N/A
38	BRN/RED	T-10 Connector, White, behind instrument panel	N/A
39	--	Not Used	--
40	--	Not Used	--
41	--	Not Used	--
42	LIL/WHT	Glow Plug Relay	N/A
43	--	Not Used	--
44	RED	Cruise Control Set Switch Signal	N/A

2000–2001 New Beetle 1.9L DFI ALH T121 Connector, *continued*

45	BLU/GRY	Cruise Control Switch Signal	N/A
46	WHT	Cruise Control Switch Signal	N/A
47	GRN/LIL	T-10 Connector, Orange, behind instrument panel	N/A
48	--	Not Used	--
49	BRN	Mass Air Flow Sensor	N/A
50	GRY/RED	Throttle Position Sensor Power	N/A
51	GRN/RED	Closed Throttle Position Switch Ground	N/A
52	BRN/BLU	Charge Air Pressure Sensor Control	N/A
53	--	Not Used	--
54	--	Not Used	--
55	--	Not Used	--
56	--	Not Used	--
57	--	Not Used	--
58	--	Not Used	--
59	--	Not Used	--
60	--	Not Used	--
61	RED/BLU	EGR Vacuum Regulator Solenoid Valve Control	N/A
62	RED/BRN	Wastegate Bypass Regulator Valve Control	N/A
63	WHT/BLU	Throttle Position Sensor Signal	N/A
64	--	Not Used	--
65	WHT/YEL	Brake Vacuum Vent Valve Switch (for cruise control)	N/A
66	WHT/RED	Clutch Vacuum Vent Valve Switch	N/A
67	--	Not Used	--
68	BRN/BLU	Mass Air Flow Sensor	N/A
69	GRY/BLU	Throttle Position Sensor Ground	N/A
70	GRY/WHT	Kick Down Switch Power	N/A
71	YEL/BLK	T-10 Connector, Orange, behind instrument panel	N/A
72	--	Not Used	--
73	GRY/GRN	T-10 Connector, Orange, behind instrument panel	N/A
74	--	Not Used	--
75	--	Not Used	--
76	--	Not Used	--
77	--	Not Used	--
78	--	Not Used	--
79	--	Not Used	--
80	--	Not Used	--
81	LIL/GRY	Change-Over Valve for Intake Manifold Flap Control	N/A
82	--	Not Used	--
83	--	Not Used	--
84	--	Not Used	--
85	--	Not Used	--
86	BRN/YEL	Ground	N/A
87	--	Not Used	--
88	--	Not Used	--
89	--	Not Used	--
90	--	Not Used	--
91	--	Not Used	--

2000–2001 New Beetle 1.9L DFI ALH T121 Connector, *continued*

92	--	Not Used	--
93	--	Not Used	--
94	--	Not Used	--
95	--	Not Used	--
96	--	Not Used	--
97	--	Not Used	--
98	--	Not Used	--
99	WHT/GRN	Modulating Piston Displacement Sensor Control	N/A
100	--	Not Used	--
101	BLU	Needle Lift Sensor Signal	N/A
102	YEL	Engine Speed Sensor Signal	N/A
103	BRN/BLU	Fuel Temperature Sensor Ground	N/A
104	BRN/GRN	Engine Coolant Temperature Sensor Signal	N/A
105	--	Not Used	--
106	GRY/GRN	Modulating Piston Displacement Sensor Signal	N/A
107	--	Not Used	--
108	LIL/WHT	Modulating Piston Displacement Sensor Control	N/A
109	GRY	Needle Lift Sensor Ground	N/A
110	WHT	Engine Speed Sensor Signal	N/A
111	YEL/RED	Fuel Temperature Sensor Power	N/A
112	BLU/BRN	Engine Coolant Temperature Sensor Ground	N/A
113	--	Not Used	--
114	BRN/BLK	Cold Start Injector Control	N/A
115	--	Not Used	--
116	BRN/RED	Quantity Adjuster Control	N/A
117	--	Not Used	--
118	--	Not Used	--
119	--	Not Used	--
120	BLK/WHT	Fuel Cut-off Valve Control	N/A
121	BRN/RED	Ground	N/A

New Beetle

2001–2002 Sedan 1.8L MFI AWP-AWV T121 Connector

ECM Pin #	Wire Color	Circuit Description (121-Pin)	Value at Hot Idle
1	BRN/RED	Ground	N/A
2	BRN/RED	Ground	N/A
3	BLK	Instrument Panel wiring harness	N/A
4	--	Not Used	--
5	BRN/BLK	Heated Oxygen Sensor (black) Power	N/A
6	--	Not Used	--
7	--	Not Used	--
8	--	Not Used	--
9	BRN/YEL	Secondary Air Injection Solenoid Valve	N/A
10	--	Not Used	--
11	--	Not Used	--
12	--	Not Used	--

2001–2002 Sedan 1.8L MFI AWP-AWV T121 Connector, *continued*

13	--	Not Used	--
14	--	Not Used	--
15	--	Not Used	--
16	--	Not Used	--
17	--	Not Used	--
18	--	Not Used	--
19	--	Not Used	--
20	--	Not Used	--
21	--	Not Used	--
22	BLU/GRN	Brake Booster Control Module	N/A
23	--	Not Used	--
24	--	Not Used	--
25	YEL/BRN	Leak Detection Pump Motor Power	N/A
26	--	Not Used	--
27	GRN	Mass Air Flow Sensor	N/A
28	BLU/LIL	T-10 Connector, White, behind instrument panel	N/A
29	BLK	Mass Air Flow Sensor	N/A
30	WHT/GRN	T-10 Connector, White, behind instrument panel	N/A
31	--	Not Used	--
32	--	Not Used	--
33	GRY/BLU	Accelerator Pedal Position Sender Ground	N/A
34	BRN/WHT	Acclcrator Pcdal Position Sender Signal	N/A
35	WHT/BLU	Throttle Position Sensor Signal	N/A
36	GRY/RED	Throttle Position Sensor Power	N/A
37	GRN/BRN	Instrument Panel Control Module	N/A
38	BLK/WHT	Cruise Control Switch Ground	N/A
39	WHT/RED	Clutch Vacuum Vent Valve Switch Signal	N/A
40	BLU/RED	T-10 Connector, Orange, behind instrument panel	N/A
41	GRN	Coolant Fan Control Module	N/A
42	--	Not Used	--
43	GRY/WHT	T-10 Connector, Orange, behind instrument panel	N/A
44	--	Not Used	--
45	--	Not Used	--
46	--	Not Used	--
47	LIL/YEL	T-10 Connector, White, behind instrument panel	N/A
48	WHT/RED	T-10 Connector, White, behind instrument panel	N/A
49	--	Not Used	--
50	--	Not Used	--
51	YEL/BLK	Heated Oxygen Sensor (black) Ground	N/A
52	YEL/BLK	Heated Oxygen Sensor (black) Signal	N/A
53	RED/LIL	Mass Air Flow Sensor	N/A
54	BLU/WHT	Vehicle Speed Signal (instrument panel)	N/A
55	WHT/YEL	Brake Vacuum Vent Valve Switch Signal	N/A
56	BLK/RED	T-10 Connector, Black, behind instrument panel	N/A
57	RED/YEL	Set Switch for Cruise Control Signal	N/A
58	ORN/BRN	ABS control module (TCM for automatic)	N/A
59	--	Not Used	--

2001–2002 Sedan 1.8L MFI AWP-AWV T121 Connector, *continued*

60	ORN/BLK	ABS control module (TCM for automatic)	N/A
61	WHT	Coolant Fan Control Module	N/A
62	RED/GRN	Fuse Box connection	N/A
63	BRN/BLK	Heated Oxygen Sensor (brown) Power	N/A
64	LIL/RED	EVAP Canister Purge Regulator Valve Signal	N/A
65	LIL/WHT	Fuel Pump Relay	N/A
66	GRY/BRN	Secondary Air Injection Pump Relay Control	N/A
67	--	Not Used	--
68	GRN	Heated Oxygen Sensor (brown) Ground	N/A
69	GRY/RED	Heated Oxygen Sensor (brown) Signal	N/A
70	GRN	Heated Oxygen Sensor (black)	N/A
71	BLK	Heated Oxygen Sensor (black)	N/A
72	GRN/WHT	Accelerator Pedal Position Sender Power	N/A
73	YEL/GRN	Throttle Position Sensor Ground	N/A
74	--	Not Used	--
75	BLU/GRY	Cruise Control Switch Signal	N/A
76	WHT	Cruise Control Switch Signal	N/A
77	ORN/BRN	Transmission Control Module (A/T only)	N/A
78	--	Not Used	--
79	ORN/BLK	Transmission Control Module (A/T only)	N/A
80	GRN/LIL	Leak Detection Pump Signal	N/A
81	GRN/WHT	T-10 Connector, Orange, for fuel gauge	N/A
82	WHT	Engine Speed Sensor Signal	N/A
83	LIL/WHT	Throttle Drive Angle Sensor Ground	N/A
84	LIL/YEL	Throttle Drive Angle Sensor Signal	N/A
85	BLU/GRN	Intake Air Temperature Sensor Control	N/A
86	LIL/YEL	Camshaft Position Sensor	N/A
87	--	Not Used	--
88	LIL/BLU	Cylinder 2 Fuel Injector Control	N/A
89	LIL/GRN	Cylinder 4 Fuel Injector Control	N/A
90	BRN	Engine Speed Sensor Signal	N/A
91	WHT/GRY	Throttle Drive Angle Sensor Power	N/A
92	BLU/WHT	Throttle Drive Angle Sensor Signal	N/A
93	GRY/YEL	Engine Coolant Temperature Sensor Control	N/A
94	BLK/LIL	Ignition Coil 4 with Power Output Stage	N/A
95	BLK/YEL	Ignition Coil 2 with Power Output Stage	N/A
96	LIL	Cylinder 1 Fuel Injector Control	N/A
97	LIL/RED	Cylinder 3 Fuel Injector Control	N/A
98	LIL/GRN	Camshaft Position Sensor	N/A
99	BLU	Knock Sensor 1 and 2 Control	N/A
100	--	Not Used	--
101	LIL/GRY	Charge Air Pressure Sensor Signal	N/A
102	BLK/BLU	Ignition Coil 1 with Power Output Stage	N/A
103	BLK/BRN	Ignition Coil 3 with Power Output Stage	N/A
104	GRN/BRN	Wastegate Bypass Regulator Valve Control	N/A
105	GRY/GRN	Recirculating Valve for the Turbocharger Control	N/A
106	GRY	Knock Sensor 1 Signal	N/A

2001–2002 Sedan 1.8L MFI AWP-AWV T121 Connector, *continued*

107	GRY	Knock Sensor 2 Signal	N/A
108	BRN/BLU	Sensor Ground	N/A
109	--	Not Used	--
110	--	Not Used	--
111	--	Not Used	--
112	--	Not Used	--
113	--	Not Used	--
114	--	Not Used	--
115	--	Not Used	--
116	--	Not Used	--
117	WHT	Throttle Drive Motor Power	N/A
118	LIL/BLK	Throttle Drive Motor Control	N/A
119	--	Not Used	--
120	--	Not Used	--
121	BLK/LIL	Ignition Coils 1-4 with Power Output Stage Ground	N/A

New Beetle

2001–2002 Sedan 2.0L MFI AVH and AZG T121 Connector

ECM Pin #	Wire Color	Circuit Description (121-Pin)	Value at Hot Idle
1	BRN/RED	Ground	N/A
2	BRN/RED	Ground	N/A
3	BLK	Instrument Panel wiring harness	N/A
4	--	Not Used	--
5	BRN/BLK	Heated Oxygen Sensor Power	N/A
6	--	Not Used	--
7	--	Not Used	--
8	--	Not Used	--
9	GRY/BRN	Secondary Air Injection Pump Relay Control	N/A
10	--	Not Used	--
11	--	Not Used	--
12	--	Not Used	--
13	--	Not Used	--
14	--	Not Used	--
15	--	Not Used	--
16	--	Not Used	--
17	--	Not Used	--
18	--	Not Used	--
19	--	Not Used	--
20	--	Not Used	--
21	LIL	ECM Power Supply Relay	N/A
22	--	Not Used	--
23	--	Not Used	--
24	--	Not Used	--
25	YEL/BRN	Leak Detection Pump Motor Power	N/A
26	LIL/GRN	Mass Air Flow Sensor	N/A

2001–2002 Sedan 2.0L MFI AVH and AZG T121 Connector, *continued*

27	GRN/RED	Mass Air Flow Sensor	N/A
28	BLU/LIL	T-10 Connector, white behind instrument panel	N/A
29	GRY/WHT	Mass Air Flow Sensor	N/A
30	--	Not Used	--
31	--	Not Used	--
32	--	Not Used	--
33	GRN/WHT	Accelerator Pedal Position Sender Ground	N/A
34	BRN/WHT	Accelerator Pedal Position Sender Signal	N/A
35	GRY/BLU	Throttle Position Sensor Signal	N/A
36	GRY/RED	Throttle Position Sensor Power	N/A
37	GRN/BRN	Instrument Panel Control Module	N/A
38	BLK/WHT	Cruise Control Switch Ground	N/A
39	WHT/RED	Clutch Vacuum Vent Valve Switch Signal	N/A
40	BLU/RED	T-10 Connector, Orange, behind instrument panel	N/A
41	GRN	T-10 Connector, Orange, behind instrument panel	N/A
42	--	Not Used	--
43	GRY/WHT	T-10 Connector, Orange, behind instrument panel	N/A
44	--	Not Used	--
45	--	Not Used	--
46	--	Not Used	--
47	--	Not Used	--
48	--	Not Used	--
49	--	Not Used	--
50	--	Not Used	--
51	ORN/LIL	Heated Oxygen Sensor Signal	N/A
52	ORN/BRN	Heated Oxygen Sensor Ground	N/A
53	RED/LIL	Mass Air Flow Sensor	N/A
54	BLU/WHT	Vehicle Speed Signal (instrument panel)	N/A
55	WHT/YEL	Brake Pedal Switch (cruise control) Control	N/A
56	BLK/RED	Brake Pedal Switch Signal	N/A
57	RED/YEL	Set Switch for Cruise Control Signal	N/A
58	ORN/BRN	ABS control module (TCM for automatic)	N/A
59	--	Not Used	--
60	ORN/BLK	ABS control module (TCM for automatic)	N/A
61	--	Not Used	--
62	RED/GRN	Fuse Box connection	N/A
63	BRN/BLK	Oxygen Sensor Power	N/A
64	LIL/RED	EVAP Canister Purge Regulator Valve Signal	N/A
65	LIL/WHT	Fuel Pump Relay	N/A
66	--	Not Used	--
67	--	Not Used	--
68	GRY/WHT	Oxygen Sensor Ground	N/A
69	GRY/RED	Oxygen Sensor Signal	N/A
70	GRN	Heated Oxygen Sensor Signal	N/A
71	BLK	Heated Oxygen Sensor Ground	N/A
72	GRN/BLU	Accelerator Pedal Position Sender Power	N/A

2001–2002 Sedan 2.0L MFI AVH and AZG T121 Connector, *continued*

73	YEL/GRN	Throttle Position Sensor Ground	N/A
74	--	Not Used	--
75	BLU/GRY	Cruise Control Switch Signal	N/A
76	WHT	Cruise Control Switch Signal	N/A
77	--	Not Used	--
78	--	Not Used	--
79	--	Not Used	--
80	GRN/LIL	Leak Detection Pump Signal	N/A
81	--	Not Used	--
82	WHT	Engine Speed Sensor Signal	N/A
83	LIL/RED	Throttle Drive Angle Sensor Ground	N/A
84	LIL/YEL	Throttle Drive Angle Sensor Signal	N/A
85	--	Not Used	--
86	GRN/LIL	Camshaft Position Sensor	N/A
87	--	Not Used	--
88	LIL/BLU	Cylinder 2 Fuel Injector Control	N/A
89	LIL/GRN	Cylinder 4 Fuel Injector Control	N/A
90	BRN	Engine Speed Sensor Signal	N/A
91	WHT	Throttle Drive Angle Sensor Power	N/A
92	LIL/WHT	Throttle Drive Angle Sensor Signal	N/A
93	BLU	Engine Coolant Temperature Sensor Control	N/A
94	LIL/RED	Ignition Coil 4 with Power Output Stage	N/A
95	LIL/BLU	Ignition Coil 2 with Power Output Stage	N/A
96	LIL	Cylinder 1 Fuel Injector Control	N/A
97	LIL/RED	Cylinder 3 Fuel Injector Control	N/A
98	GRN/YEL	Camshaft Position Sensor	N/A
99	BLU	Knock Sensor 1 and 2 Control	N/A
100	--	Not Used	--
101	--	Not Used	--
102	BLU/GRN	Ignition Coil 1 with Power Output Stage	N/A
103	BLU/YEL	Ignition Coil 3 with Power Output Stage	N/A
104	--	Not Used	--
105	--	Not Used	--
106	GRN	Knock Sensor 1 Signal	N/A
107	GRY	Knock Sensor 2 Signal	N/A
108	BRN/BLU	Sensor Ground	N/A
109	--	Not Used	--
110	--	Not Used	--
111	--	Not Used	--
112	--	Not Used	--
113	--	Not Used	--
114	--	Not Used	--
115	--	Not Used	--
116	--	Not Used	--
117	LIL	Throttle Drive Motor Power	N/A
118	LIL/WHT	Throttle Drive Motor Control	N/A

2001–2002 Sedan 2.0L MFI AVH and AZG T121 Connector, *continued*

119	--	Not Used	--
120	--	Not Used	--
121	LIL/RED	Ignition Coil (1-4) with Power Output Stage Ground	N/A

New Beetle

2002 New Beetle 1.9L DFI ALH T121 Connector

ECM Pin #	Wire Color	Circuit Description (121-Pin)	Value at Hot Idle
1	RED/LIL	Ground (wire connection in engine compartment)	N/A
2	RED/LIL	Ground (wire connection in engine compartment)	N/A
3	--	Not Used	--
4	BRN/RED	Ground	N/A
5	BRN/RED	Ground	N/A
6	ORN/BRN	T-10 Connector, White, behind instrument panel	N/A
7	ORN/WHT	T-10 Connector, White, behind instrument panel	N/A
8	--	Not Used	--
9	GRY	Transmission Control Module (A/T only)	N/A
10	--	Not Used	--
11	--	Not Used	--
12	YEL/GRN	Throttle Position Sensor Ground	N/A
13	--	Not Used	--
14	BLK/WHT	Cruise Control Switch Ground	N/A
15	--	Not Used	--
16	GRY/WHT	T-10 Connector, Orange, behind instrument panel	N/A
17	--	Not Used	--
18	BLU/YEL	Power Supply Relay	N/A
19	--	Not Used	--
20	BLU/WHT	Instrument Panel Control Module	N/A
21	BLK/BLU	Relay for Preheating Coolant (low heat output)	N/A
22	BLK/BRN	Relay for Preheating Coolant (high heat output)	N/A
23	--	Not Used	--
24	--	Not Used	--
25	--	Not Used	--
26	--	Not Used	--
27	--	Not Used	--
28	--	Not Used	--
29	GRN	Coolant Fan Control Module	N/A
30	RED/LIL	Mass Air Flow Sensor	N/A
31	LIL/RED	Charge Air Pressure Sensor Signal	N/A
32	--	Not Used	--
33	GRN/WHT	T-10 Connector, White, behind instrument panel	N/A
34	BLU/RED	T-10 Connector, Orange, behind instrument panel	N/A
35	--	Not Used	--
36	--	Not Used	--
37	BLK/LIL	Fuse Panel	N/A
38	BRN/RED	T-10 Connector, White, behind instrument panel	N/A
39	--	Not Used	--

2002 New Beetle 1.9L DFI ALH T121 Connector, *continued*

40	--	Not Used	--
41	--	Not Used	--
42	--	Not Used	--
43	--	Not Used	--
44	RED	Cruise Control Set Switch Signal	N/A
45	BLU/GRY	Cruise Control Switch Signal	N/A
46	WHT	Cruise Control Switch Signal	N/A
47	--	Not Used	--
48	--	Not Used	--
49	BRN	Mass Air Flow Sensor	N/A
50	GRY/RED	Throttle Position Sensor Power	N/A
51	GRN/RED	Closed Throttle Position Switch Ground	N/A
52	BRN/BLU	Charge Air Pressure Sensor Control	N/A
53	--	Not Used	--
54	--	Not Used	--
55	--	Not Used	--
56	--	Not Used	--
57	--	Not Used	--
58	--	Not Used	--
59	--	Not Used	--
60	--	Not Used	--
61	RED/BLU	EGR Vacuum Regulator Solenoid Valve Control	N/A
62	RED/BRN	Wastegate Bypass Regulator Valve Control	N/A
63	WHT/BLU	Throttle Position Sensor Signal	N/A
64	--	Not Used	N/A
65	WHT/YEL	Brake Vacuum Vent Valve Switch (for cruise control)	N/A
66	WHT/RED	Clutch Vacuum Vent Valve Switch (manual only)	N/A
67	--	Not Used	--
68	BRN/BLU	Mass Air Flow Sensor	N/A
69	GRY/BLU	Throttle Position Sensor Ground	N/A
70	GRY/WHT	Kick Down Switch Power	N/A
71	YEL/BLK	T-10 Connector, Orange, behind instrument panel	N/A
72	--	Not Used	--
73	GRY/GRN	T-10 Connector, Orange, behind instrument panel	N/A
74	--	Not Used	--
75	--	Not Used	--
76	--	Not Used	--
77	--	Not Used	--
78	--	Not Used	--
79	--	Not Used	--
80	--	Not Used	--
81	LIL/GRY	Change-Over Valve for Intake Manifold Flap Control	N/A
82	--	Not Used	--
83	--	Not Used	--
84	--	Not Used	--
85	--	Not Used	--
86	BRN/YEL	Ground	N/A

2002 New Beetle 1.9L DFI ALH T121 Connector, *continued*

87	--	Not Used	--
88	WHT/GRN	T-14 Connector, in plenum chamber to Fuse 10	N/A
89	--	Not Used	--
90	--	Not Used	--
91	--	Not Used	--
92	--	Not Used	--
93	--	Not Used	--
94	--	Not Used	--
95	--	Not Used	--
96	--	Not Used	--
97	--	Not Used	--
98	--	Not Used	--
99	WHT/GRN	Modulating Piston Displacement Sensor Control	N/A
100	--	Not Used	--
101	BLU	Needle Lift Sensor Signal	N/A
102	YEL	Engine Speed Sensor Signal	N/A
103	BRN/BLU	Fuel Temperature Sensor Ground	N/A
104	BRN/GRN	Engine Coolant Temperature Sensor Signal	N/A
105	--	Not Used	--
106	GRY/GRN	Modulating Piston Displacement Sensor Signal	N/A
107	--	Not Used	--
108	LIL/WHT	Modulating Piston Displacement Sensor Control	N/A
109	GRY	Needle Lift Sensor Ground	N/A
110	WHT	Engine Speed Sensor Signal	N/A
111	YEL/BLU	Fuel Temperature Sensor Power	N/A
112	BLU/BRN	Engine Coolant Temperature Sensor Ground	N/A
113	--	Not Used	--
114	BRN/BLK	Cold Start Injector Control	N/A
115	--	Not Used	--
116	BRN/RED	Quantity Adjuster Control	N/A
117	--	Not Used	--
118	--	Not Used	--
119	--	Not Used	--
120	BLK/WHT	Fuel Cut-off Valve Control	N/A
121	BRN/RED	Ground	N/A

New Beetle

2003 Sedan 2.0L MFI AVH and AZG T121 Connector

ECM Pin #	Wire Color	Circuit Description (121-Pin)	Value at Hot Idle
1	BRN/RED	Ground	N/A
2	BRN/RED	Ground	N/A
3	BLK	Instrument Panel wiring harness	N/A
4	--	Not Used	--
5	WHT	Heated Oxygen Sensor Power	N/A
6	--	Not Used	--
7	--	Not Used	--

2003 Sedan 2.0L MFI AVH and AZG T121 Connector, *continued*

8	--	Not Used	--
9	GRY/BRN	Secondary Air Injection Pump Relay Control	N/A
10	--	Not Used	--
11	--	Not Used	--
12	--	Not Used	--
13	--	Not Used	--
14	--	Not Used	--
15	--	Not Used	--
16	--	Not Used	--
17	--	Not Used	--
18	--	Not Used	--
19	--	Not Used	--
20	--	Not Used	--
21	LIL	ECM Power Supply Relay	N/A
22	BLU/GRN	Brake System Vacuum Pump Control (A/T only)	N/A
23	--	Not Used	--
24	--	Not Used	--
25	YEL/BRN	Leak Detection Pump Motor Power	N/A
26	LIL/GRN	Mass Air Flow Sensor	N/A
27	GRN/RED	Mass Air Flow Sensor	N/A
28	BLU/LIL	T-10 Connector, white behind instrument panel	N/A
29	GRY/WHT	Mass Air Flow Sensor	N/A
30	--	Not Used	--
31	--	Not Used	--
32	--	Not Used	--
33	GRN/WHT	Accelerator Pedal Position Sender Ground	N/A
34	BRN/WHT	Accelerator Pedal Position Sender Signal	N/A
35	GRY/BLU	Throttle Position Sensor Signal	N/A
36	GRY/RED	Throttle Position Sensor Power	N/A
37	GRN/BRN	Instrument Panel Control Module	N/A
38	BLK/WHT	Cruise Control Switch Ground	N/A
39	WHT/RED	Clutch Vacuum Vent Valve Switch Signal	N/A
40	BLU/RED	T-10 Connector, Orange, behind instrument panel	N/A
41	GRN	T-10 Connector, Orange, behind instrument panel	N/A
42	--	Not Used	--
43	GRY/WHT	T-10 Connector, Orange, behind instrument panel	N/A
44	--	Not Used	--
45	--	Not Used	--
46	--	Not Used	--
47	--	Not Used	--
48	--	Not Used	--
49	--	Not Used	--
50	--	Not Used	--
51	WHT	Heated Oxygen Sensor Signal	N/A
52	GRN	Heated Oxygen Sensor Ground	N/A
53	RED/LIL	Mass Air Flow Sensor	N/A
54	BLU/WHT	Vehicle Speed Signal (instrument panel)	N/A

2003 Sedan 2.0L MFI AVH and AZG T121 Connector, *continued*

55	WHT/YEL	Brake Pedal Switch (cruise control) Control	N/A
56	BLK/RED	Brake Pedal Switch Signal	N/A
57	RED/YEL	Set Switch for Cruise Control Signal	N/A
58	ORN/BRN	ABS control module (TCM for automatic)	N/A
59	--	Not Used	--
60	ORN/BLK	ABS control module (TCM for automatic)	N/A
61	--	Not Used	--
62	RED/GRN	Fuse Box connection	N/A
63	BRN	Oxygen Sensor Power	N/A
64	LIL/RED	EVAP Canister Purge Regulator Valve Signal	N/A
65	LIL/WHT	Fuel Pump Relay	N/A
66	--	Not Used	--
67	--	Not Used	--
68	BLK	Oxygen Sensor Ground	N/A
69	BLU	Oxygen Sensor Signal	N/A
70	GRN	Heated Oxygen Sensor Signal	N/A
71	BLK	Heated Oxygen Sensor Ground	N/A
72	GRN/BLU	Accelerator Pedal Position Sender Power	N/A
73	YEL/GRN	Throttle Position Sensor Ground	N/A
74	--	Not Used	--
75	BLU/GRY	Cruise Control Switch Signal	N/A
76	WHT	Cruise Control Switch Signal	N/A
77	--	Not Used	--
78	--	Not Used	--
79	--	Not Used	--
80	GRN/LIL	Leak Detection Pump Signal	N/A
81	--	Not Used	--
82	WHT	Engine Speed Sensor Signal	N/A
83	LIL/RED	Throttle Drive Angle Sensor Ground	N/A
84	LIL/YEL	Throttle Drive Angle Sensor Signal	N/A
85	--	Not Used	--
86	GRN/LIL	Camshaft Position Sensor	N/A
87	--	Not Used	--
88	LIL/BLU	Cylinder 2 Fuel Injector Control	N/A
89	LIL/GRN	Cylinder 4 Fuel Injector Control	N/A
90	BRN	Engine Speed Sensor Signal	N/A
91	WHT	Throttle Drive Angle Sensor Power	N/A
92	LIL/WHT	Throttle Drive Angle Sensor Signal	N/A
93	BLU	Engine Coolant Temperature Sensor Control	N/A
94	LIL/RED	Ignition Coil 4 with Power Output Stage	N/A
95	LIL/BLU	Ignition Coil 2 with Power Output Stage	N/A
96	LIL	Cylinder 1 Fuel Injector Control	N/A
97	LIL/RED	Cylinder 3 Fuel Injector Control	N/A
98	GRN/YEL	Camshaft Position Sensor	N/A
99	BLU	Knock Sensor 1 and 2 Control	N/A
100	--	Not Used	--
101	--	Not Used	--

2003 Sedan 2.0L MFI AVH and AZG T121 Connector, *continued*

102	BLU/GRN	Ignition Coil 1 with Power Output Stage	N/A
103	BLU/YEL	Ignition Coil 3 with Power Output Stage	N/A
104	--	Not Used	--
105	--	Not Used	--
106	GRN	Knock Sensor 1 Signal	N/A
107	GRY	Knock Sensor 2 Signal	N/A
108	BRN/BLU	Sensor Ground	N/A
109	--	Not Used	--
110	--	Not Used	--
111	--	Not Used	--
112	--	Not Used	--
113	--	Not Used	--
114	--	Not Used	--
115	--	Not Used	--
116	--	Not Used	--
117	LIL	Throttle Drive Motor Power	N/A
118	LIL/WHT	Throttle Drive Motor Control	N/A
119	--	Not Used	--
120	--	Not Used	--
121	LIL/RED	Ignition Coil (1-4) with Power Output Stage Ground	N/A

New Beetle

2003 Convertible 2.0L MFI BDC T121 Connector

ECM Pin #	Wire Color	Circuit Description (121-Pin)	Value at Hot Idle
1	BRN/RED	Ground	N/A
2	BRN/RED	Ground	N/A
3	BLK	Instrument Panel wiring harness	N/A
4	--	Not Used	--
5	WHT	Heated Oxygen Sensor Power	N/A
6	--	Not Used	--
7	--	Not Used	--
8	--	Not Used	--
9	GRY/BRN	Secondary Air Injection Solenoid Valve Signal	N/A
10	--	Not Used	--
11	--	Not Used	--
12	--	Not Used	--
13	--	Not Used	--
14	--	Not Used	--
15	--	Not Used	--
16	--	Not Used	--
17	--	Not Used	--
18	--	Not Used	--
19	--	Not Used	--
20	--	Not Used	--
21	LIL	ECM Power Supply Relay	N/A
22	BLU/GRN	Brake Booster Control Module	N/A

2003 Convertible 2.0L MFI BDC T121 Connector, *continued*

23	--	Not Used	--
24	--	Not Used	--
25	YEL/BRN	Leak Detection Pump Motor Power	N/A
26	LIL/GRN	Mass Air Flow Sensor	N/A
27	GRY/RED	Mass Air Flow Sensor	N/A
28	BLU/LIL	T-10 Connector, white, behind instrument panel	N/A
29	GRY/WHT	Mass Air Flow Sensor	N/A
30	--	Not Used	--
31	--	Not Used	--
32	--	Not Used	--
33	GRN/WHT	Accelerator Pedal Position Sender Ground	N/A
34	BRN/WHT	Accelerator Pedal Position Sender Signal	N/A
35	GRY/BLU	Throttle Position Sensor Signal	N/A
36	GRY/RED	Throttle Position Sensor Power	N/A
37	GRN/BRN	Instrument Panel Control Module (to 4/05)	N/A
38	BLK/WHT	Cruise Control Switch Ground	N/A
39	WHT/RED	Clutch Vacuum Vent Valve Switch Signal (man only)	N/A
40	BLU/RED	T-10 Connector, orange, behind instrument panel	N/A
41	GRN	Coolant Fan Control Module	N/A
42	--	Not Used	--
43	GRY/WHT	T-10 Connector, orange, behind instrument panel	N/A
44	--	Not Used	--
45	--	Not Used	--
46	--	Not Used	--
47	--	Not Used	--
48	--	Not Used	--
49	--	Not Used	--
50	--	Not Used	--
51	WHT/GRN	Heated Oxygen Sensor Signal	N/A
52	WHT/GRN	Heated Oxygen Sensor Signal	N/A
53	RED/LIL	Mass Air Flow Sensor	N/A
54	BLU/WHT	Vehicle Speed Signal (instrument panel)	N/A
55	--	Not Used	--
56	BLK/RED	T-10 Connector, black, behind instrument panel	N/A
57	RED/YEL	Set Switch for Cruise Control Signal	N/A
58	ORN/BRN	T-10 Connector, white, behind instrument panel	N/A
59	--	Not Used	--
60	ORN/BLK	T-10 Connector, white, behind instrument panel	N/A
61	--	Not Used	--
62	RED/GRN	Fuse Box connection	N/A
63	BRN/BLK	Oxygen Sensor Power	N/A
64	LIL/RED	EVAP Canister Purge Regulator Valve Signal	N/A
65	LIL/WHT	Fuel Pump Relay	N/A
66	--	Not Used	--
67	--	Not Used	--
68	BLK/BLU	Oxygen Sensor Ground	N/A
69	BLK/BLU	Oxygen Sensor Signal	N/A

2003 Convertible 2.0L MFI BDC T121 Connector, *continued*

70	GRN/BLK	Heated Oxygen Sensor Ground	N/A
71	GRN/BLK	Heated Oxygen Sensor Signal	N/A
72	GRN/BLU	Accelerator Pedal Position Sender Power	N/A
73	YEL/GRN	Throttle Position Sensor Ground	N/A
74	--	Not Used	--
75	BLU/GRY	Cruise Control Switch Signal	N/A
76	WHT	Cruise Control Switch Signal	N/A
77	--	Not Used	--
78	--	Not Used	--
79	--	Not Used	--
80	GRN/LIL	Leak Detection Pump Signal	N/A
81	--	Not Used	--
82	WHT	Engine Speed Sensor Signal	N/A
83	LIL/RED	Throttle Drive Angle Sensor Ground	N/A
84	LIL/YEL	Throttle Drive Angle Sensor Signal	N/A
85	--	Not Used	--
86	GRN/LIL	Camshaft Position Sensor	N/A
87	--	Not Used	--
88	LIL/BLU	Cylinder 2 Fuel Injector Control	N/A
89	LIL/GRN	Cylinder 4 Fuel Injector Control	N/A
90	BRN	Engine Speed Sensor Signal	N/A
91	WHT	Throttle Drive Angle Sensor Power	N/A
92	LIL/WHT	Throttle Drive Angle Sensor Signal	N/A
93	BLU	Engine Coolant Temperature Sensor Control	N/A
94	LIL/RED	Ignition Coil 4 with Power Output Stage	N/A
95	LIL/BLU	Ignition Coil 2 with Power Output Stage	N/A
96	LIL	Cylinder 1 Fuel Injector Control	N/A
97	LIL/RED	Cylinder 3 Fuel Injector Control	N/A
98	GRN/YEL	Camshaft Position Sensor	N/A
99	BLU	Knock Sensor 1 and 2 Control	N/A
100	--	Not Used	--
101	--	Not Used	--
102	BLU/GRN	Ignition Coil 1 with Power Output Stage	N/A
103	BLU/YEL	Ignition Coil 3 with Power Output Stage	N/A
104	--	Not Used	--
105	LIL/WHT	Intake Manifold Change Over Valve	N/A
106	BLK	Knock Sensor 1 Signal	N/A
107	GRY	Knock Sensor 2	N/A
108	BRN/BLU	Sensor Ground	N/A
109	--	Not Used	--
110	--	Not Used	--
111	--	Not Used	--
112	--	Not Used	--
113	--	Not Used	--
114	--	Not Used	--
115	LIL/WHT	Intake Manifold Tuning Valve 2	N/A
116	--	Not Used	--

2003 Convertible 2.0L MFI BDC T121 Connector, *continued*

117	LIL	Throttle Drive Motor Power	N/A
118	LIL/WHT	Throttle Drive Motor Control	N/A
119	--	Not Used	--
120	--	Not Used	--
121	BLK/LIL	Ground	N/A

New Beetle

2003–2004 New Beetle 1.9L DFI ALH T121 Connector

ECM Pin #	Wire Color	Circuit Description (121-Pin)	Value at Hot Idle
1	RED/LIL	Ground (wire connection in engine compartment)	N/A
2	RED/LIL	Ground (wire connection in engine compartment)	N/A
3	--	Not Used	--
4	BRN/RED	Ground	N/A
5	BRN/RED	Ground	N/A
6	ORN/BRN	T-10 Connector, White, behind instrument panel	N/A
7	ORN/WHT	T-10 Connector, White, behind instrument panel	N/A
8	--	Not Used	--
9	GRY	Transmission Control Module (A/T only)	N/A
10	--	Not Used	--
11	--	Not Used	--
12	YEL/GRN	Throttle Position Sensor Ground	N/A
13	--	Not Used	--
14	BLK/WHT	Cruise Control Switch Ground	N/A
15	--	Not Used	--
16	GRY/WHT	T-10 Connector, Orange, behind instrument panel	N/A
17	--	Not Used	--
18	BLU/YEL	Power Supply Relay	N/A
19	--	Not Used	--
20	BLU/WHT	Instrument Panel Control Module	N/A
21	BLK/BLU	Relay for Preheating Coolant (low heat output)	N/A
22	BLK/BRN	Relay for Preheating Coolant (high heat output)	N/A
23	--	Not Used	--
24	--	Not Used	--
25	--	Not Used	--
26	--	Not Used	--
27	--	Not Used	--
28	--	Not Used	--
29	GRN	Coolant Fan Control Module	N/A
30	RED/LIL	Mass Air Flow Sensor	N/A
31	LIL/RED	Charge Air Pressure Sensor Signal	N/A
32	--	Not Used	--
33	GRN/WHT	T-10 Connector, White, behind instrument panel	N/A
34	BLU/RED	T-10 Connector, Orange, behind instrument panel	N/A
35	--	Not Used	--
36	--	Not Used	--

2003–2004 New Beetle 1.9L DFI ALH T121 Connector, *continued*

37	BLK/LIL	Fuse Panel	N/A
38	BRN/RED	T-10 Connector, White, behind instrument panel	N/A
39	--	Not Used	--
40	--	Not Used	--
41	--	Not Used	--
42	--	Not Used	--
43	--	Not Used	--
44	RED	Cruise Control Set Switch Signal	N/A
45	BLU/GRY	Cruise Control Switch Signal	N/A
46	WHT	Cruise Control Switch Signal	N/A
47	GRN/LIL	T-10 Connector, White, behind instrument panel	N/A
48	--	Not Used	--
49	BRN	Mass Air Flow Sensor	N/A
50	GRY/RED	Throttle Position Sensor Power	N/A
51	GRN/RED	Closed Throttle Position Switch Ground	N/A
52	BRN/BLU	Charge Air Pressure Sensor Control	N/A
53	--	Not Used	--
54	--	Not Used	--
55	--	Not Used	--
56	--	Not Used	--
57	--	Not Used	--
58	--	Not Used	--
59	--	Not Used	--
60	--	Not Used	--
61	RED/BLU	EGR Vacuum Regulator Solenoid Valve Control	N/A
62	RED/BRN	Wastegate Bypass Regulator Valve Control	N/A
63	WHT/BLU	Throttle Position Sensor Signal	N/A
64	--	Not Used	--
65	WHT/YEL	Brake Vacuum Vent Valve Switch (for cruise control)	N/A
66	WHT/RED	Clutch Vacuum Vent Valve Switch (manual only)	N/A
67	--	Not Used	--
68	BRN/BLU	Mass Air Flow Sensor	N/A
69	GRY/BLU	Throttle Position Sensor Ground	N/A
70	GRY/WHT	Kick Down Switch Power	N/A
71	YEL/BLK	T-10 Connector, Orange, behind instrument panel	N/A
72	--	Not Used	--
73	GRY/GRN	T-10 Connector, Orange, behind instrument panel	N/A
74	--	Not Used	--
75	--	Not Used	--
76	--	Not Used	--
77	--	Not Used	--
78	--	Not Used	--
79	--	Not Used	--
80	--	Not Used	--
81	LIL/GRY	Change-Over Valve for Intake Manifold Flap Control	N/A
82	--	Not Used	--
83	--	Not Used	--

2003–2004 New Beetle 1.9L DFI ALH T121 Connector, *continued*

84	--	Not Used	--
85	--	Not Used	--
86	BRN/YEL	Ground	N/A
87	--	Not Used	--
88	WHT/GRN	T-14 Connector, in plenum chamber to Fuse 10	N/A
89	--	Not Used	--
90	--	Not Used	--
91	--	Not Used	--
92	--	Not Used	--
93	--	Not Used	--
94	--	Not Used	--
95	--	Not Used	--
96	--	Not Used	--
97	--	Not Used	--
98	--	Not Used	--
99	WHT/GRN	Modulating Piston Displacement Sensor Control	N/A
100	--	Not Used	--
101	BLU	Needle Lift Sensor Signal	N/A
102	BRN	Engine Speed Sensor Signal	N/A
103	BRN/BLU	Fuel Temperature Sensor Ground	N/A
104	BRN/GRN	Engine Coolant Temperature Sensor Signal	N/A
105	--	Not Used	--
106	GRY/GRN	Modulating Piston Displacement Sensor Signal	N/A
107	--	Not Used	--
108	LIL/WHT	Modulating Piston Displacement Sensor Control	N/A
109	GRY	Needle Lift Sensor Ground	N/A
110	WHT	Engine Speed Sensor Signal	N/A
111	YEL/BLU	Fuel Temperature Sensor Power	N/A
112	BLU/BRN	Engine Coolant Temperature Sensor Ground	N/A
113	--	Not Used	--
114	BRN/BLK	Cold Start Injector Control	N/A
115	--	Not Used	--
116	BRN/RED	Quantity Adjuster Control	N/A
117	--	Not Used	--
118	--	Not Used	--
119	--	Not Used	--
120	BLK/WHT	Fuel Cut-off Valve Control	N/A
121	BRN/RED	Ground	N/A

New Beetle

2003–2004 Sedan, 2004 Convertible (AWV) 1.8L MFI AWP and AWV T121 Connector

ECM Pin #	Wire Color	Circuit Description (121-Pin)	Value at Hot Idle
1	BRN/RED	Ground	N/A
2	BRN/RED	Ground	N/A
3	BLK	Instrument Panel wiring harness	N/A

2003–2004 Sedan, 2004 Convertible (AWV) 1.8L MFI AWP and AWV T121 Connector, *continued*

4	--	Not Used	--
5	WHT	Heated Oxygen Sensor (black) Power	N/A
6	--	Not Used	--
7	--	Not Used	--
8	--	Not Used	--
9	BRN/YEL	Secondary Air Injection Solenoid Valve	N/A
10	--	Not Used	--
11	--	Not Used	--
12	--	Not Used	--
13	--	Not Used	--
14	--	Not Used	--
15	--	Not Used	--
16	--	Not Used	--
17	--	Not Used	--
18	--	Not Used	--
19	--	Not Used	--
20	--	Not Used	--
21	--	Not Used	--
22	BLU/GRN	Brake Booster Control Module	N/A
23	--	Not Used	--
24	--	Not Used	--
25	YEL/BRN	Leak Detection Pump Motor Power	N/A
26	--	Not Used	--
27	GRY/RED	Mass Air Flow Sensor	N/A
28	BLU/LIL	T-10 Connector, White, behind instrument panel	N/A
29	GRY/WHT	Mass Air Flow Sensor	N/A
30	--	Not Used	--
31	--	Not Used	--
32	--	Not Used	--
33	GRY/BLU	Accelerator Pedal Position Sender Ground	N/A
34	BRN/WHT	Accelerator Pedal Position Sender Signal	N/A
35	WHT/BLU	Throttle Position Sensor Signal	N/A
36	GRY/RED	Throttle Position Sensor Power	N/A
37	GRN/BRN	Instrument Panel Control Module	N/A
38	BLK/WHT	Cruise Control Switch Ground	N/A
39	WHT/RED	Clutch Vacuum Vent Valve Switch Signal	N/A
40	BLU/RED	T-10 Connector, Orange, behind instrument panel	N/A
41	GRN	Coolant Fan Control Module	N/A
42	--	Not Used	--
43	GRY/WHT	T-10 Connector, Orange, behind instrument panel	N/A
44	--	Not Used	--
45	--	Not Used	--
46	--	Not Used	--
47	--	Not Used	--
48	--	Not Used	--
49	--	Not Used	--
50	--	Not Used	--

2003–2004 Sedan, 2004 Convertible (AWV) 1.8L MFI AWP and AWV T121 Connector, *continued*

51	WHT	Heated Oxygen Sensor (black) Ground	N/A
52	GRN	Heated Oxygen Sensor (black) Signal	N/A
53	RED/LIL	Mass Air Flow Sensor	N/A
54	BLU/WHT	Vehicle Speed Signal (instrument panel)	N/A
55	WHT/YEL	Brake Vacuum Vent Valve Switch Signal	N/A
56	BLK/RED	T-10 Connector, Black, behind instrument panel	N/A
57	RED/YEL	Set Switch for Cruise Control Signal	N/A
58	ORN/BRN	ABS control module (TCM for automatic)	N/A
59	--	Not Used	--
60	ORN/BLK	ABS control module (TCM for automatic)	N/A
61	WHT	Coolant Fan Control Module	N/A
62	RED/GRN	Fuse Box connection	N/A
63	GRY/YEL	Heated Oxygen Sensor (brown) Power	N/A
64	LIL/RED	EVAP Canister Purge Regulator Valve Signal	N/A
65	LIL/WHT	Fuel Pump Relay	N/A
66	GRY/BRN	Secondary Air Injection Pump Relay Control	N/A
67	--	Not Used	--
68	BLK	Heated Oxygen Sensor (brown) Ground	N/A
69	BLU	Heated Oxygen Sensor (brown) Signal	N/A
70	GRN	Heated Oxygen Sensor (black)	N/A
71	BLK	Heated Oxygen Sensor (black)	N/A
72	GRN/WHT	Accelerator Pedal Position Sender Power	N/A
73	YEL/GRN	Throttle Position Sensor Ground	N/A
74	--	Not Used	--
75	BLU/GRY	Cruise Control Switch Signal	N/A
76	WHT	Cruise Control Switch Signal	N/A
77	ORN/BRN	Transmission Control Module (A/T only)	N/A
78	--	Not Used	--
79	ORN/BLK	Transmission Control Module (A/T only)	N/A
80	GRN/LIL	Leak Detection Pump Signal	N/A
81	--	Not Used	--
82	WHT	Engine Speed Sensor Signal	N/A
83	LIL/WHT	Throttle Drive Angle Sensor Ground	N/A
84	LIL/YEL	Throttle Drive Angle Sensor Signal	N/A
85	BLU/GRN	Intake Air Temperature Sensor Control	N/A
86	GRY/YEL	Camshaft Position Sensor	N/A
87	--	Not Used	--
88	LIL/BLU	Cylinder 2 Fuel Injector Control	N/A
89	LIL/GRN	Cylinder 4 Fuel Injector Control	N/A
90	BRN	Engine Speed Sensor Signal	N/A
91	WHT	Throttle Drive Angle Sensor Power	N/A
92	LIL	Throttle Drive Angle Sensor Signal	N/A
93	BLU	Engine Coolant Temperature Sensor Signal	N/A
94	BLK/LIL	Ignition Coil 4 with Power Output Stage	N/A
95	BLK/YEL	Ignition Coil 2 with Power Output Stage	N/A
96	GRY/LIL	Cylinder 1 Fuel Injector Control	N/A
97	LIL/RED	Cylinder 3 Fuel Injector Control	N/A

2003–2004 Sedan, 2004 Convertible (AWV) 1.8L MFI AWP and AWV T121 Connector, *continued*

98	LIL/GRN	Camshaft Position Sensor	N/A
99	BLU	Knock Sensor 1 and 2 Control	N/A
100	--	Not Used	--
101	LIL/GRY	Charge Air Pressure Sensor Signal	N/A
102	BLK/BLU	Ignition Coil 1 with Power Output Stage	N/A
103	BLK/BRN	Ignition Coil 3 with Power Output Stage	N/A
104	GRN/BRN	Wastegate Bypass Regulator Valve Control	N/A
105	LIL/BLK	Recirculating Valve for the Turbocharger Control	N/A
106	GRY	Knock Sensor 1 Signal	N/A
107	GRY	Knock Sensor 2 Signal	N/A
108	BRN/BLU	Sensor Ground	N/A
109	--	Not Used	--
110	--	Not Used	--
111	--	Not Used	--
112	--	Not Used	--
113	--	Not Used	--
114	--	Not Used	--
115	GRN/WHT	Camshaft Adjustment Valve 1 Control	N/A
116	--	Not Used	--
117	WHT	Throttle Drive Motor Power	N/A
118	LIL/BLK	Throttle Drive Motor Control	N/A
119	--	Not Used	--
120	--	Not Used	--
121	BLK/LIL	Ignition Coils 1-4 with Power Output Stage Ground	N/A

New Beetle

2004 Sedan 2.0L MFI BEV T121 Connector

ECM Pin #	Wire Color	Circuit Description (121-Pin)	Value at Hot Idle
1	BRN/RED	Ground	N/A
2	BRN/RED	Ground	N/A
3	BLK	Instrument Panel wiring harness	N/A
4	--	Not Used	--
5	WHT	Heated Oxygen Sensor Power	N/A
6	--	Not Used	--
7	--	Not Used	--
8	--	Not Used	--
9	GRY/BRN	Secondary Air Injection Solenoid Valve Signal	N/A
10	--	Not Used	--
11	--	Not Used	--
12	--	Not Used	--
13	--	Not Used	--
14	--	Not Used	--
15	--	Not Used	--
16	--	Not Used	--
17	--	Not Used	--

2004 Sedan 2.0L MFI BEV T121 Connector, *continued*

18	--	Not Used	--
19	--	Not Used	--
20	--	Not Used	--
21	LIL	ECM Power Supply Relay	N/A
22	BLU/GRN	Brake Booster Control Module	N/A
23	--	Not Used	--
24	--	Not Used	--
25	YEL/BRN	Leak Detection Pump Motor Power	N/A
26	LIL/GRN	Mass Air Flow Sensor	N/A
27	GRY/RED	Mass Air Flow Sensor	N/A
28	BLU/LIL	T-10 Connector, white, behind instrument panel	N/A
29	GRY/WHT	Mass Air Flow Sensor	N/A
30	--	Not Used	--
31	--	Not Used	--
32	--	Not Used	--
33	GRN/WHT	Accelerator Pedal Position Sender Ground	N/A
34	BRN/WHT	Accelerator Pedal Position Sender Signal	N/A
35	GRY/BLU	Throttle Position Sensor Signal	N/A
36	GRY/RED	Throttle Position Sensor Power	N/A
37	GRN/BRN	Instrument Panel Control Module (to 4/05)	N/A
38	BLK/WHT	Cruise Control Switch Ground	N/A
39	WHT/RED	Clutch Vacuum Vent Valve Switch Signal (man only)	N/A
40	BLU/RED	T-10 Connector, orange, behind instrument panel	N/A
41	GRN	Coolant Fan Control Module	N/A
42	--	Not Used	--
43	GRY/WHT	T-10 Connector, orange, behind instrument panel	N/A
44	--	Not Used	--
45	--	Not Used	--
46	--	Not Used	--
47	--	Not Used	--
48	--	Not Used	--
49	--	Not Used	--
50	--	Not Used	--
51	WHT	Heated Oxygen Sensor Signal	N/A
52	GRN	Heated Oxygen Sensor Signal	N/A
53	RED/LIL	Mass Air Flow Sensor	N/A
54	BLU/WHT	Vehicle Speed Signal (instrument panel)	N/A
55	WHT/YEL	Brake Pedal Switch	N/A
56	BLK/RED	T-10 Connector, black, behind instrument panel	N/A
57	RED/YEL	Set Switch for Cruise Control Signal	N/A
58	ORN/BRN	T-10 Connector, white, behind instrument panel	N/A
59	--	Not Used	--
60	ORN/BLK	T-10 Connector, white, behind instrument panel	N/A
61	--	Not Used	--
62	RED/GRN	Fuse Box connection	N/A
63	BRN	Oxygen Sensor Power	N/A
64	LIL/RED	EVAP Canister Purge Regulator Valve Signal	N/A

2004 Sedan 2.0L MFI BEV T121 Connector, *continued*

65	LIL/WHT	Fuel Pump Relay	N/A
66	--	Not Used	--
67	--	Not Used	--
68	BLK	Oxygen Sensor Ground	N/A
69	BLU	Oxygen Sensor Signal	N/A
70	GRN	Heated Oxygen Sensor Ground	N/A
71	BLK	Heated Oxygen Sensor Signal	N/A
72	GRN/BLU	Accelerator Pedal Position Sender Power	N/A
73	YEL/GRN	Throttle Position Sensor Ground	N/A
74	--	Not Used	--
75	BLU/GRY	Cruise Control Switch Signal	N/A
76	WHT	Cruise Control Switch Signal	N/A
77	--	Not Used	--
78	--	Not Used	--
79	--	Not Used	--
80	GRN/LIL	Leak Detection Pump Signal	N/A
81	--	Not Used	--
82	WHT	Engine Speed Sensor Signal	N/A
83	LIL/RED	Throttle Drive Angle Sensor Ground	N/A
84	LIL/YEL	Throttle Drive Angle Sensor Signal	N/A
85	--	Not Used	--
86	GRN/LIL	Camshaft Position Sensor	N/A
87	--	Not Used	--
88	LIL/BLU	Cylinder 2 Fuel Injector Control	N/A
89	LIL/GRN	Cylinder 4 Fuel Injector Control	N/A
90	BRN	Engine Speed Sensor Signal	N/A
91	WHT	Throttle Drive Angle Sensor Power	N/A
92	LIL/WHT	Throttle Drive Angle Sensor Signal	N/A
93	BLU	Engine Coolant Temperature Sensor Control	N/A
94	LIL/RED	Ignition Coil 4 with Power Output Stage	N/A
95	LIL/BLU	Ignition Coil 2 with Power Output Stage	N/A
96	LIL	Cylinder 1 Fuel Injector Control	N/A
97	LIL/RED	Cylinder 3 Fuel Injector Control	N/A
98	GRN/YEL	Camshaft Position Sensor	N/A
99	BLU	Knock Sensor 1 and 2 Control	N/A
100	--	Not Used	--
101	--	Not Used	--
102	BLU/GRN	Ignition Coil 1 with Power Output Stage	N/A
103	BLU/YEL	Ignition Coil 3 with Power Output Stage	N/A
104	--	Not Used	--
105	--	Not Used	--
106	GRN	Knock Sensor 1 Signal	N/A
107	GRY	Knock Sensor 2	N/A
108	BRN/BLU	Sensor Ground	N/A
109	--	Not Used	--
110	--	Not Used	--
111	--	Not Used	--

2004 Sedan 2.0L MFI BEV T121 Connector, *continued*

112	--	Not Used	--
113	--	Not Used	--
114	--	Not Used	--
115	--	Not Used	--
116	--	Not Used	--
117	LIL	Throttle Drive Motor Power	N/A
118	LIL/WHT	Throttle Drive Motor Control	N/A
119	--	Not Used	--
120	--	Not Used	--
121	BLK/LIL	Ground	N/A

New Beetle

2004 Sedan 2.0L MFI BEV T121 Connector

ECM Pin #	Wire Color	Circuit Description (121-Pin)	Value at Hot Idle
1	BRN/RED	Ground	N/A
2	BRN/RED	Ground	N/A
3	BLK	Instrument Panel wiring harness	N/A
4	--	Not Used	--
5	WHT	Heated Oxygen Sensor Power	N/A
6	--	Not Used	--
7	--	Not Used	--
8	--	Not Used	--
9	GRY/BRN	Secondary Air Injection Solenoid Valve Signal	N/A
10	--	Not Used	--
11	--	Not Used	--
12	--	Not Used	--
13	--	Not Used	--
14	--	Not Used	--
15	--	Not Used	--
16	--	Not Used	--
17	--	Not Used	--
18	--	Not Used	--
19	--	Not Used	--
20	--	Not Used	--
21	LIL	ECM Power Supply Relay	N/A
22	BLU/GRN	Brake Booster Control Module	N/A
23	--	Not Used	--
24	--	Not Used	--
25	YEL/BRN	Leak Detection Pump Motor Power	N/A
26	LIL/GRN	Mass Air Flow Sensor	N/A
27	GRY/RED	Mass Air Flow Sensor	N/A
28	BLU/LIL	T-10 Connector, white, behind instrument panel	N/A
29	GRY/WHT	Mass Air Flow Sensor	N/A
30	--	Not Used	--
31	--	Not Used	--

2004 Sedan 2.0L MFI BEV T121 Connector, *continued*

32	--	Not Used	--
33	GRN/WHT	Accelerator Pedal Position Sender Ground	N/A
34	BRN/WHT	Accelerator Pedal Position Sender Signal	N/A
35	GRY/BLU	Throttle Position Sensor Signal	N/A
36	GRY/RED	Throttle Position Sensor Power	N/A
37	GRN/BRN	Instrument Panel Control Module (to 4/05)	N/A
38	BLK/WHT	Cruise Control Switch Ground	N/A
39	WHT/RED	Clutch Vacuum Vent Valve Switch Signal (man only)	N/A
40	BLU/RED	T-10 Connector, orange, behind instrument panel	N/A
41	GRN	Coolant Fan Control Module	N/A
42	--	Not Used	--
43	GRY/WHT	T-10 Connector, orange, behind instrument panel	N/A
44	--	Not Used	--
45	--	Not Used	--
46	--	Not Used	--
47	--	Not Used	--
48	--	Not Used	--
49	--	Not Used	--
50	--	Not Used	--
51	WHT	Heated Oxygen Sensor Signal	N/A
52	GRN	Heated Oxygen Sensor Signal	N/A
53	RED/LIL	Mass Air Flow Sensor	N/A
54	BLU/WHT	Vehicle Speed Signal (instrument panel)	N/A
55	WHT/YEL	Brake Pedal Switch	N/A
56	BLK/RED	T-10 Connector, black, behind instrument panel	N/A
57	RED/YEL	Set Switch for Cruise Control Signal	N/A
58	ORN/BRN	T-10 Connector, white, behind instrument panel	N/A
59	--	Not Used	--
60	ORN/BLK	T-10 Connector, white, behind instrument panel	N/A
61	--	Not Used	--
62	RED/GRN	Fuse Box connection	N/A
63	BRN	Oxygen Sensor Power	N/A
64	LIL/RED	EVAP Canister Purge Regulator Valve Signal	N/A
65	LIL/WHT	Fuel Pump Relay	N/A
66	--	Not Used	--
67	--	Not Used	--
68	BLK	Oxygen Sensor Ground	N/A
69	BLU	Oxygen Sensor Signal	N/A
70	GRN	Heated Oxygen Sensor Ground	N/A
71	BLK	Heated Oxygen Sensor Signal	N/A
72	GRN/BLU	Accelerator Pedal Position Sender Power	N/A
73	YEL/GRN	Throttle Position Sensor Ground	N/A
74	--	Not Used	--
75	BLU/GRY	Cruise Control Switch Signal	N/A
76	WHT	Cruise Control Switch Signal	N/A
77	--	Not Used	--
78	--	Not Used	--

2004 Sedan 2.0L MFI BEV T121 Connector, *continued*

79	--	Not Used	--
80	GRN/LIL	Leak Detection Pump Signal	N/A
81	--	Not Used	--
82	WHT	Engine Speed Sensor Signal	N/A
83	LIL/RED	Throttle Drive Angle Sensor Ground	N/A
84	LIL/YEL	Throttle Drive Angle Sensor Signal	N/A
85	--	Not Used	--
86	GRN/LIL	Camshaft Position Sensor	N/A
87	--	Not Used	--
88	LIL/BLU	Cylinder 2 Fuel Injector Control	N/A
89	LIL/GRN	Cylinder 4 Fuel Injector Control	N/A
90	BRN	Engine Speed Sensor Signal	N/A
91	WHT	Throttle Drive Angle Sensor Power	N/A
92	LIL/WHT	Throttle Drive Angle Sensor Signal	N/A
93	BLU	Engine Coolant Temperature Sensor Control	N/A
94	LIL/RED	Ignition Coil 4 with Power Output Stage	N/A
95	LIL/BLU	Ignition Coil 2 with Power Output Stage	N/A
96	LIL	Cylinder 1 Fuel Injector Control	N/A
97	LIL/RED	Cylinder 3 Fuel Injector Control	N/A
98	GRN/YEL	Camshaft Position Sensor	N/A
99	BLU	Knock Sensor 1 and 2 Control	N/A
100	--	Not Used	--
101	--	Not Used	--
102	BLU/GRN	Ignition Coil 1 with Power Output Stage	N/A
103	BLU/YEL	Ignition Coil 3 with Power Output Stage	N/A
104	--	Not Used	--
105	LIL/WHT	Intake Manifold Change Over Valve	N/A
106	GRN	Knock Sensor 1 Signal	N/A
107	GRY	Knock Sensor 2	N/A
108	BRN/BLU	Sensor Ground	N/A
109	--	Not Used	--
110	--	Not Used	--
111	--	Not Used	--
112	--	Not Used	--
113	--	Not Used	--
114	--	Not Used	--
115	LIL/WHT	Intake Manifold Tuning Valve	N/A
116	--	Not Used	--
117	LIL	Throttle Drive Motor Power	N/A
118	LIL/WHT	Throttle Drive Motor Control	N/A
119	--	Not Used	--
120	--	Not Used	--
121	BLK/LIL	Ground	N/A

New Beetle

2004–2006 Sedan 2006 Wagon 1.9L DFI BEW T94 plus T60 Connectors

ECM Pin #	Wire Color	Circuit Description (94-Pin)	Value at Hot Idle
1	BRN/RED	PVC Heating Element Power	N/A
2	BRN/RED	Ground Connection	N/A
3	RED/LIL	Wire Connection in Engine Compartment	N/A
4	BRN/RED	Intake Flap Motor	N/A
5	RED/LIL	Wire Connection in Engine Compartment	N/A
6	RED/LIL	Wire Connection in Engine Compartment	N/A
7	--	Not Used	--
8	--	Not Used	--
9	--	Not Used	--
10	BLU	Heated Oxygen Sensor Ground	N/A
11	--	Not Used	--
12	--	Not Used	--
13	--	Not Used	--
14	--	Not Used	--
15	--	Not Used	--
16	--	Not Used	--
17	GRN/RED	Closed Throttle Position Switch Ground	N/A
18	BLK/LIL	T-6 Connector, Brown, in plenum chamber	N/A
19	--	Not Used	--
20	BLK/WHT	Cruise Control Switch Ground	N/A
21	BLU/GRY	Cruise Control Switch Signal	N/A
22	--	Not Used	--
23	--	Not Used	--
24	GRY/RED	Intake Manifold Change-Over Valve Signal	N/A
25	--	Not Used	--
26	--	Not Used	--
27	--	Not Used	--
28	GRN/BRN	T-10 Connector, orange in plenum chamber	N/A
29	--	Not Used	--
30	BLU/BLK	Glow Plug Actuation Control Module	N/A
31	--	Not Used	--
32	GRY	Heated Oxygen Sensor Signal	N/A
33	--	Not Used	--
34	--	Not Used	--
35	--	Not Used	--
36	--	Not Used	--
37	--	Not Used	--
38	GRY/GRN	Intake Air Temperature Sensor Control	N/A
39	GRY/RED	Throttle Position Sensor Power	N/A
40	RED/GRN	Mass Airflow Sensor	N/A
41	--	Not Used	--
42	--	Not Used	--
43	WHT/RED	Clutch Vacuum Vent Valve Switch Control	N/A
44	WHT	Cruise Control Switch Signal	N/A

2004–2006 Sedan 2006 Wagon 1.9L DFI BEW T94 plus T60 Connectors, *continued*

45	--	Not Used	--
46	--	Not Used	--
47	BRN/WHT	Coolant Fan Control Module	N/A
48	--	Not Used	--
49	BLU/YEL	Power Supply Relay	N/A
50	GRN	A/C Connection	N/A
51	WHT	Heated Oxygen Sensor Ground	N/A
52	LIL/YEL	Fuel Pump Relay	N/A
53	--	Not Used	--
54	WHT	Heated Oxygen Sensor Ground	N/A
55	--	Not Used	--
56	--	Not Used	--
57	--	Not Used	--
58	--	Not Used	--
59	WHT/BLU	Kick Down Switch Power	N/A
60	GRN	Mass Airflow Sensor	N/A
61	GRY/WHT	Throttle Position Sensor Ground	N/A
62	LIL/RED	Charge Air Pressure Sensor	N/A
63	GRN/WHT	Glow Plug Actuation Control Module	N/A
64	BRN/RED	T-10 Connector, white in plenum chamber	N/A
65	WHT/YEL	Brake Pedal Switch Control	N/A
66	ORN/BRN	T-10 Connector, white in plenum chamber	N/A
67	--	Not Used	--
68	--	Not Used	--
69	--	Not Used	--
70	BLK/BLU	Preheating Coolant Output Relay (low heat)	N/A
71	BLK/BRN	Preheating Coolant Output Relay (high heat)	N/A
72	GRY/WHT	T-10 Connector, orange in plenum chamber	N/A
73	--	Not Used	--
74	--	Not Used	--
75	--	Not Used	--
76	BRN	Intake Air Temperature Sensor Signal	N/A
77	GRN	Heated Oxygen Sensor Signal	N/A
78	YEL/BLK	Charge Air Pressure Sensor	N/A
79	--	Not Used	--
80	--	Not Used	--
81	--	Not Used	--
82	BLK	Mass Airflow Sensor	N/A
83	GRY/BLU	Throttle Position Sensor Signal	N/A
84	YEL/GRN	Throttle Position Sensor Ground	N/A
85	BLU/RED	A/C Connection	N/A
86	--	Not Used	--
87	RED/BLK	Power Ground	N/A
88	RED	Cruise Control Set Switch Signal	N/A
89	ORN/BLK	T-10 Connector, white in plenum chamber	N/A
90	BLU/WHT	Vehicle Speed Sensor in instrument panel	N/A
91	--	Not Used	--

2004–2006 Sedan 2006 Wagon 1.9L DFI BEW T94 plus T60 Connectors, *continued*

92	--	Not Used	--
93	--	Not Used	--
94	--	Not Used	--
ECM Pin #	**Wire Color**	**Circuit Description (60-Pin)**	**Value at Hot Idle**
1	BRN/GRY	Pump/Injector Valve for Cylinder 2	N/A
2	--	Not Used	--
3	--	Not Used	--
4	--	Not Used	--
5	--	Not Used	--
6	--	Not Used	--
7	--	Not Used	--
8	--	Not Used	--
9	GRN/BLK	Turbocharger Vane Position Sensor	N/A
10	--	Not Used	--
11	--	Not Used	--
12	BLK/GRN	Camshaft Position Sensor	N/A
13	--	Not Used	--
14	--	Not Used	--
15	BRN/WHT	EGR Cooler Switch Over Valve Signal	N/A
16	--	Not Used	--
17	--	Not Used	--
18	--	Not Used	--
19	--	Not Used	--
20	--	Not Used	--
21	--	Not Used	--
22	--	Not Used	--
23	LIL/GRY	EGR Potentiometer Signal	N/A
24	LIL/BLU	EGR Potentiometer Ground	N/A
25	BLU/GRN	Intake Flap Motor	N/A
26	YEL/GRN	Turbocharger Vane Position Sensor	N/A
27	--	Not Used	--
28	BLK/BLU	Camshaft Position Sensor	N/A
29	RED/BRN	Wastegate Bypass Regulator Valve	N/A
30	--	Not Used	--
31	BLK/YEL	Ground Connection	N/A
32	BLK/YEL	Ground Connection	N/A
33	--	Not Used	--
34	--	Not Used	--
35	BLK/GRN	Fuse Connection (#37)	N/A
36	--	Not Used	--
37	--	Not Used	--
38	--	Not Used	--
39	LIL	Fuel Temperature Sensor Signal	N/A
40	BLK/YEL	Fuel Temperature Sensor Ground	N/A
41	--	Not Used	--
42	--	Not Used	--
43	WHT	Engine Speed Sensor Signal	N/A

2004–2006 Sedan 2006 Wagon 1.9L DFI BEW T94 plus T60 Connectors, *continued*

44	--	Not Used	--
45	--	Not Used	--
46	BRN/GRN	Pump/Injector Valve for Cylinder 1	N/A
47	BRN/WHT	Pump/Injector Valve for Cylinder 3	N/A
48	BRN/BLK	Pump/Injector Valve for Cylinder 4	N/A
49	RED/BLU	EGR Vacuum Regulator Solenoid Valve	N/A
50	YEL/BLK	EGR Vacuum Regulator Solenoid Motor	N/A
51	--	Not Used	--
52	BLU/BRN	Engine Coolant Temperature Sensor Ground	N/A
53	BRN/GRN	Engine Coolant Temperature Sensor Signal	N/A
54	--	Not Used	--
55	--	Not Used	--
56	--	Not Used	--
57	--	Not Used	--
58	BRN	Engine Speed Sensor Signal	N/A
59	--	Not Used	--
60	BLK/RED	Intake Flap Motor	N/A

New Beetle

2005 Sedan and Convertible 2.0L MFI BEV T121 Connector

ECM Pin #	Wire Color	Circuit Description (121-Pin)	Value at Hot Idle
1	BRN/RED	Ground	N/A
2	BRN/RED	Ground	N/A
3	BLK	Instrument Panel wiring harness	N/A
4	--	Not Used	--
5	WHT	Heated Oxygen Sensor Power	N/A
6	BLK/BRN	Oxygen Sensor Power	N/A
7	--	Not Used	--
8	--	Not Used	--
9	GRY/BRN	Secondary Air Injection Solenoid Valve Signal	N/A
10	BLU	Oxygen Sensor Ground	N/A
11	RED	Oxygen Sensor Signal	N/A
12	--	Not Used	--
13	--	Not Used	--
14	--	Not Used	--
15	--	Not Used	--
16	--	Not Used	--
17	--	Not Used	--
18	--	Not Used	--
19	--	Not Used	--
20	--	Not Used	--
21	--	Not Used	--
22	BLU/GRN	Brake Booster Control Module	N/A
23	LIL	ECM Power Supply Relay	N/A
24	--	Not Used	--
25	YEL/BRN	Leak Detection Pump Motor Power	N/A

2005 Sedan and Convertible 2.0L MFI BEV T121 Connector, *continued*

26	LIL/GRN	Mass Air Flow Sensor	N/A
27	GRY/RED	Mass Air Flow Sensor	N/A
28	BLU/LIL	T-10 Connector, white, behind instrument panel	N/A
29	GRY/WHT	Mass Air Flow Sensor	N/A
30	--	Not Used	--
31	--	Not Used	--
32	--	Not Used	--
33	GRY/BLU	Accelerator Pedal Position Sender Ground	N/A
34	BRN/WHT	Accelerator Pedal Position Sender Signal	N/A
35	WHT/BLU	Throttle Position Sensor Signal	N/A
36	GRY/RED	Throttle Position Sensor Power	N/A
37	GRN/BRN	Instrument Panel Control Module (to 4/05)	N/A
38	BLK/WHT	Cruise Control Switch Ground	N/A
39	WHT/RED	Clutch Vacuum Vent Valve Switch Signal (man only)	N/A
40	BLU/RED	T-10 Connector, orange, behind instrument panel	N/A
41	GRN	Coolant Fan Control Module	N/A
42	--	Not Used	--
43	GRY/WHT	T-10 Connector, orange, behind instrument panel	N/A
44	--	Not Used	--
45	--	Not Used	--
46	--	Not Used	--
47	--	Not Used	--
48	--	Not Used	--
49	--	Not Used	--
50	--	Not Used	--
51	BLK	Heated Oxygen Sensor Signal	N/A
52	GRN	Heated Oxygen Sensor Signal	N/A
53	RED/LIL	Mass Air Flow Sensor	N/A
54	BLU/WHT	Vehicle Speed Signal (instrument panel)	N/A
55	WHT/YEL	Brake Pedal Switch	N/A
56	BLK/RED	T-10 Connector, black, behind instrument panel	N/A
57	RED/YEL	Set Switch for Cruise Control Signal	N/A
58	ORN/BRN	T-10 Connector, white, behind instrument panel	N/A
59	--	Not Used	--
60	ORN/BLK	T-10 Connector, white, behind instrument panel	N/A
61	--	Not Used	--
62	RED/GRN	Fuse Box connection	N/A
63	--	Not Used	--
64	LIL/RED	EVAP Canister Purge Regulator Valve Signal	N/A
65	LIL/WHT	Fuel Pump Relay	N/A
66	--	Not Used	--
67	--	Not Used	--
68	--	Not Used	--
69	--	Not Used	--
70	GRN	Heated Oxygen Sensor Ground	N/A
71	WHT	Heated Oxygen Sensor Signal	N/A
72	GRN/BLU	Accelerator Pedal Position Sender Power	N/A

2005 Sedan and Convertible 2.0L MFI BEV T121 Connector, *continued*

73	YEL/GRN	Throttle Position Sensor Ground	N/A
74	--	Not Used	--
75	BLU/GRY	Cruise Control Switch Signal	N/A
76	WHT	Cruise Control Switch Signal	N/A
77	--	Not Used	--
78	--	Not Used	--
79	--	Not Used	--
80	GRN/LIL	Leak Detection Pump Signal	N/A
81	--	Not Used	--
82	WHT	Engine Speed Sensor Signal	N/A
83	LIL/RED	Throttle Drive Angle Sensor Ground	N/A
84	LIL/YEL	Throttle Drive Angle Sensor Signal	N/A
85	--	Not Used	--
86	GRN/LIL	Camshaft Position Sensor	N/A
87	--	Not Used	--
88	LIL/BLU	Cylinder 2 Fuel Injector Control	N/A
89	LIL/GRN	Cylinder 4 Fuel Injector Control	N/A
90	BRN	Engine Speed Sensor Signal	N/A
91	WHT	Throttle Drive Angle Sensor Power	N/A
92	LIL/WHT	Throttle Drive Angle Sensor Signal	N/A
93	BLU	Engine Coolant Temperature Sensor Control	N/A
94	LIL/RED	Ignition Coil 4 with Power Output Stage	N/A
95	LIL/BLU	Ignition Coil 2 with Power Output Stage	N/A
96	LIL	Cylinder 1 Fuel Injector Control	N/A
97	LIL/RED	Cylinder 3 Fuel Injector Control	N/A
98	GRN/YEL	Camshaft Position Sensor	N/A
99	BLU	Knock Sensor 1 and 2 Control	N/A
100	--	Not Used	--
101	--	Not Used	--
102	BLU/GRN	Ignition Coil 1 with Power Output Stage	N/A
103	BLU/YEL	Ignition Coil 3 with Power Output Stage	N/A
104	--	Not Used	--
105	--	Not Used	--
106	GRN	Knock Sensor 1 Signal	N/A
107	GRY	Knock Sensor 2	N/A
108	BRN/BLU	Sensor Ground	N/A
109	--	Not Used	--
110	--	Not Used	--
111	--	Not Used	--
112	--	Not Used	--
113	--	Not Used	--
114	--	Not Used	--
115	--	Not Used	--
116	--	Not Used	--
117	LIL	Throttle Drive Motor Power	N/A
118	LIL/WHT	Throttle Drive Motor Control	N/A

2005 Sedan and Convertible 2.0L MFI BEV T121 Connector, *continued*

119	--	Not Used	--
120	--	Not Used	--
121	BLK/LIL	Ground	N/A

New Beetle

2005 Sedan (BNU), Convertible (BKF) 1.8L MFI BNU and BKF T121 Connector

ECM Pin #	Wire Color	Circuit Description (121-Pin)	Value at Hot Idle
1	BRN/RED	Ground	N/A
2	BRN/RED	Ground	N/A
3		Ignition Coils with Power Output Stages 1-4 Ground	N/A
4	--	Not Used	--
5	WHT	Heated Oxygen Sensor Power	N/A
6	BLK/BRN	Heated Oxygen Sensor Power	N/A
7	--	Not Used	--
8	--	Not Used	--
9	BRN/YEL	Secondary Air Injection Solenoid Valve	N/A
10	BLU	Heated Oxygen Sensor Ground	N/A
11	RED	Heated Oxygen Sensor Signal	N/A
12	--	Not Used	--
13	--	Not Used	--
14	--	Not Used	--
15	--	Not Used	--
16	--	Not Used	--
17	--	Not Used	--
18	--	Not Used	--
19	--	Not Used	--
20	--	Not Used	--
21	--	Not Used	--
22	--	Not Used	--
23	BLK	Instrument Panel wiring harness	N/A
24	--	Not Used	--
25	YEL/BRN	Leak Detection Pump Motor Power	N/A
26	LIL/GRN	Ground	N/A
27	BRN/RED	Mass Air Flow Sensor	N/A
28	BLU/LIL	T-10 Connector, White, behind instrument panel	N/A
29	GRY/WHT	Mass Air Flow Sensor	N/A
30	--	Not Used	--
31	--	Not Used	--
32	--	Not Used	--
33	GRY/BLU	Accelerator Pedal Position Sender Ground	N/A
34	BRN/WHT	Accelerator Pedal Position Sender Signal	N/A
35	WHT/BLU	Throttle Position Sensor Signal	N/A
36	GRY/RED	Throttle Position Sensor Power	N/A
37	GRN/BRN	Instrument Panel Control Module	N/A
38	BLK/WHT	Cruise Control Switch Ground	N/A

2005 Sedan (BNU), Convertible (BKF) 1.8L MFI BNU and BKF T121 Connector, *continued*

39	WHT/RED	Clutch Vacuum Vent Valve Switch Signal	N/A
40	BLU/RED	T-10 Connector, Orange, behind instrument panel	N/A
41	GRN	Coolant Fan Control Module	N/A
42	--	Not Used	--
43	GRY/WHT	T-10 Connector, Orange, behind instrument panel	N/A
44	--	Not Used	--
45	--	Not Used	--
46	--	Not Used	--
47	--	Not Used	--
48	--	Not Used	--
49	LIL/GRY	Charge Air Pressure Sensor Signal	N/A
50	--	Not Used	--
51	WHT	Heated Oxygen Sensor Ground	N/A
52	GRN	Heated Oxygen Sensor Signal	N/A
53	RED/LIL	Mass Air Flow Sensor	N/A
54	BLU/WHT	Vehicle Speed Signal (instrument panel)	N/A
55	WHT/YEL	Brake Vacuum Vent Valve Switch Signal	N/A
56	BLK/RED	T-10 Connector, Black, behind instrument panel	N/A
57	RED/YEL	Set Switch for Cruise Control Signal	N/A
58	ORN/BRN	ABS control module (TCM for automatic)	N/A
59	--	Not Used	--
60	ORN/BLK	ABS control module (TCM for automatic)	N/A
61	WHT	Coolant Fan Control Module	N/A
62	RED/GRN	Fuse Box connection	N/A
63	GRY/YEL	Heated Oxygen Sensor Power	N/A
64	LIL/RED	EVAP Canister Purge Regulator Valve Signal	N/A
65	LIL/WHT	Fuel Pump Relay	N/A
66	--	Not Used	--
67	--	Not Used	--
68	BLK	Heated Oxygen Sensor Ground	N/A
69	BLU	Heated Oxygen Sensor Signal	N/A
70	GRN	Heated Oxygen Sensor	N/A
71	BLK	Heated Oxygen Sensor	N/A
72	GRN/WHT	Accelerator Pedal Position Sender Power	N/A
73	YEL/GRN	Throttle Position Sensor Ground	N/A
74	--	Not Used	--
75	BLU/GRY	Cruise Control Switch Signal	N/A
76	WHT	Cruise Control Switch Signal	N/A
77	ORN/BRN	Transmission Control Module (A/T only)	N/A
78	--	Not Used	--
79	ORN/BLK	Transmission Control Module (A/T only)	N/A
80	GRN/LIL	Leak Detection Pump Signal	N/A
81	--	Not Used	--
82	WHT	Engine Speed Sensor Signal	N/A
83	LIL/WHT	Throttle Drive Angle Sensor Ground	N/A
84	LIL/YEL	Throttle Drive Angle Sensor Signal	N/A
85	BLU/GRN	Intake Air Temperature Sensor Control	N/A

2005 Sedan (BNU), Convertible (BKF) 1.8L MFI BNU and BKF T121 Connector, *continued*

86	GRY/YEL	Camshaft Position Sensor	N/A
87	--	Not Used	--
88	LIL/BLU	Cylinder 2 Fuel Injector Control	N/A
89	LIL/GRN	Cylinder 4 Fuel Injector Control	N/A
90	BRN	Engine Speed Sensor Signal	N/A
91	WHT	Throttle Drive Angle Sensor Power	N/A
92	LIL	Throttle Drive Angle Sensor Signal	N/A
93	BLU	Engine Coolant Temperature Sensor Signal	N/A
94	BLK/LIL	Ignition Coil 4 with Power Output Stage	N/A
95	BLK/YEL	Ignition Coil 2 with Power Output Stage	N/A
96	GRY/LIL	Cylinder 1 Fuel Injector Control	N/A
97	LIL/RED	Cylinder 3 Fuel Injector Control	N/A
98	LIL/GRN	Camshaft Position Sensor	N/A
99	BLU	Knock Sensor 1 and 2 Control	N/A
100	--	Not Used	--
101	--	Not Used	--
102	BLK/BLU	Ignition Coil 1 with Power Output Stage	N/A
103	BLK/BRN	Ignition Coil 3 with Power Output Stage	N/A
104	GRN/BRN	Wastegate Bypass Regulator Valve Control	N/A
105	--	Not Used	--
106	GRY	Knock Sensor 1 Signal	N/A
107	GRN	Knock Sensor 2 Signal	N/A
108	BRN/BLU	Sensor Ground	N/A
109	--	Not Used	--
110	--	Not Used	--
111	--	Not Used	--
112	--	Not Used	--
113	--	Not Used	--
114	BRN/YEL	Secondary Air Injection Solenoid Valve	N/A
115	GRN/WHT	Camshaft Adjustment Valve 1 Control	N/A
116	--	Not Used	--
117	WHT	Throttle Drive Motor Power	N/A
118	LIL/BLK	Throttle Drive Motor Control	N/A
119	--	Not Used	--
120	LIL/BLK	Recirculating Valve for the Turbocharger Control	N/A
121	--	Not Used	--

New Beetle

2006 Sedan 2.5L MFI BPR T121 Connector

ECM Pin #	Wire Color	Circuit Description (121-Pin)	Value at Hot Idle
1	BRN/RED	Ground	N/A
2	BRN/RED	Ground	N/A
3	BLK/LIL	Power Supply Relay	N/A
4	--	Not Used	--
5	WHT	Heated Oxygen Sensor Power	N/A
6	BLK/BRN	Heated Oxygen Sensor Power	N/A

2006 Sedan 2.5L MFI BPR T121 Connector, *continued*

7	--	Not Used	--
8	--	Not Used	--
9	--	Not Used	--
10	BLU	Heated Oxygen Sensor Ground	N/A
11	RED	Heated Oxygen Sensor Signal	N/A
12	--	Not Used	--
13	--	Not Used	--
14	--	Not Used	--
15	--	Not Used	--
16	--	Not Used	--
17	--	Not Used	--
18	--	Not Used	--
19	--	Not Used	--
20	--	Not Used	--
21	BLK/LIL	Fuse Panel Connection	N/A
22	--	Not Used	--
23	LIL	Power Supply Relay	N/A
24	--	Not Used	--
25	YEL/BRN	Leak Detection Pump Motor Control	N/A
26	LIL/GRN	Mass Air Flow Sensor	N/A
27	GRY/RED	Mass Air Flow Sensor	N/A
28	BLU/LIL	Voltage Regulator	N/A
29	GRY/WHT	Mass Air Flow Sensor	N/A
30	RED/BLK	Ignition Starter Switch	N/A
31	BLU/BLK	Ambient Temperature Switch Signal	N/A
32	--	Not Used	--
33	GRY/BLU	Accelerator Pedal Position Sender Ground	N/A
34	BRN/WHT	Accelerator Pedal Position Sender Signal	N/A
35	WHT/BLU	Throttle Position Sensor Signal	N/A
36	GRY/RED	Throttle Position Sensor Power	N/A
37	--	Not Used	--
38	BLK/YEL	Ground on Left Headlight	N/A
39	WHT/RED	Clutch Pedal Switch Signal	N/A
40	BLU/RED	Coolant Fan Control Module	N/A
41	--	Not Used	--
42	RED/BRN	Starter Lock/Clutch Pedal Switch Relay	N/A
43	GRY/WHT	Connection (K-diagnosis wire), instrument panel	N/A
44	--	Not Used	--
45	--	Not Used	--
46	GRN/BRN	Secondary Air Injection Pump Relay	N/A
47	--	Not Used	--
48	--	Not Used	--
49	LIL/WHT	Crash Signal Relay in instrument panel	N/A
50	BLU/YEL	Ambient Temperature Switch Control	N/A
51	WHT	Heated Oxygen Sensor Signal	N/A
52	GRN	Heated Oxygen Sensor Ground	N/A
53	RED/LIL	Mass Air Flow Sensor	N/A

2006 Sedan 2.5L MFI BPR T121 Connector, *continued*

54	BLU/WHT	Vehicle Speed Signal (via the instrument panel)	N/A
55	WHT/YEL	Brake Pedal Switch Signal	N/A
56	RED/BLK	Connection in interior wiring harness	N/A
57	RED/YEL	Cruise Control Set Button Switch	N/A
58	ORN/BRN	Power Can-Bus Low Connection 1	N/A
59	--	Not Used	--
60	ORN/BLK	Power Can-Bus High Connection 1	N/A
61	WHT	High Pressure Sensor	N/A
62	RED/GRN	Power Supply Relay	N/A
63	GRY/YEL	Heated Oxygen Sensor Power	N/A
64	LIL/RED	EVAP Canister Purge Regulator Valve Signal	N/A
65	LIL/WHT	Fuel Primer Relay	N/A
66	GRN/RED	T-4 Connection	N/A
67	--	Not Used	--
68	BLK	Heated Oxygen Sensor Ground	N/A
69	BLU	Heated Oxygen Sensor Signal	N/A
70	GRN	Heated Oxygen Sensor Signal	N/A
71	BLK	Heated Oxygen Sensor Ground	N/A
72	GRN/WHT	Accelerator Pedal Position Sender Power	N/A
73	YEL/GRN	Throttle Position Sensor Ground	N/A
74	--	Not Used	--
75	BLU/GRY	Cruise Control Set Switch Signal	N/A
76	WHT	Cruise Control Set Switch Signal	N/A
77	--	Not Used	--
78	--	Not Used	--
79	--	Not Used	--
80	GRN/LIL	Leak Detection Pump Motor Signal	N/A
81	--	Not Used	--
82	LIL	Engine Speed Sensor Signal	N/A
83	BLK	Throttle Drive Angle Sensor Ground	N/A
84	LIL/YEL	Throttle Drive Angle Sensor Signal	N/A
85	--	Not Used	--
86	BLU/RED	Camshaft Position Sensor Signal	N/A
87	--	Not Used	--
88	LIL/BLU	Cylinder 4 Fuel Injector Control	N/A
89	LIL/GRY	Cylinder 5 Fuel Injector Control	N/A
90	BRN/BLU	Engine Speed Sensor Signal	N/A
91	GRN	Throttle Drive Angle Sensor Power	N/A
92	BLK/LIL	Throttle Drive Angle Sensor Signal	N/A
93	BLU/GRY	Engine Coolant Temperature Sensor Signal	N/A
94	BLK/RED	Ignition Coil 4 with Power Output Stage	N/A
95	BLK/WHT	Ignition Coil 5 with Power Output Stage	N/A
96	LIL	Cylinder 1 Fuel Injector Control	N/A
97	LIL/GRN	Cylinder 2 Fuel Injector Control	N/A
98	LIL/RED	Engine Speed Sensor Ground	N/A
99	BLU	Knock Sensor 1 and 2 Control	N/A
100	--	Not Used	--

2006 Sedan 2.5L MFI BPR T121 Connector, *continued*

101	BLU	Manifold Absolute Pressure Sensor Control	N/A
102	BLK/BLU	Ignition Coil 1 with Power Output Stage	N/A
103	BLU/YEL	Ignition Coil 2 with Power Output Stage	N/A
104	--	Not Used	--
105	--	Not Used	--
106	LIL	Knock Sensor 1 Signal	N/A
107	WHT/BRN	Knock Sensor 2 Signal	N/A
108	BRN/BLU	Camshaft Position Sensor Ground	N/A
109	--	Not Used	--
110	BLK/BRN	Ignition Coil 3 with Power Output Stage	N/A
111	--	Not Used	--
112	BLU/RED	Cylinder 3 Fuel Injector Control	N/A
113	--	Not Used	--
114	--	Not Used	--
115	GRN/YEL	Camshaft Adjustment Valve 1 Control	N/A
116	--	Not Used	--
117	LIL	Throttle Drive Motor Power	N/A
118	BRN/LIL	Throttle Drive Motor Control	N/A
119	--	Not Used	--
120	--	Not Used	--
121	BRN/BLK	Secondary Air Injection Solenoid Valve	N/A

New Beetle

2006 Sedan 2.5L MFI BPS T121 Connector

ECM Pin #	Wire Color	Circuit Description (121-Pin)	Value at Hot Idle
1	BRN/RED	Ground	N/A
2	BRN/RED	Ground	N/A
3	BLK/LIL	Power Supply Relay	N/A
4	--	Not Used	--
5	WHT	Heated Oxygen Sensor Power	N/A
6	GRY/YEL	Heated Oxygen Sensor Power	N/A
7	--	Not Used	--
8	--	Not Used	--
9	--	Not Used	--
10	BLK	Heated Oxygen Sensor Ground	N/A
11	BLU	Heated Oxygen Sensor Signal	N/A
12	--	Not Used	--
13	--	Not Used	--
14	--	Not Used	--
15	--	Not Used	--
16	--	Not Used	--
17	--	Not Used	--
18	--	Not Used	--
19	--	Not Used	--
20	--	Not Used	--

2006 Sedan 2.5L MFI BPS T121 Connector, *continued*

21	BLK/LIL	Fuse Panel Connection	N/A
22	--	Not Used	--
23	LIL	Power Supply Relay	N/A
24	--	Not Used	--
25	YEL/BRN	Leak Detection Pump Motor Control	N/A
26	LIL/GRN	Mass Air Flow Sensor	N/A
27	GRY/RED	Mass Air Flow Sensor	N/A
28	BLU/LIL	Voltage Regulator	N/A
29	GRY/WHT	Mass Air Flow Sensor	N/A
30	--	Not Used	--
31	BLU/BLK	Ambient Temperature Switch Signal	N/A
32	--	Not Used	--
33	GRY/BLU	Accelerator Pedal Position Sender Ground	N/A
34	BRN/WHT	Accelerator Pedal Position Sender Signal	N/A
35	WHT/BLU	Throttle Position Sensor Signal	N/A
36	GRY/RED	Throttle Position Sensor Power	N/A
37	--	Not Used	--
38	BLK/YEL	Ground on Left Headlight	N/A
39	WHT/RED	Clutch Pedal Switch Signal	N/A
40	BLU/RED	Coolant Fan Control Module	N/A
41	--	Not Used	--
42	--	Not Used	--
43	GRY/WHT	Connection (K-diagnosis wire), instrument panel	N/A
44	--	Not Used	--
45	--	Not Used	--
46	GRN/BRN	Secondary Air Injection Pump Relay	N/A
47	--	Not Used	--
48	--	Not Used	--
49	LIL/WHT	Crash Signal Relay in instrument panel	N/A
50	BLU/YEL	Ambient Temperature Switch Control	N/A
51	WHT	Heated Oxygen Sensor Signal	N/A
52	GRN	Heated Oxygen Sensor Ground	N/A
53	RED/LIL	Mass Air Flow Sensor	N/A
54	BLU/WHT	Vehicle Speed Signal (via the instrument panel)	N/A
55	WHT/YEL	Brake Pedal Switch Signal	N/A
56	RED/BLK	Connection in interior wiring harness	N/A
57	RED/YEL	Cruise Control Set Button Switch	N/A
58	ORN/BRN	Power Can-Bus Low Connection 1	N/A
59	--	Not Used	--
60	ORN/BLK	Power Can-Bus High Connection 1	N/A
61	WHT	High Pressure Sensor	N/A
62	RED/GRN	Power Supply Relay	N/A
63	--	Not Used	--
64	LIL/RED	EVAP Canister Purge Regulator Valve Signal	N/A
65	LIL/WHT	Fuel Primer Relay	N/A
66	GRN/RED	T-4 Connection	N/A
67	--	Not Used	--

2006 Sedan 2.5L MFI BPS T121 Connector, *continued*

68	--	Not Used	--
69	--	Not Used	--
70	GRN	Heated Oxygen Sensor Signal	N/A
71	BLK	Heated Oxygen Sensor Ground	N/A
72	GRN/WHT	Accelerator Pedal Position Sender Power	N/A
73	YEL/GRN	Throttle Position Sensor Ground	N/A
74	--	Not Used	--
75	BLU/GRY	Cruise Control Set Switch Signal	N/A
76	WHT	Cruise Control Set Switch Signal	N/A
77	--	Not Used	--
78	--	Not Used	--
79	--	Not Used	--
80	GRN/LIL	Leak Detection Pump Motor Signal	N/A
81	--	Not Used	--
82	LIL	Engine Speed Sensor Signal	N/A
83	BLK	Throttle Drive Angle Sensor Ground	N/A
84	LIL/YEL	Throttle Drive Angle Sensor Signal	N/A
85	--	Not Used	--
86	BLU/RED	Camshaft Position Sensor Signal	N/A
87	--	Not Used	--
88	LIL/BLU	Cylinder 4 Fuel Injector Control	N/A
89	LIL/GRY	Cylinder 5 Fuel Injector Control	N/A
90	BRN/BLU	Engine Speed Sensor Signal	N/A
91	GRN	Throttle Drive Angle Sensor Power	N/A
92	BLK/LIL	Throttle Drive Angle Sensor Signal	N/A
93	BLU/GRY	Engine Coolant Temperature Sensor Signal	N/A
94	BLK/RED	Ignition Coil 4 with Power Output Stage	N/A
95	BLK/WHT	Ignition Coil 5 with Power Output Stage	N/A
96	LIL	Cylinder 1 Fuel Injector Control	N/A
97	LIL/GRN	Cylinder 2 Fuel Injector Control	N/A
98	LIL/RED	Engine Speed Sensor Ground	N/A
99	BLU	Knock Sensor 1 and 2 Control	N/A
100	--	Not Used	--
101	BLU	Manifold Absolute Pressure Sensor Control	N/A
102	BLK/BLU	Ignition Coil 1 with Power Output Stage	N/A
103	BLU/YEL	Ignition Coil 2 with Power Output Stage	N/A
104	--	Not Used	--
105	--	Not Used	--
106	LIL	Knock Sensor 1 Signal	N/A
107	WHT/BRN	Knock Sensor 2 Signal	N/A
108	BRN/BLU	Camshaft Position Sensor Ground	N/A
109	--	Not Used	--
110	BLK/BRN	Ignition Coil 3 with Power Output Stage	N/A
111	--	Not Used	--
112	BLU/RED	Cylinder 3 Fuel Injector Control	N/A
113	--	Not Used	--
114	--	Not Used	--

2006 Sedan 2.5L MFI BPS T121 Connector, *continued*

115	GRN/YEL	Camshaft Adjustment Valve 1 Control	N/A
116	--	Not Used	--
117	LIL	Throttle Drive Motor Power	N/A
118	BRN/LIL	Throttle Drive Motor Control	N/A
119	--	Not Used	--
120	--	Not Used	--
121	BRN/BLK	Secondary Air Injection Solenoid Valve	N/A

Passat

1995 Sedan 2.0L MFI ABA T68 Connector

ECM Pin #	Wire Color	Circuit Description (68-Pin)	Value at Hot Idle
1	BRN/RED	Ground	N/A
2	WHT/GRY	Cylinder 2 Fuel Injector Control	N/A
3	--	Not Used	--
4	--	Not Used	--
5	YEL/BLK	Multi-Function Indicator Light (MIL)	N/A
6	YEL/BLU	Fuel Pump Relay	N/A
7	BRN/RED	Ground	N/A
8	BLK/RED	Ignition Coil Power Output Stage	N/A
9	BLK/BRN	ECM Power Supply Relay	N/A
10	--	Not Used	--
11		Not Used	--
12	--	Not Used	--
13	--	Not Used	--
14	BLU	Engine Coolant Temperature Sensor	N/A
15	LIL	EGR Temperature Sensor Control	N/A
16	BRN/RED	Mass Airflow Sensor	N/A
17	RED	Mass Airflow Sensor	N/A
18	YEL/RED	T-8 Connector (behind fuse panel) (A/T only)	N/A
19	--	Not Used	--
20	WHT	Heated Oxygen Sensor Signal	N/A
21	YEL	Data Link Connector (DLC)	N/A
22	GRN/BLK	T-44 Connector for OBD	N/A
23	RED/BLU	Control Module Display Unit in Instrument Cluster	N/A
24	WHT/BLK	Cylinder 1 Fuel Injector Control	N/A
25	WHT/GRN	Cylinder 3 Fuel Injector Control	N/A
26	WHT/BLU	Cylinder 4 Fuel Injector Control	N/A
27	WHT	Idle Air Control Valve	N/A
28	WHT	Heated Oxygen Sensor Relay	N/A
29	--	Not Used	--
30	GRN/GRY	EGR Vacuum Regulator Solenoid Valve	N/A
31	GRN/YEL	EVAP Canister Purge Regulator Valve	N/A
32	--	Not Used	--
33	BRN	Ground	N/A
34	GRY	Knock Sensor 1	N/A

1995 Sedan 2.0L MFI ABA T68 Connector, *continued*

35	--	Not Used	--
36	BLU/GRN	Intake Air Temperature Sensor Signal	N/A
37	GRN	T-2 Connector (A/C Connection)	N/A
38	BLK	Plus Connection	N/A
39	BLU	T-2 Connector (A/C Connection)	N/A
40	GRN/WHT	Throttle Position Sensor Signal	N/A
41	BLK/YEL	Throttle Position Sensor Power	N/A
42	YEL	Heated Oxygen Sensor Power	N/A
43	GRY/WHT	T-44 Connector for OBD	N/A
44	WHT/RED	Camshaft Position Sensor	N/A
45	--	Not Used	--
46	--	Not Used	--
47	--	Not Used	--
48	--	Not Used	--
49	--	Not Used	--
50	RED/GRY	Secondary AIR Valve Control	N/A
51	LIL/WHT	T-1 Connector (behind fuse panel)	N/A
52	--	Not Used	--
53	BLK/WHT	Idle Air Intake Valve Ground	N/A
54	RED/YEL	Fuel Pump Relay Signal	N/A
55	BLU/BLK	Ground	N/A
56	BRN/WHT	Ground	N/A
57	--	Not Used	N/A
58	WHT	Ground (A/T only)	N/A
59	--	Not Used	--
60	--	Not Used	--
61	--	Not Used	--
62	--	Not Used	--
63	--	Not Used	--
64	--	Not Used	--
65	WHT/YEL	T-43 Connector, Vehicle Speed Sensor	N/A
66	--	Not Used	--
67	RED	Engine Speed Sensor Signal	N/A
68	GRN	Engine Speed Sensor Signal	N/A

Passat

1996 Sedan DFI 1Z T68 Connector

ECM Pin #	Wire Color	Circuit Description (68-Pin)	Value at Hot Idle
1	BRN/RED	Power Ground	N/A
2	BLK/GRN	Power Ground	N/A
3	--	Not Used	--
4	BRN/YEL	T-8 Connector on Fuel Pump	N/A
5	BRN/YEL	Throttle Position Sensor Ground	N/A
6	BLK/BLU	Coolant Glow Plug Relay	N/A
7	GRY/GRN	Modulating Piston Displacement Sensor	N/A
8	WHT	Engine Speed Sensor Signal	N/A

1996 Sedan DFI 1Z T68 Connector, *continued*

9	LIL	Odometer Sensor	N/A
10	YEL/BLK	Catalyst 1 Temperature Sensor Signal	N/A
11	BLU	Needle Lift Sensor Signal	N/A
12	GRY	Needle Lift Sensor Ground	N/A
13	BRN/RED	Mass Air Flow Sensor Signal	N/A
14	BRN/GRN	T-24 Connector Left Cylinder Head	N/A
15	--	Not Used	--
16	--	Not Used	--
17	WHT/RED	Clutch Vacuum Vent Valve Switch Signal	N/A
18	YEL/RED	T-8 Connector	N/A
19	RED/GRN	Mass Air Flow Sensor Ground	N/A
20	--	Not Used	--
21	--	Not Used	--
22	--	Not Used	--
23	BLK/YEL	Power Supply Relay	N/A
24	BRN/RED	Power Ground	N/A
25	WHT	EGR Vacuum Regulator Solenoid Valve	N/A
26	--	Not Used	--
27	GRN/YEL	Metering Pump Motor Signal	N/A
28	GRN/BRN	T-1 Connector	N/A
29	LIL/BLK	Modulating Piston Displacement Sensor	N/A
30	BRN/YEL	Glow Plug Relay	N/A
31	BLU/YEL	T-8 Connector	N/A
32	GRY	T-8 Connector	N/A
33	BRN/BLU	Sensor Ground	N/A
34	BLU/GRY	Cruise Control Switch Signal	N/A
35	RED	Cruise Control Switch Power	N/A
36	--	Not Used	--
37	GRN	T-2 Double Connector for A/C	N/A
38	--	Not Used	--
39	BRN/RED	Power Ground (A/T only)	N/A
40	--	Not Used	--
41	--	Not Used	--
42	BLK/LIL	Power Supply Relay	N/A
43	BLU/WHT	Vehicle Speed Signal Wire Connector	N/A
44	BLK/RED	T-1 Connector	N/A
45	BLK/YEL	Fuel Injection Signal	N/A
46	BRN/RED	Sensor Ground	N/A
47	BRN/BLU	Wastegate Bypass Regulator Valve Signal	N/A
48	BLK/BRN	Coolant Glow Plug Relay	N/A
49	BRN/YEL	Ground	N/A
50	YEL/BLU	Glow Plug Relay	N/A
51	BRN/BLK	Cold Start Injector Signal	N/A
52	WHT/GRN	Modulating Piston Displacement Sensor	N/A
53	WHT/BLK	Fuel Cut-Off Valve Signal	N/A
54	--	Not Used	--
55	GRY/RED	Throttle Position Sensor Ground	N/A

1996 Sedan DFI 1Z T68 Connector, *continued*

56	LIL/BRN	Catalyst 2 Temperature Sensor Signal	N/A
57	YEL/GRN	Throttle Position Sensor Signal	N/A
58	--	Not Used	--
59	--	Not Used	--
60	--	Not Used	--
61	GRY/WHT	T-2 Data Link Double Connector	N/A
62	GRN/WHT	Kick Down Switch (A/T only)	N/A
63	YEL/BLK	Fuel Temperature Sensor	N/A
64	BLU/YEL	Intake Air Temperature Sensor	N/A
65	GRY/BLU	Closed Throttle Position Switch Signal	N/A
66	BLK/WHT	Cruise Control Switch	N/A
67	--	Not Used	--
68	BLK/YEL	Fuel Injection Signal	N/A

Passat

1996 Sedan 2.0L MFI ABA T68 Connector

ECM Pin #	Wire Color	Circuit Description (68-Pin)	Value at Hot Idle
1	BRN/RED	Power Ground	N/A
2	WHT/GRY	Cylinder 4 Fuel Injector Control	N/A
3	YEL/WHT	Generator Control	N/A
4	--	Not Used	--
5	--	Not Used	--
6	YEL/BLU	Fuel Pump Relay	N/A
7	BRN/BLK	Ground	N/A
8	BLK/RED	Ignition Coil Power Output Stage	N/A
9	--	Not Used	--
10	RED/GRN	Throttle Valve Control Module Power	N/A
11	BLU/YEL	T-8 Connector (behind fuse panel) (A/T only)	N/A
12	BRN	Oxygen Sensor Heater Control	N/A
13	WHT	Heated Oxygen Sensor Signal	N/A
14	BLU	Engine Coolant Temperature Sensor	N/A
15	LIL	EGR Temperature Sensor Control	N/A
16	BRN/BLU	Mass Airflow Sensor	N/A
17	BLU	Mass Airflow Sensor	N/A
18	YEL/RED	T-8 Connector (behind fuse panel) (A/T only)	N/A
19	--	Not Used	--
20	WHT	Heated Oxygen Sensor Signal	N/A
21	YEL	T-8 Connector (behind fuse panel) (A/T only)	N/A
22	GRN/BLK	T-44 Connector for OBD	N/A
23	BLK/YEL	Fuse Box Connection	N/A
24	GRY	Cylinder 1 Fuel Injector Control	N/A
25	GRY/GRN	Cylinder 2 Fuel Injector Control	N/A
26	GRY/RED	Cylinder 3 Fuel Injector Control	N/A
27	BLK	Throttle Valve Motor Ground	N/A
28	--	Not Used	--
29	--	Not Used	--

1996 Sedan 2.0L MFI ABA T68 Connector, *continued*

30	GRN/GRY	T-28 Connector (left cylinder head)	N/A
31	GRN/YEL	EVAP Canister Purge Regulator Valve	N/A
32	--	Not Used	--
33	BRN/BLU	Knock Sensor 1 Signal	N/A
34	GRY	Ground	N/A
35	--	Not Used	--
36	WHT/YEL	Intake Air Temperature Sensor	N/A
37	GRN	T-2 Connector (A/C Connection)	N/A
38	--	Not Used	--
39	BLU	T-2 Connector (A/C Connection)	N/A
40	RED/BLU	Throttle Valve Switch Ground	N/A
41	RED/WHT	Throttle Valve Switch Power	N/A
42	--	Not Used	--
43	GRY/WHT	Data Link Connector (DLC)	N/A
44	WHT/RED	Camshaft Position Sensor	N/A
45	--	Not Used	--
46	--	Not Used	--
47	--	Not Used	--
48	--	Not Used	--
49	--	Not Used	--
50	RED/YEL	Secondary AIR Valve Control	N/A
51	LIL/WHT	T-1 Connector (behind fuse panel)	N/A
52	--	Not Used	--
53	WHT	Throttle Valve Motor Power	N/A
54	RED	Fuel Pump Relay Signal	N/A
55	--	Not Used	--
56	BRN/BLK	Ground	N/A
57	--	Not Used	--
58	YEL	Heated Oxygen Sensor Ground	N/A
59	--	Not Used	--
60	--	Not Used	--
61	--	Not Used	--
62	RED/GRY	Throttle Valve Switch Signal	N/A
63	--	Not Used	--
64	--	Not Used	--
65	WHT/YEL	T-43 Connector, Vehicle Speed Sensor	N/A
66	BRN	Oxygen Sensor Heater Control	N/A
67	RED	Engine Speed Sensor Signal	N/A
68	GRN	Engine Speed Sensor Signal	N/A

Passat

1996–1997 Sedan 2.8L AAA T68 Connector

ECM Pin #	Wire Color	Circuit Description (68-Pin)	Value at Hot Idle
1	BRN/RED	Power Ground	N/A
2	GRY/BLU	Fuel Injector Cylinder 4	N/A
3	GRY/BLK	Fuel Injector Cylinder 5	N/A

1996–1997 Sedan 2.8L AAA T68 Connector, *continued*

4	GRY/GRN	Fuel Injector Cylinder 6	N/A
5	YEL/BLK	Relay Panel Multi Function Indicator	N/A
6	YEL/BLU	Fuel Pump Relay	N/A
7	BRN/WHT	Sensor Ground	N/A
8	BLK/BRN	Ignition Coil	N/A
9	BLK/BLU	ECM Power Supply Relay	N/A
10	WHT	Throttle Valve Control Module Sensor	N/A
11	BLU/YEL	T-8 Connector behind fuse panel (A/T only)	N/A
12	BRN	Heated Oxygen Sensor Ground	N/A
13	WHT	Heated Oxygen Sensor 2 Power	N/A
14	BLU	Engine Coolant Temperature Sensor Ground	N/A
15	LIL	EGR Temperature Sensor	N/A
16	BRN	Mass Air Flow Sensor Signal	N/A
17	BLU	Mass Air Flow Sensor Ground	N/A
18	YEL/RED	T-8 Connector behind fuse panel (A/T only)	N/A
19	--	Not Used	--
20	WHT	Heated Oxygen Sensor	N/A
21	YEL	T-8 Connector behind fuse panel (A/T only)	N/A
22	GRN/BLK	Control Module for Instrument Panel	N/A
23	BLK/YEL	ECM Power Supply Relay	N/A
24	GRY	Fuel Injector Cylinder 1	N/A
25	GRY/BLK	Fuel Injector Cylinder 2	N/A
26	GRY/RED	Fuel Injector Cylinder 3	N/A
27	BLK	Throttle Valve Control Module	N/A
28	WHT	Heated Oxygen Sensor Relay	N/A
29	--	Not Used	--
30	GRN/WHT	EGR Vacuum Regulator Valve Power	N/A
31	GRN/YEL	EVAP Canister Purge Regulator Valve	N/A
32	--	Not Used	--
33	BRN/BLU	Knock Sensor 1 Signal	N/A
34	GRN	Knock Sensor 1 Power	N/A
35	YEL/BRN	Leak Detection Pump	N/A
36	BLU/GRN	Intake Air Temperature Sensor Power	N/A
37	GRN	Double Connector (A/C connection)	N/A
38	BLK/GRN	E-6 Connector	N/A
39	BLU	Double Connector (A/C connection)	N/A
40	GRN/WHT	Throttle Valve Control Module	N/A
41	BLK/GRY	Throttle Valve Control Module Signal	N/A
42	YEL	Heated Oxygen Sensor Ground	N/A
43	GRY/WHT	Data Link Connector	N/A
44	WHT/RED	Camshaft Position Sensor	N/A
45	--	Not Used	--
46	GRN/LIL	Leak Detection Pump	N/A
47	--	Not Used	--
48	--	Not Used	--
49	RED/GRY	Secondary Air Injection Pump Relay	N/A
50	RED/YEL	Secondary Air Injection Solenoid Valve	N/A

1996–1997 Sedan 2.8L AAA T68 Connector, *continued*

51	LIL/WHT	Odometer Display	N/A
52	BLK/LIL	Ignition Coil	N/A
53	BLK/WHT	Throttle Valve Control Module Motor	N/A
54	RED	Battery Power	N/A
55	BRN/BLK	Power Ground	N/A
56	BRN/WHT	Sensor Ground	N/A
57	YEL	Knock Sensor 2 Power	N/A
58	YEL	Heated Oxygen Sensor 2 Ground	N/A
59		Not Used	--
60	BLK/BLU	Ignition Coil	N/A
61	--	Not Used	--
62	RED/GRY	Throttle Valve Control Module	N/A
63	--	Not Used	--
64	LIL/BLK	T-1 Connector	N/A
65	BLU/WHT	Odometer Display	N/A
66	BRN	Heated Oxygen Sensor 2 Signal	N/A
67	RED	Engine Speed Sensor Power	N/A
68	GRN	Engine Speed Sensor Signal	N/A

Passat

1996 Sedan DFI 1Z T68 Connector

ECM Pin #	Wire Color	Circuit Description (68-Pin)	Value at Hot Idle
1	BRN/RED	Power Ground	N/A
2	BLK/GRN	Power Ground	N/A
3	--	Not Used	--
4	BRN/YEL	T-8 Connector on Fuel Pump	N/A
5	BRN/YEL	Throttle Position Sensor Ground	N/A
6	BLK/BLU	Coolant Glow Plug Relay	N/A
7	GRY/GRN	Modulating Piston Displacement Sensor	N/A
8	WHT	Engine Speed Sensor Signal	N/A
9	LIL	Odometer Sensor	N/A
10	--	Not Used	--
11	BLU	Needle Lift Sensor Signal	N/A
12	GRY	Needle Lift Sensor Ground	N/A
13	BRN/RED	Mass Air Flow Sensor Signal	N/A
14	BRN/GRN	T-24 Connector Left Cylinder Head	N/A
15	--	Not Used	--
16	--	Not Used	--
17	WHT/RED	Clutch Vacuum Vent Valve Switch Signal	N/A
18	YEL/RED	T-8 Connector	N/A
19	RED/GRN	Mass Air Flow Sensor Ground	N/A
20	--	Not Used	--
21	--	Not Used	--
22	--	Not Used	--
23	BLK/YEL	Power Supply Relay	N/A

1996 Sedan DFI 1Z T68 Connector, *continued*

24	BRN/RED	Power Ground	N/A
25	WHT	EGR Vacuum Regulator Solenoid Valve	N/A
26	--	Not Used	--
27	GRN/YEL	Metering Pump Motor Signal	N/A
28	GRN/BRN	T-1 Connector	N/A
29	LIL/BLK	Modulating Piston Displacement Sensor	N/A
30	BRN/YEL	Glow Plug Relay	N/A
31	BLU/YEL	T-8 Connector	N/A
32	GRY	T-8 Connector	N/A
33	BRN/BLU	Sensor Ground	N/A
34	BLU/GRY	Cruise Control Switch Signal	N/A
35	RED	Cruise Control Switch Power	N/A
36	--	Not Used	--
37	GRN	T-2 Double Connector for A/C	N/A
38	--	Not Used	--
39	BRN/RED	Power Ground (A/T only)	N/A
40	--	Not Used	--
41	--	Not Used	--
42	BLK/LIL	Power Supply Relay	N/A
43	BLU/WHT	Vehicle Speed Signal Wire Connector	N/A
44	BLK/RED	T-1 Connector	N/A
45	BLK/YEL	Fuel Injection Signal	N/A
46	BRN/RED	Sensor Ground	N/A
47	BRN/BLU	Wastegate Bypass Regulator Valve Signal	N/A
48	BLK/BRN	Coolant Glow Plug Relay	N/A
49	BRN/YEL	Ground	N/A
50	YEL/BLU	Glow Plug Relay	N/A
51	BRN/BLK	Cold Start Injector Signal	N/A
52	WHT/GRN	Modulating Piston Displacement Sensor	N/A
53	WHT/BLK	Fuel Cut-Off Valve Signal	N/A
54	--	Not Used	--
55	GRY/RED	Throttle Position Sensor Ground	N/A
56	--	Not Used	--
57	YEL/GRN	Throttle Position Sensor Signal	N/A
58	--	Not Used	--
59	--	Not Used	--
60	--	Not Used	--
61	GRY/WHT	T-2 Data Link Double Connector	N/A
62	GRN/WHT	Kick Down Switch (A/T only)	N/A
63	YEL/BLK	Fuel Temperature Sensor	N/A
64	BLU/YEL	Intake Air Temperature Sensor	N/A
65	GRY/BLU	Closed Throttle Position Switch Signal	N/A
66	BLK/WHT	Cruise Control Switch	N/A
67	--	Not Used	--
68	BLK/YEL	Fuel Injection Signal	N/A

Passat

1998–1999 Sedan and Wagon 1.8L MFI AEB T80 Connector

ECM Pin #	Wire Color	Circuit Description (80-Pin)	Value at Hot Idle
1	BLK/BLU	D-52 Connection	N/A
2	BRN	Power Ground	N/A
3	RED	Battery Power	N/A
4	RED/BLU	Fuel Pump Relay	N/A
5	GRN/BLK	T-10 Controller, yellow on ECM	N/A
6	GRN/BLU	Odometer Signal	N/A
7	YEL/BLK	T-10 Controller, blue on ECM (for TCM)	N/A
8	BLK/GRY	T-10 Controller, brown on ECM	N/A
9	--	Not Used	--
10	BLK/YEL	T-10 Controller, brown on ECM	N/A
11	GRN/LIL	Camshaft Position Sensor Power	N/A
12	YEL	Mass Airflow Sensor	N/A
13	GRN	Mass Airflow Sensor	N/A
14	--	Not Used	--
15	LIL	EVAP Canister Purge Solenoid Valve Control	N/A
16	YEL/RED	Leak Detection Pump Motor Signal	N/A
17	YEL/BRN	Malfunction Indicator Light (MIL) Signal	N/A
18	YEL	Fuel Gauge Signal	N/A
19	GRN/BLK	T-10 Controller, Yellow on ECM	N/A
20	WHT/BLU	Speedometer Signal	N/A
21	BLK/LIL	Low Fuel Level Warning Light Signal	N/A
22	BRN/WHT	T-10 Controller, blue (Starting Interlock Relay)	N/A
23	LIL/BLK	T-10 Controller, blue on ECM (for TCM)	N/A
24	--	Not Used	--
25	GRY	Heated Oxygen Sensor Signal	N/A
26	BLU	Heated Oxygen Sensor Ground	N/A
27	RED/WHT	Heated Oxygen Sensor Power	N/A
28	RED/YEL	Heated Oxygen Sensor Power	N/A
29	--	Not Used	--
30	--	Not Used	--
31	--	Not Used	--
32	WHT/YEL	T-10 Controller, blue on ECM (for TCM)	N/A
33	--	Not Used	--
34	--	Not Used	--
35	--	Not Used	--
36	--	Not Used	--
37	WHT/YEL	Leak Detection Pump Signal	N/A
38	--	Not Used	--
39	--	Not Used	--
40	--	Not Used	--
41	--	Not Used	--
42	--	Not Used	--
43	WHT/BLK	T-6 Controller, red on ECM	N/A

1998–1999 Sedan and Wagon 1.8L MFI AEB T80 Connector, *continued*

44	--	Not Used	--
45	GRN/BRN	T-10 Controller, yellow on ECM (ABS Control)	N/A
46	--	Not Used	--
47	--	Not Used	--
48	--	Not Used	--
49	GRN/WHT	T-10 Controller, blue on ECM (for TCM)	N/A
50	--	Not Used	--
51	LIL	Heated Oxygen Sensor Signal	N/A
52	RED	Heated Oxygen Sensor Ground	N/A
53	BRN/GRY	Engine Coolant Temperature Sensor Control	N/A
54	BLK/RED	Intake Air Temperature Sensor Control	N/A
55	--	Not Used	N/A
56	GRY	Engine Speed Sensor Signal	N/A
57	--	Not Used	--
58	BRN/GRN	Cylinder 3 Fuel Injector Control	N/A
59	RED/LIL	Throttle Position Actuator Motor Power	N/A
60	GRN	Knock Sensor 2 Signal	N/A
61	WHT	Barometric Pressure Sensor Signal	N/A
62	LIL/GRY	Barometric Pressure Sensor Power	N/A
63	BLU	Engine Speed Sensor Signal	N/A
64	YEL/WHT	Wastegate Bypass Regulator Valve Control	N/A
65	BLK/LIL	Cylinder 4 Fuel Injector Control	N/A
66	BRN/LIL	Throttle Position Actuator Motor Control	N/A
67	GRY/WHT	Sensor Ground	N/A
68	BRN	Knock Sensor 1 Signal	N/A
69	LIL/YEL	Throttle Position Actuator Signal	N/A
70	GRY/YEL	Power Output Stage	N/A
71	GRN/WHT	Power Output Stage	N/A
72	--	Not Used	--
73	BLK/GRN	Cylinder 1 Fuel Injector Control	N/A
74	LIL/RED	Throttle Position Sensor Signal	N/A
75	WHT/YEL	Throttle Position Sensor Ground	N/A
76	GRN/GRY	Camshaft Position Sensor Signal	N/A
77	RED/GRY	Power Output Stage	N/A
78	YEL/GRY	Power Output Stage	N/A
79	--	Not Used	--
80	BRN/BLU	Cylinder 2 Fuel Injector Control	N/A

Passat

1998–1999 Sedan and Wagon 2.8L MFI AHA T80 Connector

ECM Pin #	Wire Color	Circuit Description (80-Pin)	Value at Hot Idle
1	BLK/BLU	Power Ground	N/A
2	BRN	Power Ground	N/A
3	RED	Battery Power	N/A
4	RED/BLU	Fuel Pump Relay	N/A

1998–1999 Sedan and Wagon 2.8L MFI AHA T80 Connector, *continued*

5	GRN/BLK	T-10 Controller, yellow on ECM	N/A
6	GRN/BLU	Odometer Signal	N/A
7	--	Not Used	--
8	BLK/GRY	T-10 Controller, brown on ECM	N/A
9	--	Not Used	--
10	BLK/YEL	T-10 Controller, brown on ECM	N/A
11	RED/BLK	Camshaft Position Sensor 1 and 2 Power	N/A
12	YEL	Mass Airflow Sensor	N/A
13	GRN	Mass Airflow Sensor	N/A
14	BLK	Generator	N/A
15	LIL	EVAP Canister Purge Solenoid Valve Control	N/A
16	YEL/RED	Leak Detection Pump Motor Signal	N/A
17	YEL/BRN	Malfunction Indicator Light (MIL) Signal	N/A
18	YEL	Fuel Gauge Signal	N/A
19	GRN/BLK	T-10 Controller, Yellow on ECM	N/A
20	WHT/BLU	Speedometer Signal	N/A
21	BLK/LIL	Low Fuel Level Warning Light Signal	N/A
22	--	Not Used	--
23	--	Not Used	--
24	--	Not Used	--
25	GRY	Heated Oxygen Sensor 2 Signal	N/A
26	BLU	Heated Oxygen Sensor 1 Ground	N/A
27	RED/WHT	Heated Oxygen Sensor 1 and 2 Power	N/A
28	BRN/BLK	Oxygen Sensor 1 and 2 Power	N/A
29	GRY/WHT	T-15 Connector, White on ECM (for TCM, auto only)	N/A
30	BLU/GRY	Secondary Air Injection Pump Relay Control	N/A
31	--	Not Used	--
32	--	Not Used	--
33	YEL/GRN	Secondary Air Injection Solenoid Valve	N/A
34	--	Not Used	--
35	--	Not Used	--
36	--	Not Used	--
37	WHT/YEL	Leak Detection Pump Signal	N/A
38	LIL	Oxygen Sensor 2 Signal	N/A
39	GRY	Heated Oxygen Sensor 2 Signal	N/A
40	BLU	Heated Oxygen Sensor 2 Ground	N/A
41	GRY/RED	T-15 Connector, White on ECM (for TCM, auto only)	N/A
42	--	Not Used	--
43	WHT/BLK	T-16 Controller, Data Link Connector	N/A
44	GRN/BLK	Camshaft Position Sensor 1 Signal	N/A
45	GRN/BRN	T-10 Controller, Yellow on ECM (ABS Control)	N/A
46	--	Not Used	--
47	--	Not Used	--
48	--	Not Used	--
49	--	Not Used	--
50	RED	Oxygen Sensor 2 Ground	N/A
51	LIL	Oxygen Sensor 2 Signal	N/A

1998–1999 Sedan and Wagon 2.8L MFI AHA T80 Connector, *continued*

52	RED	Oxygen Sensor 2 Ground	N/A
53	BRN/GRY	Engine Coolant Temperature Sensor	N/A
54	GRN/RED	Intake Air Temperature Sensor Control	N/A
55	BRN/GRY	Camshaft Adjustment Valves (1 and 2) Control	N/A
56	GRY	Engine Speed Sensor Signal	N/A
57	--	Not Used	--
58	BRN/GRN	Cylinder 3 Fuel Injector Control	N/A
59	RED/LIL	Throttle Position Actuator Motor Power	N/A
60	GRN	Knock Sensor 2 Signal	N/A
61	--	Not Used	--
62	GRN/LIL	Barometric Pressure Sensor Power	N/A
63	BLU	Engine Speed Sensor Signal	N/A
64	YEL/WHT	Intake Manifold Change-Over Valve Signal	N/A
65	BRN/WHT	Cylinder 4 Fuel Injector Control	N/A
66	BRN/LIL	Throttle Position Actuator Motor Control	N/A
67	GRY/WHT	Sensor Ground	N/A
68	BRN	Knock Sensor 1 Signal	N/A
69	LIL/YEL	Throttle Position Actuator Signal	N/A
70	GRN/GRY	Ignition Coil Signal	N/A
71	GRY/YEL	Ignition Coil Signal	N/A
72	BLK/RED	Cylinder 5 Fuel Injector Control	N/A
73	BLK/GRN	Cylinder 1 Fuel Injector Control	N/A
74	LIL/RED	Throttle Position Sensor Signal	N/A
75	WHT/YEL	Throttle Position Sensor Ground	N/A
76	GRN/BRN	Camshaft Position Sensor 2 Signal	N/A
77	--	Not Used	--
78	GRN/WHT	Ignition Coil Signal	N/A
79	BLK/WHT	Cylinder 6 Fuel Injector Control	N/A
80	BRN/BLU	Cylinder 2 Fuel Injector Control	N/A

Passat

2000 Sedan and Wagon 1.8L MFI ATW T121 Connector

ECM Pin #	Wire Color	Circuit Description (121-Pin)	Value at Hot Idle
1	BRN	Power Ground	N/A
2	BRN	Power Ground	N/A
3	BLK/BLU	Power Ground	N/A
4	--	Not Used	--
5	RED/WHT	HO2S Power	N/A
6	--	Not Used	--
7	--	Not Used	--
8	--	Not Used	--
9	YEL/GRN	Secondary AIR Injection Solenoid Valve Signal	N/A
10	--	Not Used	--
11	--	Not Used	--
12	--	Not Used	--

2000 Sedan and Wagon 1.8L MFI ATW T121 Connector, *continued*

13	--	Not Used	--
14	--	Not Used	--
15	--	Not Used	--
16	--	Not Used	--
17	--	Not Used	--
18	--	Not Used	--
19	--	Not Used	--
20	--	Not Used	--
21	WHT/BLK	ECM Power Supply Relay	N/A
22	--	Not Used	--
23	--	Not Used	--
24	--	Not Used	--
25	YEL/RED	Leak Detection Pump Motor Signal	N/A
26	--	Not Used	--
27	YEL	Mass Air Flow Sensor Ground	N/A
28	--	Not Used	--
29	GRN	Mass Airflow Sensor Signal	N/A
30	BLK/LIL	T15-D Connector, Red on ECM	N/A
31	--	Not Used	--
32	BLK	Sensor Ground	N/A
33	GRE/YEL	Throttle Position Sensor Ground	N/A
34	BRN/GRN	Throttle Position Sensor Signal	N/A
35	YEL/BLU	Accelerator Pedal Position Sensor Signal	N/A
36	BRN/RED	Accelerator Pedal Position Sensor Ground	N/A
37	GRN/BLU	Climatronic Control Module Signal	N/A
38	BLK/WHT	Cruise Control Switch Ground	N/A
39	RED/GRN	Clutch Vacuum Vent Valve Swith	N/A
40	BLK/YEL	Climatronic Control Module Signal	N/A
41	BLK/GRY	Climatronic Control Module Signal	N/A
42	--	Not Used	--
43	GRN/BLK	T10-D Connector, Brown on ECM	N/A
44	--	Not Used	--
45	--	Not Used	--
46	--	Not Used	--
47	YEL/BRN	T10-D Connector, Brown on ECM	N/A
48	RED/BRN	T10-D Connector, Brown on ECM	N/A
49	--	Not Used	--
50	--	Not Used	--
51	BRN	HO2S Signal	N/A
52	--	Not Used	--
53	WHT/GRN	Mass Air Flow Sensor Power	N/A
54	WHT/BLU	T10-D Connector, Brown on ECM	N/A
55	WHT/RED	Brake Vacuum Vent Valve Switch Signal	N/A
56	RED/BLK	Brake Vacuum Vent Valve Switch Signal	N/A
57	RED/YEL	Cruise Control Set Switch Signal	N/A
58	ORN/BRN	CAN-L Signal	N/A
59	--	Not Used	--

2000 Sedan and Wagon 1.8L MFI ATW T121 Connector, *continued*

60	ORN/BLK	CAN-H Signal	N/A
61	--	Not Used	--
62	RED	Battery Power	N/A
63	WHT/GRY	HO2S Power	N/A
64	LIL	EVAP Canister Purge Regulator Valve Signal	N/A
65	RED/BLU	Fuel Pump Relay Control	N/A
66	BLU/GRY	Secondary AIR Relay Control	N/A
67	LIL/BRN	Airbag Control Module Signal	N/A
68	LIL	HO2S Signal	N/A
69	RED	HO2S Ground	N/A
70	WHT	HO2S Ground	N/A
71	--	Not Used	--
72	GRY	Throttle Position Sensor Power	N/A
73	YEL/LIL	Accelerator Pedal Position Sensor Power	N/A
74	GRN/BRN	15 Pin Connector, Red on ECU	N/A
75	BLU	Cruise Control Switch Signal	N/A
76	RED/GRY	Cruise Control Switch Signal	N/A
77	--	Not Used	--
78	--	Not Used	--
79	--	Not Used	--
80	WHT/YEL	Leak Detection Pump Switch Signal	N/A
81	YEL	T10-D Connector, Brown on ECM	N/A
82	GRY	Engine Speed Sensor Signal	N/A
83	GRN/LIL	Throttle Valve Switch Signal	N/A
84	LIL/YEL	Throttle Valve Switch Ground	N/A
85	BLK/RED	Intake Air Temperature Sensor	N/A
86	GRN/GRY	Camshaft Position Sensor Signal	N/A
87	--	Not Used	--
88	GRY/LIL	Fuel Injector 4 Control	N/A
89	BRN/BLU	Fuel Injector 2 Control	N/A
90	BLU	Engine Speed Sensor Signal	N/A
91	GRY/BRN	Throttle Valve Control Module Power	N/A
92	LIL/RED	Throttle Valve Switch Signal	N/A
93	BRN/GRY	Engine Coolant Temperature Sensor Signal	N/A
94	BRN/BLK	Ignition Coil 4 Control	N/A
95	GRY/BRN	Ignition Coil 2 Control	N/A
96	WHT/GRN	Fuel Injector 1 Control	N/A
97	GRY/GRN	Fuel Injector 3 Control	N/A
98	WHT/BLK	Camshaft Position Sensor Power	N/A
99	BRN	D102 Connector 2 on the ECM Housing	N/A
100	--	Not Used	--
101	GRY/BLU	Charge Air Pressure Sensor Signal	N/A
102	GRY	Ignition Coil 1 Control	N/A
103	GRY/BLK	Ignition Coil 3 Control	N/A
104	YEL/WHT	Wastegate Bypass Regulator Valve Signal	N/A
105	BRN/BLK	Recirculating Valve for Turbocharger Signal	N/A
106	WHT	Knock Sensor 1 Signal	N/A

2000 Sedan and Wagon 1.8L MFI ATW T121 Connector, *continued*

107	YEL	Knock Sensor 2 Signal	N/A
108	BLK	Sensor Ground	N/A
109	--	Not Used	--
110	--	Not Used	--
111	--	Not Used	--
112	--	Not Used	--
113	--	Not Used	--
114	--	Not Used	--
115	--	Not Used	--
116	--	Not Used	--
117	RED/LIL	Throttle Valve Motor Power	N/A
118	BRN/LIL	Throttle Valve Motor Control	N/A
119	--	Not Used	--
120	--	Not Used	--
121	RED/GRN	Ignition Coil 1 Power	N/A

Passat

2000–2001 Sedan and Wagon 2.8L MFI ATQ T121 Connector

PCM Pin #	Wire Color	Circuit Description (121-Pin)	Value at Hot Idle
1	BRN	Power Ground	< 0.10v
2	BRN	Power Ground	< 0.10v
3	BLK/BLU	Power Ground	< 0.10v
4	RED/WHT	HO2S-11 Heater Control	Digital Signal
5	RED/WHT	HO2S-21 Heater Control	Digital Signal
6	BRN/BLK	HO2S-22 Heater Control	Digital Signal
10	VIO	HO2S-22 Ground	0.010v
11	RED	HO2S-22 Signal	0.1-0.9v
12	VIO	HO2S-21 Ground	0.010v
13	RED	HO2S-21 Signal	0.1-0.9v
25	YEL/RED	Leak Detection Pump Motor Signal	Duty Cycle
27	YEL	Mass Airflow Sensor Ground	< 0.10v
29	GRN	Mass Airflow Sensor Signal	DC Voltage
30	BLK/VIO	15 Pin Connector, RED on ECU	12-14v
32	BLK	Sensor Ground	< 0.010v
33	GRN	Throttle Position Sensor Ground	< 0.010v
34	BRN/GRN	Throttle Position Sensor Signal	0-5.1v
35	YEL/BLU	Accelerator Pedal Position Sensor Signal	0-5.1v
36	BRN/RED	Accelerator Pedal Position Sensor Ground	< 0.010v
37	GRN/BLU	Climatronic Control Module Signal	Digital Signal
38	BLK/WHT	Cruise Control Switch Ground	< 0.010v
40	BLK/YEL	Climatronic Control Module Signal	Digital Signal
41	BLK/GRY	Climatronic Control Module Signal	Digital Signal
43	GRN/WHT	T10-D Connector, Brown on ECU	12-14v
44	YEL/GRN	Secondary AIR Valve Control	12v/0v
46	BLU/GRY	Secondary AIR Relay Control	12v/0v
47	BLK/RED	T10-D Connector, Brown on ECM	12-14v

2000–2001 Sedan and Wagon 2.8L MFI ATQ T121 Connector, *continued*

48	YEL/BRN	10-D Connector, Brown on ECM	12-14v
51	GRN	HO2S-11 Ground	0.010v
53	BLU/GRN	Mass Airflow Sensor Power	12v
54	WHT/BLU	Vehicle Speed Sensor Signal	Pulse Signal
55	WHI/RED	Brake Vacuum Vent Valve Switch Signal	12v/0v
56	RED/BLK	Brake Vacuum Vent Valve Switch Signal	12v/0v
57	RED/YEL	Cruise Control Set Switch Signal	---
58	ORN/BRN	CAN-L Signal	Digital Signal
60	ORN/BLK	CAN-H Signal	Digital Signal
62	RED	Battery Power	12-14v
63	BRN/BLK	HO2S-12 Heater Control	Digital Signal
64	VIO	EVAP Canister Purge Regulator Valve Signal	12v/0v
65	RED/BLU	Fuel Pump Relay Control	12v/0v
67	VIO/BLU	Airbag Control Module Signal	Digital Signal
68	GRN	HO2S-12 Ground	0.010v
69	BLU	HO2S-12 Signal	0.010v
70	GRN	HO2S-11Signal	0.1-0.9v
72	GRN	Throttle Position Sensor Power	4.9-5.1v
73	YEL/VIO	Accelerator Pedal Position Sensor Power	4.9-5.1v
74	GRN/BRN	15 Pin Connector, Red on ECU	12-14v
75	BLU	Cruise Control Switch Signal	---
76	RED/GRY	Cruise Control Switch Signal	---
80	WHT/YEL	Leak Detection Pump Switch Signal	12v/0v
81	YEL	T10-D Connector, Brown on ECU Housing	12-14v
82	GRY	Engine Speed Sensor Signal	Pulse Signal
83	GRN/VIO	Throttle Valve Switch Signal	0v/12v
84	WHI/YEL	Throttle Valve Switch Ground	<0.01v
85	GRN/RED	Intake Air Temperature Sensor	0.4-0.9v
86	GRN/BRN	Camshaft Position Sensor Signal	Digital Signal
87	GRN/BLK	Camshaft Position Sensor Signal	Digital Signal
88	BRN/GRN	Fuel Injector 3 Control	Digital Signal
89	BLK/WHT	Fuel Injector 6 Control	Digital Signal
90	BLU	Engine Speed Sensor Signal	Pulse Signal
91	GRY/WHT	Throttle Valve Control Module Power	12v
92	VIO/RED	Throttle Valve Switch Signal	0v/12v
94	GRN/GRY	Ignition Coil Control	12v/0v
96	BLK/GRN	Fuel Injector 1 Control	Digital Signal
97	BRN/BLK	Fuel Injector 4 Control	Digital Signal
98	RED/BLK	Camshaft Position Sensor Signal	Digital Signal
99	GRN/BRN	D102 Connector 2 on the ECU Housing	12-14v
102	GRN/YEL	Ignition Coil Control	12v/0v
103	GRN/WHT	Ignition Coil Control	12v/0v
104	YEL/WHT	Intake Manifold Change-Over Valve Signal	12v/0v
106	WHT	Knock Sensor 1 Signal	AC Signal
107	YEL	Knock Sensor 2 Signal	AC Signal
108	GRY/WHT	Power Ground	< 0.1v
112	BRN/BLU	Fuel Injector 2 Control	Digital Signal

2000–2001 Sedan and Wagon 2.8L MFI ATQ T121 Connector, *continued*

113	BLK/GRN	Fuel Injector 5 Control	Digital Signal
115	BRN/GRY	Camshaft Adjustment Valve 1	12v/0v
117	RED/VIO	Throttle Valve Motor Power	Duty Cycle
118	BRN/VIO	Throttle Valve Motor Control	Duty Cycle

Passat

2001 Sedan and Wagon 1.8L MFI AUG, T121 Connector; 2001–2005 Sedan and Wagon 1.8L MFI AWM, T121 Connector

ECM Pin #	Wire Color	Circuit Description (121-Pin)	Value at Hot Idle
1	BRN	Power Ground	N/A
2	BRN	Power Ground	N/A
3	BLK/BRN	Power Ground	N/A
4	--	Not Used	--
5	RED/WHT	HO2S Power	N/A
6	--	Not Used	--
7	--	Not Used	--
8	--	Not Used	--
9	YEL/GRN	Secondary AIR Injection Solenoid Valve Signal	N/A
10	--	Not Used	--
11	--	Not Used	--
12	--	Not Used	--
13	--	Not Used	--
14	--	Not Used	--
15	--	Not Used	--
16	--	Not Used	--
17	--	Not Used	--
18	GRY/YEL	Brake Booster Pressure Sensor Signal	N/A
19	--	Not Used	--
20	--	Not Used	--
21	WHT/BLK	ECM Power Supply Relay	N/A
22	WHT/YEL	Brake Booster Relay	N/A
23	--	Not Used	--
24	--	Not Used	--
25	YEL/RED	Leak Detection Pump Motor Signal	N/A
26	--	Not Used	--
27	YEL	Mass Air Flow Sensor Ground	N/A
28	--	Not Used	--
29	GRN	Mass Airflow Sensor Signal	N/A
30	--	Not Used	--
31	--	Not Used	--
32	WHT	Sensor Ground	N/A
33	GRY/YEL	Throttle Position Sensor Ground	N/A
34	BRN/GRN	Throttle Position Sensor Signal	N/A
35	YEL/BLU	Accelerator Pedal Position Sensor Signal	N/A
36	BRN/RED	Accelerator Pedal Position Sensor Ground	N/A
37	GRN/BLU	Climatronic Control Module Signal	N/A
38	BLK/WHT	Cruise Control Switch Ground	N/A

2001 Sedan and Wagon 1.8L MFI AUG, T121 Connector; 2001–2005 Sedan and Wagon 1.8L MFI AWM, T121 Connector, *continued*

39	RED/GRN	Clutch Vacuum Vent Valve Switch (manual only)	N/A
40	--	Not Used	--
41	WHT/GRY	Climatronic Control Module Signal	N/A
42	--	Not Used	--
43	GRN/RED	T10-D Connector, Brown on ECM	N/A
44	--	Not Used	--
45	--	Not Used	--
46	--	Not Used	--
47	--	Not Used	--
48	--	Not Used	--
49	--	Not Used	--
50	--	Not Used	--
51	BRN	Heated Oxygen Sensor Signal	N/A
52	YEL/GRY	Heated Oxygen Sensor Signal	N/A
53	WHT/GRN	Mass Air Flow Sensor Power	N/A
54	WHT/BLU	T10-D Connector, Brown on ECM	N/A
55	WHT/RED	Brake Vacuum Vent Valve Switch Signal	N/A
56	RED/BLK	Brake Vacuum Vent Valve Switch Signal	N/A
57	RED/YEL	Cruise Control Set Switch Signal	N/A
58	ORN/BRN	CAN-L Signal	N/A
59	--	Not Used	--
60	ORN/BLK	CAN-H Signal	N/A
61	--	Not Used	--
62	RED	Battery Power	N/A
63	WHT/GRY	Heated Oxygen Sensor Power	N/A
64	LIL	EVAP Canister Purge Regulator Valve Signal	N/A
65	--	Not Used	--
66	BLU/GRY	Secondary AIR Relay Control	N/A
67	LIL/BLU	Airbag Control Module Signal	N/A
68	LIL	Heated Oxygen Sensor Signal	N/A
69	RED	Heated Oxygen Sensor Ground	N/A
70	WHT	Heated Oxygen Sensor Ground	N/A
71	YEL/RED	Heated Oxygen Sensor Signal	N/A
72	GRN	Throttle Position Sensor Power	N/A
73	YEL/LIL	Accelerator Pedal Position Sensor Power	N/A
74	GRN/BRN	15 Pin Connector, Red on ECU	N/A
75	BLU	Cruise Control Switch Signal	N/A
76	RED/GRY	Cruise Control Switch Signal	N/A
77	--	Not Used	--
78	--	Not Used	--
79	--	Not Used	--
80	WHT/YEL	Leak Detection Pump Switch Signal	N/A
81	--	Not Used	--
82	GRY	Engine Speed Sensor Signal	N/A
83	GRN/LIL	Throttle Valve Switch Signal	N/A
84	LIL/YEL	Throttle Valve Switch Ground	N/A
85	BLK/RED	Intake Air Temperature Sensor	N/A

2001 Sedan and Wagon 1.8L MFI AUG, T121 Connector; 2001–2005 Sedan and Wagon 1.8L MFI AWM, T121 Connector, *continued*

86	GRN/GRY	Camshaft Position Sensor Signal	N/A
87	--	Not Used	--
88	GRY/LIL	Fuel Injector 4 Control	N/A
89	BRN/BLU	Fuel Injector 2 Control	N/A
90	BLU	Engine Speed Sensor Signal	N/A
91	GRY/BRN	Throttle Valve Control Module Power	N/A
92	LIL/RED	Throttle Valve Switch Signal	N/A
93	BRN/GRY	Engine Coolant Temperature Sensor Signal	N/A
94	BRN/GRN	Ignition Coil 4 Control	N/A
95	BLU/YEL	Ignition Coil 2 Control	N/A
96	BLK/GRN	Fuel Injector 1 Control	N/A
97	GRY/GRN	Fuel Injector 3 Control	N/A
98	WHT/BLK	Camshaft Position Sensor Power	N/A
99	BRN	D102 Connector 2 on the ECM Housing	N/A
100	--	Not Used	--
101	GRY/BLU	Manifold Absolute Pressure Sensor Signal	N/A
102	BLU/RED	Ignition Coil 1 Control	N/A
103	YEL/GRN	Ignition Coil 3 Control	N/A
104	YEL/WHT	Wastegate Bypass Regulator Valve Signal	N/A
105	BRN/BLK	Recirculating Valve for Turbocharger Signal	N/A
106	WHT	Knock Sensor 1 Signal	N/A
107	YEL	Knock Sensor 2 Signal	N/A
108	BLK	Sensor Ground	N/A
109	--	Not Used	--
110	--	Not Used	--
111	--	Not Used	--
112	--	Not Used	--
113	--	Not Used	--
114	--	Not Used	--
115	GRN/YEL	Valve 1 for Camshaft Adjustment	N/A
116	--	Not Used	--
117	RED/LIL	Throttle Valve Motor Power	N/A
118	BRN/LIL	Throttle Valve Motor Control	N/A
119	--	Not Used	--
120	--	Not Used	--
121	RED/GRN	Ignition Coil 1 Power	N/A

Passat

2002–2004 Sedan and Wagon 4.0L MFI BDP T121 Connector

ECM Pin #	Wire Color	Circuit Description (121-Pin)	Value at Hot Idle
1	BRN	Power Ground	N/A
2	BRN	Power Ground	N/A
3	BLK/BLU	Fuse Box Connection	N/A
4	BRN/YEL	Heated Oxygen Sensor 2 Power	N/A
5	BRN/WHT	Heated Oxygen Sensor Power	N/A
6	GRN/LIL	Heated Oxygen Sensor Power	N/A

2002–2004 Sedan and Wagon 4.0L MFI BDP T121 Connector, T121 Connector, *continued*

7	YEL/BRN	Ignition Coil 7 with Power Output Stage	N/A
8	BRN/YEL	Ignition Coil 3 with Power Output Stage	N/A
9	--	Not Used	--
10	GRN	Heated Oxygen Sensor Ground	N/A
11	BLK	Heated Oxygen Sensor Signal	N/A
12	GRY/BLU	Heated Oxygen Sensor 2 Signal	N/A
13	WHT/GRN/BLK	Heated Oxygen Sensor 2 Ground	N/A
14	GRY/BRN	Heated Oxygen Sensor 2 Signal	N/A
15	BLK/GRN/BRN	Heated Oxygen Sensor 2 Signal	N/A
16	BLU	Knock Sensor 4 Signal	N/A
17	WHT	Engine Coolant Temperature Sensor (on radiator)	N/A
18	GRY/RED	Valve 2 for Camshaft Adjustment (exhaust)	N/A
19	BRN/BLK	Cylinder 7 Fuel Injector Control	N/A
20	--	Not Used	--
21	BLK/BLU	T-10 Connector, Orange, in engine compartment	N/A
22	GRY/BLU	Valve 1 for Camshaft Adjustment (exhaust)	N/A
23	YEL/RED	PCM Power Supply Relay Signal	N/A
24	BLK/GRN	Cylinder 3 Fuel Injector Control	N/A
25	GRY/RED	Leak Detection Pump Motor Signal	N/A
26	YEL	Mass Airflow Sensor Ground	N/A
27	GRN	Brake Booster Pressure Sensor Ground	N/A
28	BLU/GRN	Voltage Regulator	N/A
29	BLK	Mass Airflow Sensor Signal	N/A
30	--	Not Used	--
31	YEL	Knock Sensor 3 Signal	N/A
32	--	Not Used	--
33	GRY/YEL	Throttle Position Sensor Ground	N/A
34	BRN/GRN	Throttle Position Sensor Signal	N/A
35	YEL/BLU	Accelerator Pedal Position Sensor Signal	N/A
36	BRN/RED	Accelerator Pedal Position Sensor Ground	N/A
37	GRN/BLU	T-10 Connection, black, in engine compartment	N/A
38	BLK/WHT	Cruise Control Switch Ground	N/A
39	RED/GRN	Clutch Vacuum Vent Valve Switch (manual only)	N/A
40	BLK/YEL	Rear Window Defogger Switch	N/A
41	BLK/GRY	T-15 Connection, Red, in engine compartment	N/A
42	RED/BLK	Starter Ground	N/A
43	GRN/BLK	T-10 Connection, Brown, in engine compartment	N/A
44	YEL/GRN	Secondary AIR Injection Solenoid Valve Signal	N/A
45	--	Not Used	--
46	BLU/GRY	Secondary Air Injection Pump Relay	N/A
47	GRY/GRN	Auxiliary Engine Coolant Pump Relay Signal	N/A
48	--	Not Used	--
49	BLK/GRY	Ground	N/A
50	GRN/WHT	Engine Coolant Temperature Sensor (on radiator)	N/A
51	GRY/WHT	Heated Oxygen Sensor Ground	N/A
52	BLK/GRN	Heated Oxygen Sensor Signal	N/A
53	LIL/YEL	Mass Air Flow Sensor Power	N/A

2002–2004 Sedan and Wagon 4.0L MFI BDP T121 Connector, T121 Connector, *continued*

54	BLU/YEL	Camshaft Position Sensor 2 Signal	N/A
55	WHT/RED	Brake Pedal Switch Signal	N/A
56	RED/BLK	Brake Light Switch Signal	N/A
57	RED/YEL	Cruise Control Set Switch Signal	N/A
58	ORN/BRN	Can-L Low Bus Connection	N/A
59	--	Not Used	--
60	ORN/BLK	Can-L High Bus Connection	N/A
61	--	Not Used	--
62	RED	Ground	N/A
63	GRN	Heated Oxygen Sensor Power	N/A
64	LIL	EVAP Canister Purge Regulator Valve Signal	N/A
65	RED/BLU	Fuel Pump Relay	N/A
66	BLU/WHT	Coolant Fan Control Module	N/A
67	--	Not Used	--
68	GRN	Heated Oxygen Sensor Signal	N/A
69	BLK	Heated Oxygen Sensor Ground	N/A
70	WHT/GRN/BRN	Heated Oxygen Sensor Ground	N/A
71	GRY/BLK	Heated Oxygen Sensor Signal	N/A
72	GRY	Throttle Position Sensor Power	N/A
73	YEL/LIL	Accelerator Pedal Position Sensor Power	N/A
74	--	Not Used	--
75	BLU	Cruise Control Switch Signal	N/A
76	RED/GRY	Cruise Control Switch Signal	N/A
77	ORN/BRN	Can-L Low Bus Connection	N/A
78	--	Not Used	--
79	ORN/BLK	Can-L High Bus Connection	N/A
80	WHT/YEL	Leak Detection Pump Switch Signal	N/A
81	GRN/BLU	Camshaft Position Sensor 4 Signal	N/A
82	WHT/GRY	Engine Speed Sensor Signal	N/A
83	BRN/RED	Throttle Valve Switch Signal	N/A
84	BLK/BRN	Throttle Valve Switch Ground	N/A
85	GRN/BLU	Brake Booster Pressure Sensor Signal	N/A
86	GRY	Camshaft Position Sensor Signal	N/A
87	GRN/RED	Camshaft Position Sensor 3 Signal	N/A
88	BLU/BLK	Cylinder 2 Fuel Injector Control	N/A
89	BLK/YEL	Cylinder 6 Fuel Injector Control	N/A
90	BRN/WHT	Engine Speed Sensor Signal	N/A
91	GRY/WHT	Throttle Valve Control Module Power	N/A
92	LIL/RED	Throttle Valve Switch Signal	N/A
93	BRN/GRY	Engine Coolant Temperature Sensor Signal	N/A
94	GRY/BLU	Ignition Coil 2 with Power Output Stage	N/A
95	GRY/GRN	Ignition Coil 6 with Power Output Stage	N/A
96	BLK/GRY	Cylinder 1 Fuel Injector Control	N/A
97	BLK/LIL	Cylinder 5 Fuel Injector Control	N/A
98	BLK/RED	Camshaft Position Sensors 1-4 Ground	N/A
99	LIL/BLK	Knock Sensors 1-4 Ground	N/A
100	--	Not Used	--

2002–2004 Sedan and Wagon 4.0L MFI BDP T121 Connector, T121 Connector, *continued*

101	--	Not Used	--
102	GRY/RED	Ignition Coil 1 with Power Output Stage	N/A
103	GRY/LIL	Ignition Coil 5 with Power Output Stage	N/A
104	GRN/YEL	Coolant Fan 2	N/A
105	BLU/RED	Left Electro-Hydraulic Engine Mount Solenoid Valve	N/A
106	YEL	Knock Sensor 1 Signal	N/A
107	BLU	Knock Sensor 2 Signal	N/A
108	BLK	Sensor Ground	N/A
109	--	Not Used	--
110	GRY/BLK	Ignition Coil 4 with Power Output Stage	N/A
111	BLK/BLU	Ignition Coil 8 with Power Output Stage	N/A
112	BLK/WHT	Cylinder 4 Fuel Injector Control	N/A
113	BLK	Cylinder 8 Fuel Injector Control	N/A
114	WHT/YEL	Brake Booster Relay	N/A
115	BRN/GRY	Valve 1 for Camshaft Adjustment	N/A
116	GRN	Map Controlled Engine Cooling Thermostat Signal	N/A
117	BRN/LIL	Throttle Valve Motor Power	N/A
118	RED/LIL	Throttle Valve Motor Control	N/A
119	--	Not Used	--
120	BRN/BLU	Valve 2 for Camshaft Adjustment	N/A
121	--	Not Used	--

Passat

2002–2005 Sedan and Wagon 2.8L MFI ATQ T121 Connector

ECM Pin #	Wire Color	Circuit Description (121-Pin)	Value at Hot Idle
1	BRN	Power Ground	N/A
2	BRN	Power Ground	N/A
3	BLK/BLU	Power Ground	N/A
4	RED/WHT	HO2S-11 Heater Control	N/A
5	RED/WHT	HO2S-21 Heater Control	N/A
6	BRN/BLK	HO2S-22 Heater Control	N/A
7	--	Not Used	--
8	--	Not Used	--
9	--	Not Used	--
10	VIO	HO2S-22 Ground	N/A
11	RED	HO2S-22 Signal	N/A
12	VIO	HO2S-21 Ground	N/A
13	RED	HO2S-21 Signal	N/A
14	--	Not Used	--
15	--	Not Used	--
16	--	Not Used	--
17	--	Not Used	--
18	GRN/BWN	Brake Booster Pressure Sensor (A/T only)	N/A
19	--	Not Used	--
20	--	Not Used	--
21	--	Not Used	--

2002–2005 Sedan and Wagon 2.8L MFI ATQ T121 Connector, *continued*

22	--	Not Used	--
23	--	Not Used	--
24	--	Not Used	--
25	YEL/RED	Leak Detection Pump Motor Signal	N/A
26	--	Not Used	--
27	YEL	Mass Airflow Sensor Ground	N/A
28	--	Not Used	--
29	GRN	Mass Airflow Sensor Signal	N/A
30	BLK/VIO	15 Pin Connector, RED on ECM	N/A
31	--	Not Used	--
32	BLK	Sensor Ground	N/A
33	GRN	Throttle Position Sensor Ground	N/A
34	BRN/GRN	Throttle Position Sensor Signal	N/A
35	YEL/BLU	Accelerator Pedal Position Sensor Signal	N/A
36	BRN/RED	Accelerator Pedal Position Sensor Ground	N/A
37	GRN/BLU	T10-D Connector, Black on ECM	N/A
38	BLK/WHT	Cruise Control Switch Ground	N/A
39	--	Not Used	--
40	BLK/YEL	Climatronic Control Module Signal	N/A
41	BLK/GRY	Climatronic Control Module Signal	N/A
42	--	Not Used	--
43	GRN/WHT	T10-D Connector, Brown on ECM	N/A
44	YEL/GRN	Secondary AIR Valve Control	N/A
45	WHI/YEL	T15 Connector, Red on ECM (A/T only)	N/A
46	BLU/GRY	Secondary AIR Relay Control	N/A
47	BLK/RED	T10-D Connector, Brown on ECM	N/A
48	YEL/BRN	10-D Connector, Brown on ECM	N/A
49	--	Not Used	--
50	--	Not Used	--
51	GRN	HO2S-11 Ground	N/A
52	--	Not Used	--
53	BLU/GRN	Mass Airflow Sensor Power	N/A
54	WHT/BLU	Vehicle Speed Sensor Signal	N/A
55	WHI/RED	Brake Vacuum Vent Valve Switch Signal	N/A
56	RED/BLK	Brake Vacuum Vent Valve Switch Signal	N/A
57	RED/YEL	Cruise Control Set Switch Signal	N/A
58	ORN/BRN	CAN-L Signal	N/A
59	--	Not Used	--
60	ORN/BLK	CAN-H Signal	N/A
61	--	Not Used	--
62	RED	Battery Power	N/A
63	BRN/BLK	HO2S-12 Heater Control	N/A
64	VIO	EVAP Canister Purge Regulator Valve Signal	N/A
65	RED/BLU	Fuel Pump Relay Control	N/A
66	--	Not Used	--
67	VIO/BLU	Airbag Control Module Signal	N/A
68	GRN	HO2S-12 Ground	N/A

2002–2005 Sedan and Wagon 2.8L MFI ATQ T121 Connector, *continued*

69	BLU	HO2S-12 Signal	N/A
70	GRN	HO2S-11Signal	N/A
71	--	Not Used	--
72	GRN	Throttle Position Sensor Power	N/A
73	YEL/VIO	Accelerator Pedal Position Sensor Power	N/A
74	GRN/BRN	15 Pin Connector, Red on ECU	N/A
75	BLU	Cruise Control Switch Signal	N/A
76	RED/GRY	Cruise Control Switch Signal	N/A
77	--	Not Used	--
78	--	Not Used	--
79	--	Not Used	--
80	WHT/YEL	Leak Detection Pump Switch Signal	N/A
81	YEL	T10-D Connector, Brown on ECU Housing	N/A
82	GRY	Engine Speed Sensor Signal	N/A
83	GRN/VIO	Throttle Valve Switch Signal	N/A
84	WHI/YEL	Throttle Valve Switch Ground	N/A
85	GRN/RED	Intake Air Temperature Sensor	N/A
86	GRN/BRN	Camshaft Position Sensor Signal	N/A
87	GRN/BLK	Camshaft Position Sensor Signal	N/A
88	BRN/GRN	Fuel Injector 3 Control	N/A
89	BLK/WHT	Fuel Injector 6 Control	N/A
90	BLU	Engine Speed Sensor Signal	N/A
91	GRY/WHT	Throttle Valve Control Module Power	N/A
92	VIO/RED	Throttle Valve Switch Signal	N/A
93	--	Not Used	--
94	GRN/GRY	Ignition Coil Control	N/A
95	--	Not Used	--
96	BLK/GRN	Fuel Injector 1 Control	N/A
97	BRN/BLK	Fuel Injector 4 Control	N/A
98	RED/BLK	Camshaft Position Sensor Signal	N/A
99	GRN/BRN	D102 Connector 2 on the ECU Housing	N/A
100	--	Not Used	--
101	--	Not Used	--
102	GRN/YEL	Ignition Coil Control	N/A
103	GRN/WHT	Ignition Coil Control	N/A
104	YEL/WHT	Intake Manifold Change-Over Valve Signal	N/A
105	--	Not Used	--
106	WHT	Knock Sensor 1 Signal	N/A
107	YEL	Knock Sensor 2 Signal	N/A
108	GRY/WHT	Power Ground	N/A
109	--	Not Used	--
110	--	Not Used	--
111	--	Not Used	--
112	BRN/BLU	Fuel Injector 2 Control	N/A
113	BLK/GRN	Fuel Injector 5 Control	N/A
114	--	Not Used	--
115	BRN/GRY	Camshaft Adjustment Valve 1	N/A

2002–2005 Sedan and Wagon 2.8L MFI ATQ T121 Connector, *continued*

116	--	Not Used	--
117	RED/VIO	Throttle Valve Motor Power	N/A
118	BRN/VIO	Throttle Valve Motor Control	N/A
119	--	Not Used	--
120	--	Not Used	--
121	--	Not Used	--

Passat

2004–2005 Sedan and Wagon 2.0L DFI BHW T94 plus T60 Connectors

ECM Pin #	Wire Color	Circuit Description (94-Pin)	Value at Hot Idle
1	BRN	Ground	N/A
2	BRN	Ground	N/A
3	RED/GRN	Automatic Glow Time Control Module	N/A
4	BRN	Ground	N/A
5	RED/GRN	Automatic Glow Time Control Module	N/A
6	RED/GRN	Power Supply Relay	N/A
7	--	Not Used	--
8	--	Not Used	--
9	--	Not Used	--
10	GRN	Heated Oxygen Sensor Ground	N/A
11	--	Not Used	--
12	--	Not Used	--
13	--	Not Used	--
14	--	Not Used	--
15	--	Not Used	--
16	--	Not Used	--
17	BRN/GRN	Closed Throttle Position Switch Ground	N/A
18	BLK	Ignition Starter Switch	N/A
19	WHT	Additional Heater Button	N/A
20	BLK/WHT	Cruise Control Switch Ground	N/A
21	BLU	Cruise Control Switch Signal	N/A
22	--	Not Used	--
23	LIL/YEL	Turbocharger Vane Position Sensor Signal	N/A
24	--	Not Used	--
25	--	Not Used	--
26	--	Not Used	--
27	--	Not Used	--
28	--	Not Used	--
29	--	Not Used	--
30	GRN/RED	Automatic Glow Time Control Module	N/A
31	--	Not Used	--
32	WHT	Heated Oxygen Sensor Signal	N/A
33	--	Not Used	--
34	--	Not Used	--
35	--	Not Used	--
36	--	Not Used	--

2004–2005 Sedan and Wagon 2.0L DFI BHW T94 plus T60 Connectors, *continued*

37	--	Not Used	--
38	BLU/GRY	Intake Air Temperature Sensor Control	N/A
39	BRN/RED	Throttle Position Sensor Power	N/A
40	WHT/LIL	Mass Airflow Sensor	N/A
41	--	Not Used	--
42	--	Not Used	--
43	RED/GRN	Clutch Vacuum Vent Valve Switch (manual only)	N/A
44	RED/GRY	Cruise Control Switch Signal	N/A
45	--	Not Used	--
46	--	Not Used	--
47	WHT/YEL	Relay for Radiator Fan Afterrun	N/A
48	--	Not Used	--
49	BRN/LIL	Power Supply Relay Signal	N/A
50	BLK/GRY	A/C Cut-off Control Module	N/A
51	GRN/YEL	Heated Oxygen Sensor Ground	N/A
52	RED/BLU	Fuel Pump Relay Signal	N/A
53	GRY/YEL	Turbocharger Vane Position Sensor Ground	N/A
54	BRN	Heated Oxygen Sensor Ground	N/A
55	--	Not Used	--
56	--	Not Used	--
57	--	Not Used	--
58	--	Not Used	--
59	GRY	Kick Down Switch Power	N/A
60	BRN/WHT	Mass Airflow Sensor	N/A
61	GRY/YEL	Throttle Position Sensor Ground	N/A
62	LIL/BLK	Charge Air Pressure Sensor	N/A
63	RED/WHT	Automatic Glow Time Control Module	N/A
64	BLU/BLK	Generator	N/A
65	WHT/RED	Brake Pedal Switch Control	N/A
66	ORN/BRN	Connection (Low-Bus) in instrument panel	N/A
67	--	Not Used	--
68	--	Not Used	--
69	--	Not Used	--
70	WHT/YEL	Preheating Coolant Output Relay (low heat)	N/A
71	WHT/BLU	Preheating Coolant Output Relay (high heat)	N/A
72	GRN/RED	Connector (K-Diagnosis) in instrument panel	N/A
73	--	Not Used	--
74	--	Not Used	--
75	--	Not Used	--
76	BRN/BLK	Intake Air Temperature Sensor Signal	N/A
77	GRY/GRN	Heated Oxygen Sensor Signal	N/A
78	GRN/RED	Charge Air Pressure Sensor	N/A
79	--	Not Used	--
80	--	Not Used	--
81	--	Not Used	--
82	GRN/BRN	Mass Airflow Sensor	N/A
83	YEL/BLU	Throttle Position Sensor Signal	N/A

2004–2005 Sedan and Wagon 2.0L DFI BHW T94 plus T60 Connectors, *continued*

84	YEL/LIL	Throttle Position Sensor Ground	N/A
85	BLK/YEL	Rear Window Defogger Switch	N/A
86	RED/GRN	Starter	N/A
87	RED/BLK	Brake Light Switch	N/A
88	RED/YEL	Cruise Control Set Switch Signal	N/A
89	ORN/BLK	Connection (High-Bus) in instrument panel	N/A
90	WHT/BLU	Vehicle Speed Sensor in instrument panel	N/A
91	--	Not Used	--
92	--	Not Used	--
93	--	Not Used	--
94	--	Not Used	--
ECM Pin #	**Wire Color**	**Circuit Description (60-Pin)**	**Value at Hot Idle**
1	BRN/GRN	Pump/Injector Valve for Cylinder 2	N/A
2	--	Not Used	--
3	--	Not Used	--
4	--	Not Used	--
5	--	Not Used	--
6	--	Not Used	--
7	--	Not Used	--
8	--	Not Used	--
9	--	Not Used	--
10	--	Not Used	--
11	--	Not Used	--
12	RED/LIL	Camshaft Position Sensor	N/A
13	--	Not Used	--
14	--	Not Used	--
15	--	Not Used	--
16	--	Not Used	--
17	--	Not Used	--
18	--	Not Used	--
19	--	Not Used	--
20	--	Not Used	--
21	--	Not Used	--
22	--	Not Used	--
23	BLK/YEL	EGR Potentiometer Signal	N/A
24	BLK/RED	EGR Potentiometer Ground	N/A
25	WHT/YEL	Intake Flap Motor	N/A
26	BRN	EGR Potentiometer Signal	N/A
27	RED/WHT	Camshaft Position Sensor	N/A
28	LIL/BLU	Camshaft Position Sensor	N/A
29	BRN/YEL	Wastegate Bypass Regulator Valve	N/A
30	--	Not Used	--
31	GRY	Ground Connection	N/A
32	GRY/YEL	Ground Connection	N/A
33	--	Not Used	--
34	--	Not Used	--
35	--	Not Used	--

2004–2005 Sedan and Wagon 2.0L DFI BHW T94 plus T60 Connectors, *continued*

36	--	Not Used	--
37	--	Not Used	--
38	--	Not Used	--
39	GRY/GRN	Fuel Temperature Sensor Signal	N/A
40	GRY	Fuel Temperature Sensor Ground	N/A
41	--	Not Used	--
42	--	Not Used	--
43	LIL/GRN	Engine Speed Sensor Signal	N/A
44	--	Not Used	--
45	--	Not Used	--
46	BRN/WHT	Pump/Injector Valve for Cylinder 1	N/A
47	BRN/RED	Pump/Injector Valve for Cylinder 3	N/A
48	BRN/BLU	Pump/Injector Valve for Cylinder 4	N/A
49	RED/WHT	EGR Vacuum Regulator Solenoid Valve	N/A
50	BLK/BLU	EGR Vacuum Regulator Solenoid Motor	N/A
51	--	Not Used	--
52	BRN/WHT	Engine Coolant Temperature Sensor Ground	N/A
53	BRN/GRN	Engine Coolant Temperature Sensor Signal	N/A
54	--	Not Used	--
55	--	Not Used	--
56	--	Not Used	--
57	--	Not Used	--
58	GRY	Engine Speed Sensor Signal	N/A
59	--	Not Used	--
60	WHT	Intake Flap Motor	N/A

Passat

2006 Sedan and Wagon 2.0 L MFI BPY T94 plus T60 Connectors

ECM Pin #	Wire Color	Circuit Description (94-Pin)	Value at Hot Idle
1	BRN	Ground Connection 23 on main wiring harness	N/A
2	BRN	Ground Connection 23 on main wiring harness	N/A
3	RED/WHT	Plus Connection 5 on main wiring harness	N/A
4	BRN	Ground Connection 23 on main wiring harness	N/A
5	RED/WHT	Plus Connection 5 on main wiring harness	N/A
6	RED/WHT	Plus Connection 5 on main wiring harness	N/A
7	BRN/WHT	Heated Oxygen Sensor Power (behind cat. conv.)	N/A
8	WHT/YEL	Leak Detection Pump	N/A
9	--	Not Used	--
10	--	Not Used	--
11	--	Not Used	--
12	YEL	Engine Coolant Temperature Sensor (on radiator)	N/A
13	--	Not Used	--
14	--	Not Used	--
15	--	Not Used	--
16	--	Not Used	--
17	--	Not Used	--

2006 Sedan and Wagon 2.0 L MFI BPY T94 plus T60 Connectors, *continued*

18	BLK/WHT	Cruise Control Connection	N/A
19	LIL	Change Air Pressure Sensor	N/A
20	--	Not Used	--
21	--	Not Used	--
22	GRN	Mass Air Flow Sensor	N/A
23	--	Not Used	--
24	--	Not Used	--
25	BLK/RED	Connection 1 on main wiring harness	N/A
26	--	Not Used	--
27	YEL/BLU	Fuel Pump Control Module	N/A
28	YEL/LIL	Coolant Fan 2 Control	N/A
29	--	Not Used	--
30	--	Not Used	--
31	--	Not Used	--
32	GRN/YEL	Engine Control Module Power Supply Relay 2	N/A
33	RED/GRY	Recirculation Pump Relay	N/A
34	--	Not Used	--
35	--	Not Used	--
36	--	Not Used	--
37	--	Not Used	--
38	GRY	Change Air Pressure Sensor	N/A
39	--	Not Used	--
40	GRN/LIL	Leak Detection Pump	N/A
41	RED/YEL	Clutch Position Sensor (manual trans only)	N/A
42	--	Not Used	--
43	--	Not Used	--
44	--	Not Used	--
45	--	Not Used	--
46	--	Not Used	--
47	--	Not Used	--
48	--	Not Used	--
49	--	Not Used	--
50	--	Not Used	--
51	LIL/WHT	Heated Oxygen Sensor Power	N/A
52	--	Not Used	--
53	BLK	Connection 34 on main wiring harness	N/A
54	--	Not Used	--
55	--	Not Used	--
56	BLU/GRY	Accelerator Pedal Position Sensor 2	N/A
57	WHT/BLU	Accelerator Pedal Position Sensor 2	N/A
58	GRN/GRY	Throttle Position Sensor	N/A
59	--	Not Used	--
60	BLU	Heated Oxygen Sensor Signal	N/A
61	RED	Heated Oxygen Sensor Ground	N/A
62	--	Not Used	--
63	--	Not Used	--
64	BLK/BRN	Generator	N/A

2006 Sedan and Wagon 2.0 L MFI BPY T94 plus T60 Connectors, *continued*

65	--	Not Used	--
66	--	Not Used	--
67	ORN/BRN	Data Bus On-Board Diagnostic Interface	N/A
68	ORN/BLK	Data Bus On-Board Diagnostic Interface	N/A
69	BRN/LIL	Engine Control Module Power Supply Relay	N/A
70	--	Not Used	--
71	--	Not Used	--
72	--	Not Used	--
73	--	Not Used	--
74	--	Not Used	--
75	--	Not Used	--
76	GRN/RED	Heated Oxygen Sensor Ground (behind cat conv.)	N/A
77	GRN/WHT	Heated Oxygen Sensor Signal (behind cat converter)	N/A
78	LIL/BLK	Throttle Position Sensor	N/A
79	BLK/BLU	Throttle Position Sensor	N/A
80	BLK/RED	Accelerator Pedal Position Sensor 2	N/A
81	GRY/WHT	Heated Oxygen Sensor Signal	N/A
82	GRY/RED	Heated Oxygen Sensor Ground	N/A
83	--	Not Used	--
84	--	Not Used	--
85	--	Not Used	--
86	GRY/WHT	OBD Connection 1 on main wiring harness	N/A
87	BLK/WHT	Fuse 10 on Fuse Panel C	N/A
88	--	Not Used	--
89	--	Not Used	--
90	--	Not Used	--
91	--	Not Used	--
92	BLK/YEL	Engine Control Module Power Supply Relay	N/A
93	--	Not Used	--
94	--	Not Used	--
ECM Pin #	**Wire Color**	**Circuit Description (60-Pin)**	**Value at Hot Idle**
1	RED/WHT	Cylinder 2 Fuel Injector Control	N/A
2	RED/BLK	Cylinder 1 Fuel Injector Control	N/A
3	--	Not Used	--
4	LIL/WHT	Vacuum Booster Pump Valve Control	N/A
5	BLU/LIL	EVAP Canister Purge Regulator Valve 1	N/A
6	--	Not Used	--
7	LIL/GRN	Low Fuel Pressure Sensor	N/A
8	--	Not Used	--
9	RED	Knock Sensor 2 Ground	N/A
10	GRN	Engine Coolant Temperature Sensor Signal	N/A
11	--	Not Used	--
12	GRN/YEL	Throttle Valve Switch Signal	N/A
13	WHT	Intake Air Temperature Sensor Signal	N/A
14	--	Not Used	--
15	LIL	Throttle Valve Motor Power	N/A
16	RED/GRN	Cylinder 4 Fuel Injector Control	N/A

2006 Sedan and Wagon 2.0 L MFI BPY T94 plus T60 Connectors, *continued*

17	RED/LIL	Cylinder 3 Fuel Injector Control	N/A
18	--	Not Used	--
19	LIL/BRN	Fuel Pressure Regulator Valve Control	N/A
20	LIL/WHT	Camshaft Adjustment Valve 1	N/A
21	RED/YEL	Camshaft Adjustment Valve 1 (exhaust)	N/A
22	GRN	Intake Manifold Runner Position Sensor Signal	N/A
23	--	Not Used	--
24	LIL	Knock Sensor 2 Signal	N/A
25	GRY/BLU	Fuel Pressure Sensor	N/A
26	BLK/GRY	Camshaft Position Sensor Signal	N/A
27	BLU/GRN	Throttle Valve Switch Ground	N/A
28	BLK/BLU	Throttle Valve Control Module Power	N/A
29	LIL/BRN	Throttle Valve Switch Signal	N/A
30	BRN/LIL	Throttle Valve Motor Control	N/A
31	--	Not Used	--
32	BRN/LIL	Cylinder 3 Fuel Injector Ground	N/A
33	BRN/BLK	Cylinder 1 Fuel Injector Ground	N/A
34	GRY/LIL	Intake Flap Motor Power	N/A
35	--	Not Used	--
36	WHT	Engine Speed Sensor Signal	N/A
37	BLK/GRN	Low Fuel Pressure Sensor	N/A
38	--	Not Used	--
39	BRN/RED	Knock Sensor 1 Ground	N/A
40	--	Not Used	--
41	BLU/GRY	Ignition Coil 2 With Power Output Stage	N/A
42	--	Not Used	--
43	RED/GRY	Ignition Coil 3 With Power Output Stage	N/A
44	GRN/LIL	Camshaft Position Sensor Signal	N/A
45	--	Not Used	--
46	--	Not Used	--
47	BRN/WHT	Cylinder 2 Fuel Injector Ground	N/A
48	BRN/GRY	Cylinder 4 Fuel Injector Ground	N/A
49	BRN/GRN	Intake Flap Motor Ground	N/A
50	--	Not Used	--
51	BRN	Engine Speed Sensor Signal	N/A
52	BLK	Connection 2 in engine wiring harness (ground)	N/A
53	--	Not Used	--
54	WHT/BRN	Knock Sensor 1 Signal	N/A
55	--	Not Used	--
56	GRN/BRN	Ignition Coil 4 With Power Output Stage	N/A
57	--	Not Used	--
58	LIL/GRY	Ignition Coil 1 With Power Output Stage	N/A
59	--	Not Used	--
60	--	Not Used	--

Passat

2006–2007 Sedan and Wagon 3.6L MFI BLV T94 plus T60 Connectors

ECM Pin #	Wire Color	Circuit Description (94-Pin)	Value at Hot Idle
1	BRN	Ground Connection 25 on main wiring harness	N/A
2	BRN	Ground Connection 25 on main wiring harness	N/A
3	RED/WHT	Plus Connection 5 on main wiring harness	N/A
4	BRN	Ground Connection 25 on main wiring harness	N/A
5	RED/WHT	Plus Connection 5 on main wiring harness	N/A
6	RED/WHT	Plus Connection 5 on main wiring harness	N/A
7	BRN/WHT	Heated Oxygen Sensor Power (behind cat. conv.)	N/A
8	WHT/YEL	Leak Detection Pump Signal	N/A
9	--	Not Used	--
10	--	Not Used	--
11	--	Not Used	--
12	BRN	Engine Coolant Temperature Sensor (on radiator)	N/A
13	YEL/BLK	Mass Air Flow Sensor	N/A
14	--	Not Used	--
15	--	Not Used	--
16	--	Not Used	--
17	--	Not Used	--
18	BLK/WHT	Steering Column Electronic Systems Control Module	N/A
19	--	Not Used	--
20	--	Not Used	--
21	--	Not Used	--
22	YEL/GRY	Mass Air Flow Sensor	N/A
23	--	Not Used	--
24	--	Not Used	--
25	BLK/RED	Connection 1 on main wiring harness	N/A
26	YEL	Engine Coolant Temperature Sensor (on radiator)	N/A
27	YEL/BLU	Fuel Pump Control Module	N/A
28	YEL/LIL	Coolant Fan 2 Control	N/A
29	BLK/WHT	Heated Oxygen Sensor 2 Power (behind cat. conv.)	N/A
30	--	Not Used	--
31	--	Not Used	--
32	GRN	Engine Control Module Power Supply Relay 2	N/A
33	LIL/YEL	Recirculation Pump Relay	N/A
34	--	Not Used	--
35	YEL/GRY	Low Fuel Pressure Sensor	N/A
36	--	Not Used	--
37	--	Not Used	--
38	--	Not Used	--
39	--	Not Used	--
40	GRN/LIL	Leak Detection Pump Motor Ground	N/A
41	--	Not Used	--
42	YEL/RED	Intake Air Temperature Sensor	N/A
43	BLU	Heated Oxygen Sensor 3 (Bank 2) Signal	N/A
44	--	Not Used	--

2006–2007 Sedan and Wagon 3.6L MFI BLV T94 plus T60 Connectors, *continued*

45	--	Not Used	--
46	--	Not Used	--
47	--	Not Used	--
48	--	Not Used	--
49	--	Not Used	--
50	--	Not Used	--
51	LIL/WHT	Heated Oxygen Sensor Power	N/A
52	--	Not Used	--
53	BLK	Heated Oxygen Sensor 3 (Banks 1 and 2) Ground	N/A
54	WHT	Heated Oxygen Sensor 2 Ground (behind cat conv.)	N/A
55	GRN	Heated Oxygen Sensor 2 Signal (behind cat conv.)	N/A
56	GRY/BLU	Accelerator Pedal Position Sensor 2	N/A
57	BRN/BLU	Accelerator Pedal Position Sensor 2	N/A
58	BRN/WHT	Accelerator Pedal Position Sensor 2	N/A
59	RED	Heated Oxygen Sensor Signal	N/A
60	GRN	Heated Oxygen Sensor Signal	N/A
61	GRY/WHT	Heated Oxygen Sensor Ground	N/A
62	GRY/RED	Heated Oxygen Sensor Ground	N/A
63	--	Not Used	--
64	BRN/LIL	Intake Air Temperature Sensor	N/A
65	BLK/BRN	Generator	N/A
66	--	Not Used	--
67	ORN/BRN	Data Bus On-Board Diagnostic Interface	N/A
68	ORN/BLK	Data Bus On-Board Diagnostic Interface	N/A
69	BLK/GRY	Engine Control Module Power Supply Relay	N/A
70	--	Not Used	--
71	--	Not Used	--
72	--	Not Used	--
73	LIL	Heated Oxygen Sensor Power	N/A
74	--	Not Used	--
75	GRN	Heated Oxygen Sensor 3 (Bank 1) Signal	N/A
76	WHT	Heated Oxygen Sensor Ground (behind cat conv.)	N/A
77	GRN	Heated Oxygen Sensor Signal (behind cat converter)	N/A
78	GRY/RED	Throttle Position Sensor	N/A
79	BLK/BLU	Throttle Position Sensor	N/A
80	YEL/GRN	Throttle Position Sensor	N/A
81	BLK	Heated Oxygen Sensor Signal	N/A
82	GRY/RED	Heated Oxygen Sensor Ground	N/A
83	BLU	Heated Oxygen Sensor Signal	N/A
84	GRY/WHT	Heated Oxygen Sensor Ground	N/A
85	--	Not Used	--
86	GRY/WHT	OBD Connection 1 on main wiring harness	N/A
87	BLK/WHT	Plus Connection 9 in main wiring harness	N/A
88	--	Not Used	--
89	--	Not Used	--
90	--	Not Used	--
91	--	Not Used	--

2006–2007 Sedan and Wagon 3.6L MFI BLV T94 plus T60 Connectors, *continued*

92	BLK/GRN	Engine Control Power Supply Relay	N/A
93	--	Not Used	--
94	--	Not Used	--
ECM Pin #	**Wire Color**	**Circuit Description (60-Pin)**	**Value at Hot Idle**
1	YEL	Cylinder 2 Fuel Injector Control	N/A
2	WHT/LIL	Cylinder 1 Fuel Injector Control	N/A
3	BLU/WHT	Cylinder 4 Fuel Injector Control	N/A
4	YEL	Vacuum Booster Pump Valve Control	N/A
5	LIL/BLU	EVAP Canister Purge Regulator Valve 1	N/A
6	GRY/LIL	Intake Manifold Runner Control Valve	N/A
7	--	Not Used	--
8	--	Not Used	--
9	RED	Knock Sensor 2 Ground	N/A
10	GRN/BLK	Engine Coolant Temperature Sensor Signal	N/A
11	--	Not Used	N/A
12	GRN/YEL	Throttle Valve Switch Signal	N/A
13	--	Not Used	--
14	BRN/BLU	Connection 1 in engine wiring harness (ground)	N/A
15	BRN/LIL	Throttle Valve Motor Power	N/A
16	BLU/RED	Cylinder 6 Fuel Injector Control	N/A
17	GRN/BLK	Cylinder 5 Fuel Injector Control	N/A
18	RED/BLK	Cylinder 1 Fuel Injector Control	N/A
19	WHT/BLK	Fuel Pressure Regulator Valve Control	N/A
20	LIL/WHT	Camshaft Adjustment Valve 1	N/A
21	LIL/BLU	Camshaft Adjustment Valve 1 (exhaust)	N/A
22	--	Not Used	--
23	--	Not Used	--
24	LIL	Knock Sensor 2 Signal	N/A
25	YEL/RED	Fuel Pressure Sensor Signal	N/A
26	BLK/WHT	Camshaft Position Sensor Signal	N/A
27	BLU/GRN	Throttle Valve Switch Ground	N/A
28	BLK/GRN	Throttle Valve Control Module Power	N/A
29	LIL/BRN	Throttle Valve Switch Signal	N/A
30	LIL	Throttle Valve Motor Control	N/A
31	BRN/BLK	Cylinder 3 Fuel Injector Ground	N/A
32	GRY/BLK	Cylinder 5 Fuel Injector Ground	N/A
33	BLU/LIL	Cylinder 1 Fuel Injector Ground	N/A
34	--	Not Used	--
35	--	Not Used	--
36	WHT	Engine Speed Sensor Signal	N/A
37	BLK/LIL	Camshaft Position Sensor 2 Ground	N/A
38	--	Not Used	--
39	BRN/RED	Knock Sensor 1 Ground	N/A
40	--	Not Used	--
41	LIL/GRN	Ignition Coil 6 with Power Output Stage	N/A
42	LIL/YEL	Ignition Coil 4 with Power Output Stage	N/A
43	LLIL/WHT	Ignition Coil 5 with Power Output Stage	N/A

2006–2007 Sedan and Wagon 3.6L MFI BLV T94 plus T60 Connectors, *continued*

44	GRN/LIL	Camshaft Position Sensor Signal	N/A
45	BRN	Heated Oxygen Sensor 2 (Bank 2) Power	N/A
46	BLK/WHT	Cylinder 4 Fuel Injector Ground	N/A
47	BRN/YEL	Cylinder 2 Fuel Injector Ground	N/A
48	WHT/RED	Cylinder 6 Fuel Injector Ground	N/A
49	--	Not Used	--
50	--	Not Used	--
51	BRN	Engine Speed Sensor Signal	N/A
52	BLK	Connection 2 in engine wiring harness (ground)	N/A
53	--	Not Used	--
54	WHT/BRN	Knock Sensor 1 Signal	N/A
55	--	Not Used	--
56	BLU/LIL	Ignition Coil 3 with Power Output Stage	N/A
57	LIL/BLU	Ignition Coil 2 with Power Output Stage	N/A
58	LIL/GRY	Ignition Coil 1 with Power Output Stage	N/A
59	WHT/LIL	Camshaft Position Sensor 2 Signal	N/A
60	BRN/WHT	Heated Oxygen Sensor 2 (Bank 1) Power	N/A